HAESE MATHEMATICS
Specialists in mathematics publishing

Mathematics for the international student Mathematics HL (Core) third edition

David Martin
Robert Haese
Sandra Haese
Michael Haese
Mark Humphries

for use with IB Diploma Programme

MATHEMATICS FOR THE INTERNATIONAL STUDENT
Mathematics HL (Core) third edition

David Martin	B.A.,B.Sc.,M.A.,M.Ed.Admin.
Robert Haese	B.Sc.
Sandra Haese	B.Sc.
Michael Haese	B.Sc.(Hons.),Ph.D.
Mark Humphries	B.Sc.(Hons.)

Published by Haese Mathematics
152 Richmond Road, Marleston, SA 5033, AUSTRALIA
Telephone: +61 8 8210 4666, Fax: +61 8 8354 1238
Email: info@haesemathematics.com.au
Web: www.haesemathematics.com.au

National Library of Australia Card Number & ISBN 978-1-921972-11-9

First Edition	2004
Reprinted	2005 three times *(with minor corrections)*, 2006, 2007
Second Edition	2008
Reprinted	2009 *(with minor corrections)*, 2010, 2011
Third Edition	2012
Reprinted	2013 *(with minor corrections)*, 2015, 2017

Typeset in Times Roman 10.

Printed in China by Prolong Press Limited.

Acknowledgements: While every attempt has been made to trace and acknowledge copyright, the authors and publishers apologise for any accidental infringement where copyright has proved untraceable. They would be pleased to come to a suitable agreement with the rightful owner.

FOREWORD

Mathematics for the International Student: Mathematics HL has been written to embrace the syllabus for the two-year Mathematics HL Course, first examined in 2014. It is not our intention to define the course. Teachers are encouraged to use other resources. We have developed this book independently of the International Baccalaureate Organization (IBO) in consultation with many experienced teachers of IB Mathematics. The text is not endorsed by the IBO.

Syllabus references are given at the beginning of each chapter. The new edition reflects the new Mathematics HL syllabus. Discussion topics for the Theory of Knowledge have been included in this edition. See page 12 for a summary.

In response to the introduction of a calculator-free examination paper, the review sets at the end of each chapter have been categorised as 'calculator' or 'non-calculator'. Also, the final chapter contains over 200 examination-style questions, categorised as 'calculator' or 'non-calculator'. These questions should provide more difficult challenges for advanced students.

Comprehensive graphics calculator instructions for Casio fx-9860G Plus, Casio fx-CG20, TI-84 Plus and TI-$n$spire are accessible as printable pages online (see page 16) and, occasionally, where additional help may be needed, more detailed instructions are available from icons located throughout the book. The extensive use of graphics calculators and computer packages throughout the book enables students to realise the importance, application, and appropriate use of technology. No single aspect of technology has been favoured. It is as important that students work with a pen and paper as it is that they use their graphics calculator, or use a spreadsheet or graphing package on computer.

This package is language rich and technology rich. The combination of textbook and online interactive features will foster the mathematical development of students in a stimulating way. Frequent use of the online interactive features is certain to nurture a much deeper understanding and appreciation of mathematical concepts. This textbook also offers **Self Tutor** for every worked example. **Self Tutor** is accessed online – click anywhere on any worked example to hear a teacher's voice explain each step in that worked example. This is ideal for catch-up and revision, or for motivated students who want to do some independent study outside school hours.

For students who may not have a good understanding of the necessary background knowledge for this course, we have provided printable pages of information, examples, exercises, and answers online – see 'Background knowledge' (page 16).

The online interactive features allow immediate access to our own specially designed geometry software, graphing software and more. Teachers are provided with a quick and easy way to demonstrate concepts, and students can discover for themselves and re-visit when necessary.

It is not our intention that each chapter be worked through in full. Time constraints may not allow for this. Teachers must select exercises carefully, according to the abilities and prior knowledge of their students, to make the most efficient use of time and give as thorough coverage of work as possible. Investigations throughout the book will add to the discovery aspect of the course and enhance student understanding and learning.

In this changing world of mathematics education, we believe that the contextual approach shown in this book, with the associated use of technology, will enhance the students' understanding, knowledge and appreciation of mathematics, and its universal application.

We welcome your feedback.

Email: *info@haesemathematics.com.au*

Web: *www.haesemathematics.com.au*

DCM RCH SHH

PMH MAH

ACKNOWLEDGEMENTS

Cartoon artwork by John Martin. Artwork by Piotr Poturaj and Benjamin Fitzgerald.

Cover design by Piotr Poturaj.

Computer software by Thomas Jansson, Troy Cruickshank, Ashvin Narayanan, Adrian Blackburn, Edward Ross and Tim Lee.

Typeset in Australia by Charlotte Frost.

Editorial review by Catherine Quinn and Tim Knight.

Additional questions by Sriraman R. Iyer, Aditya Birla World Academy, Mumbai, India.

The authors and publishers would like to thank all those teachers who offered advice and encouragement on this book. Many of them read the page proofs and offered constructive comments and suggestions. These teachers include: Jeff Kutcher, Peter Blythe, Gail Smith, Dhruv Prajapati, Peter McCombe, and Chris Carter. To anyone we may have missed, we offer our apologies.

The publishers wish to make it clear that acknowledging these individuals does not imply any endorsement of this book by any of them, and all responsibility for the content rests with the authors and publishers.

HL & SL COMBINED CLASSES

Refer to our website www.haesemathematics.com.au for guidance in using this textbook in HL and SL combined classes.

HL OPTIONS / FURTHER MATHEMATICS

There is a separate textbook for each of the Mathematics HL Option topics:

- Topic 7 – Statistics and probability
- Topic 8 – Sets, relations and groups
- Topic 9 – Calculus
- Topic 10 – Discrete mathematics

An additional **Further Mathematics HL** textbook offers coverage of Further Mathematics topics Linear Algebra and Geometry.

ONLINE FEATURES

With the purchase of a new hard copy textbook, you will gain 27 months subscription to our online product. This subscription can be renewed for a small fee.

Students can revisit concepts taught in class and undertake their own revision and practice online.

By clicking on the relevant icon, a range of interactive features can be accessed:

- Self Tutor
- Graphics calculator instructions
- Background knowledge (as printable pages)
- Interactive links to spreadsheets, graphing and geometry software, computer demonstrations and simulations

Graphics calculator instructions: Detailed instructions are available online, as printable pages (see page 16). Click on the icon for Casio fx-9860G Plus, Casio fx-CG20, TI-84 Plus, or TI-*n*spire instructions.

COMPATIBILITY

For iPads, tablets, and other mobile devices, the interactive features may not work. However, the electronic version of the textbook and additional chapters can be viewed online using any of these devices.

REGISTERING

You will need to register to access the online features of this textbook.
Visit www.haesemathematics.com.au/register and follow the instructions. Once you have registered, you can:

- activate your digital textbook
- use your account to make additional purchases.

To activate your digital textbook, contact Haese Mathematics. On providing proof of purchase, your digital textbook will be activated. **It is important that you keep your receipt as proof of purchase.**

For general queries about registering and licence keys:

- Visit our Snowflake help page: http://snowflake.haesemathematics.com.au/help
- Contact Haese Mathematics: info@haesemathematics.com.au

ONLINE VERSION OF THE TEXTBOOK

The entire text of the book can be viewed online, allowing you to leave your textbook at school.

SELF TUTOR

Self tutor is an exciting feature of this book.

The Self Tutor icon on each worked example denotes an active online link.

Simply 'click' on the Self Tutor (or anywhere in the example box) to access the worked example, with a teacher's voice explaining each step necessary to reach the answer.

Play any line as often as you like. See how the basic processes come alive using movement and colour on the screen.

TABLE OF CONTENTS

SYMBOLS AND NOTATION USED IN THIS BOOK

$\mathbb{N}$	the set of positive integers and zero, $\{0, 1, 2, 3, ...\}$
$\mathbb{Z}$	the set of integers, $\{0, \pm1, \pm2, \pm3, ...\}$
$\mathbb{Z}^+$	the set of positive integers, $\{1, 2, 3, ...\}$
$\mathbb{Q}$	the set of rational numbers
$\mathbb{Q}^+$	the set of positive rational numbers, $\{x \mid x > 0,\ x \in \mathbb{Q}\}$
$\mathbb{R}$	the set of real numbers
$\mathbb{R}^+$	the set of positive real numbers, $\{x \mid x > 0,\ x \in \mathbb{R}\}$
$\mathbb{C}$	the set of complex numbers, $\{a + bi \mid a, b \in \mathbb{R}\}$
i	$\sqrt{-1}$
z	a complex number
z^*	the complex conjugate of z
$\|z\|$	the modulus of z
$\arg z$	the argument of z
$\mathcal{Re}\, z$	the real part of z
$\mathcal{Im}\, z$	the imaginary part of z
$\{x_1, x_2, ...\}$	the set with elements $x_1, x_2, ...$
$n(A)$	the number of elements in the finite set A
$\{x \mid ....$	the set of all x such that
$\in$	is an element of
$\notin$	is not an element of
$\varnothing$	the empty (null) set
U	the universal set
$\cup$	union
$\cap$	intersection
$\subset$	is a proper subset of
$\subseteq$	is a subset of
A'	the complement of the set A
$a^{\frac{1}{n}}, \sqrt[n]{a}$	a to the power of $\frac{1}{n}$, nth root of a (if $a \geqslant 0$ then $\sqrt[n]{a} \geqslant 0$)
$a^{\frac{1}{2}}, \sqrt{a}$	a to the power $\frac{1}{2}$, square root of a (if $a \geqslant 0$ then $\sqrt{a} \geqslant 0$)
$\|x\|$	the modulus or absolute value of x, that is $\begin{cases} x \text{ for } x \geqslant 0, & x \in \mathbb{R} \\ -x \text{ for } x < 0, & x \in \mathbb{R} \end{cases}$

$\equiv$	identity or is equivalent to
$\approx$	is approximately equal to
$>$	is greater than
$\geq$ or $\geqslant$	is greater than or equal to
$<$	is less than
$\leq$ or $\leqslant$	is less than or equal to
$\ngtr$	is not greater than
$\nless$	is not less than
$[a, b]$	the closed interval $a \leqslant x \leqslant b$
$]a, b[$	the open interval $a < x < b$
u_n	the nth term of a sequence or series
d	the common difference of an arithmetic sequence
r	the common ratio of a geometric sequence
S_n	the sum of the first n terms of a sequence, $u_1 + u_2 + ... + u_n$
S_∞ or S	the sum to infinity of a sequence, $u_1 + u_2 + ...$
$\sum\limits_{i=1}^{n} u_i$	$u_1 + u_2 + ... + u_n$
$\binom{n}{r}$	$\dfrac{n!}{r!(n-r)!}$
$f: A \to B$	f is a function under which each element of set A has an image in set B
$f: x \mapsto y$	f is a function under which x is mapped to y
$f(x)$	the image of x under the function f
f^{-1}	the inverse function of the function f
$f \circ g$	the composite function of f and g
$\lim\limits_{x \to a} f(x)$	the limit of $f(x)$ as x tends to a
$\dfrac{dy}{dx}$	the derivative of y with respect to x
$f'(x)$	the derivative of $f(x)$ with respect to x
$\dfrac{d^2y}{dx^2}$	the second derivative of y with respect to x
$f''(x)$	the second derivative of $f(x)$ with respect to x
$\dfrac{d^ny}{dx^n}$	the nth derivative of y with respect to x
$f^{(n)}(x)$	the nth derivative of $f(x)$ with respect to x
$\int y\, dx$	the indefinite integral of y with respect to x

$\int_a^b y\,dx$	the definite integral of y with respect to x between the limits $x = a$ and $x = b$
e^x	exponential function of x
$\log_a x$	logarithm to the base a of x
$\ln x$	the natural logarithm of x, $\log_e x$
sin, cos, tan	the circular functions
arcsin, arccos, arctan	the inverse circular functions
csc, sec, cot	the reciprocal circular functions
cis θ	$\cos\theta + i\sin\theta$
A(x, y)	the point A in the plane with Cartesian coordinates x and y
[AB]	the line segment with end points A and B
AB	the length of [AB]
(AB)	the line containing points A and B
$\widehat{\text{A}}$	the angle at A
$\widehat{\text{CAB}}$ or $\text{C}\widehat{\text{A}}\text{B}$	the angle between [CA] and [AB]
$\triangle$ABC	the triangle whose vertices are A, B, and C
v	the vector **v**
$\overrightarrow{\text{AB}}$	the vector represented in magnitude and direction by the directed line segment from A to B
a	the position vector $\overrightarrow{\text{OA}}$
i, **j**, **k**	unit vectors in the directions of the Cartesian coordinate axes
$\lvert$ **a** $\rvert$	the magnitude of vector **a**
$\lvert \overrightarrow{\text{AB}} \rvert$	the magnitude of $\overrightarrow{\text{AB}}$
v • **w**	the scalar product of **v** and **w**
v × **w**	the vector product of **v** and **w**
I	the identity matrix

P(A)	probability of event A
P′(A)	probability of the event "not A"
P(A | B)	probability of the event A given B
$x_1, x_2, ...$	observations of a variable
$f_1, f_2, ...$	frequencies with which the observations $x_1, x_2, x_3, ...$ occur
p_x	probability distribution function $\text{P}(X = x)$ of the discrete random variable X
$f(x)$	probability density function of the continuous random variable X
E(x)	the expected value of the random variable X
Var (X)	the variance of the random variable X
μ	population mean
σ	population standard deviation
σ^2	population variance
$\overline{x}$	sample mean
s_n^2	sample variance
s_n	standard deviation of the sample
s_{n-1}^2	unbiased estimate of the population variance
B(n, p)	binomial distribution with parameters n and p
Po(m)	Poisson distribution with mean m
N(μ, σ^2)	normal distribution with mean μ and variance σ^2
$X \sim$ B(n, p)	the random variable X has a binomial distribution with parameters n and p
$X \sim$ Po(m)	the random variable X has a Poisson distribution with mean m
$X \sim$ N(μ, σ^2)	the random variable X has a normal distribution with mean μ and variance σ^2

GREEK LETTERS

α	alpha	λ	lambda
β	beta	μ	mu
γ	gamma	π	pi
δ	delta	σ	sigma
θ	theta	ϕ	phi

THEORY OF KNOWLEDGE

Theory of Knowledge is a Core requirement in the International Baccalaureate Diploma Programme.

Students are encouraged to think critically and challenge the assumptions of knowledge. Students should be able to analyse different ways of knowing and areas of knowledge, while considering different cultural and emotional perceptions, fostering an international understanding.

The activities and discussion topics in the below table aim to help students discover and express their views on knowledge issues.

THEORY OF KNOWLEDGE

There are several theories for why one complete turn was divided into 360 degrees:

- 360 is approximately the number of days in a year.
- The Babylonians used a counting system in base 60. If they drew 6 equilateral triangles within a circle as shown, and divided each angle into 60 subdivisions, then there were 360 subdivisions in one turn. The division of an hour into 60 minutes, and a minute into 60 seconds, is from this base 60 counting system.
- 360 has 24 divisors, including every integer from 1 to 10 except 7.

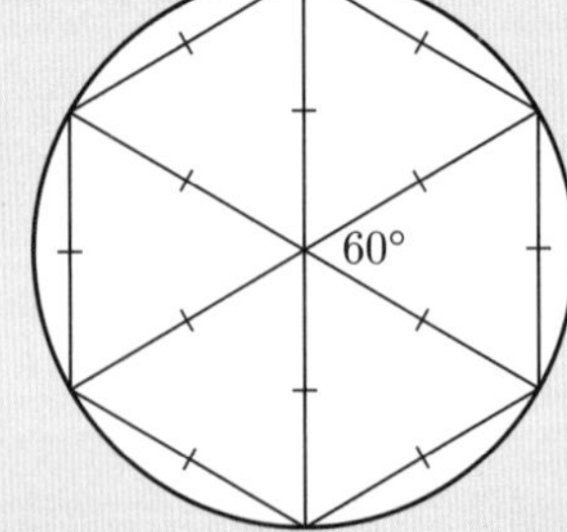

The idea of measuring an angle by the length of an arc dates to around 1400 and the Persian mathematician Al-Kashi. The concept of a radian is generally credited to Roger Cotes, however, who described it as we know it today.

1 What other measures of angle are there?

2 Which is the most *natural* unit of angle measure?

See **Chapter 10, The unit circle and radian measure**, p. 282

WRITING A MATHEMATICAL EXPLORATION

In addition to sitting examination papers, Mathematics HL students are also required to complete a **mathematical exploration**. This is a short report written by the student, based on a topic of his or her choice, and should focus on the mathematics of that topic. The mathematical exploration comprises 20% of the final mark.

The exploration should be approximately 6-12 pages long, and should be written at a level which is accessible to an audience of your peers. The exploration should also include a bibliography.

Group work should not be used for explorations. Each student's exploration is an individual piece of work.

When deciding on how to structure your exploration, you may wish to include the following sections:

Introduction: This section can be used to explain why the topic has been chosen, and to include any relevant background information.

Aim: A clear statement of intent should be given to provide perspective and direction to your exploration. This should be a short paragraph which outlines the problem or scenario under investigation.

Method and Results: This section can be used to describe the process which was followed to investigate the problem, as well as recording the unprocessed results of your investigations, in the form of a table, for example.

Analysis of Results: In this section, you should use graphs, diagrams, and calculations to analyse your results. Any graphs and diagrams should be included in the appropriate place in the report, and not attached as appendices at the end. You should also form some conjectures based on your analysis.

Conclusion: You should summarise your investigation, giving a clear response to your aim. You should also reflect on your exploration. Limitations and sources of error could be discussed, as well as potential for further exploration.

The exploration will be assessed against five assessment criteria. Refer to the Mathematics HL Subject Guide for more details.

The following two pages contain a short extract of a student's report, used with the permission of Wan Lin Oh. Please note that there is no single structure which must be followed to write a mathematical exploration. The extract displayed is only intended to illustrate some of the key features which should be included.

The electronic version of this extract contains further information, and can be accessed by clicking the icon alongside.

ELECTRONIC EXTRACT

This is an **extract** of a mathematics report used to demonstrate the components of a written report.

1. Title (and author)
A clear and concise description of the report

Population Trends in China

Written by Wan Lin Oh

2. Introduction
Outline the purpose of the task.
Include background information and definitions of key terms or variables used.

Aim

To determine the model that best fits the population of China from 1950 to 2008 by investigating different functions that best model the population of China from 1950 to 1995 (refer to *Table 1*) initially, and then re-evaluating and modifying this model to include additional data from 1983 to 2008.

Rationale

The history class had been discussing the impetus for, and the political, cultural and social implications of China's "One Child Policy", introduced in 1978 for implementation in 1979[1]. This aroused the author's curiosity about the measurable impact that the policy may have had on China's population.

Table 1: Population of China from 1950 to 1995

Year (t)	1950	1955	1960	1965	1970	1975	1980	1985	1990	1995
Population in millions (P)	554.8	609.0	657.5	729.2	830.7	927.8	998.9	1070.0	1155.3	1220.5

Choosing a model

Values from *Table 1* were used to create *Graph 1*:

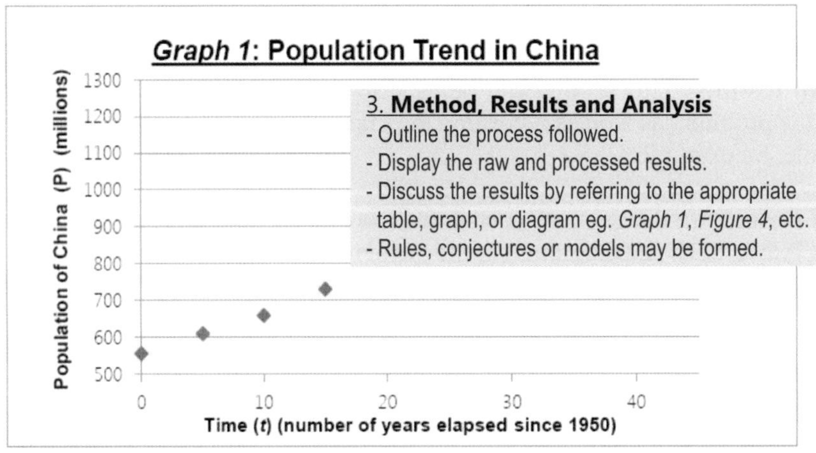

3. Method, Results and Analysis
- Outline the process followed.
- Display the raw and processed results.
- Discuss the results by referring to the appropriate table, graph, or diagram eg. *Graph 1*, *Figure 4*, etc.
- Rules, conjectures or models may be formed.

Graph 1 illustrates a positive correlation between the population of China and the number of years since 1950. This means that as time increases, the population of China also increases. *Graph 1* clearly shows that the model is not a linear function, because the graph has turning points and there is no fixed increase in corresponding to a fixed increase in P. Simple observation reveals that it is not a straight line. In addition, *Graph 1* illustrates that the function is not a power function ($P = at^b$) because it does not meet the requirements of a power function; for all positive b values, a power model must go through the origin, however *Graph 1* shows that the model's function does not pass through the origin of (0, 0).

There is a high possibility that the model could be a polynomial function because *Graph 1* indicates that there are turning point(s). A cubic and a quadratic function were then determined and compared.

Analytical Determination of Polynomial Model

As there is a high possibility that the model could be a cubic function (3rd degree polynomial function), an algebraic method can be used in order to determine the equation of the function. In order to determine this cubic equation, four points from the model will be used as there are four...

The middle section of this report has been omitted.

Conclusion

The aim of this investigation was to investigate a model that best fits the given data from 1950 to 2008. It was initially found that a 3rd degree polynomial function and an exponential function have a good possibility of fitting the given data from *Table 1* which is from year 1950 to 1995 by observing the data plots on the graph.
A cubic function (3rd degree polynomial function) was chosen eventually and consequently an algebraic method using simultaneous equations was developed to produce the equation of the function. Through this method, the equation of the cubic was deduced to be $P(t) = -0.007081t^3 + 0.5304t^2 + 5.263t + 554.8$. In addition, the use of technology was also included in this investigation to further enhance the development of the task by graphing the cubic function to determine how well the cubic function fitted the original data. The cubic graph was then compared with a quadratic function graph of $P(t) = 0.13t^2 + 8.95t + 554.8$. Ultimately, the cubic function was seen as the better fit compared to the quadratic model.

A researcher suggests that the population, P at time t can be modelled by $P(t) = \frac{K}{1+Le^{-Mt}}$. With the use of GeoGebra the parameters, K, L and M were found by trial and error to be 1590, 1.97 and 0.04 respectively. This consequently led to the equation of the logistic function of $P(t) = \frac{1590}{1+1.97e^{-0.04t}}$.

From the comparison of both the cubic and the logistic model, the cubic function was established to be a more accurate model for the given 1950 – 1995 data because the data points matched the model better, however the logistic model produced more likely values under extrapolation.
Additional data on population trends in China frc ... d by the International Monetary Fund (IMF) was given in Tabl ... graphed with the additional data points and compared. It v ... er model compared to the cubic model because it was able ... ulation of China much more precisely.
Subsequently a piecewise function was used becau ... have two distinctly different parts, each with a correspondir ... e domain $0 < t \leq 30$. The researcher's model was modified to fit the data for $30 < t \leq 58$.

4. Conclusion and Limitations
- Summarise findings in response to the stated aim including restating any rules, conjectures, or models.
- Comment on any limitations to the approach used or of the findings.
- Considerations of extensions and connections to personal/previous knowledge may also contextualise the significance of the exploration.

The piecewise function was then defined as

$$P(t) \begin{cases} -0.007081t^3 + 0.5304t^2 + 5.263t + 554.8 & 0 < t \leq 30 \\ \dfrac{1590}{1+1.97e^{-0.04t}} & 30 < t \leq 58 \end{cases}$$

This modified model matched the data points of the population of China from 1950 to 2008 closely; the model also passed through both the minimum and the maximum of the given data. In addition, the modified model exhibited good long-term behaviour and was able to predict a sensible result beyond the known values.

Limitations

In this investigation, there were several limitations that should be taken into account. Firstly, the best fit model which is the piecewise function model does not take into account the possibility of natural disasters or diseases that may occur in China in the future which will lead to a mass decrease in population. Furthermore, the model also does not consider the population pressures in China such as the one child policy. The one child policy introduced in 1978 but applied in 1979 would cause a decrease in the population in the long term. It is shown in *Graph 14* that after 1979 (P_7), the rate at which the Chinese population is increasing is slower compared to the previous years. This is because this policy leads to an increase in the abortion rate due to many families' preference for males, as males are able to take over the family name. This will consequently lead to a gender imbalance, causing a decrease in population because of the increasing difficulty for Chinese males to find partners. In addition, the model of best fit does not consider the ... countries, allowing more Chinese people to live longer, which wil ... term.

5. References and acknowledgements
A list of sources of information either footnoted on the appropriate page or given in a bibliography at the end of the report.

[1]http://geography.about.com/od/populationgeography/a/onechild.htm

BACKGROUND KNOWLEDGE

Before starting this course you can make sure that you have a good understanding of the necessary background knowledge. Click on the icon alongside to obtain a printable set of exercises and answers on this background knowledge.

BACKGROUND KNOWLEDGE

Click on the icon to access a background chapter about matrices.

MATRICES

Click on the icon to access a statistics revision chapter.

STATISTICS REVISION

Click on the icon to access printable facts about number sets.

NUMBER SETS

Click on the icon to access a printable summary of circle properties.

CIRCLE PROPERTIES

Click on the icon to access a printable summary of measurement facts.

MEASUREMENT FACTS

GRAPHICS CALCULATOR INSTRUCTIONS

Printable graphics calculator instruction booklets are available for the **Casio fx-9860G Plus**, **Casio fx-CG20**, **TI-84 Plus**, and the **TI-*n*spire**. Click on the relevant icon below.

CASIO fx-9860G Plus

CASIO fx-CG20

TI-84 Plus

TI-*n*spire

When additional calculator help may be needed, specific instructions can be printed from icons within the text.

Chapter 1

Quadratics

Syllabus reference: 2.2, 2.5, 2.6, 2.7

Contents:

OPENING PROBLEM

Abiola and Badrani are standing 40 metres apart, throwing a ball between them. When Abiola throws the ball, it travels in a smooth arc. At the time when the ball has travelled x metres horizontally towards Badrani, its height is y metres.

SIMULATION

GRAPHICS CALCULATOR INSTRUCTIONS

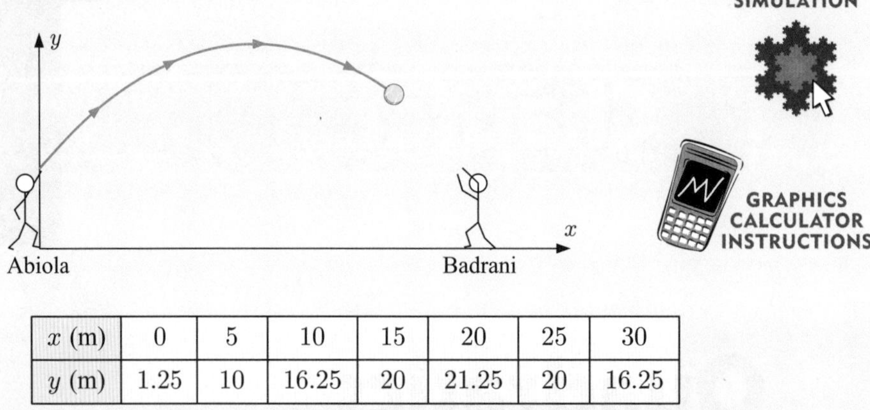

x (m)	0	5	10	15	20	25	30
y (m)	1.25	10	16.25	20	21.25	20	16.25

Things to think about:

- **a** Use technology to plot these points.
- **b** What *shape* is the graph of y against x?
- **c** What is the maximum height reached by the ball?
- **d** What *formula* gives the height of the ball when it has travelled x metres horizontally towards Badrani?
- **e** Will the ball reach Badrani before it bounces?

HISTORICAL NOTE

Galileo Galilei (1564 - 1642) was born in Pisa, Tuscany. He was a philosopher who played a significant role in the scientific revolution of that time.

Within his research he conducted a series of experiments on the paths of projectiles, attempting to find a mathematical description of falling bodies.

Two of Galileo's experiments consisted of rolling a ball down a grooved ramp that was placed at a fixed height above the floor and inclined at a fixed angle to the horizontal. In one experiment the ball left the end of the ramp and descended to the floor. In the second, a horizontal shelf was placed at the end of the ramp, and the ball travelled along this shelf before descending to the floor.

Galileo

In each experiment Galileo altered the release height h of the ball and measured the distance d the ball travelled before landing. The units of measurement were called 'punti' (points).

In both experiments Galileo found that once the ball left the ramp or shelf, its path was *parabolic* and could therefore be modelled by a *quadratic* function.

A QUADRATIC EQUATIONS

A **quadratic equation** is an equation of the form $ax^2 + bx + c = 0$ where a, b and c are constants, $a \neq 0$.

A **quadratic function** is a function of the form $y = ax^2 + bx + c$, $a \neq 0$.

Acme Leather Jacket Co. makes and sells x leather jackets each week. Their profit function is given by $P = -12.5x^2 + 550x - 2125$ dollars.

How many jackets must be made and sold each week in order to obtain a weekly profit of \$3000?

Clearly we need to solve the equation:

$$-12.5x^2 + 550x - 2125 = 3000$$

We can rearrange the equation to give

$$12.5x^2 - 550x + 5125 = 0,$$

which is of the form $ax^2 + bx + c = 0$ and is thus a quadratic equation.

SOLVING QUADRATIC EQUATIONS

To solve quadratic equations we have the following methods to choose from:

- **factorise** the quadratic and use the **Null Factor law**:

  If $ab = 0$ then $a = 0$ or $b = 0$.

- **complete the square**
- use **technology**
- use the **quadratic formula**.

The **roots** or **solutions** of $ax^2 + bx + c = 0$ are the values of x which satisfy the equation, or make it true.

For example: Consider $x^2 - 3x + 2 = 0$.

When $x = 2$, $x^2 - 3x + 2 = (2)^2 - 3(2) + 2$
$= 4 - 6 + 2$
$= 0$ ✓

So, $x = 2$ is a root of the equation $x^2 - 3x + 2 = 0$.

SOLVING BY FACTORISATION (REVISION)

Step 1: If necessary, rearrange the equation so one side is zero.

Step 2: Fully factorise the other side.

Step 3: Use the Null Factor law: If $ab = 0$ then $a = 0$ or $b = 0$.

Step 4: Solve the resulting linear equations.

Caution: Do not be tempted to divide both sides by an expression involving x.
If you do this then you may lose one of the solutions.

For example, consider $x^2 = 5x$.

Correct solution	*Incorrect solution*
$x^2 = 5x$	$x^2 = 5x$
$\therefore\ x^2 - 5x = 0$	$\therefore\ \dfrac{x^2}{x} = \dfrac{5x}{x}$
$\therefore\ x(x-5) = 0$	$\therefore\ x = 5$
$\therefore\ x = 0$ or 5	

By dividing both sides by x, we lose the solution $x = 0$.

Example 1 — Self Tutor

Solve for x: **a** $3x^2 + 5x = 0$ **b** $x^2 = 5x + 6$

a
$$3x^2 + 5x = 0$$
$$\therefore\ x(3x+5) = 0$$
$$\therefore\ x = 0 \text{ or } 3x + 5 = 0$$
$$\therefore\ x = 0 \text{ or } x = -\tfrac{5}{3}$$

b
$$x^2 = 5x + 6$$
$$\therefore\ x^2 - 5x - 6 = 0$$
$$\therefore\ (x-6)(x+1) = 0$$
$$\therefore\ x = 6 \text{ or } -1$$

Example 2 — Self Tutor

Solve for x: **a** $4x^2 + 1 = 4x$ **b** $6x^2 = 11x + 10$

a
$$4x^2 + 1 = 4x$$
$$\therefore\ 4x^2 - 4x + 1 = 0$$
$$\therefore\ (2x-1)^2 = 0$$
$$\therefore\ x = \tfrac{1}{2}$$

b
$$6x^2 = 11x + 10$$
$$\therefore\ 6x^2 - 11x - 10 = 0$$
$$\therefore\ (2x-5)(3x+2) = 0$$
$$\therefore\ x = \tfrac{5}{2} \text{ or } -\tfrac{2}{3}$$

Example 3 — Self Tutor

Solve for x: $3x + \dfrac{2}{x} = -7$

$$3x + \frac{2}{x} = -7$$
$$\therefore\ x\left(3x + \frac{2}{x}\right) = -7x \qquad \{\text{multiplying both sides by } x\}$$
$$\therefore\ 3x^2 + 2 = -7x \qquad \{\text{expanding the brackets}\}$$
$$\therefore\ 3x^2 + 7x + 2 = 0 \qquad \{\text{making the RHS } 0\}$$
$$\therefore\ (x+2)(3x+1) = 0 \qquad \{\text{factorising}\}$$
$$\therefore\ x = -2 \text{ or } -\tfrac{1}{3}$$

RHS is short for Right Hand Side.

EXERCISE 1A.1

1 Solve the following by factorisation:

a $4x^2 + 7x = 0$ **b** $3x^2 - 7x = 0$ **c** $2x^2 - 11x = 0$

d $9x = 6x^2$ **e** $x^2 - 5x + 6 = 0$ **f** $x^2 + 21 = 10x$

g $9 + x^2 = 6x$ **h** $x^2 + x = 12$ **i** $x^2 + 8x = 33$

2 Solve the following by factorisation:

a $9x^2 - 12x + 4 = 0$ **b** $2x^2 - 13x - 7 = 0$ **c** $3x^2 = 16x + 12$

d $3x^2 + 5x = 2$ **e** $2x^2 + 3 = 5x$ **f** $3x^2 + 8x + 4 = 0$

g $3x^2 = 10x + 8$ **h** $4x^2 + 4x = 3$ **i** $4x^2 = 11x + 3$

3 Solve for x:

a $(x+1)^2 = 2x^2 - 5x + 11$ **b** $5 - 4x^2 = 3(2x+1) + 2$

c $2x - \frac{1}{x} = -1$ **d** $\frac{x+3}{1-x} = -\frac{9}{x}$

SOLVING BY 'COMPLETING THE SQUARE'

Quadratics such as $x^2 + 4x + 1$ cannot be factorised by simple factorisation. An alternative way to solve equations like $x^2 + 4x + 1 = 0$ is by 'completing the square'.

Equations of the form $ax^2 + bx + c = 0$ can be converted to the form $(x+p)^2 = q$ from which the solutions are easy to obtain.

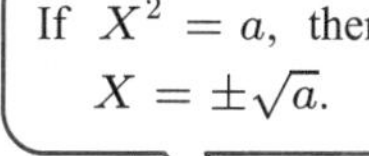
If $X^2 = a$, then $X = \pm\sqrt{a}$.

Example 4 Self Tutor

Solve exactly for x: **a** $(x+2)^2 = 7$ **b** $(x-1)^2 = -5$

a $(x+2)^2 = 7$
$\therefore\ x + 2 = \pm\sqrt{7}$
$\therefore\ x = -2 \pm \sqrt{7}$

b $(x-1)^2 = -5$
has no real solutions since the square $(x-1)^2$ cannot be negative.

Example 5 Self Tutor

Solve for exact values of x: $x^2 + 4x + 1 = 0$

$x^2 + 4x + 1 = 0$
$\therefore\ x^2 + 4x = -1$ {put the constant on the RHS}
$\therefore\ x^2 + 4x + 2^2 = -1 + 2^2$ {completing the square}
$\therefore\ (x+2)^2 = 3$ {factorising LHS}
$\therefore\ x + 2 = \pm\sqrt{3}$
$\therefore\ x = -2 \pm \sqrt{3}$

The squared number we add to both sides is $\left(\frac{\text{coefficient of } x}{2}\right)^2$

Example 6 **Self Tutor**

Solve exactly for x: $-3x^2+12x+5=0$

$-3x^2+12x+5=0$

$\therefore\ x^2-4x-\frac{5}{3}=0$ {dividing both sides by -3}

$\therefore\ x^2-4x=\frac{5}{3}$ {putting the constant on the RHS}

$\therefore\ x^2-4x+2^2=\frac{5}{3}+2^2$ {completing the square}

$\therefore\ (x-2)^2=\frac{17}{3}$ {factorising LHS}

$\therefore\ x-2=\pm\sqrt{\frac{17}{3}}$

$\therefore\ x=2\pm\sqrt{\frac{17}{3}}$

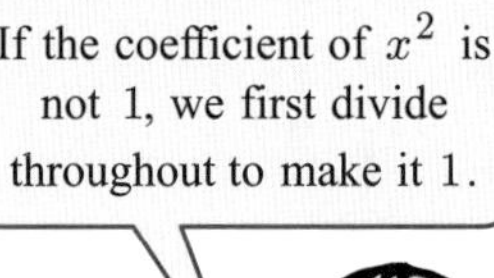

If the coefficient of x^2 is not 1, we first divide throughout to make it 1.

EXERCISE 1A.2

1 Solve exactly for x:

a $(x+5)^2=2$ b $(x+6)^2=-11$ c $(x-4)^2=8$

d $3(x-2)^2=18$ e $(2x+1)^2=3$ f $(1-3x)^2-7=0$

2 Solve exactly by completing the square:

a $x^2-4x+1=0$ b $x^2+6x+2=0$ c $x^2-14x+46=0$

d $x^2=4x+3$ e $x^2+6x+7=0$ f $x^2=2x+6$

g $x^2+6x=2$ h $x^2+10=8x$ i $x^2+6x=-11$

3 Solve exactly by completing the square:

a $2x^2+4x+1=0$ b $2x^2-10x+3=0$ c $3x^2+12x+5=0$

d $3x^2=6x+4$ e $5x^2-15x+2=0$ f $4x^2+4x=5$

SOLVING USING TECHNOLOGY

You can use your graphics calculator to solve quadratic equations.

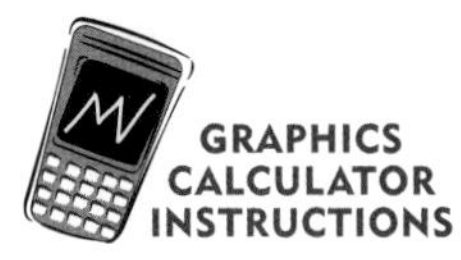

If the right hand side is zero, you can graph the expression on the left hand side. The x-intercepts of the graph will be the solutions to the quadratic.

If the right hand side is non-zero, you can either:

- rearrange the equation so the right hand side is zero, then graph the expression and find the x-intercepts, or
- graph the expressions on the left and right hand sides on the same set of axes, then find the x-coordinates of the point where they meet.

Use technology to check some of your answers to **Exercise 1A.2**.

THE QUADRATIC FORMULA

In many cases, factorising a quadratic equation or completing the square can be long or difficult. We can instead use the **quadratic formula**:

$$\text{If}\quad \boldsymbol{ax^2+bx+c=0},\quad \text{then}\quad \boldsymbol{x=\frac{-b\pm\sqrt{b^2-4ac}}{2a}}.$$

Proof:

If $ax^2+bx+c=0,\quad a\neq 0$

then $x^2+\frac{b}{a}x+\frac{c}{a}=0$ {dividing each term by a, as $a\neq 0$}

$\therefore\quad x^2+\frac{b}{a}x=-\frac{c}{a}$

$\therefore\quad x^2+\frac{b}{a}x+\left(\frac{b}{2a}\right)^2=-\frac{c}{a}+\left(\frac{b}{2a}\right)^2$ {completing the square on LHS}

$\therefore\quad \left(x+\frac{b}{2a}\right)^2=\frac{b^2-4ac}{4a^2}$ {factorising}

$\therefore\quad x+\frac{b}{2a}=\pm\sqrt{\frac{b^2-4ac}{4a^2}}$

$\therefore\quad x=\frac{-b\pm\sqrt{b^2-4ac}}{2a}$

For example, consider the Acme Leather Jacket Co. equation from page **19**.

We need to solve $12.5x^2-550x+5125=0$ for which $a=12.5$, $b=-550$, and $c=5125$.

$$\therefore\quad x=\frac{550\pm\sqrt{(-550)^2-4(12.5)(5125)}}{2(12.5)}$$

$$=\frac{550\pm\sqrt{46\,250}}{25}$$

$$\approx 30.60 \text{ or } 13.40$$

Trying to factorise this equation or using 'completing the square' would not be easy.

However, for this application the number of jackets x needs to be a whole number, so $x=13$ or 31 would produce a profit of around \$3000 each week.

HISTORICAL NOTE — THE QUADRATIC FORMULA

Thousands of years ago, people knew how to calculate the area of a shape given its side lengths. When they wanted to find the side lengths necessary to give a certain area, however, they ended up with a quadratic equation which they needed to solve.

The first known solution of a quadratic equation is written on the Berlin Papyrus from the Middle Kingdom (2160 - 1700 BC) in Egypt. By 400 BC, the Babylonians were using the method of 'completing the square'.

Pythagoras and Euclid both used geometric methods to explore the problem. Pythagoras noted that the square root wasn't always an integer, but he refused to accept that irrational solutions existed. Euclid also discovered that the square root wasn't always rational, but concluded that irrational numbers *did* exist.

A major jump forward was made in India around 700 AD, when Hindu mathematician Brahmagupta devised a general (but incomplete) solution for the quadratic equation $ax^2 + bx = c$ which was equivalent to $x = \dfrac{\sqrt{4ac+b^2}-b}{2a}$. Taking into account the sign of c, this is one of the two solutions we know today.

Brahmagupta also added *zero* to our number system!

The final, complete solution as we know it today first came around 1100 AD, by another Hindu mathematician called Baskhara. He was the first to recognise that any positive number has two square roots, which could be negative or irrational. In fact, the quadratic formula is known in some countries today as 'Baskhara's Formula'.

While the Indians had knowledge of the quadratic formula even at this early stage, it took somewhat longer for the quadratic formula to arrive in Europe.

Muhammad Al-Khwarizmi

Around 820 AD, the Islamic mathematician Muhammad bin Musa Al-Khwarizmi, who was familiar with the work of Brahmagupta, recognised that for a quadratic equation to have real solutions, the value b^2-4ac could not be negative. Al-Khwarizmi's work was brought to Europe by the Jewish mathematician and astronomer Abraham bar Hiyya (also known as Savasorda) who lived in Barcelona around 1100.

By 1545, Girolamo Cardano had blended the algebra of Al-Khwarizmi with the Euclidean geometry. His work allowed for the existence of complex or imaginary roots, as well as negative and irrational roots.

From the name Al-Khwarizmi we get the word 'algorithm'.

At the end of the 16th Century the mathematical notation and symbolism was introduced by François Viète in France.

In 1637, when René Descartes published *La Géométrie*, the quadratic formula adopted the familiar form $x = \dfrac{-b \pm \sqrt{b^2-4ac}}{2a}$.

Example 7 **Self Tutor**

Solve for x:

a $x^2 - 2x - 6 = 0$ **b** $2x^2 + 3x - 6 = 0$

a $x^2 - 2x - 6 = 0$ has $a = 1,\ b = -2,\ c = -6$

$$\therefore\ x = \frac{-(-2) \pm \sqrt{(-2)^2 - 4(1)(-6)}}{2(1)}$$

$$\therefore\ x = \frac{2 \pm \sqrt{4+24}}{2}$$

$$\therefore\ x = \frac{2 \pm \sqrt{28}}{2}$$

$$\therefore\ x = \frac{2 \pm 2\sqrt{7}}{2}$$

$$\therefore\ x = 1 \pm \sqrt{7}$$

b $2x^2 + 3x - 6 = 0$ has $a = 2,\ b = 3,\ c = -6$

$$\therefore\ x = \frac{-3 \pm \sqrt{3^2 - 4(2)(-6)}}{2(2)}$$

$$\therefore\ x = \frac{-3 \pm \sqrt{9+48}}{4}$$

$$\therefore\ x = \frac{-3 \pm \sqrt{57}}{4}$$

EXERCISE 1A.3

1 Use the quadratic formula to solve exactly for x:

a $x^2-4x-3=0$ b $x^2+6x+7=0$ c $x^2+1=4x$

d $x^2+4x=1$ e $x^2-4x+2=0$ f $2x^2-2x-3=0$

g $(3x+1)^2=-2x$ h $(x+3)(2x+1)=9$ i $x^2-2\sqrt{2}x+2=0$

2 Rearrange the following equations so they are written in the form $ax^2+bx+c=0$, then use the quadratic formula to solve exactly for x.

a $(x+2)(x-1)=2-3x$ b $(2x+1)^2=3-x$ c $(x-2)^2=1+x$

d $\dfrac{x-1}{2-x}=2x+1$ e $x-\dfrac{1}{x}=1$ f $2x-\dfrac{1}{x}=3$

B THE DISCRIMINANT OF A QUADRATIC

In the quadratic formula, the quantity b^2-4ac under the square root sign is called the **discriminant**.

The symbol **delta** Δ is used to represent the discriminant, so $\Delta=b^2-4ac$.

The quadratic formula becomes $x=\dfrac{-b\pm\sqrt{\Delta}}{2a}$ where Δ replaces b^2-4ac.

- If $\Delta>0$, $\sqrt{\Delta}$ is a positive real number, so there are **two distinct real roots**
$$x=\frac{-b+\sqrt{\Delta}}{2a} \quad\text{and}\quad x=\frac{-b-\sqrt{\Delta}}{2a}$$
- If $\Delta=0$, $x=\dfrac{-b}{2a}$ is the **only solution** (a **repeated** or **double root**)
- If $\Delta<0$, $\sqrt{\Delta}$ is not a real number and so there are **no real roots**.
- If a, b, and c are rational and Δ is a **square** then the equation has two rational roots which can be found by factorisation.

Factorisation of quadratic	*Roots of quadratic*	*Discriminant value*
two distinct linear factors	two real distinct roots	$\Delta>0$
two identical linear factors	two identical real roots (repeated)	$\Delta=0$
unable to factorise	no real roots	$\Delta<0$

Example 8 **Self Tutor**

Use the discriminant to determine the nature of the roots of:

a $2x^2-2x+3=0$ b $3x^2-4x-2=0$

a $\Delta=b^2-4ac$
$=(-2)^2-4(2)(3)$
$=-20$
Since $\Delta<0$, there are no real roots.

b $\Delta=b^2-4ac$
$=(-4)^2-4(3)(-2)$
$=40$
Since $\Delta>0$, but 40 is not a square, there are 2 distinct irrational roots.

Example 9 — Self Tutor

Consider $x^2-2x+m=0$. Find the discriminant Δ, and hence find the values of m for which the equation has:

a a repeated root **b** 2 distinct real roots **c** no real roots.

$x^2-2x+m=0$ has $a=1$, $b=-2$, and $c=m$

$$\begin{aligned}\therefore\ \Delta &= b^2-4ac\\ &= (-2)^2-4(1)(m)\\ &= 4-4m\end{aligned}$$

a For a repeated root

$$\begin{aligned}\Delta &= 0\\ \therefore\ 4-4m &= 0\\ \therefore\ 4 &= 4m\\ \therefore\ m &= 1\end{aligned}$$

b For 2 distinct real roots

$$\begin{aligned}\Delta &> 0\\ \therefore\ 4-4m &> 0\\ \therefore\ -4m &> -4\\ \therefore\ m &< 1\end{aligned}$$

c For no real roots

$$\begin{aligned}\Delta &< 0\\ \therefore\ 4-4m &< 0\\ \therefore\ -4m &< -4\\ \therefore\ m &> 1\end{aligned}$$

Example 10 — Self Tutor

Consider the equation $kx^2+(k+3)x=1$. Find the discriminant Δ and draw its sign diagram. Hence, find the value of k for which the equation has:

a two distinct real roots **b** two real roots

c a repeated root **d** no real roots.

$kx^2+(k+3)x-1=0$ has $a=k$, $b=(k+3)$, and $c=-1$

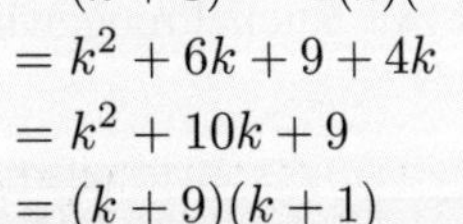

$$\begin{aligned}\therefore\ \Delta &= b^2-4ac\\ &= (k+3)^2-4(k)(-1)\\ &= k^2+6k+9+4k\\ &= k^2+10k+9\\ &= (k+9)(k+1)\end{aligned}$$

So, Δ has sign diagram:

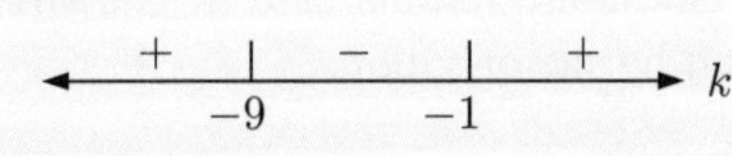

a For two distinct real roots, $\Delta>0$ $\therefore$ $k<-9$ or $k>-1$, $k\neq 0$.

b For two real roots, $\Delta\geqslant 0$ $\therefore$ $k\leqslant -9$ or $k\geqslant -1$, $k\neq 0$.

c For a repeated root, $\Delta=0$ $\therefore$ $k=-9$ or $k=-1$.

d For no real roots, $\Delta<0$ $\therefore$ $-9<k<-1$.

EXERCISE 1B

1 By using the discriminant only, state the nature of the solutions of:

a $x^2+7x-3=0$ **b** $x^2-3x+2=0$ **c** $3x^2+2x-1=0$

d $5x^2+4x-3=0$ **e** $x^2+x+5=0$ **f** $16x^2-8x+1=0$

2 By using the discriminant only, determine which of the following quadratic equations have rational roots which can be found by factorisation.

a $6x^2-5x-6=0$ **b** $2x^2-7x-5=0$ **c** $3x^2+4x+1=0$

d $6x^2-47x-8=0$ **e** $4x^2-3x+2=0$ **f** $8x^2+2x-3=0$

3 For each of the following quadratic equations, determine the discriminant Δ in simplest form and draw its sign diagram. Hence find the value(s) of m for which the equation has:

i a repeated root **ii** two distinct real roots **iii** no real roots.

a $x^2+4x+m=0$ **b** $mx^2+3x+2=0$ **c** $mx^2-3x+1=0$

4 For each of the following quadratic equations, find the discriminant Δ and hence draw its sign diagram. Find all values of k for which the equation has:

i two distinct real roots **ii** two real roots **iii** a repeated root **iv** no real roots.

a $2x^2+kx-k=0$ **b** $kx^2-2x+k=0$

c $x^2+(k+2)x+4=0$ **d** $2x^2+(k-2)x+2=0$

e $x^2+(3k-1)x+(2k+10)=0$ **f** $(k+1)x^2+kx+k=0$

C THE SUM AND PRODUCT OF THE ROOTS

If $ax^2+bx+c=0$ has roots α and β, then $\alpha+\beta=-\dfrac{b}{a}$ and $\alpha\beta=\dfrac{c}{a}$.

For example: If α and β are the roots of $2x^2-2x-1=0$ then $\alpha+\beta=1$ and $\alpha\beta=-\frac{1}{2}$.

Proof: If α and β are the roots of $ax^2+bx+c=0$,

then $ax^2+bx+c=a(x-\alpha)(x-\beta)$

$=a(x^2-[\alpha+\beta]x+\alpha\beta)$

$\therefore\ x^2+\dfrac{b}{a}x+\dfrac{c}{a}=x^2-[\alpha+\beta]x+\alpha\beta$

Equating coefficients,

$\alpha+\beta=-\dfrac{b}{a}$ and $\alpha\beta=\dfrac{c}{a}$.

Example 11 **Self Tutor**

Find the sum and product of the roots of $25x^2-20x+1=0$.
Check your answer by solving the quadratic.

If α and β are the roots then $\alpha+\beta=-\dfrac{b}{a}=\dfrac{20}{25}=\dfrac{4}{5}$

and $\alpha\beta=\dfrac{c}{a}=\dfrac{1}{25}$

Check: $25x^2-20x+1=0$ has roots

$$\frac{20\pm\sqrt{400-4(25)(1)}}{50}=\frac{20\pm\sqrt{300}}{50}=\frac{20\pm10\sqrt{3}}{50}=\frac{2\pm\sqrt{3}}{5}$$

These have sum $=\dfrac{2+\sqrt{3}}{5}+\dfrac{2-\sqrt{3}}{5}=\dfrac{4}{5}$ ✓

and product $=\left(\dfrac{2+\sqrt{3}}{5}\right)\left(\dfrac{2-\sqrt{3}}{5}\right)=\dfrac{4-3}{25}=\dfrac{1}{25}$ ✓

EXERCISE 1C

1 Find the sum and product of the roots of:

a $3x^2 - 2x + 7 = 0$ **b** $x^2 + 11x - 13 = 0$ **c** $5x^2 - 6x - 14 = 0$

2 The equation $kx^2 - (1+k)x + (3k+2) = 0$ is such that the sum of its roots is twice their product. Find the possible values of k, and the two roots in each case.

3 The quadratic equation $ax^2 - 6x + a - 2 = 0$, $a \neq 0$, has one root which is double the other.

a Let the roots be α and 2α. Hence find two equations involving α.

b Find a and the two roots of the quadratic equation.

4 The quadratic equation $kx^2 + (k-8)x + (1-k) = 0$, $k \neq 0$, has one root which is two more than the other. Find k and the two roots.

5 The roots of the equation $x^2 - 6x + 7 = 0$ are α and β.

Find the simplest quadratic equation with roots $\alpha + \dfrac{1}{\beta}$ and $\beta + \dfrac{1}{\alpha}$.

6 The roots of $2x^2 - 3x - 5 = 0$ are p and q.

Find *all* quadratic equations with roots $p^2 + q$ and $q^2 + p$.

7 $kx^2 + (k+2)x - 3 = 0$ has roots which are real and positive.

Find the possible values that k may have.

D QUADRATIC FUNCTIONS

TERMINOLOGY

The graph of a quadratic function $y = ax^2 + bx + c$, $a \neq 0$ is called a **parabola**.

The point where the graph 'turns' is called the **vertex**.

If the graph opens upwards, the vertex is the **minimum** or **minimum turning point** and the graph is **concave upwards**.

If the graph opens downwards, the vertex is the **maximum** or **maximum turning point** and the graph is **concave downwards**.

The vertical line that passes through the vertex is called the **axis of symmetry**. Every parabola is symmetrical about its axis of symmetry.

The point where the graph crosses the y-axis is the **y-intercept**.

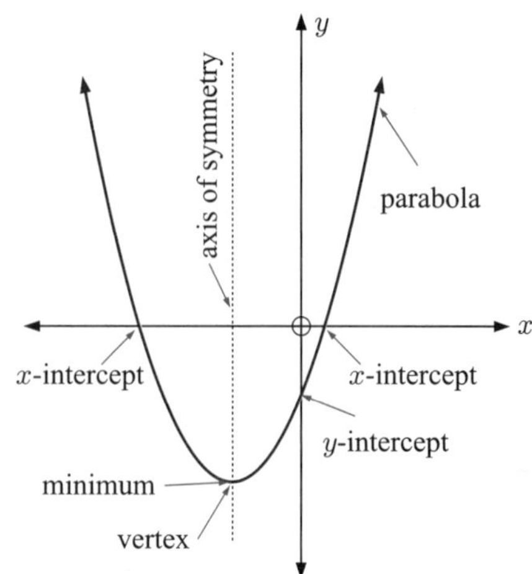

The points (if they exist) where the graph crosses the x-axis are called the **x-intercepts**. They correspond to the **roots** of the equation $y = 0$.

INVESTIGATION 1 GRAPHING $y = a(x-p)(x-q)$

What to do:

1 **a** Use technology to help you to sketch:
$y = (x-1)(x-3)$, $y = 2(x-1)(x-3)$, $y = -(x-1)(x-3)$,
$y = -3(x-1)(x-3)$, and $y = -\frac{1}{2}(x-1)(x-3)$

b Find the x-intercepts for each function in **a**.

c What is the geometrical significance of a in $y = a(x-1)(x-3)$?

2 **a** Use technology to help you to sketch:
$y = 2(x-1)(x-4)$, $y = 2(x-3)(x-5)$, $y = 2(x+1)(x-2)$,
$y = 2x(x+5)$, and $y = 2(x+2)(x+4)$

b Find the x-intercepts for each function in **a**.

c What is the geometrical significance of p and q in $y = 2(x-p)(x-q)$?

3 **a** Use technology to help you to sketch:
$y = 2(x-1)^2$, $y = 2(x-3)^2$, $y = 2(x+2)^2$, $y = 2x^2$

b Find the x-intercepts for each function in **a**.

c What is the geometrical significance of p in $y = 2(x-p)^2$?

4 Copy and complete:

- If a quadratic has the form $y = a(x-p)(x-q)$ then it ...... the x-axis at ......
- If a quadratic has the form $y = a(x-p)^2$ then it ...... the x-axis at ......

INVESTIGATION 2 GRAPHING $y = a(x-h)^2 + k$

What to do:

1 **a** Use technology to help you to sketch:
$y = (x-3)^2 + 2$, $y = 2(x-3)^2 + 2$, $y = -2(x-3)^2 + 2$,
$y = -(x-3)^2 + 2$, and $y = -\frac{1}{3}(x-3)^2 + 2$

b Find the coordinates of the vertex for each function in **a**.

c What is the geometrical significance of a in $y = a(x-3)^2 + 2$?

2 **a** Use technology to help you to sketch:
$y = 2(x-1)^2 + 3$, $y = 2(x-2)^2 + 4$, $y = 2(x-3)^2 + 1$,
$y = 2(x+1)^2 + 4$, $y = 2(x+2)^2 - 5$, and $y = 2(x+3)^2 - 2$

b Find the coordinates of the vertex for each function in **a**.

c What is the geometrical significance of h and k in $y = 2(x-h)^2 + k$?

3 Copy and complete:
If a quadratic has the form $y = a(x-h)^2 + k$ then its vertex has coordinates ......
The graph of $y = a(x-h)^2 + k$ is a ...... of the graph of $y = ax^2$ with vector ......

You should have discovered that a, the coefficient of x^2, controls the width of the graph and whether it opens upwards or downwards.

For a quadratic function $y = ax^2 + bx + c$, $a \neq 0$:

- $a > 0$ produces the shape  called concave up.

  $a < 0$ produces the shape  called concave down.
- If $-1 < a < 1$, $a \neq 0$ the graph is wider than $y = x^2$.
  If $a < -1$ or $a > 1$ the graph is narrower than $y = x^2$.

Quadratic form, $a \neq 0$	*Graph*	*Facts*
• $y = a(x-p)(x-q)$ p, q are real	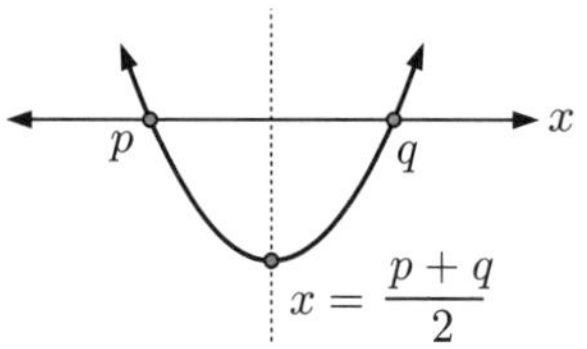	x-intercepts are p and q axis of symmetry is $x = \frac{p+q}{2}$ vertex is $\left(\frac{p+q}{2}, f(\frac{p+q}{2})\right)$
• $y = a(x-h)^2$ h is real	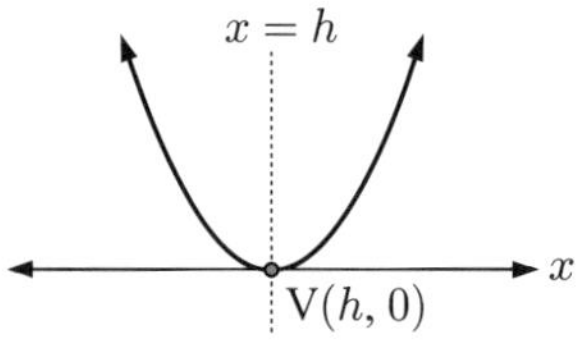	touches x-axis at h axis of symmetry is $x = h$ vertex is $(h, 0)$
• $y = a(x-h)^2 + k$	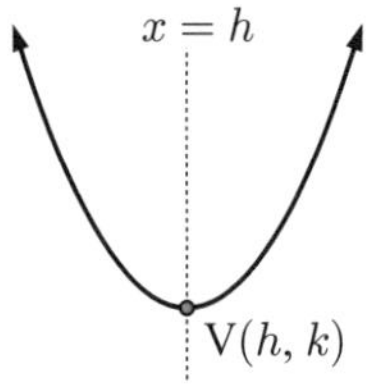	axis of symmetry is $x = h$ vertex is (h, k)
• $y = ax^2 + bx + c$	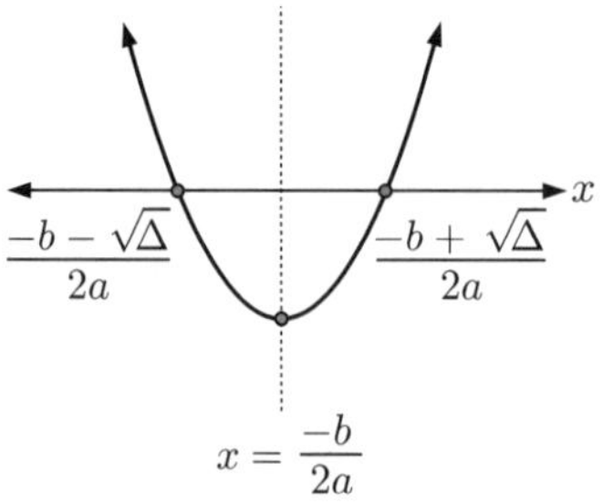	y-intercept c axis of symmetry is $x = \frac{-b}{2a}$ vertex is $\left(-\frac{b}{2a}, c - \frac{b^2}{4a}\right)$ x-intercepts for $\Delta \geqslant 0$ are $\frac{-b \pm \sqrt{\Delta}}{2a}$ where $\Delta = b^2 - 4ac$

$-\frac{b}{2a}$ is the **average** of $\frac{-b-\sqrt{\Delta}}{2a}$ and $\frac{-b+\sqrt{\Delta}}{2a}$

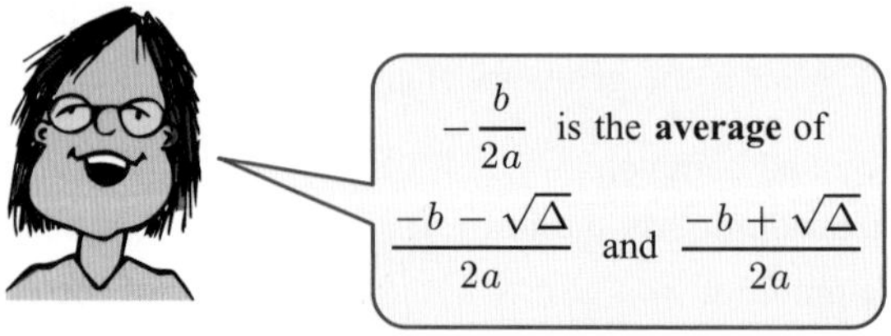

Example 12 — Self Tutor

Using axes intercepts only, sketch the graphs of:

a $y = 2(x+3)(x-1)$ **b** $y = -2(x-1)(x-2)$ **c** $y = \frac{1}{2}(x+2)^2$

a $y = 2(x+3)(x-1)$
has x-intercepts -3, 1
When $x = 0$,
$y = 2(3)(-1)$
$= -6$
$\therefore$ y-intercept is -6

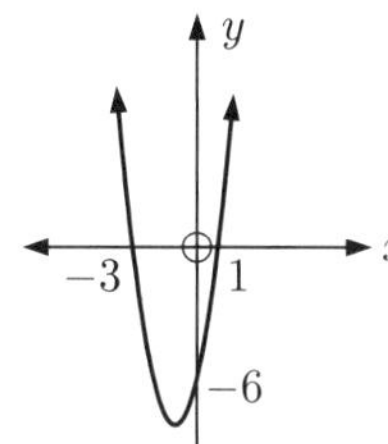

b $y = -2(x-1)(x-2)$
has x-intercepts 1, 2
When $x = 0$,
$y = -2(-1)(-2)$
$= -4$
$\therefore$ y-intercept is -4

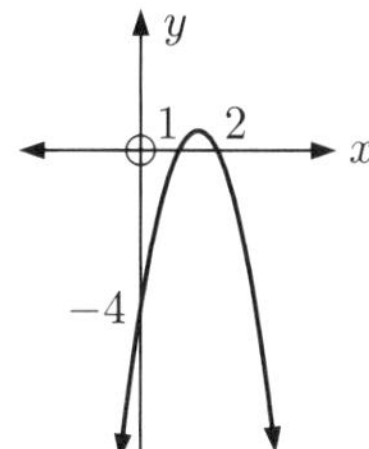

c $y = \frac{1}{2}(x+2)^2$
touches x-axis at -2
When $x = 0$,
$y = \frac{1}{2}(2)^2$
$= 2$
$\therefore$ y-intercept is 2

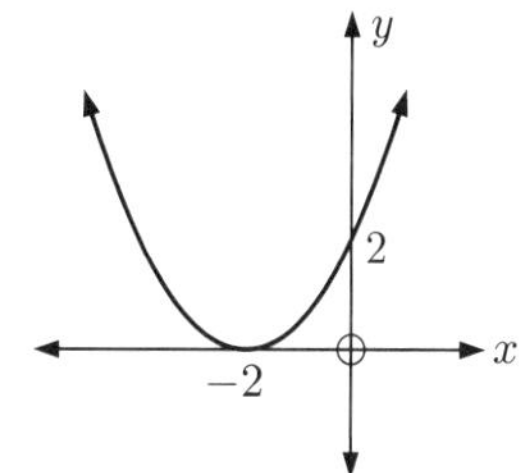

EXERCISE 1D.1

The axis of symmetry is midway between the x-intercepts.

1 Using axes intercepts only, sketch the graphs of:

a $y = (x-4)(x+2)$ **b** $y = -(x-4)(x+2)$

c $y = 2(x+3)(x+5)$ **d** $y = -3x(x+4)$

e $y = 2(x+3)^2$ **f** $y = -\frac{1}{4}(x+2)^2$

2 State the equation of the axis of symmetry for each graph in question **1**.

3 Match each quadratic function with its corresponding graph.

a $y = 2(x-1)(x-4)$ **b** $y = -(x+1)(x-4)$

c $y = (x-1)(x-4)$ **d** $y = (x+1)(x-4)$

e $y = 2(x+4)(x-1)$ **f** $y = -3(x+4)(x-1)$

g $y = -(x-1)(x-4)$ **h** $y = -3(x-1)(x-4)$

A

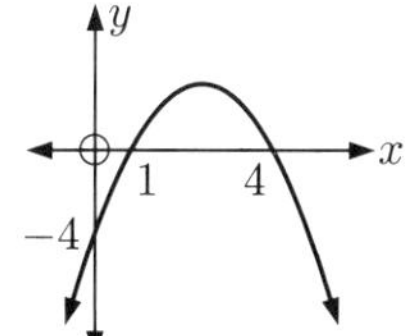

B

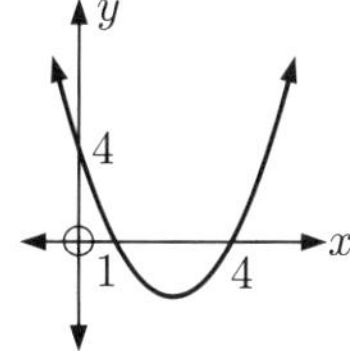

C

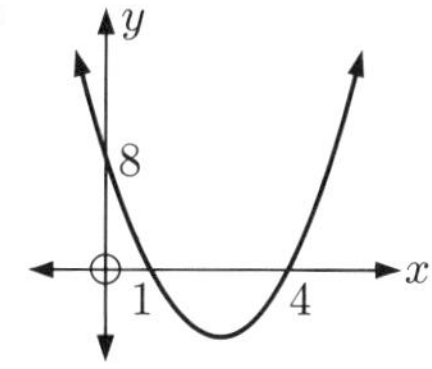

D

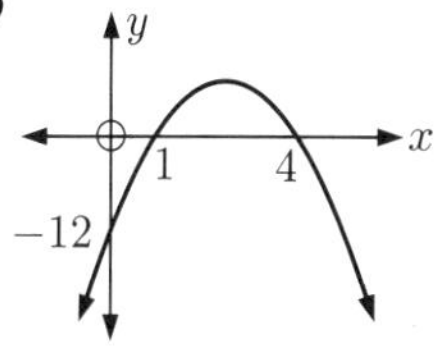

E

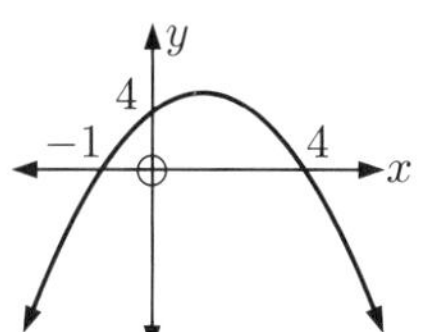

F

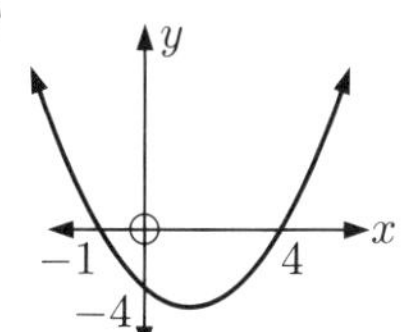

G

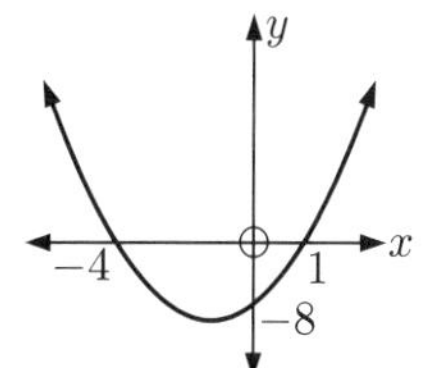

H

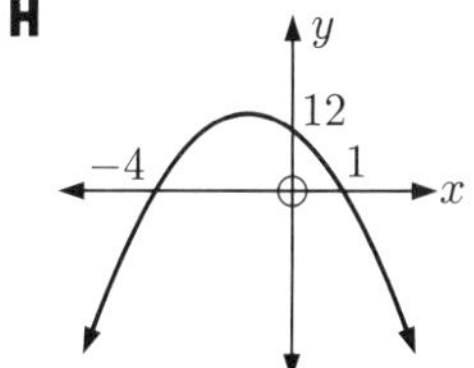

Example 13 **Self Tutor**

Use the vertex, axis of symmetry, and y-intercept to graph $y = -2(x+1)^2 + 4$.

The vertex is $(-1, 4)$.

The axis of symmetry is $x = -1$.

When $x = 0$, $y = -2(1)^2 + 4$
$= 2$

$a < 0$ so the shape is ∩

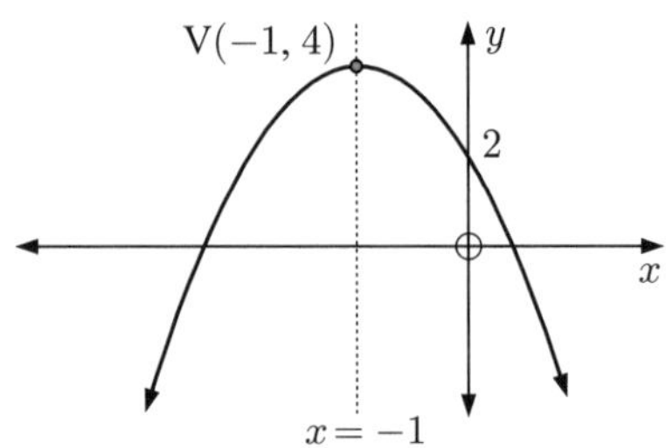

4 Use the vertex, axis of symmetry, and y-intercept to graph:

a $y = (x-1)^2 + 3$ **b** $y = 2(x+2)^2 + 1$ **c** $y = -2(x-1)^2 - 3$

d $y = \frac{1}{2}(x-3)^2 + 2$ **e** $y = -\frac{1}{3}(x-1)^2 + 4$ **f** $y = -\frac{1}{10}(x+2)^2 - 3$

5 Match each quadratic function with its corresponding graph:

a $y = -(x+1)^2 + 3$ **b** $y = -2(x-3)^2 + 2$ **c** $y = x^2 + 2$

d $y = -(x-1)^2 + 1$ **e** $y = (x-2)^2 - 2$ **f** $y = \frac{1}{3}(x+3)^2 - 3$

g $y = -x^2$ **h** $y = -\frac{1}{2}(x-1)^2 + 1$ **i** $y = 2(x+2)^2 - 1$

A

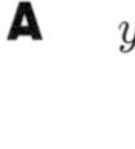

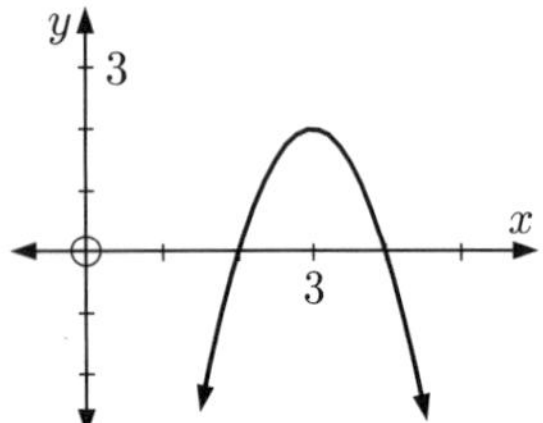

B

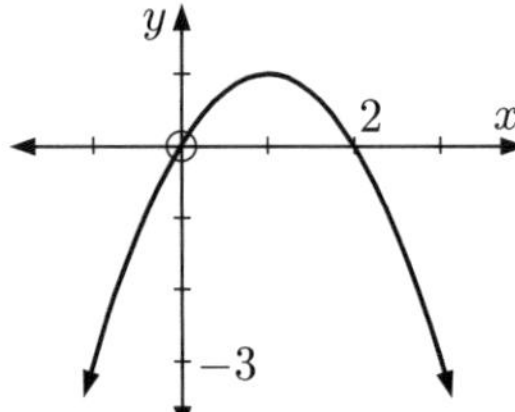

C

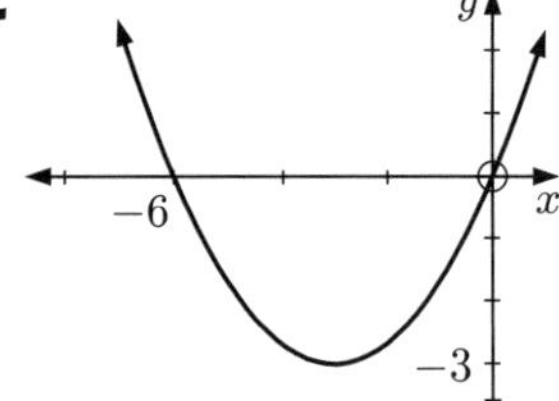

D

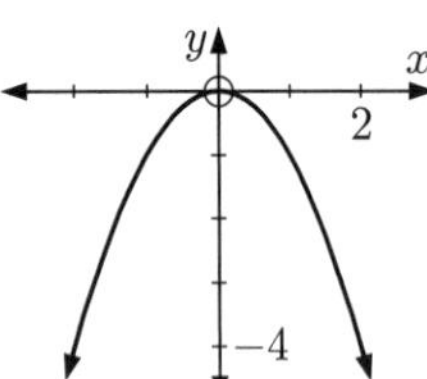

E

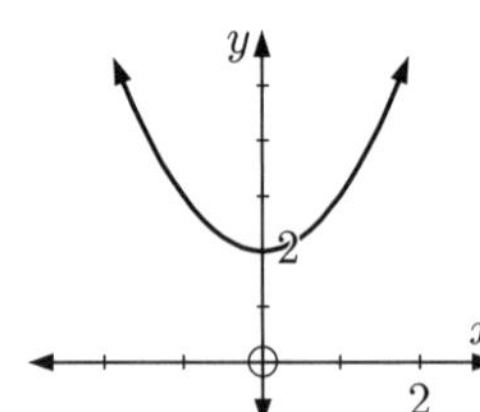

F

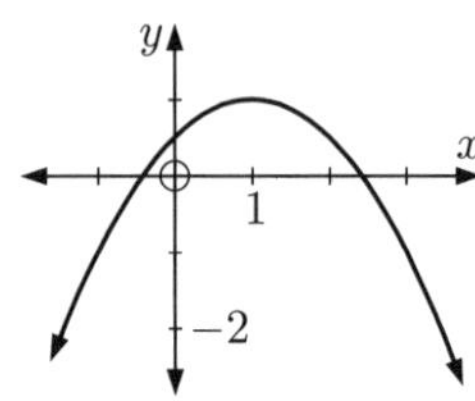

G

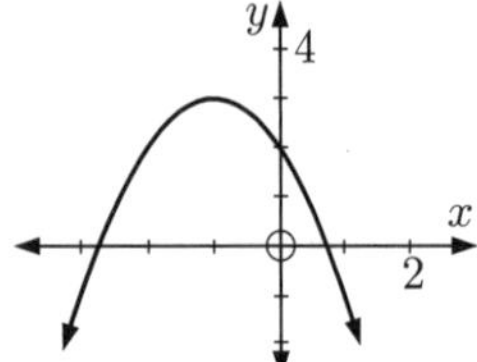

H

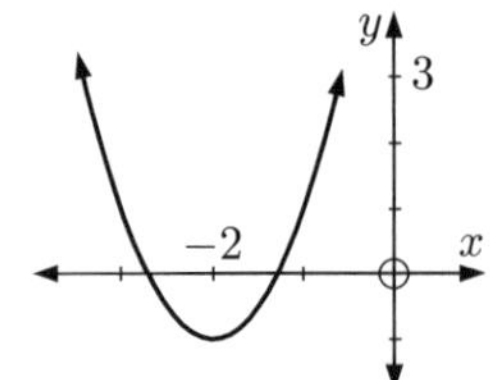

I

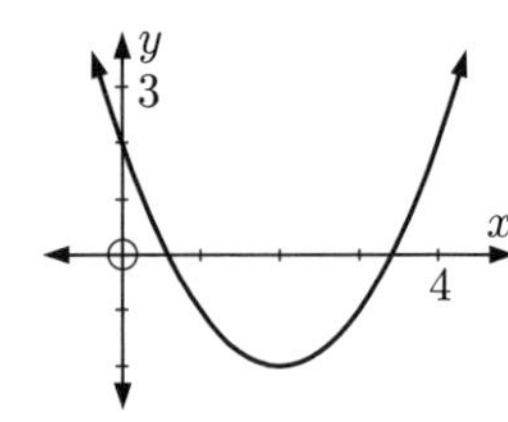

Example 14 **Self Tutor**

Determine the coordinates of the vertex of $y = 2x^2 - 8x + 1$.

$y = 2x^2 - 8x + 1$ has $a = 2$, $b = -8$, and $c = 1$

$$\therefore \frac{-b}{2a} = \frac{-(-8)}{2 \times 2} = 2$$

$\therefore$ the axis of symmetry is $x = 2$

When $x = 2$, $y = 2(2)^2 - 8(2) + 1 = -7$

$\therefore$ the vertex has coordinates $(2, -7)$.

The vertex is the **maximum** or the **minimum** depending on whether the graph is concave down or concave up.

Example 15 **Self Tutor**

Consider the quadratic $y = 2x^2 + 6x - 3$.

a State the axis of symmetry.
b Find the coordinates of the vertex.
c Find the axes intercepts.
d Hence, sketch the quadratic.

$y = 2x^2 + 6x - 3$ has $a = 2$, $b = 6$, and $c = -3$. $a > 0$ so the shape is 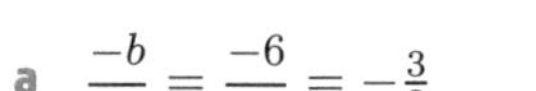

a $\frac{-b}{2a} = \frac{-6}{4} = -\frac{3}{2}$

The axis of symmetry is $x = -\frac{3}{2}$.

b When $x = -\frac{3}{2}$,

$y = 2(-\frac{3}{2})^2 + 6(-\frac{3}{2}) - 3 = -7\frac{1}{2}$

The vertex is $(-\frac{3}{2}, -7\frac{1}{2})$.

c When $x = 0$, $y = -3$

$\therefore$ y-intercept is -3.

When $y = 0$, $2x^2 + 6x - 3 = 0$

Using technology, $x \approx -3.44$ or 0.436

d

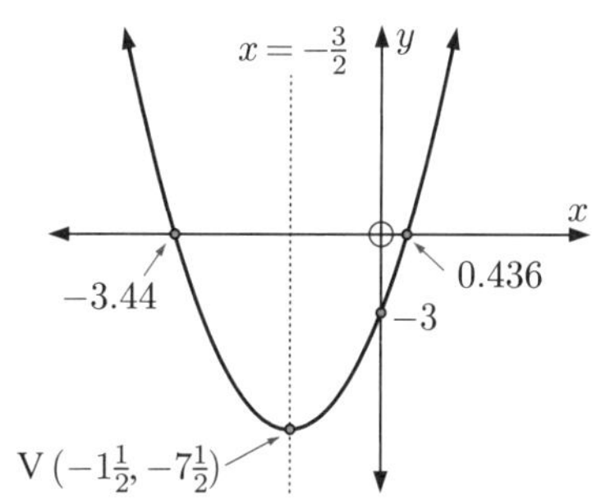

6 Locate the turning point or vertex for each of the following quadratic functions:

a $y = x^2 - 4x + 2$
b $y = 2x^2 + 4$
c $y = -3x^2 + 1$
d $y = -x^2 - 4x - 9$
e $y = 2x^2 + 6x - 1$
f $y = -\frac{1}{2}x^2 + x - 5$

7 Find the x-intercepts for:

a $y = x^2 - 9$
b $y = 2x^2 - 6$
c $y = x^2 + x - 12$
d $y = 4x - x^2$
e $y = -x^2 - 6x - 8$
f $y = -2x^2 - 4x - 2$
g $y = 4x^2 - 24x + 36$
h $y = x^2 - 4x + 1$
i $y = x^2 + 8x + 11$

8 For each of the following quadratics:

i state the axis of symmetry
ii find the coordinates of the vertex
iii find the axes intercepts, if they exist
iv sketch the quadratic.

a $y = x^2 - 2x + 5$
b $y = 2x^2 - 5x + 2$
c $y = -x^2 + 3x - 2$
d $y = -2x^2 + x + 1$
e $y = 6x - x^2$
f $y = -\frac{1}{4}x^2 + 2x + 1$

SKETCHING GRAPHS BY 'COMPLETING THE SQUARE'

If we wish to find the vertex of a quadratic given in general form $y = ax^2 + bx + c$ then one approach is to convert it to the form $y = a(x-h)^2 + k$ where we can read off the coordinates of the vertex (h, k). One way to do this is to '**complete the square**'.

Consider the simple case $y = x^2 - 4x + 1$, for which $a = 1$.

$$y = x^2 - 4x + 1$$
$$\therefore\ y = \underbrace{x^2 - 4x + 2^2}\ \underbrace{+1 - 2^2}$$
$$\therefore\ y = \quad (x-2)^2 \quad\quad -3$$

To obtain the graph of $y = x^2 - 4x + 1$ from the graph of $y = x^2$, we shift it 2 units to the right and 3 units down.

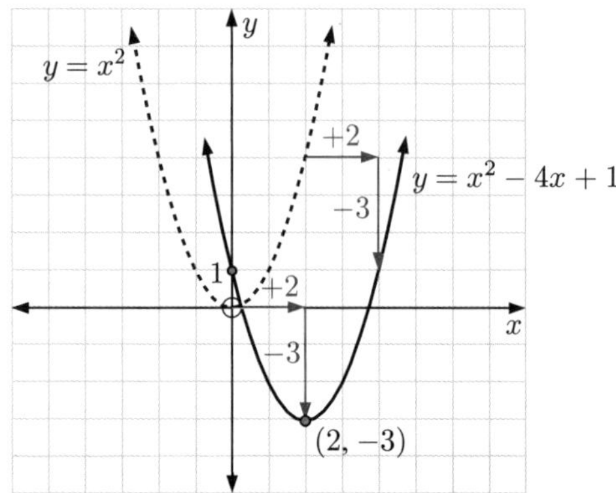

Example 16 — Self Tutor

Write $y = x^2 + 4x + 3$ in the form $y = (x-h)^2 + k$ by 'completing the square'. Hence sketch $y = x^2 + 4x + 3$, stating the coordinates of the vertex.

$$y = x^2 + 4x + 3$$
$$\therefore\ y = x^2 + 4x + 2^2 + 3 - 2^2$$
$$\therefore\ y = (x+2)^2 - 1$$

shift 2 units left   shift 1 unit down

The vertex is $(-2, -1)$ and the y-intercept is 3.

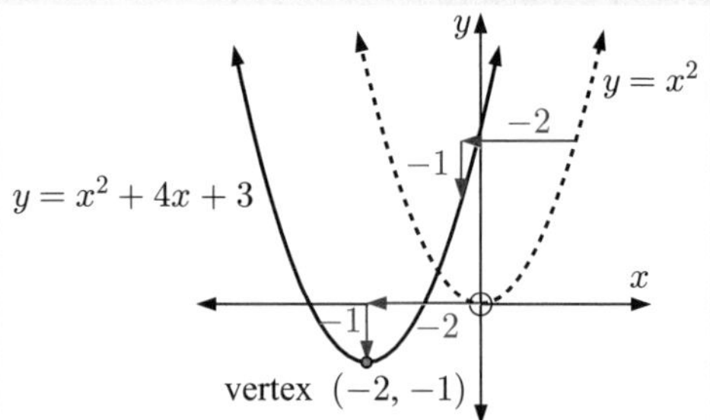

Example 17 — Self Tutor

a Convert $y = 3x^2 - 4x + 1$ to the form $y = a(x-h)^2 + k$ without technology.

b Hence, write down the coordinates of its vertex and sketch the quadratic.

a $y = 3x^2 - 4x + 1$

$= 3[x^2 - \frac{4}{3}x + \frac{1}{3}]$

$= 3[x^2 - 2(\frac{2}{3})x + (\frac{2}{3})^2 - (\frac{2}{3})^2 + \frac{1}{3}]$

$= 3[(x - \frac{2}{3})^2 - \frac{4}{9} + \frac{3}{9}]$

$= 3[(x - \frac{2}{3})^2 - \frac{1}{9}]$

$= 3(x - \frac{2}{3})^2 - \frac{1}{3}$

b The vertex is $(\frac{2}{3}, -\frac{1}{3})$ and the y-intercept is 1.

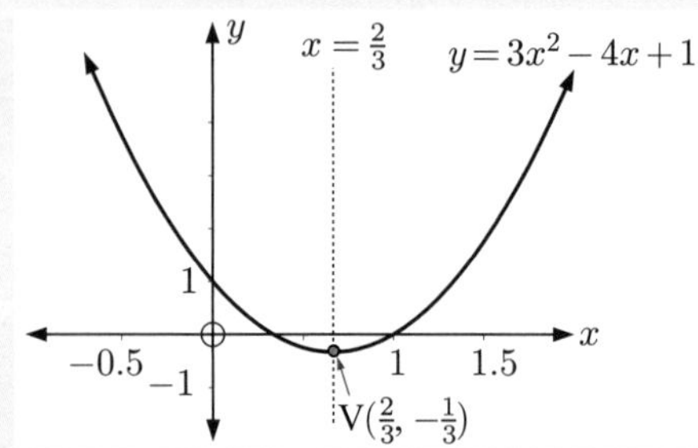

EXERCISE 1D.2

1 Write the following quadratics in the form $y = (x-h)^2 + k$ by 'completing the square'. Hence sketch each function, stating the coordinates of the vertex.

a $y = x^2 - 2x + 3$ **b** $y = x^2 + 4x - 2$ **c** $y = x^2 - 4x$
d $y = x^2 + 3x$ **e** $y = x^2 + 5x - 2$ **f** $y = x^2 - 3x + 2$
g $y = x^2 - 6x + 5$ **h** $y = x^2 + 8x - 2$ **i** $y = x^2 - 5x + 1$

2 For each of the following quadratics:

- **i** Write the quadratic in the form $y = a(x-h)^2 + k$ without using technology.
- **ii** State the coordinates of the vertex.
- **iii** Find the y-intercept.
- **iv** Sketch the graph of the quadratic.
- **v** Use technology to check your answers.

a $y = 2x^2 + 4x + 5$ **b** $y = 2x^2 - 8x + 3$
c $y = 2x^2 - 6x + 1$ **d** $y = 3x^2 - 6x + 5$
e $y = -x^2 + 4x + 2$ **f** $y = -2x^2 - 5x + 3$

ACTIVITY

Click on the icon to run a card game for quadratic functions.

THE DISCRIMINANT AND THE QUADRATIC GRAPH

The discriminant of the quadratic equation $ax^2 + bx + c = 0$ is $\Delta = b^2 - 4ac$.

We used Δ to determine the number of real roots of the equation. If they exist, these roots correspond to zeros of the quadratic $y = ax^2 + bx + c$. Δ therefore tells us about the relationship between a quadratic function and the x-axis.

The graphs of $y = x^2 - 2x + 3$, $y = x^2 - 2x + 1$, and $y = x^2 - 2x - 3$ all have the same axis of symmetry, $x = 1$.

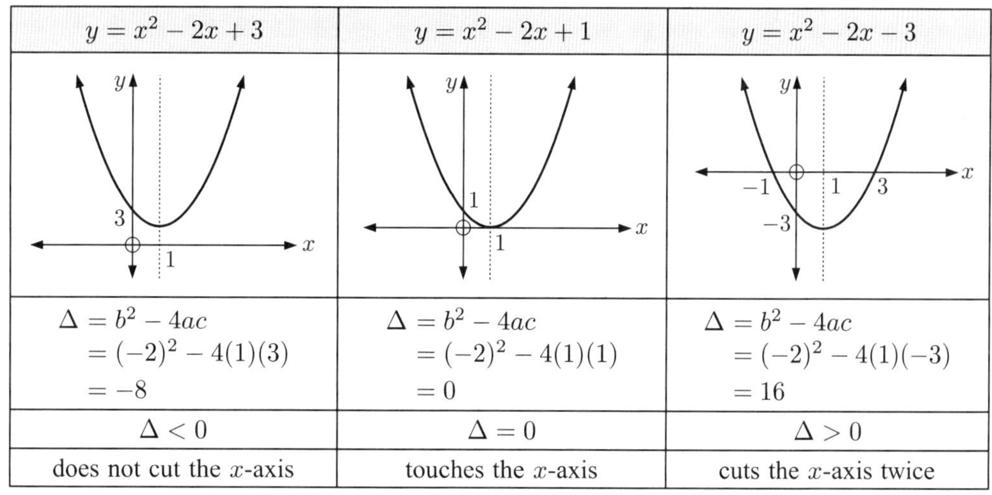

$y = x^2 - 2x + 3$	$y = x^2 - 2x + 1$	$y = x^2 - 2x - 3$
$\Delta = b^2 - 4ac$ $= (-2)^2 - 4(1)(3)$ $= -8$	$\Delta = b^2 - 4ac$ $= (-2)^2 - 4(1)(1)$ $= 0$	$\Delta = b^2 - 4ac$ $= (-2)^2 - 4(1)(-3)$ $= 16$
$\Delta < 0$	$\Delta = 0$	$\Delta > 0$
does not cut the x-axis	touches the x-axis	cuts the x-axis twice

For a quadratic function $y = ax^2 + bx + c$, we consider the discriminant $\Delta = b^2 - 4ac$.

- If $\Delta < 0$, the graph does not cut the x-axis.
- If $\Delta = 0$, the graph *touches* the x-axis.
- If $\Delta > 0$, the graph cuts the x-axis twice.

POSITIVE DEFINITE AND NEGATIVE DEFINITE QUADRATICS

Positive definite quadratics are quadratics which are positive for all values of x.
So, $ax^2 + bx + c > 0$ for all $x \in \mathbb{R}$.

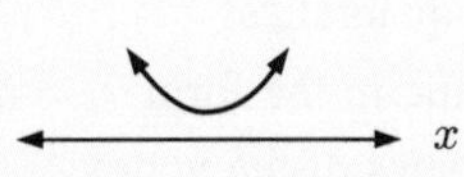

Test: A quadratic is **positive definite** $\Leftrightarrow$ $a > 0$ and $\Delta < 0$.

Negative definite quadratics are quadratics which are negative for all values of x.
So, $ax^2 + bx + c < 0$ for all $x \in \mathbb{R}$.

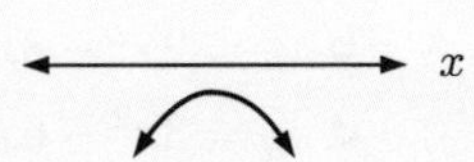

Test: A quadratic is **negative definite** $\Leftrightarrow$ $a < 0$ and $\Delta < 0$.

Example 18 **Self Tutor**

Use the discriminant to determine the relationship between the graph and the x-axis for:

a $y = x^2 + 3x + 4$ **b** $y = -2x^2 + 5x + 1$

a $a = 1, \quad b = 3, \quad c = 4$

$\therefore \Delta = b^2 - 4ac$
$= 9 - 4(1)(4)$
$= -7$

Since $\Delta < 0$, the graph does not cut the x-axis.
Since $a > 0$, the graph is concave up.

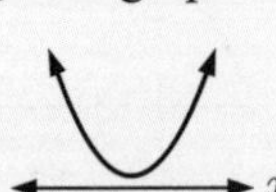

The graph is positive definite, and lies entirely above the x-axis.

b $a = -2, \quad b = 5, \quad c = 1$

$\therefore \Delta = b^2 - 4ac$
$= 25 - 4(-2)(1)$
$= 33$

Since $\Delta > 0$, the graph cuts the x-axis twice.
Since $a < 0$, the graph is concave down.

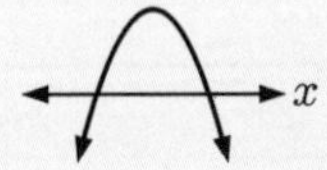

EXERCISE 1D.3

1 Use the discriminant to determine the relationship between the graph and x-axis for:

a $y = x^2 + x - 2$ **b** $y = x^2 + 7x - 2$ **c** $y = x^2 + 8x + 16$

d $y = x^2 + 4\sqrt{2}x + 8$ **e** $y = -x^2 + x + 6$ **f** $y = 9x^2 + 6x + 1$

2 Show that:

a $x^2 - 3x + 6 > 0$ for all x **b** $4x - x^2 - 6 < 0$ for all x

c $2x^2 - 4x + 7$ is positive definite **d** $-2x^2 + 3x - 4$ is negative definite.

3 Explain why $3x^2 + kx - 1$ is never positive definite for any value of k.

4 Under what conditions is $2x^2 + kx + 2$ positive definite?

E FINDING A QUADRATIC FROM ITS GRAPH

If we are given sufficient information on or about a graph we can determine the quadratic function in whatever form is required.

Example 19 Self Tutor

Find the equation of the quadratic function with graph:

a
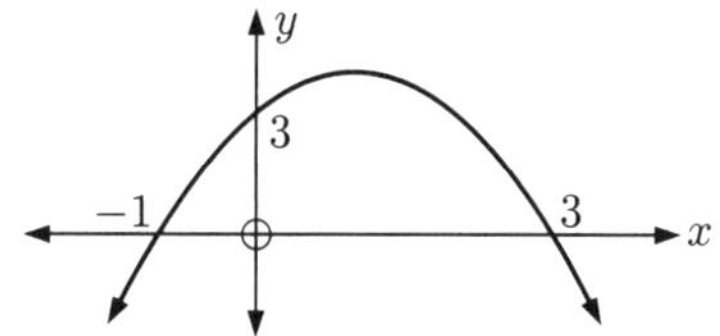

b
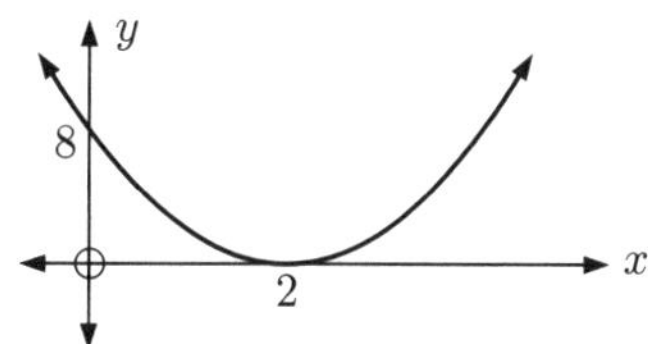

a Since the x-intercepts are -1 and 3,
$y = a(x+1)(x-3)$.
The graph is concave down, so $a < 0$.
When $x = 0$, $y = 3$
$\therefore\ 3 = a(1)(-3)$
$\therefore\ a = -1$
The quadratic function is
$y = -(x+1)(x-3)$.

b The graph touches the x-axis at $x = 2$,
so $y = a(x-2)^2$.
The graph is concave up, so $a > 0$.
When $x = 0$, $y = 8$
$\therefore\ 8 = a(-2)^2$
$\therefore\ a = 2$
The quadratic function is
$y = 2(x-2)^2$.

Example 20 Self Tutor

Find the equation of the quadratic function with graph:

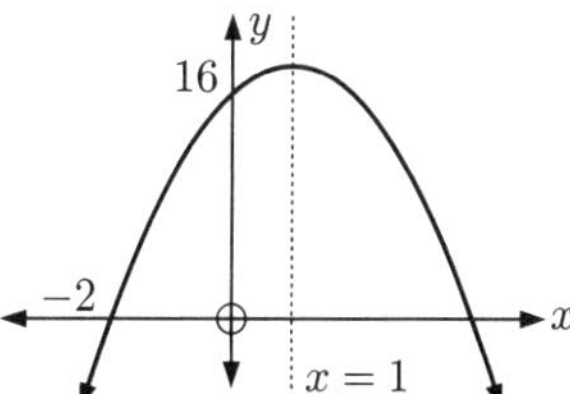

The axis of symmetry $x = 1$ lies midway between the x-intercepts.

$\therefore$ the other x-intercept is 4.

$\therefore$ the quadratic has the form

$$y = a(x+2)(x-4) \quad \text{where } a < 0$$

But when $x = 0$, $y = 16$
$\therefore\ 16 = a(2)(-4)$
$\therefore\ a = -2$

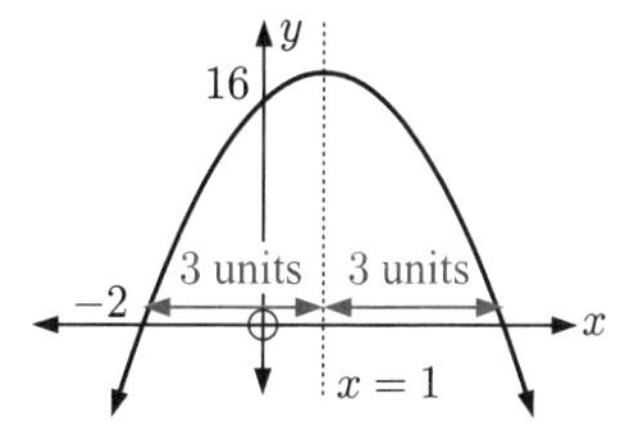

The quadratic is $y = -2(x+2)(x-4)$.

Example 21 **Self Tutor**

Find the equation of the quadratic whose graph cuts the x-axis at 4 and -3, and which passes through the point $(2, -20)$. Give your answer in the form $y = ax^2 + bx + c$.

Since the x-intercepts are 4 and -3, the quadratic has the form $y = a(x-4)(x+3)$, $a \neq 0$.

When $x = 2$, $y = -20$

$\therefore\ -20 = a(2-4)(2+3)$

$\therefore\ -20 = a(-2)(5)$

$\therefore\ a = 2$

The quadratic is $y = 2(x-4)(x+3)$

$= 2(x^2 - x - 12)$

$= 2x^2 - 2x - 24$

EXERCISE 1E

1 Find the equation of the quadratic with graph:

a

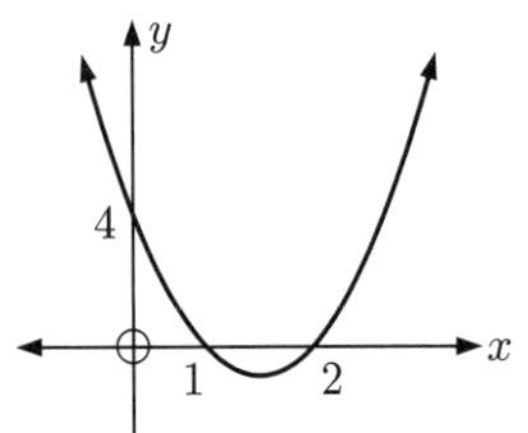

b

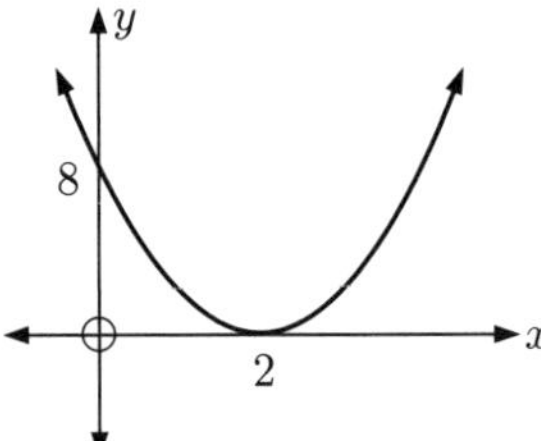

c

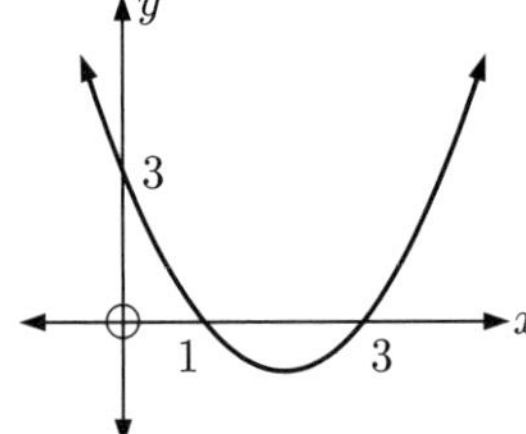

d

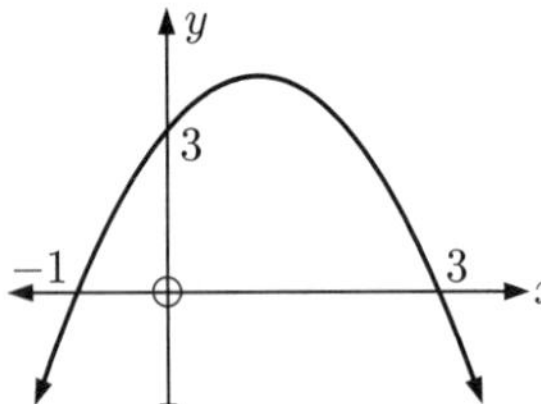

e

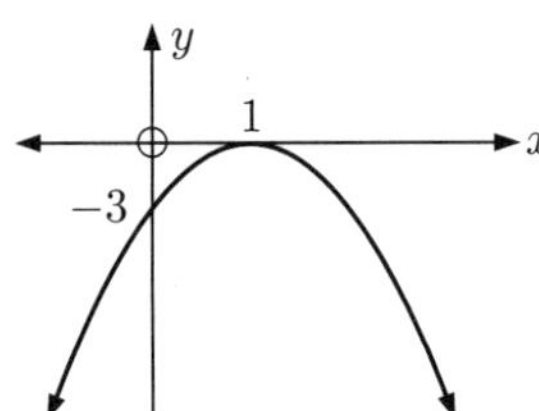

f

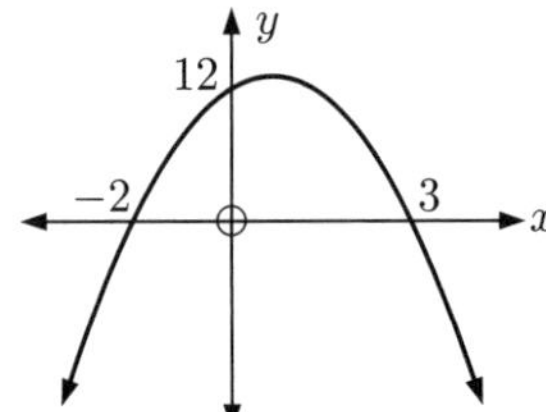

2 Find the quadratic with graph:

a

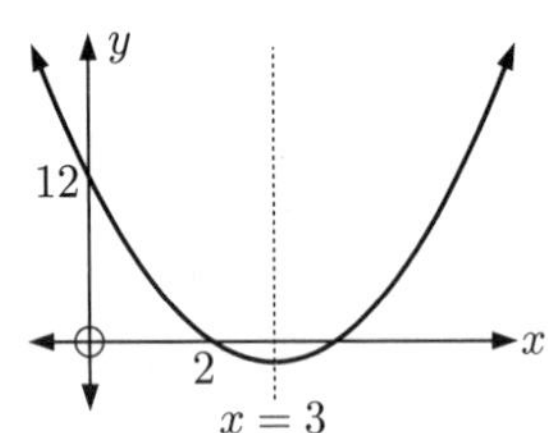

b

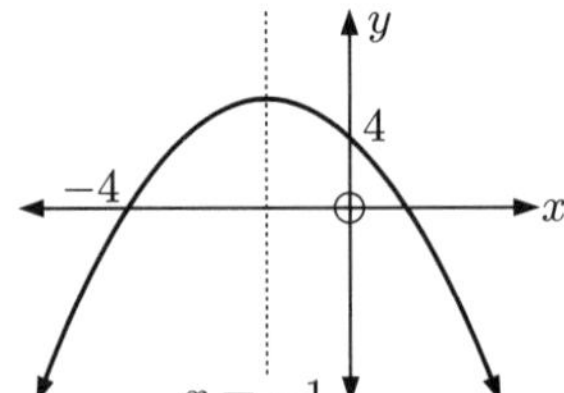

c

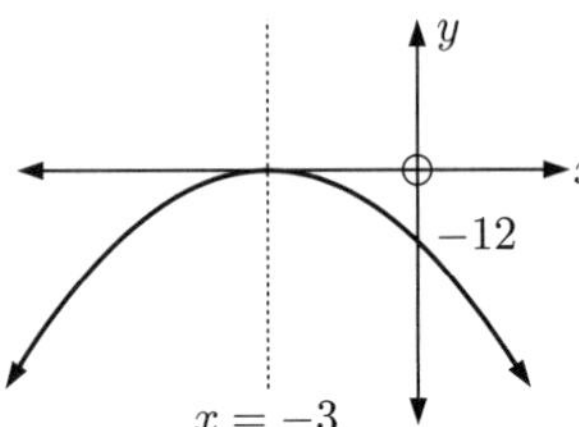

3 Find, in the form $y = ax^2 + bx + c$, the equation of the quadratic whose graph:

- **a** cuts the x-axis at 5 and 1, and passes through $(2, -9)$
- **b** cuts the x-axis at 2 and $-\frac{1}{2}$, and passes through $(3, -14)$
- **c** touches the x-axis at 3 and passes through $(-2, -25)$
- **d** touches the x-axis at -2 and passes through $(-1, 4)$
- **e** cuts the x-axis at 3, passes through $(5, 12)$ and has axis of symmetry $x = 2$
- **f** cuts the x-axis at 5, passes through $(2, 5)$ and has axis of symmetry $x = 1$.

Example 22 — Self Tutor

Find the equation of each quadratic function given its graph:

a

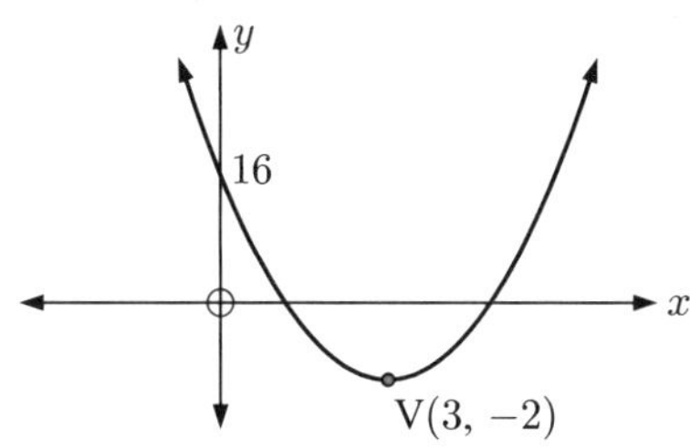

b

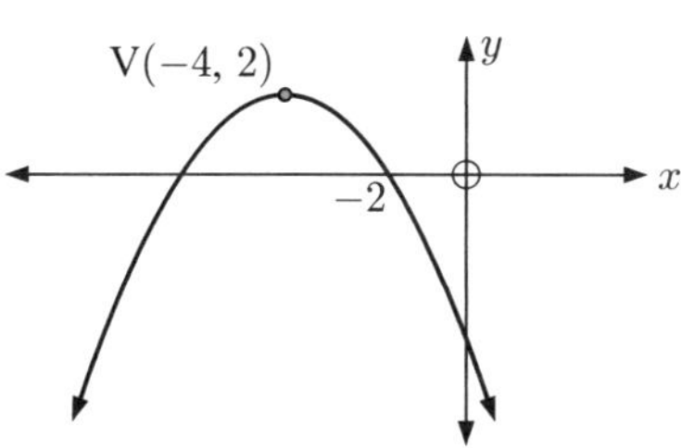

a Since the vertex is $(3, -2)$, the quadratic has the form
$y = a(x-3)^2 - 2$ where $a > 0$.
When $x = 0$, $y = 16$

$\therefore\ 16 = a(-3)^2 - 2$
$\therefore\ 16 = 9a - 2$
$\therefore\ 18 = 9a$
$\therefore\ a = 2$

The quadratic is $y = 2(x-3)^2 - 2$.

b Since the vertex is $(-4, 2)$, the quadratic has the form
$y = a(x+4)^2 + 2$ where $a < 0$.
When $x = -2$, $y = 0$

$\therefore\ 0 = a(2)^2 + 2$
$\therefore\ 4a = -2$
$\therefore\ a = -\frac{1}{2}$

The quadratic is
$y = -\frac{1}{2}(x+4)^2 + 2$.

4 If V is the vertex, find the equation of the quadratic function with graph:

a

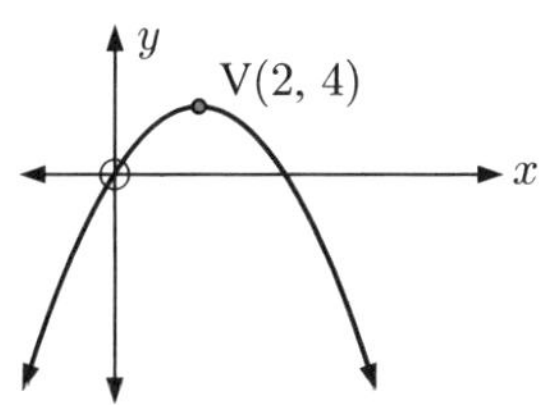

b

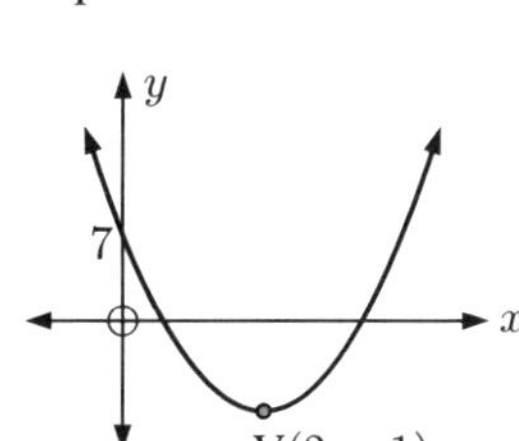

c

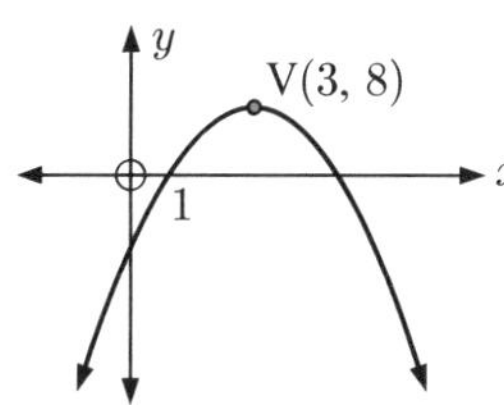

d

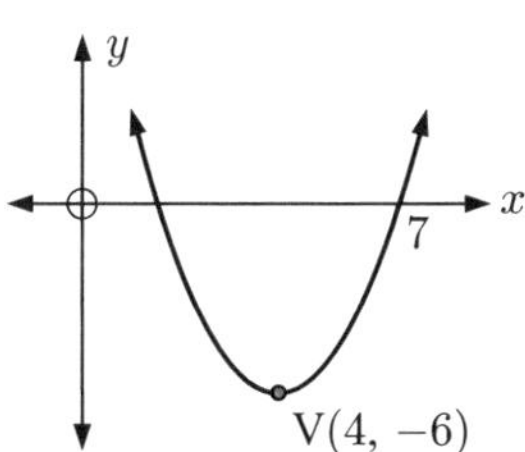

e

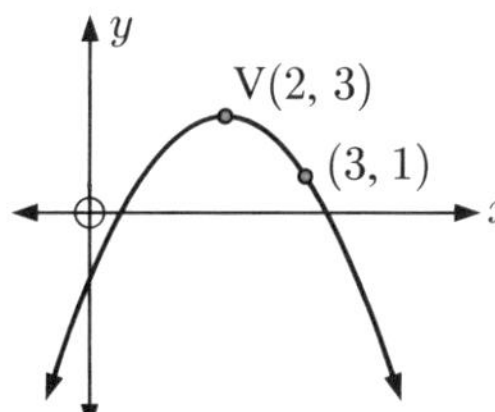

f

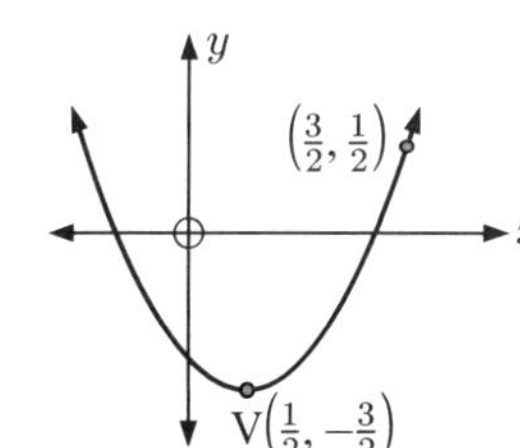

F WHERE FUNCTIONS MEET

Consider the graphs of a quadratic function and a linear function on the same set of axes.

Notice that we could have:

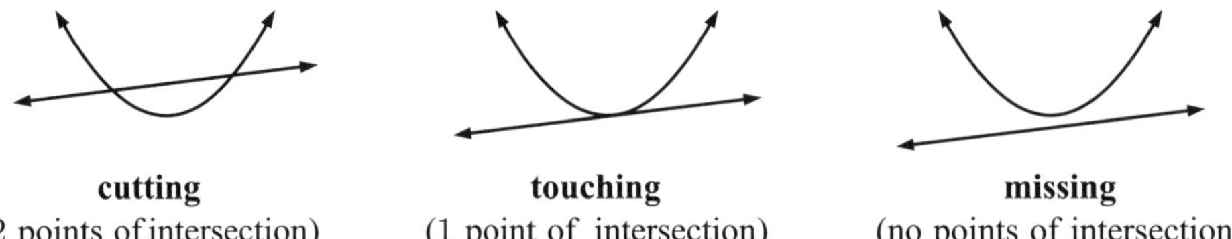

cutting	**touching**	**missing**
(2 points of intersection)	(1 point of intersection)	(no points of intersection)

If the graphs meet, the coordinates of the points of intersection of the graphs can be found by *solving the two equations simultaneously.*

Example 23 — Self Tutor

Find the coordinates of the points of intersection of the graphs with equations $y = x^2 - x - 18$ and $y = x - 3$ without technology.

$y = x^2 - x - 18$ meets $y = x - 3$ where $x^2 - x - 18 = x - 3$

$\therefore\ x^2 - 2x - 15 = 0$ {RHS = 0}

$\therefore\ (x-5)(x+3) = 0$ {factorising}

$\therefore\ x = 5$ or -3

Substituting into $y = x - 3$, when $x = 5$, $y = 2$ and when $x = -3$, $y = -6$.

$\therefore$ the graphs meet at (5, 2) and (−3, −6).

Example 24 — Self Tutor

Consider the curves $y = x^2 + 5x + 6$ and $y = 2x^2 + 2x - 4$.

a Solve for x: $x^2 + 5x + 6 = 2x^2 + 2x - 4$.

b For what values of x is $x^2 + 5x + 6 > 2x^2 + 2x - 4$?

a $x^2 + 5x + 6 = 2x^2 + 2x - 4$

$\therefore\ x^2 - 3x - 10 = 0$

$\therefore\ (x+2)(x-5) = 0$

$\therefore\ x = -2$ or 5

b

$y = 2x^2 + 2x - 4$

$y = x^2 + 5x + 6$

y, x, -2, 5

From the graphs we see that $x^2 + 5x + 6 > 2x^2 + 2x - 4$ when $-2 < x < 5$.

Example 25 | **Self Tutor**

$y = 2x + k$ is a tangent to $y = 2x^2 - 3x + 4$. Find k.

$y = 2x + k$ meets $y = 2x^2 - 3x + 4$ where

$$2x^2 - 3x + 4 = 2x + k$$

$$\therefore\ 2x^2 - 5x + (4 - k) = 0$$

Since the graphs touch, this quadratic has $\Delta = 0$

$$\therefore\ (-5)^2 - 4(2)(4-k) = 0$$

$$\therefore\ 25 - 8(4-k) = 0$$

$$\therefore\ 25 - 32 + 8k = 0$$

$$\therefore\ 8k = 7$$

$$\therefore\ k = \tfrac{7}{8}$$

A tangent *touches* a curve.

EXERCISE 1F

1 Without using technology, find the coordinates of the point(s) of intersection of:

a $y = x^2 - 2x + 8$ and $y = x + 6$

b $y = -x^2 + 3x + 9$ and $y = 2x - 3$

c $y = x^2 - 4x + 3$ and $y = 2x - 6$

d $y = -x^2 + 4x - 7$ and $y = 5x - 4$

2 a i Find where $y = x^2$ meets $y = x + 2$.

ii Solve for x: $x^2 > x + 2$.

b i Find where $y = x^2 + 2x - 3$ meets $y = x - 1$.

ii Solve for x: $x^2 + 2x - 3 > x - 1$.

c i Find where $y = 2x^2 - x + 3$ meets $y = 2 + x + x^2$.

ii Solve for x: $2x^2 - x + 3 > 2 + x + x^2$.

d i Find where $y = \frac{4}{x}$ meets $y = x + 3$.

ii Solve for x: $\frac{4}{x} > x + 3$.

3 For which value of c is the line $y = 3x + c$ a tangent to the parabola with equation $y = x^2 - 5x + 7$?

4 Find the values of m for which the lines $y = mx - 2$ are tangents to the curve with equation $y = x^2 - 4x + 2$.

5 Find the gradients of the lines with y-intercept 1 that are tangents to the curve $y = 3x^2 + 5x + 4$.

6 a For what values of c do the lines $y = x + c$ never meet the parabola with equation $y = 2x^2 - 3x - 7$?

b Choose one of the values of c found in part **a** above. Using technology, sketch the graphs to illustrate that these curves never meet.

INVESTIGATION 3 — THE PARABOLA $y^2 = 4ax$

A **parabola** is defined as the locus of all points which are equidistant from a fixed point called the *focus* and a fixed line called the *directrix*.

Suppose the focus is $F(a,\ 0)$ and the directrix is the vertical line $x = -a$.

What to do:

PRINTABLE GRAPH PAPER

1 Suggest why it is convenient to let the focus be at $(a,\ 0)$ and the directrix be the line $x = -a$.

2 Use the circular-linear graph paper provided to graph the parabola which has focus $F(2,\ 0)$ and directrix $x = -2$.

3 Use the definition given above to show that the equation of the parabola is $y^2 = 4ax$.

4 Let $y = mx + c$ be a tangent to $y^2 = 4ax$ at the point P.

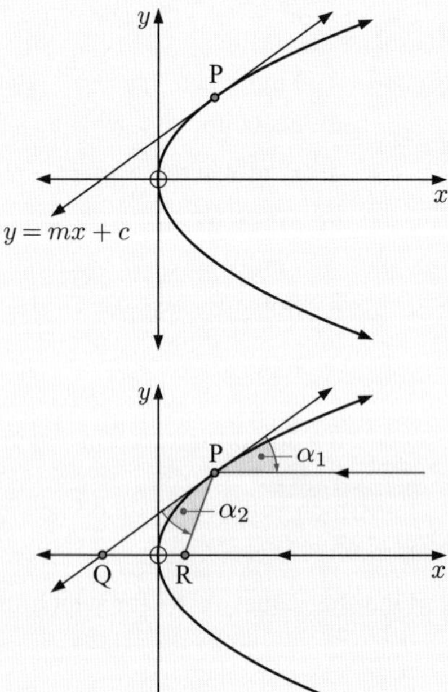

a Use quadratic theory to show that:

i $a = mc$

ii P is at $\left(\dfrac{a}{m^2},\ \dfrac{2a}{m}\right)$.

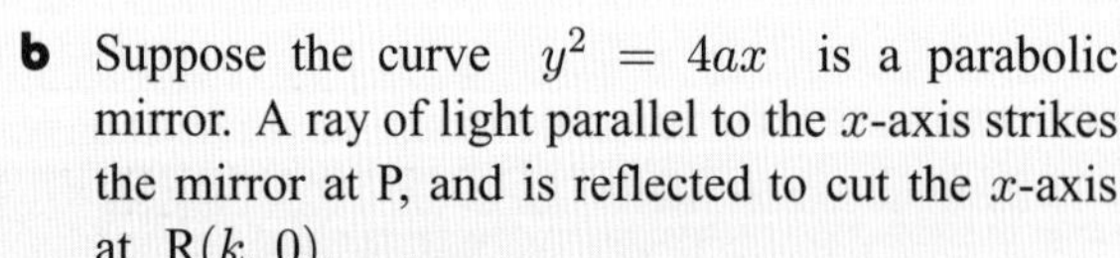

b Suppose the curve $y^2 = 4ax$ is a parabolic mirror. A ray of light parallel to the x-axis strikes the mirror at P, and is reflected to cut the x-axis at $R(k,\ 0)$.
By the principle of reflection, $\alpha_1 = \alpha_2$.

i Deduce that triangle PQR is isosceles.

ii Hence, deduce that $k = a$.

iii Clearly state the special result that follows from **4 b ii**.

5 List real life examples of where the result in **4 b iii** has been utilised.

G PROBLEM SOLVING WITH QUADRATICS

Some real world problems can be solved using a quadratic equation. We are generally only interested in any **real solutions** which result.

Any answer we obtain must be checked to see if it is reasonable. For example:

- if we are finding a length then it must be positive and we reject any negative solutions
- if we are finding 'how many people are present' then clearly the answer must be a positive integer.

We employ the following general problem solving method:

Step 1: If the information is given in words, translate it into algebra using a variable such as x for the unknown. Write down the resulting equation. Be sure to define what the variable x represents, and include units if appropriate.

Step 2: Solve the equation by a suitable method.

Step 3: Examine the solutions carefully to see if they are acceptable.

Step 4: Give your answer in a sentence.

Example 26 **Self Tutor**

A rectangle has length 3 cm longer than its width. Its area is 42 cm^2. Find its width.

If the width is x cm then the length is $(x+3)$ cm.

$\therefore\ x(x+3) = 42$ {equating areas}

$\therefore\ x^2 + 3x - 42 = 0$

$\therefore\ x \approx -8.15$ or 5.15 {using technology}

We reject the negative solution as lengths are positive.

The width is about 5.15 cm.

Example 27 **Self Tutor**

Is it possible to bend a 12 cm length of wire to form the perpendicular sides of a right angled triangle with area 20 cm^2?

Suppose the wire is bent x cm from one end.

The area $A = \frac{1}{2}x(12-x)$

$\therefore\ \frac{1}{2}x(12-x) = 20$

$\therefore\ x(12-x) = 40$

$\therefore\ 12x - x^2 - 40 = 0$

$\therefore\ x^2 - 12x + 40 = 0$

Now $\Delta = (-12)^2 - 4(1)(40)$

$= -16$ which is < 0

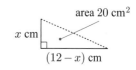

There are no real solutions, indicating this situation is **impossible**.

EXERCISE 1G

1 Two integers differ by 12 and the sum of their squares is 74. Find the integers.

2 The sum of a number and its reciprocal is $5\frac{1}{5}$. Find the number.

3 The sum of a natural number and its square is 210. Find the number.

4 The product of two consecutive odd numbers is 255. Find the numbers.

5 The length of a rectangle is 4 cm longer than its width. The rectangle has area 26 cm^2. Find its width.

6 A rectangular box has a square base with sides of length x cm. Its height is 1 cm longer than its base side length. The total surface area of the box is 240 cm^2.

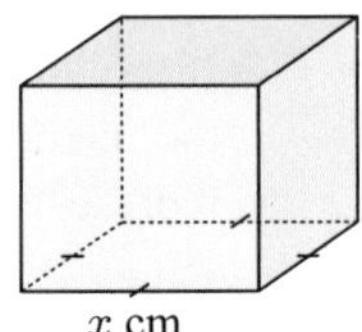

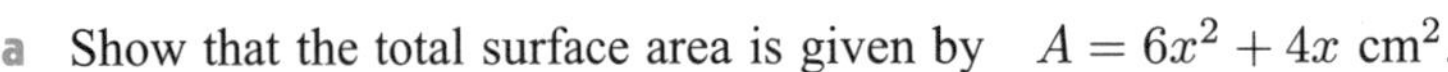

a Show that the total surface area is given by $A = 6x^2 + 4x$ cm^2.

b Find the dimensions of the box.

7 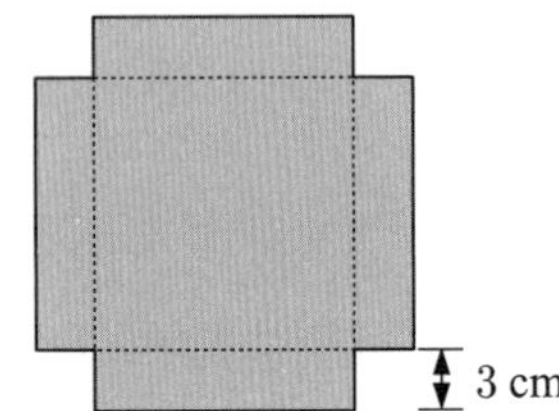

An open box can hold 80 cm^3. It is made from a square piece of tinplate with 3 cm squares cut from each of its 4 corners. Find the dimensions of the original piece of tinplate.

8 Is it possible to bend a 20 cm length of wire into the shape of a rectangle which has an area of 30 cm^2?

9 The rectangle ABCD is divided into a square and a smaller rectangle by [XY] which is parallel to its shorter sides.
The smaller rectangle BCXY is *similar* to the original rectangle, so rectangle ABCD is a **golden rectangle**.

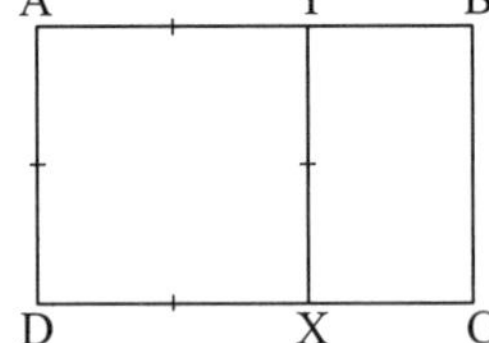

The ratio $\frac{\text{AB}}{\text{AD}}$ is called the **golden ratio**.

Show that the golden ratio is $\frac{1+\sqrt{5}}{2}$.

Hint: Let $\text{AB} = x$ units and $\text{AD} = 1$ unit.

10 Two trains travel a 160 km track each day. The express travels 10 km h^{-1} faster and takes 30 minutes less time than the normal train. Find the speed of the express.

11 Answer the **Opening Problem** on page **18**.

12 A truck carrying a wide load needs to pass through the parabolic tunnel shown. The units are metres.
The truck is 5 m high and 4 m wide.

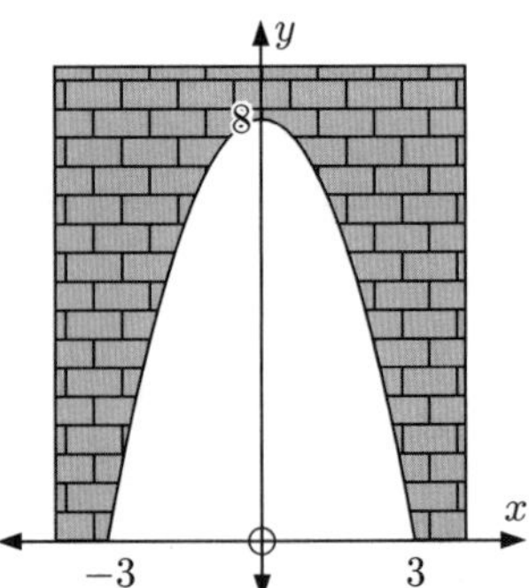

a Find the quadratic function which describes the shape of the tunnel.

b Determine whether the truck will fit.

H QUADRATIC OPTIMISATION

The process of finding the maximum or minimum value of a function is called **optimisation**.

For the quadratic function $y = ax^2 + bx + c$, we have already seen that the vertex has x-coordinate $-\frac{b}{2a}$.

- If $a > 0$, the **minimum** value of y occurs at $x = -\frac{b}{2a}$

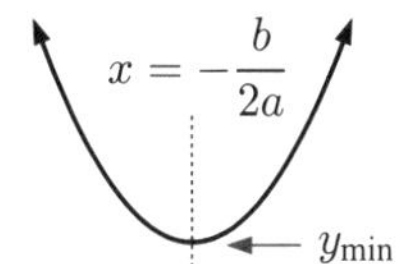

- If $a < 0$, the **maximum** value of y occurs at $x = -\frac{b}{2a}$.

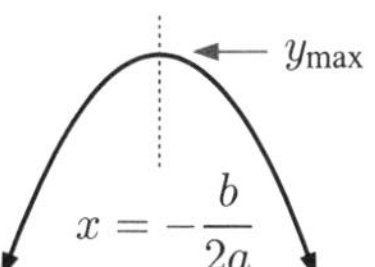

Example 28 — Self Tutor

Find the maximum or minimum value of the following quadratic functions, and the corresponding value of x:

a $y = x^2 + x - 3$ **b** $y = 3 + 3x - 2x^2$

a $y = x^2 + x - 3$ has
$a = 1$, $b = 1$, and $c = -3$.
Since $a > 0$, the shape is ∪

The minimum value occurs
when $x = \frac{-b}{2a} = -\frac{1}{2}$

and $y = (-\frac{1}{2})^2 + (-\frac{1}{2}) - 3$
$= -3\frac{1}{4}$

So, the minimum value of y is $-3\frac{1}{4}$, occurring when $x = -\frac{1}{2}$.

b $y = -2x^2 + 3x + 3$ has
$a = -2$, $b = 3$, and $c = 3$.
Since $a < 0$, the shape is ∩

The maximum value occurs
when $x = \frac{-b}{2a} = \frac{-3}{-4} = \frac{3}{4}$

and $y = -2(\frac{3}{4})^2 + 3(\frac{3}{4}) + 3$
$= 4\frac{1}{8}$

So, the maximum value of y is $4\frac{1}{8}$, occurring when $x = \frac{3}{4}$.

EXERCISE 1H

1 Find the maximum or minimum values of the following quadratic functions, and the corresponding values of x:

a $y = x^2 - 2x$ **b** $y = 4x^2 - x + 5$ **c** $y = 7x - 2x^2$

2 The profit in manufacturing x refrigerators per day, is given by the profit relation $P = -3x^2 + 240x - 800$ dollars.

a How many refrigerators should be made each day to maximise the total profit?

b What is the maximum profit?

Example 29 **Self Tutor**

A gardener has 40 m of fencing to enclose a rectangular garden plot, where one side is an existing brick wall. Suppose the two new equal sides are x m long.

a Show that the area enclosed is given by $A = x(40 - 2x)$ m^2.

b Find the dimensions of the garden of maximum area.

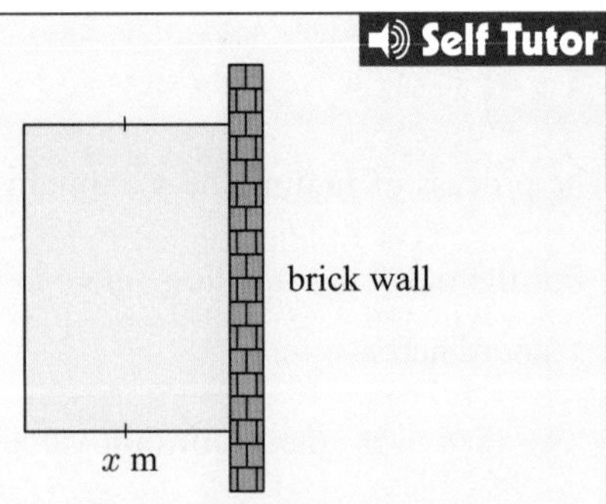

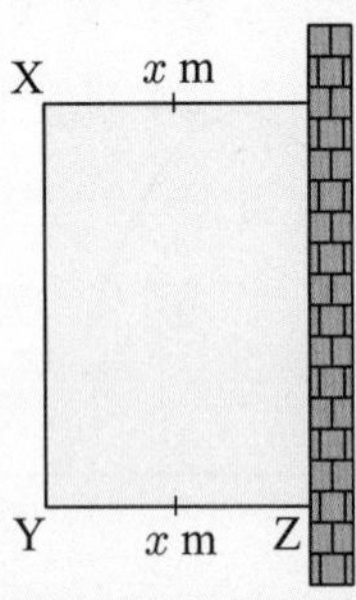

a Side [XY] has length $(40 - 2x)$ m.
Now, area = length × width
$\therefore\ A = x(40 - 2x)$ m^2

b $A = 0$ when $x = 0$ or 20.
The vertex of the function lies midway between these values, so $x = 10$.
Since $a < 0$, the shape is $\frown$

$\therefore$ the area is maximised when YZ = 10 m and XY = 20 m.

3 A rectangular plot is enclosed by 200 m of fencing and has an area of A square metres. Show that:

a $A = 100x - x^2$ where x is the length in metres of one of its sides

b the area is maximised if the rectangle is a square.

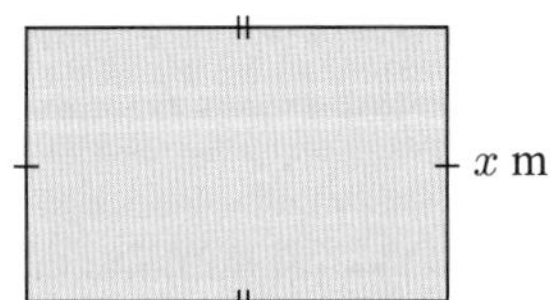

4 Three sides of a rectangular paddock are to be fenced, the fourth side being an existing straight water drain. If 1000 m of fencing is available, what dimensions should be used for the paddock so that it encloses the maximum possible area?

5 1800 m of fencing is available to fence six identical pens as shown in the diagram.

a Explain why $9x + 8y = 1800$.

b Show that the area of each pen is given by $A = -\frac{9}{8}x^2 + 225x$ m^2.

c If the area enclosed is to be maximised, what are the dimensions of each pen?

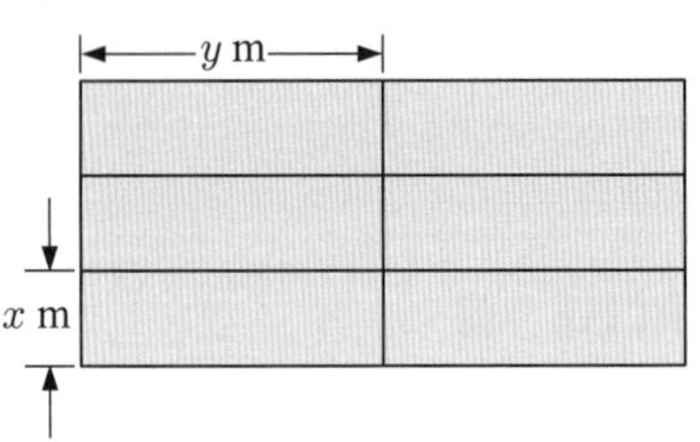

6

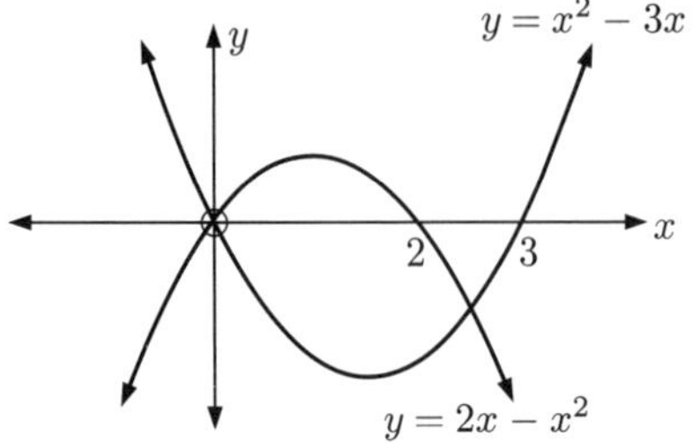

The graphs of $y = x^2 - 3x$ and $y = 2x - x^2$ are illustrated.

a Without using technology, show that the graphs meet where $x = 0$ and $x = 2\frac{1}{2}$.

b Find the maximum vertical separation between the curves for $0 \leqslant x \leqslant 2\frac{1}{2}$.

7 Infinitely many rectangles may be inscribed within the right angled triangle shown alongside. One of them is illustrated.

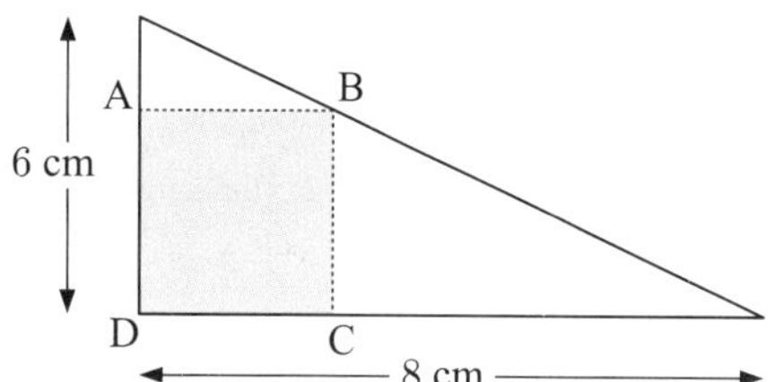

a Let AB $= x$ cm and BC $= y$ cm. Use similar triangles to find y in terms of x.

b Find the dimensions of rectangle ABCD of maximum area.

8

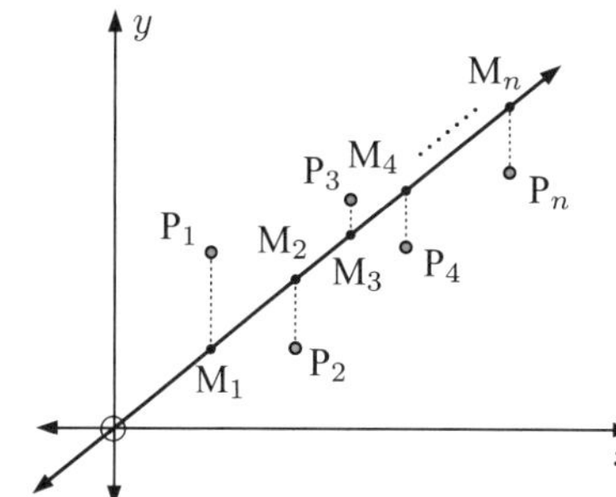

The points $P_1(a_1, b_1)$, $P(a_2, b_2)$, $P_3(a_3, b_3)$, ..., $P_n(a_n, b_n)$ are experimental data. The points are approximately linear through the origin O(0, 0).

To find the equation of the 'line of best fit' through the origin, we decide to minimise $(P_1M_1)^2 + (P_2M_2)^2 + (P_3M_3)^2 + ... + (P_nM_n)^2$ where $[P_iM_i]$ is the vertical line segment connecting each point P_i with the corresponding point M_i on the line with the same x-coordinate a_i.

Find the gradient of the 'line of best fit' in terms of a_i and b_i, $i = 1, 2, 3, 4, ..., n$.

9 Write $y = (x - a - b)(x - a + b)(x + a - b)(x + a + b)$ in expanded form and hence determine the least value of y. Assume that a and b are real constants.

10 By considering the function $y = (a_1x - b_1)^2 + (a_2x - b_2)^2$, use quadratic theory to prove the Cauchy-Schwarz inequality: $|a_1b_1 + a_2b_2| \leqslant \sqrt{a_1^2 + a_2^2}\sqrt{b_1^2 + b_2^2}$.

11 b_1, c_1, b_2, and c_2 are real, non-zero numbers such that $b_1b_2 = 2(c_1 + c_2)$. Show that at least one of the equations $x^2 + b_1x + c_1 = 0$, $x^2 + b_2x + c_2 = 0$ has two real roots.

REVIEW SET 1A — NON-CALCULATOR

1 Consider the quadratic function $y = -2(x + 2)(x - 1)$.

a State the x-intercepts.
b State the equation of the axis of symmetry.
c Find the y-intercept.
d Find the coordinates of the vertex.
e Sketch the function.

2 Solve the following equations, giving exact answers:

a $3x^2 - 12x = 0$
b $3x^2 - x - 10 = 0$
c $x^2 - 11x = 60$

3 Solve using the quadratic formula:

a $x^2 + 5x + 3 = 0$
b $3x^2 + 11x - 2 = 0$

4 Solve by 'completing the square': $x^2 + 7x - 4 = 0$

5 Use the vertex, axis of symmetry, and y-intercept to graph:

a $y = (x - 2)^2 - 4$
b $y = -\frac{1}{2}(x + 4)^2 + 6$

6 Find, in the form $y = ax^2 + bx + c$, the equation of the quadratic whose graph:

a touches the x-axis at 4 and passes through (2, 12)
b has vertex $(-4, 1)$ and passes through (1, 11).

7 Find the maximum or minimum value of the relation $y = -2x^2 + 4x + 3$ and the value of x at which this occurs.

8 The roots of $2x^2 - 3x = 4$ are α and β. Find the simplest quadratic equation which has roots $\frac{1}{\alpha}$ and $\frac{1}{\beta}$.

9 Solve the following equations:

a $x^2 + 10 = 7x$ **b** $x + \frac{12}{x} = 7$ **c** $2x^2 - 7x + 3 = 0$

10 Find the points of intersection of $y = x^2 - 3x$ and $y = 3x^2 - 5x - 24$.

11 For what values of k does the graph of $y = -2x^2 + 5x + k$ not cut the x-axis?

12 Find the values of m for which $2x^2 - 3x + m = 0$ has:

a a repeated root **b** two distinct real roots **c** no real roots.

13 The sum of a number and its reciprocal is $2\frac{1}{30}$. Find the number.

14 Show that no line with a y-intercept of (0, 10) will ever be tangential to the curve with equation $y = 3x^2 + 7x - 2$.

15 One of the roots of $kx^2 + (1 - 3k)x + (k - 6) = 0$ is the negative reciprocal of the other root. Find k and the two roots.

REVIEW SET 1B — CALCULATOR

1 Consider the quadratic function $y = 2x^2 + 6x - 3$.

a Convert it to the form $y = a(x - h)^2 + k$.

b State the coordinates of the vertex.

c Find the y-intercept.

d Sketch the graph of the function.

2 Solve:

a $(x - 2)(x + 1) = 3x - 4$ **b** $2x - \frac{1}{x} = 5$

3 Draw the graph of $y = -x^2 + 2x$.

4 Consider the quadratic function $y = -3x^2 + 8x + 7$. Find the equation of the axis of symmetry, and the coordinates of the vertex.

5 Using the discriminant only, determine the nature of the solutions of:

a $2x^2 - 5x - 7 = 0$ **b** $3x^2 - 24x + 48 = 0$

6 **a** For what values of c do the lines with equations $y = 3x + c$ intersect the parabola $y = x^2 + x - 5$ in two distinct points?

b Choose one such value of c from part **a** and find the points of intersection in this case.

7 Suppose [AB] has the same length as [CD], [BC] is 2 cm shorter than [AB], and [BE] is 7 cm in length. Find the length of [AB].

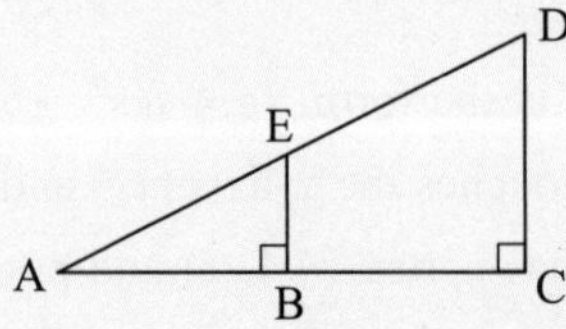

8 60 m of chicken wire is available to construct a rectangular chicken enclosure against an existing wall.

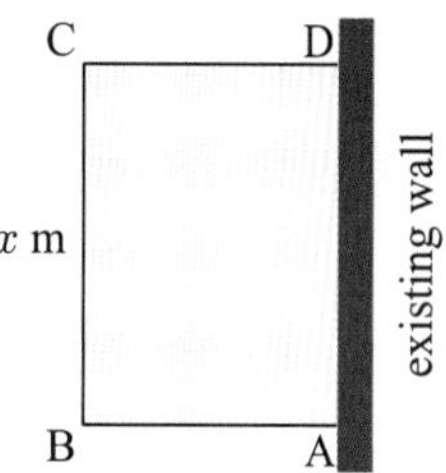

a If BC $= x$ m, show that the area of rectangle ABCD is given by $A = (30x - \frac{1}{2}x^2)$ m^2.

b Find the dimensions of the enclosure which will maximise the area enclosed.

9 Consider the quadratic function $y = 2x^2 + 4x - 1$.

a State the axis of symmetry.

b Find the coordinates of the vertex.

c Find the axes intercepts.

d Hence sketch the function.

10 An open square-based container is made by cutting 4 cm square pieces out of a piece of tinplate. If the volume of the container is 120 cm^3, find the size of the original piece of tinplate.

11 Consider $y = -x^2 - 5x + 3$ and $y = x^2 + 3x + 11$.

a Solve for x: $-x^2 - 5x + 3 = x^2 + 3x + 11$.

b Hence, or otherwise, determine the values of x for which $x^2 + 3x + 11 > -x^2 - 5x + 3$.

12 Find the maximum or minimum value of the following quadratics, and the corresponding value of x:

a $y = 3x^2 + 4x + 7$

b $y = -2x^2 - 5x + 2$

13 600 m of fencing is used to construct 6 rectangular animal pens as shown.

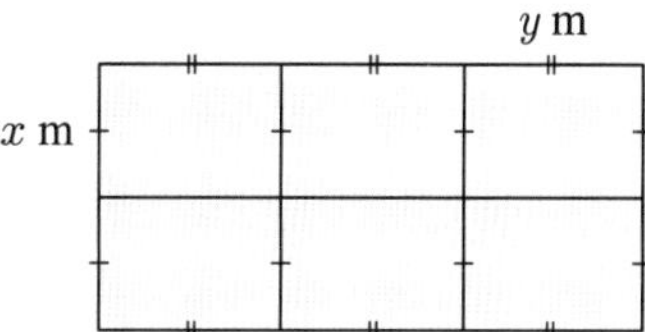

a Show that the area A of each pen is
$A = x\left(\frac{600 - 8x}{9}\right)$ m^2.

b Find the dimensions of each pen so that it has the maximum possible area.

c What is the area of each pen in this case?

14 Two different quadratic functions of the form $y = 9x^2 - kx + 4$ each touch the x-axis.

a Find the two values of k.

b Find the point of intersection of the two quadratic functions.

REVIEW SET 1C

1 Consider the quadratic function $y = \frac{1}{2}(x - 2)^2 - 4$.

a State the equation of the axis of symmetry.

b Find the coordinates of the vertex.

c Find the y-intercept.

d Sketch the function.

2 Solve the following equations:

a $x^2 - 5x - 3 = 0$

b $2x^2 - 7x - 3 = 0$

3 Solve the following using the quadratic formula:

a $x^2 - 7x + 3 = 0$ **b** $2x^2 - 5x + 4 = 0$

4 Find the equation of the quadratic function with graph:

a

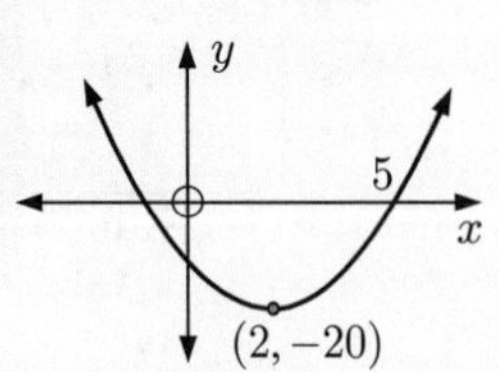

b

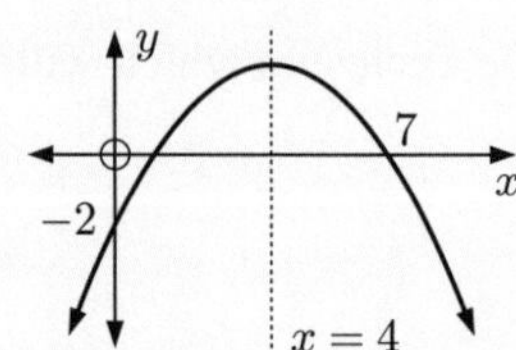

c

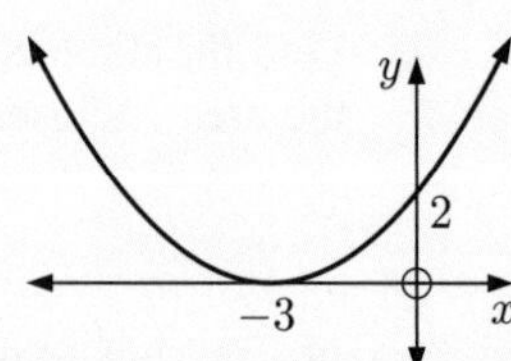

5 Use the discriminant only to find the relationship between the graph and the x-axis for:

a $y = 2x^2 + 3x - 7$ **b** $y = -3x^2 - 7x + 4$

6 Determine whether the following quadratic functions are positive definite, negative definite, or neither:

a $y = -2x^2 + 3x + 2$ **b** $y = 3x^2 + x + 11$

7 Find the equation of the quadratic function shown:

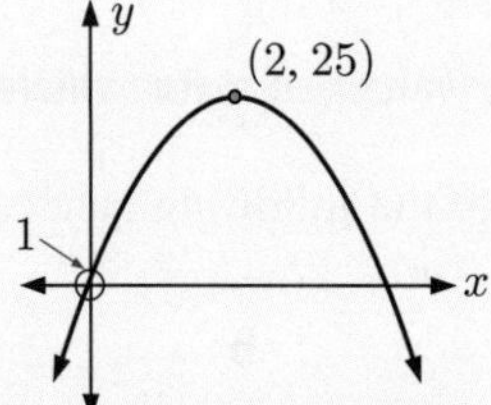

8 Find the y-intercept of the line with gradient -3 that is tangential to the parabola $y = 2x^2 - 5x + 1$.

9 For what values of k would the graph of $y = x^2 - 2x + k$ cut the x-axis twice?

10 Find the quadratic function which cuts the x-axis at 3 and -2 and which has y-intercept 24. Give your answer in the form $y = ax^2 + bx + c$.

11 For what values of m are the lines $y = mx - 10$ tangents to the parabola $y = 3x^2 + 7x + 2$?

12 $ax^2 + (3 - a)x - 4 = 0$ has roots which are real and positive. What values can a have?

13 **a** Determine the equation of:

i the quadratic function

ii the straight line.

b For what values of x is the straight line above the curve?

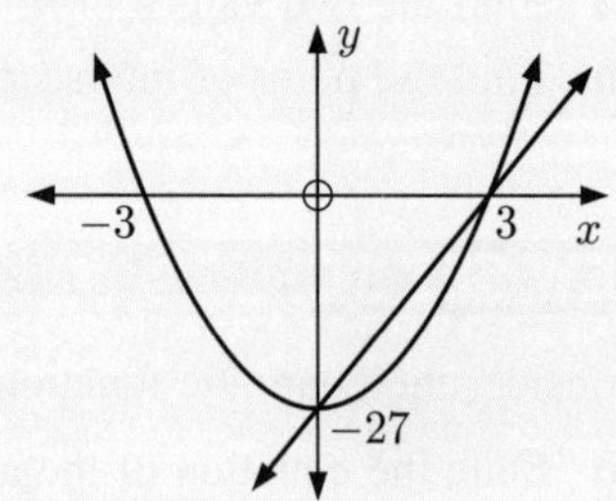

14 Show that the lines with equations $y = -5x + k$ are tangents to the parabola $y = x^2 - 3x + c$ if and only if $c - k = 1$.

15 $4x^2 - 3x - 3 = 0$ has roots p, q. Find *all* quadratic equations with roots p^3 and q^3.

Chapter 2

Functions

Syllabus reference: 2.1, 2.2, 2.4, 2.7

Contents:

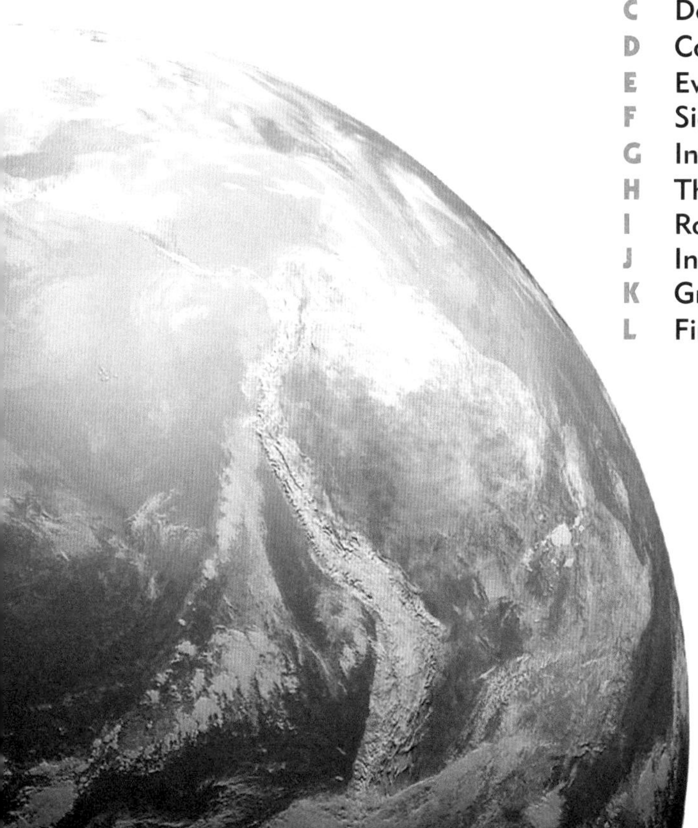

A RELATIONS AND FUNCTIONS

The charges for parking a car in a short-term car park at an airport are shown in the table below. The total charge is *dependent* on the length of time t the car is parked.

Car park charges	
Time t (hours)	*Charge*
0 - 1 hours	\$5.00
1 - 2 hours	\$9.00
2 - 3 hours	\$11.00
3 - 6 hours	\$13.00
6 - 9 hours	\$18.00
9 - 12 hours	\$22.00
12 - 24 hours	\$28.00

Looking at this table we might ask: How much would be charged for *exactly* one hour? Would it be \$5 or \$9?

To avoid confusion, we could adjust the table or draw a graph. We indicate that 2 - 3 hours really means a time over 2 hours up to and including 3 hours, by writing $2 < t \leqslant 3$ hours.

Car park charges	
Time t (hours)	*Charge*
$0 < t \leqslant 1$ hours	\$5.00
$1 < t \leqslant 2$ hours	\$9.00
$2 < t \leqslant 3$ hours	\$11.00
$3 < t \leqslant 6$ hours	\$13.00
$6 < t \leqslant 9$ hours	\$18.00
$9 < t \leqslant 12$ hours	\$22.00
$12 < t \leqslant 24$ hours	\$28.00

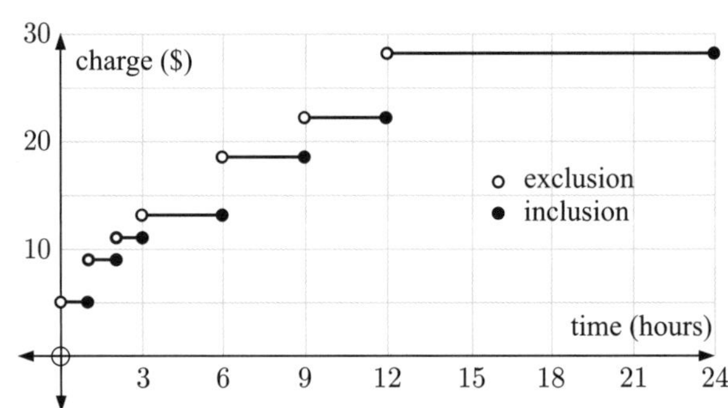

In mathematical terms, we have a relationship between two variables *time* and *charge*, so the schedule of charges is an example of a **relation**.

A relation may consist of a finite number of ordered pairs, such as $\{(1, 5), (-2, 3), (4, 3), (1, 6)\}$, or an infinite number of ordered pairs.

The parking charges example is clearly the latter, as every real value of time in the interval $0 < t \leqslant 24$ hours is represented.

The set of possible values of the variable on the horizontal axis is called the **domain** of the relation.

For example:
- the domain for the car park relation is $\{t \mid 0 < t \leqslant 24\}$
- the domain of $\{(1, 5), (-2, 3), (4, 3), (1, 6)\}$ is $\{-2, 1, 4\}$.

The set of possible values on the vertical axis is called the **range** of the relation.

For example:
- the range of the car park relation is $\{5, 9, 11, 13, 18, 22, 28\}$
- the range of $\{(1, 5), (-2, 3), (4, 3), (1, 6)\}$ is $\{3, 5, 6\}$.

We will now look at relations and functions more formally.

RELATIONS

A **relation** is any set of points which connect two variables.

A relation is often expressed in the form of an **equation** connecting the **variables** x and y. In this case the relation is a set of points (x, y) in the **Cartesian plane**.

This plane is separated into four quadrants according to the signs of x and y.

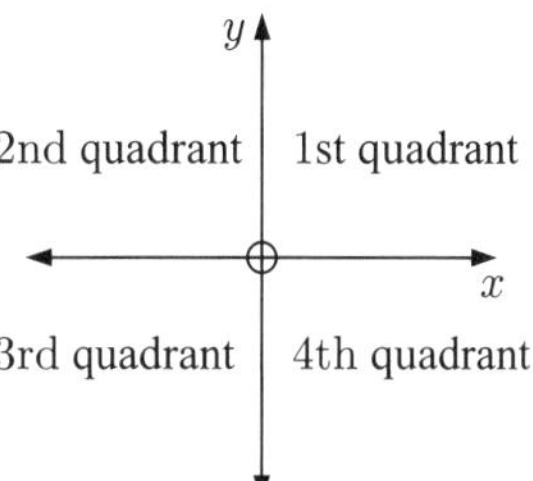

For example, $y = x + 3$ and $x = y^2$ are the equations of two relations. Each equation generates a set of ordered pairs, which we can graph:

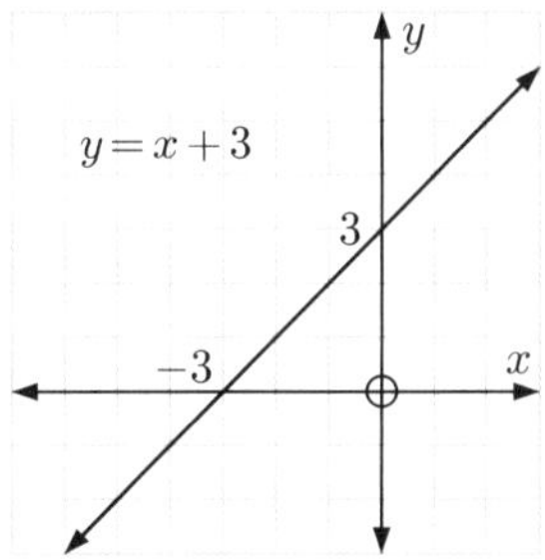

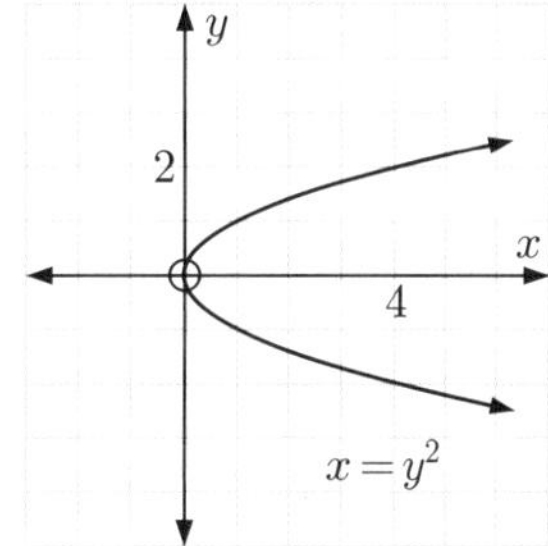

However, a relation may not be able to be defined by an equation. Below are two such examples:

(1)

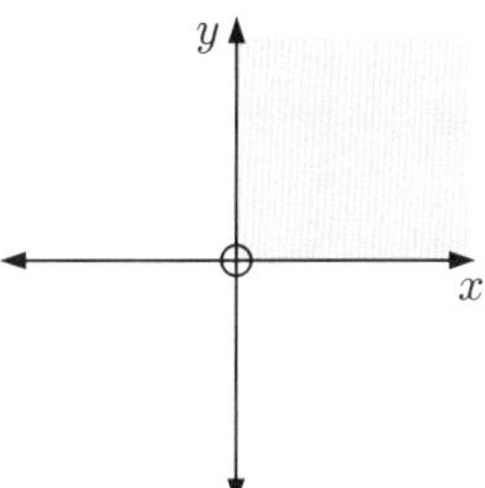

The set of all points in the first quadrant is the relation $x > 0$, $y > 0$. This is an infinite set of ordered pairs.

(2)

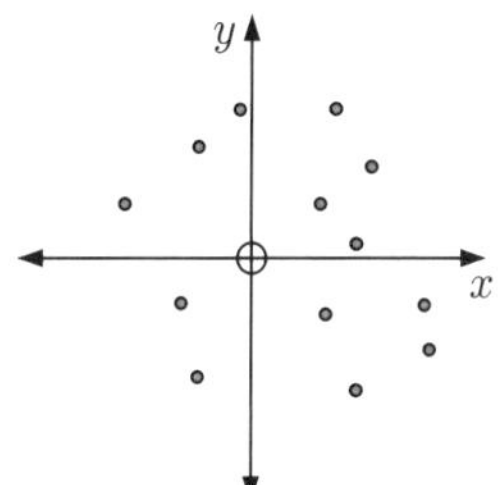

These 13 points form a relation. This is a finite set of ordered pairs.

FUNCTIONS

A **function**, sometimes called a **mapping**, is a relation in which no two different ordered pairs have the same x-coordinate or first component.

We can see from the above definition that a function is a special type of relation.

Every function is a relation, but not every relation is a function.

TESTING FOR FUNCTIONS

Algebraic Test:

If a relation is given as an equation, and the substitution of any value for x results in one and only one value of y, then the relation is a function.

For example:

- $y = 3x - 1$ is a function, since for any value of x there is only one corresponding value of y.
- $x = y^2$ is not a function, since if $x = 4$ then $y = \pm 2$. There is more than one value of y for the given value of x.

Geometric Test or Vertical Line Test:

Suppose we draw all possible vertical lines on the graph of a relation.

- If each line cuts the graph at most once, then the relation is a function.
- If at least one line cuts the graph more than once, then the relation is not a function.

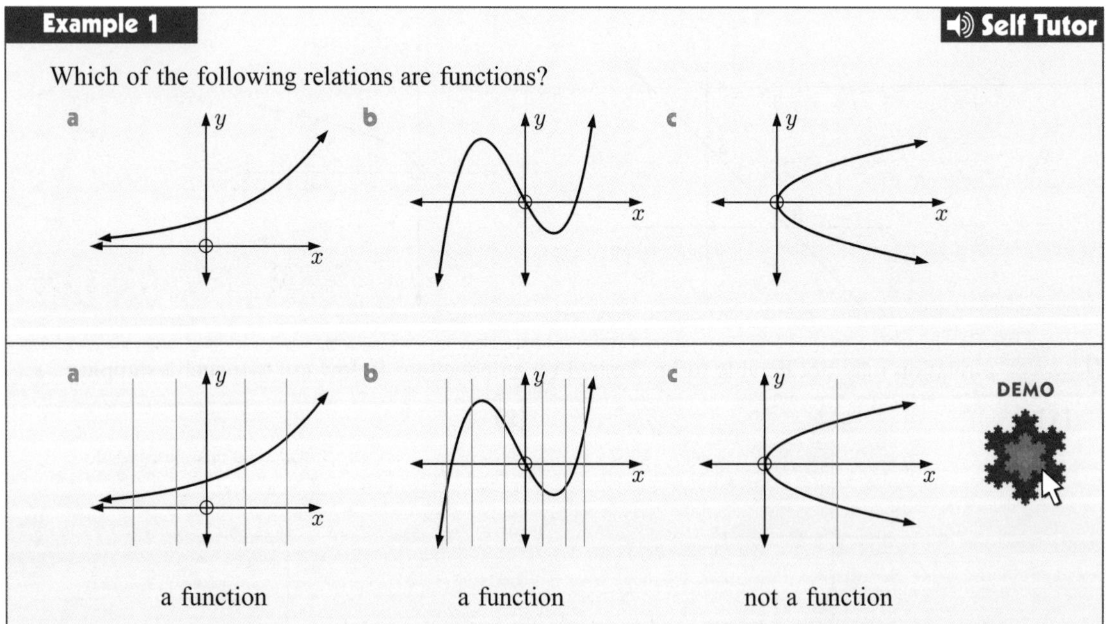

Example 1
Self Tutor
Which of the following relations are functions?
a
b
c
y
x
DEMO
a function
a function
not a function

GRAPHICAL NOTE

- If a graph contains a small **open circle** such as —o—, this point is **not included**.
- If a graph contains a small **filled-in circle** such as ——•, this point **is included**.
- If a graph contains an **arrow head** at an end such as ——→, then the graph continues indefinitely in that general direction, or the shape may repeat as it has done previously.

EXERCISE 2A

1 Which of the following sets of ordered pairs are functions? Give reasons for your answers.

a $\{(1, 3), (2, 4), (3, 5), (4, 6)\}$
b $\{(1, 3), (3, 2), (1, 7), (-1, 4)\}$
c $\{(2, -1), (2, 0), (2, 3), (2, 11)\}$
d $\{(7, 6), (5, 6), (3, 6), (-4, 6)\}$
e $\{(0, 0), (1, 0), (3, 0), (5, 0)\}$
f $\{(0, 0), (0, -2), (0, 2), (0, 4)\}$

2 Use the vertical line test to determine which of the following relations are functions:

a
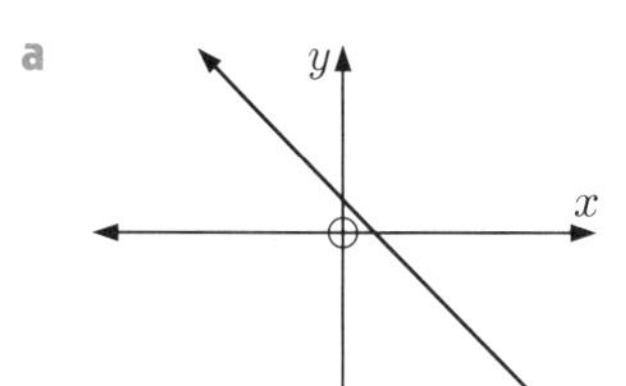

b
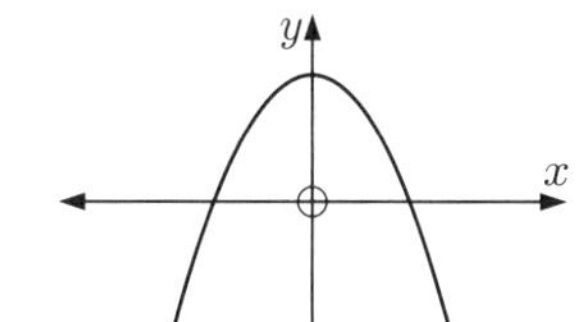

c
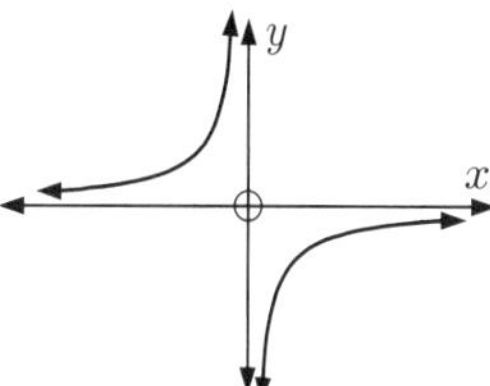

d
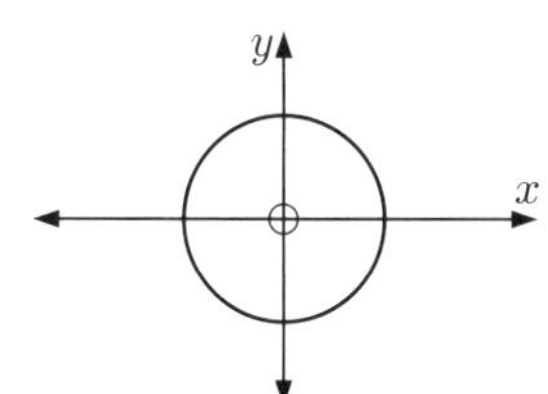

e

f
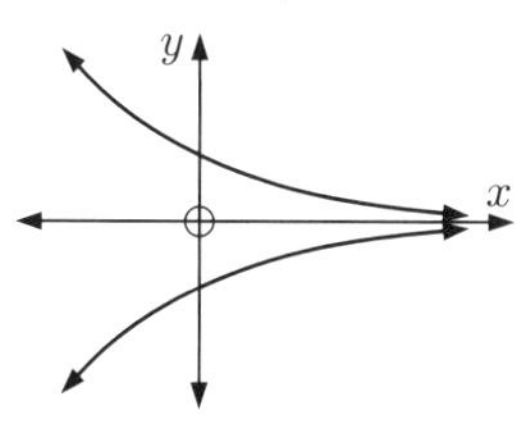

g
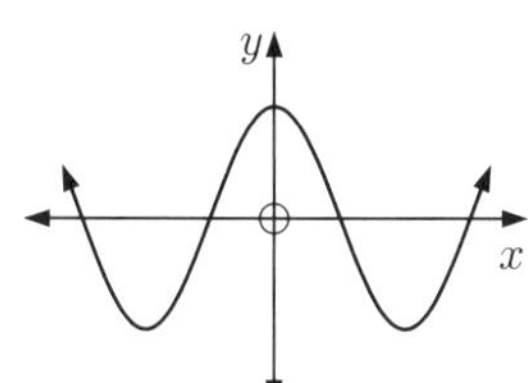

h

i
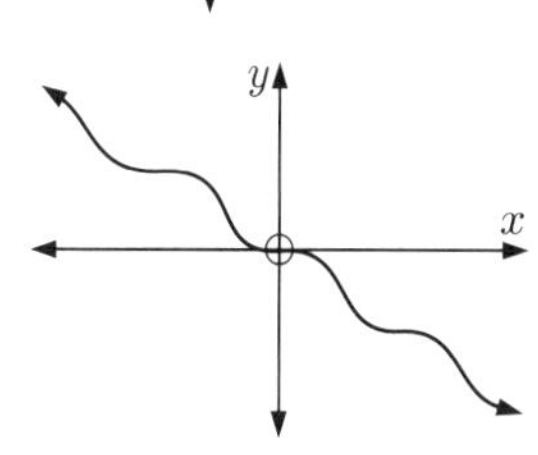

3 Will the graph of a straight line always be a function? Give evidence to support your answer.

4 Give algebraic evidence to show that the relation $x^2 + y^2 = 9$ is not a function.

B FUNCTION NOTATION

Function machines are sometimes used to illustrate how functions behave.

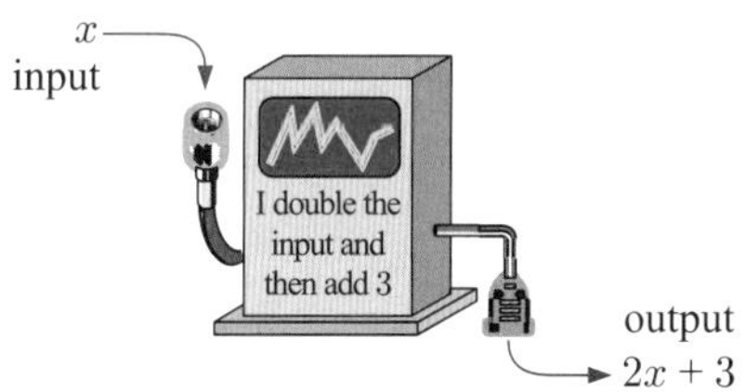

If 4 is the input fed into the machine, the output is $2(4) + 3 = 11$.

The above 'machine' has been programmed to perform a particular function.

If f is used to represent that particular function we can write:

f is the function that will convert x into $2x + 3$.

So, f would convert 2 into $2(2) + 3 = 7$ and
-4 into $2(-4) + 3 = -5$.

This function can be written as:

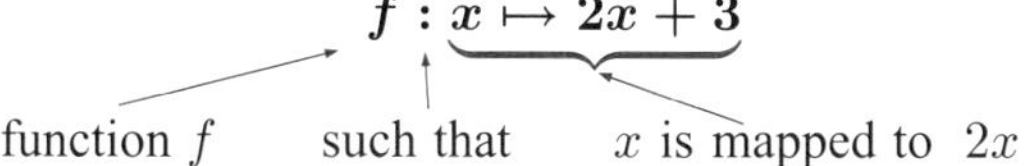

$f(x)$ is read as "f of x".

Two other equivalent forms we use are $f(x) = 2x + 3$ and $y = 2x + 3$.

$f(x)$ is the value of y for a given value of x, so $y = f(x)$.

f is the function which converts x into $f(x)$, so we write $f : x \mapsto f(x)$.

$y = f(x)$ is sometimes called the **function value** or **image** of x.

For $f(x) = 2x + 3$:

- $f(2) = 2(2) + 3 = 7$ indicates that the point (2, 7) lies on the graph of the function.
- $f(-4) = 2(-4) + 3 = -5$ indicates that the point (−4, −5) also lies on the graph.
- $f(x)$ is a **linear** function since it has the form $f(x) = ax + b$ where a, b are real constants and $a \neq 0$.

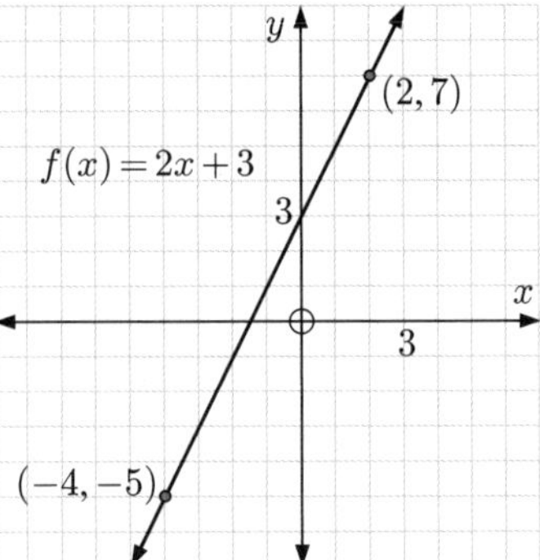

Example 2 — Self Tutor

If $f : x \mapsto 2x^2 - 3x$, find the value of: **a** $f(5)$ **b** $f(-4)$

$f(x) = 2x^2 - 3x$

a $f(5) = 2(5)^2 - 3(5)$ {replacing x with (5)}
$= 2 \times 25 - 15$
$= 35$

b $f(-4) = 2(-4)^2 - 3(-4)$ {replacing x with (−4)}
$= 2(16) + 12$
$= 44$

Example 3 — Self Tutor

If $f(x) = 5 - x - x^2$, find in simplest form: **a** $f(-x)$ **b** $f(x + 2)$

a $f(-x) = 5 - (-x) - (-x)^2$ {replacing x with $(-x)$}
$= 5 + x - x^2$

b $f(x + 2) = 5 - (x + 2) - (x + 2)^2$ {replacing x with $(x + 2)$}
$= 5 - x - 2 - [x^2 + 4x + 4]$
$= 3 - x - x^2 - 4x - 4$
$= -x^2 - 5x - 1$

EXERCISE 2B

1 If $f : x \mapsto 3x + 2$, find the value of:

a $f(0)$ **b** $f(2)$ **c** $f(-1)$ **d** $f(-5)$ **e** $f(-\frac{1}{3})$

2 If $f : x \mapsto 3x - x^2 + 2$, find the value of:

a $f(0)$ **b** $f(3)$ **c** $f(-3)$ **d** $f(-7)$ **e** $f(\frac{3}{2})$

3 If $g : x \mapsto x - \frac{4}{x}$, find the value of:

a $g(1)$ **b** $g(4)$ **c** $g(-1)$ **d** $g(-4)$ **e** $g(-\frac{1}{2})$

4 If $f(x) = 7 - 3x$, find in simplest form:

a $f(a)$ **b** $f(-a)$ **c** $f(a+3)$ **d** $f(b-1)$ **e** $f(x+2)$ **f** $f(x+h)$

5 If $F(x) = 2x^2 + 3x - 1$, find in simplest form:

a $F(x+4)$ **b** $F(2-x)$ **c** $F(-x)$ **d** $F(x^2)$ **e** $F(x^2-1)$ **f** $F(x+h)$

6 Suppose $G(x) = \dfrac{2x+3}{x-4}$.

a Evaluate: **i** $G(2)$ **ii** $G(0)$ **iii** $G(-\frac{1}{2})$

b Find a value of x such that $G(x)$ does not exist.

c Find $G(x+2)$ in simplest form.

d Find x if $G(x) = -3$.

7 f represents a function. What is the difference in meaning between f and $f(x)$?

8 The value of a photocopier t years after purchase is given by $V(t) = 9650 - 860t$ euros.

a Find $V(4)$ and state what $V(4)$ means.

b Find t when $V(t) = 5780$ and explain what this represents.

c Find the original purchase price of the photocopier.

9 On the same set of axes draw the graphs of three different functions $f(x)$ such that $f(2) = 1$ and $f(5) = 3$.

10 Find a linear function $f(x) = ax + b$ for which $f(2) = 1$ and $f(-3) = 11$.

11 Given $f(x) = ax + \dfrac{b}{x}$, $f(1) = 1$, and $f(2) = 5$, find constants a and b.

12 Given $T(x) = ax^2 + bx + c$, $T(0) = -4$, $T(1) = -2$, and $T(2) = 6$, find a, b, and c.

C DOMAIN AND RANGE

The **domain** of a relation is the set of values of x in the relation.

The **range** of a relation is the set of values of y in the relation.

NOTATION

The domain and range of a relation are often described using **set notation** or **interval notation** or a **number line graph**.

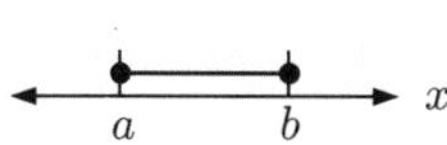

For numbers *between* a and b inclusive we write

$a \leqslant x \leqslant b$ *or* $x \in [a, b]$.

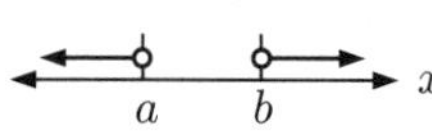

For numbers *outside* a and b we write

$x < a$ or $x > b$ *or* $x \in]-\infty, a[\cup]b, \infty[$.

Set notation	*Interval notation*	*Number line graph*	*Meaning*
$\{x \mid x \geqslant 3\}$	$x \in [3, \infty[$	3 x	the set of all x such that x is greater than or equal to 3
$\{x \mid x < 2\}$	$x \in]-\infty, 2[$	2 x	the set of all x such that x is less than 2
$\{x \mid -2 < x \leqslant 1\}$	$x \in]-2, 1]$	-2 1 x	the set of all x such that x is between -2 and 1, including 1
$\{x \mid x \leqslant 0 \ \cup \ x > 4\}$	$x \in]-\infty, 0] \ \cup \ ]4, \infty[$	0 4 x	the set of all x such that x is less than or equal to 0, or greater than 4

DOMAIN AND RANGE OF FUNCTIONS

To find the domain and range of a function, we can observe its graph.

For example:

(1)

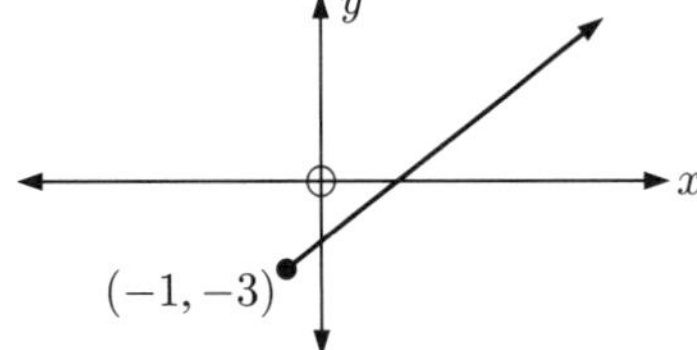

All values of $x \geqslant -1$ are included,
so the domain is $\{x \mid x \geqslant -1\}$ or $x \in [-1, \infty[$.
All values of $y \geqslant -3$ are included,
so the range is $\{y \mid y \geqslant -3\}$ or $y \in [-3, \infty[$.

(2)

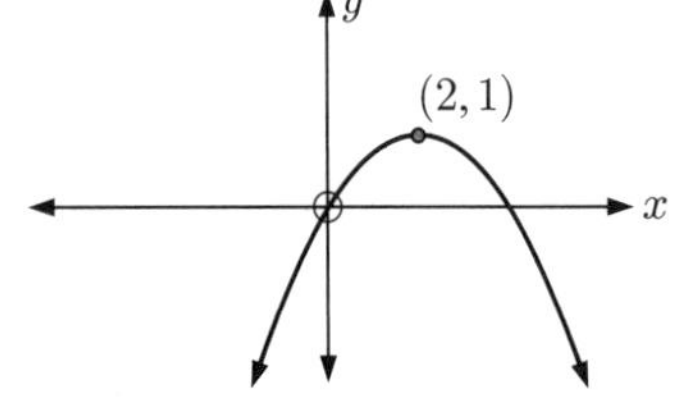

x can take any value,
so the domain is
$\{x \mid x \in \mathbb{R}\}$ or $x \in \mathbb{R}$.
y cannot be > 1,
so the range is
$\{y \mid y \leqslant 1\}$ or $y \in]-\infty, 1]$.

$x \in \mathbb{R}$ means x can be any real number.

(3)

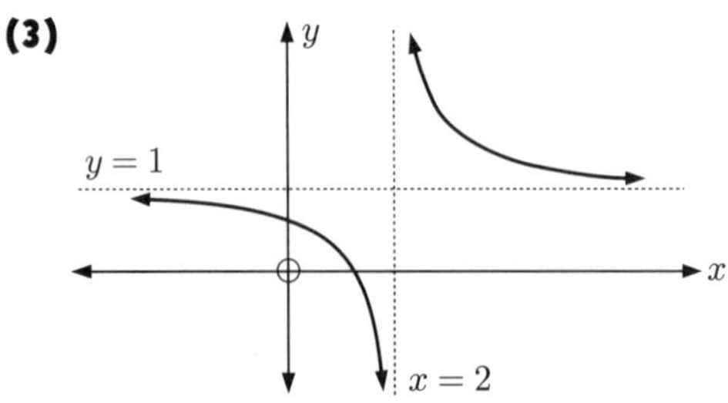

x can take all values except 2,
so the domain is $\{x \mid x \neq 2\}$ or $x \neq 2$.
y can take all values except 1,
so the range is $\{y \mid y \neq 1\}$ or $y \neq 1$.

To fully describe a function, we need both a rule *and* a domain.

For example, we can specify $f(x) = x^2$ where $x \geqslant 0$.

If a domain is not specified, we use the **natural domain**, which is the largest part of $\mathbb{R}$ for which $f(x)$ is defined.

For example, consider the domains in the table opposite:

$f(x)$	*Natural domain*
x^2	$x \in \mathbb{R}$
$\sqrt{x}$	$x \geqslant 0$
$\frac{1}{x}$	$x \neq 0$
$\frac{1}{\sqrt{x}}$	$x > 0$

Click on the icon to obtain software for finding the domain and range of different functions.

DOMAIN AND RANGE

Example 4 — Self Tutor

For each of the following graphs state the domain and range:

a

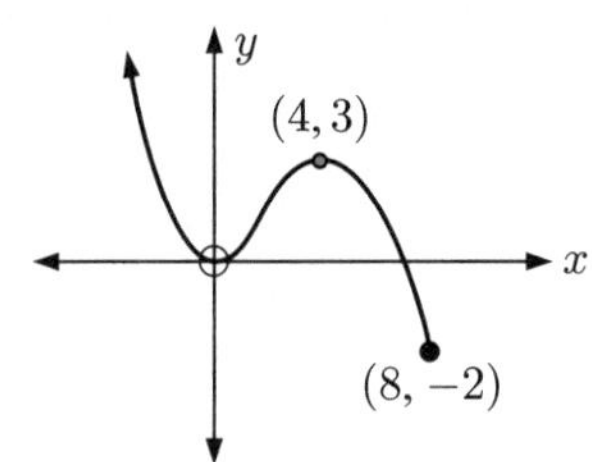

b

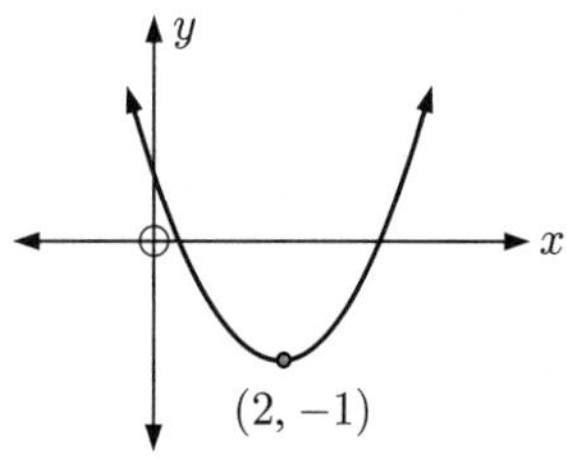

a Domain is $\{x \mid x \leqslant 8\}$
Range is $\{y \mid y \geqslant -2\}$

b Domain is $\{x \mid x \in \mathbb{R}\}$
Range is $\{y \mid y \geqslant -1\}$

EXERCISE 2C

1 For each of the following graphs, find the domain and range:

a

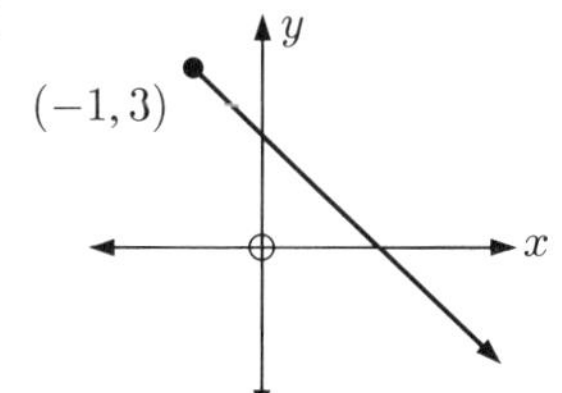

b

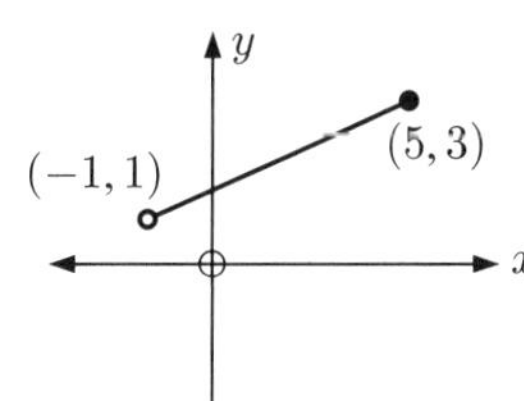

c

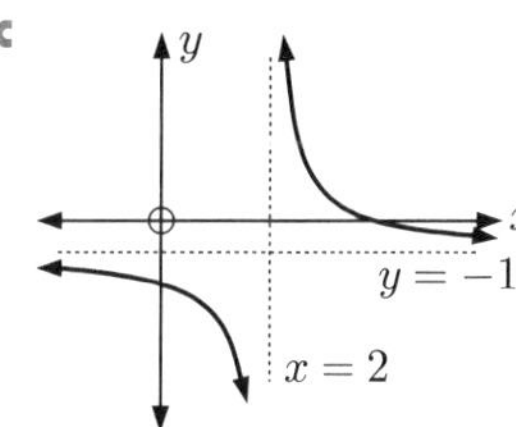

d

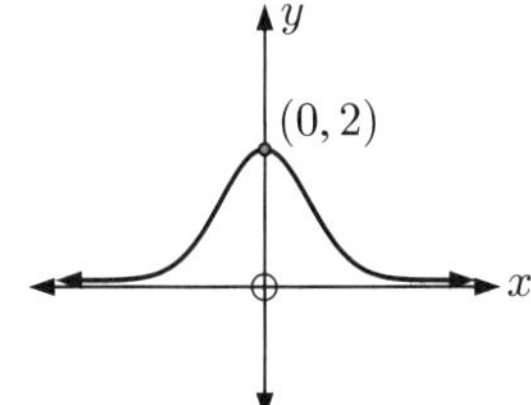

e

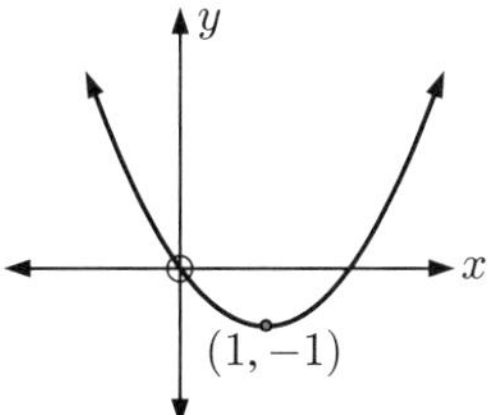

f

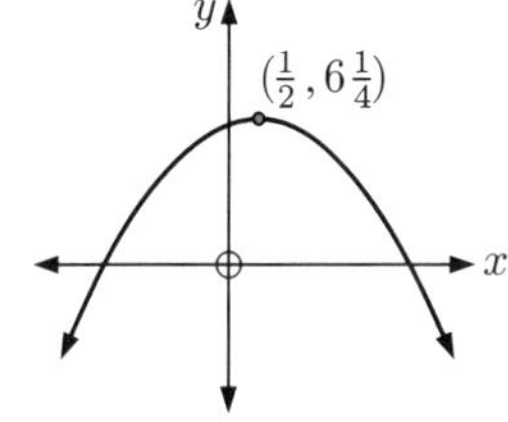

g

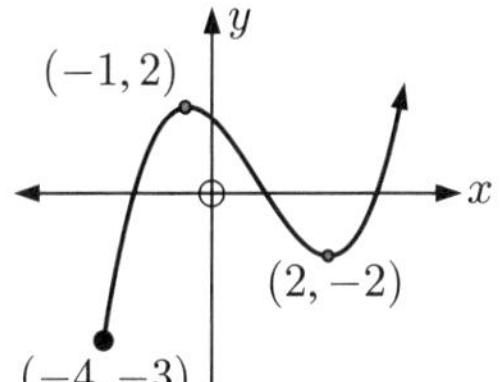

h

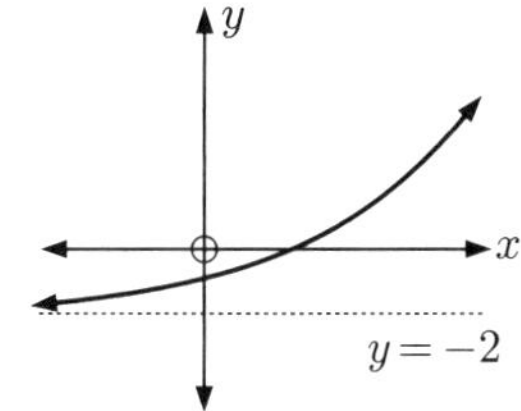

i

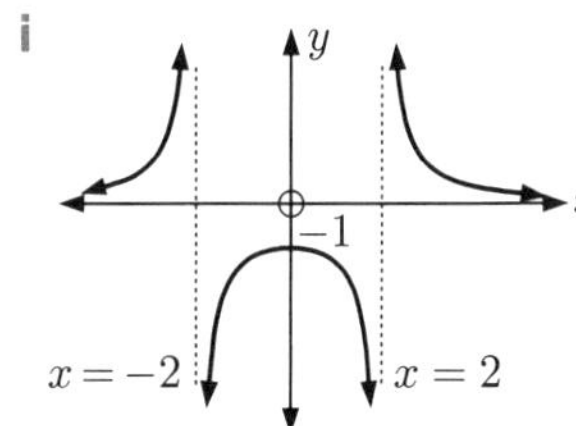

Example 5 **Self Tutor**

State the domain and range of each of the following functions:

a $f(x) = \sqrt{x-5}$ **b** $f(x) = \dfrac{1}{x-5}$ **c** $f(x) = \dfrac{1}{\sqrt{x-5}}$

a $\sqrt{x-5}$ is defined when $x - 5 \geqslant 0$
$\therefore\ x \geqslant 5$

$\therefore$ the domain is $\{x \mid x \geqslant 5\}$.
A square root cannot be negative.
$\therefore$ the range is $\{y \mid y \geqslant 0\}$.

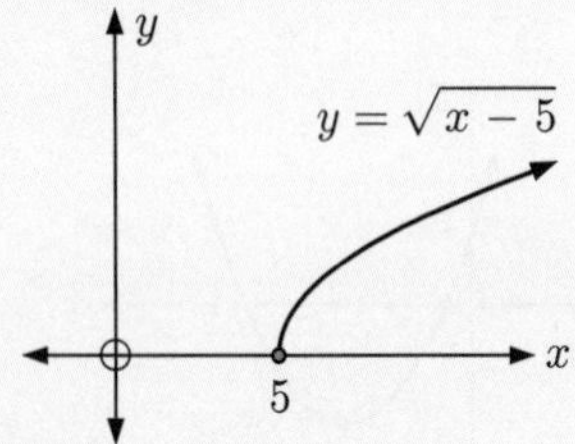

b $\dfrac{1}{x-5}$ is defined when $x - 5 \neq 0$
$\therefore\ x \neq 5$

$\therefore$ the domain is $\{x \mid x \neq 5\}$.
No matter how large or small x is,
$y = f(x)$ is never zero.
$\therefore$ the range is $\{y \mid y \neq 0\}$.

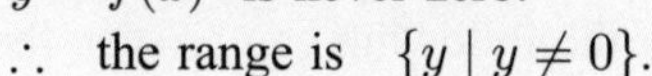

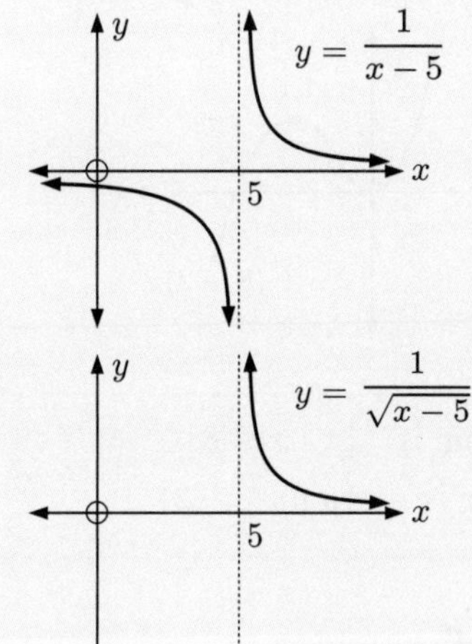

c $\dfrac{1}{\sqrt{x-5}}$ is defined when $x - 5 > 0$
$\therefore\ x > 5$

$\therefore$ the domain is $\{x \mid x > 5\}$.
$y = f(x)$ is always positive and never zero.
$\therefore$ the range is $\{y \mid y > 0\}$.

2 State the domain of each function:

a $f(x) = \sqrt{x+6}$ **b** $f : x \mapsto \dfrac{1}{x^2}$ **c** $f(x) = \dfrac{-7}{\sqrt{3-2x}}$

3 Find the domain and range of each of the following functions:

a $f : x \mapsto 2x - 1$ **b** $f(x) = 3$ **c** $f : x \mapsto \sqrt{x}$

d $f(x) = \dfrac{1}{x+1}$ **e** $f(x) = -\dfrac{1}{\sqrt{x}}$ **f** $f : x \mapsto \dfrac{1}{3-x}$

4 Use technology to help sketch graphs of the following functions.
Find the domain and range of each.

DOMAIN AND RANGE

a $f(x) = \sqrt{x-2}$ **b** $f : x \mapsto -\dfrac{1}{x^2}$ **c** $f : x \mapsto \sqrt{4-x}$

d $y = x^2 - 7x + 10$ **e** $f(x) = \sqrt{x^2+4}$ **f** $f(x) = \sqrt{x^2-4}$

g $f : x \mapsto 5x - 3x^2$ **h** $f : x \mapsto x + \dfrac{1}{x}$ **i** $y = \dfrac{x+4}{x-2}$

j $y = x^3 - 3x^2 - 9x + 10$ **k** $f : x \mapsto \dfrac{3x-9}{x^2-x-2}$ **l** $y = x^2 + x^{-2}$

m $y = x^3 + \dfrac{1}{x^3}$ **n** $f : x \mapsto x^4 + 4x^3 - 16x + 3$

5 Write down the domain and range for each of the following functions:

- **a** $\{(1, 3), (2, 5), (3, 7)\}$
- **b** $\{(-1, 3), (0, 3), (2, 5)\}$
- **c** $\{(-3, 1), (-2, 1), (-1, 1), (3, 1)\}$
- **d** $\{(x, y) \mid x^2 + y^2 = 4, \quad x \in \mathbb{Z}, \quad y \geqslant 0\}$

INVESTIGATION 1 — FLUID FILLING FUNCTIONS

When water is added at a **constant rate** to a cylindrical container, the depth of water in the container is a linear function of time. This is because the volume of water added is directly proportional to the time taken to add it. If the water was *not* added at a constant rate, depth of water over time would *not* be a linear function.

The linear depth-time graph for a cylindrical container is shown alongside.

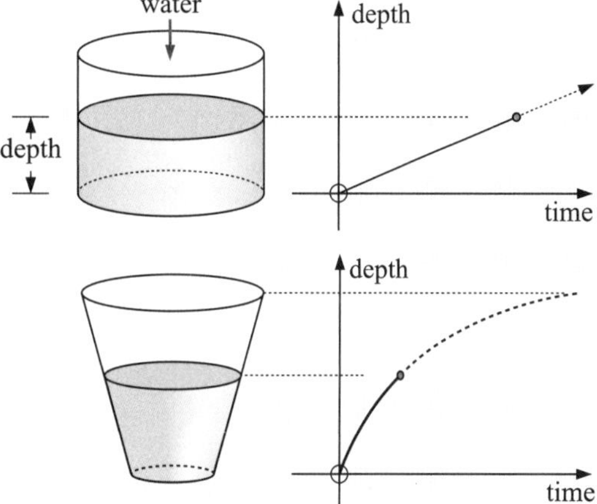

In this Investigation we explore the changes in the graph for different shaped containers such as the conical vase.

DEMO

What to do:

1 By examining the shape of each container, predict the depth-time graph when water is added at a constant rate.

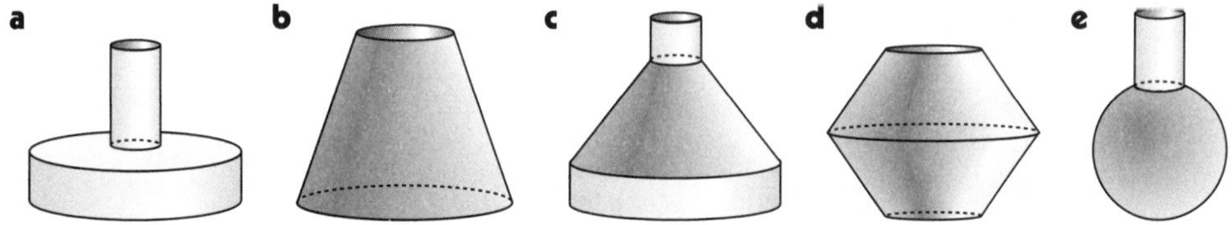

2 Use the water filling demonstration to check your answers to question **1**.

3 Write a brief report on the connection between the shape of a vessel and the corresponding shape of its depth-time graph. First examine cylindrical containers, then conical, then other shapes. Gradients of curves must be included in your report.

4 Suggest containers which would have the following depth-time graphs:

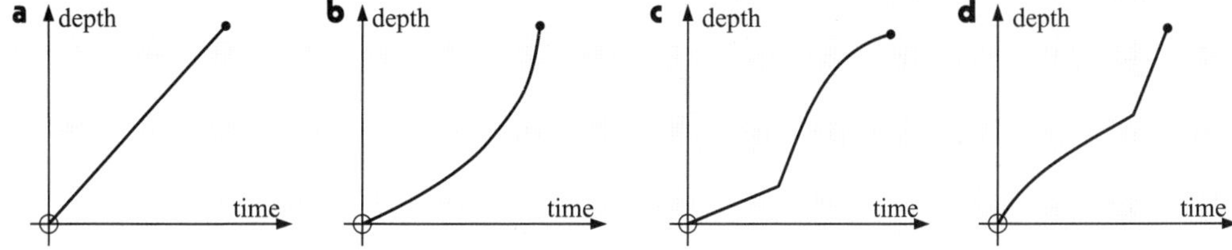

D COMPOSITE FUNCTIONS

Given functions $f : x \mapsto f(x)$ and $g : x \mapsto g(x)$, the **composite function** of f and g will convert x into $f(g(x))$.

$f \circ g$ is used to represent the composite function of f and g. It means "f following g".

$$(f \circ g)(x) = f(g(x)) \quad \text{or} \quad f \circ g : x \mapsto f(g(x)).$$

Consider $f : x \mapsto x^4$ and $g : x \mapsto 2x+3$.

$f \circ g$ means that g converts x to $2x+3$ and then f converts $(2x+3)$ to $(2x+3)^4$.

This is illustrated by the two function machines below.

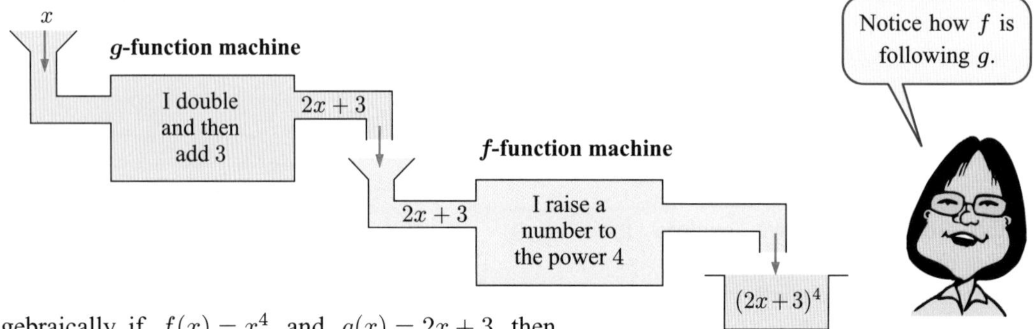

Algebraically, if $f(x) = x^4$ and $g(x) = 2x+3$ then

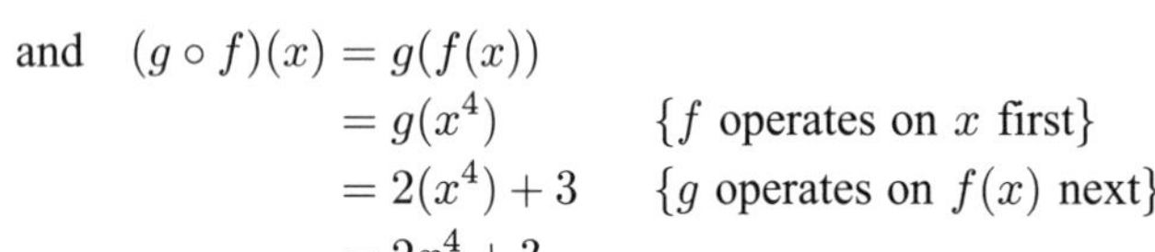

$$\begin{aligned}(f \circ g)(x) &= f(g(x)) \\ &= f(2x+3) && \{g \text{ operates on } x \text{ first}\} \\ &= (2x+3)^4 && \{f \text{ operates on } g(x) \text{ next}\}\end{aligned}$$

and

$$\begin{aligned}(g \circ f)(x) &= g(f(x)) \\ &= g(x^4) && \{f \text{ operates on } x \text{ first}\} \\ &= 2(x^4)+3 && \{g \text{ operates on } f(x) \text{ next}\} \\ &= 2x^4+3\end{aligned}$$

So, $f(g(x)) \neq g(f(x))$.

In general, $(f \circ g)(x) \neq (g \circ f)(x)$.

Example 6 **Self Tutor**

Given $f : x \mapsto 2x+1$ and $g : x \mapsto 3-4x$, find in simplest form:

a $(f \circ g)(x)$ **b** $(g \circ f)(x)$

$f(x) = 2x+1$ and $g(x) = 3-4x$

a
$$\begin{aligned}(f \circ g)(x) &= f(g(x)) \\ &= f(3-4x) \\ &= 2(3-4x)+1 \\ &= 6-8x+1 \\ &= 7-8x\end{aligned}$$

b
$$\begin{aligned}(g \circ f)(x) &= g(f(x)) \\ &= g(2x+1) \\ &= 3-4(2x+1) \\ &= 3-8x-4 \\ &= -8x-1\end{aligned}$$

In the previous Example you should have observed how we can substitute an expression into a function.

If $f(x) = 2x + 1$ then $f(\Delta) = 2(\Delta) + 1$
and so $f(3 - 4x) = 2(3 - 4x) + 1$.

Example 7 — Self Tutor

Given $f(x) = 6x - 5$ and $g(x) = x^2 + x$, determine:

a $(g \circ f)(-1)$ **b** $(f \circ f)(0)$

a $(g \circ f)(-1) = g(f(-1))$

Now $f(-1) = 6(-1) - 5$
$= -11$

$\therefore$ $(g \circ f)(-1) = g(-11)$
$= (-11)^2 + (-11)$
$= 110$

b $(f \circ f)(0) = f(f(0))$

Now $f(0) = 6(0) - 5$
$= -5$

$\therefore$ $(f \circ f)(0) = f(-5)$
$= 6(-5) - 5$
$= -35$

The domain of the composite of two functions depends on the domain of the original functions.

For example, consider $f(x) = x^2$ with domain $x \in \mathbb{R}$ and $g(x) = \sqrt{x}$ with domain $x \geqslant 0$.

$$\begin{aligned}(f \circ g)(x) &= f(g(x))\\ &= (\sqrt{x})^2\\ &= x\end{aligned}$$

The domain of $(f \circ g)(x)$ is $x \geqslant 0$, not $\mathbb{R}$, since $(f \circ g)(x)$ is defined using function $g(x)$ which has domain $x \geqslant 0$, not $\mathbb{R}$.

EXERCISE 2D

1 Given $f : x \mapsto 2x + 3$ and $g : x \mapsto 1 - x$, find in simplest form:

a $(f \circ g)(x)$ **b** $(g \circ f)(x)$ **c** $(f \circ g)(-3)$

2 Given $f(x) = \sqrt{6 - x}$ and $g(x) = 5x - 7$, find:

a $(g \circ g)(x)$ **b** $(f \circ g)(1)$ **c** $(g \circ f)(6)$

3 Given $f : x \mapsto x^2$ and $g : x \mapsto 2 - x$, find $(f \circ g)(x)$ and $(g \circ f)(x)$.
Find also the domain and range of $f \circ g$ and $g \circ f$.

4 Suppose $f : x \mapsto x^2 + 1$ and $g : x \mapsto 3 - x$.

a Find in simplest form: **i** $(f \circ g)(x)$ **ii** $(g \circ f)(x)$

b Find the value(s) of x such that $(g \circ f)(x) = f(x)$.

5 Functions f and g are defined by $f = \{(0, 3), (1, 0), (2, 1), (3, 2)\}$ and $g = \{(0, 1), (1, 2), (2, 3), (3, 0)\}$. Find $f \circ g$.

6 Functions f and g are defined by $f = \{(0, 2), (1, 5), (2, 7), (3, 9)\}$ and $g = \{(2, 2), (5, 0), (7, 1), (9, 3)\}$. Find: **a** $f \circ g$ **b** $g \circ f$.

7 Given $f(x) = \dfrac{x + 3}{x + 2}$ and $g(x) = \dfrac{x + 1}{x - 1}$, find in simplest form:

a $(f \circ g)(x)$ **b** $(g \circ f)(x)$ **c** $(g \circ g)(x)$

In each case, find the domain of the composite function.

8 Given $f(x) = \sqrt{1-x}$ and $g(x) = x^2$, find:

a $(f \circ g)(x)$ **b** the domain and range of $(f \circ g)(x)$

9 **a** If $ax + b = cx + d$ for all values of x, show that $a = c$ and $b = d$.
Hint: If it is true for all x, it is true for $x = 0$ and $x = 1$.

b Given $f(x) = 2x + 3$ and $g(x) = ax + b$ and that $(f \circ g)(x) = x$ for all values of x, deduce that $a = \frac{1}{2}$ and $b = -\frac{3}{2}$.

c Is the result in **b** true if $(g \circ f)(x) = x$ for all x?

E EVEN AND ODD FUNCTIONS

EVEN FUNCTIONS

Consider the function $f(x) = x^2$.

Now $f(3) = 3^2 = 9$
and $f(-3) = (-3)^2 = 9$
So, $f(-3) = f(3)$.

In fact, $f(-x) = f(x)$ for all x.
We know this is true since

$$\begin{aligned} f(-x) &= (-x)^2 \\ &= x^2 \\ &= f(x). \end{aligned}$$

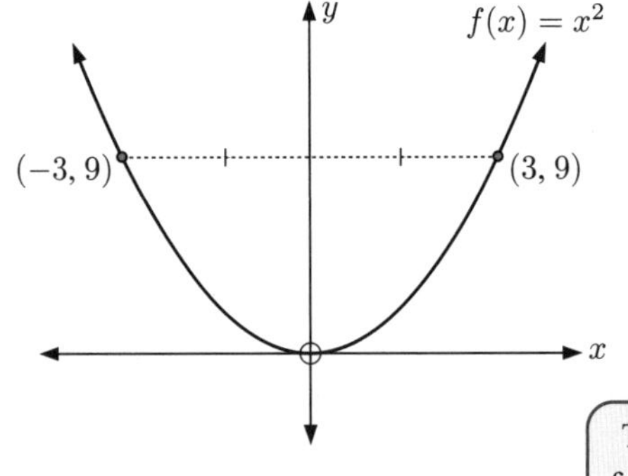

The graph of an even function is symmetrical about the y-axis.

We say that $f(x) = x^2$ is an **even function**.

A function $f(x)$ is **even** if $f(-x) = f(x)$ for all x in the domain of f.

ODD FUNCTIONS

Consider the function $f(x) = x^3$.

Now $f(2) = 2^3 = 8$
and $f(-2) = (-2)^3 = -8$
So, $f(-2) = -f(2)$.

In fact, $f(-x) = -f(x)$ for all x.
We know this is true since

$$\begin{aligned} f(-x) &= (-x)^3 \\ &= (-1)^3 x^3 \\ &= -x^3 \\ &= -f(x). \end{aligned}$$

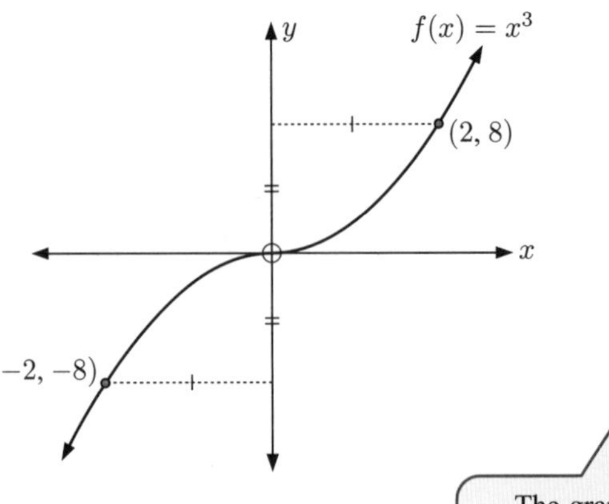

The graph of an odd function has rotational symmetry about the origin.

We say that $f(x) = x^3$ is an **odd function**.

A function $f(x)$ is **odd** if $f(-x) = -f(x)$ for all x in the domain of f.

Example 8 — Self Tutor

Determine whether the following functions are even, odd, or neither:

a $f(x)=\dfrac{1}{x}$ b $g(x)=x^4+x^2$ c $h(x)=x^2-2x+3$

a
$$f(x)=\frac{1}{x}$$
$$\therefore\ f(-x)=\frac{1}{-x}=-\frac{1}{x}=-f(x)$$
$\therefore$ $f(x)$ is an odd function.

b
$$g(x)=x^4+x^2$$
$$\therefore\ g(-x)=(-x)^4+(-x)^2=x^4+x^2=g(x)$$
$\therefore$ $g(x)$ is an even function.

c
$$h(x)=x^2-2x+3$$
$$\therefore\ h(-x)=(-x)^2-2(-x)+3=x^2+2x+3$$
which is neither $h(x)$ or $-h(x)$.
$\therefore$ $h(x)$ is neither even nor odd.

Example 9 — Self Tutor

Prove that the sum of two odd functions is also an odd function.

Suppose $h(x)=f(x)+g(x)$ where $f(x)$ and $g(x)$ are odd functions.

Now
$$h(-x)=f(-x)+g(-x)$$
$$=-f(x)-g(x)$$
$$=-(f(x)+g(x))$$
$$=-h(x)\quad\text{for all } x$$

$\therefore$ $h(x)$ is odd.

Thus the sum of two odd functions is an odd function.

EXERCISE 2E

1 Show that $f(x)=\dfrac{1}{x^2}+2$ is an even function.

2 Show that $f(x)=x^3-3x$ is an odd function.

3 Determine whether the following functions are even, odd, or neither:

a $f(x)=5x$ b $f(x)=-4x+3$ c $f(x)=\dfrac{3}{x^2-4}$

d $f(x)=2x^3-\dfrac{5}{x}$ e $f(x)=x^2+\dfrac{7}{x^2}-3$ f $f(x)=\sqrt{x}$

4 Suppose $f(x)=(2x+3)(x+a)$ where a is a constant. Find a given that f is an even function.

5 Suppose $g(x)=(x+1)\left(\dfrac{1}{x}+b\right)$ where b is a constant. Find b given that g is an odd function.

6 **a** If the quadratic function $f(x) = ax^2 + bx + c,\ a \neq 0$ is an even function, prove that $b = 0$.

b For what values of the constants is the cubic function $g(x) = ax^3 + bx^2 + cx + d,\ a \neq 0$, an odd function?

c Under what conditions is the quartic function $h(x) = ax^4 + bx^3 + cx^2 + dx + e,\ a \neq 0$, an even function?

7 Prove that:

a the sum of two even functions is an even function

b the difference between two odd functions is an odd function

c the product of two odd functions is an even function.

8 Suppose $f(x)$ is an even function and $g(x)$ is an odd function. Determine whether the function $(f \circ g)(x)$ is even, odd, or neither.

F SIGN DIAGRAMS

Sometimes we do not wish to draw a time-consuming graph of a function but wish to know when the function is positive, negative, zero, or undefined. A **sign diagram** enables us to do this and is relatively easy to construct.

For the function $f(x)$, the sign diagram consists of:

- a **horizontal line** which is really the x-axis
- **positive** $(+)$ and **negative** $(-)$ signs indicating that the graph is **above** and **below** the x-axis respectively
- the **zeros** of the function, which are the x-intercepts of the graph of $y = f(x)$, and the **roots** of the equation $f(x) = 0$
- values of x where the graph is undefined.

Consider the three functions $y = (x+2)(x-1)$, $y = -2(x-1)^2$, and $y = \dfrac{4}{x}$.

Function	$y = (x+2)(x-1)$	$y = -2(x-1)^2$	$y = \dfrac{4}{x}$
Graph	y, x, -2, 1	y, x, 1	y, x
Sign diagram	$+$ -2 $-$ 1 $+$ x	$-$ 1 $-$ x	$-$ 0 $+$ x

You should notice that:

- A sign change occurs about a zero of the function for single linear factors such as $(x+2)$ and $(x-1)$. This indicates **cutting** of the x-axis.
- No sign change occurs about a zero of the function for squared linear factors such as $(x-1)^2$. This indicates **touching** of the x-axis.
- 0 indicates that a function is **undefined** at $x = 0$.

In general:

- when a function has a linear factor with an **odd power** there is a change of sign about that zero
- when a function has a linear factor with an **even power** there is no sign change about that zero.

Example 10 **Self Tutor**

Draw sign diagrams for:

a

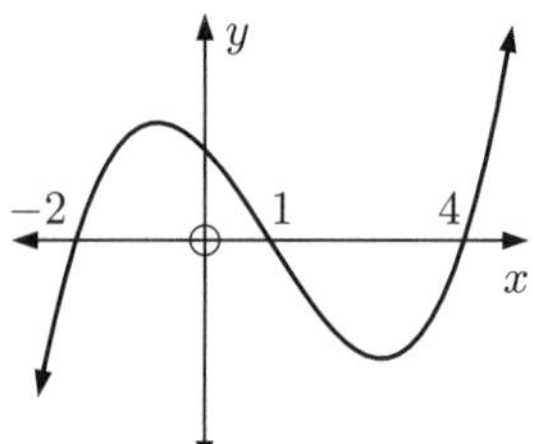

b

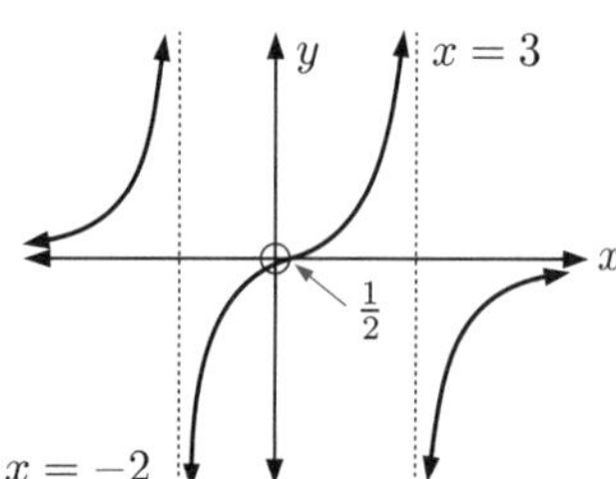

a

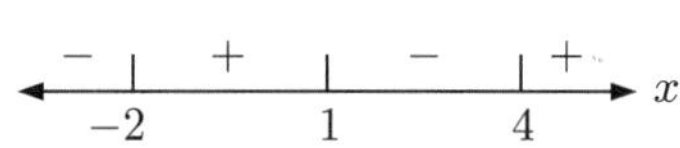

b

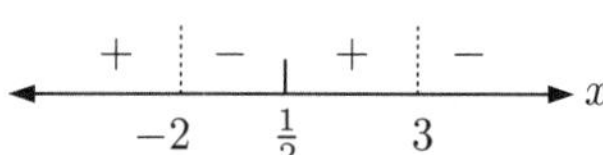

Example 11 **Self Tutor**

Draw a sign diagram for:

a $(x+3)(x-1)$

b $2(2x+5)(3-x)$

a $(x+3)(x-1)$ has zeros -3 and 1.

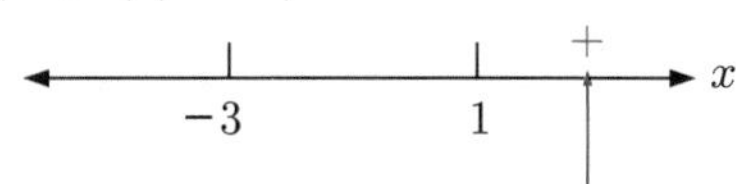

We substitute any number > 1.
When $x = 2$ we have $(5)(1) > 0$,
so we put a $+$ sign here.

As the factors are distinct and linear, the signs alternate.

b $2(2x+5)(3-x)$ has zeros $-\frac{5}{2}$ and 3.

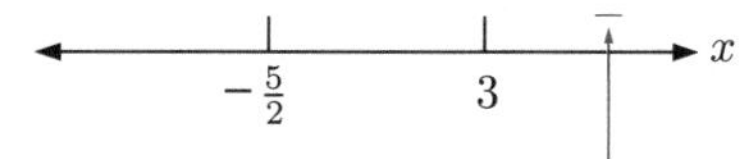

We substitute any number > 3.
When $x = 5$ we have $2(15)(-2) < 0$,
so we put a $-$ sign here.

As the factors are distinct and linear, the signs alternate.

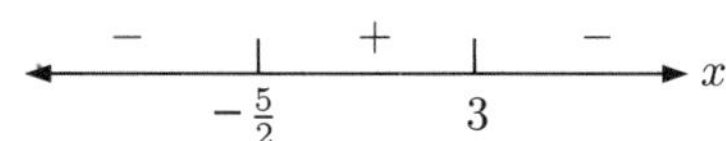

EXERCISE 2F

1 Draw sign diagrams for these graphs:

a
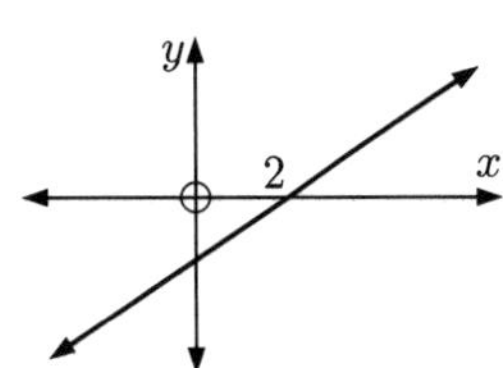

b
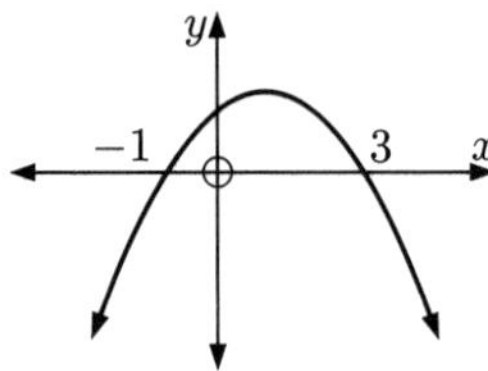

c
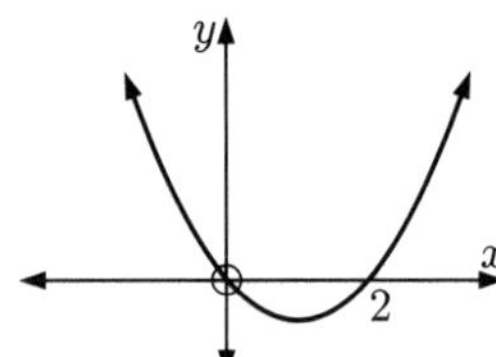

d
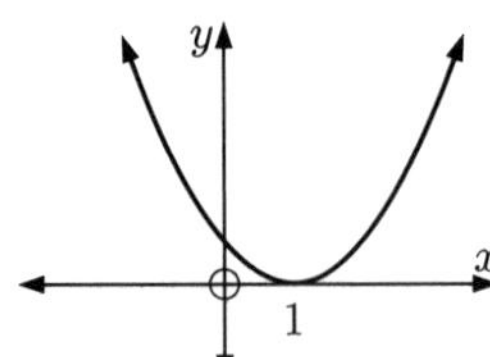

e
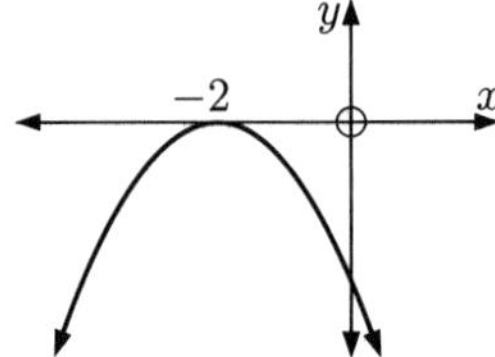

f
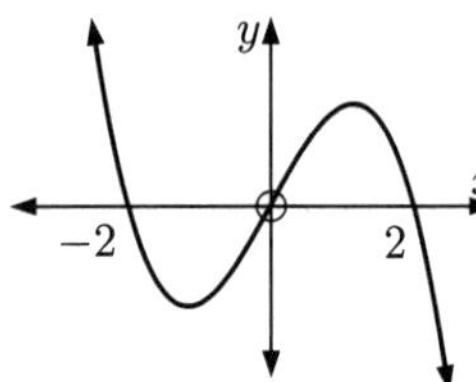

g
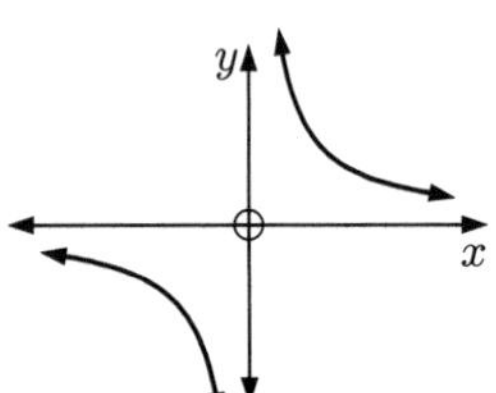

h
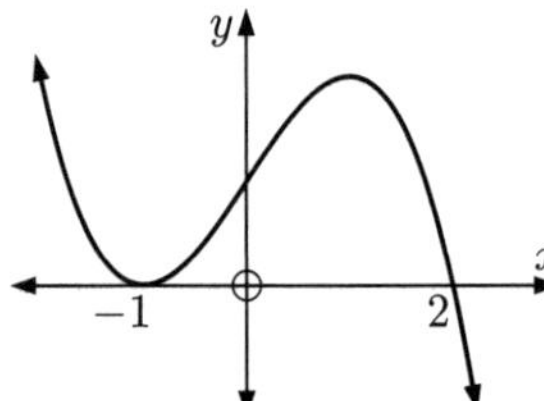

i
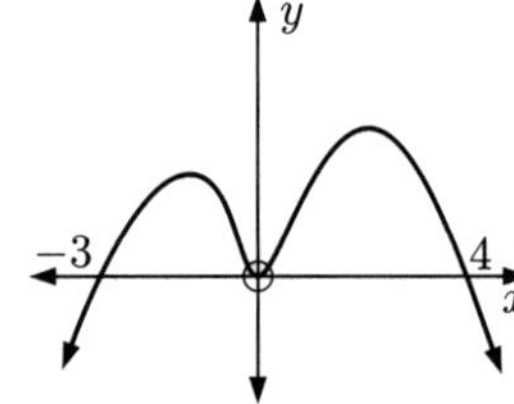

j
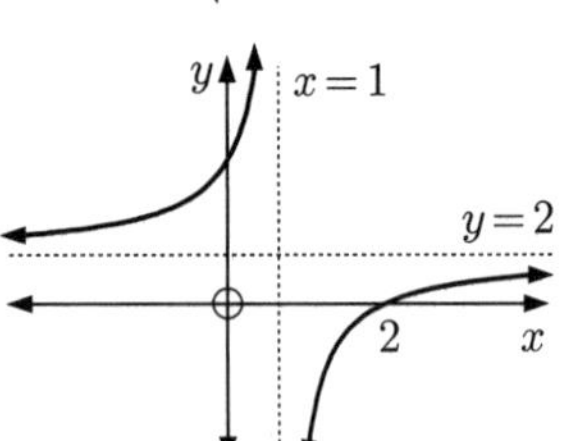

k
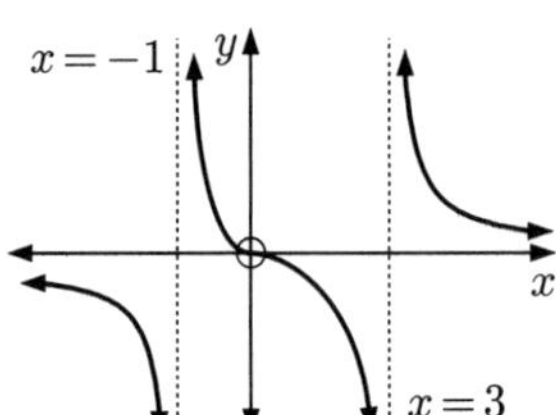

l
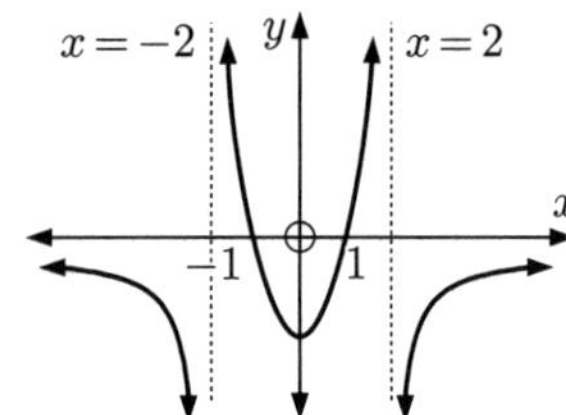

Example 12 — Self Tutor

Draw a sign diagram for: **a** $12 - 3x^2$ **b** $-4(x-3)^2$

a $12 - 3x^2 = -3(x^2 - 4)$
$= -3(x+2)(x-2)$
which has zeros -2 and 2.

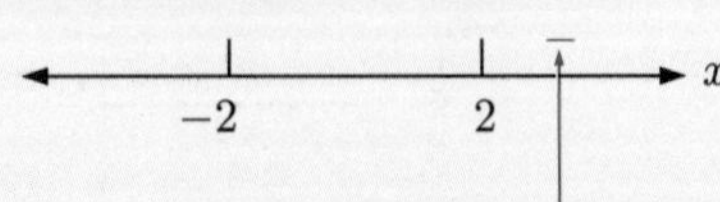

We substitute any number > 2.
When $x = 3$ we have $-3(5)(1) < 0$,
so we put a $-$ sign here.

As the factors are distinct and linear, the signs alternate.

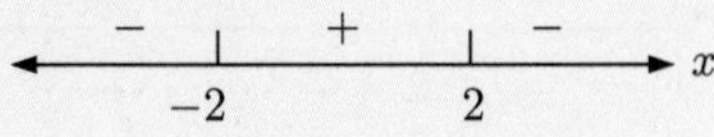

b $-4(x-3)^2$ has zero 3.

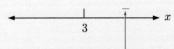

We substitute any number > 3.
When $x = 4$ we have $-4(1)^2 < 0$,
so we put a $-$ sign here.

As the linear factor is squared, the signs do not change.

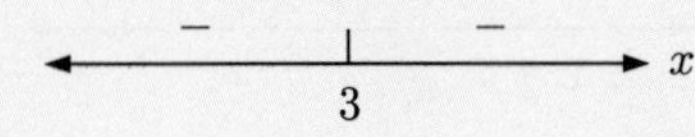

2 Draw sign diagrams for:

a $(x+4)(x-2)$
b $x(x-3)$
c $x(x+2)$
d $-(x+1)(x-3)$
e $(2x-1)(3-x)$
f $(5-x)(1-2x)$
g x^2-9
h $4-x^2$
i $5x-x^2$
j x^2-3x+2
k $2-8x^2$
l $6x^2+x-2$
m $6-16x-6x^2$
n $-2x^2+9x+5$
o $-15x^2-x+2$

3 Draw sign diagrams for:

a $(x+2)^2$
b $(x-3)^2$
c $-(x+2)^2$
d $-(x-4)^2$
e x^2-2x+1
f $-x^2+4x-4$
g $4x^2-4x+1$
h $-x^2-6x-9$
i $-4x^2+12x-9$

Example 13 **Self Tutor**

Draw a sign diagram for $\dfrac{x-1}{2x+1}$.

$\dfrac{x-1}{2x+1}$ is zero when $x=1$ and undefined when $x=-\frac{1}{2}$.

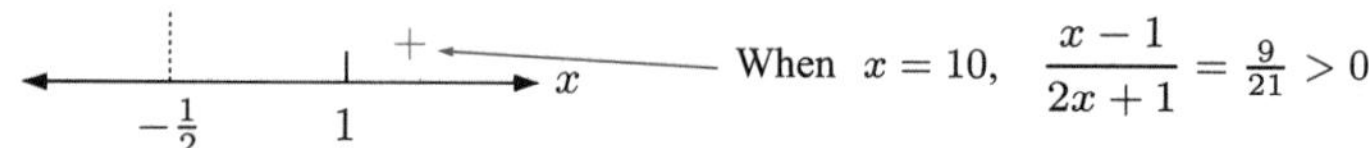

Since $(x-1)$ and $(2x+1)$ are distinct, linear factors, the signs alternate.

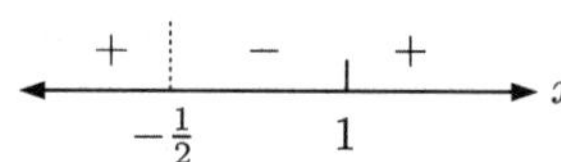

4 Draw sign diagrams for:

a $\dfrac{x+2}{x-1}$
b $\dfrac{x}{x+3}$
c $\dfrac{2x+3}{4-x}$
d $\dfrac{4x-1}{2-x}$
e $\dfrac{3x}{x-2}$
f $\dfrac{-8x}{3-x}$
g $\dfrac{(x-1)^2}{x}$
h $\dfrac{4x}{(x+1)^2}$
i $\dfrac{(x+2)(x-1)}{3-x}$
j $\dfrac{x(x-1)}{2-x}$
k $\dfrac{x^2-4}{-x}$
l $\dfrac{3-x}{2x^2-x-6}$
m $\dfrac{x^2-3}{x+1}$
n $\dfrac{x^2+1}{x}$
o $\dfrac{x^2+2x+4}{x+1}$
p $\dfrac{-(x-3)^2(x^2+2)}{x+3}$
q $\dfrac{-x^2(x+2)}{5-x}$
r $\dfrac{x^2+4}{(x-3)^2(x-1)}$
s $\dfrac{x-5}{x+1}+3$
t $\dfrac{x-2}{x+3}-4$
u $\dfrac{3x+2}{x-2}-\dfrac{x-3}{x+3}$

G INEQUALITIES (INEQUATIONS)

An **equation** is a mathematical statement that two expressions are equal.

Sometimes we have a statement that one expression is *greater than*, or else *greater than or equal to*, another. We call this an **inequation** or **inequality**.

$2x+3>11-x$ and $\dfrac{2x-1}{x}\leqslant\dfrac{x+3}{5}$ are examples of inequalities.

In this section we aim to find *all* values of the unknown for which the inequality is true.

INVESTIGATION SOLVING INEQUALITIES

Jon's method of solving $\dfrac{3x+2}{1-x}>4$ was:

If $\dfrac{3x+2}{1-x}>4$, then $3x+2>4(1-x)$
$\therefore\ 3x+2>4-4x$
$\therefore\ 7x>2$
$\therefore\ x>\frac{2}{7}$

However, Sarah pointed out that if $x=5$, $\dfrac{3x+2}{1-x}=\frac{17}{-4}=-4\frac{1}{4}$ and $-4\frac{1}{4}$ is **not** greater than 4.

They concluded that there was something wrong with Jon's method of solution.

By graphing $y=\dfrac{3x+2}{1-x}$ and $y=4$ on the same set of axes, they found that

$\dfrac{3x+2}{1-x}>4$ when $\frac{2}{7}<x<1$.

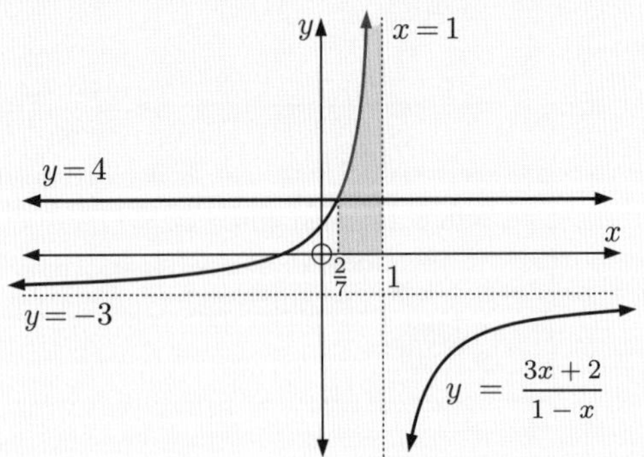

What to do:

1. At what step was Jon's method wrong?
2. Suggest an algebraic method which does give the correct answer.

From the **Investigation** above you should have concluded that multiplying both sides of an inequality by an unknown can lead to incorrect results.

We can, however, make use of the following properties:

(1) If $a>b$ and $c\in\mathbb{R}$, then $a+c>b+c$.

(2) If $a>b$ and $c>0$, then $ac>bc$.

(3) If $a>b$ and $c<0$, then $ac<bc$.

(4) If $a>b>0$, then $a^2>b^2$.

(5) If $a<b<0$, then $a^2>b^2$.

Properties **(2)** and **(3)** tell us that when we multiply an inequality by a *positive* number, the inequality sign is not changed. However, when we multiply an inequality by a *negative* number, we need to reverse the inequality sign.

For example:

$$3 < 5 \qquad\qquad\qquad 3 < 5$$
$$\therefore\ 3 \times 2 < 5 \times 2 \qquad\qquad \therefore\ 3 \times -2 > 5 \times -2$$
$$\therefore\ 6 < 10 \qquad\qquad \therefore\ -6 > -10$$

In the **Investigation**, John's method failed because he did not treat the cases where $1 - x > 0$ and $1 - x < 0$ separately.

If $1 - x > 0$ then $x < 1$

and $\dfrac{3x+2}{1-x} > 4$

$\therefore\ 3x + 2 > 4(1 - x)$

$\therefore\ 3x + 2 > 4 - 4x$

$\therefore\ 7x > 2$

$\therefore\ x > \frac{2}{7}$

This gives us $\dfrac{2}{7} < x < 1$.

If $1 - x < 0$ then $x > 1$

and $\dfrac{3x+2}{1-x} > 4$

$\therefore\ 3x + 2 < 4(1 - x)$

$\therefore\ 3x + 2 < 4 - 4x$

$\therefore\ 7x < 2$

$\therefore\ x < \frac{2}{7}$

We cannot have $x > 1$ *and* $x < \frac{2}{7}$ at the same time. We therefore have no more solutions.

The complete solution is $\dfrac{2}{7} < x < 1$.

To solve inequalities we use these steps:

- Make the RHS zero by shifting all terms to the LHS.
- Fully factorise the LHS.
- Draw a sign diagram for the LHS.
- Determine the values required from the sign diagram.

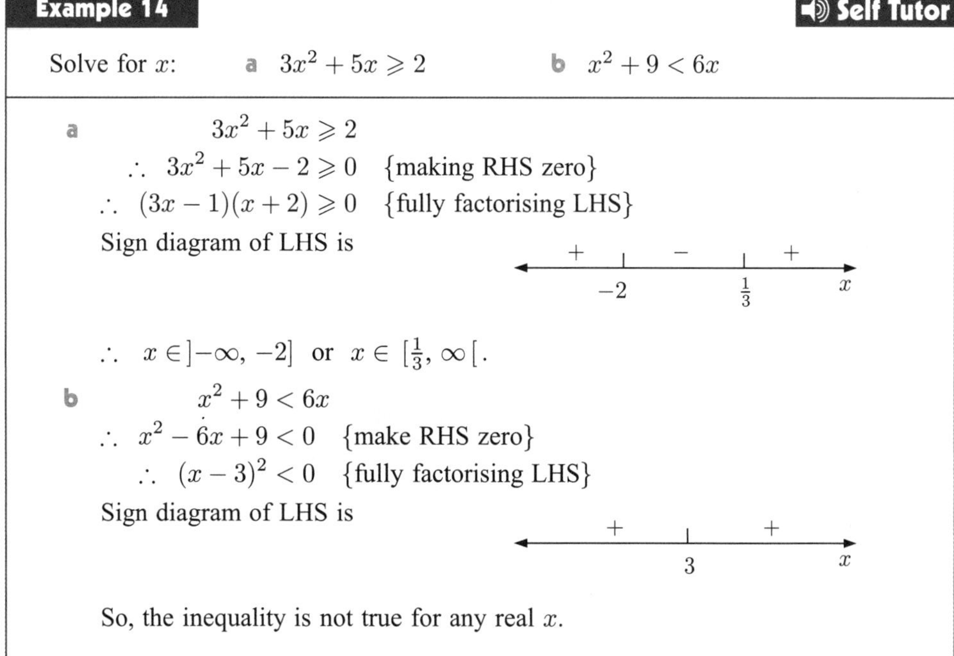

Example 14 Self Tutor

Solve for x: **a** $3x^2 + 5x \geqslant 2$ **b** $x^2 + 9 < 6x$

a $3x^2 + 5x \geqslant 2$

$\therefore\ 3x^2 + 5x - 2 \geqslant 0$ {making RHS zero}

$\therefore\ (3x - 1)(x + 2) \geqslant 0$ {fully factorising LHS}

Sign diagram of LHS is

$\therefore\ x \in]-\infty, -2]$ or $x \in [\frac{1}{3}, \infty[$.

b $x^2 + 9 < 6x$

$\therefore\ x^2 - 6x + 9 < 0$ {make RHS zero}

$\therefore\ (x - 3)^2 < 0$ {fully factorising LHS}

Sign diagram of LHS is

So, the inequality is not true for any real x.

In situations where we cannot find the sign diagram easily, we can use a graphical method.

Example 15 — Self Tutor

Solve for x: $\frac{3x+1}{x-2} \leqslant 2x - 6$.

We graph $Y_1 = \frac{3X+1}{X-2}$ and $Y_2 = 2X - 6$ on the same set of axes:

Casio fx-CG20

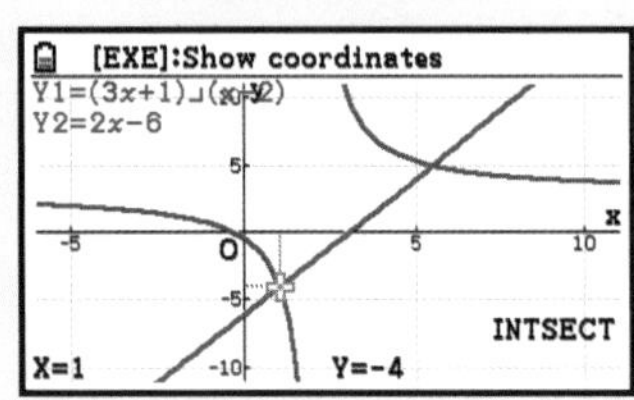

TI-84 Plus

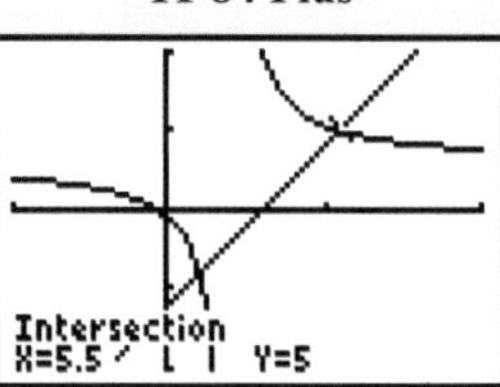

TI-*n*spire

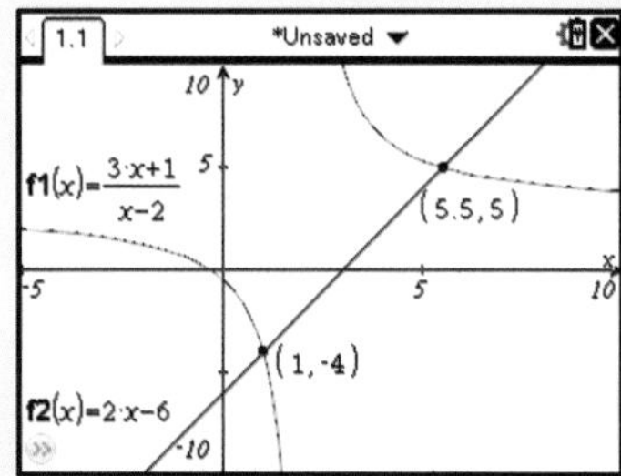

The graphs intersect at $x = 1$ and $x = 5.5$.

So, $\frac{3x+1}{x-2} \leqslant 2x - 6$ when
$1 \leqslant x < 2$ or $x \geqslant 5.5$.

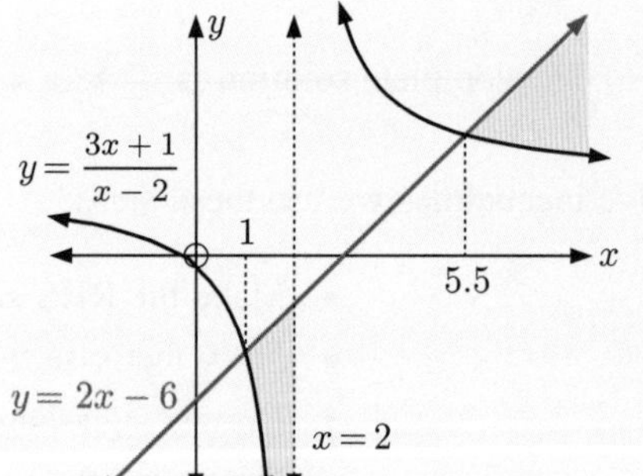

The graph of $y = \frac{3x+1}{x-2}$ is undefined when $x = 2$, so we do not include this value in our interval.

EXERCISE 2G

1 Solve for x:

a $(2 - x)(x + 3) \geqslant 0$
b $(x - 1)^2 < 0$
c $(2x + 1)(3 - x) > 0$
d $x^2 \geqslant x$
e $x^2 \geqslant 3x$
f $3x^2 + 2x < 0$
g $x^2 < 4$
h $2x^2 \geqslant 4$
i $x^2 + 4x + 4 > 0$
j $2x^2 \geqslant x + 3$
k $4x^2 - 4x + 1 < 0$
l $6x^2 + 7x < 3$
m $3x^2 > 8(x + 2)$
n $2x^2 - 4x + 2 > 0$
o $6x^2 + 1 \leqslant 5x$
p $1 + 5x < 6x^2$
q $12x^2 \geqslant 5x + 2$
r $2x^2 + 9 > 9x$

2 Solve for x:

a $\frac{x+4}{2x-1} > 0$
b $\frac{x+1}{4-x} \geqslant 0$
c $\frac{2}{2x-5} < \frac{1}{x+7}$
d $\frac{x^2+5x}{x^2-4} \leqslant 0$
e $\frac{2x-3}{x+2} < \frac{2x}{x-2}$
f $\frac{x+6}{2x-5} > 2x + 3$

3 Solve for x:

a $x^3 \geqslant x$
b $x^3 - 2x^2 + 6 < 5x$
c $x^3 - 6 \leqslant 2x - 3x^2$

H THE MODULUS FUNCTION

The **modulus** or **absolute value** of a real number x is its distance from 0 on the number line. We write the modulus of x as $|x|$.

Because the modulus is a distance, it cannot be negative.

- If $x > 0$, $|x| = x$.

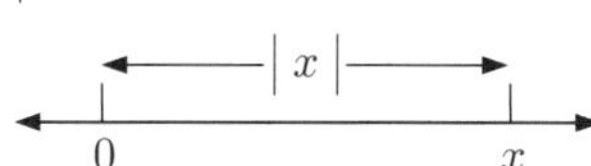

- If $x < 0$, $|x| = -x$.

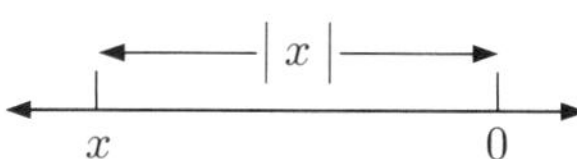

For example: The modulus of 7 is 7, which is written as $|7| = 7$.
The modulus of -5 is 5, which is written as $|-5| = 5$.

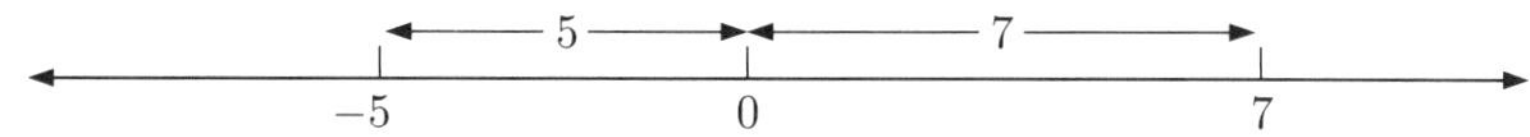

This leads us to the **algebraic definition** of modulus:

$$\text{The } \textbf{modulus of } \boldsymbol{x}, \quad |\boldsymbol{x}| = \begin{cases} \boldsymbol{x} & \textbf{if } \boldsymbol{x} \geqslant \mathbf{0} \\ -\boldsymbol{x} & \textbf{if } \boldsymbol{x} < \mathbf{0} \end{cases}$$

The relation $y = |x|$ is in fact a function. We call it the **modulus function**, and it has the graph shown.

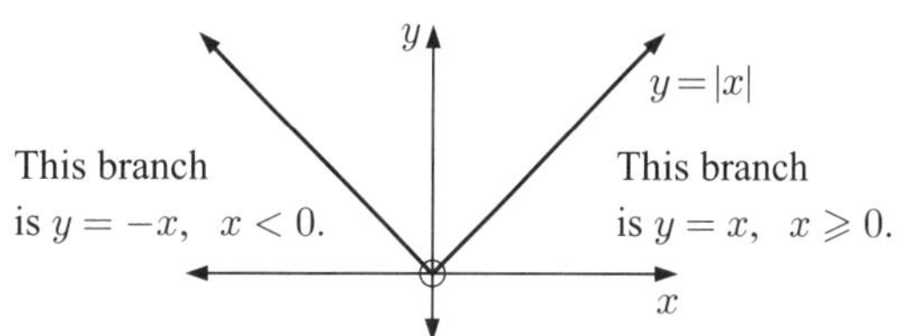

An equivalent definition of $|x|$ is: $|x| = \sqrt{x^2}$.

For example: $|7| = \sqrt{7^2} = \sqrt{49} = 7$
$|-5| = \sqrt{(-5)^2} = \sqrt{25} = 5$

Example 16 **Self Tutor**

If $a = -3$ and $b = 4$ find:

a $|7 + a|$ **b** $|ab|$ **c** $|a^2 + 2b|$

a $|7 + a|$
$= |7 + (-3)|$
$= |4|$
$= 4$

b $|ab|$
$= |(-3)(4)|$
$= |-12|$
$= 12$

c $|a^2 + 2b|$
$= |(-3)^2 + 2(4)|$
$= |9 + 8|$
$= |17|$
$= 17$

Example 17 **Self Tutor**

By replacing $|x|$ with x for $x \geqslant 0$ and $-x$ for $x < 0$, write the following functions without the modulus sign and hence graph each function:

a $f(x) = x - |x|$ **b** $f(x) = x\,|x|$

a If $x < 0$, $f(x) = x - (-x) = 2x$.
If $x \geqslant 0$, $f(x) = x - x = 0$.

So, we graph $\begin{cases} y = 2x & \text{for } x < 0 \\ y = 0 & \text{for } x \geqslant 0. \end{cases}$

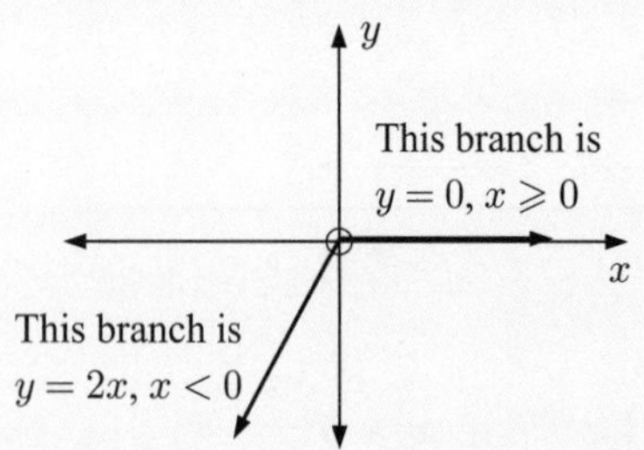

b If $x \geqslant 0$, $f(x) = x(x) = x^2$.
If $x < 0$, $f(x) = x(-x) = -x^2$.

So, we graph $\begin{cases} y = x^2 & \text{for } x \geqslant 0 \\ y = -x^2 & \text{for } x < 0. \end{cases}$

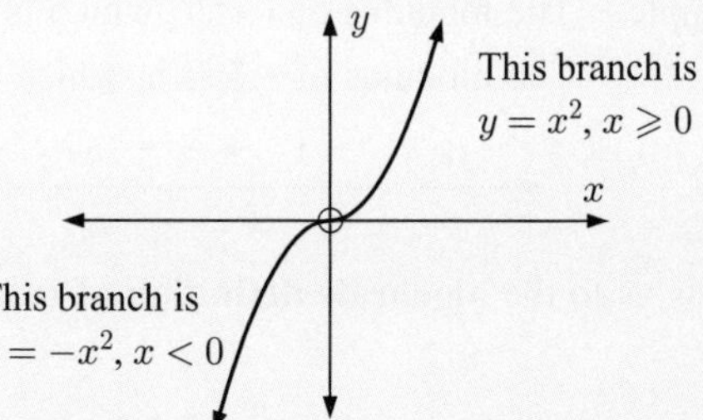

EXERCISE 2H.1

1 If $a = -2$, $b = 3$, $c = -4$ find the value of:

a $|a|$ **b** $|b|$ **c** $|a|\,|b|$ **d** $|ab|$ **e** $|a - b|$ **f** $|a| - |b|$

g $|a + b|$ **h** $|a| + |b|$ **i** $|a|^2$ **j** a^2 **k** $\left|\frac{c}{a}\right|$ **l** $\frac{|c|}{|a|}$

2 If $x = -3$, find the value of:

a $|5 - x|$ **b** $|5| - |x|$ **c** $\left|\frac{2x+1}{1-x}\right|$ **d** $|3 - 2x - x^2|$

3 **a** Copy and complete:

a	b	$\|a\| + \|b\|$	$\|a\| - \|b\|$	$\|a + b\|$	$\|a - b\|$	$\|b - a\|$
6	2					
6	-2					
-6	2					
-6	-2					

b Are the following true or false for *all* real values of a and b?

i $|a + b| = |a| + |b|$ **ii** $|a - b| = |a| - |b|$

c Use the fact that $|x| = \sqrt{x^2}$ to prove that $|a - b| = |b - a|$ for all real values of a and b.

4 **a** Copy and complete:

a	b	$\|ab\|$	$\|a\|\,\|b\|$	$\left\|\frac{a}{b}\right\|$	$\frac{\|a\|}{\|b\|}$
6	2				
6	-2				
-6	2				
-6	-2				

b Use the fact that $|x| = \sqrt{x^2}$ to prove that for all real values of a and b:

i $|ab| = |a|\,|b|$ **ii** $\left|\frac{a}{b}\right| = \frac{|a|}{|b|}, \quad b \neq 0$

5 Use $|a| = \begin{cases} a & \text{if } a \geqslant 0 \\ -a & \text{if } a < 0 \end{cases}$ to write the following functions without modulus signs.

Hence graph each function.

a $y = |x - 2|$ **b** $y = |x + 1|$ **c** $y = -|x|$

d $y = |x| + x$ **e** $y = \frac{|x|}{x}$ **f** $y = x - 2|x|$

g $y = |x| + |x - 2|$ **h** $y = |x| - |x - 1|$

MODULUS EQUATIONS AND INEQUALITIES

From the previous Exercise you should have discovered these **properties of modulus**:

- $|x| \geqslant 0$ for all x
- $|-x| = |x|$ for all x
- $|x|^2 = x^2$ for all x
- $|xy| = |x|\,|y|$ for all x and y
- $\left|\frac{x}{y}\right| = \frac{|x|}{|y|}$ for all x and y, $y \neq 0$
- $|x - y| = |y - x|$ for all x and y.

Modulus is a distance. $|x - y| = |y - x|$ says that the distance from y to x on the number line equals the distance from x to y.

It is clear that $|x| = 2$ has two solutions: $x = 2$ and $x = -2$.

If $|x| = a$ where $a > 0$, then $x = \pm a$.

We use this rule to solve equations involving the modulus function.

Example 18 **Self Tutor**

Solve for x: **a** $|2x + 3| = 7$ **b** $|3 - 2x| = -1$

a $|2x + 3| = 7$

$\therefore\ 2x + 3 = \pm 7$

$\therefore\ 2x + 3 = 7$ **or** $2x + 3 = -7$

$\therefore\ 2x = 4$ $\therefore\ 2x = -10$

$\therefore\ x = 2$ $\therefore\ x = -5$

So, $x = 2$ or -5

b $|3 - 2x| = -1$

has no solution as LHS is never negative.

Example 19 — Self Tutor

Solve for x: $\left|\dfrac{3x+2}{1-x}\right| = 4.$

If $\left|\dfrac{3x+2}{1-x}\right| = 4$ then $\dfrac{3x+2}{1-x} = \pm 4.$

$\therefore \dfrac{3x+2}{1-x} = 4$ **or** $\dfrac{3x+2}{1-x} = -4$

$\therefore 3x+2 = 4(1-x)$ $\qquad\therefore 3x+2 = -4(1-x)$

$\therefore 3x+2 = 4-4x$ $\qquad\therefore 3x+2 = -4+4x$

$\therefore 7x = 2$ $\qquad\therefore 6 = x$

$\therefore x = \frac{2}{7}$

So, $x = \frac{2}{7}$ or 6.

Also notice that: If $|x| = |b|$ then $x = \pm b.$

Example 20 — Self Tutor

Solve for x: $|x+1| = |2x-3|$

If $|x+1| = |2x-3|$, then $x+1 = \pm(2x-3)$

$\therefore x+1 = 2x-3$ **or** $x+1 = -(2x-3)$

$\therefore 4 = x$ $\qquad\therefore x+1 = -2x+3$

$\therefore 3x = 2$

$\therefore x = \frac{2}{3}$

So, $x = \frac{2}{3}$ or 4.

EXERCISE 2H.2

1 Solve for x:

a $|x| = 3$ **b** $|x| = -5$ **c** $|x| = 0$

d $|x-1| = 3$ **e** $|3-x| = 4$ **f** $|x+5| = -1$

g $|3x-2| = 1$ **h** $|3-2x| = 3$ **i** $|2-5x| = 12$

2 Solve for x:

a $\left|\dfrac{x}{x-1}\right| = 3$ **b** $\left|\dfrac{2x-1}{x+1}\right| = 5$ **c** $\left|\dfrac{x+3}{1-3x}\right| = \dfrac{1}{2}$

3 Solve for x:

a $|3x-1| = |x+2|$ **b** $|2x+5| = |1-x|$ **c** $|x+1| = |2-x|$

d $|x| = |5-x|$ **e** $|1-4x| = 2\,|x-1|$ **f** $|3x+2| = 2\,|2-x|$

4 Solve for x using:

i a graphical method **ii** an algebraic method:

a $|x+2| = 2x+1$ **b** $|2x+3| = 3\,|x|-1$ **c** $|x-2| = \frac{2}{5}x+1$

Example 21 — Self Tutor

Solve graphically: $|1-2x| > x+1$.

We draw graphs of $y = |1-2x|$ and $y = x+1$ on the same set of axes.

Casio fx-CG20

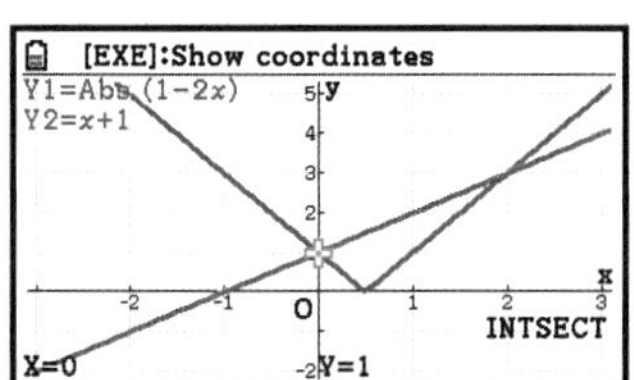

TI-84 Plus

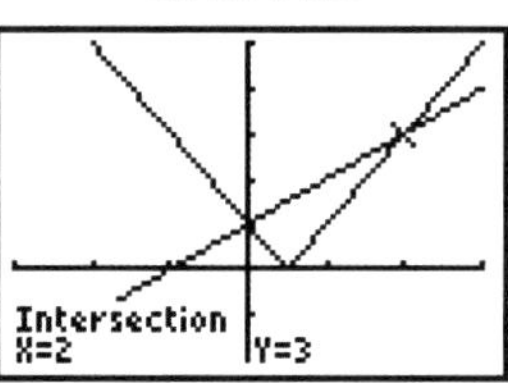

TI-*n*spire

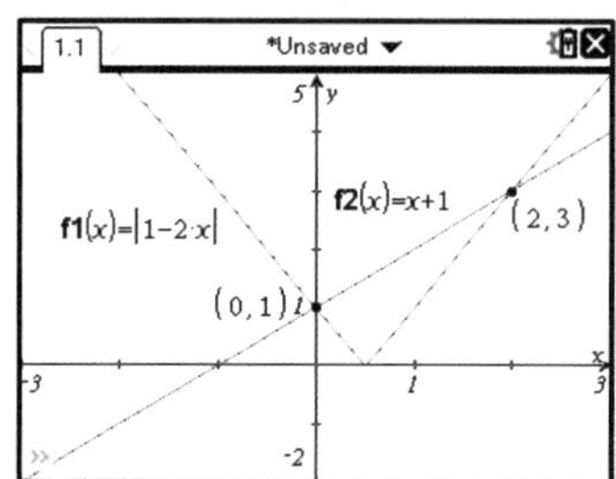

The graphs intersect at $x = 0$ and $x = 2$.

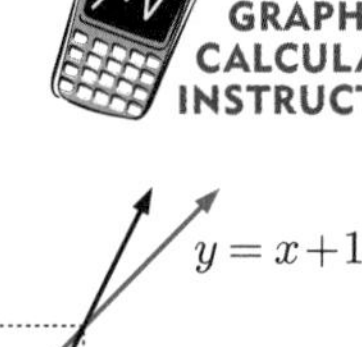

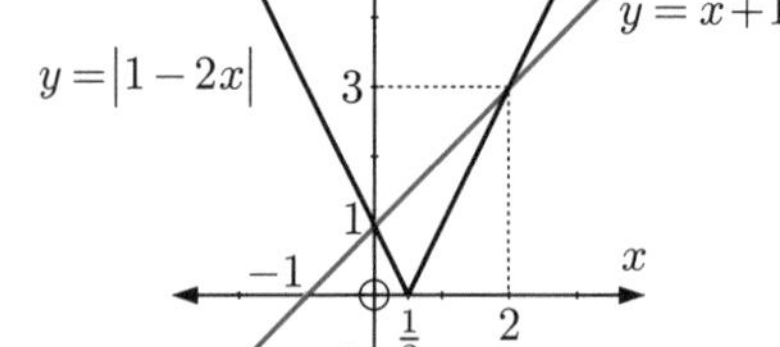

Now $|1-2x| > x+1$

when the graph of $y = |1-2x|$ lies above $y = x+1$.

$\therefore$ $x < 0$ or $x > 2$

$\therefore$ $x \in]-\infty, 0[\ \cup \ ]2, \infty[$

5 Solve for x:

a $|x| < 4$ **b** $|x| \geqslant 3$ **c** $|x+3| \leqslant 1$ **d** $|2x-1| < 3$

e $|3-4x| > 2$ **f** $|x| \geqslant |2-x|$ **g** $3|x| \leqslant |1-2x|$ **h** $\left|\dfrac{x}{x-2}\right| \geqslant 3$

i $\left|\dfrac{2x+3}{x-1}\right| \geqslant 2$ **j** $|2x-3| < x$ **k** $\left|x^2-x\right| > 2$ **l** $|x|-2 \geqslant |4-x|$

6 Graph the function $f(x) = \dfrac{|x|}{x-2}$. Hence find all values of x for which $\dfrac{|x|}{x-2} \geqslant -\frac{1}{2}$.

7 **a** Draw the graph of $y = |x+5| + |x+2| + |x| + |x-3|$.

b

P, Q, and R are factories which are respectively 5, 2, and 3 km away from factory O. A security service wants to locate its premises along (AB) so that the total length of cable to the 4 factories is a minimum.

i Suppose x is the position of the security service on (AB). Explain why the total length of cable is given by $|x+5| + |x+2| + |x| + |x-3|$.

ii Where should the security service locate its premises? What is the minimum length of cable they will need?

iii A fifth factory is located at S, 7 km to the right of O. If this factory also requires the security service, where should the security service locate its premises for minimum cable length?

8 Which of these is true? Give a proof for each answer.

a $|x+y| \leqslant |x|+|y|$ for all x, y

b $|x-y| \geqslant |x|-|y|$ for all x, y

I RATIONAL FUNCTIONS

We have seen that a linear function has the form $y = ax + b$.

When we divide a linear function by another linear function, the result is a **rational function**.

The graphs of rational functions are characterised by **asymptotes**, which are lines which the function gets closer and closer to but never reaches.

The rational functions we consider in this course can be written in the form $y = \dfrac{ax+b}{cx+d}$. These functions have asymptotes which are horizontal and vertical lines.

RECIPROCAL FUNCTIONS

A **reciprocal function** is a function of the form $x \mapsto \dfrac{k}{x}$ or $f(x) = \dfrac{k}{x}$, where $k \neq 0$ is a constant.

The simplest example of a reciprocal function is $f(x) = \dfrac{1}{x}$.

The graph of $f(x) = \dfrac{1}{x}$ is called a **rectangular hyperbola**.

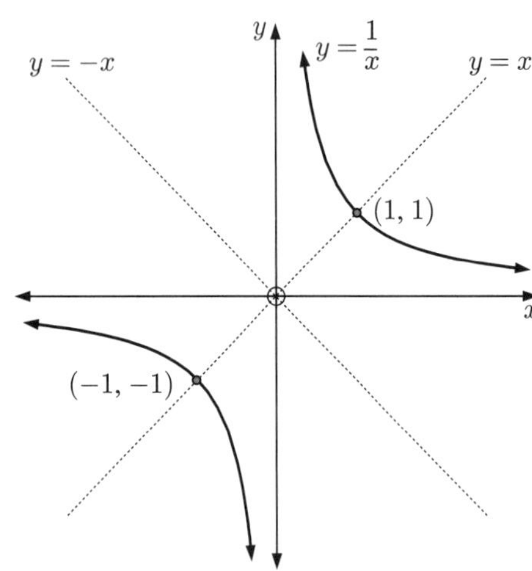

Notice that:

- The graph of $f(x) = \dfrac{1}{x}$ has two branches.
- $f(x) = \dfrac{1}{x}$ is undefined when $x = 0$
- The graph of $f(x) = \dfrac{1}{x}$ exists in the first and third quadrants only.
- $f(x) = \dfrac{1}{x}$ is symmetric about $y = x$ and about $y = -x$.
- as $x \to \infty$, $f(x) \to 0$ from above
  as $x \to -\infty$, $f(x) \to 0$ from below
  as $x \to 0$ from the right, $f(x) \to \infty$
  as $x \to 0$ from the left, $f(x) \to -\infty$
- The **asymptotes** of $f(x) = \dfrac{1}{x}$ are the x-axis and the y-axis.

GRAPHING PACKAGE

INVESTIGATION 2 RECIPROCAL FUNCTIONS

In this Investigation we explore reciprocal functions of the form $y = \frac{k}{x}$, $k \neq 0$.

DEMO

What to do:

1. Use the slider to vary the value of k for $k > 0$.
   a. Sketch the graphs of $y = \frac{1}{x}$, $y = \frac{2}{x}$, and $y = \frac{4}{x}$ on the same set of axes.
   b. Describe the effect of varying k on the graph of $y = \frac{k}{x}$.
2. Use the slider to vary the value of k for $k < 0$.
   a. Sketch the graphs of $y = -\frac{1}{x}$, $y = -\frac{2}{x}$, and $y = -\frac{4}{x}$ on the same set of axes.
   b. Describe the effect of varying k on the graph of $y = \frac{k}{x}$.

RATIONAL FUNCTIONS OF THE FORM $y = \frac{ax+b}{cx+d}$, $c \neq 0$

The graph of $f(x) = \frac{2x+1}{x-1}$ is shown below.

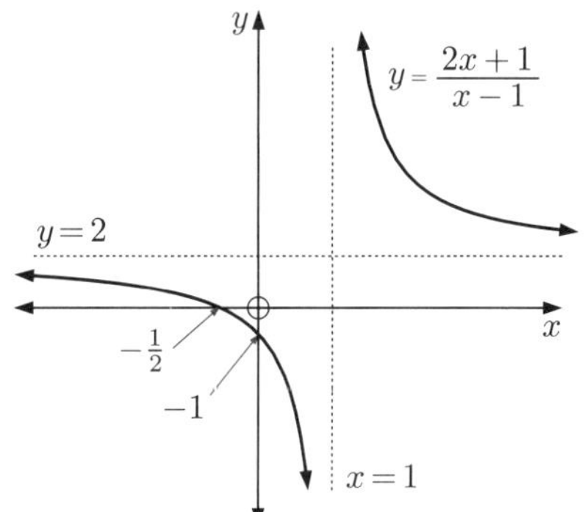

Notice that when $x = 1$, $f(x)$ is undefined.

The graph approaches the vertical line $x = 1$, so $x = 1$ is a vertical asymptote.

Notice that $f(0.999) = -2998$ and $f(1.001) = 3002$.

We can write: as $x \to 1$ from the left, $f(x) \to -\infty$
as $x \to 1$ from the right, $f(x) \to \infty$

or as $x \to 1^-$, $f(x) \to -\infty$
as $x \to 1^+$, $f(x) \to \infty$.

To determine the equation of a vertical asymptote, consider the values of x which make the function undefined.

The sign diagram of $y = \frac{2x+1}{x-1}$ is (+ | − | + on x, with critical values $-\frac{1}{2}$ and 1) and can be used to discuss the function near its vertical asymptote without having to graph the function.

The graph also approaches the horizontal line $y = 2$, so $y = 2$ is a horizontal asymptote.

Notice that $f(1000) = \frac{2001}{999} \approx 2.003$ and $f(-1000) = \frac{-1999}{-1001} \approx 1.997$

We can write: as $x \to \infty$, $y \to 2$ from above *or* as $x \to \infty$, $y \to 2^+$
as $x \to -\infty$, $y \to 2$ from below as $x \to -\infty$, $y \to 2^-$.

We can also write: as $|x| \to \infty$, $f(x) \to 2$.

This indicates that as x becomes very large (either positive or negative) the function approaches the value 2.

To determine the equation of a horizontal asymptote, we consider the behaviour of the graph of the function as $|x| \to \infty$.

In **Chapter 5** we will see how to determine the equations of the asymptotes algebraically.

INVESTIGATION 3 FINDING ASYMPTOTES

What to do:

1. Use the **graphing package** supplied or a **graphics calculator** to examine the following functions for asymptotes:

   a $y = -1 + \dfrac{3}{x-2}$ **b** $y = \dfrac{3x+1}{x+2}$ **c** $y = \dfrac{2x-9}{3-x}$

2. How can we tell directly from the function, what its vertical asymptote is?
3. Discuss whether a function can cross a vertical asymptote.

Example 22 Self Tutor

Consider the function $y = \dfrac{6}{x-2} + 4$.

a Find the asymptotes of the function. **b** Find the axes intercepts.

c Use technology to help sketch the function, including the features from **a** and **b**.

a The vertical asymptote is $x = 2$.
The horizontal asymptote is $y = 4$.

b When $y = 0$, $\dfrac{6}{x-2} = -4$

$\therefore\ -4(x-2) = 6$

$\therefore\ -4x + 8 = 6$

$\therefore\ -4x = -2$

$\therefore\ x = \frac{1}{2}$

When $x = 0$, $y = \frac{6}{-2} + 4 = 1$

So, the x-intercept is $\frac{1}{2}$ and the y-intercept is 1.

c

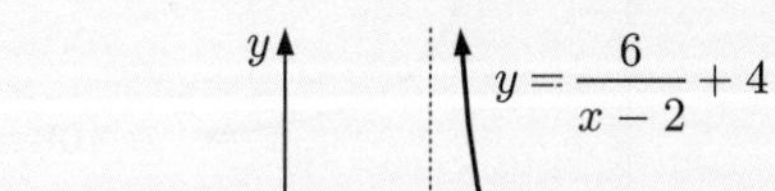

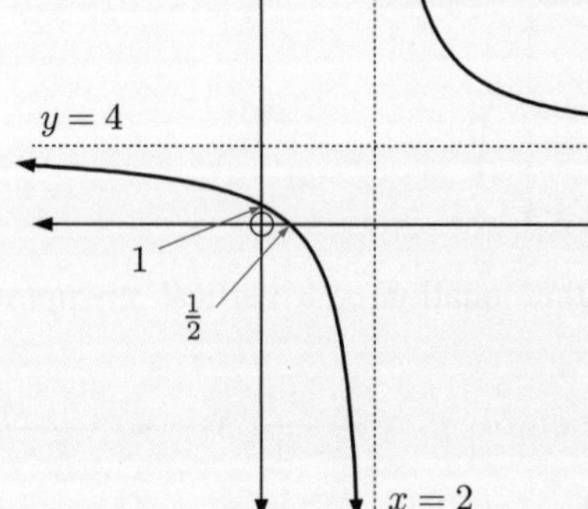

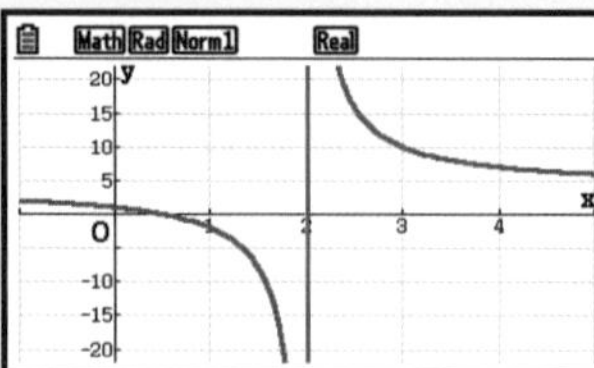

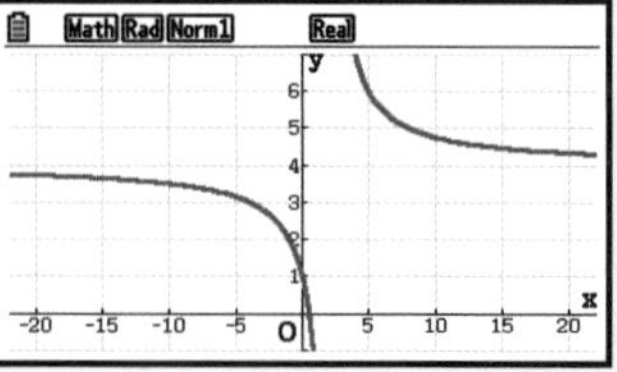

Further examples of asymptotic behaviour are seen in exponential, logarithmic, and some trigonometric functions.

EXERCISE 2I

1 For the following functions:

- **i** determine the equations of the asymptotes
- **ii** state the domain and range
- **iii** find the axes intercepts
- **iv** discuss the behaviour of the function as it approaches its asymptotes
- **v** sketch the graph of the function.

a $f : x \mapsto \dfrac{3}{x-2}$ **b** $y = 2 - \dfrac{3}{x+1}$ **c** $f : x \mapsto \dfrac{x+3}{x-2}$ **d** $f(x) = \dfrac{3x-1}{x+2}$

2 Consider the function $y = \dfrac{ax+b}{cx+d}$, where a, b, c, d are constants and $c \neq 0$.

- **a** State the domain of the function.
- **b** State the equation of the vertical asymptote.
- **c** Show that for $c \neq 0$, $\dfrac{ax+b}{cx+d} = \dfrac{a}{c} + \dfrac{b - \frac{ad}{c}}{cx+d}$.

  Hence determine the equation of the horizontal asymptote.

ACTIVITY

Click on the icon to run a card game for rational functions.

CARD GAME

J INVERSE FUNCTIONS

The operations of + and , × and :, squaring and finding the square root, are **inverse operations** as one "undoes" what the other does.

The function $y = 2x + 3$ can be "undone" by its *inverse* function $y = \dfrac{x-3}{2}$.

We can think of this as two machines. If the machines are inverses then the second machine *undoes* what the first machine does.

No matter what value of x enters the first machine, it is returned as the output from the second machine.

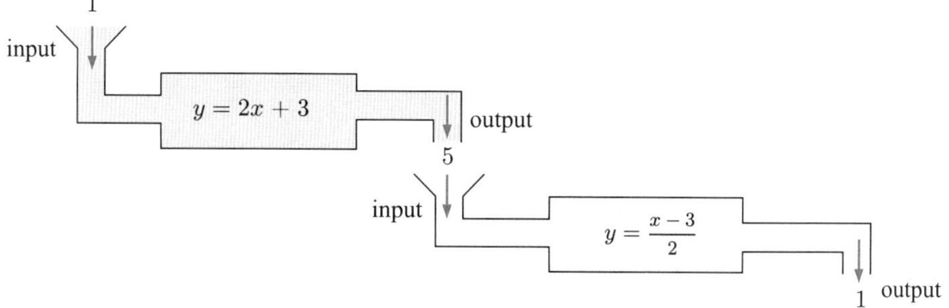

A function $y = f(x)$ *may or may not* have an inverse function. To understand which functions do have inverses, we need some more terminology.

ONE-TO-ONE AND MANY-TO-ONE FUNCTIONS

A **one-to-one** function is any function where:

- for each x there is only one value of y and
- for each y there is only one value of x.

One-to-one functions satisfy both the **vertical line test** and the **horizontal line test**.

This means that:
- no vertical line can meet the graph more than once
- no horizontal line can meet the graph more than once.

If the function $f(x)$ is **one-to-one**, it will have an inverse function which we denote $f^{-1}(x)$.

Functions that are not one-to-one are called **many-to-one**. While these functions must satisfy the '**vertical line test**' they *do not* satisfy the '**horizontal line test**'. At least one y-value has more than one corresponding x-value.

If a function $f(x)$ is **many-to-one**, it *does not* have an inverse function.

However, for a many-to-one function we can often define a new function using the same formula but with a **restricted domain** to make it a one-to-one function. This new function will have an inverse function.

PROPERTIES OF THE INVERSE FUNCTION

If $y = f(x)$ has an **inverse function**, this new function:

- is denoted $f^{-1}(x)$
- must satisfy the vertical line test
- has a graph which is the reflection of $y = f(x)$ in the line $y = x$
- satisfies $(f \circ f^{-1})(x) = x$ and $(f^{-1} \circ f)(x) = x$.

If (x, y) lies on f, then (y, x) lies on f^{-1}. Reflecting the function in the line $y = x$ has the algebraic effect of interchanging x and y.

For example, $f : y = 5x + 2$ becomes $f^{-1} : x = 5y + 2$,

which we rearrange to obtain $f^{-1} : y = \frac{x-2}{5}$.

The domain of f^{-1} is equal to the range of f.

The range of f^{-1} is equal to the domain of f.

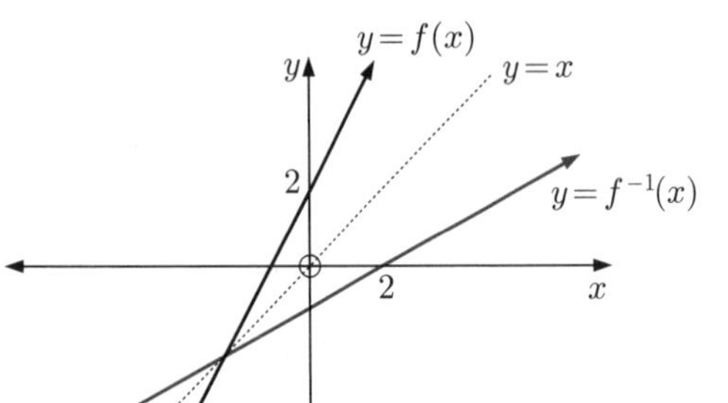

$y = f^{-1}(x)$ is the inverse of $y = f(x)$ as:

- it is also a function
- it is the reflection of $y = f(x)$ in the line $y = x$.

The parabola shown in red is the reflection of $y = f(x)$ in $y = x$, but it is *not* the inverse function of $y = f(x)$ as it fails the vertical line test.

In this case the function $y = f(x)$ does not have an inverse.

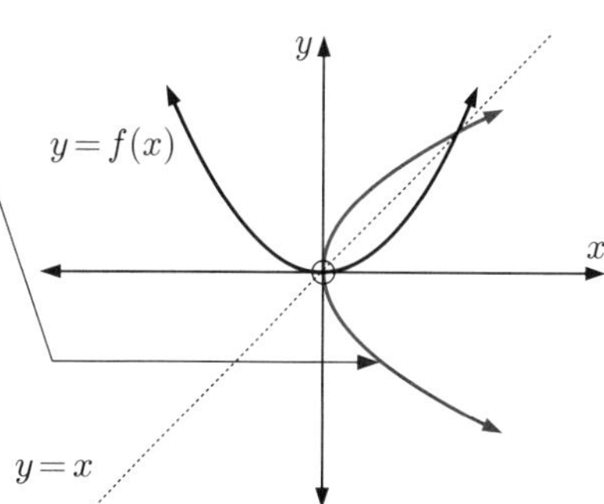

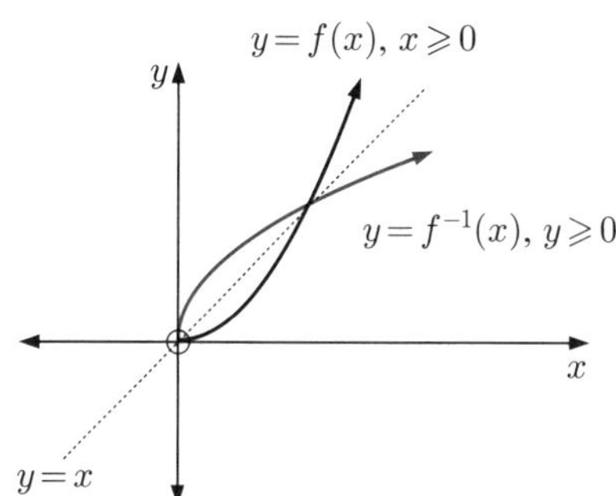

Now consider the same function $f(x)$ but with the restricted domain $x \geqslant 0$.

The function is now one-to-one, so does now have an inverse function, as shown alongside.

Example 23 **Self Tutor**

Consider $f : x \mapsto 2x + 3$.

a On the same axes, graph f and its inverse function f^{-1}.

b Find $f^{-1}(x)$ using:

  i coordinate geometry and the gradient of $y = f^{-1}(x)$ from **a**

  ii variable interchange.

c Check that $(f \circ f^{-1})(x) = (f^{-1} \circ f)(x) = x$

a $f(x) = 2x + 3$ passes through $(0, 3)$ and $(2, 7)$.

$\therefore$ $f^{-1}(x)$ passes through $(3, 0)$ and $(7, 2)$.

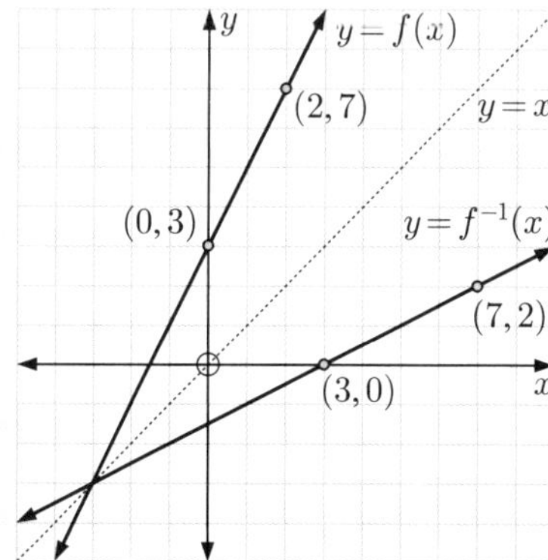

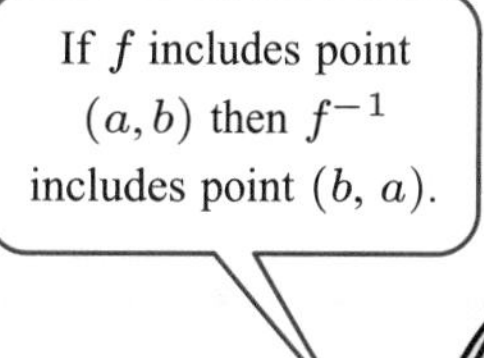

b **i** $y = f^{-1}(x)$ has gradient $\dfrac{2-0}{7-3} = \dfrac{1}{2}$

Its equation is $\dfrac{y-0}{x-3} = \dfrac{1}{2}$

$\therefore\ y = \dfrac{x-3}{2}$

$\therefore\ f^{-1}(x) = \dfrac{x-3}{2}$

ii f is $y = 2x + 3$,

$\therefore\ f^{-1}$ is $x = 2y + 3$

$\therefore\ x - 3 = 2y$

$\therefore\ \dfrac{x-3}{2} = y$

$\therefore\ f^{-1}(x) = \dfrac{x-3}{2}$

c $(f \circ f^{-1})(x)$
$= f(f^{-1}(x))$
$= f\left(\frac{x-3}{2}\right)$
$= 2\left(\frac{x-3}{2}\right) + 3$
$= x$

and

$(f^{-1} \circ f)(x)$
$= f^{-1}(f(x))$
$= f^{-1}(2x+3)$
$= \frac{(2x+3)-3}{2}$
$= \frac{2x}{2}$
$= x$

Example 24 **Self Tutor**

Consider $f : x \mapsto x^2$.

a Explain why this function does not have an inverse function.

b Does $f : x \mapsto x^2$ where $x \geqslant 0$ have an inverse function?

c Find $f^{-1}(x)$ for $f : x \mapsto x^2$, $x \geqslant 0$.

d Sketch $y = f(x)$, $y = x$, and $y = f^{-1}(x)$ for f in **b** and f^{-1} in **c**.

a $f : x \mapsto x^2$ has domain $x \in \mathbb{R}$ and is many-to-one.
It does not pass the 'horizontal line test'.

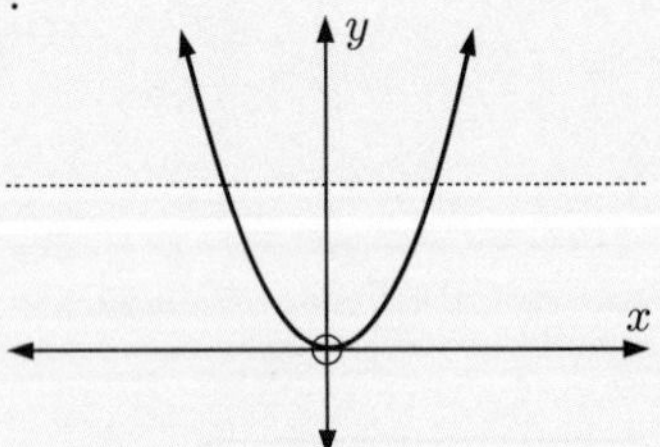

b If we restrict the domain to $x \geqslant 0$ or $x \in [0, \infty[$, the function is now one-to-one. It satisfies the 'horizontal line test' and so has an inverse function.

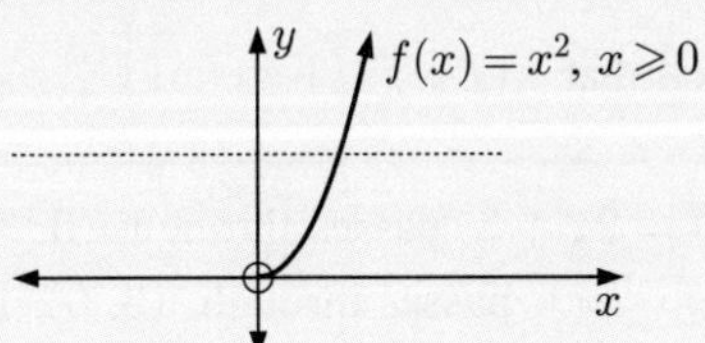

c f is defined by $y = x^2$, $x \geqslant 0$

$\therefore$ f^{-1} is defined by $x = y^2$, $y \geqslant 0$

$\therefore$ $y = \pm\sqrt{x}$, $y \geqslant 0$

$\therefore$ $y = \sqrt{x}$ {as $-\sqrt{x}$ is $\leqslant 0$}

So, $f^{-1}(x) = \sqrt{x}$

d

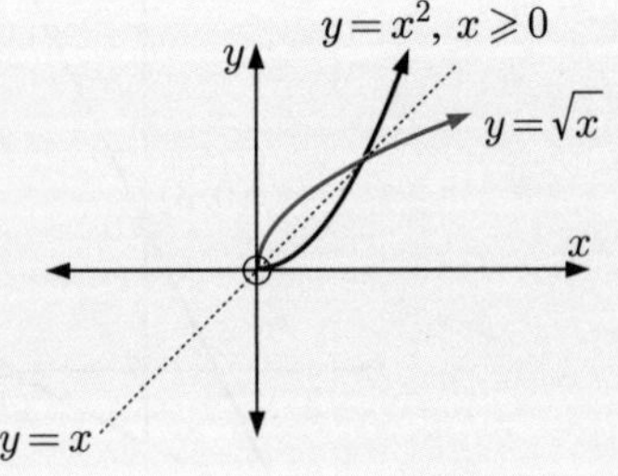

SELF-INVERSE FUNCTIONS

Any function which has an inverse, and whose graph is symmetrical about the line $y = x$, is a **self-inverse function**.

For example:

- The function $f(x) = x$ is the **identity function**, and is also a self-inverse function.
- The reciprocal function $f(x) = \frac{1}{x}$, $x \neq 0$, is also a self-inverse function, as $f = f^{-1}$.

EXERCISE 2J

1 For each of the following functions f:

- **i** On the same set of axes, graph $y = x$, $y = f(x)$, and $y = f^{-1}(x)$.
- **ii** Find $f^{-1}(x)$ using coordinate geometry and the gradient of $y = f^{-1}(x)$ from **i**.
- **iii** Find $f^{-1}(x)$ using variable interchange.

a $f : x \mapsto 3x + 1$ **b** $f : x \mapsto \dfrac{x+2}{4}$

2 For each of the following functions f:

- **i** Find $f^{-1}(x)$.
- **ii** Sketch $y = f(x)$, $y = f^{-1}(x)$, and $y = x$ on the same set of axes.
- **iii** Show that $(f^{-1} \circ f)(x) = (f \circ f^{-1})(x) = x$, the identity function.

a $f : x \mapsto 2x + 5$ **b** $f : x \mapsto \dfrac{3-2x}{4}$ **c** $f : x \mapsto x + 3$

3 Copy the graphs of the following functions and draw the graphs of $y = x$ and $y = f^{-1}(x)$ on the same set of axes.

a

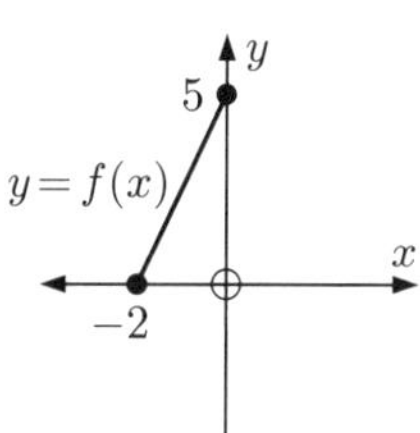

b

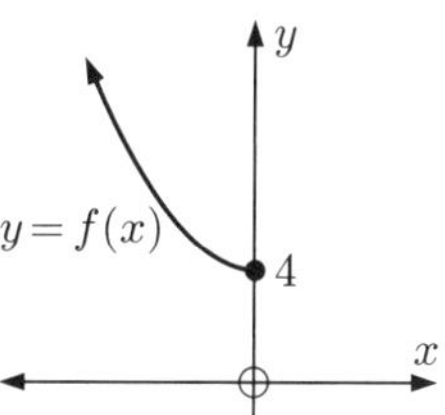

c

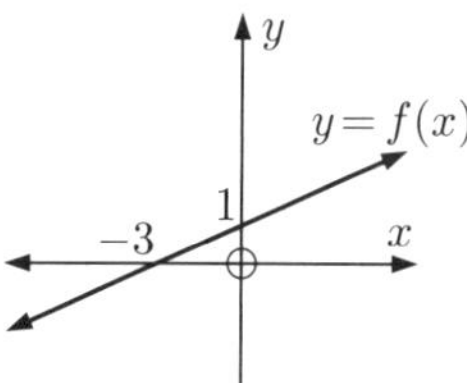

PRINTABLE GRAPHS

d

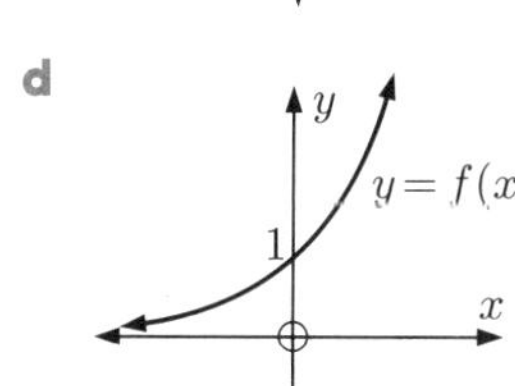

e

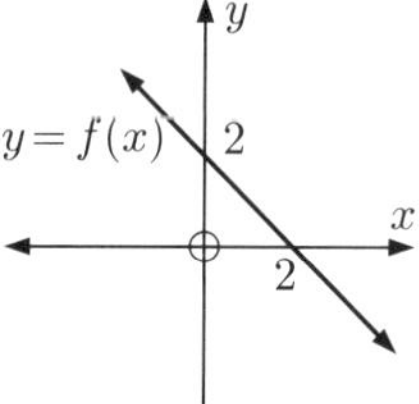

f

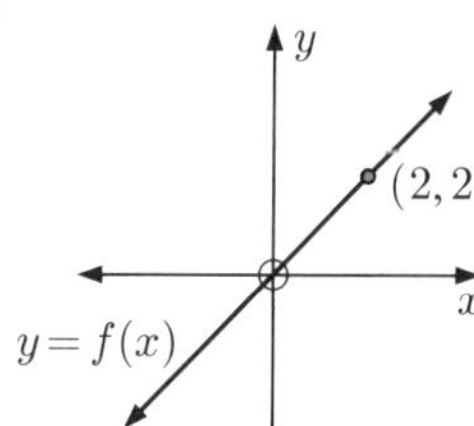

4 For the graph of $y = f(x)$ given in **3 a**, state:

a the domain of $f(x)$ **b** the range of $f(x)$

c the domain of $f^{-1}(x)$ **d** the range of $f^{-1}(x)$.

5 **a** Comment on the results from **3 e** and **f**.

b Draw a linear function that is a self-inverse function.

c Draw a rational function other than $y = \dfrac{1}{x}$, that is a self-inverse function.

A function is self-inverse if $f^{-1}(x) = f(x)$.

6 Given $f(x) = 2x - 5$, find $(f^{-1})^{-1}(x)$. What do you notice?

7 Which of the following functions have inverses? In each of these cases, write down the inverse function.

a $\{(1, 2), (2, 4), (3, 5)\}$ **b** $\{(-1, 3), (0, 2), (1, 3)\}$

c $\{(2, 1), (-1, 0), (0, 2), (1, 3)\}$ **d** $\{(-1, -1), (0, 0), (1, 1)\}$

8 **a** Sketch the graph of $f : x \mapsto x^2 - 4$ and reflect it in the line $y = x$.

b Does f have an inverse function?

c Does f with restricted domain $x \geqslant 0$ have an inverse function?

9 Sketch the graph of $f : x \mapsto x^3$ and its inverse function $f^{-1}(x)$.

10 Given $f : x \mapsto \frac{1}{x}$, $x \neq 0$, find f^{-1} algebraically and show that f is a self-inverse function.

11 Show that $f : x \mapsto \frac{3x-8}{x-3}$, $x \neq 3$ is a self-inverse function by:

a reference to its graph **b** using algebra.

12 The **horizontal line test** says:

For a function to have an inverse function, no horizontal line can cut its graph more than once.

a Explain why this is a valid test for the existence of an inverse function.

b Which of the following functions have an inverse function?

i

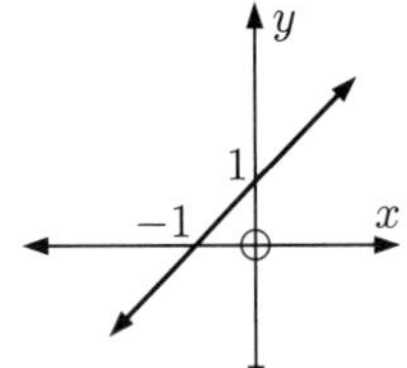

ii

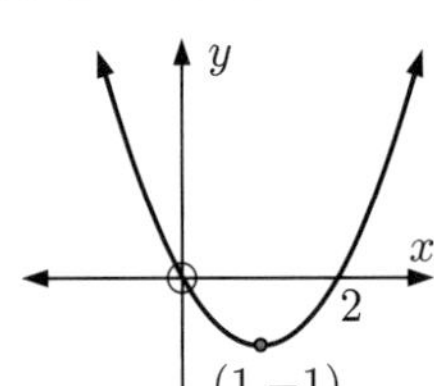

iii

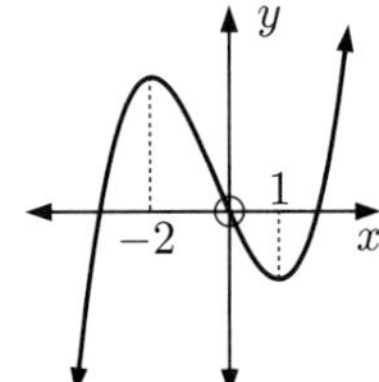

c For the functions in **b** which do not have an inverse, specify restricted domains as wide as possible such that the resulting function does have an inverse.

13 Consider $f : x \mapsto x^2$ where $x \leqslant 0$.

a Find $f^{-1}(x)$.

b Sketch $y = f(x)$, $y = x$, and $y = f^{-1}(x)$ on the same set of axes.

14 **a** Explain why $f(x) = x^2 - 4x + 3$ is a function but does not have an inverse function.

b **i** Explain why $g(x) = x^2 - 4x + 3$ where $x \geqslant 2$ has an inverse function.

ii Show that the inverse function is $g^{-1}(x) = 2 + \sqrt{1+x}$.

iii State the domain and range of:

A g **B** g^{-1}.

iv Show that $(g \circ g^{-1})(x) = (g^{-1} \circ g)(x)$, the identity function.

15 Consider $f : x \mapsto (x+1)^2 + 3$ where $x \geqslant -1$.

a Find the defining equation of f^{-1}.

b Sketch, using technology, the graphs of $y = f(x)$, $y = x$, and $y = f^{-1}(x)$.

c State the domain and range of: **i** f **ii** f^{-1}

16 Consider the functions $f : x \mapsto 2x + 5$ and $g : x \mapsto \frac{8-x}{2}$.

a Find $g^{-1}(-1)$. **b** Show that $f^{-1}(-3) - g^{-1}(6) = 0$.

c Find x such that $(f \circ g^{-1})(x) = 9$.

17 Consider the functions $f : x \mapsto 5^x$ and $g : x \mapsto \sqrt{x}$.

a Find: **i** $f(2)$ **ii** $g^{-1}(4)$. **b** Solve the equation $(g^{-1} \circ f)(x) = 25$.

18 Given $f : x \mapsto 2x$ and $g : x \mapsto 4x - 3$, show that $(f^{-1} \circ g^{-1})(x) = (g \circ f)^{-1}(x)$.

19 Which of these functions is a self-inverse function?

a $f(x) = 2x$ **b** $f(x) = x$ **c** $f(x) = -x$

d $f(x) = \frac{2}{x}$ **e** $f(x) = -\frac{6}{x}$

20

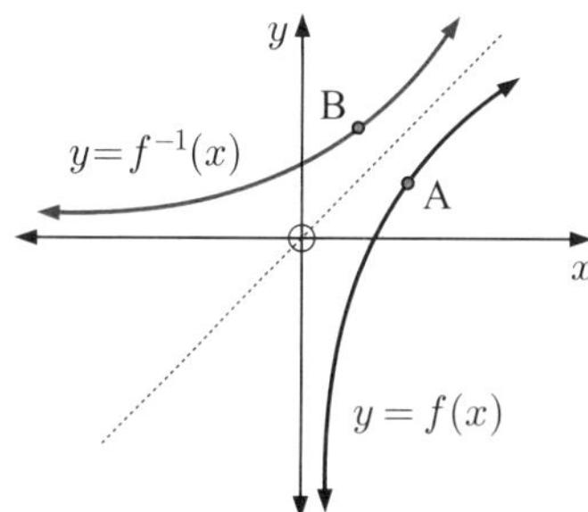

a B is the image of A under a reflection in the line $y = x$. If A is $(x, f(x))$, find the coordinates of B.

b By substituting your result from **a** into $y = f^{-1}(x)$, show that $f^{-1}(f(x)) = x$.

c Using a similar method, show that $f(f^{-1}(x)) = x$.

K GRAPHING FUNCTIONS

Drawing the graph of a function allows us to identify features such as:

- the **axes intercepts** where the graph cuts the x and y-axes
- **turning points** which could be a **local minimum** or **local maximum**
- values of x for which the function does not exist
- the presence of **asymptotes**, which are lines or curves that the graph approaches.

Many real world situations are modelled by mathematical functions which are difficult to analyse using algebra. However, we can use technology to help us graph and investigate the key features of an unfamiliar function.

If there is no domain or range specified, start with a large viewing window. This ensures we do not miss any features of the function.

Example 25 **Self Tutor**

Consider the function $y = \frac{6x^2 - 12x + 5}{x^3 + 5x - 6} + 3$. Use technology to help answer the following:

a Find the axes intercepts. **b** Find any turning points of the function.

c Find any asymptotes of the function. **d** State the domain and range of the function.

e Sketch the function, showing its key features.

a

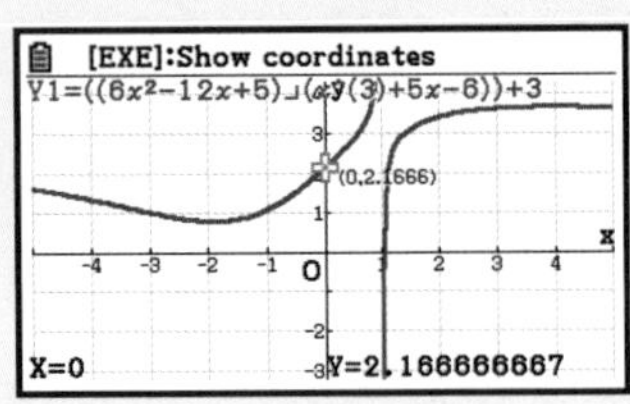

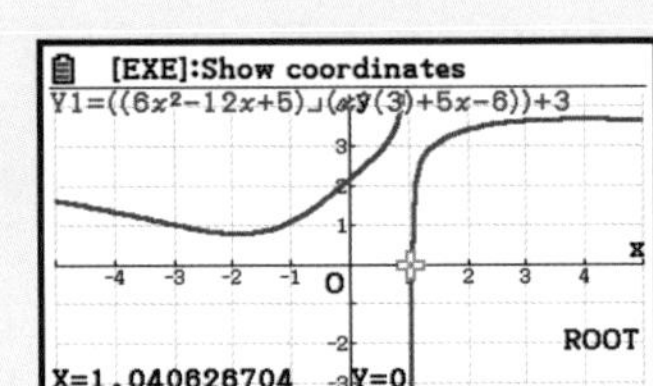

The y-intercept is 2.17. The x-intercept is 1.04.

b

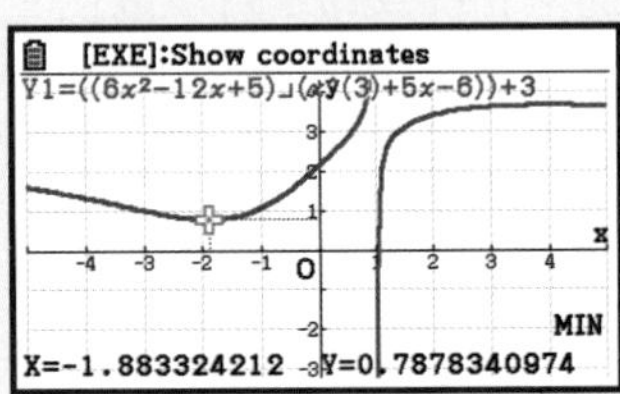

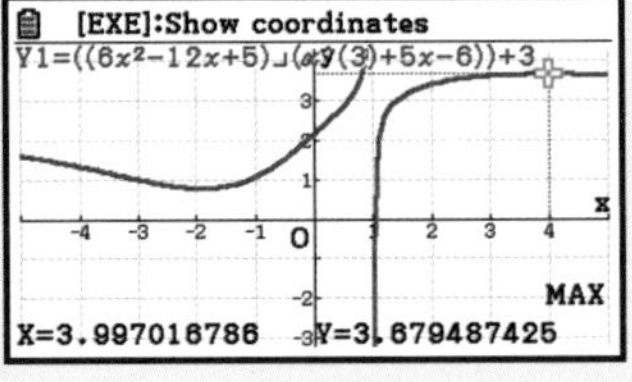

There is a local minimum at $(-1.88,\ 0.788)$, and a local maximum at $(4.00,\ 3.68)$.

c As $x \to -\infty$, $y \to 3$ from below.
As $x \to \infty$, $y \to 3$ from above.
So, the horizontal asymptote is $y = 3$.
As $x \to 1$ from the left, $y \to \infty$.
As $x \to 1$ from the right, $y \to -\infty$.
So, the vertical asymptote is $x = 1$.

d The domain is $\{x \mid x \neq 1\}$.
The range is $\{y \mid y \in \mathbb{R}\}$.

e

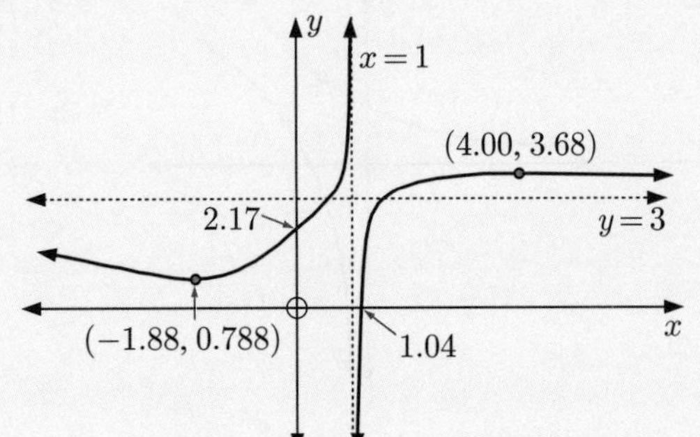

If we are asked to **sketch** a function, it will show the graph's general shape and its key features.

If we are asked to **draw** a function, it should be done more carefully to scale.

EXERCISE 2K

1 For each of the functions given, use technology to answer the following:

- **i** Find the axes intercepts.
- **ii** Find any turning points of the function.
- **iii** Find any asymptotes of the function.
- **iv** State the domain and range of the function.
- **v** Sketch the function, showing its key features.

a $y = \frac{1}{2}x(x-4)(x+3)$ **b** $y = \frac{4}{5}x^4 + 5x^3 + 5x^2 + 2$ on $-5 \leqslant x \leqslant 1$

c $y = \dfrac{x^2 - 4x - 2}{(x-4)^2}$ **d** $y = \dfrac{1 - x^2}{(x+2)^2}$ on $-5 \leqslant x \leqslant 5$

2 Find the maximum value of:

a $y = -x^4 + 2x^3 + 5x^2 + x + 2$ on the interval $0 \leqslant x \leqslant 4$

b $y = -2x^4 + 5x^2 + x + 2$ on the interval:

i $-2 \leqslant x \leqslant 2$ **ii** $-2 \leqslant x \leqslant 0$ **iii** $0 \leqslant x \leqslant 2$.

L FINDING WHERE GRAPHS MEET

Suppose we graph two functions $f(x)$ and $g(x)$ on the same set of axes. The x-coordinates of points where the graphs meet are the solutions to the equation $f(x) = g(x)$.

We can use this property to solve equations graphically, but we must make sure the graphs are drawn accurately.

Example 26 **Self Tutor**

a Draw the functions $y = \frac{1}{x}$ and $y = 2 - x$ on the same set of axes, with the domain $-4 \leqslant x \leqslant 4$.

b Hence find any solutions to the equation $2 - x = \frac{1}{x}$ on the domain $-4 \leqslant x \leqslant 4$.

c On the domain $x \in \mathbb{R}$, for what values of k does $x + \frac{1}{x} = k$ have:

i one solution **ii** no solutions **iii** two solutions?

a

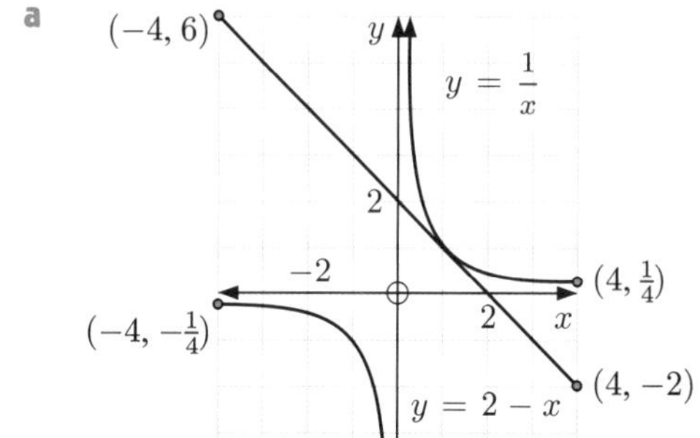

The Table menu on your calculator can be used to find endpoints.

b The graphs of $y = 2 - x$ and $y = \frac{1}{x}$ meet when $x - 1$.

$\therefore$ the solution of $2 - x = \frac{1}{x}$ is $x = 1$.

c If $x + \frac{1}{x} = k$ then $k - x = \frac{1}{x}$.

The graph of $y = k - x$ has y-intercept k and gradient -1.

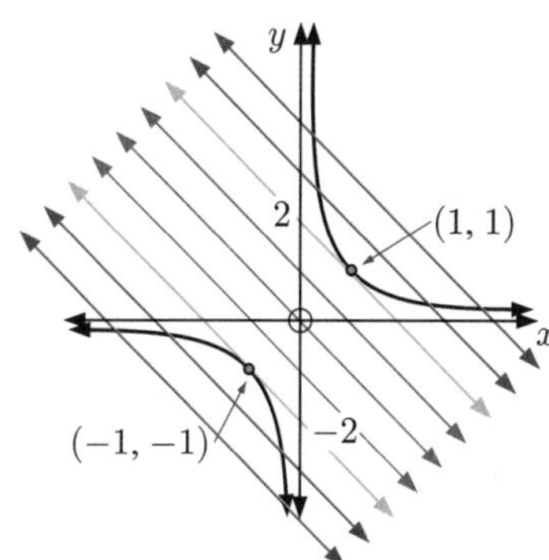

i Using **b** as a guide, we see there is one solution if $k = 2$ or -2. {green lines}

ii There are no solutions if $-2 < k < 2$. {blue lines}

iii There are two solutions if $k < -2$ or $k > 2$. {red lines}

EXERCISE 2L

1 **a** Sketch the graphs of $f(x) = x^2 - \frac{4}{x}$ and $g(x) = -x^2 + 11$ for $-6 \leqslant x \leqslant 6$.

b One of the solutions to $f(x) = g(x)$ is $x \approx 2.51$. Use technology to determine the two negative solutions.

2 Sketch the graphs of $f(x) = \sqrt{x^2 - 2x + 9}$ and $g(x) = x^2$ on the same set of axes. Hence solve $x^2 = \sqrt{x^2 - 2x + 9}$.

3 Part of the graph of $y = f(x)$ is shown alongside.

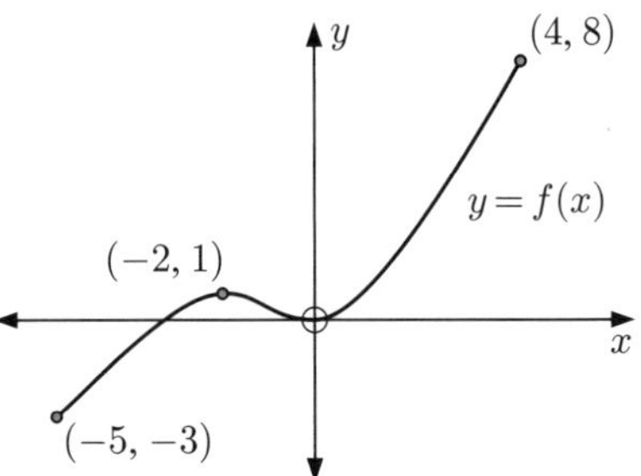

a Copy the graph, and on the same set of axes draw the graph of $g(x) = -1$, $-5 \leqslant x \leqslant 4$.

b Hence state the number of solutions of $f(x) = g(x)$ on the domain $-5 \leqslant x \leqslant 4$.

c Consider the function $h(x) = k$ on the domain $-5 \leqslant x \leqslant 4$. For what values of k does $f(x) = h(x)$ have:

i three solutions **ii** two solutions
iii one solution **iv** no solutions?

REVIEW SET 2A NON-CALCULATOR

1 For each graph, state:

i the domain **ii** the range **iii** whether the graph shows a function.

a

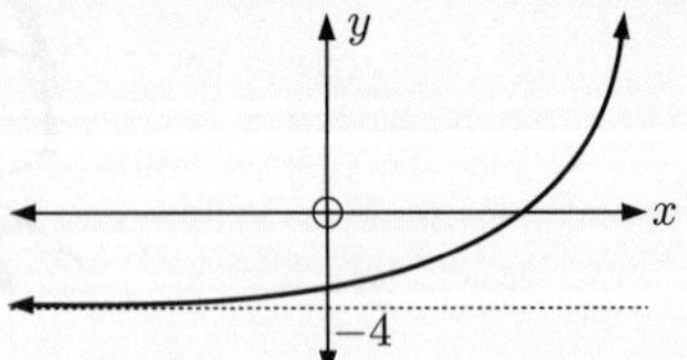

b

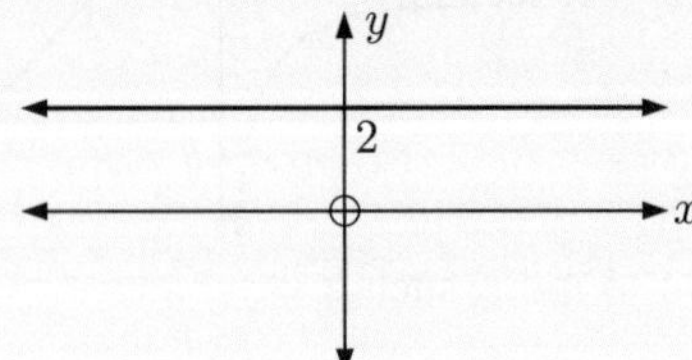

c

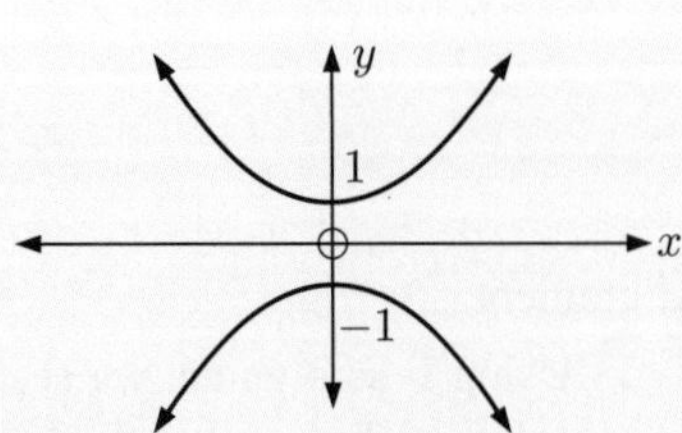

d

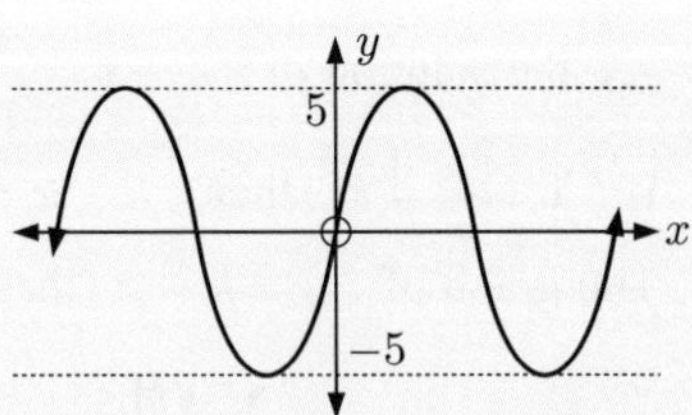

2 If $f(x) = 2x - x^2$, find: **a** $f(2)$ **b** $f(-3)$ **c** $f(-\frac{1}{2})$

3 Suppose $f(x) = ax + b$ where a and b are constants. If $f(1) = 7$ and $f(3) = -5$, find a and b.

4 State functions f and g for which:

a $f(g(x)) = \sqrt{1 - x^2}$ **b** $g(f(x)) = \left(\frac{x-2}{x+1}\right)^2$

5 Solve for x:

a $|4x - 2| = |x + 7|$ **b** $x^2 + 6x > 16$

6 If $g(x) = x^2 - 3x$, find in simplest form: **a** $g(x + 1)$ **b** $g(x^2 - 2)$

7 For each of the following graphs determine:

i the domain and range
ii the x and y-intercepts
iii whether it is a function
iv if it has an inverse function.

a

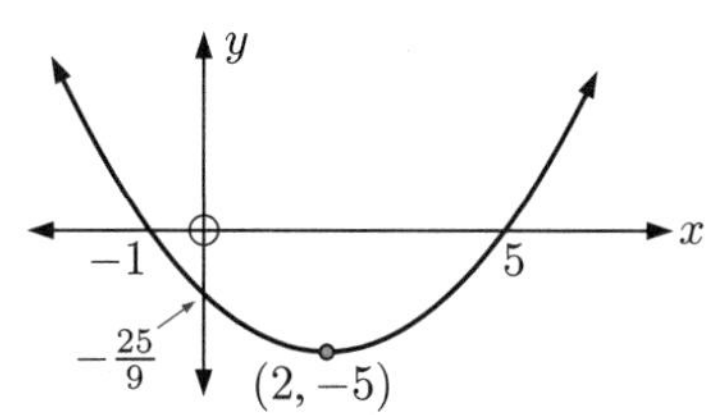

b

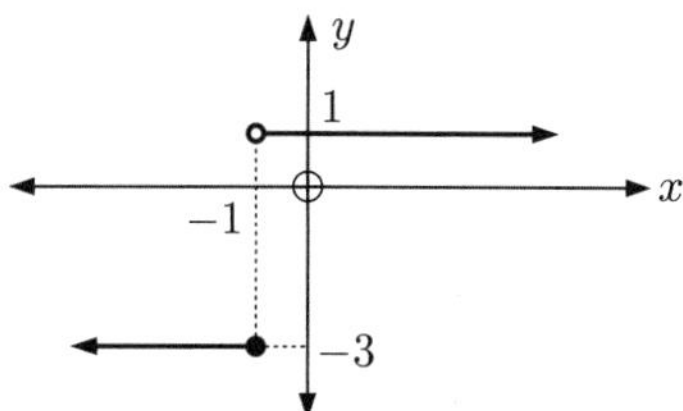

8 Determine whether the following functions are even, odd, or neither:

a $f(x) = -\dfrac{4}{x}$ **b** $f(x) = \dfrac{2x-3}{x+1}$ **c** $f(x) = \sqrt{x^2 - 5}$

9 Find $f^{-1}(x)$ given that $f(x)$ is: **a** $4x + 2$ **b** $\dfrac{3-5x}{4}$

10 Draw a sign diagram for:

a $(3x+2)(4-x)$ **b** $\dfrac{x-3}{x^2+4x+4}$

11 If $f(x) = ax + b$, $f(2) = 1$, and $f^{-1}(3) = 4$, find a and b.

12 **a** Draw a sign diagram for $\dfrac{(x+2)(x-3)}{x-1}$.

b Hence, solve for x: $\dfrac{x^2 - x - 6}{x-1} < 0$.

13 Consider $f(x) = x^2$ and $g(x) = 1 - 6x$.

a Show that $f(-3) = g(-\frac{4}{3})$. **b** Find $(f \circ g)(-2)$.

c Find x such that $g(x) = f(5)$.

14 Given $f : x \mapsto 3x + 6$ and $h : x \mapsto \dfrac{x}{3}$, show that $(f^{-1} \circ h^{-1})(x) = (h \circ f)^{-1}(x)$.

15 Suppose $h(x) = (x-4)^2 + 3$, $x \in [4, \infty[$.

a Find the defining equation of h^{-1}.

b Show that $(h \circ h^{-1})(x) = (h^{-1} \circ h)(x) = x$.

REVIEW SET 2B CALCULATOR

1 For each of the following graphs, find the domain and range:

a

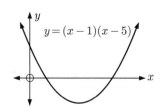

$y = (x-1)(x-5)$

b

$(1, -1)$

$x = 2$

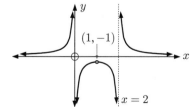

2 If $f(x) = 2x - 3$ and $g(x) = x^2 + 2$, find in simplest form:

a $(f \circ g)(x)$ **b** $(g \circ f)(x)$

3 Draw a sign diagram for: **a** $\dfrac{x^2-6x-16}{x-3}$ **b** $\dfrac{x+9}{x+5}+x$

4 Consider $f(x)=\dfrac{1}{x^2}$.

a For what value of x is $f(x)$ undefined, or not a real number?

b Sketch the graph of this function using technology.

c State the domain and range of the function.

5 Consider the function $f(x)=\dfrac{ax+3}{x-b}$.

a Find a and b given that $y=f(x)$ has asymptotes with equations $x=-1$ and $y=2$.

b Write down the domain and range of $f^{-1}(x)$.

6 For each of the following graphs, find the domain and range:

a

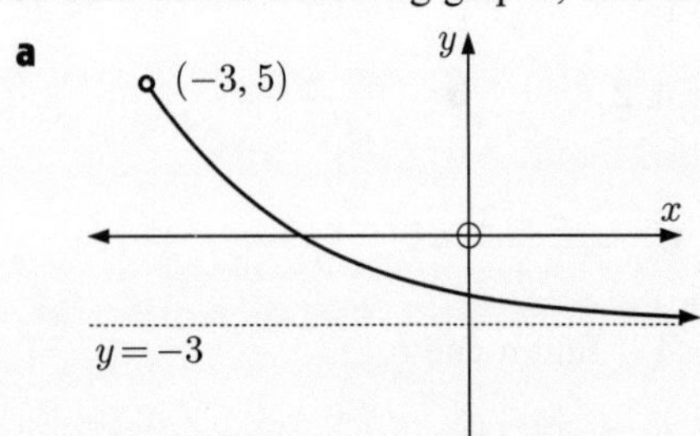

b

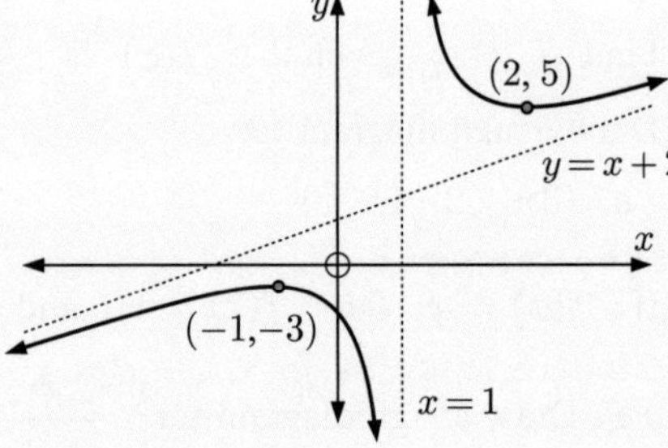

7 Solve for x: **a** $\left|\dfrac{2x+1}{x-2}\right|=3$ **b** $|3x-2|\geqslant|2x+3|$

8 Solve for x: **a** $x^2-5\leqslant 4x$ **b** $\dfrac{3}{x-1}>\dfrac{5}{2x+1}$

9 Copy the following graphs and draw the graph of each inverse function on the same set of axes:

a

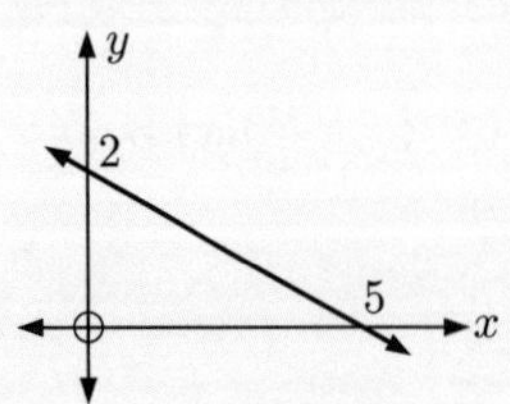

b

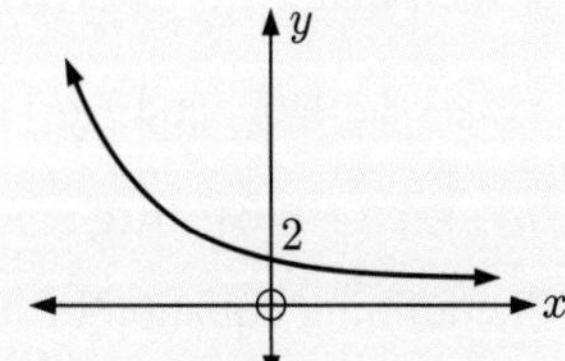

10 Consider the function $f:x\mapsto\dfrac{4x+1}{2-x}$.

a Determine the equations of the asymptotes.

b State the domain and range of the function.

c Discuss the behaviour of the function as it approaches its asymptotes.

d Determine the axes intercepts.

e Sketch the function.

11 Consider the function $f(x)=(x-3)^2+ax$ where a is a real constant. Find a given that $f(x)$ is an even function.

12 **a** Solve graphically: $|2x-6|>x+3$.

b Graph the function $f(x)=\dfrac{x}{|x|+1}$. Hence find all values of x for which $\dfrac{x}{|x|+1}\geqslant\frac{1}{3}$.

13 Consider the functions $f(x) = 3x + 1$ and $g(x) = \dfrac{2}{x}$.

a Find $(g \circ f)(x)$.

b Given $(g \circ f)(x) = -4$, solve for x.

c Let $h(x) = (g \circ f)(x)$, $x \neq -\frac{1}{3}$.

i Write down the equations of the asymptotes of $h(x)$.

ii Sketch the graph of $h(x)$ for $-3 \leqslant x \leqslant 2$.

iii State the range of $h(x)$ for the domain $-3 \leqslant x \leqslant 2$.

14 Consider $f : x \mapsto 2x - 7$.

a On the same set of axes graph $y = x$, $y = f(x)$, and $y = f^{-1}(x)$.

b Find $f^{-1}(x)$ using variable interchange.

c Show that $(f \circ f^{-1})(x) = (f^{-1} \circ f)(x) = x$, the identity function.

15 The graph of the function $f(x) = -3x^2$, $0 \leqslant x \leqslant 2$ is shown alongside.

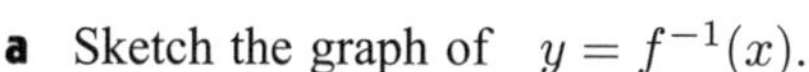

a Sketch the graph of $y = f^{-1}(x)$.

b State the range of f^{-1}.

c Solve:

i $f(x) = -10$

ii $f^{-1}(x) = 1$

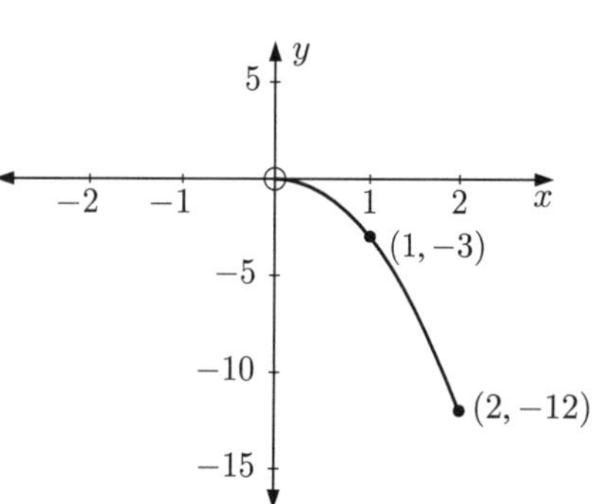

16 Consider the function $f(x) = 0.5x^3 - 4x^2 + 4x + 32$.

a Use technology to help sketch the function.

b Determine the position and nature of any turning points.

c Hence, find the maximum and minimum values of $f(x)$ on the interval $0 \leqslant x \leqslant 6$.

REVIEW SET 2C

1 For each of the following graphs, find the domain and range:

a

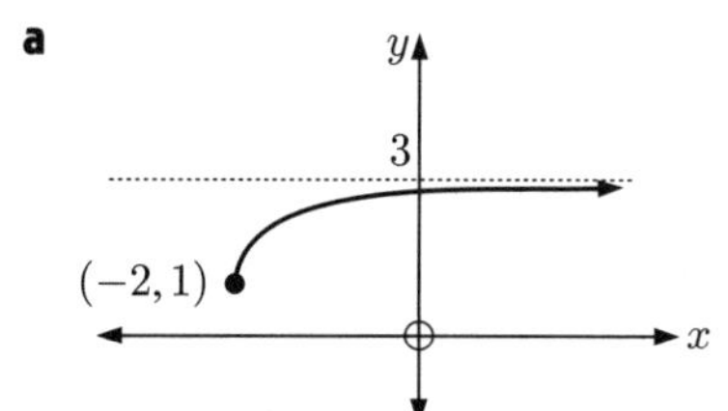

b

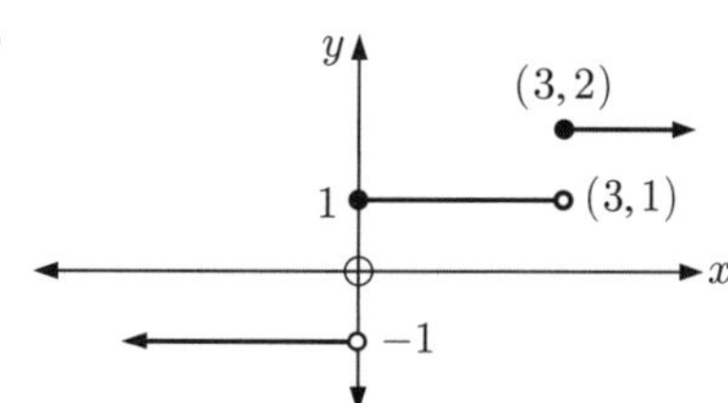

2 Given $f(x) = x^2 + 3$, find:

a $f(-3)$

b x such that $f(x) = 4$.

3 State the value(s) of x for which $f(x)$ is undefined:

a $f(x) = 10 + \dfrac{3}{2x - 1}$

b $f(x) = \sqrt{x + 7}$

4 Draw a sign diagram for:

a $f(x) = x(x+4)(3x+1)$

b $f(x) = \dfrac{-11}{(x+1)(x+8)}$

5 Given $h(x) = 7 - 3x$, find:

a $h(2x-1)$ in simplest form

b x such that $h(2x-1) = -2$.

6 Suppose $f(x) = 1 - 2x$ and $g(x) = \sqrt{x}$.

a Find in simplest form: **i** $(f \circ g)(x)$ **ii** $(g \circ f)(x)$

b State the domain and range of: **i** $(f \circ g)(x)$ **ii** $(g \circ f)(x)$

7 Suppose $f(x) = ax^2 + bx + c$. Find a, b, and c if $f(0) = 5$, $f(-2) = 21$, and $f(3) = -4$.

8 Copy the following graphs and draw the graph of each inverse function on the same set of axes:

a

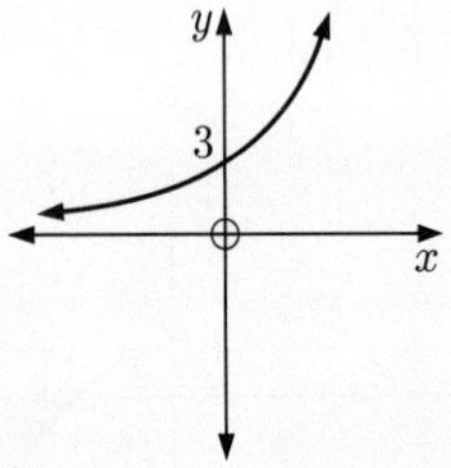

b

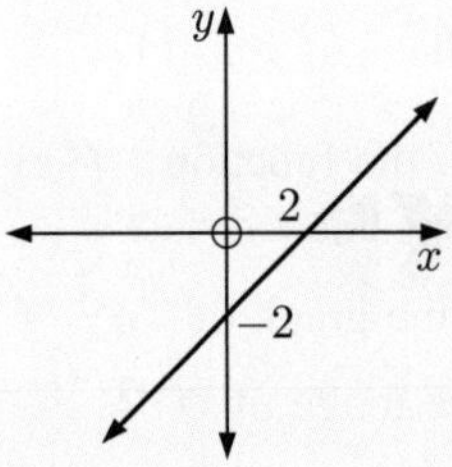

9 Suppose $f(x)$ is an odd function. Prove that $g(x) = |f(x)|$ is an even function.

10 Find the inverse function $f^{-1}(x)$ for:

a $f(x) = 7 - 4x$

b $f(x) = \dfrac{3+2x}{5}$

11 Given $f : x \mapsto 5x - 2$ and $h : x \mapsto \dfrac{3x}{4}$, show that $(f^{-1} \circ h^{-1})(x) = (h \circ f)^{-1}(x)$.

12 Solve for x:

a $2x^2 + x \leqslant 10$

b $\dfrac{x^2 - 3x - 4}{x+2} > 0$

13 Given $f(x) = 2x + 11$ and $g(x) = x^2$, find $(g \circ f^{-1})(3)$.

14 Sketch a function with domain $\{x \mid x \neq 4\}$, range $\{y \mid y \neq -1\}$, and

sign diagram 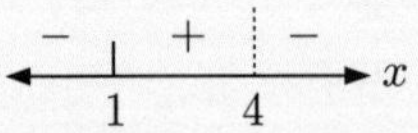 .

15 **a** Sketch the graph of $g : x \mapsto x^2 + 6x + 7$ for $x \in \,]-\infty, -3]$.

b Explain why g has an inverse function g^{-1}.

c Find algebraically, a formula for g^{-1}.

d Sketch the graph of $y = g^{-1}(x)$.

e Find the range of g.

f Find the domain and range of g^{-1}.

16 Consider the function $y = \dfrac{\sqrt{4x^2 + 5x + 8}}{x - 4} + 5$.

a Find the axes intercepts.

b Find any turning points of the function.

c Find any asymptotes of the function.

d State the domain and range of the function.

e Sketch the function, showing its key features.

Chapter 3

Exponentials

Syllabus reference: 1.2, 2.2, 2.6, 2.7, 2.8

Contents:

OPENING PROBLEM

The interior of a freezer has temperature $-10°C$. When a packet of peas is placed in the freezer, its temperature after t minutes is given by $T(t) = -10 + 32 \times 2^{-0.2t}$ °C.

Things to think about:

a What was the temperature of the packet of peas:
 i when it was first placed in the freezer
 ii after 5 minutes
 iii after 10 minutes
 iv after 15 minutes?

b What does the graph of temperature over time look like?

c According to this model, will the temperature of the packet of peas ever reach $-10°C$? Explain your answer.

We often deal with numbers that are repeatedly multiplied together. Mathematicians use **exponents**, also called **powers**, or **indices**, to construct such expressions.

Exponents have many applications in the areas of finance, engineering, physics, electronics, biology, and computer science. Problems encountered in these areas may involve situations where quantities increase or decrease over time. Such problems are often examples of **exponential growth** or **decay**.

A EXPONENTS

Rather than writing $3 \times 3 \times 3 \times 3 \times 3$, we can write this product as 3^5.

If n is a positive integer, then a^n is the product of n factors of a.

$$a^n = \underbrace{a \times a \times a \times a \times \ldots\ldots \times a}_{n \text{ factors}}$$

We say that a is the **base**, and n is the **exponent** or **index**.

$$3^5$$

base; power, index or exponent

NEGATIVE BASES

$(-1)^1 = -1$
$(-1)^2 = -1 \times -1 = 1$
$(-1)^3 = -1 \times -1 \times -1 = -1$
$(-1)^4 = -1 \times -1 \times -1 \times -1 = 1$

$(-2)^1 = -2$
$(-2)^2 = -2 \times -2 = 4$
$(-2)^3 = -2 \times -2 \times -2 = -8$
$(-2)^4 = -2 \times -2 \times -2 \times -2 = 16$

From the patterns above we can see that:

A **negative** base raised to an **odd** exponent is **negative**.
A **negative** base raised to an **even** exponent is **positive**.

CALCULATOR USE

Although different calculators vary in the appearance of keys, they all perform operations of raising to powers in a similar manner. Click on the icon for instructions for calculating exponents.

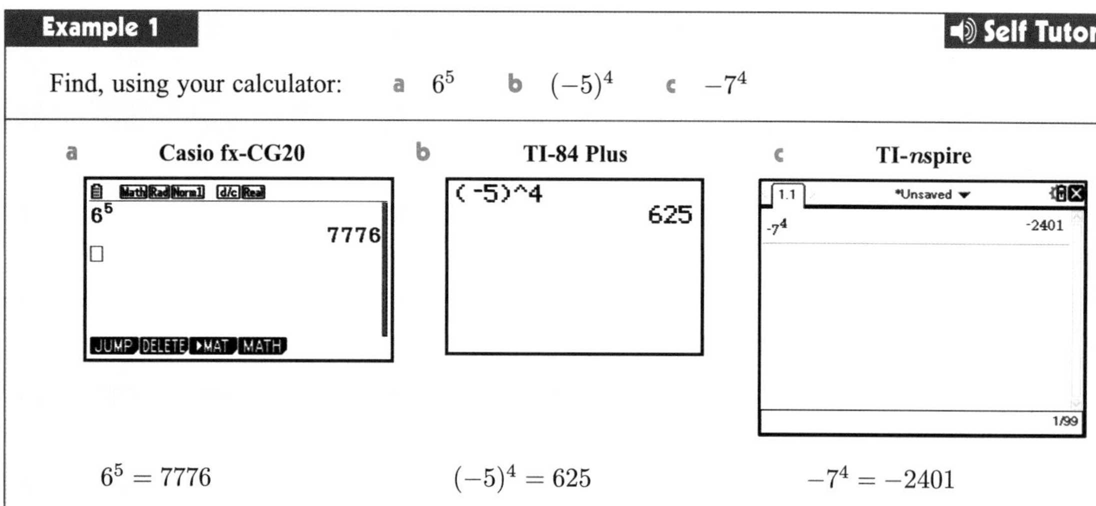

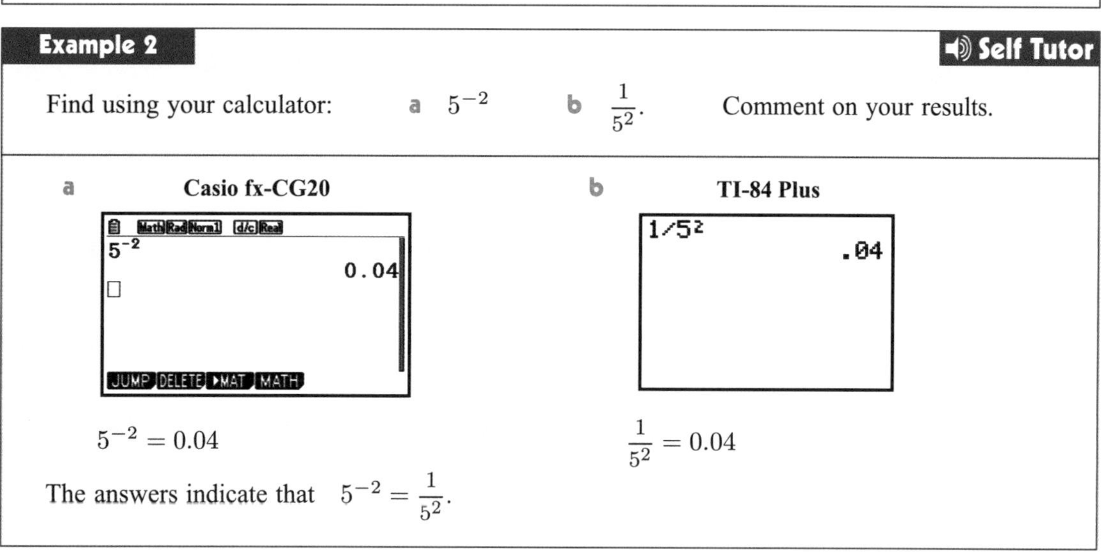

EXERCISE 3A

1 List the first six powers of:

a 2 **b** 3 **c** 4

2 Copy and complete the values of these common powers:

a $5^1 = ...$, $5^2 = ...$, $5^3 = ...$, $5^4 = ...$

b $6^1 = ...$, $6^2 = ...$, $6^3 = ...$, $6^4 = ...$

c $7^1 = ...$, $7^2 = ...$, $7^3 = ...$, $7^4 = ...$

3 Simplify, then use a calculator to check your answer:

a $(-1)^5$ **b** $(-1)^6$ **c** $(-1)^{14}$ **d** $(-1)^{19}$ **e** $(-1)^8$ **f** -1^8

g $-(-1)^8$ **h** $(-2)^5$ **i** -2^5 **j** $-(-2)^6$ **k** $(-5)^4$ **l** $-(-5)^4$

4 Use your calculator to find the value of the following, recording the entire display:

a 4^7 **b** 7^4 **c** -5^5 **d** $(-5)^5$ **e** 8^6 **f** $(-8)^6$

g -8^6 **h** 2.13^9 **i** -2.13^9 **j** $(-2.13)^9$

5 Use your calculator to find the values of the following:

a 9^{-1} b $\frac{1}{9^1}$ c 6^{-2} d $\frac{1}{6^2}$ e 3^{-4} f $\frac{1}{3^4}$

g 17^0 h $(0.366)^0$

What do you notice?

6 Consider $3^1, 3^2, 3^3, 3^4, 3^5$ .... Look for a pattern and hence find the last digit of 3^{101}.

7 What is the last digit of 7^{217}?

HISTORICAL NOTE

Nicomachus discovered an interesting number pattern involving cubes and sums of odd numbers. Nicomachus was born in Roman Syria (now Jerash, Jordan) around 100 AD. He wrote in Greek and was a Pythagorean.

$$1 = 1^3$$
$$3 + 5 = 8 = 2^3$$
$$7 + 9 + 11 = 27 = 3^3$$
$$\vdots$$

B LAWS OF EXPONENTS

The **exponent laws** for $m, n \in \mathbb{Z}$ are:

$a^m \times a^n = a^{m+n}$	To **multiply** numbers with the **same base**, keep the base and **add** the exponents.
$\frac{a^m}{a^n} = a^{m-n}, \quad a \neq 0$	To **divide** numbers with the same base, keep the base and **subtract** the exponents.
$(a^m)^n = a^{m \times n}$	When **raising** a **power** to a **power**, keep the base and **multiply** the exponents.
$(ab)^n = a^n b^n$	The power of a product is the product of the powers.
$\left(\frac{a}{b}\right)^n = \frac{a^n}{b^n}, \quad b \neq 0$	The power of a quotient is the quotient of the powers.
$a^0 = 1, \quad a \neq 0$	Any non-zero number raised to the power of zero is 1.

$a^{-n} = \frac{1}{a^n}$ and $\frac{1}{a^{-n}} = a^n$ and in particular $a^{-1} = \frac{1}{a}, \quad a \neq 0$.

Example 3 Self Tutor

Simplify using the exponent laws: a $3^5 \times 3^4$ b $\frac{5^3}{5^5}$ c $(m^4)^3$

a $3^5 \times 3^4$
$= 3^{5+4}$
$= 3^9$

b $\frac{5^3}{5^5}$
$= 5^{3-5}$
$= 5^{-2}$
$= \frac{1}{25}$

c $(m^4)^3$
$= m^{4\times 3}$
$= m^{12}$

Example 4 — Self Tutor

Write as powers of 2:

a 16 b $\frac{1}{16}$ c 1 d 4×2^n e $\dfrac{2^m}{8}$

a 16
$= 2 \times 2 \times 2 \times 2$
$= 2^4$

b $\frac{1}{16}$
$= \dfrac{1}{2^4}$
$= 2^{-4}$

c 1
$= 2^0$

d 4×2^n
$= 2^2 \times 2^n$
$= 2^{2+n}$

e $\dfrac{2^m}{8}$
$= \dfrac{2^m}{2^3}$
$= 2^{m-3}$

EXERCISE 3B

1 Simplify using the laws of exponents:

a $5^4 \times 5^7$ b $d^2 \times d^6$ c $\dfrac{k^8}{k^3}$ d $\dfrac{7^5}{7^6}$ e $(x^2)^5$ f $(3^4)^4$
g $\dfrac{p^3}{p^7}$ h $n^3 \times n^9$ i $(5^t)^3$ j $7^x \times 7^2$ k $\dfrac{10^3}{10^q}$ l $(c^4)^m$

2 Write as powers of 2:

a 4 b $\frac{1}{4}$ c 8 d $\frac{1}{8}$ e 32 f $\frac{1}{32}$
g 2 h $\frac{1}{2}$ i 64 j $\frac{1}{64}$ k 128 l $\frac{1}{128}$

3 Write as powers of 3:

a 9 b $\frac{1}{9}$ c 27 d $\frac{1}{27}$ e 3 f $\frac{1}{3}$
g 81 h $\frac{1}{81}$ i 1 j 243 k $\frac{1}{243}$

4 Write as a single power of 2:

a 2×2^a b 4×2^b c 8×2^t d $(2^{x+1})^2$ e $(2^{1-n})^{-1}$
f $\dfrac{2^c}{4}$ g $\dfrac{2^m}{2^{-m}}$ h $\dfrac{4}{2^{1-n}}$ i $\dfrac{2^{x+1}}{2^x}$ j $\dfrac{4^x}{2^{1-x}}$

5 Write as a single power of 3:

a 9×3^p b 27^a c 3×9^n d 27×3^d e 9×27^t
f $\dfrac{3^y}{3}$ g $\dfrac{3}{3^y}$ h $\dfrac{9}{27^t}$ i $\dfrac{9^a}{3^{1-a}}$ j $\dfrac{9^{n+1}}{3^{2n-1}}$

Example 5 — Self Tutor

Write in simplest form, without brackets:

a $(-3a^2)^4$ b $\left(-\dfrac{2a^2}{b}\right)^3$

a $(-3a^2)^4$
$= (-3)^4 \times (a^2)^4$
$= 81 \times a^{2\times 4}$
$= 81a^8$

b $\left(-\dfrac{2a^2}{b}\right)^3$
$= \dfrac{(-2)^3 \times (a^2)^3}{b^3}$
$= \dfrac{-8a^6}{b^3}$

6 Write without brackets:

a $(2a)^2$ b $(3b)^3$ c $(ab)^4$ d $(pq)^3$ e $\left(\frac{m}{n}\right)^2$

f $\left(\frac{a}{3}\right)^3$ g $\left(\frac{b}{c}\right)^4$ h $\left(\frac{2a}{b}\right)^0$ i $\left(\frac{m}{3n}\right)^4$ j $\left(\frac{xy}{2}\right)^3$

7 Write the following in simplest form, without brackets:

a $(-2a)^2$ b $(-6b^2)^2$ c $(-2a)^3$ d $(-3m^2n^2)^3$

e $(-2ab^4)^4$ f $\left(\frac{-2a^2}{b^2}\right)^3$ g $\left(\frac{-4a^3}{b}\right)^2$ h $\left(\frac{-3p^2}{q^3}\right)^2$

Example 6 — Self Tutor

Write without negative exponents: $\frac{a^{-3}b^2}{c^{-1}}$

$a^{-3} = \frac{1}{a^3}$ and $\frac{1}{c^{-1}} = c^1$

$\therefore \frac{a^{-3}b^2}{c^{-1}} = \frac{b^2c}{a^3}$

8 Write without negative exponents:

a ab^{-2} b $(ab)^{-2}$ c $(2ab^{-1})^2$ d $(3a^{-2}b)^2$ e $\frac{a^2b^{-1}}{c^2}$

f $\frac{a^2b^{-1}}{c^{-2}}$ g $\frac{1}{a^{-3}}$ h $\frac{a^{-2}}{b^{-3}}$ i $\frac{2a^{-1}}{d^2}$ j $\frac{12a}{m^{-3}}$

Example 7 — Self Tutor

Write $\frac{1}{2^{1-n}}$ in non-fractional form.

$$\frac{1}{2^{1-n}} = 2^{-(1-n)}$$
$$= 2^{-1+n}$$
$$= 2^{n-1}$$

9 Write in non-fractional form:

a $\frac{1}{a^n}$ b $\frac{1}{b^{-n}}$ c $\frac{1}{3^{2-n}}$ d $\frac{a^n}{b^{-m}}$ e $\frac{a^{-n}}{a^{2+n}}$

10 Simplify, giving your answers in simplest rational form:

a $\left(\frac{5}{3}\right)^0$ b $\left(\frac{7}{4}\right)^{-1}$ c $\left(\frac{1}{6}\right)^{-1}$ d $\frac{3^3}{3^0}$

e $\left(\frac{4}{3}\right)^{-2}$ f $2^1 + 2^{-1}$ g $\left(1\frac{2}{3}\right)^{-3}$ h $5^2 + 5^1 + 5^{-1}$

11 Read about Nicomachus' pattern on page **98** and find the series of odd numbers for:

a 5^3 b 7^3 c 12^3

C RATIONAL EXPONENTS

The exponent laws used previously can also be applied to **rational exponents**, or exponents which are written as a fraction.

For $a > 0$, notice that $a^{\frac{1}{2}} \times a^{\frac{1}{2}} = a^{\frac{1}{2}+\frac{1}{2}} = a^1 = a$ {exponent laws}
and $\sqrt{a} \times \sqrt{a} = a$ also.

So, $\boldsymbol{a^{\frac{1}{2}} = \sqrt{a}}$ {by direct comparison}

Likewise $a^{\frac{1}{3}} \times a^{\frac{1}{3}} \times a^{\frac{1}{3}} = a^1 = a$
and $\sqrt[3]{a} \times \sqrt[3]{a} \times \sqrt[3]{a} = a$

suggests $\boldsymbol{a^{\frac{1}{3}} = \sqrt[3]{a}}$

In general, $\boldsymbol{a^{\frac{1}{n}} = \sqrt[n]{a}}$ where $\sqrt[n]{a}$ reads 'the nth root of a', for $n \in \mathbb{Z}^+$.

We can now determine that $\sqrt[n]{a^m}$
$= (a^m)^{\frac{1}{n}}$
$= a^{\frac{m}{n}}$ $\quad \therefore \quad \boldsymbol{a^{\frac{m}{n}} = \sqrt[n]{a^m}}$ for $a > 0$, $n \in \mathbb{Z}^+$, $m \in \mathbb{Z}$

Example 8 **Self Tutor**

Write as a single power of 2: **a** $\sqrt[3]{2}$ **b** $\dfrac{1}{\sqrt{2}}$ **c** $\sqrt[5]{4}$

a $\sqrt[3]{2}$
$= 2^{\frac{1}{3}}$

b $\dfrac{1}{\sqrt{2}}$
$= \dfrac{1}{2^{\frac{1}{2}}}$
$= 2^{-\frac{1}{2}}$

c $\sqrt[5]{4}$
$= (2^2)^{\frac{1}{5}}$
$= 2^{2 \times \frac{1}{5}}$
$= 2^{\frac{2}{5}}$

EXERCISE 3C

1 Write as a single power of 2:

a $\sqrt[5]{2}$ **b** $\dfrac{1}{\sqrt[5]{2}}$ **c** $2\sqrt{2}$ **d** $4\sqrt{2}$ **e** $\dfrac{1}{\sqrt[3]{2}}$

f $2 \times \sqrt[3]{2}$ **g** $\dfrac{4}{\sqrt{2}}$ **h** $(\sqrt{2})^3$ **i** $\dfrac{1}{\sqrt[3]{16}}$ **j** $\dfrac{1}{\sqrt{8}}$

2 Write as a single power of 3:

a $\sqrt[3]{3}$ **b** $\dfrac{1}{\sqrt[3]{3}}$ **c** $\sqrt[4]{3}$ **d** $3\sqrt{3}$ **e** $\dfrac{1}{9\sqrt{3}}$

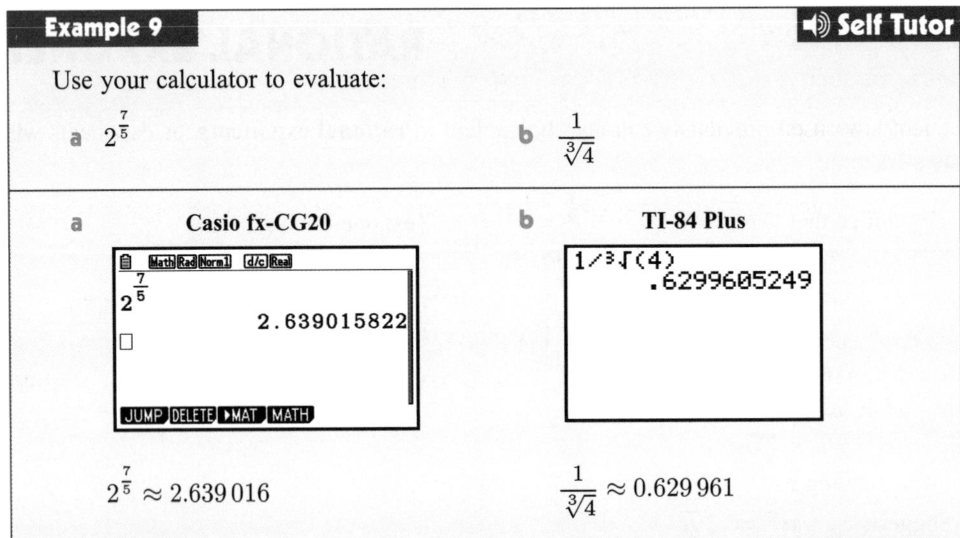

Example 9 **Self Tutor**

Use your calculator to evaluate:

a $2^{\frac{7}{5}}$ **b** $\frac{1}{\sqrt[3]{4}}$

a Casio fx-CG20

$2^{\frac{7}{5}} \approx 2.639\,016$

b TI-84 Plus

$\frac{1}{\sqrt[3]{4}} \approx 0.629\,961$

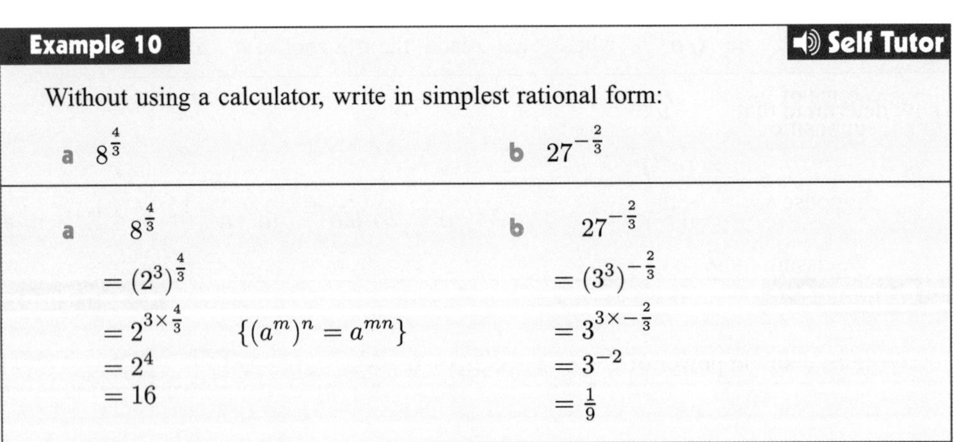

Example 10 **Self Tutor**

Without using a calculator, write in simplest rational form:

a $8^{\frac{4}{3}}$ **b** $27^{-\frac{2}{3}}$

a
$$\begin{aligned} &8^{\frac{4}{3}} \\ &= (2^3)^{\frac{4}{3}} \\ &= 2^{3\times\frac{4}{3}} \qquad \{(a^m)^n = a^{mn}\} \\ &= 2^4 \\ &= 16 \end{aligned}$$

b
$$\begin{aligned} &27^{-\frac{2}{3}} \\ &= (3^3)^{-\frac{2}{3}} \\ &= 3^{3\times-\frac{2}{3}} \\ &= 3^{-2} \\ &= \tfrac{1}{9} \end{aligned}$$

3 Write the following in the form a^x where a is a prime number and x is rational:

a $\sqrt[3]{7}$ **b** $\sqrt[4]{27}$ **c** $\sqrt[5]{16}$ **d** $\sqrt[3]{32}$ **e** $\sqrt[7]{49}$

f $\frac{1}{\sqrt[3]{7}}$ **g** $\frac{1}{\sqrt[4]{27}}$ **h** $\frac{1}{\sqrt[5]{16}}$ **i** $\frac{1}{\sqrt[3]{32}}$ **j** $\frac{1}{\sqrt[7]{49}}$

4 Use your calculator to find:

a $3^{\frac{3}{4}}$ **b** $2^{\frac{7}{8}}$ **c** $2^{-\frac{1}{3}}$ **d** $4^{-\frac{3}{5}}$ **e** $\sqrt[4]{8}$

f $\sqrt[5]{27}$ **g** $\frac{1}{\sqrt[3]{7}}$

5 Without using a calculator, write in simplest rational form:

a $4^{\frac{3}{2}}$ **b** $8^{\frac{5}{3}}$ **c** $16^{\frac{3}{4}}$ **d** $25^{\frac{3}{2}}$ **e** $32^{\frac{2}{5}}$

f $4^{-\frac{1}{2}}$ **g** $9^{-\frac{3}{2}}$ **h** $8^{-\frac{4}{3}}$ **i** $27^{-\frac{4}{3}}$ **j** $125^{-\frac{2}{3}}$

THEORY OF KNOWLEDGE

A **rational number** is a number which can be written in the form $\frac{p}{q}$ where p and q are integers, $q \neq 0$. It has been proven that a rational number has a decimal expansion which either terminates or recurs.

If we begin to write the decimal expansion of $\sqrt{2}$, there is no indication that it will terminate or recur, and we might therefore suspect that $\sqrt{2}$ is irrational.

$$1.414\,213\,562\,373\,095\,048\,801\,688\,724\,209\,698\,078\,569\,671\,875\,376\,948\,073\ldots.$$

However, we cannot *prove* that $\sqrt{2}$ is irrational by writing out its decimal expansion, as we would have to write an infinite number of decimal places. We might therefore *believe* that $\sqrt{2}$ is irrational, but it may also seem impossible to *prove* it.

1. If something has not yet been proven, does that make it untrue?
2. Is the state of an idea being true or false dependent on our ability to prove it?

In fact, we can quite easily prove that $\sqrt{2}$ is irrational by using a method called **proof by contradiction**. In this method we suppose that the opposite is true of what we want to show is true, and follow a series of logical steps until we arrive at a contradiction. The contradiction confirms that our original supposition must be false.

Proof: Suppose $\sqrt{2}$ is rational, so $\sqrt{2} = \frac{p}{q}$ for some (positive) integers p and q, $q \neq 0$.

We assume this fraction has been written in lowest terms, so p and q have no common factors.

Squaring both sides, $2 = \frac{p^2}{q^2}$

$\therefore\ p^2 = 2q^2$ .... (1)

$\therefore\ p^2$ is even, and so p must be even.

Thus $p = 2k$ for some $k \in \mathbb{Z}^+$.

Substituting into (1), $4k^2 = 2q^2$

$\therefore\ q^2 = 2k^2$

$\therefore\ q^2$ is even, and so q must be even.

Here we have a contradiction, as p and q have no common factors.

Thus our original supposition is false, and $\sqrt{2}$ is irrational.

3. Is proof by contradiction unique to mathematics, or do we use it elsewhere?

D ALGEBRAIC EXPANSION AND FACTORISATION

EXPANSION

We can use the usual expansion laws to simplify expressions containing exponents:

$$\begin{aligned} a(b+c) &= ab+ac \\ (a+b)(c+d) &= ac+ad+bc+bd \\ (a+b)(a-b) &= a^2-b^2 \\ (a+b)^2 &= a^2+2ab+b^2 \\ (a-b)^2 &= a^2-2ab+b^2 \end{aligned}$$

Example 11 — Self Tutor

Expand and simplify: $x^{-\frac{1}{2}}(x^{\frac{3}{2}}+2x^{\frac{1}{2}}-3x^{-\frac{1}{2}})$

$x^{-\frac{1}{2}}(x^{\frac{3}{2}}+2x^{\frac{1}{2}}-3x^{-\frac{1}{2}})$

$= x^{-\frac{1}{2}}\times x^{\frac{3}{2}}+x^{-\frac{1}{2}}\times 2x^{\frac{1}{2}}-x^{-\frac{1}{2}}\times 3x^{-\frac{1}{2}}$ {each term is $\times$ by $x^{-\frac{1}{2}}$}

$= x^1+2x^0-3x^{-1}$ {adding exponents}

$= x+2-\dfrac{3}{x}$

Example 12 — Self Tutor

Expand and simplify: **a** $(2^x+3)(2^x+1)$ **b** $(7^x+7^{-x})^2$

a $(2^x+3)(2^x+1)$

$= 2^x\times 2^x+2^x+3\times 2^x+3$

$= 2^{2x}+4\times 2^x+3$

$= 4^x+2^{2+x}+3$

b $(7^x+7^{-x})^2$

$= (7^x)^2+2\times 7^x\times 7^{-x}+(7^{-x})^2$

$= 7^{2x}+2\times 7^0+7^{-2x}$

$= 7^{2x}+2+7^{-2x}$

EXERCISE 3D.1

1 Expand and simplify:

a $x^2(x^3+2x^2+1)$ **b** $2^x(2^x+1)$ **c** $x^{\frac{1}{2}}(x^{\frac{1}{2}}+x^{-\frac{1}{2}})$

d $7^x(7^x+2)$ **e** $3^x(2-3^{-x})$ **f** $x^{\frac{1}{2}}(x^{\frac{3}{2}}+2x^{\frac{1}{2}}+3x^{-\frac{1}{2}})$

g $2^{-x}(2^x+5)$ **h** $5^{-x}(5^{2x}+5^x)$ **i** $x^{-\frac{1}{2}}(x^2+x+x^{\frac{1}{2}})$

2 Expand and simplify:

a $(2^x-1)(2^x+3)$ **b** $(3^x+2)(3^x+5)$ **c** $(5^x-2)(5^x-4)$

d $(2^x+3)^2$ **e** $(3^x-1)^2$ **f** $(4^x+7)^2$

g $(x^{\frac{1}{2}}+2)(x^{\frac{1}{2}}-2)$ **h** $(2^x+3)(2^x-3)$ **i** $(x^{\frac{1}{2}}+x^{-\frac{1}{2}})(x^{\frac{1}{2}}-x^{-\frac{1}{2}})$

j $(x+\frac{2}{x})^2$ **k** $(7^x-7^{-x})^2$ **l** $(5-2^{-x})^2$

FACTORISATION AND SIMPLIFICATION

Example 13 — Self Tutor

Factorise: **a** $2^{n+3} + 2^n$ **b** $2^{n+3} + 8$ **c** $2^{3n} + 2^{2n}$

a $2^{n+3} + 2^n$
$= 2^n 2^3 + 2^n$
$= 2^n(2^3 + 1)$
$= 2^n \times 9$

b $2^{n+3} + 8$
$= 2^n 2^3 + 8$
$= 8(2^n) + 8$
$= 8(2^n + 1)$

c $2^{3n} + 2^{2n}$
$= 2^{2n} 2^n + 2^{2n}$
$= 2^{2n}(2^n + 1)$

Example 14 — Self Tutor

Factorise: **a** $4^x - 9$ **b** $9^x + 4(3^x) + 4$

a $4^x - 9$
$= (2^x)^2 - 3^2$ {compare $a^2 - b^2 = (a+b)(a-b)$}
$= (2^x + 3)(2^x - 3)$

b $9^x + 4(3^x) + 4$
$= (3^x)^2 + 4(3^x) + 4$ {compare $a^2 + 4a + 4$}
$= (3^x + 2)^2$ {as $a^2 + 4a + 4 = (a+2)^2$}

EXERCISE 3D.2

1. Factorise:
   - **a** $5^{2x} + 5^x$
   - **b** $3^{n+2} + 3^n$
   - **c** $7^n + 7^{3n}$
   - **d** $5^{n+1} - 5$
   - **e** $6^{n+2} - 6$
   - **f** $4^{n+2} - 16$
2. Factorise:
   - **a** $9^x - 4$
   - **b** $4^x - 25$
   - **c** $16 - 9^x$
   - **d** $25 - 4^x$
   - **e** $9^x - 4^x$
   - **f** $4^x + 6(2^x) + 9$
   - **g** $9^x + 10(3^x) + 25$
   - **h** $4^x - 14(2^x) + 49$
   - **i** $25^x - 4(5^x) + 4$
3. Factorise:
   - **a** $4^x + 9(2^x) + 18$
   - **b** $4^x - 2^x - 20$
   - **c** $9^x + 9(3^x) + 14$
   - **d** $9^x + 4(3^x) - 5$
   - **e** $25^x + 5^x - 2$
   - **f** $49^x - 7^{x+1} + 12$

Example 15 — Self Tutor

Simplify: **a** $\dfrac{6^n}{3^n}$ **b** $\dfrac{4^n}{6^n}$

a $\dfrac{6^n}{3^n}$ or $\dfrac{6^n}{3^n}$

$= \dfrac{2^n 3^n}{3^n}$ $= \left(\frac{6}{3}\right)^n$

$= 2^n$ $= 2^n$

b $\dfrac{4^n}{6^n}$ or $\dfrac{4^n}{6^n}$

$= \dfrac{2^n 2^n}{2^n 3^n}$ $= \left(\frac{4}{6}\right)^n$

$= \dfrac{2^n}{3^n}$ $= \left(\frac{2}{3}\right)^n$

Example 16 Self Tutor

Simplify: a $\dfrac{3^n+6^n}{3^n}$ b $\dfrac{2^{m+2}-2^m}{2^m}$ c $\dfrac{2^{m+3}+2^m}{9}$

a $\dfrac{3^n+6^n}{3^n}$
$=\dfrac{3^n+2^n3^n}{3^n}$
$=\dfrac{3^n(1+2^n)}{3^n}$
$=1+2^n$

b $\dfrac{2^{m+2}-2^m}{2^m}$
$=\dfrac{2^m2^2-2^m}{2^m}$
$=\dfrac{2^m(4-1)}{2^m}$
$=3$

c $\dfrac{2^{m+3}+2^m}{9}$
$=\dfrac{2^m2^3+2^m}{9}$
$=\dfrac{2^m(8+1)}{9}$
$=2^m$

4 Simplify:

a $\dfrac{12^n}{6^n}$ b $\dfrac{20^a}{2^a}$ c $\dfrac{6^b}{2^b}$ d $\dfrac{4^n}{20^n}$

e $\dfrac{35^x}{7^x}$ f $\dfrac{6^a}{8^a}$ g $\dfrac{5^{n+1}}{5^n}$ h $\dfrac{5^{n+1}}{5}$

5 Simplify:

a $\dfrac{6^m+2^m}{2^m}$ b $\dfrac{2^n+12^n}{2^n}$ c $\dfrac{8^n+4^n}{2^n}$

d $\dfrac{12^x-3^x}{3^x}$ e $\dfrac{6^n+12^n}{1+2^n}$ f $\dfrac{5^{n+1}-5^n}{4}$

g $\dfrac{5^{n+1}-5^n}{5^n}$ h $\dfrac{4^n-2^n}{2^n}$ i $\dfrac{2^n-2^{n-1}}{2^n}$

6 Simplify:

a $2^n(n+1)+2^n(n-1)$ b $3^n\left(\dfrac{n-1}{6}\right)-3^n\left(\dfrac{n+1}{6}\right)$

E EXPONENTIAL EQUATIONS

An **exponential equation** is an equation in which the unknown occurs as part of the index or exponent.

For example: $2^x=8$ and $30\times3^x=7$ are both exponential equations.

There are a number of methods we can use to solve exponential equations. These include graphing, using technology, and by using **logarithms**, which we will study in **Chapter 4**. However, in some cases we can solve algebraically.

If the base numbers are the same, we can **equate exponents**.
If $\boldsymbol{a^x=a^k}$ then $\boldsymbol{x=k}$.

For example, if $2^x=8$ then $2^x=2^3$. Thus $x=3$, and this is the only solution.

Example 17 Self Tutor

Solve for x:

a $2^x = 16$

b $3^{x+2} = \frac{1}{27}$

a $2^x = 16$
$\therefore\ 2^x = 2^4$
$\therefore\ x = 4$

b $3^{x+2} = \frac{1}{27}$
$\therefore\ 3^{x+2} = 3^{-3}$
$\therefore\ x + 2 = -3$
$\therefore\ x = -5$

Once we have the same base we then equate the exponents.

Example 18 Self Tutor

Solve for x:

a $4^x = 8$

b $9^{x-2} = \frac{1}{3}$

a $4^x = 8$
$\therefore\ (2^2)^x = 2^3$
$\therefore\ 2^{2x} = 2^3$
$\therefore\ 2x = 3$
$\therefore\ x = \frac{3}{2}$

b $9^{x-2} = \frac{1}{3}$
$\therefore\ (3^2)^{x-2} = 3^{-1}$
$\therefore\ 3^{2(x-2)} = 3^{-1}$
$\therefore\ 2(x-2) = -1$
$\therefore\ 2x - 4 = -1$
$\therefore\ 2x = 3$
$\therefore\ x = \frac{3}{2}$

EXERCISE 3E

1 Solve for x:

a $2^x = 8$
b $5^x = 25$
c $3^x = 81$
d $7^x = 1$
e $3^x = \frac{1}{3}$
f $2^x = \sqrt{2}$
g $5^x = \frac{1}{125}$
h $4^{x+1} = 64$
i $2^{x-2} = \frac{1}{32}$
j $3^{x+1} = \frac{1}{27}$
k $7^{x+1} = 343$
l $5^{1-2x} = \frac{1}{5}$

2 Solve for x:

a $8^x = 32$
b $4^x = \frac{1}{8}$
c $9^x = \frac{1}{27}$
d $25^x = \frac{1}{5}$
e $27^x = \frac{1}{9}$
f $16^x = \frac{1}{32}$
g $4^{x+2} = 128$
h $25^{1-x} = \frac{1}{125}$
i $4^{4x-1} = \frac{1}{2}$
j $9^{x-3} = 27$
k $(\frac{1}{2})^{x+1} = 8$
l $(\frac{1}{3})^{x+2} = 9$
m $81^x = 27^{-x}$
n $(\frac{1}{4})^{1-x} = 32$
o $(\frac{1}{7})^x = 49$
p $(\frac{1}{3})^{x+1} = 243$

3 Solve for x, if possible:

a $4^{2x+1} = 8^{1-x}$
b $9^{2-x} = (\frac{1}{3})^{2x+1}$
c $2^x \times 8^{1-x} = \frac{1}{4}$

4 Solve for x:

a $3 \times 2^x = 24$
b $7 \times 2^x = 56$
c $3 \times 2^{x+1} = 24$
d $12 \times 3^{-x} = \frac{4}{3}$
e $4 \times (\frac{1}{3})^x = 36$
f $5 \times (\frac{1}{2})^x = 20$

Example 19 **Self Tutor**

Solve for x: $4^x + 2^x - 20 = 0$

$$4^x + 2^x - 20 = 0$$

$\therefore\ (2^x)^2 + 2^x - 20 = 0$ {compare $a^2 + a - 20 = 0$}

$\therefore\ (2^x - 4)(2^x + 5) = 0$ {$a^2 + a - 20 = (a-4)(a+5)$}

$\therefore\ 2^x = 4$ or $2^x = -5$

$\therefore\ 2^x = 2^2$ {2^x cannot be negative}

$\therefore\ x = 2$

5 Solve for x:

a $4^x - 6(2^x) + 8 = 0$
b $4^x - 2^x - 2 = 0$
c $9^x - 12(3^x) + 27 = 0$
d $9^x = 3^x + 6$
e $25^x - 23(5^x) - 50 = 0$
f $49^x + 1 = 2(7^x)$

Check your answers using technology. You can get instructions for doing this by clicking on the icon.

F EXPONENTIAL FUNCTIONS

We have already seen how to evaluate b^n when $n \in \mathbb{Q}$, or in other words when n is a rational number.

But what about b^n when $n \in \mathbb{R}$, so n is real but not necessarily rational?

To answer this question, we can look at graphs of exponential functions.

The most simple general **exponential function** has the form $y = b^x$ where $b > 0$, $b \neq 1$.

For example, $y = 2^x$ is an exponential function.

We construct a table of values from which we graph the function:

x	-3	-2	-1	0	1	2	3
y	$\frac{1}{8}$	$\frac{1}{4}$	$\frac{1}{2}$	1	2	4	8

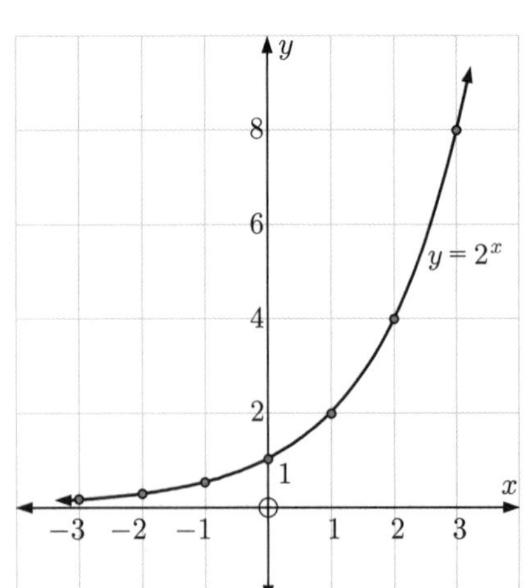

When $x = -10$, $y = 2^{-10} \approx 0.001$.
When $x = -50$, $y = 2^{-50} \approx 8.88 \times 10^{-16}$.

As x becomes large and negative, the graph of $y = 2^x$ approaches the x-axis from above but never touches it, since 2^x becomes very small but never zero.

So, as $x \to -\infty$, $y \to 0^+$.

We say that $y = 2^x$ is '**asymptotic** to the x-axis' or '$y = 0$ is a **horizontal asymptote**'.

We now have a well-defined meaning for b^n where $b, n \in \mathbb{R}$ because simple exponential functions have smooth increasing or decreasing graphs.

INVESTIGATION 1 GRAPHS OF EXPONENTIAL FUNCTIONS

In this Investigation we examine the graphs of various families of exponential functions.

DYNAMIC GRAPHING PACKAGE

Click on the icon to run the **dynamic graphing package**, or else you could use your **graphics calculator**.

What to do:

1 Explore the family of curves of the form $y = b^x$ where $b > 0$.
For example, consider $y = 2^x$, $y = 3^x$, $y = 10^x$, and $y = (1.3)^x$.

- **a** What effect does changing b have on the shape of the graph?
- **b** What is the y-intercept of each graph?
- **c** What is the horizontal asymptote of each graph?

2 Explore the family of curves of the form $y = 2^x + d$ where d is a constant.
For example, consider $y = 2^x$, $y = 2^x + 1$, and $y = 2^x - 2$.

- **a** What effect does changing d have on the position of the graph?
- **b** What effect does changing d have on the shape of the graph?
- **c** What is the horizontal asymptote of each graph?
- **d** What is the horizontal asymptote of $y = 2^x + d$?
- **e** To graph $y = 2^x + d$ from $y = 2^x$ what transformation is used?

3 Explore the family of curves of the form $y = 2^{x-c}$.
For example, consider $y = 2^x$, $y = 2^{x-1}$, $y = 2^{x+2}$, and $y = 2^{x-3}$.

- **a** What effect does changing c have on the position of the graph?
- **b** What effect does changing c have on the shape of the graph?
- **c** What is the horizontal asymptote of each graph?
- **d** To graph $y = 2^{x-c}$ from $y = 2^x$ what transformation is used?

4 Explore the relationship between $y = b^x$ and $y = b^{-x}$ where $b > 0$.
For example, consider $y = 2^x$ and $y = 2^{-x}$.

- **a** What is the y-intercept of each graph?
- **b** What is the horizontal asymptote of each graph?
- **c** What transformation moves $y = 2^x$ to $y = 2^{-x}$?

5 Explore the family of curves of the form $y = a \times 2^x$ where a is a constant.

- **a** Consider functions where $a > 0$, such as $y = 2^x$, $y = 3 \times 2^x$, and $y = \frac{1}{2} \times 2^x$.
Comment on the effect on the graph.
- **b** Consider functions where $a < 0$, such as $y = -2^x$, $y = -3 \times 2^x$, and $y = -\frac{1}{2} \times 2^x$.
Comment on the effect on the graph.
- **c** What is the horizontal asymptote of each graph? Explain your answer.

From your investigation you should have discovered that:

For the general exponential function $\boldsymbol{y = a \times b^{x-c} + d}$ where $b > 0,\ b \neq 1,\ a \neq 0$:

- b controls how steeply the graph increases or decreases
- c controls horizontal translation
- d controls vertical translation
- the equation of the horizontal asymptote is $y = d$
- if $a > 0,\ b > 1$ the function is increasing.

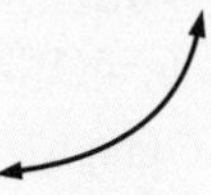

- if $a > 0,\ 0 < b < 1$ the function is decreasing.

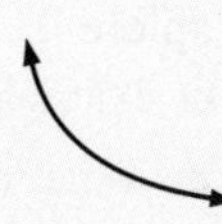

- if $a < 0,\ b > 1$ the function is decreasing.

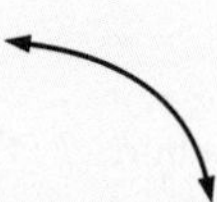

- if $a < 0,\ 0 < b < 1$ the function is increasing.

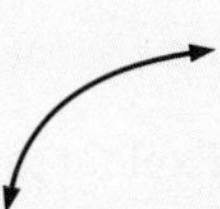

We can sketch reasonably accurate graphs of exponential functions using:

- the horizontal asymptote
- the y-intercept
- two other points, for example, when $x = 2,\ x = -2$

All exponential graphs are similar in shape and have a horizontal asymptote.

Example 20 **Self Tutor**

Sketch the graph of $y = 2^{-x} - 3$.
Hence state the domain and range of $f(x) = 2^{-x} - 3$.

For $y = 2^{-x} - 3$,
the horizontal asymptote is $y = -3$.

When $x = 0$, $y = 2^0 - 3$
$= 1 - 3$
$= -2$

$\therefore$ the y-intercept is -2.

When $x = 2$, $y = 2^{-2} - 3$
$= \frac{1}{4} - 3$
$= -2\frac{3}{4}$

When $x = -2$, $y = 2^2 - 3 = 1$

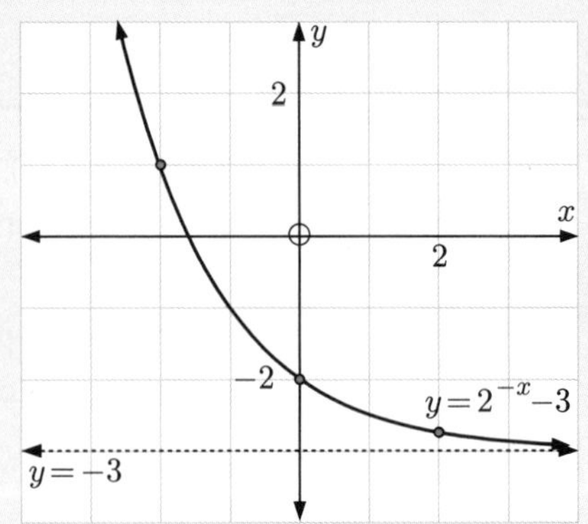

The domain is $\{x \mid x \in \mathbb{R}\}$. The range is $\{y \mid y > -3\}$.

Consider the graph of $y = 2^x$ alongside. We can use the graph to estimate:

- the value of 2^x for a given value of x, for example $2^{1.8} \approx 3.5$ {point A}
- the solutions of the exponential equation $2^x = b$, for example if $2^x = 5$ then $x \approx 2.3$ {point B}.

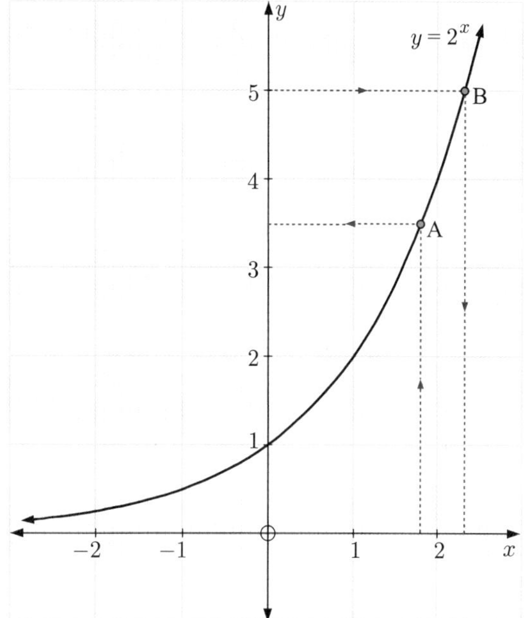

EXERCISE 3F

1 Use the graph above to estimate the value of:

a $2^{\frac{1}{2}}$ or $\sqrt{2}$ b $2^{0.8}$ c $2^{1.5}$ d $2^{-\sqrt{2}}$

2 Use the graph above to estimate the solution to:

a $2^x = 3$ b $2^x = 0.6$

3 Use the graph of $y = 2^x$ to explain why $2^x = 0$ has no solutions.

4 Draw freehand sketches of the following pairs of graphs using your observations from the previous investigation:

GRAPHING PACKAGE

a $y = 2^x$ and $y = 2^x - 2$ b $y = 2^x$ and $y = 2^{-x}$

c $y = 2^x$ and $y = 2^{x-2}$ d $y = 2^x$ and $y = 2(2^x)$

5 Draw freehand sketches of the following pairs of graphs:

a $y = 3^x$ and $y = 3^{-x}$ b $y = 3^x$ and $y = 3^x + 1$

c $y = 3^x$ and $y = -3^x$ d $y = 3^x$ and $y = 3^{x-1}$

6 For each of the functions below:

i sketch the graph of the function
ii state the domain and range
iii use your calculator to find the value of y when $x = \sqrt{2}$
iv discuss the behaviour of y as $x \to \pm\infty$
v determine the horizontal asymptotes.

a $y = 2^x + 1$ b $y = 2 - 2^x$ c $y = 2^{-x} + 3$ d $y = 3 - 2^{-x}$

G GROWTH AND DECAY

In this section we will examine situations where quantities are either increasing or decreasing exponentially. These situations are known as **growth** and **decay** modelling, and occur frequently in the world around us.

Populations of animals, people, and bacteria usually *grow* in an exponential way.

Radioactive substances, and items that depreciate in value, usually *decay* exponentially.

GROWTH

Consider a population of 100 mice which under favourable conditions is increasing by 20% each week.

To increase a quantity by 20%, we multiply it by 1.2.

If P_n is the population after n weeks, then:

$P_0 = 100$ {the *original* population}
$P_1 = P_0 \times 1.2 = 100 \times 1.2$
$P_2 = P_1 \times 1.2 = 100 \times (1.2)^2$
$P_3 = P_2 \times 1.2 = 100 \times (1.2)^3$, and so on.

From this pattern we see that $P_n = 100 \times (1.2)^n$.

So, the graph of the population is a smooth curve given by the exponential function $P_n = 100 \times (1.2)^n$.

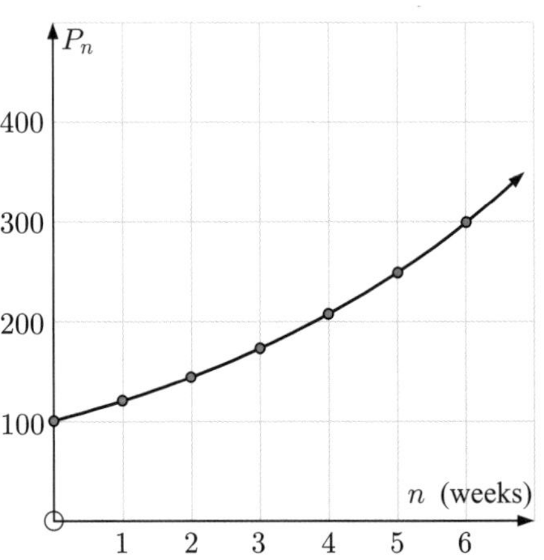

Example 21 — Self Tutor

An entomologist monitoring a grasshopper plague notices that the area affected by the grasshoppers is given by $A_n = 1000 \times 2^{0.2n}$ hectares, where n is the number of weeks after the initial observation.

a Find the original affected area.

b Find the affected area after: **i** 5 weeks **ii** 10 weeks **iii** 12 weeks.

c Draw the graph of A_n against n.

a $A_0 = 1000 \times 2^0$
$= 1000 \times 1$
$= 1000$ $\therefore$ the original affected area was 1000 ha.

b **i** $A_5 = 1000 \times 2^1$
$= 2000$
The affected area is 2000 ha.

ii $A_{10} = 1000 \times 2^2$
$= 4000$
The affected area is 4000 ha.

iii $A_{12} = 1000 \times 2^{0.2 \times 12}$
$= 1000 \times 2^{2.4}$
≈ 5280
The area affected is about 5280 ha.

c

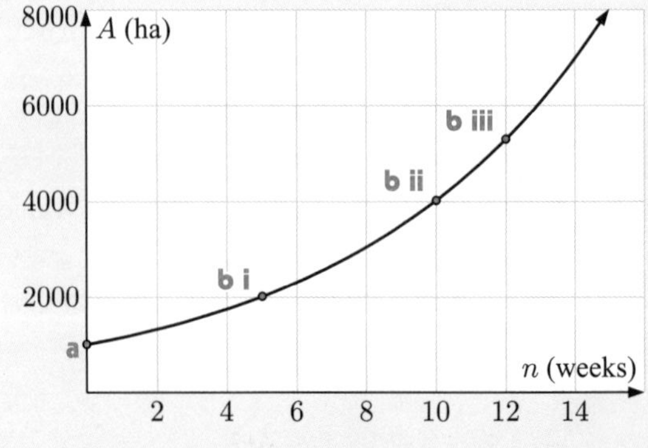

EXERCISE 3G.1

1. The weight W_t of bacteria in a culture t hours after establishment is given by $W_t = 100 \times 2^{0.1t}$ grams.
   a. Find the initial weight.
   b. Find the weight after: **i** 4 hours **ii** 10 hours **iii** 24 hours.
   c. Sketch the graph of W_t against t using the results of **a** and **b** only.
   d. Use technology to graph $\text{Y}_1 = 100 \times 2^{0.1\text{X}}$ and check your answers to **a**, **b**, and **c**.

2. A breeding program to ensure the survival of pygmy possums is established with an initial population of 50 (25 pairs). From a previous program, the expected population P_n in n years' time is given by $P_n = P_0 \times 2^{0.3n}$.
   a. What is the value of P_0?
   b. What is the expected population after: **i** 2 years **ii** 5 years **iii** 10 years?
   c. Sketch the graph of P_n against n using **a** and **b** only.
   d. Use technology to graph $\text{Y}_1 = 50 \times 2^{0.3\text{X}}$ and check your answers to **b**.

3. A species of bear is introduced to a large island off Alaska where previously there were no bears. 6 pairs of bears were introduced in 1998. It is expected that the population will increase according to $B_t = B_0 \times 2^{0.18t}$ where t is the time since the introduction.
   a. Find B_0.
   b. Find the expected bear population in 2018.
   c. Find the expected percentage increase from 2008 to 2018.

4. The speed V_t of a chemical reaction is given by $V_t = V_0 \times 2^{0.05t}$ where t is the temperature in °C.
   a. Find the reaction speed at: **i** 0°C **ii** 20°C.
   b. Find the percentage increase in reaction speed at 20°C compared with 0°C.
   c. Find $\left(\dfrac{V_{50} - V_{20}}{V_{20}}\right) \times 100\%$ and explain what this calculation means.

DECAY

Consider a radioactive substance with original weight 20 grams. It *decays* or reduces by 5% each year. The multiplier for this is 95% or 0.95.

If W_n is the weight after n years, then:

$$\begin{aligned}
W_0 &= 20 \text{ grams} \\
W_1 &= W_0 \times 0.95 = 20 \times 0.95 \text{ grams} \\
W_2 &= W_1 \times 0.95 = 20 \times (0.95)^2 \text{ grams} \\
W_3 &= W_2 \times 0.95 = 20 \times (0.95)^3 \text{ grams} \\
&\vdots \\
W_{20} &= 20 \times (0.95)^{20} \approx 7.2 \text{ grams} \\
&\vdots \\
W_{100} &= 20 \times (0.95)^{100} \approx 0.1 \text{ grams}
\end{aligned}$$

and from this pattern we see that $W_n = 20 \times (0.95)^n$.

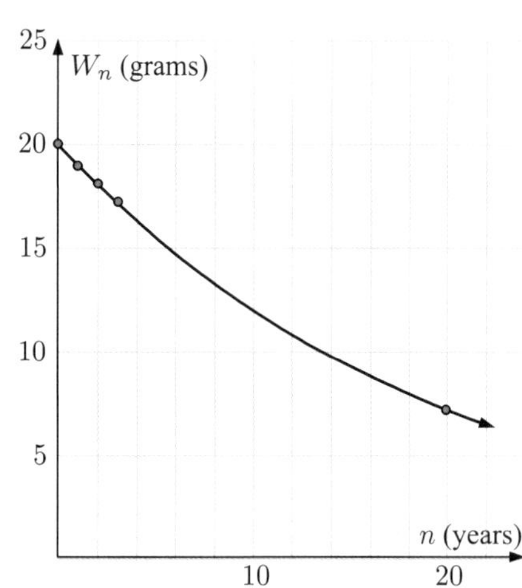

Example 22 **Self Tutor**

When a diesel-electric generator is switched off, the current dies away according to the formula $I(t) = 24 \times (0.25)^t$ amps, where t is the time in seconds after the power is cut.

a Find $I(t)$ when $t = 0$, 1, 2 and 3.

b What current flowed in the generator at the instant when it was switched off?

c Plot the graph of $I(t)$ for $t \geqslant 0$ using the information above.

d Use your graph or technology to find how long it takes for the current to reach 4 amps.

---

a $I(t) = 24 \times (0.25)^t$ amps

$I(0)$	$I(1)$	$I(2)$	$I(3)$
$= 24 \times (0.25)^0$	$= 24 \times (0.25)^1$	$= 24 \times (0.25)^2$	$= 24 \times (0.25)^3$
$= 24$ amps	$= 6$ amps	$= 1.5$ amps	$= 0.375$ amps

b When $t = 0$, $I(0) = 24$

When the generator was switched off, 24 amps of current flowed in the circuit.

c 

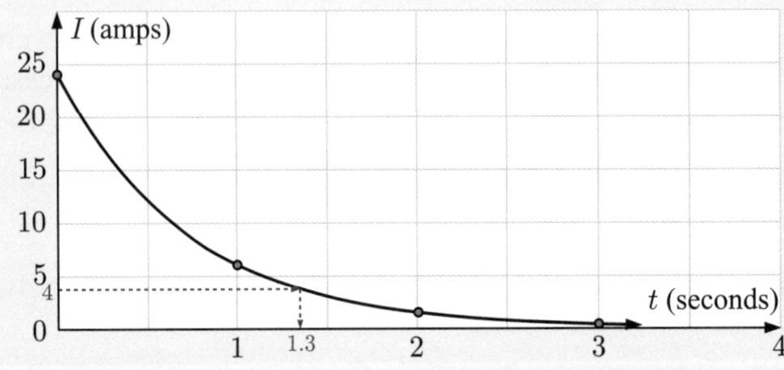

d From the graph above, the time to reach 4 amps is about 1.3 seconds. *or*

By finding the **point of intersection** of $Y_1 = 24 \times (0.25)$^X and $Y_2 = 4$ on a graphics calculator, the solution is ≈ 1.29 seconds.

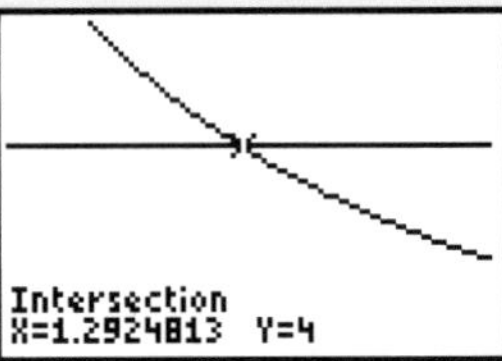

Example 23 **Self Tutor**

The weight of radioactive material remaining after t years is given by $W_t = W_0 \times 2^{-0.001t}$ grams.

a Find the original weight.

b Find the percentage remaining after 200 years.

---

a When $t = 0$, the weight remaining is $W_0 \times 2^0 = W_0$

$\therefore$ W_0 is the original weight.

b When $t = 200$,
$$W_{200} = W_0 \times 2^{-0.001 \times 200}$$
$$= W_0 \times 2^{-0.2}$$
$$\approx W_0 \times 0.8706$$
$$\approx 87.06\% \text{ of } W_0$$

After 200 years, 87.1% of the material remains.

EXERCISE 3G.2

1 The weight of a radioactive substance t years after being set aside is given by $W(t) = 250 \times (0.998)^t$ grams.

- **a** How much radioactive substance was initially set aside?
- **b** Determine the weight of the substance after:
  - **i** 400 years
  - **ii** 800 years
  - **iii** 1200 years.
- **c** Sketch the graph of $W(t)$ for $t \geqslant 0$ using **a** and **b** only.
- **d** Use your graph or graphics calculator to find how long it takes for the substance to decay to 125 grams.

2 The temperature T of a liquid which has been placed in a refrigerator is given by $T(t) = 100 \times 2^{-0.02t}$ °C where t is the time in minutes.

- **a** Find the initial temperature of the liquid.
- **b** Find the temperature after:
  - **i** 15 minutes
  - **ii** 20 minutes
  - **iii** 78 minutes.
- **c** Sketch the graph of $T(t)$ for $t \geqslant 0$ using **a** and **b** only.

3 Answer the **Opening Problem** on page **96**.

4 The weight W_t grams of radioactive substance remaining after t years is given by $W_t = 1000 \times 2^{-0.03t}$ grams.

- **a** Find the initial weight of the radioactive substance.
- **b** Find the weight remaining after:
  - **i** 10 years
  - **ii** 100 years
  - **iii** 1000 years.
- **c** Graph W_t against t using **a** and **b** only.
- **d** Use your graph or graphics calculator to find the time when 10 grams of the substance remains.
- **e** Write an expression for the amount of substance that has decayed after t years.

5 The weight W_t of a radioactive uranium-235 sample remaining after t years is given by the formula $W_t = W_0 \times 2^{-0.0002t}$ grams, $t \geqslant 0$. Find:

- **a** the original weight
- **b** the percentage weight loss after 1000 years
- **c** the time required until $\frac{1}{512}$ of the sample remains.

H THE NATURAL EXPONENTIAL e^x

We have seen that the simplest exponential functions are of the form $f(x) = b^x$ where $b > 0$, $b \neq 1$.

Graphs of some of these functions are shown alongside.

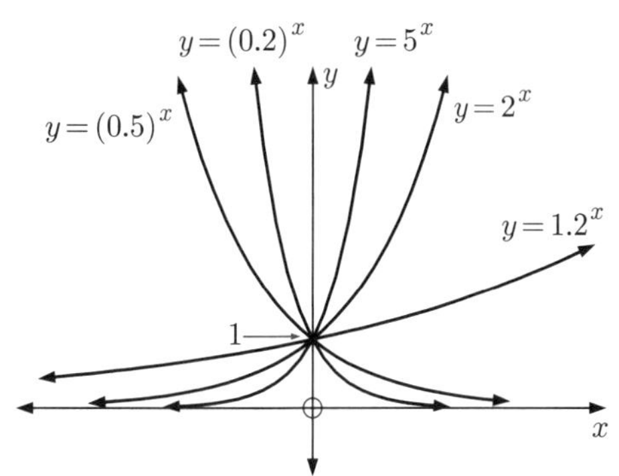

We can see that for all positive values of the base b, the graph is always positive.

Hence $b^x > 0$ for all $b > 0$.

There are an infinite number of possible choices for the base number.

However, where exponential data is examined in science, engineering, and finance, the base $e \approx 2.7183$ is commonly used.

e is a special number in mathematics. It is irrational like π, and just as π is the ratio of a circle's circumference to its diameter, e also has a physical meaning. We explore this meaning in the following Investigation.

INVESTIGATION 2 CONTINUOUS COMPOUND INTEREST

A formula for calculating the amount to which an investment grows is $u_n = u_0(1+i)^n$ where:

u_n is the final amount, u_0 is the initial amount,
i is the interest rate per compounding period,
n is the number of periods or number of times the interest is compounded.

We will investigate the final value of an investment for various values of n, and allow n to get extremely large.

What to do:

1 Suppose $1000 is invested for one year at a fixed rate of 6% per annum. Use your calculator to find the final amount or *maturing value* if the interest is paid:

a annually $(n = 1,\ i = 6\% = 0.06)$
b quarterly $(n = 4,\ i = \frac{6\%}{4} = 0.015)$
c monthly
d daily
e by the second
f by the millisecond.

2 Comment on your answers from **1**.

3 If r is the percentage rate per year, t is the number of years, and N is the number of interest payments per year, then $i = \frac{r}{N}$ and $n = Nt$.

The growth formula becomes $u_n = u_0\left(1+\frac{r}{N}\right)^{Nt}$.

If we let $a = \frac{N}{r}$, show that $u_n = u_0\left[\left(1+\frac{1}{a}\right)^a\right]^{rt}$.

4 For continuous compound growth, the number of interest payments per year N gets very large.

a Explain why a gets very large as N gets very large.

b Copy and complete the table, giving your answers as accurately as technology permits.

a	$\left(1+\frac{1}{a}\right)^a$
10	
100	
1000	
10 000	
100 000	
1 000 000	
10 000 000	

5 You should have found that for very large values of a,

$$\left(1+\frac{1}{a}\right)^a \approx 2.718\,281\,828\,459....$$

Use the e^x key of your calculator to find the value of e^1.
What do you notice?

6 For continuous growth, $u_n = u_0e^{rt}$ where u_0 is the initial amount
r is the annual percentage rate
t is the number of years

Use this formula to find the final value if $1000 is invested for 4 years at a fixed rate of 6% per annum, where the interest is calculated continuously.

From **Investigation 2** we observe that:

If interest is paid *continuously* or *instantaneously* then the formula for calculating a compounding amount $u_n = u_0(1+i)^n$ can be replaced by $u_n = u_0e^{rt}$, where r is the percentage rate per annum and t is the number of years.

HISTORICAL NOTE

The natural exponential e was first described in 1683 by Swiss mathematician Jacob Bernoulli. He discovered the number while studying compound interest, just as we did in **Investigation 2**.

The natural exponential was first called e by Swiss mathematician and physicist Leonhard Euler in a letter to the German mathematician Christian Goldbach in 1731. The number was then published with this notation in 1736.

In 1748 Euler evaluated e correct to 18 decimal places.

One may think that e was chosen because it was the first letter of Euler's name or for the word exponential, but it is likely that it was just the next vowel available since he had already used a in his work.

Leonhard Euler

EXERCISE 3H

1 Sketch, on the same set of axes, the graphs of $y = 2^x$, $y = e^x$, and $y = 3^x$. Comment on any observations.

GRAPHING PACKAGE

2 Sketch, on the same set of axes, the graphs of $y = e^x$ and $y = e^{-x}$. What is the geometric connection between these two graphs?

3 For the general exponential function $y = ae^{kx}$, what is the y-intercept?

4 Consider $y = 2e^x$.

a Explain why y can never be < 0. **b** Find y if: **i** $x = -20$ **ii** $x = 20$.

5 Find, to 3 significant figures, the value of:

a e^2 **b** e^3 **c** $e^{0.7}$ **d** $\sqrt{e}$ **e** e^{-1}

6 Write the following as powers of e:

a $\sqrt{e}$ **b** $\dfrac{1}{\sqrt{e}}$ **c** $\dfrac{1}{e^2}$ **d** $e\sqrt{e}$

7 Simplify:

a $(e^{0.36})^{\frac{t}{2}}$ **b** $(e^{0.064})^{\frac{t}{16}}$ **c** $(e^{-0.04})^{\frac{t}{8}}$ **d** $(e^{-0.836})^{\frac{t}{5}}$

8 Find, to five significant figures, the values of:

a $e^{2.31}$ **b** $e^{-2.31}$ **c** $e^{4.829}$

d $e^{-4.829}$ **e** $50e^{-0.1764}$ **f** $80e^{-0.6342}$

g $1000e^{1.2642}$ **h** $0.25e^{-3.6742}$

9 On the same set of axes, sketch and clearly label the graphs of:

$f : x \mapsto e^x$, $\quad g : x \mapsto e^{x-2}$, $\quad h : x \mapsto e^x + 3$

State the domain and range of each function.

10 On the same set of axes, sketch and clearly label the graphs of:

$f : x \mapsto e^x$, $\quad g : x \mapsto -e^x$, $\quad h : x \mapsto 10 - e^x$

State the domain and range of each function.

11 Expand and simplify:

a $(e^x + 1)^2$ **b** $(1 + e^x)(1 - e^x)$ **c** $e^x(e^{-x} - 3)$

12 The weight of bacteria in a culture is given by $W(t) = 2e^{\frac{t}{2}}$ grams where t is the time in hours after the culture was set to grow.

a Find the weight of the culture when:

i $t = 0$ **ii** $t = 30$ min **iii** $t = 1\frac{1}{2}$ hours **iv** $t = 6$ hours.

b Use **a** to sketch the graph of $W(t) = 2e^{\frac{t}{2}}$.

13 Solve for x:

a $e^x = \sqrt{e}$ **b** $e^{\frac{1}{2}x} = \frac{1}{e^2}$

14 The current flowing in an electrical circuit t seconds after it is switched off is given by $I(t) = 75e^{-0.15t}$ amps.

a What current is still flowing in the circuit after:

i 1 second **ii** 10 seconds?

b Use your graphics calculator to sketch $I(t) = 75e^{-0.15t}$ and $I = 1$.

c Hence find how long it will take for the current to fall to 1 amp.

15 Consider the function $f(x) = e^x$.

a On the same set of axes, sketch $y = f(x)$, $y = x$, and $y = f^{-1}(x)$.

b State the domain and range of f^{-1}.

ACTIVITY

Click on the icon to run a card game for exponential functions.

CARD GAME

RESEARCH — RESEARCHING e

What to do:

1. The 'bell curve' which models statistical distributions is shown alongside. Research the equation of this curve.

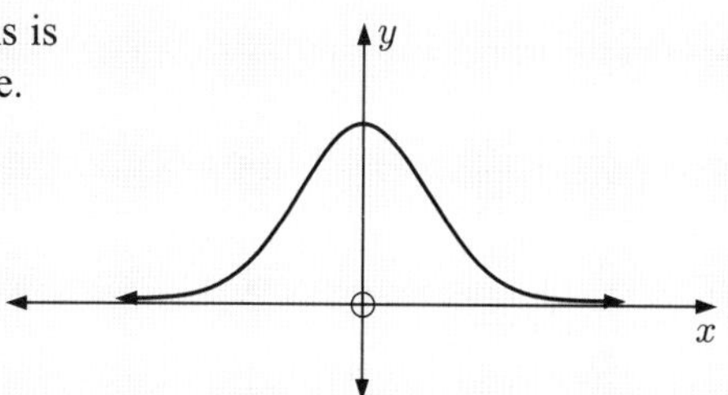

2. The function $f(x) = 1 + x + \frac{1}{2}x^2 + \frac{1}{2\times3}x^3 + \frac{1}{2\times3\times4}x^4 + ...$ has infinitely many terms. It can be shown that $f(x) = e^x$.
Check this statement by finding an approximation for $f(1)$ using its first 20 terms.

REVIEW SET 3A — NON-CALCULATOR

1. Simplify:
   a $-(-1)^{10}$ **b** $-(-3)^3$ **c** $3^0 - 3^{-1}$

2. Simplify using the laws of exponents:
   a $a^4b^5 \times a^2b^2$ **b** $6xy^5 \div 9x^2y^5$ **c** $\dfrac{5(x^2y)^2}{(5x^2)^2}$

3. Let $f(x) = 3^x$.
   a Write down the value of: **i** $f(4)$ **ii** $f(-1)$
   b Find the value of k such that $f(x+2) = k\,f(x)$, $k \in \mathbb{Z}$.

4. Write without brackets or negative exponents:
   a $x^{-2} \times x^{-3}$ **b** $2(ab)^{-2}$ **c** $2ab^{-2}$

5. Write as a single power of 3:
   a $\dfrac{27}{9^a}$ **b** $(\sqrt{3})^{1-x} \times 9^{1-2x}$

6. Evaluate:
   a $8^{\frac{2}{3}}$ **b** $27^{-\frac{2}{3}}$

7. Write without negative exponents:
   a mn^{-2} **b** $(mn)^{-3}$ **c** $\dfrac{m^2n^{-1}}{p^{-2}}$ **d** $(4m^{-1}n)^2$

8. Expand and simplify:
   a $(3 - e^x)^2$ **b** $(\sqrt{x}+2)(\sqrt{x}-2)$ **c** $2^{-x}(2^{2x} + 2^x)$

9. Find the value of x:
   a $2^{x-3} = \frac{1}{32}$ **b** $9^x = 27^{2-2x}$ **c** $e^{2x} = \dfrac{1}{\sqrt{e}}$

10 Match each equation to its corresponding graph:

a $y = -e^x$ **b** $y = 3 \times 2^x$ **c** $y = e^x + 1$ **d** $y = 3^{-x}$ **e** $y = -e^{-x}$

A

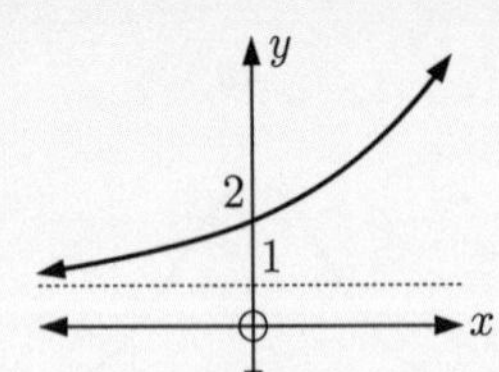

B

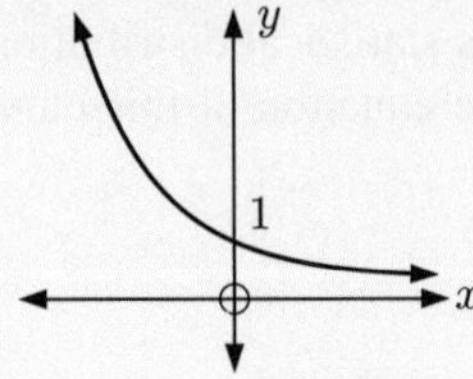

C

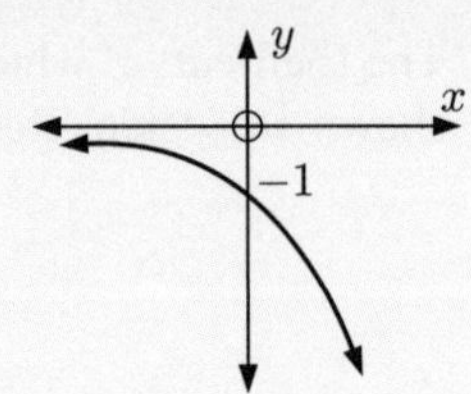

D

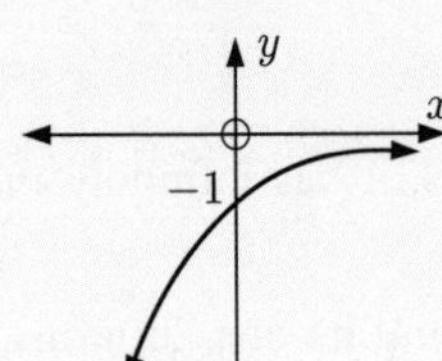

E

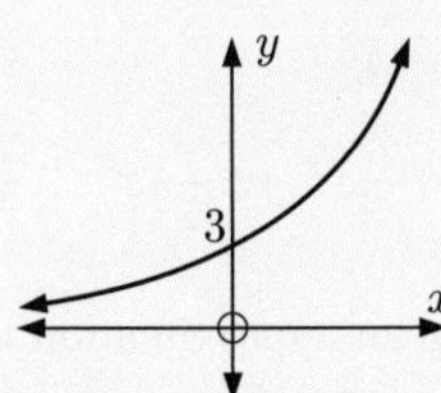

11 Suppose $y = a^x$. Express in terms of y:

a a^{2x} **b** a^{-x} **c** $\frac{1}{\sqrt{a^x}}$

REVIEW SET 3B — CALCULATOR

1 **a** Write 4×2^n as a power of 2.

b Evaluate $7^{-1} - 7^0$.

c Write $(\frac{2}{3})^{-3}$ in simplest fractional form.

d Write $\left(\frac{2a^{-1}}{b^2}\right)^2$ without negative exponents or brackets.

2 Evaluate, correct to 3 significant figures:

a $3^{\frac{3}{4}}$ **b** $27^{-\frac{1}{5}}$ **c** $\sqrt[4]{100}$

3 If $f(x) = 3 \times 2^x$, find the value of:

a $f(0)$ **b** $f(3)$ **c** $f(-2)$

4 Suppose $f(x) = 2^{-x} + 1$.

a Find $f(\frac{1}{2})$. **b** Find a such that $f(a) = 3$.

5 On the same set of axes draw the graphs of $y = 2^x$ and $y = 2^x - 4$. Include on your graph the y-intercept and the equation of the horizontal asymptote of each function.

6 The temperature of a dish t minutes after it is removed from the microwave, is given by $T = 80 \times (0.913)^t$ °C.

a Find the initial temperature of the dish.

b Find the temperature after:

i $t = 12$ **ii** $t = 24$ **iii** $t = 36$ minutes.

c Draw the graph of T against t for $t \geqslant 0$, using the above or technology.

d Hence, find the time taken for the temperature of the dish to fall to 25°C.

7 Consider $y = 3^x - 5$.

a Find y when $x = 0, \pm 1, \pm 2$. **b** Discuss y as $x \to \pm\infty$.

c Sketch the graph of $y = 3^x - 5$. **d** State the equation of any asymptote.

8 **a** On the same set of axes, sketch and clearly label the graphs of:

$f : x \mapsto e^x, \quad g : x \mapsto e^{x-1}, \quad h : x \mapsto 3 - e^x$

b State the domain and range of each function in **a**.

9 Consider $y = 3 - 2^{-x}$.

a Find y when $x = 0, \pm 1, \pm 2$. **b** Discuss y as $x \to \pm\infty$.

c Sketch the graph of $y = 3 - 2^{-x}$. **d** State the equation of any asymptote.

10 The weight of a radioactive substance after t years is given by $W = 1500 \times (0.993)^t$ grams.

a Find the original amount of radioactive material.

b Find the amount of radioactive material remaining after:

i 400 years **ii** 800 years.

c Sketch the graph of W against t, $t \geqslant 0$, using the above or technology.

d Hence, find the time taken for the weight to reduce to 100 grams.

REVIEW SET 3C

1 Given the graph of $y = 3^x$ shown, estimate solutions to the exponential equations:

a $3^x = 5$

b $3^x = \frac{1}{2}$

c $6 \times 3^x = 20$

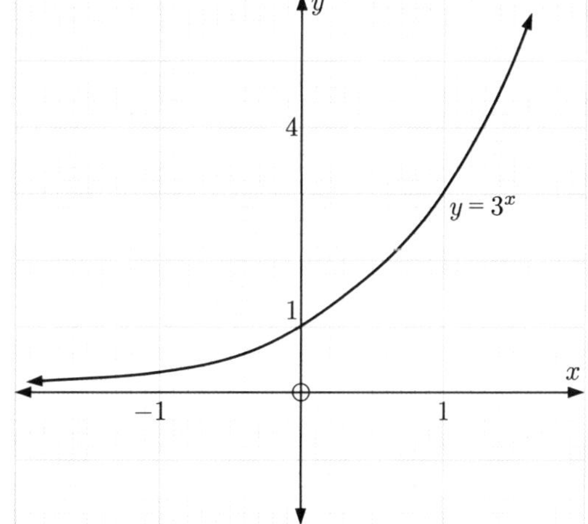

2 Simplify using the laws of exponents:

a $(a^7)^3$ **b** $pq^2 \times p^3q^4$ **c** $\dfrac{8ab^5}{2a^4b^4}$

3 Write the following as a power of 2:

a 2×2^{-4} **b** $16 \div 2^{-3}$ **c** 8^4

4 Write without brackets or negative exponents:

a b^{-3} **b** $(ab)^{-1}$ **c** ab^{-1}

5 Simplify $\dfrac{2^{x+1}}{2^{1-x}}$.

6 Write as powers of 5 in simplest form:

a 1 **b** $5\sqrt{5}$ **c** $\dfrac{1}{\sqrt[4]{5}}$ **d** 25^{a+3}

7 Expand and simplify:

a $e^x(e^{-x}+e^x)$ **b** $(2^x+5)^2$ **c** $(x^{\frac{1}{2}}-7)(x^{\frac{1}{2}}+7)$

8 Solve for x:

a $6\times 2^x = 192$ **b** $4\times(\frac{1}{3})^x = 324$

9 The point $(1, \sqrt{8})$ lies on the graph of $y=2^{kx}$. Find the value of k.

10 Solve for x without using a calculator:

a $2^{x+1}=32$ **b** $4^{x+1}=\left(\frac{1}{8}\right)^x$

11 Consider $y=2e^{-x}+1$.

a Find y when $x=0, \pm1, \pm2$.

b Discuss y as $x\to\pm\infty$.

c Sketch the graph of $y=2e^{-x}+1$.

d State the equation of any asymptote.

Chapter 4

Logarithms

Syllabus reference: 1.2, 2.2, 2.6, 2.8

Contents:

OPENING PROBLEM

In a plentiful springtime, a population of 1000 mice will double every week.

The population after t weeks is given by the exponential function $P(t) = 1000 \times 2^t$ mice.

Things to think about:

- **a** What does the graph of the population over time look like?
- **b** How long will it take for the population to reach 20 000 mice?
- **c** Can we write a function for t in terms of P, which determines the time at which the population P is reached?
- **d** What does the graph of this function look like?

A LOGARITHMS IN BASE 10

Consider the exponential function $f : x \mapsto 10^x$ or $f(x) = 10^x$.

The graph of $y = f(x)$ is shown alongside, along with its inverse function f^{-1}.

Since f is defined by $y = 10^x$,
f^{-1} is defined by $x = 10^y$.
{interchanging x and y}

y is the exponent to which the base 10 is raised in order to get x.

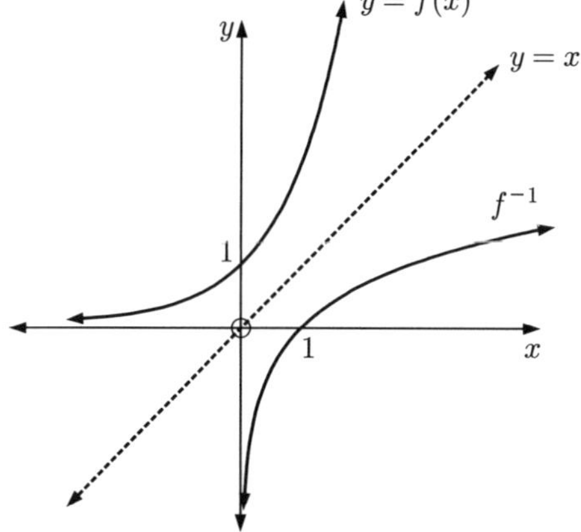

We write this as $y = \log_{10} x$ and say that y is the **logarithm in base 10, of x.**

Logarithms are thus defined to be the inverse of exponential functions:

If $f(x) = 10^x$ then $f^{-1}(x) = \log_{10} x$.

WORKING WITH LOGARITHMS

Many positive numbers can be easily written in the form 10^x.

For example:

$$\begin{aligned} 10\,000 &= 10^4 \\ 1000 &= 10^3 \\ 100 &= 10^2 \\ 10 &= 10^1 \\ 1 &= 10^0 \\ 0.1 &= 10^{-1} \\ 0.01 &= 10^{-2} \\ 0.001 &= 10^{-3} \end{aligned}$$

Numbers like $\sqrt{10}$, $10\sqrt{10}$ and $\dfrac{1}{\sqrt[5]{10}}$ can also be written in the form 10^x as follows:

$$\sqrt{10} = 10^{\frac{1}{2}} = 10^{0.5} \qquad 10\sqrt{10} = 10^1 \times 10^{0.5} = 10^{1.5} \qquad \frac{1}{\sqrt[5]{10}} = 10^{-\frac{1}{5}} = 10^{-0.2}$$

In fact, all positive numbers can be written in the form 10^x by using logarithms in base 10.

> The **logarithm in base 10** of a positive number is the power that 10 must be raised to in order to obtain the number.

For example:

- Since $1000 = 10^3$, we write $\log_{10} 1000 = 3$ or $\log 1000 = 3$.
- Since $0.01 = 10^{-2}$, we write $\log_{10}(0.01) = -2$ or $\log(0.01) = -2$.

$$\boldsymbol{a = 10^{\log a}} \quad \text{for any } a > 0.$$

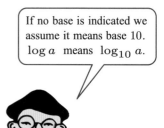

If no base is indicated we assume it means base 10. $\log a$ means $\log_{10} a$.

Notice that a must be positive since $10^x > 0$ for all $x \in \mathbb{R}$.

Notice also that $\log 1000 = \log 10^3 = 3$
and $\log 0.01 = \log 10^{-2} = -2$.

We hence conclude that $\boldsymbol{\log 10^x = x}$ for any $x \in \mathbb{R}$.

Example 1 **Self Tutor**

a Without using a calculator, find: **i** $\log 100$ **ii** $\log(\sqrt[4]{10})$.

b Check your answers using technology.

a **i** $\log 100 = \log 10^2 = 2$ **ii** $\log(\sqrt[4]{10}) = \log(10^{\frac{1}{4}}) = \frac{1}{4}$

b **i** Casio fx-CG20 **ii** TI-84 Plus

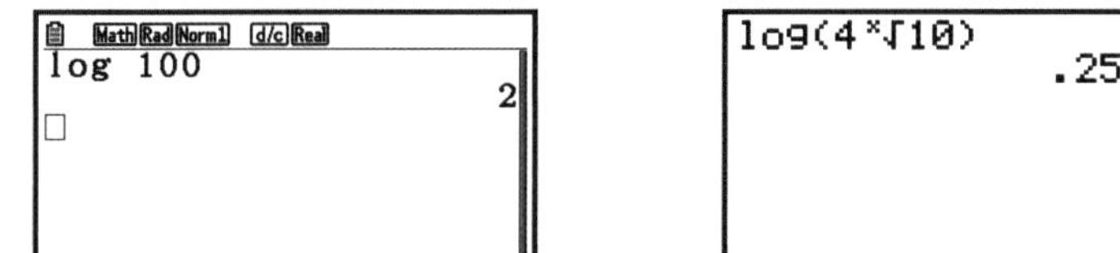

GRAPHICS CALCULATOR INSTRUCTIONS

$\log 100 = 2$ $\log(\sqrt[4]{10}) = 0.25$

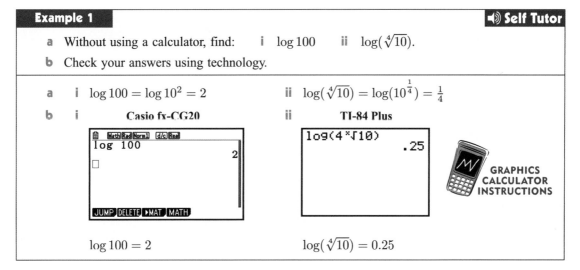

EXERCISE 4A

1 Without using a calculator, find:

a $\log 10\,000$ **b** $\log 0.001$ **c** $\log 10$ **d** $\log 1$

e $\log \sqrt{10}$ **f** $\log(\sqrt[3]{10})$ **g** $\log\left(\dfrac{1}{\sqrt[4]{10}}\right)$ **h** $\log\left(10\sqrt{10}\right)$

i $\log \sqrt[3]{100}$ **j** $\log\left(\dfrac{100}{\sqrt{10}}\right)$ **k** $\log\left(10 \times \sqrt[3]{10}\right)$ **l** $\log\left(1000\sqrt{10}\right)$

Check your answers using your calculator.

2 Simplify:

a $\log 10^n$ **b** $\log(10^a \times 100)$ **c** $\log\left(\frac{10}{10^m}\right)$ **d** $\log\left(\frac{10^a}{10^b}\right)$

Example 2 Self Tutor

Use your calculator to write the following in the form 10^x where x is correct to 4 decimal places:

a 8 **b** 800 **c** 0.08

a $8 = 10^{\log 8} \approx 10^{0.9031}$

b $800 = 10^{\log 800} \approx 10^{2.9031}$

c $0.08 = 10^{\log 0.08} \approx 10^{-1.0969}$

3 Use your calculator to write the following in the form 10^x where x is correct to 4 decimal places:

a 6 **b** 60 **c** 6000 **d** 0.6 **e** 0.006

f 15 **g** 1500 **h** 1.5 **i** 0.15 **j** 0.000 15

Example 3 Self Tutor

a Use your calculator to find: **i** $\log 2$ **ii** $\log 20$

b Explain why $\log 20 = \log 2 + 1$.

a **i** $\log 2 \approx 0.3010$
ii $\log 20 \approx 1.3010$ {calculator}

b
$$\begin{aligned}\log 20 &= \log(2 \times 10)\\ &\approx \log(10^{0.3010} \times 10^1)\\ &\approx \log 10^{1.3010} \quad \text{\{adding exponents\}}\\ &\approx 1.3010\\ &\approx \log 2 + 1\end{aligned}$$

4 **a** Use your calculator to find: **i** $\log 3$ **ii** $\log 300$

b Explain why $\log 300 = \log 3 + 2$.

5 **a** Use your calculator to find: **i** $\log 5$ **ii** $\log 0.05$

b Explain why $\log 0.05 = \log 5 - 2$.

Example 4 Self Tutor

Find x if:

a $\log x = 3$ **b** $\log x \approx -0.271$

a
$$\begin{aligned} x &= 10^{\log x}\\ \therefore\ x &= 10^3\\ \therefore\ x &= 1000\end{aligned}$$

b
$$\begin{aligned} x &= 10^{\log x}\\ \therefore\ x &\approx 10^{-0.271}\\ \therefore\ x &\approx 0.536\end{aligned}$$

6 Find x if:

a $\log x = 2$	**b** $\log x = 1$	**c** $\log x = 0$
d $\log x = -1$	**e** $\log x = \frac{1}{2}$	**f** $\log x = -\frac{1}{2}$
g $\log x = 4$	**h** $\log x = -5$	**i** $\log x \approx 0.8351$
j $\log x \approx 2.1457$	**k** $\log x \approx -1.378$	**l** $\log x \approx -3.1997$

B LOGARITHMS IN BASE a

In the previous section we defined logarithms in base 10 as the inverse of the exponential function $f(x) = 10^x$.

If $f(x) = 10^x$ then $f^{-1}(x) = \log_{10} x$.

We can use the same principle to define logarithms in other bases:

If $f(x) = a^x$ then $f^{-1}(x) = \log_a x$.

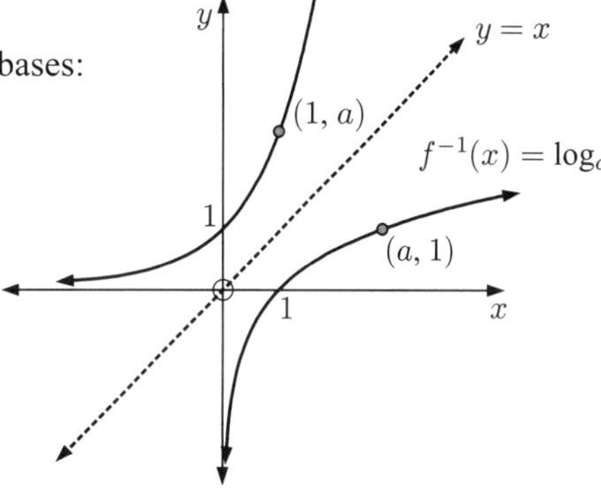

If $b = a^x$, $a \neq 1$, $a > 0$, we say that x is the **logarithm in base a, of b**, and that $b = a^x \Leftrightarrow x = \log_a b$, $b > 0$.

This means that:

"if $b = a^x$ then $x = \log_a b$, and if $x = \log_a b$ then $b = a^x$".

$b = a^x$ and $x = \log_a b$ are *equivalent* or *interchangeable* statements.

For example:
- $8 = 2^3$ means that $3 = \log_2 8$ and vice versa.
- $\log_5 25 = 2$ means that $25 = 5^2$ and vice versa.

If $y = a^x$ then $x = \log_a y$, and so $\boldsymbol{x = \log_a a^x}$.

If $x = a^y$ then $y = \log_a x$, and so $\boldsymbol{x = a^{\log_a x}}$ provided $x > 0$.

Example 5 — Self Tutor

a Write an equivalent exponential equation for $\log_{10} 1000 = 3$.

b Write an equivalent logarithmic equation for $3^4 = 81$.

a From $\log_{10} 1000 = 3$ we deduce that $10^3 = 1000$.

b From $3^4 = 81$ we deduce that $\log_3 81 = 4$.

EXERCISE 4B

1 Write an equivalent exponential equation for:

a $\log_{10} 100 = 2$ **b** $\log_{10} 10\,000 = 4$ **c** $\log_{10}(0.1) = -1$

d $\log_{10} \sqrt{10} = \frac{1}{2}$ **e** $\log_2 8 = 3$ **f** $\log_3 9 = 2$

g $\log_2(\frac{1}{4}) = -2$ **h** $\log_3 \sqrt{27} = 1.5$ **i** $\log_5(\frac{1}{\sqrt{5}}) = -\frac{1}{2}$

2 Write an equivalent logarithmic equation for:

a $2^2 = 4$ **b** $4^3 = 64$ **c** $5^2 = 25$

d $7^2 = 49$ **e** $2^6 = 64$ **f** $2^{-3} = \frac{1}{8}$

g $10^{-2} = 0.01$ **h** $2^{-1} = \frac{1}{2}$ **i** $3^{-3} = \frac{1}{27}$

Example 6 Self Tutor

Find:

a $\log_2 16$ **b** $\log_5 0.2$ **c** $\log_{10} \sqrt[5]{100}$ **d** $\log_2\left(\frac{1}{\sqrt{2}}\right)$

a $\log_2 16$
$= \log_2 2^4$
$= 4$

b $\log_5 0.2$
$= \log_5(\frac{1}{5})$
$= \log_5 5^{-1}$
$= -1$

c $\log_{10} \sqrt[5]{100}$
$= \log_{10}\left(10^2\right)^{\frac{1}{5}}$
$= \log_{10} 10^{\frac{2}{5}}$
$= \frac{2}{5}$

d $\log_2\left(\frac{1}{\sqrt{2}}\right)$
$= \log_2 2^{-\frac{1}{2}}$
$= -\frac{1}{2}$

3 Find:

a $\log_{10} 100\,000$ **b** $\log_{10}(0.01)$ **c** $\log_3 \sqrt{3}$ **d** $\log_2 8$

e $\log_2 64$ **f** $\log_2 128$ **g** $\log_5 25$ **h** $\log_5 125$

i $\log_2(0.125)$ **j** $\log_9 3$ **k** $\log_4 16$ **l** $\log_{36} 6$

m $\log_3 243$ **n** $\log_2 \sqrt[3]{2}$ **o** $\log_a a^n$ **p** $\log_8 2$

q $\log_t\left(\frac{1}{t}\right)$ **r** $\log_6 6\sqrt{6}$ **s** $\log_4 1$ **t** $\log_9 9$

4 Use your calculator to find:

a $\log_{10} 152$ **b** $\log_{10} 25$ **c** $\log_{10} 74$ **d** $\log_{10} 0.8$

5 Solve for x:

a $\log_2 x = 3$ **b** $\log_4 x = \frac{1}{2}$ **c** $\log_x 81 = 4$ **d** $\log_2(x-6) = 3$

6 Simplify:

a $\log_4 16$ **b** $\log_2 4$ **c** $\log_3\left(\frac{1}{3}\right)$ **d** $\log_{10} \sqrt[4]{1000}$

e $\log_7\left(\frac{1}{\sqrt{7}}\right)$ **f** $\log_5(25\sqrt{5})$ **g** $\log_3\left(\frac{1}{\sqrt{27}}\right)$ **h** $\log_4\left(\frac{1}{2\sqrt{2}}\right)$

i $\log_x x^2$ **j** $\log_x \sqrt{x}$ **k** $\log_m m^3$ **l** $\log_x(x\sqrt{x})$

m $\log_n\left(\frac{1}{n}\right)$ **n** $\log_a\left(\frac{1}{a^2}\right)$ **o** $\log_a\left(\frac{1}{\sqrt{a}}\right)$ **p** $\log_m \sqrt{m^5}$

THEORY OF KNOWLEDGE

Acharya Virasena was an 8th century Indian mathematician. Among other areas, he worked with the concept of *ardhaccheda*, which is how many times a number of the form 2^n can be divided by 2. The result is the integer n, and is the logarithm of the number 2^n in base 2.

In 1544, the German Michael Stifel published *Arithmetica Integra* which contains a table expressing many other integers as powers of 2. To do this, he had created an early version of a logarithmic table.

In the late 16th century, astronomers spent a large part of their working lives doing the complex and tedious calculations of spherical trigonometry needed to understand the movement of celestial bodies. In 1614, the Scottish mathematician John Napier formally proposed the idea of a logarithm, and algebraic methods for dealing with them. It was said that Napier effectively doubled the life of an astronomer by reducing the time required to do calculations.

John Napier

Just six years later, Joost Bürgi from Switzerland published a geometric approach for logarithms developed completely independently from John Napier.

1. Can anyone claim to have *invented* logarithms?
2. Can we consider the process of mathematical discovery an *evolution* of ideas?

Many areas of mathematics have been developed over centuries as several mathematicians have worked in a particular area, or taken the knowledge from one area and applied it to another field. Sometimes the process is held up because a method for solving a particular class of problem has not yet been found. In other cases, pure mathematicians have published research papers on seemingly useless mathematical ideas, which have then become vital in applications much later.

In *Everybody Counts: A report to the nation on the future of Mathematical Education* by the National Academy of Sciences (National Academy Press, 1989), there is an excellent section on the Nature of Mathematics. It includes:

> "Even the most esoteric and abstract parts of mathematics - number theory and logic, for example - are now used routinely in applications (for example, in computer science and cryptography). Fifty years ago, the leading British mathematician G.H. Hardy could boast that number theory was the most pure and least useful part of mathematics. Today, Hardy's mathematics is studied as an essential prerequisite to many applications, including control of automated systems, data transmission from remote satellites, protection of financial records, and efficient algorithms for computation."

3. Should we only study the mathematics we need to enter our chosen profession?
4. Why should we explore mathematics for its own sake, rather than to address the needs of science?

C LAWS OF LOGARITHMS

INVESTIGATION DISCOVERING THE LAWS OF LOGARITHMS

What to do:

1 Use your calculator to find:

a $\log 2 + \log 3$ **b** $\log 3 + \log 7$ **c** $\log 4 + \log 20$

d $\log 6$ **e** $\log 21$ **f** $\log 80$

From your answers, suggest a possible simplification for $\log a + \log b$.

2 Use your calculator to find:

a $\log 6 - \log 2$ **b** $\log 12 - \log 3$ **c** $\log 3 - \log 5$

d $\log 3$ **e** $\log 4$ **f** $\log(0.6)$

From your answers, suggest a possible simplification for $\log a - \log b$.

3 Use your calculator to find:

a $3\log 2$ **b** $2\log 5$ **c** $-4\log 3$

d $\log(2^3)$ **e** $\log(5^2)$ **f** $\log(3^{-4})$

From your answers, suggest a possible simplification for $n\log a$.

From the **Investigation**, you should have discovered the three important **laws of logarithms**:

If A and B are both positive then:

- $\log A + \log B = \log(AB)$
- $\log A - \log B = \log\left(\frac{A}{B}\right)$
- $n\log A = \log(A^n)$

More generally, in any base c where $c \neq 1$, $c > 0$, we have these **laws of logarithms**:

If A and B are both positive then:

- $\log_c A + \log_c B = \log_c(AB)$
- $\log_c A - \log_c B = \log_c\left(\frac{A}{B}\right)$
- $n\log_c A = \log_c(A^n)$

Proof:

- $$\begin{aligned}&\log_c(AB)\\ &= \log_c\left(c^{\log_c A} \times c^{\log_c B}\right)\\ &= \log_c\left(c^{\log_c A + \log_c B}\right)\\ &= \log_c A + \log_c B\end{aligned}$$

- $$\begin{aligned}&\log_c\left(\frac{A}{B}\right)\\ &= \log_c\left(\frac{c^{\log_c A}}{c^{\log_c B}}\right)\\ &= \log_c\left(c^{\log_c A - \log_c B}\right)\\ &= \log_c A - \log_c B\end{aligned}$$

- $\log_c(A^n)$
  $= \log_c\left((c^{\log_c A})^n\right)$
  $= \log_c\left(c^{n\log_c A}\right)$
  $= n\log_c A$

Example 7 Self Tutor

Use the laws of logarithms to write the following as a single logarithm or as an integer:

a $\log 5 + \log 3$ **b** $\log_3 24 - \log_3 8$ **c** $\log_2 5 - 1$

a $\log 5 + \log 3$
$= \log(5 \times 3)$
$= \log 15$

b $\log_3 24 - \log_3 8$
$= \log_3\left(\frac{24}{8}\right)$
$= \log_3 3$
$= 1$

c $\log_2 5 - 1$
$= \log_2 5 - \log_2 2^1$
$= \log_2\left(\frac{5}{2}\right)$

Example 8 Self Tutor

Simplify by writing as a single logarithm or as a rational number:

a $2\log 7 - 3\log 2$ **b** $2\log 3 + 3$ **c** $\frac{\log 8}{\log 4}$

a $2\log 7 - 3\log 2$
$= \log(7^2) - \log(2^3)$
$= \log 49 - \log 8$
$= \log\left(\frac{49}{8}\right)$

b $2\log 3 + 3$
$= \log(3^2) + \log(10^3)$
$= \log 9 + \log 1000$
$= \log(9000)$

c $\frac{\log 8}{\log 4} = \frac{\log 2^3}{\log 2^2}$
$= \frac{3\log 2}{2\log 2}$
$= \frac{3}{2}$

EXERCISE 4C.1

1 Write as a single logarithm or as an integer:

a $\log 8 + \log 2$ **b** $\log 4 + \log 5$ **c** $\log 40 - \log 5$
d $\log p - \log m$ **e** $\log_4 8 - \log_4 2$ **f** $\log 5 + \log(0.4)$
g $\log 2 + \log 3 + \log 4$ **h** $1 + \log_2 3$ **i** $\log 4 - 1$
j $\log 5 + \log 4 - \log 2$ **k** $2 + \log 2$ **l** $t + \log w$
m $\log_m 40 - 2$ **n** $\log_3 6 - \log_3 2 - \log_3 3$ **o** $\log 50 - 4$
p $3 - \log_5 50$ **q** $\log_5 100 - \log_5 4$ **r** $\log\left(\frac{4}{3}\right) + \log 3 + \log 7$

2 Write as a single logarithm or integer:

a $5\log 2 + \log 3$ **b** $2\log 3 + 3\log 2$ **c** $3\log 4 - \log 8$
d $2\log_3 5 - 3\log_3 2$ **e** $\frac{1}{2}\log_6 4 + \log_6 3$ **f** $\frac{1}{3}\log\left(\frac{1}{8}\right)$
g $3 - \log 2 - 2\log 5$ **h** $1 - 3\log 2 + \log 20$ **i** $2 - \frac{1}{2}\log_n 4 - \log_n 5$

3 Simplify without using a calculator:

a $\dfrac{\log 4}{\log 2}$ b $\dfrac{\log_5 27}{\log_5 9}$ c $\dfrac{\log 8}{\log 2}$ d $\dfrac{\log 3}{\log 9}$ e $\dfrac{\log_3 25}{\log_3(0.2)}$ f $\dfrac{\log_4 8}{\log_4(0.25)}$

Check your answers using a calculator.

Example 9 Self Tutor

Show that:

a $\log\left(\frac{1}{9}\right) = -2\log 3$ b $\log 500 = 3 - \log 2$

a $\log\left(\frac{1}{9}\right)$
$= \log(3^{-2})$
$= -2\log 3$

b $\log 500$
$= \log\left(\frac{1000}{2}\right)$
$= \log 1000 - \log 2$
$= \log 10^3 - \log 2$
$= 3 - \log 2$

4 Show that:

a $\log 9 = 2\log 3$ b $\log\sqrt{2} = \dfrac{1}{2}\log 2$ c $\log\left(\frac{1}{8}\right) = -3\log 2$

d $\log\left(\frac{1}{5}\right) = -\log 5$ e $\log 5 = 1 - \log 2$ f $\log 5000 = 4 - \log 2$

5 If $p = \log_b 2$, $q = \log_b 3$, and $r = \log_b 5$ write in terms of p, q, and r:

a $\log_b 6$ b $\log_b 45$ c $\log_b 108$

d $\log_b\left(\frac{5\sqrt{3}}{2}\right)$ e $\log_b\left(\frac{5}{32}\right)$ f $\log_b(0.\overline{2})$

$0.\overline{2}$ means $0.222\,222\,\ldots.$

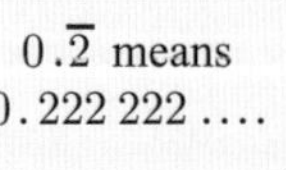

6 If $\log_2 P = x$, $\log_2 Q = y$, and $\log_2 R = z$ write in terms of x, y, and z:

a $\log_2(PR)$ b $\log_2(RQ^2)$ c $\log_2\left(\dfrac{PR}{Q}\right)$

d $\log_2(P^2\sqrt{Q})$ e $\log_2\left(\dfrac{Q^3}{\sqrt{R}}\right)$ f $\log_2\left(\dfrac{R^2\sqrt{Q}}{P^3}\right)$

7 If $\log_t M = 1.29$ and $\log_t N^2 = 1.72$ find:

a $\log_t N$ b $\log_t(MN)$ c $\log_t\left(\dfrac{N^2}{\sqrt{M}}\right)$

LOGARITHMIC EQUATIONS

We can use the laws of logarithms to write equations in a different form. This can be particularly useful if an unknown appears as an exponent.

For the logarithmic function, for every value of y, there is only one corresponding value of x. We can therefore take the logarithm of both sides of an equation without changing the solution. However, we can only do this if both sides are positive.

Example 10 — Self Tutor

Write these as logarithmic equations (in base 10):

a $y = a^2b$ **b** $y = \frac{a}{b^3}$ **c** $P = \frac{20}{\sqrt{n}}$

a
$$y = a^2b$$
$$\therefore \log y = \log(a^2b)$$
$$\therefore \log y = \log a^2 + \log b$$
$$\therefore \log y = 2\log a + \log b$$

b
$$y = \frac{a}{b^3}$$
$$\therefore \log y = \log\left(\frac{a}{b^3}\right)$$
$$\therefore \log y = \log a - \log b^3$$
$$\therefore \log y = \log a - 3\log b$$

c
$$P = \left(\frac{20}{\sqrt{n}}\right)$$
$$\therefore \log P = \log\left(\frac{20}{n^{\frac{1}{2}}}\right)$$
$$\therefore \log P = \log 20 - \log n^{\frac{1}{2}}$$
$$\therefore \log P = \log 20 - \tfrac{1}{2}\log n$$

Example 11 — Self Tutor

Write the following equations without logarithms:

a $\log A = \log b + 2\log c$ **b** $\log_2 M = 3\log_2 a - 2$

a
$$\log A = \log b + 2\log c$$
$$\therefore \log A = \log b + \log c^2$$
$$\therefore \log A = \log(bc^2)$$
$$\therefore A = bc^2$$

b
$$\log_2 M = 3\log_2 a - 2$$
$$\therefore \log_2 M = \log_2 a^3 - \log_2 2^2$$
$$\therefore \log_2 M = \log_2\left(\frac{a^3}{4}\right)$$
$$\therefore M = \frac{a^3}{4}$$

EXERCISE 4C.2

1 Write the following as logarithmic equations (in base 10), assuming all terms are positive:

a $y = 2^x$ **b** $y = 20b^3$ **c** $M = ad^4$ **d** $T = 5\sqrt{d}$

e $R = b\sqrt{l}$ **f** $Q = \frac{a}{b^n}$ **g** $y = ab^x$ **h** $F = \frac{20}{\sqrt{n}}$

i $L = \frac{ab}{c}$ **j** $N = \sqrt{\frac{a}{b}}$ **k** $S = 200 \times 2^t$ **l** $y = \frac{a^m}{b^n}$

2 Write the following equations without logarithms:

a $\log D = \log e + \log 2$ **b** $\log_a F = \log_a 5 - \log_a t$

c $\log P = \frac{1}{2}\log x$ **d** $\log_n M = 2\log_n b + \log_n c$

e $\log B = 3\log m - 2\log n$ **f** $\log N = -\frac{1}{3}\log p$

g $\log P = 3\log x + 1$ **h** $\log_a Q = 2 - \log_a x$

3 **a** Write $y = 3 \times 2^x$ as a logarithmic equation in base 2.
b Hence write x in terms of y.
c Find x when:
i $y = 3$ **ii** $y = 12$ **iii** $y = 30$

4 Solve for x:
a $\log_3 27 + \log_3(\frac{1}{3}) = \log_3 x$
b $\log_5 x = \log_5 8 - \log_5(6 - x)$
c $\log_5 125 - \log_5 \sqrt{5} = \log_5 x$
d $\log_{20} x = 1 + \log_{20} 10$
e $\log x + \log(x + 1) = \log 30$
f $\log(x + 2) - \log(x - 2) = \log 5$

D NATURAL LOGARITHMS

In **Chapter 3** we came across the **natural exponential** $e \approx 2.718\,28$.

Given the exponential function $f(x) = e^x$, the inverse function $f^{-1} = \log_e x$ is the logarithm in base e.

We use $\ln x$ to represent $\log_e x$, and call $\ln x$ the **natural logarithm** of x.

$y = \ln x$ is the reflection of $y = e^x$ in the mirror line $y = x$.

Notice that:
- $\ln 1 = \ln e^0 = 0$
- $\ln e = \ln e^1 = 1$
- $\ln e^2 = 2$
- $\ln \sqrt{e} = \ln e^{\frac{1}{2}} = \frac{1}{2}$
- $\ln\left(\frac{1}{e}\right) = \ln e^{-1} = -1$

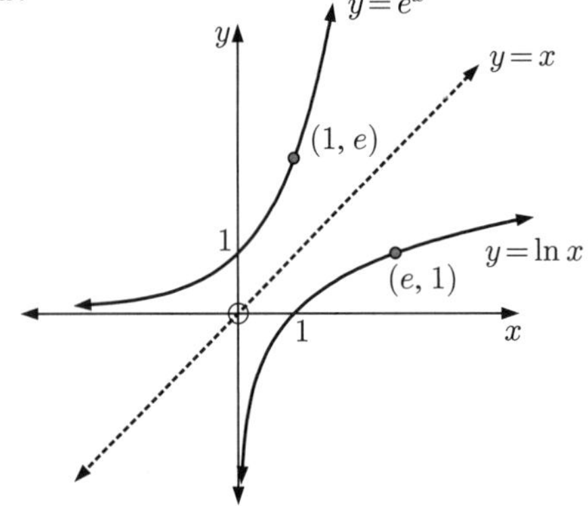

$\ln e^x = x$ and $e^{\ln x} = x$.

Since $a^x = \left(e^{\ln a}\right)^x = e^{x \ln a}$, $\quad a^x = e^{x \ln a}, \quad a > 0$.

Example 12 **Self Tutor**

Use your calculator to write the following in the form e^k where k is correct to 4 decimal places:
a 50 **b** 0.005

a 50
$= e^{\ln 50}$ {using $x = e^{\ln x}$}
$\approx e^{3.9120}$

b 0.005
$= e^{\ln 0.005}$
$\approx e^{-5.2983}$

EXERCISE 4D.1

1 Without using a calculator find:

a $\ln e^2$ b $\ln e^3$ c $\ln \sqrt{e}$ d $\ln 1$

e $\ln\left(\frac{1}{e}\right)$ f $\ln \sqrt[3]{e}$ g $\ln\left(\frac{1}{e^2}\right)$ h $\ln\left(\frac{1}{\sqrt{e}}\right)$

Check your answers using a calculator.

2 Simplify:

a $e^{\ln 3}$ b $e^{2\ln 3}$ c $e^{-\ln 5}$ d $e^{-2\ln 2}$

3 Explain why $\ln(-2)$ and $\ln 0$ cannot be found.

4 Simplify:

a $\ln e^a$ b $\ln(e \times e^a)$ c $\ln\left(e^a \times e^b\right)$ d $\ln(e^a)^b$ e $\ln\left(\frac{e^a}{e^b}\right)$

Example 13 — Self Tutor

Find x if:

a $\ln x = 2.17$ b $\ln x = -0.384$

a $\ln x = 2.17$
$\therefore\ x = e^{2.17}$
$\therefore\ x \approx 8.76$

b $\ln x = -0.384$
$\therefore\ x = e^{-0.384}$
$\therefore\ x \approx 0.681$

5 Use your calculator to write the following in the form e^k where k is correct to 4 decimal places:

a 6   b 60   c 6000   d 0.6   e 0.006

f 15   g 1500   h 1.5   i 0.15   j 0.000 15

6 Find x if:

a $\ln x = 3$ b $\ln x = 1$ c $\ln x = 0$ d $\ln x = -1$

e $\ln x = -5$ f $\ln x \approx 0.835$ g $\ln x \approx 2.145$ h $\ln x \approx -3.2971$

LAWS OF NATURAL LOGARITHMS

The laws for natural logarithms are the laws for logarithms written in base e:

For positive A and B:

- $\ln A + \ln B = \ln(AB)$
- $\ln A - \ln B = \ln\left(\frac{A}{B}\right)$
- $n \ln A = \ln(A^n)$

Example 14 **Self Tutor**

Use the laws of logarithms to write the following as a single logarithm:

a $\ln 5 + \ln 3$ **b** $\ln 24 - \ln 8$ **c** $\ln 5 - 1$

a $\ln 5 + \ln 3$
$= \ln(5 \times 3)$
$= \ln 15$

b $\ln 24 - \ln 8$
$= \ln\left(\frac{24}{8}\right)$
$= \ln 3$

c $\ln 5 - 1$
$= \ln 5 - \ln e^1$
$= \ln\left(\frac{5}{e}\right)$

Example 15 **Self Tutor**

Use the laws of logarithms to simplify:

a $2\ln 7 - 3\ln 2$ **b** $2\ln 3 + 3$

a $2\ln 7 - 3\ln 2$
$= \ln(7^2) - \ln(2^3)$
$= \ln 49 - \ln 8$
$= \ln\left(\frac{49}{8}\right)$

b $2\ln 3 + 3$
$= \ln(3^2) + \ln e^3$
$= \ln 9 + \ln e^3$
$= \ln(9e^3)$

Example 16 **Self Tutor**

Show that:

a $\ln\left(\frac{1}{9}\right) = -2\ln 3$ **b** $\ln 500 \approx 6.9078 - \ln 2$

a $\ln\left(\frac{1}{9}\right)$
$= \ln(3^{-2})$
$= -2\ln 3$

b $\ln 500 = \ln\left(\frac{1000}{2}\right)$
$= \ln 1000 - \ln 2$
$\approx 6.9078 - \ln 2$

EXERCISE 4D.2

1 Write as a single logarithm or integer:

a $\ln 15 + \ln 3$ **b** $\ln 15 - \ln 3$ **c** $\ln 20 - \ln 5$
d $\ln 4 + \ln 6$ **e** $\ln 5 + \ln(0.2)$ **f** $\ln 2 + \ln 3 + \ln 5$
g $1 + \ln 4$ **h** $\ln 6 - 1$ **i** $\ln 5 + \ln 8 - \ln 2$
j $2 + \ln 4$ **k** $\ln 20 - 2$ **l** $\ln 12 - \ln 4 - \ln 3$

2 Write in the form $\ln a$, $a \in \mathbb{Q}$:

a $5\ln 3 + \ln 4$ **b** $3\ln 2 + 2\ln 5$ **c** $3\ln 2 - \ln 8$
d $3\ln 4 - 2\ln 2$ **e** $\frac{1}{3}\ln 8 + \ln 3$ **f** $\frac{1}{3}\ln(\frac{1}{27})$
g $-\ln 2$ **h** $-\ln(\frac{1}{2})$ **i** $-2\ln(\frac{1}{4})$

3 Show that:

a $\ln 27 = 3\ln 3$ **b** $\ln\sqrt{3} = \frac{1}{2}\ln 3$ **c** $\ln(\frac{1}{16}) = -4\ln 2$
d $\ln(\frac{1}{6}) = -\ln 6$ **e** $\ln\left(\frac{1}{\sqrt{2}}\right) = -\frac{1}{2}\ln 2$ **f** $\ln\left(\frac{e}{5}\right) = 1 - \ln 5$

4 Show that:

a $\ln \sqrt[3]{5} = \frac{1}{3} \ln 5$

b $\ln(\frac{1}{32}) = -5 \ln 2$

c $\ln \left(\frac{1}{\sqrt[5]{2}} \right) = -\frac{1}{5} \ln 2$

d $\ln \left(\frac{e^2}{8} \right) = 2 - 3 \ln 2$

Example 17 **Self Tutor**

Write the following equations without logarithms:

a $\ln A = 2 \ln c + 3$

b $\ln M = 3 \ln a - 2$

a
$$\begin{aligned} \ln A &= 2 \ln c + 3 \\ \therefore \quad \ln A - 2 \ln c &= 3 \\ \therefore \quad \ln A - \ln c^2 &= 3 \\ \therefore \quad \ln \left(\frac{A}{c^2} \right) &= 3 \\ \therefore \quad \frac{A}{c^2} &= e^3 \\ \therefore \quad A &= c^2 e^3 \end{aligned}$$

b
$$\begin{aligned} \ln M &= 3 \ln a - 2 \\ \therefore \quad \ln M - 3 \ln a &= -2 \\ \therefore \quad \ln M - \ln a^3 &= -2 \\ \therefore \quad \ln \left(\frac{M}{a^3} \right) &= -2 \\ \therefore \quad \frac{M}{a^3} &= e^{-2} \\ \therefore \quad M &= \frac{a^3}{e^2} \end{aligned}$$

5 Write the following equations without logarithms, assuming all terms are positive:

a $\ln D = \ln x + 1$

b $\ln F = -\ln p + 2$

c $\ln P = \frac{1}{2} \ln x$

d $\ln M = 2 \ln y + 3$

e $\ln B = 3 \ln t - 1$

f $\ln N = -\frac{1}{3} \ln g$

g $\ln Q \approx 3 \ln x + 2.159$

h $\ln D \approx 0.4 \ln n - 0.6582$

E EXPONENTIAL EQUATIONS USING LOGARITHMS

In **Chapter 3** we found solutions to simple exponential equations where we could make equal bases and then equate exponents. However, it is not always easy to make the bases the same. In these situations we use **logarithms** to find the exact solution.

Example 18 **Self Tutor**

Consider the equation $2^x = 30$.

a Solve for x, giving an exact answer, by using: **i** base 2 **ii** base 10.

b Comment on your answers.

a **i**
$$\begin{aligned} 2^x &= 30 \\ \therefore \quad x &= \log_2 30 \end{aligned}$$

ii
$$\begin{aligned} 2^x &= 30 \\ \therefore \quad \log 2^x &= \log 30 \quad \{\text{find the logarithm of each side}\} \\ \therefore \quad x \log 2 &= \log 30 \quad \{\log(A^n) = n \log A\} \\ \therefore \quad x &= \frac{\log 30}{\log 2} \end{aligned}$$

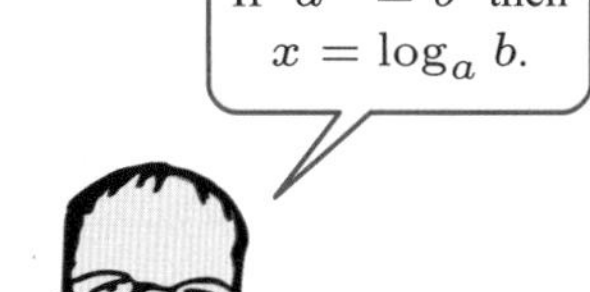

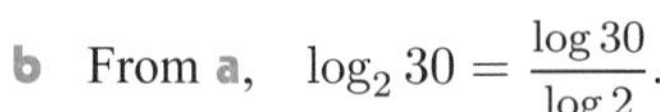

b From **a**, $\log_2 30 = \frac{\log 30}{\log 2}$.

Example 19 **Self Tutor**

Find x exactly:

a $e^x = 30$ **b** $3e^{\frac{x}{2}} = 21$

a $e^x = 30$

$\therefore \quad x = \ln 30$

b $3e^{\frac{x}{2}} = 21$

$\therefore \quad e^{\frac{x}{2}} = 7$

$\therefore \quad \frac{x}{2} = \ln 7$

$\therefore \quad x = 2\ln 7$

EXERCISE 4E

1 Solve for x, giving an exact answer in base 10:

a $2^x = 10$ **b** $3^x = 20$ **c** $4^x = 100$

d $\left(\frac{1}{2}\right)^x = 0.0625$ **e** $\left(\frac{3}{4}\right)^x = 0.1$ **f** $10^x = 0.000\,01$

2 Solve for x, giving an exact answer:

a $e^x = 10$ **b** $e^x = 1000$ **c** $2e^x = 0.3$

d $e^{\frac{x}{2}} = 5$ **e** $e^{2x} = 18$ **f** $e^{-\frac{x}{2}} = 1$

Example 20 **Self Tutor**

Consider the equation $P = 200 \times 2^{0.04t}$.

a Rearrange the equation to give t in terms of P.

b Hence find the value of t when $P = 6$.

a $P = 200 \times 2^{0.04t}$

$\therefore \quad 2^{0.04t} = \frac{P}{200}$ {dividing both sides by 200}

$\therefore \quad \log 2^{0.04t} = \log\left(\frac{P}{200}\right)$ {finding the logarithm of each side}

$\therefore \quad 0.04t \times \log 2 = \log\left(\frac{P}{200}\right)$ $\{\log(A)^n = n\log A\}$

$\therefore \quad t = \dfrac{\log\left(\frac{P}{200}\right)}{0.04 \times \log 2}$

b When $P = 6$, $t = \dfrac{\log\left(\frac{6}{200}\right)}{0.04 \times \log 2} \approx -126$

3 Consider the equation $R = 200 \times 2^{0.25t}$.

a Rearrange the equation to give t in terms of R.

b Hence find t when: **i** $R = 600$ **ii** $R = 1425$

4 Consider the equation $M = 20 \times 5^{-0.02x}$.

a Rearrange the equation to give x in terms of M.

b Hence find x when: **i** $M = 100$ **ii** $M = 232$

Example 21 — **Self Tutor**

Find algebraically the exact points of intersection of $y = e^x - 3$ and $y = 1 - 3e^{-x}$.
Check your solution using technology.

The functions meet where

$$e^x - 3 = 1 - 3e^{-x}$$
$$\therefore\ e^x - 4 + 3e^{-x} = 0$$
$$\therefore\ e^{2x} - 4e^x + 3 = 0 \quad \{\text{multiplying each term by } e^x\}$$
$$\therefore\ (e^x - 1)(e^x - 3) = 0$$
$$\therefore\ e^x = 1 \text{ or } 3$$
$$\therefore\ x = \ln 1 \text{ or } \ln 3$$
$$\therefore\ x = 0 \text{ or } \ln 3$$

When $x = 0$, $y = e^0 - 3 = -2$

When $x = \ln 3$, $y = e^{\ln 3} - 3 = 0$

$\therefore$ the functions meet at $(0, -2)$ and at $(\ln 3, 0)$.

GRAPHING PACKAGE

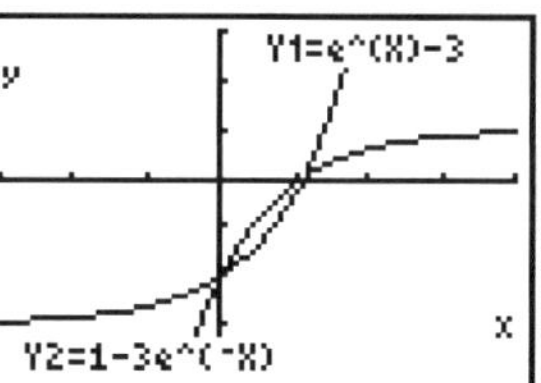

5 Solve for x, giving an exact answer:

a $4 \times 2^{-x} = 0.12$ b $300 \times 5^{0.1x} = 1000$ c $32 \times 3^{-0.25x} = 4$

d $20 \times e^{2x} = 840$ e $50 \times e^{-0.03x} = 0.05$ f $41e^{0.3x} - 27 = 0$

6 Solve for x:

a $e^{2x} = 2e^x$ b $e^x = e^{-x}$ c $e^{2x} - 5e^x + 6 = 0$

d $e^x + 2 = 3e^{-x}$ e $1 + 12e^{-x} = e^x$ f $e^x + e^{-x} = 3$

7 Find algebraically the point(s) of intersection of:

a $y = e^x$ and $y = e^{2x} - 6$ b $y = 2e^x + 1$ and $y = 7 - e^x$

c $y = 3 - e^x$ and $y = 5e^{-x} - 3$

Check your answers using technology.

F THE CHANGE OF BASE RULE

$$\boldsymbol{\log_b a = \frac{\log_c a}{\log_c b}} \quad \text{for } a, b, c > 0 \text{ and } b, c \neq 1.$$

Proof: If $\log_b a = x$, then $b^x = a$

$$\therefore\ \log_c b^x = \log_c a \quad \{\text{taking logarithms in base } c\}$$
$$\therefore\ x \log_c b = \log_c a \quad \{\text{power law of logarithms}\}$$
$$\therefore\ x = \frac{\log_c a}{\log_c b}$$
$$\therefore\ \log_b a = \frac{\log_c a}{\log_c b}$$

Example 22 **Self Tutor**

Find $\log_2 9$ by:

a letting $\log_2 9 = x$ **b** changing to base 10 **c** changing to base e.

a Let $\log_2 9 = x$

$\therefore\ 9 = 2^x$

$\therefore\ \log 2^x = \log 9$

$\therefore\ x \log 2 = \log 9$

$\therefore\ x = \dfrac{\log 9}{\log 2} \approx 3.17$

b $\log_2 9 = \dfrac{\log_{10} 9}{\log_{10} 2}$

≈ 3.17

c $\log_2 9 = \dfrac{\ln 9}{\ln 2}$

≈ 3.17

Example 23 **Self Tutor**

Solve for x exactly: $8^x - 5(4^x) = 0$

$8^x - 5(4^x) = 0$

$\therefore\ 2^{3x} - 5(2^{2x}) = 0$

$\therefore\ 2^{2x}(2^x - 5) = 0$

$\therefore\ 2^x = 5$ {as $2^{2x} > 0$ for all x}

$\therefore\ x = \log_2 5 = \dfrac{\log 5}{\log 2}$

EXERCISE 4F

1 Use the rule $\log_b a = \dfrac{\log_{10} a}{\log_{10} b}$ to find, correct to 3 significant figures:

a $\log_3 12$ **b** $\log_{\frac{1}{2}} 1250$ **c** $\log_3(0.067)$ **d** $\log_{0.4}(0.006\,984)$

2 Use the rule $\log_b a = \dfrac{\ln a}{\ln b}$ to solve, correct to 3 significant figures:

a $2^x = 0.051$ **b** $4^x = 213.8$ **c** $3^{2x+1} = 4.069$

3 Solve for x exactly:

a $25^x - 3(5^x) = 0$ **b** $8(9^x) - 3^x = 0$ **c** $2^x - 2(4^x) = 0$

4 Solve for x:

a $\log_4 x^3 + \log_2 \sqrt{x} = 8$ **b** $\log_{16} x^5 = \log_{64} 125 - \log_4 \sqrt{x}$

5 Find the exact value of x for which $4^x \times 5^{4x+3} = 10^{2x+3}$.

6 Suppose $\log_9 x + \log_{27} x = p$. Write the value of $\log_3 x + \log_{81} x$ in terms of p.

7 Without using technology, show that $2^{\frac{4}{\log_5 4} + \frac{3}{\log_7 8}} = 175$.

G GRAPHS OF LOGARITHMIC FUNCTIONS

Consider the general exponential function $f(x) = a^x$, $a > 0$, $a \neq 1$.

The graph of $y = a^x$ is:

For $0 < a < 1$:

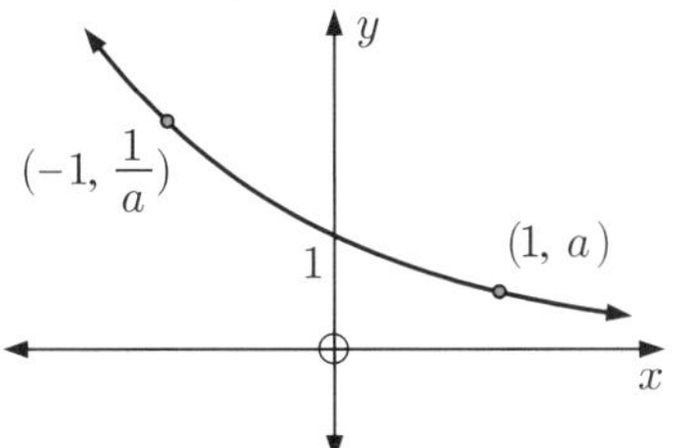

For $a > 1$:

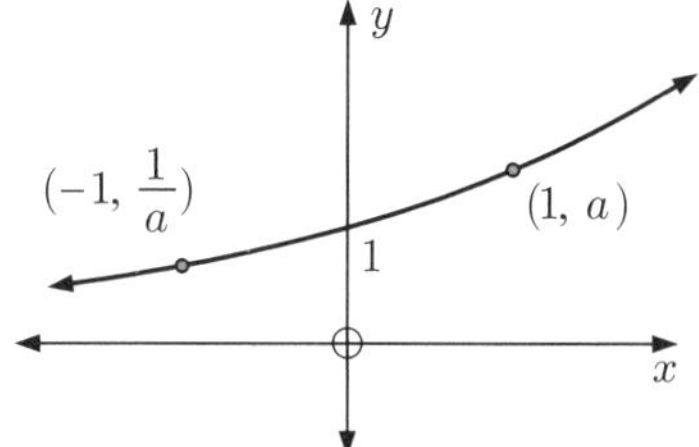

The **horizontal asymptote** for all of these functions is the x-axis $y = 0$.

The inverse function f^{-1} is given by $x = a^y$, so $y = \log_a x$.

If $\boldsymbol{f(x) = a^x}$ where $\boldsymbol{a > 0}$, $\boldsymbol{a \neq 1}$, then $\boldsymbol{f^{-1}(x) = \log_a x}$.

Since $f^{-1}(x) = \log_a x$ is an inverse function, it is a reflection of $f(x) = a^x$ in the line $y = x$. We may therefore deduce the following properties:

Function	$f(x) = a^x$	$f^{-1}(x) = \log_a x$
Domain	$\{x \mid x \in \mathbb{R}\}$	$\{x \mid x > 0\}$
Range	$\{y \mid y > 0\}$	$\{y \mid y \in \mathbb{R}\}$
Asymptote	horizontal $y = 0$	vertical $x = 0$

The graph of $y = \log_a x$ for $0 < a < 1$:

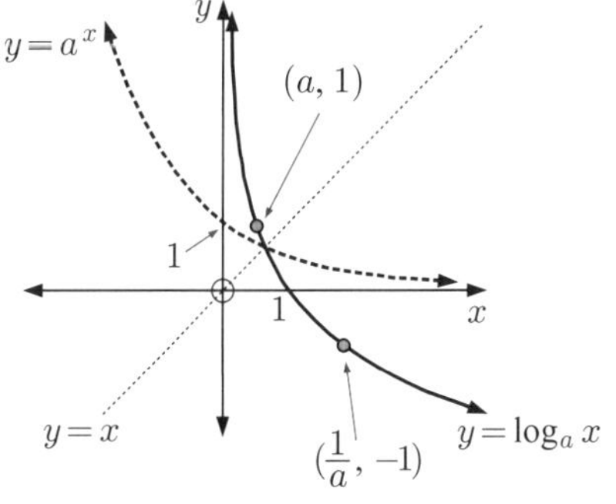

The graph of $y = \log_a x$ for $a > 1$:

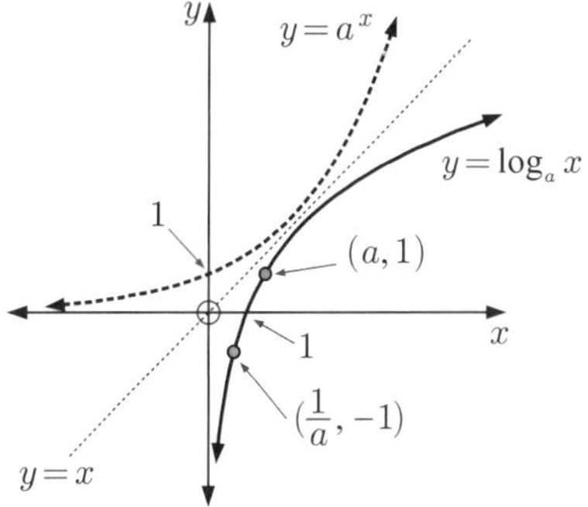

The **vertical asymptote** of $y = \log_a x$ is the y-axis $x = 0$.

For $0 < a < 1$: as $x \to \infty$, $y \to -\infty$
as $x \to 0^+$, $y \to \infty$

For $a > 1$: as $x \to \infty$, $y \to \infty$
as $x \to 0^+$, $y \to -\infty$

Since we can only find logarithms of positive numbers, the domain of $f^{-1}(x) = \log_a x$ is $\{x \mid x > 0\}$.

In general, $y = \log_a(g(x))$ is defined when $g(x) > 0$.

When graphing f, f^{-1}, and $y = x$ on your graphics calculator, it is best to set the scale so that $y = x$ makes a $45°$ angle with both axes.

Example 24 **Self Tutor**

Consider the function $f(x) = \log_2(x-1) + 1$.

a Find the domain and range of f.

b Find any asymptotes and axes intercepts.

c Sketch the graph of f showing all important features.

d Find f^{-1} and explain how to verify your answer.

a $x - 1 > 0$ when $x > 1$

So, the domain is $\{x \mid x > 1\}$ and the range is $y \in \mathbb{R}$.

b As $x \to 1$ from the right, $y \to -\infty$, so the vertical asymptote is $x = 1$.

As $x \to \infty$, $y \to \infty$.

When $x = 0$, y is undefined, so there is no y-intercept.

When $y = 0$, $\log_2(x-1) = -1$

$\therefore \quad x - 1 = 2^{-1}$

$\therefore \quad x = 1\frac{1}{2}$

So, the x-intercept is $1\frac{1}{2}$.

c To graph the function using your calculator, it may be necessary to change the base to base 10 or base e.

So, we graph $y = \dfrac{\log(x-1)}{\log 2} + 1$

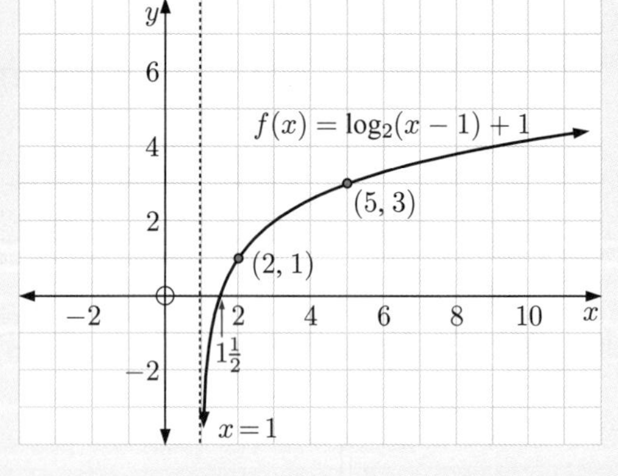

d f is defined by $y = \log_2(x-1) + 1$

$\therefore f^{-1}$ is defined by $x = \log_2(y-1) + 1$

$\therefore \quad x - 1 = \log_2(y-1)$

$\therefore \quad y - 1 = 2^{x-1}$

$\therefore \quad y = 2^{x-1} + 1$

$\therefore \quad f^{-1}(x) = 2^{x-1} + 1$

which has the horizontal asymptote $y = 1$ ✓

Its domain is $\{x \mid x \in \mathbb{R}\}$, and its range is $\{y \mid y > 1\}$.

EXERCISE 4G

1 For the following functions f:

i Find the domain and range.

ii Find any asymptotes and axes intercepts.

iii Sketch the graph of $y = f(x)$ showing all important features.

iv Solve $f(x) = -1$ algebraically and check the solution on your graph.

v Find f^{-1} and explain how to verify your answer.

a $f : x \mapsto \log_3(x+1)$

b $f : x \mapsto 1 - \log_3(x+1)$

c $f : x \mapsto \log_5(x-2) - 2$

d $f : x \mapsto 1 - \log_5(x-2)$

e $f : x \mapsto 1 - 2\log_2 x$

f $f : x \mapsto \log_2(x^2 - 3x - 4)$

Example 25 **Self Tutor**

Consider the function $f : x \mapsto e^{x-3}$.

a Find the equation defining f^{-1}.

b Sketch the graphs of f and f^{-1} on the same set of axes.

c State the domain and range of f and f^{-1}.

d Find any asymptotes and intercepts of f and f^{-1}.

a
$$f(x) = e^{x-3}$$
$$\therefore \quad f^{-1} \text{ is } x = e^{y-3}$$
$$\therefore \quad y - 3 = \ln x$$
$$\therefore \quad y = 3 + \ln x$$
So, $f^{-1}(x) = 3 + \ln x$

b

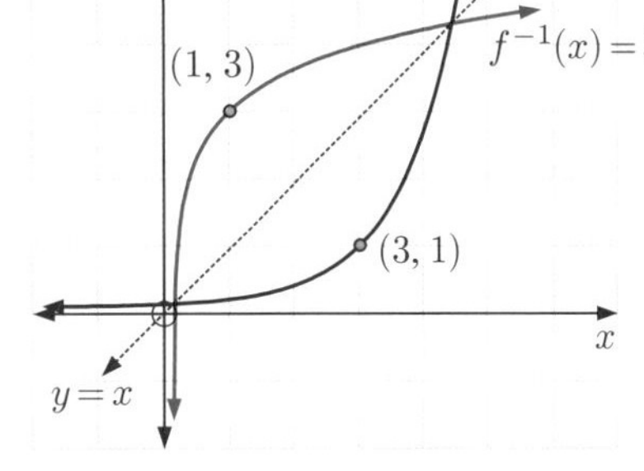

c

Function	f	f^{-1}
Domain	$x \in \mathbb{R}$	$x > 0$
Range	$y > 0$	$y \in \mathbb{R}$

d For f: Horizontal asymptote is $y = 0$, y-intercept is e^{-3}.

For f^{-1}: Vertical asymptote is $x = 0$, x-intercept is e^{-3}.

2 For the following functions f:

i Find the equation of f^{-1}.

ii Sketch the graphs of f and f^{-1} on the same set of axes.

iii State the domain and range of f and f^{-1}.

iv Find any asymptotes and intercepts of f and f^{-1}.

a $f(x) = e^x + 5$

b $f(x) = e^{x+1} - 3$

c $f(x) = \ln x - 4, \quad x > 0$

d $f(x) = \ln(x-1) + 2, \quad x > 1$

3 Given $f : x \mapsto e^{2x}$ and $g : x \mapsto 2x - 1$, find:

a $(f^{-1} \circ g)(x)$

b $(g \circ f)^{-1}(x)$

4 Consider the graphs A and B. One of them is the graph of $y = \ln x$ and the other is the graph of $y = \ln(x-2)$.

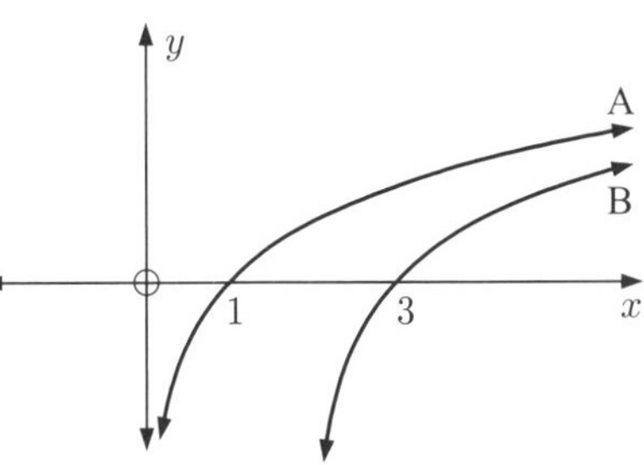

a Identify which is which. Give evidence for your answer.

b Copy the graphs onto a new set of axes and add to them the graph of $y = \ln(x+2)$.

c Find the equation of the vertical asymptote for each graph.

5 Kelly said that in order to graph $y = \ln(x^2)$, $x > 0$, you could first graph $y = \ln x$ and then double the distance of each point on the graph from the x-axis.
Is Kelly correct? Explain your answer.

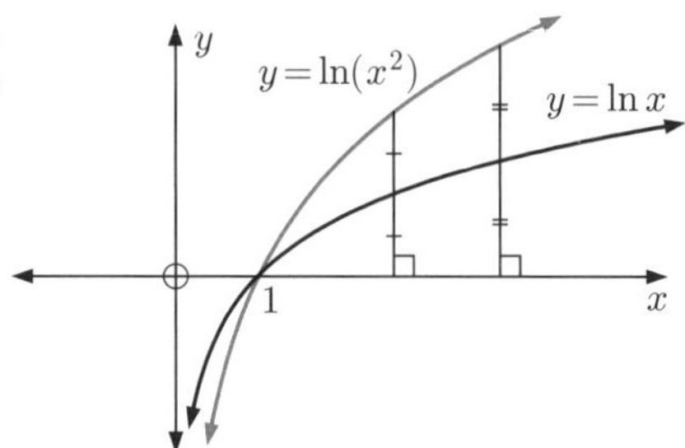

6 Consider the function $f : x \mapsto e^{x+3} + 2$.

a Find the defining equation for f^{-1}.

b Find the values of x for which:

i $f(x) < 2.1$ **ii** $f(x) < 2.01$ **iii** $f(x) < 2.001$ **iv** $f(x) < 2.0001$

Hence conjecture the horizontal asymptote for the graph of f.

c Determine the equation of the horizontal asymptote of $f(x)$ by discussing the behaviour of $f(x)$ as $x \to \pm\infty$.

d Hence, determine the vertical asymptote and the domain of f^{-1}.

Example 26

Solve for x: **a** $e^x = 2x^2 + x + 1$ **b** $e^x \geqslant 2x^2 + x + 1$

a We graph $y = e^x$ and $y = 2x^2 + x + 1$ on the same set of axes.
Using technology we find the points of intersection of the graphs.
$\therefore$ $x = 0$ and $x \approx 3.21$ are the solutions.

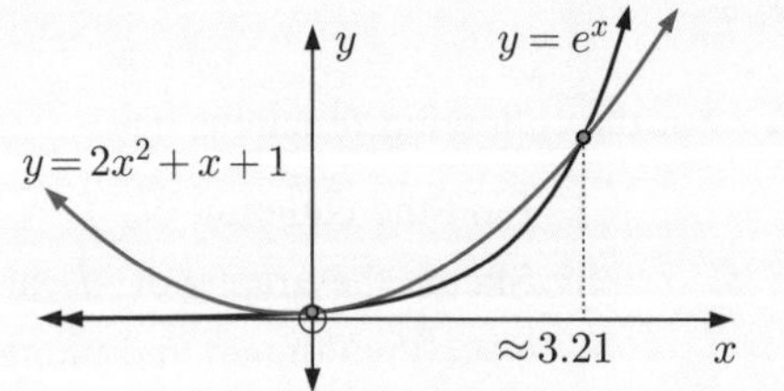

b Using the same graphs as above, we seek values of x for which the graph of $y = e^x$ either meets or is higher than the graph of $y = 2x^2 + x + 1$.
The solution is $x = 0$ or $x \geqslant 3.21$
We could also graph the solution set:

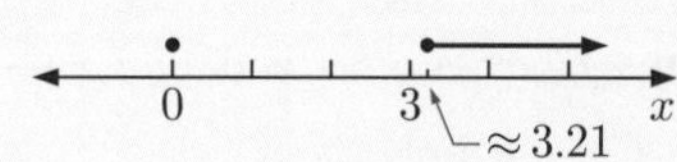

or write it as $x = 0$ or $x \in [3.21, \infty[$.

7 Solve for x:

a $x^2 > e^x$ **b** $x^3 < e^{-x}$ **c** $5 - x > \ln x$

8 State the domain of $f(x) = x^2 \ln x$. Hence find where $f(x) \leqslant 0$.

9 **a** Use technology to sketch the graph of $f(x) = \dfrac{2}{x} - e^{2x^2-x+1}$.

b State the domain and range of this function.

c Hence find all $x \in \mathbb{R}$ for which $e^{2x^2-x+1} > \dfrac{2}{x}$.

ACTIVITY

Click on the icon to obtain a card game for logarithmic functions.

CARD GAME

GROWTH AND DECAY

In **Chapter 3** we showed how exponential functions can be used to model a variety of growth and decay situations. These included the growth of populations and the decay of radioactive substances. In this section we consider more growth and decay problems, focussing particularly on how logarithms can be used in their solution.

POPULATION GROWTH

Example 27 — Self Tutor

A farmer monitoring an insect plague notices that the area affected by the insects is given by $A_n = 1000 \times 2^{0.7n}$ hectares, where n is the number of weeks after the initial observation.

a Draw an accurate graph of A_n against n and use your graph to estimate the time taken for the affected area to reach 5000 ha.

b Check your answer to **a** using logarithms and using suitable technology.

a

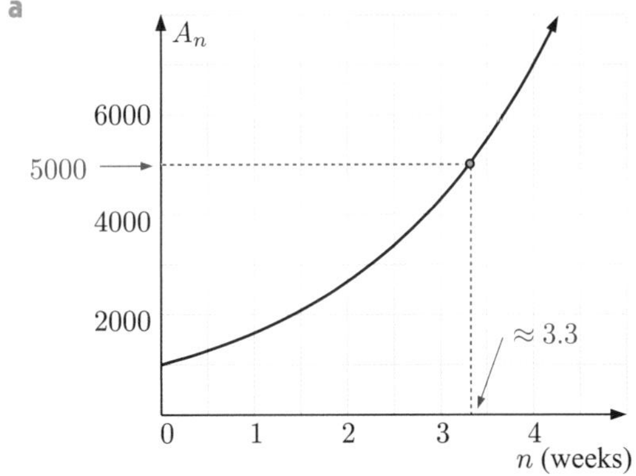

b When $A_n = 5000$,

$$1000 \times 2^{0.7n} = 5000$$
$$\therefore \quad 2^{0.7n} = 5$$
$$\therefore \quad \log 2^{0.7n} = \log 5$$
$$\therefore \quad 0.7n \log 2 = \log 5$$
$$\therefore \quad n = \frac{\log 5}{0.7 \times \log 2}$$
$$\therefore \quad n \approx 3.32$$

$\therefore$ it takes about 3 weeks and 2 days.

Using technology we find the intersection of $y = 1000 \times 2^{0.7x}$ and $y = 5000$. This confirms $n \approx 3.32$.

FINANCIAL GROWTH

Suppose an amount u_1 is invested at a fixed rate for each compounding period. In this case the value of the investment after n periods is given by $u_{n+1} = u_1 \times r^n$ where r is the multiplier corresponding to the given rate of interest. In order to find n algebraically, we need to use **logarithms**.

Example 28 — Self Tutor

Iryna has €5000 to invest in an account that pays 5.2% p.a. interest compounded annually. Find, using logarithms, how long it will take for her investment to reach €20 000.

$u_{n+1} = 20\,000$ after n years
$u_1 = 5000$
$r = 105.2\% = 1.052$

Now $u_{n+1} = u_1 \times r^n$

$\therefore\ 20\,000 = 5000 \times (1.052)^n$
$\therefore\ (1.052)^n = 4$
$\therefore\ \log(1.052)^n = \log 4$
$\therefore\ n \times \log 1.052 = \log 4$
$\therefore\ n = \dfrac{\log 4}{\log 1.052} \approx 27.3$ years

Rounding up here, it will take about 28 years to reach €20 000.

EXERCISE 4H

1 The weight W_t of bacteria in a culture t hours after establishment is given by $W_t = 20 \times 2^{0.15t}$ grams. Find, using logarithms, the time for the weight of the culture to reach:
 a 30 grams b 100 grams.

2 The mass M_t of bacteria in a culture t hours after establishment is given by $M_t = 25 \times e^{0.1t}$ grams. Show that the time required for the mass of the culture to reach 50 grams is $10 \ln 2$ hours.

3 A biologist is modelling an infestation of fire ants. He determines that the area affected by the ants is given by $A_n = 2000 \times e^{0.57n}$ hectares, where n is the number of weeks after the initial observation.
 a Draw an accurate graph of A_n against n.
 b Use your graph to estimate the time taken for the infested area to reach 10 000 ha.
 c Check your answer to **b** using: i logarithms ii suitable technology.

4 A house is expected to increase in value at an average rate of 7.5% p.a. If the house is worth £160 000 now, when would you expect it to be worth £250 000?

5 Thabo has \$10 000 to invest in an account that pays 4.8% p.a. compounded annually. How long will it take for his investment to grow to \$15 000?

6 Dien invests \$15 000 at 8.4% p.a. compounded *monthly*. He will withdraw his money when it reaches \$25 000, at which time he plans to travel. The formula $u_{n+1} = u_1 \times r^n$ can be used to model the investment, where n is the time in months.
 a Explain why $r = 1.007$. b After how many months will Dien withdraw the money?

7 The mass M_t of radioactive substance remaining after t years is given by $M_t = 1000 \times e^{-0.04t}$ grams. Find the time taken for the mass to:
 a halve b reach 25 grams c reach 1% of its original value.

8 A man jumps from an aeroplane. His speed of descent is given by $V = 50(1 - e^{-0.2t})$ m s^{-1}, where t is the time in seconds. Show that it will take $5 \ln 5$ seconds for the man's speed to reach 40 m s^{-1}.

9 Answer the **Opening Problem** on page **124**.

10 The temperature of a liquid t minutes after it is placed in a refrigerator, is given by $T = 4 + 96 \times e^{-0.03t}$ °C. Find the time required for the temperature to reach:

a 25°C b 5°C.

11 The weight of radioactive substance remaining after t years is given by $W = 1000 \times 2^{-0.04t}$ grams.

a Sketch the graph of W against t.

b Write a function for t in terms of W.

c Hence find the time required for the weight to reach:

i 20 grams ii 0.001 grams.

12 The weight of radioactive uranium remaining after t years is given by the formula $W(t) = W_0 \times 2^{-0.0002t}$ grams, $t \geqslant 0$. Find the time required for the weight to fall to:

a 25% of its original value b 0.1% of its original value.

13 The current I flowing in a transistor radio t seconds after it is switched off is given by $I = I_0 \times 2^{-0.02t}$ amps.

Show that it takes $\dfrac{50}{\log 2}$ seconds for the current to drop to 10% of its original value.

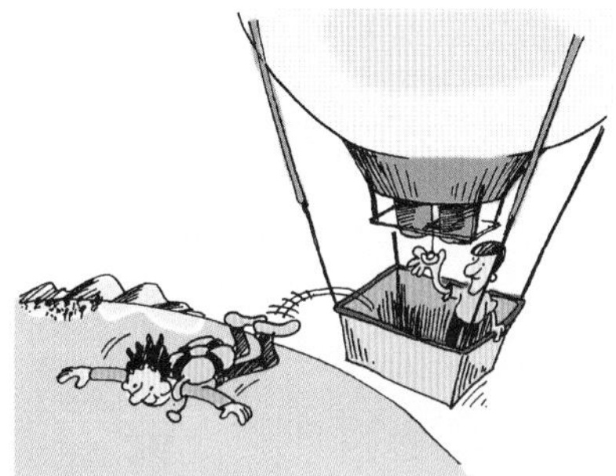

14 A parachutist jumps from the basket of a stationary hot air balloon. His speed of descent is given by $V = 60(1 - 2^{-0.2t})$ m s^{-1} where t is the time in seconds. Find the time taken for his speed to reach 50 m s^{-1}.

REVIEW SET 4A — NON-CALCULATOR

1 Find the following, showing all working.

a $\log_4 64$ b $\log_2 256$ c $\log_2(0.25)$ d $\log_{25} 5$

e $\log_8 1$ f $\log_{81} 3$ g $\log_9(0.\overline{1})$ h $\log_k \sqrt{k}$

2 Find:

a $\log \sqrt{10}$ b $\log \dfrac{1}{\sqrt[3]{10}}$ c $\log(10^a \times 10^{b+1})$

3 Simplify:

a $4\ln 2 + 2\ln 3$ b $\frac{1}{2}\ln 9 - \ln 2$ c $2\ln 5 - 1$ d $\frac{1}{4}\ln 81$

4 Find:

a $\ln(e\sqrt{e})$ b $\ln\left(\dfrac{1}{e^3}\right)$ c $\ln(e^{2x})$ d $\ln\left(\dfrac{e}{e^x}\right)$

5 Write as a single logarithm:

a $\log 16 + 2\log 3$ b $\log_2 16 - 2\log_2 3$ c $2 + \log_4 5$

6 Write as logarithmic equations:

a $P = 3 \times b^x$ b $m = \dfrac{n^3}{p^2}$

7 Show that $\log_3 7 \times 2\log_7 x = 2\log_3 x$.

8 Write the following equations without logarithms:

a $\log T = 2\log x - \log y$　　**b** $\log_2 K = \log_2 n + \frac{1}{2}\log_2 t$

9 Write in the form $a\ln k$ where a and k are positive whole numbers and k is prime:

a $\ln 32$　　**b** $\ln 125$　　**c** $\ln 729$

10 Copy and complete:

Function	$y = \log_2 x$	$y = \ln(x+5)$
Domain		
Range		

11 If $A = \log_5 2$ and $B = \log_5 3$, write in terms of A and B:

a $\log_5 36$　　**b** $\log_5 54$　　**c** $\log_5(8\sqrt{3})$　　**d** $\log_5(20.25)$　　**e** $\log_5(0.\overline{8})$

12 Solve for x:

a $3e^x - 5 = -2e^{-x}$　　**b** $2\ln x - 3\ln\left(\frac{1}{x}\right) = 10$

REVIEW SET 4B　　CALCULATOR

1 Write in the form 10^x giving x correct to 4 decimal places:

a 32　　**b** 0.0013　　**c** 8.963×10^{-5}

2 Find x if:

a $\log_2 x = -3$　　**b** $\log_5 x \approx 2.743$　　**c** $\log_3 x \approx -3.145$

3 Write the following equations without logarithms:

a $\log_2 k \approx 1.699 + x$　　**b** $\log_a Q = 3\log_a P + \log_a R$

c $\log A \approx 5\log B - 2.602$

4 Solve for x, giving exact answers:

a $5^x = 7$　　**b** $20 \times 2^{2x+1} = 640$

5 The weight of a radioactive isotope after t years is given by $W_t = 2500 \times 3^{-\frac{t}{3000}}$ grams.

a Find the initial weight of the isotope.

b Find the time taken for the isotope to reduce to 30% of its original weight.

c Find the percentage weight loss after 1500 years.

d Sketch the graph of W_t against t.

6 Show that the solution to $16^x - 5 \times 8^x = 0$ is $x = \log_2 5$.

7 Solve for x, giving exact answers:

a $\ln x = 5$　　**b** $3\ln x + 2 = 0$　　**c** $e^x = 400$

d $e^{2x+1} = 11$　　**e** $25e^{\frac{x}{2}} = 750$

8 A population of seals is given by $P_t = P_0 2^{\frac{t}{3}}$ where t is the time in years, $t \geqslant 0$.

a Find the time required for the population to double in size.

b Find the percentage increase in population during the first 4 years.

9 Consider $g : x \mapsto 2e^x - 5$.

a Find the defining equation of g^{-1}.

b Sketch the graphs of g and g^{-1} on the same set of axes.

c State the domain and range of g and g^{-1}.

d State the asymptotes and intercepts of g and g^{-1}.

10 Consider $f(x) = e^x$ and $g(x) = \ln(x+4)$, $x > -4$. Find:

a $(f \circ g)(5)$ **b** $(g \circ f)(0)$

11 **a** Sketch the graph of $f(x) = x^3 + x^2 - 6x - e^x$.

b Hence find all $x \in \mathbb{R}$ for which $e^x < x^3 + x^2 - 6x$.

REVIEW SET 4C

1 Without using a calculator, find the base 10 logarithms of:

a $\sqrt{1000}$ **b** $\dfrac{10}{\sqrt[3]{10}}$ **c** $\dfrac{10^a}{10^{-b}}$

2 Simplify:

a $e^{4\ln x}$ **b** $\ln(e^5)$ **c** $\ln(\sqrt{e})$

d $10^{\log x + \log 3}$ **e** $\ln\left(\dfrac{1}{e^x}\right)$ **f** $\dfrac{\log x^2}{\log_3 9}$

3 Write in the form e^x, where x is correct to 4 decimal places:

a 20 **b** 3000 **c** 0.075

4 Solve for x:

a $\log x = 3$ **b** $\log_3(x+2) = 1.732$ **c** $\log_2\left(\dfrac{x}{10}\right) = -0.671$

5 Write as a single logarithm:

a $\ln 60 - \ln 20$ **b** $\ln 4 + \ln 1$ **c** $\ln 200 - \ln 8 + \ln 5$

6 Write as logarithmic equations:

a $M = ab^n$ **b** $T = \dfrac{5}{\sqrt{l}}$ **c** $G = \dfrac{a^2 b}{c}$

7 Solve for x:

a $3^x = 300$ **b** $30 \times 5^{1-x} = 0.15$ **c** $3^{x+2} = 2^{1-x}$

8 Solve exactly for x:

a $e^{2x} = 3e^x$ **b** $e^{2x} - 7e^x + 12 = 0$

9 Write the following equations without logarithms:

a $\ln P = 1.5 \ln Q + \ln T$ **b** $\ln M = 1.2 - 0.5 \ln N$

10 Consider the functions $f(x) = e^{x^2} - x^6$ and $g(x) = \ln(x^2 + 1)$.

a Explain why $f(x)$ and $g(x)$ are even functions.

b Graph $y = f(x)$ and $y = g(x)$ on the domain $0 \leqslant x \leqslant 5$. Find the points of intersection of the graphs on this domain.

c Hence solve $x^6 - e^{x^2} + \ln(x^2 + 1) > 0$ for all $x \in \mathbb{R}$.

11 For the function $g : x \mapsto \log_3(x+2) - 2$:

a Find the domain and range.

b Find any asymptotes and axes intercepts for the graph of the function.

c Find the defining equation for g^{-1}. Explain how to verify your answer.

d Sketch the graphs of g, g^{-1}, and $y = x$ on the same axes.

12 The weight of a radioactive isotope remaining after t weeks is given by $W_t = 8000 \times e^{-\frac{t}{20}}$ grams. Find the time for the weight to:

a halve **b** reach 1000 g **c** reach 0.1% of its original value.

Chapter 5

Transforming functions

Syllabus reference: **2.2, 2.3**

Contents:

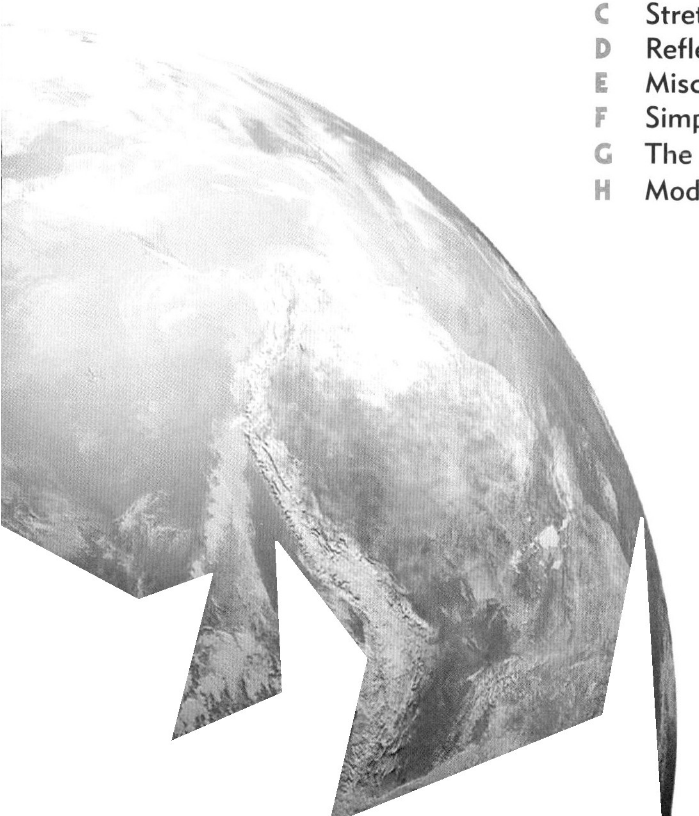

OPENING PROBLEM

Consider the function $f(x) = x^3 - 8$, whose graph is shown alongside.

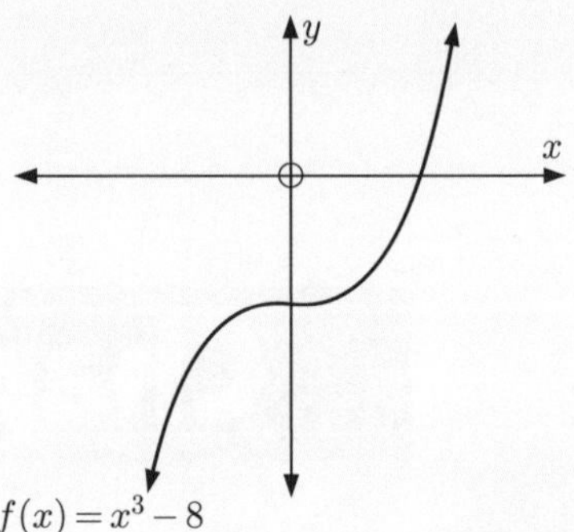

1 For each of the following functions $g(x)$, draw the graphs of $y = f(x)$ and $y = g(x)$ on the same set of axes:

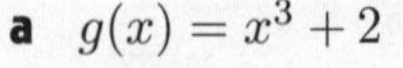

a $g(x) = x^3 + 2$ **b** $g(x) = (x-4)^3 - 8$

c $g(x) = (-x)^3 - 8$ **d** $g(x) = -(x^3 - 8)$

e $g(x) = 2(x^3 - 8)$ **f** $g(x) = \left(\frac{1}{2}x\right)^3 - 8$

g $g(x) = \left|x^3 - 8\right|$ **h** $g(x) = |x|^3 - 8$

i $g(x) = \dfrac{1}{x^3 - 8}$ **j** $g(x) = \sqrt[3]{x+8}$

GRAPHING PACKAGE

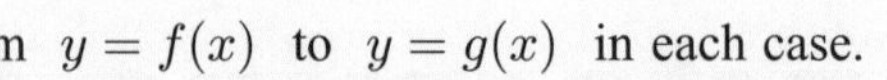

2 Describe the transformation from $y = f(x)$ to $y = g(x)$ in each case.

A TRANSFORMATION OF GRAPHS

There are several families of functions that we are already familiar with:

Name	*General form*	*Function notation*
Linear	$f(x) = ax + b, \quad a \neq 0$	$f : x \mapsto ax + b, \quad a \neq 0$
Quadratic	$f(x) = ax^2 + bx + c, \quad a \neq 0$	$f : x \mapsto ax^2 + bx + c, \quad a \neq 0$
Cubic	$f(x) = ax^3 + bx^2 + cx + d, \ a \neq 0$	$f : x \mapsto ax^3 + bx^2 + cx + d, \ a \neq 0$
Modulus	$f(x) = \|x\|$	$f : x \mapsto \|x\|$
Exponential	$f(x) = a^x, \quad a > 0, \quad a \neq 1$	$f : x \mapsto a^x, \quad a > 0, \quad a \neq 1$
Logarithmic	$f(x) = \log_e x$ or $f(x) = \ln x$	$f : x \mapsto \log_e x$ or $f : x \mapsto \ln x$
Reciprocal	$f(x) = \dfrac{k}{x}, \quad x \neq 0, \quad k \neq 0$	$f : x \mapsto \dfrac{k}{x}, \quad x \neq 0, \quad k \neq 0$

These families of functions have different and distinctive graphs. We can compare them by considering important graphical features such as axes intercepts, turning points, values of x for which the function does not exist, and asymptotes.

INVESTIGATION 1 FUNCTION FAMILIES

In this Investigation you are encouraged to use the graphing package supplied. Click on the icon to access this package.

GRAPHING PACKAGE

What to do:

1 From the menu, graph on the same set of axes:
$y = 2x + 1, \quad y = 2x + 3, \quad y = 2x - 1$
Comment on all lines of the form $y = 2x + b$.

2 From the menu, graph on the same set of axes:
$y = x + 2, \quad y = 2x + 2, \quad y = 4x + 2, \quad y = -x + 2, \quad y = -\frac{1}{2}x + 2$
Comment on all lines of the form $y = ax + 2$.

3 On the same set of axes graph:
$y = x^2, \quad y = 2x^2, \quad y = \frac{1}{2}x^2, \quad y = -x^2, \quad y = -3x^2, \quad y = -\frac{1}{5}x^2$
Comment on all functions of the form $y = ax^2, \quad a \neq 0$.

4 On the same set of axes graph:
$y = x^2, \quad y = (x-1)^2 + 2, \quad y = (x+1)^2 - 3, \quad y = (x-2)^2 - 1$
and other functions of the form $y = (x-h)^2 + k$ of your choice.
Comment on the functions of this form.

5 On the same set of axes, graph these functions:

a $y = \dfrac{1}{x}, \quad y = \dfrac{3}{x}, \quad y = \dfrac{10}{x}$

b $y = \dfrac{-1}{x}, \quad y = \dfrac{-2}{x}, \quad y = \dfrac{-5}{x}$

c $y = \dfrac{1}{x}, \quad y = \dfrac{1}{x-2}, \quad y = \dfrac{1}{x+3}$

d $y = \dfrac{1}{x}, \quad y = \dfrac{1}{x} + 2, \quad y = \dfrac{1}{x} - 2$

e $y = \dfrac{2}{x}, \quad y = \dfrac{2}{x-1} + 2, \quad y = \dfrac{2}{x+2} - 1$

Write a brief report on your discoveries.

From the **Investigation** you should have observed how different parts of a function's equation can affect its graph.

In particular, we can perform **transformations** of graphs to give the graph of a related function. These transformations include **translations**, **stretches**, and **reflections**.

In this chapter we will consider transformations of the function $y = f(x)$ into:

- $y = f(x) + b, \quad b$ is a constant
- $y = f(x - a), \quad a$ is a constant
- $y = pf(x), \quad p$ is a positive constant
- $y = f(qx), \quad q$ is a positive constant
- $y = -f(x)$
- $y = f(-x)$

When we perform a transformation on a function, a point which does not move is called an **invariant point**.

Example 1 **Self Tutor**

If $f(x) = x^2$, find in simplest form:

a $f(2x)$ **b** $f\left(\dfrac{x}{3}\right)$ **c** $2f(x) + 1$ **d** $f(x+3) - 4$

a $f(2x)$
$= (2x)^2$
$= 4x^2$

b $f\left(\dfrac{x}{3}\right)$
$= \left(\dfrac{x}{3}\right)^2$
$= \dfrac{x^2}{9}$

c $2f(x) + 1$
$= 2x^2 + 1$

d $f(x+3) - 4$
$= (x+3)^2 - 4$
$= x^2 + 6x + 9 - 4$
$= x^2 + 6x + 5$

EXERCISE 5A

1 If $f(x) = x$, find in simplest form:

a $f(2x)$ **b** $f(x) + 2$ **c** $\frac{1}{2}f(x)$ **d** $2f(x) + 3$

2 If $f(x) = x^2$, find in simplest form:

a $f(3x)$ **b** $f\left(\frac{x}{2}\right)$ **c** $3f(x)$ **d** $2f(x - 1) + 5$

3 If $f(x) = x^3$, find in simplest form:

a $f(4x)$ **b** $\frac{1}{2}f(2x)$ **c** $f(x + 1)$ **d** $2f(x + 1) - 3$

Hint: $(x + 1)^3 = x^3 + 3x^2 + 3x + 1$. See the binomial theorem in **Chapter 8**.

4 If $f(x) = 2^x$, find in simplest form:

a $f(2x)$ **b** $f(-x) + 1$ **c** $f(x - 2) + 3$ **d** $2f(x) + 3$

5 If $f(x) = \frac{1}{x}$, find in simplest form:

a $f(-x)$ **b** $f(\frac{1}{2}x)$ **c** $2f(x) + 3$ **d** $3f(x - 1) + 2$

B TRANSLATIONS

INVESTIGATION 2 TRANSLATIONS

In this Investigation we consider **translations** of the forms $y = f(x) + b$ and $y = f(x - a)$.

What to do:

GRAPHING PACKAGE

1 **a** For $f(x) = x^3$, find in simplest form:

i $f(x) + 2$ **ii** $f(x) - 3$ **iii** $f(x) + 6$

b Graph all four functions on the same set of axes.

c What effect does the constant b have when $y = f(x)$ is transformed to $y = f(x) + b$?

2 **a** For $f(x) = x^2$, find in simplest form:

i $f(x - 2)$ **ii** $f(x + 1)$ **iii** $f(x - 5)$

b Graph all four functions on the same set of axes.

c What effect does the constant a have when $y = f(x)$ is transformed to $y = f(x - a)$?

- For $y = f(x) + b$, the effect of b is to **translate** the graph **vertically** through b units.
  - If $b > 0$ it moves **upwards**.
  - If $b < 0$ it moves **downwards**.
- For $y = f(x - a)$, the effect of a is to **translate** the graph **horizontally** through a units.
  - If $a > 0$ it moves to the **right**.
  - If $a < 0$ it moves to the **left**.
- For $y = f(x - a) + b$, the graph is translated horizontally a units and vertically b units. We say it is **translated by the vector** $\binom{a}{b}$.

EXERCISE 5B

1 **a** Sketch the graph of $y = x^2$.

b On the same set of axes sketch the graphs of:

i $y = x^2 + 2$ **ii** $y = x^2 - 3$.

c What is the connection between the graphs of $y = f(x)$ and $y = f(x) + b$ if:

i $b > 0$ **ii** $b < 0$?

2 For each of the following functions f, sketch on the same set of axes the graphs of $y = f(x)$, $y = f(x) + 1$, and $y = f(x) - 2$.

a $f(x) = 2^x$ **b** $f(x) = x^3$ **c** $f(x) = \frac{1}{x}$

d $f(x) = (x-1)^2$ **e** $f(x) = |x|$

3 **a** On the same set of axes, graph $f(x) = x^2$, $y = f(x-3)$, and $y = f(x+2)$.

b What is the connection between the graphs of $y = f(x)$ and $y = f(x-a)$ if:

i $a > 0$ **ii** $a < 0$?

4 For each of the following functions f, sketch on the same set of axes the graphs of $y = f(x)$, $y = f(x-1)$, and $y = f(x+2)$.

a $f(x) = x^3$ **b** $f(x) = \ln x$ **c** $f(x) = \frac{1}{x}$

d $f(x) = (x+1)^2 + 2$ **e** $f(x) = |x|$

5 For each of the following functions f, sketch on the same set of axes the graphs of $y = f(x)$, $y = f(x-2) + 3$, and $y = f(x+1) - 4$.

a $f(x) = x^2$ **b** $f(x) = e^x$ **c** $f(x) = \frac{1}{x}$

6 Copy these functions and then draw the graph of $y = f(x-2) - 3$.

a

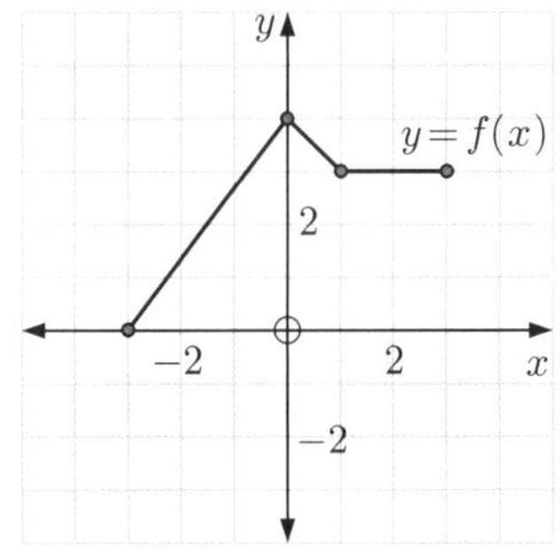

b

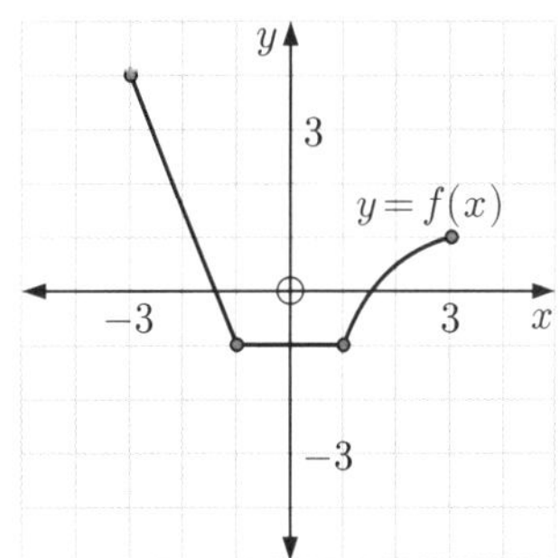

7 The graph of $f(x) = x^2 - 2x + 2$ is translated 3 units right to $g(x)$. Find $g(x)$ in the form $g(x) = ax^2 + bx + c$.

8 Suppose $f(x) = x^2$ is transformed to $g(x) = (x-3)^2 + 2$.

a Find the images of the following points on $f(x)$:

i $(0, 0)$ **ii** $(-3, 9)$ **iii** where $x = 2$

b Find the points on $f(x)$ which correspond to the following points on $g(x)$:

i $(1, 6)$ **ii** $(-2, 27)$ **iii** $(1\frac{1}{2}, 4\frac{1}{4})$

C STRETCHES

INVESTIGATION 3 STRETCHES

In this Investigation we consider **stretches** of the forms $y = pf(x)$, $p > 0$ and $y = f(qx)$, $q > 0$.

GRAPHING PACKAGE

What to do:

1. **a** For $f(x) = x + 2$, find in simplest form:
   i $3f(x)$ **ii** $\frac{1}{2}f(x)$ **iii** $5f(x)$
   b Graph all four functions on the same set of axes.
   c What effect does the constant p have when $y = f(x)$ is transformed to $y = pf(x)$, $p > 0$?

2. **a** For $f(x) = x^2$, find in simplest form:
   i $f(2x)$ **ii** $f(3x)$ **iii** $f(\frac{x}{4})$
   b Graph all four functions on the same set of axes.
   c What effect does the constant q have when $y = f(x)$ is transformed to $y = f(qx)$, $q > 0$?

- For $y = pf(x)$, $p > 0$, the effect of p is to **vertically stretch** the graph by the **scale factor** p.
  - If $p > 1$ it moves points of $y = f(x)$ **further away** from the x-axis.
  - If $0 < p < 1$ it moves points of $y = f(x)$ **closer** to the x-axis.
- For $y = f(qx)$, $q > 0$, the effect of q is to **horizontally stretch** the graph by the **scale factor** $\frac{1}{q}$.
  - If $q > 1$ it moves points of $y = f(x)$ **closer** to the y-axis.
  - If $0 < q < 1$ it moves points of $y = f(x)$ **further away** from the y-axis.

EXERCISE 5C

1. Sketch, on the same set of axes, the graphs of $y = f(x)$, $y = 2f(x)$, and $y = 3f(x)$ for each of:
   a $f(x) = x^2$ **b** $f(x) = x^3$ **c** $f(x) = e^x$
   d $f(x) = \ln x$ **e** $f(x) = \dfrac{1}{x}$ **f** $f(x) = |x|$

2. Sketch, on the same set of axes, the graphs of $y = f(x)$, $y = \frac{1}{2}f(x)$, and $y = \frac{1}{4}f(x)$ for each of:
   a $f(x) = x^2$ **b** $f(x) = x^3$ **c** $f(x) = e^x$

3. Sketch, on the same set of axes, the graphs of $y = f(x)$ and $y = f(2x)$ for each of:
   a $y = x^2$ **b** $y = (x - 1)^2$ **c** $y = (x + 3)^2$

4. Sketch, on the same set of axes, the graphs of $y = f(x)$ and $y = f(\frac{x}{2})$ for each of:
   a $y = x^2$ **b** $y = 2x$ **c** $y = (x + 2)^2$

5 Sketch, on the same set of axes, the graphs of $y = f(x)$ and $y = f(3x)$ for each of:

a $y = x$ **b** $y = x^2$ **c** $y = e^x$

6 Consider the function $f : x \mapsto x^2$.
On the same set of axes sketch the graphs of:

a $y = f(x)$, $y = 3f(x-2)+1$, and $y = 2f(x+1)-3$

b $y = f(x)$, $y = f(x-3)$, $y = f(\frac{x}{2}-3)$, $y = 2f(\frac{x}{2}-3)$, and $y = 2f(\frac{x}{2}-3)+4$

c $y = f(x)$ and $y = \frac{1}{4}f(2x+5)+1$.

7 **a** Given that the following points lie on $y = f(x)$, find the coordinates of the point each moves to under the transformation $y = 3f(2x)$:

i $(3, -5)$ **ii** $(1, 2)$ **iii** $(-2, 1)$

b Find the points on $y = f(x)$ which are moved to the following points under the transformation $y = 3f(2x)$:

i $(2, 1)$ **ii** $(-3, 2)$ **iii** $(-7, 3)$

8 The function $y = f(x)$ is transformed to the function $y = 3 + 2f(\frac{1}{2}x+1)$.

a Fully describe the transformation that maps $y = f(x)$ onto $y = 3 + 2f(\frac{1}{2}x+1)$.

b Given that the following points lie on $y = f(x)$, find the coordinates of the point each moves to under the transformation $y = 3 + 2f(\frac{1}{2}x+1)$.

i $(1, -3)$ **ii** $(2, 1)$ **iii** $(-1, -2)$

c Find the points on $y = f(x)$ which are moved to the following points under the transformation $y = 3 + 2f(\frac{1}{2}x+1)$.

i $(-2, -5)$ **ii** $(1, -1)$ **iii** $(5, 0)$

D REFLECTIONS

INVESTIGATION 4 REFLECTIONS

In this Investigation we consider **reflections** of the forms $y = -f(x)$ and $y = f(-x)$.

What to do:

GRAPHING PACKAGE

1 Consider $f(x) = x^3 - 2$.

a Find in simplest form:

i $-f(x)$ **ii** $f(-x)$

b Graph $y = f(x)$, $y = -f(x)$, and $y = f(-x)$ on the same set of axes.

2 Consider $f(x) = e^x$.

a Find in simplest form:

i $-f(x)$ **ii** $f(-x)$

b Graph $y = f(x)$, $y = -f(x)$, and $y = f(-x)$ on the same set of axes.

3 What transformation moves:

a $y = f(x)$ to $y = -f(x)$ **b** $y = f(x)$ to $y = f(-x)$?

From the **Investigation** you should have discovered that:

- For $y = -f(x)$, we **reflect** $y = f(x)$ in the ***x*-axis**.
- For $y = f(-x)$, we **reflect** $y = f(x)$ in the ***y*-axis**.

In addition, in our earlier study of functions we found that:

- For $y = f^{-1}(x)$, we **reflect** $y = f(x)$ in the line $\boldsymbol{y = x}$.

EXERCISE 5D

1 On the same set of axes, sketch the graphs of $y = f(x)$, $y = -f(x)$, and if it exists, $y = f^{-1}(x)$.

a $f(x) = 3x$ **b** $f(x) = e^x$ **c** $f(x) = x^2$

d $f(x) = \ln x$ **e** $f(x) = x^3 - 2$ **f** $f(x) = 2(x+1)^2$

2 For each of the following, find $f(-x)$. Hence graph $y = f(x)$ and $y = f(-x)$ on the same set of axes.

a $f(x) = 2x + 1$ **b** $f(x) = x^2 + 2x + 1$ **c** $f(x) = x^3$ **d** $f(x) = |x - 3|$

3 The function $f(x) = x^3 - \ln x$ is reflected in the x-axis to $g(x)$. Find $g(x)$.

4 The function $f(x) = x^4 - 2x^3 - 3x^2 + 5x - 7$ is reflected in the y-axis to $g(x)$. Find $g(x)$.

5 The function $y = f(x)$ is transformed to $g(x) = -f(x)$.

a Find the points on $g(x)$ corresponding to the following points on $f(x)$:

i (3, 0) **ii** (2, –1) **iii** (–3, 2)

b Find the points on $f(x)$ that have been transformed to the following points on $g(x)$:

i (7, –1) **ii** (–5, 0) **iii** (–3, –2)

6 The function $y = f(x)$ is transformed to $h(x) = f(-x)$.

a Find the image points on $h(x)$ for the following points on $f(x)$:

i (2, –1) **ii** (0, 3) **iii** (–1, 2) **iv** (3, 0)

b Find the points on $f(x)$ corresponding to the following points on $h(x)$:

i (5, –4) **ii** (0, 3) **iii** (2, 3) **iv** (3, 0)

7 The function $y = f(x)$ is transformed to $m(x) = f^{-1}(x)$.

a Find the image points on $m(x)$ for the following points on $f(x)$:

i (3, 1) **ii** (–2, 4) **iii** (0, –5)

b Find the points on $f(x)$ corresponding to the following points on $m(x)$:

i (–1, 1) **ii** (6, 0) **iii** (3, –2)

8 A function $f(x)$ is transformed to the function $g(x) = -f(-x)$.

a Describe the nature of the transformation.

b If (3, –7) lies on $y = f(x)$, find the transformed point on $y = g(x)$.

c Find the point on $f(x)$ that transforms to the point (–5, –1).

9 **a** Copy the graph of $y = f(x)$ alongside, then draw the graph of:

  i $y = -f(x)$ **ii** $y = f(-x)$

b Copy the graph of $y = f(x)$ alongside.

  i On the same set of axes, draw the reflection of $y = f(x)$ in the line $y = x$.

  ii Is this the graph of $y = f^{-1}(x)$?

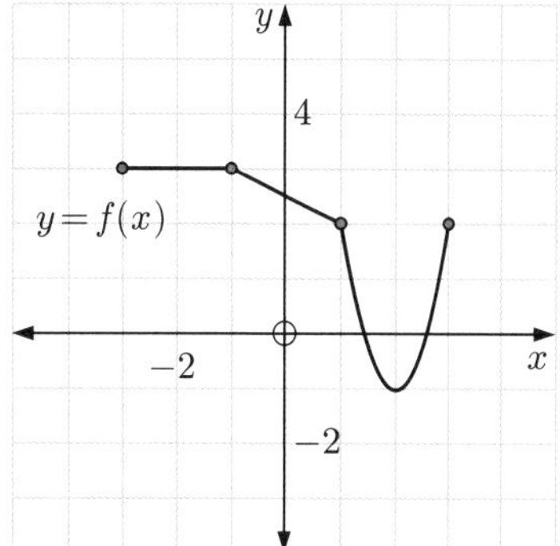

E MISCELLANEOUS TRANSFORMATIONS

A summary of all the transformations is given in the printable concept map.

CONCEPT MAP

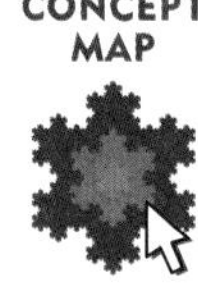

Example 2 **Self Tutor**

Consider $f(x) = \frac{1}{2}x + 1$. On separate sets of axes graph:

a $y = f(x)$ and $y = f(x + 2)$

b $y = f(x)$ and $y = f(x) + 2$

c $y = f(x)$ and $y = 2f(x)$

d $y = f(x)$ and $y = -f(x)$

a

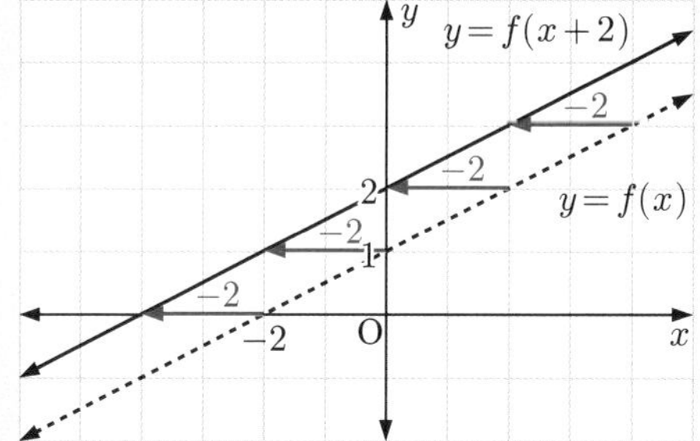

b

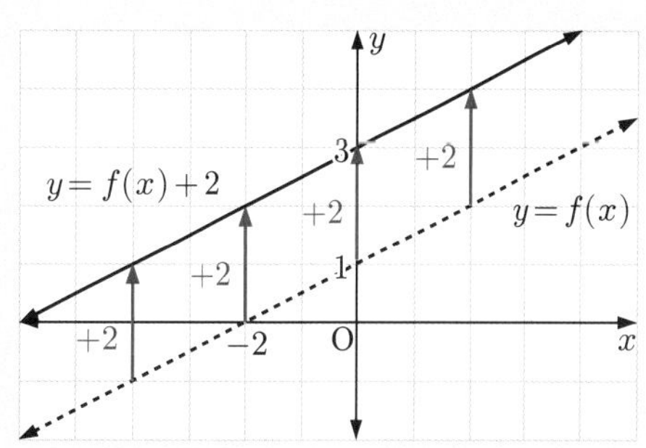

c

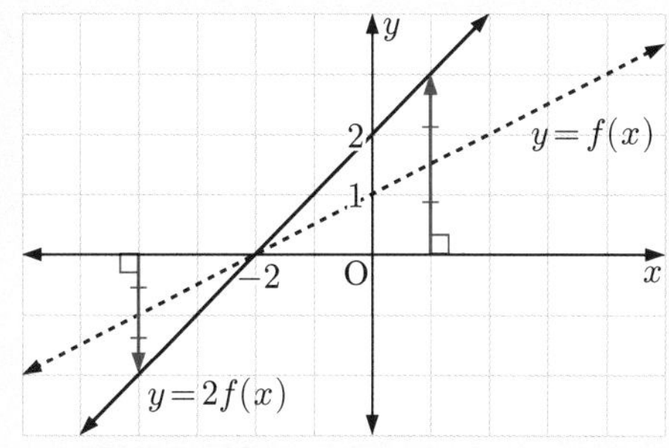

d

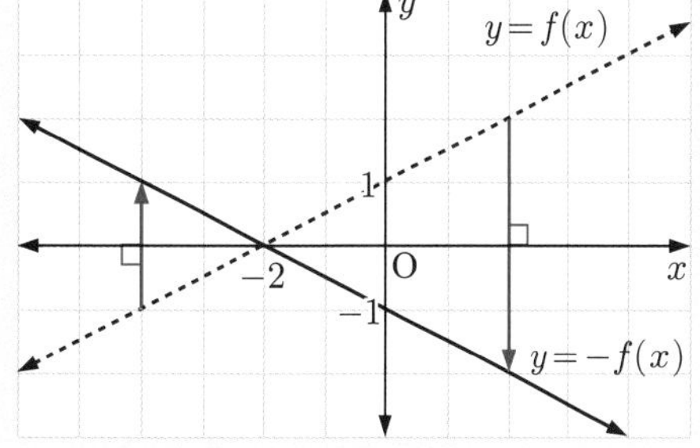

EXERCISE 5E

Invariant points do not move under a transformation.

1 Consider $f(x) = x^2 - 1$.

a Graph $y = f(x)$ and state its axes intercepts.

b Graph the functions:

i $y = f(x) + 3$ **ii** $y = f(x - 1)$

iii $y = 2f(x)$ **iv** $y = -f(x)$

c What transformation on $y = f(x)$ has occurred in each case in **b**?

d On the same set of axes graph $y = f(x)$ and $y = -2f(x)$. Describe the transformation.

e What points on $y = f(x)$ are invariant when $y = f(x)$ is transformed to $y = -2f(x)$?

2 On each of the following $f(x)$ is mapped onto $g(x)$ using a single transformation.

i Describe the transformation fully. **ii** Write $g(x)$ in terms of $f(x)$.

a

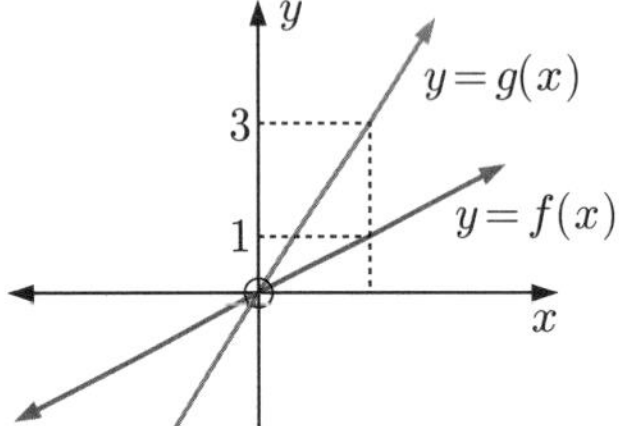

b

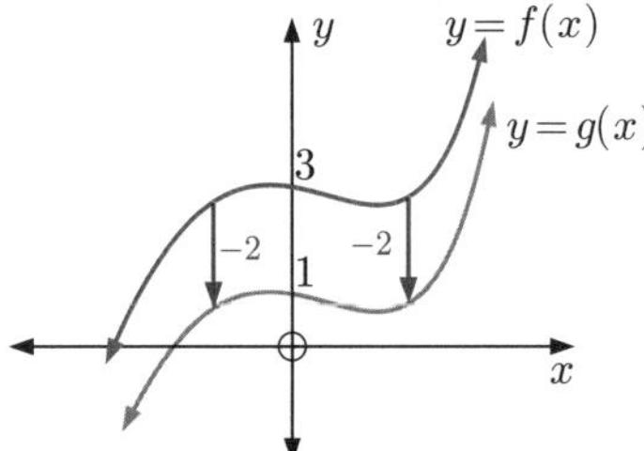

c

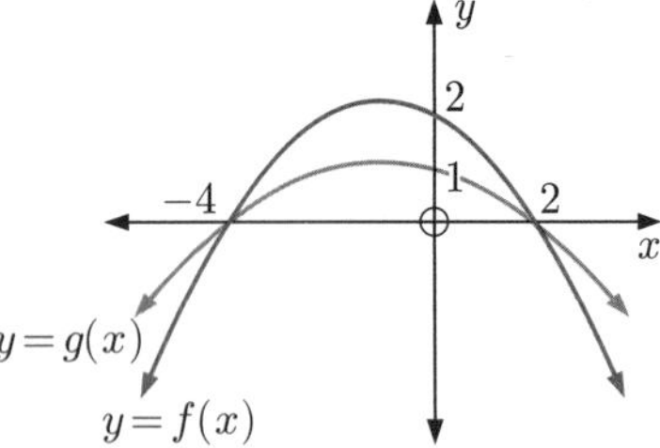

d

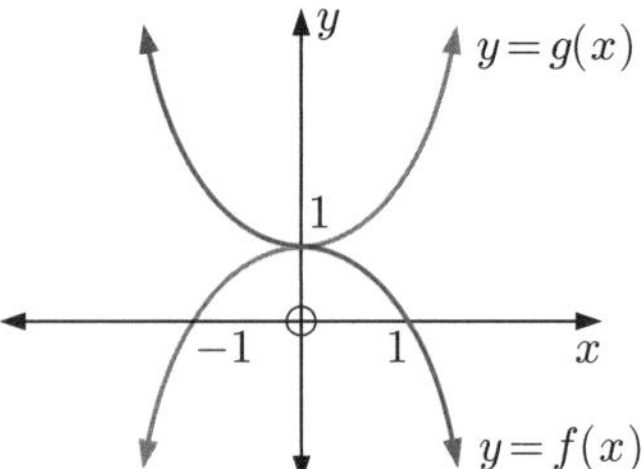

3 Copy the following graphs for $y = f(x)$ and sketch the graphs of $y = -f(x)$ on the same axes.

a

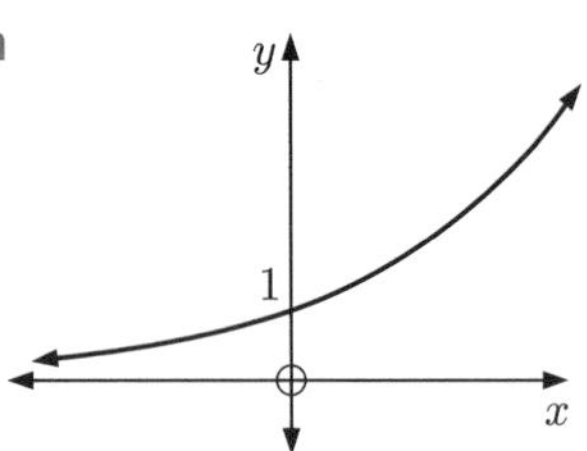

b

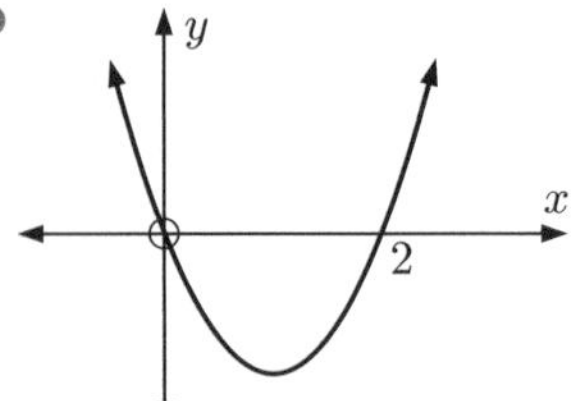

c

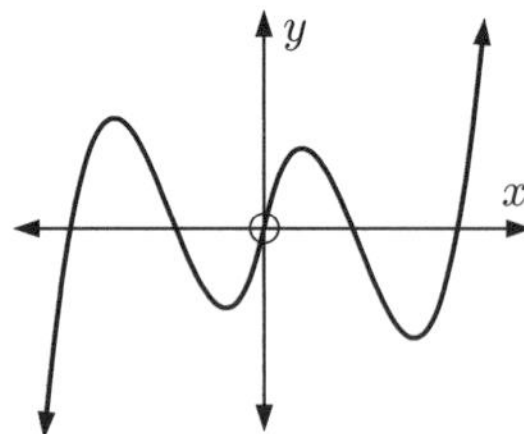

4 Given the following graphs of $y = f(x)$, sketch graphs of $y = f(-x)$:

a

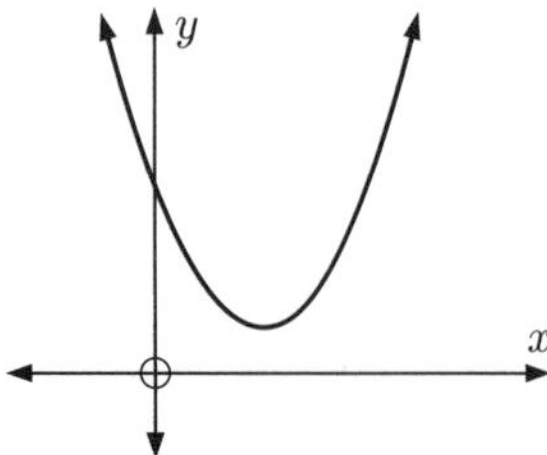

b

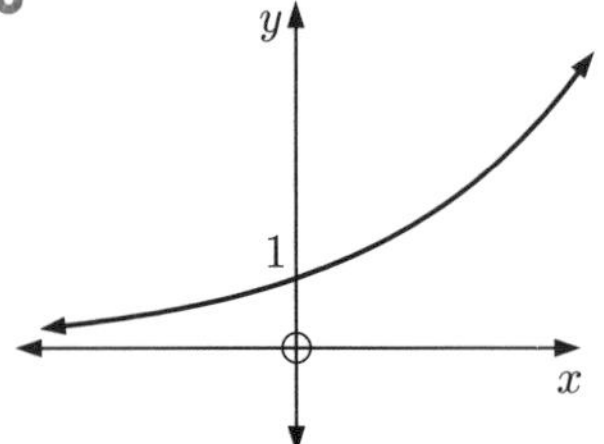

c

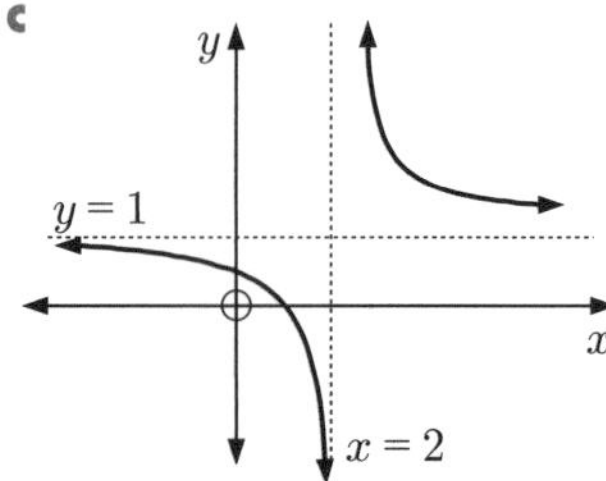

5 The scales on the graphs below are the same. Match each equation to its graph.

A $y = x^4$ **B** $y = 2x^4$ **C** $y = \frac{1}{2}x^4$ **D** $y = 6x^4$

a

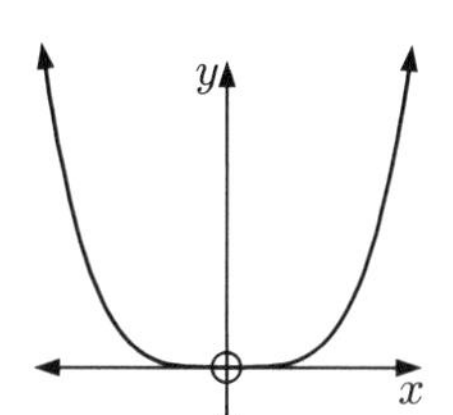

b

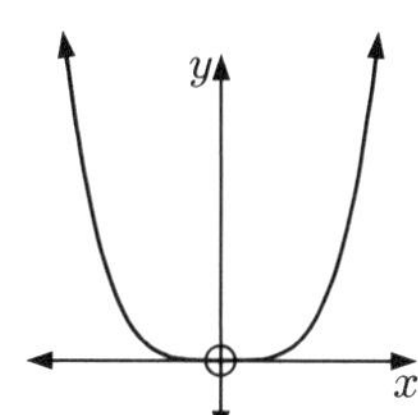

c

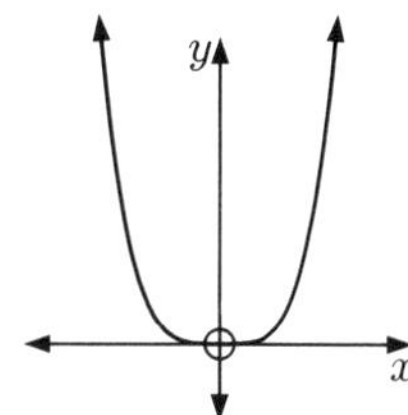

d

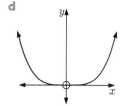

6 For the graph of $y = f(x)$ given, sketch the graph of:

a $y = 2f(x)$ **b** $y = \frac{1}{2}f(x)$

c $y = f(x+2)$ **d** $y = f(2x)$

e $y = f(\frac{1}{2}x)$

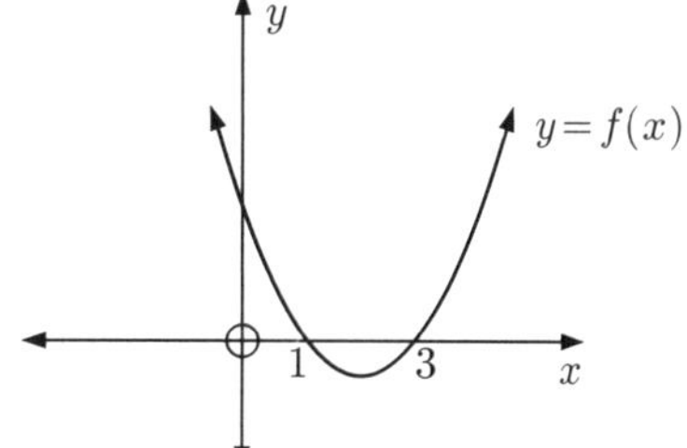

7

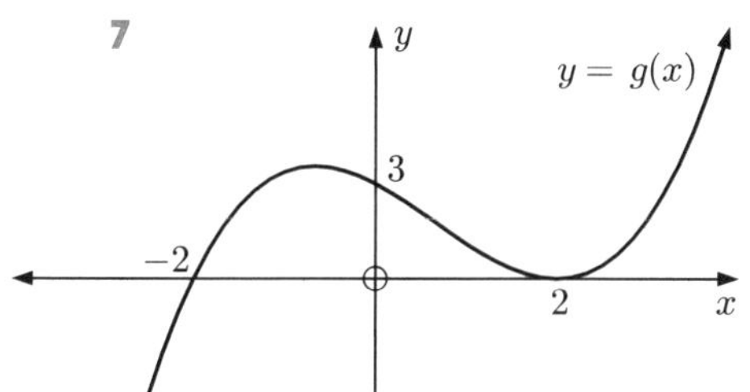

For the graph of $y = g(x)$ given, sketch the graph of:

a $y = g(x) + 2$ **b** $y = -g(x)$

c $y = g(-x)$ **d** $y = g(x+1)$

8 For the graph of $y = h(x)$ given, sketch the graph of:

a $y = h(x) + 1$ **b** $y = \frac{1}{2}h(x)$

c $y = h(-x)$ **d** $y = h\left(\frac{x}{2}\right)$

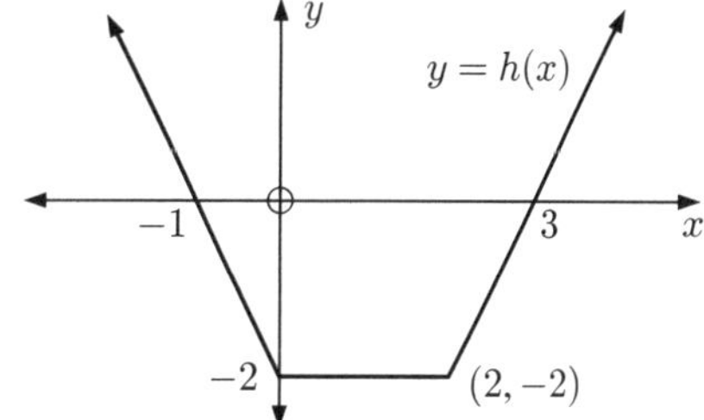

F SIMPLE RATIONAL FUNCTIONS

In **Chapter 2** we introduced rational functions and some of their properties.

A function of the form $y = \dfrac{ax+b}{cx+d}$, $x \neq -\dfrac{d}{c}$ where a, b, c, and d are constants, is called a **simple rational function**.

These functions are characterised by the presence of both a **horizontal asymptote** and a **vertical asymptote**.

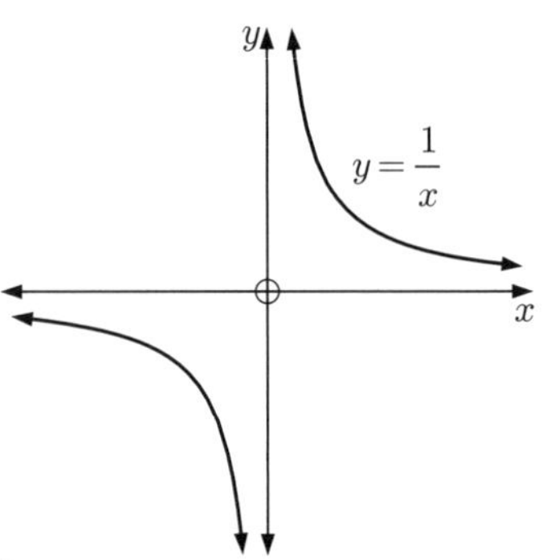

Any graph of a simple rational function can be obtained from the reciprocal function $y = \frac{1}{x}$ by a combination of transformations including:

- a **translation** (vertical and/or horizontal)
- **stretches** (vertical and/or horizontal).

For example:

- $y = \frac{k}{x}$ is a vertical stretch of $y = \frac{1}{x}$ with scale factor k.
- $y = \frac{1}{x-k}$ is a horizontal translation of $y = \frac{1}{x}$ through k units.

Example 3 **Self Tutor**

The function $g(x)$ results when $y = \frac{1}{x}$ is transformed by a vertical stretch with scale factor 2, followed by a translation of $\binom{3}{-2}$.

a Find an expression for $g(x)$.

b Find the asymptotes of $y = g(x)$.

c Sketch $y = g(x)$.

d Is $g(x)$ a self-inverse function? Explain your answer.

a Under a vertical stretch with scale factor 2, $f(x)$ becomes $2f(x)$.

$\therefore \frac{1}{x}$ becomes $2\left(\frac{1}{x}\right) = \frac{2}{x}$.

Under a translation of $\binom{3}{-2}$, $f(x)$ becomes $f(x-3)-2$.

$\therefore \frac{2}{x}$ becomes $\frac{2}{x-3} - 2$.

So, $y = \frac{1}{x}$ becomes
$$g(x) = \frac{2}{x-3} - 2 = \frac{2-2(x-3)}{x-3} = \frac{-2x+8}{x-3}$$

b The asymptotes of $y = \frac{1}{x}$ are $x = 0$ and $y = 0$.

These are unchanged by the stretch, and shifted $\binom{3}{-2}$ by the translation.

$\therefore$ the vertical asymptote is $x = 3$ and the horizontal asymptote is $y = -2$.

c

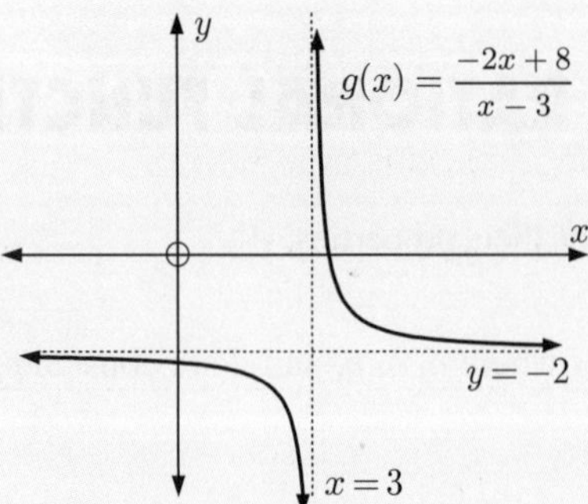

d The graph is not symmetric about $y = x$, so $g(x)$ is *not* a self-inverse function.

Example 4 — Self Tutor

Consider the function $f(x) = \dfrac{2x-6}{x+1}$.

a Find the asymptotes of $y = f(x)$.

b Discuss the behaviour of the graph near these asymptotes.

c Find the axes intercepts of $y = f(x)$.

d Sketch the graph of the function.

e Describe the transformations which transform $y = \dfrac{1}{x}$ into $y = f(x)$.

f Describe the transformations which transform $y = f(x)$ into $y = \dfrac{1}{x}$.

---

a $f(x) = \dfrac{2x-6}{x+1}$

$= \dfrac{2(x+1)-8}{x+1}$

$= \dfrac{-8}{x+1} + 2$

$y = f(x)$ is a translation of $y = \dfrac{-8}{x}$ through $\binom{-1}{2}$.

Now $y = \dfrac{-8}{x}$ has asymptotes $x = 0$ and $y = 0$.

$\therefore$ $y = f(x)$ has vertical asymptote $x = -1$ and horizontal asymptote $y = 2$.

b 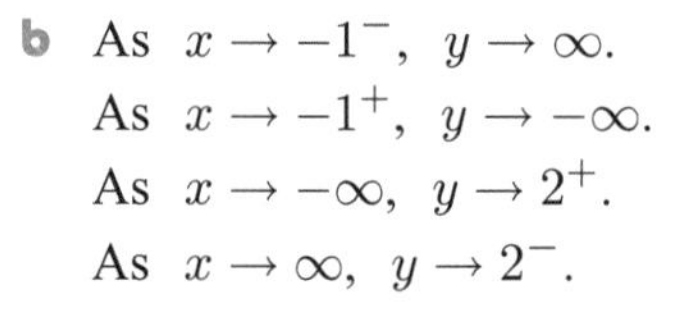

As $x \to -1^-$, $y \to \infty$.
As $x \to -1^+$, $y \to -\infty$.
As $x \to -\infty$, $y \to 2^+$.
As $x \to \infty$, $y \to 2^-$.

c 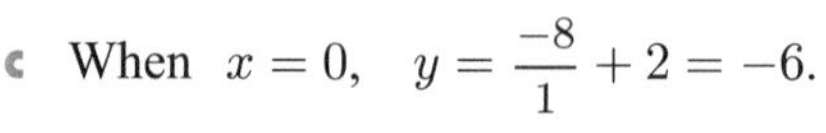

When $x = 0$, $y = \dfrac{-8}{1} + 2 = -6$.

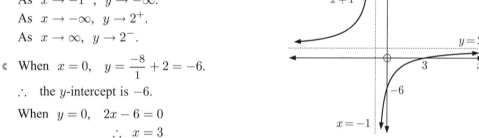

$\therefore$ the y-intercept is -6.

When $y = 0$, $2x - 6 = 0$

$\therefore$ $x = 3$

$\therefore$ the x-intercept is 3.

d 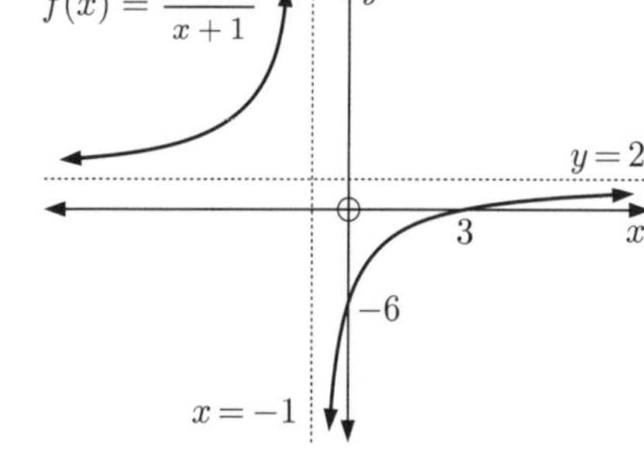

e $\dfrac{1}{x}$ becomes $\dfrac{8}{x}$ under a vertical stretch with scale factor 8.

$\dfrac{8}{x}$ becomes $\dfrac{-8}{x}$ under a reflection in the y-axis.

$-\dfrac{8}{x}$ becomes $\dfrac{-8}{x+1} + 2$ under a translation through $\binom{-1}{2}$.

So, $y = \dfrac{1}{x}$ is transformed to $y = f(x)$ under a vertical stretch with scale factor 8, followed by a reflection in the y-axis, followed by a translation through $\binom{-1}{2}$.

f To transform $y = f(x)$ into $y = \dfrac{1}{x}$, we need to reverse the process in **e**.

We need a translation through $\binom{1}{-2}$, followed by a reflection in the y-axis, followed by a vertical stretch with scale factor $\frac{1}{8}$.

EXERCISE 5F

1 Write, in the form $y = \dfrac{ax+b}{cx+d}$, the function that results when $y = \dfrac{1}{x}$ is transformed by:

a a vertical stretch with scale factor $\frac{1}{2}$

b a horizontal stretch with scale factor 3

c a horizontal translation of -3

d a vertical translation of 4.

2 The function $g(x)$ results when $y = \dfrac{1}{x}$ is transformed by a vertical stretch with scale factor 3, followed by a translation of $\binom{1}{-1}$.

a Write an expression for $g(x)$ in the form $g(x) = \dfrac{ax+b}{cx+d}$.

b Find the asymptotes of $y = g(x)$.

c State the domain and range of $g(x)$.

d Sketch $y = g(x)$.

e Is $g(x)$ a self-inverse function? Explain your answer.

3 For each of the following functions f, find:

i the asymptotes

ii how to transform $y = \dfrac{1}{x}$ into $y = f(x)$.

a $f : x \mapsto \dfrac{2x+4}{x-1}$

b $f : x \mapsto \dfrac{3x-2}{x+1}$

c $f : x \mapsto \dfrac{2x+1}{2-x}$

4 For each of the following functions $f(x)$:

i Find the asymptotes of $y = f(x)$.

ii Discuss the behaviour of the graph near these asymptotes.

iii Find the axes intercepts of $y = f(x)$.

iv Sketch the graph of $y = f(x)$.

v Describe the transformations which transform $y = \dfrac{1}{x}$ into $y = f(x)$.

vi Describe the transformations which transform $y = f(x)$ into $y = \dfrac{1}{x}$.

a $y = \dfrac{2x+3}{x+1}$

b $y = \dfrac{3}{x-2}$

c $y = \dfrac{2x-1}{3-x}$

d $y = \dfrac{5x-1}{2x+1}$

5 In order to remove noxious weeds from her property, Helga sprays with a weedicide. The chemical is slow to act, and the number of weeds per hectare remaining after t days is modelled by $N = 20 + \dfrac{100}{t+2}$ weeds/ha.

a How many weeds per ha were alive before the spraying?

b How many weeds will be alive after 8 days?

c How long will it take for the number of weeds still alive to be 40/ha?

d Sketch the graph of N against t.

e According to the model, is the spraying going to eradicate all weeds?

G THE RECIPROCAL OF A FUNCTION

For a function $f(x)$, the **reciprocal** of the function is $\frac{1}{f(x)}$.

When $y = \frac{1}{f(x)}$ is graphed from $y = f(x)$:

- the zeros of $y = f(x)$ become vertical asymptotes of $y = \frac{1}{f(x)}$
- the vertical asymptotes of $y = f(x)$ become zeros of $y = \frac{1}{f(x)}$
- the local maxima of $y = f(x)$ become local minima of $y = \frac{1}{f(x)}$
- the local minima of $y = f(x)$ become local maxima of $y = \frac{1}{f(x)}$
- when $f(x) > 0$, $\frac{1}{f(x)} > 0$ and when $f(x) < 0$, $\frac{1}{f(x)} < 0$
- when $f(x) \to 0$, $\frac{1}{f(x)} \to \pm\infty$ and when $f(x) \to \pm\infty$, $\frac{1}{f(x)} \to 0$.

Example 5 Self Tutor

Graph on the same set of axes:

a $y = x - 2$ and $y = \frac{1}{x-2}$

b $y = x^2$ and $y = \frac{1}{x^2}$

a

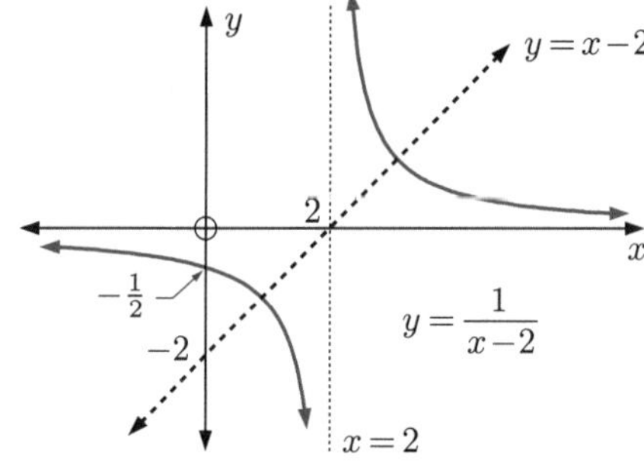

b

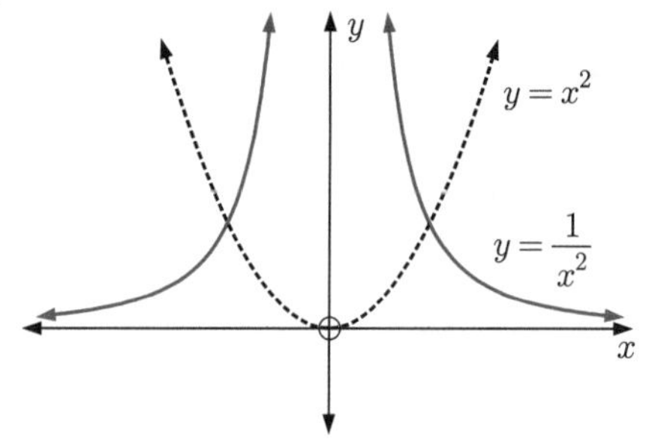

EXERCISE 5G

1 Graph on the same set of axes:

a $y = x + 3$ and $y = \frac{1}{x+3}$

b $y = -x^2$ and $y = \frac{-1}{x^2}$

c $y = \sqrt{x}$ and $y = \frac{1}{\sqrt{x}}$

d $y = (x-1)(x-3)$ and $y = \frac{1}{(x-1)(x-3)}$

2 **Invariant points** are points which do not move under a transformation.

Show that if $y = f(x)$ is transformed to $y = \frac{1}{f(x)}$, invariant points occur at $y = \pm 1$.

Check your results in question 1 for invariant points.

3 Copy the following graphs for $y = f(x)$ and on the same axes graph $y = \dfrac{1}{f(x)}$:

a

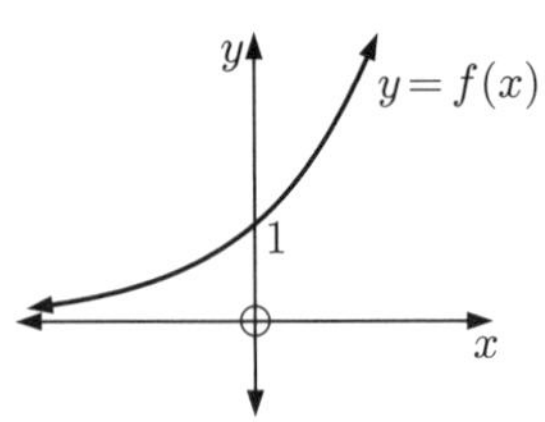

b

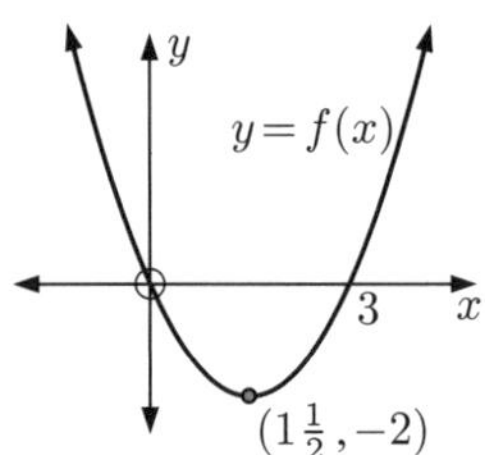

c

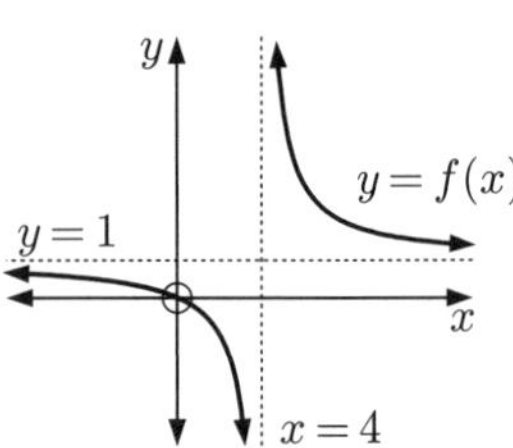

H MODULUS FUNCTIONS

We have seen that the modulus function is defined by $f : x \mapsto |x| = \begin{cases} x & \text{if } x \geqslant 0 \\ -x & \text{if } x < 0\,. \end{cases}$

To obtain the graph of $y = f(|x|)$ from the graph of $y = f(x)$:

- discard the graph for $x < 0$
- reflect the graph for $x \geqslant 0$ in the y-axis, keeping what was there
- points on the y-axis are invariant.

The modulus of the function $f(x)$ is $|f(x)| = \begin{cases} f(x) & \text{if } f(x) \geqslant 0 \\ -f(x) & \text{if } f(x) < 0\,. \end{cases}$

To obtain the graph of $y = |f(x)|$ from the graph of $y = f(x)$:

- keep the graph for $f(x) \geqslant 0$
- reflect the graph in the x-axis for $f(x) < 0$, discarding what was there
- points on the x-axis are invariant.

Example 6 **Self Tutor**

Draw the graph of $f(x) = 3x(x-2)$ and on the same set of axes draw the graph of $y = |f(x)|$ and $y = f(|x|)$.

$y = |f(x)| = \begin{cases} f(x) & \text{if } f(x) \geqslant 0 \\ -f(x) & \text{if } f(x) < 0 \end{cases}$

The graph is unchanged for $f(x) \geqslant 0$ and reflected in the x-axis for $f(x) < 0$.

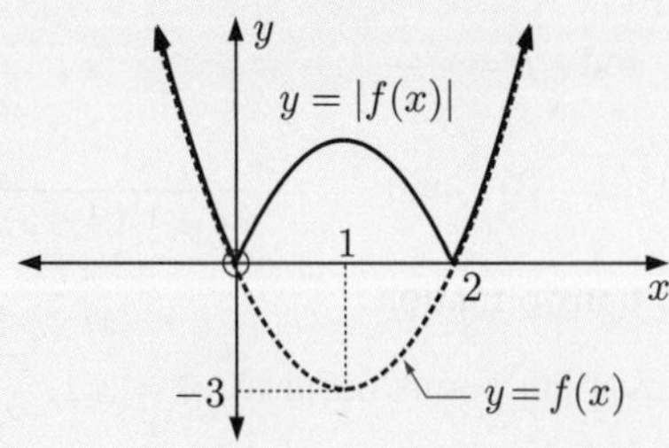

$y = f(|x|) = \begin{cases} f(x) & \text{if } x \geqslant 0 \\ f(-x) & \text{if } x < 0 \end{cases}$

The graph is unchanged for $x \geqslant 0$ and reflected in the y-axis for $x < 0$.

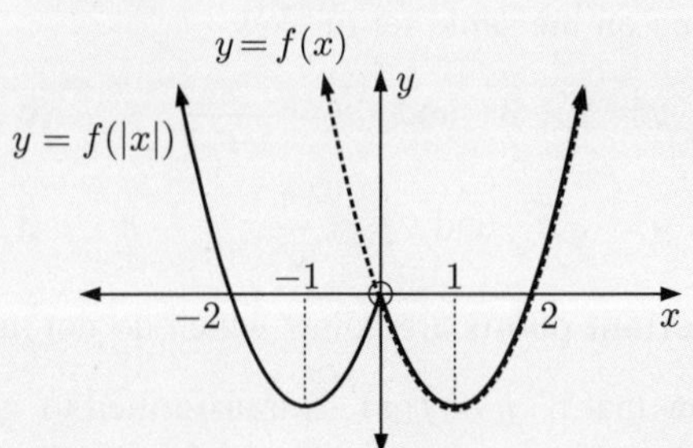

EXERCISE 5H

1 Draw $y = x(x+2)$ and on the same set of axes graph:

a $y = |f(x)|$ b $y = f(|x|)$

2 Draw $y = -x^2 + 6x - 8$ and on the same set of axes graph:

a $y = |f(x)|$ b $y = f(|x|)$

3 a Copy the following graphs for $y = f(x)$ and on the same axes graph $y = |f(x)|$:

i

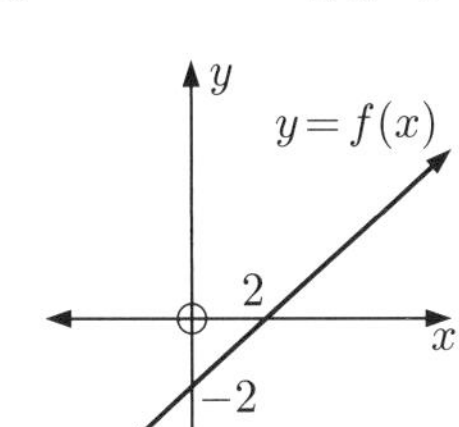

ii

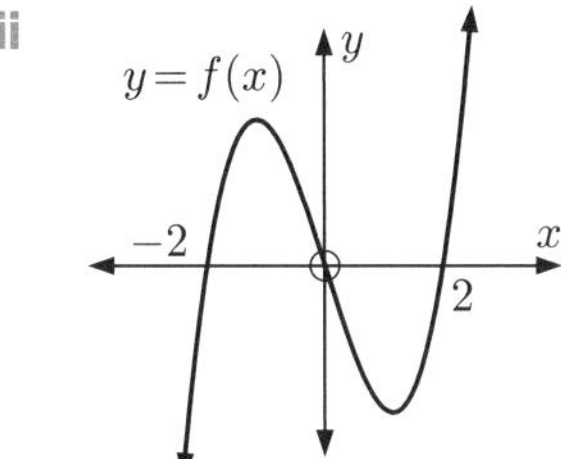

iii

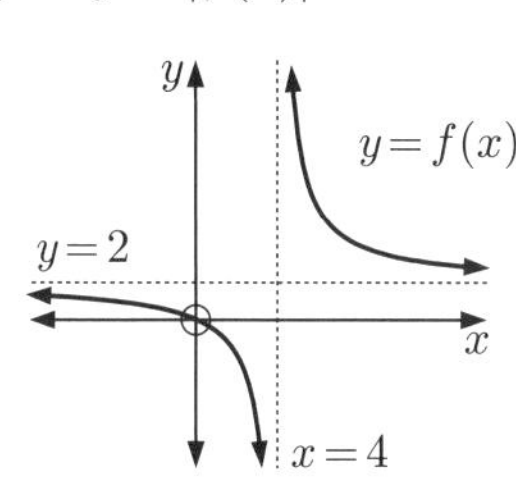

b Repeat a, but this time graph $y = f(x)$ and $y = f(|x|)$ on the same set of axes.

4 Suppose the function $f(x)$ is transformed to $|f(x)|$.
For each of the following points on $y = f(x)$, find the corresponding image point on $y = |f(x)|$:

a (3, 0) b (5, −2) c (0, 7) d (2, 2)

5 Suppose the function $f(x)$ is transformed to $f(|x|)$.

a For each of the following points on $y = f(x)$, find the image points of:

i (0, 3) ii (1, 3) iii (7, −4)

b Find the points on $y = f(x)$ that are transformed to the following points on $y = f(|x|)$:

i (0, 3) ii (−1, 3) iii (10, −8)

REVIEW SET 5A NON-CALCULATOR

1 If $f(x) = x^2 - 2x$, find in simplest form:

a $f(3)$ **b** $f(2x)$ **c** $f(-x)$ **d** $3f(x) - 2$

2 If $f(x) = 5 - x - x^2$, find in simplest form:

a $f(-1)$ **b** $f(x-1)$ **c** $f\left(\frac{x}{2}\right)$ **d** $2f(x) - f(-x)$

3 The graph of $f(x) = 3x^3 - 2x^2 + x + 2$ is translated to its image $g(x)$ by the vector $\binom{1}{-2}$.
Write the equation of $g(x)$ in the form $g(x) = ax^3 + bx^2 + cx + d$.

4 The graph of $y = f(x)$ is shown alongside.
The x-axis is a tangent to $f(x)$ at $x = a$ and $f(x)$ cuts the x-axis at $x = b$.
On the same diagram, sketch the graph of $y = f(x - c)$ where $0 < c < b - a$.
Indicate the x-intercepts of $y = f(x - c)$.

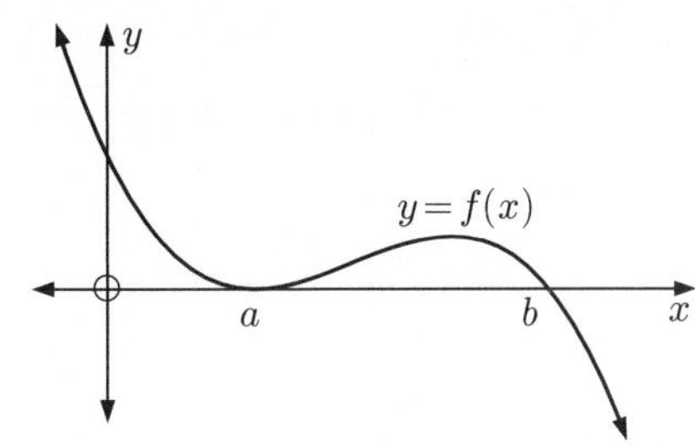

5 For the graph of $y = f(x)$ given, sketch graphs of:

a $y = f(-x)$ **b** $y = -f(x)$
c $y = f(x+2)$ **d** $y = f(x) + 2$

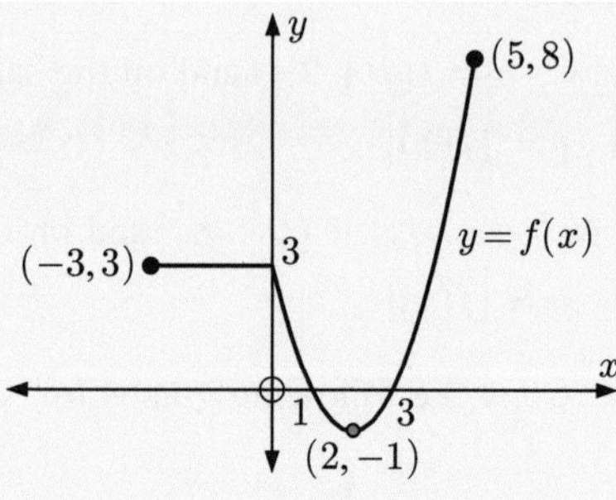

6 Consider the function $f : x \mapsto x^2$.
On the same set of axes graph:

a $y = f(x)$ **b** $y = f(x-1)$ **c** $y = 3f(x-1)$ **d** $y = 3f(x-1) + 2$

7 The graph of $y = f(x)$ is shown alongside.

a Sketch the graph of $y = g(x)$ where $g(x) = f(x+3) - 1$.

b State the equation of the vertical asymptote of $y = g(x)$.

c Identify the point A′ on the graph of $y = g(x)$ which corresponds to point A.

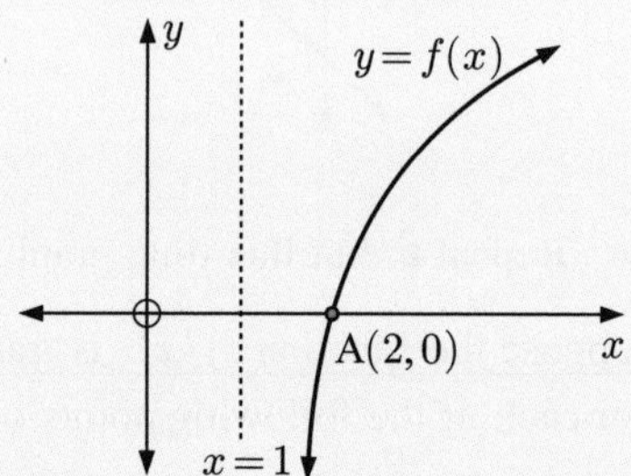

8 The graph of $y = f(x)$ is drawn alongside.

a Draw the graphs of $y = f(x)$ and $y = |f(x)|$ on the same set of axes.

b Find the y-intercept of $\dfrac{1}{f(x)}$.

c Show on the diagram the points that are invariant for the function $\dfrac{1}{f(x)}$.

d Draw the graphs of $y = f(x)$ and $y = \dfrac{1}{f(x)}$ on the same set of axes.

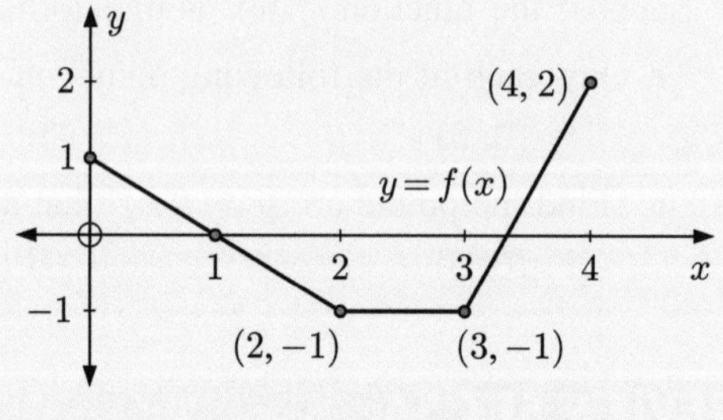

9 Let $f(x) = \dfrac{c}{x+c}$, $x \neq -c$, $c > 0$.

a On a set of axes like those shown, sketch the graph of $y = f(x)$. Label clearly any points of intersection with the axes and any asymptotes.

b On the same set of axes, sketch the graph of $y = \dfrac{1}{f(x)}$.
Label clearly any points of intersection with the axes.

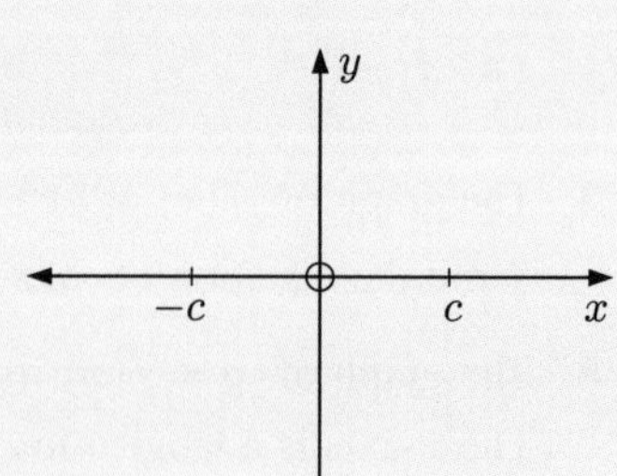

10 Consider $f(x) = x - a$ where a is a positive real number.

a Find expressions for $|f(x)|$ and $f(|x|)$.

b Sketch $y = |f(x)|$ and $y = f(|x|)$ on the same set of axes.

c Solve for x given a is a positive real number: $|x - a| = |x| - a$.

REVIEW SET 5B — CALCULATOR

1 Use your calculator to help graph $f(x) = (x+1)^2 - 4$. Include all axes intercepts, and the coordinates of the turning point of the function.

2 Consider the function $f : x \mapsto x^2$. On the same set of axes graph:

a $y = f(x)$ **b** $y = f(x+2)$ **c** $y = 2f(x+2)$ **d** $y = 2f(x+2) - 3$

3 Consider $f : x \mapsto \frac{2^x}{x}$.

a Does the function have any axes intercepts?

b Find the equations of the asymptotes of the function.

c Find any turning points of the function.

d Sketch the function for $-4 \leqslant x \leqslant 4$.

4 Consider $f(x) = 2^{-x}$.

a Use your calculator to help determine whether the following are true or false:

i As $x \to \infty$, $2^{-x} \to 0$. **ii** As $x \to -\infty$, $2^{-x} \to 0$.

iii The y-intercept is $\frac{1}{2}$. **iv** $2^{-x} > 0$ for all x.

b On the same set of axes, graph $y = f(x)$ and $y = |f(x)|$.

c Write down the equation of any asympotes of $y = |f(x)|$.

5 The graph of the function $f(x) = (x+1)^2 + 4$ is translated 2 units to the right and 4 units up.

a Find the function $g(x)$ corresponding to the translated graph.

b State the range of $f(x)$.

c State the range of $g(x)$.

6 For each of the following functions:

i Find $y = f(x)$, the result when the function is translated by $\begin{pmatrix} 1 \\ -2 \end{pmatrix}$.

ii Sketch the original function and its translated function on the same set of axes. Clearly state any asymptotes of each function.

iii State the domain and range of each function.

a $y = \frac{1}{x}$ **b** $y = 2^x$

7 Sketch the graph of $f(x) = x^2 + 1$, and on the same set of axes sketch the graphs of:

a $-f(x)$ **b** $f(2x)$ **c** $f(x) + 3$

8 Suppose $f(x) = x + 2$. The function F is obtained by stretching the function f vertically with scale factor 2, then stretching it horizontally with scale factor $\frac{1}{2}$, then translating it $\frac{1}{2}$ horizontally and -3 vertically.

a Find the function $F(x)$.

b What can be said about the point $(1, 3)$ under this transformation?

c What happens to the points $(0, 2)$ and $(-1, 1)$ under this transformation?

d Show that the points in **c** also lie on the graph of $y = F(x)$.

9 The graph of $y = f(x)$ is given.
On the same set of axes graph each pair of functions:

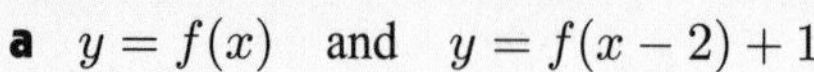

a $y = f(x)$ and $y = f(x-2)+1$

b $y = f(x)$ and $y = \dfrac{1}{f(x)}$

c $y = f(x)$ and $y = |f(x)|$

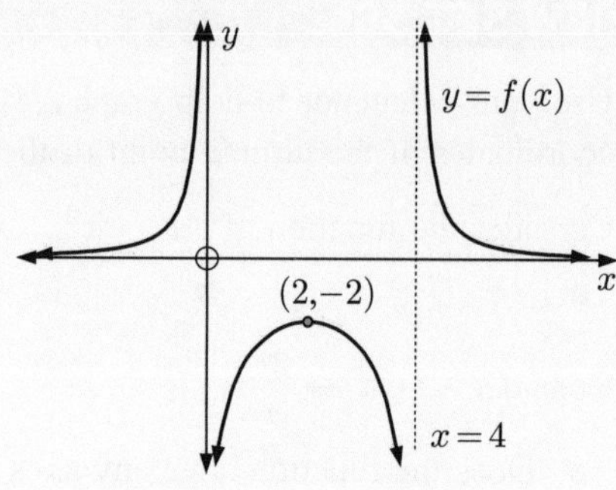

10 Consider the function $f(x) = \dfrac{2x-3}{3x+5}$.

a Find the asymptotes of $y = f(x)$.

b Discuss the behaviour of the graph near these asymptotes.

c Find the axes intercepts of $y = f(x)$.

d Sketch the graph of $y = f(x)$.

e Describe the transformations which transform $y = \dfrac{1}{x}$ into $y = f(x)$.

f Describe the transformations which transform $y = f(x)$ into $y = \dfrac{1}{x}$.

11 **a** Sketch the graph of $f(x) = -2x + 3$, clearly showing the axes intercepts.

b Find the invariant points for the graph of $y = \dfrac{1}{f(x)}$.

c State the equation of the vertical asymptote of $y = \dfrac{1}{f(x)}$ and find its y-intercept.

d Sketch the graph of $y = \dfrac{1}{f(x)}$ on the same axes as in part **a**, showing clearly the information you have found.

e On a new pair of axes, sketch the graphs of $y = |f(x)|$ and $y = f(|x|)$ showing clearly all important features.

REVIEW SET 5C

1 If $f(x) = \dfrac{4}{x}$, find in simplest form:

a $f(-4)$ **b** $f(2x)$ **c** $f\left(\dfrac{x}{2}\right)$ **d** $4f(x+2)-3$

2 Consider the graph of $y = f(x)$ shown.

a Use the graph to determine:

i the coordinates of the turning point

ii the equation of the vertical asymptote

iii the equation of the horizontal asymptote

iv the x-intercepts.

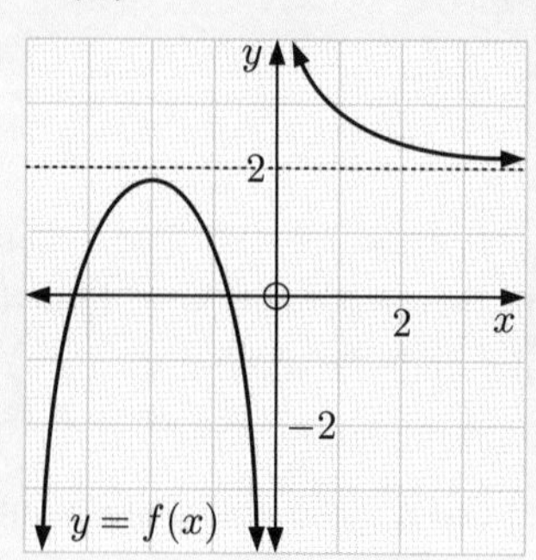

PRINTABLE GRAPH

b Graph the function $g : x \mapsto x+1$ on the same set of axes.

c Hence estimate the coordinates of the points of intersection of $y = f(x)$ and $y = g(x)$.

3 Sketch the graph of $f(x) = -x^2$, and on the same set of axes sketch the graph of:

a $y = f(-x)$ **b** $y = -f(x)$ **c** $y = f(2x)$ **d** $y = f(x-2)$

4 The graph of a cubic function $y = f(x)$ is shown alongside.

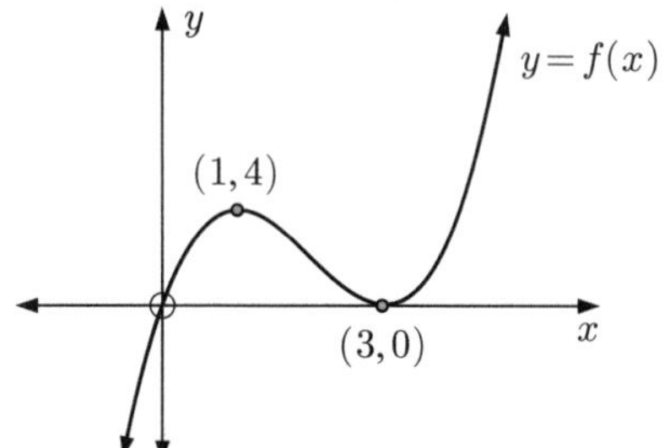

a Sketch the graph of $g(x) = -f(x-1)$.

b State the coordinates of the turning points of $y = g(x)$.

5 The graph of $f(x) = x^2$ is transformed to the graph of $g(x)$ by a reflection and a translation as illustrated. Find the formula for $g(x)$ in the form $g(x) = ax^2 + bx + c$.

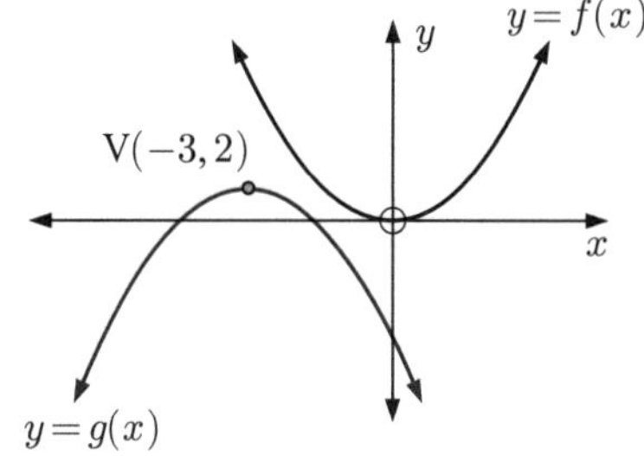

6 Given the graph of $y = f(x)$, sketch graphs of:

a $f(-x)$ **b** $f(x+1)$

c $f(x) - 3$

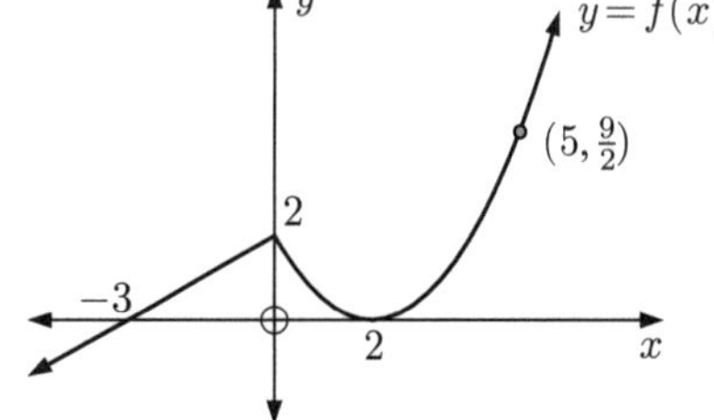

7 The graph of $f(x) = x^3 + 3x^2 - x + 4$ is translated to its image $y = g(x)$ by the vector $\binom{-1}{3}$. Write the equation of $g(x)$ in the form $g(x) = ax^3 + bx^2 + cx + d$.

8 **a** Find the equation of the line that results when the line $f(x) = 3x + 2$ is translated:

i 2 units to the left **ii** 6 units upwards.

b Show that when the linear function $f(x) = ax + b$, $a > 0$ is translated k units to the left, the resulting line is the same as when $f(x)$ is translated ka units upwards.

9 The function $f(x)$ results from transforming the function $y = \frac{1}{x}$ by a reflection in the y-axis, then a vertical stretch with scale factor 3, then a translation of $\binom{1}{1}$.

a Find an expression for $f(x)$.

b Sketch $y = f(x)$ and state its domain and range.

c Does $y = f(x)$ have an inverse function? Explain your answer.

d Is the function f a self-inverse function? Give graphical and algebraic evidence to support your answer.

10 Consider $y = \log_4 x$.

a Find the function which results from a translation of $\binom{1}{-2}$.

b Sketch the original function and the translated function on the same set of axes.

c State the asymptotes of each function.

d State the domain and range of each function.

11 The function $g(x)$ results when $y = \frac{1}{x}$ is transformed by a vertical stretch with scale factor $\frac{1}{3}$, followed by a reflection in the y-axis, followed by a translation of 2 units to the right.

a Write an expression for $g(x)$ in the form $g(x) = \frac{ax+b}{cx+d}$.

b Find the asymptotes of $y = g(x)$.

c State the domain and range of $g(x)$.

d Sketch $y = g(x)$.

Chapter 6

Complex numbers and polynomials

Syllabus reference: 1.5, 1.8, 2.5, 2.6

Contents:

A REAL QUADRATICS WITH $\Delta < 0$

In **Chapter 1**, we determined that:

If $ax^2 + bx + c = 0$, $a \neq 0$ and $a, b, c \in \mathbb{R}$, then the solutions or roots are found using the formula $x = \dfrac{-b \pm \sqrt{\Delta}}{2a}$ where $\Delta = b^2 - 4ac$ is known as the **discriminant**.

We also observed that if:

- $\Delta > 0$ we have two real *distinct* solutions
- $\Delta = 0$ we have two real *identical* solutions
- $\Delta < 0$ we have no real solutions.

However, it is in fact possible to write down two solutions for the case where $\Delta < 0$. To do this we need **imaginary numbers**.

In 1572, Rafael Bombelli defined the imaginary number $i = \sqrt{-1}$. It is called 'imaginary' because we cannot place it on a number line. With i defined, we can write down solutions for quadratic equations with $\Delta < 0$. They are called *complex* solutions because they include a real and an imaginary part.

Any number of the form $a + bi$ where a and b are real and $i = \sqrt{-1}$, is called a **complex number**.

Example 1 Self Tutor

Solve the quadratic equations: **a** $x^2 = -4$ **b** $z^2 + z + 2 = 0$

a $x^2 = -4$

$\therefore\ x = \pm\sqrt{-4}$

$\therefore\ x = \pm\sqrt{4}\sqrt{-1}$

$\therefore\ x = \pm 2i$

b $z^2 + z + 2$ has $a = 1,\ b = 1,\ c = 2$

$\therefore\ z = \dfrac{-1 \pm \sqrt{1^2 - 4(1)(2)}}{2(1)}$

$\therefore\ z = \dfrac{-1 \pm \sqrt{-7}}{2}$

$\therefore\ z = \dfrac{-1 \pm i\sqrt{7}}{2} = -\frac{1}{2} \pm \frac{\sqrt{7}}{2}i$

If the coefficient of i is a square root, we write the i first.

In **Example 1** above, notice that $\Delta < 0$ in both cases. In each case we have found two complex solutions of the form $a + bi$, where a and b are real.

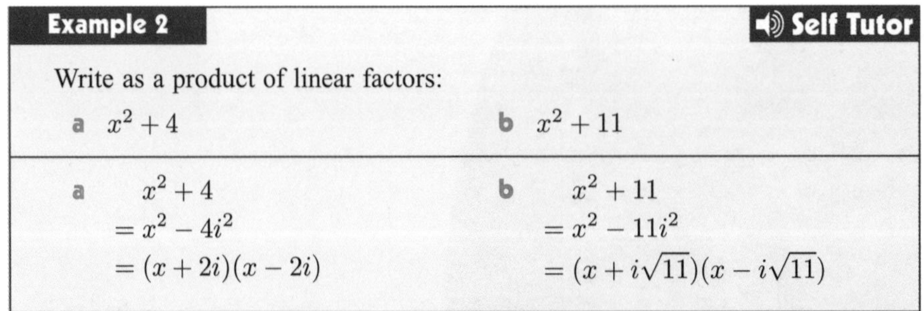

Example 2 Self Tutor

Write as a product of linear factors:

a $x^2 + 4$ **b** $x^2 + 11$

a $x^2 + 4$

$= x^2 - 4i^2$

$= (x + 2i)(x - 2i)$

b $x^2 + 11$

$= x^2 - 11i^2$

$= (x + i\sqrt{11})(x - i\sqrt{11})$

Example 3 Self Tutor

Solve for x:

a $x^2+9=0$ b $x^3+2x=0$

a
$$x^2+9=0$$
$$\therefore\ x^2-9i^2=0$$
$$\therefore\ (x+3i)(x-3i)=0$$
$$\therefore\ x=\pm 3i$$

b
$$x^3+2x=0$$
$$\therefore\ x(x^2+2)=0$$
$$\therefore\ x(x^2-2i^2)=0$$
$$\therefore\ x(x+i\sqrt{2})(x-i\sqrt{2})=0$$
$$\therefore\ x=0 \text{ or } \pm i\sqrt{2}$$

Example 4 Self Tutor

Solve for x:

a $x^2-4x+13=0$ b $x^4+x^2=6$

a
$$x^2-4x+13=0$$
$$\therefore\ x=\frac{4\pm\sqrt{16-4(1)(13)}}{2}$$
$$\therefore\ x=\frac{4\pm\sqrt{-36}}{2}$$
$$\therefore\ x=\frac{4\pm 6i}{2}$$
$$\therefore\ x=2+3i \text{ or } 2-3i$$

b
$$x^4+x^2=6$$
$$\therefore\ x^4+x^2-6=0$$
$$\therefore\ (x^2+3)(x^2-2)=0$$
$$\therefore\ (x+i\sqrt{3})(x-i\sqrt{3})(x+\sqrt{2})(x-\sqrt{2})=0$$
$$\therefore\ x=\pm i\sqrt{3} \text{ or } \pm\sqrt{2}$$

EXERCISE 6A

1 Write in terms of i:

a $\sqrt{-9}$ b $\sqrt{-64}$ c $\sqrt{-\frac{1}{4}}$ d $\sqrt{-5}$ e $\sqrt{-8}$

2 Write as a product of linear factors:

a x^2-9 b x^2+9 c x^2-7 d x^2+7
e $4x^2-1$ f $4x^2+1$ g $2x^2-9$ h $2x^2+9$
i x^3-x j x^3+x k x^4-1 l x^4-16

3 Solve for x:

a $x^2-25=0$ b $x^2+25=0$ c $x^2-5=0$ d $x^2+5=0$
e $4x^2-9=0$ f $4x^2+9=0$ g $x^3-4x=0$ h $x^3+4x=0$
i $x^3-3x=0$ j $x^3+3x=0$ k $x^4-1=0$ l $x^4=81$

4 Solve for x:

a $x^2-10x+29=0$ b $x^2+6x+25=0$ c $x^2+14x+50=0$
d $2x^2+5=6x$ e $x^2-2\sqrt{3}x+4=0$ f $2x+\dfrac{1}{x}=1$

5 Solve for x:

a $x^4+2x^2=3$ b $x^4=x^2+6$ c $x^4+5x^2=36$
d $x^4+9x^2+14=0$ e $x^4+1=2x^2$ f $x^4+2x^2+1=0$

B COMPLEX NUMBERS

Any number of the form $a + bi$ where $a, b \in \mathbb{R}$ and $i = \sqrt{-1}$, is called a **complex number**.

Notice that all real numbers are complex numbers in the special case where $b = 0$.

A complex number of the form bi where $b \in \mathbb{R}$, $b \neq 0$, is called an imaginary number or **purely imaginary**.

THE 'SUM OF TWO SQUARES'

Notice that
$$\begin{aligned} a^2 + b^2 &= a^2 - b^2 i^2 \qquad \{\text{as } i^2 = -1\} \\ &= (a + bi)(a - bi) \end{aligned}$$

Compare:
$$\begin{aligned} a^2 - b^2 &= (a + b)(a - b) \qquad \{\text{the difference of two squares factorisation}\} \\ a^2 + b^2 &= (a + bi)(a - bi) \qquad \{\text{the sum of two squares factorisation}\} \end{aligned}$$

If we write $z = a + bi$ where $a, b \in \mathbb{R}$, then:

- a is the **real part** of z and we write $\boldsymbol{a = \mathcal{R}e\,(z)}$
- b is the **imaginary part** of z and we write $\boldsymbol{b = \mathcal{I}m\,(z)}$.

For example: If $z = 2 + 3i$, then $\mathcal{R}e\,(z) = 2$ and $\mathcal{I}m\,(z) = 3$.
If $z = -\sqrt{2}i$, then $\mathcal{R}e\,(z) = 0$ and $\mathcal{I}m\,(z) = -\sqrt{2}$.

OPERATIONS WITH COMPLEX NUMBERS

Operations with complex numbers are identical to those with radicals, but with $i^2 = -1$ rather than $(\sqrt{2})^2 = 2$ or $(\sqrt{3})^2 = 3$.

For example:

- addition:
$$\begin{aligned} (2 + \sqrt{3}) + (4 + 2\sqrt{3}) &= (2 + 4) + (1 + 2)\sqrt{3} = 6 + 3\sqrt{3} \\ (2 + i) + (4 + 2i) &= (2 + 4) + (1 + 2)i = 6 + 3i \end{aligned}$$
- multiplication:
$$\begin{aligned} (2 + \sqrt{3})(4 + 2\sqrt{3}) &= 8 + 4\sqrt{3} + 4\sqrt{3} + 2(\sqrt{3})^2 = 8 + 8\sqrt{3} + 6 \\ (2 + i)(4 + 2i) &= 8 + 4i + 4i + 2i^2 = 8 + 8i - 2 \end{aligned}$$

So, we can **add**, **subtract**, **multiply**, and **divide** complex numbers in the same way we perform these operations with radicals:

$$(a + bi) + (c + di) = (a + c) + (b + d)i \qquad \textbf{addition}$$
$$(a + bi) - (c + di) = (a - c) + (b - d)i \qquad \textbf{subtraction}$$
$$(a + bi)(c + di) = ac + adi + bci + bdi^2 \qquad \textbf{multiplication}$$
$$\frac{a + bi}{c + di} = \left(\frac{a + bi}{c + di}\right)\left(\frac{c - di}{c - di}\right) = \frac{ac - adi + bci - bdi^2}{c^2 + d^2} \qquad \textbf{division}$$

Notice that in the division process, we use a multiplication technique to obtain a real number in the denominator.

Example 5 **Self Tutor**

If $z = 3 + 2i$ and $w = 4 - i$ find:

a $z + w$ **b** $z - w$ **c** zw **d** $\dfrac{z}{w}$

a $z + w$
$= (3 + 2i) + (4 - i)$
$= 7 + i$

b $z - w$
$= (3 + 2i) - (4 - i)$
$= 3 + 2i - 4 + i$
$= -1 + 3i$

c zw
$= (3 + 2i)(4 - i)$
$= 12 - 3i + 8i - 2i^2$
$= 12 + 5i + 2$
$= 14 + 5i$

d $\dfrac{z}{w} = \dfrac{3 + 2i}{4 - i}$
$= \left(\dfrac{3 + 2i}{4 - i}\right)\left(\dfrac{4 + i}{4 + i}\right)$
$= \dfrac{12 + 3i + 8i + 2i^2}{16 - i^2}$
$= \dfrac{10 + 11i}{17}$
$= \frac{10}{17} + \frac{11}{17}i$

You can use your calculator to perform operations with complex numbers.

After solving the questions in the following exercise by hand, check your answers using technology.

EXERCISE 6B.1

1 Copy and complete:

z	$\mathcal{R}e\,(z)$	$\mathcal{I}m\,(z)$	z	$\mathcal{R}e\,(z)$	$\mathcal{I}m\,(z)$
$3 + 2i$			$-3 + 4i$		
$5 - i$			$-7 - 2i$		
3			$-11i$		
0			$i\sqrt{3}$		

2 If $z = 5 - 2i$ and $w = 2 + i$, find in simplest form:

a $z + w$ **b** $2z$ **c** iw **d** $z - w$
e $2z - 3w$ **f** zw **g** w^2 **h** z^2

3 For $z = 1 + i$ and $w = -2 + 3i$, find in simplest form:

a $z + 2w$ **b** z^2 **c** z^3 **d** iz
e w^2 **f** zw **g** z^2w **h** izw

4 **a** Simplify i^n for $n = 0, 1, 2, 3, 4, 5, 6, 7, 8, 9$ and also for $n = -1, -2, -3, -4, -5$.
b Hence, simplify i^{4n+3} where n is any integer.

5 Write $(1 + i)^4$ in simplest form. Hence, find $(1 + i)^{101}$ in simplest form.

6 Suppose $z = 2 - i$ and $w = 1 + 3i$. Write in exact form $a + bi$ where $a, b \in \mathbb{R}$:

a $\dfrac{z}{w}$ **b** $\dfrac{i}{z}$ **c** $\dfrac{w}{iz}$ **d** z^{-2}

7 Simplify: **a** $\dfrac{i}{1-2i}$ **b** $\dfrac{i(2-i)}{3-2i}$ **c** $\dfrac{1}{2-i}-\dfrac{2}{2+i}$

8 If $z=2+i$ and $w=-1+2i$, find:

a $\mathcal{Im}(4z-3w)$ **b** $\mathcal{Re}(zw)$ **c** $\mathcal{Im}(iz^2)$ **d** $\mathcal{Re}\left(\dfrac{z}{w}\right)$

EQUALITY OF COMPLEX NUMBERS

Two complex numbers are **equal** when their **real parts** are equal and their **imaginary parts** are equal.

$$a+bi=c+di \quad\Leftrightarrow\quad a=c \text{ and } b=d.$$

Proof: Suppose $b\neq d$. Now if $a+bi=c+di$ where a, b, c, and d are real,

then $bi-di=c-a$

$\therefore\ i(b-d)=c-a$

$\therefore\ i=\dfrac{c-a}{b-d}$ {as $b\neq d$}

This is false as the RHS is real but the LHS is imaginary.

Thus, the supposition is false. Hence $b=d$ and furthermore $a=c$.

For the complex number $a+bi$, where a and b are real, $a+bi=0 \Leftrightarrow a=0$ and $b=0$.

Example 6 — Self Tutor

Find real numbers x and y such that:

a $(x+yi)(2-i)=-i$ **b** $(x+2i)(1-i)=5+yi$

a $(x+yi)(2-i)=-i$

$$\begin{aligned}\therefore\quad x+yi&=\frac{-i}{2-i}\\&=\left(\frac{-i}{2-i}\right)\left(\frac{2+i}{2+i}\right)\\&=\frac{-2i-i^2}{4+1}\\&=\frac{1-2i}{5}\\&=\tfrac{1}{5}-\tfrac{2}{5}i\end{aligned}$$

Equating real and imaginary parts, $x=\frac{1}{5}$ and $y=-\frac{2}{5}$.

b

$$\begin{aligned}(x+2i)(1-i)&=5+yi\\ \therefore\quad x-xi+2i+2&=5+yi\\ \therefore\quad (x+2)+(2-x)i&=5+yi\end{aligned}$$

Equating real and imaginary parts,

$\therefore\ x+2=5$ and $2-x=y$

$\therefore\ x=3$ and $y=-1$

EXERCISE 6B.2

1 Find exact real numbers x and y such that:

a $2x+3yi=-x-6i$ **b** $x^2+xi=4-2i$

c $(x+yi)(2-i)=8+i$ **d** $(3+2i)(x+yi)=-i$

2 Find exact $x,\,y\in\mathbb{R}$ such that:

a $2(x+yi)=x-yi$ **b** $(x+2i)(y-i)=-4-7i$

c $(x+i)(3-iy)=1+13i$ **d** $(x+yi)(2+i)=2x-(y+1)i$

3 The complex number z satisfies the equation $3z + 17i = iz + 11$. Write z in the form $a + bi$ where $a, b \in \mathbb{R}$ and $i = \sqrt{-1}$.

4 Express z in the form $a + bi$ where $a, b \in \mathbb{Z}$, if $z = \left(\frac{4}{1+i} + 7 - 2i\right)^2$.

5 Find the real values of m and n for which $3(m + ni) = n - 2mi - (1 - 2i)$.

6 Express $z = \frac{3i}{\sqrt{2} - i} + 1$ in the form $a + bi$ where $a, b \in \mathbb{R}$ are given exactly.

7 Suppose $(a + bi)^2 = -16 - 30i$ where $a, b \in \mathbb{R}$ and $a > 0$.
Find the possible values of a and b.

HISTORICAL NOTE

18th century mathematicians enjoyed playing with these new 'imaginary' numbers, but they were regarded as little more than interesting curiosities until the work of **Gauss** (1777 - 1855), the German mathematician, astronomer, and physicist.

Carl Friedrich Gauss

For centuries mathematicians had attempted to find a method of trisecting an angle using a compass and straight edge. Gauss put an end to this when he used complex numbers to prove the impossibility of such a construction. By his systematic use of complex numbers, he was able to convince mathematicians of their usefulness.

Early last century, the American engineer **Steinmetz** used complex numbers to solve electrical problems, illustrating that complex numbers did have a practical application.

Complex numbers are now used extensively in electronics, engineering, and physics.

COMPLEX CONJUGATES

Complex numbers $a + bi$ and $a - bi$ are called **complex conjugates**.
If $z = a + bi$ we write its conjugate as $z^* = a - bi$.

We saw on page **176** that the complex conjugate is important for division:

$\frac{z}{w} = \frac{z}{w}\frac{w^*}{w^*} = \frac{zw^*}{ww^*}$ which makes the denominator real.

Quadratics with real coefficients are called **real quadratics**. This does not necessarily mean that their zeros are real.

- If a quadratic equation has **rational coefficients** and an **irrational root** of the form $c + d\sqrt{n}$, then the **conjugate** $c - d\sqrt{n}$ is also a root of the quadratic equation.
- If a real quadratic equation has $\Delta < 0$ and $c + di$ is a complex root, then the **complex conjugate** $c - di$ is also a root.

For example:

- $x^2 - 2x + 5 = 0$ has $\Delta = (-2)^2 - 4(1)(5) = -16$
  and the solutions are $x = 1 + 2i$ and $1 - 2i$
- $x^2 + 4 = 0$ has $\Delta = 0^2 - 4(1)(4) = -16$
  and the solutions are $x = 2i$ and $-2i$

Theorem: If $c+di$ and $c-di$ are roots of a quadratic equation, then the quadratic equation is $a(x^2-2cx+(c^2+d^2))=0$ for some constant $a\neq 0$.

Proof: The sum of the roots $=2c$ and the product $=(c+di)(c-di)=c^2+d^2$

$\therefore\ x^2-(\text{sum})x+(\text{product})=0$

$\therefore\ x^2-2cx+(c^2+d^2)=0$

In general, $a(x^2-2cx+c^2+d^2)=0$ for some constant $a\neq 0$.

Notice that:

- the sum of complex conjugates $c+di$ and $c-di$ is $2c$ which is **real**
- the product is $(c+di)(c-di)=c^2+d^2$ which is also **real**.

Example 7 Self Tutor

Find all real quadratic equations having $1-2i$ as a root.

As $1-2i$ is a root, $1+2i$ is also a root.

The sum of the roots $=1-2i+1+2i$
$=2$

The product of the roots $=(1-2i)(1+2i)$
$=1+4$
$=5$

So, as $x^2-(\text{sum})x+(\text{product})=0$,
$a(x^2-2x+5)=0,\ a\neq 0$ gives all possible equations.

Example 8 Self Tutor

Find exact values for a and b if $\sqrt{2}+i$ is a root of $x^2+ax+b=0,\ a, b\in\mathbb{R}$.

Since a and b are real, the quadratic has real coefficients.

$\therefore\ \sqrt{2}-i$ is also a root.

The sum of the roots $=\sqrt{2}+i+\sqrt{2}-i=2\sqrt{2}$.

The product of the roots $=(\sqrt{2}+i)(\sqrt{2}-i)=2+1=3$.

Thus $a=-2\sqrt{2}$ and $b=3$.

EXERCISE 6B.3

1 Find all quadratic equations with real coefficients and roots of:

a $3\pm i$ b $1\pm 3i$ c $-2\pm 5i$ d $\sqrt{2}\pm i$

e $2\pm\sqrt{3}$ f 0 and $-\frac{2}{3}$ g $\pm i\sqrt{2}$ h $-6\pm i$

2 Find exact values for a and b if:

a $3+i$ is a root of $x^2+ax+b=0$, where a and b are real

b $1-\sqrt{2}$ is a root of $x^2+ax+b=0$, where a and b are rational

c $a+ai$ is a root of $x^2+4x+b=0$, where a and b are real. [Careful!]

PROPERTIES OF CONJUGATES

INVESTIGATION 1 — PROPERTIES OF CONJUGATES

The purpose of this investigation is to discover any properties that complex conjugates might have.

What to do:

1 Given $z_1 = 1 - i$ and $z_2 = 2 + i$ find:

a z_1^*	**b** z_2^*	**c** $(z_1^*)^*$	**d** $(z_2^*)^*$
e $(z_1 + z_2)^*$	**f** $z_1^* + z_2^*$	**g** $(z_1 - z_2)^*$	**h** $z_1^* - z_2^*$
i $(z_1 z_2)^*$	**j** $z_1^* z_2^*$	**k** $\left(\dfrac{z_1}{z_2}\right)^*$	**l** $\dfrac{z_1^*}{z_2^*}$
m $(z_1^2)^*$	**n** $(z_1^*)^2$	**o** $(z_2^3)^*$	**p** $(z_2^*)^3$

2 Repeat **1** with z_1 and z_2 of your choice.

3 Examine your results from **1** and **2**, and hence suggest some rules for complex conjugates.

From the **Investigation** you should have discovered the following rules for complex conjugates:

- $(z^*)^* = z$
- $(z_1 + z_2)^* = z_1^* + z_2^*$ and $(z_1 - z_2)^* = z_1^* - z_2^*$
- $(z_1 z_2)^* = z_1^* \times z_2^*$ and $\left(\dfrac{z_1}{z_2}\right)^* = \dfrac{z_1^*}{z_2^*}$, $z_2 \neq 0$
- $(z^n)^* = (z^*)^n$ for positive integers n
- $z + z^*$ and zz^* are real.

Example 9 — Self Tutor

Show that for all complex numbers z_1 and z_2:

a $(z_1 + z_2)^* = z_1^* + z_2^*$ **b** $(z_1 z_2)^* = z_1^* \times z_2^*$

a Let $z_1 = a + bi$ and $z_2 = c + di$

$\therefore$ $z_1^* = a - bi$ and $z_2^* = c - di$

Now $z_1 + z_2 = (a + c) + (b + d)i$

$$\begin{aligned}\therefore\ (z_1 + z_2)^* &= (a + c) - (b + d)i \\ &= a + c - bi - di \\ &= a - bi + c - di \\ &= z_1^* + z_2^*\end{aligned}$$

b Let $z_1 = a + bi$ and $z_2 = c + di$

$$\begin{aligned}\therefore\ z_1 z_2 &= (a + bi)(c + di) \\ &= ac + adi + bci + bdi^2 \\ &= (ac - bd) + i(ad + bc)\end{aligned}$$

$\therefore$ $(z_1 z_2)^* = (ac - bd) - i(ad + bc)$ .... (1)

Also, $z_1^* \times z_2^*$

$$\begin{aligned}&= (a - bi)(c - di) \\ &= ac - adi - bci + bdi^2 \\ &= (ac - bd) - i(ad + bc) \quad .... (2)\end{aligned}$$

From (1) and (2), $(z_1 z_2)^* = z_1^* \times z_2^*$

CONJUGATE GENERALISATIONS

Notice that $(z_1+z_2+z_3)^* = (z_1+z_2)^* + {z_3}^*$ {treating z_1+z_2 as one complex number}

$= {z_1}^* + {z_2}^* + {z_3}^*$ .... (1)

Likewise $(z_1+z_2+z_3+z_4)^* = (z_1+z_2+z_3)^* + {z_4}^*$

$= {z_1}^* + {z_2}^* + {z_3}^* + {z_4}^*$ {from (1)}

There is no reason why this process cannot continue for the conjugate of 5, 6, 7, ... complex numbers. We can therefore generalise the result:

$$(z_1+z_2+z_3+....+z_n)^* = {z_1}^* + {z_2}^* + {z_3}^* + ... + {z_n}^*$$

The process of proving the general case using the simpler cases when $n = 1, 2, 3, 4, ...$ requires **mathematical induction**. Formal proof by the Principle of Mathematical Induction is discussed in **Chapter 9**.

EXERCISE 6B.4

1 Show that $(z_1-z_2)^* = {z_1}^* - {z_2}^*$ for all complex numbers z_1 and z_2.

2 Simplify the expression $(w^*-z)^* - (w-2z^*)$ using the properties of conjugates.

3 It is known that a complex number z satisfies the equation $z^* = -z$.
Show that z is either purely imaginary or zero.

4 Suppose $z_1 = a+bi$ and $z_2 = c+di$ are complex numbers.

a Find $\frac{z_1}{z_2}$ in the form $X+iY$. **b** Show that $\left(\frac{z_1}{z_2}\right)^* = \frac{{z_1}^*}{{z_2}^*}$ for all z_1 and $z_2 \neq 0$.

5 **a** An easier way of proving $\left(\frac{z_1}{z_2}\right)^* = \frac{{z_1}^*}{{z_2}^*}$ is to start with $\left(\frac{z_1}{z_2}\right)^* \times {z_2}^*$.

Show how this can be done, remembering we have already proved that "the conjugate of a product is the product of the conjugates" in **Example 9**.

b Let $z = a+bi$ be a complex number. Prove the following:

i If $z = z^*$, then z is real. **ii** If $z^* = -z$, then z is purely imaginary or zero.

6 Prove that for all complex numbers z and w:

a $zw^* + z^*w$ is always real **b** $zw^* - z^*w$ is purely imaginary or zero.

7 **a** If $z = a+bi$, find z^2 in the form $X+iY$.

b Hence, show that $(z^2)^* = (z^*)^2$ for all complex numbers z.

c Repeat **a** and **b** but for z^3 instead of z^2.

8 Suppose $w = \frac{z-1}{z^*+1}$ where $z = a+bi$. Find the conditions under which:

a w is real **b** w is purely imaginary.

9 **a** Assuming $(z_1z_2)^* = {z_1}^*{z_2}^*$, explain why $(z_1z_2z_3)^* = {z_1}^*{z_2}^*{z_3}^*$.

b Hence show that $(z_1z_2z_3z_4)^* = {z_1}^*{z_2}^*{z_3}^*{z_4}^*$.

c Generalise your results from **a** and **b**.

d Given your generalisation in **c**, what is the result of letting all z_i values be equal to z?

C REAL POLYNOMIALS

Up to this point we have studied linear and quadratic functions at some depth, with perhaps occasional reference to cubic functions. These are part of a larger family of functions called the **polynomials**.

A **polynomial function** is a function of the form

$$P(x) = a_n x^n + a_{n-1}x^{n-1} + ... + a_2x^2 + a_1x + a_0, \quad a_1, ..., a_n \text{ constant}, \quad a_n \neq 0.$$

We say that:
- x is the **variable**
- a_0 is the **constant term**
- a_n is the **leading coefficient** and is non-zero
- a_r is the **coefficient of x^r** for $r = 0, 1, 2, ..., n$
- n is the **degree** of the polynomial, being the highest power of the variable.

In **summation notation**, we write $P(x) = \sum\limits_{r=0}^{n} a_r x^r$,

which reads: "the sum from $r = 0$ to n, of $a_r x^r$".

A **real polynomial** $P(x)$ is a polynomial for which $a_r \in \mathbb{R}$, $r = 0, 1, 2, ..., n$.

The low degree members of the polynomial family have special names, some of which you are already familiar with. For these polynomials, we commonly write their coefficients as $a, b, c, ...$

Polynomial function	*Degree*	*Name*
$ax + b, \quad a \neq 0$	1	linear
$ax^2 + bx + c, \quad a \neq 0$	2	quadratic
$ax^3 + bx^2 + cx + d, \quad a \neq 0$	3	cubic
$ax^4 + bx^3 + cx^2 + dx + e, \quad a \neq 0$	4	quartic

ADDITION AND SUBTRACTION

To **add** or **subtract** two polynomials, we collect 'like' terms.

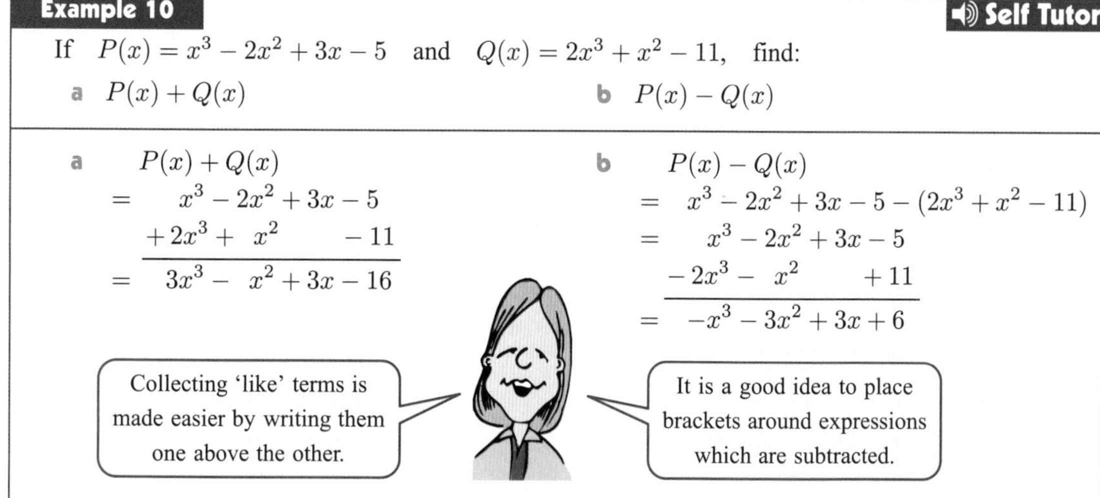

Collecting 'like' terms is made easier by writing them one above the other.
It is a good idea to place brackets around expressions which are subtracted.

Example 10 Self Tutor

If $P(x) = x^3 - 2x^2 + 3x - 5$ and $Q(x) = 2x^3 + x^2 - 11$, find:

a $P(x) + Q(x)$ **b** $P(x) - Q(x)$

a $P(x) + Q(x)$

$$\begin{array}{rl} = & x^3 - 2x^2 + 3x - 5 \\ & +2x^3 + x^2 \quad\quad - 11 \\ \hline = & 3x^3 - x^2 + 3x - 16 \end{array}$$

b $P(x) - Q(x)$

$$\begin{array}{rl} = & x^3 - 2x^2 + 3x - 5 - (2x^3 + x^2 - 11) \\ = & x^3 - 2x^2 + 3x - 5 \\ & -2x^3 - x^2 \quad\quad + 11 \\ \hline = & -x^3 - 3x^2 + 3x + 6 \end{array}$$

SCALAR MULTIPLICATION

To **multiply** a polynomial by a **scalar** (constant) we multiply each term by the scalar.

Example 11 Self Tutor

If $P(x) = x^4 - 2x^3 + 4x + 7$ find: **a** $3P(x)$ **b** $-2P(x)$

a $3P(x)$
$= 3(x^4 - 2x^3 + 4x + 7)$
$= 3x^4 - 6x^3 + 12x + 21$

b $-2P(x)$
$= -2(x^4 - 2x^3 + 4x + 7)$
$= -2x^4 + 4x^3 - 8x - 14$

POLYNOMIAL MULTIPLICATION

To **multiply** two polynomials, we multiply each term of the first polynomial by each term of the second polynomial, and then collect like terms.

Example 12 Self Tutor

If $P(x) = x^3 - 2x + 4$ and $Q(x) = 2x^2 + 3x - 5$, find $P(x)Q(x)$.

$$\begin{aligned}P(x)Q(x) &= (x^3 - 2x + 4)(2x^2 + 3x - 5)\\ &= x^3(2x^2 + 3x - 5) - 2x(2x^2 + 3x - 5) + 4(2x^2 + 3x - 5)\\ &= 2x^5 + 3x^4 - 5x^3\\ &\qquad\quad - 4x^3 - 6x^2 + 10x\\ &\qquad\qquad\quad + 8x^2 + 12x - 20\\ \hline &= 2x^5 + 3x^4 - 9x^3 + 2x^2 + 22x - 20\end{aligned}$$

SYNTHETIC MULTIPLICATION (OPTIONAL)

Polynomial multiplication can be performed using the coefficients only. We call this **synthetic multiplication**.

For example, for $(x^3 + 2x - 5)(2x + 3)$ we detach coefficients and multiply. It is different from the ordinary multiplication of large numbers because we sometimes have negative coefficients, and because we do not carry tens into the next column.

	1	0	2	−5	⟵ coefficients of $x^3 + 2x - 5$
			× 2	3	⟵ coefficients of $2x + 3$
	3	0	6	−15	
2	0	4	−10		
2	3	4	−4	−15	
x^4	x^3	x^2	x	constants	

So $(x^3 + 2x - 5)(2x + 3)$
$= 2x^4 + 3x^3 + 4x^2 - 4x - 15.$

EXERCISE 6C.1

1. If $P(x) = x^2 + 2x + 3$ and $Q(x) = 4x^2 + 5x + 6$, find in simplest form:
   a $3P(x)$ b $P(x) + Q(x)$ c $P(x) - 2Q(x)$ d $P(x)Q(x)$

2. If $f(x) = x^2 - x + 2$ and $g(x) = x^3 - 3x + 5$, find in simplest form:
   a $f(x) + g(x)$ b $g(x) - f(x)$ c $2f(x) + 3g(x)$
   d $g(x) + xf(x)$ e $f(x)\,g(x)$ f $[f(x)]^2$

3. Expand and simplify:
   a $(x^2 - 2x + 3)(2x + 1)$ b $(x - 1)^2(x^2 + 3x - 2)$ c $(x + 2)^3$
   d $(2x^2 - x + 3)^2$ e $(2x - 1)^4$ f $(3x - 2)^2(2x + 1)(x - 4)$

4. Find the following products:
   a $(2x^2 - 3x + 5)(3x - 1)$ b $(4x^2 - x + 2)(2x + 5)$
   c $(2x^2 + 3x + 2)(5 - x)$ d $(x - 2)^2(2x + 1)$
   e $(x^2 - 3x + 2)(2x^2 + 4x - 1)$ f $(3x^2 - x + 2)(5x^2 + 2x - 3)$
   g $(x^2 - x + 3)^2$ h $(2x^2 + x - 4)^2$
   i $(2x + 5)^3$ j $(x^3 + x^2 - 2)^2$

DIVISION OF POLYNOMIALS

The division of polynomials is only useful if we divide a polynomial of degree n by another of degree n or less.

DIVISION BY LINEARS

Consider $(2x^2 + 3x + 4)(x + 2) + 7$.

If we expand this expression we obtain $(2x^2 + 3x + 4)(x + 2) + 7 = 2x^3 + 7x^2 + 10x + 15$.

Dividing both sides by $(x + 2)$, we obtain

$$\begin{aligned}\frac{2x^3 + 7x^2 + 10x + 15}{x + 2} &= \frac{(2x^2 + 3x + 4)(x + 2) + 7}{x + 2} \\ &= \frac{(2x^2 + 3x + 4)(x + 2)}{x + 2} + \frac{7}{x + 2} \\ &= 2x^2 + 3x + 4 + \frac{7}{x + 2}\end{aligned}$$

where $x + 2$ is the divisor,
$2x^2 + 3x + 4$ is the quotient,
and 7 is the remainder.

If $P(x)$ is divided by $ax + b$ until a constant remainder R is obtained, then

$$\frac{P(x)}{ax + b} = Q(x) + \frac{R}{ax + b}$$

where $ax + b$ is the **divisor**, $D(x)$,
$Q(x)$ is the **quotient**,
and R is the **remainder**.

Notice that $P(x) = Q(x) \times (ax + b) + R$.

DIVISION ALGORITHM

We can divide a polynomial by another polynomial using an algorithm similar to that used for division of whole numbers:

Step 1: What do we multiply x by to get $2x^3$?
The answer is $2x^2$,
and $2x^2(x+2) = 2x^3 + 4x^2$.

Step 2: Subtract $2x^3 + 4x^2$ from $2x^3 + 7x^2$.
The answer is $3x^2$.

Step 3: Bring down the $10x$ to obtain $3x^2 + 10x$.

Return to *Step 1* with the question:
"What must we multiply x by to get $3x^2$?"
The answer is $3x$, and $3x(x+2) = 3x^2 + 6x$ ....
We continue the process until we are left with a constant.

$$\begin{array}{r|l} & \phantom{2x^3+{}}2x^2 + 3x + 4 \\ \hline x+2 & 2x^3 + 7x^2 + 10x + 15 \\ & -(2x^3 + 4x^2) \\ \hline & \phantom{2x^3+{}} 3x^2 + 10x \\ & \phantom{2x^3+{}} -(3x^2 + 6x) \\ \hline & \phantom{2x^3+7x^2+{}} 4x + 15 \\ & \phantom{2x^3+7x^2+{}} -(4x + 8) \\ \hline & 7 \end{array}$$

The division algorithm can also be performed by leaving out the variable, as shown alongside.

$$\begin{array}{r|rrrr} & & 2 & 3 & 4 \\ \hline 1 \;\; 2 & 2 & 7 & 10 & 15 \\ & -(2 & 4) & & \\ \hline & & 3 & 10 & \\ & & -(3 & 6) & \\ \hline & & & 4 & 15 \\ & & & -(4 & 8) \\ \hline & & & & 7 \end{array}$$

Either way, $\dfrac{2x^3 + 7x^2 + 10x + 15}{x+2} = 2x^2 + 3x + 4 + \dfrac{7}{x+2}$

Example 13 — Self Tutor

Find the quotient and remainder for $\dfrac{x^3 - x^2 - 3x - 5}{x-3}$.

Hence write $x^3 - x^2 - 3x - 5$ in the form $Q(x) \times (x-3) + R$.

$$\begin{array}{r|l} & \phantom{x^3-{}} x^2 + 2x + 3 \\ \hline x-3 & x^3 - x^2 - 3x - 5 \\ & -(x^3 - 3x^2) \\ \hline & \phantom{x^3-{}} 2x^2 - 3x \\ & -(2x^2 - 6x) \\ \hline & \phantom{x^3-x^2-{}} 3x - 5 \\ & -(3x - 9) \\ \hline & \phantom{x^3-x^2-3x-{}} 4 \end{array}$$

The quotient is $x^2 + 2x + 3$ and the remainder is 4.

$\therefore \dfrac{x^3 - x^2 - 3x - 5}{x-3} = x^2 + 2x + 3 + \dfrac{4}{x-3}$

$\therefore x^3 - x^2 - 3x - 5 = (x^2 + 2x + 3)(x-3) + 4.$

Check your answer by expanding the RHS.

Example 14 **Self Tutor**

Perform the division $\dfrac{x^4+2x^2-1}{x+3}$.

Hence write x^4+2x^2-1 in the form $Q(x)\times(x+3)+R$.

$$
\begin{array}{r|l}
 & x^3 - 3x^2 + 11x - 33 \\
\hline
x+3 & x^4 + 0x^3 + 2x^2 + 0x - 1 \\
 & -(x^4+3x^3) \\
 & -3x^3 + 2x^2 \\
 & -(-3x^3 - 9x^2) \\
 & 11x^2 + 0x \\
 & -(11x^2+33x) \\
 & -33x - 1 \\
 & -(-33x-99) \\
 & 98
\end{array}
$$

Notice the insertion of $0x^3$ and $0x$.

$\therefore \dfrac{x^4+2x^2-1}{x+3} = x^3-3x^2+11x-33+\dfrac{98}{x+3}$

$\therefore\ x^4+2x^2-1=(x^3-3x^2+11x-33)(x+3)+98$

EXERCISE 6C.2

1 Find the quotient and remainder for the following, and hence write the division in the form $P(x)=Q(x)\,D(x)+R$, where $D(x)$ is the divisor.

a $\dfrac{x^2+2x-3}{x+2}$ **b** $\dfrac{x^2-5x+1}{x-1}$ **c** $\dfrac{2x^3+6x^2-4x+3}{x-2}$

2 Perform the following divisions, and hence write the division in the form $P(x)=Q(x)\,D(x)+R$.

a $\dfrac{x^2-3x+6}{x-4}$ **b** $\dfrac{x^2+4x-11}{x+3}$ **c** $\dfrac{2x^2-7x+2}{x-2}$

d $\dfrac{2x^3+3x^2-3x-2}{2x+1}$ **e** $\dfrac{3x^3+11x^2+8x+7}{3x-1}$ **f** $\dfrac{2x^4-x^3-x^2+7x+4}{2x+3}$

3 Perform the divisions:

a $\dfrac{x^2+5}{x-2}$ **b** $\dfrac{2x^2+3x}{x+1}$ **c** $\dfrac{3x^2+2x-5}{x+2}$

d $\dfrac{x^3+2x^2-5x+2}{x-1}$ **e** $\dfrac{2x^3-x}{x+4}$ **f** $\dfrac{x^3+x^2-5}{x-2}$

SYNTHETIC DIVISION (OPTIONAL)

Click on the icon for an exercise involving a synthetic division process for the division of a polynomial by a linear.

PRINTABLE SECTION

DIVISION BY QUADRATICS

As with division by linears, we can use the **division algorithm** to divide polynomials by quadratics. The division process stops when the remainder has degree less than that of the divisor, so

If $P(x)$ is divided by ax^2+bx+c then

$$\frac{P(x)}{ax^2+bx+c} = Q(x) + \frac{ex+f}{ax^2+bx+c}$$

where ax^2+bx+c is the **divisor**, $Q(x)$ is the **quotient**, and $ex+f$ is the **remainder**.

The remainder will be linear if $e \neq 0$, and constant if $e = 0$.

Example 15 — Self Tutor

Find the quotient and remainder for $\dfrac{x^4+4x^3-x+1}{x^2-x+1}$.

Hence write x^4+4x^3-x+1 in the form $Q(x)\times(x^2-x+1)+R(x)$.

$$\begin{array}{r|l} & \quad x^2+5x+4 \\ \hline x^2-x+1 & x^4+4x^3+0x^2-\ \ x+1 \\ & -(x^4-\ x^3+\ x^2) \\ \hline & \qquad 5x^3-\ x^2-\ x \\ & \quad -(5x^3-5x^2+5x) \\ \hline & \qquad\qquad 4x^2-6x+1 \\ & \qquad -(4x^2-4x+4) \\ \hline & \qquad\qquad -2x-3 \end{array}$$

The quotient is x^2+5x+4 and the remainder is $-2x-3$.

$\therefore\ x^4+4x^3-x+1$
$= (x^2+5x+4)(x^2-x+1)-2x-3$

EXERCISE 6C.3

1. Find the quotient and remainder for:
   a. $\dfrac{x^3+2x^2+x-3}{x^2+x+1}$
   b. $\dfrac{3x^2-x}{x^2-1}$
   c. $\dfrac{3x^3+x-1}{x^2+1}$
   d. $\dfrac{x-4}{x^2+2x-1}$

2. Carry out the following divisions and also write each in the form $P(x)=Q(x)\,D(x)+R(x)$:
   a. $\dfrac{x^2-x+1}{x^2+x+1}$
   b. $\dfrac{x^3}{x^2+2}$
   c. $\dfrac{x^4+3x^2+x-1}{x^2-x+1}$
   d. $\dfrac{2x^3-x+6}{(x-1)^2}$
   e. $\dfrac{x^4}{(x+1)^2}$
   f. $\dfrac{x^4-2x^3+x+5}{(x-1)(x+2)}$

3. Suppose $P(x)=(x-2)(x^2+2x+3)+7$. Find the quotient and remainder when $P(x)$ is divided by $x-2$.

4. Suppose $f(x)=(x-1)(x+2)(x^2-3x+5)+15-10x$. Find the quotient and remainder when $f(x)$ is divided by x^2+x-2.

D ZEROS, ROOTS, AND FACTORS

A **zero** of a polynomial is a value of the variable which makes the polynomial equal to zero.

α is a **zero** of polynomial $P(x) \Leftrightarrow P(\alpha) = 0$.

The **roots** of a polynomial **equation** are the solutions to the equation.

α is a **root** (or **solution**) of $P(x) = 0 \Leftrightarrow P(\alpha) = 0$.

The **roots** of $P(x) = 0$ are the **zeros** of $P(x)$ and the x-intercepts of the graph of $y = P(x)$.

Consider $P(x) = x^3 + 2x^2 - 3x - 10$

$\therefore\ P(2) = 2^3 + 2(2)^2 - 3(2) - 10$
$= 8 + 8 - 6 - 10$
$= 0$

An equation has **roots**.
A polynomial has **zeros**.

This tells us:
- 2 is a zero of $x^3 + 2x^2 - 3x - 10$
- 2 is a root of $x^3 + 2x^2 - 3x - 10 = 0$
- the graph of $y = x^3 + 2x^2 - 3x - 10$ has the x-intercept 2.

If $P(x) = (x+1)(2x-1)(x+2)$, then $(x+1)$, $(2x-1)$, and $(x+2)$ are its **linear factors**.

Likewise $P(x) = (x+3)^2(2x+3)$ has been factorised into 3 linear factors, one of which is repeated.

$(x - \alpha)$ is a **factor** of the polynomial $P(x)$ $\Leftrightarrow$ there exists a polynomial $Q(x)$ such that $P(x) = (x - \alpha)Q(x)$.

Example 16 — Self Tutor

Find the zeros of: **a** $x^2 - 4x + 53$ **b** $z^3 + 3z$

a We wish to find x such that
$x^2 - 4x + 53 = 0$

$\therefore\ x = \dfrac{4 \pm \sqrt{16 - 4(1)(53)}}{2}$

$\therefore\ x = \dfrac{4 \pm \sqrt{-196}}{2} = \dfrac{4 \pm 14i}{2}$

$\therefore\ x = 2 \pm 7i$

The zeros are $2 - 7i$ and $2 + 7i$.

b We wish to find z such that

$z^3 + 3z = 0$

$\therefore\ z(z^2 + 3) = 0$

$\therefore\ z(z + i\sqrt{3})(z - i\sqrt{3}) = 0$

$\therefore\ z = 0$ or $\pm i\sqrt{3}$

The zeros are $-i\sqrt{3}$, 0, and $i\sqrt{3}$.

EXERCISE 6D.1

1 Find the zeros of:

a $2x^2 - 5x - 12$ **b** $x^2 + 6x + 10$ **c** $z^2 - 6z + 6$

d $x^3 - 4x$ **e** $z^3 + 2z$ **f** $z^4 + 4z^2 - 5$

2 Find the roots of:

a $5x^2 = 3x + 2$ **b** $(2x+1)(x^2+3) = 0$ **c** $-2z(z^2 - 2z + 2) = 0$

d $x^3 = 5x$ **e** $z^3 + 5z = 0$ **f** $z^4 = 3z^2 + 10$

Example 17 — Self Tutor

Factorise: **a** $2x^3 + 5x^2 - 3x$ **b** $z^2 + 4z + 9$

a $2x^3 + 5x^2 - 3x$
$= x(2x^2 + 5x - 3)$
$= x(2x - 1)(x + 3)$

b $z^2 + 4z + 9$ is zero when $z = \dfrac{-4 \pm \sqrt{16 - 4(1)(9)}}{2}$

$\therefore z = \dfrac{-4 \pm \sqrt{-20}}{2}$

$\therefore z = \dfrac{-4 \pm 2i\sqrt{5}}{2}$

$\therefore z = -2 \pm i\sqrt{5}$

$\therefore z^2 + 4z + 9 = (z - [-2 + i\sqrt{5}])(z - [-2 - i\sqrt{5}])$
$= (z + 2 - i\sqrt{5})(z + 2 + i\sqrt{5})$

3 Find the linear factors of:

a $2x^2 - 7x - 15$ **b** $z^2 - 6z + 16$ **c** $x^3 + 2x^2 - 4x$
d $6z^3 - z^2 - 2z$ **e** $z^4 - 6z^2 + 5$ **f** $z^4 - z^2 - 2$

4 If $P(x) = a(x - \alpha)(x - \beta)(x - \gamma)$ then α, β, and γ are its zeros.
Verify this statement by finding $P(\alpha)$, $P(\beta)$, and $P(\gamma)$.

Example 18 — Self Tutor

Find *all* cubic polynomials with zeros $\frac{1}{2}$ and $-3 \pm 2i$.

The zeros $-3 \pm 2i$ have sum $= -3 + 2i - 3 - 2i = -6$ and
product $= (-3 + 2i)(-3 - 2i) = 13$

$\therefore$ they come from the quadratic factor $x^2 + 6x + 13$

$\frac{1}{2}$ comes from the linear factor $2x - 1$.

$\therefore P(x) = a(2x - 1)(x^2 + 6x + 13), \quad a \neq 0.$

Example 19 — Self Tutor

Find *all* quartic polynomials with zeros 2, $-\frac{1}{3}$, and $-1 \pm \sqrt{5}$.

The zeros $-1 \pm \sqrt{5}$ have sum $= -1 + \sqrt{5} - 1 - \sqrt{5} = -2$ and
product $= (-1 + \sqrt{5})(-1 - \sqrt{5}) = -4$

$\therefore$ they come from the quadratic factor $x^2 + 2x - 4$.

The zeros 2 and $-\frac{1}{3}$ come from the linear factors $(x - 2)$ and $(3x + 1)$.

$\therefore P(x) = a(x - 2)(3x + 1)(x^2 + 2x - 4), \quad a \neq 0.$

5 Find *all* cubic polynomials with zeros:

a $\pm 2, 3$ **b** $-2, \pm i$ **c** $3, \ -1 \pm i$ **d** $-1, \ -2 \pm \sqrt{2}$

6 Find *all* quartic polynomials with zeros of:

a $\pm 1, \ \pm\sqrt{2}$ **b** $2, \ -1, \ \pm i\sqrt{3}$ **c** $\pm\sqrt{3}, \ 1 \pm i$ **d** $2 \pm \sqrt{5}, \ -2 \pm 3i$

POLYNOMIAL EQUALITY

Two polynomials are **equal** if and only if they have the **same degree** (order) and corresponding terms have equal coefficients.

If we know that two polynomials are **equal** then we can **equate coefficients** to find unknown coefficients.

For example, if $2x^3 + 3x^2 - 4x + 6 = ax^3 + bx^2 + cx + d$, where $a, b, c, d \in \mathbb{R}$, then $a = 2$, $b = 3$, $c = -4$, and $d = 6$.

Example 20 — Self Tutor

Find constants a, b, and c given that:

$$6x^3 + 7x^2 - 19x + 7 = (2x - 1)(ax^2 + bx + c) \quad \text{for all } x.$$

$$6x^3 + 7x^2 - 19x + 7 = (2x - 1)(ax^2 + bx + c)$$
$$\therefore\ 6x^3 + 7x^2 - 19x + 7 = 2ax^3 + 2bx^2 + 2cx - ax^2 - bx - c$$
$$\therefore\ 6x^3 + 7x^2 - 19x + 7 = 2ax^3 + (2b - a)x^2 + (2c - b)x - c$$

Since this is true for all x, we equate coefficients:

$$\therefore\ \underbrace{2a = 6}_{x^3\text{ s}} \quad \underbrace{2b - a = 7}_{x^2\text{ s}} \quad \underbrace{2c - b = -19}_{x\text{ s}} \quad \text{and} \quad \underbrace{7 = -c}_{\text{constants}}$$

$$\therefore\ a = 3 \text{ and } c = -7 \text{ and consequently } \underbrace{2b - 3 = 7 \text{ and } -14 - b = -19}_{\therefore\ b = 5 \text{ in both equations}}$$

So, $a = 3$, $b = 5$, and $c = -7$.

Example 21 — Self Tutor

Find constants a and b if $z^4 + 9 = (z^2 + az + 3)(z^2 + bz + 3)$ for all z.

$$z^4 + 9 = (z^2 + az + 3)(z^2 + bz + 3) \quad \text{for all } z$$
$$\therefore\ z^4 + 9 = z^4 + bz^3 + 3z^2 + az^3 + abz^2 + 3az + 3z^2 + 3bz + 9$$
$$\therefore\ z^4 + 9 = z^4 + (a + b)z^3 + (ab + 6)z^2 + (3a + 3b)z + 9 \quad \text{for all } z$$

Equating coefficients gives
$$\begin{cases} a + b = 0 & \text{.... (1)} \quad \{z^3\text{ s}\} \\ ab + 6 = 0 & \text{.... (2)} \quad \{z^2\text{ s}\} \\ 3a + 3b = 0 & \text{.... (3)} \quad \{z\text{ s}\} \end{cases}$$

From (1) and (3) we see that $b = -a$

$\therefore$ in (2), $a(-a) + 6 = 0$

$\therefore\ a^2 = 6$

$\therefore\ a = \pm\sqrt{6}$ and so $b = \mp\sqrt{6}$

$\therefore\ a = \sqrt{6},\ b = -\sqrt{6}$ *or* $a = -\sqrt{6},\ b = \sqrt{6}$

When simultaneously solving more equations than there are unknowns, we must check that any solutions fit **all** equations. If they do not, there are **no solutions**.

Example 22 **Self Tutor**

$(x+3)$ is a factor of $P(x) = x^3 + ax^2 - 7x + 6$. Find $a \in \mathbb{R}$ and the other factors.

Since $(x+3)$ is a factor,

The coefficient of x^3 is $1 \times 1 = 1$

This must be 2 so the constant term is $3 \times 2 = 6$

$$\begin{aligned} x^3 + ax^2 - 7x + 6 &= (x+3)(x^2 + bx + 2) \quad \text{for some constant } b \\ &= x^3 + bx^2 + 2x \\ &\qquad + 3x^2 + 3bx + 6 \\ &= x^3 + (b+3)x^2 + (3b+2)x + 6 \end{aligned}$$

Equating coefficients gives $3b + 2 = -7$ and $a = b + 3$

$\therefore$ $b = -3$ and $a = 0$

$$\begin{aligned} \therefore \quad P(x) &= (x+3)(x^2 - 3x + 2) \\ &= (x+3)(x-1)(x-2) \end{aligned}$$

The other factors are $(x-1)$ and $(x-2)$.

Example 23 **Self Tutor**

$(2x+3)$ and $(x-1)$ are factors of $2x^4 + ax^3 - 3x^2 + bx + 3$.

Find constants a and b and all zeros of the polynomial.

Since $(2x+3)$ and $(x-1)$ are factors,

The coefficient of x^4 is $2 \times 1 \times 1 = 2$

This must be -1 so the constant term is $3 \times -1 \times -1 = 3$

$$\begin{aligned} 2x^4 + ax^3 - 3x^2 + bx + 3 &= (2x+3)(x-1)(x^2 + cx - 1) \quad \text{for some } c \\ &= (2x^2 + x - 3)(x^2 + cx - 1) \\ &= 2x^4 + 2cx^3 - 2x^2 \\ &\qquad + x^3 + cx^2 - x \\ &\qquad\qquad - 3x^2 - 3cx + 3 \\ &= 2x^4 + (2c+1)x^3 + (c-5)x^2 + (-1-3c)x + 3 \end{aligned}$$

Equating coefficients gives $2c + 1 = a$, $c - 5 = -3$, and $-1 - 3c = b$

$\therefore$ $c = 2$

$\therefore$ $a = 5$ and $b = -7$

$\therefore$ $P(x) = (2x+3)(x-1)(x^2 + 2x - 1)$

Now $x^2 + 2x - 1$ has zeros $\dfrac{-2 \pm \sqrt{4 - 4(1)(-1)}}{2} = \dfrac{-2 \pm 2\sqrt{2}}{2} = -1 \pm \sqrt{2}$

$\therefore$ $P(x)$ has zeros $-\frac{3}{2}$, 1, and $-1 \pm \sqrt{2}$.

EXERCISE 6D.2

1 Find constants a, b, and c given that:

a $2x^2 + 4x + 5 = ax^2 + [2b - 6]x + c$ for all x

b $2x^3 - x^2 + 6 = (x - 1)^2(2x + a) + bx + c$ for all x.

2 Find constants a and b if:

a $z^4 + 4 = (z^2 + az + 2)(z^2 + bz + 2)$ for all z

b $2z^4 + 5z^3 + 4z^2 + 7z + 6 = (z^2 + az + 2)(2z^2 + bz + 3)$ for all z.

3 Show that $z^4 + 64$ can be factorised into two real quadratic factors of the form $z^2 + az + 8$ and $z^2 + bz + 8$, but cannot be factorised into two real quadratic factors of the form $z^2 + az + 16$ and $z^2 + bz + 4$.

4 Find real numbers a and b such that $x^4 - 4x^2 + 8x - 4 = (x^2 + ax + 2)(x^2 + bx - 2)$.
Hence solve the equation $x^4 + 8x = 4x^2 + 4$.

5 a $(2z - 3)$ is a factor of $2z^3 - z^2 + az - 3$. Find $a \in \mathbb{R}$ and all zeros of the cubic.

b $(3z + 2)$ is a factor of $3z^3 - z^2 + (a + 1)z + a$. Find $a \in \mathbb{R}$ and all the zeros of the cubic.

6 a $(2x + 1)$ and $(x - 2)$ are factors of $P(x) = 2x^4 + ax^3 + bx^2 - 12x - 8$.
Find constants a and b, and all zeros of $P(x)$.

b $(x + 3)$ and $(2x - 1)$ are factors of $2x^4 + ax^3 + bx^2 + ax + 3$.
Find constants a and b, and hence determine all zeros of the quartic.

7 a $x^3 + 3x^2 - 9x + c$, $c \in \mathbb{R}$, has two identical linear factors. Prove that c is either 5 or -27, and factorise the cubic into linear factors in each case.

b $3x^3 + 4x^2 - x + m$, $m \in \mathbb{R}$, has two identical linear factors. Find the possible values of m, and find the zeros of the polynomial in each case.

E POLYNOMIAL THEOREMS

There are many **theorems** about polynomials, some of which we look at now. Some of the theorems are true for all polynomials, while others are true only for real polynomials.

THE REMAINDER THEOREM

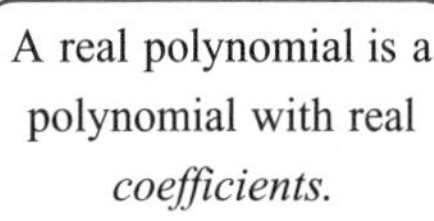

A real polynomial is a polynomial with real *coefficients*.

Consider the real cubic polynomial $P(x) = x^3 + 5x^2 - 11x + 3$.

If we divide $P(x)$ by $x - 2$, we find that

$$\frac{x^3 + 5x^2 - 11x + 3}{x - 2} = x^2 + 7x + 3 + \frac{9}{x - 2} \leftarrow \text{remainder}$$

So, when $P(x)$ is divided by $x - 2$, the remainder is 9.

Notice that $P(2) = 8 + 20 - 22 + 3$
$= 9$, which is the remainder.

By considering other examples like the one above, we formulate the **Remainder theorem**.

The Remainder Theorem

When a polynomial $P(x)$ is divided by $x - k$ until a constant remainder R is obtained, then $R = P(k)$.

Proof:

By the division algorithm, $P(x) = Q(x)(x-k) + R$
Letting $x = k$, $P(k) = Q(k) \times 0 + R$
$\therefore$ $P(k) = R$

When using the Remainder theorem, it is important to realise that the following statements are equivalent:

- $P(x) = (x-k)Q(x) + R$
- $P(k) = R$
- $P(x)$ divided by $x - k$ leaves a remainder of R.

Example 24 Self Tutor

Use the Remainder theorem to find the remainder when $x^4 - 3x^3 + x - 4$ is divided by $x + 2$.

If $P(x) = x^4 - 3x^3 + x - 4$, then

$$P(-2) = (-2)^4 - 3(-2)^3 + (-2) - 4$$
$$= 16 + 24 - 2 - 4$$
$$= 34$$

$\therefore$ when $x^4 - 3x^3 + x - 4$ is divided by $x + 2$, the remainder is 34. {Remainder theorem}

Example 25 Self Tutor

When $P(x)$ is divided by $x^2 - 3x + 7$, the quotient is $x^2 + x - 1$ and the remainder $R(x)$ is unknown.
When $P(x)$ is divided by $x - 2$ the remainder is 29. When $P(x)$ is divided by $x + 1$ the remainder is -16.
Find $R(x)$ in the form $ax + b$.

When the divisor is $x^2 - 3x + 7$, $P(x) = \underbrace{(x^2 + x - 1)}_{Q(x)}\,\underbrace{(x^2 - 3x + 7)}_{D(x)} + \underbrace{ax + b}_{R(x)}$.

Now $P(2) = 29$ {Remainder theorem}

$\therefore$ $(2^2 + 2 - 1)(2^2 - 6 + 7) + 2a + b = 29$
$\therefore$ $(5)(5) + 2a + b = 29$
$\therefore$ $2a + b = 4$ .... (1)

Also, $P(-1) = -16$ {Remainder theorem}

$\therefore$ $((-1)^2 + (-1) - 1)((-1)^2 - 3(-1) + 7) + (-a + b) = -16$
$\therefore$ $(-1)(11) - a + b = -16$
$\therefore$ $-a + b = -5$ .... (2)

Solving (1) and (2) simultaneously gives $a = 3$ and $b = -2$.

$\therefore$ $R(x) = 3x - 2$.

EXERCISE 6E.1

1 For $P(x)$ a real polynomial, write two equivalent statements for each of:

a If $P(2) = 7$, then ....

b If $P(x) = Q(x)(x+3) - 8$, then ....

c If $P(x)$ divided by $x - 5$ has a remainder of 11 then ....

2 Without performing division, find the remainder when:

a $x^3 + 2x^2 - 7x + 5$ is divided by $x - 1$

b $x^4 - 2x^2 + 3x - 1$ is divided by $x + 2$.

3 Find $a \in \mathbb{R}$ given that:

a when $x^3 - 2x + a$ is divided by $x - 2$, the remainder is 7

b when $2x^3 + x^2 + ax - 5$ is divided by $x + 1$, the remainder is -8.

4 When $x^3 + 2x^2 + ax + b$ is divided by $x - 1$ the remainder is 4, and when divided by $x + 2$ the remainder is 16. Find constants a and b.

5 $2x^n + ax^2 - 6$ leaves a remainder of -7 when divided by $x - 1$, and 129 when divided by $x + 3$. Find a and n given that $n \in \mathbb{Z}^+$.

6 When $P(z)$ is divided by $z^2 - 3z + 2$ the remainder is $4z - 7$.
Find the remainder when $P(z)$ is divided by: **a** $z - 1$ **b** $z - 2$.

7 When $P(z)$ is divided by $z + 1$ the remainder is -8, and when divided by $z - 3$ the remainder is 4. Find the remainder when $P(z)$ is divided by $(z - 3)(z + 1)$.

8 If $P(x)$ is divided by $(x - a)(x - b)$, where $a \neq b$, $a, b \in \mathbb{R}$, prove that the remainder is:
$$\left(\frac{P(b) - P(a)}{b - a}\right) \times (x - a) + P(a).$$

THE FACTOR THEOREM

For any polynomial $P(x)$, k is a zero of $P(x)$ $\Leftrightarrow$ $(x - k)$ is a factor of $P(x)$.

Proof:

k is a zero of $P(x) \Leftrightarrow P(k) = 0$	{definition of a zero}
$\Leftrightarrow R = 0$	{Remainder theorem}
$\Leftrightarrow P(x) = Q(x)(x - k)$	{division algorithm}
$\Leftrightarrow (x - k)$ is a factor of $P(x)$	{definition of a factor}

The **Factor theorem** says that if 2 is a zero of $P(x)$ then $(x - 2)$ is a factor of $P(x)$, and vice versa.

Example 26 **Self Tutor**

Find k given that $(x-2)$ is a factor of x^3+kx^2-3x+6.
Hence, fully factorise x^3+kx^2-3x+6.

Let $P(x)=x^3+kx^2-3x+6$

Since $(x-2)$ is a factor, $P(2)=0$ {Factor theorem}

$\therefore\ (2)^3+k(2)^2-3(2)+6=0$

$\therefore\ 8+4k=0$

$\therefore\ k=-2$

We now use either the division algorithm or synthetic division to find the other factors of $P(x)$:

Division algorithm *or* **Synthetic division**

$$\begin{array}{r|l} & x^2+0x-3 \\ \hline x-2 & x^3-2x^2-3x+6 \\ & -(x^3-2x^2) \\ \hline & 0x^2-3x \\ & -(0x^2-0x) \\ \hline & -3x+6 \\ & -(-3x+6) \\ \hline & 0 \end{array}$$

$$\begin{array}{c|cccc} 2 & 1 & -2 & -3 & 6 \\ & 0 & 2 & 0 & -6 \\ \hline & 1 & 0 & -3 & 0 \end{array}$$

Using either method we find that $P(x)=(x-2)(x^2-3)$
$=(x-2)(x+\sqrt{3})(x-\sqrt{3})$

EXERCISE 6E.2

1. Find the constant k and hence factorise the polynomial if:
   a. $2x^3+x^2+kx-4$ has the factor $(x+2)$
   b. $x^4-3x^3-kx^2+6x$ has the factor $(x-3)$.

2. Find constants a and b given that $2x^3+ax^2+bx+5$ has factors $(x-1)$ and $(x+5)$.

3. a. Suppose 3 is a zero of $P(z)=z^3-z^2+(k-5)z+(k^2-7)$.
   Find the possible values of $k\in\mathbb{R}$ and all the corresponding zeros of $P(z)$.
   b. Show that $(z-2)$ is a factor of $P(z)=z^3+mz^2+(3m-2)z-10m-4$ for all values of $m\in\mathbb{R}$. For what value(s) of m is $(z-2)^2$ a factor of $P(z)$?

4. a. Consider $P(x)=x^3-a^3$ where a is real.
      i. Find $P(a)$. What is the significance of this result?
      ii. Factorise x^3-a^3 as the product of a real linear and a quadratic factor.
   b. Now consider $P(x)=x^3+a^3$, where a is real.
      i. Find $P(-a)$. What is the significance of this result?
      ii. Factorise x^3+a^3 as the product of a real linear and a quadratic factor.

5 a Prove that "$x+1$ is a factor of x^n+1, $n \in \mathbb{Z}$ $\Leftrightarrow$ n is odd".
b Find the real number a such that $(x-1-a)$ is a factor of $P(x) = x^3 - 3ax - 9$.

THE FUNDAMENTAL THEOREM OF ALGEBRA

The theorems we have just seen for real polynomials can be generalised in the **Fundamental Theorem of Algebra**:

a Every polynomial of degree $n \geqslant 1$ has at least one zero which can be written in the form $a+bi$ where $a, b \in \mathbb{R}$.

b If $P(x)$ is a polynomial of degree n, then $P(x)$ has exactly n zeros, some of which may be either irrational numbers or complex numbers.

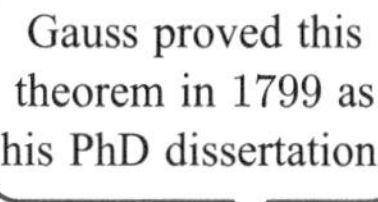
Gauss proved this
theorem in 1799 as
his PhD dissertation.

Using the Fundamental Theorem of Algebra, the following properties of **real polynomials** can be established:

- Every **real** polynomial of degree n can be factorised into n complex linear factors, some of which may be repeated.
- Every real polynomial can be expressed as a product of **real** linear and **real** irreducible quadratic factors (where $\Delta < 0$).
- Every **real** polynomial of degree n has exactly n zeros, some of which may be repeated.
- If $p+qi$ $(q \neq 0)$ is a zero of a **real** polynomial then its complex conjugate $p-qi$ is also a zero.
- Every **real** polynomial of odd degree has at least one real zero.

Example 27 **Self Tutor**

Suppose $-3+i$ is a zero of $P(x) = ax^3 + 9x^2 + ax - 30$ where a is real.

Find a and hence find all zeros of the cubic.

Since $P(x)$ is real, both $-3+i$ and $-3-i$ are zeros.

These have sum $= -6$ and product $= (-3+i)(-3-i) = 10$.

$\therefore$ the zeros $-3 \pm i$ come from the quadratic $x^2+6x+10$.

The coefficient of x^3 is a. The constant term is $10 \times (-3) = -30$.

Consequently, $ax^3 + 9x^2 + ax - 30 = (x^2+6x+10)(ax-3)$
$= ax^3 + (6a-3)x^2 + (10a-18)x - 30$

Equating coefficients of x^2 gives: $6a - 3 = 9$ $\therefore$ $a = 2$
Equating coefficients of x gives: $10a - 18 = a$ $\therefore$ $a = 2$

$\therefore$ $a = 2$ and the linear factor is $(2x-3)$

$\therefore$ the other two zeros are $-3-i$ and $\frac{3}{2}$.

Example 28 **Self Tutor**

One zero of $ax^3 + (a+1)x^2 + 10x + 15$, $a \in \mathbb{R}$, is purely imaginary.

Find a and the zeros of the polynomial.

Let the purely imaginary zero be bi, $b \neq 0$.

Since $P(x)$ is real, both bi and $-bi$ are zeros.

These have sum $= 0$ and product $= -b^2i^2 = b^2$

$\therefore$ the zeros $\pm bi$ come from the quadratic $x^2 + b^2$.

The coefficient of x^3 is a. The constant term is $b^2 \times \frac{15}{b^2} = 15$.

Consequently, $ax^3 + (a+1)x^2 + 10x + 15 = (x^2 + b^2)(ax + \frac{15}{b^2})$

$= ax^3 + \left(\frac{15}{b^2}\right)x^2 + ab^2x + 15$

Equating coefficients of x^2 gives: $a + 1 = \frac{15}{b^2}$ .... (1)

Equating coefficients of x gives: $ab^2 = 10$ .... (2)

$\therefore ab^2 + b^2 = 15$ {using (1)}

$\therefore 10 + b^2 = 15$ {using (2)}

$\therefore b^2 = 5$

$\therefore b = \pm\sqrt{5}$

In (2), since $b^2 = 5$, $5a = 10$ $\therefore$ $a = 2$.

The linear factor $ax + \frac{15}{b^2}$ becomes $2x + 3$

$\therefore$ $a = 2$ and the zeros are $\pm i\sqrt{5}$, $-\frac{3}{2}$.

EXERCISE 6E.3

1. Find all real polynomials of degree 3 with zeros $-\frac{1}{2}$ and $1 - 3i$.

2. $p(x)$ is a real cubic polynomial for which $p(1) = p(2+i) = 0$ and $p(0) = -20$. Find $p(x)$ in expanded form.

3. $2 - 3i$ is a zero of $P(z) = z^3 + pz + q$ where p and q are real.
   a. Using conjugate pairs, find p and q and the other two zeros.
   b. Check your answer by solving for p and q using $P(2 - 3i) = 0$.

4. $3 + i$ is a root of $z^4 - 2z^3 + az^2 + bz + 10 = 0$, where a and b are real. Find a and b and the other roots of the equation.

5. One zero of $P(z) = z^3 + az^2 + 3z + 9$, $a \in \mathbb{R}$, is purely imaginary. Find a and hence factorise $P(z)$ into the product of linear factors.

6. At least one zero of $P(x) = 3x^3 + kx^2 + 15x + 10$, $k \in \mathbb{R}$, is purely imaginary. Find k and hence factorise $P(x)$ into the product of linear factors.

SUM AND PRODUCT OF ROOTS THEOREM

We have seen that for the quadratic equation $ax^2 + bx + c = 0$, $a \neq 0$, the sum of the roots is $-\frac{b}{a}$ and the product of the roots is $\frac{c}{a}$.

For the polynomial equation $a_n x^n + a_{n-1}x^{n-1} + ... + a_2x^2 + a_1x + a_0 = 0$, $a_n \neq 0$

which can also be written as $\sum\limits_{r=0}^{n} a_r x^r = 0$,

the sum of the roots is $\dfrac{-a_{n-1}}{a_n}$, and the product of the roots is $\dfrac{(-1)^n a_0}{a_n}$.

We can explain this result as follows:

Consider the polynomial equation $a_n(x-\alpha_1)(x-\alpha_2)(x-\alpha_3)....(x-\alpha_n) = 0$

with roots $\alpha_1, \alpha_2, \alpha_3, ..., \alpha_n$.

Expanding the LHS we have

$$\begin{aligned}
&a_n\,\underbrace{(x-\alpha_1)(x-\alpha_2)}(x-\alpha_3)(x-\alpha_4)....(x-\alpha_n)\\
&= a_n\,\underbrace{(x^2-[\alpha_1+\alpha_2]x+(-1)^2\alpha_1\alpha_2)(x-\alpha_3)}(x-\alpha_4)....(x-\alpha_n)\\
&= a_n(x^3-[\alpha_1+\alpha_2+\alpha_3]x^2+....+(-1)^3\alpha_1\alpha_2\alpha_3)(x-\alpha_4)....(x-\alpha_n)
\end{aligned}$$

{the term of order x is no longer important as it will not contribute to either the x^{n-1} term or the constant term}

$$\vdots$$

$$\begin{aligned}
&= a_n(x^n-[\alpha_1+\alpha_2+\alpha_3+....+\alpha_n]x^{n-1}+....+(-1)^n\alpha_1\alpha_2\alpha_3....\alpha_n)\\
&= a_nx^n - a_n[\alpha_1+\alpha_2+\alpha_3+....+\alpha_n]x^{n-1}+....+(-1)^n\alpha_1\alpha_2\alpha_3....\alpha_n a_n
\end{aligned}$$

Equating coefficients,

$$a_{n-1} = -a_n[\alpha_1+\alpha_2+\alpha_3+....+\alpha_n] \quad \text{and} \quad a_0 = (-1)^n\alpha_1\alpha_2\alpha_3....\alpha_n a_n$$

$$\therefore\quad -\frac{a_{n-1}}{a_n} = \alpha_1+\alpha_2+\alpha_3+....+\alpha_n \quad \text{and} \quad \frac{(-1)^n a_0}{a_n} = \alpha_1\alpha_2\alpha_3....\alpha_n$$

Example 29 **Self Tutor**

Find the sum and product of the roots of $2x^3 - 7x^2 + 8x - 1 = 0$.

The sum of the roots $= -\dfrac{(-7)}{2} = \frac{7}{2}$.

The polynomial equation has degree 3.

$\therefore$ the product of the roots $= \dfrac{(-1)^3(-1)}{2} = \frac{1}{2}$.

Example 30 **Self Tutor**

A real polynomial has the form $P(x) = 3x^4 - 12x^3 + cx^2 + dx + e$. The graph of $y = P(x)$ has y-intercept 180. It cuts the x-axis at 2 and 6, and does not meet the x-axis anywhere else.

Suppose the other two zeros are $m \pm ni$, $n > 0$. Use the sum and product formulae to find m and n.

If the other two zeros are $m \pm ni$, the sum of the zeros is

$$2 + 6 + (m + ni) + (m - ni) = -\frac{-12}{3}$$
$$\therefore\ 2m + 8 = 4$$
$$\therefore\ m = -2$$

The constant term e is the y-intercept.

$$\therefore\ e = 180$$

The product of the zeros is $2 \times 6 \times (-2 + ni)(-2 - ni) = \frac{(-1)^4 180}{3} = 60$

$$\therefore\ 12(4 + n^2) = 60$$
$$\therefore\ 4 + n^2 = 5$$
$$\therefore\ n^2 = 1$$
$$\therefore\ n = 1 \quad \{\text{as } n > 0\}$$

So, $m = -2$ and $n = 1$.

EXERCISE 6E.4

1 Find the sum and product of the roots of:

a $2x^2 - 3x + 4 = 0$

b $3x^3 - 4x^2 + 8x - 5 = 0$

c $x^4 - x^3 + 2x^2 + 3x - 4 = 0$

d $2x^5 - 3x^4 + x^2 - 8 = 0$

e $x^7 - x^5 + 2x - 9 = 0$

f $x^6 - 1 = 0$

2 A real cubic polynomial $P(x)$ has zeros $3 \pm i\sqrt{2}$ and $\frac{2}{3}$. It has a leading coefficient of 6. Find:

a the sum and product of its zeros

b the coefficient of x^2

c the constant term.

3 A real polynomial of degree 5 has leading coefficient -1 and zeros of -2, $3 \pm i$, and $\sqrt{k} \pm 1$. The y-intercept is 18. Find:

a k

b the coefficient of x^4.

4 A real polynomial of degree 5 has leading coefficient 2 and the coefficient of x^4 is 3. When the polynomial is graphed, the y-intercept is 5, and it cuts the x-axis at $\frac{1}{2}$ and $1 \pm \sqrt{2}$ only.

Suppose the other two zeros are $m \pm ni$, $n > 0$. Use the sum and product formulae to find m and n.

5 A real quartic polynomial has leading coefficient 1 and zeros of the form $a \pm i$ and $3 \pm a$, where $a \in \mathbb{R}$. Its constant term is 25. What are the possible values that a may take?

6 $x^3 - px^2 + qx - r = 0$ has non-zero roots p, q, and r, where $p, q, r \in \mathbb{R}$.

a Show that $q = -r$ and $p = -\frac{1}{r}$.

b Hence find p, q, and r.

GRAPHING REAL POLYNOMIALS

Any polynomial with real coefficients can be graphed on the Cartesian plane.

Use of a **graphics calculator** or the **graphing package** provided will help in this section.

CUBIC POLYNOMIALS

INVESTIGATION 2 — CUBIC GRAPHS

Every real cubic polynomial can be categorised into one of four types. In each case $a \in \mathbb{R}$, $a \neq 0$, and the zeros are α, β, γ.

Type 1: Three real, distinct zeros: $P(x) = a(x-\alpha)(x-\beta)(x-\gamma)$

Type 2: Two real zeros, one repeated: $P(x) = a(x-\alpha)^2(x-\beta)$

Type 3: One real zero repeated three times: $P(x) = a(x-\alpha)^3$

Type 4: One real and two complex conjugate zeros:
$P(x) = (x-\alpha)(ax^2+bx+c)$, $\Delta = b^2 - 4ac < 0$.

What to do:

1. Experiment with the graphs of *Type 1* cubics. State the effect of changing both the size and sign of a. What is the geometrical significance of α, β, and γ?
2. Experiment with the graphs of *Type 2* cubics. What is the geometrical significance of the squared factor?
3. Experiment with the graphs of *Type 3* cubics. What is the geometrical significance of α?
4. Experiment with the graphs of *Type 4* cubics. What is the geometrical significance of α and the quadratic factor which has complex zeros?

From **Investigation 2** you should have discovered that:

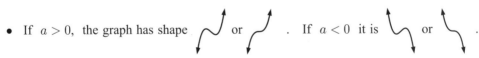

- If $a > 0$, the graph has shape or . If $a < 0$ it is or .
- All cubics are continuous smooth curves.
- Every cubic polynomial must cut the x-axis at least once, and so has at least one real zero.
- For a cubic of the form $P(x) = a(x-\alpha)(x-\beta)(x-\gamma)$, $\alpha, \beta, \gamma \in \mathbb{R}$, the graph has three distinct x-intercepts corresponding to the three distinct zeros α, β, and γ. The graph crosses over or **cuts** the x-axis at these points, as shown.

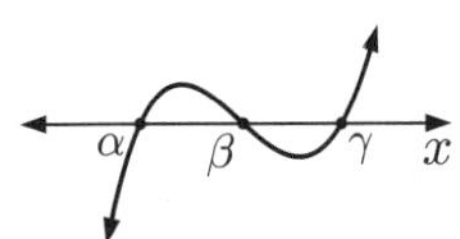

- For a cubic of the form $P(x) = a(x-\alpha)^2(x-\beta)$, α, $\beta \in \mathbb{R}$, the graph **touches** the x-axis at the repeated zero α and **cuts** it at the other x-intercept β, as shown.

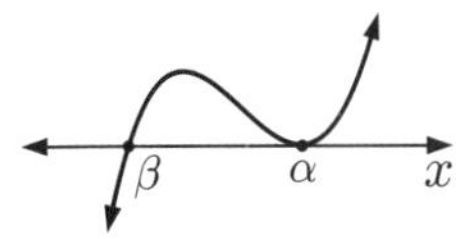

- For a cubic of the form $P(x) = a(x - \alpha)^3$, $x \in \mathbb{R}$, the graph has only one x-intercept, α. The graph is horizontal at this point, and the x-axis is a tangent to the curve even though the curve crosses over it.

- For a cubic of the form $P(x) = (x - \alpha)(ax^2 + bx + c)$ where $\Delta < 0$, there is only one x-intercept, α. The graph cuts the x-axis at this point. The other two zeros are complex and so do not appear on the graph.

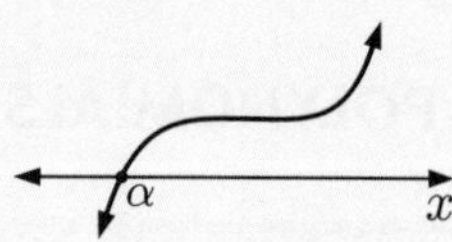

Example 31 **Self Tutor**

Find the equation of the cubic with graph:

a

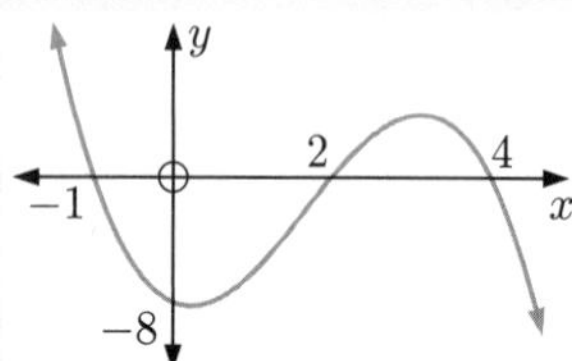

b

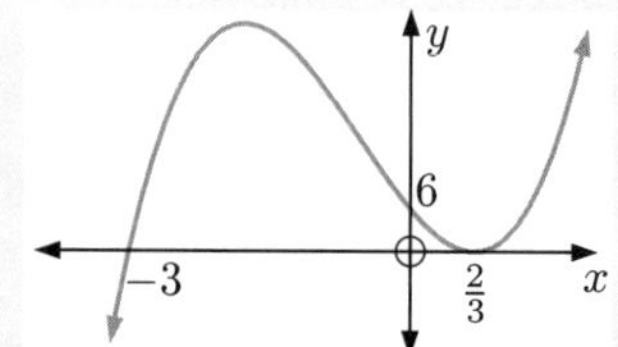

a The x-intercepts are -1, 2, 4.

$\therefore\ y = a(x+1)(x-2)(x-4)$

But when $x = 0$, $y = -8$

$\therefore\ a(1)(-2)(-4) = -8$

$\therefore\ a = -1$

So, $y = -(x+1)(x-2)(x-4)$

b The graph touches the x-axis at $\frac{2}{3}$, indicating a squared factor $(3x-2)^2$.

The other x-intercept is -3,

so $y = a(3x-2)^2(x+3)$.

But when $x = 0$, $y = 6$

$\therefore\ a(-2)^2(3) = 6$

$\therefore\ a = \frac{1}{2}$

So, $y = \frac{1}{2}(3x-2)^2(x+3)$

When determining a polynomial function from a given graph, if we are not given all the zeros, or if some of the zeros are complex, we write a factor in general form.

For example:

- If an x-intercept is not given, use $P(x) = (x-k)^2\underbrace{(ax+b)}_{\text{most general form of a linear}}$.

  Using $P(x) = a(x-k)^2(x+b)$ is more complicated.

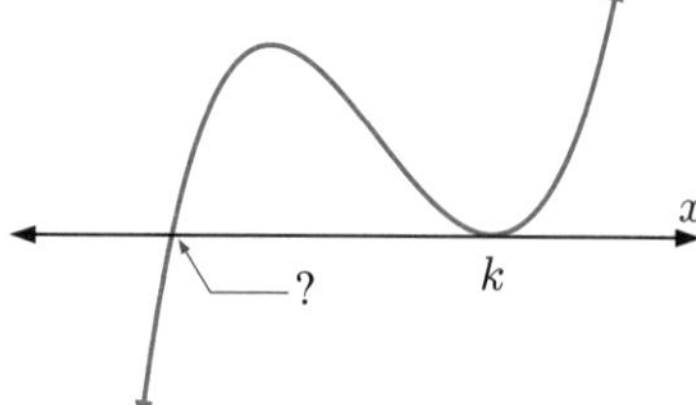

- If there is clearly only one x-intercept and that is given, use $P(x) = (x-k)\underbrace{(ax^2+bx+c)}_{\text{most general form of a quadratic}}$.

  $\Delta < 0$

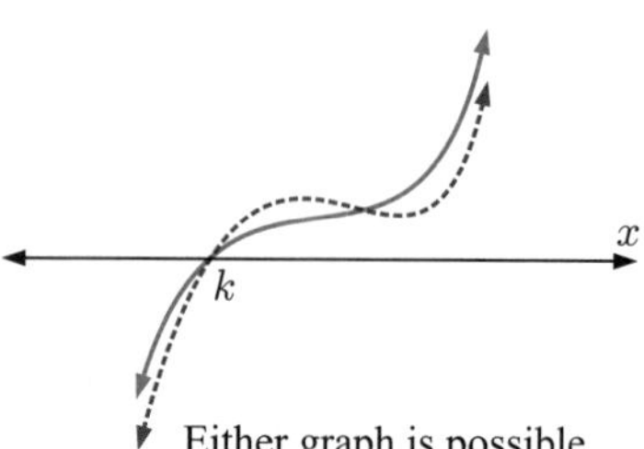

Either graph is possible.

Example 32 **Self Tutor**

Find the equation of the cubic which cuts the x-axis at 2 and -3, cuts the y-axis at -48, and which passes through the point $(1,\ -40)$.

The zeros are 2 and -3, so $y = (x-2)(x+3)(ax+b)$, $a \neq 0$.

When $x = 0$, $y = -48$

$\therefore\ (-2)(3)b = -48$

$\therefore\ b = 8$

When $x = 1$, $y = -40$

$\therefore\ (-1)(4)(a+8) = -40$

$\therefore\ a + 8 = 10$

$\therefore\ a = 2$

So, the equation is $y = (x-2)(x+3)(2x+8)$

$\therefore\ y = 2(x-2)(x+3)(x+4)$

EXERCISE 6F.1

1 For a cubic polynomial $P(x)$, state the geometrical significance of:

a a single real linear factor such as $(x-\alpha)$, $\alpha \in \mathbb{R}$

b a squared real linear factor such as $(x-\alpha)^2$, $\alpha \in \mathbb{R}$

c a cubed real linear factor such as $(x-\alpha)^3$, $\alpha \in \mathbb{R}$.

2 Find the equation of the cubic with graph:

a

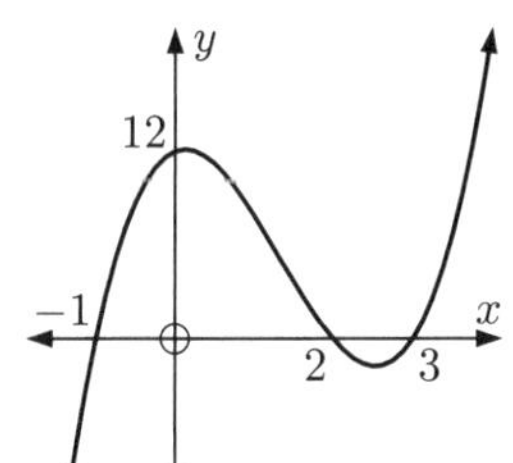

b

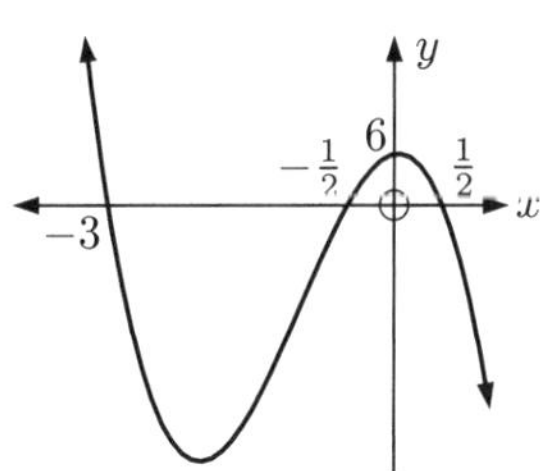

c

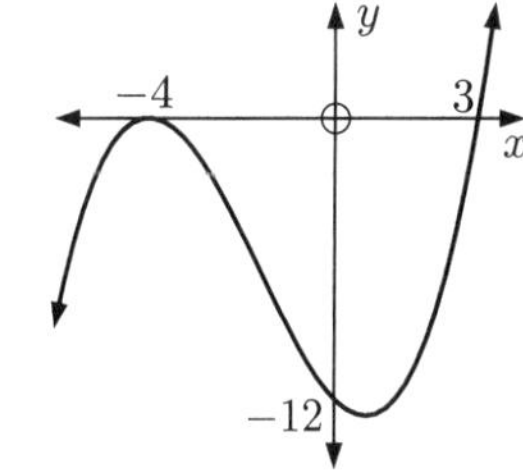

d

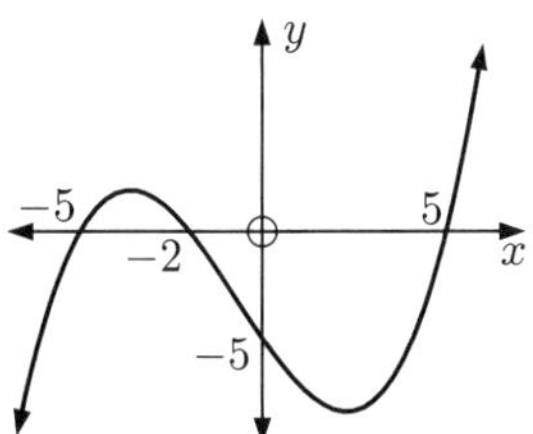

e

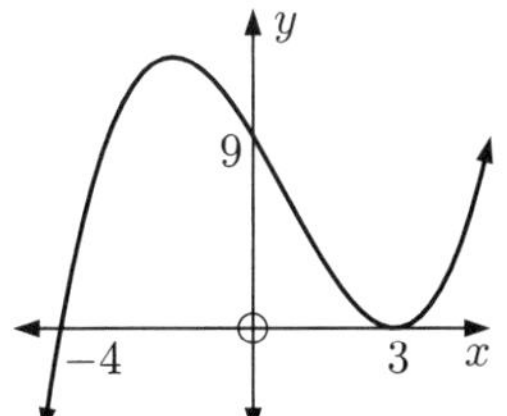

f

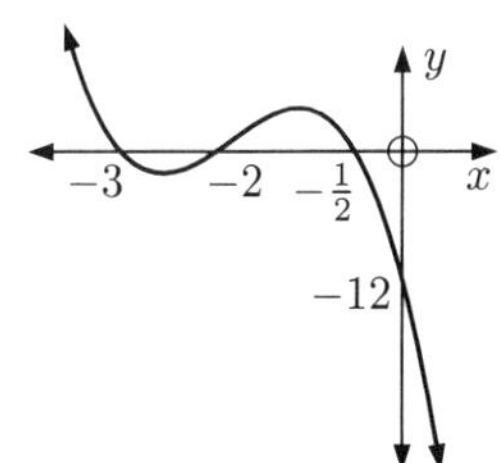

3 Find the equation of the cubic whose graph:

a cuts the x-axis at 3, 1, and -2, and passes through $(2,\ -4)$

b cuts the x-axis at -2, 0, and $\frac{1}{2}$, and passes through $(-3,\ -21)$

c touches the x-axis at 1, cuts the x-axis at -2, and passes through $(4,\ 54)$

d touches the x-axis at $-\frac{2}{3}$, cuts the x-axis at 4, and passes through $(-1,\ -5)$.

4 Match the given graphs to the corresponding cubic function:

a $y = 2(x-1)(x+2)(x+4)$
b $y = -(x+1)(x-2)(x-4)$
c $y = (x-1)(x-2)(x+4)$
d $y = -2(x-1)(x+2)(x+4)$
e $y = -(x-1)(x+2)(x+4)$
f $y = 2(x-1)(x-2)(x+4)$

A

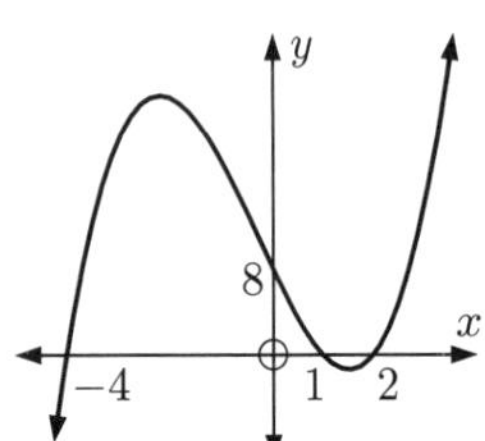

B

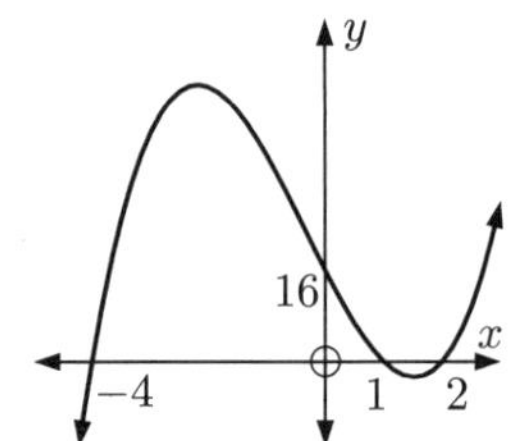

C

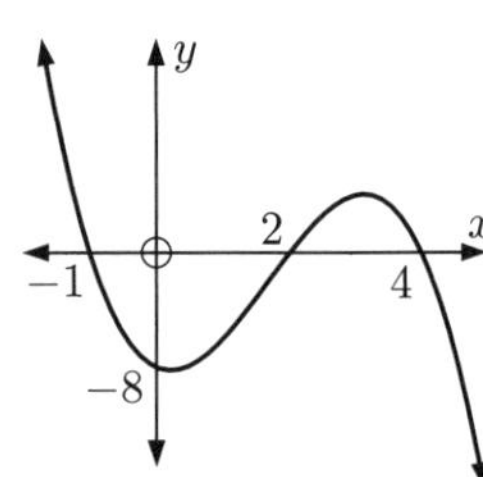

D

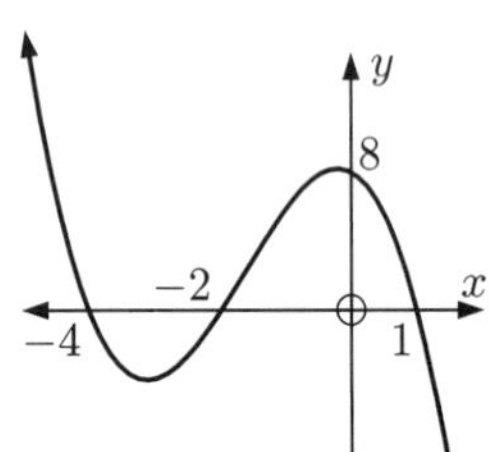

E

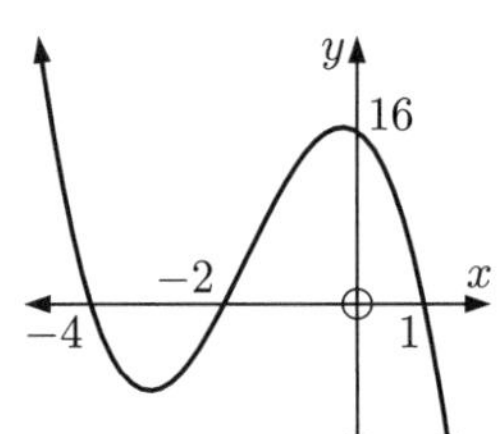

F

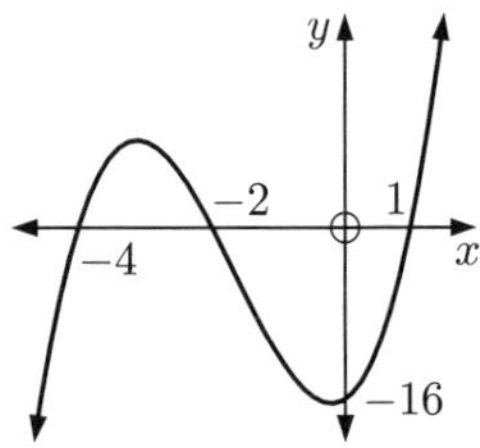

5 Find the equation of a real cubic polynomial which:

a cuts the x-axis at $\frac{1}{2}$ and -3, cuts the y-axis at 30, and passes through $(1, -20)$

b cuts the x-axis at 1, touches the x-axis at -2, and cuts the y-axis at $(0, 8)$

c cuts the x-axis at 2, cuts the y-axis at -4, and passes through $(1, -1)$ and $(-1, -21)$.

QUARTIC POLYNOMIALS

INVESTIGATION 3 — QUARTIC GRAPHS

There are considerably more possible factor types to consider for quartic functions. We will consider quartics containing certain types of factors.

What to do:

1 Experiment with the graphs of quartics which have:

a four distinct real linear factors
b a squared real linear factor and two distinct real linear factors
c two squared real linear factors
d a cubed real linear factor and one distinct real linear factor
e a real linear factor raised to the fourth power
f one real quadratic factor with $\Delta < 0$ and two real linear factors
g two real quadratic factors each with $\Delta < 0$.

GRAPHING PACKAGE

2 Summarise your observations.

From **Investigation 3** you should have discovered that:

- For a quartic polynomial in which a is the coefficient of x^4:
  - If $a > 0$ the graph opens upwards.
  - If $a < 0$ the graph opens downwards.
- If a quartic with $a > 0$ is fully factorised into real linear factors, then:
  - for a **single factor** $(x - \alpha)$, the graph **cuts** the x-axis at α

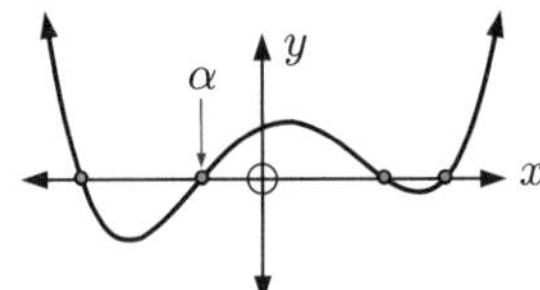

  - for a **square factor** $(x - \alpha)^2$, the graph **touches** the x-axis at α

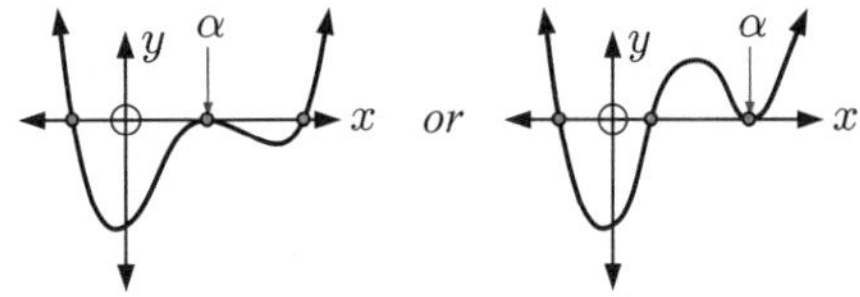

  - for a **cubed factor** $(x - \alpha)^3$, the graph **cuts** the x-axis at α and is 'flat' at α

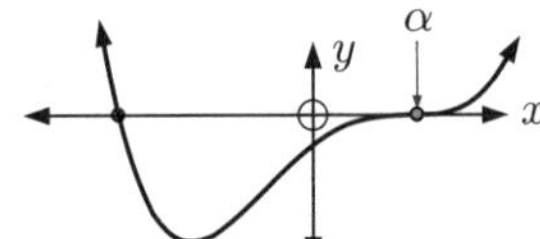

  - for a **quadruple factor** $(x - \alpha)^4$, the graph **touches** the x-axis and is 'flat' at that point.

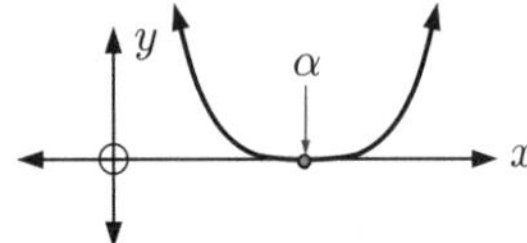

  - If a quartic with $a > 0$ has one real quadratic factor with $\Delta < 0$ we could have:

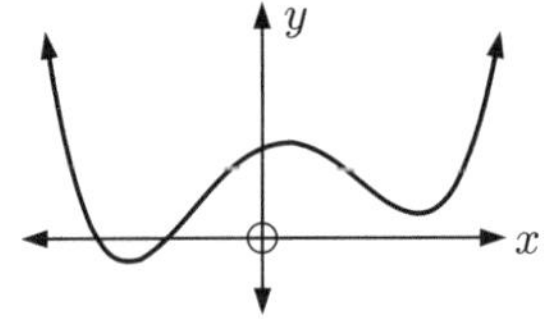

  - If a quartic with $a > 0$ has two real quadratic factors both with $\Delta < 0$ we have:

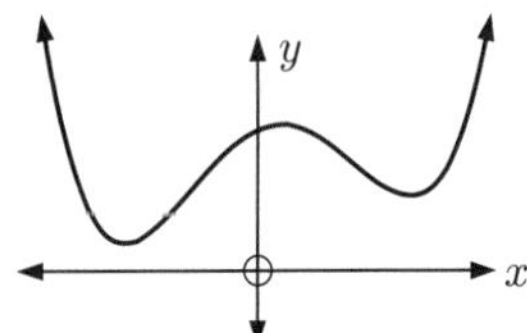

The graph does not meet the x-axis at all.

Example 33

Self Tutor

Find the equation of the quartic with graph:

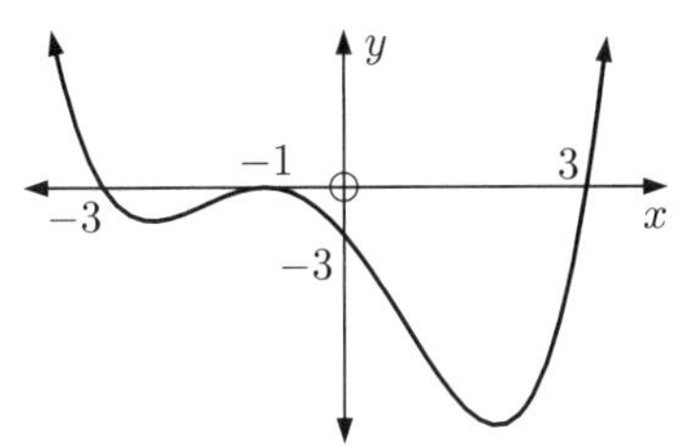

The graph touches the x-axis at -1 and cuts it at -3 and 3.

$\therefore\ y = a(x+1)^2(x+3)(x-3), \quad a \neq 0$

But when $x = 0, \quad y = -3$

$\therefore\ -3 = a(1)^2(3)(-3)$

$\therefore\ -3 = -9a$

$\therefore\ a = \frac{1}{3}$

$\therefore\ y = \frac{1}{3}(x+1)^2(x+3)(x-3)$

Example 34 **Self Tutor**

Find the quartic which touches the x-axis at 2, cuts the x-axis at -3, and also passes through $(1, -12)$ and $(3, 6)$.

The graph *touches* the x-axis at 2, so $(x-2)^2$ is a factor.
The graph *cuts* the x-axis at -3, so $(x+3)$ is a factor.
$\therefore$ $P(x) = (x-2)^2(x+3)(ax+b)$ where a and b are constants, $a \neq 0$.

Now $P(1) = -12$
$\therefore$ $(-1)^2(4)(a+b) = -12$
$\therefore$ $a + b = -3$ .... (1)

and $P(3) = 6$,
$\therefore$ $1^2(6)(3a+b) = 6$
$\therefore$ $3a + b = 1$ .... (2)

Solving (1) and (2) simultaneously gives $a = 2$, $b = -5$.
$\therefore$ $P(x) = (x-2)^2(x+3)(2x-5)$

EXERCISE 6F.2

1 Find the equation of the quartic with graph:

a
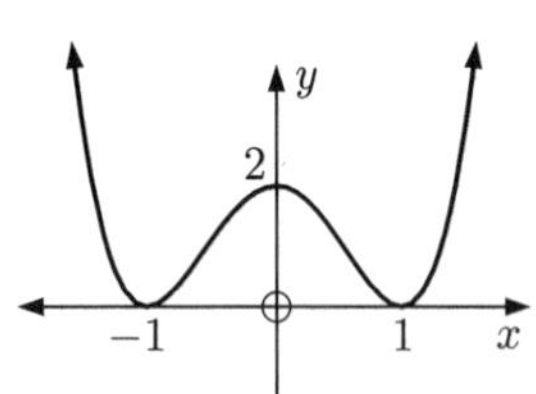

b
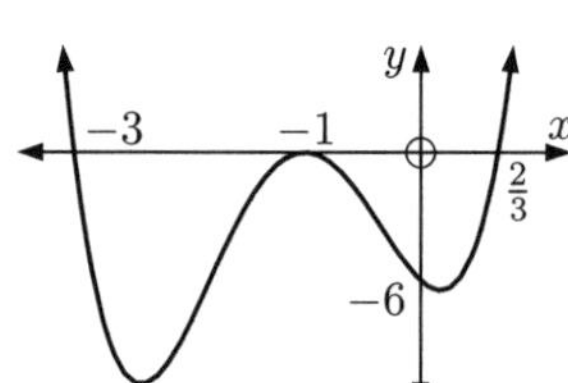

c
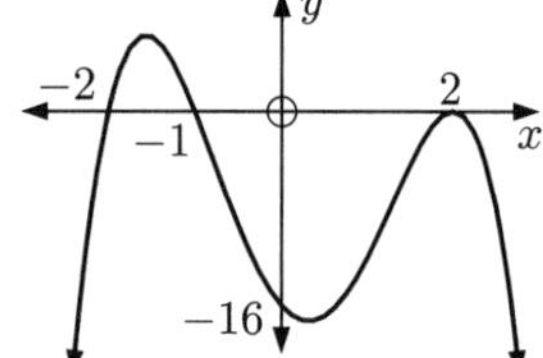

d
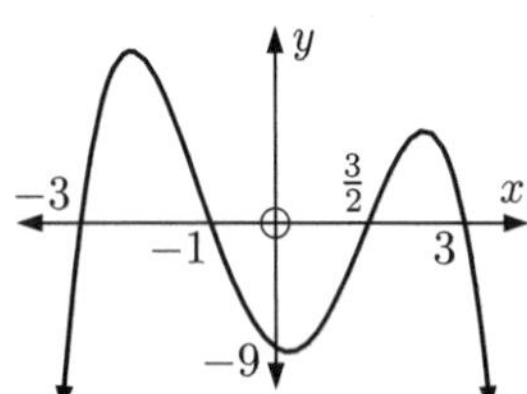

e
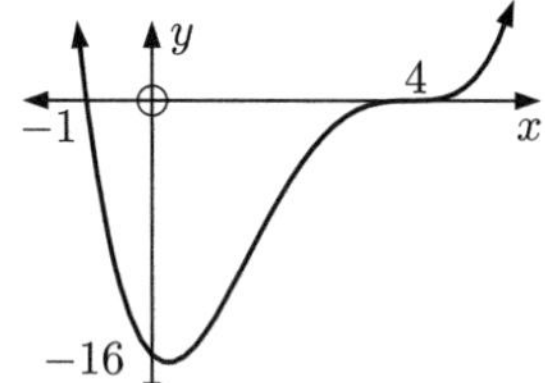

f
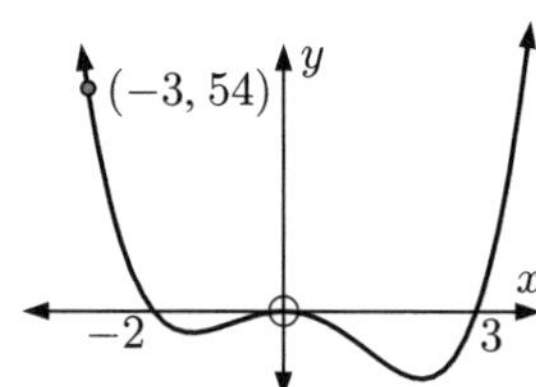

2 Match the given graphs to the corresponding quartic functions:

a $y = (x-1)^2(x+1)(x+3)$

b $y = -2(x-1)^2(x+1)(x+3)$

c $y = (x-1)(x+1)^2(x+3)$

d $y = (x-1)(x+1)^2(x-3)$

e $y = -\frac{1}{3}(x-1)(x+1)(x+3)^2$

f $y = -(x-1)(x+1)(x-3)^2$

A
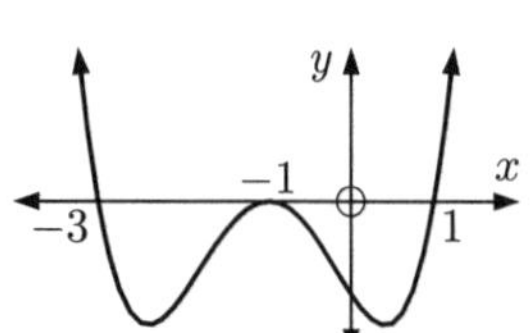

B
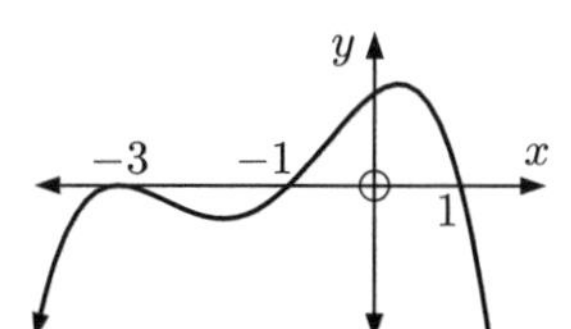

C
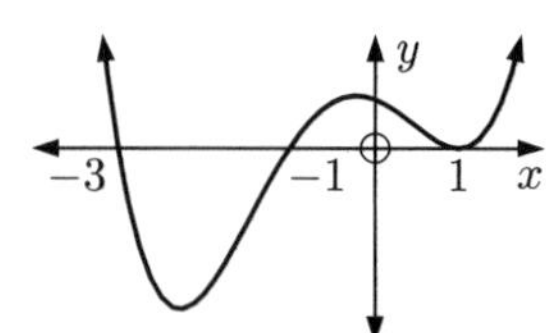

D

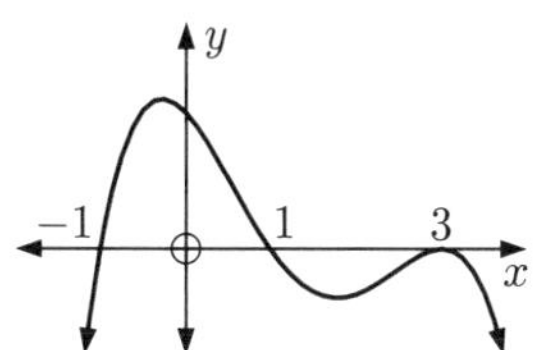

E

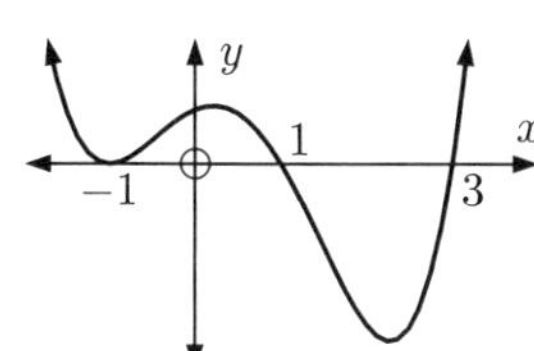

F

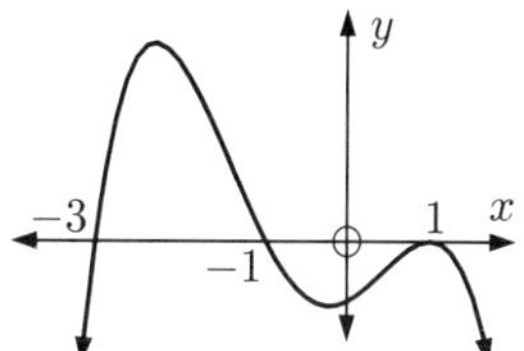

3 Find the equation of the quartic whose graph:

- **a** cuts the x-axis at -4 and $\frac{1}{2}$, touches it at 2, and passes through the point $(1, 5)$
- **b** touches the x-axis at $\frac{2}{3}$ and -3, and passes through the point $(-4, 49)$
- **c** cuts the x-axis at $\pm\frac{1}{2}$ and ± 2, and passes through the point $(1, -18)$
- **d** touches the x-axis at 1, cuts the y-axis at -1, and passes through $(-1, -4)$ and $(2, 15)$.

ACTIVITY

Click on the icon to run a card game for graphs of cubic and quartic functions.

CARD GAME

GENERAL POLYNOMIALS

We have already seen that every real cubic polynomial must cut the x-axis at least once, and so has at least one real zero.

If the exact value of the zero is difficult to find, we can use technology to help us. We can then factorise the cubic as a linear factor times a quadratic, and if necessary use the quadratic formula to find the other zeros.

This method is particularly useful if we have one rational zero and two irrational zeros that are radical conjugates.

DISCUSSION — GENERAL POLYNOMIALS

Consider the general polynomial $P(x) = a_n x^n + a_{n-1}x^{n-1} + ... + a_1 x + a_0, \quad a_n \neq 0.$

- Discuss the behaviour of the graph as $x \to -\infty$ and $x \to \infty$ depending on
  - the sign of a_n
  - whether n is odd or even.
- Under what circumstances is $P(x)$ an:
  - odd function
  - even function?

Example 35 **Self Tutor**

Find exactly the zeros of $P(x) = 3x^3 - 14x^2 + 5x + 2$.

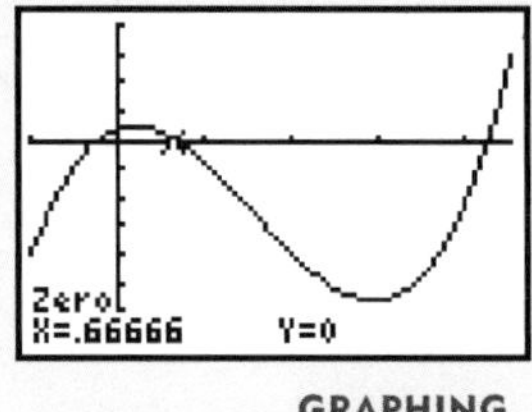

Using the calculator we search for any rational zero.

In this case, $x \approx 0.666\,667$ or $0.\overline{6}$ indicates $x = \frac{2}{3}$ is a zero.

$\therefore$ $(3x - 2)$ is a factor.

$\therefore$ $3x^3 - 14x^2 + 5x + 2 = (3x - 2)(x^2 + ax - 1)$ for some a

$= 3x^3 + (3a - 2)x^2 + (-3 - 2a)x + 2$

Equating coefficients: $3a - 2 = -14$ and $-3 - 2a = 5$

$\therefore$ $3a = -12$ and $-2a = 8$

$\therefore$ $a = -4$

$\therefore$ $P(x) = (3x - 2)(x^2 - 4x - 1)$ which has zeros $\frac{2}{3}$ and $2 \pm \sqrt{5}$ {quadratic formula}

Example 36 **Self Tutor**

Find exactly the roots of $6x^3 + 13x^2 + 20x + 3 = 0$.

Using technology, $x \approx -0.166\,666\,67 = -\frac{1}{6}$ is a root, so $(6x + 1)$ is a factor of the cubic.

$\therefore$ $(6x + 1)(x^2 + ax + 3) = 0$ for some constant a.

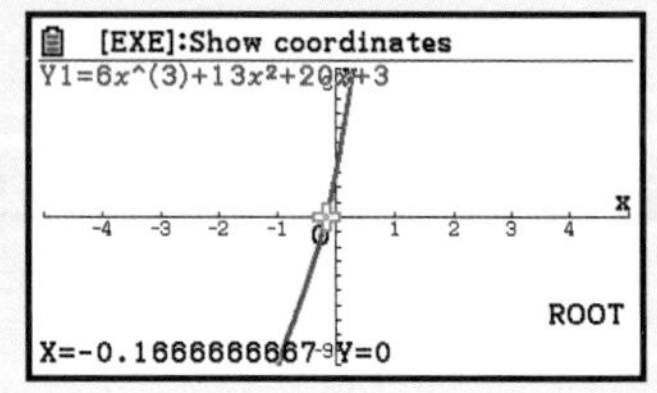

Equating coefficients of x^2: $1 + 6a = 13$

$\therefore$ $6a = 12$

$\therefore$ $a = 2$

Equating coefficients of x: $a + 18 = 20$ ✓

$\therefore$ $(6x + 1)(x^2 + 2x + 3) = 0$

$\therefore$ $x = -\frac{1}{6}$ or $-1 \pm i\sqrt{2}$ {quadratic formula}

For a quartic polynomial $P(x)$ we first need to establish if there are any x-intercepts at all. If there are not then the polynomial must have four complex zeros. If there *are* x-intercepts then we can try to identify linear or quadratic factors.

EXERCISE 6F.3

1 Find exactly all zeros of:

- **a** $x^3 - 3x^2 - 3x + 1$
- **b** $x^3 - 3x^2 + 4x - 2$
- **c** $2x^3 - 3x^2 - 4x - 35$
- **d** $2x^3 - x^2 + 20x - 10$
- **e** $4x^4 - 4x^3 - 25x^2 + x + 6$
- **f** $x^4 - 6x^3 + 22x^2 - 48x + 40$

2 Find exactly the roots of:

- **a** $x^3 + 2x^2 + 3x + 6 = 0$
- **b** $2x^3 + 3x^2 - 3x - 2 = 0$
- **c** $x^3 - 6x^2 + 12x - 8 = 0$
- **d** $2x^3 + 18 = 5x^2 + 9x$
- **e** $x^4 - x^3 - 9x^2 + 11x + 6 = 0$
- **f** $2x^4 - 13x^3 + 27x^2 = 13x + 15$

3 Factorise into linear factors with exact values:

a $x^3 - 3x^2 + 4x - 2$ **b** $x^3 + 3x^2 + 4x + 12$

c $2x^3 - 9x^2 + 6x - 1$ **d** $x^3 - 4x^2 + 9x - 10$

e $4x^3 - 8x^2 + x + 3$ **f** $3x^4 + 4x^3 + 5x^2 + 12x - 12$

g $2x^4 - 3x^3 + 5x^2 + 6x - 4$ **h** $2x^3 + 5x^2 + 8x + 20$

4 The following cubics will not factorise neatly. Find their zeros using technology.

a $x^3 + 2x^2 - 6x - 6$ **b** $x^3 + x^2 - 7x - 8$

5 A scientist is trying to design a crash test barrier with the characteristics shown graphically below.

The independent variable t is the time after impact, measured in milliseconds, such that $0 \leqslant t \leqslant 700$.

The dependent variable is the distance the barrier is depressed during the impact, measured in millimetres.

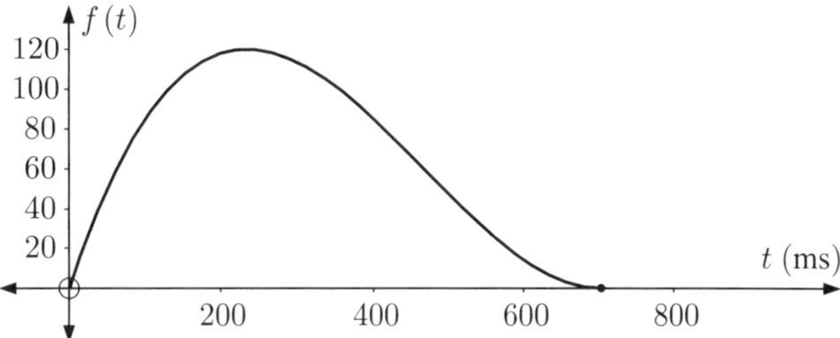

a The equation for this graph has the form $f(t) = kt(t-a)^2$, $0 \leqslant t \leqslant 700$.
Use the graph to find a. What does this value represent?

b If the ideal crash barrier is depressed by 85 mm after 100 milliseconds, find the value of k.
Hence find the equation of the graph given.

6 Last year, the volume of water in a particular reservoir could be described by the model $V(t) = -t^3 + 30t^2 - 131t + 250$ ML, where t is the time in months.

The dam authority rules that if the volume falls below 100 ML, irrigation is prohibited. During which months, if any, was irrigation prohibited in the last twelve months? Include in your answer a neat sketch of any graphs you may have used.

7 A ladder of length 10 metres is leaning against a wall so that it is just touching a cube of edge length one metre as shown.
What height up the wall does the ladder reach?

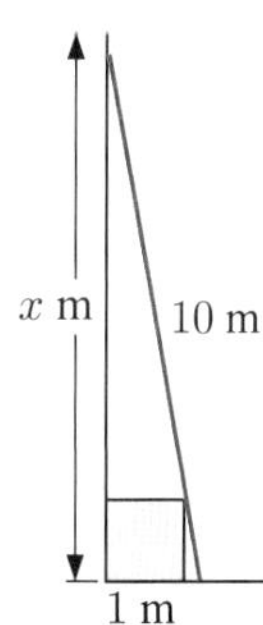

REVIEW SET 6A NON-CALCULATOR

1 Find real numbers a and b such that:

a $a + ib = 4$ **b** $(1-2i)(a+bi) = -5 - 10i$ **c** $(a+2i)(1+bi) = 17 - 19i$

2 If $z = 3 + i$ and $w = -2 - i$, find in simplest form:

a $2z - 3w$ **b** $\dfrac{z^*}{w}$ **c** z^3

3 Find exactly the real and imaginary parts of z if $z = \dfrac{3}{i+\sqrt{3}} + \sqrt{3}$.

4 Find a complex number z such that $2z - 1 = iz - i$. Write your answer in the form $z = a + bi$ where $a, b \in \mathbb{R}$.

5 Prove that $zw^* - z^*w$ is purely imaginary or zero for all complex numbers z and w.

6 Given $w = \dfrac{z+1}{z^*+1}$ where $z = a + bi$, $a, b \in \mathbb{R}$, write w in the form $x + yi$, $x, y \in \mathbb{R}$.
Hence determine the conditions under which w is purely imaginary.

7 Expand and simplify: **a** $(3x^3 + 2x - 5)(4x - 3)$ **b** $(2x^2 - x + 3)^2$

8 Carry out the following divisions: **a** $\dfrac{x^3}{x+2}$ **b** $\dfrac{x^3}{(x+2)(x+3)}$

9 Find the sum and product of the zeros of:

a $3x^4 - 4x^3 + 3x^2 + 8$ **b** $2x^6 + 2x^4 - x^3 + 7x - 10$

10 State and prove the Remainder theorem.

11 $-2 + bi$ is a solution to $z^2 + az + (3 + a) = 0$. Find constants a and b given that they are real.

12 $P(x)$ has remainder 2 when divided by $x - 3$, and remainder -13 when divided by $x + 2$. Find the remainder when $P(x)$ is divided by $x^2 - x - 6$.

13 Find *all* quartic polynomials with real, rational coefficients, having $2 - i\sqrt{3}$ and $\sqrt{2} + 1$ as two of the zeros.

14 If $f(x) = x^3 - 3x^2 - 9x + b$ has $(x - k)^2$ as a factor, show that there are two possible values of k. For each of these two values of k, find the corresponding value for b, and hence solve $f(x) = 0$.

15 Find *all* real quartic polynomials with rational coefficients, having $3 - i\sqrt{2}$ and $1 - \sqrt{2}$ as two of the zeros.

16 When $P(x) = x^n + 3x^2 + kx + 6$ is divided by $(x + 1)$ the remainder is 12. When $P(x)$ is divided by $(x - 1)$ the remainder is 8. Find k and n given that $34 < n < 38$.

17 If α and β are two of the roots of $x^3 - x + 1 = 0$, show that $\alpha\beta$ is a root of $x^3 + x^2 - 1 = 0$.
Hint: Let $x^3 - x + 1 = (x - \alpha)(x - \beta)(x - \gamma)$.

REVIEW SET 6B — CALCULATOR

1 If $z = \left(\dfrac{5}{2-i} - 3 - 2i\right)^3$, express z in the form $z = x + yi$, $x, y \in \mathbb{Z}$.

2 Without using a calculator, find z if $z^2 = 5 - 12i$. Check your answer using a calculator.

3 Prove that if z is a complex number then both $z + z^*$ and zz^* are real.

4 If $z = 4 + i$ and $w = 3 - 2i$ find $2w^* - iz$.

5 Find rationals a and b such that $\dfrac{2-3i}{2a+bi} = 3 + 2i$.

6 $a + ai$ is a root of $x^2 - 6x + b = 0$ where $a, b \in \mathbb{R}$.
Explain why b has two possible values. Find a in each case.

7 Find the remainder when $x^{47} - 3x^{26} + 5x^3 + 11$ is divided by $x + 1$.

8 Factorise $2z^3 + z^2 + 10z + 5$ as a product of linear factors with exact terms.

9 A quartic polynomial $P(x)$ has graph $y = P(x)$ which touches the x-axis at $(-2, 0)$, cuts the x-axis at $(1, 0)$, cuts the y-axis at $(0, 12)$, and passes through $(2, 80)$.
Find an expression for $P(x)$ in factored form, and hence sketch the graph of $y = P(x)$.

10 Find all zeros of $2z^4 - 5z^3 + 13z^2 - 4z - 6$.

11 Factorise $z^4 + 2z^3 - 2z^2 + 8$ into linear factors.

12 Find the general form of all real polynomials of least degree which have zeros $2 + i$ and $-1 + 3i$.

13 $3 - 2i$ is a zero of $z^4 + kz^3 + 32z + 3k - 1$, where k is real.
Find k and all zeros of the quartic.

14 Find all the zeros of the polynomial $z^4 + 2z^3 + 6z^2 + 8z + 8$ given that one of the zeros is purely imaginary.

15 When a polynomial $P(x)$ is divided by $x^2 - 3x + 2$ the remainder is $2x + 3$.
Find the remainder when $P(x)$ is divided by $x - 2$.

16 Find possible values of k if the line with equation $y = 2x + k$ does not meet the circle with equation $x^2 + y^2 + 8x - 4y + 2 = 0$.

17 $P(x) = 2x^4 - 8x^3 + ax^2 + bx - 110$, $a, b \in \mathbb{R}$, has zeros $m \pm 2i$ and $1 \pm n\sqrt{3}$ where $m, n \in \mathbb{R}$.

a Find m and n.

b Sketch the graph of $y = P(x)$.

REVIEW SET 6C

1 Find real numbers x and y such that $(3x + 2yi)(1 - i) = (3y + 1)i - x$.

2 Solve the equation: $z^2 + iz + 10 = 6z$.

3 Prove that $zw^* + z^*w$ is real for all complex numbers z and w.

4 Find real x and y such that:

a $x + iy = 0$

b $(3 - 2i)(x + i) = 17 + yi$

c $(x + iy)^2 = x - iy$

5 z and w are non-real complex numbers with the property that both $z + w$ and zw are real.
Prove that $z^* = w$.

6 Find the sum and product of the roots of:

a $2x^3 + 3x^2 - 4x + 6 = 0$

b $4x^4 = x^2 + 2x - 6$

7 Find z if $\sqrt{z} = \dfrac{2}{3 - 2i} + 2 + 5i$.

8 Find the remainder when $2x^{17} + 5x^{10} - 7x^3 + 6$ is divided by $x - 2$.

9 $5 - i$ is a zero of $2z^3 + az^2 + 62z + (a - 5)$, where a is real. Find a and the other two zeros.

10 Find, in general form, all real polynomials of least degree which have zeros:

a $i\sqrt{2}$ and $\frac{1}{2}$ **b** $1 - i$ and $-3 - i$.

11 $P(x) = 2x^3 + 7x^2 + kx - k$ is the product of 3 linear factors, 2 of which are identical.

a Show that k can take 3 distinct values.

b Write $P(x)$ as the product of linear factors for the case where k is greatest.

12 Find all roots of $2z^4 - 3z^3 + 2z^2 = 6z + 4$.

13 Suppose a and k are real. For what values of k does $z^3 + az^2 + kz + ka = 0$ have:

a one real root **b** 3 real roots?

14 $(3x + 2)$ and $(x - 2)$ are factors of $6x^3 + ax^2 - 4ax + b$. Find a and b.

15 Find the exact values of k for which the line $y = x - k$ is a tangent to the circle with equation $(x - 2)^2 + (y + 3)^2 = 4$.

16 Find the quotient and remainder when $x^4 + 3x^3 - 7x^2 + 11x - 1$ is divided by $x^2 + 2$. Hence, find constants a and b for which $x^4 + 3x^3 - 7x^2 + (2 + a)x + b$ is exactly divisible by $x^2 + 2$.

17 $P(x)$ is a real polynomial of degree 5 with leading coefficient 1 and zeros $m \pm 2i$, $1 \pm mi$, and 2 $(m \in \mathbb{R})$. The graph of $y = P(x)$ cuts the y-axis at -56.

a Show that $m = \pm\sqrt{3}$. **b** Find the coefficient of x^4.

Chapter 7

Sequences and series

Syllabus reference: 1.1

Contents:

OPENING PROBLEM THE LEGEND OF SISSA IBN DAHIR

Around 1260 AD, the Kurdish historian Ibn Khallikān recorded the following story about Sissa ibn Dahir and a chess game against the Indian King Shihram. (The story is also told in the Legend of the Ambalappuzha Paal Payasam, where the Lord Krishna takes the place of Sissa ibn Dahir, and they play a game of chess with the prize of rice grains rather than wheat.)

King Shihram was a tyrant king, and his subject Sissa ibn Dahir wanted to teach him how important all of his people were. He invented the game of chess for the king, and the king was greatly impressed. He insisted on Sissa ibn Dahir naming his reward, and the wise man asked for one grain of wheat for the first square, two grains of wheat for the second square, four grains of wheat for the third square, and so on, doubling the wheat on each successive square on the board.

The king laughed at first and agreed, for there was so little grain on the first few squares. By halfway he was surprised at the amount of grain being paid, and soon he realised his great error: that he owed more grain than there was in the world.

Things to think about:

- **a** How can we describe the number of grains of wheat for each square?
- **b** What expression gives the number of grains of wheat for the nth square?
- **c** Find the total number of grains of wheat that the king owed.

To help understand problems like the **Opening Problem**, we need to study **sequences** and their sums which are called **series**.

A NUMBER SEQUENCES

In mathematics it is important that we can:
- **recognise** a pattern in a set of numbers,
- **describe** the pattern in words, and
- **continue** the pattern.

A **number sequence** is an ordered list of numbers defined by a rule.

The numbers in the sequence are said to be its **members** or its **terms**.

A sequence which continues forever is called an **infinite sequence**.

A sequence which terminates is called a **finite sequence**.

For example, 3, 7, 11, 15, ... form an infinite number sequence.

The first term is 3, the second term is 7, the third term is 11, and so on.

We can describe this pattern in words:

"The sequence starts at 3 and each term is 4 more than the previous term."

Thus, the fifth term is 19 and the sixth term is 23.

The sequence 3, 7, 11, 15, 19, 23 which terminates with the sixth term, is a finite number sequence.

Example 1 — Self Tutor

Describe the sequence: 14, 17, 20, 23, ... and write down the next two terms.

The sequence starts at 14, and each term is 3 more than the previous term.

The next two terms are 26 and 29.

EXERCISE 7A

1. Write down the first four terms of the sequence if you start with:
   - a 4 and add 9 each time
   - b 45 and subtract 6 each time
   - c 2 and multiply by 3 each time
   - d 96 and divide by 2 each time.
2. For each of the following write a description of the sequence and find the next 2 terms:
   - a 8, 16, 24, 32, ...
   - b 2, 5, 8, 11, ...
   - c 36, 31, 26, 21, ...
   - d 96, 89, 82, 75, ...
   - e 1, 4, 16, 64, ...
   - f 2, 6, 18, 54, ...
   - g 480, 240, 120, 60, ...
   - h 243, 81, 27, 9, ...
   - i 50 000, 10 000, 2000, 400, ...
3. Describe the following number patterns and write down the next 3 terms:
   - a 1, 4, 9, 16, ...
   - b 1, 8, 27, 64, ...
   - c 2, 6, 12, 20, ...
4. Find the next two terms of:
   - a 95, 91, 87, 83, ...
   - b 5, 20, 80, 320, ...
   - c 1, 16, 81, 256, ...
   - d 2, 3, 5, 7, 11, ...
   - e 2, 4, 7, 11, ...
   - f 9, 8, 10, 7, 11, ...

B THE GENERAL TERM OF A NUMBER SEQUENCE

Sequences may be defined in one of the following ways:

- listing all terms (of a finite sequence)
- listing the first few terms and assuming that the pattern represented continues indefinitely
- giving a description in words
- using a formula which represents the **general term** or **nth term**
- using a **recurrence formula** which describes the nth term of a sequence using a formula which involves the preceding terms.

Consider the illustrated tower of bricks. The first row has three bricks, the second row has four bricks, and the third row has five bricks.

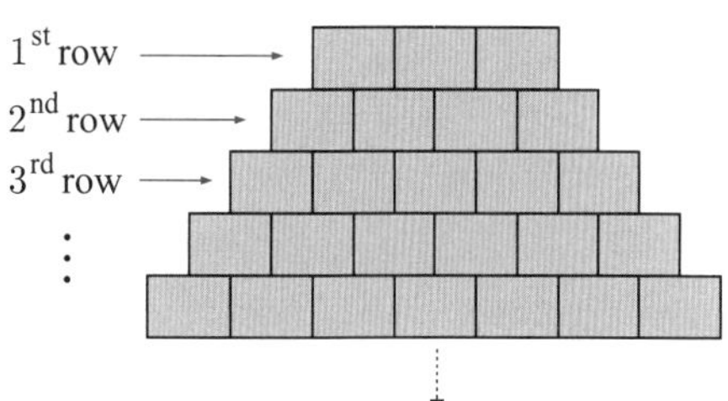

If u_n represents the number of bricks in row n (from the top) then $u_1 = 3$, $u_2 = 4$, $u_3 = 5$, $u_4 = 6$, ...

This sequence can be specified by:

- **listing terms** 3, 4, 5, 6, ...
- **using words** "The top row has three bricks, and each successive row under it has one more brick than the previous row."

- **using an explicit formula** $u_n = n + 2$ is the **general term** or **nth term** formula for $n = 1, 2, 3, 4, ...$

  Check: $u_1 = 1 + 2 = 3$ ✓ $u_2 = 2 + 2 = 4$ ✓ $u_3 = 3 + 2 = 5$ ✓

- **using a recurrence formula** The first layer has 3 bricks, and each subsequent layer has 1 more brick.

  $\therefore$ $u_1 = 3$ and $u_n = u_{n-1} + 1, \ n > 1.$

- **a pictorial or graphical representation**

  Early members of a sequence can be graphed with each represented by a dot.

  The dots *must not* be joined because n must be an integer.

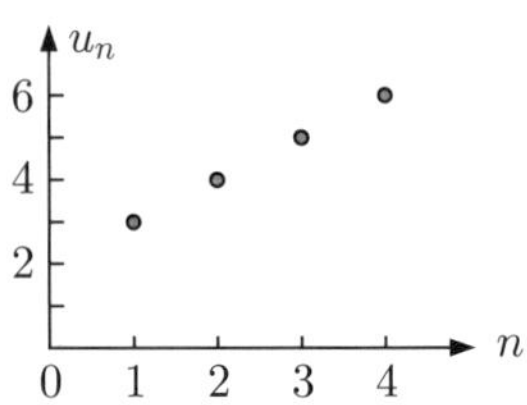

THE GENERAL TERM

The **general term** or **nth term** of a sequence is represented by a symbol with a subscript, for example u_n, T_n, t_n, or A_n. The general term is defined for $n = 1, 2, 3, 4, ...$

$\{u_n\}$ represents the sequence that can be generated by using u_n as the nth term.

The general term u_n is a function where $n \mapsto u_n$. The domain is $n \in \mathbb{Z}^+$ for an infinite sequence, or a subset of $\mathbb{Z}^+$ for a finite sequence.

For example, $\{2n + 1\}$ generates the sequence 3, 5, 7, 9, 11, ...

You can use technology to help generate sequences from a formula.

EXERCISE 7B.1

1 A sequence is defined by $u_n = 3n - 2$. Find:

a u_1 **b** u_5 **c** u_{27}

2 Consider the sequence defined by $u_n = 2n + 5$.

a Find the first four terms of the sequence. **b** Display these members on a graph.

3 Evaluate the first *five* terms of the sequence:

a $\{2n\}$ **b** $\{2n + 2\}$ **c** $\{2n - 1\}$ **d** $\{2n - 3\}$

e $\{2n + 3\}$ **f** $\{2n + 11\}$ **g** $\{3n + 1\}$ **h** $\{4n - 3\}$

4 Evaluate the first *five* terms of the sequence:

a $\{2^n\}$ **b** $\{3 \times 2^n\}$ **c** $\{6 \times (\frac{1}{2})^n\}$ **d** $\{(-2)^n\}$

5 Evaluate the first *five* terms of the sequence $\{15 - (-2)^n\}$.

RECURRENCE FORMULAE

A **recurrence formula** describes the nth term of a sequence using a formula which involves the preceding terms.

When we write a recurrence formula, we need to specify the initial terms needed for the formula to start.

For example, the recurrence formula $u_n = u_{n-1} + 3$ is by itself useless, since we do not know what the first term is.

However, if we write $u_1 = 7, \quad u_n = u_{n-1} + 3, \quad n > 1,$ then we can write

$$\begin{aligned} u_2 &= u_1 + 3 = 7 + 3 = 10 \\ u_3 &= u_2 + 3 = 10 + 3 = 13 \\ u_4 &= u_3 + 3 = 13 + 3 = 16 \\ &\vdots \end{aligned}$$

EXERCISE 7B.2

1 A sequence is defined by $u_1 = 3, \quad u_n = u_{n-1} - 4, \quad n > 1.$ Find:

a u_2 **b** u_3 **c** u_4 **d** u_5

2 The first term of a sequence is 3, and each subsequent term is double the previous one.

a Describe the sequence fully using a recurrence formula.

b Display the first 5 terms on a graph.

3 A sequence is defined by $u_1 = 1, \quad u_n = \dfrac{1}{1 + u_{n-1}}, \quad n > 1.$

a Find the next four terms of the sequence.

b Display the first 5 terms on a graph.

c Assume that the sequence terms tend to the constant value u, so that as $n \to \infty, \ u_{n-1} \to u$ and $u_n \to u$.

i Show that $u^2 + u - 1 = 0.$ **ii** Hence show that $\dfrac{1}{1+\dfrac{1}{1+\dfrac{1}{1+\dfrac{1}{1+....}}}} = \dfrac{-1+\sqrt{5}}{2}.$

4 A sequence is defined by $u_1 = 1, \quad u_n = \sqrt{\dfrac{1}{1 + u_{n-1}}}, \quad n > 1.$

a Find the next four terms of the sequence, writing your answers correct to 5 decimal places.

b Assuming that the sequence terms tend to the constant value u, show that $u^3 + u^2 - 1 = 0.$

c Hence find $\sqrt{\dfrac{1}{1+\sqrt{\dfrac{1}{1+\sqrt{\dfrac{1}{1+\sqrt{....}}}}}}}$ correct to 5 decimal places.

5 Find, correct to 5 decimal places: $\dfrac{1}{1+\left(\dfrac{1}{1+\left(\dfrac{1}{1+\left(\dfrac{1}{1+(....)}\right)^2}\right)^2}\right)^2}$

INVESTIGATION 1 THE FIBONACCI SEQUENCE

Leonardo Pisano Bigollo, known commonly as Fibonacci, was born in Pisa around 1170 AD. He is best known for the **Fibonacci sequence** which starts with 0 and 1, and then each subsequent member of the sequence is the sum of the preceding two members.

What to do:

1 Describe the Fibonacci sequence fully using a recurrence formula.

2 Write down the first 10 terms of the sequence.

3 Can the sequence be described using a formula for the general term u_n?

4 Research where the Fibonacci sequence is found in nature.

ARITHMETIC SEQUENCES

An **arithmetic sequence** is a sequence in which each term differs from the previous one by the same fixed number.

It can also be referred to as an **arithmetic progression**.

For example:

- the tower of bricks in the previous section forms an arithmetic sequence where the difference between terms is 1
- 2, 5, 8, 11, 14, ... is arithmetic as $5-2=8-5=11-8=14-11=....$
- 31, 27, 23, 19, ... is arithmetic as $27-31=23-27=19-23=....$

ALGEBRAIC DEFINITION

$\{u_n\}$ is **arithmetic** $\Leftrightarrow$ $u_{n+1}-u_n=d$ for all positive integers n where d is a constant called the **common difference**.

The symbol $\Leftrightarrow$ means 'if and only if'. It implies both:

- if $\{u_n\}$ is arithmetic then $u_{n+1}-u_n$ is a constant
- if $u_{n+1}-u_n$ is a constant then $\{u_n\}$ is arithmetic.

THE NAME 'ARITHMETIC'

If a, b, and c are any consecutive terms of an arithmetic sequence then

$$\begin{aligned} b-a &= c-b \quad \{\text{equating common differences}\} \\ \therefore\ 2b &= a+c \\ \therefore\ b &= \frac{a+c}{2} \end{aligned}$$

So, the middle term is the **arithmetic mean** of the terms on either side of it.

THE GENERAL TERM FORMULA

Suppose the first term of an arithmetic sequence is u_1 and the common difference is d.

Then $u_2=u_1+d$, $u_3=u_1+2d$, $u_4=u_1+3d$, and so on.

Hence $u_n=u_1+\underbrace{(n-1)}\,d$

term number (pointing to n in u_n)

the coefficient of d is one less than the term number

For an **arithmetic sequence** with **first term** u_1 and **common difference** d the **general term** or **nth term** is $u_n=u_1+(n-1)d$.

Example 2 | Self Tutor

Consider the sequence 2, 9, 16, 23, 30, ...

a Show that the sequence is arithmetic.

b Find a formula for the general term u_n.

c Find the 100th term of the sequence.

d Is **i** 828 **ii** 2341 a term of the sequence?

---

a $9 - 2 = 7$
$16 - 9 = 7$
$23 - 16 = 7$
$30 - 23 = 7$

The difference between successive terms is constant.
So, the sequence is arithmetic, with $u_1 = 2$ and $d = 7$.

b $u_n = u_1 + (n-1)d$
$\therefore\ u_n = 2 + 7(n-1)$
$\therefore\ u_n = 7n - 5$

c If $n = 100$,
$u_{100} = 7(100) - 5$
$= 695$

d **i** Let $u_n = 828$
$\therefore\ 7n - 5 = 828$
$\therefore\ 7n = 833$
$\therefore\ n = 119$
$\therefore$ 828 is a term of the sequence, and in fact is the 119th term.

ii Let $u_n = 2341$
$\therefore\ 7n - 5 = 2341$
$\therefore\ 7n = 2346$
$\therefore\ n = 335\frac{1}{7}$
But n must be an integer, so 2341 is not a member of the sequence.

EXERCISE 7C

1 Find the 10th term of each of the following arithmetic sequences:

a 19, 25, 31, 37, ... **b** 101, 97, 93, 89, ... **c** 8, $9\frac{1}{2}$, 11, $12\frac{1}{2}$, ...

2 Find the 15th term of each of the following arithmetic sequences:

a 31, 36, 41, 46, ... **b** 5, -3, -11, -19, ... **c** a, $a+d$, $a+2d$, $a+3d$, ...

3 Consider the sequence 6, 17, 28, 39, 50, ...

a Show that the sequence is arithmetic. **b** Find the formula for its general term.

c Find its 50th term. **d** Is 325 a member?

e Is 761 a member?

4 Consider the sequence 87, 83, 79, 75, 71, ...

a Show that the sequence is arithmetic. **b** Find the formula for its general term.

c Find the 40th term. **d** Which term of the sequence is -297?

5 A sequence is defined by $u_n = 3n - 2$.

a Prove that the sequence is arithmetic. **Hint:** Find $u_{n+1} - u_n$.

b Find u_1 and d. **c** Find the 57th term.

d What is the largest term of the sequence that is smaller than 450? Which term is this?

6 A sequence is defined by $u_n = \dfrac{71 - 7n}{2}$.

a Prove that the sequence is arithmetic. **b** Find u_1 and d. **c** Find u_{75}.

d For what values of n are the terms of the sequence less than -200?

Example 3 **Self Tutor**

Find k given that $3k+1$, k, and -3 are consecutive terms of an arithmetic sequence.

Since the terms are consecutive, $k-(3k+1) = -3-k$ {equating differences}

$$\therefore\ k-3k-1 = -3-k$$
$$\therefore\ -2k-1 = -3-k$$
$$\therefore\ -1+3 = -k+2k$$
$$\therefore\ k = 2$$

or Since the middle term is the arithmetic mean of the terms on either side of it,

$$k = \frac{(3k+1)+(-3)}{2}$$
$$\therefore\ 2k = 3k-2$$
$$\therefore\ k = 2$$

7 Find k given the consecutive arithmetic terms:

a $32, k, 3$ **b** $k, 7, 10$ **c** $k+1, 2k+1, 13$

d $k-1, 2k+3, 7-k$ **e** k, k^2, k^2+6 **f** $5, k, k^2-8$

Example 4 **Self Tutor**

Find the general term u_n for an arithmetic sequence with $u_3 = 8$ and $u_8 = -17$.

$u_3 = 8$ $\therefore$ $u_1 + 2d = 8$ .... (1) {using $u_n = u_1 + (n-1)d$}

$u_8 = -17$ $\therefore$ $u_1 + 7d = -17$ .... (2)

We now solve (1) and (2) simultaneously:

$$-u_1 - 2d = -8 \quad \text{\{multiplying both sides of (1) by } -1\}$$
$$u_1 + 7d = -17$$
$$\therefore\ 5d = -25 \quad \text{\{adding the equations\}}$$
$$\therefore\ d = -5$$

So, in (1): $u_1 + 2(-5) = 8$

$$\therefore\ u_1 - 10 = 8$$
$$\therefore\ u_1 = 18$$

Now $u_n = u_1 + (n-1)d$

$$\therefore\ u_n = 18 - 5(n-1)$$
$$\therefore\ u_n = 18 - 5n + 5$$
$$\therefore\ u_n = 23 - 5n$$

Check:

$$u_3 = 23 - 5(3) = 23 - 15 = 8\ \checkmark$$
$$u_8 = 23 - 5(8) = 23 - 40 = -17\ \checkmark$$

8 Find the general term u_n for an arithmetic sequence with:

a $u_7 = 41$ and $u_{13} = 77$ **b** $u_5 = -2$ and $u_{12} = -12\frac{1}{2}$

c seventh term 1 and fifteenth term -39

d eleventh and eighth terms being -16 and $-11\frac{1}{2}$ respectively.

Example 5 **Self Tutor**

Insert four numbers between 3 and 12 so that all six numbers are in arithmetic sequence.

Suppose the common difference is d.

$\therefore$ the numbers are $3,\ 3+d,\ 3+2d,\ 3+3d,\ 3+4d,$ and 12

$$\therefore\ 3+5d=12$$
$$\therefore\ 5d=9$$
$$\therefore\ d=\tfrac{9}{5}=1.8$$

So, the sequence is 3, 4.8, 6.6, 8.4, 10.2, 12.

9 **a** Insert three numbers between 5 and 10 so that all five numbers are in arithmetic sequence.

b Insert six numbers between -1 and 32 so that all eight numbers are in arithmetic sequence.

10 Consider the arithmetic sequence $36,\ 35\frac{1}{3},\ 34\frac{2}{3},\ ....$

a Find u_1 and d. **b** Which term of the sequence is -30?

11 An arithmetic sequence starts 23, 36, 49, 62, ... What is the first term of the sequence to exceed 100 000?

Example 6 **Self Tutor**

Ryan is a cartoonist. His comic strip has just been bought by a newspaper, so he sends them the 28 comic strips he has drawn so far. Each week after the first he mails 3 more comic strips to the newspaper.

a Find the total number of comic strips sent after 1, 2, 3, and 4 weeks.

b Show that the total number of comic strips sent after n weeks forms an arithmetic sequence.

c Find the number of comic strips sent after 15 weeks.

d When does Ryan send his 120th comic strip?

a *Week 1*: 28 comic strips
Week 2: $28+3=31$ comic strips
Week 3: $31+3=34$ comic strips
Week 4: $34+3=37$ comic strips

b Every week, Ryan sends 3 comic strips, so the difference between successive weeks is always 3. We have an arithmetic sequence with $u_1=28$ and common difference $d=3$.

c $u_n = u_1+(n-1)d$
$= 28+(n-1)\times 3$
$= 25+3n$

$\therefore\ u_{15}=25+3\times 15$
$= 70$

After 15 weeks Ryan has sent 70 comic strips.

d We want to find n such that $u_n=120$

$$\therefore\ 25+3n=120$$
$$\therefore\ 3n=95$$
$$\therefore\ n=31\tfrac{2}{3}$$

Ryan sends the 120th comic strip in the 32nd week.

12 A luxury car manufacturer sets up a factory for a new model. In the first month only 5 cars are produced. After this, 13 cars are assembled every month.

- **a** List the total number of cars that have been made in the factory by the end of each of the first six months.
- **b** Explain why the total number of cars made after n months forms an arithmetic sequence.
- **c** How many cars are made in the first year?
- **d** How long is it until the 250th car is manufactured?

13 Valéria joins a social networking website. After 1 week she has 34 online friends. At the end of 2 weeks she has 41 friends, after 3 weeks she has 48 friends, and after 4 weeks she has 55 friends.

- **a** Show that Valéria's number of friends forms an arithmetic sequence.
- **b** Assuming the pattern continues, find the number of online friends Valéria will have after 12 weeks.
- **c** After how many weeks will Valéria have 150 online friends?

14 A farmer feeds his cattle herd with hay every day in July. The amount of hay in his barn at the end of day n is given by the arithmetic sequence $u_n = 100 - 2.7n$ tonnes.

- **a** Write down the amount of hay in the barn on the first three days of July.
- **b** Find and interpret the common difference.
- **c** Find and interpret u_{25}.
- **d** How much hay is in the barn at the beginning of August?

D GEOMETRIC SEQUENCES

A sequence is **geometric** if each term can be obtained from the previous one by multiplying by the same non-zero constant.

A geometric sequence is also referred to as a **geometric progression**.

For example: 2, 10, 50, 250, ... is a geometric sequence as each term can be obtained by multiplying the previous term by 5.

Notice that $\frac{10}{2} = \frac{50}{10} = \frac{250}{50} = 5$, so each term divided by the previous one gives the same constant.

ALGEBRAIC DEFINITION

$\{u_n\}$ is **geometric** $\Leftrightarrow$ $\dfrac{u_{n+1}}{u_n} = r$ for all positive integers n where r is a constant called the **common ratio**.

For example:

- 2, 10, 50, 250, ... is geometric with $r = 5$.
- 2, -10, 50, -250, ... is geometric with $r = -5$.

THE NAME 'GEOMETRIC'

If a, b, and c are any consecutive terms of a geometric sequence then $\dfrac{b}{a} = \dfrac{c}{b}$.

$\therefore$ $b^2 = ac$ and so $b = \pm\sqrt{ac}$ where $\sqrt{ac}$ is the **geometric mean** of a and c.

THE GENERAL TERM FORMULA

Suppose the first term of a geometric sequence is u_1 and the common ratio is r.

Then $u_2 = u_1\, r$, $u_3 = u_1\, r^2$, $u_4 = u_1\, r^3$, and so on.

Hence $u_n = u_1\, r^{n-1}$

term number — The power of r is one less than the term number.

For a **geometric sequence** with **first term u_1** and **common ratio r**,
the **general term** or **nth term** is $u_n = u_1 r^{n-1}$.

Example 7 — Self Tutor

Consider the sequence $8, 4, 2, 1, \frac{1}{2}, ...$

a Show that the sequence is geometric.

b Find the general term u_n.

c Hence, find the 12th term as a fraction.

a $\dfrac{4}{8} = \frac{1}{2}$ $\quad \dfrac{2}{4} = \frac{1}{2}$ $\quad \dfrac{1}{2} = \frac{1}{2}$ $\quad \dfrac{\frac{1}{2}}{1} = \frac{1}{2}$

Consecutive terms have a common ratio of $\frac{1}{2}$.

$\therefore$ the sequence is geometric with $u_1 = 8$ and $r = \frac{1}{2}$.

b $u_n = u_1 r^{n-1}$

$\therefore$ $u_n = 8\left(\frac{1}{2}\right)^{n-1}$ *or* $u_n = 2^3 \times (2^{-1})^{n-1}$
$= 2^3 \times 2^{-n+1}$
$= 2^{3+(-n+1)}$
$= 2^{4-n}$

c $u_{12} = 8 \times \left(\frac{1}{2}\right)^{11}$
$= \frac{1}{256}$

EXERCISE 7D.1

1 For the geometric sequence with first two terms given, find b and c:

a $2, 6, b, c, ...$ b $10, 5, b, c, ...$ c $12, -6, b, c, ...$

2 Find the 6th term in each of the following geometric sequences:

a $3, 6, 12, 24, ...$ b $2, 10, 50, ...$ c $512, 256, 128, ...$

3 Find the 9th term in each of the following geometric sequences:

a $1, 3, 9, 27, ...$ b $12, 18, 27, ...$ c $\frac{1}{16}, -\frac{1}{8}, \frac{1}{4}, -\frac{1}{2}, ...$ d $a, ar, ar^2, ...$

4 a Show that the sequence $5, 10, 20, 40, ...$ is geometric.

b Find u_n and hence find the 15th term.

5 **a** Show that the sequence $12, -6, 3, -\frac{3}{2}, ...$ is geometric.

b Find u_n and hence write the 13th term as a rational number.

6 Show that the sequence $8, -6, 4.5, -3.375, ...$ is geometric. Hence find the 10th term as a decimal.

7 Show that the sequence $8, 4\sqrt{2}, 4, 2\sqrt{2}, ...$ is geometric. Hence show that the general term of the sequence is $u_n = 2^{\frac{7}{2}-\frac{1}{2}n}$.

Example 8 **Self Tutor**

$k-1$, $2k$, and $21-k$ are consecutive terms of a geometric sequence. Find k.

Since the terms are geometric, $\frac{2k}{k-1} = \frac{21-k}{2k}$ {equating the common ratio r}

$$\therefore\ 4k^2 = (21-k)(k-1)$$
$$\therefore\ 4k^2 = 21k - 21 - k^2 + k$$
$$\therefore\ 5k^2 - 22k + 21 = 0$$
$$\therefore\ (5k-7)(k-3) = 0$$
$$\therefore\ k = \tfrac{7}{5} \text{ or } 3$$

Check: If $k = \frac{7}{5}$ the terms are: $\frac{2}{5}, \frac{14}{5}, \frac{98}{5}$. ✓ $\{r = 7\}$

If $k = 3$ the terms are: 2, 6, 18. ✓ $\{r = 3\}$

8 Find k given that the following are consecutive terms of a geometric sequence:

a $7, k, 28$ **b** $k, 3k, 20-k$ **c** $k, k+8, 9k$

Example 9 **Self Tutor**

A geometric sequence has $u_2 = -6$ and $u_5 = 162$. Find its general term.

$u_2 = u_1 r = -6$ .... (1)

and $u_5 = u_1 r^4 = 162$ .... (2)

Now $\frac{u_1 r^4}{u_1 r} = \frac{162}{-6}$ {(2) ÷ (1)}

$$\therefore\ r^3 = -27$$
$$\therefore\ r = \sqrt[3]{-27}$$
$$\therefore\ r = -3$$

Using (1), $u_1(-3) = -6$

$$\therefore\ u_1 = 2$$

Thus $u_n = 2 \times (-3)^{n-1}$.

9 Find the general term u_n of the geometric sequence which has:

a $u_4 = 24$ and $u_7 = 192$

b $u_3 = 8$ and $u_6 = -1$

c $u_7 = 24$ and $u_{15} = 384$

d $u_3 = 5$ and $u_7 = \frac{5}{4}$

Example 10 — Self Tutor

Find the first term of the sequence $6, 6\sqrt{2}, 12, 12\sqrt{2}, ...$ which exceeds 1400.

The sequence is geometric with $u_1 = 6$ and $r = \sqrt{2}$

$\therefore\ u_n = 6 \times (\sqrt{2})^{n-1}$.

We need to find n such that $u_n > 1400$.

Using a graphics calculator with $Y_1 = 6 \times (\sqrt{2})^\wedge(X - 1)$, we view a *table of values*:

Casio fx-CG20

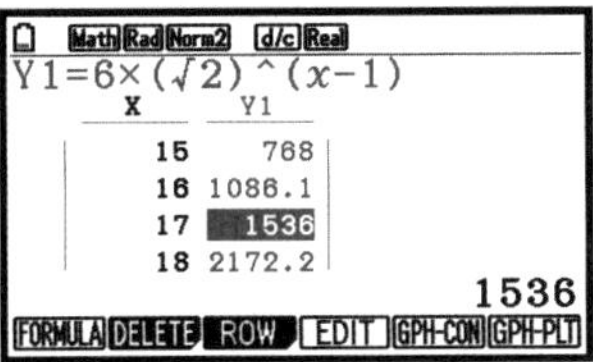

TI-84 Plus

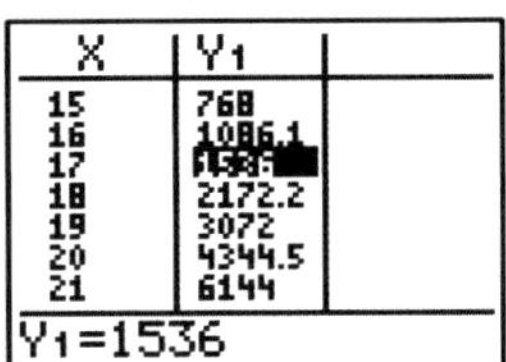

TI-*n*spire

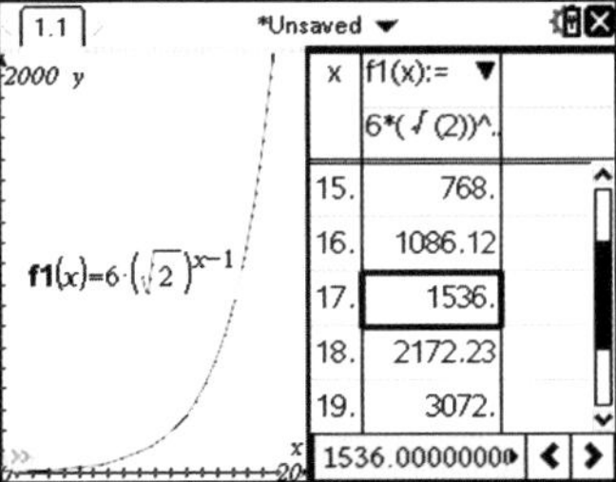

The first term to exceed 1400 is $u_{17} = 1536$.

10 **a** Find the first term of the sequence $2, 6, 18, 54, ...$ which exceeds 10 000.

b Find the first term of the sequence $4, 4\sqrt{3}, 12, 12\sqrt{3}, ...$ which exceeds 4800.

c Find the first term of the sequence $12, 6, 3, 1.5, ...$ which is less than 0.0001.

GEOMETRIC SEQUENCE PROBLEMS

Problems of **growth** and **decay** involve repeated multiplications by a constant number. We can therefore use geometric sequences to model these situations.

Example 11 — Self Tutor

The initial population of rabbits on a farm was 50.
The population increased by 7% each week.

a How many rabbits were present after:

i 15 weeks **ii** 30 weeks?

b How long would it take for the population to reach 500?

There is a fixed percentage increase each week, so the population forms a geometric sequence.

$u_1 = 50$ and $r = 1.07$

$u_2 = 50 \times 1.07^1$ = the population after 1 week

$u_3 = 50 \times 1.07^2$ = the population after 2 weeks

$\vdots$

$u_{n+1} = 50 \times 1.07^n$ = the population after n weeks.

a **i** $u_{16} = 50 \times (1.07)^{15}$
≈ 137.95
There were 138 rabbits.

ii $u_{31} = 50 \times (1.07)^{30}$
≈ 380.61
There were 381 rabbits.

b We need to solve $50 \times (1.07)^n = 500$

$$\therefore \quad (1.07)^n = 10$$

$$\therefore \quad n = \frac{\ln 10}{\ln(1.07)} \approx 34.03$$

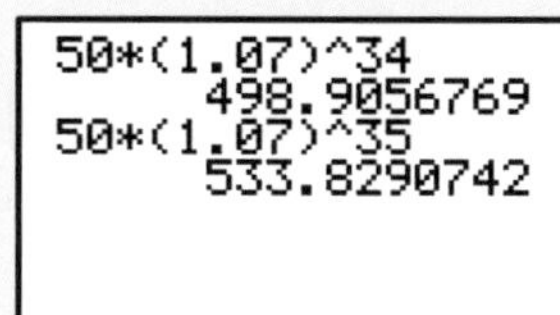

or **trial and error** on your calculator gives $n \approx 34$ weeks.

or finding the **point of intersection** of $Y_1 = 50 \times 1.07$^X and $Y_2 = 500$, the solution is ≈ 34.0 weeks.

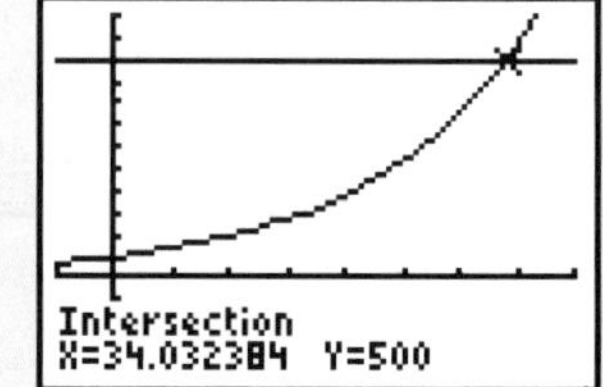

The population will reach 500 early in the 35th week.

EXERCISE 7D.2

1 A nest of ants initially contains 500 individuals. The population is increasing by 12% each week.

a How many ants will there be after:
i 10 weeks **ii** 20 weeks?

b Use technology to find how many weeks it will take for the ant population to reach 2000.

2 The animal *Eraticus* is endangered. Since 1995 there has only been one colony remaining, and in 1995 the population of the colony was 555. The population has been steadily decreasing by 4.5% per year.

a Find the population in the year 2010.

b In which year would we expect the population to have declined to 50?

3 

A herd of 32 deer is to be left unchecked on a large island off the coast of Alaska. It is estimated that the size of the herd will increase each year by 18%.

a Estimate the size of the herd after:
i 5 years **ii** 10 years.

b How long will it take for the herd size to reach 5000?

4 An endangered species of marsupials has a population of 178. However, with a successful breeding program it is expected to increase by 32% each year.

a Find the expected population size after: **i** 10 years **ii** 25 years.

b How long will it take for the population to reach 10 000?

COMPOUND INTEREST

Suppose you invest \$1000 in the bank. You leave the money in the bank for 3 years, and are paid an interest rate of 10% per annum (p.a). The interest is added to your investment each year, so the total value *increases*.

per annum means
each year.

The percentage increase each year is 10%, so at the end of the year you will have $100\% + 10\% = 110\%$ of the value at its start. This corresponds to a *multiplier* of 1.1.

After one year your investment is worth $\$1000 \times 1.1 = \$1100$.

After two years it is worth

$$\begin{aligned} &\$1100 \times 1.1 \\ &= \$1000 \times 1.1 \times 1.1 \\ &= \$1000 \times (1.1)^2 = \$1210 \end{aligned}$$

After three years it is worth

$$\begin{aligned} &\$1210 \times 1.1 \\ &= \$1000 \times (1.1)^2 \times 1.1 \\ &= \$1000 \times (1.1)^3 = \$1331 \end{aligned}$$

This suggests that if the money is left in your account for n years it would amount to $\$1000 \times (1.1)^n$.

Observe that:

$$\begin{aligned} u_1 &= \$1000 && = \text{initial investment} \\ u_2 &= u_1 \times 1.1 && = \text{amount after 1 year} \\ u_3 &= u_1 \times (1.1)^2 && = \text{amount after 2 years} \\ u_4 &= u_1 \times (1.1)^3 && = \text{amount after 3 years} \\ &\vdots \\ u_{n+1} &= u_1 \times (1.1)^n && = \text{amount after } n \text{ years} \end{aligned}$$

THE COMPOUND INTEREST FORMULA

We can calculate the value of a compounding investment using the formula

$$\boldsymbol{u_{n+1} = u_1 \times r^n}$$

where
u_1 = initial investment
r = growth multiplier for each period
n = number of compounding periods
u_{n+1} = amount after n compounding periods.

Example 12 **Self Tutor**

\$5000 is invested for 4 years at 7% p.a. compound interest, compounded annually. What will it amount to at the end of this period? Give your answer to the nearest cent.

$$\begin{aligned} u_5 &= u_1 \times r^4 && \text{is the amount after 4 years} \\ &= 5000 \times (1.07)^4 && \{\text{for a 7\% increase 100\% becomes 107\%}\} \\ &\approx 6553.98 \end{aligned}$$

The investment will amount to \$6553.98.

EXERCISE 7D.3

1. a. What will an investment of \$3000 at 10% p.a. compound interest amount to after 3 years?
   b. How much of this is interest?
2. How much compound interest is earned by investing €20 000 at 12% p.a. if the investment is over a 4 year period?
3. a. What will an investment of ¥30 000 at 10% p.a. compound interest amount to after 4 years?
   b. How much of this is interest?
4. How much compound interest is earned by investing \$80 000 at 9% p.a. if the investment is over a 3 year period?
5. What will an investment of ¥100 000 amount to after 5 years if it earns 8% p.a. compounded twice annually?
6. What will an investment of £45 000 amount to after 21 months if it earns 7.5% p.a. compounded quarterly?

Example 13 **Self Tutor**

How much should I invest now if I need a maturing value of €10 000 in 4 years' time, and I am able to invest at 8% p.a. compounded twice annually? Give your answer to the nearest cent.

The initial investment u_1 is unknown.

The investment is compounded twice annually, so the multiplier $r = 1 + \frac{0.08}{2} = 1.04$.

There are $4 \times 2 = 8$ compounding periods, so $n = 8$

$\therefore \quad u_{n+1} = u_9$

Now $u_9 = u_1 \times r^8$ $\{$using $u_{n+1} = u_1 \times r^n\}$

$\therefore \quad 10\,000 = u_1 \times (1.04)^8$

$\therefore \quad u_1 = \frac{10\,000}{(1.04)^8}$

$\therefore \quad u_1 \approx 7306.90$

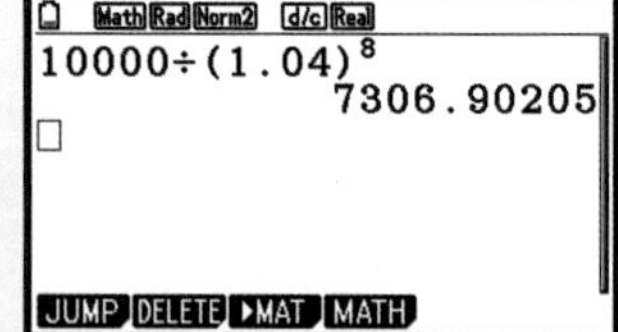

I should invest €7306.90 now.

7. How much money must be invested now if you require \$20 000 for a holiday in 4 years' time and the money can be invested at a fixed rate of 7.5% p.a. compounded annually?
8. What initial investment is required to produce a maturing amount of £15 000 in 60 months' time given a guaranteed fixed interest rate of 5.5% p.a. compounded annually?
9. How much should I invest now to yield €25 000 in 3 years' time, if the money can be invested at a fixed rate of 8% p.a. compounded quarterly?
10. What initial investment will yield ¥40 000 in 8 years' time if your money can be invested at 9% p.a. compounded monthly?

E SERIES

A **series** is the sum of the terms of a sequence.

For the **finite** sequence $\{u_n\}$ with n terms, the corresponding series is $u_1 + u_2 + u_3 + ... + u_n$.

The sum of this series is $S_n = u_1 + u_2 + u_3 + ... + u_n$ and this will always be a finite real number.

For the **infinite** sequence $\{u_n\}$, the corresponding series is $u_1 + u_2 + u_3 + ... + u_n + ...$

In many cases, the sum of an infinite series cannot be calculated. In some cases, however, it does converge to a finite number.

SIGMA NOTATION

$u_1 + u_2 + u_3 + u_4 + ... + u_n$ can be written more compactly using **sigma notation**.

The symbol $\sum$ is called **sigma**. It is the equivalent of capital S in the Greek alphabet.

We write $u_1 + u_2 + u_3 + u_4 + ... + u_n$ as $\sum\limits_{k=1}^{n} u_k$.

$\sum\limits_{k=1}^{n} u_k$ reads "the sum of all numbers of the form u_k where $k = 1, 2, 3, ...,$ up to n".

Example 14 Self Tutor

Consider the sequence 1, 4, 9, 16, 25, ...

a Write down an expression for S_n.

b Find S_n for $n = 1, 2, 3, 4,$ and 5.

a $S_n = 1^2 + 2^2 + 3^2 + 4^2 + ... + n^2$ {all terms are squares}

$= \sum\limits_{k=1}^{n} k^2$

b $S_1 = 1$

$S_2 = 1 + 4 = 5$

$S_3 = 1 + 4 + 9 = 14$

$S_4 = 1 + 4 + 9 + 16 = 30$

$S_5 = 1 + 4 + 9 + 16 + 25 = 55$

Example 15 Self Tutor

Expand and evaluate: **a** $\sum\limits_{k=1}^{7} (k+1)$ **b** $\sum\limits_{k=1}^{5} \frac{1}{2^k}$

a $\sum\limits_{k=1}^{7} (k+1)$

$= 2 + 3 + 4 + 5 + 6 + 7 + 8$

$= 35$

b $\sum\limits_{k=1}^{5} \frac{1}{2^k}$

$= \frac{1}{2} + \frac{1}{4} + \frac{1}{8} + \frac{1}{16} + \frac{1}{32}$

$= \frac{31}{32}$

You can also use technology to evaluate the sum of a series in sigma notation. Click on the icon for instructions.

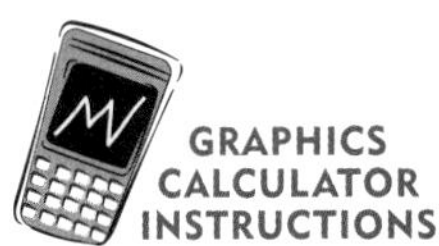

PROPERTIES OF SIGMA NOTATION

$$\sum_{k=1}^{n}(a_k+b_k)=\sum_{k=1}^{n}a_k+\sum_{k=1}^{n}b_k$$

If c is a constant, $\sum_{k=1}^{n}ca_k=c\sum_{k=1}^{n}a_k$ and $\sum_{k=1}^{n}c=cn$.

EXERCISE 7E

1 For the following sequences:

i write down an expression for S_n **ii** find S_5.

a 3, 11, 19, 27, ... **b** 42, 37, 32, 27, ... **c** 12, 6, 3, $1\frac{1}{2}$, ...

d 2, 3, $4\frac{1}{2}$, $6\frac{3}{4}$, ... **e** 1, $\frac{1}{2}$, $\frac{1}{4}$, $\frac{1}{8}$, ... **f** 1, 8, 27, 64, ...

2 Expand and evaluate:

a $\sum_{k=1}^{3}4k$ **b** $\sum_{k=1}^{6}(k+1)$ **c** $\sum_{k=1}^{4}(3k-5)$

d $\sum_{k=1}^{5}(11-2k)$ **e** $\sum_{k=1}^{7}k(k+1)$ **f** $\sum_{k=1}^{5}10\times 2^{k-1}$

3 For $u_n=3n-1$, write $u_1+u_2+u_3+....+u_{20}$ using sigma notation and evaluate the sum.

4 Show that:

a $\sum_{k=1}^{n}c=cn$ **b** $\sum_{k=1}^{n}ca_k=c\sum_{k=1}^{n}a_k$ **c** $\sum_{k=1}^{n}(a_k+b_k)=\sum_{k=1}^{n}a_k+\sum_{k=1}^{n}b_k$

5 Evaluate: **a** $\sum_{k=1}^{5}k(k+1)(k+2)$ **b** $\sum_{k=6}^{12}100\times(1.2)^{k-3}$

6 **a** Expand $\sum_{k=1}^{n}k$.

b Now write the sum with terms in the reverse order, placing each term under a term in the original expansion. Add each term with the one under it.

c Hence write an expression for the sum S_n of the first n integers.

d Hence find a and b if $\sum_{k=1}^{n}(ak+b)=8n^2+11n$ for all positive integers n.

7 **a** Explain why $\sum_{k=1}^{n}(3k^2+4k-3)=3\sum_{k=1}^{n}k^2+4\sum_{k=1}^{n}k-3n$.

b Given that $\sum_{k=1}^{n}k=\frac{n(n+1)}{2}$ and $\sum_{k=1}^{n}k^2=\frac{n(n+1)(2n+1)}{6}$,

find in simplest form $\sum_{k=1}^{n}(k+1)(k+2)$.

Check your answer in the case when $n=10$.

F ARITHMETIC SERIES

An **arithmetic series** is the sum of the terms of an arithmetic sequence.

For example: 21, 23, 25, 27, ..., 49 is a finite arithmetic sequence.

$21 + 23 + 25 + 27 + ... + 49$ is the corresponding arithmetic series.

SUM OF A FINITE ARITHMETIC SERIES

If the first term is u_1 and the common difference is d, the terms are u_1, $u_1 + d$, $u_1 + 2d$, $u_1 + 3d$, and so on.

Suppose that u_n is the final term of an arithmetic series.

So, $S_n = u_1 + (u_1 + d) + (u_1 + 2d) + ... + (u_n - 2d) + (u_n - d) + u_n$

But $S_n = u_n + (u_n - d) + (u_n - 2d) + ... + (u_1 + 2d) + (u_1 + d) + u_1$ {reversing them}

Adding these two equations vertically we get

$$2S_n = \underbrace{(u_1 + u_n) + (u_1 + u_n) + (u_1 + u_n) + ... + (u_1 + u_n) + (u_1 + u_n) + (u_1 + u_n)}_{n \text{ of these}}$$

$\therefore \quad 2S_n = n(u_1 + u_n)$

$\therefore \quad S_n = \frac{n}{2}(u_1 + u_n)$ where $u_n = u_1 + (n-1)d$

The sum of a finite arithmetic series with first term u_1, common difference d, and last term u_n, is

$$\boldsymbol{S_n = \frac{n}{2}(u_1 + u_n)} \quad \textit{or} \quad \boldsymbol{S_n = \frac{n}{2}(2u_1 + (n-1)d)}.$$

For example, from **Exercise 7E** question 6, we observe that the sum of the first n integers is

$$1 + 2 + 3 + ... + n = \frac{n(n+1)}{2} \quad \text{for } n \in \mathbb{Z}^+.$$

Example 16 Self Tutor

Find the sum of $4 + 7 + 10 + 13 + ...$ to 50 terms.

The series is arithmetic with $u_1 = 4$, $d = 3$, and $n = 50$.

Now $S_n = \frac{n}{2}(2u_1 + (n-1)d)$

$\therefore \quad S_{50} = \frac{50}{2}(2 \times 4 + 49 \times 3)$

$= 3875$

You can also use technology to evaluate series, although for some calculator models this is tedious.

Example 17 — Self Tutor

Find the sum of $-6 + 1 + 8 + 15 + ... + 141$.

The series is arithmetic with $u_1 = -6$, $d = 7$ and $u_n = 141$.

First we need to find n.

$$\begin{aligned} \text{Now} \quad u_n &= 141 \\ \therefore \quad u_1 + (n-1)d &= 141 \\ \therefore \quad -6 + 7(n-1) &= 141 \\ \therefore \quad 7(n-1) &= 147 \\ \therefore \quad n - 1 &= 21 \\ \therefore \quad n &= 22 \end{aligned}$$

$$\begin{aligned} \text{Using} \quad S_n &= \frac{n}{2}(u_1 + u_n), \\ S_{22} &= \tfrac{22}{2}(-6 + 141) \\ &= 11 \times 135 \\ &= 1485 \end{aligned}$$

EXERCISE 7F

1. Find the sum of:
   a. $3 + 7 + 11 + 15 + ...$ to 20 terms
   b. $\frac{1}{2} + 3 + 5\frac{1}{2} + 8 + ...$ to 50 terms
   c. $100 + 93 + 86 + 79 + ...$ to 40 terms
   d. $50 + 48\frac{1}{2} + 47 + 45\frac{1}{2} + ...$ to 80 terms.

2. Find the sum of:
   a. $5 + 8 + 11 + 14 + ... + 101$
   b. $50 + 49\frac{1}{2} + 49 + 48\frac{1}{2} + ... + (-20)$
   c. $8 + 10\frac{1}{2} + 13 + 15\frac{1}{2} + ... + 83$

3. Evaluate these arithmetic series:
   a. $\sum_{k=1}^{10}(2k+5)$
   b. $\sum_{k=1}^{15}(k-50)$
   c. $\sum_{k=1}^{20}\left(\frac{k+3}{2}\right)$

   Check your answers using technology.

4. An arithmetic series has seven terms. The first term is 5 and the last term is 53. Find the sum of the series.

5. An arithmetic series has eleven terms. The first term is 6 and the last term is -27. Find the sum of the series.

6. A bricklayer builds a triangular wall with layers of bricks as shown. If the bricklayer uses 171 bricks, how many layers did he build?

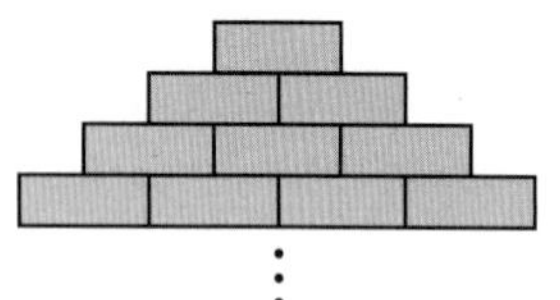

7. A soccer stadium has 25 sections of seating. Each section has 44 rows of seats, with 22 seats in the first row, 23 in the second row, 24 in the third row, and so on. How many seats are there in:
   a. row 44 of one section
   b. each section
   c. the whole stadium?

8. Find the sum of:
   a. the first 50 multiples of 11
   b. the multiples of 7 between 0 and 1000
   c. the integers between 1 and 100 which are not divisible by 3.

9 Consider the series of odd numbers $1+3+5+7+....$

  a What is the nth odd number u_n?

  b Prove that the sum of the first n odd integers is n^2.

  c Check your answer to **b** by finding S_1, S_2, S_3, and S_4.

10 Find the first two terms of an arithmetic sequence if the sixth term is 21 and the sum of the first seventeen terms is 0.

11 Three consecutive terms of an arithmetic sequence have a sum of 12 and a product of -80. Find the terms. **Hint:** Let the terms be $x-d$, x, and $x+d$.

12 Five consecutive terms of an arithmetic sequence have a sum of 40. The product of the first, middle and last terms is 224. Find the terms of the sequence.

13 Henk starts a new job selling TV sets. He sells 11 sets in the first week, 14 sets in the next, 17 sets in the next, and so on in an arithmetic sequence. In what week does Henk sell his 2000th TV set?

14 The sum of the first n terms of an arithmetic sequence is $\dfrac{n(3n+11)}{2}$.

  a Find its first two terms.

  b Find the twentieth term of the sequence.

INVESTIGATION 2 — STADIUM SEATING

A circular stadium consists of sections as illustrated, with aisles in between. The diagram shows the 13 tiers of concrete steps for the final section, Section K. Seats are placed along every concrete step, with each seat 0.45 m wide. The arc AB at the front of the first row, is 14.4 m long, while the arc CD at the back of the back row, is 20.25 m long.

1 How wide is each concrete step?

2 What is the length of the arc of the back of Row 1, Row 2, Row 3, and so on?

3 How many seats are there in Row 1, Row 2, Row 3, ..., Row 13?

4 How many sections are there in the stadium?

5 What is the total seating capacity of the stadium?

6 What is the radius r of the 'playing surface'?

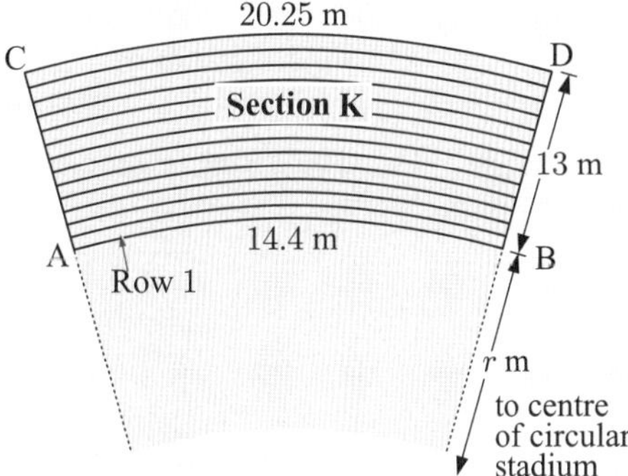

G GEOMETRIC SERIES

A **geometric series** is the sum of the terms of a geometric sequence.

For example: 1, 2, 4, 8, 16, ..., 1024 is a finite geometric sequence.

$1+2+4+8+16+....+1024$ is the corresponding finite geometric series.

If we are adding the first n terms of an infinite geometric sequence, we are then calculating a finite geometric series called the nth partial sum of the corresponding infinite series.

If we are adding all of the terms in a geometric sequence which goes on and on forever, we have an infinite geometric series.

SUM OF A FINITE GEOMETRIC SERIES

If the first term is u_1 and the common ratio is r, then the terms are: $u_1, u_1r, u_1r^2, u_1r^3, ...$

So, $S_n = u_1 + \underbrace{u_1 r}_{u_2} + \underbrace{u_1 r^2}_{u_3} + \underbrace{u_1 r^3}_{u_4} + ... + \underbrace{u_1 r^{n-2}}_{u_{n-1}} + \underbrace{u_1 r^{n-1}}_{u_n}$

For a finite geometric series with $r \neq 1$,

$$S_n = \frac{u_1(r^n - 1)}{r - 1} \quad \text{or} \quad S_n = \frac{u_1(1 - r^n)}{1 - r}.$$

Proof:

If $S_n = u_1 + u_1r + u_1r^2 + u_1r^3 + ... + u_1r^{n-2} + u_1r^{n-1}$ (*)

then $rS_n = (u_1r + u_1r^2 + u_1r^3 + u_1r^4 + ... + u_1r^{n-1}) + u_1r^n$

$\therefore\ rS_n = (S_n - u_1) + u_1r^n$ {from (*)}

$\therefore\ rS_n - S_n = u_1r^n - u_1$

$\therefore\ S_n(r - 1) = u_1(r^n - 1)$

$\therefore\ S_n = \dfrac{u_1(r^n - 1)}{r - 1}$ or $\dfrac{u_1(1 - r^n)}{1 - r}$ provided $r \neq 1$.

In the case $r = 1$ we have a sequence in which all terms are the same, and $S_n = u_1 n$.

Example 18 Self Tutor

Find the sum of $2 + 6 + 18 + 54 + ...$ to 12 terms.

The series is geometric with $u_1 = 2$, $r = 3$, and $n = 12$.

$S_n = \dfrac{u_1(r^n - 1)}{r - 1}$

$\therefore\ S_{12} = \dfrac{2(3^{12} - 1)}{3 - 1}$

$= 531\,440$

Example 19 Self Tutor

Find a formula for S_n, the sum of the first n terms of the series $9 - 3 + 1 - \frac{1}{3} + ...$

The series is geometric with $u_1 = 9$ and $r = -\frac{1}{3}$.

$S_n = \dfrac{u_1(1 - r^n)}{1 - r} = \dfrac{9(1 - (-\frac{1}{3})^n)}{\frac{4}{3}}$

$\therefore\ S_n = \frac{27}{4}(1 - (-\frac{1}{3})^n)$

This answer cannot be simplified as we do not know if n is odd or even.

EXERCISE 7G.1

1 Find the sum of the following series:

a $12+6+3+1.5+ ....$ to 10 terms
b $\sqrt{7}+7+7\sqrt{7}+49+ ....$ to 12 terms
c $6-3+1\frac{1}{2}-\frac{3}{4}+ ....$ to 15 terms
d $1-\frac{1}{\sqrt{2}}+\frac{1}{2}-\frac{1}{2\sqrt{2}}+ ....$ to 20 terms

2 Find a formula for S_n, the sum of the first n terms of the series:

a $\sqrt{3}+3+3\sqrt{3}+9+ ....$
b $12+6+3+1\frac{1}{2}+ ....$
c $0.9+0.09+0.009+0.0009+ ....$
d $20-10+5-2\frac{1}{2}+ ....$

3 A geometric sequence has partial sums $S_1=3$ and $S_2=4$.

a State the first term u_1.
b Calculate the common ratio r.
c Calculate the fifth term u_5 of the series.

4 Evaluate these geometric series:

a $\sum_{k=1}^{10} 3\times 2^{k-1}$
b $\sum_{k=1}^{12} (\frac{1}{2})^{k-2}$
c $\sum_{k=1}^{25} 6\times(-2)^k$

5 Each year a salesperson is paid a bonus of \$2000 which is banked into the same account. It earns a fixed rate of interest of 6% p.a. with interest being paid annually. The total amount in the account at the end of each year is calculated as follows:

$$A_0 = 2000$$
$$A_1 = A_0 \times 1.06 + 2000$$
$$A_2 = A_1 \times 1.06 + 2000 \quad \text{and so on.}$$

a Show that $A_2 = 2000 + 2000\times 1.06 + 2000\times(1.06)^2$.
b Show that $A_3 = 2000[1+1.06+(1.06)^2+(1.06)^3]$.
c Find the total bank balance after 10 years, assuming there are no fees or withdrawals.

6 Consider $S_n = \frac{1}{2}+\frac{1}{4}+\frac{1}{8}+\frac{1}{16}+ .... +\frac{1}{2^n}$.

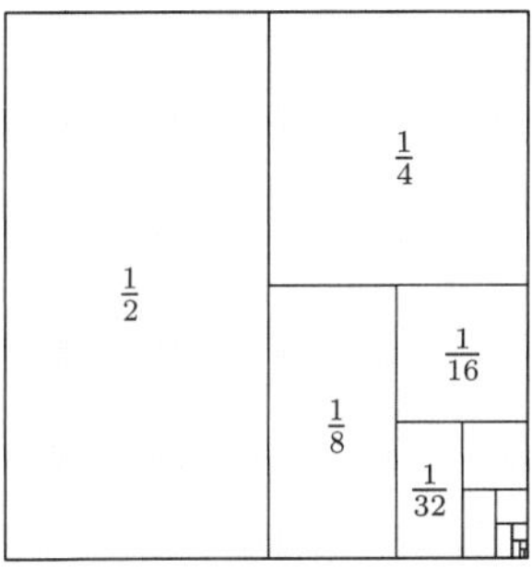

a Find S_1, S_2, S_3, S_4, and S_5 in fractional form.
b From a guess the formula for S_n.
c Find S_n using $S_n=\dfrac{u_1(1-r^n)}{1-r}$.
d Comment on S_n as n gets very large.
e What is the relationship between the given diagram and d?

7 A geometric series has second term 6. The sum of its first three terms is -14. Find its fourth term.

8 An arithmetic and a geometric sequence both have first term 1, and their second terms are equal. The 14th term of the arithmetic sequence is three times the third term of the geometric sequence. Find the twentieth term of each sequence.

9 Find n given that:
a $\sum_{k=1}^{n}(2k+3)=1517$
b $\sum_{k=1}^{n} 2\times 3^{k-1}=177\,146$

10 Suppose $u_1, u_2, ..., u_n$ is a geometric sequence with common ratio r. Show that

$$(u_1+u_2)^2+(u_2+u_3)^2+(u_3+u_4)^2+ .... +(u_{n-1}+u_n)^2=\frac{2{u_1}^2(r^{2n-1}-1)}{r-1}-(u_1^2+u_n^2).$$

11 \$8000 is borrowed over a 2-year period at a rate of 12% p.a. Quarterly repayments are made and the interest is adjusted each quarter, which means that at the end of each quarter, interest is charged on the previous balance and then the balance is reduced by the amount repaid.

There are $2 \times 4 = 8$ repayments and the interest per quarter is $\frac{12\%}{4} = 3\%$.

At the end of the first quarter the amount owed is given by $A_1 = \$8000 \times 1.03 - R$, where R is the amount of each repayment.

At the end of the second quarter, the amount owed is given by:

$$\begin{aligned} A_2 &= A_1 \times 1.03 - R \\ &= (\$8000 \times 1.03 - R) \times 1.03 - R \\ &= \$8000 \times (1.03)^2 - 1.03R - R \end{aligned}$$

- **a** Find a similar expression for the amount owed at the end of the third quarter, A_3.
- **b** Write down an expression for the amount owed at the end of the 8th quarter, A_8.
- **c** Given that $A_8 = 0$ for the loan to be fully repaid, deduce the value of R.
- **d** Suppose the amount borrowed is $\$P$, the interest rate is $r\%$ per repayment interval, and there are m repayments. Show that each repayment is $R = \dfrac{P(1+\frac{r}{100})^m \times \frac{r}{100}}{(1+\frac{r}{100})^m - 1}$ dollars.

SUM OF AN INFINITE GEOMETRIC SERIES

To examine the sum of all the terms of an infinite geometric sequence, we need to consider $S_n = \dfrac{u_1(1-r^n)}{1-r}$ when n gets very large.

If $|r| > 1$, the series is said to be **divergent** and the sum becomes infinitely large.

For example, when $r = 2$, $1 + 2 + 4 + 8 + 16 + ...$ is infinitely large.

If $|r| < 1$, or in other words $-1 < r < 1$, then as n becomes very large, r^n approaches 0.

For example, when $r = \frac{1}{2}$, $1 + \frac{1}{2} + \frac{1}{4} + \frac{1}{8} + \frac{1}{16} + ... = 2$.

This means that S_n will get closer and closer to $\dfrac{u_1}{1-r}$.

If $|r| < 1$, an infinite geometric series of the form $u_1 + u_1 r + u_1 r^2 + ... = \sum_{k=1}^{\infty} u_1 r^{k-1}$ will **converge** to the sum $\mathbf{S = \dfrac{u_1}{1-r}}$.

We call this the **limiting sum** of the series.

This result can be used to find the value of recurring decimals.

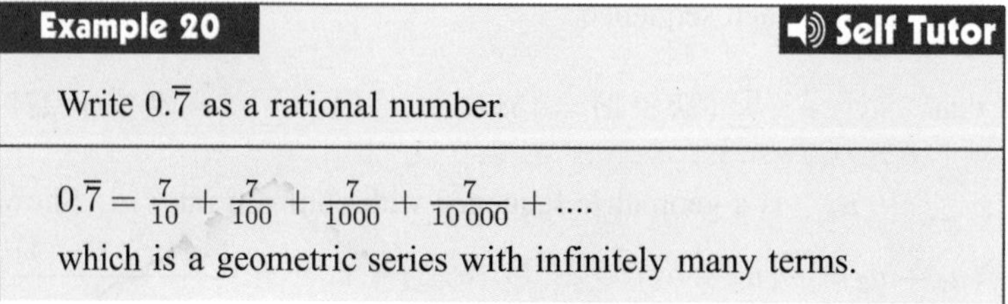

Example 20 Self Tutor

Write $0.\overline{7}$ as a rational number.

$0.\overline{7} = \frac{7}{10} + \frac{7}{100} + \frac{7}{1000} + \frac{7}{10\,000} + ...$

which is a geometric series with infinitely many terms.

In this case, $u_1 = \frac{7}{10}$ and $r = \frac{1}{10}$

$$\therefore \quad S = \frac{u_1}{1-r} = \frac{\frac{7}{10}}{1 - \frac{1}{10}} = \frac{7}{9}$$

$$\therefore \quad 0.\overline{7} = \frac{7}{9}$$

EXERCISE 7G.2

1 Consider $0.\overline{3} = \frac{3}{10} + \frac{3}{100} + \frac{3}{1000} + ...$ which is an infinite geometric series.

- **a** Find: **i** u_1 **ii** r
- **b** Using **a**, show that $0.\overline{3} = \frac{1}{3}$.

2 Write as a rational number: **a** $0.\overline{4}$ **b** $0.\overline{16}$ **c** $0.\overline{312}$

3 Use $S = \dfrac{u_1}{1-r}$ to check your answer to **Exercise 7G.1** question **6d**.

4 Find the sum of each of the following infinite geometric series:

- **a** $18 + 12 + 8 + \frac{16}{3} + ...$
- **b** $18.9 - 6.3 + 2.1 - 0.7 + ...$

5 Find each of the following:

- **a** $\displaystyle\sum_{k=1}^{\infty} \frac{3}{4^k}$
- **b** $\displaystyle\sum_{k=0}^{\infty} 6\left(-\tfrac{2}{5}\right)^k$

6 The sum of the first three terms of a convergent infinite geometric series is 19. The sum of the series is 27. Find the first term and the common ratio.

7 The second term of a convergent infinite geometric series is $\frac{8}{5}$. The sum of the series is 10. Show that there are two possible series, and find the first term and the common ratio in each case.

8

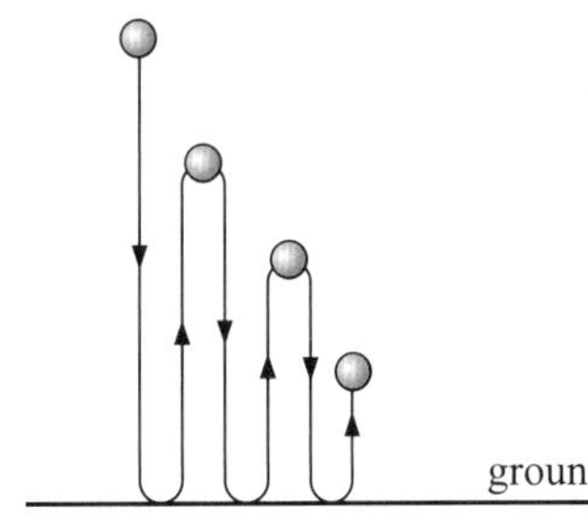

When dropped, a ball takes 1 second to hit the ground. It then takes 90% of this time to rebound to its new height, and this continues until the ball comes to rest.

- **a** Show that the total time of motion is given by $1 + 2(0.9) + 2(0.9)^2 + 2(0.9)^3 + ...$
- **b** Find S_n for the series in **a**.
- **c** How long does it take for the ball to come to rest?

Note: This diagram is inaccurate as the motion is really up and down on the same spot. It has been separated out to help us visualise what is happening.

9 Determine whether each of the following series is convergent. If so, find the sum of the series. If not, find the smallest value of n for which the sum of the first n terms of the series exceeds 100.

- **a** $18 - 9 + 4.5 - ....$
- **b** $1.2 + 1.8 + 2.7 + ...$

10 When a ball is dropped, it rebounds 75% of its height after each bounce. If the ball travels a total distance of 490 cm, from what height was the ball dropped?

11 Find x if $\displaystyle\sum_{k=1}^{\infty} \left(\frac{3x}{2}\right)^{k-1} = 4$.

ACTIVITY

Click on the icon to run a card game for sequences and series.

THEORY OF KNOWLEDGE

The German mathematician Leopold Kronecker (1823 - 1891) made important contributions in number theory and algebra. Several things are named after him, including formulae, symbols, and a theorem.

Leopold Kronecker

Kronecker made several well-known quotes, including:

"*God made integers; all else is the work of man.*"

"*A mathematical object does not exist unless it can be constructed from natural numbers in a finite number of steps.*"

1 What do you understand by the term *infinity*?

2 If the entire world were made of grains of sand, could you count them? Would the number of grains of sand be infinite?

Consider an infinite geometric series with first term u_1 and common ratio r.

If $|r| < 1$, the series will converge to the sum $S = \frac{u_1}{1-r}$.

Proof: If the first term is u_1 and the common ratio is r, the terms are u_1, $u_1 r$, $u_1 r^2$, $u_1 r^3$, and so on.

Suppose the sum of the corresponding infinite series is
$S = u_1 + u_1 r + u_1 r^2 + u_1 r^3 + u_1 r^4 + ...$ (*)

We multiply (*) by r to obtain $rS = u_1 r + u_1 r^2 + u_1 r^3 + u_1 r^4 + ...$

$\therefore \quad rS = S - u_1$ {comparing with (*)}

$\therefore \quad S(r-1) = -u_1$

$\therefore \quad S = \frac{u_1}{1-r}$ {provided $r \neq 1$}

3 Can we explain through intuition how a sum of non-zero terms, which goes on and on for ever and ever, could actually be a finite number?

In the case $r = -1$, the terms are $u_1, -u_1, u_1, -u_1, ...$
If we take partial sums of the series, the answer is always u_1 or 0.

4 What is the sum of the infinite series when $r = -1$? Is it infinite? Is it defined?

Substituting $r = -1$ into the formula above gives $S = \frac{u_1}{2}$. Could this possibly be the answer?

INVESTIGATION 3 VON KOCH'S SNOWFLAKE CURVE

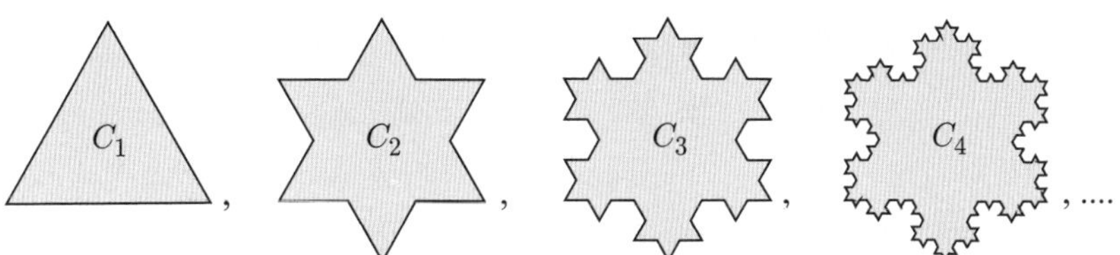

In this Investigation we consider a **limit curve** named after the Swedish mathematician Niels Fabian Helge von Koch (1870 - 1924).

To draw **Von Koch's Snowflake curve** we:

DEMO

- start with an equilateral triangle, C_1
- then divide each side into 3 equal parts
- then on each middle part draw an equilateral triangle

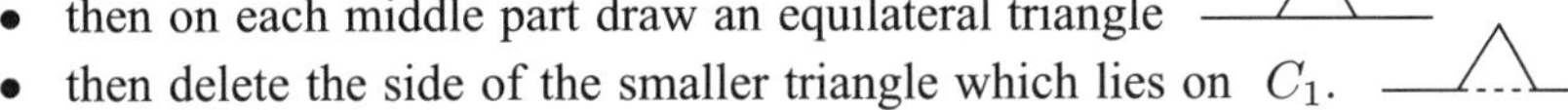

- then delete the side of the smaller triangle which lies on C_1.

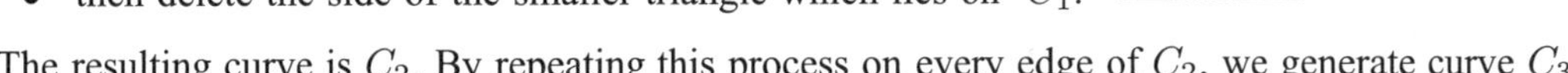

The resulting curve is C_2. By repeating this process on every edge of C_2, we generate curve C_3.

We hence obtain a sequence of special curves C_1, C_2, C_3, C_4, ... and Von Koch's curve is the limiting case when n is infinitely large.

Your task is to investigate the perimeter and area of Von Koch's curve.

What to do:

1 Suppose C_1 has a perimeter of 3 units. Find the perimeter of C_2, C_3, C_4, and C_5.

Hint: becomes so 3 parts become 4 parts.

Remembering that Von Koch's curve is C_n, where n is infinitely large, find the perimeter of Von Koch's curve.

2 Suppose the area of C_1 is 1 unit^2. Explain why the areas of C_2, C_3, C_4, and C_5 are:

$A_2 = 1 + \frac{1}{3}$ units^2 $\qquad$ $A_3 = 1 + \frac{1}{3}[1 + \frac{4}{9}]$ units^2

$A_4 = 1 + \frac{1}{3}[1 + \frac{4}{9} + (\frac{4}{9})^2]$ units^2 $\qquad$ $A_5 = 1 + \frac{1}{3}[1 + \frac{4}{9} + (\frac{4}{9})^2 + (\frac{4}{9})^3]$ units^2.

Use your calculator to find A_n where $n = 1, 2, 3, 4, 5, 6$, and 7, giving answers which are as accurate as your calculator permits.
What do you think will be the area within Von Koch's snowflake curve?

3 Is there anything remarkable about your answers to **1** and **2**?

4 Similarly, investigate the sequence of curves obtained by adding squares on successive curves from the middle third of each side. These are the curves C_1, C_2, C_3, ... shown below.

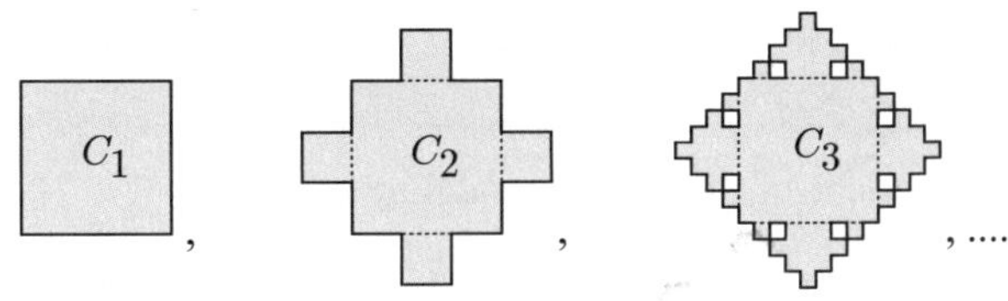

REVIEW SET 7A — NON-CALCULATOR

1 Identify the following sequences as arithmetic, geometric, or neither:

a $7, -1, -9, -17, ...$ **b** $9, 9, 9, 9, ...$ **c** $4, -2, 1, -\frac{1}{2}, ...$

d $1, 1, 2, 3, 5, 8, ...$ **e** the set of all multiples of 4 in ascending order.

2 Find k if $3k$, $k-2$, and $k+7$ are consecutive terms of an arithmetic sequence.

3 Show that $28, 23, 18, 13, ...$ is an arithmetic sequence. Hence find u_n and the sum S_n of the first n terms in simplest form.

4 Find k given that 4, k, and k^2-1 are consecutive terms of a geometric sequence.

5 Determine the general term of a geometric sequence given that its sixth term is $\frac{16}{3}$ and its tenth term is $\frac{256}{3}$.

6 Insert six numbers between 23 and 9 so that all eight numbers are in arithmetic sequence.

7 Find, in simplest form, a formula for the general term u_n of:

a $86, 83, 80, 77, ...$ **b** $\frac{3}{4}, 1, \frac{7}{6}, \frac{9}{7}, ...$ **c** $100, 90, 81, 72.9, ...$

Hint: One of these sequences is neither arithmetic nor geometric.

8 Expand and hence evaluate: **a** $\sum_{k=1}^{7} k^2$ **b** $\sum_{k=1}^{4} \frac{k+3}{k+2}$

9 Find the sum of each of the following infinite geometric series:

a $18 - 12 + 8 - ....$ **b** $8 + 4\sqrt{2} + 4 + ...$

10 A ball bounces from a height of 3 metres and returns to 80% of its previous height on each bounce. Find the total distance travelled by the ball until it stops bouncing.

11 The sum of the first n terms of an infinite sequence is $\frac{3n^2+5n}{2}$ for all $n \in \mathbb{Z}^+$.

a Find the nth term. **b** Prove that the sequence is arithmetic.

12 a, b, and c are consecutive terms of both an arithmetic and geometric sequence. What can be deduced about a, b, and c?

13 x, y, and z are consecutive terms of a geometric sequence. If $x+y+z = \frac{7}{3}$ and $x^2+y^2+z^2 = \frac{91}{9}$, find the values of x, y, and z.

14 $2x$ and $x-2$ are the first two terms of a convergent series. The sum of the series is $\frac{18}{7}$. Find x, clearly explaining why there is only one possible value.

15 a, b, and c are consecutive terms of an arithmetic sequence. Prove that the following are also consecutive terms of an arithmetic sequence:

a $b+c$, $c+a$, and $a+b$ **b** $\frac{1}{\sqrt{b}+\sqrt{c}}$, $\frac{1}{\sqrt{c}+\sqrt{a}}$, and $\frac{1}{\sqrt{a}+\sqrt{b}}$

REVIEW SET 7B — CALCULATOR

1 List the first four members of the sequences defined by:

a $u_n = 3^{n-2}$ **b** $u_n = \frac{3n+2}{n+3}$ **c** $u_n = 2^n - (-3)^n$

2 A sequence is defined by $u_n = 6(\frac{1}{2})^{n-1}$.

a Prove that the sequence is geometric. **b** Find u_1 and r.

c Find the 16th term of the sequence to 3 significant figures.

3 Consider the sequence $24, 23\frac{1}{4}, 22\frac{1}{2}, ...$

a Which term of the sequence is -36? **b** Find the value of u_{35}.

c Find S_{40}, the sum of the first 40 terms of the sequence.

4 Find the sum of:

a the first 23 terms of $3 + 9 + 15 + 21 + ...$

b the first 12 terms of $24 + 12 + 6 + 3 + ...$

5 Find the first term of the sequence $5, 10, 20, 40, ...$ which exceeds 10 000.

6 What will an investment of €6000 at 7% p.a. compound interest amount to after 5 years if the interest is compounded:

a annually **b** quarterly **c** monthly?

7 The nth term of a sequence is given by the formula $u_n = 5n - 8$.

a Find the value of u_{10}.

b Write down an expression for $u_{n+1} - u_n$ and simplify it.

c Hence explain why the sequence is arithmetic.

d Evaluate $u_{15} + u_{16} + u_{17} + ... + u_{30}$.

8 A geometric sequence has $u_6 = 24$ and $u_{11} = 768$. Determine the general term of the sequence and hence find:

a u_{17} **b** the sum of the first 15 terms.

9 Find the first term of the sequence $24, 8, \frac{8}{3}, \frac{8}{9}, ...$ which is less than 0.001.

10 **a** Determine the number of terms in the sequence $128, 64, 32, 16, ..., \frac{1}{512}$.

b Find the sum of these terms.

11 Find the sum of each of the following infinite geometric series:

a $1.21 - 1.1 + 1 - ....$ **b** $\frac{14}{3} + \frac{4}{3} + \frac{8}{21} + ...$

12 How much should be invested at a fixed rate of 9% p.a. compound interest if you need it to amount to $20 000 after 4 years with interest paid monthly?

13 In 2004 there were 3000 iguanas on a Galapagos island. Since then, the population of iguanas on the island has increased by 5% each year.

a How many iguanas were on the island in 2007?

b In what year will the population first exceed 10 000?

14 $x + 3$ and $x - 2$ are the first two terms of a geometric series. Find the values of x for which the series converges.

REVIEW SET 7C

1 A sequence is defined by $u_n = 68 - 5n$.

a Prove that the sequence is arithmetic. **b** Find u_1 and d.

c Find the 37th term of the sequence.

d State the first term of the sequence which is less than -200.

2 **a** Show that the sequence 3, 12, 48, 192, ... is geometric.

b Find u_n and hence find u_9.

3 Find the general term of the arithmetic sequence with $u_7 = 31$ and $u_{15} = -17$.
Hence, find the value of u_{34}.

4 Write using sigma notation:

a $4 + 11 + 18 + 25 + ...$ for n terms **b** $\frac{1}{4} + \frac{1}{8} + \frac{1}{16} + \frac{1}{32} + ...$ for n terms.

5 Evaluate: **a** $\sum_{k=1}^{8} \left(\frac{31 - 3k}{2}\right)$ **b** $\sum_{k=1}^{15} 50(0.8)^{k-1}$ **c** $\sum_{k=7}^{\infty} 5\left(\frac{2}{5}\right)^{k-1}$

6 How many terms of the series $11 + 16 + 21 + 26 + ...$ are needed to exceed a sum of 450?

7 £12 500 is invested in an account which pays 8.25% p.a. compounded. Find the value of the investment after 5 years if the interest is compounded:

a half-yearly **b** monthly.

8 The sum of the first two terms of an infinite geometric series is 90. The third term is 24.

a Show that there are two possible series. Find the first term and common ratio in each case.

b Show that both series converge and find their respective sums.

9 Seve is training for a long distance walk. He walks 10 km in the first week, then each week thereafter he walks an additional 500 m. If he continues this pattern for a year, how far does Seve walk:

a in the last week **b** in total?

10 **a** Under what conditions will the series $\sum_{k=1}^{\infty} 50(2x-1)^{k-1}$ converge?
Explain your answer.

b Find $\sum_{k=1}^{\infty} 50(2x-1)^{k-1}$ if $x = 0.3$.

11 a, b, c, d, and e are consecutive terms of an arithmetic sequence.
Prove that $a + e = b + d = 2c$.

12 Suppose n consecutive geometric terms are inserted between 1 and 2. Write the sum of these n terms, in terms of n.

13 An arithmetic sequence, and a geometric sequence with common ratio r, have the same first two terms. Show that the third term of the geometric sequence is $\frac{r^2}{2r-1}$ times the third term of the arithmetic sequence.

14 $11 - 2 = 9 = 3^2$ and $1111 - 22 = 1089 = 33^2$.
Show that $\underbrace{(111111....1)}_{2n \text{ 1s}} - \underbrace{(22222....2)}_{n \text{ 2s}}$ is a perfect square.

Chapter 8

Counting and the binomial expansion

Syllabus reference: 1.3

Contents:

OPENING PROBLEM

At an IB Mathematics Teachers' Conference there are 273 delegates present. The organising committee consists of 10 people.

Things to think about:

a If each committee member shakes hands with every other committee member, how many handshakes take place?
Can a 10-sided convex polygon be used to solve this problem?

b If all 273 delegates shake hands with all other delegates, how many handshakes take place now?

c If the organising committee lines up on stage to face the delegates in the audience, in how many different orders can they line up?

The **Opening Problem** is an example of a **counting** problem.

The following exercises will help us to solve counting problems without having to list and count the possibilities one by one. To do this we will examine:

- the product principle
- counting permutations
- counting combinations.

A THE PRODUCT PRINCIPLE

Suppose there are three towns A, B, and C. Four different roads could be taken from A to B, and two different roads from B to C.

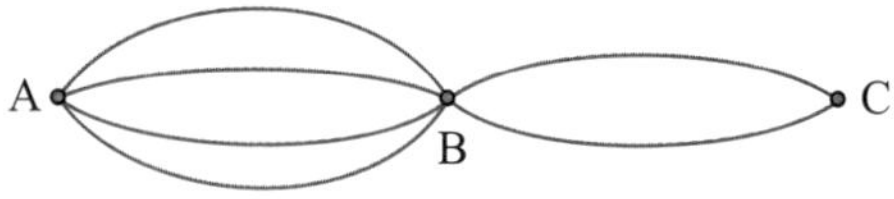

How many different pathways are there from A to C going through B?

If we take road 1, there are two alternative roads to complete our trip.

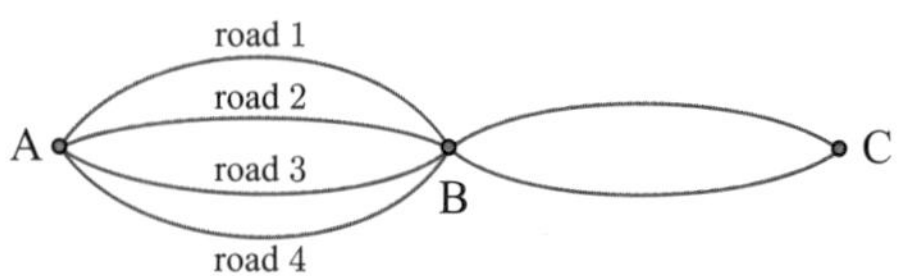

Similarly, if we take road 2, there are two alternative roads to complete our trip.

The same is true for roads 3 and 4.

So, there are $2+2+2+2=4\times 2$ different pathways from A to C going through B.

Notice that the 4 corresponds to the number of roads from A to B and the 2 corresponds to the number of roads from B to C.

THE PRODUCT PRINCIPLE

If there are m different ways of performing an operation, and for each of these there are n different ways of performing a second **independent** operation, then there are mn different ways of performing the two operations in succession.

The product principle can be extended to three or more successive independent operations.

Example 1 — **Self Tutor**

P, Q, R, and S represent where Pauline, Quentin, Reiko, and Sam live. There are two different paths from P to Q, four different paths from Q to R, and 3 different paths from R to S.

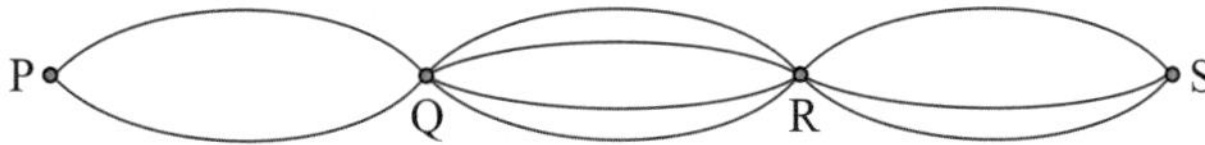

How many different pathways could Pauline take to visit Sam if she stops to see Quentin and then Reiko on the way?

The total number of different pathways $= 2 \times 4 \times 3 = 24$ {product principle}

EXERCISE 8A

1. The illustration shows the different map routes for a bus service which goes from P to S through both Q and R.
How many different routes are possible?

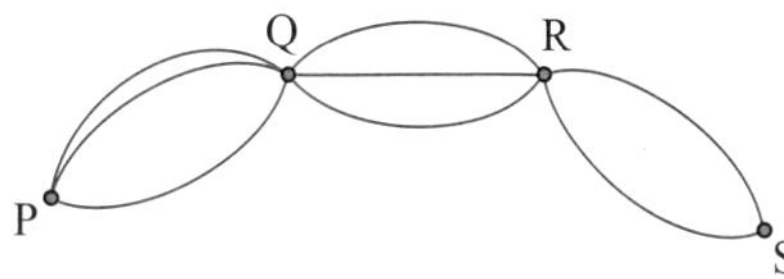

2. 
In how many ways can the vertices of a rectangle be labelled with the letters A, B, C, and D:
   a. in clockwise alphabetical order
   b. in alphabetical order
   c. in random order?

3. The wire frame shown forms the outline of a box.
An ant crawls along the wire from A to B.
How many different paths of shortest length lead from A to B?

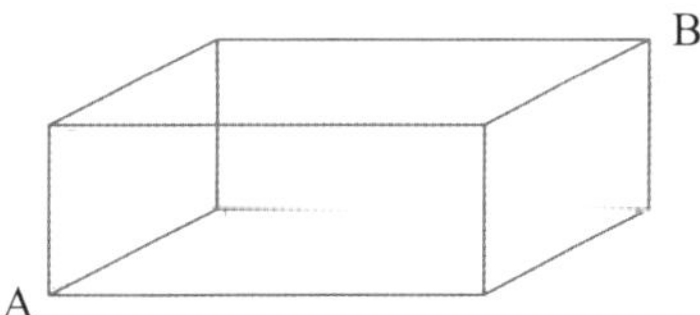

4. In how many different ways can the top two positions be filled in a table tennis competition of 7 teams?

5. A football competition is organised between 8 teams. In how many ways can the top 4 places be filled in order of premiership points obtained?

6. How many 3-digit numbers can be formed using the digits 2, 3, 4, 5, and 6:
   a. as often as desired
   b. at most once each?

7. How many different alpha-numeric plates for motor car registration can be made if the first 3 places are English alphabet letters and the remaining places are 3 digits from 0 to 9?

8. In how many ways can:
   a. 2 letters be mailed into 2 mail boxes
   b. 2 letters be mailed into 3 mail boxes
   c. 4 letters be mailed into 3 mail boxes?

B COUNTING PATHS

Consider the road system illustrated which shows the roads from P to Q.

From A to Q there are 2 paths.
From B to Q there are $3 \times 2 = 6$ paths.
From C to Q there are 3 paths.

$\therefore$ from P to Q there are $2 + 6 + 3 = 11$ paths.

Notice that:
- When going from B to G, we go from B to E **and** then from E to G. We **multiply** the possibilities.
- When going from P to Q, we must first go from P to A **or** P to B **or** P to C. We **add** the possibilities from each of these first steps.

The word **and** suggests multiplying the possibilities.
The word **or** suggests adding the possibilities.

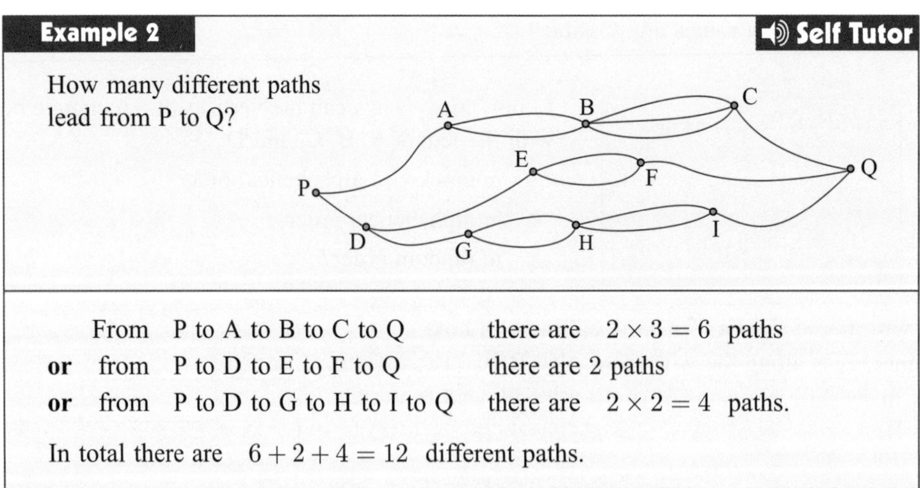

Example 2 — Self Tutor

How many different paths lead from P to Q?

From P to A to B to C to Q there are $2 \times 3 = 6$ paths
or from P to D to E to F to Q there are 2 paths
or from P to D to G to H to I to Q there are $2 \times 2 = 4$ paths.

In total there are $6 + 2 + 4 = 12$ different paths.

EXERCISE 8B

1 How many different paths lead from P to Q?

a

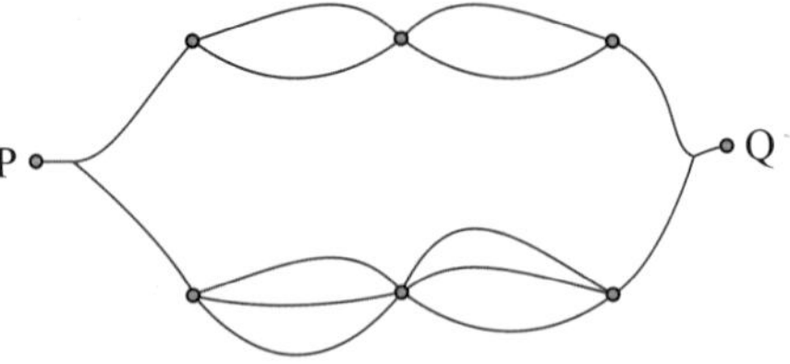

b

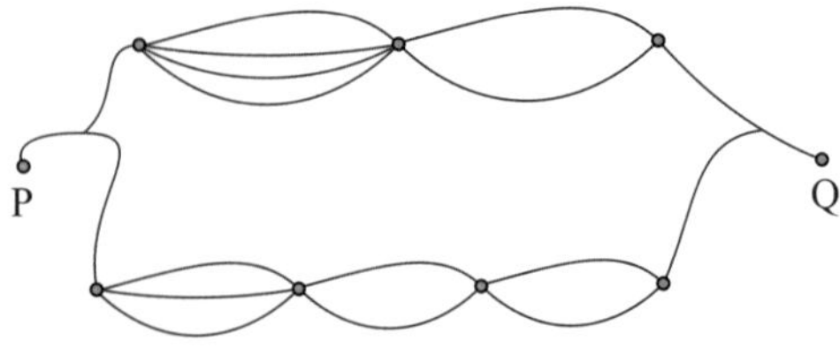

c

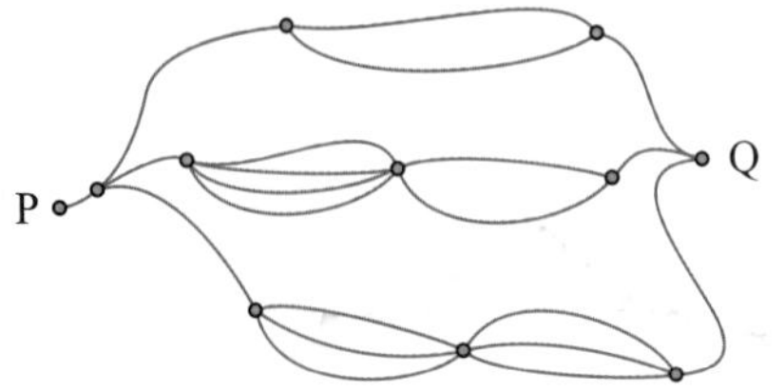

d

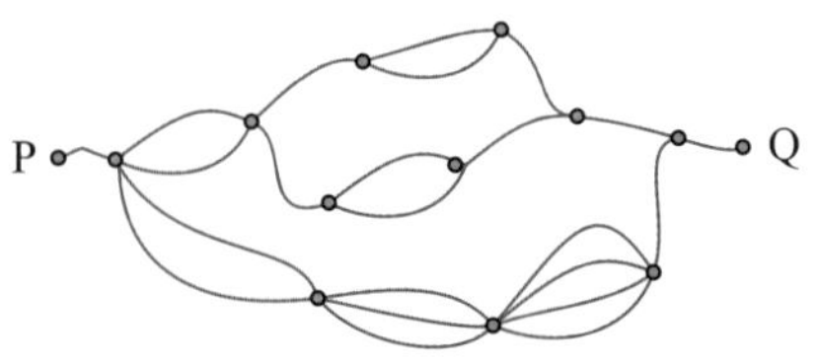

2 Katie is going on a long journey to visit her family. She lives in city A and is travelling to city E. Unfortunately there are no direct trains. However, she has the choice of several trains which stop in different cities along the way. These are illustrated in the diagram.
How many different train journeys does Katie have to choose from?

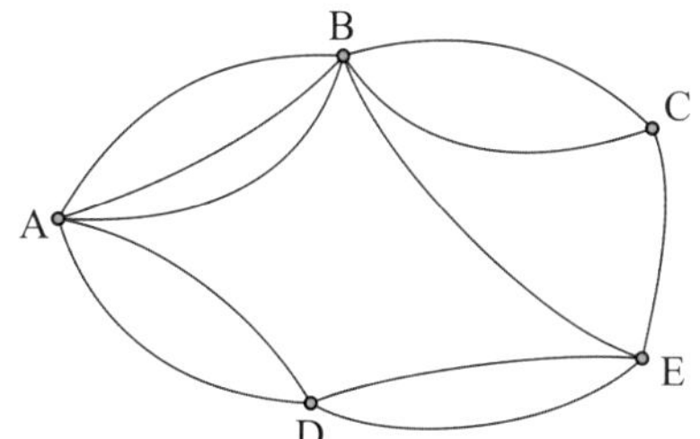

C FACTORIAL NOTATION

In problems involving counting, products of consecutive positive integers are common.

For example, $8 \times 7 \times 6$ or $6 \times 5 \times 4 \times 3 \times 2 \times 1$.

For convenience, we introduce **factorial numbers** to represent the products of consecutive positive integers.

For $n \geqslant 1$, $n!$ is the product of the first n positive integers.

$$\boldsymbol{n! = n(n-1)(n-2)(n-3).... \times 3 \times 2 \times 1}$$

$n!$ is read "n factorial".

For example, the product $6 \times 5 \times 4 \times 3 \times 2 \times 1$ can be written as $6!$.

Notice that $8 \times 7 \times 6$ can be written using factorial numbers only as

$$8 \times 7 \times 6 = \frac{8 \times 7 \times 6 \times 5 \times 4 \times 3 \times 2 \times 1}{5 \times 4 \times 3 \times 2 \times 1} = \frac{8!}{5!}$$

An alternative **recursive definition** of factorial numbers is $\boldsymbol{n! = n \times (n-1)!}$ for $n \geqslant 1$

which can be extended to $n! = n(n-1)(n-2)!$ and so on.

Using the factorial rule with $n = 1$, we have $1! = 1 \times 0!$

Therefore, for completeness we define $\boldsymbol{0! = 1}$

Example 3 — Self Tutor

What integer is equal to: **a** $4!$ **b** $\frac{5!}{3!}$ **c** $\frac{7!}{4! \times 3!}$?

a $4! = 4 \times 3 \times 2 \times 1 = 24$

b $\frac{5!}{3!} = \frac{5 \times 4 \times 3 \times 2 \times 1}{3 \times 2 \times 1} = 5 \times 4 = 20$

c $\frac{7!}{4! \times 3!} = \frac{7 \times 6 \times 5 \times 4 \times 3 \times 2 \times 1}{4 \times 3 \times 2 \times 1 \times 3 \times 2 \times 1} = 35$

EXERCISE 8C.1

1. Find $n!$ for $n = 0, 1, 2, 3, ..., 10$.
2. Simplify without using a calculator:

   a $\frac{6!}{5!}$ b $\frac{6!}{4!}$ c $\frac{6!}{7!}$ d $\frac{4!}{6!}$ e $\frac{100!}{99!}$ f $\frac{7!}{5! \times 2!}$

3. Simplify: a $\frac{n!}{(n-1)!}$ b $\frac{(n+2)!}{n!}$ c $\frac{(n+1)!}{(n-1)!}$

Example 4 — Self Tutor

Express in factorial form:

a $10 \times 9 \times 8 \times 7$ b $\frac{10 \times 9 \times 8 \times 7}{4 \times 3 \times 2 \times 1}$

a $10 \times 9 \times 8 \times 7 = \frac{10 \times 9 \times 8 \times 7 \times 6 \times 5 \times 4 \times 3 \times 2 \times 1}{6 \times 5 \times 4 \times 3 \times 2 \times 1} = \frac{10!}{6!}$

b $\frac{10 \times 9 \times 8 \times 7}{4 \times 3 \times 2 \times 1} = \frac{10 \times 9 \times 8 \times 7 \times 6 \times 5 \times 4 \times 3 \times 2 \times 1}{4 \times 3 \times 2 \times 1 \times 6 \times 5 \times 4 \times 3 \times 2 \times 1} = \frac{10!}{4! \times 6!}$

4. Express in factorial form:

   a $7 \times 6 \times 5$ b 10×9 c $11 \times 10 \times 9 \times 8 \times 7$

   d $\frac{13 \times 12 \times 11}{3 \times 2 \times 1}$ e $\frac{1}{6 \times 5 \times 4}$ f $\frac{4 \times 3 \times 2 \times 1}{20 \times 19 \times 18 \times 17}$

Example 5 — Self Tutor

Write as a product by factorising:

a $8! + 6!$ b $10! - 9! + 8!$

a $8! + 6!$
$= 8 \times 7 \times 6! + 6!$
$= 6!(8 \times 7 + 1)$
$= 6! \times 57$

b $10! - 9! + 8!$
$= 10 \times 9 \times 8! - 9 \times 8! + 8!$
$= 8!(90 - 9 + 1)$
$= 8! \times 82$

5. Write as a product by factorising:

   a $5! + 4!$ b $11! - 10!$ c $6! + 8!$ d $12! - 10!$

   e $9! + 8! + 7!$ f $7! - 6! + 8!$ g $12! - 2 \times 11!$ h $3 \times 9! + 5 \times 8!$

Example 6 — Self Tutor

Simplify $\frac{7! - 6!}{6}$ by factorising.

$\frac{7! - 6!}{6}$

$= \frac{7 \times 6! - 6!}{6}$

$= \frac{6!(\cancel{7-1})^1}{\cancel{6}_1}$

$= 6!$

6 Simplify by factorising:

a $\dfrac{12!-11!}{11}$ b $\dfrac{10!+9!}{11}$ c $\dfrac{10!-8!}{89}$ d $\dfrac{10!-9!}{9!}$

e $\dfrac{6!+5!-4!}{4!}$ f $\dfrac{n!+(n-1)!}{(n-1)!}$ g $\dfrac{n!-(n-1)!}{n-1}$ h $\dfrac{(n+2)!+(n+1)!}{n+3}$

THE BINOMIAL COEFFICIENT

The **binomial coefficient** is defined by

$$\binom{n}{r} = \underbrace{\frac{n(n-1)(n-2)\,....\,(n-r+2)(n-r+1)}{r(r-1)(r-2)\,....\,2\times 1}}_{\text{factor form}} = \underbrace{\frac{n!}{r!(n-r)!}}_{\text{factorial form}}$$

The binomial coefficient is sometimes written ${}^nC_r$ or C_r^n.

Example 7 Self Tutor

Use the formula $\binom{n}{r} = \dfrac{n!}{r!(n-r)!}$ to evaluate: a $\binom{5}{2}$ b $\binom{11}{7}$

a $\binom{5}{2} = \dfrac{5!}{2!(5-2)!}$
$= \dfrac{5!}{2!\times 3!}$
$= \dfrac{5\times 4\times 3\times 2\times 1}{2\times 1\times 3\times 2\times 1}$
$= 10$

b $\binom{11}{7} = \dfrac{11!}{7!(11-7)!}$
$= \dfrac{11!}{7!\times 4!}$
$= \dfrac{11\times 10\times 9\times 8\times 7\times 6\times 5\times 4\times 3\times 2\times 1}{7\times 6\times 5\times 4\times 3\times 2\times 1\times 4\times 3\times 2\times 1}$
$= \dfrac{7920}{24}$
$= 330$

Binomial coefficients can also be found using your graphics calculator.

EXERCISE 8C.2

1 Use the formula $\binom{n}{r} = \dfrac{n!}{r!(n-r)!}$ to evaluate:

a $\binom{3}{1}$ b $\binom{4}{2}$ c $\binom{7}{3}$ d $\binom{10}{4}$

Check your answers using technology.

2 a Use the formula $\binom{n}{r} = \dfrac{n!}{r!(n-r)!}$ to evaluate: i $\binom{8}{2}$ ii $\binom{8}{6}$

b Show that $\binom{n}{r} = \binom{n}{n-r}$ for all $n \in \mathbb{Z}^+$, $r = 0, 1, 2, ..., n$.

3 Find k if $\binom{9}{k} = 4\binom{7}{k-1}$.

D PERMUTATIONS

A **permutation** of a group of symbols is *any arrangement* of those symbols in a definite *order*.

For example, BAC is a permutation on the symbols A, B, and C in which all three of them are used. We say the symbols are "taken 3 at a time".

The set of all the different permutations on the symbols A, B, and C taken 3 at a time, is {ABC, ACB, BAC, BCA, CAB, CBA}.

Example 8 — Self Tutor

List the set of all permutations on the symbols P, Q, and R taken:

a 1 at a time **b** 2 at a time **c** 3 at a time.

a {P, Q, R}

b {PQ, QP, RP, PR, QR, RQ}

c {PQR, PRQ, QPR, QRP, RPQ, RQP}

Example 9 — Self Tutor

List all permutations on the symbols W, X, Y, and Z taken 4 at a time.

WXYZ	WXZY	WYXZ	WYZX	WZXY	WZYX
XWYZ	XWZY	XYWZ	XYZW	XZYW	XZWY
YWXZ	YWZX	YXWZ	YXZW	YZWX	YZXW
ZWXY	ZWYX	ZXWY	ZXYW	ZYWX	ZYXW

There are 24 of them.

For large numbers of symbols, listing the complete set of permutations is absurd. However, we can still count them by considering the number of options we have for filling each position.

In **Example 9** there were 4 positions to fill:

In the 1st position, any of the 4 symbols could be used, so we have 4 options.

This leaves any of 3 symbols to go in the 2nd position, which leaves any of 2 symbols to go in the 3rd position.

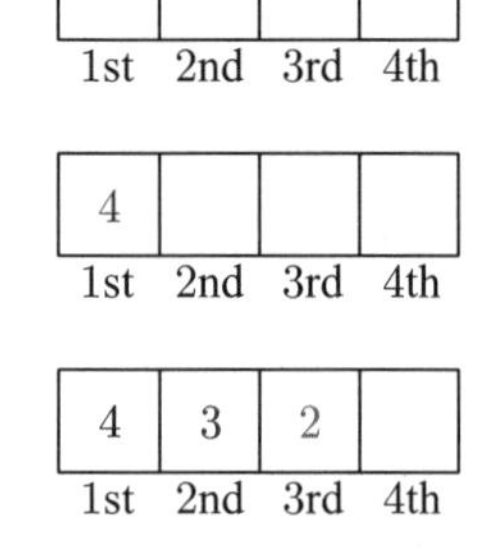

The remaining symbol must go in the 4th position.

So, the total number of permutations $= 4 \times 3 \times 2 \times 1$ {product principle}

$= 4!$

$= 24$

Example 10 Self Tutor

A chess association runs a tournament with 16 teams. In how many different ways could the top 8 positions be filled on the competition ladder?

Any of the 16 teams could fill the 'top' position.
Any of the remaining 15 teams could fill the 2nd position.
Any of the remaining 14 teams could fill the 3rd position.

⋮

Any of the remaining 9 teams could fill the 8th position.

16	15	14	13	12	11	10	9
1st	2nd	3rd	4th	5th	6th	7th	8th

The total number of permutations $= 16 \times 15 \times 14 \times 13 \times 12 \times 11 \times 10 \times 9$

$= \dfrac{16!}{8!}$

$= 518\,918\,400$

Example 11 Self Tutor

The alphabet blocks A, B, C, D, and E are placed in a row in front of you.

a How many different permutations could you have?

b How many permutations end in C?

c How many permutations have the form [...|A|...|B|...] ?

d How many begin and end with a vowel (A or E)?

a There are 5 letters taken 5 at a time.
$\therefore$ the total number of permutations $= 5 \times 4 \times 3 \times 2 \times 1 = 5! = 120$.

b

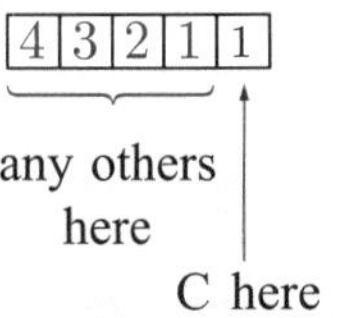

C must be in the last position. The other 4 letters could go into the remaining 4 places in 4! ways.
$\therefore$ the number of permutations $= 1 \times 4! = 24$.

c

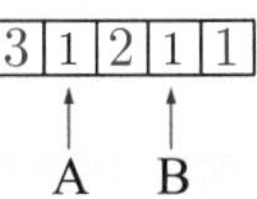

A goes into 1 place. B goes into 1 place. The remaining 3 letters go into the remaining 3 places in 3! ways.
$\therefore$ the number of permutations $= 1 \times 1 \times 3! = 6$.

d

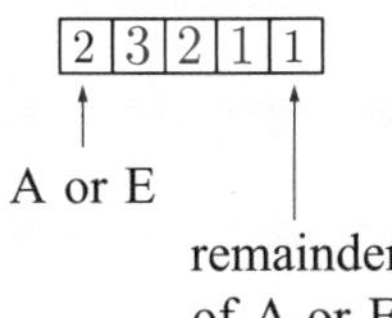

A or E could go into the 1st position, so there are two options. The other one must go into the last position.
The remaining 3 letters could go into the 3 remaining places in 3! ways.
$\therefore$ the number of permutations $= 2 \times 1 \times 3! = 12$.

EXERCISE 8D

1 List the set of all permutations on the symbols W, X, Y, and Z taken:

a 1 at a time **b** two at a time **c** three at a time.

2 List the set of all permutations on the symbols A, B, C, D, and E taken:

a 2 at a time **b** 3 at a time.

3 In how many ways can:

a 5 different books be arranged on a shelf

b 3 different paintings be chosen from a collection of 8, and hung in a row

c a signal consisting of 4 coloured flags in a row be made if there are 10 different flags to choose from?

4 Suppose you have 4 different coloured flags. How many different signals could you make using:

a 2 flags in a row **b** 3 flags in a row **c** 2 or 3 flags in a row?

5 How many different permutations on the letters A, B, C, D, E, and F are there if each letter can be used once only? How many of these:

a end in ED **b** begin with F and end with A

c begin and end with a vowel (A or E)?

6 How many 3-digit numbers can be constructed from the digits 1, 2, 3, 4, 5, 6, and 7 if each digit may be used:

a as often as desired **b** only once **c** once only and the number is odd?

7 3-digit numbers are constructed from the digits 0, 1, 2, 3, 4, 5, 6, 7, 8, and 9 using each digit at most once. How many such numbers:

a can be constructed **b** end in 5 **c** end in 0 **d** are divisible by 5?

Example 12 — Self Tutor

There are 6 different books arranged in a row on a shelf. In how many ways can two of the books, A and B, be together?

Method 1: We could have any of the following locations for A and B

A B × × × ×
B A × × × ×
× A B × × ×
× B A × × ×
× × A B × ×
× × B A × ×
× × × A B ×
× × × B A ×
× × × × A B
× × × × B A

} 10 of these

If we consider any one of these, the remaining 4 books could be placed in 4! different orderings.

$\therefore$ total number of ways $= 10 \times 4! = 240$.

Method 2: A and B can be put together in 2! ways (AB or BA).

Now consider this pairing as one book (effectively tying a string around them) which together with the other 4 books can be ordered in 5! different ways.

$\therefore$ the total number of ways $= 2! \times 5! = 240$.

8 In how many ways can 5 different books be arranged on a shelf if:

- **a** there are no restrictions
- **b** books X and Y must be together
- **c** books X and Y must not be together?

9 10 students sit in a row of 10 chairs. In how many ways can this be done if:

- **a** there are no restrictions
- **b** students A, B, and C insist on sitting together?

10 How many three-digit numbers can be made using the digits 0, 1, 3, 5, and 8 at most once each, if:

- **a** there are no restrictions
- **b** the numbers must be less than 500
- **c** the numbers must be even and greater than 300?

11 Consider the letters of the word MONDAY. How many different permutations of four letters can be chosen if:

- **a** there are no restrictions
- **b** at least one vowel (A or O) must be used
- **c** the two vowels are not together?

12 Nine boxes are each labelled with a different whole number from 1 to 9. Five people are allowed to take one box each. In how many different ways can this be done if:

- **a** there are no restrictions
- **b** the first three people decide that they will take even numbered boxes?

13 Alice has booked ten adjacent front-row seats for a basketball game for herself and nine friends.

- **a** How many different arrangements are possible if there are no restrictions?
- **b** Due to a severe snowstorm, only five of Alice's friends are able to join her for the game. In how many different ways can they be seated in the 10 seats if:
  - **i** there are no restrictions
  - **ii** any two of Alice's friends are to sit next to her?

INVESTIGATION 1 — PERMUTATIONS IN A CIRCLE

There are 6 permutations on the symbols A, B, and C **in a line**. These are:

ABC ACB BAC BCA CAB CBA.

However **in a circle** there are only 2 different permutations on these 3 symbols. They are the only possibilities with different right-hand and left-hand neighbours.

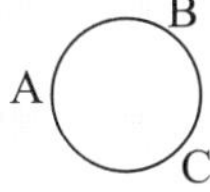

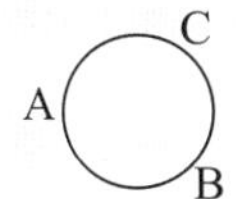

In contrast, these three diagrams show the same cyclic permutation:

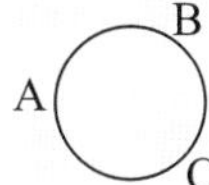

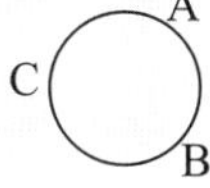

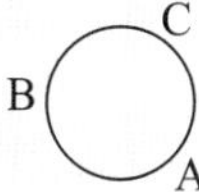

What to do:

1 Draw diagrams showing different cyclic permutations for:

- **a** one symbol: A
- **b** two symbols: A and B
- **c** three symbols: A, B, and C
- **d** four symbols: A, B, C, and D

2 Copy and complete:

Number of symbols	*Permutations in a line*	*Permutations in a circle*
1 2 3 4	 $6 = 3!$	 $2 = 2!$

3 If there are n symbols to be arranged around a circle, how many different cyclic permutations are possible?

E COMBINATIONS

A **combination** is a selection of objects *without* regard to order.

For example, the 10 different possible teams of 3 people that can be selected from A, B, C, D, and E are:

ABC ABD ABE ACD ACE ADE
BCD BCE BDE
CDE

There are 10 combinations in total.

Now given the five people A, B, C, D, and E, we know that there are $5 \times 4 \times 3 = 60$ permutations for taking three of them at a time. So why is this 6 times larger than the number of combinations?

The answer is that for the combinations, order is not important. Selecting A, B, and C for the team is the same as selecting B, C, and A. For each of the 10 possible combinations, there are $3! = 6$ ways of ordering the members of the team.

More particularly, the number of possible combinations $= \dfrac{5!}{3! \times 2!} = 10.$

number of ways to select 5 players in order (5!)

order of those in the team is not important (3!)

order of those not in the team is not important (2!)

The number of **combinations** on n distinct symbols taken r at a time is

$$C_r^n = \binom{n}{r} = \frac{n!}{r!(n-r)!}$$

Example 13 Self Tutor

How many different teams of 4 can be selected from a squad of 7 if:

a there are no restrictions **b** the teams must include the captain?

a There are 7 players up for selection and we want any 4 of them.
There are $\binom{7}{4} = 35$ possible combinations.

b The captain must be included *and* we need any 3 of the other 6.
There are $\binom{1}{1} \times \binom{6}{3} = 20$ possible combinations.

Example 14 **Self Tutor**

A committee of 4 is chosen from 7 men and 6 women. How many different committees can be chosen if:

a there are no restrictions
b there must be 2 of each sex
c there must be at least one of each sex?

a There are $7+6=13$ people up for selection and we want any 4 of them.
There are $\binom{13}{4}=715$ possible combinations.

b The 2 men can be chosen out of 7 in $\binom{7}{2}$ ways.
The 2 women can be chosen out of 6 in $\binom{6}{2}$ ways.
$\therefore$ there are $\binom{7}{2}\times\binom{6}{2}=315$ possible combinations.

c The total number of combinations
= the number with 3 men and 1 woman + the number with 2 men and 2 women
+ the number with 1 man and 3 women
$=\binom{7}{3}\times\binom{6}{1}+\binom{7}{2}\times\binom{6}{2}+\binom{7}{1}\times\binom{6}{3}$
$=665$

or The total number of combinations
$=\binom{13}{4}$ − the number with all men − the number with all women
$=\binom{13}{4}-\binom{7}{4}\times\binom{6}{0}-\binom{7}{0}\times\binom{6}{4}$
$=665$

EXERCISE 8E

1 List the different teams of 4 that can be chosen from a squad of 6 (named A, B, C, D, E, and F).
Check that the formula $\binom{n}{r}=\dfrac{n!}{r!(n-r)!}$ gives the total number of teams.

2 How many different teams of 11 can be chosen from a squad of 17?

3 Candidates for an examination are required to answer 5 questions out of 9.
a In how many ways can the questions be chosen if there are no restrictions?
b If question 1 was made compulsory, how many selections would be possible?

4 **a** How many different committees of 3 can be selected from 13 candidates?
b How many of these committees consist of the president and 2 others?

5 **a** How many different teams of 5 can be selected from a squad of 12?
b How many of these teams contain:
i the captain and vice-captain **ii** exactly one of the captain or the vice-captain?

6 A team of 9 is selected from a squad of 15. 3 particular players *must* be included, and another must be excluded because of injury. In how many ways can the team be chosen?

7 In how many ways can 4 people be selected from 10 if:
a one particular person *must* be selected
b two particular people are excluded from every selection
c one particular person is always included and two particular people are always excluded?

8 A committee of 5 is chosen from 10 men and 6 women. Determine the number of ways of selecting the committee if:

a there are no restrictions
b it must contain 3 men and 2 women
c it must contain all men
d it must contain at least 3 men
e it must contain at least one of each sex.

9 A committee of 5 is chosen from 6 doctors, 3 dentists, and 7 others.
Determine the number of ways of selecting the committee if it is to contain:

a exactly 2 doctors and 1 dentist
b exactly 2 doctors
c at least one person from either of the two given professions.

10 How many diagonals does a 20-sided convex polygon have?

11 There are 12 distinct points A, B, C, D, ..., L on a circle. Lines are drawn between each pair of points.

a How many lines: **i** are there in total **ii** pass through B?
b How many triangles: **i** are determined by the lines **ii** have one vertex B?

12 How many 4-digit numbers can be constructed for which the digits are in ascending order from left to right? You cannot start a number with 0.

13 **a** Give an example which demonstrates that:

$$\binom{5}{0}\times\binom{6}{4}+\binom{5}{1}\times\binom{6}{3}+\binom{5}{2}\times\binom{6}{2}+\binom{5}{3}\times\binom{6}{1}+\binom{5}{4}\times\binom{6}{0}=\binom{11}{4}.$$

b Copy and complete:

$$\binom{m}{0}\times\binom{n}{r}+\binom{m}{1}\times\binom{n}{r-1}+\binom{m}{2}\times\binom{n}{r-2}+\dots+\binom{m}{r-1}\times\binom{n}{1}+\binom{m}{r}\times\binom{n}{0}=\dots$$

14 In how many ways can 12 people be divided into:

a two equal groups
b three equal groups?

15 Line A contains 10 points and line B contains 7 points. If all points on line A are joined to all points on line B, determine the maximum number of points of intersection between the new lines constructed.

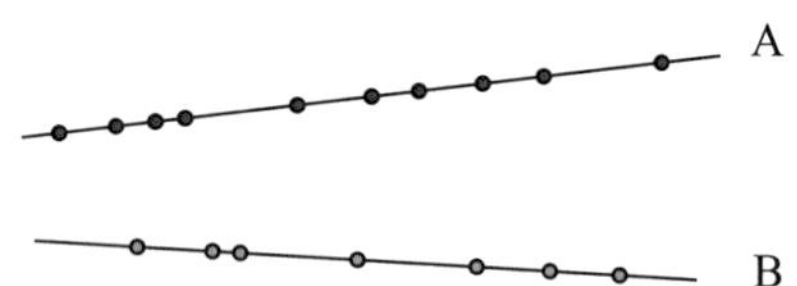

16 10 points are located on [PQ], 9 on [QR], and 8 on [RP].
All possible lines connecting these 27 points are drawn. Determine the maximum possible number of points of intersection of these lines which lie within triangle PQR.

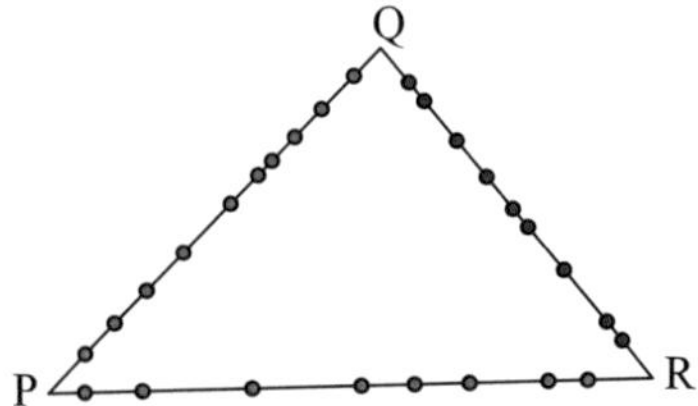

17 Answer the **Opening Problem** on page **244**.

F BINOMIAL EXPANSIONS

Consider the cube alongside, which has sides of length $(a+b)$ cm.

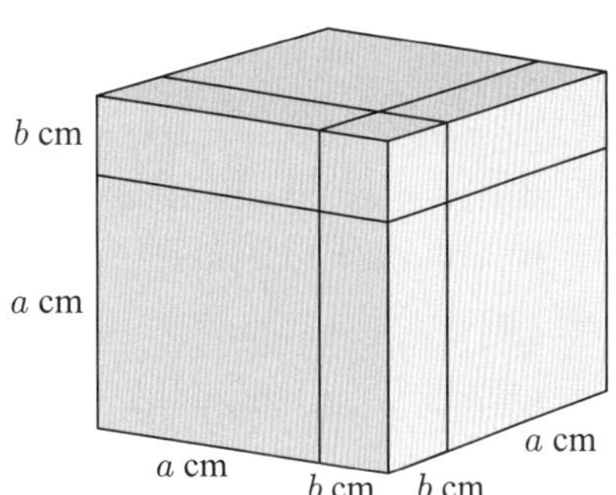

The cube has been subdivided into 8 blocks by making 3 cuts parallel to the cube's surfaces as shown.

We know that the total volume of the cube is $(a+b)^3$ cm^3. However, we can also find an expression for the cube's volume by adding the volumes of the 8 individual blocks.

We have:

1 block	$a \times a \times a$
3 blocks	$a \times a \times b$
3 blocks	$a \times b \times b$
1 block	$b \times b \times b$

$\therefore$ the cube's volume $= a^3 + 3a^2b + 3ab^2 + b^3$

$\therefore \quad (a+b)^3 = a^3 + 3a^2b + 3ab^2 + b^3$

The sum $\boldsymbol{a+b}$ is called a **binomial** as it contains two terms.

Any expression of the form $\boldsymbol{(a+b)^n}$ is called a **power of a binomial**.

All binomials raised to a power can be expanded using the same general principles. In this chapter, therefore, we consider the expansion of the general expression $(a+b)^n$ where $n \in \mathbb{N}$.

Consider the following algebraic expansions:

$(a+b)^1 = a+b$

$(a+b)^2 = a^2 + 2ab + b^2$

$$\begin{aligned}(a+b)^3 &= (a+b)(a+b)^2\\ &= (a+b)(a^2+2ab+b^2)\\ &= a^3 + 2a^2b + ab^2 + a^2b + 2ab^2 + b^3\\ &= a^3 + 3a^2b + 3ab^2 + b^3\end{aligned}$$

The **binomial expansion** of $\boldsymbol{(a+b)^2}$ is $\boldsymbol{a^2 + 2ab + b^2}$.

The **binomial expansion** of $\boldsymbol{(a+b)^3}$ is $\boldsymbol{a^3 + 3a^2b + 3ab^2 + b^3}$.

INVESTIGATION 2 THE BINOMIAL EXPANSION

What to do:

1 Expand $(a+b)^4$ in the same way as for $(a+b)^3$ above.
Hence expand $(a+b)^5$ and $(a+b)^6$.

2 The cubic expansion $(a+b)^3 = a^3 + 3a^2b + 3ab^2 + b^3$ contains 4 terms. Observe that their coefficients are: 1 3 3 1

a What happens to the powers of a and b in each term of the expansion of $(a+b)^3$?

b Does the pattern in **a** continue for the expansions of $(a+b)^4$, $(a+b)^5$, and $(a+b)^6$?

c Write down the triangle of coefficients to row 6:

$n = 1$ 1 1
$n = 2$ 1 2 1
$n = 3$ 1 3 3 1 ← row 3
⋮

3 The triangle of coefficients in **c** above is called **Pascal's triangle**. Investigate:

a the predictability of each row from the previous one

b a formula for finding the sum of the numbers in the nth row of Pascal's triangle.

4 **a** Use your results from **3** to predict the elements of the 7th row of Pascal's triangle.

b Hence write down the binomial expansion of $(a+b)^7$.

c Check your result algebraically by using $(a+b)^7 = (a+b)(a+b)^6$ and your results from **1**.

From the **Investigation** we obtained
$$\begin{aligned}(a+b)^4 &= a^4 + 4a^3b + 6a^2b^2 + 4ab^3 + b^4 \\ &= a^4 + 4a^3b^1 + 6a^2b^2 + 4a^1b^3 + b^4\end{aligned}$$

Notice that:

- As we look from left to right across the expansion, the powers of a decrease by 1, while the powers of b increase by 1.
- The sum of the powers of a and b in each term of the expansion is 4.
- The number of terms in the expansion is $4+1=5$.
- The coefficients of the terms are row 4 of Pascal's triangle.

For the expansion of $(a+b)^n$ where $n \in \mathbb{N}$:

- As we look from left to right across the expansion, the powers of a *decrease* by 1, while the powers of b *increase* by 1.
- The sum of the powers of a and b in each term of the expansion is n.
- The number of terms in the expansion is $n+1$.
- The coefficients of the terms are row n of Pascal's triangle.

In the following examples we see how the general binomial expansion $(a+b)^n$ may be put to use.

Example 15 Self Tutor

Using $(a+b)^3 = a^3 + 3a^2b + 3ab^2 + b^3$, find the binomial expansion of:

a $(2x+3)^3$ **b** $(x-5)^3$

a In the expansion of $(a+b)^3$ we substitute $a = (2x)$ and $b = (3)$.

$$\begin{aligned}\therefore\ (2x+3)^3 &= (2x)^3 + 3(2x)^2(3) + 3(2x)^1(3)^2 + (3)^3 \\ &= 8x^3 + 36x^2 + 54x + 27\end{aligned}$$

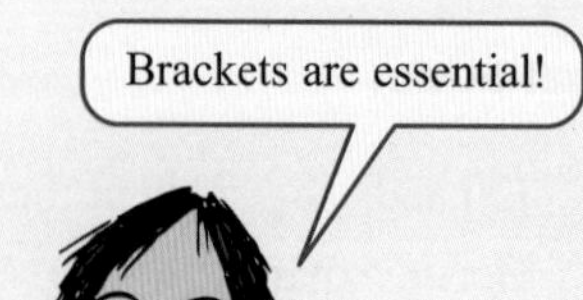
Brackets are essential!

b We substitute $a = (x)$ and $b = (-5)$

$$\begin{aligned}\therefore\ (x-5)^3 &= (x)^3 + 3(x)^2(-5) + 3(x)(-5)^2 + (-5)^3 \\ &= x^3 - 15x^2 + 75x - 125\end{aligned}$$

Example 16 **Self Tutor**

Find the:

a 5th row of Pascal's triangle **b** binomial expansion of $\left(x - \frac{2}{x}\right)^5$.

a

$$
\begin{array}{ccccccccccc}
 & & & & & 1 & & & & & \leftarrow \text{the 0th row, for } (a+b)^0 \\
 & & & & 1 & & 1 & & & & \leftarrow \text{the 1st row, for } (a+b)^1 \\
 & & & 1 & & 2 & & 1 & & & \\
 & & 1 & & 3 & & 3 & & 1 & & \\
 & 1 & & 4 & & 6 & & 4 & & 1 & \\
1 & & 5 & & 10 & & 10 & & 5 & & 1 \leftarrow \text{the 5th row, for } (a+b)^5
\end{array}
$$

b Using the coefficients obtained in **a**, $(a+b)^5 = a^5 + 5a^4b + 10a^3b^2 + 10a^2b^3 + 5ab^4 + b^5$

Letting $a = (x)$ and $b = \left(\frac{-2}{x}\right)$, we find

$$\left(x - \frac{2}{x}\right)^5 = (x)^5 + 5(x)^4\left(\frac{-2}{x}\right) + 10(x)^3\left(\frac{-2}{x}\right)^2 + 10(x)^2\left(\frac{-2}{x}\right)^3 + 5(x)\left(\frac{-2}{x}\right)^4 + \left(\frac{-2}{x}\right)^5$$

$$= x^5 - 10x^3 + 40x - \frac{80}{x} + \frac{80}{x^3} - \frac{32}{x^5}$$

EXERCISE 8F

1 Use the binomial expansion of $(a+b)^3$ to expand and simplify:

a $(p+q)^3$ **b** $(x+1)^3$ **c** $(x-3)^3$

d $(2+x)^3$ **e** $(3x-1)^3$ **f** $(2x+5)^3$

g $(2a-b)^3$ **h** $\left(3x - \frac{1}{3}\right)^3$ **i** $\left(2x + \frac{1}{x}\right)^3$

2 Use $(a+b)^4 = a^4 + 4a^3b + 6a^2b^2 + 4ab^3 + b^4$ to expand and simplify:

a $(1+x)^4$ **b** $(p-q)^4$ **c** $(x-2)^4$

d $(3-x)^4$ **e** $(1+2x)^4$ **f** $(2x-3)^4$

g $(2x+b)^4$ **h** $\left(x + \frac{1}{x}\right)^4$ **i** $\left(2x - \frac{1}{x}\right)^4$

3 Expand and simplify:

a $(x+2)^5$ **b** $(x-2y)^5$ **c** $(1+2x)^5$ **d** $\left(x - \frac{1}{x}\right)^5$

4 **a** Write down the 6th row of Pascal's triangle.

b Find the binomial expansion of:

i $(x+2)^6$ **ii** $(2x-1)^6$ **iii** $\left(x + \frac{1}{x}\right)^6$

5 Expand and simplify:

a $(1+\sqrt{2})^3$ **b** $(\sqrt{5}+2)^4$ **c** $(2-\sqrt{2})^5$

6 **a** Expand $(2+x)^6$. **b** Hence find the value of $(2.01)^6$.

7 The first two terms in a binomial expansion are: $(a+b)^3 = 8 + 12e^x + ...$

a Find a and b. **b** Hence determine the remaining two terms of the expansion.

8 Expand and simplify $(2x+3)(x+1)^4$.

9 Find the coefficient of:

a a^3b^2 in the expansion of $(3a+b)^5$

b a^3b^3 in the expansion of $(2a+3b)^6$.

G THE BINOMIAL THEOREM

In the previous section we saw how the coefficients of the binomial expansion $(a+b)^n$ can be found in the nth row of Pascal's triangle. These coefficients are in fact the **binomial coefficients** $\binom{n}{r}$ for $r = 0, 1, 2, \ldots, n$.

$$\begin{array}{ccccccccc} & & & 1 & & 1 & & & \\ & & 1 & & 2 & & 1 & & \\ & 1 & & 3 & & 3 & & 1 & \\ 1 & & 4 & & 6 & & 4 & & 1 \end{array}$$

$$\begin{array}{ccccccccc} & & & \binom{1}{0} & & \binom{1}{1} & & & \\ & & \binom{2}{0} & & \binom{2}{1} & & \binom{2}{2} & & \\ & \binom{3}{0} & & \binom{3}{1} & & \binom{3}{2} & & \binom{3}{3} & \\ \binom{4}{0} & & \binom{4}{1} & & \binom{4}{2} & & \binom{4}{3} & & \binom{4}{4} \end{array}$$

The binomial theorem states that

$$(a+b)^n = a^n + \binom{n}{1}a^{n-1}b + \ldots + \binom{n}{r}a^{n-r}b^r + \ldots + b^n$$

where $\binom{n}{r}$ is the **binomial coefficient** of $a^{n-r}b^r$ and $r = 0, 1, 2, 3, \ldots, n$.

The **general term** or $(r+1)$th term in the binomial expansion is $T_{r+1} = \binom{n}{r}a^{n-r}b^r$.

Using sigma notation we write $(a+b)^n = \sum_{r=0}^{n} \binom{n}{r}a^{n-r}b^r$.

Example 17 — Self Tutor

Write down the first three and last two terms of the expansion of $\left(2x+\frac{1}{x}\right)^{12}$.

Do not simplify your answer.

$$\left(2x+\frac{1}{x}\right)^{12} = (2x)^{12} + \binom{12}{1}(2x)^{11}\left(\frac{1}{x}\right)^1 + \binom{12}{2}(2x)^{10}\left(\frac{1}{x}\right)^2 + \ldots$$

$$\ldots + \binom{12}{11}(2x)^1\left(\frac{1}{x}\right)^{11} + \left(\frac{1}{x}\right)^{12}$$

Example 18 — Self Tutor

Find the 7th term of $\left(3x-\frac{4}{x^2}\right)^{14}$. Do not simplify your answer.

$a = (3x)$, $b = \left(\frac{-4}{x^2}\right)$, and $n = 14$

Given the general term $T_{r+1} = \binom{n}{r}a^{n-r}b^r$, we let $r = 6$

$$\therefore\ T_7 = \binom{14}{6}(3x)^8\left(\frac{-4}{x^2}\right)^6$$

EXERCISE 8G

1 Write down the first three and last two terms of the following binomial expansions. Do not simplify your answers.

a $(1+2x)^{11}$ b $\left(3x+\frac{2}{x}\right)^{15}$ c $\left(2x-\frac{3}{x}\right)^{20}$

2 Without simplifying, write down:

a the 6th term of $(2x+5)^{15}$ b the 4th term of $\left(x^2+y\right)^9$

c the 10th term of $\left(x-\frac{2}{x}\right)^{17}$ d the 9th term of $\left(2x^2-\frac{1}{x}\right)^{21}$.

Example 19 Self Tutor

In the expansion of $\left(x^2+\frac{4}{x}\right)^{12}$, find:

a the coefficient of x^6 b the constant term

$a=(x^2)$, $b=\left(\frac{4}{x}\right)$, and $n=12$

$$\therefore \text{ the general term } T_{r+1} = \binom{12}{r}(x^2)^{12-r}\left(\frac{4}{x}\right)^r = \binom{12}{r}x^{24-2r}\times\frac{4^r}{x^r} = \binom{12}{r}4^r x^{24-3r}$$

a If $24-3r=6$
then $3r=18$
$\therefore r=6$
$\therefore T_7=\binom{12}{6}4^6x^6$
$\therefore$ the coefficient of x^6 is $\binom{12}{6}4^6$ or 3 784 704.

b If $24-3r=0$
then $3r=24$
$\therefore r=8$
$\therefore T_9=\binom{12}{8}4^8x^0$
$\therefore$ the constant term is $\binom{12}{8}4^8$ or 32 440 320.

3 Consider the expansion of $(x+b)^7$.

a Write down the general term of the expansion.

b Find b given that the coefficient of x^4 is -280.

4 Find the constant term in the expansion of:

a $\left(x+\frac{2}{x^2}\right)^{15}$ b $\left(x-\frac{3}{x^2}\right)^9$.

5 a Write down the first 5 rows of Pascal's triangle.

b Find the sum of the numbers in:

i row 1 ii row 2 iii row 3 iv row 4 v row 5.

c Copy and complete: "The sum of the numbers in row n of Pascal's triangle is ......"

d Show that $(1+x)^n=\binom{n}{0}+\binom{n}{1}x+\binom{n}{2}x^2+....+\binom{n}{n-1}x^{n-1}+\binom{n}{n}x^n$.

Hence deduce that $\binom{n}{0}+\binom{n}{1}+\binom{n}{2}+....+\binom{n}{n-1}+\binom{n}{n}=2^n$.

6 Find the coefficient of:

a x^{10} in the expansion of $(3+2x^2)^{10}$

b x^3 in the expansion of $\left(2x^2-\dfrac{3}{x}\right)^6$

c x^6y^3 in the expansion of $\left(2x^2-3y\right)^6$

d x^{12} in the expansion of $\left(2x^2-\dfrac{1}{x}\right)^{12}$.

Example 20 — Self Tutor

Find the coefficient of x^5 in the expansion of $(x+3)(2x-1)^6$.

$$\begin{aligned}&(x+3)(2x-1)^6\\&=(x+3)[(2x)^6+\tbinom{6}{1}(2x)^5(-1)+\tbinom{6}{2}(2x)^4(-1)^2+....]\\&=(x+3)(2^6x^6-\tbinom{6}{1}2^5x^5+\tbinom{6}{2}2^4x^4-....)\end{aligned}$$

(1) (2)

So, the terms containing x^5 are $\binom{6}{2}2^4x^5$ from (1)

and $-3\binom{6}{1}2^5x^5$ from (2)

$\therefore$ the coefficient of x^5 is $\binom{6}{2}2^4-3\binom{6}{1}2^5=-336$

7 **a** Find the coefficient of x^5 in the expansion of $(x+2)(x^2+1)^8$.

b Find the term containing x^6 in the expansion of $(2-x)(3x+1)^9$. Simplify your answer.

8 Consider the expression $\left(x^2y-2y^2\right)^6$. Find the term in which x and y are raised to the same power.

9 **a** The third term of $(1+x)^n$ is $36x^2$. Find the fourth term.

b If $(1+kx)^n=1-12x+60x^2-....$, find the values of k and n.

10 Find a if the coefficient of x^{11} in the expansion of $\left(x^2+\dfrac{1}{ax}\right)^{10}$ is 15.

11 Expand $(1+2x-x^2)^5$ in ascending powers of x as far as the term containing x^4.

12 Show that $\binom{n}{1}=n$ and $\binom{n}{2}=\dfrac{n(n-1)}{2}$ for all $n\in\mathbb{Z}^+$, $n\geqslant 2$.

13 From the binomial expansion of $(1+x)^n$, deduce that:

a $\binom{n}{0}-\binom{n}{1}+\binom{n}{2}-\binom{n}{3}+....+(-1)^n\binom{n}{n}=0$

b $\binom{2n+1}{0}+\binom{2n+1}{1}+\binom{2n+1}{2}+....+\binom{2n+1}{n}=4^n$

14 By considering the binomial expansion of $(1+x)^n$, find $\sum\limits_{r=0}^{n}2^r\binom{n}{r}$.

15 Consider $(x^2-3x+1)^2=x^4-6x^3+11x^2-6x+1$. The sum of the coefficients in the expansion is $1-6+11-6+1=1$.
Find the sum of the coefficients in the expansion of $(x^3+2x^2+3x-7)^{100}$.

16 By considering $(1+x)^n(1+x)^n=(1+x)^{2n}$, show that
$\binom{n}{0}^2+\binom{n}{1}^2+\binom{n}{2}^2+\binom{n}{3}^2+....+\binom{n}{n}^2=\binom{2n}{n}$.

17 **a** Write down the first four and last two terms of the binomial expansion $(3+x)^n$.

b Hence find, in simplest form, the sum of the series

$$3^n + \binom{n}{1}3^{n-1} + \binom{n}{2}3^{n-2} + \binom{n}{3}3^{n-3} + ... + 3n + 1.$$

18 **a** Prove that $r\binom{n}{r} = n\binom{n-1}{r-1}$.

b Hence show that $\binom{n}{1} + 2\binom{n}{2} + 3\binom{n}{3} + 4\binom{n}{4} + ... + n\binom{n}{n} = n2^{n-1}$.

c Suppose the set of numbers $\{P_r\}$ are defined by

$$P_r = \binom{n}{r}p^r(1-p)^{n-r} \quad \text{for} \quad r = 0, 1, 2, 3, ..., n.$$

Prove that: **i** $\sum_{r=0}^{n} P_r = 1$ **ii** $\sum_{r=1}^{n} rP_r = np$

REVIEW SET 8A — NON-CALCULATOR

1 Simplify: **a** $\dfrac{n!}{(n-2)!}$ **b** $\dfrac{n!+(n+1)!}{n!}$

2 Eight people enter a room and each person shakes hands with every other person. How many hand shakes are made?

3 The letters P, Q, R, S, and T are to be arranged in a row. How many of the possible arrangements:

a end with T **b** begin with P and end with T?

4 **a** How many three digit numbers can be formed using the digits 0 to 9?

b How many of these numbers are divisible by 5?

5 The first two terms in a binomial expansion are: $(a+b)^4 = e^{4x} - 4e^{2x} + ...$

a Find a and b. **b** Copy and complete the expansion.

6 Expand and simplify $(\sqrt{3}+2)^5$ giving your answer in the form $a + b\sqrt{3}$, $a, b \in \mathbb{Z}$.

7 Find the constant term in the expansion of $\left(3x^2 + \dfrac{1}{x}\right)^8$.

8 Find c given that the expansion $(1+cx)(1+x)^4$ includes the term $22x^3$.

9 **a** Write down the first four and last two terms of the binomial expansion $(2+x)^n$.

b Hence find, in simplest form, the sum of the series

$$2^n + \binom{n}{1}2^{n-1} + \binom{n}{2}2^{n-2} + \binom{n}{3}2^{n-3} + ... + 2n + 1.$$

REVIEW SET 8B — CALCULATOR

1 Ten points are located on a 2-dimensional plane such that no three points are collinear.

a How many line segments joining two points can be drawn?

b How many different triangles can be drawn by connecting all 10 points with line segments in every possible way?

2 Use the expansion of $(4+x)^3$ to find the exact value of $(4.02)^3$.

3 Find the term independent of x in the expansion of $\left(3x - \dfrac{2}{x^2}\right)^6$.

4 Find the coefficient of x^3 in the expansion of $(x+5)^6$.

5 A team of five is chosen from six men and four women.

a How many different teams are possible with no restrictions?

b How many different teams contain at least one person of each sex?

6 A four digit number is constructed using the digits 0, 1, 2, 3, ..., 9 at most once each.

a How many numbers are possible?

b How many of these numbers are divisible by 5?

7 Use Pascal's triangle to expand $(a+b)^6$.

Hence, find the binomial expansion of: **a** $(x-3)^6$ **b** $\left(1+\frac{1}{x}\right)^6$

8 Find the coefficient of x^{-6} in the expansion of $\left(2x-\frac{3}{x^2}\right)^{12}$.

9 Find the coefficient of x^5 in the expansion of $(2x+3)(x-2)^6$.

10 Find the possible values of a if the coefficient of x^3 in $\left(2x+\frac{1}{ax^2}\right)^9$ is 288.

REVIEW SET 8C

1 Alpha-numeric number plates have two letters followed by four digits. How many plates are possible if:

a there are no restrictions **b** the first letter must be a vowel

c no letter or digit may be repeated?

2 **a** How many committees of five can be selected from eight men and seven women?

b How many of the committees contain two men and three women?

c How many of the committees contain at least one man?

3 Use the binomial expansion to find: **a** $(x-2y)^3$ **b** $(3x+2)^4$

4 Find the coefficient of x^3 in the expansion of $(2x+5)^6$.

5 Find the constant term in the expansion of $\left(2x^2-\frac{1}{x}\right)^6$.

6 The first three terms in the expansion of $(1+kx)^n$ are $1-4x+\frac{15}{2}x^2$. Find k and n.

7 Eight people enter a room and sit at random in a row of eight chairs. In how many ways can the sisters Cathy, Robyn, and Jane sit together in the row?

8 A team of eight is chosen from 11 men and 7 women. How many different teams are possible if there:

a are no restrictions **b** must be four of each sex on the team

c must be at least two women on the team?

9 Find k in the expansion $(m-2n)^{10}=m^{10}-20m^9n+km^8n^2-....+1024n^{10}$.

10 Find the possible values of q if the constant terms in the expansions of $\left(x^3+\frac{q}{x^3}\right)^8$ and $\left(x^3+\frac{q}{x^3}\right)^4$ are equal.

Chapter 9

Mathematical induction

Syllabus reference: 1.4

Contents:

OPENING PROBLEM

Pascal's triangle is named after the French mathematician **Blaise Pascal** who recorded it in 1653 in his treatise *Traité du triangle arithmétique*.

However, this triangle was studied by mathematicians long before Pascal. Indian and Persian mathematicians used it in the 10th century, as did the Chinese mathematician **Yang Hui** in the 13th century.

Yang Hui

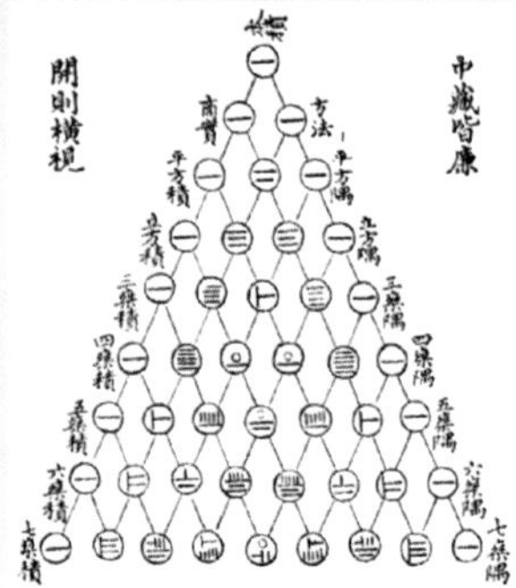

Yang Hui's Triangle

The first few rows of Pascal's triangle are shown alongside.

```
            1            ← row 0
          1   1          ← row 1
        1   2   1        ← row 2
      1   3   3   1          ⋮
    1   4   6   4   1
  1   5  10  10   5   1
1   6  15  20  15   6   1
            ⋮
```

Things to think about:

a Find the sum of the numbers in:

i row 0 **ii** row 1 **iii** row 2 **iv** row 3 **v** row 4.

b What do you suspect is the sum of the numbers in row n?

c Suppose your suggestion in **b** is true for the nth row of Pascal's triangle.

i Can you show that your suggestion must then be true for the $(n+1)$th row also?

ii How can we formally prove that your suggestion in **b** is correct?

THEORY OF KNOWLEDGE

Consider the statement: "The sum of the first n odd numbers is n^2."

How can we determine whether this statement is true? We can check that the statement is true for the first few values of n:

$$\begin{aligned}
n=1: &\quad 1=1^2\\
n=2: &\quad 1+3=4=2^2\\
n=3: &\quad 1+3+5=9=3^2\\
n=4: &\quad 1+3+5+7=16=4^2\\
n=5: &\quad 1+3+5+7+9=25=5^2
\end{aligned}$$

So, the statement is true for the first 5 values of n. But is this enough to *prove* that the statement is true in general? What if we checked the first 20 values of n, or the first 100 values of n?

1 Can we prove that a statement is true in general by checking it is true for a few specific cases?

2 How do we know when we have proven a statement to be true?

There are many **conjectures** in mathematics. These are statements that have not been proven, but are *believed* to be true.

Occasionally, incorrect statements have been thought to be true after a few specific cases were tested. The Swiss mathematician **Leonhard Euler** (1707 - 1783) stated that n^2+n+41 is a prime number for any positive integer n.

For example: when $n=1$, $n^2+n+41=43$, which is prime
when $n=2$, $n^2+n+41=47$, which is prime
when $n=3$, $n^2+n+41=53$, which is prime.

Leonhard Euler

In fact, n^2+n+41 is prime for all positive integers n from 1 to 39. However, checking $n=40$ gives a counter-example to the statement, since $40^2+40+41=41\times 41$.

3 Is it reasonable for a mathematician to assume a conjecture is true until it has been formally proven?

A THE PROCESS OF INDUCTION

The process of formulating a general result from a close examination of the simplest cases is called **mathematical induction**.

For example:

the first positive even number is	$2=2\times 1$
the second positive even number is	$4=2\times 2$
the third positive even number is	$6=2\times 3$
the fourth positive even number is	$8=2\times 4$

and from these results we *induce* that "the nth positive even number is $2\times n$ or $2n$".

The statement that "the nth positive even number is $2n$" is a summary of the observations of the simple cases $n=1, 2, 3, 4$ and is a statement which we *believe* is true. We call such a statement a **conjecture** or **proposition**.

Consider the sum of the first n odd numbers:

$$\begin{aligned}1&=1=1^2\\1+3&=4=2^2\\1+3+5&=9=3^2\\1+3+5+7&=16=4^2\\1+3+5+7+9&=25=5^2\end{aligned}$$

We may conjecture that "the sum of the first n odd numbers is n^2".

The nth odd number is $(2n-1)$, so we could also write the conjecture as:

$$1+3+5+7+9+....+(2n-1)=n^2 \quad \text{for all } n\in\mathbb{Z}^+$$

or $$\sum_{i=1}^{n}(2i-1)=n^2 \quad \text{for all } n\in\mathbb{Z}^+.$$

Example 1 Self Tutor

By examining the cases $n = 1, 2, 3, 4$, make a conjecture about the sum of
$S_n = \frac{1}{1\times2} + \frac{1}{2\times3} + \frac{1}{3\times4} + \frac{1}{4\times5} + ... + \frac{1}{n(n+1)}$.

$S_1 = \frac{1}{1\times2} = \frac{1}{2}$

$S_2 = \frac{1}{1\times2} + \frac{1}{2\times3} = \frac{1}{2} + \frac{1}{6} = \frac{2}{3}$

$S_3 = \frac{1}{1\times2} + \frac{1}{2\times3} + \frac{1}{3\times4} = \frac{2}{3} + \frac{1}{12} = \frac{3}{4}$

$S_4 = \frac{1}{1\times2} + \frac{1}{2\times3} + \frac{1}{3\times4} + \frac{1}{4\times5} = \frac{3}{4} + \frac{1}{20} = \frac{4}{5}$

If the conjecture is true then
$S_{1000} = \frac{1}{1\times2} + \frac{1}{2\times3} + ... + \frac{1}{1000\times1001}$
$= \frac{1000}{1001}$

From these results we conjecture that $S_n = \dfrac{n}{n+1}$ for all $n \in \mathbb{Z}^+$.

EXERCISE 9A

1 Copy and complete the conjecture:
The nth term of the sequence 3, 7, 11, 15, 19, ... is ____ for $n \in$ ____ .

2 Examine the following using substitutions like $n = 1, 2, 3, 4, ...$ Hence complete the proposition or conjecture.

a $3^n > 1 + 2n$ for ____

b $11^n - 1$ is divisible by ____ for ____

c $7^n + 2$ is divisible by ____ for ____

d $\left(1-\frac{1}{2}\right)\left(1-\frac{1}{3}\right)\left(1-\frac{1}{4}\right)....\left(1-\frac{1}{n+1}\right) =$ ____ for ____

3 Copy and complete each of the following conjectures, then write it using summation notation:

a $2 + 4 + 6 + 8 + 10 + ... + 2n =$ ____ for ____

b $1! + 2 \times 2! + 3 \times 3! + 4 \times 4! + ... + n \times n! =$ ____ for ____

c $\dfrac{1}{2!} + \dfrac{2}{3!} + \dfrac{3}{4!} + \dfrac{4}{5!} + ... + \dfrac{n}{(n+1)!} =$ ____ for ____

d $\dfrac{1}{2\times5} + \dfrac{1}{5\times8} + \dfrac{1}{8\times11} + ...$ to n terms $=$ ____ for ____

4 n points are placed inside a triangle.
Non-intersecting line segments are drawn connecting the 3 vertices of the triangle and the points within it, to partition the given triangle into smaller triangles.

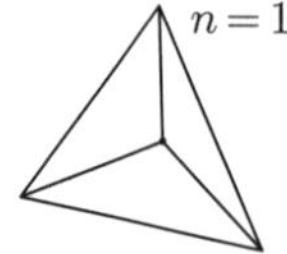

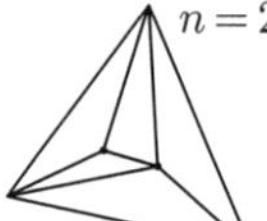

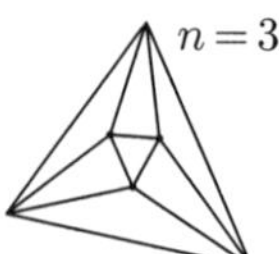

Make a proposition concerning the maximum number of triangles obtained in the general case.

5 n points are placed around a circle so that when line segments are drawn between every pair of the points, no three line segments intersect at the same point inside the circle. We consider the number of regions formed within the circle.

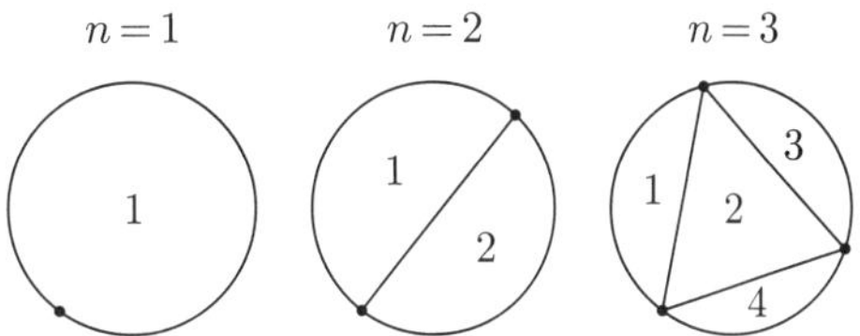

a Draw the cases $n = 4$ and $n = 5$.

b Use the cases $n = 1, 2, 3, 4, 5$ to form a conjecture about the number of regions formed in the general case.

c Draw the case $n = 6$. Do you still believe your conjecture?

B THE PRINCIPLE OF MATHEMATICAL INDUCTION

PROPOSITION NOTATION

We use P_n to represent a proposition which is defined for every integer n where $n \geqslant a$, $a \in \mathbb{Z}$.

For example, in the case of **Example 1**, our proposition P_n is:

$$\frac{1}{1\times 2}+\frac{1}{2\times 3}+\frac{1}{3\times 4}+ .... +\frac{1}{n(n+1)}=\frac{n}{n+1} \quad \text{for all } n\in\mathbb{Z}^+.$$

We see that:

- P_1 is: $\dfrac{1}{1\times 2}=\dfrac{1}{2}$
- P_2 is: $\dfrac{1}{1\times 2}+\dfrac{1}{2\times 3}=\dfrac{2}{3}$
- P_k is: $\dfrac{1}{1\times 2}+\dfrac{1}{2\times 3}+\dfrac{1}{3\times 4}+ .... +\dfrac{1}{k(k+1)}=\dfrac{k}{k+1}$.

THE PRINCIPLE OF MATHEMATICAL INDUCTION

The **principle of mathematical induction** constitutes a formal proof that a particular proposition is true.

> Suppose P_n is a proposition which is defined for every integer $n \geqslant a$, $a \in \mathbb{Z}$.
>
> If P_a is true, and if P_{k+1} is true whenever P_k is true, then P_n is true for all $n \geqslant a$.

Suppose for a given proposition that P_1 is true. If we can show that P_{k+1} is true whenever P_k is true, then the truth of P_1 implies that P_2 is true, which implies that P_3 is true, which implies that P_4 is true, and so on.

One can liken the principle of mathematical induction to the **domino effect**. We imagine an infinite set of dominoes all lined up.

Provided that:

- the first domino topples to the right, and
- the $(k+1)$th domino will topple if the kth domino topples,

then given sufficient time, every domino will topple.

DIVISIBILITY

Consider the expression $4^n + 2$ for $n = 0, 1, 2, 3, 4, 5, ...$

$4^0 + 2 = 3 \quad = 3 \times 1$

$4^1 + 2 = 6 \quad = 3 \times 2$

$4^2 + 2 = 18 \quad = 3 \times 6$

$4^3 + 2 = 66 \quad = 3 \times 22$

$4^4 + 2 = 258 = 3 \times 86$

We observe that each of the answers is divisible by 3 and so we make the conjecture

$4^n + 2$ is divisible by 3 for all $n \in \mathbb{Z}^+$.

Example 2 — Self Tutor

Prove that $4^n + 2$ is divisible by 3 for $n \in \mathbb{Z}$, $n \geqslant 0$, by using:

a the binomial expansion of $(1+3)^n$ **b** the principle of mathematical induction.

a If $n = 0$, $4^n + 2 = 4^0 + 2 = 3$ which is divisible by 3.

$$\begin{aligned} 4^n + 2 &= (1+3)^n + 2 \\ &= 1^n + \binom{n}{1}3 + \binom{n}{2}3^2 + \binom{n}{3}3^3 + ... + \binom{n}{n-1}3^{n-1} + \binom{n}{n}3^n + 2 \\ &= 3 + \binom{n}{1}3 + \binom{n}{2}3^2 + \binom{n}{3}3^3 + ... + \binom{n}{n-1}3^{n-1} + \binom{n}{n}3^n \\ &= 3\left(1 + \binom{n}{1} + \binom{n}{2}3 + \binom{n}{3}3^2 + ... + \binom{n}{n-1}3^{n-2} + \binom{n}{n}3^{n-1}\right) \end{aligned}$$

where the contents of the brackets is an integer.

$\therefore$ $4^n + 2$ is divisible by 3.

b P_n is: $4^n + 2$ is divisible by 3 for $n \in \mathbb{Z}$, $n \geqslant 0$.

Proof: (By the principle of mathematical induction)

(1) If $n = 0$, $4^0 + 2 = 3 = 1 \times 3$ $\therefore$ P_0 is true.

(2) If P_k is true, then $4^k + 2 = 3A$ where A is an integer, and $A \geqslant 1$.

$$\begin{aligned} \text{Now } 4^{k+1} + 2 &= 4^1 4^k + 2 \\ &= 4(3A - 2) + 2 \quad \{\text{using } P_k\} \\ &= 12A - 8 + 2 \\ &= 12A - 6 \\ &= 3(4A - 2) \quad \text{where } 4A - 2 \text{ is an integer as } A \in \mathbb{Z}. \end{aligned}$$

Thus $4^{k+1} + 2$ is divisible by 3 if $4^k + 2$ is divisible by 3.

Since P_0 is true, and P_{k+1} is true whenever P_k is true, then P_n is true for all $n \in \mathbb{Z}$, $n \geqslant 0$ {Principle of mathematical induction}

EXERCISE 9B.1

1 Prove that $3^n + 1$ is divisible by 2 for all integers $n \geqslant 0$, by using:

a the binomial expansion of $(1+2)^n$ **b** the principle of mathematical induction.

2 Prove that $6^n - 1$ is divisible by 5 for all integers $n \geqslant 0$, by using:

a the binomial expansion of $(5+1)^n$ **b** the principle of mathematical induction.

3 Use the principle of mathematical induction to prove that:

a $n^3 + 2n$ is divisible by 3 for all positive integers n

b $n(n^2+5)$ is divisible by 6 for all integers $n \in \mathbb{Z}^+$

c $7^n - 4^n - 3^n$ is divisible by 12 for all $n \in \mathbb{Z}^+$.

SEQUENCES AND SERIES

Example 3 — Self Tutor

a Prove that $\sum_{i=1}^{n} i^2 = \frac{n(n+1)(2n+1)}{6}$ for all $n \in \mathbb{Z}^+$.

b Find $1^2 + 2^2 + 3^2 + 4^2 + ... + 100^2$.

---

a P_n is: $1^2 + 2^2 + 3^2 + 4^2 + ... + n^2 = \frac{n(n+1)(2n+1)}{6}$ for all $n \in \mathbb{Z}^+$.

Proof: (By the principle of mathematical induction)

(1) If $n = 1$, LHS $= 1^2 = 1$ and RHS $= \frac{1 \times 2 \times 3}{6} = 1$

$\therefore$ P_1 is true.

(2) If P_k is true for $k \in \mathbb{Z}^+$, then

$$\sum_{i=1}^{k} i^2 = 1^2 + 2^2 + 3^2 + 4^2 + ... + k^2 = \frac{k(k+1)(2k+1)}{6}$$

Thus $1^2 + 2^2 + 3^2 + 4^2 + ... + k^2 + (k+1)^2$

$$= \frac{k(k+1)(2k+1)}{6} + (k+1)^2 \quad \{\text{using } P_k\}$$

$$= \frac{k(k+1)(2k+1)}{6} + (k+1)^2 \times \frac{6}{6}$$

$$= \frac{(k+1)[k(2k+1) + 6(k+1)]}{6}$$

$$= \frac{(k+1)(2k^2 + k + 6k + 6)}{6}$$

$$= \frac{(k+1)(2k^2 + 7k + 6)}{6}$$

$$= \frac{(k+1)(k+2)(2k+3)}{6}$$

$$= \frac{[k+1]([k+1]+1)(2[k+1]+1)}{6}$$

Always look for common factors.

Since P_1 is true, and P_{k+1} is true whenever P_k is true,
P_n is true for all $n \in \mathbb{Z}^+$. {Principle of mathematical induction}

b Using P_{100}, $1^2 + 2^2 + 3^2 + 4^2 + ... + 100^2 = \frac{100 \times 101 \times 201}{6} = 338\,350$

EXERCISE 9B.2

1 Use the principle of mathematical induction to prove that the following propositions are true for all positive integers n:

a $\sum_{i=1}^{n} i = \frac{n(n+1)}{2}$

b $\sum_{i=1}^{n} i(i+1) = \frac{n(n+1)(n+2)}{3}$

c $\sum_{i=1}^{n} 3i(i+4) = \frac{n(n+1)(2n+13)}{2}$

d $\sum_{i=1}^{n} i^3 = \frac{n^2(n+1)^2}{4}$

2 Prove that the sum of the first n odd numbers is n^2 using:

a the sum of an arithmetic series formula

b the principle of mathematical induction.

3 Prove that $\sum_{i=1}^{n} i \times 2^{i-1} = (n-1) \times 2^n + 1$ for all $n \in \mathbb{Z}^+$.

Example 4 — Self Tutor

Prove that $\sum_{i=1}^{n} \frac{1}{(3i-1)(3i+2)} = \frac{n}{6n+4}$ for all $n \in \mathbb{Z}^+$.

P_n is: $\frac{1}{2 \times 5} + \frac{1}{5 \times 8} + ... + \frac{1}{(3n-1)(3n+2)} = \frac{n}{6n+4}$ for all $n \in \mathbb{Z}^+$.

Proof: (By the principle of mathematical induction)

(1) If $n = 1$, LHS $= \frac{1}{2 \times 5} = \frac{1}{10}$ and RHS $= \frac{1}{6 \times 1 + 4} = \frac{1}{10}$

$\therefore$ P_1 is true.

(2) If P_k is true, then

$$\sum_{i=1}^{k} \frac{1}{(3i-1)(3i+2)} = \frac{1}{2 \times 5} + \frac{1}{5 \times 8} + ... + \frac{1}{(3k-1)(3k+2)} = \frac{k}{6k+4}$$

Now $\frac{1}{2 \times 5} + \frac{1}{5 \times 8} + ... + \frac{1}{(3k-1)(3k+2)} + \frac{1}{(3k+2)(3k+5)}$

$= \frac{k}{6k+4} + \frac{1}{(3k+2)(3k+5)}$ {using P_k}

$= \frac{k}{2(3k+2)} \times \left(\frac{3k+5}{3k+5}\right) + \frac{1}{(3k+2)(3k+5)} \times \left(\frac{2}{2}\right)$

$= \frac{3k^2+5k+2}{2(3k+2)(3k+5)}$

$= \frac{\cancel{(3k+2)}(k+1)}{2\cancel{(3k+2)}(3k+5)}$

$= \frac{k+1}{6k+10}$

$= \frac{(k+1)}{6(k+1)+4}$

Since P_1 is true, and P_{k+1} is true whenever P_k is true,
P_n is true for all $n \in \mathbb{Z}^+$. {Principle of mathematical induction}

4 **a** Prove that $\frac{1}{1 \times 2} + \frac{1}{2 \times 3} + \frac{1}{3 \times 4} + ... + \frac{1}{n(n+1)} = \frac{n}{n+1}$ for all $n \in \mathbb{Z}^+$.

b Hence find $\frac{1}{10 \times 11} + \frac{1}{11 \times 12} + \frac{1}{12 \times 13} + ... + \frac{1}{20 \times 21}$.

c Prove that $\frac{1}{1 \times 2 \times 3} + \frac{1}{2 \times 3 \times 4} + ... + \frac{1}{n(n+1)(n+2)} = \frac{n(n+3)}{4(n+1)(n+2)}$ for all $n \in \mathbb{Z}^+$.

5 **a** Prove that $1 \times 1! + 2 \times 2! + 3 \times 3! + 4 \times 4! + ... + n \times n! = (n+1)! - 1$ for all $n \in \mathbb{Z}^+$, where $n!$ is the product of the first n positive integers.

b Prove that $\frac{1}{2!} + \frac{2}{3!} + \frac{3}{4!} + \frac{4}{5!} + ... + \frac{n}{(n+1)!} = \frac{(n+1)! - 1}{(n+1)!}$ for all $n \in \mathbb{Z}^+$.

c Hence find the sum $\frac{1}{2!} + \frac{2}{3!} + \frac{3}{4!} + \frac{4}{5!} + ... + \frac{9}{10!}$ in rational form.

6 Prove that $n + 2(n-1) + 3(n-2) + ... + (n-2)3 + (n-1)2 + n = \frac{n(n+1)(n+2)}{6}$ for all integers $n \geqslant 1$.

Hint: $1 \times 6 + 2 \times 5 + 3 \times 4 + 4 \times 3 + 5 \times 2 + 6 \times 1$
$= 1 \times 5 + 2 \times 4 + 3 \times 3 + 4 \times 2 + 5 \times 1 + (1 + 2 + 3 + 4 + 5 + 6)$

Example 5 — Self Tutor

A sequence is defined by $u_1 = 1$ and $u_{n+1} = 2u_n + 1$ for all $n \in \mathbb{Z}^+$.

Prove that $u_n = 2^n - 1$ for all $n \in \mathbb{Z}^+$.

P_n is: if $u_1 = 1$ and $u_{n+1} = 2u_n + 1$ for all $n \in \mathbb{Z}^+$, then $u_n = 2^n - 1$.

Proof: (By the principle of mathematical induction)

(1) If $n = 1$, $u_1 = 2^1 - 1 = 2 - 1 = 1$ $\therefore$ P_1 is true.

(2) If P_k is true, then $u_k = 2^k - 1$

$$\begin{aligned} \text{Now } u_{k+1} &= 2u_k + 1 \\ &= 2(2^k - 1) + 1 \quad \{\text{using } P_k\} \\ &= 2^{k+1} - 2 + 1 \\ &= 2^{k+1} - 1 \end{aligned}$$

$\therefore$ P_{k+1} is also true.

Since P_1 is true, and P_{k+1} is true whenever P_k is true,
then P_n is true for all $n \in \mathbb{Z}^+$ {Principle of mathematical induction}

7 Use the principle of mathematical induction to prove these propositions:

a If a sequence $\{u_n\}$ is defined by $u_1 = 5$ and $u_{n+1} = u_n + 8n + 5$ for all $n \in \mathbb{Z}^+$, then $u_n = 4n^2 + n$.

b If the first term of a sequence is 1, and subsequent terms are defined by the recursion formula $u_{n+1} = 2 + 3u_n$, then $u_n = 2(3^{n-1}) - 1$.

c If a sequence is defined by $u_1 = 2$ and $u_{n+1} = \frac{u_n}{2(n+1)}$ for all $n \in \mathbb{Z}^+$, then $u_n = \frac{2^{2-n}}{n!}$.

d If a sequence is defined by $u_1 = 1$ and $u_{n+1} = u_n + (-1)^n(n+1)^2$ for all $n \in \mathbb{Z}^+$, then $u_n = \frac{(-1)^{n-1}n(n+1)}{2}$.

8 A sequence is defined by $u_1 = 1$ and $u_{n+1} = u_n + (2n+1)$ for all $n \in \mathbb{Z}^+$.

a By finding u_n for $n = 2$, 3, and 4, conjecture a formula for u_n in terms of n only.

b Prove that your conjecture is true using the principle of mathematical induction.

9 A sequence is defined by $u_1 = \frac{1}{3}$ and $u_{n+1} = u_n + \frac{1}{(2n+1)(2n+3)}$ for all $n \in \mathbb{Z}^+$.

a By finding u_n for $n = 2$, 3, and 4, conjecture a formula for u_n in terms of n only.

b Prove that your conjecture is true using the principle of mathematical induction.

10 $(2+\sqrt{3})^n = A_n + B_n\sqrt{3}$ for all $n \in \mathbb{Z}^+$, where A_n and B_n are integers.

- **a** Find A_n and B_n for $n = 1, 2, 3$, and 4.
- **b** Without using induction, show that $A_{n+1} = 2A_n + 3B_n$ and $B_{n+1} = A_n + 2B_n$.
- **c** Calculate $(A_n)^2 - 3(B_n)^2$ for $n = 1, 2, 3$, and 4 and hence conjecture a result.
- **d** Prove that your conjecture is true.

11 Prove that $u_n = \dfrac{2^n - (-1)^n}{3}$ is an odd number for all $n \in \mathbb{Z}^+$.

12 Sometimes, in order to prove the next inductive step, we require the previous *two* cases. Another form of the principle of mathematical induction is therefore:

> Suppose P_n is a proposition defined for all $n \in \mathbb{Z}^+$.
> If P_1 and P_2 are true, and also P_{k+2} is true whenever P_k and P_{k+1} are true, then P_n is true for all $n \in \mathbb{Z}^+$.

Use this form to prove that:

- **a** If a sequence u_n is defined by $u_1 = 11$, $u_2 = 37$, and $u_{n+2} = 5u_{n+1} - 6u_n$ for all $n \in \mathbb{Z}^+$, then $u_n = 5(3^n) - 2^{n+1}$.
- **b** If $u_n = (3+\sqrt{5})^n + (3-\sqrt{5})^n$ where $n \in \mathbb{Z}^+$, then u_n is a multiple of 2^n.
  Hint: First find a and b such that $u_{n+2} = au_{n+1} + bu_n$.

OTHER APPLICATIONS

Proof by the principle of mathematical induction is used in several other areas of mathematics, including:

- inequalities
- products
- geometrical generalisations
- complex numbers.
- differential calculus

You will find some proofs with these topics in the appropriate chapters later in the book.

Example 6 Self Tutor

Prove that a convex n-sided polygon has $\frac{1}{2}n(n-3)$ diagonals for all $n \geqslant 3$.

P_n is: A convex n-sided polygon has $\frac{1}{2}n(n-3)$ diagonals for all $n \geqslant 3$.

Proof: (By the principle of mathematical induction)

(1) If $n = 3$ we have a triangle with 0 diagonals,
and $\frac{1}{2} \times 3 \times 0 = 0$
$\therefore$ P_3 is true.

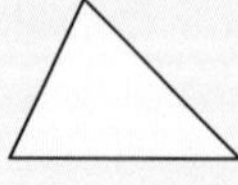

(2) If P_k is true, a convex k-sided polygon has $\frac{1}{2}k(k-3)$ diagonals.
We label the vertices
$1, 2, 3, 4, 5, \ldots, k-1, k$
and $k+1$ as an additional vertex.

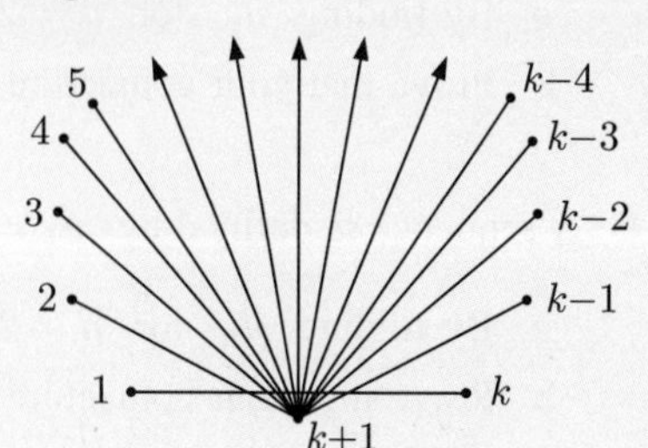

The number of diagonals when $n = k+1$ is
the number of diagonals when $n = k$, $+ \underbrace{k-2}$ $(+1)$ ← the line from vertex 1 to vertex k was previously a side and is now a diagonal

↑ the number of diagonals from vertex $k+1$ to the vertices 2, 3, 4, 5, ..., $k-1$

$\therefore$ the number of diagonals when $n = k+1$ is

$$\begin{aligned}
&\tfrac{1}{2}k(k-3) + k - 1 \\
&= \tfrac{1}{2}k(k-3) + \tfrac{2}{2}(k-1) \\
&= \tfrac{1}{2}[k^2 - 3k + 2k - 2] \\
&= \tfrac{1}{2}[k^2 - k - 2] \\
&= \tfrac{1}{2}(k+1)(k-2) \\
&= \tfrac{1}{2}[k+1]([k+1] - 3)
\end{aligned}$$

Try this process for
$k = 4, 5, 6, ...$

Since P_3 is true, and P_{k+1} is true whenever P_k is true, then P_n is true for all $n \geqslant 3$ {Principle of mathematical induction}

EXERCISE 9B.3

1 Use the principle of mathematical induction to prove the following propositions:

a $\left(1-\frac{1}{2}\right)\left(1-\frac{1}{3}\right)\left(1-\frac{1}{4}\right)\left(1-\frac{1}{5}\right)....\left(1-\frac{1}{n+1}\right) = \frac{1}{n+1}$, $n \in \mathbb{Z}^+$.

Hint: The LHS $\neq \sum_{i=1}^{n} \left(1 - \frac{1}{i+1}\right)$ since it is a product not a sum.

b If n straight lines are drawn on a plane such that each line intersects every other line and no three lines have a common point of intersection, then the plane is divided into $\dfrac{n(n+1)}{2} + 1$ regions.

c If n points are placed inside a triangle, and non-intersecting lines are drawn connecting the 3 vertices of the triangle and the points within it to partition the triangle into smaller triangles, then the maximum number of triangles resulting is $2n+1$.

d $\left(1-\frac{1}{2^2}\right)\left(1-\frac{1}{3^2}\right)\left(1-\frac{1}{4^2}\right)....\left(1-\frac{1}{n^2}\right) = \dfrac{n+1}{2n}$ for all integers $n \geqslant 2$.

2 Use the principle of mathematical induction to prove the following propositions:

a $3^n \geqslant 1 + 2n$ for all $n \in \mathbb{N}$

b $n! \geqslant 2^n$ for all $n \in \mathbb{Z}$, $n \geqslant 4$

c $8^n \geqslant n^3$ for all $n \in \mathbb{Z}^+$

d $(1-h)^n \leqslant \dfrac{1}{1+nh}$ for $0 \leqslant h \leqslant 1$ and all $n \in \mathbb{Z}^+$.

3 Use the principle of mathematical induction to prove the following propositions. You may use the results from **Chapter 6** that for all complex numbers z_1 and z_2, $(z_1+z_2)^* = {z_1}^* + {z_2}^*$ and $(z_1 z_2)^* = {z_1}^* {z_2}^*$.

a $(z_1 + z_2 + ... + z_n)^* = {z_1}^* + {z_2}^* + ... + z_n^*$ for all $n \in \mathbb{Z}^+$ and complex $z_1, z_2, ..., z_n$

b $(z_1 z_2 .... z_n)^* = {z_1}^* {z_2}^* .... z_n^*$ for all $n \in \mathbb{Z}^+$ and complex $z_1, z_2, ..., z_n$

c $(z^n)^* = (z^*)^n$ for all $n \in \mathbb{Z}^+$ and complex z.

INVESTIGATION 1 SEQUENCES, SERIES, AND INDUCTION

This Investigation involves the principle of mathematical induction as well as concepts from sequences, series, and counting.

The sequence of numbers $\{u_n\}$ is defined by $u_n = n \times n!$.

What to do:

1. Let $S_n = u_1 + u_2 + u_3 + ... + u_n$. Investigate S_n for several different values of n.
2. Hence conjecture an expression for S_n.
3. Prove your conjecture to be true using the principle of mathematical induction.
4. Show that u_n can be written as $(n+1)! - n!$ and hence devise an alternative direct proof of your conjecture for S_n.
5. Let $C_n = u_n + u_{n+1}$. Write an expression for C_n in factorial notation, and simplify it.
6. Let $T_n = C_1 + C_2 + C_3 + ... + C_n$. Find T_n for $n = 1, 2, 3, 4$, and 5.
7. Conjecture an expression for T_n.
8. Prove your conjecture for T_n by any method.

INVESTIGATION 2 EQUIVALENT STATEMENTS

Already in the course we have seen *equivalent statements* which are connected by "if and only if".

For example, if the statements A and B are equivalent we write $A \Leftrightarrow B$.

In order to prove that two statements are equivalent, we need to prove that *both*:

- if A is true then B is also true, written $A \Rightarrow B$
- if B is true then A is also true, written $B \Rightarrow A$

For example, consider the equivalent statements for $n \in \mathbb{Z}^+$:

$$(n+2)^2 - n^2 \text{ is a multiple of 8} \quad \Leftrightarrow \quad n \text{ is odd}$$

Proof: If $(n+2)^2 - n^2$ is a multiple of 8 then

$n^2 + 4n + 4 - n^2 = 8k$ for some integer k

$\therefore \; 4n + 4 = 8k$

$\therefore \; 4(n+1) = 8k$

$\therefore \; n + 1 = 2k$

$\therefore \; n = 2k - 1$

$\therefore \; n$ is odd.

If n is odd then $n = 2k - 1$ for some integer k

$\therefore \; n + 1 = 2k$

$\therefore \; 4(n+1) = 8k$

$\therefore \; 4n + 4 = 8k$

$\therefore \; n^2 + 4n + 4 - n^2 = 8k$

$\therefore \; (n+2)^2 - n^2$ is a multiple of 8

We need to show that $A \Rightarrow B$ *and* that $B \Rightarrow A$.

What to do:

1. Prove using the method of mathematical induction that $(n+2)^2 - n^2$ is a multiple of 8 for all odd integers n.
2. Prove that for all $x \in \mathbb{Z}^+$, x is not divisible by 3 $\Leftrightarrow$ $x^2 - 1$ is divisible by 3.
3. Can you prove using the method of mathematical induction that $x^2 - 1$ is divisible by 3 for all positive integers x that are *not* divisible by 3?

REVIEW SET 9A

Prove the following propositions using the principle of mathematical induction:

1. $\sum\limits_{i=1}^{n} (2i-1) = n^2$ for all $n \in \mathbb{Z}^+$.
2. $7^n + 2$ is divisible by 3 for all $n \in \mathbb{Z}^+$.
3. $\sum\limits_{i=1}^{n} i(i+1)(i+2) = \dfrac{n(n+1)(n+2)(n+3)}{4}$ for all $n \in \mathbb{Z}^+$.
4. $1 + r + r^2 + r^3 + r^4 + ... + r^{n-1} = \dfrac{1-r^n}{1-r}$ for all $n \in \mathbb{Z}^+$, provided $r \neq 1$.
5. $5^{2n} - 1$ is divisible by 24 for all $n \in \mathbb{Z}^+$.
6. $5^n \geqslant 1 + 4n$ for all $n \in \mathbb{Z}^+$.
7. If $u_1 = 1$ and $u_{n+1} = 3u_n + 2^n$, then $u_n = 3^n - 2^n$ for all $n \in \mathbb{Z}^+$.

REVIEW SET 9B

Prove the following propositions using the principle of mathematical induction:

1. $\sum\limits_{i=1}^{n} (2i-1)^2 = \dfrac{n(2n+1)(2n-1)}{3}$ for all $n \in \mathbb{Z}^+$.
2. $3^{2n+2} - 8n - 9$ is divisible by 64 for all positive integers n.
3. $\sum\limits_{i=1}^{n} (2i+1)2^{i-1} = 1 + (2n-1) \times 2^n$ for all positive integers n.
4. $5^n + 3$ is divisible by 4 for all integers $n \geqslant 0$.
5. $\sum\limits_{i=1}^{n} i(i+1)^2 = \dfrac{n(n+1)(n+2)(3n+5)}{12}$ for all positive integers n.
6. $5^n + 3^n \geqslant 2^{2n+1}$ for all $n \in \mathbb{Z}^+$.
7. If $u_1 = 9$ and $u_{n+1} = 2u_n + 3(5^n)$, then $u_n = 2^{n+1} + 5^n$ for all $n \in \mathbb{Z}^+$.

REVIEW SET 9C

Prove the following propositions using the principle of mathematical induction:

1 $\sum_{i=1}^{n} i(i+2) = \dfrac{n(n+1)(2n+7)}{6}$ for all $n \in \mathbb{Z}^+$.

2 $7^n - 1$ is divisible by 6 for all $n \in \mathbb{Z}^+$.

3 $\sum_{i=1}^{n} (2i-1)^3 = n^2(2n^2-1)$ for all positive integers $n \geqslant 1$.

4 $3^n - 1 - 2n$ is divisible by 4 for all non-negative integers n.

5 $\sum_{i=1}^{n} \dfrac{1}{(2i-1)(2i+1)} = \dfrac{n}{2n+1}$ for all positive integers n.

6 If $u_1 = 5$ and $u_{n+1} = 2u_n - 3(-1)^n$, then $u_n = 3(2^n) + (-1)^n$ for all $n \in \mathbb{Z}^+$.

7 $\sqrt[n]{n!} \leqslant \dfrac{n+1}{2}$ for all $n \in \mathbb{Z}^+$.

Chapter 10

The unit circle and radian measure

Syllabus reference: 3.1, 3.2

Contents:

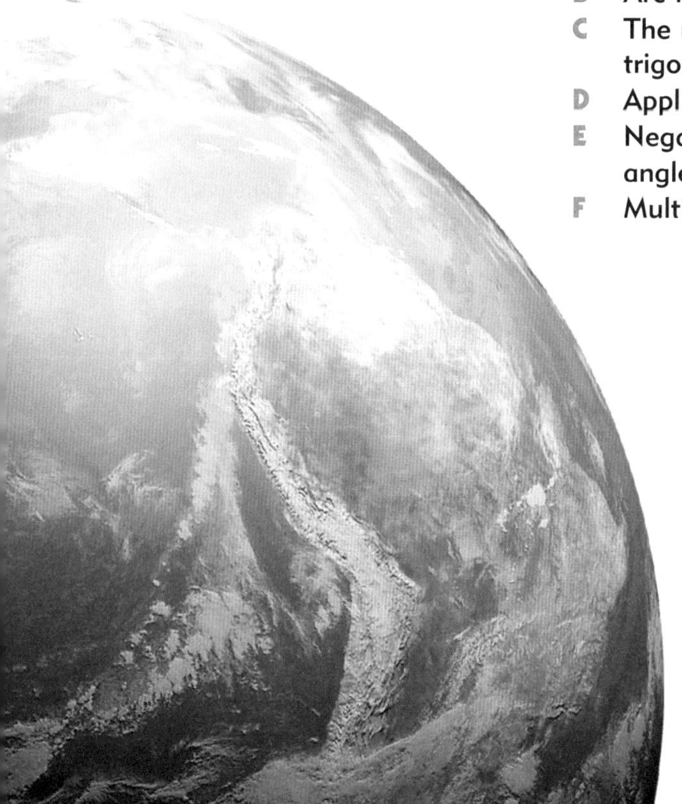

OPENING PROBLEM

Consider an equilateral triangle with sides 2 cm long. Altitude [AN] bisects side [BC] and the vertical angle BAC.

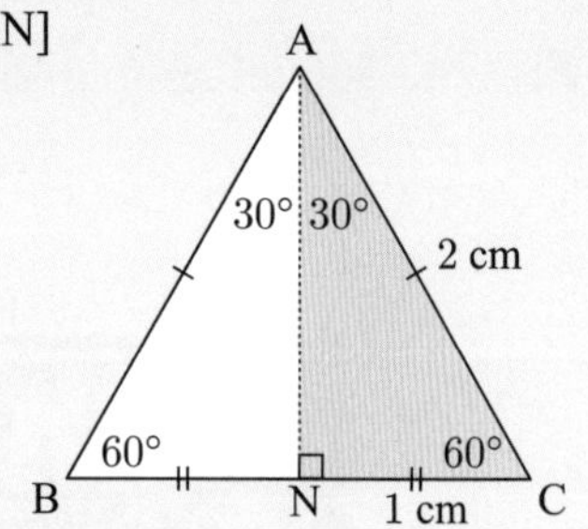

Things to think about:

a Can you use this figure to explain why $\sin 30° = \frac{1}{2}$?

b Use your calculator to find the values of $\sin 30°$, $\sin 150°$, $\sin 390°$, $\sin 1110°$, and $\sin(-330°)$.
Can you explain your results even though the angles are not between $0°$ and $90°$?

A RADIAN MEASURE

DEGREE MEASUREMENT OF ANGLES

We have seen previously that one full revolution makes an angle of $360°$, and the angle on a straight line is $180°$. Hence, one **degree**, $1°$, can be defined as $\frac{1}{360}$th of one full revolution. This measure of angle is commonly used by surveyors and architects.

For greater accuracy we define one **minute**, $1'$, as $\frac{1}{60}$th of one degree and one **second**, $1''$, as $\frac{1}{60}$th of one minute. Obviously a minute and a second are very small angles.

Most graphics calculators can convert fractions of angles measured in degrees into minutes and seconds. This is also useful for converting fractions of hours into minutes and seconds for time measurement, as one minute is $\frac{1}{60}$th of one hour, and one second is $\frac{1}{60}$th of one minute.

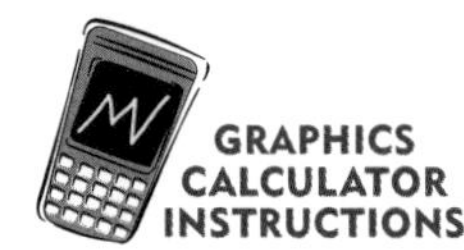

RADIAN MEASUREMENT OF ANGLES

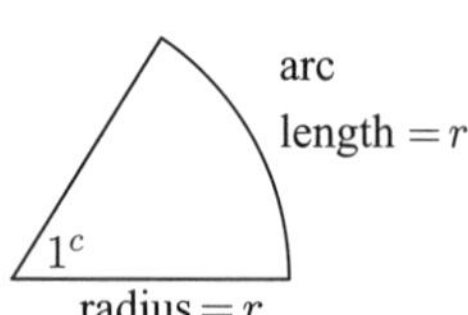

An angle is said to have a measure of 1 **radian** (1^c) if it is subtended at the centre of a circle by an arc equal in length to the radius.

The symbol 'c' is used for radian measure but is usually omitted. By contrast, the degree symbol is *always* used when the measure of an angle is given in degrees.

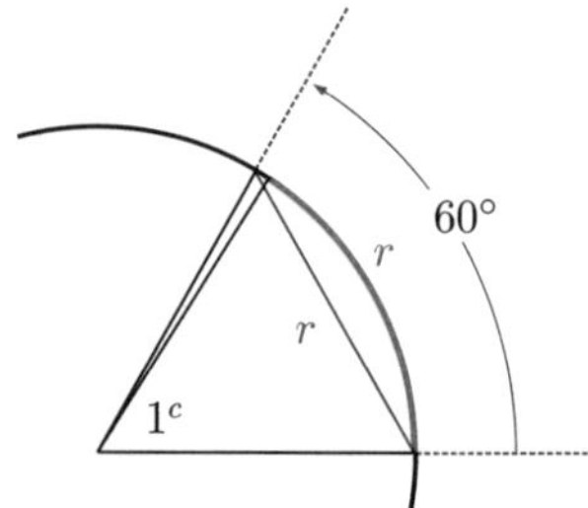

From the diagram to the right, it can be seen that 1^c is slightly smaller than $60°$. In fact, $1^c \approx 57.3°$.

The word 'radian' is an abbreviation of 'radial angle'.

DEGREE-RADIAN CONVERSIONS

If the radius of a circle is r, then an arc of length πr, or half the circumference, will subtend an angle of π radians.

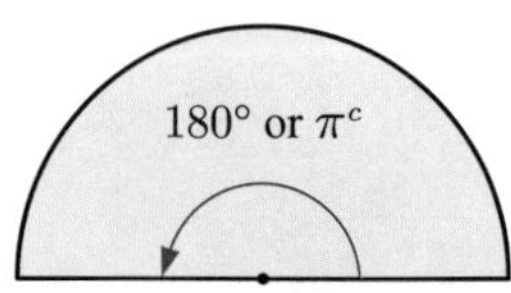

Therefore, 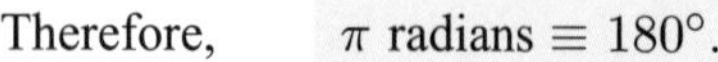 π radians $\equiv 180°$.

So, $1^c = \left(\frac{180}{\pi}\right)^\circ \approx 57.3^\circ$ and $1^\circ = \left(\frac{\pi}{180}\right)^c \approx 0.0175^c$.

To convert from degrees to radians, we multiply by $\frac{\pi}{180}$.

To convert from radians to degrees, we multiply by $\frac{180}{\pi}$.

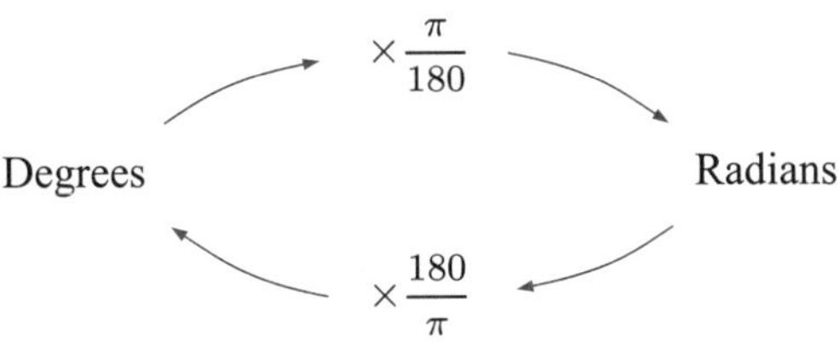

We indicate degrees with a small °.
To indicate radians we use a small ^{c} or else use no symbol at all.

Example 1 **Self Tutor**

Convert 45° to radians in terms of π.

$45^\circ = (45 \times \frac{\pi}{180})$ radians
$= \frac{\pi}{4}$ radians

or

$180^\circ = \pi$ radians
$\therefore \left(\frac{180}{4}\right)^\circ = \frac{\pi}{4}$ radians
$\therefore 45^\circ = \frac{\pi}{4}$ radians

Angles in radians may be expressed either in terms of π or as decimals.

Example 2 **Self Tutor**

Convert 126.5° to radians.

126.5°
$= (126.5 \times \frac{\pi}{180})$ radians
≈ 2.21 radians

EXERCISE 10A

1 Convert to radians, in terms of π:

a 90°	**b** 60°	**c** 30°	**d** 18°	**e** 9°
f 135°	**g** 225°	**h** 270°	**i** 360°	**j** 720°
k 315°	**l** 540°	**m** 36°	**n** 80°	**o** 230°

2 Convert to radians, correct to 3 significant figures:

a 36.7°	**b** 137.2°	**c** 317.9°	**d** 219.6°	**e** 396.7°

Example 3 **Self Tutor**

Convert $\frac{5\pi}{6}$ to degrees.

$\frac{5\pi}{6}$
$= \left(\frac{5\pi}{6} \times \frac{180}{\pi}\right)^\circ$
$= 150^\circ$

3 Convert the following radian measures to degrees:

a $\frac{\pi}{5}$ **b** $\frac{3\pi}{5}$ **c** $\frac{3\pi}{4}$ **d** $\frac{\pi}{18}$ **e** $\frac{\pi}{9}$

f $\frac{7\pi}{9}$ **g** $\frac{\pi}{10}$ **h** $\frac{3\pi}{20}$ **i** $\frac{7\pi}{6}$ **j** $\frac{\pi}{8}$

Example 4 **Self Tutor**

Convert 0.638 radians to degrees.

0.638 radians

$= (0.638 \times \frac{180}{\pi})^\circ$

$\approx 36.6^\circ$

4 Convert the following radians to degrees. Give your answers correct to 2 decimal places.

a 2 **b** 1.53 **c** 0.867 **d** 3.179 **e** 5.267

5 Copy and complete, giving answers in terms of π:

a

Degrees	0	45	90	135	180	225	270	315	360
Radians									

b

Degrees	0	30	60	90	120	150	180	210	240	270	300	330	360
Radians													

THEORY OF KNOWLEDGE

There are several theories for why one complete turn was divided into 360 degrees:

- 360 is approximately the number of days in a year.
- The Babylonians used a counting system in base 60. If they drew 6 equilateral triangles within a circle as shown, and divided each angle into 60 subdivisions, then there were 360 subdivisions in one turn. The division of an hour into 60 minutes, and a minute into 60 seconds, is from this base 60 counting system.
- 360 has 24 divisors, including every integer from 1 to 10 except 7.

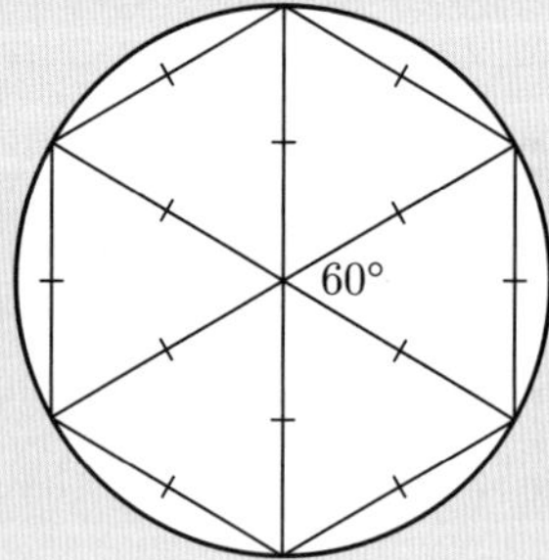

The idea of measuring an angle by the length of an arc dates to around 1400 and the Persian mathematician Al-Kashi. The concept of a radian is generally credited to Roger Cotes, however, who described it as we know it today.

1 What other measures of angle are there?

2 Which is the most *natural* unit of angle measure?

B ARC LENGTH AND SECTOR AREA

You should be familiar with these terms relating to the parts of a circle:

sector
arc (part of circle)
centre
chord
radius
segment

An arc, sector, or segment is described as:

- **minor** if it involves less than half the circle
- **major** if it involves more than half the circle.

For example:

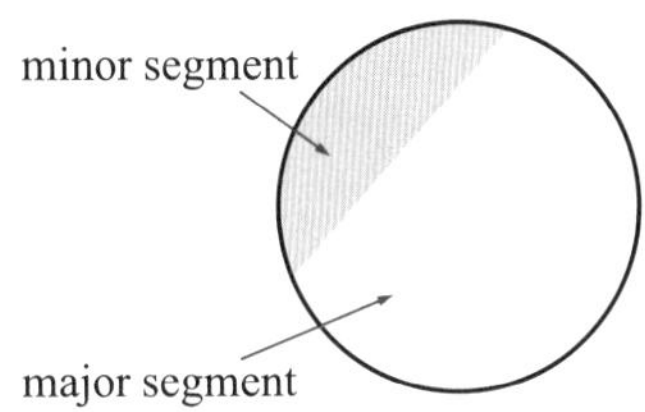

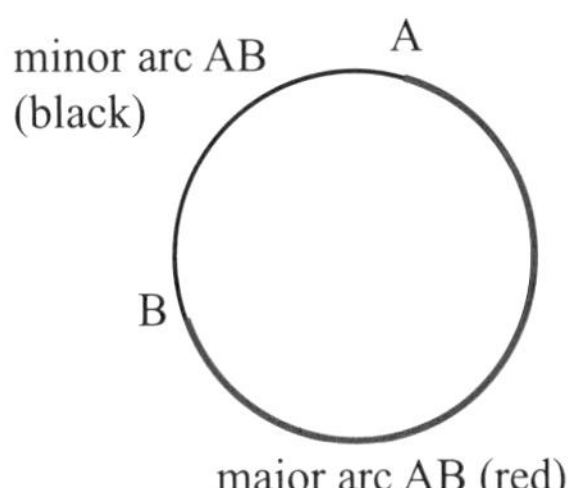

ARC LENGTH

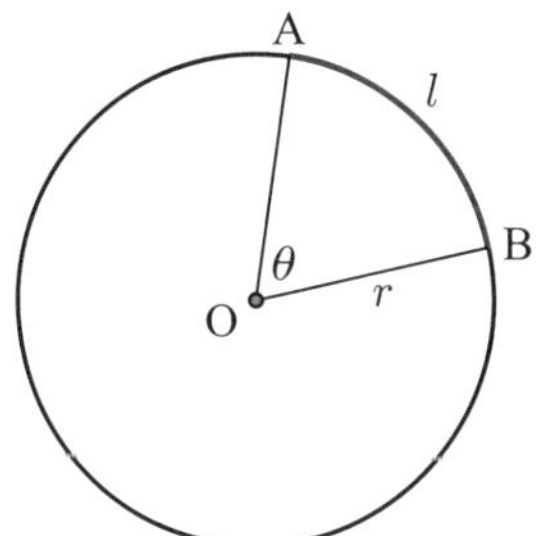

In the diagram, the **arc length** AB is l. Angle θ is measured in **radians**. We use a ratio to obtain:

$$\frac{\text{arc length}}{\text{circumference}} = \frac{\theta}{2\pi}$$

$$\therefore \quad \frac{l}{2\pi r} = \frac{\theta}{2\pi}$$

$$\therefore \quad l = \theta r$$

For θ in radians, arc length $\boldsymbol{l = \theta r}$.

For θ in degrees, arc length $\boldsymbol{l = \frac{\theta}{360} \times 2\pi r}$.

AREA OF SECTOR

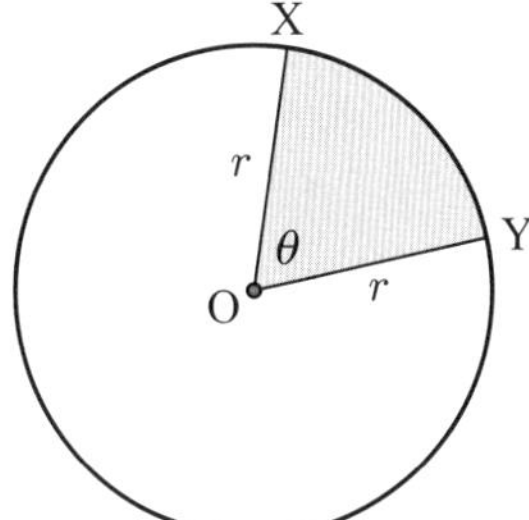

In the diagram, the area of minor sector XOY is shaded. θ is measured in **radians**. We use a ratio to obtain:

$$\frac{\text{area of sector}}{\text{area of circle}} = \frac{\theta}{2\pi}$$

$$\therefore \quad \frac{A}{\pi r^2} = \frac{\theta}{2\pi}$$

$$\therefore \quad A = \tfrac{1}{2}\theta r^2$$

For θ in radians, area of sector $\boldsymbol{A = \frac{1}{2}\theta r^2}$.

For θ in degrees, area of sector $\boldsymbol{A = \frac{\theta}{360} \times \pi r^2}$.

Example 5 — Self Tutor

A sector has radius 12 cm and angle 3 radians. Find its:

a arc length **b** area

a arc length $= \theta r$
$= 3 \times 12$
$= 36$ cm

b area $= \frac{1}{2}\theta r^2$
$= \frac{1}{2} \times 3 \times 12^2$
$= 216\text{ cm}^2$

EXERCISE 10B

1 Use radians to find the arc length and area of a sector of a circle of:

a radius 9 cm and angle $\frac{7\pi}{4}$ **b** radius 4.93 cm and angle 4.67 radians.

2 A sector has an angle of $107.9°$ and an arc length of 5.92 m. Find its:

a radius **b** area.

3 A sector has an angle of 1.19 radians and an area of 20.8 cm^2. Find its:

a radius **b** perimeter.

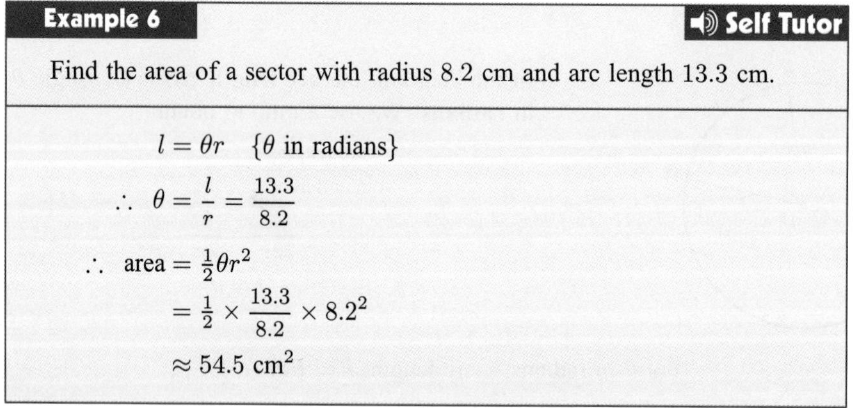

Example 6 — Self Tutor

Find the area of a sector with radius 8.2 cm and arc length 13.3 cm.

$l = \theta r$ {θ in radians}

$\therefore \theta = \frac{l}{r} = \frac{13.3}{8.2}$

$\therefore$ area $= \frac{1}{2}\theta r^2$
$= \frac{1}{2} \times \frac{13.3}{8.2} \times 8.2^2$
$\approx 54.5\text{ cm}^2$

4 Find, in radians, the angle of a sector of:

a radius 4.3 m and arc length 2.95 m **b** radius 10 cm and area 30 cm^2.

5 Find θ (in radians) for each of the following, and hence find the area of each figure:

a

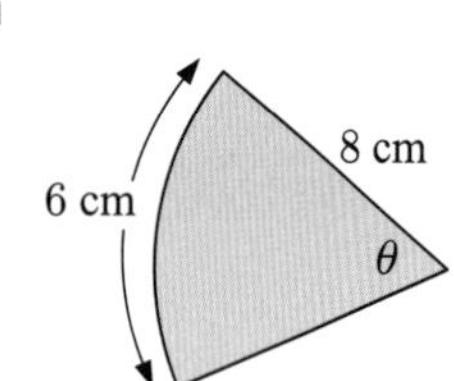

b

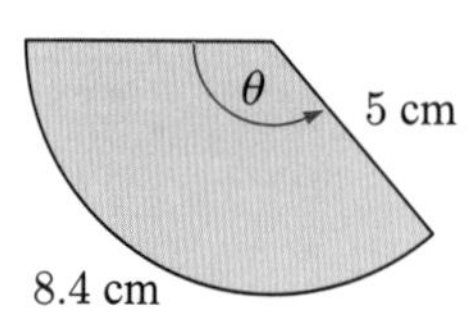

c

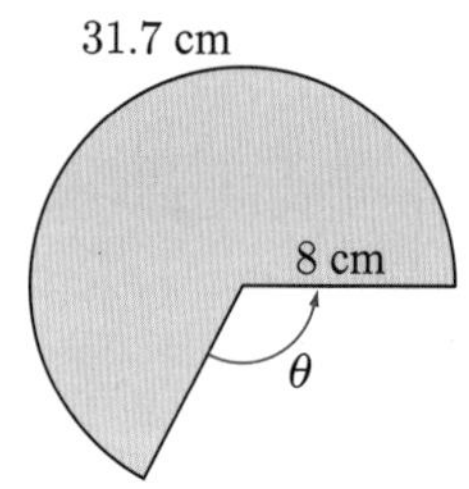

6 Find the arc length and area of a sector of radius 5 cm and angle 2 radians.

7 If a sector has radius $2x$ cm and arc length x cm, show that its area is $x^2\text{ cm}^2$.

8 The cone

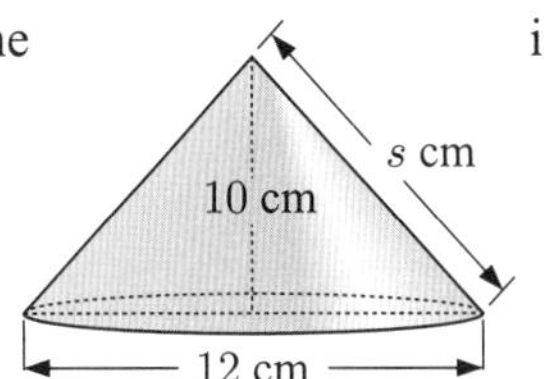

is made from this sector:

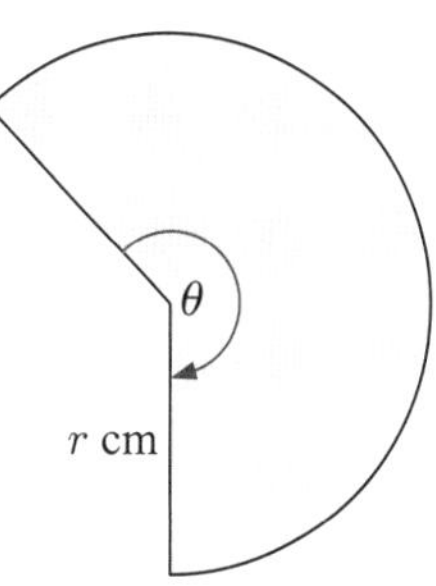

Find correct to 3 significant figures:

- **a** the slant length s cm
- **b** the value of r
- **c** the arc length of the sector
- **d** the sector angle θ in radians.

9 The end wall of a building has the shape illustrated, where the centre of arc AB is at C. Find:

- **a** α to 4 significant figures
- **b** θ to 4 significant figures
- **c** the area of the wall.

10 [AT] is a tangent to the given circle. OA = 13 cm and the circle has radius 5 cm. Find the perimeter of the shaded region.

11 A **nautical mile** (nmi) is the distance on the Earth's surface that subtends an angle of 1 minute (or $\frac{1}{60}$ of a degree) of the Great Circle arc measured from the centre of the Earth.
A **knot** is a speed of 1 nautical mile per hour.

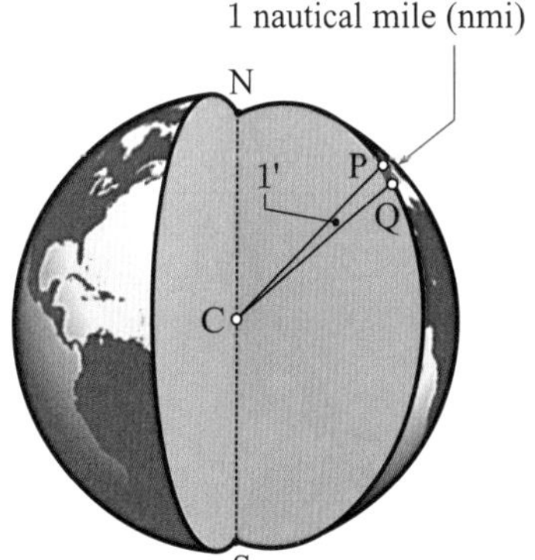

- **a** Given that the radius of the Earth is 6370 km, show that 1 nmi is approximately equal to 1.853 km.
- **b** Calculate how long it would take a plane to fly from Perth to Adelaide (a distance of 2130 km) if the plane can fly at 480 knots.

12 A sheep is tethered to a post which is 6 m from a long fence. The length of rope is 9 m. Find the area which the sheep can feed on.

13 A belt fits tightly around two pulleys of radii 4 cm and 6 cm respectively. The distance between their centres is 20 cm.

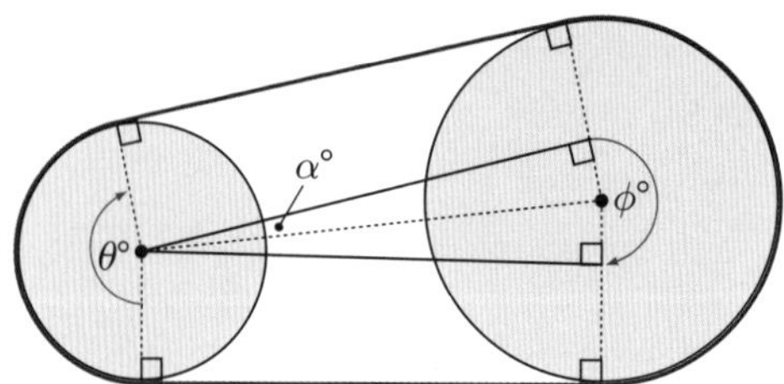

Find, correct to 4 significant figures:

a α **b** θ **c** ϕ **d** the length of the belt.

THE UNIT CIRCLE AND THE TRIGONOMETRIC RATIOS

The **unit circle** is the circle with centre $(0, 0)$ and radius 1 unit.

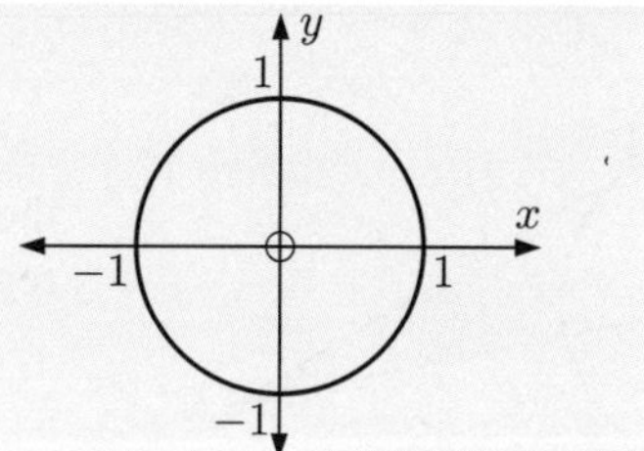

Suppose $P(x, y)$ is any point on the circle with centre $(0, 0)$ and radius r units.

$$\text{Since} \quad OP = r,$$
$$\sqrt{(x-0)^2 + (y-0)^2} = r \qquad \{\text{distance formula}\}$$
$$\therefore \quad x^2 + y^2 = r^2$$

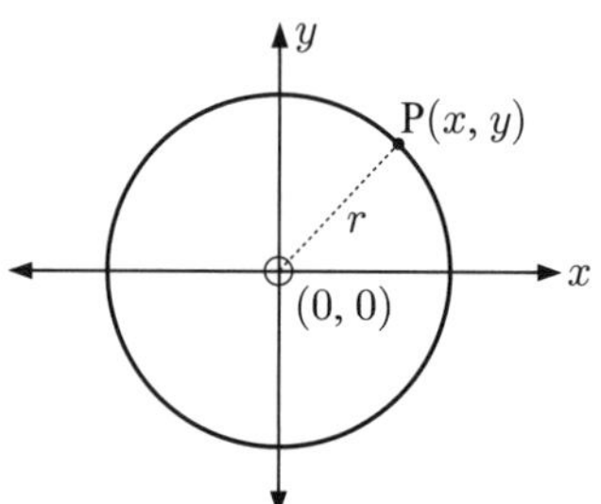

$\boldsymbol{x^2 + y^2 = r^2}$ is the equation of a circle with centre $(0, 0)$ and radius r.
The equation of the **unit circle** is $x^2 + y^2 = 1$.

ANGLE MEASUREMENT

Suppose P lies anywhere on the unit circle and A is $(1, 0)$.
Let θ be the angle measured from [OA] on the positive x-axis.

θ is **positive** for anticlockwise rotations and **negative** for clockwise rotations.

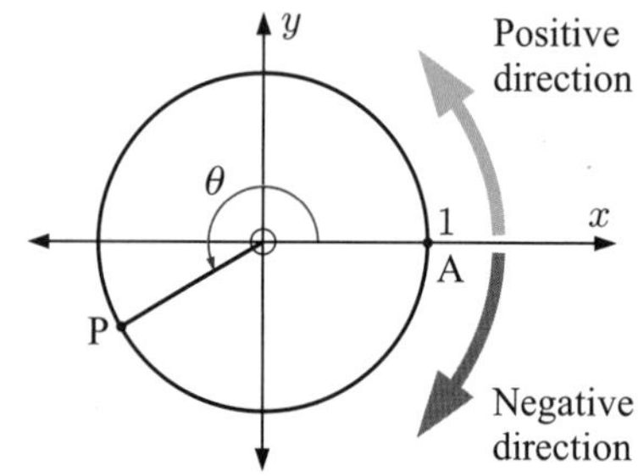

For example, $\theta = 210° = \frac{7\pi}{6}$
and $\phi = -150° = -\frac{5\pi}{6}$.

You can explore angle measurement further by clicking on the icon.

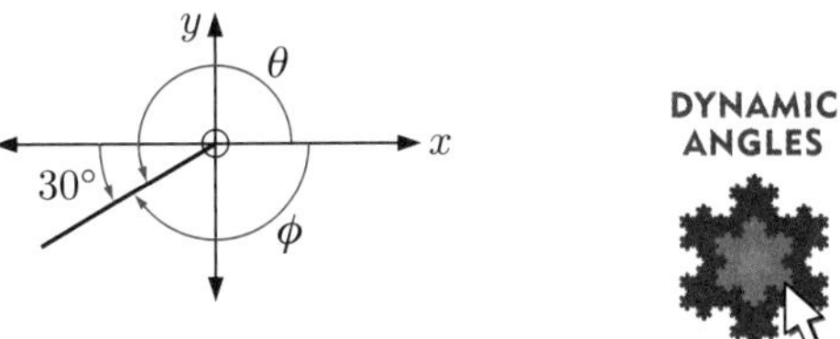

DEFINITION OF SINE AND COSINE

Consider a point $P(a, b)$ which lies on the unit circle in the first quadrant. [OP] makes an angle θ with the x-axis as shown.

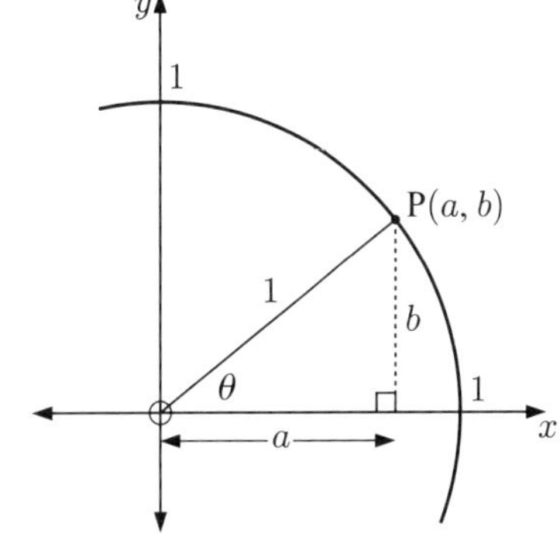

Using right angled triangle trigonometry:

$$\cos\theta = \frac{\text{ADJ}}{\text{HYP}} = \frac{a}{1} = a$$

$$\sin\theta = \frac{\text{OPP}}{\text{HYP}} = \frac{b}{1} = b$$

$$\tan\theta = \frac{\text{OPP}}{\text{ADJ}} = \frac{b}{a} = \frac{\sin\theta}{\cos\theta}$$

In general, for a point P anywhere on the unit circle:

- $\boldsymbol{\cos\theta}$ is the x-coordinate of P
- $\boldsymbol{\sin\theta}$ is the y-coordinate of P

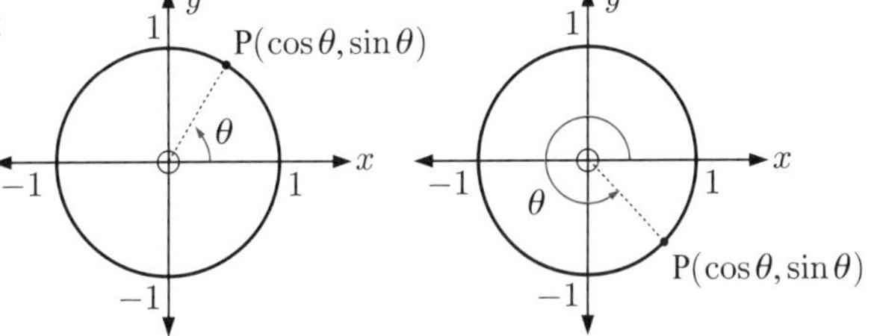

We can hence find the coordinates of any point on the unit circle with given angle θ measured from the positive x-axis.

Since the unit circle has equation $x^2 + y^2 = 1$, $(\cos\theta)^2 + (\sin\theta)^2 = 1$ for all θ.

We commonly write this as $\boldsymbol{\cos^2\theta + \sin^2\theta = 1}$.

For all points on the unit circle, $-1 \leqslant x \leqslant 1$ and $-1 \leqslant y \leqslant 1$.

So, $\boldsymbol{-1 \leqslant \cos\theta \leqslant 1}$ and $\boldsymbol{-1 \leqslant \sin\theta \leqslant 1}$ for all θ.

DEFINITION OF TANGENT

Suppose we extend [OP] to meet the tangent from $A(1, 0)$. We let the intersection between these lines be point Q. Note that as P moves, so does Q.

The position of Q relative to A is defined as the **tangent function**.

Notice that $\triangle$s ONP and OAQ are equiangular and therefore similar.

Consequently $\frac{\text{AQ}}{\text{OA}} = \frac{\text{NP}}{\text{ON}}$ and hence $\frac{\text{AQ}}{1} = \frac{\sin\theta}{\cos\theta}$.

Under the definition that $\text{AQ} = \tan\theta$, $\boldsymbol{\tan\theta = \frac{\sin\theta}{\cos\theta}}$.

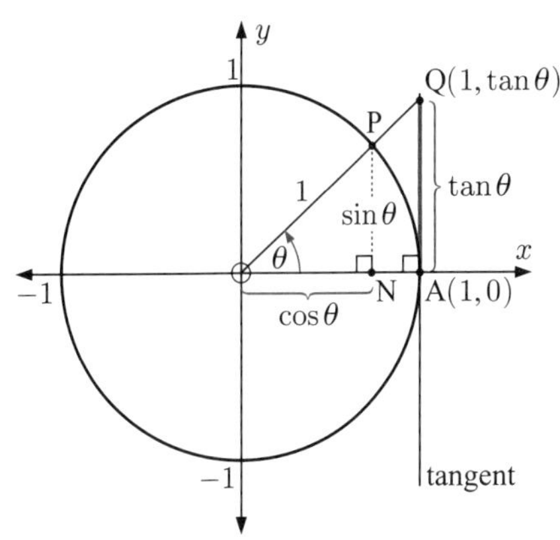

INVESTIGATION 1 THE TRIGONOMETRIC RATIOS

In this Investigation we explore the signs of the trigonometric ratios in each quadrant of the unit circle.

THE UNIT CIRCLE

What to do:

1. Click on the icon to run the Unit Circle software.
   Drag the point P slowly around the circle.
   Note the *sign* of each trigonometric ratio in each quadrant.

Quadrant	$\cos\theta$	$\sin\theta$	$\tan\theta$
1	positive		
2			
3			
4			

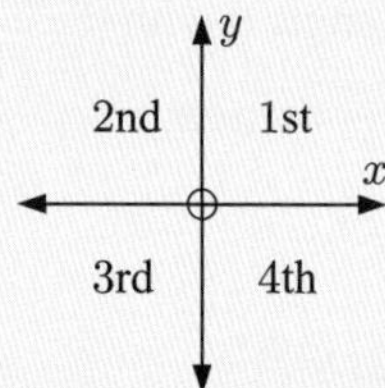

2. Hence note down the trigonometric ratios which are *positive* in each quadrant.

From the **Investigation** you should have discovered that:

- $\sin\theta$, $\cos\theta$, and $\tan\theta$ are positive in quadrant 1
- only $\sin\theta$ is positive in quadrant 2
- only $\tan\theta$ is positive in quadrant 3
- only $\cos\theta$ is positive in quadrant 4.

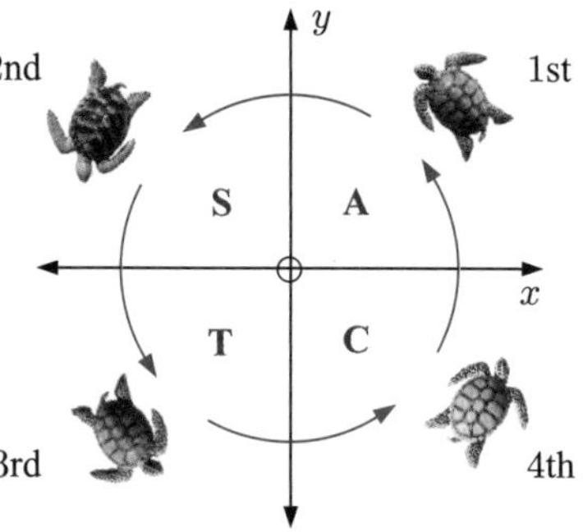

We can use a letter to show which trigonometric ratios are positive in each quadrant. The A stands for *all* of the ratios.

You might like to remember them using

All **S**illy **T**urtles **C**rawl.

Example 7 — Self Tutor

Use a unit circle diagram to find the values of $\cos(-270°)$ and $\sin(-270°)$.

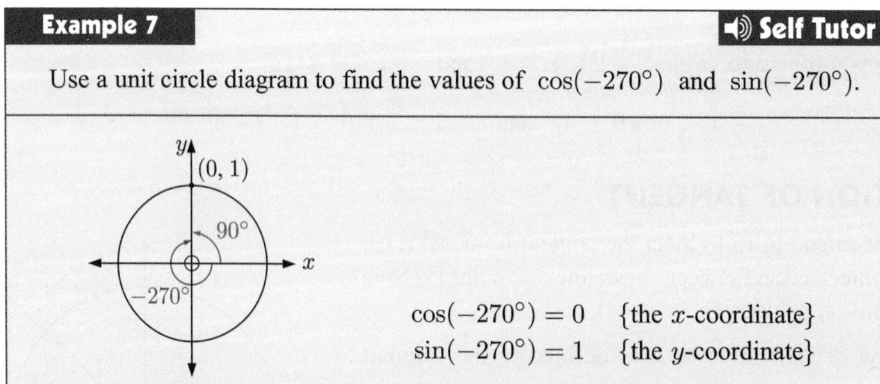

$\cos(-270°) = 0$ {the x-coordinate}

$\sin(-270°) = 1$ {the y-coordinate}

We can use the unit circle to find various relationships between cosine and sine.

Example 8 — Self Tutor

Use the unit circle to show that $\cos\left(\frac{\pi}{2}+\theta\right) = -\sin\theta$ and $\sin\left(\frac{\pi}{2}+\theta\right) = \cos\theta$ for $0 < \theta < \pi$.

Let point $B(\cos\theta, \sin\theta)$ be on the unit circle such that [OB] makes an angle of θ with the positive x-axis.

Let point P be on the unit circle such that [OP] makes an angle of $\left(\frac{\pi}{2}+\theta\right)$ with the positive x-axis.

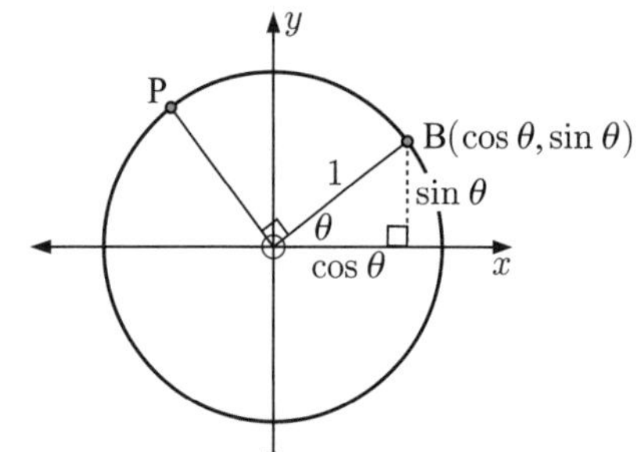

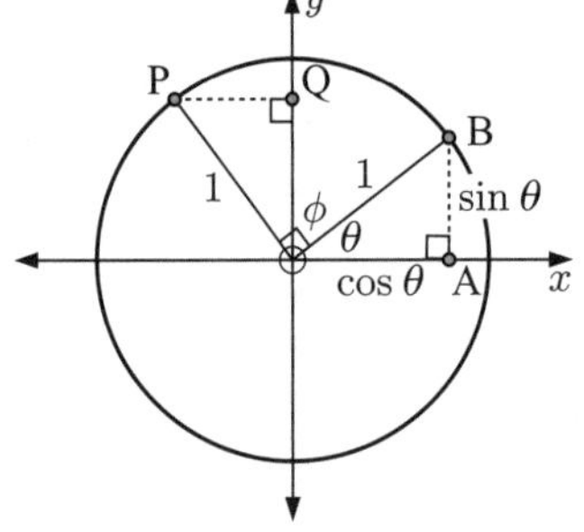

In the diagram alongside, $\phi = \frac{\pi}{2} - \theta$

$\therefore \ \widehat{POQ} = \theta$

In $\triangle$s POQ and BOA:
- $OP = OB$
- $\widehat{POQ} = \widehat{BOA}$
- $\widehat{PQO} = \widehat{BAO}$

$\therefore$ $\triangle$s POQ and BOA are congruent. {AAcorrS}

$\therefore \ PQ = AB = \sin\theta$

and $OQ = OA = \cos\theta$

$\therefore$ P has coordinates $(-\sin\theta, \cos\theta)$.

But P has coordinates $\left(\cos\left(\frac{\pi}{2}+\theta\right), \sin\left(\frac{\pi}{2}+\theta\right)\right)$.

$\therefore \ \cos\left(\frac{\pi}{2}+\theta\right) = -\sin\theta$ and $\sin\left(\frac{\pi}{2}+\theta\right) = \cos\theta$

In order to prove that $\cos\left(\frac{\pi}{2}+\theta\right) = -\sin\theta$ and $\sin\left(\frac{\pi}{2}+\theta\right) = \cos\theta$ for all θ, we would need to consider each quadrant separately.

PERIODICITY OF TRIGONOMETRIC RATIOS

Since there are 2π radians in a full revolution, if we add any integer multiple of 2π to θ (in radians) then the position of P on the unit circle is unchanged.

For θ in radians and $k \in \mathbb{Z}$,

$$\cos(\theta + 2k\pi) = \cos\theta \ \text{ and } \ \sin(\theta + 2k\pi) = \sin\theta.$$

We notice that for any point $(\cos\theta, \sin\theta)$ on the unit circle, the point directly opposite is $(-\cos\theta, -\sin\theta)$.

$$\therefore \ \cos(\theta+\pi) = -\cos\theta$$
$$\sin(\theta+\pi) = -\sin\theta$$
$$\text{and } \ \tan(\theta+\pi) = \tan\theta$$

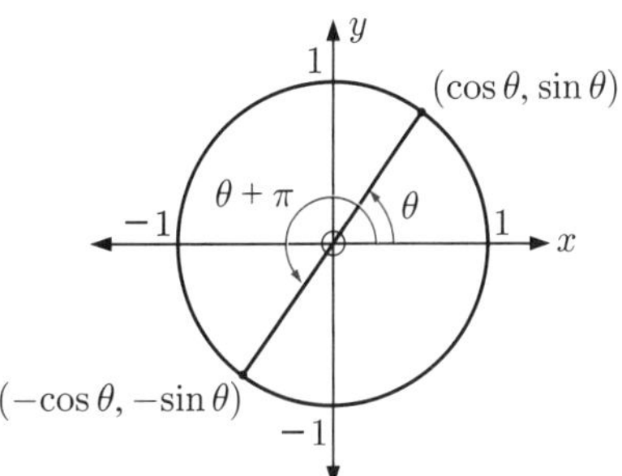

For θ in radians and $k \in \mathbb{Z}$, $\boldsymbol{\tan(\theta + k\pi) = \tan\theta}$.

This **periodic** feature is an important property of the trigonometric functions.

EXERCISE 10C

1 For each unit circle illustrated:

   i state the exact coordinates of points A, B, and C in terms of sine and cosine

   ii use your calculator to give the coordinates of A, B, and C correct to 3 significant figures.

a

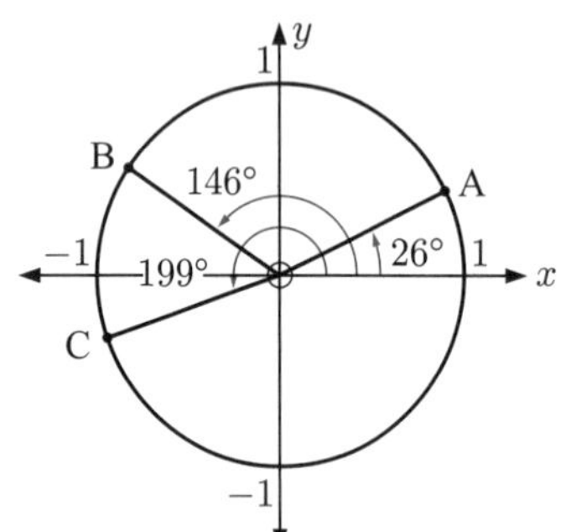

b

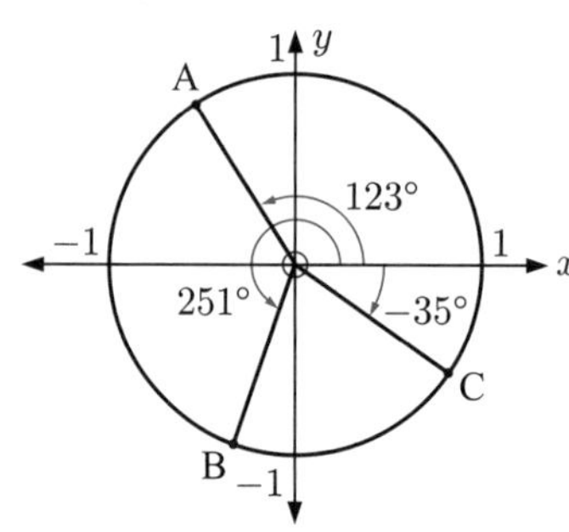

2 With the aid of a unit circle, complete the following table:

θ (degrees)	$0°$	$90°$	$180°$	$270°$	$360°$	$450°$
θ (radians)						
sine						
cosine						
tangent						

3 **a** Use your calculator to evaluate: **i** $\frac{1}{\sqrt{2}}$ **ii** $\frac{\sqrt{3}}{2}$

   b Copy and complete the following table. Use your calculator to evaluate the trigonometric ratios, then **a** to write them exactly.

θ (degrees)	$30°$	$45°$	$60°$	$135°$	$150°$	$240°$	$315°$
θ (radians)							
sine							
cosine							
tangent							

4 **a** Copy and complete:

Quadrant	*Degree measure*	*Radian measure*	$\cos\theta$	$\sin\theta$	$\tan\theta$
1	$0° < \theta < 90°$	$0 < \theta < \frac{\pi}{2}$	positive	positive	
2					
3					
4					

   b In which quadrants are the following true?

   i $\cos\theta$ is positive.

   ii $\cos\theta$ is negative.

   iii $\cos\theta$ and $\sin\theta$ are both negative.

   iv $\cos\theta$ is negative and $\sin\theta$ is positive.

5 **a** Use your calculator to evaluate:

   i $\sin 100°$ **ii** $\sin 80°$ **iii** $\sin 120°$ **iv** $\sin 60°$

   v $\sin 150°$ **vi** $\sin 30°$ **vii** $\sin 45°$ **viii** $\sin 135°$

b Use the results from **a** to copy and complete:
$\sin(180° - \theta) = ...$

c Justify your answer using the diagram alongside:

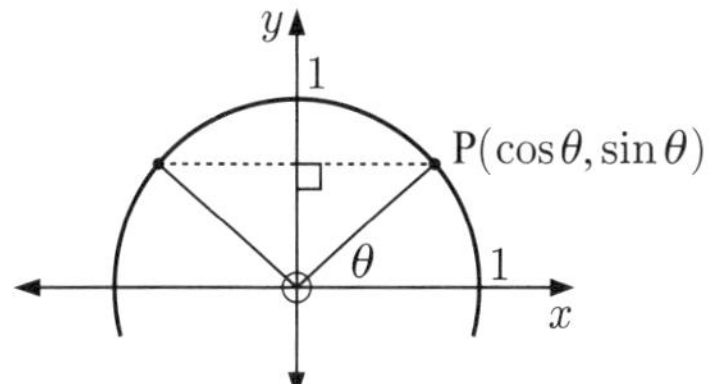

d Find the obtuse angle with the same sine as:
i $45°$ **ii** $51°$ **iii** $\frac{\pi}{3}$ **iv** $\frac{\pi}{6}$

6 **a** Use your calculator to evaluate:
i $\cos 70°$ **ii** $\cos 110°$ **iii** $\cos 60°$ **iv** $\cos 120°$
v $\cos 25°$ **vi** $\cos 155°$ **vii** $\cos 80°$ **viii** $\cos 100°$

b Use the results from **a** to copy and complete:
$\cos(180° - \theta) = ...$

c Justify your answer using the diagram alongside:

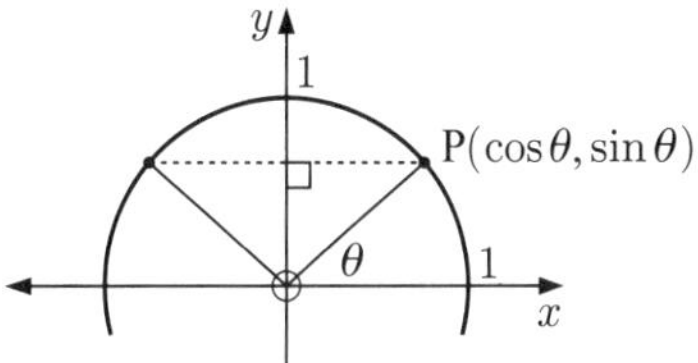

d Find the obtuse angle which has the negative cosine of:
i $40°$ **ii** $19°$ **iii** $\frac{\pi}{5}$ **iv** $\frac{2\pi}{5}$

7 Use the unit circle to show that, for $0 < \theta < \frac{\pi}{2}$:

a $\cos(\pi + \theta) = -\cos\theta$ and $\sin(\pi + \theta) = -\sin\theta$

b $\cos\left(\frac{3\pi}{2} + \theta\right) = \sin\theta$ and $\sin\left(\frac{3\pi}{2} + \theta\right) = -\cos\theta$

8 Without using your calculator, find:

a $\sin 137°$ if $\sin 43° \approx 0.6820$
b $\sin 59°$ if $\sin 121° \approx 0.8572$
c $\cos 143°$ if $\cos 37° \approx 0.7986$
d $\cos 24°$ if $\cos 156° \approx -0.9135$
e $\sin 115°$ if $\sin 65° \approx 0.9063$
f $\cos 132°$ if $\cos 48° \approx 0.6691$

9 **a** If $A\widehat{O}P = B\widehat{O}Q = \theta$, what is the measure of $A\widehat{O}Q$?

b Copy and complete:
[OQ] is a reflection of [OP] in the ......
and so Q has coordinates ......

c What trigonometric formulae can be deduced from **a** and **b**?

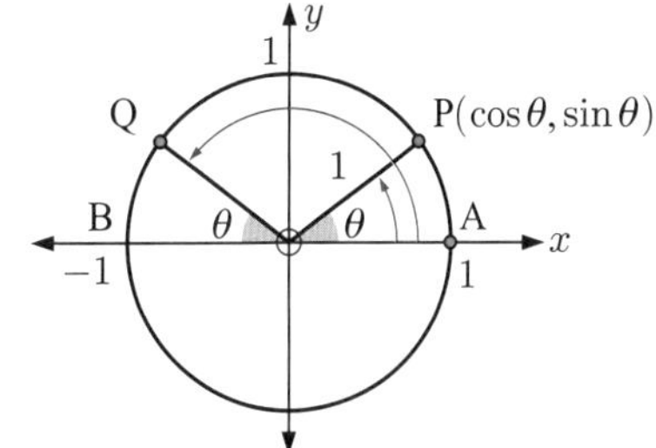

10 **a** Copy and complete:

θ^c	$\sin\theta$	$\sin(-\theta)$	$\cos\theta$	$\cos(-\theta)$
0.75				
1.772				
3.414				
6.25				
-1.17				

b What trigonometric formulae can be deduced from your results in **a**?

c 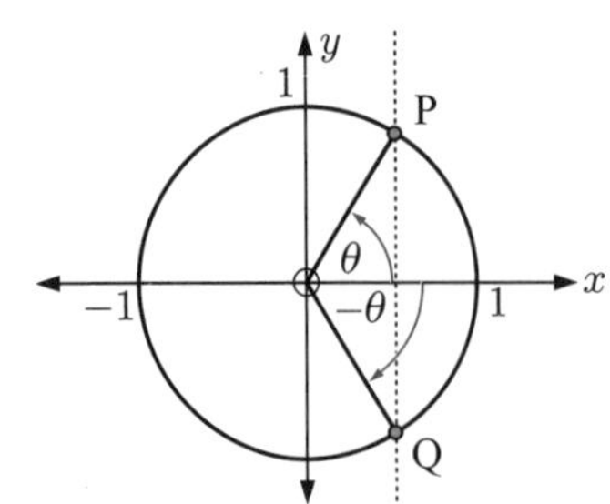

The coordinates of P in the figure are $(\cos\theta, \sin\theta)$.

i By finding the coordinates of Q in terms of θ in *two different* ways, prove your formulae in **b**.

ii Hence explain why $\cos(2\pi - \theta) = \cos\theta$.

D APPLICATIONS OF THE UNIT CIRCLE

The identity $\cos^2\theta + \sin^2\theta = 1$ is essential for finding trigonometric ratios.

Example 9 **Self Tutor**

Find exactly the possible values of $\cos\theta$ for $\sin\theta = \frac{2}{3}$. Illustrate your answers.

$\cos^2\theta + \sin^2\theta = 1$

$\therefore \cos^2\theta + \left(\frac{2}{3}\right)^2 = 1$

$\therefore \cos^2\theta = \frac{5}{9}$

$\therefore \cos\theta = \pm\frac{\sqrt{5}}{3}$

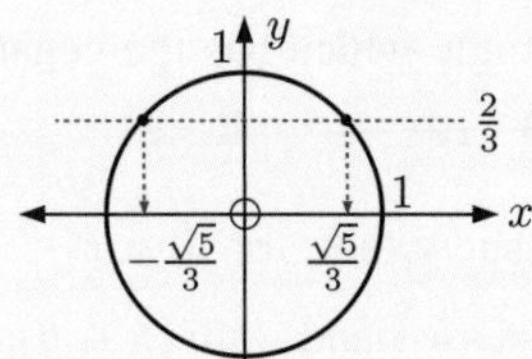

EXERCISE 10D.1

1 Find the possible exact values of $\cos\theta$ for:

a $\sin\theta = \frac{1}{2}$ b $\sin\theta = -\frac{1}{3}$ c $\sin\theta = 0$ d $\sin\theta = -1$

2 Find the possible exact values of $\sin\theta$ for:

a $\cos\theta = \frac{4}{5}$ b $\cos\theta = -\frac{3}{4}$ c $\cos\theta = 1$ d $\cos\theta = 0$

Example 10 **Self Tutor**

If $\sin\theta = -\frac{3}{4}$ and $\pi < \theta < \frac{3\pi}{2}$, find $\cos\theta$ and $\tan\theta$. Give exact values.

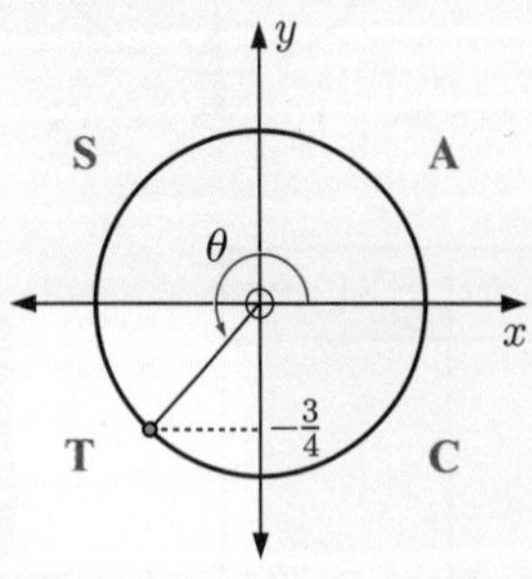

Now $\cos^2\theta + \sin^2\theta = 1$

$\therefore \cos^2\theta + \frac{9}{16} = 1$

$\therefore \cos^2\theta = \frac{7}{16}$

$\therefore \cos\theta = \pm\frac{\sqrt{7}}{4}$

But $\pi < \theta < \frac{3\pi}{2}$, so θ is a quadrant 3 angle

$\therefore \cos\theta$ is negative.

$\therefore \cos\theta = -\frac{\sqrt{7}}{4}$

and $\tan\theta = \dfrac{\sin\theta}{\cos\theta} = \dfrac{-\frac{3}{4}}{-\frac{\sqrt{7}}{4}} = \dfrac{3}{\sqrt{7}}$

3 Without using a calculator, find:

a $\sin\theta$ if $\cos\theta = \frac{2}{3}$ and $0 < \theta < \frac{\pi}{2}$

b $\cos\theta$ if $\sin\theta = \frac{2}{5}$ and $\frac{\pi}{2} < \theta < \pi$

c $\cos\theta$ if $\sin\theta = -\frac{3}{5}$ and $\frac{3\pi}{2} < \theta < 2\pi$

d $\sin\theta$ if $\cos\theta = -\frac{5}{13}$ and $\pi < \theta < \frac{3\pi}{2}$.

Example 11 — Self Tutor

If $\tan\theta = -2$ and $\frac{3\pi}{2} < \theta < 2\pi$, find $\sin\theta$ and $\cos\theta$. Give exact answers.

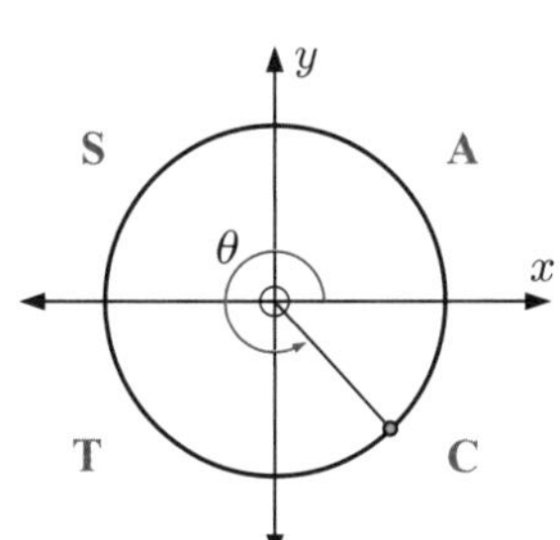

$$\tan\theta = \frac{\sin\theta}{\cos\theta} = -2$$

$$\therefore \sin\theta = -2\cos\theta$$

Now $\sin^2\theta + \cos^2\theta = 1$

$\therefore (-2\cos\theta)^2 + \cos^2\theta = 1$

$\therefore 4\cos^2\theta + \cos^2\theta = 1$

$\therefore 5\cos^2\theta = 1$

$\therefore \cos\theta = \pm\frac{1}{\sqrt{5}}$

But $\frac{3\pi}{2} < \theta < 2\pi$, so θ is a quadrant 4 angle.

$\therefore$ $\cos\theta$ is positive and $\sin\theta$ is negative.

$\therefore$ $\cos\theta = \frac{1}{\sqrt{5}}$ and $\sin\theta = -\frac{2}{\sqrt{5}}$.

4 **a** If $\sin x = \frac{1}{3}$ and $\frac{\pi}{2} < x < \pi$, find $\tan x$ exactly.

b If $\cos x = \frac{1}{5}$ and $\frac{3\pi}{2} < x < 2\pi$, find $\tan x$ exactly.

c If $\sin x = -\frac{1}{\sqrt{3}}$ and $\pi < x < \frac{3\pi}{2}$, find $\tan x$ exactly.

d If $\cos x = -\frac{3}{4}$ and $\frac{\pi}{2} < x < \pi$, find $\tan x$ exactly.

5 Find exact values for $\sin x$ and $\cos x$ given that:

a $\tan x = \frac{2}{3}$ and $0 < x < \frac{\pi}{2}$

b $\tan x = -\frac{4}{3}$ and $\frac{\pi}{2} < x < \pi$

c $\tan x = \frac{\sqrt{5}}{3}$ and $\pi < x < \frac{3\pi}{2}$

d $\tan x = -\frac{12}{5}$ and $\frac{3\pi}{2} < x < 2\pi$

6 Suppose $\tan x = k$ where k is a constant and $\pi < x < \frac{3\pi}{2}$. Write expressions for $\sin x$ and $\cos x$ in terms of k.

FINDING ANGLES WITH PARTICULAR TRIGONOMETRIC RATIOS

From **Exercise 10C** you should have discovered that:

For θ in radians:

- $\boldsymbol{\sin(\pi - \theta) = \sin\theta}$
- $\boldsymbol{\cos(\pi - \theta) = -\cos\theta}$
- $\boldsymbol{\cos(2\pi - \theta) = \cos\theta}$

We need results such as these, and also the periodicity of the trigonometric ratios, to find angles which have a particular sine, cosine, or tangent.

Example 12 **Self Tutor**

Find the two angles θ on the unit circle, with $0 \leqslant \theta \leqslant 2\pi$, such that:

a $\cos\theta = \frac{1}{3}$ **b** $\sin\theta = \frac{3}{4}$ **c** $\tan\theta = 2$

a Using technology, $\cos^{-1}(\frac{1}{3}) \approx 1.23$

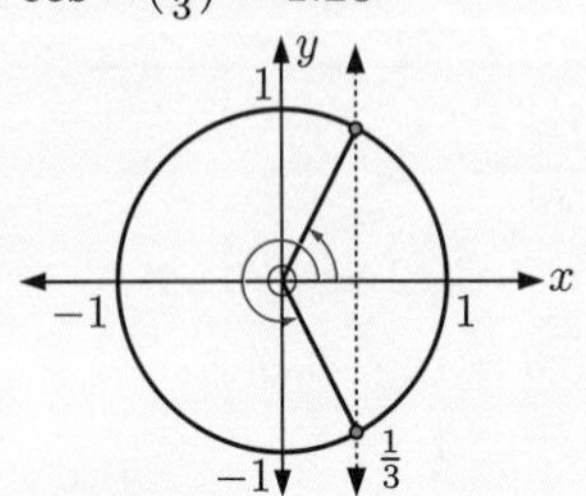

$\therefore \ \theta \approx 1.23$ or $2\pi - 1.23$

$\therefore \ \theta \approx 1.23$ or 5.05

b Using technology, $\sin^{-1}(\frac{3}{4}) \approx 0.848$

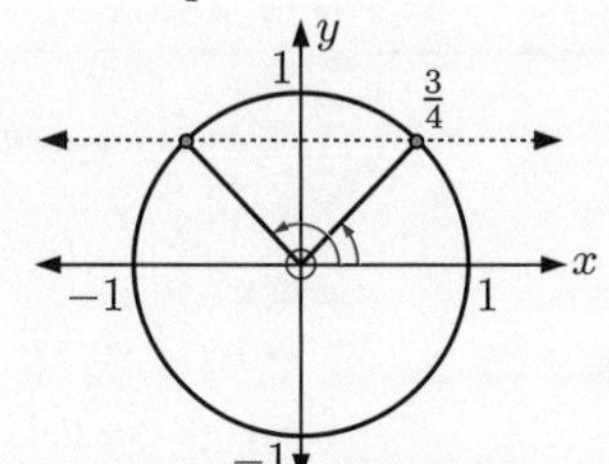

$\therefore \ \theta \approx 0.848$ or $\pi - 0.848$

$\therefore \ \theta \approx 0.848$ or 2.29

c Using technology, $\tan^{-1}(2) \approx 1.11$

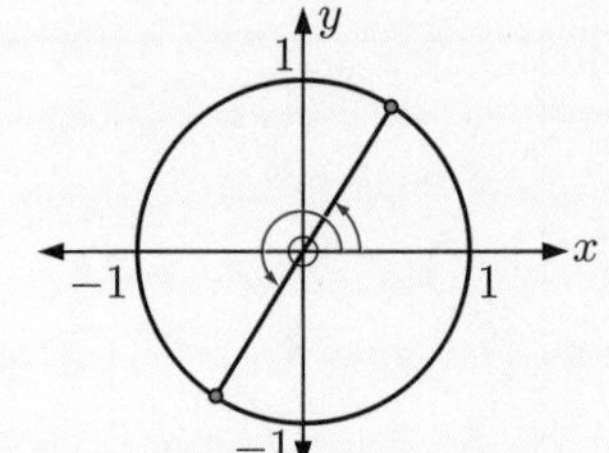

$\therefore \ \theta \approx 1.11$ or $\pi + 1.11$

$\therefore \ \theta \approx 1.11$ or 4.25

Example 13 **Self Tutor**

Find two angles θ on the unit circle, with $0 \leqslant \theta \leqslant 2\pi$, such that:

a $\sin\theta = -0.4$ **b** $\cos\theta = -\frac{2}{3}$ **c** $\tan\theta = -\frac{1}{3}$

a Using technology, $\sin^{-1}(-0.4) \approx -0.412$

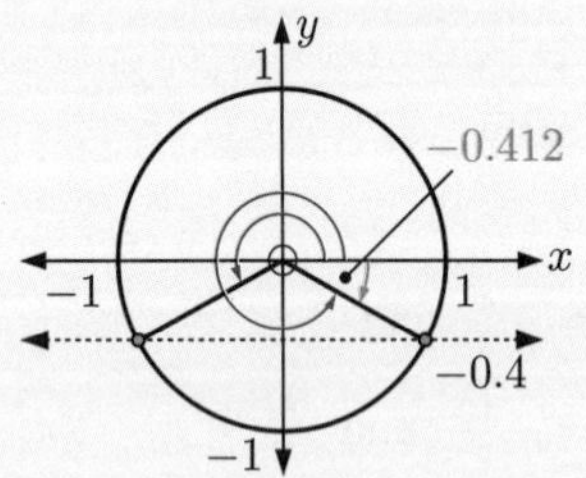

But $0 \leqslant \theta \leqslant 2\pi$

$\therefore \ \theta \approx \pi + 0.412$ or $2\pi - 0.412$

$\therefore \ \theta \approx 3.55$ or 5.87

b Using technology, $\cos^{-1}(-\frac{2}{3}) \approx 2.30$

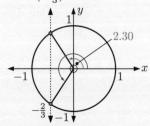

But $0 \leqslant \theta \leqslant 2\pi$

$\therefore \ \theta \approx 2.30$ or $2\pi - 2.30$

$\therefore \ \theta \approx 2.30$ or 3.98

c Using technology, $\tan^{-1}(-\frac{1}{3}) \approx -0.322$

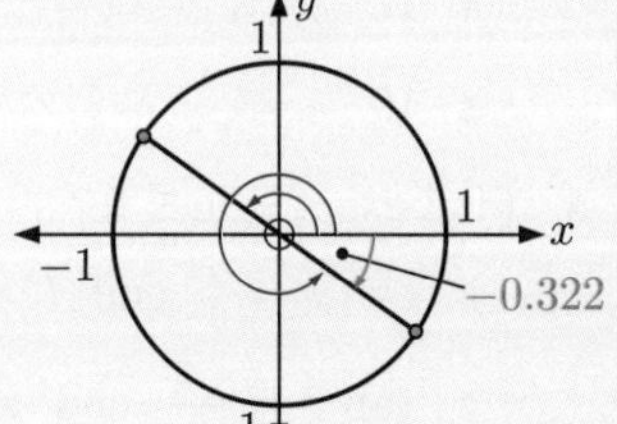

But $0 \leqslant \theta \leqslant 2\pi$

$\therefore \ \theta \approx \pi - 0.322$ or $2\pi - 0.322$

$\therefore \ \theta \approx 2.82$ or 5.96

The green arrow shows the angle that your calculator gives.

EXERCISE 10D.2

1 Find two angles θ on the unit circle, with $0 \leqslant \theta \leqslant 2\pi$, such that:

a $\tan\theta = 4$ **b** $\cos\theta = 0.83$ **c** $\sin\theta = \frac{3}{5}$

d $\cos\theta = 0$ **e** $\tan\theta = 1.2$ **f** $\cos\theta = 0.7816$

g $\sin\theta = \frac{1}{11}$ **h** $\tan\theta = 20.2$ **i** $\sin\theta = \frac{39}{40}$

2 Find two angles θ on the unit circle, with $0 \leqslant \theta \leqslant 2\pi$, such that:

a $\cos\theta = -\frac{1}{4}$ **b** $\sin\theta = 0$ **c** $\tan\theta = -3.1$

d $\sin\theta = -0.421$ **e** $\tan\theta = -6.67$ **f** $\cos\theta = -\frac{2}{17}$

g $\tan\theta = -\sqrt{5}$ **h** $\cos\theta = \frac{-1}{\sqrt{3}}$ **i** $\sin\theta = -\frac{\sqrt{2}}{\sqrt{5}}$

NEGATIVE AND COMPLEMENTARY ANGLE FORMULAE

For any given angle θ:

- the **negative** of θ is $-\theta$
- the **complement** of θ is $\frac{\pi}{2} - \theta$.

INVESTIGATION 2 NEGATIVE AND COMPLEMENTARY ANGLES

The purpose of this Investigation is to discover relationships (if they exist) between $\cos(-\theta)$, $\sin(-\theta)$, $\cos\left(\frac{\pi}{2} - \theta\right)$, $\sin\left(\frac{\pi}{2} - \theta\right)$, $\cos\theta$, and $\sin\theta$.

What to do:

1 Copy and complete, adding angles of your choice to the table:

θ	$\sin\theta$	$\cos\theta$	$\sin(-\theta)$	$\cos(-\theta)$	$\sin\left(\frac{\pi}{2} - \theta\right)$	$\cos\left(\frac{\pi}{2} - \theta\right)$
2.67						
0.642						
$\frac{\pi}{6}$						

2 Hence predict formulae for $\sin(-\theta)$, $\cos(-\theta)$, $\sin\left(\frac{\pi}{2} - \theta\right)$, and $\cos\left(\frac{\pi}{2} - \theta\right)$.

NEGATIVE ANGLE FORMULAE

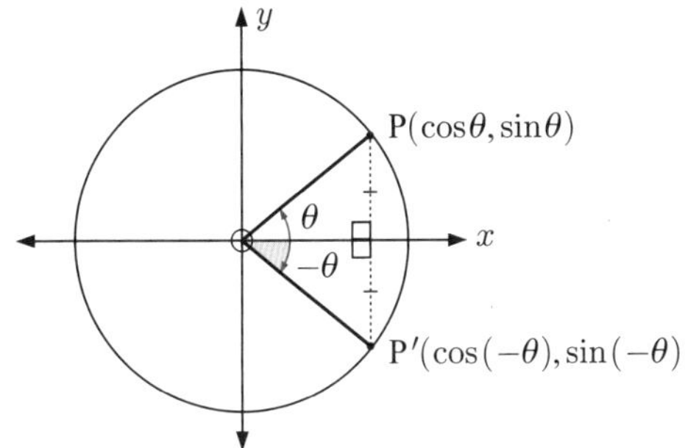

In the diagram, P′ is the reflection of P in the x-axis. It therefore corresponds to the angle $-\theta$.

P and P′ have the same x-coordinate, but their y-coordinates are negatives.

Hence $\cos(-\theta) = \cos\theta$ and $\sin(-\theta) = -\sin\theta$.

Using these results,

$$\tan(-\theta) = \frac{\sin(-\theta)}{\cos(-\theta)} = \frac{-\sin\theta}{\cos\theta} = -\tan\theta$$

We have hence deduced the **negative angle formulae**:

$$\cos(-\theta) = \cos\theta \qquad \sin(-\theta) = -\sin\theta \qquad \tan(-\theta) = -\tan\theta$$

COMPLEMENTARY ANGLE FORMULAE

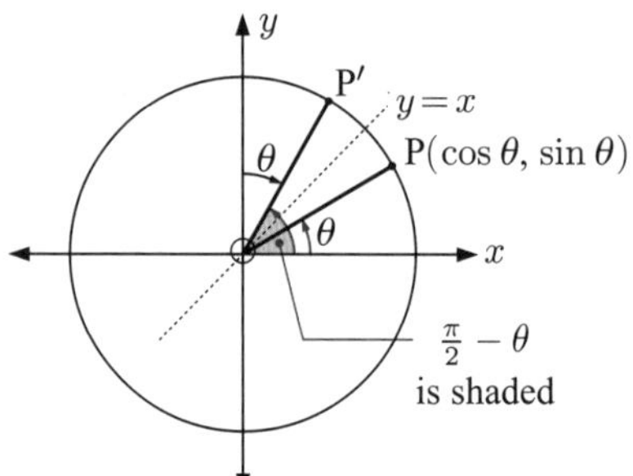

Consider P′ on the unit circle which corresponds to the angle $\frac{\pi}{2} - \theta$.

$\therefore$ P′ is $\left(\cos\left(\frac{\pi}{2} - \theta\right), \sin\left(\frac{\pi}{2} - \theta\right)\right)$ .... (1)

But P′ is the image of P under a reflection in the line $y = x$.

$\therefore$ P′ is $(\sin\theta, \cos\theta)$ .... (2)

Comparing (1) and (2) gives the **complementary angle formulae**:

$$\cos\left(\tfrac{\pi}{2} - \theta\right) = \sin\theta \qquad \sin\left(\tfrac{\pi}{2} - \theta\right) = \cos\theta$$

Example 14 — Self Tutor

Simplify:

a $2\sin(-\theta) + 3\sin\theta$ **b** $2\cos\theta + \cos(-\theta)$ **c** $3\sin\left(\frac{\pi}{2} - \theta\right) + 2\cos\theta$

a $2\sin(-\theta) + 3\sin\theta$
$= -2\sin\theta + 3\sin\theta$
$= \sin\theta$

b $2\cos\theta + \cos(-\theta)$
$= 2\cos\theta + \cos\theta$
$= 3\cos\theta$

c $3\sin\left(\frac{\pi}{2} - \theta\right) + 2\cos\theta$
$= 3\cos\theta + 2\cos\theta$
$= 5\cos\theta$

EXERCISE 10E

1 Simplify:

a $\sin\theta + \sin(-\theta)$ **b** $\tan(-\theta) - \tan\theta$ **c** $2\cos\theta + \cos(-\theta)$
d $3\sin\theta - \sin(-\theta)$ **e** $\cos^2(-\alpha)$ **f** $\sin^2(-\alpha)$
g $\cos(-\alpha)\cos\alpha - \sin(-\alpha)\sin\alpha$

2 Simplify:

a $2\sin\theta - \cos(90° - \theta)$ **b** $\sin(-\theta) - \cos(90° - \theta)$ **c** $\sin(90° - \theta) - \cos\theta$
d $3\cos(-\theta) - 4\sin\left(\frac{\pi}{2} - \theta\right)$ **e** $3\cos\theta + \sin\left(\frac{\pi}{2} - \theta\right)$ **f** $\cos\left(\frac{\pi}{2} - \theta\right) + 4\sin\theta$

3 Explain why $\sin(\theta - \phi) = -\sin(\phi - \theta)$, $\cos(\theta - \phi) = \cos(\phi - \theta)$.

4 Simplify:

a $\dfrac{\sin\theta}{\cos\theta}$ **b** $\dfrac{\sin(-\theta)}{\cos(-\theta)}$ **c** $\dfrac{\sin\left(\frac{\pi}{2} - \theta\right)}{\cos\theta}$

d $\dfrac{-\sin(-\theta)}{\cos\theta}$ **e** $\dfrac{\cos\left(\frac{\pi}{2} - \theta\right)}{\sin\left(\frac{\pi}{2} - \theta\right)}$ **f** $\dfrac{\cos\left(\frac{\pi}{2} - \theta\right)}{\cos\theta}$

INVESTIGATION 3 PARAMETRIC EQUATIONS

Usually we write functions in the form $y = f(x)$.

For example: $y = 3x + 7$, $y = x^2 - 6x + 8$, $y = \sin x$

However, sometimes it is useful to express **both** x and y in terms of another variable t, called the **parameter**. In this case we say we have **parametric equations**.

What to do:

1 **a** Use your graphics calculator to plot $\{(x, y) \mid x = \cos t,\ y = \sin t,\ 0° \leqslant t \leqslant 360°\}$. Use the same scale on both axes.
Note: Your calculator will need to be set to degrees.
b Describe the resulting graph. Is it the graph of a function?
c Evaluate $x^2 + y^2$. Hence determine the equation of this graph in terms of x and y only.

2 Use your graphics calculator to plot:
a $\{(x, y) \mid x = 2\cos t,\ y = \sin(2t),\ 0° \leqslant t \leqslant 360°\}$
b $\{(x, y) \mid x = 2\cos t,\ y = 2\sin(3t),\ 0° \leqslant t \leqslant 360°\}$
c $\{(x, y) \mid x = 2\cos t,\ y = \cos t - \sin t,\ 0° \leqslant t \leqslant 360°\}$
d $\{(x, y) \mid x = \cos^2 t + \sin 2t,\ y = \cos t,\ 0° \leqslant t \leqslant 360°\}$
e $\{(x, y) \mid x = \cos^3 t,\ y = \sin t,\ 0° \leqslant t \leqslant 360°\}$

F MULTIPLES OF $\frac{\pi}{6}$ AND $\frac{\pi}{4}$

Angles which are multiples of $\frac{\pi}{6}$ and $\frac{\pi}{4}$ occur frequently, so it is important for us to write their trigonometric ratios exactly.

MULTIPLES OF $\frac{\pi}{4}$ OR 45°

Triangle OBP is isosceles as angle OPB also measures 45°.

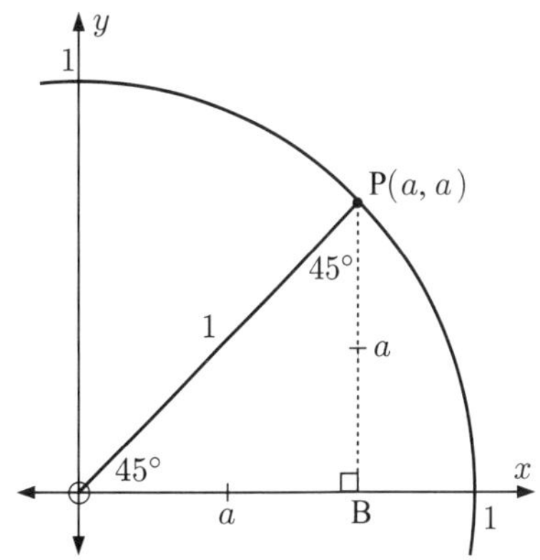

Letting OB = BP = a,

$a^2 + a^2 = 1^2$ {Pythagoras}

$\therefore\ 2a^2 = 1$

$\therefore\ a^2 = \frac{1}{2}$

$\therefore\ a = \frac{1}{\sqrt{2}}$ as $a > 0$

So, P is $\left(\frac{1}{\sqrt{2}}, \frac{1}{\sqrt{2}}\right)$ where $\frac{1}{\sqrt{2}} \approx 0.707$.

$$\cos\frac{\pi}{4} = \frac{1}{\sqrt{2}} \quad \text{and} \quad \sin\frac{\pi}{4} = \frac{1}{\sqrt{2}}$$

You should remember these values. If you forget, draw a right angled isosceles triangle with side length 1.

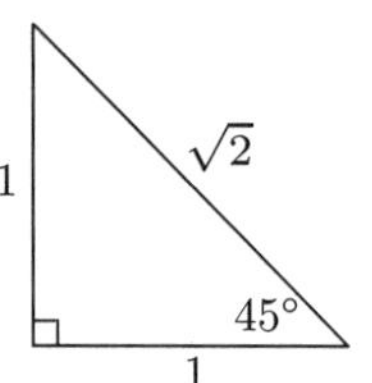

For **multiples of $\frac{\pi}{4}$**, the number $\frac{1}{\sqrt{2}}$ is the important thing to remember. The signs of the coordinates are determined by which quadrant the angle is in.

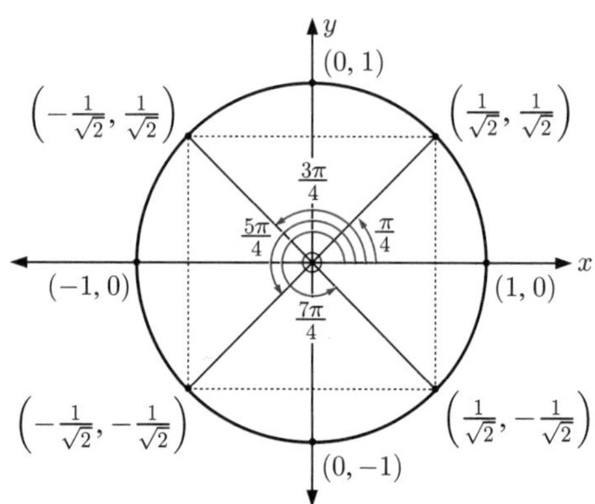

MULTIPLES OF $\frac{\pi}{6}$ OR 30°

Since OA = OP, triangle OAP is isosceles.

The remaining angles are therefore also 60°, and so triangle AOP is equilateral.

The altitude [PN] bisects base [OA],

so $\text{ON} = \frac{1}{2}$

If P is $(\frac{1}{2}, k)$, then $(\frac{1}{2})^2 + k^2 = 1$ {Pythagoras}

$\therefore\ k^2 = \frac{3}{4}$

$\therefore\ k = \frac{\sqrt{3}}{2}$ {as $k > 0$}

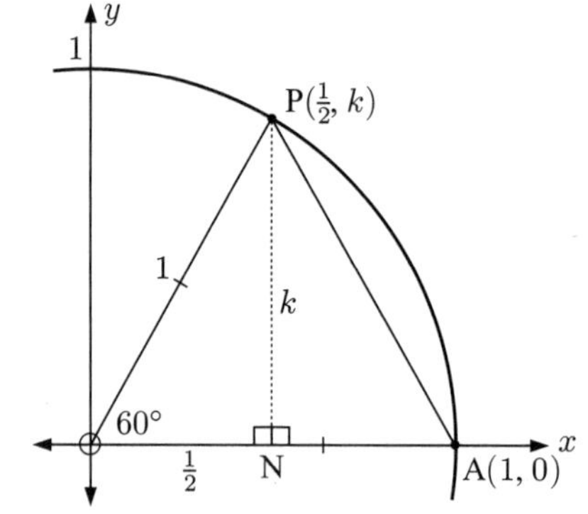

So, P is $(\frac{1}{2}, \frac{\sqrt{3}}{2})$ where $\frac{\sqrt{3}}{2} \approx 0.866$.

$$\cos\frac{\pi}{3} = \frac{1}{2} \quad \text{and} \quad \sin\frac{\pi}{3} = \frac{\sqrt{3}}{2}$$

Now $\widehat{NPO} = \frac{\pi}{6} = 30°$.

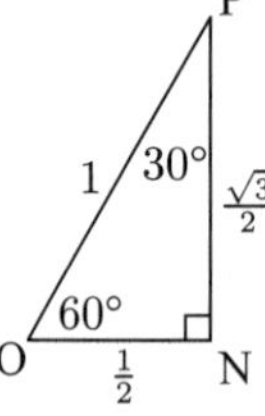

Hence $$\cos\frac{\pi}{6} = \frac{\sqrt{3}}{2} \quad \text{and} \quad \sin\frac{\pi}{6} = \frac{1}{2}$$

You should remember these values. If you forget, divide in two an equilateral triangle with side length 2.

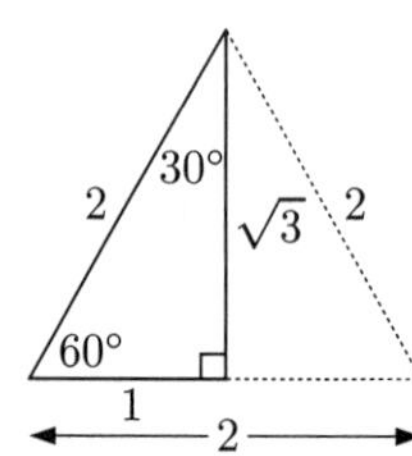

For **multiples of $\frac{\pi}{6}$**, the numbers $\frac{1}{2}$ and $\frac{\sqrt{3}}{2}$ are important. The exact coordinates of each point are found by symmetry.

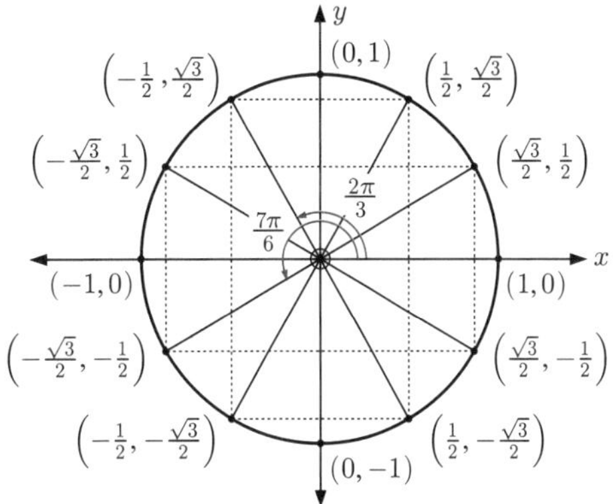

SUMMARY

- For **multiples of $\frac{\pi}{2}$**, the coordinates of the points on the unit circle involve 0 and ± 1.
- For *other* **multiples of $\frac{\pi}{4}$**, the coordinates involve $\pm\frac{1}{\sqrt{2}}$.
- For *other* **multiples of $\frac{\pi}{6}$**, the coordinates involve $\pm\frac{1}{2}$ and $\pm\frac{\sqrt{3}}{2}$.

You should be able to use this summary to find the trigonometric ratios for angles which are multiples of $\frac{\pi}{6}$ and $\frac{\pi}{4}$.

For example, consider the angles:

- $225° = \frac{5\pi}{4}$

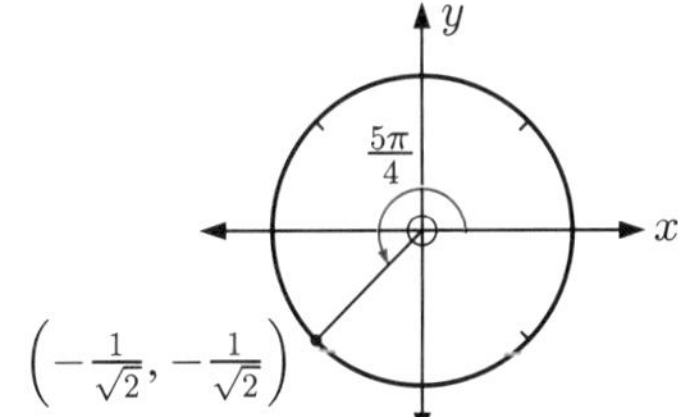

$\frac{5\pi}{4}$ is in quadrant 3, so the signs are both negative and both have $\frac{1}{\sqrt{2}}$ size.

- $300° = \frac{5\pi}{3}$ which is a multiple of $\frac{\pi}{6}$.

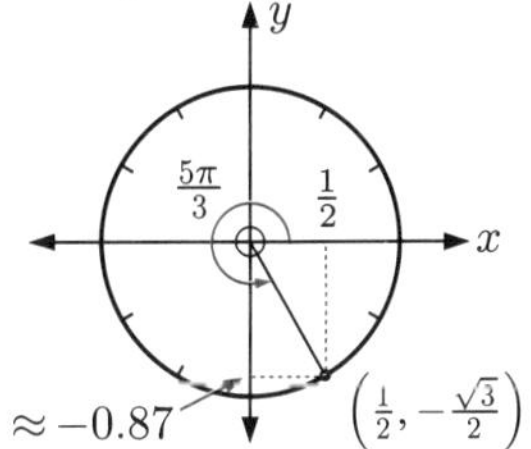

$\frac{5\pi}{3}$ is in quadrant 4, so the signs are $(+, -)$ and from the diagram the x-value is $\frac{1}{2}$.

Example 15 **Self Tutor**

Find the exact values of $\sin\alpha$, $\cos\alpha$, and $\tan\alpha$ for $\alpha = \frac{3\pi}{4}$.

$\left(-\frac{1}{\sqrt{2}}, \frac{1}{\sqrt{2}}\right)$ ≈ 0.7 $\frac{3\pi}{4}$ ≈ -0.7

$\sin\left(\frac{3\pi}{4}\right) = \frac{1}{\sqrt{2}}$

$\cos\left(\frac{3\pi}{4}\right) = -\frac{1}{\sqrt{2}}$

$\tan\left(\frac{3\pi}{4}\right) = -1$

Example 16 — Self Tutor

Find the exact values of $\sin\frac{4\pi}{3}$, $\cos\frac{4\pi}{3}$, and $\tan\frac{4\pi}{3}$.

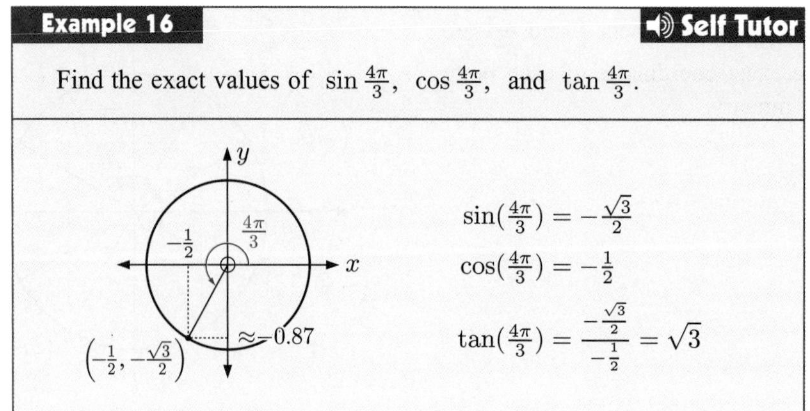

$\sin(\frac{4\pi}{3}) = -\frac{\sqrt{3}}{2}$

$\cos(\frac{4\pi}{3}) = -\frac{1}{2}$

$\tan(\frac{4\pi}{3}) = \dfrac{-\frac{\sqrt{3}}{2}}{-\frac{1}{2}} = \sqrt{3}$

EXERCISE 10F

1 Use a unit circle diagram to find exact values for $\sin\theta$, $\cos\theta$, and $\tan\theta$ for θ equal to:

a $\frac{\pi}{4}$ **b** $\frac{3\pi}{4}$ **c** $\frac{7\pi}{4}$ **d** π **e** $-\frac{3\pi}{4}$

2 Use a unit circle diagram to find exact values for $\sin\beta$, $\cos\beta$, and $\tan\beta$ for β equal to:

a $\frac{\pi}{6}$ **b** $\frac{2\pi}{3}$ **c** $\frac{7\pi}{6}$ **d** $\frac{5\pi}{3}$ **e** $\frac{11\pi}{6}$

3 Find the exact values of:

a $\cos 120°$, $\sin 120°$, and $\tan 120°$ **b** $\cos(-45°)$, $\sin(-45°)$, and $\tan(-45°)$

4 **a** Find the exact values of $\cos 90°$ and $\sin 90°$.

b What can you say about $\tan 90°$?

Example 17 — Self Tutor

Without using a calculator, show that $8\sin(\frac{\pi}{3})\cos(\frac{5\pi}{6}) = -6$.

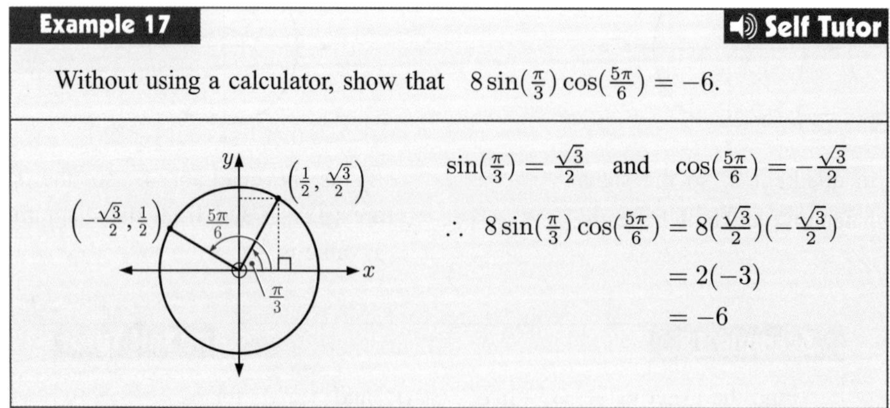

$\sin(\frac{\pi}{3}) = \frac{\sqrt{3}}{2}$ and $\cos(\frac{5\pi}{6}) = -\frac{\sqrt{3}}{2}$

$$\begin{aligned} \therefore\ 8\sin(\tfrac{\pi}{3})\cos(\tfrac{5\pi}{6}) &= 8(\tfrac{\sqrt{3}}{2})(-\tfrac{\sqrt{3}}{2}) \\ &= 2(-3) \\ &= -6 \end{aligned}$$

5 Without using a calculator, evaluate:

a $\sin^2 60°$ **b** $\sin 30° \cos 60°$ **c** $4\sin 60° \cos 30°$

d $1 - \cos^2(\frac{\pi}{6})$ **e** $\sin^2(\frac{2\pi}{3}) - 1$ **f** $\cos^2(\frac{\pi}{4}) - \sin(\frac{7\pi}{6})$

g $\sin(\frac{3\pi}{4}) - \cos(\frac{5\pi}{4})$ **h** $1 - 2\sin^2(\frac{7\pi}{6})$ **i** $\cos^2(\frac{5\pi}{6}) - \sin^2(\frac{5\pi}{6})$

j $\tan^2(\frac{\pi}{3}) - 2\sin^2(\frac{\pi}{4})$ **k** $2\tan(-\frac{5\pi}{4}) - \sin(\frac{3\pi}{2})$ **l** $\dfrac{2\tan 150°}{1 - \tan^2 150°}$

Check all answers using your calculator.

Example 18 **Self Tutor**

Find all angles $0 \leqslant \theta \leqslant 2\pi$ with a cosine of $\frac{1}{2}$.

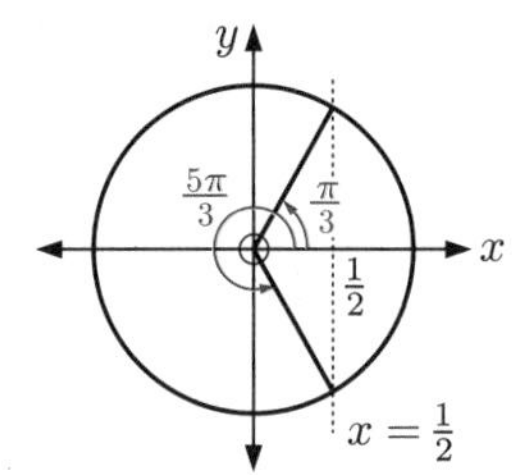

Since the cosine is $\frac{1}{2}$, we draw the vertical line $x = \frac{1}{2}$.

Because $\frac{1}{2}$ is involved, we know the required angles are multiples of $\frac{\pi}{6}$.

They are $\frac{\pi}{3}$ and $\frac{5\pi}{3}$.

6 Find all angles between $0°$ and $360°$ with:

a a sine of $\frac{1}{2}$ **b** a sine of $\frac{\sqrt{3}}{2}$ **c** a cosine of $\frac{1}{\sqrt{2}}$

d a cosine of $-\frac{1}{2}$ **e** a cosine of $-\frac{1}{\sqrt{2}}$ **f** a sine of $-\frac{\sqrt{3}}{2}$

7 Find all angles between 0 and 2π (inclusive) which have:

a a tangent of 1 **b** a tangent of -1 **c** a tangent of $\sqrt{3}$

d a tangent of 0 **e** a tangent of $\frac{1}{\sqrt{3}}$ **f** a tangent of $-\sqrt{3}$

8 Find all angles between 0 and 4π with:

a a cosine of $\frac{\sqrt{3}}{2}$ **b** a sine of $-\frac{1}{2}$ **c** a sine of -1

9 Find θ if $0 \leqslant \theta \leqslant 2\pi$ and:

a $\cos\theta = \frac{1}{2}$ **b** $\sin\theta = \frac{\sqrt{3}}{2}$ **c** $\cos\theta = -1$ **d** $\sin\theta = 1$

e $\cos\theta = -\frac{1}{\sqrt{2}}$ **f** $\sin^2\theta = 1$ **g** $\cos^2\theta = 1$ **h** $\cos^2\theta = \frac{1}{2}$

i $\tan\theta = -\frac{1}{\sqrt{3}}$ **j** $\tan^2\theta = 3$

10 Find *all* values of θ for which $\tan\theta$ is: **a** zero **b** undefined.

REVIEW SET 10A NON-CALCULATOR

1 Convert these to radians in terms of π:

a $120°$ **b** $225°$ **c** $150°$ **d** $540°$

2 Find the acute angles that would have the same:

a sine as $\frac{2\pi}{3}$ **b** sine as $165°$ **c** cosine as $276°$.

3 Find:

a $\sin 159°$ if $\sin 21° \approx 0.358$ **b** $\cos 92°$ if $\cos 88° \approx 0.035$

c $\cos 75°$ if $\cos 105° \approx -0.259$ **d** $\sin(-133°)$ if $\sin 47° \approx 0.731$.

4 Use a unit circle diagram to find:

a $\cos 360°$ and $\sin 360°$ **b** $\cos(-\pi)$ and $\sin(-\pi)$.

5 Explain how to use the unit circle to find θ when $\cos\theta = -\sin\theta$, $0 \leqslant \theta \leqslant 2\pi$.

6 Find exact values for $\sin\theta$, $\cos\theta$, and $\tan\theta$ for θ equal to:

a $\frac{2\pi}{3}$ **b** $\frac{8\pi}{3}$

7 If $\sin x = -\frac{1}{4}$ and $\pi < x < \frac{3\pi}{2}$, find $\tan x$ exactly.

8 If $\cos\theta = \frac{3}{4}$ find the possible values of $\sin\theta$.

9 Evaluate:

a $2\sin(\frac{\pi}{3})\cos(\frac{\pi}{3})$ **b** $\tan^2(\frac{\pi}{4}) - 1$ **c** $\cos^2(\frac{\pi}{6}) - \sin^2(\frac{\pi}{6})$

10 Given $\tan x = -\frac{3}{2}$ and $\frac{3\pi}{2} < x < 2\pi$, find: **a** $\sin x$ **b** $\cos x$.

11

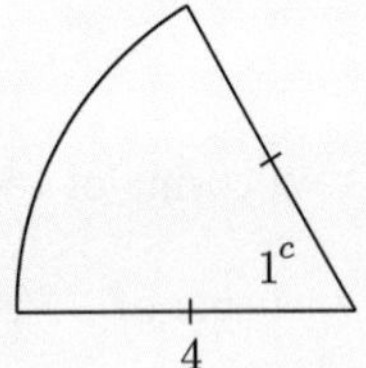

Find the perimeter and area of the sector.

12 Suppose $\cos\theta = \frac{\sqrt{11}}{\sqrt{17}}$ and θ is acute. Find the exact value of $\tan\theta$.

13 Simplify:

a $\cos\left(\frac{\pi}{2} - \theta\right) - \sin\theta$ **b** $\cos(-\theta)\tan\theta$ **c** $\sin(-\alpha)\cos\left(\alpha - \frac{\pi}{2}\right)$

REVIEW SET 10B — CALCULATOR

1 Determine the coordinates of the point on the unit circle corresponding to an angle of:

a 320° **b** 163°

2 Convert to radians to 4 significant figures:

a 71° **b** 124.6° **c** −142°

3 Convert these radian measurements to degrees, to 2 decimal places:

a 3 **b** 1.46 **c** 0.435 **d** −5.271

4 Determine the area of a sector of angle $\frac{5\pi}{12}$ and radius 13 cm.

5 Find the coordinates of the points M, N, and P on the unit circle.

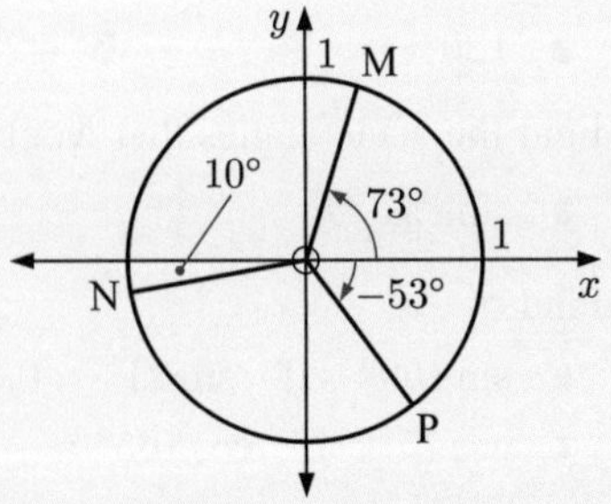

6 Find the angle [OA] makes with the positive x-axis if the x-coordinate of the point A on the unit circle is −0.222 .

7 Find all angles between $0°$ and $360°$ which have:

a a cosine of $-\frac{\sqrt{3}}{2}$ **b** a sine of $\frac{1}{\sqrt{2}}$ **c** a tangent of $-\sqrt{3}$

8 Find θ for $0 \leqslant \theta \leqslant 2\pi$ if:

a $\cos\theta = -1$ **b** $\sin^2\theta = \frac{3}{4}$

9 Find the obtuse angles which have the same:

a sine as $47°$ **b** sine as $\frac{\pi}{15}$ **c** cosine as $186°$

10 Find the perimeter and area of a sector of radius 11 cm and angle $63°$.

11 Find the radius and area of a sector of perimeter 36 cm with an angle of $\frac{2\pi}{3}$.

12 Find two angles on the unit circle with $0 \leqslant \theta \leqslant 2\pi$, such that:

a $\cos\theta = \frac{2}{3}$ **b** $\sin\theta = -\frac{1}{4}$ **c** $\tan\theta = 3$

REVIEW SET 10C

1 Convert these radian measurements to degrees:

a $\frac{2\pi}{5}$ **b** $\frac{5\pi}{4}$ **c** $\frac{7\pi}{9}$ **d** $\frac{11\pi}{6}$

2 Illustrate the regions where $\sin\theta$ and $\cos\theta$ have the same sign.

3 Use a unit circle diagram to find:

a $\cos(\frac{3\pi}{2})$ and $\sin(\frac{3\pi}{2})$ **b** $\cos(-\frac{\pi}{2})$ and $\sin(-\frac{\pi}{2})$

4 Suppose $m = \sin p$, where p is acute. Write an expression in terms of m for:

a $\sin(\pi - p)$ **b** $\sin(p + 2\pi)$ **c** $\cos p$ **d** $\tan p$

5

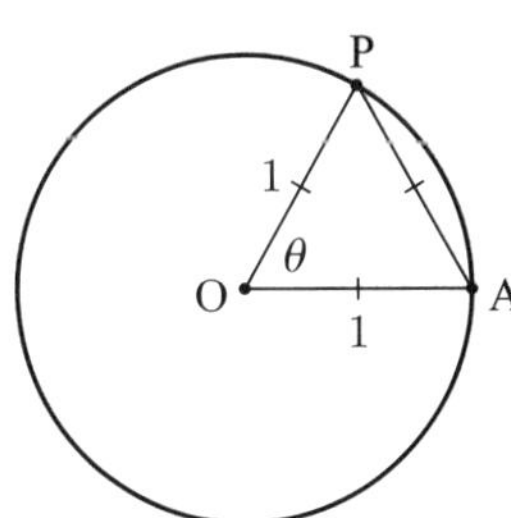

a State the value of θ in:

i degrees **ii** radians.

b State the arc length AP.

c State the area of the minor *sector* OAP.

6 Without a calculator, evaluate $\tan^2(\frac{2\pi}{3})$.

7 Show that $\cos(\frac{3\pi}{4}) - \sin(\frac{3\pi}{4}) = -\sqrt{2}$.

8 If $\cos\theta = -\frac{3}{4}$, $\frac{\pi}{2} < \theta < \pi$ find the exact value of:

a $\sin\theta$ **b** $\tan\theta$ **c** $\sin(\theta + \pi)$

9 Without using a calculator, evaluate:

a $\tan^2 60° - \sin^2 45°$ **b** $\cos^2(\frac{\pi}{4}) + \sin(\frac{\pi}{2})$ **c** $\cos(\frac{5\pi}{3}) - \tan(\frac{5\pi}{4})$

10 Simplify:

a $\sin(\pi - \theta) - \sin\theta$ **b** $\sin\left(\frac{\pi}{2} - \theta\right) - 2\cos\theta$

11 Use a unit circle diagram to show that
$\cos\left(\frac{\pi}{2}+\theta\right) = -\sin\theta$ for $\frac{\pi}{2} < \theta < \pi$.

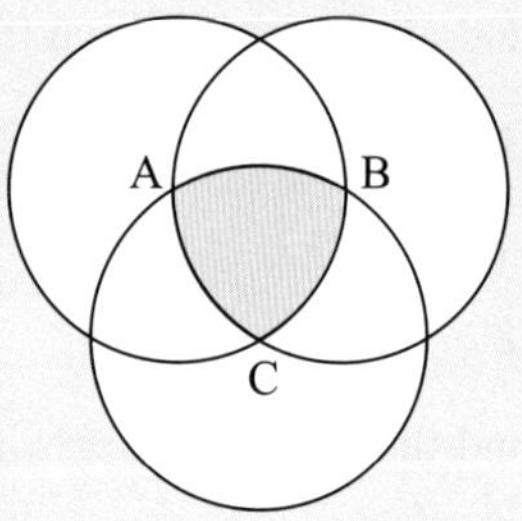

12 Three circles with radius r are drawn as shown, each with its centre on the circumference of the other two circles. A, B and C are the centres of the three circles. Prove that an expression for the area of the shaded region is $A = \frac{r^2}{2}(\pi - \sqrt{3})$.

Chapter 11

Non-right angled triangle trigonometry

Syllabus reference: 3.7

Contents:

OPENING PROBLEM

A triangular sail is to be cut from a section of cloth. Two of the sides must have lengths 4 m and 6 m as illustrated. The total area for the sail must be 11.6 m^2, the maximum allowed for the boat to race in its class.

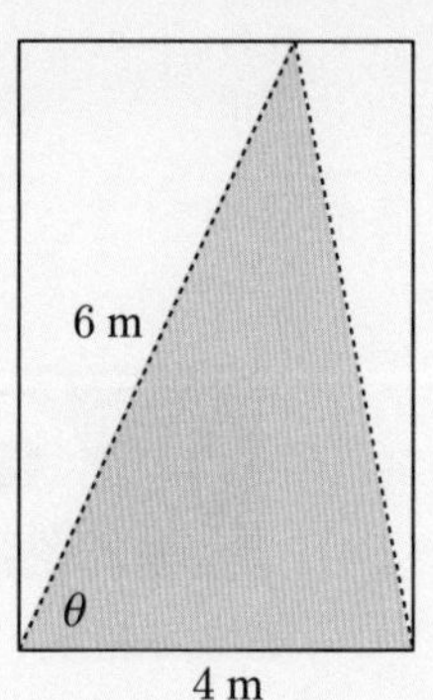

Things to think about:

- **a** Can you find the size of the angle θ between the two sides of given length?
- **b** Can you find the length of the third side of the sail?

A AREAS OF TRIANGLES

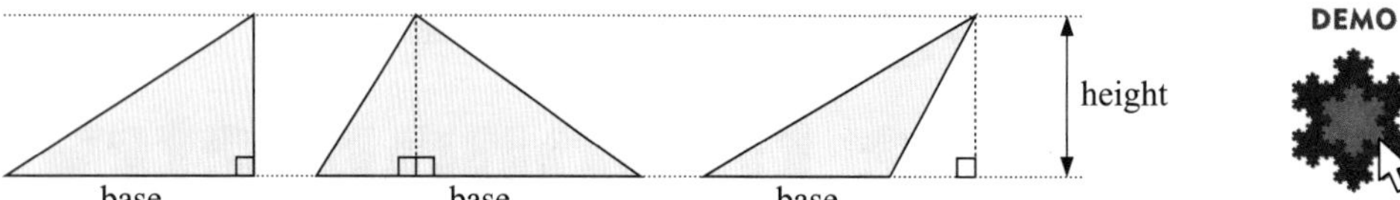

If we know the base and height measurements of a triangle we can calculate the area using **area** $= \frac{1}{2}$ **base** $\times$ **height**.

However, cases arise where we do not know the height but we can use trigonometry to calculate the area.

These cases are:

- knowing two sides and the **included angle** between them

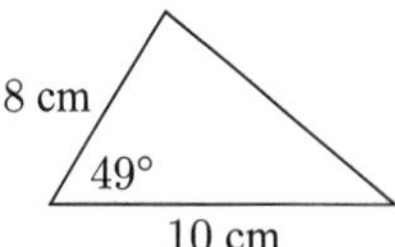

- knowing all three sides

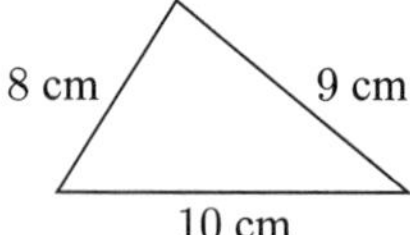

USING THE INCLUDED ANGLE

Triangle ABC has angles of size A, B, and C. The sides opposite these angles are labelled a, b, and c respectively.

Using trigonometry, we can develop an alternative area formula that does not depend on a perpendicular height.

Any triangle that is not right angled must be either acute or obtuse. We will consider both cases:

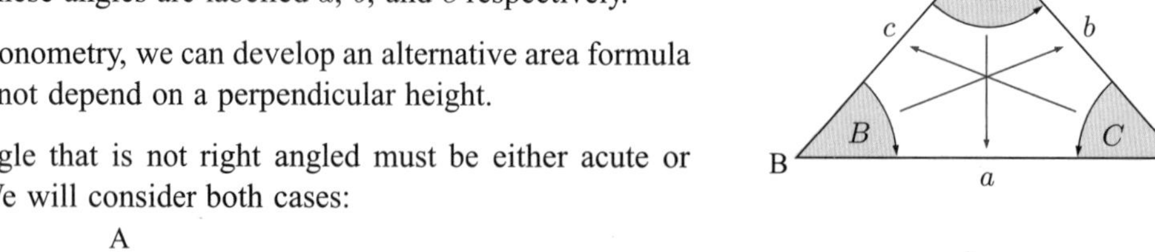

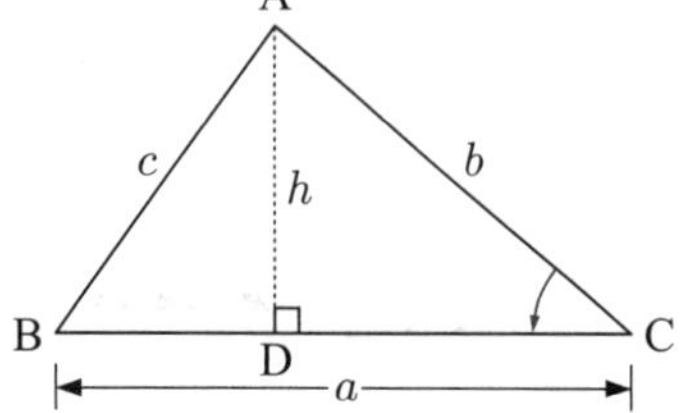

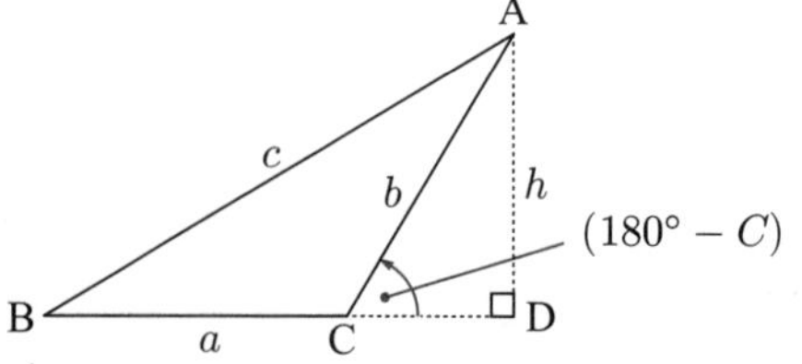

In both triangles the altitude h is constructed from A to D on [BC] (extended if necessary).

$$\sin C = \frac{h}{b} \qquad \therefore \quad h = b\sin C$$

$$\sin(180° - C) = \frac{h}{b}$$
$$\therefore \quad h = b\sin(180° - C)$$
But $\sin(180° - C) = \sin C$
$$\therefore \quad h = b\sin C$$

So, since area $= \frac{1}{2}ah$, we now have **Area** $= \frac{1}{2}\boldsymbol{ab}\,\mathbf{sin}\,\boldsymbol{C}$.

Using different altitudes we can show that the area is also $\frac{1}{2}bc\sin A$ or $\frac{1}{2}ac\sin B$.

Given the lengths of two sides of a triangle, and the size of the included angle between them, the area of the triangle is

half of the product of two sides and the sine of the included angle.

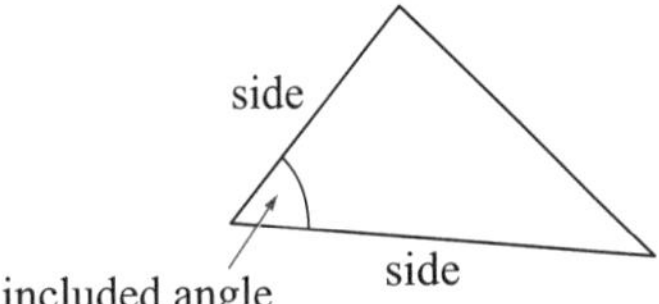

Example 1 — Self Tutor

Find the area of triangle ABC.

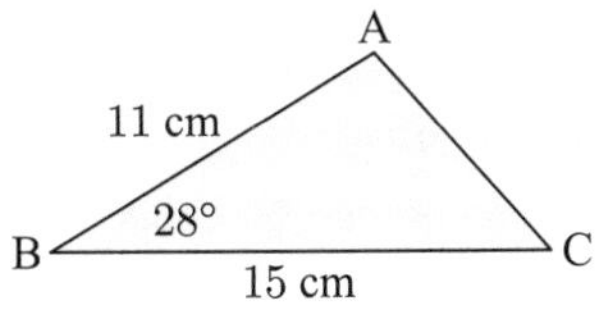

Area $= \frac{1}{2}ac\sin B$
$= \frac{1}{2} \times 15 \times 11 \times \sin 28°$
$\approx 38.7 \text{ cm}^2$

If we rearrange the area formula to find the included angle between two sides, we need to use the **inverse** sine function denoted $\sin^{-1}$. For help with this and the other inverse trigonometric functions you should consult the Background Knowledge chapter online.

Example 2 — Self Tutor

A triangle has sides of length 10 cm and 11 cm and an area of 50 cm^2. Determine the two possible measures of the included angle. Give your answers accurate to 1 decimal place.

If the included angle is θ, then $\frac{1}{2} \times 10 \times 11 \times \sin\theta = 50$
$$\therefore \quad \sin\theta = \frac{50}{55}$$

Now $\sin^{-1}(\frac{50}{55}) \approx 65.4°$
$\therefore \quad \theta \approx 65.4°$ or $180° - 65.4°$
$\therefore \quad \theta \approx 65.4°$ or $114.6°$

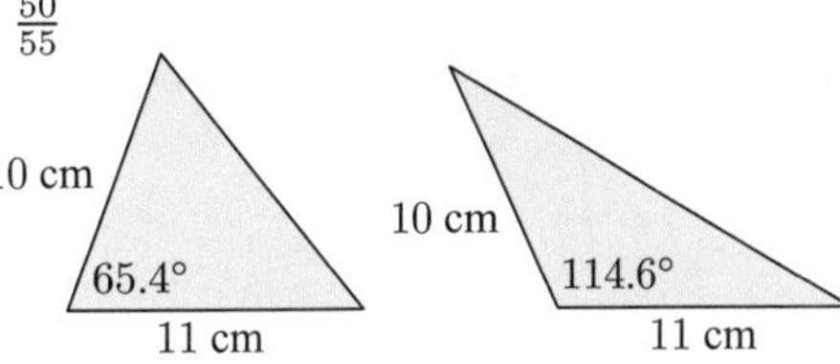

The two different possible angles are 65.4° and 114.6°.

EXERCISE 11A

1 Find the area of:

a

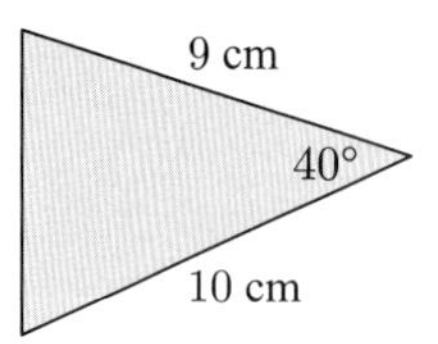

b

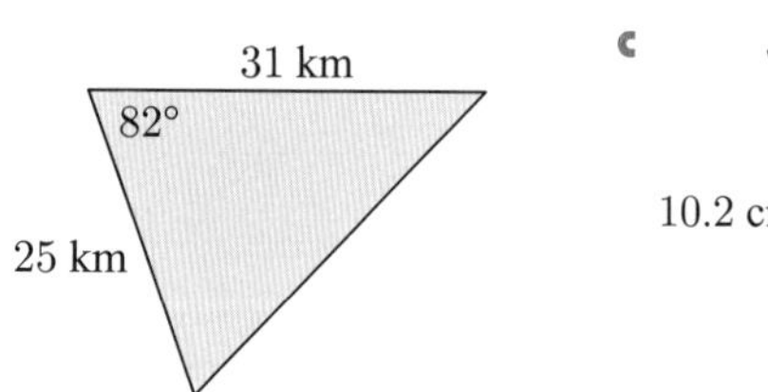

c

10.2 cm

$\frac{2\pi}{3}$

6.4 cm

2 If triangle ABC has area 150 cm^2, find the value of x:

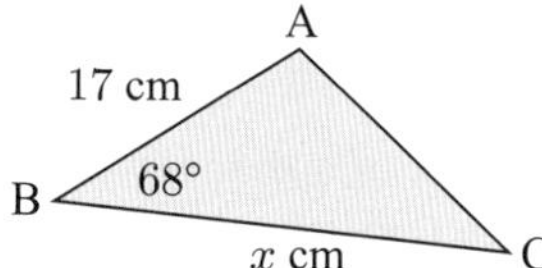

3 A parallelogram has two adjacent sides with lengths 4 cm and 6 cm respectively. If the included angle measures 52°, find the area of the parallelogram.

4 A rhombus has sides of length 12 cm and an angle of 72°. Find its area.

5 Find the area of a regular hexagon with sides of length 12 cm.

6 A rhombus has an area of 50 cm^2 and an internal angle of size 63°. Find the length of its sides.

7 A regular pentagonal garden plot has centre of symmetry O and an area of 338 m^2. Find the distance OA.

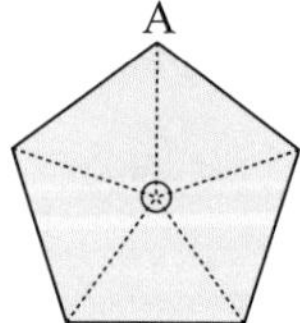

8 Find the possible values of the included angle of a triangle with:

a sides of length 5 cm and 8 cm, and area 15 cm^2

b sides of length 45 km and 53 km, and area 800 km^2.

9 The Australian 50 cent coin has the shape of a regular dodecagon, which is a polygon with 12 sides.
Eight of these 50 cent coins will fit exactly on an Australian $10 note as shown. What fraction of the $10 note is *not* covered?

10 Find the shaded area in:

a

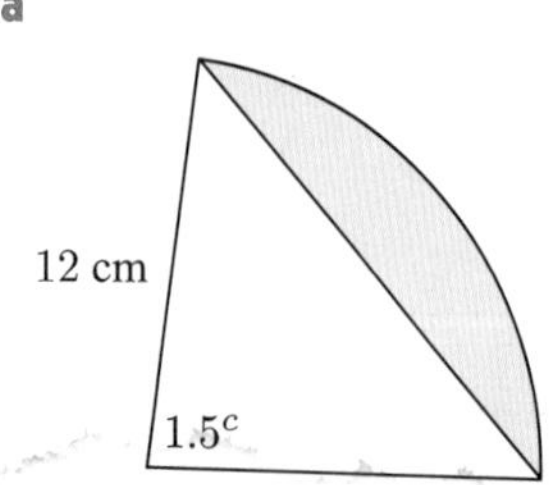

b

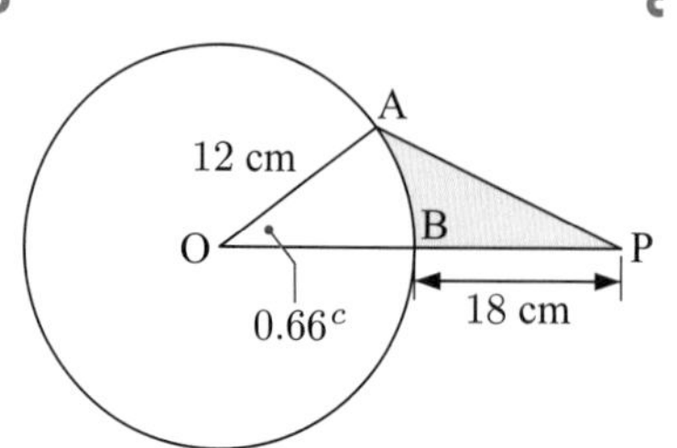

c

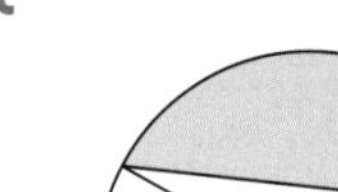

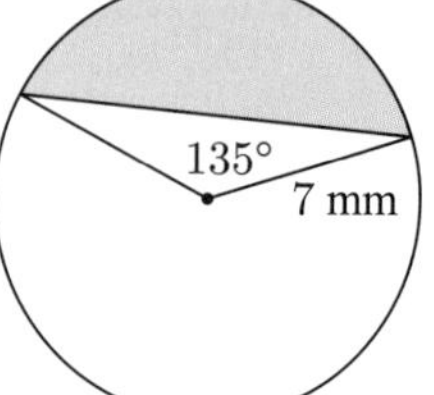

11 ADB is an arc of the circle with centre C and radius 7.3 cm. AEB is an arc of the circle with centre F and radius 8.7 cm.
Find the shaded area.

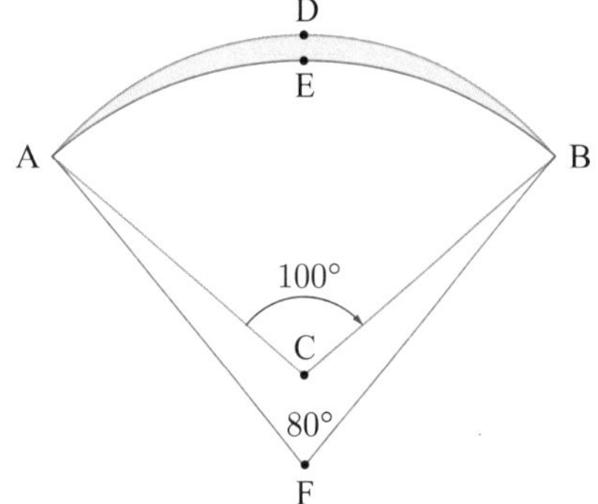

INVESTIGATION 1 HERON'S FORMULA

In his book *Metrica* written about 60 AD, **Heron** (or **Hero**) **of Alexandria** proved the following formula for the area of a triangle with side lengths a, b, and c:

$$A = \sqrt{s(s-a)(s-b)(s-c)}$$

where $s = \frac{a+b+c}{2}$

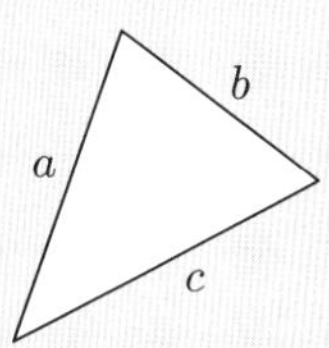

What to do:

1 Find the area of the right angled triangle with sides 3 cm, 4 cm, and 5 cm:

a without using Heron's formula

b using Heron's formula.

2 Find the area of a triangle with sides of length:

a 6 cm, 8 cm, and 12 cm

b 7.2 cm, 8.9 cm, and 9.7 cm.

3 Research Heron's proof of his theorem using **cyclic quadrilaterals**.

A more common modern proof of Heron's formula is to use the **cosine rule**, which we see in the next section.

B THE COSINE RULE

The **cosine rule** involves the sides and angles of a triangle.

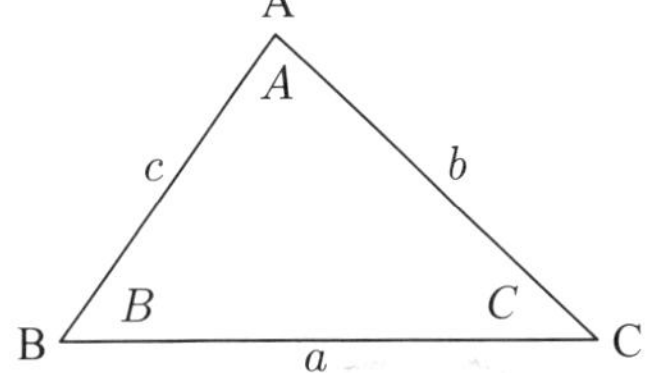

In any $\triangle$ABC:

$$\begin{aligned} & \boldsymbol{a^2 = b^2 + c^2 - 2bc\cos A} \\ \textit{or} \quad & \boldsymbol{b^2 = a^2 + c^2 - 2ac\cos B} \\ \textit{or} \quad & \boldsymbol{c^2 = a^2 + b^2 - 2ab\cos C} \end{aligned}$$

We will develop the first formula for both an acute and an obtuse triangle.

Proof:

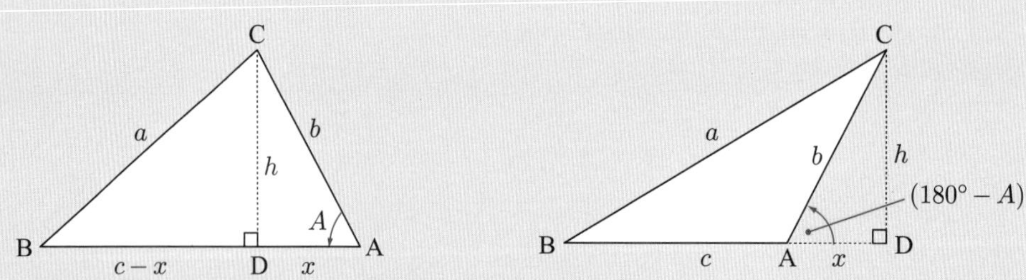

In both triangles draw the altitude from C down to [AB] (extended if necessary), meeting it at D.

Let $\text{AD} = x$ and let $\text{CD} = h$.

Apply the theorem of Pythagoras in $\triangle$BCD:

$$a^2 = h^2 + (c - x)^2 \qquad\qquad a^2 = h^2 + (c + x)^2$$

$$\therefore\ a^2 = h^2 + c^2 - 2cx + x^2 \qquad\qquad \therefore\ a^2 = h^2 + c^2 + 2cx + x^2$$

In both cases, applying Pythagoras to $\triangle$ADC gives $h^2 + x^2 = b^2$.

$\therefore\ h^2 = b^2 - x^2$, and we substitute this into the equations above.

$$\therefore\ a^2 = b^2 + c^2 - 2cx \qquad\qquad \therefore\ a^2 = b^2 + c^2 + 2cx$$

In $\triangle$ADC, $\cos A = \dfrac{x}{b}$

$\therefore\ b\cos A = x$

$\therefore\ a^2 = b^2 + c^2 - 2bc\cos A$

$\cos(180° - A) = \dfrac{x}{b}$

$\therefore\ -\cos A = \dfrac{x}{b}$

$\therefore\ -b\cos A = x$

$\therefore\ a^2 = b^2 + c^2 - 2bc\cos A$

The other variations of the cosine rule could be developed by rearranging the vertices of $\triangle$ABC.

Note that if $A = 90°$ then $\cos A = 0$ and $a^2 = b^2 + c^2 - 2bc\cos A$ reduces to $a^2 = b^2 + c^2$, which is the Pythagorean Rule.

The **cosine rule** can be used to solve problems involving triangles given:

- **two sides** and an **included angle**
- **three sides**.

If we are given **two sides** and a **non-included angle**, then when we try to find the third side we obtain a quadratic equation. This is an *ambiguous* case where there may be two plausible solutions. We may not be able to solve for the length uniquely if there are two positive, plausible solutions to the quadratic equation.

Example 3 — Self Tutor

Find, correct to 2 decimal places, the length of [BC].

B
11 cm
42°
A
C
13 cm

By the cosine rule:

$\text{BC}^2 = 11^2 + 13^2 - 2 \times 11 \times 13 \times \cos 42°$

$\therefore\ \text{BC} \approx \sqrt{(11^2 + 13^2 - 2 \times 11 \times 13 \times \cos 42°)}$

$\therefore\ \text{BC} \approx 8.801$

$\therefore$ [BC] is 8.80 cm in length.

Rearrangement of the original cosine rule formulae can be used for finding angles if we know all three sides. The formulae for finding the angles are:

$$\cos A = \frac{b^2 + c^2 - a^2}{2bc} \qquad \cos B = \frac{c^2 + a^2 - b^2}{2ca} \qquad \cos C = \frac{a^2 + b^2 - c^2}{2ab}$$

We then need to use the **inverse** cosine function $\cos^{-1}$ to evaluate the angle.

Example 4 **Self Tutor**

In triangle ABC, AB = 7 cm, BC = 5 cm, and CA = 8 cm.

a Find the measure of $B\widehat{C}A$.

b Find the exact area of triangle ABC.

a

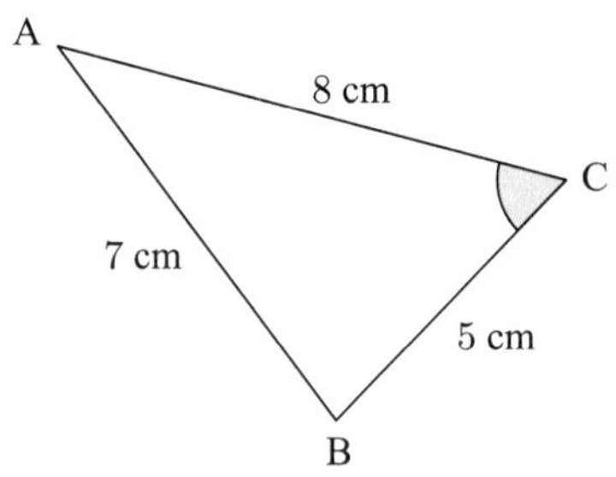

By the cosine rule:

$\cos C = \frac{(5^2 + 8^2 - 7^2)}{(2 \times 5 \times 8)}$

$\therefore \ C = \cos^{-1}\left(\frac{5^2 + 8^2 - 7^2}{2 \times 5 \times 8}\right)$

$\therefore \ C = \cos^{-1}(\frac{1}{2})$

$\therefore \ C = 60°$

So, $B\widehat{C}A$ measures 60°.

b The area of $\triangle ABC = \frac{1}{2} \times 8 \times 5 \times \sin 60°$

$= 20 \times \frac{\sqrt{3}}{2} \qquad \{\sin 60° = \frac{\sqrt{3}}{2}\}$

$= 10\sqrt{3} \text{ cm}^2$

EXERCISE 11B

1 Find the length of the remaining side in the given triangle:

a

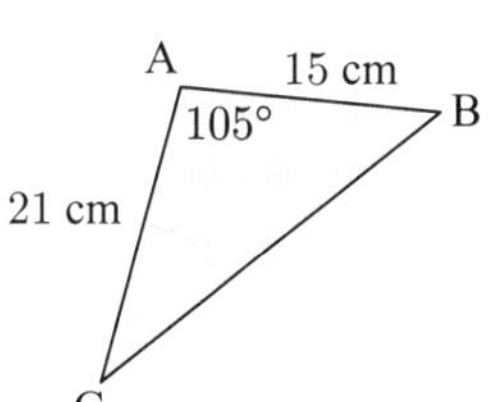

b

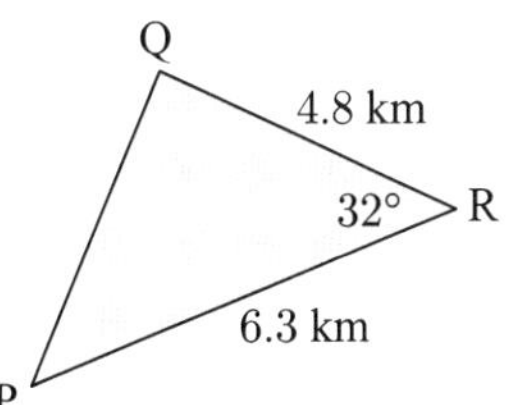

c

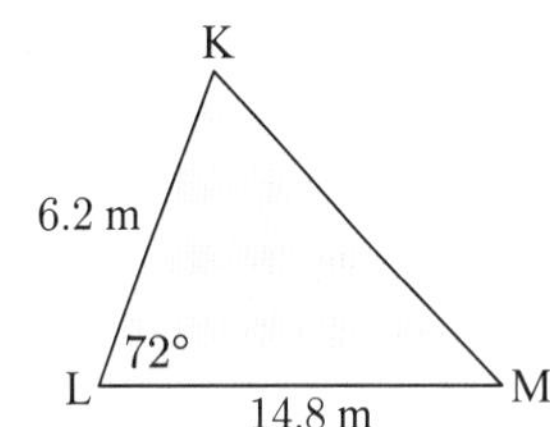

2 Find the measure of all angles of:

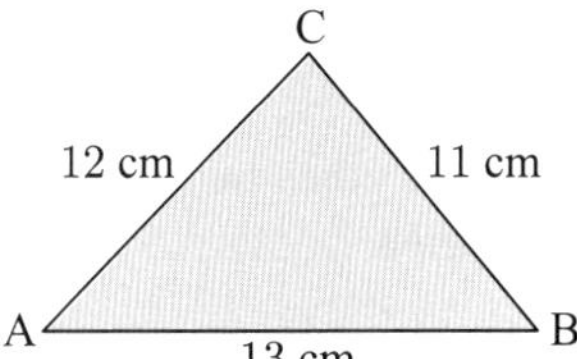

3 **a** Find the measure of obtuse $P\widehat{Q}R$.

b Hence find the area of $\triangle PQR$.

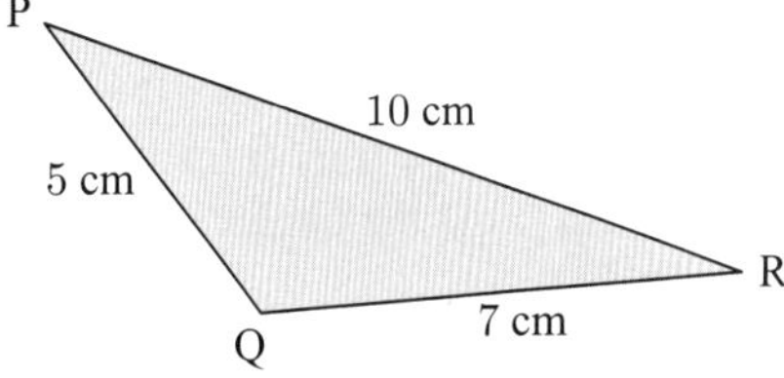

4 **a** Find the smallest angle of a triangle with sides 11 cm, 13 cm, and 17 cm.

b Find the largest angle of a triangle with sides 4 cm, 7 cm, and 9 cm.

The smallest angle is opposite the shortest side.

5 **a** Find $\cos\theta$ but not θ.

b Find the value of x.

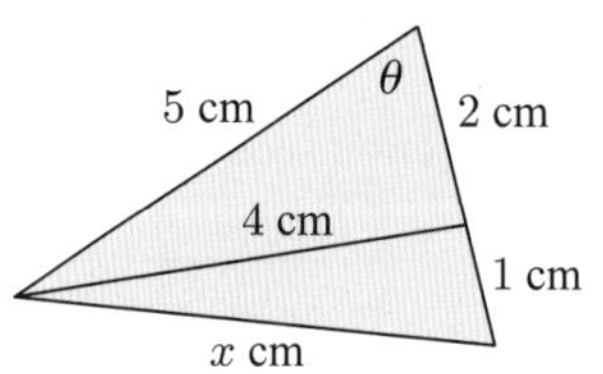

6 Find the exact value of x in each of the following diagrams:

a

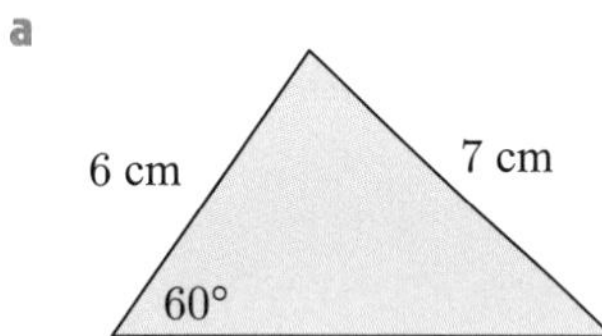

b

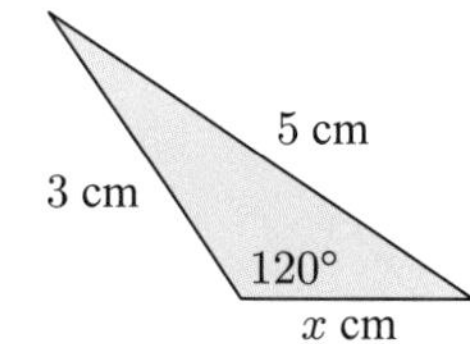

c

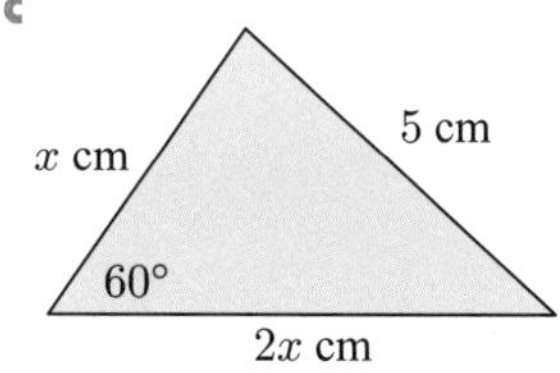

7 Solve the **Opening Problem** on page **306**.

8 Find x in each of the following diagrams:

a

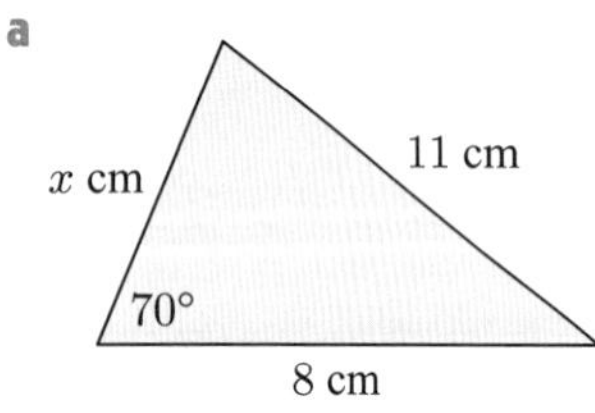

b

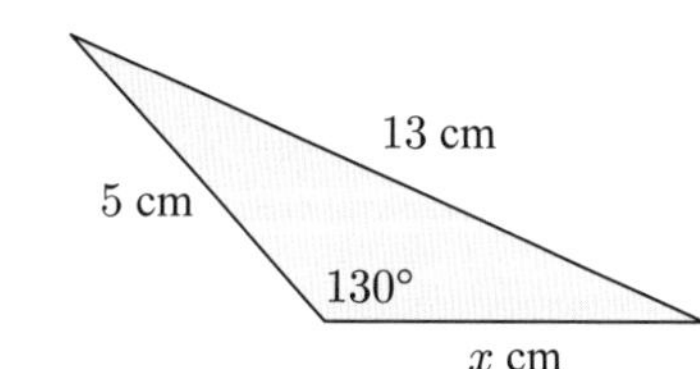

9 Show that there are two plausible values for x in this triangle:

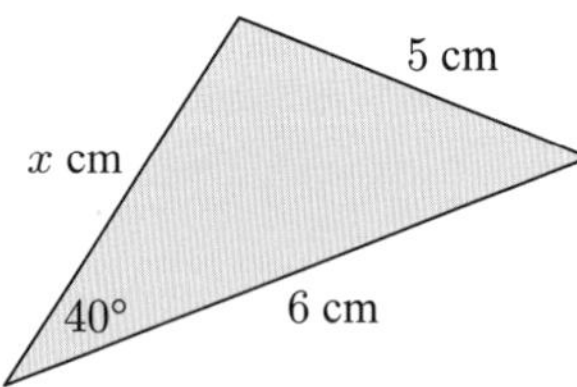

10 In the diagram alongside, $\cos\theta = -\frac{1}{5}$.

a Find x.

b Hence find the exact area of the triangle.

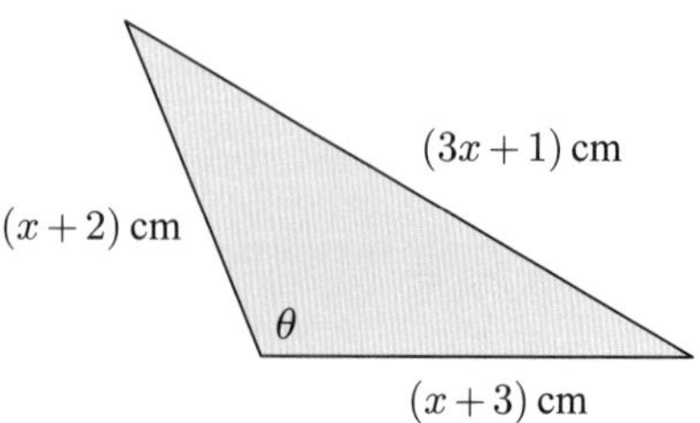

11 The parallel sides of a trapezium have lengths 5 cm and 8 cm. The other two sides have lengths 6 cm and 4 cm. Find the angles of the trapezium, to the nearest degree.

THE SINE RULE

The **sine rule** is a set of equations which connects the lengths of the sides of any triangle with the sines of the angles of the triangle. The triangle does not have to be right angled for the sine rule to be used.

In any triangle ABC with sides a, b, and c units in length, and opposite angles A, B, and C respectively,

$$\frac{\sin A}{a} = \frac{\sin B}{b} = \frac{\sin C}{c} \quad \text{or} \quad \frac{a}{\sin A} = \frac{b}{\sin B} = \frac{c}{\sin C}.$$

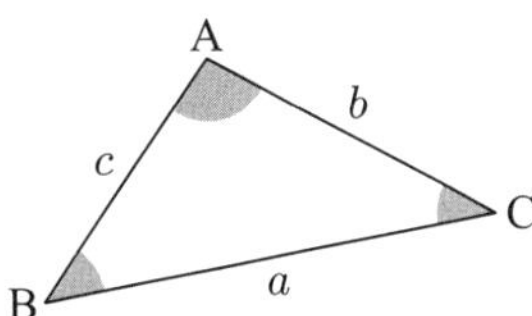

Proof: The area of any triangle ABC is given by $\frac{1}{2}bc\sin A = \frac{1}{2}ac\sin B = \frac{1}{2}ab\sin C$.

Dividing each expression by $\frac{1}{2}abc$ gives $\frac{\sin A}{a} = \frac{\sin B}{b} = \frac{\sin C}{c}$.

The sine rule is used to solve problems involving triangles, given:

- **two angles** and **one side**
- **two sides** and a **non-included angle**.

FINDING SIDE LENGTHS

Example 5 — Self Tutor

Find the length of [AC] correct to two decimal places.

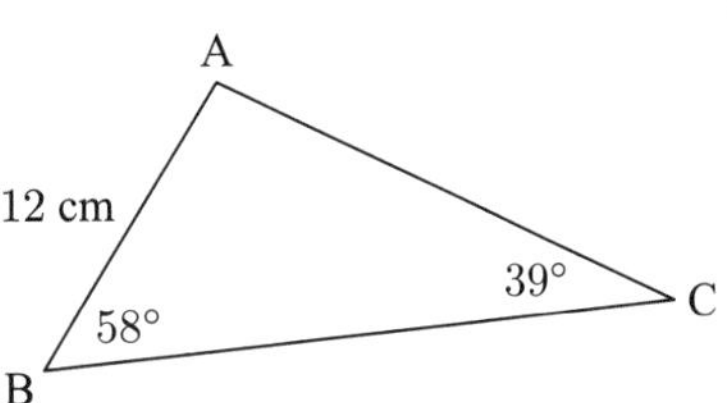

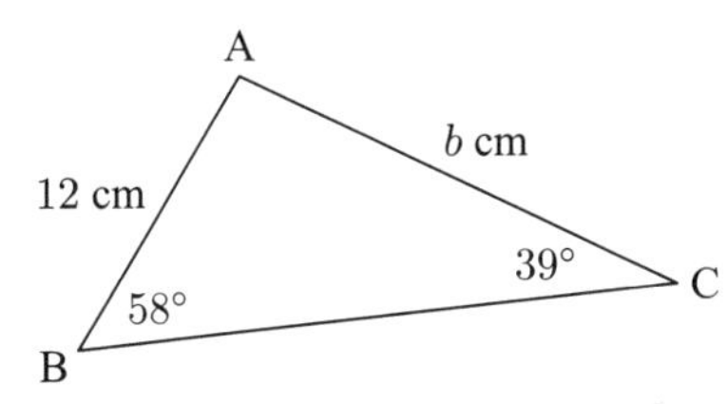

Using the sine rule, $\frac{b}{\sin 58^\circ} = \frac{12}{\sin 39^\circ}$

$\therefore\ b = \frac{12 \times \sin 58^\circ}{\sin 39^\circ}$

$\therefore\ b \approx 16.170\,74$

$\therefore$ [AC] is about 16.17 cm long.

EXERCISE 11C.1

1 Find the value of x:

a 23 cm, x cm, 37°, 48°

b

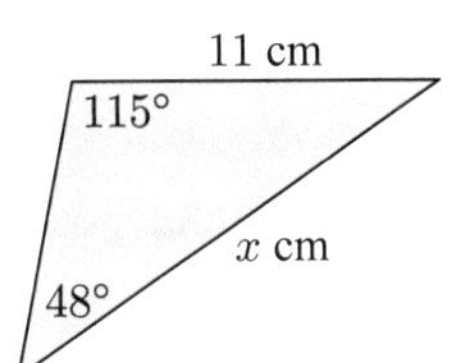

c

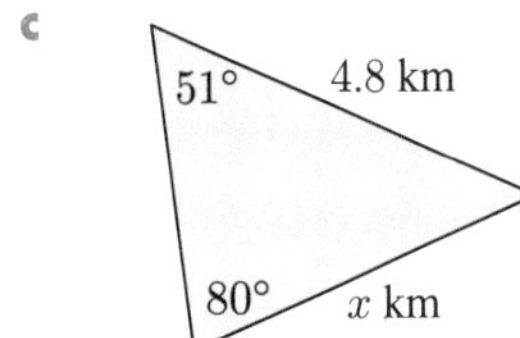

2 Consider triangle ABC.

a Given $A = 63°$, $B = 49°$, and $b = 18$ cm, find a.

b Given $A = 82°$, $C = 25°$, and $c = 34$ cm, find b.

c Given $B = 21°$, $C = 48°$, and $a = 6.4$ cm, find c.

FINDING ANGLES

The problem of finding angles using the sine rule is more complicated because there may be two possible answers. For example, if $\sin\theta = \frac{\sqrt{3}}{2}$ then $\theta = 60°$ or $120°$. We call this situation an **ambiguous case**.

DEMO

You can click on the icon to obtain an interactive demonstration of the ambiguous case, or else you can work through the following investigation.

INVESTIGATION 2 THE AMBIGUOUS CASE

You will need a blank sheet of paper, a ruler, a protractor, and a compass for the tasks that follow. In each task you will be required to construct triangles from given information. You could also do this using a computer package such as 'The Geometer's Sketchpad'.

What to do:

Task 1: Draw AB = 10 cm. At A construct an angle of 30°. Using B as the centre, draw an arc of a circle of radius 6 cm. Let C denote the point where the arc intersects the ray from A. How many different possible points C are there, and therefore how many different triangles ABC may be constructed?

Task 2: As before, draw AB = 10 cm and construct a 30° angle at A. This time draw an arc of radius 5 cm centred at B. How many different triangles are possible?

Task 3: Repeat, but this time draw an arc of radius 3 cm centred at B. How many different triangles are possible?

Task 4: Repeat, but this time draw an arc of radius 12 cm centred at B. How many different triangles are possible now?

You should have discovered that when you are given two sides and a non-included angle there are a number of different possibilities. You could get two triangles, one triangle, or it may be impossible to draw any triangles at all for some given dimensions.

Now consider the calculations involved in each of the cases of the investigation.

Task 1: Given: $c = 10$ cm, $a = 6$ cm, $A = 30°$

$$\frac{\sin C}{c} = \frac{\sin A}{a}$$

$$\therefore \sin C = \frac{c \sin A}{a}$$

$$\therefore \sin C = \frac{10 \times \sin 30°}{6} \approx 0.8333$$

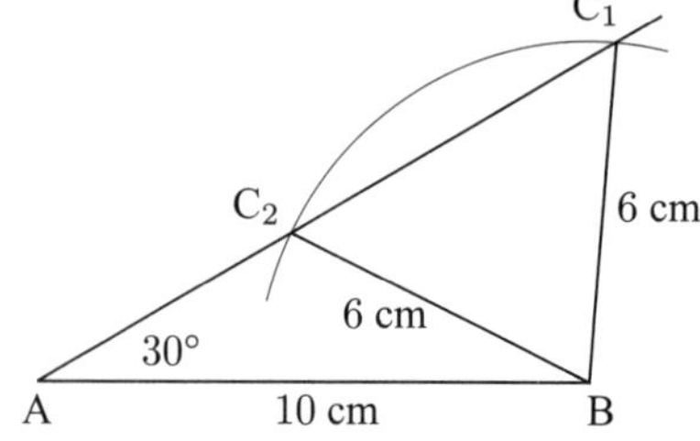

Because $\sin\theta = \sin(180° - \theta)$ there are two possible angles:
$C \approx 56.44°$ *or* $180° - 56.44° = 123.56°$

Task 2: Given: $c = 10$ cm, $a = 5$ cm, $A = 30°$

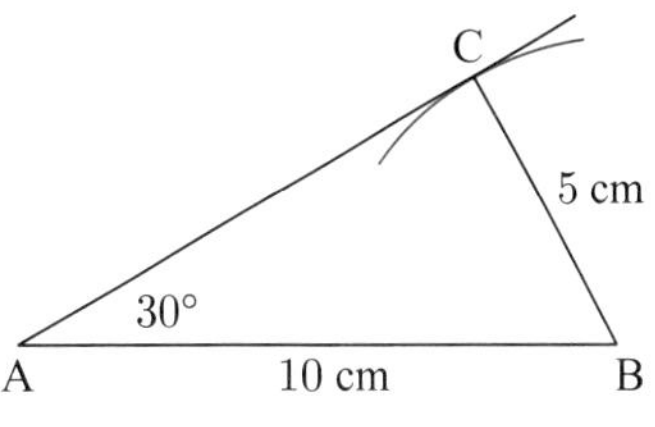

$$\frac{\sin C}{c} = \frac{\sin A}{a}$$

$$\therefore \quad \sin C = \frac{c \sin A}{a}$$

$$\therefore \quad \sin C = \frac{10 \times \sin 30°}{5} = 1$$

There is only one possible solution for C in the range from $0°$ to $180°$, and that is $C = 90°$. Only one triangle is therefore possible. Complete the solution of the triangle yourself.

Task 3: Given: $c = 10$ cm, $a = 3$ cm, $A = 30°$

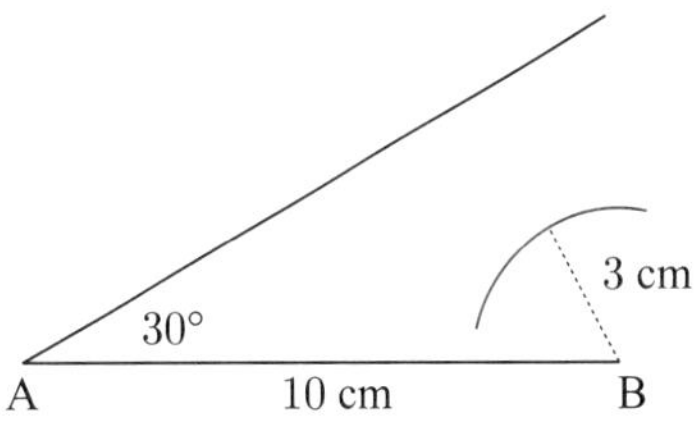

$$\frac{\sin C}{c} = \frac{\sin A}{a}$$

$$\therefore \quad \sin C = \frac{c \sin A}{a}$$

$$\therefore \quad \sin C = \frac{10 \times \sin 30°}{3} \approx 1.6667$$

There is no angle that has a sine value > 1. Therefore there is *no solution* for this given data, and no triangles can be drawn to match the information given.

Task 4: Given: $c = 10$ cm, $a = 12$ cm, $A = 30°$

$$\frac{\sin C}{c} = \frac{\sin A}{a}$$

$$\therefore \quad \sin C = \frac{c \sin A}{a}$$

$$\therefore \quad \sin C = \frac{10 \times \sin 30°}{12} \approx 0.4167$$

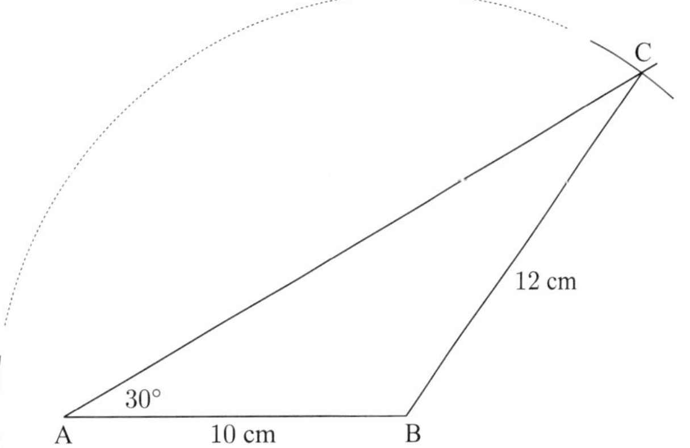

Two angles have a sine ratio of 0.4167:

$C \approx 24.62°$ *or*
$180° - 24.62° = 155.38°$

However, in this case only one of these two angles is valid. If $A = 30°$ then C cannot possibly equal $155.38°$ because $30° + 155.38° > 180°$.

Therefore, there is only one possible solution, $C \approx 24.62°$.

Conclusion: Each situation using the sine rule with two sides and a non-included angle must be examined very carefully.

Example 6 Self Tutor

Find the measure of angle C in triangle ABC if AC = 7 cm, AB = 11 cm, and angle B measures 25°.

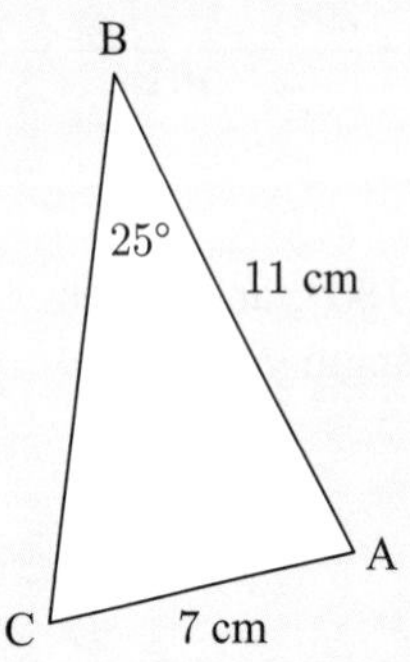

$$\frac{\sin C}{c} = \frac{\sin B}{b} \quad \{\text{sine rule}\}$$

$$\therefore \frac{\sin C}{11} = \frac{\sin 25°}{7}$$

$$\therefore \sin C = \frac{11 \times \sin 25°}{7}$$

$$\therefore C = \sin^{-1}\left(\frac{11 \times \sin 25°}{7}\right) \quad \text{or its supplement } \{\text{as } C \text{ may be obtuse}\}$$

$$\therefore C \approx 41.6° \text{ or } 180° - 41.6°$$

$$\therefore C \approx 41.6° \text{ or } 138.4°$$

$\therefore$ C measures 41.6° if angle C is acute, or 138.4° if angle C is obtuse.

In this case there is insufficient information to determine the actual shape of the triangle. There are two possible triangles.

Sometimes there is information in the question which enables us to **reject** one of the answers.

Example 7 Self Tutor

Find the measure of angle L in triangle KLM given that angle K measures 56°, LM = 16.8 m, and KM = 13.5 m.

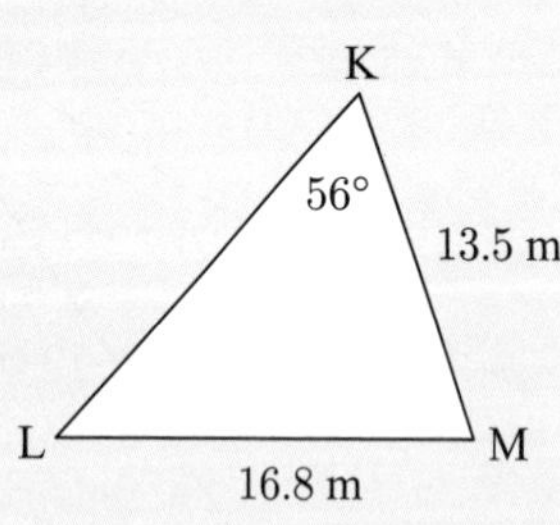

$$\frac{\sin L}{13.5} = \frac{\sin 56°}{16.8} \quad \{\text{by the sine rule}\}$$

$$\therefore \sin L = \frac{13.5 \times \sin 56°}{16.8}$$

$$\therefore L = \sin^{-1}\left(\frac{13.5 \times \sin 56°}{16.8}\right) \quad \text{or its supplement}$$

$$\therefore L \approx 41.8° \text{ or } 180° - 41.8°$$

$$\therefore L \approx 41.8° \text{ or } 138.2°$$

We reject $L \approx 138.2°$, since $138.2° + 56° > 180°$ which is impossible in a triangle.

$\therefore$ $L \approx 41.8°$, a unique solution in this case.

EXERCISE 11C.2

1 Triangle ABC has angle $B = 40°$, $b = 8$ cm, and $c = 11$ cm. Find the two possible measures of angle C.

2 Consider triangle ABC.

a Given $a = 14.6$ cm, $b = 17.4$ cm, and $A\widehat{B}C = 65°$, find the measure of $B\widehat{A}C$.

b Given $b = 43.8$ cm, $c = 31.4$ cm, and $A\widehat{C}B = 43°$, find the measure of $A\widehat{B}C$.

c Given $a = 6.5$ km, $c = 4.8$ km, and $B\widehat{A}C = 71°$, find the measure of $A\widehat{C}B$.

3 Is it possible to have a triangle with the measurements shown? Explain your answer.

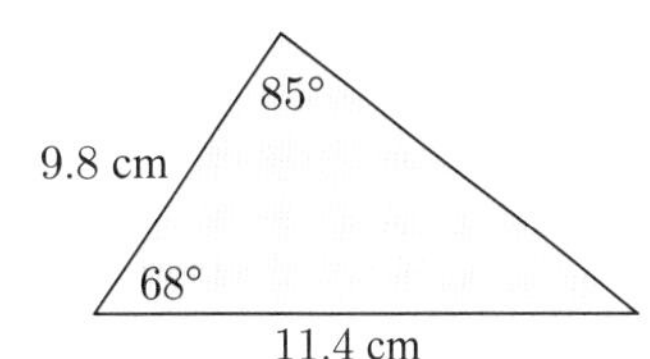

4 Given AD = 20 cm, find the magnitude of $\widehat{ABC}$ and hence the length BD.

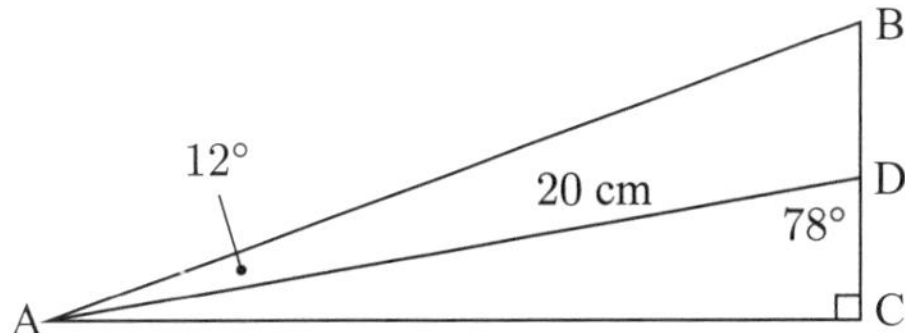

5 Find x and y in the given figure.

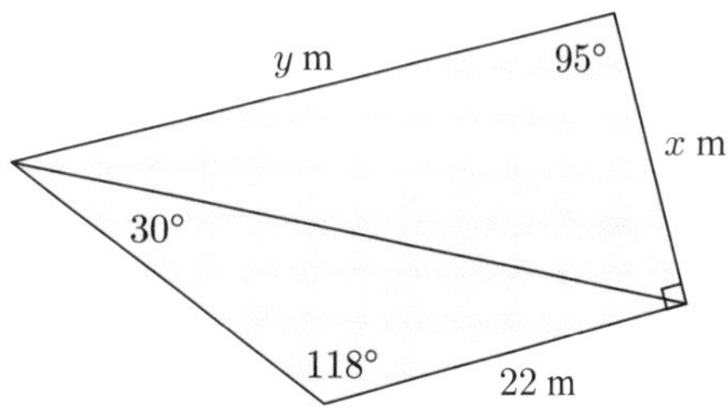

6 Triangle ABC has $\widehat{A} = 58°$, AB = 5.1 cm, AC = 8 cm, and BC = 6.84 cm.

a Find $\widehat{B}$ correct to the nearest degree using the sine rule.

b Find $\widehat{B}$ correct to the nearest degree using the cosine rule.

c Copy and complete: "When faced with using either the sine rule or the cosine rule it is better to use the ...... as it avoids ......"

7 In triangle ABC, $\widehat{ABC} = 30°$, AC = 9 cm, and AB = 7 cm. Find the area of the triangle.

8 In the parallelogram ABCD, AB = 67 mm, AD = 35 mm, and $\widehat{BAD} = 47°$. Calculate $\widehat{BAC}$.

9 Find the exact value of x, giving your answer in the form $a + b\sqrt{2}$ where $a, b \in \mathbb{Q}$.

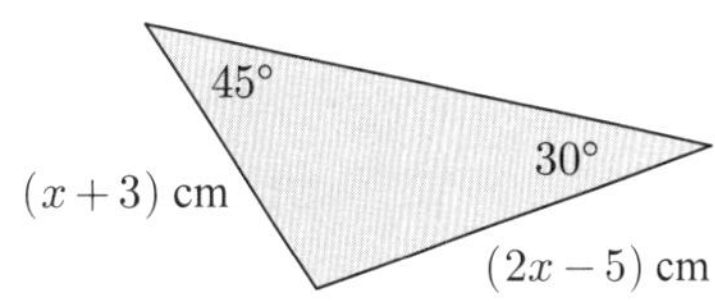

D USING THE SINE AND COSINE RULES

If we are given a problem involving a triangle, we must first decide which rule is best to use.

If the triangle is right angled then the trigonometric ratios or Pythagoras' Theorem can be used. For some problems we can add an extra line or two to the diagram to create a right angled triangle.

However, if we do not have a right angled triangle then we usually have to choose between the sine and cosine rules. In these cases the following checklist may be helpful:

Use the **cosine rule** when given:

- three sides
- two sides and an included angle.

Use the **sine rule** when given:

- one side and two angles
- two sides and a non-included angle, but beware of an *ambiguous case* which can occur when the smaller of the two given sides is opposite the given angle.

Example 8 — Self Tutor

The angles of elevation to the top of a mountain are measured from two beacons A and B at sea.

The measurements are shown on the diagram.

If the beacons are 1473 m apart, how high is the mountain?

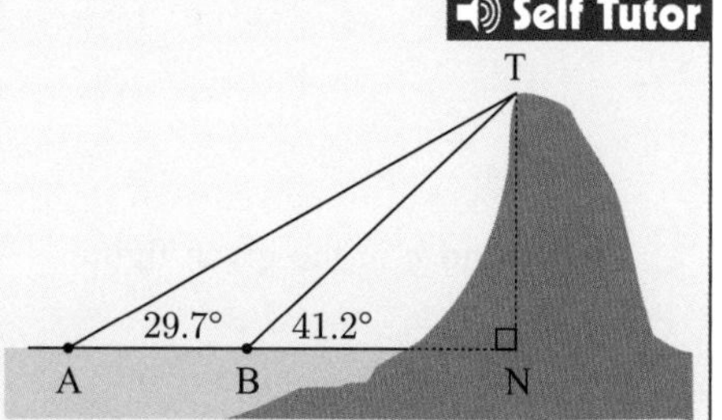

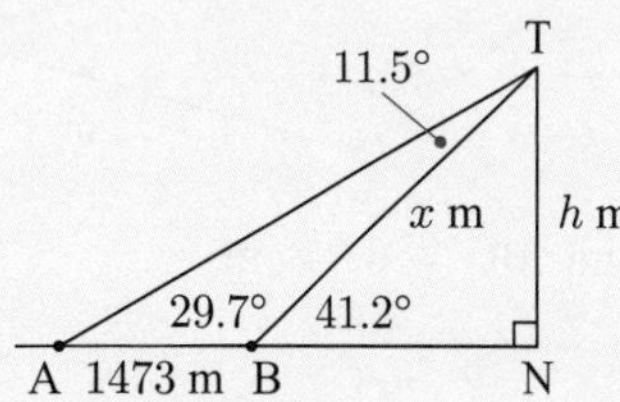

Let BT be x m and NT be h m.

$\widehat{ATB} = 41.2° - 29.7°$ {exterior angle of $\triangle$}

$= 11.5°$

We find x in $\triangle$ABT using the sine rule:

$$\frac{x}{\sin 29.7°} = \frac{1473}{\sin 11.5°}$$

$$\therefore\ x = \frac{1473}{\sin 11.5°} \times \sin 29.7°$$

$$\approx 3660.62$$

Now, in $\triangle$BNT, $\sin 41.2° = \dfrac{h}{x} \approx \dfrac{h}{3660.62}$

$$\therefore\ h \approx \sin 41.2° \times 3660.62$$

$$\approx 2410$$

The mountain is about 2410 m high.

EXERCISE 11D

1 Rodrigo wishes to determine the height of a flagpole. He takes a sighting to the top of the flagpole from point P. He then moves further away from the flagpole by 20 metres to point Q and takes a second sighting. The information is shown in the diagram alongside. How high is the flagpole?

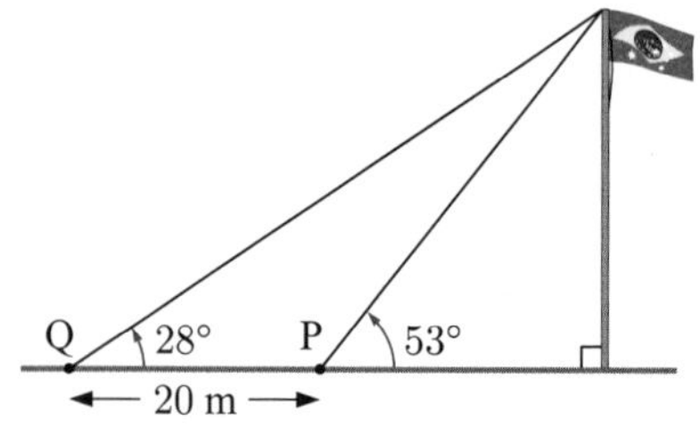

2 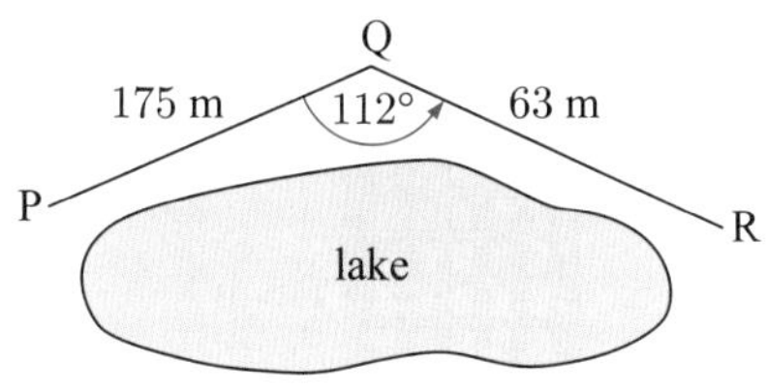

To get from P to R, a park ranger had to walk along a path to Q and then to R as shown.
What is the distance in a straight line from P to R?

3 A golfer played his tee shot a distance of 220 m to point A. He then played a 165 m six iron to the green. If the distance from tee to green is 340 m, determine the angle the golfer was off line with his tee shot.

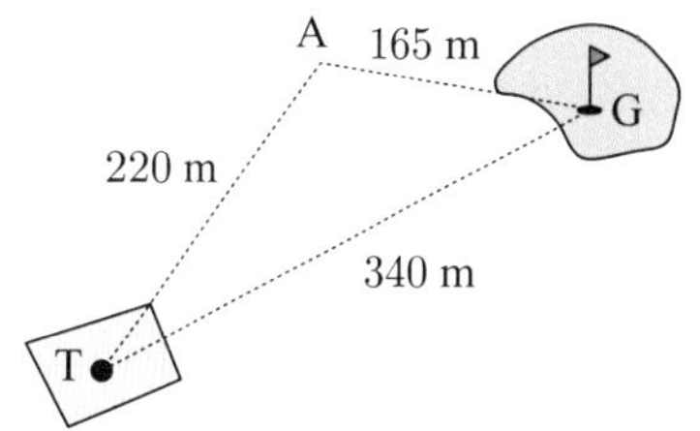

4 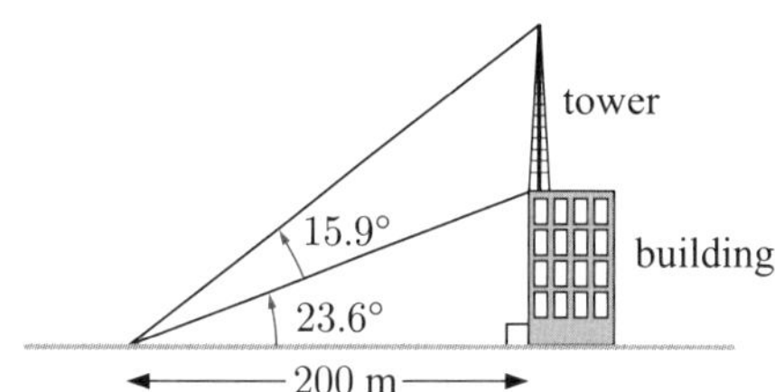

A communications tower is constructed on top of a building as shown. Find the height of the tower.

5 Hikers Ritva and Esko leave point P at the same time. Ritva walks 4 km on a bearing of 040°, then a further 6 km on a bearing of 155°.
Esko hikes directly from P to the camp site.

a How far does Esko hike?

b In which direction does Esko hike?

c Ritva hikes at 10 km h^{-1} and Esko hikes at 6 km h^{-1}.

- **i** Who will arrive at the camp site first?
- **ii** How long will this person need to wait before the other person arrives?

d On what bearing should the hikers walk from the camp site to return to P?

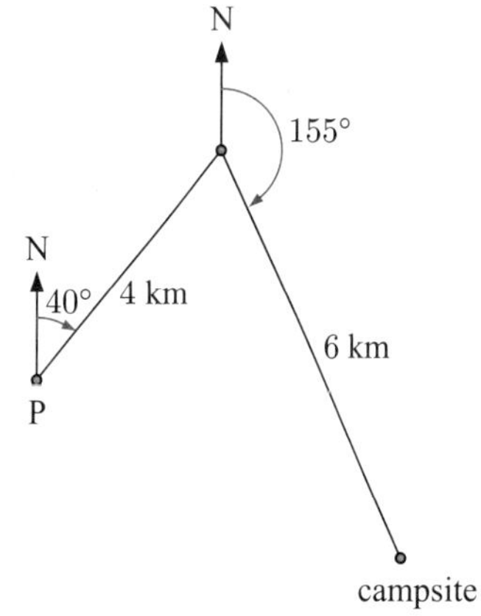

6 A football goal is 5 metres wide. When a player is 26 metres from one goal post and 23 metres from the other, he shoots for goal. What is the angle of view of the goals that the player sees?

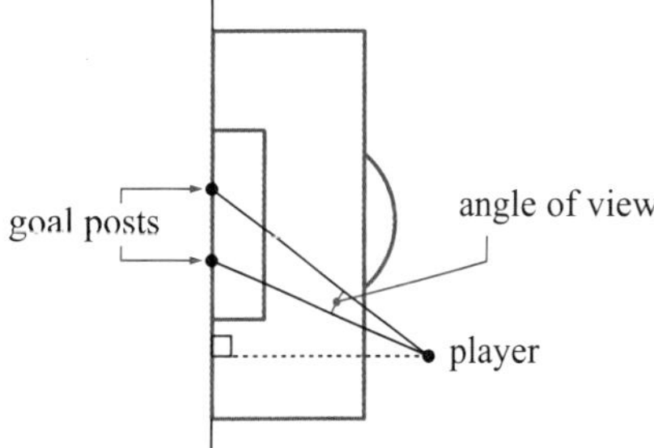

7 A tower 42 metres high stands on top of a hill. From a point some distance from the base of the hill, the angle of elevation to the top of the tower is 13.2° and the angle of elevation to the bottom of the tower is 8.3°. Find the height of the hill.

8 From the foot of a building I have to look 22° upwards to sight the top of a tree. From the top of the building, 150 metres above ground level, I have to look down at an angle of 50° below the horizontal to sight the tree top.

a How high is the tree?

b How far from the building is this tree?

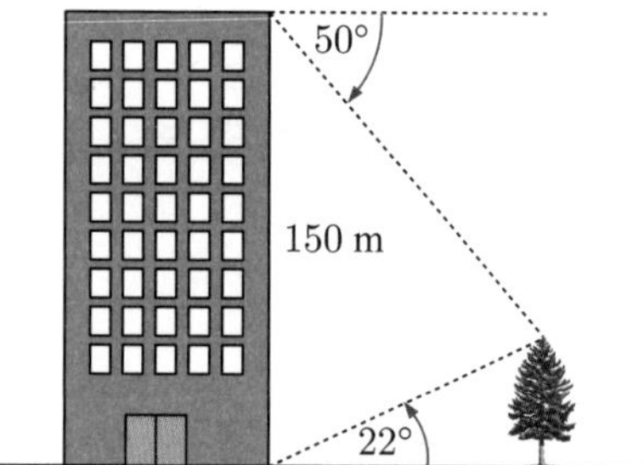

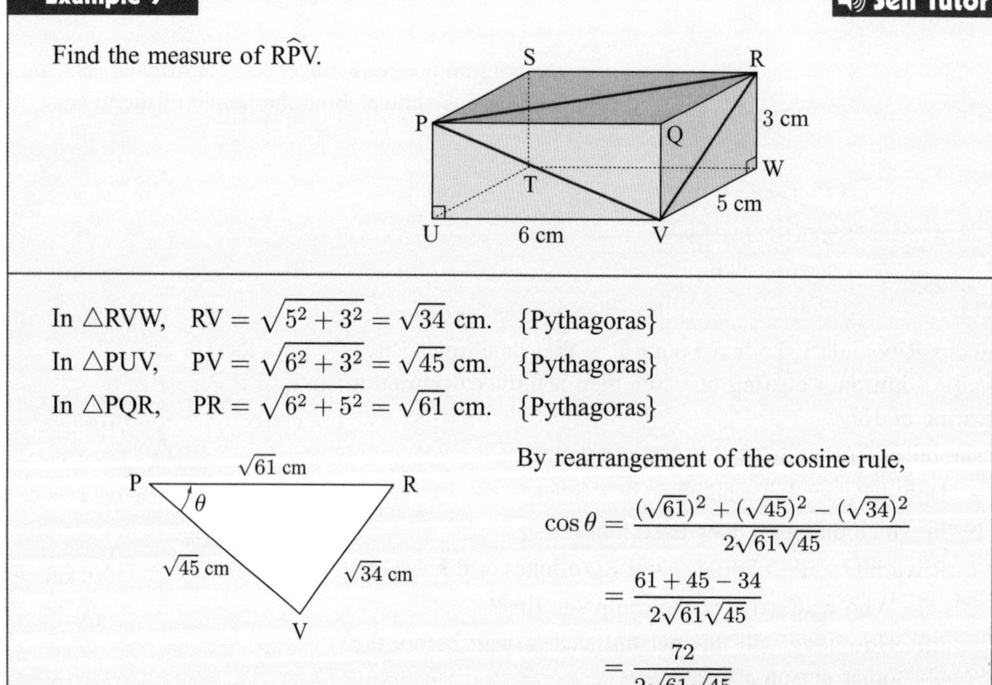

Example 9 **Self Tutor**

Find the measure of $\widehat{RPV}$.

In $\triangle$RVW, $RV = \sqrt{5^2 + 3^2} = \sqrt{34}$ cm. {Pythagoras}

In $\triangle$PUV, $PV = \sqrt{6^2 + 3^2} = \sqrt{45}$ cm. {Pythagoras}

In $\triangle$PQR, $PR = \sqrt{6^2 + 5^2} = \sqrt{61}$ cm. {Pythagoras}

By rearrangement of the cosine rule,

$$\cos\theta = \frac{(\sqrt{61})^2 + (\sqrt{45})^2 - (\sqrt{34})^2}{2\sqrt{61}\sqrt{45}}$$

$$= \frac{61 + 45 - 34}{2\sqrt{61}\sqrt{45}}$$

$$= \frac{72}{2\sqrt{61}\sqrt{45}}$$

$$\therefore\ \theta = \cos^{-1}\left(\frac{36}{\sqrt{61}\sqrt{45}}\right) \approx 46.6^\circ$$

$\therefore$ $\widehat{RPV}$ measures about 46.6°.

9 Find the measure of $\widehat{PQR}$ in the rectangular box shown.

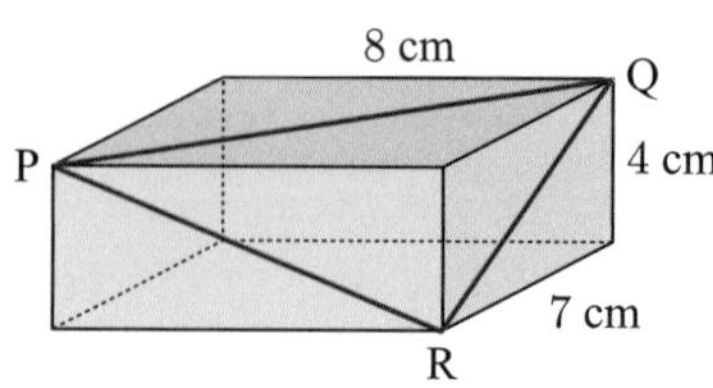

10 Two observation posts A and B are 12 km apart. A third observation post C is located such that $\widehat{CAB}$ is 42° and $\widehat{CBA}$ is 67°. Find the distance of C from A and from B.

11 Stan and Olga are considering buying a sheep farm. A surveyor has supplied them with the given accurate sketch. Find the area of the property, giving your answer in:

a km^2 **b** hectares.

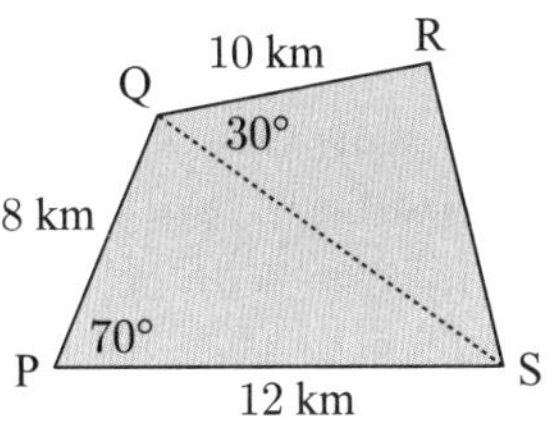

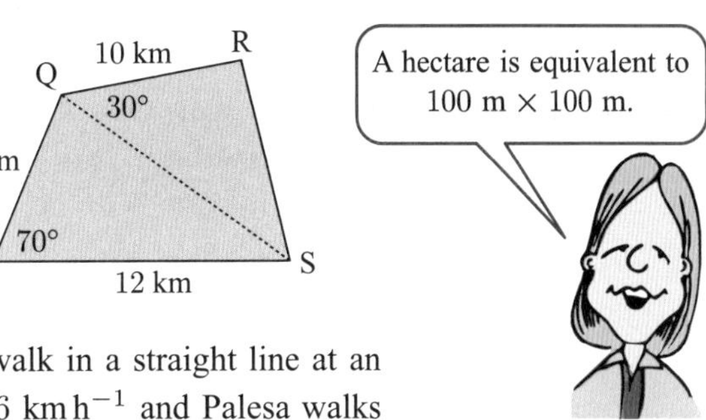

12 Thabo and Palesa start at point A. They each walk in a straight line at an angle of 120° to one another. Thabo walks at 6 km h^{-1} and Palesa walks at 8 km h^{-1}. How far apart are they after 45 minutes?

13 The cross-section design of the kerbing for a driverless-bus roadway is shown opposite. The metal strip is inlaid into the concrete and is used to control the direction and speed of the bus. Find the width of the metal strip.

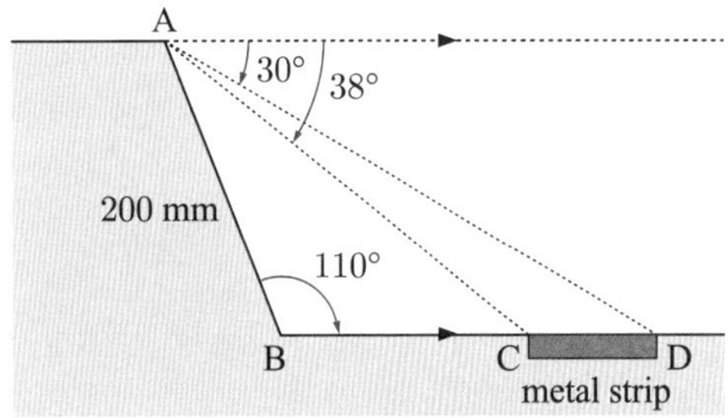

14 An orienteer runs for $4\frac{1}{2}$ km, then turns through an angle of 32° and runs for another 6 km. How far is she from her starting point?

15 Sam and Markus are standing on level ground 100 metres apart. A large tree is due north of Markus and on a bearing of 065° from Sam. The top of the tree appears at an angle of elevation of 25° to Sam and 15° to Markus. Find the height of the tree.

16 A helicopter A observes two ships B and C. B is 23.8 km from the helicopter and C is 31.9 km from it. The angle of view $B\widehat{A}C$ from the helicopter to B and C, is 83.6°. How far are the ships apart?

REVIEW SET 11A NON-CALCULATOR

1 Determine the area of the triangle.

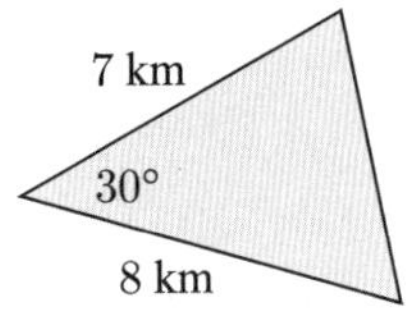

2 You are given enough details of a triangle so that you could use either the cosine rule or the sine rule to find an unknown. Which rule should you use? Explain your answer.

3 Kady was asked to draw the illustrated triangle exactly.

a Use the cosine rule to find x.

b What should Kady's response be?

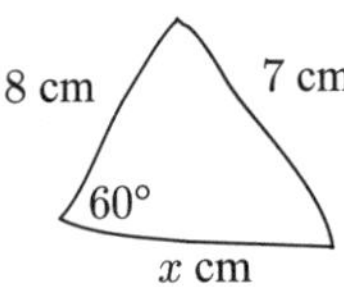

4 A triangle has sides of length 7 cm and 13 cm, and its area is 42 cm^2. Find the sine of the included angle.

5 Consider the kite ABCD alongside:

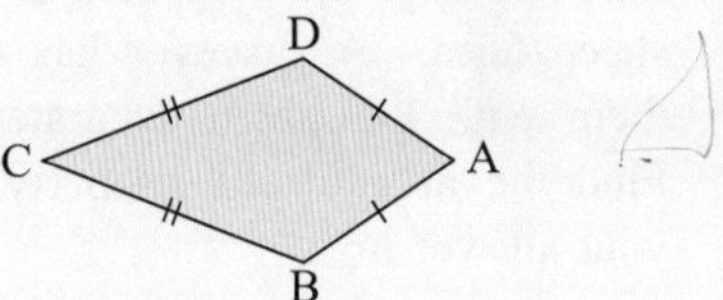

a Use the cosine rule to show that $\widehat{ADC} = \widehat{ABC}$.

b Use the sine rule to show that $\widehat{DAC} = \widehat{BAC}$.

6

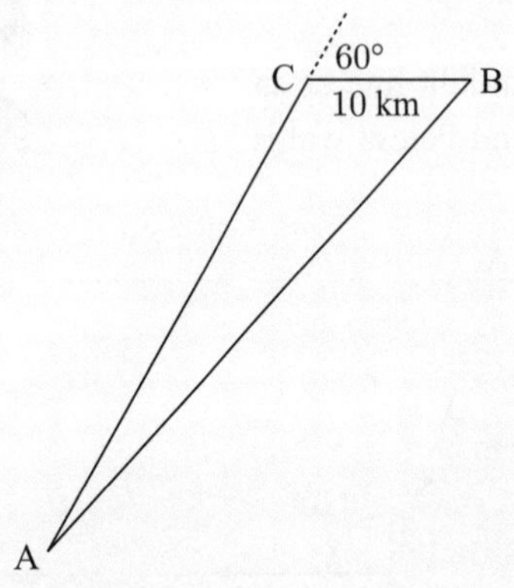

A boat is meant to be sailing directly from A to B. However, it travels in a straight line to C before the captain realises he is off course. The boat is turned through an angle of 60°, then travels another 10 km to B. The trip would have been 4 km shorter if the boat had gone straight from A to B. How far did the boat travel?

7 Show that the yellow shaded area is given by $A = \frac{49}{2}\left(\frac{13\pi}{18} - \sin(\frac{13\pi}{18})\right)$.

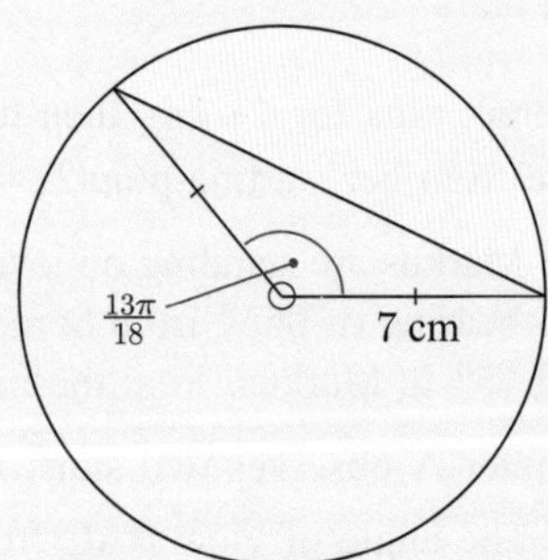

8 In triangle ABC, AB = 5 m, AC = d m, BC = x m, and $\widehat{ABC}$ measures 20°.

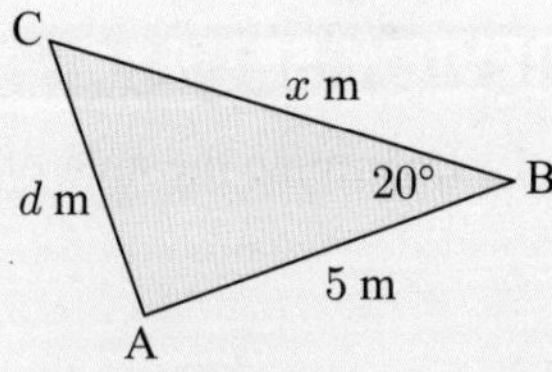

a Find an expression for d^2 in terms of x.

b For $d > 0$, d is minimised when d^2 is minimised. Use this fact to find the exact value of x which minimises d.

c Hence, show that d is minimised when $\widehat{BCA}$ is a right angle.

9 **a** For the quadratic function $y = -x^2 + 12x - 20$, find the maximum or minimum value and the corresponding value of x.

b In triangle ABC, AB = y, BC = x, AC = 8, and the perimeter of the triangle is 20.

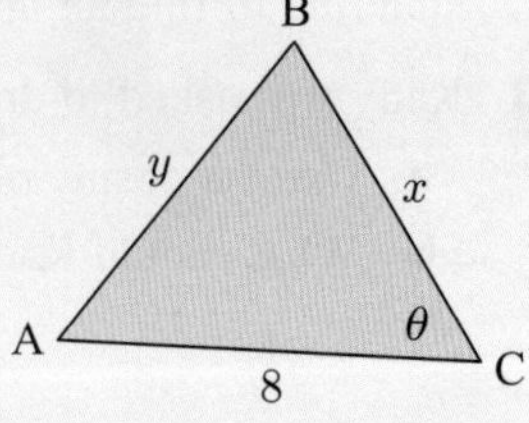

i Write y in terms of x.

ii Use the cosine rule to write y^2 in terms of x and $\cos\theta$.

iii Hence show that $\cos\theta = \dfrac{3x - 10}{2x}$.

c If the area of the triangle is A, show that $A^2 = 20(-x^2 + 12x - 20)$.

d Hence, find the maximum area of the triangle, and the triangle's shape when this occurs.

REVIEW SET 11B — CALCULATOR

1 Determine the value of x:

a

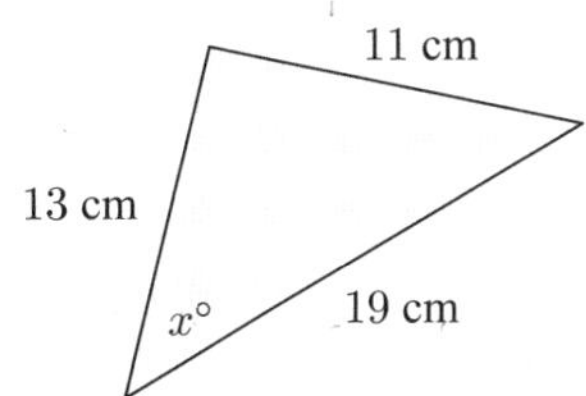

b

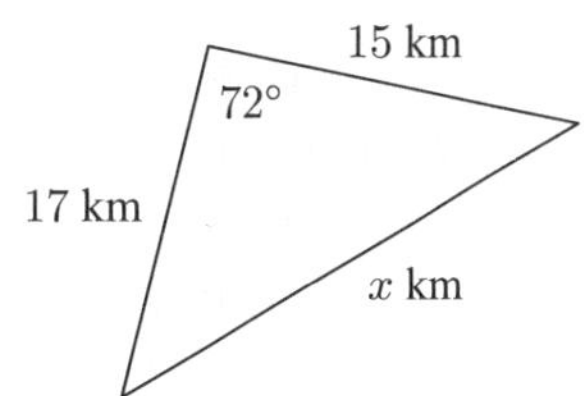

2 Find the unknown side and angles:

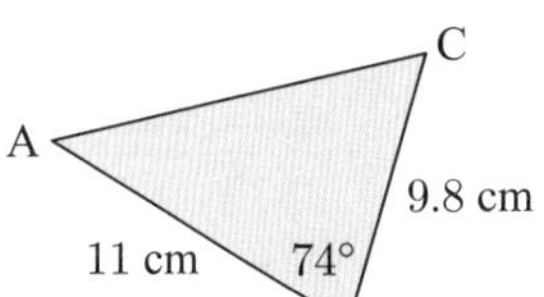

3 Find the area of quadrilateral ABCD:

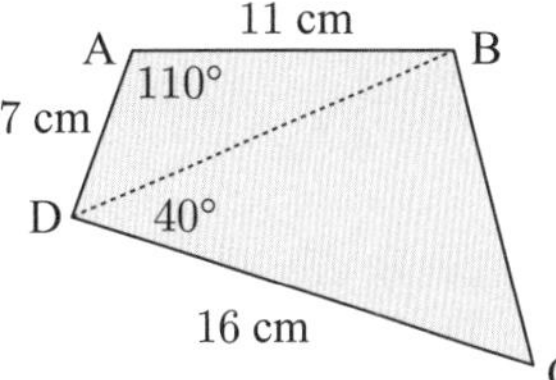

4 A vertical tree is growing on the side of a hill with gradient 10° to the horizontal. From a point 50 m downhill from the tree, the angle of elevation to the top of the tree is 18°. Find the height of the tree.

5

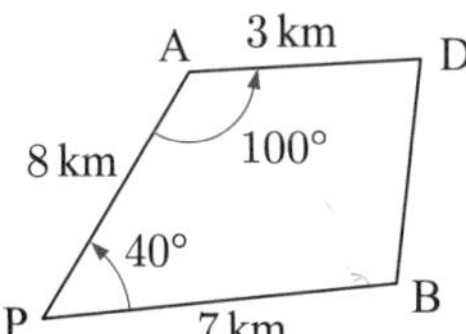

Hikers Andrew and Brett take separate trails from their starting point P to get to their destination D. They walk at an angle of 40° apart from each other as shown, and stop to eat their lunch at positions A and B respectively. How far does Brett still have to walk to reach the destination?

6 At 1 pm, runners A and B start out from the same point. Runner A runs at 14 km h^{-1} on the bearing 025°. Runner B runs at 12 km h^{-1} on the bearing 097°.

a At what time will A and B be 20 km apart?

b Find the bearing of B from A at this time.

7 From point A, the angle of elevation to the top of a tall building is 20°. On walking 80 m towards the building, the angle of elevation is now 23°. How tall is the building?

8 Peter, Sue, and Alix are sea-kayaking. Peter is 430 m from Sue on a bearing of 113°. Alix is on a bearing of 210° and is 310 m from Sue. Find the distance and bearing of Peter from Alix.

REVIEW SET 11C

1 Find the value of x:

a

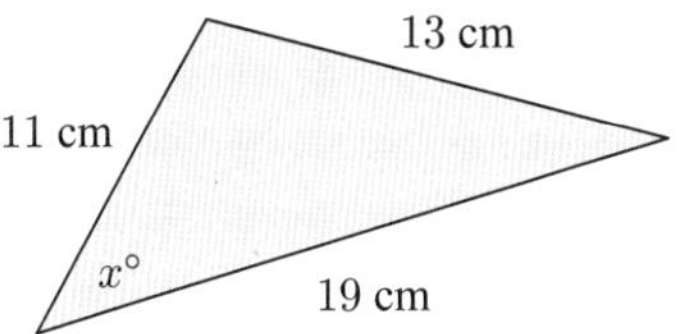

b

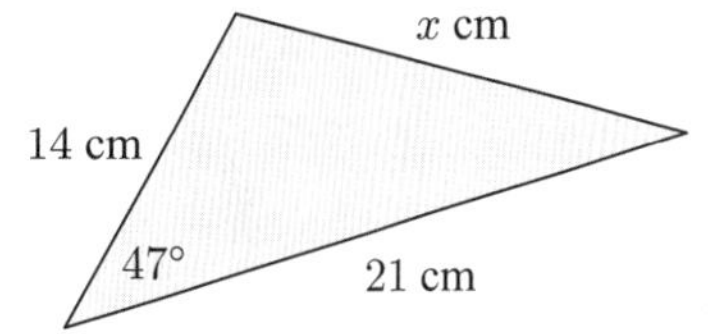

2 The triangle ABC has area 80 cm^2.

a Find the value of x.

b Hence find the length AC.

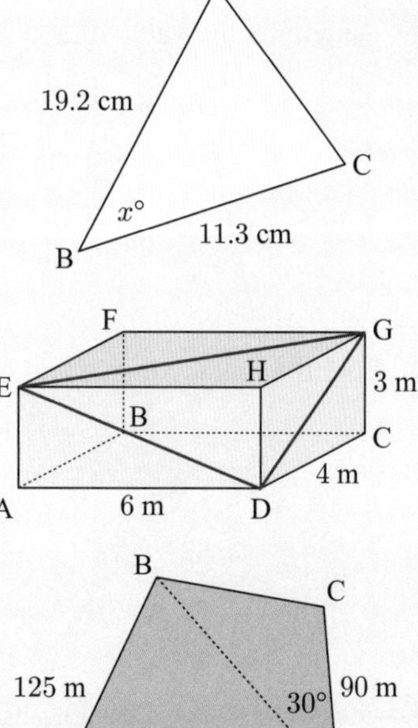

3 Find the measure of $\text{E}\widehat{\text{D}}\text{G}$:

4 Anke and Lucas are considering buying a block of land. The land agent supplies them with the given accurate sketch. Find the area of the property, giving your answer in:

a m^2 **b** hectares.

5 A family in Germany drives at 140 km h^{-1} for 45 minutes on the bearing 032°, and then 180 km h^{-1} for 40 minutes on the bearing 317°. Find the distance and bearing of the car from its starting point.

6 Soil contractor Frank was given the following dimensions over the telephone:

The triangular garden plot ABC has $\text{C}\widehat{\text{A}}\text{B}$ measuring 44°, [AC] is 8 m long, and [BC] is 6 m long. Soil to a depth of 10 cm is required.

a Explain why Frank needs extra information from his client.

b What is the maximum volume of soil needed if his client is unable to supply the necessary information?

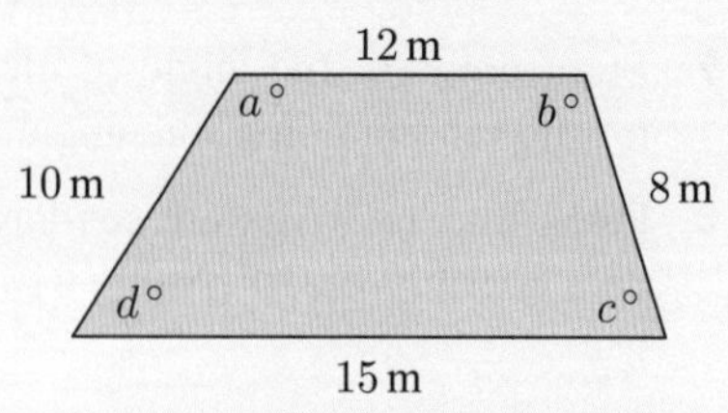

7 **a** Explain why $\cos(180° - \theta) = -\cos\theta$.

b In the given quadrilateral, $b + d = 180$.

i Show that $300\cos d° - 192\cos b° = 117$.

ii Find the values of b and d.

iii Hence, find the values of a and c.

8 Quadrilateral PQRS has been divided into two triangles by the diagonal [QS]. PQ = 3, QR = 7, PS = 6, $\text{Q}\widehat{\text{P}}\text{S} = \phi$, and $\text{Q}\widehat{\text{R}}\text{S} = 32°$.

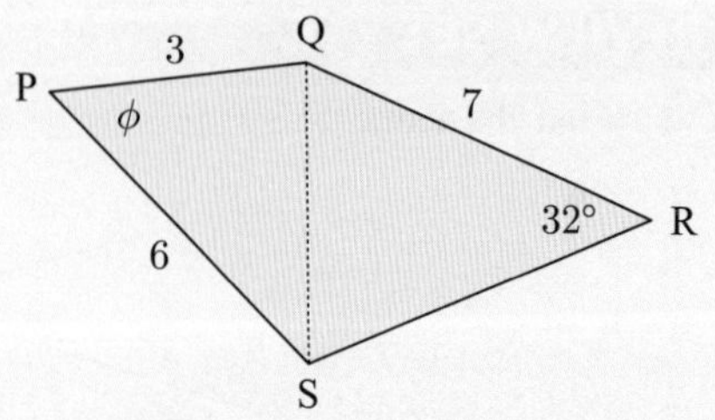

a Find QS in terms of $\cos\phi$.

b Suppose $\phi = 50°$ and $\text{R}\widehat{\text{S}}\text{Q}$ is acute.

i Find the measure of $\text{R}\widehat{\text{S}}\text{Q}$.

ii Find the perimeter of the quadrilateral.

iii Find the area of the quadrilateral.

Chapter 12

Trigonometric functions

Syllabus reference: 3.2, 3.4, 3.5

Contents:

OPENING PROBLEM

A Ferris wheel rotates at a constant speed. The wheel's radius is 10 m and the bottom of the wheel is 2 m above ground level. From a point in front of the wheel, Andrew is watching a green light on the perimeter of the wheel. Andrew notices that the green light moves in a circle. He estimates how high the light is above ground level at two second intervals and draws a scatter diagram of his results.

Things to think about:

- **a** What will Andrew's scatter diagram look like?
- **b** What function can be used to model the data?
- **c** How could this function be used to find:
  - **i** the light's position at any point in time
  - **ii** the times when the light is at its maximum and minimum heights?
- **d** What part of the function indicates the time for one full revolution of the wheel?

Click on the icon to visit a simulation of the Ferris wheel. You will be able to view the light from:

DEMO

- in front of the wheel
- a side-on position
- above the wheel.

You can then observe the graph of the green light's position as the wheel rotates at a constant rate.

A PERIODIC BEHAVIOUR

Periodic phenomena occur all the time in the physical world. For example, in:

- seasonal variations in our climate
- variations in average maximum and minimum monthly temperatures
- the number of daylight hours at a particular location
- tidal variations in the depth of water in a harbour
- the phases of the moon
- animal populations.

These phenomena illustrate variable behaviour which is repeated over time. The repetition may be called **periodic**, **oscillatory**, or **cyclic** in different situations.

In this chapter we will see how trigonometric functions can be used to model periodic phenomena.

OBSERVING PERIODIC BEHAVIOUR

The table below shows the mean monthly maximum temperature for Cape Town, South Africa.

Month	Jan	Feb	Mar	Apr	May	Jun	Jul	Aug	Sep	Oct	Nov	Dec
Temp. T (°C)	28	27	$25\frac{1}{2}$	22	$18\frac{1}{2}$	16	15	16	18	$21\frac{1}{2}$	24	26

On the scatter diagram alongside we plot the temperature T on the vertical axis. We assign January as $t = 1$ month, February as $t = 2$ months, and so on for the 12 months of the year.

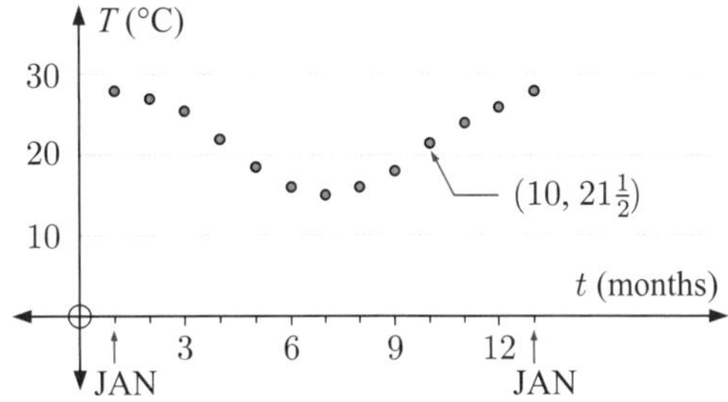

The temperature shows a variation from an average of 28°C in January through a range of values across the months. The cycle will approximately repeat itself for each subsequent 12 month period. By the end of the chapter we will be able to establish a **periodic function** which approximately fits this set of points.

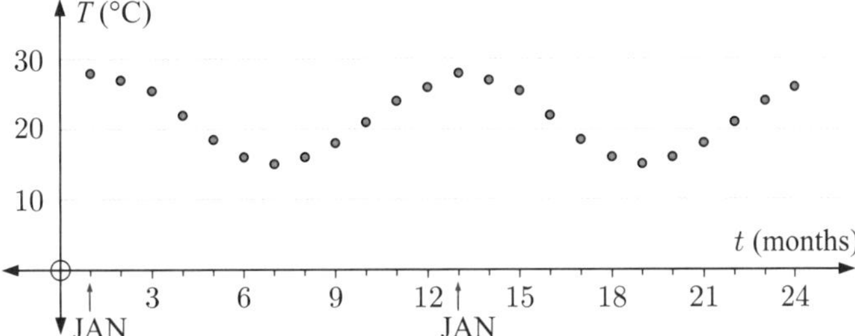

HISTORICAL NOTE

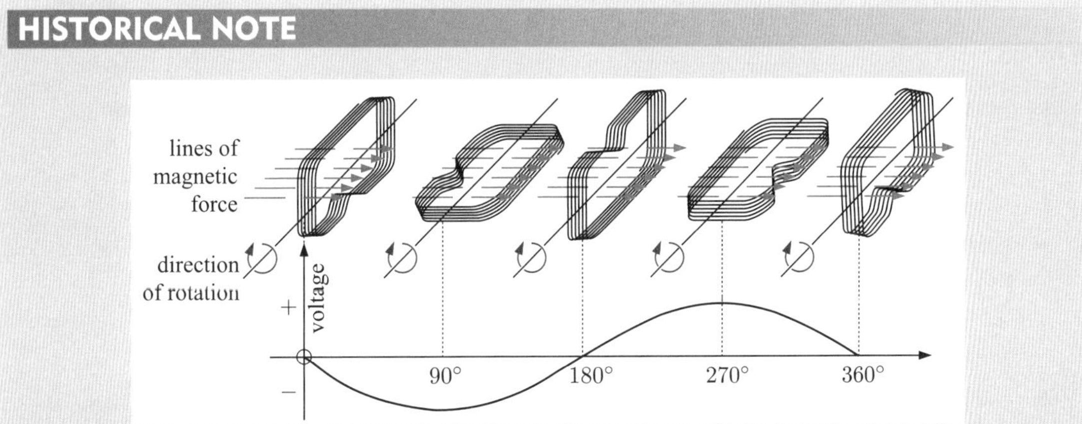

In 1831 **Michael Faraday** discovered that an electric current was generated by rotating a coil of wire in a magnetic field. The electric current produced showed a voltage which varied between positive and negative values as the coil rotated through 360°.

Graphs with this basic shape, where the cycle is repeated over and over, are called **sine waves**.

GATHERING PERIODIC DATA

Data on a number of periodic phenomena can be found online or in other publications. For example:

- Maximum and minimum monthly temperatures can be found at www.weatherbase.com
- Tidal details can be obtained from daily newspapers or internet sites such as http://tidesandcurrents.noaa.gov or http://www.bom.gov.au/oceanography

TERMINOLOGY USED TO DESCRIBE PERIODICITY

A **periodic function** is one which repeats itself over and over in a horizontal direction, in intervals of the same length.

The **period** of a periodic function is the length of one repetition or cycle.

$f(x)$ is a periodic function with period $p \Leftrightarrow \boldsymbol{f(x+p)=f(x)}$ for all x, and p is the smallest positive value for this to be true.

A **cycloid** is an example of a periodic function. It is the curve traced out by a point on a circle as the circle rolls across a flat surface in a straight line.

DEMO

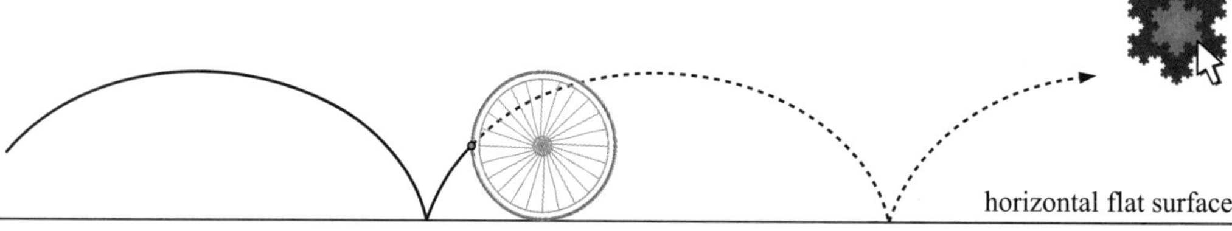

Use a **graphing package** to examine the function $f(x) = x - [x]$ where $[x]$ is "the largest integer less than or equal to x".
Is $f(x)$ periodic? What is its period?

GRAPHING PACKAGE

WAVES

In this course we are mainly concerned with periodic phenomena which show a wave pattern:

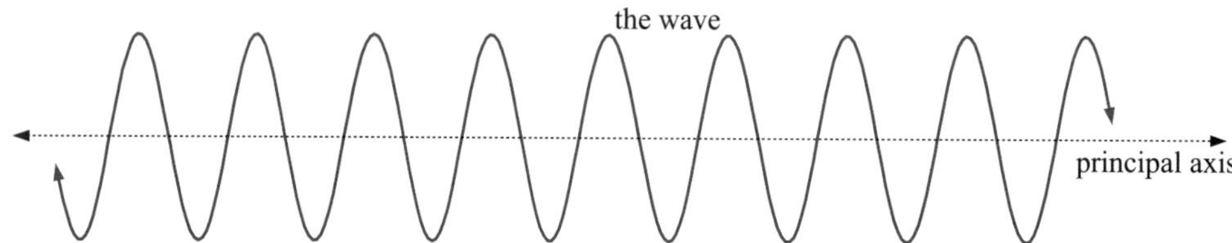

The wave oscillates about a horizontal line called the **principal axis** or **mean line** which has equation $y = \dfrac{\textbf{max} + \textbf{min}}{\textbf{2}}$.

A **maximum point** occurs at the top of a crest, and a **minimum point** at the bottom of a trough.

The **amplitude** is the distance between a maximum (or minimum) point and the principal axis.

$$\textbf{amplitude} = \frac{\textbf{max} - \textbf{min}}{\textbf{2}}$$

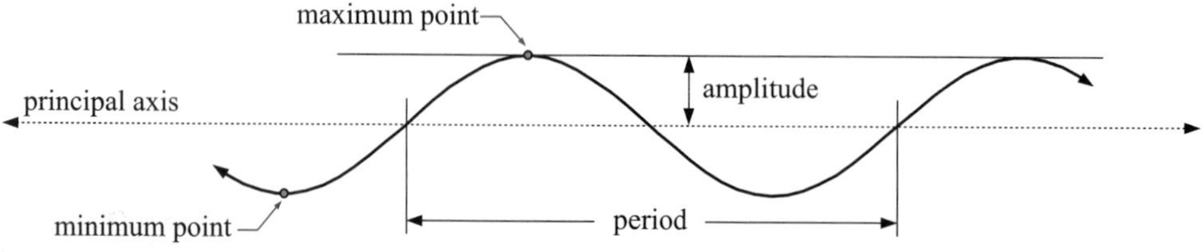

EXERCISE 12A

1 Which of these graphs show periodic behaviour?

a
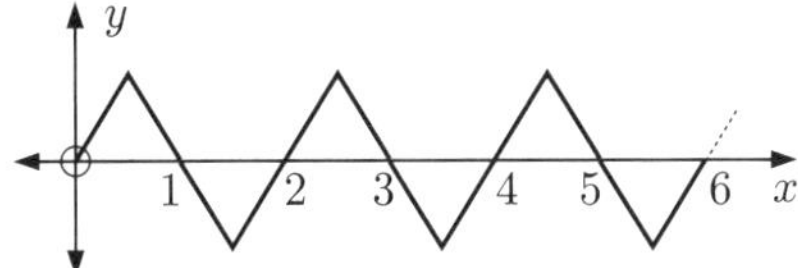

b
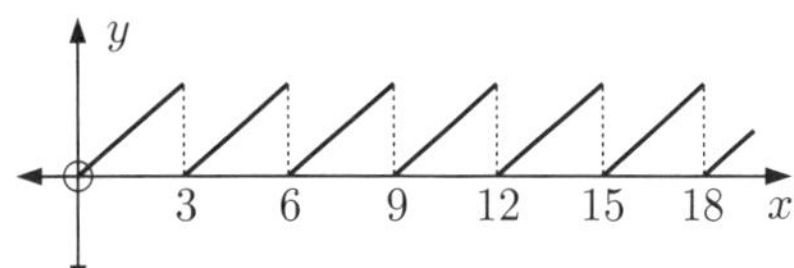

c
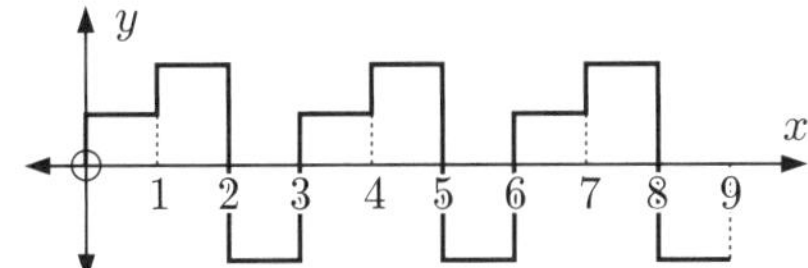

d
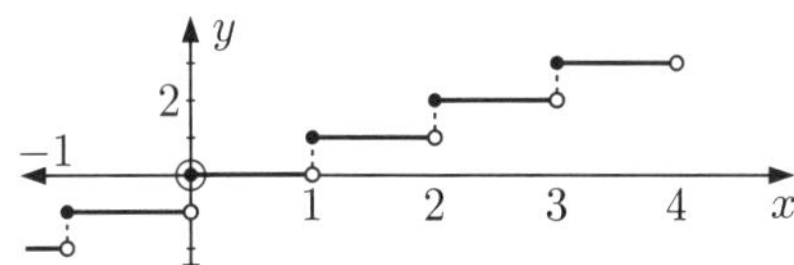

e
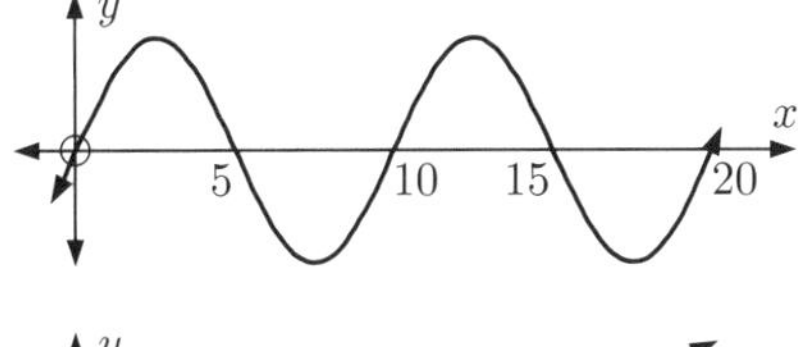

f
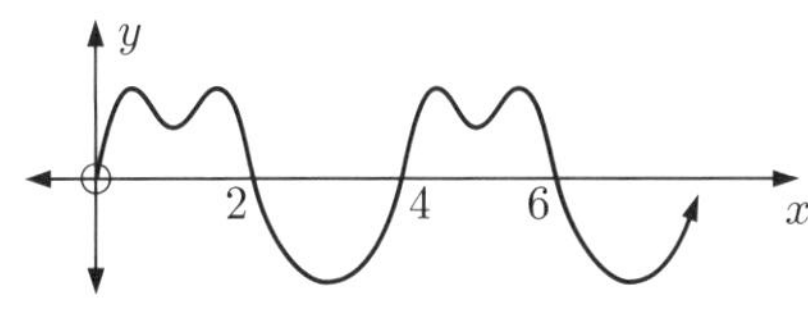

g
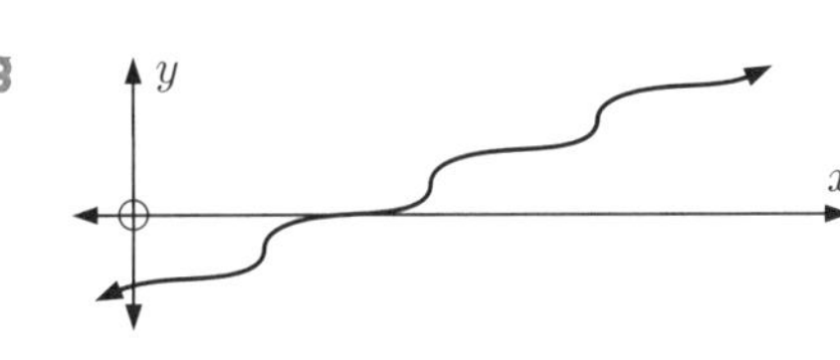

h
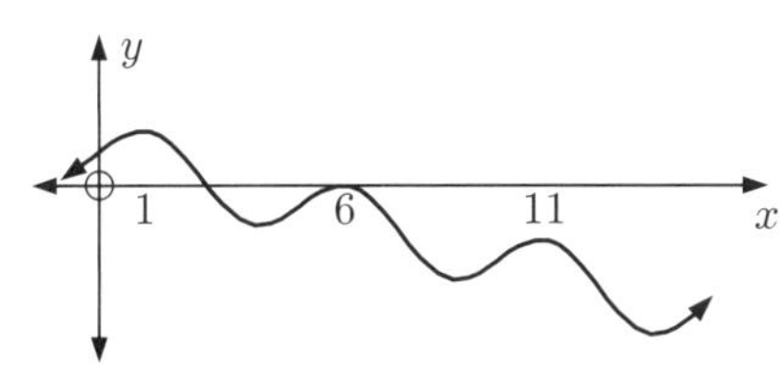

2 The following tabled values show the height above the ground of a point on a bicycle wheel as the bicycle is wheeled along a flat surface.

Distance travelled (cm)	0	20	40	60	80	100	120	140	160	180	200
Height above ground (cm)	0	6	23	42	57	64	59	43	23	7	1

Distance travelled (cm)	220	240	260	280	300	320	340	360	380	400
Height above ground (cm)	5	27	40	55	63	60	44	24	9	3

a Plot the graph of height against distance.

b Is it reasonable to fit a curve to this data, or should we leave it as discrete points?

c Is the data periodic? If so, estimate:

i the equation of the principal axis

ii the maximum value

iii the period

iv the amplitude.

3 Draw a scatter diagram for each set of data below. Is there any evidence to suggest the data is periodic?

a

x	0	1	2	3	4	5	6	7	8	9	10	11	12
y	0	1	1.4	1	0	-1	-1.4	-1	0	1	1.4	1	0

b

x	0	2	3	4	5	6	7	8	9	10	12
y	0	4.7	3.4	1.7	2.1	5.2	8.9	10.9	10.2	8.4	10.4

THEORY OF KNOWLEDGE

In mathematics we clearly define terms so there is no misunderstanding of their exact meaning.

We can understand the need for specific definitions by considering integers and rational numbers:

- 2 is an integer, and is also a rational number since $2 = \frac{4}{2}$.
- $\frac{4}{2}$ is a rational number, and is also an integer since $\frac{4}{2} = 2$.
- $\frac{4}{3}$ is a rational number, but is *not* an integer.

Symbols are frequently used in mathematics to take the place of phrases. For example:

- $=$ is read as "is equal to"
- $\sum$ is read as "the sum of all"
- $\in$ is read as "is an element of" or "is in".

1 Is mathematics a language?

2 Why is it important that mathematicians use the same notation?

3 Does a mathematical argument need to read like a good piece of English?

The word *similar* is used in mathematics to describe two figures which are in proportion. This is different from how *similar* is used in everyday speech.

Likewise the words *function*, *domain*, *range*, *period*, and *wave* all have different or more specific mathematical meanings.

4 What is the difference between *equal*, *equivalent*, and *the same*?

5 Are there any words which we use only in mathematics? What does this tell us about the nature of mathematics and the world around us?

B THE SINE FUNCTION

In previous studies of trigonometry we have only considered static situations where an angle is fixed. However, when an object moves around a circle, the situation is dynamic. The angle θ between the radius [OP] and the positive x-axis continually changes with time.

Consider again the **Opening Problem** in which a Ferris wheel of radius 10 m revolves at constant speed. We let P represent the green light on the wheel.

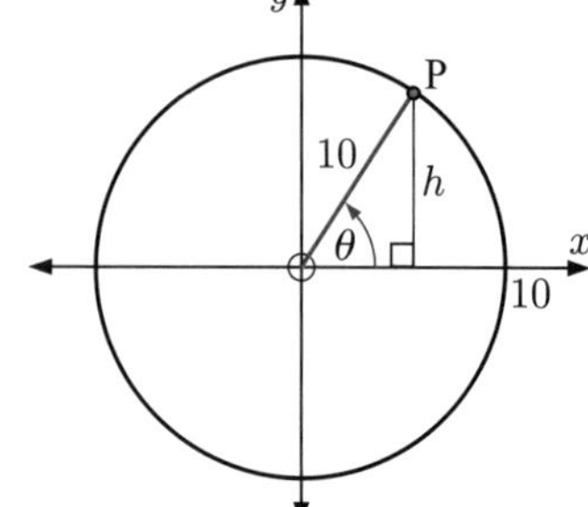

The height of P relative to the x-axis can be determined using right angled triangle trigonometry.

$$\sin\theta = \frac{h}{10}, \quad \text{so} \quad h = 10\sin\theta.$$

As time goes by, θ changes and so does h.

So, we can write h as a function of θ, or alternatively we can write h as a function of time t.

Suppose the Ferris wheel observed by Andrew takes 100 seconds for a full revolution. The graph below shows the height of the light above or below the principal axis against the time in seconds.

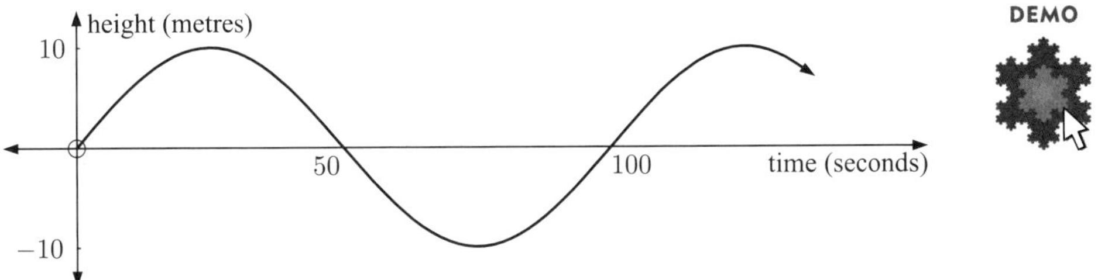

We observe that the amplitude is 10 metres and the period is 100 seconds.

THE BASIC SINE CURVE

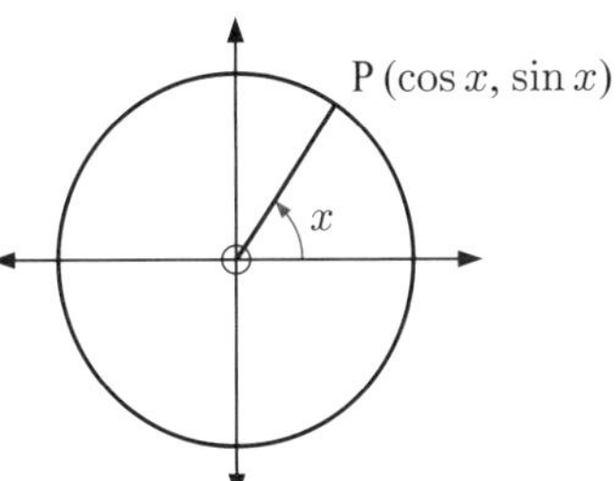

Suppose point P moves around the unit circle so the angle [OP] makes with the positive horizontal axis is x. In this case P has coordinates $(\cos x, \sin x)$.

If we project the values of $\sin x$ from the unit circle to a set of axes alongside, we can obtain the graph of $y = \sin x$.

Note carefully that x on the unit circle diagram is an *angle*, and becomes the horizontal coordinate of the sine function.

Unless indicated otherwise, you should assume that x is measured in radians. Degrees are only included on this graph for the sake of completeness.

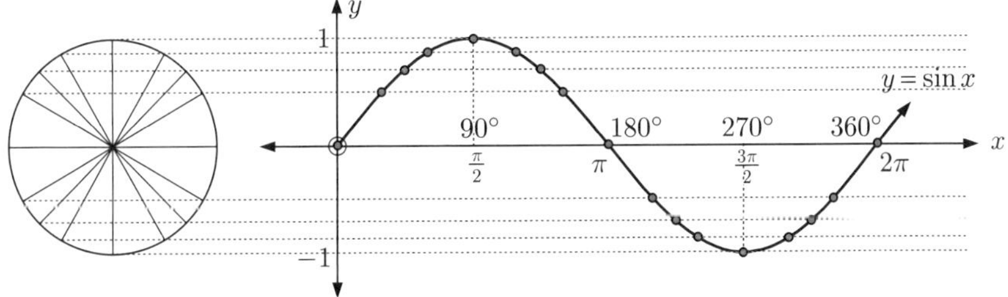

Click on the icon to generate the sine function for yourself.

You should observe that the sine function can be continued beyond $0 \leqslant x \leqslant 2\pi$ in either direction.

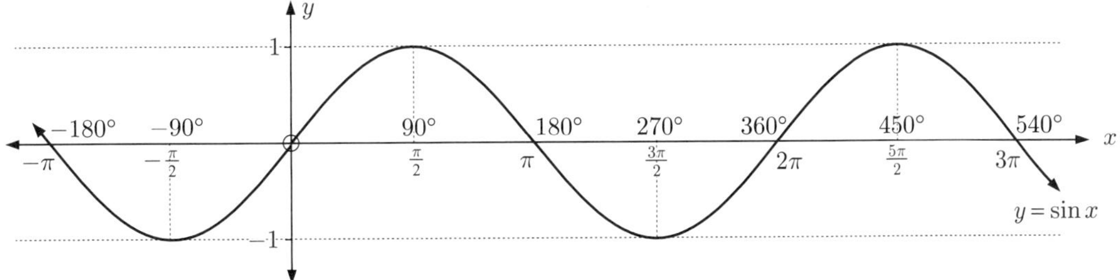

The unit circle repeats itself after one full revolution, so its *period* is 2π.

The *maximum* value is 1 and the *minimum* is -1, as $-1 \leqslant y \leqslant 1$ on the unit circle.

The *amplitude* is 1.

TRANSFORMATIONS OF THE SINE CURVE

In the investigations that follow, we will consider applying transformations to the sine curve $y = \sin x$. Using the transformations we learnt in **Chapter 5**, we can generate the curve for the general sine function $y = a\sin(b(x-c)) + d$.

INVESTIGATION 1 THE FAMILY $y = a\sin x,\ a \neq 0$

Click on the icon to explore the family $y = a\sin x$, $a \neq 0$.

DYNAMIC SINE FUNCTION

Notice that x is measured in radians.

What to do:

1 Use the slider to vary the value of a. Observe the changes to the graph of the function.

2 Use the software to help complete the table:

a	*Function*	*Maximum*	*Minimum*	*Period*	*Amplitude*
1	$y = \sin x$	1	-1	2π	1
2	$y = 2\sin x$				
0.5	$y = 0.5\sin x$				
-1	$y = -\sin x$				
a	$y = a\sin x$				

3 How does a affect the function $y = a\sin x$?

4 State the amplitude of:

a $y = 3\sin x$ **b** $y = \sqrt{7}\sin x$ **c** $y = -2\sin x$

INVESTIGATION 2 THE FAMILY $y = \sin bx,\ b > 0$

Click on the icon to explore the family $y = \sin bx$, $b > 0$.

DYNAMIC SINE FUNCTION

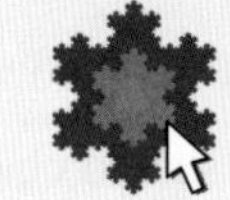

What to do:

1 Use the slider to vary the value of b. Observe the changes to the graph of the function.

2 Use the software to help complete the table:

b	*Function*	*Maximum*	*Minimum*	*Period*	*Amplitude*
1	$y = \sin x$	1	-1	2π	1
2	$y = \sin 2x$				
$\frac{1}{2}$	$y = \sin(\frac{1}{2}x)$				
b	$y = \sin bx$				

3 How does b affect the function $y = \sin bx$?

4 State the period of:

a $y = \sin 3x$ **b** $y = \sin(\frac{1}{3}x)$ **c** $y = \sin(1.2x)$ **d** $y = \sin bx$

From the previous **Investigations** you should have found:

Family $y = a\sin x$, $a \neq 0$

- a affects the amplitude of the graph; amplitude $= |a|$
- The graph is a vertical stretch of $y = \sin x$ with scale factor $|a|$.
- If $a < 0$, the graph of $y = \sin x$ is also reflected in the x-axis.

Family $y = \sin bx$, $b > 0$

- The graph is a horizontal stretch of $y = \sin x$ with scale factor $\frac{1}{b}$.
- period $= \frac{2\pi}{b}$

$|a|$ is the modulus of a. It is the *size* of a, and cannot be negative.

Example 1 — Self Tutor

Without using technology, sketch the graphs of:

a $y = 2\sin x$ b $y = -2\sin x$ for $0 \leqslant x \leqslant 2\pi$.

a This is a vertical stretch of $y = \sin x$ with scale factor 2.
The amplitude is 2 and the period is 2π.

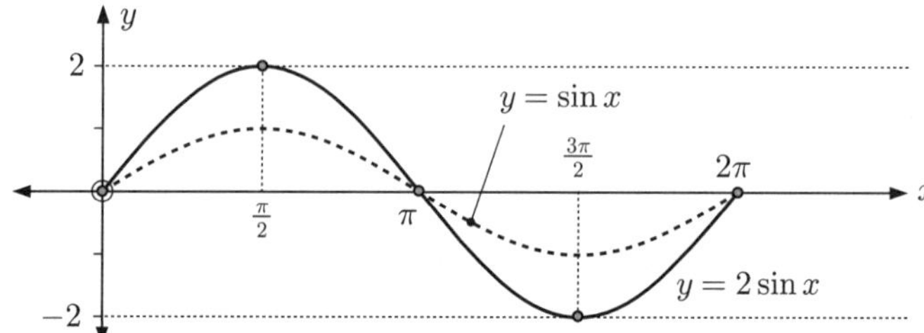

b The amplitude is 2 and the period is 2π. It is the reflection of $y = 2\sin x$ in the x-axis.

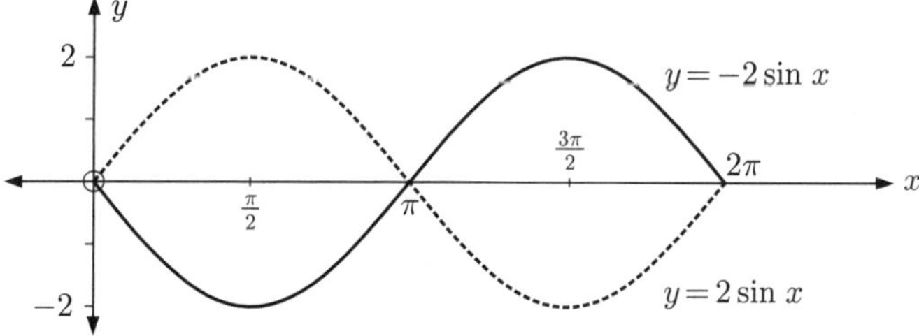

Example 2 — Self Tutor

Without using technology, sketch the graph of $y = \sin 2x$ for $0 \leqslant x \leqslant 2\pi$.

This is a horizontal stretch of $y = \sin x$ with scale factor $\frac{1}{2}$.

The period is $\frac{2\pi}{2} = \pi$, so the maximum values are π units apart.

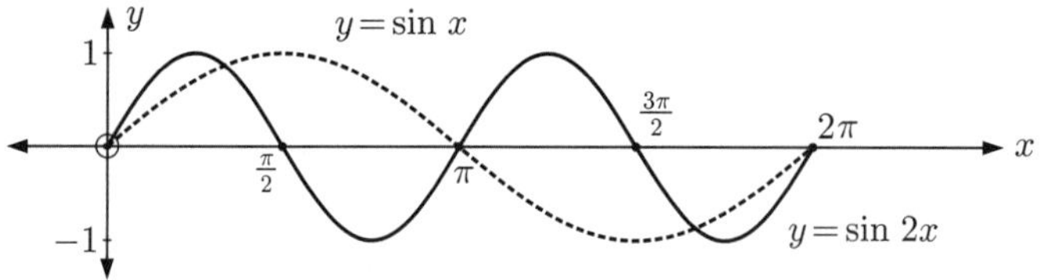

Since $\sin 2x$ has half the period of $\sin x$, the first maximum is at $\frac{\pi}{4}$ not $\frac{\pi}{2}$.

EXERCISE 12B.1

1 Without using technology, sketch the graphs of the following for $0 \leqslant x \leqslant 2\pi$:

a $y = 3\sin x$ **b** $y = -3\sin x$ **c** $y = \frac{3}{2}\sin x$ **d** $y = -\frac{3}{2}\sin x$

2 Without using technology, sketch the graphs of the following for $0 \leqslant x \leqslant 3\pi$:

a $y = \sin 3x$ **b** $y = \sin\left(\frac{x}{2}\right)$ **c** $y = \sin(-2x)$

3 State the period of:

a $y = \sin 4x$ **b** $y = \sin(-4x)$ **c** $y = \sin\left(\frac{x}{3}\right)$ **d** $y = \sin(0.6x)$

4 Find b given that the function $y = \sin bx$, $b > 0$ has period:

a 5π **b** $\frac{2\pi}{3}$ **c** 12π **d** 4 **e** 100

5 Use a graphics calculator or graphing package to help you graph for $0° \leqslant x \leqslant 720°$:

a $y = 2\sin x + \sin 2x$ **b** $y = \sin x + \sin 2x + \sin 3x$ **c** $y = \dfrac{1}{\sin x}$

6 **a** Use a graphing package or graphics calculator to graph for $-4\pi \leqslant x \leqslant 4\pi$:

i $f(x) = \sin x + \dfrac{\sin 3x}{3} + \dfrac{\sin 5x}{5}$

ii $f(x) = \sin x + \dfrac{\sin 3x}{3} + \dfrac{\sin 5x}{5} + \dfrac{\sin 7x}{7} + \dfrac{\sin 9x}{9} + \dfrac{\sin 11x}{11}$

b Predict the graph of $f(x) = \sin x + \dfrac{\sin 3x}{3} + \dfrac{\sin 5x}{5} + \dfrac{\sin 7x}{7} + ... + \dfrac{\sin 1001x}{1001}$

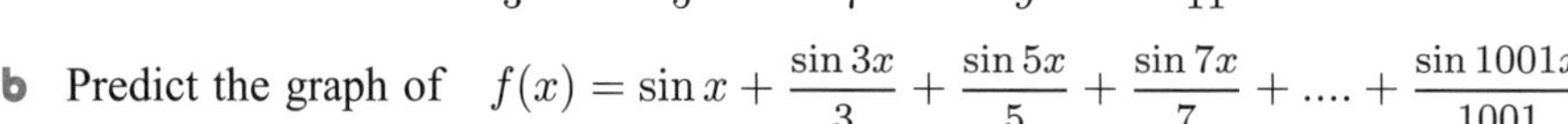

INVESTIGATION 3 THE FAMILIES $y = \sin(x - c)$ AND $y = \sin x + d$

Click on the icon to explore the families $y = \sin(x - c)$ and $y = \sin x + d$.

What to do:

1 Use the slider to vary the value of c. Observe the changes to the graph of the function, and complete the table:

c	*Function*	*Maximum*	*Minimum*	*Period*	*Amplitude*
0	$y = \sin x$	1	-1	2π	1
-2	$y = \sin(x - 2)$				
2	$y = \sin(x + 2)$				
$-\frac{\pi}{3}$	$y = \sin(x - \frac{\pi}{3})$				
c	$y = \sin(x - c)$				

2 What transformation moves $y = \sin x$ to $y = \sin(x - c)$?

3 Return the value of c to zero, and now vary the value of d. Observe the changes to the graph of the function, and complete the table:

d	*Function*	*Maximum*	*Minimum*	*Period*	*Amplitude*
0	$y = \sin x$	1	-1	2π	1
3	$y = \sin x + 3$				
-2	$y = \sin x - 2$				
d	$y = \sin x + d$				

4 What transformation moves $y = \sin x$ to $y = \sin x + d$?

5 What transformation moves $y = \sin x$ to $y = \sin(x - c) + d$?

From **Investigation 3** we observe that:

- $y = \sin(x - c)$ is a **horizontal translation** of $y = \sin x$ through c units.
- $y = \sin x + d$ is a **vertical translation** of $y = \sin x$ through d units.
- $y = \sin(x - c) + d$ is a **translation** of $y = \sin x$ through vector $\binom{c}{d}$.

Example 3 Self Tutor

On the same set of axes graph for $0 \leqslant x \leqslant 4\pi$:

a $y = \sin x$ and $y = \sin(x - 1)$ **b** $y = \sin x$ and $y = \sin x - 1$

a This is a horizontal translation of $y = \sin x$ to the right by 1 unit.

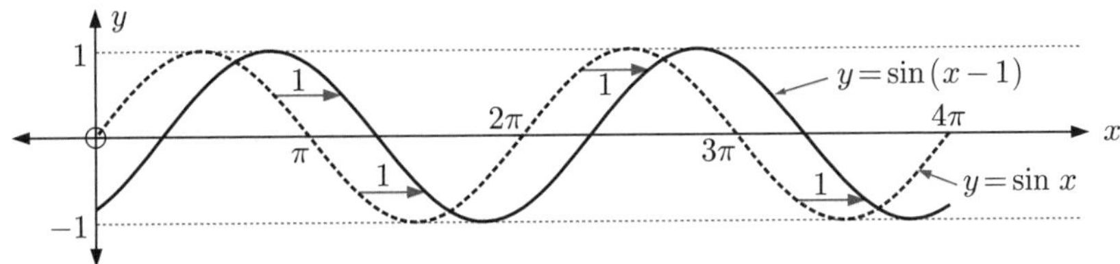

b This is a vertical translation of $y = \sin x$ downwards by 1 unit.

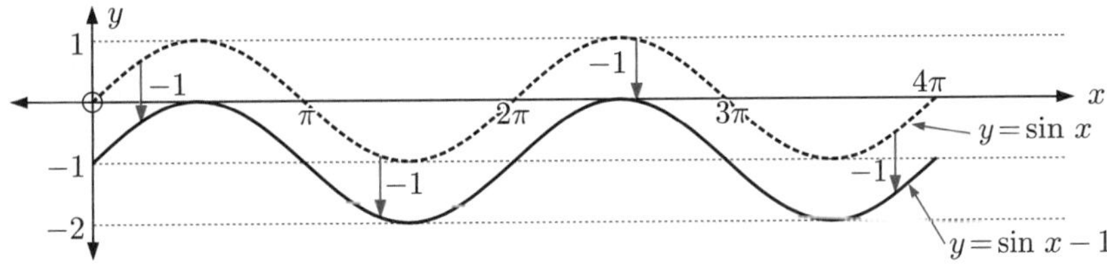

THE GENERAL SINE FUNCTION

The **general sine function** is

$$\boldsymbol{y = a\sin(b(x - c)) + d} \quad \text{where } b > 0.$$

affects **amplitude** (a), affects **period** (b), affects **horizontal translation** (c), affects **vertical translation** (d)

The **principal axis** of the general sine function is $\boldsymbol{y = d}$.

The **period** of the general sine function is $\dfrac{\boldsymbol{2\pi}}{\boldsymbol{b}}$.

The **amplitude** of the general sine function is $|\boldsymbol{a}|$.

For example, $y = 2\sin(3(x - \frac{\pi}{4})) + 1$ is a translation of $y = 2\sin 3x$ with translation vector $\begin{pmatrix} \frac{\pi}{4} \\ 1 \end{pmatrix}$.

Starting with $y = \sin x$ we would:

- double the amplitude to produce $y = 2\sin x$, then
- divide the period by 3 to produce $y = 2\sin 3x$, then
- translate by $\begin{pmatrix} \frac{\pi}{4} \\ 1 \end{pmatrix}$ to produce $y = 2\sin(3(x - \frac{\pi}{4})) + 1$.

EXERCISE 12B.2

1 Sketch the graphs of the following for $0 \leqslant x \leqslant 4\pi$:

a $y = \sin x - 2$ **b** $y = \sin(x - 2)$ **c** $y = \sin(x + 2)$

d $y = \sin x + 2$ **e** $y = \sin(x + \frac{\pi}{4})$ **f** $y = \sin(x - \frac{\pi}{6}) + 1$

Check your answers using technology.

GRAPHING PACKAGE

2 State the period of:

a $y = \sin 5t$ **b** $y = \sin\left(\frac{t}{4}\right)$ **c** $y = \sin(-2t)$

3 Find b in $y = \sin bx$ if $b > 0$ and the period is:

a 3π **b** $\frac{\pi}{10}$ **c** 100π **d** 50

4 State the transformations which map:

a $y = \sin x$ onto $y = \sin x - 1$ **b** $y = \sin x$ onto $y = \sin(x - \frac{\pi}{4})$

c $y = \sin x$ onto $y = 2\sin x$ **d** $y = \sin x$ onto $y = \sin 4x$

e $y = \sin x$ onto $y = \frac{1}{2}\sin x$ **f** $y = \sin x$ onto $y = \sin\left(\frac{x}{4}\right)$

g $y = \sin x$ onto $y = -\sin x$ **h** $y = \sin x$ onto $y = -3 + \sin(x + 2)$

i $y = \sin x$ onto $y = 2\sin 3x$ **j** $y = \sin x$ onto $y = \sin(x - \frac{\pi}{3}) + 2$

C MODELLING USING SINE FUNCTIONS

When patterns of variation can be identified and quantified using a formula or equation, predictions may be made about behaviour in the future. Examples of this include tidal movement which can be predicted many months ahead, and the date of a future full moon.

In this section we use sine functions to model periodic biological and physical phenomena.

MEAN MONTHLY TEMPERATURE

Consider again the mean monthly maximum temperature for Cape Town over a 12 month period:

Month	Jan	Feb	Mar	Apr	May	Jun	Jul	Aug	Sep	Oct	Nov	Dec
Temp. T (°C)	28	27	$25\frac{1}{2}$	22	$18\frac{1}{2}$	16	15	16	18	$21\frac{1}{2}$	24	26

The graph over a two year period is shown below:

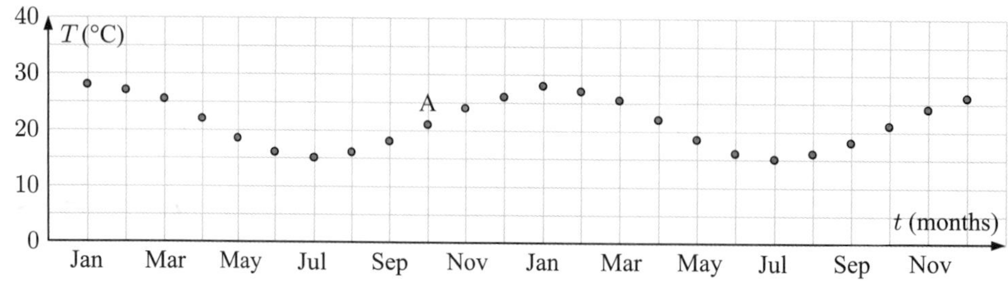

We attempt to model this data using the general sine function $y = a\sin(b(x-c)) + d$,
or in this case $T = a\sin(b(t-c)) + d$.

The period is 12 months, so $\frac{2\pi}{b} = 12$ and $\therefore$ $b = \frac{\pi}{6}$.

The amplitude $= \frac{\text{max} - \text{min}}{2} \approx \frac{28-15}{2} \approx 6.5$, so $a \approx 6.5$.

The principal axis is midway between the maximum and minimum, so $d \approx \frac{28+15}{2} \approx 21.5$.

So, the model is $T \approx 6.5\sin(\frac{\pi}{6}(t-c)) + 21.5$ for some constant c.

On the original graph, point A lies on the principal axis, and is the first point shown at which we are starting a new period. Since A is at (10, 21.5), $c = 10$.

The model is therefore $T \approx 6.5\sin(\frac{\pi}{6}(t-10)) + 21.5$, and we can superimpose it on the original data as follows.

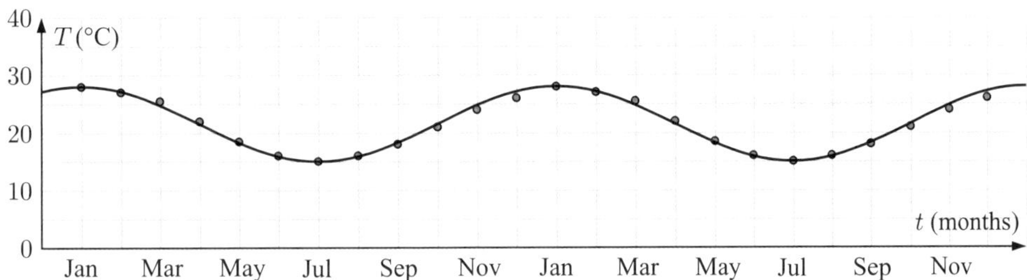

TIDAL MODELS

The tides at Juneau, Alaska were recorded over a two day period. The results are shown in the table opposite:

Day 1	high tide	1:18 pm
	low tide	6:46 am, 7:13 pm
Day 2	high tide	1:31 am, 2:09 pm
	low tide	7:30 am, 7:57 pm

Suppose high tide corresponds to height 1 and low tide to height -1.

Plotting these times with t being the time after midnight before the first low tide, we get:

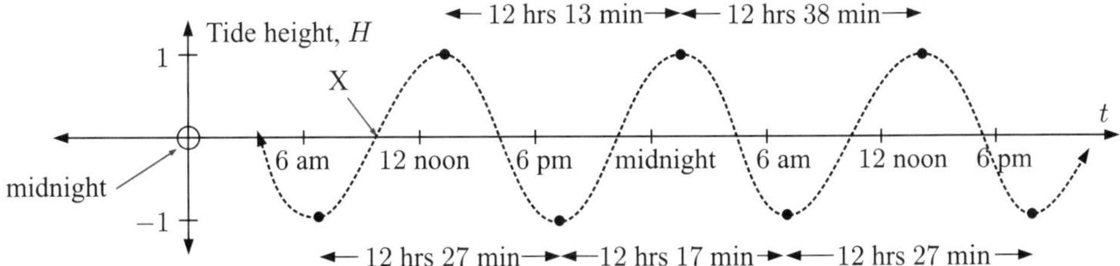

We attempt to model this periodic data using $H = a\sin(b(t-c)) + d$.

The principal axis is $H = 0$, so $d = 0$.

The amplitude is 1, so $a = 1$.

The graph shows that the 'average' period is about 12 hours 24 min $\approx$ 12.4 hours.

But the period is $\frac{2\pi}{b}$, so $\frac{2\pi}{b} \approx 12.4$ and $\therefore$ $b \approx \frac{2\pi}{12.4} \approx 0.507$.

The model is now $H \approx \sin(0.507(t - c))$ for some constant c.

We find point X which is midway between the *first minimum* 6:46 am and the *following maximum* 1:18 pm. Its x-coordinate is $\frac{6.77 + 13.3}{2} \approx 10.0$, so $c \approx 10.0$.

So, the model is $H \approx \sin(0.507(t - 10.0))$.

Below is our original graph of seven plotted points and our model which attempts to fit them.

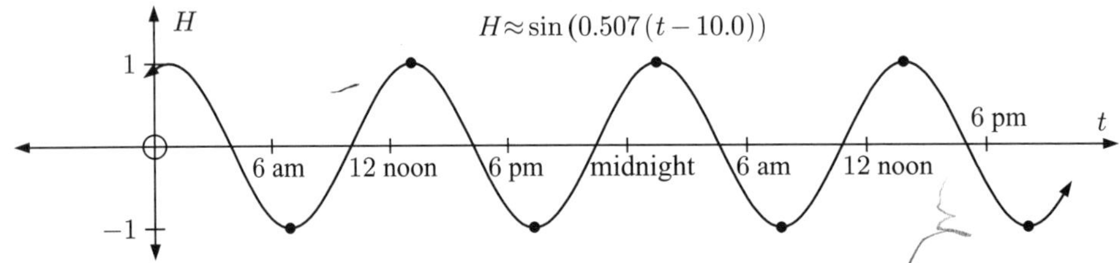

Use your **graphics calculator** to check this result.
The times must be given in hours after midnight, so

the low tide at 6:46 am is $(6.77, -1)$,

the high tide at 1:18 pm is $(13.3, 1)$, and so on.

EXERCISE 12C

1 Below is a table which shows the mean monthly maximum temperatures for a city in Greece.

Month	Jan	Feb	Mar	Apr	May	Jun	July	Aug	Sept	Oct	Nov	Dec
Temperature (°C)	15	14	15	18	21	25	27	26	24	20	18	16

a Use a sine function of the form $T \approx a\sin(b(t-c))+d$ to model the data. Find good estimates of the constants a, b, c and d without using technology. Use Jan $\equiv 1$, Feb $\equiv 2$, and so on.

b Use technology to check your answer to **a**. How well does your model fit?

2 The data in the table shows the mean monthly temperatures for Christchurch, New Zealand.

Month	Jan	Feb	Mar	Apr	May	Jun	July	Aug	Sept	Oct	Nov	Dec
Temperature (°C)	15	16	$14\frac{1}{2}$	12	10	$7\frac{1}{2}$	7	$7\frac{1}{2}$	$8\frac{1}{2}$	$10\frac{1}{2}$	$12\frac{1}{2}$	14

a Find a sine model for this data in the form $T \approx a\sin(b(t-c))+d$. Assume Jan $\equiv 1$, Feb $\equiv 2$, and so on. Do not use technology.

b Use technology to check your answer to **a**.

3 Some of the largest tides in the world are observed in Canada's Bay of Fundy. The difference between high and low tides is 14 metres, and the average time difference between high tides is about 12.4 hours.

a Find a sine model for the height of the tide H in terms of the time t.

b Sketch the graph of the model over one period.

4 At the Mawson base in Antarctica, the mean monthly temperatures for the last 30 years are:

Month	Jan	Feb	Mar	Apr	May	Jun	July	Aug	Sept	Oct	Nov	Dec
Temperature (°C)	0	0	-4	-9	-14	-17	-18	-19	-17	-13	-6	-2

a Find a sine model for this data without using technology. Use Jan $\equiv 1$, Feb $\equiv 2$, and so on.

b How appropriate is the model?

5 Revisit the **Opening Problem** on page **326**.
The wheel takes 100 seconds to complete one revolution. Find the sine model which gives the height of the light above the ground at any point in time. Assume that at time $t = 0$, the light is at its lowest point.

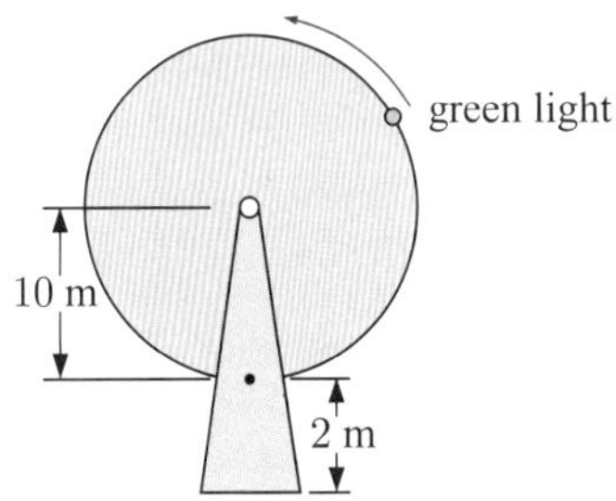

D THE COSINE FUNCTION

We return to the Ferris wheel and now view the movement of the green light from above.

Now $\cos\theta = \dfrac{d}{10}$ so $d = 10\cos\theta$

The graph being generated over time is therefore a **cosine function**.

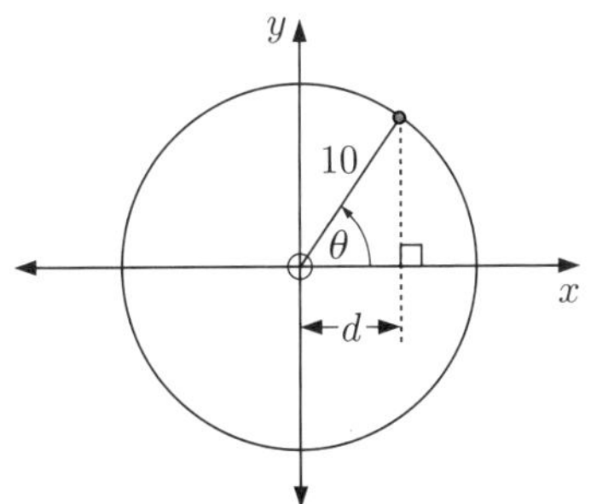

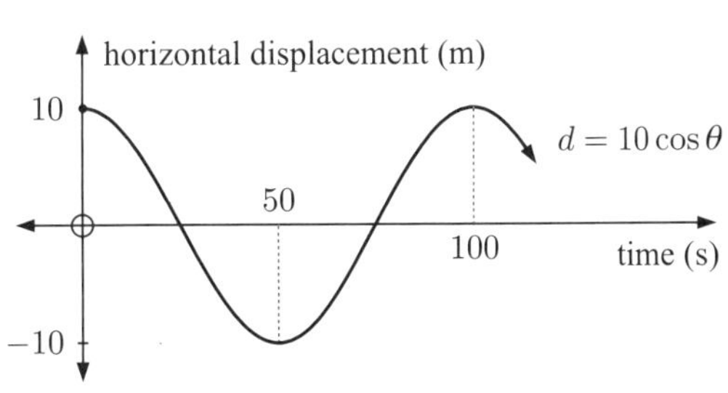

DEMO

The cosine curve $y = \cos x$, like the sine curve $y = \sin x$, has a **period** of 2π, an **amplitude** of 1, and its **range** is $-1 \leqslant y \leqslant 1$.

Use your graphics calculator or graphing package to check these features.

GRAPHING PACKAGE

Now view the relationship between the sine and cosine functions.

You should observe that $y = \cos x$ and $y = \sin x$ are identical in shape, but the cosine function is $\frac{\pi}{2}$ units left of the sine function under a horizontal translation.

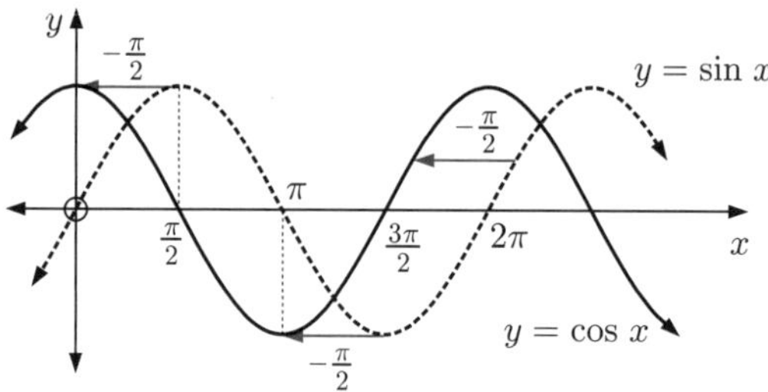

This suggests that $\cos x = \sin\left(x + \frac{\pi}{2}\right)$.

Use your graphing package or graphics calculator to check this by graphing $y = \cos x$ and $y = \sin\left(x + \frac{\pi}{2}\right)$ on the same set of axes.

GRAPHING PACKAGE

THE GENERAL COSINE FUNCTION

The **general cosine function** is $y = a\cos(b(x - c)) + d$ where $a \neq 0,\ b > 0$.

Since the cosine function is a horizontal translation of the sine function, the constants a, b, c, and d have the same effects as for the general sine function. Click on the icon to check this.

DYNAMIC COSINE FUNCTION

Example 4 **Self Tutor**

On the same set of axes graph $y = \cos x$ and $y = \cos\left(x - \frac{\pi}{3}\right)$ for $-2\pi \leqslant x \leqslant 2\pi$.

$y = \cos\left(x - \frac{\pi}{3}\right)$ is a horizontal translation of $y = \cos x$ through $\frac{\pi}{3}$ units to the right.

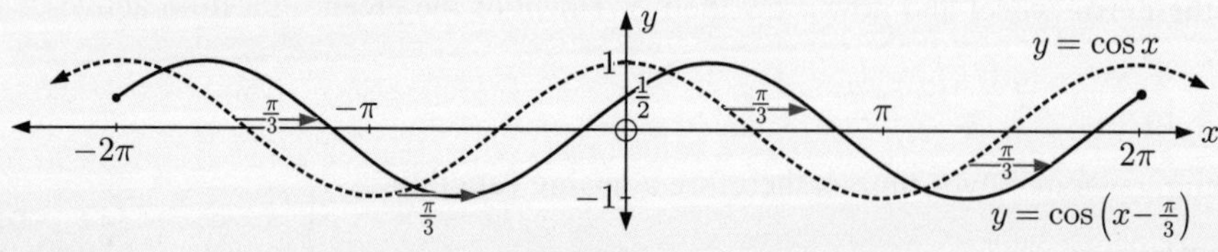

Example 5 **Self Tutor**

Without using technology, sketch the graph of $y = 3\cos 2x$ for $0 \leqslant x \leqslant 2\pi$.

Notice that $a = 3$, so the amplitude is $|3| = 3$.

$b = 2$, so the period is $\frac{2\pi}{b} = \frac{2\pi}{2} = \pi$.

To obtain this from $y = \cos x$, we have a vertical stretch with scale factor 3 followed by a horizontal stretch with scale factor $\frac{1}{2}$, as the period has been halved.

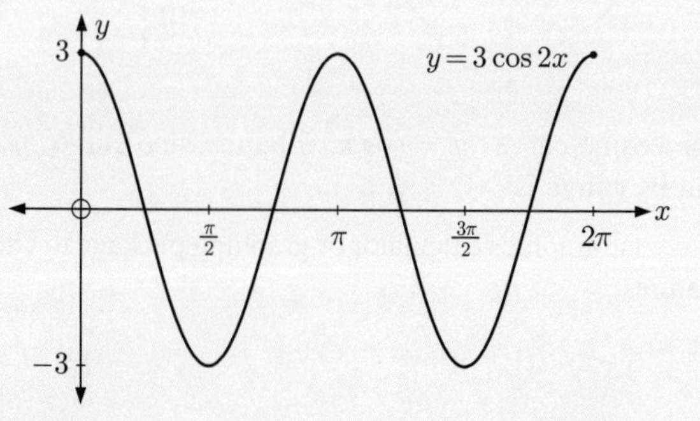

EXERCISE 12D

1. Given the graph of $y = \cos x$, sketch the graphs of:

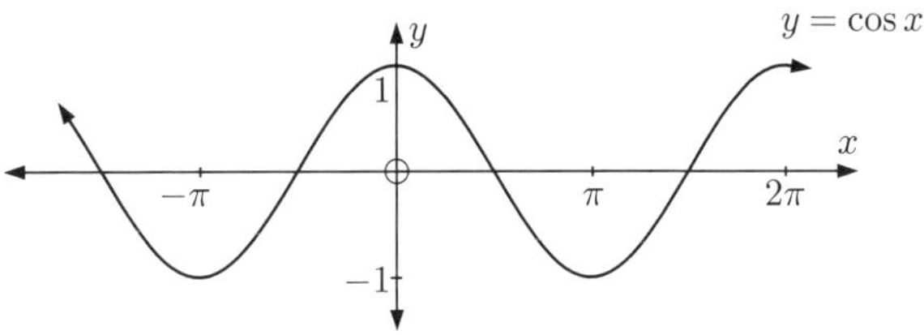

a $y = \cos x + 2$
b $y = \cos x - 1$
c $y = \cos\left(x - \frac{\pi}{4}\right)$
d $y = \cos\left(x + \frac{\pi}{6}\right)$
e $y = \frac{2}{3}\cos x$
f $y = \frac{3}{2}\cos x$
g $y = -\cos x$
h $y = \cos\left(x - \frac{\pi}{6}\right) + 1$
i $y = \cos\left(x + \frac{\pi}{4}\right) - 1$
j $y = \cos 2x$
k $y = \cos\left(\frac{x}{2}\right)$
l $y = 3\cos 2x$

2. Without graphing them, state the periods of:

a $y = \cos 3x$
b $y = \cos\left(\frac{x}{3}\right)$
c $y = \cos\left(\frac{\pi}{50}x\right)$

3. The general cosine function is $y = a\cos(b(x - c)) + d$.
State the geometrical significance of a, b, c, and d.

4. Find the cosine function shown in the graph:

a
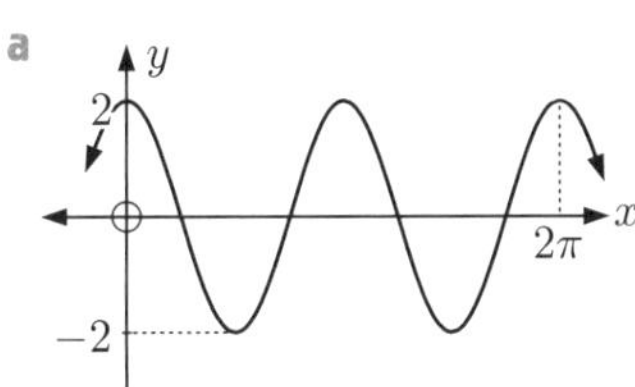

b
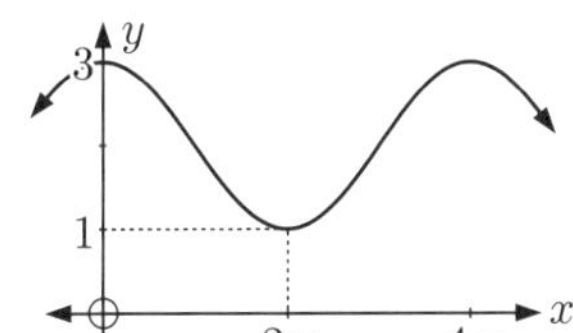

c
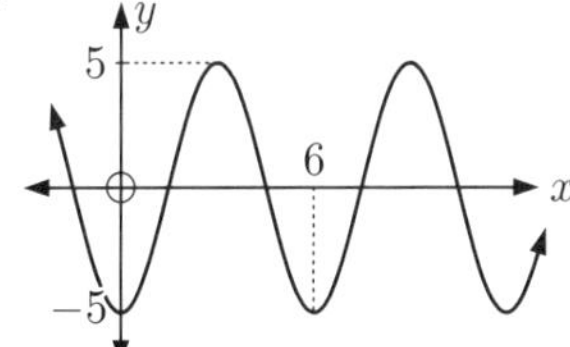

E THE TANGENT FUNCTION

We have seen that if $P(\cos\theta, \sin\theta)$ is a point which is free to move around the unit circle, and if [OP] is extended to meet the tangent at $A(1, 0)$, the intersection between these lines occurs at $Q(1, \tan\theta)$.

This enables us to define the **tangent function**

$$\tan\theta = \frac{\sin\theta}{\cos\theta}.$$

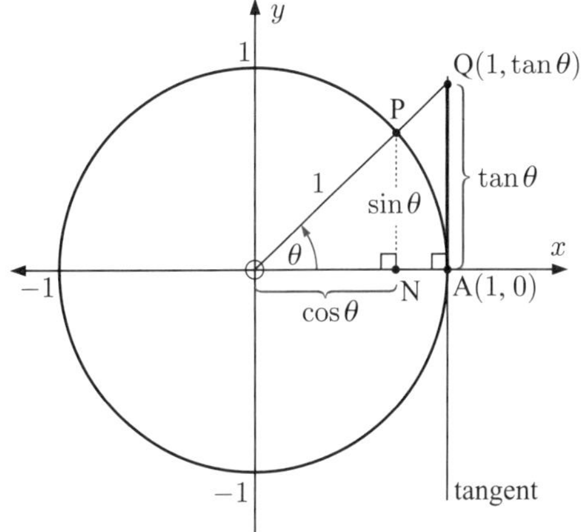

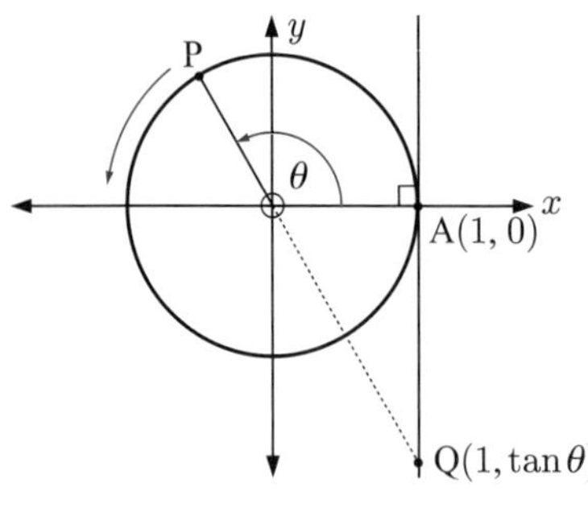

For θ in quadrant 2, $\sin\theta$ is positive and $\cos\theta$ is negative and so $\tan\theta = \dfrac{\sin\theta}{\cos\theta}$ is negative.

As before, [OP] is extended to meet the tangent at A at Q(1, $\tan\theta$).

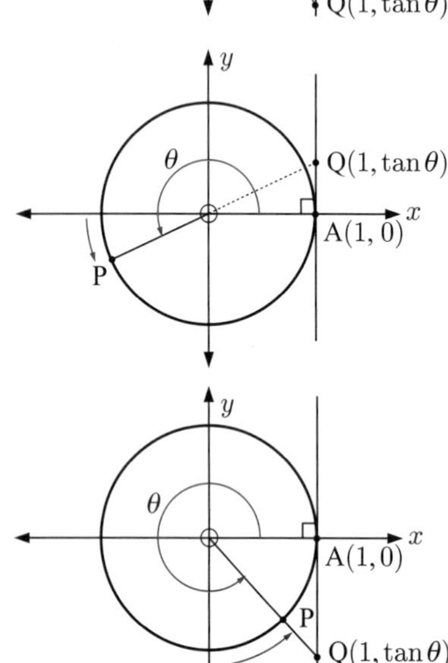

For θ in quadrant 3, $\sin\theta$ and $\cos\theta$ are both negative and so $\tan\theta$ is positive. This is clearly demonstrated as Q is back above the x-axis.

For θ in quadrant 4, $\sin\theta$ is negative and $\cos\theta$ is positive. $\tan\theta$ is again negative.

DISCUSSION

What happens to $\tan\theta$ when P is at (0, 1) and (0, -1)?

THE GRAPH OF $y = \tan x$

Since $\tan x = \dfrac{\sin x}{\cos x}$, $\tan x$ will be undefined whenever $\cos x = 0$.

The zeros of the function $y = \cos x$ correspond to vertical asymptotes of the function $y = \tan x$.

The graph of $y = \tan x$ is

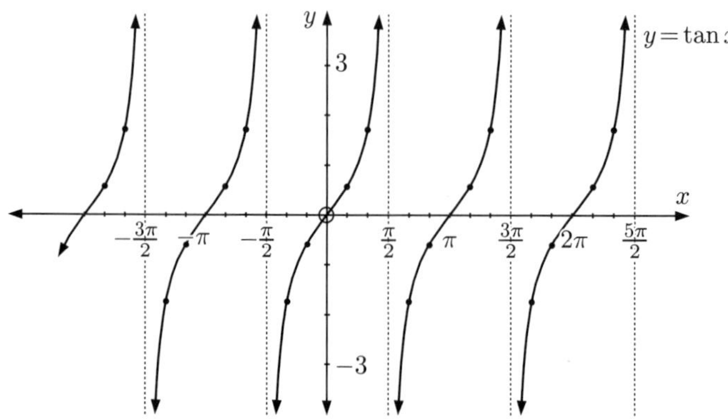

We observe that $y = \tan x$ has:
- **period** of π
- **range** $y \in \mathbb{R}$
- **vertical asymptotes** $x = \frac{\pi}{2} + k\pi$ for all $k \in \mathbb{Z}$.

Click on the icon to explore how the tangent function is produced from the unit circle.

TANGENT FUNCTION

DISCUSSION

- Discuss how to find the x-intercepts of $y = \tan x$.
- What must $\tan(x - \pi)$ simplify to?
- How many solutions does the equation $\tan x = 2$ have?

THE GENERAL TANGENT FUNCTION

The **general tangent function** is $y = a\tan(b(x - c)) + d$, $a \neq 0$, $b > 0$.

- The **principal axis** is $y = d$.
- The **period** of this function is $\frac{\pi}{b}$.
- The **amplitude** of this function is undefined.
- There are infinitely many vertical asymptotes.

Click on the icon to explore the properties of this function.

DYNAMIC TANGENT FUNCTION

Example 6 — Self Tutor

Without using technology, sketch the graph of $y = \tan(x + \frac{\pi}{4})$ for $0 \leqslant x \leqslant 3\pi$.

$y = \tan(x + \frac{\pi}{4})$ is a horizontal translation of $y = \tan x$ through $-\frac{\pi}{4}$

$y = \tan x$ has vertical asymptotes $x = \frac{\pi}{2}$, $x = \frac{3\pi}{2}$, and $x = \frac{5\pi}{2}$

Its x-axis intercepts are 0, π, 2π, and 3π.

$\therefore$ $y = \tan(x + \frac{\pi}{4})$ has vertical asymptotes $x = \frac{\pi}{4}$, $x = \frac{5\pi}{4}$, $x = \frac{9\pi}{4}$, and x-intercepts $\frac{3\pi}{4}$, $\frac{7\pi}{4}$, and $\frac{11\pi}{4}$.

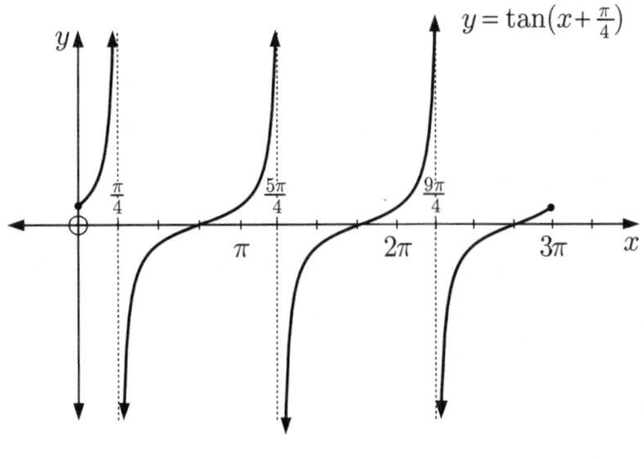

Example 7 **Self Tutor**

Without using technology, sketch the graph of $y = \tan 2x$ for $-\pi \leqslant x \leqslant \pi$.

$y = \tan 2x$ is a horizontal stretch of $y = \tan x$ with scale factor $\frac{1}{2}$.

Since $b = 2$, the period is $\frac{\pi}{2}$.

The vertical asymptotes are $x = \pm\frac{\pi}{4}$, $x = \pm\frac{3\pi}{4}$, and the x-axis intercepts are at $0, \pm\frac{\pi}{2}, \pm\pi$.

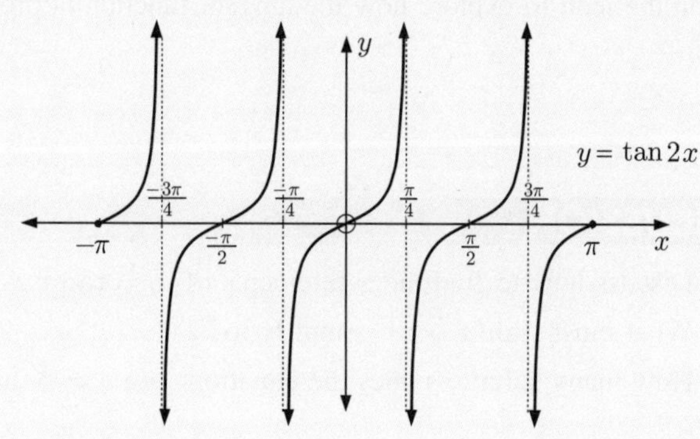

EXERCISE 12E

1 a *Sketch* the following functions for $0 \leqslant x \leqslant 3\pi$:
   i $y = \tan(x - \frac{\pi}{2})$ ii $y = -\tan x$ iii $y = \tan 3x$

   b Use technology to check your answers to a.
   Look in particular for asymptotes and the x-intercepts.

GRAPHING PACKAGE

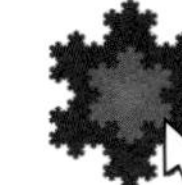

2 Describe the transformation(s) which moves the first curve to the second curve:
   a $y = \tan x$ to $y = \tan(x - 1) + 2$ b $y = \tan x$ to $y = -\tan x$
   c $y = \tan x$ to $y = 2\tan\left(\frac{x}{2}\right)$

3 State the period of:
   a $y = \tan x$ b $y = \tan 3x$ c $y = \tan nx$, $n \neq 0$

GRAPHING PACKAGE

F GENERAL TRIGONOMETRIC FUNCTIONS

In the previous sections we have explored properties of the sine, cosine, and tangent functions, and observed how they can be transformed into more general trigonometric functions.

The following tables summarise our observations:

FEATURES OF CIRCULAR FUNCTIONS

Function	*Sketch for* $0 \leqslant x \leqslant 2\pi$	*Period*	*Amplitude*	*Domain*	*Range*
$y = \sin x$	y, 1, −1, x	2π	1	$x \in \mathbb{R}$	$-1 \leqslant y \leqslant 1$

Function	*Sketch for* $0 \leqslant x \leqslant 2\pi$	*Period*	*Amplitude*	*Domain*	*Range*
$y = \cos x$		2π	1	$x \in \mathbb{R}$	$-1 \leqslant y \leqslant 1$
$y = \tan x$		π	undefined	$x \neq \pm\frac{\pi}{2}, \pm\frac{3\pi}{2}, ...$	$y \in \mathbb{R}$

GENERAL TRIGONOMETRIC FUNCTIONS				
General function	a affects vertical stretch	$b > 0$ affects horizontal stretch	c affects horizontal translation	d affects vertical translation
$y = a\sin(b(x-c)) + d$ $y = a\cos(b(x-c)) + d$	amplitude $= \lvert a \rvert$	period $= \dfrac{2\pi}{b}$	• $c > 0$ moves the graph right • $c < 0$ moves the graph left	• $d > 0$ moves the graph up • $d < 0$ moves the graph down • principal axis is $y = d$
$y = a\tan(b(x-c)) + d$	amplitude undefined	period $= \dfrac{\pi}{b}$		

EXERCISE 12F

1 State the amplitude, where appropriate, of:

a $y = \sin 4x$ **b** $y = 2\tan(\frac{x}{2})$ **c** $y = -\cos(3(x - \frac{\pi}{4}))$

2 State the period of:

a $y = -\tan x$ **b** $y = \cos(\frac{x}{3}) - 1$ **c** $y = \sin(2(x - \frac{\pi}{4}))$

3 Find b given:

a $y = \sin bx$ has period 2π **b** $y = \cos bx$ has period $\frac{2\pi}{3}$

c $y = \tan bx$ has period $\frac{\pi}{2}$ **d** $y = \sin bx$ has period 4

4 Sketch the graphs of these functions for $0 \leqslant x \leqslant 2\pi$:

a $y = \frac{2}{3}\cos x$ **b** $y = \sin x + 1$ **c** $y = \tan(x + \frac{\pi}{2})$

d $y = 3\cos 2x$ **e** $y = \sin(x + \frac{\pi}{4}) - 1$ **f** $y = \tan x - 2$

5 State the maximum and minimum values, where appropriate, of:

a $y = -\sin 5x$ **b** $y = 3\cos x$ **c** $y = 2\tan x$

d $y = -\cos 2x + 3$ **e** $y = 1 + 2\sin x$ **f** $y = \sin(x - \frac{\pi}{2}) - 3$

6 State the transformation(s) which map(s):

a $y = \sin x$ onto $y = \frac{1}{2}\sin x$

b $y = \cos x$ onto $y = \cos(\frac{x}{4})$

c $y = \sin x$ onto $y = -\sin x$

d $y = \cos x$ onto $y = \cos x - 2$

e $y = \tan x$ onto $y = \tan(x + \frac{\pi}{4})$

f $y = \sin x$ onto $y = \sin(-x)$

7 Find m and n given the following graph is of the function $y = m\sin x + n$.

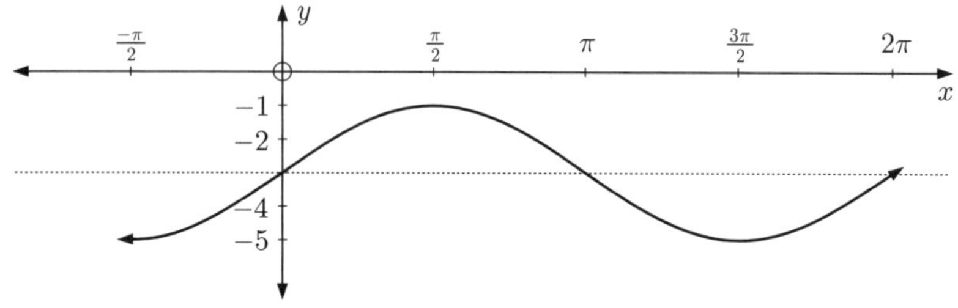

8 Find p and q given the following graph is of the function $y = \tan pt + q$.

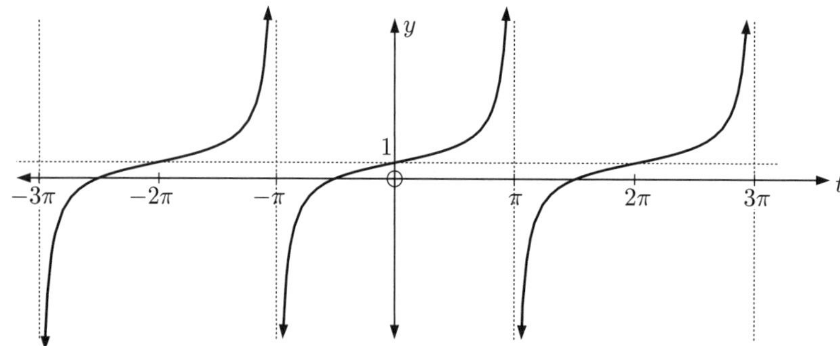

ACTIVITY

Click on the icon to run a card game for trigonometric functions.

CARD GAME

G RECIPROCAL TRIGONOMETRIC FUNCTIONS

We define the reciprocal trigonometric functions cosec x, secant x, and cotangent x as:

$$\csc x = \frac{1}{\sin x}, \quad \sec x = \frac{1}{\cos x}, \quad \text{and} \quad \cot x = \frac{1}{\tan x} = \frac{\cos x}{\sin x}$$

Using these definitions we can derive the identities:

$$\tan^2 x + 1 = \sec^2 x \quad \text{and} \quad 1 + \cot^2 x = \csc^2 x$$

Proof: Using $\sin^2 x + \cos^2 x = 1$,

$$\frac{\sin^2 x}{\cos^2 x} + \frac{\cos^2 x}{\cos^2 x} = \frac{1}{\cos^2 x} \quad \{\text{dividing each term by } \cos^2 x\}$$

$$\therefore \tan^2 x + 1 = \sec^2 x$$

Also using $\sin^2 x + \cos^2 x = 1$,

$$\frac{\sin^2 x}{\sin^2 x} + \frac{\cos^2 x}{\sin^2 x} = \frac{1}{\sin^2 x} \quad \{\text{dividing each term by } \sin^2 x\}$$

$$\therefore \; 1 + \cot^2 x = \csc^2 x$$

Example 8 — Self Tutor

Using the graph of $y = \sin x$, sketch the graph of $y = \dfrac{1}{\sin x} = \csc x$ for $x \in [-2\pi, 2\pi]$.

Check your answer using technology.

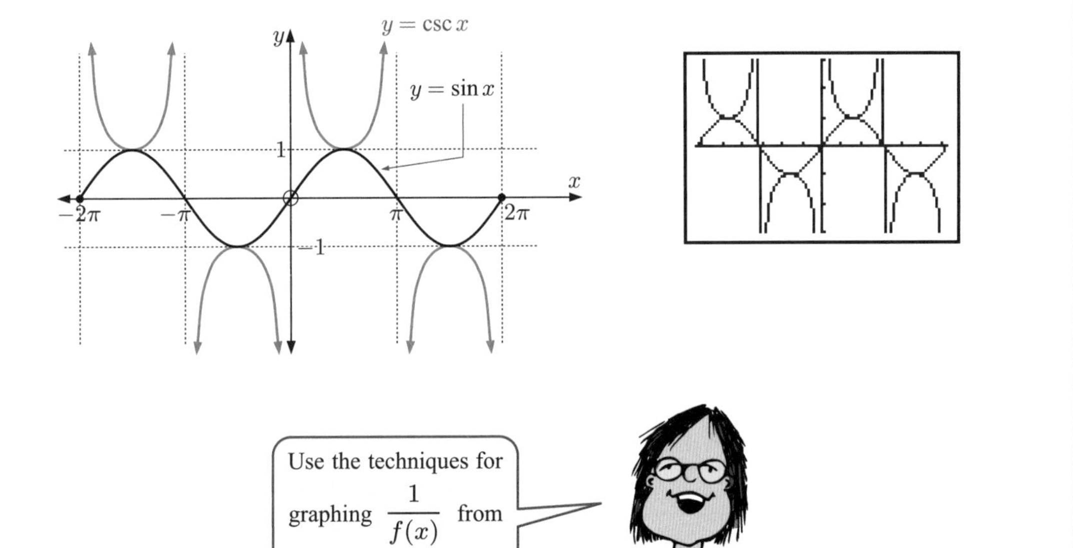

Use the techniques for graphing $\dfrac{1}{f(x)}$ from **Chapter 5**.

EXERCISE 12G

1 Without using a calculator, find:

a $\csc\left(\frac{\pi}{3}\right)$ b $\cot\left(\frac{2\pi}{3}\right)$ c $\sec\left(\frac{5\pi}{6}\right)$ d $\cot(\pi)$

2 Without using a calculator, find $\csc x$, $\sec x$, and $\cot x$ given:

a $\sin x = \frac{3}{5}$, $0 \leqslant x \leqslant \frac{\pi}{2}$ b $\cos x = \frac{2}{3}$, $\frac{3\pi}{2} < x < 2\pi$

3 Find the other *five* trigonometric ratios if:

a $\cos x = \frac{3}{4}$ and $\frac{3\pi}{2} < x < 2\pi$ b $\sin x = -\frac{2}{3}$ and $\pi < x < \frac{3\pi}{2}$

c $\sec x = 2\frac{1}{2}$ and $0 < x < \frac{\pi}{2}$ d $\csc x = 2$ and $\frac{\pi}{2} < x < \pi$

e $\tan \beta = \frac{1}{2}$ and $\pi < \beta < \frac{3\pi}{2}$ f $\cot \theta = \frac{4}{3}$ and $\pi < \theta < \frac{3\pi}{2}$

4 Simplify:

a $\tan x \cot x$ b $\sin x \csc x$ c $\csc x \cot x$

d $\sin x \cot x$ e $\dfrac{\cot x}{\csc x}$ f $\dfrac{2\sin x \cot x + 3\cos x}{\cot x}$

5 Using the graph of $y = \cos x$, sketch the graph of $y = \dfrac{1}{\cos x} = \sec x$ for $x \in [-2\pi, 2\pi]$. Check your answer using technology.

6 Using the graph of $y = \tan x$, sketch the graph of $y = \dfrac{1}{\tan x} = \cot x$ for $x \in [-2\pi, 2\pi]$.

7 Use technology to sketch $y = \sec x$ and $y = \csc\left(x + \frac{\pi}{2}\right)$ on the same set of axes for $x \in [-2\pi, 2\pi]$. Explain your answer.

H INVERSE TRIGONOMETRIC FUNCTIONS

In many problems we need to know what angle results in a particular trigonometric ratio. We have already seen this for right angled triangle problems and the cosine and sine rules.

For example:

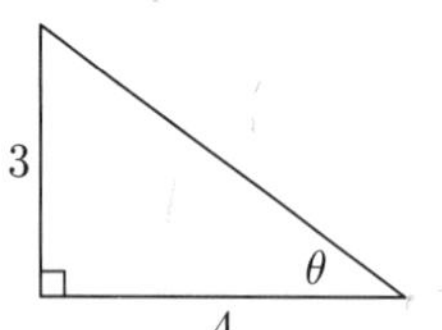

$$\tan\theta = \frac{\text{OPP}}{\text{ADJ}} = \frac{3}{4}$$
$$\therefore \quad \theta = \tan^{-1}(\tfrac{3}{4})$$
$$\therefore \quad \theta \approx 36.9^\circ$$

Rather than writing the inverse trigonometric functions as $\sin^{-1} x$, $\cos^{-1} x$, and $\tan^{-1} x$, which can be confused with reciprocals, mathematicians more formally refer to these functions as arcsine x, arccosine x, and arctangent x.

Since $\sin x$, $\cos x$, and $\tan x$ are all many-to-one functions, their domains must be restricted in order for them to have inverse functions. The inverse functions are therefore defined as follows:

Function	*Definition*	*Range*
$y = \arcsin x$	$x = \sin y, \quad -1 \leqslant x \leqslant 1$	$-\frac{\pi}{2} \leqslant y \leqslant \frac{\pi}{2}$
$y = \arccos x$	$x = \cos y, \quad -1 \leqslant x \leqslant 1$	$0 \leqslant y \leqslant \pi$
$y = \arctan x$	$x = \tan y, \quad x \in \mathbb{R}$	$-\frac{\pi}{2} < y < \frac{\pi}{2}$

The graphs of these functions are illustrated below as the inverse functions of $\sin x$, $\cos x$, and $\tan x$ on restricted domains which are the ranges of $\arcsin x$, $\arccos x$, and $\arctan x$ respectively.

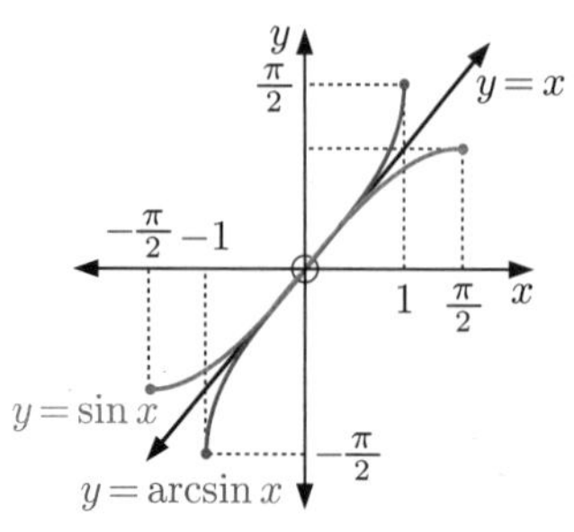

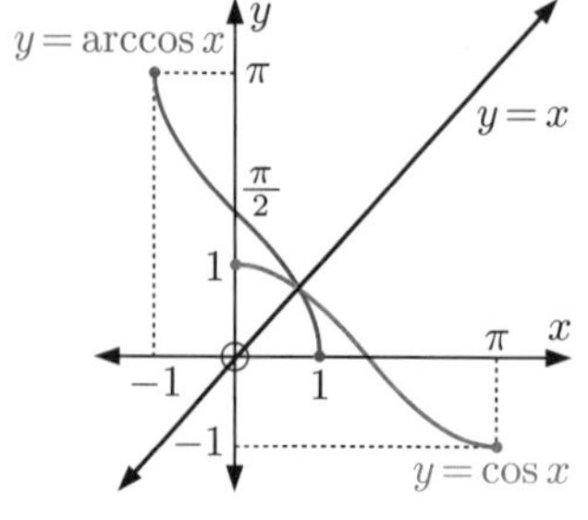

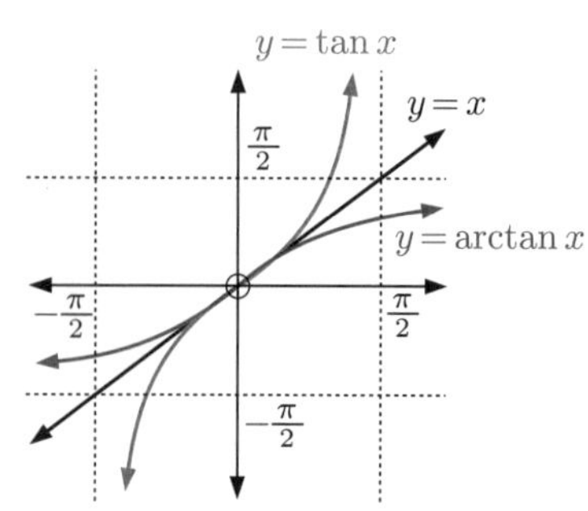

EXERCISE 12H

1 Copy and complete the table below, giving the restricted domain and range for each trigonometric function so that its inverse function exists:

Function	*Restricted domain*	*Restricted range*	*Inverse function*	*Domain*	*Range*
$y = \sin x$			$y = \arcsin x$		
$y = \cos x$			$y = \arccos x$		
$y = \tan x$			$y = \arctan x$		

2 Find, giving your answer in radians:

a $\arccos(1)$
b $\arcsin(-1)$
c $\arctan(1)$
d $\arctan(-1)$
e $\arcsin\left(\frac{1}{2}\right)$
f $\arccos\left(\frac{-\sqrt{3}}{2}\right)$
g $\arctan\left(\sqrt{3}\right)$
h $\arccos\left(-\frac{1}{\sqrt{2}}\right)$
i $\arctan\left(-\frac{1}{\sqrt{3}}\right)$
j $\sin^{-1}(-0.767)$
k $\cos^{-1}(0.327)$
l $\tan^{-1}(-50)$

3 Find the invariant point for the inverse transformation from:

a $y = \sin x$ to $y = \arcsin x$
b $y = \tan x$ to $y = \arctan x$
c $y = \cos x$ to $y = \arccos x$.

4 a State the equations of the asymptotes of $y = \arctan x$.

b Do the functions $y = \arcsin x$ and $y = \arccos x$ have vertical asymptotes? Explain your answer.

5 Simplify:

a $\arcsin(\sin \frac{\pi}{3})$
b $\arccos(\cos\left(-\frac{\pi}{6}\right))$
c $\tan(\arctan(0.3))$
d $\cos(\arccos(-\frac{1}{2}))$
e $\arctan(\tan \pi)$
f $\arcsin(\sin \frac{4\pi}{3})$

Remember to think about domain and range.

INVESTIGATION 4

Carl Friedrich Gauss used his **Gaussian hypergeometric series** to analyse the continued fraction:

$$\cfrac{x}{1+\cfrac{(1x)^2}{3+\cfrac{(2x)^2}{5+\cfrac{(3x)^2}{7+\cfrac{(4x)^2}{9+\dots}}}}}$$

What to do:

1 Evaluate the fraction with $x = 1$ for as many levels as necessary for the answer to be accurate to 5 decimal places. You may wish to use a spreadsheet.

2 Compare your result with $\arctan 1$.

3 Compare the continued fraction and $\arctan x$ for another value of x of your choosing.

REVIEW SET 12A — NON-CALCULATOR

1 Which of the following graphs display periodic behaviour?

a

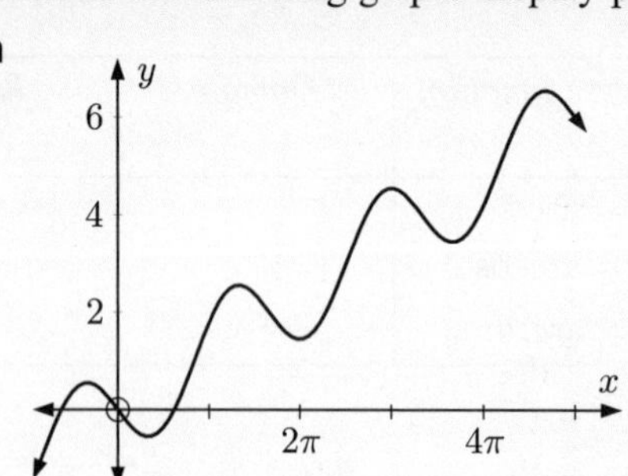

b

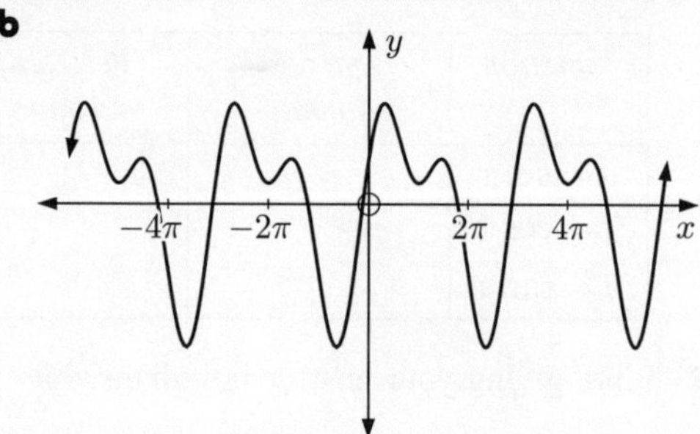

2 Draw the graph of $y = 4\sin x$ for $0 \leqslant x \leqslant 2\pi$.

3 State the minimum and maximum values of:

a $1 + \sin x$ **b** $-2\cos 3x$

4 State the period of:

a $y = 4\sin(\frac{x}{5})$ **b** $y = -2\cos(4x)$ **c** $y = 4\cos(\frac{x}{2}) + 4$ **d** $y = \frac{1}{2}\tan(3x)$

5 Complete the table:

Function	*Period*	*Amplitude*	*Domain*	*Range*
$y = -3\sin(\frac{x}{4}) + 1$				
$y = \tan 2x$				
$y = 3\cos \pi x$				

6 Find the cosine function represented in each of the following graphs:

a

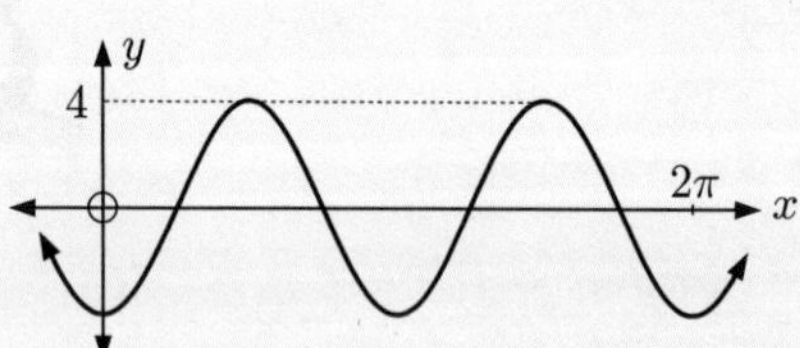

b

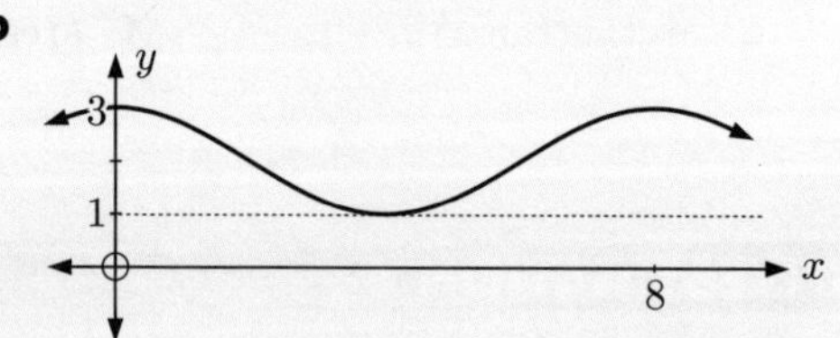

7 State the transformations which map:

a $y = \sin x$ onto $y = 3\sin(2x)$ **b** $y = \cos x$ onto $y = \cos\left(x - \frac{\pi}{3}\right) - 1$

8 Find the remaining five trigonometric ratios from sin, cos, tan, csc, sec, and cot, if:

a $\cos x = \frac{1}{3}$ and $0 < x < \pi$ **b** $\tan x = \frac{4}{5}$ and $\pi < x < 2\pi$.

9 Simplify:

a $\arctan(\tan(-0.5))$ **b** $\arcsin(\sin(-\frac{\pi}{6}))$ **c** $\arccos(\cos 2\pi)$

REVIEW SET 12B — CALCULATOR

1 For each set of data below, draw a scatter diagram and state if the data exhibits approximately periodic behaviour.

a

x	0	1	2	3	4	5	6	7	8	9	10	11	12
y	2.7	0.8	−1.7	−3	−2.1	0.3	2.5	2.9	1.3	−1.3	−2.9	−2.5	−0.3

b

x	0	1	2	3	4	5	6	7	8	9
y	5	3.5	6	−1.5	4	−2.5	−0.8	0.9	2.6	4.3

2 Draw the graph of $y = \sin 3x$ for $0 \leqslant x \leqslant 2\pi$.

3 State the period of: **a** $y = 4\sin(\frac{x}{3})$ **b** $y = -2\tan 4x$

4 Draw the graph of $y = 0.6\cos(2.3x)$ for $0 \leqslant x \leqslant 5$.

5 A robot on Mars records the temperature every Mars day. A summary series, showing every one hundredth Mars day, is shown in the table below.

Number of Mars days	0	100	200	300	400	500	600	700	800	900	1000	1100	1200	1300
Temp. (°C)	−43	−15	−5	−21	−59	−79	−68	−50	−27	−8	−15	−70	−78	−68

a Find the maximum and minimum temperatures recorded by the robot.

b Find a sine model for the temperature T in terms of the number of Mars days n.

c Use this information to estimate the length of a Mars year.

6 State the minimum and maximum values of:

a $y = 5\sin x - 3$ **b** $y = \frac{1}{3}\cos x + 1$

7 State the transformations which map:

a $y = \tan x$ onto $y = -\tan(2x)$ **b** $y = \sin x$ onto $y = 2\sin\left(\frac{x}{2} - \frac{\pi}{4}\right) + \frac{1}{2}$

8 **a** Sketch the graphs of $y = \sec x$ and $y = \csc x$ on the same set of axes for $-2\pi \leqslant x \leqslant 2\pi$.

b State a transformation which maps $y = \sec x$ onto $y = \csc x$ for all $x \in \mathbb{R}$.

9 **a** Sketch the graphs of $y = \arcsin x$ and $y = \arccos x$ on the same set of axes.

b State the domain and range of each function.

c State the transformations which map $y = \arcsin x$ onto $y = \arccos x$.

REVIEW SET 12C

1 Find b given that the function $y = \sin bx$, $b > 0$ has period:

a 6π **b** $\frac{\pi}{12}$ **c** 9

2 **a** Without using technology, draw the graph of $f(x) = \sin(x - \frac{\pi}{3}) + 2$ for $0 \leqslant x \leqslant 2\pi$.

b For what values of k will $f(x) = k$ have solutions?

3 Consider the graph alongside.

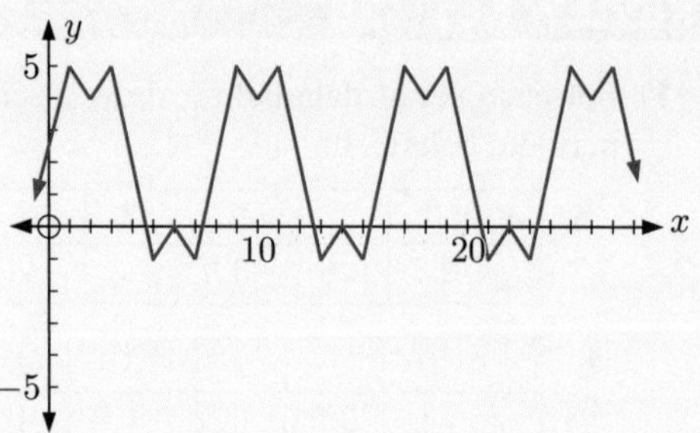

a Explain why this graph shows periodic behaviour.

b State:

i the period

ii the maximum value

iii the minimum value

4 On the same set of axes, for the domain $0 \leqslant x \leqslant 2\pi$, sketch:

a $y = \cos x$ and $y = \cos x - 3$

b $y = \cos x$ and $y = \cos(x - \frac{\pi}{4})$

c $y = \cos x$ and $y = 3\cos 2x$

d $y = \cos x$ and $y = 2\cos(x - \frac{\pi}{3}) + 3$

5 The table below gives the mean monthly maximum temperature for Perth Airport in Western Australia.

Month	Jan	Feb	Mar	Apr	May	Jun	Jul	Aug	Sep	Oct	Nov	Dec
Temp. (°C)	31.5	31.8	29.5	25.4	21.5	18.8	17.7	18.3	20.1	22.4	25.5	28.8

a A sine function of the form $T \approx a\sin(b(t - c)) + d$ is used to model the data. Find good estimates of the constants a, b, c, and d without using technology. Use Jan $\equiv 1$, Feb $\equiv 2$, and so on.

b Check your answer to **a** using technology. How well does your model fit?

6 State the transformations which map:

a $y = \cos x$ onto $y = \cos(x - \frac{\pi}{3}) + 1$

b $y = \tan x$ onto $y = -2\tan x$

c $y = \sin x$ onto $y = \sin(3x)$

7 Find the function represented in each of the following graphs:

a

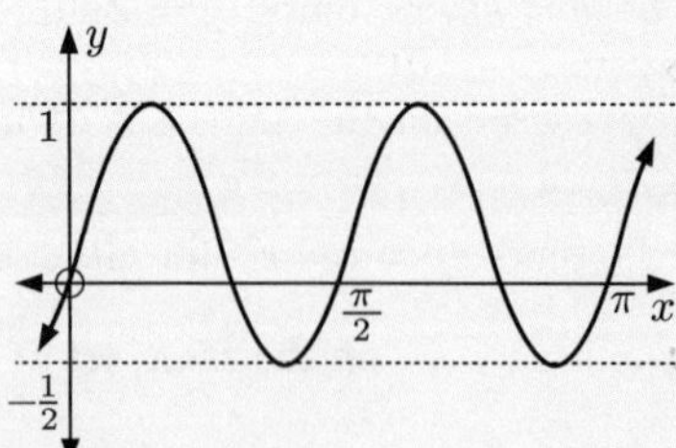

b

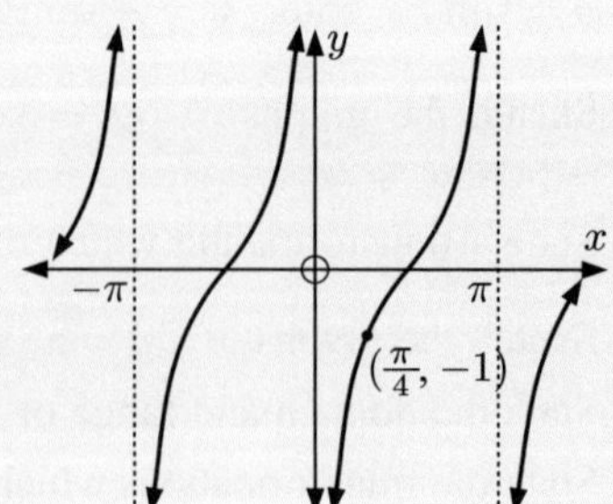

8 Simplify:

a $\csc x \tan x$

b $\dfrac{\tan x}{\sec x}$

c $\sec x - \tan x \sin x$

9 **a** For what restricted domain of $y = \tan x$, is $y = \arctan x$ the inverse function?

b Sketch $y = \tan x$ for this domain, and $y = \arctan x$, on the same set of axes.

Chapter 13

Trigonometric equations and identities

Syllabus reference: 3.3, 3.6

Contents:

OPENING PROBLEM

Andrew is watching a Ferris wheel rotate at constant speed. There are many lights around the Ferris wheel, and Andrew watches a green light closely. The height of the green light after t seconds is given by $H(t) = 10\sin(\frac{\pi}{50}(t-25)) + 12$ metres.

Things to think about:

a At what height will the green light be after 50 seconds?

b How long does it take for the wheel to complete a full circle?

c At what times in the first three minutes will the green light be 16 metres above the ground?

A TRIGONOMETRIC EQUATIONS

Linear equations such as $2x+3=11$ have exactly one solution. Quadratic equations of the form $ax^2+bx+c=0,\ a\neq 0$ have at most two real solutions.

Trigonometric equations generally have infinitely many solutions unless a restricted domain such as $0 \leqslant x \leqslant 3\pi$ is given.

For example, in the **Opening Problem**, the green light will be 16 metres above the ground when $10\sin(\frac{\pi}{50}(t-25)) + 12 = 16$ metres.

This is a trigonometric equation, and it has infinitely many solutions provided the wheel keeps rotating. For this reason we need to specify if we are interested in the first three minutes of its rotation, which is when $0 \leqslant t \leqslant 180$.

We will examine solving trigonometric equations using:

- pre-prepared graphs
- technology
- algebra.

GRAPHICAL SOLUTION OF TRIGONOMETRIC EQUATIONS

Sometimes simple trigonometric graphs are available on grid paper. In such cases we can estimate solutions straight from the graph.

Example 1 **Self Tutor**

Solve $\cos x = 0.4$ for $0 \leqslant x \leqslant 10$ radians using the graph of $y = \cos x$.

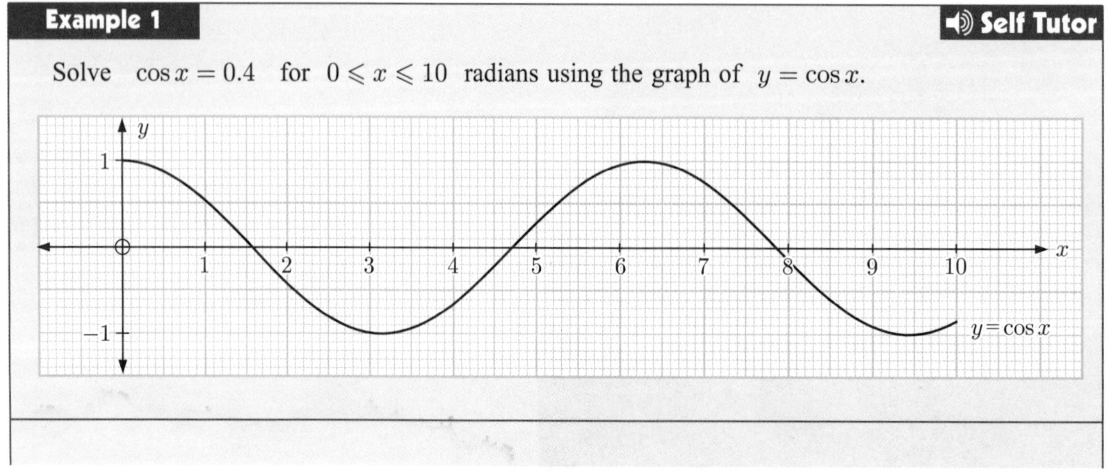

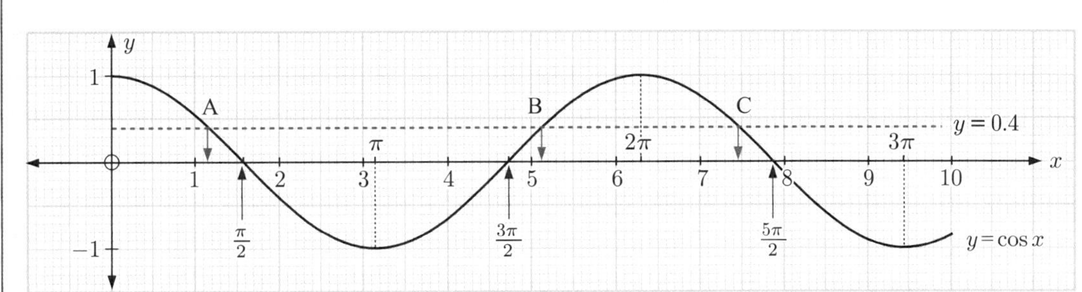

$y = 0.4$ meets $y = \cos x$ at A, B, and C. Hence $x \approx 1.2, 5.1$, or 7.4.

The solutions of $\cos x = 0.4$ for $0 \leqslant x \leqslant 10$ radians are 1.2, 5.1, and 7.4.

EXERCISE 13A.1

1

Use the graph of $y = \sin x$ to find, correct to 1 decimal place, the solutions of:

a $\sin x = 0.3$ for $0 \leqslant x \leqslant 15$

b $\sin x = -0.4$ for $5 \leqslant x \leqslant 15$.

2

Use the graph of $y = \cos x$ to find, correct to 1 decimal place, the solutions of:

a $\cos x = 0.4$ for $0 \leqslant x \leqslant 10$

b $\cos x = -0.3$ for $4 \leqslant x \leqslant 12$.

3

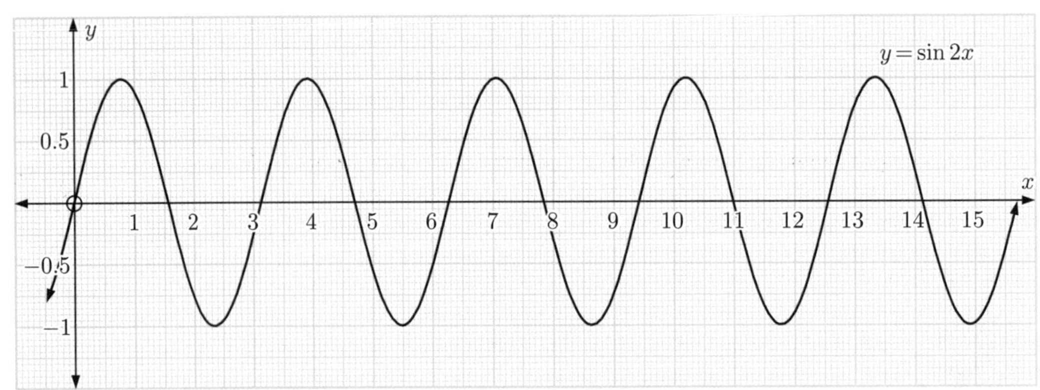

Use the graph of $y = \sin 2x$ to find, correct to 1 decimal place, the solutions of:

a $\sin 2x = 0.7$ for $0 \leqslant x \leqslant 16$

b $\sin 2x = -0.3$ for $0 \leqslant x \leqslant 16$.

4

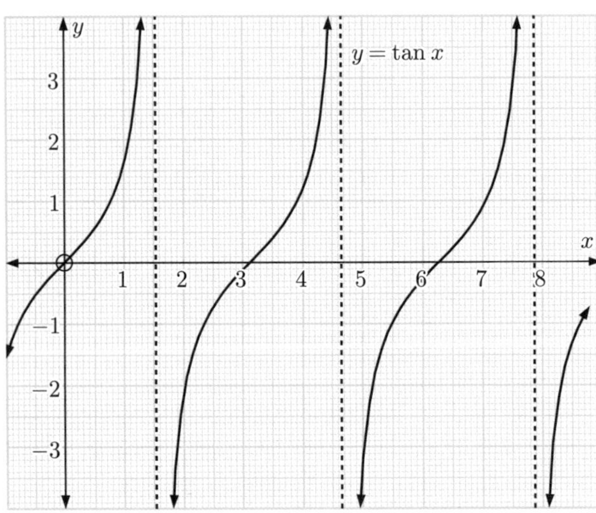

The graph of $y = \tan x$ is illustrated.

a Use the graph to estimate: i $\tan 1$ ii $\tan 2.3$

Check your answers with a calculator.

b Find, correct to 1 decimal place, the solutions of:

i $\tan x = 2$ for $0 \leqslant x \leqslant 8$

ii $\tan x = -1.4$ for $2 \leqslant x \leqslant 7$.

SOLVING TRIGONOMETRIC EQUATIONS USING TECHNOLOGY

Trigonometric equations may be solved using either a **graphing package** or a **graphics calculator**.

GRAPHING PACKAGE

When using a graphics calculator make sure that the **mode** is set to **radians**.

Example 2 **Self Tutor**

Solve $2\sin x - \cos x = 4 - x$ for $0 \leqslant x \leqslant 2\pi$.

We graph the functions $Y_1 = 2\sin X - \cos X$ and $Y_2 = 4 - X$ on the same set of axes.

We need to use **window** settings just larger than the domain.

In this case, $X\text{min} = -\frac{\pi}{6}$ $X\text{max} = \frac{13\pi}{6}$ $X\text{scale} = \frac{\pi}{6}$

GRAPHICS CALCULATOR INSTRUCTIONS

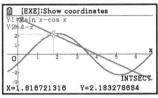

Casio fx-CG20 **TI-84 Plus** **TI-*n*spire**

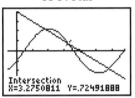

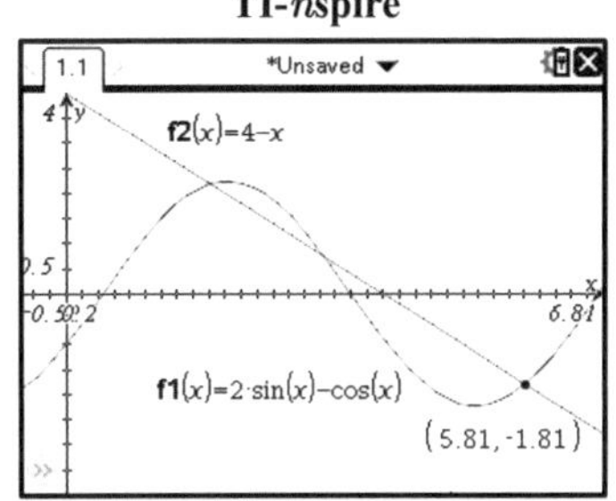

The solutions are $x \approx 1.82$, 3.28, and 5.81.

EXERCISE 13A.2

1. Solve for x on the domain $0 < x < 12$:
   a. $\sin x = 0.431$
   b. $\cos x = -0.814$
   c. $3\tan x - 2 = 0$
2. Solve for x on the domain $-5 \leqslant x \leqslant 5$:
   a. $5\cos x - 4 = 0$
   b. $2\tan x + 13 = 0$
   c. $8\sin x + 3 = 0$
3. Solve each of the following for $0 \leqslant x \leqslant 2\pi$:
   a. $\sin(x + 2) = 0.0652$
   b. $\sin^2 x + \sin x - 1 = 0$
   c. $x\tan\left(\frac{x^2}{10}\right) = x^2 - 6x + 1$
   d. $2\sin(2x)\cos x - \ln x$
4. Solve for x: $\cos(x - 1) + \sin(x + 1) = 6x + 5x^2 - x^3$, $-2 \leqslant x \leqslant 6$.

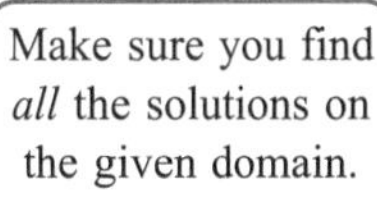

Make sure you find *all* the solutions on the given domain.

SOLVING TRIGONOMETRIC EQUATIONS USING ALGEBRA

Using a calculator we get approximate decimal or **numerical** solutions to trigonometric equations.

Sometimes exact solutions are needed in terms of π, and these arise when the solutions are multiples of $\frac{\pi}{6}$ or $\frac{\pi}{4}$. Exact solutions obtained using algebra are called **analytical** solutions.

We use the periodicity of the trigonometric functions to give us all solutions in the required domain. Remember that $\sin x$ and $\cos x$ both have period 2π, and $\tan x$ has period π.

For example, consider $\sin x = 1$. We know from the unit circle that a solution is $x = \frac{\pi}{2}$. However, since the period of $\sin x$ is 2π, there are infinitely many solutions spaced 2π apart.

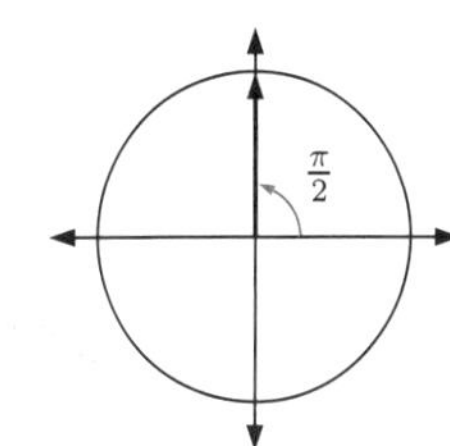

In general, $x = \frac{\pi}{2} + k2\pi$ is a solution for any $k \in \mathbb{Z}$.

In this course we will be solving equations on a fixed domain. This means there will be a finite number of solutions.

Reminder:

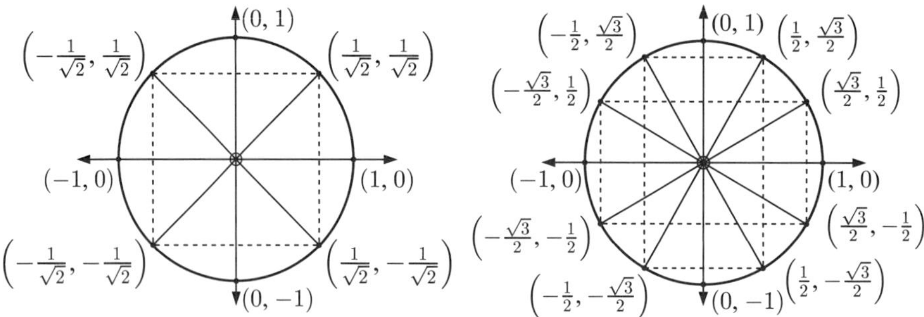

Example 3 **Self Tutor**

Solve for x: $2\sin x - 1 = 0, \quad 0 \leqslant x \leqslant \pi$

$2\sin x - 1 = 0$

$\therefore \sin x = \frac{1}{2}$

There are two points on the unit circle with sine $\frac{1}{2}$.

They correspond to angles $\frac{\pi}{6}$ and $\frac{5\pi}{6}$.

These are the only solutions in the domain $0 \leqslant x \leqslant \pi$, so $x = \frac{\pi}{6}$ or $\frac{5\pi}{6}$.

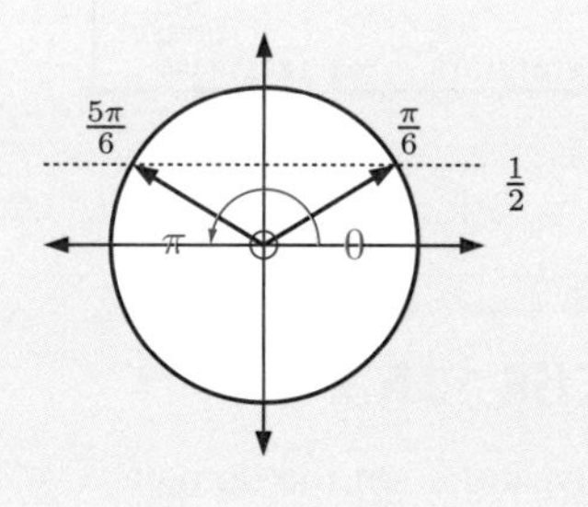

Since the tangent function is periodic with period π we see that $\tan(x + \pi) = \tan x$ for all values of x. This means that equal tan values are π units apart.

Example 4 **Self Tutor**

Solve $\tan x + \sqrt{3} = 0$ for $0 < x < 4\pi$.

$\tan x + \sqrt{3} = 0$

$\therefore \tan x = -\sqrt{3}$

There are two points on the unit circle with tangent $-\sqrt{3}$.

They correspond to angles $\frac{2\pi}{3}$ and $\frac{5\pi}{3}$.

For the domain $0 < x < 4\pi$ we have 4 solutions:

$x = \frac{2\pi}{3}, \frac{5\pi}{3}, \frac{8\pi}{3}$ or $\frac{11\pi}{3}$.

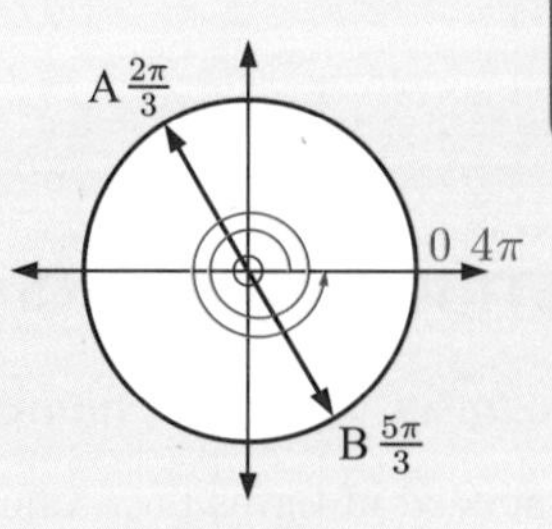

Start at angle 0 and work around to 4π, noting down the angle every time you reach points A and B.

EXERCISE 13A.3

1 Solve for x on the domain $0 \leqslant x \leqslant 4\pi$:

a $2\cos x - 1 = 0$ b $\sqrt{2}\sin x = 1$ c $\tan x = 1$

2 Solve for x on the domain $-2\pi \leqslant x \leqslant 2\pi$:

a $2\sin x - \sqrt{3} = 0$ b $\sqrt{2}\cos x + 1 = 0$ c $\tan x = -1$

Example 5 **Self Tutor**

Solve exactly for $0 \leqslant x \leqslant 3\pi$:

a $\sin x = -\frac{1}{2}$ b $\sin 2x = -\frac{1}{2}$ c $\sin(x - \frac{\pi}{6}) = -\frac{1}{2}$

The three equations all have the form $\sin\theta = -\frac{1}{2}$.

There are two points on the unit circle with sine $-\frac{1}{2}$.

They correspond to angles $\frac{7\pi}{6}$ and $\frac{11\pi}{6}$.

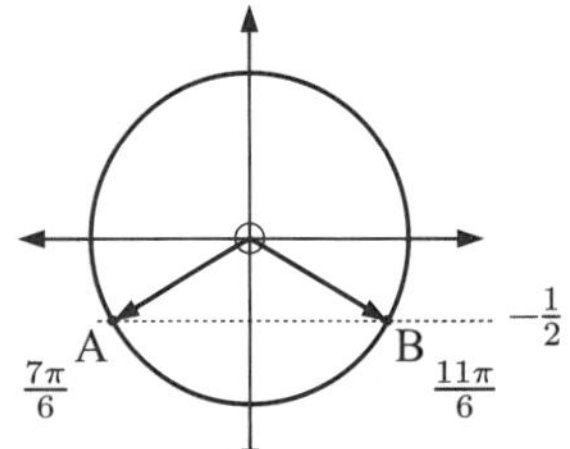

a In this case θ is simply x, so we have the domain $0 \leqslant x \leqslant 3\pi$.
The only solutions for this domain are $x = \frac{7\pi}{6}$ or $\frac{11\pi}{6}$.

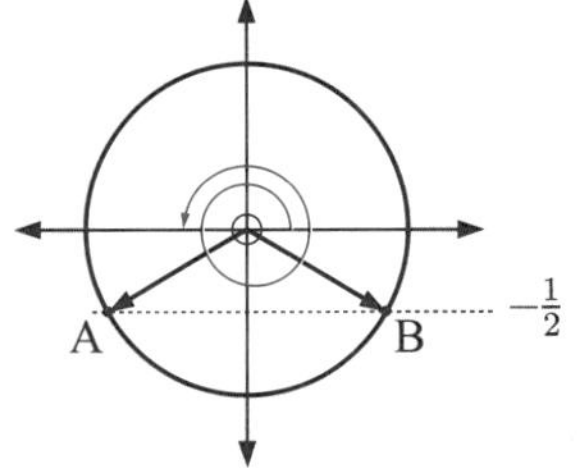

b In this case θ is $2x$.
If $0 \leqslant x \leqslant 3\pi$ then $0 \leqslant 2x \leqslant 6\pi$.
$\therefore\ 2x = \frac{7\pi}{6}, \frac{11\pi}{6}, \frac{19\pi}{6}, \frac{23\pi}{6}, \frac{31\pi}{6}$, or $\frac{35\pi}{6}$
$\therefore\ x = \frac{7\pi}{12}, \frac{11\pi}{12}, \frac{19\pi}{12}, \frac{23\pi}{12}, \frac{31\pi}{12}$, or $\frac{35\pi}{12}$

c In this case θ is $x - \frac{\pi}{6}$.
If $0 \leqslant x \leqslant 3\pi$ then $-\frac{\pi}{6} \leqslant x - \frac{\pi}{6} \leqslant \frac{17\pi}{6}$.

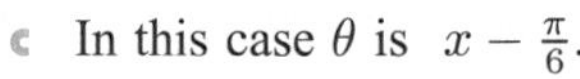

$\therefore\ x - \frac{\pi}{6} = -\frac{\pi}{6}, \frac{7\pi}{6}$, or $\frac{11\pi}{6}$
$\therefore\ x = 0, \frac{4\pi}{3}$, or 2π

Start at $-\frac{\pi}{6}$ and work around to $\frac{17\pi}{6}$, noting down the angle every time you reach points A and B.

3 If $0 \leqslant x \leqslant 2\pi$, what are the possible values of:

a $2x$ b $\frac{x}{3}$ c $x + \frac{\pi}{2}$ d $x - \frac{\pi}{6}$ e $2(x - \frac{\pi}{4})$ f $-x$

4 If $-\pi \leqslant x \leqslant \pi$, what are the possible values of:

a $3x$ b $\frac{x}{4}$ c $x - \frac{\pi}{2}$ d $2x + \frac{\pi}{2}$ e $-2x$ f $\pi - x$

5 Solve exactly for $0 \leqslant x \leqslant 3\pi$:

a $\cos x = \frac{1}{2}$ b $\cos 2x = \frac{1}{2}$ c $\cos(x + \frac{\pi}{3}) = \frac{1}{2}$

Example 6 — Self Tutor

Find exact solutions of $\sqrt{2}\cos(x - \frac{3\pi}{4}) + 1 = 0$ for $0 \leqslant x \leqslant 6\pi$.

Rearranging $\sqrt{2}\cos(x - \frac{3\pi}{4}) + 1 = 0$, we find $\cos(x - \frac{3\pi}{4}) = -\frac{1}{\sqrt{2}}$.

We recognise $\frac{1}{\sqrt{2}}$ as a special fraction for multiples of $\frac{\pi}{4}$, and we identify two points on the unit circle with cosine $-\frac{1}{\sqrt{2}}$.

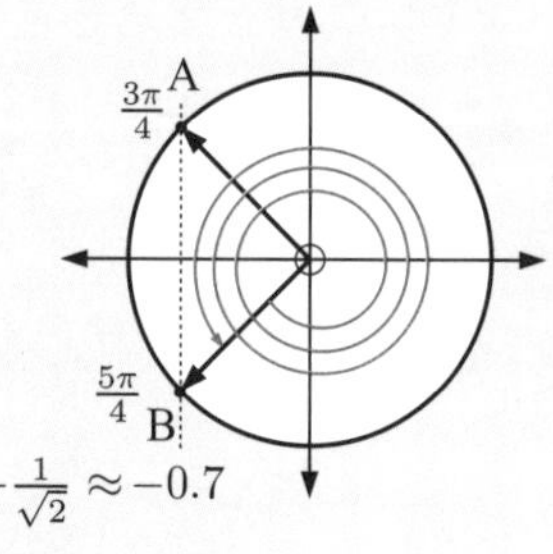

Start at $-\frac{3\pi}{4}$ and work around to $\frac{21\pi}{4}$, noting down the angle every time you reach points A and B.

Since $0 \leqslant x \leqslant 6\pi$,

$-\frac{3\pi}{4} \leqslant x - \frac{3\pi}{4} \leqslant \frac{21\pi}{4}$

So, $x - \frac{3\pi}{4} = -\frac{3\pi}{4}, \frac{3\pi}{4}, \frac{5\pi}{4}, \frac{11\pi}{4}, \frac{13\pi}{4}, \frac{19\pi}{4}$, or $\frac{21\pi}{4}$

$\therefore\ x = 0, \frac{3\pi}{2}, 2\pi, \frac{7\pi}{2}, 4\pi, \frac{11\pi}{2}$, or 6π

6 Find the exact solutions of:

a $\cos x = -\frac{1}{2},\ 0 \leqslant x \leqslant 5\pi$

b $2\sin x - 1 = 0,\ -360° \leqslant x \leqslant 360°$

c $2\cos x + \sqrt{3} = 0,\ 0 \leqslant x \leqslant 3\pi$

d $\cos(x - \frac{2\pi}{3}) = \frac{1}{2},\ -2\pi \leqslant x \leqslant 2\pi$

e $2\sin(x + \frac{\pi}{3}) = 1,\ -3\pi \leqslant x \leqslant 3\pi$

f $\sqrt{2}\sin(x - \frac{\pi}{4}) + 1 = 0,\ 0 \leqslant x \leqslant 3\pi$

g $3\cos 2x + 3 = 0,\ 0 \leqslant x \leqslant 3\pi$

h $4\cos 3x + 2 = 0,\ -\pi \leqslant x \leqslant \pi$

i $\sin(4(x - \frac{\pi}{4})) = 0,\ 0 \leqslant x \leqslant \pi$

j $2\sin(2(x - \frac{\pi}{3})) = -\sqrt{3},\ 0 \leqslant x \leqslant 2\pi$

7 Solve for x exactly where $x \in [0, 2\pi]$:

a $\sec x = 2$ **b** $\csc x = -\sqrt{2}$ **c** $\sqrt{3}\sec 2x = -2$ **d** $\csc\left(x + \frac{\pi}{6}\right) + \sqrt{2} = 0$

Example 7 — Self Tutor

Find exact solutions of $\tan(2x - \frac{\pi}{3}) = 1$ for $-\pi \leqslant x \leqslant \pi$.

There are two points on the unit circle which have tangent 1.

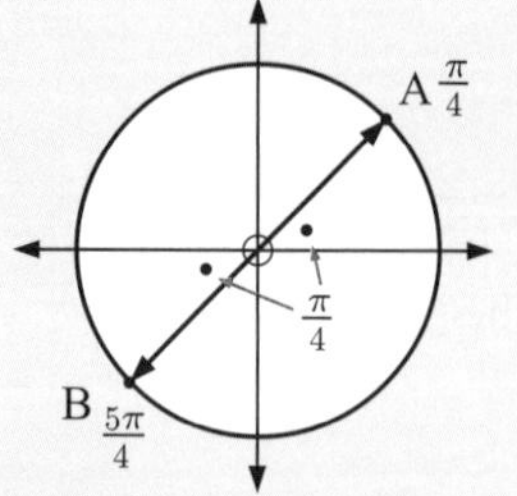

Start at $-\frac{7\pi}{3}$ and work around to $\frac{5\pi}{3}$, noting down the angle every time you reach points A and B.

Since $-\pi \leqslant x \leqslant \pi$,

$-2\pi \leqslant 2x \leqslant 2\pi$

$\therefore\ -\frac{7\pi}{3} \leqslant 2x - \frac{\pi}{3} \leqslant \frac{5\pi}{3}$

So, $2x - \frac{\pi}{3} = -\frac{7\pi}{4}, -\frac{3\pi}{4}, \frac{\pi}{4}$, or $\frac{5\pi}{4}$

$\therefore\ 2x = -\frac{17\pi}{12}, -\frac{5\pi}{12}, \frac{7\pi}{12}$, or $\frac{19\pi}{12}$

$\therefore\ x = -\frac{17\pi}{24}, -\frac{5\pi}{24}, \frac{7\pi}{24}$, or $\frac{19\pi}{24}$

8 Find the exact solutions of $\tan x = \sqrt{3}$ for $0 \leqslant x \leqslant 2\pi$. Hence solve the following equations for $0 \leqslant x \leqslant 2\pi$:

a $\tan(x - \frac{\pi}{6}) = \sqrt{3}$ **b** $\tan 4x = \sqrt{3}$ **c** $\tan^2 x = 3$

9 Find the exact solutions of the following for $0 \leqslant x \leqslant 2\pi$:

a $\cot x + 1 = 0$ **b** $\cot\left(2x - \frac{\pi}{4}\right) - \sqrt{3} = 0$

10 Find exactly the zeros of:

a $y = \sin 2x$ for $0° \leqslant x \leqslant 180°$ **b** $y = \sin(x - \frac{\pi}{4})$ for $0 \leqslant x \leqslant 3\pi$

Example 8 Self Tutor

Find the exact solutions of $\sqrt{3}\sin x = \cos x$ for $0° \leqslant x \leqslant 360°$.

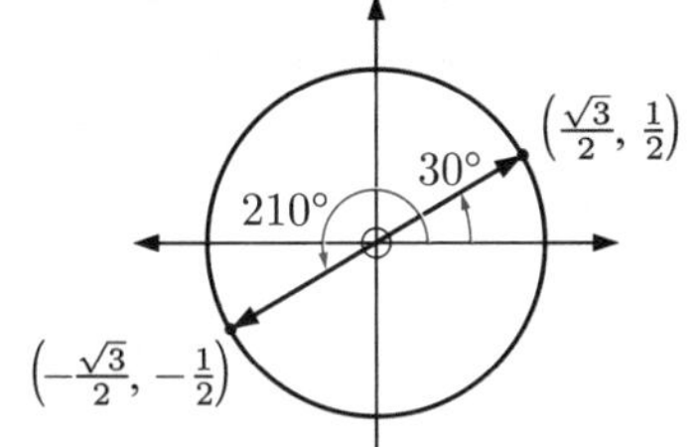

$\sqrt{3}\sin x = \cos x$

$\therefore \quad \frac{\sin x}{\cos x} = \frac{1}{\sqrt{3}}$ {dividing both sides by $\sqrt{3}\cos x$}

$\therefore \quad \tan x = \frac{1}{\sqrt{3}}$

$\therefore \quad x = 30°$ or $210°$

11 **a** Use your graphics calculator to sketch the graphs of $y = \sin x$ and $y = \cos x$ on the same set of axes on the domain $0 \leqslant x \leqslant 2\pi$.

b Find the x-coordinates of the points of intersection of the two graphs.

12 Find the exact solutions to these equations for $0 \leqslant x \leqslant 2\pi$:

a $\sin x = -\cos x$ **b** $\sin(3x) = \cos(3x)$ **c** $\sin(2x) = \sqrt{3}\cos(2x)$

Check your answers using a graphics calculator by finding the points of intersection of the appropriate graphs.

Example 9 Self Tutor

Find, where possible, the exact solutions of:

a $\arctan x = \frac{\pi}{3}$ **b** $\arccos(x - 1) = \frac{2\pi}{3}$ **c** $\arcsin x = \frac{2\pi}{3}$

a The range of $y = \arctan x$ is $-\frac{\pi}{2} < y < \frac{\pi}{2}$. $\frac{\pi}{3}$ is within the range.

$\therefore \quad x = \tan\frac{\pi}{3} = \sqrt{3}$

b The range of $y = \arccos(x - 1)$ is $0 \leqslant y \leqslant \pi$. $\frac{2\pi}{3}$ is within the range.

$\therefore \quad x - 1 = \cos(\frac{2\pi}{3})$

$\therefore \quad x - 1 = -\frac{1}{2}$

$\therefore \quad x = \frac{1}{2}$

c The range of $y = \arcsin x$ is $-\frac{\pi}{2} \leqslant y \leqslant \frac{\pi}{2}$, and $\frac{2\pi}{3}$ is outside this range.

$\therefore$ no solution exists, even though we can find $\sin(\frac{2\pi}{3})$.

13 Find, where possible, the exact solutions of:

a $\arctan x = \frac{\pi}{4}$ **b** $\arcsin x = -\frac{\pi}{3}$ **c** $\arccos x = \frac{3\pi}{4}$

d $\arcsin(x + 1) = \frac{\pi}{6}$ **e** $\arccos x = -\frac{\pi}{4}$ **f** $\arctan(x - \sqrt{3}) = -\frac{\pi}{3}$

B USING TRIGONOMETRIC MODELS

Having discussed the solution of trigonometric equations, we can now put into use the trigonometric model from **Chapter 12**.

Example 10 Self Tutor

The height of the tide above mean sea level on January 24th at Cape Town is modelled approximately by $h(t) = 3\sin(\frac{\pi t}{6})$ metres where t is the number of hours after midnight.

a Graph $y = h(t)$ for $0 \leqslant t \leqslant 24$.

b When is high tide and what is the maximum height?

c What is the height of the tide at 2 pm?

d A ship can cross the harbour provided the tide is at least 2 m above mean sea level. When is crossing possible on January 24th?

a $h(0) = 0$

$h(t) = 3\sin(\frac{\pi t}{6})$ has period $= \dfrac{2\pi}{\frac{\pi}{6}} = 2\pi \times \frac{6}{\pi} = 12$ hours

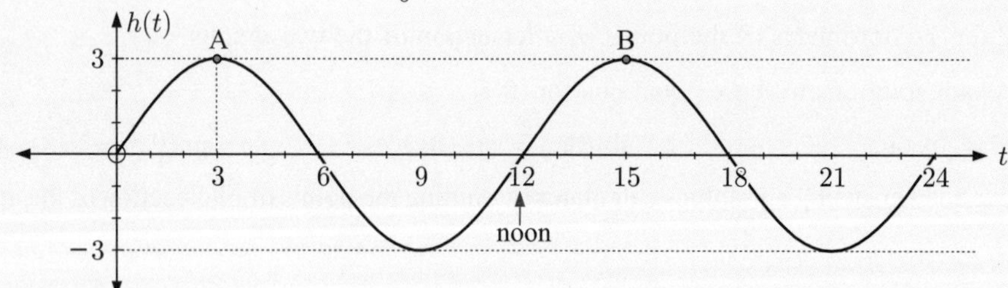

b High tide is at 3 am and 3 pm. The maximum height is 3 m above the mean as seen at points A and B.

c At 2 pm, $t = 14$ and $h(14) = 3\sin(\frac{14\pi}{6}) \approx 2.60$ m.

So, the tide is 2.6 m above the mean.

d

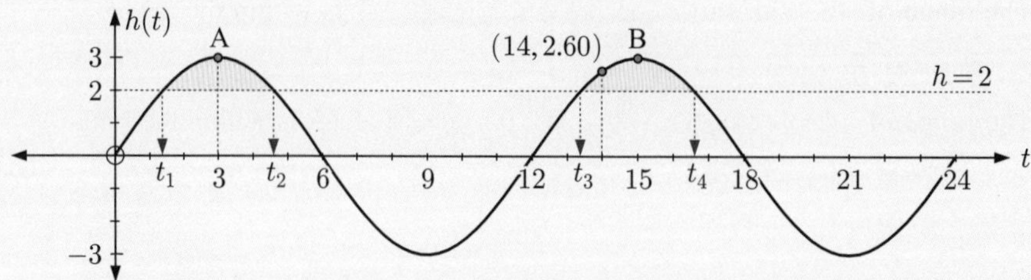

We need to solve $h(t) = 2$, so $3\sin(\frac{\pi t}{6}) = 2$.

Using a graphics calculator with $\text{Y}_1 = 3\sin(\frac{\pi \text{X}}{6})$ and $\text{Y}_2 = 2$

we obtain $t_1 \approx 1.39$, $t_2 \approx 4.61$, $t_3 \approx 13.39$, $t_4 \approx 16.61$

Now 1.39 hours = 1 hour 23 minutes, and so on.

So, the ship can cross between 1:23 am and 4:37 am or 1:23 pm and 4:37 pm.

EXERCISE 13B

1. Answer the **Opening Problem** on page **354**.

2. The population of grasshoppers after t weeks where $0 \leqslant t \leqslant 12$ is estimated by $P(t) = 7500 + 3000\sin(\frac{\pi t}{8})$.
   a. Find: i the initial estimate ii the estimate after 5 weeks.
   b. What is the greatest population size over this interval and when does it occur?
   c. When is the population: i 9000 ii 6000?
   d. During what time interval(s) does the population size exceed 10 000?

3. The model for the height of a light on a certain Ferris wheel is $H(t) = 20 - 19\sin(\frac{2\pi t}{3})$, where H is the height in metres above the ground, and t is in minutes.
   a. Where is the light at time $t = 0$?
   b. At what time is the light at its lowest in the first revolution of the wheel?
   c. How long does the wheel take to complete one revolution?
   d. Sketch the graph of the function $H(t)$ over one revolution.

4. The population of water buffalo is given by $P(t) = 400 + 250\sin(\frac{\pi t}{2})$ where t is the number of years since the first estimate was made.
   a. What was the initial estimate?
   b. What was the population size after:
      i 6 months ii two years?
   c. Find $P(1)$. What is the significance of this value?
   d. Find the smallest population size and when it first occurred.
   e. Find the first time when the herd exceeded 500.

5. A paint spot X lies on the outer rim of the wheel of a paddle-steamer.
   The wheel has radius 3 m. It rotates anticlockwise at a constant rate, and X is seen entering the water every 4 seconds.
   H is the distance of X above the bottom of the boat. At time $t = 0$, X is at its highest point.

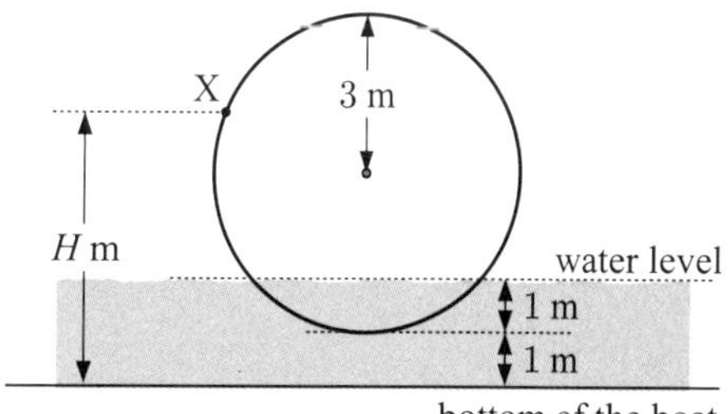

   a. Find a cosine model for H in the form $H(t) = a\cos(b(t + c)) + d$.
   b. At what time t does X first enter the water?

6. Over a 28 day period, the cost per litre of petrol was modelled by $C(t) = 9.2\sin(\frac{\pi}{7}(t - 4)) + 107.8$ cents L^{-1}.
   a. True or false?
      i "The cost per litre oscillates about 107.8 cents with maximum price \$1.17 per litre."
      ii "Every 14 days, the cycle repeats itself."
   b. What was the cost of petrol on day 7, to the nearest tenth of a cent per litre?
   c. On which days was the petrol priced at \$1.10 per litre?
   d. What was the minimum cost per litre and when did it occur?

C TRIGONOMETRIC RELATIONSHIPS

There are a vast number of trigonometric relationships. However, we only need to remember a few because we can obtain the rest by rearrangement or substitution.

SIMPLIFYING TRIGONOMETRIC EXPRESSIONS

For any given angle θ, $\sin\theta$ and $\cos\theta$ are real numbers. $\tan\theta$ is also real whenever it is defined. The algebra of trigonometry is therefore identical to the algebra of real numbers.

An expression like $2\sin\theta + 3\sin\theta$ compares with $2x + 3x$ when we wish to do simplification, and so $2\sin\theta + 3\sin\theta = 5\sin\theta$.

To simplify complicated trigonometric expressions, we often use the identities:

$$\boldsymbol{\sin^2\theta + \cos^2\theta = 1}$$

$$\boldsymbol{\tan\theta = \frac{\sin\theta}{\cos\theta}}$$

$$\boldsymbol{\tan^2\theta + 1 = \sec^2\theta}$$

$$\boldsymbol{1 + \cot^2\theta = \csc^2\theta}$$

We can also use rearrangements of these formulae, such as:

$$\sin^2\theta = 1 - \cos^2\theta \qquad \tan^2\theta = \sec^2\theta - 1$$
$$\cos^2\theta = 1 - \sin^2\theta \qquad \cot^2\theta = \csc^2\theta - 1$$

Example 11 — Self Tutor

Simplify: **a** $3\cos\theta + 4\cos\theta$ **b** $\tan\alpha - 3\tan\alpha$

a $3\cos\theta + 4\cos\theta = 7\cos\theta$
{compare with $3x + 4x = 7x$}

b $\tan\alpha - 3\tan\alpha = -2\tan\alpha$
{compare with $x - 3x = -2x$}

EXERCISE 13C.1

1 Simplify:

a $\sin\theta + \sin\theta$ **b** $2\cos\theta + \cos\theta$ **c** $3\sin\theta - \sin\theta$
d $3\sin\theta - 2\sin\theta$ **e** $\tan\theta - 3\tan\theta$ **f** $2\cos^2\theta - 5\cos^2\theta$

Example 12 — Self Tutor

Simplify: **a** $2 - 2\sin^2\theta$ **b** $\cos^2\theta\sin\theta + \sin^3\theta$

a
$2 - 2\sin^2\theta$
$= 2(1 - \sin^2\theta)$
$= 2\cos^2\theta$
{as $\cos^2\theta + \sin^2\theta = 1$}

b
$\cos^2\theta\sin\theta + \sin^3\theta$
$= \sin\theta(\cos^2\theta + \sin^2\theta)$
$= \sin\theta \times 1$
$= \sin\theta$

Example 13 **Self Tutor**

Expand and simplify: $(\cos\theta - \sin\theta)^2$

$$\begin{aligned}&(\cos\theta - \sin\theta)^2\\ &= \cos^2\theta - 2\cos\theta\sin\theta + \sin^2\theta \qquad \{\text{using } (a-b)^2 = a^2 - 2ab + b^2\}\\ &= \cos^2\theta + \sin^2\theta - 2\cos\theta\sin\theta\\ &= 1 - 2\cos\theta\sin\theta\end{aligned}$$

2 Simplify:

- **a** $3\sin^2\theta + 3\cos^2\theta$
- **b** $-2\sin^2\theta - 2\cos^2\theta$
- **c** $-\cos^2\theta - \sin^2\theta$
- **d** $3 - 3\sin^2\theta$
- **e** $4 - 4\cos^2\theta$
- **f** $\cos^3\theta + \cos\theta\sin^2\theta$
- **g** $\cos^2\theta - 1$
- **h** $\sin^2\theta - 1$
- **i** $2\cos^2\theta - 2$
- **j** $\dfrac{1-\sin^2\theta}{\cos^2\theta}$
- **k** $\dfrac{1-\cos^2\theta}{\sin\theta}$
- **l** $\dfrac{\cos^2\theta - 1}{-\sin\theta}$

3 Simplify:

- **a** $3\tan x - \dfrac{\sin x}{\cos x}$
- **b** $\dfrac{\sin^2 x}{\cos^2 x}$
- **c** $\tan x\cos x$
- **d** $\dfrac{\sin x}{\tan x}$
- **e** $3\sin x + 2\cos x\tan x$
- **f** $\dfrac{2\tan x}{\sin x}$

4 Expand and simplify if possible:

- **a** $(1+\sin\theta)^2$
- **b** $(\sin\alpha - 2)^2$
- **c** $(\tan\alpha - 1)^2$
- **d** $(\sin\alpha + \cos\alpha)^2$
- **e** $(\sin\beta - \cos\beta)^2$
- **f** $-(2-\cos\alpha)^2$

5 Simplify:

- **a** $1 - \sec^2\beta$
- **b** $\dfrac{\tan^2\theta(\cot^2\theta + 1)}{\tan^2\theta + 1}$
- **c** $\cos^2\alpha(\sec^2\alpha - 1)$
- **d** $(\sin x + \tan x)(\sin x - \tan x)$
- **e** $(2\sin\theta + 3\cos\theta)^2 + (3\sin\theta - 2\cos\theta)^2$
- **f** $(1+\csc\theta)(\sin\theta - \sin^2\theta)$
- **g** $\sec A - \sin A\tan A - \cos A$

FACTORISING TRIGONOMETRIC EXPRESSIONS

Example 14 **Self Tutor**

Factorise:

a $\cos^2\alpha - \sin^2\alpha$ **b** $\tan^2\theta - 3\tan\theta + 2$

a $$\begin{aligned}&\cos^2\alpha - \sin^2\alpha\\ &= (\cos\alpha + \sin\alpha)(\cos\alpha - \sin\alpha) \qquad \{a^2 - b^2 = (a+b)(a-b)\}\end{aligned}$$

b $$\begin{aligned}&\tan^2\theta - 3\tan\theta + 2\\ &= (\tan\theta - 2)(\tan\theta - 1) \qquad \{x^2 - 3x + 2 = (x-2)(x-1)\}\end{aligned}$$

Example 15 **Self Tutor**

Simplify:

a $\dfrac{2-2\cos^2\theta}{1+\cos\theta}$ **b** $\dfrac{\cos\theta-\sin\theta}{\cos^2\theta-\sin^2\theta}$

a $\dfrac{2-2\cos^2\theta}{1+\cos\theta}$

$=\dfrac{2(1-\cos^2\theta)}{1+\cos\theta}$

$=\dfrac{2(1+\cos\theta)(1-\cos\theta)}{(1+\cos\theta)}$

$=2(1-\cos\theta)$

b $\dfrac{\cos\theta-\sin\theta}{\cos^2\theta-\sin^2\theta}$

$=\dfrac{(\cos\theta-\sin\theta)}{(\cos\theta+\sin\theta)(\cos\theta-\sin\theta)}$

$=\dfrac{1}{\cos\theta+\sin\theta}$

EXERCISE 13C.2

1 Factorise:

a $1-\sin^2\theta$ **b** $\sin^2\alpha-\cos^2\alpha$ **c** $\tan^2\alpha-1$

d $2\sin^2\beta-\sin\beta$ **e** $2\cos\phi+3\cos^2\phi$ **f** $3\sin^2\theta-6\sin\theta$

g $\tan^2\theta+5\tan\theta+6$ **h** $2\cos^2\theta+7\cos\theta+3$ **i** $6\cos^2\alpha-\cos\alpha-1$

j $3\tan^2\alpha-2\tan\alpha$ **k** $\sec^2\beta-\csc^2\beta$ **l** $2\cot^2x-3\cot x+1$

m $2\sin^2x+7\sin x\cos x+3\cos^2x$

2 Simplify:

a $\dfrac{1-\sin^2\alpha}{1-\sin\alpha}$ **b** $\dfrac{\tan^2\beta-1}{\tan\beta+1}$ **c** $\dfrac{\cos^2\phi-\sin^2\phi}{\cos\phi+\sin\phi}$

d $\dfrac{\cos^2\phi-\sin^2\phi}{\cos\phi-\sin\phi}$ **e** $\dfrac{\sin\alpha+\cos\alpha}{\sin^2\alpha-\cos^2\alpha}$ **f** $\dfrac{3-3\sin^2\theta}{6\cos\theta}$

g $1-\dfrac{\cos^2\theta}{1+\sin\theta}$ **h** $\dfrac{1+\cot\theta}{\csc\theta}-\dfrac{\sec\theta}{\tan\theta+\cot\theta}$ **i** $\dfrac{\tan^2\theta}{\sec\theta-1}$

3 Show that:

a $(\cos\theta+\sin\theta)^2+(\cos\theta-\sin\theta)^2=2$ **b** $(2\sin\theta+3\cos\theta)^2+(3\sin\theta-2\cos\theta)^2=13$

c $(1-\cos\theta)\left(1+\dfrac{1}{\cos\theta}\right)=\tan\theta\sin\theta$ **d** $\left(1+\dfrac{1}{\sin\theta}\right)(\sin\theta-\sin^2\theta)=\cos^2\theta$

e $\sec A-\cos A=\tan A\sin A$ **f** $\dfrac{\cos\theta}{1-\sin\theta}=\sec\theta+\tan\theta$

g $\dfrac{\cos\alpha}{1-\tan\alpha}+\dfrac{\sin\alpha}{1-\cot\alpha}=\sin\alpha+\cos\alpha$ **h** $\dfrac{\sin\theta}{1+\cos\theta}+\dfrac{1+\cos\theta}{\sin\theta}=2\csc\theta$

i $\dfrac{\sin\theta}{1-\cos\theta}-\dfrac{\sin\theta}{1+\cos\theta}=2\cot\theta$ **j** $\dfrac{1}{1-\sin\theta}+\dfrac{1}{1+\sin\theta}=2\sec^2\theta$

Use a graphing package to check these simplifications by graphing each function on the same set of axes.

D DOUBLE ANGLE FORMULAE

INVESTIGATION 1 DOUBLE ANGLE FORMULAE

What to do:

1 Copy and complete, using angles of your choice as well:

θ	$\sin 2\theta$	$2\sin\theta$	$2\sin\theta\cos\theta$	$\cos 2\theta$	$2\cos\theta$	$\cos^2\theta - \sin^2\theta$
0.631						
57.81°						
-3.697						

2 Write down any discoveries from your table of values.

3 In the diagram alongside, the semi-circle has radius 1 unit, and $\widehat{PAB} = \theta$.

$\widehat{APO} = \theta$ {$\triangle$AOP is isosceles}

$\widehat{PON} = 2\theta$ {exterior angle of a triangle}

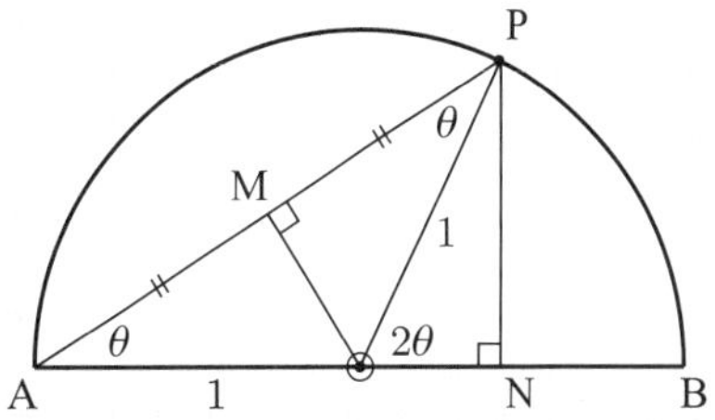

P
M
θ
1
θ
2θ
A
1
N
B

a Find in terms of θ, the lengths of:

i [OM] **ii** [AM]

iii [ON] **iv** [PN]

b Use $\triangle$ANP and the lengths in **a** to show that:

i $\cos\theta = \dfrac{\sin 2\theta}{2\sin\theta}$ **ii** $\cos\theta = \dfrac{1+\cos 2\theta}{2\cos\theta}$

c Hence deduce that:

i $\sin 2\theta = 2\sin\theta\cos\theta$ **ii** $\cos 2\theta = 2\cos^2\theta - 1$

From the **Investigation** you should have deduced two of the **double angle formulae**:
$\sin 2\theta = 2\sin\theta\cos\theta$ and $\cos 2\theta = 2\cos^2\theta - 1$.

These formulae are in fact true for all angles θ.

Using $\cos^2\theta = 1 - \sin^2\theta$, we find $\cos 2\theta = \cos^2\theta - \sin^2\theta$
and $\cos 2\theta = 1 - 2\sin^2\theta$.

Now
$$\tan 2\theta = \frac{\sin 2\theta}{\cos 2\theta} = \frac{2\sin\theta\cos\theta}{\cos^2\theta - \sin^2\theta} = \frac{\frac{2\sin\theta\cos\theta}{\cos^2\theta}}{\frac{\cos^2\theta}{\cos^2\theta} - \frac{\sin^2\theta}{\cos^2\theta}} = \frac{2\tan\theta}{1-\tan^2\theta}$$

So, the **double angle formulae** are:

$$\sin 2\theta = 2\sin\theta\cos\theta$$
$$\cos 2\theta = \cos^2\theta - \sin^2\theta = 1 - 2\sin^2\theta = 2\cos^2\theta - 1$$
$$\tan 2\theta = \frac{2\tan\theta}{1-\tan^2\theta}$$

GRAPHING PACKAGE

Example 16 Self Tutor

Given that $\sin\alpha = \frac{3}{5}$ and $\cos\alpha = -\frac{4}{5}$ find:

a $\sin 2\alpha$ **b** $\cos 2\alpha$

a $\sin 2\alpha$
$= 2\sin\alpha\cos\alpha$
$= 2(\frac{3}{5})(-\frac{4}{5})$
$= -\frac{24}{25}$

b $\cos 2\alpha$
$= \cos^2\alpha - \sin^2\alpha$
$= (-\frac{4}{5})^2 - (\frac{3}{5})^2$
$= \frac{7}{25}$

Example 17 Self Tutor

If $\sin\alpha = \frac{5}{13}$ where $\frac{\pi}{2} < \alpha < \pi$, find the exact value of: **a** $\sin 2\alpha$ **b** $\tan 2\alpha$.

α is in quadrant 2, so $\cos\alpha$ is negative.

Now $\cos^2\alpha + \sin^2\alpha = 1$
$\therefore \cos^2\alpha + \frac{25}{169} = 1$
$\therefore \cos^2\alpha = \frac{144}{169}$
$\therefore \cos\alpha = \pm\frac{12}{13}$
$\therefore \cos\alpha = -\frac{12}{13}$

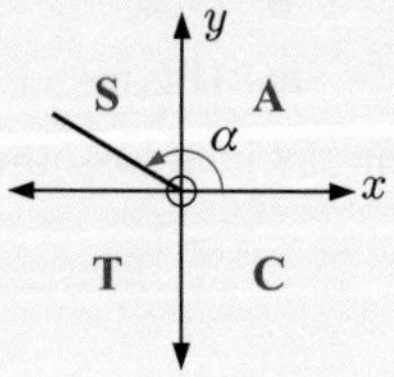

a $\sin 2\alpha = 2\sin\alpha\cos\alpha$
$\therefore \sin 2\alpha = 2(\frac{5}{13})(-\frac{12}{13})$
$= -\frac{120}{169}$

b $\tan\alpha = \frac{\sin\alpha}{\cos\alpha} = -\frac{5}{12}$
$\therefore \tan 2\alpha = \frac{2\tan\alpha}{1-\tan^2\alpha}$
$= \frac{2\left(-\frac{5}{12}\right)}{1-\left(-\frac{5}{12}\right)^2}$
$= \frac{-\frac{5}{6}}{1-\frac{25}{144}}$
$= -\frac{120}{119}$

EXERCISE 13D

1 If $\sin\theta = \frac{4}{5}$ and $\cos\theta = \frac{3}{5}$ find the exact values of:

a $\sin 2\theta$ **b** $\cos 2\theta$ **c** $\tan 2\theta$

2 **a** If $\cos A = \frac{1}{3}$, find $\cos 2A$. **b** If $\sin\phi = -\frac{2}{3}$, find $\cos 2\phi$.

3 If $\sin\alpha = -\frac{2}{3}$ where $\pi < \alpha < \frac{3\pi}{2}$, find the exact value of:

a $\cos\alpha$ **b** $\sin 2\alpha$

4 If $\cos\beta = \frac{2}{5}$ where $270° < \beta < 360°$, find the exact value of:

a $\sin\beta$ **b** $\sin 2\beta$

Example 18 **Self Tutor**

If α is acute and $\cos 2\alpha = \frac{3}{4}$ find the exact values of: **a** $\cos\alpha$ **b** $\sin\alpha$.

a
$$\cos 2\alpha = 2\cos^2\alpha - 1$$
$$\therefore \frac{3}{4} = 2\cos^2\alpha - 1$$
$$\therefore \cos^2\alpha = \frac{7}{8}$$
$$\therefore \cos\alpha = \pm\frac{\sqrt{7}}{2\sqrt{2}}$$
$$\therefore \cos\alpha = \frac{\sqrt{7}}{2\sqrt{2}}$$
{as α is acute, $\cos\alpha > 0$}

b
$$\sin\alpha = \sqrt{1-\cos^2\alpha}$$
{as α is acute, $\sin\alpha > 0$}
$$\therefore \sin\alpha = \sqrt{1-\frac{7}{8}}$$
$$\therefore \sin\alpha = \sqrt{\frac{1}{8}}$$
$$\therefore \sin\alpha = \frac{1}{2\sqrt{2}}$$

5 If α is acute and $\cos 2\alpha = -\frac{7}{9}$, find without a calculator: **a** $\cos\alpha$ **b** $\sin\alpha$.

6 Find the exact value of $\tan A$ if:

a $\tan 2A = \frac{21}{20}$ and A is obtuse. **b** $\tan 2A = -\frac{12}{5}$ and A is acute.

7 Find the exact value of $\tan\left(\frac{\pi}{8}\right)$.

Example 19 **Self Tutor**

Use an appropriate 'double angle formula' to simplify:

a $3\sin\theta\cos\theta$ **b** $4\cos^2 2B - 2$

a
$$3\sin\theta\cos\theta$$
$$= \frac{3}{2}(2\sin\theta\cos\theta)$$
$$= \frac{3}{2}\sin 2\theta$$

b
$$4\cos^2 2B - 2$$
$$= 2(2\cos^2 2B - 1)$$
$$= 2\cos 2(2B)$$
$$= 2\cos 4B$$

8 Find the exact value of $\left[\cos(\frac{\pi}{12}) + \sin(\frac{\pi}{12})\right]^2$.

9 Use an appropriate 'double angle' formula to simplify:

a $2\sin\alpha\cos\alpha$ **b** $4\cos\alpha\sin\alpha$ **c** $\sin\alpha\cos\alpha$

d $2\cos^2\beta - 1$ **e** $1 - 2\cos^2\phi$ **f** $1 - 2\sin^2 N$

g $2\sin^2 M - 1$ **h** $\cos^2\alpha - \sin^2\alpha$ **i** $\sin^2\alpha - \cos^2\alpha$

j $2\sin 2A\cos 2A$ **k** $2\cos 3\alpha\sin 3\alpha$ **l** $2\cos^2 4\theta - 1$

m $1 - 2\cos^2 3\beta$ **n** $1 - 2\sin^2 5\alpha$ **o** $2\sin^2 3D - 1$

p $\cos^2 2A - \sin^2 2A$ **q** $\cos^2(\frac{\alpha}{2}) - \sin^2(\frac{\alpha}{2})$ **r** $2\sin^2 3P - 2\cos^2 3P$

10 Show that:

a $(\sin\theta + \cos\theta)^2 = 1 + \sin 2\theta$ **b** $\cos^4\theta - \sin^4\theta = \cos 2\theta$

GRAPHING PACKAGE

11 Solve exactly for x where $0 \leqslant x \leqslant 2\pi$:

a $\sin 2x + \sin x = 0$

b $\sin 2x - 2\cos x = 0$

c $\sin 2x + 3\sin x = 0$

12 Use the double angle formula to show that:

a $\sin^2\theta = \frac{1}{2} - \frac{1}{2}\cos 2\theta$

b $\cos^2\theta = \frac{1}{2} + \frac{1}{2}\cos 2\theta$

13 Find the exact value of $\cos A$ in the diagram:

a

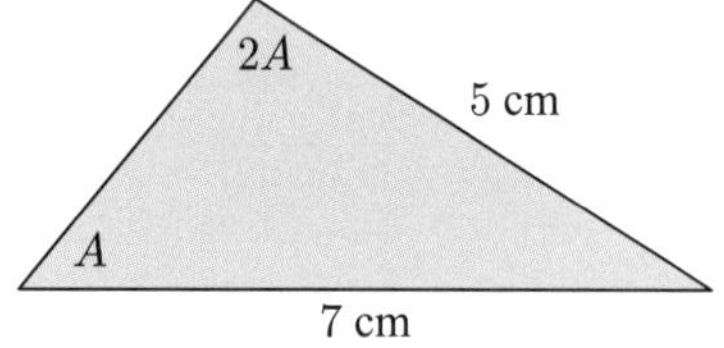

b

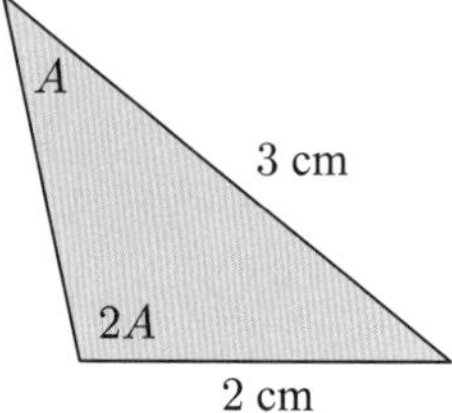

14 Prove the identities:

a $\dfrac{\sin 2\theta}{1 - \cos 2\theta} = \cot\theta$

b $\dfrac{\sin\theta + \sin 2\theta}{1 + \cos\theta + \cos 2\theta} = \tan\theta$

15 Prove:

a $\dfrac{\sin 2\theta}{1 + \cos 2\theta} - \tan\theta = 0$

b $\csc 2\theta = \tan\theta + \cot 2\theta$

c $\dfrac{\sin 2\theta}{\sin\theta} - \dfrac{\cos 2\theta}{\cos\theta} = \sec\theta$

E COMPOUND ANGLE FORMULAE

INVESTIGATION 2 COMPOUND ANGLE FORMULAE

What to do:

1 Copy and complete for angles A and B in radians or degrees. Include some angles of your own choosing.

A	B	$\cos A$	$\cos B$	$\cos(A-B)$	$\cos A - \cos B$	$\cos A\cos B + \sin A\sin B$
47°	24°					
138°	49°					
3^c	2^c					
$\vdots$	$\vdots$					

2 Write down any discoveries from your table of values.

3 Copy and complete for four pairs of angles A and B of your own choosing:

A	B	$\sin A$	$\sin B$	$\sin(A+B)$	$\sin A + \sin B$	$\sin A\cos B + \cos A\sin B$
$\vdots$	$\vdots$					

4 Write down any discoveries from your table of values.

If A and B are **any** two angles then:

$$\mathbf{\cos(A \pm B) = \cos A \cos B \mp \sin A \sin B}$$
$$\mathbf{\sin(A \pm B) = \sin A \cos B \pm \cos A \sin B}$$
$$\mathbf{\tan(A \pm B) = \frac{\tan A \pm \tan B}{1 \mp \tan A \tan B}}$$

These are known as the **compound angle formulae**. There are many ways of establishing them, but many are unsatisfactory as the arguments limit the angles A and B to being acute.

Proof:

Consider P$(\cos A, \sin A)$ and Q$(\cos B, \sin B)$ as any two points on the unit circle, as shown.

Angle POQ is $A - B$.

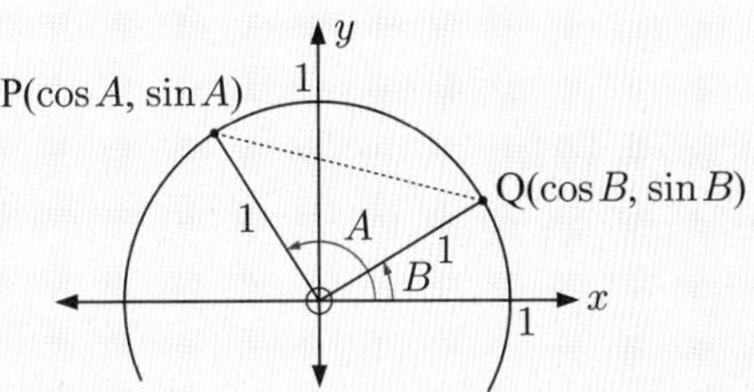

Using the distance formula:

$$\text{PQ} = \sqrt{(\cos A - \cos B)^2 + (\sin A - \sin B)^2}$$
$$\begin{aligned}\therefore\ (\text{PQ})^2 &= \cos^2 A - 2\cos A\cos B + \cos^2 B + \sin^2 A - 2\sin A\sin B + \sin^2 B\\ &= (\cos^2 A + \sin^2 A) + (\cos^2 B + \sin^2 B) - 2(\cos A\cos B + \sin A\sin B)\\ &= 2 - 2(\cos A\cos B + \sin A\sin B) \quad .... (1)\end{aligned}$$

But, by the *cosine rule* in $\triangle$POQ,

$$\begin{aligned}(\text{PQ})^2 &= 1^2 + 1^2 - 2(1)(1)\cos(A - B)\\ &= 2 - 2\cos(A - B) \quad .... (2)\end{aligned}$$
$$\therefore\ \cos(A - B) = \cos A\cos B + \sin A\sin B \qquad \{\text{comparing (1) and (2)}\}$$

From this formula the other formulae can be established:

$$\begin{aligned}\cos(A + B) &= \cos(A - (-B))\\ &= \cos A\cos(-B) + \sin A\sin(-B)\\ &= \cos A\cos B + \sin A(-\sin B) \qquad \{\cos(-\theta) = \cos\theta \text{ and } \sin(-\theta) = -\sin\theta\}\\ &= \cos A\cos B - \sin A\sin B\end{aligned}$$

Also

$$\begin{aligned}&\sin(A - B)\\ &= \cos(\tfrac{\pi}{2} - (A - B))\\ &= \cos((\tfrac{\pi}{2} - A) + B)\\ &= \cos(\tfrac{\pi}{2} - A)\cos B - \sin(\tfrac{\pi}{2} - A)\sin B\\ &= \sin A\cos B - \cos A\sin B\end{aligned}$$

$$\begin{aligned}&\sin(A + B)\\ &= \sin(A - (-B))\\ &= \sin A\cos(-B) - \cos A\sin(-B)\\ &= \sin A\cos B - \cos A(-\sin B)\\ &= \sin A\cos B + \cos A\sin B\end{aligned}$$

$$\begin{aligned}&\tan(A + B)\\ &= \frac{\sin(A + B)}{\cos(A + B)}\\ &= \frac{\sin A\cos B + \cos A\sin B}{\cos A\cos B - \sin A\sin B}\\ &= \frac{\frac{\sin A\cos B}{\cos A\cos B} + \frac{\cos A\sin B}{\cos A\cos B}}{\frac{\cos A\cos B}{\cos A\cos B} - \frac{\sin A\sin B}{\cos A\cos B}}\\ &= \frac{\tan A + \tan B}{1 - \tan A\tan B}\end{aligned}$$

$$\begin{aligned}&\tan(A - B)\\ &= \tan(A + (-B))\\ &= \frac{\tan A + \tan(-B)}{1 - \tan A\tan(-B)}\\ &= \frac{\tan A - \tan B}{1 + \tan A\tan B}\end{aligned}$$
$$\{\tan(-B) = -\tan B\}$$

Example 20 **Self Tutor**

Expand and simplify $\sin(270° + \alpha)$.

$\sin(270° + \alpha)$
$= \sin 270° \cos\alpha + \cos 270° \sin\alpha$ {compound angle formula}
$= -1 \times \cos\alpha + 0 \times \sin\alpha$
$= -\cos\alpha$

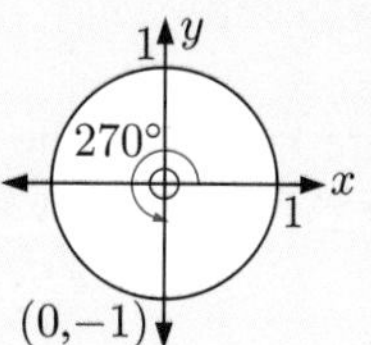

EXERCISE 13E

1 Expand and simplify:

a $\sin(90° + \theta)$
b $\cos(90° + \theta)$
c $\sin(180° - \theta)$
d $\cos(\pi + \alpha)$
e $\sin(2\pi - A)$
f $\cos\left(\frac{3\pi}{2} - \theta\right)$
g $\tan\left(\frac{\pi}{4} + \theta\right)$
h $\tan\left(\theta - \frac{3\pi}{4}\right)$
i $\tan(\pi + \theta)$

2 Expand, then simplify and write your answer in the form $A\sin\theta + B\cos\theta$:

a $\sin\left(\theta + \frac{\pi}{3}\right)$
b $\cos\left(\frac{2\pi}{3} - \theta\right)$
c $\cos\left(\theta + \frac{\pi}{4}\right)$
d $\sin\left(\frac{\pi}{6} - \theta\right)$

Example 21 **Self Tutor**

Simplify $\cos 3\theta \cos\theta - \sin 3\theta \sin\theta$.

$\cos 3\theta \cos\theta - \sin 3\theta \sin\theta$
$= \cos(3\theta + \theta)$ {compound angle formula in reverse}
$= \cos 4\theta$

3 Simplify using the compound angle formulae in reverse:

a $\cos 2\theta \cos\theta + \sin 2\theta \sin\theta$
b $\sin 2A \cos A + \cos 2A \sin A$
c $\cos A \sin B - \sin A \cos B$
d $\sin\alpha \sin\beta + \cos\alpha \cos\beta$
e $\sin\phi \sin\theta - \cos\phi \cos\theta$
f $2\sin\alpha \cos\beta - 2\cos\alpha \sin\beta$
g $\dfrac{\tan 2\theta - \tan\theta}{1 + \tan 2\theta \tan\theta}$
h $\dfrac{\tan 2A + \tan A}{1 - \tan 2A \tan A}$

4 Use the compound angle formulae to prove the double angle formulae:

a $\sin 2\theta = 2\sin\theta\cos\theta$
b $\cos 2\theta = \cos^2\theta - \sin^2\theta$
c $\tan 2\theta = \dfrac{2\tan\theta}{1 - \tan^2\theta}$

5 Simplify using the compound angle formulae:

a $\cos(\alpha + \beta)\cos(\alpha - \beta) - \sin(\alpha + \beta)\sin(\alpha - \beta)$
b $\sin(\theta - 2\phi)\cos(\theta + \phi) - \cos(\theta - 2\phi)\sin(\theta + \phi)$
c $\cos\alpha\cos(\beta - \alpha) - \sin\alpha\sin(\beta - \alpha)$

Example 22 **Self Tutor**

Without using your calculator, show that $\sin 75° = \frac{\sqrt{6}+\sqrt{2}}{4}$.

$$\begin{aligned}\sin 75° &= \sin(45° + 30°)\\ &= \sin 45° \cos 30° + \cos 45° \sin 30°\\ &= (\tfrac{1}{\sqrt{2}})(\tfrac{\sqrt{3}}{2}) + (\tfrac{1}{\sqrt{2}})(\tfrac{1}{2})\\ &= \left(\frac{\sqrt{3}+1}{2\sqrt{2}}\right)\frac{\sqrt{2}}{\sqrt{2}}\\ &= \frac{\sqrt{6}+\sqrt{2}}{4}\end{aligned}$$

6 Without using your calculator, show that:

a $\cos 75° = \frac{\sqrt{6}-\sqrt{2}}{4}$ **b** $\sin 105° = \frac{\sqrt{6}+\sqrt{2}}{4}$ **c** $\cos(\frac{13\pi}{12}) = \frac{-\sqrt{6}-\sqrt{2}}{4}$

7 Find the exact value of: **a** $\tan(\frac{5\pi}{12})$ **b** $\tan 105°$

8 If $\tan A = \frac{2}{3}$ and $\tan B = -\frac{1}{5}$, find the exact value of $\tan(A+B)$.

9 If $\tan A = \frac{3}{4}$, find $\tan\left(A + \frac{\pi}{4}\right)$.

10 Simplify:

a $\tan\left(A + \frac{\pi}{4}\right)\tan\left(A - \frac{\pi}{4}\right)$ **b** $\dfrac{\tan(A+B) + \tan(A-B)}{1 - \tan(A+B)\tan(A-B)}$

11 If $\sin A = -\frac{1}{3}$, $\pi \leqslant A \leqslant \frac{3\pi}{2}$ and $\cos B = \frac{1}{\sqrt{5}}$, $0 \leqslant B \leqslant \frac{\pi}{2}$, find:

a $\tan(A+B)$ **b** $\tan 2A$.

12 Simplify, giving your answer exactly: $\dfrac{\tan 80° \quad \tan 20°}{1 + \tan 80° \tan 20°}$

13 If $\tan(A+B) = \frac{3}{5}$ and $\tan B = \frac{2}{3}$, find the exact value of $\tan A$.

14 Find $\tan A$ if $\tan(A-B)\tan(A+B) = 1$.

15 Find the exact value of $\tan\alpha$:

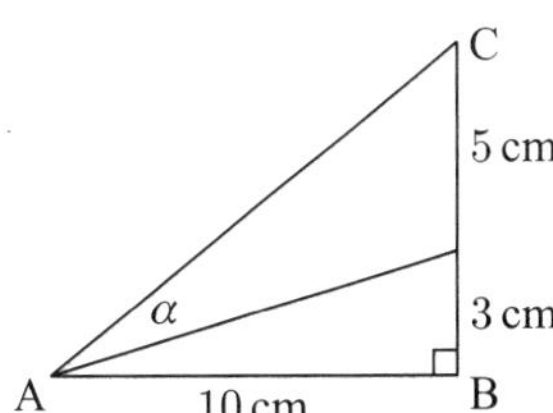

16 Find the exact value of the tangent of the acute angle between two lines if their gradients are $\frac{1}{2}$ and $\frac{2}{3}$.

17 Express $\tan(A+B+C)$ in terms of $\tan A$, $\tan B$, and $\tan C$.
Hence show that if A, B, and C are the angles of a triangle, then
$$\tan A + \tan B + \tan C = \tan A \tan B \tan C.$$

18 Show that:

a $\sqrt{2}\cos\left(\theta+\frac{\pi}{4}\right)=\cos\theta-\sin\theta$

b $2\cos\left(\theta-\frac{\pi}{3}\right)=\cos\theta+\sqrt{3}\sin\theta$

c $\cos(\alpha+\beta)-\cos(\alpha-\beta)=-2\sin\alpha\sin\beta$

d $\cos(\alpha+\beta)\cos(\alpha-\beta)=\cos^2\alpha-\sin^2\beta.$

19 Prove that, in the given figure, $\alpha+\beta=\frac{\pi}{4}$.

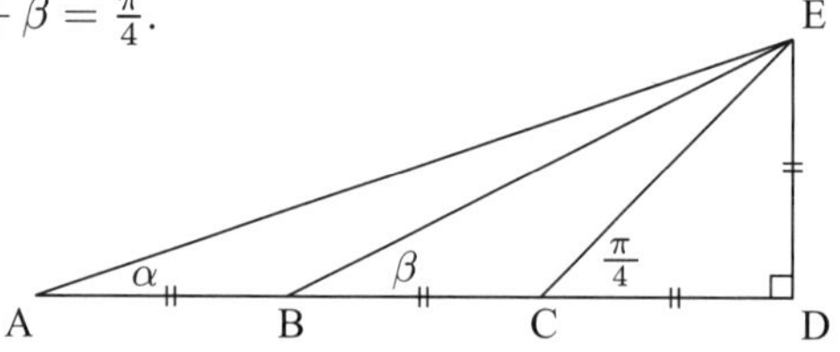

Example 23 — Self Tutor

Suppose $\sin x-\sqrt{3}\cos x=k\cos(x+b)$ for $k>0$ and $0<b<2\pi$. Find k and b.

$$\begin{aligned}\sin x-\sqrt{3}\cos x&=k\cos(x+b)\\&=k[\cos x\cos b-\sin x\sin b]\\&=k\cos x\cos b-k\sin x\sin b\end{aligned}$$

Equating coefficients of $\cos x$ and $\sin x$,

$k\cos b=-\sqrt{3}$ .... (1) and $-k\sin b=1$ .... (2)

$\therefore\ k^2\cos^2 b=3$ and $k^2\sin^2 b=1$ {squaring both sides}

$\therefore\ k^2(\cos^2 b+\sin^2 b)=4$ {adding the two equations}

$\therefore\ k^2=4$

$\therefore\ k=2$ {since $k>0$}

Substituting $k=2$ into (1) gives $\cos b=-\frac{\sqrt{3}}{2}$

and into (2) gives $\sin b=-\frac{1}{2}$

$\therefore\ b=\frac{7\pi}{6}$

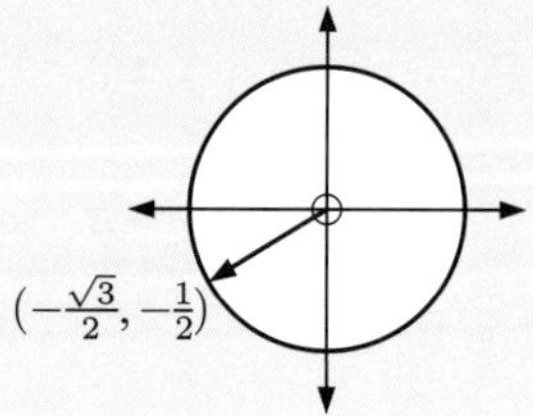

20 Suppose $\sqrt{3}\sin x+\cos x=k\sin(x+b)$ for $k>0$ and $0<b<2\pi$. Find k and b.

21 **a** Prove that $\cos 3\theta=4\cos^3\theta-3\cos\theta$ by replacing 3θ by $(2\theta+\theta)$.

b Hence, solve the equation $8\cos^3\theta-6\cos\theta+1=0$ for $\theta\in[-\pi,\pi]$.

22 **a** Write $\sin 3\theta$ in the form $a\sin^3\theta+b\sin\theta$ where $a, b\in\mathbb{Z}$.

b Hence, solve the equation $\sin 3\theta=\sin\theta$ for $\theta\in[0, 3\pi]$.

23 Use the basic definition of periodicity to show algebraically that the period of $f(x)=\sin(nx)$ is $\dfrac{2\pi}{n}$, for all $n>0$.

24 **a** Write $2\cos x-5\sin x$ in the form $k\cos(x+b)$ for $k>0$, $0<b<2\pi$.

b Hence solve the equation $2\cos x-5\sin x=-2$ for $0\leqslant x\leqslant\pi$.

c Given that $t=\tan\left(\frac{x}{2}\right)$, prove that $\sin x=\dfrac{2t}{1+t^2}$ and $\cos x=\dfrac{1-t^2}{1+t^2}$.

d Hence solve $2\cos x-5\sin x=-2$ for $0\leqslant x\leqslant\pi$ using **c**.

25 Use $\tan(\theta-\phi)=\dfrac{\tan\theta-\tan\phi}{1+\tan\theta\tan\phi}$ to show that $\arctan(5)-\arctan\left(\frac{2}{3}\right)=\frac{\pi}{4}$.

26 Without using technology, show that:

a $\arctan\left(\frac{1}{5}\right)+\arctan\left(\frac{2}{3}\right)=\frac{\pi}{4}$

b $\arctan\left(\frac{4}{3}\right)=2\arctan\left(\frac{1}{2}\right)$

27 Find the exact value of $4\arctan\left(\frac{1}{5}\right)-\arctan\left(\frac{1}{239}\right)$.

28 a Show that $\sin(A+B)+\sin(A-B)=2\sin A\cos B$.

b Hence show that $\sin A\cos B=\frac{1}{2}\sin(A+B)+\frac{1}{2}\sin(A-B)$.

c Hence write the following as sums:

i $\sin 3\theta\cos\theta$

ii $\sin 6\alpha\cos\alpha$

iii $2\sin 5\beta\cos\beta$

iv $4\cos\theta\sin 4\theta$

v $6\cos 4\alpha\sin 3\alpha$

vi $\frac{1}{3}\cos 5A\sin 3A$

29 a Show that $\cos(A+B)+\cos(A-B)=2\cos A\cos B$.

b Hence show that $\cos A\cos B=\frac{1}{2}\cos(A+B)+\frac{1}{2}\cos(A-B)$.

c Hence write the following as a *sum* of cosines:

i $\cos 4\theta\cos\theta$

ii $\cos 7\alpha\cos\alpha$

iii $2\cos 3\beta\cos\beta$

iv $6\cos x\cos 7x$

v $3\cos P\cos 4P$

vi $\frac{1}{4}\cos 4x\cos 2x$

30 a Show that $\cos(A-B)-\cos(A+B)=2\sin A\sin B$.

b Hence show that $\sin A\sin B=\frac{1}{2}\cos(A-B)-\frac{1}{2}\cos(A+B)$.

c Hence write the following as a *difference* of cosines:

i $\sin 3\theta\sin\theta$

ii $\sin 6\alpha\sin\alpha$

iii $2\sin 5\beta\sin\beta$

iv $4\sin\theta\sin 4\theta$

v $10\sin 2A\sin 8A$

vi $\frac{1}{5}\sin 3M\sin 7M$

31 The **products to sums formulae** are:

$$\sin A\cos B=\tfrac{1}{2}\sin(A+B)+\tfrac{1}{2}\sin(A-B) \quad .... (1)$$

$$\cos A\cos B=\tfrac{1}{2}\cos(A+B)+\tfrac{1}{2}\cos(A-B) \quad .... (2)$$

$$\sin A\sin B=\tfrac{1}{2}\cos(A-B)-\tfrac{1}{2}\cos(A+B) \quad .... (3)$$

a What formulae result if we replace B by A in each of these formulae?

b Suppose $A+B=S$ and $A-B=D$.

i Show that $A=\dfrac{S+D}{2}$ and $B=\dfrac{S-D}{2}$.

ii Using the substitutions $A+B=S$ and $A-B=D$, show that equation (1) becomes $\sin S+\sin D=2\sin\left(\frac{S+D}{2}\right)\cos\left(\frac{S-D}{2}\right)$ .... (4)

iii By replacing D by $(-D)$ in (4), show that $\sin S-\sin D=2\cos\left(\frac{S+D}{2}\right)\sin\left(\frac{S-D}{2}\right)$.

c What formula results when the substitution $A=\frac{S+D}{2}$ and $B=\frac{S-D}{2}$ is made into (2)?

d What formula results when the substitution $A=\frac{S+D}{2}$ and $B=\frac{S-D}{2}$ is made into (3)?

32 The **factor formulae** are:

$$\sin S+\sin D=2\sin\left(\tfrac{S+D}{2}\right)\cos\left(\tfrac{S-D}{2}\right) \qquad \cos S+\cos D=2\cos\left(\tfrac{S+D}{2}\right)\cos\left(\tfrac{S-D}{2}\right)$$

$$\sin S-\sin D=2\cos\left(\tfrac{S+D}{2}\right)\sin\left(\tfrac{S-D}{2}\right) \qquad \cos S-\cos D=-2\sin\left(\tfrac{S+D}{2}\right)\sin\left(\tfrac{S-D}{2}\right)$$

Use these formulae to convert the following to products:

a $\sin 5x+\sin x$

b $\cos 8A+\cos 2A$

c $\cos 3\alpha-\cos\alpha$

d $\sin 5\theta-\sin 3\theta$

e $\cos 7\alpha-\cos\alpha$

f $\sin 3\alpha+\sin 7\alpha$

g $\cos 2B-\cos 4B$

h $\sin(x+h)-\sin x$

i $\cos(x+h)-\cos x$

F TRIGONOMETRIC EQUATIONS IN QUADRATIC FORM

Sometimes we may be given trigonometric equations in quadratic form.

For example, $2\sin^2 x + \sin x = 0$ and $2\cos^2 x + \cos x - 1 = 0$ are clearly quadratic equations where the variables are $\sin x$ and $\cos x$ respectively.

These equations can be factorised by quadratic factorisation and then solved for x.

Example 24 Self Tutor

Solve for x on $0 \leqslant x \leqslant 2\pi$, giving your answers as exact values:

a $2\sin^2 x + \sin x = 0$ **b** $2\cos^2 x + \cos x - 1 = 0$

a
$$2\sin^2 x + \sin x = 0$$
$$\therefore \sin x(2\sin x + 1) = 0$$
$$\therefore \sin x = 0 \text{ or } -\tfrac{1}{2}$$

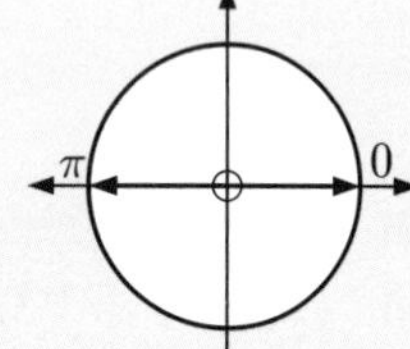

$\sin x = 0$ when
$x = 0$, π, or 2π

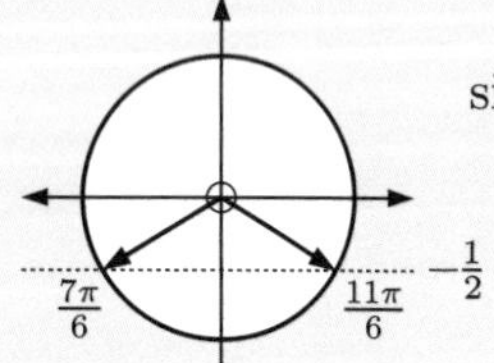

$\sin x = -\frac{1}{2}$ when
$x = \frac{7\pi}{6}$ or $\frac{11\pi}{6}$

The solutions are:
$x = 0$, π, $\frac{7\pi}{6}$, $\frac{11\pi}{6}$, or 2π.

b

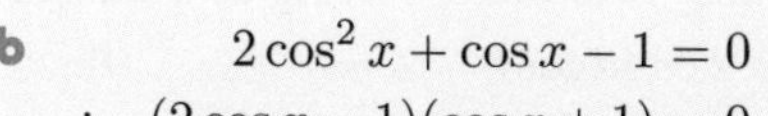

$$2\cos^2 x + \cos x - 1 = 0$$
$$\therefore (2\cos x - 1)(\cos x + 1) = 0$$
$$\therefore \cos x = \tfrac{1}{2} \text{ or } -1$$

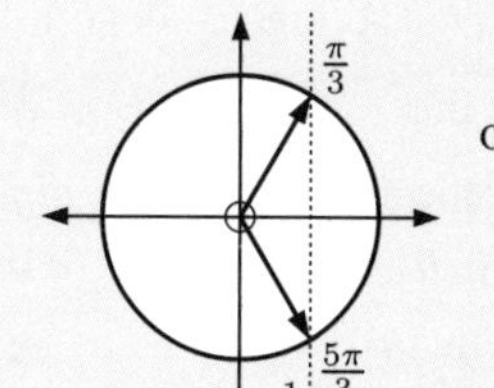

$\cos x = \frac{1}{2}$ when
$x = \frac{\pi}{3}$ or $\frac{5\pi}{3}$

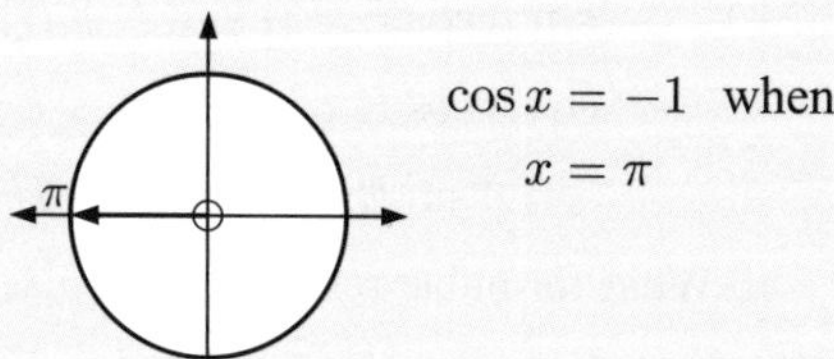

$\cos x = -1$ when
$x = \pi$

The solutions are:
$x = \frac{\pi}{3}$, π, or $\frac{5\pi}{3}$.

EXERCISE 13F

1 Solve for $0 \leqslant x \leqslant 2\pi$ giving your answers as exact values:

a $2\sin^2 x + \sin x = 0$ **b** $2\cos^2 x = \cos x$ **c** $2\cos^2 x + \cos x - 1 = 0$

d $2\sin^2 x + 3\sin x + 1 = 0$ **e** $\sin^2 x = 2 - \cos x$ **f** $3\tan x = \cot x$

2 Solve for $0 \leqslant x \leqslant 2\pi$ giving your answers as exact values:

a $\cos 2x - \cos x = 0$ **b** $\cos 2x + 3\cos x = 1$ **c** $\cos 2x + \sin x = 0$

d $\sin 4x = \sin 2x$ **e** $\sin x + \cos x = \sqrt{2}$ **f** $2\cos^2 x = 3\sin x$

3 Solve for $x \in [-\pi, \pi]$, giving exact answers:

a $2\sin x + \csc x = 3$ **b** $\sin 2x + \cos x - 2\sin x - 1 = 0$

c $\tan^4 x - 2\tan^2 x - 3 = 0$

4 Solve for $x \in [0, 2\pi]$:

a $2\cos^2 x = \sin x$ b $\cos 2x + 5\sin x = 0$ c $2\tan^2 x + 3\sec^2 x = 7$

G TRIGONOMETRIC SERIES AND PRODUCTS

A **trigonometric series** is the *addition* of a sequence of trigonometric expressions.

For example, $\sin x + \sin 3x + \sin 5x + \sin 7x$ is a trigonometric series.

A **trigonometric product** is the *product* of a sequence of trigonometric expressions.

For example, $\cos x \times \cos 2x \times \cos 3x \times \cos 4x$ is a trigonometric product.

If the trigonometric expressions follow a pattern, we may be able to simplify the series or product. We can use the principle of mathematical induction to prove the result.

Example 25 Self Tutor

Use the principle of mathematical induction to prove that:

$$2\cos 2x + 2\cos 4x + 2\cos 6x + ... + 2\cos 2nx = \frac{\sin[(2n+1)x]}{\sin x} - 1 \quad \text{for all } n \in \mathbb{Z}^+.$$

P_n is $2\cos 2x + 2\cos 4x + 2\cos 6x + ... + 2\cos 2nx = \dfrac{\sin[(2n+1)x]}{\sin x} - 1$ for all $n \in \mathbb{Z}^+$.

(1) If $n = 1$, LHS $= 2\cos 2x$ and

$$\begin{aligned}
\text{RHS} &= \frac{\sin 3x}{\sin x} - 1 \\
&= \frac{\sin(2x + x)}{\sin x} - 1 \\
&= \frac{\sin 2x \cos x + \cos 2x \sin x}{\sin x} - 1 \\
&= \frac{(2\cancel{\sin x}\cos x)\cos x + \cos 2x\cancel{\sin x}}{\cancel{\sin x}} - 1 \\
&= 2\cos^2 x + \cos 2x - 1 \\
&= (2\cos^2 x - 1) + \cos 2x \\
&= 2\cos 2x \qquad \therefore\ P_1 \text{ is true.}
\end{aligned}$$

(2) If P_k is true, then $2\cos 2x + 2\cos 4x + 2\cos 6x + ... + 2\cos 2kx = \dfrac{\sin[(2k+1)x]}{\sin x} - 1$

$\therefore$ $2\cos 2x + 2\cos 4x + 2\cos 6x + ... + 2\cos 2kx + 2\cos[2(k+1)x]$

$$= \left(\frac{\sin[(2k+1)x]}{\sin x} - 1\right) + 2\cos[2(k+1)x]$$

$$= \frac{\sin[(2k+1)x] + 2\sin x\cos[2(k+1)x]}{\sin x} - 1$$

$$= \frac{\sin[(2k+1)x] + \sin[x + 2(k+1)x] + \sin[x - 2(k+1)x]}{\sin x} - 1$$

$\{2\sin A\cos B = \sin(A+B) + \sin(A-B)\}$

$$= \frac{\sin[(2k+1)x] + \sin[(2(k+1)+1)x] + \sin[-(2k+1)x]}{\sin x} - 1$$

$$= \frac{\cancel{\sin[(2k+1)x]} + \sin[(2(k+1)+1)x] - \cancel{\sin[(2k+1)x]}}{\sin x} - 1$$

$$= \frac{\sin[(2(k+1)+1)x]}{\sin x} - 1$$

Since P_1 is true, and P_{k+1} is true whenever P_k is true, P_n is true for all $n \in \mathbb{Z}^+$.

{Principle of mathematical induction}

EXERCISE 13G

1 **a** Simplify $1 + \sin x + \sin^2 x + \sin^3 x + \sin^4 x + ... + \sin^{n-1} x$.

b Find the sum of the infinite series $1 + \sin x + \sin^2 x + \sin^3 x + ...$

c Find the value of $x \in [0, 2\pi]$ such that the series in **b** has sum $\frac{2}{3}$.

2 We know that $2\sin x \cos x = \sin 2x$.

a Show that:

i $2\sin x(\cos x + \cos 3x) = \sin 4x$

ii $2\sin x(\cos x + \cos 3x + \cos 5x) = \sin 6x$

b Predict what the following would simplify to:

i $2\sin x(\cos x + \cos 3x + \cos 5x + \cos 7x)$

ii $\cos x + \cos 3x + \cos 5x + ... + \cos 19x$ (10 terms)

iii $\cos x + \cos 3x + \cos 5x + ... + \cos[(2n-1)x]$ (n terms)

c Use the principle of mathematical induction to prove that:

$$\cos\theta + \cos 3\theta + \cos 5\theta + ... + \cos(2n-1)\theta = \frac{\sin 2n\theta}{2\sin\theta} \quad \text{for all } n \in \mathbb{Z}^+.$$

3 From $\sin 2x = 2\sin x \cos x$ we observe that $\sin x \cos x = \frac{\sin 2x}{2} = \frac{\sin(2^1 x)}{2^1}$.

a Prove that:

i $\sin x \cos x \cos 2x = \frac{\sin(2^2 x)}{2^2}$

ii $\sin x \cos x \cos 2x \cos 4x = \frac{\sin(2^3 x)}{2^3}$

b Assuming the pattern in **a** continues, simplify:

i $\sin x \cos x \cos 2x \cos 4x \cos 8x$

ii $\sin x \cos x \cos 2x .... \cos 32x$

c Generalise the results from **a** and **b**. Prove your generalisation using the principle of mathematical induction.

4 **a** Use the principle of mathematical induction to prove that:

$$\sin\theta + \sin 3\theta + \sin 5\theta + ... + \sin(2n-1)\theta = \frac{1-\cos 2n\theta}{2\sin\theta} \quad \text{for all positive integers } n.$$

b Hence, find the value of $\sin\frac{\pi}{7} + \sin\frac{3\pi}{7} + \sin\frac{5\pi}{7} + \sin\pi + \sin\frac{9\pi}{7} + \sin\frac{11\pi}{7} + \sin\frac{13\pi}{7}$.

5 Use the principle of mathematical induction to prove that:

$$\cos x \times \cos 2x \times \cos 4x \times \cos 8x \times .... \times \cos(2^{n-1}x) = \frac{\sin(2^n x)}{2^n \times \sin x} \quad \text{for all } n \in \mathbb{Z}^+.$$

6 Use the principle of mathematical induction to prove that:

$$\cos^2\theta + \cos^2 2\theta + \cos^2 3\theta + \cos^2 4\theta + ... + \cos^2(n\theta) = \frac{1}{2}\left[n + \frac{\cos[(n+1)\theta]\sin n\theta}{\sin\theta}\right] \quad \text{for all } n \in \mathbb{Z}^+.$$

THEORY OF KNOWLEDGE

Trigonometry appears to be one of the most useful disciplines of mathematics, having great importance in building and engineering. Its study has been driven by the need to solve real world problems throughout history.

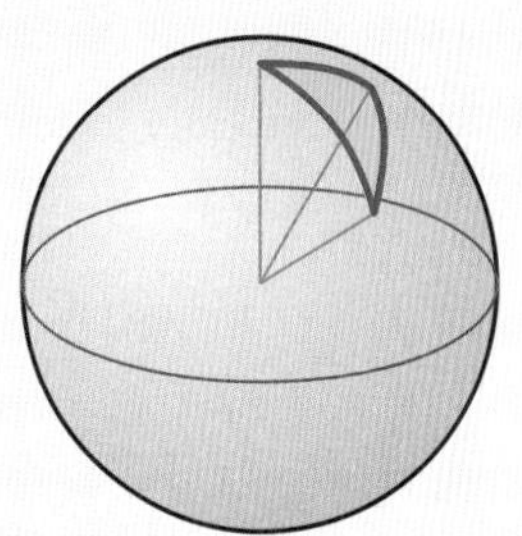

The study of trigonometry began when Greek, Babylonian, and Arabic astronomers needed to calculate the positions of stars and planets. These early mathematicians considered the trigonometry of spherical triangles, which are triangles on the surface of a sphere.

Trigonometric functions were developed by Hipparchus around 140 BC, and then by Ptolemy and Menelaus around 100 AD.

Around 500 AD, Hindu mathematicians published a table called the *Aryabhata*. It was a table of lengths of half chords, which are the lengths $AM = r\sin x$ in the diagram. This is trigonometry of triangles in a plane, as we study in schools today.

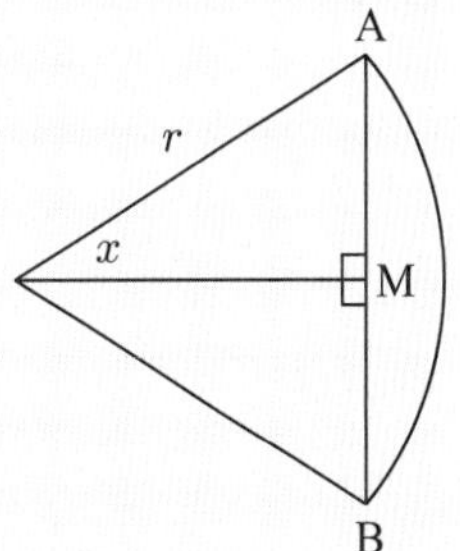

1. How do society and culture affect mathematical knowledge?
2. Should congruence and similarity, or the work of Pythagoras, be considered part of modern trigonometry?
3. Is the angle sum of a triangle always equal to 180°?

REVIEW SET 13A — NON-CALCULATOR

1

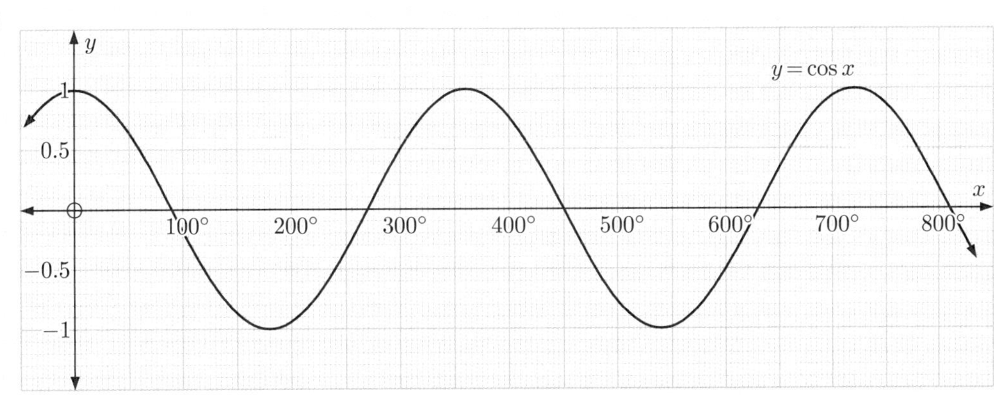

Use the graph of $y = \cos x$ to find the solutions of:

a $\cos x = -0.4, \quad 0 \leqslant x \leqslant 800°$ **b** $\cos x = 0.9, \quad 0 \leqslant x \leqslant 600°$

2 Solve in terms of π:

a $2\sin x = -1$ for $0 \leqslant x \leqslant 4\pi$ **b** $\sqrt{2}\sin x - 1 = 0$ for $-2\pi \leqslant x \leqslant 2\pi$

3 Find the x-intercepts of:

a $y = 2\sin 3x + \sqrt{3}$ for $0 \leqslant x \leqslant 2\pi$ **b** $y = \sqrt{2}\sin\left(x + \frac{\pi}{4}\right)$ for $0 \leqslant x \leqslant 3\pi$

4 Solve algebraically in terms of π:

a $\cot x = \sqrt{3}$ for $x \in [0, 2\pi]$ **b** $\sec^2 x = \tan x + 1$ for $x \in [0, 2\pi]$

5 Simplify: **a** $\cos\left(\frac{3\pi}{2} - \theta\right)$ **b** $\sin\left(\theta + \frac{\pi}{2}\right)$

6 Simplify: **a** $\dfrac{1 - \cos^2\theta}{1 + \cos\theta}$ **b** $\dfrac{\sin\alpha - \cos\alpha}{\sin^2\alpha - \cos^2\alpha}$ **c** $\dfrac{4\sin^2\alpha - 4}{8\cos\alpha}$

7 If $\sin\alpha = -\frac{3}{4}$, $\pi \leqslant \alpha \leqslant \frac{3\pi}{2}$, find the exact value of:

a $\cos\alpha$ **b** $\sin 2\alpha$ **c** $\cos 2\alpha$ **d** $\tan 2\alpha$

8 Show that $\dfrac{\sin 2\alpha - \sin\alpha}{\cos 2\alpha - \cos\alpha + 1}$ simplifies to $\tan\alpha$.

9 Find the exact value of: **a** $\cos(165°)$ **b** $\tan(\frac{\pi}{12})$

10 Solve for x in $[0, 2\pi]$: **a** $2\cos 2x + 1 = 0$ **b** $\sin 2x = -\sqrt{3}\cos 2x$

11

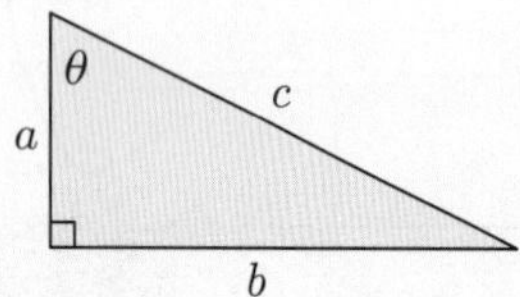

Prove that: **a** $\sin 2\theta = \dfrac{2ab}{c^2}$

b $\cos 2\theta = \dfrac{a^2 - b^2}{c^2}$

12 Find the exact solutions of:

a $\sqrt{2}\cos\left(x + \frac{\pi}{4}\right) - 1 = 0$, $x \in [0, 4\pi]$ **b** $\tan 2x - \sqrt{3} = 0$, $x \in [0, 2\pi]$

13

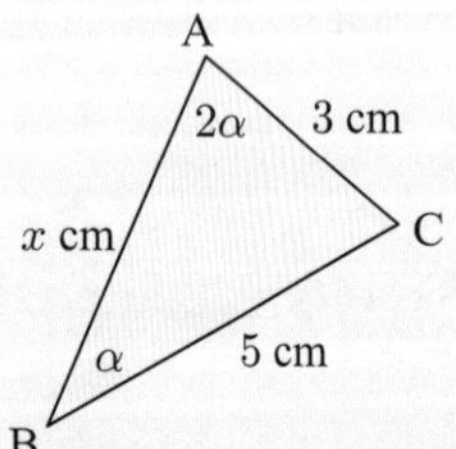

a Show that $\cos\alpha = \frac{5}{6}$.

b Show that x is a solution of $3x^2 - 25x + 48 = 0$.

c Find x by solving the equation in **b**.

REVIEW SET 13B — CALCULATOR

1 Solve for $0 \leqslant x \leqslant 8$: **a** $\sin x = 0.382$ **b** $\tan(\frac{x}{2}) = -0.458$

2 Solve:

a $\cos x = 0.4379$ for $0 \leqslant x \leqslant 10$ **b** $\cos(x - 2.4) = -0.6014$ for $0 \leqslant x \leqslant 6$

3 If $\sin A = \frac{5}{13}$ and $\cos A = \frac{12}{13}$, find: **a** $\sin 2A$ **b** $\cos 2A$ **c** $\tan 2A$

4 **a** Solve for $0 \leqslant x \leqslant 10$:

i $\tan x = 4$ **ii** $\tan(\frac{x}{4}) = 4$ **iii** $\tan(x - 1.5) = 4$

b Find exact solutions for x given $-\pi \leqslant x \leqslant \pi$:

i $\tan(x + \frac{\pi}{6}) = -\sqrt{3}$ **ii** $\tan 2x = -\sqrt{3}$ **iii** $\tan^2 x - 3 = 0$

c Solve $3\tan(x - 1.2) = -2$ for $0 \leqslant x \leqslant 10$.

5 Show that:

- **a** $\sqrt{2}\cos\left(\theta+\frac{\pi}{4}\right)=\cos\theta-\sin\theta$
- **b** $\cos\alpha\cos(\beta-\alpha)-\sin\alpha\sin(\beta-\alpha)=\cos\beta$

6 If $\cos x=-\frac{3}{4}$ and $\pi<x<\frac{3\pi}{2}$ find the exact value of $\sin\left(\frac{x}{2}\right)$.

7 Solve for $0\leqslant x\leqslant 2\pi$:

- **a** $\cos x=0.3$
- **b** $2\sin(3x)=\sqrt{2}$
- **c** $43+8\sin x=50.1$

8 An ecologist studying a species of water beetle estimates the population of a colony over an eight week period. If t is the number of weeks after the initial estimate is made, then the population in thousands can be modelled by $P(t)=5+2\sin(\frac{\pi t}{3})$ where $0\leqslant t\leqslant 8$.

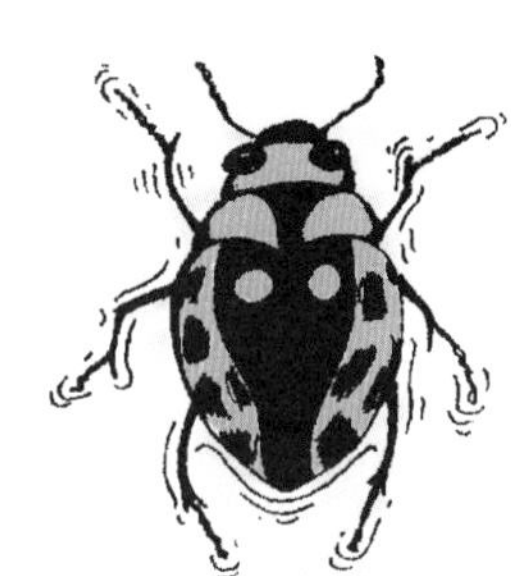

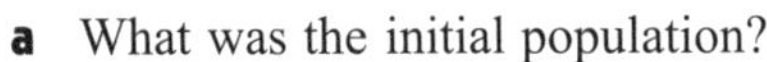

- **a** What was the initial population?
- **b** What were the smallest and largest populations?
- **c** During what time interval(s) did the population exceed 6000?

9 Solve for x: $3\cos x+\sin 2x=1$ for $0\leqslant x\leqslant 10$.

10 Write $3\sin x+4\cos x$ in the form $k\cos(x+b)$, where $k>0$ and $0<b<2\pi$.

11 From ground level, a shooter is aiming at targets on a vertical brick wall. At the current angle of elevation of his rifle, he will hit a target 20 m above ground level. If he doubles the angle of elevation of the rifle, he will hit a target 45 m above ground level. How far is the shooter from the wall?

12 Find exactly the length of BC:

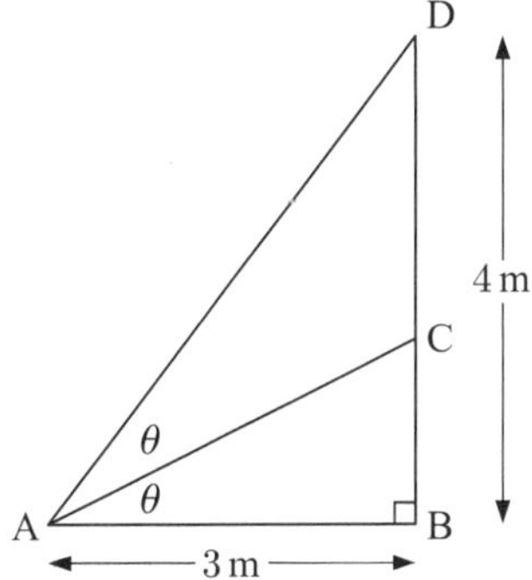

13 In an industrial city, the amount of pollution in the air becomes greater during the working week when factories are operating, and lessens over the weekend. The number of milligrams of pollutants in a cubic metre of air is given by

$P(t)=40+12\sin\frac{2\pi}{7}(t-\frac{37}{12})$

where t is the number of days after midnight on Saturday night.

- **a** What is the minimum level of pollution?
- **b** At what time during the week does this minimum level occur?

REVIEW SET 13C

1 Consider $y = \sin(\frac{x}{3})$ on the domain $-7 \leqslant x \leqslant 7$. Use the graph to solve, correct to 1 decimal place: **a** $\sin(\frac{x}{3}) = -0.9$ **b** $\sin(\frac{x}{3}) = \frac{1}{4}$

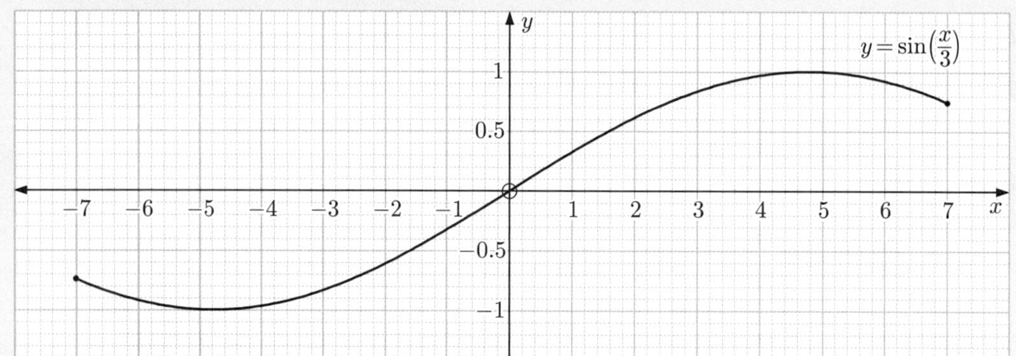

2 Solve algebraically for $0 \leqslant x \leqslant 2\pi$, giving answers in terms of π:

a $\sin^2 x - \sin x - 2 = 0$ **b** $4\sin^2 x = 1$

3 Find the exact solutions of:

a $\tan(x - \frac{\pi}{3}) = \frac{1}{\sqrt{3}}$, $0 \leqslant x \leqslant 4\pi$ **b** $\cos(x + \frac{2\pi}{3}) = \frac{1}{2}$, $-2\pi \leqslant x \leqslant 2\pi$

4 Simplify:

a $\cos^3\theta + \sin^2\theta\cos\theta$ **b** $\dfrac{\cos^2\theta - 1}{\sin\theta}$ **c** $5 - 5\sin^2\theta$

d $\dfrac{\sin^2\theta - 1}{\cos\theta}$ **e** $\dfrac{\tan\theta + \cot\theta}{\sec\theta}$ **f** $\cos^2\theta(\tan\theta + 1)^2 - 1$

5 If $\tan 2\alpha = \frac{4}{3}$ for $\alpha \in \,]0, \frac{\pi}{2}[$, find the exact value of $\sin\alpha$ without using a calculator.

6 Simplify: **a** $(2\sin\alpha - 1)^2$ **b** $(\cos\alpha - \sin\alpha)^2$

7 Show that:

a $\dfrac{\cos\theta}{1 + \sin\theta} + \dfrac{1 + \sin\theta}{\cos\theta} = 2\sec\theta$ **b** $\left(1 + \dfrac{1}{\cos\theta}\right)\left(\cos\theta - \cos^2\theta\right) = \sin^2\theta$

8 Solve exactly: **a** $\arcsin x = \frac{\pi}{3}$ **b** $\arctan(x - 2) = \frac{\pi}{6}$

9 Show that $\sin\left(\frac{\pi}{8}\right) = \frac{1}{2}\sqrt{2 - \sqrt{2}}$ using a suitable double angle formula.

10 If α and β are the other angles of a right angled triangle, show that $\sin 2\alpha = \sin 2\beta$.

11 Prove that: **a** $(\sin\theta + \cos\theta)^2 = 1 + \sin 2\theta$ **b** $\csc 2x + \cot 2x = \cot x$

12 Use the principle of mathematical induction to prove that:

$$\sin^2\theta + \sin^2 2\theta + \sin^2 3\theta + ... + \sin^2(n\theta) = \frac{1}{2}\left[n - \frac{\cos[(n+1)\theta]\sin n\theta}{\sin\theta}\right] \text{ for all } n \in \mathbb{Z}^+.$$

13 **a** Show that $\cos(\alpha - \beta) - \cos(\alpha + \beta) = 2\sin\alpha\sin\beta$.

b Hence show that $\sin\alpha\sin\beta = \frac{1}{2}(\cos(\alpha - \beta) - \cos(\alpha + \beta))$.

c Hence show that $\sin[(k+1)\theta]\sin\frac{\theta}{2} + \sin\frac{k\theta}{2}\sin\frac{(k+1)\theta}{2} = \sin\frac{(k+1)\theta}{2}\sin\frac{(k+2)\theta}{2}$.

d Use the principle of mathematical induction to prove that:

$$\sin\theta + \sin 2\theta + \sin 3\theta + ... + \sin(n\theta) = \frac{\sin\left[\frac{1}{2}(n+1)\theta\right]\sin\left(\frac{1}{2}n\theta\right)}{\sin\left(\frac{1}{2}\theta\right)} \text{ for all } n \in \mathbb{Z}^+.$$

Chapter 14

Vectors

Syllabus reference: 4.1, 4.2, 4.5

Contents:

OPENING PROBLEM

An aeroplane in calm conditions is flying at 800 $km\,h^{-1}$ due east. A cold wind suddenly blows from the south-west at 35 $km\,h^{-1}$, pushing the aeroplane slightly off course.

Things to think about:

- **a** How can we illustrate the plane's movement and the wind using a scale diagram?
- **b** What operation do we need to perform to find the effect of the wind on the aeroplane?
- **c** Can you use a scale diagram to determine the resulting speed and direction of the aeroplane?

A VECTORS AND SCALARS

In the **Opening Problem**, the effect of the wind on the aeroplane is determined by both its speed *and* its direction. The effect would be different if the wind was blowing against the aeroplane rather than from behind it.

Quantities which have only magnitude are called **scalars**.

Quantities which have both magnitude and direction are called **vectors**.

The *speed* of the plane is a scalar. It describes its size or strength.

The *velocity* of the plane is a vector. It includes both its speed and also its direction.

Other examples of vector quantities are:

- acceleration
- force
- displacement
- momentum

For example, farmer Giles needs to remove a fence post. He starts by pushing on the post sideways to loosen the ground. Giles has a choice of how hard to push the post and in which direction. The force he applies is therefore a vector.

DIRECTED LINE SEGMENT REPRESENTATION

We can represent a vector quantity using a **directed line segment** or **arrow**.

The **length of the arrow** represents the size or magnitude of the quantity, and the **arrowhead** shows its direction.

For example, if farmer Giles pushes the post with a force of 50 Newtons (N) to the north-east, we can draw a scale diagram of the force relative to the north line.

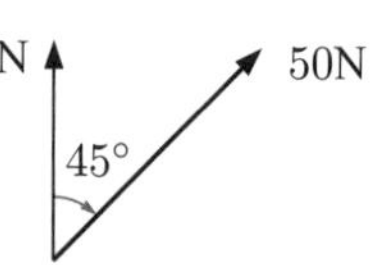

Scale: 1 cm represents 25 N

Example 1 **Self Tutor**

Draw a scale diagram to represent a force of 40 Newtons in a north-easterly direction.

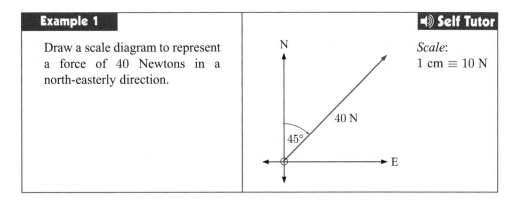

N, 40 N, 45°, E

Scale:
1 cm $\equiv$ 10 N

EXERCISE 14A.1

1 Using a scale of 1 cm represents 10 units, sketch a vector to represent:

- **a** 30 Newtons in a south-easterly direction
- **b** 25 m s^{-1} in a northerly direction
- **c** an excavator digging a tunnel at a rate of 30 cm min^{-1} at an angle of 30° to the ground
- **d** an aeroplane taking off at an angle of 10° to the runway with a speed of 50 m s^{-1}.

2 If ⟶ represents a velocity of 50 m s^{-1} due east, draw a directed line segment representing a velocity of:

- **a** 100 m s^{-1} due west
- **b** 75 m s^{-1} north-east.

3 Draw a scale diagram to represent the following vectors:

- **a** a force of 30 Newtons in the NW direction
- **b** a velocity of 36 m s^{-1} vertically downwards
- **c** a displacement of 4 units at an angle of 15° to the positive x-axis
- **d** an aeroplane taking off at an angle of 8° to the runway at a speed of 150 km h^{-1}.

VECTOR NOTATION

Consider the vector from the origin O to the point A. We call this the **position vector** of point A.

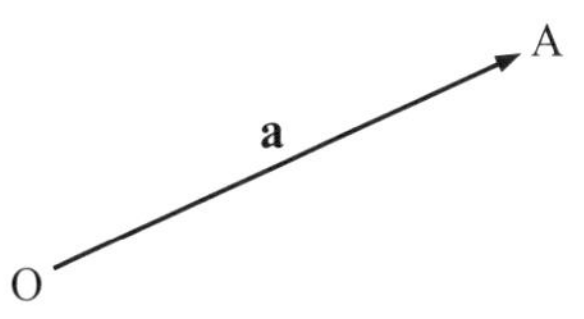

- This **position vector** could be represented by

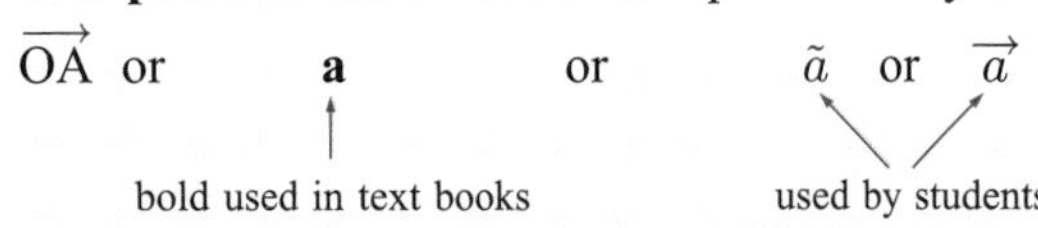

- The **magnitude** or **length** could be represented by
$|\overrightarrow{OA}|$ or OA or $|\mathbf{a}|$ or $|\tilde{a}|$ or $|\vec{a}|$

For

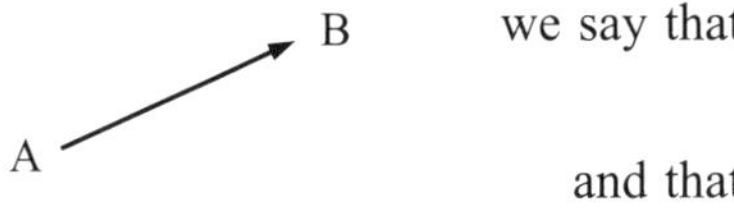

we say that $\overrightarrow{AB}$ is the vector which **originates** at A and **terminates** at B,

and that $\overrightarrow{AB}$ is the **position vector** of B relative to A.

GEOMETRIC VECTOR EQUALITY

Two vectors are **equal** if they have the same magnitude and direction.

Equal vectors are **parallel** and in the same direction, and are **equal in length**. The arrows that represent them are translations of one another.

We can draw a vector with given magnitude and direction from *any* point, so we consider vectors to be **free**. They are sometimes referred to as **free vectors**.

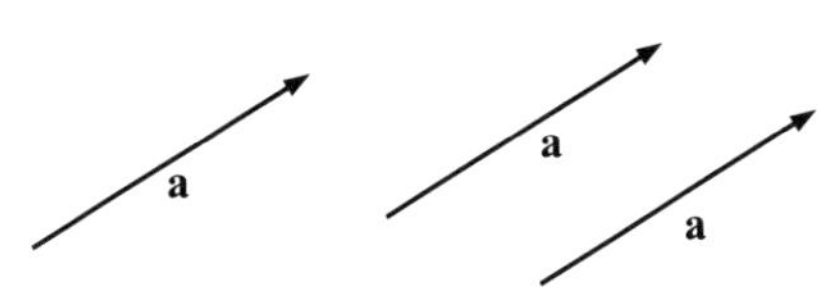

GEOMETRIC NEGATIVE VECTORS

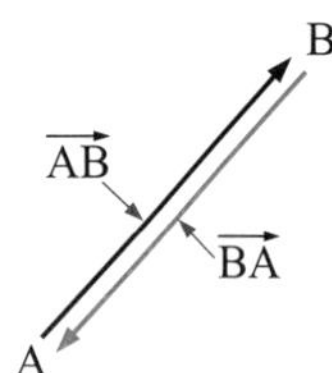

$\overrightarrow{AB}$ and $\overrightarrow{BA}$ have the same length, but they have opposite directions.

We say that $\overrightarrow{BA}$ is the negative of $\overrightarrow{AB}$ and write $\overrightarrow{BA} = -\overrightarrow{AB}$.

$\mathbf{a}$ and $-\mathbf{a}$ are parallel and equal in length, but opposite in direction.

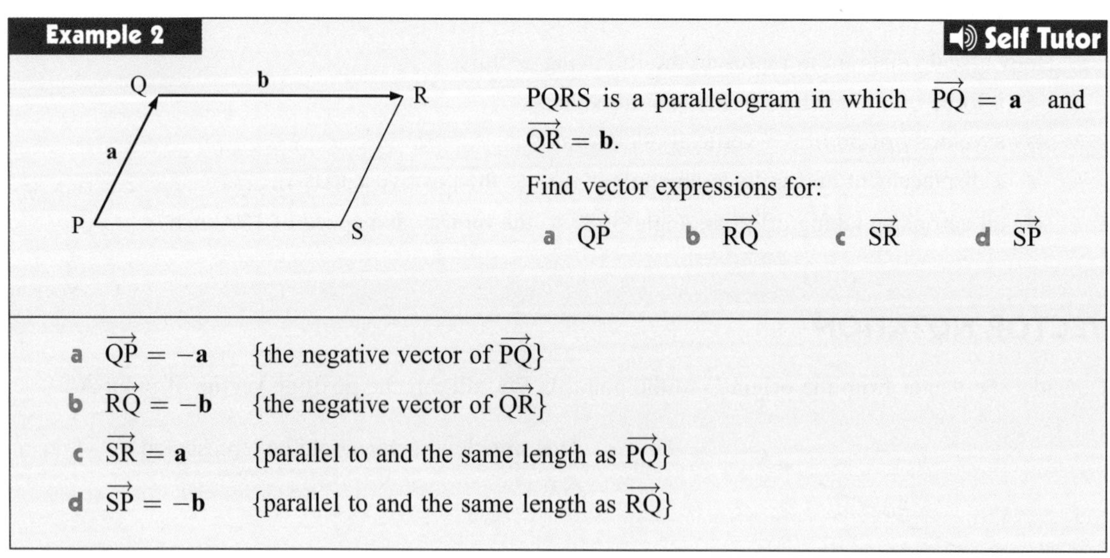

Example 2 Self Tutor

PQRS is a parallelogram in which $\overrightarrow{PQ} = \mathbf{a}$ and $\overrightarrow{QR} = \mathbf{b}$.

Find vector expressions for:

a $\overrightarrow{QP}$ **b** $\overrightarrow{RQ}$ **c** $\overrightarrow{SR}$ **d** $\overrightarrow{SP}$

a $\overrightarrow{QP} = -\mathbf{a}$ {the negative vector of $\overrightarrow{PQ}$}

b $\overrightarrow{RQ} = -\mathbf{b}$ {the negative vector of $\overrightarrow{QR}$}

c $\overrightarrow{SR} = \mathbf{a}$ {parallel to and the same length as $\overrightarrow{PQ}$}

d $\overrightarrow{SP} = -\mathbf{b}$ {parallel to and the same length as $\overrightarrow{RQ}$}

EXERCISE 14A.2

1 State the vectors which are:

- **a** equal in magnitude
- **b** parallel
- **c** in the same direction
- **d** equal
- **e** negatives of one another.

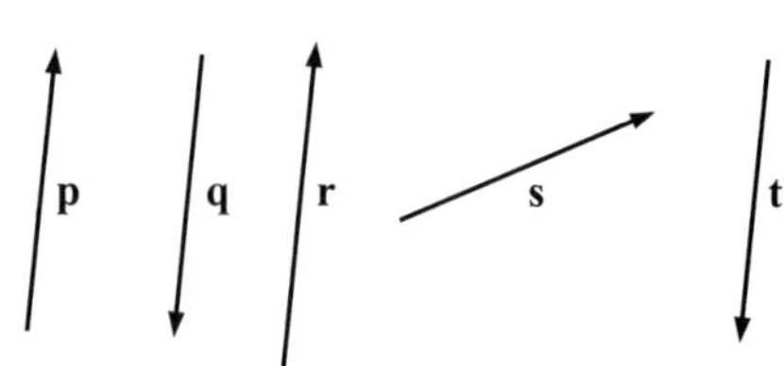

2 The figure alongside consists of two equilateral triangles. A, B, and C lie on a straight line. $\overrightarrow{AB} = \mathbf{p}$, $\overrightarrow{AE} = \mathbf{q}$, and $\overrightarrow{DC} = \mathbf{r}$.

Which of the following statements are true?

a $\overrightarrow{EB} = \mathbf{r}$ **b** $|\mathbf{p}| = |\mathbf{q}|$ **c** $\overrightarrow{BC} = \mathbf{r}$

d $\overrightarrow{DB} = \mathbf{q}$ **e** $\overrightarrow{ED} = \mathbf{p}$ **f** $\mathbf{p} = \mathbf{q}$

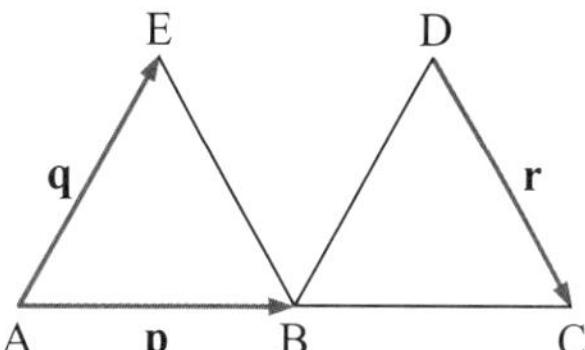

3

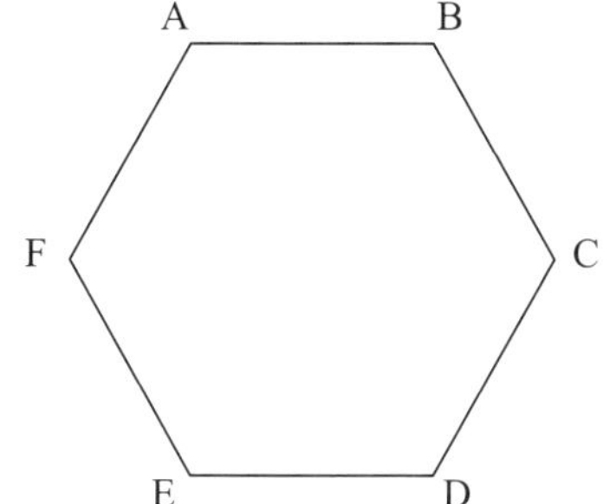

ABCDEF is a regular hexagon.

a Write down the vector which:

i originates at B and terminates at C

ii is equal to $\overrightarrow{AB}$.

b Write down *all* vectors which:

i are the negative of $\overrightarrow{EF}$

ii have the same length as $\overrightarrow{ED}$.

c Write down a vector which is parallel to $\overrightarrow{AB}$ and twice its length.

DISCUSSION

- Could we have a zero vector?
- What would its length be?
- What would its direction be?

B GEOMETRIC OPERATIONS WITH VECTORS

In previous years we have often used vectors for problems involving distances and directions. The vectors in this case are **displacements**.

A typical problem could be:

A runner runs east for 4 km and then south for 2 km.

How far is she from her starting point and in what direction?

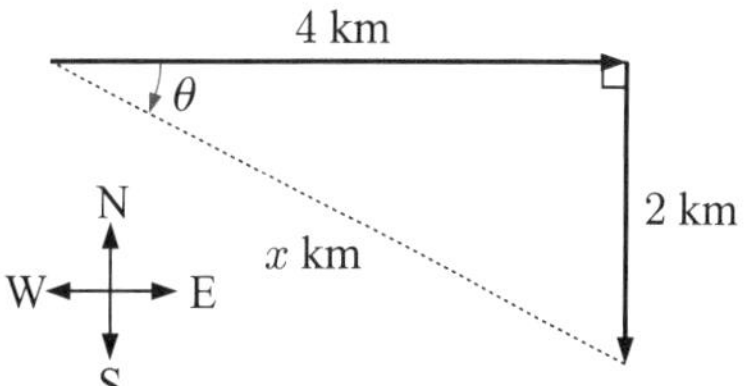

In problems like these we use trigonometry and Pythagoras' theorem to find the unknown lengths and angles.

GEOMETRIC VECTOR ADDITION

Suppose we have three towns P, Q, and R.

A trip from P to Q followed by a trip from Q to R has the same origin and destination as a trip from P to R.

This can be expressed in vector form as the sum $\overrightarrow{PQ} + \overrightarrow{QR} = \overrightarrow{PR}$.

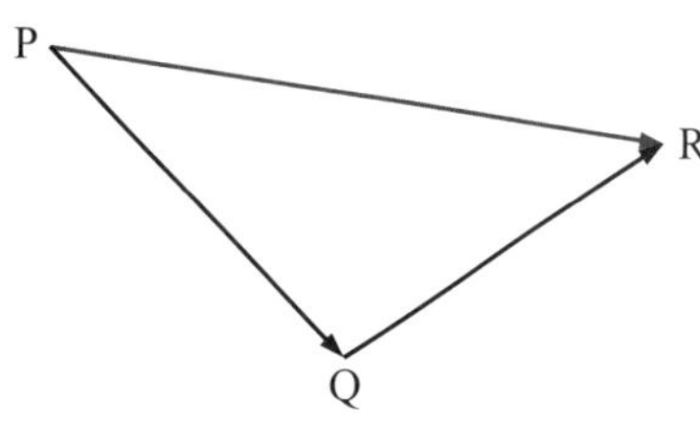

This triangular diagram could take all sorts of shapes, but in each case the sum will be true. For example:

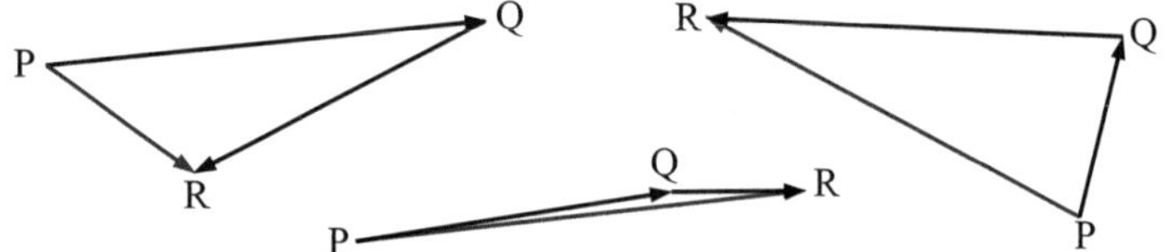

After considering diagrams like those above, we can now define vector addition geometrically:

To construct $\mathbf{a} + \mathbf{b}$:

Step 1: Draw $\mathbf{a}$.

Step 2: At the arrowhead end of $\mathbf{a}$, draw $\mathbf{b}$.

Step 3: Join the beginning of $\mathbf{a}$ to the arrowhead end of $\mathbf{b}$. This is vector $\mathbf{a} + \mathbf{b}$.

DEMO

Example 3 — Self Tutor

Given $\mathbf{a}$ and $\mathbf{b}$ as shown, construct $\mathbf{a} + \mathbf{b}$.

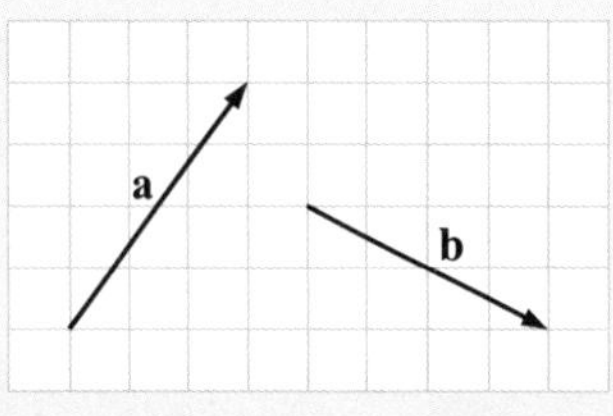

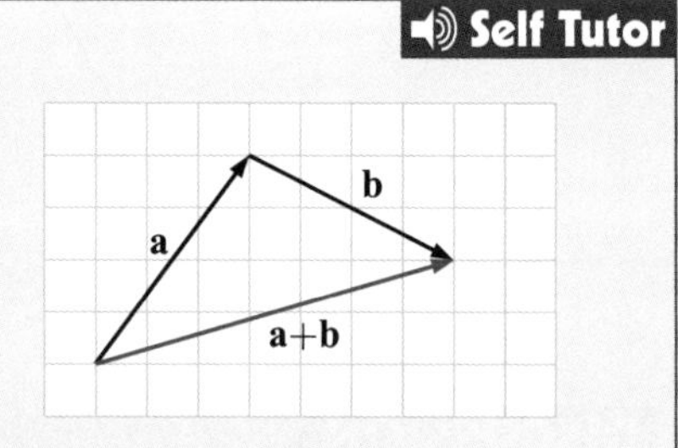

THE ZERO VECTOR

Having defined vector addition, we are now able to state that:

The **zero vector 0** is a vector of length 0.

For any vector $\mathbf{a}$: $\mathbf{a} + \mathbf{0} = \mathbf{0} + \mathbf{a} = \mathbf{a}$

$\mathbf{a} + (-\mathbf{a}) = (-\mathbf{a}) + \mathbf{a} = \mathbf{0}$.

When we write the zero vector by hand, we usually write $\overrightarrow{0}$.

Example 4 — Self Tutor

Find a single vector which is equal to:

a $\overrightarrow{BC} + \overrightarrow{CA}$

b $\overrightarrow{BA} + \overrightarrow{AE} + \overrightarrow{EC}$

c $\overrightarrow{AB} + \overrightarrow{BC} + \overrightarrow{CA}$

d $\overrightarrow{AB} + \overrightarrow{BC} + \overrightarrow{CD} + \overrightarrow{DE}$

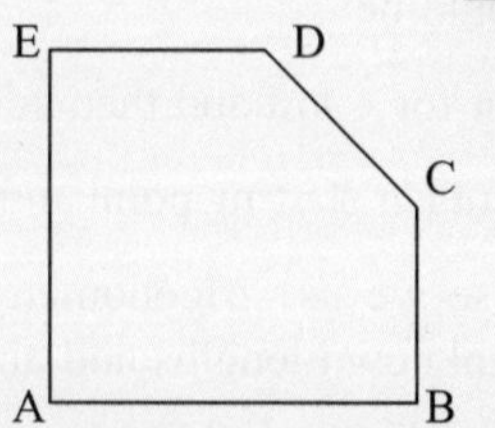

a $\overrightarrow{BC} + \overrightarrow{CA} = \overrightarrow{BA}$ {as shown}

b $\overrightarrow{BA} + \overrightarrow{AE} + \overrightarrow{EC} = \overrightarrow{BC}$

c $\overrightarrow{AB} + \overrightarrow{BC} + \overrightarrow{CA} = \overrightarrow{AA} = \mathbf{0}$

d $\overrightarrow{AB} + \overrightarrow{BC} + \overrightarrow{CD} + \overrightarrow{DE} = \overrightarrow{AE}$

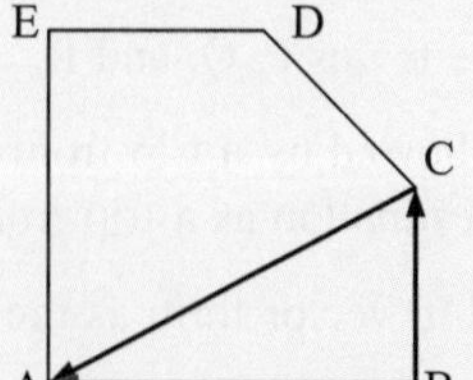

EXERCISE 14B.1

1 Use the given vectors **p** and **q** to construct $\mathbf{p}+\mathbf{q}$:

a
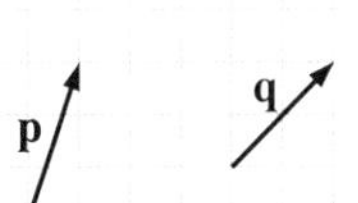

b

c
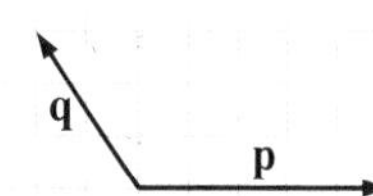

d
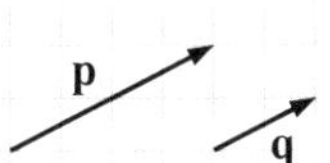

e
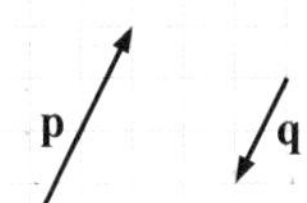

f
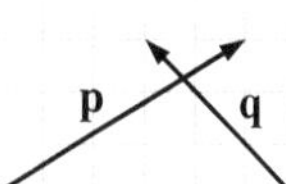

2 Find a single vector which is equal to:

a $\overrightarrow{AB}+\overrightarrow{BC}$ b $\overrightarrow{BC}+\overrightarrow{CD}$ c $\overrightarrow{AB}+\overrightarrow{BA}$

d $\overrightarrow{AB}+\overrightarrow{BC}+\overrightarrow{CD}$ e $\overrightarrow{AC}+\overrightarrow{CB}+\overrightarrow{BD}$ f $\overrightarrow{BC}+\overrightarrow{CA}+\overrightarrow{AB}$

3 a Given
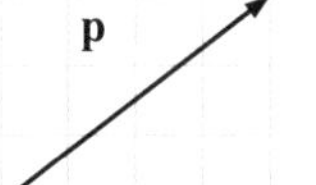

and q use vector diagrams to find:

i $\mathbf{p}+\mathbf{q}$ ii $\mathbf{q}+\mathbf{p}$.

b For any two vectors **p** and **q**, is $\mathbf{p}+\mathbf{q}=\mathbf{q}+\mathbf{p}$?

4 Consider:
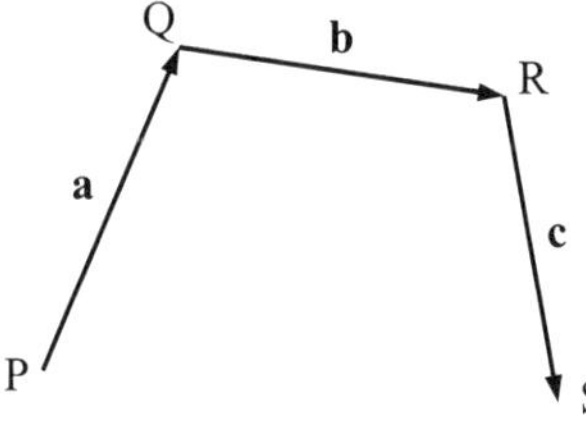

One way of finding $\overrightarrow{PS}$ is:

$$\overrightarrow{PS}=\overrightarrow{PR}+\overrightarrow{RS}$$
$$=(\mathbf{a}+\mathbf{b})+\mathbf{c}.$$

Use the diagram to show that $(\mathbf{a}+\mathbf{b})+\mathbf{c}=\mathbf{a}+(\mathbf{b}+\mathbf{c})$.

5 Answer the **Opening Problem** on page **384**.

GEOMETRIC VECTOR SUBTRACTION

To subtract one vector from another, we simply **add its negative**. $\mathbf{a}-\mathbf{b}=\mathbf{a}+(-\mathbf{b})$

For example,

given
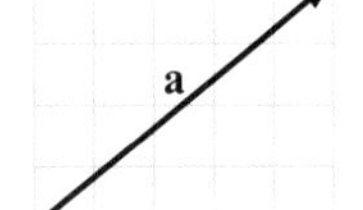

and
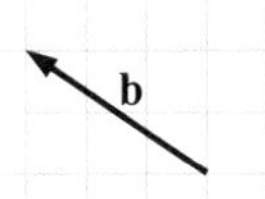

then
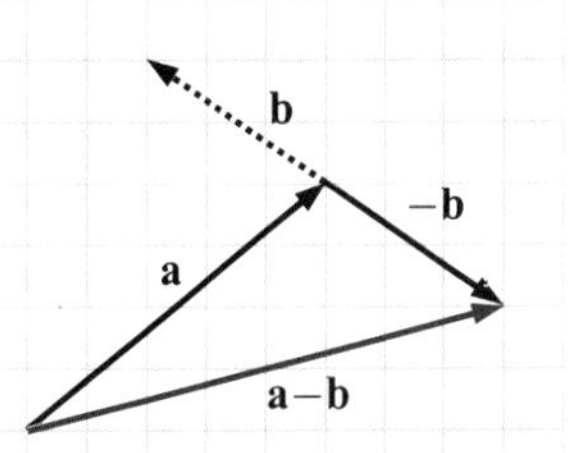

Example 5 Self Tutor

For **r**, **s**, and **t** shown, find geometrically:

a $\mathbf{r} - \mathbf{s}$

b $\mathbf{s} - \mathbf{t} - \mathbf{r}$

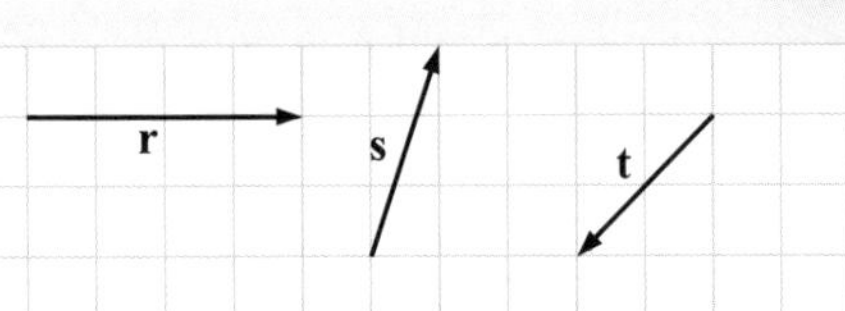

a

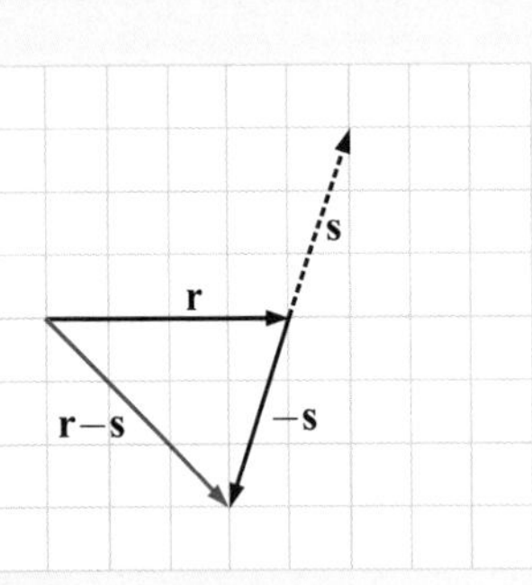

b

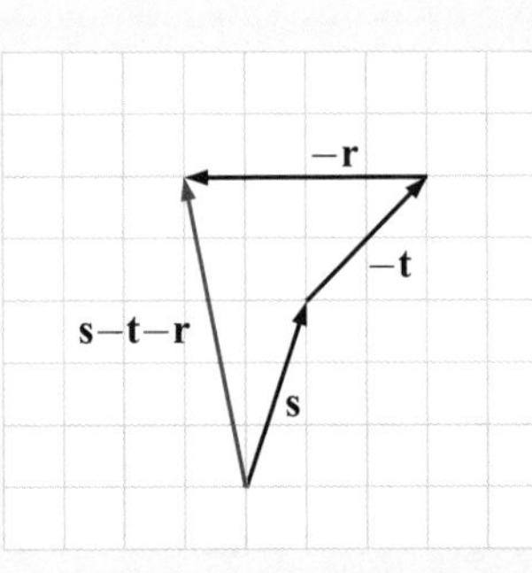

Example 6 Self Tutor

For points A, B, C, and D, simplify the following vector expressions:

a $\overrightarrow{AB} - \overrightarrow{CB}$ **b** $\overrightarrow{AC} - \overrightarrow{BC} - \overrightarrow{DB}$

a $\overrightarrow{AB} - \overrightarrow{CB}$

$= \overrightarrow{AB} + \overrightarrow{BC}$ $\{$as $\overrightarrow{BC} = -\overrightarrow{CB}\}$

$= \overrightarrow{AC}$

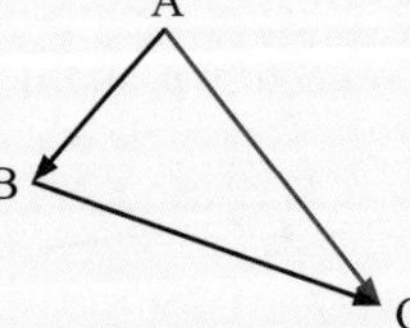

b $\overrightarrow{AC} - \overrightarrow{BC} - \overrightarrow{DB}$

$= \overrightarrow{AC} + \overrightarrow{CB} + \overrightarrow{BD}$

$= \overrightarrow{AD}$

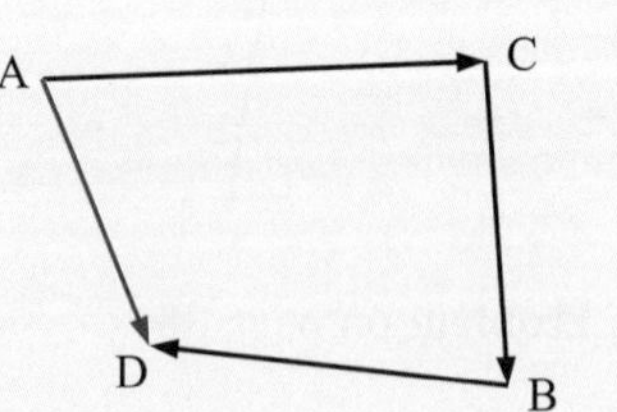

EXERCISE 14B.2

1 For the following vectors **p** and **q**, show how to construct $\mathbf{p} - \mathbf{q}$:

a

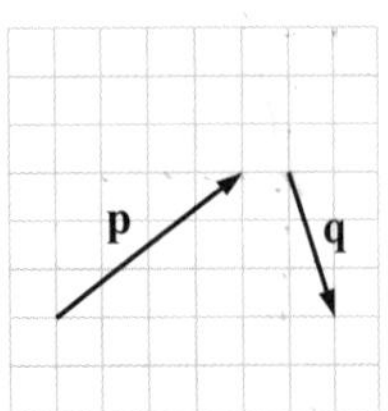

b

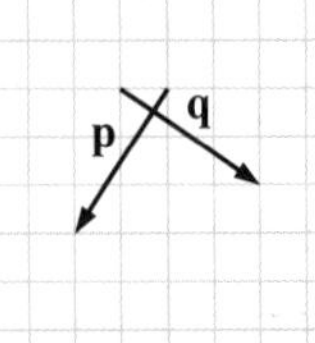

c

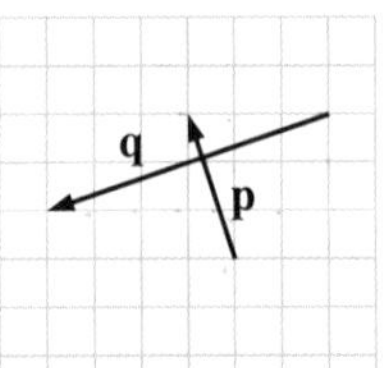

d

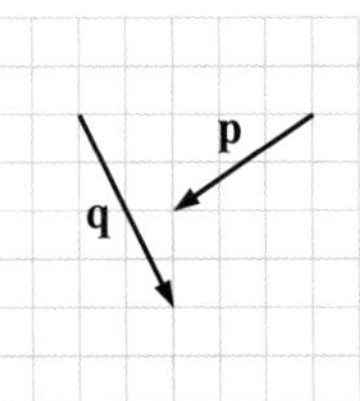

2 For the vectors illustrated, show how to construct:

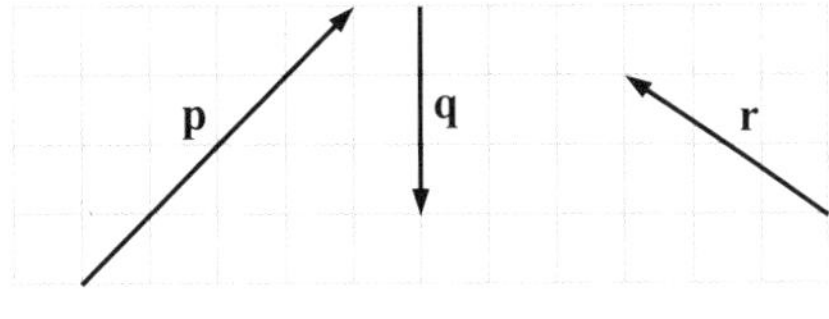

a $\mathbf{p} + \mathbf{q} - \mathbf{r}$ **b** $\mathbf{p} - \mathbf{q} - \mathbf{r}$ **c** $\mathbf{r} - \mathbf{q} - \mathbf{p}$

3 For points A, B, C, and D, simplify the following vector expressions:

a $\overrightarrow{AC} + \overrightarrow{CB}$ **b** $\overrightarrow{AD} - \overrightarrow{BD}$ **c** $\overrightarrow{AC} + \overrightarrow{CA}$

d $\overrightarrow{AB} + \overrightarrow{BC} + \overrightarrow{CD}$ **e** $\overrightarrow{BA} - \overrightarrow{CA} + \overrightarrow{CB}$ **f** $\overrightarrow{AB} - \overrightarrow{CB} - \overrightarrow{DC}$

VECTOR EQUATIONS

Whenever we have vectors which form a closed polygon, we can write a **vector equation** which relates the variables.

The vector equation can usually be written in several ways, but they are all equivalent.

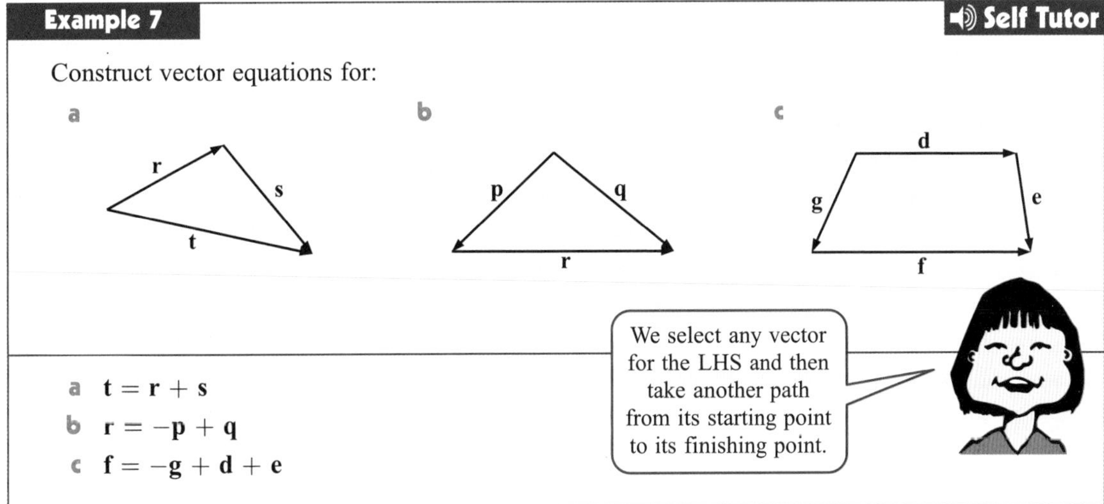

EXERCISE 14B.3

1 Construct vector equations for:

a

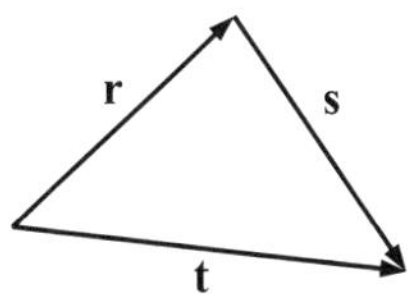

b

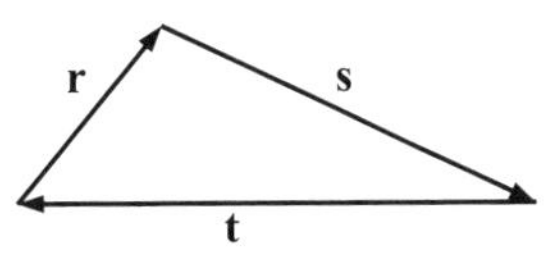

c

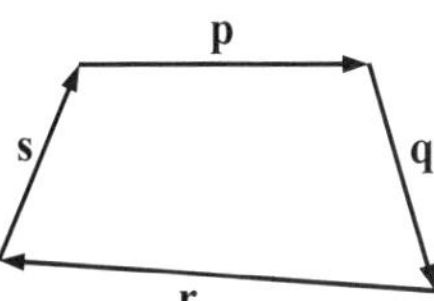

d

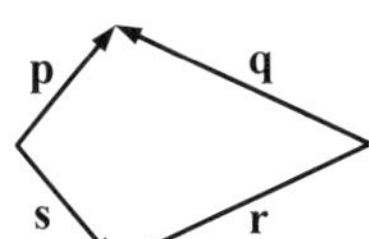

e

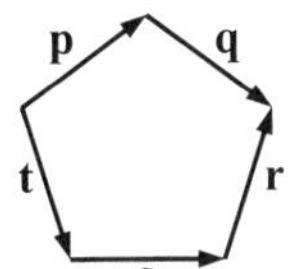

f

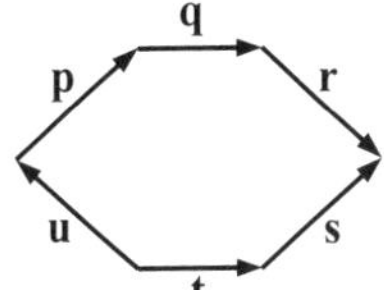

Example 8 **Self Tutor**

Find, in terms of **r**, **s**, and **t**:

a $\overrightarrow{RS}$ **b** $\overrightarrow{SR}$ **c** $\overrightarrow{ST}$

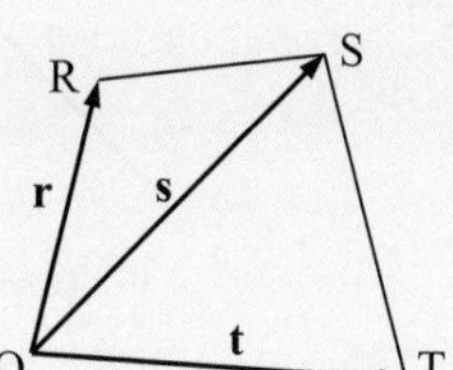

a $\overrightarrow{RS}$
$= \overrightarrow{RO} + \overrightarrow{OS}$
$= -\overrightarrow{OR} + \overrightarrow{OS}$
$= -\mathbf{r} + \mathbf{s}$
$= \mathbf{s} - \mathbf{r}$

b $\overrightarrow{SR}$
$= \overrightarrow{SO} + \overrightarrow{OR}$
$= -\overrightarrow{OS} + \overrightarrow{OR}$
$= -\mathbf{s} + \mathbf{r}$
$= \mathbf{r} - \mathbf{s}$

c $\overrightarrow{ST}$
$= \overrightarrow{SO} + \overrightarrow{OT}$
$= -\overrightarrow{OS} + \overrightarrow{OT}$
$= -\mathbf{s} + \mathbf{t}$
$= \mathbf{t} - \mathbf{s}$

2 **a** Find, in terms of **r**, **s**, and **t**:
i $\overrightarrow{OB}$ **ii** $\overrightarrow{CA}$ **iii** $\overrightarrow{OC}$

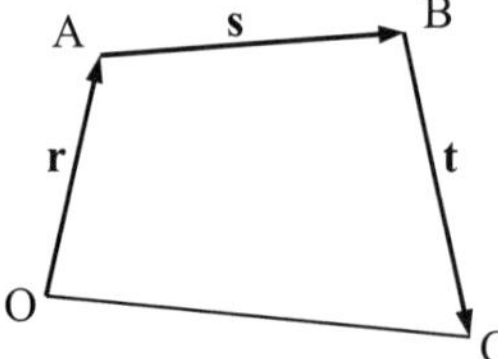

b Find, in terms of **p**, **q**, and **r**:
i $\overrightarrow{AD}$ **ii** $\overrightarrow{BC}$ **iii** $\overrightarrow{AC}$

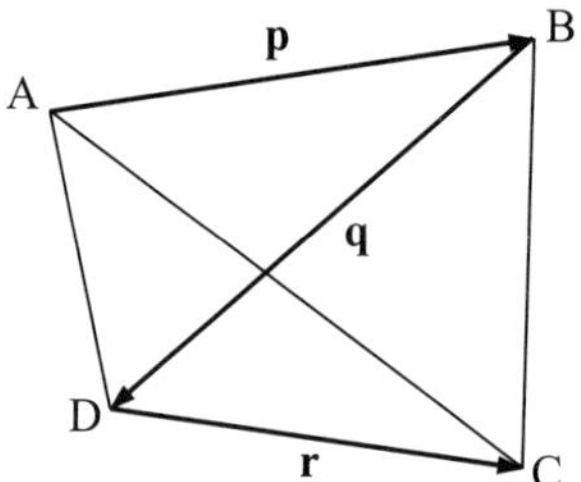

GEOMETRIC SCALAR MULTIPLICATION

A **scalar** is a non-vector quantity. It has a size but no direction.

We can multiply vectors by scalars such as 2 and -3, or in fact any $k \in \mathbb{R}$.

If **a** is a vector, we define $2\mathbf{a} = \mathbf{a} + \mathbf{a}$ and $3\mathbf{a} = \mathbf{a} + \mathbf{a} + \mathbf{a}$
so $-3\mathbf{a} = 3(-\mathbf{a}) = (-\mathbf{a}) + (-\mathbf{a}) + (-\mathbf{a})$.

If **a** is **a** then **a** **a** 2**a** **a** **a** **a** 3**a** $-$**a** $-$**a** $-$**a** -3**a**

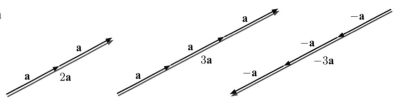

So, $2\mathbf{a}$ is in the same direction as **a** but is twice as long as **a**
$3\mathbf{a}$ is in the same direction as **a** but is three times longer than **a**
$-3\mathbf{a}$ has the opposite direction to **a** and is three times longer than **a**.

If **a** is a vector and k is a scalar, then $k\mathbf{a}$ is also a vector and we are performing **scalar multiplication**.

VECTOR SCALAR MULTIPLICATION

If $k > 0$, $k\mathbf{a}$ and $\mathbf{a}$ have the same direction.
If $k < 0$, $k\mathbf{a}$ and $\mathbf{a}$ have opposite directions.
If $k = 0$, $k\mathbf{a} = \mathbf{0}$, the zero vector.

Example 9 Self Tutor

Given vectors 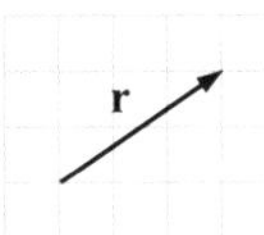

and 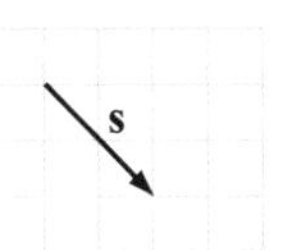

,

construct geometrically: **a** $2\mathbf{r} + \mathbf{s}$ **b** $\mathbf{r} - 3\mathbf{s}$

a

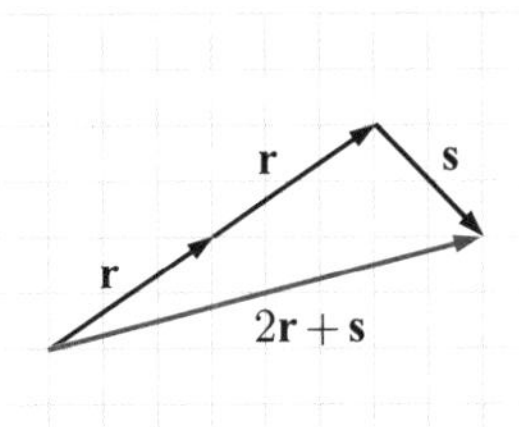

b

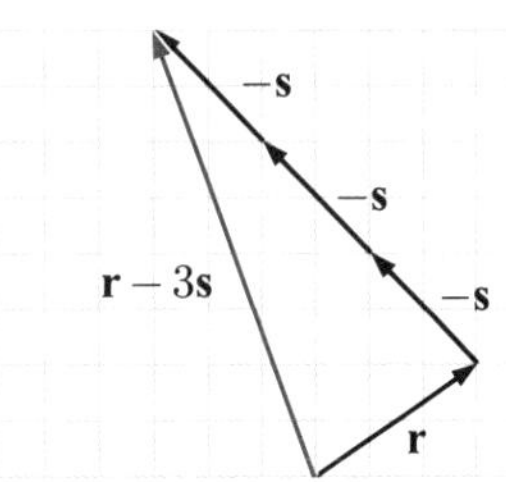

Example 10 Self Tutor

Sketch vectors **p** and **q** if: **a** $\mathbf{p} = 3\mathbf{q}$ **b** $\mathbf{p} = -\frac{1}{2}\mathbf{q}$

Suppose **q** is: 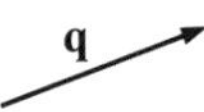

a

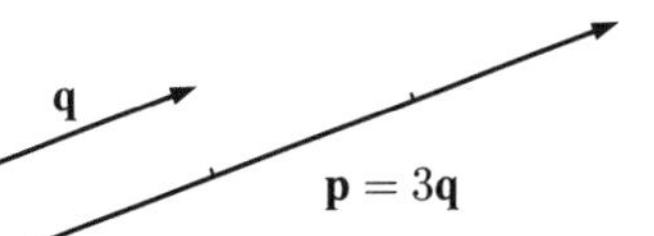

b

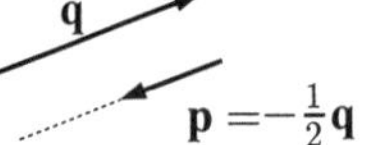

EXERCISE 14B.4

1 Given vectors 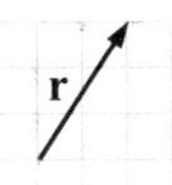

and 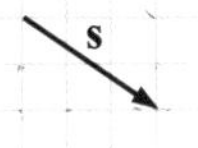

, construct geometrically:

a $-\mathbf{r}$ **b** $2\mathbf{s}$ **c** $\frac{1}{2}\mathbf{r}$ **d** $-\frac{3}{2}\mathbf{s}$

e $2\mathbf{r} - \mathbf{s}$ **f** $2\mathbf{r} + 3\mathbf{s}$ **g** $\frac{1}{2}\mathbf{r} + 2\mathbf{s}$ **h** $\frac{1}{2}(\mathbf{r} + 3\mathbf{s})$

2 Sketch vectors **p** and **q** if:

a $\mathbf{p} = \mathbf{q}$ **b** $\mathbf{p} = -\mathbf{q}$ **c** $\mathbf{p} = 2\mathbf{q}$ **d** $\mathbf{p} = \frac{1}{3}\mathbf{q}$ **e** $\mathbf{p} = -3\mathbf{q}$

3 **a** Copy this diagram and on it mark the points:

i X such that $\overrightarrow{MX} = \overrightarrow{MN} + \overrightarrow{MP}$

ii Y such that $\overrightarrow{MY} = \overrightarrow{MN} - \overrightarrow{MP}$

iii Z such that $\overrightarrow{PZ} = 2\overrightarrow{PM}$

b What type of figure is MNYZ?

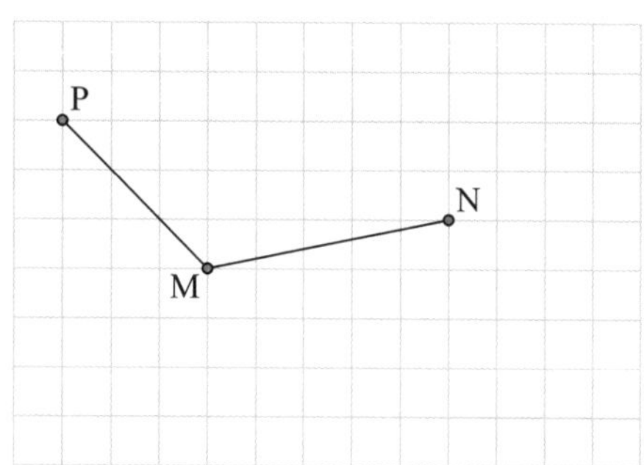

4

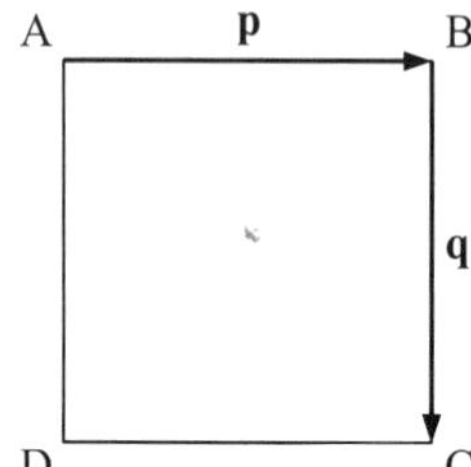

ABCD is a square. Its diagonals [AC] and [BD] intersect at M. If $\overrightarrow{AB} = \mathbf{p}$ and $\overrightarrow{BC} = \mathbf{q}$, find in terms of $\mathbf{p}$ and $\mathbf{q}$:

a $\overrightarrow{CD}$ **b** $\overrightarrow{AC}$ **c** $\overrightarrow{AM}$ **d** $\overrightarrow{BM}$

5

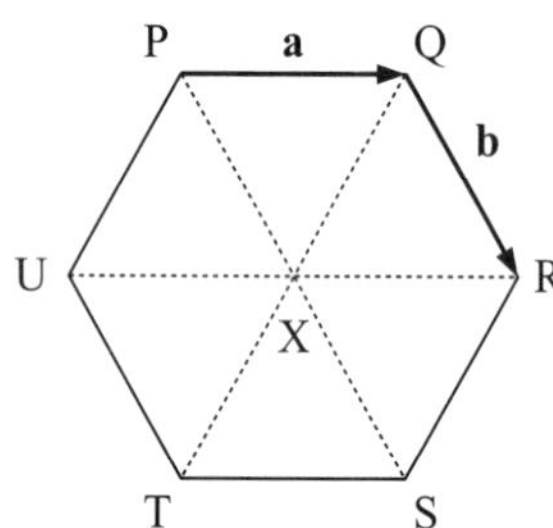

PQRSTU is a regular hexagon.

If $\overrightarrow{PQ} = \mathbf{a}$ and $\overrightarrow{QR} = \mathbf{b}$, find in terms of $\mathbf{a}$ and $\mathbf{b}$:

a $\overrightarrow{PX}$ **b** $\overrightarrow{PS}$ **c** $\overrightarrow{QX}$ **d** $\overrightarrow{RS}$

C VECTORS IN THE PLANE

When we plot points in the Cartesian plane, we move first in the x-direction and then in the y-direction.

For example, to plot the point P(2, 5), we start at the origin, move 2 units in the x-direction, and then 5 units in the y-direction.

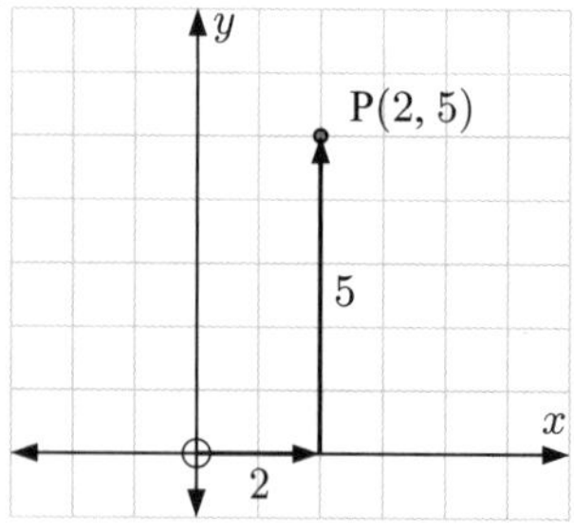

In transformation geometry, translating a point a units in the x-direction and b units in the y-direction can be achieved using the translation vector $\begin{pmatrix} a \\ b \end{pmatrix}$.

So, the vector from O to P is $\overrightarrow{OP} = \begin{pmatrix} 2 \\ 5 \end{pmatrix}$.

Suppose that $\mathbf{i} = \begin{pmatrix} 1 \\ 0 \end{pmatrix}$ is a translation 1 unit in the positive x-direction

and that $\mathbf{j} = \begin{pmatrix} 0 \\ 1 \end{pmatrix}$ is a translation 1 unit in the positive y-direction.

i and **j** are examples of **unit vectors** because they have length 1.

We can see that moving from O to P is equivalent to two lots of **i** plus 5 lots of **j**.

$$\overrightarrow{OP} = 2\mathbf{i} + 5\mathbf{j}$$

$$\therefore \begin{pmatrix} 2 \\ 5 \end{pmatrix} = 2\begin{pmatrix} 1 \\ 0 \end{pmatrix} + 5\begin{pmatrix} 0 \\ 1 \end{pmatrix}$$

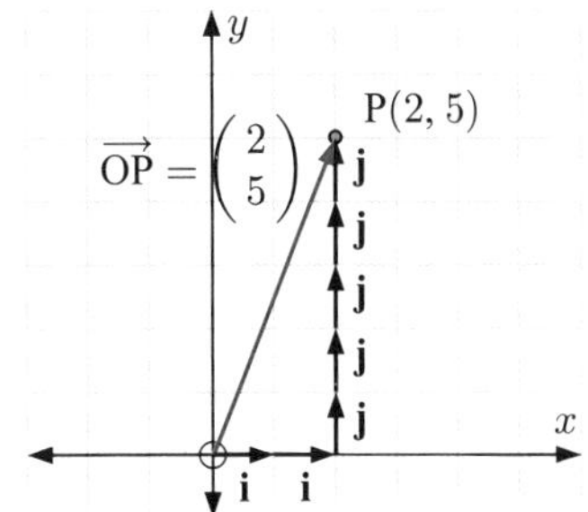

The point P(x, y) has **position vector** $\overrightarrow{OP} = \begin{pmatrix} x \\ y \end{pmatrix} = x\mathbf{i} + y\mathbf{j}$

component form   unit vector form

$\mathbf{i} = \begin{pmatrix} 1 \\ 0 \end{pmatrix}$ is the **base unit vector** in the x-direction.

$\mathbf{j} = \begin{pmatrix} 0 \\ 1 \end{pmatrix}$ is the **base unit vector** in the y-direction.

The set of vectors $\{\mathbf{i}, \mathbf{j}\} = \left\{\begin{pmatrix} 1 \\ 0 \end{pmatrix}, \begin{pmatrix} 0 \\ 1 \end{pmatrix}\right\}$ is the **standard basis** for the 2-dimensional (x, y) coordinate system.

All vectors in the plane can be described in terms of the base unit vectors **i** and **j**.

For example: $\mathbf{a} = 3\mathbf{i} - \mathbf{j}$
$\mathbf{b} = -4\mathbf{i} + 3\mathbf{j}$

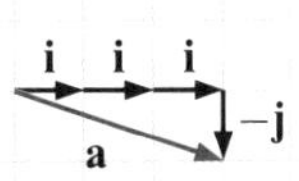

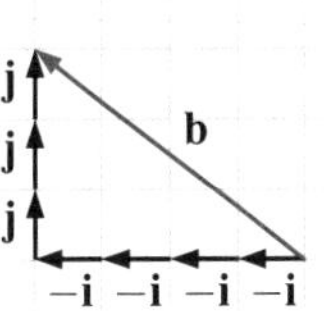

Two vectors are **equal** if their components are equal.

Example 11 — Self Tutor

a Write $\overrightarrow{OA}$ and $\overrightarrow{CB}$ in component form and in unit vector form.

b Comment on your answers in **a**.

a $\overrightarrow{OA} = \begin{pmatrix} 3 \\ 1 \end{pmatrix} = 3\mathbf{i} + \mathbf{j}$ $\overrightarrow{CB} = \begin{pmatrix} 3 \\ 1 \end{pmatrix} = 3\mathbf{i} + \mathbf{j}$

b The vectors $\overrightarrow{OA}$ and $\overrightarrow{CB}$ are equal.

EXERCISE 14C

1 Write the illustrated vectors in component form and in unit vector form:

a 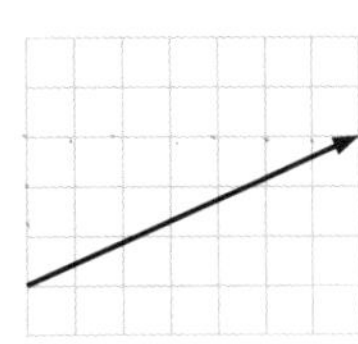

b 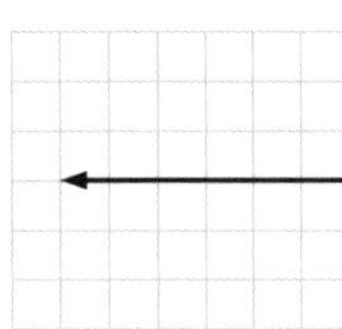

c 

d 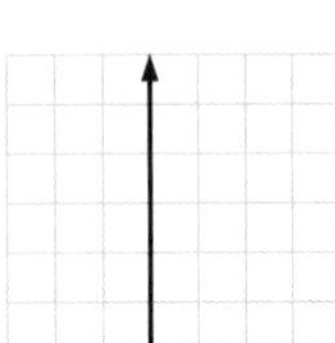

e 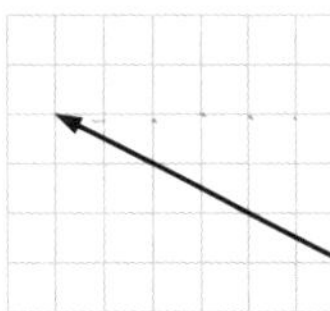

f 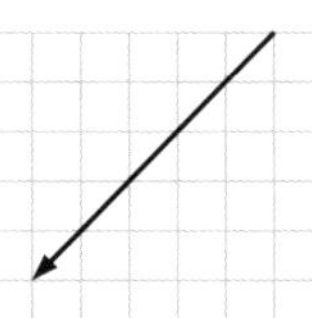

2 Write each vector in unit vector form, and illustrate it using an arrow diagram:

a $\begin{pmatrix} 3 \\ 4 \end{pmatrix}$ **b** $\begin{pmatrix} 2 \\ 0 \end{pmatrix}$ **c** $\begin{pmatrix} 2 \\ -5 \end{pmatrix}$ **d** $\begin{pmatrix} -1 \\ -3 \end{pmatrix}$

3 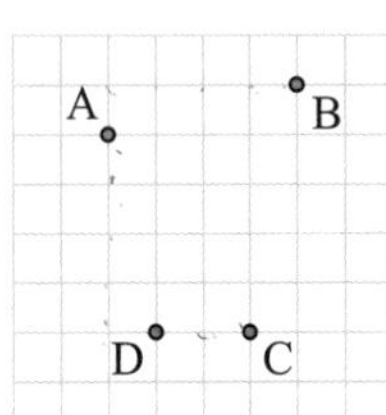

Find in component form and in unit vector form:

a $\overrightarrow{BA}$ **b** $\overrightarrow{BC}$ **c** $\overrightarrow{DC}$
d $\overrightarrow{AC}$ **e** $\overrightarrow{CA}$ **f** $\overrightarrow{DB}$

4 Write in component form and illustrate using a directed line segment:

a $\mathbf{i} + 2\mathbf{j}$ **b** $-\mathbf{i} + 3\mathbf{j}$ **c** $-5\mathbf{j}$ **d** $4\mathbf{i} - 2\mathbf{j}$

5 Write the zero vector **0** in component form.

D THE MAGNITUDE OF A VECTOR

Consider vector $\mathbf{v} = \begin{pmatrix} 2 \\ 3 \end{pmatrix} = 2\mathbf{i} + 3\mathbf{j}$.

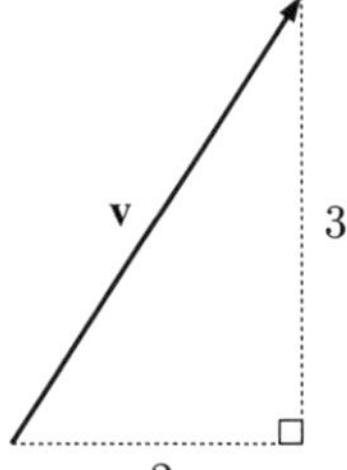

The **magnitude** or **length** of **v** is represented by $|\mathbf{v}|$.

By Pythagoras, $|\mathbf{v}|^2 = 2^2 + 3^2 = 4 + 9 = 13$

$\therefore \ |\mathbf{v}| = \sqrt{13}$ units {since $|\mathbf{v}| > 0$}

If $\mathbf{v} = \begin{pmatrix} v_1 \\ v_2 \end{pmatrix} = v_1\mathbf{i} + v_2\mathbf{j}$, the **magnitude** or **length** of **v** is $|\mathbf{v}| = \sqrt{{v_1}^2 + {v_2}^2}$.

Example 12 — Self Tutor

If $\mathbf{p} = \begin{pmatrix} 3 \\ -5 \end{pmatrix}$ and $\mathbf{q} = 2\mathbf{i} - 5\mathbf{j}$ find: **a** $|\mathbf{p}|$ **b** $|\mathbf{q}|$

a $\mathbf{p} = \begin{pmatrix} 3 \\ -5 \end{pmatrix}$

$\therefore\ |\mathbf{p}| = \sqrt{3^2 + (-5)^2}$

$= \sqrt{34}$ units

b As $2\mathbf{i} - 5\mathbf{j} = \begin{pmatrix} 2 \\ -5 \end{pmatrix}$,

$|\mathbf{q}| = \sqrt{2^2 + (-5)^2}$

$= \sqrt{29}$ units.

UNIT VECTORS

A **unit vector** is any vector which has a length of one unit.

$\mathbf{i} = \begin{pmatrix} 1 \\ 0 \end{pmatrix}$ and $\mathbf{j} = \begin{pmatrix} 0 \\ 1 \end{pmatrix}$ are the base unit vectors in the positive x and y-directions respectively.

Example 13 — Self Tutor

Find k given that $\begin{pmatrix} -\frac{1}{3} \\ k \end{pmatrix}$ is a unit vector.

Since $\begin{pmatrix} -\frac{1}{3} \\ k \end{pmatrix}$ is a unit vector, $\sqrt{(-\frac{1}{3})^2 + k^2} = 1$

$\therefore\ \sqrt{\frac{1}{9} + k^2} = 1$

$\therefore\ \frac{1}{9} + k^2 = 1$ {squaring both sides}

$\therefore\ k^2 = \frac{8}{9}$

$\therefore\ k = \pm\frac{\sqrt{8}}{3}$

EXERCISE 14D

1 Find the magnitude of:

a $\begin{pmatrix} 3 \\ 4 \end{pmatrix}$ **b** $\begin{pmatrix} -4 \\ 3 \end{pmatrix}$ **c** $\begin{pmatrix} 2 \\ 0 \end{pmatrix}$ **d** $\begin{pmatrix} -2 \\ 2 \end{pmatrix}$ **e** $\begin{pmatrix} 0 \\ -3 \end{pmatrix}$

2 Find the length of:

a $\mathbf{i} + \mathbf{j}$ **b** $5\mathbf{i} - 12\mathbf{j}$ **c** $-\mathbf{i} + 4\mathbf{j}$ **d** $3\mathbf{i}$ **e** $k\mathbf{j}$

3 Which of the following are unit vectors?

a $\begin{pmatrix} 0 \\ -1 \end{pmatrix}$ **b** $\begin{pmatrix} -\frac{1}{\sqrt{2}} \\ \frac{1}{\sqrt{2}} \end{pmatrix}$ **c** $\begin{pmatrix} \frac{2}{3} \\ \frac{1}{3} \end{pmatrix}$ **d** $\begin{pmatrix} -\frac{3}{5} \\ -\frac{4}{5} \end{pmatrix}$ **e** $\begin{pmatrix} \frac{2}{7} \\ -\frac{5}{7} \end{pmatrix}$

4 Find k for the unit vectors:

a $\begin{pmatrix} 0 \\ k \end{pmatrix}$ b $\begin{pmatrix} k \\ 0 \end{pmatrix}$ c $\begin{pmatrix} k \\ 1 \end{pmatrix}$ d $\begin{pmatrix} k \\ k \end{pmatrix}$ e $\begin{pmatrix} \frac{1}{2} \\ k \end{pmatrix}$

5 Given $\mathbf{v} = \begin{pmatrix} 8 \\ p \end{pmatrix}$ and $|\mathbf{v}| = \sqrt{73}$ units, find the possible values of p.

E OPERATIONS WITH PLANE VECTORS

ALGEBRAIC VECTOR ADDITION

Consider adding vectors $\mathbf{a} = \begin{pmatrix} a_1 \\ a_2 \end{pmatrix}$ and $\mathbf{b} = \begin{pmatrix} b_1 \\ b_2 \end{pmatrix}$.

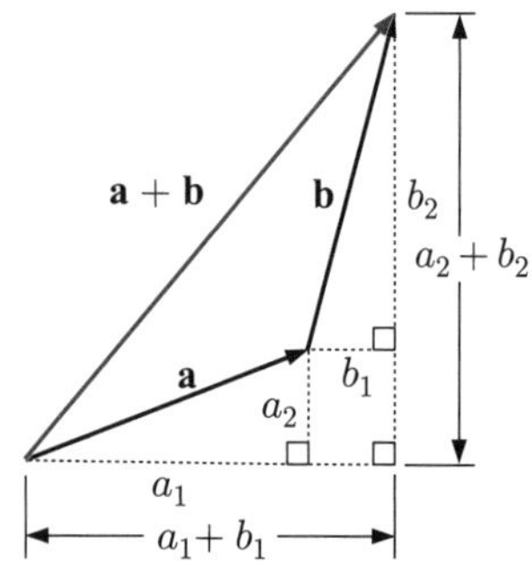

Notice that:

- the horizontal step for $\mathbf{a} + \mathbf{b}$ is $a_1 + b_1$
- the vertical step for $\mathbf{a} + \mathbf{b}$ is $a_2 + b_2$.

If $\mathbf{a} = \begin{pmatrix} a_1 \\ a_2 \end{pmatrix}$ and $\mathbf{b} = \begin{pmatrix} b_1 \\ b_2 \end{pmatrix}$ then $\mathbf{a} + \mathbf{b} = \begin{pmatrix} a_1 + b_1 \\ a_2 + b_2 \end{pmatrix}$.

Example 14 **Self Tutor**

If $\mathbf{a} = \begin{pmatrix} 1 \\ -3 \end{pmatrix}$ and $\mathbf{b} = \begin{pmatrix} 4 \\ 7 \end{pmatrix}$, find $\mathbf{a} + \mathbf{b}$. Check your answer graphically.

$$\begin{aligned} \mathbf{a} + \mathbf{b} &= \begin{pmatrix} 1 \\ -3 \end{pmatrix} + \begin{pmatrix} 4 \\ 7 \end{pmatrix} \\ &= \begin{pmatrix} 1+4 \\ -3+7 \end{pmatrix} \\ &= \begin{pmatrix} 5 \\ 4 \end{pmatrix} \end{aligned}$$

Graphical check:

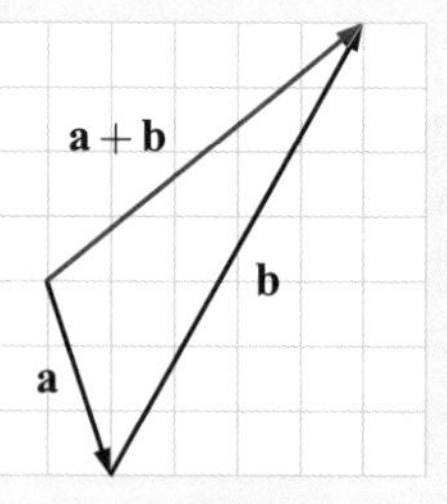

ALGEBRAIC NEGATIVE VECTORS

In the diagram we see the vector $\mathbf{a} = \begin{pmatrix} 2 \\ 3 \end{pmatrix}$

and its negative $-\mathbf{a} = \begin{pmatrix} -2 \\ -3 \end{pmatrix}$.

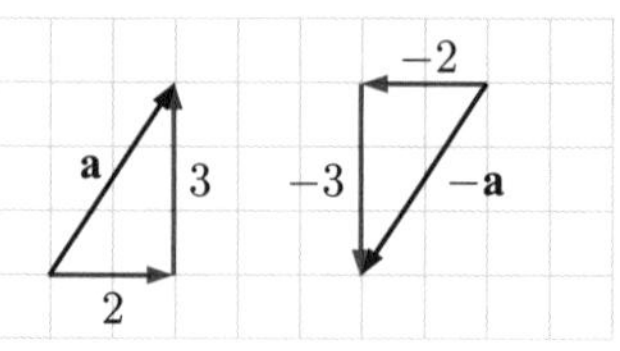

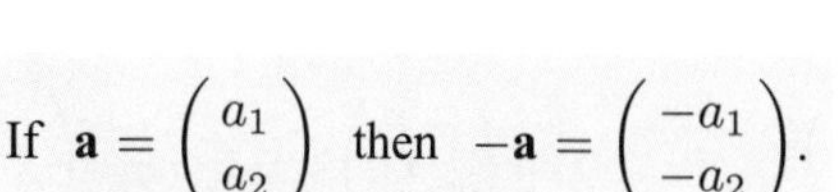

If $\mathbf{a} = \begin{pmatrix} a_1 \\ a_2 \end{pmatrix}$ then $-\mathbf{a} = \begin{pmatrix} -a_1 \\ -a_2 \end{pmatrix}$.

ALGEBRAIC VECTOR SUBTRACTION

To subtract one vector from another, we simply **add its negative**.

So, if $\mathbf{a} = \begin{pmatrix} a_1 \\ a_2 \end{pmatrix}$ and $\mathbf{b} = \begin{pmatrix} b_1 \\ b_2 \end{pmatrix}$

then $\mathbf{a} - \mathbf{b} = \mathbf{a} + (-\mathbf{b})$

$= \begin{pmatrix} a_1 \\ a_2 \end{pmatrix} + \begin{pmatrix} -b_1 \\ -b_2 \end{pmatrix}$

$= \begin{pmatrix} a_1 - b_1 \\ a_2 - b_2 \end{pmatrix}$

a

b

−b

a − b

If $\mathbf{a} = \begin{pmatrix} a_1 \\ a_2 \end{pmatrix}$ and $\mathbf{b} = \begin{pmatrix} b_1 \\ b_2 \end{pmatrix}$, then $\mathbf{a} - \mathbf{b} = \begin{pmatrix} a_1 - b_1 \\ a_2 - b_2 \end{pmatrix}$.

Example 15 **Self Tutor**

Given $\mathbf{p} = \begin{pmatrix} 3 \\ -2 \end{pmatrix}$, $\mathbf{q} = \begin{pmatrix} 1 \\ 4 \end{pmatrix}$, and $\mathbf{r} = \begin{pmatrix} -2 \\ -5 \end{pmatrix}$, find:

a $\mathbf{q} - \mathbf{p}$

b $\mathbf{p} - \mathbf{q} - \mathbf{r}$

a $\mathbf{q} - \mathbf{p}$

$= \begin{pmatrix} 1 \\ 4 \end{pmatrix} - \begin{pmatrix} 3 \\ -2 \end{pmatrix}$

$= \begin{pmatrix} 1-3 \\ 4+2 \end{pmatrix}$

$= \begin{pmatrix} -2 \\ 6 \end{pmatrix}$

b $\mathbf{p} - \mathbf{q} - \mathbf{r}$

$= \begin{pmatrix} 3 \\ -2 \end{pmatrix} - \begin{pmatrix} 1 \\ 4 \end{pmatrix} - \begin{pmatrix} -2 \\ -5 \end{pmatrix}$

$= \begin{pmatrix} 3-1+2 \\ -2-4+5 \end{pmatrix}$

$= \begin{pmatrix} 4 \\ -1 \end{pmatrix}$

ALGEBRAIC SCALAR MULTIPLICATION

We have already seen a geometric approach for integer scalar multiplication:

a

a a

2a

Consider $\mathbf{a} = \begin{pmatrix} 1 \\ 3 \end{pmatrix}$.

$2\mathbf{a} = \mathbf{a} + \mathbf{a} = \begin{pmatrix} 1 \\ 3 \end{pmatrix} + \begin{pmatrix} 1 \\ 3 \end{pmatrix} = \begin{pmatrix} 2 \\ 6 \end{pmatrix} = \begin{pmatrix} 2 \times 1 \\ 2 \times 3 \end{pmatrix}$

$3\mathbf{a} = \mathbf{a} + \mathbf{a} + \mathbf{a} = \begin{pmatrix} 1 \\ 3 \end{pmatrix} + \begin{pmatrix} 1 \\ 3 \end{pmatrix} + \begin{pmatrix} 1 \\ 3 \end{pmatrix} = \begin{pmatrix} 3 \\ 9 \end{pmatrix} = \begin{pmatrix} 3 \times 1 \\ 3 \times 3 \end{pmatrix}$

If k is any scalar and $\mathbf{v} = \begin{pmatrix} v_1 \\ v_2 \end{pmatrix}$, then $k\mathbf{v} = \begin{pmatrix} kv_1 \\ kv_2 \end{pmatrix}$.

Notice that:

- $(-1)\mathbf{v} = \begin{pmatrix} (-1)\,v_1 \\ (-1)\,v_2 \end{pmatrix} = \begin{pmatrix} -v_1 \\ -v_2 \end{pmatrix} = -\mathbf{v}$
- $(0)\mathbf{v} = \begin{pmatrix} (0)\,v_1 \\ (0)\,v_2 \end{pmatrix} = \begin{pmatrix} 0 \\ 0 \end{pmatrix} = \mathbf{0}$

Example 16 — Self Tutor

For $\mathbf{p}=\begin{pmatrix}4\\1\end{pmatrix}$, $\mathbf{q}=\begin{pmatrix}2\\-3\end{pmatrix}$ find: **a** $3\mathbf{q}$ **b** $\mathbf{p}+2\mathbf{q}$ **c** $\frac{1}{2}\mathbf{p}-3\mathbf{q}$

a $3\mathbf{q}$
$=3\begin{pmatrix}2\\-3\end{pmatrix}$
$=\begin{pmatrix}6\\-9\end{pmatrix}$

b $\mathbf{p}+2\mathbf{q}$
$=\begin{pmatrix}4\\1\end{pmatrix}+2\begin{pmatrix}2\\-3\end{pmatrix}$
$=\begin{pmatrix}4+2(2)\\1+2(-3)\end{pmatrix}$
$=\begin{pmatrix}8\\-5\end{pmatrix}$

c $\frac{1}{2}\mathbf{p}-3\mathbf{q}$
$=\frac{1}{2}\begin{pmatrix}4\\1\end{pmatrix}-3\begin{pmatrix}2\\-3\end{pmatrix}$
$=\begin{pmatrix}\frac{1}{2}(4)-3(2)\\\frac{1}{2}(1)-3(-3)\end{pmatrix}$
$=\begin{pmatrix}-4\\9\frac{1}{2}\end{pmatrix}$

Example 17 — Self Tutor

If $\mathbf{p}=3\mathbf{i}-5\mathbf{j}$ and $\mathbf{q}=-\mathbf{i}-2\mathbf{j}$, find $|\mathbf{p}-2\mathbf{q}|$.

$\mathbf{p}-2\mathbf{q}=3\mathbf{i}-5\mathbf{j}-2(-\mathbf{i}-2\mathbf{j})$
$=3\mathbf{i}-5\mathbf{j}+2\mathbf{i}+4\mathbf{j}$
$=5\mathbf{i}-\mathbf{j}$

$\therefore\ |\mathbf{p}-2\mathbf{q}|=\sqrt{5^2+(-1)^2}$
$=\sqrt{26}$ units

VECTOR RACE GAME

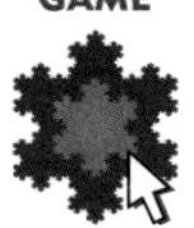

EXERCISE 14E

1 If $\mathbf{a}=\begin{pmatrix}-3\\2\end{pmatrix}$, $\mathbf{b}=\begin{pmatrix}1\\4\end{pmatrix}$, and $\mathbf{c}=\begin{pmatrix}-2\\-5\end{pmatrix}$ find:

a $\mathbf{a}+\mathbf{b}$ **b** $\mathbf{b}+\mathbf{a}$ **c** $\mathbf{b}+\mathbf{c}$ **d** $\mathbf{c}+\mathbf{b}$
e $\mathbf{a}+\mathbf{c}$ **f** $\mathbf{c}+\mathbf{a}$ **g** $\mathbf{a}+\mathbf{a}$ **h** $\mathbf{b}+\mathbf{a}+\mathbf{c}$

2 Given $\mathbf{p}=\begin{pmatrix}-4\\2\end{pmatrix}$, $\mathbf{q}=\begin{pmatrix}-1\\-5\end{pmatrix}$, and $\mathbf{r}=\begin{pmatrix}3\\-2\end{pmatrix}$ find:

a $\mathbf{p}-\mathbf{q}$ **b** $\mathbf{q}-\mathbf{r}$ **c** $\mathbf{p}+\mathbf{q}-\mathbf{r}$
d $\mathbf{p}-\mathbf{q}-\mathbf{r}$ **e** $\mathbf{q}-\mathbf{r}-\mathbf{p}$ **f** $\mathbf{r}+\mathbf{q}-\mathbf{p}$

3 Consider $\mathbf{a}=\begin{pmatrix}a_1\\a_2\end{pmatrix}$.

a Use vector addition to show that $\mathbf{a}+\mathbf{0}=\mathbf{a}$.
b Use vector subtraction to show that $\mathbf{a}-\mathbf{a}=\mathbf{0}$.

4 For $\mathbf{p}=\begin{pmatrix}1\\5\end{pmatrix}$, $\mathbf{q}=\begin{pmatrix}-2\\4\end{pmatrix}$, and $\mathbf{r}=\begin{pmatrix}-3\\-1\end{pmatrix}$ find:

a $-3\mathbf{p}$ **b** $\frac{1}{2}\mathbf{q}$ **c** $2\mathbf{p}+\mathbf{q}$ **d** $\mathbf{p}-2\mathbf{q}$
e $\mathbf{p}-\frac{1}{2}\mathbf{r}$ **f** $2\mathbf{p}+3\mathbf{r}$ **g** $2\mathbf{q}-3\mathbf{r}$ **h** $2\mathbf{p}-\mathbf{q}+\frac{1}{3}\mathbf{r}$

5 Consider $\mathbf{p} = \begin{pmatrix} 1 \\ 1 \end{pmatrix}$ and $\mathbf{q} = \begin{pmatrix} 2 \\ -1 \end{pmatrix}$. Find geometrically and then comment on the results:

a $\mathbf{p}+\mathbf{p}+\mathbf{q}+\mathbf{q}+\mathbf{q}$ **b** $\mathbf{p}+\mathbf{q}+\mathbf{p}+\mathbf{q}+\mathbf{q}$ **c** $\mathbf{q}+\mathbf{p}+\mathbf{q}+\mathbf{p}+\mathbf{q}$

6 For $\mathbf{r} = \begin{pmatrix} 2 \\ 3 \end{pmatrix}$ and $\mathbf{s} = \begin{pmatrix} -1 \\ 4 \end{pmatrix}$ find:

a $|\mathbf{r}|$ **b** $|\mathbf{s}|$ **c** $|\mathbf{r}+\mathbf{s}|$ **d** $|\mathbf{r}-\mathbf{s}|$ **e** $|\mathbf{s}-2\mathbf{r}|$

7 If $\mathbf{p} = \begin{pmatrix} 1 \\ 3 \end{pmatrix}$ and $\mathbf{q} = \begin{pmatrix} -2 \\ 4 \end{pmatrix}$ find:

a $|\mathbf{p}|$ **b** $|2\mathbf{p}|$ **c** $|-2\mathbf{p}|$ **d** $|3\mathbf{p}|$ **e** $|-3\mathbf{p}|$
f $|\mathbf{q}|$ **g** $|4\mathbf{q}|$ **h** $|-4\mathbf{q}|$ **i** $|\frac{1}{2}\mathbf{q}|$ **j** $|-\frac{1}{2}\mathbf{q}|$

8 Suppose $\mathbf{x} = \begin{pmatrix} x_1 \\ x_2 \end{pmatrix}$ and $\mathbf{a} = \begin{pmatrix} a_1 \\ a_2 \end{pmatrix}$. Show by equating components, that if $k\mathbf{x} = \mathbf{a}$ then $\mathbf{x} = \frac{1}{k}\mathbf{a}$.

9 From your answers in **7**, you should have noticed that $|k\mathbf{v}| = |k||\mathbf{v}|$.
So, (the length of $k\mathbf{v}$) = (the modulus of k) $\times$ (the length of $\mathbf{v}$).

By letting $\mathbf{v} = \begin{pmatrix} v_1 \\ v_2 \end{pmatrix}$, prove that $|k\mathbf{v}| = |k||\mathbf{v}|$.

F THE VECTOR BETWEEN TWO POINTS

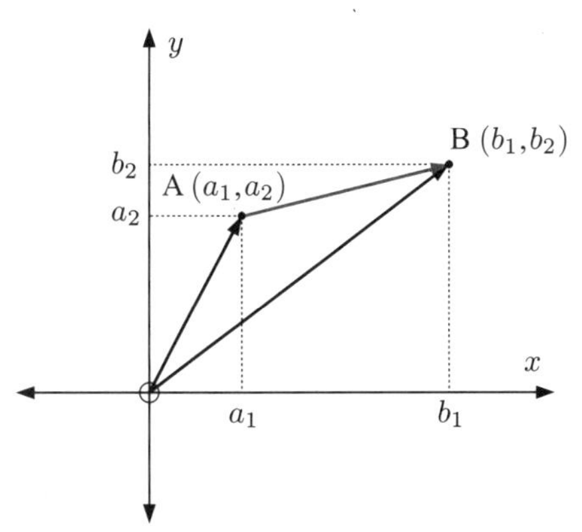

In the diagram, point A has position vector $\overrightarrow{OA} = \begin{pmatrix} a_1 \\ a_2 \end{pmatrix}$,

and point B has position vector $\overrightarrow{OB} = \begin{pmatrix} b_1 \\ b_2 \end{pmatrix}$.

$$\begin{aligned} \therefore \quad \overrightarrow{AB} &= \overrightarrow{AO} + \overrightarrow{OB} \\ &= -\overrightarrow{OA} + \overrightarrow{OB} \\ &= \overrightarrow{OB} - \overrightarrow{OA} \\ &= \begin{pmatrix} b_1 \\ b_2 \end{pmatrix} - \begin{pmatrix} a_1 \\ a_2 \end{pmatrix} \\ &= \begin{pmatrix} b_1 - a_1 \\ b_2 - a_2 \end{pmatrix} \end{aligned}$$

The **position vector of B relative to A** is $\overrightarrow{AB} = \overrightarrow{OB} - \overrightarrow{OA} = \begin{pmatrix} b_1 - a_1 \\ b_2 - a_2 \end{pmatrix}$.

We can also observe this in terms of transformations.

In translating point A to point B in the diagram, the translation vector is $\begin{pmatrix} b_1 - a_1 \\ b_2 - a_2 \end{pmatrix}$.

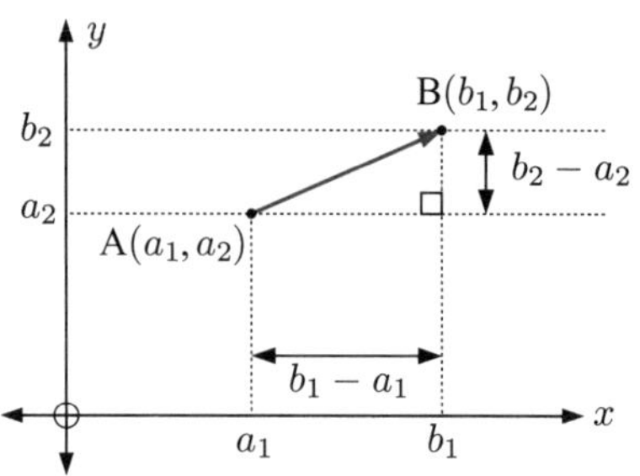

In general, for two points A and B with position vectors **a** and **b** respectively, we observe

$$\overrightarrow{AB} = -\mathbf{a} + \mathbf{b} = \mathbf{b} - \mathbf{a} = \begin{pmatrix} b_1 - a_1 \\ b_2 - a_2 \end{pmatrix}$$

and

$$\overrightarrow{BA} = -\mathbf{b} + \mathbf{a} = \mathbf{a} - \mathbf{b} = \begin{pmatrix} a_1 - b_1 \\ a_2 - b_2 \end{pmatrix}$$

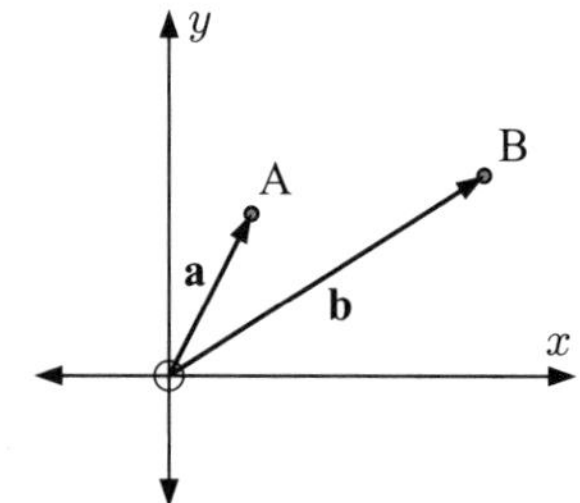

Example 18 **Self Tutor**

Given points A(−1, 2), B(3, 4), and C(4, −5), find the position vector of:

a B from O **b** B from A **c** A from C

a The position vector of B relative to O is $\overrightarrow{OB} = \begin{pmatrix} 3-0 \\ 4-0 \end{pmatrix} = \begin{pmatrix} 3 \\ 4 \end{pmatrix}$.

b The position vector of B relative to A is $\overrightarrow{AB} = \begin{pmatrix} 3--1 \\ 4-2 \end{pmatrix} = \begin{pmatrix} 4 \\ 2 \end{pmatrix}$.

c The position vector of A relative to C is $\overrightarrow{CA} = \begin{pmatrix} -1-4 \\ 2--5 \end{pmatrix} = \begin{pmatrix} -5 \\ 7 \end{pmatrix}$.

Example 19 **Self Tutor**

[AB] is the diameter of a circle with centre C(−1, 2). If B is (3, 1), find:

a $\overrightarrow{BC}$ **b** the coordinates of A.

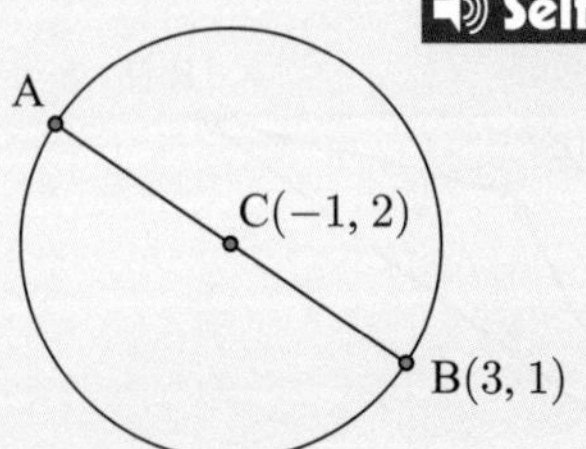

a $\overrightarrow{BC} = \begin{pmatrix} -1-3 \\ 2-1 \end{pmatrix} = \begin{pmatrix} -4 \\ 1 \end{pmatrix}$

b If A has coordinates (a, b), then $\overrightarrow{CA} = \begin{pmatrix} a-(-1) \\ b-2 \end{pmatrix} = \begin{pmatrix} a+1 \\ b-2 \end{pmatrix}$

But $\overrightarrow{CA} = \overrightarrow{BC}$, so $\begin{pmatrix} a+1 \\ b-2 \end{pmatrix} = \begin{pmatrix} -4 \\ 1 \end{pmatrix}$

$\therefore\ a+1 = -4$ and $b-2 = 1$

$\therefore\ a = -5$ and $b = 3$

$\therefore$ A is $(-5, 3)$.

EXERCISE 14F

1 Find $\overrightarrow{AB}$ given:

a A(2, 3) and B(4, 7)
b A(3, −1) and B(1, 4)
c A(−2, 7) and B(1, 4)
d B(3, 0) and A(2, 5)
e B(6, −1) and A(0, 4)
f B(0, 0) and A(−1, −3)

2 Consider the point A(1, 4). Find the coordinates of:

a B given $\overrightarrow{AB} = \begin{pmatrix} 3 \\ -2 \end{pmatrix}$
b C given $\overrightarrow{CA} = \begin{pmatrix} -1 \\ 2 \end{pmatrix}$.

3 [PQ] is the diameter of a circle with centre C.

a Find $\overrightarrow{PC}$.
b Hence find the coordinates of Q.

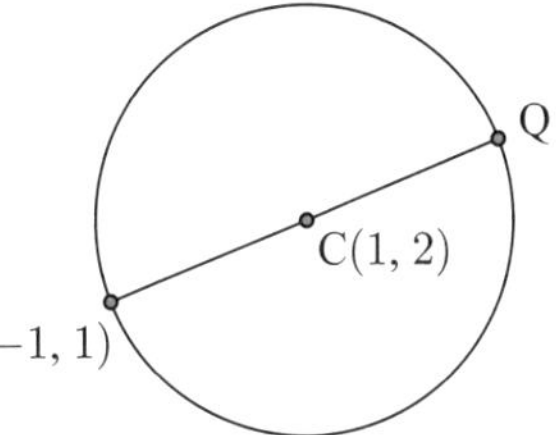

4 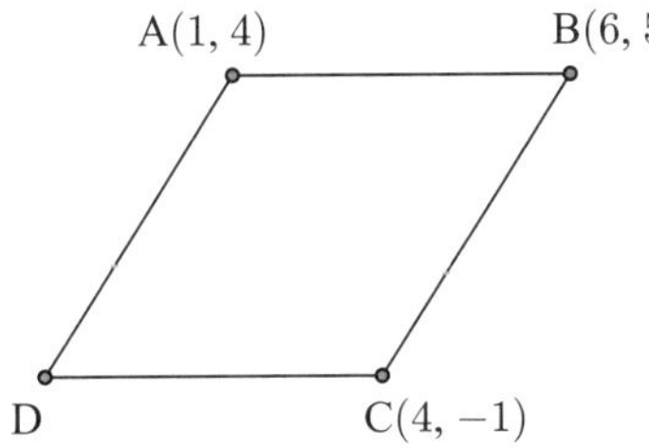

ABCD is a parallelogram.

a Find $\overrightarrow{AB}$.
b Find $\overrightarrow{CD}$.
c Hence find the coordinates of D.

5 A(−1, 3) and B(3, k) are two points which are 5 units apart.

a Find $\overrightarrow{AB}$ and $|\overrightarrow{AB}|$.
b Hence, find the two possible values of k.
c Show, by illustration, why k should have two possible values.

6 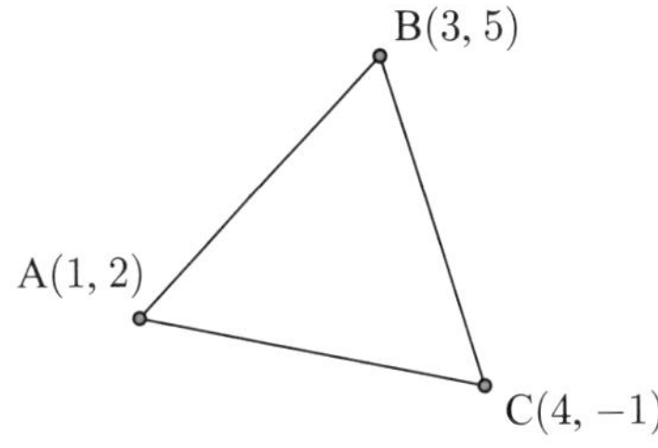

a Find $\overrightarrow{AB}$ and $\overrightarrow{AC}$.
b Explain why $\overrightarrow{BC} = -\overrightarrow{AB} + \overrightarrow{AC}$.
c Hence find $\overrightarrow{BC}$.
d Check your answer to c by direct evaluation.

7 **a** Given $\overrightarrow{BA} = \begin{pmatrix} 2 \\ -3 \end{pmatrix}$ and $\overrightarrow{BC} = \begin{pmatrix} -3 \\ 1 \end{pmatrix}$, find $\overrightarrow{AC}$.

b Given $\overrightarrow{AB} = \begin{pmatrix} -1 \\ 3 \end{pmatrix}$ and $\overrightarrow{CA} = \begin{pmatrix} 2 \\ -1 \end{pmatrix}$, find $\overrightarrow{CB}$.

c Given $\overrightarrow{PQ} = \begin{pmatrix} -1 \\ 4 \end{pmatrix}$, $\overrightarrow{RQ} = \begin{pmatrix} 2 \\ 1 \end{pmatrix}$, and $\overrightarrow{RS} = \begin{pmatrix} -3 \\ 2 \end{pmatrix}$, find $\overrightarrow{SP}$.

8

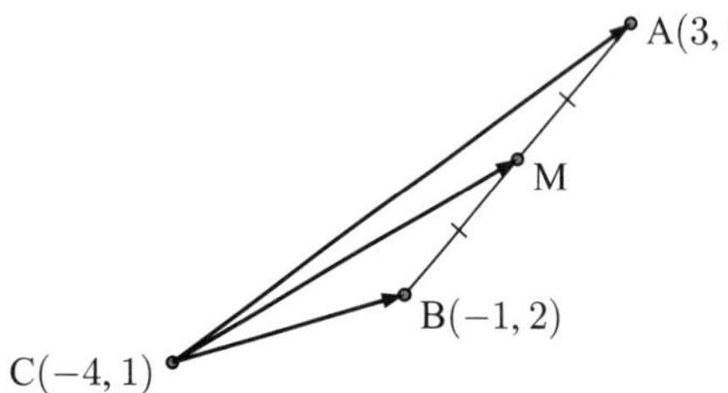

a Find the coordinates of M.

b Find vectors $\overrightarrow{CA}$, $\overrightarrow{CM}$, and $\overrightarrow{CB}$.

c Verify that $\overrightarrow{CM} = \frac{1}{2}\overrightarrow{CA} + \frac{1}{2}\overrightarrow{CB}$.

G VECTORS IN SPACE

To specify points in **3-dimensional space** we need a point of reference O, called the **origin**.

Through O we draw 3 **mutually perpendicular** lines and call them the X, Y, and Z-axes. We often think of the YZ-plane as the plane of the page, with the X-axis coming directly out of the page. However, we cannot of course draw this.

In the diagram alongside the **coordinate planes** divide space into 8 regions, with each pair of planes intersecting on the axes.

The **positive direction** of each axis is a solid line whereas the **negative direction** is 'dashed'.

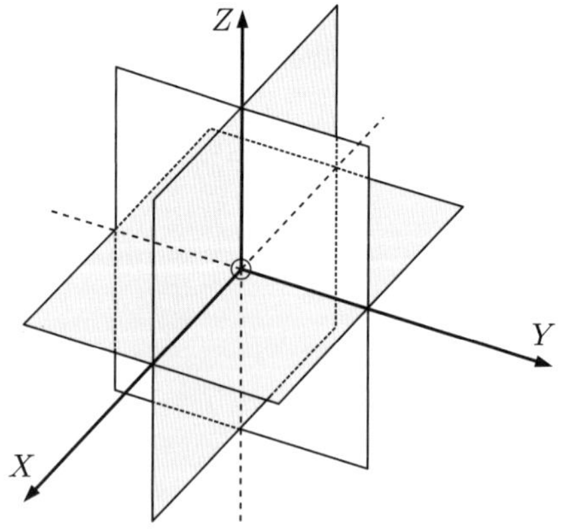

Any point P in space can be specified by an **ordered triple** of numbers (x, y, z) where x, y, and z are the **steps** in the X, Y, and Z directions from the origin O, to P.

The **position vector** of P is

$$\overrightarrow{OP} = \begin{pmatrix} x \\ y \\ z \end{pmatrix} = x\mathbf{i} + y\mathbf{j} + z\mathbf{k}$$

where $\mathbf{i} = \begin{pmatrix} 1 \\ 0 \\ 0 \end{pmatrix}$, $\mathbf{j} = \begin{pmatrix} 0 \\ 1 \\ 0 \end{pmatrix}$, and $\mathbf{k} = \begin{pmatrix} 0 \\ 0 \\ 1 \end{pmatrix}$

are the **base unit vectors** in the X, Y, and Z directions respectively.

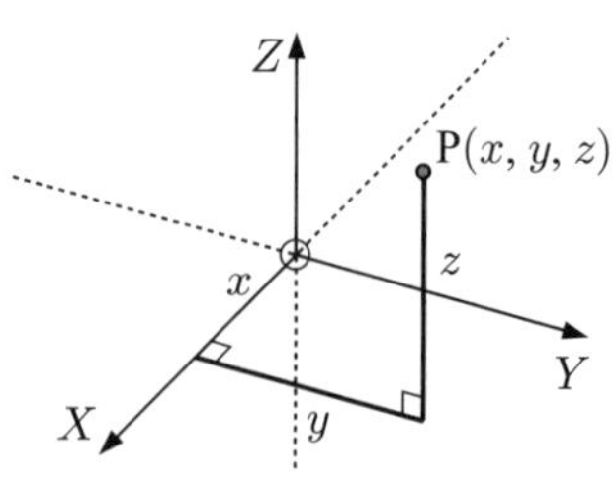

The set of vectors $\{\mathbf{i}, \mathbf{j}, \mathbf{k}\} = \left\{ \begin{pmatrix} 1 \\ 0 \\ 0 \end{pmatrix}, \begin{pmatrix} 0 \\ 1 \\ 0 \end{pmatrix}, \begin{pmatrix} 0 \\ 0 \\ 1 \end{pmatrix} \right\}$

is the **standard basis** for the 3-dimensional (x, y, z) coordinate system.

To help us visualise the 3-D position of a point on our 2-D paper, it is useful to complete a rectangular prism or box with the origin O as one vertex, the axes as sides adjacent to it, and P being the vertex opposite O.

3-D POINT PLOTTER

Example 20 Self Tutor

Illustrate the points:

a A(0, 2, 0) **b** B(3, 0, 2) **c** C(−1, 2, 3)

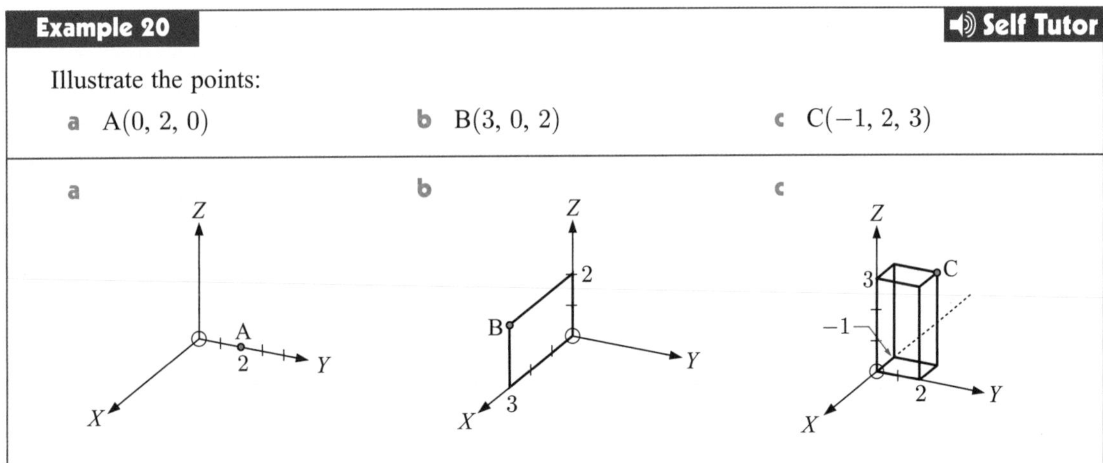

THE MAGNITUDE OF A VECTOR

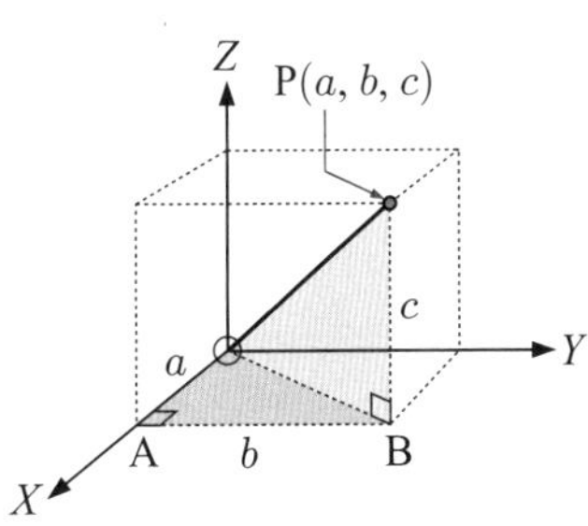

Triangle OAB is right angled at A

$\therefore \quad \text{OB}^2 = a^2 + b^2 \quad .... \; (1) \quad \{\text{Pythagoras}\}$

Triangle OBP is right angled at B

$\therefore \quad \text{OP}^2 = \text{OB}^2 + c^2 \quad \{\text{Pythagoras}\}$

$\therefore \quad \text{OP}^2 = a^2 + b^2 + c^2 \quad \{\text{using (1)}\}$

$\therefore \quad \text{OP} = \sqrt{a^2 + b^2 + c^2}$

The magnitude or length of the vector $\mathbf{v} = \begin{pmatrix} v_1 \\ v_2 \\ v_3 \end{pmatrix}$ is $|\mathbf{v}| = \sqrt{{v_1}^2 + {v_2}^2 + {v_3}^2}$.

THE VECTOR BETWEEN TWO POINTS

If $A(x_1, y_1, z_1)$ and $B(x_2, y_2, z_2)$ are two points in space then:

$$\overrightarrow{AB} = \overrightarrow{OB} - \overrightarrow{OA} = \begin{pmatrix} x_2 - x_1 \\ y_2 - y_1 \\ z_2 - z_1 \end{pmatrix} \begin{matrix} \leftarrow x\text{-step} \\ \leftarrow y\text{-step} \\ \leftarrow z\text{-step} \end{matrix}$$

$\overrightarrow{AB}$ is called the 'vector AB' or the **'position vector of B relative to A'**.

The magnitude of $\overrightarrow{AB}$ is $AB = \sqrt{(x_2 - x_1)^2 + (y_2 - y_1)^2 + (z_2 - z_1)^2}$ which is the distance between the points A and B.

Example 21 — Self Tutor

If P is $(-3, 1, 2)$ and Q is $(1, -1, 3)$, find: **a** $\overrightarrow{OP}$ **b** $\overrightarrow{PQ}$ **c** $|\overrightarrow{PQ}|$

a $\overrightarrow{OP} = \begin{pmatrix} -3 \\ 1 \\ 2 \end{pmatrix}$

b $\overrightarrow{PQ} = \begin{pmatrix} 1-(-3) \\ -1-1 \\ 3-2 \end{pmatrix} = \begin{pmatrix} 4 \\ -2 \\ 1 \end{pmatrix}$

c $|\overrightarrow{PQ}| = \sqrt{4^2 + (-2)^2 + 1^2}$
$= \sqrt{21}$ units

VECTOR EQUALITY

Two vectors are **equal** if they have the same magnitude and direction.

If $\mathbf{a} = \begin{pmatrix} a_1 \\ a_2 \\ a_3 \end{pmatrix}$ and $\mathbf{b} = \begin{pmatrix} b_1 \\ b_2 \\ b_3 \end{pmatrix}$, then $\mathbf{a} = \mathbf{b} \Leftrightarrow a_1 = b_1,\ a_2 = b_2,\ a_3 = b_3$.

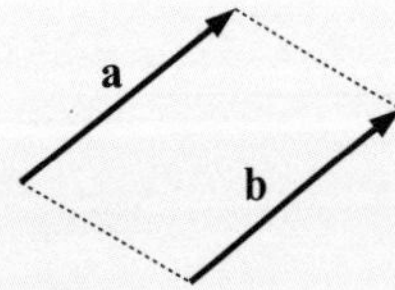

If **a** and **b** do not coincide, then they are opposite sides of a parallelogram, and lie in the same plane.

Example 22 — Self Tutor

ABCD is a parallelogram. A is $(-1, 2, 1)$, B is $(2, 0, -1)$, and D is $(3, 1, 4)$. Find the coordinates of C.

Let C be (a, b, c).

[AB] is parallel to [DC], and they have the same length, so $\overrightarrow{DC} = \overrightarrow{AB}$

A(−1, 2, 1) B(2, 0, −1) C(a, b, c) D(3, 1, 4)

$\therefore \begin{pmatrix} a-3 \\ b-1 \\ c-4 \end{pmatrix} = \begin{pmatrix} 3 \\ -2 \\ -2 \end{pmatrix}$

$\therefore \quad a-3=3, \quad b-1=-2, \quad \text{and} \quad c-4=-2$

$\therefore \quad a=6, \quad b=-1, \quad \text{and} \quad c=2$

So, C is $(6, -1, 2)$.

Check: Midpoint of [DB] is $\left(\frac{3+2}{2}, \frac{1+0}{2}, \frac{4+-1}{2}\right)$ which is $\left(\frac{5}{2}, \frac{1}{2}, \frac{3}{2}\right)$.

Midpoint of [AC] is $\left(\frac{-1+6}{2}, \frac{2+-1}{2}, \frac{1+2}{2}\right)$ which is $\left(\frac{5}{2}, \frac{1}{2}, \frac{3}{2}\right)$.

The midpoints are the same, so the diagonals of the parallelogram bisect. ✓

EXERCISE 14G

1 Consider the point $T(3, -1, 4)$.

- **a** Draw a diagram to locate the position of T in space.
- **b** Find $\overrightarrow{OT}$.
- **c** How far is it from O to T?

2 Illustrate P and find its distance from the origin O:

- **a** $P(0, 0, -3)$
- **b** $P(0, -1, 2)$
- **c** $P(3, 1, 4)$
- **d** $P(-1, -2, 3)$

3 Given $A(-3, 1, 2)$ and $B(1, 0, -1)$ find:

- **a** $\overrightarrow{AB}$ and $\overrightarrow{BA}$
- **b** the lengths $|\overrightarrow{AB}|$ and $|\overrightarrow{BA}|$.

4 Given $A(3, 1, 0)$ and $B(-1, 1, 2)$ find $\overrightarrow{OA}$, $\overrightarrow{OB}$, and $\overrightarrow{AB}$.

5 Given $M(4, -2, -1)$ and $N(-1, 2, 0)$ find:

- **a** the position vector of M relative to N
- **b** the position vector of N relative to M
- **c** the distance between M and N.

6 Consider $A(-1, 2, 5)$, $B(2, 0, 3)$, and $C(-3, 1, 0)$.

- **a** Find the position vector $\overrightarrow{OA}$ and its length OA.
- **b** Find the position vector $\overrightarrow{AB}$ and its length AB.
- **c** Find the position vector $\overrightarrow{AC}$ and its length AC.
- **d** Find the position vector $\overrightarrow{CB}$ and its length CB.
- **e** Hence classify triangle ABC.

7 Find the shortest distance from $Q(3, 1, -2)$ to:

- **a** the Y-axis
- **b** the origin
- **c** the YOZ plane.

8 Show that $P(0, 4, 4)$, $Q(2, 6, 5)$, and $R(1, 4, 3)$ are vertices of an isosceles triangle.

9 Use side lengths to classify triangle ABC given the coordinates:

- **a** $A(0, 0, 3)$, $B(2, 8, 1)$, and $C(-9, 6, 18)$
- **b** $A(1, 0, -3)$, $B(2, 2, 0)$, and $C(4, 6, 6)$.

10 The vertices of triangle ABC are $A(5, 6, -2)$, $B(6, 12, 9)$, and $C(2, 4, 2)$.

- **a** Use distances to show that the triangle is right angled.
- **b** Hence find the area of the triangle.

11 A sphere has centre $C(-1, 2, 4)$ and diameter [AB] where A is $(-2, 1, 3)$.
Find the coordinates of B and the radius of the sphere.

12 **a** State the coordinates of any general point A on the Y-axis.

b Use **a** and the diagram opposite to find the coordinates of two points on the Y-axis which are $\sqrt{14}$ units from B$(-1, -1, 2)$.

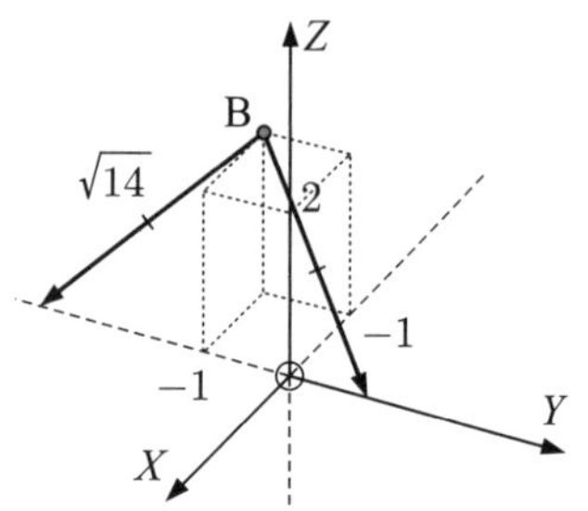

13 Find a, b, and c if:

a $\begin{pmatrix} a-4 \\ b-3 \\ c+2 \end{pmatrix} = \begin{pmatrix} 1 \\ 3 \\ -4 \end{pmatrix}$ **b** $\begin{pmatrix} a-5 \\ b-2 \\ c+3 \end{pmatrix} = \begin{pmatrix} 3-a \\ 2-b \\ 5-c \end{pmatrix}$

14 Find k given the unit vector:

a $\begin{pmatrix} -\frac{1}{2} \\ k \\ \frac{1}{4} \end{pmatrix}$ **b** $\begin{pmatrix} k \\ \frac{2}{3} \\ -\frac{1}{3} \end{pmatrix}$

15 A$(-1, 3, 4)$, B$(2, 5, -1)$, C$(-1, 2, -2)$, and D(r, s, t) are four points in space.

Find r, s, and t if: **a** $\overrightarrow{AC} = \overrightarrow{BD}$ **b** $\overrightarrow{AB} = \overrightarrow{DC}$

16 A quadrilateral has vertices A$(1, 2, 3)$, B$(3, -3, 2)$, C$(7, -4, 5)$, and D$(5, 1, 6)$.

a Find $\overrightarrow{AB}$ and $\overrightarrow{DC}$.

b What can be deduced about the quadrilateral ABCD?

17 PQRS is a parallelogram. P is $(-1, 2, 3)$, Q is $(1, -2, 5)$, and R is $(0, 4, -1)$.

a Use vectors to find the coordinates of S.

b Use midpoints of diagonals to check your answer.

H OPERATIONS WITH VECTORS IN SPACE

The rules for algebra with vectors readily extend from 2-D to 3-D:

If $\mathbf{a} = \begin{pmatrix} a_1 \\ a_2 \\ a_3 \end{pmatrix}$ and $\mathbf{b} = \begin{pmatrix} b_1 \\ b_2 \\ b_3 \end{pmatrix}$ then $\mathbf{a}+\mathbf{b} = \begin{pmatrix} a_1+b_1 \\ a_2+b_2 \\ a_3+b_3 \end{pmatrix}$, $\mathbf{a}-\mathbf{b} = \begin{pmatrix} a_1-b_1 \\ a_2-b_2 \\ a_3-b_3 \end{pmatrix}$,

and $k\mathbf{a} = \begin{pmatrix} ka_1 \\ ka_2 \\ ka_3 \end{pmatrix}$ for any scalar k.

INVESTIGATION — PROPERTIES OF VECTORS IN SPACE

There are several properties or rules which are valid for the addition of real numbers. For example, we know that $a+b=b+a$.

Our task is to identify some similar properties of vectors.

What to do:

1 Use general vectors $\mathbf{a}=\begin{pmatrix}a_1\\a_2\\a_3\end{pmatrix}$, $\mathbf{b}=\begin{pmatrix}b_1\\b_2\\b_3\end{pmatrix}$, and $\mathbf{c}=\begin{pmatrix}c_1\\c_2\\c_3\end{pmatrix}$ to find:

a $\mathbf{a}+\mathbf{b}$ and $\mathbf{b}+\mathbf{a}$
b $\mathbf{a}+\mathbf{0}$
c $\mathbf{a}+(-\mathbf{a})$ and $(-\mathbf{a}+\mathbf{a})$
d $(\mathbf{a}+\mathbf{b})+\mathbf{c}$ and $\mathbf{a}+(\mathbf{b}+\mathbf{c})$

2 Summarise your observations from **1**. Do they match the rules for real numbers?

3 Prove that for scalar k and vectors $\mathbf{a}=\begin{pmatrix}a_1\\a_2\\a_3\end{pmatrix}$ and $\mathbf{b}=\begin{pmatrix}b_1\\b_2\\b_3\end{pmatrix}$:

a $|k\mathbf{a}|=|k|\,|\mathbf{a}|$
b $k(\mathbf{a}+\mathbf{b})=k\mathbf{a}+k\mathbf{b}$

From the **Investigation** you should have found that for vectors $\mathbf{a}$, $\mathbf{b}$, $\mathbf{c}$ and $k\in\mathbb{R}$:

- $\mathbf{a}+\mathbf{b}=\mathbf{b}+\mathbf{a}$ {commutative property}
- $(\mathbf{a}+\mathbf{b})+\mathbf{c}=\mathbf{a}+(\mathbf{b}+\mathbf{c})$ {associative property}
- $\mathbf{a}+\mathbf{0}=\mathbf{0}+\mathbf{a}=\mathbf{a}$ {additive identity}
- $\mathbf{a}+(-\mathbf{a})=(-\mathbf{a})+\mathbf{a}=\mathbf{0}$ {additive inverse}
- $|k\mathbf{a}|=|k|\,|\mathbf{a}|$ where $k\mathbf{a}$ is parallel to $\mathbf{a}$
  (length of $k\mathbf{a}$ = modulus of k × length of $\mathbf{a}$)
- $k(\mathbf{a}+\mathbf{b})=k\mathbf{a}+k\mathbf{b}$ {distributive property}

The rules for solving vector equations are similar to those for solving real number equations. However, there is no such thing as dividing a vector by a scalar. Instead, we multiply by the reciprocal scalar.

For example, if $2\mathbf{x}=\mathbf{a}$ then $\mathbf{x}=\frac{1}{2}\mathbf{a}$ and *not* $\frac{\mathbf{a}}{2}$.

$\frac{\mathbf{a}}{2}$ has no meaning in vector algebra.

Two useful rules are:

- if $\mathbf{x}+\mathbf{a}=\mathbf{b}$ then $\mathbf{x}=\mathbf{b}-\mathbf{a}$
- if $k\mathbf{x}=\mathbf{a}$ then $\mathbf{x}=\frac{1}{k}\mathbf{a}$ $(k\neq 0)$

Proof:

- If $\mathbf{x}+\mathbf{a}=\mathbf{b}$
  then $\mathbf{x}+\mathbf{a}+(-\mathbf{a})=\mathbf{b}+(-\mathbf{a})$
  $\therefore\ \mathbf{x}+\mathbf{0}=\mathbf{b}-\mathbf{a}$
  $\therefore\ \mathbf{x}=\mathbf{b}-\mathbf{a}$
- If $k\mathbf{x}=\mathbf{a}$
  then $\frac{1}{k}(k\mathbf{x})=\frac{1}{k}\mathbf{a}$
  $\therefore\ 1\mathbf{x}=\frac{1}{k}\mathbf{a}$
  $\therefore\ \mathbf{x}=\frac{1}{k}\mathbf{a}$

Example 23 Self Tutor

Solve for $\mathbf{x}$: **a** $3\mathbf{x} - \mathbf{r} = \mathbf{s}$ **b** $\mathbf{c} - 2\mathbf{x} = \mathbf{d}$

a $3\mathbf{x} - \mathbf{r} = \mathbf{s}$
$\therefore\ 3\mathbf{x} = \mathbf{s} + \mathbf{r}$
$\therefore\ \mathbf{x} = \frac{1}{3}(\mathbf{s} + \mathbf{r})$

b $\mathbf{c} - 2\mathbf{x} = \mathbf{d}$
$\therefore\ \mathbf{c} - \mathbf{d} = 2\mathbf{x}$
$\therefore\ \frac{1}{2}(\mathbf{c} - \mathbf{d}) = \mathbf{x}$

Example 24 Self Tutor

If $\mathbf{a} = \begin{pmatrix} -1 \\ 3 \\ 2 \end{pmatrix}$, find $|2\mathbf{a}|$.

$$\begin{aligned} |2\mathbf{a}| &= 2|\mathbf{a}| \\ &= 2\sqrt{(-1)^2 + 3^2 + 2^2} \\ &= 2\sqrt{1 + 9 + 4} \\ &= 2\sqrt{14} \text{ units} \end{aligned}$$

Example 25 Self Tutor

Find the coordinates of C and D:

B(−1,−2,2)
D
C
A(−2,−5,3)

$$\overrightarrow{AB} = \begin{pmatrix} -1-(-2) \\ -2-(-5) \\ 2-3 \end{pmatrix} = \begin{pmatrix} 1 \\ 3 \\ -1 \end{pmatrix}$$

$$\begin{aligned} \overrightarrow{OC} &= \overrightarrow{OA} + \overrightarrow{AC} \\ &= \overrightarrow{OA} + 2\overrightarrow{AB} \\ &= \begin{pmatrix} -2 \\ -5 \\ 3 \end{pmatrix} + \begin{pmatrix} 2 \\ 6 \\ -2 \end{pmatrix} = \begin{pmatrix} 0 \\ 1 \\ 1 \end{pmatrix} \end{aligned}$$

$\therefore$ C is $(0, 1, 1)$

$$\begin{aligned} \overrightarrow{OD} &= \overrightarrow{OA} + \overrightarrow{AD} \\ &= \overrightarrow{OA} + 3\overrightarrow{AB} \\ &= \begin{pmatrix} -2 \\ -5 \\ 3 \end{pmatrix} + \begin{pmatrix} 3 \\ 9 \\ -3 \end{pmatrix} = \begin{pmatrix} 1 \\ 4 \\ 0 \end{pmatrix} \end{aligned}$$

$\therefore$ D is $(1, 4, 0)$

EXERCISE 14H

1 Solve the following vector equations for $\mathbf{x}$:

a $2\mathbf{x} = \mathbf{q}$ **b** $\frac{1}{2}\mathbf{x} = \mathbf{n}$ **c** $-3\mathbf{x} = \mathbf{p}$

d $\mathbf{q} + 2\mathbf{x} = \mathbf{r}$ **e** $4\mathbf{s} - 5\mathbf{x} = \mathbf{t}$ **f** $4\mathbf{m} - \frac{1}{3}\mathbf{x} = \mathbf{n}$

2 Suppose $\mathbf{a} = \begin{pmatrix} -1 \\ 2 \\ 3 \end{pmatrix}$ and $\mathbf{b} = \begin{pmatrix} 2 \\ -2 \\ 1 \end{pmatrix}$. Find $\mathbf{x}$ if:

a $2\mathbf{a} + \mathbf{x} = \mathbf{b}$ **b** $3\mathbf{x} - \mathbf{a} = 2\mathbf{b}$ **c** $2\mathbf{b} - 2\mathbf{x} = -\mathbf{a}$

3 If $\overrightarrow{OA} = \begin{pmatrix} -2 \\ -1 \\ 1 \end{pmatrix}$ and $\overrightarrow{OB} = \begin{pmatrix} 1 \\ 3 \\ -1 \end{pmatrix}$, find $\overrightarrow{AB}$ and hence the distance from A to B.

4 For A(−1, 3, 2) and B(3, −2, 1) find:

a $\overrightarrow{AB}$ in terms of **i**, **j**, and **k**

b the magnitude of $\overrightarrow{AB}$

5 If $\mathbf{a} = \begin{pmatrix} 1 \\ 0 \\ 3 \end{pmatrix}$ and $\mathbf{b} = \begin{pmatrix} -2 \\ 1 \\ 1 \end{pmatrix}$, find:

a $|\mathbf{a}|$
b $|\mathbf{b}|$
c $2|\mathbf{a}|$
d $|2\mathbf{a}|$
e $-3|\mathbf{b}|$
f $|-3\mathbf{b}|$
g $|\mathbf{a}+\mathbf{b}|$
h $|\mathbf{a}-\mathbf{b}|$

6 If $\overrightarrow{AB} = \mathbf{i} - \mathbf{j} + \mathbf{k}$ and $\overrightarrow{BC} = -2\mathbf{i} + \mathbf{j} - 3\mathbf{k}$ find $\overrightarrow{AC}$ in terms of **i**, **j**, and **k**.

7 Consider the points A(2, 1, −2), B(0, 3, −4), C(1, −2, 1), and D(−2, −3, 2).
Deduce that $\overrightarrow{BD} = 2\overrightarrow{AC}$.

8 Find the coordinates of C, D, and E.

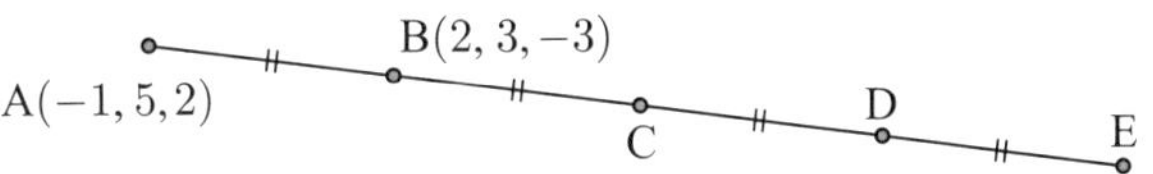

9 Use vectors to determine whether ABCD is a parallelogram:

a A(3, −1), B(4, 2), C(−1, 4), and D(−2, 1)

b A(5, 0, 3), B(−1, 2, 4), C(4, −3, 6), and D(10, −5, 5)

c A(2, −3, 2), B(1, 4, −1), C(−2, 6, −2), and D(−1, −1, 2).

10 Use vector methods to find the remaining vertex of:

a
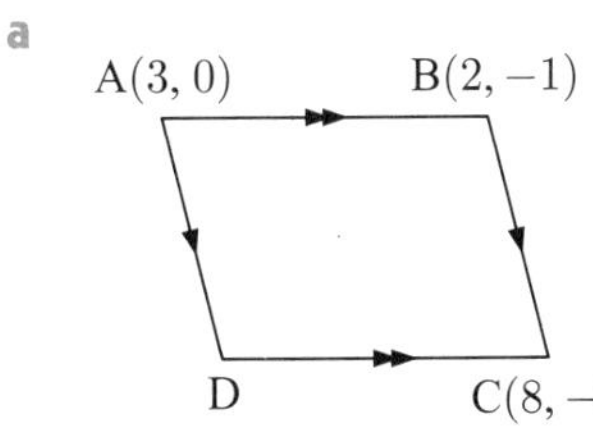

b
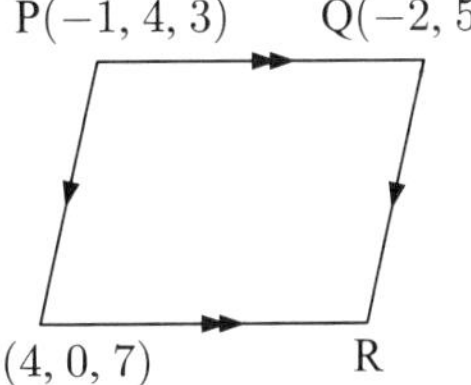

c
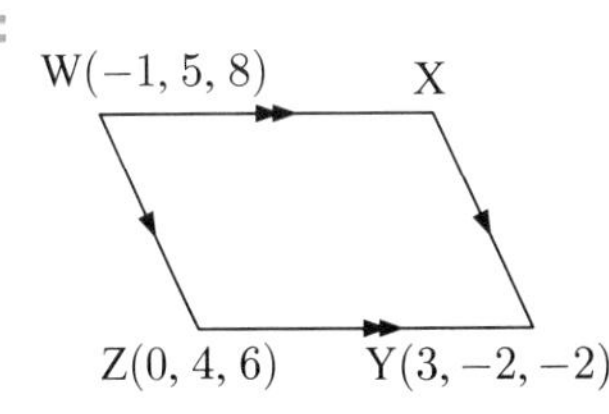

11
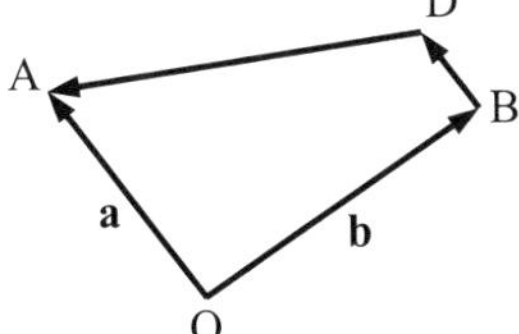

In the given figure [BD] is parallel to [OA] and half its length. Find, in terms of **a** and **b**, vector expressions for:

a $\overrightarrow{BD}$
b $\overrightarrow{AB}$
c $\overrightarrow{BA}$
d $\overrightarrow{OD}$
e $\overrightarrow{AD}$
f $\overrightarrow{DA}$

12 If $\overrightarrow{AB} = \begin{pmatrix} -1 \\ 3 \\ 2 \end{pmatrix}$, $\overrightarrow{AC} = \begin{pmatrix} 2 \\ -1 \\ 4 \end{pmatrix}$, and $\overrightarrow{BD} = \begin{pmatrix} 0 \\ 2 \\ -3 \end{pmatrix}$, find:

a $\overrightarrow{AD}$
b $\overrightarrow{CB}$
c $\overrightarrow{CD}$

13 For $\mathbf{a}=\begin{pmatrix}2\\-1\\1\end{pmatrix}$, $\mathbf{b}=\begin{pmatrix}1\\2\\-3\end{pmatrix}$, and $\mathbf{c}=\begin{pmatrix}0\\1\\-3\end{pmatrix}$, find:

a $\mathbf{a}+\mathbf{b}$ **b** $\mathbf{a}-\mathbf{b}$ **c** $\mathbf{b}+2\mathbf{c}$

d $\mathbf{c}-\frac{1}{2}\mathbf{a}$ **e** $\mathbf{a}-\mathbf{b}-\mathbf{c}$ **f** $2\mathbf{b}-\mathbf{c}+\mathbf{a}$

14 If $\mathbf{a}=\begin{pmatrix}-1\\1\\3\end{pmatrix}$, $\mathbf{b}=\begin{pmatrix}1\\-3\\2\end{pmatrix}$, and $\mathbf{c}=\begin{pmatrix}-2\\2\\4\end{pmatrix}$, find:

a $|\mathbf{a}|$ **b** $|\mathbf{b}|$ **c** $|\mathbf{b}+\mathbf{c}|$

d $|\mathbf{a}-\mathbf{c}|$ **e** $|\mathbf{a}|\,\mathbf{b}$ **f** $\frac{1}{|\mathbf{a}|}\mathbf{a}$

15 Find scalars a, b, and c:

a $2\begin{pmatrix}1\\0\\3a\end{pmatrix}=\begin{pmatrix}b\\c-1\\2\end{pmatrix}$ **b** $a\begin{pmatrix}1\\1\\0\end{pmatrix}+b\begin{pmatrix}2\\0\\-1\end{pmatrix}+c\begin{pmatrix}0\\1\\1\end{pmatrix}=\begin{pmatrix}-1\\3\\3\end{pmatrix}$

c $a\begin{pmatrix}2\\-3\\1\end{pmatrix}+b\begin{pmatrix}1\\7\\2\end{pmatrix}=\begin{pmatrix}7\\-19\\2\end{pmatrix}$

I PARALLELISM

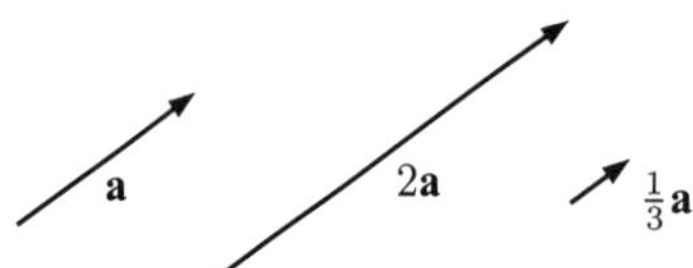

are parallel vectors of different length.

Two non-zero vectors are **parallel** if and only if one is a scalar multiple of the other.

Given any non-zero vector $\mathbf{v}$ and non-zero scalar k, the vector $k\mathbf{v}$ is parallel to $\mathbf{v}$.

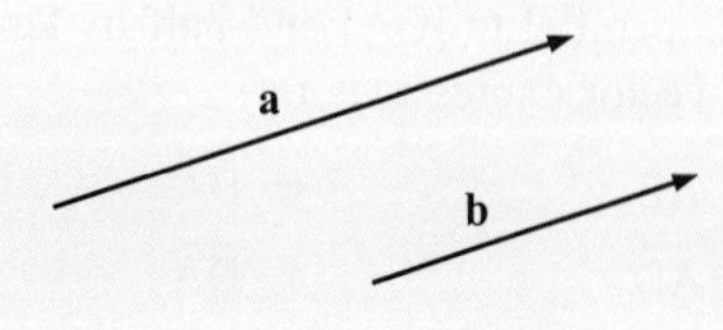

- If $\mathbf{a}$ is parallel to $\mathbf{b}$, then there exists a scalar k such that $\mathbf{a}=k\mathbf{b}$.
- If $\mathbf{a}=k\mathbf{b}$ for some scalar k, then
  - $\mathbf{a}$ is parallel to $\mathbf{b}$, and
  - $|\mathbf{a}|=|k|\,|\mathbf{b}|$.

$|k|$ is the modulus of k, whereas $|\mathbf{a}|$ is the length of vector $\mathbf{a}$.

Example 26 — Self Tutor

Find r and s given that $\mathbf{a} = \begin{pmatrix} 2 \\ -1 \\ r \end{pmatrix}$ is parallel to $\mathbf{b} = \begin{pmatrix} s \\ 2 \\ -3 \end{pmatrix}$.

Since $\mathbf{a}$ and $\mathbf{b}$ are parallel, $\mathbf{a} = k\mathbf{b}$ for some scalar k.

$$\therefore \begin{pmatrix} 2 \\ -1 \\ r \end{pmatrix} = k \begin{pmatrix} s \\ 2 \\ -3 \end{pmatrix}$$

$\therefore\ 2 = ks,\quad -1 = 2k$ and $r = -3k$

Consequently, $k = -\frac{1}{2}$ and $\therefore\ 2 = -\frac{1}{2}s$ and $r = -3(-\frac{1}{2})$

$\therefore\ r = \frac{3}{2}$ and $s = -4$

UNIT VECTORS

Given a non-zero vector $\mathbf{v}$, its magnitude $|\mathbf{v}|$ is a scalar quantity.

If we multiply $\mathbf{v}$ by the scalar $\dfrac{1}{|\mathbf{v}|}$, we obtain the parallel vector $\dfrac{1}{|\mathbf{v}|}\mathbf{v}$.

The length of this vector is $\left|\dfrac{1}{|\mathbf{v}|}\right| |\mathbf{v}| = \dfrac{|\mathbf{v}|}{|\mathbf{v}|} = 1$, so $\dfrac{1}{|\mathbf{v}|}\mathbf{v}$ is a unit vector in the direction of $\mathbf{v}$.

- A unit vector in the direction of $\mathbf{v}$ is $\dfrac{1}{|\mathbf{v}|}\mathbf{v}$.
- A vector $\mathbf{b}$ of length k in the same direction as $\mathbf{a}$ is $\mathbf{b} = \dfrac{k}{|\mathbf{a}|}\mathbf{a}$.
- A vector $\mathbf{b}$ of length k which is *parallel to* $\mathbf{a}$ could be $\mathbf{b} = \pm\dfrac{k}{|\mathbf{a}|}\mathbf{a}$.

Example 27 — Self Tutor

If $\mathbf{a} = 3\mathbf{i} - \mathbf{j}$ find:

a a unit vector in the direction of $\mathbf{a}$

b a vector of length 4 units in the direction of $\mathbf{a}$

c vectors of length 4 units which are parallel to $\mathbf{a}$.

a $|\mathbf{a}| = \sqrt{3^2 + (-1)^2} = \sqrt{9+1} = \sqrt{10}$ units

$\therefore$ the unit vector is $\frac{1}{\sqrt{10}}(3\mathbf{i} - \mathbf{j}) = \frac{3}{\sqrt{10}}\mathbf{i} - \frac{1}{\sqrt{10}}\mathbf{j}$

b A vector of length 4 units in the direction of $\mathbf{a}$ is $\frac{4}{\sqrt{10}}(3\mathbf{i} - \mathbf{j}) = \frac{12}{\sqrt{10}}\mathbf{i} - \frac{4}{\sqrt{10}}\mathbf{j}$

c The vectors of length 4 units which are parallel to $\mathbf{a}$ are

$\frac{12}{\sqrt{10}}\mathbf{i} - \frac{4}{\sqrt{10}}\mathbf{j}$ and $-\frac{12}{\sqrt{10}}\mathbf{i} + \frac{4}{\sqrt{10}}\mathbf{j}$.

Example 28 Self Tutor

Find a vector $\mathbf{b}$ of length 7 in the opposite direction to the vector $\mathbf{a} = \begin{pmatrix} 2 \\ -1 \\ 1 \end{pmatrix}$.

The unit vector in the direction of $\mathbf{a}$ is $\dfrac{1}{|\,\mathbf{a}\,|}\mathbf{a} = \dfrac{1}{\sqrt{4+1+1}}\begin{pmatrix} 2 \\ -1 \\ 1 \end{pmatrix} = \frac{1}{\sqrt{6}}\begin{pmatrix} 2 \\ -1 \\ 1 \end{pmatrix}$.

We multiply this unit vector by -7. The negative reverses the direction and the 7 gives the required length.

Thus $\mathbf{b} = -\frac{7}{\sqrt{6}}\begin{pmatrix} 2 \\ -1 \\ 1 \end{pmatrix}$.

Check that $|\,\mathbf{b}\,| = 7$.

COLLINEAR POINTS

Three or more points are said to be **collinear** if they lie on the same straight line.

A, B, and C are **collinear** if $\overrightarrow{AB} = k\overrightarrow{BC}$ for some scalar k.

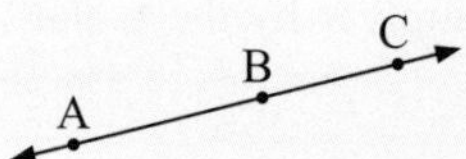

Example 29 Self Tutor

Prove that A(-1, 2, 3), B(4, 0, -1), and C(14, -4, -9) are collinear.

$\overrightarrow{AB} = \begin{pmatrix} 5 \\ -2 \\ -4 \end{pmatrix}$ and $\overrightarrow{BC} = \begin{pmatrix} 10 \\ -4 \\ -8 \end{pmatrix} = 2\begin{pmatrix} 5 \\ -2 \\ -4 \end{pmatrix}$

$\therefore$ $\overrightarrow{BC} = 2\overrightarrow{AB}$

$\therefore$ [BC] is parallel to [AB].

Since B is common to both, A, B, and C are collinear.

EXERCISE 14I

1 $\mathbf{a} = \begin{pmatrix} 2 \\ -1 \\ 3 \end{pmatrix}$ and $\mathbf{b} = \begin{pmatrix} -6 \\ r \\ s \end{pmatrix}$ are parallel. Find r and s.

2 Find scalars a and b given that $\begin{pmatrix} 3 \\ -1 \\ 2 \end{pmatrix}$ and $\begin{pmatrix} a \\ 2 \\ b \end{pmatrix}$ are parallel.

3 What can be deduced from the following?

a $\overrightarrow{AB} = 3\overrightarrow{CD}$ **b** $\overrightarrow{RS} = -\frac{1}{2}\overrightarrow{KL}$ **c** $\overrightarrow{AB} = 2\overrightarrow{BC}$

4 The position vectors of P, Q, R, and S are $\begin{pmatrix}3\\2\\-1\end{pmatrix}$, $\begin{pmatrix}1\\4\\-3\end{pmatrix}$, $\begin{pmatrix}2\\-1\\2\end{pmatrix}$, and $\begin{pmatrix}-1\\-2\\3\end{pmatrix}$ respectively.

a Deduce that [PR] and [QS] are parallel.

b What is the relationship between the lengths of [PR] and [QS]?

5 If $\mathbf{a} = \begin{pmatrix}2\\4\end{pmatrix}$, write down the vector:

a in the same direction as **a** and twice its length

b in the opposite direction to **a** and half its length.

6 Find the unit vector in the direction of:

a $\mathbf{i} + 2\mathbf{j}$ b $2\mathbf{i} - 3\mathbf{k}$ c $2\mathbf{i} - 2\mathbf{j} + \mathbf{k}$

7 Find a vector **v** which has:

a the same direction as $\begin{pmatrix}2\\-1\end{pmatrix}$ and length 3 units

b the opposite direction to $\begin{pmatrix}-1\\-4\end{pmatrix}$ and length 2 units.

8 A is (3, 2) and point B is 4 units from A in the direction $\begin{pmatrix}1\\-1\end{pmatrix}$.

a Find $\overrightarrow{AB}$.

b Find $\overrightarrow{OB}$ using $\overrightarrow{OB} = \overrightarrow{OA} + \overrightarrow{AB}$.

c Hence deduce the coordinates of B.

9 a Find vectors of length 1 unit which are parallel to $\mathbf{a} = \begin{pmatrix}2\\-1\\-2\end{pmatrix}$.

b Find vectors of length 2 units which are parallel to $\mathbf{b} = \begin{pmatrix}-2\\-1\\2\end{pmatrix}$.

10 Find a vector **b** in:

a the same direction as $\begin{pmatrix}-1\\4\\1\end{pmatrix}$ and with length 6 units

b the opposite direction to $\begin{pmatrix}-1\\-2\\-2\end{pmatrix}$ and with length 5 units.

11 a Prove that A(−2, 1, 4), B(4, 3, 0), and C(19, 8, −10) are collinear.

b Prove that P(2, 1, 1), Q(5, −5, −2), and R(−1, 7, 4) are collinear.

c A(2, −3, 4), B(11, −9, 7), and C(−13, a, b) are collinear. Find a and b.

d K(1, −1, 0), L(4, −3, 7), and M(a, 2, b) are collinear. Find a and b.

12 The **triangle inequality** states that:
"In any triangle, the sum of any two sides must always be greater than the third side."

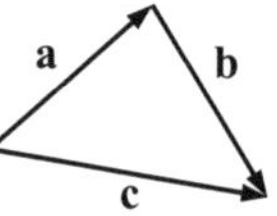

Prove that $|\mathbf{a} + \mathbf{b}| \leqslant |\mathbf{a}| + |\mathbf{b}|$ using a geometrical argument.

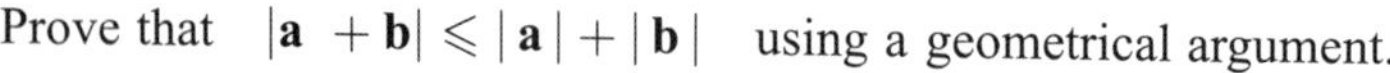

Hint: Consider:
- **a** is not parallel to **b** and use the triangle inequality
- **a** and **b** parallel
- any other cases.

J THE SCALAR PRODUCT OF TWO VECTORS

For ordinary numbers a and b we can write the product of a and b as ab or $a \times b$. There is only one interpretation for this product, so we can use power notation $a^2 = a \times a$, $a^3 = a \times a \times a$, and so on as shorthand.

However, there are *two* different types of product involving two vectors. These are:

- The **scalar product** of 2 vectors, which results in a **scalar** answer and has the notation $\mathbf{v} \bullet \mathbf{w}$ (read "**v** dot **w**").
- The **vector product** of 2 vectors, which results in a **vector** answer and has the notation $\mathbf{v} \times \mathbf{w}$ (read "**v** cross **w**").

Consequently, for vector **v**, $\mathbf{v}^2$ or $(\mathbf{v})^2$ has no meaning and is not used, as it is not clear which of the vector products it would refer to.

SCALAR PRODUCT

The **scalar product** of two vectors is also known as the **dot product** or **inner product**.

If $\mathbf{v} = \begin{pmatrix} v_1 \\ v_2 \\ v_3 \end{pmatrix}$ and $\mathbf{w} = \begin{pmatrix} w_1 \\ w_2 \\ w_3 \end{pmatrix}$, the **scalar product** of **v** and **w** is defined as

$\mathbf{v} \bullet \mathbf{w} = v_1w_1 + v_2w_2 + v_3w_3$.

ANGLE BETWEEN VECTORS

Consider the vectors v $\mathbf{v} = \begin{pmatrix} v_1 \\ v_2 \\ v_3 \end{pmatrix}$ and w $\mathbf{w} = \begin{pmatrix} w_1 \\ w_2 \\ w_3 \end{pmatrix}$.

We translate one of the vectors so that they both originate from the same point.

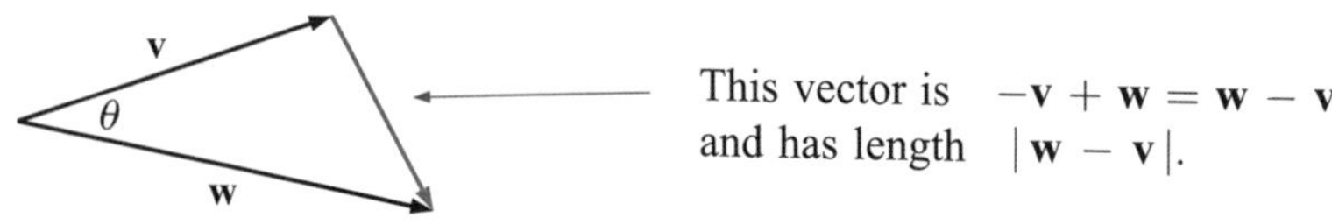

This vector is $-\mathbf{v} + \mathbf{w} = \mathbf{w} - \mathbf{v}$ and has length $|\mathbf{w} - \mathbf{v}|$.

Using the cosine rule, $|\mathbf{w} - \mathbf{v}|^2 = |\mathbf{v}|^2 + |\mathbf{w}|^2 - 2|\mathbf{v}||\mathbf{w}|\cos\theta$

But $\mathbf{w} - \mathbf{v} = \begin{pmatrix} w_1 \\ w_2 \\ w_3 \end{pmatrix} - \begin{pmatrix} v_1 \\ v_2 \\ v_3 \end{pmatrix} = \begin{pmatrix} w_1 - v_1 \\ w_2 - v_2 \\ w_3 - v_3 \end{pmatrix}$

$$\therefore \quad (w_1-v_1)^2+(w_2-v_2)^2+(w_3-v_3)^2 = v_1^2+v_2^2+v_3^2+w_1^2+w_2^2+w_3^2-2|\mathbf{v}||\mathbf{w}|\cos\theta$$

$$\therefore \quad v_1w_1+v_2w_2+v_3w_3 = |\mathbf{v}||\mathbf{w}|\cos\theta$$

$$\therefore \quad \mathbf{v} \bullet \mathbf{w} = |\mathbf{v}||\mathbf{w}|\cos\theta$$

The angle θ between two vectors $\mathbf{v}$ and $\mathbf{w}$ can be found using

$$\cos\theta = \frac{\mathbf{v} \bullet \mathbf{w}}{|\mathbf{v}||\mathbf{w}|}$$

ALGEBRAIC PROPERTIES OF THE SCALAR PRODUCT

The scalar product has the following algebraic properties for both 2-D and 3-D vectors:

- $\mathbf{v} \bullet \mathbf{w} = \mathbf{w} \bullet \mathbf{v}$
- $\mathbf{v} \bullet \mathbf{v} = |\mathbf{v}|^2$
- $\mathbf{v} \bullet (\mathbf{w}+\mathbf{x}) = \mathbf{v} \bullet \mathbf{w} + \mathbf{v} \bullet \mathbf{x}$
- $(\mathbf{v}+\mathbf{w}) \bullet (\mathbf{x}+\mathbf{y}) = \mathbf{v} \bullet \mathbf{x} + \mathbf{v} \bullet \mathbf{y} + \mathbf{w} \bullet \mathbf{x} + \mathbf{w} \bullet \mathbf{y}$

These properties are proven by using general vectors such as: $\mathbf{v} = \begin{pmatrix} v_1 \\ v_2 \\ v_3 \end{pmatrix}$ and $\mathbf{w} = \begin{pmatrix} w_1 \\ w_2 \\ w_3 \end{pmatrix}$.

Be careful not to confuse the **scalar product**, which is the product of two vectors to give a scalar answer, with **scalar multiplication**, which is the product of a scalar and a vector to give a parallel vector. They are quite different.

GEOMETRIC PROPERTIES OF THE SCALAR PRODUCT

- For non-zero vectors $\mathbf{v}$ and $\mathbf{w}$:
  $\mathbf{v} \bullet \mathbf{w} = 0 \;\Leftrightarrow\;$ $\mathbf{v}$ and $\mathbf{w}$ are **perpendicular** or **orthogonal**.
- $|\mathbf{v} \bullet \mathbf{w}| = |\mathbf{v}||\mathbf{w}| \;\Leftrightarrow\;$ $\mathbf{v}$ and $\mathbf{w}$ are non-zero **parallel vectors**.
- If θ is the angle between vectors $\mathbf{v}$ and $\mathbf{w}$ then: $\mathbf{v} \bullet \mathbf{w} = |\mathbf{v}||\mathbf{w}|\cos\theta$
  If θ is acute, $\cos\theta > 0$ and so $\mathbf{v} \bullet \mathbf{w} > 0$
  If θ is obtuse, $\cos\theta < 0$ and so $\mathbf{v} \bullet \mathbf{w} < 0$.
  The angle between two vectors is always taken as the angle θ such that $0° \leqslant \theta \leqslant 180°$, rather than reflex angle α.

The first two of these results can be demonstrated as follows:

If $\mathbf{v}$ is perpendicular to $\mathbf{w}$ then $\theta = 90°$.

$$\begin{aligned}\therefore \quad \mathbf{v} \bullet \mathbf{w} &= |\mathbf{v}||\mathbf{w}|\cos\theta \\ &= |\mathbf{v}||\mathbf{w}|\cos 90° \\ &= 0\end{aligned}$$

If $\mathbf{v}$ is parallel to $\mathbf{w}$ then $\theta = 0°$ or $180°$.

$$\begin{aligned}\therefore \quad \mathbf{v} \bullet \mathbf{w} &= |\mathbf{v}||\mathbf{w}|\cos\theta \\ &= |\mathbf{v}||\mathbf{w}|\cos 0° \text{ or } |\mathbf{v}||\mathbf{w}|\cos 180° \\ &= \pm|\mathbf{v}||\mathbf{w}|\end{aligned}$$

$$\therefore \quad |\mathbf{v} \bullet \mathbf{w}| = |\mathbf{v}||\mathbf{w}|$$

To formally prove these results we must also show that their converses are true.

DISCUSSION

a

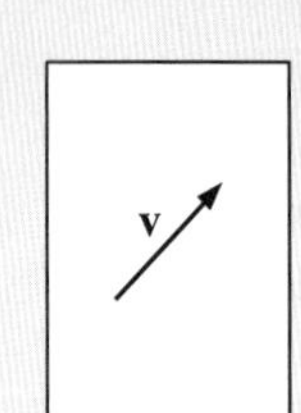

Elaine has drawn a vector $\mathbf{v}$ on a *plane* which is a sheet of paper. It is therefore a 2-dimensional vector.

i How many vectors can she draw which are *perpendicular* to $\mathbf{v}$?

ii Are all of these vectors parallel?

b

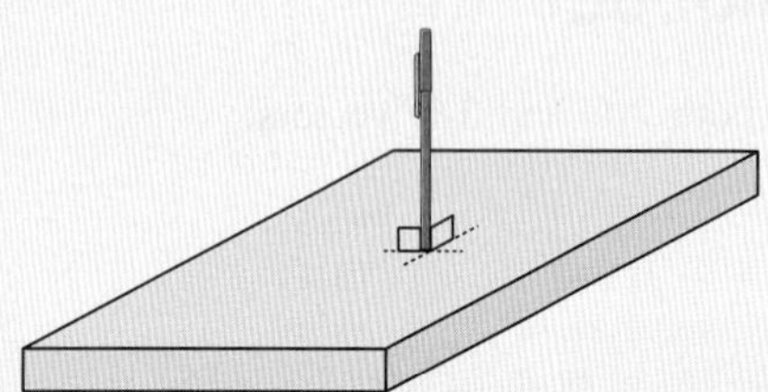

Edward is thinking about vectors in *space*. These are 3-dimensional vectors. He is holding his pen vertically on his desk to represent a vector $\mathbf{w}$.

i How many vectors are there which are *perpendicular* to $\mathbf{w}$?

ii If Edward was to draw a vector (in pencil) on his desk, would it be perpendicular to $\mathbf{w}$?

iii Are all of the vectors which are perpendicular to $\mathbf{w}$, parallel to one another?

Example 30 **Self Tutor**

If $\mathbf{p} = \begin{pmatrix} 2 \\ 3 \\ -1 \end{pmatrix}$ and $\mathbf{q} = \begin{pmatrix} -1 \\ 0 \\ 2 \end{pmatrix}$, find:

a $\mathbf{p} \bullet \mathbf{q}$ **b** the angle between $\mathbf{p}$ and $\mathbf{q}$.

a

$$\begin{aligned} &\mathbf{p} \bullet \mathbf{q} \\ &= \begin{pmatrix} 2 \\ 3 \\ -1 \end{pmatrix} \bullet \begin{pmatrix} -1 \\ 0 \\ 2 \end{pmatrix} \\ &= 2(-1) + 3(0) + (-1)2 \\ &= -2 + 0 - 2 \\ &= -4 \end{aligned}$$

b

$$\begin{aligned} \mathbf{p} \bullet \mathbf{q} &= |\mathbf{p}||\mathbf{q}| \cos\theta \\ \therefore \cos\theta &= \frac{\mathbf{p} \bullet \mathbf{q}}{|\mathbf{p}||\mathbf{q}|} \\ &= \frac{-4}{\sqrt{4+9+1}\sqrt{1+0+4}} \\ &= \frac{-4}{\sqrt{70}} \\ \therefore \theta &= \arccos\left(\frac{-4}{\sqrt{70}}\right) \approx 119^\circ \end{aligned}$$

Since $\mathbf{p} \bullet \mathbf{q} < 0$ the angle is obtuse.

EXERCISE 14J

1 For $\mathbf{p} = \begin{pmatrix} 3 \\ 2 \end{pmatrix}$, $\mathbf{q} = \begin{pmatrix} -1 \\ 5 \end{pmatrix}$, and $\mathbf{r} = \begin{pmatrix} -2 \\ 4 \end{pmatrix}$, find:

a $\mathbf{q} \bullet \mathbf{p}$ **b** $\mathbf{q} \bullet \mathbf{r}$ **c** $\mathbf{q} \bullet (\mathbf{p} + \mathbf{r})$ **d** $3\mathbf{r} \bullet \mathbf{q}$

e $2\mathbf{p} \bullet 2\mathbf{p}$ **f** $\mathbf{i} \bullet \mathbf{p}$ **g** $\mathbf{q} \bullet \mathbf{j}$ **h** $\mathbf{i} \bullet \mathbf{i}$

2 For $\mathbf{a} = \begin{pmatrix} 2 \\ 1 \\ 3 \end{pmatrix}$, $\mathbf{b} = \begin{pmatrix} -1 \\ 1 \\ 1 \end{pmatrix}$, and $\mathbf{c} = \begin{pmatrix} 0 \\ -1 \\ 1 \end{pmatrix}$, find:

a $\mathbf{a} \bullet \mathbf{b}$ **b** $\mathbf{b} \bullet \mathbf{a}$ **c** $|\mathbf{a}|^2$

d $\mathbf{a} \bullet \mathbf{a}$ **e** $\mathbf{a} \bullet (\mathbf{b} + \mathbf{c})$ **f** $\mathbf{a} \bullet \mathbf{b} + \mathbf{a} \bullet \mathbf{c}$

3 If $\mathbf{p} = \begin{pmatrix} 3 \\ -1 \\ 2 \end{pmatrix}$ and $\mathbf{q} = \begin{pmatrix} -2 \\ 1 \\ 3 \end{pmatrix}$, find: **a** $\mathbf{p} \bullet \mathbf{q}$ **b** the angle between **p** and **q**.

4 Find the angle between **m** and **n** if:

a $\mathbf{m} = \begin{pmatrix} 2 \\ -1 \\ -1 \end{pmatrix}$ and $\mathbf{n} = \begin{pmatrix} -1 \\ 3 \\ 2 \end{pmatrix}$ **b** $\mathbf{m} = 2\mathbf{j} - \mathbf{k}$ and $\mathbf{n} = \mathbf{i} + 2\mathbf{k}$

5 Find: **a** $(\mathbf{i} + \mathbf{j} - \mathbf{k}) \bullet (2\mathbf{j} + \mathbf{k})$ **b** $\mathbf{i} \bullet \mathbf{i}$ **c** $\mathbf{i} \bullet \mathbf{j}$

6 Find $\mathbf{p} \bullet \mathbf{q}$ if:

a $|\mathbf{p}| = 2,\ |\mathbf{q}| = 5,\ \theta = 60°$ **b** $|\mathbf{p}| = 6,\ |\mathbf{q}| = 3,\ \theta = 120°$

7 **a** Suppose $|\mathbf{v}| = 3$ and $|\mathbf{w}| = 4$. State the possible values of $\mathbf{v} \bullet \mathbf{w}$ if **v** and **w** are:

i parallel **ii** at 60° to each other.

b Suppose $\mathbf{a} \bullet \mathbf{b} = -12$ and **b** is a unit vector.

i Explain why **a** and **b** are not perpendicular.

ii Find $|\mathbf{a}|$ if **a** and **b** are parallel.

c Suppose $|\mathbf{c}| = |\mathbf{d}| = \sqrt{5}$. What can be deduced about **c** and **d** if:

i $\mathbf{c} \bullet \mathbf{d} = 5$ **ii** $\mathbf{c} \bullet \mathbf{d} = -5$?

8 In the given figure:

a State the coordinates of P.

b Find $\overrightarrow{BP}$ and $\overrightarrow{AP}$.

c Find $\overrightarrow{AP} \bullet \overrightarrow{BP}$ using **b**.

d What property of a semi-circle has been deduced in **c**?

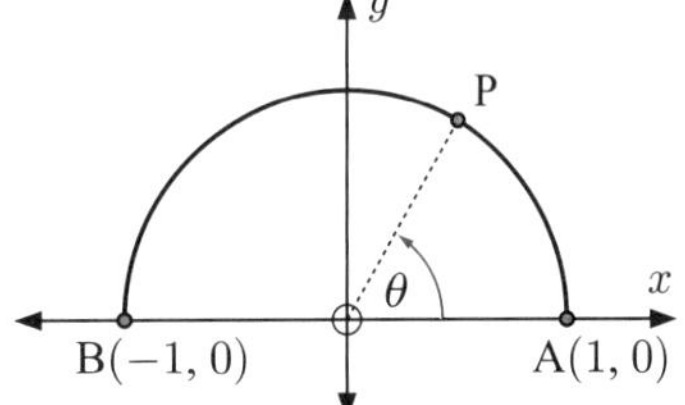

9 Use $\mathbf{a} = \begin{pmatrix} a_1 \\ a_2 \\ a_3 \end{pmatrix}$, $\mathbf{b} = \begin{pmatrix} b_1 \\ b_2 \\ b_3 \end{pmatrix}$, and $\mathbf{c} = \begin{pmatrix} c_1 \\ c_2 \\ c_3 \end{pmatrix}$ to prove that $\mathbf{a} \bullet (\mathbf{b} + \mathbf{c}) = \mathbf{a} \bullet \mathbf{b} + \mathbf{a} \bullet \mathbf{c}$.

Hence, prove that $(\mathbf{a} + \mathbf{b}) \bullet (\mathbf{c} + \mathbf{d}) = \mathbf{a} \bullet \mathbf{c} + \mathbf{a} \bullet \mathbf{d} + \mathbf{b} \bullet \mathbf{c} + \mathbf{b} \bullet \mathbf{d}$.

Example 31 **Self Tutor**

Find t such that $\mathbf{a} = \begin{pmatrix} -1 \\ 5 \end{pmatrix}$ and $\mathbf{b} = \begin{pmatrix} 2 \\ t \end{pmatrix}$ are perpendicular.

Since **a** and **b** are perpendicular, $\mathbf{a} \bullet \mathbf{b} = 0$

$$\therefore \begin{pmatrix} -1 \\ 5 \end{pmatrix} \bullet \begin{pmatrix} 2 \\ t \end{pmatrix} = 0$$

$$\therefore (-1)(2) + 5t = 0$$

$$\therefore -2 + 5t = 0$$

$$\therefore 5t = 2$$

$$\therefore t = \tfrac{2}{5}$$

If two vectors are perpendicular then their scalar product is **zero**.

10 Find t if the given pair of vectors are:

i perpendicular **ii** parallel.

a $\mathbf{p}=\begin{pmatrix}3\\t\end{pmatrix}$ and $\mathbf{q}=\begin{pmatrix}-2\\1\end{pmatrix}$ **b** $\mathbf{r}=\begin{pmatrix}t\\t+2\end{pmatrix}$ and $\mathbf{s}=\begin{pmatrix}3\\-4\end{pmatrix}$

c $\mathbf{a}=\begin{pmatrix}t\\t+2\end{pmatrix}$ and $\mathbf{b}=\begin{pmatrix}2-3t\\t\end{pmatrix}$

11 **a** Show that $\begin{pmatrix}1\\1\\5\end{pmatrix}$ and $\begin{pmatrix}2\\3\\-1\end{pmatrix}$ are perpendicular.

b Show that $\mathbf{a}=\begin{pmatrix}3\\1\\2\end{pmatrix}$, $\mathbf{b}=\begin{pmatrix}-1\\1\\1\end{pmatrix}$, and $\mathbf{c}=\begin{pmatrix}1\\5\\-4\end{pmatrix}$ are mutually perpendicular.

c Find t if the following vectors are perpendicular:

i $\mathbf{a}=\begin{pmatrix}3\\-1\\t\end{pmatrix}$ and $\mathbf{b}=\begin{pmatrix}2t\\-3\\-4\end{pmatrix}$ **ii** $\begin{pmatrix}3\\t\\-2\end{pmatrix}$ and $\begin{pmatrix}1-t\\-3\\4\end{pmatrix}$.

Vectors are *mutually perpendicular* if each one is perpendicular to all the others.

Example 32 **Self Tutor**

Consider the points A(2, 1), B(6, −1), and C(5, −3). Use a scalar product to check if triangle ABC is right angled. If it is, state the right angle.

$\overrightarrow{AB}=\begin{pmatrix}4\\-2\end{pmatrix}$, $\overrightarrow{BC}=\begin{pmatrix}-1\\-2\end{pmatrix}$, and $\overrightarrow{AC}=\begin{pmatrix}3\\-4\end{pmatrix}$.

$\overrightarrow{AB}\bullet\overrightarrow{BC}=4(-1)+(-2)(-2)=-4+4=0.$

$\therefore$ $\overrightarrow{AB}\perp\overrightarrow{BC}$ and so triangle ABC is right angled at B.

We do not need to check $\overrightarrow{AB}\bullet\overrightarrow{AC}$ and $\overrightarrow{BC}\bullet\overrightarrow{AC}$ because a triangle cannot have more than one right angle.

12 Use a scalar product to check if triangle ABC is right angled. If it is, state the right angle.

a A(−2, 1), B(−2, 5), and C(3, 1) **b** A(4, 7), B(1, 2), and C(−1, 6)

c A(2, −2), B(5, 7), and C(−1, −1) **d** A(10, 1), B(5, 2), and C(7, 4)

13 Consider triangle ABC in which A is (5, 1, 2), B is (6, −1, 0), and C is (3, 2, 0). Using scalar product only, show that the triangle is right angled.

14 A(2, 4, 2), B(−1, 2, 3), C(−3, 3, 6), and D(0, 5, 5) are vertices of a quadrilateral.

a Prove that ABCD is a parallelogram.

b Find $|\overrightarrow{AB}|$ and $|\overrightarrow{BC}|$. What can be said about ABCD?

c Find $\overrightarrow{AC}\bullet\overrightarrow{BD}$. Describe what property of ABCD you have shown.

Example 33 **Self Tutor**

Find the form of all vectors which are perpendicular to $\begin{pmatrix} 3 \\ 4 \end{pmatrix}$. Are all of the vectors parallel?

$\begin{pmatrix} 3 \\ 4 \end{pmatrix} \bullet \begin{pmatrix} -4 \\ 3 \end{pmatrix} = -12 + 12 = 0$

$\therefore$ $\begin{pmatrix} -4 \\ 3 \end{pmatrix}$ is a vector perpendicular to $\begin{pmatrix} 3 \\ 4 \end{pmatrix}$.

The required vectors have the form $k\begin{pmatrix} -4 \\ 3 \end{pmatrix}$, $k \neq 0$. All of these vectors are parallel.

15 Find the form of all vectors which are perpendicular to:

a $\begin{pmatrix} 5 \\ 2 \end{pmatrix}$ **b** $\begin{pmatrix} -1 \\ -2 \end{pmatrix}$ **c** $\begin{pmatrix} 3 \\ -1 \end{pmatrix}$ **d** $\begin{pmatrix} -4 \\ 3 \end{pmatrix}$ **e** $\begin{pmatrix} 2 \\ 0 \end{pmatrix}$

16 Find any two vectors which are not parallel, but which are *both* perpendicular to $\begin{pmatrix} 1 \\ 2 \\ -1 \end{pmatrix}$.

17 Find the angle ABC of triangle ABC for A(3, 0, 1), B(−3, 1, 2), and C(−2, 1, −1).

Hint: To find the angle at B, use $\overrightarrow{BA}$ and $\overrightarrow{BC}$.
What angle is found if $\overrightarrow{BA}$ and $\overrightarrow{CB}$ are used?

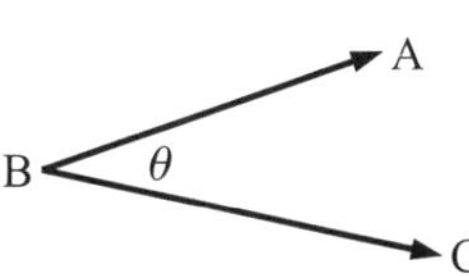

Example 34 **Self Tutor**

Use vector methods to determine the measure of $A\widehat{B}C$.

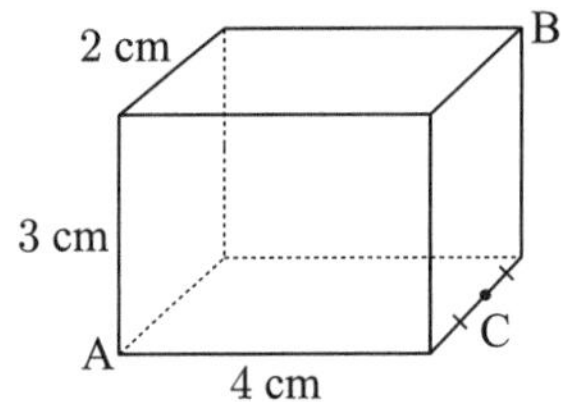

The vectors used must both be away from B (or both towards B). If this is not done you will be finding the exterior angle at B.

Placing the coordinate axes as illustrated,
A is (2, 0, 0), B is (0, 4, 3), and C is (1, 4, 0).

$\therefore$ $\overrightarrow{BA} = \begin{pmatrix} 2 \\ -4 \\ -3 \end{pmatrix}$ and $\overrightarrow{BC} = \begin{pmatrix} 1 \\ 0 \\ -3 \end{pmatrix}$

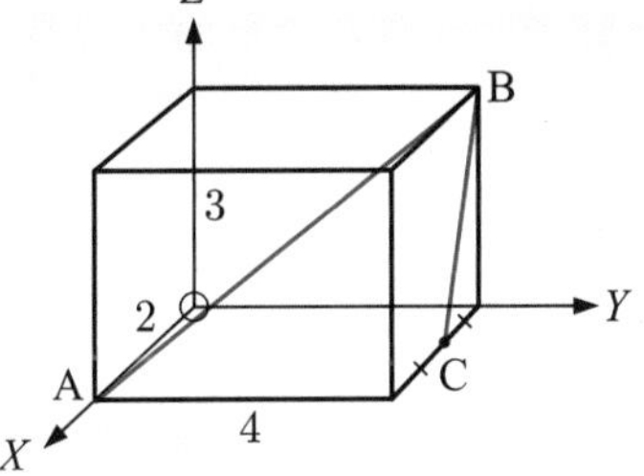

$$\cos(A\widehat{B}C) = \frac{\overrightarrow{BA} \bullet \overrightarrow{BC}}{|\overrightarrow{BA}|\,|\overrightarrow{BC}|}$$

$$= \frac{2(1) + (-4)(0) + (-3)(-3)}{\sqrt{4+16+9}\sqrt{1+0+9}}$$

$$= \frac{11}{\sqrt{290}}$$

$$\therefore\ A\widehat{B}C = \arccos\left(\frac{11}{\sqrt{290}}\right) \approx 49.8°$$

18 The cube alongside has sides of length 2 cm. Find, using vector methods, the measure of:

a $\widehat{ABS}$ **b** $\widehat{RBP}$ **c** $\widehat{PBS}$

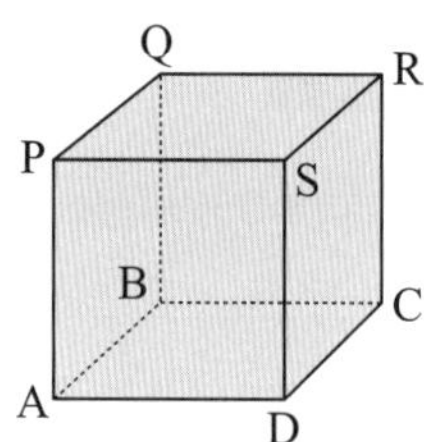

19 [KL], [LM], and [LX] are 8, 5, and 3 units long respectively. P is the midpoint of [KL]. Find, using vector methods, the measure of:

a $\widehat{YNX}$ **b** $\widehat{YNP}$

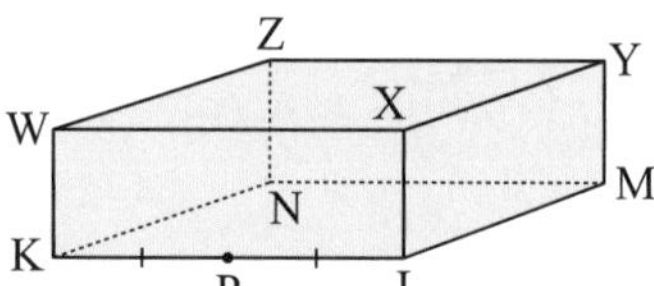

20 Consider tetrahedron ABCD.

a Find the coordinates of M.

b Find the measure of $\widehat{DMA}$.

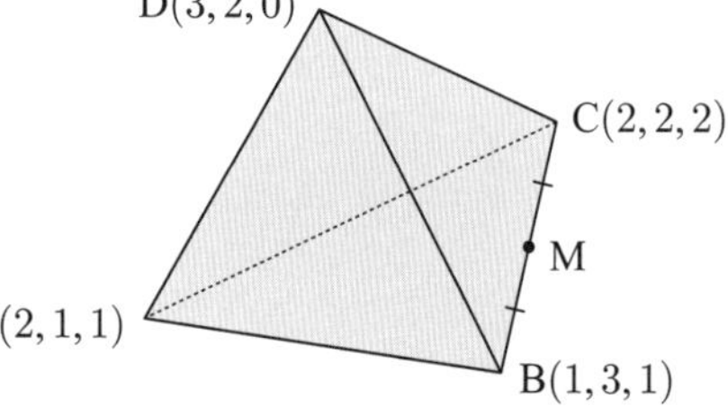

21 **a** Find t if $2\mathbf{i} + t\mathbf{j} + (t-2)\mathbf{k}$ and $t\mathbf{i} + 3\mathbf{j} + t\mathbf{k}$ are perpendicular.

b Find r, s, and t if $\mathbf{a} = \begin{pmatrix} 1 \\ 2 \\ 3 \end{pmatrix}$, $\mathbf{b} = \begin{pmatrix} 2 \\ 2 \\ r \end{pmatrix}$, and $\mathbf{c} = \begin{pmatrix} s \\ t \\ 1 \end{pmatrix}$ are mutually perpendicular.

22 Find the angle made by: **a** $\mathbf{i}$ and $\begin{pmatrix} 1 \\ 2 \\ 3 \end{pmatrix}$ **b** $\mathbf{j}$ and $\begin{pmatrix} -1 \\ 1 \\ 3 \end{pmatrix}$.

23 Find three vectors $\mathbf{a}$, $\mathbf{b}$, and $\mathbf{c}$ such that $\mathbf{a} \neq \mathbf{0}$ and $\mathbf{a} \bullet \mathbf{b} = \mathbf{a} \bullet \mathbf{c}$ but $\mathbf{b} \neq \mathbf{c}$.

24 Show, using $|\mathbf{x}|^2 = \mathbf{x} \bullet \mathbf{x}$, that:

a $|\mathbf{a}+\mathbf{b}|^2 + |\mathbf{a}-\mathbf{b}|^2 = 2|\mathbf{a}|^2 + 2|\mathbf{b}|^2$ **b** $|\mathbf{a}+\mathbf{b}|^2 - |\mathbf{a}-\mathbf{b}|^2 = 4\,\mathbf{a} \bullet \mathbf{b}$

25 $\mathbf{a}$ and $\mathbf{b}$ are the position vectors of two distinct points A and B, neither of which is the origin. Show that if $|\mathbf{a}+\mathbf{b}| = |\mathbf{a}-\mathbf{b}|$ then $\mathbf{a}$ is perpendicular to $\mathbf{b}$, using:

a a vector algebraic method **b** a geometric argument.

26 If $|\mathbf{a}| = 3$ and $|\mathbf{b}| = 4$, find $(\mathbf{a}+\mathbf{b}) \bullet (\mathbf{a}-\mathbf{b})$.

27 Explain why $\mathbf{a} \bullet \mathbf{b} \bullet \mathbf{c}$ is meaningless.

K THE VECTOR PRODUCT OF TWO VECTORS

We have seen how the scalar product of two vectors results in a **scalar**.

The second form of product between two vectors is the **vector product** or **vector cross product**, and this results in a **vector**.

The **vector product** arises when we attempt to find a vector which is *perpendicular to two other known vectors.*

Suppose $\mathbf{x} = \begin{pmatrix} x \\ y \\ z \end{pmatrix}$ is perpendicular to both $\mathbf{a} = \begin{pmatrix} a_1 \\ a_2 \\ a_3 \end{pmatrix}$ and $\mathbf{b} = \begin{pmatrix} b_1 \\ b_2 \\ b_3 \end{pmatrix}$.

$$\therefore \begin{cases} a_1x + a_2y + a_3z = 0 \\ b_1x + b_2y + b_3z = 0 \end{cases} \quad \{\text{as dot products are zero}\}$$

$$\therefore \begin{cases} a_1x + a_2y = -a_3z & \text{.... (1)} \\ b_1x + b_2y = -b_3z & \text{.... (2)} \end{cases}$$

We will now try to solve these two equations to get expressions for x and y in terms of z.

To eliminate x, we multiply (1) by $-b_1$ and (2) by a_1.

$$\begin{aligned} -a_1b_1x - a_2b_1y &= a_3b_1z && \{-b_1 \times (1)\} \\ a_1b_1x + a_1b_2y &= -a_1b_3z && \{a_1 \times (2)\} \end{aligned}$$

Adding these gives $\quad (a_1b_2 - a_2b_1)y = (a_3b_1 - a_1b_3)z$

$\therefore \quad y = (a_3b_1 - a_1b_3)t$ and $z = (a_1b_2 - a_2b_1)t$ for any non-zero t.

Substituting into (1), $\quad a_1x + a_2(a_3b_1 - a_1b_3)t = -a_3(a_1b_2 - a_2b_1)t$

$$\begin{aligned} \therefore \quad a_1x &= (-a_1a_3b_2 + a_2a_3b_1 - a_2a_3b_1 + a_1a_2b_3)t \\ \therefore \quad a_1x &= a_1(a_2b_3 - a_3b_2)t \\ \therefore \quad x &= (a_2b_3 - a_3b_2)t \end{aligned}$$

The simplest vector perpendicular to both **a** and **b** is obtained by letting $t = 1$. In this case we find
$\mathbf{x} = \begin{pmatrix} a_2b_3 - a_3b_2 \\ a_3b_1 - a_1b_3 \\ a_1b_2 - a_2b_1 \end{pmatrix}$, which we call the vector cross product of **a** and **b**.

The **vector cross product** of vectors $\mathbf{a} = \begin{pmatrix} a_1 \\ a_2 \\ a_3 \end{pmatrix}$ and $\mathbf{b} = \begin{pmatrix} b_1 \\ b_2 \\ b_3 \end{pmatrix}$ is $\mathbf{a} \times \mathbf{b} = \begin{pmatrix} a_2b_3 - a_3b_2 \\ a_3b_1 - a_1b_3 \\ a_1b_2 - a_2b_1 \end{pmatrix}$.

Rather than remembering this formula, mathematicians commonly write the vector cross product as:

The signs alternate $+$, $-$, $+$.

$$\mathbf{a} \times \mathbf{b} = \begin{vmatrix} \mathbf{i} & \mathbf{j} & \mathbf{k} \\ a_1 & a_2 & a_3 \\ b_1 & b_2 & b_3 \end{vmatrix} = \begin{vmatrix} a_2 & a_3 \\ b_2 & b_3 \end{vmatrix}\mathbf{i} - \begin{vmatrix} a_1 & a_3 \\ b_1 & b_3 \end{vmatrix}\mathbf{j} + \begin{vmatrix} a_1 & a_2 \\ b_1 & b_2 \end{vmatrix}\mathbf{k}$$

From this form: Cover up the top row and **i** column. Cover up the top row and **j** column. Cover up the top row and **k** column.

Be careful to get the sign of the middle term correct!

where $\begin{vmatrix} a_2 & a_3 \\ b_2 & b_3 \end{vmatrix} = (a_2b_3 - a_3b_2)$ is the product of the main (red) diagonal minus the product of the other diagonal (green)

$$\begin{vmatrix} a_2 & a_3 \\ b_2 & b_3 \end{vmatrix}$$

We find that $\mathbf{a} \times \mathbf{b} = (a_2b_3 - a_3b_2)\mathbf{i} - (a_1b_3 - a_3b_1)\mathbf{j} + (a_1b_2 - a_2b_1)\mathbf{k}$
$= (a_2b_3 - a_3b_2)\mathbf{i} + (a_3b_1 - a_1b_3)\mathbf{j} + (a_1b_2 - a_2b_1)\mathbf{k}$

ALGEBRAIC PROPERTIES OF THE VECTOR CROSS PRODUCT

The vector cross product has the following algebraic properties for non-zero 3-dimensional vectors **a**, **b**, **c**, and **d**:

- $\mathbf{a} \times \mathbf{b}$ is a vector which is perpendicular to both **a** and **b**.
- $\mathbf{a} \times \mathbf{a} = \mathbf{0}$ for all **a**.
- $\mathbf{a} \times \mathbf{b} = -\mathbf{b} \times \mathbf{a}$ for all **a** and **b**.
  This means that $\mathbf{a} \times \mathbf{b}$ and $\mathbf{b} \times \mathbf{a}$ have the same length but opposite direction.
- $\mathbf{a} \bullet (\mathbf{b} \times \mathbf{c})$ is called the **scalar triple product**.
- $\mathbf{a} \times (\mathbf{b} + \mathbf{c}) = (\mathbf{a} \times \mathbf{b}) + (\mathbf{a} \times \mathbf{c})$
- $(\mathbf{a} + \mathbf{b}) \times (\mathbf{c} + \mathbf{d}) = (\mathbf{a} \times \mathbf{c}) + (\mathbf{a} \times \mathbf{d}) + (\mathbf{b} \times \mathbf{c}) + (\mathbf{b} \times \mathbf{d})$.

You will prove or verify these results in the next **Exercise**.

Example 35 Self Tutor

If $\mathbf{a} = \begin{pmatrix} 2 \\ 3 \\ -1 \end{pmatrix}$ and $\mathbf{b} = \begin{pmatrix} -1 \\ 2 \\ 4 \end{pmatrix}$, find $\mathbf{a} \times \mathbf{b}$.

After finding $\mathbf{a} \times \mathbf{b}$, check that your answer is perpendicular to both **a** and **b**.

$$\mathbf{a} \times \mathbf{b} = \begin{vmatrix} \mathbf{i} & \mathbf{j} & \mathbf{k} \\ 2 & 3 & -1 \\ -1 & 2 & 4 \end{vmatrix}$$

$$= \begin{vmatrix} 3 & -1 \\ 2 & 4 \end{vmatrix} \mathbf{i} - \begin{vmatrix} 2 & -1 \\ -1 & 4 \end{vmatrix} \mathbf{j} + \begin{vmatrix} 2 & 3 \\ -1 & 2 \end{vmatrix} \mathbf{k}$$

$$= (3 \times 4 - (-1) \times 2)\mathbf{i} - (2 \times 4 - (-1) \times (-1))\mathbf{j} + (2 \times 2 - 3 \times (-1))\mathbf{k}$$

$$= \begin{pmatrix} 14 \\ -7 \\ 7 \end{pmatrix}$$

Example 36 Self Tutor

For $\mathbf{a} = \begin{pmatrix} 2 \\ 1 \\ -1 \end{pmatrix}$, $\mathbf{b} = \begin{pmatrix} 1 \\ 2 \\ 3 \end{pmatrix}$, and $\mathbf{c} = \begin{pmatrix} 2 \\ 0 \\ 4 \end{pmatrix}$, find:

a $\mathbf{b} \times \mathbf{c}$

b $\mathbf{a} \bullet (\mathbf{b} \times \mathbf{c})$

a $$\mathbf{b} \times \mathbf{c} = \begin{vmatrix} \mathbf{i} & \mathbf{j} & \mathbf{k} \\ 1 & 2 & 3 \\ 2 & 0 & 4 \end{vmatrix}$$

$$= \begin{vmatrix} 2 & 3 \\ 0 & 4 \end{vmatrix} \mathbf{i} - \begin{vmatrix} 1 & 3 \\ 2 & 4 \end{vmatrix} \mathbf{j} + \begin{vmatrix} 1 & 2 \\ 2 & 0 \end{vmatrix} \mathbf{k}$$

$$= \begin{pmatrix} 8 \\ 2 \\ -4 \end{pmatrix}$$

b $$\mathbf{a} \bullet (\mathbf{b} \times \mathbf{c}) = \begin{pmatrix} 2 \\ 1 \\ -1 \end{pmatrix} \bullet \begin{pmatrix} 8 \\ 2 \\ -4 \end{pmatrix}$$

$$= 16 + 2 + 4$$

$$= 22$$

EXERCISE 14K.1

1 Calculate:

a $\begin{pmatrix} 2 \\ -3 \\ 1 \end{pmatrix} \times \begin{pmatrix} 1 \\ 4 \\ -2 \end{pmatrix}$

b $\begin{pmatrix} -1 \\ 0 \\ 2 \end{pmatrix} \times \begin{pmatrix} 3 \\ -1 \\ -2 \end{pmatrix}$

c $(\mathbf{i} + \mathbf{j} - 2\mathbf{k}) \times (\mathbf{i} - \mathbf{k})$

d $(2\mathbf{i} - \mathbf{k}) \times (\mathbf{j} + 3\mathbf{k})$

2 Suppose $\mathbf{a} = \begin{pmatrix} 1 \\ 2 \\ 3 \end{pmatrix}$ and $\mathbf{b} = \begin{pmatrix} -1 \\ 3 \\ -1 \end{pmatrix}$.

a Find $\mathbf{a} \times \mathbf{b}$.

b Hence determine $\mathbf{a} \bullet (\mathbf{a} \times \mathbf{b})$ and $\mathbf{b} \bullet (\mathbf{a} \times \mathbf{b})$.

c Explain your results.

3 $\mathbf{i}$, $\mathbf{j}$, and $\mathbf{k}$ are the unit vectors parallel to the coordinate axes.

a Find $\mathbf{i} \times \mathbf{i}$, $\mathbf{j} \times \mathbf{j}$, and $\mathbf{k} \times \mathbf{k}$. Comment on your results.

b Find: i $\mathbf{i} \times \mathbf{j}$ and $\mathbf{j} \times \mathbf{i}$ ii $\mathbf{j} \times \mathbf{k}$ and $\mathbf{k} \times \mathbf{j}$ iii $\mathbf{i} \times \mathbf{k}$ and $\mathbf{k} \times \mathbf{i}$.
Comment on your results.

4 Using $\mathbf{a} = \begin{pmatrix} a_1 \\ a_2 \\ a_3 \end{pmatrix}$ and $\mathbf{b} = \begin{pmatrix} b_1 \\ b_2 \\ b_3 \end{pmatrix}$, prove that:

a $\mathbf{a} \times \mathbf{a} = \mathbf{0}$ for all 3-dimensional vectors $\mathbf{a}$

b $\mathbf{a} \times \mathbf{b} = -(\mathbf{b} \times \mathbf{a})$ for all 3-dimensional vectors $\mathbf{a}$ and $\mathbf{b}$.

5 Suppose $\mathbf{a} = \begin{pmatrix} 1 \\ 3 \\ 2 \end{pmatrix}$, $\mathbf{b} = \begin{pmatrix} 2 \\ -1 \\ 1 \end{pmatrix}$, and $\mathbf{c} = \begin{pmatrix} 0 \\ 1 \\ -2 \end{pmatrix}$. Find:

a $\mathbf{b} \times \mathbf{c}$

b $\mathbf{a} \bullet (\mathbf{b} \times \mathbf{c})$

6 Suppose $\mathbf{a} = \mathbf{i} + 2\mathbf{k}$, $\mathbf{b} = -\mathbf{j} + \mathbf{k}$, and $\mathbf{c} = 2\mathbf{i} - \mathbf{k}$. Find:

a $\mathbf{a} \times \mathbf{b}$

b $\mathbf{a} \times \mathbf{c}$

c $(\mathbf{a} \times \mathbf{b}) + (\mathbf{a} \times \mathbf{c})$

d $\mathbf{a} \times (\mathbf{b} + \mathbf{c})$

Check that $\mathbf{a} \times (\mathbf{b} + \mathbf{c}) = (\mathbf{a} \times \mathbf{b}) + (\mathbf{a} \times \mathbf{c})$ for other vectors $\mathbf{a}$, $\mathbf{b}$, and $\mathbf{c}$ of your choosing.

7 Prove that $\mathbf{a} \times (\mathbf{b} + \mathbf{c}) = \mathbf{a} \times \mathbf{b} + \mathbf{a} \times \mathbf{c}$ using
$\mathbf{a} = \begin{pmatrix} a_1 \\ a_2 \\ a_3 \end{pmatrix}$, $\mathbf{b} = \begin{pmatrix} b_1 \\ b_2 \\ b_3 \end{pmatrix}$, and $\mathbf{c} = \begin{pmatrix} c_1 \\ c_2 \\ c_3 \end{pmatrix}$.

8 Use $\mathbf{a} \times (\mathbf{b} + \mathbf{c}) = (\mathbf{a} \times \mathbf{b}) + (\mathbf{a} \times \mathbf{c})$ to prove that
$(\mathbf{a} + \mathbf{b}) \times (\mathbf{c} + \mathbf{d}) = (\mathbf{a} \times \mathbf{c}) + (\mathbf{a} \times \mathbf{d}) + (\mathbf{b} \times \mathbf{c}) + (\mathbf{b} \times \mathbf{d})$.
Note that since $\mathbf{x} \times \mathbf{y} = -(\mathbf{y} \times \mathbf{x})$, the order of the vectors must be maintained.

9 Use the properties of vector cross product to simplify:

a $\mathbf{a} \times (\mathbf{a} + \mathbf{b})$

b $(\mathbf{a} + \mathbf{b}) \times (\mathbf{a} + \mathbf{b})$

c $(\mathbf{a} + \mathbf{b}) \times (\mathbf{a} - \mathbf{b})$

d $2\mathbf{a} \bullet (\mathbf{a} \times \mathbf{b})$

Example 37 **Self Tutor**

Find *all* vectors perpendicular to both $\mathbf{a} = \begin{pmatrix} 1 \\ 2 \\ -1 \end{pmatrix}$ and $\mathbf{b} = \begin{pmatrix} 1 \\ 0 \\ -3 \end{pmatrix}$.

$$\mathbf{a} \times \mathbf{b} = \begin{vmatrix} \mathbf{i} & \mathbf{j} & \mathbf{k} \\ 1 & 2 & -1 \\ 1 & 0 & -3 \end{vmatrix} = \begin{vmatrix} 2 & -1 \\ 0 & -3 \end{vmatrix} \mathbf{i} - \begin{vmatrix} 1 & -1 \\ 1 & -3 \end{vmatrix} \mathbf{j} + \begin{vmatrix} 1 & 2 \\ 1 & 0 \end{vmatrix} \mathbf{k}$$

$$= -6\mathbf{i} + 2\mathbf{j} - 2\mathbf{k}$$

$$= -2(3\mathbf{i} - \mathbf{j} + \mathbf{k})$$

The vectors have the form $k \begin{pmatrix} 3 \\ -1 \\ 1 \end{pmatrix}$ where k is any non-zero real number.

10 Find *all* vectors perpendicular to both:

a $\begin{pmatrix} 2 \\ -1 \\ 3 \end{pmatrix}$ and $\begin{pmatrix} 1 \\ 1 \\ 1 \end{pmatrix}$

b $\begin{pmatrix} -1 \\ 3 \\ 4 \end{pmatrix}$ and $\begin{pmatrix} 5 \\ 0 \\ 2 \end{pmatrix}$

c $\mathbf{i} + \mathbf{j}$ and $\mathbf{i} - \mathbf{j} - \mathbf{k}$

d $\mathbf{i} - \mathbf{j} - \mathbf{k}$ and $2\mathbf{i} + 2\mathbf{j} - 3\mathbf{k}$.

11 Find all vectors perpendicular to both $\mathbf{a} = \begin{pmatrix} 2 \\ 3 \\ -1 \end{pmatrix}$ and $\mathbf{b} = \begin{pmatrix} 1 \\ -2 \\ 2 \end{pmatrix}$.

Hence find *two* vectors of length 5 units which are perpendicular to both **a** and **b**.

Example 38 **Self Tutor**

Find a vector which is perpendicular to the plane passing through the points A(1, −1, 2), B(3, 1, 0), and C(−1, 2, −3).

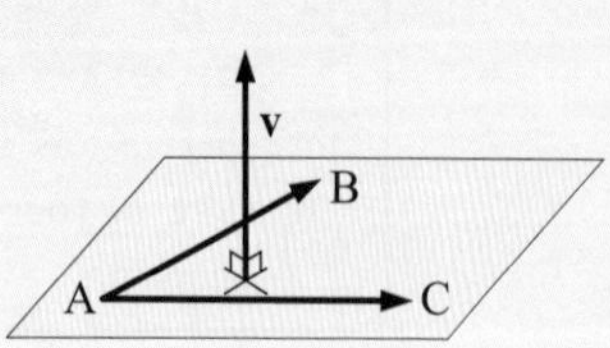

$\overrightarrow{AB} = \begin{pmatrix} 2 \\ 2 \\ -2 \end{pmatrix}$ and $\overrightarrow{AC} = \begin{pmatrix} -2 \\ 3 \\ -5 \end{pmatrix}$

The vector **v** must be perpendicular to both $\overrightarrow{AB}$ and $\overrightarrow{AC}$.

$$\therefore\ \mathbf{v} = \begin{vmatrix} \mathbf{i} & \mathbf{j} & \mathbf{k} \\ 2 & 2 & -2 \\ -2 & 3 & -5 \end{vmatrix} = \begin{vmatrix} 2 & -2 \\ 3 & -5 \end{vmatrix} \mathbf{i} - \begin{vmatrix} 2 & -2 \\ -2 & -5 \end{vmatrix} \mathbf{j} + \begin{vmatrix} 2 & 2 \\ -2 & 3 \end{vmatrix} \mathbf{k}$$

$$= -4\mathbf{i} + 14\mathbf{j} + 10\mathbf{k}$$

$$= -2(2\mathbf{i} - 7\mathbf{j} - 5\mathbf{k})$$

Any non-zero multiple of $\begin{pmatrix} 2 \\ -7 \\ -5 \end{pmatrix}$ will be perpendicular to the plane.

12 Find a vector which is perpendicular to the plane passing through the points:

a A(1, 3, 2), B(0, 2, −5), and C(3, 1, −4)

b P(2, 0, −1), Q(0, 1, 3), and R(1, −1, 1).

DIRECTION OF $\mathbf{a} \times \mathbf{b}$

We have already observed that $\mathbf{a} \times \mathbf{b} = -(\mathbf{b} \times \mathbf{a})$, so $\mathbf{a} \times \mathbf{b}$ and $\mathbf{b} \times \mathbf{a}$ are in opposite directions.

However, what is the direction of each?

In the last **Exercise**, we saw that $\mathbf{i} \times \mathbf{j} = \mathbf{k}$ and $\mathbf{j} \times \mathbf{i} = -\mathbf{k}$.

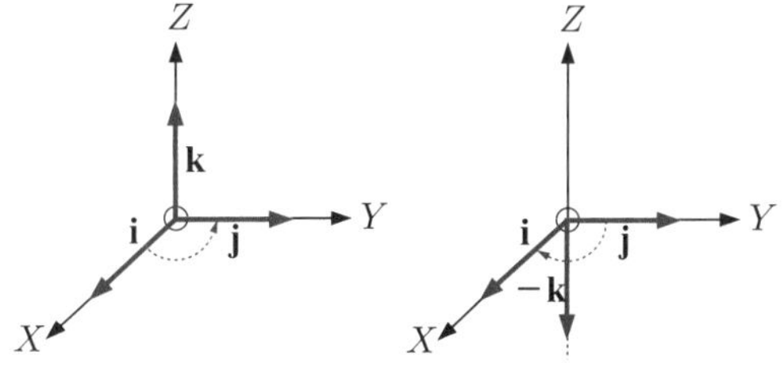

In general, the **direction** of $\mathbf{x} \times \mathbf{y}$ is determined by the **right hand rule**:

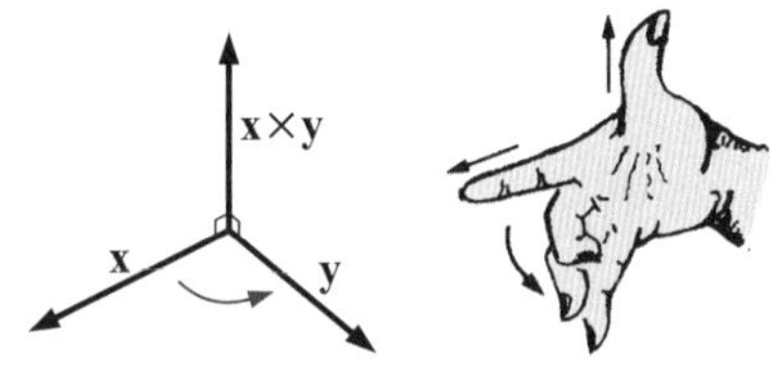

If the fingers on your right hand turn from $\mathbf{x}$ to $\mathbf{y}$, then your thumb points in the direction of $\mathbf{x} \times \mathbf{y}$.

THE LENGTH OF $\mathbf{a} \times \mathbf{b}$

Using $\mathbf{a} \times \mathbf{b} = \begin{pmatrix} a_2b_3 - a_3b_2 \\ a_3b_1 - a_1b_3 \\ a_1b_2 - a_2b_1 \end{pmatrix}$,

$$|\mathbf{a} \times \mathbf{b}| = \sqrt{(a_2b_3 - a_3b_2)^2 + (a_3b_1 - a_1b_3)^2 + (a_1b_2 - a_2b_1)^2}$$

However, another very useful form of the length of $\mathbf{a} \times \mathbf{b}$ exists. This is:

$$|\mathbf{a} \times \mathbf{b}| = |\mathbf{a}||\mathbf{b}|\sin\theta \quad \text{where } \theta \text{ is the angle between } \mathbf{a} \text{ and } \mathbf{b}.$$

Proof:

$$\begin{aligned}
|\mathbf{a}|^2|\mathbf{b}|^2\sin^2\theta &= |\mathbf{a}|^2|\mathbf{b}|^2(1-\cos^2\theta) \\
&= |\mathbf{a}|^2|\mathbf{b}|^2 - |\mathbf{a}|^2|\mathbf{b}|^2\cos^2\theta \\
&= |\mathbf{a}|^2|\mathbf{b}|^2 - (\mathbf{a} \bullet \mathbf{b})^2 \\
&= (a_1^2 + a_2^2 + a_3^2)(b_1^2 + b_2^2 + b_3^2) - (a_1b_1 + a_2b_2 + a_3b_3)^2 \\
&\qquad \text{which on expanding and then factorising gives} \\
&= (a_2b_3 - a_3b_2)^2 + (a_3b_1 - a_1b_3)^2 + (a_1b_2 - a_2b_1)^2 \\
&= |\mathbf{a} \times \mathbf{b}|^2 \\
\therefore\ |\mathbf{a} \times \mathbf{b}| &= |\mathbf{a}||\mathbf{b}|\sin\theta \qquad \{\text{as } \sin\theta > 0\}
\end{aligned}$$

The immediate consequences of this result are:

- If $\mathbf{u}$ is a **unit vector** in the direction of $\mathbf{a} \times \mathbf{b}$ then $\mathbf{a} \times \mathbf{b} = |\mathbf{a}||\mathbf{b}|\sin\theta\,\mathbf{u}$. In some texts this is the **geometric definition** of $\mathbf{a} \times \mathbf{b}$.
- If $\mathbf{a}$ and $\mathbf{b}$ are non-zero vectors, then $\mathbf{a} \times \mathbf{b} = \mathbf{0} \;\Leftrightarrow\; \mathbf{a}$ is parallel to $\mathbf{b}$.

EXERCISE 14K.2

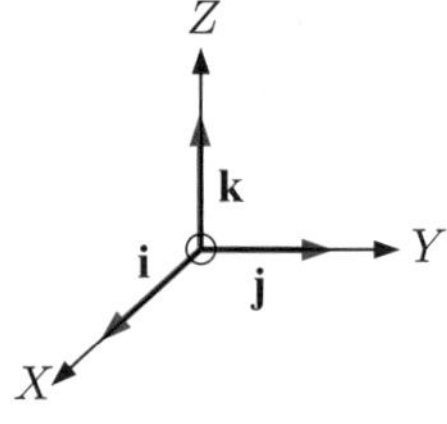

1
  a Find $\mathbf{i} \times \mathbf{k}$ and $\mathbf{k} \times \mathbf{i}$ using the original definition of $\mathbf{a} \times \mathbf{b}$.
  b Check that the **right hand rule** correctly gives the direction of $\mathbf{i} \times \mathbf{k}$ and $\mathbf{k} \times \mathbf{i}$.
  c Check that $\mathbf{a} \times \mathbf{b} = |\mathbf{a}||\mathbf{b}|\sin\theta\,\mathbf{u}$, where $\mathbf{u}$ is the unit vector in the direction of $\mathbf{a} \times \mathbf{b}$, is true for $\mathbf{i} \times \mathbf{k}$ and $\mathbf{k} \times \mathbf{i}$.

2 Consider $\mathbf{a} = \begin{pmatrix} 2 \\ -1 \\ 3 \end{pmatrix}$ and $\mathbf{b} = \begin{pmatrix} 1 \\ 0 \\ -1 \end{pmatrix}$.
  a Find $\mathbf{a} \bullet \mathbf{b}$ and $\mathbf{a} \times \mathbf{b}$.
  b Find $\cos\theta$ using $\mathbf{a} \bullet \mathbf{b} = |\mathbf{a}||\mathbf{b}|\cos\theta$.
  c Find $\sin\theta$ using $\sin^2\theta + \cos^2\theta = 1$.
  d Find $\sin\theta$ using $|\mathbf{a} \times \mathbf{b}| = |\mathbf{a}||\mathbf{b}|\sin\theta$.

3 Prove the property:
"If $\mathbf{a}$ and $\mathbf{b}$ are non-zero vectors then $\mathbf{a} \times \mathbf{b} = \mathbf{0} \Leftrightarrow \mathbf{a}$ is parallel to $\mathbf{b}$."

4 Consider the points A(2, 3, −1) and B(−1, 1, 2).

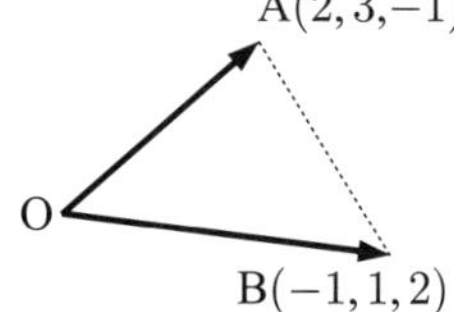

  a Find:
    i $\overrightarrow{OA}$ and $\overrightarrow{OB}$ ii $\overrightarrow{OA} \times \overrightarrow{OB}$ iii $|\overrightarrow{OA} \times \overrightarrow{OB}|$
  b Explain why the area of triangle OAB is $\frac{1}{2}|\overrightarrow{OA} \times \overrightarrow{OB}|$.

5 A, B, and C are 3 distinct points with non-zero position vectors $\mathbf{a}$, $\mathbf{b}$, and $\mathbf{c}$ respectively.
  a If $\mathbf{a} \times \mathbf{c} = \mathbf{b} \times \mathbf{c}$, what can be deduced about $\overrightarrow{OC}$ and $\overrightarrow{AB}$?
  b If $\mathbf{a} + \mathbf{b} + \mathbf{c} = \mathbf{0}$, what relationship exists between $\mathbf{a} \times \mathbf{b}$ and $\mathbf{b} \times \mathbf{c}$?
  c If $\mathbf{c} \neq \mathbf{0}$ and $\mathbf{b} \times \mathbf{c} = \mathbf{c} \times \mathbf{a}$, prove that $\mathbf{a} + \mathbf{b} = k\mathbf{c}$ for some scalar k.

REVIEW SET 14A NON-CALCULATOR

1 Using a scale of 1 cm represents 10 units, sketch a vector to represent:
  a an aeroplane taking off at an angle of 8° to a runway with a speed of 60 m s^{-1}
  b a displacement of 45 m in a north-easterly direction.

2 Simplify: **a** $\overrightarrow{AB} - \overrightarrow{CB}$ **b** $\overrightarrow{AB} + \overrightarrow{BC} - \overrightarrow{DC}$.

3 Construct vector equations for:

a

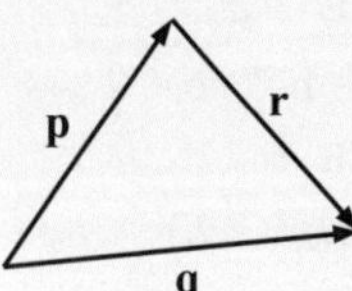

b

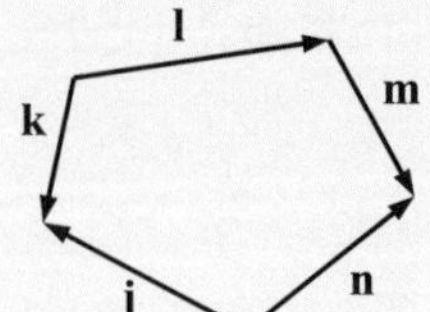

4 If $\overrightarrow{PQ} = \begin{pmatrix} -4 \\ 1 \end{pmatrix}$, $\overrightarrow{RQ} = \begin{pmatrix} -1 \\ 2 \end{pmatrix}$, and $\overrightarrow{RS} = \begin{pmatrix} 2 \\ -3 \end{pmatrix}$, find $\overrightarrow{SP}$.

5

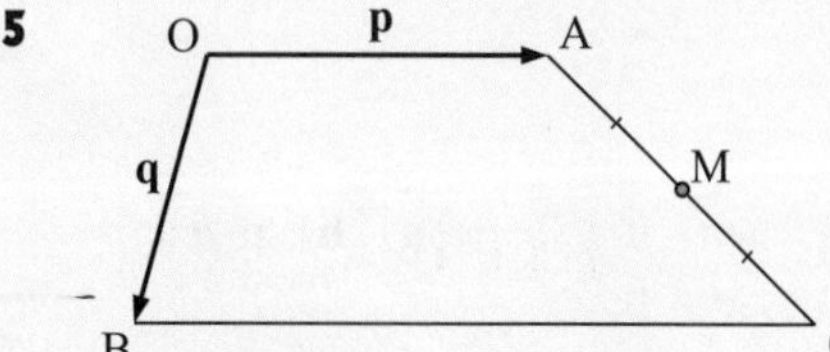

[BC] is parallel to [OA] and is twice its length. Find, in terms of $\mathbf{p}$ and $\mathbf{q}$, vector expressions for:
  a $\overrightarrow{AC}$ **b** $\overrightarrow{OM}$.

6 Find m and n if $\begin{pmatrix} 3 \\ m \\ n \end{pmatrix}$ and $\begin{pmatrix} -12 \\ -20 \\ 2 \end{pmatrix}$ are parallel vectors.

7 If $\overrightarrow{AB} = \begin{pmatrix} 2 \\ -7 \\ 4 \end{pmatrix}$ and $\overrightarrow{AC} = \begin{pmatrix} -6 \\ 1 \\ -3 \end{pmatrix}$, find $\overrightarrow{CB}$.

8 If $\mathbf{p} = \begin{pmatrix} 3 \\ -2 \end{pmatrix}$, $\mathbf{q} = \begin{pmatrix} -1 \\ 5 \end{pmatrix}$, and $\mathbf{r} = \begin{pmatrix} -3 \\ 4 \end{pmatrix}$, find: **a** $\mathbf{p} \bullet \mathbf{q}$ **b** $\mathbf{q} \bullet (\mathbf{p} - \mathbf{r})$

9 Consider points X(−2, 5), Y(3, 4), W(−3, −1), and Z(4, 10). Use vectors to show that WYZX is a parallelogram.

10 Consider points A(2, 3), B(−1, 4), and C(3, k). Find k if $\widehat{BAC}$ is a right angle.

11 Explain why:

a $\mathbf{a} \bullet \mathbf{b} \bullet \mathbf{c}$ is meaningless **b** the expression $\mathbf{a} \bullet \mathbf{b} \times \mathbf{c}$ does not need brackets.

12 Find all vectors which are perpendicular to the vector $\begin{pmatrix} -4 \\ 5 \end{pmatrix}$.

13 In this question you may **not** assume any diagonal properties of parallelograms.

OABC is a parallelogram with $\overrightarrow{OA} = \mathbf{p}$ and $\overrightarrow{OC} = \mathbf{q}$. M is the midpoint of [AC].

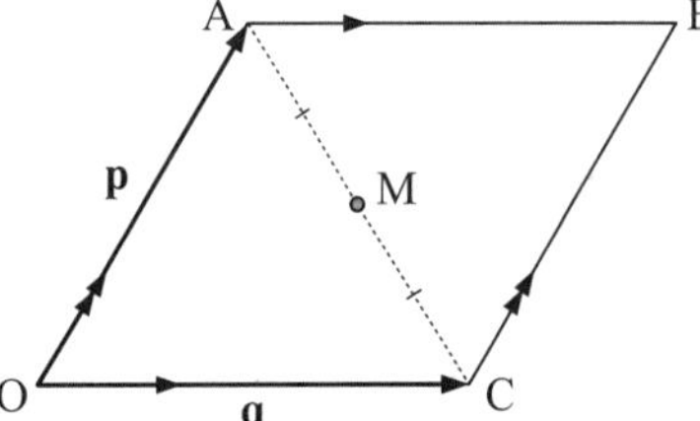

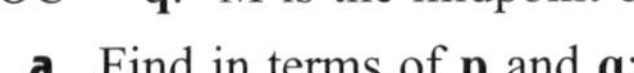

a Find in terms of $\mathbf{p}$ and $\mathbf{q}$:

i $\overrightarrow{OB}$ **ii** $\overrightarrow{OM}$

b Hence show that O, M, and B are collinear, and that M is the midpoint of [OB].

14 Find the values of k such that the following are unit vectors: **a** $\begin{pmatrix} \frac{4}{7} \\ \frac{1}{k} \end{pmatrix}$ **b** $\begin{pmatrix} k \\ k \end{pmatrix}$

15 Suppose $|\mathbf{a}| = 2$, $|\mathbf{b}| = 4$, and $|\mathbf{c}| = 5$.

Find: **a** $\mathbf{a} \bullet \mathbf{b}$

b $\mathbf{b} \bullet \mathbf{c}$

c $\mathbf{a} \bullet \mathbf{c}$

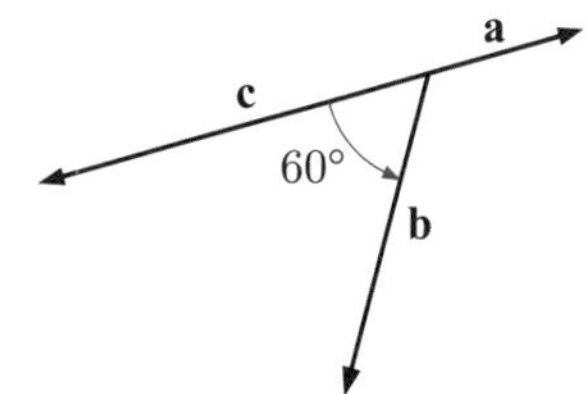

16 Find a and b if J(−4, 1, 3), K(2, −2, 0), and L(a, b, 2) are collinear.

17 Given $|\mathbf{u}| = 3$, $|\mathbf{v}| = 5$, and $\mathbf{u} \times \mathbf{v} = \begin{pmatrix} 1 \\ -3 \\ -4 \end{pmatrix}$, find the possible values of $\mathbf{u} \bullet \mathbf{v}$.

18 [AB] and [CD] are diameters of a circle with centre O.

a If $\overrightarrow{OC} = \mathbf{q}$ and $\overrightarrow{OB} = \mathbf{r}$, find:

i $\overrightarrow{DB}$ in terms of $\mathbf{q}$ and $\mathbf{r}$ **ii** $\overrightarrow{AC}$ in terms of $\mathbf{q}$ and $\mathbf{r}$.

b What can be deduced about [DB] and [AC]?

19 **a** Find t given that $\begin{pmatrix} 2-t \\ 3 \\ t \end{pmatrix}$ and $\begin{pmatrix} t \\ 4 \\ t+1 \end{pmatrix}$ are perpendicular.

b Show that K(4, 3, −1), L(−3, 4, 2), and M(2, 1, −2) are vertices of a right angled triangle.

REVIEW SET 14B — CALCULATOR

1 Copy the given vectors and find geometrically:

a $\mathbf{x} + \mathbf{y}$ **b** $\mathbf{y} - 2\mathbf{x}$

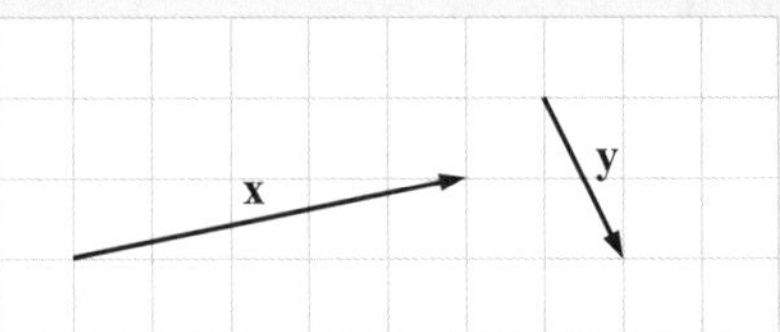

2 Show that A(−2, −1, 3), B(4, 0, −1), and C(−2, 1, −4) are vertices of an isosceles triangle.

3 If $\mathbf{r} = \begin{pmatrix} 4 \\ 1 \end{pmatrix}$ and $\mathbf{s} = \begin{pmatrix} -3 \\ 2 \end{pmatrix}$ find: **a** $|\mathbf{s}|$ **b** $|\mathbf{r} + \mathbf{s}|$ **c** $|2\mathbf{s} - \mathbf{r}|$

4 Find scalars r and s such that $r\begin{pmatrix} -2 \\ 1 \end{pmatrix} + s\begin{pmatrix} 3 \\ -4 \end{pmatrix} = \begin{pmatrix} 13 \\ -24 \end{pmatrix}$.

5 Given P(2, 3, −1) and Q(−4, 4, 2), find:

a $\overrightarrow{PQ}$ **b** the distance between P and Q **c** the midpoint of [PQ].

6 If A(4, 2, −1), B(−1, 5, 2), C(3, −3, c) are vertices of triangle ABC which is right angled at B, find the value of c.

7 Suppose $\mathbf{a} = \begin{pmatrix} 2 \\ -3 \\ 1 \end{pmatrix}$ and $\mathbf{b} = \begin{pmatrix} -1 \\ 2 \\ 3 \end{pmatrix}$. Find $\mathbf{x}$ given $\mathbf{a} - 3\mathbf{x} = \mathbf{b}$.

8 Find the angle between the vectors $\mathbf{a} = 3\mathbf{i} + \mathbf{j} - 2\mathbf{k}$ and $\mathbf{b} = 2\mathbf{i} + 5\mathbf{j} + \mathbf{k}$.

9 Find two points on the Z-axis which are 6 units from P(−4, 2, 5).

10 Determine all possible values of t if $\begin{pmatrix} 3 \\ 3-2t \end{pmatrix}$ and $\begin{pmatrix} t^2+t \\ -2 \end{pmatrix}$ are perpendicular.

11 Prove that P(−6, 8, 2), Q(4, 6, 8), and R(19, 3, 17) are collinear.

12 If $\mathbf{u} = \begin{pmatrix} -4 \\ 2 \\ 1 \end{pmatrix}$ and $\mathbf{v} = \begin{pmatrix} -1 \\ 3 \\ -2 \end{pmatrix}$, find: **a** $\mathbf{u} \bullet \mathbf{v}$ **b** the angle between $\mathbf{u}$ and $\mathbf{v}$.

13 [AP] and [BQ] are altitudes of triangle ABC.

Let $\overrightarrow{OA} = \mathbf{p}$, $\overrightarrow{OB} = \mathbf{q}$, and $\overrightarrow{OC} = \mathbf{r}$.

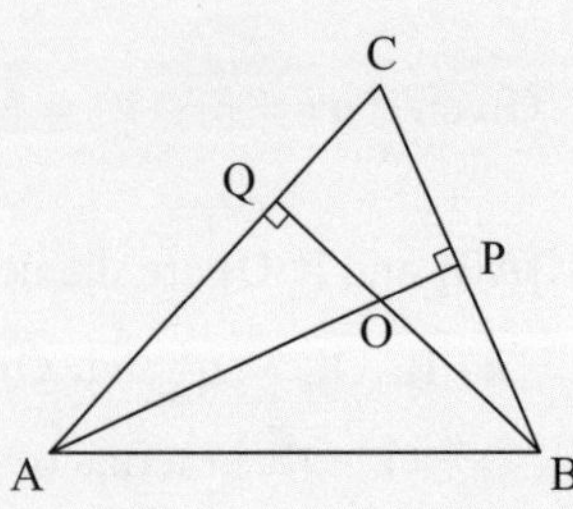

a Find vector expressions for $\overrightarrow{AC}$ and $\overrightarrow{BC}$ in terms of $\mathbf{p}$, $\mathbf{q}$, and $\mathbf{r}$.

b Using the property $\mathbf{a} \bullet (\mathbf{b} - \mathbf{c}) = \mathbf{a} \bullet \mathbf{b} - \mathbf{a} \bullet \mathbf{c}$, deduce that $\mathbf{q} \bullet \mathbf{r} = \mathbf{p} \bullet \mathbf{q} = \mathbf{p} \bullet \mathbf{r}$.

c Hence prove that [OC] is perpendicular to [AB].

14 Find *two* vectors of length 4 units which are parallel to $3\mathbf{i} - 2\mathbf{j} + \mathbf{k}$.

15 Find the measure of $D\widehat{M}C$.

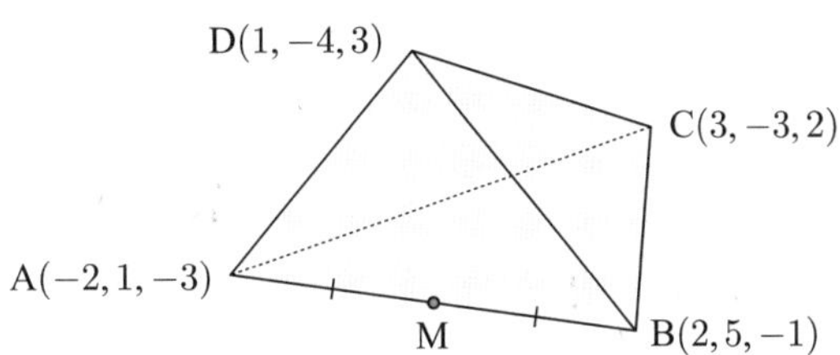

16 Find all vectors perpendicular to both $\begin{pmatrix} 2 \\ -1 \\ -2 \end{pmatrix}$ and $\begin{pmatrix} 1 \\ 1 \\ 2 \end{pmatrix}$.

17 **a** Find k given that $\begin{pmatrix} k \\ \frac{1}{\sqrt{2}} \\ -k \end{pmatrix}$ is a unit vector.

b Find the vector which is 5 units long and has the opposite direction to $\begin{pmatrix} 3 \\ 2 \\ -1 \end{pmatrix}$.

18 Find the angle between $\begin{pmatrix} 2 \\ -4 \\ 3 \end{pmatrix}$ and $\begin{pmatrix} -1 \\ 1 \\ 3 \end{pmatrix}$.

19 Determine the measure of $Q\widehat{D}M$ given that M is the midpoint of [PS].

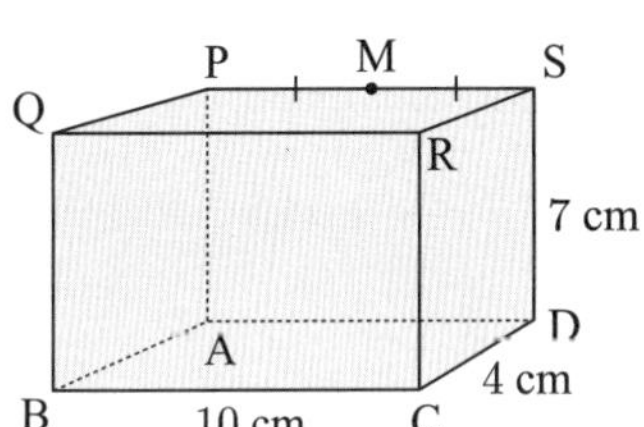

REVIEW SET 14C

1 Find a single vector which is equal to: **a** $\overrightarrow{PR} + \overrightarrow{RQ}$ **b** $\overrightarrow{PS} + \overrightarrow{SQ} + \overrightarrow{QR}$

2 For $\mathbf{m} = \begin{pmatrix} 6 \\ -3 \\ 1 \end{pmatrix}$, $\mathbf{n} = \begin{pmatrix} 2 \\ 3 \\ -4 \end{pmatrix}$, and $\mathbf{p} = \begin{pmatrix} -1 \\ 3 \\ 6 \end{pmatrix}$, find:

a $\mathbf{m} - \mathbf{n} + \mathbf{p}$ **b** $2\mathbf{n} - 3\mathbf{p}$ **c** $|\mathbf{m} + \mathbf{p}|$

3 What geometrical facts can be deduced from the equations:

a $\overrightarrow{AB} = \frac{1}{2}\overrightarrow{CD}$ **b** $\overrightarrow{AB} = 2\overrightarrow{AC}$?

4 Given P(2, −5, 6) and Q(−1, 7, 9), find:

a the position vector of Q relative to P **b** the distance from P to Q

c the distance from P to the X-axis.

5 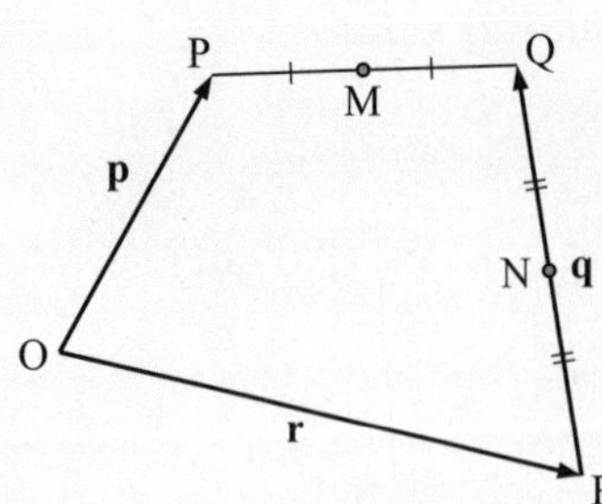

In the figure alongside, $\overrightarrow{OP} = \mathbf{p}$, $\overrightarrow{OR} = \mathbf{r}$, and $\overrightarrow{RQ} = \mathbf{q}$. M and N are the midpoints of [PQ] and [QR] respectively.
Find, in terms of **p**, **q**, and **r**:

a $\overrightarrow{OQ}$ **b** $\overrightarrow{PQ}$ **c** $\overrightarrow{ON}$ **d** $\overrightarrow{MN}$

6 Suppose $\mathbf{p} = \begin{pmatrix} -3 \\ 1 \\ 2 \end{pmatrix}$, $\mathbf{q} = \begin{pmatrix} 2 \\ -4 \\ 1 \end{pmatrix}$, and $\mathbf{r} = \begin{pmatrix} 3 \\ 2 \\ 0 \end{pmatrix}$. Find **x** if: **a** $\mathbf{p} - 3\mathbf{x} = \mathbf{0}$ **b** $2\mathbf{q} - \mathbf{x} = \mathbf{r}$

7 Suppose $|\mathbf{v}| = 3$ and $|\mathbf{w}| = 2$. If **v** is parallel to **w**, what values might $\mathbf{v} \bullet \mathbf{w}$ take?

8 Find a unit vector which is parallel to $\mathbf{i} + r\mathbf{j} + 2\mathbf{k}$ and perpendicular to $2\mathbf{i} + 2\mathbf{j} - \mathbf{k}$.

9 Find t if $\begin{pmatrix} -4 \\ t+2 \\ t \end{pmatrix}$ and $\begin{pmatrix} t \\ 1+t \\ -3 \end{pmatrix}$ are perpendicular vectors.

10 Find all angles of the triangle with vertices K(3, 1, 4), L(−2, 1, 3), and M(4, 1, 3).

11 Find k if the following are unit vectors: **a** $\begin{pmatrix} \frac{5}{13} \\ k \end{pmatrix}$ **b** $\begin{pmatrix} k \\ k \\ k \end{pmatrix}$

12 Use vector methods to find the measure of $\widehat{GAC}$ in the rectangular box alongside.

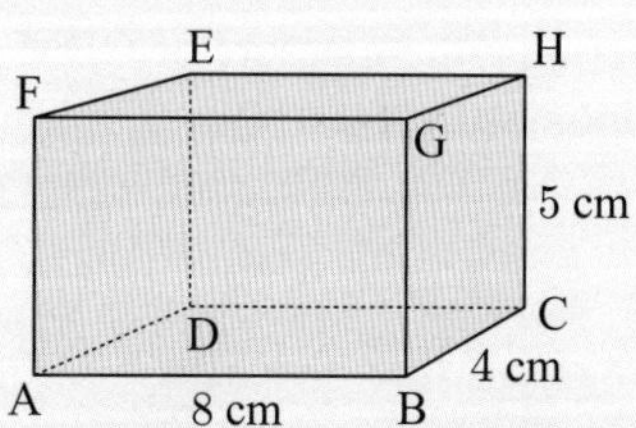

13 Using $\mathbf{p} = \begin{pmatrix} 3 \\ -2 \end{pmatrix}$, $\mathbf{q} = \begin{pmatrix} -2 \\ 5 \end{pmatrix}$, and $\mathbf{r} = \begin{pmatrix} 1 \\ -3 \end{pmatrix}$, verify that:

$\mathbf{p} \bullet (\mathbf{q} - \mathbf{r}) = \mathbf{p} \bullet \mathbf{q} - \mathbf{p} \bullet \mathbf{r}$.

14 P(−1, 2, 3) and Q(4, 0, −1) are two points in space. Find:

a $\overrightarrow{PQ}$ **b** the angle that $\overrightarrow{PQ}$ makes with the X-axis.

15 Suppose $\overrightarrow{OM} = \begin{pmatrix} -2 \\ 4 \end{pmatrix}$, $\overrightarrow{MP} = \begin{pmatrix} 5 \\ -1 \end{pmatrix}$, $\overrightarrow{MP} \bullet \overrightarrow{PT} = 0$, and $|\overrightarrow{MP}| = |\overrightarrow{PT}|$.

Write down the two possible position vectors $\overrightarrow{OT}$.

16 Given $\mathbf{p} = 2\mathbf{i} - \mathbf{j} + 4\mathbf{k}$ and $\mathbf{q} = -\mathbf{i} - 4\mathbf{j} + 2\mathbf{k}$, find the angle between **p** and **q**.

17 Suppose $\mathbf{u} = 2\mathbf{i} + \mathbf{j}$, $\mathbf{v} = 3\mathbf{j}$, and θ is the acute angle between **u** and **v**.
Find the exact value of $\sin\theta$.

18 Find *two* vectors of length 3 units which are perpendicular to both $-\mathbf{i} + 3\mathbf{k}$ and $2\mathbf{i} - \mathbf{j} + \mathbf{k}$.

Chapter 15

Vector applications

Syllabus reference: 4.3, 4.4, 4.5, 4.6, 4.7

Contents:

OPENING PROBLEM

A yacht club is situated at (0, 0) and at 12:00 noon a yacht is at point A(2, 20). The yacht is moving with constant speed in the straight path shown in the diagram. The grid intervals are kilometres.

At 1:00 pm the yacht is at (6, 17).
At 2:00 pm it is at (10, 14).

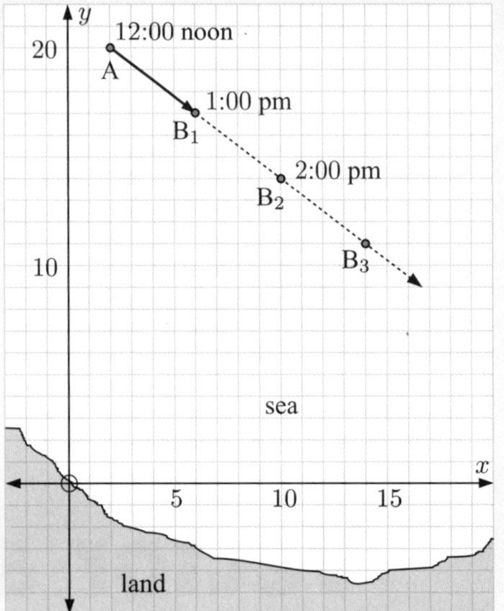

What to do:

a Find the position vectors of:
  i A **ii** B_1 **iii** B_2

b Find $\overrightarrow{AB_1}$. Explain what it means.

c **i** How far does the yacht travel in one hour?
  ii What is its speed?

d Find $\overrightarrow{AB_2}$, $\overrightarrow{AB_3}$, and $\overrightarrow{AB_4}$. Hence write $\overrightarrow{AB_t}$ in terms of $\overrightarrow{AB_1}$.

e What is represented by the vector $\overrightarrow{OB_2}$?

A PROBLEMS INVOLVING VECTOR OPERATIONS

When we apply vectors to problems in the real world, we often consider the combined effect when vectors are added together. This sum is called the **resultant vector**.

We have an example of vector addition when two tug boats are used to pull a ship into port. If the tugs tow with forces $\mathbf{F}_1$ and $\mathbf{F}_2$ then the resultant force is $\mathbf{F}_1 + \mathbf{F}_2$.

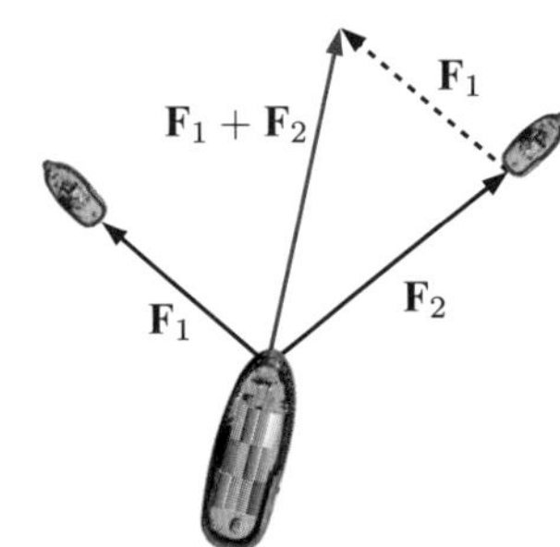
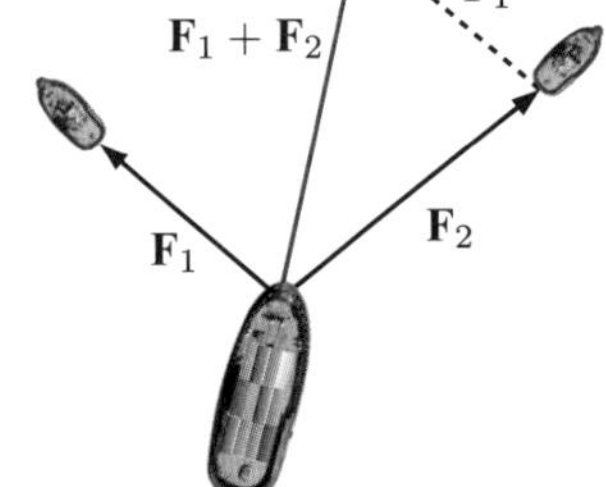

Example 1 — Self Tutor

In still water, Jacques can swim at 1.5 m s^{-1}. Jacques is at point A on the edge of a canal, and considers point B directly opposite. A current is flowing from the left at a constant speed of 0.5 m s^{-1}.

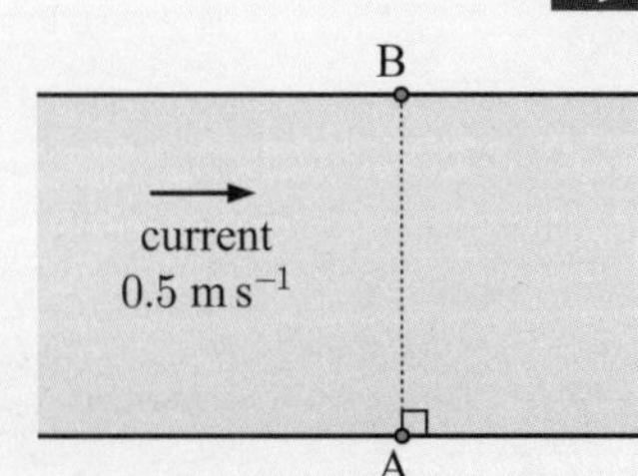

a If Jacques dives in straight towards B, and swims without allowing for the current, what will his actual speed and direction be?

b Jacques wants to swim directly across the canal to point B.

  i At what angle should Jacques *aim* to swim in order that the current will correct his direction?

  ii What will Jacques' actual speed be?

Suppose **c** is the current's velocity vector,
s is the velocity vector Jacques would have if the water was still, and
$\mathbf{f} = \mathbf{c} + \mathbf{s}$ is Jacques' resultant velocity vector.

a Jacques aims directly across the river, but the current takes him downstream to the right.

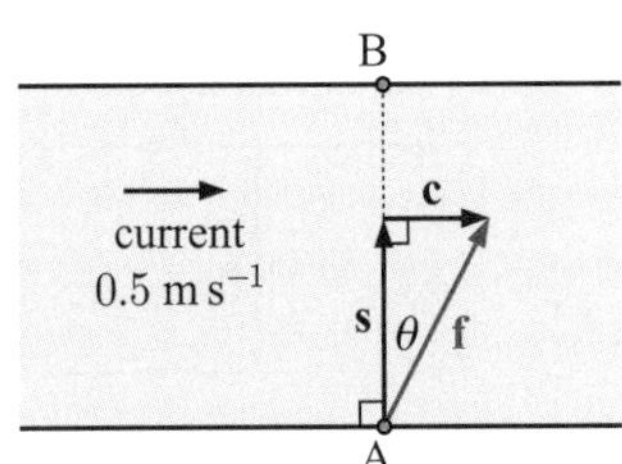

$$|\mathbf{f}|^2 = |\mathbf{c}|^2 + |\mathbf{s}|^2$$
$$= 0.5^2 + 1.5^2$$
$$= 2.5$$
$$\therefore \ |\mathbf{f}| \approx 1.58$$

$$\tan\theta = \frac{0.5}{1.5}$$
$$\therefore \ \theta \approx 18.4°$$

Jacques has an actual speed of approximately 1.58 m s^{-1} and his direction of motion is approximately 18.4° to the right of his intended line.

b Jacques needs to aim to the left of B so the current will correct his direction.

i $\sin\phi = \frac{0.5}{1.5}$

$\therefore \ \phi \approx 19.5°$

Jacques needs to aim approximately 19.5° to the left of B.

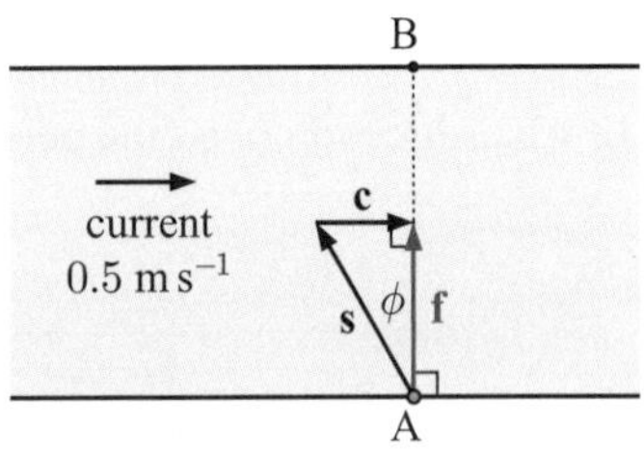

ii $|\mathbf{f}|^2 + |\mathbf{c}|^2 = |\mathbf{s}|^2$

$\therefore \ |\mathbf{f}|^2 + 0.5^2 = 1.5^2$

$\therefore \ |\mathbf{f}|^2 = 2$

$\therefore \ |\mathbf{f}| \approx 1.41$

In these conditions, Jacques' actual speed towards B is approximately 1.41 m s^{-1}.

Another example of vector addition is when an aircraft is affected by wind. A pilot needs to know how to compensate for the wind, especially during take-off and landing.

SIMULATION

EXERCISE 15A

1 An athlete can normally run with constant speed 6 m s^{-1}. Using a vector diagram to illustrate each situation, find the athlete's speed if:

  a he is assisted by a wind of 1 m s^{-1} from directly behind him

  b he runs into a head wind of 1 m s^{-1}.

2 In still water, Mary can swim at 1.2 m s^{-1}. She is standing at point P on the edge of a canal, directly opposite point Q. The water is flowing to the right at a constant speed of 0.6 m s^{-1}.

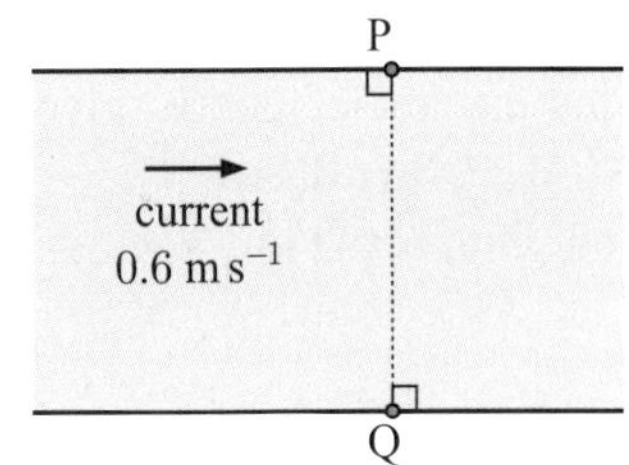

  a If Mary tries to swim directly from P to Q without allowing for the current, what will her actual velocity be?

  b Mary wants to swim directly across the canal to point Q.

    i At what angle should she *aim* to swim in order that the current corrects her direction?

    ii What will Mary's actual speed be?

3 A boat needs to travel south at a speed of 20 km h^{-1}. However a constant current of 6 km h^{-1} is flowing from the south-east. Use vectors to find:

- **a** the equivalent speed in still water for the boat to achieve the actual speed of 20 km h^{-1}
- **b** the direction in which the boat must head to compensate for the current.

4 As part of an endurance race, Stephanie needs to swim from X to Y across a wide river.
Stephanie swims at 1.8 m s^{-1} in still water.
The river flows with a consistent current of 0.3 m s^{-1} as shown.

- **a** Find the distance from X to Y.
- **b** In which direction should Stephanie *aim* so that the current will push her onto a path directly towards Y?

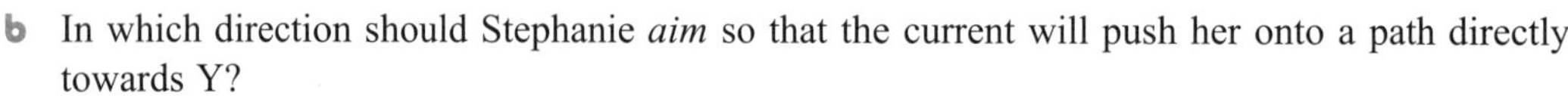

- **c** Find the time Stephanie will take to cross the river.

5 An aeroplane needs to fly due east from one city to another at a speed of 400 km h^{-1}. However, a 50 km h^{-1} wind blows constantly from the north-east.

- **a** How does the wind affect the speed of the aeroplane?
- **b** In what direction must the aeroplane head to compensate for the wind?

B AREA

The vector cross product can be used to find the areas of triangles and parallelograms defined by two vectors.

TRIANGLES

If a triangle has defining vectors **a** and **b** then its area is $\frac{1}{2}|\mathbf{a}\times\mathbf{b}|$ units^2.

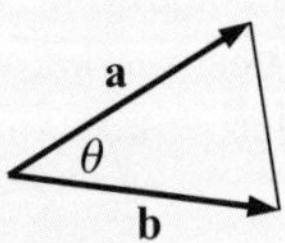

Proof: Area $= \frac{1}{2} \times$ product of two sides $\times$ sine of included angle

$$= \tfrac{1}{2}|\mathbf{a}||\mathbf{b}|\sin\theta$$

$$= \tfrac{1}{2}|\mathbf{a}\times\mathbf{b}|$$

Example 2 **Self Tutor**

Find the area of $\triangle$ABC given A$(-1, 2, 3)$, B$(2, 1, 4)$, and C$(0, 5, -1)$.

$$\overrightarrow{AB}\times\overrightarrow{AC} = \begin{vmatrix} \mathbf{i} & \mathbf{j} & \mathbf{k} \\ 3 & -1 & 1 \\ 1 & 3 & -4 \end{vmatrix}$$

$$= \begin{vmatrix} -1 & 1 \\ 3 & -4 \end{vmatrix}\mathbf{i} - \begin{vmatrix} 3 & 1 \\ 1 & -4 \end{vmatrix}\mathbf{j} + \begin{vmatrix} 3 & -1 \\ 1 & 3 \end{vmatrix}\mathbf{k}$$

$$= \mathbf{i} + 13\mathbf{j} + 10\mathbf{k}$$

$$\therefore \text{ area} = \tfrac{1}{2}|\overrightarrow{AB}\times\overrightarrow{AC}| = \tfrac{1}{2}\sqrt{1+169+100}$$

$$= \tfrac{1}{2}\sqrt{270} \text{ units}^2$$

PARALLELOGRAMS

If a parallelogram has defining vectors $\mathbf{a}$ and $\mathbf{b}$ then its area is $|\mathbf{a} \times \mathbf{b}|$ units^2.

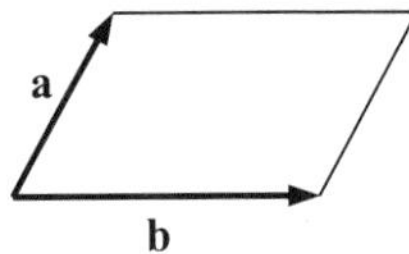

The proof follows directly from that of a triangle, as the parallelogram consists of two congruent triangles with defining vectors $\mathbf{a}$ and $\mathbf{b}$.

EXERCISE 15B

1 Calculate the area of triangle ABC given:
  a A(2, 1, 1), B(4, 3, 0), and C(1, 3, −2)
  b A(0, 0, 0), B(−1, 2, 3), and C(1, 2, 6)
  c A(1, 3, 2), B(2, −1, 0), and C(1, 10, 6)

2 Calculate the area of parallelogram ABCD given A(−1, 2, 2), B(2, −1, 4), and C(0, 1, 0).

3 ABCD is a parallelogram with vertices A(−1, 3, 2), B(2, 0, 4), and C(−1, −2, 5). Find:
  a the coordinates of D
  b the area of the parallelogram.

4 A(−1, 1, 2), B(2, 0, 1), and C(k, 2, −1) are three points in space. Find k if the area of triangle ABC is $\sqrt{88}$ units^2.

5 A, B, and C are three points with position vectors $\mathbf{a}$, $\mathbf{b}$, and $\mathbf{c}$ respectively. Find a formula for S, the total surface area of the tetrahedron OABC.

C LINES IN 2-D AND 3-D

In both 2-D and 3-D geometry we can determine the **equation of a line** using its **direction** and any **fixed point** on the line.

Suppose a line passes through a fixed point A with position vector $\mathbf{a}$, and that the line is parallel to the vector $\mathbf{b}$.

DEMO

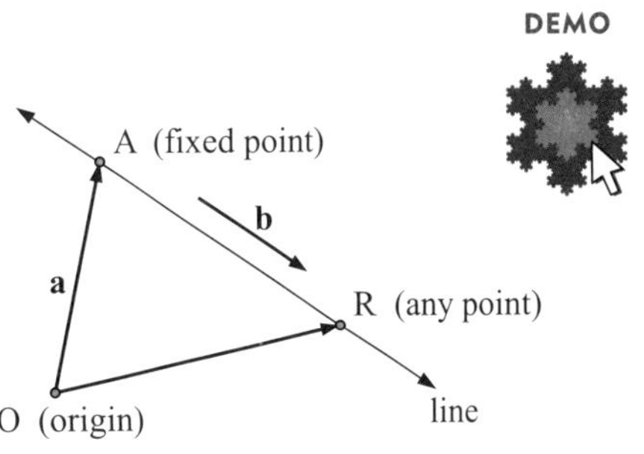

Consider a point R on the line so that $\overrightarrow{OR} = \mathbf{r}$.

By vector addition, $\overrightarrow{OR} = \overrightarrow{OA} + \overrightarrow{AR}$
$\therefore \ \mathbf{r} = \mathbf{a} + \overrightarrow{AR}$.

Since $\overrightarrow{AR}$ is parallel to $\mathbf{b}$,

$\overrightarrow{AR} = \lambda\mathbf{b}$ for some scalar $\lambda \in \mathbb{R}$
$\therefore \ \mathbf{r} = \mathbf{a} + \lambda\mathbf{b}$

So, $\mathbf{r} = \mathbf{a} + \lambda\mathbf{b}, \ \lambda \in \mathbb{R}$ is the **vector equation** of the line.

LINES IN 2-D

- In 2-D we are dealing with a **line in a plane**.
- $\begin{pmatrix} x \\ y \end{pmatrix} = \begin{pmatrix} a_1 \\ a_2 \end{pmatrix} + \lambda \begin{pmatrix} b_1 \\ b_2 \end{pmatrix}$ is the **vector equation** of the line

  where $R(x, y)$ is any point on the line,

  $A(a_1, a_2)$ is a known fixed point on the line,

  and $\mathbf{b} = \begin{pmatrix} b_1 \\ b_2 \end{pmatrix}$ is the **direction vector** of the line.

- The gradient of the line is $m = \dfrac{b_2}{b_1}$.
- Since $\begin{pmatrix} x \\ y \end{pmatrix} = \begin{pmatrix} a_1 + \lambda b_1 \\ a_2 + \lambda b_2 \end{pmatrix}$, we can write the **parametric equations** of the line as

  $\begin{cases} x = a_1 + \lambda b_1 \\ y = a_2 + \lambda b_2 \end{cases}$ where $\lambda \in \mathbb{R}$ is the **parameter**.

  Each point on the line corresponds to exactly one value of λ.

- We can convert these equations into Cartesian form by equating λ values.

  Using $\lambda = \dfrac{x - a_1}{b_1} = \dfrac{y - a_2}{b_2}$ we obtain $b_2x - b_1y = b_2a_1 - b_1a_2$

  which is the **Cartesian equation** of the line.

Example 3 — Self Tutor

A line passes through the point $A(1, 5)$ and has direction vector $\begin{pmatrix} 3 \\ 2 \end{pmatrix}$.

Describe the line using:

a a vector equation **b** parametric equations **c** a Cartesian equation.

a The vector equation is $\mathbf{r} = \mathbf{a} + \lambda\mathbf{b}$ where

$\mathbf{a} = \overrightarrow{OA} = \begin{pmatrix} 1 \\ 5 \end{pmatrix}$ and $\mathbf{b} = \begin{pmatrix} 3 \\ 2 \end{pmatrix}$

$\therefore \begin{pmatrix} x \\ y \end{pmatrix} = \begin{pmatrix} 1 \\ 5 \end{pmatrix} + \lambda \begin{pmatrix} 3 \\ 2 \end{pmatrix}, \quad \lambda \in \mathbb{R}$

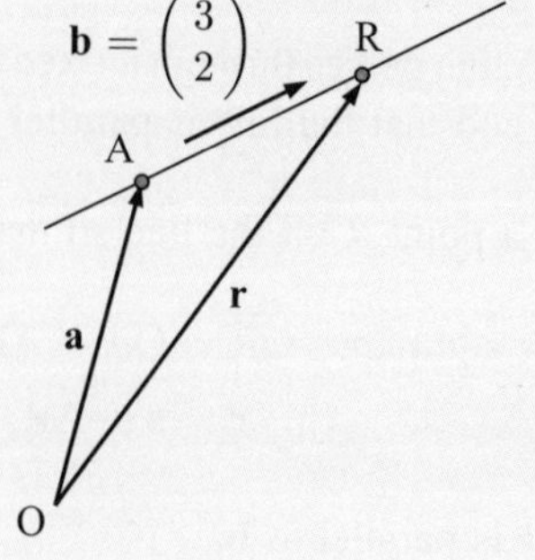

b $x = 1 + 3\lambda$ and $y = 5 + 2\lambda, \quad \lambda \in \mathbb{R}$

c Now $\lambda = \dfrac{x-1}{3} = \dfrac{y-5}{2}$

$\therefore 2x - 2 = 3y - 15$

$\therefore 2x - 3y = -13$

LINES IN 3-D

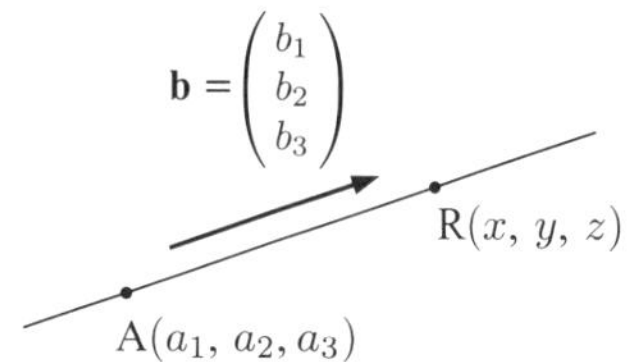

- In 3-D we are dealing with a **line in space**.
- $\begin{pmatrix} x \\ y \\ z \end{pmatrix} = \begin{pmatrix} a_1 \\ a_2 \\ a_3 \end{pmatrix} + \lambda \begin{pmatrix} b_1 \\ b_2 \\ b_3 \end{pmatrix}$ is the **vector equation** of the line

  where $\text{R}(x,\ y,\ z)$ is any point on the line,
  $\text{A}(a_1,\ a_2,\ a_3)$ is the known or fixed point on the line,
  and $\mathbf{b} = \begin{pmatrix} b_1 \\ b_2 \\ b_3 \end{pmatrix}$ is the **direction vector** of the line.

- The **parametric equations** of the line are:

  $x = a_1 + \lambda b_1$
  $y = a_2 + \lambda b_2$ where $\lambda \in \mathbb{R}$ is called the **parameter**.
  $z = a_3 + \lambda b_3$

  Each point on the line corresponds to exactly one value of λ.

- By equating λ values, we obtain the **Cartesian equations** of the line $\lambda = \dfrac{x - a_1}{b_1} = \dfrac{y - a_2}{b_2} = \dfrac{z - a_3}{b_3}$.

We do not talk about the **gradient** of a line in 3-D.
We describe its direction only by its **direction vector**.

Example 4 — Self Tutor

Find a vector equation and the corresponding parametric equations of the line through $(1, -2, 3)$ in the direction $4\mathbf{i} + 5\mathbf{j} - 6\mathbf{k}$.

The vector equation is $\mathbf{r} = \mathbf{a} + \lambda\mathbf{b}$

$$\therefore \begin{pmatrix} x \\ y \\ z \end{pmatrix} = \begin{pmatrix} 1 \\ -2 \\ 3 \end{pmatrix} + \lambda \begin{pmatrix} 4 \\ 5 \\ -6 \end{pmatrix}, \quad \lambda \in \mathbb{R}.$$

The parametric equations are:

$$x = 1 + 4\lambda, \quad y = -2 + 5\lambda, \quad z = 3 - 6\lambda, \quad \lambda \in \mathbb{R}.$$

NON-UNIQUENESS OF THE VECTOR EQUATION OF A LINE

Consider the line passing through $(5, 4)$ and $(7, 3)$. When writing the equation of the line, we could use either point to give the position vector $\mathbf{a}$.

Similarly, we could use the direction vector $\begin{pmatrix} 2 \\ -1 \end{pmatrix}$, but we could also use $\begin{pmatrix} -2 \\ 1 \end{pmatrix}$ or indeed any non-zero scalar multiple of these vectors.

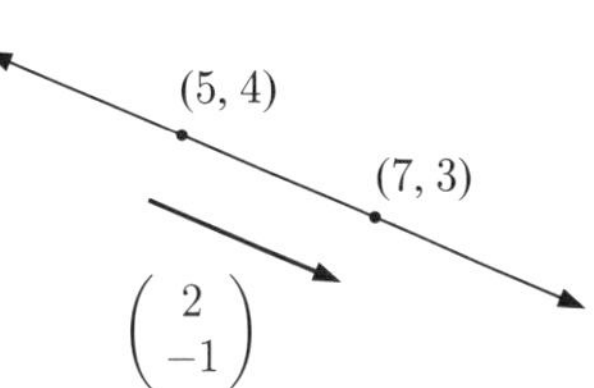

We could thus write the equation of the line as

$$\mathbf{x} = \begin{pmatrix} 5 \\ 4 \end{pmatrix} + \lambda \begin{pmatrix} 2 \\ -1 \end{pmatrix}, \quad \lambda \in \mathbb{R} \quad \text{or} \quad \mathbf{x} = \begin{pmatrix} 7 \\ 3 \end{pmatrix} + \mu \begin{pmatrix} -2 \\ 1 \end{pmatrix}, \quad \mu \in \mathbb{R} \quad \text{and so on.}$$

Notice how we use different parameters λ and μ when we write these equations. This is because the parameters are clearly not the same: when $\lambda = 0$, we have the point (5, 4)
when $\mu = 0$, we have the point (7, 3).

In fact, the parameters are related by $\mu = 1 - \lambda$.

Example 5 **Self Tutor**

Find parametric equations of the line through A(2, –1, 4) and B(–1, 0, 2).

We require a direction vector for the line, either $\overrightarrow{AB}$ or $\overrightarrow{BA}$.

$$\overrightarrow{AB} = \begin{pmatrix} -1-2 \\ 0--1 \\ 2-4 \end{pmatrix} = \begin{pmatrix} -3 \\ 1 \\ -2 \end{pmatrix}$$

Using point A, the equations are: $x = 2 - 3\lambda,\ y = -1 + \lambda,\ z = 4 - 2\lambda,\ \lambda \in \mathbb{R}$

or using the point B, the equations are: $x = -1 - 3\mu,\ y = \mu,\ z = 2 - 2\mu,\ \mu \in \mathbb{R}$.

EXERCISE 15C

1 Describe each of the following lines using:

i a vector equation **ii** parametric equations **iii** a Cartesian equation

a a line with direction $\begin{pmatrix} 1 \\ 4 \end{pmatrix}$ which passes through (3, –4)

b a line passing through (5, 2) which is perpendicular to $\begin{pmatrix} 5 \\ 2 \end{pmatrix}$

c a line parallel to $3\mathbf{i} + 7\mathbf{j}$ which cuts the x-axis at –6

d a line passing through (–1, 11) and (–3, 12).

2 A line passes through (–1, 4) with direction vector $\begin{pmatrix} 2 \\ -1 \end{pmatrix}$.

a Write parametric equations for the line using the parameter λ.

b Find the points on the line for which $\lambda = 0, 1, 3, -1$, and -4.

3 **a** Does (3, –2) lie on the line with vector equation $\mathbf{r} = \begin{pmatrix} 2 \\ 1 \end{pmatrix} + \lambda \begin{pmatrix} 1 \\ -3 \end{pmatrix}$?

b $(k, 4)$ lies on the line with parametric equations $x = 1 - 2\lambda,\ y = 1 + \lambda$. Find k.

4 Line L has vector equation $\mathbf{r} = \begin{pmatrix} 1 \\ 5 \end{pmatrix} + \lambda \begin{pmatrix} -1 \\ 3 \end{pmatrix}$.

a Locate the point on the line corresponding to $\lambda = 1$.

b Explain why the direction of the line could also be described by $\begin{pmatrix} 1 \\ -3 \end{pmatrix}$.

c Use your answers to **a** and **b** to write an alternative vector equation for line L.

5 Describe each of the following lines using:

i a vector equation **ii** parametric equations **iii** a Cartesian equation.

a a line parallel to $\begin{pmatrix} 2 \\ 1 \\ 3 \end{pmatrix}$ which passes through $(1, 3, -7)$

b a line which passes through $(0, 1, 2)$ with direction vector $\mathbf{i} + \mathbf{j} - 2\mathbf{k}$

c a line parallel to the X-axis which passes through $(-2, 2, 1)$

d a line parallel to $2\mathbf{i} - \mathbf{j} + 3\mathbf{k}$ which passes through $(0, 2, -1)$

e a line perpendicular to the XOY plane which passes through $(3, 2, -1)$.

6 Find the vector equation of the line which passes through:

a $\mathrm{A}(1, 2, 1)$ and $\mathrm{B}(-1, 3, 2)$ **b** $\mathrm{C}(0, 1, 3)$ and $\mathrm{D}(3, 1, -1)$

c $\mathrm{E}(1, 2, 5)$ and $\mathrm{F}(1, -1, 5)$ **d** $\mathrm{G}(0, 1, -1)$ and $\mathrm{H}(5, -1, 3)$

7 Find the direction vector of the line:

a $\begin{pmatrix} 4 \\ -1 \\ 1 \end{pmatrix} + \lambda \begin{pmatrix} -2 \\ 0 \\ 3 \end{pmatrix}$ **b** $x = 5 - t, \quad y = 1 + t, \quad z = 2 - 3t$

c $\dfrac{x-2}{3} = \dfrac{y+1}{2} = z - 1$ **d** $\dfrac{1-x}{2} = \dfrac{y}{4} = \dfrac{z-3}{3}$

8 Find the coordinates of the point where the line with parametric equations $x = 1 - \lambda$, $y = 3 + \lambda$, and $z = 3 - 2\lambda$ meets:

a the XOY plane **b** the YOZ plane **c** the XOZ plane.

9 The parametric equations of a line are $x = x_0 + \lambda l$, $y = y_0 + \lambda m$, $z = z_0 + \lambda n$ for $\lambda \in \mathbb{R}$.

a Find the coordinates of the point on the line corresponding to $\lambda = 0$.

b Find the direction vector of the line.

c Write the Cartesian equation for the line.

10 Find points on the line with parametric equations $x = 2 - \lambda$, $y = 3 + 2\lambda$, and $z = 1 + \lambda$ which are $5\sqrt{3}$ units from the point $(1, 0, -2)$.

11 The perpendicular from a point to a line minimises the distance from the point to that line. Use quadratic theory to find the coordinates of the foot of the perpendicular:

a from $(1, 1, 2)$ to the line with equations $x = 1 + \lambda$, $y = 2 - \lambda$, $z = 3 + \lambda$

b from $(2, 1, 3)$ to the line with vector equation $\begin{pmatrix} x \\ y \\ z \end{pmatrix} = \begin{pmatrix} 1 \\ 2 \\ 0 \end{pmatrix} + \mu \begin{pmatrix} 1 \\ -1 \\ 2 \end{pmatrix}$.

D THE ANGLE BETWEEN TWO LINES

In **Chapter 14** we saw that the angle between two vectors is measured in the range $0° \leqslant \theta \leqslant 180°$. We used the formula $\cos\theta = \dfrac{\mathbf{b}_1 \bullet \mathbf{b}_2}{|\mathbf{b}_1||\mathbf{b}_2|}$.

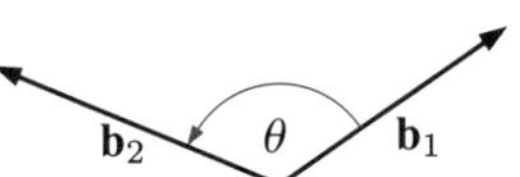

In the case of lines which continue infinitely in both directions, we agree to talk about the *acute* angle between them. For an acute angle, $\cos\theta > 0$, so we use the formula

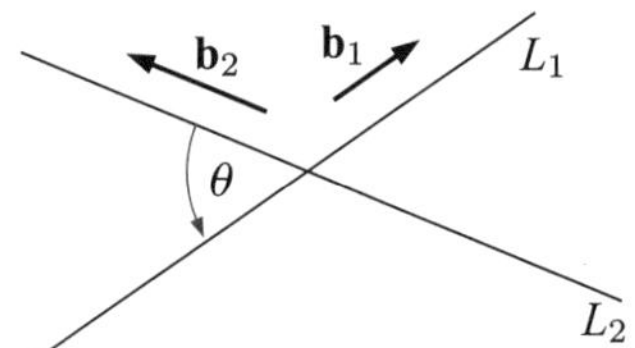

$$\cos\theta = \frac{|\mathbf{b}_1 \bullet \mathbf{b}_2|}{|\mathbf{b}_1||\mathbf{b}_2|}$$

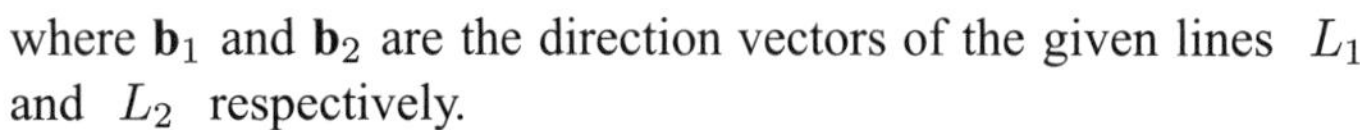

where $\mathbf{b}_1$ and $\mathbf{b}_2$ are the direction vectors of the given lines L_1 and L_2 respectively.

Example 6 **Self Tutor**

Find the angle between the lines L_1: $x = 2 - 3t$, $y = -1 + t$ and L_2: $x = 1 + 2s$, $y = -4 + 3s$.

$$\mathbf{b}_1 = \begin{pmatrix} -3 \\ 1 \end{pmatrix}, \quad \mathbf{b}_2 = \begin{pmatrix} 2 \\ 3 \end{pmatrix}$$

$$\therefore \cos\theta = \frac{|-6+3|}{\sqrt{10}\sqrt{13}}$$

$$\therefore \cos\theta \approx 0.2631$$

$$\therefore \theta \approx 74.7° \quad (1.30 \text{ radians})$$

Consider two lines $\mathbf{r}_1 = \mathbf{a}_1 + \lambda\mathbf{b}_1$ and $\mathbf{r}_2 = \mathbf{a}_2 + \mu\mathbf{b}_2$.

- $\mathbf{r}_1$ and $\mathbf{r}_2$ are **parallel** if $\mathbf{b}_1 = k\mathbf{b}_2$ for some scalar k. We write $\mathbf{r}_1 \parallel \mathbf{r}_2$.
- $\mathbf{r}_1$ and $\mathbf{r}_2$ are **perpendicular** if $\mathbf{b}_1 \bullet \mathbf{b}_2 = 0$. We write $\mathbf{r}_1 \perp \mathbf{r}_2$.

Example 7 **Self Tutor**

Find the angle between the lines:

$$L_1\text{: } \mathbf{r}_1 = \begin{pmatrix} 2 \\ -1 \\ 4 \end{pmatrix} + \lambda \begin{pmatrix} -3 \\ 1 \\ -2 \end{pmatrix} \quad \text{and} \quad L_2\text{: } \frac{1-x}{3} = y = \frac{2-z}{2}.$$

$$\mathbf{b}_1 = \begin{pmatrix} -3 \\ 1 \\ -2 \end{pmatrix} \quad \text{and} \quad \mathbf{b}_2 = \begin{pmatrix} -3 \\ 1 \\ -2 \end{pmatrix}$$

Since $\mathbf{b}_1 = \mathbf{b}_2$, the lines are parallel and the angle between them is $0°$.

EXERCISE 15D

1 Find the angle between the lines L_1: $x = -4 + 12t$, $y = 3 + 5t$
and L_2: $x = 3s$, $y = -6 - 4s$

2 Consider the lines L_1: $x = 2 + 5p$, $y = 19 - 2p$
L_2: $x = 3 + 4r$, $y = 7 + 10r$

Show that the lines are perpendicular.

3 The line L_1 passes through $(-6, 3)$ and is parallel to $\begin{pmatrix} 4 \\ -3 \end{pmatrix}$.

The line L_2 has direction $5\mathbf{i} + 4\mathbf{j}$ and cuts the y-axis at $(0, 8)$.
Find the acute angle between the lines.

4 **a** Find the angle between the lines:

$$L_1: \quad \frac{x-8}{3} = \frac{9-y}{16} = \frac{z-10}{7} \quad \text{and} \quad L_2: \quad \mathbf{r}_2 = \begin{pmatrix} 15 \\ 29 \\ 5 \end{pmatrix} + t\begin{pmatrix} 3 \\ 8 \\ -5 \end{pmatrix}.$$

b A third line L_3 is perpendicular to L_1 and has direction vector $\begin{pmatrix} 0 \\ -3 \\ x \end{pmatrix}$.

Find the value of x.

Example 8 **Self Tutor**

Find the measure of the acute angle between the lines $2x + y = 5$ and $3x - 2y = 8$.

$2x + y = 5$ has gradient $-\frac{2}{1}$ and $\therefore$ direction vector $\begin{pmatrix} 1 \\ -2 \end{pmatrix}$ which we call $\mathbf{a}$.

$3x - 2y = 8$ has gradient $\frac{3}{2}$ and $\therefore$ direction vector $\begin{pmatrix} 2 \\ 3 \end{pmatrix}$ which we call $\mathbf{b}$.

If the angle between the lines is θ, then

$$\cos\theta = \frac{|\mathbf{a} \bullet \mathbf{b}|}{|\mathbf{a}|\,|\mathbf{b}|}$$
$$= \frac{|(1 \times 2) + (-2 \times 3)|}{\sqrt{1+4}\sqrt{4+9}}$$
$$= \frac{4}{\sqrt{5}\sqrt{13}}$$

$$\therefore \quad \theta = \arccos\left(\frac{4}{\sqrt{65}}\right) \approx 60.3^\circ$$

If a line has gradient $\frac{b}{a}$, it has direction vector $\begin{pmatrix} a \\ b \end{pmatrix}$.

5 Find the measure of the angle between the lines:

a $x - y = 3$ and $3x + 2y = 11$
b $y = x + 2$ and $y = 1 - 3x$
c $y + x = 7$ and $x - 3y + 2 = 0$
d $y = 2 - x$ and $x - 2y = 7$

E CONSTANT VELOCITY PROBLEMS

Consider again the yacht in the **Opening Problem**.

The **initial position** of the yacht is given by the position vector $\mathbf{a} = \begin{pmatrix} 2 \\ 20 \end{pmatrix}$.

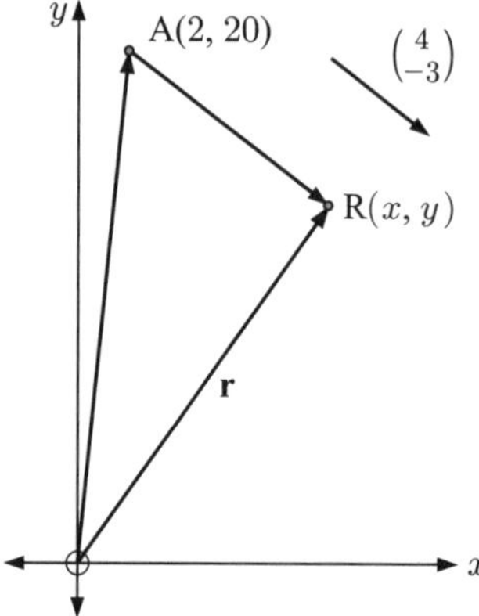

In the hour from 1:00 pm until 2:00 pm, the yacht travels $\begin{pmatrix} 4 \\ -3 \end{pmatrix}$.

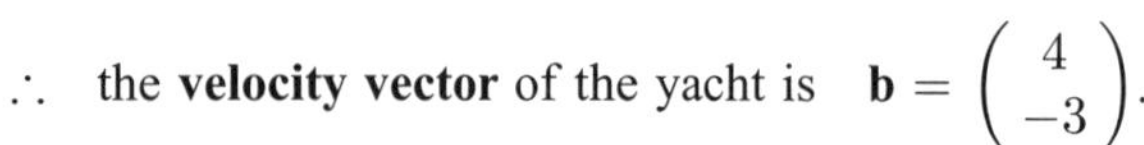

$\therefore$ the **velocity vector** of the yacht is $\mathbf{b} = \begin{pmatrix} 4 \\ -3 \end{pmatrix}$.

Suppose that t hours after leaving A, the yacht is at R(x, y).

$$\overrightarrow{OR} = \overrightarrow{OA} + \overrightarrow{AR}$$

$$\therefore \mathbf{r} = \begin{pmatrix} 2 \\ 20 \end{pmatrix} + t\begin{pmatrix} 4 \\ -3 \end{pmatrix} \quad \text{for } t \geqslant 0$$

$$\therefore \begin{pmatrix} x \\ y \end{pmatrix} = \begin{pmatrix} 2 \\ 20 \end{pmatrix} + t\begin{pmatrix} 4 \\ -3 \end{pmatrix}$$ is the **vector equation** of the yacht's path.

In velocity problems the parameter corresponds to time.

If an object has initial position vector **a** and moves with constant velocity **b**, its position at time t is given by

$$\mathbf{r} = \mathbf{a} + t\mathbf{b} \quad \text{for } t \geqslant 0.$$

The **speed** of the object is $|\mathbf{b}|$.

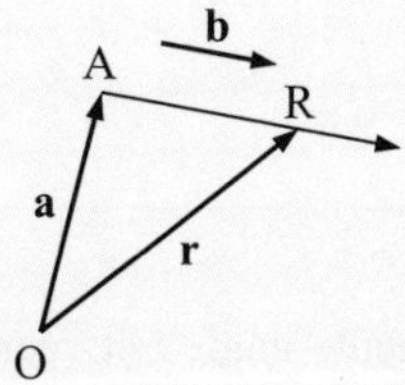

Example 9 — Self Tutor

$\begin{pmatrix} x \\ y \end{pmatrix} = \begin{pmatrix} 1 \\ 9 \end{pmatrix} + t\begin{pmatrix} 3 \\ -4 \end{pmatrix}$ is the vector equation of the path of an object.

The time t is in seconds, $t \geqslant 0$. The distance units are metres.

a Find the object's initial position.

b Plot the path of the object for $t = 0, 1, 2, 3$.

c Find the velocity vector of the object.

d Find the object's speed.

e If the object continues in the same direction but increases its speed to 30 m s^{-1}, state its new velocity vector.

---

a At $t = 0$, $\begin{pmatrix} x \\ y \end{pmatrix} = \begin{pmatrix} 1 \\ 9 \end{pmatrix}$

$\therefore$ the object is at (1, 9).

b

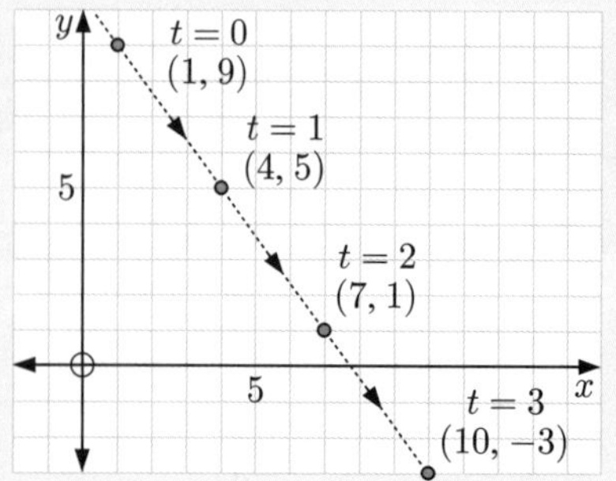

c The velocity vector is $\begin{pmatrix} 3 \\ -4 \end{pmatrix}$.

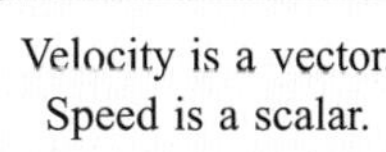

d The speed is $\left|\begin{pmatrix} 3 \\ -4 \end{pmatrix}\right| = \sqrt{3^2 + (-4)^2}$
$= 5 \text{ m s}^{-1}$.

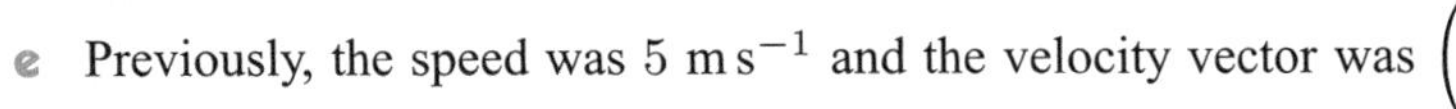

e Previously, the speed was 5 m s^{-1} and the velocity vector was $\begin{pmatrix} 3 \\ -4 \end{pmatrix}$.

$\therefore$ the new velocity vector is $6\begin{pmatrix} 3 \\ -4 \end{pmatrix} = \begin{pmatrix} 18 \\ -24 \end{pmatrix}$.

Example 10 **Self Tutor**

An object is initially at (5, 10) and moves with velocity vector $3\mathbf{i} - \mathbf{j}$ metres per minute. Find:

a the position of the object at time t minutes
b the speed of the object
c the position of the object at $t = 3$ minutes
d the time when the object is due east of (0, 0).

a

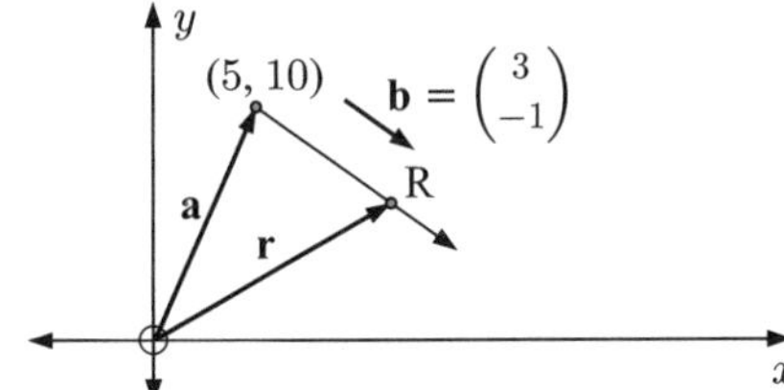

$$\mathbf{r} = \mathbf{a} + t\mathbf{b}$$

$$\therefore \begin{pmatrix} x \\ y \end{pmatrix} = \begin{pmatrix} 5 \\ 10 \end{pmatrix} + t\begin{pmatrix} 3 \\ -1 \end{pmatrix}, \quad t \in \mathbb{R}$$

$$\therefore \begin{pmatrix} x \\ y \end{pmatrix} = \begin{pmatrix} 5 + 3t \\ 10 - t \end{pmatrix}$$

After t minutes, the object is at $(5+3t, 10-t)$.

b The speed of the object is $|\mathbf{b}| = \sqrt{3^2 + (-1)^2} = \sqrt{10}$ metres per minute.

c At $t = 3$ minutes, $5 + 3t = 14$ and $10 - t = 7$. The object is at (14, 7).

d When the object is due east of (0, 0), y must be zero.

$\therefore 10 - t = 0$

$\therefore t = 10$

The object is due east of (0, 0) after 10 minutes.

EXERCISE 15E

1 A particle at $P(x(t), y(t))$ moves such that $x(t) = 1 + 2t$ and $y(t) = 2 - 5t$, $t \geqslant 0$. The distances are in centimetres and t is in seconds.

a Find the initial position of P.
b Illustrate the initial part of the motion of P where $t = 0, 1, 2, 3$.
c Find the velocity vector of P.
d Find the speed of P.

2 **a** Find the vector equation of a boat initially at (2, 3), which travels with velocity vector $\begin{pmatrix} 4 \\ -5 \end{pmatrix}$. The grid units are kilometres and the time is in hours.

b Locate the boat's position after 90 minutes.

c How long will it take for the boat to reach the point $(5, -0.75)$?

3 A remote controlled toy car is initially at the point $(-3, -2)$. It moves with constant velocity $2\mathbf{i} + 4\mathbf{j}$. The distance units are centimetres, and the time is in seconds.

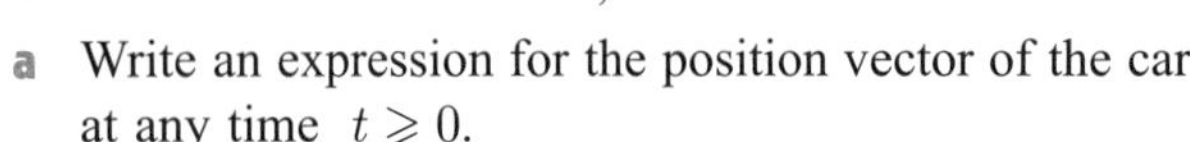

a Write an expression for the position vector of the car at any time $t \geqslant 0$.

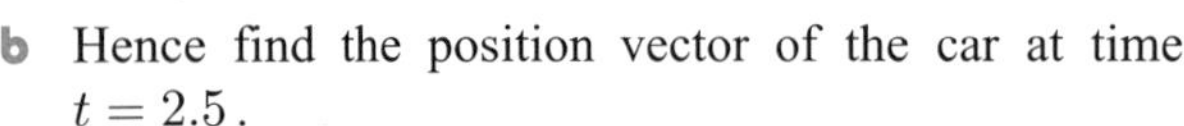

b Hence find the position vector of the car at time $t = 2.5$.

c Find when the car is **i** due north **ii** due west of the observation point (0, 0).

d Plot the car's positions at times $t = 0, \frac{1}{2}, 1, 1\frac{1}{2}, 2, 2\frac{1}{2}, ...$

4 Each of the following vector equations represents the path of a moving object. t is measured in seconds and $t \geqslant 0$. Distances are measured in metres. In each case, find:

i the initial position **ii** the velocity vector **iii** the speed of the object.

a $\begin{pmatrix} x \\ y \end{pmatrix} = \begin{pmatrix} -4 \\ 3 \end{pmatrix} + t\begin{pmatrix} 12 \\ 5 \end{pmatrix}$

b $x = 3 + 2t, \quad y = -t, \quad z = 4 - 2t$

5 Find the velocity vector of a speed boat moving parallel to:

a $\begin{pmatrix} 4 \\ -3 \end{pmatrix}$ with a speed of 150 km h^{-1}

b $2\mathbf{i} + \mathbf{j}$ with a speed of 50 km h^{-1}.

6 Find the velocity vector of a swooping eagle moving in the direction $-2\mathbf{i} + 5\mathbf{j} - 14\mathbf{k}$ with a speed of 90 km h^{-1}.

7 Yacht A moves according to $x(t) = 4 + t, \ y(t) = 5 - 2t$ where the distance units are kilometres and the time units are hours. Yacht B moves according to $x(t) = 1 + 2t, \ y(t) = -8 + t, \ t \geqslant 0$.

a Find the initial position of each yacht.

b Find the velocity vector of each yacht.

c Show that the speed of each yacht is constant, and state these speeds.

d Verify algebraically that the paths of the yachts are at right angles to each other.

8 Submarine P is at $(-5, 4)$ and fires a torpedo with velocity vector $\begin{pmatrix} 3 \\ -1 \end{pmatrix}$ at 1:34 pm.

Submarine Q is at (15, 7) and a minutes later fires a torpedo with velocity vector $\begin{pmatrix} -4 \\ -3 \end{pmatrix}$.

Distances are measured in kilometres and time is in minutes.

a Show that the position of P's torpedo can be written as $P(x_1(t), \ y_1(t))$ where $x_1(t) = -5 + 3t$ and $y_1(t) = 4 - t$.

b What is the speed of P's torpedo?

c Show that the position of Q's torpedo can be written as $Q(x_2(t), \ y_2(t))$ where $x_2(t) = 15 - 4(t - a)$ and $y_2(t) = 7 - 3(t - a)$.

d Q's torpedo is successful in knocking out P's torpedo. At what time did Q fire its torpedo and at what time did the explosion occur?

9 A helicopter at $A(6,\ 9,\ 3)$ moves with constant velocity in a straight line. 10 minutes later it is at $B(3,\ 10,\ 2.5)$. Distances are in kilometres.

a Find $\overrightarrow{AB}$.

b Find the helicopter's speed.

c Determine the equation of the straight line path of the helicopter.

d The helicopter is travelling directly towards its helipad, which has z-coordinate 0. Find the total time taken for the helicopter to land.

F THE SHORTEST DISTANCE FROM A LINE TO A POINT

A ship R sails through point A in the direction **b** and continues past a port P. At what time will the ship be closest to the port?

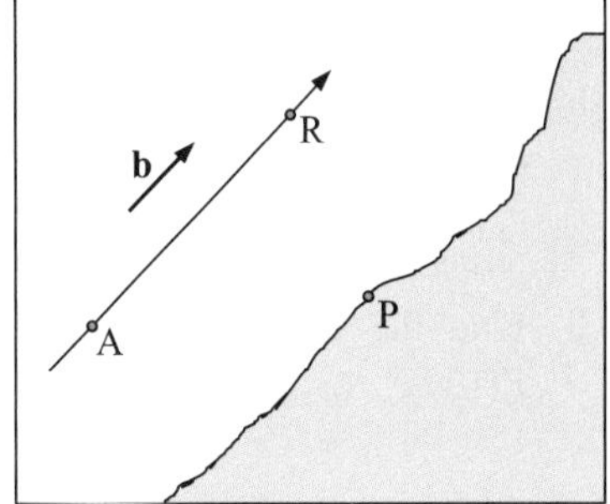

The ship is closest when [PR] is perpendicular to [AR].

$\therefore\ \overrightarrow{PR} \bullet \mathbf{b} = 0$

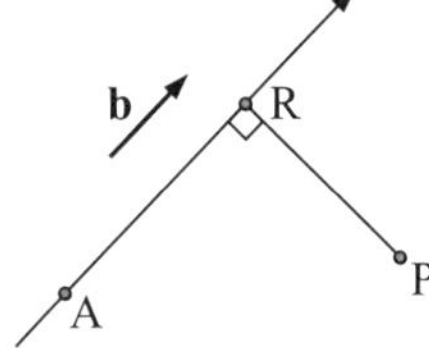

In this situation, point R is called the **foot of the perpendicular from P to the line**.

Example 11 **Self Tutor**

A line has vector equation $\begin{pmatrix} x \\ y \end{pmatrix} = \begin{pmatrix} 1 \\ 2 \end{pmatrix} + t\begin{pmatrix} 3 \\ -1 \end{pmatrix},\ t \in \mathbb{R}$. Let P be the point $(5,\ -1)$.

Find exactly the shortest distance from P to the line.

Let N be the point on the line closest to P.

N has coordinates $(1+3t,\ 2-t)$ for some t, and $\overrightarrow{PN}$ is $\begin{pmatrix} 1+3t-5 \\ 2-t-(-1) \end{pmatrix} = \begin{pmatrix} 3t-4 \\ 3-t \end{pmatrix}$.

Now $\overrightarrow{PN} \bullet \begin{pmatrix} 3 \\ -1 \end{pmatrix} = 0$ {as $\begin{pmatrix} 3 \\ -1 \end{pmatrix}$ is the direction vector of the line}

$\therefore\ \begin{pmatrix} 3t-4 \\ 3-t \end{pmatrix} \bullet \begin{pmatrix} 3 \\ -1 \end{pmatrix} = 0$

$\therefore\ 3(3t-4) - (3-t) = 0$

$\therefore\ 9t - 12 - 3 + t = 0$

$\therefore\ 10t = 15$

$\therefore\ t = \frac{15}{10} = \frac{3}{2}$

N$(1+3t,\ 2-t)$

$\begin{pmatrix} 3 \\ -1 \end{pmatrix}$

P$(5,\ -1)$

Thus $\overrightarrow{PN} = \begin{pmatrix} \frac{9}{2}-4 \\ 3-\frac{3}{2} \end{pmatrix} = \begin{pmatrix} \frac{1}{2} \\ \frac{3}{2} \end{pmatrix} = \frac{1}{2}\begin{pmatrix} 1 \\ 3 \end{pmatrix}$

and $|\overrightarrow{PN}| = \frac{1}{2}\left|\begin{pmatrix} 1 \\ 3 \end{pmatrix}\right| = \frac{1}{2}\sqrt{1^2+3^2} = \frac{\sqrt{10}}{2}$ units

EXERCISE 15F

1 Find the shortest distance from:

a P(3, 2) to the line with parametric equations $x = 2 + t, \quad y = 3 + 2t, \quad t \in \mathbb{R}$

b Q(−1, 1) to the line with parametric equations $x = t, \quad y = 1 - t, \quad t \in \mathbb{R}$

c R(−3, −1) to the line $\begin{pmatrix} x \\ y \end{pmatrix} = \begin{pmatrix} 2 \\ 3 \end{pmatrix} + s\begin{pmatrix} 1 \\ -1 \end{pmatrix}, \quad s \in \mathbb{R}$

d S(5, −2) to the line $\begin{pmatrix} x \\ y \end{pmatrix} = \begin{pmatrix} 2 \\ 5 \end{pmatrix} + t\begin{pmatrix} 3 \\ -7 \end{pmatrix}, \quad t \in \mathbb{R}$.

Example 12 — Self Tutor

On the map shown, distances are measured in kilometres.

Ship R is moving in the direction $\begin{pmatrix} 3 \\ 4 \end{pmatrix}$ at 10 km h^{-1}.

a Write an expression for the position of the ship in terms of t, the number of hours after leaving port A.

b Find the time when the ship is closest to port P(10, 2).

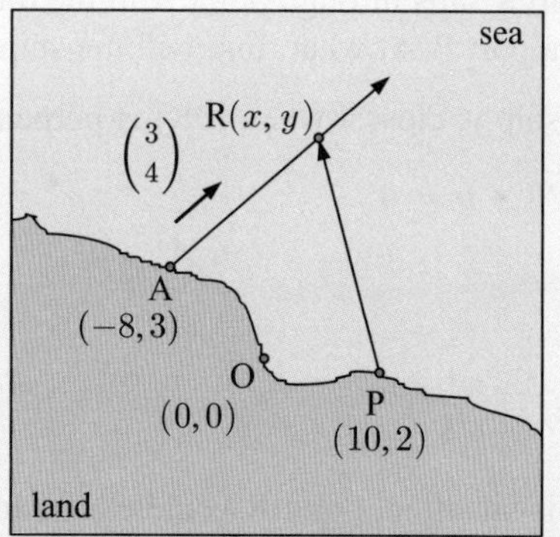

a $\left|\begin{pmatrix} 3 \\ 4 \end{pmatrix}\right| = \sqrt{3^2 + 4^2} = 5$

Since the speed is 10 km h^{-1}, the ship's velocity vector must be $2\begin{pmatrix} 3 \\ 4 \end{pmatrix} = \begin{pmatrix} 6 \\ 8 \end{pmatrix}$.

Now $\overrightarrow{OR} = \overrightarrow{OA} + \overrightarrow{AR}$

$\therefore \begin{pmatrix} x \\ y \end{pmatrix} = \begin{pmatrix} -8 \\ 3 \end{pmatrix} + t\begin{pmatrix} 6 \\ 8 \end{pmatrix}$

$\therefore$ the position of ship R is $(-8 + 6t, 3 + 8t)$.

b The ship is closest to P when $\overrightarrow{PR} \perp \begin{pmatrix} 3 \\ 4 \end{pmatrix}$

$$\therefore \overrightarrow{PR} \bullet \begin{pmatrix} 3 \\ 4 \end{pmatrix} = 0$$

$$\therefore \begin{pmatrix} -8 + 6t - 10 \\ 3 + 8t - 2 \end{pmatrix} \bullet \begin{pmatrix} 3 \\ 4 \end{pmatrix} = 0$$

$$\therefore 3(6t - 18) + 4(1 + 8t) = 0$$

$$\therefore 18t - 54 + 4 + 32t = 0$$

$$\therefore 50t - 50 = 0$$

$$\therefore t = 1$$

So, the ship is closest to port P one hour after leaving A.

2 An ocean liner is at $(6, -6)$, cruising at 10 km h^{-1} in the direction $\begin{pmatrix} -3 \\ 4 \end{pmatrix}$.

A fishing boat is anchored at $(0, 0)$. Distances are in kilometres.

- **a** Find, in terms of $\mathbf{i}$ and $\mathbf{j}$, the initial position vector of the liner from the fishing boat.
- **b** Write an expression for the position vector of the liner at any time t hours after it has sailed from $(6, -6)$.
- **c** When will the liner be due east of the fishing boat?
- **d** Find the time and position of the liner when it is nearest to the fishing boat.

3 Let $\mathbf{i}$ represent a displacement 1 km due east and $\mathbf{j}$ represent a displacement 1 km due north.

The control tower of an airport is at $(0, 0)$. Aircraft within 100 km of $(0, 0)$ are visible on the radar screen at the control tower.

At 12:00 noon an aircraft is 200 km east and 100 km north of the control tower. It is flying parallel to the vector $\mathbf{b} = -3\mathbf{i} - \mathbf{j}$ with a speed of $40\sqrt{10}\text{ km h}^{-1}$.

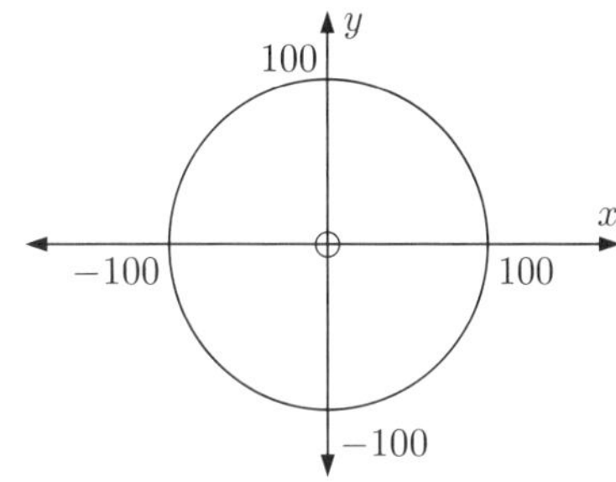

- **a** Write down the velocity vector of the aircraft.
- **b** Write a vector equation for the path of the aircraft using t to represent the time in hours that have elapsed since 12:00 noon.
- **c** Find the position of the aircraft at 1:00 pm.
- **d** Show that the aircraft first becomes visible on the radar screen at 1:00 pm.
- **e** Find the time when the aircraft is closest to the control tower, and find the distance between the aircraft and the control tower at this time.
- **f** At what time will the aircraft disappear from the radar screen?

4 The diagram shows a railway track that has equation $2x + 3y = 36$.

The axes represent two long country roads.

All distances are in kilometres.

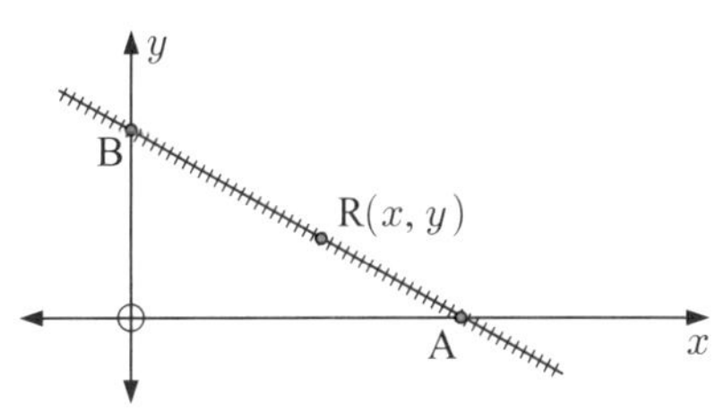

- **a** Find the coordinates of A and B.
- **b** $R(x, y)$ is a point on the railway track. Express the coordinates of point R in terms of x only.
- **c** Some railway workers have set up a base camp at $P(4, 0)$. Find $\overrightarrow{PR}$ and $\overrightarrow{AB}$.
- **d** Find the coordinates of the point on the railway track that is closest to P. Hence, find the shortest distance from the base camp to the railway track.

5 Boat A's position is given by $x(t) = 3 - t$, $y(t) = 2t - 4$, and boat B's position is given by $x(t) = 4 - 3t$, $y(t) = 3 - 2t$. The distance units are kilometres, and the time units are hours.

- **a** Find the initial position of each boat.
- **b** Find the velocity vector of each boat.
- **c** What is the acute angle between the paths of the boats?
- **d** At what time are the boats closest to each other?

Example 13 **Self Tutor**

Consider the point P$(-1, 2, 3)$ and the line with parametric equations $x = 1+2t$, $y = -4+3t$, $z = 3+t$.

a Find the coordinates of the foot of the perpendicular from P to the line.

b Find the shortest distance from the point to the line.

a The line has direction vector $\mathbf{b} = \begin{pmatrix} 2 \\ 3 \\ 1 \end{pmatrix}$.

Let A$(1+2t, -4+3t, 3+t)$ be any point on the given line.

$$\therefore \overrightarrow{PA} = \begin{pmatrix} 1+2t-(-1) \\ -4+3t-2 \\ 3+t-3 \end{pmatrix} = \begin{pmatrix} 2+2t \\ -6+3t \\ t \end{pmatrix}$$

b

A$(1+2t, -4+3t, 3+t)$

P$(-1, 2, 3)$

If A is the closest point on the line to P, then $\overrightarrow{PA}$ and **b** are perpendicular.

$$\therefore \overrightarrow{PA} \bullet \mathbf{b} = 0$$

$$\therefore \begin{pmatrix} 2+2t \\ -6+3t \\ t \end{pmatrix} \bullet \begin{pmatrix} 2 \\ 3 \\ 1 \end{pmatrix} = 0$$

$$\therefore 2(2+2t) + 3(-6+3t) + 1(t) = 0$$

$$\therefore 4+4t-18+9t+t = 0$$

$$\therefore 14t = 14$$

$$\therefore t = 1$$

Substituting $t = 1$ into the parametric equations, we obtain the foot of the perpendicular $(3, -1, 4)$.

b When $t = 1$, $\overrightarrow{PA} = \begin{pmatrix} 2+2 \\ -6+3 \\ 1 \end{pmatrix} = \begin{pmatrix} 4 \\ -3 \\ 1 \end{pmatrix}$

$$\therefore \text{PA} = \sqrt{4^2 + (-3)^2 + 1^2} = \sqrt{26} \text{ units}$$

$\therefore$ the shortest distance from P to the line is $\sqrt{26}$ units.

6 **a** Find the coordinates of the foot of the perpendicular from $(3, 0, -1)$ to the line with parametric equations $x = 2+3t$, $y = -1+2t$, $z = 4+t$.

b Find the shortest distance from the line to the point.

7 **a** Find the coordinates of the foot of the perpendicular from $(1, 1, 3)$ to the line with vector equation $\begin{pmatrix} x \\ y \\ z \end{pmatrix} = \begin{pmatrix} 1 \\ -1 \\ 2 \end{pmatrix} + t\begin{pmatrix} 2 \\ 3 \\ 1 \end{pmatrix}$.

b Find the shortest distance from the point to the line.

G INTERSECTING LINES

Vector equations of two intersecting lines can be **solved simultaneously** to find the point where the lines meet.

Example 14 Self Tutor

Line 1 has vector equation $\begin{pmatrix} x \\ y \end{pmatrix} = \begin{pmatrix} -2 \\ 1 \end{pmatrix} + s\begin{pmatrix} 3 \\ 2 \end{pmatrix}$ and

line 2 has vector equation $\begin{pmatrix} x \\ y \end{pmatrix} = \begin{pmatrix} 15 \\ 5 \end{pmatrix} + t\begin{pmatrix} -4 \\ 1 \end{pmatrix}$, where s and t are scalars.

Use vector methods to find where the two lines meet.

The lines meet where $\begin{pmatrix} -2 \\ 1 \end{pmatrix} + s\begin{pmatrix} 3 \\ 2 \end{pmatrix} = \begin{pmatrix} 15 \\ 5 \end{pmatrix} + t\begin{pmatrix} -4 \\ 1 \end{pmatrix}$

$\therefore\ -2 + 3s = 15 - 4t$ and $1 + 2s = 5 + t$

$\therefore\ 3s + 4t = 17$ .... (1) and $2s - t = 4$ .... (2)

$$\begin{aligned} 3s + 4t &= 17 \quad \{(1)\} \\ 8s - 4t &= 16 \quad \{(2) \times 4\} \\ \hline \therefore\ 11s \quad &= 33 \\ \therefore\ s &= 3 \end{aligned}$$

Using (2), $2(3) - t = 4$

$\therefore\ t = 2$

Using line 1, $\begin{pmatrix} x \\ y \end{pmatrix} = \begin{pmatrix} -2 \\ 1 \end{pmatrix} + 3\begin{pmatrix} 3 \\ 2 \end{pmatrix} = \begin{pmatrix} 7 \\ 7 \end{pmatrix}$

Checking in line 2, $\begin{pmatrix} x \\ y \end{pmatrix} = \begin{pmatrix} 15 \\ 5 \end{pmatrix} + 2\begin{pmatrix} -4 \\ 1 \end{pmatrix} = \begin{pmatrix} 7 \\ 7 \end{pmatrix}$

$\therefore$ the lines meet at (7, 7).

EXERCISE 15G

1 Triangle ABC is formed by three lines:

Line 1 (AB) is $\begin{pmatrix} x \\ y \end{pmatrix} = \begin{pmatrix} -1 \\ 6 \end{pmatrix} + r\begin{pmatrix} 3 \\ -2 \end{pmatrix}$, line 2 (AC) is $\begin{pmatrix} x \\ y \end{pmatrix} = \begin{pmatrix} 0 \\ 2 \end{pmatrix} + s\begin{pmatrix} 1 \\ 1 \end{pmatrix}$

and line 3 (BC) is $\begin{pmatrix} x \\ y \end{pmatrix} = \begin{pmatrix} 10 \\ -3 \end{pmatrix} + t\begin{pmatrix} -2 \\ 3 \end{pmatrix}$ where r, s, and t are scalars.

a Draw the three lines accurately on a grid.

b Hence, find the coordinates of A, B and C.

c Prove that $\triangle$ABC is isosceles.

d Use vector methods to *check* your answers to **b**.

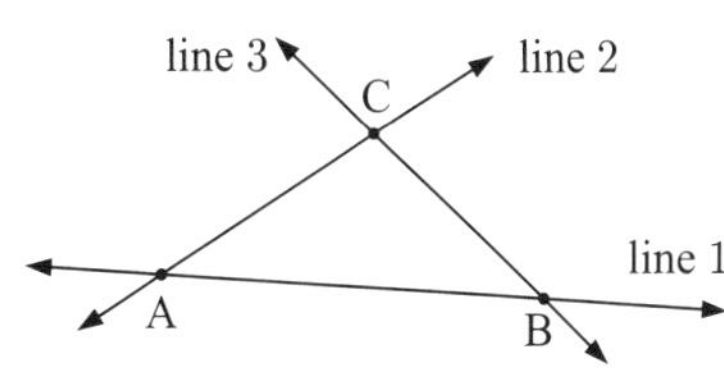

2 A parallelogram is defined by four lines as follows:

Line (AB) is $\begin{pmatrix} x \\ y \end{pmatrix} = \begin{pmatrix} -4 \\ 6 \end{pmatrix} + r\begin{pmatrix} 7 \\ 3 \end{pmatrix}$. Line (AD) is $\begin{pmatrix} x \\ y \end{pmatrix} = \begin{pmatrix} -4 \\ 6 \end{pmatrix} + s\begin{pmatrix} 1 \\ 2 \end{pmatrix}$.

Line (CD) is $\begin{pmatrix} x \\ y \end{pmatrix} = \begin{pmatrix} 22 \\ 25 \end{pmatrix} + t\begin{pmatrix} -7 \\ -3 \end{pmatrix}$. Line (CB) is $\begin{pmatrix} x \\ y \end{pmatrix} = \begin{pmatrix} 22 \\ 25 \end{pmatrix} + u\begin{pmatrix} -1 \\ -2 \end{pmatrix}$.

r, s, t, and u are scalars.

- **a** Draw an accurate sketch of the four lines and the parallelogram formed by them. Label the vertices.
- **b** From your diagram find the coordinates of A, B, C, and D.
- **c** Use vector methods to confirm your answers to **b**.

3 An isosceles triangle ABC is formed by these lines:

(AB) is $\begin{pmatrix} x \\ y \end{pmatrix} = \begin{pmatrix} 0 \\ 2 \end{pmatrix} + r\begin{pmatrix} 2 \\ 1 \end{pmatrix}$, (BC) is $\begin{pmatrix} x \\ y \end{pmatrix} = \begin{pmatrix} 8 \\ 6 \end{pmatrix} + s\begin{pmatrix} -1 \\ -2 \end{pmatrix}$,

and (AC) is $\begin{pmatrix} x \\ y \end{pmatrix} = \begin{pmatrix} 0 \\ 5 \end{pmatrix} + t\begin{pmatrix} 1 \\ -1 \end{pmatrix}$, where r, s and t are scalars.

- **a** Use vector methods to find the coordinates of A, B, and C.
- **b** Which two sides of the triangle are equal in length? Find all side lengths.

4 (QP) is $\begin{pmatrix} x \\ y \end{pmatrix} = \begin{pmatrix} 3 \\ -1 \end{pmatrix} + r\begin{pmatrix} 14 \\ 10 \end{pmatrix}$, (QR) is $\begin{pmatrix} x \\ y \end{pmatrix} = \begin{pmatrix} 3 \\ -1 \end{pmatrix} + s\begin{pmatrix} 17 \\ -9 \end{pmatrix}$, and

(PR) is $\begin{pmatrix} x \\ y \end{pmatrix} = \begin{pmatrix} 0 \\ 18 \end{pmatrix} + t\begin{pmatrix} 5 \\ -7 \end{pmatrix}$, where r, s, and t are scalars.

- **a** Use vector methods to find the coordinates of P, Q, and R.
- **b** Find vectors $\overrightarrow{PQ}$ and $\overrightarrow{PR}$ and evaluate $\overrightarrow{PQ} \bullet \overrightarrow{PR}$.
- **c** Hence, find the size of $Q\widehat{P}R$.
- **d** Find the area of $\triangle$PQR.

5 Quadrilateral ABCD is formed by these lines:

(AB) is $\begin{pmatrix} x \\ y \end{pmatrix} = \begin{pmatrix} 2 \\ 5 \end{pmatrix} + r\begin{pmatrix} 4 \\ 1 \end{pmatrix}$, (BC) is $\begin{pmatrix} x \\ y \end{pmatrix} = \begin{pmatrix} 18 \\ 9 \end{pmatrix} + s\begin{pmatrix} -8 \\ 32 \end{pmatrix}$,

(CD) is $\begin{pmatrix} x \\ y \end{pmatrix} = \begin{pmatrix} 14 \\ 25 \end{pmatrix} + t\begin{pmatrix} -8 \\ -2 \end{pmatrix}$, and (AD) is $\begin{pmatrix} x \\ y \end{pmatrix} = \begin{pmatrix} 3 \\ 1 \end{pmatrix} + u\begin{pmatrix} -3 \\ 12 \end{pmatrix}$,

where r, s, t, and u are scalars.

- **a** Use vector methods to find the coordinates of A, B, C, and D.
- **b** Write down vectors $\overrightarrow{AC}$ and $\overrightarrow{DB}$ and hence find:
  - **i** $|\overrightarrow{AC}|$
  - **ii** $|\overrightarrow{DB}|$
  - **iii** $\overrightarrow{AC} \bullet \overrightarrow{DB}$
- **c** What do the answers to **b** tell you about quadrilateral ABCD?

THEORY OF KNOWLEDGE

In his 1827 book entitled *The Barycentric Calculus*, German mathematician August Möbius demonstrated the use of directed line segments in projective geometry. He discussed the addition and scalar multiplication of directed line segments, but he did not name them vectors.

Gauss, Hamilton, Grassmann, Laplace, and Lagrange all helped in the evolution of vectors and vector analysis as we know them today. In particular, in 1843 Sir William Hamilton defined a *quaternion* as the quotient of two directed line segments.

Sir William Hamilton

The American mathematician Josaiah Willard Gibbs (1839 - 1903) began teaching a course on vector analysis at Yale University in 1879. He published *Elements of Vector Analysis* in two halves, and in 1901 his work was summarised by his student Edwin Wilson in the book *Vector Analysis*.

Oliver Heaviside (1850 - 1925) developed vector analysis separately from Gibbs, receiving a copy of Gibbs' work later, in 1888. Heaviside wrote:

> "... the invention of quaternions must be regarded as a most remarkable feat of human ingenuity. Vector analysis, without quaternions, could have been found by any mathematician by carefully examining the mechanics of the Cartesian mathematics; but to find out quaternions required a genius."

1. Are geometry and algebra two separate domains of knowledge?
2. When Gibbs and Heaviside developed vector analysis independently, who can claim to be the founder of this field?
3. Does mathematics *evolve*?

H RELATIONSHIPS BETWEEN LINES

We have just seen how the intersection of two lines can be found by vector methods. We have also seen how to determine whether two lines are parallel.

We can now summarise the possible relationships between lines in 2 and 3 dimensions.

LINE CLASSIFICATION IN 2 DIMENSIONS

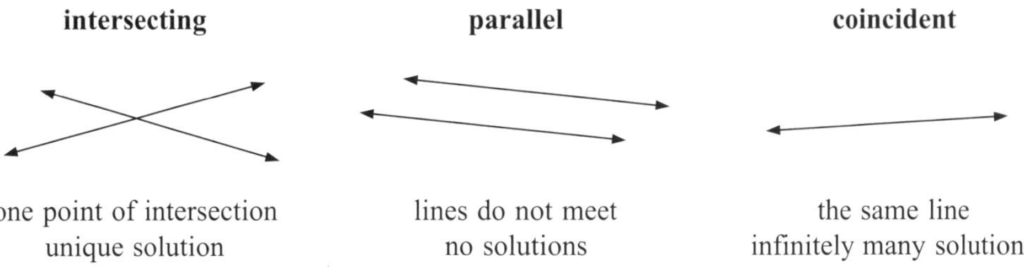

In order to classify lines in 2 dimensions, we can solve the equations of the lines simultaneously. We will do this with **row operations**, since we will also use this technique later for dealing with planes.

The system of equations $\begin{cases} 2x + y = -1 \\ x - 3y = 17 \end{cases}$ is called a 2×2 system, because there are **2 equations** in **2 unknowns**.

In the method of 'elimination' used to solve these equations, we observe that the following operations produce equations with the same solutions as the original pair.

- The equations can be interchanged without affecting the solutions.

  For example, $\begin{cases} 2x + y = -1 \\ x - 3y = 17 \end{cases}$ has the same solutions as $\begin{cases} x - 3y = 17 \\ 2x + y = -1 \end{cases}$.

- An equation can be replaced by a non-zero multiple of itself.

  For example, $2x + y = -1$ could be replaced by $-6x - 3y = 3$ {multiplying each term by -3}.

- Any equation can be replaced by a multiple of itself plus (or minus) a multiple of another equation.

  For example, suppose we replace the second equation by "twice the second equation minus the first equation".

  In this case we have:
  $$\begin{array}{r} 2x - 6y = 34 \\ -(2x + y = -1) \\ \hline -7y = 35 \end{array}$$

  so $\begin{cases} 2x + y = -1 \\ x - 3y = 17 \end{cases}$ becomes $\begin{cases} 2x + y = -1 \\ -7y = 35 \end{cases}$.

AUGMENTED MATRICES

Instead of writing $\begin{cases} 2x + y = -1 \\ x - 3y = 17 \end{cases}$ we detach the coefficients and write the system as the **augmented matrix** $\left[\begin{array}{cc|c} 2 & 1 & -1 \\ 1 & -3 & 17 \end{array}\right]$.

In this form we can use **elementary row operations** equivalent to the three legitimate operations with equations. We can hence:

- interchange rows
- replace any row by a non-zero multiple of itself
- replace any row by itself plus (or minus) a multiple of another row.

This process of solving a system using elementary row operations is known as **row reduction**. Our aim is to perform row operations until there are all zeros in the bottom left hand corner. At this time the system is said to be in **echelon form**.

In practice, we can combine several elementary row operations in one step. You will see this in the **Example** which follows.

Example 15 | Self Tutor

Use row operations to solve: $\begin{cases} 2x+3y=4 \\ 5x+4y=17 \end{cases}$

In augmented matrix form the system is:

$$\left[\begin{array}{cc|c} 2 & 3 & 4 \\ 5 & 4 & 17 \end{array}\right]$$

$$\sim \left[\begin{array}{cc|c} 2 & 3 & 4 \\ 0 & 7 & -14 \end{array}\right] \longleftarrow R_2 \to 5R_1 - 2R_2$$

$$\begin{array}{rrr} 10 & 15 & 20 \\ -10 & -8 & -34 \\ \hline 0 & 7 & -14 \end{array}$$

$\sim$ is read as "which has the same solution as"

From row 2, $7y = -14$

$\therefore\ y = -2$

Back substituting into the first equation, we have $2x + 3(-2) = 4$

$\therefore\ 2x - 6 = 4$

$\therefore\ 2x = 10$

$\therefore\ x = 5$

Don't forget to check your solution.

So, the solution is $x = 5$, $y = -2$.

EXERCISE 15H.1

1 Solve by row reduction:

a $x - 2y = 8$, $4x + y = 5$

b $4x + 5y = 21$, $5x - 3y = -20$

c $3x + y = -10$, $2x + 5y = -24$

2 By inspection, classify the following pairs of equations as either intersecting, parallel, or coincident lines:

a $x - 3y = 2$, $3x + y = 8$

b $x + y = 7$, $3x + 3y = 1$

c $4x - y = 8$, $y = 2$

d $x - 2y = 4$, $2x - 4y = 8$

e $5x - 11y = 2$, $6x + y = 8$

f $3x - 4y = 5$, $-3x + 4y = 2$

3 Consider the equation pair $\begin{cases} x+2y=3 \\ 2x+4y=6 \end{cases}$.

a Explain why there are infinitely many solutions, giving geometric evidence.

b Explain why the second equation can be ignored when finding all solutions.

c Give all solutions in the form: i $x = t$, $y = ...$ ii $y = s$, $x = ...$

4 a Use row reduction on the system $\begin{cases} 2x+3y=5 \\ 2x+3y=11 \end{cases}$ to reduce the system to $\left[\begin{array}{cc|c} 2 & 3 & 5 \\ 0 & 0 & 6 \end{array}\right]$.

b What does the second row indicate?

c Explain this result geometrically.

5 a Use row reduction on the system $\begin{cases} 2x+3y=5 \\ 4x+6y=10 \end{cases}$ to reduce the system to $\left[\begin{array}{cc|c} 2 & 3 & 5 \\ 0 & 0 & 0 \end{array}\right]$.

b Explain this result geometrically.

Example 16 — Self Tutor

Use row operations to solve $\begin{cases} x + 3y = 5 \\ 4x + 12y = k \end{cases}$ where k is a constant.

Give a geometric interpretation of the solutions.

In augmented matrix form, the system is:

$$\left[\begin{array}{cc|c} 1 & 3 & 5 \\ 4 & 12 & k \end{array}\right]$$

$$\sim \left[\begin{array}{cc|c} 1 & 3 & 5 \\ 0 & 0 & k-20 \end{array}\right] \longleftarrow R_2 \to R_2 - 4R_1$$

$$\begin{array}{ccc} 4 & 12 & k \\ -4 & -12 & -20 \\ \hline 0 & 0 & k-20 \end{array}$$

The second equation actually reads $0x + 0y = k - 20$

If $k \neq 20$ we have $0 =$ a non-zero number, which is absurd. So there are no solutions.

In this case the equations represent parallel lines.

If $k = 20$ we have $0 = 0$, which is true for all x and y.

This means that all solutions come from $x + 3y = 5$ alone.

Letting $y = t$, $x = 5 - 3t$ for all values of t.

$\therefore$ there are infinitely many solutions of the form $x = 5 - 3t$, $y = t$, $t \in \mathbb{R}$.

In this case the lines are coincident.

6 **a** By using augmented matrices, show that $\begin{cases} 3x - y = 2 \\ 6x - 2y = 4 \end{cases}$ has infinitely many solutions of the form $x = t$, $y = 3t - 2$.

b Discuss the solutions to $\begin{cases} 3x - y = 2 \\ 6x - 2y = k \end{cases}$ where k can take any real value.

7 Consider $\begin{cases} 3x - y = 8 \\ 6x - 2y = k \end{cases}$ where k is any real number.

a Use row operations to reduce the system to: $\left[\begin{array}{cc|c} 3 & -1 & 8 \\ 0 & 0 & .... \end{array}\right]$

b **i** For what value of k is there infinitely many solutions?

ii What form do the infinite number of solutions have?

c **i** When does the system have no solutions?

ii Explain this result geometrically.

8 Consider the system $\begin{cases} 4x + 8y = 1 \\ 2x - ay = 11 \end{cases}$.

a Use row operations to reduce the system to: $\left[\begin{array}{cc|c} 4 & 8 & 1 \\ 0 & .... & .... \end{array}\right]$

b For what values of a does the system have a unique solution?

c Show that the unique solution is $x = \dfrac{a + 88}{4a + 16}$, $y = \dfrac{-21}{2a + 8}$ for these values of a.

d What is the solution in all other cases?

9 **a** Use row operations to find the values of m when the system $\begin{cases} mx + 2y = 6 \\ 2x + my = 6 \end{cases}$ has a unique solution.

b Find the unique solution.

c Discuss the solutions in the other *two* cases.

LINE CLASSIFICATION IN 3 DIMENSIONS

Lines are **coplanar** if they lie in the same plane.
If the lines are not coplanar then they are **skew**.

- If the lines are **coplanar**, they may be intersecting, parallel, or coincident.
- If the lines are **parallel**, the angle between them is $0°$.
- If the lines are **intersecting**, the angle between them is θ, as shown.

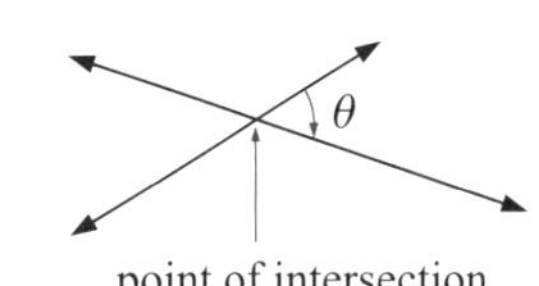

- If the lines are **skew**, we suppose one line is translated to intersect with the other. The angle between the original lines is defined as the angle between the intersecting lines, which is the angle θ.

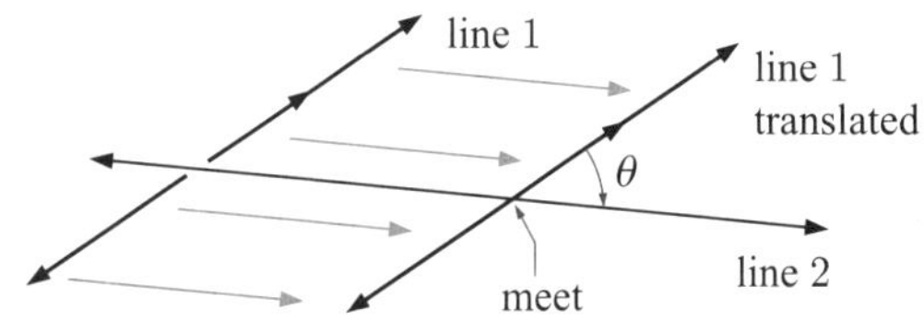

Example 17 — Self Tutor

3D GRAPHING

Line 1 has equations $x = -1 + 2s$, $y = 1 - 2s$, and $z = 1 + 4s$.
Line 2 has equations $x = 1 - t$, $y = t$, and $z = 3 - 2t$.
Show that the lines are parallel but not coincident.

Line 1 is $\begin{pmatrix} x \\ y \\ z \end{pmatrix} = \begin{pmatrix} -1 \\ 1 \\ 1 \end{pmatrix} + s\begin{pmatrix} 2 \\ -2 \\ 4 \end{pmatrix}$ with direction vector $\begin{pmatrix} 2 \\ -2 \\ 4 \end{pmatrix}$.

Line 2 is $\begin{pmatrix} x \\ y \\ z \end{pmatrix} = \begin{pmatrix} 1 \\ 0 \\ 3 \end{pmatrix} + t\begin{pmatrix} -1 \\ 1 \\ -2 \end{pmatrix}$ with direction vector $\begin{pmatrix} -1 \\ 1 \\ -2 \end{pmatrix}$.

Since $\begin{pmatrix} 2 \\ -2 \\ 4 \end{pmatrix} = -2\begin{pmatrix} -1 \\ 1 \\ -2 \end{pmatrix}$, the lines are parallel.

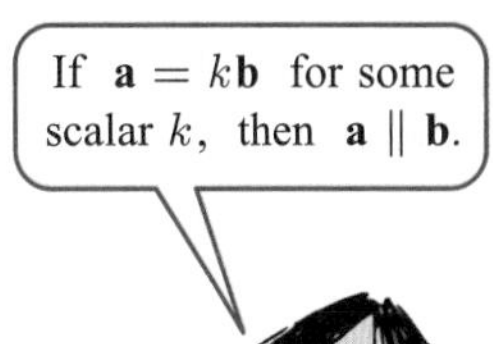

If $\mathbf{a} = k\mathbf{b}$ for some scalar k, then $\mathbf{a} \parallel \mathbf{b}$.

When $s = 0$, the point on line 1 is $(-1, 1, 1)$.

For line 2, $y = t$, so the unique point on line 2 with y-coordinate 1 is the point where $t = 1$. This point is $(0, 1, 1)$.

Since $(0, 1, 1) \neq (-1, 1, 1)$, the lines are not coincident.

Example 18

Self Tutor

Line 1 has equations $x = -1 + 2s$, $y = 1 - 2s$, and $z = 1 + 4s$.

Line 2 has equations $x = 1 - t$, $y = t$, and $z = 3 - 2t$.

Line 3 has equations $x = 1 + 2u$, $y = -1 - u$, and $z = 4 + 3u$.

a Show that line 2 and line 3 intersect and find the angle between them.

b Show that line 1 and line 3 are skew.

3D GRAPHING

a Equating x, y, and z values in lines 2 and 3 gives

$$1 - t = 1 + 2u \qquad t = -1 - u \quad \text{and} \quad 3 - 2t = 4 + 3u$$
$$\therefore\ t = -2u \qquad \therefore\ t = -1 - u \quad \text{and} \quad 3u + 2t = -1 \quad .... (1)$$

Solving these we get $-2u = -1 - u$

$\therefore\ -u = -1$

$\therefore\ u = 1$ and so $t = -2$

Checking in (1): $3u + 2t = 3(1) + 2(-2) = 3 - 4 = -1$ ✓

$\therefore$ $u = 1$, $t = -2$ satisfies all three equations, a *common solution.*

Using $u = 1$, lines 2 and 3 meet at $(1 + 2(1), -1 - (1), 4 + 3(1))$ which is $(3, -2, 7)$.

Direction vectors for lines 2 and 3 are $\mathbf{a} = \begin{pmatrix} -1 \\ 1 \\ -2 \end{pmatrix}$ and $\mathbf{b} = \begin{pmatrix} 2 \\ -1 \\ 3 \end{pmatrix}$ respectively.

If θ is the acute angle between $\mathbf{a}$ and $\mathbf{b}$, then

$$\cos\theta = \frac{|\mathbf{a} \bullet \mathbf{b}|}{|\mathbf{a}|\,|\mathbf{b}|} = \frac{|-2-1-6|}{\sqrt{1+1+4}\sqrt{4+1+9}} = \frac{9}{\sqrt{84}}$$

$$\therefore\ \theta \approx 10.89^\circ$$

$\therefore$ the angle between lines 2 and 3 is about 10.9°.

b Equating x, y, and z values in lines 1 and 3 gives

$$-1 + 2s = 1 + 2u \qquad 1 - 2s = -1 - u \quad \text{and} \quad 1 + 4s = 4 + 3u$$
$$\therefore\ 2s - 2u = 2 \qquad \therefore\ -2s + u = -2 \quad \text{and} \quad 4s - 3u = 3 \quad .... (2)$$

Solving these we get $2s - 2u = 2$

$\therefore\ -2s + u = -2$

$\therefore\ -u = 0$ {adding them}

$\therefore$ $2s = 2$ and so $s = 1$

Checking in (2), $4s - 3u = 4(1) - 3(0) = 4 \neq 3$

So, there is no simultaneous solution to all 3 equations.

$\therefore$ the lines do not intersect.

The direction vector for line 1 is $\begin{pmatrix} 2 \\ -2 \\ 4 \end{pmatrix}$ and $\begin{pmatrix} 2 \\ -2 \\ 4 \end{pmatrix} \neq k \begin{pmatrix} 2 \\ -1 \\ 3 \end{pmatrix}$ for any $k \in \mathbb{R}$.

$\therefore$ lines 1 and 3 are not parallel.

Since they do not intersect and are not parallel, they are skew.

EXERCISE 15H.2

1 Classify the following line pairs as either parallel, intersecting, coincident, or skew. In each case find the measure of the acute angle between them.

  a $x = 1 + 2t, \quad y = 2 - t, \quad z = 3 + t$ and $x = -2 + 3s, \quad y = 3 - s, \quad z = 1 + 2s$
  b $x = -1 + 2\lambda, \quad y = 2 - 12\lambda, \quad z = 4 + 12\lambda$ and $x = 4\mu - 3, \quad y = 3\mu + 2, \quad z = -\mu - 1$
  c $x = 6t, \quad y = 3 + 8t, \quad z = -1 + 2t$ and $x = 2 + 3s, \quad y = 4s, \quad z = 1 + s$
  d $x = 2 - y = z + 2$ and $x = 1 + 3s, \quad y = -2 - 2s, \quad z = 2s + \frac{1}{2}$
  e $x = 1 + \lambda, \quad y = 2 - \lambda, \quad z = 3 + 2\lambda$ and $x = 2 + 3\mu, \quad y = 3 - 2\mu, \quad z = \mu - 5$
  f $x = 1 - 2t, \quad y = 8 + t, \quad z = 5$ and $x = 2 + 4s, \quad y = -1 - 2s, \quad z = 3$
  g $x = 1 + 2\lambda, \quad y = -\lambda, \quad z = 1 + 3\lambda$ and $x = 3 - 4\mu, \quad y = -1 + 2\mu, \quad z = 4 - 6\mu$

2 Discuss the relationship between:

Line 1: $\mathbf{r}_1 = \begin{pmatrix} 3 \\ -1 \\ 2 \end{pmatrix} + \lambda \begin{pmatrix} 1 \\ -2 \\ -1 \end{pmatrix}$

Line 2: $x = 1 - 2\mu, \quad y = 4\mu, \quad z = -1 + 2\mu$

Line 3: $x = \dfrac{y-1}{2} = z - 1$

INVESTIGATION MOTION OF PARTICLES IN SPACE

In this Investigation we consider the motion of two particles in space, and whether they will collide.

What to do:

1 Consider the two lines:

$$L_1: \begin{pmatrix} 5 \\ 4 \\ 4 \end{pmatrix} + t \begin{pmatrix} -2 \\ -3 \\ 4 \end{pmatrix} \quad \text{and} \quad L_2: \begin{pmatrix} 1 \\ -4 \\ 14 \end{pmatrix} + u \begin{pmatrix} 0 \\ -2 \\ 2 \end{pmatrix}$$

Show that these lines intersect, and give the coordinates of the point of intersection.

2 Suppose we have two particles moving in space.
Particle A's position after t seconds is given by
$$x(t) = 5 - 2t, \quad y(t) = 4 - 3t, \quad z(t) = 4 + 4t.$$
Particle B's position after t seconds is given by
$$x(t) = 1, \quad y(t) = -4 - 2t, \quad z(t) = 14 + 2t.$$
All distance units are metres.

  a Find the initial position of each particle.
  b Find the velocity vector of each particle.
  c Will the particles collide? Explain why or why not.

3 Suppose D is the distance between particles A and B at time t.

  a Write an equation connecting D and t.
  b Hence find the minimum distance between the particles, and the time when this occurs.

PLANES

A **plane** is a flat surface that extends forever, and which has zero thickness.

It fits into a scheme which starts with a point and which builds to three dimensional space.

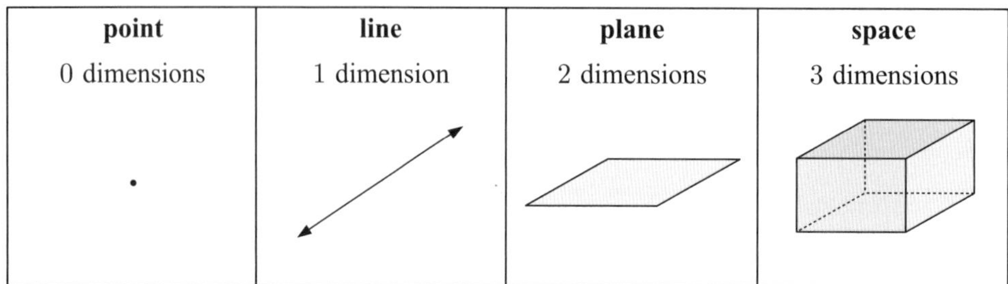

To find the equation of a plane, we need to know a point on the plane and also its orientation in space.

A vector is parallel to a plane if, by travelling along the vector, we remain on the plane.

For example, if A and B both lie on a plane then $\mathbf{a} = \overrightarrow{AB}$ is parallel to the plane.

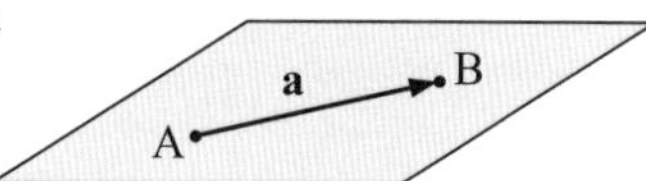

The orientation of a plane cannot be given by a single vector *parallel* to the plane because infinitely many planes of different orientation are parallel to a single direction vector.

To define the orientation of a plane uniquely, we require either:

- two non-parallel vectors which are both parallel to the plane, or
- one vector perpendicular or *normal* to the plane.

Suppose the point $R(x, y, z)$ lies on the plane containing a known point $A(a_1, a_2, a_3)$ and two non-parallel vectors $\mathbf{b} = \begin{pmatrix} b_1 \\ b_2 \\ b_3 \end{pmatrix}$ and $\mathbf{c} = \begin{pmatrix} c_1 \\ c_2 \\ c_3 \end{pmatrix}$.

$$\overrightarrow{AR} = \lambda\mathbf{b} + \mu\mathbf{c} \quad \text{for some scalars } \lambda \text{ and } \mu$$

$$\therefore \quad \overrightarrow{OR} - \overrightarrow{OA} = \lambda\mathbf{b} + \mu\mathbf{c}$$

$$\therefore \quad \overrightarrow{OR} = \overrightarrow{OA} + \lambda\mathbf{b} + \mu\mathbf{c}$$

$\mathbf{r} = \mathbf{a} + \lambda\mathbf{b} + \mu\mathbf{c}$ is the **vector equation of a plane**
where **r** is the position vector of any point on the plane,
a is the position vector of a known point $A(a_1, a_2, a_3)$ on the plane
b and **c** are any two non-parallel vectors that are parallel to the plane
$\lambda, \mu \in \mathbb{R}$ are two independent parameters.

Alternatively, if the two non-parallel vectors **b** and **c** are parallel to the plane, then the **normal vector** $\mathbf{n} = \mathbf{b} \times \mathbf{c}$ is perpendicular to any vector or line in or parallel to the plane.

Suppose $R(x, y, z)$ lies on the plane containing a known point $A(x_1, y_1, z_1)$, and which has normal vector $\mathbf{n} = \begin{pmatrix} a \\ b \\ c \end{pmatrix}$.

Now $\overrightarrow{AR}$ is perpendicular to $\mathbf{n}$

$$\therefore \quad \mathbf{n} \bullet \overrightarrow{AR} = 0$$

$$\therefore \quad \begin{pmatrix} a \\ b \\ c \end{pmatrix} \bullet \begin{pmatrix} x - x_1 \\ y - y_1 \\ z - z_1 \end{pmatrix} = 0$$

$$\therefore \quad a(x - x_1) + b(y - y_1) + c(z - z_1) = 0$$

$$\therefore \quad ax + by + cz = ax_1 + by_1 + cz_1 \quad \text{where the RHS is a constant.}$$

$\mathbf{n} = \begin{pmatrix} a \\ b \\ c \end{pmatrix}$

$R(x, y, z)$

$A(x_1, y_1, z_1)$

$\mathbf{n} \bullet \overrightarrow{AR} = 0$ is another form of the **vector equation of the plane**.

It could also be written as $\mathbf{n} \bullet (\mathbf{r} - \mathbf{a}) = 0$
or $\mathbf{r} \bullet \mathbf{n} = \mathbf{a} \bullet \mathbf{n}$.

If a plane has normal vector $\mathbf{n} = \begin{pmatrix} a \\ b \\ c \end{pmatrix}$ and passes through (x_1, y_1, z_1) then it has equation $ax + by + cz = ax_1 + by_1 + cz_1 = d$, where d is a constant.

This is the **Cartesian equation of the plane**.

Example 19 **Self Tutor**

Find the equation of the plane with normal vector $\begin{pmatrix} 1 \\ 2 \\ 3 \end{pmatrix}$ and which contains the point $(-1, 2, 4)$.

$\mathbf{n} = \begin{pmatrix} 1 \\ 2 \\ 3 \end{pmatrix}$, and the point $(-1, 2, 4)$ lies on the plane.

$\therefore$ the equation is $x + 2y + 3z = (-1) + 2(2) + 3(4)$
which is $x + 2y + 3z = 15$

EXERCISE 15I

1. Find the equation of the plane:

   a. with normal vector $\begin{pmatrix} 2 \\ -1 \\ 3 \end{pmatrix}$ and which passes through $(-1, 2, 4)$

   b. perpendicular to the line connecting $A(2, 3, 1)$ and $B(5, 7, 2)$, and which passes through A

   c. containing $A(3, 2, 1)$ and the line $x = 1 + t$, $y = 2 - t$, $z = 3 + 2t$.

2 State the normal vector to the plane with equation:

a $2x+3y-z=8$ **b** $3x-y=11$ **c** $z=2$ **d** $x=0$

3 Find the equation of the:

a XOZ-plane **b** plane perpendicular to the Z-axis, which passes through $(2,\ -1,\ 4)$.

Example 20 **Self Tutor**

Find the equation of the plane through A$(-1,\ 2,\ 0)$, B$(3,\ 1,\ 1)$, and C$(1,\ 0,\ 3)$:

a in vector form **b** in Cartesian form.

a $\overrightarrow{AB} = \begin{pmatrix} 4 \\ -1 \\ 1 \end{pmatrix}$ and $\overrightarrow{CB} = \begin{pmatrix} 2 \\ 1 \\ -2 \end{pmatrix}$ are two non-parallel vectors in the plane.

Using C as the known (fixed) point on the plane,

$$\mathbf{r} = \begin{pmatrix} x \\ y \\ z \end{pmatrix} = \begin{pmatrix} 1 \\ 0 \\ 3 \end{pmatrix} + \lambda \begin{pmatrix} 4 \\ -1 \\ 1 \end{pmatrix} + \mu \begin{pmatrix} 2 \\ 1 \\ -2 \end{pmatrix}, \quad \lambda,\ \mu \in \mathbb{R}.$$

b If $\mathbf{n}$ is the normal vector, then $\mathbf{n} = \overrightarrow{AB} \times \overrightarrow{AC} = \begin{pmatrix} 4 \\ -1 \\ 1 \end{pmatrix} \times \begin{pmatrix} 2 \\ -2 \\ 3 \end{pmatrix}$

$$\therefore\ \mathbf{n} = \begin{vmatrix} \mathbf{i} & \mathbf{j} & \mathbf{k} \\ 4 & -1 & 1 \\ 2 & -2 & 3 \end{vmatrix}$$

$$= \begin{vmatrix} -1 & 1 \\ -2 & 3 \end{vmatrix} \mathbf{i} - \begin{vmatrix} 4 & 1 \\ 2 & 3 \end{vmatrix} \mathbf{j} + \begin{vmatrix} 4 & -1 \\ 2 & -2 \end{vmatrix} \mathbf{k}$$

$$= -1\mathbf{i} - 10\mathbf{j} - 6\mathbf{k}$$

$$= \begin{pmatrix} -1 \\ -10 \\ -6 \end{pmatrix} \text{ or } -\begin{pmatrix} 1 \\ 10 \\ 6 \end{pmatrix}$$

Check that all 3 points satisfy this equation.

Thus the plane has equation

$x+10y+6z = (-1)+10(2)+6(0)$ {using point A}

$\therefore\ x+10y+6z = 19$

4 Find the equation of the plane containing the following points, in:

i vector form **ii** Cartesian form.

a A$(0,\ 2,\ 6)$, B$(1,\ 3,\ 2)$, and C$(-1,\ 2,\ 4)$ **b** A$(3,\ 1,\ 2)$, B$(0,\ 4,\ 0)$, and C$(0,\ 0,\ 1)$

c A$(2,\ 0,\ 3)$, B$(0,\ -1,\ 2)$, and C$(4,\ -3,\ 0)$.

5 Find the equations of the *line*:

a passing through $(1,\ -2,\ 0)$ and normal to the plane $x-3y+4z=8$

b passing through $(3,\ 4,\ -1)$ and normal to the plane $x-y-2z=11$.

Example 21 Self Tutor

Find the parametric equations of the line through A$(-1, 2, 3)$ and B$(2, 0, -3)$.
Hence find where this line meets the plane with equation $x - 2y + 3z = 26$.

$\overrightarrow{AB} = \begin{pmatrix} 3 \\ -2 \\ -6 \end{pmatrix}$ so (AB) has parametric equations
$x = -1 + 3\lambda, \quad y = 2 - 2\lambda, \quad z = 3 - 6\lambda$ .... $(*)$

This line meets the plane $x - 2y + 3z = 26$ where

$$\begin{aligned} -1 + 3\lambda - 2(2 - 2\lambda) + 3(3 - 6\lambda) &= 26 \\ \therefore \quad 4 - 11\lambda &= 26 \\ \therefore \quad -11\lambda &= 22 \\ \therefore \quad \lambda &= -2 \end{aligned}$$

Substituting $\lambda = -2$ into $*$, the line meets the plane at $(-7, 6, 15)$.

6 Find the parametric equations of the line through A$(2, -1, 3)$ and B$(1, 2, 0)$.
Hence find where this line meets the plane with equation $x + 2y - z = 5$.

7 **a** Find the parametric equations of the line through P$(1, -2, 4)$ and Q$(2, 0, -1)$.

b Hence find where this line meets:

i the YOZ-plane

ii the plane with equation $y + z = 2$

iii the line with Cartesian equation $\dfrac{x-3}{2} = \dfrac{y+2}{3} = \dfrac{z-30}{-1}$.

Example 22 Self Tutor

Suppose N is the foot of the normal from A$(2, -1, 3)$ to the plane $x - y + 2z = 27$. Find the coordinates of N, and hence find the shortest distance from A to the plane.

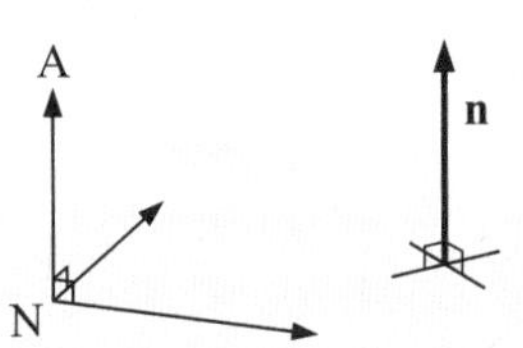

$x - y + 2z = 27$ has normal vector $\begin{pmatrix} 1 \\ -1 \\ 2 \end{pmatrix}$

$\therefore$ the parametric equations of (AN) are

$x = 2 + t, \quad y = -1 - t, \quad z = 3 + 2t.$

This line meets the plane $x - y + 2z = 27$ where

$$\begin{aligned} 2 + t - (-1 - t) + 2(3 + 2t) &= 27 \\ \therefore \quad 2 + t + 1 + t + 6 + 4t &= 27 \\ \therefore \quad 6t + 9 &= 27 \\ \therefore \quad 6t &= 18 \\ \therefore \quad t &= 3 \end{aligned}$$

$\therefore$ N is $(5, -4, 9)$.

The shortest distance, $\text{AN} = \sqrt{(5-2)^2 + (-4 - -1)^2 + (9-3)^2}$
$= \sqrt{54}$ units.

8 In each of the following cases, find the foot of the normal from A to the given plane, and hence find the shortest distance from A to the plane:

a A(1, 0, 2), $2x + y - 2z + 11 = 0$ **b** A(2, −1, 3), $x - y + 3z = -10$

c A(1, −4, −3), $4x - y - 2z = 8$

9 Find the coordinates of the mirror image of A(3, 1, 2) when reflected in the plane $x+2y+z=1$.

10 Does the line passing through (3, 4, −1) and which is normal to $x + 4y - z = -2$, intersect any of the coordinate axes?

11 Find the equation of the plane through A(1, 2, 3) and B(0, −1, 2) which is parallel to:

a the X-axis **b** the Y-axis **c** the Z-axis.

12 Show that the lines $x - 1 = \frac{y-2}{2} = z + 3$ and $x + 1 = y - 3 = 2z + 5$ are coplanar, and find the equation of the plane which contains them.

13 A(1, 2, k) lies on the plane $x + 2y - 2z = 8$. Find:

a the value of k

b the coordinates of B such that (AB) is normal to the plane, and B is 6 units from it.

Example 23 Self Tutor

Suppose N is the foot of the normal from A(2, −1, 3) to the plane with equation $\mathbf{r} = \mathbf{i} + 3\mathbf{k} + \lambda(4\mathbf{i} - \mathbf{j} + \mathbf{k}) + \mu(2\mathbf{i} + \mathbf{j} - 2\mathbf{k})$, $\lambda, \mu \in \mathbb{R}$.
Find the coordinates of N.

The normal to the plane has direction vector given by

$$(4\mathbf{i} - \mathbf{j} + \mathbf{k}) \times (2\mathbf{i} + \mathbf{j} - 2\mathbf{k}) = \begin{vmatrix} \mathbf{i} & \mathbf{j} & \mathbf{k} \\ 4 & -1 & 1 \\ 2 & 1 & -2 \end{vmatrix}$$

$$= \begin{vmatrix} -1 & 1 \\ 1 & -2 \end{vmatrix}\mathbf{i} - \begin{vmatrix} 4 & 1 \\ 2 & -2 \end{vmatrix}\mathbf{j} + \begin{vmatrix} 4 & -1 \\ 2 & 1 \end{vmatrix}\mathbf{k} = \begin{pmatrix} 1 \\ 10 \\ 6 \end{pmatrix}$$

The equation of the normal through A is $\begin{pmatrix} x \\ y \\ z \end{pmatrix} = \begin{pmatrix} 2 \\ -1 \\ 3 \end{pmatrix} + t\begin{pmatrix} 1 \\ 10 \\ 6 \end{pmatrix}$

so N must have coordinates of the form $(2 + t, -1 + 10t, 3 + 6t)$.

But N lies in the plane, so $\begin{pmatrix} 2+t \\ -1+10t \\ 3+6t \end{pmatrix} = \begin{pmatrix} 1 \\ 0 \\ 3 \end{pmatrix} + \lambda\begin{pmatrix} 4 \\ -1 \\ 1 \end{pmatrix} + \mu\begin{pmatrix} 2 \\ 1 \\ -2 \end{pmatrix}$

$$\therefore \begin{cases} 2 + t = 1 + 4\lambda + 2\mu \\ -1 + 10t = -\lambda + \mu \\ 3 + 6t = 3 + \lambda - 2\mu \end{cases} \quad \text{which simplifies to} \quad \begin{cases} 4\lambda + 2\mu - t = 1 \\ -\lambda + \mu - 10t = -1 \\ \lambda - 2\mu - 6t = 0 \end{cases}$$

Solving simultaneously with technology gives $\lambda = \frac{40}{137}$, $\mu = \frac{-7}{137}$, $t = \frac{9}{137}$

$\therefore$ N is the point $(2\frac{9}{137}, -\frac{47}{137}, 3\frac{54}{137})$

Check by substituting for λ and μ in the equation of the plane.

14 In each of the following cases, find the foot of the normal from A to the given plane, and hence find the shortest distance from A to the plane:

 a A(3, 2, 1), $\mathbf{r} = 3\mathbf{i} + \mathbf{j} + 2\mathbf{k} + \lambda(2\mathbf{i} + \mathbf{j} + \mathbf{k}) + \mu(4\mathbf{i} + 2\mathbf{j} - 2\mathbf{k})$

 b A(1, 0, −2), $\mathbf{r} = \mathbf{i} - \mathbf{j} + \mathbf{k} + \lambda(3\mathbf{i} - \mathbf{j} + 2\mathbf{k}) + \mu(-\mathbf{i} + \mathbf{j} - \mathbf{k})$

15 Suppose **b** and **c** are non-parallel vectors which lie in the same plane. The normal vector to the plane is $\mathbf{n} = \mathbf{b} \times \mathbf{c}$. Prove that $\mathbf{n} \perp (\lambda\mathbf{b} + \mu\mathbf{c})$ for all $\lambda, \mu \in \mathbb{R}$ except when $\lambda = \mu = 0$.

16 Suppose Q is any point in the plane $Ax + By + Cz + D = 0$. d is the distance from $P(x_1, y_1, z_1)$ to the given plane.

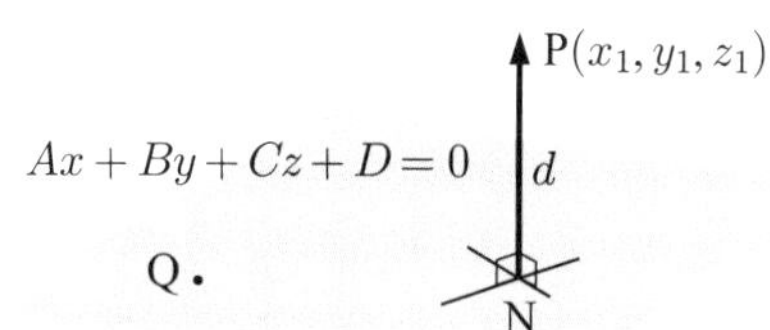

 a Explain why $d = \dfrac{|\overrightarrow{QP} \bullet \mathbf{n}|}{|\mathbf{n}|}$.

 b Hence, show that $d = \dfrac{|Ax_1 + By_1 + Cz_1 + D|}{\sqrt{A^2 + B^2 + C^2}}$.

 c Check your answers to question **8** using the formula in **b**.

 d Find the distance from:

  i $(0, 0, 0)$ to $x + 2y - z = 10$ **ii** $(1, -3, 2)$ to $x + y - z = 2$.

17 Find the distance between the parallel planes:

 a $x + y + 2z = 4$ and $2x + 2y + 4z + 11 = 0$

 b $ax + by + cz + d_1 = 0$ and $ax + by + cz + d_2 = 0$.

18 Show that the line $x = 2 + t$, $y = -1 + 2t$, $z = -3t$ is parallel to the plane $11x - 4y + z = 0$, and find its distance from the plane.

19 Find the equations of the two planes which are parallel to $2x - y + 2z = 5$ and 2 units from it.

J ANGLES IN SPACE

THE ANGLE BETWEEN A LINE AND A PLANE

Suppose a line has direction vector **d** and a plane has normal vector **n**. We allow **n** to intersect the line, making an angle of θ with it. The required acute angle is ϕ and

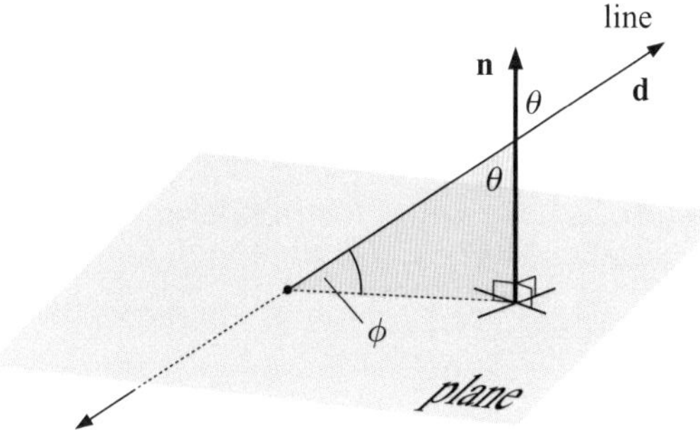

$$\sin\phi = \cos\theta = \frac{|\mathbf{n} \bullet \mathbf{d}|}{|\mathbf{n}||\mathbf{d}|}$$

$$\therefore \quad \phi = \arcsin\left(\frac{|\mathbf{n} \bullet \mathbf{d}|}{|\mathbf{n}||\mathbf{d}|}\right)$$

Example 24 **Self Tutor**

Find the angle between the plane $x + 2y - z = 8$ and the line with equations $x = t,\quad y = 1 - t,\quad z = 3 + 2t$.

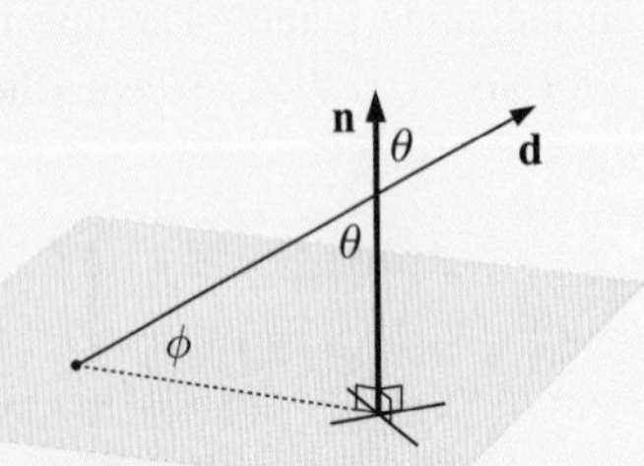

$\mathbf{n} = \begin{pmatrix} 1 \\ 2 \\ -1 \end{pmatrix}$ and $\mathbf{d} = \begin{pmatrix} 1 \\ -1 \\ 2 \end{pmatrix}$

The angle between the plane and the line is

$$\begin{aligned} \phi &= \arcsin\left(\frac{|1 - 2 - 2|}{\sqrt{1+4+1}\sqrt{1+1+4}}\right) \\ &= \arcsin\left(\frac{3}{\sqrt{6}\sqrt{6}}\right) \\ &= \arcsin\left(\tfrac{1}{2}\right) = 30^\circ \end{aligned}$$

THE ANGLE BETWEEN TWO PLANES

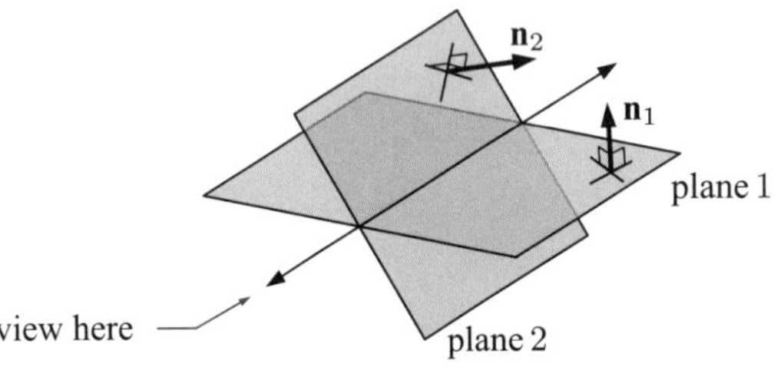

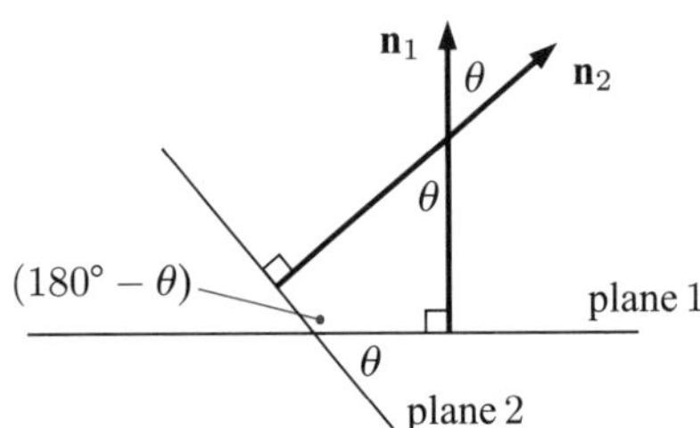

The cosine of the acute angle between the two normals is $\cos\theta = \dfrac{|\,\mathbf{n}_1 \bullet \mathbf{n}_2\,|}{|\,\mathbf{n}_1\,|\,|\,\mathbf{n}_2\,|}$.

If two planes have normal vectors $\mathbf{n}_1$ and $\mathbf{n}_2$, and θ is the acute angle between them, then $\theta = \arccos\left(\dfrac{|\,\mathbf{n}_1 \bullet \mathbf{n}_2\,|}{|\,\mathbf{n}_1\,|\,|\,\mathbf{n}_2\,|}\right)$.

Example 25 **Self Tutor**

Find the angle between the planes with equations $x + y - z = 8$ and $2x - y + 3z = -1$.

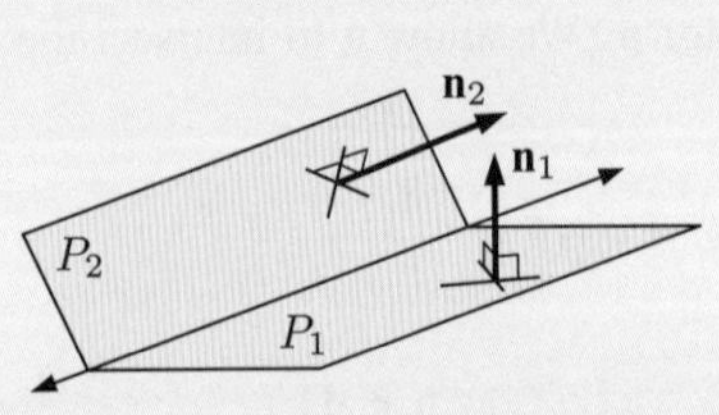

$x + y - z = 8$ has normal vector $\mathbf{n}_1 = \begin{pmatrix} 1 \\ 1 \\ -1 \end{pmatrix}$ and

$2x - y + 3z = -1$ has normal vector $\mathbf{n}_2 = \begin{pmatrix} 2 \\ -1 \\ 3 \end{pmatrix}$.

The acute angle between the planes is

$$\begin{aligned} \theta &= \arccos\left(\frac{|\,\mathbf{n}_1 \bullet \mathbf{n}_2\,|}{|\,\mathbf{n}_1\,|\,|\,\mathbf{n}_2\,|}\right) \\ &= \arccos\left(\frac{|2 + -1 + -3|}{\sqrt{1+1+1}\sqrt{4+1+9}}\right) \\ &= \arccos\left(\frac{2}{\sqrt{42}}\right) \approx 72.0^\circ \end{aligned}$$

EXERCISE 15J

1 Find the angle between:

a the plane $x - y + z = 5$ and the line $\dfrac{x-1}{4} = \dfrac{y+1}{3} = z + 2$

b the plane $2x - y + z = 8$ and the line $x = t + 1$, $y = -1 + 3t$, $z = t$

c the plane $3x + 4y - z = -4$ and the line $x - 4 = 3 - y = 2(z+1)$

d the plane $\mathbf{r}_p = 2\mathbf{i} - \mathbf{j} + \mathbf{k} + \lambda(3\mathbf{i} - 4\mathbf{j} - \mathbf{k}) + \mu(\mathbf{i} + \mathbf{j} - 2\mathbf{k})$ and the line $\mathbf{r}_l = 3\mathbf{i} + 2\mathbf{j} - \mathbf{k} + t(\mathbf{i} - \mathbf{j} + \mathbf{k})$.

2 Find the acute angle between the planes with equations:

a $2x - y + z = 3$
$x + 3y + 2z = 8$

b $x - y + 3z = 2$
$3x + y - z = -5$

c $3x - y + z = -11$
$2x + 4y - z = 2$

d $\mathbf{r}_1 = 3\mathbf{i} + 2\mathbf{j} - \mathbf{k} - \lambda(\mathbf{i} - \mathbf{j} + \mathbf{k}) + \mu(2\mathbf{i} - 4\mathbf{j} + 3\mathbf{k})$
$\mathbf{r}_2 = \mathbf{i} + \mathbf{j} - \mathbf{k} - \lambda(2\mathbf{i} + \mathbf{j} + \mathbf{k}) + \mu(\mathbf{i} + \mathbf{j} + \mathbf{k})$

e $3x - 4y + z = -2$ and $\mathbf{r} = \begin{pmatrix} 2 \\ -1 \\ 1 \end{pmatrix} + \lambda\begin{pmatrix} 3 \\ -1 \\ 0 \end{pmatrix} + \mu\begin{pmatrix} 2 \\ 1 \\ 1 \end{pmatrix}$

K INTERSECTING PLANES

- **Two planes** in space could have any of the following three arrangements:

(1) intersecting

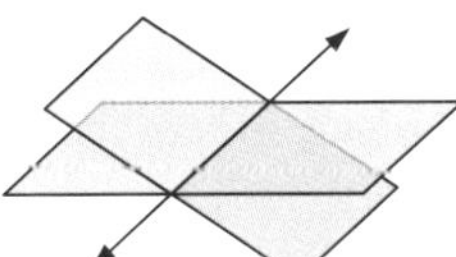

(2) parallel

(3) coincident

$P_1 = P_2$

- **Three planes** in space could have any of the following eight arrangements:

(1) all coincident

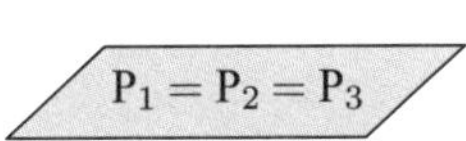

(2) two coincident and one intersecting

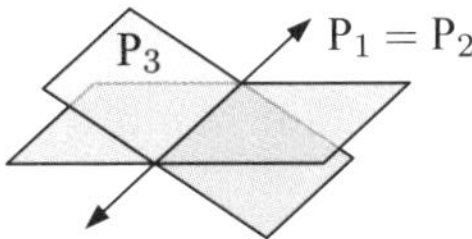

(3) two coincident and one parallel

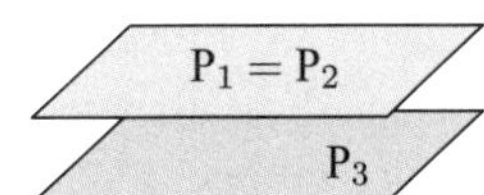

(4) two parallel and one intersecting

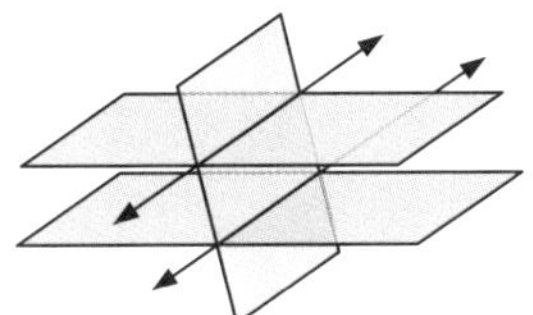

(5) all three parallel

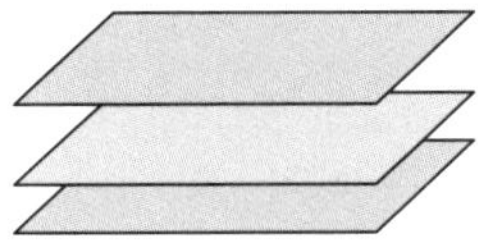

(6) all meet at the one point

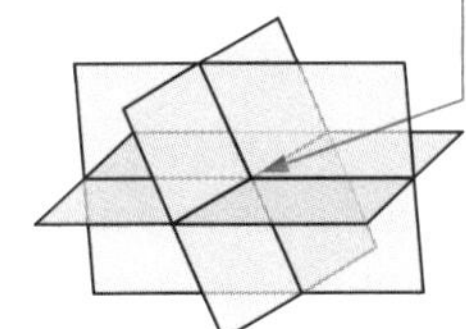

(7) all meet in a common line

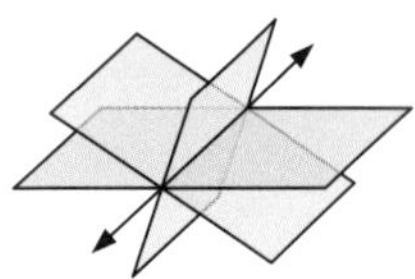

(8) the line of intersection of any two is parallel to the third plane.

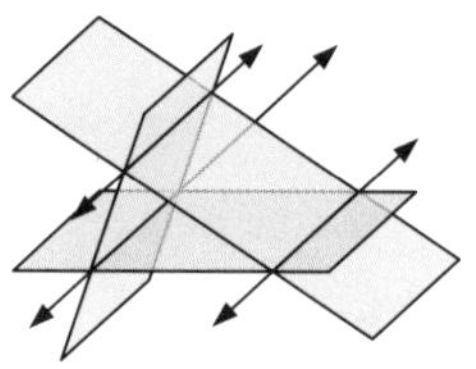

Click on the icon to explore these cases in three dimensions.

To find how planes intersect, we need to solve a system of linear equations.
We can do this using technology or using **row operations**.

USING ROW OPERATIONS TO SOLVE A 3×3 SYSTEM

A general 3×3 system in variables x, y, and z has the form $\begin{cases} a_1x + b_1y + c_1z = d_1 \\ a_2x + b_2y + c_2z = d_2 \\ a_3x + b_3y + c_3z = d_3 \end{cases}$

where the coefficients of x, y, and z are constants.

$\left[\begin{array}{ccc|c} a_1 & b_1 & c_1 & d_1 \\ a_2 & b_2 & c_2 & d_2 \\ a_3 & b_3 & c_3 & d_3 \end{array}\right]$ is the system's **augmented matrix form** which we need to reduce to **echelon form**: $\left[\begin{array}{ccc|c} a & b & c & d \\ 0 & e & f & g \\ 0 & 0 & h & i \end{array}\right]$ using **row operations**.

Notice the creation of a **triangle of zeros** in the **bottom left hand corner**.

In this form we can easily solve the system because the last row is really $hz = i$.

- If $h \neq 0$ we can determine z uniquely using $z = \frac{i}{h}$, and likewise y and x from the other two rows. Thus we arrive at a **unique solution**.
- If $h = 0$ and $i \neq 0$, the last row reads $0 \times z = i$ where $i \neq 0$ which is absurd. Hence, there is **no solution** and we say that the system is **inconsistent**.
- If $h = 0$ and $i = 0$, the last row is all zeros. Consequently, there are **infinitely many** solutions of the form $x = p + kt$, $y = q + lt$, and $z = t$ where $t \in \mathbb{R}$.

Note that the parametric representation of infinite solutions in terms of the parameter t is not unique.

This particular form assumes you are eliminating x and y to get z from Row 3. It may be easier to eliminate, for example, x and z to get y.

Example 26 **Self Tutor**

Solve the system: $\begin{cases} x + 3y - z = 15 \\ 2x + y + z = 7 \\ x - y - 2z = 0 \end{cases}$.

Interpret the result geometrically.

In augmented matrix form, the system is:

$$\left[\begin{array}{ccc|c} 1 & 3 & -1 & 15 \\ 2 & 1 & 1 & 7 \\ 1 & -1 & -2 & 0 \end{array}\right]$$

$$\sim \left[\begin{array}{ccc|c} 1 & 3 & -1 & 15 \\ 0 & -5 & 3 & -23 \\ 0 & -4 & -1 & -15 \end{array}\right] \quad \begin{array}{l} R_2 \to R_2 - 2R_1 \\ R_3 \to R_3 - R_1 \end{array}$$

$$\sim \left[\begin{array}{ccc|c} 1 & 3 & -1 & 15 \\ 0 & -5 & 3 & -23 \\ 0 & 0 & -17 & 17 \end{array}\right] \quad R_3 \to 5R_3 - 4R_2$$

$$\begin{array}{cccc} 2 & 1 & 1 & 7 \\ -2 & -6 & 2 & -30 \\ \hline 0 & -5 & 3 & -23 \end{array} \qquad \begin{array}{cccc} 1 & -1 & -2 & 0 \\ -1 & -3 & 1 & -15 \\ \hline 0 & -4 & -1 & -15 \end{array} \qquad \begin{array}{cccc} 0 & -20 & -5 & -75 \\ 0 & 20 & -12 & 92 \\ \hline 0 & 0 & -17 & 17 \end{array}$$

The last row gives $-17z = 17$

$\therefore\ z = -1$

Using the final row 2 $-5y + 3z = -23$

we get $-5y - 3 = -23$

$\therefore\ -5y = -20$

$\therefore\ y = 4$

Using the final row 1 $x + 3y - z = 15$

we get $x + 12 + 1 = 15$

$\therefore\ x = 2$

Thus we have a unique solution $x = 2,\ y = 4,\ z = -1.$

The system represents three planes that meet at the single point $(2, 4, -1)$. {case **(6)**}

Example 27 **Self Tutor**

Solve the system: $\left\{\begin{array}{r} 2x - y + z = 5 \\ x + y - z = 2 \\ 3x - 3y + 3z = 8 \end{array}\right.$. Interpret the result geometrically.

In augmented matrix form, the system is:

$$\left[\begin{array}{ccc|c} 1 & 1 & -1 & 2 \\ 2 & -1 & 1 & 5 \\ 3 & -3 & 3 & 8 \end{array}\right]$$

Notice the swapping $R_1 \leftrightarrow R_2$

$$\sim \left[\begin{array}{ccc|c} 1 & 1 & -1 & 2 \\ 0 & -3 & 3 & 1 \\ 0 & -6 & 6 & 2 \end{array}\right] \quad \begin{array}{l} R_2 \to R_2 - 2R_1 \\ R_3 \to R_3 - 3R_1 \end{array}$$

$$\sim \left[\begin{array}{ccc|c} 1 & 1 & -1 & 2 \\ 0 & -3 & 3 & 1 \\ 0 & 0 & 0 & 0 \end{array}\right] \quad R_3 \to R_3 - 2R_2$$

$$\begin{array}{cccc} 2 & -1 & 1 & 5 \\ -2 & -2 & 2 & -4 \\ \hline 0 & -3 & 3 & 1 \end{array} \qquad \begin{array}{cccc} 3 & -3 & 3 & 8 \\ -3 & -3 & 3 & -6 \\ \hline 0 & -6 & 6 & 2 \end{array} \qquad \begin{array}{cccc} 0 & -6 & 6 & 2 \\ 0 & 6 & -6 & -2 \\ \hline 0 & 0 & 0 & 0 \end{array}$$

The row of zeros indicates infinitely many solutions.

If we let $z = t$ in row 2, then $-3y + 3t = 1$

$\therefore\ -3y = 1 - 3t$

$\therefore\ y = \dfrac{1 - 3t}{-3} = -\frac{1}{3} + t$

Thus in row 1, $x+(-\frac{1}{3}+t)-t=2$

$\therefore\ x-\frac{1}{3}=2$ and so $x=\frac{7}{3}$

$\therefore$ the solutions have the form: $x=\frac{7}{3},\ \ y=-\frac{1}{3}+t,\ \ z=t,$ where $t\in\mathbb{R}$.

The system represents three planes that intersect in a line. {case **(7)**}

EXERCISE 15K

1 Solve the following systems using row reduction:

a $\begin{cases} x-2y+5z=1 \\ 2x-4y+8z=2 \\ -3x+6y+7z=-3 \end{cases}$ **b** $\begin{cases} x+4y+11z=7 \\ x+6y+17z=9 \\ x+4y+8z=4 \end{cases}$

c $\begin{cases} 2x-y+3z=17 \\ 2x-2y-5z=4 \\ 3x+2y+2z=10 \end{cases}$ **d** $\begin{cases} 2x+3y+4z=1 \\ 5x+6y+7z=2 \\ 8x+9y+10z=4 \end{cases}$

2 **a** How many scenarios are possible when solving simultaneously: $\begin{cases} a_1x+b_1y+c_1z=d_1 \\ a_2x+b_2y+c_2z=d_2 \end{cases}$?

b Under what conditions will the planes in **a** be: **i** parallel **ii** coincident?

c Solve the following using elementary row operations, and interpret each system of equations geometrically:

i $\begin{cases} x-3y+2z=8 \\ 3x-9y+2z=4 \end{cases}$ **ii** $\begin{cases} 2x+y+z=5 \\ x-y+z=3 \end{cases}$ **iii** $\begin{cases} x+2y-3z=6 \\ 3x+6y-9z=18 \end{cases}$

3 Discuss the possible solutions of the following systems where k is a real number, and interpret the solutions geometrically.

a $\begin{cases} x+2y-z=6 \\ 2x+4y+kz=12 \end{cases}$ **b** $\begin{cases} x-y+3z=8 \\ 2x-2y+6z=k \end{cases}$

4 For each of the eight possible geometric scenarios of three planes in space, comment on the solutions to the corresponding system of linear equations.

For example, $P_1=P_2=P_3$ has infinitely many solutions where x, y, and z are expressed in terms of two parameters s and t.

5 Solve the following systems using row operations. In each case state the geometric meaning of your solution:

a $\begin{cases} x+y-z=-5 \\ x-y+2z=11 \\ 4x+y-5z=-18 \end{cases}$ **b** $\begin{cases} x-y+2z=1 \\ 2x+y-z=8 \\ 5x-2y+5z=11 \end{cases}$ **c** $\begin{cases} x+2y-z=8 \\ 2x-y-z=5 \\ 3x-4y-z=2 \end{cases}$

d $\begin{cases} x-y+z=8 \\ 2x-2y+2z=11 \\ x+3y-z=-2 \end{cases}$ **e** $\begin{cases} x+y-2z=1 \\ x-y+z=4 \\ 3x+3y-6z=3 \end{cases}$ **f** $\begin{cases} x-y-z=5 \\ x+y+z=1 \\ 5x-y+2z=17 \end{cases}$

Example 28 **Self Tutor**

Consider the system $\begin{cases} x - 2y - z = -1 \\ 2x + y + 3z = 13 \\ x + 8y + 9z = a \end{cases}$ where a takes all real values.

a Use row operations to reduce the system to echelon form.

b When does the system have no solutions? Interpret the result geometrically.

c When does the system have infinitely many solutions? Find the solutions in this case, and interpret them geometrically.

---

a In augmented matrix form, the system is:

$$\left[\begin{array}{ccc|c} 1 & -2 & -1 & -1 \\ 2 & 1 & 3 & 13 \\ 1 & 8 & 9 & a \end{array}\right]$$

$$\begin{array}{rrrr} 2 & 1 & 3 & 13 \\ -2 & 4 & 2 & 2 \\ \hline 0 & 5 & 5 & 15 \end{array}$$

$$\sim \left[\begin{array}{ccc|c} 1 & -2 & -1 & -1 \\ 0 & 5 & 5 & 15 \\ 0 & 10 & 10 & a+1 \end{array}\right] \begin{array}{l} \\ \longleftarrow R_2 \to R_2 - 2R_1 \\ \longleftarrow R_3 \to R_3 - R_1 \end{array}$$

$$\begin{array}{rrrr} 1 & 8 & 9 & a \\ -1 & 2 & 1 & 1 \\ \hline 0 & 10 & 10 & a+1 \end{array}$$

$$\sim \left[\begin{array}{ccc|c} 1 & -2 & -1 & -1 \\ 0 & 5 & 5 & 15 \\ 0 & 0 & 0 & a-29 \end{array}\right] \begin{array}{l} \\ \\ \longleftarrow R_3 \to R_3 - 2R_2 \end{array}$$

$$\begin{array}{rrrr} 0 & 10 & 10 & a+1 \\ 0 & -10 & -10 & -30 \\ \hline 0 & 0 & 0 & a-29 \end{array}$$

b If $a \neq 29$, we have an inconsistent system as zero = non-zero, and therefore no solutions. In this case we have three planes with no common point of intersection.

c If $a = 29$, the last row is all zeros, indicating infinitely many solutions.

Letting $z = t$ in row 2 gives $5y + 5t = 15$

$\therefore\ y = 3 - t$

Using row 1, $x - 2y - z = -1$

$\therefore\ x - 2(3 - t) - t = -1$

$\therefore\ x - 6 + 2t - t = -1$

$\therefore\ x = 5 - t$

We have infinitely many solutions, in the form:

$x = 5 - t,\quad y = 3 - t,\quad z = t,\quad$ where $t \in \mathbb{R}$.

In this case we have three planes which meet in a line.

6 Solve the system of equations $\begin{cases} x - y + 3z = 1 \\ 2x - 3y - z = 3 \\ 3x - 5y - 5z = k \end{cases}$ where k is a real number.

State the geometrical meaning of the different solutions.

7 A system of equations is $\begin{cases} x + 3y + 3z = a - 1 \\ 2x - y + z = 7 \\ 3x - 5y + az = 16 \end{cases}$.

a Reduce the system to echelon form using row operations.

b Show that if $a = -1$ the system has infinitely many solutions, and find their form. Interpret this result geometrically.

c If $a \neq -1$, find the unique solution in terms of a. Interpret this result geometrically.

8 Reduce the system of equations: $\begin{cases} 2x + y - z = 3 \\ mx - 2y + z = 1 \\ x + 2y + mz = -1 \end{cases}$ to a form in which the solutions may be determined for all real values of m.

a Show that the system has no solutions for one value of m. Interpret this result geometrically.

b Show that there are infinitely many solutions for another value of m. Interpret this result geometrically.

c i For what values of m does the system have a unique solution? Interpret this result geometrically.

ii Show that the unique solution is $x = \dfrac{7}{m+5}$, $y = \dfrac{3(m-2)}{m+5}$, $z = \dfrac{-7}{m+5}$.

9 Find if and where the following planes meet:

P_1: $\mathbf{r}_1 = 2\mathbf{i} - \mathbf{j} + \lambda(3\mathbf{i} + \mathbf{k}) + \mu(\mathbf{i} + \mathbf{j} - \mathbf{k})$ P_2: $\mathbf{r}_2 = 3\mathbf{i} - \mathbf{j} + 3\mathbf{k} + r(2\mathbf{i} - \mathbf{k}) + s(\mathbf{i} + \mathbf{j})$

P_3: $\mathbf{r}_3 = \begin{pmatrix} 2 \\ -1 \\ 2 \end{pmatrix} + t\begin{pmatrix} 1 \\ -1 \\ 0 \end{pmatrix} - u\begin{pmatrix} 0 \\ -1 \\ 2 \end{pmatrix}$

REVIEW SET 15A NON-CALCULATOR

1 For the line that passes through $(-6, 3)$ with direction $\begin{pmatrix} 4 \\ -3 \end{pmatrix}$, write down the corresponding:

a vector equation b parametric equations c Cartesian equation.

2 $(-3, m)$ lies on the line with vector equation $\begin{pmatrix} x \\ y \end{pmatrix} = \begin{pmatrix} 18 \\ -2 \end{pmatrix} + t\begin{pmatrix} -7 \\ 4 \end{pmatrix}$. Find m.

3 Line L has equation $\mathbf{r} = \begin{pmatrix} 3 \\ -3 \end{pmatrix} + t\begin{pmatrix} 2 \\ 5 \end{pmatrix}$.

a Locate the point on the line corresponding to $t = 1$.

b Explain why the direction of the line could also be described by $\begin{pmatrix} 4 \\ 10 \end{pmatrix}$.

c Use your answers to **a** and **b** to write an alternative vector equation for line L.

4 P(2, 0, 1), Q(3, 4, −2), and R(−1, 3, 2) are three points in space.

a Find parametric equations of line (PQ).

b Show that if $\theta = \widehat{PQR}$, then $\cos\theta = \dfrac{20}{\sqrt{26}\sqrt{33}}$.

5 Triangle ABC is formed by three lines:

Line (AB) is $\begin{pmatrix} x \\ y \end{pmatrix} = \begin{pmatrix} 4 \\ -1 \end{pmatrix} + t\begin{pmatrix} 1 \\ 3 \end{pmatrix}$. Line (BC) is $\begin{pmatrix} x \\ y \end{pmatrix} = \begin{pmatrix} 7 \\ 4 \end{pmatrix} + s\begin{pmatrix} 1 \\ -1 \end{pmatrix}$.

Line (AC) is $\begin{pmatrix} x \\ y \end{pmatrix} = \begin{pmatrix} -1 \\ 0 \end{pmatrix} + u\begin{pmatrix} 3 \\ 1 \end{pmatrix}$. s, t, and u are scalars.

a Use vector methods to find the coordinates of A, B, and C.

b Find $|\overrightarrow{AB}|$, $|\overrightarrow{BC}|$, and $|\overrightarrow{AC}|$.

c Classify triangle ABC.

6 **a** Consider two unit vectors **a** and **b**. Prove that the vector $\mathbf{a}+\mathbf{b}$ bisects the angle between vector **a** and vector **b**.

b Consider the points $H(9, 5, -5)$, $J(7, 3, -4)$, and $K(1, 0, 2)$.

Find the equation of the line L that passes through J and bisects $\widehat{HJK}$.

c Find the coordinates of the point where L meets (HK).

7 Suppose A is $(3, 2, -1)$ and B is $(-1, 2, 4)$.

a Write down a vector equation of the line through A and B.

b Find the equation of the plane through B with normal $\overrightarrow{AB}$.

c Find *two* points on (AB) which are $2\sqrt{41}$ units from A.

8 For $C(-3, 2, -1)$ and $D(0, 1, -4)$, find the coordinates of the point where the line passing through C and D meets the plane with equation $2x - y + z = 3$.

9 Suppose $\overrightarrow{OA} = \mathbf{a}$, $\overrightarrow{OB} = \mathbf{b}$, $|\mathbf{a}| = 3$, $|\mathbf{b}| = \sqrt{7}$, and $\mathbf{a} \times \mathbf{b} = \mathbf{i} + 2\mathbf{j} - 3\mathbf{k}$. Find:

a $\mathbf{a} \bullet \mathbf{b}$

b the area of triangle OAB.

10 **a** How far is $X(-1, 1, 3)$ from the plane $x - 2y - 2z = 8$?

b Find the coordinates of the foot of the perpendicular from $Q(-1, 2, 3)$ to the line $2 - x = y - 3 = -\frac{1}{2}z$.

11 **a** Find if possible the point where the line through $L(1, 0, 1)$ and $M(-1, 2, -1)$ meets the plane with equation $x - 2y - 3z = 14$.

b Find the shortest distance from L to the plane.

12 The equations of two lines are: L_1: $x = 3t - 4$, $y = t + 2$, $z = 2t - 1$

L_2: $x = \dfrac{y-5}{2} = \dfrac{-z-1}{2}$.

a Find the point of intersection of L_1 and the plane $2x + y - z = 2$.

b Find the point of intersection of L_1 and L_2.

c Find the equation of the plane that contains L_1 and L_2.

13 Show that the line $x - 1 = \dfrac{y+2}{2} = \dfrac{z-3}{4}$ is parallel to the plane $6x + 7y - 5z = 8$ and find the distance between them.

14 $x^2 + y^2 + z^2 = 26$ is the equation of a sphere with centre $(0, 0, 0)$ and radius $\sqrt{26}$ units. Find the point(s) where the line through $(3, -1, -2)$ and $(5, 3, -4)$ meets the sphere.

15 When an archer fires an arrow, he is suddenly aware of a breeze which pushes his shot off-target. The speed of the shot $|\mathbf{v}|$ is *not* affected by the wind, but the arrow's flight is $2°$ off-line.

a Draw a vector diagram to represent the situation.

b Hence explain why:

i the breeze must be $91°$ to the intended direction of the arrow

ii the speed of the breeze must be $2|\mathbf{v}|\sin 1°$.

16 In the figure ABCD is a parallelogram. X is the midpoint of BC, and Y is on [AX] such that $\overrightarrow{AY} = 2\overrightarrow{YX}$.

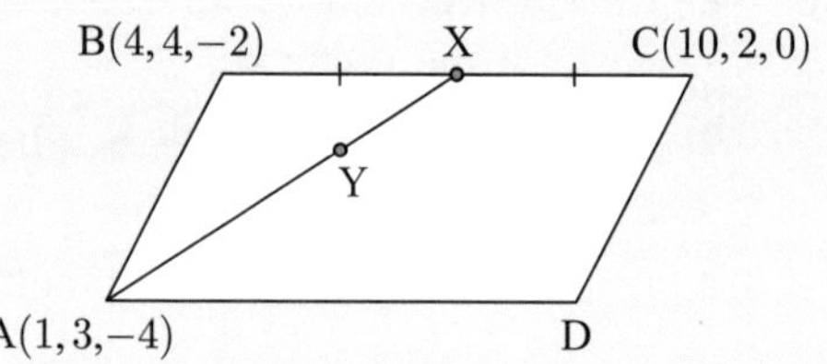

a Find the coordinates of X and D.

b Find the coordinates of Y.

c Show that B, Y, and D are collinear.

17 Solve the system $\begin{cases} x - y + z = 5 \\ 2x + y - z = -1 \\ 7x + 2y + kz = -k \end{cases}$ for any real number k

using row operations.

Give geometric interpretations of your results.

REVIEW SET 15B — CALCULATOR

1 Find the vector equation of the line which cuts the y-axis at (0, 8) and has direction $5\mathbf{i} + 4\mathbf{j}$.

2 A yacht is sailing with constant speed $5\sqrt{10}$ km h^{-1} in the direction $-\mathbf{i} - 3\mathbf{j}$. Initially it is at point $(-6, 10)$. A beacon is at (0, 0) at the centre of a tiny atoll. Distances are in kilometres.

a Find in terms of **i** and **j** :

i the initial position vector of the yacht

ii the direction vector of the yacht

iii the position vector of the yacht at any time t hours, $t \geqslant 0$.

b Find the time when the yacht is closest to the beacon.

c If there is a reef of radius 8 km around the atoll, will the yacht hit the reef?

3 Write down **i** a vector equation **ii** parametric equations for the line passing through:

a $(2, -3)$ with direction $\begin{pmatrix} 4 \\ -1 \end{pmatrix}$

b $(-1, 6, 3)$ and $(5, -2, 0)$.

4 A small plane can fly at 350 km h^{-1} in still conditions. Its pilot needs to fly due north, but needs to deal with a 70 km h^{-1} wind from the east.

a In what direction should the pilot face the plane in order that his resultant velocity is due north?

b What will the speed of the plane be?

5 Find the angle between line L_1 passing through (0, 3) and $(5, -2)$, and line L_2 passing through $(-2, 4)$ and $(-6, 7)$.

6 Submarine X23 is at (2, 4). It fires a torpedo with velocity vector $\begin{pmatrix} 1 \\ -3 \end{pmatrix}$ at exactly 2:17 pm.

Submarine Y18 is at (11, 3). It fires a torpedo with velocity vector $\begin{pmatrix} -1 \\ a \end{pmatrix}$ at 2:19 pm to intercept the torpedo from X23. Distance units are kilometres. t is in minutes.

a Find $x_1(t)$ and $y_1(t)$ for the torpedo fired from submarine X23.

b Find $x_2(t)$ and $y_2(t)$ for the torpedo fired from submarine Y18.

c At what time does the interception occur?

d What was the direction and speed of the interception torpedo?

7 Suppose P_1 is the plane $2x - y - 2z = 9$ and P_2 is the plane $x + y + 2z = 1$.
L is the line with parametric equations $x = t$, $y = 2t - 1$, $z = 3 - t$.
Find the acute angle between: **a** L and P_1 **b** P_1 and P_2.

8 Consider the lines L_1: $\dfrac{x-8}{3} = \dfrac{y+9}{-16} = \dfrac{z-10}{7}$ and L_2: $x = 15 + 3t$, $y = 29 + 8t$, $z = 5 - 5t$.

a Show that the lines are skew. **b** Find the acute angle between them.

c Line L_3 is a translation of L_1 which intersects L_2. Find the equation of the plane containing L_2 and L_3.

d Find the shortest distance between them.

9 **a** Find the equation of the plane through A$(-1, 2, 3)$, B$(1, 0, -1)$, and C$(0, -1, 5)$.

b If X is $(3, 2, 4)$, find the angle that (AX) makes with this plane.

10 **a** Find all vectors of length 3 units which are normal to the plane $x - y + z = 6$.

b Find a unit vector parallel to $\mathbf{i} + r\mathbf{j} + 3\mathbf{k}$ and perpendicular to $2\mathbf{i} - \mathbf{j} + 2\mathbf{k}$.

c The distance from A$(-1, 2, 3)$ to the plane with equation $2x - y + 2z = k$ is 3 units. Find k.

11 Find the angle between the lines with equations $4x - 5y = 11$ and $2x + 3y = 7$.

12 Consider A$(2, -1, 3)$ and B$(0, 1, -1)$.

a Find the vector equation of the line through A and B.

b Hence find the coordinates of C on (AB) which is 2 units from A.

13 Find the angle between the plane $2x + 2y - z = 3$ and the line
$x = t - 1$, $y = -2t + 4$, $z = -t + 3$.

14 Let $\mathbf{r} = 2\mathbf{i} - 2\mathbf{j} - \mathbf{k}$, $\mathbf{s} = 2\mathbf{i} + \mathbf{j} + 2\mathbf{k}$, $\mathbf{t} = \mathbf{i} + 2\mathbf{j} - \mathbf{k}$, be the position vectors of the points R, S, and T, respectively. Find the area of the triangle RST.

15 Classify the following line pairs as either parallel, intersecting, or skew. In each case find the measure of the acute angle between them.

a $x = 2 + t$, $y = -1 + 2t$, $z = 3 - t$ and $x = -8 + 4s$, $y = s$, $z = 7 - 2s$

b $x = 3 + t$, $y = 5 - 2t$, $z = -1 + 3t$ and $x = 2 - s$, $y = 1 + 3s$, $z = 4 + s$

16 $\mathbf{p} = \begin{pmatrix} 1 \\ -1 \\ 2 \end{pmatrix}$ and $\mathbf{q} = \begin{pmatrix} 2 \\ 3 \\ -1 \end{pmatrix}$

a Find $\mathbf{p} \times \mathbf{q}$.

b Find m if $\mathbf{p} \times \mathbf{q}$ is perpendicular to the line L with equation $\mathbf{r} = \begin{pmatrix} 1 \\ -2 \\ 3 \end{pmatrix} + \lambda \begin{pmatrix} 2 \\ 1 \\ m \end{pmatrix}$.

c Hence find the equation of the plane P containing L which is perpendicular to $\mathbf{p} \times \mathbf{q}$.

d Find t if the point A$(4, t, 2)$ lies on the plane P.

e For the value of t found in **d**, if B is the point $(6, -3, 5)$, find the exact value of the sine of the angle between (AB) and the plane P.

17 **a** Show that the plane $2x + y + z = 5$ contains the line
L_1: $x = -2t + 2$, $y = t$, $z = 3t + 1$, $t \in \mathbb{R}$.

b For what values of k does the plane $x + ky + z = 3$ contain L_1?

c *Hence* find the values of p and q for which the following system of equations has an infinite number of solutions. Clearly explain your reasoning.

$$\begin{cases} 2x + y + z = 5 \\ x - y + z = 3 \\ -2x + py + 2z = q \end{cases}$$

18 Consider the system $\begin{cases} x - 3y + 2z = -5 \\ 3x + y + (2-k)z = 10 \\ -2x + 6y + kz = 5 \end{cases}$ where k can take any real value.

a Reduce the system to echelon form.

b For what value of k does the system have no solutions? Interpret this result geometrically.

c **i** For what value(s) of k does the system have a unique solution?

ii Find the unique solution in terms of k, and interpret the result geometrically.

iii Find the unique solution when $k = 1$.

REVIEW SET 15C

1 Find the velocity vector of an object moving in the direction $3\mathbf{i} - \mathbf{j}$ with speed 20 km h^{-1}.

2 A moving particle has coordinates $P(x(t), y(t))$ where $x(t) = -4 + 8t$ and $y(t) = 3 + 6t$. The distance units are metres, and $t \geqslant 0$ is the time in seconds. Find the:

a initial position of the particle

b position of the particle after 4 seconds

c particle's velocity vector

d speed of the particle.

3 Trapezium KLMN is formed by the following lines:

(KL) is $\begin{pmatrix} x \\ y \end{pmatrix} = \begin{pmatrix} 2 \\ 19 \end{pmatrix} + p\begin{pmatrix} 5 \\ -2 \end{pmatrix}$. (ML) is $\begin{pmatrix} x \\ y \end{pmatrix} = \begin{pmatrix} 33 \\ -5 \end{pmatrix} + q\begin{pmatrix} -11 \\ 16 \end{pmatrix}$.

(NK) is $\begin{pmatrix} x \\ y \end{pmatrix} = \begin{pmatrix} 3 \\ 7 \end{pmatrix} + r\begin{pmatrix} 4 \\ 10 \end{pmatrix}$. (MN) is $\begin{pmatrix} x \\ y \end{pmatrix} = \begin{pmatrix} 43 \\ -9 \end{pmatrix} + s\begin{pmatrix} -5 \\ 2 \end{pmatrix}$.

p, q, r, and s are scalars.

a Which two lines are parallel? Explain your answer.

b Which lines are perpendicular? Explain your answer.

c Use vector methods to find the coordinates of K, L, M, and N.

d Calculate the area of trapezium KLMN.

4 Find the angle between the lines:
L_1: $x = 1 - 4t$, $y = 3t$ and L_2: $x = 2 + 5s$, $y = 5 - 12s$.

5 Consider A(3, −1, 1) and B(0, 2, −2).

a Find $|\overrightarrow{AB}|$.

b Show that the line passing through A and B can be described by
$\mathbf{r} = 2\mathbf{j} - 2\mathbf{k} + \lambda(-\mathbf{i} + \mathbf{j} - \mathbf{k})$ where λ is a scalar.

c Find the angle between (AB) and the line with vector equation $t(\mathbf{i} + \mathbf{j} + \mathbf{k})$.

6 Let $\mathbf{i}$ represent a displacement 1 km due east and $\mathbf{j}$ represent a displacement 1 km due north. Road A passes through $(-9, 2)$ and $(15, -16)$. Road B passes through $(6, -18)$ and $(21, 18)$.

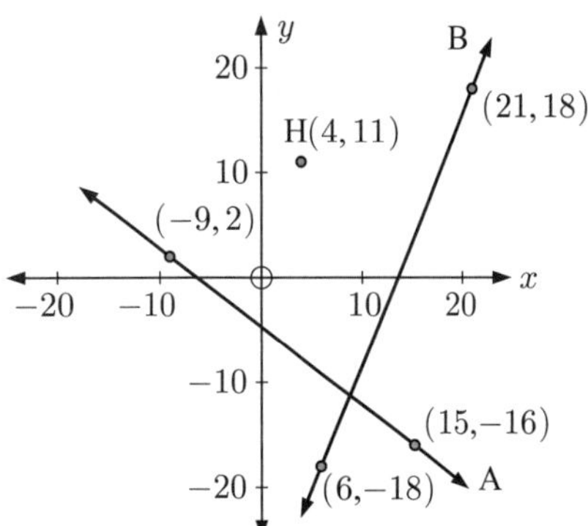

a Find a vector equation for each of the roads.

b An injured hiker is at $(4, 11)$, and needs to travel the shortest possible distance to a road. Towards which road should he head, and how far will he need to walk to reach this road?

7 Given the points $A(4, 2, -1)$, $B(2, 1, 5)$, and $C(9, 4, 1)$:

a Show that $\overrightarrow{AB}$ is perpendicular to $\overrightarrow{AC}$.

b Find the equation of the line through: **i** A and B **ii** A and C.

8 The triangle with vertices $P(-1, 2, 1)$, $Q(0, 1, 4)$, and $R(a, -1, -2)$ has area $\sqrt{118}$ units^2. Find a.

9 Consider $A(-1, 2, 3)$, $B(2, 0, -1)$, and $C(-3, 2, -4)$.

a Find the equation of the plane defined by A, B, and C.

b Find the measure of $C\widehat{A}B$.

c $D(r, 1, -r)$ is a point such that $B\widehat{D}C$ is a right angle. Find r.

10 Given $A(-1, 2, 3)$, $B(1, 0, -1)$, and $C(1, 3, 0)$, find:

a the normal vector to the plane containing A, B, and C

b D, the fourth vertex of parallelogram ACBD

c the area of parallelogram ACBD

d the coordinates of the foot of the perpendicular from C to the line AB.

11 $P(2, 0, 1)$, $Q(3, 4, -2)$, and $R(-1, 3, 2)$ are three points in space. Find:

a $\overrightarrow{PQ}$, $|\overrightarrow{PQ}|$, and $\overrightarrow{QR}$

b the parametric equations of (PQ)

c a vector equation of the plane PQR.

12 Given the point $A(-1, 3, 2)$, the plane $2x - y + 2z = 8$, and the line defined by $x = 7 - 2t$, $y = -6 + t$, $z = 1 + 5t$, find:

a the distance from A to the plane

b the coordinates of the point on the plane nearest to A

c the shortest distance from A to the line.

13 **a** Find the equation of the plane through $A(-1, 0, 2)$, $B(0, -1, 1)$, and $C(1, 2, -1)$.

b Find the equation of the line, in parametric form, which passes through the origin and is normal to the plane in **a**.

c Find the point where the line in **b** intersects the plane in **a**.

14 Consider the lines with equations $\dfrac{x-3}{2} = y - 4 = \dfrac{z+1}{-2}$ and $x = -1 + 3t, \quad y = 2 + 2t, \quad z = 3 - t.$

a Are the lines parallel, intersecting, or skew? Justify your answer.

b Determine the acute angle between the lines.

15 Line 1 has equation $\dfrac{x-8}{3} = \dfrac{y+9}{-16} = \dfrac{z-10}{7}.$

Line 2 has vector equation $\begin{pmatrix} x \\ y \\ z \end{pmatrix} = \begin{pmatrix} 15 \\ 29 \\ 5 \end{pmatrix} + \lambda \begin{pmatrix} 3 \\ 8 \\ -5 \end{pmatrix}.$

a Show that lines 1 and 2 are skew.

b Line 3 is a translation of line 1 which intersects line 2. Find the equation of the plane containing lines 2 and 3.

c Hence find the shortest distance between lines 1 and 2.

d Find the coordinates of the points where the common perpendicular meets the lines 1 and 2.

16 Lines L_1 and L_2 are defined by

$$L_1: \ \mathbf{r} = \begin{pmatrix} 3 \\ -2 \\ -2 \end{pmatrix} + s \begin{pmatrix} -1 \\ 1 \\ 2 \end{pmatrix} \quad \text{and} \quad L_2: \ \mathbf{r} = \begin{pmatrix} 3 \\ 0 \\ -1 \end{pmatrix} + t \begin{pmatrix} -1 \\ -1 \\ 1 \end{pmatrix}.$$

a Find the coordinates of A, the point of intersection of the lines.

b Show that the point B(0, −3, 2) lies on the line L_2.

c Find the equation of the line BC given that C(3, −2, −2) lies on L_1.

d Find the equation of the plane containing A, B, and C.

e Find the area of triangle ABC.

f Show that the point D(9, −4, 2) lies on the normal to the plane passing through C.

17 Three planes have the equations given below:

Plane A: $x + 3y + 2z = 5$

Plane B: $2x + y + 9z = 20$

Plane C: $x - y + 6z = 8$

a Show that plane A and plane B intersect in a line L_1.

b Show that plane B and plane C intersect in a line L_2.

c Show that plane A and plane C intersect in a line L_3.

d Show that L_1, L_2, and L_3 are parallel but not coincident.

e What does this mean geometrically?

Chapter 16

Complex numbers

Syllabus reference: 1.6, 1.7

Contents:

HISTORICAL NOTE **EULER'S BEAUTIFUL EQUATION**

One of the most remarkable results in mathematics is known as **Euler's beautiful equation** $e^{i\pi} = -1$ named after **Leonhard Euler**.

It is called beautiful because it links together three great constants of mathematics: Euler's constant e; the imaginary number i; and the ratio of a circle's circumference to its diameter, π.

The beautiful equation comes from Euler's more general result: $e^{i\theta} = \cos\theta + i\sin\theta, \quad \theta \in \mathbb{R}$.

We will see in this chapter the importance and usefulness of this formula. However, the concept of a real number raised to an imaginary power is completely abstract.

Harvard lecturer **Benjamin Pierce** said of $e^{i\pi} = -1$,

> "Gentlemen, that is surely true, it is absolutely paradoxical; we cannot understand it, and we don't know what it means, but we have proved it, and therefore we know it must be the truth."

Having now studied both trigonometry and vectors, we can return to study complex numbers in greater detail.

A COMPLEX NUMBERS AS 2-D VECTORS

In **Chapter 6** we saw that a complex number can be written in Cartesian form as $z = a + bi$ where $a = \mathcal{R}e(z)$ and $b = \mathcal{I}m(z)$ are both real numbers.

There is a one-to-one relationship between any complex number $a + bi$ and the point (a, b) in the Cartesian plane.

We can therefore plot any complex number on a plane as a unique ordered number pair. We refer to the plane as the **complex plane** or the **Argand plane**.

On the complex plane, the x-axis is called the **real axis** and the y-axis is called the **imaginary axis**.

- All **real** numbers with $b = 0$ lie on the real axis.
- All **purely imaginary** numbers with $a = 0$ lie on the imaginary axis.
- The origin $(0, 0)$ lies on both axes and it corresponds to $z = 0$, a real number.
- Complex numbers that are neither real nor purely imaginary (a and b are not both 0) lie in one of the four quadrants.

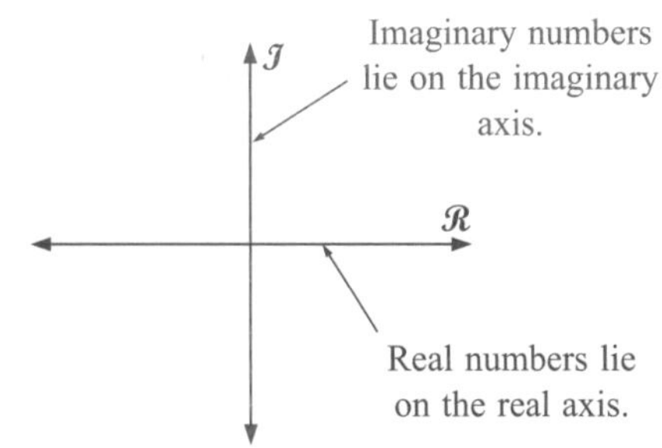

We saw in **Chapter 14** that any point P on the Cartesian plane can be represented by a vector. The position vector for the point P is $\overrightarrow{OP}$. In the same way we can illustrate complex numbers using vectors on the Argand plane. We call this an **Argand diagram**.

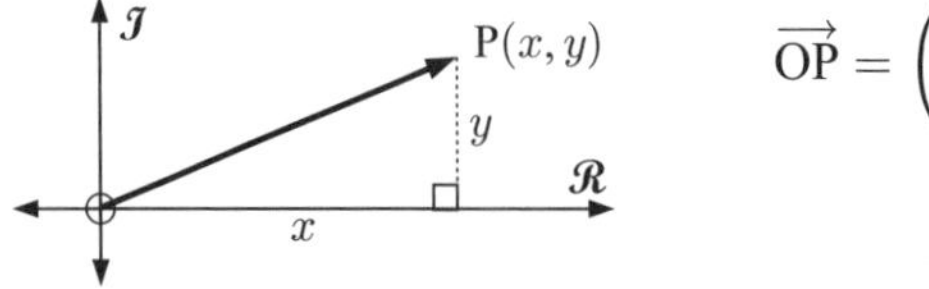

$\overrightarrow{OP} = \begin{pmatrix} x \\ y \end{pmatrix}$ represents $x + yi$

For example:

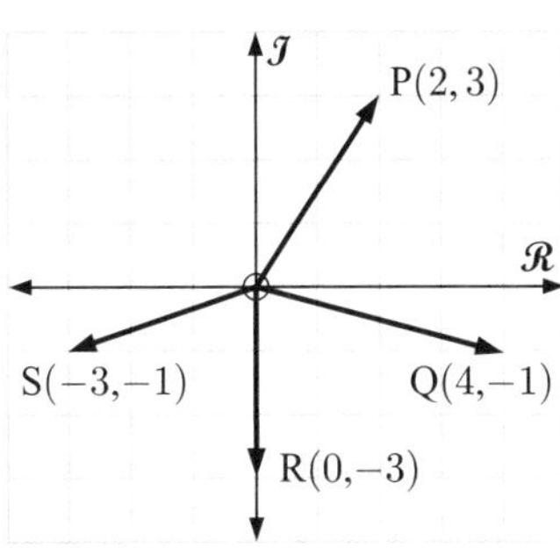

$\overrightarrow{OP} = \begin{pmatrix} 2 \\ 3 \end{pmatrix}$ represents $2 + 3i$

$\overrightarrow{OQ} = \begin{pmatrix} 4 \\ -1 \end{pmatrix}$ represents $4 - i$

$\overrightarrow{OR} = \begin{pmatrix} 0 \\ -3 \end{pmatrix}$ represents $-3i$

$\overrightarrow{OS} = \begin{pmatrix} -3 \\ -1 \end{pmatrix}$ represents $-3 - i$

Example 1 **Self Tutor**

Illustrate the positions of the following complex numbers on an Argand diagram:

$z_1 = 3,\quad z_2 = 4 + 3i,$
$z_3 = 5i,\quad z_4 = -4 + 2i,$
$z_5 = -3 - i,\quad z_6 = -2i.$

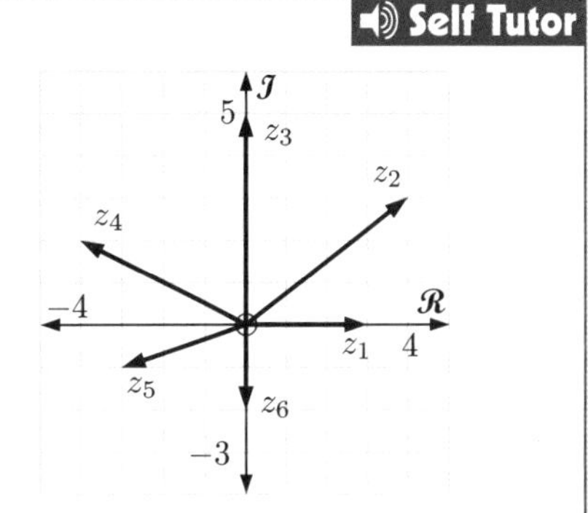

We can apply the vector operations of addition, subtraction, and scalar multiplication to give the correct answers for these operations with complex numbers.

Example 2 **Self Tutor**

Suppose $z_1 = 3 + i$ and $z_2 = 1 - 4i$. Find using both algebra and vectors:

a $z_1 + z_2$ **b** $z_1 - z_2$

a $z_1 + z_2$
$= 3 + i + 1 - 4i$
$= 4 - 3i$

b $z_1 - z_2$
$= 3 + i - (1 - 4i)$
$= 3 + i - 1 + 4i$
$= 2 + 5i$

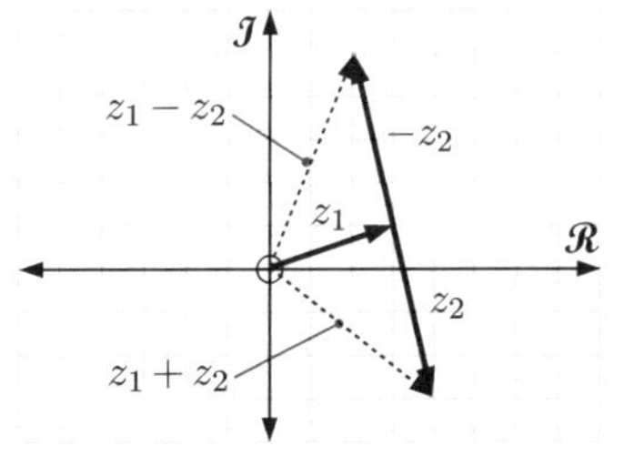

Example 3 **Self Tutor**

Suppose $z = 1 + 2i$ and $w = 3 - i$. Find using both algebra and vectors:

a $2z + w$ **b** $z - 2w$

a $2z + w$
$= 2(1 + 2i) + 3 - i$
$= 2 + 4i + 3 - i$
$= 5 + 3i$

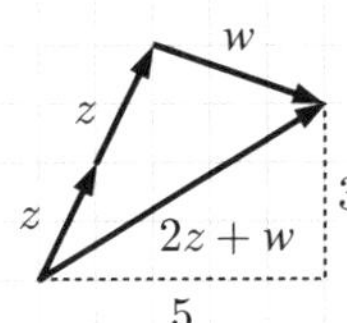

b $z - 2w$
$= 1 + 2i - 2(3 - i)$
$= 1 + 2i - 6 + 2i$
$= -5 + 4i$

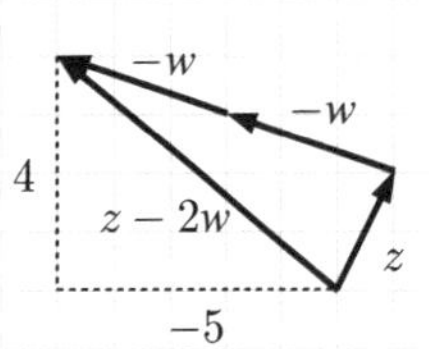

REPRESENTING CONJUGATES

If $z = x + iy$, then $z^* = x - iy$.

This means that if $\overrightarrow{OP} = \begin{pmatrix} x \\ y \end{pmatrix}$ represents z, then $\overrightarrow{OP'} = \begin{pmatrix} x \\ -y \end{pmatrix}$ represents z^*.

For example:

$\overrightarrow{OP}$ represents $2 + 4i$ and
$\overrightarrow{OP'}$ represents $2 - 4i$.

$\overrightarrow{OQ}$ represents $-3i$ and
$\overrightarrow{OQ'}$ represents $3i$.

$\overrightarrow{OR}$ represents $-3 + 2i$ and
$\overrightarrow{OR'}$ represents $-3 - 2i$.

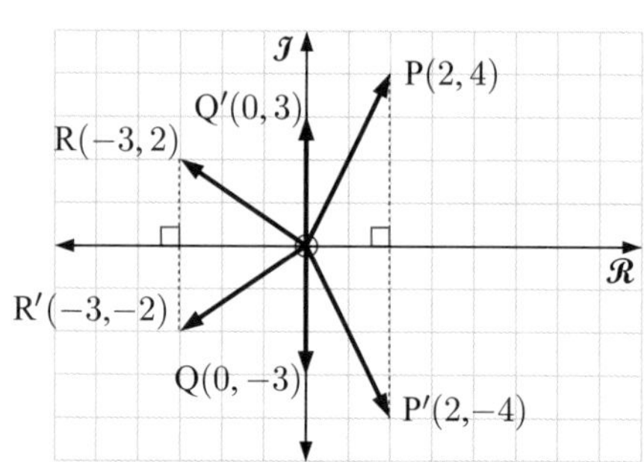

If z is $\overrightarrow{OP}$, its conjugate z^* is a reflection of $\overrightarrow{OP}$ in the real axis.

EXERCISE 16A

1 On an Argand diagram, illustrate the complex numbers:

a $z_1 = 5$ **b** $z_2 = -1 + 2i$ **c** $z_3 = -6 - 2i$
d $z_4 = -6i$ **e** $z_5 = 2 - i$ **f** $z_6 = 4i$

2 Suppose $z = 1 + 2i$ and $w = 3 - i$. Find using both algebra and vectors:

a $z + w$ **b** $z - w$ **c** $2z - w$ **d** $w - 3z$

3 Suppose $z_1 = 4 - i$ and $z_2 = 2 + 3i$. Find using both algebra and vectors:

a $z_1 + 1$ **b** $z_1 + 2i$ **c** $z_2 + \frac{1}{2}z_1$ **d** $\dfrac{z_1 + 4}{2}$

4 Suppose z is a complex number. Explain, with illustration, how to find geometrically:

a $3z$ **b** $-2z$ **c** z^* **d** $3i - z$
e $2 - z$ **f** $z^* + i$ **g** $\dfrac{z+2}{3}$ **h** $\dfrac{z-4}{2}$

5 Show on an Argand diagram:

a $z = 3 + 2i$ and its conjugate $z^* = 3 - 2i$
b $z = -2 + 5i$ and its conjugate $z^* = -2 - 5i$

6 Suppose $z = 2 - i$. From the diagram we can see that $z + z^* = 4$, which is real.
Explain, with illustration, why $z + z^*$ is always real for any complex number z.

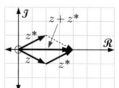

7 Explain, with illustration, why $z - z^*$ is always purely imaginary or zero. What distinguishes these two cases?

8 Explain using an Argand diagram why for all complex numbers z and w:

a $(z + w)^* = z^* + w^*$ **b** $(z - w)^* = z^* - w^*$

DISCUSSION

We have seen that for two vectors **a** and **b**, there are two types of vector product:

- the scalar dot product $\mathbf{a} \bullet \mathbf{b}$
- the vector cross product $\mathbf{a} \times \mathbf{b}$.

Are either of these products useful when considering the product of two complex numbers?

Are there any other tools we have seen that can help with the product of two complex numbers?

B MODULUS

We know that vectors have magnitude and direction, so we can also attribute a magnitude and direction to complex numbers. The magnitude of a complex number is called its **modulus**, and its direction is called its **argument**.

The **modulus** of the complex number $z = a + bi$ is the length of the vector $\begin{pmatrix} a \\ b \end{pmatrix}$, which is the real number $|z| = \sqrt{a^2 + b^2}$.

Notice that if $z = a + bi$ then $|z|$ gives the distance of the point (a, b) from the origin. This is consistent with the definition in **Chapter 2** of $|x|$ for $x \in \mathbb{R}$. We stated there that $|x|$ is the distance of real number x from the origin O.

For example, consider the complex number $z = 3 + 2i$.

The distance from O to P is its modulus, $|z|$.

$\therefore \quad |z| = \sqrt{3^2 + 2^2} \quad$ {Pythagoras}

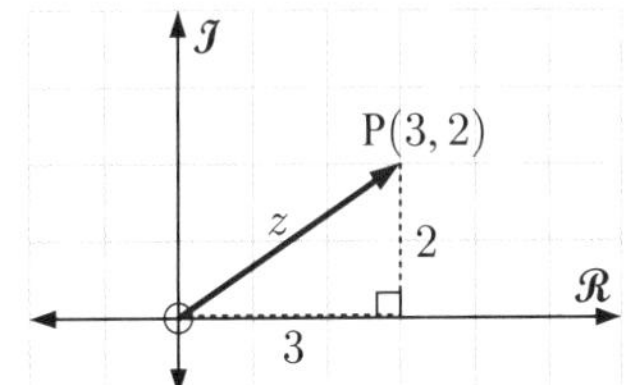

Example 4 — Self Tutor

Find $|z|$ for z equal to:

a $3 + 2i$ **b** $3 - 2i$ **c** $-3 - 2i$

a $|z| = \sqrt{3^2 + 2^2} = \sqrt{13}$

b $|z| = \sqrt{3^2 + (-2)^2} = \sqrt{9 + 4} = \sqrt{13}$

c $|z| = \sqrt{(-3)^2 + (-2)^2} = \sqrt{9 + 4} = \sqrt{13}$

PROPERTIES OF MODULUS

- $|z^*| = |z|$
- $|z|^2 = zz^*$
- $|z_1 z_2| = |z_1||z_2|$ and $\left|\dfrac{z_1}{z_2}\right| = \dfrac{|z_1|}{|z_2|}$ provided $z_2 \neq 0$
- $|z_1 z_2 z_3 .... z_n| = |z_1||z_2||z_3| .... |z_n|$ and $|z^n| = |z|^n$ for $n \in \mathbb{Z}^+$.

EXERCISE 16B.1

1 Find $|z|$ for z equal to:

a $3-4i$ **b** $5+12i$ **c** $-8+2i$ **d** $3i$ **e** -4

2 If $z=2+i$ and $w=-1+3i$ find:

a $|z|$ **b** $|z^*|$ **c** $|z^*|^2$
d zz^* **e** $|zw|$ **f** $|z||w|$
g $\left|\frac{z}{w}\right|$ **h** $\frac{|z|}{|w|}$ **i** $|z^2|$
j $|z|^2$ **k** $|z^3|$ **l** $|z|^3$

Use your answers to check the properties of modulus.

Example 5 **Self Tutor**

Prove that $|z_1z_2|=|z_1||z_2|$ for all complex numbers z_1 and z_2.

Let $z_1=a+bi$ and $z_2=c+di$ where a, b, c, and d are real.

$\therefore\ z_1z_2=(a+bi)(c+di)=[ac-bd]+i[ad+bc]$

Thus
$$\begin{aligned}|z_1z_2|&=\sqrt{(ac-bd)^2+(ad+bc)^2}\\&=\sqrt{a^2c^2-\cancel{2abcd}+b^2d^2+a^2d^2+\cancel{2abcd}+b^2c^2}\\&=\sqrt{a^2(c^2+d^2)+b^2(c^2+d^2)}\\&=\sqrt{(c^2+d^2)(a^2+b^2)}\\&=\sqrt{a^2+b^2}\times\sqrt{c^2+d^2}\\&=|z_1||z_2|\end{aligned}$$

3 Prove that for all complex numbers z:

a $|z^*|=|z|$ **b** $|z|^2=zz^*$

4 Find $|z|$ given:

a $z=\cos\theta+i\sin\theta$ **b** $z=r(\cos\theta+i\sin\theta),\ \ r\in\mathbb{R}$

5 Prove that $\left|\frac{z}{w}\right|=\frac{|z|}{|w|}$ for all complex numbers z and w, $w\neq 0$.

6 **a** Use the result $|z_1z_2|=|z_1||z_2|$ to show that:

i $|z_1z_2z_3|=|z_1||z_2||z_3|$ **ii** $|z_1z_2z_3z_4|=|z_1||z_2||z_3||z_4|$

b Generalise your results from **a**.

c Use the principle of mathematical induction to prove your conjecture in **b**.

d Hence prove that $|z^n|=|z|^n$ for all $n\in\mathbb{Z}^+$.

e Hence find $|z^{20}|$ for $z=1-i\sqrt{3}$.

7 Given $|z|=3$, use the rules $|zw|=|z||w|$ and $\left|\frac{z}{w}\right|=\frac{|z|}{|w|}$ to find:

a $|2z|$ **b** $|-3z|$ **c** $|(1+2i)z|$
d $|iz|$ **e** $\left|\frac{1}{z}\right|$ **f** $\left|\frac{2i}{z^2}\right|$

8 Suppose $w = \frac{z+1}{z-1}$ where $z = a + bi$, $a, b \in \mathbb{R}$.

a Write w in the form $X + Yi$ where X and Y involve a and b.

b If $|z| = 1$, find $\mathcal{Re}(w)$.

Example 6 — Self Tutor

Find $|z|$ given that $5|z-1| = |z-25|$ where z is a complex number.

$$
\begin{aligned}
5|z-1| &= |z-25| \\
\therefore\ 25|z-1|^2 &= |z-25|^2 \\
\therefore\ 25(z-1)(z-1)^* &= (z-25)(z-25)^* \quad \{\text{as } zz^* = |z|^2\} \\
\therefore\ 25(z-1)(z^*-1) &= (z-25)(z^*-25) \quad \{\text{as } (z-w)^* = z^* - w^*\} \\
\therefore\ 25zz^* - 25z - 25z^* + 25 &= zz^* - 25z - 25z^* + 625 \\
\therefore\ 24zz^* &= 600 \\
\therefore\ zz^* &= 25 \\
\therefore\ |z|^2 &= 25 \\
\therefore\ |z| &= 5 \quad \{\text{as } |z| > 0\}
\end{aligned}
$$

9 Suppose z is a complex number. Find $|z|$ if:

a $|z+9| = 3|z+1|$

b $\left|\frac{z+4}{z+1}\right| = 2$

10 Suppose z and w are non-zero complex numbers, and $|z+w| = |z-w|$.

Show that $\frac{z}{z^*} = -\frac{w}{w^*}$.

DISTANCE IN THE NUMBER PLANE

Suppose P_1 and P_2 are two points in the complex plane which correspond to the complex numbers z_1 and z_2.

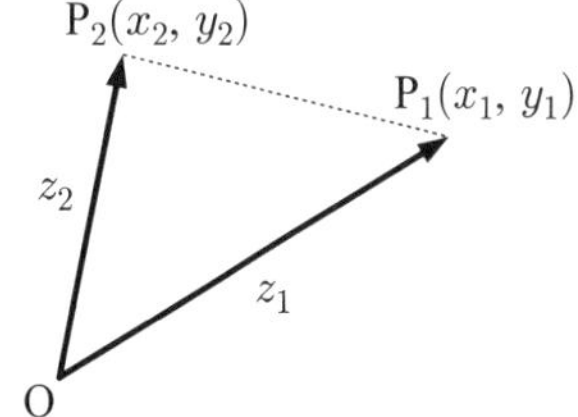

Now
$$
\begin{aligned}
|z_1 - z_2| &= |(x_1 + y_1 i) - (x_2 + y_2 i)| \\
&= |(x_1 - x_2) + (y_1 - y_2)i| \\
&= \sqrt{(x_1 - x_2)^2 + (y_1 - y_2)^2}
\end{aligned}
$$
which we recognise as the distance between P_1 and P_2.

Alternatively,
$$
\begin{aligned}
\overrightarrow{P_2P_1} &= \overrightarrow{P_2O} + \overrightarrow{OP_1} \\
&= -z_2 + z_1 \\
&= z_1 - z_2 \\
\therefore\ |z_1 - z_2| &= |\overrightarrow{P_2P_1}|
\end{aligned}
$$
= distance between P_1 and P_2.

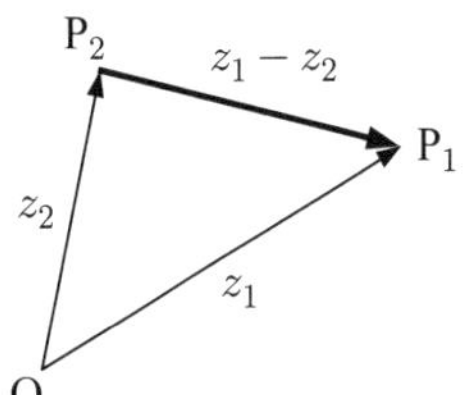

If $z_1 \equiv \overrightarrow{OP_1}$ and $z_2 \equiv \overrightarrow{OP_2}$ then $|z_1 - z_2|$ is the distance between the points P_1 and P_2.

CONNECTION TO COORDINATE GEOMETRY

There is a clear connection between complex numbers, vector geometry, and coordinate geometry.

For example, the parallelogram shown is formed by complex numbers w and z on the Argand plane.

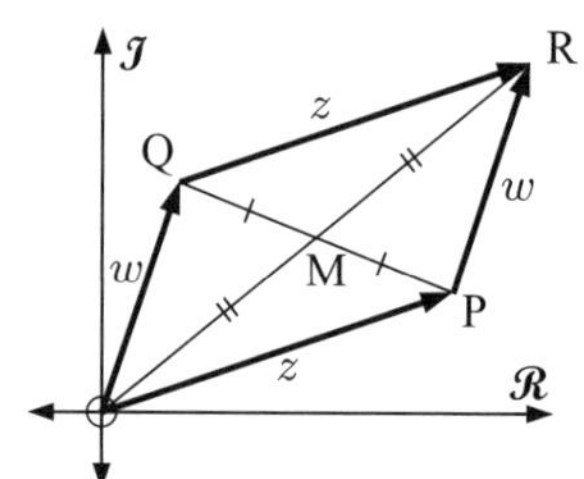

The diagonals of the parallelogram are $\overrightarrow{OR}$ and $\overrightarrow{PQ}$.

Since these diagonals bisect each other, and $\overrightarrow{OR} \equiv w + z$,

$$\overrightarrow{OM} \equiv \frac{w+z}{2}.$$

Example 7 — Self Tutor

P(2, 3) and Q(6, 1) are two points on the Cartesian plane. Use complex numbers to find:

a distance PQ **b** the midpoint of [PQ].

a If $z = 2 + 3i$ and $w = 6 + i$

then $z - w = 2 + 3i - 6 - i$
$= -4 + 2i$

$\therefore\ |z - w| = \sqrt{(-4)^2 + 2^2} = \sqrt{20}$

$\therefore$ PQ $= \sqrt{20}$ units

b $\dfrac{z+w}{2} = \dfrac{2 + 3i + 6 + i}{2} = 4 + 2i$

$\therefore$ the midpoint of [PQ] is (4, 2).

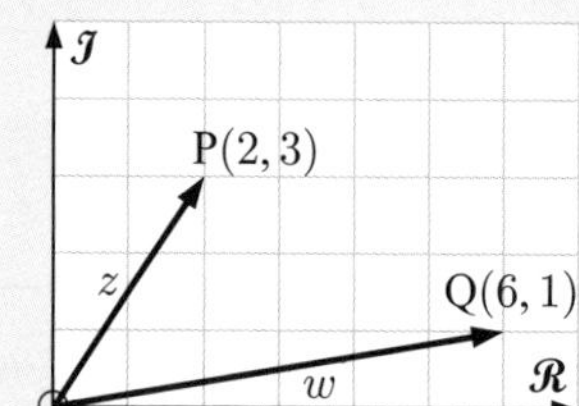

Example 8 — Self Tutor

What transformation moves z to iz?

If $z = x + iy$, then $iz = i(x + iy)$
$= xi + i^2y$
$= -y + xi$

If $z \equiv \text{P}(x, y)$ then $iz \equiv \text{P}'(-y, x)$

We notice that $|z| = \sqrt{x^2 + y^2}$

and $|iz| = \sqrt{(-y)^2 + x^2}$
$= \sqrt{x^2 + y^2}$

$\therefore$ OP$'$ = OP

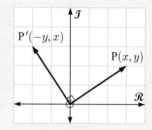

So, z moves to iz under an anti-clockwise rotation of $\frac{\pi}{2}$ about O.

EXERCISE 16B.2

1 Use complex numbers to find **i** distance AB **ii** the midpoint of [AB] for:

a A(3, 6) and B(−1, 2) **b** A(−4, 7) and B(1, −3).

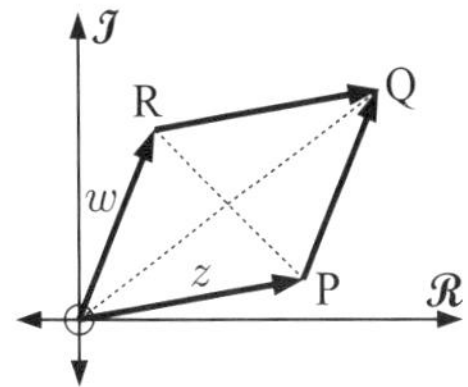

2 OPQR is a parallelogram. $\overrightarrow{OP}$ represents z and $\overrightarrow{OR}$ represents w, where z and w are complex numbers.

a Find, in terms of z and w:

i 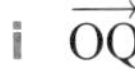 $\overrightarrow{OQ}$ **ii** $\overrightarrow{PR}$

b Explain from triangle OPQ, why $|z+w| \leqslant |z| + |w|$. Discuss the case of equality.

c Explain from triangle OPR, why $|z-w| \geqslant |w| - |z|$. Discuss the case of equality.

3 If $z_1 \equiv \overrightarrow{OP_1}$ and $z_2 \equiv \overrightarrow{OP_2}$, explain why $z_1 - z_2 = \overrightarrow{P_2P_1}$.

4 Find the transformation which moves:

a z to z^* **b** z to $-z$ **c** z to $-z^*$ **d** z to $-iz$.

5 Find the complex number z that satisfies the equation $\dfrac{50}{z^*} - \dfrac{10}{z} = 2 + 9i$ given $|z| = 2\sqrt{10}$.

C ARGUMENT AND POLAR FORM

The direction of the vector $\begin{pmatrix} a \\ b \end{pmatrix}$ can be described by its angle from the positive real axis.

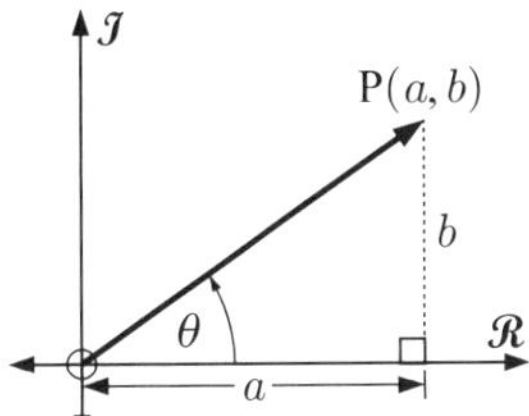

Suppose the complex number $z = a + bi$ is represented by the vector $\overrightarrow{OP}$, and that $\overrightarrow{OP}$ makes an angle of θ with the **positive real axis**.

The angle θ between $\overrightarrow{OP}$ and the positive real axis is called the **argument** of z, or simply $\arg z$.

To avoid confusion with infinitely many possibilities for θ which are 2π apart, we choose the domain of $\arg z$ to be $\theta \in\,]-\pi, \pi\,]$. This guarantees that $z \mapsto \arg z = \theta$ is a function.

Real numbers have argument of 0 or π.
Purely imaginary numbers have argument of $\frac{\pi}{2}$ *or* $-\frac{\pi}{2}$.

POLAR FORM

We have seen that the **Cartesian form** of a complex number is $z = a + bi$. However, we can also use a **polar form** which is based on the modulus and argument of z.

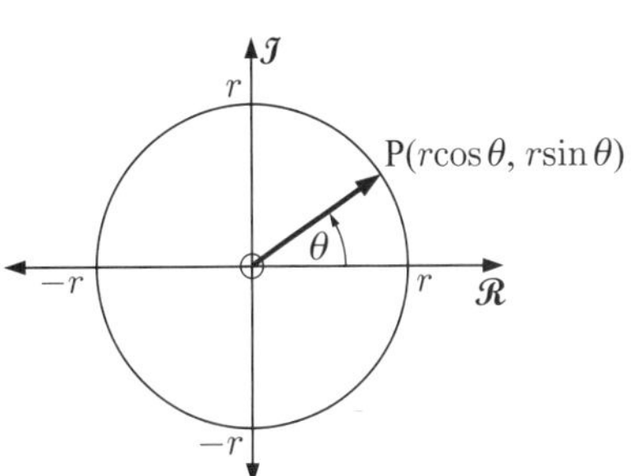

Any point P which lies on a circle with centre O(0, 0) and radius r, has Cartesian coordinates $(r\cos\theta, r\sin\theta)$.

$\therefore$ on the Argand plane, the complex number represented by $\overrightarrow{OP}$ is $z = r\cos\theta + ir\sin\theta = r(\cos\theta + i\sin\theta)$.

But $r = |z|$ and if we define $\textbf{cis}\,\boldsymbol{\theta} = \boldsymbol{\cos\theta + i\sin\theta}$

then $z = |z| \text{ cis } \theta$.

Any complex number z has **Cartesian form** $z = x + yi$ or **polar form** $z = |\,z\,|$ cis θ where $|\,z\,|$ is the **modulus** of z, θ is the **argument** of z, and cis $\theta = \cos\theta + i\sin\theta$.

We will soon see that polar form is extremely powerful for dealing with multiplication and division of complex numbers, as well as quickly finding powers and roots of numbers.

The conjugate of z has the same length as z, and its argument is $(-\theta)$.

If $z = |\,z\,|$ cis θ, then $z^* = |\,z\,|$ cis $(-\theta)$.

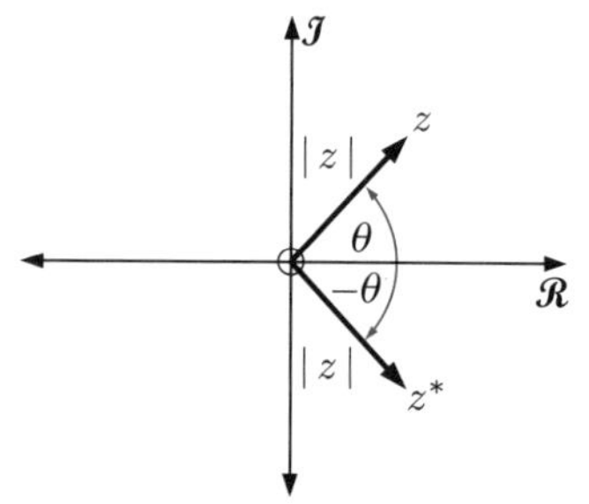

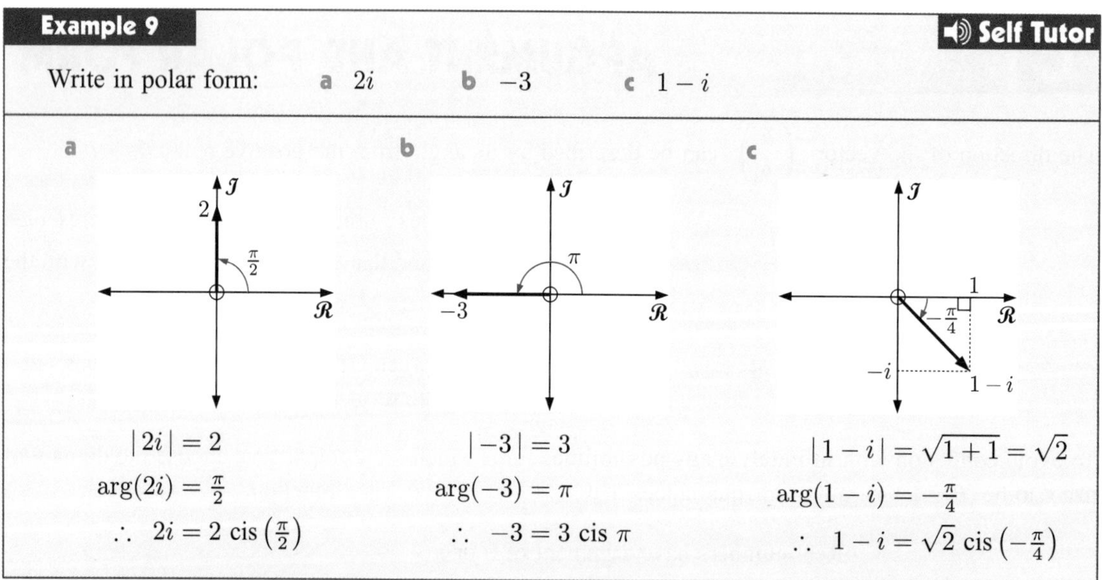

Example 9 — Self Tutor

Write in polar form: **a** $2i$ **b** -3 **c** $1-i$

a

$|\,2i\,| = 2$

$\arg(2i) = \frac{\pi}{2}$

$\therefore\ 2i = 2 \text{ cis } \left(\frac{\pi}{2}\right)$

b

$|-3| = 3$

$\arg(-3) = \pi$

$\therefore\ -3 = 3 \text{ cis } \pi$

c

$|1-i| = \sqrt{1+1} = \sqrt{2}$

$\arg(1-i) = -\frac{\pi}{4}$

$\therefore\ 1 - i = \sqrt{2} \text{ cis } \left(-\frac{\pi}{4}\right)$

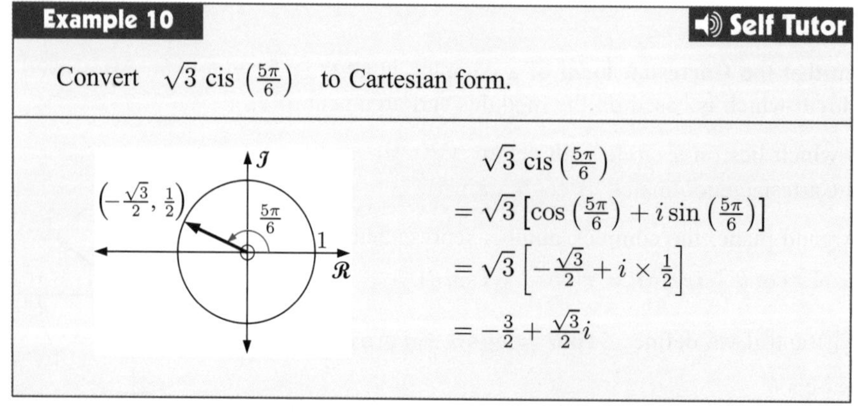

Example 10 — Self Tutor

Convert $\sqrt{3} \text{ cis } \left(\frac{5\pi}{6}\right)$ to Cartesian form.

$\sqrt{3} \text{ cis } \left(\frac{5\pi}{6}\right)$

$= \sqrt{3}\left[\cos\left(\frac{5\pi}{6}\right) + i\sin\left(\frac{5\pi}{6}\right)\right]$

$= \sqrt{3}\left[-\frac{\sqrt{3}}{2} + i \times \frac{1}{2}\right]$

$= -\frac{3}{2} + \frac{\sqrt{3}}{2}i$

EXERCISE 16C.1

1 Write in polar form:

- **a** 4
- **b** $4i$
- **c** -6
- **d** $-3i$
- **e** $1+i$
- **f** $2-2i$
- **g** $-\sqrt{3}+i$
- **h** $2\sqrt{3}+2i$

2 What complex number cannot be written in polar form? Explain your answer.

3 Convert $k+ki$ to polar form.
Hint: You must consider $k>0$, $k=0$, and $k<0$ separately.

4 Convert to Cartesian form without using a calculator:

- **a** $2 \text{ cis } \left(\frac{\pi}{2}\right)$
- **b** $8 \text{ cis } \left(\frac{\pi}{4}\right)$
- **c** $4 \text{ cis } \left(\frac{\pi}{6}\right)$
- **d** $\sqrt{2} \text{ cis } \left(-\frac{\pi}{4}\right)$
- **e** $\sqrt{3} \text{ cis } \left(\frac{2\pi}{3}\right)$
- **f** $5 \text{ cis } \pi$

5
- **a** Find the value of cis 0.
- **b** Find $|\text{cis } \theta|$.
- **c** Show that $\text{cis } \alpha \text{ cis } \beta = \text{cis } (\alpha+\beta)$.

PROPERTIES OF cis θ

Three useful properties of cis θ are:

- $\text{cis } \theta \times \text{cis } \phi = \text{cis } (\theta+\phi)$
- $\dfrac{\text{cis } \theta}{\text{cis } \phi} = \text{cis } (\theta-\phi)$
- $\text{cis } (\theta+k2\pi) = \text{cis } \theta$ for all $k \in \mathbb{Z}$.

The first two of these are similiar to the exponent laws: $a^\theta a^\phi = a^{\theta+\phi}$ and $\dfrac{a^\theta}{a^\phi} = a^{\theta-\phi}$.

Proof:

- $$\begin{aligned}\text{cis } \theta \times \text{cis } \phi &= (\cos\theta + i\sin\theta)(\cos\phi + i\sin\phi)\\ &= [\cos\theta\cos\phi - \sin\theta\sin\phi] + i[\sin\theta\cos\phi + \cos\theta\sin\phi]\\ &= \cos(\theta+\phi) + i\sin(\theta+\phi) \quad \{\text{compound angle identities}\}\\ &= \text{cis } (\theta+\phi)\end{aligned}$$

- $$\begin{aligned}\frac{\text{cis } \theta}{\text{cis } \phi} &= \frac{\text{cis } \theta}{\text{cis } \phi} \times \frac{\text{cis } (-\phi)}{\text{cis } (-\phi)}\\ &= \frac{\text{cis } (\theta-\phi)}{\text{cis } 0}\\ &= \text{cis } (\theta-\phi) \quad \{\text{as cis } 0 = 1\}\end{aligned}$$

- $$\begin{aligned}&\text{cis } (\theta+2k\pi)\\ &= \text{cis } \theta \times \text{cis } (k2\pi)\\ &= \text{cis } \theta \times 1 \quad \{\text{since } k \in \mathbb{Z}\}\\ &= \text{cis } \theta\end{aligned}$$

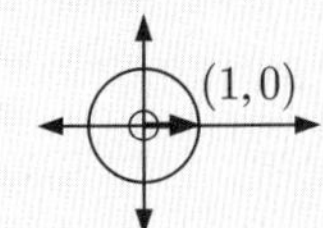

Example 11 Self Tutor

Use the properties of cis to simplify: **a** $\text{cis}\left(\frac{\pi}{5}\right) \text{cis}\left(\frac{3\pi}{10}\right)$ **b** $\dfrac{\text{cis}\left(\frac{\pi}{5}\right)}{\text{cis}\left(\frac{7\pi}{10}\right)}$

a $\text{cis}\left(\frac{\pi}{5}\right) \text{cis}\left(\frac{3\pi}{10}\right)$
$= \text{cis}\left(\frac{\pi}{5} + \frac{3\pi}{10}\right)$
$= \text{cis}\left(\frac{5\pi}{10}\right)$
$= \text{cis}\left(\frac{\pi}{2}\right)$
$= \cos\left(\frac{\pi}{2}\right) + i\sin\left(\frac{\pi}{2}\right)$
$= 0 + i(1)$
$= i$

b $\dfrac{\text{cis}\left(\frac{\pi}{5}\right)}{\text{cis}\left(\frac{7\pi}{10}\right)}$
$= \text{cis}\left(\frac{\pi}{5} - \frac{7\pi}{10}\right)$
$= \text{cis}\left(-\frac{\pi}{2}\right)$
$= \cos\left(-\frac{\pi}{2}\right) + i\sin\left(-\frac{\pi}{2}\right)$
$= 0 + i(-1)$
$= -i$

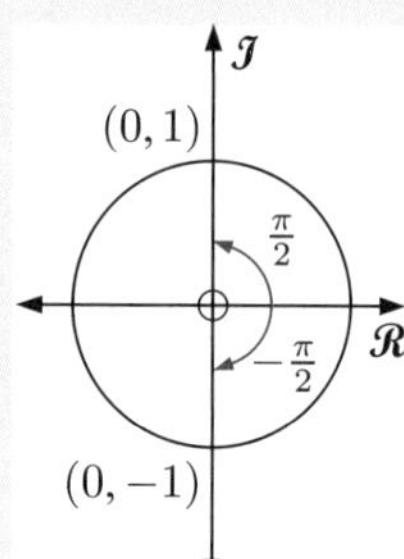

Example 12 Self Tutor

Simplify $\text{cis}\left(\frac{107\pi}{6}\right)$.

$\frac{107\pi}{6} = 17\frac{5}{6}\pi = 18\pi - \frac{\pi}{6}$

$\therefore \text{cis}\left(\frac{107\pi}{6}\right) = \text{cis}\left(18\pi - \frac{\pi}{6}\right)$
$= \text{cis}\left(-\frac{\pi}{6}\right)$ $\{\text{cis}\,(\theta + k2\pi) = \text{cis}\,\theta\}$
$= \frac{\sqrt{3}}{2} - \frac{1}{2}i$

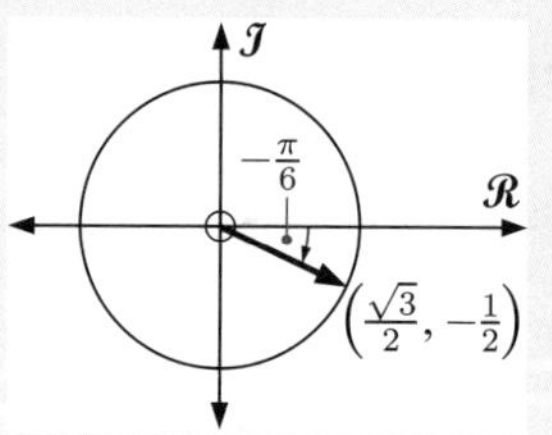

EXERCISE 16C.2

1 Use the properties of cis to simplify:

a $\text{cis}\,\theta\,\text{cis}\,2\theta$ **b** $\dfrac{\text{cis}\,3\theta}{\text{cis}\,\theta}$ **c** $[\text{cis}\,\theta]^3$

d $\text{cis}\left(\frac{\pi}{18}\right) \text{cis}\left(\frac{\pi}{9}\right)$ **e** $2\,\text{cis}\left(\frac{\pi}{12}\right) \text{cis}\left(\frac{\pi}{6}\right)$ **f** $2\,\text{cis}\left(\frac{2\pi}{5}\right) \times 4\,\text{cis}\left(\frac{8\pi}{5}\right)$

g $\dfrac{4\,\text{cis}\left(\frac{\pi}{12}\right)}{2\,\text{cis}\left(\frac{7\pi}{12}\right)}$ **h** $\dfrac{\sqrt{32}\,\text{cis}\left(\frac{\pi}{8}\right)}{\sqrt{2}\,\text{cis}\left(-\frac{7\pi}{8}\right)}$ **i** $\left[\sqrt{2}\,\text{cis}\left(\frac{\pi}{8}\right)\right]^4$

2 Use the property $\text{cis}\,(\theta + k2\pi) = \text{cis}\,\theta$ to evaluate:

a $\text{cis}\,17\pi$ **b** $\text{cis}\,(-37\pi)$ **c** $\text{cis}\left(\frac{91\pi}{3}\right)$

3 If $z = 2\,\text{cis}\,\theta$:

a What is $|z|$ and $\arg z$?
b Write z^* in polar form.
c Write $-z$ in polar form.
d Write $-z^*$ in polar form.

$-2\,\text{cis}\,\theta$ is *not* in polar form since the modulus of a complex number is a length and therefore positive.

Example 13 — Self Tutor

Use the properties of cis θ to find the transformation which moves z to iz.

Let $z = r \text{ cis } \theta$ and use $i = 1 \text{ cis } \left(\frac{\pi}{2}\right)$.

$\therefore \; iz = r \text{ cis } \theta \times \text{cis } \left(\frac{\pi}{2}\right)$

$\qquad = r \text{ cis } \left(\theta + \frac{\pi}{2}\right)$

$\therefore \; z$ has been rotated anti-clockwise by $\frac{\pi}{2}$ about O.

4
 a Write i in polar form.
 b Suppose $z = r \text{ cis } \theta$ is any complex number. Write iz in polar form.
 c What transformation maps z onto $-iz$?

5
 a Write in polar form: i $\cos\theta - i\sin\theta$ ii $\sin\theta - i\cos\theta$
 b Copy and complete: "If $z = r \text{ cis } \theta$ then $z^* = ...$ in polar form."

MULTIPLICATION OF COMPLEX NUMBERS

Now $zw = |z| \text{ cis } \theta \times |w| \text{ cis } \phi$

$\qquad = \underbrace{|z||w|}_{\text{non-negative}} \text{ cis } (\theta + \phi)$ {property of cis}

$\therefore \; |zw| = |z||w|$ {the non-negative number multiplied by cis (....)}

and $\arg(zw) = \theta + \phi = \arg z + \arg w$.

If a complex number z is multiplied by $r \text{ cis } \theta$ then its modulus is *multiplied* by r and its argument is *increased* by θ.

Example 14 — Self Tutor

a Write $z = 1 + \sqrt{3}i$ in polar form.
b Hence multiply z by $2 \text{ cis } \left(\frac{\pi}{6}\right)$.
c Illustrate what has happened on an Argand diagram.
d What transformations have taken place when multiplying z by $2 \text{ cis } \left(\frac{\pi}{6}\right)$?

a

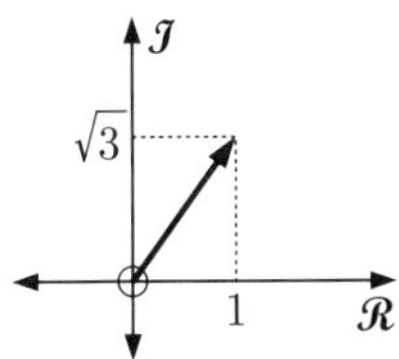

If $z = 1 + \sqrt{3}i$, then $|z| = \sqrt{1^2 + (\sqrt{3})^2} = 2$

$\therefore \; z = 2\left(\frac{1}{2} + \frac{\sqrt{3}}{2}i\right)$

$\therefore \; z = 2\left(\cos\left(\frac{\pi}{3}\right) + i\sin\left(\frac{\pi}{3}\right)\right)$

$\therefore \; z = 2 \text{ cis } \left(\frac{\pi}{3}\right)$

b $(1+\sqrt{3}i) \times 2 \text{ cis}\left(\frac{\pi}{6}\right) = 2 \text{ cis}\left(\frac{\pi}{3}\right) \times 2 \text{ cis}\left(\frac{\pi}{6}\right)$
$= 4 \text{ cis}\left(\frac{\pi}{3}+\frac{\pi}{6}\right)$
$= 4 \text{ cis}\left(\frac{\pi}{2}\right)$
$= 4(0+1i)$
$= 4i$

c

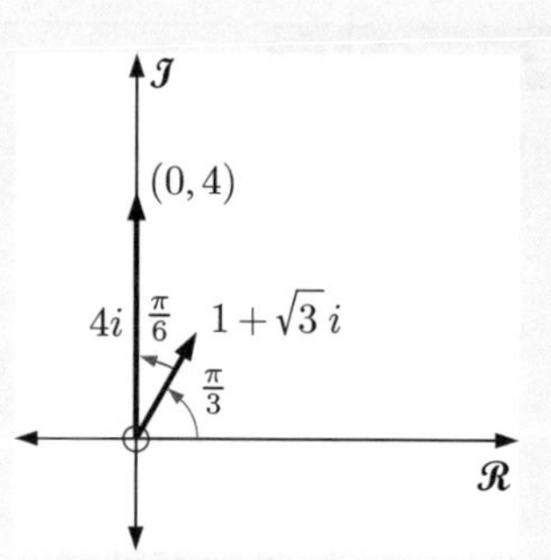

d When z was multiplied by $2 \text{ cis}\left(\frac{\pi}{6}\right)$, its modulus has been doubled, and it has been rotated anti-clockwise through $\frac{\pi}{6}$.

EXERCISE 16C.3

1 **a** Write $z = 2 - 2i$ in polar form. **b** Hence multiply z by $\frac{1}{3} \text{ cis}\left(\frac{\pi}{3}\right)$.

c Illustrate what has happened on an Argand diagram.

d What transformations have taken place when multiplying z by $\frac{1}{3} \text{ cis}\left(\frac{\pi}{3}\right)$?

2 **a** Write $z = -\sqrt{3} + i$ in polar form. **b** Hence multiply z by $\frac{1}{2} \text{ cis}\left(-\frac{\pi}{3}\right)$.

c Illustrate what has happened on an Argand diagram.

d What transformations have taken place when multiplying z by $\frac{1}{2} \text{ cis}\left(-\frac{\pi}{3}\right)$?

Example 15 — Self Tutor

Use complex numbers to write $\cos\left(\frac{7\pi}{12}\right)$ and $\sin\left(\frac{7\pi}{12}\right)$ in simplest surd form.

$\cos\left(\frac{7\pi}{12}\right) + i\sin\left(\frac{7\pi}{12}\right)$
$= \text{cis}\left(\frac{7\pi}{12}\right)$
$= \text{cis}\left(\frac{3\pi}{12}+\frac{4\pi}{12}\right)$
$= \text{cis}\left(\frac{\pi}{4}\right) \times \text{cis}\left(\frac{\pi}{3}\right)$ $\quad \{\text{cis}\,(\theta+\phi) = \text{cis}\,\theta \times \text{cis}\,\phi\}$
$= \left(\frac{1}{\sqrt{2}} + \frac{1}{\sqrt{2}}i\right)\left(\frac{1}{2} + \frac{\sqrt{3}}{2}i\right)$
$= \left(\frac{1}{2\sqrt{2}} - \frac{\sqrt{3}}{2\sqrt{2}}\right) + i\left(\frac{\sqrt{3}}{2\sqrt{2}} + \frac{1}{2\sqrt{2}}\right)$

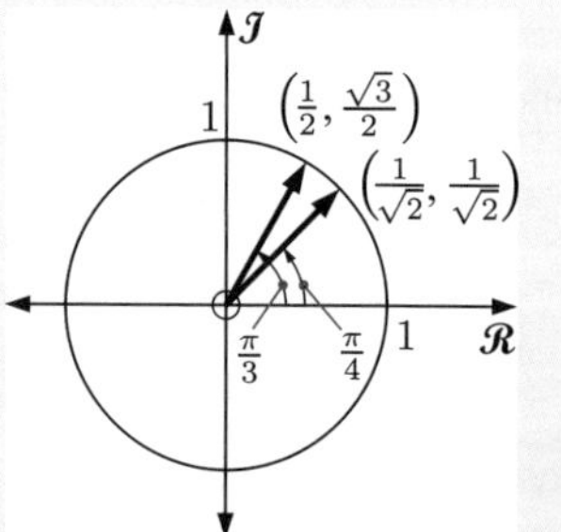

Equating real parts: $\quad \cos\left(\frac{7\pi}{12}\right) = \left(\frac{1-\sqrt{3}}{2\sqrt{2}}\right) \times \frac{\sqrt{2}}{\sqrt{2}} = \frac{\sqrt{2}-\sqrt{6}}{4}$

Equating imaginary parts: $\quad \sin\left(\frac{7\pi}{12}\right) = \frac{\sqrt{3}+1}{2\sqrt{2}} = \frac{\sqrt{6}+\sqrt{2}}{4}$

3 Use complex number methods to find, in simplest surd form:

a $\cos\left(\frac{\pi}{12}\right)$ and $\sin\left(\frac{\pi}{12}\right)$ **b** $\cos\left(\frac{11\pi}{12}\right)$ and $\sin\left(\frac{11\pi}{12}\right)$

4 Use the principle of mathematical induction to prove that $\arg(z^n) = n\arg z$ for all $n \in \mathbb{Z}^+$.

5 Use polar form to establish:

$$\left|\frac{z}{w}\right| = \frac{|z|}{|w|} \quad \text{and} \quad \arg\left(\frac{z}{w}\right) = \arg z - \arg w, \quad \text{provided } w \neq 0.$$

Example 16 Self Tutor

Suppose $z = \sqrt{2}\text{ cis }\theta$ where θ is obtuse. Find the modulus and argument of:

a $2z$ **b** $\frac{z}{i}$ **c** $(1-i)z$

a $2z = 2\sqrt{2}\text{ cis }\theta$

$\therefore\ |2z| = 2\sqrt{2}$ and $\arg(2z) = \theta$

b $i = \text{cis}\left(\frac{\pi}{2}\right)$

$\therefore\ \frac{z}{i} = \frac{\sqrt{2}\text{ cis }\theta}{\text{cis}\left(\frac{\pi}{2}\right)}$

$= \sqrt{2}\text{ cis}\left(\theta - \frac{\pi}{2}\right)$

$\therefore\ \left|\frac{z}{i}\right| = \sqrt{2}$ and $\arg\left(\frac{z}{i}\right) = \theta - \frac{\pi}{2}$ {since θ is obtuse, $\left(\theta - \frac{\pi}{2}\right) \in\ ]-\pi, \pi]$}

c

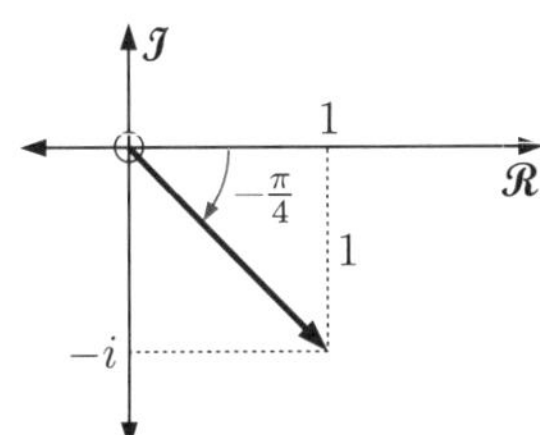

$1 - i = \sqrt{2}\text{ cis}\left(-\frac{\pi}{4}\right)$

$\therefore\ (1-i)z = \sqrt{2}\text{ cis}\left(-\frac{\pi}{4}\right) \times \sqrt{2}\text{ cis }\theta$

$= 2\text{ cis}\left(-\frac{\pi}{4} + \theta\right)$

$\therefore\ |(1-i)z| = 2$ and $\arg((1-i)z) = \theta - \frac{\pi}{4}$

6 Suppose $z = 3\text{ cis }\theta$ where θ is acute. Determine the modulus and argument of:

a $-z$ **b** z^* **c** iz **d** $(1+i)z$ **e** $\frac{z}{i}$ **f** $\frac{z}{1-i}$

Example 17 Self Tutor

Suppose $z = \text{cis }\phi$ where ϕ is acute. Find the modulus and argument of $z + 1$.

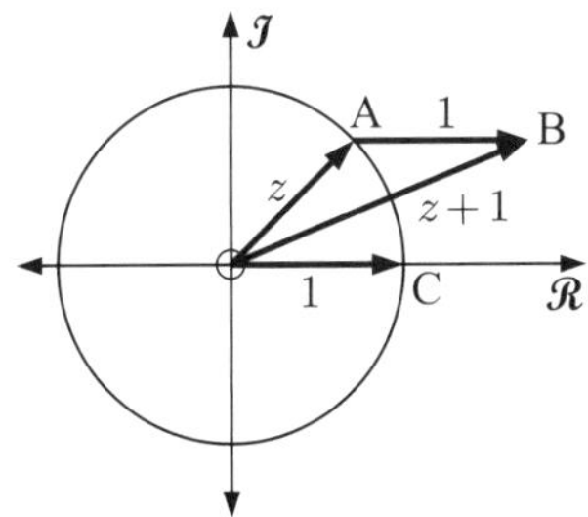

$|z| = 1$, so z ends on the unit circle.

$z + 1$ is the vector $\overrightarrow{\text{OB}}$ shown.

OABC is a rhombus.

$\therefore\ \arg(z+1) = \frac{\phi}{2}$

{diagonals bisect the angles of the rhombus}

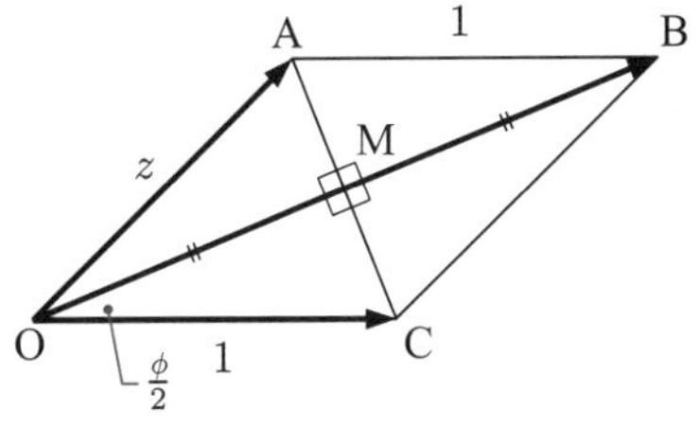

Also $\cos\left(\frac{\phi}{2}\right) = \frac{\text{OM}}{1}$

$\therefore\ \text{OM} = \cos\left(\frac{\phi}{2}\right)$

$\therefore\ \text{OB} = 2\cos\left(\frac{\phi}{2}\right)$

$\therefore\ |z+1| = 2\cos\left(\frac{\phi}{2}\right)$

7 **a** If $z = \text{cis}\,\phi$ where ϕ is acute, determine the modulus and argument of $z-1$.

b Hence write $z-1$ in polar form.

c Hence write $(z-1)^*$ in polar form.

8 ABC is an equilateral triangle. Suppose z_1 represents $\overrightarrow{OA}$, z_2 represents $\overrightarrow{OB}$, and z_3 represents $\overrightarrow{OC}$.

a Explain what vectors represent $z_2 - z_1$ and $z_3 - z_2$.

b Find $\left|\dfrac{z_2 - z_1}{z_3 - z_2}\right|$.

c Determine $\arg\left(\dfrac{z_2 - z_1}{z_3 - z_2}\right)$.

d Hence find the value of $\left(\dfrac{z_2 - z_1}{z_3 - z_2}\right)^3$.

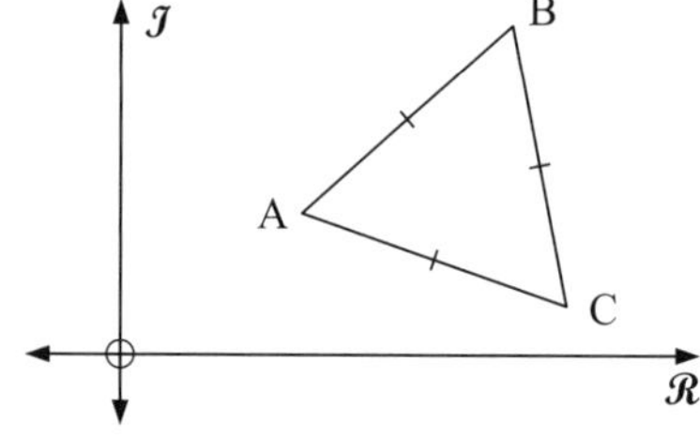

9 Suppose $z = a - i$ where a is real, and $\arg z = -\frac{5\pi}{6}$. Find the exact value of a.

FURTHER CONVERSION BETWEEN CARTESIAN AND POLAR FORMS

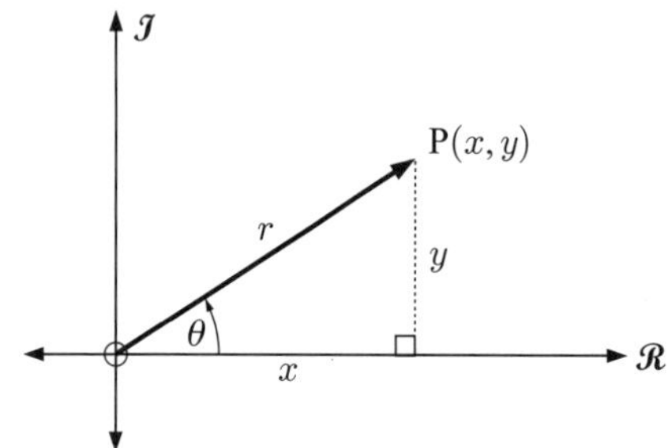

If $z = \underbrace{x + iy}_{\text{Cartesian form}} = \underbrace{r\ \text{cis}\,\theta}_{\text{Polar form}}$ then:

- $r = \sqrt{x^2 + y^2}$
- $\tan\theta = \dfrac{y}{x}$
- $\cos\theta = \dfrac{x}{\sqrt{x^2+y^2}}$
- $\sin\theta = \dfrac{y}{\sqrt{x^2+y^2}}$

POLAR TO CARTESIAN

$$\begin{aligned} z &= 2\ \text{cis}\left(\tfrac{\pi}{8}\right) \\ &= 2\left(\cos\left(\tfrac{\pi}{8}\right) + i\sin\left(\tfrac{\pi}{8}\right)\right) \\ &\approx 1.85 + 0.765i \end{aligned}$$

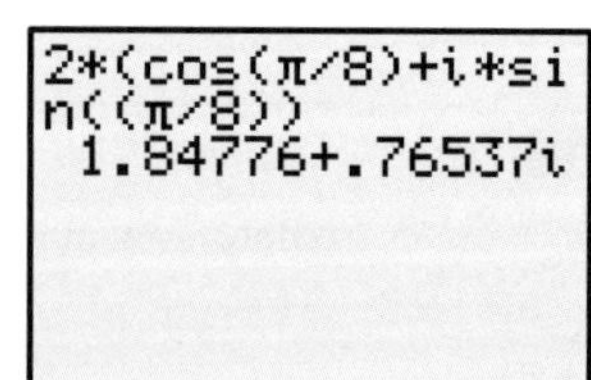

CARTESIAN TO POLAR

$z = -3 + i$ has $r = \sqrt{(-3)^2 + 1^2} = \sqrt{10}$

$\therefore\ \underbrace{\cos\theta = \dfrac{-3}{\sqrt{10}} \quad \text{and} \quad \sin\theta = \dfrac{1}{\sqrt{10}}}$

$\therefore\ \theta = \pi - \arcsin\left(\frac{1}{\sqrt{10}}\right)$ {quadrant 2}

$\therefore\ -3 + i \approx \sqrt{10}\ \text{cis}\,(2.82)$

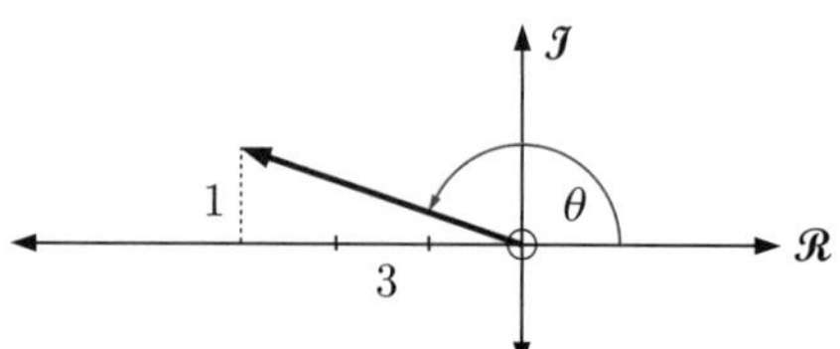

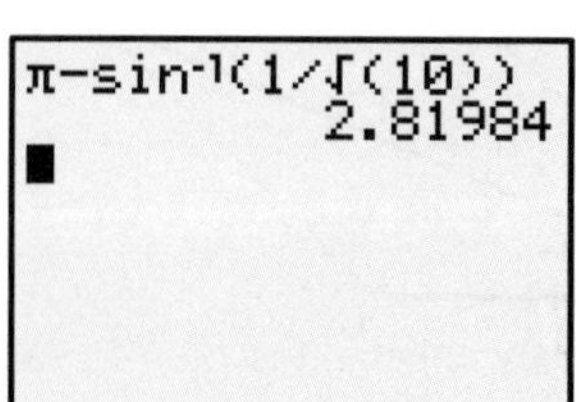

EXERCISE 16C.4

1 Use your calculator to convert to Cartesian form:

a $\sqrt{3}\text{ cis }(2.5187)$ **b** $\sqrt{11}\text{ cis }\left(-\frac{3\pi}{8}\right)$ **c** $2.836\,49\text{ cis }(-2.684\,32)$

2 Use your calculator to convert to polar form:

a $3-4i$ **b** $-5-12i$ **c** $-11.6814+13.2697i$

3 Add the following using $a+bi$ surd form and convert your answer to polar form:

a $3\text{ cis }\left(\frac{\pi}{4}\right)+\text{cis }\left(\frac{-3\pi}{4}\right)$ **b** $2\text{ cis }\left(\frac{2\pi}{3}\right)+5\text{ cis }\left(\frac{-2\pi}{3}\right)$

4 Use the sum and product of roots to find the real quadratic equations with roots of:

a $2\text{ cis }\left(\frac{2\pi}{3}\right),\ 2\text{ cis }\left(\frac{4\pi}{3}\right)$ **b** $\sqrt{2}\text{ cis }\left(\frac{\pi}{4}\right),\ \sqrt{2}\text{ cis }\left(\frac{-\pi}{4}\right)$

D EULER'S FORM

Leonhard Euler proved that the complex number $z=|z|\text{ cis }\theta$ can also be written in the form $z=|z|e^{i\theta}$.

To achieve this he proved that $e^{i\theta}=\cos\theta+i\sin\theta$.

For example, consider the complex number $z=1+i$.

$|z|=\sqrt{2}$ and $\theta=\frac{\pi}{4}$,

$\therefore\ 1+i=\sqrt{2}\text{ cis }\left(\frac{\pi}{4}\right)=\sqrt{2}e^{i\frac{\pi}{4}}$

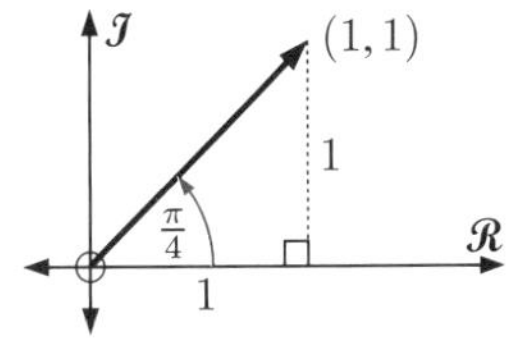

So, $\sqrt{2}\text{ cis }\left(\frac{\pi}{4}\right)$ is the **polar form** of $1+i$.

and $\sqrt{2}e^{i\frac{\pi}{4}}$ is the **Euler form** of $1+i$.

We cannot formally prove the identity $e^{i\theta}=\cos\theta+i\sin\theta$ until we have studied calculus. However, in the following investigation we put together our knowledge of exponentials and complex numbers to justify why it is reasonable.

INVESTIGATION EULER'S FORM

In this Investigation we consider what happens when a complex number z is multiplied by a complex number in Euler's form. The aim is to explore the relation $e^{i\theta}=\cos\theta+i\sin\theta$ and provide insight into its meaning.

What to do:

1 On the Cartesian plane, the equation of the unit circle is $x^2+y^2=1$.

a State the equation of the unit circle on the Argand plane.

b Write, in the form $x+yi$, the complex number on the unit circle whose corresponding vector makes an angle of θ with the positive x-axis.

2 In **Chapter 3** we studied continuous compound interest. We saw that for 100% compound growth with interest calculated in n subintervals, the value after each subinterval is generated using the sequence $1, \left(1+\frac{1}{n}\right), \left(1+\frac{1}{n}\right)^2, ..., \left(1+\frac{1}{n}\right)^n$. We saw that as $n \to \infty$, $\left(1+\frac{1}{n}\right)^n \to e^1$.

Now consider the sequence $1, \left(1+\frac{i}{n}\right), \left(1+\frac{i}{n}\right)^2, ..., \left(1+\frac{i}{n}\right)^n$ which has $n+1$ terms.

We now have *imaginary* compound growth with interest calculated in n subintervals.

a For each of the following cases, find the terms of the sequence, then plot them on an Argand diagram and connect them with straight line segments.

i $n=1$ **ii** $n=2$ **iii** $n=3$ **iv** $n=4$ **v** $n=5$

b Click on the icon and use the software to plot the terms of the sequence for values of n up to 500.

DEMO

i What do you notice about the curve being generated as n gets very large?

ii Find $\cos 1^c$ and $\sin 1^c$.

iii Locate $\left(1+\frac{i}{500}\right)^{500}$.

iv Copy and complete: As $n \to \infty$, $\left(1+\frac{i}{n}\right)^n \to ...... + ...\, i = e^i$. {Euler}

3 In **Chapter 3** we also saw that if interest is paid continuously, then the compounded amount after 1 year is the initial amount multiplied by e^R, where R is the interest rate.

Now consider the sequence $1, \left(1+\frac{i\theta}{n}\right), \left(1+\frac{i\theta}{n}\right)^2, ..., \left(1+\frac{i\theta}{n}\right)^n$ which has $n+1$ terms.

Click on the icon to run the software.

a Set $\theta = 2$. Plot the terms of the sequence for values of n up to 500.

i What do you notice about the curve being generated as n gets very large?

ii Find $\cos 2^c$ and $\sin 2^c$.

iii Copy and complete: As $n \to \infty$, $\left(1+\frac{2i}{n}\right)^n \to ...... + ...\, i = e^{2i}$. {Euler}

b Set $\theta = \frac{\pi}{2}$. Plot the terms of the sequence for values of n up to 500.

i Copy and complete: As $n \to \infty$, $\left(1+\frac{i\frac{\pi}{2}}{n}\right)^n \to ...... + ...\, i = e^{i\frac{\pi}{2}}$. {Euler}

ii Comment on your answer with respect to the result in **Examples 8** and **13**.

4 Copy and complete:

a The effect of *real* compound growth is to increase the size or modulus of a quantity. So, when we multiply a complex number z by e^r, the vector representing z is ...... with scale factor .... This means the ...... of z is multiplied by e^r.

b The effect of *imaginary* compound growth is to rotate the quantity in the complex plane. So, when we multiply a complex number z by $\text{cis}\,\theta = e^{\cdots}$, z is ...... anti-clockwise through angle ...... about the origin O. This means the ...... of z is increased by θ.

5 Consider $z = |z|\,e^{i\theta}$ and $w = |w|\,e^{i\phi}$.
Explain the result when z is multiplied by w.

Example 18 — Self Tutor

Evaluate: **a** $e^{-i\frac{\pi}{4}}$ **b** i^{-i}

a $e^{-i\frac{\pi}{4}}$
$= \cos\left(-\frac{\pi}{4}\right) + i\sin\left(-\frac{\pi}{4}\right)$
$= \frac{1}{\sqrt{2}} - i\frac{1}{\sqrt{2}}$

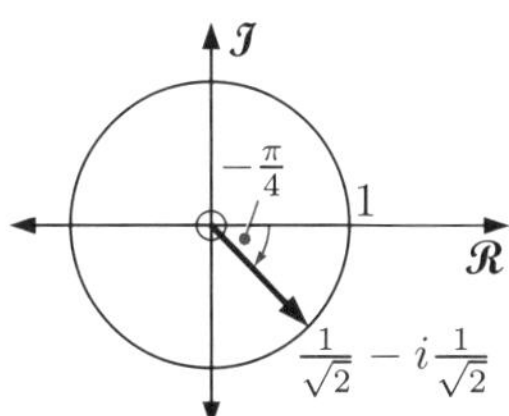

b Now $i = 0 + 1i$
$= \cos\left(\frac{\pi}{2}\right) + i\sin\left(\frac{\pi}{2}\right)$
$= e^{i\frac{\pi}{2}}$
$\therefore\ i^{-i} = \left(e^{i\frac{\pi}{2}}\right)^{-i}$
$= e^{-i^2\frac{\pi}{2}}$
$= e^{\frac{\pi}{2}}$

EXERCISE 16D

1 Use $e^{i\theta} = \cos\theta + i\sin\theta$ to find the value of:

a $e^{i\pi}$ **b** $e^{i\frac{\pi}{3}}$ **c** $e^{-i\frac{\pi}{2}}$

2 Use $\text{cis}\ \theta = e^{i\theta}$ to prove that:

a $\text{cis}\ \theta\ \text{cis}\ \phi = \text{cis}(\theta+\phi)$ **b** $\dfrac{\text{cis}\ \theta}{\text{cis}\ \phi} = \text{cis}\ (\theta-\phi)$

3 Suppose $z = \text{cis}\ \theta$, find the argument of:

a $\sqrt{z}$ **b** iz **c** $-iz^2$ **d** $\dfrac{i}{z}$

4 Evaluate:

a e^i **b** 3^i **c** i^i **d** $\left(\left(i^i\right)^i\right)^2$

E DE MOIVRE'S THEOREM

Polar form enables us to easily calculate powers of complex numbers.

For example, if $z = |z|\ \text{cis}\ \theta$

then $z^2 = |z|\ \text{cis}\ \theta \times |z|\ \text{cis}\ \theta$
$= |z|^2\ \text{cis}\ (\theta+\theta)$
$= |z|^2\ \text{cis}\ 2\theta$

and $z^3 = z^2 z$
$= |z|^2\ \text{cis}\ 2\theta \times |z|\ \text{cis}\ \theta$
$= |z|^3\ \text{cis}\ (2\theta+\theta)$
$= |z|^3\ \text{cis}\ 3\theta$

The generalisation of this process is **De Moivre's Theorem**:

$$(|z|\ \textbf{cis}\ \theta)^n = |z|^n\ \textbf{cis}\ n\theta \quad \text{for all rational } n.$$

Proof for $n \in \mathbb{Z}^+$: {By the principle of mathematical induction}

P_n is: "$(|z| \text{ cis } \theta)^n = |z|^n \text{ cis } n\theta$" for all $n \in \mathbb{Z}^+$.

(1) If $n = 1$, then $(|z| \text{ cis } \theta)^1 = |z| \text{ cis } \theta$ $\quad \therefore$ P_1 is true.

(2) If P_k is true, then $(|z| \text{ cis } \theta)^k = |z|^k \text{ cis } k\theta$

Thus
$$\begin{aligned}(|z| \text{ cis } \theta)^{k+1} &= (|z| \text{ cis } \theta)^k \times |z| \text{ cis } \theta && \{\text{exponent law}\}\\ &= |z|^k \text{ cis } k\theta \times |z| \text{ cis } \theta && \{\text{using } P_k\}\\ &= |z|^{k+1} \text{ cis } (k\theta + \theta) && \{\text{exponent law and cis property}\}\\ &= |z|^{k+1} \text{ cis } (k+1)\theta\end{aligned}$$

Since P_1 is true, and P_{k+1} is true whenever P_k is true,
P_n is true for all $n \in \mathbb{Z}^+$. {Principle of mathematical induction}

We also observe that $\text{cis } (-n\theta) = \text{cis } (0 - n\theta) = \dfrac{\text{cis } 0}{\text{cis } n\theta}$ $\quad \left\{\text{as } \dfrac{\text{cis } \theta}{\text{cis } \phi} = \text{cis } (\theta - \phi)\right\}$

$$\begin{aligned}\therefore \text{ cis } (-n\theta) &= \frac{1}{\text{cis } n\theta} && \{\text{as cis } 0 = 1\}\\ &= \frac{1}{[\text{cis } \theta]^n} && \{\text{for } n \in \mathbb{Z}^+\}\\ &= [\text{cis } \theta]^{-n} && \text{so the theorem is true for all } n \in \mathbb{Z}\end{aligned}$$

Also, $\left[\text{cis}\left(\frac{\theta}{n}\right)\right]^n = \text{cis}\left(n\left(\frac{\theta}{n}\right)\right) = \text{cis } \theta$

$\therefore \quad [\text{cis } \theta]^{\frac{1}{n}} = \text{cis}\left(\frac{\theta}{n}\right)$

$\therefore \quad (\text{cis } \theta)^{\frac{m}{n}} = \left((\text{cis } \theta)^{\frac{1}{n}}\right)^m = \left(\text{cis}\left(\frac{\theta}{n}\right)\right)^m = \text{cis}\left(\frac{m}{n}\theta\right)$ $\quad$ so the theorem is true for all $n \in \mathbb{Q}$.

Example 19 **Self Tutor**

Find the exact value of $\left(\sqrt{3} + i\right)^8$ using De Moivre's theorem.
Check your answer by calculator.

$\sqrt{3} + i$ has modulus $\sqrt{\left(\sqrt{3}\right)^2 + 1^2} = \sqrt{4} = 2$

$$\begin{aligned}\therefore \sqrt{3} + i &= 2\left(\tfrac{\sqrt{3}}{2} + \tfrac{1}{2}i\right)\\ &= 2 \text{ cis}\left(\tfrac{\pi}{6}\right)\end{aligned}$$

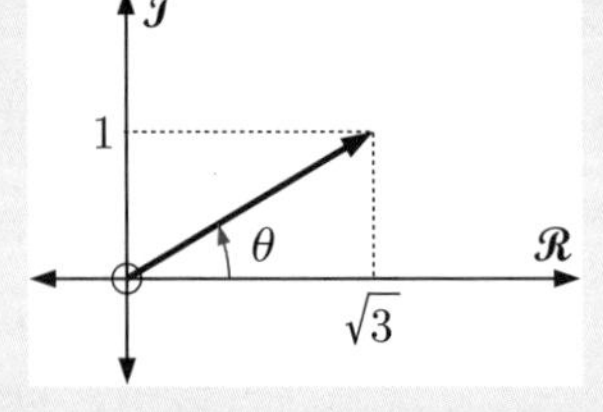

$$\begin{aligned}\therefore \left(\sqrt{3} + i\right)^8 &= \left(2 \text{ cis}\left(\tfrac{\pi}{6}\right)\right)^8\\ &= 2^8 \text{ cis}\left(\tfrac{8\pi}{6}\right)\\ &= 2^8 \text{ cis}\left(\tfrac{4\pi}{3}\right)\\ &= 2^8\left(-\tfrac{1}{2} - \tfrac{\sqrt{3}}{2}i\right)\\ &= -128 - 128\sqrt{3}i\end{aligned}$$

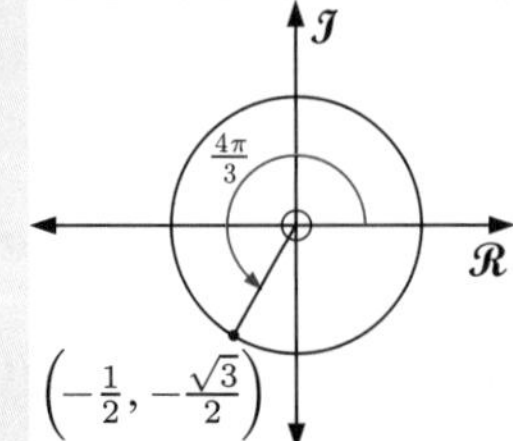

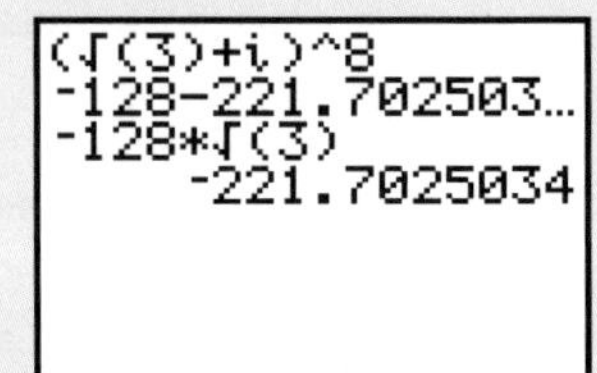

EXERCISE 16E

1 Prove De Moivre's theorem using Euler's form and the laws of exponents.

2 Use De Moivre's theorem to simplify:

a $\left(\sqrt{2}\text{ cis }\left(\frac{\pi}{5}\right)\right)^{10}$ **b** $\left(\text{cis }\left(\frac{\pi}{12}\right)\right)^{36}$ **c** $\left(\sqrt{2}\text{ cis }\left(\frac{\pi}{8}\right)\right)^{12}$

d $\sqrt{5\text{ cis }\left(\frac{\pi}{7}\right)}$ **e** $\sqrt[3]{8\text{ cis }\left(\frac{\pi}{2}\right)}$ **f** $\left(8\text{ cis }\left(\frac{\pi}{5}\right)\right)^{\frac{5}{3}}$

3 Use De Moivre's theorem to find the exact value of:

a $(1+i)^{15}$ **b** $\left(1-i\sqrt{3}\right)^{11}$ **c** $\left(\sqrt{2}-i\sqrt{2}\right)^{-19}$

d $(-1+i)^{-11}$ **e** $\left(\sqrt{3}-i\right)^{\frac{1}{2}}$ **f** $\left(2+2i\sqrt{3}\right)^{-\frac{5}{2}}$

Check your answers using technology.

4 Suppose $z = |\,z\,|\text{ cis }\theta$ where $-\pi < \theta \leqslant \pi$.

a Use De Moivre's theorem to find $\sqrt{z}$ in terms of $|\,z\,|$ and θ.

b What restrictions apply to $\phi = \arg(\sqrt{z})$?

c True or false? "$\sqrt{z}$ has a non-negative real part."

5 Show that $\cos\theta - i\sin\theta = \text{cis }(-\theta)$. Hence, simplify $(\cos\theta - i\sin\theta)^{-3}$.

6 **a** Write $z = 1+i$ in polar form and hence write z^n in polar form.

b Find all values of n for which:

i z^n is real **ii** z^n is purely imaginary.

7 Suppose $|\,z\,| = 2$ and $\arg z = \theta$. Determine the modulus and argument of:

a z^3 **b** iz^2 **c** $\dfrac{1}{z}$ **d** $-\dfrac{i}{z^2}$

8 If $z = \text{cis }\theta$, prove that $\dfrac{z^2-1}{z^2+1} = i\tan\theta$.

Example 20 **Self Tutor**

By considering $\cos 2\theta + i\sin 2\theta$, deduce the double angle formulae for $\cos 2\theta$ and $\sin 2\theta$.

$$\begin{aligned}\cos 2\theta + i\sin 2\theta &= \text{cis } 2\theta \\ &= [\text{cis } \theta]^2 \qquad \{\text{De Moivre's theorem}\} \\ &= [\cos\theta + i\sin\theta]^2 \\ &= [\cos^2\theta - \sin^2\theta] + i\,[2\sin\theta\cos\theta]\end{aligned}$$

Equating imaginary parts, $\sin 2\theta = 2\sin\theta\cos\theta$

Equating real parts, $\cos 2\theta = \cos^2\theta - \sin^2\theta$

9 **a** Use complex number methods to deduce that:

i $\cos 3\theta = 4\cos^3\theta - 3\cos\theta$ **ii** $\sin 3\theta = 3\sin\theta - 4\sin^3\theta$

b Hence find $\tan 3\theta$ in terms of $\tan\theta$ only.

c Hence solve the equations:

i $4x^3 - 3x = -\frac{1}{\sqrt{2}}$ **ii** $x^3 - 3\sqrt{3}x^2 - 3x + \sqrt{3} = 0$

10 Points A, B, and C form an isosceles triangle with a right angle at B. Suppose A, B, and C are represented by the complex numbers z_1, z_2, and z_3 respectively.

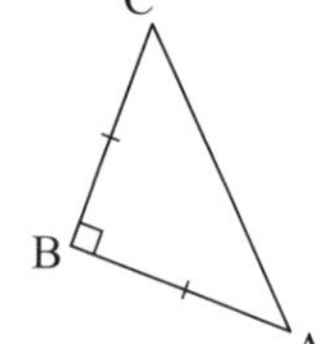

a Show that $(z_1 - z_2)^2 = -(z_3 - z_2)^2$.

b If ABCD forms a square, what complex number represents the point D? Give your answer in terms of z_1, z_2, and z_3.

11 Find a formula for:

a $\cos 4\theta$ in terms of $\cos\theta$ **b** $\sin 4\theta$ in terms of $\cos\theta$ and $\sin\theta$.

12 **a** **i** If $z = \text{cis}\ \theta$, prove that $z^n + \dfrac{1}{z^n} = 2\cos n\theta$.

ii Hence explain why $z + \dfrac{1}{z} = 2\cos\theta$.

iii Use the binomial theorem to expand $\left(z + \dfrac{1}{z}\right)^3$, and simplify your result.

iv Hence show that $\cos^3\theta = \frac{1}{4}\cos 3\theta + \frac{3}{4}\cos\theta$.

v Hence show that the exact value of $\cos^3\left(\frac{13\pi}{12}\right)$ is $\frac{-5\sqrt{2}-3\sqrt{6}}{16}$. **Hint:** $\frac{13\pi}{12} = \frac{3\pi}{4} + \frac{\pi}{3}$

b Show that if $z = \text{cis}\ \theta$ then $z^n - \dfrac{1}{z^n} = 2i\sin n\theta$.

Hence show that $\sin^3\theta = \frac{3}{4}\sin\theta - \frac{1}{4}\sin 3\theta$.

c Hence show that $\sin^3\theta\cos^3\theta = \frac{1}{32}(3\sin 2\theta - \sin 6\theta)$.

F ROOTS OF COMPLEX NUMBERS

With De Moivre's theorem to help us, we can now find roots of complex numbers.

SOLVING $z^n = c$

Consider an equation of the form $z^n = c$ where n is a positive integer and c is a complex number.

The **nth roots of the complex number c** are the n solutions of $z^n = c$.

- There are **exactly** n nth roots of c.
- If $c \in \mathbb{R}$, the complex roots must occur in conjugate pairs.
- If $c \notin \mathbb{R}$, the complex roots do not all occur in conjugate pairs.
- The roots of z^n will all have the same modulus which is $|c|^{\frac{1}{n}}$.
- On an Argand diagram, the roots all lie on a circle with radius $|c|^{\frac{1}{n}}$.
- The roots on the circle $r = |c|^{\frac{1}{n}}$ will be equally spaced around the circle.
  If you join all the points you will get a geometric shape that is a regular polygon.
  For example, $n = 3$ (equilateral $\triangle$), $n = 4$ (square)
  $n = 5$ (regular pentagon)
  $n = 6$ (regular hexagon) etc.

For example, the 4th roots of $2i$ are the four solutions of $z^4 = 2i$.

The roots of a complex number may be found by factorisation, but this is usually very difficult. It is therefore desirable to have an alternative method such as the **nth roots method** presented in the following example.

Example 21 — Self Tutor

Find the four 4th roots of 1 by:

a factorisation

b the 'nth roots method'.

The 4th roots of 1 are the 4 solutions of $z^4 = 1$.

a By factorisation,
$$z^4 = 1$$
$$\therefore\ z^4 - 1 = 0$$
$$\therefore\ (z^2+1)(z^2-1) = 0$$
$$\therefore\ (z+i)(z-i)(z+1)(z-1) = 0$$
$$\therefore\ z = \pm i \text{ or } \pm 1$$

b By the 'nth roots method',

$\mathcal{I}$, $\mathcal{R}$, 1

$$z^4 = 1$$
$$\therefore\ z^4 = 1 \text{ cis } (0 + k2\pi) \quad \text{where } k \in \mathbb{Z} \quad \{\text{polar form}\}$$
$$\therefore\ z = [\text{cis } (k2\pi)]^{\frac{1}{4}}$$
$$\therefore\ z = \text{cis}\left(\tfrac{k2\pi}{4}\right) \quad \{\text{De Moivre}\}$$
$$\therefore\ z = \text{cis}\left(\tfrac{k\pi}{2}\right)$$
$$\therefore\ z = \text{cis } 0,\ \text{cis}\left(\tfrac{\pi}{2}\right),\ \text{cis } \pi,\ \text{cis}\left(\tfrac{3\pi}{2}\right)$$
$\{$letting $k = 0, 1, 2, 3\}$
$$\therefore\ z = 1, i, -1, -i$$

The substitution of $k = 0, 1, 2, 3$ to find the 4 roots could be done using any 4 consecutive integers for k.

EXERCISE 16F.1

1 Find the three cube roots of 1 using: a factorisation b the 'nth roots method'.

2 Solve for z: a $z^3 = -8i$ b $z^3 = -27i$

3 Find the three cube roots of -1, and display them on an Argand diagram.

4 Solve for z: a $z^4 = 16$ b $z^4 = -16$

5 Find the four fourth roots of $-i$, and display them on an Argand diagram.

6 Solve the following and display the roots on an Argand diagram:

a $z^3 = 2 + 2i$
b $z^3 = -2 + 2i$
c $z^2 = \frac{1}{2} + \frac{\sqrt{3}}{2}i$
d $z^4 = \sqrt{3} + i$
e $z^5 = -4 - 4i$
f $z^3 = -2\sqrt{3} - 2i$

Example 22 — Self Tutor

a Find the fourth roots of -4 in the form $a+bi$.

b Hence write z^4+4 as a product of real quadratic factors.

a The fourth roots of -4 are the solutions of

$z^4=-4$

$\therefore\ z^4=4\text{ cis }(\pi+k2\pi)$ where $k\in\mathbb{Z}$

$\therefore\ z=[4\text{ cis }(\pi+k2\pi)]^{\frac{1}{4}}$

$\therefore\ z=4^{\frac{1}{4}}\text{ cis}\left(\frac{\pi+k2\pi}{4}\right)$ {De Moivre}

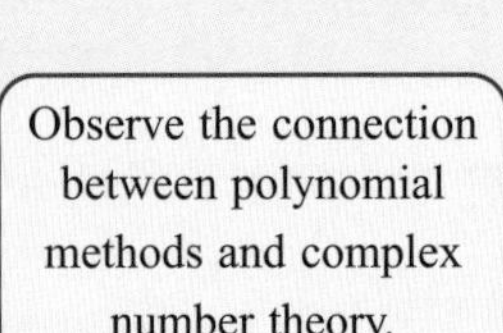

$\therefore\ z=2^{\frac{1}{2}}\text{ cis}\left(\frac{\pi}{4}\right),\ 2^{\frac{1}{2}}\text{ cis}\left(\frac{3\pi}{4}\right),\ 2^{\frac{1}{2}}\text{ cis}\left(\frac{5\pi}{4}\right),\ 2^{\frac{1}{2}}\text{ cis}\left(\frac{7\pi}{4}\right)$ {letting $k=0, 1, 2, 3$}

$\therefore\ z=\sqrt{2}\left(\frac{1}{\sqrt{2}}+\frac{1}{\sqrt{2}}i\right),\ \sqrt{2}\left(-\frac{1}{\sqrt{2}}+\frac{1}{\sqrt{2}}i\right),\ \sqrt{2}\left(-\frac{1}{\sqrt{2}}-\frac{1}{\sqrt{2}}i\right),\ \sqrt{2}\left(\frac{1}{\sqrt{2}}-\frac{1}{\sqrt{2}}i\right)$

$\therefore\ z=1+i,\quad -1+i,\quad -1-i,\quad 1-i$

b The roots $1\pm i$ have sum $=2$ and product $=(1+i)(1-i)=2$

$\therefore$ they come from the quadratic factor z^2-2z+2.

The roots $-1\pm i$ have sum $=-2$ and product $=(-1+i)(-1-i)=2$

$\therefore$ they come from the quadratic factor z^2+2z+2.

Thus $z^4+4=(z^2-2z+2)(z^2+2z+2)$

Observe the connection between polynomial methods and complex number theory.

7 **a** Find the four solutions of $z^4+1=0$ in the form $a+bi$, and display them on an Argand diagram.

b Hence write z^4+1 as the product of two real quadratic factors.

8 Consider $z=\dfrac{\left(\frac{\sqrt{3}}{2}-\frac{1}{2}i\right)^2}{\left(\cos(\frac{\pi}{10})-i\sin(\frac{\pi}{10})\right)^5\left(\cos(\frac{\pi}{30})+i\sin(\frac{\pi}{30})\right)^{25}}$

a Using polar form and De Moivre's theorem, find the *modulus* and *argument* of z.

b Hence show that z is a cube root of 1.

c Without using a calculator, show that $(1-2z)(2z^2-1)$ is a real number.

9 **a** Write $-16i$ in polar form.

b Let z be the fourth root of $-16i$ which lies in the second quadrant. Express z exactly in:

i polar form **ii** Cartesian form.

THE nTH ROOTS OF UNITY

The **nth roots of unity** are the solutions of $z^n=1$.

Example 23 **Self Tutor**

Find the three cube roots of unity and display them on an Argand diagram. If w is the root with smallest positive argument, show that the roots are 1, w, and w^2 and that $1 + w + w^2 = 0$.

The cube roots of unity are the solutions of $z^3 = 1$.

But $1 = \text{cis } 0 = \text{cis } (0 + k2\pi)$ for all $k \in \mathbb{Z}$

$\therefore\ z^3 = \text{cis } (k2\pi)$

$\therefore\ z = [\text{cis } (k2\pi)]^{\frac{1}{3}}$ {De Moivre's theorem}

$\therefore\ z = \text{cis } \left(\frac{k2\pi}{3}\right)$

$\therefore\ z = \text{cis } 0,\ \text{cis } \left(\frac{2\pi}{3}\right),\ \text{cis } \left(\frac{4\pi}{3}\right)$ {letting $k = 0, 1, 2$}

$\therefore\ z = 1,\ -\frac{1}{2} + \frac{\sqrt{3}}{2}i,\ -\frac{1}{2} - \frac{\sqrt{3}}{2}i$

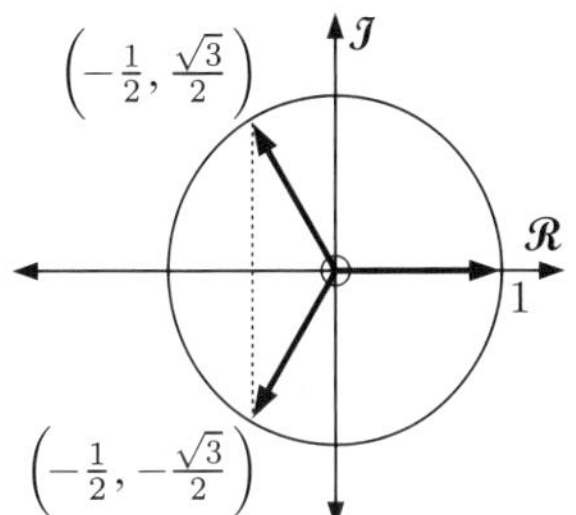

$w = \text{cis } \left(\frac{2\pi}{3}\right)$ and $w^2 = \left[\text{cis } \left(\frac{2\pi}{3}\right)\right]^2 = \text{cis } \left(\frac{4\pi}{3}\right)$

$\therefore$ the roots are 1, w, and w^2 where $w = \text{cis } \left(\frac{2\pi}{3}\right)$

and $1 + w + w^2 = 1 + \left(-\frac{1}{2} + \frac{\sqrt{3}}{2}i\right) + \left(-\frac{1}{2} - \frac{\sqrt{3}}{2}i\right) = 0$

EXERCISE 16F.2

1. In **Example 23** we showed that the cube roots of 1 are 1, w, w^2 where $w = \text{cis } \left(\frac{2\pi}{3}\right)$.
   a. Use this fact to solve the following equations, giving your answers in terms of w:
      i. $(z + 3)^3 = 1$
      ii. $(z - 1)^3 = 8$
      iii. $(2z - 1)^3 = -1$
   b. Show by vector addition that $1 + w + w^2 = 0$.

2. In **Example 21** we showed that the four fourth roots of unity were $1, i, -1, -i$.
   a. Show that the four fourth roots of unity can be written in the form $1, w, w^2, w^3$ where $w = \text{cis } \left(\frac{\pi}{2}\right)$.
   b. Hence show that $1 + w + w^2 + w^3 = 0$.
   c. Check the result in **b** by vector addition.

3. a. Find the 5 fifth roots of unity and display them on an Argand diagram.
   b. If w is the root with smallest positive argument, show that the roots are $1, w, w^2, w^3$, and w^4.
   c. Simplify $(1+w+w^2+w^3+w^4)(1-w)$ and hence show that $1+w+w^2+w^3+w^4 = 0$.
   d. Show by vector addition that $1 + w + w^2 + w^3 + w^4 = 0$.

4. a. Show that $w = \text{cis } \left(\frac{2\pi}{n}\right)$ is the nth root of unity with the smallest positive argument.
   b. Hence show that:
      i. the n roots of $z^n = 1$ are $1, w, w^2, w^3, ..., w^{n-1}$
      ii. $1 + w + w^2 + w^3 + ... + w^{n-1} = 0$

5. Show that for any complex number α, the sum of the n zeros of $z^n = \alpha$ is 0.

G MISCELLANEOUS PROBLEMS

The following questions combine complex number theory with topics covered in previous chapters.

EXERCISE 16G

1. Solve for z: $z^2 - (2+i)z + (3+i) = 0$
2. Find the Cartesian equation for the locus of points $P(x, y)$ if $z = x + iy$ and:
   a. $z^* = -iz$
   b. $\arg(z - i) = \frac{\pi}{6}$
   c. $|z+3| + |z-3| = 8$
3. Use the binomial expansion of $(1+i)^{2n}$ to prove that:
   $\binom{2n}{0} - \binom{2n}{2} + \binom{2n}{4} - \binom{2n}{6} + \ldots + (-1)^n \binom{2n}{2n} = 2^n \cos\left(\frac{n\pi}{2}\right), \quad n \in \mathbb{Z}^+$.
4. By considering $1 + \text{cis}\,\theta + \text{cis}\,2\theta + \text{cis}\,3\theta + \ldots + \text{cis}\,n\theta$ as a geometric series, find $\sum_{r=0}^{n} \cos r\theta$.
5. Prove that $1 + \text{cis}\,\theta = 2\cos\left(\frac{\theta}{2}\right) \text{cis}\left(\frac{\theta}{2}\right)$. Hence determine the sum of the series $\sum_{r=0}^{n} \binom{n}{r} \cos(r\theta)$.

REVIEW SET 16A — NON-CALCULATOR

1. Find the real and imaginary parts of $\left(i - \sqrt{3}\right)^5$.
2. Find the Cartesian equation for the locus of points $P(x, y)$ if $z = x + iy$ and:
   a. $|z - i| = |z + 1 + i|$
   b. $z^* - iz = 0$
3. Find $|z|$ if z is a complex number and $|z + 16| = 4|z + 1|$.
4. Find a single transformation which maps z onto: a. z^* b. $-z$ c. iz
5. z and w are non-real complex numbers such that both $z + w$ and zw are real. Prove that $z^* = w$.
6. If $(x + iy)^n = X + Yi$ where n is a positive integer, show that $X^2 + Y^2 = (x^2 + y^2)^n$.
7. Prove that $|z - w|^2 + |z + w|^2 = 2\left(|z|^2 + |w|^2\right)$.
8. $z^5 = 1$ has roots $1, \alpha, \alpha^2, \alpha^3$, and α^4 where $\alpha = \text{cis}\left(\frac{2\pi}{5}\right)$.
   a. Prove that $1 + \alpha + \alpha^2 + \alpha^3 + \alpha^4 = 0$.
   b. Solve $\left(\frac{z+2}{z-1}\right)^5 = 1$, giving your answer in terms of α.
9. If $z \neq 0$ and $\left|\frac{z+1}{z-1}\right| = 1$, prove that z is purely imaginary.
10. If $z = \text{cis}\,\phi$ and $w = \frac{1+z}{1+z^*}$, show that $w = \text{cis}\,\phi$ also.
11. Find the cube roots of $-64i$, giving your answers in the form $a + bi$ where $a, b \in \mathbb{R}$.
12. If $z = \text{cis}\,\theta$, find the modulus and argument of: a. $(2z)^{-1}$ b. $1 - z$
13. Write $-1 + i\sqrt{3}$ in polar form. Hence find the values of m for which $\left(-1 + i\sqrt{3}\right)^m$ is real.
14. Prove that $\text{cis}\,\theta + \text{cis}\,\phi = 2\cos\left(\frac{\theta-\phi}{2}\right) \text{cis}\left(\frac{\theta+\phi}{2}\right)$. Hence show that $\left(\frac{z+1}{z-1}\right)^5 = 1$ has solutions of the form $z = i\cot\left(\frac{n\pi}{5}\right)$ for $n = 1, 2, 3$, and 4.

REVIEW SET 16B CALCULATOR

1 Let $z_1 = \cos\left(\frac{\pi}{6}\right) + i\sin\left(\frac{\pi}{6}\right)$ and $z_2 = \cos\left(\frac{\pi}{4}\right) + i\sin\left(\frac{\pi}{4}\right)$.

Express $\left(\frac{z_1}{z_2}\right)^3$ in the form $z = a + bi$.

2 If $z = 4 + i$ and $w = 2 - 3i$, find:

a $2w^* - iz$ **b** $|w - z^*|$ **c** $|z^{10}|$ **d** $\arg(w - z)$

3 Find rational numbers a and b such that $\dfrac{2 - 3i}{2a + bi} = 3 + 2i$.

4 Find the Cartesian equation for the locus of points $P(x, y)$ if $z = x + iy$ and:

a $\arg(z - i) = \frac{\pi}{2}$ **b** $\left|\dfrac{z+2}{z-2}\right| = 2$

5 Write $2 - 2\sqrt{3}i$ in polar form. Hence find all values of n for which $\left(2 - 2\sqrt{3}i\right)^n$ is real.

6 Determine the cube roots of -27.

7 If $z = 4 \operatorname{cis} \theta$, find the modulus and argument of:

a z^3 **b** $\dfrac{1}{z}$ **c** iz^*

8 Prove: **a** $\arg(z^n) = n \arg z$ for all complex numbers z and rational n

b $\left(\dfrac{z}{w}\right)^* = \dfrac{z^*}{w^*}$ for all z and for all $w \neq 0$.

9 Points A and B are the representations in the complex plane of the numbers $z = 2 - 2i$ and $w = -1 - \sqrt{3}i$ respectively.

a Given the origin O, find $\widehat{AOB}$ in terms of π.

b Calculate the argument of zw in terms of π.

10 Illustrate the region defined by $\{z:\ 2 \leqslant |z| \leqslant 5 \text{ and } -\frac{\pi}{4} < \arg z \leqslant \frac{\pi}{2}\}$.
Show clearly all included boundary points.

11 Use polar form to deduce that $\left|\dfrac{1}{z}\right| = \dfrac{1}{|z|}$ for $z \neq 0$ and $\arg\left(\dfrac{1}{z}\right) = -\arg z$.

12 If $z = \operatorname{cis} \alpha$, write $1 + z$ in polar form. Hence determine the modulus and argument of $1 + z$.

Hint: $\sin\theta = 2\sin\left(\frac{\theta}{2}\right)\cos\left(\frac{\theta}{2}\right)$

13 $\triangle P_1P_2P_3$ is an equilateral triangle.

O is an origin such that $\overrightarrow{OP_1} \equiv z_1$, $\overrightarrow{OP_2} \equiv z_2$, and $\overrightarrow{OP_3} \equiv z_3$.

Suppose $\arg(z_2 - z_1) = \alpha$.

a Show that $\arg(z_3 - z_2) = \alpha - \frac{2\pi}{3}$.

b Find the modulus and argument of $\dfrac{z_2 - z_1}{z_3 - z_2}$.

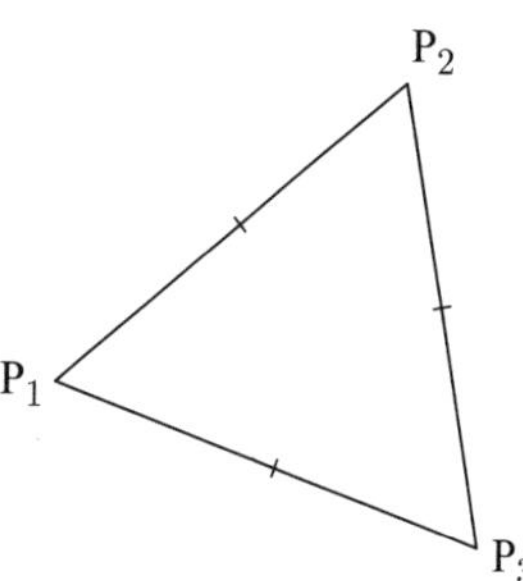

14 State the five fifth roots of unity, and hence solve:

a $(2z - 1)^5 = 32$ **b** $z^5 + 5z^4 + 10z^3 + 10z^2 + 5z = 0$ **c** $(z + 1)^5 = (z - 1)^5$

REVIEW SET 16C

1 Write in polar form: **a** $-5i$ **b** $2-2i\sqrt{3}$ **c** $k-ki$ where $k<0$

2 Suppose $z=(1+bi)^2$ where b is real and positive. Find the exact value of b if $\arg z=\frac{\pi}{3}$.

3 **a** Prove that $\text{cis}\,\theta\times\text{cis}\,\phi=\text{cis}\,(\theta+\phi)$.

b If $z=2\sqrt{2}\,\text{cis}\,\alpha$, write $(1-i)z$ in polar form. Hence find $\arg[(1-i)z]$.

4 $z_1\equiv\overrightarrow{\text{OA}}$ and $z_2\equiv\overrightarrow{\text{OB}}$ represent two sides of a right angled isosceles triangle OAB.

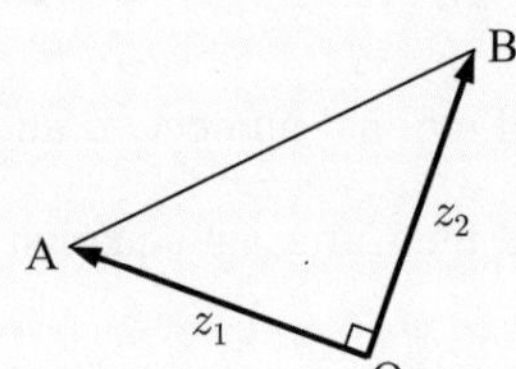

a Determine the modulus and argument of $\dfrac{{z_1}^2}{{z_2}^2}$.

b Hence, deduce that ${z_1}^2+{z_2}^2=0$.

5 Let $z=\sqrt[4]{a}\left(\cos\left(\frac{\pi}{6}\right)+i\sin\left(\frac{\pi}{6}\right)\right)$ and $w=\sqrt[4]{b}\left(\cos\left(\frac{\pi}{4}\right)-i\sin\left(\frac{\pi}{4}\right)\right)$.

Find, in terms of a and b, the real and imaginary parts of $\left(\dfrac{z}{w}\right)^4$.

6 **a** List the five fifth roots of 1 in terms of w, where $w=\text{cis}\left(\frac{2\pi}{5}\right)$. Display these roots on an Argand diagram.

b By considering the factorisation of z^5-1 in *two different ways*, show that: $z^4+z^3+z^2+z+1=(z-w)(z-w^2)(z-w^3)(z-w^4)$.

c Hence, find the value of $(2-w)(2-w^2)(2-w^3)(2-w^4)$.

7 Find the cube roots of $-8i$, giving your answers in the form $a+bi$ where a and b do not involve trigonometric ratios.

8 Write each of the following in the form $(\text{cis}\,\theta)^n$:

a $\cos 3\theta+i\sin 3\theta$ **b** $\dfrac{1}{\cos 2\theta+i\sin 2\theta}$ **c** $\cos\theta-i\sin\theta$

9 Determine the fifth roots of $2+2i$.

10 If $z+\dfrac{1}{z}$ is real, prove that either $|z|=1$ or z is real.

11 If $z=\text{cis}\,\theta$, prove that:

a $|z|=1$ **b** $z^*=\dfrac{1}{z}$ **c** $\sin^4\theta=\frac{1}{8}(\cos 4\theta-4\cos 2\theta+3)$

12 Suppose w is the root of $z^5=1$ with the smallest positive argument. Find real quadratic equations with roots: **a** w and w^4 **b** $(w+w^4)$ and (w^2+w^3).

13 If $|z+w|=|z-w|$, prove that $\arg z$ and $\arg w$ differ by $\frac{\pi}{2}$.

14 The complex number z is a root of the equation $|z|=|z+4|$.

a Show that the real part of z is -2.

b Let v and w be two possible values of z such that $|z|=4$. v is in the 2nd quadrant.

i Sketch the points that represent v and w on an Argand diagram.

ii Show that $\arg v=\frac{2\pi}{3}$. **iii** Find $\arg w$ where $-\pi<\arg w\leqslant\pi$.

iv Find $\arg\left(\dfrac{v^m w}{i}\right)$ in terms of m and π.

v Hence find a value of m for which $\dfrac{v^m w}{i}$ is a real number.

Chapter 17

Introduction to differential calculus

Syllabus reference: 6.1

Contents:

OPENING PROBLEM

In a BASE jumping competition from the Petronas Towers in Kuala Lumpur, the altitude of a professional jumper in the first 3 seconds is given by $f(t) = 452 - 4.8t^2$ metres, where $0 \leqslant t \leqslant 3$ seconds.

Things to think about:

a What will a graph of the altitude of the jumper in the first 3 seconds look like?

b Does the jumper travel with constant speed?

c Can you find the speed of the jumper when:

- **i** $t = 0$ seconds
- **ii** $t = 1$ second
- **iii** $t = 2$ seconds
- **iv** $t = 3$ seconds?

Calculus is a major branch of mathematics which builds on algebra, trigonometry, and analytic geometry. It has widespread applications in science, engineering, and financial mathematics.

The study of calculus is divided into two fields, **differential calculus** and **integral calculus**. These fields are linked by the **Fundamental Theorem of Calculus** which we will study later in the course.

HISTORICAL NOTE

Calculus is a Latin word meaning 'pebble'. Ancient Romans used stones for counting.

The history of calculus begins with the **Egyptian Moscow papyrus** from about 1850 BC.

The Greek mathematicians **Democritus**, **Zeno of Elea**, **Antiphon**, and **Eudoxes** studied **infinitesimals**, dividing objects into an infinite number of pieces in order to calculate the area of regions, and volume of solids.

Archimedes of Syracuse was the first to find the tangent to a curve other than a circle. His methods were the foundation of modern calculus developed almost 2000 years later.

Archimedes

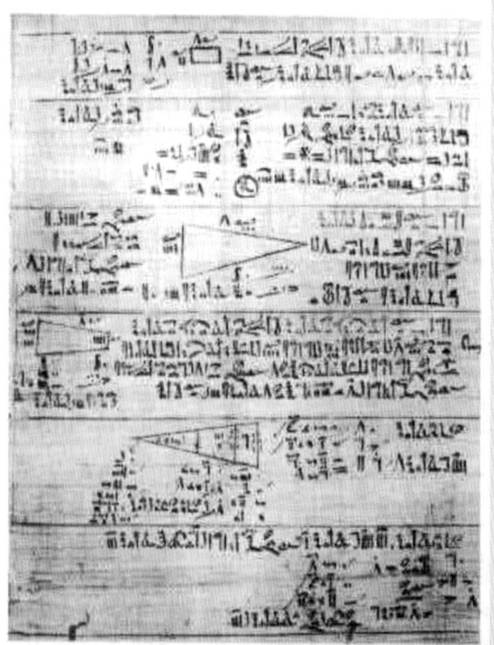

Egyptian Moscow papyrus

LIMITS

The concept of a **limit** is essential to differential calculus. We will see that calculating limits is necessary for finding the gradient of a tangent to a curve at any point on the curve.

Consider the following table of values for $f(x) = x^2$ where x is less than 2 but increasing and getting closer and closer to 2:

x	1	1.9	1.99	1.999	1.9999
$f(x)$	1	3.61	3.9601	3.996 00	3.999 60

We say that as x approaches 2 from the left, $f(x)$ approaches 4 from below.

We can construct a similar table of values where x is greater than 2 but decreasing and getting closer and closer to 2:

x	3	2.1	2.01	2.001	2.0001
$f(x)$	9	4.41	4.0401	4.004 00	4.000 40

We say that as x approaches 2 from the right, $f(x)$ approaches 4 from above.

So, as x approaches 2 from either direction, $f(x)$ approaches a limit of 4. We write this as $\lim\limits_{x \to 2} x^2 = 4$.

INFORMAL DEFINITION OF A LIMIT

The following definition of a limit is informal but adequate for the purposes of this course:

If $f(x)$ can be made as close as we like to some real number A by making x sufficiently close to (but not equal to) a, then we say that $f(x)$ has a **limit** of A as x approaches a, and we write

$$\lim_{x \to a} f(x) = A.$$

In this case, $f(x)$ is said to **converge** to A as x approaches a.

Notice that the limit is defined for x close to but *not equal to* a. Whether the function f is defined or not at $x = a$ is not important to the definition of the limit of f as x approaches a. What *is* important is the behaviour of the function as x gets *very close to* a.

For example, if $f(x) = \dfrac{5x + x^2}{x}$ and we wish to find the limit as $x \to 0$, it is tempting for us to simply substitute $x = 0$ into $f(x)$. However, in doing this, not only do we get the meaningless value of $\frac{0}{0}$, but also we destroy the basic limit method.

Observe that if $f(x) = \dfrac{5x + x^2}{x} = \dfrac{x(5 + x)}{x}$

$$\text{then} \quad f(x) = \begin{cases} 5 + x \ \text{ if } \ x \neq 0 \\ \text{is undefined if } \ x = 0. \end{cases}$$

The graph of $y = f(x)$ is shown alongside. It is the straight line $y = x + 5$ with the point (0, 5) missing, called a **point of discontinuity** of the function.

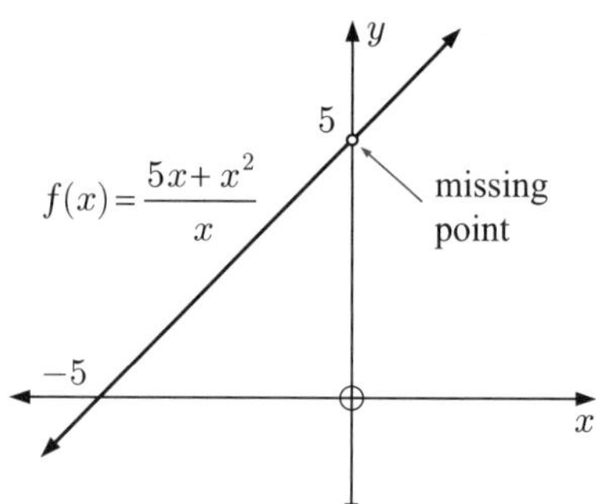

However, even though this point is missing, the *limit* of $f(x)$ as x approaches 0 does exist. In particular, as $x \to 0$ from either direction, $f(x) \to 5$.

We write $\lim\limits_{x \to 0} \dfrac{5x + x^2}{x} = 5$ which reads:

"the limit as x approaches 0, of $f(x) = \dfrac{5x + x^2}{x}$, is 5".

A function f is said to be **continuous** at $x = a$ if and only if three conditions are all satisfied:

1 $f(a)$ is defined **2** $\lim\limits_{x \to a} f(x)$ exists **3** $f(a) = \lim\limits_{x \to a} f(x)$

In practice we do not need to graph functions each time to determine limits, and most can be found algebraically.

Example 1 **Self Tutor**

Evaluate: **a** $\lim\limits_{x \to 2} x^2$ **b** $\lim\limits_{x \to 0} \dfrac{x^2 + 3x}{x}$ **c** $\lim\limits_{x \to 3} \dfrac{x^2 - 9}{x - 3}$

a x^2 can be made as close as we like to 4 by making x sufficiently close to 2.

$\therefore \lim\limits_{x \to 2} x^2 = 4.$

b
$$\lim_{x \to 0} \frac{x^2 + 3x}{x}$$
$$= \lim_{x \to 0} \frac{\cancel{x}(x + 3)}{\cancel{x}} \quad \text{since } x \neq 0$$
$$= \lim_{x \to 0} (x + 3)$$
$$= 3$$

c
$$\lim_{x \to 3} \frac{x^2 - 9}{x - 3}$$
$$= \lim_{x \to 3} \frac{(x + 3)\cancel{(x - 3)}}{\cancel{x - 3}} \quad \text{since } x \neq 3$$
$$= \lim_{x \to 3} (x + 3)$$
$$= 6$$

RULES FOR LIMITS

If $f(x)$ and $g(x)$ are functions such that $\lim\limits_{x \to a} f(x)$ and $\lim\limits_{x \to a} g(x)$ exist, and c is a constant:

- $\lim\limits_{x \to a} c = c$
- $\lim\limits_{x \to a} c\,f(x) = c \lim\limits_{x \to a} f(x)$
- $\lim\limits_{x \to a} [f(x) \pm g(x)] = \lim\limits_{x \to a} f(x) \pm \lim\limits_{x \to a} g(x)$
- $\lim\limits_{x \to a} [f(x)\,g(x)] = \lim\limits_{x \to a} f(x) \times \lim\limits_{x \to a} g(x)$
- $\lim\limits_{x \to a} \dfrac{f(x)}{g(x)} = \lim\limits_{x \to a} f(x) \div \lim\limits_{x \to a} g(x)$ provided $\lim\limits_{x \to a} g(x) \neq 0$

Example 2 — Self Tutor

Use the rules to evaluate the following limits. Explain what the results mean.

a $\lim_{x \to 3} (x+2)(x-1)$ **b** $\lim_{x \to 1} \frac{x^2+2}{x-2}$

a As $x \to 3$, $x+2 \to 5$ and $x-1 \to 2$

$\therefore \lim_{x \to 3} (x+2)(x-1) = 5 \times 2 = 10$

So, as $x \to 3$, $(x+2)(x-1)$ *converges* to 10.

b As $x \to 1$, $x^2+2 \to 3$ and $x-2 \to -1$

$\therefore \lim_{x \to 1} \frac{x^2+2}{x-2} = \frac{3}{-1} = -3$

So, as $x \to 1$, $\frac{x^2+2}{x-2}$ *converges* to -3.

EXERCISE 17A

1 Evaluate:

a $\lim_{x \to 3} (x+4)$ **b** $\lim_{x \to -1} (5-2x)$ **c** $\lim_{x \to 4} (3x-1)$

d $\lim_{x \to 2} (5x^2-3x+2)$ **e** $\lim_{h \to 0} h^2(1-h)$ **f** $\lim_{x \to 0} (x^2+5)$

2 Evaluate:

a $\lim_{x \to 0} 5$ **b** $\lim_{h \to 2} 7$ **c** $\lim_{x \to 0} c$, c a constant

3 Evaluate:

a $\lim_{x \to 1} \frac{x^2-3x}{x}$ **b** $\lim_{h \to 2} \frac{h^2+5h}{h}$ **c** $\lim_{x \to 0} \frac{x-1}{x+1}$ **d** $\lim_{x \to 0} \frac{x}{x}$

4 At what values of x are the following functions *not* continuous?
Explain your answer in each case.

a $f(x) = \frac{1}{x}$ **b** $f(x) = \frac{x^2-x}{x}$

5 Evaluate the following limits:

a $\lim_{x \to 0} \frac{x^2-3x}{x}$ **b** $\lim_{x \to 0} \frac{x^2+5x}{x}$ **c** $\lim_{x \to 0} \frac{2x^2-x}{x}$

d $\lim_{h \to 0} \frac{2h^2+6h}{h}$ **e** $\lim_{h \to 0} \frac{3h^2-4h}{h}$ **f** $\lim_{h \to 0} \frac{h^3-8h}{h}$

g $\lim_{x \to 1} \frac{x^2-x}{x-1}$ **h** $\lim_{x \to 2} \frac{x^2-2x}{x-2}$ **i** $\lim_{x \to 3} \frac{x^2-x-6}{x-3}$

B LIMITS AT INFINITY

We can use the idea of limits to discuss the behaviour of functions for extreme values of x.

We write $x \to \infty$ to mean when x becomes as large as we like and positive,

and $x \to -\infty$ to mean when x becomes as large as we like and negative.

We read $x \to \infty$ as "x tends to plus infinity" and $x \to -\infty$ as "x tends to minus infinity".

Notice that as $x \to \infty$, $1 < x < x^2 < x^3 < ....$ and as x gets very large, the value of $\frac{1}{x}$ gets very small. In fact, we can make $\frac{1}{x}$ as close to 0 as we like by making x large enough. This means that $\lim\limits_{x \to \infty} \frac{1}{x} = 0$ even though $\frac{1}{x}$ never actually reaches 0.

Similarly, $$\begin{aligned}\lim_{x \to \infty} \frac{x}{x^2} &= \lim_{x \to \infty} \frac{1}{x} \times \frac{x}{x} \\ &= \lim_{x \to \infty} \frac{1}{x} \qquad \{\text{since } x \neq 0\} \\ &= 0\end{aligned}$$

Example 3 **Self Tutor**

Evaluate: **a** $\lim\limits_{x \to \infty} \frac{2x+3}{x-4}$ **b** $\lim\limits_{x \to \infty} \frac{x^2-3x+2}{1-x^2}$

a $$\begin{aligned}&\lim_{x \to \infty} \frac{2x+3}{x-4} \\ &= \lim_{x \to \infty} \frac{2+\frac{3}{x}}{1-\frac{4}{x}} \qquad \{\text{dividing each term in both numerator and denominator by } x\} \\ &= \frac{2}{1} \qquad \{\text{as } x \to \infty,\ \frac{3}{x} \to 0 \text{ and } \frac{4}{x} \to 0\} \\ &= 2\end{aligned}$$

b $$\begin{aligned}&\lim_{x \to \infty} \frac{x^2-3x+2}{1-x^2} \\ &= \lim_{x \to \infty} \frac{1-\frac{3}{x}+\frac{2}{x^2}}{\frac{1}{x^2}-1} \qquad \{\text{dividing each term by } x^2\} \\ &= \frac{1}{-1} \qquad \{\text{as } x \to \infty,\ \frac{3}{x} \to 0,\ \frac{2}{x^2} \to 0, \text{ and } \frac{1}{x^2} \to 0\} \\ &= -1\end{aligned}$$

EXERCISE 17B.1

1 Examine $\lim\limits_{x \to \infty} \frac{1}{x^2}$.

2 Evaluate:

a $\lim\limits_{x \to \infty} \frac{3x-2}{x+1}$ **b** $\lim\limits_{x \to \infty} \frac{1-2x}{3x+2}$ **c** $\lim\limits_{x \to \infty} \frac{x}{1-x}$

d $\lim\limits_{x \to \infty} \frac{x^2+3}{x^2-1}$ **e** $\lim\limits_{x \to \infty} \frac{x^2-2x+4}{x^2+x-1}$

INVESTIGATION 1 LIMITS IN NUMBER SEQUENCES

The sequence 0.3, 0.33, 0.333, ... can be defined by the general term $x_n = 0.333....3$ where there are n 3s after the decimal point, $n \in \mathbb{Z}^+$.

What to do:

1 Copy and complete the table alongside:

n	x_n	$3x_n$	$1 - 3x_n$
1			
2			
3			
4			
5			
10			

2 Consider x_{100} which contains 100 3s. In the number $(1-3x_{100})$, how many 0s are there between the decimal point and the 1?

3 In the limit as n tends to infinity, x_n contains an increasingly large number of 3s. In the number $(1 - 3x_n)$, how many 0s will there be before the 1?

4 Using your answer to **3**, state $\lim_{n \to \infty} 1 - 3x_n$.

5 Hence state $\lim_{n \to \infty} x_n$, which is the exact value of $0.\overline{3}$.

ASYMPTOTES

In **Chapter 2** we studied **rational functions**, and saw how they are characterised by the presence of **asymptotes**.

Consider the function $f(x) = \dfrac{2x+3}{x-4}$ which has domain $\{x \mid x \neq 4, \ x \in \mathbb{R}\}$.

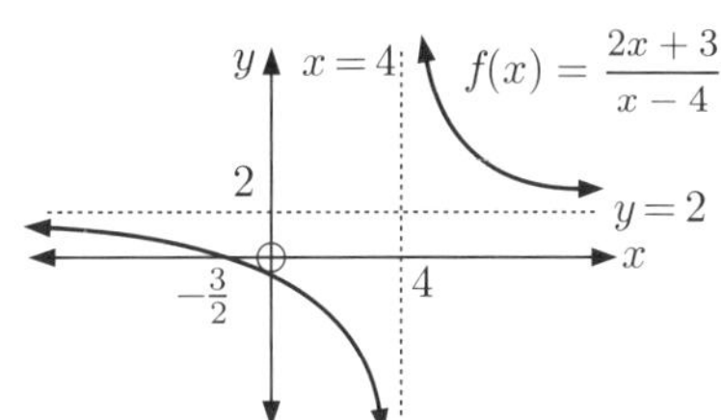

From the graph of $y = f(x)$, we can see the function has a vertical asymptote $x = 4$, and a horizontal asymptote $y = 2$.

Both of these asymptotes can be described in terms of limits:

As $x \to 4^-$, $f(x) \to -\infty$
As $x \to 4^+$, $f(x) \to +\infty$
As $x \to -\infty$, $f(x) \to 2^-$
As $x \to +\infty$, $f(x) \to 2^+$

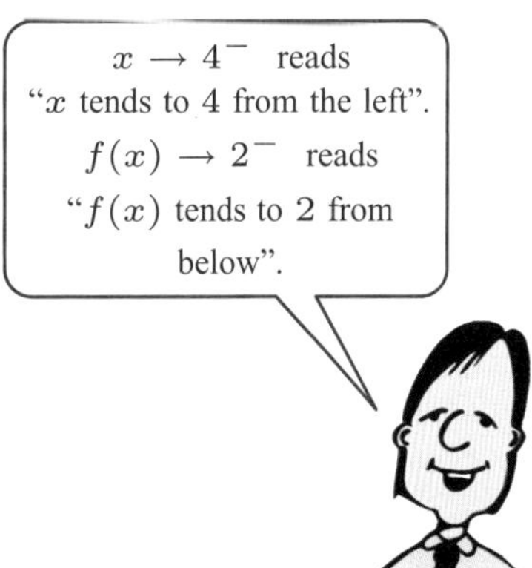

$x \longrightarrow 4^-$ reads "x tends to 4 from the left".
$f(x) \longrightarrow 2^-$ reads "$f(x)$ tends to 2 from below".

Since $f(x)$ converges to a finite value as $x \to -\infty$, we see $\lim_{x \to -\infty} f(x) = 2$.

Since $f(x)$ converges to a finite value as $x \to +\infty$, we see $\lim_{x \to \infty} f(x) = 2$.

This matches what we see algebraically, since

$$\lim_{x\to\infty} \frac{2x+3}{x-4} = \lim_{x\to\infty} \frac{2+\frac{3}{x}}{1-\frac{4}{x}} = 2$$

and

$$\lim_{x\to-\infty} \frac{2x+3}{x-4} = \lim_{x\to-\infty} \frac{2+\frac{3}{x}}{1-\frac{4}{x}} = 2$$

Example 4 **Self Tutor**

a Discuss the behaviour of $f(x) = \dfrac{2-x}{1+x}$ near its asymptotes, and hence deduce their equations.

b State the values of $\lim\limits_{x \to -\infty} f(x)$ and $\lim\limits_{x \to \infty} f(x)$ if they exist.

a As $x \to -1^-$, $f(x) \to -\infty$
As $x \to -1^+$, $f(x) \to \infty$
As $x \to -\infty$, $f(x) \to -1^-$
As $x \to \infty$, $f(x) \to -1^+$
The vertical asymptote is $x = -1$.
The horizontal asymptote is $y = -1$.

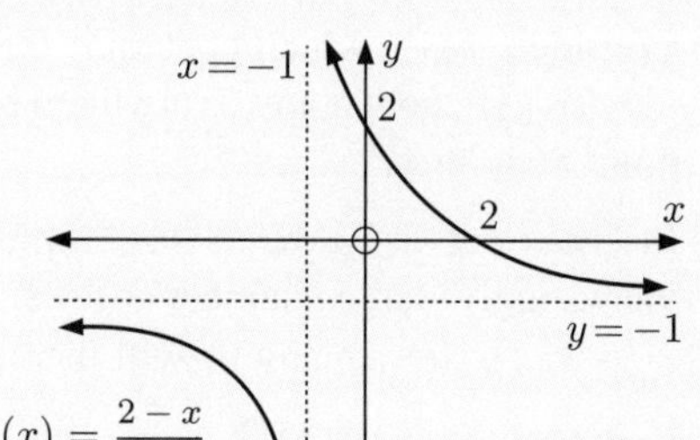

b $\lim\limits_{x \to -\infty} f(x) = -1$ and $\lim\limits_{x \to \infty} f(x) = -1$.

Example 5 **Self Tutor**

Find, if possible, $\lim\limits_{x \to -\infty} (3 - e^{-x})$ and $\lim\limits_{x \to \infty} (3 - e^{-x})$.

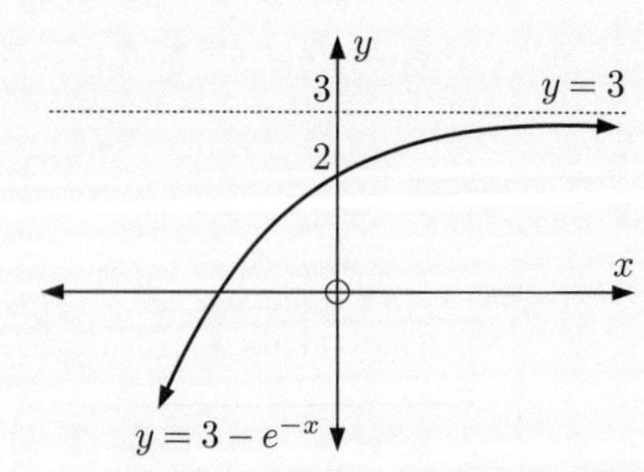

As $x \to -\infty$, $3 - e^{-x} \to -\infty$.

Since $3 - e^{-x}$ does not approach a finite value, $\lim\limits_{x \to -\infty} (3 - e^{-x})$ does not exist.

As $x \to \infty$, $3 - e^{-x} \to 3^-$

$\therefore \lim\limits_{x \to \infty} (3 - e^{-x}) = 3$

EXERCISE 17B.2

1 For each of the following functions:

i discuss the behaviour near the asymptotes and hence deduce their equations

ii state the values of $\lim\limits_{x \to -\infty} f(x)$ and $\lim\limits_{x \to \infty} f(x)$.

a $f(x) = \dfrac{1}{x}$ **b** $f(x) = \dfrac{3x-2}{x+3}$ **c** $f(x) = \dfrac{1-2x}{3x+2}$

d $f(x) = \dfrac{x}{1-x}$ **e** $f(x) = \dfrac{x^2-1}{x^2+1}$ **f** $f(x) = \dfrac{x}{x^2+1}$

2 **a** Sketch the graph of $y = e^x - 6$.

b Hence discuss the value and geometric interpretation of:

i $\lim\limits_{x \to -\infty} (e^x - 6)$ **ii** $\lim\limits_{x \to \infty} (e^x - 6)$

3 Find, if possible, $\lim\limits_{x \to -\infty} (2e^{-x} - 3)$ and $\lim\limits_{x \to \infty} (2e^{-x} - 3)$.

4 Determine the asymptotes of the graphs of the following functions, and discuss the behaviour of the graphs near these asymptotes:

a $f(x) = \ln x$ **b** $f(x) = e^{x - \frac{1}{x}}$

5 If a graph approaches a straight line which is neither vertical nor horizontal, we call it an **oblique asymptote**.

Determine the asymptotes of the graphs of the following functions, and discuss the behaviour of the graphs near these asymptotes:

a $f(x) = x + \ln x$ **b** $f(x) = e^x - x$ **c** $f(x) = \dfrac{x^3 - 2}{x^2 + 1}$ **d** $f(x) = (x - 2)e^{-x}$

C TRIGONOMETRIC LIMITS

INVESTIGATION 2 EXAMINING $\dfrac{\sin\theta}{\theta}$ NEAR $\theta = 0$

This Investigation examines $\dfrac{\sin\theta}{\theta}$ when θ is close to 0 and θ is in radians.

We consider this ratio graphically, numerically, and geometrically.

What to do:

1 Show that $f(\theta) = \dfrac{\sin\theta}{\theta}$ is an even function. What does this mean graphically?

2 Since $\dfrac{\sin\theta}{\theta}$ is even we need only examine $\dfrac{\sin\theta}{\theta}$ for positive θ.

a Is $f(\theta) = \dfrac{\sin\theta}{\theta}$ continuous at $\theta = 0$?

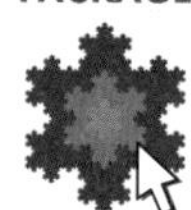

b Graph $y = \dfrac{\sin\theta}{\theta}$ for $-\frac{\pi}{2} \leqslant \theta \leqslant \frac{\pi}{2}$ using a graphics calculator or graphing package.

c Explain using your graph why $\displaystyle\lim_{\theta \to 0} \frac{\sin\theta}{\theta} = 1$.

3 **a** Copy and complete the given table, using your calculator.

b Construct a table for negative values of θ which approach 0.

c Use your tables to discuss $\displaystyle\lim_{\theta \to 0} \frac{\sin\theta}{\theta}$.

θ	$\sin\theta$	$\dfrac{\sin\theta}{\theta}$
1		
0.5		
0.1		
0.01		
0.001		
⋮		

4 **a** Explain why the area of the shaded segment is $A = \frac{1}{2}r^2(\theta - \sin\theta)$.

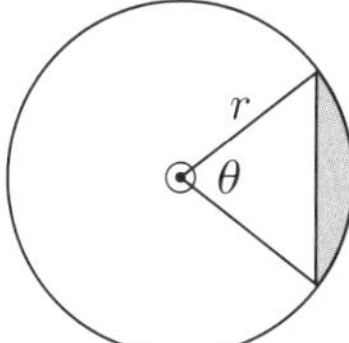

b Use the given figure to show that for θ in radians, as $\theta \to 0^+$, $\dfrac{\sin\theta}{\theta} \to 1$.

5 Repeat the Investigation to find $\displaystyle\lim_{\theta \to 0} \frac{\sin\theta}{\theta}$ for θ in degrees rather than radians.

If θ is in radians, then $\lim\limits_{\theta \to 0} \dfrac{\sin\theta}{\theta} = 1$.

Proof:

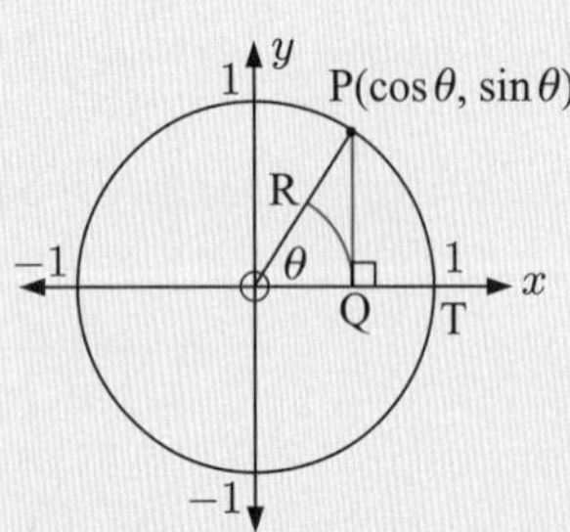

Suppose $P(\cos\theta, \sin\theta)$ lies on the unit circle in the first quadrant.

[PQ] is drawn perpendicular to the x-axis, and arc QR with centre O is drawn. Now,

area of sector OQR $\leqslant$ area $\triangle$OQP $\leqslant$ area sector OTP

$\therefore \ \frac{1}{2}(\text{OQ})^2 \times \theta \leqslant \frac{1}{2}(\text{OQ})(\text{PQ}) \leqslant \frac{1}{2}(\text{OT})^2 \times \theta$

$\therefore \ \frac{1}{2}\theta\cos^2\theta \leqslant \frac{1}{2}\cos\theta\sin\theta \leqslant \frac{1}{2}\theta$

$\therefore \ \cos\theta \leqslant \dfrac{\sin\theta}{\theta} \leqslant \dfrac{1}{\cos\theta}$

{dividing throughout by $\frac{1}{2}\theta\cos\theta$, which is > 0}

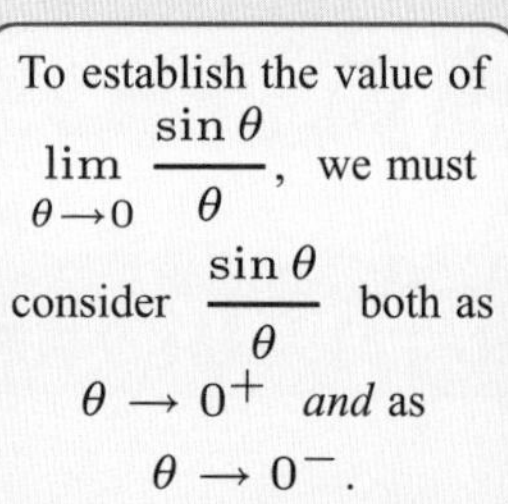

To establish the value of $\lim\limits_{\theta \to 0} \dfrac{\sin\theta}{\theta}$, we must consider $\dfrac{\sin\theta}{\theta}$ both as $\theta \to 0^+$ *and* as $\theta \to 0^-$.

Now as $\theta \to 0$, both $\cos\theta \to 1$ and $\dfrac{1}{\cos\theta} \to 1$

$\therefore$ as $\theta \to 0^+$, $\dfrac{\sin\theta}{\theta} \to 1$.

But $\dfrac{\sin\theta}{\theta}$ is an even function, so as $\theta \to 0^-$, $\dfrac{\sin\theta}{\theta} \to 1$ also.

Thus $\lim\limits_{\theta \to 0} \dfrac{\sin\theta}{\theta} = 1$.

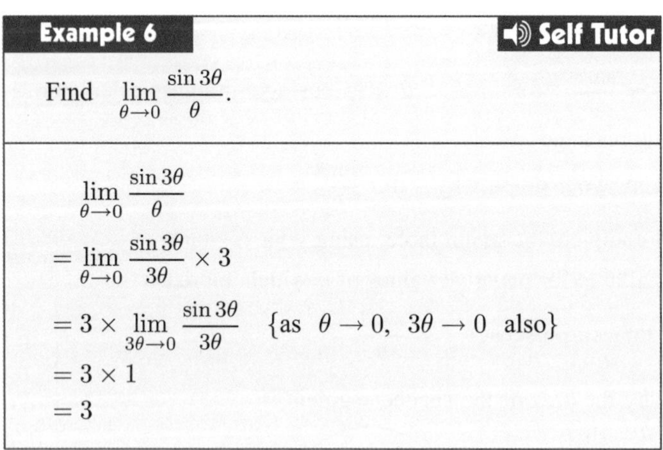

Example 6 — Self Tutor

Find $\lim\limits_{\theta \to 0} \dfrac{\sin 3\theta}{\theta}$.

$\lim\limits_{\theta \to 0} \dfrac{\sin 3\theta}{\theta}$

$= \lim\limits_{\theta \to 0} \dfrac{\sin 3\theta}{3\theta} \times 3$

$= 3 \times \lim\limits_{3\theta \to 0} \dfrac{\sin 3\theta}{3\theta}$ {as $\theta \to 0$, $3\theta \to 0$ also}

$= 3 \times 1$

$= 3$

EXERCISE 17C

1 Find:

a $\lim\limits_{\theta \to 0} \dfrac{\sin 2\theta}{\theta}$ **b** $\lim\limits_{\theta \to 0} \dfrac{\theta}{\sin\theta}$ **c** $\lim\limits_{\theta \to 0} \dfrac{\tan\theta}{\theta}$

d $\lim\limits_{\theta \to 0} \dfrac{\sin\theta\sin 4\theta}{\theta^2}$ **e** $\lim\limits_{h \to 0} \dfrac{\sin\left(\frac{h}{2}\right)\cos h}{h}$ **f** $\lim\limits_{n \to \infty} n\sin\left(\frac{2\pi}{n}\right)$

2 A circle of radius r contains n congruent isosceles triangles, all with apex O and with base vertices on the circle as shown.

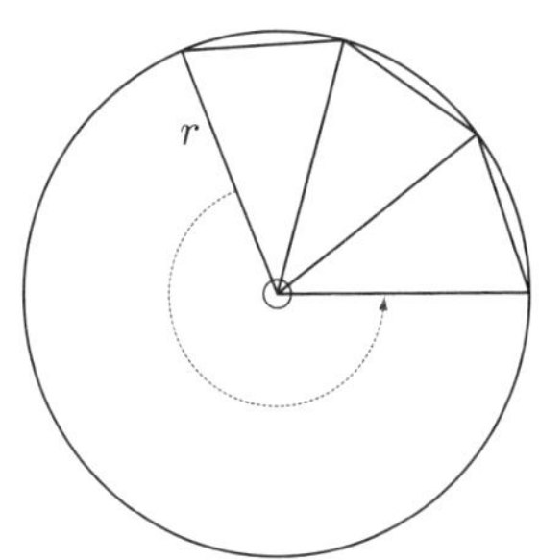

a Explain why the sum of the areas of the triangles is $S_n = \frac{1}{2}nr^2 \sin\left(\frac{2\pi}{n}\right)$.

b Find $\lim\limits_{n\to\infty} S_n$ **i** geometrically **ii** algebraically.

c What can be deduced from **b**?

3 **a** Show that $\cos(A+B) - \cos(A-B) = -2\sin A \sin B$.

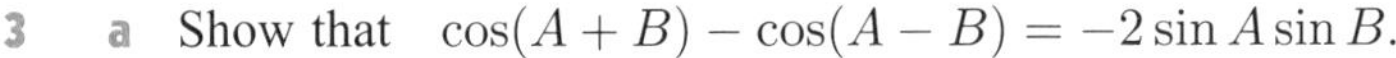

b Suppose $A + B = S$ and $A - B = D$.

Show that $\cos S - \cos D = -2\sin\left(\frac{S+D}{2}\right)\sin\left(\frac{S-D}{2}\right)$.

c Hence find $\lim\limits_{h\to 0} \dfrac{\cos(x+h) - \cos x}{h}$.

THEORY OF KNOWLEDGE

The Greek philosopher Zeno of Elea lived in what is now southern Italy, in the 5th century BC. He is most famous for his paradoxes, which were recorded in Aristotle's work *Physics*.

The arrow paradox

"*If everything when it occupies an equal space is at rest, and if that which is in locomotion is always occupying such a space at any moment, the flying arrow is therefore motionless.*"

This argument says that if we fix an instant in time, an arrow appears motionless. Consequently, how is it that the arrow actually moves?

The dichotomy paradox

"*That which is in locomotion must arrive at the half-way stage before it arrives at the goal.*"

If an object is to move a fixed distance then it must travel half that distance. Before it can travel a half the distance, it must travel a half *that* distance. With this process continuing indefinitely, motion is impossible.

Achilles and the tortoise

"*In a race, the quickest runner can never overtake the slowest, since the pursuer must first reach the point whence the pursued started, so that the slower must always hold a lead.*"

According to this principle, the athlete Achilles will never be able to catch the slow tortoise!

1. A paradox is a logical argument that leads to a contradiction or a situation which defies logic or reason. Can a paradox be the truth?
2. Are Zeno's paradoxes really paradoxes?
3. Are the three paradoxes essentially the same?
4. We know from experience that things *do* move, and that Achilles *would* catch the tortoise. Does that mean that logic has failed?
5. What do Zeno's paradoxes have to do with limits?

D RATES OF CHANGE

A **rate** is a comparison between two quantities with different units.

We often judge performances by rates. For example:

- Sir Donald Bradman's average batting rate at Test cricket level was 99.94 *runs per innings*.
- Michael Jordan's average basketball scoring rate was 30.1 *points per game*.
- Rangi's average typing rate is 63 *words per minute* with an error rate of 2.3 *errors per page*.

Speed is a commonly used rate. It is the rate of change in distance per unit of time.
We are familiar with the formula:

$$\textbf{average speed} = \frac{\textbf{distance travelled}}{\textbf{time taken}}$$

However, if a car has an average speed of 60 km h^{-1} for a journey, it does not mean that the car travels at exactly 60 km h^{-1} for the whole time.

In fact, the speed will probably vary continuously throughout the journey.

So, how can we calculate the car's speed at any particular time?

Suppose we are given a graph of the car's distance travelled against time taken. If this graph is a straight line, then we know the speed is constant and is given by the *gradient* of the line.

If the graph is a curve, then the car's instantaneous speed is given by the *gradient of the tangent* to the curve at that time.

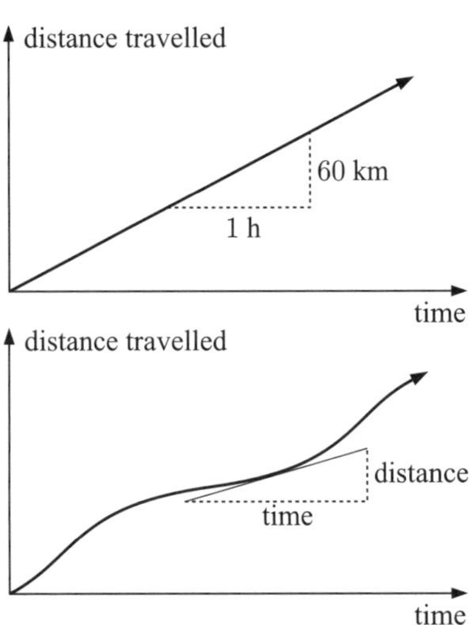

The **instantaneous rate of change** of a dependent variable with respect to the independent variable at a particular instant, is given by the **gradient of the tangent** to the graph at that point.

HISTORICAL NOTE

The modern study of **differential calculus** originated in the 17th century with the work of **Sir Isaac Newton** and **Gottfried Wilhelm Leibniz**. They developed the necessary theory while attempting to find algebraic methods for solving problems dealing with the **gradients of tangents** to curves, and finding the **rate of change** in one variable with respect to another.

Isaac Newton 1642 – 1727

Gottfried Leibniz 1646 – 1716

INVESTIGATION 3 INSTANTANEOUS SPEED

A ball bearing is dropped from the top of a tall building. The distance D it has fallen after t seconds is recorded, and the following graph of distance against time obtained.

We choose a fixed point F on the curve when $t = 2$ seconds. We then choose another point M on the curve, and draw in the line segment or **chord** [FM] between the two points. To start with, we let M be the point when $t = 4$ seconds.

D, t, 20, 40, 60, 80, 2, 4, 6
M(4, 80)
chord
curve
F(2, 20)

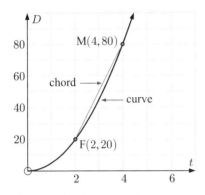

The *average* speed in the time interval $2 \leqslant t \leqslant 4$ is

$$= \frac{\text{distance travelled}}{\text{time taken}}$$

$$= \frac{(80 - 20)\text{ m}}{(4 - 2)\text{ s}}$$

$$= \tfrac{60}{2}\text{ m s}^{-1}$$

$$= 30\text{ m s}^{-1}$$

In this Investigation we will try to measure the *instantaneous* speed of the ball when $t = 2$ seconds.

What to do:

DEMO

1 Click on the icon to start the demonstration.
F is the point where $t = 2$ seconds, and M is another point on the curve.
To start with, M is at $t = 4$ seconds.
The number in the box marked *gradient* is the gradient of the chord [FM]. This is the *average speed* of the ball bearing in the interval from F to M. For M at $t = 4$ seconds, you should see the average speed is 30 m s^{-1}.

2 Click on M and drag it slowly towards F. Copy and complete the table alongside with the gradient of the chord [FM] for M being the points on the curve at the given varying times t.

t	gradient of [FM]
3	
2.5	
2.1	
2.01	

3 Observe what happens as M reaches F. Explain why this is so.

4 For $t = 2$ seconds, what do you suspect will be the instantaneous speed of the ball bearing?

5 Move M to the origin, and then slide it towards F from the left. Copy and complete the table with the gradient of the chord [FM] for various times t.

t	gradient of [FM]
0	
1.5	
1.9	
1.99	

6 Do your results agree with those in **4**?

THE TANGENT TO A CURVE

A **chord** of a curve is a straight line segment which joins any two points on the curve.

The gradient of the chord [AB] measures the average rate of change of the function values for the given change in x-values.

A **tangent** is a straight line which *touches* a curve at a single point.

The gradient of the tangent at point A measures the instantaneous rate of change of the function at point A.

As B approaches A, the limit of the gradient of the chord [AB] will be the gradient of the tangent at A.

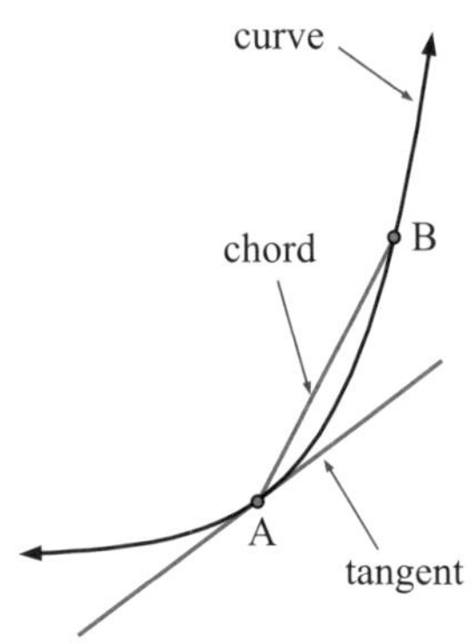

The **gradient of the tangent** to $y = f(x)$ at $x = a$ is the instantaneous rate of change in $f(x)$ with respect to x at that point.

INVESTIGATION 4 — THE GRADIENT OF A TANGENT

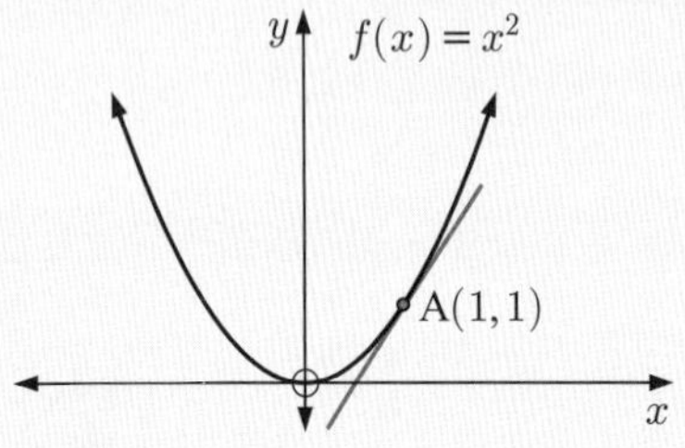

Given a curve $f(x)$, we wish to find the gradient of the tangent at the point $(a, f(a))$.

In this Investigation we will find the gradient of the tangent to $f(x) = x^2$ at the point A(1, 1).

DEMO

What to do:

1 Suppose B lies on $f(x) = x^2$ and B has coordinates (x, x^2).

a Show that the chord [AB] has gradient $\dfrac{x^2 - 1}{x - 1}$.

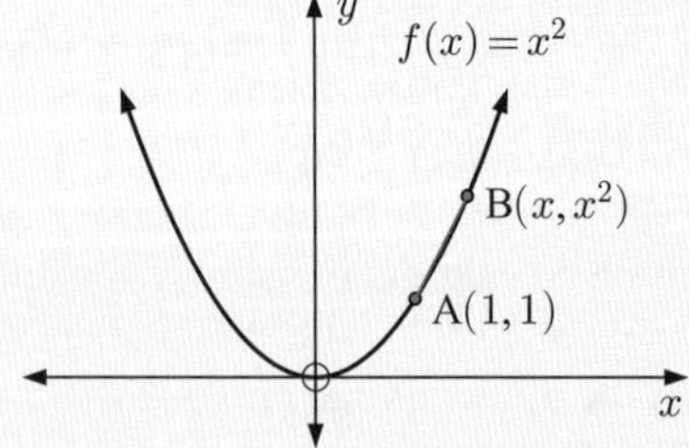

b Copy and complete the table alongside:

c Comment on the gradient of [AB] as x gets closer to 1.

x	Point B	gradient of [AB]
5	(5, 25)	6
3		
2		
1.5		
1.1		
1.01		
1.001		

2 Repeat the process letting x get closer to 1, but from the left of A. Use the points where $x = 0$, 0.8, 0.9, 0.99, and 0.999.

3 Click on the icon to view a demonstration of the process.

4 What do you suspect is the gradient of the tangent at A?

Fortunately we do not have to use a graph and table of values each time we wish to find the gradient of a tangent. Instead we can use an algebraic and geometric approach which involves **limits**.

From **Investigation 4**, the gradient of [AB] $= \dfrac{x^2-1}{x-1}$.

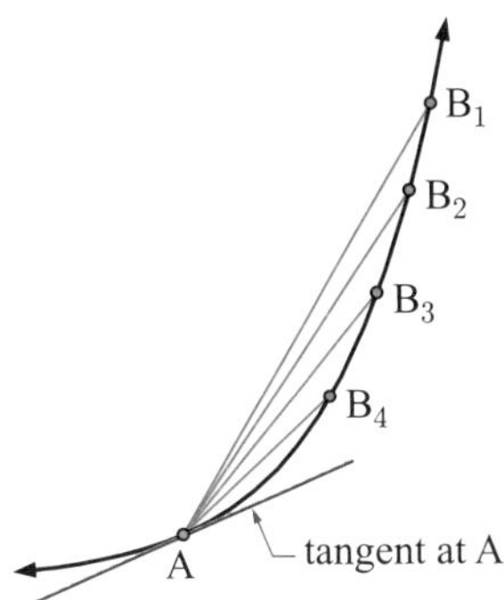

As B approaches A, $x \to 1$ and
the gradient of [AB] $\to$ the gradient of the tangent at A.

So, the gradient of the tangent at the point A is

$$\begin{aligned} m_T &= \lim_{x\to 1} \frac{x^2-1}{x-1} \\ &= \lim_{x\to 1} \frac{(x+1)(x-1)}{x-1} \\ &= \lim_{x\to 1} (x+1) \quad \text{since } x \neq 1 \\ &= 2 \end{aligned}$$

As B approaches A, the gradient of [AB] approaches or **converges** to 2.

EXERCISE 17D

1. Use the method in **Investigation 3** to answer the **Opening Problem** on page **508**.

2. a. Use the method in **Investigation 4** to find the gradient of the tangent to $y = x^2$ at the point (2, 4).

   b. Evaluate $\lim\limits_{x\to 2} \dfrac{x^2-4}{x-2}$, and provide a geometric interpretation of this result.

E THE DERIVATIVE FUNCTION

For a non-linear function with equation $y = f(x)$, the gradients of the tangents at various points are different.

Our task is to determine a **gradient function** which gives the gradient of the tangent to $y = f(x)$ at $x = a$, for any point a in the domain of f.

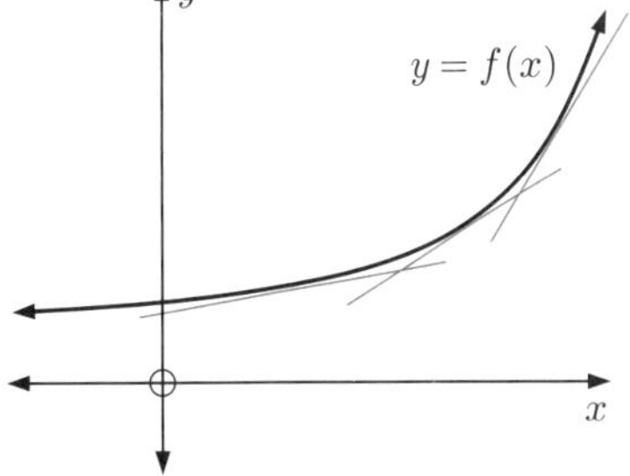

The gradient function of $y = f(x)$ is called its **derivative function** and is labelled $f'(x)$.

We read the derivative function as "eff dashed x".

The value of $f'(a)$ is the gradient of the tangent to $y = f(x)$ at the point where $x = a$.

Example 7 **Self Tutor**

For the given graph, find $f'(4)$ and $f(4)$.

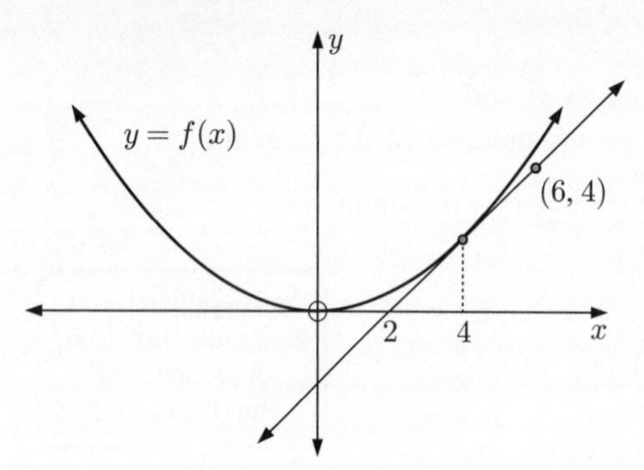

The graph shows the tangent to the curve $y = f(x)$ at the point where $x = 4$.

The tangent passes through (2, 0) and (6, 4), so its gradient is $f'(4) = \frac{4-0}{6-2} = 1$.

The equation of the tangent is $\frac{y-0}{x-2} = 1$

$\therefore \quad y = x - 2$

When $x = 4$, $y = 2$, so the point of contact between the tangent and the curve is (4, 2).

$\therefore \quad f(4) = 2$

EXERCISE 17E

1 Using the graph below, find:

a $f(2)$ **b** $f'(2)$

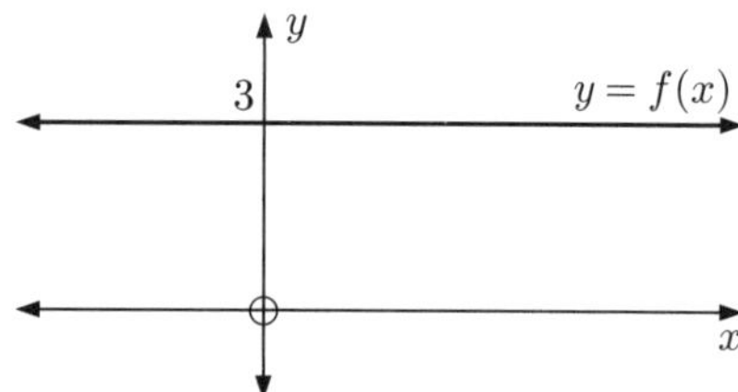

2 Using the graph below, find:

a $f(0)$ **b** $f'(0)$

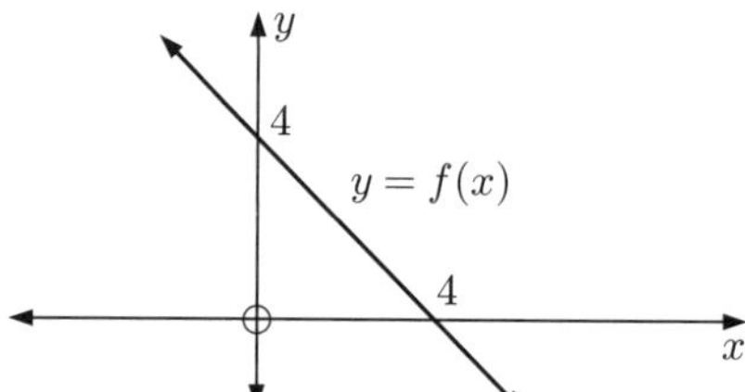

3 Consider the graph alongside. Find $f(2)$ and $f'(2)$.

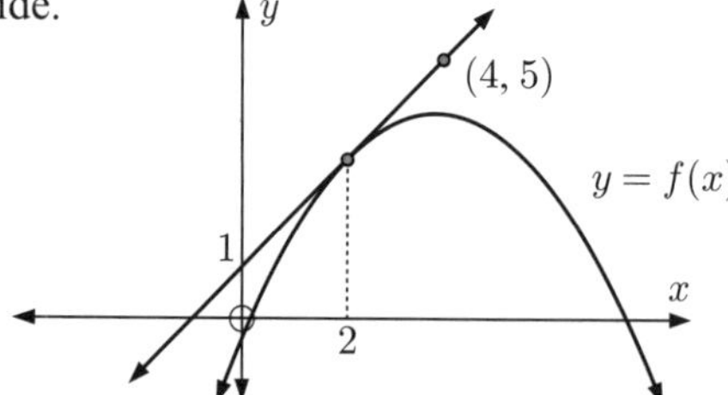

INVESTIGATION 5 GRADIENT FUNCTIONS

The software online can be used to find the gradient of the tangent to a function $f(x)$ at any point. By sliding the point along the graph we can observe the changing gradient of the tangent. We can hence generate the gradient function $f'(x)$.

GRADIENT FUNCTIONS

What to do:

1 Consider the functions $f(x) = 0$, $f(x) = 2$, and $f(x) = 4$.

a For each of these functions, what is the gradient?

b Is the gradient constant for all values of x?

2 Consider the function $f(x) = mx + c$.

a State the gradient of the function.

b Is the gradient constant for all values of x?

c Use the software to graph the following functions and observe the gradient function $f'(x)$. Hence verify that your answer in **b** is correct.

i $f(x) = x - 1$ **ii** $f(x) = 3x + 2$ **iii** $f(x) = -2x + 1$

3 **a** Observe the function $f(x) = x^2$ using the software. What *type* of function is the gradient function $f'(x)$?

b Observe the following quadratic functions using the software:

i $f(x) = x^2 + x - 2$ **ii** $f(x) = 2x^2 - 3$

iii $f(x) = -x^2 + 2x - 1$ **iv** $f(x) = -3x^2 - 3x + 6$

c What *type* of function is each of the gradient functions $f'(x)$ in **b**?

4 **a** Observe the function $f(x) = \ln x$ using the software.

b What *type* of function is the gradient function $f'(x)$?

c What is the *domain* of the gradient function $f'(x)$?

5 **a** Observe the function $f(x) = e^x$ using the software.

b What is the gradient function $f'(x)$?

F DIFFERENTIATION FROM FIRST PRINCIPLES

Consider a general function $y = f(x)$ where A is the point $(x, f(x))$ and B is the point $(x + h, f(x + h))$.

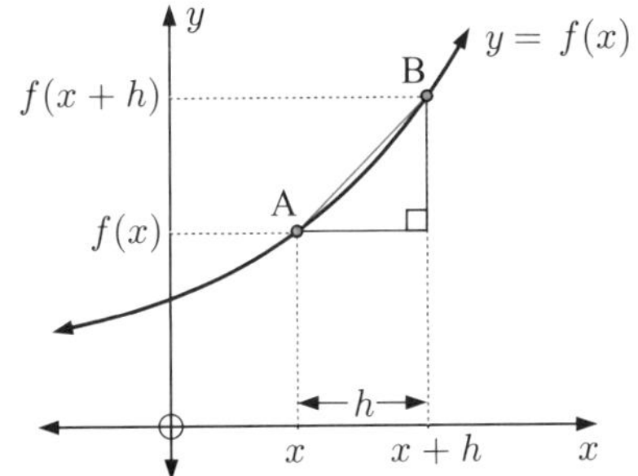

The chord [AB] has gradient $= \dfrac{f(x+h) - f(x)}{x + h - x}$

$= \dfrac{f(x+h) - f(x)}{h}$

If we let B approach A, then the gradient of [AB] approaches the gradient of the tangent at A.

So, the gradient of the tangent at the variable point $(x, f(x))$ is $\lim\limits_{h \to 0} \dfrac{f(x+h) - f(x)}{h}$.

This formula gives the gradient of the tangent to the curve $y = f(x)$ at the point $(x, f(x))$ for any value of x for which this limit exists. Since there is at most one value of the gradient for each value of x, the formula is actually a function.

The **derivative function** or simply **derivative** of $y = f(x)$ is defined as

$$f'(x) = \lim_{h \to 0} \frac{f(x+h) - f(x)}{h}$$

When we evaluate this limit to find a derivative function, we say we are **differentiating from first principles**.

Example 8 — Self Tutor

Use the definition of $f'(x)$ to find the gradient function of $f(x) = x^2$.

$$\begin{aligned} f'(x) &= \lim_{h\to 0} \frac{f(x+h)-f(x)}{h} \\ &= \lim_{h\to 0} \frac{(x+h)^2 - x^2}{h} \\ &= \lim_{h\to 0} \frac{\cancel{x^2} + 2hx + h^2 - \cancel{x^2}}{h} \\ &= \lim_{h\to 0} \frac{\cancel{h}(2x+h)}{\cancel{h}} \\ &= \lim_{h\to 0} (2x+h) \qquad \{\text{as } h \neq 0\} \\ &= 2x \end{aligned}$$

ALTERNATIVE NOTATION

If we are given a function $f(x)$ then $f'(x)$ represents the derivative function.

If we are given y in terms of x then y' or $\frac{dy}{dx}$ are commonly used to represent the derivative.

$\frac{dy}{dx}$ reads "dee y by dee x" or "the derivative of y with respect to x".

$\frac{dy}{dx}$ is **not a fraction**. However, the notation $\frac{dy}{dx}$ is a result of taking the limit of a fraction. If we replace h by δx and $f(x+h) - f(x)$ by δy, then

$$f'(x) = \lim_{h\to 0} \frac{f(x+h)-f(x)}{h} \quad \text{becomes}$$

$$\begin{aligned} f'(x) &= \lim_{\delta x\to 0} \frac{\delta y}{\delta x} \\ &= \frac{dy}{dx}. \end{aligned}$$

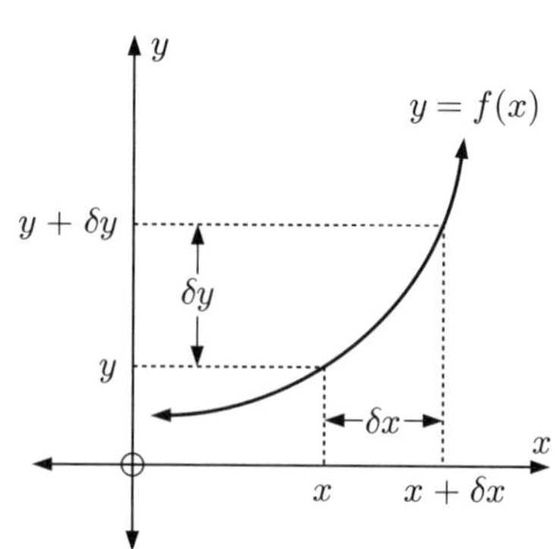

THE DERIVATIVE WHEN $x = a$

The gradient of the tangent to $y = f(x)$ at the point where $x = a$ is denoted $f'(a)$, where

$$f'(a) = \lim_{h\to 0} \frac{f(a+h)-f(a)}{h}$$

Example 9 **Self Tutor**

Use the first principles formula $f'(a) = \lim_{h \to 0} \frac{f(a+h) - f(a)}{h}$ to find the instantaneous rate of change in $f(x) = x^2 + 2x$ at the point where $x = 5$.

$f(5) = 5^2 + 2(5) = 35$

$$\therefore \quad f'(5) = \lim_{h \to 0} \frac{f(5+h) - f(5)}{h}$$

$$= \lim_{h \to 0} \frac{(5+h)^2 + 2(5+h) - 35}{h}$$

$$= \lim_{h \to 0} \frac{\cancel{25} + 10h + h^2 + \cancel{10} + 2h - \cancel{35}}{h}$$

$$= \lim_{h \to 0} \frac{h^2 + 12h}{h}$$

$$= \lim_{h \to 0} \frac{\cancel{h}(h+12)}{\cancel{h}_1} \quad \{\text{as } h \neq 0\}$$

$$= 12$$

$\therefore$ the instantaneous rate of change in $f(x)$ at $x = 5$ is 12.

You can also find the gradient of the tangent at a given point on a function using your graphics calculator. Instructions for doing this can be found online.

EXERCISE 17F

1 **a** Find, from first principles, the gradient function of $f(x)$ where $f(x)$ is:

i x **ii** 5
iii x^3 **iv** x^4

b Hence predict a formula for $f'(x)$ where $f(x) = x^n$, $n \in \mathbb{N}$.

Remember the binomial expansions.

2 Find $f'(x)$ from first principles, given that $f(x)$ is:

a $2x + 5$ **b** $x^2 - 3x$ **c** $-x^2 + 5x - 3$

3 Find $\frac{dy}{dx}$ from first principles given:

a $y = 4 - x$ **b** $y = 2x^2 + x - 1$ **c** $y = x^3 - 2x^2 + 3$

4 Use the first principles formula $f'(a) = \lim_{h \to 0} \frac{f(a+h) - f(a)}{h}$ to find:

a $f'(2)$ for $f(x) = x^3$ **b** $f'(3)$ for $f(x) = x^4$.

5 Use the first principles formula to find the gradient of the tangent to:

a $f(x) = 3x + 5$ at $x = -2$

b $f(x) = 5 - 2x^2$ at $x = 3$

c $f(x) = x^2 + 3x - 4$ at $x = 3$

d $f(x) = 5 - 2x - 3x^2$ at $x = -2$

6 **a** Given $y = x^3 - 3x$, find $\frac{dy}{dx}$ from first principles.

b *Hence* find the points on the graph at which the tangent has zero gradient.

7 Find $\frac{dy}{dx}$ from first principles given:

a $y = \frac{4}{x}$

b $y = \frac{4x+1}{x-2}$

8 Use the first principles formula to find the gradient of the tangent to:

a $f(x) = \frac{1}{x^2}$ at $x = 3$

b $f(x) = \frac{3x}{x^2+1}$ at $x = -4$

c $f(x) = \sqrt{x}$ at $x = 4$

d $f(x) = \frac{1}{\sqrt{x}}$ at $x = 1$

REVIEW SET 17A — NON-CALCULATOR

1 Evaluate:

a $\lim_{x\to 1} (6x - 7)$

b $\lim_{h\to 0} \frac{2h^2 - h}{h}$

c $\lim_{x\to 4} \frac{x^2 - 16}{x - 4}$

2 Find, from first principles, the derivative of:

a $f(x) = x^2 + 2x$

b $y = 4 - 3x^2$

3 Determine the asymptotes of the graphs of the following functions, and discuss the behaviour of the graphs near these asymptotes:

a $f(x) = e^{x-2} - 3$

b $f(x) = \ln(x^2 + 3)$

c $f(x) = \ln(-x) + 2$

4 Find:

a $\lim_{\theta\to 0} \frac{\sin 4\theta}{\theta}$

b $\lim_{\theta\to 0} \frac{2\theta}{\sin 3\theta}$

c $\lim_{n\to\infty} n \sin\left(\frac{\pi}{n}\right)$

5 Given $f(x) = 5x - x^2$, find $f'(1)$ from first principles.

6 In the **Opening Problem**, the altitude of the jumper is given by $f(t) = 452 - 4.8t^2$ metres, where $0 \leqslant t \leqslant 3$ seconds.

a Find $f'(t) = \lim_{h\to 0} \frac{f(t+h) - f(t)}{h}$.

b *Hence* find the speed of the jumper when $t = 2$ seconds.

REVIEW SET 17B — CALCULATOR

1 **a** Discuss the behaviour of $y = \frac{x-7}{3x+2}$ near its asymptotes. Hence deduce their equations.

b State the values of $\lim_{x\to -\infty} \frac{x-7}{3x+2}$ and $\lim_{x\to\infty} \frac{x-7}{3x+2}$.

2 **a** Sketch the graph of $y = 2 - e^{x+1}$.

b Hence find, if possible, $\lim\limits_{x \to -\infty} (2 - e^{x+1})$ and $\lim\limits_{x \to \infty} (2 - e^{x+1})$.

c State the equation of the asymptote of $y = 2 - e^{x+1}$.

3 At what values of x are the following functions *not* continuous? Explain your answer in each case.

a $f(x) = \ln(x^2)$ **b** $f(x) = \dfrac{x^2 - 1}{1 - x}$

4 Find $\lim\limits_{h \to 0} \dfrac{2\cos\left(x + \frac{h}{2}\right)\sin\left(\frac{h}{2}\right)}{h}$. State any assumptions made in finding your answer.

5 Consider $f(x) = 2x^2$.

a Show that $\dfrac{f(x+h) - f(x)}{h} = 4x + 2h$ provided $h \neq 0$.

b Hence evaluate $\dfrac{f(3+h) - f(3)}{h}$ when:

i $h = 0.1$ **ii** $h = 0.01$ **iii** in the limit as h approaches zero.

c Give a geometric interpretation of your result from **b**.

6 The horizontal asymptote of the function $y = \dfrac{2x+3}{4-x}$ is $y = -2$.

Justify this statement using a limit argument.

REVIEW SET 17C

1 Evaluate the limits:

a $\lim\limits_{h \to 0} \dfrac{h^3 - 3h}{h}$ **b** $\lim\limits_{x \to 1} \dfrac{3x^2 - 3x}{x - 1}$ **c** $\lim\limits_{x \to 2} \dfrac{x^2 - 3x + 2}{2 - x}$

2 **a** Sketch the graph of $y = \dfrac{2+x}{x-4}$.

b Discuss the behaviour of the graph near its asymptotes, and hence deduce their equations.

c State the values of $\lim\limits_{x \to -\infty} \dfrac{2+x}{x-4}$ and $\lim\limits_{x \to \infty} \dfrac{2+x}{x-4}$.

3 Given $f(x) = x^4 - 2x$, find $f'(1)$ from first principles.

4 **a** Show that $\sin(A+B) - \sin(A-B) = 2\cos A \sin B$.

b Suppose $A + B = S$ and $A - B = D$.
Show that $\sin S - \sin D = 2\cos\left(\frac{S+D}{2}\right)\sin\left(\frac{S-D}{2}\right)$.

c Hence, find $\lim\limits_{h \to 0} \dfrac{\sin(x+h) - \sin x}{h}$.

d Explain the significance of this result.

5 **a** Given $y = 2x^2 - 1$, find $\dfrac{dy}{dx}$ from first principles.

b Hence state the gradient of the tangent to $y = 2x^2 - 1$ at the point where $x = 4$.

c For what value of x is the gradient of the tangent to $y = 2x^2 - 1$ equal to -12?

6 **a** Prove that for θ in radians:

i If $\theta > 0$ then $\lim\limits_{\theta \to 0} \dfrac{\sin\theta}{\theta} = 1$.

ii If $\theta < 0$ then $\lim\limits_{\theta \to 0} \dfrac{\sin\theta}{\theta} = 1$ also.

b Show that the area of the shaded segment of a circle of radius r, is given by $\frac{1}{2}r^2(\theta - \sin\theta)$, where θ is in radians.

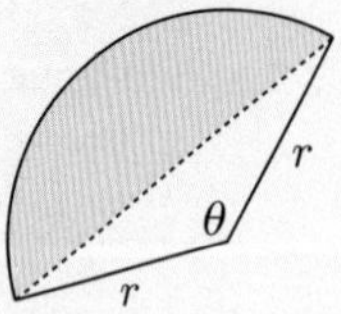

c Hence give geometric evidence that **a i** is true.

Chapter 18

Rules of differentiation

Syllabus reference: 6.1, 6.2

Contents:

OPENING PROBLEM

Consider the curve $y = x^2$.

In the previous chapter we found that the gradient function of this curve is $\frac{dy}{dx} = 2x$.

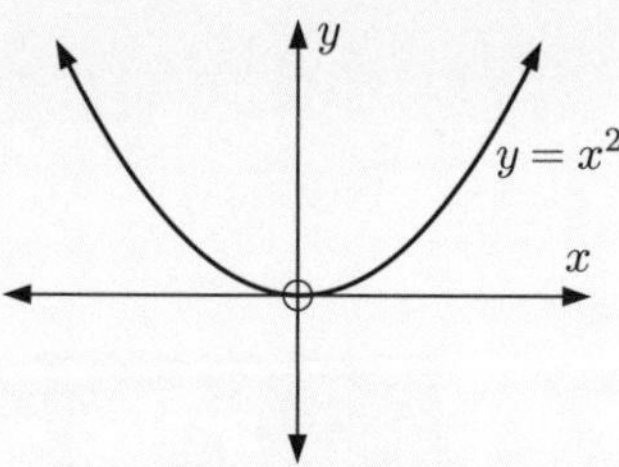

Things to think about:

a Consider the transformation of $y = x^2$ onto $y = x^2 + 3$.
  i What transformation has taken place?
  ii For a given value of x, has the gradient of the tangent to the function changed?
  iii What is the gradient function of $y = x^2 + 3$?

b Consider the transformation of $y = x^2$ onto $y = (x - 2)^2$.
  i What transformation has taken place?
  ii How does the gradient function of $y = (x - 2)^2$ relate to the gradient function of $y = x^2$?
  iii Can you write down the gradient function of $y = (x - 2)^2$?

DEMO

A SIMPLE RULES OF DIFFERENTIATION

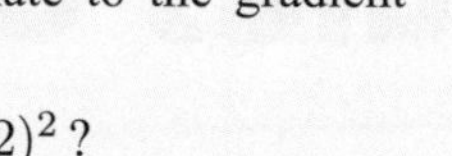

Differentiation is the process of finding a derivative or gradient function.

Given a function $f(x)$, we obtain $f'(x)$ by **differentiating with respect to** the variable x.

There are a number of rules associated with differentiation. These rules can be used to differentiate more complicated functions without having to use first principles.

INVESTIGATION 1 SIMPLE RULES OF DIFFERENTIATION

In this Investigation we attempt to differentiate functions of the form x^n, cx^n where c is a constant, and functions which are a sum or difference of polynomial terms of the form cx^n.

What to do:

1 Differentiate from first principles: **a** x^2 **b** x^3 **c** x^4

2 Consider the binomial expansion:

$$(x+h)^n = \binom{n}{0}x^n + \binom{n}{1}x^{n-1}h + \binom{n}{2}x^{n-2}h^2 + ... + \binom{n}{n}h^n$$
$$= x^n + nx^{n-1}h + \binom{n}{2}x^{n-2}h^2 + ... + h^n$$

Use the first principles formula $f'(x) = \lim_{h \to 0} \frac{f(x+h) - f(x)}{h}$ to find the derivative of $f(x) = x^n$ for $x \in \mathbb{N}$.

3 **a** Find, from first principles, the derivatives of: **i** $4x^2$ **ii** $2x^3$
  b By comparison with **1**, copy and complete: "If $f(x) = cx^n$, then $f'(x) = ...$"

4 **a** Use first principles to find $f'(x)$ for:

i $f(x) = x^2 + 3x$ **ii** $f(x) = x^3 - 2x^2$

b Copy and complete: "If $f(x) = u(x) + v(x)$ then $f'(x) = ...$"

The rules you found in the **Investigation** are much more general than the cases you just considered.

For example, if $f(x) = x^n$ then $f'(x) = nx^{n-1}$ is true not just for all $n \in \mathbb{N}$, but actually for all $n \in \mathbb{R}$.

We can summarise the following rules:

$f(x)$	$f'(x)$	*Name of rule*
c (a constant)	0	**differentiating a constant**
x^n	nx^{n-1}	**power rule**
$c\,u(x)$	$c\,u'(x)$	**scalar multiplication rule**
$u(x) + v(x)$	$u'(x) + v'(x)$	**addition rule**

The last two rules can be proved using the first principles definition of $f'(x)$.

- If $f(x) = c\,u(x)$ where c is a constant, then $f'(x) = c\,u'(x)$.

Proof:

$$\begin{aligned} f'(x) &= \lim_{h\to 0} \frac{f(x+h) - f(x)}{h} \\ &= \lim_{h\to 0} \frac{c\,u(x+h) - c\,u(x)}{h} \\ &= \lim_{h\to 0} c\left[\frac{u(x+h) - u(x)}{h}\right] \\ &= c \lim_{h\to 0} \frac{u(x+h) - u(x)}{h} \\ &= c\,u'(x) \end{aligned}$$

- If $f(x) = u(x) + v(x)$ then $f'(x) = u'(x) + v'(x)$

Proof:

$$\begin{aligned} f'(x) &= \lim_{h\to 0} \frac{f(x+h) - f(x)}{h} \\ &= \lim_{h\to 0} \left(\frac{u(x+h) + v(x+h) - [u(x) + v(x)]}{h}\right) \\ &= \lim_{h\to 0} \left(\frac{u(x+h) - u(x) + v(x+h) - v(x)}{h}\right) \\ &= \lim_{h\to 0} \frac{u(x+h) - u(x)}{h} + \lim_{h\to 0} \frac{v(x+h) - v(x)}{h} \\ &= u'(x) + v'(x) \end{aligned}$$

Using the rules we have now developed we can differentiate sums of powers of x.

For example, if $f(x) = 3x^4 + 2x^3 - 5x^2 + 7x + 6$ then

$$\begin{aligned} f'(x) &= 3(4x^3) + 2(3x^2) - 5(2x) + 7(1) + 0 \\ &= 12x^3 + 6x^2 - 10x + 7 \end{aligned}$$

Example 1 — Self Tutor

If $y = 3x^2 - 4x$, find $\dfrac{dy}{dx}$ and interpret its meaning.

As $y = 3x^2 - 4x$, $\dfrac{dy}{dx} = 6x - 4$.

$\dfrac{dy}{dx}$ is:
- the gradient function or derivative of $y = 3x^2 - 4x$ from which the gradient of the tangent at any point on the curve can be found
- the instantaneous rate of change of y with respect to x.

Example 2 — Self Tutor

Find $f'(x)$ for $f(x)$ equal to: **a** $5x^3 + 6x^2 - 3x + 2$ **b** $7x - \dfrac{4}{x} + \dfrac{3}{x^3}$

a
$$\begin{aligned} f(x) &= 5x^3 + 6x^2 - 3x + 2 \\ \therefore\ f'(x) &= 5(3x^2) + 6(2x) - 3(1) \\ &= 15x^2 + 12x - 3 \end{aligned}$$

b
$$\begin{aligned} f(x) &= 7x - \frac{4}{x} + \frac{3}{x^3} \\ &= 7x - 4x^{-1} + 3x^{-3} \\ \therefore\ f'(x) &= 7(1) - 4(-1x^{-2}) + 3(-3x^{-4}) \\ &= 7 + 4x^{-2} - 9x^{-4} \\ &= 7 + \frac{4}{x^2} - \frac{9}{x^4} \end{aligned}$$

Remember that $\dfrac{1}{x^n} = x^{-n}$.

Example 3 — Self Tutor

Find the gradient function of $y = x^2 - \dfrac{4}{x}$ and hence find the gradient of the tangent to the function at the point where $x = 2$.

$$\begin{aligned} y &= x^2 - \frac{4}{x} \\ &= x^2 - 4x^{-1} \end{aligned}$$

$$\begin{aligned} \therefore\ \frac{dy}{dx} &= 2x - 4(-1x^{-2}) \\ &= 2x + 4x^{-2} \\ &= 2x + \frac{4}{x^2} \end{aligned}$$

When $x = 2$, $\dfrac{dy}{dx} = 4 + 1 = 5$.

So, the tangent has gradient 5.

You can also use your graphics calculator to evaluate the gradient of the tangent to a function at a given point.

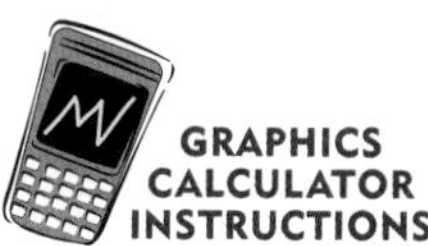

Example 4 **Self Tutor**

Find the gradient function for each of the following:

a $f(x) = 3\sqrt{x} + \frac{2}{x}$

b $g(x) = x^2 - \frac{4}{\sqrt{x}}$

a
$$\begin{aligned} f(x) &= 3\sqrt{x} + \frac{2}{x} \\ &= 3x^{\frac{1}{2}} + 2x^{-1} \\ \therefore\ f'(x) &= 3(\tfrac{1}{2}x^{-\frac{1}{2}}) + 2(-1x^{-2}) \\ &= \tfrac{3}{2}x^{-\frac{1}{2}} - 2x^{-2} \\ &= \frac{3}{2\sqrt{x}} - \frac{2}{x^2} \end{aligned}$$

b
$$\begin{aligned} g(x) &= x^2 - \frac{4}{\sqrt{x}} \\ &= x^2 - 4x^{-\frac{1}{2}} \\ \therefore\ g'(x) &= 2x - 4(-\tfrac{1}{2}x^{-\frac{3}{2}}) \\ &= 2x + 2x^{-\frac{3}{2}} \\ &= 2x + \frac{2}{x\sqrt{x}} \end{aligned}$$

EXERCISE 18A

1 Find $f'(x)$ given that $f(x)$ is:

a x^3 **b** $2x^3$ **c** $7x^2$ **d** $6\sqrt{x}$

e $3\sqrt[3]{x}$ **f** $x^2 + x$ **g** $4 - 2x^2$ **h** $x^2 + 3x - 5$

i $\frac{1}{2}x^4 - 6x^2$ **j** $\dfrac{3x - 6}{x}$ **k** $\dfrac{2x - 3}{x^2}$ **l** $\dfrac{x^3 + 5}{x}$

m $\dfrac{x^3 + x - 3}{x}$ **n** $\dfrac{1}{\sqrt{x}}$ **o** $(2x - 1)^2$ **p** $(x + 2)^3$

2 Find $\dfrac{dy}{dx}$ for:

a $y = 2.5x^3 - 1.4x^2 - 1.3$ **b** $y = \pi x^2$ **c** $y = \dfrac{1}{5x^2}$

d $y = 100x$ **e** $y = 10(x + 1)$ **f** $y = 4\pi x^3$

3 Differentiate with respect to x:

a $6x + 2$ **b** $x\sqrt{x}$ **c** $(5 - x)^2$ **d** $\dfrac{6x^2 - 9x^4}{3x}$

e $(x + 1)(x - 2)$ **f** $\dfrac{1}{x^2} + 6\sqrt{x}$ **g** $4x - \dfrac{1}{4x}$ **h** $x(x + 1)(2x - 5)$

4 Find the gradient of the tangent to:

a $y = x^2$ at $x = 2$

b $y = \dfrac{8}{x^2}$ at the point $(9, \frac{8}{81})$

c $y = 2x^2 - 3x + 7$ at $x = -1$

d $y = \dfrac{2x^2 - 5}{x}$ at the point $(2, \frac{3}{2})$

e $y = \dfrac{x^2 - 4}{x^2}$ at the point $(4, \frac{3}{4})$

f $y = \dfrac{x^3 - 4x - 8}{x^2}$ at $x = -1$

Check your answers using technology.

5 Suppose $f(x) = x^2 + (b + 1)x + 2c$, $f(2) = 4$, and $f'(-1) = 2$.
Find the constants b and c.

6 Find the gradient function of $f(x)$ where $f(x)$ is:

a $4\sqrt{x}+x$ **b** $\sqrt[3]{x}$ **c** $-\dfrac{2}{\sqrt{x}}$ **d** $2x-\sqrt{x}$

e $\dfrac{4}{\sqrt{x}}-5$ **f** $3x^2-x\sqrt{x}$ **g** $\dfrac{5}{x^2\sqrt{x}}$ **h** $2x-\dfrac{3}{x\sqrt{x}}$

7 **a** If $y=4x-\dfrac{3}{x}$, find $\dfrac{dy}{dx}$ and interpret its meaning.

b The position of a car moving along a straight road is given by $S=2t^2+4t$ metres where t is the time in seconds. Find $\dfrac{dS}{dt}$ and interpret its meaning.

c The cost of producing x toasters each week is given by $C=1785+3x+0.002x^2$ dollars. Find $\dfrac{dC}{dx}$ and interpret its meaning.

B THE CHAIN RULE

In **Chapter 2** we defined the **composite** of two functions g and f as $(g\circ f)(x)$ or $g(f(x))$.

We can often write complicated functions as the composite of two or more simpler functions.

For example $y=(x^2+3x)^4$ could be rewritten as $y=u^4$ where $u=x^2+3x$, or as $y=g(f(x))$ where $g(x)=x^4$ and $f(x)=x^2+3x$.

Example 5 Self Tutor

Find: **a** $g(f(x))$ if $g(x)=\sqrt{x}$ and $f(x)=2-3x$

b $g(x)$ and $f(x)$ such that $g(f(x))=\dfrac{1}{x-x^2}$.

a $g(f(x))$
$=g(2-3x)$
$=\sqrt{2-3x}$

b $g(f(x))=\dfrac{1}{x-x^2}=\dfrac{1}{f(x)}$

$\therefore\ g(x)=\dfrac{1}{x}$ and $f(x)=x-x^2$

There are several possible answers for **b**.

EXERCISE 18B.1

1 Find $g(f(x))$ if:

a $g(x)=x^2$ and $f(x)=2x+7$ **b** $g(x)=2x+7$ and $f(x)=x^2$

c $g(x)=\sqrt{x}$ and $f(x)=3-4x$ **d** $g(x)=3-4x$ and $f(x)=\sqrt{x}$

e $g(x)=\dfrac{2}{x}$ and $f(x)=x^2+3$ **f** $g(x)=x^2+3$ and $f(x)=\dfrac{2}{x}$

2 Find $g(x)$ and $f(x)$ such that $g(f(x))$ is:

a $(3x+10)^3$ **b** $\dfrac{1}{2x+4}$ **c** $\sqrt{x^2-3x}$ **d** $\dfrac{10}{(3x-x^2)^3}$

DERIVATIVES OF COMPOSITE FUNCTIONS

The reason we are interested in writing complicated functions as composite functions is to make finding derivatives easier.

INVESTIGATION 2 DIFFERENTIATING COMPOSITE FUNCTIONS

The purpose of this Investigation is to learn how to differentiate composite functions.

Based on the rule "if $y = x^n$ then $\frac{dy}{dx} = nx^{n-1}$", we might suspect that if $y = (2x+1)^2$ then $\frac{dy}{dx} = 2(2x+1)^1$. But is this so?

What to do:

1 Expand $y = (2x+1)^2$ and hence find $\frac{dy}{dx}$. How does this compare with $2(2x+1)^1$?

2 Expand $y = (3x+1)^2$ and hence find $\frac{dy}{dx}$. How does this compare with $2(3x+1)^1$?

3 Expand $y = (ax+1)^2$ where a is a constant, and hence find $\frac{dy}{dx}$. How does this compare with $2(ax+1)^1$?

4 Suppose $y = u^2$.

a Find $\frac{dy}{du}$.

b Now suppose $u = ax + 1$, so $y = (ax+1)^2$.

i Find $\frac{du}{dx}$.

ii Write $\frac{dy}{du}$ from **a** in terms of x.

iii *Hence* find $\frac{dy}{du} \times \frac{du}{dx}$.

iv Compare your answer to the result in **3**.

c If $y = u^2$ where u is a function of x, what do you suspect $\frac{dy}{dx}$ will be equal to?

5 Expand $y = (x^2+3x)^2$ and hence find $\frac{dy}{dx}$.

Does your answer agree with the rule you suggested in **4 c**?

6 Consider $y = (2x+1)^3$.

a Expand the brackets and then find $\frac{dy}{dx}$.

b If we let $u = 2x+1$, then $y = u^3$.

i Find $\frac{du}{dx}$.

ii Find $\frac{dy}{du}$, and write it in terms of x.

iii Hence find $\frac{dy}{du} \times \frac{du}{dx}$.

iv Compare your answer to the result in **a**.

7 Copy and complete: "If y is a function of u, and u is a function of x, then $\frac{dy}{dx} = ...$"

THE CHAIN RULE

$$\text{If } \boldsymbol{y = g(u)} \text{ where } \boldsymbol{u = f(x)} \text{ then } \boldsymbol{\frac{dy}{dx} = \frac{dy}{du}\frac{du}{dx}}.$$

This rule is extremely important and enables us to differentiate complicated functions much faster. For example, for any function $f(x)$:

$$\text{If } \boldsymbol{y = [f(x)]^n} \text{ then } \boldsymbol{\frac{dy}{dx} = n[f(x)]^{n-1} \times f'(x)}.$$

Proof: Consider $y = g(u)$ where $u = f(x)$.

For a small change of δx in x, there is a small change of $f(x+\delta x) - f(x) = \delta u$ in u and a small change of δy in y.

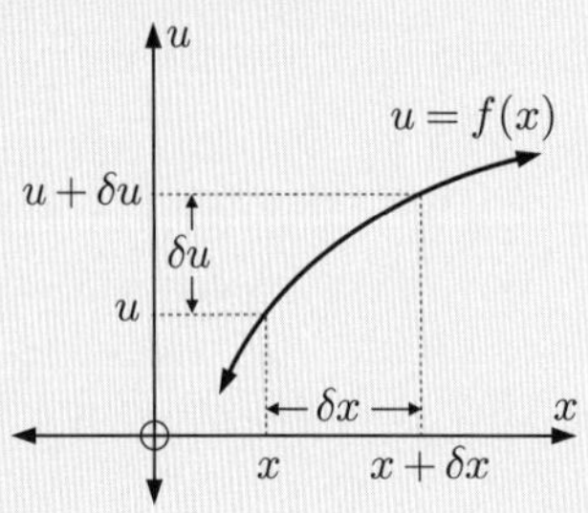

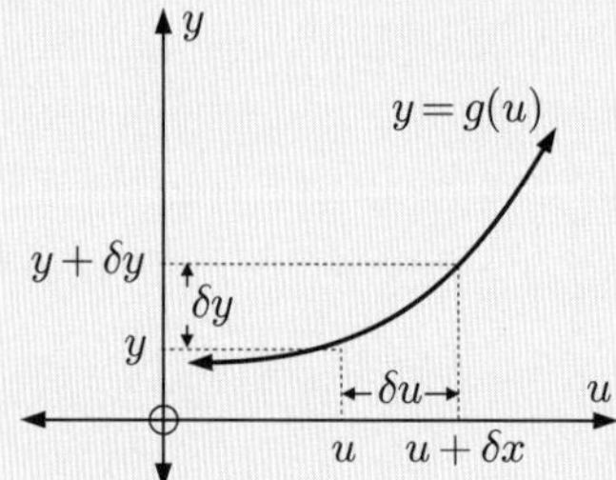

Now $\dfrac{\delta y}{\delta x} = \dfrac{\delta y}{\delta u} \times \dfrac{\delta u}{\delta x}$ {fraction multiplication}

As $\delta x \to 0$, $\delta u \to 0$ also.

$\therefore \displaystyle\lim_{\delta x \to 0} \frac{\delta y}{\delta x} = \lim_{\delta u \to 0} \frac{\delta y}{\delta u} \times \lim_{\delta x \to 0} \frac{\delta u}{\delta x}$ {limit rule}

$\therefore \dfrac{dy}{dx} = \dfrac{dy}{du}\dfrac{du}{dx}$

We can compare this proof with the limit definition of the derivative function $f'(x) = \displaystyle\lim_{h \to 0} \frac{f(x+h) - f(x)}{h}$.

If we let $h = \delta x$ and $f(x+h) - f(x) = \delta y$, then $f'(x) = \dfrac{dy}{dx} = \displaystyle\lim_{\delta x \to 0} \frac{\delta y}{\delta x}$.

Example 6 **Self Tutor**

Find $\dfrac{dy}{dx}$ if:

a $y = (x^2 - 2x)^4$

b $y = \dfrac{4}{\sqrt{1-2x}}$

a

$y = (x^2 - 2x)^4$

$\therefore y = u^4$ where $u = x^2 - 2x$

Now $\dfrac{dy}{dx} = \dfrac{dy}{du}\dfrac{du}{dx}$ {chain rule}

$= 4u^3(2x - 2)$

$= 4(x^2 - 2x)^3(2x - 2)$

b

$y = \dfrac{4}{\sqrt{1-2x}}$

$\therefore y = 4u^{-\frac{1}{2}}$ where $u = 1 - 2x$

Now $\dfrac{dy}{dx} = \dfrac{dy}{du}\dfrac{du}{dx}$ {chain rule}

$= 4 \times (-\frac{1}{2}u^{-\frac{3}{2}}) \times (-2)$

$= 4u^{-\frac{3}{2}}$

$= 4(1 - 2x)^{-\frac{3}{2}}$

The brackets around $2x - 2$ are essential.

EXERCISE 18B.2

1 Write in the form au^n, clearly stating what u is:

a $\frac{1}{(2x-1)^2}$ **b** $\sqrt{x^2-3x}$ **c** $\frac{2}{\sqrt{2-x^2}}$

d $\sqrt[3]{x^3-x^2}$ **e** $\frac{4}{(3-x)^3}$ **f** $\frac{10}{x^2-3}$

2 Find the gradient function $\frac{dy}{dx}$ for:

a $y=(4x-5)^2$ **b** $y=\frac{1}{5-2x}$ **c** $y=\sqrt{3x-x^2}$

d $y=(1-3x)^4$ **e** $y=6(5-x)^3$ **f** $y=\sqrt[3]{2x^3-x^2}$

g $y=\frac{6}{(5x-4)^2}$ **h** $y=\frac{4}{3x-x^2}$ **i** $y=2\left(x^2-\frac{2}{x}\right)^3$

3 Find the gradient of the tangent to:

a $y=\sqrt{1-x^2}$ at $x=\frac{1}{2}$ **b** $y=(3x+2)^6$ at $x=-1$

c $y=\frac{1}{(2x-1)^4}$ at $x=1$ **d** $y=6\times\sqrt[3]{1-2x}$ at $x=0$

e $y=\frac{4}{x+2\sqrt{x}}$ at $x=4$ **f** $y=\left(x+\frac{1}{x}\right)^3$ at $x=1$.

Check your answers using technology.

4 The gradient function of $f(x)=(2x-b)^a$ is $f'(x)=24x^2-24x+6$.
Find the constants a and b.

5 Suppose $y=\frac{a}{\sqrt{1+bx}}$ where a and b are constants. When $x=3$, $y=1$ and $\frac{dy}{dx}=-\frac{1}{8}$.
Find a and b.

6 If $y=x^3$ then $x=y^{\frac{1}{3}}$.

a Find $\frac{dy}{dx}$ and $\frac{dx}{dy}$ and hence show that $\frac{dy}{dx}\times\frac{dx}{dy}=1$.

b Explain why $\frac{dy}{dx}\times\frac{dx}{dy}=1$ whenever these derivatives exist for any general function $y=f(x)$.

C THE PRODUCT RULE

We have seen the addition rule:

$$\text{If } f(x)=u(x)+v(x) \text{ then } f'(x)=u'(x)+v'(x).$$

We now consider the case $f(x)=u(x)\,v(x)$. Is $f'(x)=u'(x)\,v'(x)$?

In other words, does the derivative of a product of two functions equal the product of the derivatives of the two functions?

INVESTIGATION 3 — THE PRODUCT RULE

Suppose $u(x)$ and $v(x)$ are two functions of x, and that $f(x) = u(x)\,v(x)$ is the product of these functions.

The purpose of this Investigation is to find a rule for determining $f'(x)$.

What to do:

1 Suppose $u(x) = x$ and $v(x) = x$, so $f(x) = x^2$.

- **a** Find $f'(x)$ by direct differentiation.
- **b** Find $u'(x)$ and $v'(x)$.
- **c** Does $f'(x) = u'(x)\,v'(x)$?

2 Suppose $u(x) = x$ and $v(x) = \sqrt{x}$, so $f(x) = x\sqrt{x} = x^{\frac{3}{2}}$.

- **a** Find $f'(x)$ by direct differentiation.
- **b** Find $u'(x)$ and $v'(x)$.
- **c** Does $f'(x) = u'(x)\,v'(x)$?

3 Copy and complete the following table, finding $f'(x)$ by direct differentiation.

$f(x)$	$f'(x)$	$u(x)$	$v(x)$	$u'(x)$	$v'(x)$	$u'(x)\,v(x) + u(x)\,v'(x)$
x^2		x	x			
$x^{\frac{3}{2}}$		x	$\sqrt{x}$			
$x(x+1)$		x	$x+1$			
$(x-1)(2-x^2)$		$x-1$	$2-x^2$			

4 Copy and complete:
"If $u(x)$ and $v(x)$ are two functions of x and $f(x) = u(x)\,v(x)$, then $f'(x) = ...$"

THE PRODUCT RULE

If $\boldsymbol{f(x) = u(x)\,v(x)}$, then $\boldsymbol{f'(x) = u'(x)\,v(x) + u(x)\,v'(x)}$.

Alternatively, if $y = uv$ where u and v are functions of x, then $\boldsymbol{\dfrac{dy}{dx} = \dfrac{du}{dx}\,v + u\,\dfrac{dv}{dx}}$.

Proof: Let $y = u(x)v(x)$. Suppose there is a small change of δx in x which causes corresponding changes of δu in u, δv in v, and δy in y.

Since $y = uv$, $\quad y + \delta y = (u + \delta u)(v + \delta v)$

$$\therefore \; y + \delta y = uv + (\delta u)v + u(\delta v) + \delta u \delta v$$

$$\therefore \; \delta y = (\delta u)v + u(\delta v) + \delta u \delta v$$

$$\therefore \; \frac{\delta y}{\delta x} = \left(\frac{\delta u}{\delta x}\right) v + u\left(\frac{\delta v}{\delta x}\right) + \left(\frac{\delta u}{\delta x}\right)\delta v \qquad \{\text{dividing each term by } \delta x\}$$

$$\therefore \; \lim_{\delta x \to 0} \frac{\delta y}{\delta x} = \left(\lim_{\delta x \to 0} \frac{\delta u}{\delta x}\right) v + u\left(\lim_{\delta x \to 0} \frac{\delta v}{\delta x}\right) + 0 \qquad \{\text{as } \delta x \to 0,\ \delta v \to 0 \text{ also}\}$$

$$\therefore \; \frac{dy}{dx} = \frac{du}{dx}\,v + u\,\frac{dv}{dx}$$

Example 7 — Self Tutor

Find $\frac{dy}{dx}$ if: **a** $y = \sqrt{x}(2x+1)^3$ **b** $y = x^2(x^2-2x)^4$

a $y = \sqrt{x}(2x+1)^3$ is the product of $u = x^{\frac{1}{2}}$ and $v = (2x+1)^3$

$\therefore\ u' = \frac{1}{2}x^{-\frac{1}{2}}$ and $v' = 3(2x+1)^2 \times 2$ {chain rule}

$= 6(2x+1)^2$

Now $\frac{dy}{dx} = u'v + uv'$ {product rule}

$= \frac{1}{2}x^{-\frac{1}{2}}(2x+1)^3 + x^{\frac{1}{2}} \times 6(2x+1)^2$

$= \frac{1}{2}x^{-\frac{1}{2}}(2x+1)^3 + 6x^{\frac{1}{2}}(2x+1)^2$

b $y = x^2(x^2-2x)^4$ is the product of $u = x^2$ and $v = (x^2-2x)^4$

$\therefore\ u' = 2x$ and $v' = 4(x^2-2x)^3(2x-2)$ {chain rule}

Now $\frac{dy}{dx} = u'v + uv'$ {product rule}

$= 2x(x^2-2x)^4 + x^2 \times 4(x^2-2x)^3(2x-2)$

$= 2x(x^2-2x)^4 + 4x^2(x^2-2x)^3(2x-2)$

EXERCISE 18C

1 Use the product rule to differentiate:

a $f(x) = x(x-1)$ **b** $f(x) = 2x(x+1)$ **c** $f(x) = x^2\sqrt{x+1}$

2 Find $\frac{dy}{dx}$ using the product rule:

a $y = x^2(2x-1)$ **b** $y = 4x(2x+1)^3$ **c** $y = x^2\sqrt{3-x}$

d $y = \sqrt{x}(x-3)^2$ **e** $y = 5x^2(3x^2-1)^2$ **f** $y = \sqrt{x}(x-x^2)^3$

3 Find the gradient of the tangent to:

a $y = x^4(1-2x)^2$ at $x = -1$ **b** $y = \sqrt{x}(x^2-x+1)^2$ at $x = 4$

c $y = x\sqrt{1-2x}$ at $x = -4$ **d** $y = x^3\sqrt{5-x^2}$ at $x = 1$

Check your answers using technology.

4 Consider $y = \sqrt{x}(3-x)^2$.

a Show that $\frac{dy}{dx} = \frac{(3-x)(3-5x)}{2\sqrt{x}}$.

b Find the x-coordinates of all points on $y = \sqrt{x}(3-x)^2$ where the tangent is horizontal.

c For what values of x is $\frac{dy}{dx}$ undefined?

d Are there any values of x for which y is defined but $\frac{dy}{dx}$ is not?

e What is the graphical significance of your answer in **d**?

5 Suppose $y = -2x^2(x+4)$. For what values of x does $\frac{dy}{dx} = 10$?

D THE QUOTIENT RULE

Expressions like $\frac{x^2+1}{2x-5}$, $\frac{\sqrt{x}}{1-3x}$, and $\frac{x^3}{(x-x^2)^4}$ are called **quotients** because they represent the division of one function by another.

Quotient functions have the form $Q(x) = \frac{u(x)}{v(x)}$.

$$\begin{aligned}
\text{Notice that}\quad u(x) &= Q(x)\,v(x) \\
\therefore\quad u'(x) &= Q'(x)\,v(x) + Q(x)\,v'(x) \qquad \{\text{product rule}\} \\
\therefore\quad u'(x) - Q(x)\,v'(x) &= Q'(x)\,v(x) \\
\therefore\quad Q'(x)\,v(x) &= u'(x) - \frac{u(x)}{v(x)}\,v'(x) \\
\therefore\quad Q'(x)\,v(x) &= \frac{u'(x)\,v(x) - u(x)\,v'(x)}{v(x)} \\
\therefore\quad Q'(x) &= \frac{u'(x)\,v(x) - u(x)\,v'(x)}{[v(x)]^2} \quad \text{when this exists.}
\end{aligned}$$

THE QUOTIENT RULE

If $\boldsymbol{Q(x) = \frac{u(x)}{v(x)}}$ then $\boldsymbol{Q'(x) = \frac{u'(x)\,v(x) - u(x)\,v'(x)}{[v(x)]^2}}$

Alternatively, if $\boldsymbol{y = \frac{u}{v}}$ where u and v are functions of x, then $\boldsymbol{\frac{dy}{dx} = \frac{u'v - uv'}{v^2}}$.

Example 8 **Self Tutor**

Use the quotient rule to find $\frac{dy}{dx}$ if: **a** $y = \frac{1+3x}{x^2+1}$ **b** $y = \frac{\sqrt{x}}{(1-2x)^2}$

a $y = \frac{1+3x}{x^2+1}$ is a quotient with $u = 1+3x$ and $v = x^2+1$

$\therefore$ $u' = 3$ and $v' = 2x$

$$\begin{aligned}
\text{Now}\quad \frac{dy}{dx} &= \frac{u'v - uv'}{v^2} \qquad \{\text{quotient rule}\} \\
&= \frac{3(x^2+1) - (1+3x)2x}{(x^2+1)^2} \\
&= \frac{3x^2+3-2x-6x^2}{(x^2+1)^2} \\
&= \frac{3-2x-3x^2}{(x^2+1)^2}
\end{aligned}$$

b $y = \frac{\sqrt{x}}{(1-2x)^2}$ is a quotient with $u = x^{\frac{1}{2}}$ and $v = (1-2x)^2$

$\therefore$ $u' = \frac{1}{2}x^{-\frac{1}{2}}$ and $v' = 2(1-2x)^1 \times (-2)$ {chain rule}

$= -4(1-2x)$

$$\text{Now}\quad \frac{dy}{dx} = \frac{u'v - uv'}{v^2} \qquad \{\text{quotient rule}\}$$

$$= \frac{\frac{1}{2}x^{-\frac{1}{2}}(1-2x)^2 - x^{\frac{1}{2}} \times (-4(1-2x))}{(1-2x)^4}$$

$$= \frac{\frac{1}{2}x^{-\frac{1}{2}}(1-2x)^2 + 4x^{\frac{1}{2}}(1-2x)}{(1-2x)^4}$$

$$= \frac{\cancel{(1-2x)}\left[\frac{1-2x}{2\sqrt{x}} + 4\sqrt{x}\left(\frac{2\sqrt{x}}{2\sqrt{x}}\right)\right]}{(1-2x)^{\cancel{4}3}} \qquad \{\text{look for common factors}\}$$

$$= \frac{1-2x+8x}{2\sqrt{x}(1-2x)^3}$$

$$= \frac{6x+1}{2\sqrt{x}(1-2x)^3}$$

Note: Simplification of $\frac{dy}{dx}$ as in the above example is often time consuming and unnecessary, especially if you simply want to find the gradient of a tangent at a given point. In such cases you can substitute a value for x without simplifying the derivative function first.

EXERCISE 18D

1 Use the quotient rule to find $\frac{dy}{dx}$ if:

a $y = \frac{1+3x}{2-x}$ b $y = \frac{x^2}{2x+1}$ c $y = \frac{x}{x^2-3}$

d $y = \frac{\sqrt{x}}{1-2x}$ e $y = \frac{x^2-3}{3x-x^2}$ f $y = \frac{x}{\sqrt{1-3x}}$

2 Find the gradient of the tangent to:

a $y = \frac{x}{1-2x}$ at $x = 1$ b $y = \frac{x^3}{x^2+1}$ at $x = -1$

c $y = \frac{\sqrt{x}}{2x+1}$ at $x = 4$ d $y = \frac{x^2}{\sqrt{x^2+5}}$ at $x = -2$

Check your answers using technology.

3 a If $y = \frac{2\sqrt{x}}{1-x}$, show that $\frac{dy}{dx} = \frac{x+1}{\sqrt{x}(1-x)^2}$.

b For what values of x is $\frac{dy}{dx}$ i zero ii undefined?

4 a If $y = \frac{x^2-3x+1}{x+2}$, show that $\frac{dy}{dx} = \frac{x^2+4x-7}{(x+2)^2}$.

b For what values of x is $\frac{dy}{dx}$ i zero ii undefined?

c What is the graphical significance of your answers in **b**?

E IMPLICIT DIFFERENTIATION

For relations such as $y^3 + 3xy^2 - xy + 11 = 0$ it is often difficult or impossible to write y as a function of x. Such relationships between x and y are called **implicit relations**.

Although we cannot write $y = f(x)$ and find a derivative function $f'(x)$, we can still find $\frac{dy}{dx}$, the rate of change in y with respect to x.

Consider the circle with centre $(0, 0)$ and radius 2.

The equation of the circle is $x^2 + y^2 = 4$.

Method 1:

Suppose $A(x, y)$ lies on the circle.

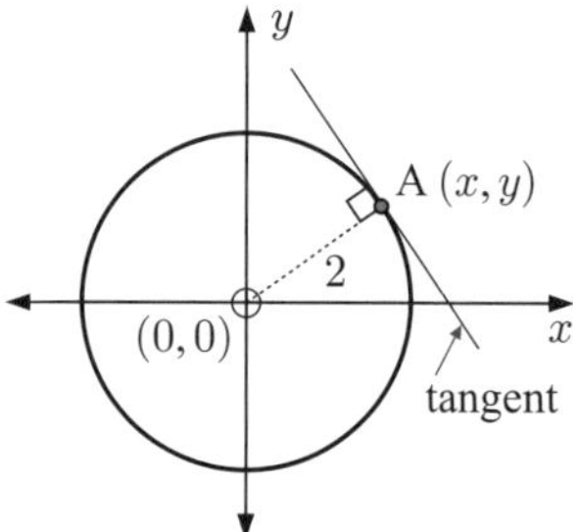

The radius [OA] has gradient $= \frac{y\text{-step}}{x\text{-step}} = \frac{y-0}{x-0} = \frac{y}{x}$

$\therefore$ the tangent at A has gradient $= -\frac{x}{y}$ {the negative reciprocal}

Thus for all points (x, y) on the circle, $\frac{dy}{dx} = -\frac{x}{y}$.

So, in this case we have found $\frac{dy}{dx}$ by using a circle property.

Method 2:

Another way of finding $\frac{dy}{dx}$ for a circle is to split the relation into two parts.

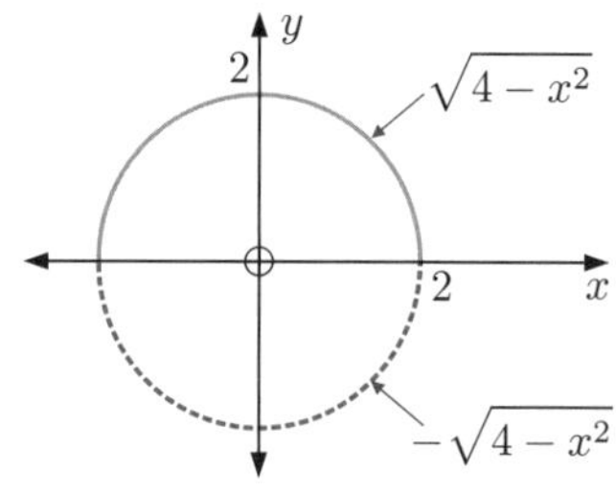

If $x^2 + y^2 = 4$, then $y^2 = 4 - x^2$

and so $y = \pm\sqrt{4 - x^2}$.

For the (yellow) top half:

$y = \sqrt{4 - x^2} = (4 - x^2)^{\frac{1}{2}}$

$$\therefore \frac{dy}{dx} = \frac{1}{2}(4 - x^2)^{-\frac{1}{2}} \times (-2x) = \frac{-x}{\sqrt{4-x^2}} = -\frac{x}{y}$$

For the (green) bottom half:

$y = -\sqrt{4 - x^2} = -(4 - x^2)^{\frac{1}{2}}$

$$\therefore \frac{dy}{dx} = -\frac{1}{2}(4 - x^2)^{-\frac{1}{2}} \times (-2x) = \frac{x}{\sqrt{4-x^2}} = \frac{x}{-y}$$

So, in both cases $\frac{dy}{dx} = -\frac{x}{y}$.

IMPLICIT DIFFERENTIATION

A more formal method for solving this type of problem is **implicit differentiation**.

If we are given an implicit relation between y and x and we want $\frac{dy}{dx}$, we can differentiate both sides of the equation with respect to x, applying the chain, product, quotient, and any other rules as appropriate. This will generate terms containing $\frac{dy}{dx}$, which we then proceed to make the subject of the equation.

For example, if $x^2 + y^2 = 4$, then $\frac{d}{dx}\left(x^2+y^2\right) = \frac{d}{dx}(4)$

$$\therefore \quad 2x + 2y\frac{dy}{dx} = 0$$

$$\therefore \quad \frac{dy}{dx} = -\frac{x}{y}$$

A useful property using the chain rule is that $\boldsymbol{\frac{d}{dx}\left(y^n\right) = ny^{n-1}\frac{dy}{dx}}$.

Example 9 Self Tutor

If y is a function of x, find:

a $\frac{d}{dx}\left(y^3\right)$ **b** $\frac{d}{dx}\left(\frac{1}{y}\right)$ **c** $\frac{d}{dx}\left(xy^2\right)$

a $\frac{d}{dx}\left(y^3\right)$

$= 3y^2\frac{dy}{dx}$

b $\frac{d}{dx}\left(\frac{1}{y}\right)$

$= \frac{d}{dx}\left(y^{-1}\right)$

$= -y^{-2}\frac{dy}{dx}$

$= -\frac{1}{y^2}\frac{dy}{dx}$

c $\frac{d}{dx}\left(xy^2\right)$

$= 1 \times y^2 + x \times 2y\frac{dy}{dx}$ {product rule}

$= y^2 + 2xy\frac{dy}{dx}$

Example 10 Self Tutor

Find $\frac{dy}{dx}$ if: **a** $x^2 + y^3 = 8$ **b** $x + x^2y + y^3 = 100$

a
$$x^2 + y^3 = 8$$
$$\therefore \quad \frac{d}{dx}\left(x^2\right) + \frac{d}{dx}\left(y^3\right) = \frac{d}{dx}(8)$$
$$\therefore \quad 2x + 3y^2\frac{dy}{dx} = 0$$
$$\therefore \quad \frac{dy}{dx} = \frac{-2x}{3y^2}$$

b
$$x + x^2y + y^3 = 100$$
$$\therefore \quad \frac{d}{dx}(x) + \frac{d}{dx}\left(x^2y\right) + \frac{d}{dx}\left(y^3\right) = \frac{d}{dx}(100)$$
$$\therefore \quad 1 + \underbrace{\left[2xy + x^2\frac{dy}{dx}\right]}_{\text{\{product rule\}}} + 3y^2\frac{dy}{dx} = 0$$
$$\therefore \quad \left(x^2+3y^2\right)\frac{dy}{dx} = -1 - 2xy$$
$$\therefore \quad \frac{dy}{dx} = \frac{-1-2xy}{x^2+3y^2}$$

Example 11 — Self Tutor

Find the gradient of the tangent to $x^2 + y^3 = 5$ at the point where $x = 2$.

First we find $\frac{dy}{dx}$: $\quad 2x + 3y^2\frac{dy}{dx} = 0$ {implicit differentiation}

$$\therefore\ 3y^2\frac{dy}{dx} = -2x$$

$$\therefore\ \frac{dy}{dx} = \frac{-2x}{3y^2}$$

When $x = 2$, $\quad 4 + y^3 = 5$

$$\therefore\ y = 1$$

So, when $x = 2$, $\quad \frac{dy}{dx} = \frac{-2(2)}{3(1)^2} = -\frac{4}{3}$

$\therefore$ the gradient of the tangent at $x = 2$ is $-\frac{4}{3}$.

EXERCISE 18E

1 If y is a function of x, find:

a $\frac{d}{dx}(2y)$ **b** $\frac{d}{dx}(-3y)$ **c** $\frac{d}{dx}(y^3)$ **d** $\frac{d}{dx}\left(\frac{1}{y}\right)$ **e** $\frac{d}{dx}(y^4)$

f $\frac{d}{dx}(\sqrt{y})$ **g** $\frac{d}{dx}\left(\frac{1}{y^2}\right)$ **h** $\frac{d}{dx}(xy)$ **i** $\frac{d}{dx}(x^2y)$ **j** $\frac{d}{dx}(xy^2)$

2 Find $\frac{dy}{dx}$ if:

a $x^2 + y^2 = 25$ **b** $x^2 + 3y^2 = 9$ **c** $y^2 - x^2 = 8$

d $x^2 - y^3 = 10$ **e** $x^2 + xy = 4$ **f** $x^3 - 2xy = 5$

3 Find the gradient of the tangent to:

a $x + y^3 = 4y$ at $y = 1$ **b** $x + y = 8xy$ at $x = \frac{1}{2}$

F DERIVATIVES OF EXPONENTIAL FUNCTIONS

In **Chapter 3** we saw that the simplest **exponential functions** have the form $f(x) = b^x$ where b is any positive constant, $b \neq 1$.

The graphs of all members of the exponential family $f(x) = b^x$ have the following properties:

- pass through the point $(0, 1)$
- asymptotic to the x-axis at one end
- lie above the x-axis for all x.

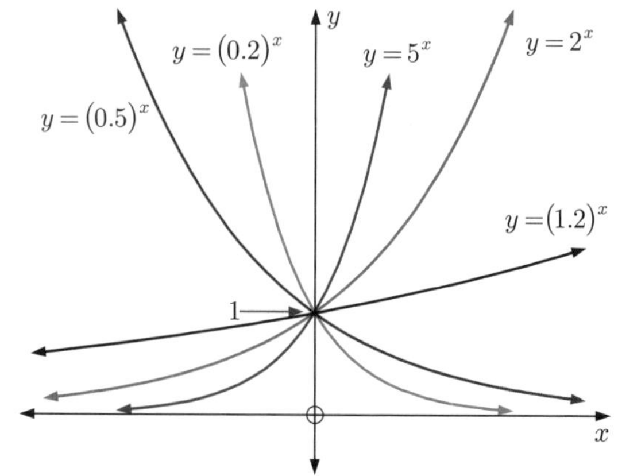

INVESTIGATION 4 THE DERIVATIVE OF $y = b^x$

This Investigation could be done by using a **graphics calculator** or by clicking on the icon.

The purpose of this Investigation is to observe the nature of the derivatives of $f(x) = b^x$ for various values of b.

What to do:

1 Use your graphics calculator or the software provided to help fill in the table for $y = 2^x$:

x	y	$\frac{dy}{dx}$	$\frac{dy}{dx} \div y$
0			
0.5			
1			
1.5			
2			

CALCULUS DEMO

2 Repeat **1** for the following functions:

a $y = 3^x$ **b** $y = 5^x$ **c** $y = (0.5)^x$

3 Use your observations from **1** and **2** to write a statement about the derivative of the general exponential $y = b^x$ for $b > 0$, $b \neq 1$.

From the **Investigation** you should have discovered that:

$$\text{If } f(x) = b^x \text{ then } f'(x) = f'(0) \times b^x.$$

Proof:

If $f(x) = b^x$,

then $f'(x) = \lim\limits_{h \to 0} \dfrac{b^{x+h} - b^x}{h}$ {first principles definition of the derivative}

$= \lim\limits_{h \to 0} \dfrac{b^x(b^h - 1)}{h}$

$= b^x \times \left(\lim\limits_{h \to 0} \dfrac{b^h - 1}{h} \right)$ {as b^x is independent of h}

But $f'(0) = \lim\limits_{h \to 0} \dfrac{f(0+h) - f(0)}{h}$

$= \lim\limits_{h \to 0} \dfrac{b^h - 1}{h}$

$\therefore f'(x) = b^x \times f'(0)$

Given this result, if we can find a value of b such that $f'(0) = 1$, then we will have found *a function which is its own derivative*!

INVESTIGATION 5 SOLVING FOR b IF $f(x) = b^x$ AND $f'(x) = b^x$

Click on the icon to graph $f(x) = b^x$ and its derivative function $y = f'(x)$. **DEMO**

Experiment with different values of b until the graphs of $f(x) = b^x$ and $y = f'(x)$ appear the same.

Estimate the corresponding value of b to 3 decimal places.

You should have discovered that $f(x) = f'(x) = b^x$ when $b \approx 2.718$.

To find this value of b more accurately we return to the algebraic approach:

We have already shown that if $f(x) = b^x$ then $f'(x) = b^x \left(\lim\limits_{h \to 0} \dfrac{b^h - 1}{h} \right)$.

So if $f'(x) = b^x$ then we require $\lim\limits_{h \to 0} \dfrac{b^h - 1}{h} = 1$.

$$\therefore \lim_{h \to 0} b^h = \lim_{h \to 0} (1 + h)$$

Letting $h = \dfrac{1}{n}$, we notice that $\dfrac{1}{n} \to 0$ if $n \to \infty$

$$\therefore \lim_{n \to \infty} b^{\frac{1}{n}} = \lim_{n \to \infty} \left(1 + \frac{1}{n}\right)$$

$$\therefore b = \lim_{n \to \infty} \left(1 + \frac{1}{n}\right)^n \quad \text{if this limit exists}$$

We have in fact already seen this limit in compound interest.

We found that as $n \to \infty$,

$$\left(1 + \frac{1}{n}\right)^n \to 2.718\,281\,828\,459\,045\,235\ldots.$$

and this irrational number is the natural exponential e.

We now have: **If $f(x) = e^x$ then $f'(x) = e^x$.**

e^x is sometimes written as $\exp(x)$.
For example, $\exp(1 - x) = e^{1-x}$.

PROPERTIES OF $y = e^x$

$\dfrac{dy}{dx} = e^x = y$

As $x \to \infty$, $y \to \infty$ very rapidly, and so $\dfrac{dy}{dx} \to \infty$.

This means that the gradient of the curve is very large for large values of x. The curve increases in steepness as x gets larger.

As $x \to -\infty$, $y \to 0$ but never reaches 0, and so $\dfrac{dy}{dx} \to 0$ also.

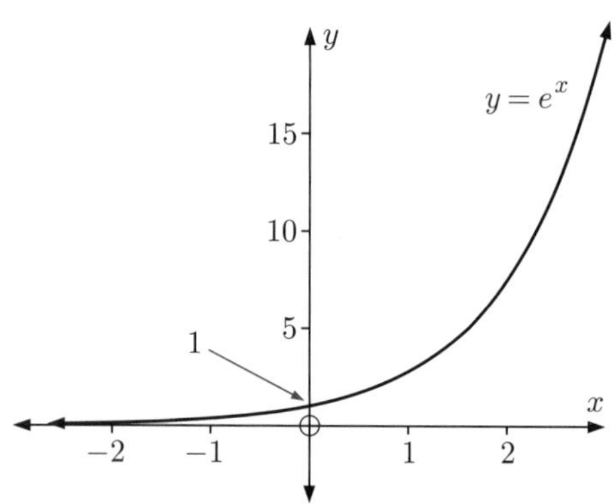

For large negative x, $f(x) = e^x$ approaches the asymptote $y = 0$.

$e^x > 0$ for all x, so the range of $f(x) = e^x$ is $\mathbb{R}^+$.

THE DERIVATIVE OF $e^{f(x)}$

The functions e^{-x}, e^{2x+3}, and e^{-x^2} all have the form $e^{f(x)}$.

Since $e^x > 0$ for all x, $e^{f(x)} > 0$ for all x, no matter what the function $f(x)$.

Suppose $y = e^{f(x)} = e^u$ where $u = f(x)$.

Now $\frac{dy}{dx} = \frac{dy}{du}\frac{du}{dx}$ {chain rule}

$= e^u \frac{du}{dx}$

$= e^{f(x)} \times f'(x)$

Function	*Derivative*
e^x	e^x
$e^{f(x)}$	$e^{f(x)} \times f'(x)$

Example 12 Self Tutor

Find the gradient function for y equal to:

a $2e^x + e^{-3x}$ **b** x^2e^{-x} **c** $\frac{e^{2x}}{x}$

a If $y = 2e^x + e^{-3x}$ then $\frac{dy}{dx} = 2e^x + e^{-3x}(-3)$

$= 2e^x - 3e^{-3x}$

b If $y = x^2e^{-x}$ then $\frac{dy}{dx} = 2xe^{-x} + x^2e^{-x}(-1)$ {product rule}

$= 2xe^{-x} - x^2e^{-x}$

c If $y = \frac{e^{2x}}{x}$ then $\frac{dy}{dx} = \frac{e^{2x}(2)x - e^{2x}(1)}{x^2}$ {quotient rule}

$= \frac{e^{2x}(2x-1)}{x^2}$

Example 13 Self Tutor

Find the gradient function for y equal to: **a** $(e^x - 1)^3$ **b** $\frac{1}{\sqrt{2e^{-x}+1}}$

a $y = (e^x - 1)^3$

$= u^3$ where $u = e^x - 1$

$\frac{dy}{dx} = \frac{dy}{du}\frac{du}{dx}$ {chain rule}

$= 3u^2 \frac{du}{dx}$

$= 3(e^x - 1)^2 \times e^x$

$= 3e^x(e^x - 1)^2$

b $y = (2e^{-x} + 1)^{-\frac{1}{2}}$

$= u^{-\frac{1}{2}}$ where $u = 2e^{-x} + 1$

$\frac{dy}{dx} = \frac{dy}{du}\frac{du}{dx}$ {chain rule}

$= -\frac{1}{2}u^{-\frac{3}{2}}\frac{du}{dx}$

$= -\frac{1}{2}(2e^{-x} + 1)^{-\frac{3}{2}} \times 2e^{-x}(-1)$

$= e^{-x}(2e^{-x} + 1)^{-\frac{3}{2}}$

THE DERIVATIVE OF a^x, $a > 0$

Consider $y = a^x$ where $a > 0$

$$= \left(e^{\ln a}\right)^x = e^{(\ln a)x}$$

$$\therefore \quad \frac{dy}{dx} = e^{(\ln a)x} \times \ln a$$

$$= a^x \ln a$$

If $y = a^x$ where $a > 0$, then $\dfrac{dy}{dx} = a^x \ln a$.

EXERCISE 18F

1 Find the gradient function for $f(x)$ equal to:

a e^{4x} **b** $e^x + 3$ **c** e^{-2x} **d** $e^{\frac{x}{2}}$

e $2e^{-\frac{x}{2}}$ **f** $1 - 2e^{-x}$ **g** $4e^{\frac{x}{2}} - 3e^{-x}$ **h** $\dfrac{e^x + e^{-x}}{2}$

i e^{-x^2} **j** $e^{\frac{1}{x}}$ **k** $10(1 + e^{2x})$ **l** $20(1 - e^{-2x})$

m e^{2x+1} **n** $e^{\frac{x}{4}}$ **o** e^{1-2x^2} **p** $e^{-0.02x}$

2 Find the derivative of:

a xe^x **b** x^3e^{-x} **c** $\dfrac{e^x}{x}$ **d** $\dfrac{x}{e^x}$

e x^2e^{3x} **f** $\dfrac{e^x}{\sqrt{x}}$ **g** $\sqrt{x}e^{-x}$ **h** $\dfrac{e^x + 2}{e^{-x} + 1}$

3 Find the gradient of the tangent to:

a $y = (e^x + 2)^4$ at $x = 0$ **b** $y = \dfrac{1}{2 - e^{-x}}$ at $x = 0$

c $y = \sqrt{e^{2x} + 10}$ at $x = \ln 3$

Check your answers using technology.

4 Find the gradient function for $f(x)$ equal to:

a $\dfrac{1}{(1 - e^{3x})^2}$ **b** $\dfrac{1}{\sqrt{1 - e^{-x}}}$ **c** $x\sqrt{1 - 2e^{-x}}$

5 Given $f(x) = e^{kx} + x$ and $f'(0) = -8$, find k.

6 **a** By substituting $e^{\ln 2}$ for 2 in $y = 2^x$, find $\dfrac{dy}{dx}$.

b Show that if $y = b^x$ where $b > 0$, $b \neq 1$, then $\dfrac{dy}{dx} = b^x \times \ln b$.

7 The tangent to $f(x) = x^2e^{-x}$ at point P is horizontal. Find the possible coordinates of P.

8 Find $\dfrac{dy}{dx}$ for:

a $y = 2^x$ **b** $y = 5^x$ **c** $y = x2^x$

d $y = x^3 6^{-x}$ **e** $y = \dfrac{2^x}{x}$ **f** $y = \dfrac{x}{3^x}$

9 Find $\dfrac{dy}{dx}$ if $x^3e^{3y} + 4x^2y^3 = 27e^{-2x}$.

G DERIVATIVES OF LOGARITHMIC FUNCTIONS

INVESTIGATION 6 THE DERIVATIVE OF $\ln x$

If $y = \ln x$, what is the gradient function?

CALCULUS DEMO

What to do:

1. Click on the icon to see the graph of $y = \ln x$. Observe the gradient function being drawn as the point moves from left to right along the graph.
2. Predict a formula for the gradient function of $y = \ln x$.
3. Find the gradient of the tangent to $y = \ln x$ for $x = 0.25, 0.5, 1, 2, 3, 4$, and 5. Do your results confirm your prediction in **2**?

From the **Investigation** you should have observed:

$$\text{If } y = \ln x \text{ then } \frac{dy}{dx} = \frac{1}{x}.$$

Proof: If $y = \ln x$ then $x = e^y$

Using implicit differentiation with respect to x, $\quad 1 = e^y \frac{dy}{dx}$ {chain rule}

$\therefore \; 1 = x \frac{dy}{dx}$ {as $e^y = x$}

$\therefore \; \frac{1}{x} = \frac{dy}{dx}$

THE DERIVATIVE OF $\ln f(x)$

Suppose $y = \ln f(x)$

$\therefore \; y = \ln u$ where $u = f(x)$.

Now $\frac{dy}{dx} = \frac{dy}{du}\frac{du}{dx}$ {chain rule}

$\therefore \; \frac{dy}{dx} = \frac{1}{u}\frac{du}{dx}$

$= \frac{f'(x)}{f(x)}$

Function	*Derivative*
$\ln x$	$\frac{1}{x}$
$\ln f(x)$	$\frac{f'(x)}{f(x)}$

The laws of logarithms can help us to differentiate some logarithmic functions more easily.

For $a > 0$, $b > 0$, $n \in \mathbb{R}$: $\quad \boldsymbol{\ln(ab) = \ln a + \ln b}$

$$\boldsymbol{\ln\left(\frac{a}{b}\right) = \ln a - \ln b}$$

$$\boldsymbol{\ln(a^n) = n \ln a}$$

Example 14 Self Tutor

Find the gradient function of:

a $y = \ln(kx)$, k a constant **b** $y = \ln(1-3x)$ **c** $y = x^3 \ln x$

a $y = \ln(kx)$

$\therefore \dfrac{dy}{dx} = \dfrac{k}{kx}$

$= \dfrac{1}{x}$

b $y = \ln(1-3x)$

$\therefore \dfrac{dy}{dx} = \dfrac{-3}{1-3x}$

$= \dfrac{3}{3x-1}$

c $y = x^3 \ln x$

$\therefore \dfrac{dy}{dx} = 3x^2 \ln x + x^3 \left(\dfrac{1}{x}\right)$ {product rule}

$= 3x^2 \ln x + x^2$

$= x^2(3\ln x + 1)$

$\ln(kx) = \ln k + \ln x$
$= \ln x +$ constant
so $\ln(kx)$ and $\ln x$ both have derivative $\dfrac{1}{x}$.

Example 15 Self Tutor

Differentiate with respect to x:

a $y = \ln(xe^{-x})$ **b** $y = \ln\left[\dfrac{x^2}{(x+2)(x-3)}\right]$

a $y = \ln(xe^{-x})$

$= \ln x + \ln e^{-x}$ $\{\ln(ab) = \ln a + \ln b\}$

$= \ln x - x$ $\{\ln e^a = a\}$

$\therefore \dfrac{dy}{dx} = \dfrac{1}{x} - 1$

b $y = \ln\left[\dfrac{x^2}{(x+2)(x-3)}\right]$

$= \ln x^2 - \ln[(x+2)(x-3)]$ $\{\ln\left(\frac{a}{b}\right) = \ln a - \ln b\}$

$= 2\ln x - [\ln(x+2) + \ln(x-3)]$

$= 2\ln x - \ln(x+2) - \ln(x-3)$

$\therefore \dfrac{dy}{dx} = \dfrac{2}{x} - \dfrac{1}{x+2} - \dfrac{1}{x-3}$

A derivative function will only be valid on *at most* the domain of the original function.

EXERCISE 18G

1 Find the gradient function of:

a $y = \ln(7x)$ **b** $y = \ln(2x+1)$ **c** $y = \ln(x - x^2)$

d $y = 3 - 2\ln x$ **e** $y = x^2 \ln x$ **f** $y = \dfrac{\ln x}{2x}$

g $y = e^x \ln x$ **h** $y = (\ln x)^2$ **i** $y = \sqrt{\ln x}$

j $y = e^{-x} \ln x$ **k** $y = \sqrt{x}\ln(2x)$ **l** $y = \dfrac{2\sqrt{x}}{\ln x}$

m $y = 3 - 4\ln(1-x)$ **n** $y = x\ln(x^2+1)$

2 Find $\frac{dy}{dx}$ for:

a $y = x\ln 5$ **b** $y = \ln(x^3)$ **c** $y = \ln(x^4 + x)$

d $y = \ln(10 - 5x)$ **e** $y = [\ln(2x+1)]^3$ **f** $y = \frac{\ln(4x)}{x}$

g $y = \ln\left(\frac{1}{x}\right)$ **h** $y = \ln(\ln x)$ **i** $y = \frac{1}{\ln x}$

3 Use the logarithm laws to help differentiate with respect to x:

a $y = \ln\sqrt{1-2x}$ **b** $y = \ln\left(\frac{1}{2x+3}\right)$ **c** $y = \ln\left(e^x\sqrt{x}\right)$

d $y = \ln\left(x\sqrt{2-x}\right)$ **e** $y = \ln\left(\frac{x+3}{x-1}\right)$ **f** $y = \ln\left(\frac{x^2}{3-x}\right)$

g $f(x) = \ln\left((3x-4)^3\right)$ **h** $f(x) = \ln\left(x(x^2+1)\right)$ **i** $f(x) = \ln\left(\frac{x^2+2x}{x-5}\right)$

4 Find the gradient of the tangent to $y = x\ln x$ at the point where $x = e$.

5 Suppose $f(x) = a\ln(2x+b)$ where $f(e) = 3$ and $f'(e) = \frac{6}{e}$. Find the constants a and b.

6 Find $\frac{dy}{dx}$ for: **a** $y = \log_2 x$ **b** $y = \log_{10} x$ **c** $y = x\log_3 x$

7 Find $\frac{da}{db}$ if $e^{2a}\ln b^2 - a^3b + \ln(ab) = 21$.

H DERIVATIVES OF TRIGONOMETRIC FUNCTIONS

In **Chapter 12** we saw that sine and cosine curves arise naturally from motion in a circle.

Click on the icon to observe the motion of point P around the unit circle. Observe the graphs of P's height relative to the x-axis, and then P's horizontal displacement from the y-axis. The resulting graphs are those of $y = \sin t$ and $y = \cos t$.

Suppose P moves anticlockwise around the unit circle with constant linear speed of 1 unit per second.

After 2π seconds, P will travel 2π units which is one full revolution, and thus through 2π radians.

So, after t seconds P will travel through t radians, and at time t, P is at $(\cos t, \sin t)$.

The **angular velocity** of P is the rate of change of $A\widehat{O}P$ with respect to time.

Angular velocity is only meaningful in motion along a circular or elliptical arc.

For the example above, the angular velocity of P is $\frac{d\theta}{dt}$ and $\frac{d\theta}{dt} = 1$ radian per second.

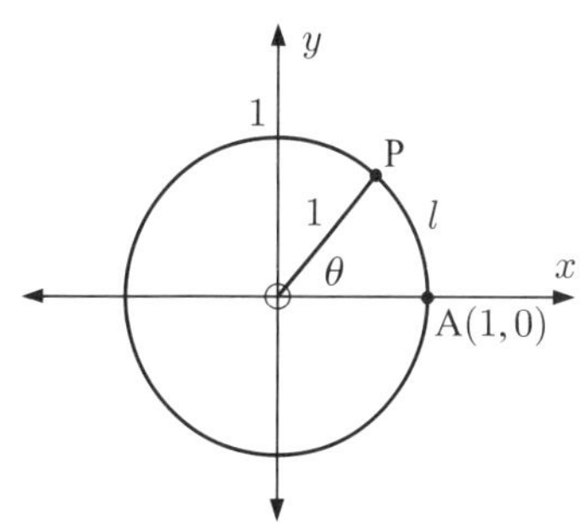

If we let l be the arc length AP, the **linear speed** of P is $\frac{dl}{dt}$, the rate of change of l with respect to time.

For the example above, $l = \theta r = \theta \times 1 = \theta$ and $\frac{dl}{dt} = 1$ unit per second.

INVESTIGATION 7 DERIVATIVES OF $\sin x$ AND $\cos x$

Our aim is to use a computer demonstration to investigate the derivatives of $\sin x$ and $\cos x$.

What to do:

1 Click on the icon to observe the graph of $y = \sin x$. A tangent with x-step of length 1 unit moves across the curve, and its y-step is translated onto the gradient graph. Predict the derivative of the function $y = \sin x$.

DERIVATIVES DEMO

2 Repeat the process in **1** for the graph of $y = \cos x$. Hence predict the derivative of the function $y = \cos x$.

From the **Investigation** you should have deduced that for t in radians,

$$\frac{d}{dt}(\sin t) = \cos t \quad \text{and} \quad \frac{d}{dt}(\cos t) = -\sin t.$$

We will now show these derivatives using first principles. To do this we make use of two results from **Chapter 17**:

- If θ is in radians, then $\lim\limits_{\theta \to 0} \dfrac{\sin\theta}{\theta} = 1$
- $\sin S - \sin D = 2\cos\left(\frac{S+D}{2}\right)\sin\left(\frac{S-D}{2}\right)$

THE DERIVATIVE OF $\sin x$

Consider $f(x) = \sin x$.

Now
$$\begin{aligned} f'(x) &= \lim_{h\to 0} \frac{f(x+h)-f(x)}{h} \\ &= \lim_{h\to 0} \frac{\sin(x+h)-\sin x}{h} \\ &= \lim_{h\to 0} \frac{2\cos\left(\frac{x+h+x}{2}\right)\sin\left(\frac{x+h-x}{2}\right)}{h} && \{\text{identity above}\} \\ &= \lim_{h\to 0} \frac{2\cos\left(x+\frac{h}{2}\right)\sin\left(\frac{h}{2}\right)}{h} \\ &= \lim_{h\to 0} \frac{2\cos\left(x+\frac{h}{2}\right)}{2} \times \frac{\sin\left(\frac{h}{2}\right)}{\frac{h}{2}} \\ &= \cos x \times 1 && \left\{\text{as } h \to 0,\ \tfrac{h}{2} \to 0,\ \frac{\sin\left(\frac{h}{2}\right)}{\frac{h}{2}} \to 1\right\} \\ &= \cos x \end{aligned}$$

If $f(x) = \sin x$ where x is in radians, then $f'(x) = \cos x$.

THE DERIVATIVE OF $\cos x$

Consider $y = \cos x = \sin\left(\frac{\pi}{2} - x\right)$

$\therefore$ $y = \sin u$ where $u = \frac{\pi}{2} - x$

Now
$$\begin{aligned} \frac{dy}{dx} &= \frac{dy}{du}\frac{du}{dx} && \{\text{chain rule}\} \\ &= \cos u \times (-1) \\ &= -\cos\left(\tfrac{\pi}{2} - x\right) \\ &= -\sin x \end{aligned}$$

If $f(x) = \cos x$ where x is in radians, then $f'(x) = -\sin x$.

THE DERIVATIVE OF $\tan x$

Consider $y = \tan x = \dfrac{\sin x}{\cos x}$

$$\therefore \frac{dy}{dx} = \frac{\cos x \cos x - \sin x(-\sin x)}{(\cos x)^2} \quad \{\text{quotient rule}\}$$
$$= \frac{\cos^2 x + \sin^2 x}{\cos^2 x}$$
$$= \frac{1}{\cos^2 x}$$
$$= \sec^2 x$$

Summary: For x in radians:

Function	*Derivative*
$\sin x$	$\cos x$
$\cos x$	$-\sin x$
$\tan x$	$\sec^2 x$

THE DERIVATIVES OF $\sin[f(x)]$, $\cos[f(x)]$, AND $\tan[f(x)]$

Suppose $y = \sin[f(x)]$

If we let $u = f(x)$, then $y = \sin u$.

But $\dfrac{dy}{dx} = \dfrac{dy}{du}\dfrac{du}{dx}$ {chain rule}

$$\therefore \frac{dy}{dx} = \cos u \times f'(x)$$
$$= \cos[f(x)] \times f'(x)$$

We can perform the same procedure for $\cos[f(x)]$ and $\tan[f(x)]$, giving the results shown:

Function	*Derivative*
$\sin[f(x)]$	$\cos[f(x)]\, f'(x)$
$\cos[f(x)]$	$-\sin[f(x)]\, f'(x)$
$\tan[f(x)]$	$\sec^2[f(x)]\, f'(x)$

Example 16 **Self Tutor**

Differentiate with respect to x:

a $x \sin x$

b $4\tan^2(3x)$

a Suppose $y = x \sin x$

Using the product rule,

$$\frac{dy}{dx} = (1)\sin x + (x)\cos x$$
$$= \sin x + x\cos x$$

b Suppose $y = 4\tan^2(3x)$

$$= 4[\tan(3x)]^2$$
$$= 4u^2 \text{ where } u = \tan(3x)$$
$$\frac{dy}{dx} = \frac{dy}{du}\frac{du}{dx} \quad \{\text{chain rule}\}$$
$$\therefore \frac{dy}{dx} = 8u \times \frac{du}{dx}$$
$$= 8\tan(3x) \times \frac{3}{\cos^2(3x)}$$
$$= \frac{24\sin(3x)}{\cos^3(3x)}$$

THE DERIVATIVES OF $\csc x$, $\sec x$, AND $\cot x$

Suppose $y = \csc x = \dfrac{1}{\sin x}$

If $u = \sin x$ then $y = u^{-1}$ and $\dfrac{du}{dx} = \cos x$

$$\begin{aligned}\therefore \quad \frac{dy}{dx} &= -1u^{-2}\frac{du}{dx} \qquad \{\text{chain rule}\}\\ &= \frac{-1}{(\sin x)^2}\cos x\\ &= -\frac{1}{\sin x}\frac{\cos x}{\sin x}\\ &= -\csc x \cot x\end{aligned}$$

Likewise:

- if $y = \sec x$, then $\dfrac{dy}{dx} = \sec x \tan x$
- if $y = \cot x$, then $\dfrac{dy}{dx} = -\csc^2 x$

Summary: For x in radians:

Function	*Derivative*
$\csc x$	$-\csc x \cot x$
$\sec x$	$\sec x \tan x$
$\cot x$	$-\csc^2 x$

Example 17 **Self Tutor**

Find $\dfrac{dy}{dx}$ for: **a** $y = \csc(3x)$ **b** $y = \sqrt{\cot\left(\frac{x}{2}\right)}$

a

$$\begin{aligned} y &= \csc(3x)\\ \therefore \quad \frac{dy}{dx} &= -\csc(3x)\cot(3x)\frac{d}{dx}(3x)\\ &= -3\csc(3x)\cot(3x)\end{aligned}$$

b

$$\begin{aligned} y &= \left(\cot\left(\tfrac{x}{2}\right)\right)^{\frac{1}{2}}\\ \therefore \quad \frac{dy}{dx} &= \tfrac{1}{2}\left(\cot\left(\tfrac{x}{2}\right)\right)^{-\frac{1}{2}} \times -\csc^2\left(\tfrac{x}{2}\right) \times \tfrac{1}{2}\\ &= \frac{-\csc^2\left(\frac{x}{2}\right)}{4\sqrt{\cot\left(\frac{x}{2}\right)}}\end{aligned}$$

EXERCISE 18H

1 Find $\dfrac{dy}{dx}$ for:

a $y = \sin(2x)$ **b** $y = \sin x + \cos x$ **c** $y = \cos(3x) - \sin x$

d $y = \sin(x+1)$ **e** $y = \cos(3-2x)$ **f** $y = \tan(5x)$

g $y = \sin(\frac{x}{2}) - 3\cos x$ **h** $y = 3\tan(\pi x)$ **i** $y = 4\sin x - \cos(2x)$

2 Differentiate with respect to x:

a $x^2 + \cos x$ **b** $\tan x - 3\sin x$ **c** $e^x \cos x$

d $e^{-x}\sin x$ **e** $\ln(\sin x)$ **f** $e^{2x}\tan x$

g $\sin(3x)$ **h** $\cos(\frac{x}{2})$ **i** $3\tan(2x)$

j $x\cos x$ **k** $\dfrac{\sin x}{x}$ **l** $x\tan x$

3 Prove that for x in radians:

a $\dfrac{d}{dx}(\sec x) = \sec x \tan x$ **b** $\dfrac{d}{dx}(\cot x) = -\csc^2 x$

4 Differentiate with respect to x:

a $\sin(x^2)$ **b** $\cos(\sqrt{x})$ **c** $\sqrt{\cos x}$
d $\sin^2 x$ **e** $\cos^3 x$ **f** $\cos x \sin(2x)$
g $\cos(\cos x)$ **h** $\cos^3(4x)$ **i** $\csc(4x)$
j $\sec(2x)$ **k** $\dfrac{2}{\sin^2(2x)}$ **l** $8\cot^3\left(\frac{x}{2}\right)$

5 Find the gradient of the tangent to:

a $f(x) = \sin^3 x$ at the point where $x = \frac{2\pi}{3}$
b $f(x) = \cos x \sin x$ at the point where $x = \frac{\pi}{4}$.

6 Find $\dfrac{dy}{dx}$ for:

a $y = x \sec x$ **b** $y = e^x \cot x$ **c** $y = 4\sec(2x)$
d $y = e^{-x}\cot\left(\frac{x}{2}\right)$ **e** $y = x^2 \csc x$ **f** $y = x\sqrt{\csc x}$
g $y = \ln(\sec x)$ **h** $y = x\csc(x^2)$ **i** $y = \dfrac{\cot x}{\sqrt{x}}$

I DERIVATIVES OF INVERSE TRIGONOMETRIC FUNCTIONS

We have seen that $f(x) = \sin x$, $x \in [-\frac{\pi}{2}, \frac{\pi}{2}]$ is a one-one function and has the inverse function $f^{-1}(x) = \arcsin x$ or $\sin^{-1} x$.

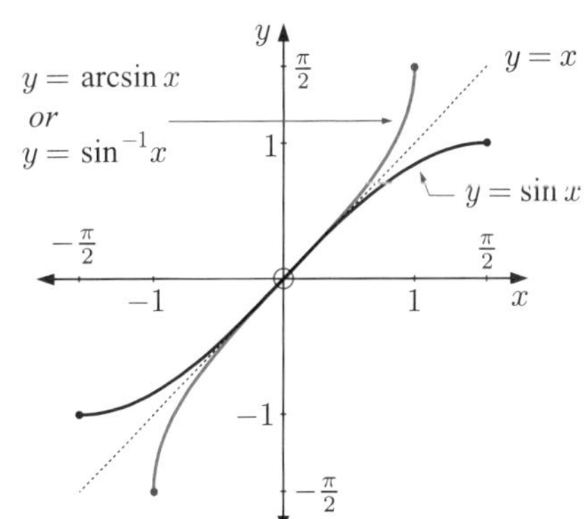

Reminder:

- $\sin^{-1} x$ **is not** $\dfrac{1}{\sin x}$ or $\csc x$
- $\sin^{-1} x$ is the **inverse** function of $y = \sin x$.
- $\csc x$ is the **reciprocal** function of $y = \sin x$.

Consider the following differentiation of $y = \arcsin x$:

If $y = \arcsin x$ then $x = \sin y$

$$\therefore \frac{dx}{dy} = \cos y = \sqrt{1 - \sin^2 y}$$

$$\therefore \frac{dx}{dy} = \sqrt{1 - x^2}$$

From the chain rule, $\dfrac{dy}{dx}\dfrac{dx}{dy} = \dfrac{dy}{dy} = 1$, so $\dfrac{dy}{dx}$ and $\dfrac{dx}{dy}$ are reciprocals.

$$\therefore \frac{dy}{dx} = \frac{1}{\sqrt{1-x^2}}, \quad x \in \,]-1, 1[$$

If $y = \arcsin x$, then $\dfrac{dy}{dx} = \dfrac{1}{\sqrt{1-x^2}}$, $x \in \,]-1, 1[$

$f(x) = \cos x$, $x \in [0, \pi]$ has inverse function $f^{-1}(x) = \arccos x$ or $\cos^{-1} x$.

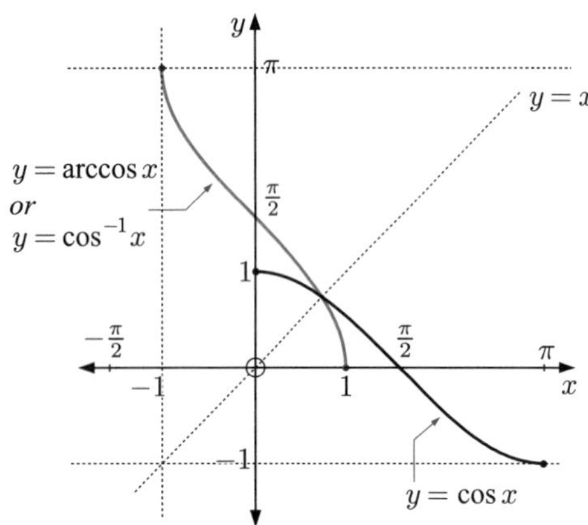

If $y = \arccos x$, then
$\frac{dy}{dx} = \frac{-1}{\sqrt{1-x^2}}$, $x \in \left]-1, 1\right[$

$f(x) = \tan x$, $x \in \left]-\frac{\pi}{2}, \frac{\pi}{2}\right[$ has inverse function $f^{-1}(x) = \arctan x$ or $\tan^{-1} x$.

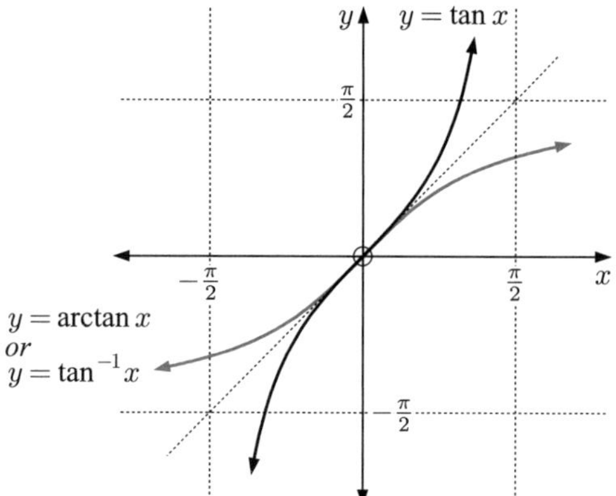

If $y = \arctan x$, then
$\frac{dy}{dx} = \frac{1}{1+x^2}$, $x \in \mathbb{R}$

EXERCISE 18I

1 If $y = \arccos x$, show that $\dfrac{dy}{dx} = \dfrac{-1}{\sqrt{1-x^2}}$, $x \in \left]-1, 1\right[$.

2 If $y = \arctan x$, show that $\dfrac{dy}{dx} = \dfrac{1}{1+x^2}$, $x \in \mathbb{R}$.

3 Find $\dfrac{dy}{dx}$ for:

a $y = \arctan(2x)$ **b** $y = \arccos(3x)$ **c** $y = \arcsin\left(\frac{x}{4}\right)$

d $y = \arccos\left(\frac{x}{5}\right)$ **e** $y = \arctan(x^2)$ **f** $y = \arccos(\sin x)$

4 Find $\dfrac{dy}{dx}$ for:

a $y = x \arcsin x$ **b** $y = e^x \arccos x$ **c** $y = e^{-x} \arctan x$

5 **a** For constant $a \in \mathbb{R}$, $a > 0$, prove that if $y = \arcsin\left(\frac{x}{a}\right)$, then $\dfrac{dy}{dx} = \dfrac{1}{\sqrt{a^2 - x^2}}$ for $x \in \left]-a, a\right[$.

b For constant $a \in \mathbb{R}$, $a > 0$, prove that if $y = \arctan\left(\frac{x}{a}\right)$, then $\dfrac{dy}{dx} = \dfrac{a}{a^2 + x^2}$ for $x \in \mathbb{R}$.

c For constant $a \in \mathbb{R}$, $a > 0$, if $y = \arccos\left(\frac{x}{a}\right)$, find $\dfrac{dy}{dx}$.

J SECOND AND HIGHER DERIVATIVES

Given a function $f(x)$, the derivative $f'(x)$ is known as the **first derivative**.

The **second derivative** of $f(x)$ is the derivative of $f'(x)$, or **the derivative of the first derivative**.

We use $f''(x)$ or y'' or $\frac{d^2y}{dx^2}$ to represent the second derivative.

$f''(x)$ reads "*f double dashed x*".

$\frac{d^2y}{dx^2} = \frac{d}{dx}\left(\frac{dy}{dx}\right)$ reads "*dee two y by dee x squared*".

Example 18 — Self Tutor

Find $f''(x)$ given that $f(x) = x^3 - \frac{3}{x}$.

$$\begin{aligned}
\text{Now} \quad f(x) &= x^3 - 3x^{-1} \\
\therefore \quad f'(x) &= 3x^2 + 3x^{-2} \\
\therefore \quad f''(x) &= 6x - 6x^{-3} \\
&= 6x - \frac{6}{x^3}
\end{aligned}$$

We can continue to differentiate to obtain higher derivatives.

The **nth derivative of y with respect to x** is obtained by differentiating $y = f(x)$ n times.
We use the notation $f^{(n)}(x)$ or $\frac{d^ny}{dx^n}$ for the nth derivative.

Example 19 — Self Tutor

Given $f(x) = \cos 2x$, show that $f^{(3)}(\frac{\pi}{8}) = 4\sqrt{2}$.

$$\begin{aligned}
f(x) &= \cos 2x \\
\therefore \quad f'(x) &= -2\sin 2x \\
\therefore \quad f''(x) &= -4\cos 2x \\
\therefore \quad f^{(3)}(x) &= 8\sin 2x \\
\therefore \quad f^{(3)}(\tfrac{\pi}{8}) &= 8\sin(2 \times \tfrac{\pi}{8}) \\
&= 8\sin\tfrac{\pi}{4} \\
&= 8 \times \tfrac{1}{\sqrt{2}} \\
&= 4\sqrt{2}
\end{aligned}$$

EXERCISE 18J

1 Find $f''(x)$ given that:

a $f(x) = 3x^2 - 6x + 2$ **b** $f(x) = \frac{2}{\sqrt{x}} - 1$ **c** $f(x) = 2x^3 - 3x^2 - x + 5$

d $f(x) = \frac{2-3x}{x^2}$ **e** $f(x) = (1-2x)^3$ **f** $f(x) = \frac{x+2}{2x-1}$

2 Find $\frac{d^2y}{dx^2}$ given that:

a $y = x - x^3$ **b** $y = x^2 - \frac{5}{x^2}$ **c** $y = 2 - \frac{3}{\sqrt{x}}$

d $y = \frac{4-x}{x}$ **e** $y = (x^2 - 3x)^3$ **f** $y = x^2 - x + \frac{1}{1-x}$

3 Given $f(x) = x^3 - 2x + 5$, find:

a $f(2)$ **b** $f'(2)$ **c** $f''(2)$ **d** $f^{(3)}(2)$

4 Suppose $y = Ae^{bx}$ where A and b are constants.

Prove by mathematical induction that $\frac{d^ny}{dx^n} = b^n y$, for $n \in \mathbb{Z}^+$.

5 Find x when $f''(x) = 0$ for:

a $f(x) = 2x^3 - 6x^2 + 5x + 1$ **b** $f(x) = \frac{x}{x^2+2}$

6 Consider the function $f(x) = 2x^3 - x$.

Complete the following table by indicating whether $f(x)$, $f'(x)$, and $f''(x)$ are positive (+), negative (−), or zero (0) at the given values of x.

x	-1	0	1
$f(x)$	−		
$f'(x)$			
$f''(x)$			

7 Given $f(x) = \frac{2}{3}\sin 3x$, show that $f^{(3)}(\frac{2\pi}{9}) = 9$.

8 Suppose $f(x) = 2\sin^3 x - 3\sin x$.

a Show that $f'(x) = -3\cos x \cos 2x$. **b** Find $f''(x)$.

9 Find $\frac{d^2y}{dx^2}$ given:

a $y = -\ln x$ **b** $y = x\ln x$ **c** $y = (\ln x)^2$

10 Given $f(x) = x^2 - \frac{1}{x}$, find:

a $f(1)$ **b** $f'(1)$ **c** $f''(1)$ **d** $f^{(3)}(1)$

11 If $y = 2e^{3x} + 5e^{4x}$, show that $\frac{d^2y}{dx^2} - 7\frac{dy}{dx} + 12y = 0$.

12 If $y = \sin(2x+3)$, show that $\frac{d^2y}{dx^2} + 4y = 0$.

13 If $y = \sin x$, show that $\frac{d^4y}{dx^4} = y$.

14 If $y = 2\sin x + 3\cos x$, show that $y'' + y = 0$ where y'' represents $\frac{d^2y}{dx^2}$.

15 Find $\frac{d^2y}{dx^2}$ for each of the following implicit relations:

a $x^2 + y^2 = 25$ **b** $x^2 - y^2 = 10$ **c** $x^3 + 2xy = 4$

16 Suppose $3V^2 + 2q = 2Vq$. Find: **a** $\frac{dV}{dq}$ **b** $\frac{d^2q}{dV^2}$

17 Consider the function $f(x) = e^{ax}(x+1)$, $a \in \mathbb{R}$. Show that:

a $f'(x) = e^{ax}(a[x+1]+1)$

b $f''(x) = ae^{ax}(a[x+1]+2)$

c if $f^{(k)}(x) = a^{k-1}e^{ax}(a[x+1]+k)$, $k \in \mathbb{Z}$, then $f^{(k+1)}(x) = a^k e^{ax}(a[x+1]+[k+1])$.

18 Consider the function $f(x) = e^{-x}(x+2)$.

a Find: **i** $f'(x)$ **ii** $f''(x)$ **iii** $f'''(x)$ **iv** $f^{(4)}(x)$.

b Conjecture a formula for $f^{(n)}(x)$, $n \in \mathbb{Z}^+$.

c Use the principle of mathematical induction to prove your conjecture.

REVIEW SET 18A NON-CALCULATOR

1 If $f(x) = 7 + x - 3x^2$, find:

a $f(3)$ **b** $f'(3)$ **c** $f''(3)$.

2 Find $\frac{dy}{dx}$ for:

a $y = 3x^2 - x^4$ **b** $y = \frac{x^3 - x}{x^2}$

3 At what point on the curve $f(x) = \frac{x}{\sqrt{x^2+1}}$ does the tangent have gradient 1?

4 Find $\frac{dy}{dx}$ if:

a $y = e^{x^3+2}$ **b** $y = \ln\left(\frac{x+3}{x^2}\right)$ **c** $\ln(2y+1) = xe^y$

5 Given $y = 3e^x - e^{-x}$, show that $\frac{d^2y}{dx^2} = y$.

6 Differentiate with respect to x:

a $\sin(5x)\ln(x)$ **b** $\sin(x)\cos(2x)$ **c** $e^{-2x}\tan x$

7 Find the gradient of the tangent to $y = \sin^2 x$ at the point where $x = \frac{\pi}{3}$.

8 Find $\frac{dy}{dx}$ and $\frac{d^2y}{dx^2}$ given $x^2 + 2xy + y^2 = 4$.

9 Determine the derivative with respect to t of:

a $M = (t^2+3)^4$ **b** $A = \frac{\sqrt{t+5}}{t^2}$

10 Use the rules of differentiation to find $\dfrac{dy}{dx}$ for:

a $y = \dfrac{4}{\sqrt{x}} - 3x$ **b** $y = \sqrt{x^2 - 3x}$

11 Find $f''(2)$ for:

a $f(x) = 3x^2 - \dfrac{1}{x}$ **b** $f(x) = \sqrt{x}$

12 Given $y = (1 - \frac{1}{3}x)^3$, show that $\dfrac{d^3y}{dx^3} = -\frac{2}{9}$.

13 For $y = \dfrac{1}{2x+1}$, prove that $\dfrac{d^ny}{dx^n} = \dfrac{(-2)^n n!}{(2x+1)^{n+1}}$ for all $n \in \mathbb{Z}^+$.

REVIEW SET 18B CALCULATOR

1 Differentiate with respect to x:

a $5x - 3x^{-1}$ **b** $(3x^2 + x)^4$ **c** $(x^2 + 1)(1 - x^2)^3$

2 Find all points on the curve $y = 2x^3 + 3x^2 - 10x + 3$ where the gradient of the tangent is 2.

3 If $y = \sqrt{5 - 4x}$, find:

a $\dfrac{dy}{dx}$ **b** $\dfrac{d^2y}{dx^2}$ **c** $\dfrac{d^3y}{dx^3}$

4 Consider the curves $y = e^{x-1} + 1$ and $y = 3 - e^{1-x}$.

a Sketch the curves on the same set of axes.

b Find the point of intersection of the two curves.

c Show that the tangents to each curve at this point have the same gradient.

d Comment on the significance of this result.

5 Find $\dfrac{dy}{dx}$ if:

a $y = \ln(x^3 - 3x)$ **b** $y = \dfrac{e^x}{x^2}$ **c** $e^{x+y} = \ln(y^2 + 1)$

6 Find x if $f''(x) = 0$ and $f(x) = 2x^4 - 4x^3 - 9x^2 + 4x + 7$.

7 If $f(x) = x - \cos x$, find:

a $f(\pi)$ **b** $f'(\frac{\pi}{2})$ **c** $f''(\frac{3\pi}{4})$

8 Given that a and b are constants, differentiate $y = 3\sin bx - a\cos 2x$ with respect to x.

Find a and b if $y + \dfrac{d^2y}{dx^2} = 6\cos 2x$.

9 Differentiate with respect to x:

a $10x - \sin(10x)$ **b** $\ln\left(\dfrac{1}{\cos x}\right)$ **c** $\sin(5x)\ln(2x)$

10 Differentiate with respect to x:

a $f(x) = \dfrac{(x+3)^3}{\sqrt{x}}$ **b** $f(x) = x^4\sqrt{x^2 + 3}$

11 Find $\frac{dy}{dx}$ for:

a $y = \frac{x}{\sqrt{\sec x}}$ **b** $y = e^x \cot(2x)$ **c** $y = \arccos\left(\frac{x}{2}\right)$

12 The curve $f(x) = 2x^3 + Ax + B$ has a tangent with gradient 10 at the point $(-2, 33)$. Find the values of A and B.

13 Find $\frac{dy}{dx}$ given $x^2 - 3y^2 = 0$. Explain your answer.

REVIEW SET 18C

1 Differentiate with respect to x:

a $y = x^3\sqrt{1-x^2}$ **b** $y = \frac{x^2 - 3x}{\sqrt{x+1}}$

2 Find $\frac{d^2y}{dx^2}$ for:

a $y = 3x^4 - \frac{2}{x}$ **b** $y = x^3 - x + \frac{1}{\sqrt{x}}$

3 Find all points on the curve $y = xe^x$ where the gradient of the tangent is $2e$.

4 Differentiate with respect to x:

a $f(x) = \ln(e^x + 3)$ **b** $f(x) = \ln\left[\frac{(x+2)^3}{x}\right]$ **c** $f(x) = x^{x^2}$

5 Given $y = \left(x - \frac{1}{x}\right)^4$, find $\frac{dy}{dx}$ when $x = 1$.

6 **a** Find $f'(x)$ and $f''(x)$ for $f(x) = \sqrt{x}\cos(4x)$.

b Hence find $f'\left(\frac{\pi}{16}\right)$ and $f''\left(\frac{\pi}{8}\right)$.

7 Suppose $y = 3\sin 2x + 2\cos 2x$. Show that $4y + \frac{d^2y}{dx^2} = 0$.

8 Consider $f(x) = \frac{6x}{3+x^2}$. Find the value(s) of x when:

a $f(x) = -\frac{1}{2}$ **b** $f'(x) = 0$ **c** $f''(x) = 0$

9 The function f is defined by $f : x \mapsto -10\sin 2x \cos 2x$, $0 \leqslant x \leqslant \pi$.

a Write down an expression for $f(x)$ in the form $k\sin 4x$.

b Solve $f'(x) = 0$, giving exact answers.

10 Given the curve $e^x y - xy^2 = 1$, find:

a $\frac{dy}{dx}$ **b** the gradient of the curve at $x = 0$.

11 Prove using the principle of mathematical induction that if $y = x^n$, $n \in \mathbb{Z}^+$, then $\frac{dy}{dx} = nx^{n-1}$. You may assume the product rule of differentiation.

12 Use the product rule for differentiation to prove that:

a If $y = uv$ where u and v are functions of x, then

$$\frac{d^2y}{dx^2} = \left(\frac{d^2u}{dx^2}\right)v + 2\frac{du}{dx}\frac{dv}{dx} + u\left(\frac{d^2v}{dx^2}\right).$$

b If $y = uvw$ where u, v, and w are functions of x, then

$$\frac{dy}{dx} = \frac{du}{dx}vw + u\frac{dv}{dx}w + uv\frac{dw}{dx}.$$

13 Consider the function $f(x) = xe^{ax}$, where a is a constant.

a Find $f^{(n)}(x)$ for $n = 1, 2, 3$, and 4.

b Conjecture a formula for $f^{(n)}(x)$, $n \in \mathbb{Z}^+$.

c Use the principle of mathematical induction to prove your conjecture.

Chapter 19

Properties of curves

Syllabus reference: 6.1, 6.3

Contents:

OPENING PROBLEM

The curve $y = x^3 - 4x$ is shown on the graph alongside.

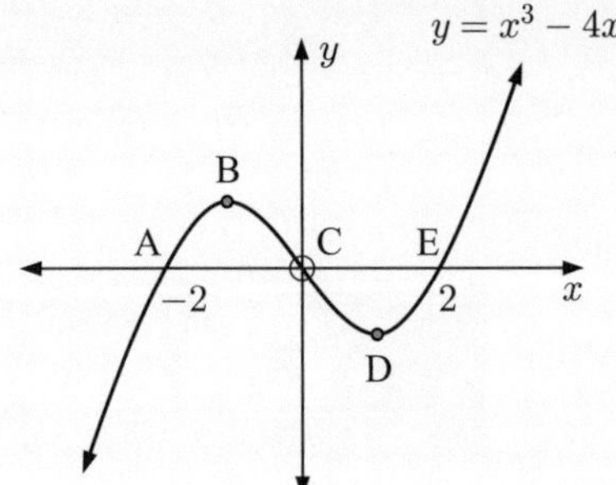

Things to think about:

a At which points is the tangent to the curve horizontal?

b On which intervals is the curve:
 i increasing **ii** decreasing?

c At which point does the curve change its *shape*? How can we use calculus to find this point?

In the previous chapter we saw how to differentiate many types of functions. In this chapter we will learn how to use derivatives to find:

- tangents and normals to curves
- intervals where a function is increasing or decreasing
- turning points which are local minima and maxima
- points where a function changes its shape.

A TANGENTS AND NORMALS

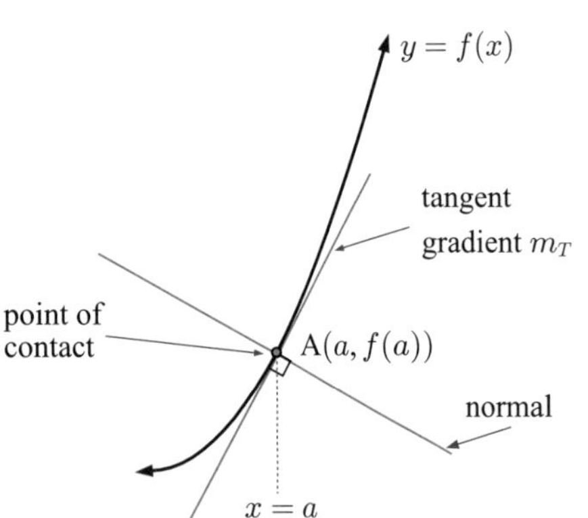

Consider a curve $y = f(x)$.

If A is the point with x-coordinate a, then the gradient of the tangent to the curve at this point is $f'(a) = m_T$.

The equation of the tangent is

$$\frac{y - f(a)}{x - a} = f'(a)$$

$$\textit{or} \quad y - f(a) = f'(a)(x - a).$$

A **normal** to a curve is a line which is perpendicular to the tangent at the point of contact.

The gradients of perpendicular lines are negative reciprocals of each other, so:

The gradient of the **normal** to the curve at $x = a$ is $m_N = -\dfrac{1}{f'(a)}$.

For example:

If $f(x) = x^2$ then $f'(x) = 2x$.

At $x = 2$, $m_T = f'(2) = 4$ and $m_N = -\dfrac{1}{f'(2)} = -\frac{1}{4}$.

So, at $x = 2$ the tangent has gradient 4 and the normal has gradient $-\frac{1}{4}$.

Since $f(2) = 4$, the tangent has equation $y - 4 = 4(x - 2)$ or $y = 4x - 4$

and the normal has equation $y - 4 = -\frac{1}{4}(x - 2)$ or $y = -\frac{1}{4}x + \frac{9}{2}$.

Reminder: If a line has gradient $\frac{4}{5}$ say, and passes through $(2, -3)$ say, another quick way to write down its equation is $4x - 5y = 4(2) - 5(-3)$ or $4x - 5y = 23$.

If the gradient was $-\frac{4}{5}$, we would have:

$4x + 5y = 4(2) + 5(-3)$ or $4x + 5y = -7$.

You can also find the equations of tangents at a given point using your graphics calculator.

Example 1 **Self Tutor**

Find the equation of the tangent to $f(x) = x^2 + 1$ at the point where $x = 1$.

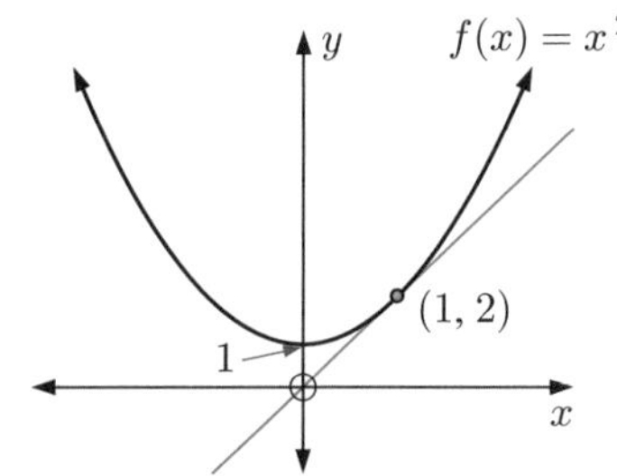

Since $f(1) = 1 + 1 = 2$, the point of contact is $(1, 2)$.

Now $f'(x) = 2x$, so $m_T = f'(1) = 2$

$\therefore$ the tangent has equation $\dfrac{y-2}{x-1} = 2$

which is $y - 2 = 2x - 2$

or $y = 2x$.

EXERCISE 19A

1. Find the equation of the tangent to:
   - **a** $y = x - 2x^2 + 3$ at $x = 2$
   - **b** $y = \sqrt{x} + 1$ at $x = 4$
   - **c** $y = x^3 - 5x$ at $x = 1$
   - **d** $y = \dfrac{4}{\sqrt{x}}$ at $(1, 4)$
   - **e** $y = \dfrac{3}{x} - \dfrac{1}{x^2}$ at $(-1, -4)$
   - **f** $y = 3x^2 - \dfrac{1}{x}$ at $x = -1$.

   Check your answers using technology.

Example 2 **Self Tutor**

Find the equation of the normal to $y = \dfrac{8}{\sqrt{x}}$ at the point where $x = 4$.

When $x = 4$, $y = \dfrac{8}{\sqrt{4}} = \dfrac{8}{2} = 4$. So, the point of contact is $(4, 4)$.

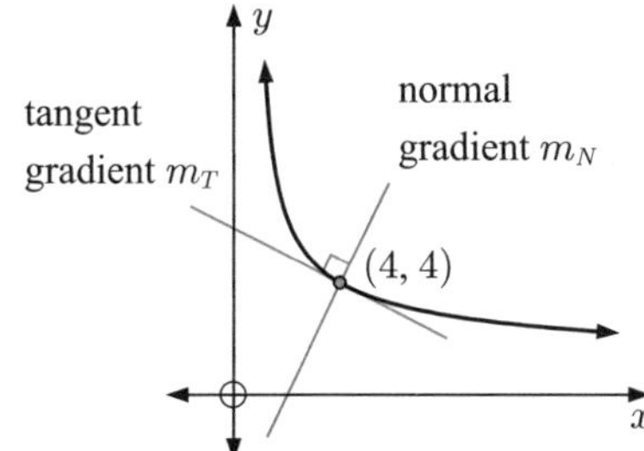

Now as $y = 8x^{-\frac{1}{2}}$, $\dfrac{dy}{dx} = -4x^{-\frac{3}{2}}$

$\therefore$ when $x = 4$, $m_T = -4 \times 4^{-\frac{3}{2}} = -\frac{1}{2}$

$\therefore$ the normal at $(4, 4)$ has gradient $m_N = \frac{2}{1}$.

$\therefore$ the equation of the normal is

$2x - 1y = 2(4) - 1(4)$

or $2x - y = 4$

2 Find the equation of the normal to:

a $y = x^2$ at the point $(3, 9)$

b $y = x^3 - 5x + 2$ at $x = -2$

c $y = \dfrac{5}{\sqrt{x}} - \sqrt{x}$ at the point $(1, 4)$

d $y = 8\sqrt{x} - \dfrac{1}{x^2}$ at $x = 1$.

Example 3 — Self Tutor

Find the equations of any horizontal tangents to $y = x^3 - 12x + 2$.

Since $y = x^3 - 12x + 2$, $\dfrac{dy}{dx} = 3x^2 - 12$

Casio fx-CG20

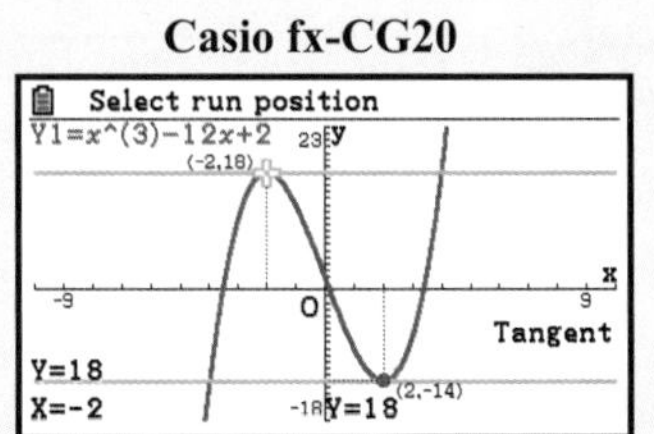

Horizontal tangents have gradient 0,

so $3x^2 - 12 = 0$

$\therefore\ 3(x^2 - 4) = 0$

$\therefore\ 3(x + 2)(x - 2) = 0$

$\therefore\ x = -2$ or 2

When $x = 2$, $y = 8 - 24 + 2 = -14$

When $x = -2$, $y = -8 + 24 + 2 = 18$

$\therefore$ the points of contact are $(2, -14)$ and $(-2, 18)$

$\therefore$ the tangents are $y = -14$ and $y = 18$.

3 **a** Find the equations of the horizontal tangents to $y = 2x^3 + 3x^2 - 12x + 1$.

b Find the points of contact where horizontal tangents meet the curve $y = 2\sqrt{x} + \dfrac{1}{\sqrt{x}}$.

c Find k if the tangent to $y = 2x^3 + kx^2 - 3$ at the point where $x = 2$ has gradient 4.

d Find the equation of another tangent to $y = 1 - 3x + 12x^2 - 8x^3$ which is parallel to the tangent at $(1, 2)$.

4 **a** Consider the curve $y = x^2 + ax + b$ where a and b are constants. The tangent to this curve at the point where $x = 1$ is $2x + y = 6$. Find the values of a and b.

b Consider the curve $y = a\sqrt{x} + \dfrac{b}{\sqrt{x}}$ where a and b are constants. The normal to this curve at the point where $x = 4$ is $4x + y = 22$. Find the values of a and b.

c Show that the equation of the tangent to $y = 2x^2 - 1$ at the point where $x = a$, is $4ax - y = 2a^2 + 1$.

5 Find the equation of the tangent to:

a $y = \sqrt{2x + 1}$ at $x = 4$

b $y = \dfrac{1}{2 - x}$ at $x = -1$

c $f(x) = \dfrac{x}{1 - 3x}$ at $(-1, -\frac{1}{4})$

d $f(x) = \dfrac{x^2}{1 - x}$ at $(2, -4)$.

6 Find the equation of the normal to:

a $y = \dfrac{1}{(x^2 + 1)^2}$ at $(1, \frac{1}{4})$

b $y = \dfrac{1}{\sqrt{3 - 2x}}$ at $x = -3$

c $f(x) = \sqrt{x}(1 - x)^2$ at $x = 4$

d $f(x) = \dfrac{x^2 - 1}{2x + 3}$ at $x = -1$.

7 Consider the curve $y = a\sqrt{1-bx}$ where a and b are constants. The tangent to this curve at the point where $x = -1$ is $3x + y = 5$. Find the values of a and b.

8 Consider $f : x \mapsto \dfrac{x}{\sqrt{2-x}}$.

- **a** State the domain of f.
- **b** Show that $f'(x) = \dfrac{4-x}{2(2-x)^{\frac{3}{2}}}$.
- **c** Find the equation of the normal to $y = f(x)$ at the point where $f(x) = -1$.

Example 4 **Self Tutor**

Show that the equation of the tangent to $y = \ln x$ at the point where $y = -1$ is $y = ex - 2$.

When $y = -1$, $\ln x = -1$

$\therefore\ x = e^{-1} = \frac{1}{e}$

$\therefore$ the point of contact is $\left(\frac{1}{e}, -1\right)$.

Now $f(x) = \ln x$ has derivative $f'(x) = \dfrac{1}{x}$

$\therefore$ the tangent at $\left(\frac{1}{e}, -1\right)$ has gradient $\dfrac{1}{\frac{1}{e}} = e$

$\therefore$ the tangent has equation $\dfrac{y - -1}{x - \frac{1}{e}} = e$

which is $y + 1 = e(x - \frac{1}{e})$

or $y = ex - 2$

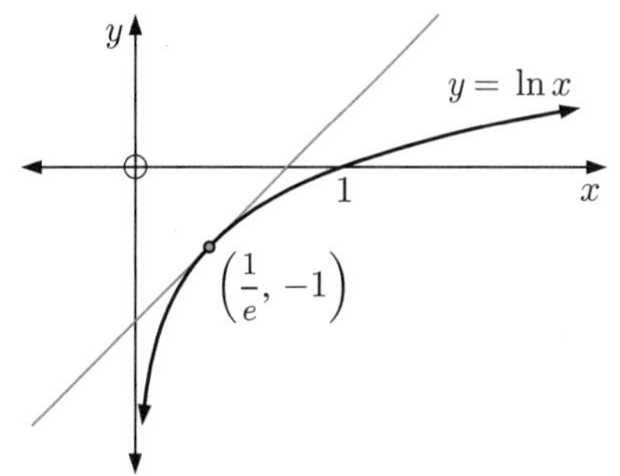

9 Find the equation of:

- **a** the tangent to the function $f : x \mapsto e^{-x}$ at the point where $x = 1$
- **b** the tangent to $y = \ln(2 - x)$ at the point where $x = -1$
- **c** the normal to $y = \ln\sqrt{x}$ at the point where $y = -1$.

Example 5 **Self Tutor**

Find the equation of the tangent to $y = \tan x$ at the point where $x = \frac{\pi}{4}$.

When $x = \frac{\pi}{4}$, $y = \tan\frac{\pi}{4} = 1$

$\therefore$ the point of contact is $(\frac{\pi}{4}, 1)$.

Now $f(x) = \tan x$ has derivative $f'(x) = \dfrac{1}{\cos^2 x}$

$\therefore$ the tangent at $(\frac{\pi}{4}, 1)$ has gradient $\dfrac{1}{\cos^2\frac{\pi}{4}} = \dfrac{1}{\left(\frac{1}{\sqrt{2}}\right)^2} = 2$

$\therefore$ the tangent has equation $\dfrac{y-1}{x-\frac{\pi}{4}} = 2$

which is $y - 1 = 2x - \frac{\pi}{2}$

or $y = 2x + (1 - \frac{\pi}{2})$

10 Show that the curve with equation $y = \dfrac{\cos x}{1 + \sin x}$ does not have any horizontal tangents.

11 Find the equation of:

a the tangent to $y = \sin x$ at the origin

b the tangent to $y = \tan x$ at the origin

c the normal to $y = \cos x$ at the point where $x = \frac{\pi}{6}$

d the normal to $y = \dfrac{1}{\sin(2x)}$ at the point where $x = \frac{\pi}{4}$.

12 Find the equation of the tangent to:

a $y = \sec x$ at $x = \frac{\pi}{4}$

b $y = \cot\left(\frac{x}{2}\right)$ at $x = \frac{\pi}{3}$

13 Find the equation of the normal to:

a $y = \csc x$ at $x = \frac{\pi}{6}$

b $y = \sqrt{\sec(\frac{x}{3})}$ at $x = \pi$

Example 6 **Self Tutor**

Find the coordinates of the point(s) where the tangent to $y = x^3 + x + 2$ at (1, 4) meets the curve again.

Let $f(x) = x^3 + x + 2$

$\therefore\ f'(x) = 3x^2 + 1$

$\therefore\ f'(1) = 3 + 1 = 4$

$\therefore$ the equation of the tangent at (1, 4) is $4x - y = 4(1) - 4$

or $y = 4x$.

Casio fx-CG20

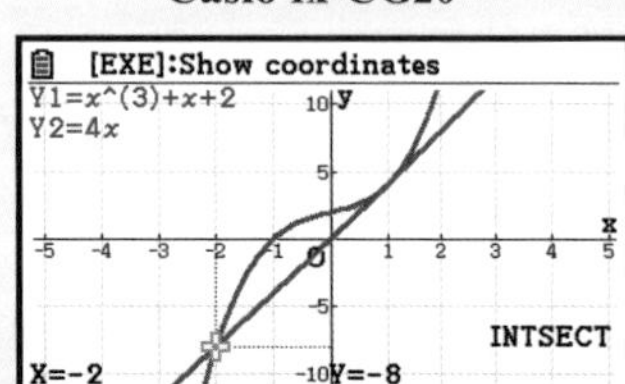

TI-84 Plus

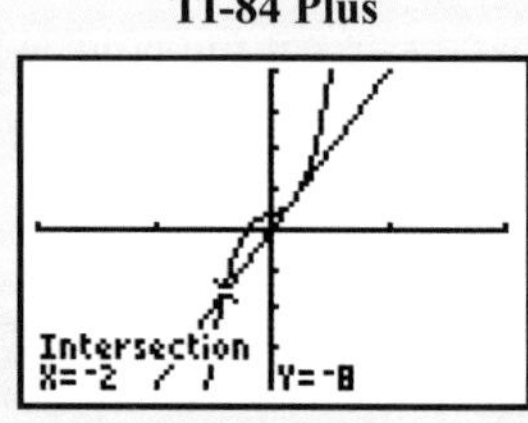

TI-*n*spire

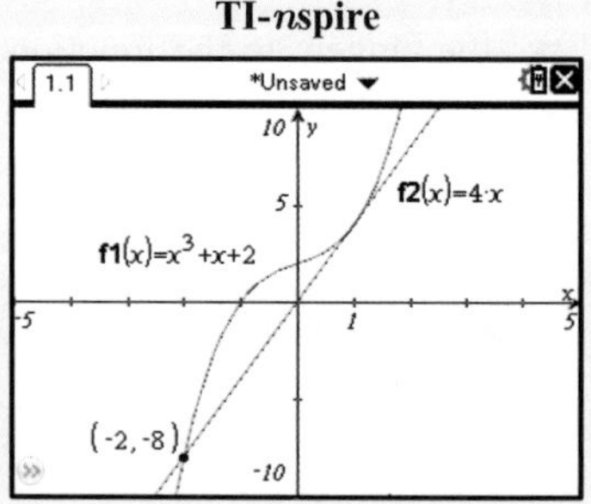

Using technology, $y = 4x$ meets $y = x^3 + x + 2$ at $(-2, -8)$.

14 **a** Find where the tangent to the curve $y = x^3$ at the point where $x = 2$, meets the curve again.

b Find where the tangent to the curve $y = -x^3 + 2x^2 + 1$ at the point where $x = -1$, meets the curve again.

15 Consider the function $f(x) = x^2 + \dfrac{4}{x^2}$.

a Find $f'(x)$.

b Find the values of x at which the tangent to the curve is horizontal.

c Show that the tangents at these points are the same line.

16 The tangent to $y = x^2e^x$ at $x = 1$ cuts the x and y-axes at A and B respectively. Find the coordinates of A and B.

Example 7 | **Self Tutor**

Find the equations of the tangents to $y = x^2$ from the external point (2, 3).

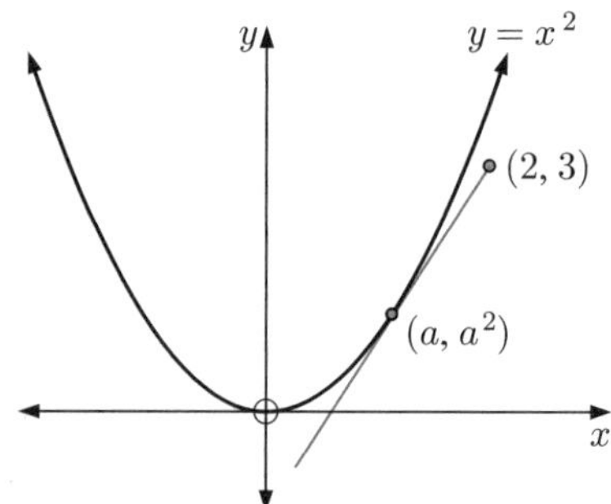

Let (a, a^2) be a general point on $f(x) = x^2$.

Now $f'(x) = 2x$, so $f'(a) = 2a$

$\therefore$ the equation of the tangent at (a, a^2) is

$$\frac{y - a^2}{x - a} = 2a$$

or $y - a^2 = 2ax - 2a^2$

or $y = 2ax - a^2$

Thus the tangents which pass through (2, 3) satisfy

$$3 = 2a(2) - a^2$$
$$\therefore \quad a^2 - 4a + 3 = 0$$
$$\therefore \quad (a-1)(a-3) = 0$$
$$\therefore \quad a = 1 \text{ or } 3$$

$\therefore$ exactly two tangents pass through the external point (2, 3).

If $a = 1$, the tangent has equation $y = 2x - 1$ with point of contact (1, 1).

If $a = 3$, the tangent has equation $y = 6x - 9$ with point of contact (3, 9).

17 **a** Find the equation of the tangent to $y = x^2 - x + 9$ at the point where $x = a$. Hence, find the equations of the two tangents from (0, 0) to the curve. State the coordinates of the points of contact.

b Find the equations of the tangents to $y = x^3$ from the external point $(-2, 0)$.

c Find the equation of the normal to $y = \sqrt{x}$ from the external point (4, 0).
Hint: There is no normal at the point where $x = 0$, as this is the endpoint of the function.

18 Find the equation of the tangent to $y = e^x$ at the point where $x = a$.
Hence, find the equation of the tangent to $y = e^x$ which passes through the origin.

19 Consider $f(x) = \dfrac{8}{x^2}$.

a Sketch the graph of the function.

b Find the equation of the tangent at the point where $x = a$.

c If the tangent in **b** cuts the x-axis at A and the y-axis at B, find the coordinates of A and B.

d Find the area of triangle OAB and discuss the area of the triangle as $a \to \infty$.

20 The graphs of $y = \sqrt{x + a}$ and $y = \sqrt{2x - x^2}$ have the same gradient at their point of intersection.
Find a and the point of intersection.

21 Suppose P is $(-2, 3)$ and Q is $(6, -3)$. The line (PQ) is a tangent to $y = \dfrac{b}{(x+1)^2}$. Find b.

22 Find, correct to 2 decimal places, the angle between the tangents to $y = 3e^{-x}$ and $y = 2 + e^x$ at their point of intersection.

23 A quadratic of the form $y = ax^2$, $a > 0$, touches the logarithmic function $y = \ln x$ as shown.

  a If the x-coordinate of the point of contact is b, explain why $ab^2 = \ln b$ and $2ab = \dfrac{1}{b}$.

  b Deduce that the point of contact is $(\sqrt{e}, \frac{1}{2})$.

  c Find the value of a.

  d Find the equation of the common tangent.

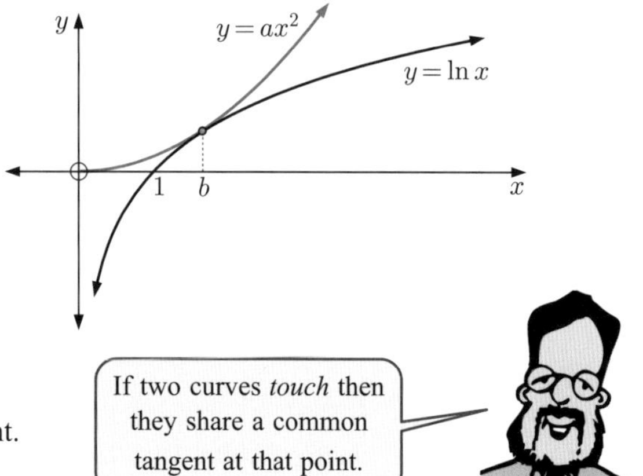
y
y = ax²
y = ln x
1
b
x
If two curves *touch* then they share a common tangent at that point.

24 **a** Find the tangents to the unit circle $x^2 + y^2 = 1$ at the points where $x = \frac{1}{2}$.

  b Find the point on the x-axis where these tangents intersect.

B INCREASING AND DECREASING FUNCTIONS

The concepts of increasing and decreasing are closely linked to **intervals** or subsets of a function's domain.

We commonly use the algebraic notation shown in the table to describe subsets of the real numbers corresponding to intervals of the real number line.

Algebraic form	*Geometric form*
$x \geqslant 2$	2 x
$x > 2$	2 x
$x \leqslant 4$	4 x
$x < 4$	4 x
$2 \leqslant x \leqslant 4$	2 4 x
$2 \leqslant x < 4$	2 4 x

Suppose S is an interval in the domain of $f(x)$, so $f(x)$ is defined for all x in S.

- $f(x)$ is **increasing** on S $\Leftrightarrow$ $f(a) \leqslant f(b)$ for all $a, b \in S$ such that $a < b$.
- $f(x)$ is **decreasing** on S $\Leftrightarrow$ $f(a) \geqslant f(b)$ for all $a, b \in S$ such that $a < b$.

For example:

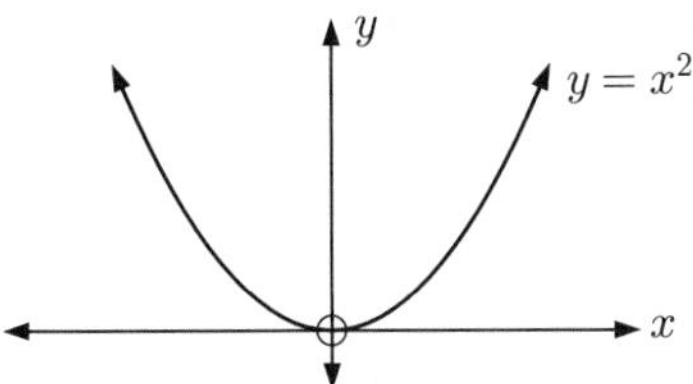

$y = x^2$ is decreasing for $x \leqslant 0$ and increasing for $x \geqslant 0$.

People often get confused about the point $x = 0$. They wonder how the curve can be both increasing and decreasing at the same point when it is clear that the tangent is horizontal. The answer is that increasing and decreasing are associated with *intervals*, not particular values for x. We must clearly state that $y = x^2$ is decreasing *on the interval* $x \leqslant 0$ and increasing *on the interval* $x \geqslant 0$.

We can determine intervals where a curve is increasing or decreasing by considering $f'(x)$ on the interval in question. For most functions that we deal with in this course:

- $f(x)$ is **increasing** on S $\Leftrightarrow$ $f'(x) \geqslant 0$ for all x in S
- $f(x)$ is **strictly increasing** on S $\Leftrightarrow$ $f'(x) > 0$ for all x in S
- $f(x)$ is **decreasing** on S $\Leftrightarrow$ $f'(x) \leqslant 0$ for all x in S
- $f(x)$ is **strictly decreasing** on S $\Leftrightarrow$ $f'(x) < 0$ for all x in S.

MONOTONICITY

Many functions are either increasing or decreasing for all $x \in \mathbb{R}$. We say these functions are **monotone increasing** or **monotone decreasing**.

For example:

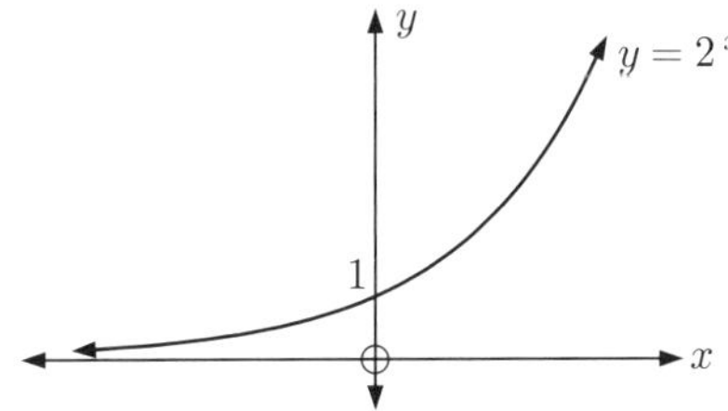

$y = 2^x$ is strictly increasing for all x.

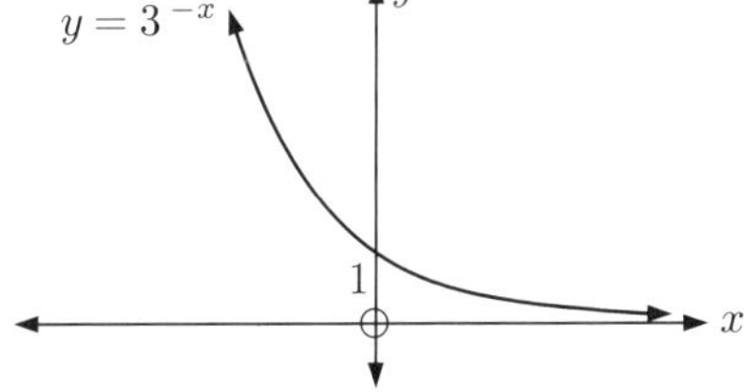

$y = 3^{-x}$ is strictly decreasing for all x.

The word "strictly" is not required for this course, but it is useful for understanding. It allows us to make statements like:

- for a **strictly increasing** function, an increase in x produces an increase in y
- for a **strictly decreasing** function, an increase in x produces a decrease in y.

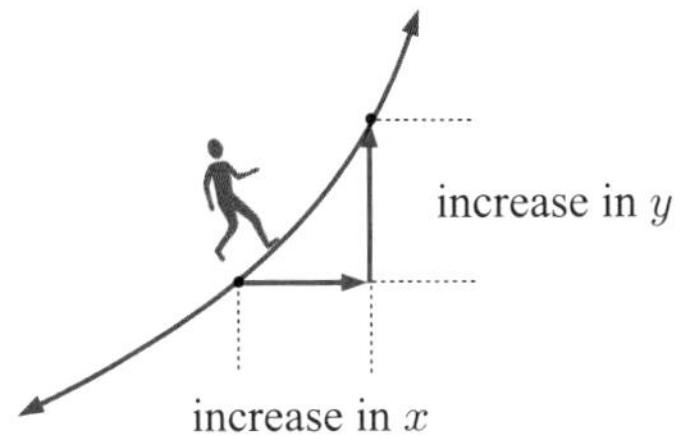

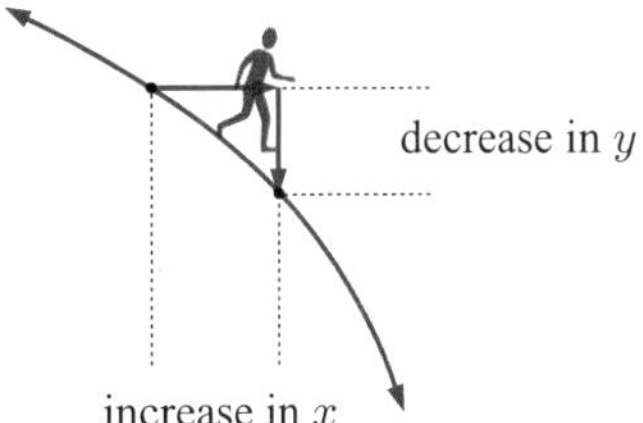

Example 8 **Self Tutor**

Find intervals where $f(x)$ is:

a increasing

b decreasing.

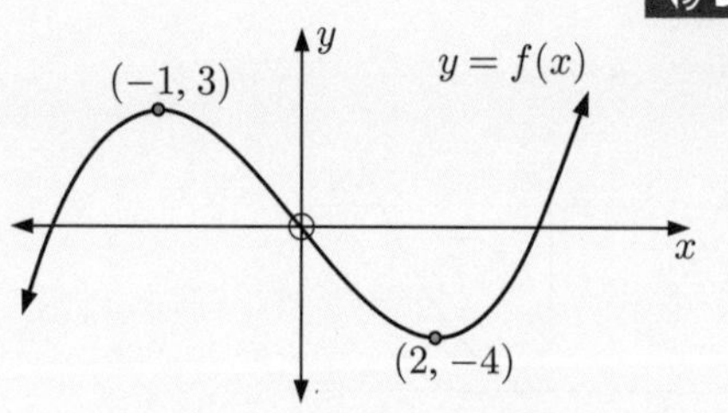

a $f(x)$ is increasing for $x \leqslant -1$ and for $x \geqslant 2$ since $f'(x) \geqslant 0$ on these intervals.

b $f(x)$ is decreasing for $-1 \leqslant x \leqslant 2$.

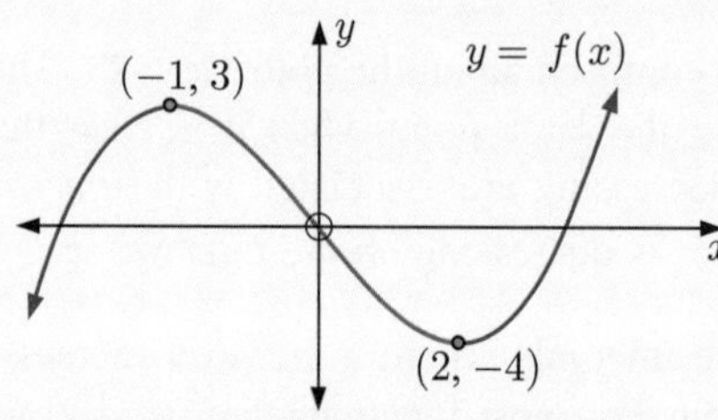

Sign diagrams for the derivative are extremely useful for determining intervals where a function is increasing or decreasing. Consider the following examples:

$f(x) = x^2$

DEMO

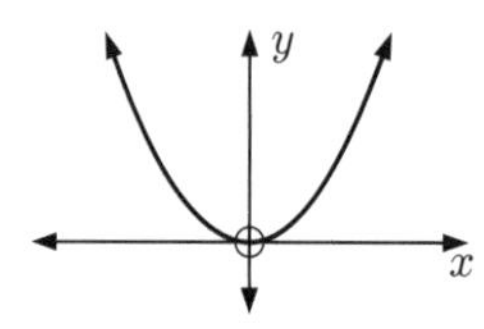

$f'(x) = 2x$ which has sign diagram

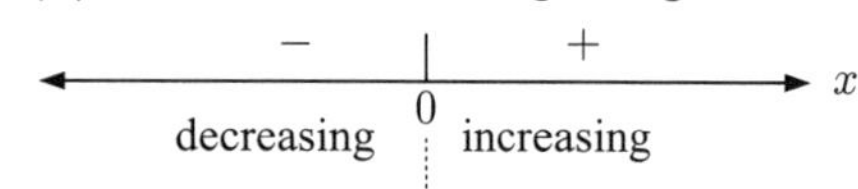

$\therefore \ f(x) = x^2$ is $\begin{cases} \text{decreasing for } x \leqslant 0 \\ \text{increasing for } x \geqslant 0. \end{cases}$

$f(x) = -x^2$

DEMO

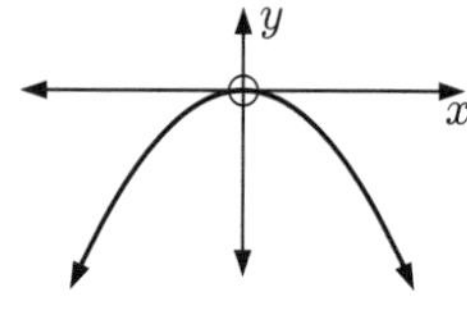

$f'(x) = -2x$ which has sign diagram

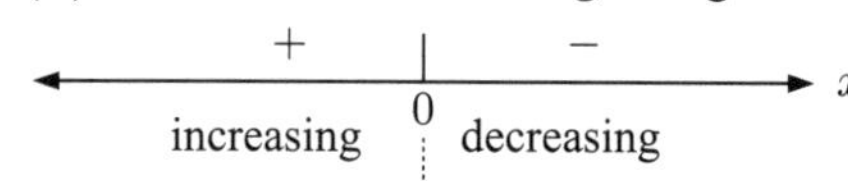

$\therefore \ f(x) = -x^2$ is $\begin{cases} \text{increasing for } x \leqslant 0 \\ \text{decreasing for } x \geqslant 0. \end{cases}$

$f(x) = x^3$

DEMO

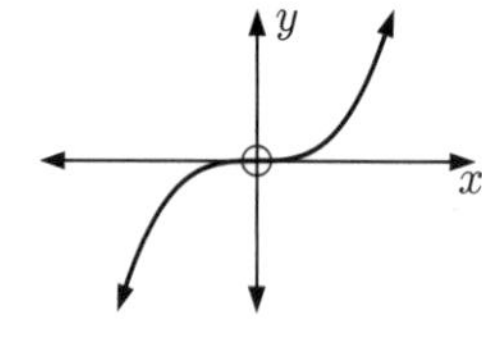

$f'(x) = 3x^2$ which has sign diagram

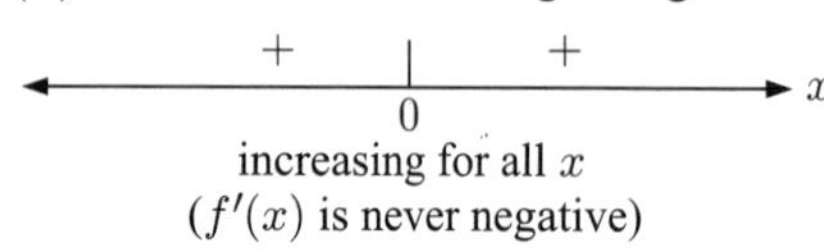

$\therefore \ f(x) = x^3$ is monotone increasing.

$f(x) = x^3 - 3x + 4$

DEMO

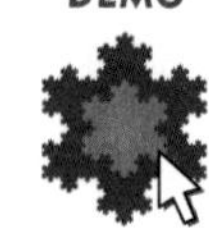

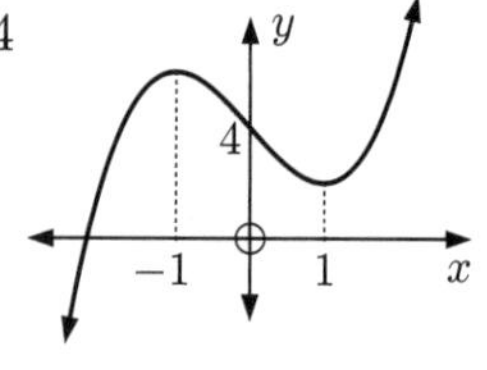

$$\begin{aligned} f'(x) &= 3x^2 - 3 \\ &= 3(x^2 - 1) \\ &= 3(x+1)(x-1) \end{aligned}$$

which has sign diagram

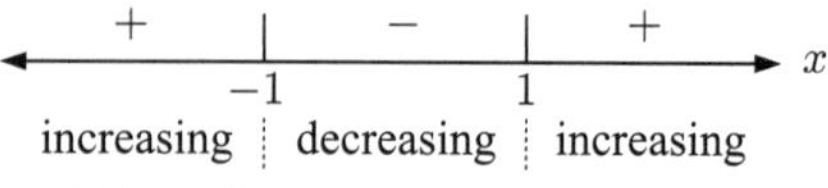

$\therefore \ f(x) = x^3 - 3x + 4$ is

$\begin{cases} \text{increasing for } x \leqslant -1 \text{ and for } x \geqslant 1 \\ \text{decreasing for } -1 \leqslant x \leqslant 1. \end{cases}$

Example 9 — Self Tutor

Find the intervals where the following functions are increasing or decreasing:

a $f(x) = -x^3 + 3x^2 + 5$ **b** $f(x) = 3x^4 - 8x^3 + 2$

a $f(x) = -x^3 + 3x^2 + 5$

$\therefore\ f'(x) = -3x^2 + 6x$

$= -3x(x-2)$

which has sign diagram

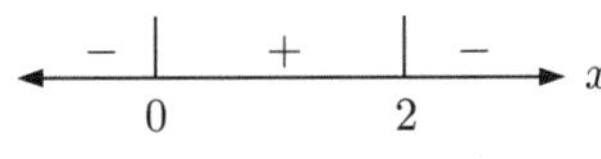

TI-84 Plus

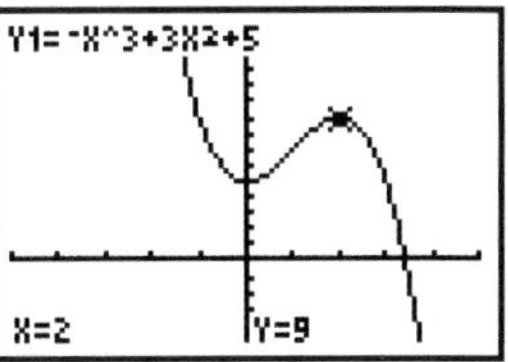

So, $f(x)$ is decreasing for $x \leqslant 0$ and for $x \geqslant 2$, and increasing for $0 \leqslant x \leqslant 2$.

b $f(x) = 3x^4 - 8x^3 + 2$

$\therefore\ f'(x) = 12x^3 - 24x^2$

$= 12x^2(x-2)$

which has sign diagram

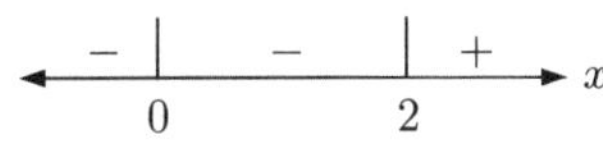

Casio fx-CG20

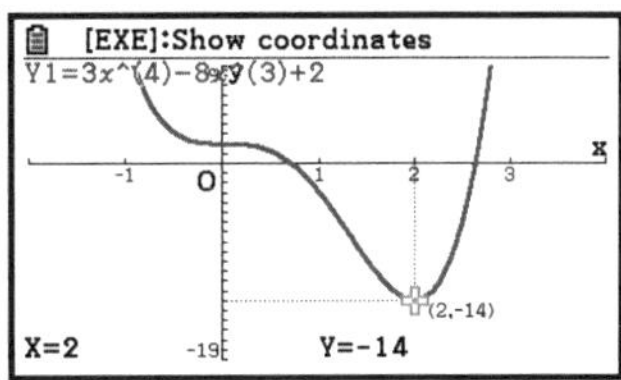

So, $f(x)$ is decreasing for $x \leqslant 2$, and increasing for $x \geqslant 2$.

Remember that $f(x)$ must be defined for all x on an interval before we can classify the function as increasing or decreasing on that interval. We need to take care with vertical asymptotes and other values for x where the function is not defined.

EXERCISE 19B

1 Write down the intervals where the graphs are: **i** increasing **ii** decreasing.

a

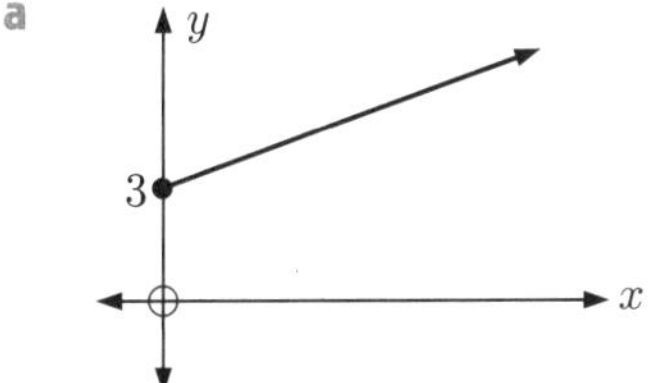

b

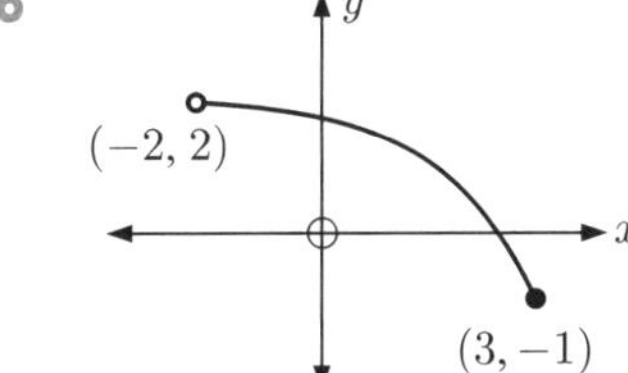

c

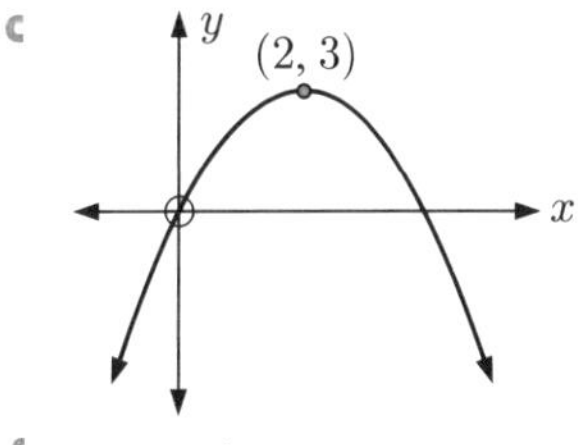

d

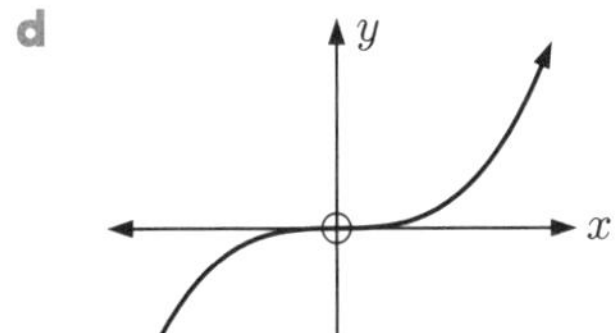

e

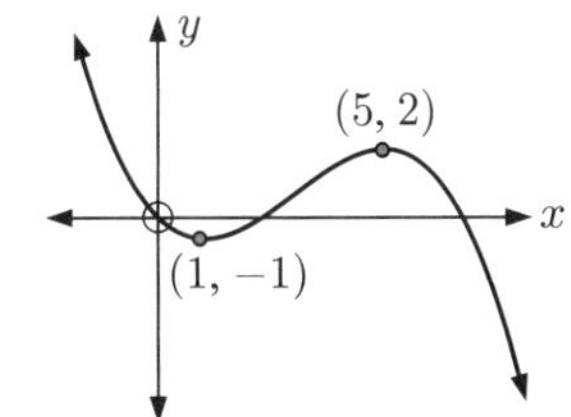

f

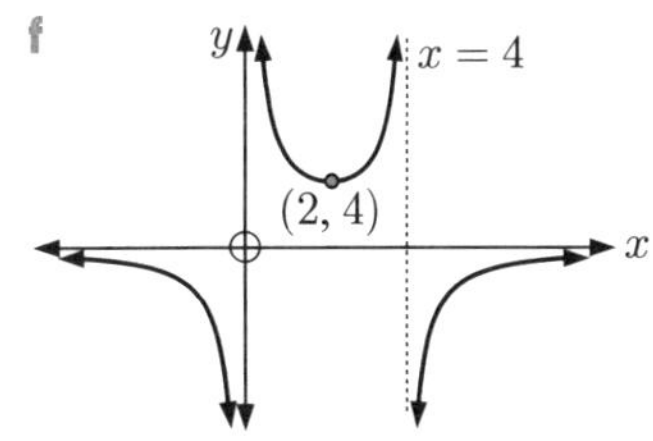

2 Find the intervals where $f(x)$ is increasing or decreasing:

a $f(x) = x^2$ **b** $f(x) = -x^3$ **c** $f(x) = \sqrt{x}$ **d** $f(x) = 2x^2 + 3x - 4$

e $f(x) = \dfrac{2}{\sqrt{x}}$ **f** $f(x) = x^3 - 6x^2$ **g** $f(x) = e^x$ **h** $f(x) = \ln x$

i $f(x) = -2x^3 + 4x$ **j** $f(x) = 3 + e^{-x}$ **k** $f(x) = xe^x$ **l** $f(x) = x - 2\sqrt{x}$

3 Find the intervals where $f(x)$ is increasing or decreasing:

a $f(x) = -4x^3 + 15x^2 + 18x + 3$

b $f(x) = 3x^4 - 16x^3 + 24x^2 - 2$

c $f(x) = 2x^3 + 9x^2 + 6x - 7$

d $f(x) = x^3 - 6x^2 + 3x - 1$

Example 10 **Self Tutor**

Consider $f(x) = \dfrac{2x-3}{x^2+2x-3}$.

a Show that $f'(x) = \dfrac{-2x(x-3)}{(x+3)^2(x-1)^2}$ and draw its sign diagram.

b Hence, find the intervals where $y = f(x)$ is increasing or decreasing.

a
$$f(x) = \frac{2x-3}{x^2+2x-3}$$
$$\therefore\ f'(x) = \frac{2(x^2+2x-3) - (2x-3)(2x+2)}{(x^2+2x-3)^2} \qquad \{\text{quotient rule}\}$$
$$= \frac{2x^2+4x-6-[4x^2-2x-6]}{((x-1)(x+3))^2}$$
$$= \frac{-2x^2+6x}{(x-1)^2(x+3)^2}$$
$$= \frac{-2x(x-3)}{(x-1)^2(x+3)^2}$$
which has sign diagram:

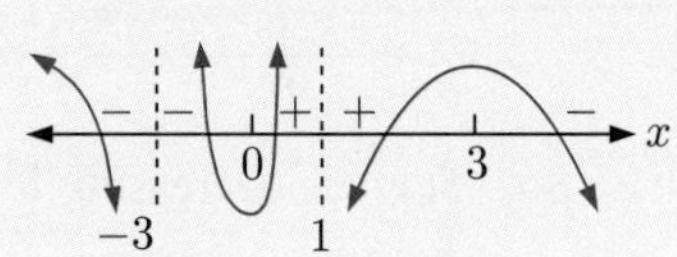

b $f(x)$ is increasing for $0 \leqslant x < 1$
and for $1 < x \leqslant 3$.

$f(x)$ is decreasing for $x < -3$
and for $-3 < x \leqslant 0$
and for $x \geqslant 3$.

TI-*n*spire

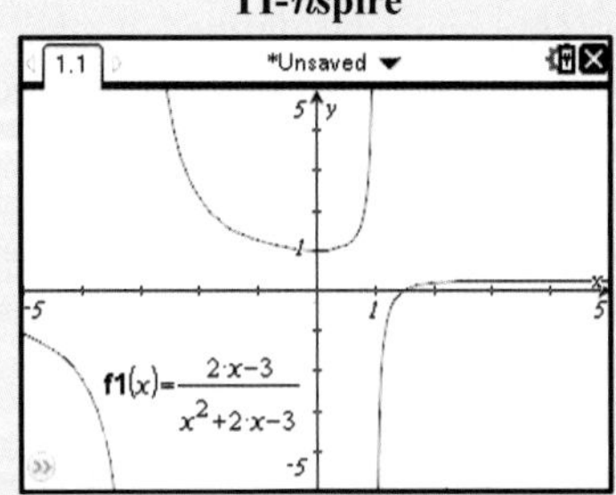

4 Consider $f(x) = \dfrac{4x}{x^2+1}$.

a Show that $f'(x) = \dfrac{-4(x+1)(x-1)}{(x^2+1)^2}$ and draw its sign diagram.

b Hence, find intervals where $y = f(x)$ is increasing or decreasing.

5 Consider $f(x) = \dfrac{4x}{(x-1)^2}$.

a Show that $f'(x) = \dfrac{-4(x+1)}{(x-1)^3}$ and draw its sign diagram.

b Hence, find intervals where $y = f(x)$ is increasing or decreasing.

6 Consider $f(x) = \dfrac{-x^2+4x-7}{x-1}$.

a Show that $f'(x) = \dfrac{-(x+1)(x-3)}{(x-1)^2}$ and draw its sign diagram.

b Hence, find intervals where $y = f(x)$ is increasing or decreasing.

7 Discuss intervals where the following functions are increasing or decreasing:

a $y = \arccos x$ b $y = \arcsin x$ c $y = \arctan x$

d $y = \cos x$ e $y = \sin x$ f $y = \tan x$

8 Find intervals where $f(x)$ is increasing or decreasing if:

a $f(x) = \dfrac{x^3}{x^2 - 1}$ b $f(x) = e^{-x^2}$ c $f(x) = x^2 + \dfrac{4}{x-1}$

d $f(x) = \dfrac{e^{-x}}{x}$ e $f(x) = x + \cos x$

C STATIONARY POINTS

A **stationary point** of a function is a point where $f'(x) = 0$.

It could be a local maximum, local minimum, or stationary inflection.

TURNING POINTS (MAXIMA AND MINIMA)

Consider the following graph which has a restricted domain of $-5 \leqslant x \leqslant 6$.

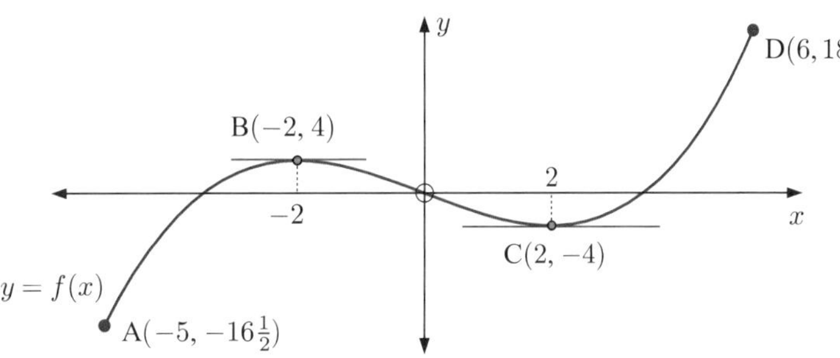

A is a **global minimum**. It is the minimum value of y on the entire domain.

B is a **local maximum**. It is a turning point where $f'(x) = 0$ and the curve has shape ∩.

C is a **local minimum**. It is a turning point where $f'(x) = 0$ and the curve has shape ∪.

D is a **global maximum**. It is the maximum value of y on the entire domain.

For many functions, a local maximum or minimum is also the global maximum or minimum.

For example, for $y = x^2$ the point (0, 0) is a local minimum and is also the global minimum.

STATIONARY POINTS OF INFLECTION

It is not always true that whenever we find a value of x where $f'(x) = 0$, we have a local maximum or minimum.

For example, $f(x) = x^3$ has $f'(x) = 3x^2$,
so $f'(x) = 0$ when $x = 0$.

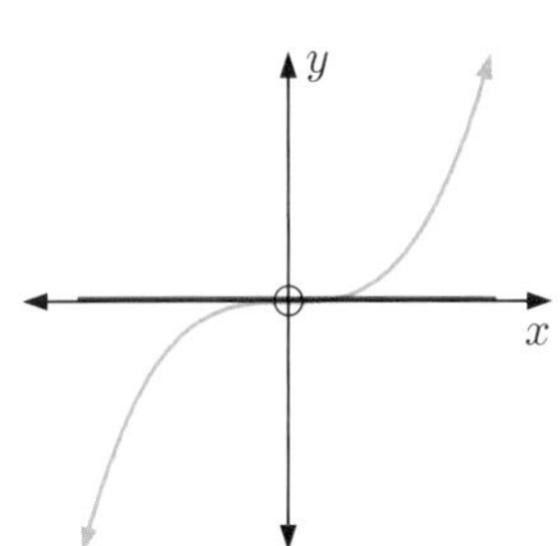

The x-axis is a tangent to the curve which actually crosses over the curve at O(0, 0). This tangent is horizontal but O(0, 0) is neither a local maximum nor a local minimum.

It is called a **stationary inflection** (or **inflexion**) as the curve changes its curvature or shape.

SIGN DIAGRAMS

Consider the graph alongside.

The sign diagram of its gradient function is shown directly beneath it.

We can use the sign diagram to describe the stationary points of the function.

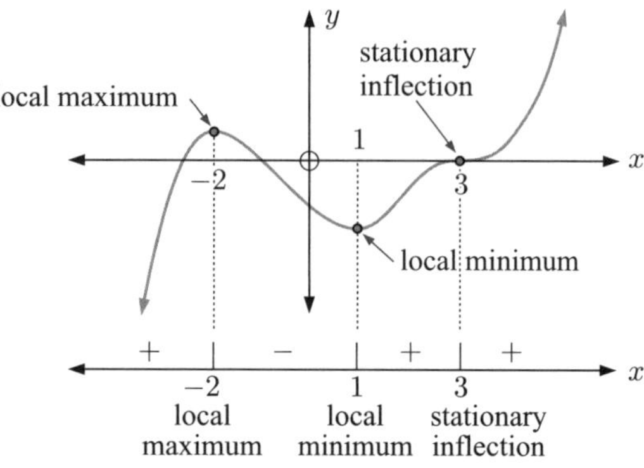

Stationary point where $f'(a) = 0$	*Sign diagram of* $f'(x)$ *near* $x = a$	*Shape of curve near* $x = a$
local maximum	$+$ | $-$ at a, x	$x = a$
local minimum	$-$ | $+$ at a, x	$x = a$
stationary inflection	$+$ | $+$ at a, x or $-$ | $-$ at a, x	$x = a$ or $x = a$

Example 11 **Self Tutor**

Find and classify all stationary points of $f(x) = x^3 - 3x^2 - 9x + 5$.

$$f(x) = x^3 - 3x^2 - 9x + 5$$

$$\therefore \quad f'(x) = 3x^2 - 6x - 9$$

$$= 3(x^2 - 2x - 3)$$

$$= 3(x - 3)(x + 1) \quad \text{which has sign diagram:}$$

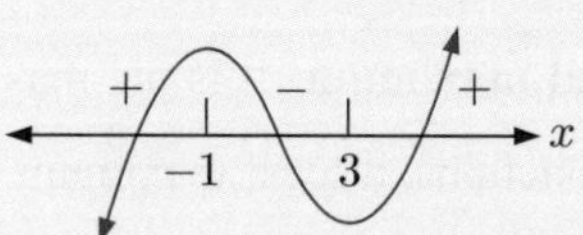

So, we have a local maximum at $x = -1$ and a local minimum at $x = 3$.

$$f(-1) = (-1)^3 - 3(-1)^2 - 9(-1) + 5 = 10$$

$$f(3) = 3^3 - 3 \times 3^2 - 9 \times 3 + 5 = -22$$

TI-84 Plus

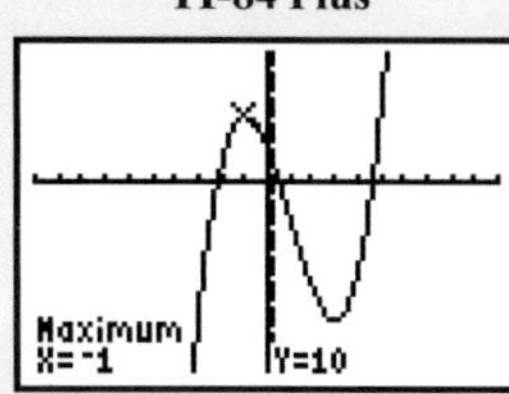

There is a local maximum at $(-1, 10)$.

Casio fx-CG20

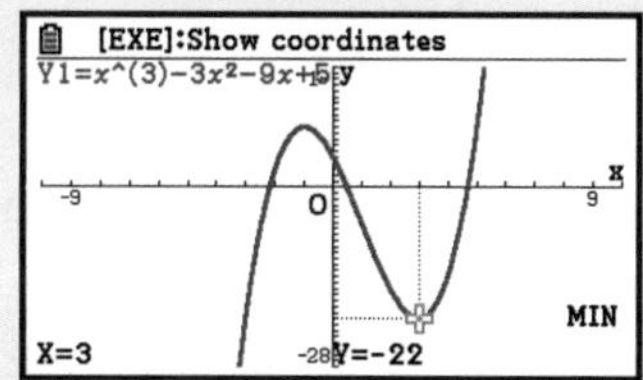

There is a local minimum at $(3, -22)$.

Example 12 **Self Tutor**

Find the exact position and nature of the stationary point of $y = (x-2)e^{-x}$.

$\dfrac{dy}{dx} = (1)e^{-x} + (x-2)e^{-x}(-1)$ {product rule}

$= e^{-x}(1-(x-2))$

$= \dfrac{3-x}{e^x}$ where e^x is positive for all x.

So, $\dfrac{dy}{dx} = 0$ when $x = 3$.

The sign diagram of $\dfrac{dy}{dx}$ is:

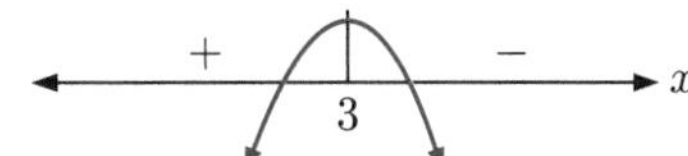

$\therefore$ at $x = 3$ we have a local maximum.

But when $x = 3$, $y = (1)e^{-3} = \dfrac{1}{e^3}$

$\therefore$ the local maximum is at $(3, \frac{1}{e^3})$.

If we are asked to find the greatest or least value of a function on an interval, then we must also check the value of the function at the endpoints. We seek the *global* maximum or minimum on the given domain.

Example 13 **Self Tutor**

Find the greatest and least value of $y = x^3 - 6x^2 + 5$ on the interval $-2 \leqslant x \leqslant 5$.

Now $\dfrac{dy}{dx} = 3x^2 - 12x$

$= 3x(x-4)$

$\therefore \dfrac{dy}{dx} = 0$ when $x = 0$ or 4.

The sign diagram of $\dfrac{dy}{dx}$ is:

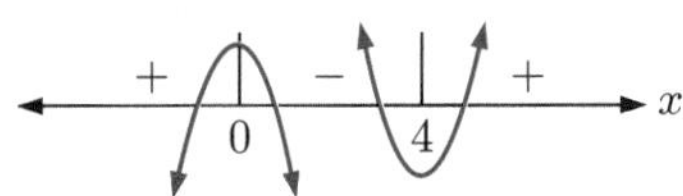

TI-nspire

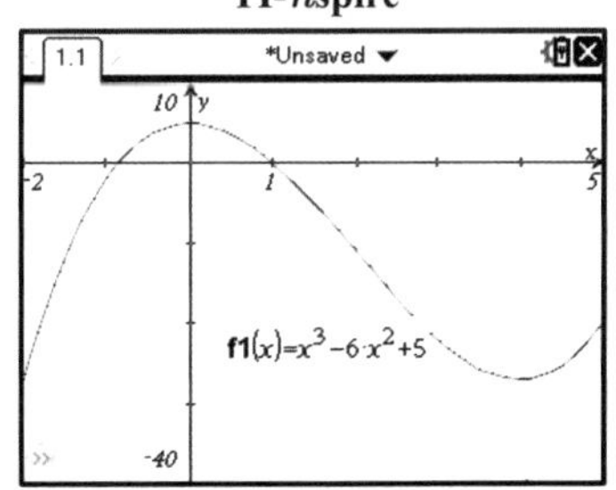

$\therefore$ there is a local maximum at $x = 0$, and a local minimum at $x = 4$.

Critical value (x)	$f(x)$
-2 (end point)	-27
0 (local max)	5
4 (local min)	-27
5 (end point)	-20

The greatest of these values is 5 when $x = 0$.

The least of these values is -27 when $x = -2$ and when $x = 4$.

EXERCISE 19C

1 The tangents at points A, B and C are horizontal.

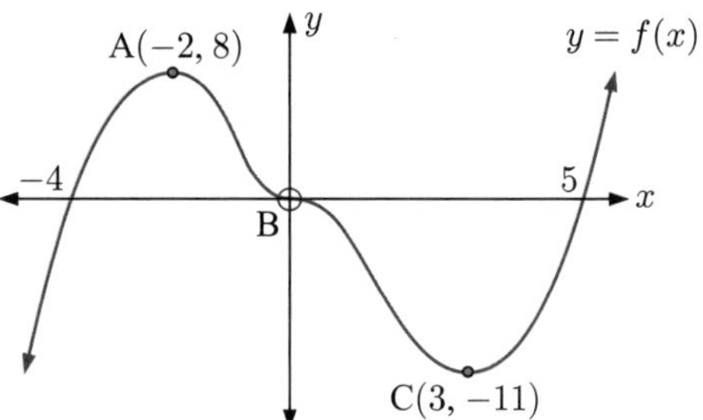

a Classify points A, B and C.

b Draw a sign diagram for the gradient function $f'(x)$ for all x.

c State intervals where $y = f(x)$ is:

i increasing **ii** decreasing.

d Draw a sign diagram for $f(x)$ for all x.

2 For each of the following functions, find and classify any stationary points. Sketch the function, showing all important features.

a $f(x) = x^2 - 2$ **b** $f(x) = x^3 + 1$ **c** $f(x) = x^3 - 3x + 2$

d $f(x) = x^4 - 2x^2$ **e** $f(x) = \sqrt{x} + 2$ **f** $f(x) = x^3 - 6x^2 + 12x + 1$

g $f(x) = x - \sqrt{x}$ **h** $f(x) = 1 - x\sqrt{x}$ **i** $f(x) = x^4 - 6x^2 + 8x - 3$

j $f(x) = x^4 - 2x^2 - 8$

3 At what value of x does the quadratic function $f(x) = ax^2 + bx + c$, $a \neq 0$, have a stationary point? Under what conditions is the stationary point a local maximum or a local minimum?

4 Find the position and nature of the stationary point(s) of:

a $y = xe^{-x}$ **b** $y = x^2e^x$ **c** $y = \dfrac{e^x}{x}$ **d** $y = e^{-x}(x + 2)$

5 $f(x) = 2x^3 + ax^2 - 24x + 1$ has a local maximum at $x = -4$. Find a.

6 $f(x) = x^3 + ax + b$ has a stationary point at $(-2, 3)$.

a Find the values of a and b.

b Find the position and nature of all stationary points.

7 Consider $f(x) = x \ln x$.

a For what values of x is $f(x)$ defined?

b Show that the global minimum value of $f(x)$ is $-\dfrac{1}{e}$.

8 For each of the following, determine the position and nature of the stationary points on the interval $0 \leqslant x \leqslant 2\pi$, then show them on a graph of the function.

a $f(x) = \sin x$ **b** $f(x) = \cos(2x)$ **c** $f(x) = \sin^2 x$

d $f(x) = e^{\sin x}$ **e** $f(x) = \sin(2x) + 2\cos x$

9 The cubic polynomial $P(x) = ax^3 + bx^2 + cx + d$ touches the line with equation $y = 9x + 2$ at the point $(0, 2)$, and has a stationary point at $(-1, -7)$. Find $P(x)$.

10 Find the greatest and least value of:

a $x^3 - 12x - 2$ for $-3 \leqslant x \leqslant 5$ **b** $4 - 3x^2 + x^3$ for $-2 \leqslant x \leqslant 3$

11 Show that $y = 4e^{-x}\sin x$ has a local maximum when $x = \frac{\pi}{4}$.

12 $f(t) = ate^{bt^2}$ has a maximum value of 1 when $t = 2$. Find constants a and b.

13 Prove that $\dfrac{\ln x}{x} \leqslant \dfrac{1}{e}$ for all $x > 0$.

14 Consider the function $f(x) = x - \ln x$.

- **a** Show that $y = f(x)$ has a local minimum and that this is the only turning point.
- **b** Hence prove that $\ln x \leqslant x - 1$ for all $x > 0$.

15 Consider the function $f(x) = \sec x$.

- **a** For what values of x is $f(x)$ undefined on the interval $0 \leqslant x \leqslant 2\pi$?
- **b** Find the position and nature of any stationary points on the interval $0 \leqslant x \leqslant 2\pi$.
- **c** Prove that $f(x+2\pi) = f(x)$ for all $x \in \mathbb{R}$ where $f(x)$ is defined. Explain the geometrical significance of this result.
- **d** Sketch the graph of $y = \sec x$, $x \in [-\frac{\pi}{2}, \frac{5\pi}{2}]$ and show its stationary points.

INFLECTIONS AND SHAPE

When a curve, or part of a curve, has shape:

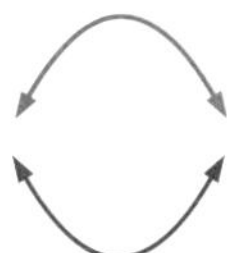

we say that the shape is **concave downwards**

we say that the shape is **concave upwards**.

TEST FOR SHAPE

Consider the **concave downwards** curve:

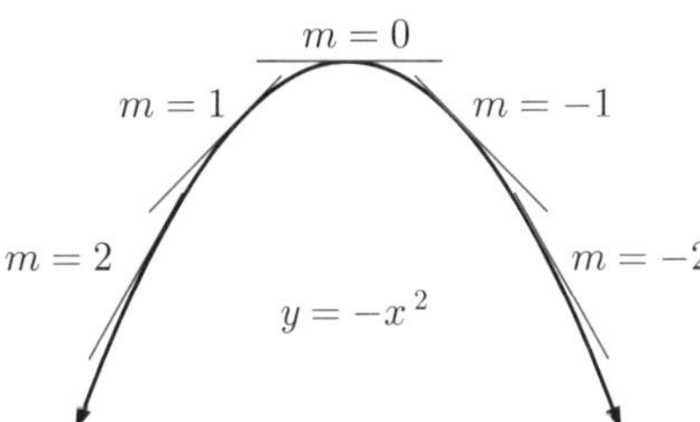

Wherever we are on the curve, as x increases, the gradient of the tangent decreases.

$\therefore$ $f'(x)$ is decreasing

$\therefore$ its derivative is negative, which means $f''(x) < 0$.

Likewise, if the curve is **concave upwards**:

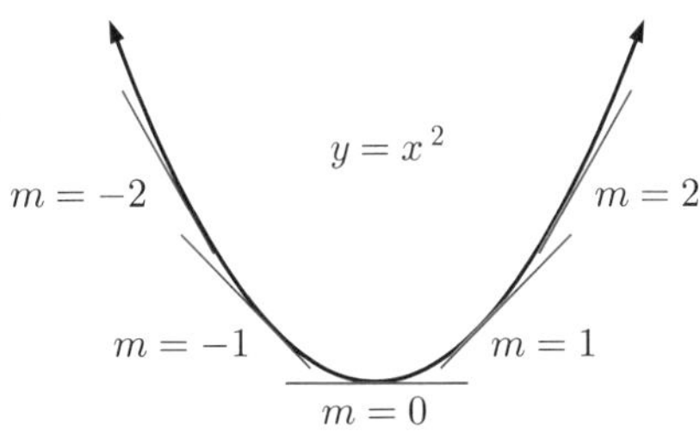

Wherever we are on the curve, as x increases, the gradient of the tangent increases.

$\therefore$ $f'(x)$ is increasing

$\therefore$ its derivative is positive, which means $f''(x) > 0$.

POINTS OF INFLECTION (INFLEXION)

A **point of inflection** is a point on a curve at which there is a change of **curvature** or shape.

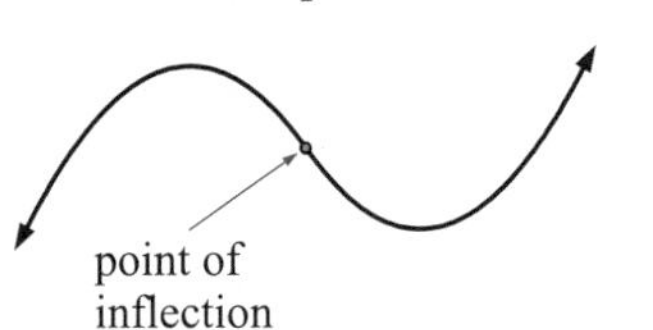

or

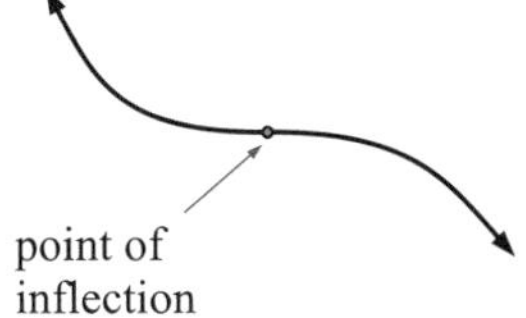

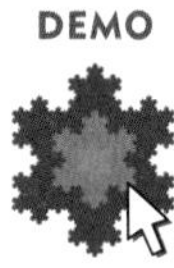

If the tangent at a point of inflection is horizontal then this point is a **horizontal** or **stationary inflection**.

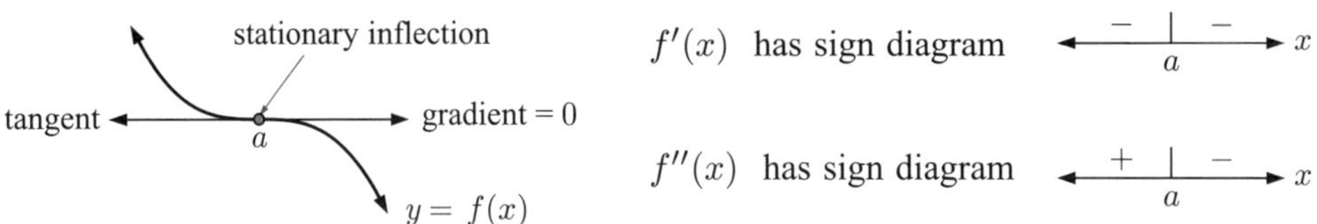

If the tangent at a point of inflection is not horizontal, we have a **non-horizontal** or **non-stationary inflection**.

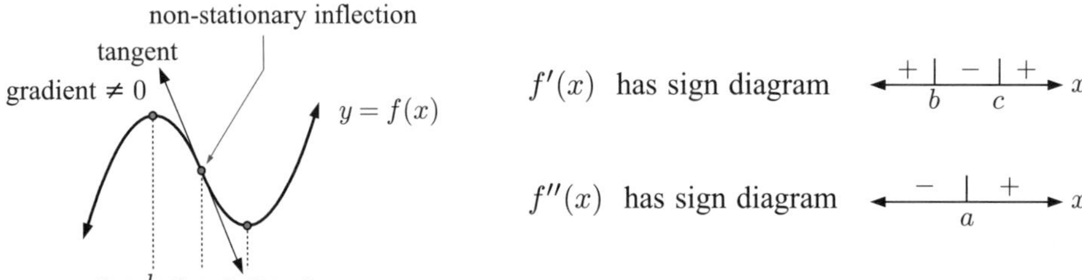

The tangent at the point of inflection, also called the **inflecting tangent**, crosses the curve at that point.

There is a **point of inflection** at $x = a$ if $f''(a) = 0$ **and** the sign of $f''(x)$ changes at $x = a$.

The point of inflection corresponds to a change in **curvature**.

In the vicinity of a, $f''(x)$ has sign diagram either $+ \mid -$ (at a) or $- \mid +$ (at a)

Observe that if $f(x) = x^4$ then $f'(x) = 4x^3$
and $f''(x) = 12x^2$ and $f''(x)$ has sign diagram $+ \mid +$ (at 0)

Although $f''(0) = 0$ we do not have a point of inflection at $(0, 0)$ because the sign of $f''(x)$ does not change at $x = 0$.
In fact, the graph of $f(x) = x^4$ is:

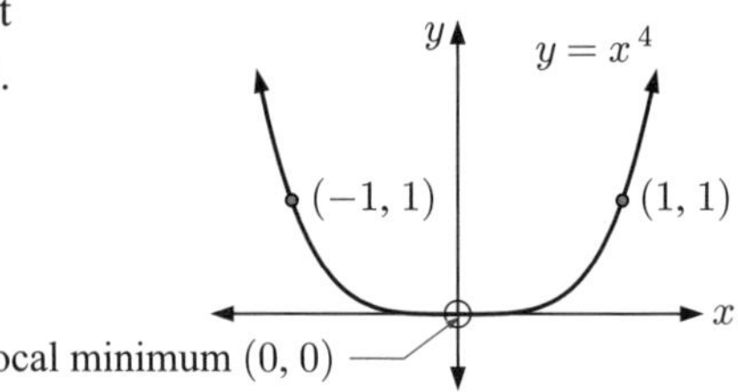

SUMMARY

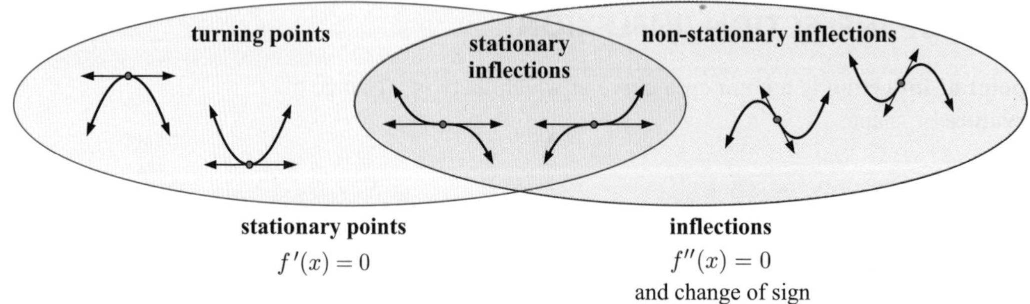

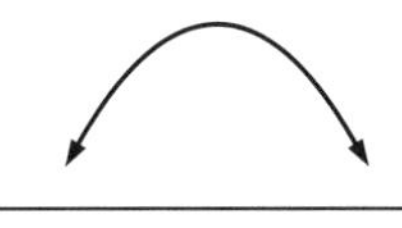	If a curve is **concave downwards** on an interval S then $f''(x) \leqslant 0$ for all x in S.
	If a curve is **concave upwards** on an interval S then $f''(x) \geqslant 0$ for all x in S.

If $f''(x)$ changes sign at $x = a$, and $f''(a) = 0$, then we have a

- **stationary inflection** if $f'(a) = 0$
- **non-stationary inflection** if $f'(a) \neq 0$.

Click on the demo icon to examine some standard functions for turning points, points of inflection, and intervals where the function is increasing, decreasing, and concave up or down.

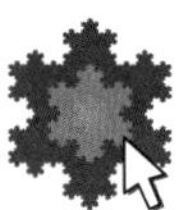

Example 14 **Self Tutor**

Find and classify all points of inflection of $f(x) = x^4 - 4x^3 + 5$.

$f(x) = x^4 - 4x^3 + 5$

$\therefore \ f'(x) = 4x^3 - 12x^2 = 4x^2(x-3)$

$\therefore \ f''(x) = 12x^2 - 24x$

$= 12x(x-2)$

$\therefore \ f''(x) = 0$ when $x = 0$ or 2

$f'(x)$: $-$ | 0 | $-$ | 3 | $+$

$f''(x)$: $+$ (concave up) | 0 | $-$ (concave down) | 2 | $+$ (concave up)

Since the signs of $f''(x)$ change about $x = 0$ and $x = 2$, these two points are points of inflection.

Now $f'(0) = 0$, $f'(2) = 32 - 48 \neq 0$

and $f(0) = 5$, $f(2) = 16 - 32 + 5 = -11$

Thus $(0, 5)$ is a stationary inflection, and $(2, -11)$ is a non-stationary inflection.

EXERCISE 19D.1

1 a In the diagram shown, B and D are stationary points, and C is a point of inflection. Complete the table by indicating whether each value is zero, positive, or negative:

Point	$f(x)$	$f'(x)$	$f''(x)$
A	$+$		
B			
C			
D		0	
E			

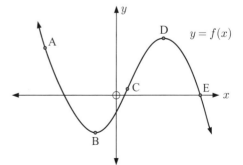

b Describe the turning points of $y = f(x)$.

c Describe the inflection point of $y = f(x)$.

2 Find and classify all points of inflection of:

a $f(x) = x^2 + 3$

b $f(x) = 2 - x^3$

c $f(x) = x^3 - 6x^2 + 9x + 1$

d $f(x) = -3x^4 - 8x^3 + 2$

e $f(x) = 3 - \dfrac{1}{\sqrt{x}}$

f $f(x) = x^3 + 6x^2 + 12x + 5$

Example 15 **Self Tutor**

Consider $f(x) = 3x^4 - 16x^3 + 24x^2 - 9$.

a Find and classify all points where $f'(x) = 0$.

b Find and classify all points of inflection.

c Find intervals where the function is increasing or decreasing.

d Find intervals where the function is concave up or down.

e Sketch the function showing the features you have found.

---

a $f(x) = 3x^4 - 16x^3 + 24x^2 - 9$

$\therefore \quad f'(x) = 12x^3 - 48x^2 + 48x$

$= 12x(x^2 - 4x + 4)$

$= 12x(x-2)^2$

$\therefore$ $f'(x)$ has sign diagram:

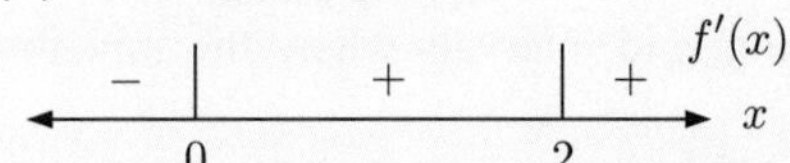

Now $f(0) = -9$ and $f(2) = 7$

$\therefore$ $(0, -9)$ is a local minimum and $(2, 7)$ is a stationary inflection.

b $f''(x) = 36x^2 - 96x + 48$

$= 12(3x^2 - 8x + 4)$

$= 12(x-2)(3x-2)$

$\therefore$ $f''(x)$ has sign diagram:

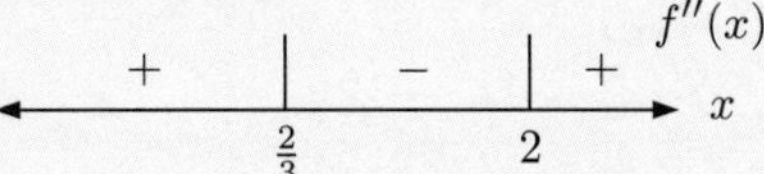

Now $f(\frac{2}{3}) \approx -2.48$

$\therefore$ $(2, 7)$ is a stationary inflection and $(\frac{2}{3}, -2.48)$ is a non-stationary inflection.

c $f(x)$ is decreasing for $x \leqslant 0$

$f(x)$ is increasing for $x \geqslant 0$.

d $f(x)$ is concave up for $x \leqslant \frac{2}{3}$ and $x \geqslant 2$

$f(x)$ is concave down for $\frac{2}{3} \leqslant x \leqslant 2$.

e

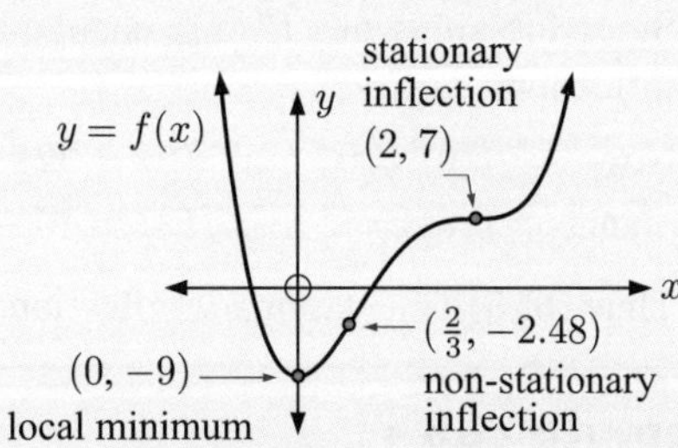

3 For each of the following functions:

i Find and classify all points where $f'(x) = 0$

ii Find and classify all points of inflection

iii Find intervals where the function is increasing or decreasing

iv Find intervals where the function is concave up or down

v Sketch the function showing the features you have found.

a $f(x) = x^2$

b $f(x) = x^3$

c $f(x) = \sqrt{x}$

d $f(x) = x^3 - 3x^2 - 24x + 1$

e $f(x) = 3x^4 + 4x^3 - 2$

f $f(x) = (x-1)^4$

g $f(x) = x^4 - 4x^2 + 3$

h $f(x) = 3 - \dfrac{4}{\sqrt{x}}$

Example 16 — **Self Tutor**

Consider the function $y = 2 - e^{-x}$.

a Find the x-intercept.

b Find the y-intercept.

c Show algebraically that the function is increasing for all x.

d Show algebraically that the function is concave down for all x.

e Use technology to help graph $y = 2 - e^{-x}$.

f Explain why $y = 2$ is a horizontal asymptote.

---

a The x-intercept occurs when $y = 0$

$\therefore\ e^{-x} = 2$

$\therefore\ -x = \ln 2$

$\therefore\ x = -\ln 2$

$\therefore$ the x-intercept is $-\ln 2 \approx -0.693$

b The y-intercept occurs when $x = 0$

$\therefore\ y = 2 - e^0 = 2 - 1 = 1$

c $\dfrac{dy}{dx} = 0 - e^{-x}(-1) = e^{-x} = \dfrac{1}{e^x}$

Now $e^x > 0$ for all x,

so $\dfrac{dy}{dx} > 0$ for all x.

$\therefore$ the function is increasing for all x.

d $\dfrac{d^2y}{dx^2} = e^{-x}(-1)$

$= \dfrac{-1}{e^x}$ which is < 0 for all x.

$\therefore$ the function is concave down for all x.

e

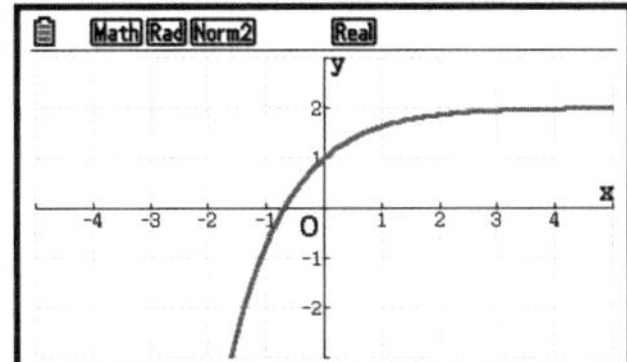

TI-84 Plus

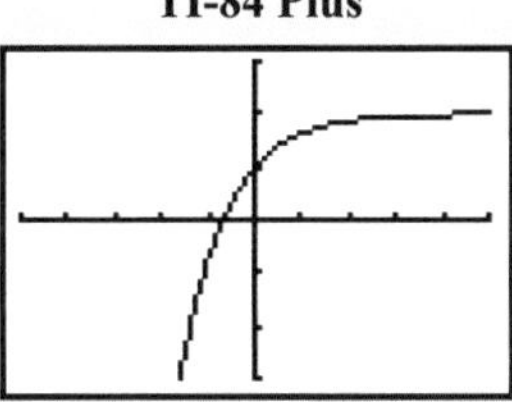

TI-*n*spire

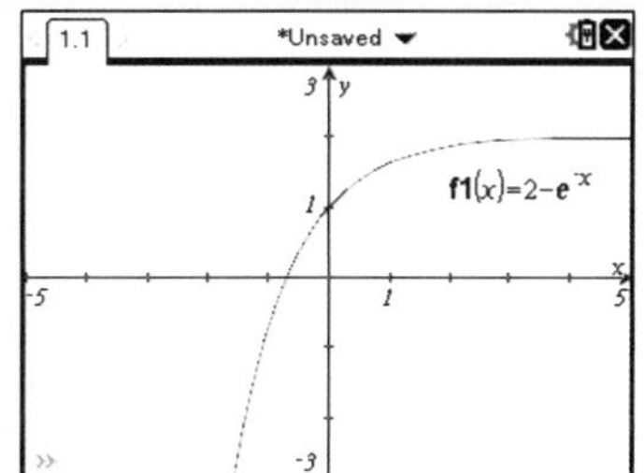

f As $x \to \infty$, $e^x \to \infty$ and $e^{-x} \to 0$

$\therefore\ y \to 2^-$

Hence, the horizontal asymptote is $y = 2$.

4 The function $f(x) = e^{2x} - 3$ cuts the x-axis at A and the y-axis at B.

a Find the coordinates of A and B.

b Show algebraically that the function is increasing for all x.

c Find $f''(x)$ and hence explain why $f(x)$ is concave up for all x.

d Use technology to help graph $y = e^{2x} - 3$.

e Explain why $y = -3$ is a horizontal asymptote.

5 Suppose $f(x) = e^x - 3$ and $g(x) = 3 - 5e^{-x}$ where $-3 \leqslant x \leqslant 4$.

a Find the x and y-intercepts of both functions.

b Discuss $f(x)$ and $g(x)$ as $x \to \infty$ and as $x \to -\infty$.

c Find algebraically the point(s) of intersection of the functions.

d Sketch the graph of both functions on the same set of axes. Show all important features on your graph.

6 The function $y = e^x - 3e^{-x}$ cuts the x-axis at P and the y-axis at Q.

a Determine the coordinates of P and Q.

b Prove that the function is increasing for all x.

c Show that $\frac{d^2y}{dx^2} = y$. What can be deduced about the concavity of the function above and below the x-axis?

d Use technology to help graph $y = e^x - 3e^{-x}$.
Show the features of **a**, **b**, and **c** on the graph.

7 Consider $f(x) = \ln(2x - 1) - 3$.

a Find the x-intercept.

b Can $f(0)$ be found? What is the significance of this result?

c Find the gradient of the tangent to the curve at $x = 1$.

d Find the domain of f.

e Find $f''(x)$ and hence explain why $f(x)$ is concave down for all x in the domain of f.

f Graph the function, showing the features you have found.

8 Consider $f(x) = \ln x$.

a For what values of x is $f(x)$ defined?

b Find the signs of $f'(x)$ and $f''(x)$ and comment on the geometrical significance of each.

c Sketch the graph of $f(x) = \ln x$ and find the equation of the normal at the point where $y = 1$.

9 Consider the function $f(x) = \frac{e^x}{x}$.

a Does the graph of $y = f(x)$ have any x or y-intercepts?

b Discuss $f(x)$ as $x \to \infty$ and as $x \to -\infty$.

c Find and classify any stationary points of $y = f(x)$.

d Find the intervals where $f(x)$ is: **i** concave up **ii** concave down.

e Sketch the graph of $y = f(x)$ showing all important features.

f Find the equation of the tangent to $f(x) = \frac{e^x}{x}$ at the point where $x = -1$.

10 A function commonly used in statistics is the *normal distribution function* $f(x) = \frac{1}{\sqrt{2\pi}}e^{-\frac{1}{2}x^2}$.

a Find the stationary points of the function and find the intervals where the function is increasing and decreasing.

b Find all points of inflection.

c Discuss $f(x)$ as $x \to \infty$ and as $x \to -\infty$.

d Sketch the graph of $y = f(x)$ showing all important features.

11 Consider the function $y = 4^x - 2^x$:

a Find the axes intercepts.

b Discuss the graph as $x \to \infty$ and as $x \to -\infty$.

c Find the position and nature of any stationary points.

d Discuss the concavity of the function.

e Sketch the function, showing the features you have found.

12 Consider the **surge function** $f(t) = Ate^{-bt}$, $t \geqslant 0$, where A and b are positive constants.

  a Prove that it has:

    i a local maximum at $t = \dfrac{1}{b}$ **ii** a point of inflection at $t = \dfrac{2}{b}$.

  b Sketch the function, showing the features you have found.

13 Consider the **logistic function** $f(t) = \dfrac{C}{1 + Ae^{-bt}}$, $t \geqslant 0$, where A, b, and C are positive constants.

  a Find the y-intercept.

  b Prove that:

    i $y = C$ is its horizontal asymptote

    ii if $A > 1$, there is a point of inflection with y-coordinate $\dfrac{C}{2}$.

  c Sketch the function, showing the features you have found.

ZEROS OF $f'(x)$ AND $f''(x)$

Suppose $y = f(x)$ has an inflection point at $x = a$.

$\therefore$ $f''(a) = 0$ and thus the derivative function $f'(x)$ has a stationary point at $x = a$.

Since $f''(x)$ changes sign at $x = a$, the stationary point of $f'(x)$ is a local maximum or local minimum.

If, in addition, $f'(a) = 0$, then the local maximum or local minimum of $f'(x)$ lies on the x-axis, and we know $y = f(x)$ has a stationary inflection point at $x = a$. Otherwise, $y = f(x)$ has a non-stationary inflection point at $x = a$.

Example 17

Using the graph of $y = f(x)$ alongside, sketch the graphs of $y = f'(x)$ and $y = f''(x)$.

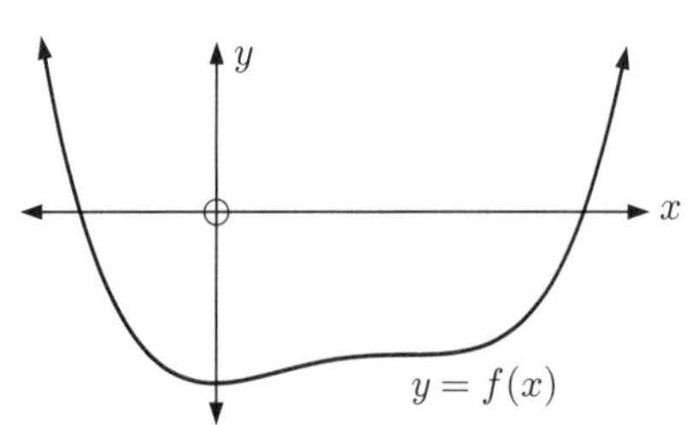

The local minimum corresponds to $f'(x) = 0$ and $f''(x) \neq 0$.

The non-stationary point of inflection corresponds to $f'(x) \neq 0$ and $f''(x) = 0$.

The stationary point of inflection corresponds to $f'(x) = 0$ and $f''(x) = 0$.

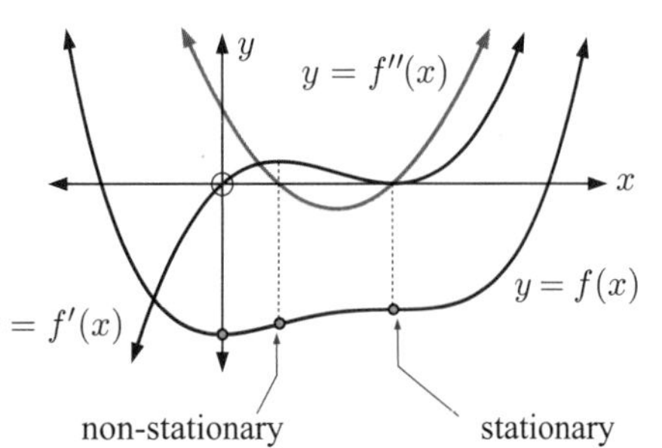

Example 18 **Self Tutor**

The graph alongside shows a gradient function $y = f'(x)$.

Sketch a graph which could be $y = f(x)$, showing clearly the x-values corresponding to all stationary points and points of inflection.

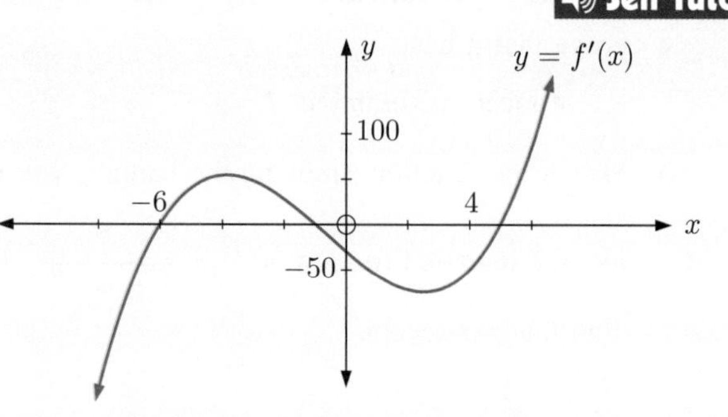

Sign diagram of $f'(x)$ is:

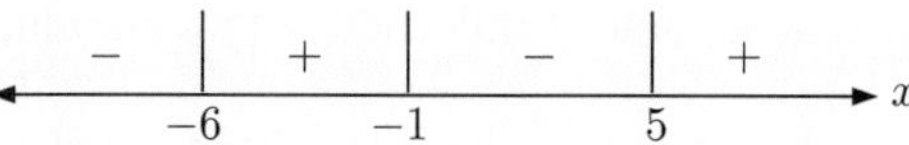

$f'(x)$ is a maximum when $x = -4$ and a minimum when $x \approx 2\frac{1}{2}$.

At these points $f''(x) = 0$ but $f'(x) \neq 0$, so they correspond to non-stationary points of inflection.

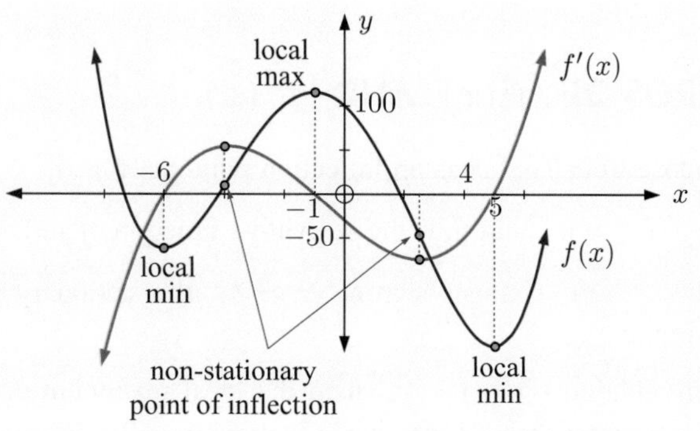

EXERCISE 19D.2

1 Using the graphs of $y = f(x)$ below, sketch the graphs of $y = f'(x)$ and $y = f''(x)$. Show clearly the axes intercepts and turning points.

a

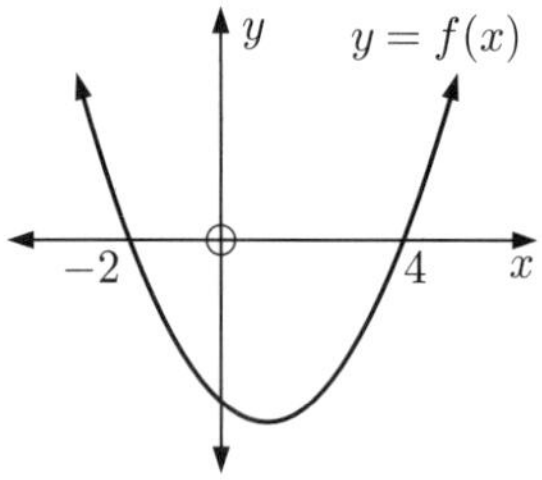

b

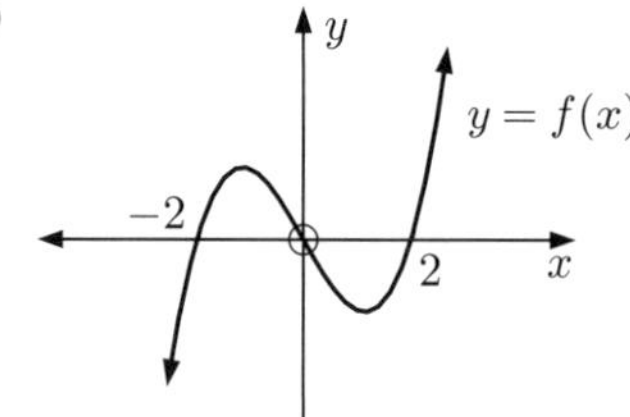

c

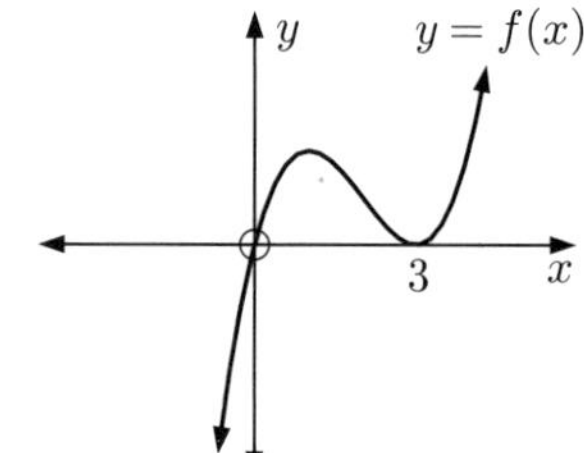

2 For the graphs of $y = f'(x)$ below, sketch a graph which could be $y = f(x)$. Show clearly the location of any stationary points and points of inflection.

a

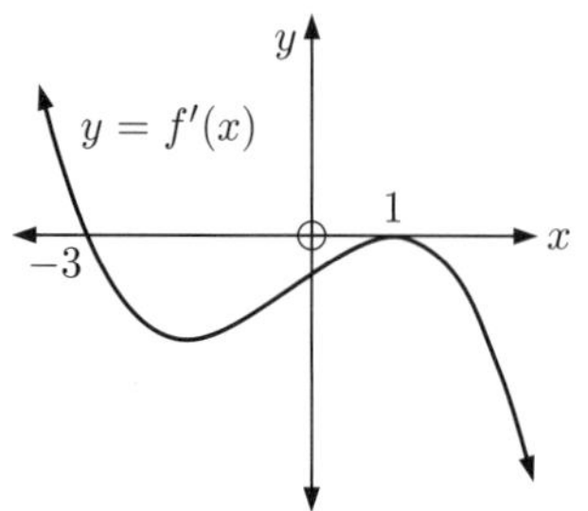

b

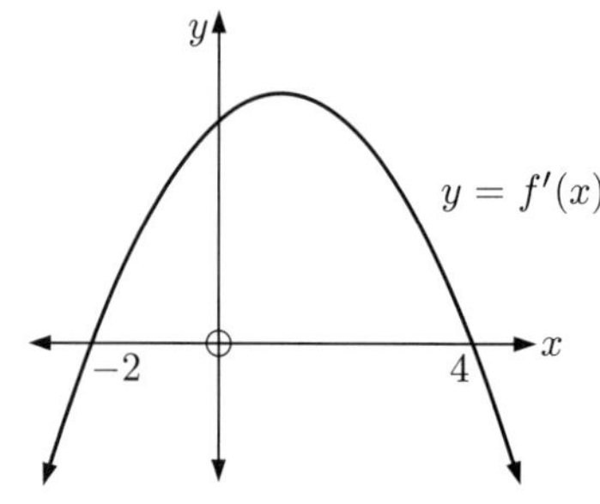

ACTIVITY

Click on the icon to run a card game on curve properties.

CARD GAME

REVIEW SET 19A — NON-CALCULATOR

1 Find the equation of the tangent to $y = -2x^2$ at the point where $x = -1$.

2 Find the equation of the normal to $y = \dfrac{1-2x}{x^2}$ at the point where $x = 1$.

3 Consider the function $f(x) = \dfrac{3x-2}{x+3}$.

- **a** State the equation of the vertical asymptote.
- **b** Find the axes intercepts.
- **c** Find $f'(x)$ and draw its sign diagram.
- **d** Does the function have any stationary points?

4 Find the equation of the normal to $y = e^{-x^2}$ at the point where $x = 1$.

5 Show that the equation of the tangent to $y = x\tan x$ at $x = \frac{\pi}{4}$ is $(2+\pi)x - 2y = \dfrac{\pi^2}{4}$.

6 The tangent to $y = \dfrac{ax+b}{\sqrt{x}}$ at $x = 1$ is $2x - y = 1$. Find a and b.

7 Show that the equation of the tangent to $f(x) = 4\ln(2x)$ at the point P$(1,\ 4\ln 2)$ is given by $y = 4x + 4\ln 2 - 4$.

8 Consider the function $f(x) = \dfrac{e^x}{x-1}$.

- **a** Find the y-intercept of the function.
- **b** For what values of x is $f(x)$ defined?
- **c** Find the signs of $f'(x)$ and $f''(x)$ and comment on the geometrical significance of each.
- **d** Sketch the graph of $y = f(x)$.
- **e** Find the equation of the tangent at the point where $x = 2$.

9 The line through A$(2,\ 4)$ and B$(0,\ 8)$ is a tangent to $y = \dfrac{a}{(x+2)^2}$. Find a.

10 Find the coordinates of P and Q if (PQ) is the tangent to $y = \dfrac{5}{\sqrt{x}}$ at $(1,\ 5)$.

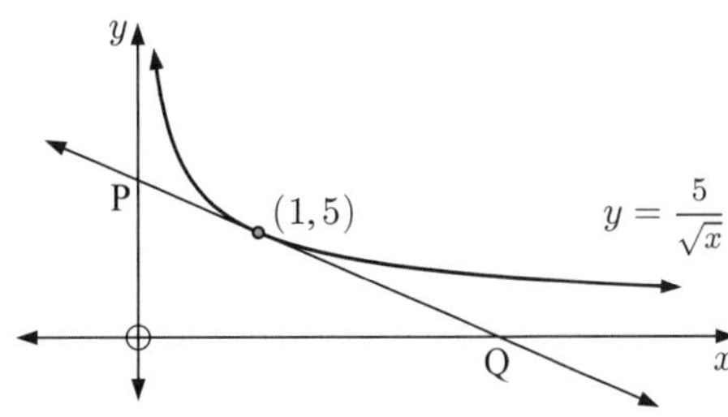

11 Given the graph of $y = f'(x)$ drawn alongside, sketch a possible curve for $y = f(x)$. Show clearly any turning points and points of inflection.

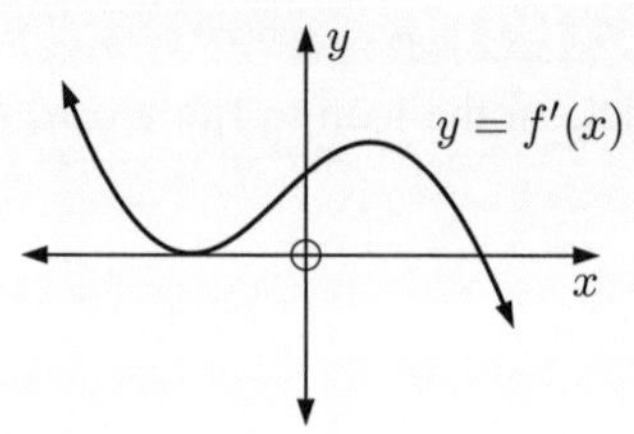

12 Find the equation of the tangent to $y = \ln(x^2 + 3)$ at the point where $x = 0$.

13 Find the equation of:

a the tangent to $y = \sec x$ at the point where $x = \frac{\pi}{3}$

b the normal to $y = \arctan x$ at the point where $x = \sqrt{3}$.

14 Prove that every normal to the unit circle $x^2 + y^2 = 1$ passes through the origin.

REVIEW SET 19B CALCULATOR

1 Determine the equation of any horizontal tangents to the curve with equation $y = x^3 - 3x^2 - 9x + 2$.

2 The tangent to $y = x^2\sqrt{1-x}$ at $x = -3$ cuts the axes at points A and B. Determine the area of triangle OAB.

3 Suppose $f(x) = x^3 + ax$, $a < 0$ has a turning point when $x = \sqrt{2}$.

a Find a.

b Find the position and nature of all stationary points of $y = f(x)$.

c Sketch the graph of $y = f(x)$.

4 At the point where $x = 0$, the tangent to $f(x) = e^{4x} + px + q$ has equation $y = 5x - 7$. Find p and q.

5 Find where the tangent to $y = 2x^3 + 4x - 1$ at $(1, 5)$ cuts the curve again.

6 Find a given that the tangent to $y = \dfrac{4}{(ax+1)^2}$ at $x = 0$ passes through $(1, 0)$.

7 Consider the function $f(x) = e^x - x$.

a Find and classify any stationary points of $y = f(x)$.

b Discuss what happens to $f(x)$ as $x \to \infty$.

c Find $f''(x)$ and draw its sign diagram. Give a geometrical interpretation for the sign of $f''(x)$.

d Sketch the graph of $y = f(x)$.

e Deduce that $e^x \geqslant x + 1$ for all x.

8 Find the equation of the normal to $y = \dfrac{x+1}{x^2-2}$ at the point where $x = 1$.

9 Show that $y = 2 - \dfrac{7}{1+2x}$ has no horizontal tangents.

10 Find the equation of the quadratic function $g(x)$ where $y = g(x)$ is the parabola shown. Give your answer in the form $g(x) = ax^2 + bx + c$.

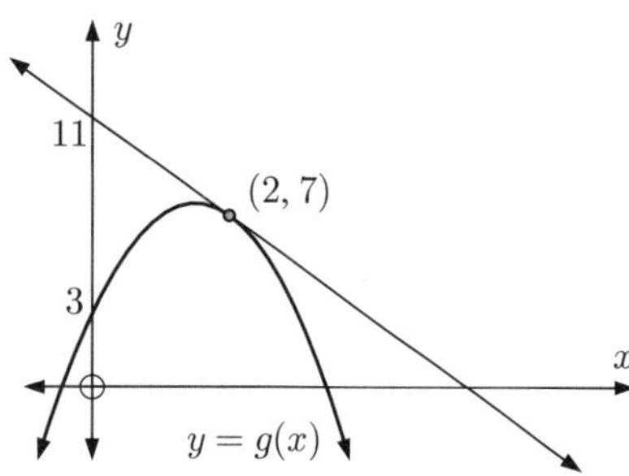

11 Consider $f(x) = \sqrt{\cos x}$ for $0 \leqslant x \leqslant 2\pi$.

a For what values of x in this interval is $f(x)$ defined?

b Find $f'(x)$ and hence find intervals where $f(x)$ is increasing and decreasing.

c Sketch the graph of $y = f(x)$ on $0 \leqslant x \leqslant 2\pi$.

12 The graph of $y = f(x)$ is given. On the same axes sketch the graph of $y = f'(x)$.

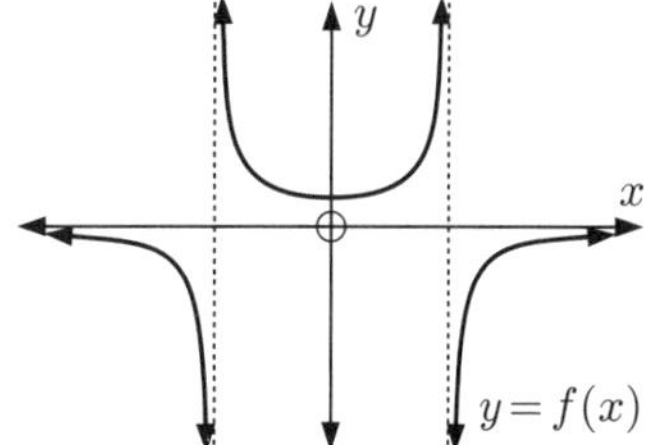

13 **a** Sketch the graph of $x \mapsto \frac{4}{x}$ for $x > 0$.

b Find the equation of the tangent to the function at the point where $x = k$, $k > 0$.

c If the tangent in **b** cuts the x-axis at A and the y-axis at B, find the coordinates of A and B.

d What can be deduced about the area of triangle OAB?

e Find k if the normal to the curve at $x = k$ passes through the point $(1, 1)$.

14 $y = \frac{x}{\sqrt{1-x}}$ has a tangent with equation $5x + by = a$ at the point where $x = -3$. Find the values of a and b.

REVIEW SET 19C

1 Find the equation of the normal to $y = \frac{1}{\sqrt{x}}$ at the point where $x = 4$.

2 $y = f(x)$ is the parabola shown.

a Find $f(3)$ and $f'(3)$.

b Hence find $f(x)$ in the form $f(x) = ax^2 + bx + c$.

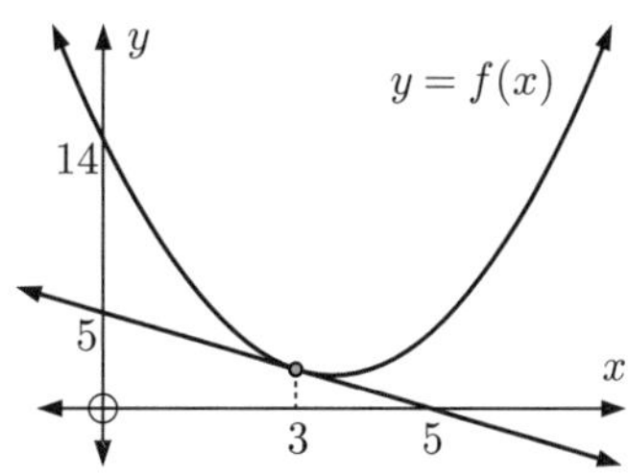

3 $y = 2x$ is a tangent to the curve $y = x^3 + ax + b$ at $x = 1$. Find a and b.

4 The tangent to $y = x^3 + ax^2 - 4x + 3$ at $x = 1$ is parallel to the line $y = 3x$.

- **a** Find the value of a and the equation of the tangent at $x = 1$.
- **b** Where does the tangent cut the curve again?

5 Find where the tangent to $y = \ln(x^4 + 3)$ at $x = 1$ cuts the y-axis.

6 Consider the function $f(x) = 2x^3 - 3x^2 - 36x + 7$.

- **a** Find and classify all stationary points and points of inflection.
- **b** Find intervals where the function is increasing and decreasing.
- **c** Find intervals where the function is concave up or down.
- **d** Sketch the graph of $y = f(x)$ showing all important features.

7 The normal to $f(x) = \dfrac{3x}{1+x}$ at $(2, 2)$ cuts the axes at B and C. Find the length of [BC].

8 Consider the function $f(x) = x^3 - 4x^2 + 4x$.

- **a** Find all axes intercepts.
- **b** Find and classify all stationary points and points of inflection.
- **c** Sketch the graph of $y = f(x)$ showing features from **a** and **b**.

9 Find the equation of:

- **a** the tangent to $y = \dfrac{1}{\sin x}$ at the point where $x = \frac{\pi}{3}$
- **b** the normal to $y = \cos(\frac{x}{2})$ at the point where $x = \frac{\pi}{2}$.

10 The curve $f(x) = 3x^3 + ax^2 + b$ has tangent with gradient 0 at the point $(-2, 14)$. Find a and b and hence $f''(-2)$.

11 Show that the curves whose equations are $y = \sqrt{3x+1}$ and $y = \sqrt{5x - x^2}$ have a common tangent at their point of intersection. Find the equation of this common tangent.

12 Consider the function $f(x) = x + \ln x$.

- **a** Find the values of x for which $f(x)$ is defined.
- **b** Find the signs of $f'(x)$ and $f''(x)$ and comment on the geometrical significance of each.
- **c** Sketch the graph of $y = f(x)$.
- **d** Find the equation of the normal at the point where $x = 1$.

13

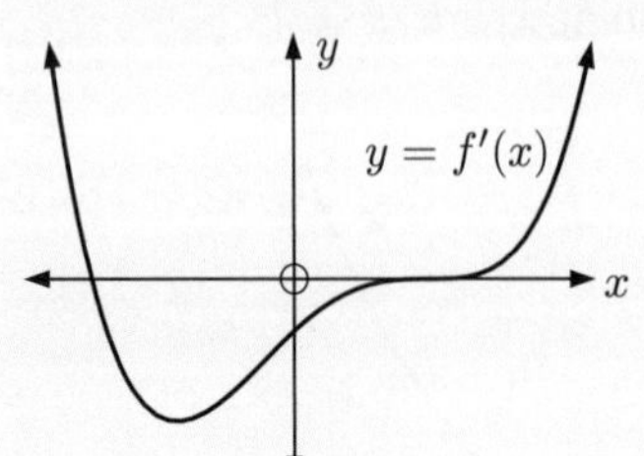

The graph of $y = f'(x)$ is drawn.
On the same axes clearly draw a possible graph of $y = f(x)$. Show all turning points and points of inflection.

14 If $f(x) = \arcsin x + \arccos x$, find $f'(x)$. What can we conclude about $f(x)$?

15

- **a** Find the equations of the tangents to the circle $x^2 + y^2 = 4$ at the points where $y = 1$.
- **b** Find the point of intersection of these tangents.

Chapter 20

Applications of differential calculus

Syllabus reference: 6.2, 6.3, 6.6

Contents:

OPENING PROBLEM

Michael rides up a hill and down the other side to his friend's house. The dots on the graph show Michael's position at various times t.

$t = 0$ $t = 5$ $t = 10$ $t = 15$ $t = 17$ $t = 19$

Michael's place   friend's house

DEMO

The distance Michael has travelled at various times is given by the function $s(t) = 1.2t^3 - 30t^2 + 285t$ metres for $0 \leqslant t \leqslant 19$ minutes.

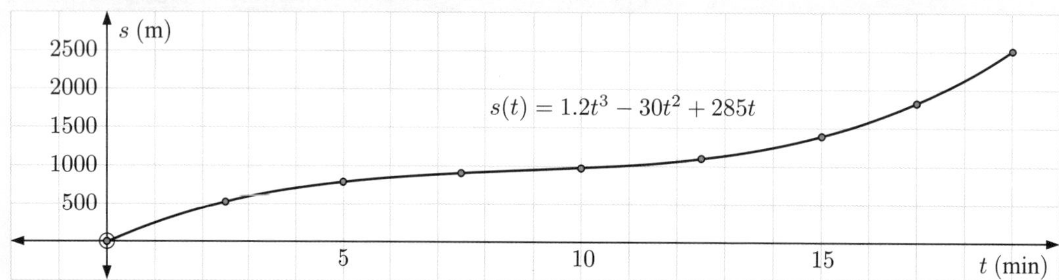

Things to think about:

a Explain why $s(t)$ should be an increasing function.

b Can you find a function for Michael's *speed* at any time t?

c Michael's *acceleration* is the rate at which his speed is changing with respect to time. How can we interpret $s''(t)$?

d Can you find Michael's speed and acceleration at the time $t = 15$ minutes?

e At what point do you think the hill was steepest? How far had Michael travelled to this point?

We saw in the previous chapter some of the curve properties that can be analysed using calculus. In this chapter we look at applying these techniques in real world problems of:

- kinematics (motion problems of displacement, velocity, and acceleration)
- rates of change
- optimisation (maxima and minima).

A KINEMATICS

In the **Opening Problem** we are dealing with the movement of Michael riding his bicycle. We do not know the direction Michael is travelling, so we talk simply about the *distance* he has travelled and his *speed*.

For problems of **motion in a straight line**, we can include the direction the object is travelling along the line. We therefore can talk about *displacement* and *velocity*.

DISPLACEMENT

Suppose an object P moves along a straight line so that its position s from an origin O is given as some function of time t. We write $s = s(t)$ where $t \geqslant 0$.

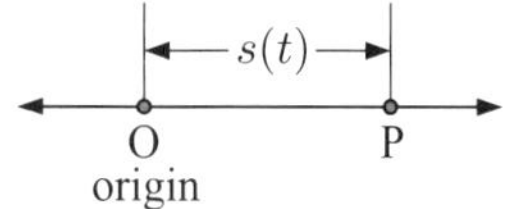

$s(t)$ is a **displacement function** and for any value of t it gives the displacement from O.

$s(t)$ is a vector quantity. Its magnitude is the distance from O, and its sign indicates the direction from O.

On the horizontal axis through O:

- if $s(t) > 0$, P is located to the **right of O**
- if $s(t) = 0$, P is located **at O**
- if $s(t) < 0$, P is located to the **left of O**.

MOTION GRAPHS

Consider $s(t) = t^2 + 2t - 3$ cm.

$s(0) = -3$ cm, $s(1) = 0$ cm, $s(2) = 5$ cm, $s(3) = 12$ cm, $s(4) = 21$ cm.

To appreciate the motion of P we draw a **motion graph**. You can also view the motion by clicking on the icon.

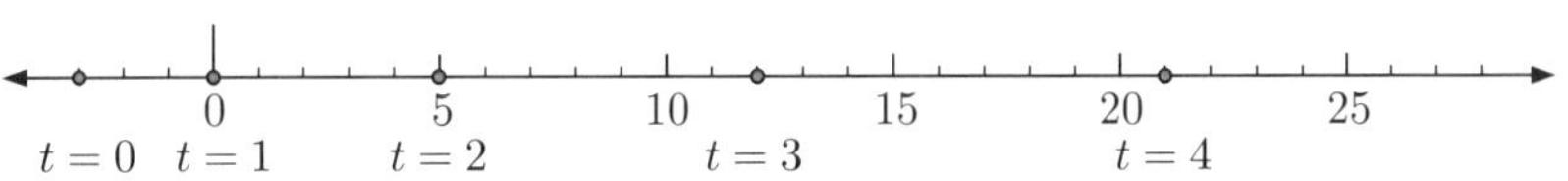

DEMO

Fully animated, we not only get a good idea of the position of P, but also of what is happening to its velocity and acceleration.

VELOCITY

The **average velocity** of an object moving in a straight line in the time interval from $t = t_1$ to $t = t_2$ is the ratio of the change in displacement to the time taken.

If $s(t)$ is the displacement function then **average velocity** $= \dfrac{s(t_2) - s(t_1)}{t_2 - t_1}$.

On a graph of $s(t)$ against t for the time interval from $t = t_1$ to $t = t_2$, the average velocity is the gradient of a chord through the points $(t_1, s(t_1))$ and $(t_2, s(t_2))$.

In **Chapter 17** we established that the instantaneous rate of change of a quantity is given by its derivative.

If $s(t)$ is the displacement function of an object moving in a straight line, then $v(t) = s'(t) = \lim\limits_{h \to 0} \dfrac{s(t+h) - s(t)}{h}$ is the **instantaneous velocity** or **velocity function** of the object at time t.

On a graph of $s(t)$ against t, the instantaneous velocity at a particular time is the gradient of the tangent to the graph at that point.

ACCELERATION

If an object moves in a straight line with velocity function $v(t)$ then:

- the **average acceleration** for the time interval from $t = t_1$ to $t = t_2$ is the ratio of the change in velocity to the time taken

$$\textbf{average acceleration} = \frac{v(t_2) - v(t_1)}{t_2 - t_1}$$

- the **instantaneous acceleration** at time t is $a(t) = v'(t) = \lim\limits_{h \to 0} \dfrac{v(t+h) - v(t)}{h}$.

UNITS

Each time we differentiate with respect to time t, we calculate a rate per unit of time. So, for a displacement in metres and time in seconds:

- the units of velocity are m s^{-1}
- the units of acceleration are m s^{-2}.

Example 1 — Self Tutor

A particle moves in a straight line with displacement from O given by $s(t) = 3t - t^2$ metres at time t seconds. Find:

a the average velocity for the time interval from $t = 2$ to $t = 5$ seconds

b the average velocity for the time interval from $t = 2$ to $t = 2 + h$ seconds

c $\lim\limits_{h \to 0} \dfrac{s(2+h) - s(2)}{h}$ and comment on its significance.

a average velocity

$$= \frac{s(5) - s(2)}{5 - 2}$$
$$= \frac{(15 - 25) - (6 - 4)}{3}$$
$$= \frac{-10 - 2}{3}$$
$$= -4 \text{ m s}^{-1}$$

b average velocity

$$= \frac{s(2+h) - s(2)}{2 + h - 2}$$
$$= \frac{3(2+h) - (2+h)^2 - 2}{h}$$
$$= \frac{6 + 3h - 4 - 4h - h^2 - 2}{h}$$
$$= \frac{-h - h^2}{h}$$
$$= -1 - h \ \text{ m s}^{-1} \quad \text{provided } h \neq 0$$

c
$$\lim_{h \to 0} \frac{s(2+h) - s(2)}{h}$$
$$= \lim_{h \to 0} (-1 - h) \qquad \{\text{since } h \neq 0\}$$
$$= -1 \text{ m s}^{-1}$$

This is the instantaneous velocity of the particle at time $t = 2$ seconds.

EXERCISE 20A.1

1 A particle P moves in a straight line with displacement function $s(t) = t^2 + 3t - 2$ metres, where $t \geqslant 0$, t in seconds.

a Find the average velocity from $t = 1$ to $t = 3$ seconds.

b Find the average velocity from $t = 1$ to $t = 1 + h$ seconds.

c Find the value of $\lim\limits_{h \to 0} \dfrac{s(1+h) - s(1)}{h}$ and comment on its significance.

d Find the average velocity from time t to time $t+h$ seconds and interpret $\lim\limits_{h \to 0} \dfrac{s(t+h) - s(t)}{h}$.

2 A particle P moves in a straight line with displacement function $s(t) = 5 - 2t^2$ cm, where $t \geqslant 0$, t in seconds.

a Find the average velocity from $t = 2$ to $t = 5$ seconds.

b Find the average velocity from $t = 2$ to $t = 2 + h$ seconds.

c Find the value of $\lim\limits_{h \to 0} \dfrac{s(2+h) - s(2)}{h}$ and state the meaning of this value.

d Interpret $\lim\limits_{h \to 0} \dfrac{s(t+h) - s(t)}{h}$.

3 A particle moves in a straight line with velocity function $v(t) = 2\sqrt{t} + 3$ cm s^{-1}, $t \geqslant 0$.

a Find the average acceleration from $t = 1$ to $t = 4$ seconds.

b Find the average acceleration from $t = 1$ to $t = 1 + h$ seconds.

c Find the value of $\lim\limits_{h \to 0} \dfrac{v(1+h) - v(1)}{h}$. Interpret this value.

d Interpret $\lim\limits_{h \to 0} \dfrac{v(t+h) - v(t)}{h}$.

4 An object moves in a straight line with displacement function $s(t)$ and velocity function $v(t)$, $t \geqslant 0$. State the meaning of:

a $\lim\limits_{h \to 0} \dfrac{s(4+h) - s(4)}{h}$

b $\lim\limits_{h \to 0} \dfrac{v(4+h) - v(4)}{h}$

VELOCITY AND ACCELERATION FUNCTIONS

If a particle P moves in a straight line and its position is given by the displacement function $s(t)$, $t \geqslant 0$, then:

- the **velocity** of P at time t is given by $\boldsymbol{v(t) = s'(t)}$
- the **acceleration** of P at time t is given by $\boldsymbol{a(t) = v'(t) = s''(t)}$
- $s(0)$, $v(0)$, and $a(0)$ give us the position, velocity, and acceleration of the particle at time $t = 0$, and these are called the **initial conditions**.

$$\underset{\text{displacement}}{s(t)} \xrightarrow{\text{differentiate}} \underset{\text{velocity}}{v(t) = \frac{ds}{dt}} \xrightarrow{\text{differentiate}} \underset{\text{acceleration}}{a(t) = \frac{dv}{dt} = \frac{d^2s}{dt^2}}$$

SIGN INTERPRETATION

Suppose a particle P moves in a straight line with displacement function $s(t)$ relative to an origin O. Its velocity function is $v(t)$ and its acceleration function is $a(t)$.

We can use **sign diagrams** to interpret:

- where the particle is located relative to O
- the direction of motion and where a change of direction occurs
- when the particle's velocity is increasing or decreasing.

SIGNS OF $s(t)$:

$s(t)$	*Interpretation*
$=0$	P is at O
>0	P is located to the right of O
<0	P is located to the left of O

SIGNS OF $v(t)$:

$v(t)$	*Interpretation*
$=0$	P is instantaneously at rest
>0	P is moving to the right
<0	P is moving to the left

$$v(t) = \lim_{h \to 0} \frac{s(t+h) - s(t)}{h}$$

If $v(t) > 0$ then $s(t+h) - s(t) > 0$

$\therefore$ $s(t+h) > s(t)$

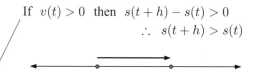

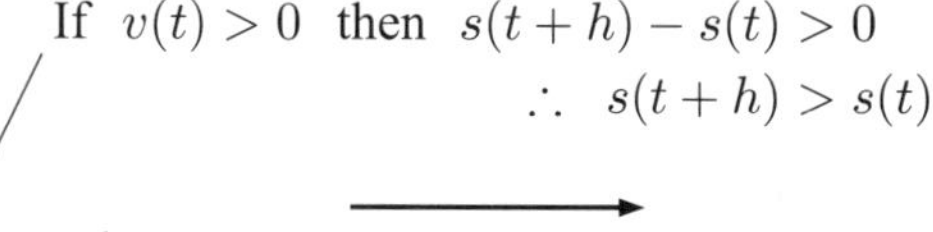

$\therefore$ P is moving to the right.

SIGNS OF $a(t)$:

$a(t)$	*Interpretation*
>0	velocity is increasing
<0	velocity is decreasing
$=0$	velocity may be a maximum or minimum or possibly constant

A useful table:

Phrase used in a question	t	s	v	a
initial conditions	0			
at the origin		0		
stationary			0	
reverses			0	
maximum or minimum displacement			0	
constant velocity				0
maximum or minimum velocity				0

When a particle reverses direction, its velocity must change sign.
This corresponds to a local maximum or local minimum distance from the origin O.

We need a sign diagram of a to determine if the velocity of the point is a local maximum or minimum.

SPEED

As we have seen, velocities have size (magnitude) and sign (direction). In contrast, speed simply measures *how fast* something is travelling, regardless of the direction of travel. Speed is a *scalar* quantity which has size but no sign. Speed cannot be negative.

The **speed** at any instant is the magnitude of the object's velocity.
If $S(t)$ represents speed then $S = |v|$.

Be careful not to confuse speed $S(t)$ with displacement $s(t)$.

To determine when the speed $S(t)$ of an object P with displacement $s(t)$ is increasing or decreasing, we use a **sign test**.

- If the signs of $v(t)$ and $a(t)$ are the same (both positive or both negative), then the speed of P is increasing.
- If the signs of $v(t)$ and $a(t)$ are opposite, then the speed of P is decreasing.

We prove *the first* of these as follows:

Proof: Let $S = |v|$ be the speed of P at any instant, so $S = \begin{cases} v & \text{if } v \geqslant 0 \\ -v & \text{if } v < 0. \end{cases}$

Case 1: If $v > 0$, $S = v$ and $\therefore \frac{dS}{dt} = \frac{dv}{dt} = a(t)$

If $a(t) > 0$ then $\frac{dS}{dt} > 0$ which implies that S is increasing.

Case 2: If $v < 0$, $S = -v$ and $\therefore \frac{dS}{dt} = -\frac{dv}{dt} = -a(t)$

If $a(t) < 0$ then $\frac{dS}{dt} > 0$ which also implies that S is increasing.

Thus if $v(t)$ and $a(t)$ have the same sign then the speed of P is increasing.

INVESTIGATION — DISPLACEMENT, VELOCITY, AND ACCELERATION GRAPHS

In this Investigation we examine the motion of a projectile which is fired in a vertical direction. The projectile is affected by gravity, which is responsible for the projectile's constant acceleration.

MOTION DEMO

We then extend the Investigation to consider other cases of motion in a straight line.

What to do:

1. Click on the icon to examine vertical projectile motion.
   Observe first the displacement along the line, then look at the velocity which is the rate of change in displacement. When is the velocity positive and when is it negative?
2. Examine the following graphs and comment on their shapes:
   - *displacement* v *time*
   - *velocity* v *time*
   - *acceleration* v *time*
3. Pick from the menu or construct functions of your own choosing to investigate the relationship between displacement, velocity, and acceleration.

You are encouraged to use the motion demo above to help answer questions in the following exercise.

Example 2 **Self Tutor**

A particle moves in a straight line with position relative to O given by $s(t) = t^3 - 3t + 1$ cm, where t is the time in seconds, $t \geqslant 0$.

a Find expressions for the particle's velocity and acceleration, and draw sign diagrams for each of them.

b Find the initial conditions and hence describe the motion at this instant.

c Describe the motion of the particle at $t = 2$ seconds.

d Find the position of the particle when changes in direction occur.

e Draw a motion diagram for the particle.

f For what time interval is the particle's speed increasing?

g What is the total distance travelled in the time from $t = 0$ to $t = 2$ seconds?

---

a $s(t) = t^3 - 3t + 1$ cm

$\therefore\ v(t) = 3t^2 - 3$ $\{$as $v(t) = s'(t)\}$

$= 3(t^2 - 1)$

$= 3(t + 1)(t - 1)$ cm s^{-1}

which has sign diagram:

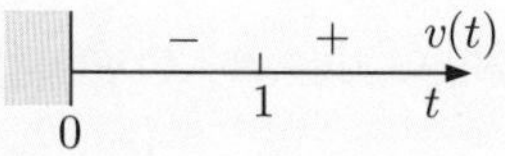

and $a(t) = 6t$ cm s^{-2} $\{$as $a(t) = v'(t)\}$

which has sign diagram:

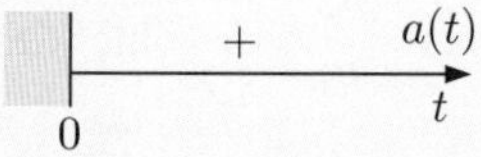

b When $t = 0$, $s(0) = 1$ cm

$v(0) = -3$ cm s^{-1}

$a(0) = 0$ cm s^{-2}

$\therefore$ the particle is 1 cm to the right of O, moving to the left at a speed of 3 cm s^{-1}.

c When $t = 2$, $s(2) = 8 - 6 + 1 = 3$ cm

$v(2) = 12 - 3 = 9$ cm s^{-1}

$a(2) = 12$ cm s^{-2}

$\therefore$ the particle is 3 cm to the right of O, moving to the right at a speed of 9 cm s^{-1}.

Since a and v have the same sign, the speed of the particle is increasing.

d Since $v(t)$ changes sign when $t = 1$, a change of direction occurs at this instant.

$s(1) = 1 - 3 + 1 = -1$, so the particle changes direction when it is 1 cm to the left of O.

e

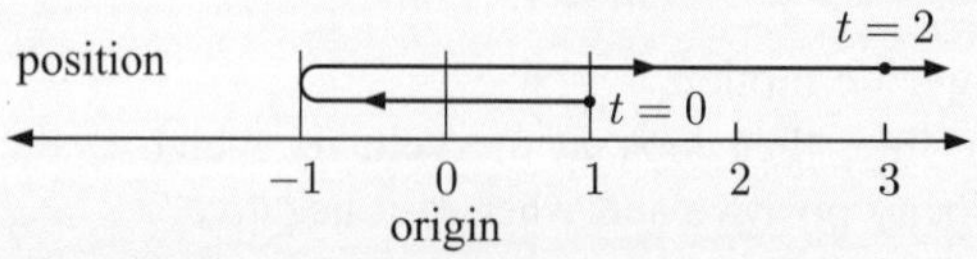

origin

The motion is actually **on the line**, not above it as shown.

As $t \to \infty$, $s(t) \to \infty$ and $v(t) \to \infty$.

f Speed is increasing when $v(t)$ and $a(t)$ have the same sign. This is for $t \geqslant 1$.

g Total distance travelled $= 2 + 4 = 6$ cm.

$\therefore$ the stationary point at $t = -1$ is not required.

In later chapters on integral calculus we will see another technique for finding the distances travelled and displacement over time.

EXERCISE 20A.2

1 An object moves in a straight line with position given by $s(t) = t^2 - 4t + 3$ cm from O, where t is in seconds, $t \geqslant 0$.

a Find expressions for the object's velocity and acceleration, and draw sign diagrams for each function.

b Find the initial conditions and explain what is happening to the object at that instant.

c Describe the motion of the object at time $t = 2$ seconds.

d At what time does the object reverse direction? Find the position of the object at this instant.

e Draw a motion diagram for the object.

f For what time intervals is the speed of the object decreasing?

2 A stone is projected vertically so that its position above ground level after t seconds is given by $s(t) = 98t - 4.9t^2$ metres, $t \geqslant 0$.

a Find the velocity and acceleration functions for the stone and draw sign diagrams for each function.

b Find the initial position and velocity of the stone.

c Describe the stone's motion at times $t = 5$ and $t = 12$ seconds.

d Find the maximum height reached by the stone.

e Find the time taken for the stone to hit the ground.

3 When a ball is thrown, its height above the ground is given by $s(t) = 1.2 + 28.1t - 4.9t^2$ metres where t is the time in seconds.

a From what distance above the ground was the ball released?

b Find $s'(t)$ and state what it represents.

c Find t when $s'(t) = 0$. What is the significance of this result?

d What is the maximum height reached by the ball?

e Find the ball's speed: i when released ii at $t = 2$ s iii at $t = 5$ s.
State the significance of the sign of the derivative $s'(t)$.

f How long will it take for the ball to hit the ground?

g What is the significance of $s''(t)$?

4 A shell is accidentally fired vertically from a mortar at ground level and reaches the ground again after 14.2 seconds. Its height above the ground at time t seconds is given by $s(t) = bt - 4.9t^2$ metres where b is constant.

a Show that the initial velocity of the shell is b m s^{-1} upwards.

b Find the initial velocity of the shell.

5 A particle moves in a straight line with displacement function $s(t) = 12t - 2t^3 - 1$ centimetres where t is in seconds, $t \geqslant 0$.

a Find velocity and acceleration functions for the particle's motion.

b Find the initial conditions and interpret their meaning.

c Find the times and positions when the particle reverses direction.

d At what times is the particle's: i speed increasing ii velocity increasing?

6 The position of a particle moving along the x-axis is given by $x(t) = t^3 - 9t^2 + 24t$ metres where t is in seconds, $t \geqslant 0$.

a Draw sign diagrams for the particle's velocity and acceleration functions.

b Find the position of the particle at the times when it reverses direction, and hence draw a motion diagram for the particle.

c At what times is the particle's:

i speed decreasing **ii** velocity decreasing?

d Find the total distance travelled by the particle in the first 5 seconds of motion.

When finding the total distance travelled, always look for direction reversals first.

7 A particle P moves in a straight line with displacement function $s(t) = 100t + 200e^{-\frac{t}{5}}$ cm, where t is the time in seconds, $t \geqslant 0$.

a Find the velocity and acceleration functions.

b Find the initial position, velocity, and acceleration of P.

c Discuss the velocity of P as $t \to \infty$.

d Sketch the graph of the velocity function.

e Find when the velocity of P is 80 cm per second.

8 A particle P moves along the x-axis with position given by $x(t) = 1 - 2\cos t$ cm where t is the time in seconds.

a State the initial position, velocity, and acceleration of P.

b Describe the motion when $t = \frac{\pi}{4}$ seconds.

c Find the times when the particle reverses direction on $0 < t < 2\pi$, and find the position of the particle at these instants.

d When is the particle's speed increasing on $0 \leqslant t \leqslant 2\pi$?

9 In an experiment, an object is fired vertically from the earth's surface. From the results, a two-dimensional graph of the position $s(t)$ metres above the earth's surface is plotted, where t is the time in seconds. It is noted that the graph is *parabolic*.

Assuming a constant gravitational acceleration g and an initial velocity of $v(0)$, show that:

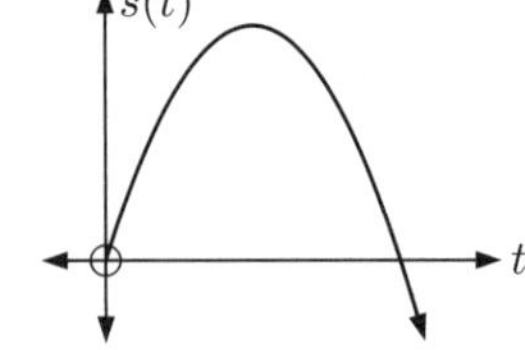

a $v(t) = v(0) + gt$ **b** $s(t) = v(0) \times t + \frac{1}{2}gt^2$.

Hint: Assume that $s(t) = at^2 + bt + c$.

10 The velocity of an object after t seconds, $t \geqslant 0$, is given by $v = 25te^{-2t}$ cm s^{-1}.

a Sketch the velocity function.

b Show that the object's acceleration at time t is given by $a = 25(1 - 2t)e^{-2t}$ cm s^{-2}.

c When is the velocity increasing?

B RATES OF CHANGE

We have seen previously that if $s(t)$ is a displacement function then $s'(t)$ or $\frac{ds}{dt}$ is the instantaneous rate of change in displacement with respect to time, which we call velocity.

There are countless examples in the real world where quantities vary with time, or with respect to some other variable.

For example:
- temperature varies continuously
- the height of a tree varies as it grows
- the prices of stocks and shares vary with each day's trading.

We have already seen that if $y = f(x)$ then $f'(x)$ or $\frac{dy}{dx}$ is the gradient of the tangent to $y = f(x)$ at the given point.

$\frac{dy}{dx}$ gives the **rate of change in y with respect to x.**

We can therefore use the derivative of a function to tell us the **rate** at which something is happening.

For example:

- $\frac{dH}{dt}$ or $H'(t)$ could be the instantaneous rate of ascent of a person in a Ferris wheel.
  It might have units metres per second or m s^{-1}.
- $\frac{dC}{dt}$ or $C'(t)$ could be a person's instantaneous rate of change in lung capacity.
  It might have units litres per second or L s^{-1}.

Example 3 **Self Tutor**

According to a psychologist, the ability of a person to understand spatial concepts is given by $A = \frac{1}{3}\sqrt{t}$ where t is the age in years, $5 \leqslant t \leqslant 18$.

a Find the rate of improvement in ability to understand spatial concepts when a person is:
  i 9 years old   ii 16 years old.

b Show that $\frac{dA}{dt} > 0$ for $5 \leqslant t \leqslant 18$. Comment on the significance of this result.

c Show that $\frac{d^2A}{dt^2} < 0$ for $5 \leqslant t \leqslant 18$. Comment on the significance of this result.

---

a $A = \frac{1}{3}\sqrt{t} = \frac{1}{3}t^{\frac{1}{2}}$ $\quad \therefore \quad \frac{dA}{dt} = \frac{1}{6}t^{-\frac{1}{2}} = \frac{1}{6\sqrt{t}}$

i When $t = 9$, $\frac{dA}{dt} = \frac{1}{18}$

$\therefore$ the rate of improvement is $\frac{1}{18}$ units per year for a 9 year old.

ii When $t = 16$, $\frac{dA}{dt} = \frac{1}{24}$

$\therefore$ the rate of improvement is $\frac{1}{24}$ units per year for a 16 year old.

b Since $\sqrt{t}$ is never negative, $\frac{1}{6\sqrt{t}}$ is never negative

$\therefore \quad \frac{dA}{dt} > 0$ for all $5 \leqslant t \leqslant 18$.

This means that the ability to understand spatial concepts increases with age.

c $\dfrac{dA}{dt} = \frac{1}{6}t^{-\frac{1}{2}}$

$\therefore \quad \dfrac{d^2A}{dt^2} = -\frac{1}{12}t^{-\frac{3}{2}} = -\dfrac{1}{12t\sqrt{t}}$

$\therefore \quad \dfrac{d^2A}{dt^2} < 0$ for all $5 \leqslant t \leqslant 18$.

This means that while the ability to understand spatial concepts increases with age, the rate of increase slows down with age.

You are encouraged to use technology to graph each function you need to consider. This is often useful in interpreting results.

EXERCISE 20B

1 The estimated future profits of a small business are given by $P(t) = 2t^2 - 12t + 118$ thousand dollars, where t is the time in years from now.

- **a** What is the current annual profit?
- **b** Find $\dfrac{dP}{dt}$ and state its units.
- **c** Explain the significance of $\dfrac{dP}{dt}$.
- **d** For what values of t will the profit:
  - **i** decrease
  - **ii** increase

  on the previous year?
- **e** What is the minimum profit and when does it occur?
- **f** Find $\dfrac{dP}{dt}$ when $t = 4$, 10 and 25. What do these figures represent?

2 Water is draining from a swimming pool. The remaining volume of water after t minutes is $V = 200(50 - t)^2$ m^3. Find:

- **a** the average rate at which the water leaves the pool in the first 5 minutes
- **b** the instantaneous rate at which the water is leaving at $t = 5$ minutes.

3 The quantity of a chemical in human skin which is responsible for its 'elasticity' is given by $Q = 100 - 10\sqrt{t}$ where t is the age of a person in years.

- **a** Find Q at: **i** $t = 0$ **ii** $t = 25$ **iii** $t = 100$ years.
- **b** At what rate is the quantity of the chemical changing at the age of:
  - **i** 25 years
  - **ii** 50 years?
- **c** Show that the rate at which the skin loses the chemical is decreasing for all $t > 0$.

4 The height of *pinus radiata*, grown in ideal conditions, is given by $H = 20 - \dfrac{97.5}{t+5}$ metres, where t is the number of years after the tree was planted from an established seedling.

- **a** How high was the tree at the time of its planting?
- **b** Find the height of the tree after 4, 8, and 12 years.
- **c** Find the rate at which the tree is growing after 0, 5, and 10 years.
- **d** Show that $\dfrac{dH}{dt} > 0$ for all $t \geqslant 0$. What is the significance of this result?

Example 4 **Self Tutor**

The cost in dollars of producing x items in a factory each day is given by

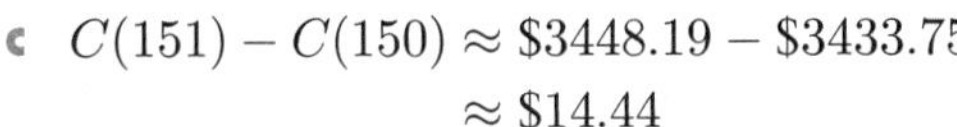

$$C(x) = \underbrace{0.000\,13x^3 + 0.002x^2}_{\text{labour}} + \underbrace{5x}_{\text{raw materials}} + \underbrace{2200}_{\text{fixed costs}}$$

a Find $C'(x)$, which is called the marginal cost function.

b Find the marginal cost when 150 items are produced. Interpret this result.

c Find $C(151) - C(150)$. Compare this with the answer in **b**.

---

a The marginal cost function is
$C'(x) = 0.000\,39x^2 + 0.004x + 5$ dollars per item.

b $C'(150) = \$14.38$
This is the rate at which the costs are increasing with respect to the production level x when 150 items are made per day.
It gives an estimate of the cost of making the 151st item each day.

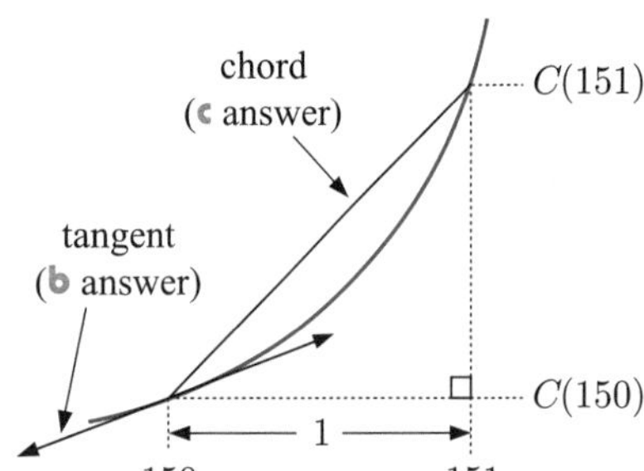

c $C(151) - C(150) \approx \$3448.19 - \$3433.75$
$\approx \$14.44$

This is the actual cost of making the 151st item each day, so the answer in **b** gives a good estimate.

5 Seablue make denim jeans. The cost model for making x pairs per day is
$C(x) = 0.0003x^3 + 0.02x^2 + 4x + 2250$ dollars.

a Find the marginal cost function $C'(x)$.

b Find $C'(220)$. What does it estimate?

c Find $C(221) - C(220)$. What does this represent?

d Find $C''(x)$ and the value of x when $C''(x) = 0$. What is the significance of this point?

6 The total cost of running a train from Paris to Marseille is given by $C(v) = \frac{1}{5}v^2 + \frac{200\,000}{v}$ euros where v is the average speed of the train in km h^{-1}.

a Find the total cost of the journey if the average speed is:

i 50 km h^{-1} **ii** 100 km h^{-1}.

b Find the rate of change in the cost of running the train at speeds of:

i 30 km h^{-1} **ii** 90 km h^{-1}.

c At what speed will the cost be a minimum?

7 A tank contains 50 000 litres of water. The tap is left fully on and all the water drains from the tank in 80 minutes. The volume of water remaining in the tank after t minutes is given by
$V = 50\,000\left(1 - \frac{t}{80}\right)^2$ litres where $0 \leqslant t \leqslant 80$.

a Find $\frac{dV}{dt}$ and draw the graph of $\frac{dV}{dt}$ against t.

b At what time was the outflow fastest?

c Show that $\frac{d^2V}{dt^2}$ is always constant and positive. Interpret this result.

8 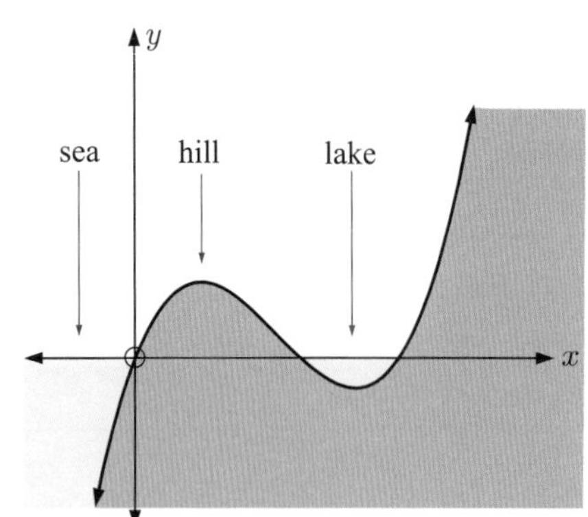

Alongside is a land and sea profile where the x-axis is sea level. The function $y = \frac{1}{10}x(x-2)(x-3)$ km gives the height of the land or sea bed relative to sea level at distance x km from the shore line.

a Find where the lake is located relative to the shore line of the sea.

b Find $\frac{dy}{dx}$ and interpret its value when $x = \frac{1}{2}$ km and when $x = 1\frac{1}{2}$ km.

c Find the deepest point of the lake and the depth at this point.

9 A fish farm grows and harvests barramundi in a large dam. The population of fish after t years is given by the function $P(t)$.

The rate of change in the population $\frac{dP}{dt}$ is modelled by $\frac{dP}{dt} = aP\left(1 - \frac{P}{b}\right) - \left(\frac{c}{100}\right)P$ where a, b, and c are known constants. a is the birth rate of the barramundi, b is the maximum carrying capacity of the dam, and c is the percentage of barramundi that are harvested each year.

a Explain why the fish population is stable when $\frac{dP}{dt} = 0$.

b If the birth rate is 6%, the maximum carrying capacity is 24 000, and 5% are harvested each year, find the stable population.

c If the harvest rate changes to 4%, what will the stable population increase to?

10 A radioactive substance decays according to the formula $W = 20e^{-kt}$ grams where t is the time in hours.

a Find k given that after 50 hours the weight is 10 grams.

b Find the weight of radioactive substance present:

  i initially **ii** after 24 hours **iii** after 1 week.

c How long will it take for the weight to reach 1 gram?

d Find the rate of radioactive decay at: **i** $t = 100$ hours **ii** $t = 1000$ hours.

e Show that $\frac{dW}{dt}$ is proportional to the weight of substance remaining.

11 The temperature of a liquid after being placed in a refrigerator is given by $T = 5 + 95e^{-kt}$ °C where k is a positive constant and t is the time in minutes.

a Find k if the temperature of the liquid is 20°C after 15 minutes.

b What was the temperature of the liquid when it was first placed in the refrigerator?

c Show that $\frac{dT}{dt} = c(T - 5)$ for some constant c. Find the value of c.

d At what rate is the temperature changing at:

  i $t = 0$ mins **ii** $t = 10$ mins **iii** $t = 20$ mins?

12 The height of a certain species of shrub t years after it is planted is given by $H(t) = 20\ln(3t+2) + 30$ cm, $t \geqslant 0$.

- **a** How high was the shrub when it was planted?
- **b** How long will it take for the shrub to reach a height of 1 m?
- **c** At what rate is the shrub's height changing:
  - **i** 3 years after being planted
  - **ii** 10 years after being planted?

13 In the conversion of sugar solution to alcohol, the chemical reaction obeys the law $A = s(1-e^{-kt})$, $t \geqslant 0$ where t is the number of hours after the reaction commenced, s is the original sugar concentration (%), and A is the alcohol produced, in litres.

- **a** Find A when $t = 0$.
- **b** Suppose $s = 10$ and $A = 5$ after 3 hours.
  - **i** Find k.
  - **ii** Find the speed of the reaction at time 5 hours.
- **c** Show that the speed of the reaction is proportional to $A - s$.

14 The number of bees in a hive after t months is modelled by $B(t) = \dfrac{C}{1 + 0.5e^{-1.73t}}$.

- **a** Find the initial bee population in terms of C.
- **b** Find the percentage increase in the population after 1 month.
- **c** Is there a limit to the population size? If so, what is it?
- **d** After 2 months the bee population is 4500. Find the original population size.
- **e** Find $B'(t)$ and use it to explain why the population is increasing over time.
- **f** Sketch the graph of $B(t)$.

Example 5 **Self Tutor**

Find the rate of change in the area of triangle ABC as θ changes, at the time when $\theta = 60°$.

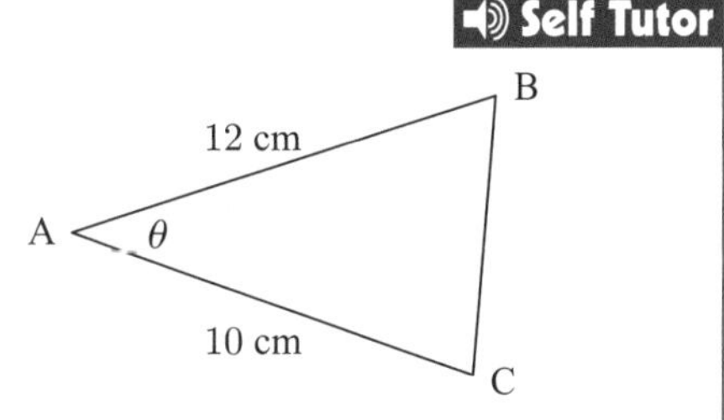

Area $A = \frac{1}{2} \times 10 \times 12 \times \sin\theta$ $\quad \{\text{Area} = \frac{1}{2}bc\sin A\}$

$\therefore\ A = 60\sin\theta$ cm^2

$\therefore\ \dfrac{dA}{d\theta} = 60\cos\theta$

When $\theta = \frac{\pi}{3}$, $\cos\theta = \frac{1}{2}$

$\therefore\ \dfrac{dA}{d\theta} = 30$ cm^2 per radian

θ must be in **radians** so the dimensions are correct.

15 Find exactly the rate of change in the area of triangle PQR as θ changes, at the time when $\theta = 45°$.

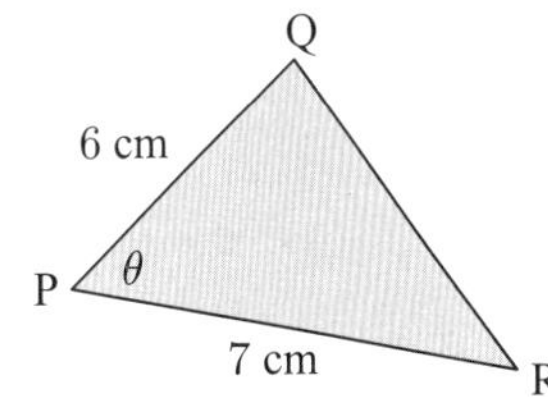

16 On the Indonesian coast, the depth of water at time t hours after midnight is given by $d = 9.3 + 6.8\cos(0.507t)$ metres.

a Find the rate of change in the depth of water at 8:00 am.

b Is the tide rising or falling at this time?

17 The voltage in a circuit is given by $V(t) = 340\sin(100\pi t)$ where t is the time in seconds. At what rate is the voltage changing:

a when $t = 0.01$ **b** when $V(t)$ is a maximum?

18 A piston is operated by rod [AP] attached to a flywheel of radius 1 m. AP = 2 m. P has coordinates $(\cos t, \sin t)$ and point A is $(-x, 0)$.

a Show that $x = \sqrt{4 - \sin^2 t} - \cos t$.

b Find the rate at which x is changing at the instant when:

i $t = 0$ **ii** $t = \frac{\pi}{2}$ **iii** $t = \frac{2\pi}{3}$

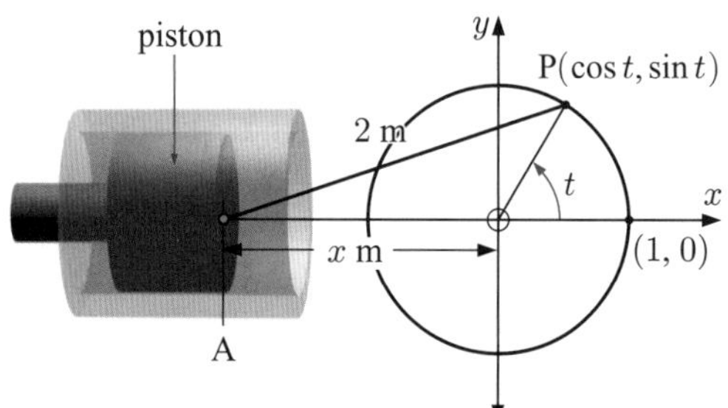

C OPTIMISATION

There are many problems for which we need to find the **maximum** or **minimum** value of a function. The solution is often referred to as the **optimum** solution and the process is called **optimisation**.

We can find optimum solutions in several ways:

- using technology to graph the function and search for the maximum or minimum value
- using analytical methods such as the formula $x = -\dfrac{b}{2a}$ for the vertex of a parabola
- using differential calculus to locate the turning points of a function.

These last two methods are useful especially when exact solutions are required.

WARNING

The maximum or minimum value does not always occur when the first derivative is zero. It is essential to also examine the values of the function at the endpoint(s) of the interval under consideration for global maxima and minima.

For example:

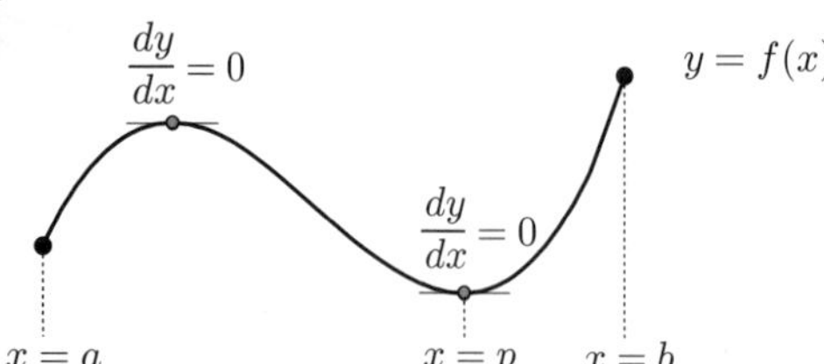

The maximum value of y occurs at the endpoint $x = b$. The minimum value of y occurs at the local minimum $x = p$.

TESTING OPTIMAL SOLUTIONS

If one is trying to optimise a function $f(x)$ and we find values of x such that $f'(x) = 0$, there are several tests we can use to see whether we have a maximum or a minimum solution:

SIGN DIAGRAM TEST

If, near to $x = a$ where $f'(a) = 0$ the sign diagram is:

- 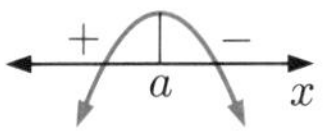

we have a **local maximum**
- 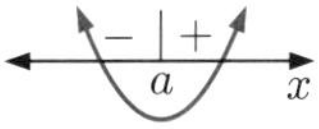

we have a **local minimum**.

SECOND DERIVATIVE TEST

At $x = a$ where $f'(a) = 0$:

- If $\dfrac{d^2y}{dx^2} < 0$ we have  shape, which indicates we have a **local maximum**.
- If $\dfrac{d^2y}{dx^2} > 0$ we have  shape, which indicates we have a **local minimum**.

GRAPHICAL TEST

If the graph of $y = f(x)$ shows:

-  we have a **local maximum**
-  we have a **local minimum**.

OPTIMISATION PROBLEM SOLVING METHOD

Step 1: Draw a large, clear diagram of the situation.

Step 2: Construct a formula with the variable to be **optimised** as the subject. It should be written in terms of **one** convenient **variable**, for example x. You should write down what domain restrictions there are on x.

Step 3: Find the **first derivative** and find the values of x which make the first derivative **zero**.

Step 4: For a restricted domain such as $a \leqslant x \leqslant b$, the maximum or minimum may occur either when the derivative is zero, at an endpoint, or at a point where the derivative is not defined. Show using the **sign diagram test**, the **second derivative test**, or the **graphical test**, that you have a maximum or a minimum.

Step 5: Write your answer in a sentence, making sure you specifically answer the question.

Example 6 **Self Tutor**

A rectangular cake dish is made by cutting out squares from the corners of a 25 cm by 40 cm rectangle of tin-plate, and then folding the metal to form the container.

What size squares must be cut out to produce the cake dish of maximum volume?

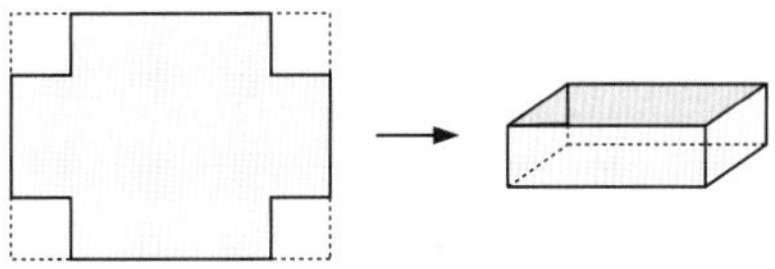

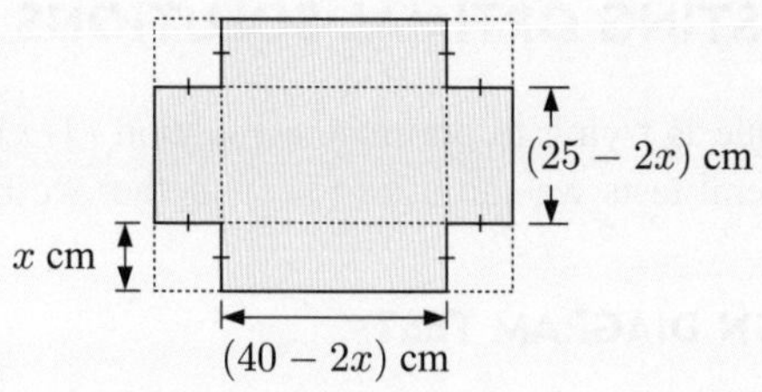

Step 1: Let x cm be the side lengths of the squares that are cut out.

Step 2: Volume = length × width × depth

$$= (40-2x)(25-2x)x$$
$$= (1000 - 80x - 50x + 4x^2)x$$
$$= 1000x - 130x^2 + 4x^3 \text{ cm}^3$$

Since the side lengths must be positive, $x > 0$ and $25 - 2x > 0$.

$\therefore \quad 0 < x < 12.5$

Step 3:

$$\frac{dV}{dx} = 12x^2 - 260x + 1000$$
$$= 4(3x^2 - 65x + 250)$$
$$= 4(3x - 50)(x - 5)$$

$\therefore \quad \frac{dV}{dx} = 0 \quad$ when $\quad x = \frac{50}{3} \quad$ or $\quad x = 5$

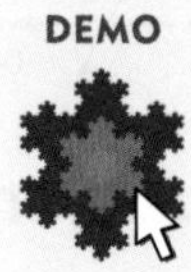

Step 4: **Sign diagram test** *or* **Second derivative test**

$\frac{dV}{dx}$ has sign diagram:

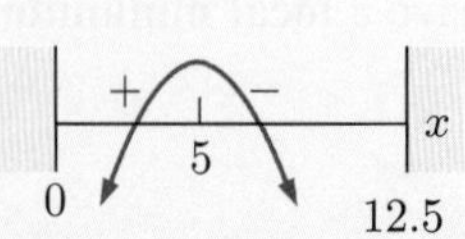

$$\frac{d^2V}{dx^2} = 24x - 260$$

When $x = 5$, $\quad \frac{d^2V}{dx^2} = -140$ which is < 0

$\therefore$ the shape is ∩ and we have a local maximum.

Step 5: The maximum volume is obtained when $x = 5$, which is when 5 cm squares are cut from the corners.

Example 7 **Self Tutor**

A 4 litre container must have a square base, vertical sides, and an open top. Find the most economical shape which minimises the surface area of material needed.

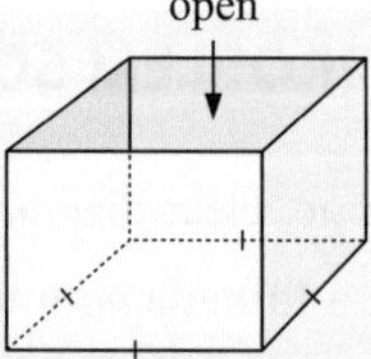

Step 1:

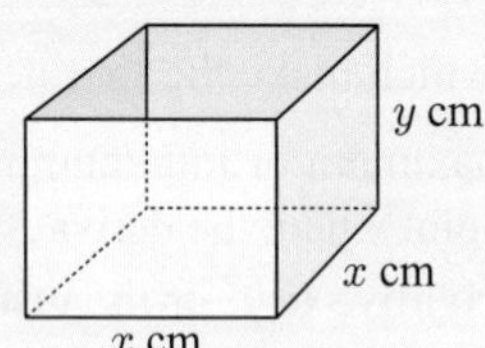

Let the base lengths be x cm and the depth be y cm.

The volume $\quad V = \text{length} \times \text{width} \times \text{depth}$

$\therefore \quad V = x^2 y$

$\therefore \quad 4000 = x^2 y \quad .... (1) \quad \{1 \text{ litre} \equiv 1000 \text{ cm}^3\}$

Step 2: The total surface area $\quad A = \text{area of base} + 4(\text{area of one side})$

$$= x^2 + 4xy$$
$$= x^2 + 4x\left(\frac{4000}{x^2}\right) \qquad \{\text{using (1)}\}$$

$\therefore \quad A(x) = x^2 + 16\,000x^{-1} \quad$ where $\quad x > 0$

Step 3: $\quad A'(x) = 2x - 16\,000x^{-2}$

$\therefore \quad A'(x) = 0 \quad$ when $\quad 2x = 16\,000x^{-2}$

$\therefore \quad 2x^3 = 16\,000$

$\therefore \quad x = \sqrt[3]{8000} = 20$

Step 4: **Sign diagram test** *or* **Second derivative test**

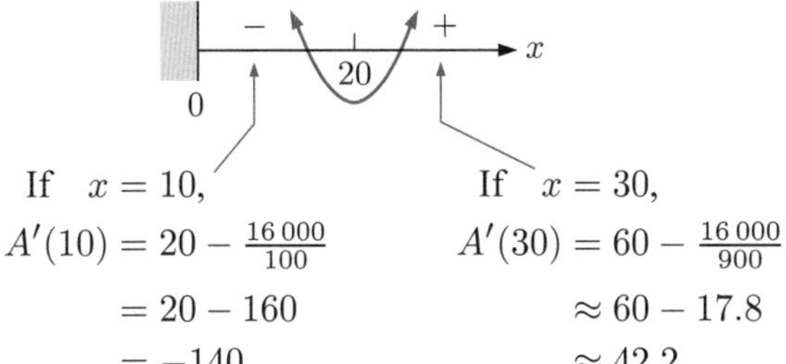

If $x = 10$,
$A'(10) = 20 - \frac{16\,000}{100}$
$= 20 - 160$
$= -140$

If $x = 30$,
$A'(30) = 60 - \frac{16\,000}{900}$
$\approx 60 - 17.8$
≈ 42.2

$A''(x) = 2 + 32\,000x^{-3}$
$= 2 + \dfrac{32\,000}{x^3}$

which is always positive as $x^3 > 0$ for all $x > 0$.

The minimum material is used to make the container when $x = 20$ and $y = \dfrac{4000}{20^2} = 10$.

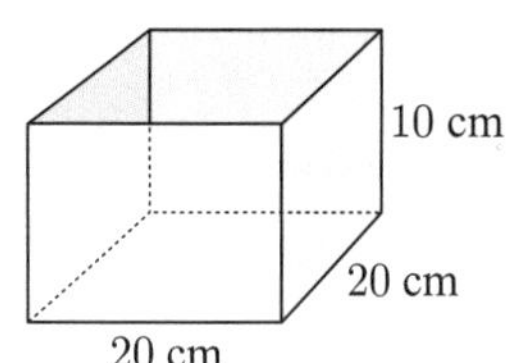

Step 5: The most economical shape has a square base 20 cm × 20 cm, and height 10 cm.

Sometimes the variable to be optimised is in the form of a single square root function. In these situations it is convenient to square the function and use the fact that if $A(x) > 0$ for all x in the interval under consideration, then the optimum value of $A(x)$ occurs at the same value of x as the optimum value of $[A(x)]^2$.

Example 8 **Self Tutor**

An animal enclosure is a right angled triangle with one side being a drain. The farmer has 300 m of fencing available for the other two sides, [AB] and [BC].

a If AB $= x$ m, show that AC $= \sqrt{90\,000 - 600x}$.

b Find the maximum possible area of the triangular enclosure.

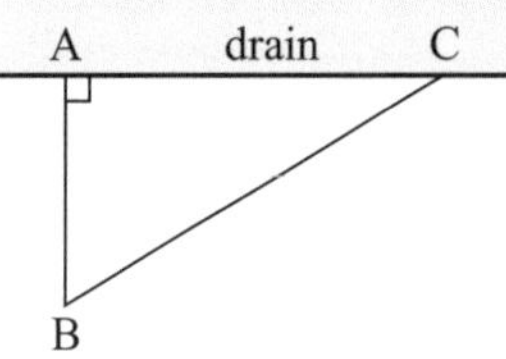

a $(AC)^2 + x^2 = (300 - x)^2$ {Pythagoras}

$\therefore (AC)^2 = 90\,000 - 600x + x^2 - x^2$

$= 90\,000 - 600x$

$\therefore AC = \sqrt{90\,000 - 600x}, \quad 0 < x < 300$

b The area of triangle ABC is

$A(x) = \frac{1}{2}(\text{base} \times \text{altitude})$

$= \frac{1}{2}(AC \times x)$

$= \frac{1}{2}x\sqrt{90\,000 - 600x}$

$\therefore [A(x)]^2 = \dfrac{x^2}{4}(90\,000 - 600x)$

$= 22\,500x^2 - 150x^3$

$\therefore \dfrac{d}{dx}[A(x)]^2 = 45\,000x - 450x^2$

$= 450x(100 - x)$

$\therefore \dfrac{d}{dx}[A(x)]^2$ has sign diagram:

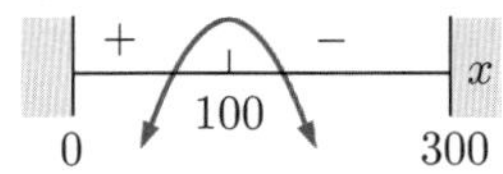

$A(x)$ is maximised when $x = 100$

$\therefore A_{\max} = \frac{1}{2}(100)\sqrt{90\,000 - 60\,000} \approx 8660 \text{ m}^2$

The maximum area is about 8660 m^2 when the enclosure has the side lengths shown.

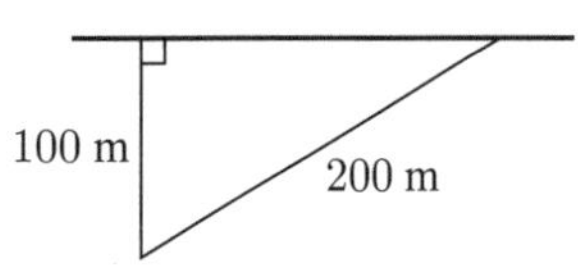

Use **calculus techniques** to answer the following problems.

GRAPHING PACKAGE

In cases where finding the zeros of the derivatives is difficult you may use a **graphics calculator** or **graphing package** to help you.

EXERCISE 20C

1 When a manufacturer makes x items per day, the cost function is $C(x) = 720 + 4x + 0.02x^2$ dollars and the price function is $p(x) = 15 - 0.002x$ dollars per item. Find the production level that will maximise profits.

2 A duck farmer wishes to build a rectangular enclosure of area 100 m^2. The farmer must purchase wire netting for three of the sides as the fourth side is an existing fence. Naturally, the farmer wishes to minimise the length (and therefore cost) of fencing required to complete the job.

a If the shorter sides have length x m, show that the required length of wire netting to be purchased is $L = 2x + \dfrac{100}{x}$.

b Use *technology* to help you sketch the graph of $y = 2x + \dfrac{100}{x}$.

c Find the minimum value of L and the corresponding value of x when this occurs.

d Sketch the optimum situation showing all dimensions.

3 A manufacturer can produce x fittings per day where $0 \leqslant x \leqslant 10\,000$. The production costs are:

- €1000 per day for the workers
- €2 per day per fitting
- €$\dfrac{5000}{x}$ per day for running costs and maintenance.

How many fittings should be produced daily to minimise the total production costs?

4 The total cost of producing x blankets per day is $\frac{1}{4}x^2 + 8x + 20$ dollars, and for this production level each blanket may be sold for $(23 - \frac{1}{2}x)$ dollars.

How many blankets should be produced per day to maximise the total profit?

5 The cost of running a boat is £$\left(\dfrac{v^2}{10} + 22\right)$ per hour, where v km h^{-1} is the speed of the boat.

Find the speed which will minimise the total cost per kilometre.

6 A psychologist claims that the ability A to memorise simple facts during infancy years can be calculated using the formula $A(t) = t\ln t + 1$ where $0 < t \leqslant 5$, t being the age of the child in years.

a At what age is the child's memorising ability a minimum?

b Sketch the graph of $A(t)$ for $0 < t \leqslant 5$.

7 A manufacturing company makes door hinges. They have a standing order filled by producing 50 each hour, but production of more than 150 per hour is useless as they will not sell. The cost function for making x hinges per hour is:

$$C(x) = 0.0007x^3 - 0.1796x^2 + 14.663x + 160 \text{ dollars where } 50 \leqslant x \leqslant 150.$$

Find the minimum and maximum hourly costs, and the production levels when each occurs.

8 A manufacturer of electric kettles performs a cost control study. They discover that to produce x kettles per day, the cost per kettle is given by $C(x) = 4\ln x + \left(\dfrac{30 - x}{10}\right)^2$ dollars with a minimum production capacity of 10 kettles per day.

How many kettles should be manufactured to keep the cost per kettle to a minimum?

9 Radioactive waste is to be disposed of in fully enclosed lead boxes of inner volume 200 cm^3. The base of the box has dimensions in the ratio $2:1$.

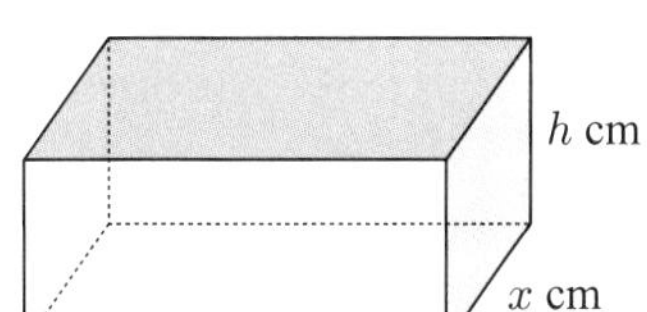

a What is the inner length of the box?

b Explain why $x^2h = 100$.

c Explain why the inner surface area of the box is given by $A(x) = 4x^2 + \frac{600}{x}$ cm^2.

d Use *technology* to help sketch the graph of $y = 4x^2 + \frac{600}{x}$.

e Find the minimum inner surface area of the box and the corresponding value of x.

f Sketch the optimum box shape showing all dimensions.

10 Infinitely many rectangles which sit on the x-axis can be inscribed under the curve $y = e^{-x^2}$. Determine the coordinates of C such that rectangle ABCD has maximum area.

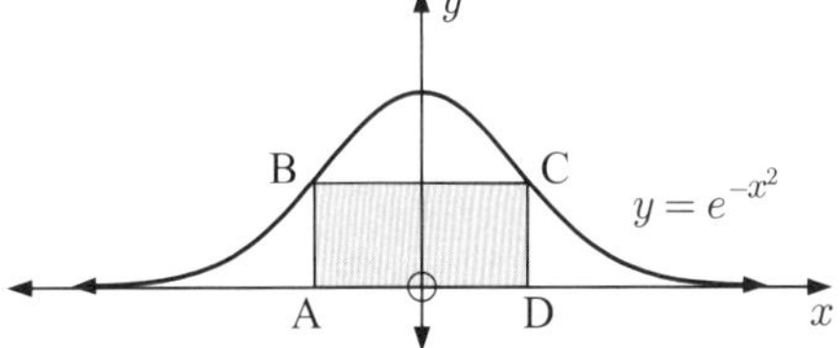

11 Consider the manufacture of cylindrical tin cans of 1 L capacity, where the cost of the metal used is to be minimised.

a Explain why the height h is given by $h = \frac{1000}{\pi r^2}$ cm.

b Show that the total surface area A is given by $A = 2\pi r^2 + \frac{2000}{r}$ cm^2.

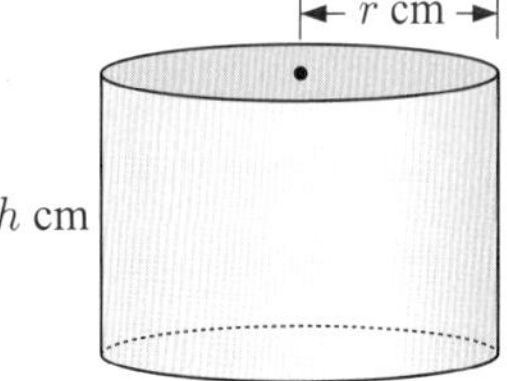

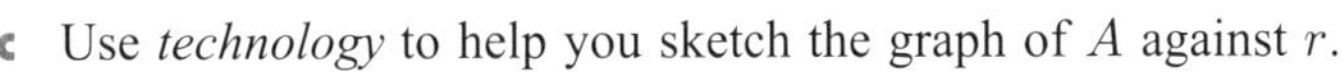

c Use *technology* to help you sketch the graph of A against r.

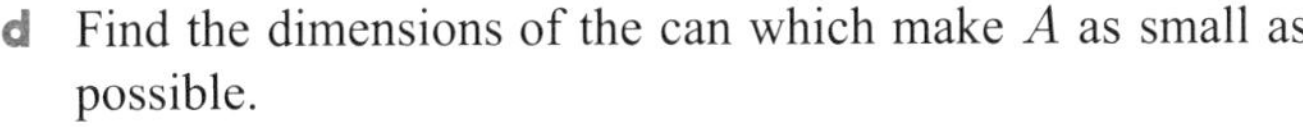

d Find the dimensions of the can which make A as small as possible.

e Sketch the can of smallest surface area.

12 A circular piece of tinplate of radius 10 cm has 3 segments removed as illustrated. The angle θ is measured in radians.

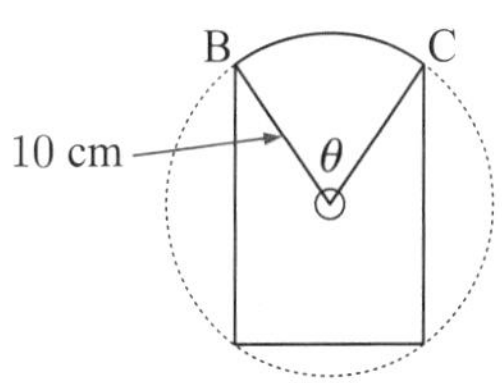

a Show that the remaining area is given by $A = 50(\theta + 3\sin\theta)$ cm^2.

b Find θ such that the area A is a maximum, and also the area A in this case.

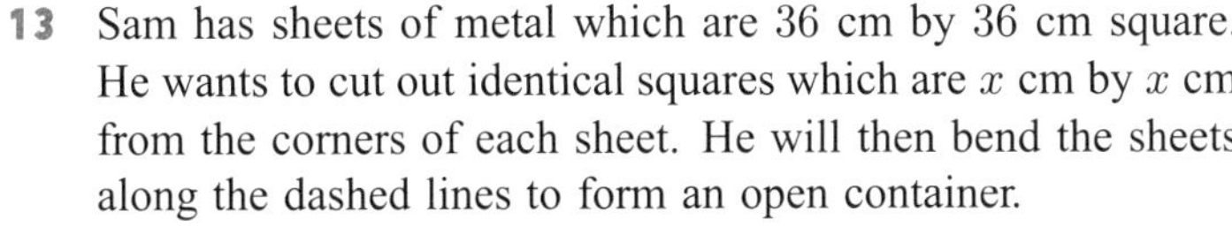

13 Sam has sheets of metal which are 36 cm by 36 cm square. He wants to cut out identical squares which are x cm by x cm from the corners of each sheet. He will then bend the sheets along the dashed lines to form an open container.

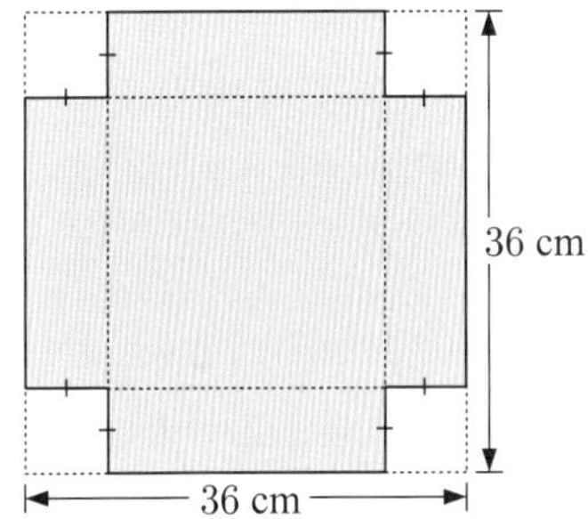

a Show that the volume of the container is given by $V(x) = x(36 - 2x)^2$ cm^3.

b What sized squares should be cut out to produce the container of greatest capacity?

c Suppose the initial square was a cm by a cm. Prove that the volume of the container will be maximised when squares of side length $\frac{a}{6}$ cm are cut from each corner.

14 An athletics track has two 'straights' of length l m, and two semi-circular ends of radius x m. The perimeter of the track is 400 m.

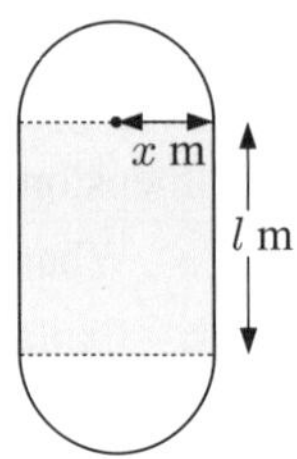

- **a** Show that $l = 200 - \pi x$ and write down the possible values that x may have.
- **b** What values of l and x maximise the shaded rectangle inside the track? What is this maximum area?

15 A small population of wasps is observed. After t weeks the population is modelled by $P(t) = \dfrac{50\,000}{1 + 1000e^{-0.5t}}$ wasps, where $0 \leqslant t \leqslant 25$.
Find when the wasp population is growing fastest.

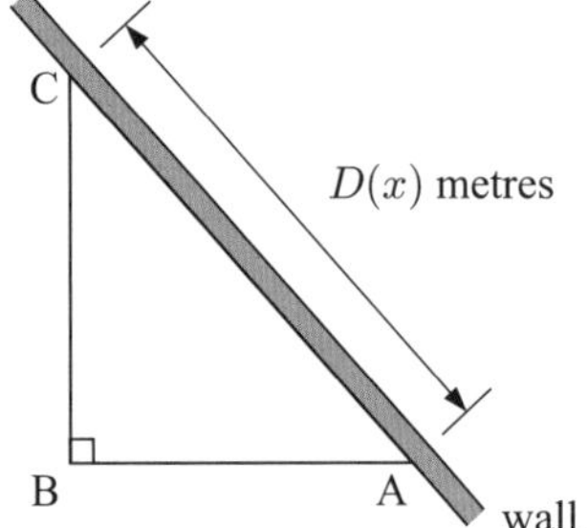

16 When a new pain killing injection is administered, the effect is modelled by $E(t) = 750te^{-1.5t}$ units, where $t \geqslant 0$ is the time in hours after the injection.
At what time is the drug most effective?

17 A right angled triangular pen is made from 24 m of fencing, all used for sides [AB] and [BC]. Side [AC] is an existing brick wall.

- **a** If AB $= x$ m, find $D(x)$ in terms of x.
- **b** Find $\dfrac{d[D(x)]^2}{dx}$ and hence draw its sign diagram.
- **c** Find the smallest possible value of $D(x)$ and the design of the pen in this case.

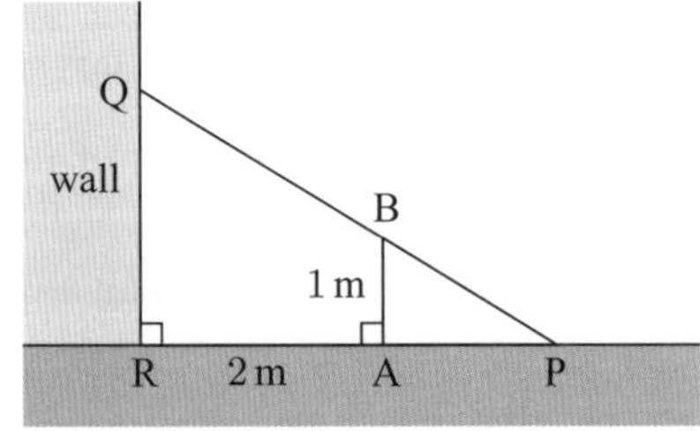

18 [AB] is a 1 m high fence which is 2 m from a vertical wall [RQ]. An extension ladder [PQ] rests on the fence so that it touches the ground at P and the wall at Q.

- **a** If AP $= x$ m, find QR in terms of x.
- **b** If the ladder has length L m, show that $[L(x)]^2 = (x+2)^2\left(1 + \dfrac{1}{x^2}\right)$.
- **c** Show that $\dfrac{d[L(x)]^2}{dx} = 0$ only when $x = \sqrt[3]{2}$.
- **d** Find, correct to the nearest centimetre, the shortest length of the extension ladder. You must prove that this length is the shortest.

19 A symmetrical gutter is made from a sheet of metal 30 cm wide by bending it twice as shown.

- **a** Deduce that the cross-sectional area of the gutter is given by $A = 100\cos\theta(1 + \sin\theta)$.
- **b** Show that $\dfrac{dA}{d\theta} = 0$ when $\sin\theta = \frac{1}{2}$ or -1.
- **c** For what value of θ does the gutter have maximum carrying capacity? Find the cross-sectional area for this value of θ.

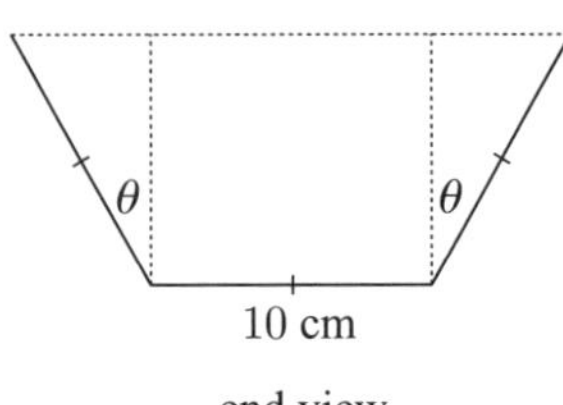

20 A sector of radius 10 cm and angle $\theta°$ is bent to form a conical cup as shown.

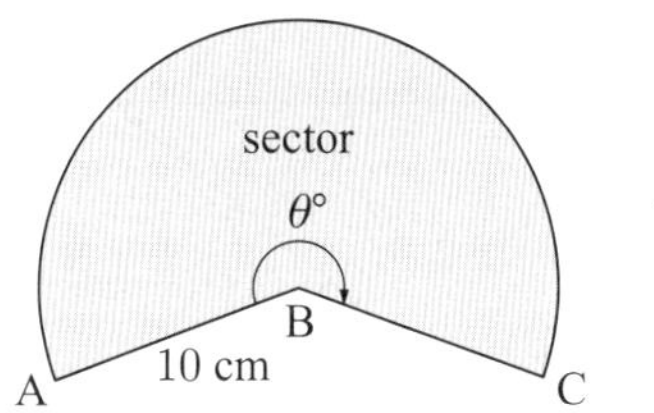

becomes

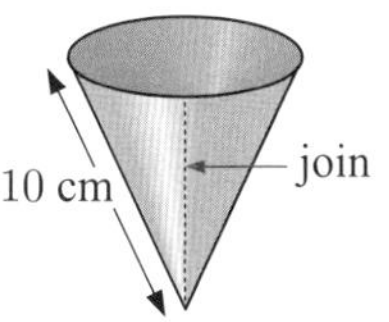

when edges [AB] and [CB] are joined with tape.

Suppose the resulting cone has base radius r cm and height h cm.

a Show using the sector that arc AC $= \frac{\theta\pi}{18}$.

b Explain why $r = \frac{\theta}{36}$.

c Show that $h = \sqrt{100 - \left(\frac{\theta}{36}\right)^2}$.

d Find the cone's capacity V in terms of θ only.

e Use technology to sketch the graph of $V(\theta)$.

f Find θ when $V(\theta)$ is a maximum.

21 At 1:00 pm a ship A leaves port P. It sails in the direction 030° at 12 km h^{-1}. At the same time, ship B is 100 km due east of P, and is sailing at 8 km h^{-1} towards P.

a Show that the distance $D(t)$ between the two ships is given by $D(t) = \sqrt{304t^2 - 2800t + 10\,000}$ km, where t is the number of hours after 1:00 pm.

b Find the minimum value of $[D(t)]^2$ for all $t \geqslant 0$.

c At what time, to the nearest minute, are the ships closest?

22 B is a row boat 5 km out at sea from A. [AC] is a straight sandy beach, 6 km long. Peter can row the boat at 8 km h^{-1} and run along the beach at 17 km h^{-1}. Suppose Peter rows directly from B to point X on [AC] such that AX $= x$ km.

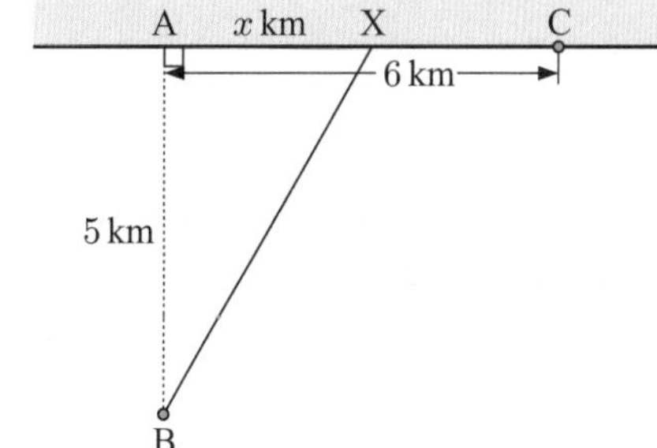

a Explain why $0 \leqslant x \leqslant 6$.

b Show that the *total time* Peter takes to row to X and then run along the beach to C, is given by $T(x) = \frac{\sqrt{x^2+25}}{8} + \frac{6-x}{17}$ hours, $0 \leqslant x \leqslant 6$.

c Find x such that $\frac{dT}{dx} = 0$. Explain the significance of this value.

23 A pumphouse is to be placed at some point X along a river.
Two pipelines will then connect the pumphouse to homesteads A and B.
How far from M should point X be so that the total length of pipeline is minimised?

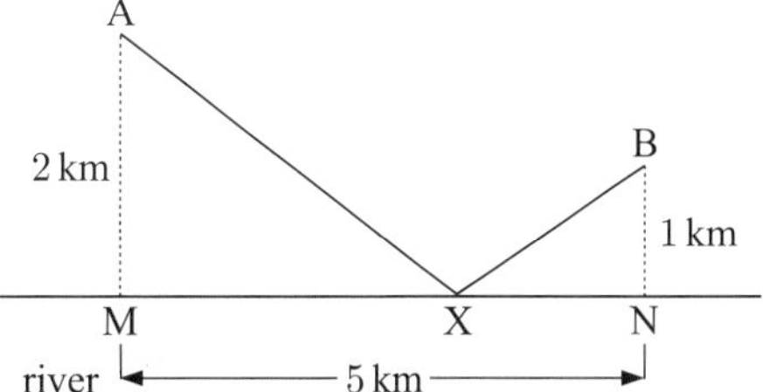

24 Two lamps have intensities 40 and 5 candle-power respectively. They are placed 6 m apart. If the intensity of illumination I at any point is directly proportional to the power of the source, and inversely proportional to the square of the distance from the source, find the darkest point on the line joining the two lamps.

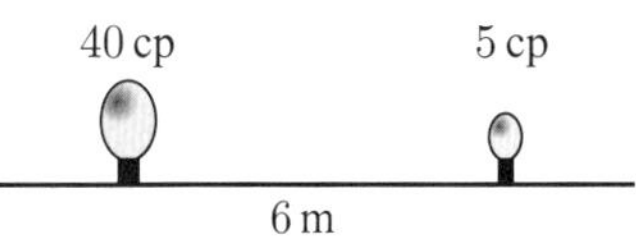

Example 9 **Self Tutor**

Two corridors meet at right angles and are 2 m and 3 m wide respectively. θ is the angle marked on the given figure. [AB] is a thin metal tube which must be kept horizontal and cannot be bent as it moves around the corner from one corridor to the other.

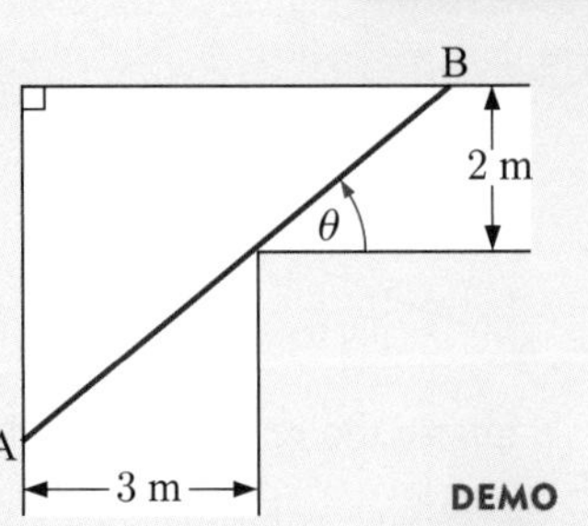

a Show that the length AB is given by $L = \dfrac{3}{\cos\theta} + \dfrac{2}{\sin\theta}$.

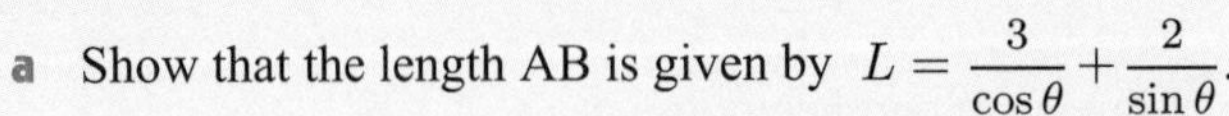

b Show that $\dfrac{dL}{d\theta} = 0$ when $\theta = \tan^{-1}\left(\sqrt[3]{\frac{2}{3}}\right) \approx 41.1°$.

c Find L when $\theta = \tan^{-1}\left(\sqrt[3]{\frac{2}{3}}\right)$ and comment on the significance of this value.

a $\cos\theta = \dfrac{3}{a}$ and $\sin\theta = \dfrac{2}{b}$

$\therefore\ a = \dfrac{3}{\cos\theta}$ and $b = \dfrac{2}{\sin\theta}$

$\therefore\ L = a + b = \dfrac{3}{\cos\theta} + \dfrac{2}{\sin\theta}$

b $L = 3[\cos\theta]^{-1} + 2[\sin\theta]^{-1}$

$$\therefore\ \frac{dL}{d\theta} = -3[\cos\theta]^{-2}(-\sin\theta) - 2[\sin\theta]^{-2}\cos\theta$$

$$= \frac{3\sin\theta}{\cos^2\theta} - \frac{2\cos\theta}{\sin^2\theta}$$

$$= \frac{3\sin^3\theta - 2\cos^3\theta}{\cos^2\theta\sin^2\theta}$$

Thus $\dfrac{dL}{d\theta} = 0$ when $3\sin^3\theta = 2\cos^3\theta$

$\therefore\ \tan^3\theta = \frac{2}{3}$

$\therefore\ \tan\theta = \sqrt[3]{\frac{2}{3}}$

$\therefore\ \theta = \tan^{-1}\left(\sqrt[3]{\frac{2}{3}}\right) \approx 41.1°$

c *Sign diagram* of $\dfrac{dL}{d\theta}$:

When $\theta = 30°$, $\dfrac{dL}{d\theta} \approx -4.93 < 0$

When $\theta = 60°$, $\dfrac{dL}{d\theta} \approx 9.06 > 0$

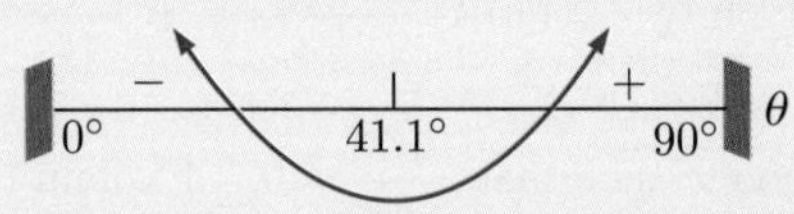

Thus, AB is minimised when $\theta \approx 41.1°$. At this time $L \approx 7.02$ metres. Ignoring the width of the rod, the greatest length of rod able to be horizontally carried around the corner is 7.02 m.

25 In a hospital, two corridors 4 m wide and 3 m wide meet at right angles. What is the maximum possible length of an X-ray screen which can be carried upright around the corner?

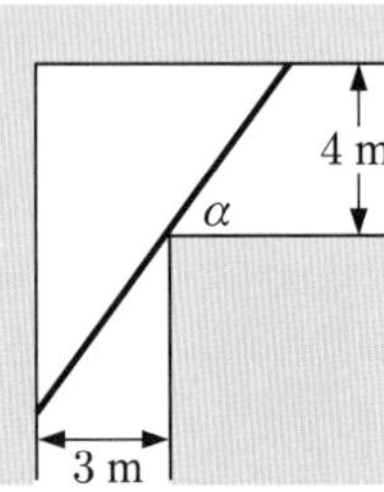

26 Fence [AB] is 2 m high and is 2 m from a house. [XY] is a ladder which touches the ground at X, the house at Y, and the fence at B.

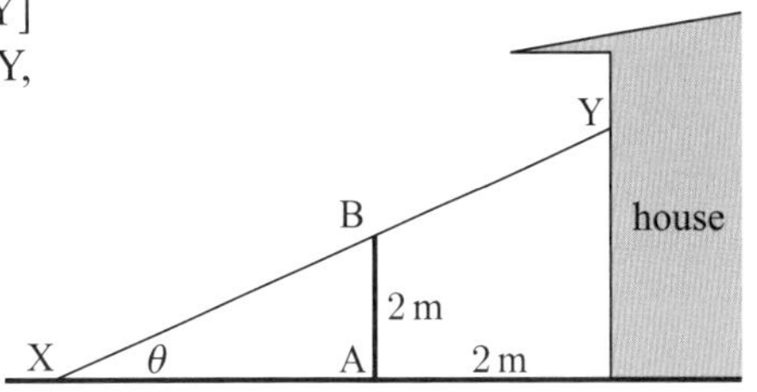

a If L is the length of [XY], show that $L = 2\sec\theta + 2\csc\theta$.

b Show that $\dfrac{dL}{d\theta} = \dfrac{2\sin^3\theta - 2\cos^3\theta}{\sin^2\theta\cos^2\theta}$.

c Find the length of the shortest ladder [XY] which touches at X, B, and Y.

27

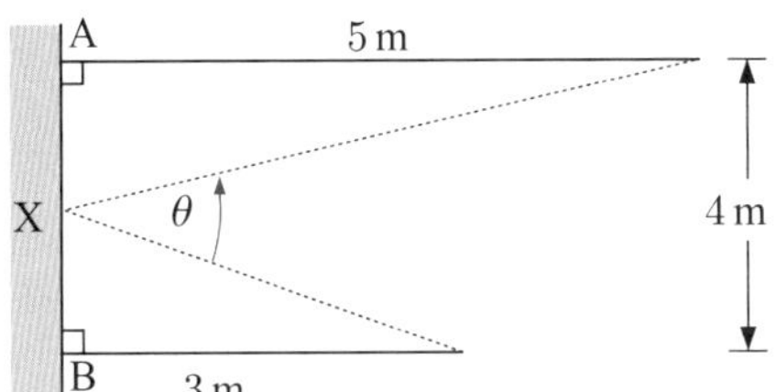

How far should X be from A for angle θ to be maximised?

28 Hieu can row a boat across a circular lake of radius 2 km at 3 km h^{-1}. He can walk around the edge of the lake at 5 km h^{-1}.
What is the longest possible time Hieu could take to get from P to R by rowing from P to Q and then walking from Q to R?

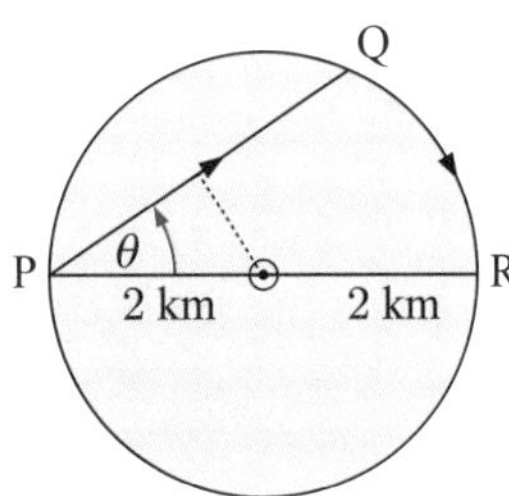

29 Sonia approaches a painting which has its bottom edge 2 m above eye level and its top edge 3 m above eye level.

a Given α and θ as shown in the diagram, find $\tan\alpha$ and $\tan(\alpha+\theta)$.

b Find θ in terms of x only.
Hint: $\theta = (\alpha+\theta) - \alpha$.

c Show that $\dfrac{d\theta}{dx} = \dfrac{2}{x^2+4} - \dfrac{3}{x^2+9}$ and hence find x when $\dfrac{d\theta}{dx} = 0$.

d Interpret the result you have found in **c**.

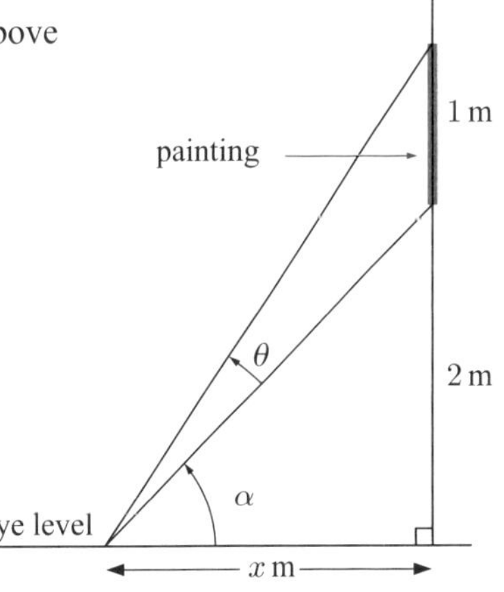

30

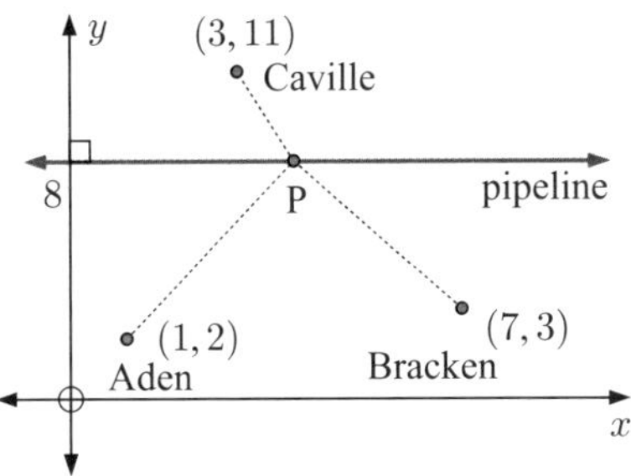

Three towns and their grid references are marked on the diagram alongside. A pumping station is to be located at P on the pipeline, to pump water to the three towns. The grid units are kilometres.
Exactly where should P be located so that the total length of the pipelines to Aden, Bracken, and Caville is minimised? What is the shortest total length of pipe required?

31 A, B, and C are computers which are networked to a printer P. Where should P be located so that the total cable length $AP + BP + CP$ is a minimum?

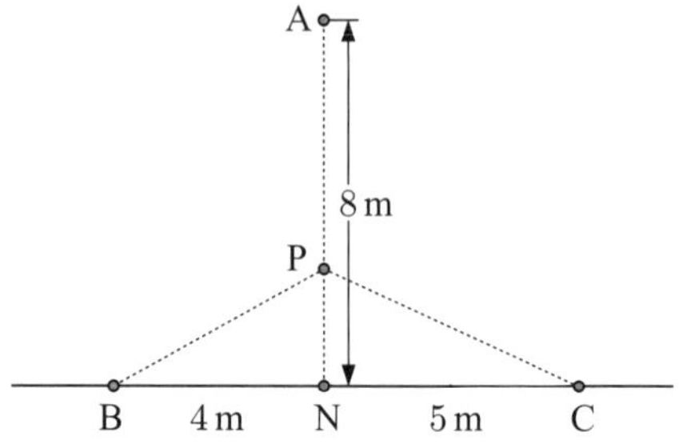

32

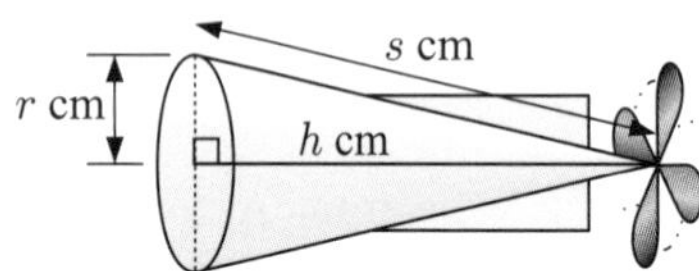

The back end of a guided long range torpedo is to be conical with slant edge s cm, and will be filled with fuel.

Find the ratio of $s : r$ such that the fuel carrying capacity is maximised.

33 A company constructs rectangular seating arrangements for pop concerts on sports grounds. The oval shown has equation $\frac{x^2}{a^2} + \frac{y^2}{b^2} = 1$ where a and b are the lengths of the semi-major and semi-minor axes.

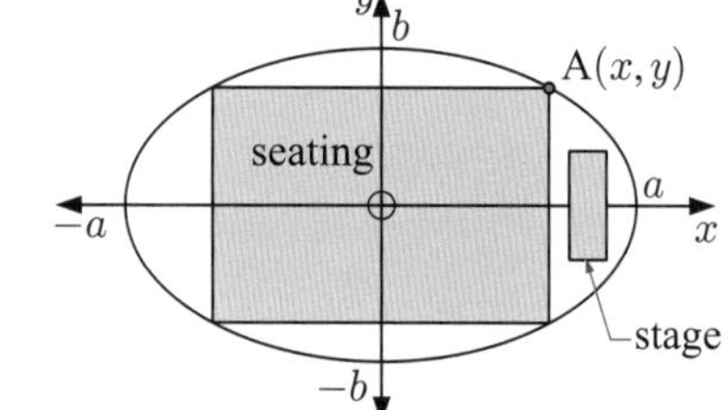

a Show that $y = \frac{b}{a}\sqrt{a^2 - x^2}$ for A as shown.

b Show that the seating area is given by $A(x) = \frac{4bx}{a}\sqrt{a^2 - x^2}$.

c Prove that the seating area is maximised when $x = \frac{a}{\sqrt{2}}$.

d Given that the area of the ellipse is πab, what percentage of the ground is occupied by the seats in the optimum case?

THEORY OF KNOWLEDGE

"Aristotle is recognized as the inventor of scientific method because of his refined analysis of logical implications contained in demonstrative discourse, which goes well beyond natural logic and does not owe anything to the ones who philosophized before him."

– Riccardo Pozzo

A **scientific method** of inquiry for investigating phenomena has been applied in varying degrees throughout the course of history. The first formal statement of such a method was made by René Descartes in his *Discourse on the Method* published in 1637. This work is perhaps best known for Descartes' quote, "*Je pense, donc je suis*" which means "I think, therefore I am". In 1644 in his *Principles of Philosophy* he published the same quote in Latin: "*Cogito ergo sum*".

The scientific method involves a series of steps:

Step 1: asking a question (how, when, why, ...)
Step 2: conducting appropriate research
Step 3: constructing a hypothesis, or possible explanation why things are so
Step 4: testing the hypothesis by a fair experiment
Step 5: analysing the results
Step 6: drawing a conclusion
Step 7: communicating your findings

Snell's law states the relationship between the angles of incidence and refraction when a ray of light passes from one homogeneous medium to another.

It was first discovered in 984 AD by the Persian scientist Ibn Sahl, who was studying the shape of lenses. However, it is named after Willebrord Snellius, one of those who rediscovered the law in the Renaissance. The law was published by Descartes in the *Discourse on the Method.*

In the figure alongside, a ray passes from A to B via point X. The refractive indices of the two media are n and m. The angle of incidence is α and the angle of refraction is β.

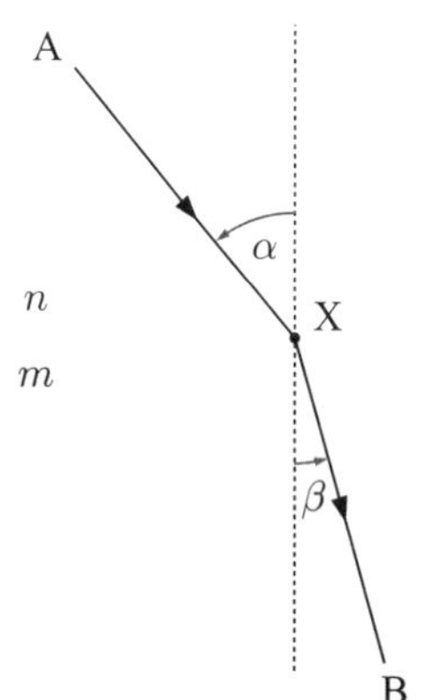

Snell's law is: $\boldsymbol{n \sin \alpha = m \sin \beta}$.

The law follows from Fermat's principle of least time. It gives the path of least time for the ray travelling from A to B.

1 Is optimisation unique to mathematics?

2 How does mathematics fit into the scientific method?

3 Does mathematics have a prescribed method of its own?

4 Is mathematics a science?

D RELATED RATES

A 5 m ladder rests against a vertical wall at point B. Its feet are at point A on horizontal ground.

The ladder slips and slides down the wall.

Click on the icon to view the motion of the sliding ladder.

DEMO

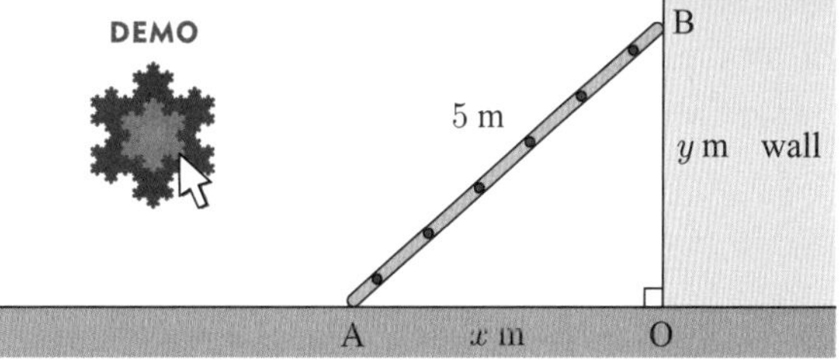

The following diagram shows the positions of the ladder at certain instances.

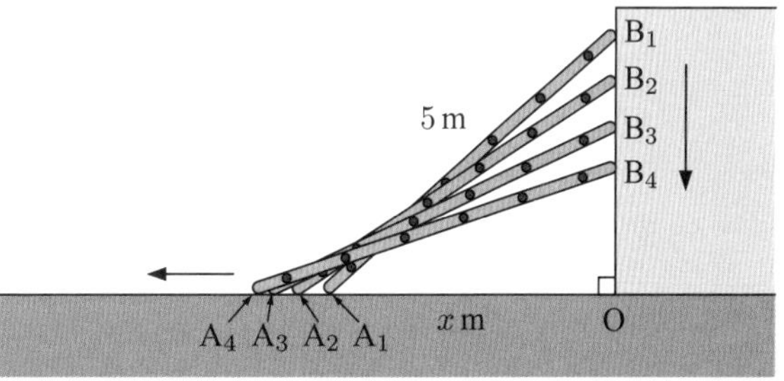

If AO $= x$ m and OB $= y$ m,
then $x^2 + y^2 = 5^2$. {Pythagoras}

Differentiating this equation with respect to time t gives $2x\dfrac{dx}{dt} + 2y\dfrac{dy}{dt} = 0$

or $x\dfrac{dx}{dt} + y\dfrac{dy}{dt} = 0$.

This equation is called a **differential equation** and describes the motion of the ladder at any instant.

$\dfrac{dx}{dt}$ is the rate of change in x with respect to time t, and is the speed of A relative to point O.

$\dfrac{dx}{dt}$ is *positive* as x is increasing.

$\dfrac{dy}{dt}$ is the rate of change in y with respect to time t, and is the speed at which B moves downwards.

$\dfrac{dy}{dt}$ is *negative* as y is decreasing.

Problems involving differential equations where one of the variables is time t are called **related rates** problems.

The method for solving related rates problems is:

Step 1: Draw a large, clear **diagram** of the situation. Sometimes two or more diagrams are necessary.

Step 2: Write down the information, label the diagram(s), and make sure you distinguish between the **variables** and the **constants**.

Step 3: Write an **equation** connecting the variables. You will often need to use:

- Pythagoras' theorem
- similar triangles where corresponding sides are in proportion
- right angled triangle trigonometry
- sine and cosine rules.

Step 4: **Differentiate** the equation with respect to t to obtain a **differential equation**.

Step 5: Solve for the **particular case** which is some instant in time.

Warning:

We **must not** substitute values for the particular case too early. Otherwise we will incorrectly treat variables as constants. The differential equation in fully generalised form must be established first.

Example 10 — Self Tutor

A 5 m long ladder rests against a vertical wall with its feet on horizontal ground. The feet on the ground slip, and at the instant when they are 3 m from the wall, they are moving at 10 m s^{-1}.

At what speed is the other end of the ladder moving at this instant?

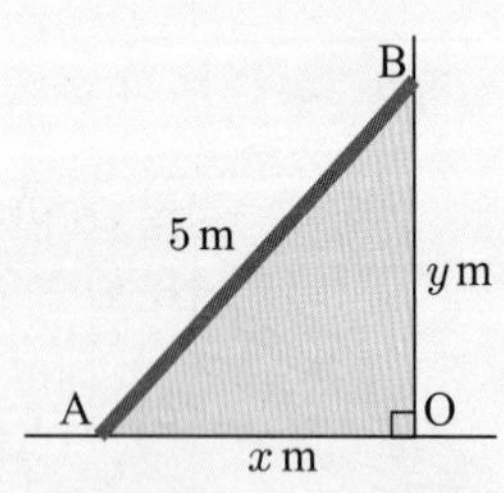

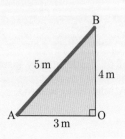

Let OA $= x$ m and OB $= y$ m

$\therefore\ x^2 + y^2 = 5^2$ {Pythagoras}

Differentiating with respect to t gives

$$2x\frac{dx}{dt} + 2y\frac{dy}{dt} = 0$$

$$\therefore\ x\frac{dx}{dt} + y\frac{dy}{dt} = 0$$

Particular case:

At the instant when $\frac{dx}{dt} = 10$ m s^{-1},

$$\therefore\ 3(10) + 4\frac{dy}{dt} = 0$$

$$\therefore\ \frac{dy}{dt} = -\tfrac{15}{2} = -7.5 \text{ m s}^{-1}$$

Thus OB is decreasing at 7.5 m s^{-1}.

$\therefore$ the other end of the ladder is moving down the wall at 7.5 m s^{-1} at that instant.

We must differentiate **before** we substitute values for the particular case. Otherwise we will incorrectly treat the variables as constants.

Example 11 Self Tutor

A cube is expanding so its volume increases at a constant rate of 10 $cm^3\,s^{-1}$. Find the rate of change in its total surface area, at the instant when its sides are 20 cm long.

Let x cm be the lengths of the sides of the cube, so the surface area $A = 6x^2$ cm^2 and the volume $V = x^3$ cm^3.

$\therefore \quad \frac{dA}{dt} = 12x\frac{dx}{dt}$ and $\frac{dV}{dt} = 3x^2\frac{dx}{dt}$

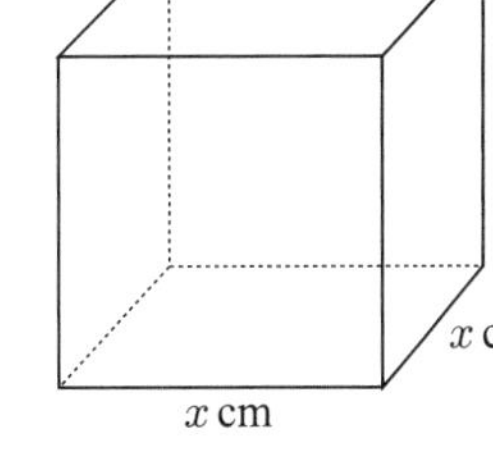

Particular case:

At the instant when $x = 20$, $\frac{dV}{dt} = 10$

$\therefore \quad 10 = 3 \times 20^2 \times \frac{dx}{dt}$

$\therefore \quad \frac{dx}{dt} = \frac{10}{1200} = \frac{1}{120}$ $cm\,s^{-1}$

Thus $\frac{dA}{dt} = 12 \times 20 \times \frac{1}{120}$ $cm^2\,s^{-1}$

$= 2$ $cm^2\,s^{-1}$

$\therefore$ the surface area is increasing at 2 $cm^2\,s^{-1}$.

Example 12 Self Tutor

Triangle ABC is right angled at A, and AB = 20 cm. $\widehat{ABC}$ increases at a constant rate of 1° per minute. At what rate is BC changing at the instant when $\widehat{ABC}$ measures 30°?

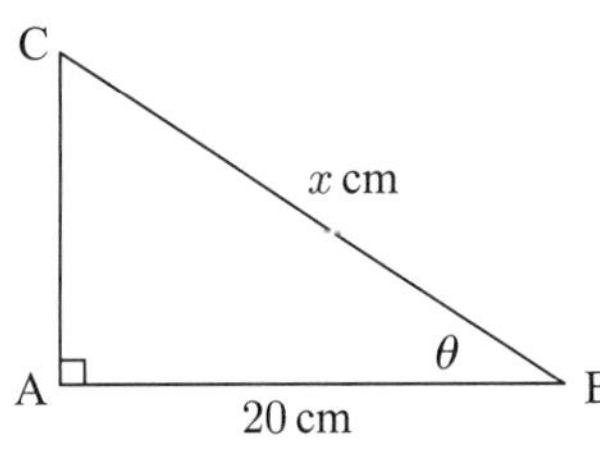

Let $\widehat{ABC} = \theta$ and BC $= x$ cm

Now $\cos\theta = \frac{20}{x} = 20x^{-1}$

$\therefore \quad -\sin\theta\,\frac{d\theta}{dt} = -20x^{-2}\,\frac{dx}{dt}$

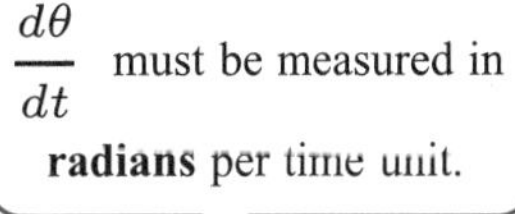

$\frac{d\theta}{dt}$ must be measured in **radians** per time unit.

Particular case:

When $\theta = 30°$, $\cos 30° = \frac{20}{x}$

$\therefore \quad \frac{\sqrt{3}}{2} = \frac{20}{x}$

$\therefore \quad x = \frac{40}{\sqrt{3}}$

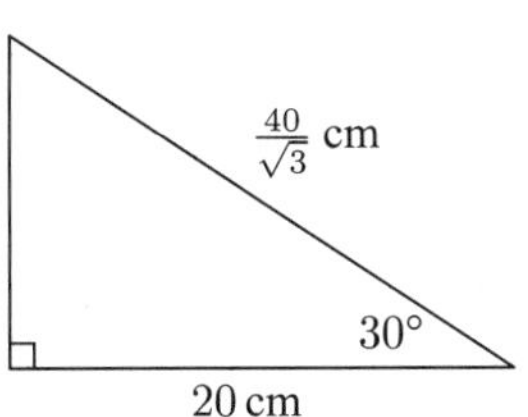

Also, $\frac{d\theta}{dt} = 1°$ per min

$= \frac{\pi}{180}$ radians per min

Thus $-\sin 30° \times \frac{\pi}{180} = -20 \times \frac{3}{1600} \times \frac{dx}{dt}$

$\therefore \quad -\frac{1}{2} \times \frac{\pi}{180} = -\frac{3}{80}\,\frac{dx}{dt}$

$\therefore \quad \frac{dx}{dt} = \frac{\pi}{360} \times \frac{80}{3}$ cm per min

≈ 0.2327 cm per min

$\therefore$ BC is increasing at approximately 0.233 cm per min.

EXERCISE 20D

1 a and b are variables related by the equation $ab^3 = 40$. At the instant when $a = 5$, b is increasing at 1 unit per second. What is happening to a at this instant?

2 The length of a rectangle is decreasing at 1 cm per minute. However, the area of the rectangle remains constant at 100 cm^2. At what rate is the breadth increasing at the instant when the rectangle is a square?

3 A stone is thrown into a lake and a circular ripple moves out at a constant speed of 1 m s^{-1}. Find the rate at which the circle's area is increasing at the instant when:

a $t = 2$ seconds **b** $t = 4$ seconds.

4 Air is pumped into a spherical weather balloon at a constant rate of 6π m^3 per minute. Find the rate of change in its surface area at the instant when the radius of the balloon is 2 m.

5 For a given mass of gas in a piston, $pV^{1.5} = 400$ where p is the pressure in N m^{-2}, and V is the volume in m^3.

Suppose the pressure increases at a constant rate of 3 N m^{-2} per minute. Find the rate at which the volume is changing at the instant when the pressure is 50 N m^{-2}.

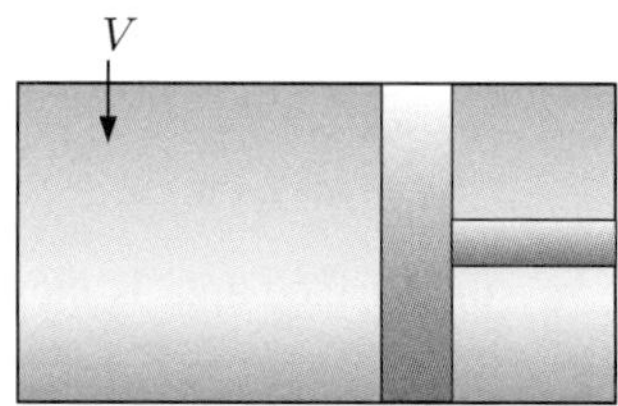

6 Wheat runs from a hole in a silo at a constant rate and forms a conical heap whose base radius is treble its height. After 1 minute, the height of the heap is 20 cm. Find the rate at which the height is rising at this instant.

7 A trough of length 6 m has a uniform cross-section which is an equilateral triangle with sides of length 1 m. Water leaks from the bottom of the trough at a constant rate of 0.1 m^3/min.

Find the rate at which the water level is falling at the instant when the water is 20 cm deep.

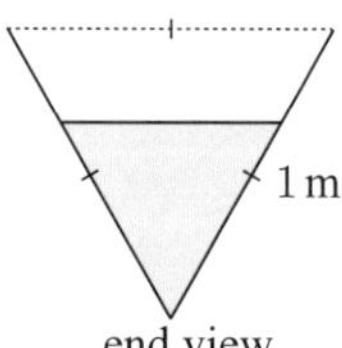

8 Two jet aeroplanes fly on parallel courses which are 12 km apart. Their air speeds are 200 m s^{-1} and 250 m s^{-1} respectively. How fast is the distance between them changing at the instant when the slower jet is 5 km ahead of the faster one?

9 A ground-level floodlight located 40 m from the foot of a building shines in the direction of the building. A 2 m tall person walks directly from the floodlight towards the building at 1 m s^{-1}. How fast is the person's shadow on the building shortening at the instant when the person is:

a 20 m from the building

b 10 m from the building?

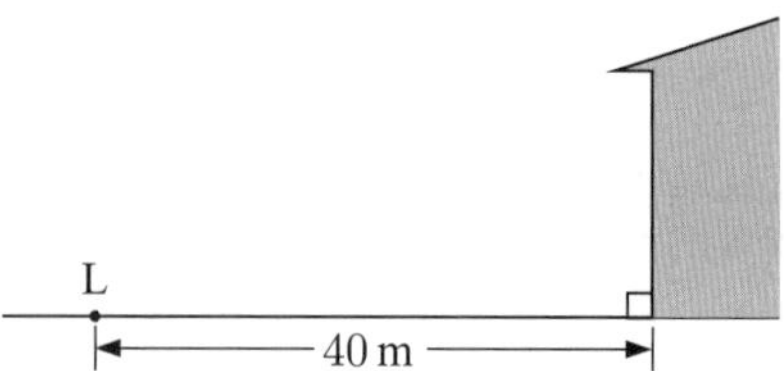

10 A right angled triangle ABC has a fixed hypotenuse [AC] of length 10 cm, and side [AB] increases in length at 0.1 cm s^{-1}. At what rate is $C\widehat{A}B$ decreasing at the instant when the triangle is isosceles?

11 An aeroplane passes directly overhead then flies horizontally away from an observer at an altitude of 5000 m and air speed of 200 m s^{-1}. At what rate is its angle of elevation to the observer changing at the instant when the angle of elevation is:

a 60° **b** 30°?

12 Rectangle PQRS has [PQ] of fixed length 20 cm, and [QR] increases in length at a constant rate of 2 cm s^{-1}. At what rate is the acute angle between the diagonals of the rectangle changing at the instant when [QR] is 15 cm long?

13 Triangle PQR is right angled at Q, and [PQ] is 6 cm long. [QR] increases in length at 2 cm per minute. Find the rate of change in $Q\widehat{P}R$ at the instant when [QR] is 8 cm long.

14 Two cyclists A and B leave X simultaneously at 120° to one another, with constant speeds of 12 m s^{-1} and 16 m s^{-1} respectively. Find the rate at which the distance between them is changing after 2 minutes.

15 AOB is a fixed diameter of a circle of radius 5 cm. Point P moves around the circle at a constant rate of 1 revolution in 10 seconds. Find the rate at which the distance AP is changing at the instant when:

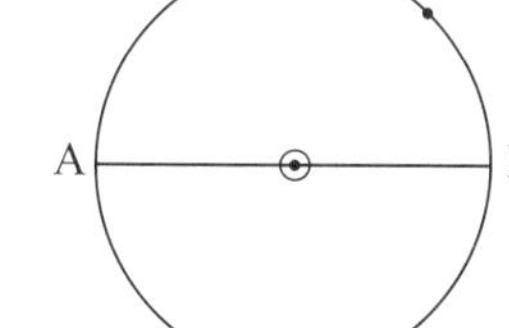

a AP = 5 cm and increasing

b P is at B.

16

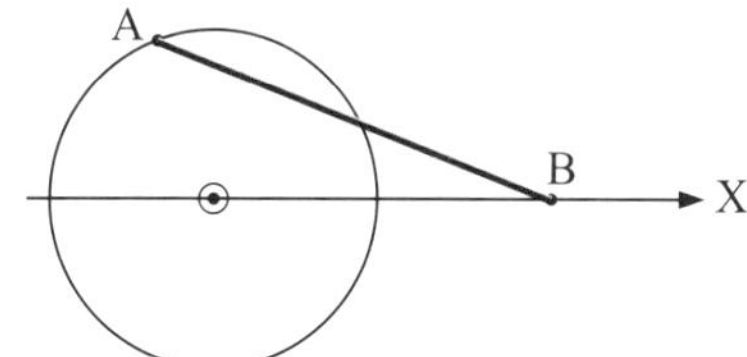

Shaft [AB] is 30 cm long and is attached to a flywheel at A. B is confined to motion along [OX]. The radius of the wheel is 15 cm, and the wheel rotates clockwise at 100 revolutions per second. Find the rate of change in $A\widehat{B}O$ when $A\widehat{O}X$ is:

a 120° **b** 180°

17 A farmer has a water trough of length 8 m which has a semi-circular cross-section of diameter 1 m. Water is pumped into the trough at a constant rate of 0.1 m^3 per minute.

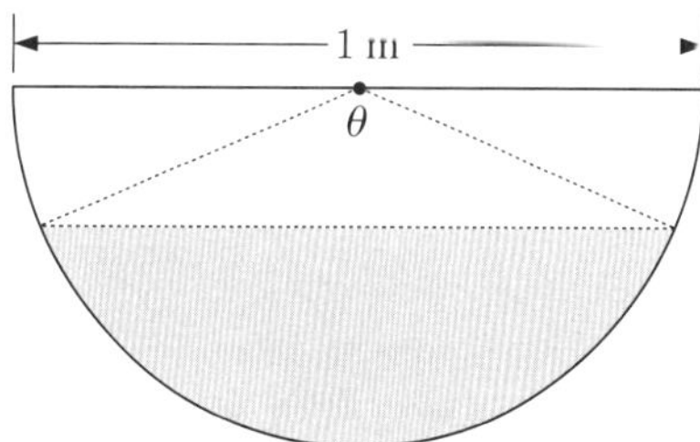

a Show that the volume of water in the trough is given by $V = \theta - \sin\theta$, where θ is the angle illustrated (in radians).

b Find the rate at which the water level is rising at the instant when the water is 25 cm deep.

Hint: First find $\frac{d\theta}{dt}$ and then find $\frac{dh}{dt}$ at the given instant.

REVIEW SET 20A NON-CALCULATOR

1 A particle moves in a straight line along the x-axis with position given by $x(t) = 3 + \sin(2t)$ cm after t seconds.

a Find the initial position, velocity, and acceleration of the particle.

b Find the times when the particle changes direction during $0 \leqslant t \leqslant \pi$ seconds.

c Find the total distance travelled by the particle in the first π seconds.

2 A particle P moves in a straight line with position relative to the origin O given by $s(t) = 2t^3 - 9t^2 + 12t - 5$ cm, where t is the time in seconds, $t \geqslant 0$.

a Find expressions for the particle's velocity and acceleration and draw sign diagrams for each of them.

b Find the initial conditions.

c Describe the motion of the particle at time $t = 2$ seconds.

d Find the times and positions where the particle changes direction.

e Draw a diagram to illustrate the motion of P.

f Determine the time intervals when the particle's speed is increasing.

3 Rectangle ABCD is inscribed within the parabola $y = k - x^2$ and the x-axis, as shown.

a If $\text{OD} = x$, show that the rectangle ABCD has area function $A(x) = 2kx - 2x^3$.

b If the area of ABCD is a maximum when $\text{AD} = 2\sqrt{3}$, find k.

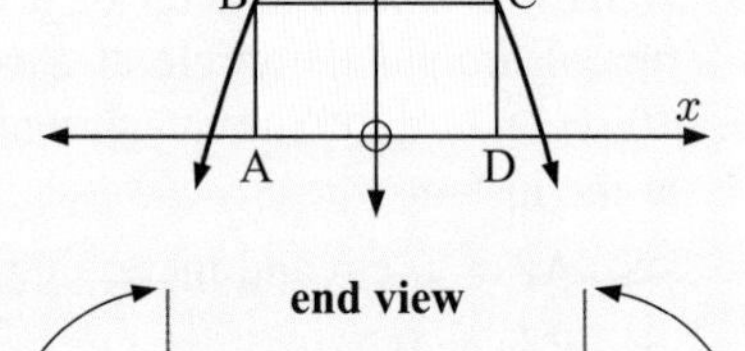

4 A rectangular gutter is formed by bending a 24 cm wide sheet of metal as shown.

Where must the bends be made in order to maximise the capacity of the gutter?

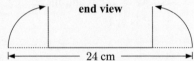

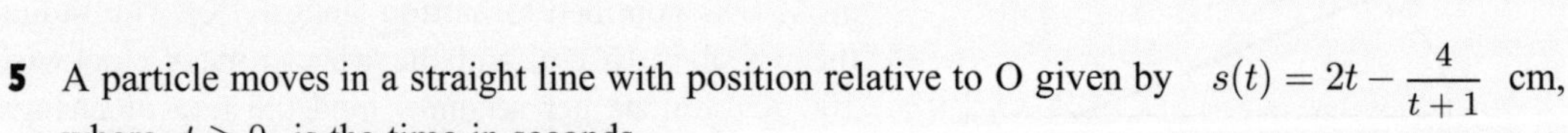

5 A particle moves in a straight line with position relative to O given by $s(t) = 2t - \dfrac{4}{t+1}$ cm, where $t \geqslant 0$ is the time in seconds.

a Find velocity and acceleration functions for the particle's motion and draw sign diagrams for each of them.

b Describe the motion of the particle at $t = 1$ second.

c Does the particle ever change direction? If so, where and when does it do this?

d Draw a diagram to illustrate the motion of the particle.

e Find the time intervals when the:

i velocity is increasing **ii** speed is increasing.

6 A rectangular sheet of tin-plate is $2k$ cm by k cm. Four squares, each with sides x cm, are cut from its corners. The remainder is bent into the shape of an open rectangular container. Find the value of x which will maximise the capacity of the container.

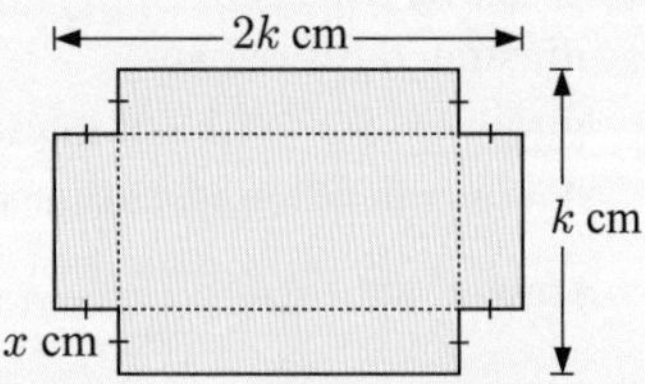

7 A cork bobs up and down in a bucket of water such that the distance from the centre of the cork to the bottom of the bucket is given by $s(t) = 30 + \cos(\pi t)$ cm, $t \geqslant 0$ seconds.

a Find the cork's velocity at times $t = 0, \frac{1}{2}, 1, 1\frac{1}{2}, 2$ s.

b Find the time intervals when the cork is falling.

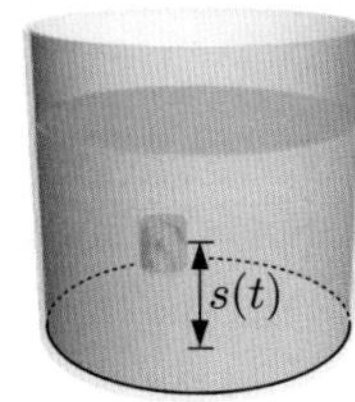

8 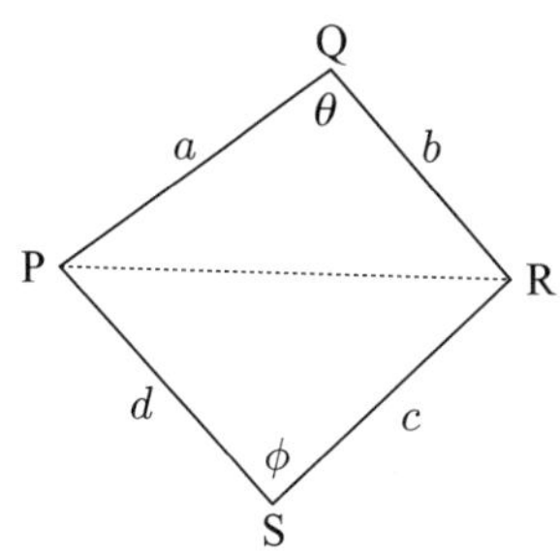

Four straight sticks of fixed lengths a, b, c, and d are hinged together at P, Q, R, and S.

a Use the cosine rule to find an equation which connects a, b, c, d, $\cos\theta$, and $\cos\phi$.

b Hence show that $\dfrac{d\theta}{d\phi} = \dfrac{cd\sin\phi}{ab\sin\theta}$.

c Hence show that the area of quadrilateral PQRS is a maximum when PQRS is a cyclic quadrilateral.

9 The graph of $y = ae^{-x}$ for $a > 0$ is shown.
P is a moving point on the graph, and A and B lie on the axes as shown so that OAPB is a rectangle.
As P moves along the curve, the rectangle constantly changes shape.
Find the x-coordinate of P, in terms of a, such that the rectangle OAPB has minimum perimeter.

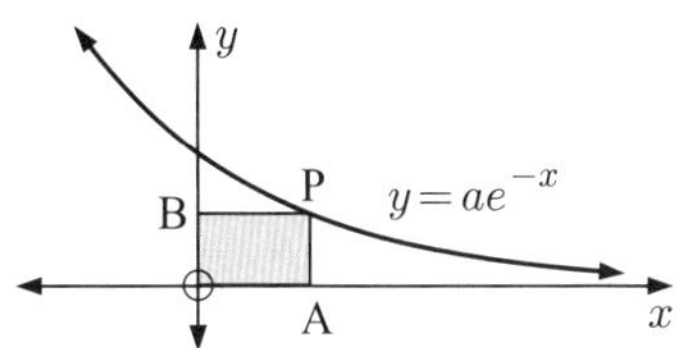

10 A and B are two houses directly opposite one another and 1 km from a straight road [CD]. MC is 3 km and C is a house at the roadside.
A power unit is to be located on [DC] at P such that PA + PB + PC is minimised. This ensures that the cost of trenching and cable will be as small as possible.

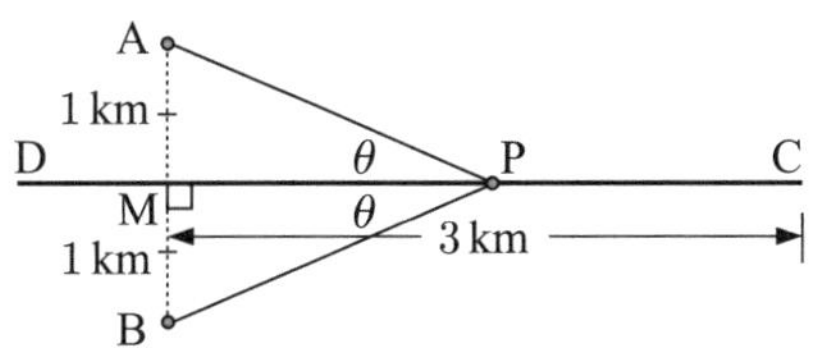

a What cable length would be required if P is at: **i** M **ii** C?

b Show that if $\theta = \widehat{APM} = \widehat{BPM}$, then the length of cable is given by
$L(\theta) = 2\csc\theta + 3 - \cot\theta$ metres.

c Show that $\dfrac{dL}{d\theta} = \dfrac{1 - 2\cos\theta}{\sin^2\theta}$ and hence show that the minimum length of cable required is $(3 + \sqrt{3})$ km.

REVIEW SET 20B — CALCULATOR

1 The height of a tree t years after it was planted is given by $H(t) = 60 + 40\ln(2t + 1)$ cm, $t \geqslant 0$.

a How high was the tree when it was planted?

b How long does it take for the tree to reach: **i** 150 cm **ii** 300 cm?

c At what rate is the tree's height increasing after: **i** 2 years **ii** 20 years?

2 A particle P moves in a straight line with position given by $s(t) = 80e^{-\frac{t}{10}} - 40t$ m where t is the time in seconds, $t \geqslant 0$.

a Find the velocity and acceleration functions.

b Find the initial position, velocity, and acceleration of P.

c Discuss the velocity of P as $t \to \infty$.

d Sketch the graph of the velocity function.

e Find the exact time when the velocity is -44 m s^{-1}.

3 The cost per hour of running a freight train is given by $C(v) = \frac{v^2}{30} + \frac{9000}{v}$ dollars where v is the average speed of the train in km h^{-1}.

a Find the cost of running the train for:

i two hours at 45 km h^{-1} **ii** 5 hours at 64 km h^{-1}.

b Find the rate of change in the hourly cost of running the train at speeds of:

i 50 km h^{-1} **ii** 66 km h^{-1}.

c At what speed will the cost per hour be a minimum?

4 A particle moves along the x-axis with position relative to origin O given by $x(t) = 3t - \sqrt{t+1}$ cm, where t is the time in seconds, $t \geqslant 0$.

a Find expressions for the particle's velocity and acceleration at any time t, and draw sign diagrams for each function.

b Find the initial conditions and hence describe the motion at that instant.

c Describe the motion of the particle at $t = 8$ seconds.

d Find the time and position when the particle reverses direction.

e Determine the time interval when the particle's speed is decreasing.

5 The value of a car t years after its purchase is given by $V = 20\,000e^{-0.4t}$ dollars. Calculate:

a the purchase price of the car

b the rate of decrease of the value of the car 10 years after it was purchased.

6 When a shirt maker sells x shirts per day, their income is given by

$$I(x) = 200 \ln \left(1 + \frac{x}{100}\right) + 1000 \text{ dollars.}$$

The manufacturing costs are determined by the cost function

$$C(x) = (x - 100)^2 + 200 \text{ dollars.}$$

How many shirts should be sold daily to maximise profits? What is the maximum daily profit?

7 A 200 m fence is placed around a lawn which has the shape of a rectangle with a semi-circle on one of its sides.

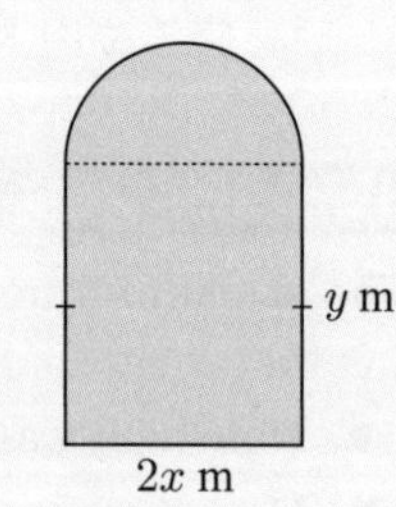

a Using the dimensions shown on the figure, show that $y = 100 - x - \frac{\pi}{2}x$.

b Find the area of the lawn A in terms of x only.

c Find the dimensions of the lawn if it has the maximum possible area.

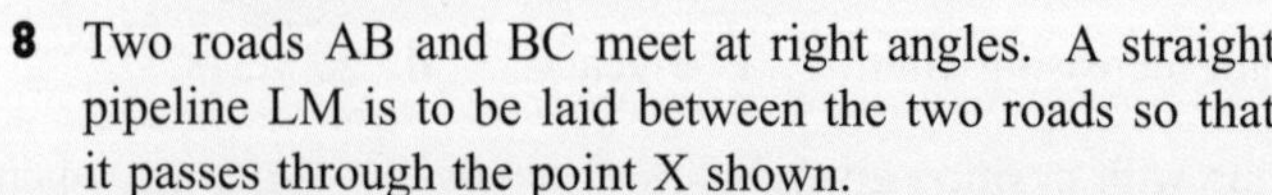

8 Two roads AB and BC meet at right angles. A straight pipeline LM is to be laid between the two roads so that it passes through the point X shown.

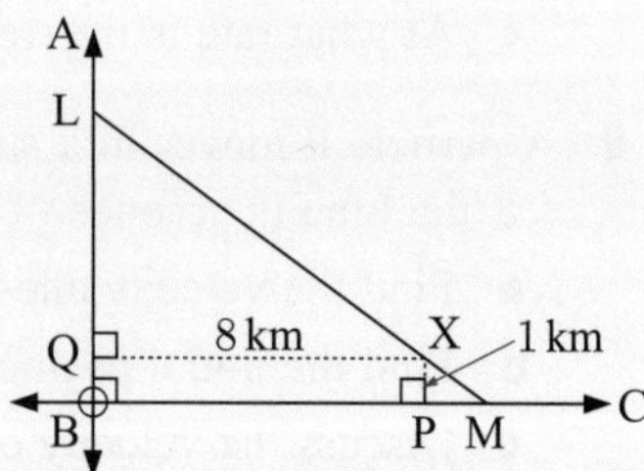

a If PM $= x$ km, find LQ in terms of x.

b Hence show that the length of the pipeline is given by $L(x) = \sqrt{x^2+1}\left(1 + \frac{8}{x}\right)$ km.

c Find $\frac{d[L(x)]^2}{dx}$.

d Hence find the shortest possible length for the pipeline.

9 The point $(3\cos\theta,\ 2\sin\theta)$ lies on a curve, $\theta \in [0,\ 2\pi]$.

a Find the equation of the curve in Cartesian form. **b** Find $\dfrac{dy}{dx}$ in terms of θ.

c A tangent to the curve meets the x-axis at A and the y-axis at B.
Find the smallest area of triangle OAB and the values of θ when it occurs.

10 A man on a jetty pulls a boat directly towards him so the rope is coming in at a rate of 20 metres per minute. The rope is attached to the boat 1 m above water level and the man's hands are 6 m above the water level. How fast is the boat approaching the jetty at the instant when it is 15 m from the jetty?

11 Water exits a conical tank at a constant rate of 0.2 m^3/minute. If the surface of the water has radius r:

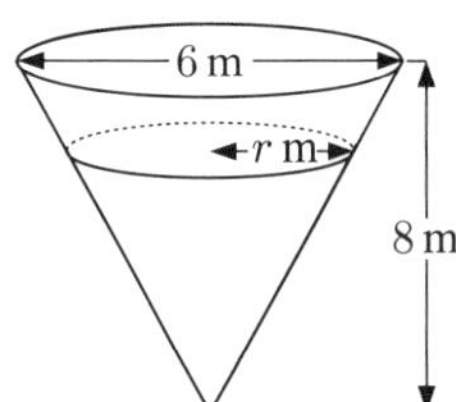

a find $V(r)$, the volume of the water remaining

b find the rate at which the surface radius is changing at the instant when the water is 5 m deep.

REVIEW SET 20C

1 A manufacturer of open steel boxes has to make one with a square base and a capacity of 1 m^3. The steel costs \$2 per square metre.

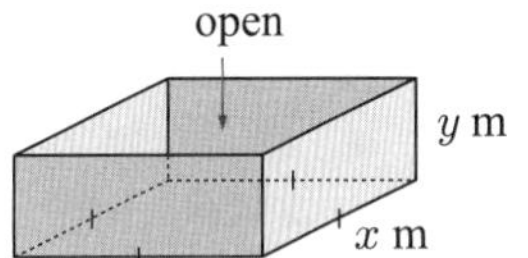

a If the base measures x m by x m and the height is y m, find y in terms of x.

b Hence, show that the total cost of the steel is $C(x) = 2x^2 + \dfrac{8}{x}$ dollars.

c Find the dimensions of the box which would cost the least in steel to make.

2 A particle P moves in a straight line with position from O given by $s(t) = 15t - \dfrac{60}{(t+1)^2}$ cm, where t is the time in seconds, $t \geqslant 0$.

a Find velocity and acceleration functions for P's motion.

b Describe the motion of P at $t = 3$ seconds.

c For what values of t is the particle's speed increasing?

3 Infinitely many rectangles can be inscribed under the curve $y = e^{-2x}$ as shown.
Determine the coordinates of A such that the rectangle OBAC has maximum area.

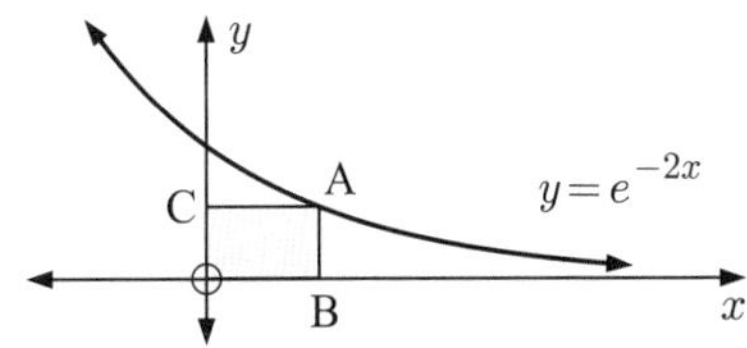

4 The height of a tree t years after it is planted is given by $H(t) = 6\left(1 - \dfrac{2}{t+3}\right)$ metres, $t \geqslant 0$.

a How high was the tree when it was planted?

b Determine the height of the tree after $t = 3$, 6 and 9 years.

c Find the rate at which the tree is growing at $t = 0$, 3, 6 and 9 years.

d Show that $H'(t) > 0$ and explain the significance of this result.

e Sketch the graph of $H(t)$ against t.

5 A particle P moves in a straight line with position given by $s(t) = 25t - 10\ln t$ cm, $t \geqslant 1$, where t is the time in minutes.

a Find the velocity and acceleration functions.

b Find the position, velocity, and acceleration when $t = e$ minutes.

c Discuss the velocity as $t \to \infty$.

d Sketch the graph of the velocity function.

e Find when the velocity of P is 20 cm per minute.

6 A triangular pen is enclosed by two fences [AB] and [BC], each of length 50 m, with the river being the third side.

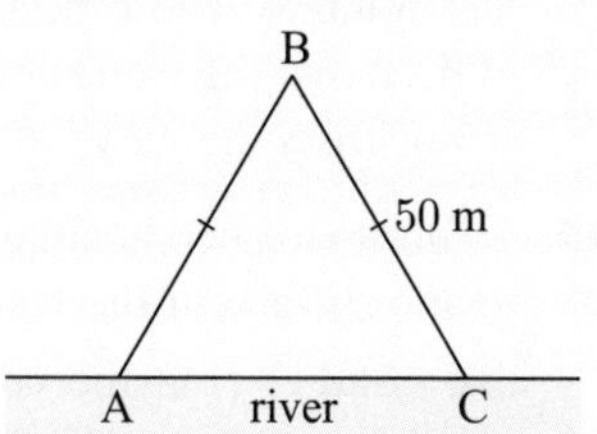

a If $AC = 2x$ m, show that the area of triangle ABC is $A(x) = x\sqrt{2500 - x^2}$ m^2.

b Find $\dfrac{d[A(x)]^2}{dx}$.

c Hence find x such that the area is a maximum.

7 A light bulb hangs from the ceiling at height h metres above the floor, directly above point N. At any point A on the floor which is x metres from the light bulb, the illumination I is given by $I = \dfrac{\sqrt{8}\cos\theta}{x^2}$ units.

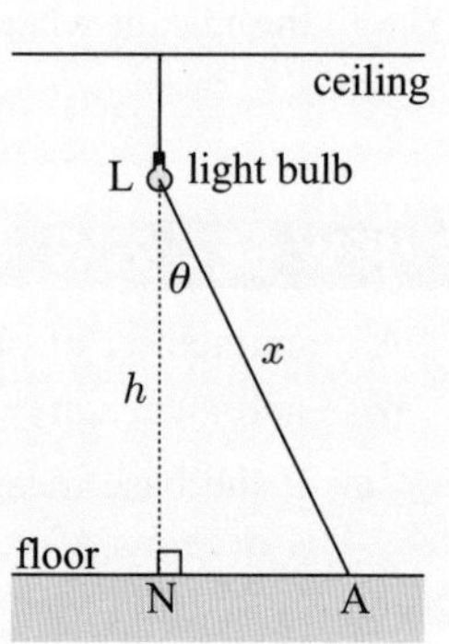

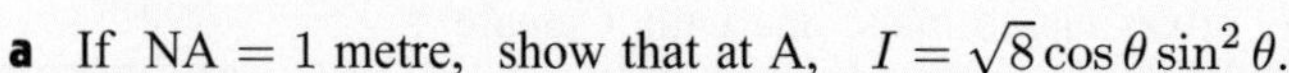

a If NA = 1 metre, show that at A, $I = \sqrt{8}\cos\theta\sin^2\theta$.

b The light bulb may be lifted or lowered to change the intensity at A. Assuming NA = 1 metre, find the height the bulb should be above the floor for greatest illumination at A.

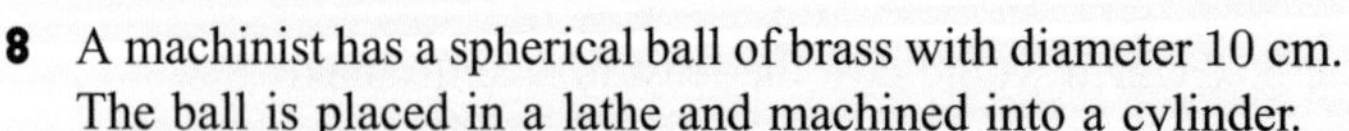

8 A machinist has a spherical ball of brass with diameter 10 cm. The ball is placed in a lathe and machined into a cylinder.

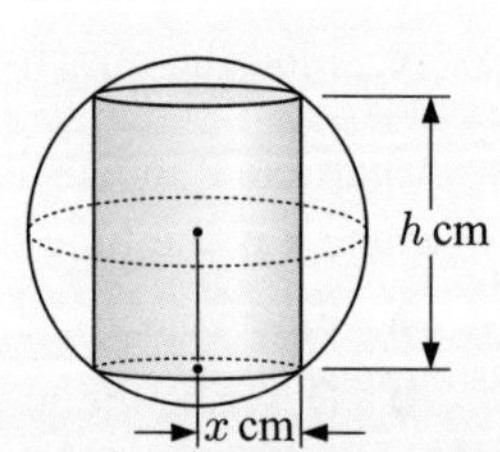

a If the cylinder has radius x cm, show that the cylinder's volume is given by $V(x) = \pi x^2\sqrt{100 - 4x^2}$ cm^3.

b Hence find the dimensions of the cylinder of largest volume which can be made.

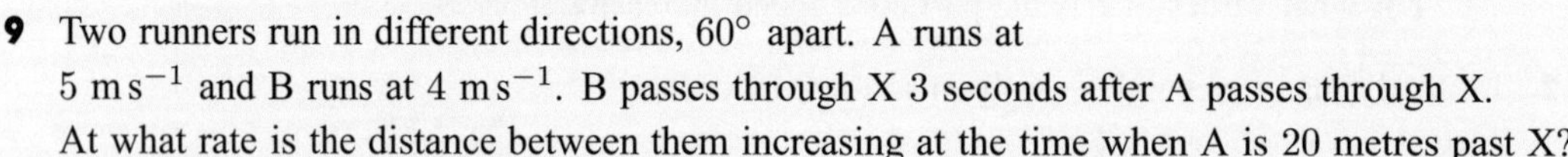

9 Two runners run in different directions, $60°$ apart. A runs at $5\ \text{m s}^{-1}$ and B runs at $4\ \text{m s}^{-1}$. B passes through X 3 seconds after A passes through X. At what rate is the distance between them increasing at the time when A is 20 metres past X?

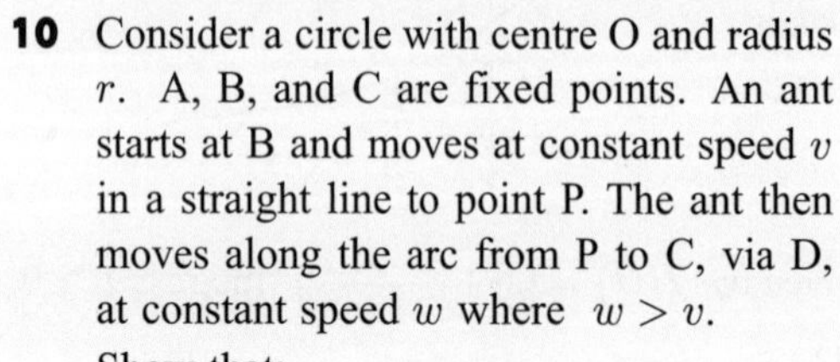

10 Consider a circle with centre O and radius r. A, B, and C are fixed points. An ant starts at B and moves at constant speed v in a straight line to point P. The ant then moves along the arc from P to C, via D, at constant speed w where $w > v$.
Show that:

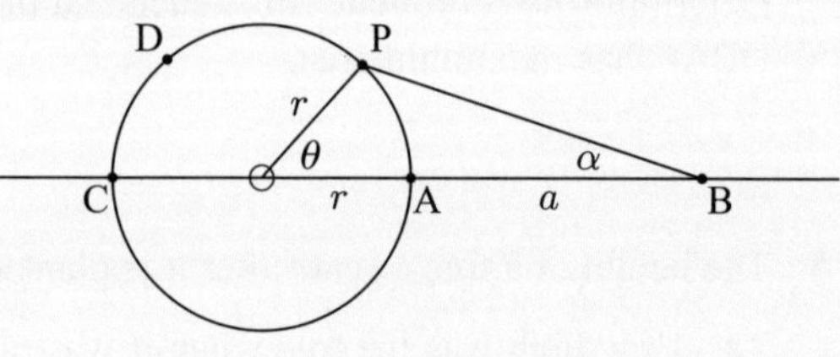

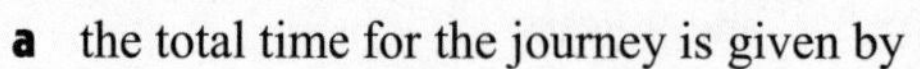

a the total time for the journey is given by $T = \dfrac{\sqrt{r^2 + (a+r)^2 - 2r(a+r)\cos\theta}}{v} + \dfrac{r(\pi - \theta)}{w}$.

b $\dfrac{dT}{d\theta} = \dfrac{a+r}{v}\left(\sin\alpha - \dfrac{rv}{(a+r)w}\right)$.

c T is minimised when $\sin\alpha = \dfrac{rv}{(a+r)w}$.

Chapter 21

Integration

Syllabus reference: 6.4, 6.5, 6.7

Contents:

OPENING PROBLEM

The function $f(x) = x^2 + 1$ lies above the x-axis for all $x \in \mathbb{R}$.

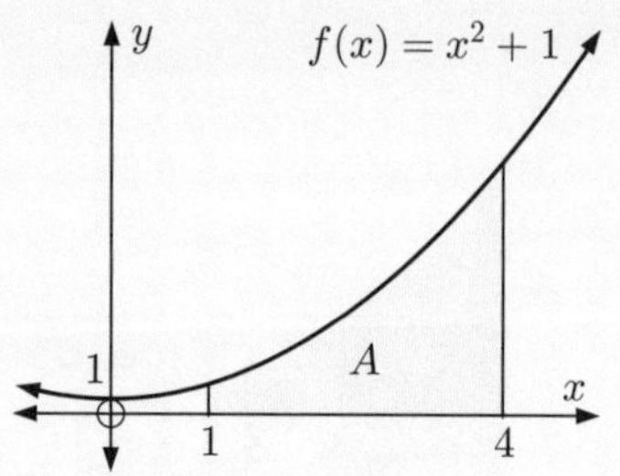

Things to think about:

a How can we calculate the shaded area A, which is the area under the curve for $1 \leqslant x \leqslant 4$?

b What function has $x^2 + 1$ as its derivative?

In the previous chapters we used differential calculus to find the derivatives of many types of functions. We also used it in problem solving, in particular to find the gradients of graphs and rates of changes, and to solve optimisation problems.

In this chapter we consider **integral calculus**. This involves **antidifferentiation** which is the reverse process of differentiation. Integral calculus also has many useful applications, including:

- finding areas of shapes with curved boundaries
- finding volumes of revolution
- finding distances travelled from velocity functions
- solving problems in economics, biology, and statistics
- solving differential equations.

A THE AREA UNDER A CURVE

The task of finding the area under a curve has been important to mathematicians for thousands of years. In the history of mathematics it was fundamental to the development of integral calculus. We will therefore begin our study by calculating the area under a curve using the same methods as the ancient mathematicians.

UPPER AND LOWER RECTANGLES

Consider the function $f(x) = x^2 + 1$.

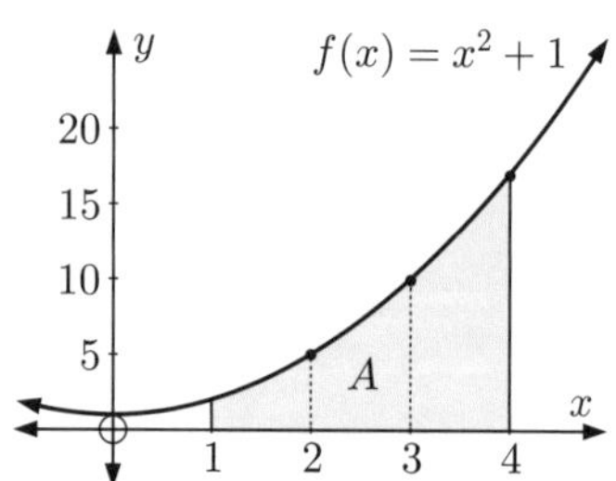

We wish to estimate the area A enclosed by $y = f(x)$, the x-axis, and the vertical lines $x = 1$ and $x = 4$.

Suppose we divide the interval $1 \leqslant x \leqslant 4$ into three strips of width 1 unit as shown. We obtain three subintervals of equal width.

The diagram alongside shows **upper rectangles**, which are rectangles with top edges at the maximum value of the curve on that subinterval.

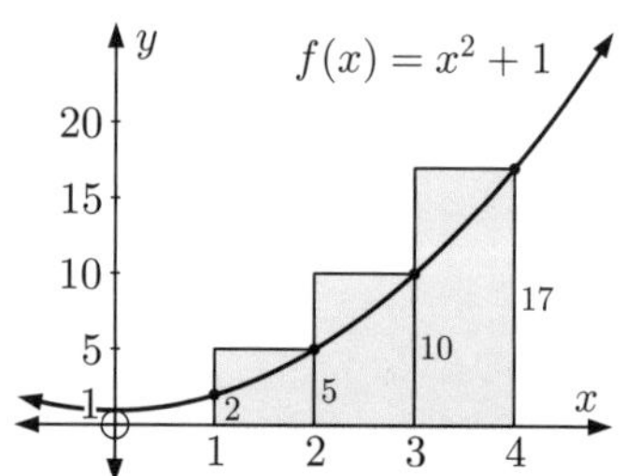

The area of the upper rectangles,

$$\begin{aligned} A_U &= 1 \times f(2) + 1 \times f(3) + 1 \times f(4) \\ &= 5 + 10 + 17 \\ &= 32 \text{ units}^2 \end{aligned}$$

The next diagram shows **lower rectangles**, which are rectangles with top edges at the minimum value of the curve on that subinterval.

The area of the lower rectangles,

$$\begin{aligned} A_L &= 1 \times f(1) + 1 \times f(2) + 1 \times f(3) \\ &= 2 + 5 + 10 \\ &= 17 \text{ units}^2 \end{aligned}$$

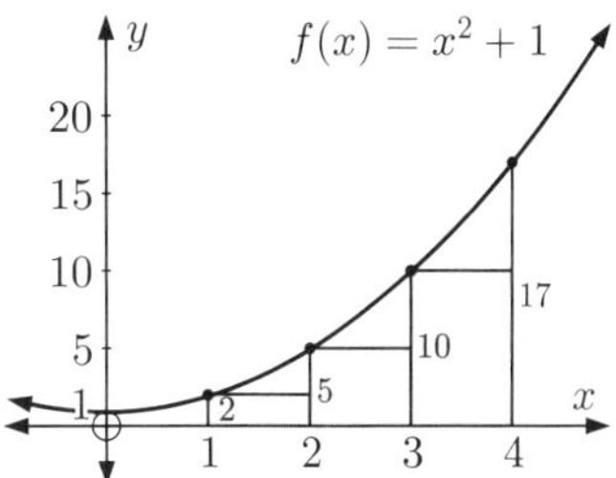

Now clearly $A_L < A < A_U$, so the area A lies between 17 units^2 and 32 units^2.

If the interval $1 \leqslant x \leqslant 4$ was divided into 6 subintervals, each of width $\frac{1}{2}$, then

$$\begin{aligned} A_U &= \tfrac{1}{2}f(1\tfrac{1}{2}) + \tfrac{1}{2}f(2) + \tfrac{1}{2}f(2\tfrac{1}{2}) + \tfrac{1}{2}f(3) + \tfrac{1}{2}f(3\tfrac{1}{2}) + \tfrac{1}{2}f(4) \\ &= \tfrac{1}{2}(\tfrac{13}{4} + 5 + \tfrac{29}{4} + 10 + \tfrac{53}{4} + 17) \\ &= 27.875 \text{ units}^2 \end{aligned}$$

$$\begin{aligned} \text{and} \quad A_L &= \tfrac{1}{2}f(1) + \tfrac{1}{2}f(1\tfrac{1}{2}) + \tfrac{1}{2}f(2) + \tfrac{1}{2}f(2\tfrac{1}{2}) + \tfrac{1}{2}f(3) + \tfrac{1}{2}f(3\tfrac{1}{2}) \\ &= \tfrac{1}{2}(2 + \tfrac{13}{4} + 5 + \tfrac{29}{4} + 10 + \tfrac{53}{4}) \\ &= 20.375 \text{ units}^2 \end{aligned}$$

From this refinement we conclude that the area A lies between 20.375 and 27.875 units^2.

As we create more subintervals, the estimates A_L and A_U will become more and more accurate. In fact, as the subinterval width is reduced further and further, both A_L and A_U will **converge** to A.

We illustrate this process by estimating the area A between the graph of $y = x^2$ and the x-axis for $0 \leqslant x \leqslant 1$.

This example is of historical interest. **Archimedes** (287 - 212 BC) found the exact area. In an article that contains 24 propositions, he developed the essential theory for what is now known as integral calculus.

Consider $f(x) = x^2$ and divide the interval $0 \leqslant x \leqslant 1$ into 4 subintervals of equal width.

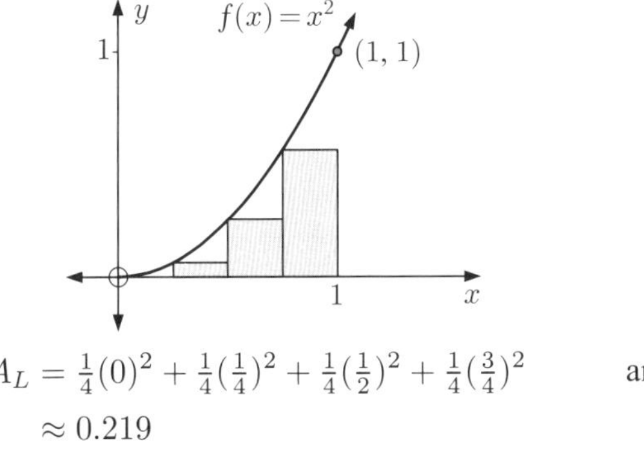

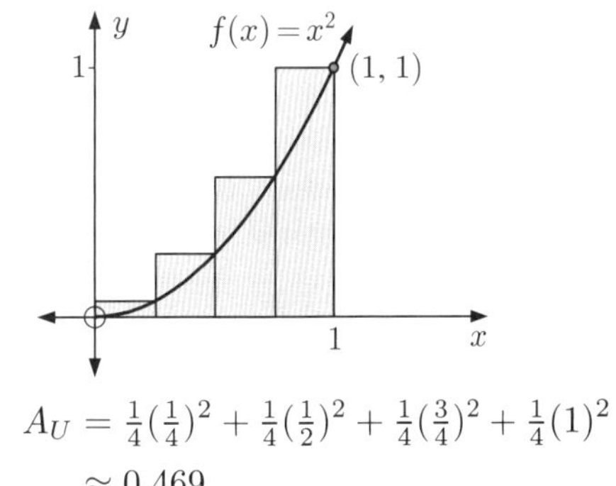

$$\begin{aligned} A_L &= \tfrac{1}{4}(0)^2 + \tfrac{1}{4}(\tfrac{1}{4})^2 + \tfrac{1}{4}(\tfrac{1}{2})^2 + \tfrac{1}{4}(\tfrac{3}{4})^2 \\ &\approx 0.219 \end{aligned}$$

and

$$\begin{aligned} A_U &= \tfrac{1}{4}(\tfrac{1}{4})^2 + \tfrac{1}{4}(\tfrac{1}{2})^2 + \tfrac{1}{4}(\tfrac{3}{4})^2 + \tfrac{1}{4}(1)^2 \\ &\approx 0.469 \end{aligned}$$

Now suppose there are n subintervals between $x = 0$ and $x = 1$, each of width $\dfrac{1}{n}$.

We can use technology to help calculate A_L and A_U for large values of n.

Click on the appropriate icon to access our **area finder** software or instructions for the procedure on a **graphics calculator**.

AREA FINDER

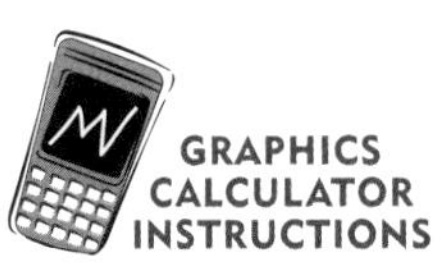

The table alongside summarises the results you should obtain for $n = 4$, 10, 25, and 50.

The exact value of A is in fact $\frac{1}{3}$, as we will find later in the chapter. Notice how both A_L and A_U are converging to this value as n increases.

n	A_L	A_U	*Average*
4	0.218 75	0.468 75	0.343 75
10	0.285 00	0.385 00	0.335 00
25	0.313 60	0.353 60	0.333 60
50	0.323 40	0.343 40	0.333 40

EXERCISE 21A.1

1 Consider the area between $y = x$ and the x-axis from $x = 0$ to $x = 1$.

a Divide the interval into 5 strips of equal width, then estimate the area using:

i upper rectangles **ii** lower rectangles.

b Calculate the actual area and compare it with your answers in **a**.

2 Consider the area between $y = \dfrac{1}{x}$ and the x-axis from $x = 2$ to $x = 4$. Divide the interval into 6 strips of equal width, then estimate the area using:

a upper rectangles **b** lower rectangles.

3 Use rectangles to find lower and upper sums for the area between the graph of $y = x^2$ and the x-axis for $1 \leqslant x \leqslant 2$. Use $n = 10$, 25, 50, 100, and 500. Give your answers to 4 decimal places. As n gets larger, both A_L and A_U converge to the same number which is a simple fraction. What is it?

4 **a** Use lower and upper sums to estimate the area between each of the following functions and the x-axis for $0 \leqslant x \leqslant 1$. Use values of $n = 5$, 10, 50, 100, 500, 1000, and 10 000. Give your answer to 5 decimal places in each case.

i $y = x^3$ **ii** $y = x$ **iii** $y = x^{\frac{1}{2}}$ **iv** $y = x^{\frac{1}{3}}$

b For each case in **a**, A_L and A_U converge to the same number which is a simple fraction. What fractions are they?

c Using your answer to **b**, predict the area between the graph of $y = x^a$ and the x-axis for $0 \leqslant x \leqslant 1$ and any number $a > 0$.

5 Consider the quarter circle of centre $(0, 0)$ and radius 2 units illustrated.

Its area is $\frac{1}{4}$(full circle of radius 2)

$= \frac{1}{4} \times \pi \times 2^2$

$= \pi$ units^2

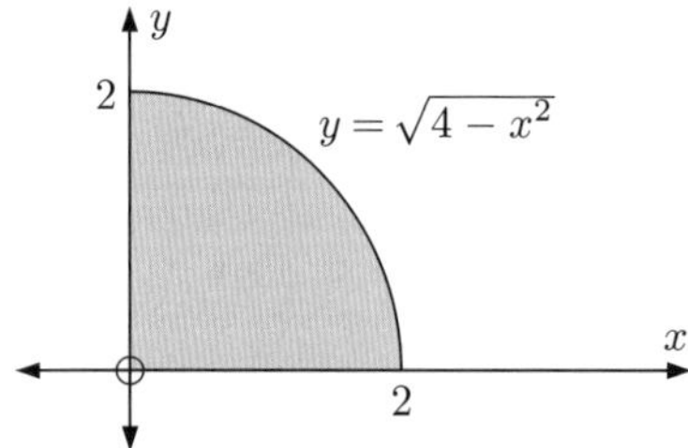

a Estimate the area using lower and upper rectangles for $n = 10$, 50, 100, 200, 1000, and 10 000. Hence, find rational bounds for π.

b Archimedes found the famous approximation $3\frac{10}{71} < \pi < 3\frac{1}{7}$.

For what value of n is your estimate for π better than that of Archimedes?

THE DEFINITE INTEGRAL

Consider the lower and upper rectangle sums for a function which is positive and increasing on the interval $a \leqslant x \leqslant b$.

We divide the interval into n subintervals, each of width $w = \dfrac{b-a}{n}$.

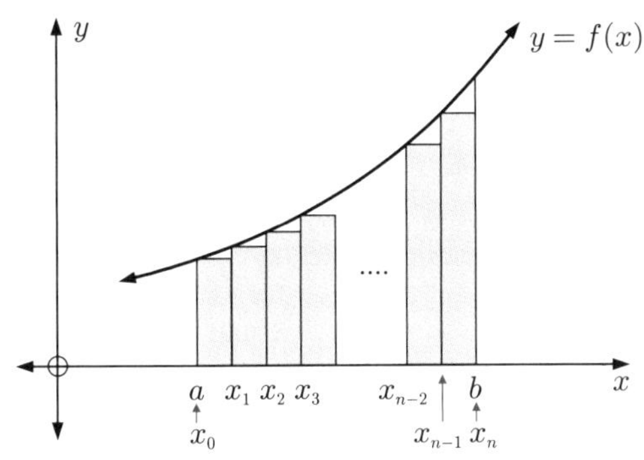

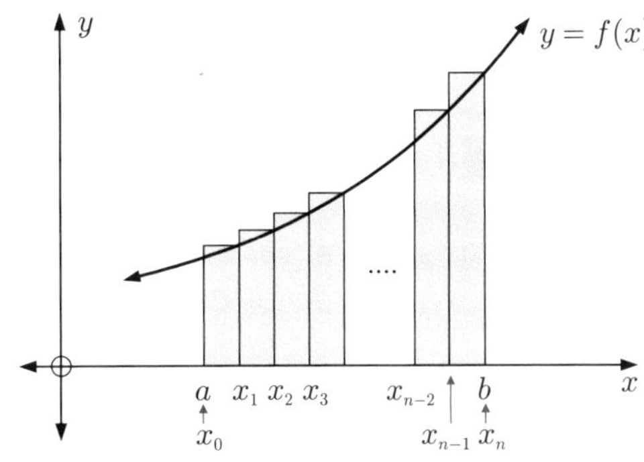

Since the function is increasing,

$$A_L = w\,f(x_0) + w\,f(x_1) + w\,f(x_2) + ... + w\,f(x_{n-2}) + w\,f(x_{n-1}) = w\sum_{i=0}^{n-1} f(x_i)$$

and $$A_U = w\,f(x_1) + w\,f(x_2) + w\,f(x_3) + ... + w\,f(x_{n-1}) + w\,f(x_n) = w\sum_{i=1}^{n} f(x_i)$$

Notice that
$$A_U - A_L = w\,(f(x_n) - f(x_0))$$
$$= \frac{1}{n}(b-a)\,(f(b) - f(a))$$

$$\therefore \quad \lim_{n\to\infty}(A_U - A_L) = 0 \qquad \{\text{since } \lim_{n\to\infty}\frac{1}{n} = 0\}$$

$$\therefore \quad \lim_{n\to\infty} A_L = \lim_{n\to\infty} A_U \qquad \{\text{when both limits exist}\}$$

$\lim_{n\to\infty}$ means we have infinitely many subintervals.

$\therefore$ since $A_L < A < A_U$ for all values of n, it follows that $\lim_{n\to\infty} A_L = A = \lim_{n\to\infty} A_U$

This fact is true for all positive continuous functions on an interval $a \leqslant x \leqslant b$.

The value A is known as the "**definite integral** of $f(x)$ from a to b", written $A = \displaystyle\int_a^b f(x)\,dx$.

If $f(x) \geqslant 0$ for all $a \leqslant x \leqslant b$, then

$\displaystyle\int_a^b f(x)\,dx$ is equal to the shaded area.

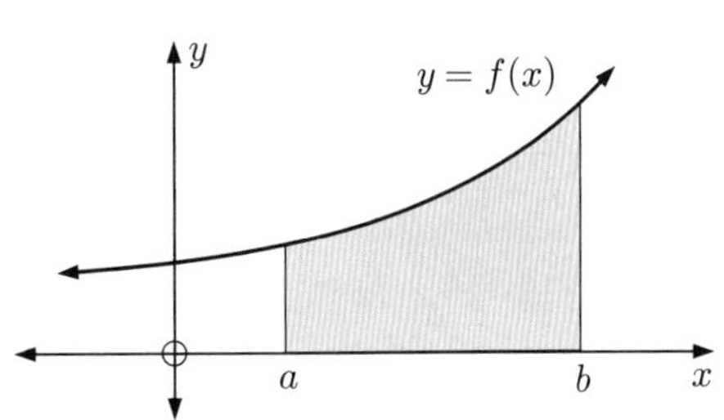

HISTORICAL NOTE

The word **integration** means "*to put together into a whole*". An **integral** is the "whole" produced from integration, since the areas $f(x_i) \times w$ of the thin rectangular strips are put together into one whole area.

The symbol $\int$ is called an **integral sign**. In the time of **Newton** and **Leibniz** it was the stretched out letter s, but it is no longer part of the alphabet.

Example 1

Self Tutor

a Sketch the graph of $y = x^4$ for $0 \leqslant x \leqslant 1$. Shade the area described by $\int_0^1 x^4\, dx$.

b Use technology to calculate the lower and upper rectangle sums for n equal subintervals where $n = 5$, 10, 50, 100, and 500.

c Hence find $\int_0^1 x^4\, dx$ to 2 significant figures.

AREA FINDER

a

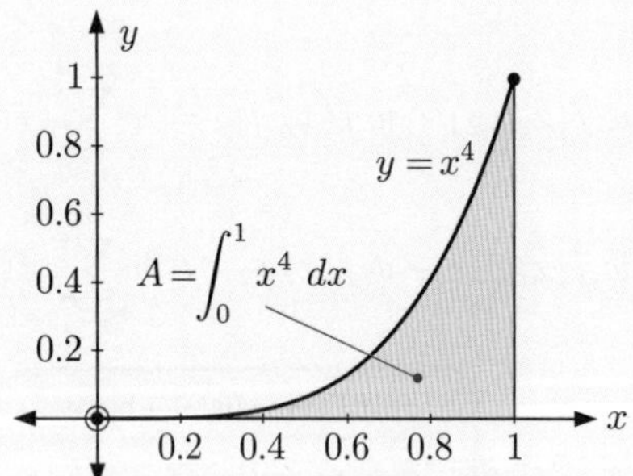

b

n	A_L	A_U
5	0.1133	0.3133
10	0.1533	0.2533
50	0.1901	0.2101
100	0.1950	0.2050
500	0.1990	0.2010

c When $n = 500$, $A_L \approx A_U \approx 0.20$, to 2 significant figures.

$\therefore$ since $A_L < \int_0^1 x^4\, dx < A_U$, $\int_0^1 x^4\, dx \approx 0.20$

EXERCISE 21A.2

1 **a** Sketch the graph of $y = \sqrt{x}$ for $0 \leqslant x \leqslant 1$.

Shade the area described by $\int_0^1 \sqrt{x}\, dx$.

b Find the lower and upper rectangle sums for $n = 5$, 10, 50, 100, and 500.

c Hence find $\int_0^1 \sqrt{x}\, dx$ to 2 significant figures.

2 **a** Sketch the graph of $y = \sqrt{1+x^3}$ and the x-axis for $0 \leqslant x \leqslant 2$.

b Write expressions for the lower and upper rectangle sums using n subintervals, $n \in \mathbb{N}$.

c Find the lower and upper rectangle sums for $n = 50$, 100, and 500.

d Hence estimate $\int_0^2 \sqrt{1+x^3}\, dx$.

Example 2 **Self Tutor**

Use graphical evidence and known area facts to find: **a** $\int_0^2 (2x+1)\,dx$ **b** $\int_0^1 \sqrt{1-x^2}\,dx$

a

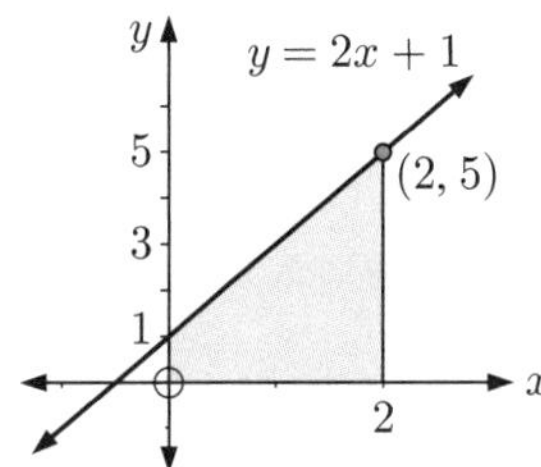

$\int_0^2 (2x+1)\,dx$
$=$ shaded area
$= \left(\frac{1+5}{2}\right) \times 2$
$= 6$

b If $y = \sqrt{1-x^2}$ then $y^2 = 1 - x^2$ and so $x^2 + y^2 = 1$ which is the equation of the unit circle. $y = \sqrt{1-x^2}$ is the upper half.

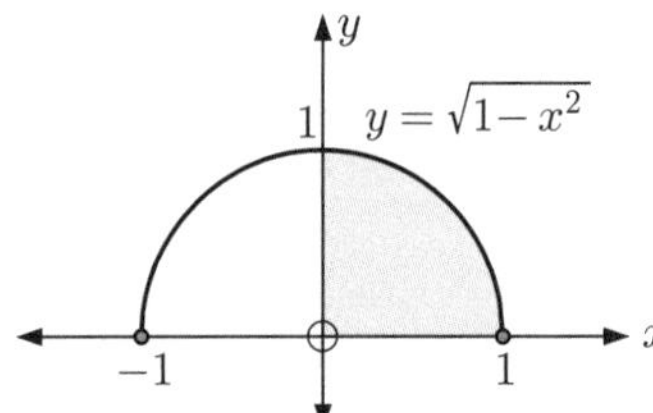

$\int_0^1 \sqrt{1-x^2}\,dx$
$=$ shaded area
$= \frac{1}{4}(\pi r^2)$ where $r = 1$
$= \frac{\pi}{4}$

3 Use graphical evidence and known area facts to find:

a $\int_1^3 (1+4x)\,dx$ **b** $\int_{-1}^2 (2-x)\,dx$ **c** $\int_{-2}^2 \sqrt{4-x^2}\,dx$

INVESTIGATION 1 ESTIMATING $\int_{-3}^3 e^{-\frac{x^2}{2}}\,dx$

The integral $\int_{-3}^3 e^{-\frac{x^2}{2}}\,dx$ is of considerable interest to statisticians.

In this Investigation we shall estimate the value of this integral using upper and lower rectangular sums for $n = 4500$. We will perform the integration in sections.

What to do:

1 Sketch the graph of $y = e^{-\frac{x^2}{2}}$ for $-3 \leqslant x \leqslant 3$.

2 Calculate the upper and lower rectangular sums for the interval $0 \leqslant x \leqslant 3$ using $n = 2250$.

GRAPHICS CALCULATOR INSTRUCTIONS

3 Use the fact that the function $y = e^{-\frac{x^2}{2}}$ is symmetric to find upper and lower rectangular sums for $-3 \leqslant x \leqslant 0$ for $n = 2250$.

4 Use your results of **2** and **3** to estimate $\int_{-3}^3 e^{-\frac{x^2}{2}}\,dx$.

How accurate is your estimate compared with $\sqrt{2\pi}$?

AREA FINDER

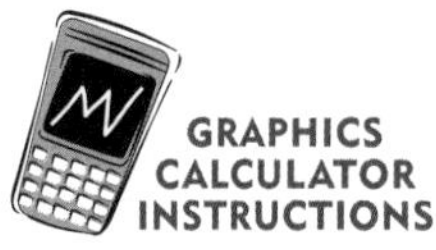

B ANTIDIFFERENTIATION

In many problems in calculus we know the rate of change of one variable with respect to another, but we do not have a formula which relates the variables. In other words, we know $\frac{dy}{dx}$, but we need to know y in terms of x.

Examples of problems we need to solve include:

- The gradient function $f'(x)$ of a curve is $2x+3$ and the curve passes through the origin. What is the function $y = f(x)$?
- The rate of change in temperature is $\frac{dT}{dt} = 10e^{-t}$ °C per minute where $t \geqslant 0$. What is the temperature function given that initially the temperature was 11°C?

The process of finding y from $\frac{dy}{dx}$, or $f(x)$ from $f'(x)$, is the reverse process of differentiation. We call it **antidifferentiation** or **integration**.

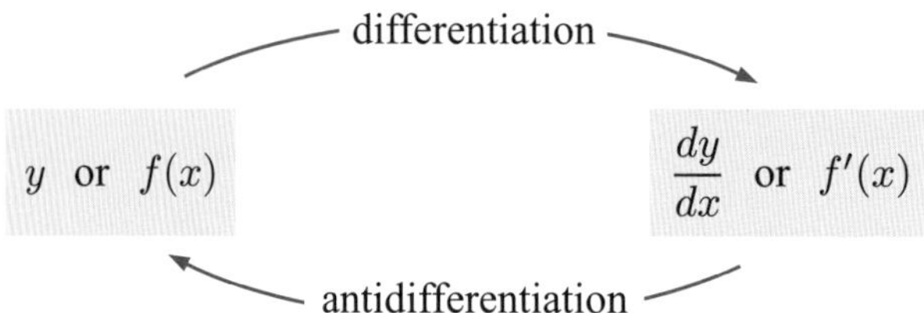

Consider $\frac{dy}{dx} = x^2$.

From our work on differentiation, we know that when we differentiate power functions the index reduces by 1. We hence know that y must involve x^3.

Now if $y = x^3$ then $\frac{dy}{dx} = 3x^2$, so if we start with $y = \frac{1}{3}x^3$ then $\frac{dy}{dx} = x^2$.

However, for all of the cases $y = \frac{1}{3}x^3 + 2$, $y = \frac{1}{3}x^3 + 100$, and $y = \frac{1}{3}x^3 - 7$, we find that $\frac{dy}{dx} = x^2$.

In fact, there are infinitely many functions of the form $y = \frac{1}{3}x^3 + c$ where c is an arbitrary constant which will give $\frac{dy}{dx} = x^2$. Ignoring the arbitrary constant, we say that $\frac{1}{3}x^3$ is the **antiderivative** of x^2. It is the simplest function which, when differentiated, gives x^2.

If $F(x)$ is a function where $F'(x) = f(x)$ we say that:

- the **derivative** of $F(x)$ is $f(x)$ and
- the **antiderivative** of $f(x)$ is $F(x)$.

Example 3 **Self Tutor**

Find the antiderivative of: **a** x^3 **b** e^{2x} **c** $\frac{1}{\sqrt{x}}$

a $\frac{d}{dx}\left(x^4\right) = 4x^3$

$\therefore \frac{d}{dx}\left(\frac{1}{4}x^4\right) = x^3$

$\therefore$ the antiderivative of x^3 is $\frac{1}{4}x^4$.

b $\frac{d}{dx}\left(e^{2x}\right) = e^{2x} \times 2$

$\therefore \frac{d}{dx}\left(\frac{1}{2}e^{2x}\right) = \frac{1}{2} \times e^{2x} \times 2 = e^{2x}$

$\therefore$ the antiderivative of e^{2x} is $\frac{1}{2}e^{2x}$.

c $\frac{1}{\sqrt{x}} = x^{-\frac{1}{2}}$

Now $\frac{d}{dx}(x^{\frac{1}{2}}) = \frac{1}{2}x^{-\frac{1}{2}}$

$\therefore \frac{d}{dx}(2x^{\frac{1}{2}}) = 2(\frac{1}{2})x^{-\frac{1}{2}} = x^{-\frac{1}{2}}$

$\therefore$ the antiderivative of $\frac{1}{\sqrt{x}}$ is $2\sqrt{x}$.

EXERCISE 21B

1 **a** Find the antiderivative of:

i x **ii** x^2 **iii** x^5 **iv** x^{-2} **v** x^{-4} **vi** $x^{\frac{1}{3}}$ **vii** $x^{-\frac{1}{2}}$

b Predict a general rule for the antiderivative of x^n, for $n \neq -1$.

2 **a** Find the antiderivative of:

i e^{2x} **ii** e^{5x} **iii** $e^{\frac{1}{2}x}$ **iv** $e^{0.01x}$ **v** $e^{\pi x}$ **vi** $e^{\frac{x}{3}}$

b Predict a general rule for the antiderivative of e^{kx} where k is a constant.

3 Find the antiderivative of:

a $6x^2 + 4x$ by first differentiating $x^3 + x^2$

b e^{3x+1} by first differentiating e^{3x+1}

c $\sqrt{x}$ by first differentiating $x\sqrt{x}$

d $(2x+1)^3$ by first differentiating $(2x+1)^4$.

C THE FUNDAMENTAL THEOREM OF CALCULUS

Sir Isaac Newton and **Gottfried Wilhelm Leibniz** showed the link between differential calculus and the definite integral or limit of an area sum we saw in **Section A**. This link is called the **fundamental theorem of calculus**. The beauty of this theorem is that it enables us to evaluate complicated summations.

We have already observed that:

If $f(x)$ is a continuous positive function on an interval $a \leqslant x \leqslant b$ then the area under the curve between $x = a$ and $x = b$ is $\int_a^b f(x)\, dx$.

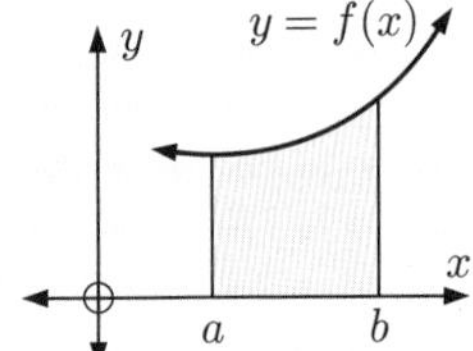

INVESTIGATION 2 THE AREA FUNCTION

Consider the constant function $f(x) = 5$.

We wish to find an **area function** which will give the area under the function between $x = a$ and some other value of x which we will call t.

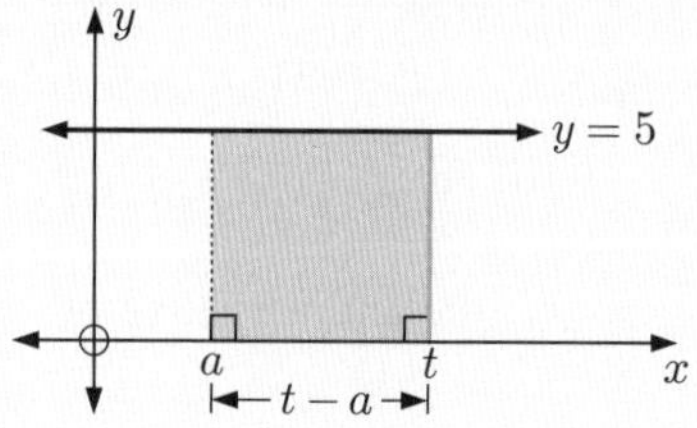

The area function is $A(t) = \int_a^t 5\, dx$

$= \text{shaded area in graph}$

$= (t - a)5$

$= 5t - 5a$

$\therefore$ we can write $A(t)$ in the form $F(t) - F(a)$ where $F(t) = 5t$ or equivalently, $F(x) = 5x$.

What to do:

1. What is the derivative $F'(x)$ of the function $F(x) = 5x$? How does this relate to the function $f(x)$?

2. Consider the simplest linear function $f(x) = x$.
The corresponding area function is

$$A(t) = \int_a^t x\, dx$$

$$= \text{shaded area in graph}$$

$$= \left(\frac{t + a}{2}\right)(t - a)$$

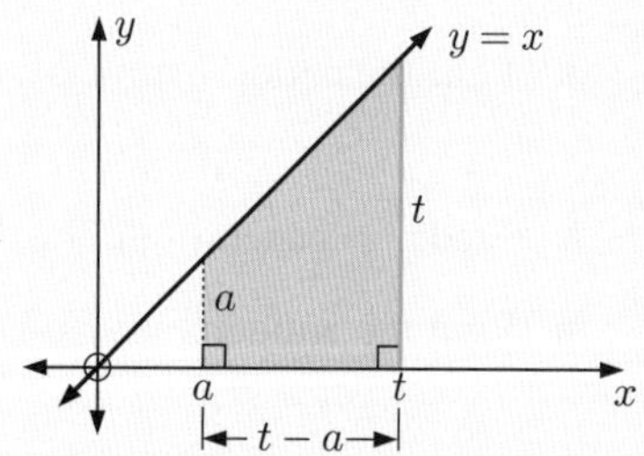

   a Write $A(t)$ in the form $F(t) - F(a)$.

   b What is the derivative $F'(x)$? How does it relate to the function $f(x)$?

3. Consider $f(x) = 2x + 3$. The corresponding area function is

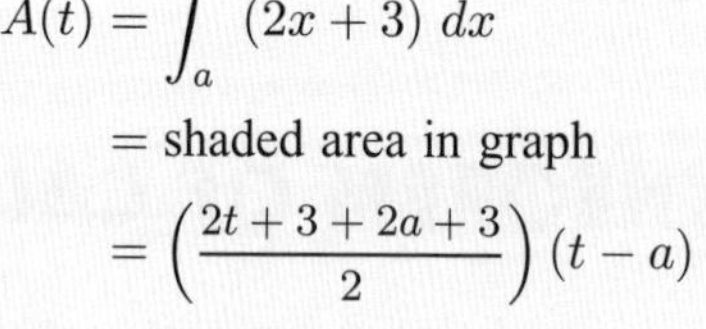

$$A(t) = \int_a^t (2x + 3)\, dx$$

$$= \text{shaded area in graph}$$

$$= \left(\frac{2t + 3 + 2a + 3}{2}\right)(t - a)$$

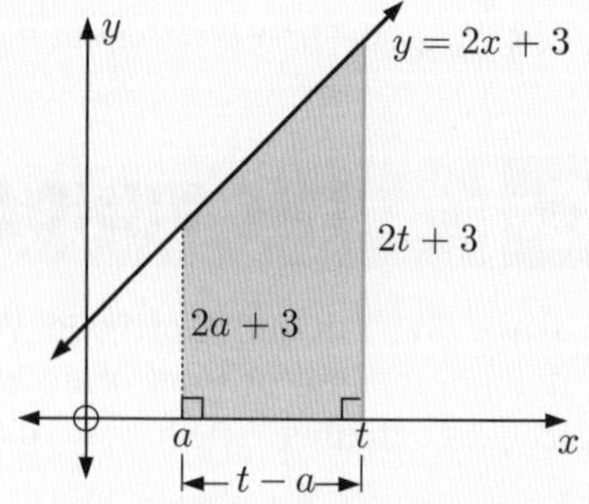

   a Write $A(t)$ in the form $F(t) - F(a)$.

   b What is the derivative $F'(x)$?
   How does it relate to the function $f(x)$?

4. Repeat the procedure in **2** and **3** to find area functions for:

   a $f(x) = \frac{1}{2}x + 3$ **b** $f(x) = 5 - 2x$

   Do your results fit with your earlier observations?

5. If $f(x) = 3x^2 + 4x + 5$, predict what $F(x)$ would be without performing the algebraic procedure.

From the **Investigation** you should have discovered that, for $f(x) \geqslant 0$,

$$\int_a^t f(x)\,dx = F(t) - F(a) \quad \text{where} \quad F'(x) = f(x). \quad F(x) \text{ is the } \textbf{antiderivative} \text{ of } f(x).$$

The following argument shows why this is true for all functions $f(x) \geqslant 0$.

Consider a function $y = f(x)$ which has antiderivative $F(x)$ and an area function $A(t) = \int_a^t f(x)\,dx$ which is the area from $x = a$ to $x = t$.

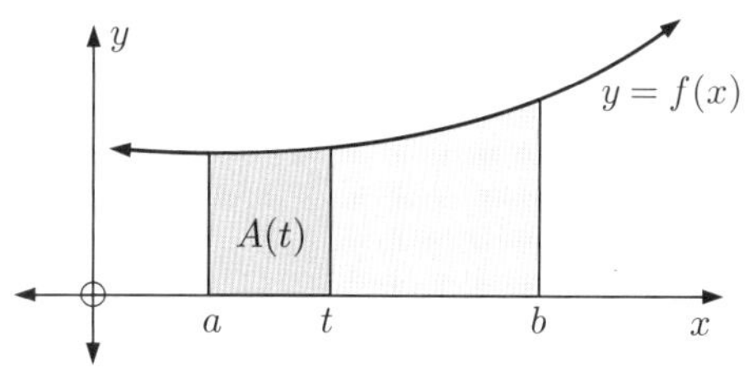

$A(t)$ is clearly an increasing function and

$A(a) = 0$ .... (1)

Consider the narrow strip between $x = t$ and $x = t + h$. The area of this strip is $A(t+h) - A(t)$, but we also know it must lie between a lower and upper rectangle on the interval $t \leqslant x \leqslant t+h$ of width h.

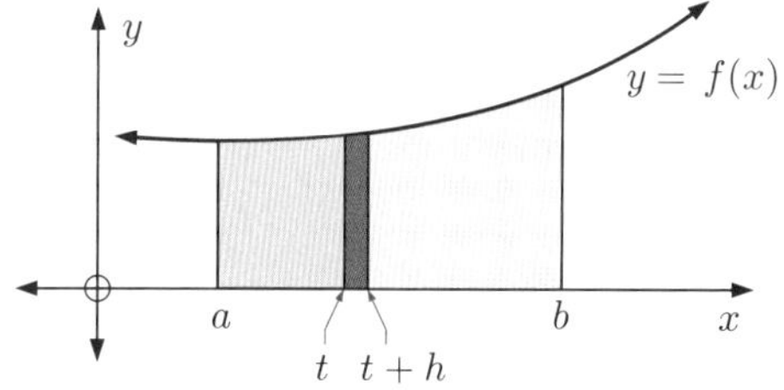

$$\text{area of smaller rectangle} \leqslant A(t+h) - A(t) \leqslant \text{area of larger rectangle}$$

If $f(x)$ is increasing on this interval then

$$hf(t) \leqslant A(t+h) - A(t) \leqslant hf(t+h)$$

$$\therefore \quad f(t) \leqslant \frac{A(t+h) - A(t)}{h} \leqslant f(t+h)$$

Equivalently, if $f(x)$ is decreasing on this interval then $f(t+h) \leqslant \dfrac{A(t+h) - A(t)}{h} \leqslant f(t)$.

Taking the limit as $h \to 0$ gives $f(t) \leqslant A'(t) \leqslant f(t)$

$$\therefore \quad A'(t) = f(t)$$

The area function $A(t)$ must only differ from the antiderivative of $f(t)$ by a constant.

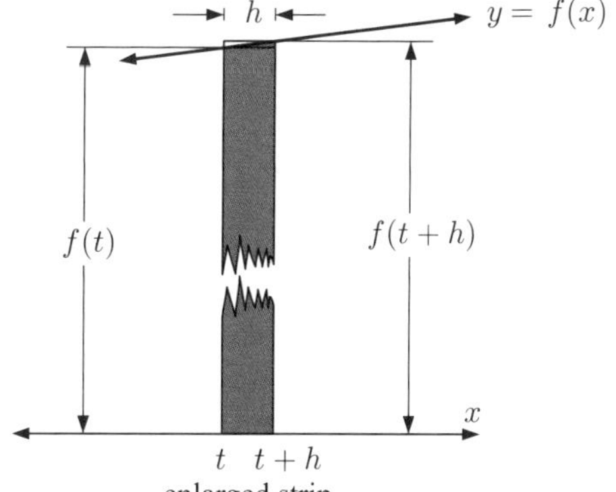

enlarged strip

$\therefore \quad A(t) = F(t) + c$

Letting $t = a$, $A(a) = F(a) + c$

But from (1), $A(a) = 0$, so $c = -F(a)$

$\therefore \quad A(t) = F(t) - F(a)$

Letting $t = b$, $\displaystyle\int_a^b f(x)\,dx = F(b) - F(a)$

This result is in fact true for all continuous functions $f(x)$.

THE FUNDAMENTAL THEOREM OF CALCULUS

For a continuous function $f(x)$ with antiderivative $F(x)$, $\displaystyle\int_a^b \boldsymbol{f(x)\,dx = F(b) - F(a)}$.

Instructions for evaluating definite integrals on your calculator can be found by clicking on the icon.

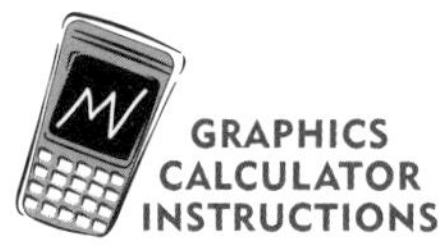

PROPERTIES OF DEFINITE INTEGRALS

The following properties of definite integrals can all be deduced from the fundamental theorem of calculus:

- $\int_a^a f(x)\,dx = 0$
- $\int_a^b c\,dx = c(b-a)$ {c is a constant}
- $\int_b^a f(x)\,dx = -\int_a^b f(x)\,dx$
- $\int_a^b c\,f(x)\,dx = c\int_a^b f(x)\,dx$
- $\int_a^b f(x)\,dx + \int_b^c f(x)\,dx = \int_a^c f(x)\,dx$
- $\int_a^b [f(x) \pm g(x)]dx = \int_a^b f(x)\,dx \pm \int_a^b g(x)\,dx$

Example proof:

$$\int_a^b f(x)\,dx + \int_b^c f(x)\,dx$$
$$= F(b) - F(a) + F(c) - F(b)$$
$$= F(c) - F(a)$$
$$= \int_a^c f(x)\,dx$$

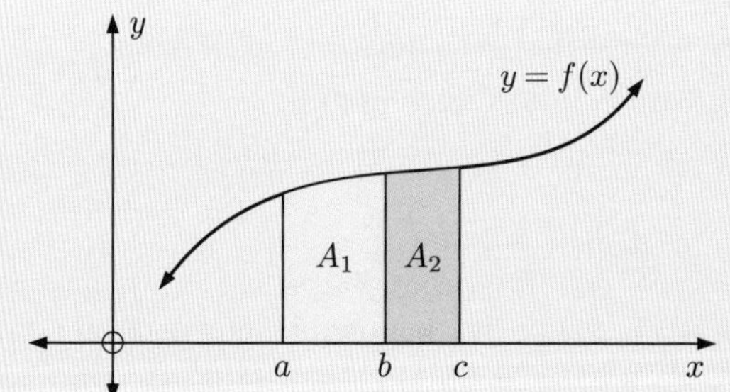

For the case where $a \leqslant b \leqslant c$ and $f(x) \geqslant 0$ for $a \leqslant x \leqslant c$,

$$\int_a^b f(x)\,dx + \int_b^c f(x)\,dx = A_1 + A_2 = \int_a^c f(x)\,dx$$

Example 4 **Self Tutor**

Use the fundamental theorem of calculus to find the area between:

a the x-axis and $y = x^2$ from $x = 0$ to $x = 1$

b the x-axis and $y = \sqrt{x}$ from $x = 1$ to $x = 9$.

a

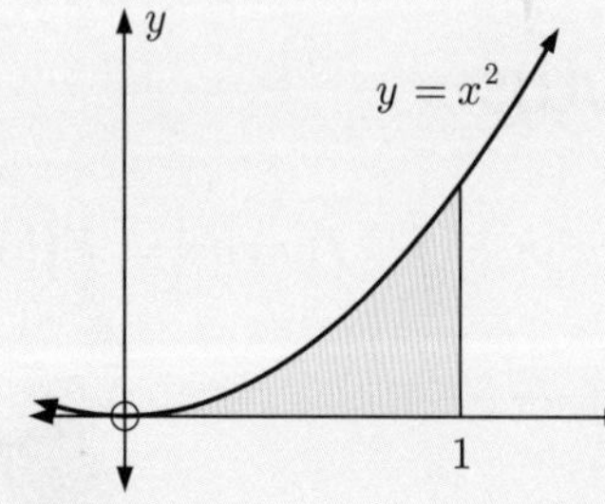

$f(x) = x^2$ has antiderivative $F(x) = \frac{x^3}{3}$

$\therefore$ the area $= \int_0^1 x^2\,dx$

$= F(1) - F(0)$

$= \frac{1}{3} - 0$

$= \frac{1}{3}$ units^2

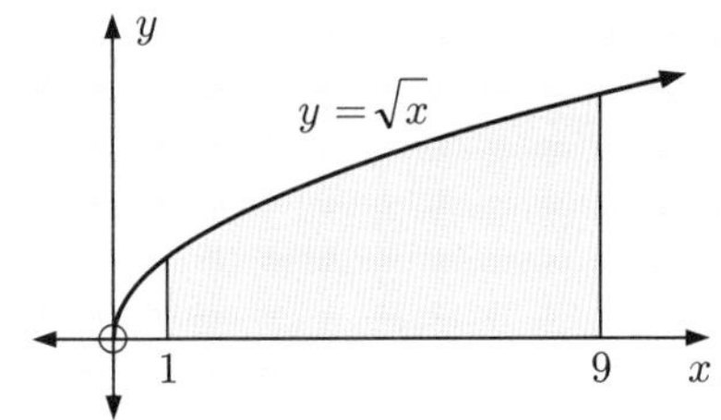

Casio fx-CG20

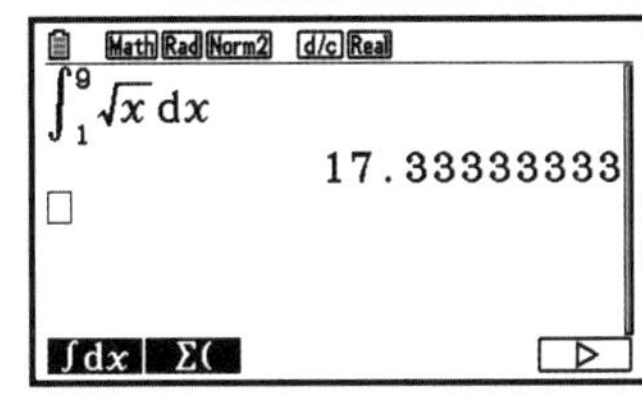

$f(x) = \sqrt{x} = x^{\frac{1}{2}}$ has antiderivative

$F(x) = \dfrac{x^{\frac{3}{2}}}{\frac{3}{2}} = \frac{2}{3}x\sqrt{x}$

$$\begin{aligned}\therefore \text{ the area } &= \int_1^9 x^{\frac{1}{2}}\,dx \\ &= F(9) - F(1) \\ &= \tfrac{2}{3} \times 27 - \tfrac{2}{3} \times 1 \\ &= 17\tfrac{1}{3} \text{ units}^2\end{aligned}$$

EXERCISE 21C

1 Use the fundamental theorem of calculus to find the area between:

- **a** the x-axis and $y = x^3$ from $x = 0$ to $x = 1$
- **b** the x-axis and $y = x^2$ from $x = 1$ to $x = 2$
- **c** the x-axis and $y = \sqrt{x}$ from $x = 0$ to $x = 1$.

Check your answers using technology.

2 Use the fundamental theorem of calculus to show that:

- **a** $\displaystyle\int_a^a f(x)\,dx = 0$ and explain the result graphically
- **b** $\displaystyle\int_a^b c\,dx = c(b-a)$ where c is a constant
- **c** $\displaystyle\int_b^a f(x)\,dx = -\int_a^b f(x)\,dx$
- **d** $\displaystyle\int_a^b c\,f(x)\,dx = c\int_a^b f(x)\,dx$ where c is a constant
- **e** $\displaystyle\int_a^b [f(x)+g(x)]\,dx = \int_a^b f(x)\,dx + \int_a^b g(x)\,dx$

3 Use the fundamental theorem of calculus to find the area between the x-axis and:

- **a** $y = x^3$ from $x = 1$ to $x = 2$
- **b** $y = x^2 + 3x + 2$ from $x = 1$ to $x = 3$
- **c** $y = \sqrt{x}$ from $x = 1$ to $x = 2$
- **d** $y = e^x$ from $x = 0$ to $x = 1.5$
- **e** $y = \dfrac{1}{\sqrt{x}}$ from $x = 1$ to $x = 4$
- **f** $y = x^3 + 2x^2 + 7x + 4$ from $x = 1$ to $x = 1.25$

4 Use technology to find the area between the x-axis and $y=\sqrt{9-x^2}$ from $x=0$ to $x=3$.
Check your answer by direct calculation of the area.

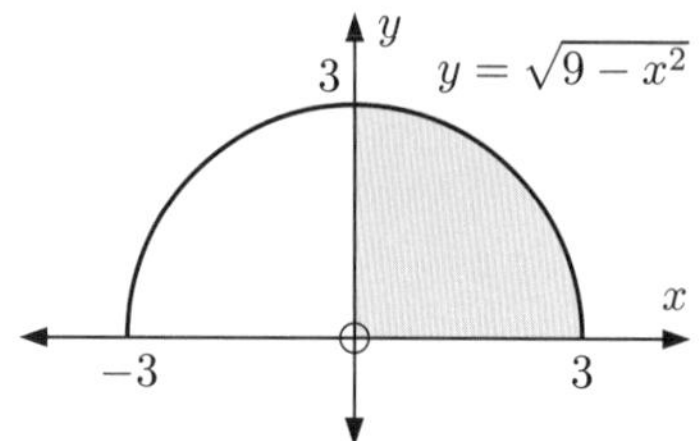

5 **a** Use the fundamental theorem of calculus to show that
$$\int_a^b (-f(x))\,dx = -\int_a^b f(x)\,dx$$

b Hence show that if $f(x)\leqslant 0$ for all x on $a\leqslant x\leqslant b$ then the shaded area $= -\int_a^b f(x)\,dx$.

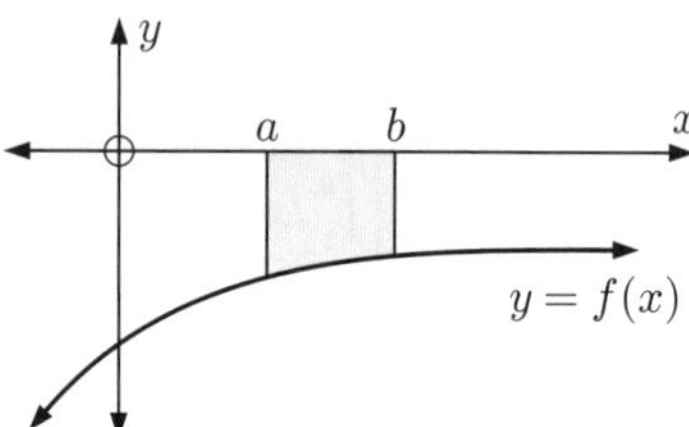

c Calculate the following integrals, and give graphical interpretations of your answers:

i $\int_0^1 (-x^2)\,dx$ **ii** $\int_0^1 (x^2-x)\,dx$ **iii** $\int_{-2}^0 3x\,dx$

d Use graphical evidence and known area facts to find $\int_0^2 \left(-\sqrt{4-x^2}\right)dx$.

D INTEGRATION

Earlier we showed that the **antiderivative** of x^2 is $\frac{1}{3}x^3$.

We showed that any function of the form $\frac{1}{3}x^3+c$ where c is a constant, has derivative x^2.

We say that the **indefinite integral** or **integral** of x^2 is $\frac{1}{3}x^3+c$, and write $\int x^2\,dx = \frac{1}{3}x^3+c$.

We read this as "the integral of x^2 with respect to x is $\frac{1}{3}x^3+c$, where c is a constant".

If $F'(x)=f(x)$ then $\int f(x)\,dx = F(x)+c$.

This process is known as **indefinite integration**. It is indefinite because it is not being applied to a particular interval.

DISCOVERING INTEGRALS

Since integration is the reverse process of differentiation we can sometimes discover integrals by differentiation. For example:

- if $F(x)=x^4$ then $F'(x)=4x^3$
  $\therefore \int 4x^3\,dx = x^4+c$
- if $F(x)=\sqrt{x}=x^{\frac{1}{2}}$ then $F'(x)=\frac{1}{2}x^{-\frac{1}{2}}=\frac{1}{2\sqrt{x}}$
  $\therefore \int \frac{1}{2\sqrt{x}}dx = \sqrt{x}+c$

The following rules may prove useful:

- Any constant may be written in front of the integral sign.

$$\int k\,f(x)\,dx = k\int f(x)\,dx, \quad k \text{ is a constant}$$

Proof: Consider differentiating $kF(x)$ where $F'(x) = f(x)$.

$$\frac{d}{dx}(k\,F(x)) = k\,F'(x) = k\,f(x)$$

$$\therefore \int k\,f(x)\,dx = k\,F(x)$$

$$= k\int f(x)\,dx$$

- The integral of a sum is the sum of the separate integrals. This rule enables us to integrate term by term.

$$\int [f(x) + g(x)]\,dx = \int f(x)\,dx + \int g(x)\,dx$$

Example 5 — Self Tutor

If $y = x^4 + 2x^3$, find $\frac{dy}{dx}$. Hence find $\int (2x^3 + 3x^2)\,dx$.

If $y = x^4 + 2x^3$ then $\frac{dy}{dx} = 4x^3 + 6x^2$

$$\therefore \int (4x^3 + 6x^2)\,dx = x^4 + 2x^3 + c$$

$$\therefore \int 2(2x^3 + 3x^2)\,dx = x^4 + 2x^3 + c$$

$$\therefore 2\int (2x^3 + 3x^2)\,dx = x^4 + 2x^3 + c$$

$$\therefore \int (2x^3 + 3x^2)\,dx = \tfrac{1}{2}x^4 + x^3 + c$$

c represents a general constant, so is simply any value $c \in \mathbb{R}$. Instead of writing $\frac{c}{2}$, we can therefore still write just c.

EXERCISE 21D

1. If $y = x^7$, find $\frac{dy}{dx}$. Hence find $\int x^6\,dx$.

2. If $y = x^3 + x^2$, find $\frac{dy}{dx}$. Hence find $\int (3x^2 + 2x)\,dx$.

3. If $y = e^{2x+1}$, find $\frac{dy}{dx}$. Hence find $\int e^{2x+1}\,dx$.

4. If $y = (2x + 1)^4$ find $\frac{dy}{dx}$. Hence find $\int (2x + 1)^3\,dx$.

Example 6 — Self Tutor

Suppose $y = \sqrt{5x-1}$.

a Find $\frac{dy}{dx}$. **b** Hence find $\int \frac{1}{\sqrt{5x-1}}\,dx$.

a
$$\begin{aligned} y &= \sqrt{5x-1} \\ &= (5x-1)^{\frac{1}{2}} \\ \therefore\ \frac{dy}{dx} &= \tfrac{1}{2}(5x-1)^{-\frac{1}{2}}(5) \quad \{\text{chain rule}\} \\ &= \frac{5}{2\sqrt{5x-1}} \end{aligned}$$

b Using **a**, $\int \frac{5}{2\sqrt{5x-1}}\,dx = \sqrt{5x-1} + c$

$$\therefore\ \tfrac{5}{2}\int \frac{1}{\sqrt{5x-1}}\,dx = \sqrt{5x-1} + c$$

$$\therefore\ \int \frac{1}{\sqrt{5x-1}}\,dx = \tfrac{2}{5}\sqrt{5x-1} + c$$

5 If $y = x\sqrt{x}$, find $\frac{dy}{dx}$. Hence find $\int \sqrt{x}\,dx$.

6 If $y = \frac{1}{\sqrt{x}}$, find $\frac{dy}{dx}$. Hence find $\int \frac{1}{x\sqrt{x}}\,dx$.

7 If $y = \cos 2x$, find $\frac{dy}{dx}$. Hence find $\int \sin 2x\,dx$.

8 If $y = \sin(1-5x)$, find $\frac{dy}{dx}$. Hence find $\int \cos(1-5x)\,dx$.

9 By considering $\frac{d}{dx}(x^2-x)^3$, find $\int (2x-1)(x^2-x)^2\,dx$.

10 Prove the rule $\int [f(x)+g(x)]\,dx = \int f(x)\,dx + \int g(x)\,dx$.

11 Find $\frac{dy}{dx}$ if $y = \sqrt{1-4x}$. Hence find $\int \frac{1}{\sqrt{1-4x}}\,dx$.

12 By considering $\frac{d}{dx}\ln(5-3x+x^2)$, find $\int \frac{4x-6}{5-3x+x^2}\,dx$.

13 Find $\frac{d}{dx}(\csc x)$. Hence find $\int \frac{\cos x}{\sin^2 x}\,dx$.

14 By considering $\frac{d}{dx}(\arctan x)$, find $\int \frac{-3}{x^2+1}\,dx$.

15 By considering $\frac{d}{dx}(2^x)$, find $\int 2^x\,dx$. **Hint:** $2^x = \left(e^{\ln 2}\right)^x$.

16 By considering $\frac{d}{dx}(x\ln x)$, find $\int \ln x\,dx$.

We can check that an integral is correct by differentiating the answer. It should give us the **integrand**, the function we originally integrated.

E RULES FOR INTEGRATION

In earlier chapters we developed a set of rules to help us differentiate functions more efficiently:

Function	*Derivative*	*Name*
c, a constant	0	
$mx + c$, m and c are constants	m	
x^n	nx^{n-1}	**power rule**
$cu(x)$	$cu'(x)$	**scalar multiplication rule**
$u(x) + v(x)$	$u'(x) + v'(x)$	**addition rule**
$u(x)v(x)$	$u'(x)v(x) + u(x)v'(x)$	**product rule**
$\frac{u(x)}{v(x)}$	$\frac{u'(x)v(x) - u(x)v'(x)}{[v(x)]^2}$	**quotient rule**
$y = f(u)$ where $u = u(x)$	$\frac{dy}{dx} = \frac{dy}{du}\frac{du}{dx}$	**chain rule**
e^x	e^x	
$e^{f(x)}$	$e^{f(x)}f'(x)$	
$\ln x$	$\frac{1}{x}$	
$\ln f(x)$	$\frac{f'(x)}{f(x)}$	
$[f(x)]^n$	$n[f(x)]^{n-1}f'(x)$	
$\sin x$	$\cos x$	
$\cos x$	$-\sin x$	
$\tan x$	$\sec^2 x$	
$\csc x$	$-\csc x \cot x$	
$\sec x$	$\sec x \tan x$	
$\cot x$	$-\csc^2 x$	
$\arcsin x$	$\frac{1}{\sqrt{1-x^2}}$	
$\arccos x$	$\frac{-1}{\sqrt{1-x^2}}$	
$\arctan x$	$\frac{1}{1+x^2}$	

These rules or combinations of them can be used to differentiate all of the functions we consider in this course.

However, the task of finding **antiderivatives** is not so easy, and cannot be written as a simple list of rules as we did above. In fact, huge books of different types of functions and their integrals have been written. Fortunately our course is restricted to a few special cases.

RULES FOR INTEGRATION

For k a constant, $\frac{d}{dx}(kx+c)=k$ $\quad\therefore\quad$ $\displaystyle\int k\,dx = kx + c$

If $n \neq -1$, $\frac{d}{dx}\left(\frac{x^{n+1}}{n+1}+c\right)=\frac{(n+1)x^n}{n+1}=x^n$ $\quad\therefore\quad$ $\displaystyle\int x^n\,dx = \frac{x^{n+1}}{n+1}+c,\ n \neq -1$

$\frac{d}{dx}(e^x+c)=e^x$ $\quad\therefore\quad$ $\displaystyle\int e^x\,dx = e^x + c$

If $x>0$, $\frac{d}{dx}(\ln x+c)=\frac{1}{x}$

If $x<0$, $\frac{d}{dx}(\ln(-x)+c)=\frac{-1}{-x}=\frac{1}{x}$ $\quad\therefore\quad$ $\displaystyle\int \frac{1}{x}\,dx = \ln|x| + c$

$\frac{d}{dx}(\sin x+c)=\cos x$ $\quad\therefore\quad$ $\displaystyle\int \cos x\,dx = \sin x + c$

$\frac{d}{dx}(-\cos x+c)=\sin x$ $\quad\therefore\quad$ $\displaystyle\int \sin x\,dx = -\cos x + c$

$\frac{d}{dx}(\tan x+c)=\sec^2 x$ $\quad\therefore\quad$ $\displaystyle\int \sec^2 x\,dx = \tan x + c$

Function	*Integral*
k, a constant	$kx+c$
$x^n,\ n \neq -1$	$\frac{x^{n+1}}{n+1}+c$
e^x	e^x+c
$\frac{1}{x}$	$\ln\|x\|+c$
$\cos x$	$\sin x + c$
$\sin x$	$-\cos x + c$
$\sec^2 x$	$\tan x + c$

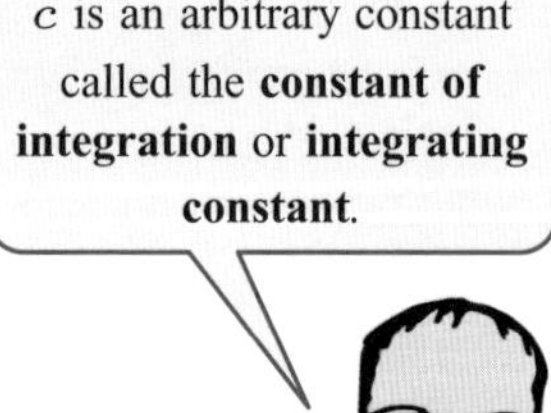

c is an arbitrary constant called the **constant of integration** or **integrating constant**.

Remember that you can always check your integration by differentiating the resulting function.

Example 7 — Self Tutor

Find:

a $\displaystyle\int (x^3 - 2x^2 + 5)\,dx$

b $\displaystyle\int \left(\frac{1}{x^3} - \sqrt{x}\right)dx$

a $\displaystyle\int (x^3 - 2x^2 + 5)\,dx$

$= \frac{x^4}{4} - \frac{2x^3}{3} + 5x + c$

b $\displaystyle\int \left(\frac{1}{x^3} - \sqrt{x}\right)dx$

$= \displaystyle\int (x^{-3} - x^{\frac{1}{2}})\,dx$

$= \frac{x^{-2}}{-2} - \frac{x^{\frac{3}{2}}}{\frac{3}{2}} + c$

$= -\frac{1}{2x^2} - \frac{2}{3}x^{\frac{3}{2}} + c$

Example 8 Self Tutor

Integrate with respect to x:

a $2\sin x - \cos x$ b $-\frac{2}{x} + 3e^x$ c $3\sec^2 x - x^2$

a $\int (2\sin x - \cos x)\,dx$
$= 2(-\cos x) - \sin x + c$
$= -2\cos x - \sin x + c$

b $\int \left(-\frac{2}{x} + 3e^x\right) dx$
$= -2\ln|x| + 3e^x + c$

c $\int \left(3\sec^2 x - x^2\right) dx$
$= 3\tan x - \frac{1}{3}x^3 + c$

There is no product or quotient rule for integration. Consequently we often have to carry out multiplication or division before we integrate.

Example 9 Self Tutor

Find: a $\int \left(3x + \frac{2}{x}\right)^2 dx$ b $\int \left(\frac{x^2-2}{\sqrt{x}}\right) dx$

a $\int \left(3x + \frac{2}{x}\right)^2 dx$
$= \int \left(9x^2 + 12 + \frac{4}{x^2}\right) dx$
$= \int (9x^2 + 12 + 4x^{-2})\,dx$
$= \frac{9x^3}{3} + 12x + \frac{4x^{-1}}{-1} + c$
$= 3x^3 + 12x - \frac{4}{x} + c$

We expand the brackets and simplify to a form that can be integrated.

b $\int \left(\frac{x^2-2}{\sqrt{x}}\right) dx$
$= \int \left(\frac{x^2}{\sqrt{x}} - \frac{2}{\sqrt{x}}\right) dx$
$= \int (x^{\frac{3}{2}} - 2x^{-\frac{1}{2}})\,dx$
$= \frac{x^{\frac{5}{2}}}{\frac{5}{2}} - \frac{2x^{\frac{1}{2}}}{\frac{1}{2}} + c$
$= \frac{2}{5}x^2\sqrt{x} - 4\sqrt{x} + c$

EXERCISE 21E.1

1 Find:

a $\int (x^4 - x^2 - x + 2)\,dx$ b $\int (\sqrt{x} + e^x)\,dx$ c $\int \left(3e^x - \frac{1}{x}\right) dx$

d $\int \left(x\sqrt{x} - \frac{2}{x}\right) dx$ e $\int \left(\frac{1}{x\sqrt{x}} + \frac{4}{x}\right) dx$ f $\int \left(\frac{1}{2}x^3 - x^4 + x^{\frac{1}{3}}\right) dx$

g $\int \left(x^2 + \frac{3}{x}\right) dx$ h $\int \left(\frac{1}{2x} + x^2 - e^x\right) dx$ i $\int \left(5e^x + \frac{1}{3}x^3 - \frac{4}{x}\right) dx$

2 Integrate with respect to x:

a $3\sin x - 2$ b $4x - 2\cos x$ c $\sin x - 2\cos x + e^x$

d $x^2\sqrt{x} - 10\sin x$ e $\frac{x(x-1)}{3} + \cos x$ f $-\sin x + 2\sqrt{x}$

3 Find:

a $\int (x^2 + 3x - 2)\, dx$

b $\int \left(\sqrt{x} - \frac{1}{\sqrt{x}}\right) dx$

c $\int \left(2e^x - \frac{1}{x^2}\right) dx$

d $\int \frac{1 - 4x}{x\sqrt{x}}\, dx$

e $\int (2x + 1)^2\, dx$

f $\int \frac{x^2 + x - 3}{x}\, dx$

g $\int \frac{2x - 1}{\sqrt{x}}\, dx$

h $\int \frac{x^2 - 4x + 10}{x^2\sqrt{x}}\, dx$

i $\int (x + 1)^3\, dx$

4 Find:

a $\int \left(\sqrt{x} + \frac{1}{2}\cos x\right) dx$

b $\int \left(2e^t - 4\sin t\right) dt$

c $\int \left(3\cos t - \frac{1}{t}\right) dt$

d $\int \left(\sec^2 x + 2\sin x\right) dx$

e $\int (\theta - \sin\theta)\, d\theta$

f $\int \left(\frac{2}{\theta} - \sec^2\theta\right) d\theta$

5 Find y if:

a $\frac{dy}{dx} = 6$

b $\frac{dy}{dx} = 4x^2$

c $\frac{dy}{dx} = 5\sqrt{x} - x^2$

d $\frac{dy}{dx} = \frac{1}{x^2}$

e $\frac{dy}{dx} = 2e^x - 5$

f $\frac{dy}{dx} = 4x^3 + 3x^2$

6 Integrate with respect to x:

a $(1 - 2x)^2$

b $\sqrt{x} - \frac{2}{\sqrt{x}}$

c $\frac{x^2 + 2x - 5}{x^2}$

7 Find $f(x)$ if:

a $f'(x) = x^3 - 5\sqrt{x} + 3$

b $f'(x) = 2\sqrt{x}(1 - 3x)$

c $f'(x) = 3e^x - \frac{4}{x}$

Example 10 **Self Tutor**

Find $\frac{d}{dx}(x\sin x)$ and hence determine $\int x\cos x\, dx$.

$\frac{d}{dx}(x\sin x) = (1)\sin x + (x)\cos x$ {product rule of differentiation}

$= \sin x + x\cos x$

Thus $\int (\sin x + x\cos x)\, dx = x\sin x + c$ {antidifferentiation}

$\therefore \int \sin x\, dx + \int x\cos x\, dx = x\sin x + c$

$\therefore -\cos x + \int x\cos x\, dx = x\sin x + c$

$\therefore \int x\cos x\, dx = x\sin x + \cos x + c$

8 Find $\frac{d}{dx}(e^x\sin x)$ and hence find $\int e^x(\sin x + \cos x)\, dx$.

9 Find $\frac{d}{dx}(e^{-x}\sin x)$ and hence find $\int \frac{\cos x - \sin x}{e^x}\, dx$.

10 Find $\frac{d}{dx}(x\cos x)$ and hence find $\int x\sin x\,dx$.

11 Find $\frac{d}{dx}(\sec x)$ and hence find $\int \tan x \sec x\,dx$.

12 Find $\frac{d}{dx}(x - 3\arctan x)$ and hence determine $\int \frac{x^2-2}{x^2+1}\,dx$.

13 **a** Find $\frac{d}{dx}(\arccos x)$ and $\frac{d}{dx}(\arcsin x)$.

b Hence find *two* expressions for $\int \frac{1}{\sqrt{1-x^2}}\,dx$.

c Explain your answer.

PARTICULAR VALUES

We can find the constant of integration c if we are given a particular value of the function.

Example 11 **Self Tutor**

Find $f(x)$ given that $f'(x) = x^3 - 2x^2 + 3$ and $f(0) = 2$.

Since $f'(x) = x^3 - 2x^2 + 3$,

$$f(x) = \int (x^3 - 2x^2 + 3)\,dx$$

$$\therefore\ f(x) = \frac{x^4}{4} - \frac{2x^3}{3} + 3x + c$$

But $f(0) = 2$, so $c = 2$

Thus $f(x) = \frac{x^4}{4} - \frac{2x^3}{3} + 3x + 2$

Example 12 **Self Tutor**

Find $f(x)$ given that $f'(x) = 2\sin x - \sqrt{x}$ and $f(0) = 4$.

$$f(x) = \int \left(2\sin x - x^{\frac{1}{2}}\right) dx$$

$$\therefore\ f(x) = 2 \times (-\cos x) - \frac{x^{\frac{3}{2}}}{\frac{3}{2}} + c$$

$$\therefore\ f(x) = -2\cos x - \tfrac{2}{3}x^{\frac{3}{2}} + c$$

But $f(0) = 4$, so $-2\cos 0 - 0 + c = 4$

$\therefore\ c = 6$

Thus $f(x) = -2\cos x - \tfrac{2}{3}x^{\frac{3}{2}} + 6$.

If we are given the second derivative we need to integrate twice to find the function. This creates two integrating constants, so we need two other facts about the curve in order to determine these constants.

Example 13 — Self Tutor

Find $f(x)$ given that $f''(x) = 12x^2 - 4$, $f'(0) = -1$, and $f(1) = 4$.

If $f''(x) = 12x^2 - 4$

then $f'(x) = \frac{12x^3}{3} - 4x + c$ {integrating with respect to x}

$\therefore$ $f'(x) = 4x^3 - 4x + c$

But $f'(0) = -1$, so $c = -1$

Thus $f'(x) = 4x^3 - 4x - 1$

$\therefore$ $f(x) = \frac{4x^4}{4} - \frac{4x^2}{2} - x + d$ {integrating again}

$\therefore$ $f(x) = x^4 - 2x^2 - x + d$

But $f(1) = 4$, so $1 - 2 - 1 + d = 4$ and hence $d = 6$

Thus $f(x) = x^4 - 2x^2 - x + 6$

EXERCISE 21E.2

1 Find $f(x)$ given that:

a $f'(x) = 2x - 1$ and $f(0) = 3$

b $f'(x) = 3x^2 + 2x$ and $f(2) = 5$

c $f'(x) = e^x + \frac{1}{\sqrt{x}}$ and $f(1) = 1$

d $f'(x) = x - \frac{2}{\sqrt{x}}$ and $f(1) = 2$

2 Find $f(x)$ given that:

a $f'(x) = x^2 - 4\cos x$ and $f(0) = 3$

b $f'(x) = 2\cos x - 3\sin x$ and $f\left(\frac{\pi}{4}\right) = \frac{1}{\sqrt{2}}$

c $f'(x) = \sqrt{x} - 2\sec^2 x$ and $f(\pi) = 0$

3 Find $f(x)$ given that:

a $f''(x) = 2x + 1$, $f'(1) = 3$, and $f(2) = 7$

b $f''(x) = 15\sqrt{x} + \frac{3}{\sqrt{x}}$, $f'(1) = 12$, and $f(0) = 5$

c $f''(x) = \cos x$, $f'(\frac{\pi}{2}) = 0$, and $f(0) = 3$

d $f''(x) = 2x$ and the points (1, 0) and (0, 5) lie on the curve.

F INTEGRATING $f(ax + b)$

In this section we deal with integrals of functions which are composite with the linear function $ax + b$.

Notice that $\frac{d}{dx}\left(\frac{1}{a}e^{ax+b}\right) = \frac{1}{a}e^{ax+b} \times a = e^{ax+b}$

$$\therefore \quad \int e^{ax+b}\,dx = \frac{1}{a}e^{ax+b} + c \quad \text{for } a \neq 0$$

For $n \neq -1$, $\dfrac{d}{dx}\left(\dfrac{1}{a(n+1)}(ax+b)^{n+1}\right) = \dfrac{1}{a(n+1)}(n+1)(ax+b)^n \times a,$

$$= (ax+b)^n$$

$$\therefore \quad \boldsymbol{\int (ax+b)^n \, dx = \frac{1}{a}\frac{(ax+b)^{n+1}}{(n+1)} + c \quad \text{for } n \neq -1}$$

Also, $\dfrac{d}{dx}\left(\dfrac{1}{a}\ln(ax+b)\right) = \dfrac{1}{a}\left(\dfrac{a}{ax+b}\right) = \dfrac{1}{ax+b}$ for $ax+b>0$, $a \neq 0$

$$\therefore \quad \int \frac{1}{ax+b}\, dx = \frac{1}{a}\ln(ax+b) + c \quad \text{for} \quad ax+b>0, \quad a \neq 0$$

In fact, $$\boldsymbol{\int \frac{1}{ax+b}\, dx = \frac{1}{a}\ln|ax+b| + c \quad \text{for } a \neq 0.}$$

We can perform the same process for the circular functions:

$$\frac{d}{dx}(\sin(ax+b)) = a\cos(ax+b)$$

$$\therefore \quad \int a\cos(ax+b)\, dx = \sin(ax+b) + c$$

$$\therefore \quad a\int \cos(ax+b)\, dx = \sin(ax+b) + c$$

So, $$\boldsymbol{\int \cos(ax+b)\, dx = \frac{1}{a}\sin(ax+b) + c \quad \text{for } a \neq 0.}$$

Likewise we can show $$\boldsymbol{\int \sin(ax+b)\, dx = -\frac{1}{a}\cos(ax+b) + c \quad \text{for } a \neq 0.}$$

and $$\boldsymbol{\int \sec^2(ax+b)\, dx = \frac{1}{a}\tan(ax+b) + c \quad \text{for } a \neq 0.}$$

For a, b constants with $a \neq 0$, we have:

Function	*Integral*
e^{ax+b}	$\frac{1}{a}e^{ax+b} + c$
$(ax+b)^n,\ n \neq -1$	$\frac{1}{a}\frac{(ax+b)^{n+1}}{n+1} + c$
$\frac{1}{ax+b}$	$\frac{1}{a}\ln\|ax+b\| + c$
$\cos(ax+b)$	$\frac{1}{a}\sin(ax+b) + c$
$\sin(ax+b)$	$-\frac{1}{a}\cos(ax+b) + c$
$\sec^2(ax+b)$	$\frac{1}{a}\tan(ax+b) + c$

Example 14 **Self Tutor**

Find: **a** $\int (2x+3)^4\,dx$ **b** $\int \frac{1}{\sqrt{1-2x}}\,dx$

a $\int (2x+3)^4\,dx$

$= \frac{1}{2} \times \frac{(2x+3)^5}{5} + c$

$= \frac{1}{10}(2x+3)^5 + c$

b $\int \frac{1}{\sqrt{1-2x}}\,dx$

$= \int (1-2x)^{-\frac{1}{2}}\,dx$

$= \frac{1}{-2} \times \frac{(1-2x)^{\frac{1}{2}}}{\frac{1}{2}} + c$

$= -\sqrt{1-2x} + c$

Example 15 **Self Tutor**

Find: **a** $\int (2e^{2x} - e^{-3x})\,dx$ **b** $\int \frac{4}{1-2x}\,dx$

a $\int (2e^{2x} - e^{-3x})\,dx$

$= 2(\frac{1}{2})e^{2x} - (\frac{1}{-3})e^{-3x} + c$

$= e^{2x} + \frac{1}{3}e^{-3x} + c$

b $\int \frac{4}{1-2x}\,dx$

$= 4\int \frac{1}{1-2x}\,dx$

$= 4\left(\frac{1}{-2}\right)\ln|1-2x| + c$

$= -2\ln|1-2x| + c$

Example 16 **Self Tutor**

Integrate with respect to x: $2\sin(3x) + \cos(4x+\pi)$

$\int (2\sin(3x) + \cos(4x+\pi))\,dx$

$= 2 \times \frac{1}{3}(-\cos(3x)) + \frac{1}{4}\sin(4x+\pi) + c$

$= -\frac{2}{3}\cos(3x) + \frac{1}{4}\sin(4x+\pi) + c$

Integrals involving $\sin^2(ax+b)$ and $\cos^2(ax+b)$ can be found by first using the identities

$$\sin^2\theta = \frac{1}{2} - \frac{1}{2}\cos(2\theta) \quad \text{or} \quad \cos^2\theta = \frac{1}{2} + \frac{1}{2}\cos(2\theta).$$

For example, we would substitute

$$\sin^2(3x - \tfrac{\pi}{2}) = \tfrac{1}{2} - \tfrac{1}{2}\cos(6x-\pi) \quad \text{or} \quad \cos^2(\tfrac{x}{2}) = \tfrac{1}{2} + \tfrac{1}{2}\cos x$$

$$= \tfrac{1}{2} + \tfrac{1}{2}\cos(6x)$$

Example 17 **Self Tutor**

Integrate $(2-\sin x)^2$ with respect to x.

$$\begin{aligned}&\int (2-\sin x)^2\,dx\\ &= \int (4-4\sin x+\sin^2 x)\,dx\\ &= \int \left(4-4\sin x+\tfrac{1}{2}-\tfrac{1}{2}\cos(2x)\right)\,dx\\ &= \int \left(\tfrac{9}{2}-4\sin x-\tfrac{1}{2}\cos(2x)\right)\,dx\\ &= \tfrac{9}{2}x+4\cos x-\tfrac{1}{2}\times\tfrac{1}{2}\sin(2x)+c\\ &= \tfrac{9}{2}x+4\cos x-\tfrac{1}{4}\sin(2x)+c\end{aligned}$$

EXERCISE 21F

1 Find:

a $\int (2x+5)^3\,dx$ **b** $\int \frac{1}{(3-2x)^2}\,dx$ **c** $\int \frac{4}{(2x-1)^4}\,dx$

d $\int (4x-3)^7\,dx$ **e** $\int \sqrt{3x-4}\,dx$ **f** $\int \frac{10}{\sqrt{1-5x}}\,dx$

g $\int 3(1-x)^4\,dx$ **h** $\int \frac{4}{\sqrt{3-4x}}\,dx$

2 Integrate with respect to x:

a $\sin(3x)$ **b** $2\cos(-4x)+1$ **c** $\sec^2(2x)$

d $3\cos\left(\frac{x}{2}\right)$ **e** $3\sin(2x)-e^{-x}$ **f** $e^{2x}-2\sec^2\left(\frac{x}{2}\right)$

g $2\sin\left(2x+\frac{\pi}{6}\right)$ **h** $-3\cos\left(\frac{\pi}{4}-x\right)$ **i** $4\sec^2\left(\frac{\pi}{3}-2x\right)$

j $\cos(2x)+\sin(2x)$ **k** $2\sin(3x)+5\cos(4x)$ **l** $\frac{1}{2}\cos(8x)-3\sin x$

3 **a** Find $y=f(x)$ given $\frac{dy}{dx}=\sqrt{2x-7}$ and that $y=11$ when $x=8$.

b The function $f(x)$ has gradient function $f'(x)=\frac{4}{\sqrt{1-x}}$, and the curve $y=f(x)$ passes through the point $(-3, -11)$.
Find the point on the graph of $y=f(x)$ with x-coordinate -8.

4 Integrate with respect to x:

a $\cos^2 x$ **b** $\sin^2 x$

c $1+\cos^2(2x)$ **d** $3-\sin^2(3x)$

e $\frac{1}{2}\cos^2(4x)$ **f** $(1+\cos x)^2$

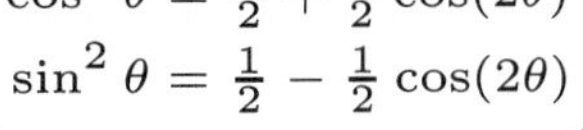
Use the identities
$\cos^2\theta=\frac{1}{2}+\frac{1}{2}\cos(2\theta)$
$\sin^2\theta=\frac{1}{2}-\frac{1}{2}\cos(2\theta)$

5 Find:

a $\int 3(2x-1)^2\,dx$ b $\int (x^2-x)^2\,dx$ c $\int (1-3x)^3\,dx$

d $\int (1-x^2)^2\,dx$ e $\int 4\sqrt{5-x}\,dx$ f $\int (x^2+1)^3\,dx$

6 a Use the identity $\cos^2\theta = \frac{1}{2}+\frac{1}{2}\cos(2\theta)$ to show that $\cos^4 x = \frac{1}{8}\cos(4x)+\frac{1}{2}\cos(2x)+\frac{3}{8}$.

b Hence find $\int \cos^4 x\,dx$.

7 Find:

a $\int \left(2e^x+5e^{2x}\right)dx$ b $\int \left(3e^{5x-2}\right)dx$ c $\int \left(e^{7-3x}\right)dx$

d $\int \frac{1}{2x-1}\,dx$ e $\int \frac{5}{1-3x}\,dx$ f $\int \left(e^{-x}-\frac{4}{2x+1}\right)dx$

g $\int \left(e^x+e^{-x}\right)^2 dx$ h $\int \left(e^{-x}+2\right)^2 dx$ i $\int \left(x-\frac{5}{1-x}\right)dx$

8 Find y given that:

a $\frac{dy}{dx}=(1-e^x)^2$ b $\frac{dy}{dx}=1-2x+\frac{3}{x+2}$ c $\frac{dy}{dx}=e^{-2x}+\frac{4}{2x-1}$

9 To find $\int \frac{1}{4x}\,dx$, Tracy's answer was $\int \frac{1}{4x}\,dx = \frac{1}{4}\ln|4x|+c$.

Nadine's answer was $\int \frac{1}{4x}\,dx = \frac{1}{4}\int \frac{1}{x}\,dx = \frac{1}{4}\ln|x|+c$.

Which of them has found the correct answer? Prove your statement.

10 Suppose $f'(x)=p\sin\left(\frac{1}{2}x\right)$ where p is a constant. $f(0)=1$ and $f(2\pi)=0$. Find p and hence $f(x)$.

11 Consider a function g such that $g''(x)=-\sin 2x$.
Show that the gradients of the tangents to $y=g(x)$ when $x=\pi$ and $x=-\pi$ are equal.

12 a Find $f(x)$ given $f'(x)=2e^{-2x}$ and $f(0)=3$.

b Find $f(x)$ given $f'(x)=2x-\frac{2}{1-x}$ and $f(-1)=3$.

c A curve has gradient function $\sqrt{x}+\frac{1}{2}e^{-4x}$ and passes through $(1, 0)$. Find the equation of the function.

13 Show that $(\sin x+\cos x)^2 = 1+\sin 2x$ and hence determine $\int (\sin x+\cos x)^2\,dx$.

14 Show that $(\cos x+1)^2=\frac{1}{2}\cos 2x+2\cos x+\frac{3}{2}$ and hence determine $\int (\cos x+1)^2\,dx$.

15 Show that $\frac{3}{x+2}-\frac{1}{x-2}=\frac{2x-8}{x^2-4}$ and hence find $\int \frac{2x-8}{x^2-4}\,dx$.

16 Show that $\frac{1}{2x-1}-\frac{1}{2x+1}=\frac{2}{4x^2-1}$ and hence find $\int \frac{2}{4x^2-1}\,dx$.

INVESTIGATION 3 EULER'S FORM

In our study of complex numbers we used Euler's form $e^{ix} = \cos x + i\sin x$. We will now prove this result using calculus.

What to do:

1 Differentiate e^{ix} with respect to x.

2 We know that every complex number can be written in polar form.
Therefore suppose $e^{ix} = r\,\text{cis}\,\theta = r\cos\theta + ir\sin\theta$ where r and θ are dependent on x.

a Use **1** to show that the derivative of e^{ix} with respect to x is $ir\cos\theta - r\sin\theta$.

b Show that the derivative of $r\cos\theta + ir\sin\theta$ with respect to x is
$$\left(\cos\theta\,\frac{dr}{dx} - r\sin\theta\,\frac{d\theta}{dx}\right) + i\left(\sin\theta\,\frac{dr}{dx} + r\cos\theta\,\frac{d\theta}{dx}\right).$$

3 By comparing the derivatives obtained in **2**, explain why $\frac{dr}{dx} = 0$ and $\frac{d\theta}{dx} = 1$.

4 We know that $e^{i0} = 1$. $\therefore$ when $x = 0$, $r = 1$ and $\theta = 0$.

a Integrate $\frac{dr}{dx}$ with respect to x. Use $r(0) = 1$ to show that $r(x) = 1$ for all x.

b Integrate $\frac{d\theta}{dx}$ with respect to x. Use $\theta(0) = 0$ to show that $\theta(x) = x$ for all x.

5 Hence explain why $e^{ix} = \cos x + i\sin x$.

G INTEGRATION BY SUBSTITUTION

Consider the integral $\int (x^2+3x)^4(2x+3)\,dx$.

The function we are integrating is the product of two expressions:

- The first expression $(x^2+3x)^4$ has the form $f(u(x))$ where $f(u) = u^4$ and $u(x) = x^2+3x$.
- The second expression is $(2x+3)$, and we notice that this is also $u'(x)$.

So, we can write the integral in the form $\int f(u(x))\,u'(x)\,dx$ where $f(u) = u^4$ and $u(x) = x^2+3x$.

We can integrate functions of this form using the theorem $\int f(u)\,\frac{du}{dx}\,dx = \int f(u)\,du$

Proof: Suppose $F(u)$ is the antiderivative of $f(u)$, so $\frac{dF}{du} = f(u)$

$$\therefore \int f(u)\,du = F(u) + c \quad .... (1)$$

But $\frac{dF}{dx} = \frac{dF}{du}\frac{du}{dx}$ {chain rule}

$$= f(u)\,\frac{du}{dx}$$

$$\therefore \int f(u)\,\frac{du}{dx}\,dx = F(u) + c = \int f(u)\,du \quad \text{\{from (1)\}}$$

Using the theorem,

$$\int (x^2+3x)^4(2x+3)\,dx$$
$$= \int u^4 \frac{du}{dx}\,dx \qquad \left\{u = x^2+3x, \quad \frac{du}{dx} = 2x+3\right\}$$
$$= \int u^4\,du \qquad \left\{\text{replacing } \frac{du}{dx}\,dx \text{ by } du\right\}$$
$$= \frac{u^5}{5} + c$$
$$= \tfrac{1}{5}(x^2+3x)^5 + c$$

We use $u(x)$ in our solution, but we give our answer in terms of x, since the original integral was with respect to x.

Example 18 — Self Tutor

Use substitution to find: $\int \sqrt{x^3+2x}\,(3x^2+2)\,dx$

$$\int \sqrt{x^3+2x}\,(3x^2+2)\,dx$$
$$= \int \sqrt{u}\,\frac{du}{dx}\,dx \qquad \left\{u = x^3+2x, \quad \frac{du}{dx} = 3x^2+2\right\}$$
$$= \int u^{\frac{1}{2}}\,du$$
$$= \frac{u^{\frac{3}{2}}}{\frac{3}{2}} + c$$
$$= \tfrac{2}{3}(x^3+2x)^{\frac{3}{2}} + c$$

Example 19 — Self Tutor

Use substitution to find: **a** $\int \frac{3x^2+2}{x^3+2x}\,dx$ **b** $\int xe^{1-x^2}\,dx$

a
$$\int \frac{3x^2+2}{x^3+2x}\,dx$$
$$= \int \frac{1}{x^3+2x}(3x^2+2)\,dx$$
$$= \int \frac{1}{u}\frac{du}{dx}\,dx \qquad \left\{u = x^3+2x, \quad \frac{du}{dx} = 3x^2+2\right\}$$
$$= \int \frac{1}{u}\,du$$
$$= \ln|u| + c$$
$$= \ln\left|x^3+2x\right| + c$$

b
$$\int xe^{1-x^2}\,dx$$
$$= -\tfrac{1}{2}\int (-2x)e^{1-x^2}\,dx$$
$$= -\tfrac{1}{2}\int e^u \frac{du}{dx}\,dx \qquad \left\{u = 1-x^2, \quad \frac{du}{dx} = -2x\right\}$$
$$= -\tfrac{1}{2}\int e^u\,du$$
$$= -\tfrac{1}{2}e^u + c$$
$$= -\tfrac{1}{2}e^{1-x^2} + c$$

Example 20 — Self Tutor

Integrate with respect to x: **a** $\cos^3 x \sin x$ **b** $\dfrac{\cos x}{\sin x}$

a
$$\int \cos^3 x \sin x \, dx$$
$$= \int (\cos x)^3 \sin x \, dx$$
$$= \int u^3 \left(-\frac{du}{dx}\right) dx \qquad \left\{u = \cos x, \ \frac{du}{dx} = -\sin x\right\}$$
$$= -\int u^3 \, du$$
$$= -\frac{u^4}{4} + c$$
$$= -\tfrac{1}{4}\cos^4 x + c$$

b
$$\int \frac{\cos x}{\sin x} \, dx$$
$$= \int \frac{1}{u}\frac{du}{dx} \, dx \qquad \left\{u = \sin x, \ \frac{du}{dx} = \cos x\right\}$$
$$= \int \frac{1}{u} \, du$$
$$= \ln|u| + c$$
$$= \ln|\sin x| + c$$

Note: The substitutions we make need to be chosen with care.

For example, in **Example 20** part **b**, if we let $u = \cos x$, $\dfrac{du}{dx} = -\sin x$ then we obtain $\displaystyle\int \frac{\cos x}{\sin x}\,dx = \int \frac{u}{-\frac{du}{dx}}\,dx$. This is not in the correct form to apply our theorem, which tells us we have made the wrong substitution and we need to try another.

Example 21 — Self Tutor

Find $\displaystyle\int \sin^3 x \, dx$.

$$\int \sin^3 x \, dx$$
$$= \int \sin^2 x \sin x \, dx$$
$$= \int \left(1 - \cos^2 x\right) \sin x \, dx$$
$$= \int \left(1 - u^2\right)\left(-\frac{du}{dx}\right) dx \qquad \left\{u = \cos x, \ \frac{du}{dx} = -\sin x\right\}$$
$$= \int (u^2 - 1) \, du$$
$$= \frac{u^3}{3} - u + c$$
$$= \tfrac{1}{3}\cos^3 x - \cos x + c$$

EXERCISE 21G.1

1 Use the substitutions given to perform the integration:

a $\displaystyle\int 3x^2(x^3+1)^4 \, dx$ using $u(x) = x^3 + 1$

b $\displaystyle\int x^2 e^{x^3+1} \, dx$ using $u(x) = x^3 + 1$

c $\displaystyle\int \sin^4 x \cos x \, dx$ using $u(x) = \sin x$

d $\displaystyle\int \frac{e^{\frac{x-1}{x}}}{x^2} \, dx$ using $u(x) = \dfrac{x-1}{x}$

2 Integrate by substitution:

a $4x^3(2+x^4)^3$ b $\dfrac{2x}{\sqrt{x^2+3}}$ c $\dfrac{x}{(1-x^2)^5}$

d $\sqrt{x^3+x}\,(3x^2+1)$ e $(x^3+2x+1)^4(3x^2+2)$ f $\dfrac{x^2}{(3x^3-1)^4}$

g $\dfrac{x+2}{(x^2+4x-3)^2}$ h $x^4(x+1)^4(2x+1)$

3 Find:

a $\displaystyle\int -2e^{1-2x}\,dx$ b $\displaystyle\int 2xe^{x^2}\,dx$ c $\displaystyle\int \frac{e^{\sqrt{x}}}{\sqrt{x}}\,dx$ d $\displaystyle\int (2x-1)e^{x-x^2}\,dx$

4 Find:

a $\displaystyle\int \frac{2x}{x^2+1}\,dx$ b $\displaystyle\int \frac{x}{2-x^2}\,dx$ c $\displaystyle\int \frac{2x-3}{x^2-3x}\,dx$

d $\displaystyle\int \frac{6x^2-2}{x^3-x}\,dx$ e $\displaystyle\int \frac{4x-10}{5x-x^2}\,dx$ f $\displaystyle\int \frac{1-x^2}{x^3-3x}\,dx$

5 Integrate with respect to x:

a $x^2(3-x^3)^2$ b $\dfrac{4}{x\ln x}$ c $x\sqrt{1-x^2}$

d xe^{1-x^2} e $\dfrac{1-3x^2}{x^3-x}$ f $\dfrac{(\ln x)^3}{x}$

6 Integrate with respect to x:

a $\sin^4 x\cos x$ b $\dfrac{\sin x}{\sqrt{\cos x}}$ c $\tan x$ d $\sqrt{\sin x}\cos x$

e $\dfrac{\cos x}{(2+\sin x)^2}$ f $\dfrac{\sin x}{\cos^3 x}$ g $\dfrac{\sin x}{1-\cos x}$ h $\dfrac{\cos(2x)}{\sin(2x)-3}$

i $x\sin(x^2)$ j $\dfrac{\sin^3 x}{\cos^5 x}$ k $\csc^3(2x)\cot(2x)$ l $\cos^3 x$

7 Find:

a $\displaystyle\int \sin^5 x\,dx$ b $\displaystyle\int \sin^4 x\cos^3 x\,dx$ c $\displaystyle\int \sin^3(2x)\cos(2x)\,dx$

8 Find $f(x)$ if $f'(x)$ is:

a $\sin x e^{\cos x}$ b $\dfrac{\sin x+\cos x}{\sin x-\cos x}$ c $\dfrac{e^{\tan x}}{\cos^2 x}$

9 Find:

a $\displaystyle\int \cot x\,dx$ b $\displaystyle\int \cot(3x)\,dx$ c $\displaystyle\int \csc^2 x\,dx$

d $\displaystyle\int \sec x\tan x\,dx$ e $\displaystyle\int \csc x\cot x\,dx$ f $\displaystyle\int \tan(3x)\sec(3x)\,dx$

g $\displaystyle\int \csc\left(\tfrac{x}{2}\right)\cot\left(\tfrac{x}{2}\right)\,dx$ h $\displaystyle\int \sec^3 x\sin x\,dx$ i $\displaystyle\int \frac{\csc^2 x}{\sqrt{\cot x}}\,dx$

10 Find:

a $\int \frac{x\ln(x^2+7)}{x^2+7}\,dx$ using the substitution $u = \ln(x^2+7)$

b $\int x^2\sqrt{x-16}\,dx$ using the substitution $u = x-16$

c $\int \frac{1}{36+4x^2}\,dx$ using the substitution $x = 3\tan\theta$

d $\int \frac{\sqrt{x-1}}{x}\,dx$ using the substitution $u = \sqrt{x-1}$

e $\int \frac{\sqrt{4x^2-1}}{5x}\,dx$ using the substitution $x = \frac{1}{2}\sec\theta$.

11 a Find $\frac{d}{dx}(\arcsin x)$ and hence find $\int \frac{1}{\sqrt{1-x^2}}\,dx$.

b By substituting $\theta = \frac{x}{a}$, find $\int \frac{1}{\sqrt{a^2-x^2}}\,dx$.

c State the domain on which the integral in **b** is defined.

d Hence find:

i $\int \frac{4}{\sqrt{1-x^2}}\,dx$ ii $\int \frac{3}{\sqrt{4-x^2}}\,dx$ iii $\int \frac{1}{\sqrt{1-4x^2}}\,dx$ iv $\int \frac{2}{\sqrt{4-9x^2}}\,dx$

12 a Find $\frac{d}{dx}(\arctan x)$ and hence find $\int \frac{1}{x^2+1}\,dx$.

b By substituting $\theta = \frac{x}{a}$, find $\int \frac{1}{x^2+a^2}\,dx$.

c State the domain on which the integral in **b** is defined.

d Hence find:

i $\int \frac{1}{x^2+16}\,dx$ ii $\int \frac{1}{4x^2+1}\,dx$ iii $\int \frac{1}{4+2x^2}\,dx$ iv $\int \frac{5}{9+4x^2}\,dx$

13 Suppose ξ is a constant. Use the substitution $u = \sqrt{\frac{x-\xi}{x+\xi}}$ to find the following integrals, giving your answers in simplest form:

a $\int \frac{x-\xi}{(x+\xi)^3}\,dx$ b $\int (x-\xi)^{\frac{1}{2}}\,(x+\xi)^{-\frac{5}{2}}\,dx$

14 Find $\int \frac{1}{\sqrt{x}+\sqrt[3]{x}}\,dx$ using the substitution $u = x^{\frac{1}{6}}$.

CHOOSING WHAT TO SUBSTITUTE

Here are some suggestions of substitutions to help integrate more difficult functions.

Note that these substitutions may not always lead to success, so sometimes other substitutions will be needed.

With practice you will develop a feeling for which substitution is best in a given situation.

When a function contains	*Try substituting*
$\sqrt{f(x)}$	$u = f(x)$
$\ln x$	$u = \ln x$
$\sqrt{a^2-x^2}$	$x = a\sin\theta$
x^2+a^2 *or* $\sqrt{x^2+a^2}$	$x = a\tan\theta$
$\sqrt{x^2-a^2}$	$x = a\sec\theta$

Example 22 **Self Tutor**

Find $\int x\sqrt{x+2}\,dx$.

Let $u = x+2$ $\quad \therefore \frac{du}{dx} = 1$

$$\therefore \int x\sqrt{x+2}\,dx = \int (u-2)\sqrt{u}\,\frac{du}{dx}\,dx$$
$$= \int \left(u^{\frac{3}{2}} - 2u^{\frac{1}{2}}\right) du$$
$$= \frac{u^{\frac{5}{2}}}{\frac{5}{2}} - \frac{2u^{\frac{3}{2}}}{\frac{3}{2}} + c$$
$$= \tfrac{2}{5}(x+2)^{\frac{5}{2}} - \tfrac{4}{3}(x+2)^{\frac{3}{2}} + c$$

Example 23 **Self Tutor**

Find $\int \frac{\sqrt{x^2-9}}{x}\,dx$.

Let $x = 3\sec\theta$ $\quad \therefore \frac{dx}{d\theta} = 3\sec\theta\tan\theta$

$$\therefore \int \frac{\sqrt{x^2-9}}{x}\,dx = \int \frac{\sqrt{9\sec^2\theta - 9}}{3\sec\theta}\,3\sec\theta\tan\theta\,d\theta$$
$$= \int 3\sqrt{\sec^2\theta - 1}\,\tan\theta\,d\theta$$
$$= \int 3\tan^2\theta\,d\theta$$
$$= \int (3\sec^2\theta - 3)\,d\theta$$
$$= 3\tan\theta - 3\theta + c$$

Since $\sec\theta = \frac{x}{3}$, $\cos\theta = \frac{3}{x}$

$\therefore \theta = \arccos\left(\frac{3}{x}\right)$

and $\tan\theta = \frac{\sqrt{x^2-9}}{3}$

So, $\int \frac{\sqrt{x^2-9}}{x}\,dx = \sqrt{x^2-9} - 3\arccos\left(\frac{3}{x}\right) + c$

EXERCISE 21G.2

1 Find, using a suitable substitution:

a $\int x\sqrt{x-3}\,dx$ **b** $\int x^2\sqrt{x+1}\,dx$ **c** $\int x^3\sqrt{3-x^2}\,dx$ **d** $\int t^3\sqrt{t^2+2}\,dt$

2 Integrate with respect to x:

a $\dfrac{x^2}{9+x^2}$ **b** $\dfrac{x^2}{\sqrt{1-x^2}}$ **c** $\dfrac{2x}{x^2+9}$ **d** $\dfrac{4\ln x}{x\left(1+[\ln x]^2\right)}$

e $\dfrac{\sqrt{x^2-4}}{x}$ **f** $\sin x\cos 2x$ **g** $\dfrac{1}{\sqrt{9-4x^2}}$ **h** $\dfrac{x^3}{1+x^2}$

i $\dfrac{1}{x\left(9+4[\ln x]^2\right)}$ **j** $\dfrac{1}{x(x^2+16)}$ **k** $\dfrac{1}{x^2\sqrt{16-x^2}}$ **l** $x^2\sqrt{4-x^2}$

H INTEGRATION BY PARTS

The method of **integration by parts** comes from the product rule of differentiation. It allows us to integrate a function which is written as a product.

Since $\dfrac{d}{dx}(uv)=u'v+uv'$, $\displaystyle\int(u'v+uv')\,dx=uv$

$$\therefore\ \int u'v\,dx+\int uv'\,dx=uv$$

$$\therefore\ \int uv'\,dx=uv-\int u'v\,dx$$

So, providing $\displaystyle\int u'v\,dx$ can be easily found, we can find $\displaystyle\int uv'\,dx$ using

$$\boldsymbol{\int uv'\,dx=uv-\int u'v\,dx}$$

Example 24 **Self Tutor**

Find: **a** $\displaystyle\int xe^{-x}\,dx$ **b** $\displaystyle\int x\cos x\,dx$

a $u=x$ $\quad v'=e^{-x}$

$u'=1$ $\quad v=-e^{-x}$

$$\therefore\ \int xe^{-x}\,dx=-xe^{-x}-\int(-e^{-x})\,dx$$
$$=-xe^{-x}+(-e^{-x})+c$$
$$=-e^{-x}(x+1)+c$$

Check:

$\dfrac{d}{dx}(-e^{-x}(x+1)+c)$

$=e^{-x}(x+1)+-e^{-x}(1)+0$

$=xe^{-x}+\cancel{e^{-x}}-\cancel{e^{-x}}$

$=xe^{-x}$ ✓

b $u=x$ $\quad v'=\cos x$

$u'=1$ $\quad v=\sin x$

$$\therefore\ \int x\cos x\,dx=x\sin x-\int\sin x\,dx$$
$$=x\sin x-(-\cos x)+c$$
$$=x\sin x+\cos x+c$$

Check:

$\dfrac{d}{dx}(x\sin x+\cos x+c)$

$=1\times\sin x+x\cos x-\sin x$

$=\cancel{\sin x}+x\cos x-\cancel{\sin x}$

$=x\cos x$ ✓

EXERCISE 21H

1 Use integration by parts to find the integral of the following functions with respect to x:

a xe^x **b** $x\sin x$ **c** $x^2 \ln x$
d $x\sin 3x$ **e** $x\cos 2x$ **f** $x\sec^2 x$

When using 'integration by parts' the function u should be easy to differentiate and the function v' should be easy to integrate.

2 **a** Find $\int \ln x\, dx$ by first writing $\ln x$ as $1 \times \ln x$.

b Hence find $\int (\ln x)^2\, dx$.

3 Find $\int \arctan x\, dx$ by first writing $\arctan x$ as $1 \times \arctan x$.

Example 25 **Self Tutor**

Find $\int e^x \sin x\, dx$.

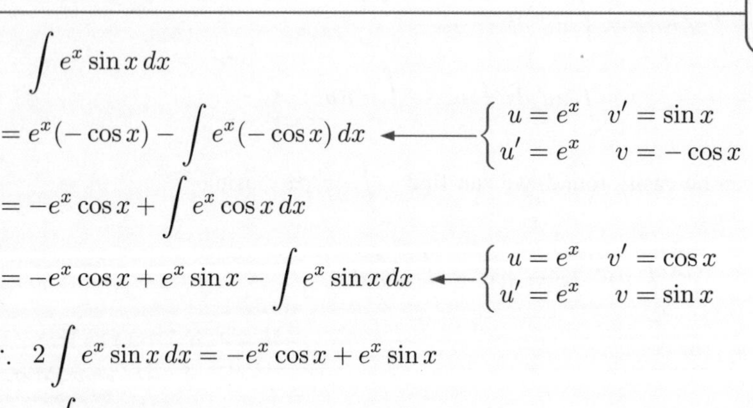

$\int e^x \sin x\, dx$

$= e^x(-\cos x) - \int e^x(-\cos x)\, dx$ ⟵ $\begin{cases} u = e^x & v' = \sin x \\ u' = e^x & v = -\cos x \end{cases}$

$= -e^x \cos x + \int e^x \cos x\, dx$

$= -e^x \cos x + e^x \sin x - \int e^x \sin x\, dx$ ⟵ $\begin{cases} u = e^x & v' = \cos x \\ u' = e^x & v = \sin x \end{cases}$

$\therefore\ 2\int e^x \sin x\, dx = -e^x \cos x + e^x \sin x$

$\therefore\ \int e^x \sin x\, dx = \frac{1}{2}e^x(\sin x - \cos x) + c$

Sometimes we need to use integration by parts *twice* in order to find an integral.

4 Integrate with respect to x:

a $x^2 e^{-x}$ **b** $e^x \cos x$ **c** $e^{-x}\sin x$ **d** $x^2 \sin x$

5 **a** Use integration by parts to find $\int u^2 e^u\, du$.

b Hence find $\int (\ln x)^2\, dx$ using the substitution $u = \ln x$.

6 **a** Use integration by parts to find $\int u \sin u\, du$.

b Hence find $\int \sin\sqrt{2x}\, dx$ using the substitution $u^2 = 2x$.

7 Find $\int \cos\sqrt{3x}\, dx$ using the substitution $u^2 = 3x$.

I MISCELLANEOUS INTEGRATION

We have now practised several integration techniques with clues given as to what method to use. In this section we attempt to find integrals without clues.

EXERCISE 21I

1 Integrate with respect to x:

a $\dfrac{e^x + e^{-x}}{e^x - e^{-x}}$ b 7^x c $(3x+5)^5$ d $\dfrac{\sin x}{2-\cos x}$

e $x\sec^2 x$ f $\cot 2x$ g $x(x+3)^3$ h $\dfrac{(x+1)^3}{x}$

2 Integrate with respect to x:

a $x^2 e^{-x}$ b $x\sqrt{1-x}$ c $x^2\sqrt{1-x^2}$ d $\dfrac{3}{x\sqrt{x^2-4}}$

e $x^2\sqrt{x-3}$ f $\tan^3 x$ g $\dfrac{\ln(x+2)}{(x+2)^2}$ h $\dfrac{1}{x^2+2x+3}$

3 Integrate with respect to x:

a $\dfrac{1}{x^2+9}$ b $\dfrac{4}{\sqrt{x}\sqrt{1-x}}$ c $\ln(2x)$ d $e^{-x}\cos x$

e $\dfrac{1}{x(1+x^2)}$ f $\dfrac{\arctan x}{1+x^2}$ g $\sqrt{9-x^2}$ h $\dfrac{(\ln x)^2}{x^2}$

i $\dfrac{x}{\sqrt{x-3}}$ j $\sin 4x \cos x$ k $\dfrac{2x+3}{x^2-2x+5}$ l $\cos^3 x$

m $\dfrac{x+4}{x^2+4}$ n $\dfrac{1-2x}{\sqrt{4-x^2}}$ o $\dfrac{x^3}{(2-x)^3}$ p $\sin^5 x\cos^5 x$

J DEFINITE INTEGRALS

Earlier we saw the **fundamental theorem of calculus**:

If $F(x)$ is the antiderivative of $f(x)$ where $f(x)$ is continuous on the interval $a \leqslant x \leqslant b$, then the **definite integral** of $f(x)$ on this interval is $\displaystyle\int_a^b \boldsymbol{f(x)\,dx = F(b) - F(a)}$.

$\displaystyle\int_a^b f(x)\,dx$ reads "the integral from $x = a$ to $x = b$ of $f(x)$ with respect to x"
or "the integral from a to b of $f(x)$ with respect to x".

It is called a **definite** integral because there are lower and upper limits for the integration, and it therefore results in a numerical answer.

When calculating definite integrals we can omit the constant of integration c as this will always cancel out in the subtraction process.

It is common to write $F(b) - F(a)$ as $[F(x)]_a^b$, and so

$$\int_a^b f(x)\,dx = [F(x)]_a^b = F(b) - F(a)$$

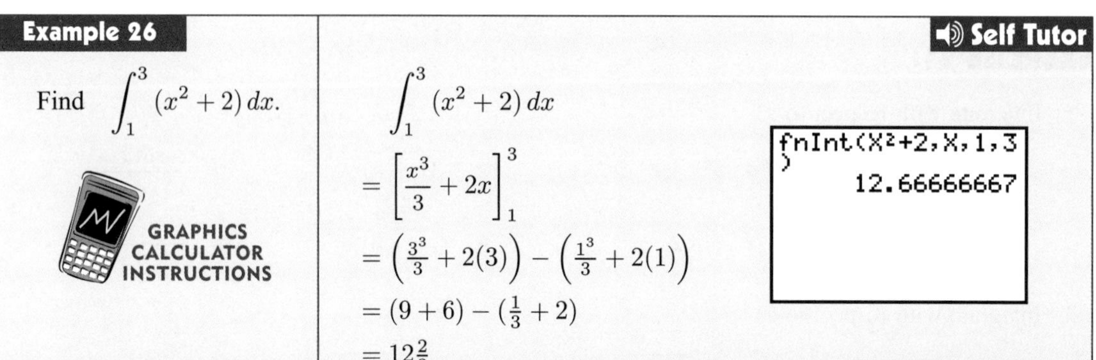

Example 26 Self Tutor

Find $\int_1^3 (x^2+2)\,dx$.

GRAPHICS CALCULATOR INSTRUCTIONS

$$\int_1^3 (x^2+2)\,dx$$
$$= \left[\frac{x^3}{3} + 2x\right]_1^3$$
$$= \left(\frac{3^3}{3} + 2(3)\right) - \left(\frac{1^3}{3} + 2(1)\right)$$
$$= (9+6) - (\tfrac{1}{3}+2)$$
$$= 12\tfrac{2}{3}$$

```
fnInt(X²+2,X,1,3
)
          12.66666667
```

Example 27 Self Tutor

Evaluate: **a** $\int_0^{\frac{\pi}{3}} \sin x\,dx$ **b** $\int_1^4 \left(2x + \frac{3}{x}\right) dx$

a
$$\int_0^{\frac{\pi}{3}} \sin x\,dx$$
$$= [-\cos x]_0^{\frac{\pi}{3}}$$
$$= (-\cos\tfrac{\pi}{3}) - (-\cos 0)$$
$$= -\tfrac{1}{2} + 1$$
$$= \tfrac{1}{2}$$

b
$$\int_1^4 \left(2x + \frac{3}{x}\right) dx$$
$$= \left[x^2 + 3\ln x\right]_1^4 \quad \{\text{since } x > 0\}$$
$$= (16 + 3\ln 4) - (1 + 3\ln 1)$$
$$= 15 + 6\ln 2$$

EXERCISE 21J.1

1 Evaluate the following and check with your graphics calculator:

a $\int_0^1 x^3\,dx$ **b** $\int_0^2 (x^2 - x)\,dx$ **c** $\int_0^1 e^x\,dx$

d $\int_1^4 \left(x - \frac{3}{\sqrt{x}}\right) dx$ **e** $\int_4^9 \frac{x-3}{\sqrt{x}}\,dx$ **f** $\int_1^3 \frac{1}{x}\,dx$

g $\int_1^2 (e^{-x}+1)^2\,dx$ **h** $\int_2^6 \frac{1}{\sqrt{2x-3}}\,dx$ **i** $\int_0^1 e^{1-x}\,dx$

2 Evaluate:

a $\int_0^{\frac{\pi}{6}} \cos x\,dx$ **b** $\int_{\frac{\pi}{3}}^{\frac{\pi}{2}} \sin x\,dx$ **c** $\int_{\frac{\pi}{4}}^{\frac{\pi}{3}} \sec^2 x\,dx$

d $\int_0^{\frac{\pi}{6}} \sin(3x)\,dx$ **e** $\int_0^{\frac{\pi}{4}} \cos^2 x\,dx$ **f** $\int_0^{\frac{\pi}{2}} \sin^2 x\,dx$

3 Show that $\dfrac{4x+1}{x-1}$ may be written in the form $4 + \dfrac{5}{x-1}$.

Hence show that $\displaystyle\int_3^5 \frac{4x+1}{x-1}\,dx = 8 + 5\ln 2$.

4 Find m such that:

a $\displaystyle\int_m^{-2} \frac{1}{4-x}\,dx = \ln\tfrac{3}{2}$ **b** $\displaystyle\int_m^{2m} (2x-1)\,dx = 4$

5 Evaluate $\displaystyle\int_{-1}^{1} e^{-x^2}$ using technology.

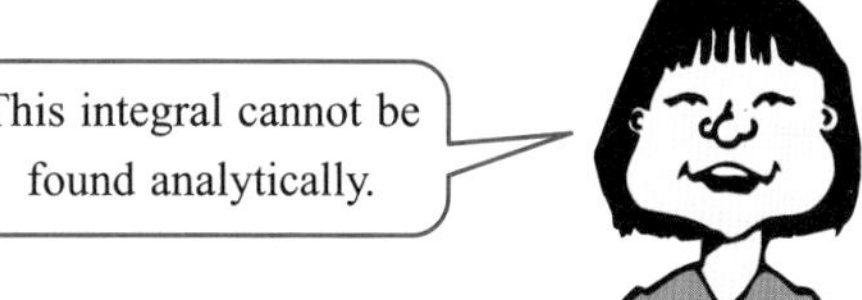
This integral cannot be found analytically.

Example 28 **Self Tutor**

Evaluate $\displaystyle\int_2^5 xe^x\,dx$ using integration by parts. Check your answer using technology.

$u = x \qquad v' = e^x$

$u' = 1 \qquad v = e^x$

$$\therefore \int xe^x\,dx = xe^x - \int e^x\,dx$$
$$= xe^x - e^x + c$$

$$\therefore \int_2^5 xe^x\,dx = [xe^x - e^x]_2^5$$
$$= (5e^5 - e^5) - (2e^2 - e^2)$$
$$= 4e^5 - e^2$$
$$\approx 586.3$$

GRAPHICS CALCULATOR INSTRUCTIONS

Casio fx-CG20

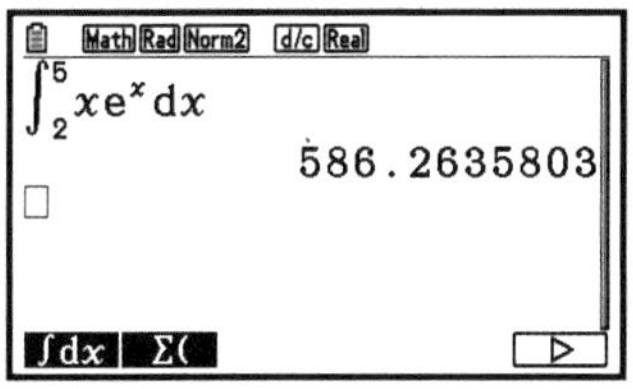

$\int_2^5 xe^x\,dx$ 586.2635803

TI-84 Plus

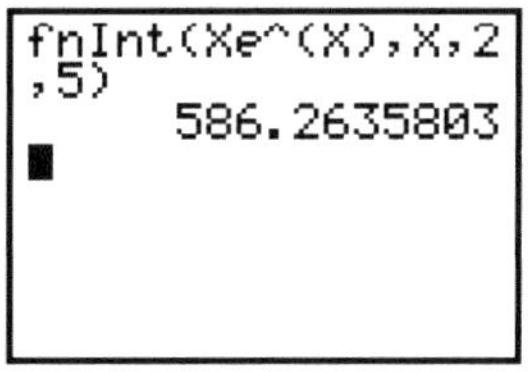

```
fnInt(Xe^(X),X,2
,5)
        586.2635803
```

TI-*n*spire

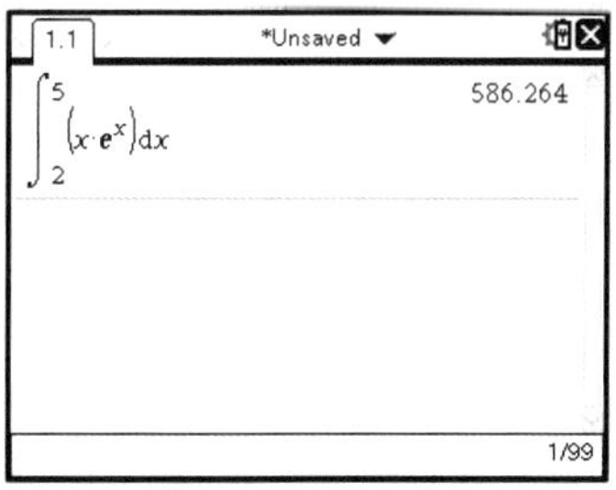

$\int_2^5 (x\cdot e^x)dx$ 586.264

6 Evaluate the following integrals using integration by parts.
Check your answers using technology.

a $\displaystyle\int_0^1 -xe^{-x}\,dx$ **b** $\displaystyle\int_0^{\frac{\pi}{2}} x\sin x\,dx$ **c** $\displaystyle\int_1^3 \ln x\,dx$

DEFINITE INTEGRALS INVOLVING SUBSTITUTION

When we solve a definite integral by substitution, we need to make sure the endpoints are converted to the new variable.

Example 29 **Self Tutor**

Evaluate: **a** $\displaystyle\int_2^3 \frac{x}{x^2-1}\,dx$ **b** $\displaystyle\int_0^1 \frac{6x}{(x^2+1)^3}\,dx$

a Let $u = x^2 - 1$ $\therefore \dfrac{du}{dx} = 2x$

When $x = 2$, $u = 2^2 - 1 = 3$

When $x = 3$, $u = 3^2 - 1 = 8$

$$\therefore \int_2^3 \frac{x}{x^2-1}\,dx = \int_2^3 \frac{1}{u}\left(\tfrac{1}{2}\frac{du}{dx}\right)dx$$
$$= \tfrac{1}{2}\int_3^8 \frac{1}{u}\,du$$
$$= \tfrac{1}{2}\left[\ln|u|\right]_3^8$$
$$= \tfrac{1}{2}(\ln 8 - \ln 3)$$
$$= \tfrac{1}{2}\ln\left(\tfrac{8}{3}\right)$$

b Let $u = x^2 + 1$ $\therefore \dfrac{du}{dx} = 2x$

When $x = 0$, $u = 1$

When $x = 1$, $u = 2$

$$\therefore \int_0^1 \frac{6x}{(x^2+1)^3}\,dx = \int_0^1 \frac{1}{u^3}\left(3\frac{du}{dx}\right)dx$$
$$= 3\int_1^2 u^{-3}\,du$$
$$= 3\left[\frac{u^{-2}}{-2}\right]_1^2$$
$$= 3\left(\frac{2^{-2}}{-2} - \frac{1^{-2}}{-2}\right)$$
$$= \tfrac{9}{8}$$

Example 30 **Self Tutor**

Evaluate: $\displaystyle\int_{\frac{\pi}{6}}^{\frac{\pi}{2}} \sqrt{\sin x}\cos x\,dx$

Let $u = \sin x$ $\therefore \dfrac{du}{dx} = \cos x$

When $x = \frac{\pi}{2}$, $u = \sin\frac{\pi}{2} = 1$

When $x = \frac{\pi}{6}$, $u = \sin\frac{\pi}{6} = \frac{1}{2}$

$$\therefore \int_{\frac{\pi}{6}}^{\frac{\pi}{2}} \sqrt{\sin x}\cos x\,dx$$
$$= \int_{\frac{\pi}{6}}^{\frac{\pi}{2}} \sqrt{u}\,\frac{du}{dx}\,dx$$
$$= \int_{\frac{1}{2}}^{1} u^{\frac{1}{2}}\,du$$
$$= \left[\frac{u^{\frac{3}{2}}}{\frac{3}{2}}\right]_{\frac{1}{2}}^{1}$$
$$= \tfrac{2}{3}(1)^{\frac{3}{2}} - \tfrac{2}{3}\left(\tfrac{1}{2}\right)^{\frac{3}{2}}$$
$$= \tfrac{2}{3} - \frac{1}{3\sqrt{2}}$$

Casio fx-CG20

```
∫(π÷6 → π÷2) √sin x cos xdx
                 0.4309644063
2÷3-1÷(3√2)
                 0.4309644063
```

TI-84 Plus

```
fnInt(√(sin(X))*cos(X),X,π/6,π/2)
                 .4309644063
2/3-1/(3√(2))
                 .4309644063
```

TI-nspire

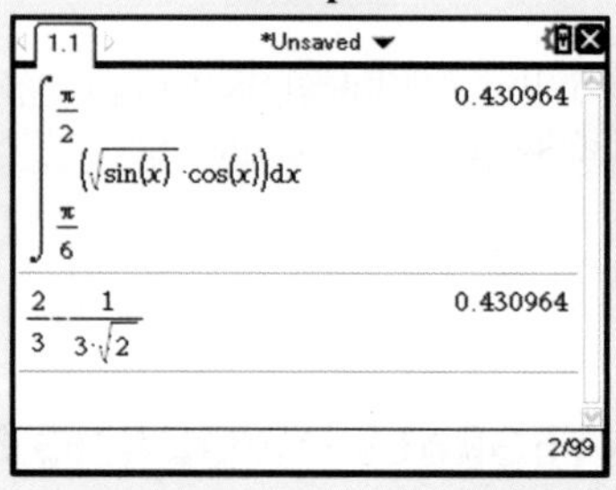

EXERCISE 21J.2

1 Evaluate the following and check with your graphics calculator:

a $\displaystyle\int_1^2 \frac{x}{(x^2+2)^2}\,dx$ **b** $\displaystyle\int_0^1 x^2 e^{x^3+1}\,dx$ **c** $\displaystyle\int_0^3 x\sqrt{x^2+16}\,dx$

d $\displaystyle\int_1^2 xe^{-2x^2}\,dx$ **e** $\displaystyle\int_2^3 \frac{x}{2-x^2}\,dx$ **f** $\displaystyle\int_1^2 \frac{\ln x}{x}\,dx$

g $\displaystyle\int_0^1 \frac{1-3x^2}{1-x^3+x}\,dx$ **h** $\displaystyle\int_2^4 \frac{6x^2-4x+4}{x^3-x^2+2x}\,dx$

2 Evaluate:

a $\displaystyle\int_0^{\frac{\pi}{3}} \frac{\sin x}{\sqrt{\cos x}}\,dx$ **b** $\displaystyle\int_0^{\frac{\pi}{6}} \sin^2 x\cos x\,dx$ **c** $\displaystyle\int_0^{\frac{\pi}{4}} \tan x\,dx$

d $\displaystyle\int_{\frac{\pi}{6}}^{\frac{\pi}{2}} \cot x\,dx$ **e** $\displaystyle\int_0^{\frac{\pi}{6}} \frac{\cos x}{1-\sin x}\,dx$ **f** $\displaystyle\int_0^{\frac{\pi}{4}} \sec^2 x\tan^3 x\,dx$

3 Evaluate for $n \in \mathbb{Z}$: $\displaystyle\int_0^1 (x^2+2x)^n\,(x+1)\,dx$

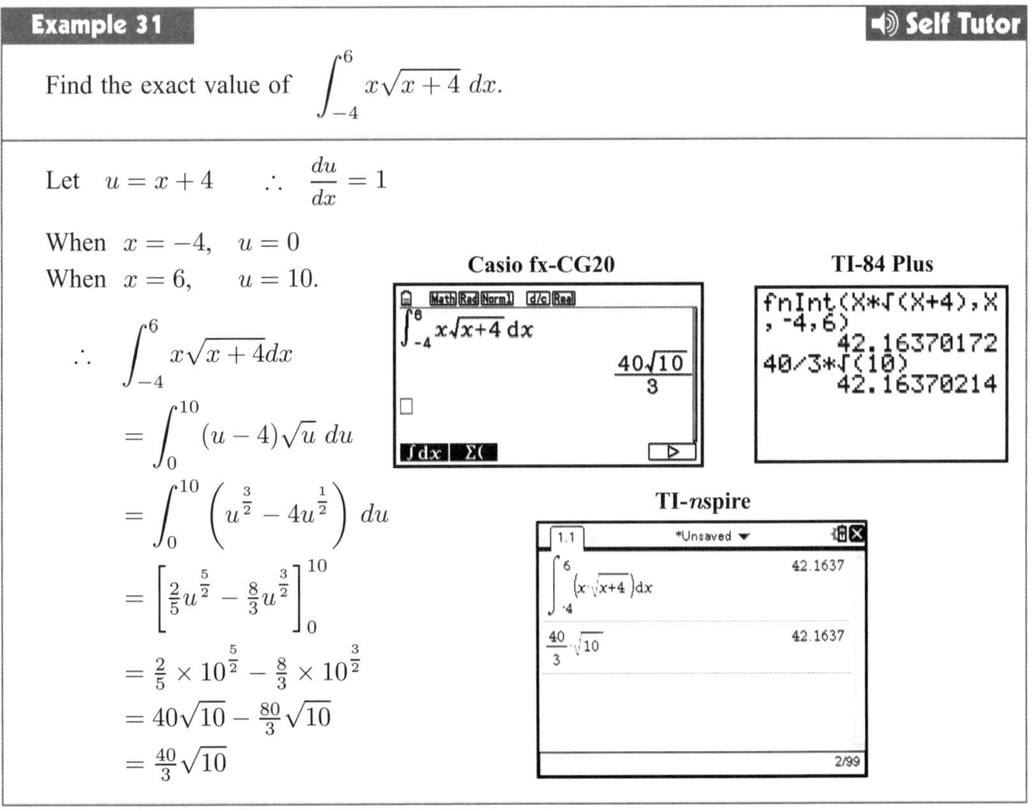

Example 31 Self Tutor

Find the exact value of $\displaystyle\int_{-4}^{6} x\sqrt{x+4}\,dx.$

Let $u = x+4$ $\quad\therefore\ \dfrac{du}{dx} = 1$

When $x=-4,\quad u=0$
When $x=6,\quad u=10.$

$$\therefore \int_{-4}^{6} x\sqrt{x+4}\,dx$$
$$= \int_0^{10} (u-4)\sqrt{u}\,du$$
$$= \int_0^{10} \left(u^{\frac{3}{2}} - 4u^{\frac{1}{2}}\right) du$$
$$= \left[\tfrac{2}{5}u^{\frac{5}{2}} - \tfrac{8}{3}u^{\frac{3}{2}}\right]_0^{10}$$
$$= \tfrac{2}{5}\times 10^{\frac{5}{2}} - \tfrac{8}{3}\times 10^{\frac{3}{2}}$$
$$= 40\sqrt{10} - \tfrac{80}{3}\sqrt{10}$$
$$= \tfrac{40}{3}\sqrt{10}$$

4 Find the exact value of:

a $\displaystyle\int_3^4 x\sqrt{x-1}\,dx$ **b** $\displaystyle\int_0^3 x\sqrt{x+6}\,dx$ **c** $\displaystyle\int_2^5 x^2\sqrt{x-2}\,dx$

Check each answer using technology.

PROPERTIES OF DEFINITE INTEGRALS

Earlier in the chapter we proved the following properties of definite integrals using the fundamental theorem of calculus:

- $\int_a^b f(x)\,dx = -\int_b^a f(x)\,dx$
- $\int_a^b cf(x)\,dx = c\int_a^b f(x)\,dx$, c is any constant
- $\int_a^b f(x)\,dx + \int_b^c f(x)\,dx = \int_a^c f(x)\,dx$
- $\int_a^b [f(x) + g(x)]\,dx = \int_a^b f(x)\,dx + \int_a^b g(x)\,dx$

EXERCISE 21J.3

Use questions **1** to **4** to check the properties of definite integrals.

1 Find: **a** $\int_1^4 \sqrt{x}\,dx$ and $\int_1^4 (-\sqrt{x})\,dx$ **b** $\int_0^1 x^7\,dx$ and $\int_0^1 (-x^7)\,dx$

2 Find: **a** $\int_0^1 x^2\,dx$ **b** $\int_1^2 x^2\,dx$ **c** $\int_0^2 x^2\,dx$ **d** $\int_0^1 3x^2\,dx$

3 Find: **a** $\int_0^2 (x^3 - 4x)\,dx$ **b** $\int_2^3 (x^3 - 4x)\,dx$ **c** $\int_0^3 (x^3 - 4x)\,dx$

4 Find: **a** $\int_0^1 x^2\,dx$ **b** $\int_0^1 \sqrt{x}\,dx$ **c** $\int_0^1 (x^2 + \sqrt{x})\,dx$

5 Evaluate the following integrals using area interpretation:

a $\int_0^3 f(x)\,dx$ **b** $\int_3^7 f(x)\,dx$

c $\int_2^4 f(x)\,dx$ **d** $\int_0^7 f(x)\,dx$

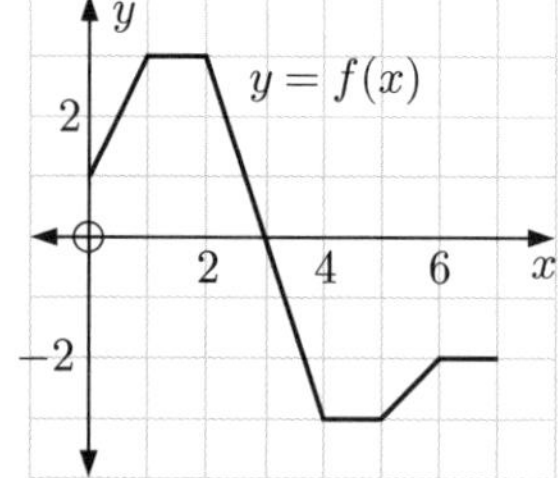

6 Evaluate the following integrals using area interpretation:

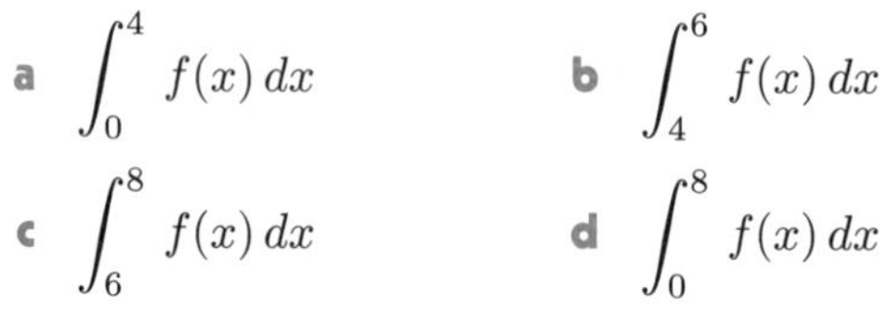

a $\int_0^4 f(x)\,dx$ **b** $\int_4^6 f(x)\,dx$

c $\int_6^8 f(x)\,dx$ **d** $\int_0^8 f(x)\,dx$

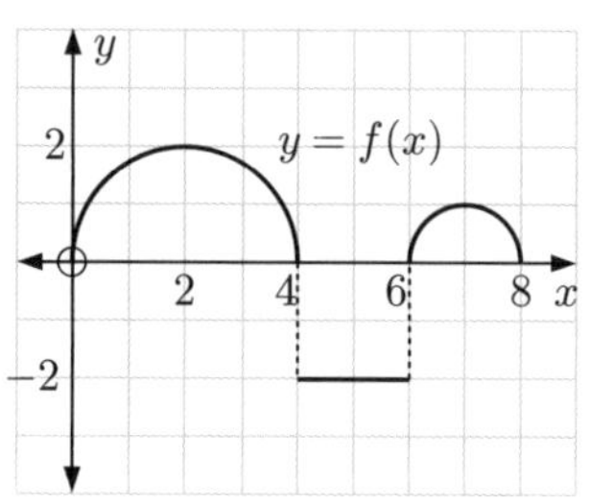

7 Write as a single integral:

a $\int_2^4 f(x)\,dx + \int_4^7 f(x)\,dx$ **b** $\int_1^3 g(x)\,dx + \int_3^8 g(x)\,dx + \int_8^9 g(x)\,dx$

8 **a** If $\int_1^3 f(x)\,dx = 2$ and $\int_1^6 f(x)\,dx = -3$, find $\int_3^6 f(x)\,dx$.

b If $\int_0^2 f(x)\,dx = 5$, $\int_4^6 f(x)\,dx = -2$, and $\int_0^6 f(x)\,dx = 7$, find $\int_2^4 f(x)\,dx$.

9 Given that $\int_{-1}^1 f(x)\,dx = -4$, determine the value of:

a $\int_1^{-1} f(x)\,dx$ **b** $\int_{-1}^1 (2 + f(x))\,dx$ **c** $\int_{-1}^1 2f(x)\,dx$

d k such that $\int_{-1}^1 kf(x)\,dx = 7$

10 If $g(2) = 4$ and $g(3) = 5$, calculate $\int_2^3 \left(g'(x) - 1\right) dx$.

REVIEW SET 21A NON-CALCULATOR

1 The graph of $y = f(x)$ is illustrated:
Evaluate the following using area interpretation:

a $\int_0^4 f(x)\,dx$ **b** $\int_4^6 f(x)\,dx$

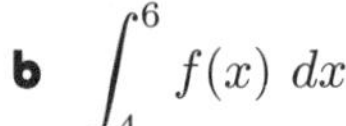

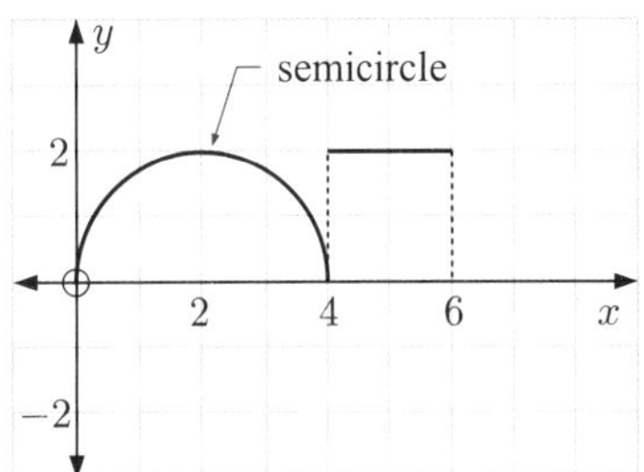

2 Integrate with respect to x:

a $\dfrac{4}{\sqrt{x}}$ **b** $\dfrac{3}{1-2x}$ **c** $\sin(4x-5)$ **d** e^{4-3x}

3 Find the exact value of:

a $\int_{-5}^{-1} \sqrt{1-3x}\,dx$ **b** $\int_0^{\frac{\pi}{2}} \cos\left(\frac{x}{2}\right) dx$ **c** $\int_0^1 \dfrac{4x^2}{(x^3+2)^3}\,dx$

4 By differentiating $y = \sqrt{x^2-4}$, find $\int \dfrac{x}{\sqrt{x^2-4}}\,dx$.

5 Find the values of b such that $\int_0^b \cos x\,dx = \dfrac{1}{\sqrt{2}}$, $0 < b < \pi$.

6 Find: **a** $\int (2-\cos x)^2\,dx$ **b** $\int x\cos(x^2)\,dx$.

7 By differentiating $(3x^2+x)^3$, find $\int (3x^2+x)^2(6x+1)\,dx$.

8 If $\int_1^4 f(x)\,dx = 3$, determine:

a $\int_1^4 (f(x)+1)\,dx$ **b** $\int_1^2 f(x)\,dx - \int_4^2 f(x)\,dx$

9 Integrate with respect to x:

a $\sin^7 x \cos x$ **b** $\tan(2x)$ **c** $e^{\sin x} \cos x$

10 Given that $f''(x) = 2\sin(2x)$, $f'(\frac{\pi}{2}) = 0$, and $f(0) = 3$, find the exact value of $f(\frac{\pi}{2})$.

11 Find exactly:

a $\int_0^{\frac{\pi}{6}} 4\sin^2\left(\frac{x}{2}\right) dx$ **b** $\int_{\frac{\pi}{6}}^{\frac{\pi}{2}} \cot\theta \, d\theta$ **c** $\int_{\frac{\pi}{4}}^{\frac{\pi}{3}} \frac{\sec^2 x}{\tan x} dx$

12 Find $\int x^2\sqrt{4-x}\, dx$.

13 Use integration by parts to find $\int \arctan x \, dx$. Check your answer using differentiation.

14 **a** Find $\int 2x(x^2+1)^3 \, dx$ using the substitution $u(x) = x^2 + 1$.

b *Hence* evaluate:

i $\int_0^1 2x(x^2+1)^3 \, dx$ **ii** $\int_{-1}^2 -x(1+x^2)^3 \, dx$

15 **a** Find constants A, B, and C such that $\frac{4}{x(1-x^2)} = \frac{A}{x} + \frac{B}{x+1} + \frac{C}{x-1}$.

b Hence find $\int \frac{4}{x(1-x^2)} dx$.

c Find in simplest form, the exact value of $\int_2^4 \frac{4}{x(1-x^2)} dx$.

REVIEW SET 21B CALCULATOR

1 **a** Sketch the region between the curve $y = \frac{4}{1+x^2}$ and the x-axis for $0 \leqslant x \leqslant 1$.
Divide the interval into 5 equal parts and display the 5 upper and lower rectangles.

b Find the lower and upper rectangle sums for $n = 5$, 50, 100, and 500.

c Give your best estimate for $\int_0^1 \frac{4}{1+x^2} dx$ and compare this answer with π.

2 Find y if:

a $\frac{dy}{dx} = (x^2-1)^2$ **b** $\frac{dy}{dx} = 400 - 20e^{-\frac{x}{2}}$ **c** $\frac{dy}{dx} = x(x^2-1)^2$

3 Evaluate correct to 4 significant figures:

a $\int_{-2}^0 4e^{-x^2} dx$ **b** $\int_0^1 \frac{10x}{\sqrt{3x+1}} dx$

4 Find $\frac{d}{dx}(\ln x)^2$ and hence find $\int \frac{\ln x}{x} dx$.

5 A curve $y = f(x)$ has $f''(x) = 18x + 10$. Find $f(x)$ if $f(0) = -1$ and $f(1) = 13$.

6 If $\int_0^a e^{1-2x}\,dx = \frac{e}{4}$, find a in the form $\ln k$.

7 Find the following integrals exactly, then check your answers using technology.

a $\int_3^4 \frac{1}{\sqrt{2x+1}}\,dx$ **b** $\int_0^1 x^2 e^{x+1}\,dx$

8 Suppose $f''(x) = 3x^2 + 2x$ and $f(0) = f(2) = 3$. Find:

a $f(x)$ **b** the equation of the normal to $y = f(x)$ at $x = 2$.

9 **a** Find $(e^x + 2)^3$ using the binomial expansion.

b Hence find the exact value of $\int_0^1 (e^x + 2)^3\,dx$. Check your answer using technology.

10 Integrate by substitution: $\int_{\frac{\pi}{4}}^{\frac{\pi}{3}} \sin^5 x \cos x\,dx$. Check your answer using technology.

11 Suppose $f''(x) = 4x^2 - 3$, $f'(0) = 6$, and $f(2) = 3$. Find $f(3)$.

12 Find the derivative of $x \tan x$ and hence determine $\int x \sec^2 x\,dx$.

13 **a** Write $\frac{4x-3}{2x+1}$ in the form $A + \frac{B}{2x+1}$. **b** Hence evaluate $\int_0^2 \frac{4x-3}{2x+1}\,dx$.

14 Integrate with respect to x:

a $e^{-x}\cos x$ **b** $x^2 e^x$ **c** $\frac{x^3}{\sqrt{9-x^2}}$

15 **a** Find $\int \frac{1}{x+2}\,dx - \int \frac{2}{x-1}\,dx$. **b** Hence find $\int \frac{x+5}{(x+2)(x-1)}\,dx$.

REVIEW SET 21C

1 Find:

a $\int \left(2e^{-x} - \frac{1}{x} + 3\right) dx$ **b** $\int \left(\sqrt{x} - \frac{1}{\sqrt{x}}\right)^2 dx$ **c** $\int \left(3 + e^{2x-1}\right)^2 dx$

2 Given that $f'(x) = x^2 - 3x + 2$ and $f(1) = 3$, find $f(x)$.

3 Find the exact value of:

a $\int_2^3 \frac{1}{\sqrt{3x-4}}\,dx$ **b** $\int_0^{\frac{\pi}{3}} \cos^2\left(\frac{x}{2}\right) dx$ **c** $\int_0^{\frac{\pi}{4}} \tan x\,dx$

4 Find $\frac{d}{dx}(e^{-2x}\sin x)$ and hence find $\int_0^{\frac{\pi}{2}} \left[e^{-2x}(\cos x - 2\sin x)\right] dx$

5 Find $\int (2x+3)^n\,dx$ for all integers n.

6 A function has gradient function $2\sqrt{x}+\dfrac{a}{\sqrt{x}}$ and passes through the points (0, 2) and (1, 4).

Find a and hence explain why the function $y=f(x)$ has no stationary points.

7 $\displaystyle\int_a^{2a}(x^2+ax+2)\,dx=\frac{73a}{2}$. Find a.

8

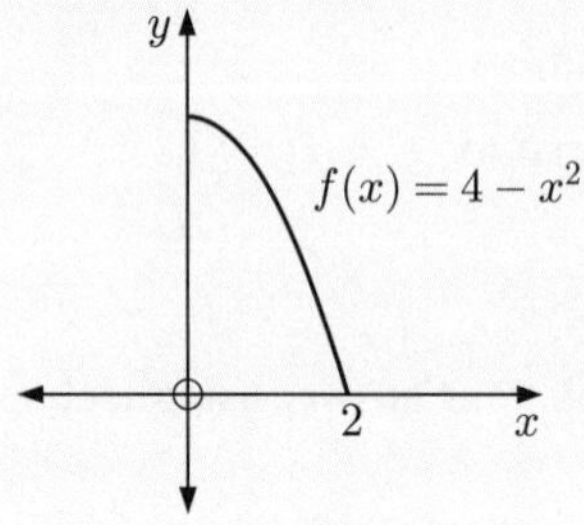

a Use *four* upper and lower rectangles to find rational numbers A and B such that:

$$A<\int_0^2(4-x^2)\,dx<B.$$

b Hence, find a good estimate for

$$\int_0^2(4-x^2)\,dx.$$

9 Integrate by substitution:

a $\displaystyle\int\frac{2x}{\sqrt{x^2-5}}\,dx$ **b** $\displaystyle\int\frac{\sin x}{\cos^4 x}\,dx$ **c** $\displaystyle\int_1^2 3x^2\sqrt{x^3-1}\,dx$

10 Differentiate $\ln(\sec x)$ with respect to x, for x such that $\sec x>0$.
What integral can be deduced from this derivative?

11 Find:

a $\displaystyle\int\frac{5}{\sqrt{9-x^2}}\,dx$ **b** $\displaystyle\int\frac{1}{9+4x^2}\,dx$ **c** $\displaystyle\int_7^{10}x\sqrt{x-5}\,dx$

12 Suppose $\dfrac{x^3-3x+2}{x-2}=Ax^2+Bx+C+\dfrac{D}{x-2}$. Find A, B, C, and D using the division algorithm. Hence find $\displaystyle\int\frac{x^3-3x+2}{x-2}\,dx$.

13 Find:

a $\displaystyle\int x\cos x\,dx$ **b** $\displaystyle\int\frac{\sqrt{x^2-4}}{x}\,dx$

14 Find $\dfrac{d}{dx}\left(\dfrac{e^{1-x}}{x^2}\right)$. Hence evaluate $\displaystyle\int_1^2\frac{e^{1-x}(x+2)}{x^3}\,dx$ exactly.

15 Find $\displaystyle\int\frac{\sin x}{\sqrt{\cos^n x}}\,dx$. Comment on the existence of this integral.

Chapter 22

Applications of integration

Syllabus reference: 6.5, 6.6

Contents:

OPENING PROBLEM

A wooden bowl is made in the shape of a *paraboloid.*

We start with the curve $y = 4\sqrt{x}$ for $0 \leqslant x \leqslant 4$, then rotate this curve through 360° around the x-axis.

DEMO

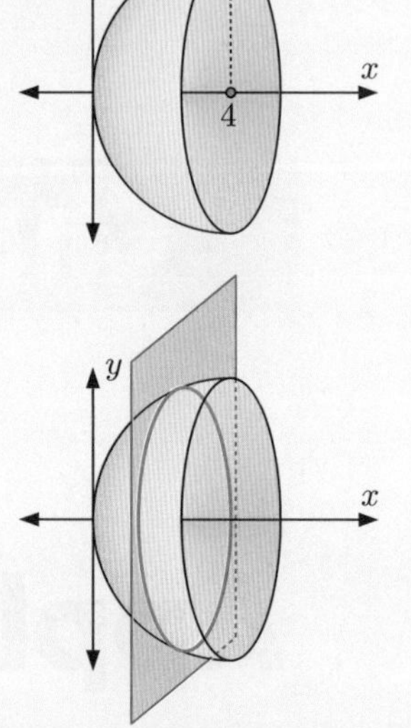

Things to think about:

a If we take a vertical slice of the bowl, what shape do we obtain?

b Can you explain why the capacity of the bowl is given by $\int_0^4 \pi(4\sqrt{x})^2 \, dx$?

c Hence find the capacity of the bowl.

We have already seen how definite integrals can be related to the areas between functions and the x-axis. In this chapter we explore this relationship further, and consider other applications of integral calculus such as kinematics and volumes of solids of revolution.

A THE AREA UNDER A CURVE

Following the work of Newton and Leibniz, integration was rigorously formalised using limits by the German mathematician **Bernhard Riemann** (1826 - 1866).

In **Chapter 21** we established a definite integral for the area under a curve. It is called the **Riemann integral** in his honour.

Bernhard Riemann

If $f(x)$ is positive and continuous on the interval $a \leqslant x \leqslant b$, then the area bounded by $y = f(x)$, the x-axis, and the vertical lines $x = a$ and $x = b$, is given by $A = \int_a^b f(x)\, dx$ or $\int_a^b y \, dx$.

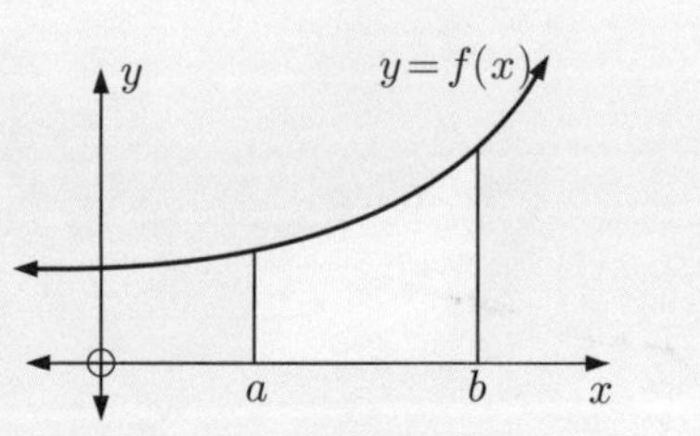

We can apply a similar area integral for curves which can be expressed in the form $x = f(y)$.

If $f(y)$ is positive and continuous on the interval $a \leqslant y \leqslant b$, then the area bounded by $x = f(y)$, the y-axis, and the horizontal lines $y = a$ and $y = b$, is given by $A = \int_a^b f(y)\, dy$.

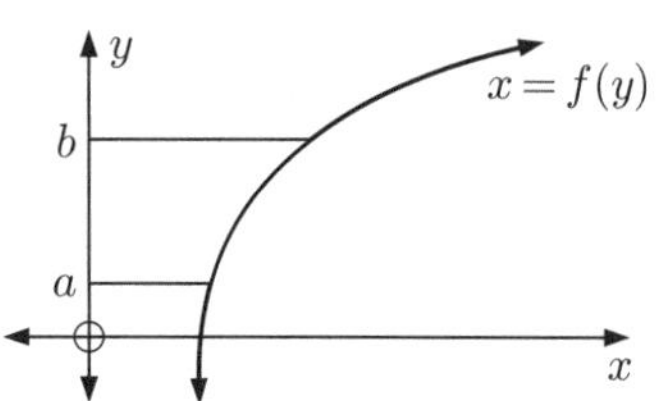

Example 1 — Self Tutor

Find the area of the region enclosed by $y = 2x$, the x-axis, $x = 0$, and $x = 4$ by using:

a a geometric argument **b** integration.

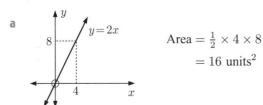

a

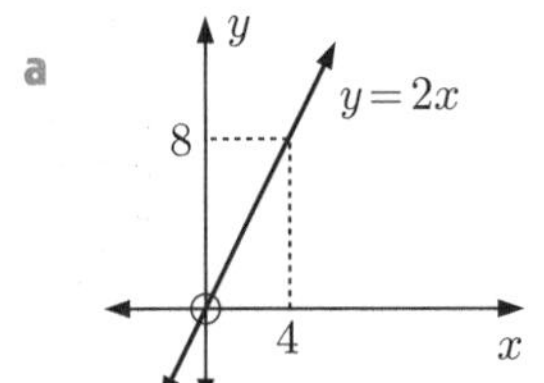

$$\text{Area} = \tfrac{1}{2} \times 4 \times 8$$
$$= 16 \text{ units}^2$$

b
$$\text{Area} = \int_0^4 2x\, dx$$
$$= \left[x^2\right]_0^4$$
$$= 4^2 - 0^2$$
$$= 16 \text{ units}^2$$

Example 2 — Self Tutor

Find the area of the region enclosed by $y = x^2 + 1$, the x-axis, $x = 1$, and $x = 2$.

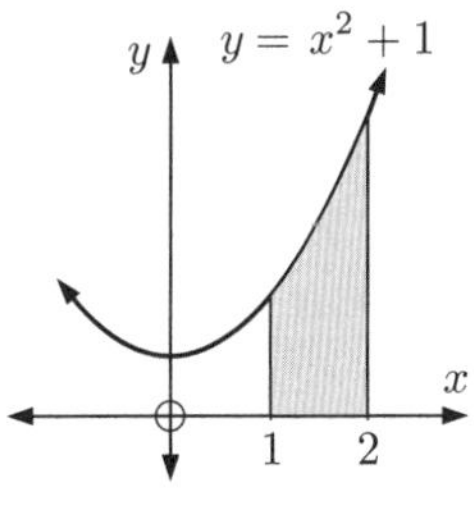

$$\text{Area} = \int_1^2 (x^2 + 1)\, dx$$
$$= \left[\frac{x^3}{3} + x\right]_1^2$$
$$= \left(\tfrac{8}{3} + 2\right) - \left(\tfrac{1}{3} + 1\right)$$
$$= 3\tfrac{1}{3} \text{ units}^2$$

It is helpful to sketch the region.

We can check this result using a graphics calculator or graphing package.

GRAPHING PACKAGE

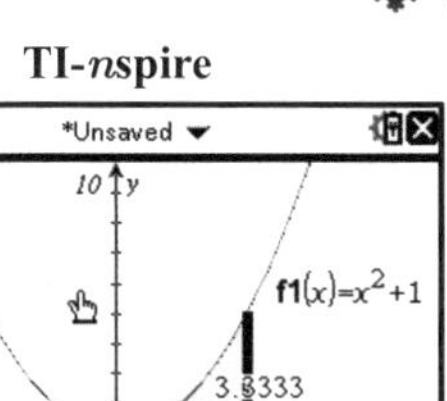

Casio fx-CG20

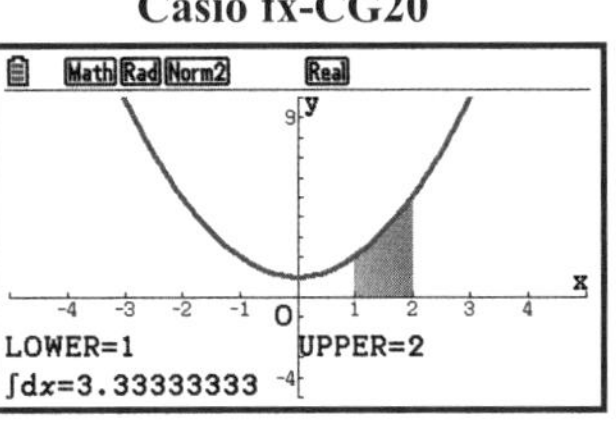

TI-84 Plus

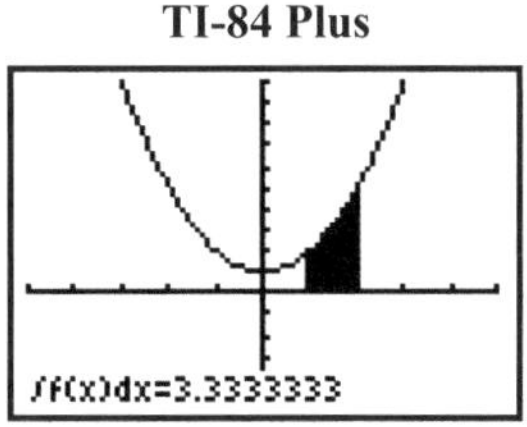

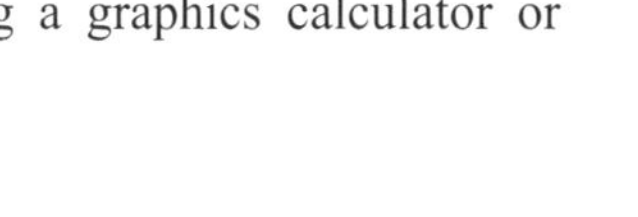

TI-nspire

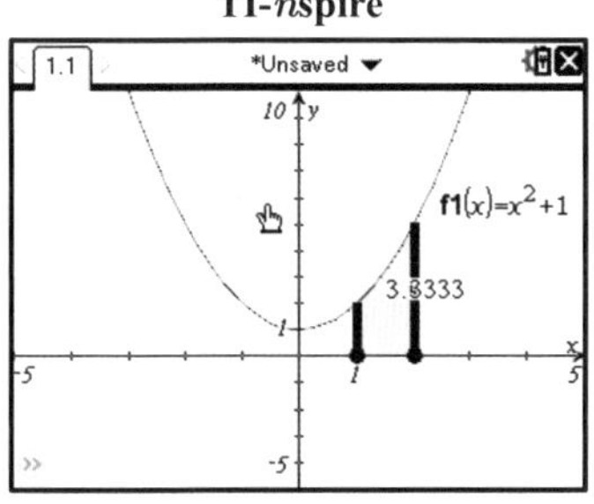

EXERCISE 22A

1 Find the area of each of the regions described below by using:

i a geometric argument **ii** integration

a $y = 5$, the x-axis, $x = -6$, and $x = 0$

b $y = x$, the x-axis, $x = 4$, and $x = 5$

c $y = -3x$, the x-axis, $x = -3$, and $x = 0$

d $y = -x$, the x-axis, $x = 0$, and $x = 2$

2 Find the exact value of the area of the region bounded by:

a $y = x^2$, the x-axis, and $x = 1$

b $y = \sin x$, the x-axis, $x = 0$, and $x = \pi$

c $y = x^3$, the x-axis, $x = 1$, and $x = 4$

d $y = e^x$, the x-axis, the y-axis, and $x = 1$

e the x-axis and the part of $y = 6 + x - x^2$ above the x-axis

f the axes and $y = \sqrt{9 - x}$

g $y = \dfrac{1}{x}$, the x-axis, $x = 1$, and $x = 4$

h $y = 2 - \dfrac{1}{\sqrt{x}}$, the x-axis, and $x = 4$

i $y = e^x + e^{-x}$, the x-axis, $x = -1$, and $x = 1$

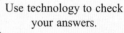

3 Find the exact value of the area of the region bounded by:

a $x = y^2 + 1$, the y-axis, and the lines $y = 1$ and $y = 2$

b $x = \sqrt{y + 5}$, the y-axis, and the lines $y = -1$ and $y = 4$

Example 3 **Self Tutor**

Find the area enclosed by one arch of the curve $y = \sin 2x$ and the x-axis.

The period of $y = \sin 2x$ is $\frac{2\pi}{2} = \pi$, so the first positive x-intercept is $\frac{\pi}{2}$.

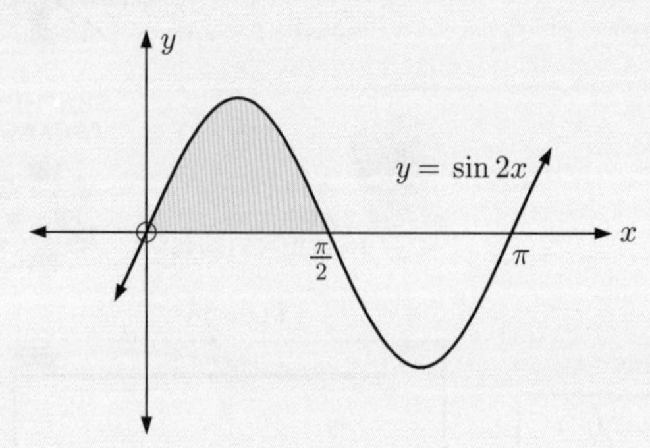

$$\begin{aligned}
\text{The required area} &= \int_0^{\frac{\pi}{2}} \sin 2x \, dx \\
&= \left[\tfrac{1}{2}(-\cos 2x)\right]_0^{\frac{\pi}{2}} \\
&= -\tfrac{1}{2}\left[\cos 2x\right]_0^{\frac{\pi}{2}} \\
&= -\tfrac{1}{2}(\cos \pi - \cos 0) \\
&= 1 \text{ unit}^2
\end{aligned}$$

4 Find the area enclosed by one arch of the curve $y = \cos 3x$ and the x-axis.

5 **a** Show that the area enclosed by $y = \sin x$ and the x-axis from $x = 0$ to $x = \pi$ is 2 units^2.

b Find the area enclosed by $y = \sin^2 x$ and the x-axis from $x = 0$ to $x = \pi$.

6 Use $\int \ln x \, dx = x \ln x - x + c$ to find the exact area of the region bounded by $y = \ln x$, the x-axis, and $x = 4$. Use technology to check your answer.

7 **a** Use integration by parts to find $\int x \sin x \, dx$.

b Hence find the exact area of the region bounded by $y = x \sin x$, the x-axis, $x = 1$, and $x = \frac{\pi}{2}$. Check your answer using technology.

INVESTIGATION 1 $\int_a^b f(x)\,dx$ AND AREAS

Does $\int_a^b f(x)\,dx$ always give us an area?

What to do:

1. Find $\int_0^1 x^3\,dx$ and $\int_{-1}^1 x^3\,dx$.
2. Using a graph, explain why the first integral in **1** gives an area, whereas the second integral does not.
3. Find $\int_{-1}^0 x^3\,dx$ and explain why the answer is negative.
4. Show that $\int_{-1}^0 x^3\,dx + \int_0^1 x^3\,dx = \int_{-1}^1 x^3\,dx$.
5. Find $\int_0^{-1} x^3\,dx$ and interpret its meaning.
6. Suppose $f(x)$ is a function such that $f(x) \leqslant 0$ for all $a \leqslant x \leqslant b$. Suggest an expression for the area between the curve and the function for $a \leqslant x \leqslant b$.

B THE AREA BETWEEN TWO FUNCTIONS

If two functions $f(x)$ and $g(x)$ intersect at $x = a$ and $x = b$, and $f(x) \geqslant g(x)$ for all $a \leqslant x \leqslant b$, then the area of the shaded region between their points of intersection is given by

$$A = \int_a^b [f(x) - g(x)]\,dx.$$

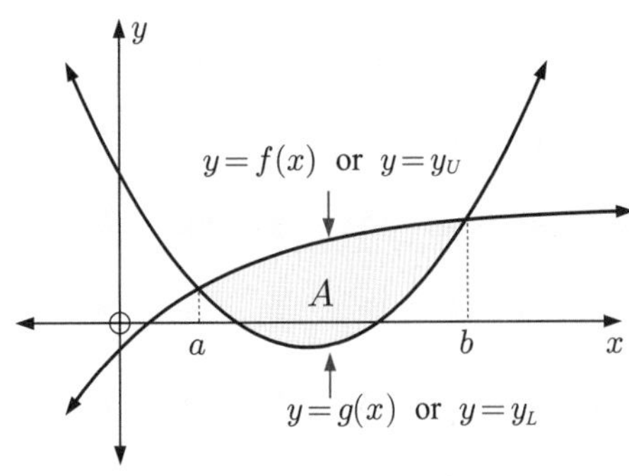

Alternatively, if the upper and lower functions are $y = y_U$ and $y = y_L$ respectively, then the area is

$$A = \int_a^b [y_U - y_L]\,dx.$$

Proof: If we translate each curve vertically through $\binom{0}{k}$ until it is completely above the x-axis, the area does not change.

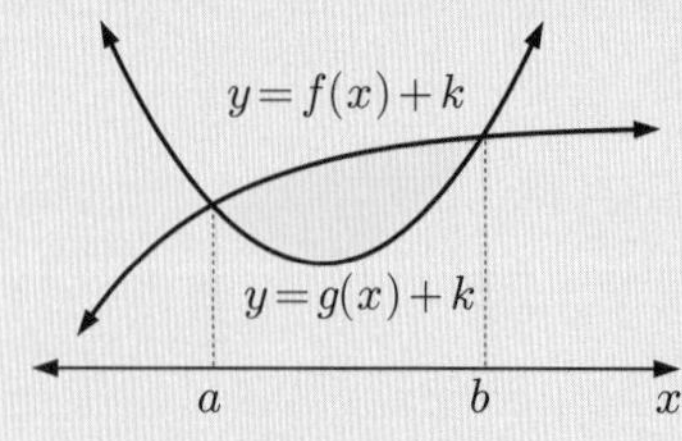

Area of shaded region

$$= \int_a^b [f(x)+k]\,dx - \int_a^b [g(x)+k]\,dx$$

$$= \int_a^b [f(x) - g(x)]\,dx$$

We can see immediately that if $f(x)$ is the axis $f(x) = 0$ then the enclosed area is $\int_a^b [-g(x)]\,dx$

or $-\int_a^b g(x)\,dx$.

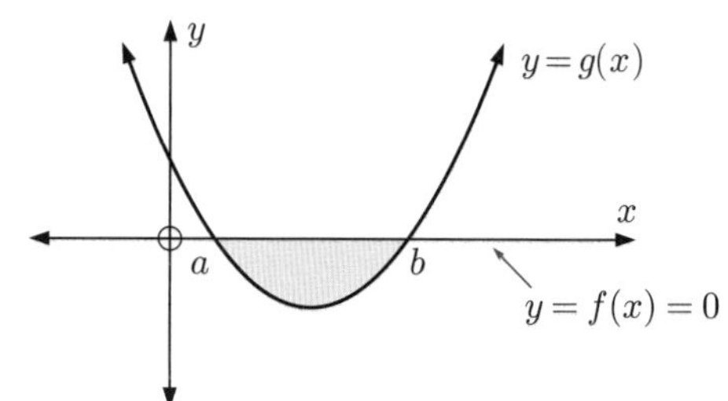

Example 4 **Self Tutor**

Use $\int_a^b [y_U - y_L]\,dx$ to find the area bounded by the x-axis and $y = x^2 - 2x$.

The curve cuts the x-axis when $y = 0$

$\therefore\ x^2 - 2x = 0$

$\therefore\ x(x-2) = 0$

$\therefore\ x = 0$ or 2

$\therefore$ the x-intercepts are 0 and 2.

$$\text{Area} = \int_0^2 [y_U - y_L]\,dx$$

$$= \int_0^2 [0 - (x^2 - 2x)]\,dx$$

$$= \int_0^2 (2x - x^2)\,dx$$

$$= \left[x^2 - \frac{x^3}{3}\right]_0^2$$

$$= \left(4 - \tfrac{8}{3}\right) - (0)$$

$\therefore$ the area is $\frac{4}{3}$ units^2.

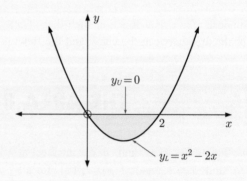

```
fnInt(2X-X²,X,0,
2)
        1.333333333
```

Example 5 **Self Tutor**

Find the area of the region enclosed by $y = x + 2$ and $y = x^2 + x - 2$.

$y = x + 2$ meets $y = x^2 + x - 2$

where $x^2 + x - 2 = x + 2$

$\therefore \quad x^2 - 4 = 0$

$\therefore \quad (x+2)(x-2) = 0$

$\therefore \quad x = \pm 2$

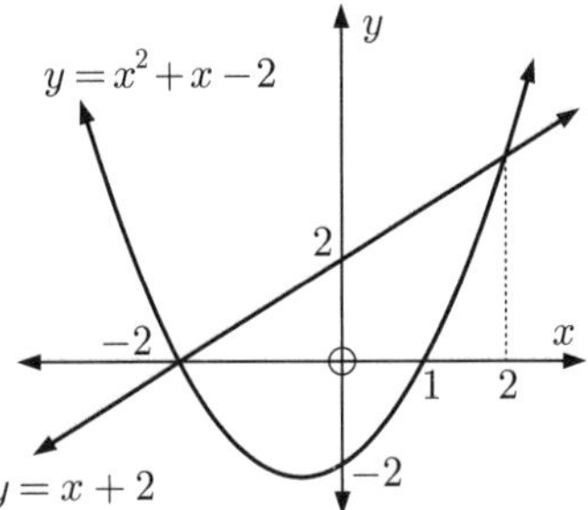

$$\begin{aligned} \text{Area} &= \int_{-2}^{2} [y_U - y_L]\, dx \\ &= \int_{-2}^{2} [(x+2) - (x^2 + x - 2)]\, dx \\ &= \int_{-2}^{2} (4 - x^2)\, dx \\ &= \left[4x - \frac{x^3}{3}\right]_{-2}^{2} \\ &= \left(8 - \tfrac{8}{3}\right) - \left(-8 + \tfrac{8}{3}\right) \\ &= 10\tfrac{2}{3} \text{ units}^2 \end{aligned}$$

$\therefore$ the area is $10\frac{2}{3}$ units^2.

Example 6 **Self Tutor**

Find the total area of the regions contained by $y = f(x)$ and the x-axis for $f(x) = x^3 + 2x^2 - 3x$.

$$\begin{aligned} f(x) &= x^3 + 2x^2 - 3x \\ &= x(x^2 + 2x - 3) \\ &= x(x-1)(x+3) \end{aligned}$$

$\therefore \quad y = f(x)$ cuts the x-axis at 0, 1, and -3.

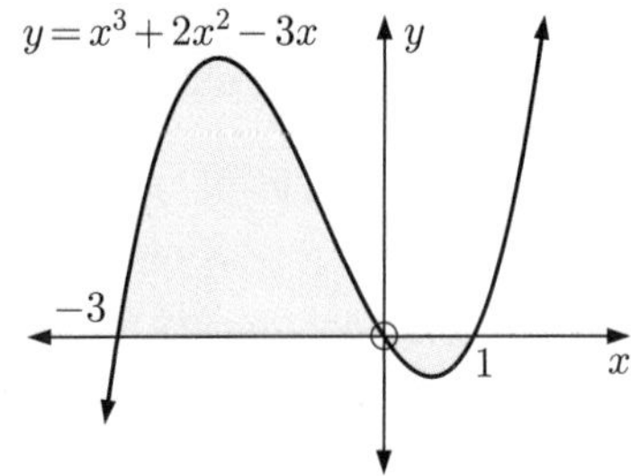

Total area

$$\begin{aligned} &= \int_{-3}^{0} (x^3 + 2x^2 - 3x)\, dx - \int_{0}^{1} (x^3 + 2x^2 - 3x)\, dx \\ &= \left[\frac{x^4}{4} + \frac{2x^3}{3} - \frac{3x^2}{2}\right]_{-3}^{0} - \left[\frac{x^4}{4} + \frac{2x^3}{3} - \frac{3x^2}{2}\right]_{0}^{1} \\ &= \left(0 - -11\tfrac{1}{4}\right) - \left(-\tfrac{7}{12} - 0\right) \\ &= 11\tfrac{5}{6} \text{ units}^2 \end{aligned}$$

For cases where there is more than one region contained between two functions, we can also use the property:

The area between the functions $f(x)$ and $g(x)$ on the interval $a \leqslant x \leqslant b$ is
$$A = \int_a^b |f(x) - g(x)|\, dx.$$

The modulus ensures the two components of the area are added together.

For example, the area in **Example 6** may be found using technology using

area $= \displaystyle\int_{-3}^{1} \left|x^3 + 2x^2 - 3x\right| \, dx$.

Casio fx-CG20

LOWER=-3 UPPER=1
∫dx=11.8333333

TI-84 Plus

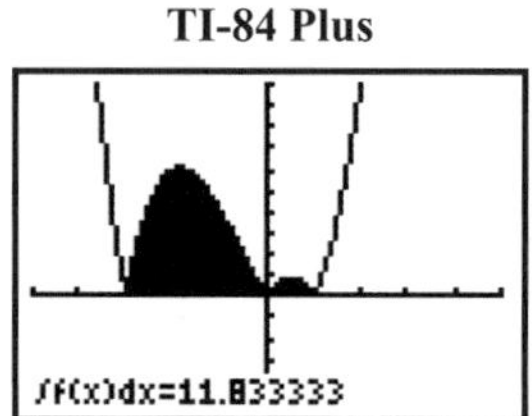

TI-nspire

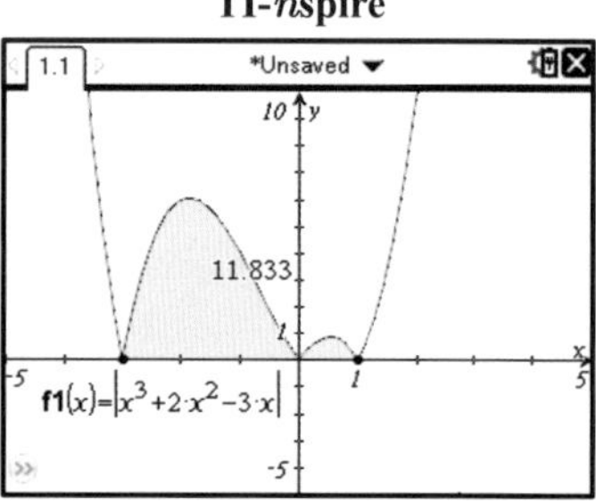

EXERCISE 22B

1 Find the exact value of the area bounded by:

  a the x-axis and $y = x^2 + x - 2$

  b the x-axis, $y = e^{-x} - 1$, and $x = 2$

  c $y = \dfrac{1}{x}$, the x-axis, $x = -1$, and $x = -3$

  d the x-axis and the part of $y = 3x^2 - 8x + 4$ below the x-axis

  e $y = \cos x$, the x-axis, $x = \frac{\pi}{2}$, and $x = \frac{3\pi}{2}$

  f $y = x^3 - 4x$, the x-axis, $x = 1$, and $x = 2$

  g $y = \sin x - 1$, the x-axis, $x = 0$, and $x = \frac{\pi}{2}$

2 Find the area of the region enclosed by $y = x^2 - 2x$ and $y = 3$.

3 Consider the graphs of $y = x - 3$ and $y = x^2 - 3x$.

  a Sketch the graphs on the same set of axes.

  b Find the coordinates of the points where the graphs meet.

  c Find the area of the region enclosed by the two graphs.

4 Determine the area of the region enclosed by $y = \sqrt{x}$ and $y = x^2$.

5 **a** On the same set of axes, graph $y = e^x - 1$ and $y = 2 - 2e^{-x}$, showing axes intercepts and asymptotes.

  b Find algebraically the points of intersection of $y = e^x - 1$ and $y = 2 - 2e^{-x}$.

  c Find the area of the region enclosed by the two curves.

6 Determine exactly the area of the region bounded by $y = 2e^x$, $y = e^{2x}$, and $x = 0$.

7 **a** On the same set of axes, draw the graphs of the functions $y = 2x$ and $y = 4x^2$. Determine exactly the area of the region enclosed by these functions.

  b On the same set of axes, draw the graphs of the relations $y = 2x$ and $y^2 = 4x$. Determine exactly the area of the region enclosed by these relations.

8 Use $\int \ln y \, dy = y \ln y - y + c$ to find the exact area of the region bounded by $y = e^x$, the y-axis, and the lines $y = 2$ and $y = 3$.

9 A region with $x \geqslant 0$ has boundaries defined by $y = \sin x$, $y = \cos x$, and the y-axis. Find the area of the region.

10 The graph alongside shows $y = \tan x$ for $-\frac{\pi}{2} < x < \frac{\pi}{2}$.
A is the point on the graph with y-coordinate 1.

a Find the coordinates of A.

b Find the shaded area.

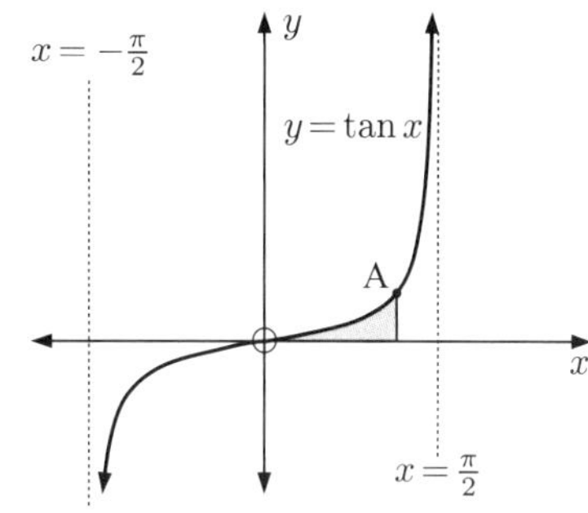

11 Sketch the circle with equation $x^2 + y^2 = 9$.

a Explain why the upper half of the circle has equation $y = \sqrt{9 - x^2}$.

b Hence, determine $\int_0^3 \sqrt{9 - x^2} \, dx$ without actually integrating the function.

c Check your answer using technology.

12 Find the area enclosed by the function $y = f(x)$ and the x-axis for:

a $f(x) = x^3 - 9x$ **b** $f(x) = -x(x-2)(x-4)$ **c** $f(x) = x^4 - 5x^2 + 4$.

13 The illustrated curves are those of $y = \sin x$ and $y = \sin(2x)$.

a Identify each curve.

b Find algebraically the coordinates of A.

c Find the total area enclosed by C_1 and C_2 for $0 \leqslant x \leqslant \pi$.

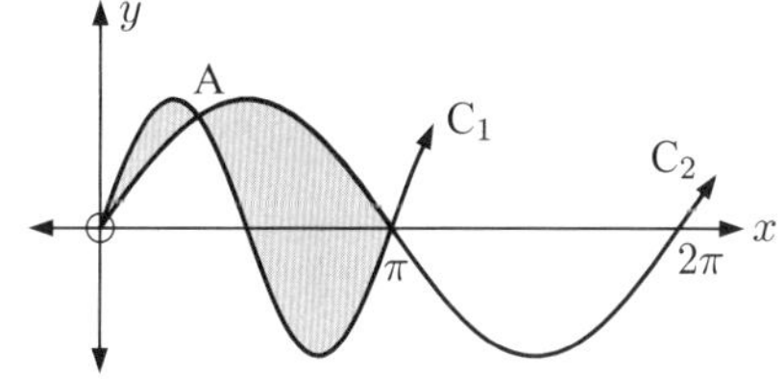

14

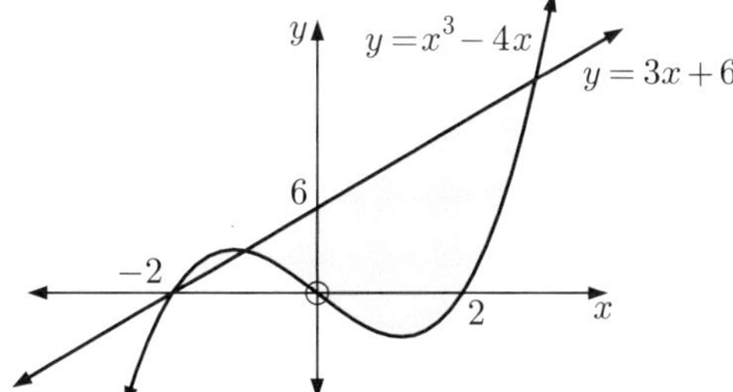

a Write the shaded area as:

i the sum of two definite integrals

ii a single definite integral involving modulus.

b Find the total shaded area.

15 Find the area enclosed by:

a $y = x^3 - 5x$ and $y = 2x^2 - 6$

b $y = -x^3 + 3x^2 + 6x - 8$ and $y = 5x - 5$

c $y = 2x^3 - 3x^2 + 18$ and $y = x^3 + 10x - 6$.

16 **a** Explain why the total area shaded is *not* equal to $\int_1^7 f(x)\,dx$.

b Write an expression for the total shaded area in terms of integrals.

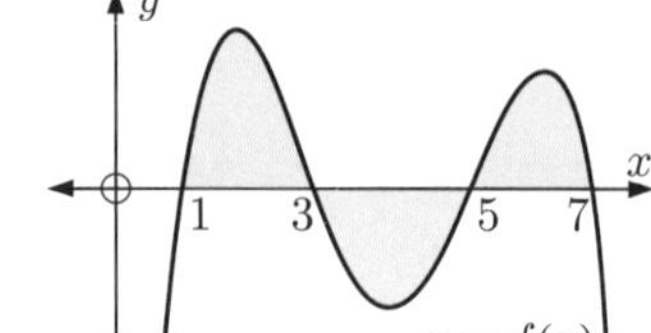

17

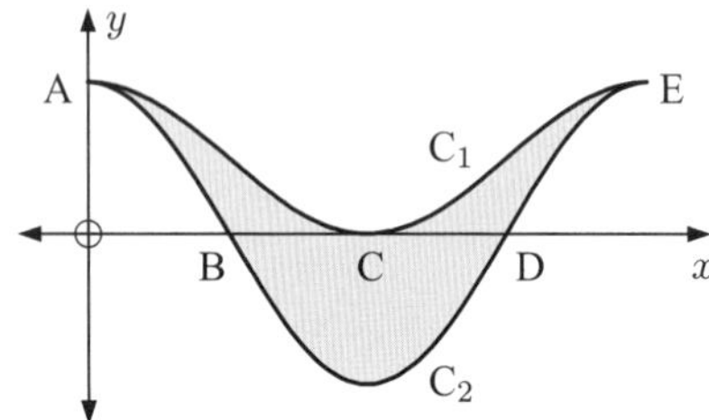

The illustrated curves are $y = \cos(2x)$ and $y = \cos^2 x$.

a Identify each curve as C_1 or C_2.

b Determine the coordinates of A, B, C, D, and E.

c Show that the area of the shaded region is $\frac{\pi}{2}$ units^2.

18 Find, correct to 3 significant figures, the areas of the regions enclosed by the curves:

a $y = e^{-x^2}$ and $y = x^2 - 1$

b $y = x^x$ and $y = 4x - \frac{1}{10}x^4$

19 **a** The shaded area is 0.2 units^2. Find k, correct to 4 decimal places.

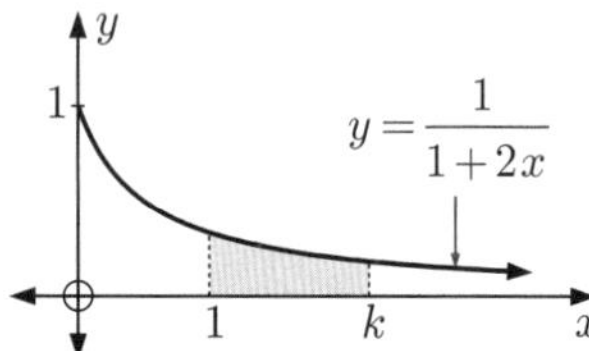

b The shaded area is 1 unit^2. Find b, correct to 4 decimal places.

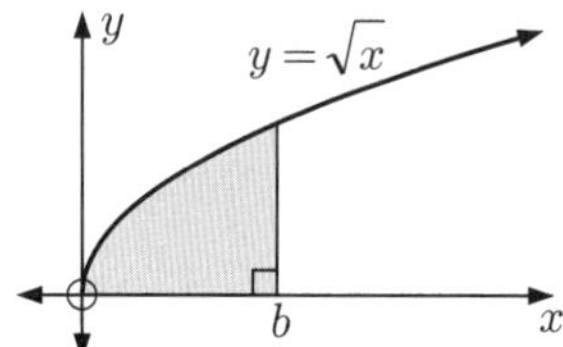

20 **a** The shaded area is 2.4 units^2. Find k, correct to 4 decimal places.

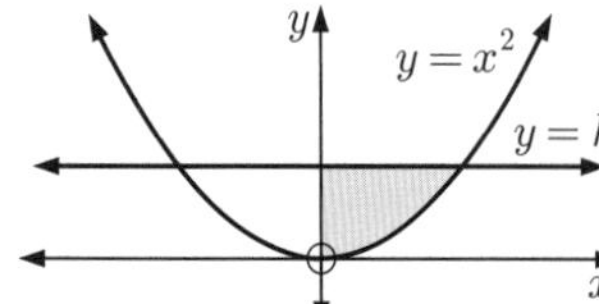

b The shaded area is $6a$ units^2. Find the exact value of a.

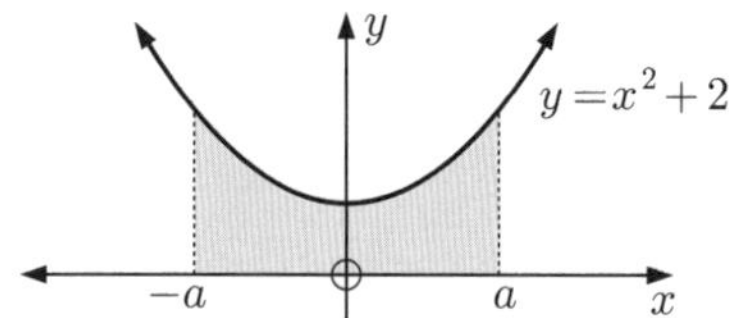

21 **a** Sketch the graph of $y = \dfrac{1}{\sqrt{1 - x^2}}$.

b Explain algebraically why the function

i is symmetrical about the y-axis

ii has domain $x \in \left]-1, 1\right[$.

c Find the exact area enclosed by the function and the x-axis, the y-axis, and the line $x = \frac{1}{2}$.

INVESTIGATION 2 NUMERICAL INTEGRATION

There are many functions that do not have indefinite integrals. However, definite integrals can still be determined by **numerical methods**. An example of this is the upper and lower rectangles we used in **Chapter 21**.

A slightly more accurate method is the **midpoint rule** in which we take the height of each rectangle to be the value of the function at the midpoint of the subinterval.

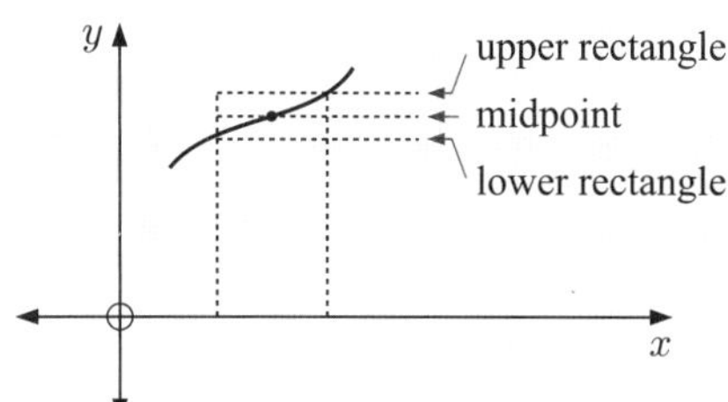

For example, consider the area between $f(x) = \sin(x^2)$ and the x-axis from $x = 0$ to $x = 1$.

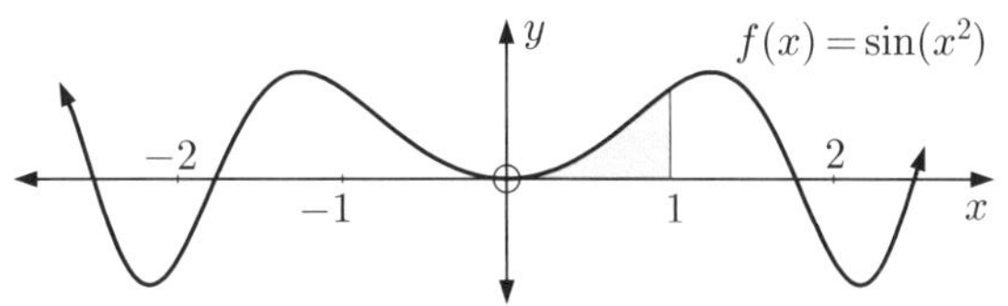

$f(x)$ is an even function and does not have an indefinite integral, so a numerical method is essential for $\int_0^1 \sin(x^2)\,dx$ to be evaluated.

What to do:

1 Suppose the interval from 0 to 1 is divided into 10 equal subintervals of width 0.1. Using the midpoint rule, the shaded area is

$$\int_0^1 \sin(x^2)\,dx \approx [f(0.05) + f(0.15) + f(0.25) + ... + f(0.95)]\,\delta x$$

where $\delta x = 0.1$ is the subinterval width.

Use this formula to estimate the integral to 3 decimal places.

2 Click on the area finder icon. Use the software to estimate the area for $n = 10, 100, 1000, 10\,000$.

Compare the results with those given by your graphics calculator.

AREA FINDER

3 Use the midpoint rule to estimate:

a $\int_0^2 \sin(x^2)\,dx$

b the area enclosed between $y = \sin(x^2)$, the x-axis, and the vertical line $x = 2$.

C KINEMATICS

DISTANCES FROM VELOCITY GRAPHS

Suppose a car travels at a constant positive velocity of 60 km h^{-1} for 15 minutes.

We know the distance travelled = speed × time

$$= 60 \text{ km h}^{-1} \times \tfrac{1}{4} \text{ h}$$
$$= 15 \text{ km.}$$

When we graph *speed* against *time*, the graph is a horizontal line, and we can see that the distance travelled is the area shaded.

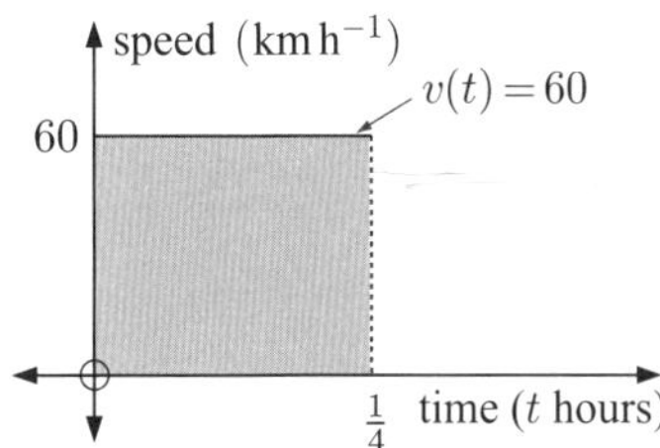

So, the distance travelled can also be found by the definite integral $\int_0^{\frac{1}{4}} 60\,dt = 15$ km.

Now suppose the speed decreases at a constant rate so that the car, initially travelling at 60 km h^{-1}, stops in 6 minutes or $\frac{1}{10}$ hour.

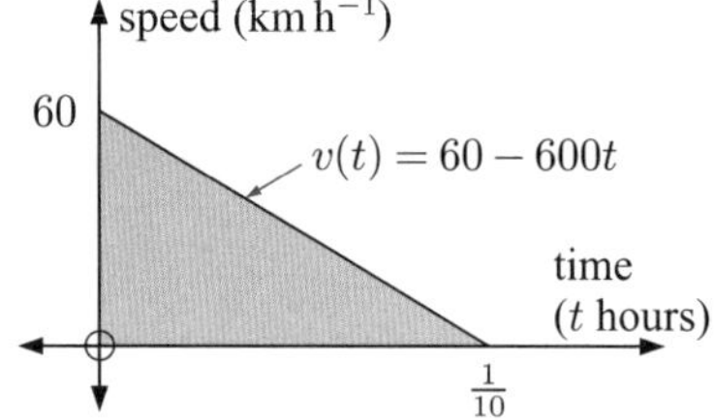

In this case the *average* speed is 30 km h^{-1}, so the distance travelled $= 30$ km h^{-1} $\times \frac{1}{10}$ h $= 3$ km

But the triangle has area $= \frac{1}{2} \times \text{base} \times \text{altitude}$

$= \frac{1}{2} \times \frac{1}{10} \times 60 = 3$

So, once again the shaded area gives us the distance travelled, and we can find it using the definite integral $\int_0^{\frac{1}{10}} (60 - 600t)\,dt = 3$.

These results suggest that distance travelled $= \int_{t_1}^{t_2} v(t)\,dt$ provided we do not change direction.

If we have a change of direction within the time interval then the velocity will change sign. We therefore need to add the components of area above and below the t-axis to find the total distance travelled.

For a velocity-time function $v(t)$ where $v(t) \geqslant 0$ on the interval $t_1 \leqslant t \leqslant t_2$,

distance travelled $= \int_{t_1}^{t_2} v(t)\,dt$.

For a velocity-time function $v(t)$ where $v(t) \leqslant 0$ on the interval $t_1 \leqslant t \leqslant t_2$,

distance travelled $= -\int_{t_1}^{t_2} v(t)\,dt$.

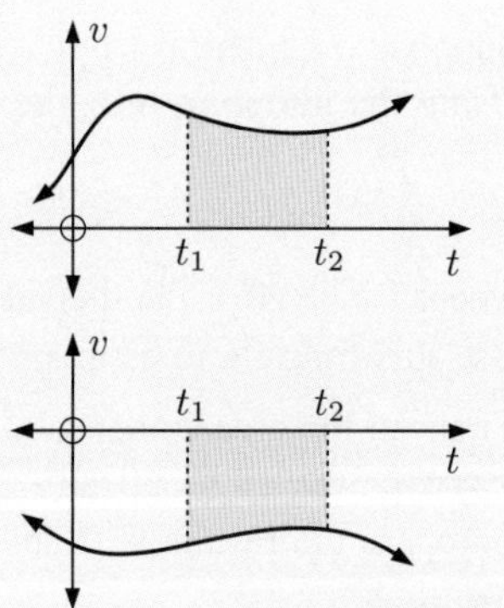

Example 7 **Self Tutor**

The velocity-time graph for a train journey is illustrated in the graph alongside. Find the total distance travelled by the train.

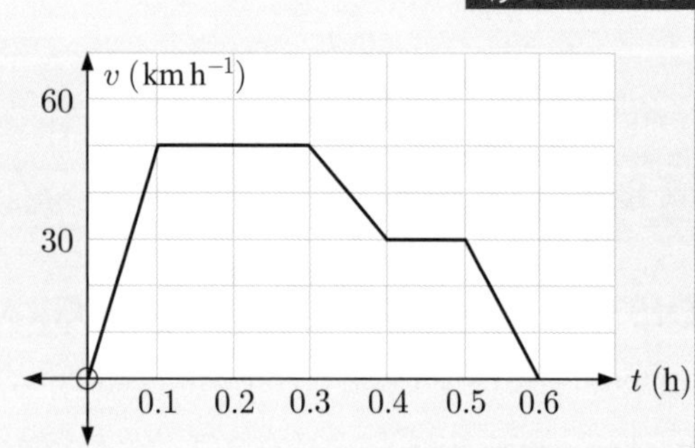

Total distance travelled

$=$ total area under the graph

$=$ area A + area B + area C + area D + area E

$= \frac{1}{2}(0.1)50 + (0.2)50 + \left(\frac{50+30}{2}\right)(0.1) + (0.1)30 + \frac{1}{2}(0.1)30$

$= 2.5 + 10 + 4 + 3 + 1.5$

$= 21$ km

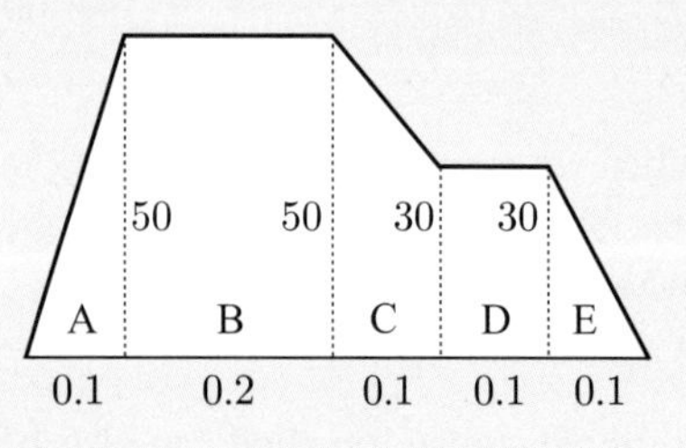

EXERCISE 22C.1

1. A runner has the velocity-time graph shown. Find the total distance travelled by the runner.

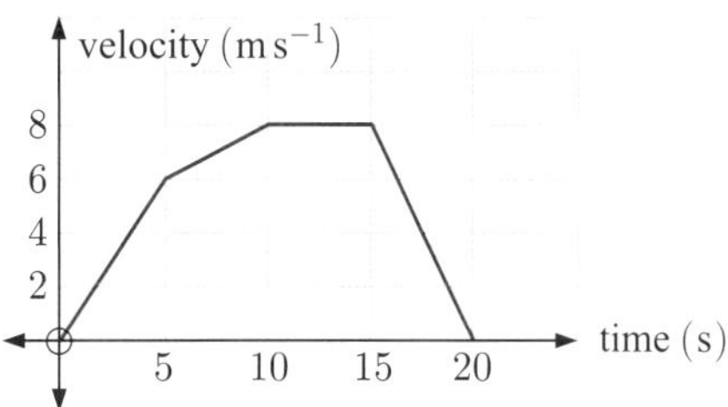

2. 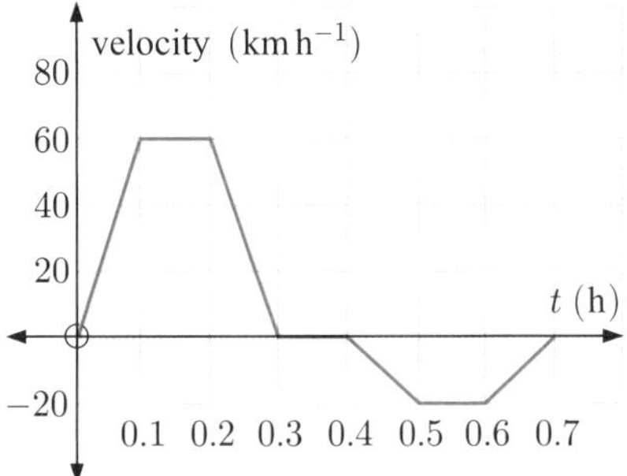

A car travels along a straight road with the velocity-time function illustrated.

   a. What is the significance of the graph:
      i. above the t-axis
      ii. below the t-axis?
   b. Find the total distance travelled by the car.
   c. Find the final displacement of the car from its starting point.

3. A cyclist rides off from rest, accelerating at a constant rate for 3 minutes until she reaches 40 $km\,h^{-1}$. She then maintains a constant speed for 4 minutes until reaching a hill. She slows down at a constant rate over one minute to 30 $km\,h^{-1}$, then continues at this rate for 10 minutes. At the top of the hill she reduces her speed uniformly and is stationary 2 minutes later.

   a. Draw a graph to show the cyclist's motion.
   b. How far has the cyclist travelled?

DISPLACEMENT AND VELOCITY FUNCTIONS

In this section we are concerned with **motion in a straight line**.

For some displacement function $s(t)$, the velocity function is $v(t) = s'(t)$.

So, given a velocity function we can determine the displacement function by the integral

$$s(t) = \int v(t)\, dt$$

The constant of integration determines the **initial position** on the line where the object begins.

Using the displacement function we can determine the change in displacement in a time interval $t_1 \leqslant t \leqslant t_2$ using the integral:

$$\textbf{Displacement} = s(t_2) - s(t_1) = \int_{t_1}^{t_2} v(t)\, dt$$

TOTAL DISTANCE TRAVELLED

To determine the total distance travelled in a time interval $t_1 \leqslant t \leqslant t_2$, we need to account for any changes of direction in the motion.

To find the total distance travelled given a velocity function $v(t) = s'(t)$ on $t_1 \leqslant t \leqslant t_2$:
- Draw a sign diagram for $v(t)$ so we can determine any changes of direction.
- Determine $s(t)$ by integration, including a constant of integration.
- Find $s(t_1)$ and $s(t_2)$. Also find $s(t)$ at each time the direction changes.
- Draw a motion diagram.
- Determine the total distance travelled from the motion diagram.

Using technology, we can simply use

$$\textbf{total distance travelled} \text{ from } t = t_1 \text{ to } t = t_2 = \int_{t_1}^{t_2} |v(t)|\, dt$$

VELOCITY AND ACCELERATION FUNCTIONS

We know that the acceleration function is the derivative of the velocity function, so $a(t) = v'(t)$.

So, given an acceleration function, we can determine the velocity function by integration:

$$v(t) = \int a(t)\, dt.$$

SUMMARY

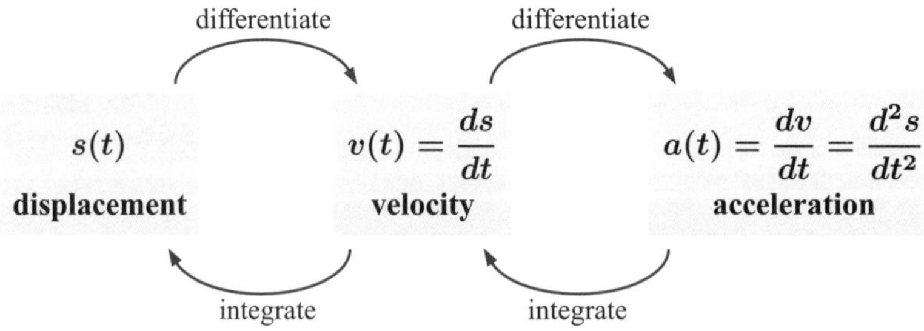

Example 8 Self Tutor

A particle P moves in a straight line with velocity function $v(t) = t^2 - 3t + 2$ m s^{-1}. Answer the following without using a calculator:

a How far does P travel in the first 4 seconds of motion?

b Find the displacement of P after 4 seconds.

a $v(t) = s'(t) = t^2 - 3t + 2$
$= (t-1)(t-2)$

$\therefore$ the sign diagram of v is: (0) $+$ (1) $-$ (2) $+$ t

Since the signs change, P reverses direction at $t = 1$ and $t = 2$ seconds.

Now $s(t) = \int (t^2 - 3t + 2)\, dt = \frac{t^3}{3} - \frac{3t^2}{2} + 2t + c$

Hence $s(0) = c$ $\qquad s(1) = \frac{1}{3} - \frac{3}{2} + 2 + c = c + \frac{5}{6}$

$s(2) = \frac{8}{3} - 6 + 4 + c = c + \frac{2}{3}$ $\qquad s(4) = \frac{64}{3} - 24 + 8 + c = c + 5\frac{1}{3}$

Motion diagram:

c $\quad c+\frac{2}{3}$ $\quad c+\frac{5}{6}$ $\quad c+5\frac{1}{3}$

$\therefore$ total distance travelled $= (c + \frac{5}{6} - c) + (c + \frac{5}{6} - [c + \frac{2}{3}]) + (c + 5\frac{1}{3} - [c + \frac{2}{3}])$

$= \frac{5}{6} + \frac{5}{6} - \frac{2}{3} + 5\frac{1}{3} - \frac{2}{3}$

$= 5\frac{2}{3}$ m

b Displacement = final position – original position

$= s(4) - s(0)$

$= c + 5\frac{1}{3} - c$

$= 5\frac{1}{3}$ m

So, the displacement is $5\frac{1}{3}$ m to the right.

EXERCISE 22C.2

1 A particle has velocity function $v(t) = 1 - 2t$ cm s^{-1} as it moves in a straight line. The particle is initially 2 cm to the right of O.

 a Write a formula for the displacement function $s(t)$.
 b Find the total distance travelled in the first second of motion.
 c Find the displacement of the particle at the end of one second.

2 Particle P is initially at the origin O. It moves with the velocity function $v(t) = t^2 - t - 2$ cm s^{-1}.

 a Write a formula for the displacement function $s(t)$.
 b Find the total distance travelled in the first 3 seconds of motion.
 c Find the displacement of the particle at the end of three seconds.

3 The velocity of a moving object is given by $v(t) = 32 + 4t$ m s^{-1}.

 a If $s = 16$ m when $t = 0$ seconds, find the displacement function.
 b Explain why the displacement of the object and its total distance travelled in the interval $0 \leqslant t \leqslant t_1$, can both be represented by the definite integral $\int_0^{t_1} (32 + 4t)\, dt$.
 c Show that the object is travelling with constant acceleration.

4 An object has velocity function $v(t) = \cos(2t)$ m s^{-1}. If $s(\frac{\pi}{4}) = 1$ m, determine $s(\frac{\pi}{3})$ exactly.

5 A particle moves along the x-axis with velocity function $x'(t) = 16t - 4t^3$ units per second. Find the total distance travelled in the time interval:

 a $0 \leqslant t \leqslant 3$ seconds $\qquad$ **b** $1 \leqslant t \leqslant 3$ seconds.

6 A particle moves in a straight line with velocity function $v(t) = \cos t$ m s^{-1}.

 a Show that the particle oscillates between two points.
 b Find the distance between the two points in **a**.

7 The velocity of a particle travelling in a straight line is given by $v(t) = 50 - 10e^{-0.5t}$ m s^{-1}, where $t \geqslant 0$, t in seconds.

- **a** State the initial velocity of the particle.
- **b** Find the velocity of the particle after 3 seconds.
- **c** How long will it take for the particle's velocity to increase to 45 m s^{-1}?
- **d** Discuss $v(t)$ as $t \to \infty$.
- **e** Show that the particle's acceleration is always positive.
- **f** Draw the graph of $v(t)$ against t.
- **g** Find the total distance travelled by the particle in the first 3 seconds of motion.

Example 9 **Self Tutor**

A particle is initially at the origin and moving to the right at 5 cm s^{-1}. It accelerates with time according to $a(t) = 4 - 2t$ cm s^{-2}. Answer the following without using a calculator:

- **a** Find the velocity function of the particle, and sketch its graph for $0 \leqslant t \leqslant 6$ s.
- **b** For the first 6 seconds of motion, determine the:
  - **i** displacement of the particle
  - **ii** total distance travelled.

a $v(t) = \int a(t)\,dt = \int (4 - 2t)\,dt$

$= 4t - t^2 + c$

But $v(0) = 5$, so $c = 5$

$\therefore\ v(t) = -t^2 + 4t + 5$ cm s^{-1}

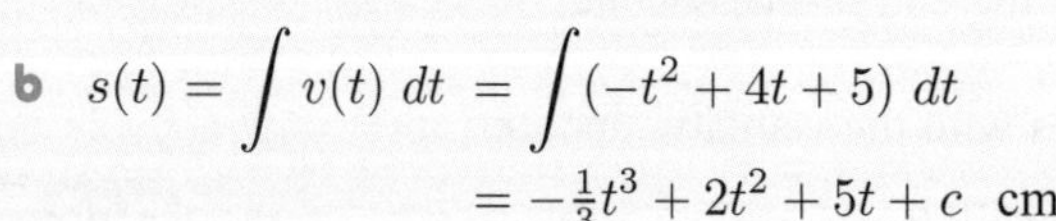

b $s(t) = \int v(t)\,dt = \int (-t^2 + 4t + 5)\,dt$

$= -\frac{1}{3}t^3 + 2t^2 + 5t + c$ cm

But $s(0) = 0$, so $c = 0$

$\therefore\ s(t) = -\frac{1}{3}t^3 + 2t^2 + 5t$ cm

i Displacement $= s(6) - s(0)$

$= -\frac{1}{3}(6)^3 + 2(6)^2 + 5(6)$

$= 30$ cm

ii The particle changes direction when $t = 5$ s.

Now $s(5) = -\frac{1}{3}(5)^3 + 2(5)^2 + 5(5)$

$= 33\frac{1}{3}$ cm

Motion diagram:

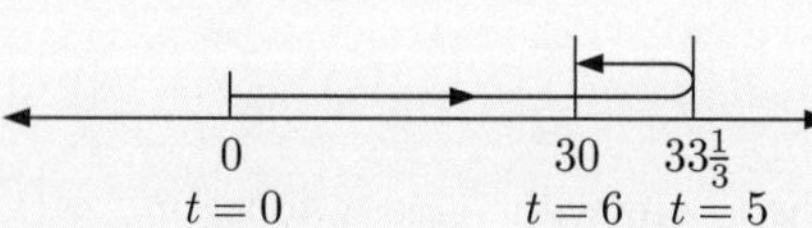

$\therefore$ the total distance travelled $= 33\frac{1}{3} + 3\frac{1}{3}$

$= 36\frac{2}{3}$ cm

Using technology, we can check that $\int_0^6 \left|-t^2 + 4t + 5\right|\,dt = 36\frac{2}{3}$

8 A particle is initially stationary at the origin. It accelerates according to the function $a(t) = \dfrac{-1}{(t+1)^2}$ m s^{-2}.

- **a** Find the velocity function $v(t)$ for the particle.
- **b** Find the displacement function $s(t)$ for the particle.
- **c** Describe the motion of the particle at the time $t = 2$ seconds.

9 A train moves along a straight track with acceleration $(\frac{t}{10} - 3)$ m s^{-2}. The initial velocity of the train is 45 m s^{-1}.

- **a** Determine the velocity function $v(t)$.
- **b** Evaluate $\displaystyle\int_0^{60} v(t)\, dt$ and explain what this value represents.

10 An object has initial velocity 20 m s^{-1} as it moves in a straight line with acceleration function $4e^{-\frac{t}{20}}$ m s^{-2}.

- **a** Show that as t increases the object approaches a limiting velocity.
- **b** Find the total distance travelled in the first 10 seconds of motion.

D PROBLEM SOLVING BY INTEGRATION

When we studied differential calculus, we saw how to find the rate of change of a function by differentiation.

In practical situations it is sometimes easier to measure the rate of change of a variable, and use integration to find a function for the quantity concerned. For example, we can measure the rate at which electricity is used, and then integrate to find the total amount used over time.

Example 10 **Self Tutor**

The marginal cost of producing x urns per week is given by
$\dfrac{dC}{dx} = 2.15 - 0.02x + 0.000\,36x^2$ dollars per urn provided $0 \leqslant x \leqslant 120$.
The initial costs before production starts are \$185.
Find the total cost of producing 100 urns per day.

The marginal cost is $\dfrac{dC}{dx} = 2.15 - 0.02x + 0.000\,36x^2$ dollars per urn

$$\therefore\ C(x) = \int (2.15 - 0.02x + 0.000\,36x^2)\, dx$$
$$= 2.15x - 0.02\,\frac{x^2}{2} + 0.000\,36\,\frac{x^3}{3} + c \text{ dollars}$$
$$= 2.15x - 0.01x^2 + 0.000\,12x^3 + c \text{ dollars}$$

But $C(0) = 185$ dollars $\therefore\ c = 185$

$$\therefore\ C(x) = 2.15x - 0.01x^2 + 0.000\,12x^3 + 185 \text{ dollars}$$
$$\therefore\ C(100) = 2.15(100) - 0.01(100)^2 + 0.000\,12(100)^3 + 185 \text{ dollars}$$
$$= 420 \text{ dollars}$$

$\therefore$ the total cost is \$420.

EXERCISE 22D

1 The marginal cost per day of producing x gadgets is $C'(x) = 3.15 + 0.004x$ euros per gadget. Find the total cost of daily production of 800 gadgets given that the fixed costs before production commences are €450 per day.

2 The marginal profit for producing x dinner plates per week is given by $P'(x) = 15 - 0.03x$ dollars per plate. If no plates are made then a loss of \$650 each week occurs.

 a Find the profit function $P(x)$.

 b What is the maximum profit and when does it occur?

 c What production levels enable a profit to be made?

3 Jon needs to bulk-up for the football season. His rate of energy intake t days after starting his weight gain program is given by $E'(t) = 350(80 + 0.15t)^{0.8} - 120(80 + 0.15t)$ calories per day. Find Jon's total energy needs over the first week of the program.

Example 11 **Self Tutor**

A metal tube has an annulus cross-section as shown. The outer radius is 4 cm and the inner radius is 2 cm. Within the tube, water is maintained at a temperature of 100°C. Within the metal the temperature drops off from inside to outside according to $\frac{dT}{dx} = -\frac{10}{x}$, where x is the distance from the central axis and $2 \leqslant x \leqslant 4$.

Find the temperature of the outer surface of the tube.

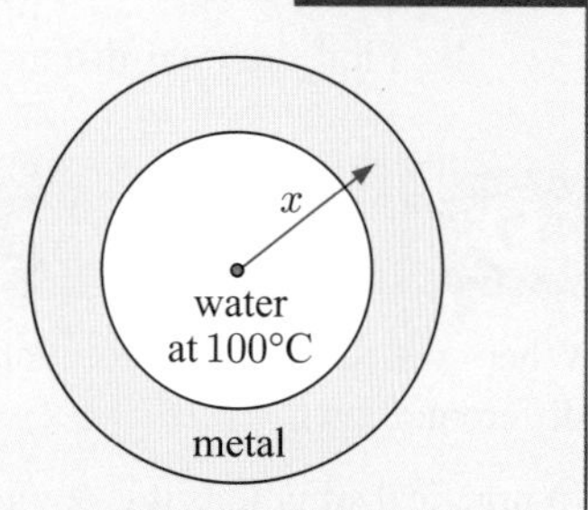

tube cross-section

Since $\frac{dT}{dx} = \frac{-10}{x}$, $\quad T = \int \frac{-10}{x}\,dx$

$\therefore\ T = -10\ln|x| + c$

But when $x = 2$, $T = 100$

$\therefore\ 100 = -10\ln 2 + c$

$\therefore\ c = 100 + 10\ln 2$

Thus $T = -10\ln x + 100 + 10\ln 2$ $\quad \{\text{as } x > 0,\ |x| = x\}$

$\therefore\ T = 100 + 10\ln\left(\frac{2}{x}\right)$

When $x = 4$, $T = 100 + 10\ln\left(\frac{1}{2}\right) \approx 93.1$

$\therefore$ the outer surface temperature is 93.1°C.

4 The tube cross-section shown has inner radius 3 cm and outer radius 6 cm. Within the tube, water is maintained at a temperature of 100°C. Within the metal the temperature falls off at the rate $\frac{dT}{dx} = \frac{-20}{x^{0.63}}$, where x is the distance from the central axis and $3 \leqslant x \leqslant 6$.

Find the temperature of the outer surface of the tube.

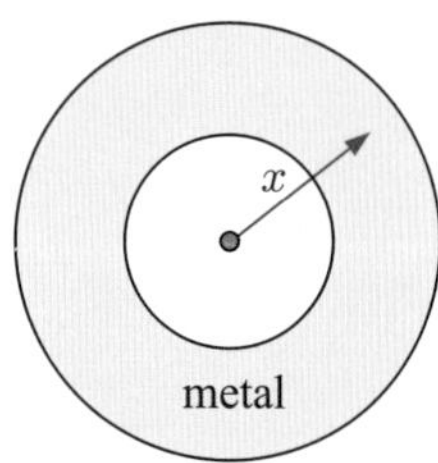

5 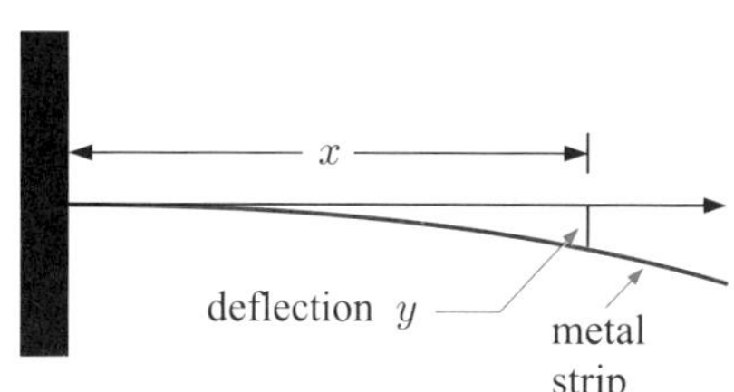

A thin horizontal metal strip of length 1 m is deflected by y m at the distance of x m from its fixed end.

It is known that $\frac{d^2y}{dx^2} = -\frac{1}{10}(1-x)^2$.

a Find y and $\frac{dy}{dx}$ when $x = 0$.

b Find the function $y(x)$ which measures the deflection from the horizontal at any point on the metal strip.

c Determine the greatest deflection of the metal strip.

6 A wooden plank 4 m long is supported only at its ends, O and P. The plank sags under its own weight by y m at the distance x m from O.

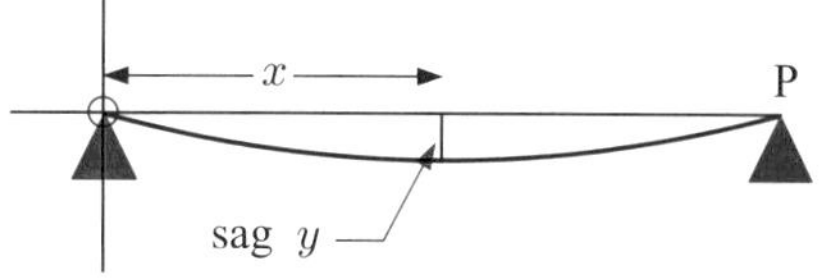

The differential equation $\frac{d^2y}{dx^2} = 0.01\left(2x - \frac{x^2}{2}\right)$

relates the variables x and y, for $0 \leqslant x \leqslant 4$.

a Find the function $y(x)$ which measures the sag from the horizontal at any point along the plank.

b Find the maximum sag from the horizontal.

c Find the sag 1 m away from P.

d Find the angle the plank makes with the horizontal at the point 1 m from P.

7 A contractor digs cylindrical wells to a depth of h metres. He estimates that the cost of digging at the depth x metres is $\left(\frac{1}{2}x^2 + 4\right)$ dollars per m^3 of earth and rock extracted.

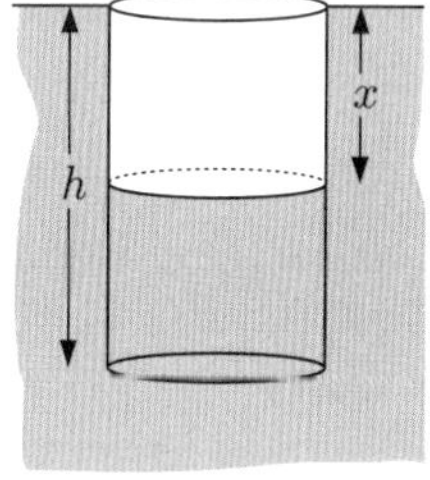

Show that the total cost of digging a well of radius r m is given by $C(h) = \pi r^2\left(\frac{h^3 + 24h}{6}\right) + C_0$ dollars.

Hint: $\frac{dC}{dx} = \frac{dC}{dV}\frac{dV}{dx}$

8 The length of a continuous function $y = f(x)$ on $a \leqslant x \leqslant b$ is given by

$L(x) = \int_a^b \sqrt{1 + \left(\frac{dy}{dx}\right)^2}\, dx$.

Find, correct to 5 decimal places, the length of $y = \sin x$ on $0 \leqslant x \leqslant \pi$.

9 A farmer with a large property plans a rectangular fruit orchard. One boundary will be an irrigation canal, and he has 4 km of fencing to fence the remaining three sides.

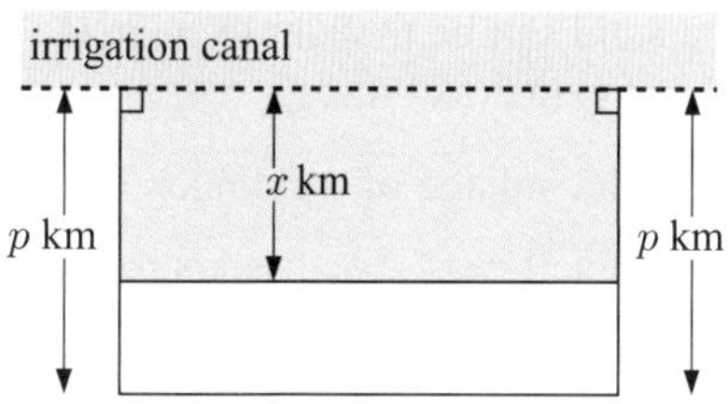

The farmer knows that the yield per unit of area is proportional to $\frac{1}{\sqrt{x+4}}$, where x is the distance in kilometres away from the canal.

Suppose the yield from the field is Y, the area of the orchard is A km^2, and the two vertical fences are p km long.

a Explain why $\frac{dY}{dA} = \frac{k}{\sqrt{x+4}}$ where k is a constant.

b Show that $\dfrac{dY}{dx} = \dfrac{k(4-2p)}{\sqrt{x+4}}$ by using the chain rule.

c Explain why $Y = \displaystyle\int_0^p \frac{k(4-2p)}{\sqrt{x+4}}\,dx$.

d Show that $Y = 4k(2-p)[\sqrt{p+4}-2]$.

e What dimensions should the orchard be, to maximise the yield?

E SOLIDS OF REVOLUTION

Consider the curve $y = f(x)$ for $a \leqslant x \leqslant b$.

If the shaded area below is **revolved about the x-axis** through $360°$ or 2π, a 3-dimensional solid will be formed. Such a solid is called a **solid of revolution**.

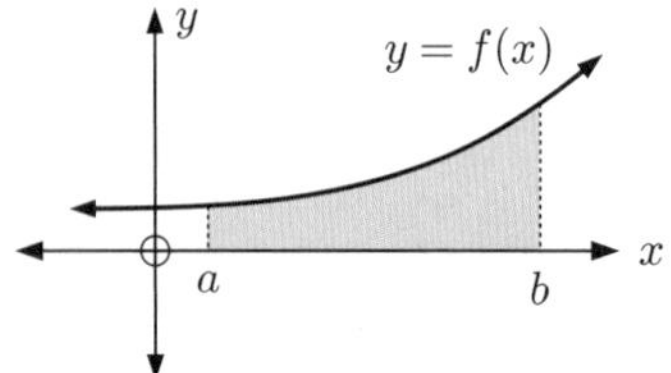

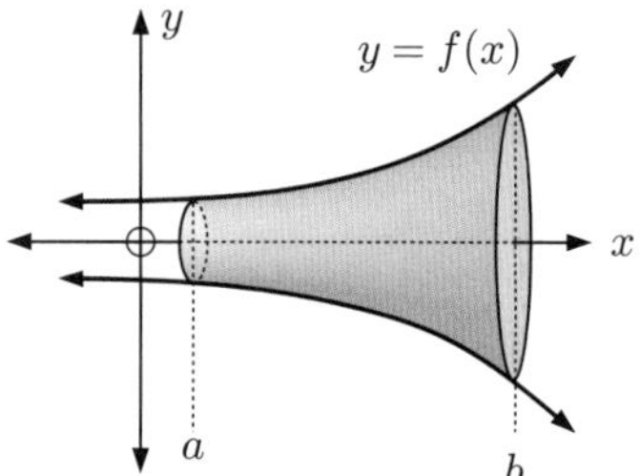

DEMO

For a curve defined by $x = g(y)$ on $a \leqslant y \leqslant b$, we can also define a solid of revolution. In this case the shaded area is **revolved about the y-axis** through $360°$ or 2π.

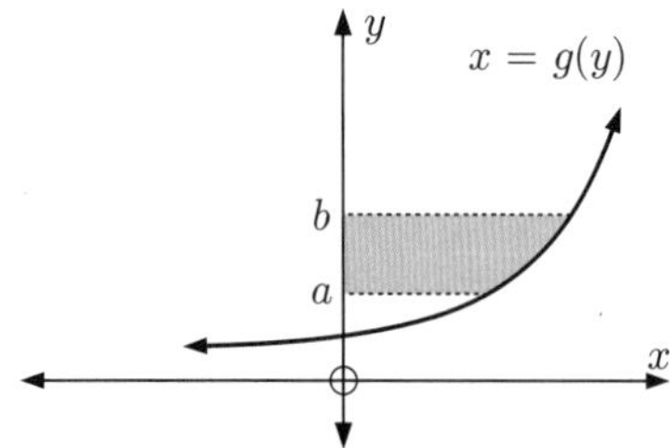

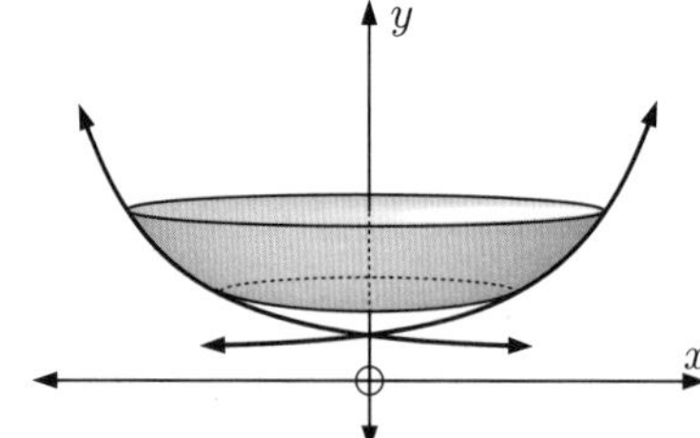

DEMO

VOLUME OF REVOLUTION

We can use integration to find volumes of revolution between $x = a$ and $x = b$.

The solid can be thought to be made up of an infinite number of thin cylindrical discs.

Since the volume of a cylinder $= \pi r^2 h$:

- the left-most disc has approximate volume $\pi[f(a)]^2 h$
- the right-most disc has approximate volume $\pi[f(b)]^2 h$
- the middle disc has approximate volume $\pi[f(x)]^2 h$.

As there are infinitely many discs, we let $h \to 0$.

We obtain $V = \displaystyle\lim_{h\to 0} \sum_{x=a}^{x=b} \pi[f(x)]^2\,h = \int_a^b \pi[f(x)]^2\,dx = \pi\int_a^b y^2\,dx.$

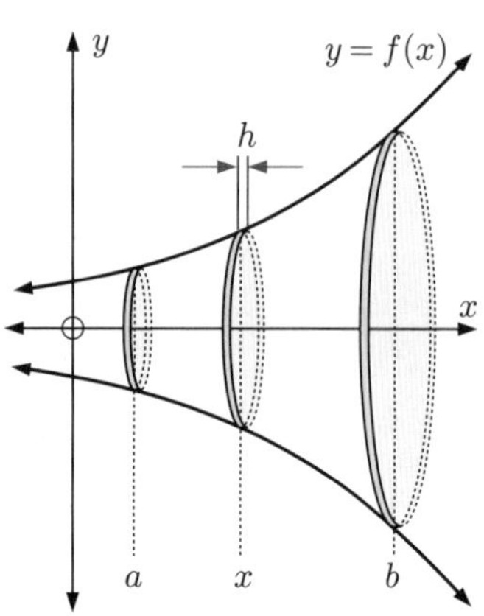

When the region enclosed by $y = f(x)$, the x-axis, and the vertical lines $x = a$ and $x = b$ is revolved through 2π about the x-axis to generate a solid, the volume of the solid is given by

Volume of revolution $= \pi \int_a^b y^2\, dx$.

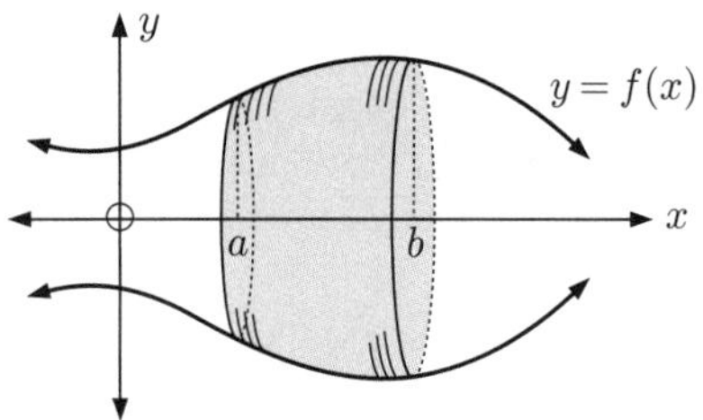

Using the same limit method we can derive a formula for the volume of a solid of revolution formed by rotation about the y-axis.

When the region enclosed by $x = f(y)$, the y-axis, and the horizontal lines $y = a$ and $y = b$ is revolved through 2π about the y-axis to generate a solid, the volume of the solid is given by

Volume of revolution $= \pi \int_a^b x^2\, dy$.

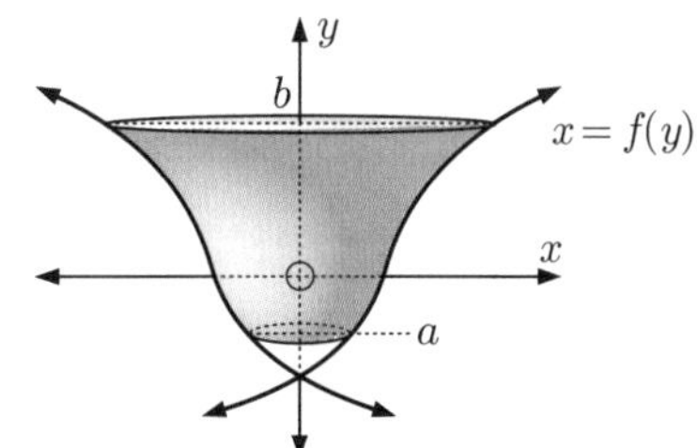

Example 12 **Self Tutor**

Use integration to find the volume of the solid generated when the line $y = x$ for $1 \leqslant x \leqslant 4$ is revolved through 2π around the x-axis.

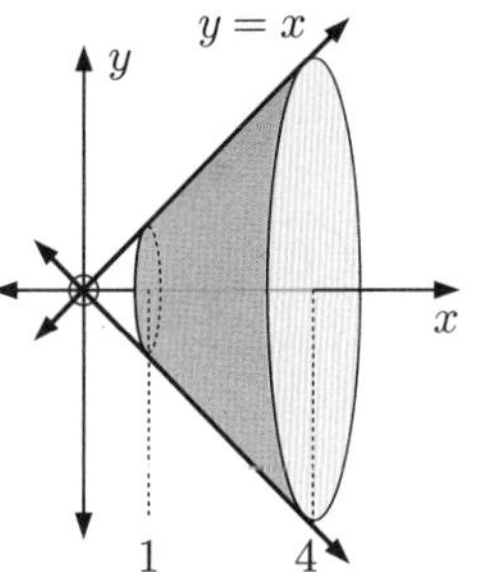

$$\begin{aligned}\text{Volume of revolution} &= \pi \int_a^b y^2\, dx \\ &= \pi \int_1^4 x^2\, dx \\ &= \pi \left[\frac{x^3}{3}\right]_1^4 \\ &= \pi\left(\tfrac{64}{3} - \tfrac{1}{3}\right) \\ &= 21\pi \text{ cubic units}\end{aligned}$$

The volume of a cone of height h and base radius r is

$V_{\text{cone}} = \frac{1}{3}\pi r^2 h$

So, in this example

$$\begin{aligned}V &= \tfrac{1}{3}\pi 4^2(4) - \tfrac{1}{3}\pi 1^2(1) \\ &= \tfrac{64\pi}{3} - \tfrac{\pi}{3} \\ &= 21\pi \quad \checkmark\end{aligned}$$

Example 13 **Self Tutor**

Find the volume of the solid formed when the graph of the function $y = x^2$ for $0 \leqslant x \leqslant 5$ is revolved through 2π about the x-axis.

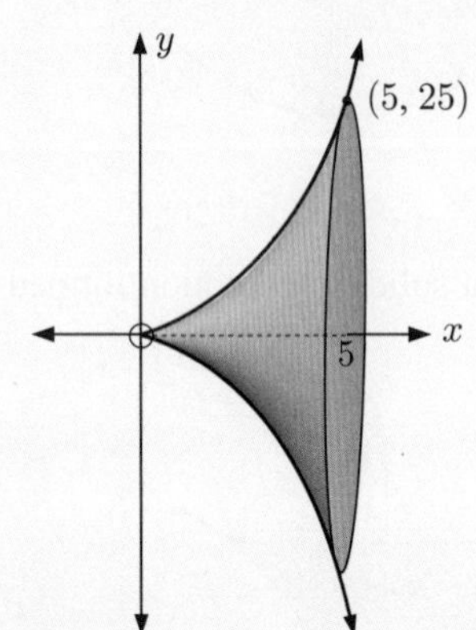

$$\text{Volume of revolution} = \pi \int_a^b y^2\, dx$$
$$= \pi \int_0^5 (x^2)^2\, dx$$
$$= \pi \int_0^5 x^4\, dx$$
$$= \pi \left[\frac{x^5}{5}\right]_0^5$$
$$= \pi(625 - 0)$$
$$= 625\pi \text{ cubic units}$$

EXERCISE 22E.1

1 Find the volume of the solid formed when the following are revolved through 2π about the x-axis:

a $y = 2x$ for $0 \leqslant x \leqslant 3$

b $y = \sqrt{x}$ for $0 \leqslant x \leqslant 4$

c $y = x^3$ for $1 \leqslant x \leqslant 2$

d $y = x^{\frac{3}{2}}$ for $1 \leqslant x \leqslant 4$

e $y = x^2$ for $2 \leqslant x \leqslant 4$

f $y = \sqrt{25 - x^2}$ for $0 \leqslant x \leqslant 5$

g $y = \dfrac{1}{x-1}$ for $2 \leqslant x \leqslant 3$

h $y = x + \dfrac{1}{x}$ for $1 \leqslant x \leqslant 3$

2 Use technology to find, correct to 3 significant figures, the volume of the solid of revolution formed when these functions are rotated through $360°$ about the x-axis:

a $y = \dfrac{x^3}{x^2+1}$ for $1 \leqslant x \leqslant 3$

b $y = e^{\sin x}$ for $0 \leqslant x \leqslant 2$.

3 Find the volume of revolution when the shaded region is revolved through 2π about the x-axis.

a
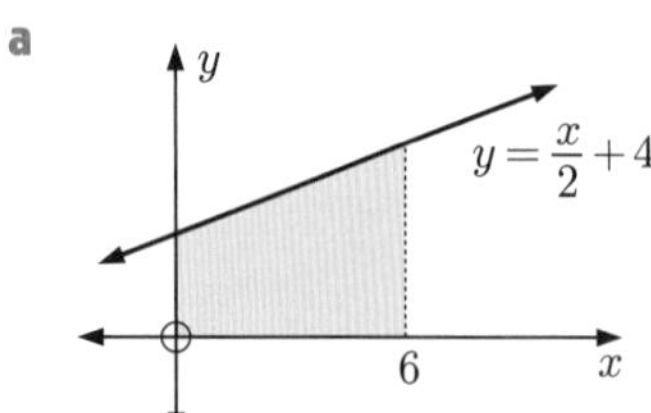

b
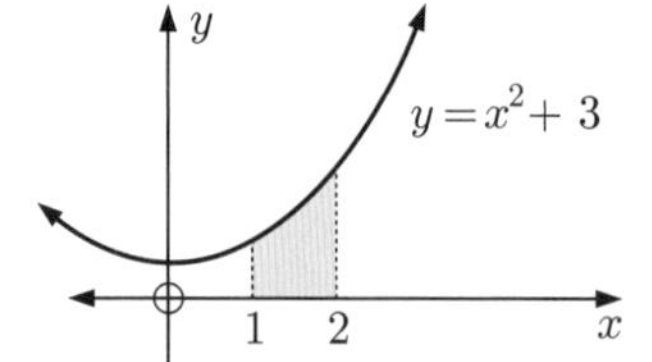

c
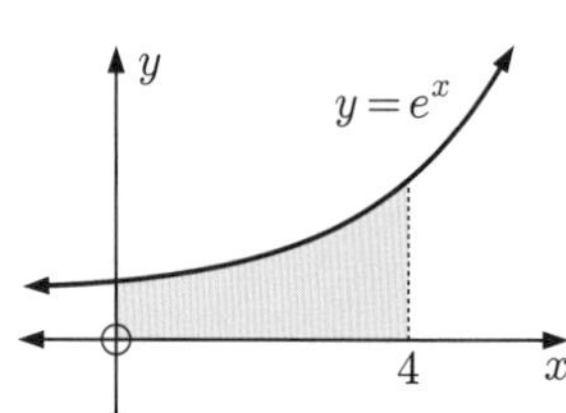

4 Answer the **Opening Problem** on page **672**.

5
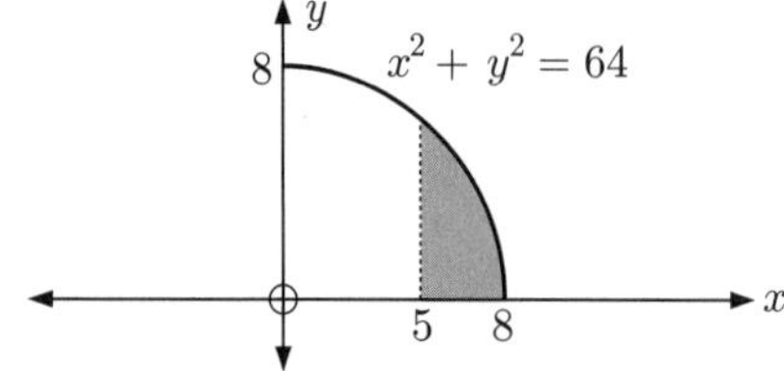

The shaded region is rotated through $360°$ about the x-axis.

a Find the volume of revolution.

b A hemispherical bowl of radius 8 cm contains water to a depth of 3 cm.
What is the volume of water?

6 a What is the name of the solid of revolution when the shaded region is revolved about the x-axis?

b Find the equation of the line segment [AB] in the form $y = ax + b$.

c Find a formula for the volume of the solid using $V = \pi \int_a^b y^2\,dx$.

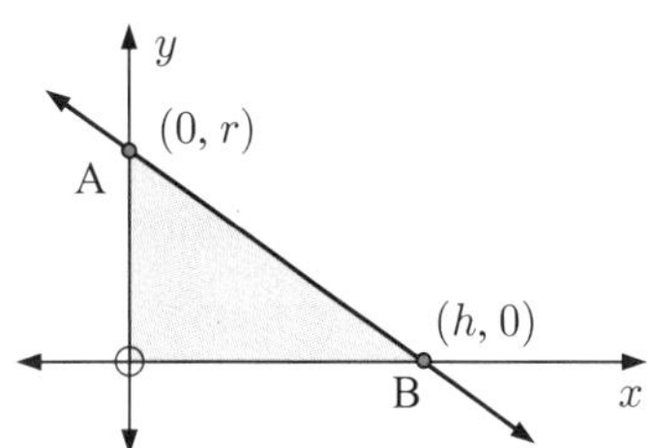

7 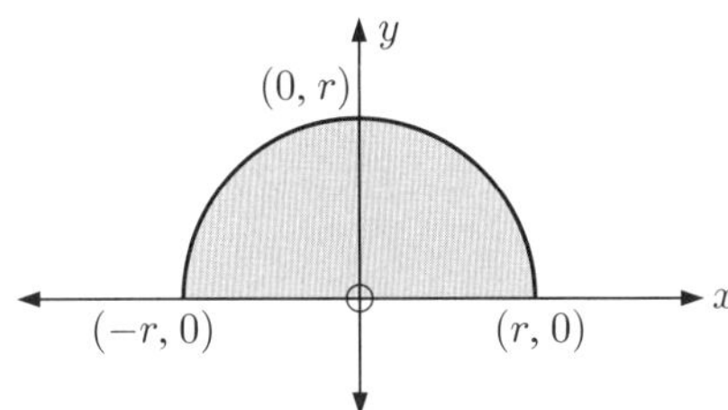

A circle with centre (0, 0) and radius r units has equation $x^2 + y^2 = r^2$.

a If the shaded region is revolved about the x-axis, what solid is formed?

b Use integration to show that the volume of revolution is $\frac{4}{3}\pi r^3$.

Example 14 **Self Tutor**

The graph of $y = \ln x$, $x \in [1, e]$ is revolved through 2π about the y-axis. Find the volume of revolution.

When $x = 1$, $y = 0$
When $x = e$, $y = 1$

$\therefore$ we rotate the function for $y \in [0, 1]$.

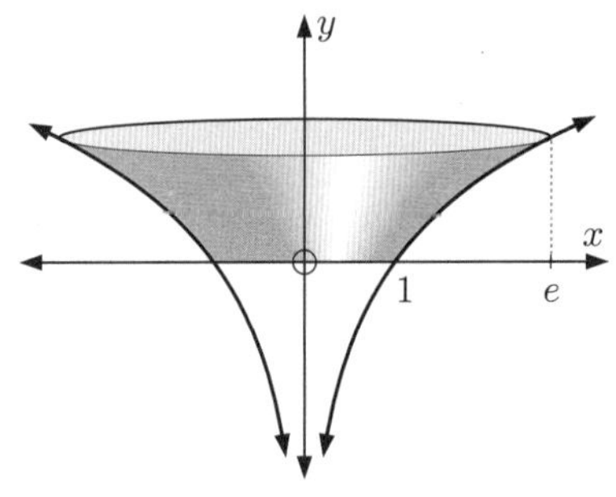

$\therefore$ Volume

$$= \pi \int_a^b x^2\,dy$$

$$= \pi \int_0^1 (e^y)^2\,dy$$

$$= \pi \int_0^1 e^{2y}\,dy$$

$$= \pi \left[\tfrac{1}{2}e^{2y}\right]_0^1$$

$$= \pi \left[\tfrac{1}{2}e^2 - \tfrac{1}{2}e^0\right]$$

$$= \tfrac{\pi}{2}(e^2 - 1) \text{ units}^3$$

8 Find the volumes of the solids formed when the following are revolved through 2π about the y-axis:

a $y = x^2$ between $y = 0$ and $y = 4$

b $y = \sqrt{x}$ between $y = 1$ and $y = 4$

c $y = \ln x$ between $y = 0$ and $y = 2$

d $y = \sqrt{x - 2}$ between $x = 2$ and $x = 11$

e $y = (x - 1)^3$ between $x = 1$ and $x = 3$.

9 Find exactly the volume of the solid of revolution formed by revolving the relation $\dfrac{x^2}{9} + \dfrac{y^2}{16} = 1$, $x \geqslant 0$ through $360°$ about the y-axis.

Example 15 — Self Tutor

One arch of $y = \sin x$ is revolved through $360°$ about the x-axis.

Using the identity $\sin^2 x = \frac{1}{2} - \frac{1}{2}\cos(2x)$ to help you, find the volume of revolution.

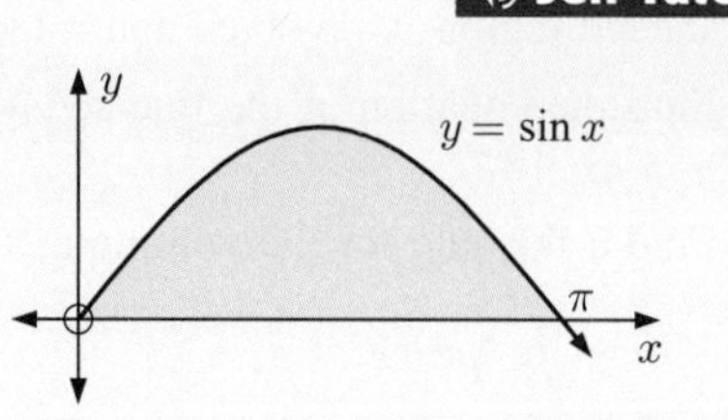

$$\begin{aligned}
\text{Volume} &= \pi \int_a^b y^2\, dx \\
&= \pi \int_0^{\pi} \sin^2 x\, dx \\
&= \pi \int_0^{\pi} \left[\tfrac{1}{2} - \tfrac{1}{2}\cos(2x)\right] dx \\
&= \pi \left[\frac{x}{2} - \tfrac{1}{2}\left(\tfrac{1}{2}\right)\sin(2x)\right]_0^{\pi} \\
&= \pi \left[\left(\tfrac{\pi}{2} - \tfrac{1}{4}\sin(2\pi)\right) - \left(0 - \tfrac{1}{4}\sin 0\right)\right] \\
&= \pi \times \tfrac{\pi}{2} \\
&= \tfrac{\pi^2}{2} \text{ units}^3
\end{aligned}$$

10 Find the volume of revolution when the following regions are revolved through 2π about the x-axis:

- **a** $y = \cos x$ for $0 \leqslant x \leqslant \frac{\pi}{2}$
- **b** $y = \cos(2x)$ for $0 \leqslant x \leqslant \frac{\pi}{4}$
- **c** $y = \sqrt{\sin x}$ for $0 \leqslant x \leqslant \pi$
- **d** $y = \dfrac{1}{\cos x}$ for $0 \leqslant x \leqslant \frac{\pi}{3}$
- **e** $y = \sec(3x)$ for $x \in [0, \frac{\pi}{12}]$
- **f** $y = \tan\left(\frac{x}{2}\right)$ for $x \in [0, \frac{\pi}{2}]$

11 **a** Sketch the graph of $y = \sin x + \cos x$ for $0 \leqslant x \leqslant \frac{\pi}{2}$.

b Hence, find the volume of revolution of the shape bounded by $y = \sin x + \cos x$, the x-axis, $x = 0$, and $x = \frac{\pi}{4}$ when it is rotated through 2π about the x-axis.

12 **a** Sketch the graph of $y = 4\sin(2x)$ for $0 \leqslant x \leqslant \frac{\pi}{4}$.

b Hence, find the volume of revolution of the shape bounded by $y = 4\sin(2x)$, the x-axis, $x = 0$, and $x = \frac{\pi}{4}$ when it is rotated through 2π about the x-axis.

VOLUMES FOR TWO DEFINING FUNCTIONS

Consider the circle with centre (0, 3) and radius 1 unit.

When this circle is revolved about the x-axis, we obtain a doughnut shape or *torus*.

DEMO

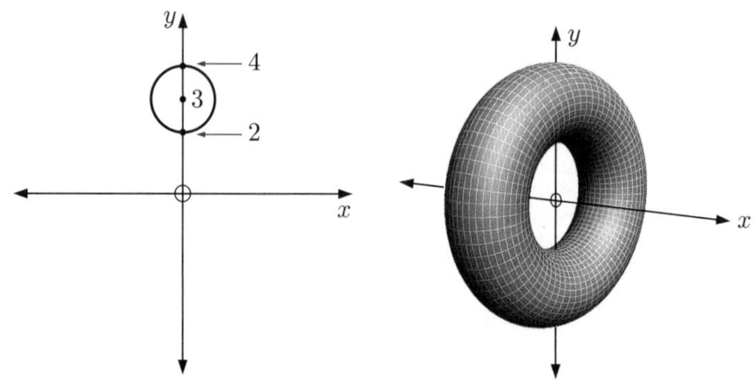

Suppose $y_U = f(x)$ and $y_L = g(x)$, where $f(x) \geqslant g(x)$ for all $a \leqslant x \leqslant b$.

If the region bounded by the upper function $y_U = f(x)$ and the lower function $y_L = g(x)$, and the lines $x = a$, $x = b$ is revolved about the x-axis, then its volume of revolution is given by:

$$V = \pi \int_a^b \left([f(x)]^2 - [g(x)]^2 \right) dx \quad \text{or} \quad V = \pi \int_a^b \left({y_U}^2 - {y_L}^2 \right) dx$$

Rotating about the x-axis gives

y
$y = f(x)$ or y_U
$y = g(x)$ or y_L
a b x

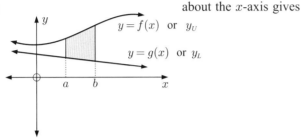

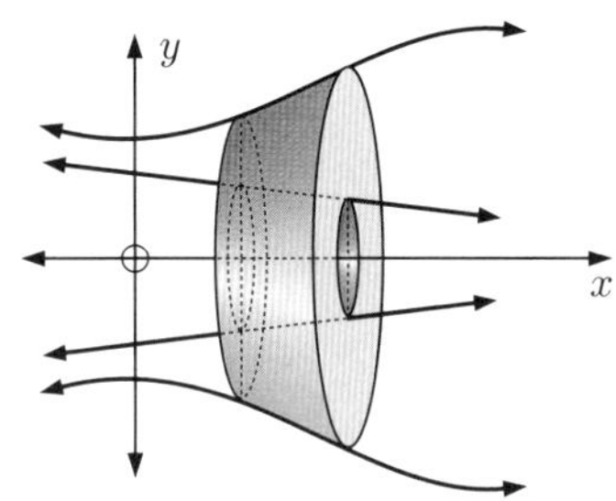

Example 16 — Self Tutor

Find the volume of revolution generated by revolving the region between $y = x^2$ and $y = \sqrt{x}$ about the x-axis.

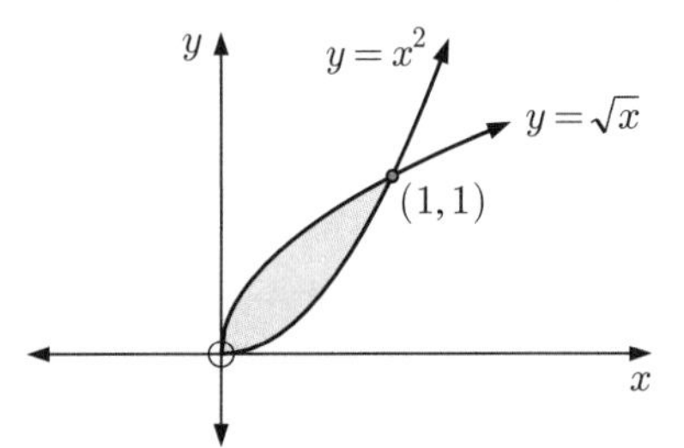

$$\begin{aligned}
\text{Volume} &= \pi \int_0^1 \left({y_U}^2 - {y_L}^2 \right) dx \\
&= \pi \int_0^1 \left(\left(\sqrt{x}\right)^2 - (x^2)^2 \right) dx \\
&= \pi \int_0^1 (x - x^4)\, dx \\
&= \pi \left[\frac{x^2}{2} - \frac{x^5}{5} \right]_0^1 \\
&= \pi \left(\left(\tfrac{1}{2} - \tfrac{1}{5} \right) - (0) \right) \\
&= \tfrac{3\pi}{10} \text{ units}^3
\end{aligned}$$

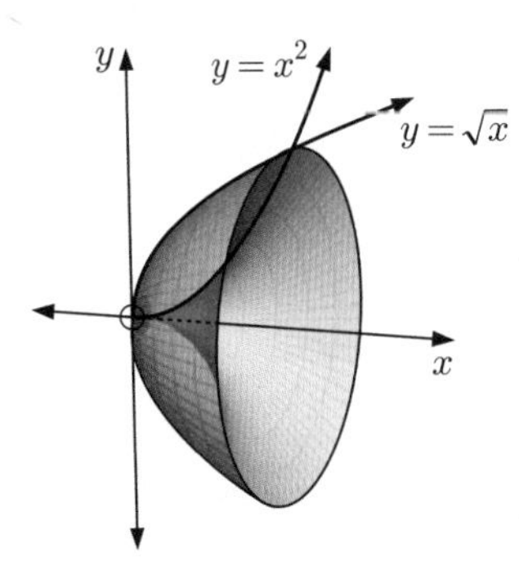

EXERCISE 22E.2

1 The shaded region between $y = 4 - x^2$ and $y = 3$ is revolved about the x-axis.

  a What are the coordinates of A and B?

  b Find the volume of revolution.

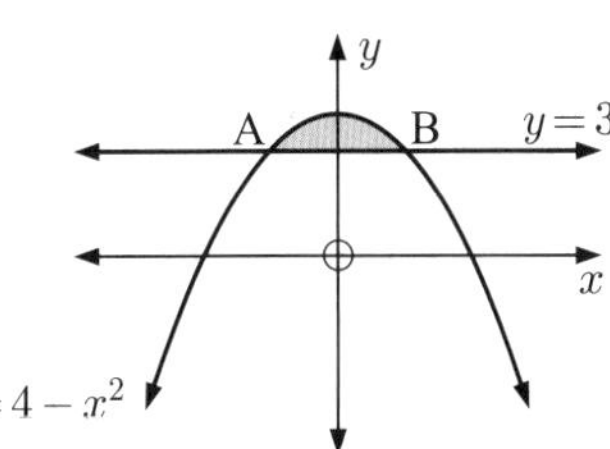

2 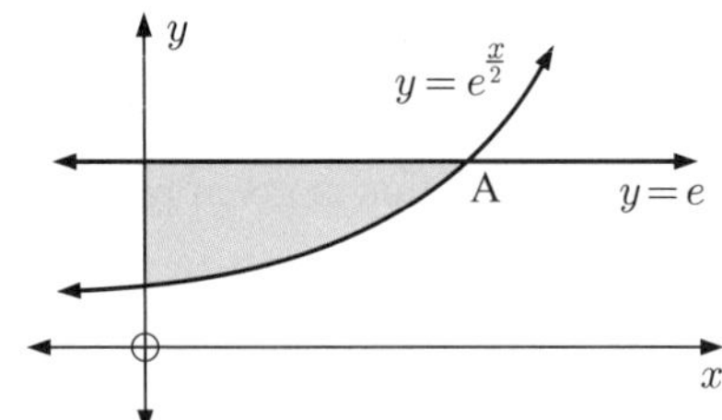

The shaded region is revolved about the x-axis.

a Find the coordinates of A.

b Find the volume of revolution.

3 The shaded region between $y = x$, $y = \dfrac{1}{x}$, and $x = 2$ is revolved about the x-axis.

a Find the coordinates of A.

b Find the volume of revolution.

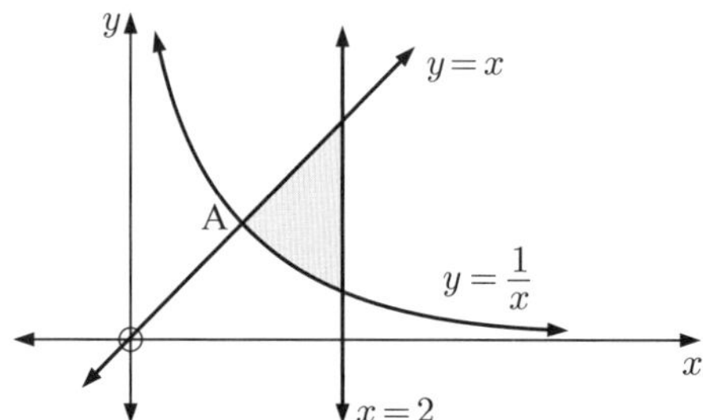

4 **a** Sketch $y = x^2 - 4x + 6$ and $x + y = 6$ on the same set of axes. Shade the area enclosed by these functions.

b Find exactly the volume of the solid of revolution generated by rotating the region enclosed by $y = x^2 - 4x + 6$ and $x + y = 6$ through $360°$ about the x-axis.

5 The shaded region is revolved about the x-axis.

a State the coordinates of A.

b Find the volume of revolution.

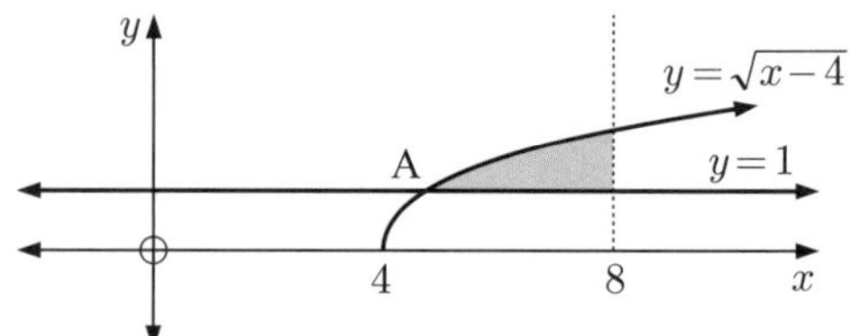

6 The illustrated circle has equation $x^2 + (y - 3)^2 = 4$.

a Show that $y = 3 \pm \sqrt{4 - x^2}$.

b Illustrate which part of the circle is represented by $y = 3 + \sqrt{4 - x^2}$ and which part by $y = 3 - \sqrt{4 - x^2}$.

c Find the volume of the solid of revolution generated by revolving the shaded region through 2π about the x-axis.

Hint: Substitute $x = 2\sin u$ to evaluate the integral. Use your calculator to check your answer.

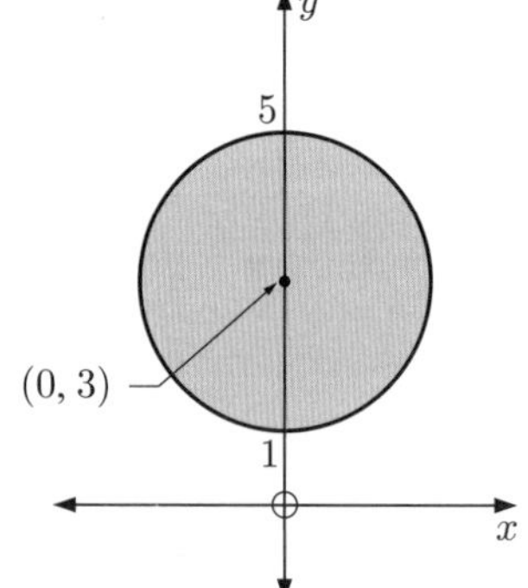

7 A chord of the circle with equation $x^2 + y^2 = r^2$ is parallel to the y-axis, and has length equal to the radius of the circle. A solid of revolution is generated by revolving the minor segment cut off by the chord through $360°$ about the y-axis.

Prove that the volume of the solid formed is given by $V = \dfrac{\pi r^3}{6}$.

8 A circle has equation $x^2 + y^2 = r^2$ where $r > 3$. A minor segment is cut off by a chord of length 6 units drawn parallel to the y-axis.

Show that the volume of the solid of revolution formed by rotating the segment through $360°$ about the y-axis, is independent of the value of r.

9 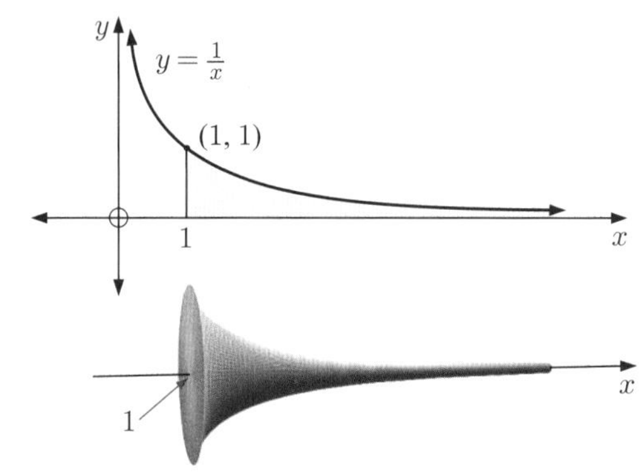

Prove that the shaded area from $x = 1$ to infinity is infinite whereas its volume of revolution is finite.

REVIEW SET 22A NON-CALCULATOR

1 Write an expression for the total shaded area.

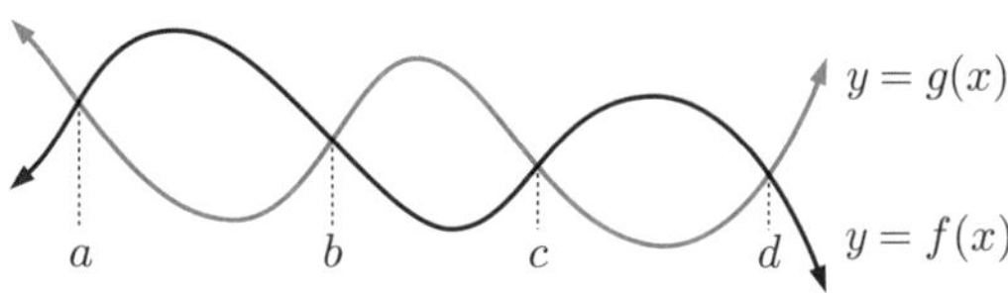

2 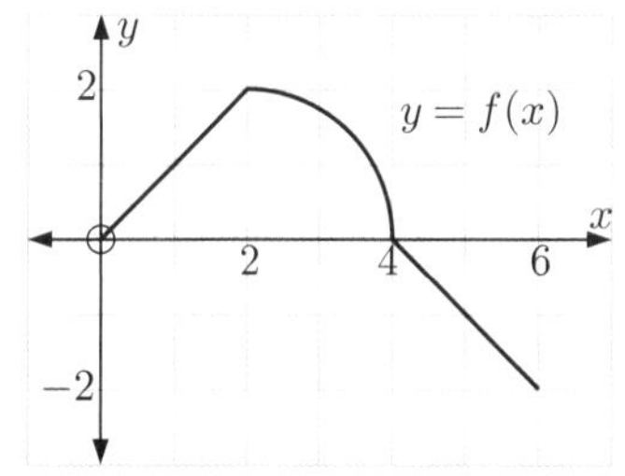

Find: **a** $\int_0^4 f(x)\,dx$ **b** $\int_4^6 f(x)\,dx$ **c** $\int_0^6 f(x)\,dx$

3 Does $\int_{-1}^{3} f(x)\,dx$ represent the area of the shaded region? Explain your answer briefly.

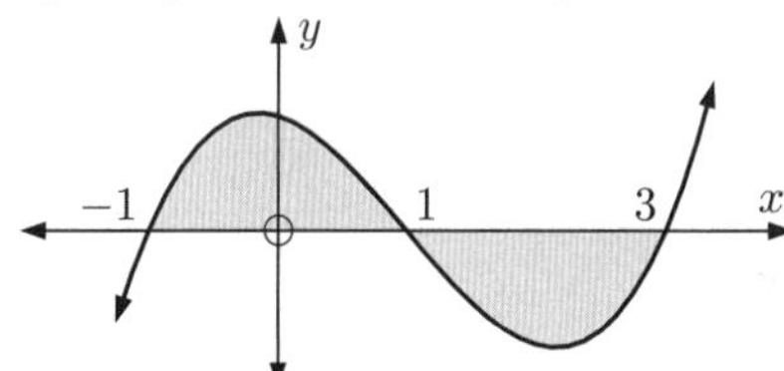

4 Determine k if the enclosed region has area $5\frac{1}{3}$ units^2.

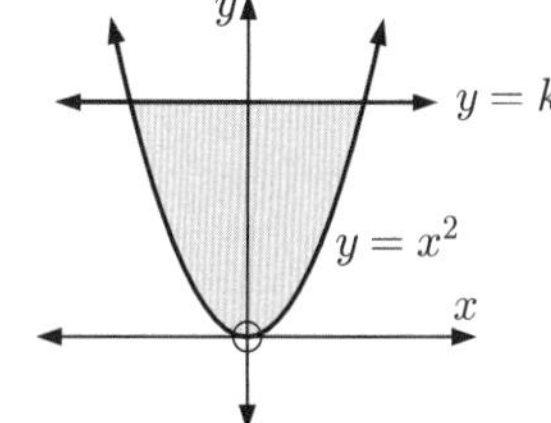

5 By appealing only to geometrical evidence, explain why

$$\int_0^1 e^x\,dx + \int_1^e \ln x\,dx = e.$$

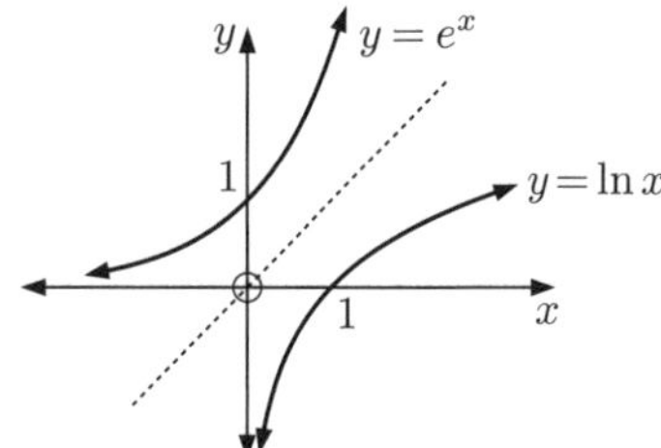

6 Find the area of the region enclosed by $y = x^2 + 4x + 1$ and $y = 3x + 3$.

7 Determine the area enclosed by the axes and $y = 4e^x - 1$.

8 A particle moves in a straight line with velocity $v(t) = t^2 - 6t + 8\ \text{m s}^{-1}$, for $t \geqslant 0$ seconds.

a Draw a sign diagram for $v(t)$.

b Describe what happens to the particle in the first 5 seconds of motion.

c After 5 seconds, how far is the particle from its original position?

d Find the total distance travelled in the first 5 seconds of motion.

9 Find the volume of the solid of revolution formed when the following are revolved through $360°$ about the x-axis:

a $y = x$ between $x = 4$ and $x = 10$

b $y = x + 1$ between $x = 4$ and $x = 10$

c $y = \sqrt{\sin x}$ between $x = 0$ and $x = \pi$

d $y = 1 - \cos x$ between $x = 0$ and $x = \frac{\pi}{2}$.

10 A cantilever of length L m is deflected by y m at the distance x m from the fixed end. The variables are connected by $\dfrac{d^2y}{dx^2} = k(L - x)^2$ where k is the proportionality constant.

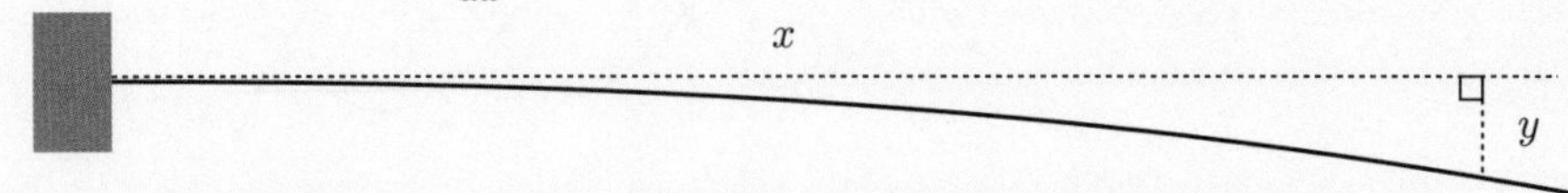

Find the greatest deflection of the cantilever in terms of k and L.

11

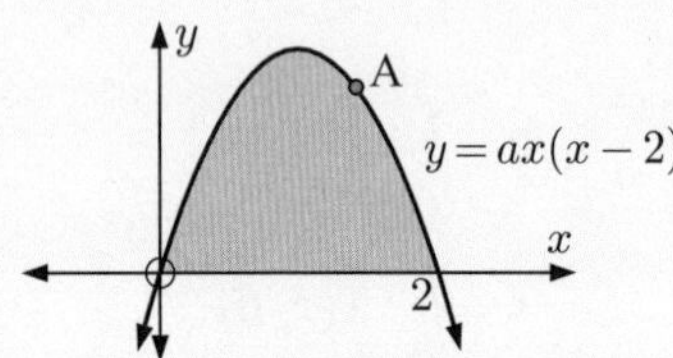

a Find a given that the shaded area is 4 units^2.

b Find the x-coordinate of A if [OA] divides the shaded region into equal areas.

12 Consider the given function $y = f(x)$, $0 \leqslant x \leqslant 6$.

a Show that A has equation $y = \sqrt{4x - x^2}$, $0 \leqslant x \leqslant 4$.

b Show that B has equation $y = -\sqrt{10x - x^2 - 24}$, $4 \leqslant x \leqslant 6$.

c Use the properties of circles to find $\displaystyle\int_0^4 \sqrt{4x - x^2}\, dx$ and $\displaystyle\int_4^6 -\sqrt{10x - x^2 - 24}\, dx$.

d Find $\displaystyle\int_0^6 f(x)\, dx$.

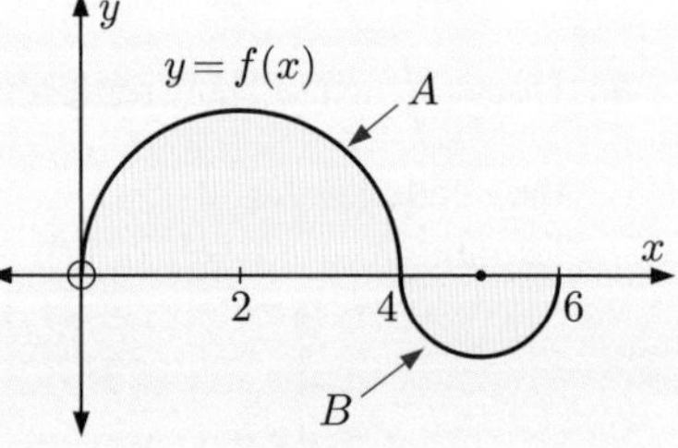

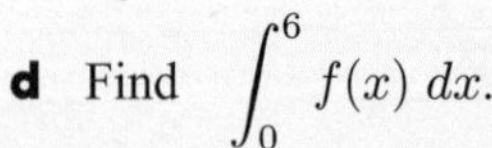

13 Find the volume of revolution when $y = \csc x$ is rotated through $360°$ about the x-axis for $\frac{\pi}{4} \leqslant x \leqslant \frac{3\pi}{4}$.

14

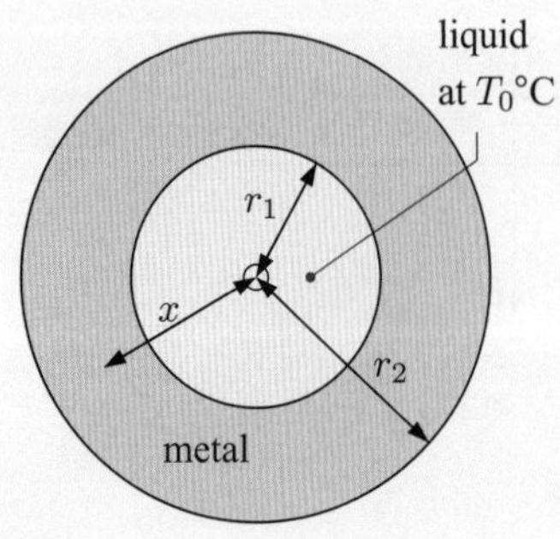

A metal tube has an annulus cross-section with radii r_1 and r_2 as shown.

Within the tube a liquid is maintained at temperature T_0 °C.

Within the metal, the temperature drops from inside to outside according to $\dfrac{dT}{dx} = \dfrac{k}{x}$ where k is a negative constant and x is the distance from the central axis.

Show that the outer surface has temperature $T_0 + k \ln\left(\dfrac{r_2}{r_1}\right)$.

15 Show that the volume of revolution generated by rotating the shaded region through $360°$ about the x-axis is $\frac{256}{15}\pi$ units^3.

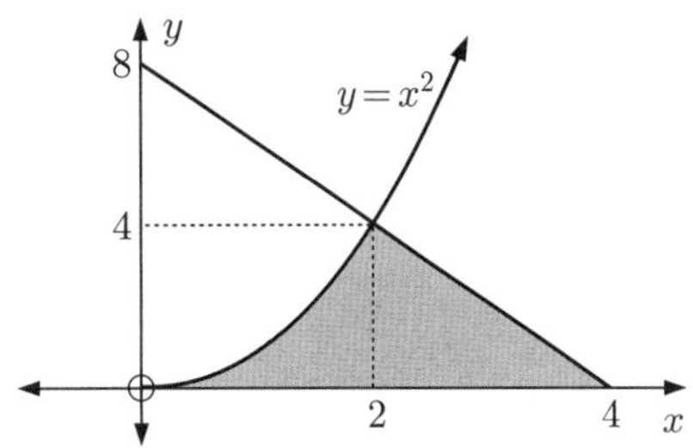

16 Find the volume of the solid of revolution formed when the following are rotated through 2π about the y-axis:

a $x = y^2$ between $y = 1$ and $y = 2$

b $y = \sqrt[3]{x^2}$ between $y = 2$ and $y = 3$.

REVIEW SET 22B — CALCULATOR

1 A particle moves in a straight line with velocity $v(t) = 2t - 3t^2$ m s^{-1}, for $t \geqslant 0$ seconds.

a Find a formula for the acceleration function $a(t)$.

b Find a formula for the displacement function $s(t)$.

c Find the change in displacement after two seconds.

2 Consider $f(x) = \dfrac{x}{1+x^2}$.

a Find the position and nature of all turning points of $y = f(x)$.

b Discuss $f(x)$ as $x \to \infty$ and as $x \to -\infty$.

c Sketch the graph of $y = f(x)$.

d Find, using technology, the area enclosed by $y = f(x)$, the x-axis, and the vertical line $x = -2$.

3 A particle moves in a straight line with velocity given by $v(t) = \sin t$ m s^{-1}, where $t \geqslant 0$ seconds. Find the total distance travelled by the particle in the first 4 seconds of motion.

4 A boat travelling in a straight line has its engine turned off at time $t = 0$. Its velocity at time t seconds thereafter is given by $v(t) = \dfrac{100}{(t+2)^2}$ m s^{-1}.

a Find the initial velocity of the boat, and its velocity after 3 seconds.

b Discuss $v(t)$ as $t \to \infty$.

c Sketch the graph of $v(t)$ against t.

d Find how long it takes for the boat to travel 30 metres from when the engine is turned off.

e Find the acceleration of the boat at any time t.

f Show that $\dfrac{dv}{dt} = -kv^{\frac{3}{2}}$, and find the value of the constant k.

5 The figure shows the graphs of $y = \cos(2x)$ and $y = e^{3x}$ for $-\pi \leqslant x \leqslant \frac{\pi}{2}$.

Find correct to 4 decimal places:

a the x-coordinates of their points of intersection

b the area of the shaded region.

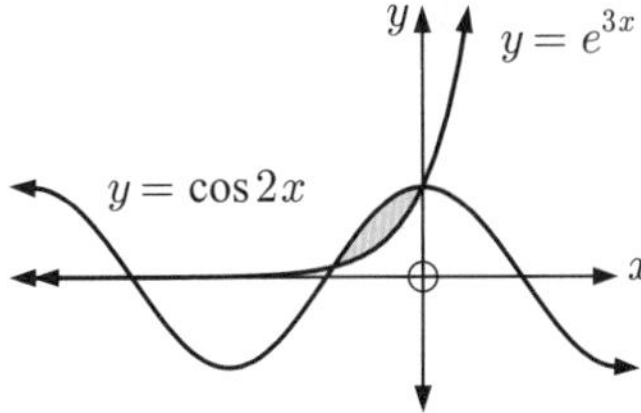

6 **a** Draw the graphs of $y^2 = x - 1$ and $y = x - 3$ on the same set of axes.

b Find the coordinates of the points where the graphs meet.

c Find the area enclosed by the graphs.

7 The shaded region has area $\frac{1}{2}$ unit^2. Find the value of m.

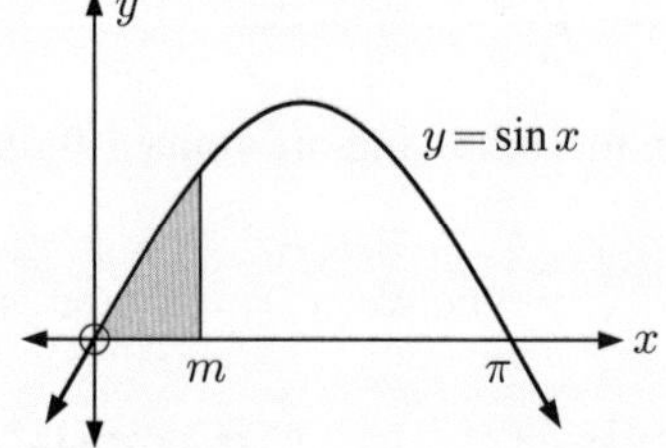

8 Find, correct to 4 decimal places:

a the value of a

b the area of the shaded region.

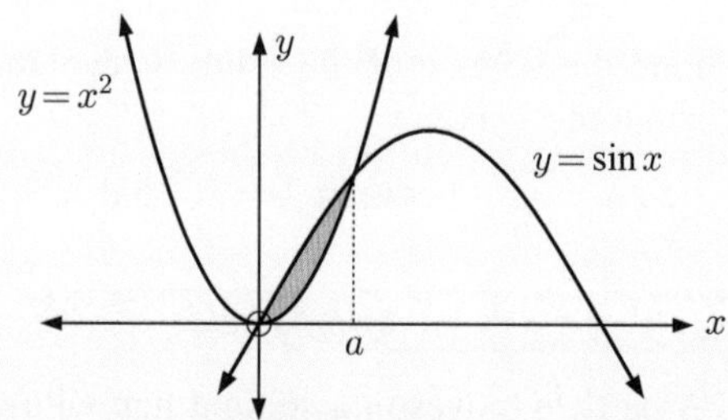

9 Find the volume of the solid of revolution formed when the shaded region is revolved through 360° about the x-axis:

a

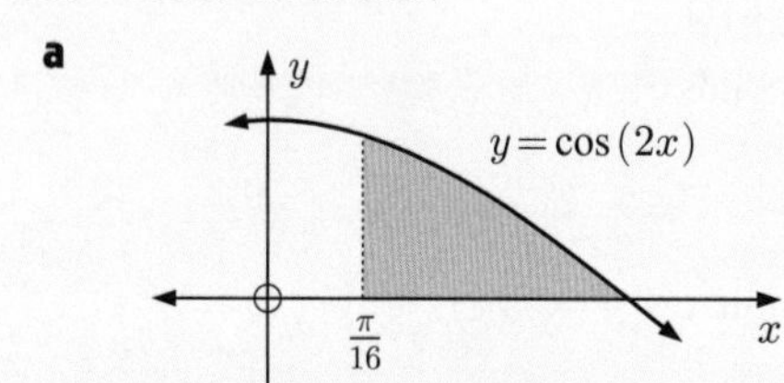

b

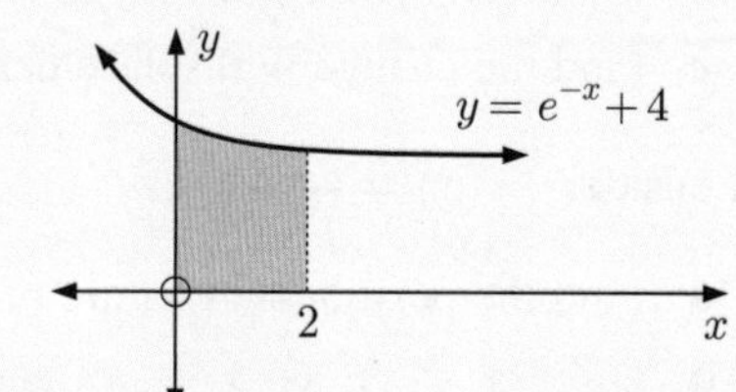

10 Find the area enclosed by $y = 2x^3 - 9x$ and $y = 3x^2 - 10$.

11 Consider $f(x) = 2 - \sec^2 x$ on $[-4, 4]$.

a Use technology to help sketch the graph of the function.

b Find the equations of the function's vertical asymptotes.

c Find the axes intercepts.

d Find the area of the region bounded by one arch of the function and the x-axis.

12 Determine the area enclosed by the y-axis, the line $y = 3$, and the curve $x = \ln\left(\frac{y+3}{2}\right)$.

13 **a** The graph of $y = \sin x$ is drawn alongside for $0 \leqslant x \leqslant \pi$.

Use the graph to explain why

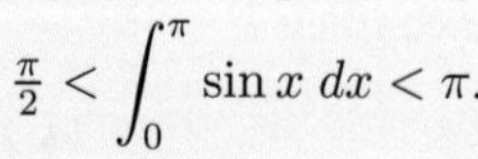

$$\frac{\pi}{2} < \int_0^{\pi} \sin x \, dx < \pi.$$

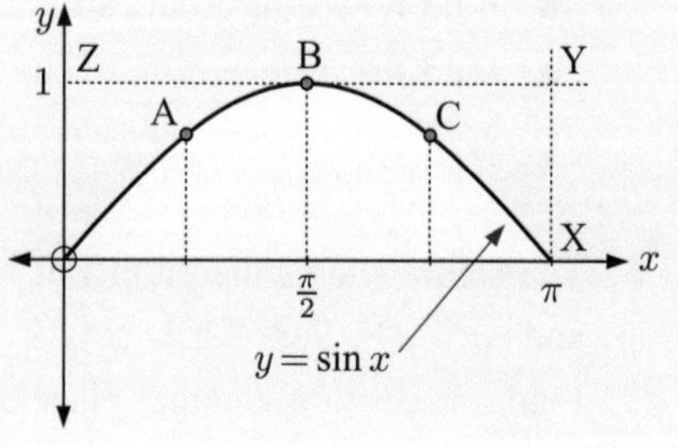

b A is $\left(\frac{\pi}{4}, \frac{1}{\sqrt{2}}\right)$ and C is $\left(\frac{3\pi}{4}, \frac{1}{\sqrt{2}}\right)$.

Use the diagram to show that the area under the illustrated arch of $y = \sin x$ is greater than $\frac{\pi}{4}(1 + \sqrt{2})$ units^2.

c Find exactly the area under one arch of $y = \sin x$.

14 **a** Sketch the region bounded by $y = x^3 + 2$, the y-axis, and the horizontal lines $y = 3$ and $y = 6$.

b Write x in the form $f(y)$.

c Find the area of the region graphed in **a**.

15 Find the area of the region enclosed by $y = x^3 + x^2 + 2x + 6$ and $y = 7x^2 - x - 4$.

16 Without actually integrating $\sin^3 x$, prove that $\int_0^\pi \sin^3 x\, dx < 4$.

Hint: Graph $y = \sin^3 x$ for $0 \leqslant x \leqslant \pi$.

REVIEW SET 22C

1 At time $t = 0$ a particle passes through the origin with velocity 27 cm s^{-1}. Its acceleration t seconds later is $6t - 30$ cm s^{-2}.

a Write an expression for the particle's velocity.

b Write an integral which represents the displacement from the origin after 6 seconds. Hence calculate this displacement.

2 **a** Sketch the graphs of $y = \sin^2 x$ and $y = \sin x$ on the same set of axes for $0 \leqslant x \leqslant \pi$.

b Find the exact value of the area enclosed by these curves for $0 \leqslant x \leqslant \frac{\pi}{2}$.

3 Find a given that the area of the region between $y = e^x$ and the x-axis from $x = 0$ to $x = a$ is 2 units^2. Hence determine b such that the area of the region from $x = a$ to $x = b$ is also 2 units^2.

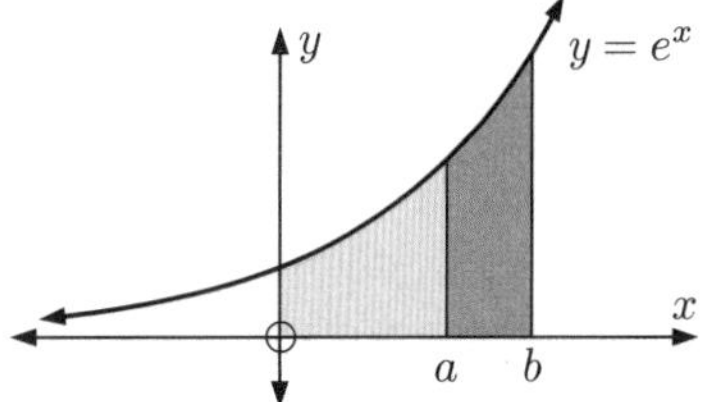

4 Determine the area of the region enclosed by $y = x$, $y = \sin x$, and $x = \pi$.

5 Determine the area enclosed by $y = \frac{2}{\pi}x$ and $y = \sin x$.

6 OABC is a rectangle and the two shaded regions are equal in area. Find k.

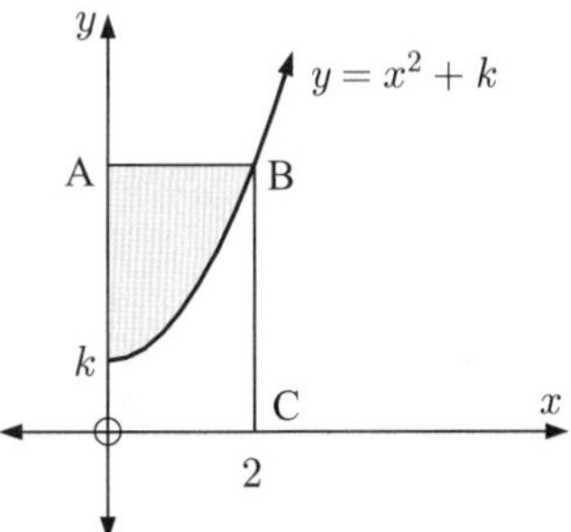

7 Determine $\int_0^2 \sqrt{4 - x^2}\, dx$ using graphical evidence only.

8 Find the volume of the solid of revolution formed when the following are rotated about the x-axis:

a $y = \sin x$ between $x = 0$ and $x = \pi$

b $y = \sqrt{9 - x^2}$ between $x = 0$ and $x = 3$.

9 Determine the total finite area enclosed by $y = x^3$ and $y = 7x^2 - 10x$.

10 Find the exact area of the region enclosed by $y = \tan x$, the x-axis, and the vertical line $x = \frac{\pi}{3}$.

11 The current $I(t)$ milliamps in a circuit falls off in accordance with $\dfrac{dI}{dt} = \dfrac{-100}{t^2}$ where t is the time in seconds, $t \geqslant 0.2$.
When $t = 2$ seconds, the current is measured to be 150 milliamps.

a Find a formula for the current at any time $t \geqslant 0.2$.

b Hence find:

i the current after 20 seconds **ii** what happens to the current as $t \to \infty$.

12 The ellipse shown has equation $\dfrac{x^2}{16} + \dfrac{y^2}{4} = 1$.

a Copy the graph and shade the area represented by

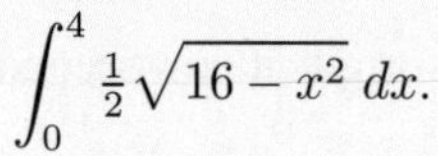

$\displaystyle\int_0^4 \tfrac{1}{2}\sqrt{16 - x^2}\, dx$.

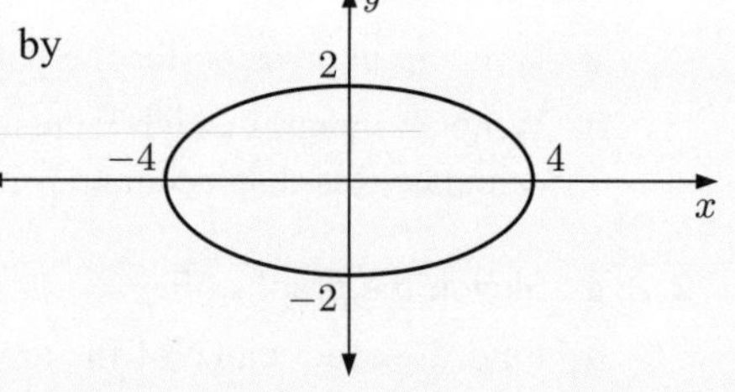

b Explain using the graph why

$$8 < \int_0^4 \sqrt{16 - x^2}\, dx < 16.$$

13 **a** Use $V = \frac{1}{3}\pi r^2 h$ to find the volume of this cone.

b Check your answer to **a** by defining an appropriate solid of revolution and calculating its volume.

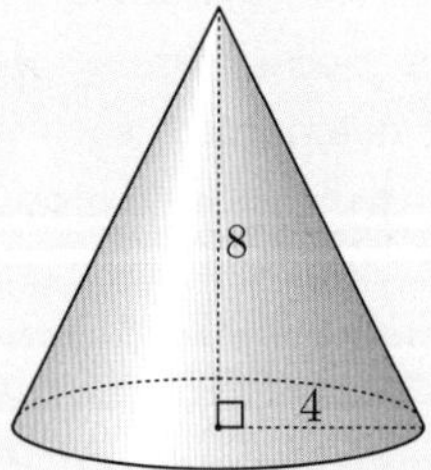

14 **a** Find $\dfrac{d}{dx}[\ln(\tan x + \sec x)]$ and hence find $\displaystyle\int \sec x\, dx$.

b Consider $x \mapsto \sec(2x)$ for $x \in [0, \pi]$.

i Sketch the graph of the function on the given domain.

ii Find the area of the region bounded by $y = \sec(2x)$, the y-axis, and the line $y = 3$.

15

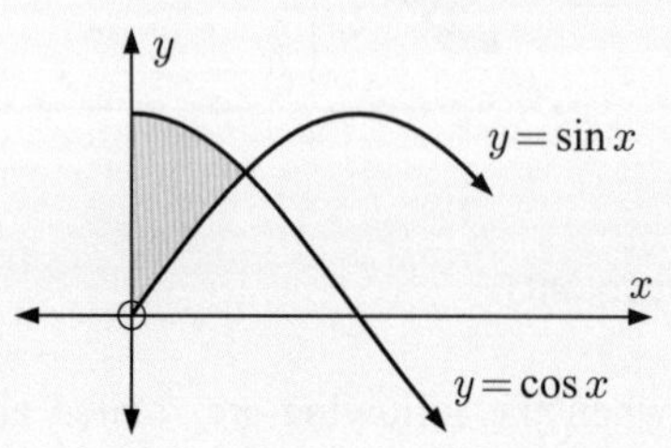

Find the volume of the solid of revolution obtained when the shaded region is rotated through 360° about the x-axis:

16 Find the volume enclosed when $y = x^3$, $0 \leqslant y \leqslant 8$ is revolved through 2π about the y-axis.

Chapter 23

Descriptive statistics

Syllabus reference: 5.1

Contents:

OPENING PROBLEM

A farmer wanted to investigate the effect of a new organic fertiliser on his crops of peas. He divided a small garden into two equal plots and planted many peas in each. Both plots were treated the same except that fertiliser was used on one but not the other.

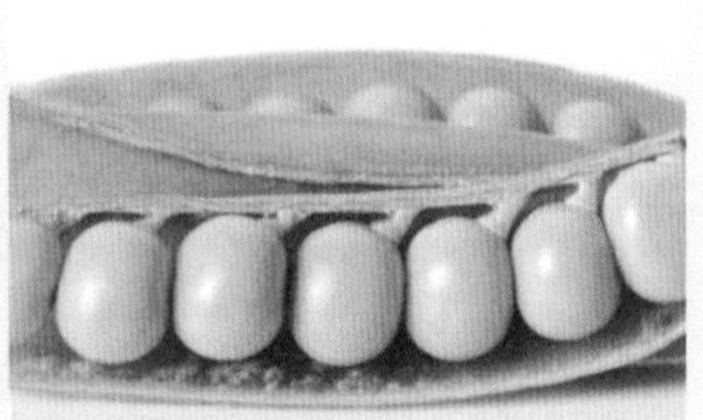

A random sample of 150 pods was harvested from each plot at the same time, and the number of peas in each pod was counted. The results were:

Without fertiliser

4 6 5 6 5 6 4 6 4 9 5 3 6 8 5 4 6 8 6 5 6 7 4 6 5 2 8 6 5 6 5 5 5 4 4 4 6 7 5 6 7 5 5 6
4 8 5 3 7 5 3 6 4 7 5 6 5 7 5 7 6 7 5 4 7 5 5 5 6 6 5 6 7 5 8 6 8 6 7 6 6 3 7 6 8 3 3 4
4 7 6 5 6 4 5 7 3 7 7 6 7 7 4 6 6 5 6 7 6 3 4 6 6 3 7 6 7 6 8 6 6 6 6 4 7 6 6 5 3 8 6 7
6 8 6 7 6 6 6 8 4 4 8 6 6 2 6 5 7 3

With fertiliser

6 7 7 4 9 5 5 5 8 9 8 9 7 7 5 8 7 6 6 7 9 7 7 7 8 9 3 7 4 8 5 10 8 6 7 6 7 5 6 8 7 9 4
4 9 6 8 5 8 7 7 4 7 8 10 6 10 7 7 7 9 7 7 8 6 8 6 8 7 4 8 6 8 7 3 8 7 6 9 7 6 9 7 6 8 3
9 5 7 6 8 7 9 7 8 4 8 7 7 7 6 6 8 6 3 8 5 8 7 6 7 4 9 6 6 6 8 4 7 8 9 7 7 4 7 5 7 4 7 6
4 6 7 7 6 7 8 7 6 6 7 8 6 7 10 5 13 4 7 11

Things to think about:

- Can you state clearly the problem that the farmer wants to solve?
- How has the farmer tried to make a fair comparison?
- How could the farmer make sure that his selection was random?
- What is the best way of organising this data?
- What are suitable methods of displaying the data?
- Are there any abnormally high or low results and how should they be treated?
- How can we best describe the most typical pod size?
- How can we best describe the spread of possible pod sizes?
- Can the farmer make a reasonable conclusion from his investigation?

A KEY STATISTICAL CONCEPTS

In statistics we collect information about a group of individuals, then analyse this information to draw conclusions about those individuals.

You should already be familiar with these words which are commonly used in statistics:

- **Population** - an entire collection of individuals about which we want to draw conclusions
- **Census** - the collection of information from the **whole population**
- **Sample** - a subset of the population which should be chosen at **random** to avoid **bias** in the results
- **Survey** - the collection of information from a **sample**

- **Data** - information about individuals in a population
- **Categorical variable** - describes a particular quality or characteristic which can be divided into categories
- **Numerical variable** - describes a characteristic which has a numerical value that can be counted or measured
- **Parameter** - a numerical quantity measuring some aspect of a population
- **Statistic** - a quantity calculated from data gathered from a sample, usually used to estimate a population parameter
- **Distribution** - the pattern of variation of data, which may be described as:

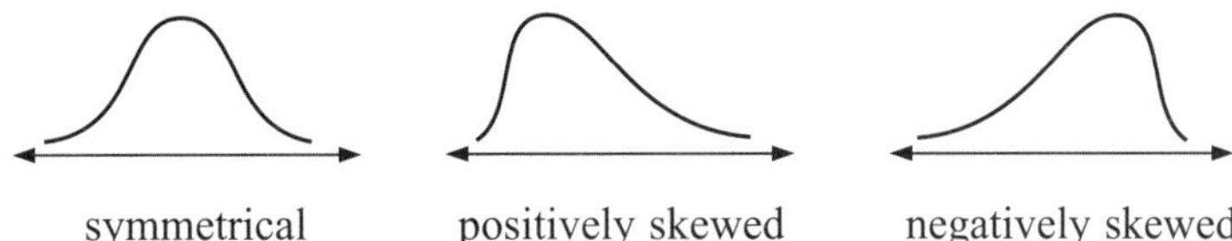

symmetrical   positively skewed   negatively skewed

- **Outliers** - data values that are either much larger or much smaller than the general body of data; they should be included in analysis *unless* they are the result of human or other known error

NUMERICAL VARIABLES

There are two types of numerical variable we will deal with in this course:

A **discrete numerical variable** takes exact number values and is often a result of **counting**.

Examples of discrete numerical variables are:

- *The number of people in a car*: the variable could take the values 1, 2, 3, ...
- *The score out of 20 for a test*: the variable could take the values 0, 1, 2, 3, ..., 20.

A **continuous variable** takes numerical values within a certain continuous range. It is usually a result of **measuring**.

Examples of continuous numerical variables are:

- *The height of Year 11 students*: the variable can take any value from about 140 cm to 200 cm.
- *The speed of cars on a stretch of highway*: the variable can take any value from 0 $\text{km}\,\text{h}^{-1}$ to the fastest speed that a car can travel, but is most likely to be in the range 60 $\text{km}\,\text{h}^{-1}$ to 160 $\text{km}\,\text{h}^{-1}$.

FREQUENCY TABLES

One of the simplest ways to organise data is using a frequency table.

For example, consider the data set:

1 3 1 2 4   2 4 1 5 3   1 3 2 2 4
1 3 4 1 2   3 2 4 1 3   2 1 2 5 2

A **tally** is used to count the number of 1s, 2s, 3s, and so on. As we read the data from left to right, we place a vertical stroke in the tally column. We use 𝍸 to represent 5.

The **frequency** column summarises the number of each particular data value.

The **relative frequency** column measures the percentage of the total number of data values that are in each group.

Value	*Tally*	*Frequency* (f)	*% relative frequency*
1	~~IIII~~ III	8	26.7 ← $\frac{8}{30} \times 100\%$
2	~~IIII~~ IIII	9	30.0
3	~~IIII~~ I	6	20.0
4	~~IIII~~	5	16.7
5	II	2	6.7
	Totals	30	100

DISPLAY OF NUMERICAL DATA

From previous courses you should be familiar with **column graphs** used to display **discrete** numerical variables.

When data is recorded for a **continuous** variable there are likely to be many different values. We organise the data in a frequency table by grouping it into **class intervals** of **equal width**.

A special type of graph called a **frequency histogram** or just **histogram** is used to display the data. This is similar to a column graph but, to account for the continuous nature of the variable, a number line is used for the horizontal axis and the 'columns' are joined together.

Column graphs and frequency histograms both have the following features:

- The frequency of occurrence is on the vertical axis.
- The range of scores is on the horizontal axis.
- The column widths are equal and the column height represents frequency.

The **modal class**, or class of values that appears most often, is easy to identify from the highest column of the frequency histogram.

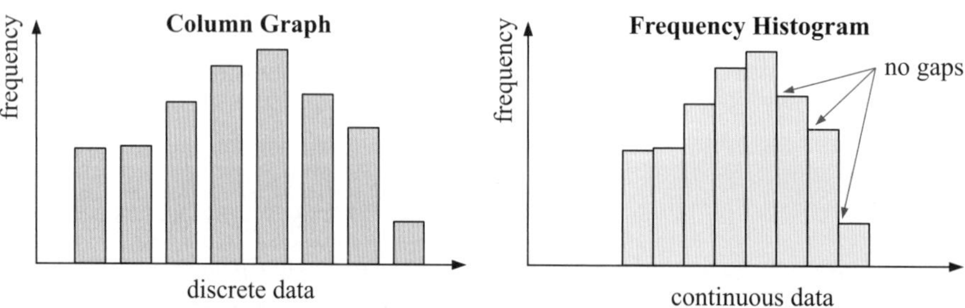

CASE STUDY — DRIVING A GOLF BALL

While attending a golf championship, I measured how far Ethan, a professional golfer, hit 30 drives on the practice fairway. The results are given below in metres:

244.6	245.1	248.0	248.8	250.0	251.1
251.2	253.9	254.5	254.6	255.9	257.0
260.6	262.8	262.9	263.1	263.2	264.3
264.4	265.0	265.5	265.6	266.5	267.4
269.7	270.5	270.7	272.9	275.6	277.5

To organise the data, we sort it into **groups** in a frequency table.

When forming groups, we find the lowest and highest values, and then choose a group width so that there are about 6 to 12 groups. In this case the lowest value is 244.6 m and the highest is 277.5 m. If we choose a group width of 5 m, we obtain eight groups of equal width between values 240 m and 280 m, which cover all of the data values.

Suppose d is the length of a drive. The first group $240 \leqslant d < 245$ includes any data value which is at least 240 m but less than 245 m. The group $260 \leqslant d < 265$ includes data which is at least 260 m but < 265 m. We use this technique to create eight groups into which all data values will fall.

Ethan's 30 drives

Distance (m)	*Tally*	*Frequency* (f)	% *relative frequency*
$240 \leqslant d < 245$	|	1	3.3 ← $\approx \frac{1}{30} \times 100$
$245 \leqslant d < 250$	|||	3	10.0 ← $= \frac{3}{30} \times 100$
$250 \leqslant d < 255$	~~||||~~ |	6	20.0
$255 \leqslant d < 260$	||	2	6.7
$260 \leqslant d < 265$	~~||||~~ ||	7	23.3
$265 \leqslant d < 270$	~~||||~~ |	6	20.0
$270 \leqslant d < 275$	|||	3	10.0
$275 \leqslant d < 280$	||	2	6.7
	Totals	30	100.0

From this table we can draw both a **frequency histogram** and a **relative frequency histogram**:

Frequency histogram

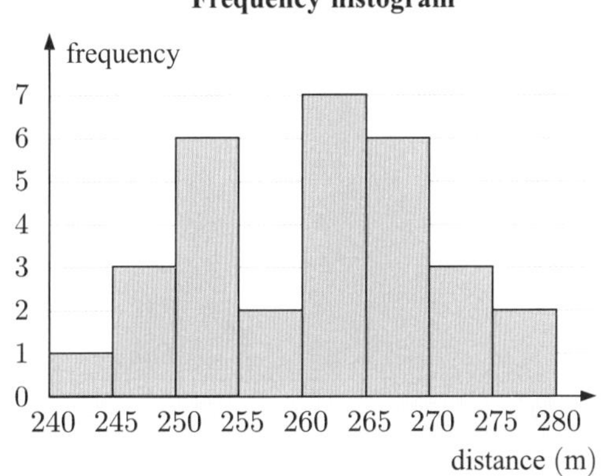

Relative frequency histogram

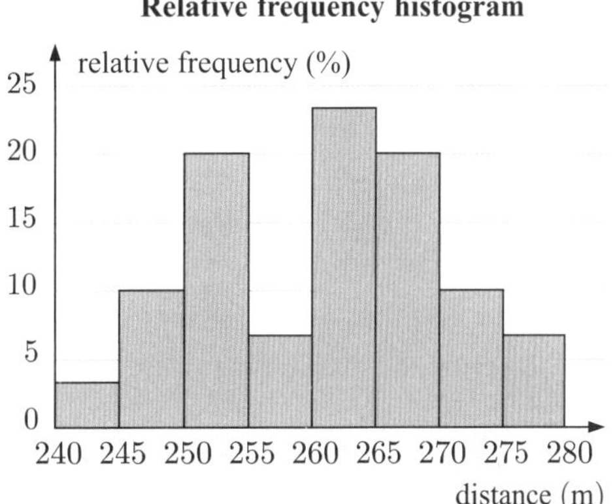

We can see that the modal class is $260 \leqslant d < 265$.

The advantage of the relative frequency histogram is that we can easily compare it with other distributions with different numbers of data values. Using percentages allows for a fair comparison.

Each graph should have a title and its axes should be labelled.

Example 1 | **Self Tutor**

A sample of 20 juvenile lobsters is randomly selected from a tank containing several hundred. Each lobster is measured for length (in cm) and the results are:

4.9, 5.6, 7.2, 6.7, 3.1, 4.6, 6.0, 5.0, 3.7, 7.3,
6.0, 5.4, 4.2, 6.6, 4.7, 5.8, 4.4, 3.6, 4.2, 5.4

a Organise the data using a frequency table, and hence graph the data.

b State the modal class for the data.

c Describe the distribution of the data.

a The variable 'the length of a lobster' is *continuous* even though lengths have been rounded to the nearest mm.

The shortest length is 3.1 cm and the longest is 7.3 cm, so we will use class intervals of length 1 cm.

Length l (cm)	*Frequency*
$3 \leqslant l < 4$	3
$4 \leqslant l < 5$	6
$5 \leqslant l < 6$	5
$6 \leqslant l < 7$	4
$7 \leqslant l < 8$	2

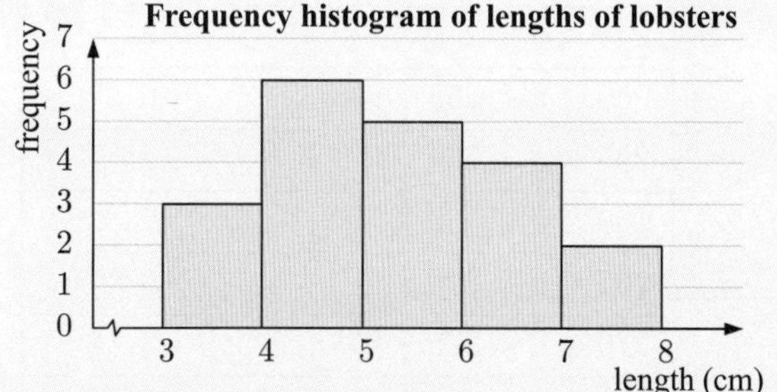

b The modal class is $4 \leqslant l < 5$ cm as this occurred most frequently.

c The data is positively skewed.

EXERCISE 23A

1 A frequency table for the heights of a basketball squad is given alongside.

Height (cm)	*Frequency*
$170 \leqslant H < 175$	1
$175 \leqslant H < 180$	8
$180 \leqslant H < 185$	9
$185 \leqslant H < 190$	11
$190 \leqslant H < 195$	9
$195 \leqslant H < 200$	3
$200 \leqslant H < 205$	3

a Explain why 'height' is a continuous variable.

b Construct a frequency histogram for the data. The axes should be clearly marked and labelled, and the graph should have a title.

c What is the modal class? Explain what this means.

d Describe the distribution of the data.

2 A school has conducted a survey of 60 students to investigate the time it takes for them to travel to school. The following data gives the travel times to the nearest minute:

12 15 16 8 10 17 25 34 42 18 24 18 45 33 38
45 40 3 20 12 10 10 27 16 37 45 15 16 26 32
35 8 14 18 15 27 19 32 6 12 14 20 10 16 14
28 31 21 25 8 32 46 14 15 20 18 8 10 25 22

a Is travel time a discrete or continuous variable?

b Copy and complete the frequency table shown.

c Describe the distribution of the data.

d What is the travelling time modal class?

Time (min)	*Tally*	*Frequency*
0 - 9		
10 - 19		
20 - 29		
30 - 39		
40 - 49		

3 For the following data, state whether a frequency histogram or a column graph should be used, and draw the appropriate graph.

a The number of matches in 30 match boxes:

Number of matches per box	47	49	50	51	52	53	55
Frequency	1	1	9	12	4	2	1

b The heights of 25 hockey players (to the nearest cm):

Height (cm)	120 - 129	130 - 139	140 - 149	150 - 159	160 - 169
Frequency	1	2	7	14	1

4 A plant inspector takes a random sample of six month old seedlings from a nursery and measures their heights to the nearest mm.

The results are shown in the frequency histogram.

a How many of the seedlings are 400 mm or more in height?

b What percentage of the seedlings are between 349 and 400 mm?

c The total number of seedlings in the nursery is 1462. Estimate the number of seedlings which measure:

i less than 400 mm **ii** between 374 and 425 mm.

Frequency histogram of heights of seedlings

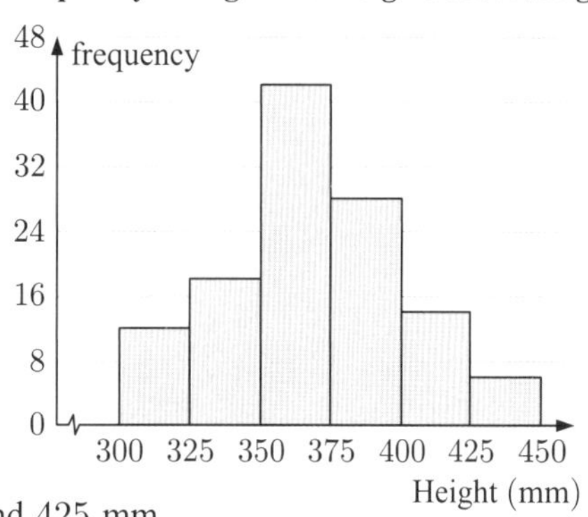

B MEASURING THE CENTRE OF DATA

We can get a better understanding of a data set if we can locate the **middle** or **centre** of the data, and also get an indication of its **spread** or dispersion. Knowing one of these without the other is often of little use.

There are *three statistics* that are used to measure the **centre** of a data set. These are the **mode**, the **mean**, and the **median**.

THE MODE

For discrete numerical data, the **mode** is the most frequently occurring value in the data set.

For continuous numerical data, we cannot talk about a mode in this way because no two data values will be *exactly* equal. Instead we talk about a **modal class**, which is the class or group that occurs most frequently.

THE MEAN

The **mean** of a data set is the statistical name for the arithmetic average.

$$\textbf{mean} = \frac{\textbf{sum of all data values}}{\textbf{the number of data values}}$$

The mean gives us a single number which indicates a centre of the data set. It is usually not a member of the data set.

For example, a mean test mark of 73% tells us that there are several marks below 73% and several above it. 73% is at the centre, but it is not always the case that one of the students scored 73%.

Suppose x is a numerical variable. We let:

x_i be the ith data value

n be the number of data values in the sample or population

$\overline{x}$ represent the mean of a **sample**, so

$$\overline{x} = \frac{x_1 + x_2 + x_3 + ... + x_n}{n} = \frac{\sum_{i=1}^{n} x_i}{n}$$

μ represent the mean of a **population**, so $\mu = \dfrac{\sum_{i=1}^{n} x_i}{n}$.

In many cases we do not have data from all of the members of a population, so the exact value of μ is often unknown. Instead we collect data from a sample of the population, and use the mean of the sample $\overline{x}$ as an approximation for μ.

Important: For examination purposes, all data will be treated as a population. We therefore use the statistic μ for the mean.

THE MEDIAN

The **median** is the *middle value* of an ordered data set.

An ordered data set is obtained by listing the data from smallest to largest value.

The median splits the data in halves. Half of the data are less than or equal to the median, and half are greater than or equal to it.

For example, if the median mark for a test is 73% then you know that half the class scored less than or equal to 73% and half scored greater than or equal to 73%.

For an **odd number** of data, the median is one of the original data values.

For an **even number** of data, the median is the average of the two middle values, and hence may not be in the original data set.

If there are n data values listed in order from smallest to largest, the median is the $\left(\dfrac{n+1}{2}\right)$th data value.

For example:

If $n = 13$, $\dfrac{13+1}{2} = 7$, so the median is the 7th ordered data value.

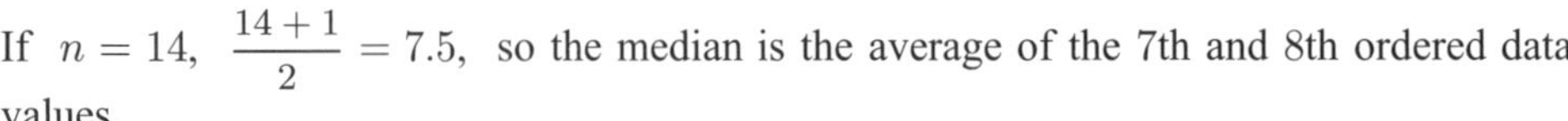

If $n = 14$, $\dfrac{14+1}{2} = 7.5$, so the median is the average of the 7th and 8th ordered data values.

THE MERITS OF THE MEAN AND MEDIAN AS MEASURES OF CENTRE

The **median** is the only measure of centre that will locate the true centre regardless of the data set's features. It is unaffected by the presence of extreme values. It is called a *resistant* measure of centre.

The **mean** is an accurate measure of centre if the distribution is symmetrical or approximately symmetrical. If it is not, then unbalanced high or low values will *drag* the mean toward them and cause it to be an inaccurate measure of the centre. It is called a *non-resistant* measure of centre because it is influenced by all data values in the set. *If it is considered inaccurate, it should not be used in discussion.*

THE RELATIONSHIP BETWEEN THE MEAN AND THE MEDIAN FOR DIFFERENT DISTRIBUTIONS

For distributions that are **symmetric** about the centre, the mean and median will be approximately equal.

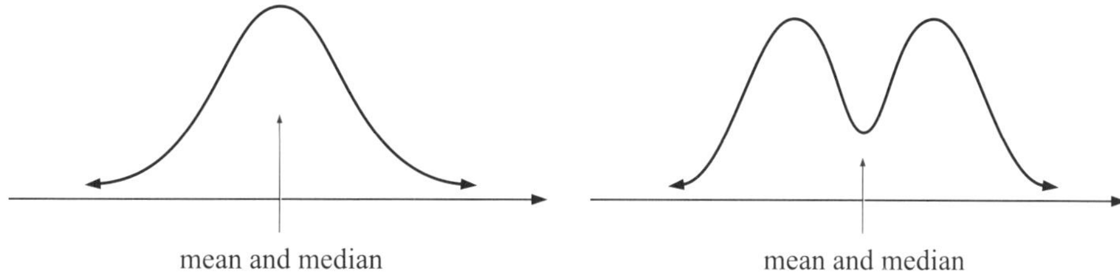

If the data set has symmetry, both the mean and the median should accurately measure the centre of the distribution.

If the data set is not symmetric, it may be positively or negatively skewed:

positively skewed distribution **negatively skewed distribution**

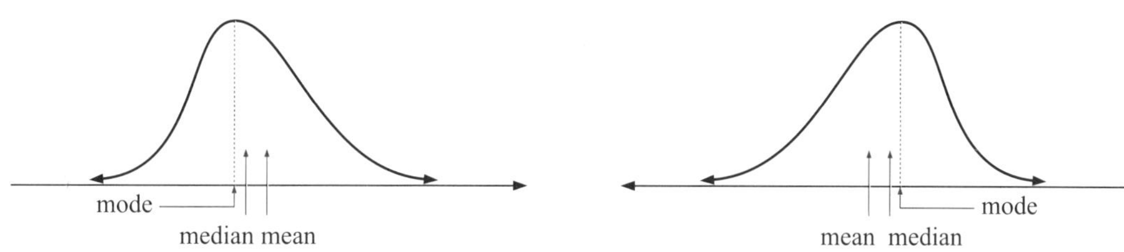

Notice that the mean and median are clearly different for these skewed distributions.

INVESTIGATION 1 MERITS OF THE MEAN AND MEDIAN

For this investigation you can either use the software provided online or your graphics calculator.

STATISTICS PACKAGE

Consider the data gained from Ethan, a professional golfer. The data was as follows:

244.6	245.1	248.0	248.8	250.0	251.1	251.2	253.9	254.5	254.6
255.9	257.0	260.6	262.8	262.9	263.1	263.2	264.3	264.4	265.0
265.5	265.6	266.5	267.4	269.7	270.5	270.7	272.9	275.6	277.5

GRAPHICS CALCULATOR INSTRUCTIONS

What to do:

1. Enter the data into your graphics calculator as a list, or use the **statistics package** supplied.
   - **a** Produce a frequency histogram of the data. Set the X values from 240 to 280 with an increment of 5. Set the Y values from 0 to 30.
   - **b** Comment on the shape of the distribution.
   - **c** Find the mean and median of the data.
   - **d** Compare the mean and the median. Is the mean an accurate measure of the centre?
2. Since we have continuous numerical data, we have a modal class rather than an individual mode.
   - **a** What is the modal class?
   - **b** What would the modal class be if our intervals were 2 m wide starting at 240 m?

3 Now suppose Ethan had hit a few very bad drives. Let us say that his three shortest drives were very short!

- **a** Change the three shortest drives to 82.1 m, 103.2 m, and 111.1 m.
- **b** Repeat **1 a**, **b**, **c**, and **d** but set the X values from 75 to 300 with an increment of 25 for the frequency histogram.
- **c** Describe the distribution as symmetric, positively skewed, or negatively skewed.
- **d** What effect have the changed values had on the mean and median as measures of the centre of the data?

4 What would have happened if Ethan had hit a few really long balls in addition to the very bad ones? Let us imagine that the longest balls he hit were very long indeed!

- **a** Change the three longest drives to 403.9 m, 415.5 m, and 420.0 m.
- **b** Repeat **1 a**, **b**, **c**, and **d** but set the X values from 50 to 450 with an increment of 50 for the frequency histogram.
- **c** Describe the distribution as symmetric, positively skewed, or negatively skewed.
- **d** What effect have the changed values had on the mean and median as measures of the centre of the data?

5 While collecting the data from Ethan, I decided I would also hit 30 golf balls with my driver. The relative frequency histogram alongside shows the results. The distribution is clearly positively skewed.
Discuss the merits of the median and mean as measures of the centre of this distribution.

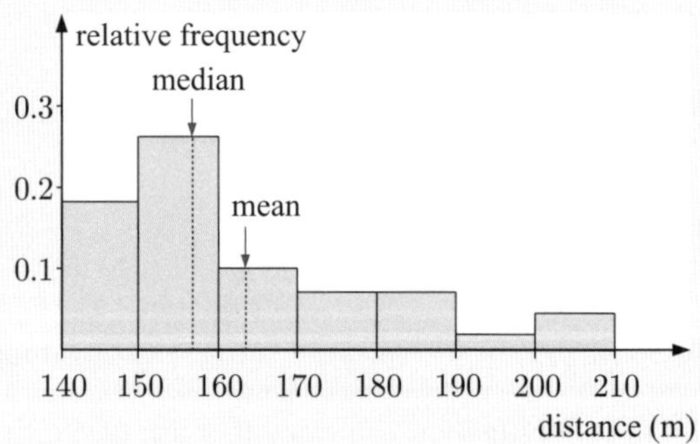

UNGROUPED DATA

Example 2 — Self Tutor

The number of faulty products returned to an electrical goods store over a 21 day period is:

3 4 4 9 8 8 6 4 7 9 1 3 5 3 5 9 8 6 3 7 1

For this data set, find the: **a** mean **b** median **c** mode.

a $\text{mean} = \dfrac{3+4+4+....+3+7+1}{21}$ ← sum of the data / ← 21 data values

$= \frac{113}{21}$

≈ 5.38 faulty products

b As $n = 21$, $\dfrac{n+1}{2} = 11$

The ordered data set is: ~~1 1 3 3 3 3 4 4 4 5~~ 5 ~~6 6 7 7 8 8 8 9 9 9~~

↑ 11th value

$\therefore$ median = 5 faulty products

c 3 is the score which occurs the most often, so the mode is 3 faulty products.

For the faulty products data in **Example 2**, how are the measures of the middle affected if on the 22nd day the number of faulty products was 9?

We expect the mean to rise as the new data value is greater than the old mean.

In fact, the new mean $= \frac{113+9}{22} = \frac{122}{22} \approx 5.55$ faulty products.

As $n = 22$, $\frac{n+1}{2} = 11.5$

The new ordered data set would be: $1\ 1\ 3\ 3\ 3\ 3\ 4\ 4\ 4\ 5\ \underbrace{5\ 6}_{\text{two middle scores}}\ 6\ 7\ 7\ 8\ 8\ 8\ 9\ 9\ 9\ 9$

The new median $= \frac{5+6}{2} = 5.5$ faulty products.

This new data set has *two* modes, which are 3 and 9 faulty products. We say that the data set is **bimodal**.

Note:
- If a data set has three or more modes, we do not use the mode as a measure of the middle or centre of the data values.
- Consider the data: 4 2 5 6 7 4 5 3 5 4 7 6 3 5 8 6 5.

  The dot plot of this data is:

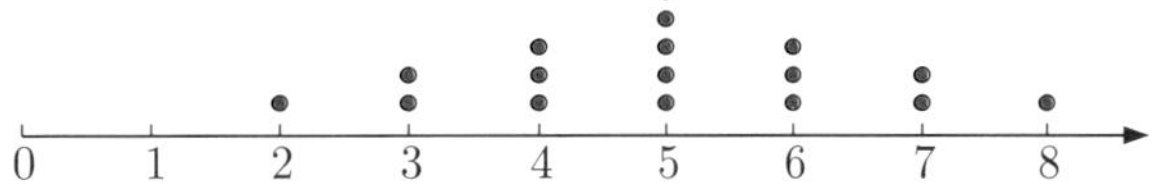

  For this data the mean, median, and mode are all 5.

  For a *symmetrical distribution* of data, the mean, mode, and median will all be equal.

  However, equal or approximately equal values of the mean, mode, and median does not necessarily indicate a distribution is symmetric.

Example 3 — Self Tutor

If 6 people have a mean mass of 53.7 kg, find their total mass.

$$\frac{\text{sum of masses}}{6} = 53.7 \text{ kg}$$

$$\therefore \text{ sum of masses} = 53.7 \times 6$$

$$\therefore \text{ the total mass} = 322.2 \text{ kg.}$$

EXERCISE 23B.1

1 For each of the following data sets, find the: **i** mean **ii** median **iii** mode.

  a 2, 3, 3, 3, 4, 4, 4, 5, 5, 5, 5, 6, 6, 6, 6, 6, 7, 7, 8, 8, 8, 9, 9

  b 10, 12, 12, 15, 15, 16, 16, 17, 18, 18, 18, 18, 19, 20, 21

  c 22.4, 24.6, 21.8, 26.4, 24.9, 25.0, 23.5, 26.1, 25.3, 29.5, 23.5

  Check your answers using the statistics features on your graphics calculator.

2 Consider the following two data sets:

  Data set A: 3, 4, 4, 5, 6, 6, 7, 7, 7, 8, 8, 9, 10

  Data set B: 3, 4, 4, 5, 6, 6, 7, 7, 7, 8, 8, 9, 15

  a Find the mean for both data set A and data set B.

  b Find the median of both data set A and data set B.

  c Explain why the mean of data set A is less than the mean of data set B.

  d Explain why the median of data set A is the same as the median of data set B.

3 The annual salaries of ten office workers are:

$23 000, $46 000, $23 000, $38 000, $24 000,
$23 000, $23 000, $38 000, $23 000, $32 000

a Find the mean, median, and modal salaries of this group.

b Explain why the mode is an unsatisfactory measure of the middle in this case.

c Is the median a satisfactory measure of the middle of this data set?

4 The following raw data is the daily rainfall (to the nearest millimetre) for the month of July 2007 in the desert:

3, 1, 0, 0, 0, 0, 0, 2, 0, 0, 3, 0, 0, 0, 7, 1, 1, 0, 3, 8, 0, 0, 0, 42, 21, 3, 0, 3, 1, 0, 0

a Find the mean, median, and mode for the data.

b Explain why the median is not the most suitable measure of centre for this set of data.

c Explain why the mode is not the most suitable measure of centre for this set of data.

d Are there any outliers in this data set?

e On some occasions outliers are removed because they must be due to errors in observation or calculation. If the outliers in the data set were accurately found, should they be removed before finding the measures of the middle?

5 A basketball team scored 43, 55, 41, and 37 points in their first four matches.

a Find the mean number of points scored for the first four matches.

b What score will the team need to shoot in the next match so that they maintain the same mean score?

c The team scores only 25 points in the fifth match.

i Find the mean number of points scored for the five matches.

ii The team then scores 41 points in their sixth and final match. Will this increase or decrease their previous mean score? What is the mean score for all six matches?

6 This year, the mean monthly sales for a clothing store have been $15 467. Calculate the total sales for the store for the year.

7 While on an outback safari, Bill drove an average of 262 km per day for a period of 12 days. How far did Bill drive in total while on safari?

8 The table alongside compares the mass at birth of some guinea pigs with their mass when they were two weeks old.

a What was the mean birth mass?

b What was the mean mass after two weeks?

c What was the mean increase over the two weeks?

Guinea Pig	*Mass (g) at birth*	*Mass (g) at 2 weeks*
A	75	210
B	70	200
C	80	200
D	70	220
E	74	215
F	60	200
G	55	206
H	83	230

9 Given $\mu = 11.6$ and $n = 10$, calculate $\sum_{i=1}^{10} x_i$.

10 Towards the end of season, a netballer had played 14 matches and had thrown an average of 16.5 goals per game. In the final two matches of the season she threw 21 goals and 24 goals. Find the netballer's average for the whole season.

11 The selling prices of the last 10 houses sold in a certain district were as follows:

$146 400, $127 600, $211 000, $192 500,
$256 400, $132 400, $148 000, $129 500,
$131 400, $162 500

a Calculate the mean and median selling prices and comment on the results.

b Which measure would you use if you were:

i a vendor wanting to sell your house

ii looking to buy a house in the district?

12 Find x if 5, 9, 11, 12, 13, 14, 17, and x have a mean of 12.

13 Find a given that 3, 0, a, a, 4, a, 6, a, and 3 have a mean of 4.

14 Over the complete assessment period, Aruna averaged 35 out of a possible 40 marks for her maths tests. However, when checking her files, she could only find 7 of the 8 tests. For these she scored 29, 36, 32, 38, 35, 34, and 39. How many marks out of 40 did she score for the eighth test?

15 A sample of 10 measurements has a mean of 15.7 and a sample of 20 measurements has a mean of 14.3. Find the mean of all 30 measurements.

16 The mean and median of a set of 9 measurements are both 12. Seven of the measurements are 7, 9, 11, 13, 14, 17, and 19. Find the other two measurements.

17 Jana took seven spelling tests, each with ten words, but she could only find the results of five of them. These were 9, 5, 7, 9, and 10. She asked her teacher for the other two results and the teacher said that the mode of her scores was 9 and the mean was 8. Given that Jana remembers she only got one 10, find the two missing results.

MEASURES OF THE CENTRE FROM OTHER SOURCES

When the same data values appear several times we often summarise the data in a frequency table.

Consider the data in the given table:

Data value (x)	*Frequency* (f)	*Product* (fx)
3	1	$1 \times 3 = 3$
4	1	$1 \times 4 = 4$
5	3	$3 \times 5 = 15$
6	7	$7 \times 6 = 42$
7	15	$15 \times 7 = 105$
8	8	$8 \times 8 = 64$
9	5	$5 \times 9 = 45$
Total	$\sum f = 40$	$\sum fx = 278$

We can find the measures of the centre directly from the table.

The mode

The data value 7 has the highest frequency.

The mode is therefore 7.

The mean

Adding a 'Product' column to the table helps to add all scores.

For example, there are 15 data of value 7 and these add to $7 \times 15 = 105$.

Remembering that the mean $= \dfrac{\text{sum of all data values}}{\text{the number of data values}}$, we find

$$\mu = \frac{f_1x_1 + f_2x_2 + f_3x_3 + ... + f_kx_k}{f_1 + f_2 + f_3 + ... + f_k}$$ where k is the number of *different* data values.

$$= \frac{\sum_{i=1}^{k} f_ix_i}{n}$$

This formula is often abbreviated as $\mu = \dfrac{\sum fx}{n}$.

In this case the mean $= \dfrac{278}{40} = 6.95$.

The median

Since $\dfrac{n+1}{2} = \dfrac{41}{2} = 20.5$, the median is the average of the 20th and 21st data values, when they are listed in order.

In the table, the blue numbers show us accumulated frequency values, or the **cumulative frequency**.

We can see that the 20th and 21st data values (in order) are both 7s.

Data value	*Frequency*	*Cumulative frequency*
3	1	1 ⟵ one number is 3
4	1	2 ⟵ two numbers are 4 or less
5	3	5 ⟵ five numbers are 5 or less
6	7	12 ⟵ 12 numbers are 6 or less
7	15	27 ⟵ 27 numbers are 7 or less
8	8	35 ⟵ 35 numbers are 8 or less
9	5	40 ⟵ all numbers are 9 or less
Total	40	

$\therefore$ the median $= \dfrac{7+7}{2} = 7$

Notice that we have a skewed distribution even though the mean, median and mode are nearly equal. So, we must be careful if we use measures of the middle to call a distribution symmetric.

Example 4 Self Tutor

The table below shows the number of aces served by tennis players in their first sets of a tournament.

Number of aces	1	2	3	4	5	6
Frequency	4	11	18	13	7	2

Determine the: **a** mean **b** median **c** mode for this data.

Number of aces (x)	*Frequency* (f)	*Product* (fx)	*Cumulative frequency*
1	4	4	4
2	11	22	15
3	18	54	33
4	13	52	46
5	7	35	53
6	2	12	55
Total	$n = 55$	$\sum fx = 179$	

a $\mu = \dfrac{\sum fx}{n}$

$= \dfrac{179}{55}$

≈ 3.25 aces

In this case $\dfrac{\sum fx}{n}$ is short for $\dfrac{\sum_{i=1}^{6} f_i x_i}{n}$.

b $\dfrac{n+1}{2} = 28$, so the median is the 28th data value. From the cumulative frequency column, the data values 16 to 33 are 3 aces.

$\therefore$ the 28th data value is 3 aces.

$\therefore$ the median is 3 aces.

c Looking down the frequency column, the highest frequency is 18. This corresponds to 3 aces, so the mode is 3 aces.

EXERCISE 23B.2

1 The table alongside shows the results when 3 coins were tossed simultaneously 30 times.

Number of heads	*Frequency*
0	4
1	12
2	11
3	3
Total	30

Calculate the:

a mode **b** median **c** mean.

Check your answers using the statistics features of your graphics calculator.

2 The following frequency table records the number of phone calls made in a day by 50 fifteen-year-olds.

Number of phone calls	*Frequency*
0	5
1	8
2	13
3	8
4	6
5	3
6	3
7	2
8	1
11	1

a For this data, find the:

i mean **ii** median **iii** mode.

b Construct a column graph for the data and show the position of the mean, median, and mode on the horizontal axis.

c Describe the distribution of the data.

d Why is the mean larger than the median for this data?

e Which measure of centre would be the most suitable for this data set?

3 A company claims that their match boxes contain, on average, 50 matches per box. On doing a survey, the Consumer Protection Society recorded the following results:

Number in a box	*Frequency*
47	5
48	4
49	11
50	6
51	3
52	1
Total	30

a Calculate the:

i mode **ii** median **iii** mean

b Do the results of this survey support the company's claim?

c In a court for 'false advertising', the company won their case against the Consumer Protection Society. Suggest how they did this.

4 Families at a school in Australia were surveyed, and the number of children in each family recorded. The results of the survey are shown alongside.

Number of children	*Frequency*
1	5
2	28
3	15
4	8
5	2
6	1
Total	59

a Calculate the:

i mean **ii** mode **iii** median.

b The average Australian family has 2.2 children. How does this school compare to the national average?

c The data set is skewed. Is the skewness positive or negative?

d How has the skewness of the data affected the measures of the centre of the data set?

5 The frequency column graph gives the value of donations for the Heart Foundation collected in a particular street.

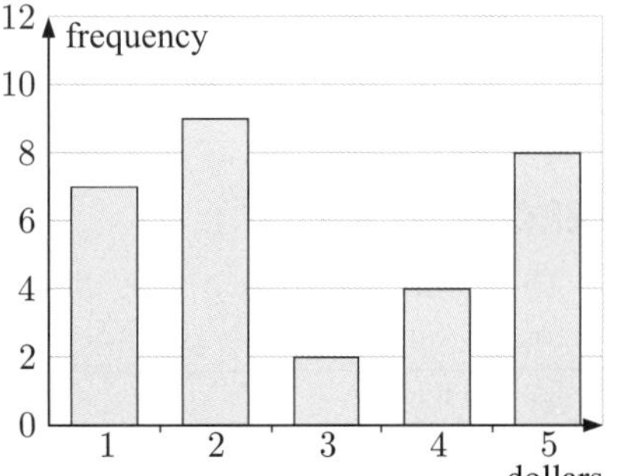

a Construct a frequency table from the graph.

b Determine the total number of donations.

c Find the: **i** mean **ii** mode **iii** median of the donations.

d Which of the measures of central tendency can be found easily using the graph only?

6 The table shows the IB mathematics scores for a class. A pass is considered to be a score of 4 or more.

Score	1	2	3	4	5	6	7
Number of students	0	2	3	5	x	4	1

a Given the mean score was 4.45, find the value of x.

b Find the percentage of students who passed.

7 Revisit the **Opening Problem** on page **704**.

a Use a frequency table for the *Without fertiliser* data to find the:

i mean **ii** mode **iii** median number of peas per pod.

b Use a frequency table for the *With fertiliser* data to find the:

i mean **ii** mode **iii** median number of peas per pod.

c Which of the measures of the centre is appropriate to use in a report on this data?

d Has the application of fertiliser significantly improved the number of peas per pod?

8 Out of 31 measurements, 15 are below 10 cm and 12 are above 11 cm. Find the median if the other 4 measurements are 10.1 cm, 10.4 cm, 10.7 cm, and 10.9 cm.

9 Two brands of toothpicks claim that their boxes contain an average of 50 toothpicks per box. In a survey the Consumer Protection Society (C.P.S.) recorded the following results:

Brand A

Number in a box	46	47	48	49	50	51	52	53	55
Frequency	1	2	3	7	10	20	15	3	1

Brand B

Number in a box	48	49	50	51	52	53	54
Frequency	3	17	30	7	2	1	1

a Find the average contents of Brands A and B toothpick boxes.

b Would it be fair for the C.P.S. to prosecute the manufacturers of either brand, based on these statistics?

10 In an office of 20 people there are only 4 salary levels paid:

€50 000 (1 person), €42 000 (3 people), €35 000 (6 people), €28 000 (10 people).

a Calculate: **i** the median salary **ii** the modal salary **iii** the mean salary.

b Which measure of central tendency might be used by the boss who is against a pay rise for the other employees?

GROUPED DATA

When information has been gathered in groups or classes, we use the **midpoint** or **mid-interval value** to represent all scores within that interval.

We are assuming that the scores within each class are evenly distributed throughout that interval. The mean calculated is an **approximation** of the true value, and we cannot do better than this without knowing each individual data value.

INVESTIGATION 2 — MID-INTERVAL VALUES

When mid-interval values are used to represent all scores within that interval, what effect will this have on estimating the mean of the grouped data?

The table alongside summarises the marks received by students for a physics examination out of 50. The exact results for each student have been lost.

Marks	*Frequency*
0 - 9	2
10 - 19	31
20 - 29	73
30 - 39	85
40 - 49	28

What to do:

1 Suppose that all of the students scored the lowest possible result in their class interval, so 2 students scored 0, 31 students scored 10, and so on.
Calculate the mean of these results, and hence complete:
"The mean score of students in the physics examination must be *at least* ...."

2 Now suppose that all of the students scored the highest possible result in their class interval.
Calculate the mean of these results, and hence complete:
"The mean score of students in the physics examination must be *at most* ...."

3 We now have two extreme values between which the actual mean must lie.
Now suppose that all of the students scored the mid-interval value in their class interval. We assume that 2 students scored 4.5, 31 students scored 14.5, and so on.

a Calculate the mean of these results.

b How does this result compare with lower and upper limits found in **1** and **2**?

c Copy and complete:
"The mean score of the students in the physics examination was approximately ...."

The publishers acknowledge the late Mr Jim Russell, General Features for the reproduction of this cartoon.

Example 5 **Self Tutor**

Estimate the mean of the following *ages of bus drivers* data, to the nearest year:

Age (yrs)	21 - 25	26 - 30	31 - 35	36 - 40	41 - 45	46 - 50	51 - 55
Frequency	11	14	32	27	29	17	7

Age (yrs)	*Frequency* (f)	*Midpoint* (x)	$f\,x$
21 - 25	11	23	253
26 - 30	14	28	392
31 - 35	32	33	1056
36 - 40	27	38	1026
41 - 45	29	43	1247
46 - 50	17	48	816
51 - 55	7	53	371
Total	$n = 137$		$\sum f x = 5161$

$$\mu = \frac{\sum f\,x}{n} = \frac{5161}{137} \approx 37.7$$

$\therefore$ the mean age of the drivers is about 38 years.

EXERCISE 23B.3

1 50 students sit a mathematics test. Estimate the mean score given these results:

Score	0 - 9	10 - 19	20 - 29	30 - 39	40 - 49
Frequency	2	5	7	27	9

Check your answers using your calculator.

2 The table shows the petrol sales in one day by a number of city service stations.

  a How many service stations were involved in the survey?

  b Estimate the total amount of petrol sold for the day by the service stations.

  c Find the approximate mean sales of petrol for the day.

Litres (L)	*Frequency*
$2000 < L \leqslant 3000$	4
$3000 < L \leqslant 4000$	4
$4000 < L \leqslant 5000$	9
$5000 < L \leqslant 6000$	14
$6000 < L \leqslant 7000$	23
$7000 < L \leqslant 8000$	16

3 This frequency histogram illustrates the results of an aptitude test given to a group of people seeking positions in a company.

  a How many people sat for the test?

  b Estimate the mean score for the test.

  c What fraction of the people scored less than 100 for the test?

  d What percentage of the people scored more than 130 for the test?

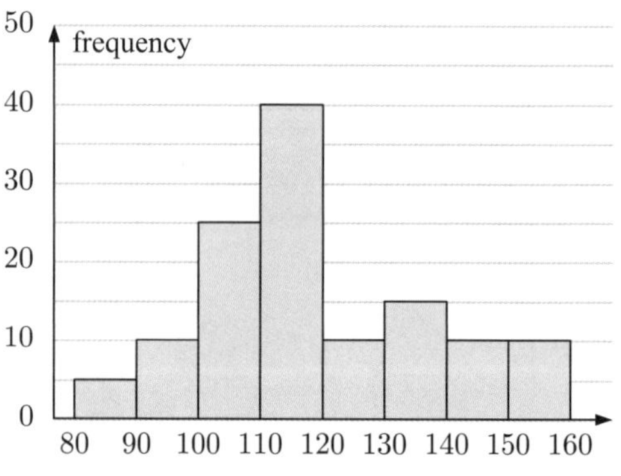

C VARIANCE AND STANDARD DEVIATION

To accurately describe a distribution we need to measure both its **centre** and its **spread** or **dispersion**.

The distributions shown have the same mean, but clearly they have different spreads. The A distribution has most scores close to the mean whereas the C distribution has the greatest spread.

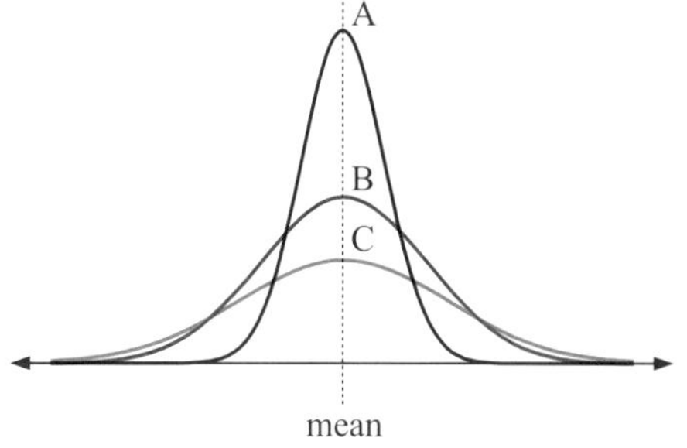

SIMPLE MEASURES OF SPREAD

In previous years you should have studied the **range** and **interquartile range** of a data set, which are two simple measures of its spread. These concepts led on to **boxplots** to display data, and **cumulative frequency graphs** to show **percentiles**.

You can revise this work by clicking on the icon.

VARIANCE AND STANDARD DEVIATION

The problem with using the range and the IQR as measures of spread or dispersion of scores is that both of them only use two values in their calculation. Some data sets can therefore have their spread characteristics hidden when the range or IQR are quoted. We therefore turn to the **variance** and **standard deviation** of a data set.

Consider a data set of n values: $x_1, x_2, x_3, x_4, ..., x_n,$ with mean $\overline{x}$.

$x_i - \overline{x}$ measures how far x_i deviates from the mean, so one might suspect that the mean of the deviations $\frac{1}{n}\sum_{i-1}^{n}(x_i - \overline{x})$ would give a good measure of spread. However, this value turns out to always be zero.

Instead, we define the **variance** of n data values to be $\sigma^2 = \dfrac{\sum_{i=1}^{n}(x_i - \overline{x})^2}{n}$.

Notice in this formula that:

- $(x_i - \overline{x})^2$ is also a measure of how far x_i deviates from $\overline{x}$. However, the square ensures that each term in the sum is positive, which is why the sum turns out not to be zero.
- If $\sum_{i=1}^{n}(x_i - \overline{x})^2$ is small, it will indicate that most of the data values are close to $\overline{x}$.
- Dividing by n gives an indication of how far, on average, the data is from the mean.

For a data set of n values, $\sigma = \sqrt{\dfrac{\sum_{i=1}^{n}(x_i - \overline{x})^2}{n}}$ is called the **standard deviation**.

The square root in the standard deviation is used to correct the units. For example, if x_i is the weight of a student in kg, σ^2 would be in kg^2. For this reason the standard deviation is more frequently quoted than the variance.

For the purpose of this course, we assume the formulae for variance and standard deviation of a whole population are the same as those for a sample.

The **variance** of a population of size n is $\sigma^2 = \dfrac{\sum\limits_{i=1}^{n}(x_i - \mu)^2}{n} = \dfrac{\sum\limits_{i=1}^{n} x_i^{\,2}}{n} - \mu^2$.

The **standard deviation** of a population of size n is $\sigma = \sqrt{\dfrac{\sum\limits_{i=1}^{n}(x_i - \mu)^2}{n}}$.

The standard deviation is a **non-resistant** measure of spread. This is due to its dependence on the mean of the sample and because extreme data values will give large values for $(x_i - \mu)^2$. It is only a useful measure if the distribution is approximately symmetrical. However, the standard deviation is particularly useful when the data from which it came is **normally distributed**. This will be discussed in detail in **Chapter 26**.

The IQR and percentiles are more appropriate tools for measuring spread if the distribution is considerably skewed.

INVESTIGATION 3 — STANDARD DEVIATION

A group of 5 students is chosen from each of three schools, to test their ability to solve puzzles. The 15 students are each given a series of puzzles and two hours to solve as many as they can individually. The results were:

School A: 7, 7, 7, 7, 7
School B: 5, 6, 7, 8, 9
School C: 3, 5, 7, 9, 11

What to do:

1. Show that the mean and median for each school is 7.
2. Given the mean $\mu = 7$ for each group, complete a table like the one following, for each school:

School ....

Score (x_i)	*Deviation* $(x_i - \mu)$	*Square of deviation* $(x_i - \mu)^2$
Sum		

3. Calculate the variance $\dfrac{\sum(x_i - \mu)^2}{n}$ and standard deviation $\sqrt{\dfrac{\sum(x_i - \mu)^2}{n}}$ for each group.

   Check your results match the following table:

School	*Mean*	*Variance*	*Standard deviation*
A	7	0	0
B	7	2	$\sqrt{2}$
C	7	8	$\sqrt{8}$

4. Use the table above to compare the performances of the different schools.
5. A group of 5 students from a higher year level at school C are given the same test. They each score 2 more than the students in the lower year group, so their scores are: 5, 7, 9, 11, 13.
   - **a** Find the mean, variance, and standard deviation for this set.
   - **b** Comment on the effect of adding 2 to each member of a data set.

6 A group of 5 teachers from B decide to show their students how clever they are. They complete twice as many puzzles as each of their students, so their scores are: 10, 12, 14, 16, 18.

  a Find the mean, variance, and standard deviation for this set.

  b Comment on the effect of doubling each member of a data set.

Example 6 — **Self Tutor**

A library surveys 20 borrowers each day from Monday to Friday, and records the number who are not satisfied with the range of reading material. The results are: 3 7 6 8 11.

The following year the library receives a grant that enables the purchase of a large number of books. The survey is then repeated and the results are: 2 3 5 4 6.

a Find the mean and standard deviation for each survey using:

  i the formula **ii** technology.

b What do these statistics tell us?

a **i** *Survey 1*

x	$x-\mu$	$(x-\mu)^2$
3	-4	16
7	0	0
6	-1	1
8	1	1
11	4	16
35	*Total*	34

$\therefore \ \mu = \frac{35}{5} = 7$

$$\sigma = \sqrt{\frac{\sum(x-\mu)^2}{n}} = \sqrt{\frac{34}{5}} \approx 2.61$$

Survey 2

x	$x-\mu$	$(x-\mu)^2$
2	-2	4
3	-1	1
5	1	1
4	0	0
6	2	4
20	*Total*	10

$\therefore \ \mu = \frac{20}{5} = 4$

$$\sigma = \sqrt{\frac{\sum(x-\mu)^2}{n}} = \sqrt{\frac{10}{5}} \approx 1.41$$

ii **Casio fx-CG20**

1-Variable
x̄ =7
Σx =35
Σx² =279
σx =2.60768096
sx =2.91547594
n =5

TI-84 Plus

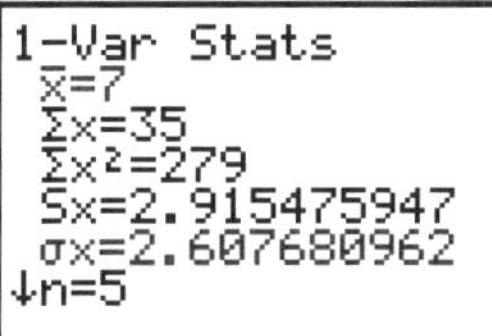

TI-nspire

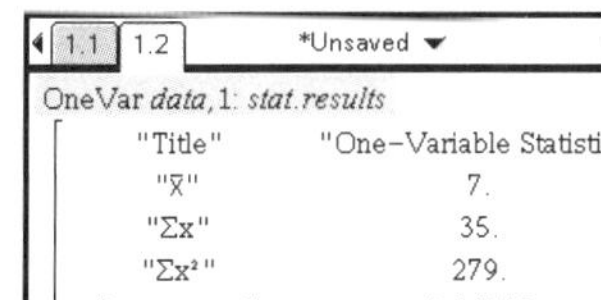

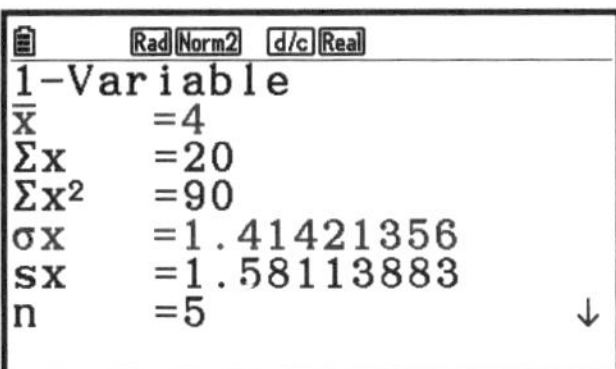

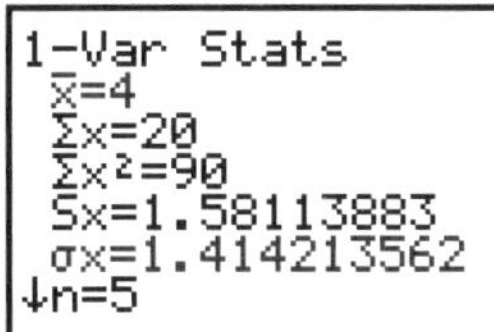

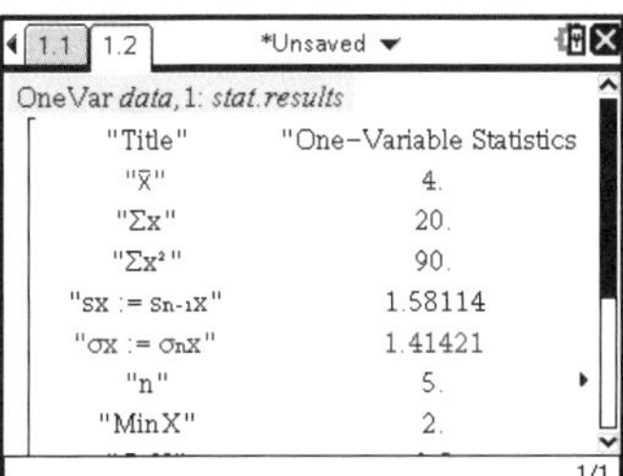

b The second survey shows that the number of dissatisfied borrowers has almost halved and there is less variability in the number of dissatisfied borrowers.

EXERCISE 23C.1

1 The column graphs show two distributions:

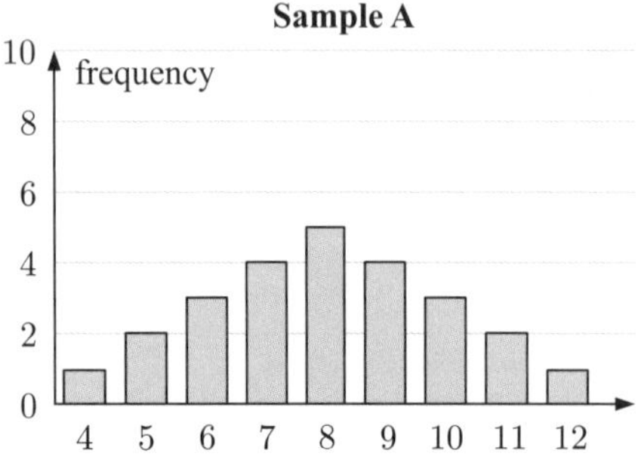

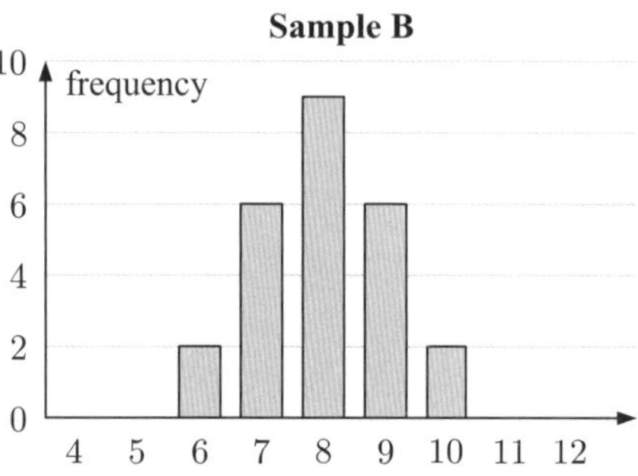

- **a** By looking at the graphs, which distribution appears to have wider spread?
- **b** Find the mean of each sample.
- **c** Find the standard deviation for each sample. Comment on your answers.
- **d** The range and interquartile range for the two samples are given in the table. In what way does the standard deviation give a better description of how the data is distributed?

Sample	*Range*	*IQR*
A	8	3
B	4	2

2 The number of points scored by Andrew and Brad in the last 8 basketball matches are tabulated below.

Points by Andrew	23	17	31	25	25	19	28	32
Points by Brad	9	29	41	26	14	44	38	43

- **a** Find the mean and standard deviation of the number of points scored by each player.
- **b** Which of the two players is more consistent?

3 Two baseball coaches compare the number of runs scored by their teams in their last ten matches:

Rockets	0	10	1	9	11	0	8	5	6	7
Bullets	4	3	4	1	4	11	7	6	12	5

- **a** Show that each team has the same mean and range of runs scored.
- **b** Which team's performance do you suspect is more variable over the period?
- **c** Check your answer to **b** by finding the standard deviation for each distribution.
- **d** Does the range or the standard deviation give a better indication of variability?

4 A manufacturer of soft drinks employs a statistician for quality control. He needs to check that 375 mL of drink goes into each can, but realises the machine which fills the cans will slightly vary each delivery.

- **a** Would you expect the standard deviation for the whole production run to be the same for one day as it is for one week? Explain your answer.
- **b** If samples of 125 cans are taken each day, what measure would be used to:
  - **i** check that an average of 375 mL of drink goes into each can
  - **ii** check the variability of the volume of drink going into each can?
- **c** What is the significance of a low standard deviation in this case?

5 The weights in kg of seven footballers are: 79, 64, 59, 71, 68, 68, 74.

a Find the mean and standard deviation for this group.

b When measured five years later, each footballer's weight had increased by exactly 10 kg. Find the new mean and standard deviation.

c Comment on your results in general terms.

6 The weights of ten young turkeys to the nearest 0.1 kg are:
0.8, 1.1, 1.2, 0.9, 1.2, 1.2, 0.9, 0.7, 1.0, 1.1

a Find the mean and standard deviation for the weights of the turkeys.

b After being fed a special diet for one month, the weights of the turkeys doubled. Find the new mean and standard deviation.

c Comment on your results.

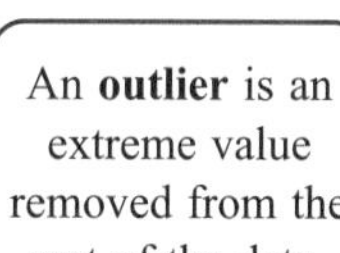

An **outlier** is an extreme value removed from the rest of the data.

7 The following table shows the decrease in cholesterol levels in 6 volunteers after a two week trial of special diet and exercise.

Volunteer	A	B	C	D	E	F
Decrease in cholesterol	0.8	0.6	0.7	0.8	0.4	2.8

a Find the standard deviation of the data.

b Recalculate the standard deviation with the outlier removed.

c Discuss the effect of an extreme value on the standard deviation.

8 A set of 8 integers $\{1, 3, 5, 7, 4, 5, p, q\}$ has a mean of 5 and a variance of 5.25. Find p and q given that $p < q$.

9 A set of 10 integers $\{3, 9, 5, 5, 6, 4, a, 6, b, 8\}$ has a mean of 6 and a variance of 3.2. Find a and b given that $a > b$.

10 **a** Prove that $\sum_{i=1}^{n}(x_i - \mu)^2 = \sum_{i=1}^{n}\left(x_i^{\,2}\right) - n\mu^2$.

b Find the mean of the data set $\{x_1, x_2, ..., x_{25}\}$ given that $\sum_{i=1}^{25} x_i^{\,2} = 2568.25$ and the standard deviation is 5.2.

STANDARD DEVIATION FOR GROUPED DATA

Suppose there are n scores in a set of grouped data with mean μ.

If the different data values are $x_1, x_2, ..., x_k$, with corresponding frequencies $f_1, f_2, ..., f_k$, then:

- the variance is given by $\boldsymbol{\sigma^2 = \dfrac{\sum_{i=1}^{k} f_i(x_i - \mu)^2}{n} = \dfrac{\sum_{i=1}^{k} f_i x_i^{\,2}}{n} - \mu^2}$
- the standard deviation is given by $\boldsymbol{\sigma = \sqrt{\dfrac{\sum_{i=1}^{k} f_i(x_i - \mu)^2}{n}}}$.

Example 7 — Self Tutor

Find the standard deviation of the distribution using:

a the standard deviation formula **b** technology.

GRAPHICS CALCULATOR INSTRUCTIONS

Score	1	2	3	4	5
Frequency	1	2	4	2	1

a

x	f	fx	$x-\mu$	$(x-\mu)^2$	$f(x-\mu)^2$
1	1	1	-2	4	4
2	2	4	-1	1	2
3	4	12	0	0	0
4	2	8	1	1	2
5	1	5	2	4	4
Total	10	30			12

$$\mu = \frac{\sum fx}{n} = \frac{30}{10} = 3$$

$$\sigma = \sqrt{\frac{\sum f(x-\mu)^2}{n}}$$

$$= \sqrt{\frac{12}{10}} \approx 1.10$$

b

Casio fx-CG20

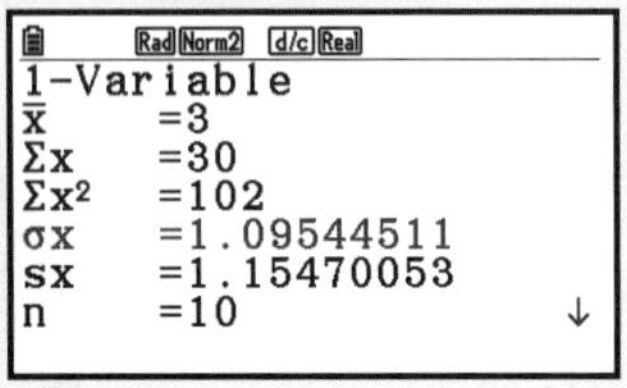

```
1-Variable
x̄    =3
Σx   =30
Σx²  =102
σx   =1.09544511
sx   =1.15470053
n    =10
```

TI-84 Plus

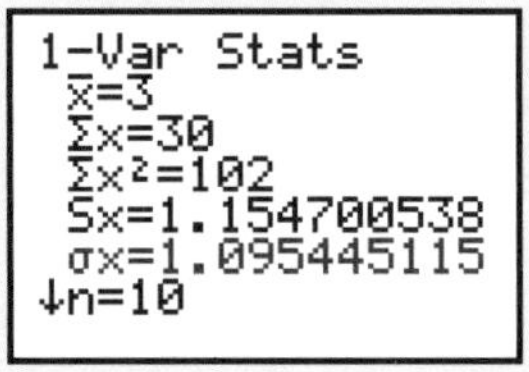

```
1-Var Stats
x̄=3
Σx=30
Σx²=102
Sx=1.154700538
σx=1.095445115
↓n=10
```

TI-nspire

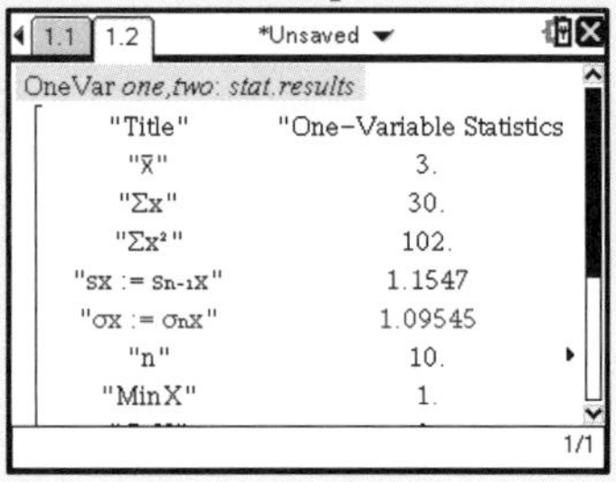

```
OneVar one,two: stat.results
"Title"        "One-Variable Statistics"
"x̄"            3.
"Σx"           30.
"Σx²"          102.
"sx := sn-1x"  1.1547
"σx := σnx"    1.09545
"n"            10.
"MinX"         1.
```

For continuous data, or data that has been grouped in classes, we use the midpoint of the interval to represent all data in that interval.

EXERCISE 23C.2

1 Below is a sample of family sizes taken at random from people in a city.

Number of children, x	0	1	2	3	4	5	6	7
Frequency, f	14	18	13	5	3	2	2	1

Find the sample mean and standard deviation.

2 The table shows the ages of squash players at the Junior National Squash Championship.

Age	11	12	13	14	15	16	17	18
Frequency	2	1	4	5	6	4	2	1

Find the mean and standard deviation of the ages.

3 The numbers of toothpicks in 48 boxes were counted and the results tabulated.

Number of toothpicks	33	35	36	37	38	39	40
Frequency	1	5	7	13	12	8	2

Find the mean and standard deviation of the number of toothpicks in the boxes.

4 The lengths of 30 12-day old babies were measured, and the following data obtained:

Estimate the mean length and the standard deviation of the lengths.

Length (cm)	*Frequency*
$40 \leqslant L < 42$	1
$42 \leqslant L < 44$	1
$44 \leqslant L < 46$	3
$46 \leqslant L < 48$	7
$48 \leqslant L < 50$	11
$50 \leqslant L < 52$	5
$52 \leqslant L < 54$	2

5 The weekly wages (in dollars) of 200 steel workers are given in the table.

Estimate the mean and the standard deviation of the wages.

Wage ($)	*Number of workers*
360 - 369.99	17
370 - 379.99	38
380 - 389.99	47
390 - 399.99	57
400 - 409.99	18
410 - 419.99	10
420 - 429.99	10
430 - 439.99	3

THEORY OF KNOWLEDGE

Statistics are often used to give the reader a misleading impression of what the data actually means. In some cases this happens by accident through mistakes in the statistical process. Often, however, it is done deliberately in an attempt to persuade the reader to believe something.

Even simple things like the display of data can be done so as to create a false impression. For example, the two graphs below show the profits of a company for the first four months of the year.

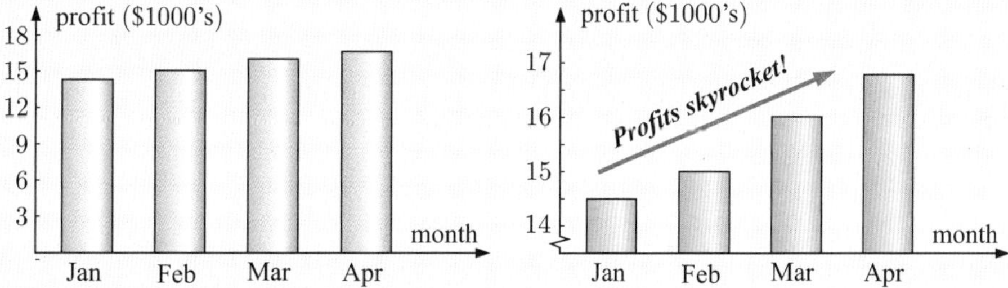

Both graphs accurately display the data, but on one graph the vertical axis has a break in its scale which can give the impression that the increase in profits is much larger than it really is. The comment 'Profits skyrocket!' encourages the reader to come to that conclusion without looking at the data more carefully.

1 Given that the data is presented with mathematical accuracy in both graphs, would you say the author in the second case has lied?

When data is collected by sampling, the choice of a biased sample can be used to give misleading results. There is also the question of whether outliers should be considered as genuine data, or ignored and left out of statistical analysis.

2 In what other ways can statistics be used to deliberately mislead the target audience?

The use of statistics in science and medicine has been widely debated, as companies employ scientific 'experts' to back their claims. For example, in the multi-billion dollar tobacco industry, huge volumes of data have been collected which claim that smoking leads to cancer and other harmful effects. However, the industry has sponsored other studies which deny these claims.

There are many scientific articles and books which discuss the uses and misuses of statistics. For example:

- *Surgeons General's reports on smoking and cancer: uses and misuses of statistics and of science*, R J Hickey and I E Allen, Public Health Rep. 1983 Sep-Oct; **98**(5): 410-411.
- *Misusage of Statistics in Medical Researches*, I Ercan, B Yazici, Y Yang, G Ozkaya, S Cangur, B Ediz, I Kan, 2007, European Journal of General Medicine, **4**(3),127-133.
- *Sex, Drugs, and Body Counts: The Politics of Numbers in Global Crime and Conflict*, P Andreas and K M Greenhill, 2010, Cornell University Press.

3 Can we trust statistical results published in the media and in scientific journals?

4 What role does ethics have to play in mathematics?

REVIEW SET 23A — NON-CALCULATOR

1 The data supplied below is the diameter (in cm) of a number of bacteria colonies as measured by a microbiologist 12 hours after seeding.

0.4 2.1 3.4 3.9 4.7 3.7 0.8 3.6 4.1 4.9 2.5 3.1 1.5 2.6 4.0
1.3 3.5 0.9 1.5 4.2 3.5 2.1 3.0 1.7 3.6 2.8 3.7 2.8 3.2 3.3

a Find the **i** median **ii** range of the data.

b Group the data in 5 groups and display it using a frequency histogram.

c Comment on the skewness of the data.

2 The data set 4, 6, 9, a, 3, b has a mean and mode of 6. Find the values of a and b given $a > b$.

3 The histograms alongside show the times for the 100 metre freestyle recorded by members of a swimming squad.

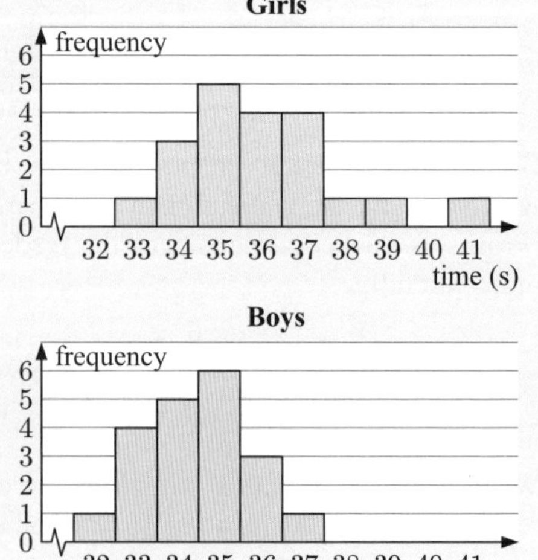

a Copy and complete:

Distribution	*Girls*	*Boys*
shape		
median		
mean		
modal class		

b Discuss the distributions of times for the boys and girls. What conclusion can you make?

4 Find k given that 2, 5, k, k, 3, k, 7, and 4 have a mean of 6.

5 Consider the column graph alongside.

a Find the:

i mode

ii median

iii mean of the data.

b Describe the distribution of the data.

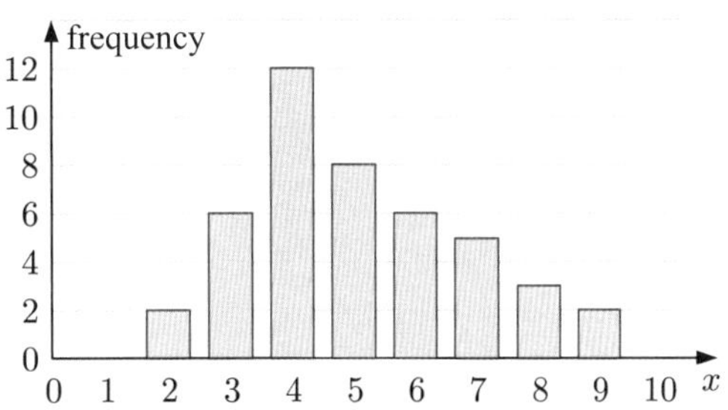

6 The table alongside shows the number of appointments at a dentist surgery over 54 days.

a Draw a column graph to display the data.

b Describe the distribution of the data.

c Find the:

i mode **ii** mean of the data.

Number of appointments	*Frequency*
2	1
3	2
4	5
5	7
6	11
7	15
8	9
9	4

7 Suppose a, b, c, d, and e have a mean of 8. Find the mean of $10 - a$, $10 - b$, $20 - c$, $20 - d$, and $50 - e$.

8 The data set $\{12, 13, 8, 10, 14, 7, a, b\}$ has mean of 10, and a variance of 8.5. Find a and b given that $a < b$.

9 Consider the data set $\{8, 11, 12, 9, a\}$.

a Find the mean of the data set in terms of a.

b Given that the variance of the data set is 6, find the possible values of a.

10 Prove that the variance of any set of five consecutive integers is 2.

REVIEW SET 23B — CALCULATOR

1 The data below shows the distance in metres that Thabiso threw a baseball:

71.2	65.1	68.0	71.1	74.6	68.8	83.2	85.0	74.5	87.4
84.3	77.0	82.8	84.4	80.6	75.9	89.7	83.2	97.5	82.9
90.5	85.5	90.7	92.9	95.6	85.5	64.6	73.9	80.0	86.5

a Determine the highest and lowest value for the data set.

b Choose between 6 and 12 groups into which all the data values can be placed.

c Prepare a frequency distribution table.

d Draw a frequency histogram for the data.

e Determine the: **i** mean **ii** median.

2 Consider the data set: $k - 2$, k, $k + 3$, $k + 3$.

a Show that the mean of the data set is equal to $k + 1$.

b Suppose each number in the data set is increased by 2. Find the new mean of the data set in terms of k.

3 Consider the following distribution of continuous grouped data:

Scores	$0 \leqslant x < 10$	$10 \leqslant x < 20$	$20 \leqslant x < 30$	$30 \leqslant x < 40$	$40 \leqslant x < 50$
Frequency	1	13	27	17	2

Estimate the mean and standard deviation of the data.

4 The number of litres of petrol purchased by a group of motor vehicle drivers is shown alongside. Estimate the mean and standard deviation of the number of litres purchased.

Litres	*Number of vehicles*
$15 \leqslant l < 20$	5
$20 \leqslant l < 25$	13
$25 \leqslant l < 30$	17
$30 \leqslant l < 35$	29
$35 \leqslant l < 40$	27
$40 \leqslant l < 45$	18
$45 \leqslant l < 50$	7

5 The table alongside shows the number of matches in a sample of boxes.

Number	47	48	49	50	51	52
Frequency	21	29	35	42	18	31

a Find the mean and standard deviation for this data.

b Does this result justify a claim that the average number of matches per box is 50?

6 Katie loves cats. She visits every house in her street to find out how many cats live there. The responses are given below:

Number of cats	*Frequency*
0	36
1	9
2	11
3	5
4	1
5	1

a Draw a graph to display this data.

b Describe the distribution.

c Find the: **i** mode **ii** mean **iii** median.

d Which of the measures of centre is most appropriate for this data? Explain your answer.

7 The table shows the length of time cars spent in a particular parking lot in one day.

a How many cars parked in the parking lot?

b Estimate the mean and standard deviation of the data.

Time (t hours)	*Frequency*
$0 < t \leqslant 1$	32
$1 < t \leqslant 2$	85
$2 < t \leqslant 3$	123
$3 < t \leqslant 4$	97
$4 < t \leqslant 5$	62
$5 < t \leqslant 6$	27

8 Consider the data set: 42 58 74 62 51 45 73 54 66 84.
Find the: **a** mean **b** median **c** standard deviation

9 Friends Kevin and Felicity each completed a set of 20 crossword puzzles. The time taken, in minutes, to complete each puzzle is shown below.

Kevin

37	53	47	33	39
49	37	48	32	36
39	42	34	29	52
48	33	56	39	41

Felicity

33	36	41	26	52
38	49	57	39	44
48	25	34	27	53
38	34	35	50	31

a Find the mean of each data set.
b Find the standard deviation of each data set.
c Who generally solved the puzzles faster?
d Who was more consistent?

10 A set of 20 data values $\{x_1, x_2, ..., x_{20}\}$ has $\sum_{i=1}^{20} x_i^2 = 2872$ and $\mu = 11$. Find the variance of the data set.

REVIEW SET 23C

1 The winning margin in 100 basketball games was recorded. The results are given alongside:

a Is the winning margin discrete or continuous?
b Draw an appropriate graph to represent this information.
c Can you calculate the mean winning margin exactly? Explain your answer.

Margin (points)	*Frequency*
1 - 10	13
11 - 20	35
21 - 30	27
31 - 40	18
41 - 50	7

2 The following distribution has a mean score of 5.7:

a Find the value of x.
b Hence find the variance of the distribution.

Score	2	5	x	$x+6$
Frequency	3	2	4	1

3 The table alongside shows the number of customers visiting a supermarket on various days.
Estimate the mean number of customers per day.

Number of customers	*Frequency*
250 - 299	14
300 - 349	34
350 - 399	68
400 - 449	72
450 - 499	54
500 - 549	23
550 - 599	7

4 The data set 33, 18, 25, 40, 36, 41, m, n has a mode of 36 and a mean of 32. Find m and n given that $m < n$.

5 Consider the data set $\{m, m+4, m-2, m+1, m+6, m-3\}$.

a Find, in terms of m, the median and mean of the data set.
b Find the variance of the data set.

6 The weekly supermarket bills for a number of families was observed and recorded in the table given.
Estimate the mean bill and the standard deviation of the bills.

Bill (€)	*Number of families*
70 - 79.99	27
80 - 89.99	32
90 - 99.99	48
100 - 109.99	25
110 - 119.99	37
120 - 129.99	21
130 - 139.99	18
140 - 149.99	7

7 Pratik is a quality control officer for a biscuit company. He needs to check that 250 g of biscuits go into each packet, but realises that the weight in each packet will vary slightly.

a Would you expect the standard deviation for the whole population to be the same for one day as it is for one week? Explain your answer.

b If a sample of 100 packets is measured each day, what measure would be used to check:

i that an average of 250 g of biscuits goes into each packet

ii the variability of the mass going into each packet?

c Explain the significance of a low standard deviation in this case.

8 A group of students were asked how long they slept for last night. The results are shown in the table.
Estimate the mean and standard deviation of the data.

Time (t hours)	*Frequency*
$6 \leqslant t < 7$	5
$7 \leqslant t < 8$	19
$8 \leqslant t < 9$	38
$9 \leqslant t < 10$	22
$10 \leqslant t < 11$	6

9 Roger and Clinton play golf together every Saturday. Their scores for the past 30 weeks are given below.

Roger

78 74 82 85 79 73
92 79 88 77 85 87
82 96 90 80 82 88
74 89 93 91 85 78
75 94 79 94 90 85

Clinton

77 79 82 84 75 75
73 78 71 83 85 72
77 76 78 75 74 81
80 77 74 71 72 75
79 81 72 78 76 73

a Find the mean and standard deviation of each data set.

b Which player generally has the lower score?

c Which player has the greater variation in their scores?

Chapter 24

Probability

Syllabus reference: 5.2, 5.3, 5.4

Contents:

OPENING PROBLEM

In the late 17th century, English mathematicians compiled and analysed mortality tables which showed the number of people who died at different ages. From these tables they could estimate the probability that a person would be alive at a future date. This led to the establishment of the first life-insurance company in 1699.

Life Insurance Companies use statistics on **life expectancy** and **death rates** to work out the premiums to charge people who insure with them.

The **life table** shown is from Australia. It shows the number of people out of 100 000 births who survive to different ages, and the expected years of remaining life at each age.

For example, we can see that out of 100 000 births, 98 052 males are expected to survive to the age of 20, and from that age the survivors are expected to live a further 54.35 years.

LIFE TABLE					
Male			Female		
Age	*Number surviving*	*Expected remaining life*	Age	*Number surviving*	*Expected remaining life*
0	100 000	73.03	0	100 000	79.46
5	98 809	68.90	5	99 307	75.15
10	98 698	63.97	10	99 125	70.22
15	98 555	59.06	15	98 956	65.27
20	98 052	54.35	20	98 758	60.40
25	97 325	49.74	25	98 516	55.54
30	96 688	45.05	30	98 278	50.67
35	96 080	40.32	35	98 002	45.80
40	95 366	35.60	40	97 615	40.97
45	94 323	30.95	45	96 997	36.22
50	92 709	26.45	50	95 945	31.59
55	89 891	22.20	55	94 285	27.10
60	85 198	18.27	60	91 774	22.76
65	78 123	14.69	65	87 923	18.64
70	67 798	11.52	70	81 924	14.81
75	53 942	8.82	75	72 656	11.36
80	37 532	6.56	80	58 966	8.38
85	20 998	4.79	85	40 842	5.97
90	8416	3.49	90	21 404	4.12
95	2098	2.68	95	7004	3.00
99	482	2.23	99	1953	2.36

Things to think about:

a Can you use the life table to estimate how many years you can expect to live?

b Can you estimate the chance that a new-born boy or girl reaches the age of 15?

c Can the table be used to estimate the chance that:

- **i** a 15 year old boy *will* reach age 75
- **ii** a 15 year old girl *will not* reach age 75?

d An insurance company sells policies to people to insure them against death over a 30-year period. If the person dies during this period, the beneficiaries receive the agreed payout figure. Why are such policies cheaper to take out for a 20 year old than for a 50 year old?

e How many of your classmates would you expect to be alive and able to attend a 30 year class reunion?

In the field of **probability theory** we use mathematics to describe the **chance** or **likelihood** of an event happening.

We apply probability theory in physical and biological sciences, economics, politics, sport, life insurance, quality control, production planning, and a host of other areas.

We assign to every event a number which lies between 0 and 1 inclusive. We call this number a **probability**.

> An **impossible** event which has 0% chance of happening is assigned a probability of 0.
>
> A **certain** event which has 100% chance of happening is assigned a probability of 1.
>
> All other events can be assigned a probability between 0 and 1.

The number line below shows how we could interpret different probabilities:

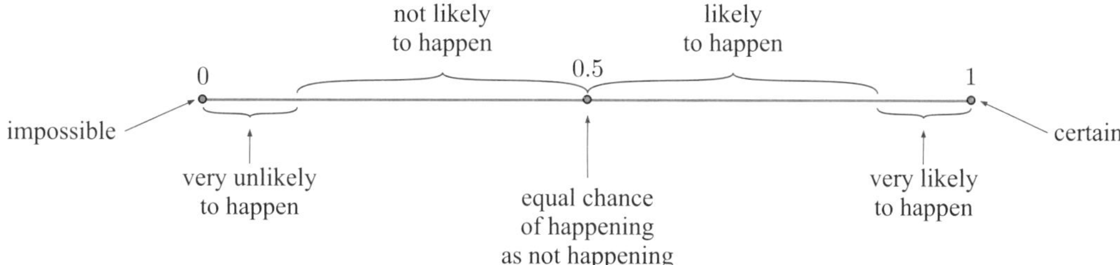

The assigning of probabilities is usually based on either:

- observing the results of an experiment (experimental probability), or
- using arguments of symmetry (theoretical probability).

A EXPERIMENTAL PROBABILITY

In experiments involving chance we use the following terms to talk about what we are doing and the results we obtain:

- The **number of trials** is the total number of times the experiment is repeated.
- The **outcomes** are the different results possible for one trial of the experiment.
- The **frequency** of a particular outcome is the number of times that this outcome is observed.
- The **relative frequency** of an outcome is the frequency of that outcome expressed as a fraction or percentage of the total number of trials.

For example, when a small plastic cone was tossed into the air 279 times it fell on its *side* 183 times and on its *base* 96 times.

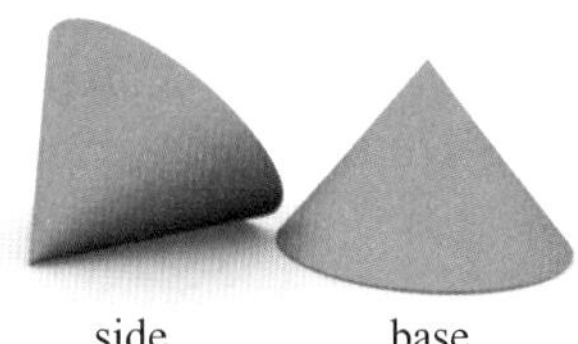

We say:

- the number of trials is 279
- the outcomes are *side* and *base*
- the frequencies of *side* and *base* are 183 and 96 respectively
- the relative frequencies of *side* and *base* are $\frac{183}{279} \approx 0.656$ and $\frac{96}{279} \approx 0.344$ respectively.

  In the absence of any further data, the relative frequency of each event is our best estimate of the probability of that event occurring.

Experimental probability = relative frequency.

In this case, we write

Experimental P(side) = the experimental probability the cone will land on its side when tossed
≈ 0.656

Experimental P(base) = the experimental probability the cone will land on its base when tossed
≈ 0.344

The larger the number of trials, the more confident we are that the estimated probability will be accurate.

INVESTIGATION 1 — TOSSING DRAWING PINS

If a drawing pin tossed in the air finishes  we say it has finished on its *back*. If it finishes  we say it has finished on its *side*.

If two drawing pins are tossed simultaneously, the possible results are:

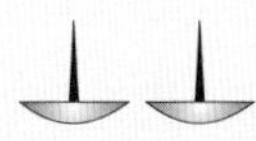
two backs

back and side

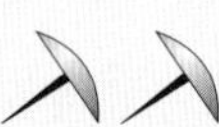
two sides

What to do:

1. Obtain two drawing pins of the same shape and size. Toss the pair 80 times and record the outcomes in a table.
2. Obtain relative frequencies (experimental probabilities) for each of the three outcomes.
3. Pool your results with four other people and so obtain experimental probabilities from 400 tosses. The other people must have pins with the same shape.
4. Which gives the more reliable probability estimates, your results or the whole group's? Explain your answer.
5. Keep your results as they may be useful later in this chapter.

In some situations, for example in the **Investigation** above, experimentation is the only way of obtaining probabilities.

EXERCISE 24A

1. When a batch of 145 paper clips was dropped onto 6 cm by 6 cm squared paper, it was observed that 113 fell completely inside squares and 32 finished up on the grid lines. Find, to 2 decimal places, the experimental probability of a clip falling:

   a inside a square b on a line.

2.

Length	*Frequency*
0 - 19	17
20 - 39	38
40 - 59	19
60+	4

Jose surveyed the length of TV commercials (in seconds). Find, to 3 decimal places, the experimental probability that a randomly chosen TV commercial will last:

   a 20 to 39 seconds b more than a minute
   c between 20 and 59 seconds (inclusive).

3 Betul records the number of phone calls she receives over a period of consecutive days.

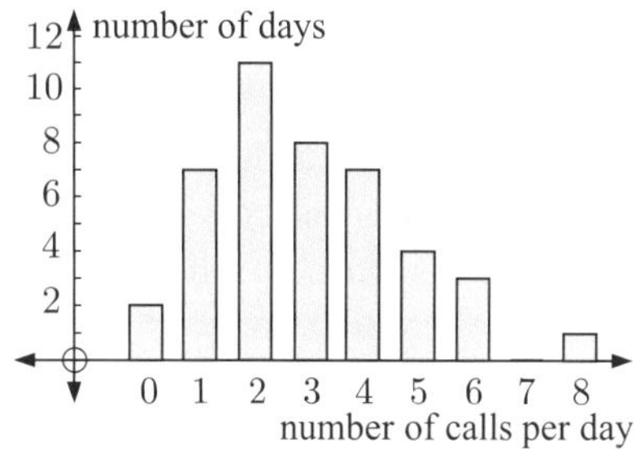

a For how many days did the survey last?

b Estimate Betul's chance of receiving:

i no phone calls in a day

ii 5 or more phone calls in a day

iii less than 3 phone calls in a day.

4 Pat does a lot of travelling in her car, and she keeps records on how often she fills her car with petrol. The table alongside shows the frequencies of the number of days between refills. Estimate the likelihood that:

a there is a four day gap between refills

b there is at least a four day gap between refills.

Days between refills	*Frequency*
1	37
2	81
3	48
4	17
5	6
6	1

INVESTIGATION 2 COIN TOSSING EXPERIMENTS

The coins of most currencies have two distinct faces, usually referred to as 'heads' and 'tails'. When we toss a coin in the air, we expect it to finish on a head or tail with equal likelihood.

In this Investigation the coins do not have to be all the same type.

What to do:

1 Toss *one coin* 40 times. Record the number of heads in each trial, in a table:

Result	*Tally*	*Frequency*	*Relative frequency*
1 head			
no heads			

2 Toss *two coins* 60 times. Record the number of heads in each trial, in a table:

Result	*Tally*	*Frequency*	*Relative frequency*
2 heads			
1 head			
no heads			

3 Toss *three coins* 80 times. Record the number of heads in each trial, in a table:

Result	*Tally*	*Frequency*	*Relative frequency*
3 heads			
2 heads			
1 head			
no heads			

4 Share your results to **1**, **2**, and **3** with several other students. Comment on any similarities and differences.

5 Pool your results and find new relative frequencies for tossing one coin, two coins, and three coins.

6 Click on the icon to examine a coin tossing simulation.
Set it to toss one coin 10 000 times.
Run the simulation ten times, each time recording the relative frequency for each possible result. Comment on these results. Do your results agree with what you expected?

COIN TOSSING

7 Experiment with the simulation for *two coins* and then *three coins*.

From the previous **Investigation** you should have observed that, when tossing two coins, there are roughly twice as many 'one head' results as there are 'no heads' or 'two heads'.

The explanation for this is best seen using two different coins where you could get:

two heads one head one head no heads

We should expect the ratio two heads : one head : no heads to be 1 : 2 : 1. However, due to chance, there will be variations from this when we look at experimental results.

INVESTIGATION 3 DICE ROLLING EXPERIMENTS

You will need: At least one normal six-sided die with numbers 1 to 6 on its faces. Several dice would be useful to speed up the experimentation.

WORKSHEET

What to do:

1 List the possible outcomes for the uppermost face when the die is rolled.

2 Suppose the die is rolled 60 times.
Copy and complete the following table of your **expected results**:

Outcomes	*Expected frequency*	*Expected relative frequency*
1		
2		
⋮		
6		

3 Roll the die 60 times. Record the results in a table like the one shown:

Outcome	*Tally*	*Frequency*	*Relative frequency*
1			
2			
⋮			
6			
	Total	60	1

4 Pool as much data as you can with other students.

  a Look at similarities and differences from one set to another.

  b Summarise the overall pooled data in one table.

  c Compare your results with your expectation in **2**.

SIMULATION

5 Use the die rolling simulation online to roll the die 10 000 times. Repeat this 10 times. On each occasion, record your results in a table like that in **3**. Do your results further confirm your expected results?

6 The different possible outcomes when a pair of dice is rolled are shown alongside.

There are 36 possible outcomes.

Notice that the three outcomes {1, 3}, {2, 2}, and {3, 1}, give a sum of 4.

We observe that the possible sums of the dice are 2, 3, ..., 12.

Using the illustration above, copy and complete the table of **expected results**:

Sum of dice	2	3	4	5	6	7	8	9	10	11	12
Fraction of possible outcomes with this sum			$\frac{3}{36}$								
Fraction as decimal			0.083								

7 If a pair of dice is rolled 360 times, how many of each result (2, 3, 4, ..., 12) would you expect to get? Extend the table in **6** by adding another row and writing your **expected frequencies** within it.

8 Toss two dice 360 times. Record the *sum of the two numbers* for each toss in a table.

WORKSHEET

Sum	*Tally*	*Frequency*	*Relative frequency*
2			
3			
4			
⋮			
12			
	Total	360	1

9 Pool as much data as you can with other students and find the overall relative frequency of each sum.

10 Use the two dice simulation online to roll the pair of dice 10 000 times. Repeat this 10 times and on each occasion record your results in a table like that in **8**. Are your results consistent with your expectations?

SIMULATION

B SAMPLE SPACE

A **sample space** U is the set of all possible outcomes of an experiment.
It is also referred to as the **universal set** U.

There are a variety of ways of representing or illustrating sample spaces, including:

- lists
- tables of outcomes
- 2-dimensional grids
- Venn diagrams
- tree diagrams

We will use tables and Venn diagrams later in the chapter.

LISTING OUTCOMES

Example 1 Self Tutor

List the sample space of possible outcomes for:

a tossing a coin

b rolling a normal die.

a When a coin is tossed, there are two possible outcomes.
$\therefore$ sample space = {H, T}

b When a die is rolled, there are 6 possible outcomes.
$\therefore$ sample space = {1, 2, 3, 4, 5, 6}

2-DIMENSIONAL GRIDS

When an experiment involves more than one operation we can still use listing to illustrate the sample space. However, a grid is often more efficient.

Example 2 Self Tutor

Use a 2-dimensional grid to illustrate the possible outcomes when 2 coins are tossed.

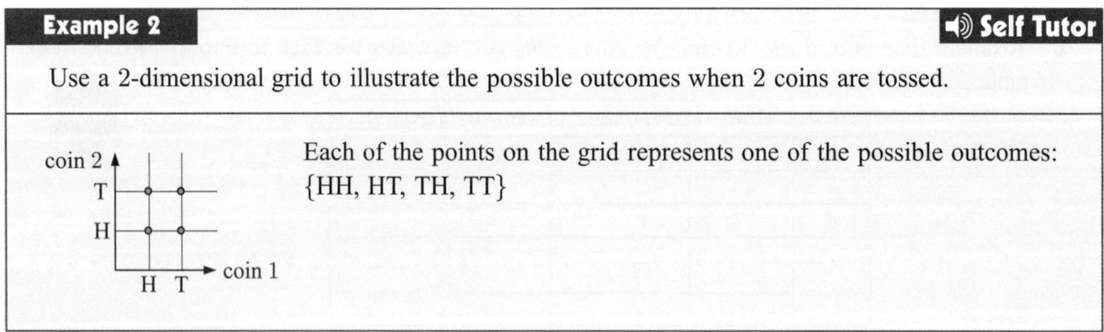

Each of the points on the grid represents one of the possible outcomes:
{HH, HT, TH, TT}

TREE DIAGRAMS

The sample space in **Example 2** could also be represented by a tree diagram. The advantage of tree diagrams is that they can be used when more than two operations are involved.

Example 3 Self Tutor

Illustrate, using a tree diagram, the possible outcomes when:

a tossing two coins

b drawing two marbles from a bag containing many red, green, and yellow marbles.

a

coin 1	coin 2	possible outcomes
H	H	HH
	T	HT
T	H	TH
	T	TT

Each 'branch' gives a different outcome.
The sample space is {HH, HT, TH, TT}.

b

marble 1	marble 2	possible outcomes
R	R	RR
	G	RG
	Y	RY
G	R	GR
	G	GG
	Y	GY
Y	R	YR
	G	YG
	Y	YY

EXERCISE 24B

1. List the sample space for the following:
   a. twirling a square spinner labelled A, B, C, D
   b. the sexes of a 2-child family
   c. the order in which 4 blocks A, B, C and D can be lined up
   d. the 8 different 3-child families.

2. Illustrate on a 2-dimensional grid the sample space for:
   a. rolling a die and tossing a coin simultaneously
   b. rolling two dice
   c. rolling a die and spinning a spinner with sides A, B, C, D
   d. twirling two square spinners, one labelled A, B, C, D and the other 1, 2, 3, 4.

3. Illustrate on a tree diagram the sample space for:
   a. tossing a 5-cent and a 10-cent coin simultaneously
   b. tossing a coin and twirling an equilateral triangular spinner labelled A, B, and C
   c. twirling two equilateral triangular spinners labelled 1, 2, and 3, and X, Y, and Z
   d. drawing two tickets from a hat containing a large number of pink, blue, and white tickets.

C THEORETICAL PROBABILITY

Consider the **octagonal spinner** alongside.

Since the spinner is symmetrical, when it is spun the arrowed marker could finish with equal likelihood on each of the sections marked 1 to 8.

The likelihood of obtaining the outcome 4 would be:

1 chance in 8, $\frac{1}{8}$, $12\frac{1}{2}\%$, or 0.125.

This is a **mathematical** or **theoretical** probability and is based on what we theoretically expect to occur. It is the chance of that event occurring in any trial of the experiment.

If we are interested in the event of getting a result of *6 or more* from one spin of the octagonal spinner, there are three favourable results (6, 7, or 8) out of the eight possible results. Since each of these is equally likely to occur, $P(6 \text{ or more}) = \frac{3}{8}$.

We read $\frac{3}{8}$ as
"3 chances in 8".

In general, for an event A containing **equally likely** possible results, the probability of A occurring is

$$P(A) = \frac{\textbf{the number of members of the event } A}{\textbf{the total number of possible outcomes}} = \frac{n(A)}{n(U)}.$$

Example 4 — Self Tutor

A ticket is *randomly selected* from a basket containing 3 green, 4 yellow, and 5 blue tickets. Determine the probability of getting:

- **a** a green ticket
- **b** a green or yellow ticket
- **c** an orange ticket
- **d** a green, yellow, or blue ticket.

There are $3+4+5=12$ tickets which could be selected with equal chance.

a $P(G) = \frac{3}{12} = \frac{1}{4}$

b $P(G \text{ or } Y) = \frac{3+4}{12} = \frac{7}{12}$

c $P(O) = \frac{0}{12} = 0$

d $P(G, Y, \text{ or } B) = \frac{3+4+5}{12} = 1$

From **Example 4**, notice that:

- In **c** an orange result cannot occur. The calculated probability is 0 because the event has *no chance of occurring.*
- In **d** the outcome of a green, yellow, or blue is certain to occur. It is 100% likely so the theoretical probability is 1.

Events which have *no chance of occurring* or probability 0, or are *certain to occur* or probability 1, are two extremes.

For any event A, the probability $P(A)$ of A occurring satisfies $\mathbf{0 \leqslant P(A) \leqslant 1}$.

Example 5 — Self Tutor

An ordinary 6-sided die is rolled once. Determine the chance of:

- **a** getting a 6
- **b** not getting a 6
- **c** getting a 1 or 2
- **d** not getting a 1 or 2

The sample space of possible outcomes is $\{1, 2, 3, 4, 5, 6\}$.

a $P(6) = \frac{1}{6}$

b $P(\text{not a } 6) = P(1, 2, 3, 4, \text{ or } 5) = \frac{5}{6}$

c $P(1 \text{ or } 2) = \frac{2}{6} = \frac{1}{3}$

d $P(\text{not a 1 or } 2) = P(3, 4, 5, \text{ or } 6) = \frac{4}{6} = \frac{2}{3}$

COMPLEMENTARY EVENTS

In **Example 5** notice that $P(6) + P(\text{not a } 6) = 1$ and that

$$P(1 \text{ or } 2) + P(\text{not a } 1 \text{ or } 2) = 1.$$

This is no surprise as *a* 6 and *not a* 6 are **complementary events**. It is certain that one of the events will occur, and impossible for both of them to occur at the same time.

> Two events are **complementary** if exactly one of the events *must* occur.
> If A is an event, then A' is the complementary event of A, or 'not A'.
>
> $$\mathbf{P(A) + P(A') = 1}$$

EXERCISE 24C.1

1 A marble is randomly selected from a box containing 5 green, 3 red, and 7 blue marbles. Determine the probability that the marble is:

- a red
- b green
- c blue
- d not red
- e neither green nor blue
- f green or red.

2 A carton of a dozen eggs contains eight brown eggs. The rest are white.

- a How many white eggs are there in the carton?
- b Find the probability that an egg selected at random is:
  - i brown
  - ii white.

3 A dart board has 36 sectors labelled 1 to 36. Determine the probability that a dart thrown at the centre of the board will hit a sector labelled with:

- a a multiple of 4
- b a number between 6 and 9 inclusive
- c a number greater than 20
- d 9
- e a multiple of 13
- f an odd number that is a multiple of 3
- g a multiple of both 4 and 6
- h a multiple of 4 or 6, or both.

4 What is the probability that a randomly chosen person has his or her next birthday:

- a on a Tuesday
- b on a weekend
- c in July
- d in January or February?

5
- a List the six different orders in which Antti, Kai, and Neda may sit in a row.
- b If the three of them sit randomly in a row, determine the probability that:
  - i Antti sits in the middle
  - ii Antti sits at the left end
  - iii Antti does not sit at the right end
  - iv Kai and Neda are seated together.

6
- a List the 8 possible 3-child families according to the gender of the children. For example, GGB means "*the first is a girl, the second is a girl, the third is a boy*".
- b Assuming that each of these is equally likely to occur, determine the probability that a randomly selected 3-child family consists of:
  - i all boys
  - ii all girls
  - iii boy then girl then girl
  - iv two girls and a boy
  - v a girl for the eldest
  - vi at least one boy.

7 a List, in systematic order, the 24 different orders in which four people A, B, C, and D may sit in a row.

b Determine the probability that when the four people sit at random in a row:

i A sits on one of the end seats
ii B sits on one of the two middle seats
iii A and B are seated together
iv A, B, and C are seated together, not necessarily in that order.

USING GRIDS TO FIND PROBABILITIES

Two-dimensional grids can give us excellent visual displays of sample spaces. We can use them to count favourable outcomes and so calculate probabilities.

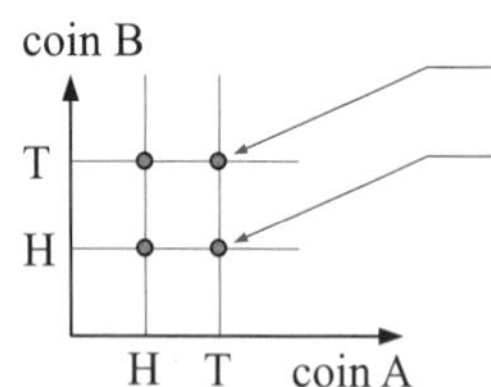

This point represents 'a tail from coin A' and 'a tail from coin B'.

This point represents 'a tail from coin A' and 'a head from coin B'.

There are four members of the sample space.

Example 6 **Self Tutor**

Use a two-dimensional grid to illustrate the sample space for tossing a coin and rolling a die simultaneously. From this grid determine the probability of:

a tossing a head
b getting a tail and a 5
c getting a tail or a 5.

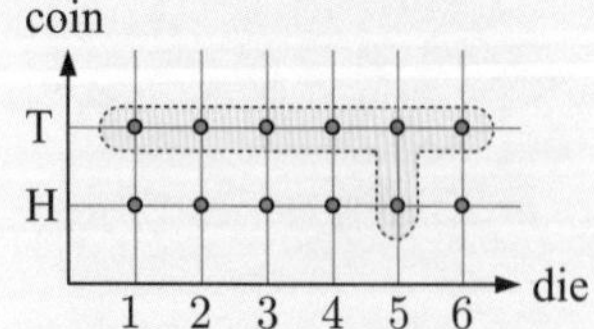

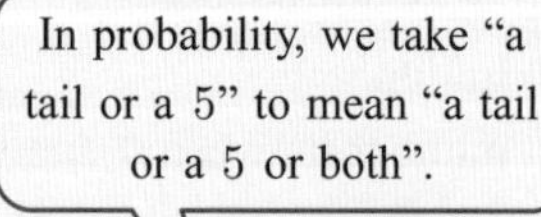

In probability, we take "a tail or a 5" to mean "a tail or a 5 or both".

There are 12 members in the sample space.

a $P(\text{head}) = \frac{6}{12} = \frac{1}{2}$

b $P(\text{tail and a '5'}) = \frac{1}{12}$

c $P(\text{tail or a '5'}) = \frac{7}{12}$ {the enclosed points}

EXERCISE 24C.2

1 Draw the grid of the sample space when a 5-cent and a 10-cent coin are tossed simultaneously. Hence determine the probability of getting:

a two heads
b two tails
c exactly one head
d at least one head.

2 A coin and a pentagonal spinner with sectors 1, 2, 3, 4, and 5 are tossed and spun respectively.

a Draw a grid to illustrate the sample space of possible outcomes.

b How many outcomes are possible?

c Use your grid to determine the chance of getting:
 i a tail and a 3
 ii a head and an even number
 iii an odd number
 iv a head or a 5.

"A head or a 5" means "a head or a 5, or both".

3 A pair of dice is rolled. The 36 different possible results are illustrated in the 2-dimensional grid.
Use the grid to determine the probability of getting:

a two 3s
b a 5 and a 6
c a 5 or a 6 (or both)
d at least one 6
e exactly one 6
f no sixes
g a sum of 7
h a sum greater than 8
i a sum of 7 or 11
j a sum of no more than 8.

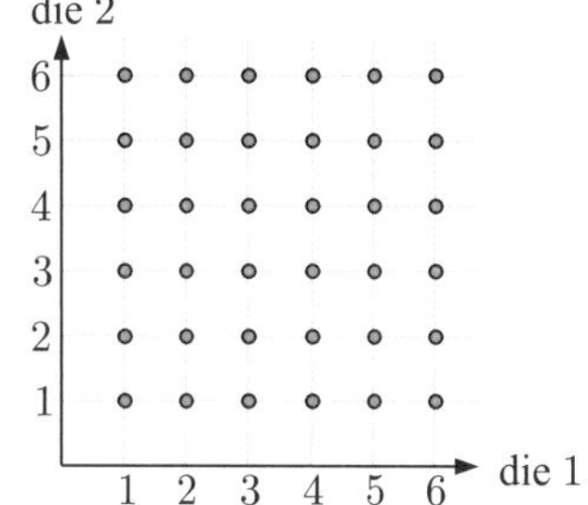

DISCUSSION

Three children have been experimenting with a coin, tossing it in the air and recording the outcomes. They have done this 10 times and have recorded 10 tails. Before the next toss they make the following statements:

Jackson: "It's got to be a head next time!"

Sally: "No, it always has an equal chance of being a head or a tail. The coin cannot remember what the outcomes have been."

Amy: "Actually, I think it will probably be a tail again, because I think the coin must be biased. It might be weighted so it is more likely to give a tail."

Discuss the statements of each child. Who do you think is correct?

D TABLES OF OUTCOMES

Tables of outcomes are tables which compare two categorical variables. They usually result from a survey.

For example, a group of teachers was asked which mode of transport they used to travel to school. Their responses are summarised in the table below. The variables are *gender* and *mode of transport*.

	Car	Bicycle	Bus
Male	37	10	10
Female	30	5	13

← 13 female teachers catch the bus to school.

In the following example we will see how these tables can be used to estimate probabilities. To help us, we extend the table to include totals in each row and column.

Example 7 — Self Tutor

People exiting a new ride at a theme park were asked whether they liked or disliked the ride. The results are shown in the table alongside.

	Child	Adult
Liked the ride	55	28
Disliked the ride	17	30

Use this table to estimate the probability that a randomly selected person who went on the ride:

a liked the ride

b is a child *and* disliked the ride

c is an adult *or* disliked the ride

d liked the ride, given that he or she is a child

e is an adult, given that he or she disliked the ride.

We extend the table to include totals:

	Child	Adult	Total
Liked the ride	55	28	83
Disliked the ride	17	30	47
Total	72	58	130

a 83 out of the 130 people surveyed liked the ride.

$\therefore$ P(liked the ride) $\approx \frac{83}{130} \approx 0.638$

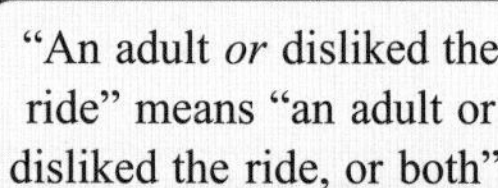
"An adult *or* disliked the ride" means "an adult or disliked the ride, or both".

b 17 of the 130 people surveyed are children who disliked the ride.

$\therefore$ P(child *and* disliked the ride) $\approx \frac{17}{130} \approx 0.131$

c $28 + 30 + 17 = 75$ of the 130 people are adults or disliked the ride.

$\therefore$ P(adults *or* disliked the ride) $\approx \frac{75}{130} \approx 0.577$

d Of the 72 children, 55 liked the ride.

$\therefore$ P(liked the ride given that he or she is a child) $\approx \frac{55}{72} \approx 0.764$

e Of the 47 people who disliked the ride, 30 were adults.

$\therefore$ P(adult given that he or she disliked the ride) $\approx \frac{30}{47} \approx 0.638$

EXERCISE 24D

1 A sample of adults in a suburb were surveyed about their current employment status and their level of education. The results are summarised in the table below.

	Employed	Unemployed
Attended university	225	34
Did not attend university	197	81

Estimate the probability that the next randomly chosen adult:

a attended university

b did not attend university and is currently employed

c is unemployed

d is employed, given that they attended university

e attended university, given that they are unemployed.

2 The types of ticket used to gain access to a basketball match were recorded as people entered the stadium. The results are shown alongside.

	Adult	Child
Season ticket holder	1824	779
Not a season ticket holder	3247	1660

a What was the total attendance for the match?

b One person is randomly selected to sit on the home team's bench. Find the probability that the person selected:

i is a child **ii** is not a season ticket holder

iii is an adult season ticket holder.

3 A small hotel in London has kept a record of all the room bookings made for the year, categorised by season and booking type. Find the probability that a randomly selected booking was:

	Single	Double	Family
Peak season	125	220	98
Off-peak season	248	192	152

a in the peak season **b** a single room in the off-peak season

c a single or a double room **d** a family room, given that it was in the off-peak season

e in the peak season, given that it was not a single room.

E COMPOUND EVENTS

Suppose box X contains 2 blue and 2 green balls, and box Y contains 1 white and 3 red balls. A ball is randomly selected from each of the boxes. Determine the probability of getting "a blue ball from X *and* a red ball from Y".

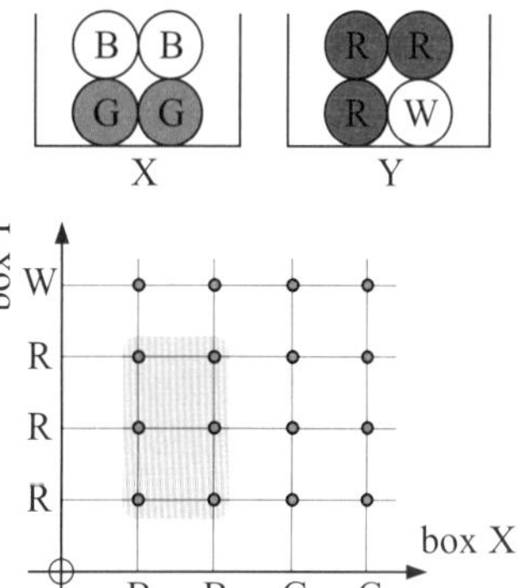

By illustrating the sample space on the two-dimensional grid shown, we can see that 6 of the 16 possibilities are blue from X and red from Y. Each of the outcomes is equally likely, so

$$\text{P(blue from X \textbf{and} red from Y)} = \tfrac{6}{16}.$$

In this section we look for a quicker method for finding the probability in such a situation.

INVESTIGATION 4 PROBABILITIES OF COMPOUND EVENTS

Suppose a coin is tossed and a die is rolled at the same time. The result of the coin toss will be called outcome A, and the result of the die roll will be outcome B.

What to do:

1 Copy and complete, using a 2-dimensional grid if necessary:

	P(A and B)	P(A)	P(B)
P(a head and a 4)			
P(a head and an odd number)			
P(a tail and a number larger than 1)			
P(a tail and a number less than 3)			

2 What is the connection between P(A and B), P(A), and P(B)?

INVESTIGATION 5 REVISITING DRAWING PINS

We cannot find by theoretical argument the probability that a drawing pin will land on its back . We can only find this probability by experimentation.

When tossing *two* drawing pins, can we still use the rule for compound events:

P(back *and* back) $=$ P(back) $\times$ P(back)?

What to do:

1 From **Investigation 1** on page **736**, what is your estimate of P(back *and* back)?

2 **a** Count the number of drawing pins in a full packet. They must be identical to each other and the same ones that you used in **Investigation 1**.

b Drop the whole packet onto a solid surface and count the number of *backs* and *sides*. Repeat this several times. Pool results with others and finally estimate P(back).

c Hence find P(back) $\times$ P(back) using **2 b**.

3 Is P(back *and* back) $\approx$ P(back) $\times$ P(back)?

From **Investigations 4** and **5**, it seems that if A and B are two events for which the occurrence of each one does not affect the occurrence of the other, then $\text{P}(A \textbf{ and } B) = \text{P}(A) \times \text{P}(B)$.

Before we can formalise this as a rule, however, we need to distinguish between **independent** and **dependent** events.

INDEPENDENT EVENTS

Events are **independent** if the occurrence of either of them does not affect the probability that the others occur.

Consider again the example on the previous page. Suppose we happen to choose a blue ball from box X. This does not affect the outcome when we choose a ball from box Y. The probability of selecting a red ball from box Y is $\frac{3}{4}$ regardless of which colour ball is selected from box X.

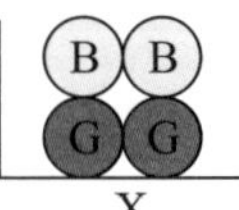

X

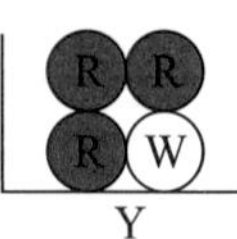

Y

So, the two events "a blue ball from X" and "a red ball from Y" are independent.

If A and B are **independent events** then $\text{P}(A \textbf{ and } B) = \text{P}(A) \times \text{P}(B)$.

This rule can be extended for any number of independent events.

For example: If A, B, and C are all **independent events**, then $\text{P}(A \textbf{ and } B \textbf{ and } C) = \text{P}(A) \times \text{P}(B) \times \text{P}(C)$.

Example 8 **Self Tutor**

A coin and a die are tossed simultaneously. Determine the probability of getting a head and a 3 without using a grid.

P(a head and a 3) $=$ P(H) $\times$ P(3) {events are independent}

$= \frac{1}{2} \times \frac{1}{6} = \frac{1}{12}$

EXERCISE 24E.1

1. At a mountain village in Papua New Guinea it rains on average 6 days a week. Determine the probability that it rains on:
   a. any one day
   b. two successive days
   c. three successive days.

2. A coin is tossed 3 times. Determine the probability of getting the following sequences of results:
   a. head then head then head
   b. tail then head then tail.

3. A school has two photocopiers. On any one day, machine A has an 8% chance of malfunctioning and machine B has a 12% chance of malfunctioning. Determine the probability that on any one day both machines will:
   a. malfunction
   b. work effectively.

4. A couple would like 4 children, none of whom will be adopted. They will be disappointed if the children are not born in the order boy, girl, boy, girl. Determine the probability that they will be:
   a. happy with the order of arrival
   b. unhappy with the order of arrival.

5. Two marksmen fire at a target simultaneously. Jiri hits the target 70% of the time and Benita hits it 80% of the time. Determine the probability that:
   a. they both hit the target
   b. they both miss the target
   c. Jiri hits but Benita misses
   d. Benita hits but Jiri misses.

6. 

   An archer always hits a circular target with each arrow fired, and hits the bullseye on average 2 out of every 5 shots. If 3 arrows are fired at the target, determine the probability that the bullseye is hit:
   a. every time
   b. the first two times, but not on the third shot
   c. on no occasion.

DEPENDENT EVENTS

Suppose a hat contains 5 red and 3 blue tickets. One ticket is randomly chosen, its colour is noted, and it is then put aside and so *not* put back in the hat. A second ticket is then randomly selected. What is the chance that it is red?

If the first ticket was red, P(second is red) $= \frac{4}{7}$ ← 4 reds remaining, ← 7 to choose from

If the first ticket was blue, P(second is red) $= \frac{5}{7}$ ← 5 reds remaining, ← 7 to choose from

So, the probability of the second ticket being red *depends* on what colour the first ticket was. We therefore have **dependent events**.

Two or more events are **dependent** if they are **not independent**.

Events are **dependent** if the occurrence of one of the events *does affect* the occurrence of the other event.

If A and B are dependent events then

$$\text{P}(A \textbf{ then } B) = \text{P}(A) \times \text{P}(B \text{ given that } A \text{ has occurred}).$$

Example 9 — Self Tutor

A box contains 4 red and 2 yellow tickets. Two tickets are randomly selected from the box one by one *without* replacement. Find the probability that:

a both are red **b** the first is red and the second is yellow.

a P(both red)

= P(first selected is red *and* second is red)

= P(first selected is red) × P(second is red given that the first is red)

$= \frac{4}{6} \times \frac{3}{5}$ ← If a red is drawn first, 3 reds remain out of a total of 5. ($\frac{4}{6}$: 4 reds out of a total of 6 tickets)

$= \frac{2}{5}$

b P(first is red *and* second is yellow)

= P(first is red) × P(second is yellow given that the first is red)

$= \frac{4}{6} \times \frac{2}{5}$ ← If a red is drawn first, 2 yellows remain out of a total of 5. ($\frac{4}{6}$: 4 reds out of a total of 6 tickets)

$= \frac{4}{15}$

Example 10 — Self Tutor

A hat contains 20 tickets numbered 1, 2, 3, ..., 20. If 3 tickets are drawn from the hat without replacement, determine the probability that they are all prime numbers.

In each fraction the numerator is the number of outcomes in the event. The denominator is the total number of possible outcomes.

{2, 3, 5, 7, 11, 13, 17, 19} are primes.

∴ there are 20 numbers of which 8 are primes.

∴ P(3 primes)

= P(1st drawn is prime *and* 2nd is prime *and* 3rd is prime)

$= \frac{8}{20}$ ← 8 primes out of 20 numbers

$\times \frac{7}{19}$ ← 7 primes out of 19 numbers after a successful first draw

$\times \frac{6}{18}$ ← 6 primes out of 18 numbers after two successful draws

≈ 0.0491

EXERCISE 24E.2

1 A bin contains 12 identically shaped chocolates of which 8 are strawberry creams. If 3 chocolates are selected simultaneously from the bin, determine the probability that:

  a they are all strawberry creams
  b none of them are strawberry creams.

Drawing three chocolates *simultaneously* implies there is no replacement.

2 A box contains 7 red and 3 green balls. Two balls are drawn one after another from the box without replacement. Determine the probability that:

  a both are red
  b the first is green and the second is red
  c a green and a red are obtained.

3 A lottery has 100 tickets which are placed in a barrel. Three tickets are drawn at random from the barrel, without replacement, to decide 3 prizes. If John has 3 tickets in the lottery, determine his probability of winning:

  a first prize
  b first and second prize
  c all 3 prizes
  d none of the prizes.

4 A hat contains 7 names of players in a tennis squad including the captain and the vice captain. If a team of 3 is chosen at random by drawing the names from the hat, determine the probability that it does *not* contain:

  a the captain
  b the captain or the vice captain.

5 Two students are chosen at random from a group of two girls and five boys, all of different ages. Find the probability that the two students chosen will be:

  a two boys
  b the eldest two students.

F TREE DIAGRAMS

Tree diagrams can be used to illustrate sample spaces if the possible outcomes are not too numerous. Once the sample space is illustrated, the tree diagram can be used for determining probabilities.

Consider two archers firing simultaneously at a target. These are independent events.

Li has probability $\frac{3}{4}$ of hitting a target and Yuka has probability $\frac{4}{5}$.

The tree diagram for this information is:

H = hit M = miss

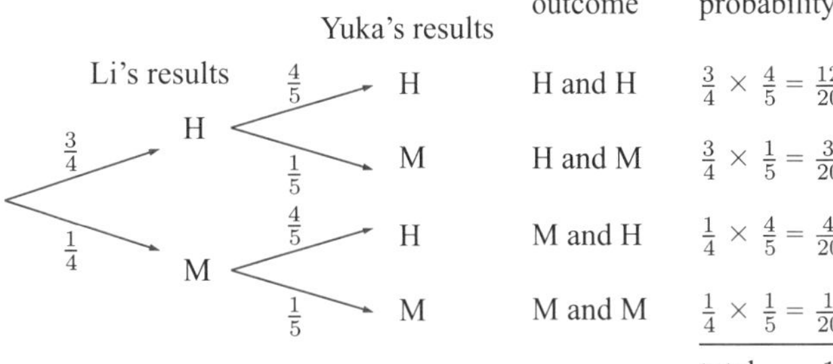

Notice that:

- The probabilities for hitting and missing are marked on the branches.
- There are *four* alternative branches, each showing a particular outcome.
- All outcomes are represented.
- The probability of each outcome is obtained by **multiplying** the probabilities along its branch.

Example 11 Self Tutor

Carl is not having much luck lately. His car will only start 80% of the time and his motorbike will only start 60% of the time.

a Draw a tree diagram to illustrate this situation.

b Use the tree diagram to determine the chance that:

i both will start **ii** Carl can only use his car.

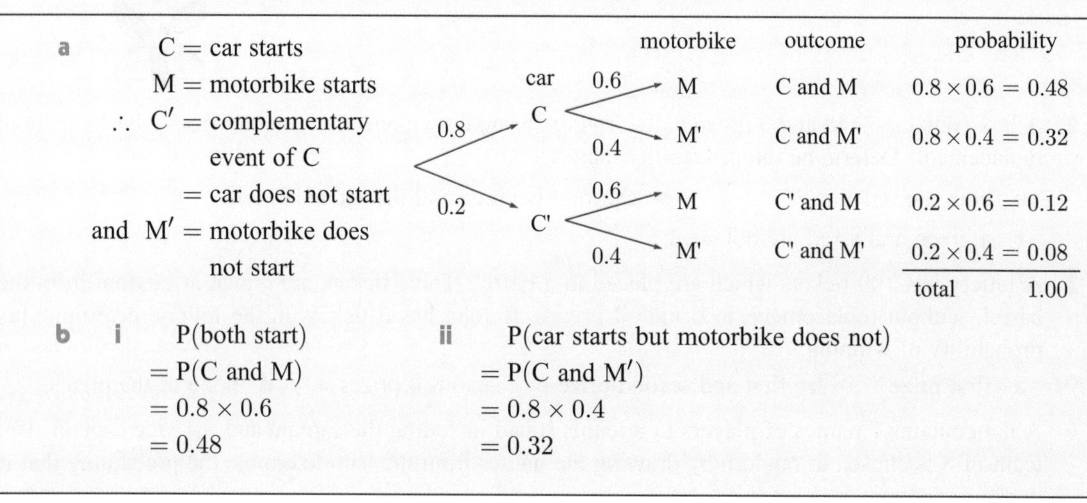

a C = car starts

M = motorbike starts

$\therefore$ C′ = complementary event of C = car does not start

and M′ = motorbike does not start

b **i** P(both start)
= P(C and M)
$= 0.8 \times 0.6$
$= 0.48$

ii P(car starts but motorbike does not)
= P(C and M′)
$= 0.8 \times 0.4$
$= 0.32$

If there is more than one outcome that results in an event occurring, then we need to **add** the probabilities of these outcomes.

Example 12

Two boxes each contain 6 petunia plants that are not yet flowering. Box A contains 2 plants that will have purple flowers and 4 plants that will have white flowers. Box B contains 5 plants that will have purple flowers and 1 plant that will have white flowers. A box is selected by tossing a coin, and one plant is removed at random from it. Determine the probability that it will have purple flowers.

Box A

P	W	W
W	P	W

Box B

P	P	P
W	P	P

box, flower: A ($\frac{1}{2}$): P ($\frac{2}{6}$) ✓, W ($\frac{4}{6}$); B ($\frac{1}{2}$): P ($\frac{5}{6}$) ✓, W ($\frac{1}{6}$)

P(purple flowers)

$= \text{P}(A \text{ and } P) + \text{P}(B \text{ and } P)$

$= \frac{1}{2} \times \frac{2}{6} + \frac{1}{2} \times \frac{5}{6}$ {branches marked ✓}

$= \frac{7}{12}$

EXERCISE 24F

1 Of the students in a class playing musical instruments, 60% are female. 20% of the females and 30% of the males play the violin.

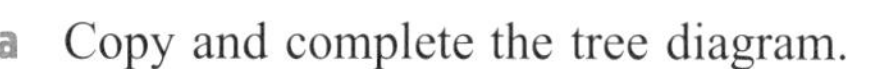

- **a** Copy and complete the tree diagram.
- **b** What is the probability that a randomly selected student:
  - **i** is male and does not play the violin
  - **ii** plays the violin?

0.6
F
violin
not violin
M
violin
not violin

2

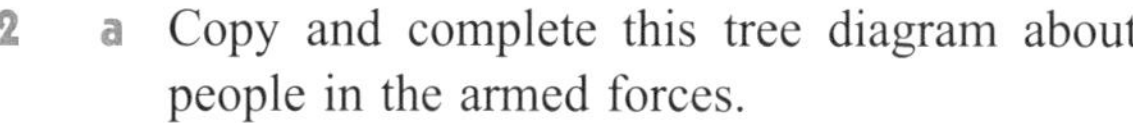

- **a** Copy and complete this tree diagram about people in the armed forces.
- **b** What is the probability that a member of the armed forces:
  - **i** is an officer
  - **ii** is not an officer in the navy
  - **iii** is not an army or air force officer?

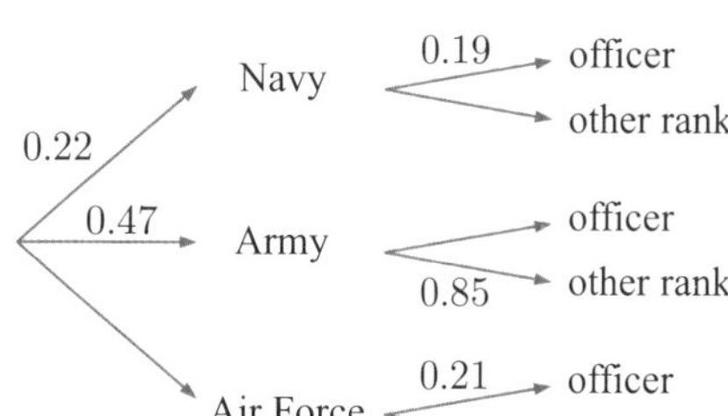

3 Suppose this spinner is spun twice.

- **a** Copy and complete the branches on the tree diagram shown.

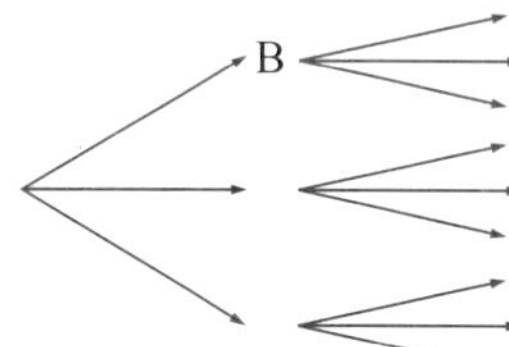

- **b** Find the probability that black appears on both spins.
- **c** Find the probability that yellow appears on both spins.
- **d** Find the probability that different colours appear on the two spins.
- **e** Find the probability that black appears on either spin.

4 The probability of rain tomorrow is estimated to be $\frac{1}{5}$. If it does rain, Mudlark will start favourite in the horse race, with probability $\frac{1}{2}$ of winning. If it is fine, he only has a 1 in 20 chance of winning. Display the sample space of possible results of the horse race on a tree diagram. Hence determine the probability that Mudlark will win tomorrow.

5 Machine A makes 40% of the bottles produced at a factory. Machine B makes the rest. Machine A spoils 5% of its product, while Machine B spoils only 2%. Using an appropriate tree diagram, determine the probability that the next bottle inspected at this factory is spoiled.

6 Jar A contains 2 white and 3 red discs. Jar B contains 3 white and 1 red disc. A jar is chosen at random by the flip of a coin, and one disc is taken at random from it. Determine the probability that the disc is red.

7 The English Premier League consists of 20 teams. Tottenham is currently in 8th place on the table. It has 20% chance of winning and 50% chance of losing against any team placed above it. If a team is placed below it, Tottenham has a 50% chance of winning and a 30% chance of losing. Find the probability that Tottenham will draw its next game.

8 Three bags contain different numbers of blue and red marbles.
A bag is selected using a die which has three A faces, two B faces, and one C face. One marble is then selected randomly from the bag.
Determine the probability that the marble is:

a blue **b** red.

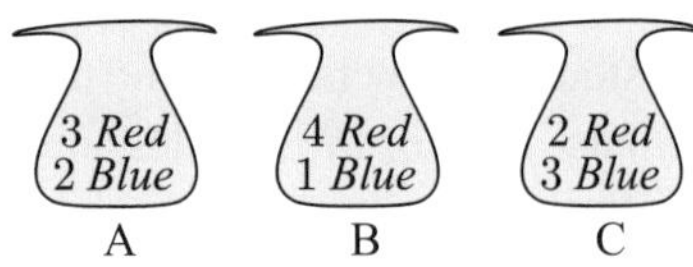

G SAMPLING WITH AND WITHOUT REPLACEMENT

Suppose we have a large group of objects. If we select one of the objects at random and inspect it for particular features, then this process is known as **sampling**.

If the object is put back in the group before an object is chosen again, we call it **sampling with replacement**.

If the object is put to one side, we call it **sampling without replacement**.

Sampling is commonly used in the quality control of industrial processes.

Sometimes the inspection process makes it impossible to return the object to the large group. For example:

- To see if a chocolate is hard or soft-centred, we need to bite it or squeeze it.
- To see if an egg contains one or two yolks, we need to break it open.
- To see if an object is correctly made, we may need to pull it apart.

Consider a box containing 3 red, 2 blue, and 1 yellow marble. If we sample two marbles, we can do this either:

- **with replacement** of the first before the second is drawn, or
- **without replacement** of the first before the second is drawn.

Examine how the tree diagrams differ:

With replacement

1st		2nd	
R ($\frac{3}{6}$)	$\frac{3}{6}$	R	
	$\frac{2}{6}$	B	(✓)
	$\frac{1}{6}$	Y	(✓)
B ($\frac{2}{6}$)	$\frac{3}{6}$	R	(✓)
	$\frac{2}{6}$	B	
	$\frac{1}{6}$	Y	(✓)
Y ($\frac{1}{6}$)	$\frac{3}{6}$	R	(✓)
	$\frac{2}{6}$	B	(✓)
	$\frac{1}{6}$	Y	

Without replacement

1st		2nd	
R ($\frac{3}{6}$)	$\frac{2}{5}$	R	
	$\frac{2}{5}$	B	(×)
	$\frac{1}{5}$	Y	(×)
B ($\frac{2}{6}$)	$\frac{3}{5}$	R	(×)
	$\frac{1}{5}$	B	
	$\frac{1}{5}$	Y	(×)
Y ($\frac{1}{6}$)	$\frac{3}{5}$	R	(×)
	$\frac{2}{5}$	B	(×)
		can't have YY	

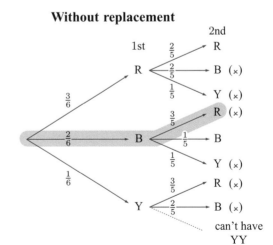

The highlighted branch represents a blue marble with the first draw and a red marble with the second draw. We write this as BR.

Notice that:

- with replacement (independent events), P(two reds) $= \frac{3}{6} \times \frac{3}{6} = \frac{1}{4}$
- without replacement (dependent events), P(two reds) $= \frac{3}{6} \times \frac{2}{5} = \frac{1}{5}$

Example 13 — Self Tutor

A box contains 3 red, 2 blue and 1 yellow marble. Find the probability of getting two different colours:

a if replacement occurs **b** if replacement does not occur.

To answer this question we use the tree diagram on page **754**.

a P(two different colours)
$= \text{P(RB or RY or BR or BY or YR or YB)}$ {ticked ones ✓}
$= \frac{3}{6}\times\frac{2}{6}+\frac{3}{6}\times\frac{1}{6}+\frac{2}{6}\times\frac{3}{6}+\frac{2}{6}\times\frac{1}{6}+\frac{1}{6}\times\frac{3}{6}+\frac{1}{6}\times\frac{2}{6}$
$= \frac{11}{18}$

b P(two different colours)
$= \text{P(RB or RY or BR or BY or YR or YB)}$ {crossed ones ×}
$= \frac{3}{6}\times\frac{2}{5}+\frac{3}{6}\times\frac{1}{5}+\frac{2}{6}\times\frac{3}{5}+\frac{2}{6}\times\frac{1}{5}+\frac{1}{6}\times\frac{3}{5}+\frac{1}{6}\times\frac{2}{5}$
$= \frac{11}{15}$

In **b**, "2 different colours" and "2 same colours" are complementary events.
$\therefore$ P(2 different colours)
$= 1 - \text{P(2 the same)}$
$= 1 - \text{P(RR or BB)}$
$= 1 - (\frac{3}{6}\times\frac{2}{5}+\frac{2}{6}\times\frac{1}{5})$
$= \frac{11}{15}$

Example 14 — Self Tutor

A bag contains 5 red and 3 blue marbles. Two marbles are drawn simultaneously from the bag. Determine the probability that at least one is red.

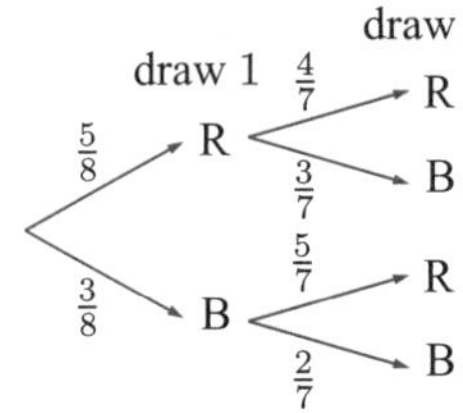

P(at least one red)
$= \text{P(RR or RB or BR)}$
$= \frac{5}{8}\times\frac{4}{7}+\frac{5}{8}\times\frac{3}{7}+\frac{3}{8}\times\frac{5}{7}$
$= \frac{20+15+15}{56}$
$= \frac{25}{28}$

Drawing *simultaneously* is the same as sampling *without* replacement.

Alternatively, P(at least one red)
$= 1 - \text{P(no reds)}$ {complementary events}
$= 1 - \text{P(BB)}$ and so on.

EXERCISE 24G

1 Two marbles are drawn in succession from a box containing 2 purple and 5 green marbles. Determine the probability that the two marbles are different colours if:

a the first is replaced **b** the first is *not* replaced.

2 5 tickets numbered 1, 2, 3, 4, and 5 are placed in a bag. Two are taken from the bag without replacement.

a Complete the tree diagram by writing the probabilities on the branches.

b Determine the probability that:

i both are odd
ii both are even
iii one is odd and the other even.

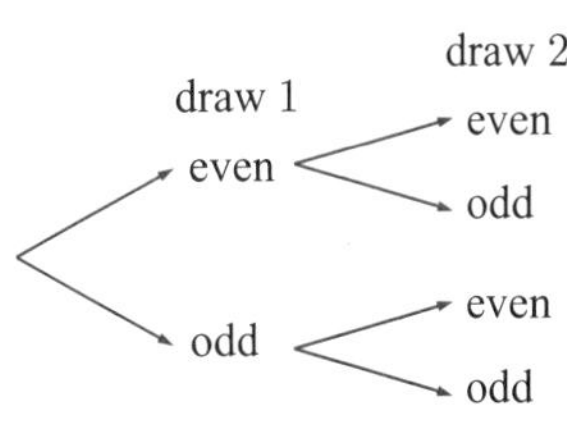

3 A die has 4 faces showing A, and 2 faces showing B. Jar A contains 3 red and 2 green tickets. Jar B contains 3 red and 7 green tickets.

A roll of the die is used to select either jar A or jar B. Once a jar has been selected, two tickets are randomly selected from it without replacement. Determine the probability that:

a both are green **b** they are different in colour.

4 Marie has a bag of sweets which are all identical in shape. The bag contains 6 orange drops and 4 lemon drops. She selects one sweet at random, eats it, and then takes another at random.

a Determine the probability that:

i both sweets are orange drops

ii both sweets are lemon drops

iii the first is an orange drop and the second is a lemon drop

iv the first is a lemon drop and the second is an orange drop.

b Add your answers in **a**. Explain why the total must be 1.

5 A board game uses the spinner shown. If the first spin is red, then the spinner is spun a second time.

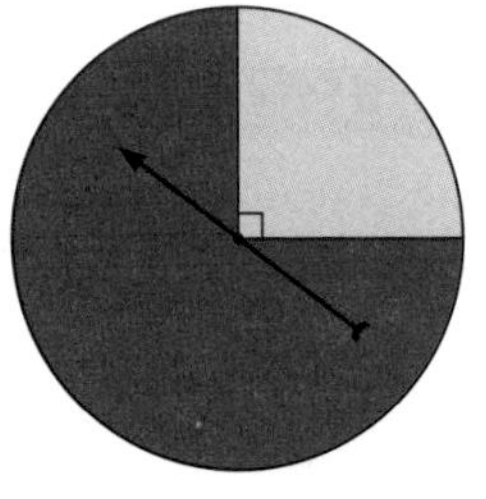

a Complete the tree diagram by labelling each branch with its probability.

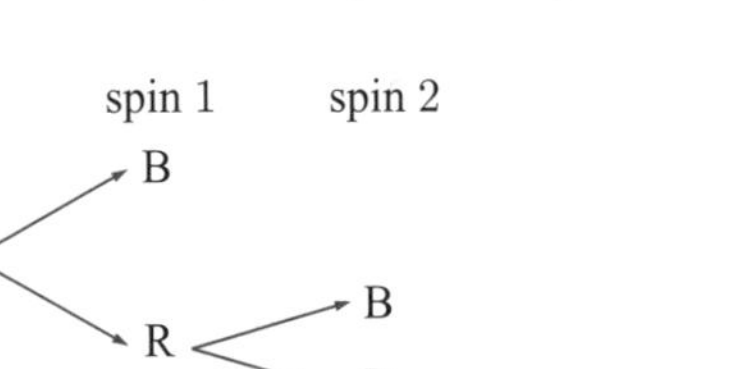

b Calculate the probability that the end result is blue.

6 A bag contains four red and two blue marbles. Three marbles are selected simultaneously. Determine the probablity that:

a all are red **b** only two are red **c** at least two are red.

7 Bag A contains 3 red and 2 white marbles. Bag B contains 4 red and 3 white marbles. One marble is randomly selected from A and its colour noted. If it is red, 2 reds are added to bag B. If it is white, 2 whites are added to bag B. A marble is then selected from bag B. What is the chance that the marble selected is white?

8 A man holds two tickets in a 100-ticket lottery in which there are two winning tickets chosen. If no replacement occurs, determine the probability that he will win:

a both prizes **b** neither prize **c** at least one prize.

9 A container holds 3 red, 7 white, and 2 black balls. A ball is chosen at random from the container and is not replaced. A second ball is then chosen. Find the probability of choosing one white and one black ball in any order.

10 A bag contains 7 yellow and n blue markers.

Two markers are selected at random, without replacement. The probability that both markers are yellow is $\frac{3}{13}$. How many blue markers are there in the bag?

H SETS AND VENN DIAGRAMS

Venn diagrams are a useful way of representing the events in a sample space. These diagrams usually consist of a rectangle which represents the complete sample space U, and circles within it which represent particular events.

Venn diagrams can be used to solve certain types of probability questions and also to establish a number of probability laws.

When we roll an ordinary die, the sample space or universal set is $U = \{1, 2, 3, 4, 5, 6\}$.

If the event A is "*a number less than 3*", then there are two outcomes which satisfy event A. We can write $A = \{1, 2\}$.

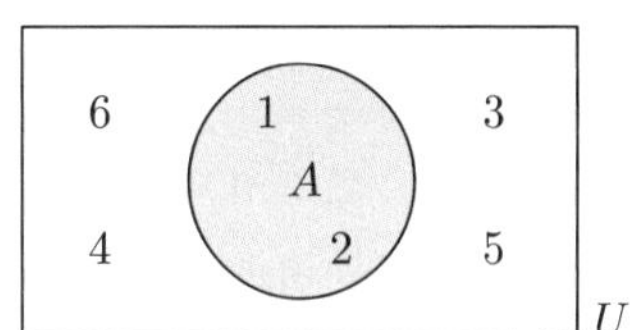

The Venn diagram alongside illustrates the event A within the universal set U.

$n(U) = 6$ and $n(A) = 2$, so $\text{P}(A) = \dfrac{n(A)}{n(U)} = \frac{2}{6} = \frac{1}{3}$.

SET NOTATION

- 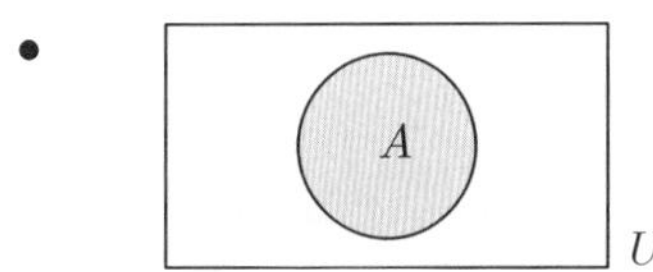

  The **universal set** or **sample space** U is represented by a rectangle.
  A set within the universal set is usually represented by a circle.

- A' (shaded green) is the **complement** of A (shaded purple).
  A' represents the non-occurrence of A, so $\text{P}(A) + \text{P}(A') = 1$.

  If $U = \{1, 2, 3, 4, 5, 6, 7\}$ and $A = \{2, 4, 6\}$ then $A' = \{1, 3, 5, 7\}$.
- $x \in A$ reads 'x is in A' and means that x is an element of the set A.
- $n(A)$ reads 'the number of elements in set A'.
- $A \cap B$ denotes the **intersection** of sets A and B. This set contains all elements common to **both** sets.

  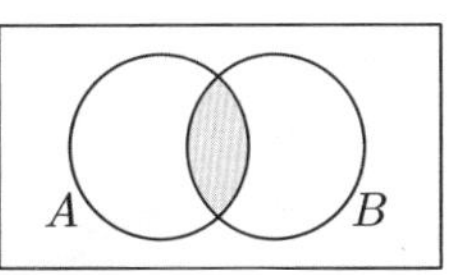

  $A \cap B$ is shaded in purple.

  $A \cap B = \{x \mid x \in A \textbf{ and } x \in B\}$

- $A \cup B$ denotes the **union** of sets A and B. This set contains all elements belonging to A **or** B **or both** A and B.

  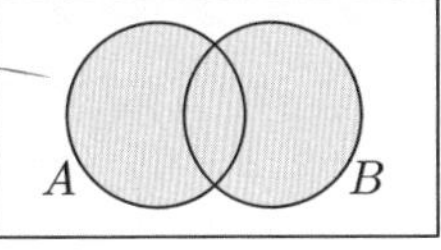

  $A \cup B$ is shaded in purple.

  $A \cup B = \{x \mid x \in A \textbf{ or } x \in B \textbf{ or } x \in A \cap B\}$

- **Disjoint sets** are sets which do not have elements in common.

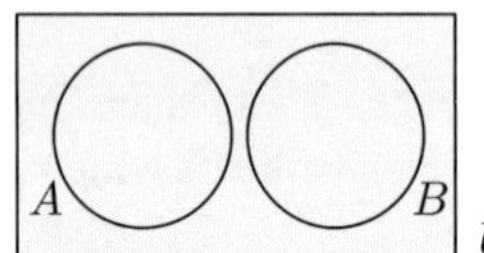

These two sets are disjoint.

$A \cap B = \varnothing$ where $\varnothing$ denotes the **empty set**.

A and B are said to be **mutually exclusive**.

Example 15 Self Tutor

Suppose A is the set of all factors of 24 and B is the set of all factors of 28.
$U = \{x \mid 1 \leqslant x \leqslant 28, \quad x \in \mathbb{Z}\}$

a Find $A \cap B$. **b** Illustrate A and B on a Venn diagram. **c** Find $A \cup B$.

$A = \{1, 2, 3, 4, 6, 8, 12, 24\}$ and $B = \{1, 2, 4, 7, 14, 28\}$

a $A \cap B =$ the set of factors common to both 24 **and** 28 $= \{1, 2, 4\}$

b

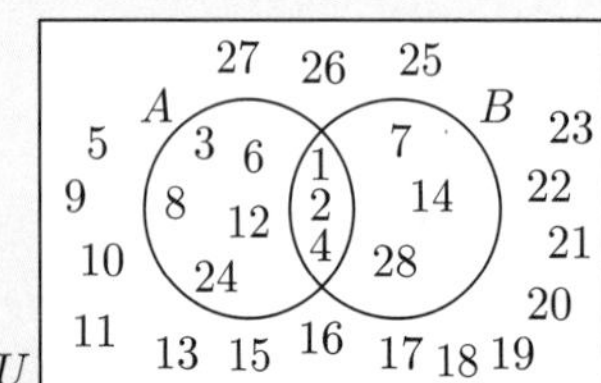

c $A \cup B$
$=$ the set of factors of 24 **or** 28 (or both)
$= \{1, 2, 3, 4, 6, 7, 8, 12, 14, 24, 28\}$

Example 16 Self Tutor

On separate Venn diagrams containing two events A and B that intersect, shade the region representing:

a in A but not in B **b** neither in A nor B.

Shading is shown in green.

a

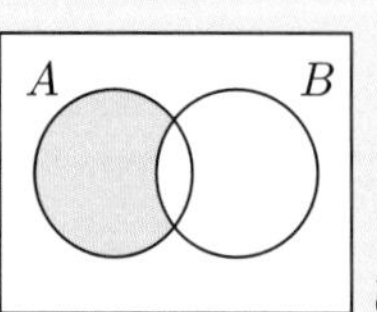

b

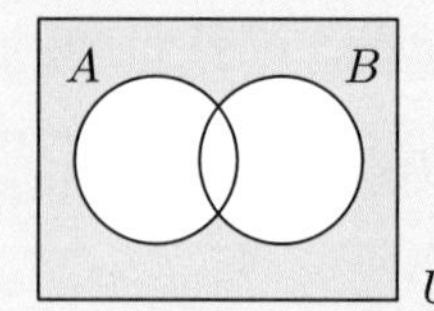

To practise shading the regions represented by various subsets, click on the icon.

DEMO

EXERCISE 24H.1

1 Let A be the set of all factors of 6, B be the set of all positive even integers < 11, and $U = \{x \mid 1 \leqslant x \leqslant 10, \ x \in \mathbb{Z}\}$

a Describe A and B using set notation.

b Illustrate A and B on a Venn diagram.

c Find: **i** $n(A)$ **ii** $A \cup B$ **iii** $A \cap B$.

2 On separate Venn diagrams containing two events A and B that intersect, shade the region representing:

- **a** in A
- **b** in B
- **c** in both A and B
- **d** in A or B
- **e** in B but not in A
- **f** in exactly one of A or B.

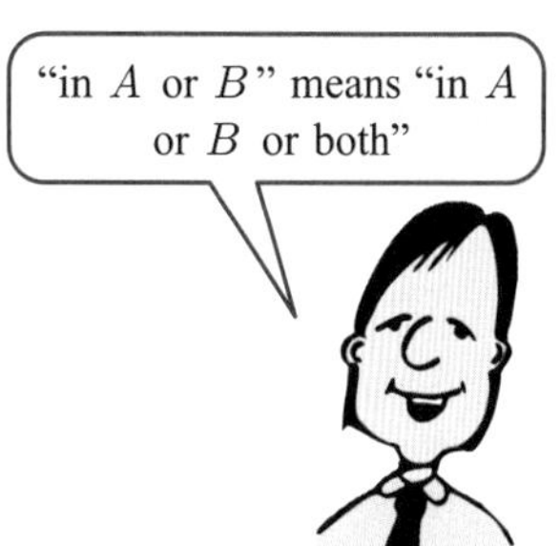

"in A or B" means "in A or B or both"

3 If A and B are two non-disjoint sets, shade the region of a Venn diagram representing:

- **a** A'
- **b** $A' \cap B$
- **c** $A \cup B'$
- **d** $A' \cap B'$

4 The diagram alongside is the most general case for three events in the same sample space U.
On separate Venn diagram sketches, shade:

- **a** A
- **b** B'
- **c** $B \cap C$
- **d** $A \cup C$
- **e** $A \cap B \cap C$
- **f** $(A \cup B) \cap C$

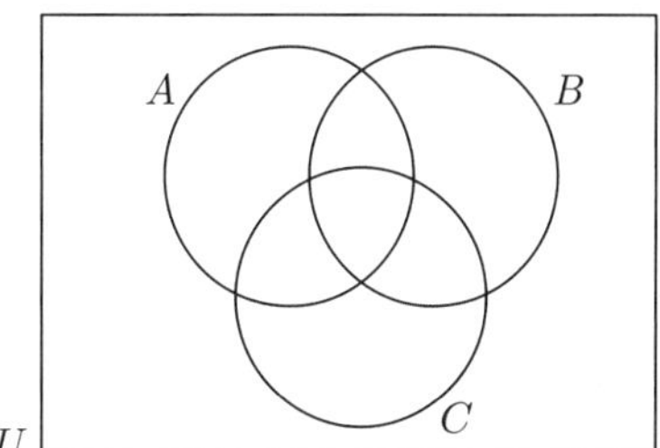

Example 17 **Self Tutor**

If the Venn diagram alongside illustrates the number of people in a sporting club who play tennis (T) and hockey (H), determine the number of people:

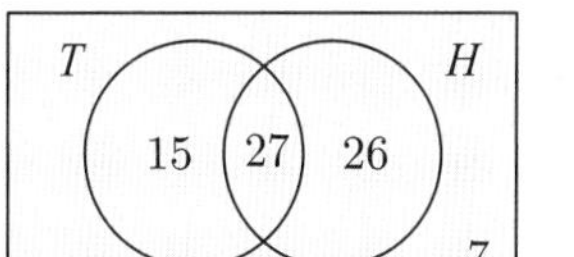

- **a** in the club
- **b** who play hockey
- **c** who play both sports
- **d** who play neither sport
- **e** who play at least one sport.

a Number in the club
$= 15 + 27 + 26 + 7 = 75$

b Number who play hockey
$= 27 + 26 = 53$

c Number who play both sports $= 27$

d Number who play neither sport $= 7$

e Number who play at least one sport
$= 15 + 27 + 26 = 68$

5 The Venn diagram alongside illustrates the number of students in a particular class who study Chemistry (C) and History (H). Determine the number of students:

- **a** in the class
- **b** who study both subjects
- **c** who study at least one of the subjects
- **d** who only study Chemistry.

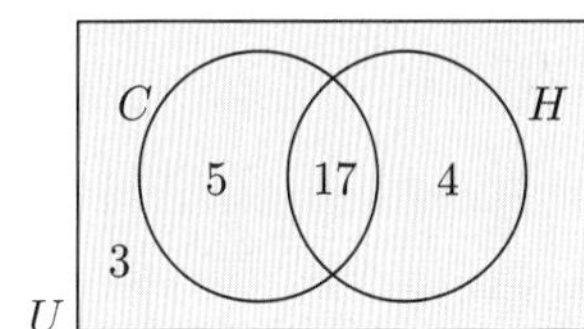

Example 18 Self Tutor

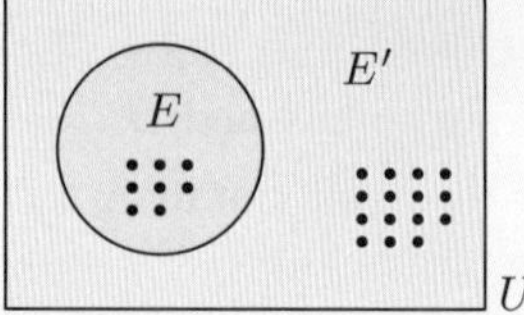

The Venn diagram alongside represents the set U of all children in a class. Each dot represents a student. The event E shows all those students with blue eyes. Determine the probability that a randomly selected child:

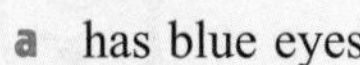

a has blue eyes **b** does not have blue eyes.

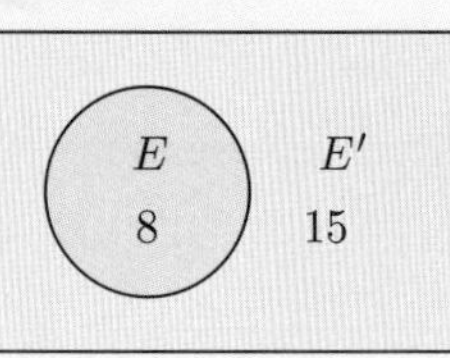

$n(U) = 23, \quad n(E) = 8$

a $\text{P(blue eyes)} = \dfrac{n(E)}{n(U)} = \frac{8}{23}$

b $\text{P(not blue eyes)} = \dfrac{n(E')}{n(U)} = \frac{15}{23}$

or $\text{P(not blue)} = 1 - \text{P(blue eyes)}$

$= 1 - \frac{8}{23} = \frac{15}{23}$

Example 19 Self Tutor

In a class of 30 students, 19 study Physics, 17 study Chemistry, and 15 study both of these subjects. Display this information on a Venn diagram and hence determine the probability that a randomly selected class member studies:

a both subjects

b at least one of the subjects

c Physics but not Chemistry

d exactly one of the subjects

e neither subject

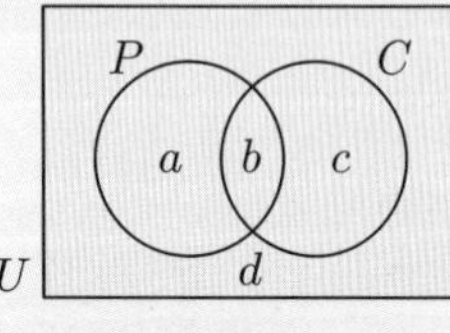

Let P represent the event of 'studying Physics' and C represent the event of 'studying Chemistry'.

Now

$$\begin{aligned} a + b &= 19 && \{\text{as 19 study Physics}\} \\ b + c &= 17 && \{\text{as 17 study Chemistry}\} \\ b &= 15 && \{\text{as 15 study both}\} \\ a + b + c + d &= 30 && \{\text{as there are 30 in the class}\} \end{aligned}$$

$\therefore \quad b = 15, \ a = 4, \ c = 2, \ d = 9.$

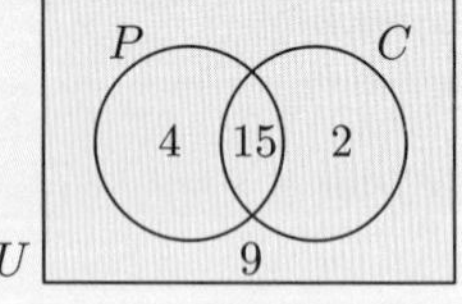

a P(studies both)

$= \frac{15}{30}$ or $\frac{1}{2}$

b P(studies at least one subject)

$= \frac{4+15+2}{30}$

$= \frac{7}{10}$

c $\text{P}(P \text{ but not } C)$

$= \frac{4}{30}$

$= \frac{2}{15}$

d P(studies exactly one)

$= \frac{4+2}{30}$

$= \frac{1}{5}$

e P(studies neither)

$= \frac{9}{30}$

$= \frac{3}{10}$

6 In a survey at an alpine resort, people were asked whether they liked skiing (S) or snowboarding (B). Use the Venn diagram to determine the number of people:

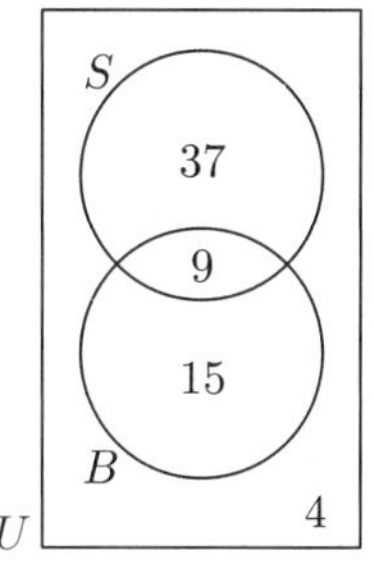

a in the survey
b who liked both activities
c who liked neither activity
d who liked exactly one activity.

7 For two events A and B we are given that $n(A \cap B) = 5$, $n(A) = 11$, $n(A \cup B) = 12$, and $n(B') = 8$.

a Complete the Venn diagram:

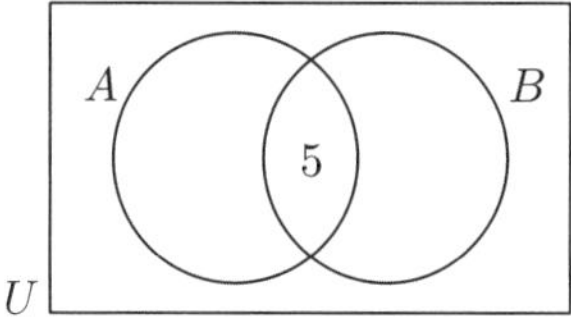

b Hence find:
i $P(A \cup B)$ ii $P(A')$

8 In a class of 40 students, 19 play tennis, 20 play netball, and 8 play neither of these sports. A student is randomly chosen from the class. Determine the probability that the student:

a plays tennis
b does not play netball
c plays at least one of the sports
d plays one and only one of the sports
e plays netball but not tennis.

9 50 married men were asked whether they gave their wife flowers or chocolates for her last birthday. The results were: 31 gave chocolates, 12 gave flowers, and 5 gave both chocolates and flowers. If one of the married men was chosen at random, determine the probability that he gave his wife:

a chocolates or flowers
b chocolates but not flowers
c neither chocolates nor flowers.

10 The medical records for a class of 30 children showed that 24 had previously had measles, 12 had previously had measles and mumps, and 26 had previously had at least one of measles or mumps. If one child from the class is selected at random, determine the probability that he or she has had:

a mumps
b mumps but not measles
c neither mumps nor measles.

11 The Venn diagram opposite indicates the types of program a group of 40 individuals watched on television last night.

M represents movies, S represents sport, and D represents drama.

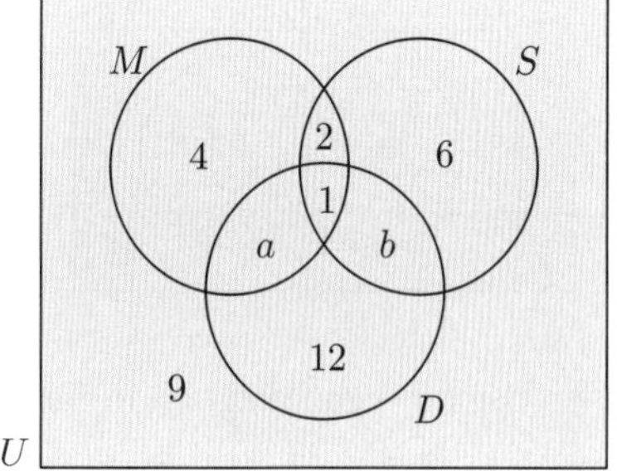

a Given that 10 people watched a movie last night, calculate a and b.

b Find the probability that one of these individuals, selected at random, watched:
i sport
ii drama and sport
iii a movie but not sport
iv drama but not a movie
v drama or a movie.

USING VENN DIAGRAMS TO VERIFY SET IDENTITIES

Example 20 **Self Tutor**

Verify that $(A \cup B)' = A' \cap B'$.

this shaded region is $(A \cup B)$

this shaded region is $(A \cup B)'$

represents A'

represents B'

represents $A' \cap B'$

$(A \cup B)'$ and $A' \cap B'$ are represented by the same regions, verifying that $(A \cup B)' = A' \cap B'$.

EXERCISE 24H.2

1 By drawing appropriate Venn diagrams, verify that:

  a $(A \cap B)' = A' \cup B'$ **b** $A \cup (B \cap C) = (A \cup B) \cap (A \cup C)$

  c $A \cap (B \cup C) = (A \cap B) \cup (A \cap C)$

2 Suppose $S = \{x \mid x \text{ is a positive integer} < 100\}$.
Let $A = \{\text{multiples of 7 in } S\}$ and $B = \{\text{multiples of 5 in } S\}$.

  a How many elements are there in: **i** A **ii** B **iii** $A \cap B$ **iv** $A \cup B$?

  b If $n(E)$ represents the number of elements in set E, verify that
  $n(A \cup B) = n(A) + n(B) - n(A \cap B)$.

  c Use the figure alongside to establish that
  $n(A \cup B) = n(A) + n(B) - n(A \cap B)$
  for all sets A and B in a universal set U.

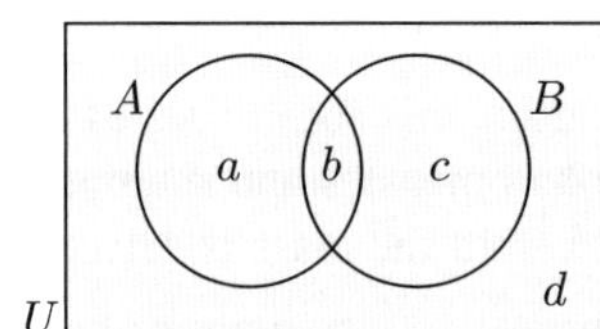

3

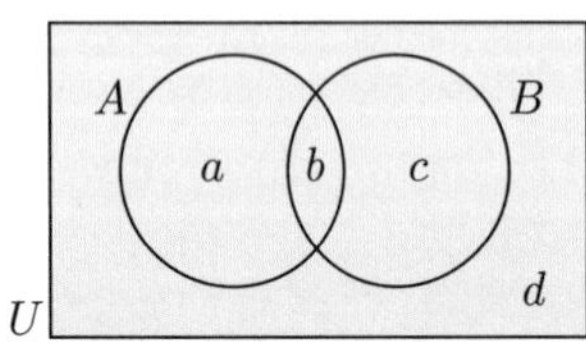

From the Venn diagram, $\text{P}(A) = \dfrac{a+b}{a+b+c+d}$.

  a Use the Venn diagram to find:
  i $\text{P}(B)$ **ii** $\text{P}(A \cap B)$ **iii** $\text{P}(A \cup B)$ **iv** $\text{P}(A) + \text{P}(B) - \text{P}(A \cap B)$

  b What is the connection between $\text{P}(A \cup B)$ and $\text{P}(A) + \text{P}(B) - \text{P}(A \cap B)$?

I LAWS OF PROBABILITY

THE ADDITION LAW

In the previous exercise we demonstrated the **addition law of probability**:

For two events A and B, $\quad P(A \cup B) = P(A) + P(B) - P(A \cap B)$

which means: $\quad$ P(**either** A **or** B **or** both) = P(A) + P(B) − P(**both** A **and** B).

Example 21 — Self Tutor

If $P(A) = 0.6$, $P(A \cup B) = 0.7$, and $P(A \cap B) = 0.3$, find $P(B)$.

$P(A \cup B) = P(A) + P(B) - P(A \cap B)$

$\therefore \quad 0.7 = 0.6 + P(B) - 0.3$

$\therefore \quad P(B) = 0.4$

or

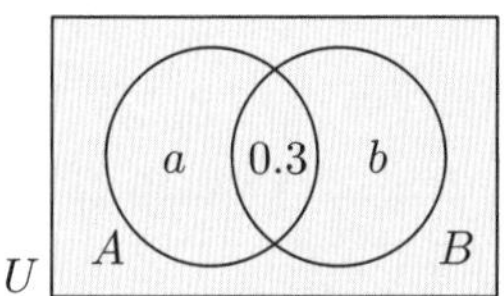

Using a Venn diagram with the probabilities on it,

$a + 0.3 = 0.6 \quad$ and $\quad a + b + 0.3 = 0.7$

$\therefore \quad a = 0.3 \qquad \therefore \quad a + b = 0.4$

$\therefore \quad 0.3 + b = 0.4$

$\therefore \quad b = 0.1$

$\therefore \quad P(B) = 0.3 + b = 0.4$

MUTUALLY EXCLUSIVE OR DISJOINT EVENTS

If A and B are **mutually exclusive** events then $P(A \cap B) = 0$ and so the addition law becomes $P(A \cup B) = P(A) + P(B)$.

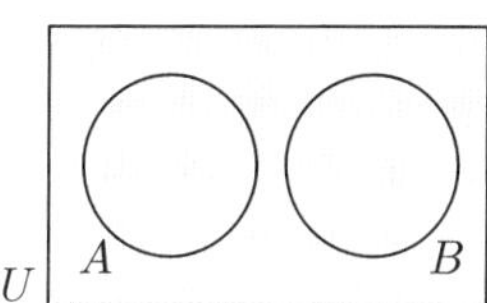

Example 22 — Self Tutor

A chocolate is randomly selected from a box which contains 6 chocolates with hard centres and 12 chocolates with soft centres.

Let H be the event that a randomly selected chocolate from the box has a hard centre, and S be the event that a randomly selected chocolate from the box has a soft centre.

a Are the events H and S mutually exclusive?

b Find: **i** $P(H)$ **ii** $P(S)$ **iii** $P(H \cap S)$ **iv** $P(H \cup S)$.

a Chocolates cannot have both a hard and a soft centre.

$\therefore$ H and S are mutually exclusive.

b **i** $P(H) = \frac{6}{18} = \frac{1}{3}$ **ii** $P(S) = \frac{12}{18} = \frac{2}{3}$

iii $P(H \cap S) = 0$ {a chocolate cannot have a hard *and* a soft centre}

iv $P(H \cup S) = \frac{18}{18} = 1$ {a chocolate must have a hard *or* a soft centre}

CONDITIONAL PROBABILITY

Given two events A and B, the **conditional probability of A given B** is the probability that A occurs given that B has already occurred.

The conditional probability is written $A \mid B$ and read as "A given B".

Example 23 — Self Tutor

In a class of 25 students, 14 like pizza and 16 like iced coffee. One student likes neither and 6 students like both. One student is randomly selected from the class. What is the probability that the student:

a likes pizza

b likes pizza given that he or she likes iced coffee?

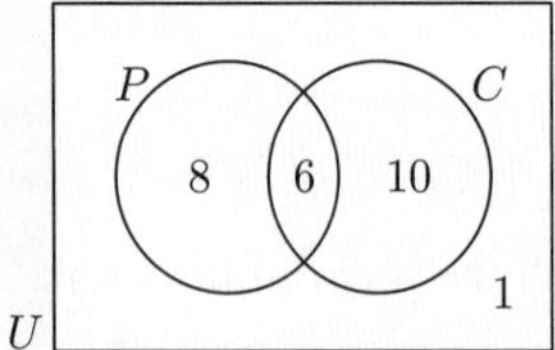

The Venn diagram of the situation is shown.

a Of the 25 students, 14 like pizza.

$\therefore \quad \text{P(pizza)} = \frac{14}{25}$

b Of the 16 who like iced coffee, 6 like pizza.

$\therefore \quad \text{P(pizza} \mid \text{iced coffee)} = \frac{6}{16}$

If A and B are events then $\mathbf{P(A \mid B) = \dfrac{P(A \cap B)}{P(B)}}$.

Proof:

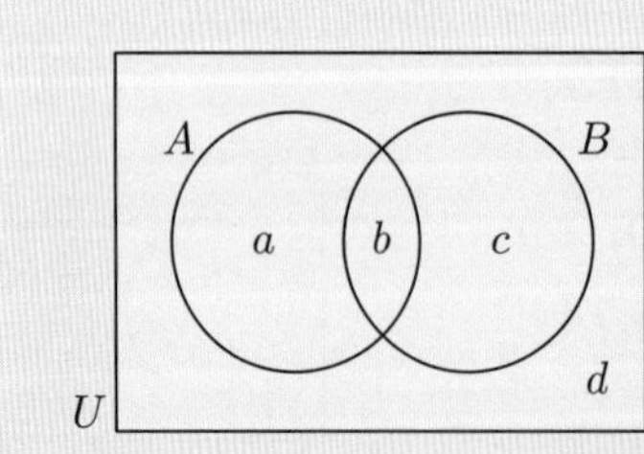

$$\text{P}(A \mid B) = \frac{b}{b+c} \quad \{\text{Venn diagram}\}$$

$$= \frac{\frac{b}{(a+b+c+d)}}{\frac{b+c}{(a+b+c+d)}}$$

$$= \frac{\text{P}(A \cap B)}{\text{P}(B)}$$

It follows that:

$$\mathbf{P(A \cap B) = P(A \mid B)\,P(B)}$$
$$\mathbf{P(A \cap B) = P(B \mid A)\,P(A).}$$

EXERCISE 24I

1 If $\text{P}(A) = 0.4$, $\text{P}(A \cup B) = 0.9$, and $\text{P}(A \cap B) = 0.1$, find $\text{P}(B)$.

2 If $\text{P}(X) = 0.6$, $\text{P}(Y) = 0.5$, and $\text{P}(X \cup Y) = 0.9$, find $\text{P}(X \cap Y)$.

3 A and B are mutually exclusive events.

If $\text{P}(B) = 0.45$ and $\text{P}(A \cup B) = 0.8$, find $\text{P}(A)$.

Example 24 — Self Tutor

In a class of 40 students, 34 like bananas, 22 like pineapple, and 2 dislike both fruits. A student is randomly selected. Find the probability that the student:

- **a** likes both fruits
- **b** likes at least one fruit
- **c** likes bananas given that he or she likes pineapple
- **d** dislikes pineapple given that he or she likes bananas.

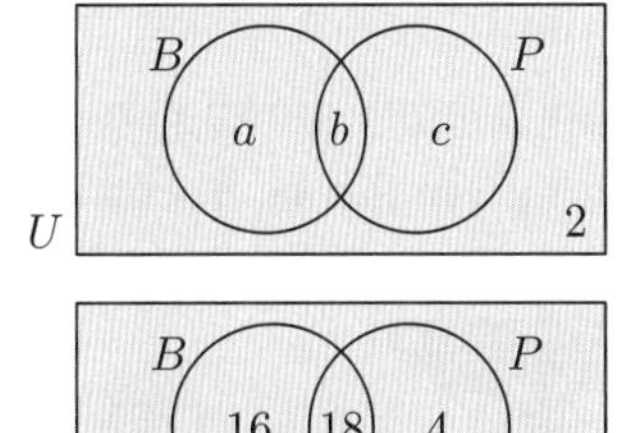

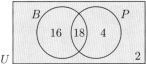

B represents students who like bananas.
P represents students who like pineapple.

We are given that $a + b = 34$
$b + c = 22$
$a + b + c = 38$

$\therefore\ c = 38 - 34 = 4$ and so $b = 18$
and $a = 16$

a P(likes both) $= \frac{18}{40}$ $= \frac{9}{20}$

b P(likes at least one) $= \frac{38}{40}$ $= \frac{19}{20}$

c $\text{P}(B \mid P)$ $= \frac{18}{22}$ $= \frac{9}{11}$

d $\text{P}(P' \mid B)$ $= \frac{16}{34}$ $= \frac{8}{17}$

Example 25 — Self Tutor

The top shelf in a cupboard contains 3 cans of pumpkin soup and 2 cans of chicken soup. The bottom shelf contains 4 cans of pumpkin soup and 1 can of chicken soup.

Lukas is twice as likely to take a can from the bottom shelf as he is from the top shelf. Suppose Lukas takes one can of soup without looking at the label. Determine the probability that it:

- **a** is chicken
- **b** was taken from top shelf given that it is chicken.

T represents the top shelf.
B represents the bottom shelf.
P represents the pumpkin soup.
C represents the chicken soup.

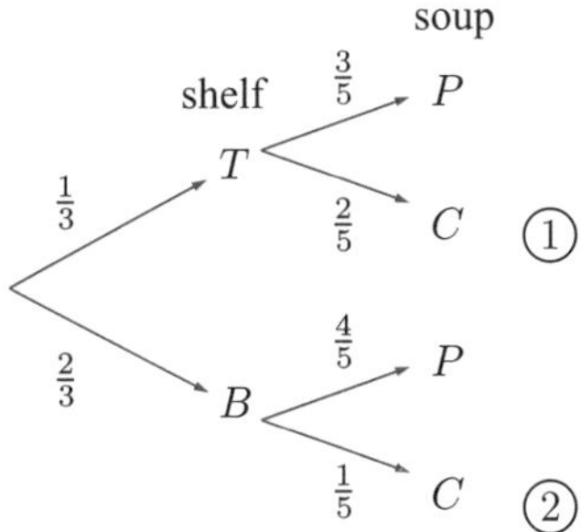

a P(soup is chicken)

$= \frac{1}{3} \times \frac{2}{5} + \frac{2}{3} \times \frac{1}{5}$ {paths ① and ②}

$= \frac{4}{15}$

b P(top shelf | chicken)

$= \dfrac{\text{P(top shelf and chicken)}}{\text{P(chicken)}}$

$= \dfrac{\frac{1}{3} \times \frac{2}{5}}{\frac{4}{15}}$ ← path ① (numerator); ← from **a** (denominator)

$= \frac{1}{2}$

4 In a group of 50 students, 40 study Mathematics, 32 study Physics, and each student studies at least one of these subjects.

a Use a Venn diagram to find how many students study both subjects.

b If a student from this group is randomly selected, find the probability that he or she:

i studies Mathematics but not Physics

ii studies Physics given that he or she studies Mathematics.

5 In a group of 40 boys, 23 have dark hair, 18 have brown eyes, and 26 have dark hair, brown eyes, or both. One of the boys is selected at random. Determine the probability that he has:

a dark hair and brown eyes

b neither dark hair nor brown eyes

c dark hair but not brown eyes

d brown eyes given that he has dark hair.

6 50 students went bushwalking. 23 were sunburnt, 22 were bitten by ants, and 5 were both sunburnt and bitten by ants. Determine the probability that a randomly selected student:

a escaped being bitten

b was bitten or sunburnt (or both)

c was neither bitten nor sunburnt

d was bitten, given that he or she was sunburnt

e was sunburnt, given that he or she was not bitten.

7 400 families were surveyed. It was found that 90% had a TV set and 60% had a computer. Every family had at least one of these items. One of these families is randomly selected, and it is found that it has a computer. Find the probability that it also has a TV set.

8 In a certain town three newspapers are published. 20% of the population read A, 16% read B, 14% read C, 8% read A and B, 5% read A and C, 4% read B and C, and 2% read all 3 newspapers. A person is selected at random. Use a Venn diagram to help determine the probability that the person reads:

a none of the papers

b at least one of the papers

c exactly one of the papers

d A or B (or both)

e A, given that the person reads at least one paper

f C, given that the person reads either A or B or both.

9 Urn A contains 2 red and 3 blue marbles, and urn B contains 4 red and 1 blue marble. Peter selects an urn by tossing a coin, and takes a marble from that urn.

a Determine the probability that it is red.

b Given that the marble is red, what is the probability that it came from B?

10 The probability that Greta's mother takes her shopping is $\frac{2}{5}$. When Greta goes shopping with her mother she gets an icecream 70% of the time. When Greta does not go shopping with her mother she gets an icecream 30% of the time.

Determine the probability that:

a Greta's mother buys her an icecream when shopping

b Greta went shopping with her mother, given that her mother buys her an icecream.

11 On a given day, machine A has a 10% chance of malfunctioning and machine B has a 7% chance of the same. Given that at least one of the machines malfunctioned today, what is the chance that machine B malfunctioned?

12 On any day, the probability that a boy eats his prepared lunch is 0.5. The probability that his sister eats her lunch is 0.6. The probability that the girl eats her lunch given that the boy eats his is 0.9. Determine the probability that:

a both eat their lunch
b the boy eats his lunch given that the girl eats hers
c at least one of them eats their lunch.

13 The probability that a randomly selected person has cancer is 0.02. The probability that he or she reacts positively to a test which detects cancer is 0.95 if he or she has cancer, and 0.03 if he or she does not. Determine the probability that a randomly tested person:

a reacts positively
b has cancer given that he or she reacts positively.

14 A double-headed, a double-tailed, and an ordinary coin are placed in a tin can. One of the coins is randomly chosen without identifying it. The coin is tossed and falls "heads". Determine the probability that the coin is the "double-header".

J INDEPENDENT EVENTS

A and B are **independent events** if the occurrence of each one of them does not affect the probability that the other occurs.

This means that $\text{P}(A \mid B) = \text{P}(A \mid B') = \text{P}(A)$.

Using $\text{P}(A \cap B) = \text{P}(A \mid B)\,\text{P}(B)$ we see that

A and B are **independent events** $\Leftrightarrow$ $\text{P}(A \cap B) = \text{P}(A)\,\text{P}(B)$

which is the result we saw earlier.

Example 26 **Self Tutor**

When two coins are tossed, A is the event of getting 2 heads. When a die is rolled, B is the event of getting a 5 or 6. Show that A and B are independent events.

$\text{P}(A) = \frac{1}{4}$ and $\text{P}(B) = \frac{2}{6}$.

Therefore, $\text{P}(A)\,\text{P}(B) = \frac{1}{4} \times \frac{2}{6} = \frac{1}{12}$

$\text{P}(A \cap B)$
$= \text{P}(2 \text{ heads } \textbf{and} \text{ a } 5 \text{ or a } 6)$
$= \frac{2}{24}$
$= \frac{1}{12}$

die: 1, 2, 3, 4, 5, 6
2 coins: HH, HT, TH, TT

Since $\text{P}(A \cap B) = \text{P}(A)\,\text{P}(B)$, the events A and B are independent.

Example 27 **Self Tutor**

Suppose $P(A) = \frac{1}{2}$, $P(B) = \frac{1}{3}$, and $P(A \cup B) = p$. Find p if:

a A and B are mutually exclusive **b** A and B are independent.

a If A and B are mutually exclusive, $A \cap B = \varnothing$ and so $P(A \cap B) = 0$

But $P(A \cup B) = P(A) + P(B) - P(A \cap B)$

$\therefore \quad p = \frac{1}{2} + \frac{1}{3} - 0 = \frac{5}{6}$

b If A and B are independent, $P(A \cap B) = P(A)\,P(B) = \frac{1}{2} \times \frac{1}{3} = \frac{1}{6}$

$\therefore \quad P(A \cup B) = \frac{1}{2} + \frac{1}{3} - \frac{1}{6}$

$\therefore \quad p = \frac{2}{3}$

Example 28 **Self Tutor**

Suppose $P(A) = \frac{2}{5}$, $P(B \mid A) = \frac{1}{3}$, and $P(B \mid A') = \frac{1}{4}$. Find: **a** $P(B)$ **b** $P(A \cap B')$

$P(B \mid A) = \dfrac{P(B \cap A)}{P(A)}$ so $P(B \cap A) = P(B \mid A)\,P(A) = \frac{1}{3} \times \frac{2}{5} = \frac{2}{15}$

Similarly, $P(B \cap A') = P(B \mid A')\,P(A') = \frac{1}{4} \times \frac{3}{5} = \frac{3}{20}$

$\therefore$ the Venn diagram is:

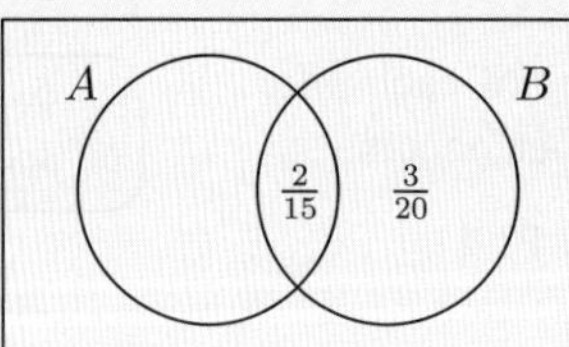

a $P(B) = \frac{2}{15} + \frac{3}{20} = \frac{17}{60}$

b $P(A \cap B') = P(A) - P(A \cap B)$

$= \frac{2}{5} - \frac{2}{15}$

$= \frac{4}{15}$

EXERCISE 24J

1 If $P(R) = 0.4$, $P(S) = 0.5$, and $P(R \cup S) = 0.7$, are R and S independent events?

2 If $P(A) = \frac{2}{5}$, $P(B) = \frac{1}{3}$, and $P(A \cup B) = \frac{1}{2}$, find:

a $P(A \cap B)$ **b** $P(B \mid A)$ **c** $P(A \mid B)$

Are A and B independent events?

3 If $P(X) = 0.5$, $P(Y) = 0.7$, and X and Y are independent events, determine the probability of the occurrence of:

a both X and Y **b** X or Y or both **c** neither X nor Y

d X but not Y **e** X given that Y occurs.

4 The probabilities that students A, B, and C can solve a particular problem are $\frac{3}{5}$, $\frac{2}{3}$, and $\frac{1}{2}$ respectively. If they all try, determine the probability that at least one of the group solves the problem.

5 **a** Find the probability of getting at least one six when a die is rolled 3 times.

b Find the smallest n such that P(at least one 6 in n rolls of a die) $> 99\%$.

6 A and B are independent events. Prove that A' and B' are also independent events.

7 Suppose $P(A \cap B) = 0.1$ and $P(A \cap B') = 0.4$. Find $P(A \cup B')$ given that A and B are independent.

8 Suppose $P(C) = \frac{9}{20}$, $P(C \mid D') = \frac{3}{7}$, and $P(C \mid D) = \frac{6}{13}$.

  a Find: **i** $P(D)$ **ii** $P(C' \cup D')$

  b Are C and D independent events? Give a reason for your answer.

9 Ruba and Hania play a game in which they take it in turns to select a card, with replacement, from a well-shuffled pack of 52 playing cards. The first person to select an ace wins the game. Ruba has the first turn.

  a **i** Find the probability that Ruba wins on her third turn.

  ii Show that the probability that Ruba wins prior to her $(n+1)$th turn is $\frac{13}{25}\left(1 - (\frac{12}{13})^{2n}\right)$.

  iii Hence, find the probability that Ruba wins the game.

  b If Ruba and Hania play this game seven times, find the probability that Ruba will win more games than Hania.

10 Robot 3PCO stands on the edge of a cliff and takes random steps either towards or away from the cliff's edge.
The probability of 3PCO stepping away from the edge is $\frac{3}{5}$, and towards the edge is $\frac{2}{5}$.
Find the probability 3PCO does not step over the cliff in his first four steps.

K PROBABILITIES USING PERMUTATIONS AND COMBINATIONS

Permutations and **combinations** can sometimes be used to find probabilities of various events. They are particularly useful if the sample size is large.

Remember that:

- **permutations** involve the **ordering** of objects or things
- **combinations** involve **selections** such as committees or teams.

Example 29 **Self Tutor**

A squad of 13 players includes 4 brothers. A team of 7 is randomly selected by drawing names from a hat. Determine the probability that the team contains:

a all the brothers **b** at least 2 of the brothers.

There are $\binom{13}{7}$ different teams of 7 that can be chosen from 13 people.

a Of these teams, $\binom{4}{4}\binom{9}{3}$ contain all 4 brothers and any 3 others.

$$\therefore \text{ P(team contains all the brothers)} = \frac{\binom{4}{4}\binom{9}{3}}{\binom{13}{7}} \approx 0.0490$$

b
$$\begin{aligned}\text{P(at least 2 brothers)} &= \text{P(2 brothers or 3 brothers or 4 brothers)}\\ &= \frac{\binom{4}{2}\binom{9}{5}}{\binom{13}{7}} + \frac{\binom{4}{3}\binom{9}{4}}{\binom{13}{7}} + \frac{\binom{4}{4}\binom{9}{3}}{\binom{13}{7}}\\ &\approx 0.783\end{aligned}$$

EXERCISE 24K

1 A committee of 4 is chosen from 11 people by random selection. Two sisters were amongst the group from which the selection was made. Find the probability that both sisters are chosen for the committee.

2 4 alphabet blocks D, A, I, and S are placed at random in a row. What is the likelihood that they spell out either AIDS or SAID?

3 A team of 7 is randomly chosen from a squad of 12 including the club captain and vice captain. Determine the probability that both the captain and vice-captain are chosen.

4 Of the 22 people on board a plane, 3 are professional golfers. If the plane crashes and 4 people are killed, determine the chance that all three golfers survive.

5 5 boys sit at random on 5 seats in a row. Determine the probability that the two friends Keong and James sit:

a at the ends of the row **b** together.

6 A committee of 5 is randomly selected from 9 men and 7 women. Determine the likelihood that it consists of:

a all men **b** at least 3 men **c** at least one of each sex.

7 6 people including friends A, B, and C are randomly seated on a row of 6 chairs. Determine the likelihood that A, B, and C are seated together.

8 A school committee of 7 is chosen at random from 11 senior students and 3 junior students. Find the probability that:

a only senior students are chosen **b** all three junior students are chosen.

BAYES' THEOREM

Suppose a sample space U is partitioned into two mutually exclusive regions by an event A and its complement A'.

We can show this on a Venn diagram as

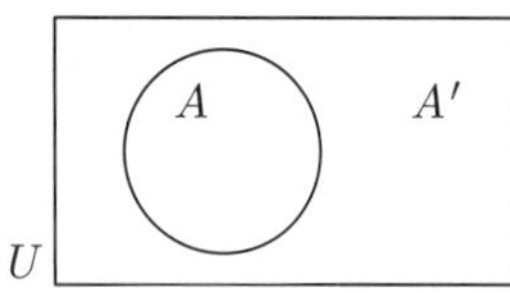

or

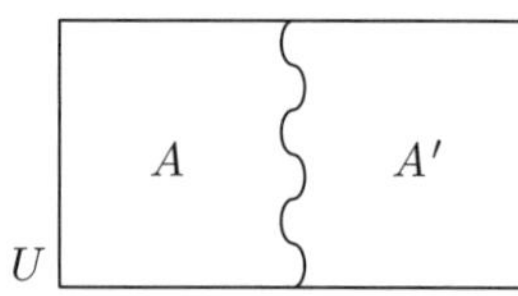

Now consider another event B in the sample space U. We can show this on a Venn diagram as

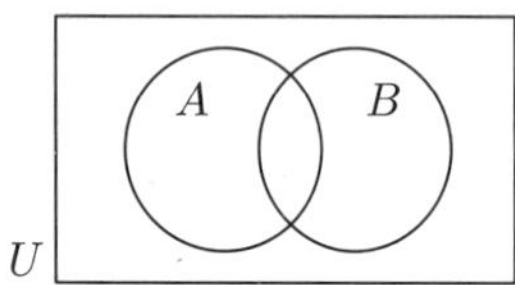

or

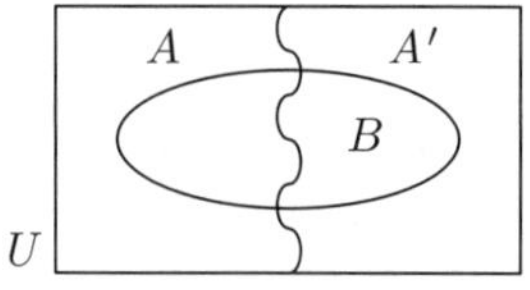

Bayes' theorem states that $\mathbf{P}(A \mid B) = \dfrac{\mathbf{P}(B \mid A)\, \mathbf{P}(A)}{\mathbf{P}(B)}$

where $\text{P}(B) = \text{P}(B \mid A)\,\text{P}(A) + \text{P}(B \mid A')\,\text{P}(A')$.

Proof:

$$P(A \mid B) = \frac{P(A \cap B)}{P(B)} = \frac{P(B \mid A)\,P(A)}{P(B)}$$

where $P(B) = P(B \cap A) + P(B \cap A')$
$= P(B \mid A)\,P(A) + P(B \mid A')\,P(A')$

Example 30 — Self Tutor

A can contains 4 blue and 2 green marbles. One marble is randomly drawn from the can without replacement and its colour is noted. A second marble is then drawn. Find the probability that:

a the second marble is blue

b the first marble was green given that the second marble is blue.

Let A be the event that the first marble is green.

Let B be the event that the second marble is blue.

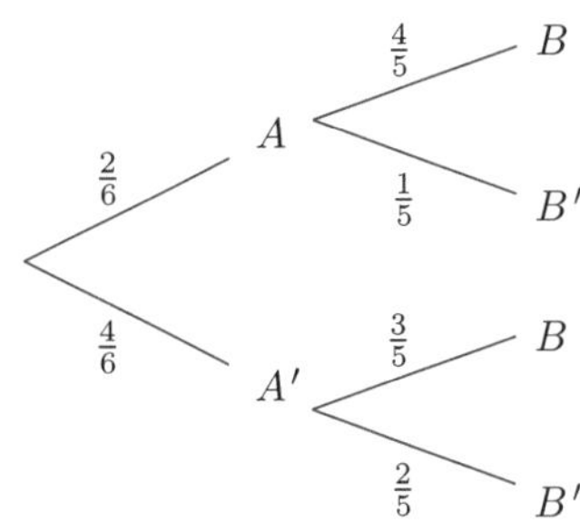

a P(second marble is blue)

$= P(B)$

$= P(B \mid A)\,P(A) + P(B \mid A')\,P(A')$

$= \frac{4}{5} \times \frac{2}{6} + \frac{3}{5} \times \frac{4}{6}$

$= \frac{2}{3}$

b P(first was green | second is blue)

$= P(A \mid B)$

$= \frac{P(B \mid A)\,P(A)}{P(B)}$ {Bayes' theorem}

$= \frac{\frac{4}{5} \times \frac{2}{6}}{\frac{2}{3}}$ {using a}

$= \frac{2}{5}$

EXERCISE 24L

1 Coffee machines A and B produce coffee in identically shaped plastic cups. A produces 65% of the coffee sold each day, and B produces the remainder. A underfills a cup 4% of the time, while B underfills a cup 5% of the time.

 a A cup of coffee is chosen at random. Find the probability that it is underfilled.

 b A cup of coffee is randomly chosen and is found to be underfilled. Find the probability that it came from A.

2 54% of the students at a university are female. 8% of the male students are colour-blind and 2% of the female students are colour-blind.

 a A randomly chosen student is colour-blind. Find the probability that the student is male.

 b A randomly chosen student is not colour-blind. Find the probability that the student is female.

3 A marble is randomly chosen from a can containing 3 red and 5 blue marbles. It is replaced by two marbles of the other colour. Another marble is then randomly chosen from the can. If the marbles chosen are the same colour, what is the probability that they are both blue?

4 35% of the animals in a deer herd carry the TPC gene. 58% of these deer also carry the SD gene, while 23% of the deer without the TPC gene carry the SD gene. If a deer is randomly chosen and is found to carry the SD gene, what is the probability that it does not carry the TPC gene?

5 A new blood test has been designed to detect a form of cancer. The probability that the test correctly identifies someone with the cancer is 0.97, and the probability that the test correctly identifies someone without the cancer is 0.93. Approximately 0.1% of the general population are known to contract this cancer.

When a patient has a blood test, the test results are positive for the cancer. Find the probability that the patient actually has the cancer.

6 A man drives his car to work 80% of the time. The remainder of the time he rides his bicycle. When he rides his bicycle to work he is late 25% of the time, whereas when he drives his car to work he is late 15% of the time. On a particular day, the man arrives early. Find the probability that he rode his bicycle to work that day.

7 The probabilities that Hiran's mother and father will be alive after ten years are 0.99 and 0.98 respectively. What is the probability that if only one of them is alive after ten years, it is his mother?

8 A manufacturer produces drink bottles. He uses 2 machines which produce 60% and 40% of the bottles respectively. 3% of the bottles made by the first machine are defective and 5% of the bottles made by the second machine are defective. Find the probability that a defective bottle came from:

a the first machine **b** the second machine.

9

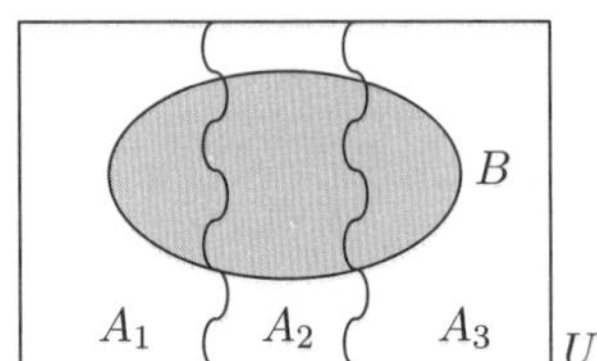

A sample space U is partitioned into three by the mutually exclusive events A_1, A_2, and A_3.
The sample space also contains another event B.

a Show that $\mathrm{P}(B) = \mathrm{P}(B \mid A_1)\,\mathrm{P}(A_1) + \mathrm{P}(B \mid A_2)\,\mathrm{P}(A_2) + \mathrm{P}(B \mid A_3)\,\mathrm{P}(A_3)$.

b Hence show that Bayes' theorem for the case of three partitions is

$$\mathrm{P}(A_i \mid B) = \frac{\mathrm{P}(B \mid A_i)\,\mathrm{P}(A_i)}{\mathrm{P}(B)}, \quad i \in \{1, 2, 3\} \quad \text{where} \quad \mathrm{P}(B) = \sum_{j=1}^{3} \mathrm{P}(B \mid A_j)\,\mathrm{P}(A_j).$$

10 A newspaper printer has three presses A, B, and C which print 30%, 40%, and 30% of daily production respectively. Due to the age of the machines and other problems, the presses will produce streaks on their output 3%, 5%, and 7% of the time, respectively.

a Find the probability that a randomly chosen newspaper does not have streaks.

b If a randomly chosen newspaper does not have streaks, find the probability that it was printed by press A.

c If a randomly chosen newspaper has streaks, find the probability that it was printed by either press A or C.

11 12% of the over-60 population of Agento have lung cancer. Of those with lung cancer, 50% were heavy smokers, 40% were moderate smokers, and 10% were non-smokers. Of those without lung cancer, 5% were heavy smokers, 15% were moderate smokers, and 80% were non-smokers.
A member of the over-60 population of Agento is chosen at random. Find the probability that the person:

- **a** was a heavy smoker
- **b** has lung cancer given the person was a moderate smoker
- **c** has lung cancer given the person was a non-smoker.

THEORY OF KNOWLEDGE

Modern probability theory began in 1653 when gambler Chevalier de Mere contacted mathematician **Blaise Pascal** with a problem on how to divide the stakes when a gambling game is interrupted during play. Pascal involved **Pierre de Fermat**, a lawyer and amateur mathematician, and together they solved the problem. In the process they laid the foundations upon which the laws of probability were formed.

Blaise Pascal

Pierre de Fermat

Agner Krarup Erlang

Applications of probability are now found from quantum physics to medicine and industry.

The first research paper on **queueing theory** was published in 1909 by the Danish engineer **Agner Krarup Erlang** who worked for the Copenhagen Telephone Exchange. In the last hundred years this theory has become an integral part of the huge global telecommunications industry, but it is equally applicable to modelling car traffic right down to queues at your local supermarket.

Statistics and probability are used extensively to predict the behaviour of the global stock market. For example, American mathematician **Edward Oakley Thorp** developed and applied hedge fund techniques for the financial markets in the 1960s.

On the level of an individual investor, money is put into the stock market if there is a good probability that the value of the shares will increase. This investment has risk, however, as witnessed recently with the Global Financial Crisis of 2008 - 2009.

1. In what ways can mathematics model the world without using functions?
2. How does a knowledge of probability theory affect decisions we make?
3. Do ethics play a role in the use of mathematics?

REVIEW SET 24A NON-CALCULATOR

1 List the different orders in which 4 people A, B, C, and D could line up. If they line up at random, determine the probability that:

a A is next to C **b** there is exactly one person between A and C.

2 Given $P(A) = m$ is the probability of event A occurring in any given trial:

a Write $P(A')$ in terms of m.

b State the range of possible values of m.

c Suppose two trials are performed independently. Find, in terms of m, the probability of A occurring:

i exactly once **ii** at least once.

3 A coin is tossed and a square spinner labelled A, B, C, D, is twirled. Determine the probability of obtaining:

a a head and consonant **b** a tail and C **c** a tail or a vowel or both.

4 The probability that a man will be alive in 25 years is $\frac{3}{5}$, and the probability that his wife will be alive is $\frac{2}{3}$. Determine the probability that in 25 years:

a both will be alive **b** at least one will be alive **c** only the wife will be alive.

5 Given $P(Y) = 0.35$ and $P(X \cup Y) = 0.8$, and that X and Y are mutually exclusive events, find:

a $P(X \cap Y)$ **b** $P(X)$

c the probability that X occurs or Y occurs, but not both X and Y.

6 What is meant by: **a** independent events **b** mutually exclusive events?

7 Graph the sample space of all possible outcomes when a pair of dice is rolled. Hence determine the probability of getting:

a a sum of 7 or 11 **b** a sum of at least 8.

8 In a group of 40 students, 22 study Economics, 25 study Law, and 3 study neither of these subjects. Determine the probability that a randomly chosen student studies:

a both Economics and Law **b** at least one of these subjects

c Economics given that he or she studies Law.

9 The probability that a particular salesman will leave his sunglasses behind in any store is $\frac{1}{5}$. Suppose the salesman visits two stores in succession and leaves his sunglasses behind in one of them. What is the probability that the salesman left his sunglasses in the first store?

10 Each time Mae and Ravi play chess, Mae has probability $\frac{4}{5}$ of winning. If they play 5 games, determine the probability that:

a Mae wins 3 of the games **b** Mae wins either 4 or 5 of the games.

11 Suppose $P(X' \mid Y) = \frac{2}{3}$, $P(Y) = \frac{5}{6}$, and $X' \cap Y' = \varnothing$. Find $P(X)$.

12 The diagram alongside shows an electrical circuit with switches. The probability that any switch is open is $\frac{1}{3}$. Determine the probability that the current flows from A to B.

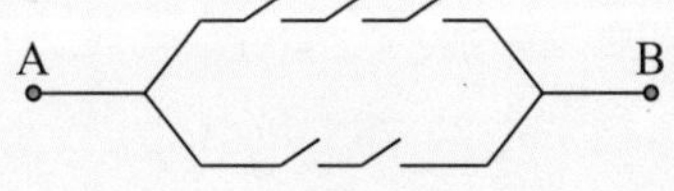

13 One letter is randomly selected from each of the names JONES, PETERS, and EVANS. Find the probability that:

a the three letters are the same　　**b** exactly two of the letters are the same.

REVIEW SET 24B　　CALCULATOR

1 Niklas and Rolf play tennis with the winner being the first to win two sets. Niklas has a 40% chance of beating Rolf in any set. Draw a tree diagram showing the possible outcomes and hence determine the probability that Niklas will win the match.

2 If I buy 4 tickets in a 500 ticket lottery, and the prizes are drawn without replacement, determine the probability that I will win:

a the first 3 prizes　　**b** at least one of the first 3 prizes.

3 The students in a school are all vaccinated against measles. 48% of the students are males, of whom 16% have an allergic reaction to the vaccine. 35% of the girls also have an allergic reaction. A student is randomly chosen from the school. Find the probability that the student:

a has an allergic reaction

b is female given that a reaction occurs.

4 On any one day it could rain with 25% chance and be windy with 36% chance.

a Draw a tree diagram showing the possibilities with regard to wind and rain on a particular day.

b Hence determine the probability that on a particular day there will be:

i rain and wind　　**ii** rain or wind or both.

c What assumption have you made in your answers?

5 A, B, and C have 10%, 20%, and 30% chance of independently solving a certain maths problem. If they all try independently of one another, what is the probability that at least one of them will solve the problem?

6 Two events are defined such that $\text{P}(A) = 0.11$ and $\text{P}(B) = 0.7$. $n(B) = 14$.

a Calculate: **i** $\text{P}(A')$　**ii** $n(U)$

b If A and B are independent events, find: **i** $\text{P}(A \cap B)$　**ii** $\text{P}(A \mid B)$

c If instead, A and B are mutually exclusive events, find $\text{P}(A \cup B)$.

7 Let C be the event that "a person has a cat" and D be the event that "a person has a dog". $\text{P}(C) = \frac{3}{7}$, $\text{P}(D \mid C') = \frac{2}{5}$, and $\text{P}(D' \mid C) = \frac{3}{4}$.

a Copy and complete the tree diagram by marking a probability on each branch.

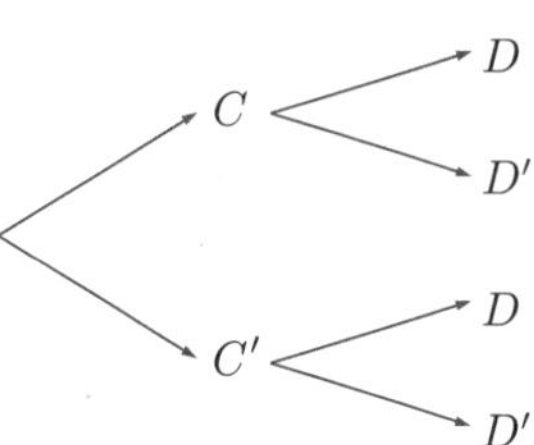

b If a person is chosen at random, find the probability that the person has:

i a cat and a dog

ii at least one pet (cat or dog).

8 A survey of 200 people included 90 females. It found that 60 people smoked, 40 of whom were male.

a Use the given information to complete the table:

	Female	*Male*	*Total*
Smoker			
Non-smoker			
Total			

b A person is selected at random. Find the probability that this person is:

i a female non-smoker

ii a male given the person was a non-smoker.

c If two people from the survey are selected at random, calculate the probability that:

i both of them are non-smoking females

ii one is a smoker and the other is a non-smoker.

9 In a certain class, 91% of the students passed Mathematics and 88% of the students passed Chemistry. 85% of students passed both Mathematics and Chemistry.

a Show that the events of passing Mathematics and passing Chemistry are *not* independent.

b A randomly selected student passed Chemistry. Find the probability that this student did not pass Mathematics.

10 A group of ten students includes three from Year 12 and four from Year 11. The principal calls a meeting with five students randomly selected from the group. Calculate the probability that exactly two Year 12 and two Year 11 students are called to the meeting.

11 A person with a university degree has a 0.33 chance of getting an executive position. A person without a university degree has a 0.17 chance of getting an executive position. If 78% of all applicants for an executive position have a university degree, find the probability that the successful applicant does not have one.

12 A team of five is randomly chosen from six doctors and four dentists. Determine the probability that it consists of:

a all doctors **b** at least two doctors.

13 With each pregnancy, a particular woman will give birth to either a single baby or twins. There is a 15% chance of having twins during each pregnancy. Suppose that after 2 pregnancies she has given birth to 3 children. Find the probability that she had twins first.

14 Four different numbers are randomly chosen from the set $S = \{1, 2, 3, 4, 5, ..., 10\}$. X is the second largest of the numbers selected.
Determine the probability that X is: **a** 2 **b** 7 **c** 9.

REVIEW SET 24C

1 Systematically list the possible sexes of a 4-child family. Hence determine the probability that a randomly selected 4-child family has two children of each sex.

2 A bag contains 3 red, 4 yellow and 5 blue marbles. Two marbles are randomly selected from the bag without replacement. What is the probability that:

a both are blue **b** they have the same colour

c at least one is red **d** exactly one is yellow?

3 A class contains 25 students. 13 play tennis, 14 play volleyball, and 1 plays neither of these sports.

a A student is randomly selected from the class. Determine the probability that the student:

i plays both tennis and volleyball **ii** plays at least one of these sports

iii plays volleyball given that he or she does not play tennis.

b Three students are randomly selected from the class. Determine the probability that:

i none of these students play tennis

ii at least one of these students plays tennis.

4 An urn contains three red balls and six blue balls.

a A ball is drawn at random and found to be blue. What is the probability that a second draw with no replacement will also produce a blue ball?

b Two balls are drawn without replacement and the second is found to be red. What is the probability that the first ball was also red?

c Based on the toss of a coin, either a red ball or a blue ball is added to the urn. A ball is then drawn at random and found to be blue. Find the probability the added ball was red.

5 A and B are independent events where $\mathrm{P}(A) = 0.8$ and $\mathrm{P}(B) = 0.65$.

Determine: **a** $\mathrm{P}(A \cup B)$ **b** $\mathrm{P}(A \mid B)$ **c** $\mathrm{P}(A' \mid B')$ **d** $\mathrm{P}(B \mid A)$.

6 A school photocopier has a 95% chance of working on any particular day. Find the probability that it will be working on at least one of the next two days.

7 Jon goes cycling on three random mornings of each week. When he goes cycling he has eggs for breakfast 70% of the time. When he does not go cycling he has eggs for breakfast 25% of the time. Determine the probability that Jon:

a has eggs for breakfast **b** goes cycling given that he has eggs for breakfast.

8 A survey of 50 men and 50 women was conducted to see how many people prefer coffee or tea. It was found that 15 men and 24 women prefer tea.

a Display this information in a two-way table.

b Let C represent the people who prefer coffee and M represent the men. Hence complete the Venn diagram.

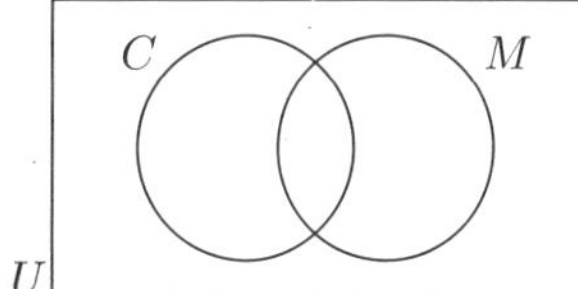

c Calculate: **i** $\mathrm{P}(C')$ **ii** $\mathrm{P}(M \mid C)$

9 T and M are events such that $n(U) = 30$, $n(T) = 10$, $n(M) = 17$, and $n((T \cup M)') = 5$.

a Draw a Venn diagram to display this information.

b Hence find: **i** $\mathrm{P}(T \cap M)$ **ii** $\mathrm{P}((T \cap M) \mid M)$

10 Answer the questions in the **Opening Problem** on page **734**.

11 The independent probabilities that 3 components of a TV set will need replacing within one year are $\frac{1}{20}$, $\frac{1}{50}$, and $\frac{1}{100}$ respectively. Calculate the probability that there will need to be a replacement of:

a at least one component within a year **b** exactly one component within a year.

12 When Peter plays John at tennis, the probability that Peter wins his service game is p and the probability that John wins his service game is q, where $p > q$ and $p+q > 1$. Which is more likely:

A Peter will win at least two consecutive games out of 3 when he serves first

B Peter will win at least two consecutive games out of 3 when John serves first?

13 Two different numbers were chosen at random from the digits 1 to 9 inclusive. It was observed that their sum was even. Determine the probability that both numbers were odd.

14 Using a 52 card pack, a 'royal flush' consists of the 10, J, Q, K, A of one suit. Find the probability of dealing:

a a royal flush in any order

b a royal flush in the order 10, J, Q, K, A.

Chapter 25

Discrete random variables

Syllabus reference: 5.7, 5.8

Contents:

OPENING PROBLEM

It is known that 3.2% of the pens manufactured in a factory are faulty.

The factory quality controller randomly tests 120 pens, checking them one at a time, and replacing them into the batch before the next one is chosen.

Things to think about:

a What will the probability be that:

 i all are faulty **ii** none are faulty?

b How can we find the probability that:

 i less than 2 are faulty **ii** between 5 and 10 (inclusive) are faulty?

A DISCRETE RANDOM VARIABLES

RANDOM VARIABLES

In previous work on probability we have often described events by using words. However, if possible, it is far more convenient to use numbers.

A **random variable** represents in number form the possible outcomes which could occur for some random experiment.

A **discrete random variable** X has a set of distinct possible values. In this course you will consider only a finite number of outcomes, so we label them $x_1, x_2, x_3, ..., x_n$.

For example, X could be:

- the number of houses in your suburb which have a 'power safety switch'
- the number of new bicycles sold each year by a bicycle store
- the number of defective light bulbs in the purchase order of a city store.

To determine the value of a discrete random variable we need to **count**.

A **continuous random variable** X could take possible values in some interval on the number line.

For example, X could be:

- the heights of men, which would all lie in the interval $50 < X < 250$ cm
- the volume of water in a rainwater tank, which could lie in the interval $0 < X < 100$ m^3.

To determine the value of a continuous random variable we need to **measure**.

PROBABILITY DISTRIBUTIONS

For any random variable there is a corresponding **probability distribution** which describes the probability that the variable will take any particular value.

The probability that the variable X takes value x is denoted $P(X = x)$.

For example, when tossing two coins, the random variable X could be 0 heads, 1 head, or 2 heads, so $X = 0$, 1, or 2. The associated probability distribution is $P(X = 0) = \frac{1}{4}$, $P(X = 1) = \frac{1}{2}$, and $P(X = 2) = \frac{1}{4}$.

Spike graph

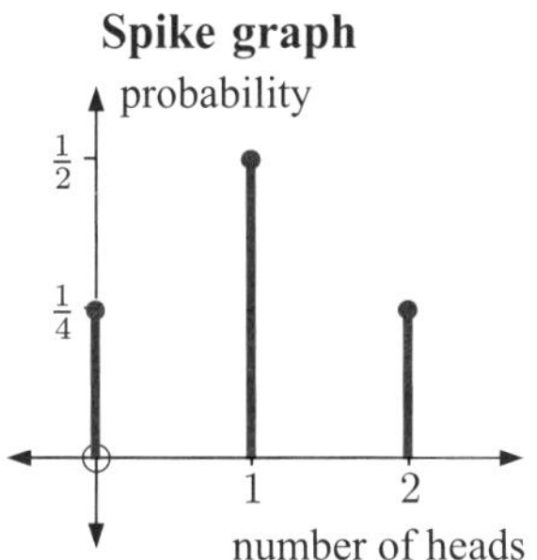

or

Probability column graph

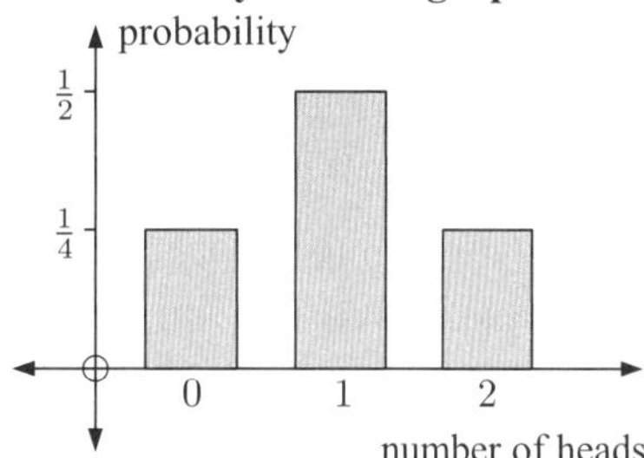

Example 1 **Self Tutor**

A supermarket has three checkout points A, B, and C. A government inspector checks the weighing scales for accuracy at each checkout. If a weighing scale is accurate then Y is recorded, and if not, N is recorded. Suppose the random variable X is the number of accurate weighing scales at the supermarket.

a List the possible outcomes.

b Describe, using X, the events of there being:

i one accurate scale   ii at least one accurate scale.

a Possible outcomes:

A	B	C	X
N	N	N	0
Y	N	N	1
N	Y	N	1
N	N	Y	1
N	Y	Y	2
Y	N	Y	2
Y	Y	N	2
Y	Y	Y	3

b i $X = 1$

ii $X = 1, 2,$ or 3

EXERCISE 25A

1 Classify the following random variables as continuous or discrete:

a the quantity of fat in a sausage
b the mark out of 50 for a geography test
c the weight of a seventeen year old student
d the volume of water in a cup of coffee
e the number of trout in a lake
f the number of hairs on a cat
g the length of hairs on a horse
h the height of a sky-scraper.

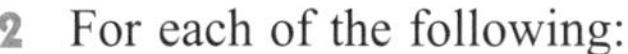

2 For each of the following:

i identify the random variable being considered
ii give possible values for the random variable
iii indicate whether the variable is continuous or discrete.

a To measure the rainfall over a 24-hour period in Singapore, the height of water collected in a rain gauge (up to 400 mm) is used.

b To investigate the stopping distance for a tyre with a new tread pattern, a braking experiment is carried out.

c To check the reliability of a new type of light switch, switches are repeatedly turned off and on until they fail.

3 A supermarket has four checkouts A, B, C, and D. Management checks the weighing devices at each checkout. If a weighing device is accurate a Y is recorded; otherwise, N is recorded. The random variable being considered is the number of weighing devices which are accurate.

a Suppose X is the random variable. What values can X have?

b Tabulate the possible outcomes and the corresponding values for X.

c Describe, using X, the events of:

i exactly two devices being accurate ii at least two devices being accurate.

4 Consider tossing three coins simultaneously. The random variable under consideration is the number of heads that could result.

a List the possible values of X.

b Tabulate the possible outcomes and the corresponding values of X.

c Find the values of $P(X = x)$, the probability of each x value occurring.

d Graph the probability distribution $P(X = x)$ against x as a probability column graph.

B DISCRETE PROBABILITY DISTRIBUTIONS

We saw in **Chapter 24** that probabilities may be assigned to events in a number of ways.

For example:

- we can conduct experiments where we perform trials many times over until a pattern emerges
- we can use symmetry to say the chances of a coin being a head or a tail are both $\frac{1}{2}$, or an ordinary die showing a particular number is $\frac{1}{6}$
- we can evaluate the form of tennis players to predict their chances in an upcoming match.

Whichever way probabilities are assigned, they must satisfy the following rule:

If X is a random variable with sample space $\{x_1, x_2, x_3, ..., x_n\}$ and corresponding probabilities $\{p_1, p_2, p_3, ..., p_n\}$ so that $P(X = x_i) = p_i$, $i = 1, ..., n$, then:

- $0 \leqslant p_i \leqslant 1$ for all $i = 1$ to n
- $\sum_{i=1}^{n} p_i = p_1 + p_2 + p_3 + ... + p_n = 1$
- $\{p_1, ..., p_n\}$ describes the **probability distribution** of X.

The **mode** of the distribution is the most frequently occurring value of the variable. This is the data value x_i whose probability p_i is the highest.

The **median** of the distribution corresponds to the 50th percentile. If the elements of the sample space $\{x_1, x_2, x_3, ..., x_n\}$ are listed in ascending order, it will be the value x_j when the cumulative sum $p_1 + p_2 + ... + p_j$ reaches 0.5.

For example, when a coin is tossed twice, the number of heads X has sample space $\{0, 1, 2\}$ with corresponding probabilities $\{\frac{1}{4}, \frac{1}{2}, \frac{1}{4}\}$. We see that $0 \leqslant p_i \leqslant 1$ for each value of i, and that the probabilities add up to 1.

The **probability distribution** of a **discrete random variable** can be given:

- in table form
- in graphical form
- in function form as a **probability distribution function** or **probability mass function** $P(x) = \text{P}(X = x)$. The domain of the probability mass function is the set of possible values of the variable, and the range is the set of values in the probability distribution of the variable.

When we consider discrete random variables, we often look at intervals of values which the variable may take. We need to be careful about whether the end points of an interval are included. For example, consider these intervals:

Notation	*Statement*
$\text{P}(X = 3)$	the probability that X equals 3
$\text{P}(X < 3)$	the probability that X is less than 3
$\text{P}(X \leqslant 3)$	the probability that X is at most 3
$\text{P}(X > 3)$	the probability that X is more than 3
$\text{P}(X \geqslant 3)$	the probability that X is at least 3
$\text{P}(3 < X < 7)$	the probability that X is between 3 and 7
$\text{P}(3 \leqslant X \leqslant 7)$	the probability that X is at least 3 but no more than 7
$\text{P}(3 < X \leqslant 7)$	the probability that X is more than 3 but no more than 7
$\text{P}(3 \leqslant X < 7)$	the probability that X is at least 3 but less than 7

Example 2 **Self Tutor**

A magazine store recorded the number of magazines purchased by its customers in one week. 23% purchased one magazine, 38% purchased two, 21% purchased three, 13% purchased four, and 5% purchased five.

a What is the random variable?

b Make a probability table for the random variable.

c Graph the probability distribution using a spike graph.

d State the mode and median of the distribution.

a The random variable X is the number of magazines sold.
So, $X = 1, 2, 3, 4,$ or 5.

b

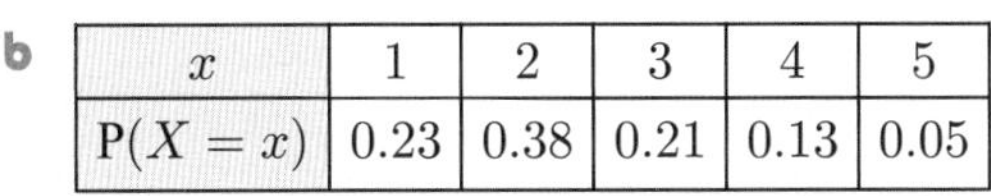

x	1	2	3	4	5
$\text{P}(X = x)$	0.23	0.38	0.21	0.13	0.05

c

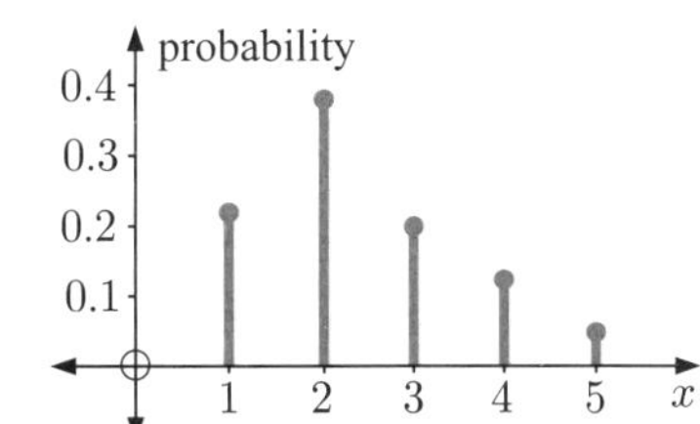

d Customers are most likely to buy 2 magazines, so this is the mode of the distribution.

$p_1 = 0.23$

$p_1 + p_2 = 0.23 + 0.38 = 0.61$

Since $p_1 + p_2 \geqslant 0.5$, the median is 2 magazines.

Example 3 — Self Tutor

Show that the following are probability distribution functions:

a $P(x) = \dfrac{x^2+1}{34}, \quad x = 1, 2, 3, 4$ **b** $P(x) = \binom{3}{x}(0.6)^x(0.4)^{3-x}, \quad x = 0, 1, 2, 3$

a $P(1) = \frac{2}{34}$ $P(2) = \frac{5}{34}$ $P(3) = \frac{10}{34}$ $P(4) = \frac{17}{34}$

All of these obey $0 \leqslant P(x_i) \leqslant 1$, and $\sum P(x_i) = \frac{2}{34} + \frac{5}{34} + \frac{10}{34} + \frac{17}{34} = 1$

$\therefore$ $P(x)$ is a probability distribution function.

b For $P(x) = \binom{3}{x}(0.6)^x(0.4)^{3-x}$,

$P(0) = \binom{3}{0}(0.6)^0(0.4)^3 = 1 \times 1 \times (0.4)^3 = 0.064$

$P(1) = \binom{3}{1}(0.6)^1(0.4)^2 = 3 \times (0.6) \times (0.4)^2 = 0.288$

$P(2) = \binom{3}{2}(0.6)^2(0.4)^1 = 3 \times (0.6)^2 \times (0.4) = 0.432$

$P(3) = \binom{3}{3}(0.6)^3(0.4)^0 = 1 \times (0.6)^3 \times 1 = 0.216$

Total 1.000

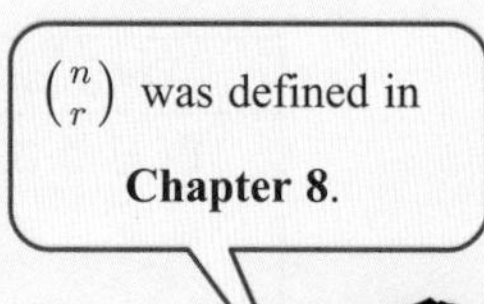
$\binom{n}{r}$ was defined in
Chapter 8.

All probabilities lie between 0 and 1, and $\sum P(x_i) = 1$.

$\therefore$ $P(x)$ is a probability distribution function.

EXERCISE 25B

1 Find k in these probability distributions:

a

x	0	1	2
$P(X = x)$	0.3	k	0.5

b

x	0	1	2	3
$P(X = x)$	k	$2k$	$3k$	k

2 The probabilities of Jason scoring home runs in each game during his baseball career are given in the following table. X is the number of home runs per game.

x	0	1	2	3	4	5
$P(x)$	a	0.3333	0.1088	0.0084	0.0007	0.0000

a State the value of $P(2)$.

b Find the value of a. Explain what this number means.

c Find the value of $P(1) + P(2) + P(3) + P(4) + P(5)$. Explain what this means.

d Draw a probability distribution spike graph of $P(x)$ against x.

e State the mode and median of the distribution.

3 Explain why the following are not valid probability distribution functions:

a

x	0	1	2	3
$P(x)$	0.2	0.3	0.4	0.2

b

x	2	3	4	5
$P(x)$	0.3	0.4	0.5	-0.2

4 Sally's number of hits in each softball match has the following probability distribution:

x	0	1	2	3	4	5
$P(X = x)$	0.07	0.14	k	0.46	0.08	0.02

a State clearly what the random variable represents.

b Find: **i** k **ii** $P(X \geqslant 2)$ **iii** $P(1 \leqslant X \leqslant 3)$

c Find the mode and median number of hits.

5 A die is rolled twice.

  a Draw a grid which shows the sample space.

  b Suppose X denotes the sum of the results for the two rolls.

    i Find the probability distribution of X.

    ii Draw a probability distribution column graph for X.

    iii State the mode and median of X.

6 Find k for the following probability distributions:

  a $P(x) = k(x+2)$ for $x = 1, 2, 3$ **b** $P(x) = \dfrac{k}{x+1}$ for $x = 0, 1, 2, 3$.

7 A discrete random variable X has probability distribution given by $\text{P}(X = x) = k\left(\frac{1}{3}\right)^x\left(\frac{2}{3}\right)^{4-x}$ where $x = 0, 1, 2, 3, 4$.

  a Find $\text{P}(X = x)$ for $x = 0, 1, 2, 3$ and 4. **b** Find k and hence find $\text{P}(X \geqslant 2)$.

8 Electrical components are produced and packed into boxes of 10. It is known that 4% of the components produced are faulty. The random variable X denotes the number of faulty items in the box, and has probability distribution function

$$\text{P}(X = x) = \binom{10}{x}(0.04)^x(0.96)^{10-x}, \quad x = 0, 1, 2, ..., 10.$$

Find the probability that a randomly selected box:

  a will not contain a faulty component **b** will contain at least one faulty component.

Example 4 — Self Tutor

Two marbles are randomly selected without replacement from a bag containing 4 red and 2 blue marbles. Let X denote the number of red marbles selected.

  a Find the probability distribution of X.

  b Illustrate the probability distribution using a spike graph.

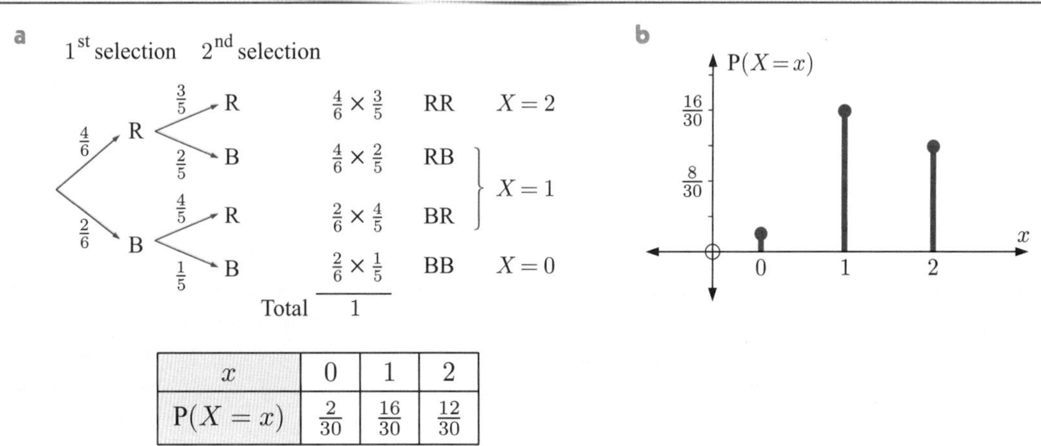

x	0	1	2
$\text{P}(X = x)$	$\frac{2}{30}$	$\frac{16}{30}$	$\frac{12}{30}$

9 A bag contains 5 blue and 3 green tickets and a number of tickets are randomly selected without replacement. We let X denote the number of blue tickets selected.

Find the probability distribution of X if:

  a two tickets are randomly selected **b** three tickets are randomly selected.

10 When a pair of dice is rolled, D denotes the sum of the top faces.

- **a** Display the possible results in a table.
- **b** Find $P(D = 7)$.
- **c** Find the probability distribution of D.
- **d** Find $P(D \geqslant 8 \mid D \geqslant 6)$.

11 When a pair of dice is rolled, N denotes the difference between the numbers on the top faces.

- **a** Display the possible results in a table.
- **b** Construct a probability distribution table for the possible values of N.
- **c** Find: **i** $P(N = 3)$ **ii** $P(N \geqslant 3 \mid N \geqslant 1)$

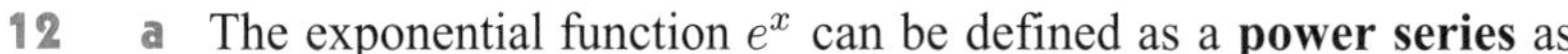

12 **a** The exponential function e^x can be defined as a **power series** as

$$e^x = \sum_{n=0}^{\infty} \frac{x^n}{n!} = 1 + x + \frac{x^2}{2!} + \frac{x^3}{3!} + ...$$

Use this definition to evaluate $\sum_{n=0}^{\infty} \frac{(0.2)^n e^{-0.2}}{n!}$.

b Suppose X is the number of cars that pass a shop during the period from 3:00 pm to 3:03 pm. The probability distribution for X is given by

$$P(X = x) = \frac{(0.2)^x e^{-0.2}}{x!} \quad \text{where } x = 0, 1, 2, 3, ...$$

- **i** Explain why this is a valid probability distribution.
- **ii** Evaluate $P(X = 0)$, $P(X = 1)$, and $P(X = 2)$.
- **iii** Find the probability that at least three cars will pass the shop in the given period.

C EXPECTATION

Consider the following problem:

A die is to be rolled 120 times. On how many occasions should we *expect* the result to be a "six"?

In order to answer this question we must first consider all possible outcomes of rolling the die. The possibilities are 1, 2, 3, 4, 5, and 6, and each of these is equally likely to occur.

Therefore, we would expect $\frac{1}{6}$ of them to be a "six".

$\frac{1}{6}$ of 120 is 20, so we expect 20 of the 120 rolls of the die to yield a "six".

However, this does not mean that we *will* get 20 sixes when we roll a die 120 times.

If there are n trials of an experiment, and an event has probability p of occurring in each of the trials, then the number of times we **expect** the event to occur is np.

We can also talk about the *expected* or *mean* outcome from one trial of an experiment.

Suppose we have 10 counters. One shows the number 1, four show the number 2, three show 3, and the other two show 4. One counter is to be randomly selected from a hat.

Suppose we perform this experiment 10 times. We can summarise the possible results in a table:

Outcome x_i	1	2	3	4
Probability p_i	$\frac{1}{10}$	$\frac{4}{10}$	$\frac{3}{10}$	$\frac{2}{10}$
Expected frequency f_i *from* 10 *trials*	1	4	3	2

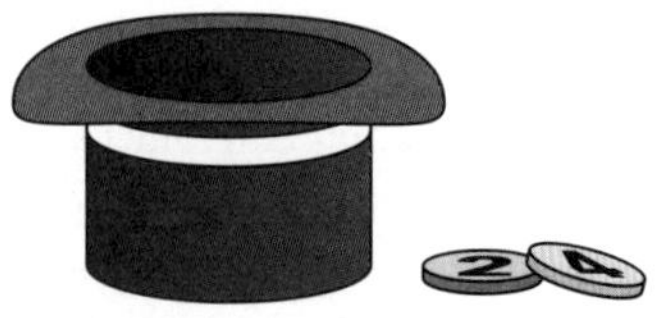

If the actual results match the expected frequencies, then the mean result will be

$$\begin{aligned}\mu &= \frac{\sum f_i x_i}{n} \\ &= \frac{1\times 1 + 2\times 4 + 3\times 3 + 4\times 2}{1+4+3+2} \\ &= 1\times \tfrac{1}{10} + 2\times \tfrac{4}{10} + 3\times \tfrac{3}{10} + 4\times \tfrac{2}{10} \\ &= \sum x_i p_i\end{aligned}$$

In the general case, suppose a random variable has k possible values $x_1, x_2, x_3, ..., x_k$
with probabilities $p_1, p_2, p_3, ..., p_k$.

From n trials, the expected frequencies are $f_1, f_2, f_3, ..., f_k$, where $n = \sum f_i$.

The population mean
$$\begin{aligned}\mu &= \frac{\sum f_i x_i}{n} \quad \{\text{mean for tabled values}\} \\ &= \frac{f_1x_1 + f_2x_2 + f_3x_3 + ... + f_kx_k}{n} \\ &= x_1\left(\frac{f_1}{n}\right) + x_2\left(\frac{f_2}{n}\right) + x_3\left(\frac{f_3}{n}\right) + ... + x_k\left(\frac{f_k}{n}\right) \\ &= x_1p_1 + x_2p_2 + x_3p_3 + ... + x_kp_k \\ &= \sum x_i p_i\end{aligned}$$

The expected outcome for the random variable X is the **mean** result μ.

The **expectation** of the random variable X is

$$\mathbf{E}(\boldsymbol{X}) = \boldsymbol{\mu} = \sum_{i=1}^{n} \boldsymbol{x_i p_i} \quad \text{or} \quad \sum_{i=1}^{n} \boldsymbol{x_i}\, \mathbf{P}(\boldsymbol{X} = \boldsymbol{x_i})$$

Example 5 — Self Tutor

Each time a footballer kicks for goal he has a $\frac{3}{4}$ chance of being successful.
In a particular game he has 12 kicks for goal. How many goals would you expect him to kick?

$p = \text{P(goal)} = \frac{3}{4}$ $\quad\therefore$ the expected number of goals is
$$\begin{aligned}E &= np \\ &= 12 \times \tfrac{3}{4} \\ &= 9\end{aligned}$$

Example 6 — Self Tutor

Find the mean of the magazine store data in **Example 2**. Explain what it means.

The probability table is:

x_i	1	2	3	4	5
p_i	0.23	0.38	0.21	0.13	0.05

Now
$$\begin{aligned}\mu &= \sum x_i p_i \\ &= 1(0.23) + 2(0.38) + 3(0.21) + 4(0.13) + 5(0.05) \\ &= 2.39\end{aligned}$$

In the long run, the average number of magazines purchased per customer is 2.39 .

FAIR GAMES

In gambling, we say that the **expected gain** of the player from each game is the expected return or payout from the game, less the amount it cost them to play.

The game will be **fair** if the expected gain is zero.

Suppose X represents the gain of a player from each game.
The game is **fair** if $\text{E}(X) = 0$.

Example 7 — Self Tutor

In a game of chance, a player spins a square spinner labelled 1, 2, 3, 4. The player wins the amount of money shown in the table alongside, depending on which number comes up. Determine:

Number	1	2	3	4
Winnings	\$1	\$2	\$5	\$8

a the expected return for one spin of the spinner

b the expected *gain* of the player if it costs \$5 to play each game

c whether you would recommend playing this game.

a Let Y denote the return or payout from each spin.
As each outcome is equally likely, the probability for each outcome is $\frac{1}{4}$
$\therefore$ expected return $= \text{E}(Y) = \frac{1}{4} \times 1 + \frac{1}{4} \times 2 + \frac{1}{4} \times 5 + \frac{1}{4} \times 8 = \$4$.

b Let X denote the *gain* of the player from each game.
Since it costs \$5 to play the game, the expected gain $= \text{E}(X) = \text{E}(Y) - \$5$
$= \$4 - \$5$
$= -\$1$

c Since $\text{E}(X) \neq 0$, the game is not fair. In particular, since $\text{E}(X) = -\$1$, we expect the player to lose \$1 on average with each spin. We would not recommend that a person play the game.

EXERCISE 25C

1 In a particular region, the probability that it will rain on any one day is 0.28. On how many days of the year would you expect it to rain?

2 **a** If 3 coins are tossed, what is the chance that they all fall heads?

b If the 3 coins are tossed 200 times, on how many occasions would you expect them all to fall heads?

3 If two dice are rolled simultaneously 180 times, on how many occasions would you expect to get a double?

4 A single coin is tossed once. If a head appears you win \$2, and if a tail appears you lose \$1. How much would you expect to win when playing this game three times?

5 During the snow season there is a $\frac{3}{7}$ probability of snow falling on any particular day. If Udo skis for five weeks, on how many days could he expect to see snow falling?

6 A goalkeeper has probability $\frac{3}{10}$ of saving a penalty attempt. How many goals would he expect to save from 90 attempts?

7 In a random survey of her electorate, politician A discovered the residents' voting intentions in relation to herself and her two opponents B and C. The results are indicated alongside:

A	B	C
165	87	48

a Estimate the probability that a randomly chosen voter in the electorate will vote for:
i A **ii** B **iii** C.

b If there are 7500 people in the electorate, how many of these would you expect to vote for:
i A **ii** B **iii** C?

8 A charity fundraiser gets a licence to run the following gambling game: A die is rolled and the returns to the player are given in the 'pay table' alongside. To play the game costs \$4. A result of getting a 6 wins \$10, so in fact you are ahead by \$6 if you get a 6 on the first roll.

Result	*Wins*
6	\$10
4, 5	\$4
1, 2, 3	\$1

a What are your chances of playing one game and winning:
i \$10 **ii** \$4 **iii** \$1?

b Your expected return from throwing a 6 is $\frac{1}{6} \times \$10$. What is your expected return from throwing:
i a 4 or 5 **ii** a 1, 2, or 3 **iii** a 1, 2, 3, 4, 5, or 6?

c What is your overall expected result at the end of one game?

d What is your overall expected result at the end of 100 games?

9 A person rolls a normal six-sided die and wins the number of euros shown on the face.

a Find the expected return from one roll of the die.

b Find the expected *gain* of the player if it costs €4 to play the game. Would you advise the person to play several games?

c Suppose it costs €k to play the game. What value(s) of k will result in:
i a fair game **ii** a profit being made by the vendor?

10 A person plays a game with a pair of coins. If two heads appear then £10 is won. If a head and a tail appear then £3 is won. If two tails appear then £5 is lost.

a How much would a person expect to win playing this game once?

b If the organiser of the game wishes to make an average of £1 per game, how much should the organiser charge people per game to play?

11 At a charity event there is a money-raising game involving a pair of ordinary dice. The game costs \$$a$ to play. When the two dice are rolled, their sum is described by the variable X. The organisers decide that a sum which is less than 4 or between 7 and 9 inclusive gives a loss of \$$\frac{a}{3}$, a result between 4 and 6 inclusive gives a return of \$7, and a result of 10 or more gives a return of \$21.

a Determine $P(X \leqslant 3)$, $P(4 \leqslant X \leqslant 6)$, $P(7 \leqslant X \leqslant 9)$, and $P(X \geqslant 10)$.

b Show that the expected gain of a player is given by $\frac{1}{6}(35 - 7a)$ dollars.

c What value would a need to have for the game to be 'fair'?

d Explain why the organisers would not let a be 4.

e If the organisers let a be 6 and the game was played 2406 times, estimate the amount of money raised by this game.

D VARIANCE AND STANDARD DEVIATION

Consider again having 10 counters, one which shows 1, four which show 2, three which show 3, and two which show 4. One counter is to be randomly selected from a hat. If we perform this experiment 10 times, the possible results are:

Outcome x_i	1	2	3	4
Probability p_i	$\frac{1}{10}$	$\frac{4}{10}$	$\frac{3}{10}$	$\frac{2}{10}$
Expected frequency f_i *from* 10 *trials*	1	4	3	2

We have already seen that the mean or expected outcome is $\mu = \dfrac{\sum f_i x_i}{n} = \sum x_i p_i$.

The variance of the results is
$$\begin{aligned}\sigma^2 &= \frac{\sum f_i(x_i-\mu)^2}{n}\\ &= \frac{1(x_1-\mu)^2}{10} + \frac{4(x_2-\mu)^2}{10} + \frac{3(x_3-\mu)^2}{10} + \frac{2(x_4-\mu)^2}{10}\\ &= \tfrac{1}{10}(x_1-\mu)^2 + \tfrac{4}{10}(x_2-\mu)^2 + \tfrac{3}{10}(x_3-\mu)^2 + \tfrac{2}{10}(x_4-\mu)^2\\ &= \sum(x_i-\mu)^2 p_i\end{aligned}$$

Now consider the general case, where a random variable has k possible values $x_1, x_2, x_3, ..., x_k$ with probabilities $p_1, p_2, p_3, ..., p_k$.

From n trials, the expected frequencies are $f_1, f_2, f_3, ..., f_k$, where $n = \sum f_i$.

The variance
$$\begin{aligned}\sigma^2 &= \frac{\sum f_i(x_i-\mu)^2}{n}\\ &= \frac{f_1(x_1-\mu)^2}{n} + \frac{f_2(x_2-\mu)^2}{n} + \frac{f_3(x_3-\mu)^2}{n} + ... + \frac{f_k(x_k-\mu)^2}{n}\\ &= p_1(x_1-\mu)^2 + p_2(x_2-\mu)^2 + p_3(x_3-\mu)^2 + ... + p_k(x_k-\mu)^2\\ &= \sum(x_i-\mu)^2 p_i\end{aligned}$$

If a discrete random variable has k possible values $x_1, x_2, x_3, ..., x_k$ with probabilities $p_1, p_2, p_3, ..., p_k$

then:
- the population **mean** or **expectation** is $\mathbf{E}(X) = \mu = \sum x_i p_i$
- the population **variance** is $\mathbf{Var}(X) = \sigma^2 = \sum(x_i-\mu)^2 p_i = \mathbf{E}(X-\mu)^2$
- the population **standard deviation** is $\sigma = \sqrt{\sum(x_i-\mu)^2 p_i}$.

Example 8 — Self Tutor

Find the standard deviation for the distribution:

x_i	1	2	3	4	5
p_i	0.23	0.38	0.21	0.13	0.05

The mean $\mu = \sum x_i p_i = 2.39$ See **Example 6**.

The standard deviation

$$\begin{aligned}\sigma &= \sqrt{\sum(x_i-\mu)^2 p_i}\\ &= \sqrt{(1-2.39)^2 \times 0.23 + (2-2.39)^2 \times 0.38 + ... + (5-2.39)^2 \times 0.05}\\ &\approx 1.122\end{aligned}$$

An alternative formula for the population standard deviation is $\boldsymbol{\sigma = \sqrt{\sum x_i{}^2 p_i - \mu^2}}$.

For example, for the roll of a die:

$$\mu = \sum x_i p_i = 1(\tfrac{1}{6}) + 2(\tfrac{1}{6}) + 3(\tfrac{1}{6}) + 4(\tfrac{1}{6}) + 5(\tfrac{1}{6}) + 6(\tfrac{1}{6}) = 3.5$$

and $\sigma^2 = \sum x_i{}^2 p_i - \mu^2 = 1^2(\tfrac{1}{6}) + 2^2(\tfrac{1}{6}) + 3^2(\tfrac{1}{6}) + 4^2(\tfrac{1}{6}) + 5^2(\tfrac{1}{6}) + 6^2(\tfrac{1}{6}) - (3.5)^2$

$$\approx 2.92$$

Consequently, $\sigma \approx 1.71$.

You can check these results using your calculator by generating 800 random digits from 1 to 6. The mean and standard deviation of the results should be a good approximation of the theoretical values above.

EXERCISE 25D

1 A country exports crayfish to overseas markets. The buyers are prepared to pay high prices when the crayfish arrive still alive.

If X is the number of deaths per dozen crayfish, the probability distribution for X is given by:

x	0	1	2	3	4	5	> 5
$P(X = x)$	0.54	0.26	0.15	k	0.01	0.01	0.00

a Find k.

b Over a long period, what is the mean number of deaths per dozen crayfish?

c Find the standard deviation σ for the probability distribution.

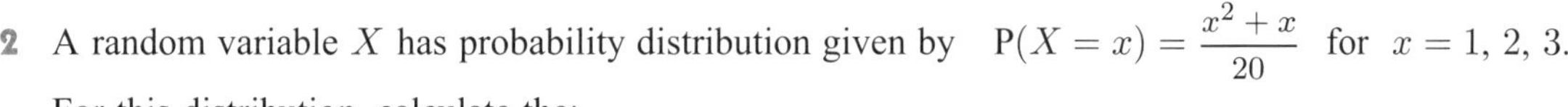

2 A random variable X has probability distribution given by $P(X = x) = \dfrac{x^2 + x}{20}$ for $x = 1, 2, 3$.

For this distribution, calculate the:

a mode b median c mean μ d standard deviation σ.

3 A random variable X has probability distribution given by

$$P(x) = C_x^3 (0.4)^x (0.6)^{3-x} \quad \text{for} \quad x = 0, 1, 2, 3.$$

a Find $P(x)$ for $x = 0, 1, 2$, and 3, and display the results in table form.

b Find the mean and standard deviation for the distribution.

4 Use $\sigma^2 = \sum (x_i - \mu)^2 p_i$ to show that $\sigma^2 = \sum x_i{}^2 p_i - \mu^2$.

5 A random variable X has the probability distribution shown.

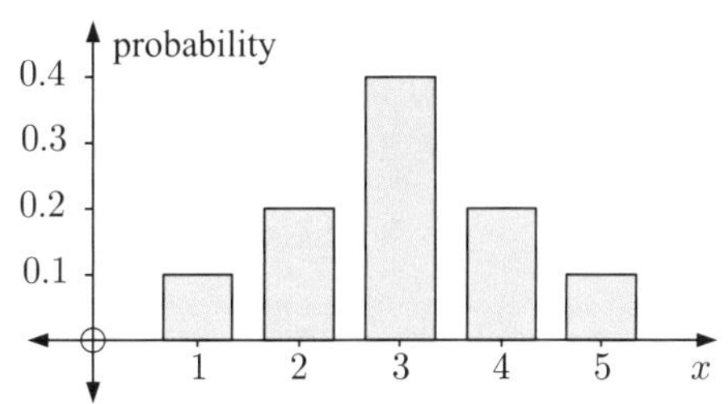

a Copy and complete:

x_i	1	2	3	4	5
$P(x_i)$					

b Find the mean μ and standard deviation σ for the distribution.

c Determine i $P(\mu - \sigma < x < \mu + \sigma)$ ii $P(\mu - 2\sigma < x < \mu + 2\sigma)$.

6 An insurance policy covers a \$20 000 sapphire ring against theft and loss. If the ring is stolen then the insurance company will pay the policy owner in full. If it is lost then they will pay the owner \$8000. From past experience, the insurance company knows that the probability of theft is 0.0025, and of loss is 0.03. How much should the company charge to cover the ring in order that their expected return is \$100?

7 When a pair of dice is rolled, the random variable M is the larger of the two numbers that are shown uppermost, or the value of a single die if a double is thrown.

  a In table form, record the probability distribution of M.

  b For the M-distribution find the:

    i mode   **ii** median   **iii** mean   **iv** standard deviation.

8 A uniform distribution has $P(x_1) = P(x_2) = P(x_3) = ...$ Give *two* examples of a uniform distribution.

E PROPERTIES OF E(X) AND Var(X)

PROPERTIES OF E(X)

If E(X) is the expected value of the random variable X, then:

- $\mathrm{E}(k) = k$ for any constant k
- $\mathrm{E}(kX) = k\,\mathrm{E}(X)$ for any constant k
- $\mathrm{E}(A(X) + B(X)) = \mathrm{E}(A(X)) + \mathrm{E}(B(X))$ for functions A and B.

The expectation of a sum is the sum of the individual expectations.

For example, these properties enable us to deduce that:

- $\mathrm{E}(5) = 5$
- $\mathrm{E}(3X) = 3\,\mathrm{E}(X)$
- $\mathrm{E}(X^2 + 2X + 3) = \mathrm{E}(X^2) + 2\,\mathrm{E}(X) + 3$

PROPERTY OF Var(X)

$$\mathbf{Var}(X) = \mathbf{E}(X^2) - (\mathbf{E}(X))^2 \quad \text{or} \quad \mathbf{Var}(X) = \mathbf{E}(X^2) - \mu^2$$

Proof:

$$\begin{aligned}\mathrm{Var}(X) &= \mathrm{E}(X-\mu)^2 \\ &= \mathrm{E}(X^2 - 2\mu X + \mu^2) \\ &= \mathrm{E}(X^2) - 2\mu\,\mathrm{E}(X) + \mu^2 \qquad \{\text{properties of } \mathrm{E}(X)\} \\ &= \mathrm{E}(X^2) - 2\mu^2 + \mu^2 \\ &= \mathrm{E}(X^2) - \mu^2\end{aligned}$$

Example 9 **Self Tutor**

Suppose X has the probability distribution:

x	1	2	3	4
p_x	0.1	0.3	0.4	0.2

Find:

a the mean of X **b** the variance of X **c** the standard deviation of X.

a $\text{E}(X) = \sum x_i p_i = 1(0.1) + 2(0.3) + 3(0.4) + 4(0.2)$

$\therefore \text{E}(X) = 2.7$ so $\mu = 2.7$

b $\text{E}(X^2) = \sum {x_i}^2 p_i = 1^2(0.1) + 2^2(0.3) + 3^2(0.4) + 4^2(0.2) = 8.1$

$\therefore \text{Var}(X) = \text{E}(X^2) - (\text{E}(X))^2$
$= 8.1 - 2.7^2$
$= 0.81$

c $\sigma = \sqrt{\text{Var}(X)} = 0.9$

INVESTIGATION 1 $\text{E}(aX+b)$ AND $\text{Var}(aX+b)$

The purpose of this Investigation is to discover a relationship between $\text{E}(aX+b)$ and $\text{E}(X)$ and also $\text{Var}(aX+b)$ and $\text{Var}(X)$.

What to do:

1 Consider the X-distribution: 1, 2, 3, 4, 5, each occurring with equal probability.

a Find $\text{E}(X)$ and $\text{Var}(X)$.

b If $Y = 2X + 3$, find the Y-distribution.
Hence find $\text{E}(2X+3)$ and $\text{Var}(2X+3)$.

c Repeat **b** for:

i $Y = 3X - 2$ **ii** $Y = -2X + 5$ **iii** $Y = \dfrac{X+1}{2}$.

2 Make up your own sample distribution for a random variable X and repeat **1**.

3 Record all your results in table form for both distributions.

4 State the relationship between:
- $\text{E}(X)$ and $\text{E}(aX+b)$
- $\text{Var}(X)$ and $\text{Var}(aX+b)$.

From the **Investigation** you should have discovered that

$$\mathbf{E(aX+b) = a\,E(X) + b} \quad \text{and} \quad \mathbf{Var(aX+b) = a^2 Var(X)}.$$

These results will be formally proved in the **Exercise** which follows.

Example 10 **Self Tutor**

X is distributed with mean 8.1 and standard deviation 2.37. If $Y = 4X - 7$, find the mean and standard deviation of the Y-distribution.

$\text{E}(X) = 8.1$ and $\text{Var}(X) = 2.37^2$

$$\begin{aligned}&\text{E}(4X - 7)\\&= 4\text{E}(X) - 7\\&= 4(8.1) - 7\\&= 25.4\end{aligned}$$

$$\begin{aligned}&\text{Var}(4X - 7)\\&= 4^2\text{Var}(X)\\&= 4^2 \times 2.37^2\end{aligned}$$

For the Y-distribution, the mean is 25.4 and the standard deviation is $4 \times 2.37 = 9.48$.

EXERCISE 25E

1 Suppose X has the probability distribution:

x	2	3	4	5	6
p_x	0.3	0.3	0.2	0.1	0.1

Find:

a the mean of X **b** the variance of X **c** the standard deviation of X.

2 Suppose X has the probability distribution:

x	5	6	7	8
p_x	0.2	k	0.4	0.1

Find:

a the value of k **b** the mean of X **c** the variance of X.

3 Suppose X has the probability distribution:

x	1	2	3	4
p_x	0.4	0.3	0.2	0.1

Find:

a $\text{E}(X)$ **b** $\text{E}(X^2)$ **c** $\text{Var}(X)$ **d** σ

e $\text{E}(X+1)$ **f** $\text{Var}(X+1)$ **g** $\text{E}(2X^2 + 3X - 7)$

4 Suppose X has the probability distribution:

x	1	2	3	4
p_x	0.2	a	0.3	b

a Given that $\text{E}(X) = 2.8$, find a and b.

b Hence show that $\text{Var}(X) = 1.26$.

5 Suppose X is the number of marsupials entering a park at night. It is suspected that X has a probability distribution of the form $\text{P}(X = x) = a(x^2 - 8x)$ where $X = 0, 1, 2, 3, ..., 8$.

a Find the constant a.

b Find the expected number of marsupials entering the park on a given night.

c Find the standard deviation of X.

6 An unbiased coin is tossed four times. Suppose X is the number of heads which appear.

a Find the probability distribution of X.

b Find: **i** the mean of X **ii** the standard deviation of X.

7 A box contains 10 almonds, two of which are bitter and the remainder are normal. Brit randomly selects three almonds without replacement. Let X be the random variable for the number of bitter almonds Brit selects.

  a Find the probability distribution of X.

  b Find: **i** the mean of X **ii** the standard deviation of X.

8 The probability distribution of the discrete random variable Y is:

Y	-1	0	1	2
$\mathrm{P}(Y = y)$	0.1	a	0.3	b

  a Given $\mathrm{E}(Y) = 0.9$, find a and b. **b** Hence find $\mathrm{Var}(Y)$.

9 The score X obtained by rolling a biased pentagonal die has the probability distribution:

X	1	2	3	4	5
$\mathrm{P}(X = x)$	$\frac{1}{6}$	$\frac{1}{3}$	$\frac{1}{12}$	a	$\frac{1}{6}$

  a Find a.

  b Hence find: **i** $\mathrm{E}(X)$ and $\mathrm{Var}(X)$
  ii $\mathrm{E}(2X)$ and $\mathrm{Var}(2X)$
  iii $\mathrm{E}(X-1)$ and $\mathrm{Var}(X-1)$.

10 X is distributed with mean 6 and standard deviation 2. If $Y = 2X + 5$, find the mean and standard deviation of the Y-distribution.

11 **a** Use the properties of $\mathrm{E}(X)$ to prove that $\mathrm{E}(aX + b) = a\,\mathrm{E}(X) + b$.

  b The mean of an X-distribution is 3. Find the mean of the Y-distribution where:

  i $Y = 3X + 4$ **ii** $Y = -2X + 1$ **iii** $Y = \dfrac{4X - 2}{3}$

12 X is a random variable with mean 5 and standard deviation 2.
Find $\mathrm{E}(Y)$ and $\mathrm{Var}(Y)$ for:

  a $Y = 2X + 3$ **b** $Y = -2X + 3$ **c** $Y = \dfrac{X - 5}{2}$

13 Suppose $Y = 2X + 3$ where X is a random variable.
Find in terms of $\mathrm{E}(X)$ and $\mathrm{E}(X^2)$:

  a $\mathrm{E}(Y)$ **b** $\mathrm{E}(Y^2)$ **c** $\mathrm{Var}(Y)$

14 Using $\mathrm{Var}(X) = \mathrm{E}(X^2) - (\mathrm{E}(X))^2$, prove that $\mathrm{Var}(aX + b) = a^2\mathrm{Var}(X)$.

F THE BINOMIAL DISTRIBUTION

Thus far in the chapter we have considered properties of general discrete random variables.

We now examine a special type of discrete random variable which is applied to **sampling with replacement**. The probability distribution associated with this variable is the **binomial probability distribution**.

For sampling without replacement the hypergeometric probability distribution is the model used, but that distribution is not part of this course.

BINOMIAL EXPERIMENTS

Consider an experiment for which there are two possible results: **success** if some event occurs, or **failure** if the event does not occur.

If we repeat this experiment in a number of **independent trials**, we call it a **binomial experiment**.

The probability of a success p must be constant for all trials. Since success and failure are complementary events, the probability of a failure is $1-p$ and is constant for all trials.

The random variable X is the total number of successes in n trials.

INVESTIGATION 2 SAMPLING SIMULATION

When balls enter the 'sorting' chamber shown they hit a metal rod and may go left or right. This movement continues as the balls fall from one level of rods to the next. The balls finally come to rest in collection chambers at the bottom of the sorter.

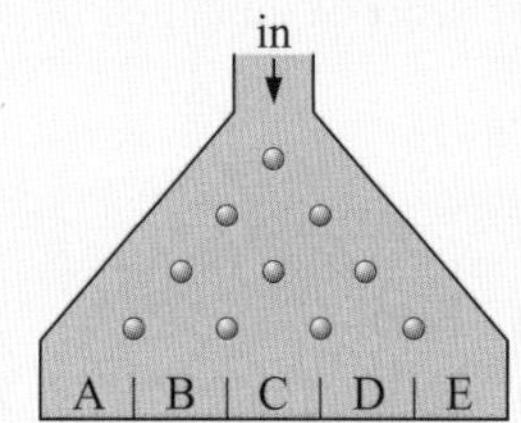

This sorter looks very much like a tree diagram rotated through $90°$.

Click on the icon to open the simulation. Notice that the sliding bar will alter the probabilities of balls going to the left or right at each rod.

What to do:

1 To simulate the results of tossing two coins, set the bar to 50% and the sorter as shown.
Run the simulation 200 times and repeat this four more times. Record each set of results.
What do you notice about the results?

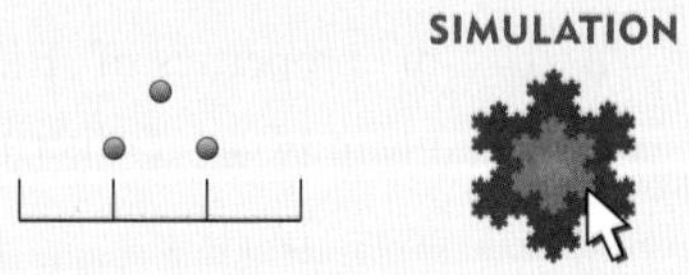

2 A bag contains 7 blue and 3 red marbles. Two marbles are randomly selected from the bag, the first being *replaced* before the second is drawn.
Since $\text{P(blue)} = \frac{7}{10} = 70\%$, set the bar to 70%.

a Run the simulation a large number of times. Hence estimate the probability of getting:
i two blues **ii** one blue **iii** no blues.

b The following tree diagram shows the theoretical probabilities for the different outcomes:

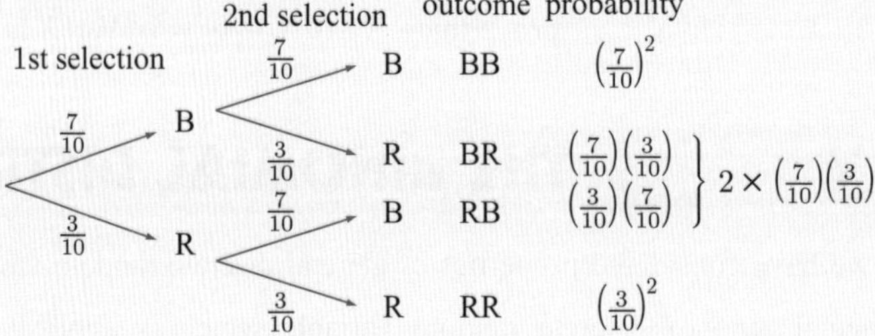

i Do the theoretical probabilities agree with the experimental results above?

ii Write down the algebraic expansion of $(a+b)^2$.

iii Substitute $a=\frac{7}{10}$ and $b=\frac{3}{10}$ in the $(a+b)^2$ expansion. What do you notice?

3 From the bag of 7 blue and 3 red marbles, *three* marbles are randomly selected *with replacement*. Set the sorter to 3 levels and the bar to 70%.

a Run the simulation many times to obtain experimental probabilities of getting:

i three blues **ii** two blues **iii** one blue **iv** no blues.

b Use a tree diagram showing 1st selection, 2nd selection, and 3rd selection to find theoretical probabilities for this experiment.

c Show that $(a+b)^3 = a^3 + 3a^2b + 3ab^2 + b^3$. Substitute $a = \frac{7}{10}$ and $b = \frac{3}{10}$ and compare your results with **a** and **b**.

THE BINOMIAL PROBABILITY DISTRIBUTION

Suppose a spinner has three blue edges and one white edge. Clearly, for each spin we will get either a blue or a white.

The chance of finishing on blue is $\frac{3}{4}$ and on white is $\frac{1}{4}$.

If we call a blue result a 'success' and a white result a 'failure', then we have a binomial experiment.

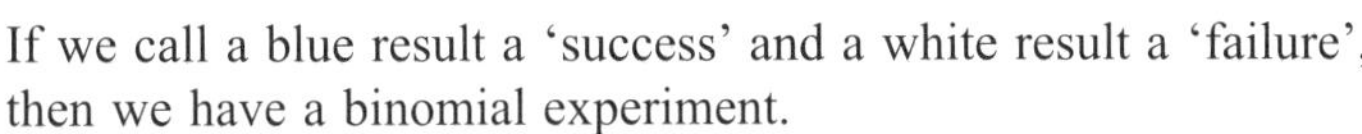

We let p be the probability of getting a blue, so $p = \frac{3}{4}$. The probability of getting a white is $1 - p = \frac{1}{4}$.

Consider twirling the spinner $n = 3$ times.

Let the random variable X be the number of 'successes' or blue results, so $X = 0$, 1, 2, or 3.

$$\begin{aligned} P(X=0) &= P(\text{none are blue}) \\ &= \tfrac{1}{4} \times \tfrac{1}{4} \times \tfrac{1}{4} \\ &= \left(\tfrac{1}{4}\right)^3 \end{aligned}$$

$$\begin{aligned} P(X=1) &= P(\text{1 blue and 2 white}) \\ &= P(\text{BWW or WBW or WWB}) \\ &= \left(\tfrac{3}{4}\right)\left(\tfrac{1}{4}\right)^2 \times 3 \quad \{\text{the 3 branches } \checkmark\} \end{aligned}$$

$$\begin{aligned} P(X=2) &= P(\text{2 blue and 1 white}) \\ &= P(\text{BBW or BWB or WBB}) \\ &= \left(\tfrac{3}{4}\right)^2\left(\tfrac{1}{4}\right) \times 3 \end{aligned}$$

$$\begin{aligned} P(X=3) &= P(\text{3 blues}) \\ &= \left(\tfrac{3}{4}\right)^3 \end{aligned}$$

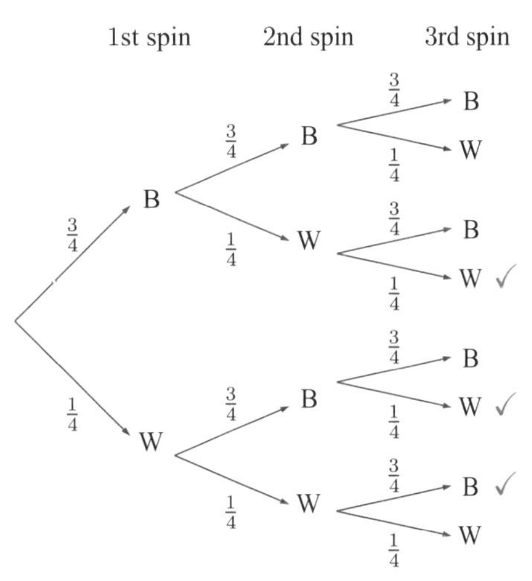

The coloured factor 3 is the number of ways of getting one success in three trials, which is $\binom{3}{1}$.

Notice that:

$$\begin{aligned} P(X=0) &= \left(\tfrac{1}{4}\right)^3 &&= \tbinom{3}{0}\left(\tfrac{3}{4}\right)^0\left(\tfrac{1}{4}\right)^3 \approx 0.0156 \\ P(X=1) &= 3\left(\tfrac{1}{4}\right)^2\left(\tfrac{3}{4}\right)^1 &&= \tbinom{3}{1}\left(\tfrac{3}{4}\right)^1\left(\tfrac{1}{4}\right)^2 \approx 0.1406 \\ P(X=2) &= 3\left(\tfrac{1}{4}\right)^1\left(\tfrac{3}{4}\right)^2 &&= \tbinom{3}{2}\left(\tfrac{3}{4}\right)^2\left(\tfrac{1}{4}\right)^1 \approx 0.4219 \\ P(X=3) &= \left(\tfrac{3}{4}\right)^3 &&= \tbinom{3}{3}\left(\tfrac{3}{4}\right)^3\left(\tfrac{1}{4}\right)^0 \approx 0.4219 \end{aligned}$$

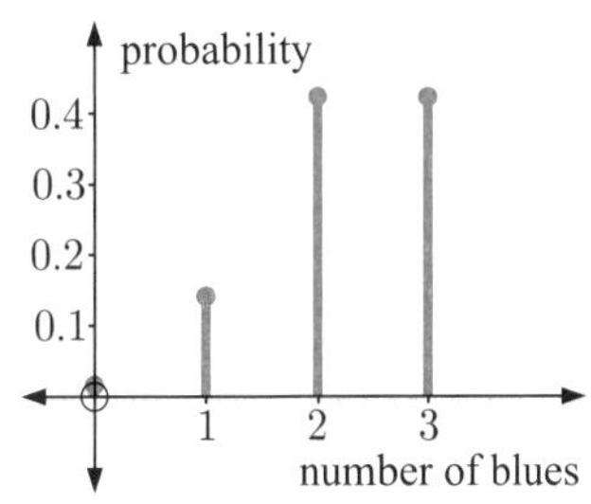

This suggests that $\mathrm{P}(X=x)=\binom{3}{x}\left(\frac{3}{4}\right)^x\left(\frac{1}{4}\right)^{3-x}$ where $x=0, 1, 2, 3$.

The sum of the probabilities $\mathrm{P}(X=0)+\mathrm{P}(X=1)+\mathrm{P}(X=2)+\mathrm{P}(X=3)$

$$=\left(\tfrac{1}{4}\right)^3+3\left(\tfrac{1}{4}\right)^2\left(\tfrac{3}{4}\right)+3\left(\tfrac{1}{4}\right)\left(\tfrac{3}{4}\right)^2+\left(\tfrac{3}{4}\right)^3$$

which is the binomial expansion for $\left(\frac{1}{4}+\frac{3}{4}\right)^3$, and $\left(\frac{1}{4}+\frac{3}{4}\right)^3=1^3=1$.

Example 11 **Self Tutor**

a Expand $\left(\frac{9}{10}+\frac{1}{10}\right)^5$.

b An archer has a 90% chance of hitting a target with each arrow. If 5 arrows are fired, determine the chance of hitting the target:

i twice only ii at most 3 times.

a $\left(\frac{9}{10}+\frac{1}{10}\right)^5$

$$=\left(\tfrac{9}{10}\right)^5+5\left(\tfrac{9}{10}\right)^4\left(\tfrac{1}{10}\right)+10\left(\tfrac{9}{10}\right)^3\left(\tfrac{1}{10}\right)^2+10\left(\tfrac{9}{10}\right)^2\left(\tfrac{1}{10}\right)^3+5\left(\tfrac{9}{10}\right)\left(\tfrac{1}{10}\right)^4+\left(\tfrac{1}{10}\right)^5$$

b The probability of success with each arrow is $p=\frac{9}{10}$.
Let X be the number of arrows that hit the target.
The expansion in **a** gives the probability distribution for X.

$\left(\frac{9}{10}\right)^5$	$+5\left(\frac{9}{10}\right)^4\left(\frac{1}{10}\right)$	$+10\left(\frac{9}{10}\right)^3\left(\frac{1}{10}\right)^2$	$+10\left(\frac{9}{10}\right)^2\left(\frac{1}{10}\right)^3$	$+5\left(\frac{9}{10}\right)\left(\frac{1}{10}\right)^4+$	$\left(\frac{1}{10}\right)^5$
$\mathrm{P}(X=5)$	$\mathrm{P}(X=4)$	$\mathrm{P}(X=3)$	$\mathrm{P}(X=2)$	$\mathrm{P}(X=1)$	$\mathrm{P}(X=0)$
5 hits	4 hits 1 miss	3 hits 2 misses	2 hits 3 misses	1 hit 4 misses	5 misses

i P(hits twice only)
$=\mathrm{P}(X=2)$
$=10\left(\frac{9}{10}\right)^2\left(\frac{1}{10}\right)^3$
$=0.0081$

ii P(hits at most 3 times)

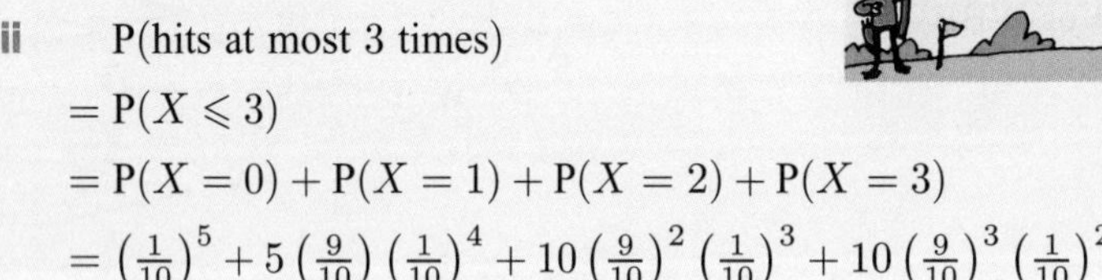

$=\mathrm{P}(X\leqslant 3)$
$=\mathrm{P}(X=0)+\mathrm{P}(X=1)+\mathrm{P}(X=2)+\mathrm{P}(X=3)$
$=\left(\frac{1}{10}\right)^5+5\left(\frac{9}{10}\right)\left(\frac{1}{10}\right)^4+10\left(\frac{9}{10}\right)^2\left(\frac{1}{10}\right)^3+10\left(\frac{9}{10}\right)^3\left(\frac{1}{10}\right)^2$
≈ 0.0815

EXERCISE 25F.1

1 a Expand $(p+q)^4$.

b If a coin is tossed *four* times, what is the probability of getting 3 heads?

2 a Expand $(p+q)^5$.

b If *five* coins are tossed simultaneously, what is the probability of getting:

i 4 heads and 1 tail in any order ii 2 heads and 3 tails in any order

iii 4 heads and 1 tail in that order?

3 **a** Expand $\left(\frac{2}{3}+\frac{1}{3}\right)^4$.

b Four chocolates are selected at random, with replacement, from a box which contains strawberry creams and almond centres in the ratio $2:1$. Find the probability of getting:

i all strawberry creams **ii** two of each type

iii at least 2 strawberry creams.

4 **a** Expand $\left(\frac{3}{4}+\frac{1}{4}\right)^5$.

b In New Zealand in 1946 there were two different coins of value two shillings. These were 'normal' kiwis and 'flat back' kiwis, in the ratio $3:1$. From a very large batch of 1946 two shilling coins, five were selected at random with replacement. Find the probability that:

i two were 'flat backs'

ii at least 3 were 'flat backs'

iii at most 3 were 'normal' kiwis.

5 When rifle shooter Huy fires a shot, he hits the target 80% of the time. If Huy fires 4 shots at the target, determine the probability that he has:

a 2 hits and 2 misses in any order **b** at least 2 hits.

THE BINOMIAL PROBABILITY DISTRIBUTION FUNCTION

Consider a binomial experiment for which p is the probability of a *success* and $1-p$ is the probability of a *failure*.

If there are n independent trials then the probability that there are r *successes* and $n-r$ *failures* is $\text{P}(X=r)=\binom{n}{r}p^r(1-p)^{n-r}$ where $r=0, 1, 2, 3, 4, ..., n$.

$\text{P}(X=r)$ is the **binomial probability distribution function**.

The **expected** or **mean** outcome of the experiment is $\mu=\text{E}(X)=np$.

If X is the random variable of a binomial experiment with parameters n and p, then we write $\boldsymbol{X\sim \text{B}(n, p)}$ where $\sim$ reads "*is distributed as*".

We can quickly calculate binomial probabilities using a graphics calculator.

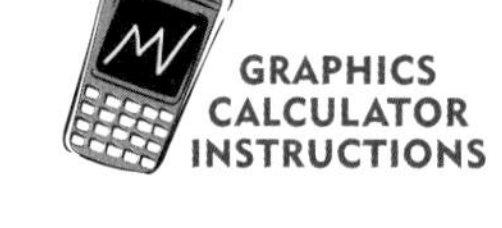

For example:

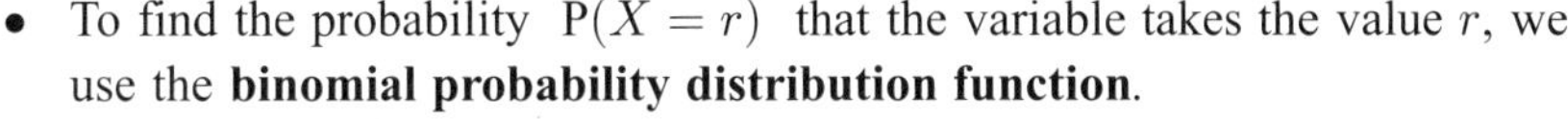

- To find the probability $\text{P}(X=r)$ that the variable takes the value r, we use the **binomial probability distribution function**.
- To find the probability that the variable takes a *range* of values, such as $\text{P}(X\leqslant r)$ or $\text{P}(X\geqslant r)$, we use the **binomial cumulative distribution function**.

Some calculator models, such as the **TI-84 Plus**, only allow you to calculate $\text{P}(X\leqslant r)$. To find the probability $\text{P}(X\geqslant r)$ for these models, it is often easiest to find the complement $\text{P}(X\leqslant r-1)$ and use $\text{P}(X\geqslant r)=1-\text{P}(X\leqslant r-1)$.

Example 12 **Self Tutor**

72% of union members are in favour of a certain change to their conditions of employment. A random sample of five members is taken. Find:

a the probability that three members are in favour of the change in conditions

b the probability that at least three members are in favour of the changed conditions

c the expected number of members in the sample that are in favour of the change.

Let X denote the number of members in the sample in favour of the changes.

$n = 5$, so $X = 0, 1, 2, 3, 4$, or 5, and $p = 72\% = 0.72$

$\therefore \quad X \sim \text{B}(5, 0.72)$.

a $\text{P}(X = 3) = \binom{5}{3}(0.72)^3(0.28)^2$

≈ 0.293

Casio fx-CG20

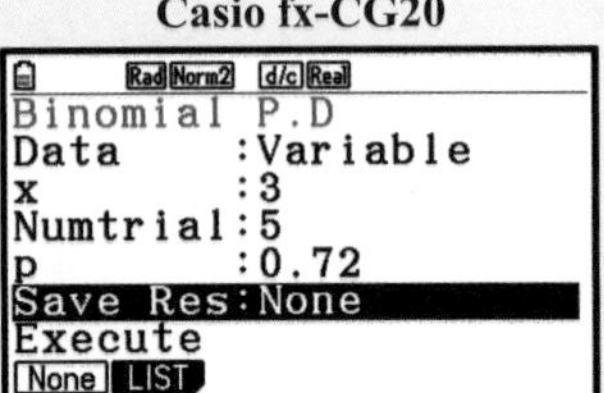

TI-84 Plus

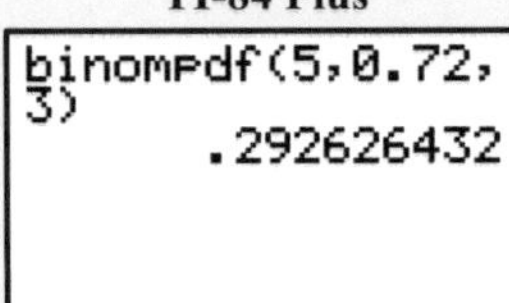

TI-*n*spire

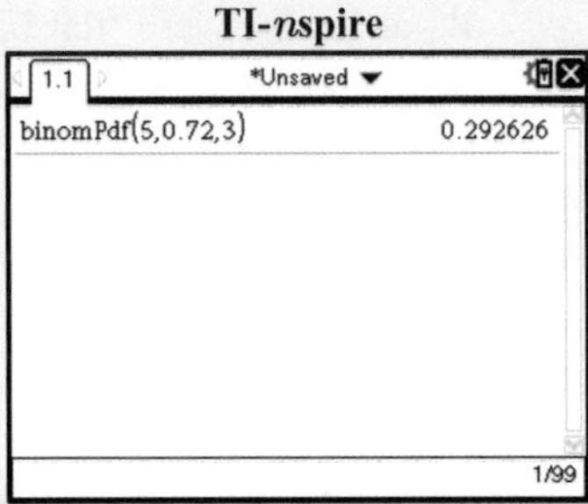

b $\text{P}(X \geqslant 3) \approx 0.862$

Casio fx-CG20

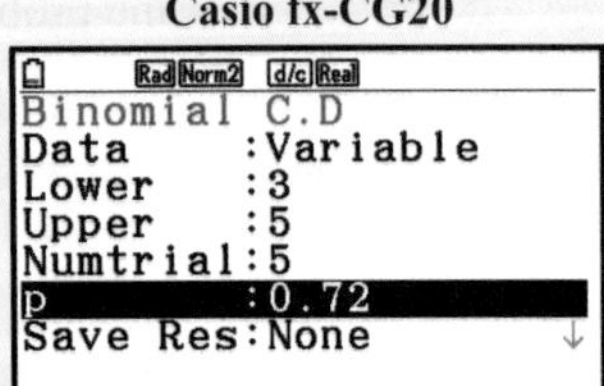

TI-84 Plus

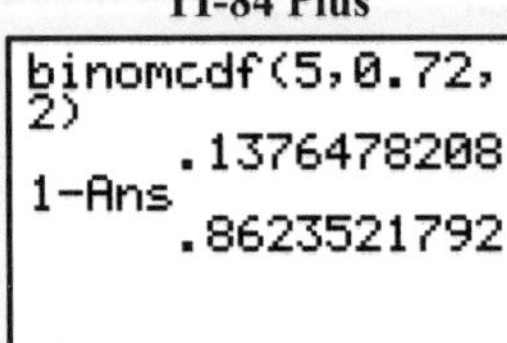

TI-*n*spire

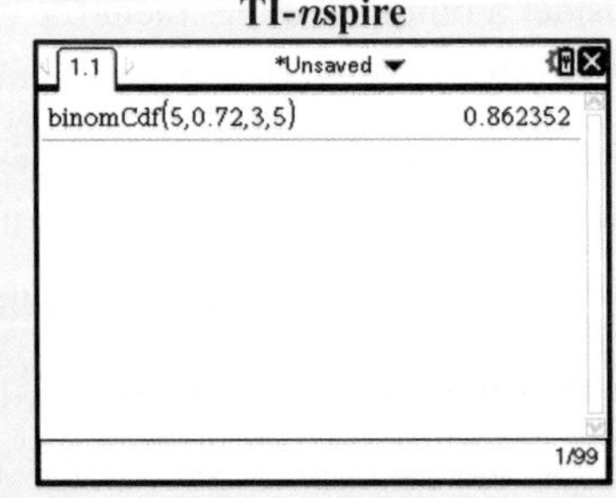

c $\text{E}(X) = np = 5 \times 0.72 = 3.6$ members

EXERCISE 25F.2

1 For which of these probability experiments does the binomial distribution apply? Justify your answers, using a full sentence.

a A coin is thrown 100 times. The variable is the number of heads.

b One hundred coins are each thrown once. The variable is the number of heads.

c A box contains 5 blue and 3 red marbles. I draw out 5 marbles, replacing the marble each time. The variable is the number of red marbles drawn.

d A box contains 5 blue and 3 red marbles. I draw out 5 marbles without replacement. The variable is the number of red marbles drawn.

e A large bin contains ten thousand bolts, 1% of which are faulty. I draw a sample of 10 bolts from the bin. The variable is the number of faulty bolts.

2 5% of electric light bulbs are defective at manufacture. If 6 bulbs are tested at random with each one being replaced before the next is chosen, determine the probability that:

a two are defective **b** at least one is defective.

3 In a multiple choice test there are 10 questions. Each question has 5 choices, one of which is correct. If 70% is the pass mark and Raj, who knows absolutely nothing about the subject, guesses each answer at random, determine the probability that he will pass.

4 At a manufacturing plant, 35% of the employees work night-shift. If 7 employees are each selected from the entire group at random, find the probability that:

a exactly 3 of them work night-shift **b** less than 4 of them work night-shift

c at least 4 of them work night-shift.

5 Records show that 6% of the items assembled on a production line are faulty.
A random sample of 12 items is selected with replacement. Find the probability that:

a none will be faulty **b** at most one will be faulty

c at least two will be faulty **d** less than four will be faulty.

6 There is a 5% chance that any apple in a crate will have a blemish. If a random sample of 25 apples is taken with replacement, find:

a the probability that exactly 2 of these have blemishes

b the probability that at least one has a blemish

c the expected number of apples that will have a blemish.

7 The local bus service does not have a good reputation. The 8 am bus will run late on average two days out of every five. For any week of the year taken at random, find the probability of the 8 am bus being on time:

a all 7 days **b** only on Monday **c** on any 6 days **d** on at least 4 days.

8 An infectious flu virus is spreading through a school. The probability of a randomly selected student having the flu next week is 0.3.

a Mr C has a class of 25 students.

  i Calculate the probability that 2 or more students will have the flu next week.

  ii If more than 20% of the students have the flu next week, a class test will have to be cancelled. What is the probability that the test will be cancelled?

b If the school has 350 students, find the expected number that will have the flu next week.

9 During a season, a basketball player has a 94% success rate in shooting from the free throw line. In one match the basketballer has 20 shots from the free throw line.

a Find the probability that the basketballer is successful on:

  i all 20 throws **ii** at least 18 throws.

b Find the expected number of successful throws for the match.

10 Martina beats Jelena in 2 games out of 3 at tennis. What is the probability that Jelena wins a set of tennis 6 games to 4?

Hint: What is the score after 9 games?

11 How many ordinary dice are needed for there to be a better than an even chance of at least one six when they are thrown together?

12 If a fair coin is tossed 200 times, find the probability of obtaining:

- **a** between 90 and 110 (inclusive) heads
- **b** more than 95 but less than 105 heads.

13 A new drug has 75% probability of curing a patient within one week. If 38 patients are treated using this drug, what is the probability that between 24 and 31 patients (inclusive) will be cured within a week?

14 **a** Suppose $P(x) = \binom{n}{x} p^{n-x}(1-p)^x$ where $x = 0, 1, 2, 3, ..., n$.

Prove that $P(x+1) = \left(\dfrac{n-x}{x+1}\right)\left(\dfrac{1-p}{p}\right)P(x)$ where $P(0) = p^n$.

b If $n = 5$ and $p = \frac{1}{2}$, find $P(0), P(1), P(2), ..., P(5)$.

THE MEAN AND STANDARD DEVIATION OF A BINOMIAL DISTRIBUTION

We have already seen that the expected outcome for a binomial experiment is the mean value $\mu = np$.

INVESTIGATION 3 THE MEAN AND STANDARD DEVIATION OF A BINOMIAL DISTRIBUTION

In this Investigation we use a calculator to calculate the mean and standard deviation of a number of binomial distributions. A spreadsheet can also be used to speed up the process and handle a larger number of examples.

We will first calculate the mean and standard deviation for the variable $X \sim B(30, 0.25)$.

What to do:

1 Enter the possible values for X from $x = 0$ to $x = 30$ into **List 1**, and their corresponding binomial probabilities $P(X = x)$ into **List 2**.

2 Draw the scatter diagram of **List 1** against **List 2**.

3 Calculate the descriptive statistics for the distribution.

```
1-Var Stats
x̄=7.5        ← μ
Σx=7.5
Σx²=61.875
Sx=
σx=2.371708245   ← σ
↓n=1
```

4 Copy and complete the following table by repeating steps **1** to **3** for the remaining values of n and p.

	$p = 0.1$	$p = 0.25$	$p = 0.5$	$p = 0.7$
$n = 10$				
$n = 30$		$\mu = 7.5$ $\sigma \approx 2.3717$		
$n = 50$				

5 Compare your values with the formulae $\mu = np$ and $\sigma = \sqrt{np(1-p)}$.

From this **Investigation** you should have observed the following results which are true in general:

Suppose X is a binomial random variable with parameters n and p, so $X \sim \text{B}(n, p)$.

- The **mean** of X is $\boldsymbol{\mu = np}$.
- The **standard deviation** of X is $\boldsymbol{\sigma = \sqrt{np(1-p)}}$.
- The **variance** of X is $\boldsymbol{\sigma^2 = np(1-p)}$.

For the case $n = 1$, the number of successes could be 0 or 1.

x_i	0	1
p_i	$1-p$	p

$$\begin{aligned}\text{Now } \mu &= \sum p_i x_i \\ &= (1-p)(0) + p(1) \\ &= p\end{aligned}$$

$$\begin{aligned}\text{and } \sigma^2 &= \sum x_i{}^2 p_i - \mu^2 \\ &= [(0)^2(1-p) + (1)^2 p] - p^2 \\ &= p - p^2 \\ &= p(1-p) \\ \therefore \sigma &= \sqrt{p(1-p)}\end{aligned}$$

For the case $n = 2$, the number of successes could be 0, 1, or 2.

$$\begin{aligned}P(0) &= \binom{2}{0} p^0 (1-p)^2 = (1-p)^2 \\ P(1) &= \binom{2}{1} p^1 (1-p)^1 = 2p(1-p) \\ P(2) &= \binom{2}{2} p^2 (1-p)^0 = p^2\end{aligned}$$

x_i	0	1	2
p_i	$(1-p)^2$	$2p(1-p)$	p^2

$$\begin{aligned}\text{Now } \mu &= \sum p_i x_i \\ &= (1-p)^2(0) + 2p(1-p)(1) + p^2(2) \\ &= 2p(1-p) + 2p^2 \\ &= 2p(1-p+p) \\ &= 2p\end{aligned}$$

$$\begin{aligned}\text{and } \sigma^2 &= \sum x_i{}^2 p_i - \mu^2 \\ &= [0^2 \times (1-p)^2 + 1^2 \times 2p(1-p) + 2^2 \times p^2] - (2p)^2 \\ &= 2p(1-p) + 4p^2 - 4p^2 \\ &= 2p(1-p) \\ \therefore \sigma &= \sqrt{2p(1-p)}\end{aligned}$$

Example 13 **Self Tutor**

A fair die is rolled twelve times and X is the number of sixes that could result. Find the mean and standard deviation of the X-distribution.

This is a binomial distribution with $n = 12$ and $p = \frac{1}{6}$, so $X \sim \text{B}(12, \frac{1}{6})$.

$$\begin{aligned}\text{So, } \mu &= np \\ &= 12 \times \tfrac{1}{6} \\ &= 2\end{aligned}$$

$$\begin{aligned}\text{and } \sigma &= \sqrt{np(1-p)} \\ &= \sqrt{12 \times \tfrac{1}{6} \times \tfrac{5}{6}} \\ &\approx 1.291\end{aligned}$$

We expect a six to be rolled 2 times, with standard deviation 1.291.

EXERCISE 25F.3

1 Suppose $X \sim B(6, p)$. For each of the following cases:

- **i** find the mean and standard deviation of the X-distribution
- **ii** graph the distribution using a column graph
- **iii** comment on the shape of the distribution.

a $p = 0.5$ **b** $p = 0.2$ **c** $p = 0.8$

2 A coin is tossed 10 times and X is the number of heads which occur. Find the mean and variance of the X-distribution.

3 Suppose $X \sim B(3, p)$, so $P(x) = \binom{3}{x} p^x (1-p)^{3-x}$.

a Find $P(0)$, $P(1)$, $P(2)$, and $P(3)$.
Display your results in a table like the one opposite.

x_i	0	1	2	3
p_i				

b Use $\mu = \sum p_i x_i$ to show that $\mu = 3p$.

c Use $\sigma^2 = \sum {x_i}^2 p_i - \mu^2$ to show that $\sigma = \sqrt{3p(1-p)}$.

Example 14 Self Tutor

5% of a batch of batteries are defective. A random sample of 80 batteries is taken with replacement. Find the mean and standard deviation of the number of defective batteries in the sample.

This is a binomial sampling situation with $n = 80$, $p = 5\% = \frac{1}{20}$.

If X is the random variable for the number of defectives, then $X \sim B(80, \frac{1}{20})$.

So, $\mu = np = 80 \times \frac{1}{20} = 4$ and $\sigma = \sqrt{np(1-p)} = \sqrt{80 \times \frac{1}{20} \times \frac{19}{20}} \approx 1.949$

We expect a defective battery 4 times, with standard deviation 1.949.

4 Bolts produced by a machine vary in quality. The probability that a given bolt is defective is 0.04. Random samples of 30 bolts are taken from the week's production.

a If X is the number of defective bolts in a sample, find the mean and standard deviation of the X-distribution.

b If Y is the number of non-defective bolts in a sample, find the mean and standard deviation of the Y-distribution.

5 A city restaurant knows that 13% of reservations are not honoured, which means the group does not arrive. Suppose the restaurant receives 30 reservations. Let X be the random variable of the number of groups that do not arrive. Find the mean and standard deviation of the X-distribution.

G THE POISSON DISTRIBUTION

The **Poisson random variable** was first introduced by the French mathematician **Siméon-Denis Poisson** (1781 - 1840). He discovered it as a limit of the binomial distribution as the number of trials $n \to \infty$.

Whereas the binomial distribution $B(n, p)$ is used to determine the probability of obtaining a certain number of successes in a given number of independent trials, the **Poisson distribution** is used to determine the probability of obtaining a certain number of successes within a certain interval (of time or space).

Examples are:

- the number of incoming telephone calls per hour
- the number of misprints on a typical page of a book
- the number of fish caught in a large lake per day
- the number of car accidents on a given road per month.

The probability distribution function for the discrete Poisson random variable is

$$P(x) = P(X = x) = \frac{m^x e^{-m}}{x!} \quad \text{for} \quad x = 0, 1, 2, 3, 4, 5, ...$$

where m is the **parameter** of the distribution.

PROOF

INVESTIGATION 4 POISSON MEAN AND VARIANCE

In this Investigation you should discover the mean and variance of the Poisson distribution.

What to do:

1 Show that $f(x) = e^x$ satisfies the differential equation $f'(x) = f(x)$ where $f(0) = 1$.

2 Consider $f(x) = 1 + \frac{x}{1!} + \frac{x^2}{2!} + \frac{x^3}{3!} + ... + \frac{x^n}{n!} + ...$ which is an infinite power series.

a Check that $f(0) = 1$ and find $f'(x)$. What do you notice?

b From the result of question **1**, what can be deduced about $f(x)$?

c Check your result in **b** by calculating the sum of the first 12 terms of the power series for $x = 1$.

3 By observing that $\frac{2x}{1!} = \frac{x}{1!} + \frac{x}{1!}$, $\frac{3x^2}{2!} = \frac{x^2}{2!} + \frac{x^2}{1!}$, ...,

show that $1 + \frac{2x}{1!} + \frac{3x^2}{2!} + \frac{4x^3}{3!} + ... = e^x(1 + x)$.

4 If $p_x = P(X = x) = \frac{m^x e^{-m}}{x!}$ for $x = 0, 1, 2, 3, ...,$ show that:

a $\sum_{x=0}^{\infty} p_x = 1$ **b** $E(X) = m$ **c** $Var(X) = m$

If X is a Poisson discrete random variable with parameter μ then $\mu = m$ and $\sigma^2 = m$.

Since $\mu = \sigma^2 = m$, we can describe a Poisson distribution using the single parameter m.

We can therefore denote the distribution simply by $Po(m)$.

If X is a Poisson random variable then we write $X \sim Po(m)$ where $m = \mu = \sigma^2$.

Conditions for a distribution to be Poisson:

1. The average number of occurrences μ is constant for each interval. It should be equally likely that the event occurs in one specific interval as in any other.
2. The probability of more than one occurrence in a given interval is very small. The typical number of occurrences in a given interval should be much less than is theoretically possible, say about 10% or less.
3. The number of occurrences in disjoint intervals are independent of each other.

Consider the following example:

When Sandra proof read 80 pages of a text book she observed the following distribution for X, the number of errors per page:

X	0	1	2	3	4	5	6	7	8	9	10
Frequency	3	11	16	18	15	9	5	1	1	0	1

The mean for this data is $\mu = \dfrac{\sum f_i x_i}{n} = \dfrac{257}{80} = 3.2125$.

Using the Poisson model with $m = 3.2125$, the expected frequencies are:

x_i	0	1	2	3	4	5	6	7	8	9	10
f_i	3.22	10.3	16.6	17.8	14.3	9.18	4.92	2.26	0.906	0.323	0.104

The expected frequencies are close to the observed frequencies, which suggests the Poisson model is a good model for representing this distribution.

Further evidence is that

$$\begin{aligned}\sigma^2 &= \sum x_i p_i^2 - \mu^2 \\ &= 0^2(\tfrac{3}{80}) + 1^2(\tfrac{11}{80}) + 2^2(\tfrac{16}{80}) + ... + 10^2(\tfrac{1}{80}) - 3.2125^2 \\ &\approx 3.367\end{aligned}$$

which is fairly close to $m = 3.2125$.

There is a formal statistic test for establishing whether a Poisson distribution is appropriate. This test is covered in the **Statistics Option Topic**.

EXERCISE 25G

1. Between 9:00 am and 9:15 am on Fridays, Sven's Florist Shop receives the X-distribution of phone calls shown.

X	0	1	2	3	4	5	6
Frequency	12	18	12	6	3	0	1

   a. Find the mean of the X-distribution.
   b. Compare the actual data with that generated by a Poisson model.

2. The Poisson random variable X has standard deviation 2.67.
   a. What is its mean?
   b. What is its probability generating function?
   c. Find:
      i. $P(X = 2)$ ii. $P(X \leqslant 3)$ iii. $P(X \geqslant 5)$ iv. $P(X \geqslant 3 \mid X \geqslant 1)$

3 One gram of radioactive substance is positioned so that each emission of an alpha-particle will flash on a screen. The emissions over 500 periods of 10 seconds duration are summarised in the following table.

Number per period	0	1	2	3	4	5	6	7
Frequency	91	156	132	75	33	9	3	1

- **a** Find the mean of the distribution.
- **b** Fit a Poisson model to the data and compare the actual data to that from the model.
- **c** Find the standard deviation of the distribution. How close is it to the square root of the mean found in **a**?

4 Top Cars rents cars to tourists. They have four cars which are hired out on a daily basis. The number of requests each day has a Poisson distribution with mean 3. Determine the probability that:

- **a** none of the cars is rented
- **b** at least 3 of the cars are rented
- **c** some requests will have to be refused
- **d** all of the cars are hired out given that at least two are.

5 Consider a random variable $X \sim \text{Po}(m)$.

- **a** Find m given that $\text{P}(X = 1) + \text{P}(X = 2) = \text{P}(X = 3)$.
- **b** If $m = 2.7$, find: **i** $\text{P}(X \geqslant 3)$ **ii** $\text{P}(X \leqslant 4 \mid X \geqslant 2)$.

6 Wind tunnel experiments on a new aerofoil indicated there was a 98% chance of the aerofoil not disintegrating at maximum airspeed. In a sample of 100 aerofoils, use the Poisson distribution to determine the probability that:

- **a** only one
- **b** only two
- **c** at most 2 aerofoils will disintegrate.

7 Road safety figures for a large city show that any driver has a 0.02% chance of being killed each time he or she drives a car.

- **a** Use the Poisson approximation to the binomial distribution to find the probability that a driver can use a car 10 times a week for a year and survive.
- **b** If this data does not change, for how many years can you drive in this city and still have a better than even chance of surviving?

8 A supplier of clothing materials looks for flaws before selling the materials to customers. The number of flaws follows a Poisson distribution with a mean of 1.7 flaws per metre.

- **a** Find the probability that there are exactly 3 flaws in 1 metre of material.
- **b** Determine the probability that there is at least one flaw in 2 metres of material.
- **c** Find the modal value of this Poisson distribution.

9 The random variable Y has a Poisson distribution with mean m, and satisfies

$$\text{P}(Y = 3) = \text{P}(Y = 1) + 2\text{P}(Y = 2).$$

- **a** Find the value of m correct to 4 decimal places.
- **b** Hence find: **i** $\text{P}(1 < Y < 5)$ **ii** $\text{P}(2 \leqslant Y \leqslant 6 \mid Y \geqslant 4)$

10 The random variable U has a Poisson distribution with mean x. Let y be the probability that U takes one of the values 0, 1, or 2.

- **a** Write down an expression for y as a function of x.
- **b** Sketch the graph of y for $0 \leqslant x \leqslant 3$.
- **c** Use calculus to show that as the mean increases, $\text{P}(U \leqslant 2)$ decreases.

REVIEW SET 25A — NON-CALCULATOR

1 $P(X = x) = \frac{a}{x^2+1}$, $x = 0, 1, 2, 3$ is a probability distribution function.

a Find a. **b** Find $P(X \geqslant 1)$.

2 A random sample of 120 toothbrushes is taken with replacement from a very large batch where 4% are known to be defective. Find the mean number and standard deviation of defectives in the sample.

3 A random variable X has the probability distribution function $P(x)$ described in the table.

x	0	1	2	3	4
$P(x)$	0.10	0.30	0.45	0.10	k

a Find k. **b** Find $P(X \geqslant 3)$.

c Find the expectation $E(X)$ for the distribution.

d Find the standard deviation σ for the distribution.

4 **a** Expand $\left(\frac{3}{5}+\frac{2}{5}\right)^4$.

b A tin contains 20 pens of which 12 have blue ink. Four pens are randomly selected, with replacement, from the tin. Find the probability that:

i two of them have blue ink **ii** at most two have blue ink.

5 Three green balls and two yellow balls are placed in a hat. When two balls are randomly drawn without replacement, X is the number of green balls drawn. Find:

a $P(X = 0)$ **b** $P(X = 1)$ **c** $P(X = 2)$ **d** $E(X)$

6 Lakshmi rolls a normal six-sided die. She wins twice the number of pounds as the number shown on the face.

a How much does Lakshmi expect to win from one roll of the die?

b If it costs £8 to play the game, would you advise Lakshmi to play several games? Explain your answer.

7 A binomial distribution has probability distribution function $P(X = x) = k\left(\frac{1}{3}\right)^x\left(\frac{2}{3}\right)^{7-x}$ where $x = 0, 1, 2, 3, ..., 7$.

a Write k in the form $\binom{n}{r}$. **b** Find the mean and variance of the distribution.

8 **a** Expand $\left(\frac{4}{5}+\frac{1}{5}\right)^5$.

b With every attempt, Jack has an 80% chance of kicking a goal. In one quarter of a match he has 5 kicks for goal.

Determine the probability that he scores:

i 3 goals then misses twice

ii 3 goals and misses twice.

9 At a social club fundraiser there is a dice game where, on a single roll of a six-sided die, the following payouts are made:

\$2 for an odd number, \$3 for a 2, \$6 for a 4, and \$9 for a 6.

a Find the expected return for a single roll of the die.

b If the club charges \$5 for each roll, and 75 people play the game once each, how much money will the club expect to make?

10 Consider the two spinners illustrated:

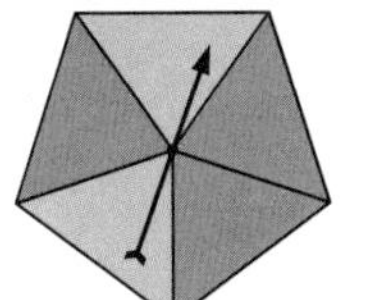

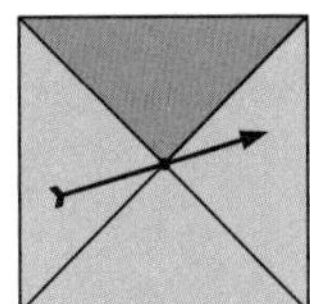

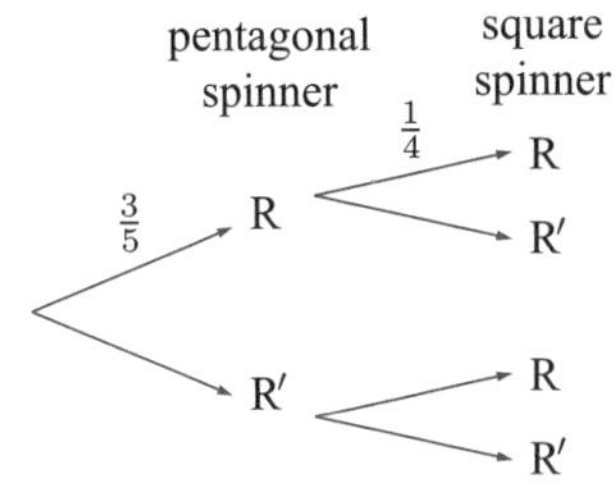

a Copy and complete the tree diagram which shows all possible results when the two are spun together.

b Calculate the probability that exactly one red will occur.

c The pair of spinners is now spun 10 times and X is the number of times that exactly one red occurs.

i Write down expressions for $P(X = 1)$ and $P(X = 9)$.

ii Hence determine which of these outcomes is more likely.

11 A biased tetrahedral die has the numbers 6, 12, and 24 marked on three of its faces. The fourth number is x. The table shows the probability of each of the numbers occurring if the die is rolled once.

Number	6	12	x	24
Probability	$\frac{1}{3}$	$\frac{1}{6}$	$\frac{1}{4}$	y

a Find y, the probability of obtaining the number 24.

b Find the fourth number if the average result when rolling the die once is 14.

c Find the median and modal score for one roll of this die.

12 The random variable X has mean μ and standard deviation σ.
Prove that the random variable $Y = aX + b$ has mean $a\mu + b$ and standard deviation $|a|\sigma$.

REVIEW SET 25B — CALCULATOR

1 A discrete random variable X has probability distribution function $P(x) = k\left(\frac{3}{4}\right)^x\left(\frac{1}{4}\right)^{3-x}$ where $x = 0, 1, 2, 3$ and k is a constant. Find:

a k **b** $P(X \geqslant 1)$ **c** $E(X)$

d the standard deviation of the distribution.

2 A manufacturer finds that 18% of the items produced from its assembly lines are defective. During a floor inspection, the manufacturer randomly selects ten items with replacement. Find the probability that the manufacturer finds:

a one defective **b** two defective **c** at least two defective items.

3 From data over the last fifteen years it is known that the chance of a netballer with a knee injury needing major knee surgery in any one season is 0.0132. In 2007 there were 487 knee injuries in netball games throughout the country. Find the expected number of major knee surgeries required.

4 24% of visitors to a museum make voluntary donations. On a certain day the museum has 175 visitors. Find:

a the expected number of donations

b the probability that less than 40 visitors make a donation.

5 An X-ray has probability of 0.96 of showing a fracture in the arm. If four different X-rays are taken of a particular fracture, find the probability that:

- **a** all four show the fracture
- **b** the fracture does not show up
- **c** at least three X-rays show the fracture
- **d** only one X-ray shows the fracture.

6 A school basketball team has 8 players, each of whom has a 75% chance of turning up for any given game. The team needs at least 5 players to avoid forfeiting the game.

- **a** Find the probability that for a randomly chosen game, the team will:
  - **i** have all of its players
  - **ii** have to forfeit the game.
- **b** The team plays 30 games for the season. How many games would you expect the team to forfeit?

7 The binomial distribution $X \sim \text{B}(n, p)$ has mean 30 and variance 22.5.

- **a** Find n and p.
- **b** Hence find: **i** $\text{P}(X = 25)$ **ii** $\text{P}(X \geqslant 25)$ **iii** $\text{P}(15 \leqslant X \leqslant 25)$

8 The discrete random variable X has the probability distribution function

$$\text{P}(X = x) = a\left(\frac{5}{6}\right)^x \quad \text{for } x = 0, 1, 2, 3, ...$$

Find the value of a.

9 When X plays Y at table tennis, we know from past experience that X wins 3 sets in every 5 played.

- **a** If they play 6 sets, write down the probability generator.
- **b** Hence determine the probability that:
  - **i** Y wins 3 of them
  - **ii** Y wins at least 5 of them.

10 One glass blower was known to break one out of every 200 objects he attempts. A second glass blower was known to break three out of every 200 objects she attempts. During one period, the first glass blower produced 20 objects and the second glass blower produced 40 objects. Use the Poisson distribution to find the probability that the glass blowers broke 2 or more objects between them.

11 For a given binomial random variable X with 7 independent trials, we know that

$$\text{P}(X = 3) = 0.226\,89.$$

- **a** Find the smallest possible value of p, the probability of obtaining a success in one trial.
- **b** Hence calculate the probability of getting at most 4 successes in 10 trials.

12 A discrete random variable has the probability distribution function

$$\text{P}(X = x) = k(x + x^{-1}) \quad \text{where} \quad x = 1, 2, 3, 4. \quad \text{Find:}$$

- **a** the exact value of k
- **b** $\text{E}(X)$ and $\text{Var}(X)$
- **c** the median and mode of X.

13 The random variable Y has a Poisson distribution with $\text{P}(Y > 3) \approx 0.033\,768\,97$. Find $\text{P}(Y < 3)$.

REVIEW SET 25C

1 Find k for the following probability distribution functions:

a $P(X = x) = \frac{k}{2x}$, $x = 1, 2, 3$

b

x	0	1	2	3
$P(x)$	$\frac{k}{2}$	0.2	k^2	0.3

2 A random variable X has probability distribution function $P(X = x) = \binom{4}{x}\left(\frac{1}{2}\right)^x\left(\frac{1}{2}\right)^{4-x}$ for $x = 0, 1, 2, 3, 4$.

a Find $P(X = x)$ for $x = 0, 1, 2, 3, 4$.
b Find the mean μ for the distribution.
c Find the standard deviation σ for the distribution.

3 A die is biased such that the probability of obtaining a 6 is $\frac{2}{5}$. The die is rolled 1200 times. Let X be the number of sixes obtained. Find the mean and standard deviation of X.

4 Only 40% of young trees that are planted will survive the first year. The Botanical Gardens buys five young trees. Assuming independence, find the probability that during the first year:

a exactly one tree will survive
b at most one tree will survive
c at least one tree will survive.

5 In a game, the numbers from 1 to 20 are written on tickets and placed in a bag. A player draws out a number at random. He or she wins \$3 if the number is even, \$6 if the number is a square number, and \$9 if the number is both even and square.

a Calculate the probability that the player wins: **i** \$3 **ii** \$6 **iii** \$9
b How much should be charged to play the game so that it is a fair game?

6 A fair die is rolled 360 times. Find the probability that:

a less than 50 results are a 6
b between 55 and 65 results (inclusive) are a 6.

7 An unbiased coin is tossed n times. Find the smallest value of n for which the probability of getting at least two heads is greater than 99%.

8 A hot water unit relies on 20 solar components for its power and will operate provided at least one of its 20 components is working. The probability that an individual solar component will fail in a year is 0.85, and the failure of each individual component is independent of the others.

a Find the probability that the hot water unit will fail within one year.
b Find the smallest number of solar components required to ensure that a hot water service like this one is operating at the end of one year with a probability of at least 0.98.

9 A dart thrower has a one in three chance of hitting the correct number with any throw. He throws 5 darts at the board and X is the number of successful hits.

a Find the probability generator for X.
b Calculate the probability of the thrower scoring an odd number of successful hits given that he has at least two successful hits.

10 A Poisson random variable X is such that $P(X = 1) = P(2 \leqslant X \leqslant 4)$.

a Find the mean and standard deviation of: **i** X **ii** $Y = \frac{X+1}{2}$.
b Find $P(X \geqslant 2)$.

11 A Poisson random variable X satisfies the rule $5\text{Var}(X) = 2[\text{E}(X)]^2 - 12$.

a Find the mean of X.

b Find $\text{P}(X < 3)$.

12 The random variable X has a binomial distribution for which $\text{P}(X > 2) \approx 0.070\,198$ for 10 independent trials. Find $\text{P}(X < 2)$.

13 During peak period, customers arrive at random at a fish and chip shop at the rate of 20 customers every 15 minutes.

a Find the probability that during peak period, 15 customers will arrive in the next quarter of an hour.

b If the probability that more than 10 customers will arrive at the fish and chip shop during 10 minutes of peak period is greater than 80%, the manager will employ an extra shop assistant. Will the manager hire an extra shop assistant?

Chapter 26

Continuous random variables

Syllabus reference: 5.5, 5.7

Contents:

OPENING PROBLEM

A salmon breeder catches hundreds of adult fish. He records their weights in a frequency table with class intervals $3 \leqslant w < 3.1$ kg, $3.1 \leqslant w < 3.2$ kg, $3.2 \leqslant w < 3.3$ kg, and so on.

The mean weight is 4.73 kg, and the standard deviation is 0.53 kg.

A frequency histogram of the data is bell-shaped and symmetric about the mean.

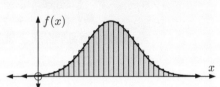

Things to think about:

a Can we use the mean and standard deviation only to find the proportion of salmon whose weight is:

  i greater than 6 kg **ii** between 4 kg and 6 kg?

b How can we find the weight:

  i which 90% of salmon weigh less than **ii** which 25% of salmon weigh more than?

A CONTINUOUS RANDOM VARIABLES

In the previous chapter we looked at discrete random variables and examined binomial probability distributions where the random variable X could take the non-negative integer values $x = 0, 1, 2, 3, 4, ..., n$ for some finite $n \in \mathbb{N}$.

For a **continuous random variable** X, x can take any real value.

We use a function called a **probability density function** to define the probability distribution.

INVESTIGATION 1 PROBABILITY DENSITY FUNCTIONS

In statistics we can use a **cumulative frequency graph** to help find the percentage of data values above or below a certain value.

For example, the graph alongside shows the cumulative frequency, expressed as a percentile of all the data, for different values of the variable X.

We see that $0.7 = 70\%$ of the data is less than 5.

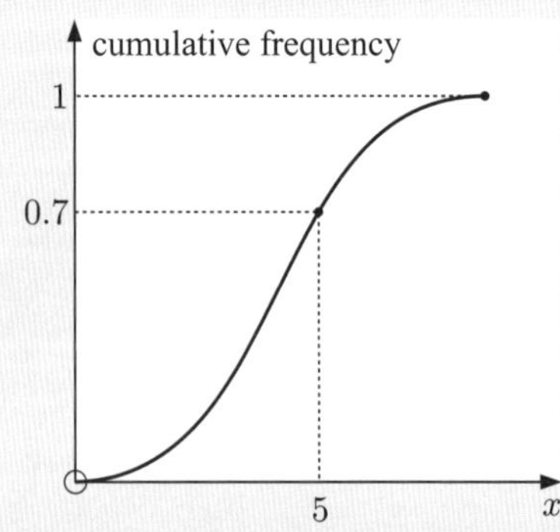

What to do:

1 **a** Explain why the graph of a cumulative frequency function is always increasing.

b Find the range of a cumulative frequency function if the cumulative frequency is expressed as a percentile of all the data.

c If the cumulative frequency graph becomes very steep over a particular interval, what does this indicate about the number of data values in that interval?

2 Now consider the **gradient function** of the cumulative frequency function. Explain why:

- **a** the gradient function is always positive
- **b** a high value for the gradient function at a given value of x indicates a high frequency of data values around that value of x
- **c** the gradient function gives the **probability density function** of the data.

So, if we *differentiate* the cumulative frequency function, we obtain the probability density function.

Reversing this process, if we know the probability density function of a data set, then we can *integrate* it to obtain the cumulative frequency function.

3 The students at a school were asked how far they live from their local library. The probability density function for the distances is $f(x) = \frac{1}{2}x^2e^{-x}$, $0 < x < 12$.

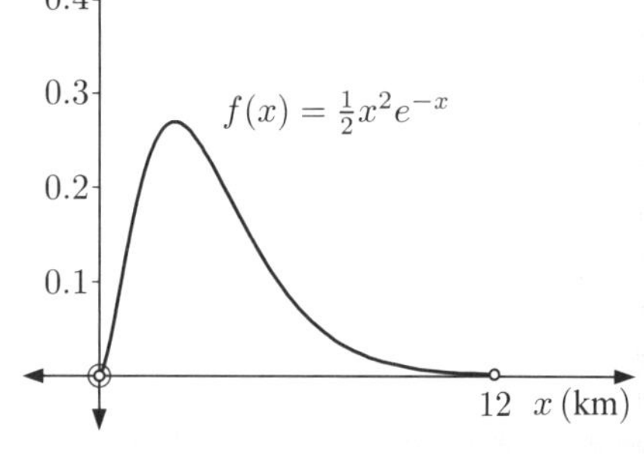

- **a** Explain the significance of the domain $0 < x < 12$.
- **b** Use integration by parts to show that the cumulative frequency function is $F(x) = -\frac{1}{2}x^2e^{-x} - xe^{-x} - e^{-x} + c$, where c is a constant.
- **c** **i** Explain why $F(0) = 0$. **ii** Hence, find c.
- **d** State the domain of $F(x)$.
- **e** **i** Find $F(3)$, and interpret your result.
  ii Find $\int_0^3 f(x)\,dx$. Compare your answer to **e i**.
- **f** Use technology to sketch the graph of $F(x)$.
- **g** Predict the value of $\int_0^{12} f(x)\,dx$. Use technology to check your answer.

For a continuous random variable X, the **probability density function** is a function $f(x)$ such that $f(x) \geqslant 0$ on its entire domain.

If the domain of the function is $a \leqslant x \leqslant b$, then $\int_a^b f(x)\,dx = 1$.

The probability that X lies in the interval $c \leqslant X \leqslant d$ is

$$\mathbf{P}(c \leqslant X \leqslant d) = \int_c^d f(x)\,dx.$$

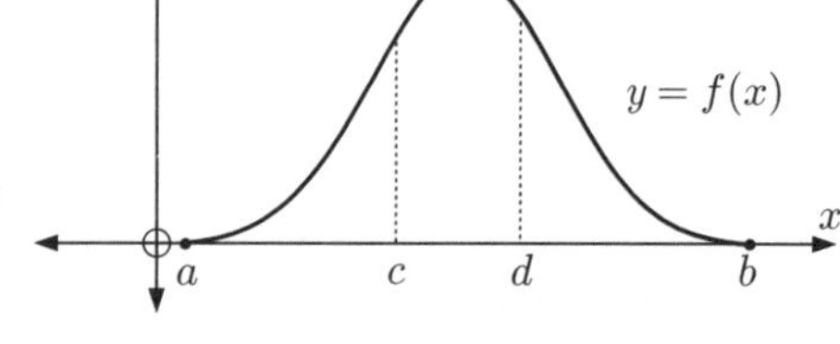

For a continuous variable X, the probability that X is *exactly* equal to a particular value is zero. So, $\text{P}(X = x) = 0$ for all x.

For example, the probability that an egg will weigh *exactly* 72.9 g is zero.

If you were to weigh an egg on scales that weigh to the nearest 0.1 g, a reading of 72.9 g means the weight lies somewhere between 72.85 g and 72.95 g. No matter how accurate your scales are, you can only ever know the weight of an egg within a range.

So, for a continuous variable we can only talk about the probability that an event lies in an **interval**.

A consequence of this is that $P(c \leqslant X \leqslant d) = P(c < X \leqslant d) = P(c \leqslant X < d) = P(c < X < d)$.

This would not be true if X was discrete.

PROPERTIES OF PROBABILITY DENSITY FUNCTIONS

For a continuous probability density function:

- The **mode** is the value of x at the maximum value of $f(x)$ on $[a, b]$.
- The **median** m is the solution for m of the equation $\int_a^m f(x)\,dx = \frac{1}{2}$.
- The **mean** μ or $E(X)$ is defined as $\mu = \int_a^b x\,f(x)\,dx$.
- The **variance** $\text{Var}(X) = E(X^2) - \{E(X)\}^2 = \int_a^b x^2\,f(x)\,dx - \mu^2$.

We can compare these with the properties of probability distribution functions for discrete variables as follows:

Discrete random variable

- $\mu = E(X) = \sum x p_x$
- $\sigma^2 = \text{Var}(X) = E(X-\mu)^2$
 $= \sum (x-\mu)^2 p_x$
 $= \sum x^2 p_x - \mu^2$
 $= E(X^2) - \{E(X)\}^2$

Continuous random variable

- $\mu = E(X) = \int x\,f(x)\,dx$
- $\sigma^2 = \text{Var}(X) = E(X-\mu)^2$
 $= \int (x-\mu)^2\,f(x)\,dx$
 $= \int x^2\,f(x)\,dx - \mu^2$
 $= E(X^2) - \{E(X)\}^2$

Example 1 — Self Tutor

Consider the probability density function $f(x) = \frac{1}{2}x, \quad 0 \leqslant x \leqslant 2$.

a Check that $f(x)$ is able to be a probability density function.

b For this distribution find the:

 i mode **ii** median **iii** mean.

c Find $\text{Var}(X)$ and σ.

a

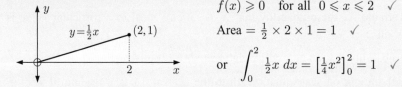

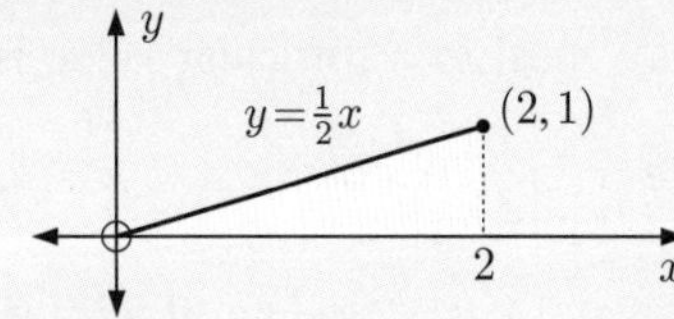

$f(x) \geqslant 0$ for all $0 \leqslant x \leqslant 2$ ✓

Area $= \frac{1}{2} \times 2 \times 1 = 1$ ✓

or $\int_0^2 \frac{1}{2}x\,dx = \left[\frac{1}{4}x^2\right]_0^2 = 1$ ✓

b **i** The maximum value of $f(x)$ is when $x = 2$
$\therefore$ the mode $= 2$

ii The median is the solution of

$$\int_0^m \tfrac{1}{2}x\,dx = \tfrac{1}{2}$$

$$\therefore \left[\tfrac{1}{4}x^2\right]_0^m = \tfrac{1}{2}$$

$$\therefore \frac{m^2}{4} = \tfrac{1}{2}$$

$$\therefore m^2 = 2$$

$$\therefore m = \sqrt{2} \quad \{\text{as } m \in [0, 2]\}$$

iii $$\mu = \int_0^2 x\,f(x)\,dx = \int_0^2 \tfrac{1}{2}x^2\,dx = \left[\tfrac{1}{6}x^3\right]_0^2 = 1\tfrac{1}{3}$$

c $$\mathrm{E}(X^2) = \int_0^2 x^2\,f(x)\,dx = \int_0^2 \tfrac{1}{2}x^3\,dx = \left[\tfrac{1}{8}x^4\right]_0^2 = 2$$

$$\therefore \mathrm{Var}(X) = 2 - (1\tfrac{1}{3})^2 = \tfrac{2}{9}$$

and $$\sigma = \sqrt{\mathrm{Var}(X)} = \tfrac{\sqrt{2}}{3}$$

EXERCISE 26A

1 $f(x) = ax(x-4)$, $0 \leqslant x \leqslant 4$ is a continuous probability density function.

a Find a. **b** Sketch the graph of $y = f(x)$.

c For this distribution find the:

i mean **ii** mode **iii** median **iv** variance.

2 $f(x) = -0.2x(x-b)$, $0 \leqslant x \leqslant b$ is a probability density function. Find:

a b **b** the mean **c** the variance.

3 $f(x) = ke^{-x}$, $0 \leqslant x \leqslant 3$ is a probability density function.

a Find k correct to 4 decimal places. **b** Find the median of the distribution.

4 $f(x) = kx^2(x-6)$, $0 \leqslant x \leqslant 5$ is a probability density function. Find:

a k **b** the mode **c** the median **d** the mean **e** the variance.

5 The probability density function for the random variable Y is $f(y) = 5 - 12y$, $0 \leqslant y \leqslant k$.

a Under what conditions can Y be a continuous random variable?

b Without using calculus, find k.

c Notice that $\displaystyle\int_0^{\frac{1}{2}} (5-12y)\,dy = 1$. Explain why $k \neq \frac{1}{2}$ despite this result.

d Find the mean and median value of Y.

6 The probability density function for the random variable X is $f(x) = k$, $a \leqslant x \leqslant b$.

a Find k in terms of a and b.

b Calculate the mean, median, and mode of X.

c Calculate $\mathrm{Var}(X)$ and the standard deviation of X.

7 The continuous random variable X has the probability density function $f(x) = 2e^{-2x}$, $x \geqslant 0$. Calculate:

a the median of X **b** the mode of X.

8 A continuous random variable X has probability density function $f(x) = 6\cos 3x$, $0 \leqslant x \leqslant a$. Find:

a a **b** the mean **c** the 20th percentile

d the standard deviation of X.

9 The continuous random variable X has the probability density function $f(x) = ax^4$, $0 \leqslant x \leqslant k$. Given that $P(X \leqslant \frac{2}{3}) = \frac{1}{243}$, find a and k.

10 The time taken X, in hours, to perform a particular task has the probability density function:

$$f(x) = \begin{cases} \frac{125}{18}x^2 & 0 \leqslant x < 0.6 \\ \dfrac{9}{10x^2} & 0.6 \leqslant x \leqslant 0.9 \\ 0 & \text{otherwise.} \end{cases}$$

a Sketch the graph of this function.

b Show that $f(x)$ is a well defined probability density function.

c Find the mean, median, and mode of X.

d Find the variance and standard deviation of X.

e Find $P(0.3 < X < 0.7)$ and interpret your answer.

B THE NORMAL DISTRIBUTION

The normal distribution is the most important distribution for a continuous random variable. Many naturally occurring phenomena have a distribution that is normal, or approximately normal. Some examples are:

- physical attributes of a population such as height, weight, and arm length
- crop yields
- scores for tests taken by a large population

Once a normal model has been established, we can use it to make predictions about a distribution and to answer other relevant questions.

HOW A NORMAL DISTRIBUTION ARISES

Consider the oranges picked from an orange tree. They do not all have the same weight. The variation may be due to several factors, including:

- genetics
- different times when the flowers were fertilised
- different amounts of sunlight reaching the leaves and fruit
- different weather conditions such as the prevailing winds.

The result is that most of the fruit will have weights close to the mean, while fewer oranges will be *much* heavier or *much* lighter.

This results in a **bell-shaped distribution** for the weight of oranges in a crop, which is symmetric about the mean.

A TYPICAL NORMAL DISTRIBUTION

A large sample of cockle shells was collected and the maximum width of each shell was measured. Click on the video clip icon to see how a histogram of the data is built up. Then click on the demo icon to observe the effect of changing the class interval lengths for normally distributed data.

DEMO

THE NORMAL PROBABILITY DENSITY FUNCTION

If X is **normally distributed** then its **probability density function** is

$$f(x) = \frac{1}{\sigma\sqrt{2\pi}}\, e^{-\frac{1}{2}\left(\frac{x-\mu}{\sigma}\right)^2} \quad \text{for} \quad -\infty < x < \infty$$

where μ is the mean and σ^2 is the variance of the distribution.

We write $\boldsymbol{X \sim \mathrm{N}(\mu, \sigma^2)}$.

- The curve $y = f(x)$, which is called a **normal curve**, is symmetrical about the vertical line $x = \mu$.
- As $x \to \pm\infty$ the normal curve approaches its asymptote, the x-axis.
- $f(x) > 0$ for all x.
- Since the total probability must be 1, $\displaystyle\int_{-\infty}^{\infty} f(x)\, dx = 1$.

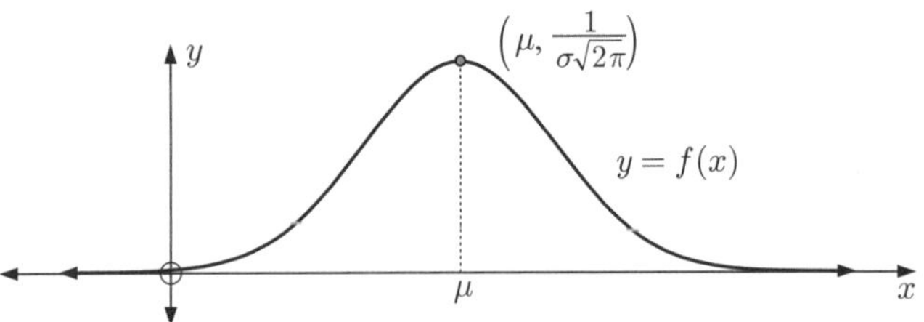

- More scores are distributed closer to the mean than further away. This results in the typical **bell shape**.

Click on the icon to explore the normal probability density function and how it changes when μ and σ are altered.

DEMO

THE GEOMETRIC SIGNIFICANCE OF μ AND σ

Differentiating $f(x) = \dfrac{1}{\sigma\sqrt{2\pi}}\, e^{-\frac{1}{2}\left(\frac{x-\mu}{\sigma}\right)^2}$

we obtain $f'(x) = \dfrac{-1}{\sigma^2\sqrt{2\pi}} \left(\dfrac{x-\mu}{\sigma}\right) e^{-\frac{1}{2}\left(\frac{x-\mu}{\sigma}\right)^2}$

$\therefore$ $f'(x) = 0$ only when $x = \mu$, and this corresponds to the point on the graph when $f(x)$ is a maximum.

Differentiating again, we obtain $f''(x) = \dfrac{-1}{\sigma^2\sqrt{2\pi}}\, e^{-\frac{1}{2}\left(\frac{x-\mu}{\sigma}\right)^2}\left[\dfrac{1}{\sigma} - \dfrac{(x-\mu)^2}{\sigma^3}\right]$

$\therefore \quad f''(x) = 0 \quad \text{when} \quad \dfrac{(x-\mu)^2}{\sigma^3} = \dfrac{1}{\sigma}$

$\therefore \quad (x-\mu)^2 = \sigma^2$

$\therefore \quad x - \mu = \pm\sigma$

$\therefore \quad x = \mu \pm \sigma$

So, the points of inflection are at $x = \mu + \sigma$ and $x = \mu - \sigma$.

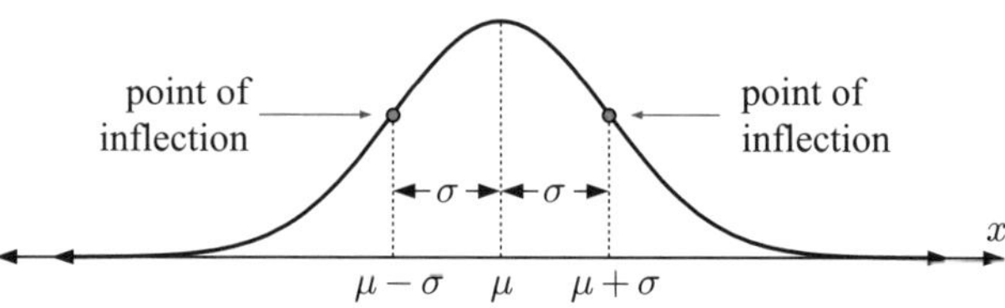

For a normal curve, the standard deviation is uniquely determined as the horizontal distance from the line of symmetry $x = \mu$ to a point of inflection.

INVESTIGATION 2 STANDARD DEVIATION

The purpose of this Investigation is to find the proportions of normal distribution data which lie within σ, 2σ, and 3σ of the mean.

What to do:

1 Click on the icon to start the demonstration in Microsoft® Excel.

DEMO

2 Take a random sample of size $n = 1000$ from a normal distribution.

3 Find the sample mean $\overline{x}$ and standard deviation s.

4 Find:

a $\overline{x} - s$ and $\overline{x} + s$ **b** $\overline{x} - 2s$ and $\overline{x} + 2s$ **c** $\overline{x} - 3s$ and $\overline{x} + 3s$

5 Count all values between:

a $\overline{x} - s$ and $\overline{x} + s$ **b** $\overline{x} - 2s$ and $\overline{x} + 2s$ **c** $\overline{x} - 3s$ and $\overline{x} + 3s$

6 Determine the percentage of data values in these intervals.

7 Repeat the procedure several times. Hence suggest the proportions of normal distribution data which lie within:

a σ **b** 2σ **c** 3σ from the mean.

HISTORICAL NOTE

The normal distribution was first characterised by **Carl Friedrich Gauss** in 1809 as a way to rationalize his **method of least squares** for linear regression.

Marquis de Laplace was the first to calculate $\displaystyle\int_{-\infty}^{\infty} e^{-x^2}\,dx = \sqrt{\pi}$. This led to the correct *normalization constant* $\dfrac{1}{\sigma\sqrt{2\pi}}$ being used for the normal distribution, ensuring $\displaystyle\int_{-\infty}^{\infty} f(x)\,dx = 1$.

For a normal distribution with mean μ and standard deviation σ, the percentage breakdown of where the random variable could lie is shown below:

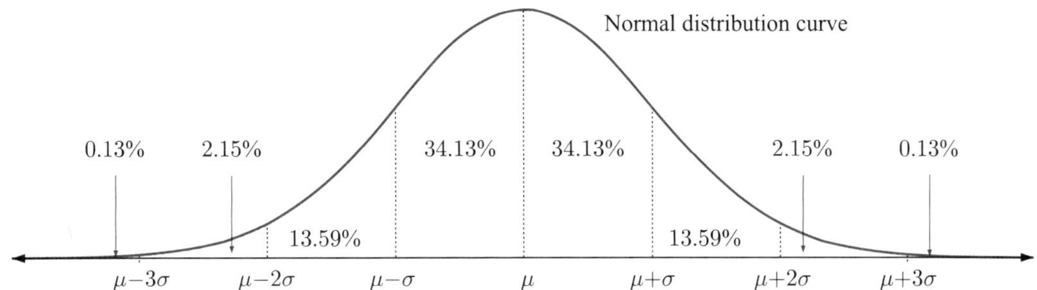

Notice that:
- $\approx 68.26\%$ of values lie between $\mu - \sigma$ and $\mu + \sigma$
- $\approx 95.44\%$ of values lie between $\mu - 2\sigma$ and $\mu + 2\sigma$
- $\approx 99.74\%$ of values lie between $\mu - 3\sigma$ and $\mu + 3\sigma$.

Example 2 **Self Tutor**

The chest measurements of 18 year old male footballers are normally distributed with a mean of 95 cm and a standard deviation of 8 cm.

a Find the percentage of footballers with chest measurements between:
  i 87 cm and 103 cm **ii** 103 cm and 111 cm

b Find the probability that the chest measurement of a randomly chosen footballer is between 87 cm and 111 cm.

c Find the value of k such that approximately 16% of chest measurements are below k cm.

a **i** We need the percentage between $\mu - \sigma$ and $\mu + \sigma$.
$\therefore$ about 68.3% of footballers have a chest measurement between 87 cm and 103 cm.

ii We need the percentage between $\mu + \sigma$ and $\mu + 2\sigma$.
$\therefore$ about 13.6% of footballers have a chest measurement between 103 cm and 111 cm.

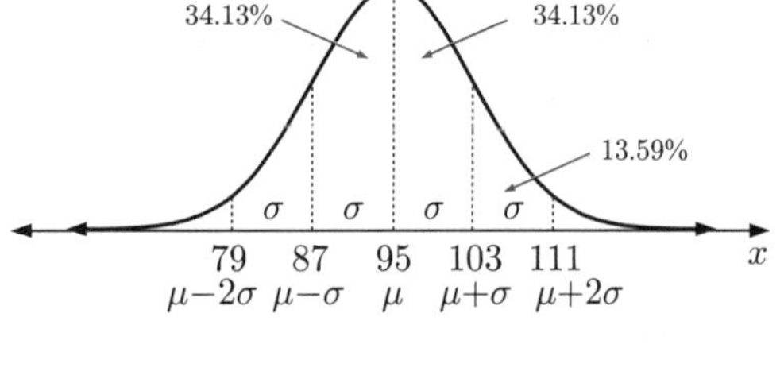

b We need the percentage between $\mu - \sigma$ and $\mu + 2\sigma$.

This is $2(34.13\%) + 13.59\%$
$\approx 81.9\%$.

So, the probability is ≈ 0.819.

c Approximately 16% of data lies more than one standard deviation lower than the mean.

$\therefore$ k is σ below the mean μ

$\therefore$ $k = 95 - 8$
$= 87$ cm

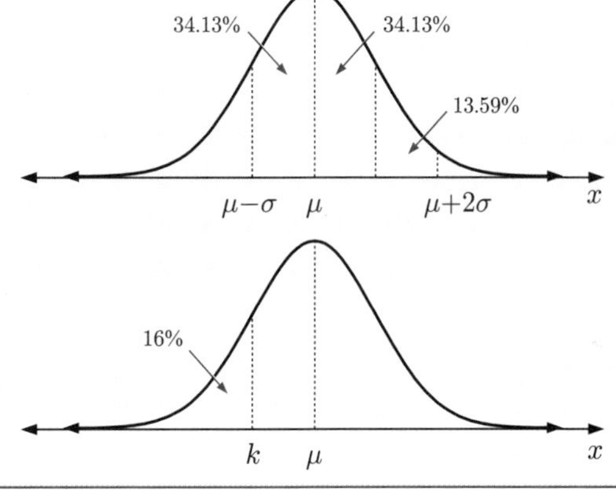

EXERCISE 26B

1 Draw each of the following normal distributions accurately on one set of axes.

Distribution	*Mean* (mL)	*Standard deviation* (mL)
A	25	5
B	30	2
C	21	10

2 Explain why it is likely that the distributions of the following variables will be normal:

- a the volume of soft drink in cans
- b the diameter of bolts immediately after manufacture.

3 State the probability that a randomly selected, normally distributed value lies between:

- a σ below the mean and σ above the mean
- b the mean and the value 2σ above the mean.

4 The mean height of players in a basketball competition is 184 cm. If the standard deviation is 5 cm, what percentage of them are likely to be:

- a taller than 189 cm
- b taller than 179 cm
- c between 174 cm and 199 cm
- d over 199 cm tall?

5 The mean average rainfall of Claudona for August is 48 mm with a standard deviation of 6 mm. Over a 20 year period, how many times would you expect there to be less than 42 mm of rainfall during August in Claudona?

6 Two hundred lifesavers competed in a swimming race. The mean time was 10 minutes 30 seconds, and the standard deviation was 15 seconds. Estimate the number of competitors who:

- a took longer than 11 minutes
- b took less than 10 minutes 15 seconds
- c completed the race in a time between 10 min 15 s and 10 min 45 s.

7 The weights of babies born at Prince Louis Maternity Hospital last year averaged 3.0 kg with a standard deviation of 200 grams. If there were 545 babies born at this hospital last year, estimate the number that weighed:

- a less than 3.2 kg
- b between 2.8 kg and 3.4 kg.

8 Given $X \sim N(3, 0.1^2)$, find:

- a the probability that a randomly selected value lies within 2 standard deviations of the mean
- b the value of X which is 1 standard deviation below the mean.

9 The weights of Jason's oranges are normally distributed. 84% of the crop weigh more than 152 grams and 16% weigh more than 200 grams.

- a Find μ and σ for the crop.
- b What percentage of the oranges weigh between 152 grams and 224 grams?

10 The height of male students in a university is normally distributed with mean 170 cm and standard deviation 8 cm.

- a Find the percentage of male students whose height is:
  - i between 162 cm and 170 cm
  - ii between 170 cm and 186 cm.

b Find the probability that a randomly chosen student from this group has a height:

- **i** between 178 cm and 186 cm
- **ii** less than 162 cm
- **iii** less than 154 cm
- **iv** greater than 162 cm.

c Find the value of k such that 16% of the students are taller than k cm.

11 The heights of 13 year old boys are normally distributed. 97.72% of them are above 131 cm and 2.28% are above 179 cm.

a Find μ and σ for the height distribution.

b A 13 year old boy is randomly chosen. What is the probability that his height lies between 143 cm and 191 cm?

12 When a specific variety of radish is grown without fertiliser, the weights of the radishes produced are normally distributed with mean 40 g and standard deviation 10 g.
When the same variety of radish is grown in the same way but with fertiliser added, the weights of the radishes produced are also normally distributed, but with mean 140 g and standard deviation 40 g.
Determine the proportion of radishes grown:

a without fertiliser with weights less than 50 grams

b with fertiliser with weights less than 60 grams

c **i** with and **ii** without fertiliser, which have weights between 20 and 60 g

d **i** with and **ii** without fertiliser, which have weights greater than 60 g.

13 A bottle filling machine fills an average of 20 000 bottles a day with a standard deviation of 2000. Assuming that production is normally distributed and the year comprises 260 working days, calculate the approximate number of working days on which:

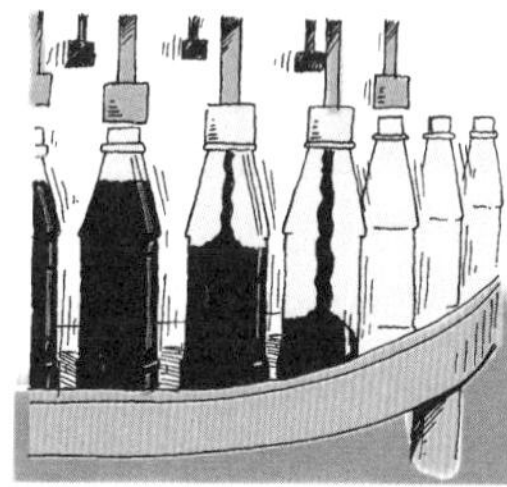

a under 18 000 bottles are filled

b over 16 000 bottles are filled

c between 18 000 and 24 000 bottles (inclusive) are filled.

C PROBABILITIES USING A CALCULATOR

Using the properties of the normal probability density function, we have considered probabilities in regions of width σ either side of the mean.

To find probabilities more generally we use technology.

Suppose $X \sim \text{N}(10,\ 2.3^2)$, so X is normally distributed with mean 10 and standard deviation 2.3.

To find $\text{P}(8 \leqslant X \leqslant 11)$, we need to calculate $\int_8^{11} f(x)\,dx$, where $f(x)$ is the probability density function of the normal curve.

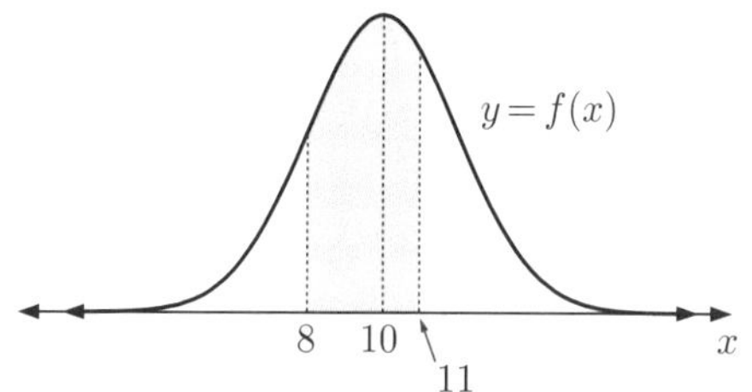

Click on the icon to find instructions for these processes.

Example 3 — Self Tutor

If $X \sim N(10, 2.3^2)$, find these probabilities:

a $P(8 \leqslant X \leqslant 11)$ **b** $P(X \leqslant 12)$ **c** $P(X > 9)$. Illustrate your results.

X is normally distributed with mean 10 and standard deviation 2.3.

Using technology:

a

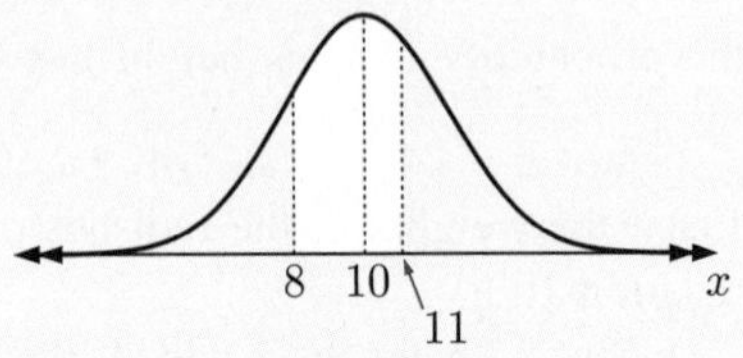

$P(8 \leqslant X \leqslant 11) \approx 0.476$

b

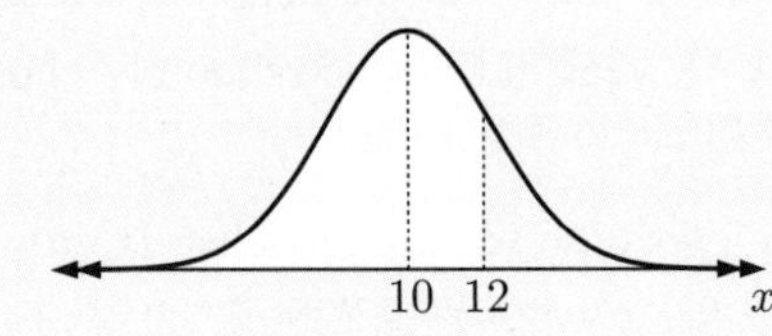

$P(X \leqslant 12) \approx 0.808$

c

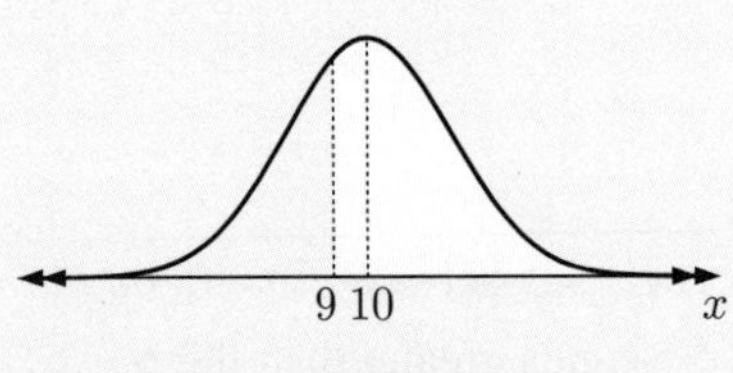

$P(X > 9) \approx 0.668$

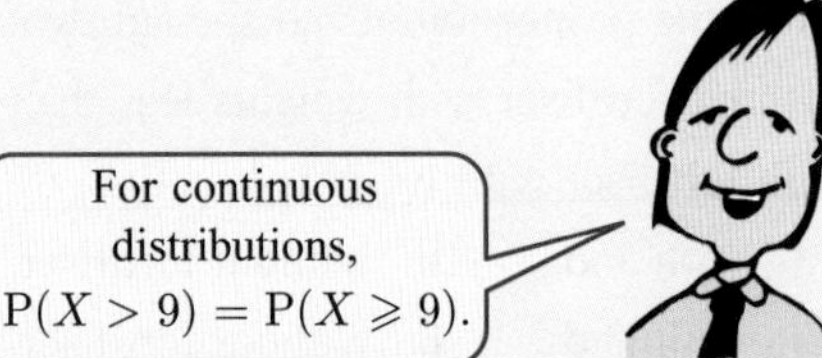

Example 4 — Self Tutor

In 1972 the heights of rugby players were approximately normally distributed with mean 179 cm and standard deviation 7 cm. Find the probability that a randomly selected player in 1972 was:

a at least 175 cm tall **b** between 170 cm and 190 cm.

If X is the height of a player then X is normally distributed with $\mu = 179$, $\sigma = 7$.

Using technology:

a

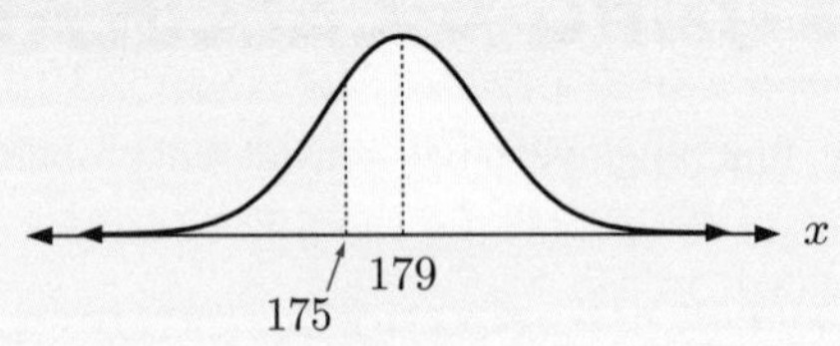

$P(X \geqslant 175) \approx 0.716$

b

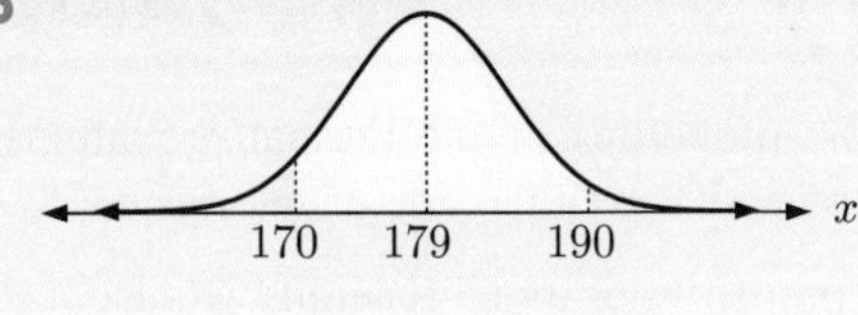

$P(170 < X < 190) \approx 0.843$

EXERCISE 26C

1 X is a random variable that is distributed normally with mean 70 and standard deviation 4. Find:

a $P(70 \leqslant X \leqslant 74)$ **b** $P(68 \leqslant X \leqslant 72)$ **c** $P(X \leqslant 65)$

2 X is a random variable that is distributed normally with mean 60 and standard deviation 5. Find:

a $P(60 \leqslant X \leqslant 65)$ **b** $P(62 \leqslant X \leqslant 67)$ **c** $P(X \geqslant 64)$

d $P(X \leqslant 68)$ **e** $P(X \leqslant 61)$ **f** $P(57.5 \leqslant X \leqslant 62.5)$

3 A machine produces metal bolts. The lengths of these bolts have a normal distribution with mean 19.8 cm and standard deviation 0.3 cm. If a bolt is selected at random from the machine, find the probability that it will have a length between 19.7 cm and 20 cm.

4 Max's customers put money for charity into a collection box on the front counter of his shop. The average weekly collection is approximately normally distributed with mean \$40 and standard deviation \$6. What proportion of weeks would he expect to collect:

a between \$30.00 and \$50.00 **b** at least \$50.00?

5 Eels are washed onto a beach after a storm. Their lengths have a normal distribution with mean 41 cm and variance 11 cm^2.

a If an eel is randomly selected, find the probability that it is at least 50 cm long.

b Find the proportion of eels measuring between 40 cm and 50 cm long.

c How many eels from a sample of 200 would you expect to measure at least 45 cm in length?

6 The speed of cars passing the supermarket is normally distributed with mean 56.3 km h^{-1} and standard deviation 7.4 km h^{-1}. Find the probability that a randomly selected car has speed:

a between 60 and 75 km h^{-1} **b** at most 70 km h^{-1}

c at least 60 km h^{-1}.

Example 5 **Self Tutor**

The time taken by students to complete a puzzle is normally distributed with mean 28.3 minutes and standard deviation 3.6 minutes. Calculate the probability that:

a a randomly selected student took at least 30 minutes to complete the puzzle

b out of 10 randomly selected students, at most half of them took at least 30 minutes to complete the puzzle.

---

a Let X denote the time for a student to complete the puzzle.

$X \sim \text{N}(28.3, 3.6^2)$

$\therefore$ $\text{P}(X \geqslant 30) \approx 0.318\,38$ {using technology}

≈ 0.318

b Let Y denote the number of students who took at least 30 minutes to complete the puzzle.

Then $Y \sim \text{B}(10, 0.318\,38)$

$\therefore$ $\text{P}(Y \leqslant 5) \approx 0.938$ {using technology}

7 Apples from a grower's crop were normally distributed with mean 173 grams and standard deviation 34 grams. Apples weighing less than 130 grams were too small to sell.

- **a** Find the proportion of apples from this crop which were too small to sell.
- **b** Find the probability that in a picker's basket of 100 apples, up to 10 apples were too small to sell.

8 People found to have high blood pressure are started on a course of tablets and have their blood pressure checked at the end of 4 weeks. The drop in blood pressure over the period is normally distributed with mean 5.9 units and standard deviation 1.9 units.

- **a** Find the proportion of people who show a drop of more than 4 units.
- **b** Eight people from the large population taking the course of tablets are selected at random. Find the probability that more than five of them will show a drop in blood pressure of more than 4 units.

D THE STANDARD NORMAL DISTRIBUTION (Z-DISTRIBUTION)

Every normal X-distribution can be **transformed** into the **standard normal distribution** or **Z-distribution** using the transformation $z = \frac{x - \mu}{\sigma}$.

INVESTIGATION 3 PROPERTIES OF $z = \frac{x - \mu}{\sigma}$

Suppose a random variable X is **normally distributed** with mean μ and standard deviation σ.

For each value of x we can calculate a **z-score** using the algebraic transformation $z = \frac{x - \mu}{\sigma}$.

What to do:

1 Consider the x-values: 1, 2, 2, 3, 3, 3, 3, 4, 4, 4, 4, 4, 5, 5, 5, 5, 6, 6, 7.

- **a** Draw a graph of the distribution to check that it is approximately normal.
- **b** Find the mean μ and standard deviation σ for the distribution of x-values.
- **c** Use the transformation $z = \frac{x - \mu}{\sigma}$ to convert each x-value into a z-value.
- **d** Find the mean and standard deviation for the distribution of z-values.

2 Click on the icon to load a large sample drawn from a normal population. By clicking appropriately we can repeat the four steps in **1**.

3 Summarise your findings.

DEMO

The Z-distribution has mean 0 and standard deviation 1.

In fact, $Z \sim N(0, 1)$.

No matter what the parameters μ and σ of the X-distribution are, we always end up with the same Z-distribution $Z \sim N(0, 1)$.

THE Z-TRANSFORMATION

To explain how this works, remember that a normal X-distribution with mean μ and standard deviation σ has probability density function

$$f(x) = \frac{1}{\sigma\sqrt{2\pi}}\, e^{-\frac{1}{2}\left(\frac{x-\mu}{\sigma}\right)^2}.$$

Consider first the normal distribution $\text{N}(0,\ 1)$ with $\mu = 0$, $\sigma = 1$, and probability density function

$$f(x) = \frac{1}{\sqrt{2\pi}}\, e^{-\frac{1}{2}x^2}.$$

With $\mu = 0$ and $\sigma = 1$, the transformation $z = \dfrac{x-\mu}{\sigma}$ is simply $z = x$.

So, the distribution $\text{N}(0,\ 1)$ is unchanged under the transformation, and the probability density function for the Z-distribution is

$$f(z) = \frac{1}{\sqrt{2\pi}}\, e^{-\frac{1}{2}z^2}.$$

So what happens with other X-distributions?

$\text{N}(0,\ 1)$ is symmetric about the mean $x = 0$, with points of inflection at $x = \pm 1$.

To transform $\text{N}(0,\ 1)$ into $\text{N}(\mu,\ \sigma^2)$ we need:

- a horizontal stretch with scale factor σ to shift the inflection points to $x = \pm\sigma$
- a horizontal translation μ units to the right to shift the mean to $x = \mu$.

This gives us $f\left(\frac{x-\mu}{\sigma}\right) = \frac{1}{\sqrt{2\pi}}\, e^{-\frac{1}{2}\left(\frac{x-\mu}{\sigma}\right)^2}$.

However, when we made the horizontal stretch, we changed the area under the curve. We know this needs to be 1 for a probability density function. We therefore also need:

- a vertical stretch with scale factor $\dfrac{1}{\sigma}$ to ensure that $\displaystyle\int_{-\infty}^{\infty} f(x)\, dx = 1$.

DEMO

We now have the probability density function $f(x) = \dfrac{1}{\sigma\sqrt{2\pi}}\, e^{-\frac{1}{2}\left(\frac{x-\mu}{\sigma}\right)^2}$.

The Z-transformation does this in reverse, so no matter what normal distribution we start with, we end up with the standard normal distribution function $f(z)$.

HISTORICAL NOTE

Notice that the normal distribution function $f(x)$ has two parameters μ and σ, whereas the standard normal distribution function $f(z)$ has no parameters. This means that a unique table of values can be constructed for $f(z)$.

Before graphics calculators and computer packages, it was impossible to calculate probabilities for a general normal distribution $\text{N}(\mu,\ \sigma^2)$.

Instead, all data was transformed using the Z-transformation, and the standard normal distribution table was consulted for the required probability values.

THE SIGNIFICANCE OF THE Z-DISTRIBUTION

The **probability density function** for the Z-distribution is $f(z) = \frac{1}{\sqrt{2\pi}} e^{-\frac{1}{2}z^2}$, $-\infty < z < \infty$.

If x is an observation from a normal distribution with mean μ and standard deviation σ, the **z-score** of x, $z = \dfrac{x-\mu}{\sigma}$, is the number of standard deviations x is from the mean.

For example:

- if $z = 1.84$, then x is 1.84 standard deviations to the right of the mean
- if $z = -0.273$, then x is 0.273 standard deviations to the left of the mean.

This diagram shows how the z-score is related to a general normal curve:

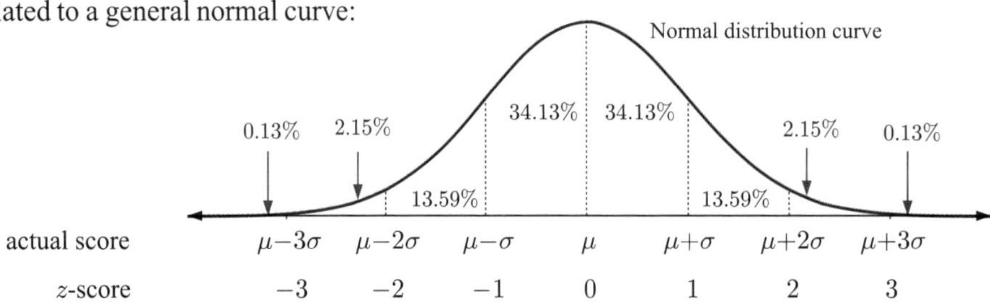

This means that for $X \sim N(\mu, \sigma^2)$ and $Z \sim N(0, 1^2)$ and for any $x_1, x_2 \in \mathbb{R}$, $x_1 < x_2$, with corresponding z-values $z_1 = \dfrac{x_1-\mu}{\sigma}$ and $z_2 = \dfrac{x_2-\mu}{\sigma}$, then:

- $P(X \geqslant x_1) = P(Z \geqslant z_1)$

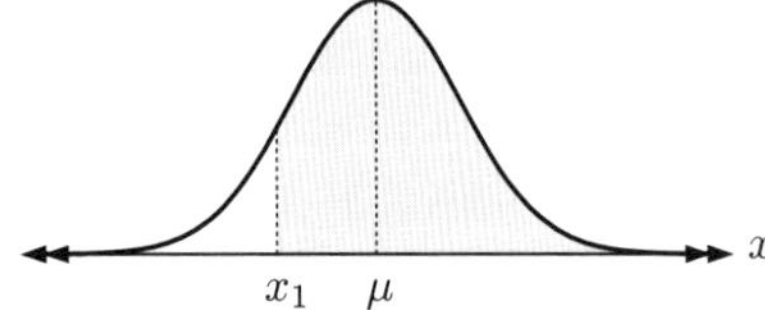

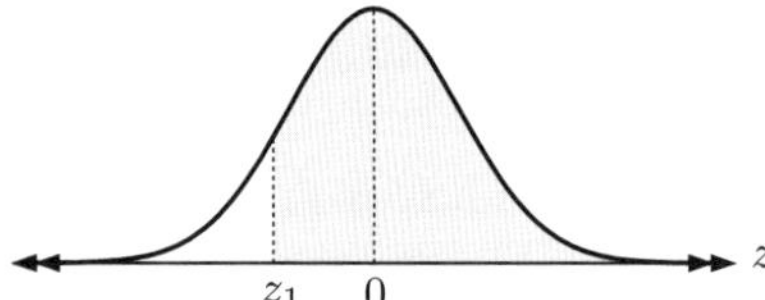

- $P(X \leqslant x_1) = P(Z \leqslant z_1)$

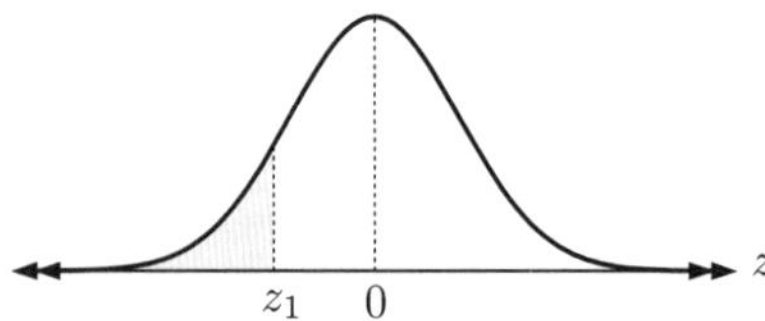

- $P(x_1 \leqslant X \leqslant x_2) = P(z_1 \leqslant Z \leqslant z_2)$

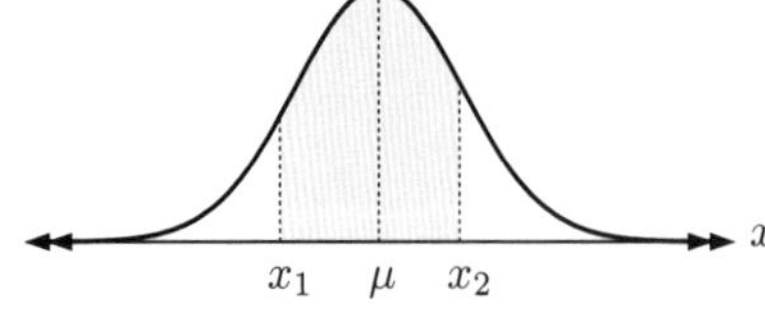

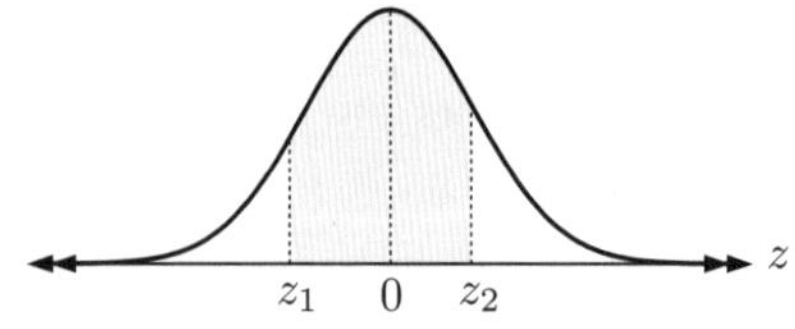

z-scores are particularly useful when comparing two populations with different μ and σ. However, these comparisons will only be reasonable if both distributions are approximately normal.

Example 6 Self Tutor

Kelly scored 73% in History where the class mean was 68% and the standard deviation was 10.2%. In Mathematics she scored 66%, where the class mean was 62% and the standard deviation was 6.8%.

In which subject did Kelly perform better compared with the rest of her class?

Assume the scores for both subjects were normally distributed.

Kelly's z-score for History $= \frac{73-68}{10.2} \approx 0.490$

Kelly's z-score for Mathematics $= \frac{66-62}{6.8} \approx 0.588$

So, Kelly's result in Mathematics was 0.588 standard deviations above the mean, whereas her result in History was 0.490 standard deviations above the mean.

$\therefore$ Kelly's result in Mathematics was better compared to her class, even though her percentage was lower.

EXERCISE 26D

1 The table shows Emma's midyear exam results. The exam results for each subject are normally distributed with the mean μ and standard deviation σ shown in the table.

Subject	*Emma's score*	μ	σ
English	48	40	4.4
Mandarin	81	60	9
Geography	84	55	18
Biology	68	50	20
Maths	84	50	15

a Find the z-score for each of Emma's subjects.

b Arrange Emma's subjects from 'best' to 'worst' in terms of the z-scores.

2 The table alongside shows Sergio's results in his final examinations, along with the class means and standard deviations.

	Sergio	μ	σ
Physics	73%	78%	10.8%
Chemistry	77%	72%	11.6%
Mathematics	76%	74%	10.1%
German	91%	86%	9.6%
Biology	58%	62%	5.2%

a Find Sergio's Z-value for each subject.

b Arrange Sergio's performances in each subject in order from 'best' to 'worst'.

3 Consider the normal distribution probabilities:

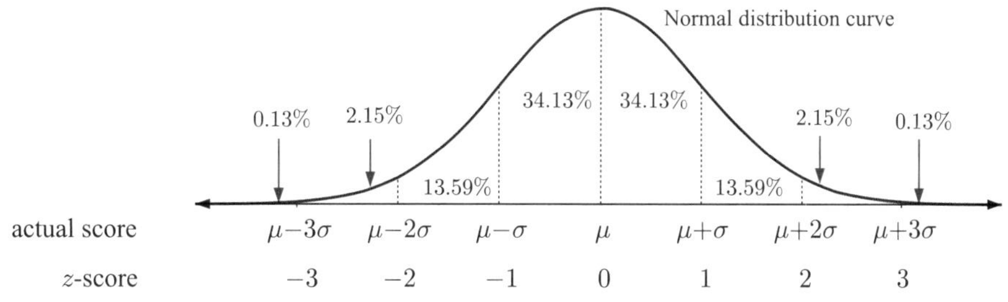

Use the diagram to calculate the following probabilities. In each case sketch the Z-distribution and shade in the region of interest.

a $P(-1 < Z < 1)$
b $P(-1 \leqslant Z \leqslant 3)$
c $P(-1 < Z < 0)$
d $P(Z < 2)$
e $P(-1 < Z)$
f $P(Z \geqslant 1)$

4 For a random variable X, the mean is μ and standard deviation is σ.

Using properties of $\mathrm{E}(X)$ and $\mathrm{Var}(X)$ find: **a** $\mathrm{E}\left(\dfrac{X-\mu}{\sigma}\right)$ **b** $\mathrm{Var}\left(\dfrac{X-\mu}{\sigma}\right)$.

5 Given $X \sim \mathrm{N}(\mu, \sigma^2)$ and $Z \sim \mathrm{N}(0, 1)$, determine the values of a and b such that:

a $\mathrm{P}(\mu - \sigma < X < \mu + 2\sigma) = \mathrm{P}(a < Z < b)$

b $\mathrm{P}(\mu - 0.5\sigma < X < \mu) = \mathrm{P}(a < Z < b)$

c $\mathrm{P}(0 \leqslant Z \leqslant 3) = \mathrm{P}(\mu - a\sigma \leqslant X \leqslant \mu + b\sigma)$

Example 7 **Self Tutor**

Use technology to illustrate and calculate:

a $\mathrm{P}(-0.41 \leqslant Z \leqslant 0.67)$ **b** $\mathrm{P}(Z \leqslant 1.5)$ **c** $\mathrm{P}(Z > 0.84)$

$Z \sim \mathrm{N}(0, 1)$

a

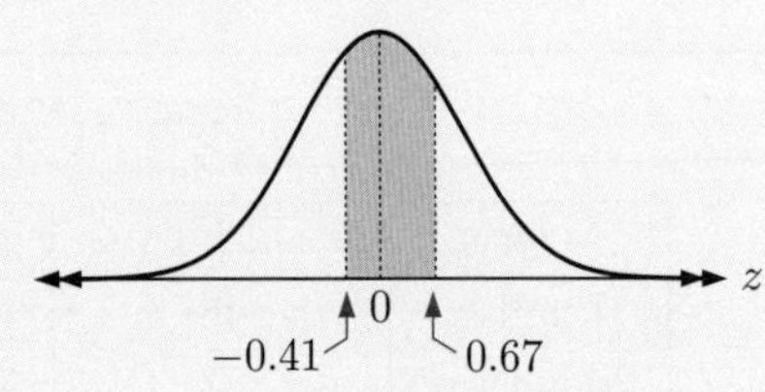

$\mathrm{P}(-0.41 \leqslant Z \leqslant 0.67) \approx 0.408$

b

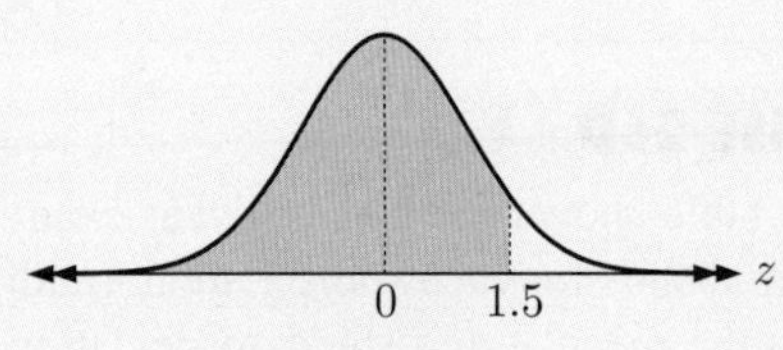

$\mathrm{P}(Z \leqslant 1.5) \approx 0.933$

c

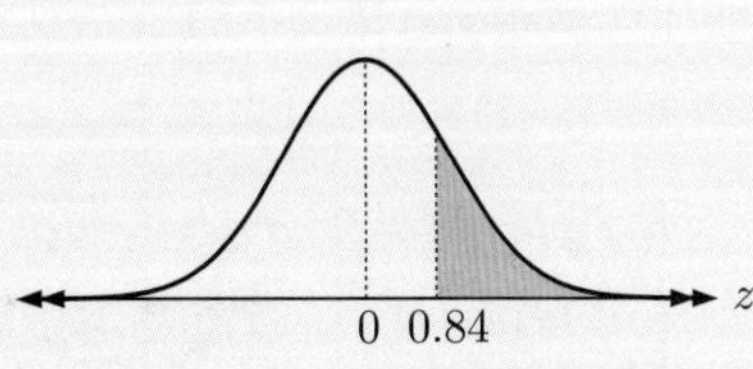

$\mathrm{P}(Z > 0.84) \approx 0.200$

6 If Z is the standard normal distribution, find the following probabilities using technology. In each case sketch the regions.

a $\mathrm{P}(-0.86 \leqslant Z \leqslant 0.32)$ **b** $\mathrm{P}(-2.3 \leqslant Z \leqslant 1.5)$ **c** $\mathrm{P}(Z \leqslant 1.2)$

d $\mathrm{P}(Z \leqslant -0.53)$ **e** $\mathrm{P}(Z \geqslant 1.3)$ **f** $\mathrm{P}(Z \geqslant -1.4)$

g $\mathrm{P}(Z > 4)$ **h** $\mathrm{P}(-0.5 < Z < 0.5)$ **i** $\mathrm{P}(-1.960 \leqslant Z \leqslant 1.960)$

7 Suppose the variable X is normally distributed with mean $\mu = 58.3$ and standard deviation $\sigma = 8.96$.

a Let the z-score of $x = 50.6$ be z_1 and the z-score of $x = 68.9$ be z_2.

i Calculate z_1 and z_2. **ii** Find $\mathrm{P}(z_1 \leqslant Z \leqslant z_2)$.

b Check your answer by calculating $\mathrm{P}(50.6 \leqslant X \leqslant 68.9)$ directly from your calculator.

E QUANTILES OR k-VALUES

Consider a population of crabs where the length of a shell, X mm, is normally distributed with mean 70 mm and standard deviation 10 mm.

A biologist wants to protect the population by allowing only the largest 5% of crabs to be harvested. He therefore asks the question: "95% of the crabs have lengths less than what?".

To answer this question we need to find k such that $P(X \leqslant k) = 0.95$.

The number k is known as a **quantile**, and in this case the 95% quantile.

When finding quantiles we are given a probability and are asked to calculate the corresponding measurement. This is the *inverse* of finding probabilities, and we use the **inverse normal function** on our calculator.

Example 8 Self Tutor

If $X \sim N(23.6, 3.1^2)$, find k for which $P(X < k) = 0.95$.

X has mean 23.6 and standard deviation 3.1.

If $P(X < k) = 0.95$ then

$k \approx 28.7$

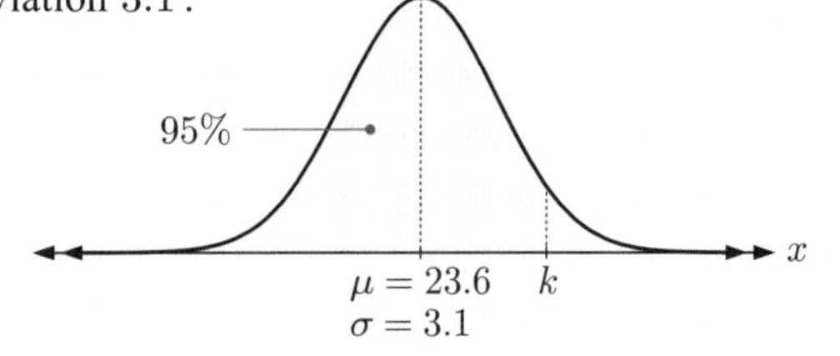

Example 9 Self Tutor

If Z is the standard normal distribution, find k such that $P(Z > k) = 0.73$. Interpret your result.

If $P(Z > k) = 0.73$ then

$P(Z \leqslant k) = 0.27$

$\therefore \; k \approx -0.613$

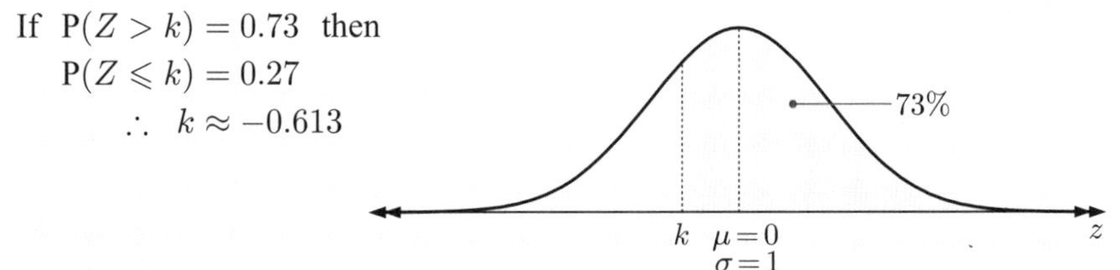

This means 73% of the Z-distribution values are more than -0.613.

EXERCISE 26E.1

1 Suppose Z is the standard normal distribution. Illustrate with a sketch and find k such that:

a $P(Z \leqslant k) = 0.81$
b $P(Z \leqslant k) = 0.58$
c $P(Z \leqslant k) = 0.17$
d $P(Z \geqslant k) = 0.95$
e $P(Z \geqslant k) = 0.90$

2 $X \sim \mathrm{N}(20, 3^2)$. Illustrate with a sketch and find k such that:

a $\mathrm{P}(X \leqslant k) = 0.348$ **b** $\mathrm{P}(X \leqslant k) = 0.878$ **c** $\mathrm{P}(X \leqslant k) = 0.5$

3 Suppose $X \sim \mathrm{N}(38.7, 8.2^2)$. Illustrate with a sketch and find k such that:

a $\mathrm{P}(X \leqslant k) = 0.9$ **b** $\mathrm{P}(X \geqslant k) = 0.8$

4 Given that $X \sim \mathrm{N}(23, 5^2)$, find a such that:

a $\mathrm{P}(X < a) = 0.378$ **b** $\mathrm{P}(X \geqslant a) = 0.592$

c $\mathrm{P}(23 - a < X < 23 + a) = 0.427$

Example 10 **Self Tutor**

A university professor determines that no more than 80% of this year's History candidates should pass the final examination. The examination results were approximately normally distributed with mean 62 and standard deviation 13. Find the lowest score necessary to pass the examination.

Let X denote the final examination result, so $X \sim \mathrm{N}(62, 13^2)$.

We need to find k such that

$\mathrm{P}(X \geqslant k) = 0.8$

$\therefore$ $\mathrm{P}(X < k) = 0.2$

$\therefore$ $k \approx 51.059$

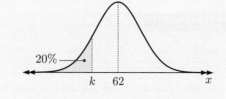

So, the minimum pass mark is 52.

5 The students of Class X sat a Physics test. The average score was 46 with a standard deviation of 25. The teacher decided to award an A to the top 7% of the students in the class. Assuming that the scores were normally distributed, find the lowest score that would achieve an A.

6 The length of fish from a particular species is normally distributed with mean 35 cm and standard deviation 8 cm. The fisheries department has decided that the smallest 10% of the fish are not to be harvested. What is the size of the smallest fish that can be harvested?

7 The length of a screw produced by a machine is normally distributed with mean 75 mm and standard deviation 0.1 mm. If a screw is too long it is automatically rejected. If 1% of screws are rejected, what is the length of the smallest screw to be rejected?

8 Pedro is studying Algebra and Geometry. He sits for the mid-year exams in each subject.

Pedro's Algebra mark is 56%, and the class mean and standard deviation are 50.2% and 15.8% respectively.

In Geometry he is told that the class mean and standard deviation are 58.7% and 18.7% respectively.

What percentage does Pedro need to have scored in Geometry, to have an equivalent result to his Algebra mark?

9 The volume of cool drink in a bottle filled by a machine is normally distributed with mean 503 mL and standard deviation 0.5 mL. 1% of the bottles are rejected because they are underfilled, and 2% are rejected because they are overfilled; otherwise they are kept for retail. What range of volumes is in the bottles that are kept?

THE IMPORTANCE OF QUANTILES

For some questions we **must** convert to z-scores in order to find the answer.

We always need to convert to z-scores if we are trying to find an unknown mean μ or standard deviation σ.

Example 11 **Self Tutor**

An adult scallop population is known to be normally distributed with a standard deviation of 5.9 g. If 15% of scallops weigh less than 58.2 g, find the mean weight of the population.

Let the mean weight of the population be μ g.

If X g denotes the weight of an adult scallop, then $X \sim \text{N}(\mu, 5.9^2)$.

As we do not know μ we cannot use the invNorm directly, but we can convert to z-scores and use the properties of $\text{N}(0, 1^2)$.

Now $\quad \text{P}(X \leqslant 58.2) = 0.15$

$\therefore \quad \text{P}\left(Z \leqslant \frac{58.2-\mu}{5.9}\right) = 0.15$

Using invNorm for $\text{N}(0, 1^2)$,

$\frac{58.2-\mu}{5.9} \approx -1.0364$

$\therefore \quad 58.2 - \mu \approx -6.1$

$\mu \approx 64.3$

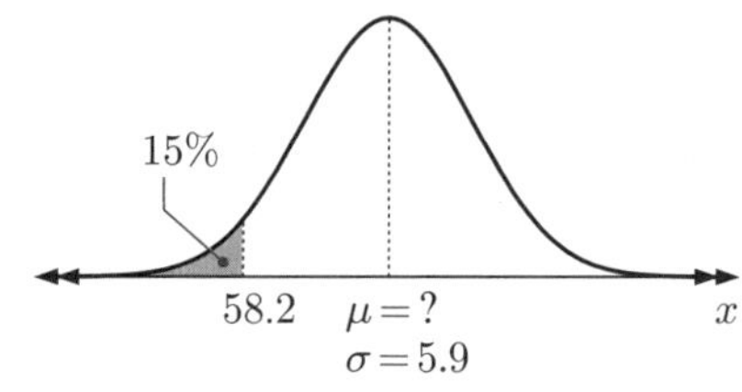

So, the mean weight is 64.3 g.

Example 12 **Self Tutor**

Find the mean and standard deviation of a normally distributed random variable X if $\text{P}(X \leqslant 20) = 0.1$ and $\text{P}(X \geqslant 29) = 0.15$.

$X \sim \text{N}(\mu, \sigma^2)$ where we have to find μ and σ.

We start by finding z_1 and z_2 which correspond to $x_1 = 20$ and $x_2 = 29$.

0.1 0.15

$x_1 = 20$ μ $x_2 = 29$

z_1 z_2

Now $\quad \text{P}(X \leqslant x_1) = 0.1$

$\therefore \quad \text{P}(Z \leqslant \frac{20-\mu}{\sigma}) = 0.1$

$\therefore \quad z_1 = \frac{20-\mu}{\sigma} \approx -1.282$

$\therefore \quad 20 - \mu \approx -1.282\sigma \quad .... (1)$

and $\quad \text{P}(X \leqslant x_2) = 0.85$

$\therefore \quad \text{P}(Z \leqslant \frac{29-\mu}{\sigma}) = 0.85$

$\therefore \quad z_2 = \frac{29-\mu}{\sigma} \approx 1.036$

$\therefore \quad 29 - \mu \approx 1.036\sigma \quad .... (2)$

Solving (1) and (2) simultaneously we get $\mu \approx 25.0$ and $\sigma \approx 3.88$.

EXERCISE 26E.2

1 The IQs of students at school are normally distributed with a standard deviation of 15. If 20% of students have an IQ higher than 125, find the mean IQ of students at school.

2 The distances an athlete jumps are normally distributed with mean 5.2 m. If 15% of the jumps by this athlete are less than 5 m, what is the standard deviation?

3 The weekly income of a bakery is normally distributed with a mean of \$6100. If 85% of the time the weekly income exceeds \$6000, what is the standard deviation?

4 The arrival times of buses at a depot are normally distributed with standard deviation 5 minutes. If 10% of the buses arrive before 3:55 pm, find the mean arrival time of buses at the depot.

5 Find the mean and standard deviation of a normally distributed random variable X if $\text{P}(X \geqslant 35) = 0.32$ and $\text{P}(X \leqslant 8) = 0.26$.

6
 a Find the mean and standard deviation of a normally distributed random variable X, given that $\text{P}(X \geqslant 80) = 0.1$ and $\text{P}(X \leqslant 30) = 0.15$.
 b In a Mathematics examination it was found that 10% of the students scored at least 80, and no more than 15% scored under 30. Assuming the scores are normally distributed, what proportion of students scored more than 50?

7 The diameters of pistons manufactured by a company are normally distributed. Only those pistons whose diameters lie between 3.994 cm and 4.006 cm are acceptable.
 a Find the mean and the standard deviation of the distribution if 4% of the pistons are rejected as being too small, and 5% are rejected as being too large.
 b Determine the probability that the diameter of a randomly chosen piston lies between 3.997 cm and 4.003 cm.

8 Circular metal tokens are used to operate a washing machine in a laundromat. The diameters of the tokens are normally distributed, and only tokens with diameters between 1.94 and 2.06 cm will operate the machine.
 a Find the mean and standard deviation of the distribution given that 2% of the tokens are too small, and 3% are too large.
 b Find the probability that at most one token out of a randomly selected sample of 20 will not operate the machine.

ACTIVITY

Click on the icon to obtain a card game for the normal distribution.

CARD GAME

REVIEW SET 26A NON-CALCULATOR

1 The average height of 17 year old boys is normally distributed with mean 179 cm and standard deviation 8 cm. Calculate the percentage of 17 year old boys whose heights are:

a more than 195 cm **b** between 163 cm and 195 cm

c between 171 cm and 187 cm.

2 The contents of cans of a certain brand of soft drink are normally distributed with mean 377 mL and standard deviation 4.2 mL.

a Find the percentage of cans with contents:

i less than 368.6 mL **ii** between 372.8 mL and 389.6 mL.

b Find the probability that a randomly selected can contains between 377 mL and 381.2 mL.

3 The edible part of a batch of Coffin Bay oysters is normally distributed with mean 38.6 grams and standard deviation 6.3 grams.

Let the random variable X be the mass of a Coffin Bay oyster.

a Find a if $P(38.6 - a \leqslant X \leqslant 38.6 + a) = 0.6826$.

b Find b if $P(X \geqslant b) = 0.8413$.

4 A random variable X has probability density function $f(x) = a(x+1)x(x-1)(x-2)$ for $0 < x < 1$.

a Show that $a = \frac{30}{11}$. **b** Find the mode of X.

c Show that $f(\frac{1}{2} - x) = f(\frac{1}{2} + x)$. **d** Hence find the median of X.

5 The results of a test are normally distributed. Harri gained a z-score equal to -2.

a Interpret this z-score with regard to the mean and standard deviation of the test scores.

b What proportion of students obtained a better score than Harri?

c The mean test score was 151 and Harri's actual score was 117. Find the standard deviation of the test scores.

6 The continuous random variable Z is distributed such that $Z \sim N(0, 1)$.
Find the value of k if $P(-k \leqslant Z \leqslant k) = 0.95$.

7 A continuous random variable X has the probability density function
$f(x) = ax(4 - x^2), \quad 0 \leqslant x \leqslant 2$.

a Find a. **b** Find the mode of X.

c Show that the median of X is $\sqrt{4 - 2\sqrt{2}}$. **d** Find the mean of X.

8 The distance that a 15 year old boy can throw a tennis ball is normally distributed with mean 35 m and standard deviation 4 m.

The distance that a 10 year old boy can throw a tennis ball is normally distributed with mean 25 m and standard deviation 3 m.

Jarrod is 15 years old and can throw a tennis ball 41 m. How far does his 10 year old brother Paul need to throw a tennis ball to perform as well as Jarrod?

9 State the probability that a randomly selected, normally distributed value lies between:

a σ above the mean and 2σ above the mean **b** the mean and σ above the mean.

10 

A bottle shop sells on average 2500 bottles per day with a standard deviation of 300 bottles. Assuming that the number of bottles sold per day is normally distributed, calculate the percentage of days when:

a less than 1900 bottles are sold
b more than 2200 bottles are sold
c between 2200 and 3100 bottles are sold.

11 The continuous random variable X has probability density function $f(x) = 2e^{-x}$, $0 \leqslant x \leqslant k$.

a Find the exact value of k.
b Find the probability that X lies between $\ln \frac{4}{3}$ and $\ln \frac{5}{3}$.
c Find the mean of X.
d Show that the variance of X is $1 - 2(\ln 2)^2$.

REVIEW SET 26B — CALCULATOR

1 The mean and standard deviation of a normal distribution are 150 and 12 respectively. What percentage of values lie between:

a 138 and 162 **b** 126 and 174 **c** 126 and 162 **d** 162 and 174?

2 The arm lengths of 18 year old females are normally distributed with mean 64 cm and standard deviation 4 cm.

a Find the percentage of 18 year old females whose arm lengths are:
i between 60 cm and 72 cm **ii** greater than 60 cm.
b Find the probability that a randomly chosen 18 year old female has an arm length in the range 56 cm to 64 cm.
c The arm lengths of 70% of the 18 year old females are more than x cm. Find the value of x.

3 The length of steel rods produced by a machine is normally distributed with a standard deviation of 3 mm. It is found that 2% of all rods are less than 25 mm long. Find the mean length of rods produced by the machine.

4 The continuous random variable Z is distributed such that $Z \sim \mathrm{N}(0, 1)$.
Find the value of k if $\mathrm{P}(|Z| > k) = 0.376$.

5 $f(x) = ax(x-3)$, $0 \leqslant x \leqslant 2$ is a continuous probability density function.

a Find a. **b** Sketch the graph of $y = f(x)$.
c For this distribution, find the:
i mean **ii** mode **iii** median **iv** variance.
d Find $\mathrm{P}(1 \leqslant x \leqslant 2)$.

6 The distribution curve shown corresponds to $X \sim \mathrm{N}(\mu, \sigma^2)$.
Area A = Area $B = 0.2$.

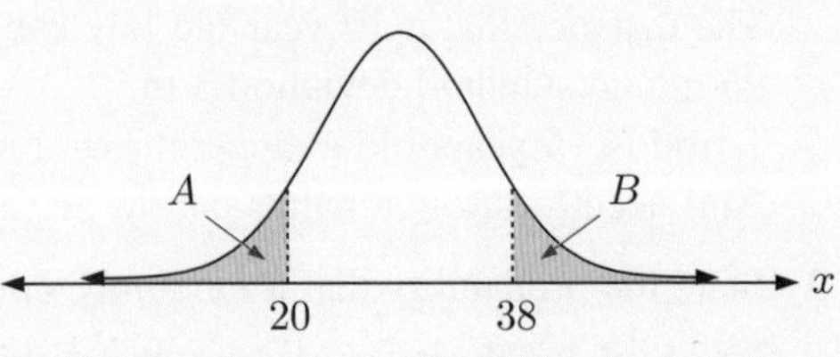

a Find μ and σ.
b Calculate:
i $\mathrm{P}(X \leqslant 35)$ **ii** $\mathrm{P}(23 \leqslant X \leqslant 30)$

7 Let X be the weight in grams of bags of sugar filled by a machine. Bags less than 500 grams are considered underweight.
Suppose that $X \sim \text{N}(503, 2^2)$.

a What proportion of bags are underweight?

b If a quality inspector randomly selects 20 bags, what is the probability that at most 2 bags are underweight?

8 The marks of 2376 candidates in an IB examination are normally distributed with mean 49 marks and variance 225.

a If the pass mark is 45, estimate the number of candidates who passed the examination.

b If the top 7% of the candidates are awarded a '7', find the minimum mark required to obtain a '7'.

c Find the interquartile range of the distribution of marks obtained.

9 The life of a Xenon-brand battery is normally distributed with mean 33.2 weeks and standard deviation 2.8 weeks.

a Find the probability that a randomly selected battery will last at least 35 weeks.

b For how many weeks can the manufacturer expect the batteries to last before 8% of them fail?

10 The random variable X is normally distributed with $\text{P}(X \leqslant 30) = 0.0832$ and $\text{P}(X \geqslant 90) = 0.101$.

a Find the mean μ and standard deviation σ.

b Hence find $\text{P}(-7 \leqslant X - \mu \leqslant 7)$.

11 Kerry's marks for an English essay and a Chemistry test were 26 out of 40, and 82% respectively.

a Explain briefly why the information given is not sufficient to determine whether Kerry's results are better in English than in Chemistry.

b Suppose that the marks of all students in the English essay and the Chemistry test were normally distributed as $\text{N}(22, 4^2)$ and $\text{N}(75, 7^2)$ respectively. Use this information to determine which of Kerry's two marks is better.

REVIEW SET 26C

1 The middle 68% of a normal distribution lies between 16.2 and 21.4.

a What is the mean and standard deviation of the distribution?

b Over what range of values would you expect the middle 95% of the data to spread?

2 A random variable X is normally distributed with mean 20.5 and standard deviation 4.3. Find:

a $\text{P}(X \geqslant 22)$ **b** $\text{P}(18 \leqslant X \leqslant 22)$ **c** k such that $\text{P}(X \leqslant k) = 0.3$.

3 X is a continuous random variable where $X \sim \text{N}(\mu, 2^2)$.
Find $\text{P}(-0.524 < X - \mu < 0.524)$.

4 The lengths of metal rods produced in a manufacturing process are normally distributed with mean μ cm and standard deviation 6 cm. 5.63% of the rods have length greater than 89.52 cm. Find the mean, median, and modal length of the metal rods.

5 The random variable T represents the lifetime in years of a component of a solar cell. Its probability density function is $F(t) = 0.4e^{-0.4t}, \ t \geqslant 0$.

a Find the probability that this component of the solar cell fails within 1 year.

b Each solar cell has 5 of these components which operate independently of each other. The cell will work provided at least 3 of the components continue to work. Find the probability that a solar cell will still operate after 1 year.

6 The curve shown is the probability density function for a normally distributed random variable X. Its mean is 50, and $\mathrm{P}(X < 90) \approx 0.975$. Find the shaded area.

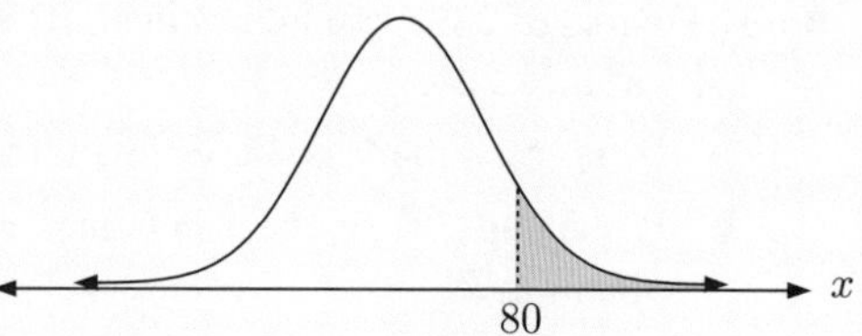

7 The weight of an apple in an apple harvest is normally distributed with mean 300 grams and standard deviation 50 grams. Only apples with weights between 250 and 350 grams are considered fit for sale.

a Find the proportion of apples fit for sale.

b In a sample of 100 apples, what is the probability that at least 75 are fit for sale?

8 It is claimed that the continuous random variable X has probability density function $f(x) = \dfrac{4}{1+x^2}, \quad 0 \leqslant x \leqslant 1$.

a Show that this is not possible.

b Use your working from **a** to find an exact value of k for which $F(x) = k\,f(x)$ would be a well-defined probability density function.

c Hence, find the exact values of the mean and variance of X.

Hint: $\dfrac{x^2}{1+x^2} = 1 - \dfrac{1}{1+x^2}$.

9 A factory has a machine designed to fill bottles of drink with volume 375 mL of liquid. It is found that the average amount of drink in each bottle is 376 mL, and that 2.3% of the drink bottles have a volume smaller than 375 mL. Assuming that the amount of drink in each bottle is normally distributed, find the standard deviation.

10 The height of an 18 year old boy is normally distributed with mean 187 cm. Fifteen percent of 18 year old boys have heights greater than 193 cm. Find the probability that two 18 year old boys, chosen at random, will have heights greater than 185 cm.

11 The continuous random variable X has probability density function defined by:

$$f(x) = \begin{cases} \dfrac{x}{5} & \text{for } 0 \leqslant x < 2 \\[2mm] \dfrac{8}{5x^2} & \text{for } 2 \leqslant x \leqslant k \\[2mm] 0 & \text{elsewhere.} \end{cases}$$

a Find the value of k.

b Find the exact value of the median of X.

c Find the mean and variance of X.

Chapter 27

Miscellaneous questions

Contents:

A NON-CALCULATOR QUESTIONS

EXERCISE 27A

1 a Evaluate $(1-i)^2$ and hence simplify $(1-i)^{4n}$.
  b Evaluate $(1-i)^{16}$.
  c Hence find *two* solutions of $z^{16} = 256$. Give clear reasons for your answers.

2 The sum of the first n terms of a series is given by $S_n = n^3 + 2n - 1$.
Write an expression for u_n, the nth term of the series.

3 Solve for x: $\dfrac{3x-1}{|x+1|} > 2$.

4 Let $z = \dfrac{-1+i\sqrt{3}}{4}$ and $w = \dfrac{\sqrt{2}+i\sqrt{2}}{4}$.
  a Write z and w in the form $r(\cos\theta + i\sin\theta)$ where $0 \leqslant \theta \leqslant \pi$.
  b Show that $zw = \frac{1}{4}\left(\cos\left(\frac{11\pi}{12}\right) + i\sin\left(\frac{11\pi}{12}\right)\right)$.
  c Express zw in the form $a+ib$, and hence find the exact values of $\cos\left(\frac{11\pi}{12}\right)$ and $\sin\left(\frac{11\pi}{12}\right)$.

5 Suppose $y = mx + c$ is a tangent to $y^2 = 4x$. Show that $c = \dfrac{1}{m}$, and that the coordinates of the point of contact are $\left(\dfrac{1}{m^2}, \dfrac{2}{m}\right)$.

6 Solve for x: $\sin^2 x + \sin x - 2 = 0$, $-2\pi \leqslant x \leqslant 2\pi$.

7 Suppose $f : x \mapsto \ln x$ and $g : x \mapsto 3 + x$. Find: a $f^{-1}(2) \times g^{-1}(2)$ b $(f \circ g)^{-1}(2)$.

8 Given an angle θ where $\sin\theta = -\frac{5}{13}$ and $-\frac{\pi}{2} < \theta < 0$, find the exact values of:
  a $\cos\theta$ b $\tan\theta$ c $\sin 2\theta$ d $\sec 2\theta$.

9 Solve $\sqrt{3}\cos x \csc x + 1 = 0$ for $0 \leqslant x \leqslant 2\pi$.

10 The number of snails in a garden plot can be modelled by a Poisson distribution with standard deviation d. The chance of finding exactly 8 snails is half that of finding exactly 7 snails. Find d.

11 The equation of line L is $r = 2\mathbf{i} - 3\mathbf{j} + \mathbf{k} + \lambda(-\mathbf{i} + \mathbf{j} - \mathbf{k})$, $\lambda \in \mathbb{R}$. Find the coordinates of the point on L that is nearest to the origin.

12 The function $f(x)$ has the following properties: $f'(x) > 0$ and $f''(x) < 0$ for all x, $f(2) = 1$, and $f'(2) = 2$.
  a Find the equation of the tangent to $f(x)$ at $x = 2$, and sketch it on a graph.
  b Sketch a possible graph of $y = f(x)$ on the same set of axes.
  c Explain why $f(x)$ has exactly one zero.
  d Estimate an interval in which the zero of $f(x)$ lies.

13 Prove by induction that $2n^3 - 3n^2 + n + 31 \geqslant 0$ for all $n \in \mathbb{Z}$, $n \geqslant -2$.

14 Use the method of integration by parts to find $\displaystyle\int x \arctan x \, dx$.

Check your answer using differentiation.

15 Solve for x:

a $\log_2(x^2 - 2x + 1) = 1 + \log_2(x - 1)$ **b** $3^{2x+1} = 5(3^x) + 2$

16 Prove by induction that $\sum_{r=1}^{n} r3^r = \frac{3}{4}[(2n-1)3^n + 1]$ for all $n \in \mathbb{Z}^+$.

17 Show that the equation of the tangent to the ellipse with equation $\frac{x^2}{a^2} + \frac{y^2}{b^2} = 1$ at the point $\mathrm{P}(x_1, y_1)$ is $\left(\frac{x_1}{a^2}\right)x + \left(\frac{y_1}{b^2}\right)y = 1$.

18

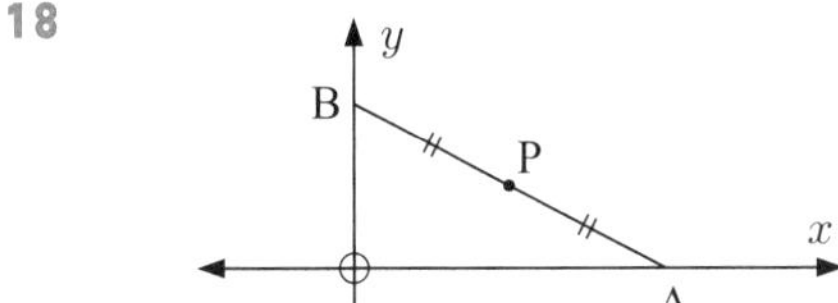

[AB] is a thin metal rod of fixed length, and P is its centre. A is free to move on the x-axis and B is free to move on the y-axis.

What path or *locus* is traced out by point P as the rod moves to all possible places?

19 **a** Find constants A, B, and C such that $\frac{x+5}{(x^2+5)(1-x)} = \frac{Ax+B}{x^2+5} + \frac{C}{x-1}$.

b Hence, find the exact value of $\int_2^4 \frac{x+5}{(x^2+5)(1-x)}\,dx$.

20 **a** If $z + \frac{1}{z}$ is real, prove that either $|z| = 1$ or z is real.

b If $|z+w| = |z-w|$ prove that $\arg z$ and $\arg w$ differ by $\frac{\pi}{2}$.

c If $z = r \operatorname{cis} \theta$, write z^4, $\frac{1}{z}$, and iz^* in polar form.

21 **a** Simplify: **i** $(A \cup B) \cap A'$ **ii** $(A \cap B) \cup (A' \cap B)$.

b Verify that $(A \cap B) \cup C = (A \cup C) \cap (B \cup C)$.

c Prove that if A and B are independent events, then so are:

i A' and B' **ii** A and B'.

22 Find: **a** $\int \frac{x}{\sqrt{1-x^2}}\,dx$ **b** $\int \frac{1+x}{1+x^2}\,dx$ **c** $\int \frac{1}{\sqrt{1-x^2}}\,dx$

23 A flagpole is erected at point A. The top of the flagpole is point B. At point C, due west of A, the angle of elevation to B is α. At point D, due south of A, the angle of elevation to B is β.

Point E is due south of C and due west of D. Show that the angle of elevation from E to B is $\operatorname{arccot}\left(\sqrt{\cot^2\alpha + \cot^2\beta}\right)$.

24 Evaluate:

a $\frac{1}{1+\sqrt{2}} + \frac{1}{\sqrt{2}+\sqrt{3}} + \frac{1}{\sqrt{3}+\sqrt{4}} + ... + \frac{1}{\sqrt{99}+\sqrt{100}}$

b $\frac{1}{1+\sqrt{2}} + \frac{1}{\sqrt{2}+\sqrt{3}} + \frac{1}{\sqrt{3}+\sqrt{4}} + ... + \frac{1}{\sqrt{n}+\sqrt{n+1}}$, $n \in \mathbb{Z}^+$.

25 The three numbers x, y, and z are such that $x > y > z > 0$.

Show that if $\frac{1}{x}$, $\frac{1}{y}$, and $\frac{1}{z}$ are in arithmetic progression, then $x - z$, y, and $x - y + z$ could be the lengths of the sides of a right angled triangle.

26 A rectangle is divided by m lines parallel to one pair of opposite sides and n lines parallel to the other pair. How many rectangles are there in the figure obtained?

27 Two different numbers are randomly chosen from the set $\{1, 2, 3, 4, 5, ..., n\}$, where n is a multiple of four. Determine the probability that one of the numbers is four times larger than the other.

28 Use the cosine rule and the given kite to show that $\sin^2\theta = \frac{1}{2} - \frac{1}{2}\cos 2\theta$ and $\cos^2\theta = \frac{1}{2} + \frac{1}{2}\cos 2\theta$.

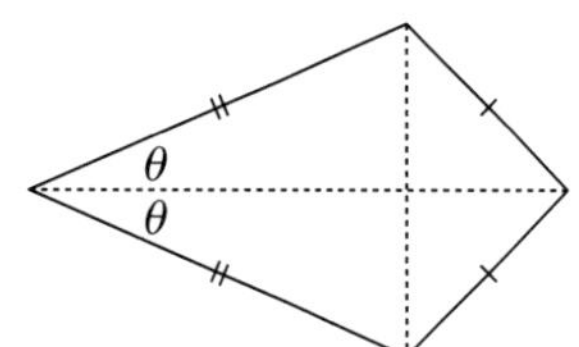

29 Show that $\tan\theta = 3\tan\alpha$.

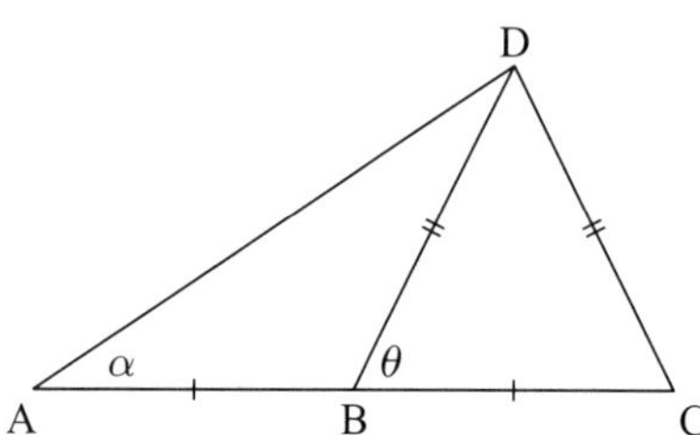

30 Solve for x:

a $8^{2x+3} = 4\sqrt[3]{2}$ **b** $3^{2x+1} + 8(3^x) = 3$ **c** $\ln(\ln x) = 1$

d $\log_{\frac{1}{9}} x = \log_9 5$

31 By considering $\frac{d}{dx}(\tan^3 x)$, find $\int \sec^4 x\, dx$.

32 What can be deduced if $A \cap B$ and $A \cup B$ are independent events?

33 Write $(3 - i\sqrt{2})^4$ in the form $x + y\sqrt{2}i$ where $x, y \in \mathbb{Z}$.

34 Solve the equation $\sin\theta\cos\theta = \frac{1}{4}$ for the interval $\theta \in [-\pi, \pi]$.

35 Find real numbers a and b such that $z^3 + az^2 + bz + 15 = 0$ has a root $2 + i$.

36 Solve for x:

a $(0.5)^{x+1} > 0.125$ **b** $\left(\frac{2}{3}\right)^x > \left(\frac{3}{2}\right)^{x-1}$ **c** $4^x + 2^{x+3} < 48$

37 Two people agree to meet each other at the corner of two city streets between 1 pm and 2 pm. However, neither will wait for the other for more than 30 minutes. If each person is equally likely to arrive at any time during the one hour period, determine the probability that they will in fact meet.

38

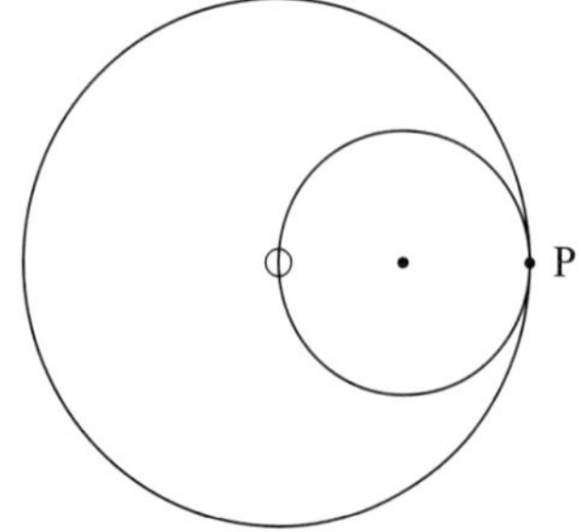

A circle is centred at the origin O. A second circle has half the diameter of the original circle and touches it internally. P is a fixed point on the smaller circle as shown, and lies on the x-axis.

The smaller circle now rolls around the inside of the larger one without slipping.

Show that for all positions of the smaller circle, P remains on the x-axis.

39 f is defined by $x \mapsto \ln(x(x-2))$.

a State the domain of f. **b** Find $f'(x)$.

c Find the equation of the tangent to $y = f(x)$ at the point where $x = 3$.

40 **a** The complex number z has argument θ. Show that iz has argument $\theta + \frac{\pi}{2}$.

b On an Argand plane, points P, Q, and R represent the complex numbers z_1, z_2, and z_3 respectively. If $i(z_3 - z_2) = z_1 - z_2$, what can be deduced about triangle PQR?

41 By considering the identity $(1+k)^n = (1+k)^2(1+k)^{n-2}$, deduce that $\binom{n}{r} = \binom{n-2}{r} + 2\binom{n-2}{r-1} + \binom{n-2}{r-2}$.

42 If $x^2 + y^2 = 52xy$, and $0 < y < x$, show that $\log\left(\frac{x-y}{5}\right) = \frac{1}{2}(\log x + \log 2y)$.

43 Find $\frac{dy}{dx}$ in terms of x and y, if $\sin(xy) + y^2 = x$.

44 Use the figure alongside to show that $\cos 36° = \frac{1+\sqrt{5}}{4}$.

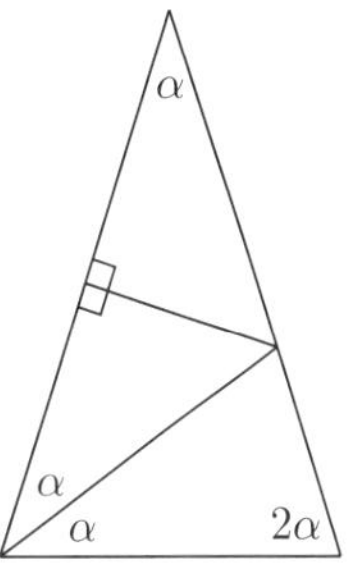

45 $x^n + ax^2 - 6$ leaves a remainder of -3 when divided by $(x-1)$, and a remainder of -15 when divided by $(x+3)$. Find the values of a and n.

46 Hat 1 contains three green and four blue tickets. Hat 2 contains four green and three blue tickets. One ticket is randomly selected from each hat.

a Find the probability that the tickets have the same colour.

b Given that the tickets have different colours, what is the probability that the green ticket came from Hat 2?

47 The point A$(-2, 3)$ lies on the graph of $y = f(x)$. Give the coordinates of the point that A moves to under the transformation:

a $y = f(x-2) + 1$ **b** $y = 2f(x-2)$ **c** $y = -|f(x)| - 2$

d $y = f(2x-3)$ **e** $y = \frac{1}{f(x)}$ **f** $y = f^{-1}(x)$

48 $z = re^{i\theta}$, $r > 0$, is a non-zero complex number such that $z + \frac{1}{z} = a + bi$, $a, b \in \mathbb{R}$.

a Find expressions for a and b in terms of r and θ.

b Hence, find all complex numbers z such that $z + \frac{1}{z}$ is real.

49 A sequence u_n is defined by $u_1 = u_2 = 1$ and $u_{n+2} = u_{n+1} + u_n$ for all $n \in \mathbb{Z}^+$. Prove by induction that $u_n \leqslant 2^n$ for all $n \in \mathbb{Z}^+$.

50 Use the principle of mathematical induction to prove that

$$\left(1 - \frac{1}{2^2}\right)\left(1 - \frac{1}{3^2}\right)\left(1 - \frac{1}{4^2}\right) .... \left(1 - \frac{1}{n^2}\right) = \frac{n+1}{2n} \quad \text{for } n \in \mathbb{Z}^+,\ n \geqslant 2.$$

51 **a** Find the turning points of $y = x^3 - 12x^2 + 45x$. Hence graph the function.

b If $x^3 - 12x^2 + 45x = k$ has three distinct real roots, what values can k have?

52 Triangle ABC has perimeter 20 cm.

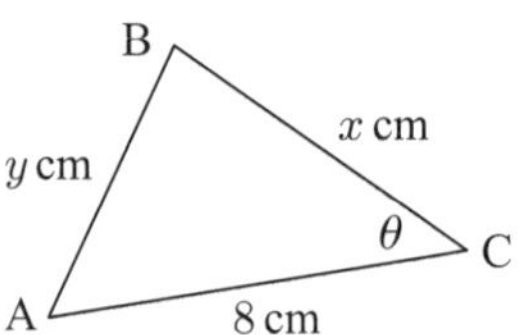

a Find y^2 in terms of x and θ and hence write $\cos\theta$ in terms of x only.

b If the triangle has area A, show that $A^2 = -20(x^2 - 12x + 20)$.

c Find the maximum area of the triangle. Comment on the triangle's shape when its area is a maximum.

53 $\frac{1}{k}$, k, k^2+1 are respectively the 3rd, 4th, and 6th terms of an arithmetic sequence.
Given that $k \in \mathbb{Q}$, find:

a k **b** the general term u_n.

54 Let $\mathbf{r} = 2\mathbf{i} - 2\mathbf{j} + \mathbf{k}$, $\mathbf{s} = 3\mathbf{i} + \mathbf{j} + 2\mathbf{k}$, and $\mathbf{t} = \mathbf{i} + 2\mathbf{j} - \mathbf{k}$ be the position vectors of the points R, S, and T respectively. Find the area of triangle RST.

55 Given that $x = \log_3 y^2$, express $\log_y 81$ in terms of x.

56 Find a trigonometric equation of the form $y = a\sin(b(x+c)) + d$ that represents the following graph with the information given below.
You may assume that (3, -5) is a minimum point and (6, -1) lies on the principal axis.

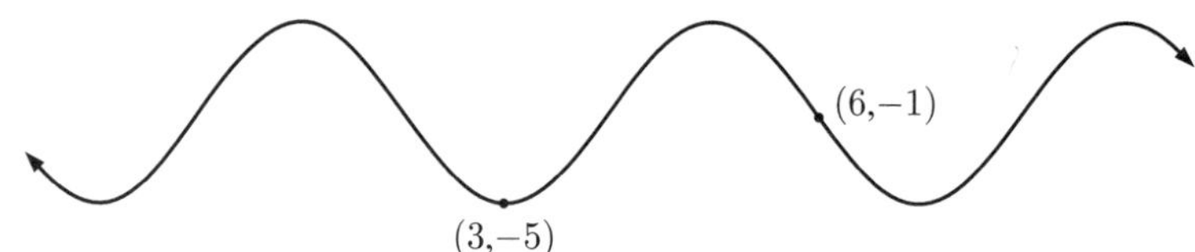

57 Two events A and B are independent with $P(A \mid B) = \frac{1}{4}$ and $P(B \mid A) = \frac{2}{5}$.
Find $P(A \cup B')$.

58 Find z in the form $a + bi$ if $z^2 = 1 + i + \dfrac{58}{9(3-7i)}$.

59 Solve simultaneously: $4^x = 8^y$ and $9^y = \dfrac{243}{3^x}$.

60 Determine the domain of $f(x) = \arccos(1 + x - x^2)$, and find $f'(x)$.

61 If $\sum\limits_{n=1}^{m} f(n) = m^3 + 3m$, find $f(n)$.

62 A normally distributed random variable X has a mean of 90. Given that $P(X < 85) \approx 0.16$:

a find $P(90 < X < 95)$

b estimate the standard deviation for the random variable X.

63 Find the equations of the asymptotes of $y = \dfrac{\tan x}{\sin(2x) + 1}$ where $-\pi \leqslant x \leqslant \pi$.

64 **a** Given $\mathbf{a} = \mathbf{i} + \mathbf{j} - 3\mathbf{k}$ and $\mathbf{b} = \mathbf{j} + 2\mathbf{k}$, find $\mathbf{a} \times \mathbf{b}$.

b Find a vector of length 5 units which is perpendicular to both $\mathbf{a}$ and $\mathbf{b}$.

65 If $z = \cos\theta + i\sin\theta$ where $0 < \theta < \frac{\pi}{4}$, find the modulus and argument of $1 - z^2$.

66 Find $\displaystyle\int x^2 \sin x \, dx$.

67 If $f(2x+3) = 5x - 7$, find $f^{-1}(x)$.

68 Solve for x: $\log_x 4 + \log_2 x = 3$

69 Prove that the roots of $(m-1)x^2 + x - m = 0$ are real and positive for all $0 < m < 1$.

70 **a** Show that $\sin 15° = \dfrac{\sqrt{6}-\sqrt{2}}{4}$.

b Hence, find the exact value of $\cos^2 165° + \cos^2 285°$.

71 **a** Find $\dfrac{dy}{dx}$ if $x^2 - 3xy + y^2 = 7$.

b Hence find the coordinates of all points on the curve where the gradient of the tangent is $\frac{2}{3}$.

72 A and B are two events such that $\text{P}(A) = \frac{1}{3}$ and $\text{P}(B) = \frac{2}{7}$.

a Find $\text{P}(A \cup B)$ if A and B are: **i** mutually exclusive **ii** independent.

b Find $\text{P}(A \mid B)$ if $\text{P}(A \cup B) = \frac{3}{7}$.

73 **a** Express $1+i$ and $\sqrt{3}-i$ in the form $re^{i\theta}$.

Hence write $z = \dfrac{-1-i}{\sqrt{3}-i}$ in the form $re^{i\theta}$.

b Find the smallest positive integer n such that z^n is a real number.

74 Consider the vectors $\mathbf{p} = \begin{pmatrix} 1 \\ 2 \\ -2 \end{pmatrix}$ and $\mathbf{q} = \begin{pmatrix} -t \\ 1+t \\ 2t \end{pmatrix}$. Find t such that:

a $\mathbf{p}$ and $\mathbf{q}$ are perpendicular **b** $\mathbf{p}$ and $\mathbf{q}$ are parallel.

75 Determine the sequence of transformations which transform the function $f(x) = 3x^2 - 12x + 5$ to $g(x) = -3x^2 + 18x - 10$.

76 Find the area of the region bounded by the curve $y = \tan^2 x + 2\sin^2 x$, the x-axis, and the line $x = \frac{\pi}{4}$.

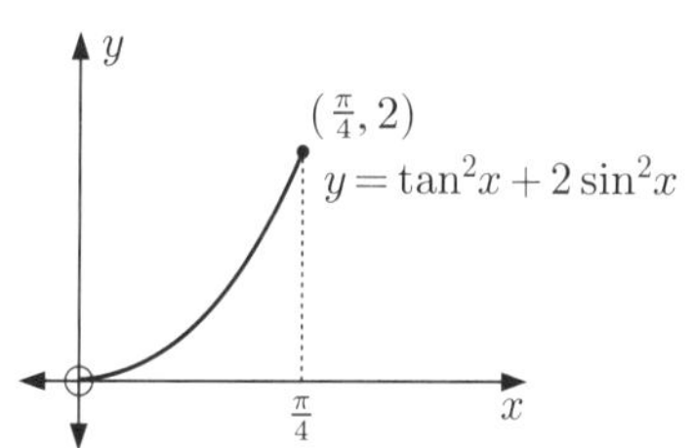

77 The graph of a quartic polynomial $y = f(x)$ cuts the x-axis at $x = -3$ and $-\frac{1}{4}$, and touches it at $x = \frac{3}{2}$. The y-intercept is 9. Find $f(x)$.

78 Find exact solutions for the following:

a $|1 - 4x| > \frac{1}{3}|2x - 1|$ **b** $\dfrac{x-2}{6-5x-x^2} \leqslant 0$

79 For $-\pi \leqslant x \leqslant \pi$, find the exact solutions of $3\sec 2x = \cot 2x + 3\tan 2x$.

80 Find $\dfrac{dy}{dx}$ if $e^{xy} + xy^2 - \sin y = 2$.

81 Suppose $z = x + 2i$ and $u = 3 + iy$ where $x, y \in \mathbb{R}$. Find the smallest positive value of x for which $\dfrac{z+u}{z-u}$ is purely imaginary.

82 Find the area of the region enclosed by the graph of $y = \dfrac{\tan x}{\cos(2x)+1}$, the x-axis, and the line $x = \frac{\pi}{3}$.

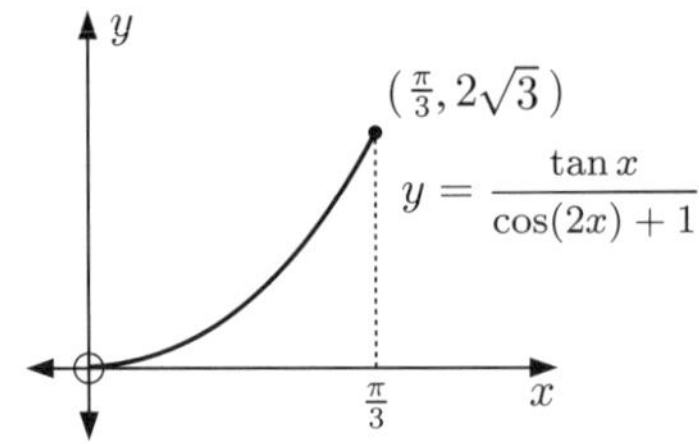

83 The scores a, b, 6, 13, and 7 where $b > a$, have a mean and variance of 8. Find the values of a and b.

84 The graph of $y = f(x)$ for $-9 \leqslant x \leqslant 9$ is shown alongside.
The function has vertical asymptotes $x = 2$ and $x = -3$, and the horizontal asymptote $y = 2$.
Copy and sketch the graph of $y = \dfrac{1}{f(x)}$, indicating clearly the axes intercepts and all asymptotes.

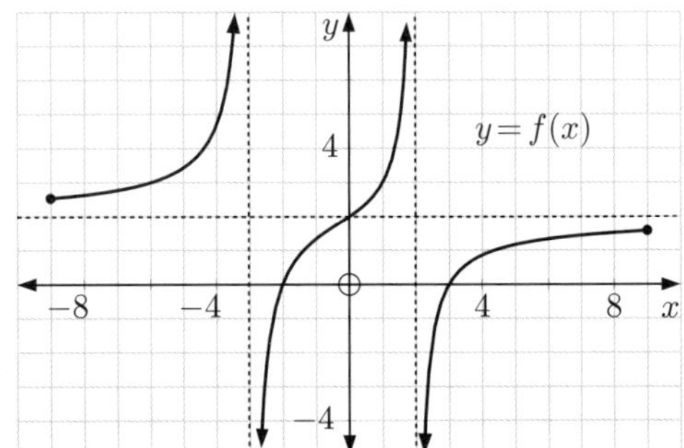

85 The height of a cone is always twice the radius of its base. The volume of the cone is increasing at a constant rate of 5 $\text{cm}^3\,\text{s}^{-1}$. Find the rate of change in the radius when the height is 20 cm.

86 The line passing through the points A(0, 5, 6) and B(4, 1, −2), and the line $\mathbf{r} = \begin{pmatrix} a \\ 3 \\ 2 \end{pmatrix} + s\begin{pmatrix} 2 \\ -1 \\ 1 \end{pmatrix}$, are coplanar. Find the value of a.

87 Find the coordinates of the stationary points on the curve $y = \dfrac{\sin x}{\tan x + 1}$ where $-\pi \leqslant x \leqslant \frac{\pi}{2}$.

88 A discrete random variable X has the probability distribution function $\text{P}(X = x) = a\left(\frac{2}{5}\right)^x$, $x = 0, 1, 2, 3, ...$ Find the value of a.

89 Solve exactly for x: $4\sin x = \sqrt{3}\csc x + 2 - 2\sqrt{3}$, $0 \leqslant x \leqslant 2\pi$.

90 Consider the following system of linear equations in which p and q are constants:

$$\begin{cases} x - 2y + 3z = 1 \\ x + py + 2z = 0 \\ -2x + p^2y - 4z = q \end{cases}$$

a Write this system of equations in augmented matrix form.

b Show, using clearly defined row operations, that this augmented matrix can be reduced to: $\left(\begin{array}{ccc|c} 1 & -2 & 3 & 1 \\ 0 & p+2 & -1 & -1 \\ 0 & 0 & p & p+q \end{array}\right)$.

c What values can p and q take when the system has:

i a unique solution **ii** no solutions **iii** infinite solutions?

Explain the geometric significance of each case.

d Specify the infinite solutions in parametric form.

91 The real polynomial $P(z)$ of degree 4 has one complex zero of the form $1-2i$, and another of the form ai, where $a \neq 0$, $a \in \mathbb{R}$.
Find $P(z)$ if $P(0) = 10$ and the coefficient of z^4 is 1. Leave your answer in factorised form.

92 Given that A is an acute angle and $\tan 2A = \frac{3}{2}$, find the exact value of $\tan A$.

93 Let $\mathbf{a} = 3\mathbf{i} + 2\mathbf{j} - \mathbf{k}$, $\mathbf{b} = \mathbf{i} + \mathbf{j} - \mathbf{k}$, and $\mathbf{c} = 2\mathbf{i} - \mathbf{j} + \mathbf{k}$.

- **a** Show that $\mathbf{b} \times \mathbf{c} = -3\mathbf{j} - 3\mathbf{k}$.
- **b** Verify for the given vectors that $\mathbf{a} \times (\mathbf{b} \times \mathbf{c}) = \mathbf{b}\,(\mathbf{a} \bullet \mathbf{c}) - \mathbf{c}\,(\mathbf{a} \bullet \mathbf{b})$.

94
- **a** Simplify $(-1 + i\sqrt{2})^3$.
- **b** Write $5 + i\sqrt{2}$ in the form $a^3 \operatorname{cis} \theta$, stating the exact values of a and θ.
- **c** Find the exact solutions of $z^3 = 5 + i\sqrt{2}$.
- **d** Hence, show that $\arctan\left(\frac{\sqrt{2}}{5}\right) + 2\pi = 3\arccos\left(\frac{-1}{\sqrt{3}}\right)$.

95 Let P be a point on the circle $x^2 + y^2 = a^2$, where $a \in \mathbb{R}$, $a > 0$, such that P lies in the first quadrant.

- **a** Write down the coordinates of P in terms of $x > 0$.
- **b** Let $A(-a, 0)$ and $B(a, 0)$ be the points where the circle meets the x-axis.
  - **i** Find vectors $\overrightarrow{AP}$, $\overrightarrow{AB}$, and $\overrightarrow{OP}$.
  - **ii** Find $\cos(P\widehat{A}B)$ and $\cos(P\widehat{O}B)$ in terms of x and a.
- **c** Verify that $\cos(P\widehat{O}B) = 2\cos^2(P\widehat{A}B) - 1$. Which property of circles have you verified?

96 Prove by induction that $x^n - y^n$ has a factor of $x - y$ for all $n \in \mathbb{Z}^+$.

97 Let θ be the angle between unit vectors $\mathbf{u}$ and $\mathbf{v}$, where $0 \leqslant \theta \leqslant \pi$.

- **a** Express $|\mathbf{u} - \mathbf{v}|$ and $|\mathbf{u} + \mathbf{v}|$ in terms of θ.
- **b** Hence determine the value of $\cos\theta$ such that $|\mathbf{u} + \mathbf{v}| = 5\,|\mathbf{u} - \mathbf{v}|$.

98 The diagram shows a sector POR of a circle with radius 1 unit and centre O. The angle $P\widehat{O}R = \theta$, and the line segments [PQ], [P_1Q_1], [P_2Q_2], [P_3Q_3], ... are all perpendicular to [OR].

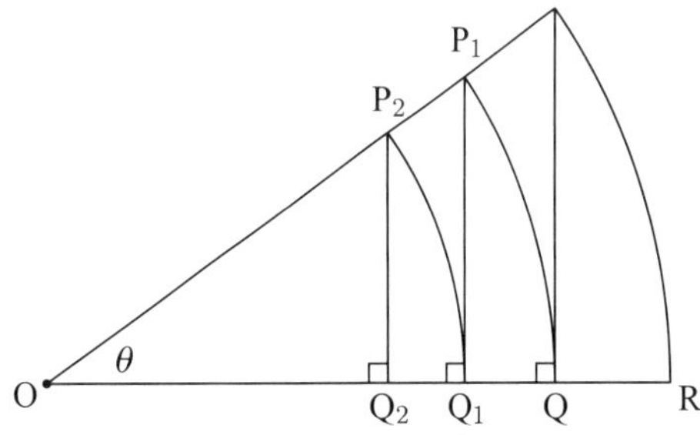

Calculate, in terms of θ, the sum to infinity of the lengths $PQ + P_1Q_1 + P_2Q_2 + P_3Q_3 + ...$

99 $x^2 + ax + bc = 0$ and $x^2 + bx + ca = 0$ where $a \neq 0$, $b \neq 0$, $c \neq 0$, have a single common root. Prove that the other roots satisfy $x^2 + cx + ab = 0$.

100 The illustrated ellipse has equation $\dfrac{x^2}{a^2}+\dfrac{y^2}{b^2}=1$.

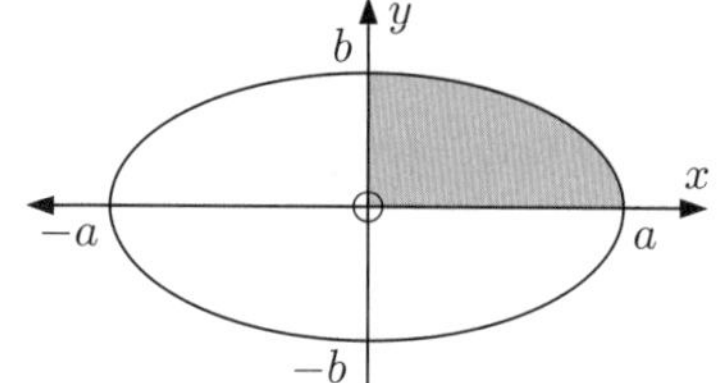

a Show that the shaded region has area given by $\dfrac{b}{a}\displaystyle\int_0^a \sqrt{a^2-x^2}\,dx$.

b Find the area of the ellipse in terms of a and b.

c An ellipsoid is obtained by rotating the ellipse through $360°$ about the x-axis.
Prove that the volume of the ellipsoid is given by $V=\frac{4}{3}\pi ab^2$.

101 Consider the following quadratic function where $a_i,\ b_i \in \mathbb{R}$:

$$f(x)=(a_1x-b_1)^2+(a_2x-b_2)^2+(a_3x-b_3)^2+\ldots.+(a_nx-b_n)^2.$$

Use quadratic theory to prove the 'Cauchy-Schwartz inequality':

$$\left(\sum_{i=1}^{n} {a_i}^2\right)\left(\sum_{i=1}^{n} {b_i}^2\right) \geqslant \left(\sum_{i=1}^{n} a_ib_i\right)^2.$$

102 **a** Assuming Pascal's rule $\binom{n}{r}+\binom{n}{r+1}=\binom{n+1}{r+1}$,
prove that $(1+x)^n=1+\binom{n}{1}x+\binom{n}{2}x^2+\ldots.+\binom{n}{n}x^n$ for all $n\in\mathbb{Z}^+$.

b Hence establish the binomial expansion for $(a+b)^n$, by letting $x=\dfrac{b}{a}$.

103 Solve simultaneously: $x=16y$ and $\log_y x-\log_x y=\frac{8}{3}$.

104 **a** Find the cube roots of $-2-2i$.

b Display the cube roots of $-2-2i$ on an Argand diagram.

c If the cube roots are α_1, α_2, and α_3, show that $\alpha_1+\alpha_2+\alpha_3=0$.

d Using an algebraic argument, prove that if β is any complex number then the sum of the zeros of $z^n=\beta$ is 0.

105 Find all values of m for which the quartic equation $x^4-(3m+2)x^2+m^2=0$ has 4 real roots in arithmetic progression.

106 Find $\arctan\left(\frac{1}{7}\right)+2\arctan\left(\frac{1}{3}\right)$.

107 Let P be the plane which intersects the coordinate axes at $x=a$, $y=b$, and $z=c$, where a, b, c are non-zero. Use vector techniques to derive the equation of P as $\dfrac{x}{a}+\dfrac{y}{b}+\dfrac{z}{c}=1$.

108 $(1+x)^n=\binom{n}{0}+\binom{n}{1}x+\binom{n}{2}x^2+\binom{n}{3}x^3+\ldots.+\binom{n}{n}x^n$ for all $n\in\mathbb{Z}^+$. Prove that:

a $\binom{n}{1}+2\binom{n}{2}+3\binom{n}{3}+\ldots.+n\binom{n}{n}=n\,2^{n-1}$

b $\binom{n}{0}+2\binom{n}{1}+3\binom{n}{2}+\ldots.+(n+1)\binom{n}{n}=(n+2)2^{n-1}$

c $\binom{n}{0}+\frac{1}{2}\binom{n}{1}+\frac{1}{3}\binom{n}{2}+\ldots.+\dfrac{1}{n+1}\binom{n}{n}=\dfrac{2^{n+1}-1}{n+1}$.

109 Over 2000 years ago, **Heron** or **Hero** discovered a formula for finding the area of a triangle with sides a, b, and c. It is $A=\sqrt{s(s-a)(s-b)(s-c)}$ where $2s=a+b+c$.
Prove Heron's formula.

110 The random variable X has a Poisson distribution such that $P(X = 2) - P(X = 1) = 3\,P(X = 0)$. Find, in surd form, the standard deviation σ of the distribution.

111 Consider the series $\dfrac{1}{1 \times 3} + \dfrac{1}{2 \times 4} + \dfrac{1}{3 \times 5} + ... + \dfrac{1}{n(n+2)}$.

- **a** Find constants A and B such that $\dfrac{1}{n(n+2)} = \dfrac{A}{n} + \dfrac{B}{n+2}$.
- **b** Hence show that the sum of the series is $\frac{3}{4} - \dfrac{1}{2n+2} - \dfrac{1}{2n+4}$.
- **c** Find $\sum\limits_{r=1}^{\infty} \dfrac{1}{r(r+2)}$.
- **d** Prove your answer to **b** using mathematical induction.

112
- **a** Given that $y = \ln(\tan x)$, $x \in \,]0, \frac{\pi}{2}[$, show that $\dfrac{dy}{dx} = k\csc(2x)$ for some constant k.
- **b** The graph of $y = \csc(2x)$ is illustrated on the interval $]0, \frac{\pi}{2}[$.
  Find the area of the shaded region. Give your answer in the form $a \ln b$ where $a \in \mathbb{Q}$ and $b \in \mathbb{Z}^+$.

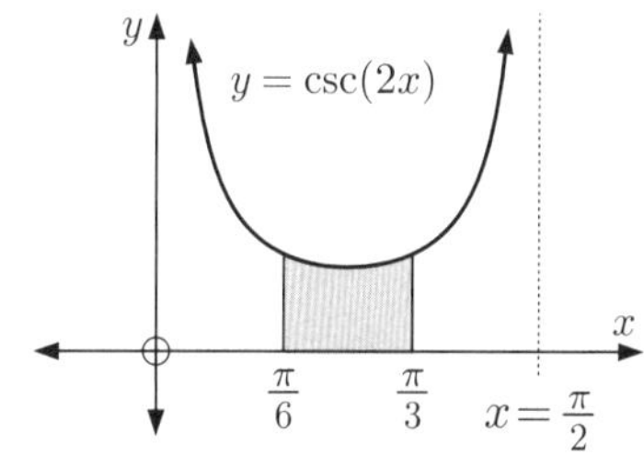

113 ABC is an equilateral triangle with sides of length $2k$. P is any point within the triangle. [PX], [PY], and [PZ] are altitudes from P to the sides [AB], [BC], and [CA] respectively.

- **a** By letting $P\widehat{C}Z$ be θ, show that PX + PY + PZ is constant for all positions of P.
- **b** Check that your solution to **a** is correct when P is at A.
- **c** Prove that the result from **a** is true using areas of triangles only.

114 Find the sum of the series:

- **a** $1 + a\cos\theta + a^2\cos 2\theta + a^3\cos 3\theta + ... + a^n\cos n\theta$
- **b** $a\sin\theta + a^2\sin 2\theta + a^3\sin 3\theta + ... + a^n\sin n\theta$ for $n \in \mathbb{Z}^+$.

115 Find $\displaystyle\int \frac{x}{1+\sqrt{x+2}}\,dx$.

116 While driving, Bernard passes through n intersections which are independently controlled by traffic lights. Each set of lights has probability p of stopping him.

- **a** Find the probability that Bernard will be stopped at least once.
- **b** Suppose A_k is the event that Bernard is stopped at exactly k intersections, and B_k is the event that Bernard is stopped at at least k intersections.
  Write down the conditional probability $P(A_k \mid B_k)$.
- **c** Find p such that:
  - **i** A_1 and B_1 are independent
  - **ii** $P(A_2 \mid B_2) = P(A_1)$ and $n = 2$.

117
- **a** Use complex number methods to show that $\tan 4\theta = \dfrac{4\tan\theta - 4\tan^3\theta}{1 - 6\tan^2\theta + \tan^4\theta}$.
- **b** Hence, find the roots of the equation $x^4 + 4x^3 - 6x^2 - 4x + 1 = 0$.

118 Consider the system of linear equations: $\begin{cases} 2x - y + 3z = 4 \\ 2x + y + (a+3)z = 10 - a \\ 4x + 6y + (a^2+6)z = a^2 \end{cases}$

a Write the system in an augmented matrix, and use row operations to reduce the system to echelon form.

b Find the value(s) of a for which the system has:

i no solutions **ii** infinitely many solutions, and find the form of these solutions

iii a unique solution, and find the solution in the case where $a = 2$.

In each case explain the geometric significance of your answers.

119 Let $f(x) = xe^{1-2x^2}$.

a Find $f'(x)$ and $f''(x)$.

b Find the coordinates of the stationary points of $y = f(x)$, and determine their nature.

c Find the x-coordinates of the points of inflection of the function.

d Discuss the behaviour of the function as $x \to \pm\infty$.

e Sketch the graph of the function.

f The region bounded by $y = f(x)$, the x-axis, and the line $x = k$, $k > 0$, has area $\frac{1}{4}(e-1)$ units^2. Find the value of k.

120 **a** Assuming the identities

$\cos S + \cos D = 2\cos\left(\frac{S+D}{2}\right)\cos\left(\frac{S-D}{2}\right)$ and $\sin S + \sin D = 2\sin\left(\frac{S+D}{2}\right)\cos\left(\frac{S-D}{2}\right)$,

prove that $\text{cis}\,\theta + \text{cis}\,\phi = 2\cos\left(\frac{\theta-\phi}{2}\right)\text{cis}\left(\frac{\theta+\phi}{2}\right)$.

b Hence find the modulus and argument of $\text{cis}\,\theta + \text{cis}\,\phi$.

c Show that your answers in **b** are correct using a geometrical argument.

d Prove that the solutions of $\left(\frac{z+1}{z-1}\right)^5 = 1$ are $z = -i\cot\left(\frac{k\pi}{5}\right)$, $k = 1, 2, 3, 4$.

121 Use the substitution $u = \ln x$ to find $\int \sin(\ln x)\,dx$ for $x > 0$.

122 A real polynomial $P(x) = x^4 + ax^3 + bx^2 + cx - 10$ has two integer zeros p and q, and a complex zero $1 + ki$, $k \in \mathbb{Z}$.

a Use the zero $1 + ki$ to write an expression for a real quadratic factor of $P(x)$.

b State all possible values of k.

c Use p and q to write another expression for a real quadratic factor of $P(x)$. Hence list all possible values of pq.

d Given that $p + q = -1$, show that there is only one possible value for pq. Hence find all zeros of $P(x)$.

123 Alongside is a sketch of the gradient function of $y = f(x)$.

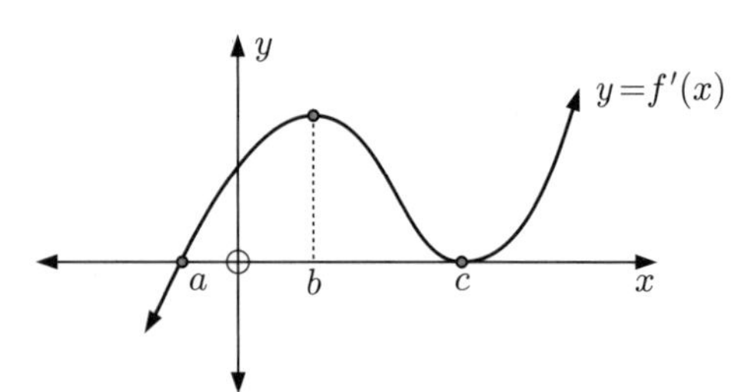

a Sketch a possible solution curve for $y = f(x)$, showing as many features as possible.

b Let $y = f_1(x)$ be the solution curve you have drawn. Write down the form of all possible solution functions $f(x)$ in terms of $f_1(x)$.

124 An infinite number of circles are drawn in a sector of a circle of radius 10 cm and angle $\alpha = \frac{\pi}{3}$ as shown.

a Find the total area of this infinite series of circles.

b Write an expression for the total area of all circles for a general angle α such that $0 \leqslant \alpha \leqslant \frac{\pi}{2}$.

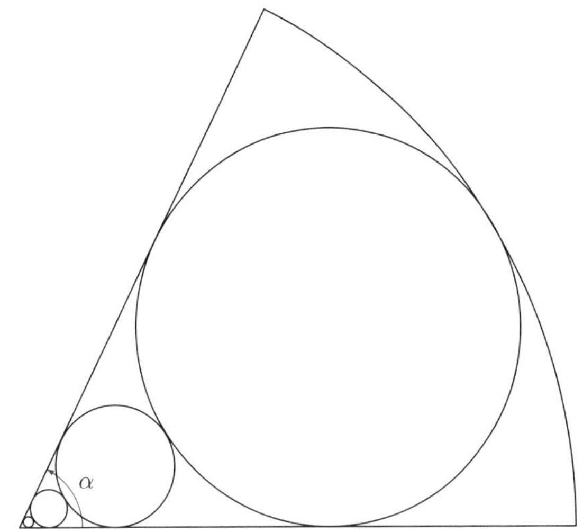

125 The tangent to the curve $y = f(x)$ at the point $A(x, y)$ meets the x-axis at the point $B(x - \frac{1}{2}, 0)$. The curve meets the y-axis at the point $C(0, \frac{1}{e})$.

a Show that at any point on the curve, $\dfrac{dx}{dy} = \dfrac{1}{2y}$.

b Hence find the equation of the curve.

126 **a** Prove that in any triangle with angles A, B, and C:

i $\sin 2A + \sin 2B + \sin 2C = 4 \sin A \sin B \sin C$

ii $\tan A + \tan B + \tan C = \tan A \tan B \tan C$.

b Consider *any* three angles A, B, and C for which $\tan A + \tan B + \tan C = \tan A \tan B \tan C$. What can be deduced about $A + B + C$?

127 **a** Show that $\sqrt{14 - 4\sqrt{6}}$ cannot be written in the form $a + b\sqrt{6}$ where $a, b \in \mathbb{Z}$.

b Write $\sqrt{14 - 4\sqrt{6}}$ in the form $a\sqrt{m} + b\sqrt{n}$ where $a, b, m, n \in \mathbb{Z}$.

128 **a** The acute angled triangle ABC has vertices on a circle with radius r.

Show that the area of the triangle is given by $\dfrac{abc}{4r}$.

b In triangle ABC it is known that $\sin A = \cos B + \cos C$.
Show that the triangle is right angled.

129 Let L_1 and L_2 be parallel lines. Let P be a fixed point on L_1 and let Q be any point on L_2. Let $\mathbf{v}$ be the direction vector of L_1.

Show that the distance between the lines L_1 and L_2 is given by $\dfrac{|\overrightarrow{QP} \times \mathbf{v}|}{|\mathbf{v}|}$.

130 **a** Find constants P and Q such that $\dfrac{1}{a^2 - x^2} = \dfrac{P}{a - x} + \dfrac{Q}{a + x}$.

b Hence show that $\displaystyle\int \frac{1}{a^2 - x^2}\, dx = \frac{1}{2a} \ln \left| \frac{a + x}{a - x} \right| + c$.

c Check **b** using differentiation.

131 The non-zero vectors $\mathbf{a}$ and $\mathbf{b}$ are not parallel. Find constants x and y if vectors $(2 - x)\,\mathbf{a} + y\mathbf{b}$ and $y\mathbf{a} + (x - 3)\,\mathbf{b}$ are equal.

132 R and Q are two fixed points on either side of line segment [AB]. P is free to move on the line segment so that the angles θ and ϕ vary. a and b are the distances of Q and R respectively from [AB].

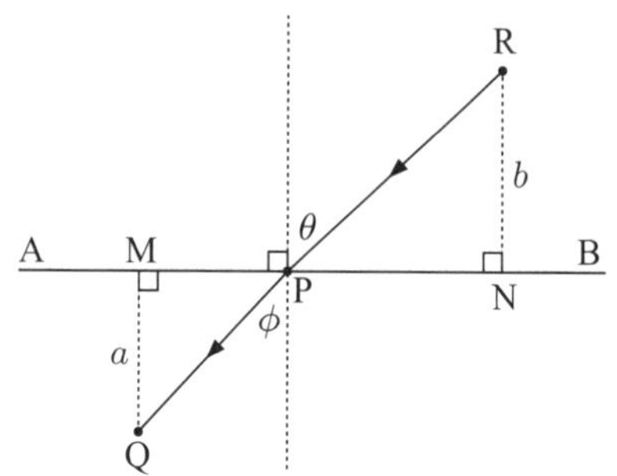

a Show that for all positions of P, $\dfrac{d\phi}{d\theta} = \dfrac{-a\cos^2\phi}{b\cos^2\theta}$.

b A particle moves from R to P with constant speed v_1, and from P to Q with constant speed v_2.
Deduce that the time taken to go from R to P to Q is a minimum when $\dfrac{\sin\theta}{\sin\phi} = \dfrac{v_1}{v_2}$.

133 A game is played in which the wheel shown is spun by first the player and then the game operator.
The player wins $\$a$ if their spin is higher than the operator's.
It costs $\$k$ to play the game.
If the game is fair, find an expression linking a and k.

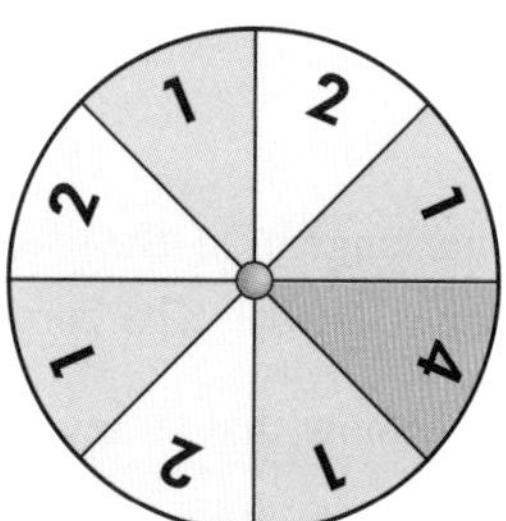

B CALCULATOR QUESTIONS

EXERCISE 27B

1 x and y satisfy the equations $x^2 + 3xy + 9 = 0$ and $y^2 + x - 1 = 0$.
Solve these equations simultaneously for x, given that x is real.

2 Each summer, 10% of the trees on a certain plantation die out, and each winter, workmen plant 100 new trees. At the end of the winter in 1980 there were 1200 trees in the plantation.

a How many living trees were there at the end of winter in 1970?

b What will happen to the number of trees in the plantation during the 21st century assuming the conditions remain unchanged?

3 Two marksmen, A and B, fire simultaneously at a target. A is twice as likely to hit the target as B, and the probability that the target is hit is $\frac{1}{2}$. Find the probability of A hitting the target.

4 Prove by induction that for all $n \in \mathbb{Z}^+$,

$$\frac{1}{a(a+1)} + \frac{1}{(a+1)(a+2)} + \frac{1}{(a+2)(a+3)} + ... + \frac{1}{(a+n-1)(a+n)} = \frac{n}{a(a+n)}.$$

5 z and w are two complex numbers such that $2z + w = i$ and $z - 3w = 7 - 10i$.
Find $z + w$ in the form $a + bi$, where $a, b \in \mathbb{Z}$.

6 $x^2 + b_1x + c_1 = 0$ and $x^2 + b_2x + c_2 = 0$ are two quadratic equations where $b_1b_2 = 2(c_1 + c_2)$.
Prove that at least one of the equations has real roots.

7 Find α:

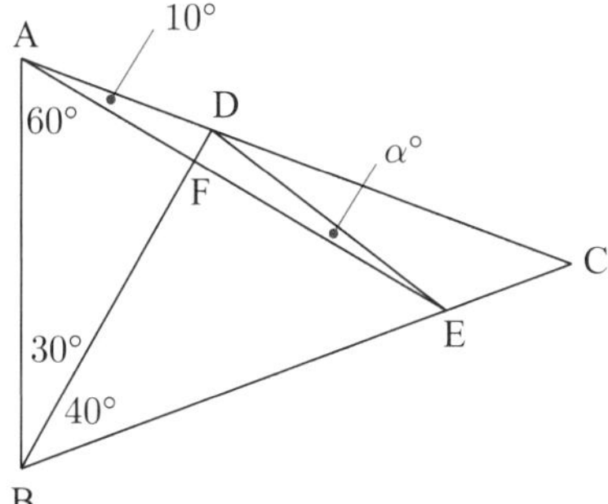

8 The line L with direction vector $\mathbf{v}$, passes through points A and Q. Q is the point on L which is closest to point P.

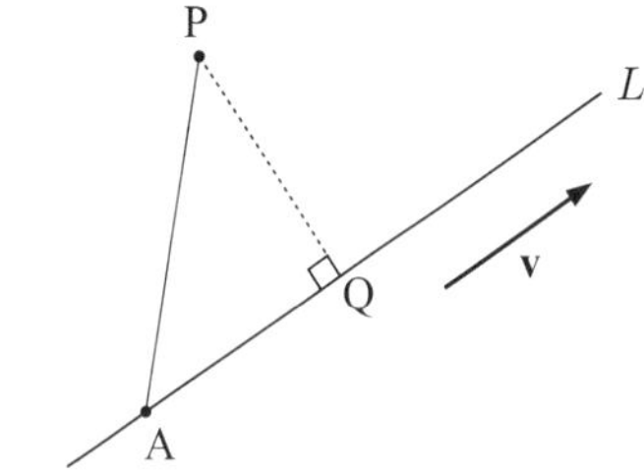

a Prove that $\text{PQ} = \dfrac{|\overrightarrow{\text{AP}} \times \mathbf{v}|}{|\mathbf{v}|}$.

b Hence, find the shortest distance from $(2, -1, 3)$ to the line $\begin{pmatrix} x \\ y \\ z \end{pmatrix} = \begin{pmatrix} -1 \\ 1 \\ 2 \end{pmatrix} + \lambda \begin{pmatrix} 3 \\ -1 \\ 1 \end{pmatrix}$.

9 A mountain is perfectly conical in shape. The base is a circle of radius 2 km, and the steepest slopes leading up to the top T are 3 km long.

From the southernmost point A on the base, a path leads up around the side of the mountain to B, the point on the northern slope which is 1.5 km up the slope from the base C. A and C are diametrically opposite.

If the path leading from A to B is the shortest possible distance from A to B around the mountainside, find the length of this path.

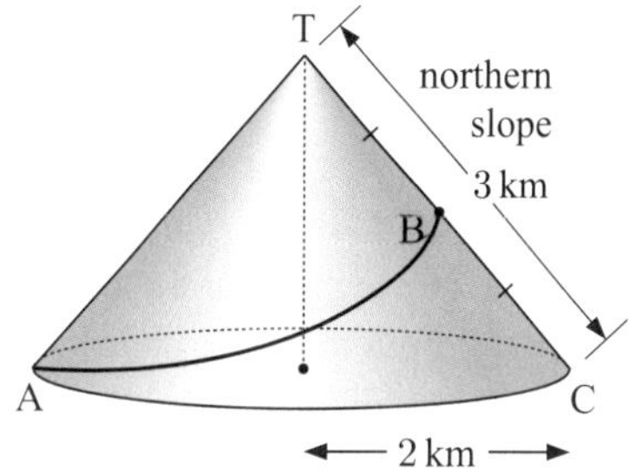

10 Show that the equation $\log_3(x-k) + \log_3(x+2) = 1$ has a real solution for every real value of k.

11 a Find the general term u_n of the sequence: $\dfrac{1}{\sin\theta} - \sin\theta,\ \cos\theta,\ \sin\theta,\ \dfrac{1}{\cos\theta} - \cos\theta,\ ....$

b Find an equation connecting consecutive terms of the sequence:
$1, \cos\theta, \cos^3\theta, \cos^7\theta, \cos^{15}\theta, ...$

12 A hundred seeds are planted in ten rows of ten seeds per row. Assuming that each seed independently germinates with probability $\frac{1}{2}$, find the probability that the row with the maximum number of germinations contains at least 8 seedlings.

13 For a continuous function defined on the interval $[a, b]$, the length of the curve can be found using $L = \displaystyle\int_a^b \sqrt{1 + [f'(x)]^2}\, dx$. Find the length of:

a $y = x^2$ on the interval $[0, 1]$

b $y = \sin x$ on the interval $[0, \pi]$.

14 When $P(x)$ is divided by $(x-a)^2$, prove that the remainder is $P'(a)(x-a) + P(a)$ where $P'(x)$ is the derivative of $P(x)$.

15 **a** Write $-8i$ in polar form.

b Hence find the three cube roots of $-8i$, calling them z_1, z_2, and z_3.

c Illustrate the roots from **b** on an Argand diagram.

d Show that $z_1^2 = z_2 z_3$ where z_1 is any one of the three cube roots.

e Find the product of the three cube roots.

16 A lampshade is a truncated cone open at the bottom and top.
Find the pattern needed to make this lampshade from a flat sheet of material.

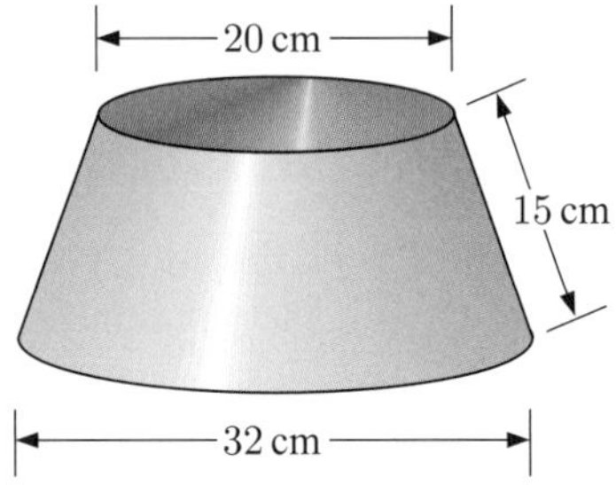

17 A normally distributed random variable X has a mean of 90. Given that $\text{P}(X < 88) \approx 0.289\,25$, find:

a the standard deviation of X, to 5 decimal places

b the probability that a randomly chosen score is *either* greater than 91 *or* less than 89.

18 If $f : x \mapsto 2x + 1$ and $g : x \mapsto \dfrac{x+1}{x-2}$, find: **a** $(f \circ g)(x)$ **b** $g^{-1}(x)$.

19 [AB] represents a painting on a wall.
AB = 2 m and BC = 1 m.
The angle of view observed by a girl between the top and bottom of the painting is 30°.
How far is the girl from the wall?

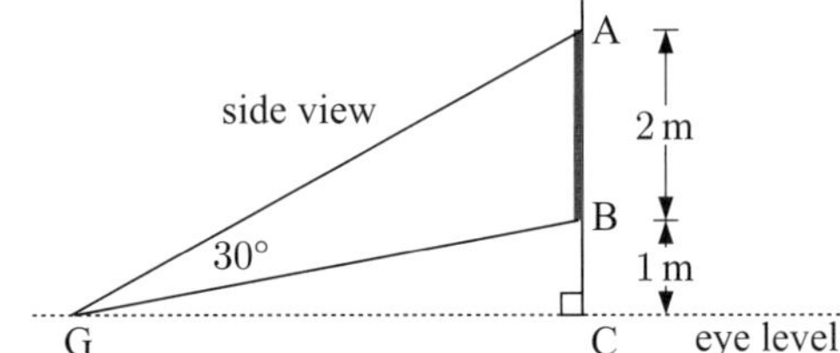

20 Find x in terms of a: $\log_a(x+2) = \log_a x + 2$ and $a > 1$.

21 Use vector methods to prove that joining the midpoints of the sides of a rhombus produces a rectangle.

22 The diagram shows a simple electrical network.

Each symbol —✓•— represents a switch.

All four switches operate independently, and the probability of each one of them being closed is p.

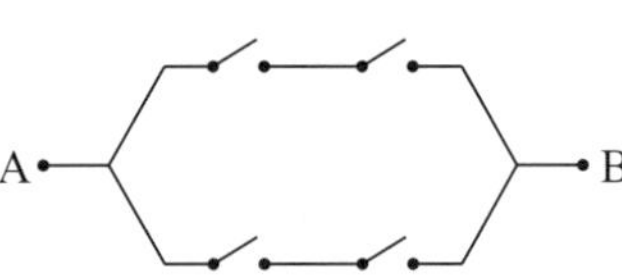

a In terms of p, find the probability that current flows from A to B.

b Find the least value of p for which the probability of current flow is more than 0.5.

23 Let $h(x) = x^3 - 6tx^2 + 11t^2x - 6t^3$ where $t \in \mathbb{R}$.

a Show that t is a zero of $h(x)$.

b Factorise $h(x)$ as a product of linear factors.

c Hence or otherwise, find the coordinates of the points where the graphs of $y = x^3 + 6x^2$ and $y = -6 - 11x$ meet.

24 A Year 12 class completed a calculus test with the following results:

$\sum_{i=1}^{25} x_i = 1650$ and $\sum_{i=1}^{25} x_i^2 = 115\,492$, where x_i denotes the percentage result of the ith student in the class. For this data, find the:

a mean **b** variance.

25 ABC is an equilateral triangle with sides 10 cm long. P is the point within the triangle which is 5 cm from A and 6 cm from B. How far is P from C?

26 Suppose $w = \dfrac{z-1}{z^* + 1}$ where $z = a + bi$ and z^* is the complex conjugate of z.

a Write w in the form $x + yi$.

b Hence determine the conditions under which w is purely imaginary.

27 Factorise $f(x) = 2x^3 - x^2 - 8x - 5$, and hence find the values of x for which $f(x) \geqslant 0$.

28 In an International school there are 78 students preparing for the IB Diploma. 38 of these students are male, and 17 of these males are studying Mathematics at the higher level. Of the female students, 25 are not studying Mathematics at the higher level.

A student is selected at random and found to be studying Mathematics at the higher level. Find the probability that this student is male.

29 Suppose that for all $n \in \mathbb{Z}^+$, $(2 - \sqrt{3})^n = a_n - b_n\sqrt{3}$ where a_n and b_n are integers.

a Show that $a_{n+1} = 2a_n + 3b_n$ and $b_{n+1} = a_n + 2b_n$.

b Calculate $a_n^2 - 3b_n^2$ for $n = 1$, 2, and 3.

c What do you propose from **b**?

d Prove your proposition from **c**.

30

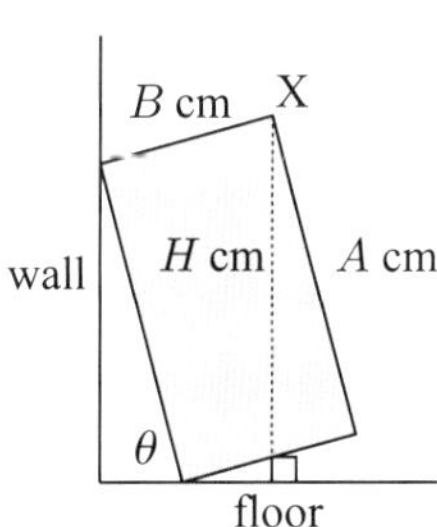

An A cm by B cm rectangular refrigerator leans at an angle of θ to the floor against a wall.

a Find H in terms of A, B, and θ.

b Use the figure to prove that

$A\sin\theta + B\cos\theta \leqslant \sqrt{A^2 + B^2}$, with equality when $\tan\theta = \dfrac{A}{B}$.

31 Find the coefficient of:

a x^{12} in the expansion of $\left(2x^3 - \dfrac{1}{2x}\right)^8$ **b** x^2 in the expansion of $(1 + 2x)^5(2 - x)^6$

c x^3 in the expansion of $(1 + 2x - 3x^2)^4$.

32 A company manufactures computer chips, and it is known that 3% of them are faulty. In a batch of 500 such chips, find the probability that between 1 and 2 percent (inclusive) of the chips are faulty.

33 The real polynomial $p(x) = x^3 + (5 + 4a)x + 5a$ has a zero $-2 + i$.

a Find a real quadratic factor of $p(x)$.

b Hence, find the value of a and the real zero of $p(x)$.

34 Solve the equation: $\binom{n}{3} = 3\binom{n-1}{2} - \binom{n-1}{1}$.

35 In a game, a player rolls a biased tetrahedral (four-faced) die. The probability of obtaining each possible score is shown in the table.

Score	1	2	3	4
Probability	$\frac{1}{12}$	k	$\frac{1}{4}$	$\frac{1}{3}$

a Find the value of k.

b Let the random variable X denote the number of 2s that occur when the die is rolled 2400 times. Calculate the exact mean and standard deviation of X.

36 The first 3 terms of a geometric sequence have a sum of 39. If the middle term is increased by $66\frac{2}{3}\%$, the first three terms now form an arithmetic sequence. Find the smallest possible value of the first term.

37 The points A$(-1, 0)$, B$(1, 0)$, and C$(0, -0.5)$ lie at the axes intercepts of $y = f(x)$.
On the same set of axes, sketch the following graphs. In each case, explain what has happened to the points A, B, and C.

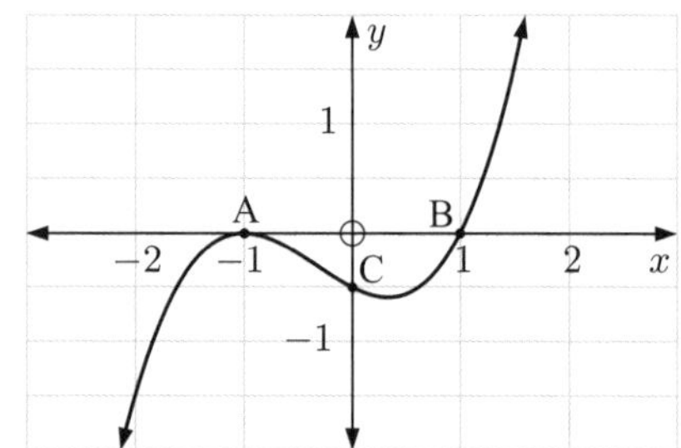

a $y = f(x+1) - 1$ **b** $y = -2f(x-1)$

c $y = |\, f(x) \,|$ **d** $y = \dfrac{1}{f(x)}$

38 A factory manufactures rope, and the rope has an average of 0.7 flaws per 100 metres. It is known that the number of flaws produced in the rope follows a Poisson distribution.

a Determine the probability that there will be exactly 2 flaws in 200 metres of rope.

b Find the probability that there will be at least 2 flaws in 400 metres of rope.

39 In the given figure, ABCD is a parallelogram. X is the midpoint of [BC], and Y lies on [AX] such that AY : YX = 2 : 1. The coordinates of A, B, and C are $(1, 3, -4)$, $(4, 4, -2)$, and $(10, 2, 0)$ respectively.

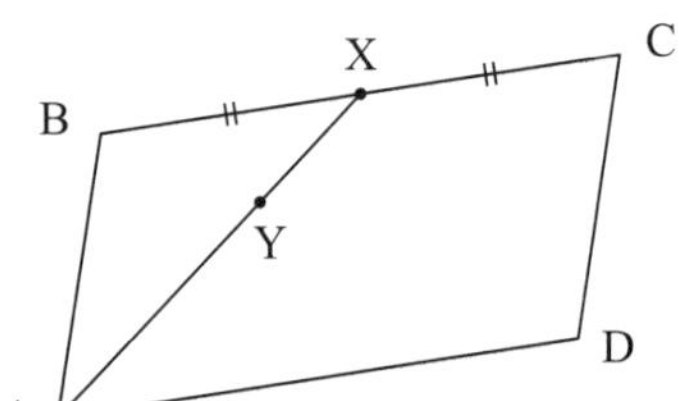

a Find the coordinates of D, X, and Y.

b Prove that B, D, and Y are collinear.

40 Solve $\dfrac{dy}{dx} = \cos^2 x$ given that $y(0) = 4$.

41 A random variable X is normally distributed with standard deviation 2.83. Find the probability that a randomly selected score from X will differ from the mean by less than 4.

42

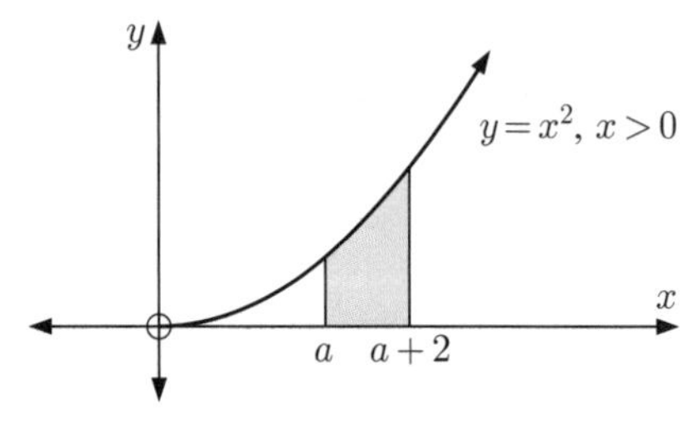

Find a given that the shaded region has area $5\frac{1}{6}$ units^2.

43 If $1 - 2i$ is a zero of $P(x) = x^4 + 11x^2 - 10x + 50$, find all the other zeros.

44 A club has n female members and n male members. A committee of *three* members is to be randomly chosen, and must contain more females than males.

- **a** How many committees consist of 2 females and 1 male?
- **b** How many committees consist of 3 females?
- **c** Use **a** and **b** to deduce that $n\binom{n}{2}+\binom{n}{3}=\frac{1}{2}\binom{2n}{3}$.
- **d** Suppose the club consists of 12 people, and that Mr and Mrs Jones are both members. Find the probability that a randomly selected committee contains:
  - **i** Mrs Jones
  - **ii** Mr Jones, given that it contains Mrs Jones.

45 The Ferris wheel at the Royal Show turns one full circle every minute. The lowest point is 1 metre from the ground, and the highest point is 25 metres above the ground.

- **a** When sitting on the Ferris wheel, your height above ground level after t seconds is given by $h(t)=a+b\sin(c(t-d))$ metres. Find the values of a, b, c, and d given that you start your ride by entering your seat at the lowest point.
- **b** If the motor driving the Ferris wheel broke down after 91 seconds, how high up would you be while waiting to be rescued?

46 The ratio of the zeros of x^2+ax+b is $2:1$. Find a relationship between a and b.

47 Find the exact value of the volume of the solid formed when the region enclosed by $y=xe^{x^3}$, the x-axis, and the line $x=1$, is rotated through 360° about the x-axis.

48 The equations of two lines are:

$$L_1\colon\ \mathbf{r}=\begin{pmatrix}-4\\2\\-1\end{pmatrix}+\lambda\begin{pmatrix}3\\1\\2\end{pmatrix},\ \lambda\in\mathbb{R},\quad\text{and}\quad L_2\colon\ x=\frac{y-5}{2}=\frac{-z-1}{2}.$$

- **a** Determine the point of intersection of L_1 with the plane $2x+y-z=2$.
- **b** Explain why L_1 and L_2 are not parallel.
- **c** Find the point of intersection of L_1 and L_2.
- **d** Find the equation of the plane that contains L_1 and L_2.

49 The lifetime n (in years) of a panel in a solar powered calculator is given by the probability density function $f(n)=0.6e^{-0.6n}$, $n\geqslant 0$.

- **a** Find the probability that a randomly chosen panel will last for at least one year.
- **b** A solar calculator has 8 of these panels, each of which operates independently of each other. The calculator will continue to operate provided at least one of the panels is operating. Find the probability that a randomly chosen calculator fails within one year.

50 Suppose A and B are events such that $P(A)=0.3+x$, $P(B)=0.2+x$, and $P(A\cap B)=x$. Find the possible values of x if A and B are:

- **a** mutually exclusive events
- **b** independent events.

51 The real quadratic function $f(x)$ has a zero of $3+2i$, and a y-intercept of -13. Write the function in the form:

- **a** $f(x)=ax^2+bx+c$
- **b** $f(x)=a(x-h)^2+k$.

52 Suppose $f(x) = 2\tan(3(x-1)) + 4$, $x \in [-1, 1]$. Find:

- **a** the period of $y = f(x)$
- **b** the equations of any asymptotes
- **c** the transformations that transform $y = \tan x$ into $y = f(x)$
- **d** the domain and range of $y = f(x)$.

53 The velocity of a particle travelling in a straight line is given by $v = \cos\left(\frac{1}{3}t\right)$ cm s^{-1}. Find the distance travelled by this particle in the first 10π seconds of motion.

54 Year 12 students at a government school can choose from 16 subjects to study for their Certificate. Seven of these subjects are in group I, six are in group II, and the other three are in group III. Students must study six subjects to qualify for the Certificate. How many combinations of subjects are possible if:

- **a** there are no restrictions
- **b** students must choose 2 subjects from groups I and II, and the remaining subjects could be from any group
- **c** French (a group I subject) is compulsory, and they must choose at least one subject from group III?

55
- **a** Show that the plane $2x + y + z = 5$ contains the line L_1: $x = -2t + 2$, $y = t$, $z = 3t + 1$, $t \in \mathbb{R}$.
- **b** For what values of k does the plane $x + ky + z = 3$ contain L_1?
- **c** Without using row operations, find the values of p and q for which the following system of equations has an infinite number of solutions. Clearly explain your reasoning.

$$\begin{cases} 2x + y + z = 5 \\ x - y + z = 3 \\ 2x + py + 2z = q \end{cases}$$

- **d** Check your result using row operations.

56 The sum of an infinite geometric series is 49 and the second term of the series is 10. Find the possible values for the sum of the first three terms of the series.

57 The average number of amoebas in 50 mL of pond water is 20.

- **a** Assuming that the number of amoebas in pond water follows a Poisson distribution, find the probability that no more than 5 amoebas are present in 10 mL of randomly sampled pond water.
- **b** If a researcher collected 10 mL of pond water each weekday over 4 weeks (20 samples in all), find the probability that the researcher collected no more than 5 amoebas on more than 10 occasions in that 4 week period.

58 The function f is defined by $f : x \mapsto e^{\sin^2 x}$, $x \in [0, \pi]$.

- **a** Use calculus to find the exact value(s) of x for which $f(x)$ has a maximum value.
- **b** Find $f''(x)$.
- **c** Find the point(s) of inflection in the given domain.

59 When a real cubic polynomial $P(x)$ is divided by $x(2x-3)$, the remainder is $ax + b$.

- **a** If the quotient is the same as the remainder, write down an expression for $P(x)$.
- **b** Prove that $(2x-1)$ and $(x-1)$ are both factors of $P(x)$.
- **c** Find an expression for $P(x)$ given that it has y-intercept 7, and that it passes through the point $(2, 39)$.

60 Find a if $a > 0$ and $\displaystyle\int_0^a \frac{x}{x^2+1}\,dx = 3$.

61 One of the five fifth roots of the complex number $a + bi$ is $1 + i$.
Without finding a and b, find the other four roots in polar form.

62 Find the acute angle between the plane $2x + 2y - z = 3$ and the line
$x = \lambda - 1$, $y = -2\lambda + 4$, $z = -\lambda + 3$.

63 **a** Simplify $\sin(2\arcsin x)$. **b** Find $\displaystyle\int \sin(2\arcsin x)\,dx$.

c Hence find $\displaystyle\int_0^1 \sin(2\arcsin x)\,dx$.

64 The lines $\mathbf{r} = \begin{pmatrix} 3 \\ -2 \\ 2 \end{pmatrix} + \lambda \begin{pmatrix} a \\ -1 \\ 2 \end{pmatrix}$ and $\dfrac{x-4}{2} = 1 - y = \dfrac{z+2}{3}$ intersect at point P.

a Find the value of a and hence find the coordinates of P.
b Find the acute angle between the two lines.
c Find the equation of the plane which contains the two lines.

65 The position of an object travelling along the x-axis at time t seconds is given by
$s(t) = t^3 - 7t^2 + 10t + 14$ metres, where $t \geqslant 0$.

a At which times is the object stationary?
b Find the total distance travelled by the object in the first five seconds.
c **i** When is the object's acceleration zero?
ii Describe what is happening to the object's speed and velocity at this time.

66 Find, to 3 significant figures, the area of the region enclosed by the graphs of $y = xe^{\sin x}$ and $y = x^2 - 4x + 6$.

67 $P(z) = z^3 + az^2 + bz + c$ where $a, b, c \in \mathbb{R}$.
Two of the roots of $P(z)$ are -2 and $-3 + 2i$.

a Find a, b, and c. **b** Find the values of z for which $P(z) \geqslant 0$.

68 Let $f(x) = x \tan\sqrt{1 - x^2}$, $-1 \leqslant x \leqslant 1$.

a Sketch the graph of $y = f(x)$.
b Find the volume of the solid formed when the region bounded by $y = f(x)$, $x = 0$, and $x = 1$ is revolved through 2π about the x-axis.

69 Prove that $3(5^{2n+1}) + 2^{3n+1}$ is divisible by 17 for all $n \in \mathbb{Z}^+$.

70 **a** If $x = a^{\frac{1}{3}} + b^{\frac{1}{3}}$, show that $x^3 = 3(ab)^{\frac{1}{3}}x + (a + b)$.
b Hence, find all real solutions of the equation $x^3 = 6x + 6$.

71 A function f with positive values has the property that its value at any point is three times the gradient of the tangent at that point. Given that $f(0) = 3$, find the function f.

72 **a** I wish to borrow $20 000 for 10 years at 12% p.a. where the interest is compounded quarterly. I intend to pay off the loan in quarterly instalments. How much do I need to pay back each quarter?

b Find a formula for calculating the repayments $\$R$ if the total amount borrowed is $\$P$, for n years, at $r\%$ p.a. interest compounding m times per annum, and there are m equal payments at equal intervals each year.

73 **a** Schools A and B *each* choose 11 members to go into a squad of 22. A combined team of 11 must be chosen to be sent away interstate.
In how many ways can a team of 11 be selected and a captain be chosen if the captain must come from A?

b Generalise the solution from **a** to show that:
$1\binom{n}{1}^2 + 2\binom{n}{2}^2 + 3\binom{n}{3}^2 + ... + n\binom{n}{n}^2 = n\binom{2n-1}{n-1}$.

74 Consider the curve $y = xe^{2x}$.

a Find the exact value of $k \in \mathbb{R}$ such that $y = k$ is a horizontal tangent to the curve.

b For which values of $k \in \mathbb{R}$ does the line $y = k$ meet the curve in:

i exactly one point **ii** two distinct points **iii** no points?

c Now consider the family of curves $y = xe^{ax}$, $a \in \mathbb{R}$, $a > 0$.

i Show that $y = x$ is a tangent to all such curves and find the point of contact.

ii Find the equation of the normal to $y = xe^{ax}$, $a \in \mathbb{R}$, $a > 0$, when $x = 0$ and find the acute angle this normal makes with the x-axis.

75 Consider a randomly chosen n child family, where $n > 1$. Let A be the event that the family has at most one boy, and B be the event that every child in the family is of the same sex. For what values of n are the events A and B independent?

76 z is a complex number where $|z| = 1$ and $\arg z \in [0, \frac{\pi}{2}]$.

Given that $\arg\left(\frac{z}{z+2}\right) = \frac{\pi}{4}$, find $|z+2|$.

77 A quadratic equation $ax^2 + bx + c = 0$ is copied by a typist. However, the coefficients a, b, and c are blurred, and she can only see that they are integers of one digit. If she chooses a random digit for each coefficient, what is the probability that the equation she types has real roots?

78 In triangle ABC, the angle at A is double the angle at B.
If AC = 5 cm and BC = 6 cm, find:

a the cosine of the angle at B **b** the length of [AB] using the cosine rule.

79 Prove that $\dfrac{1}{\sin 2x} + \dfrac{1}{\sin 4x} + ... + \dfrac{1}{\sin(2^n x)} = \cot x - \cot(2^n x)$ for all $n \in \mathbb{Z}^+$.

80 Let $I_n = \displaystyle\int_0^{\frac{\pi}{2}} x^n \cos x \, dx$.

a Calculate $I_0 = \displaystyle\int_0^{\frac{\pi}{2}} \cos x \, dx$. **b** Calculate $I_1 = \displaystyle\int_0^{\frac{\pi}{2}} x \cos x \, dx$.

c Show that for $n \geqslant 2$, $I_n = \left(\frac{\pi}{2}\right)^n - n(n-1)I_{n-2}$.

d Hence find $\displaystyle\int_0^{\frac{\pi}{2}} x^3 \cos x \, dx$.

81 **a** Use complex number methods to prove that $\cos^3\theta = \frac{3}{4}\cos\theta + \frac{1}{4}\cos 3\theta$.

b Solve the equation $x^3 - 3x + 1 = 0$ by letting $x = \dfrac{1}{m}\cos\theta$.

82 Points P and Q are 3 metres apart, and free to move on the coordinate axes. N is the foot of the perpendicular from the origin to the line segment [PQ]. [PQ] makes an angle of θ with the y-axis.

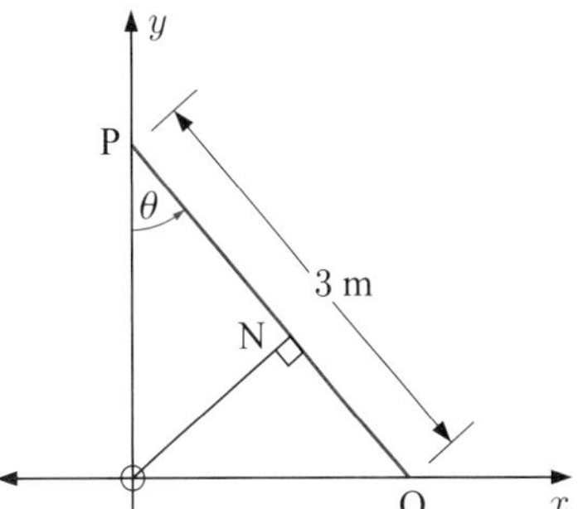

a Show that N is at $(3\sin\theta\cos^2\theta,\ 3\sin^2\theta\cos\theta)$.

b Sketch the curve defined by:
$x = 3\sin\theta\cos^2\theta, \quad y = 3\sin^2\theta\cos\theta$.

83 An arithmetic sequence has first term u_1 and common difference $d \neq 0$. Its distinct terms u_{m+1}, u_{n+1}, and u_{p+1} are consecutive terms of a geometric sequence.

a Find $\dfrac{d}{u_1}$.

b Prove that if $\dfrac{2mp}{m+p} = n$ then $\dfrac{d}{u_1}$ is negative for all n.

84 Suppose line L_1 is defined by $\mathbf{r} = \begin{pmatrix} 8 \\ -13 \\ -3 \end{pmatrix} + \lambda \begin{pmatrix} 3 \\ -5 \\ -2 \end{pmatrix}$ and line L_2 is defined by $\dfrac{x+10}{6} = \dfrac{y-7}{-5} = \dfrac{z-11}{-5}$.

a Find the coordinates of A, the point of intersection of lines L_1 and L_2.

b Find the coordinates of B, where L_1 meets the plane $3x + 2y - z = -2$.

c The point $C(p,\ 0,\ q)$ lies on the plane in **b**.

Find the possible values of p if the area of triangle ABC is $\frac{\sqrt{3}}{2}$ units^2.

85 Suppose e^x can be written as the infinite series $e^x = a_0 + a_1x + a_2x^2 + ... + a_nx^n + ...$

a Show that $a_0 = 1$, $a_1 = 1$, $a_2 = \frac{1}{2!}$, $a_3 = \frac{1}{3!}$, ...

b Hence conjecture an infinite geometric series representation for e^x.

c Check your answer to **b** using the substitution $x = 1$.

86 If two vectors **a** and **b** are such that $|\mathbf{a}| = 7$, $|\mathbf{b}| = 16$, and $|\mathbf{a} - \mathbf{b}| = 20$, find $|\mathbf{a} + \mathbf{b}|$ to 3 significant figures.

87 α and β are two of the roots of $x^3 + ax^2 + bx + c = 0$.
Prove that $\alpha\beta$ is a root of $x^3 - bx^2 + acx - c^2 = 0$.

88 The graph shows the relation $(x^2 + y^2)^2 = x^2 - y^2$.

a Find the axes intercepts for this relation.

b Find the coordinates of four points on the graph at which the tangent is horizontal.

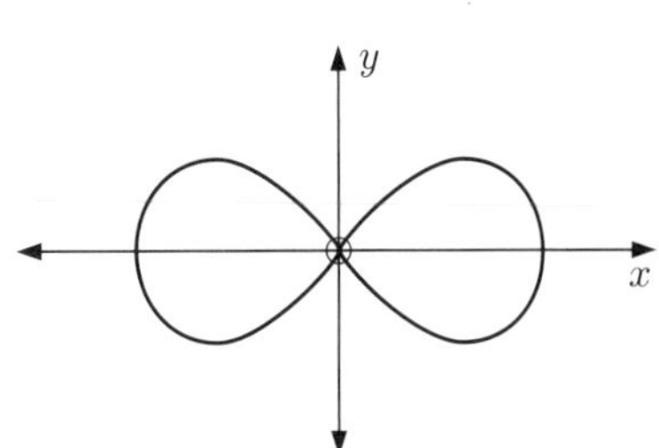

89 **a** Use integration by parts to find $\int \ln x \, dx$. Check that your answer is correct by differentiation.

b The continuous random variable X has the probability density function $f(x) = \ln x$, $1 \leqslant x \leqslant k$. Find the *exact* value of k.

c Find the *median* value of the random variable X.

90 At A on the surface of the Earth, a rocket is launched vertically upwards. After t hours it is at R, h km above the surface. B is the horizon seen from R.

Suppose the Earth has radius r km, $\widehat{BOR}$ is θ, and arc AB is y km long.

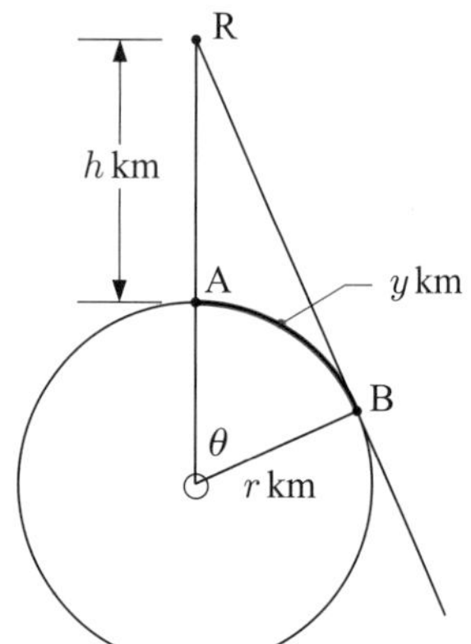

a Show that $\dfrac{dy}{dt} = \dfrac{\cos^2\theta}{\sin\theta}\dfrac{dh}{dt}$.

b If the velocity of the rocket after t hours is given by $r\sin t$, $t \in [0, \pi]$, find the height of the rocket at $t = \frac{\pi}{2}$ hours.

c If $r \approx 6000$, find the rate at which arc AB is changing at the instant when $t = \frac{\pi}{2}$.

91 **a** The hyperbolic tangent function $\text{arctanh}(z)$ is $\text{arctanh}(z) = \frac{1}{2}[\ln(1+z) - \ln(1-z)]$.

Show that the derivative of $\text{arctanh}(z)$ is $\dfrac{1}{1-z^2}$.

b A function $f(x)$ has derivative $f'(x) = x\left(1-x^2\right)^{-\frac{3}{2}}$.

The volume of revolution bounded by $y = f(x)$, the x-axis, $x = 0$, and $x = \frac{1}{2}$, when rotated 2π around the x-axis, is 14.589 units. Find $f(x)$.

92 **a** Let $f(x) = x - 4\sqrt{2}\sqrt{x}$.

- **i** Write down the domain of f.
- **ii** Find the minimum value of f on its domain, and find the coordinates of the point where f attains its minimum value.
- **iii** Sketch $y = f(x)$ and $y = f'(x)$ on the same graph.

b For two numbers $a, b \geqslant 0$, the arithmetic mean $\geqslant$ geometric mean.

This means that $\dfrac{a+b}{2} \geqslant \sqrt{ab}$.

- **i** Prove this result by considering $\left(\sqrt{a} - \sqrt{b}\right)^2$.
- **ii** For a suitable choice of a and b, establish an alternative proof for the result of **a ii**.

93 **a** Find a and k if the line L_1 given by $\mathbf{r} = \begin{pmatrix} 1 \\ -1 \\ 2 \end{pmatrix} + \lambda \begin{pmatrix} 1 \\ a \\ -1 \end{pmatrix}$ lies on the plane P_1 with equation $3x - ky + z = 3$.

b Show that the plane P_2 with equation $2x - y - 4z = 9$ is perpendicular to P_1.

c Find the equation of L_2, the line of intersection of P_1 and P_2.

d Find the point of intersection of L_1 and L_2.

e Find the angle between the lines L_1 and L_2.

94 There are 12 students in a school's Hungarian class. Being well-mannered, they line up in a single file to enter the class.

a How many orders are possible?

b How many orders are there if:

i Irena and Eva are among the last four in the line

ii Istvan is between Paul and Laszlo and they are all together

iii Istvan is between Paul and Laszlo but they are not necessarily together

iv there are exactly three students between Annabelle and Holly?

c Once inside, the class is split into 3 groups of four students each for a vocabulary quiz. In how many ways can this be done if:

i there are no restrictions **ii** Ben and Marton must be in the same group?

95 A particle moves in a straight line with displacement from point O given by the function $s(t)$. The acceleration of the particle is $a(t)$ and its velocity is $v(t)$ where $a = \frac{1}{2}v^2$ and $v(0) = -1$.

a Show that $\dfrac{dt}{dv} = \dfrac{2}{v^2}$.

b Hence find $v(t)$.

c Find the distance travelled in the first 2 seconds of motion.

96 Prove that

$$n^2 \times 1 + (n-1)^2 \times 2 + (n-2)^2 \times 3 + (n-3)^2 \times 4 + ... + 2^2 \times (n-1) + 1^2 \times n = \frac{n(n+1)^2(n+2)}{12}$$

for all positive integers n.

97 A ball of icecream with initial radius 8 cm takes 5 minutes to melt. During this time, its radius decreases at a constant rate.

a Find the rate of change of volume of the icecream ball 2.5 minutes after the icecream begins to melt.

b Find the average rate of change of volume for the last 4 minutes of melting time.

98 Suppose 3, $\log_y x$, $3\log_2 y$, and $7\log_x z$ are consecutive terms of an arithmetic sequence.

a Prove that $x^{18} = y^{21} = z^{28}$.

b Find x in terms of: **i** y **ii** z.

99 Consider the function $f(x) = \cos(2x)e^{\cos x + \sin x}$.

a State the period of $f(x)$.

b Sketch the function for the interval $0 \leqslant x \leqslant 4\pi$.

c Find:

i $\displaystyle\int f(x)dx$ using the substitution $u = e^{\cos x + \sin x}$

ii the first positive x-intercept of the function.

d Hence find the area enclosed by the function and the x-axis from $x = 0$ until the first positive x-intercept.

100 A manufacturer constructs symmetric eggcups from wood using a lathe. He programs his machinery

with the functions $\begin{cases} y_1 = 2 + 5e^{\frac{-2}{x^2+0.5}} \\ y_2 = \dfrac{4\ln x}{\ln 2} \\ y_3 = 5.2\,. \end{cases}$ All units are in centimetres.

a **i** Sketch and shade the region R bounded by the lines $x = 0$, $y = 0$, and the curves y_1, y_2, and y_3.

ii Find the exact value of the y-intercept of the curve y_1.

b **i** If $y = 2 + 5e^{\frac{-2}{x^2+0.5}}$ and $x \geqslant 0$, show that $x = \sqrt{\dfrac{-2}{\ln\left(\frac{y-2}{5}\right)} - 0.5}\,.$

ii If $y = \dfrac{4\ln x}{\ln 2}$ for $x > 0$, show that $x = 2^{\frac{y}{4}}$.

c Each eggcup corresponds to the solid obtained by rotating the region R about the y-axis.

i Write down an expression involving integrals which gives the volume of wood in an eggcup.

ii Calculate the volume of wood in an eggcup.

d Let $f_2(x) = \dfrac{4\ln x}{\ln 2}$ and $g_2(x) = 2^{\frac{x}{4}}$.

i Sketch $y = f_2(x)$, $y = g_2(x)$, and $y = x$ on the same axes.

ii Show that $(f_2 \circ g_2)(x) = x$ for all $x \in \mathbb{R}$, and show that $(g_2 \circ f_2)(x) = x$ for all $x > 0$.

iii What is the relationship between $f_2(x)$ and $g_2(x)$?

e Let $f_1(x) = 2 + 5e^{\frac{-2}{x^2+0.5}}$, where $x \geqslant 0$.

i Write down a function $g_1(x)$ for which $(f_1 \circ g_1)(x) = x$.

ii State the domain of the function $g_1(x)$.

ANSWERS

EXERCISE 1A.1

1 a $x = 0, -\frac{7}{4}$ b $x = 0, \frac{7}{3}$ c $x = 0, \frac{11}{2}$
d $x = 0, \frac{3}{2}$ e $x = 3, 2$ f $x = 3, 7$
g $x = 3$ h $x = -4, 3$ i $x = -11, 3$

2 a $x = \frac{2}{3}$ b $x = -\frac{1}{2}, 7$ c $x = -\frac{2}{3}, 6$
d $x = \frac{1}{3}, -2$ e $x = \frac{3}{2}, 1$ f $x = -\frac{2}{3}, -2$
g $x = -\frac{2}{3}, 4$ h $x = \frac{1}{2}, -\frac{3}{2}$ i $x = -\frac{1}{4}, 3$

3 a $x = 2, 5$ b $x = 0, -\frac{3}{2}$ c $x = \frac{1}{2}, -1$ d $x = 3$

EXERCISE 1A.2

1 a $x = -5 \pm \sqrt{2}$ b no real solns c $x = 4 \pm 2\sqrt{2}$
d $x = 2 \pm \sqrt{6}$ e $x = -\frac{1}{2} \pm \frac{1}{2}\sqrt{3}$ f $x = \frac{1}{3} \pm \frac{\sqrt{7}}{3}$

2 a $x = 2 \pm \sqrt{3}$ b $x = -3 \pm \sqrt{7}$ c $x = 7 \pm \sqrt{3}$
d $x = 2 \pm \sqrt{7}$ e $x = -3 \pm \sqrt{2}$ f $x = 1 \pm \sqrt{7}$
g $x = -3 \pm \sqrt{11}$ h $x = 4 \pm \sqrt{6}$ i no real solns

3 a $x = -1 \pm \frac{1}{\sqrt{2}}$ b $x = \frac{5}{2} \pm \frac{\sqrt{19}}{2}$ c $x = -2 \pm \sqrt{\frac{7}{3}}$
d $x = 1 \pm \sqrt{\frac{7}{3}}$ e $x = \frac{3}{2} \pm \sqrt{\frac{37}{20}}$ f $x = -\frac{1}{2} \pm \frac{\sqrt{6}}{2}$

EXERCISE 1A.3

1 a $x = 2 \pm \sqrt{7}$ b $x = -3 \pm \sqrt{2}$ c $x = 2 \pm \sqrt{3}$
d $x = -2 \pm \sqrt{5}$ e $x = 2 \pm \sqrt{2}$ f $x = \frac{1}{2} \pm \frac{1}{2}\sqrt{7}$
g $x = -\frac{4}{9} \pm \frac{\sqrt{7}}{9}$ h $x = -\frac{7}{4} \pm \frac{\sqrt{97}}{4}$ i $x = \sqrt{2}$

2 a $x = -2 \pm 2\sqrt{2}$ b $x = -\frac{5}{8} \pm \frac{\sqrt{57}}{8}$ c $x = \frac{5}{2} \pm \frac{\sqrt{13}}{2}$
d $x = \frac{1}{2} \pm \frac{1}{2}\sqrt{7}$ e $x = \frac{1}{2} \pm \frac{\sqrt{5}}{2}$ f $x = \frac{3}{4} \pm \frac{\sqrt{17}}{4}$

EXERCISE 1B

1 a 2 real distinct solutions b 2 distinct rational solutions
c 2 distinct rational solutions d 2 real distinct solutions
e no real solutions f a repeated solution

2 a, c, d, f

3 a $\Delta = 16 - 4m$

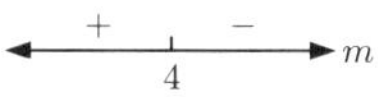

i $m = 4$ ii $m < 4$ iii $m > 4$

b $\Delta = 9 - 8m$

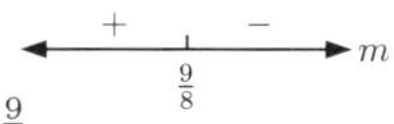

i $m = \frac{9}{8}$ ii $m < \frac{9}{8}$ iii $m > \frac{9}{8}$

c $\Delta = 9 - 4m$

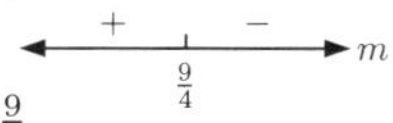

i $m = \frac{9}{4}$ ii $m < \frac{9}{4}$ iii $m > \frac{9}{4}$

4 a $\Delta = k^2 + 8k$

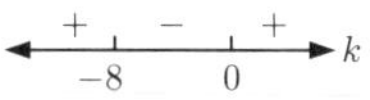

i $k < -8$ or $k > 0$ ii $k \leqslant -8$ or $k \geqslant 0$
iii $k = -8$ or 0 iv $-8 < k < 0$

b $\Delta = 4 - 4k^2$

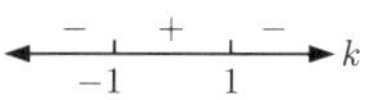

i $-1 < k < 1$ ii $-1 \leqslant k \leqslant 1$
iii $k = \pm 1$ iv $k < -1$ or $k > 1$

c $\Delta = k^2 + 4k - 12$

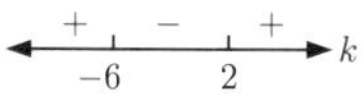

i $k < -6$ or $k > 2$ ii $k \leqslant -6$ or $k \geqslant 2$
iii $k = -6$ or 2 iv $-6 < k < 2$

d $\Delta = k^2 - 4k - 12$

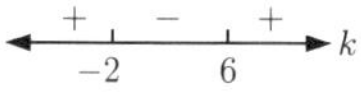

i $k < -2$ or $k > 6$ ii $k \leqslant -2$ or $k \geqslant 6$
iii $k = -2$ or 6 iv $-2 < k < 6$

e $\Delta = 9k^2 - 14k - 39$

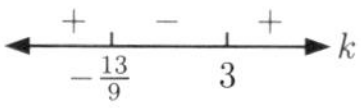

i $k < -\frac{13}{9}$ or $k > 3$ ii $k \leqslant -\frac{13}{9}$ or $k \geqslant 3$
iii $k = -\frac{13}{9}$ or 3 iv $-\frac{13}{9} < k < 3$

f $\Delta = -3k^2 - 4k$

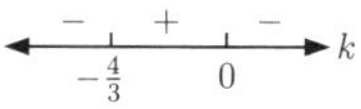

i $-\frac{4}{3} < k < 0$ ii $-\frac{4}{3} \leqslant k \leqslant 0$
iii $k = -\frac{4}{3}$ or 0 iv $k < -\frac{4}{3}$ or $k > 0$

EXERCISE 1C

1 a sum $= \frac{2}{3}$, product $= \frac{7}{3}$ b sum $= -11$, product $= -13$
c sum $= \frac{6}{5}$, product $= -\frac{14}{5}$

2 $k = -\frac{3}{5}$, roots are -1 and $\frac{1}{3}$

3 a $3\alpha = \dfrac{6}{a}$, $2\alpha^2 = \dfrac{a-2}{a}$
b $a = 4$, roots are $\frac{1}{2}$ and 1 or $a = -2$, roots are -1 and -2

4 $k = 4$, roots are $-\frac{1}{2}$ and $\frac{3}{2}$ or $k = 16$, roots are $-\frac{5}{4}$ and $\frac{3}{4}$

5 $7x^2 - 48x + 64 = 0$ 6 $a(8x^2 - 70x + 147) = 0$, $a \neq 0$

7 $-8 + \sqrt{60} \leqslant k < 0$

EXERCISE 1D.1

1 a $y = (x - 4)(x + 2)$ b $y = -(x - 4)(x + 2)$

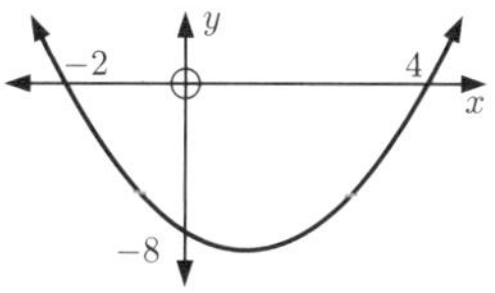

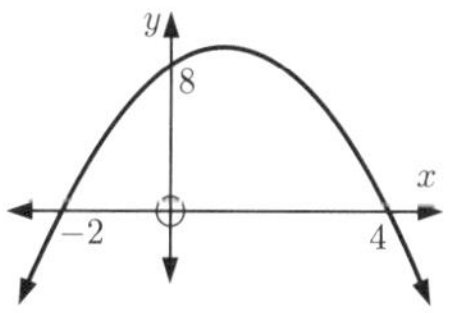

c $y = 2(x + 3)(x + 5)$ d $y = -3x(x + 4)$

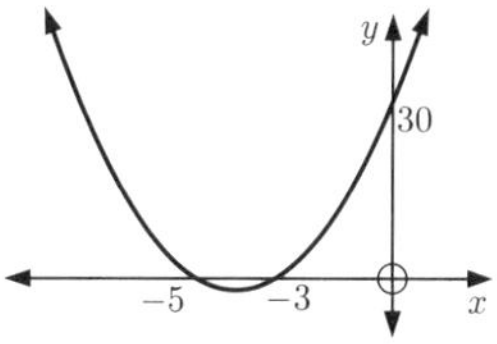

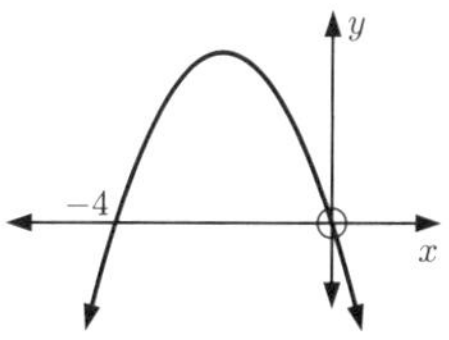

e $y = 2(x + 3)^2$ f $y = -\frac{1}{4}(x + 2)^2$

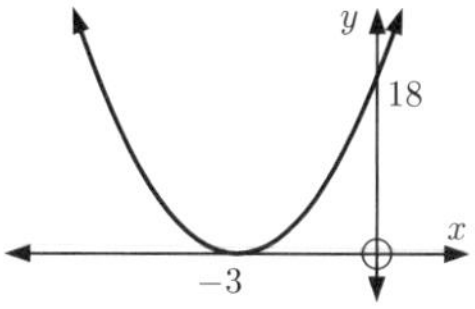

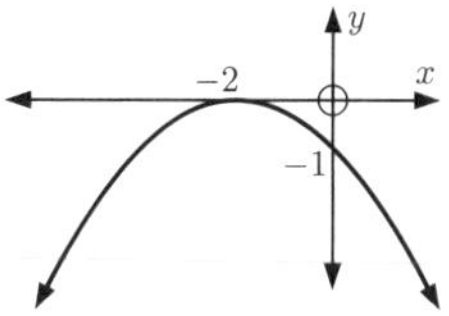

2 a $x = 1$ b $x = 1$ c $x = -4$ d $x = -2$
e $x = -3$ f $x = -2$

3 a **C** b **E** c **B** d **F** e **G** f **H**
g **A** h **D**

4 **a**

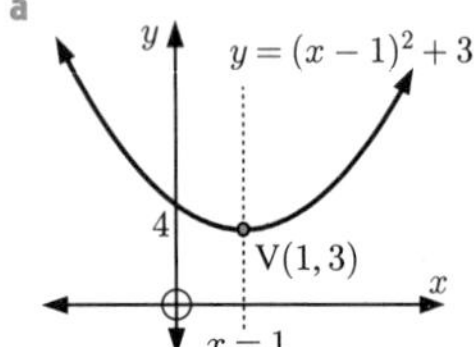

b

$y=2(x+2)^2+1$, $x=-2$, 9, V(−2, 1)

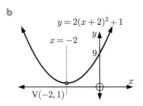

c

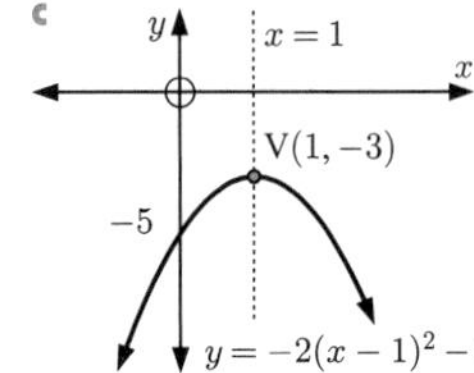

d

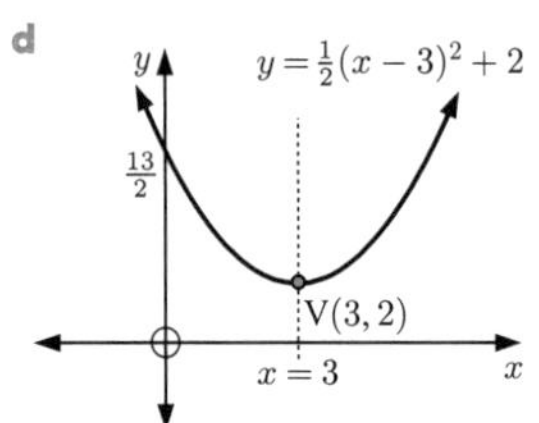

e

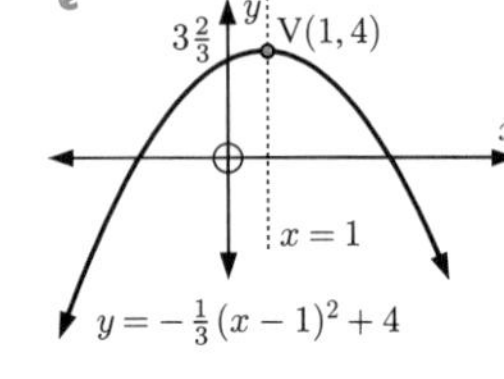

f

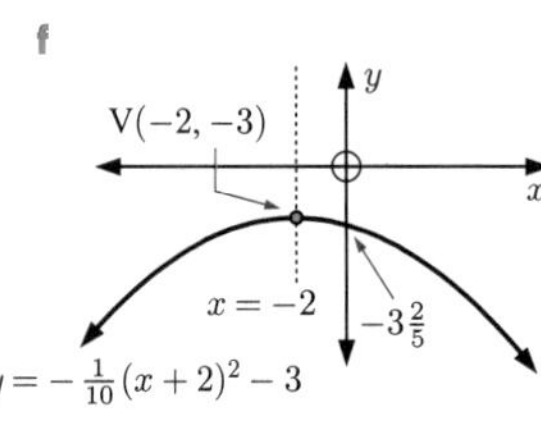

5 **a** **G** **b** **A** **c** **E** **d** **B** **e** **I** **f** **C**
g **D** **h** **F** **i** **H**

6 **a** (2, −2) **b** (0, 4) **c** (0, 1)
d (−2, −5) **e** $(-\frac{3}{2}, -\frac{11}{2})$ **f** $(1, -\frac{9}{2})$

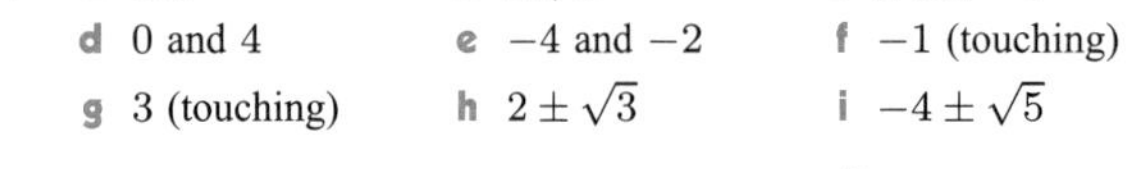

7 **a** ±3 **b** $\pm\sqrt{3}$ **c** 3 and −4
d 0 and 4 **e** −4 and −2 **f** −1 (touching)
g 3 (touching) **h** $2\pm\sqrt{3}$ **i** $-4\pm\sqrt{5}$

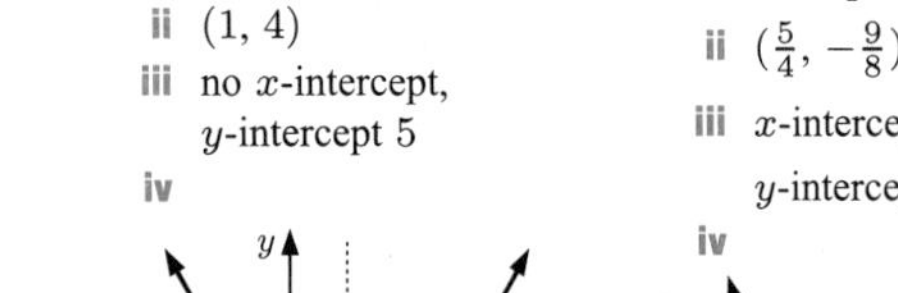

8 **a** **i** $x=1$
ii (1, 4)
iii no x-intercept, y-intercept 5
iv

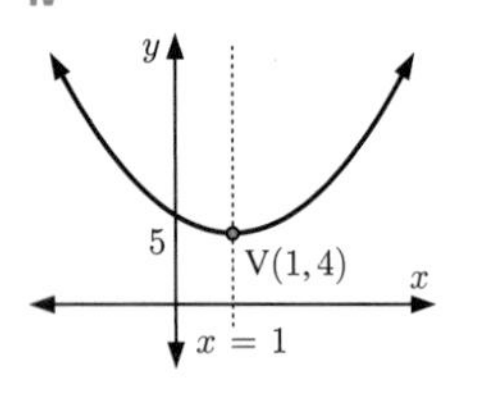

b **i** $x=\frac{5}{4}$
ii $(\frac{5}{4}, -\frac{9}{8})$
iii x-intercepts $\frac{1}{2}$, 2, y-intercept 2
iv

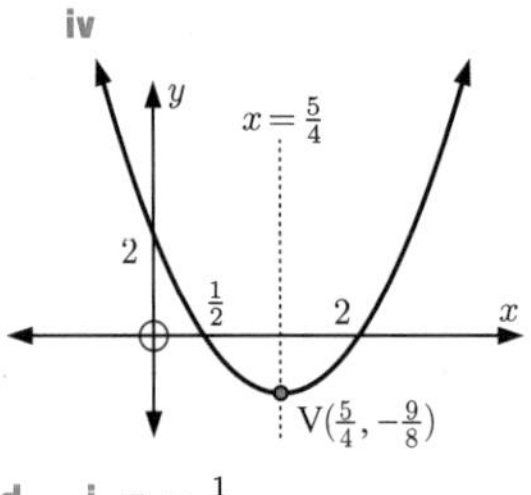

c **i** $x=\frac{3}{2}$
ii $(\frac{3}{2}, \frac{1}{4})$
iii x-intercepts 1, 2, y-intercept −2
iv

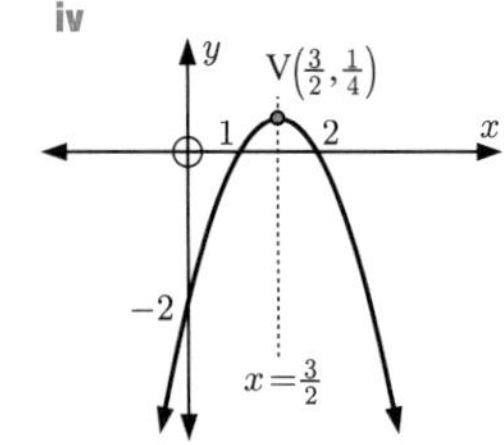

d **i** $x=\frac{1}{4}$
ii $(\frac{1}{4}, \frac{9}{8})$
iii x-intercepts $-\frac{1}{2}$, 1, y-intercept 1
iv

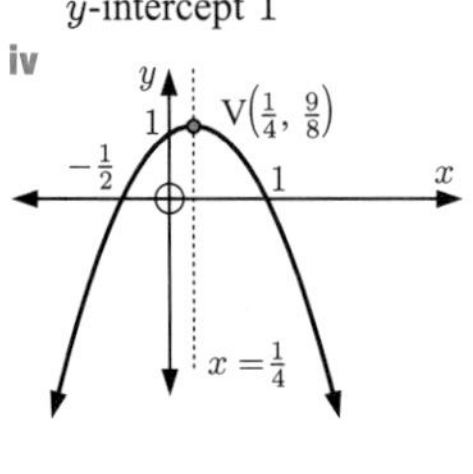

e **i** $x=3$ **ii** (3, 9)
iii x-intercepts 0, 6, y-intercept 0
iv

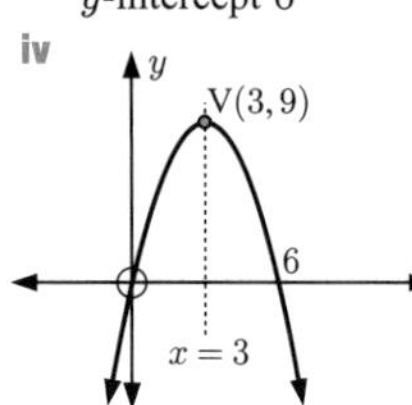

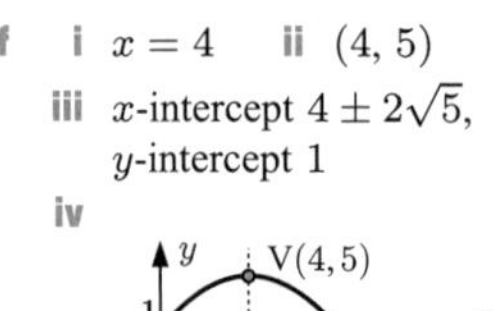

f **i** $x=4$ **ii** (4, 5)
iii x-intercept $4\pm2\sqrt{5}$, y-intercept 1
iv

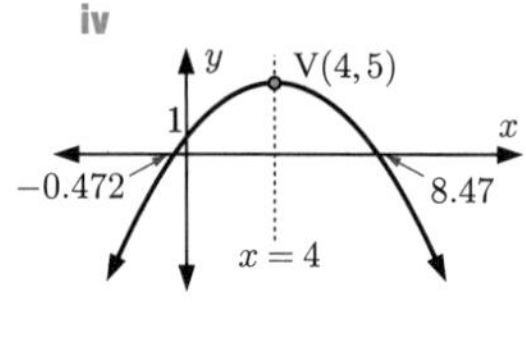

EXERCISE 1D.2

1 **a** $y=(x-1)^2+2$

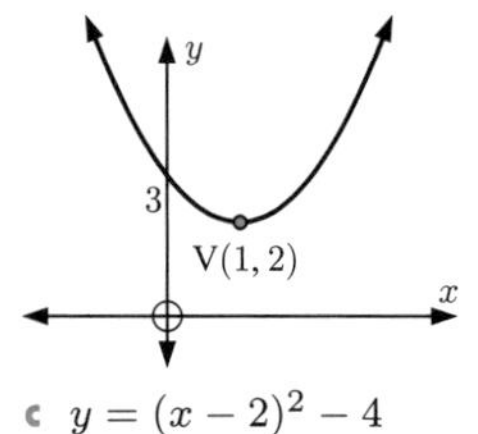

b $y=(x+2)^2-6$

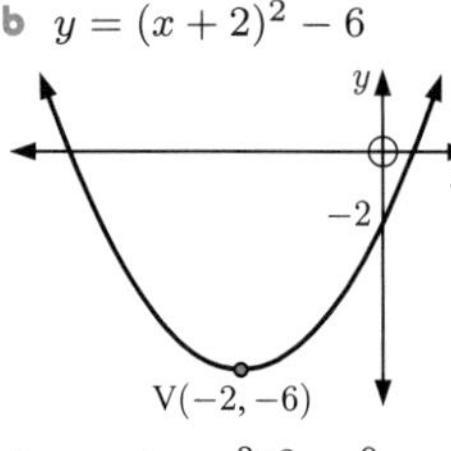

c $y=(x-2)^2-4$

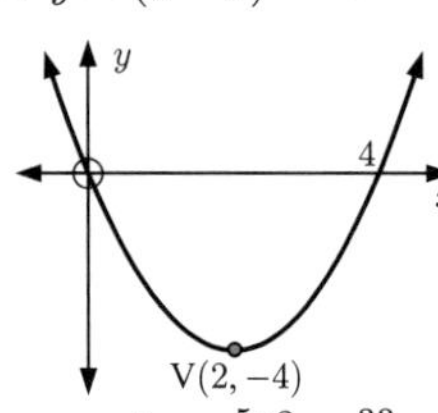

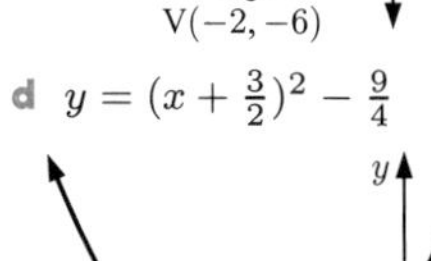

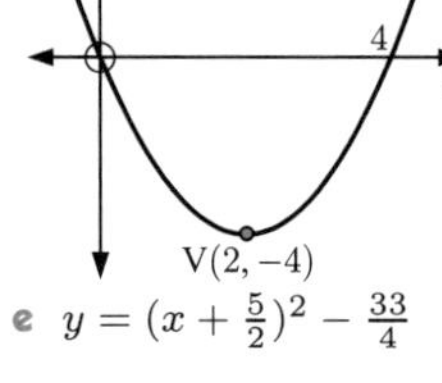

d $y=(x+\frac{3}{2})^2-\frac{9}{4}$

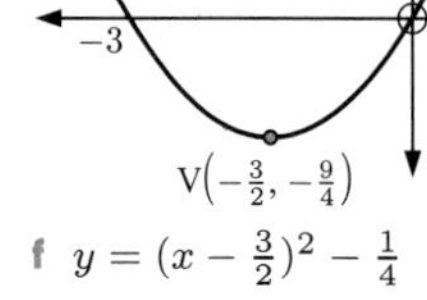

e $y=(x+\frac{5}{2})^2-\frac{33}{4}$

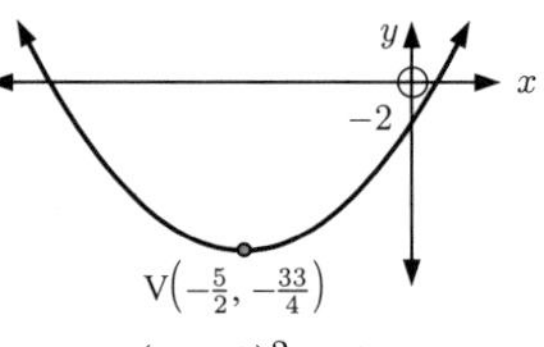

f $y=(x-\frac{3}{2})^2-\frac{1}{4}$

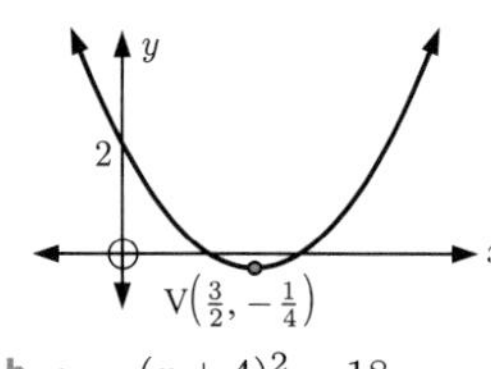

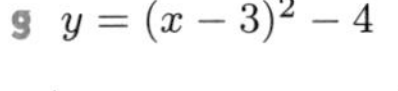

g $y=(x-3)^2-4$

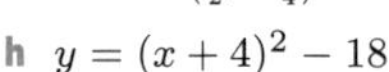

h $y=(x+4)^2-18$

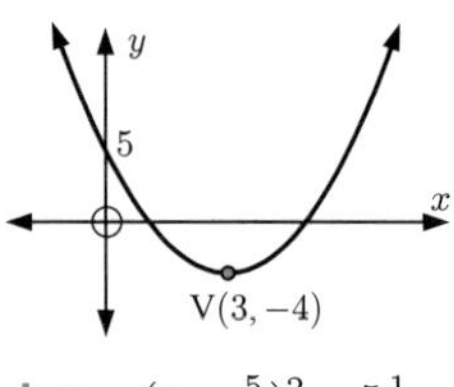

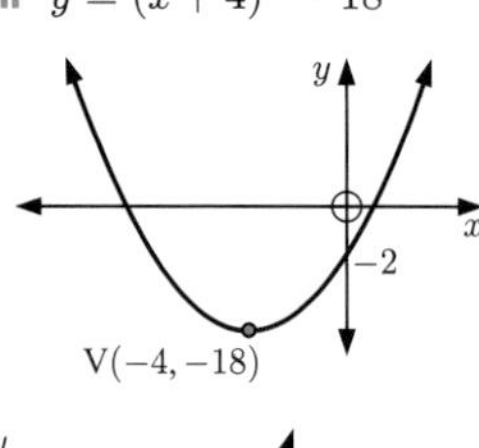

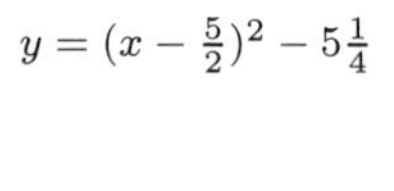

i $y=(x-\frac{5}{2})^2-5\frac{1}{4}$

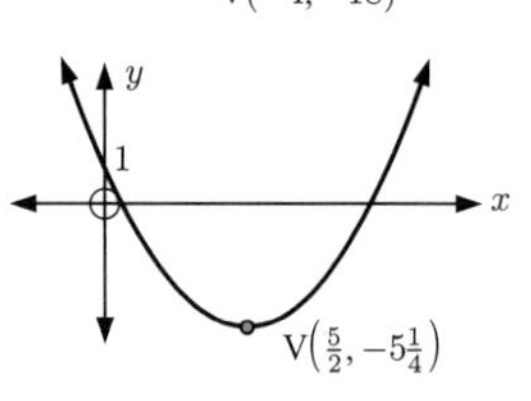

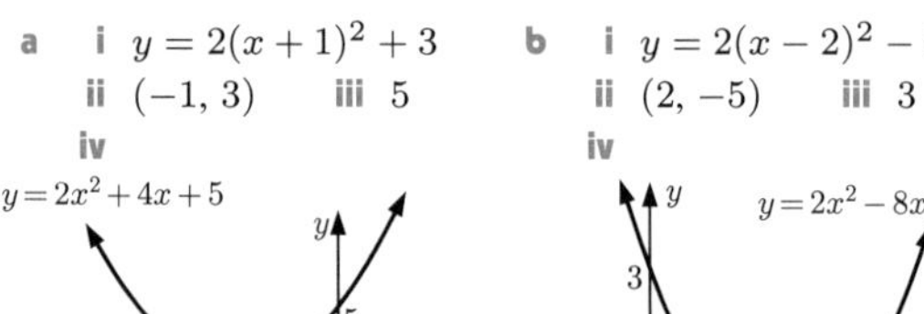

2 **a** **i** $y=2(x+1)^2+3$
ii (−1, 3) **iii** 5
iv

b **i** $y=2(x-2)^2-5$
ii (2, −5) **iii** 3
iv

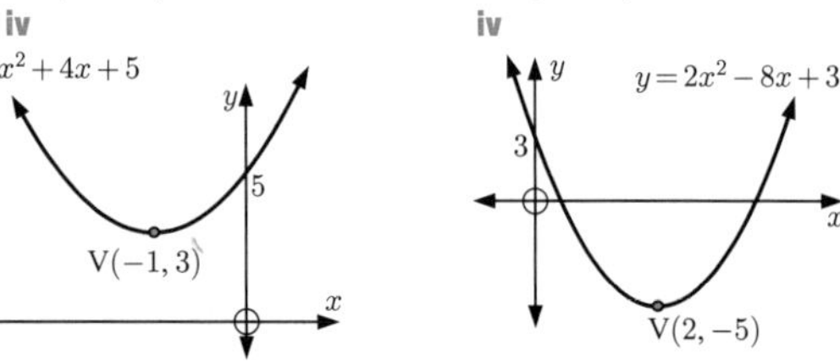

c i $y = 2(x-\frac{3}{2})^2 - \frac{7}{2}$
ii $(\frac{3}{2}, -\frac{7}{2})$ iii 1
iv

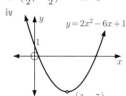

d i $y = 3(x-1)^2 + 2$
ii $(1, 2)$ iii 5
iv

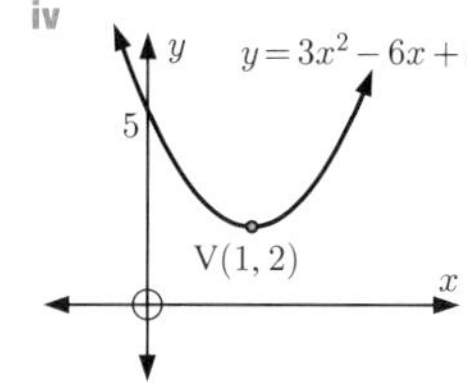

e i $y = -(x-2)^2 + 6$
ii $(2, 6)$ iii 2
iv

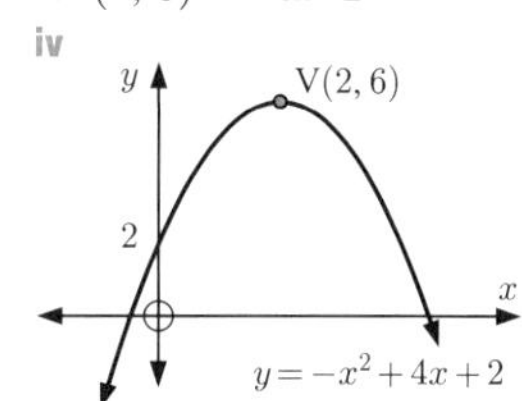

f i $y = -2(x+\frac{5}{4})^2 + \frac{49}{8}$
ii $(-\frac{5}{4}, \frac{49}{8})$ iii 3
iv

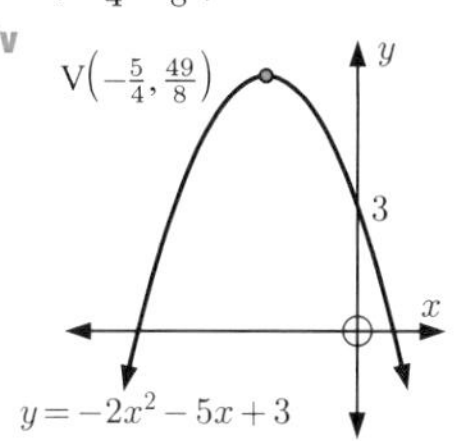

EXERCISE 1D.3

1 a cuts x-axis twice, concave up
b cuts x-axis twice, concave up
c touches x-axis, concave up
d touches x-axis, concave up
e cuts x-axis twice, concave down
f touches x-axis, concave up

2 a $a = 1$ which is > 0 and $\Delta = -15$ which is < 0
b $a = -1$ which is < 0 and $\Delta = -8$ which is < 0
c $a = 2$ which is > 0 and $\Delta = -40$ which is < 0
d $a = -2$ which is < 0 and $\Delta = -23$ which is < 0

3 $a = 3$ which is > 0 and $\Delta = k^2 + 12$ which is always > 0 {as $k^2 > 0$ for all k}

4 $a = 2$ which is > 0 and $\Delta = k^2 - 16$ $\therefore$ positive definite when $k^2 < 16$ so $-4 < k < 4$

EXERCISE 1E

1 a $y = 2(x-1)(x-2)$ b $y = 2(x-2)^2$
c $y = (x-1)(x-3)$ d $y = -(x-3)(x+1)$
e $y = -3(x-1)^2$ f $y = -2(x+2)(x-3)$

2 a $y = \frac{3}{2}(x-2)(x-4)$ b $y = -\frac{1}{2}(x+4)(x-2)$
c $y = -\frac{4}{3}(x+3)^2$

3 a $y = 3x^2 - 18x + 15$ b $y = -4x^2 + 6x + 4$
c $y = -x^2 + 6x - 9$ d $y = 4x^2 + 16x + 16$
e $y = \frac{3}{2}x^2 - 6x + \frac{9}{2}$ f $y = -\frac{1}{3}x^2 + \frac{2}{3}x + 5$

4 a $y = -(x-2)^2 + 4$ b $y = 2(x-2)^2 - 1$
c $y = -2(x-3)^2 + 8$ d $y = \frac{2}{3}(x-4)^2 - 6$
e $y = -2(x-2)^2 + 3$ f $y = 2(x-\frac{1}{2})^2 - \frac{3}{2}$

EXERCISE 1F

1 a $(1, 7)$ and $(2, 8)$ b $(4, 5)$ and $(-3, -9)$
c $(3, 0)$ (touching) d graphs do not meet

2 a i $(2, 4)$ and $(-1, 1)$ ii $x < -1$ or $x > 2$
b i $(1, 0)$ and $(-2, -3)$ ii $x \nless -2$ or $x > 1$
c i $(1, 4)$ touches ii $x \neq 1$
d i $(1, 4)$ and $(-4, -1)$ ii $x < -4$ or $0 < x < 1$

3 $c = -9$ 4 $m = 0$ or -8 5 -1 or 11

6 a $c < -9$
b e.g., if $c = -10$

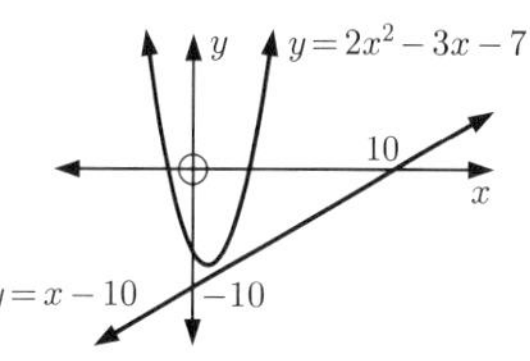

EXERCISE 1G

1 7 and -5 or -7 and 5 2 5 or $\frac{1}{5}$ 3 14
4 15 and 17 or -15 and -17 5 3.48 cm
6 b 6 cm by 6 cm by 7 cm 7 11.2 cm square
8 no 10 61.8 km h^{-1}
11 a

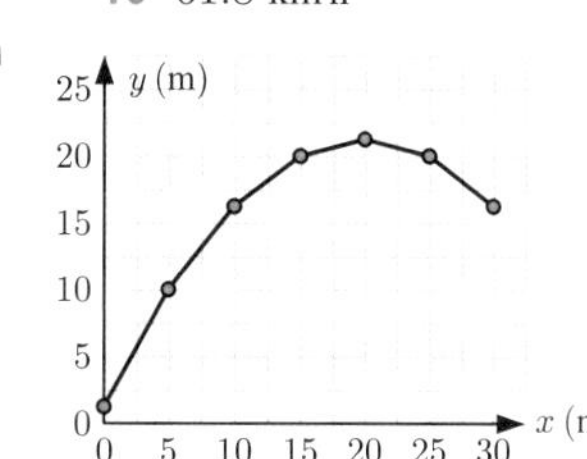

b parabolic c 21.25 m d $y = -0.05x^2 + 2x + 1.25$
e Yes, when $x = 40$, $y = 1.25$

12 a $y = -\frac{8}{9}x^2 + 8$
b No, tunnel is only 3.67 m wide 5 m above ground level.

EXERCISE 1H

1 a min. -1, when $x = 1$ b min. $4\frac{15}{16}$, when $x = \frac{1}{8}$
c max. $6\frac{1}{8}$, when $x = \frac{7}{4}$

2 a 40 refrigerators b $4000

4 500 m by 250 m

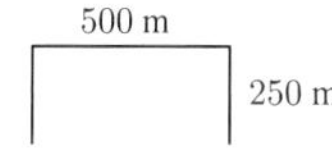

5 c 100 m by 112.5 m 6 b $3\frac{1}{8}$ units

7 a $y = 6 - \frac{3}{4}x$ b 3 cm by 4 cm

8 $m = \dfrac{a_1b_1 + a_2b_2 + ... + a_nb_n}{a_1^2 + a_2^2 + ... + a_n^2}$

9 $y = x^4 - 2(a^2+b^2)x^2 + (a^2-b^2)^2$,
least value $= -4a^2b^2$

REVIEW SET 1A

1 a -2 and 1
b $x = -\frac{1}{2}$
c 4
d $(-\frac{1}{2}, \frac{9}{2})$
e

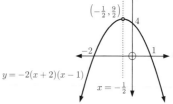

2 a $x = 0$ or 4 b $x = -\frac{5}{3}$ or 2 c $x = 15$ or -4

3 a $x = -\frac{5}{2} \pm \frac{\sqrt{13}}{2}$ b $x = \frac{-11 \pm \sqrt{145}}{6}$

4 $x = -\frac{7}{2} \pm \frac{\sqrt{65}}{2}$

5 **a** 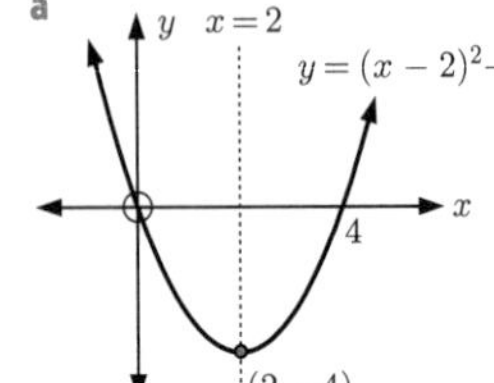

b 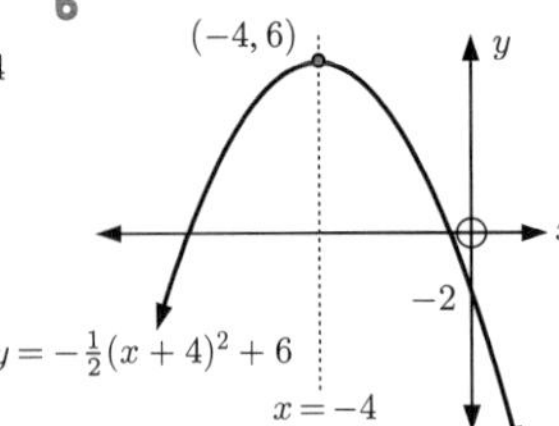

6 **a** $y = 3x^2 - 24x + 48$ **b** $y = \frac{2}{5}x^2 + \frac{16}{5}x + \frac{37}{5}$

7 $a = -2$ which is < 0 $\therefore$ ∩ a max.
max. $= 5$ when $x = 1$

8 $4x^2 + 3x - 2 = 0$

9 **a** $x = 5$ or 2 **b** $x = 3$ or 4 **c** $x = \frac{1}{2}$ or 3

10 $(4, 4)$ and $(-3, 18)$ **11** $k < -3\frac{1}{8}$

12 **a** $m = \frac{9}{8}$ **b** $m < \frac{9}{8}$ **c** $m > \frac{9}{8}$

13 $\frac{6}{5}$ or $\frac{5}{6}$ **15** $k = 3$, roots are $-\frac{1}{3}$ and 3

REVIEW SET 1B

1 **a** $y = 2(x + \frac{3}{2})^2 - \frac{15}{2}$
b $(-\frac{3}{2}, -\frac{15}{2})$
c -3
d 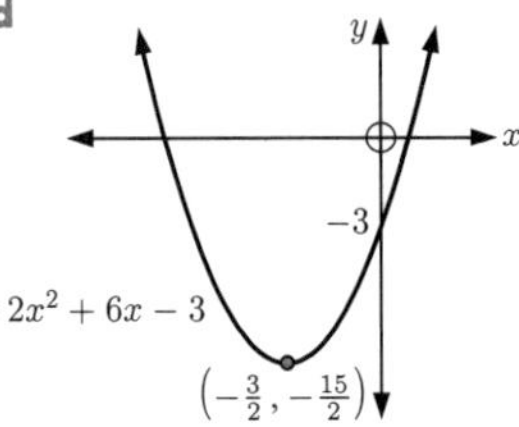

2 **a** $x \approx 0.586$ or 3.414 **b** $x \approx -0.186$ or 2.686

3 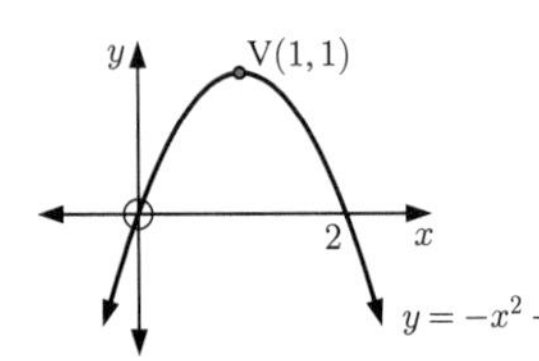

4 $x = \frac{4}{3}$, $V(\frac{4}{3}, 12\frac{1}{3})$

5 **a** two distinct rational solutions **b** a repeated solution

6 **a** $c > -6$ **b** e.g., $c = -2$, $(-1, -5)$ and $(3, 7)$

7 12.9 cm **8** **b** 15 m by 30 m

9 **a** $x = -1$
b $(-1, -3)$
c y-intercept -1,
x-ints. $-1 \pm \frac{1}{2}\sqrt{6}$
d 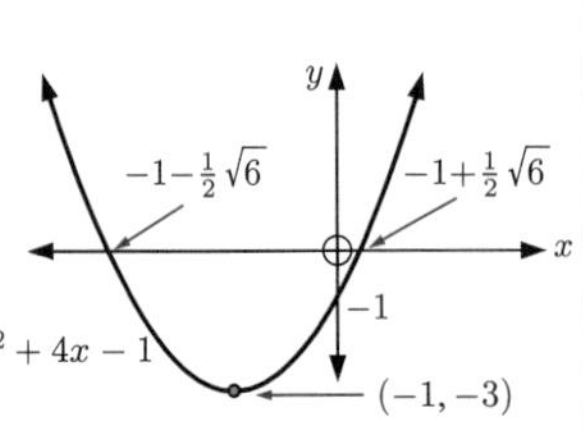

10 ≈ 13.5 cm by 13.5 cm **11** **a** $x = -2$ **b** $x \neq -2$

12 **a** min. $= 5\frac{2}{3}$ when $x = -\frac{2}{3}$
b max. $= 5\frac{1}{8}$ when $x = -\frac{5}{4}$

13 **b** $37\frac{1}{2}$ m by $33\frac{1}{3}$ m **c** 1250 m^2

14 **a** $k = 12$ or -12 **b** $(0, 4)$

REVIEW SET 1C

1 **a** $x = 2$
b $(2, -4)$
c -2
d 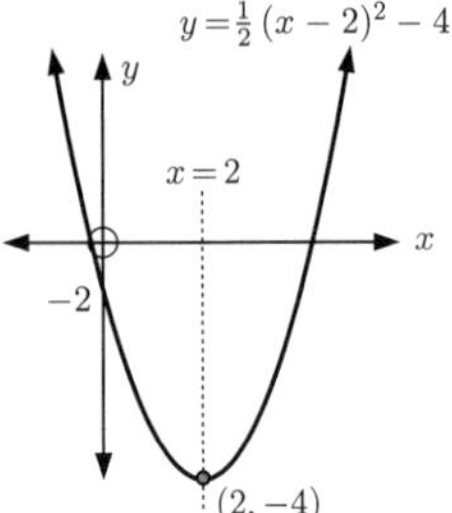

2 **a** $x = \frac{5}{2} \pm \frac{\sqrt{37}}{2}$ **b** $x = \frac{7}{4} \pm \frac{\sqrt{73}}{4}$

3 **a** $x = \frac{7}{2} \pm \frac{\sqrt{37}}{2}$ **b** no real roots

4 **a** $y = \frac{20}{9}(x - 2)^2 - 20$ **b** $y = -\frac{2}{7}(x - 1)(x - 7)$
c $y = \frac{2}{9}(x + 3)^2$

5 **a** graph cuts x-axis twice ∪ **b** graph cuts x-axis twice ∩

6 **a** neither **b** positive definite

7 $y = -6(x - 2)^2 + 25$ **8** y-intercept $= \frac{1}{2}$

9 $k < 1$ **10** $y = -4x^2 + 4x + 24$

11 $m = -5$ or 19 **12** $a \leqslant -9$, $-1 \leqslant a < 0$

13 **a** **i** $y = 3(x - 3)(x + 3)$ **ii** $y = 9x - 27$
b $0 < x < 3$

15 $a(64x^2 - 135x - 27) = 0$, $a \neq 0$

EXERCISE 2A

1 a, d, e **2** a, b, c, e, g, i **3** No, for example $x = 1$

4 No, for example $(0, 3)$ and $(0, -3)$ satisfy the relation.

EXERCISE 2B

1 **a** 2 **b** 8 **c** -1 **d** -13 **e** 1

2 **a** 2 **b** 2 **c** -16 **d** -68 **e** $\frac{17}{4}$

3 **a** -3 **b** 3 **c** 3 **d** -3 **e** $\frac{15}{2}$

4 **a** $7 - 3a$ **b** $7 + 3a$ **c** $-3a - 2$ **d** $10 - 3b$
e $1 - 3x$ **f** $7 - 3x - 3h$

5 **a** $2x^2 + 19x + 43$ **b** $2x^2 - 11x + 13$
c $2x^2 - 3x - 1$ **d** $2x^4 + 3x^2 - 1$
e $2x^4 - x^2 - 2$ **f** $2x^2 + (4h + 3)x + 2h^2 + 3h - 1$

6 **a** **i** $-\frac{7}{2}$ **ii** $-\frac{3}{4}$ **iii** $-\frac{4}{9}$
b $x = 4$ **c** $\frac{2x + 7}{x - 2}$ **d** $x = \frac{9}{5}$

7 f is the function which converts x into $f(x)$ whereas $f(x)$ is the value of the function for any value of x.

8 **a** 6210 euros, value after 4 years
b $t = 4.5$ years, the time for the photocopier to reach a value of 5780 euros.
c 9650 euros

9 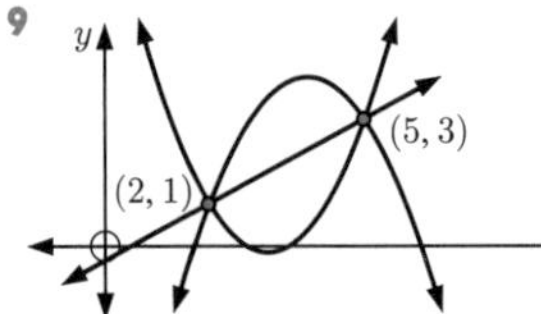

10 $f(x) = -2x + 5$

11 $a = 3$, $b = -2$

12 $a = 3$, $b = -1$, $c = -4$

EXERCISE 2C

1. a Domain = $\{x \mid x \geqslant -1\}$, Range = $\{y \mid y \leqslant 3\}$
   b Domain = $\{x \mid -1 < x \leqslant 5\}$, Range = $\{y \mid 1 < y \leqslant 3\}$
   c Domain = $\{x \mid x \neq 2\}$, Range = $\{y \mid y \neq -1\}$
   d Domain = $\{x \mid x \in \mathbb{R}\}$, Range = $\{y \mid 0 < y \leqslant 2\}$
   e Domain = $\{x \mid x \in \mathbb{R}\}$, Range = $\{y \mid y \geqslant -1\}$
   f Domain = $\{x \mid x \in \mathbb{R}\}$, Range = $\{y \mid y \leqslant \frac{25}{4}\}$
   g Domain = $\{x \mid x \geqslant -4\}$, Range = $\{y \mid y \geqslant -3\}$
   h Domain = $\{x \mid x \in \mathbb{R}\}$, Range = $\{y \mid y > -2\}$
   i Domain = $\{x \mid x \neq \pm 2\}$,
   Range = $\{y \mid y \leqslant -1 \text{ or } y > 0\}$
2. a Domain = $\{x \mid x \geqslant -6\}$ b Domain = $\{x \mid x \neq 0\}$
   c Domain = $\{x \mid x < \frac{3}{2}\}$
3. a Domain = $\{x \mid x \in \mathbb{R}\}$, Range = $\{y \mid y \in \mathbb{R}\}$
   b Domain = $\{x \mid x \in \mathbb{R}\}$, Range = $\{3\}$
   c Domain = $\{x \mid x \geqslant 0\}$, Range = $\{y \mid y \geqslant 0\}$
   d Domain = $\{x \mid x \neq -1\}$, Range = $\{y \mid y \neq 0\}$
   e Domain = $\{x \mid x > 0\}$, Range = $\{y \mid y < 0\}$
   f Domain = $\{x \mid x \neq 3\}$, Range = $\{y \mid y \neq 0\}$
4. a Domain = $\{x \mid x \geqslant 2\}$, Range = $\{y \mid y \geqslant 0\}$
   b Domain = $\{x \mid x \neq 0\}$, Range = $\{y \mid y < 0\}$
   c Domain = $\{x \mid x \leqslant 4\}$, Range = $\{y \mid y \geqslant 0\}$
   d Domain = $\{x \mid x \in \mathbb{R}\}$, Range = $\{y \mid y \geqslant -2\frac{1}{4}\}$
   e Domain = $\{x \mid x \in \mathbb{R}\}$, Range = $\{y \mid y \geqslant 2\}$
   f Domain = $\{x \mid x \leqslant -2 \text{ or } x \geqslant 2\}$, Range = $\{y \mid y \geqslant 0\}$
   g Domain = $\{x \mid x \in \mathbb{R}\}$, Range = $\{y \mid y \leqslant \frac{25}{12}\}$
   h Domain = $\{x \mid x \neq 0\}$, Range = $\{y \mid y \leqslant -2 \text{ or } y \geqslant 2\}$
   i Domain = $\{x \mid x \neq 2\}$, Range = $\{y \mid y \neq 1\}$
   j Domain = $\{x \mid x \in \mathbb{R}\}$, Range = $\{y \mid y \in \mathbb{R}\}$
   k Domain = $\{x \mid x \neq -1, \text{ and } x \neq 2\}$,
   Range = $\{y \mid y \leqslant \frac{1}{3} \text{ or } y \geqslant 3\}$
   l Domain = $\{x \mid x \neq 0\}$, Range = $\{y \mid y \geqslant 2\}$
   m Domain = $\{x \mid x \neq 0\}$, Range = $\{y \mid y \leqslant -2 \text{ or } y \geqslant 2\}$
   n Domain = $\{x \mid x \in \mathbb{R}\}$, Range = $\{y \mid y \geqslant -8\}$
5. a Domain = $\{1, 2, 3\}$, Range = $\{3, 5, 7\}$
   b Domain = $\{-1, 0, 2\}$, Range = $\{3, 5\}$
   c Domain = $\{-3, -2, -1, 3\}$, Range = $\{1\}$
   d Domain = $\{-2, -1, 0, 1, 2\}$, Range = $\{0, \sqrt{3}, 2\}$

EXERCISE 2D

1. a $5 - 2x$ b $-2x - 2$ c 11
2. a $25x - 42$ b $\sqrt{8}$ c -7
3. $(f \circ g)(x) = (2 - x)^2$,
   Domain = $\{x \mid x \in \mathbb{R}\}$,
   Range = $\{y \mid y \geqslant 0\}$
   $(g \circ f)(x) = 2 - x^2$,
   Domain = $\{x \mid x \in \mathbb{R}\}$,
   Range = $\{y \mid y \leqslant 2\}$
4. a i $x^2 - 6x + 10$ ii $2 - x^2$ b $x = \pm\frac{1}{\sqrt{2}}$
5. $f \circ g = \{(0, 0), (1, 1), (2, 2), (3, 3)\}$
6. a $f \circ g = \{(2, 7), (5, 2), (7, 5), (9, 9)\}$
   b $g \circ f = \{(0, 2), (1, 0), (2, 1), (3, 3)\}$
7. a $(f \circ g)(x) = \dfrac{4x - 2}{3x - 1}$, $x \neq 1$, Domain = $\{x \mid x \neq \frac{1}{3} \text{ or } 1\}$
   b $(g \circ f)(x) = 2x + 5$, $x \neq -2$, Domain = $\{x \mid x \neq -2\}$
   c $(g \circ g)(x) = x$, $x \neq 1$, Domain = $\{x \mid x \neq 1\}$
8. a $(f \circ g)(x) = \sqrt{1 - x^2}$
   b Domain = $\{x \mid -1 \leqslant x \leqslant 1\}$, Range = $\{y \mid 0 \leqslant y \leqslant 1\}$
9. a Let $x = 0$, $\therefore$ $b = d$ and so
   $ax + b = cx + b$
   $\therefore$ $ax = cx$ for all x
   Let $x = 1$, $\therefore$ $a = c$
   b $(f \circ g)(x) = [2a]x + [2b + 3] = 1x + 0$ for all x
   $\therefore$ $2a = 1$ and $2b + 3 = 0$
   c Yes, $\{(g \circ f)(x) = [2a]x + [3a + b]\}$

EXERCISE 2E

3. a odd b neither c even d odd
   e even f neither
4. $a = -\frac{3}{2}$
5. $b = -1$
6. b $b = 0$, $d = 0$ c $b = 0$, $d = 0$
8. even

EXERCISE 2F

1. a $-$ | 2 | $+$ (x)
   b $-$ | -1 | $+$ | 3 | $-$ (x)
   c $+$ | 0 | $-$ | 2 | $+$ (x)
   d $+$ | 1 | $+$ (x)
   e $-$ | -2 | $-$ (x)
   f $+$ | -2 | $-$ | 0 | $+$ | 2 | $-$ (x)
   g $-$ ¦ 0 ¦ $+$ (x)
   h $+$ | -1 | $+$ | 2 | $-$ (x)
   i $-$ | -3 | $+$ | 0 | $+$ | 4 | $-$ (x)
   j $+$ ¦ 1 ¦ $-$ | 2 | $+$ (x)
   k $-$ ¦ -1 ¦ $+$ | 0 | $-$ ¦ 3 ¦ $+$ (x)
   l $-$ ¦ -2 ¦ $+$ | -1 | $-$ | 1 | $+$ ¦ 2 ¦ $-$ (x)
2. a $+$ | -4 | $-$ | 2 | $+$ (x)
   b $+$ | 0 | $-$ | 3 | $+$ (x)
   c $+$ | -2 | $-$ | 0 | $+$ (x)
   d $-$ | -1 | $+$ | 3 | $-$ (x)
   e $-$ | $\frac{1}{2}$ | $+$ | 3 | $-$ (x)
   f $+$ | $\frac{1}{2}$ | $-$ | 5 | $+$ (x)
   g $+$ | -3 | $-$ | 3 | $+$ (x)
   h $-$ | -2 | $+$ | 2 | $-$ (x)
   i $-$ | 0 | $+$ | 5 | $-$ (x)
   j $+$ | 1 | $-$ | 2 | $+$ (x)
   k $-$ | $-\frac{1}{2}$ | $+$ | $\frac{1}{2}$ | $-$ (x)
   l $+$ | $-\frac{2}{3}$ | $-$ | $\frac{1}{2}$ | $+$ (x)
   m $-$ | -3 | $+$ | $\frac{1}{3}$ | $-$ (x)
   n $-$ | $-\frac{1}{2}$ | $+$ | 5 | $-$ (x)
   o $-$ | $-\frac{2}{5}$ | $+$ | $\frac{1}{3}$ | $-$ (x)
3. a $+$ | -2 | $+$ (x)
   b $+$ | 3 | $+$ (x)
   c $-$ | -2 | $-$ (x)
   d $-$ | 4 | $-$ (x)

e $+ \;|\; 1 \;|\; +$ (x) **f** $- \;|\; 2 \;|\; -$ (x)

g $+ \;|\; \frac{1}{2} \;|\; +$ (x) **h** $- \;|\; -3 \;|\; -$ (x)

i $- \;|\; \frac{3}{2} \;|\; -$ (x)

4 **a** $+ \;|\; -2 \;|\; - \;|\; 1 \;|\; +$ (x) **b** $+ \;|\; -3 \;|\; - \;|\; 0 \;|\; +$ (x)

c $- \;|\; -\frac{3}{2} \;|\; + \;|\; 4 \;|\; -$ (x) **d** $- \;|\; \frac{1}{4} \;|\; + \;|\; 2 \;|\; -$ (x)

e $+ \;|\; 0 \;|\; - \;|\; 2 \;|\; +$ (x) **f** $+ \;|\; 0 \;|\; - \;|\; 3 \;|\; +$ (x)

g $- \;|\; 0 \;|\; + \;|\; 1 \;|\; +$ (x) **h** $- \;|\; -1 \;|\; - \;|\; 0 \;|\; +$ (x)

i $+ \;|\; -2 \;|\; - \;|\; 1 \;|\; + \;|\; 3 \;|\; -$ (x) **j** $+ \;|\; 0 \;|\; - \;|\; 1 \;|\; + \;|\; 2 \;|\; -$ (x)

k $+ \;|\; -2 \;|\; - \;|\; 0 \;|\; + \;|\; 2 \;|\; -$ (x) **l** $+ \;|\; -\frac{3}{2} \;|\; - \;|\; 2 \;|\; + \;|\; 3 \;|\; -$ (x)

m $- \;|\; -\sqrt{3} \;|\; + \;|\; -1 \;|\; - \;|\; \sqrt{3} \;|\; +$ (x) **n** $- \;|\; 0 \;|\; +$ (x)

o $- \;|\; -1 \;|\; +$ (x) **p** $+ \;|\; -3 \;|\; - \;|\; 3 \;|\; -$ (x)

q $+ \;|\; -2 \;|\; - \;|\; 0 \;|\; - \;|\; 5 \;|\; +$ (x) **r** $- \;|\; 1 \;|\; + \;|\; 3 \;|\; +$ (x)

s $+ \;|\; -1 \;|\; - \;|\; \frac{1}{2} \;|\; +$ (x) **t** $- \;|\; -\frac{14}{3} \;|\; + \;|\; -3 \;|\; -$ (x)

u $+ \;|\; -8 \;|\; - \;|\; -3 \;|\; + \;|\; 0 \;|\; - \;|\; 2 \;|\; +$ (x)

EXERCISE 2G

1 **a** $x \in [-3, 2]$ **b** no solutions **c** $x \in \left]-\frac{1}{2}, 3\right[$
d $x \in \left]-\infty, 0\right]$ or $[1, \infty[$ **e** $x \in \left]-\infty, 0\right]$ or $[3, \infty[$
f $x \in \left]-\frac{2}{3}, 0\right[$ **g** $x \in]-2, 2[$
h $x \in]-\infty, -\sqrt{2}]$ or $[\sqrt{2}, \infty[$ **i** $x \neq -2$
j $x \in]-\infty, -1]$ or $[\frac{3}{2}, \infty[$ **k** no solutions
l $x \in \left]-\frac{3}{2}, \frac{1}{3}\right[$ **m** $x \in \left]-\infty, -\frac{4}{3}\right[$ or $]4, \infty[$
n $x \neq 1$ **o** $x \in \left[\frac{1}{3}, \frac{1}{2}\right]$
p $x \in \left]-\infty, -\frac{1}{6}\right[$ or $]1, \infty[$
q $x \in \left]-\infty, -\frac{1}{4}\right]$ or $\left[\frac{2}{3}, \infty\right[$
r $x \in \left]-\infty, \frac{3}{2}\right[$ or $]3, \infty[$

2 **a** $x < -4,\ x > \frac{1}{2}$ **b** $-1 \leqslant x < 4$
c $-7 < x < 2.5$ **d** $-5 \leqslant x < -2,\ 0 \leqslant x < 2$
e $-2 < x < 0.545,\ x > 2$ **f** $x < -1.75,\ 2.5 < x < 3$

3 **a** $-1 \leqslant x \leqslant 0,\ x \geqslant 1$ **b** $x < -2,\ 1 < x < 3$
c $x \leqslant -3,\ -1.41 \leqslant x \leqslant 1.41$

EXERCISE 2H.1

1 **a** 2 **b** 3 **c** 6 **d** 6 **e** 5 **f** −1 **g** 1
h 5 **i** 4 **j** 4 **k** 2 **l** 2

2 **a** 8 **b** 2 **c** $\frac{5}{4}$ **d** 0

3 **a**

a	b	$\|a\|+\|b\|$	$\|a\|-\|b\|$	$\|a+b\|$	$\|a-b\|$	$\|b-a\|$
6	2	8	4	8	4	4
6	−2	8	4	4	8	8
−6	2	8	4	4	8	8
−6	−2	8	4	8	4	4

b **i** false **ii** false

4 **a**

a	b	$\|ab\|$	$\|a\|\,\|b\|$	$\left\|\frac{a}{b}\right\|$	$\frac{\|a\|}{\|b\|}$
6	2	12	12	3	3
6	−2	12	12	3	3
−6	2	12	12	3	3
−6	−2	12	12	3	3

5 **a** $y = |x-2|$ **b** $y = |x+1|$

$y = \begin{cases} x-2, & x \geqslant 2 \\ 2-x, & x < 2 \end{cases}$ $\qquad y = \begin{cases} x+1, & x \geqslant -1 \\ -x-1, & x < -1 \end{cases}$

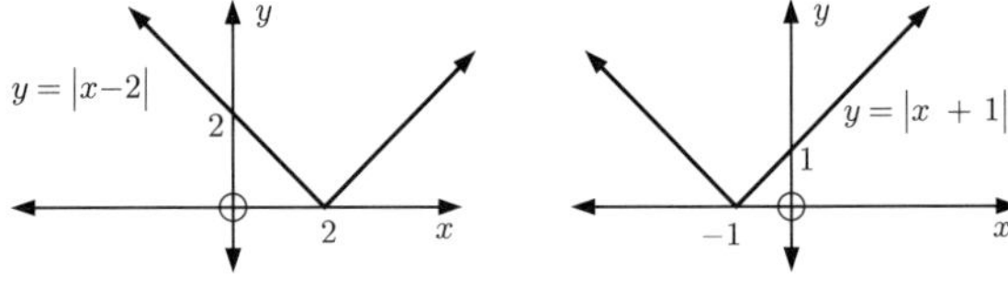

c $y = \begin{cases} -x, & x \geqslant 0 \\ x, & x < 0 \end{cases}$ **d** $y = \begin{cases} 2x, & x \geqslant 0 \\ 0, & x < 0 \end{cases}$

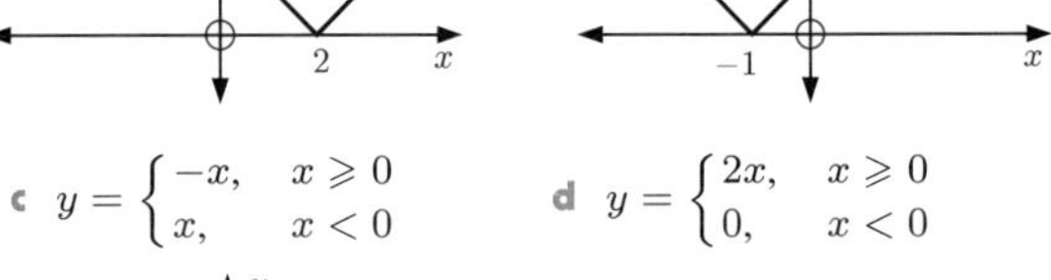

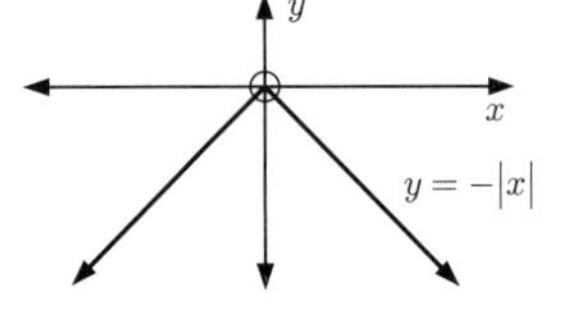

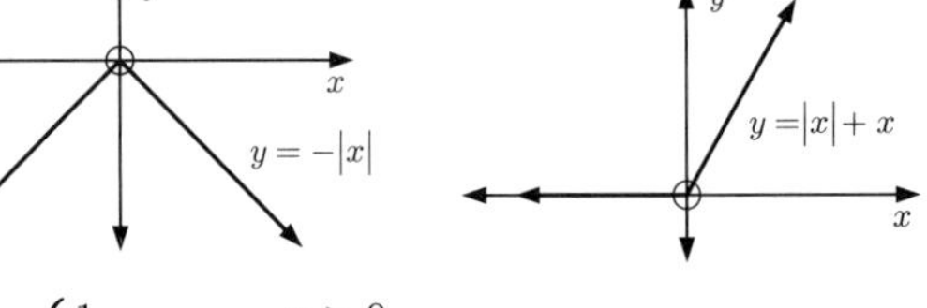

e $y = \begin{cases} 1, & x > 0 \\ \text{undefined}, & x = 0 \\ -1, & x < 0 \end{cases}$ **f** $y = \begin{cases} -x, & x \geqslant 0 \\ 3x, & x < 0 \end{cases}$

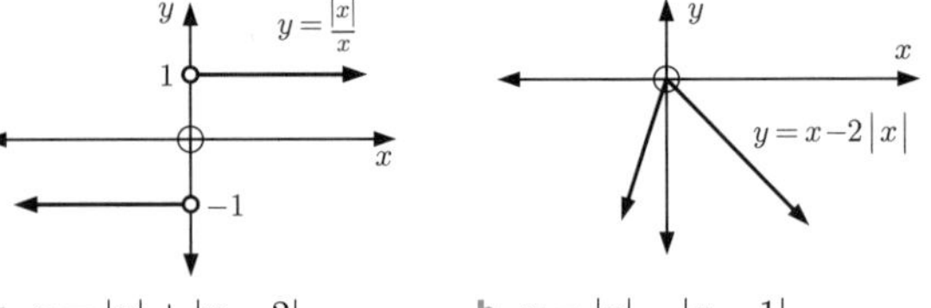

g $y = |x| + |x-2|$ **h** $y = |x| - |x-1|$

$y = \begin{cases} 2x-2, & x \geqslant 2 \\ 2, & 0 \leqslant x < 2 \\ 2-2x, & x < 0 \end{cases}$ $\qquad y = \begin{cases} 1, & x \geqslant 1 \\ 2x-1, & 0 \leqslant x < 1 \\ -1, & x < 0 \end{cases}$

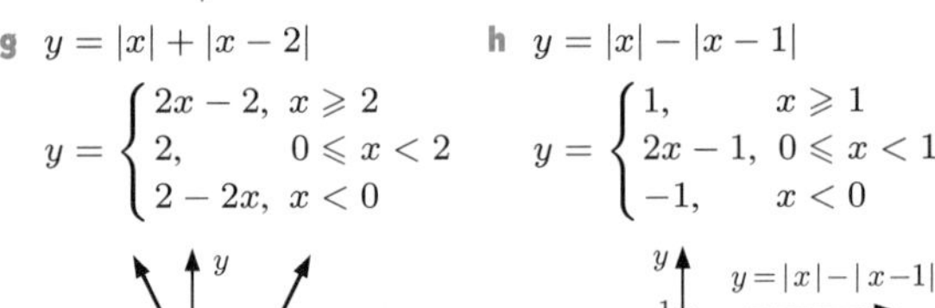

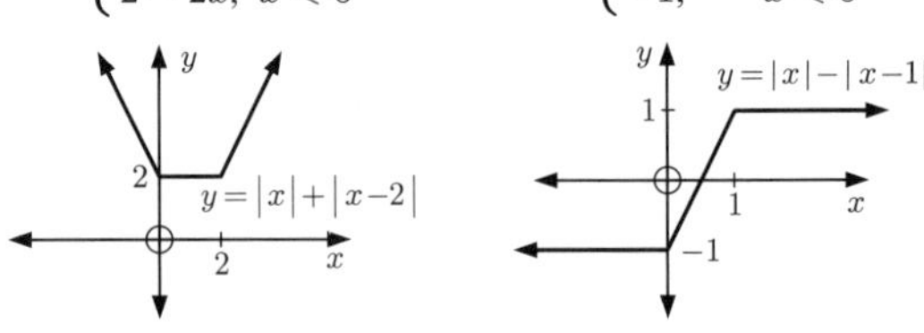

EXERCISE 2H.2

1 a $x = \pm 3$ b no solution c $x = 0$
d $x = 4$ or -2 e $x = -1$ or 7 f no solution
g $x = 1$ or $\frac{1}{3}$ h $x = 0$ or 3 i $x = -2$ or $\frac{14}{5}$

2 a $x = \frac{3}{2}$ or $\frac{3}{4}$ b $x = -2$ or $-\frac{4}{7}$ c $x = -1$ or 7

3 a $x = -\frac{1}{4}$ or $\frac{3}{2}$ b $x = -6$ or $-\frac{4}{3}$ c $x = \frac{1}{2}$
d $x = \frac{5}{2}$ e $x = \pm\frac{1}{2}$ f $x = -6$ or $\frac{2}{5}$

4 a $x = 1$ b $x = -\frac{4}{5}$ or 4 c $x = \frac{5}{7}$ or 5

5 a $x \in \,]-4, 4[$ b $x \in \,]-\infty, -3]$ or $[3, \infty[$
c $x \in [-4, -2]$ d $x \in \,]-1, 2[$
e $x \in \,]-\infty, \frac{1}{4}[$ or $]\frac{5}{4}, \infty[$ f $x \in [1, \infty[$
g $x \in [-1, \frac{1}{5}]$ h $x \in [\frac{3}{2}, 2[$ or $x \in \,]2, 3]$
i $x \in [-\frac{1}{4}, 1[$ or $x \in \,]1, \infty[$ j $x \in \,]1, 3[$
k $x \in \,]-\infty, -1[$ or $x \in \,]2, \infty[$ l $x \in [3, \infty[$

6

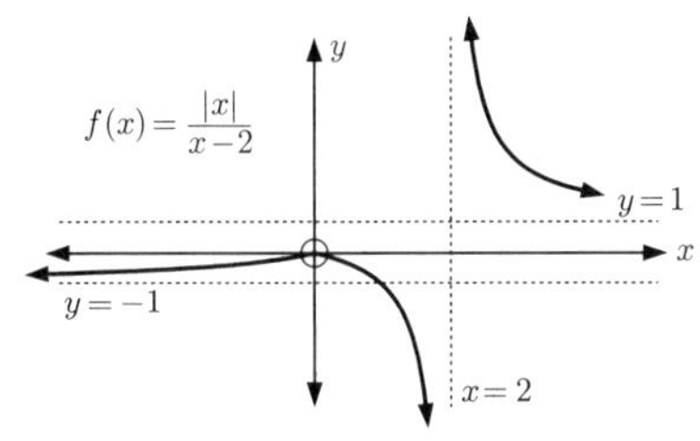

$x \in [-2, \frac{2}{3}]$ or $x \in \,]2, \infty[$

7 a

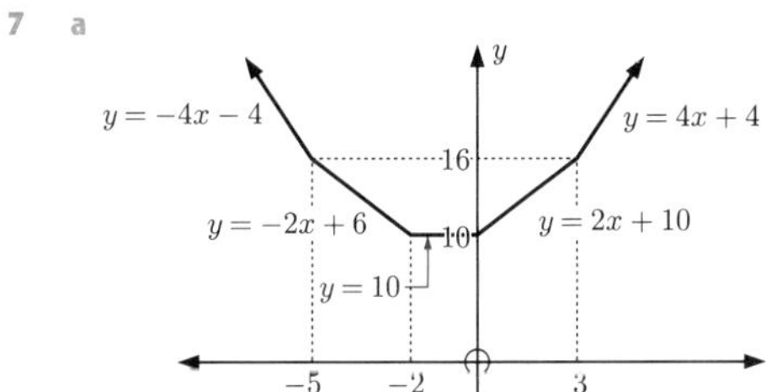

b ii Anywhere between O and Q inclusive, the minimum length of cable is 10 km.
iii At O, minimum length of cable is 17 km.

8 a True b True

EXERCISE 2I

1 a i vertical asymptote $x = 2$, horizontal asymptote $y = 0$
ii Domain $= \{x \mid x \neq 2\}$, Range $= \{y \mid y \neq 0\}$
iii no x-intercept, y-intercept $-\frac{3}{2}$
iv as $x \to 2^-$, $y \to -\infty$ as $x \to \infty$, $y \to 0^+$
as $x \to 2^+$, $y \to \infty$ as $x \to -\infty$, $y \to 0^-$
v

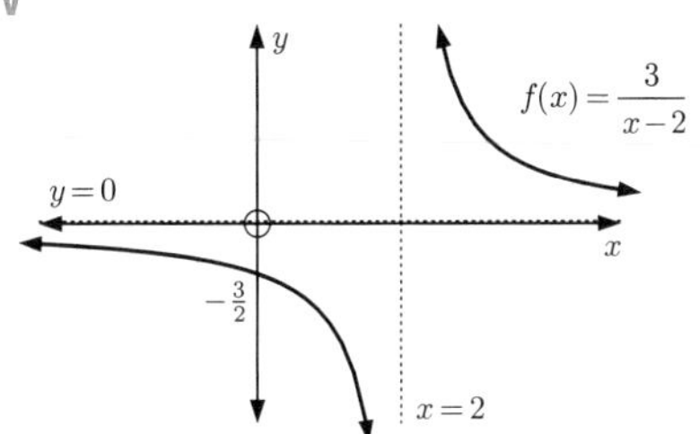

b i vertical asymptote $x = -1$, horizontal asymptote $y = 2$
ii Domain $= \{x \mid x \neq -1\}$, Range $= \{y \mid y \neq 2\}$
iii x-intercept $\frac{1}{2}$, y-intercept -1
iv as $x \to -1^-$, $y \to \infty$ as $x \to \infty$, $y \to 2^-$
as $x \to -1^+$, $y \to -\infty$ as $x \to -\infty$, $y \to 2^+$
v

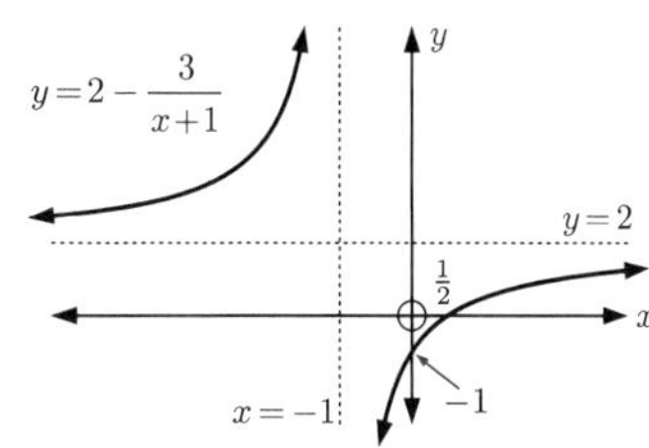

c i vertical asymptote $x = 2$, horizontal asymptote $y = 1$
ii Domain $= \{x \mid x \neq 2\}$, Range $= \{y \mid y \neq 1\}$
iii x-intercept -3, y-intercept $-\frac{3}{2}$
iv as $x \to 2^-$, $y \to -\infty$ as $x \to \infty$, $y \to 1^+$
as $x \to 2^+$, $y \to \infty$ as $x \to -\infty$, $y \to 1^-$
v

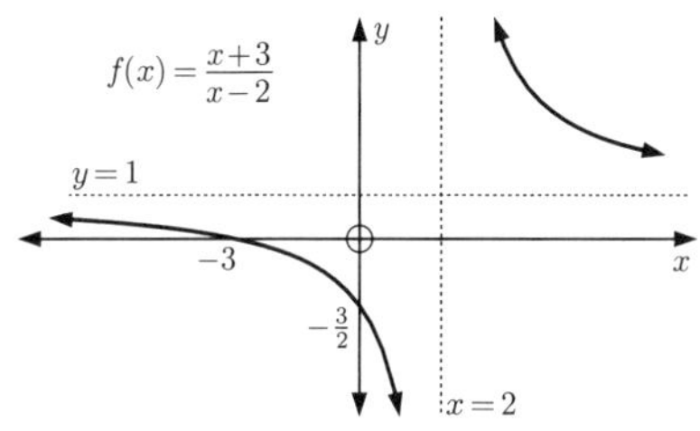

d i vertical asymptote $x = -2$, horizontal asymptote $y = 3$
ii Domain $= \{x \mid x \neq -2\}$, Range $= \{y \mid y \neq 3\}$
iii x-intercept $\frac{1}{3}$, y-intercept $-\frac{1}{2}$
iv as $x \to -2^-$, $y \to \infty$ as $x \to \infty$, $y \to 3^-$
as $x \to -2^+$, $y \to -\infty$ as $x \to -\infty$, $y \to 3^+$
v

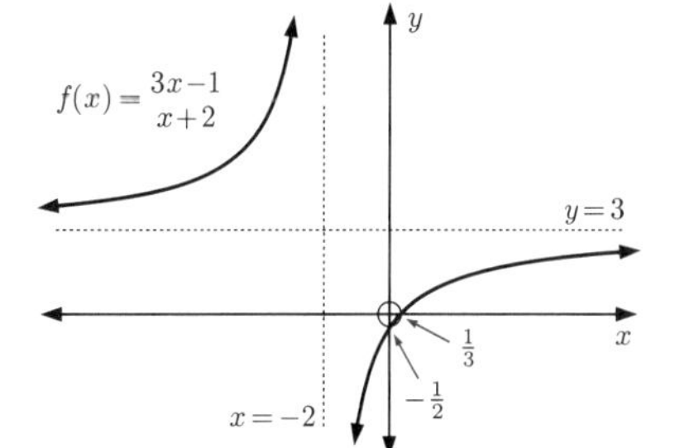

2 a Domain $= \{x \mid x \neq -\frac{d}{c}\}$
b Vertical asymptote $x = -\frac{d}{c}$
c Horizontal asymptote $y = \frac{a}{c}$

EXERCISE 2J

1 a i

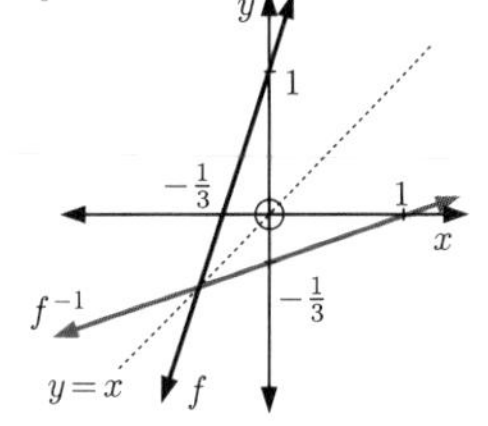

ii, iii $f^{-1}(x) = \dfrac{x-1}{3}$

b i

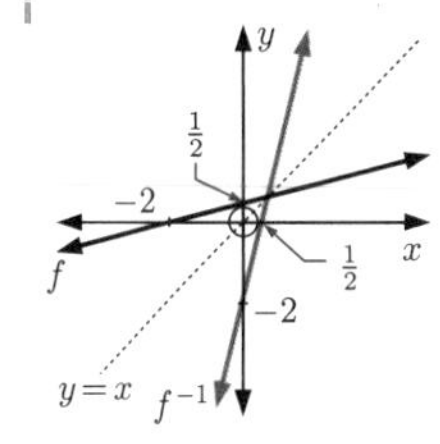

ii, iii $f^{-1}(x) = 4x - 2$

2 **a** **i** 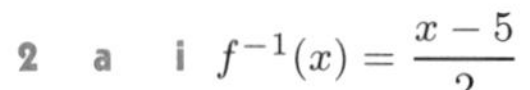 $f^{-1}(x) = \dfrac{x-5}{2}$

ii

b **i** $f^{-1}(x) = -2x + \frac{3}{2}$

ii

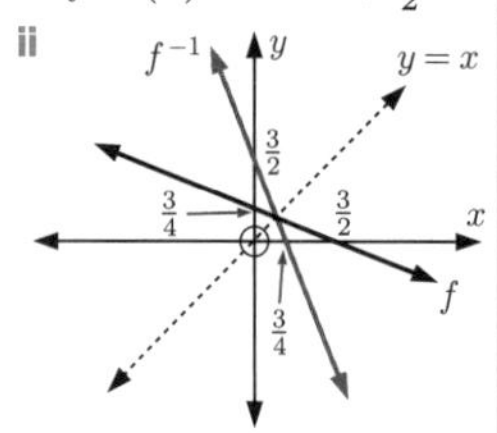

c **i** $f^{-1}(x) = x - 3$ **ii**

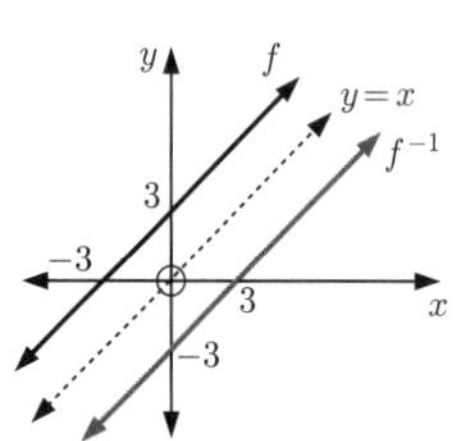

3 **a**

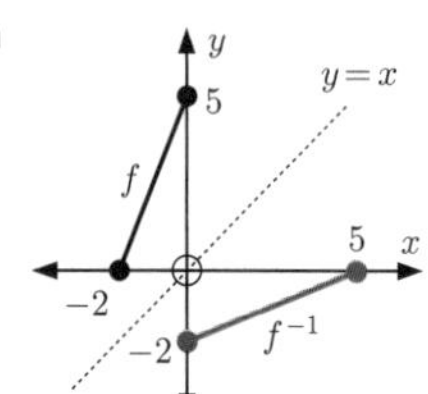

b

c

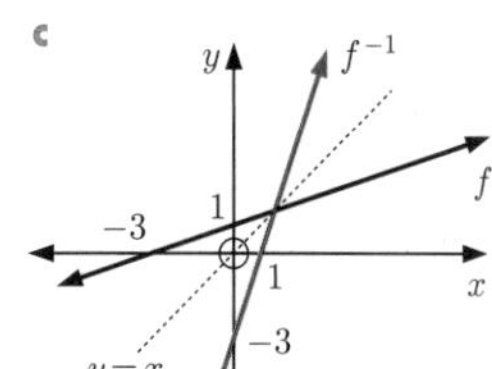

d

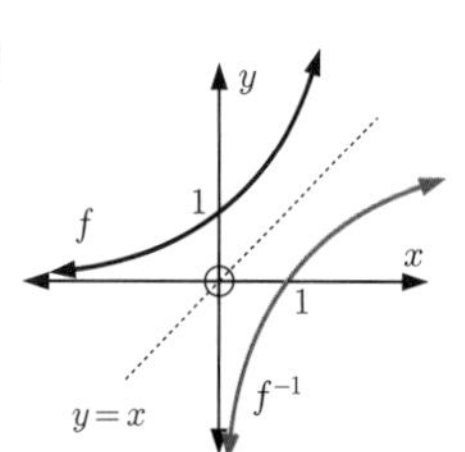

e

f

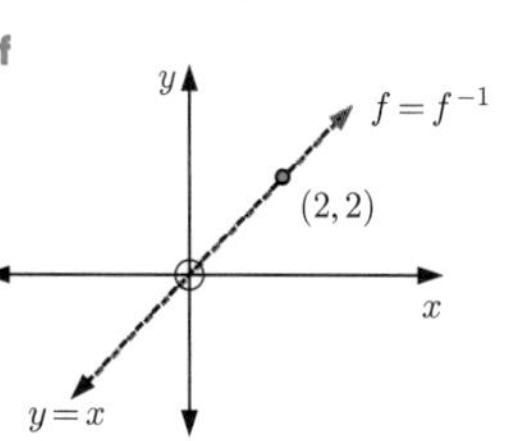

4 **a** $\{x \mid -2 \leqslant x \leqslant 0\}$ **b** $\{y \mid 0 \leqslant y \leqslant 5\}$

c $\{x \mid 0 \leqslant x \leqslant 5\}$ **d** $\{y \mid -2 \leqslant y \leqslant 0\}$

5 **a** f and f^{-1} are the same. They are self-inverse functions.

b For example, any linear function with slope $= -1$.

c For example, any of $y = \dfrac{a}{x}$, $a \in \mathbb{R}$, $a \neq 0, 1$.

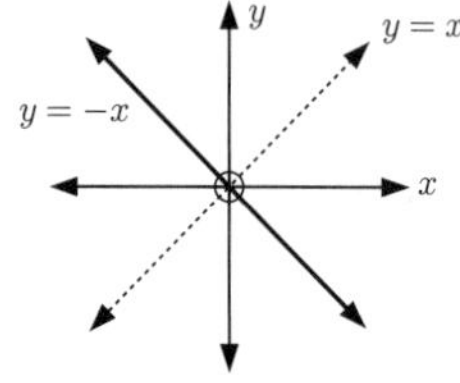

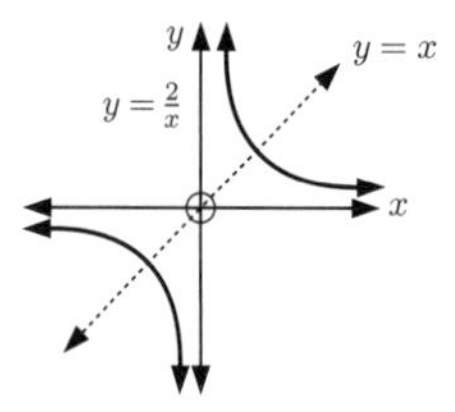

6 $f(x)$ is the same as $(f^{-1})^{-1}(x)$

7 **a** $\{(2, 1), (4, 2), (5, 3)\}$ **b** inverse does not exist

c $\{(0, -1), (1, 2), (2, 0), (3, 1)\}$

d $\{(-1, -1), (0, 0), (1, 1)\}$

8 **a**

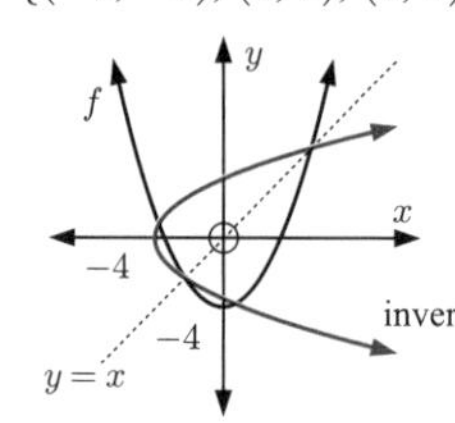

9

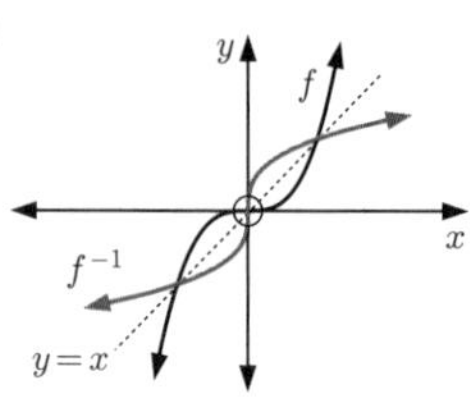

b No

c Yes, it is $y = \sqrt{x+4}$

10 $f^{-1}(x) = \dfrac{1}{x}$ and $f(x) = \dfrac{1}{x}$ $\therefore$ $f = f^{-1}$

$\therefore$ f is a self-inverse function

11 **a** $y = \dfrac{3x-8}{x-3}$ is symmetrical about $y = x$

$\therefore$ f is a self-inverse function.

b $f^{-1}(x) = \dfrac{3x-8}{x-3}$ and $f(x) = \dfrac{3x-8}{x-3}$

$\therefore$ $f = f^{-1}$ $\therefore$ f is a self-inverse function

12 **b** **i** is the only one

c **i** Domain $= \{x \mid x \leqslant 1\}$ **ii** Domain $= \{x \mid x \geqslant 1\}$

13 **a** $f^{-1}(x) = -\sqrt{x}$ **b**

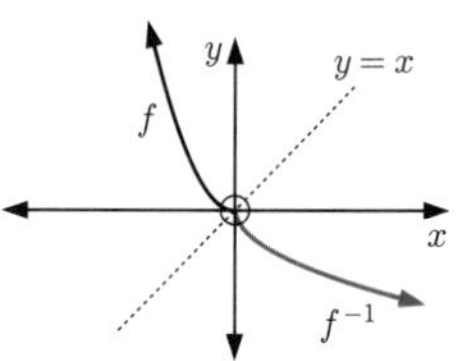

14 **a**

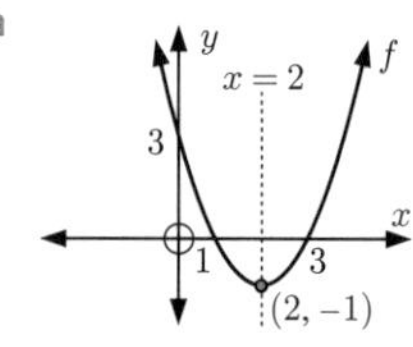

A horizontal line above the vertex cuts the graph **twice**. So, it does not have an inverse.

b **i** For $x \geqslant 2$, all horizontal lines cut 0 or once only, $\therefore$ has an inverse.

ii **Hint:** Inverse is $x = y^2 - 4y + 3$ for $y \geqslant 2$

iii **A** Domain $= \{x \mid x \geqslant 2\}$, Range $= \{y \mid y \geqslant -1\}$

B Domain $= \{x \mid x \geqslant -1\}$, Range $= \{y \mid y \geqslant 2\}$

iv **Hint:** Find $(g \circ g^{-1})(x)$ and $(g^{-1} \circ g)(x)$ and show that they both equal x.

15 **a** $f^{-1}(x) = \sqrt{x-3} - 1$, $x \geqslant 3$

b

c **i** Domain $= \{x \mid x \geqslant -1\}$
Range $= \{y \mid y \geqslant 3\}$

ii Domain $= \{x \mid x \geqslant 3\}$
Range $= \{y \mid y \geqslant -1\}$

16 **a** 10 **c** $x = 3$ **17** **a** **i** 25 **ii** 16 **b** $x = 1$

18 $(f^{-1} \circ g^{-1})(x) = \dfrac{x+3}{8}$ and $(g \circ f)^{-1}(x) = \dfrac{x+3}{8}$

19 **a** Is not **b** Is **c** Is **d** Is **e** Is

20 **a** B is $(f(x), x)$

EXERCISE 2K

1 a i x-intercepts -3, 0, and 4, y-intercept 0
ii max. $(-1.69, 6.30)$, min. $(2.36, -10.4)$
iii no asymptotes
iv Domain $= \{x \mid x \in \mathbb{R}\}$, Range $= \{y \mid y \in \mathbb{R}\}$
v

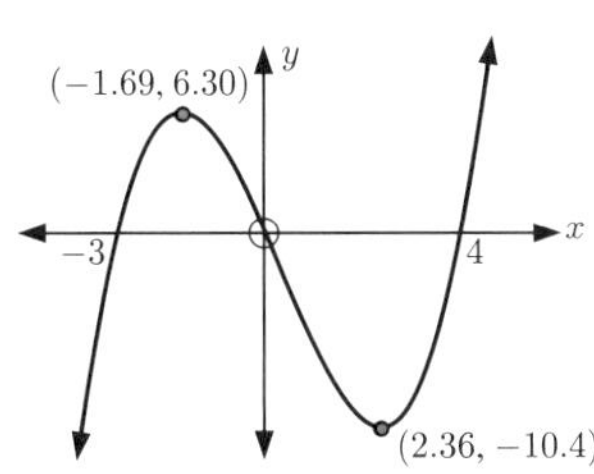

b i x-intercepts -4.97 and -1.55, y-intercept 2
ii local minima at $(-3.88, -33.5)$, $(0, 2)$
local maximum at $(-0.805, 2.97)$
iii no asymptotes
iv Domain $= \{x \mid -5 \leqslant x \leqslant 1\}$,
Range $= \{y \mid -33.5 \leqslant y \leqslant 12.8\}$
v

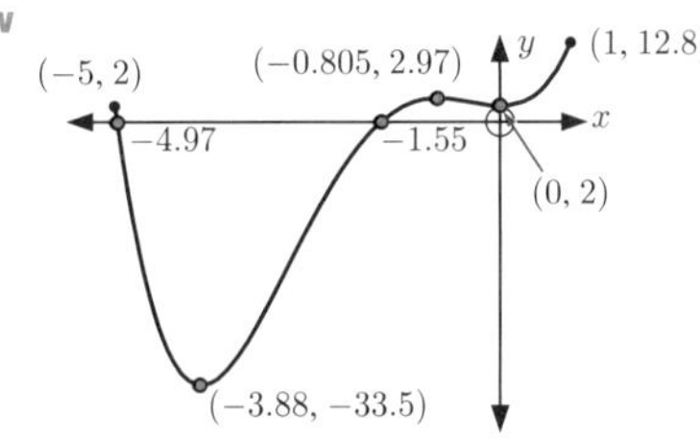

c i x-intercepts -0.449 and 4.45, y-intercept -0.125
ii local maximum at $(5, 3)$
iii vertical asymptote of $x = 4$,
horizontal asymptote of $y = 1$
iv Domain $= \{x \mid x \neq 4\}$, Range $= \{y \mid y \leqslant 3\}$
v

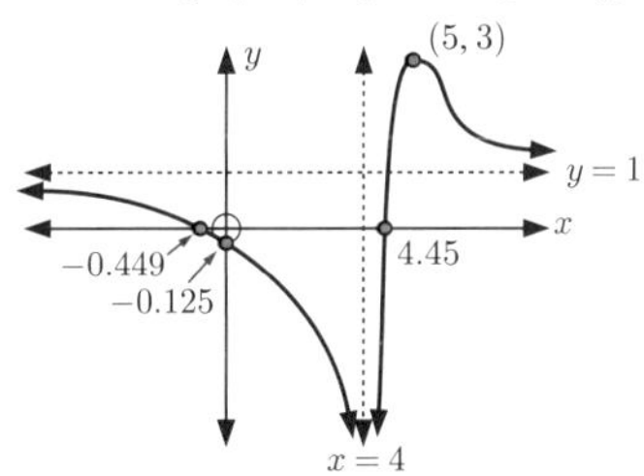

d i x-intercepts -1 and 1, y-intercept 0.25
ii local maximum at $(-0.5, 0.333)$
iii vertical asymptote of $x = -2$,
no horizontal asymptote
iv Domain $= \{x \mid -5 \leqslant x \leqslant 5,\ x \neq 2\}$,
Range $= \{y \mid y \leqslant 0.333\}$
v

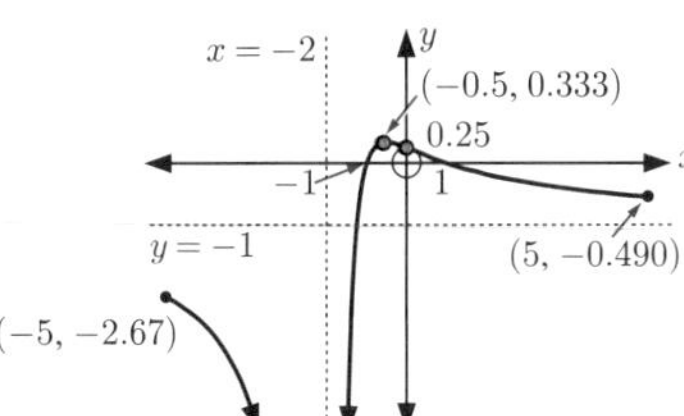

2 a max. 28.0
b i max. 6.27 ii max. 4.03 iii max. 6.27

EXERCISE 2L

1 a

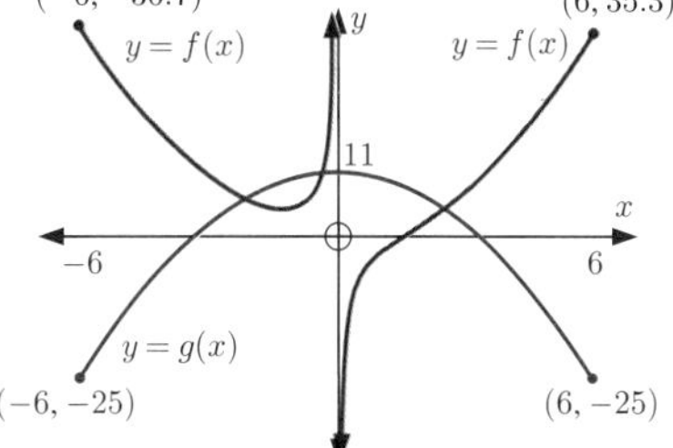

b $x = -0.373$ and -2.14

2

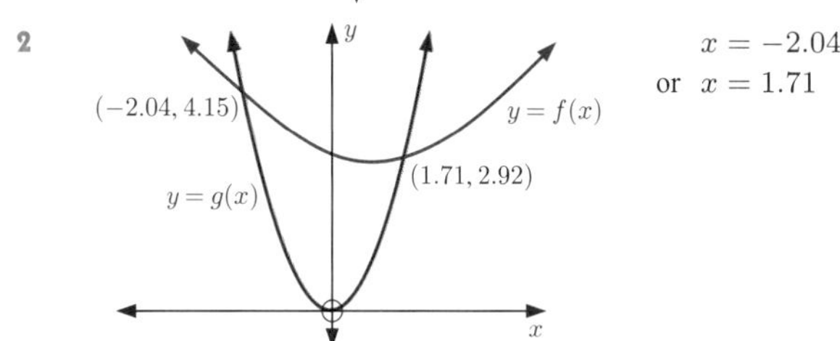

$x = -2.04$ or $x = 1.71$

3 a

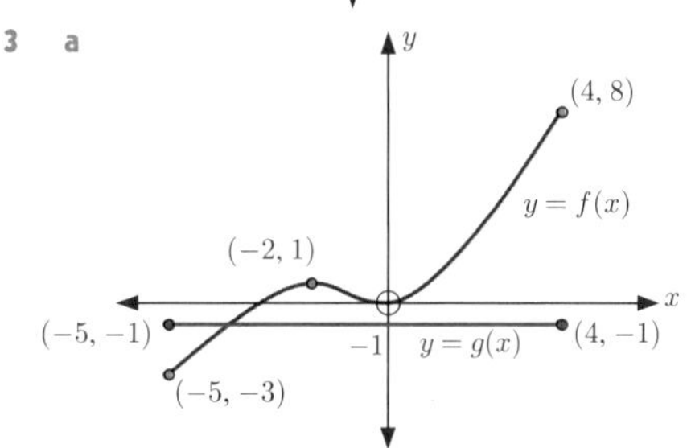

b one solution
c i $0 < k < 1$ ii $k = 0$ or 1
iii $-3 \leqslant k < 0$ or $1 < k \leqslant 8$ iv no solutions

REVIEW SET 2A

1 a i Domain $= \{x \mid x \in \mathbb{R}\}$ ii Range $= \{y \mid y > -4\}$
iii Yes
b i Domain $= \{x \mid x \in \mathbb{R}\}$ ii Range $= \{2\}$ iii Yes
c i Domain $= \{x \mid x \in \mathbb{R}\}$
ii Range $= \{y \mid y \leqslant -1$ or $y \geqslant 1\}$ iii No
d i Domain $= \{x \mid x \in \mathbb{R}\}$
ii Range $= \{y \mid -5 \leqslant y \leqslant 5\}$ iii Yes

2 a 0 b -15 c $-\frac{5}{4}$ 3 $a = -6$, $b = 13$

4 a $f(x) = \sqrt{x}$, $g(x) = 1 - x^2$
b $g(x) = x^2$, $f(x) = \dfrac{x-2}{x+1}$

5 a $x = -1$ or 3 b $x \in \,]-\infty, -8[$ or $]2, \infty[$

6 a $x^2 - x - 2$ b $x^4 - 7x^2 + 10$

7 a i Domain $= \{x \mid x \in \mathbb{R}\}$, Range $= \{y \mid y \geqslant -5\}$
ii x-int -1, 5, y-int $-\frac{25}{9}$ iii is a function iv No
b i Domain $= \{x \mid x \in \mathbb{R}\}$, Range $= \{y \mid y = 1$ or $-3\}$
ii no x-intercepts, y-intercept 1 iii is a function
iv No

8 a odd b neither c even

9 a $f^{-1}(x) = \dfrac{x-2}{4}$ b $f^{-1}(x) = \dfrac{3-4x}{5}$

10 **a** sign diagram: $-$ | $+$ | $-$ at $-\frac{2}{3}$, 4 (x) **b** sign diagram: $-$ ¦ $-$ | $+$ at -2, 3 (x)

11 $a = 1,\ b = -1$

12 **a** sign diagram: $-$ | $+$ ¦ $-$ | $+$ at -2, 1, 3 (x) **b** $x \in\ ]-\infty, -2[\ \text{ or }\ ]1, 3[$

13 **a** $f(-3) = (-3)^2 = 9$ **b** 169 **c** $x = -4$
$g(-\frac{4}{3}) = 1 - 6(-\frac{4}{3}) = 9$

14 $(f^{-1} \circ h^{-1})(x) = (h \circ f)^{-1}(x) = x - 2$

15 **a** $h^{-1}(x) = 4 + \sqrt{x-3},\quad x \geqslant 3$

REVIEW SET 2B

1 **a** Domain $= \{x \mid x \in \mathbb{R}\}$, Range $= \{y \mid y \geqslant -4\}$
b Domain $= \{x \mid x \neq 0,\ x \neq 2\}$,
Range $= \{y \mid y \leqslant -1 \text{ or } y > 0\}$

2 **a** $2x^2 + 1$ **b** $4x^2 - 12x + 11$

3 **a** sign diagram: $-$ | $+$ ¦ $-$ | $+$ at -2, 3, 8 (x) **b** sign diagram: $-$ ¦ $+$ | $+$ at -5, -3 (x)

4 **a** $x = 0$ **b**

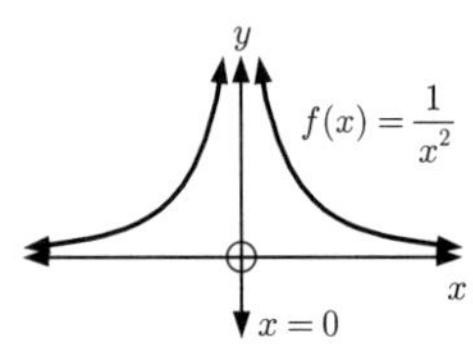

c Domain $= \{x \mid x \neq 0\}$, Range $= \{y \mid y > 0\}$

5 **a** $a = 2,\ b = -1$
b Domain $= \{x \mid x \neq 2\}$, Range $= \{y \mid y \neq -1\}$

6 **a** Domain $= \{x \mid x > -3\}$, Range $= \{y \mid -3 < y < 5\}$
b Domain $= \{x \mid x \neq 1\}$, Range $= \{y \mid y \leqslant -3 \text{ or } y \geqslant 5\}$

7 **a** $x = 1$ or 7 **b** $x \in\ ]-\infty, -\frac{1}{5}]\ \text{ or }\ [5, \infty[$

8 **a** $x \in [-1, 5]$ **b** $x \in\ ]-8, -\frac{1}{2}[\ \text{ or }\ ]1, \infty[$

9 **a**

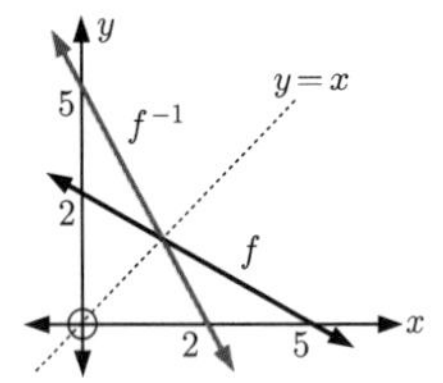

b

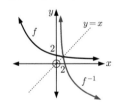

10 **a** vertical asymptote $x = 2$, horizontal asymptote $y = -4$
b Domain $= \{x \mid x \neq 2\}$, Range $= \{y \mid y \neq -4\}$
c as $x \to 2^-$, $y \to \infty$ as $x \to \infty$, $y \to -4^-$
as $x \to 2^+$, $y \to -\infty$ as $x \to -\infty$, $y \to -4^+$
d x-intercept $-\frac{1}{4}$, y-intercept $\frac{1}{2}$
e

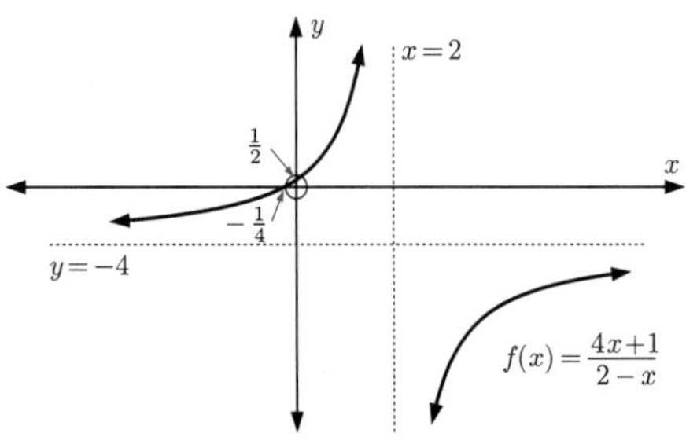

11 $a = 6$

12 **a** $x \in\ ]-\infty, 1[\ \text{ or }\ ]9, \infty[$
b

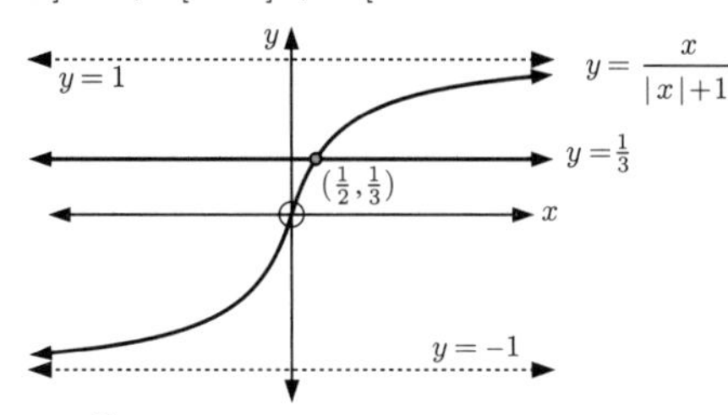

$\therefore\ \dfrac{x}{|x|+1} \geqslant \frac{1}{3}$ for $x \in [\frac{1}{2}, \infty[$

13 **a** $(g \circ f)(x) = \dfrac{2}{3x+1}$ **b** $x = -\frac{1}{2}$
c **i** vertical asymptote $x = -\frac{1}{3}$,
horizontal asymptote $y = 0$
ii

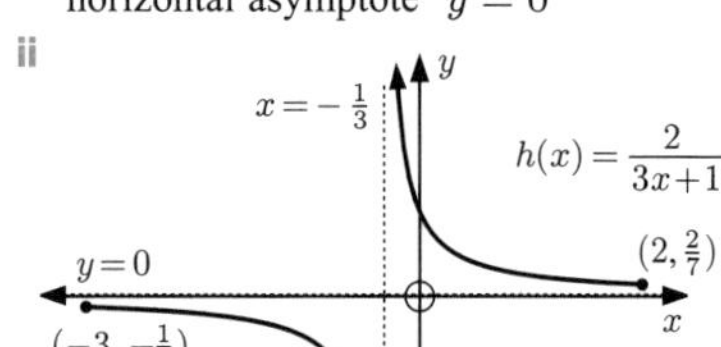

iii Range $= \{y \mid y \leqslant -\frac{1}{4} \text{ or } y \geqslant \frac{2}{7}\}$

14 **a**

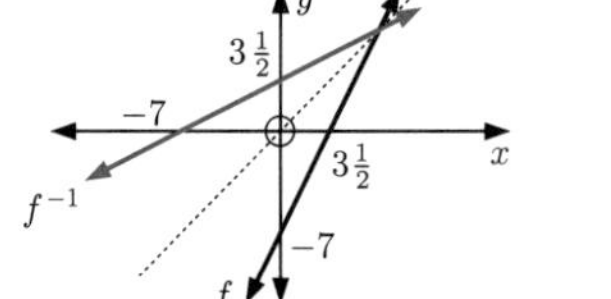

b $f^{-1}(x) = \dfrac{x+7}{2}$

15 **a**

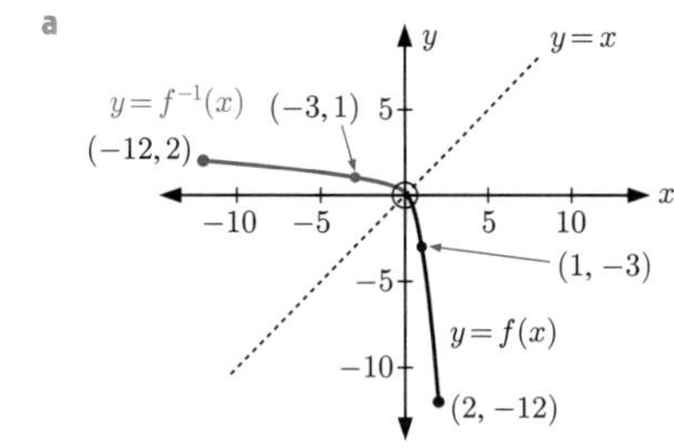

b Range $= \{y \mid 0 \leqslant y \leqslant 2\}$
c **i** $x \approx 1.83$ **ii** $x = -3$

16 **a**

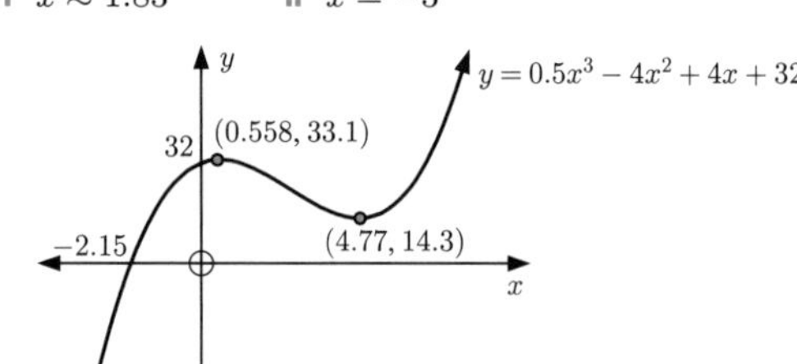

b local maximum at $(0.558, 33.1)$
local minimum at $(4.77, 14.3)$
c maximum is 33.1, minimum is 14.3

REVIEW SET 2C

1 a Domain $= \{x \mid x \geqslant -2\}$, Range $= \{y \mid 1 \leqslant y < 3\}$
  b Domain $= \{x \in \mathbb{R}\}$, Range $= \{-1, 1, 2\}$

2 a 12 b $x = \pm 1$

3 a $x = \frac{1}{2}$ b $x < -7$

4 a $-$ $+$ $-$ $+$; -4, $-\frac{1}{3}$, 0, x
  b $-$ $+$ $-$; -8, -1, x

5 a $10 - 6x$ b $x = 2$

6 a i $1 - 2\sqrt{x}$ ii $\sqrt{1 - 2x}$
  b i Domain $= \{x \mid x \geqslant 0\}$, Range $= \{y \mid y \leqslant 1\}$
   ii Domain $= \{x \mid x \leqslant 0.5\}$, Range $= \{y \mid y \geqslant 0\}$

7 $a = 1$, $b = -6$, $c = 5$

8 a 

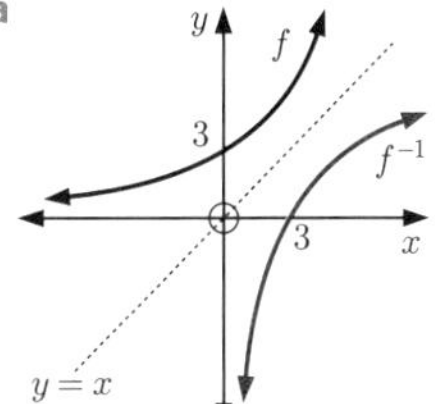

b 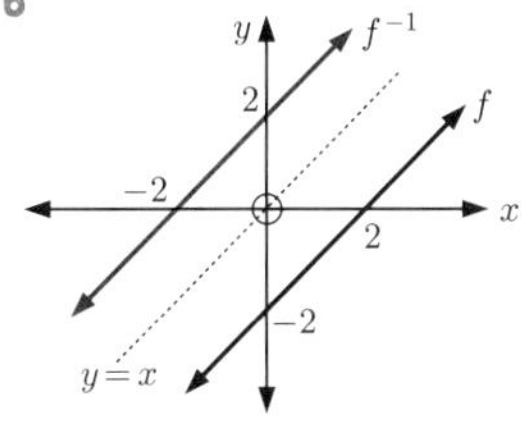

10 a $f^{-1}(x) = \dfrac{7 - x}{4}$ b $f^{-1}(x) = \dfrac{5x - 3}{2}$

11 $(f^{-1} \circ h^{-1})(x) = (h \circ f)^{-1}(x) = \dfrac{4x + 6}{15}$

12 a $x \in [-\frac{5}{2}, 2]$ b $x \in]-2, -1[$ or $]4, \infty[$

13 16

14 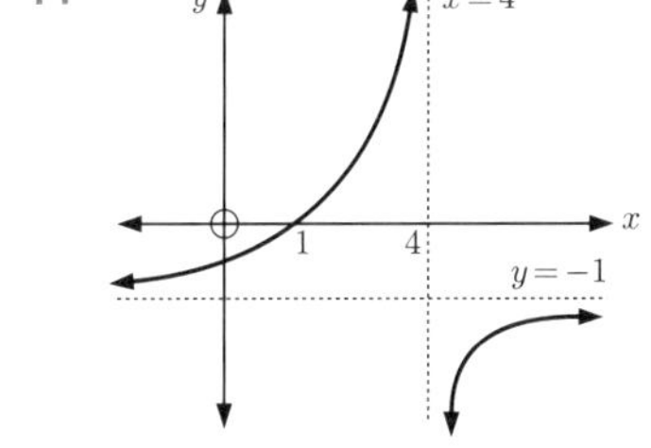

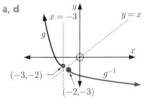

15 a, d $x = -3$, $y = x$, g, x, $(-3, -2)$, g^{-1}, $(-2, -3)$
  b Any horizontal line cuts the graph at most once.
  c $g(x) = -3 - \sqrt{x + 2}$, $x \geqslant -2$
  e Range of g $\{y \mid y \geqslant -2\}$
  f Domain of g^{-1} $\{x \mid x \geqslant -2\}$,
   Range of g^{-1} $\{y \mid y \leqslant -3\}$

16 a x-intercept 2.61, y-intercept 4.29
  b local maximum at $(-0.973, 4.47)$
  c vertical asymptote $x = 4$,
   horizontal asymptotes $y = 3$, $y = 7$
  d Domain $= \{x \mid x \neq 4\}$, Range $= \{y \mid y \leqslant 4.47,\ y > 7\}$
  e 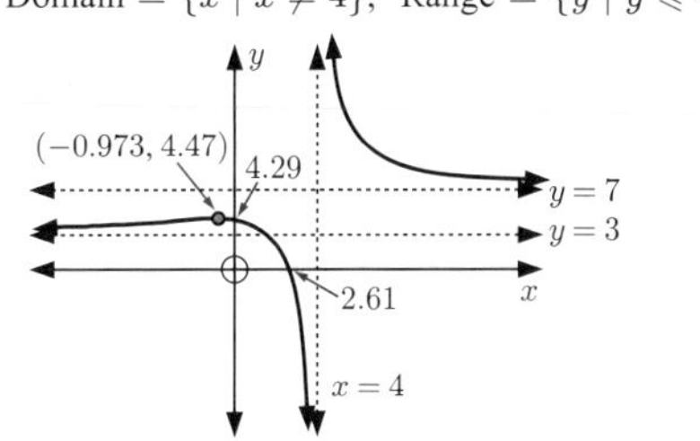

EXERCISE 3A

1 a $2^1 = 2$, $2^2 = 4$, $2^3 = 8$, $2^4 = 16$, $2^5 = 32$, $2^6 = 64$
  b $3^1 = 3$, $3^2 = 9$, $3^3 = 27$, $3^4 = 81$, $3^5 = 243$, $3^6 = 729$
  c $4^1 = 4$, $4^2 = 16$, $4^3 = 64$, $4^4 = 256$, $4^5 = 1024$, $4^6 = 4096$

2 a $5^1 = 5$, $5^2 = 25$, $5^3 = 125$, $5^4 = 625$
  b $6^1 = 6$, $6^2 = 36$, $6^3 = 216$, $6^4 = 1296$
  c $7^1 = 7$, $7^2 = 49$, $7^3 = 343$, $7^4 = 2401$

3 a -1 b 1 c 1 d -1 e 1 f -1
  g -1 h -32 i -32 j -64 k 625 l -625

4 a 16 384 b 2401 c -3125 d -3125
  e 262 144 f 262 144 g $-262\,144$
  h 902.436 039 6 i $-902.436\,039\,6$
  j $-902.436\,039\,6$

5 a $0.\overline{1}$ b $0.\overline{1}$ c $0.02\overline{7}$ d $0.02\overline{7}$
  e 0.012 345 679 f 0.012 345 679 g 1 h 1

Notice that $a^{-n} = \dfrac{1}{a^n}$

6 3 7 7

EXERCISE 3B

1 a 5^{11} b d^8 c k^5 d $\frac{1}{7}$ e x^{10} f 3^{16}
  g p^{-4} h n^{12} i 5^{3t} j 7^{x+2} k 10^{3-q} l c^{4m}

2 a 2^2 b 2^{-2} c 2^3 d 2^{-3} e 2^5 f 2^{-5}
  g 2^1 h 2^{-1} i 2^6 j 2^{-6} k 2^7 l 2^{-7}

3 a 3^2 b 3^{-2} c 3^3 d 3^{-3} e 3^1 f 3^{-1}
  g 3^4 h 3^{-4} i 3^0 j 3^5 k 3^{-5}

4 a 2^{a+1} b 2^{b+2} c 2^{t+3} d 2^{2x+2} e 2^{n-1}
  f 2^{c-2} g 2^{2m} h 2^{n+1} i 2^1 j 2^{3x-1}

5 a 3^{p+2} b 3^{3a} c 3^{2n+1} d 3^{d+3} e 3^{3t+2}
  f 3^{y-1} g 3^{1-y} h 3^{2-3t} i 3^{3a-1} j 3^3

6 a $4a^2$ b $27b^3$ c a^4b^4 d p^3q^3 e $\dfrac{m^2}{n^2}$
  f $\dfrac{a^3}{27}$ g $\dfrac{b^4}{c^4}$ h $1,\ b \neq 0$ i $\dfrac{m^4}{81n^4}$ j $\dfrac{x^3y^3}{8}$

7 a $4a^2$ b $36b^4$ c $-8a^3$ d $-27m^6n^6$
  e $16a^4b^{16}$ f $\dfrac{-8a^6}{b^6}$ g $\dfrac{16a^6}{b^2}$ h $\dfrac{9p^4}{q^6}$

8 a $\dfrac{a}{b^2}$ b $\dfrac{1}{a^2b^2}$ c $\dfrac{4a^2}{b^2}$ d $\dfrac{9b^2}{a^4}$ e $\dfrac{a^2}{bc^2}$
  f $\dfrac{a^2c^2}{b}$ g a^3 h $\dfrac{b^3}{a^2}$ i $\dfrac{2}{ad^2}$ j $12am^3$

9 a a^{-n} b b^n c 3^{n-2} d a^nb^m e a^{-2n-2}

10 a 1 b $\frac{4}{7}$ c 6 d 27 e $\frac{9}{16}$ f $\frac{5}{2}$
  g $\frac{27}{125}$ h $\frac{151}{5}$

11 a $5^3 = 21 + 23 + 25 + 27 + 29$
  b $7^3 = 43 + 45 + 47 + 49 + 51 + 53 + 55$
  c $12^3 = 133 + 135 + 137 + 139 + 141 + 143 + 145 + 147 + 149 + 151 + 153 + 155$

EXERCISE 3C

1 a $2^{\frac{1}{5}}$ b $2^{-\frac{1}{5}}$ c $2^{\frac{3}{2}}$ d $2^{\frac{5}{2}}$ e $2^{-\frac{1}{3}}$
  f $2^{\frac{4}{3}}$ g $2^{\frac{3}{2}}$ h $2^{\frac{3}{2}}$ i $2^{-\frac{4}{3}}$ j $2^{-\frac{3}{2}}$

2 a $3^{\frac{1}{3}}$ b $3^{-\frac{1}{3}}$ c $3^{\frac{1}{4}}$ d $3^{\frac{3}{2}}$ e $3^{-\frac{5}{2}}$

3 **a** $7^{\frac{1}{3}}$ **b** $3^{\frac{3}{4}}$ **c** $2^{\frac{4}{5}}$ **d** $2^{\frac{5}{3}}$ **e** $7^{\frac{2}{7}}$
f $7^{-\frac{1}{3}}$ **g** $3^{-\frac{3}{4}}$ **h** $2^{-\frac{4}{5}}$ **i** $2^{-\frac{5}{3}}$ **j** $7^{-\frac{2}{7}}$

4 **a** 2.28 **b** 1.83 **c** 0.794 **d** 0.435 **e** 1.68
f 1.93 **g** 0.523

5 **a** 8 **b** 32 **c** 8 **d** 125 **e** 4
f $\frac{1}{2}$ **g** $\frac{1}{27}$ **h** $\frac{1}{16}$ **i** $\frac{1}{81}$ **j** $\frac{1}{25}$

EXERCISE 3D.1

1 **a** $x^5 + 2x^4 + x^2$ **b** $4^x + 2^x$ **c** $x + 1$
d $49^x + 2(7^x)$ **e** $2(3^x) - 1$ **f** $x^2 + 2x + 3$
g $1 + 5(2^{-x})$ **h** $5^x + 1$ **i** $x^{\frac{3}{2}} + x^{\frac{1}{2}} + 1$

2 **a** $4^x + 2^{x+1} - 3$ **b** $9^x + 7(3^x) + 10$
c $25^x - 6(5^x) + 8$ **d** $4^x + 6(2^x) + 9$
e $9^x - 2(3^x) + 1$ **f** $16^x + 14(4^x) + 49$
g $x - 4$ **h** $4^x - 9$ **i** $x - x^{-1}$ **j** $x^2 + 4 + \dfrac{4}{x^2}$
k $7^{2x} - 2 + 7^{-2x}$ **l** $25 - 10(2^{-x}) + 4^{-x}$

EXERCISE 3D.2

1 **a** $5^x(5^x + 1)$ **b** $10(3^n)$ **c** $7^n(1 + 7^{2n})$
d $5(5^n - 1)$ **e** $6(6^{n+1} - 1)$ **f** $16(4^n - 1)$

2 **a** $(3^x+2)(3^x-2)$ **b** $(2^x+5)(2^x-5)$ **c** $(4+3^x)(4-3^x)$
d $(5+2^x)(5-2^x)$ **e** $(3^x+2^x)(3^x-2^x)$ **f** $(2^x+3)^2$
g $(3^x+5)^2$ **h** $(2^x-7)^2$ **i** $(5^x-2)^2$

3 **a** $(2^x+3)(2^x+6)$ **b** $(2^x+4)(2^x-5)$
c $(3^x+2)(3^x+7)$ **d** $(3^x+5)(3^x-1)$
e $(5^x+2)(5^x-1)$ **f** $(7^x-4)(7^x-3)$

4 **a** 2^n **b** 10^a **c** 3^b **d** $\dfrac{1}{5^n}$ **e** 5^x
f $(\frac{3}{4})^a$ **g** 5 **h** 5^n

5 **a** $3^m + 1$ **b** $1 + 6^n$ **c** $4^n + 2^n$ **d** $4^x - 1$
e 6^n **f** 5^n **g** 4 **h** $2^n - 1$ **i** $\frac{1}{2}$

6 **a** $n\,2^{n+1}$ **b** -3^{n-1}

EXERCISE 3E

1 **a** $x = 3$ **b** $x = 2$ **c** $x = 4$ **d** $x = 0$
e $x = -1$ **f** $x = \frac{1}{2}$ **g** $x = -3$ **h** $x = 2$
i $x = -3$ **j** $x = -4$ **k** $x = 2$ **l** $x = 1$

2 **a** $x = \frac{5}{3}$ **b** $x = -\frac{3}{2}$ **c** $x = -\frac{3}{2}$ **d** $x = -\frac{1}{2}$
e $x = -\frac{2}{3}$ **f** $x = -\frac{5}{4}$ **g** $x = \frac{3}{2}$ **h** $x = \frac{5}{2}$
i $x = \frac{1}{8}$ **j** $x = \frac{9}{2}$ **k** $x = -4$ **l** $x = -4$
m $x = 0$ **n** $x = \frac{7}{2}$ **o** $x = -2$ **p** $x = -6$

3 **a** $x = \frac{1}{7}$ **b** has no solutions **c** $x = 2\frac{1}{2}$

4 **a** $x = 3$ **b** $x = 3$ **c** $x = 2$
d $x = 2$ **e** $x = -2$ **f** $x = -2$

5 **a** $x = 1$ or 2 **b** $x = 1$ **c** $x = 1$ or 2
d $x = 1$ **e** $x = 2$ **f** $x = 0$

EXERCISE 3F

1 **a** 1.4 **b** 1.7 **c** 2.8 **d** 0.4

2 **a** $x \approx 1.6$ **b** $x \approx -0.7$

3 $y = 2^x$ has a horizontal asymptote of $y = 0$

4 **a** **b**

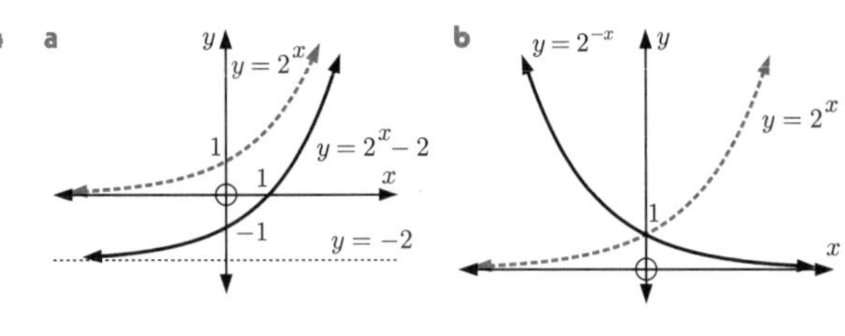

c **d**

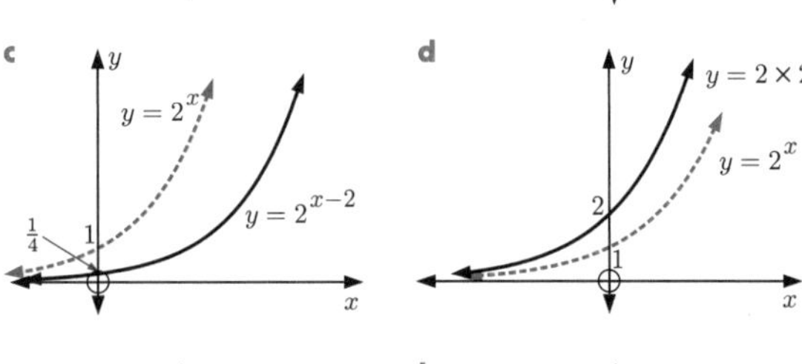

5 **a** **b** **c** **d**

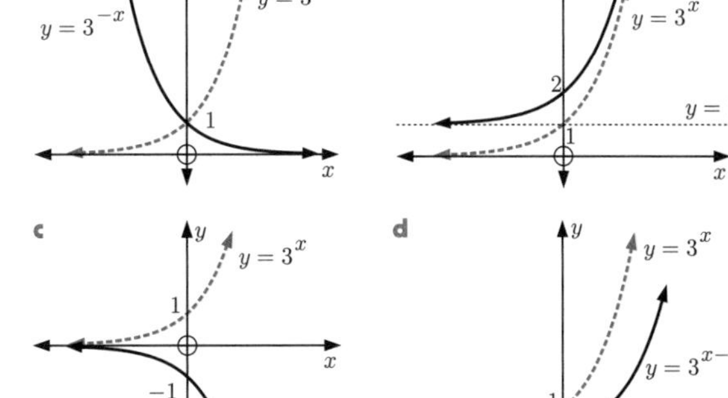

6 **a** **i**

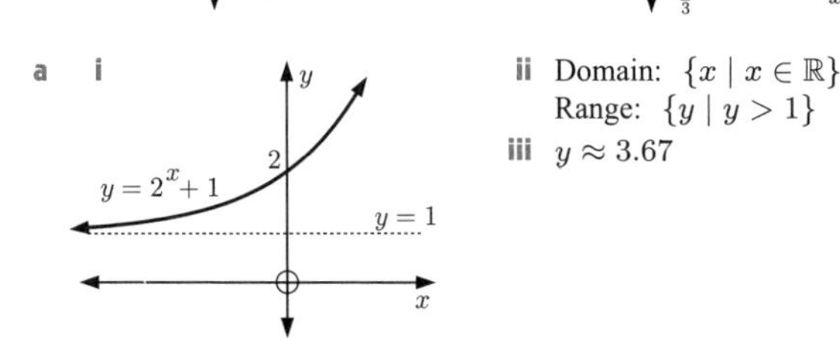

ii Domain: $\{x \mid x \in \mathbb{R}\}$
Range: $\{y \mid y > 1\}$
iii $y \approx 3.67$

iv As $x \to \infty$, $y \to \infty$
As $x \to -\infty$, $y \to 1$ from above
v $y = 1$

b **i**

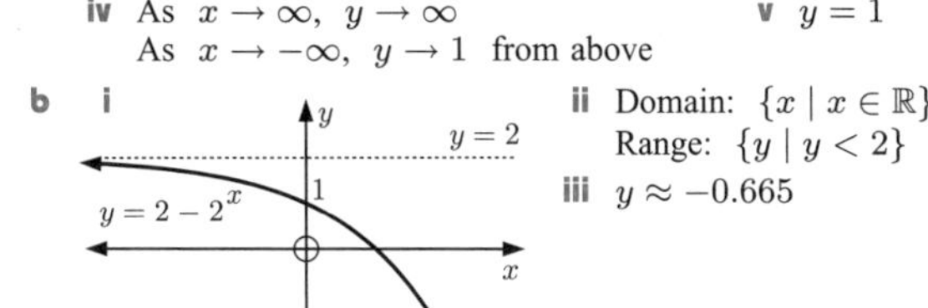

ii Domain: $\{x \mid x \in \mathbb{R}\}$
Range: $\{y \mid y < 2\}$
iii $y \approx -0.665$

iv As $x \to \infty$, $y \to -\infty$
As $x \to -\infty$, $y \to 2$ from below
v $y = 2$

c **i**

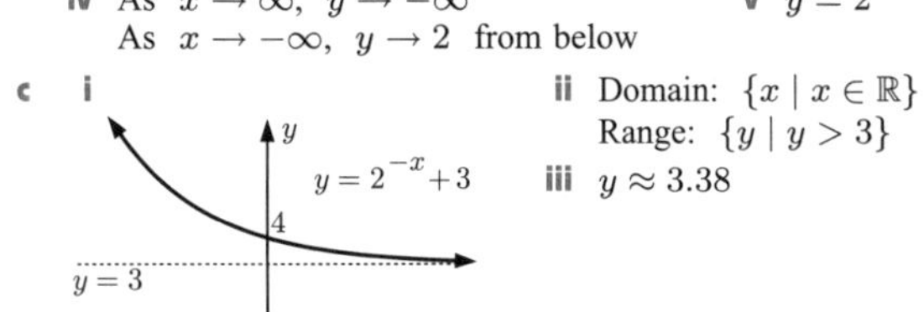

ii Domain: $\{x \mid x \in \mathbb{R}\}$
Range: $\{y \mid y > 3\}$
iii $y \approx 3.38$

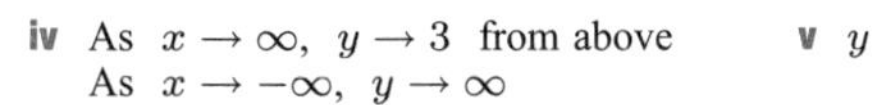

iv As $x \to \infty$, $y \to 3$ from above
As $x \to -\infty$, $y \to \infty$
v $y = 3$

d **i**

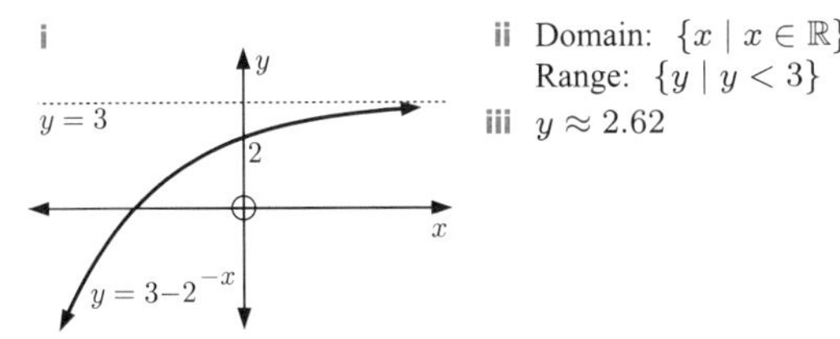

ii Domain: $\{x \mid x \in \mathbb{R}\}$
Range: $\{y \mid y < 3\}$

iii $y \approx 2.62$

iv As $x \to \infty$, $y \to 3$ from below
As $x \to -\infty$, $y \to -\infty$

v $y = 3$

EXERCISE 3G.1

1 **a** 100 grams

b **i** 132 g **ii** 200 g **iii** 528 g

c

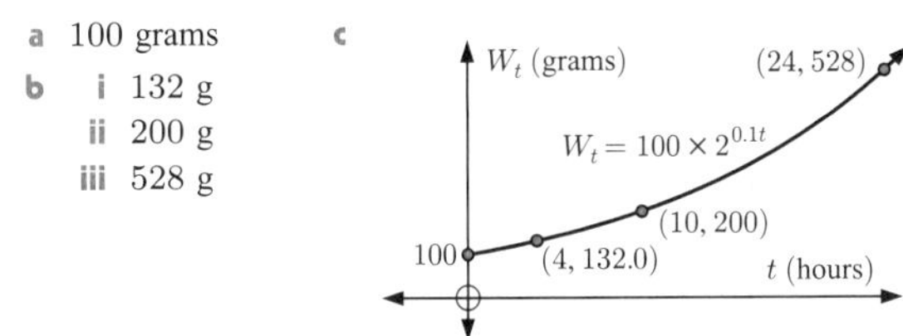

2 **a** 50

b

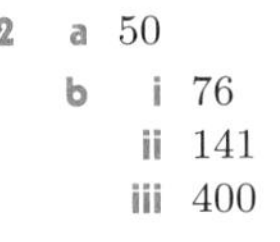
i 76 **ii** 141 **iii** 400

c

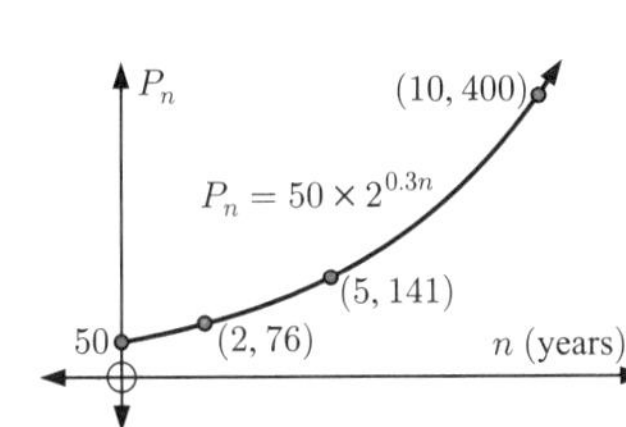

3 **a** 12 bears **b** 146 bears **c** 248% increase

4 **a** **i** V_0 **ii** $2V_0$ **b** 100%

c 183% increase, it is the percentage increase at 50°C compared with 20°C

EXERCISE 3G.2

1 **a** 250 g **b** **i** 112 g **ii** 50.4 g **iii** 22.6 g

c **d** $\approx$ 346 years

2 **a** 100°C

b

i 81.2°C **ii** 75.8°C **iii** 33.9°C

c

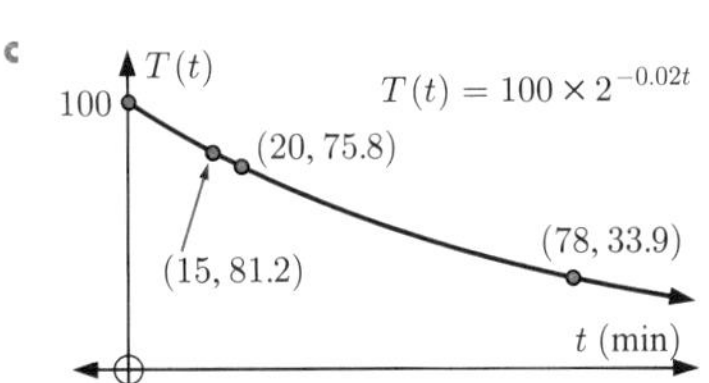

3 **a** **i** 22°C

ii 6°C **iii** −2°C **iv** −6°C

b

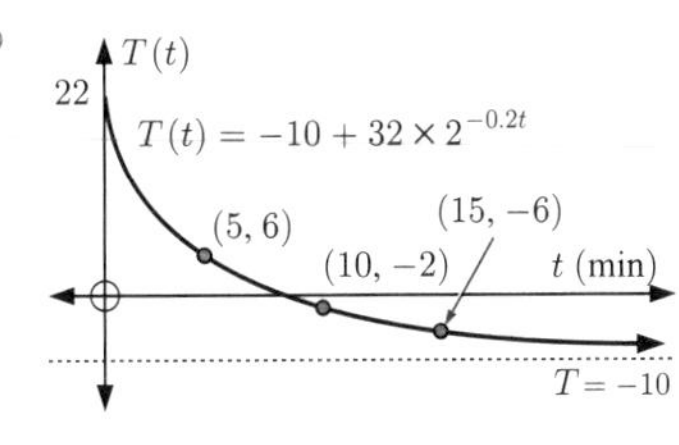

c Never, since $32 \times 2^{-0.2t}$ is always > 0

4 **a** 1000 g

b **i** 812 g **ii** 125 g **iii** 9.31×10^{-7} g

c

d 221 years

e $1000(1 - 2^{-0.03t})$ g

5 **a** W_0 **b** 12.9% **c** 45 000 years

EXERCISE 3H

1 The graph of $y = e^x$ lies between $y = 2^x$ and $y = 3^x$.

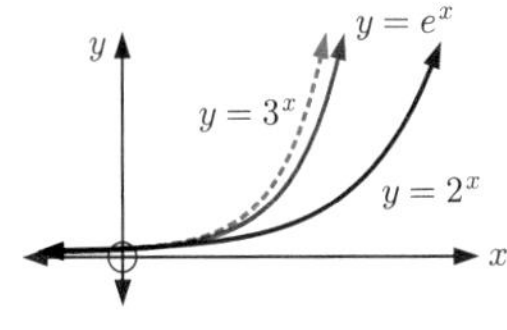

2 One is the other reflected in the y-axis.

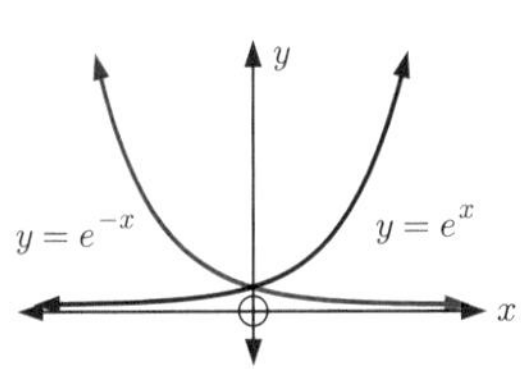

3 a

4 **a** $e^x > 0$ for all x

b **i** 0.000 000 004 12 **ii** 970 000 000

5 **a** ≈ 7.39 **b** ≈ 20.1 **c** ≈ 2.01 **d** ≈ 1.65 **e** ≈ 0.368

6 **a** $e^{\frac{1}{2}}$ **b** $e^{-\frac{1}{2}}$ **c** e^{-2} **d** $e^{\frac{3}{2}}$

7 **a** $e^{0.18t}$ **b** $e^{0.004t}$ **c** $e^{-0.005t}$ **d** $\approx e^{-0.167t}$

8 **a** 10.074 **b** 0.099 261 **c** 125.09 **d** 0.007 994 5 **e** 41.914 **f** 42.429 **g** 3540.3 **h** 0.006 342 4

9

Domain of f, g, and h is $\{x \mid x \in \mathbb{R}\}$
Range of f is $\{y \mid y > 0\}$, Range of g is $\{y \mid y > 0\}$
Range of h is $\{y \mid y > 3\}$

10

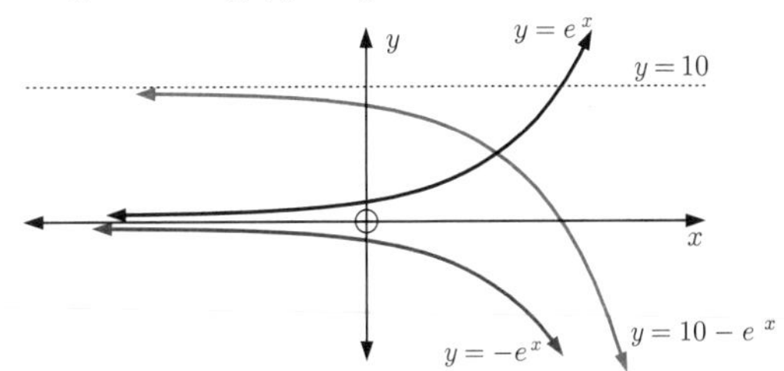

Domain of f, g, and h is $\{x \mid x \in \mathbb{R}\}$
Range of f is $\{y \mid y > 0\}$, Range of g is $\{y \mid y < 0\}$
Range of h is $\{y \mid y < 10\}$

11 **a** $e^{2x} + 2e^x + 1$ **b** $1 - e^{2x}$ **c** $1 - 3e^x$

12 **a** **i** 2 g **ii** 2.57 g **iii** 4.23 g **iv** 40.2 g **b**

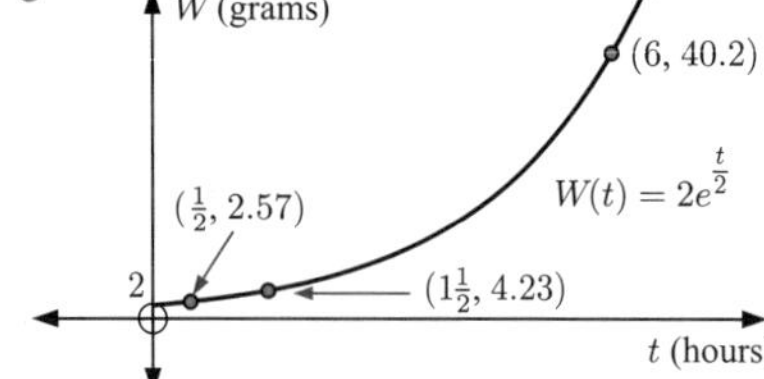

13 **a** $x = \frac{1}{2}$ **b** $x = -4$

14 **a** **i** 64.6 amps **ii** 16.7 amps **c** 28.8 seconds **b**

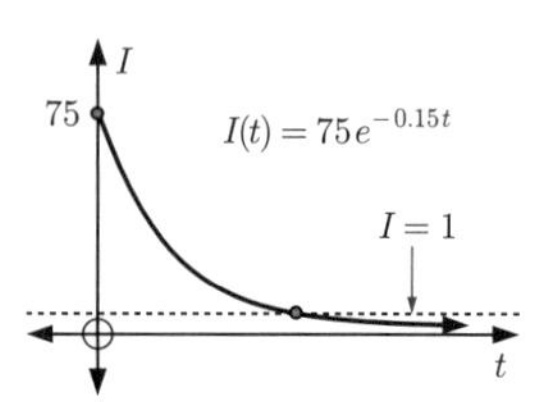

15 **a** $f^{-1}(x) = \log_e x$

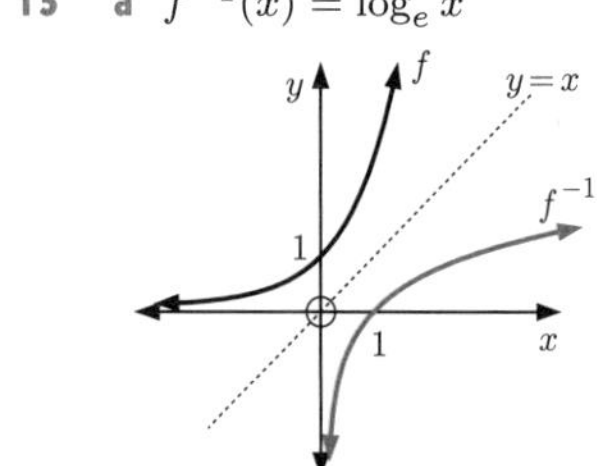

b Domain of f^{-1} is $\{x \mid x > 0\}$, Range of f^{-1} is $\{y \mid y \in \mathbb{R}\}$

REVIEW SET 3A

1 **a** -1 **b** 27 **c** $\frac{2}{3}$ **2** **a** a^6b^7 **b** $\dfrac{2}{3x}$ **c** $\dfrac{y^2}{5}$

3 **a** **i** 81 **ii** $\frac{1}{3}$ **b** $k = 9$

4 **a** $\dfrac{1}{x^5}$ **b** $\dfrac{2}{a^2b^2}$ **c** $\dfrac{2a}{b^2}$ **5** **a** 3^{3-2a} **b** $3^{\frac{5}{2}-\frac{9}{2}x}$

6 **a** 4 **b** $\frac{1}{9}$ **7** **a** $\dfrac{m}{n^2}$ **b** $\dfrac{1}{m^3n^3}$ **c** $\dfrac{m^2p^2}{n}$ **d** $\dfrac{16n^2}{m^2}$

8 **a** $9 - 6e^x + e^{2x}$ **b** $x - 4$ **c** $2^x + 1$

9 **a** $x = -2$ **b** $x = \frac{3}{4}$ **c** $x = -\frac{1}{4}$

10 **a** **C** **b** **E** **c** **A** **d** **B** **e** **D**

11 **a** y^2 **b** y^{-1} **c** $\dfrac{1}{\sqrt{y}}$ (or $y^{-\frac{1}{2}}$)

REVIEW SET 3B

1 **a** 2^{n+2} **b** $-\frac{6}{7}$ **c** $3\frac{3}{8}$ **d** $\dfrac{4}{a^2b^4}$

2 **a** 2.28 **b** 0.517 **c** 3.16 **3** **a** 3 **b** 24 **c** $\frac{3}{4}$

4 **a** $\frac{1}{\sqrt{2}} + 1 \approx 1.71$ **b** $a = -1$

5

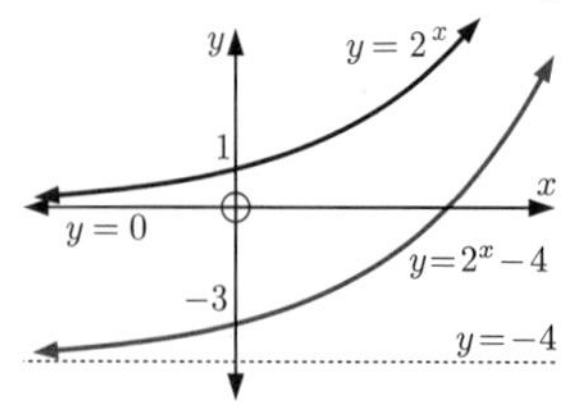

6 **a** 80°C **b** **i** 26.8°C **ii** 9.00°C **iii** 3.02°C **d** $\approx$ 12.8 min **c**

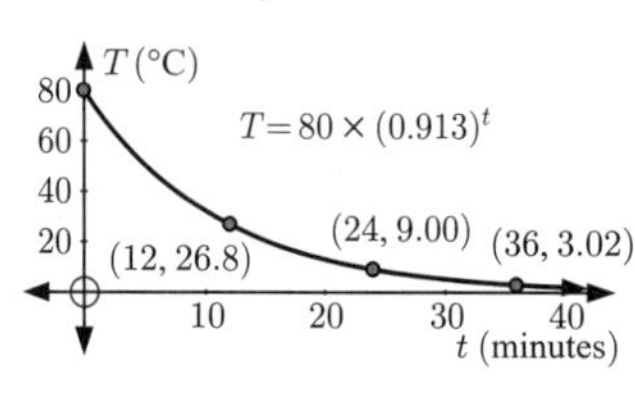

7 **a**

x	-2	-1	0	1	2
y	$-4\frac{8}{9}$	$-4\frac{2}{3}$	-4	-2	4

b as $x \to \infty$, $y \to \infty$; as $x \to -\infty$, $y \to -5$ (above)

c

d $y = -5$

8 **a**

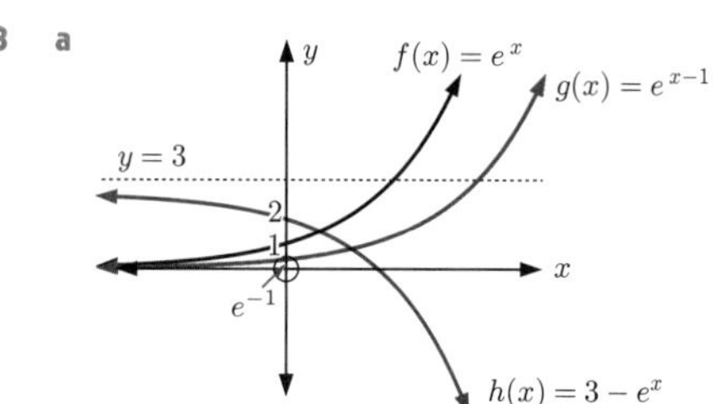

b Domain of f, g, and h is $\{x \mid x \in \mathbb{R}\}$
Range of f is $\{y \mid y > 0\}$, Range of g is $\{y \mid y > 0\}$,
Range of h is $\{y \mid y < 3\}$

9 **a**

x	-2	-1	0	1	2
y	-1	1	2	$2\frac{1}{2}$	$2\frac{3}{4}$

b as $x \to \infty$, $y \to 3$ (below); as $x \to -\infty$, $y \to -\infty$

c

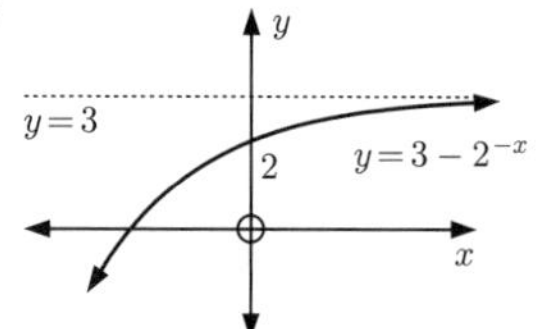

d $y = 3$

10 **a** 1500 g **b** **i** 90.3 g **ii** 5.44 g **d** 386 years **c**

REVIEW SET 3C

1 **a** $x \approx 1.5$ **b** $x \approx -0.6$ **c** $x \approx 1.1$

2 **a** a^{21} **b** p^4q^6 **c** $\dfrac{4b}{a^3}$

3 **a** 2^{-3} **b** 2^7 **c** 2^{12} **4** **a** $\dfrac{1}{b^3}$ **b** $\dfrac{1}{ab}$ **c** $\dfrac{a}{b}$

5 2^{2x} **6** **a** 5^0 **b** $5^{\frac{3}{2}}$ **c** $5^{-\frac{1}{4}}$ **d** 5^{2a+6}

7 **a** $1 + e^{2x}$ **b** $2^{2x} + 10(2^x) + 25$ **c** $x - 49$

8 **a** $x = 5$ **b** $x = -4$ **9** $k = \frac{3}{2}$

10 **a** $x = 4$ **b** $x = -\frac{2}{5}$

11 **a**

x	-2	-1	0	1	2
y	15.8	6.44	3	1.74	1.27

b as $x \to \infty$, $y \to 1$ (above); as $x \to -\infty$, $y \to \infty$

c

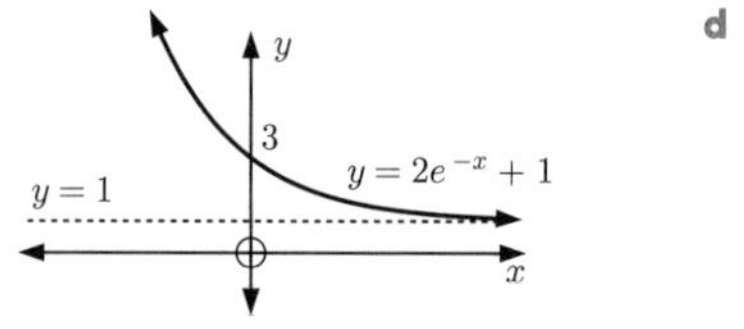

d $y = 1$

EXERCISE 4A

1 a 4 b -3 c 1 d 0 e $\frac{1}{2}$ f $\frac{1}{3}$ g $-\frac{1}{4}$ h $1\frac{1}{2}$ i $\frac{2}{3}$ j $1\frac{1}{2}$ k $1\frac{1}{3}$ l $3\frac{1}{2}$

2 a n b $a+2$ c $1-m$ d $a-b$

3 a $10^{0.7782}$ b $10^{1.7782}$ c $10^{3.7782}$ d $10^{-0.2218}$ e $10^{-2.2218}$ f $10^{1.1761}$ g $10^{3.1761}$ h $10^{0.1761}$ i $10^{-0.8239}$ j $10^{-3.8239}$

4 a i 0.477 ii 2.477 b $\log 300 = \log(3 \times 10^2)$

5 a i 0.699 ii -1.301 b $\log 0.05 = \log(5 \times 10^{-2})$

6 a $x = 100$ b $x = 10$ c $x = 1$ d $x = \frac{1}{10}$ e $x = 10^{\frac{1}{2}}$ f $x = 10^{-\frac{1}{2}}$ g $x = 10\,000$ h $x = 0.000\,01$ i $x \approx 6.84$ j $x \approx 140$ k $x \approx 0.0419$ l $x \approx 0.000\,631$

EXERCISE 4B

1 a $10^2 = 100$ b $10^4 = 10\,000$ c $10^{-1} = 0.1$ d $10^{\frac{1}{2}} = \sqrt{10}$ e $2^3 = 8$ f $3^2 = 9$ g $2^{-2} = \frac{1}{4}$ h $3^{1.5} = \sqrt{27}$ i $5^{-\frac{1}{2}} = \frac{1}{\sqrt{5}}$

2 a $\log_2 4 = 2$ b $\log_4 64 = 3$ c $\log_5 25 = 2$ d $\log_7 49 = 2$ e $\log_2 64 = 6$ f $\log_2(\frac{1}{8}) = -3$ g $\log_{10} 0.01 = -2$ h $\log_2(\frac{1}{2}) = -1$ i $\log_3(\frac{1}{27}) = -3$

3 a 5 b -2 c $\frac{1}{2}$ d 3 e 6 f 7 g 2 h 3 i -3 j $\frac{1}{2}$ k 2 l $\frac{1}{2}$ m 5 n $\frac{1}{3}$ o n, $a > 0$ p $\frac{1}{3}$ q -1, $t > 0$ r $\frac{3}{2}$ s 0 t 1

4 a ≈ 2.18 b ≈ 1.40 c ≈ 1.87 d ≈ -0.0969

5 a $x = 8$ b $x = 2$ c $x = 3$ d $x = 14$

6 a 2 b 2 c -1 d $\frac{3}{4}$ e $-\frac{1}{2}$ f $\frac{5}{2}$ g $-\frac{3}{2}$ h $-\frac{3}{4}$ i 2, $x > 0$ j $\frac{1}{2}$, $x > 0$ k 3, $m > 0$ l $\frac{3}{2}$, $x > 0$ m -1, $n > 0$ n -2, $a > 0$ o $-\frac{1}{2}$, $a > 0$ p $\frac{5}{2}$, $m > 0$

EXERCISE 4C.1

1 a $\log 16$ b $\log 20$ c $\log 8$ d $\log \dfrac{p}{m}$ e 1 f $\log 2$ g $\log 24$ h $\log_2 6$ i $\log 0.4$ j 1 k $\log 200$ l $\log(10^t \times w)$ m $\log_m \left(\dfrac{40}{m^2}\right)$ n 0 o $\log(0.005)$ p $\log_5(\frac{5}{2})$ q 2 r $\log 28$

2 a $\log 96$ b $\log 72$ c $\log 8$ d $\log_3(\frac{25}{8})$ e 1 f $\log \frac{1}{2}$ g $\log 20$ h $\log 25$ i $\log_n \left(\dfrac{n^2}{10}\right)$

3 a 2 b $\frac{3}{2}$ c 3 d $\frac{1}{2}$ e -2 f $-\frac{3}{2}$

4 For example, for a, $\log 9 = \log 3^2 = 2\log 3$

5 a $p+q$ b $2q+r$ c $2p+3q$ d $r+\frac{1}{2}q-p$ e $r-5p$ f $p-2q$

6 a $x+z$ b $z+2y$ c $x+z-y$ d $2x+\frac{1}{2}y$ e $3y-\frac{1}{2}z$ f $2z+\frac{1}{2}y-3x$

7 a 0.86 b 2.15 c 1.075

EXERCISE 4C.2

1 a $\log y = x\log 2$ b $\log y \approx 1.30 + 3\log b$ c $\log M = \log a + 4\log d$ d $\log T \approx 0.699 + \frac{1}{2}\log d$ e $\log R = \log b + \frac{1}{2}\log l$ f $\log Q = \log a - n\log b$ g $\log y = \log a + x\log b$ h $\log F \approx 1.30 - \frac{1}{2}\log n$ i $\log L = \log a + \log b - \log c$ j $\log N = \frac{1}{2}\log a - \frac{1}{2}\log b$ k $\log S \approx 2.30 + t\log 2$ l $\log y = m\log a - n\log b$

2 a $D = 2e$ b $F = \dfrac{5}{t}$ c $P = \sqrt{x}$ d $M = b^2c$ e $B = \dfrac{m^3}{n^2}$ f $N = \dfrac{1}{\sqrt[3]{p}}$ g $P = 10x^3$ h $Q = \dfrac{a^2}{x}$

3 a $\log_2 y = \log_2 3 + x$ b $x = \log_2\left(\dfrac{y}{3}\right)$ c i $x = 0$ ii $x = 2$ iii $x \approx 3.32$

4 a $x = 9$ b $x = 2$ or 4 c $x = 25\sqrt{5}$ d $x = 200$ e $x = 5$ f $x = 3$

EXERCISE 4D.1

1 a 2 b 3 c $\frac{1}{2}$ d 0 e -1 f $\frac{1}{3}$ g -2 h $-\frac{1}{2}$

2 a 3 b 9 c $\frac{1}{5}$ d $\frac{1}{4}$

3 x does not exist such that $e^x = -2$ or 0

4 a a b $a+1$ c $a+b$ d ab e $a-b$

5 a $e^{1.7918}$ b $e^{4.0943}$ c $e^{8.6995}$ d $e^{-0.5108}$ e $e^{-5.1160}$ f $e^{2.7081}$ g $e^{7.3132}$ h $e^{0.4055}$ i $e^{-1.8971}$ j $e^{-8.8049}$

6 a $x \approx 20.1$ b $x \approx 2.72$ c $x = 1$ d $x \approx 0.368$ e $x \approx 0.006\,74$ f $x \approx 2.30$ g $x \approx 8.54$ h $x \approx 0.037\,0$

EXERCISE 4D.2

1 a $\ln 45$ b $\ln 5$ c $\ln 4$ d $\ln 24$ e $\ln 1 = 0$ f $\ln 30$ g $\ln(4e)$ h $\ln\left(\frac{6}{e}\right)$ i $\ln 20$ j $\ln(4e^2)$ k $\ln\left(\frac{20}{e^2}\right)$ l $\ln 1 = 0$

2 a $\ln 972$ b $\ln 200$ c $\ln 1 = 0$ d $\ln 16$ e $\ln 6$ f $\ln\left(\frac{1}{3}\right)$ g $\ln\left(\frac{1}{2}\right)$ h $\ln 2$ i $\ln 16$

3 For example, for a, $\ln 27 = \ln 3^3 = 3\ln 3$

4 **Hint:** In d, $\ln\left(\frac{e^2}{8}\right) = \ln e^2 - \ln 2^3$

5 a $D = ex$ b $F = \dfrac{e^2}{p}$ c $P = \sqrt{x}$ d $M = e^3y^2$ e $B = \dfrac{t^3}{e}$ f $N = \dfrac{1}{\sqrt[3]{g}}$ g $Q \approx 8.66x^3$ h $D \approx 0.518n^{0.4}$

EXERCISE 4E

1 a $x = \dfrac{1}{\log 2}$ b $x = \dfrac{\log 20}{\log 3}$ c $x = \dfrac{2}{\log 4}$ d $x = 4$ e $x = -\dfrac{1}{\log(\frac{3}{4})}$ f $x = -5$

2 a $x = \ln 10$ b $x = \ln 1000$ c $x = \ln 0.15$ d $x = 2\ln 5$ e $x = \frac{1}{2}\ln 18$ f $x = 0$

3 a $t = \dfrac{\log R - \log 200}{0.25\log 2}$ b i $t \approx 6.34$ ii $t \approx 11.3$

4 **a** $x = \dfrac{\log M - \log 20}{-0.02 \log 5}$ **b** **i** $x = -50$ **ii** $x \approx -76.1$

5 **a** $x = -\dfrac{\log(0.03)}{\log 2}$ **b** $x = \dfrac{10 \log\left(\frac{10}{3}\right)}{\log 5}$

c $x = \dfrac{-4 \log\left(\frac{1}{8}\right)}{\log 3}$ **d** $x = \frac{1}{2} \ln 42$

e $x = -\frac{100}{3} \ln(0.001)$ **f** $x = \frac{10}{3} \ln\left(\frac{27}{41}\right)$

6 **a** $x = \ln 2$ **b** $x = 0$ **c** $x = \ln 2$ or $\ln 3$ **d** $x = 0$

e $x = \ln 4$ **f** $x = \ln\left(\frac{3+\sqrt{5}}{2}\right)$ or $\ln\left(\frac{3-\sqrt{5}}{2}\right)$

7 **a** $(\ln 3, 3)$ **b** $(\ln 2, 5)$ **c** $(0, 2)$ and $(\ln 5, -2)$

EXERCISE 4F

1 **a** ≈ 2.26 **b** ≈ -10.3 **c** ≈ -2.46 **d** ≈ 5.42

2 **a** $x \approx -4.29$ **b** $x \approx 3.87$ **c** $x \approx 0.139$

3 **a** $x = \dfrac{\log 3}{\log 5}$ **b** $x = \dfrac{\log\left(\frac{1}{8}\right)}{\log 3}$ **c** $x = -1$

4 **a** $x = 16$ **b** $x = \sqrt[3]{5} \approx 1.71$

5 $x = \dfrac{\log 8}{\log 25}$ or $\log_{25} 8$ **6** $\frac{3}{2}p$

EXERCISE 4G

1 **a** **i** Domain is $\{x \mid x > -1\}$, Range is $\{y \mid y \in \mathbb{R}\}$
ii VA is $x = -1$, x and y-intercepts 0
iii

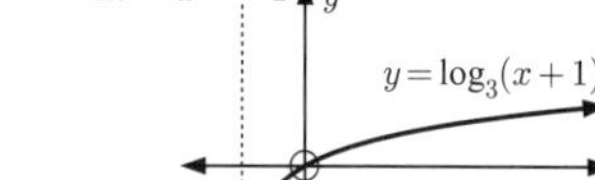

iv $x = -\frac{2}{3}$
v $f^{-1}(x) = 3^x - 1$

b **i** Domain is $\{x \mid x > -1\}$, Range is $\{y \mid y \in \mathbb{R}\}$
ii VA is $x = -1$, x-intercept 2, y-intercept 1
iii

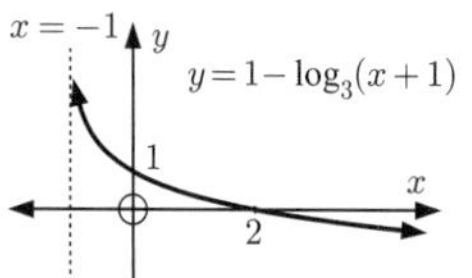

iv $x = 8$
v $f^{-1}(x) = 3^{1-x} - 1$

c **i** Domain is $\{x \mid x > 2\}$, Range is

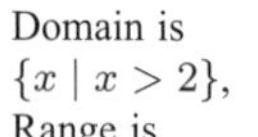

$\{y \mid y \in \mathbb{R}\}$
ii VA is $x = 2$, x-intercept 27, no y-intercept
iii $y = \log_5(x-2) - 2$; 27; $x = 2$
iv $x = 7$
v $f^{-1}(x) = 5^{2+x} + 2$

d **i** Domain is $\{x \mid x > 2\}$, Range is $\{y \mid y \in \mathbb{R}\}$
ii VA is $x = 2$, x-intercept 7, no y-intercept
iii $y = 1 - \log_5(x-2)$; 7; $x = 2$
iv $x = 27$
v $f^{-1}(x) = 5^{1-x} + 2$

e **i** Domain is $\{x \mid x > 0\}$, Range is $\{y \mid y \in \mathbb{R}\}$
ii VA is $x = 0$, x-intercept $\sqrt{2}$, no y-intercept
iii

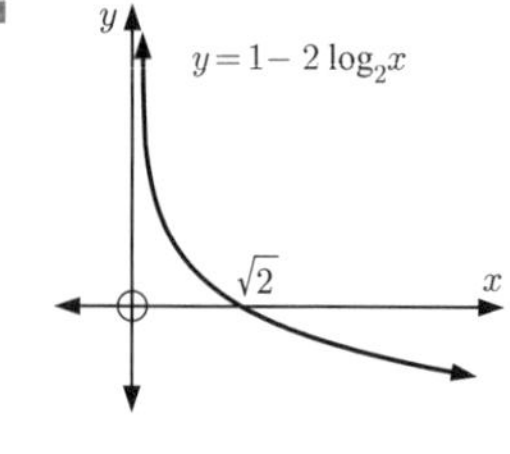

iv $x = 2$
v $f^{-1}(x) = 2^{\frac{1-x}{2}}$

f **i** Domain is $\{x \mid x < -1$ or $x > 4\}$, Range is $\{y \mid y \in \mathbb{R}\}$
ii VA $x = 4$, $x = -1$, x-ints. 4.19, -1.19, no y-intercept
iii $y = \log_2(x^2 - 3x - 4)$; $x = -1$; $x = 4$; -1.19; 4.19
iv $x = -1.10$ and 4.10
v if $x > 4$, $f^{-1}(x) = \dfrac{3 + \sqrt{25 + 2^{x+2}}}{2}$

if $x < -1$, $f^{-1}(x) = \dfrac{3 - \sqrt{25 + 2^{x+2}}}{2}$

2 **a** **i** $f^{-1}(x) = \ln(x - 5)$
ii $f(x) = e^x + 5$; 6; $x = 5$; $y = 5$; 6; $y = x$; f^{-1}
iii Domain of f is $\{x \mid x \in \mathbb{R}\}$, Range is $\{y \mid y > 5\}$
Domain of f^{-1} is $\{x \mid x > 5\}$, Range is $\{y \mid y \in \mathbb{R}\}$
iv f has a HA $y = 5$, f has y-int 6
f^{-1} has a VA $x = 5$, f^{-1} has x-int 6

b **i** $f^{-1}(x) = \ln(x + 3) - 1$
ii $x = -3$; f^{-1}; $f(x) = e^{x+1} - 3$; $y = -3$; $y = x$
iii Domain of f is $\{x \mid x \in \mathbb{R}\}$, Range is $\{y \mid y > -3\}$
Domain of f^{-1} is $\{x \mid x > -3\}$, Range is $\{y \mid y \in \mathbb{R}\}$
iv f has a HA $y = -3$, x-int $\ln 3 - 1$, y-int $e - 3$
f^{-1} has a VA $x = -3$, x-int $e - 3$, y-int $\ln 3 - 1$

c **i** $f^{-1}(x) = e^{x+4}$
ii

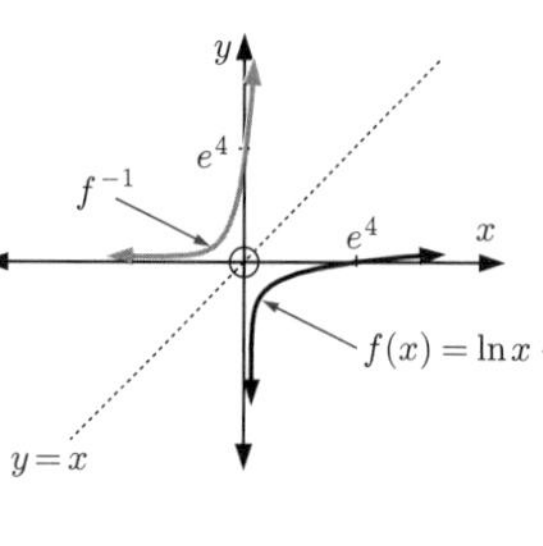

iii Domain of f is $\{x \mid x > 0\}$, Range of f is $\{y \mid y \in \mathbb{R}\}$
Domain of f^{-1} is $\{x \mid x \in \mathbb{R}\}$, Range is

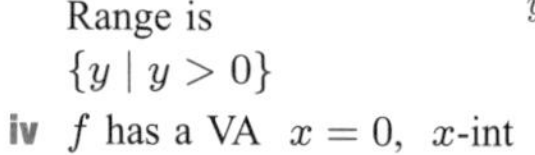

$\{y \mid y > 0\}$
iv f has a VA $x = 0$, x-int e^4
f^{-1} has a HA $y = 0$, y-int e^4

d **i** $f^{-1}(x) = 1 + e^{x-2}$ **ii**

y $x=1$ $y=x$ f^{-1} $y=1$ x $f(x) = \ln(x-1)+2$

iii Domain of f is $\{x \mid x > 1\}$, Range is $\{y \mid y \in \mathbb{R}\}$
Domain of f^{-1} is $\{x \mid x \in \mathbb{R}\}$, Range is $\{y \mid y > 1\}$

iv f has a VA $x = 1$, x-int $1 + e^{-2}$
f^{-1} has a HA $y = 1$, y-int $1 + e^{-2}$

3 $f^{-1}(x) = \frac{1}{2}\ln x$
- **a** $(f^{-1} \circ g)(x) = \frac{1}{2}\ln(2x-1)$
- **b** $(g \circ f)^{-1}(x) = \frac{1}{2}\ln\left(\frac{x+1}{2}\right)$

4 **a** A is $y = \ln x$ as its x-intercept is 1 **b**

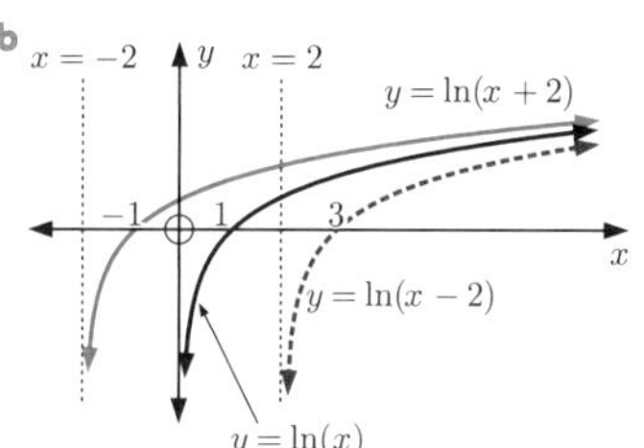

c $y = \ln x$ has VA $x = 0$
$y = \ln(x-2)$ has VA $x = 2$
$y = \ln(x+2)$ has VA $x = -2$

5 $y = \ln(x^2) = 2\ln x$, so she is correct.
This is because the y-values are twice as large for $y = \ln(x^2)$ as they are for $y = \ln x$.

6 **a** $f^{-1} : x \mapsto \ln(x-2) - 3$
- **b** **i** $x < -5.30$ **ii** $x < -7.61$ **iii** $x < -9.91$
  iv $x < -12.2$ Conjecture HA is $y = 2$
- **c** as $x \to \infty$, $f(x) \to \infty$,
  as $x \to -\infty$, $e^{x+3} \to 0$ and $f(x) \to 2$
  $\therefore$ HA is $y = 2$
- **d** VA of f^{-1} is $x = 2$, Domain of f^{-1} is $\{x \mid x > 2\}$

7 **a** $x < -0.703$ **b** $x < 0.773$
c $0 < x < 3.69$ as $\ln x$ is only defined for $x > 0$

8 Domain is $x \in \,]0, \infty[$, $f(x) \leqslant 0$ for $x \in \,]0, 1]$

9 **a**

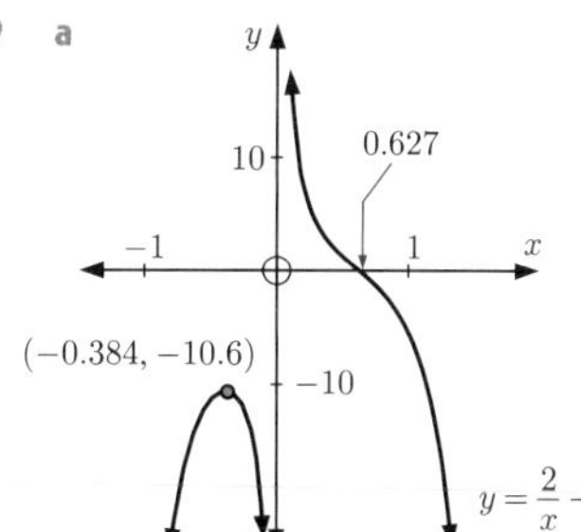

b Domain is $\{x \mid x \in \mathbb{R},\ x \neq 0\}$
Range is $\{y \mid y \in \mathbb{R}\}$

c For $x \in \,]-\infty, 0[$ or $]0.627, \infty[$

EXERCISE 4H

1 **a** 3.90 h **b** 15.5 h

3 **a, b** see graph below **c** **i** $n \approx 2.82$ weeks

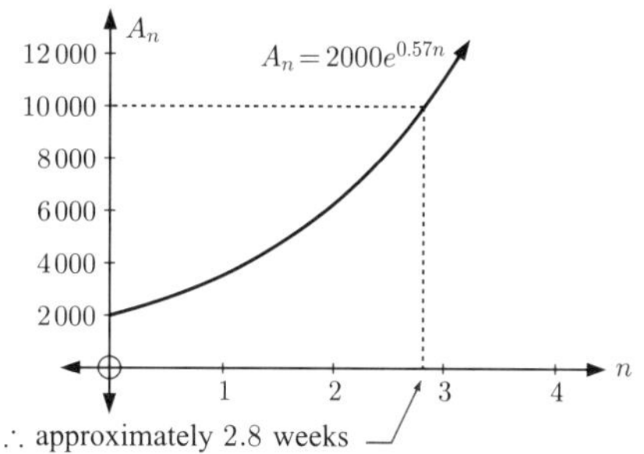

4 In 6.17 years, or 6 years 62 days

5 9 years

6 **a** $\frac{8.4\%}{12} = 0.7\% = 0.007$, $r = 1 + 0.007 = 1.007$
b after 74 months

7 **a** 17.3 years **b** 92.2 years **c** 115 years

8 **Hint:** Set $V = 40$, solve for t.

9 **a** **d**

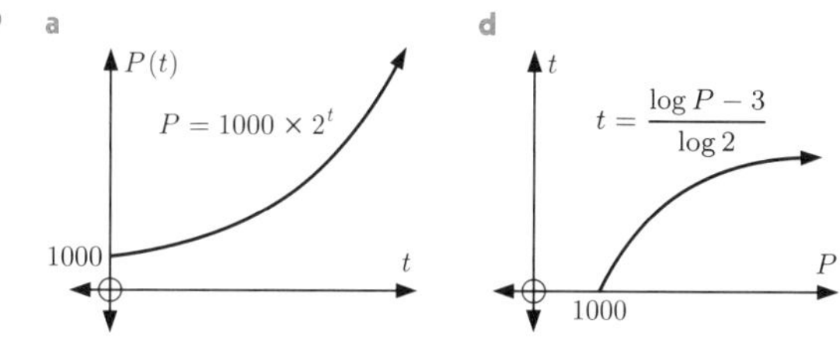

b ≈ 4.32 weeks **c** $t = \frac{\log P - 3}{\log 2}$

10 **a** 50.7 min **b** 152 min

11 **a**

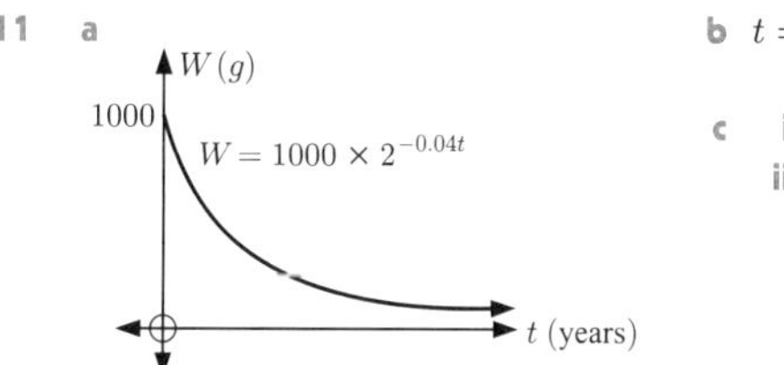

b $t = \frac{3 - \log W}{0.04 \log 2}$

c **i** $t \approx 141$ years **ii** $t \approx 498$ years

12 **a** 10 000 years **b** 49 800 years

13 **Hint:** $t = \frac{-50\log(0.1)}{\log 2}$ **14** 12.9 seconds

REVIEW SET 4A

1 **a** 3 **b** 8 **c** -2 **d** $\frac{1}{2}$ **e** 0
f $\frac{1}{4}$ **g** -1 **h** $\frac{1}{2}$, $k > 0$

2 **a** $\frac{1}{2}$ **b** $-\frac{1}{3}$ **c** $a + b + 1$

3 **a** $\ln 144$ **b** $\ln\left(\frac{3}{2}\right)$ **c** $\ln\left(\frac{25}{e}\right)$ **d** $\ln 3$

4 **a** $\frac{3}{2}$ **b** -3 **c** $2x$ **d** $1 - x$

5 **a** $\log 144$ **b** $\log_2\left(\frac{16}{9}\right)$ **c** $\log_4 80$

6 **a** $\log P = \log 3 + x \log b$ **b** $\log m = 3\log n - 2\log p$

7 **Hint:** Use change of base rule.

8 **a** $T = \frac{x^2}{y}$ **b** $K = n\sqrt{t}$

9 **a** $5\ln 2$ **b** $3\ln 5$ **c** $6\ln 3$

10

Function	$y = \log_2 x$	$y = \ln(x+5)$
Domain	$x > 0$	$x > -5$
Range	$y \in \mathbb{R}$	$y \in \mathbb{R}$

11 a $2A + 2B$ b $A + 3B$ c $3A + \frac{1}{2}B$
d $4B - 2A$ e $3A - 2B$

12 a $x = 0$ or $\ln\left(\frac{2}{3}\right)$ b $x = e^2$

REVIEW SET 4B

1 a $\approx 10^{1.5051}$ b $\approx 10^{-2.8861}$ c $\approx 10^{-4.0475}$

2 a $x = \frac{1}{8}$ b $x \approx 82.7$ c $x \approx 0.0316$

3 a $k \approx 3.25 \times 2^x$ b $Q = P^3R$ c $A \approx \dfrac{B^5}{400}$

4 a $x = \dfrac{\log 7}{\log 5}$ b $x = 2$

5 a 2500 g
b 3290 years
c 42.3%
d

$W_t = 2500 \times 3^{-\frac{t}{3000}}$

6 **Hint:** $2^{4x} - 5 \times 2^{3x} = 0$

7 a $x = e^5$ b $x = e^{-\frac{2}{3}}$ c $x = \ln 400$
d $x = \dfrac{\ln 11 - 1}{2}$ e $x = 2\ln 30$

8 a 3 years b 152%

9 a $g^{-1}(x)$

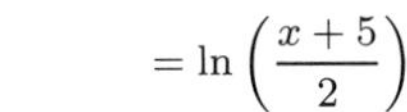

$= \ln\left(\dfrac{x+5}{2}\right)$

b

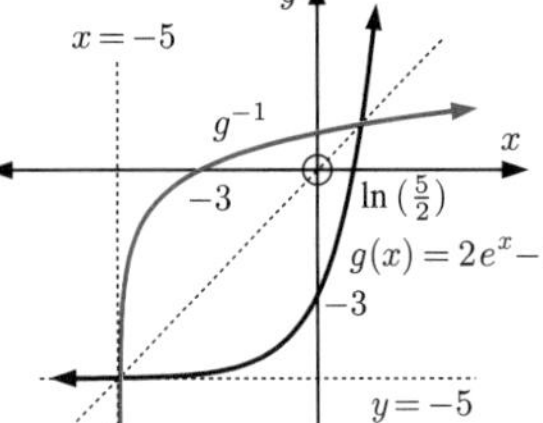

c Domain of g is $\{x \mid x \in \mathbb{R}\}$,
Range is $\{y \mid y > -5\}$
Domain of g^{-1} is $\{x \mid x > -5\}$,
Range is $\{y \mid y \in \mathbb{R}\}$

d g has horizontal asymptote $y = -5$,
x-intercept is $\ln\left(\frac{5}{2}\right) \approx 0.916$, y-intercept is -3
g^{-1} has vertical asymptote $x = -5$,
x-intercept is -3, y-intercept is ≈ 0.916

10 a 9 b $\ln 5$

11 a

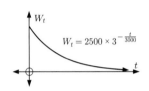

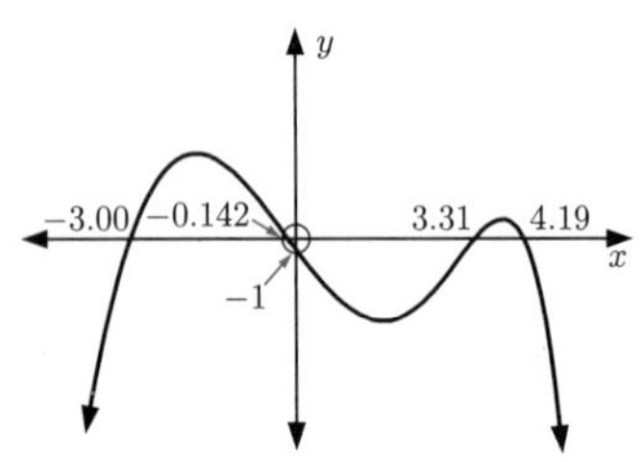

b $-3.00 < x < -0.142$, $3.31 < x < 4.19$

REVIEW SET 4C

1 a $\frac{3}{2}$ b $\frac{2}{3}$ c $a + b$

2 a x^4 b 5 c $\frac{1}{2}$ d $3x$ e $-x$ f $\log x$

3 a $\approx e^{2.9957}$ b $\approx e^{8.0064}$ c $\approx e^{-2.5903}$

4 a $x = 1000$ b $x \approx 4.70$ c $x \approx 6.28$

5 a $\ln 3$ b $\ln 4$ c $\ln 125$

6 a $\log M = \log a + n \log b$ b $\log T = \log 5 - \frac{1}{2}\log l$
c $\log G = 2\log a + \log b - \log c$

7 a $x \approx 5.19$ b $x \approx 4.29$ c $x \approx -0.839$

8 a $x = \ln 3$ b $x = \ln 3$ or $\ln 4$

9 a $P = TQ^{1.5}$ b $M = \dfrac{e^{1.2}}{\sqrt{N}}$

10 a $f(-x) = e^{(-x)^2} - (-x)^6 = e^{x^2} - x^6 = f(x)$
$g(-x) = \ln((-x)^2 + 1) = \ln(x^2 + 1) = g(x)$
$\therefore$ both even

b

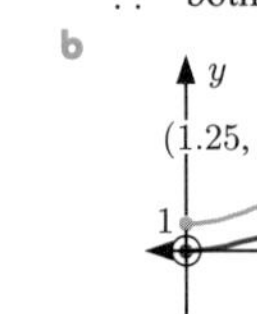

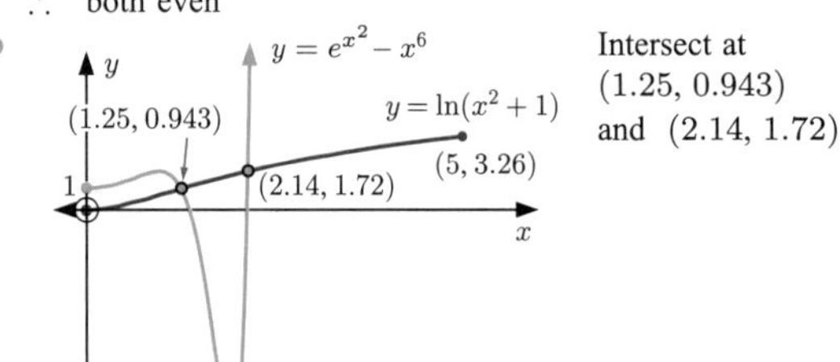

Intersect at (1.25, 0.943) and (2.14, 1.72).

c $1.25 < x < 2.14$

11 a Domain is $\{x \mid x > -2\}$
Range is $\{y \mid y \in \mathbb{R}\}$
b VA is $x = -2$,
x-intercept is 7,
y-intercept is ≈ -1.37
c $g^{-1}(x) = 3^{x+2} - 2$
d

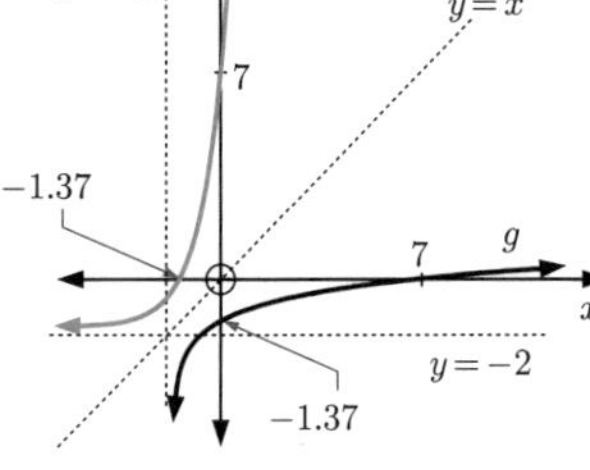

12 a 13.9 weeks b 41.6 weeks c 138 weeks

EXERCISE 5A

1 a $2x$ b $x + 2$ c $\dfrac{x}{2}$ d $2x + 3$

2 a $9x^2$ b $\dfrac{x^2}{4}$ c $3x^2$ d $2x^2 - 4x + 7$

3 a $64x^3$ b $4x^3$ c $x^3 + 3x^2 + 3x + 1$
d $2x^3 + 6x^2 + 6x - 1$

4 a 4^x b $2^{-x} + 1$ c $2^{x-2} + 3$ d $2^{x+1} + 3$

5 a $-\dfrac{1}{x}$ b $\dfrac{2}{x}$ c $\dfrac{2 + 3x}{x}$ d $\dfrac{2x + 1}{x - 1}$

EXERCISE 5B

1 a, b

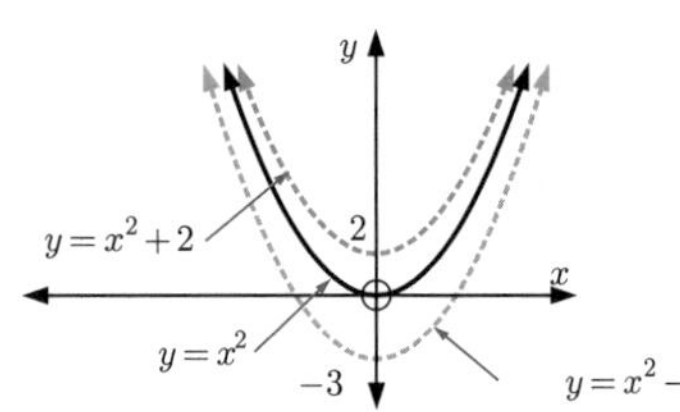

c i If $b > 0$, the function is translated vertically upwards through b units.
ii If $b < 0$, the function is translated vertically downwards $|b|$ units.

2 **a**

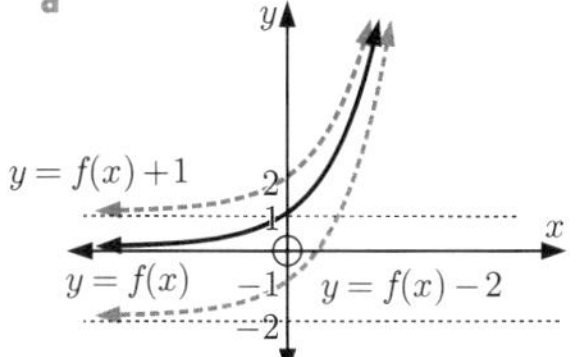

b

y
y = f(x)
y = f(x) + 1
1
x
−2
y = f(x) − 2

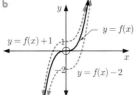

c

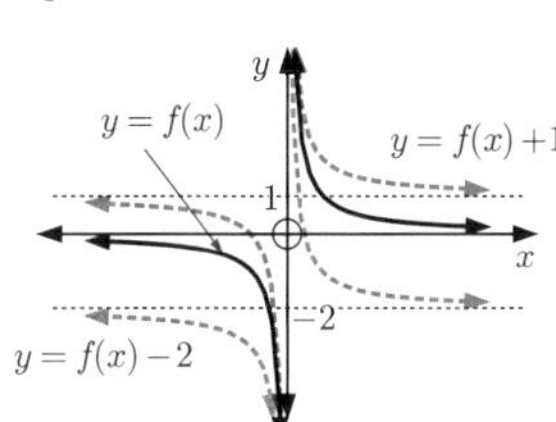

d

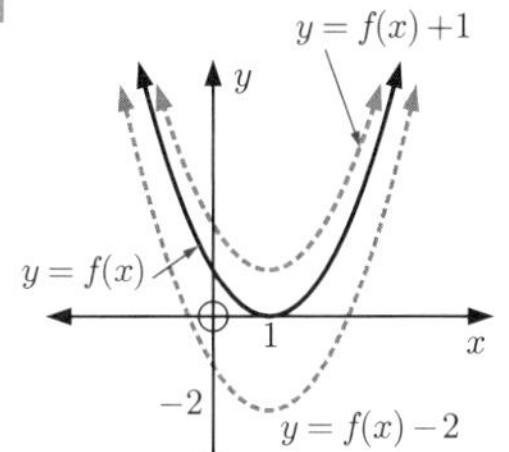

e

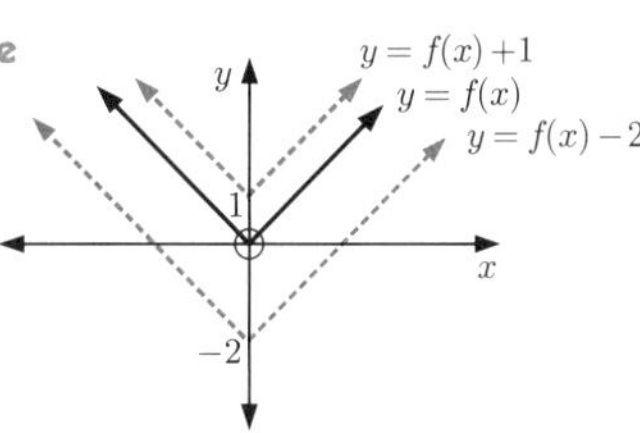

3 **a**

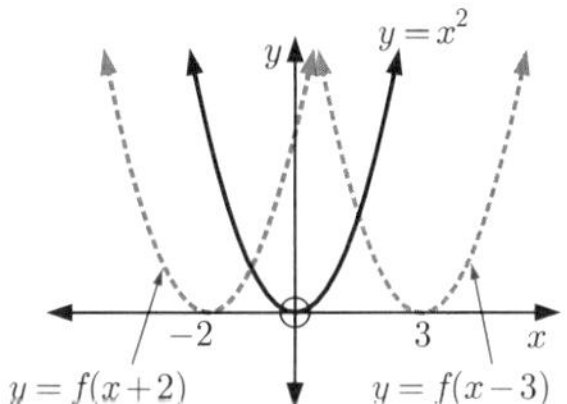

b **i** If $a > 0$, the graph is translated a units right.
ii If $a < 0$, the graph is translated $|a|$ units left.

4 **a**

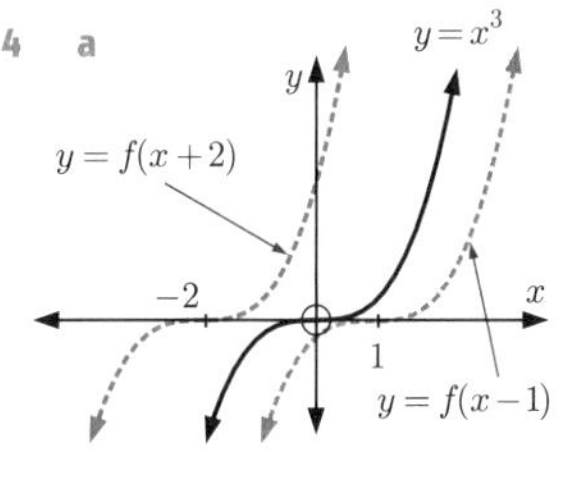

b

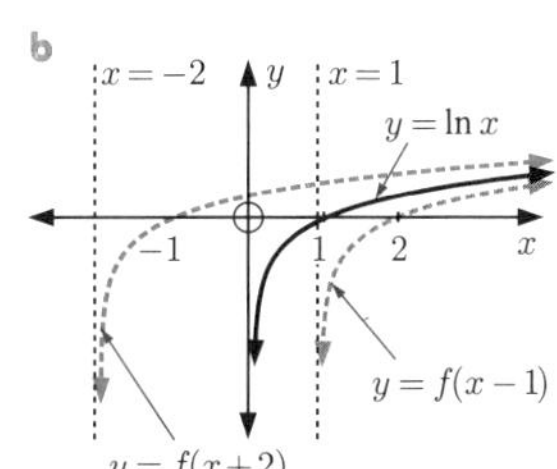

c **d**

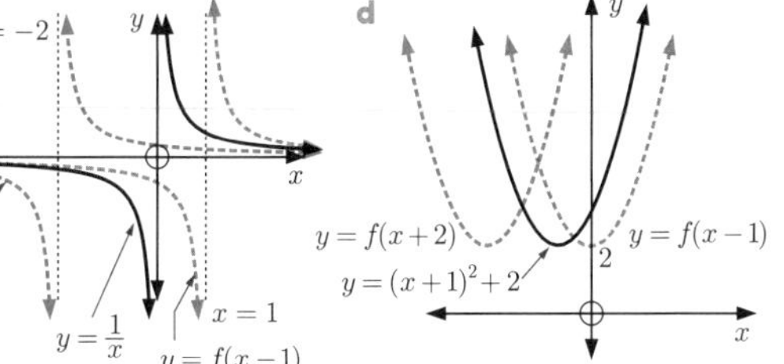

e

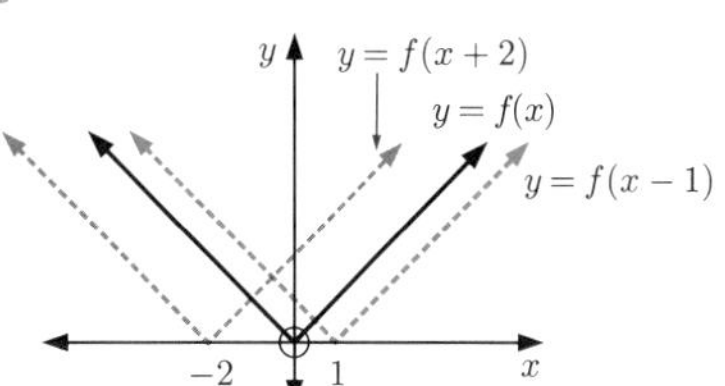

5 **a**

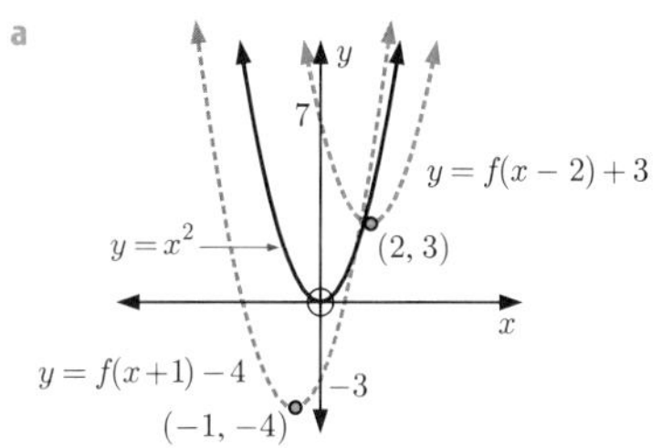

b

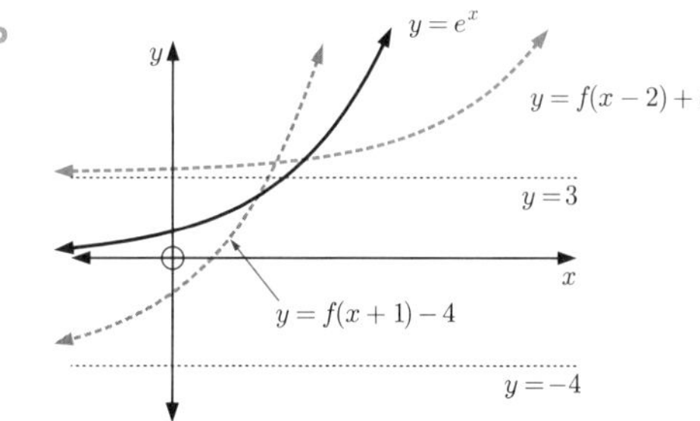

c

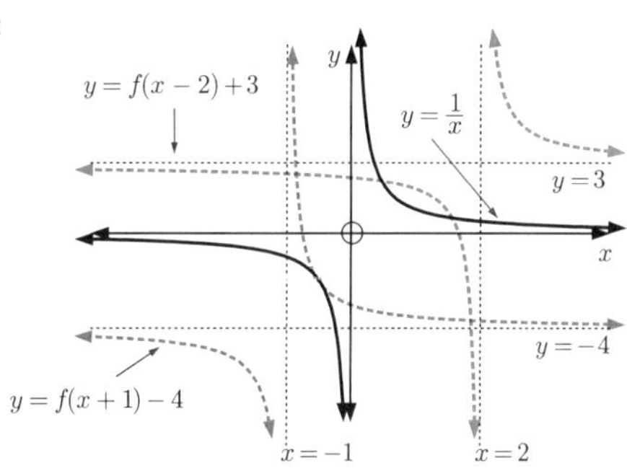

6 A translation of $\begin{pmatrix} 2 \\ -3 \end{pmatrix}$.

a

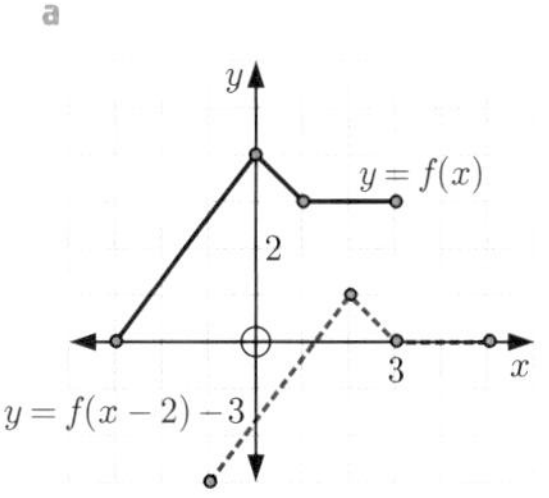

b

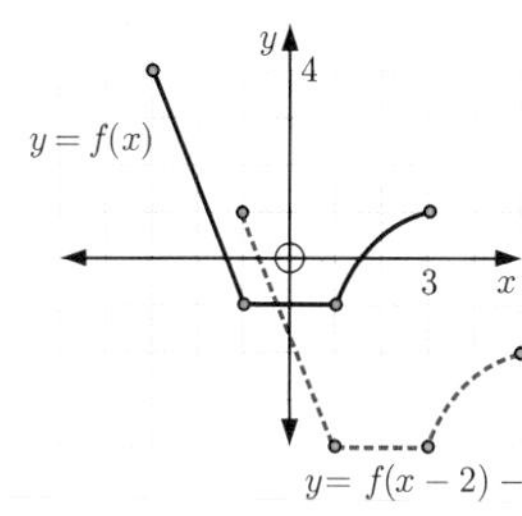

7 $g(x) = x^2 - 8x + 17$

8 **a** **i** $(3, 2)$ **ii** $(0, 11)$ **iii** $(5, 6)$
b **i** $(-2, 4)$ **ii** $(-5, 25)$ **iii** $(-1\frac{1}{2}, 2\frac{1}{4})$

EXERCISE 5C

1 **a**

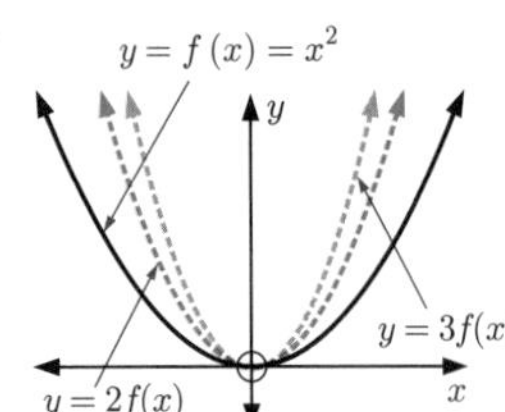

b

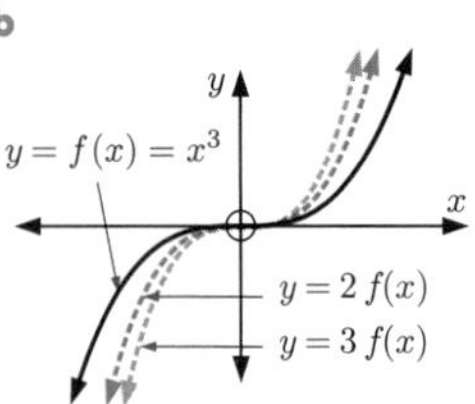

c

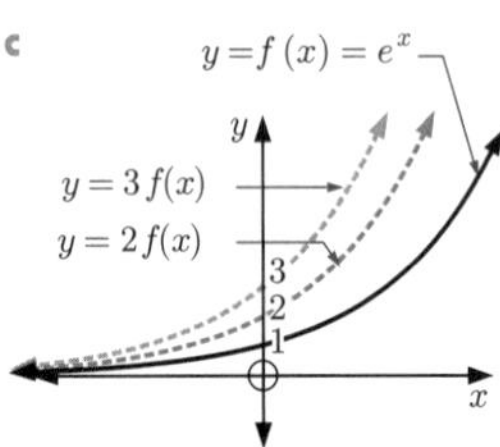

d

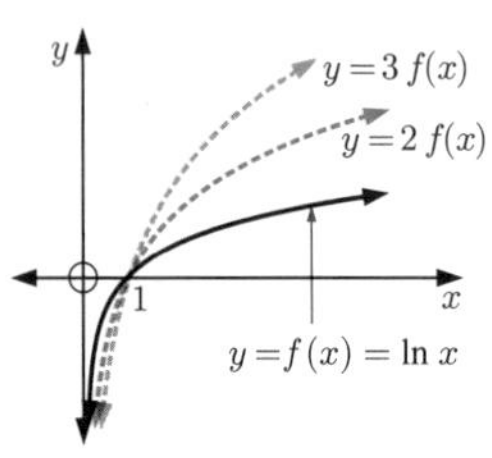

e

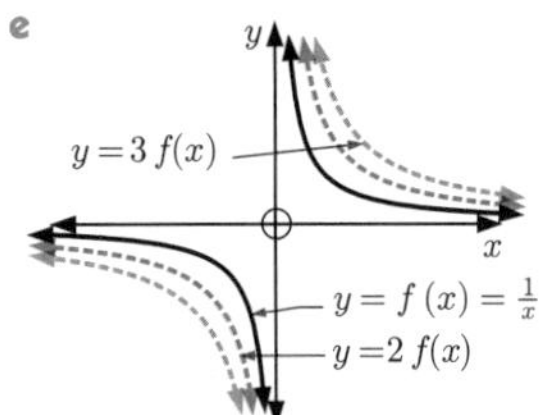

f

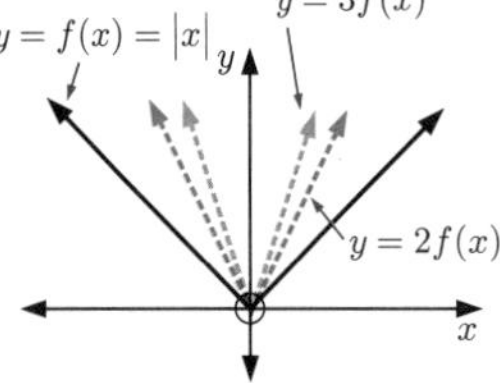

2 **a**

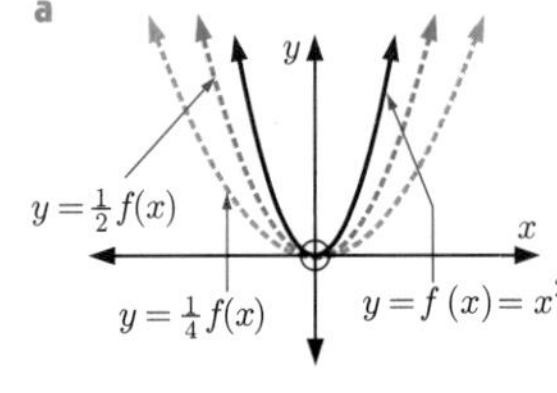

b

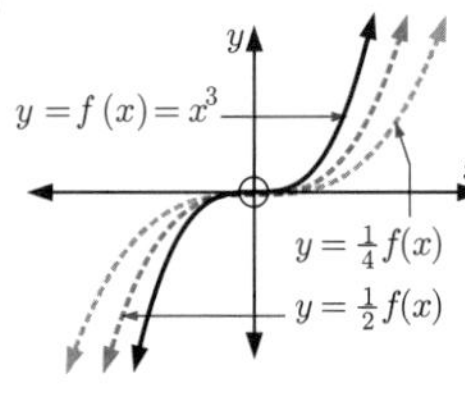

c

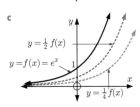

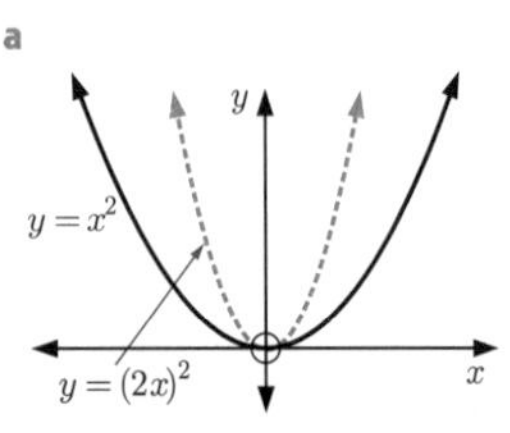

3 **a**

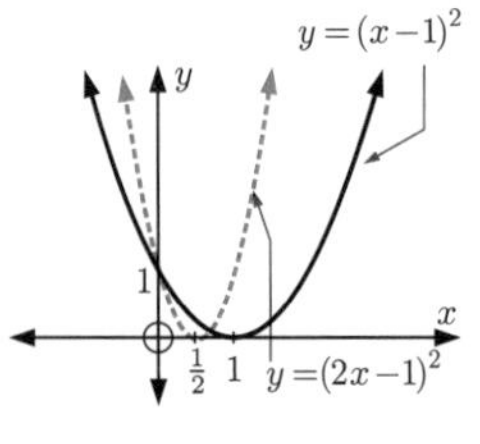

b

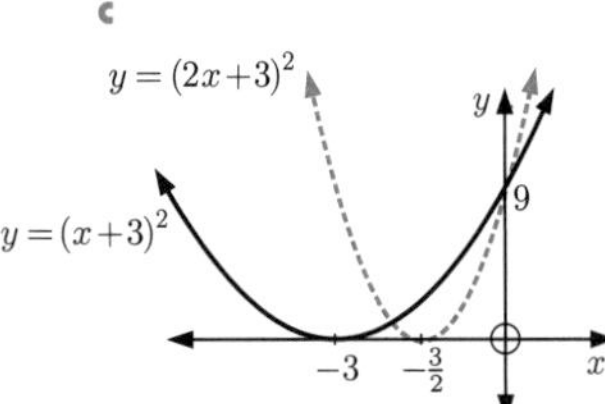

c

4 **a**

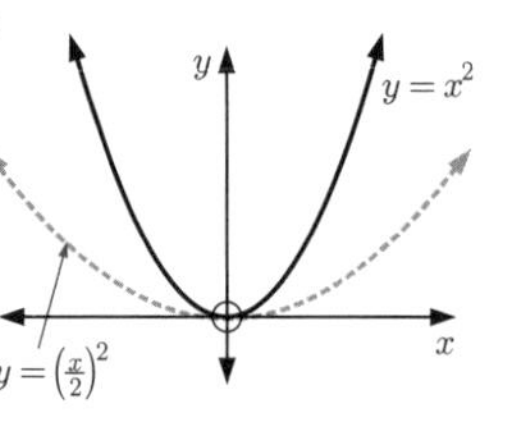

b

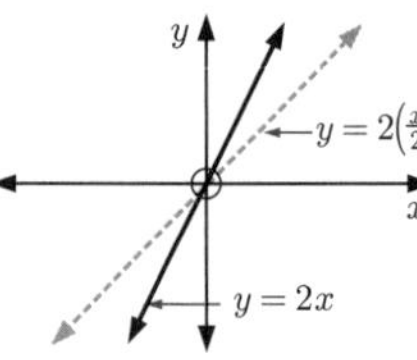

c

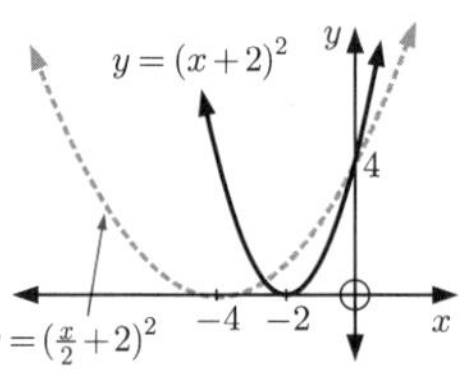

5 **a**

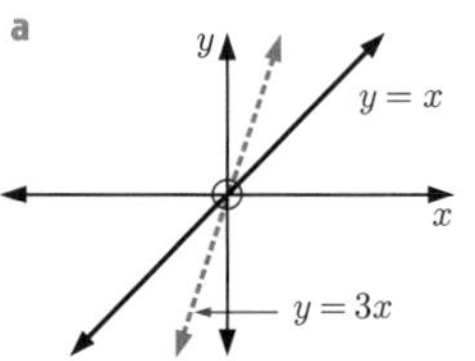

b

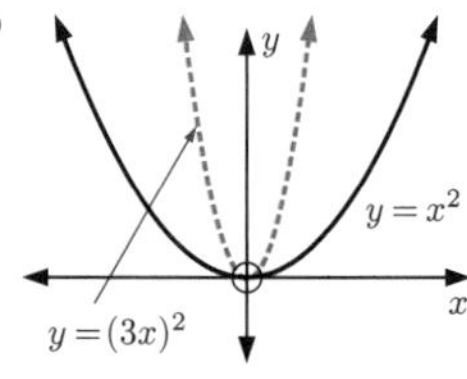

c

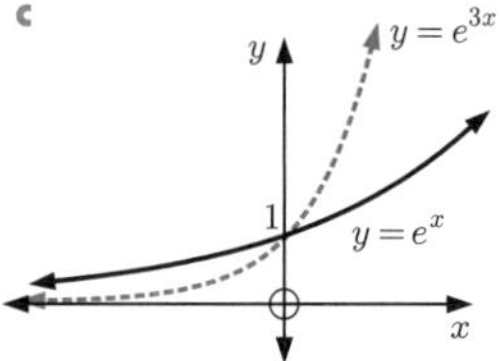

6 **a**

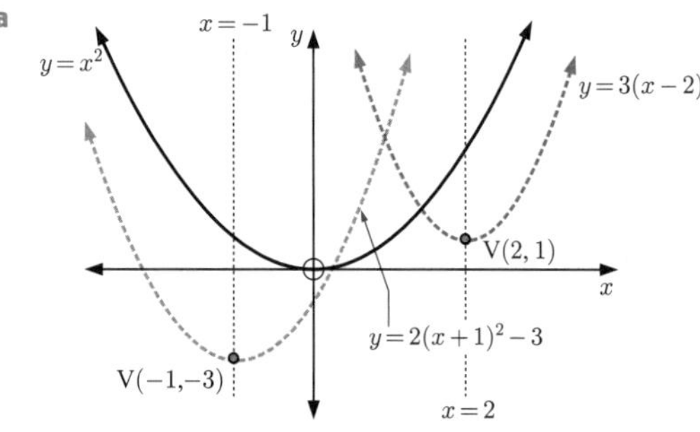

b

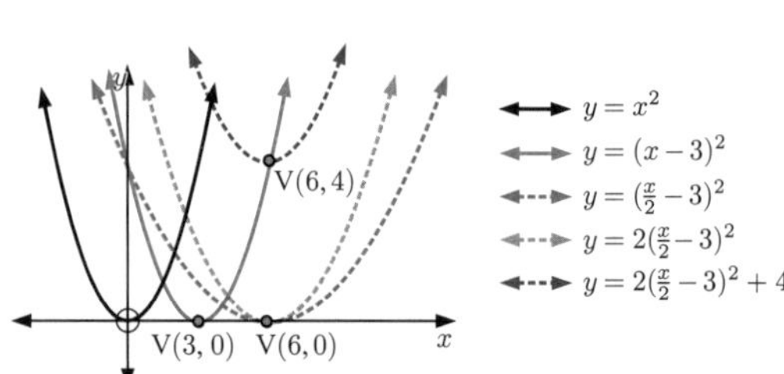

c

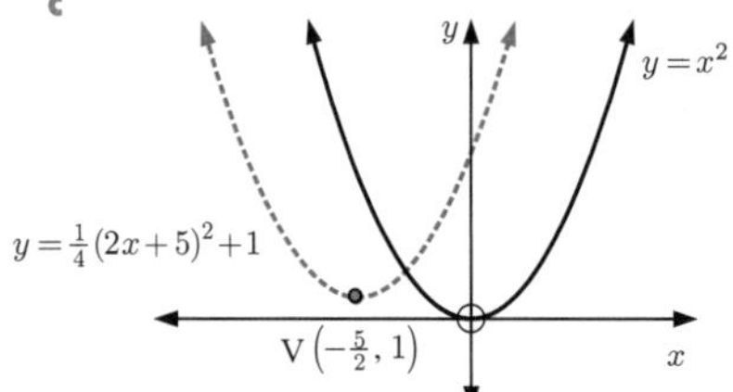

7 **a** **i** $(\frac{3}{2}, -15)$ **ii** $(\frac{1}{2}, 6)$ **iii** $(-1, 3)$

b **i** $(4, \frac{1}{3})$ **ii** $(-6, \frac{2}{3})$ **iii** $(-14, 1)$

8 **a** $f(x)$ is translated horizontally 1 unit left, then horizontally stretched by a factor of 2, then vertically stretched by a factor of 2, then translated 3 units upwards.

b **i** $(0, -3)$ **ii** $(2, 5)$ **iii** $(-4, -1)$

c **i** $(0, -4)$ **ii** $(\frac{3}{2}, -2)$ **iii** $(\frac{7}{2}, -\frac{3}{2})$

EXERCISE 5D

1 **a**

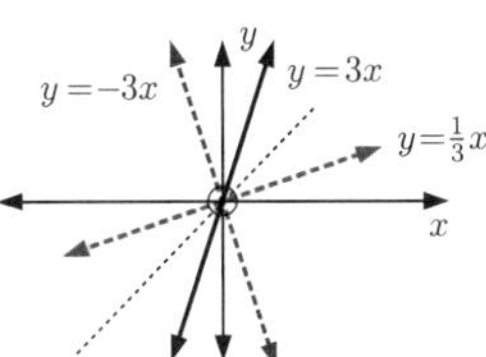

b

c

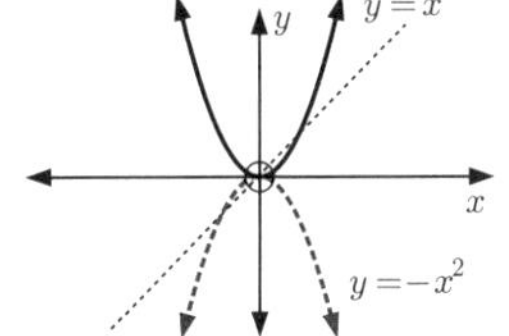

$y = x^2$ does not have an inverse.

d

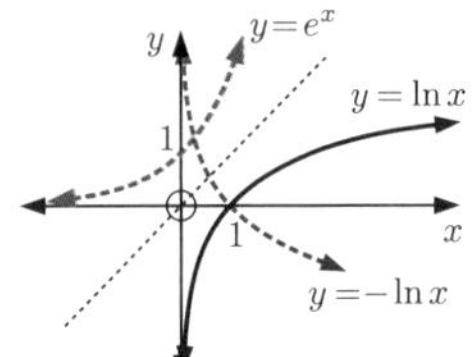

e

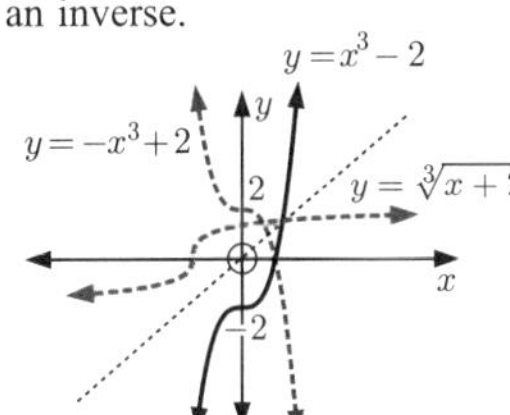

f

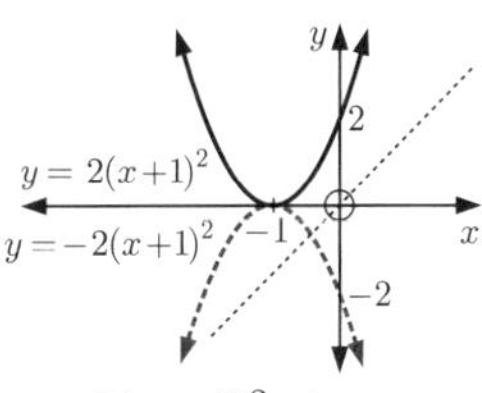

$y = 2(x + 1)^2$ does not have an inverse.

2 **a** $f(-x) = -2x + 1$

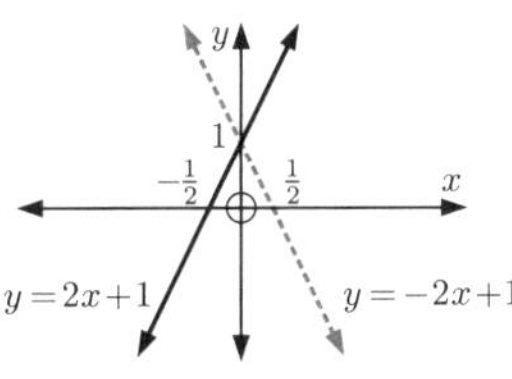

b $f(-x) = x^2 - 2x + 1$

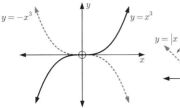

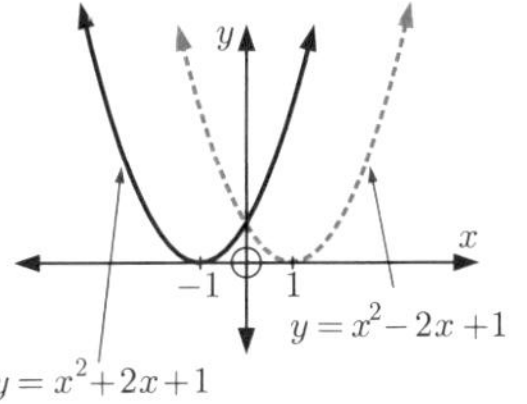

c $f(-x) = -x^3$

d $f(-x) = |-x - 3|$
$= |x + 3|$

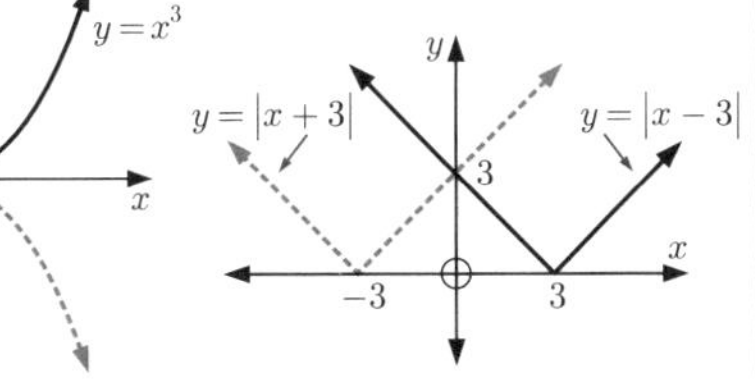

3 $g(x) = \ln x - x^3$ **4** $g(x) = x^4 + 2x^3 - 3x^2 - 5x - 7$

5 **a** **i** $(3, 0)$ **ii** $(2, 1)$ **iii** $(-3, -2)$

b **i** $(7, 1)$ **ii** $(-5, 0)$ **iii** $(-3, 2)$

6 **a** **i** $(-2, -1)$ **ii** $(0, 3)$ **iii** $(1, 2)$ **iv** $(-3, 0)$

b **i** $(-5, -4)$ **ii** $(0, 3)$ **iii** $(-2, 3)$ **iv** $(-3, 0)$

7 **a** **i** $(1, 3)$ **ii** $(4, -2)$ **iii** $(-5, 0)$

b **i** $(1, -1)$ **ii** $(0, 6)$ **iii** $(-2, 3)$

8 **a** A rotation about the origin through $180°$. **b** $(-3, 7)$

c $(5, 1)$

9 **a**

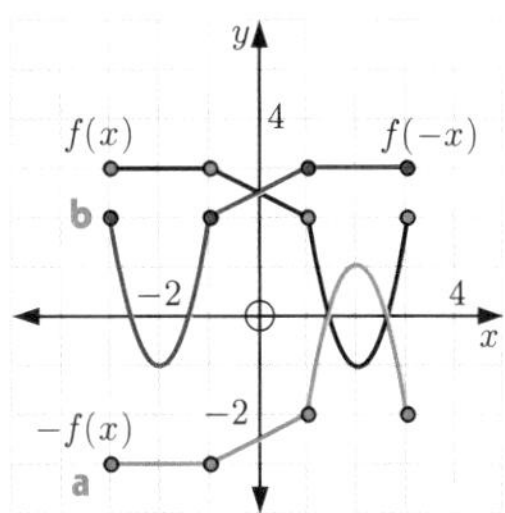

b **i**

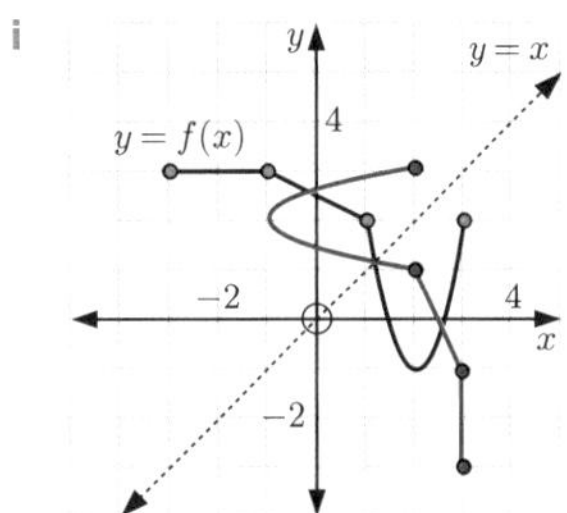

ii No, this is not a function.

EXERCISE 5E

1 **a** x-intercepts are ± 1, y-intercept is -1

b **i, ii**

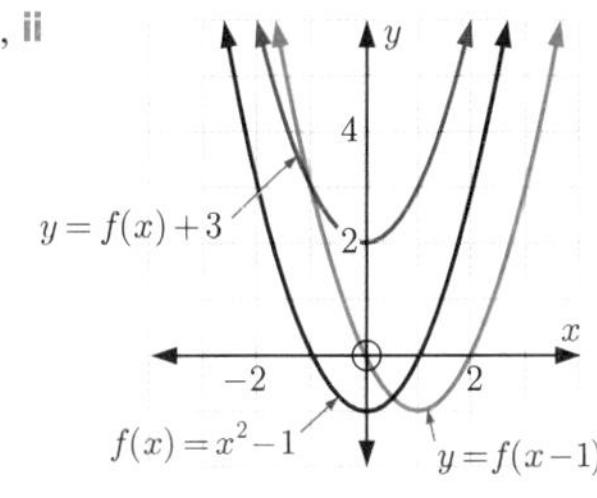

iii, iv

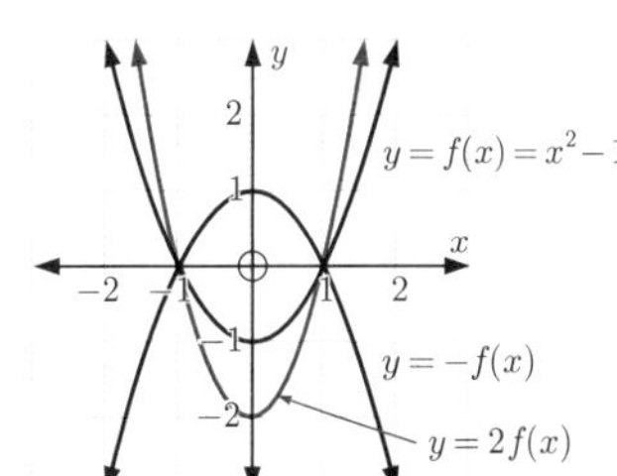

c **i** a vertical translation of 3 units upwards

ii a horizontal translation of 1 unit to the right

iii a vertical stretch with scale factor 2

iv a reflection in the x-axis

d

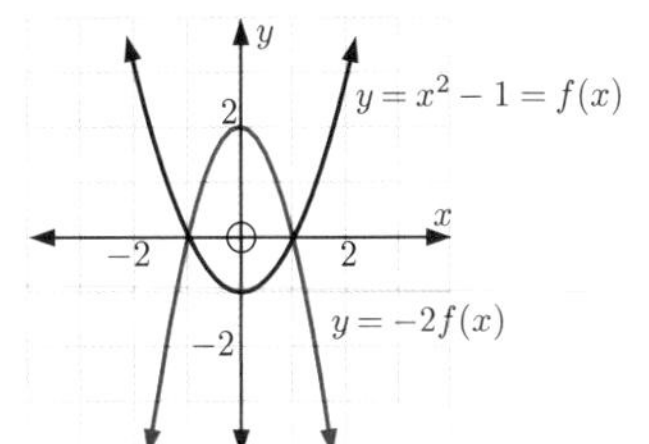

A reflection in the x-axis, then a vertical stretch with scale factor 2.

e $(-1, 0)$ and $(1, 0)$

2 **a** **i** A vertical stretch with scale factor 3. **ii** $g(x) = 3f(x)$

b **i** A vertical translation of $\begin{pmatrix} 0 \\ -2 \end{pmatrix}$. **ii** $g(x) = f(x) - 2$

c **i** A vertical stretch with scale factor $\frac{1}{2}$. **ii** $g(x) = \frac{1}{2}f(x)$

d **i** A reflection in the y-axis. **ii** $g(x) = f(-x)$

3 **a**

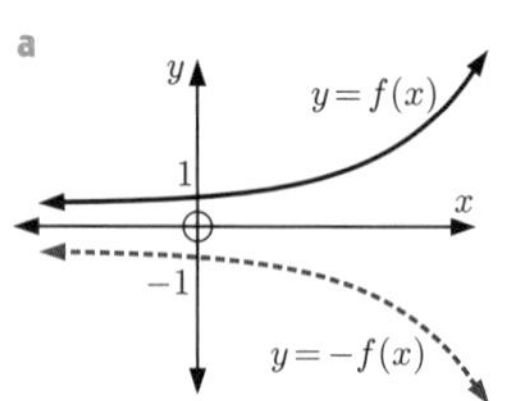

b

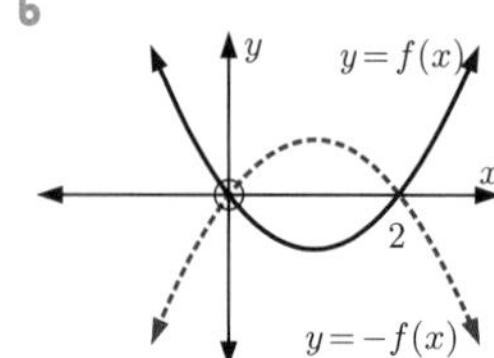

c

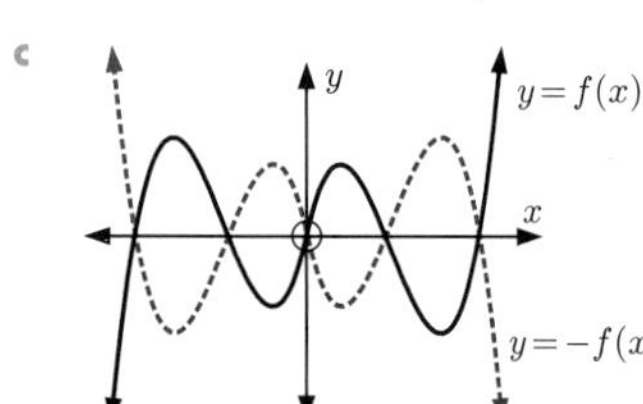

4 **a**

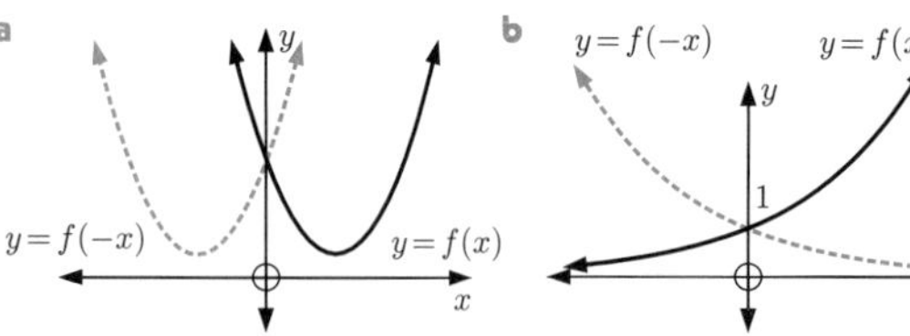

b (see graph above)

c

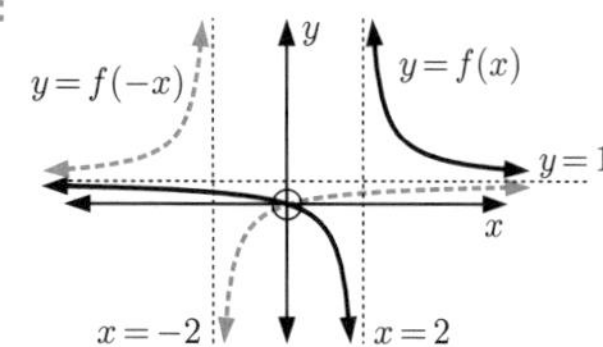

5 **a** **A** **b** **B** **c** **D** **d** **C**

6

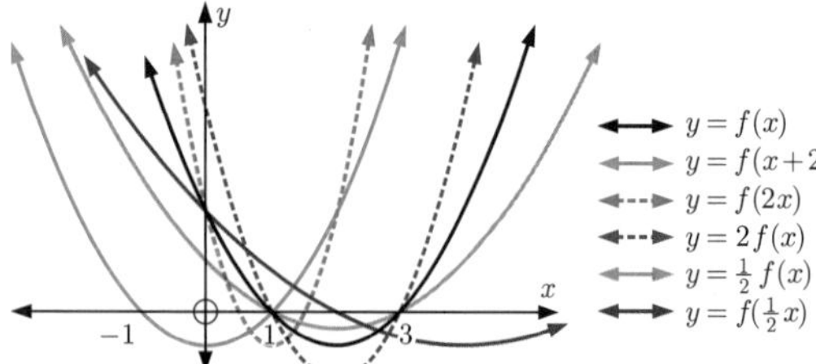

7

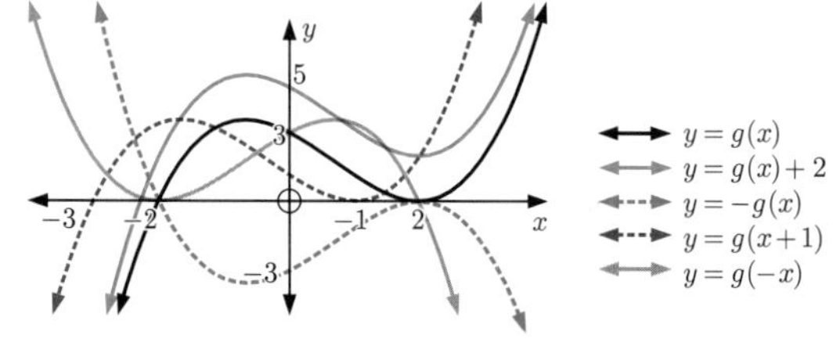

8

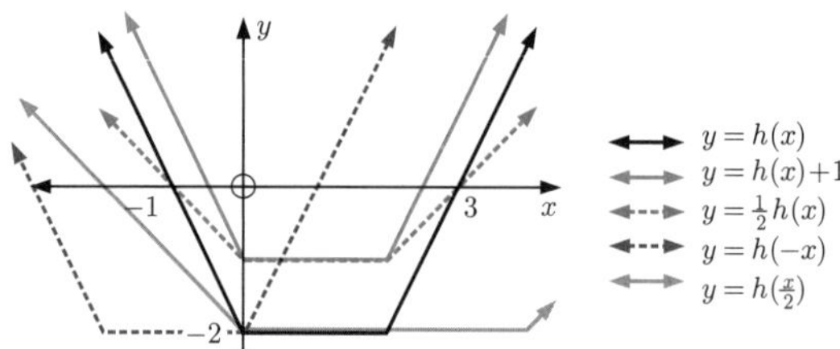

EXERCISE 5F

1 **a** $y = \dfrac{1}{2x}$ **b** $y = \dfrac{3}{x}$ **c** $y = \dfrac{1}{x+3}$

d $y = 4 + \dfrac{1}{x} = \dfrac{4x+1}{x}$

2 **a** $g(x) = \dfrac{3}{x-1} - 1 = \dfrac{-x+4}{x-1}$

b vertical asymptote $x = 1$, horizontal asymptote $y = -1$

c Domain $= \{x \mid x \neq 1\}$, Range $= \{y \mid y \neq -1\}$

d

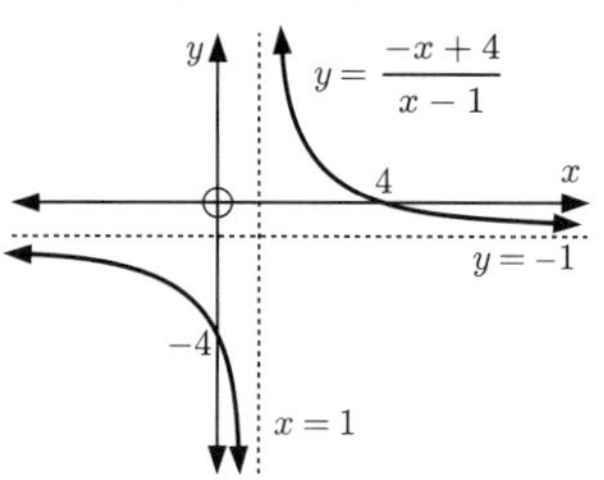

e No, the graph is not symmetric about the line $y = x$.

3 **a** **i** VA is $x = 1$, HA is $y = 2$

ii A vertical stretch with scale factor 6, then a translation of $\begin{pmatrix} 1 \\ 2 \end{pmatrix}$.

b **i** VA is $x = -1$, HA is $y = 3$

ii A vertical stretch with scale factor 5, reflection in the x-axis, then a translation of $\begin{pmatrix} -1 \\ 3 \end{pmatrix}$.

c **i** VA is $x = 2$, HA is $y = -2$

ii A vertical stretch with scale factor 5, reflection in the x-axis, then a translation of $\begin{pmatrix} 2 \\ -2 \end{pmatrix}$.

4 **a** **i** VA is $x = -1$, HA is $y = 2$

ii as $x \to -1^-$, $y \to -\infty$
as $x \to -1^+$, $y \to \infty$
as $x \to -\infty$, $y \to 2^-$
as $x \to \infty$, $y \to 2^+$

iii x-intercept is $-\frac{3}{2}$, y-intercept is 3

iv

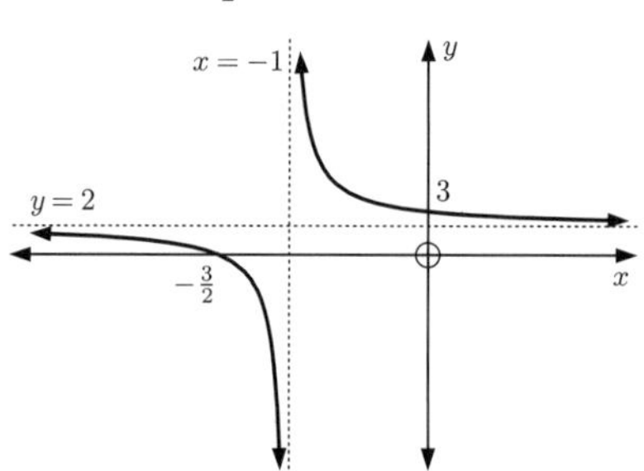

v Translate $\begin{pmatrix} -1 \\ 2 \end{pmatrix}$. **vi** Translate $\begin{pmatrix} 1 \\ -2 \end{pmatrix}$.

b **i** VA is $x = 2$, HA is $y = 0$

ii as $x \to 2^-$, $y \to -\infty$
as $x \to 2^+$, $y \to \infty$
as $x \to \infty$, $y \to 0^+$
as $x \to -\infty$, $y \to 0^-$

iii no x-intercept, y-intercept is $-1\frac{1}{2}$

iv

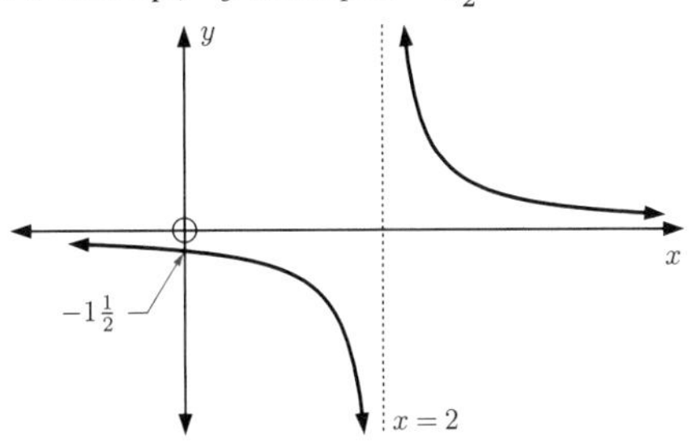

v Vertical stretch with scale factor 3, then translate $\binom{2}{0}$.

vi Translate $\binom{-2}{0}$, then vertical stretch with scale factor $\frac{1}{3}$.

c **i** VA is $x = 3$, HA is $y = -2$

ii as $x \to 3^-$, $y \to \infty$
as $x \to 3^+$, $y \to -\infty$
as $x \to -\infty$, $y \to -2^+$
as $x \to \infty$, $y \to -2^-$

iii x-intercept is $\frac{1}{2}$, y-intercept is $-\frac{1}{3}$

iv

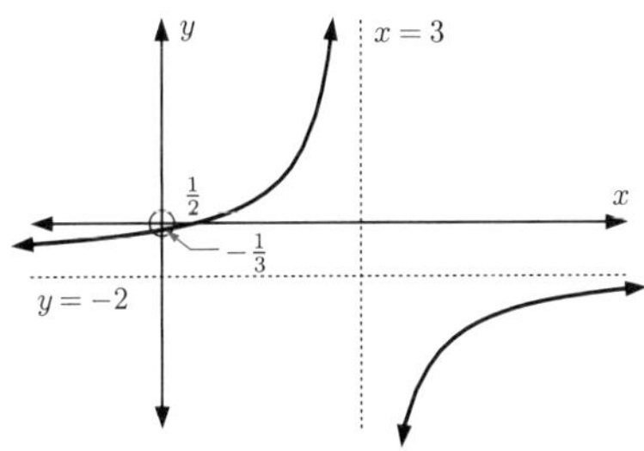

v Vertical stretch with scale factor 5, reflect in x-axis, then translate $\binom{3}{-2}$.

vi Translate $\binom{-3}{2}$, reflect in x-axis, then vertical stretch with scale factor $\frac{1}{5}$.

d **i** VA is $x = -\frac{1}{2}$, HA is $y = 2\frac{1}{2}$

ii as $x \to -\frac{1}{2}^-$, $y \to \infty$
as $x \to -\frac{1}{2}^+$, $y \to -\infty$
as $x \to \infty$, $y \to 2\frac{1}{2}^-$
as $x \to -\infty$, $y \to 2\frac{1}{2}^+$

iii x-intercept is $\frac{1}{5}$, y-intercept is -1

iv

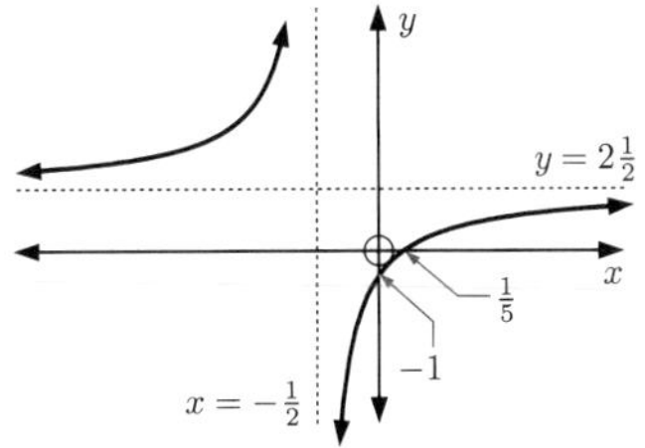

v Vertical stretch with scale factor $\frac{7}{4}$, reflect in x-axis, then translate $\begin{pmatrix} -\frac{1}{2} \\ \frac{5}{2} \end{pmatrix}$.

vi Translate $\begin{pmatrix} \frac{1}{2} \\ -\frac{5}{2} \end{pmatrix}$, reflect in x-axis, then vertical stretch with scale factor $\frac{4}{7}$.

5 **a** 70 weeds/ha **b** 30 weeds/ha **c** 3 days

d

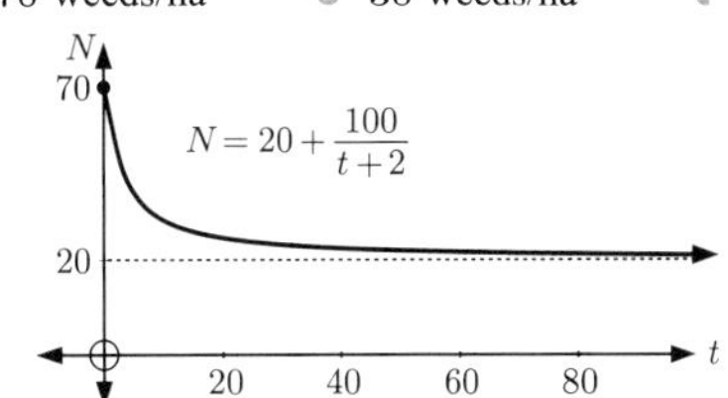

e No, the number of weeds/ha will approach 20 (from above), so at least 20 weeds will remain.

EXERCISE 5G

1 **a**

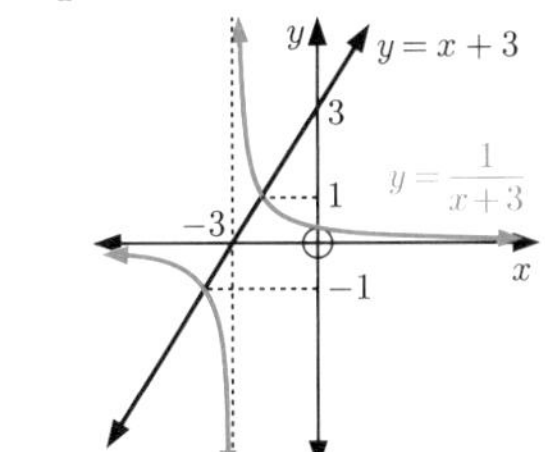

b

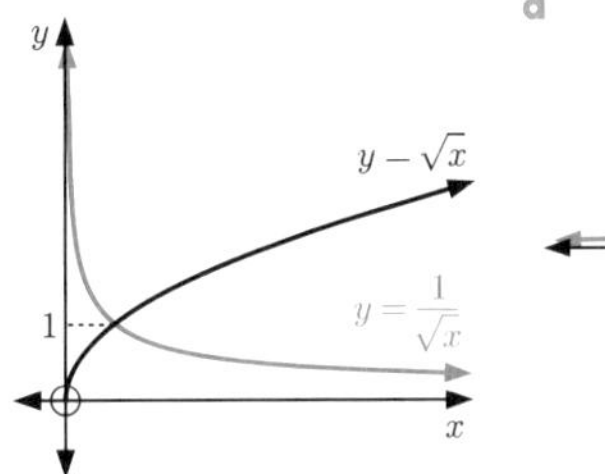

c

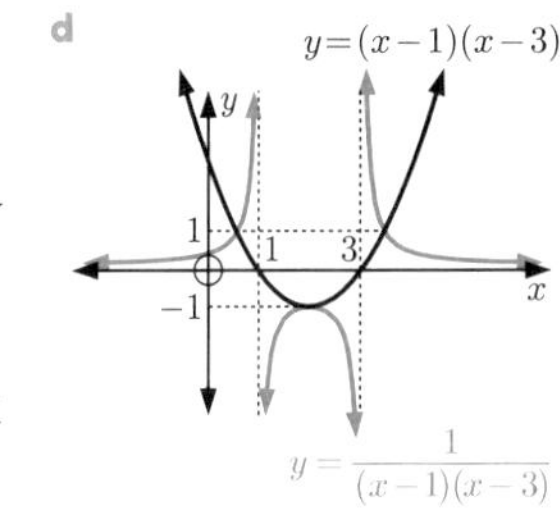

d

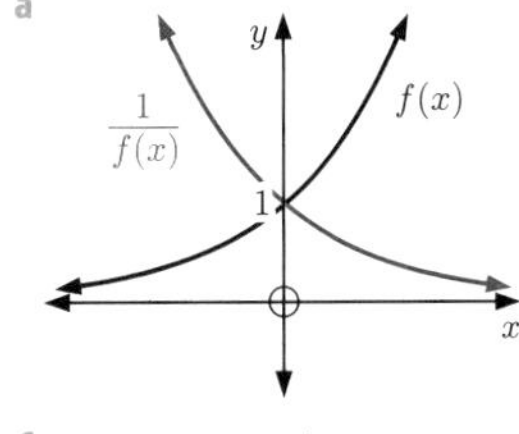

2 **a** invariant points are $(-2, 1)$ and $(-4, -1)$
b invariant points are $(-1, -1)$ and $(1, -1)$
c invariant point is $(1, 1)$
d invariant points are $(0.586, 1)$, $(2, -1)$, and $(3.41, 1)$

3 **a**

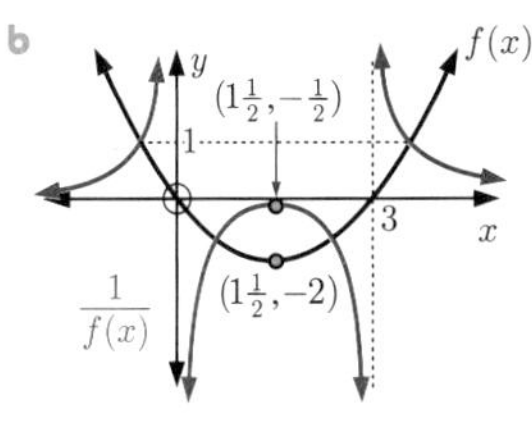

b

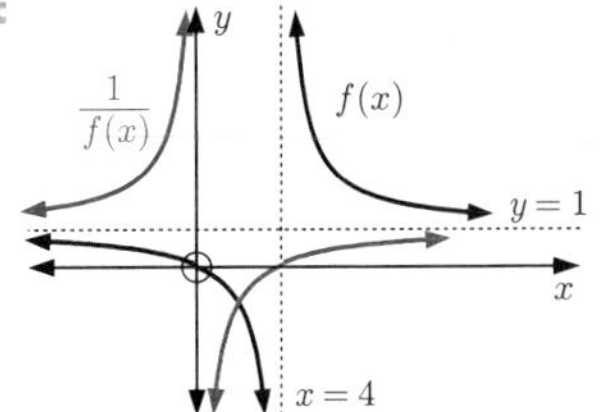

c

EXERCISE 5H

1 **a**

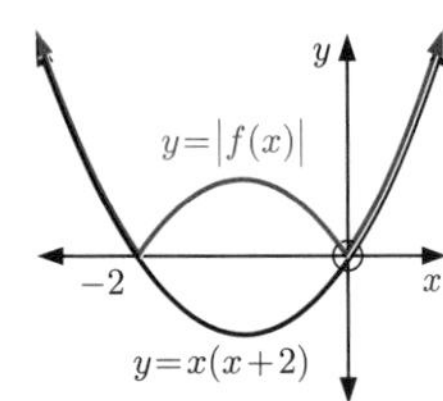

b

2 **a**

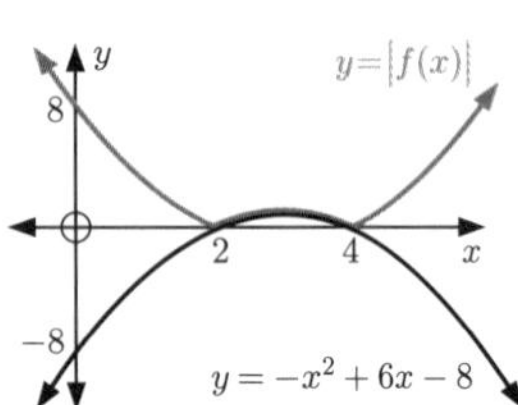

b

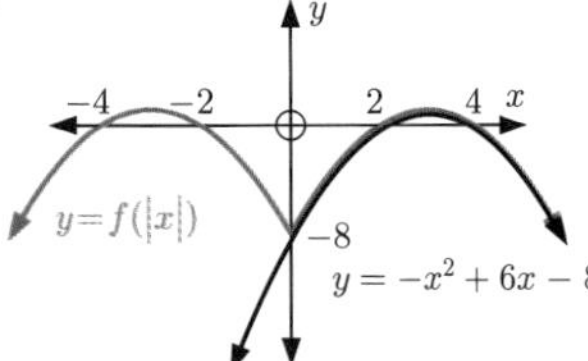

3 **a** **i**

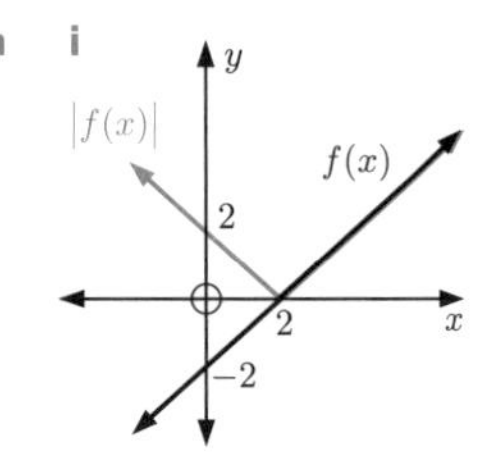

ii

iii

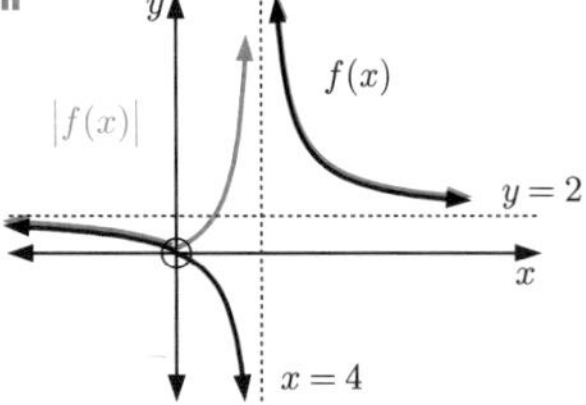

b **i**

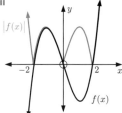

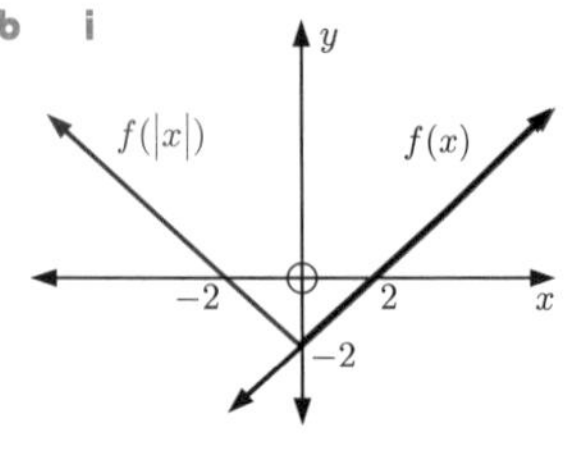

ii

iii

4 **a** (3, 0) **b** (5, 2) **c** (0, 7) **d** (2, 2)

5 **a** **i** (0, 3) **ii** (1, 3) and (−1, 3)
iii (7, −4) and (−7, −4)
b **i** (0, 3) **ii** (1, 3) **iii** (10, −8)

REVIEW SET 5A

1 **a** 3 **b** $4x^2 - 4x$ **c** $x^2 + 2x$ **d** $3x^2 - 6x - 2$

2 **a** 5 **b** $-x^2+x+5$ **c** $5-\frac{1}{2}x-\frac{1}{4}x^2$ **d** $-x^2-3x+5$

3 $g(x) = 3x^3 - 11x^2 + 14x - 6$

4

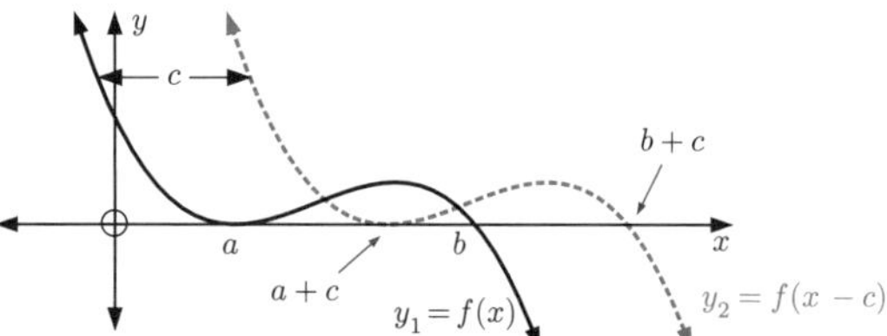

5

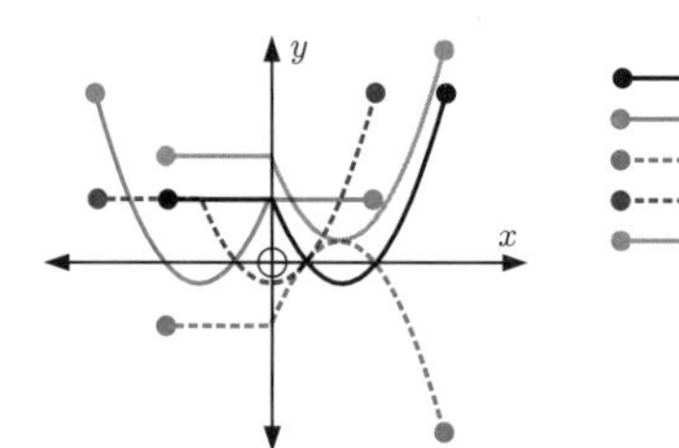

6

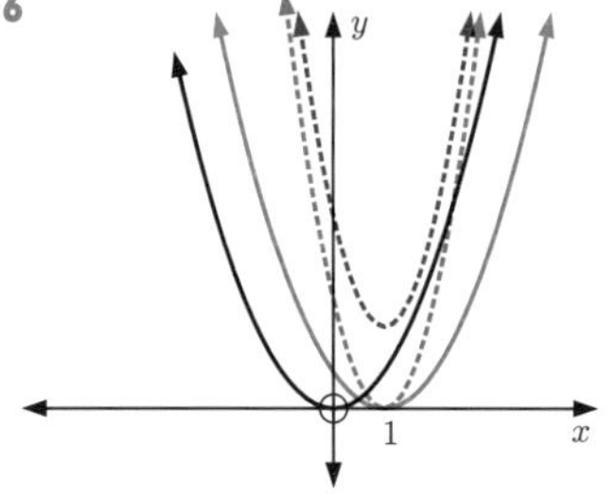

7 **a**

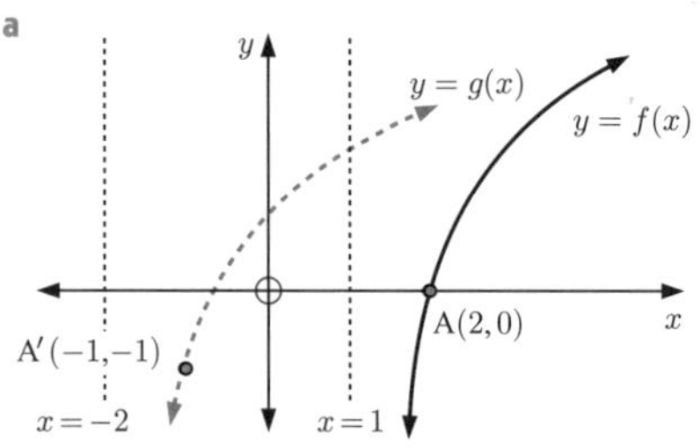

b $x = -2$ **c** A′(−1, −1)

8 **a**

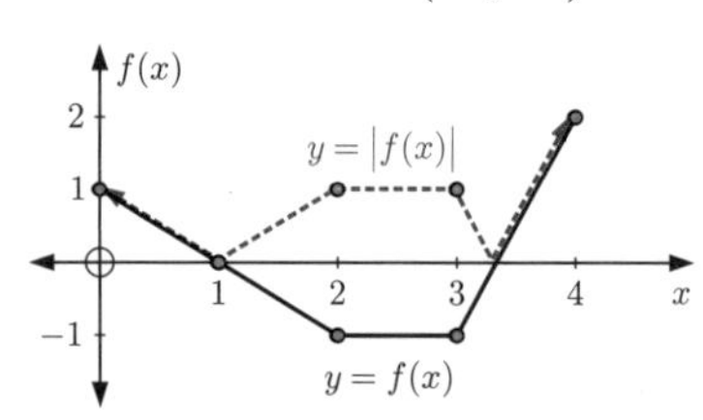

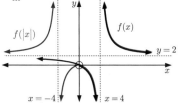

b 1 **c** (0, 1), $(\frac{11}{3}, 1)$, all points on $y = -1$, $x \in [2, 3]$

d

$f(x)$, $(\frac{11}{3}, 1)$, $(4, 2)$, $(4, \frac{1}{2})$, $y = f(x)$, $y = \frac{1}{f(x)}$, $x = 1$, $x = 3\frac{1}{3}$

9 a, b

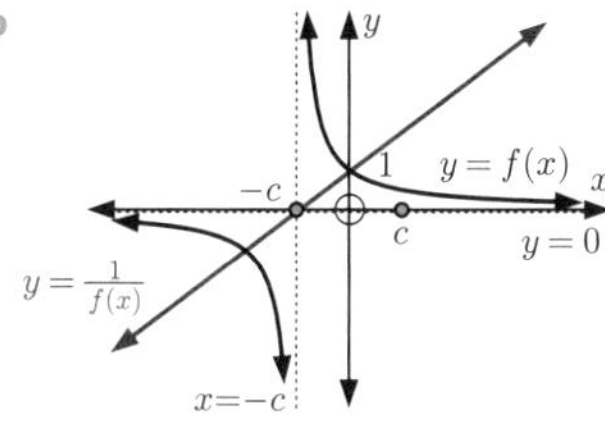

10 a $|f(x)| = |x - a|$, $f(|x|) = |x| - a$

b

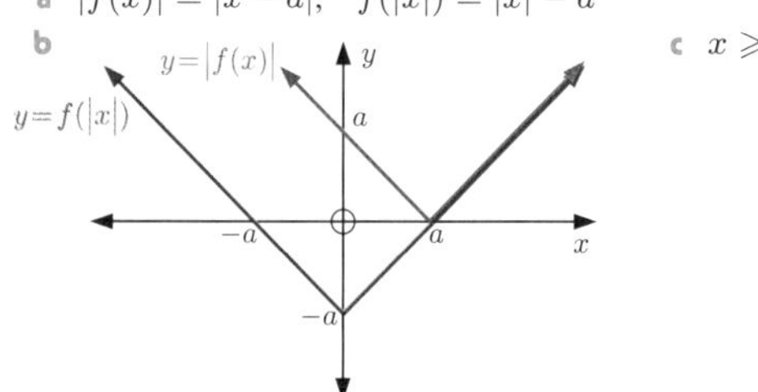

c $x \geqslant a$

REVIEW SET 5B

1

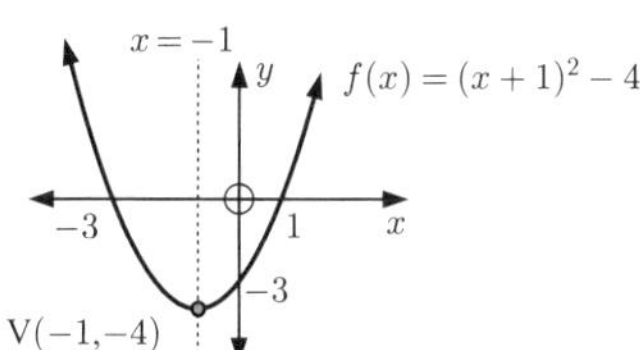

2

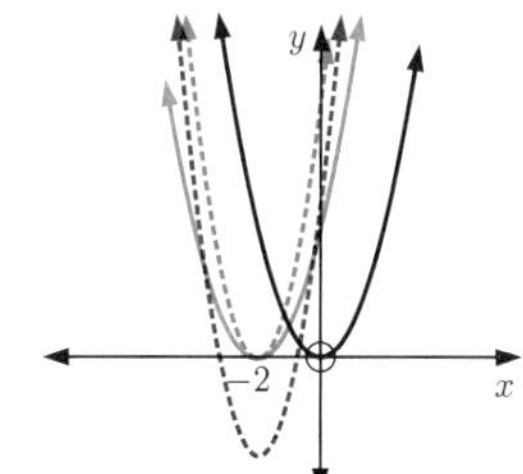

$y = f(x) = x^2$
$y = f(x+2)$
$y = 2\,f(x+2)$
$y = 2\,f(x+2) - 3$

3 a no

b horizontal asymptote $y = 0$, vertical asymptote $x = 0$

c min. turning point (1.44, 1.88)

d

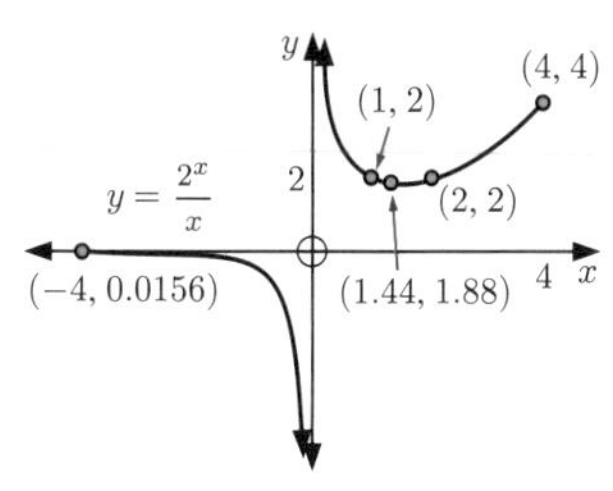

4 a i true
ii false
iii false
iv true

b

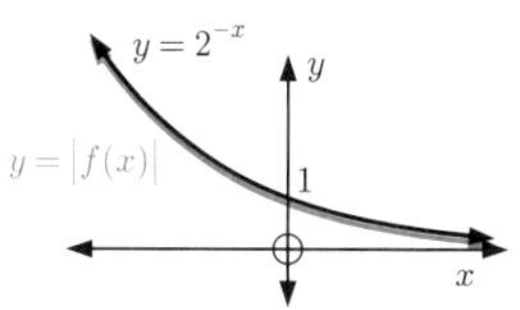

c horizontal asymptote $y = 0$

5 a $g(x) = (x-1)^2 + 8$ b $\{y \mid y \geqslant 4\}$ c $\{y \mid y \geqslant 8\}$

6 a i $y = \dfrac{1}{x-1} - 2$

ii

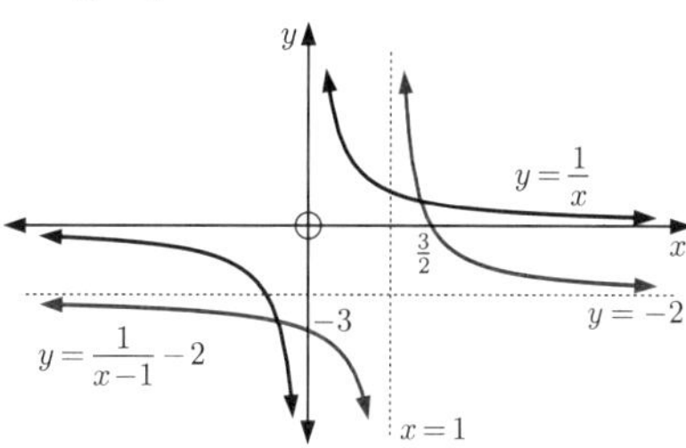

For $y = \dfrac{1}{x}$, VA is $x = 0$, HA is $y = 0$

For $y = \dfrac{1}{x-1} - 2$, VA is $x = 1$, HA is $y = -2$

iii For $y = \dfrac{1}{x}$, domain is $\{x \mid x \neq 0\}$, range is $\{y \mid y \neq 0\}$

For $y = \dfrac{1}{x-1} - 2$, domain is $\{x \mid x \neq 1\}$, range is $\{y \mid y \neq -2\}$

b i $y = 2^{x-1} - 2$

ii

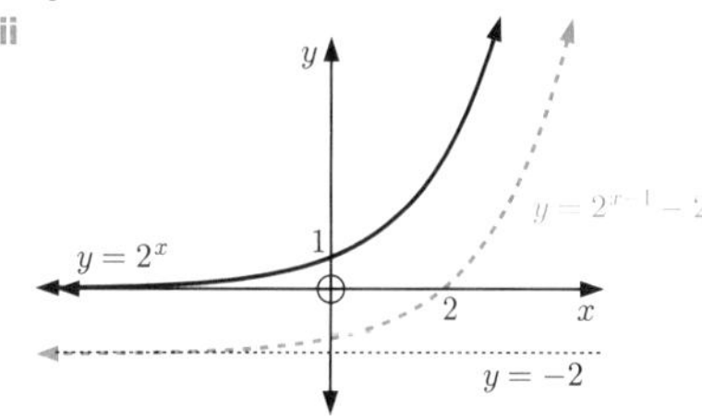

For $y = 2^x$, HA is $y = 0$, no VA
For $y = 2^{x-1} - 2$, HA is $y = -2$, no VA

iii For $y = 2^x$, domain is $\{x \mid x \in \mathbb{R}\}$, range is $\{y \mid y > 0\}$

For $y = 2^{x-1} - 2$, domain is $\{x \mid x \in \mathbb{R}\}$, range is $\{y \mid y > -2\}$

7

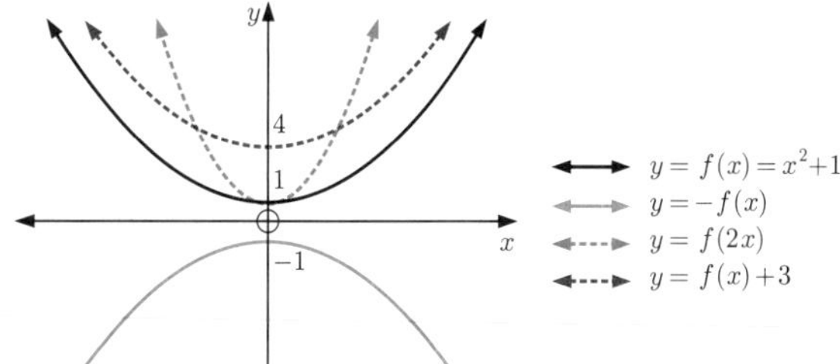

8 a $F(x) = 4x - 1$ b It is invariant

c $(0, 2) \to (\frac{1}{2}, 1)$ and $(-1, 1) \to (0, -1)$

d $F(\frac{1}{2}) = 1$ and $F(0) = -1$

9 **a**

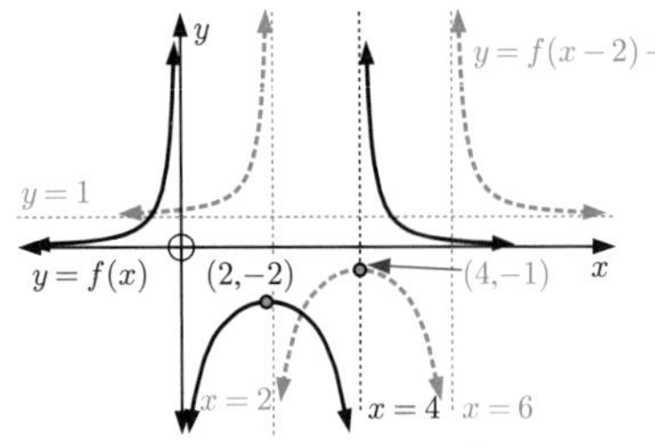

b

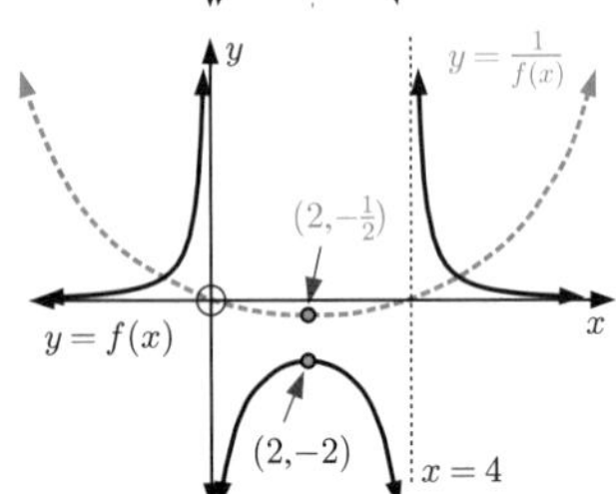

c

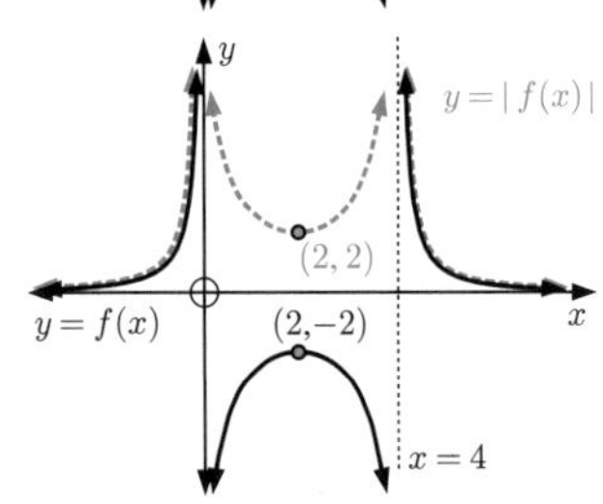

10 **a** vertical asymptote $x = -\frac{5}{3}$
horizontal asymptote $y = \frac{2}{3}$

b as $x \to -\frac{5}{3}^-$, $y \to \infty$
as $x \to -\frac{5}{3}^+$, $y \to -\infty$
$x \to -\infty$, as $y \to \frac{2}{3}^+$
$x \to \infty$, as $y \to \frac{2}{3}^-$

c y-intercept is $-\frac{3}{5}$, x-intercept is $\frac{3}{2}$

d

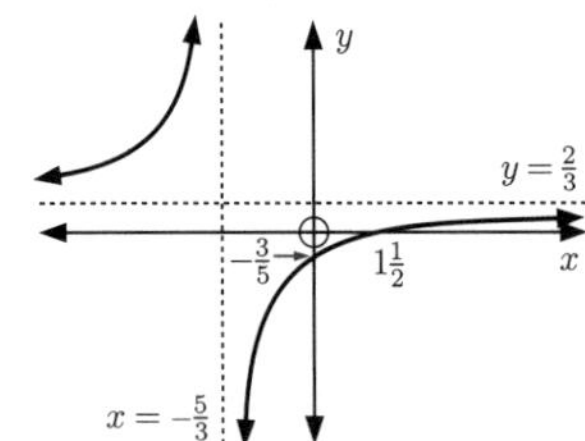

e Vertical stretch with scale factor $\frac{19}{9}$, reflection in the x-axis, then translate $\begin{pmatrix} -\frac{5}{3} \\ \frac{2}{3} \end{pmatrix}$.

f Translate $\begin{pmatrix} \frac{5}{3} \\ -\frac{2}{3} \end{pmatrix}$, reflect in x-axis, then vertical stretch with scale factor $\frac{9}{19}$.

11 **a, d**

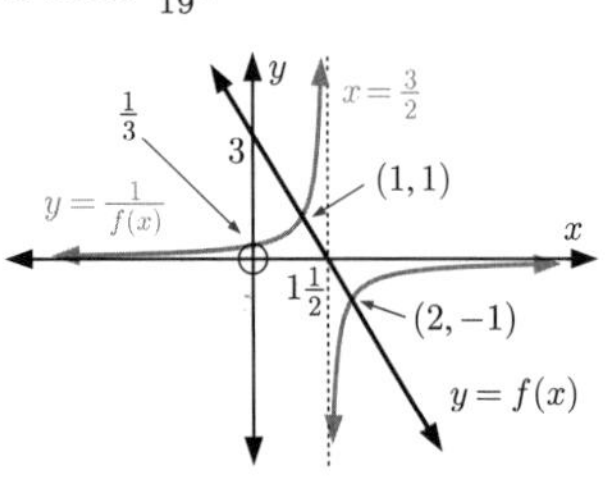

b (1, 1) and (2, −1) **c** VA is $x = \frac{3}{2}$, y-int is $\frac{1}{3}$

e

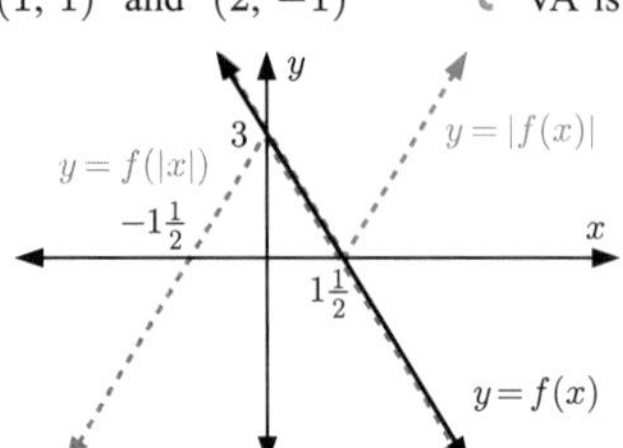

REVIEW SET 5C

1 **a** -1 **b** $\dfrac{2}{x}$ **c** $\dfrac{8}{x}$ **d** $\dfrac{10-3x}{x+2}$

2 **a** **i** $(-2, 1.8)$ **ii** $x = 0$ **iii** $y = 2$
iv x-intercepts -3.2 and -0.75

b

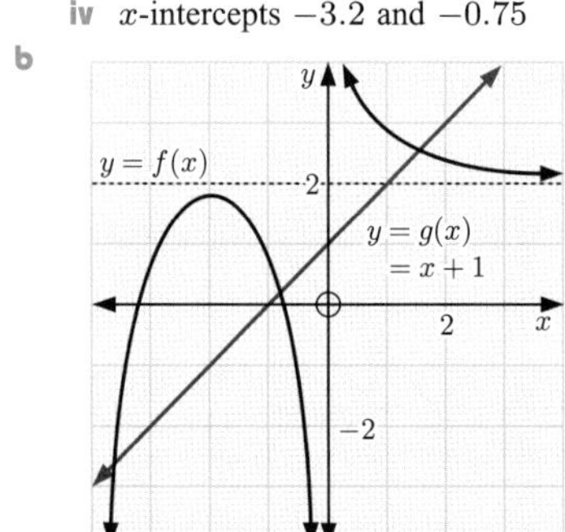

c $(-3.65, -2.65)$, $(-0.8, 0.2)$, and $(1.55, 2.55)$

3

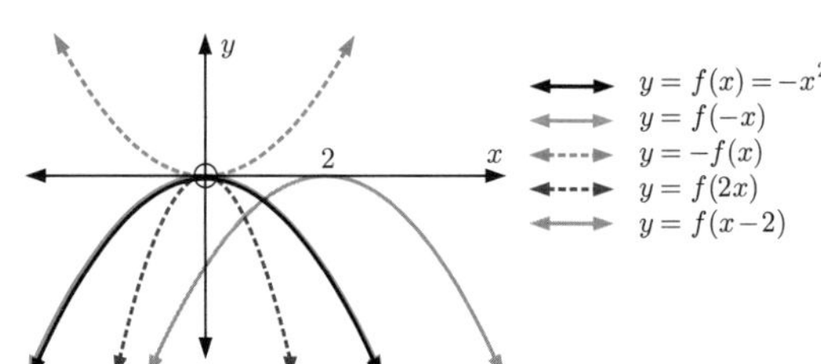

4 **a**

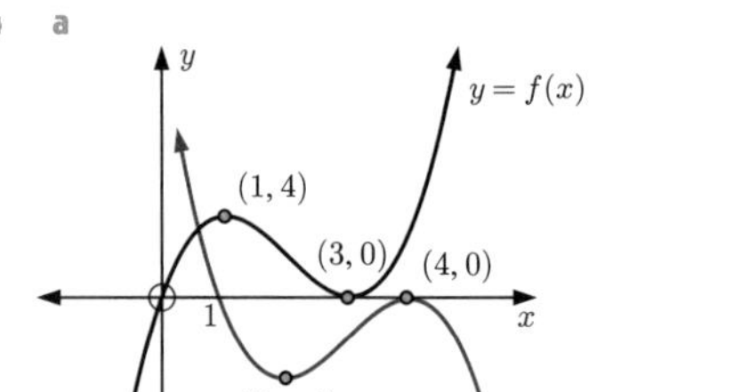

b (2, −4) and (4, 0)

5 $g(x) = -x^2 - 6x - 7$

6

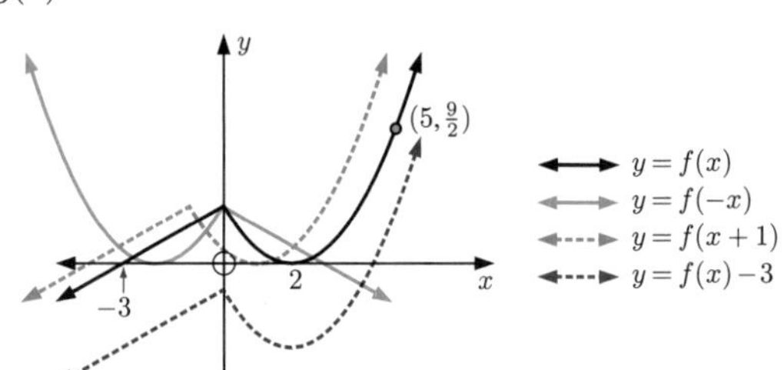

7 $g(x) = x^3 + 6x^2 + 8x + 10$

8 a i $y = 3x + 8$ ii $y = 3x + 8$

b $f(x+k) = a(x+k) + b = ax + b + ka = f(x) + ka$

9 a $f(x) = \frac{-3}{x-1} + 1$ b

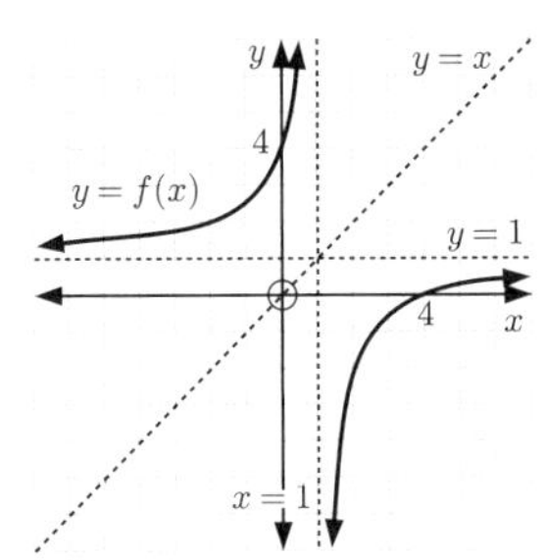

Domain of $f(x)$ is $\{x \mid x \neq 1\}$
Range of $f(x)$ is $\{y \mid y \neq 1\}$

c Yes, since it is a one-to-one function (passes both the vertical and horizontal line tests).

d Yes, since $f^{-1}(x) = f(x) = \frac{-3}{x-1} + 1$. Also, the graph of $f(x)$ is symmetrical about the line $y = x$.

10 a $y = \log_4(x-1) - 2$

b

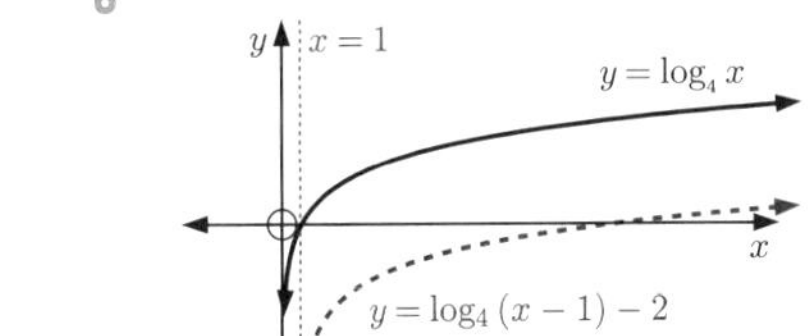

c For $y = \log_4 x$, VA is $x = 0$, no HA
For $y = \log_4(x-1) - 2$, VA is $x = 1$, no HA

d For $y = \log_4 x$, Domain is $\{x \mid x > 0\}$,
Range is $\{y \in \mathbb{R}\}$
For $y = \log_4(x-1) - 2$, Domain is $\{x \mid x > 1\}$,
Range is $\{y \in \mathbb{R}\}$

11 a $g(x) = \frac{-1}{3x-6}$

b HA is $y = 0$, VA is $x = 2$

c Domain $= \{x \mid x \neq 2\}$, Range $= \{y \mid y \neq 0\}$

d

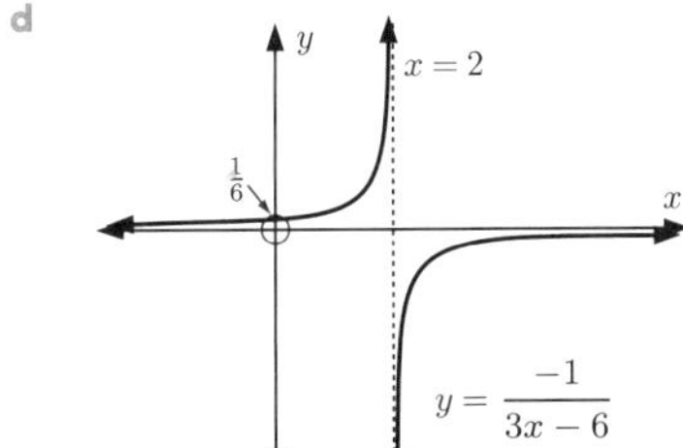

EXERCISE 6A

1 a $3i$ b $8i$ c $\frac{1}{2}i$ d $i\sqrt{5}$ e $i\sqrt{8}$

2 a $(x+3)(x-3)$ b $(x+3i)(x-3i)$
c $(x+\sqrt{7})(x-\sqrt{7})$ d $(x+i\sqrt{7})(x-i\sqrt{7})$
e $(2x+1)(2x-1)$ f $(2x+i)(2x-i)$
g $(\sqrt{2}x+3)(\sqrt{2}x-3)$ h $(\sqrt{2}x+3i)(\sqrt{2}x-3i)$
i $x(x+1)(x-1)$ j $x(x+i)(x-i)$
k $(x+1)(x-1)(x+i)(x-i)$
l $(x+2)(x-2)(x+2i)(x-2i)$

3 a $x = \pm 5$ b $x = \pm 5i$ c $x = \pm\sqrt{5}$
d $x = \pm i\sqrt{5}$ e $x = \pm\frac{3}{2}$ f $x = \pm\frac{3}{2}i$
g $x = 0,\ x = \pm 2$ h $x = 0,\ x = \pm 2i$
i $x = 0,\ x = \pm\sqrt{3}$ j $x = 0,\ x = \pm i\sqrt{3}$
k $x = \pm 1,\ x = \pm i$ l $x = \pm 3,\ x = \pm 3i$

4 a $x = 5 \pm 2i$ b $x = -3 \pm 4i$ c $x = -7 \pm i$
d $x = \frac{3}{2} \pm \frac{1}{2}i$ e $x = \sqrt{3} \pm i$ f $x = \frac{1}{4} \pm \frac{i\sqrt{7}}{4}$

5 a $x = \pm i\sqrt{3}$ or ± 1 b $x = \pm\sqrt{3}$ or $\pm i\sqrt{2}$
c $x = \pm 3i$ or ± 2 d $x = \pm i\sqrt{7}$ or $\pm i\sqrt{2}$
e $x = \pm 1$ f $x = \pm i$

EXERCISE 6B.1

1

z	$\mathcal{R}e\,(z)$	$\mathcal{I}m\,(z)$	z	$\mathcal{R}e\,(z)$	$\mathcal{I}m\,(z)$
$3+2i$	3	2	$-3+4i$	-3	4
$5-i$	5	-1	$-7-2i$	-7	-2
3	3	0	$-11i$	0	-11
0	0	0	$i\sqrt{3}$	0	$\sqrt{3}$

2 a $7-i$ b $10-4i$ c $-1+2i$ d $3-3i$
e $4-7i$ f $12+i$ g $3+4i$ h $21-20i$

3 a $-3+7i$ b $2i$ c $-2+2i$ d $-1+i$
e $-5-12i$ f $-5+i$ g $-6-4i$ h $-1-5i$

4 a $i^0 = 1,\ i^1 = i,\ i^2 = -1,\ i^3 = -i,\ i^4 = 1,\ i^5 = i,$
$i^6 = -1,\ i^7 = -i,\ i^8 = 1,\ i^9 = i,\ i^{-1} = -i,$
$i^{-2} = -1,\ i^{-3} = i,\ i^{-4} = 1,\ i^{-5} = -i$
b $i^{4n+3} = -i$

5 $(1+i)^4 = -4$, $(1+i)^{101} = -2^{50}(1+i)$

6 a $-\frac{1}{10} - \frac{7}{10}i$ b $-\frac{1}{5} + \frac{2}{5}i$ c $\frac{7}{5} + \frac{1}{5}i$ d $\frac{3}{25} + \frac{4}{25}i$

7 a $-\frac{2}{5} + \frac{1}{5}i$ b $-\frac{1}{13} + \frac{8}{13}i$ c $-\frac{2}{5} + \frac{3}{5}i$

8 a -2 b -4 c 3 d 0

EXERCISE 6B.2

1 a $x = 0,\ y = -2$ b $x = -2$
c $x = 3,\ y = 2$ d $x = -\frac{2}{13},\ y = -\frac{3}{13}$

2 a $x = 0,\ y = 0$ b $x = 3,\ y = -2$ or $x = 4,\ y = -\frac{3}{2}$
c $x = 2,\ y = -5$ or $x = -\frac{5}{3},\ y = 6$ d $x = -1,\ y = 0$

3 $z = 5 - 4i$ **4** $z = 65 - 72i$ **5** $m = -\frac{1}{11},\ n = \frac{8}{11}$

6 $z = i\sqrt{2}$ **7** $a = 3,\ b = -5$

EXERCISE 6B.3

1 a $a(x^2 - 6x + 10) = 0,\ a \neq 0$
b $a(x^2 - 2x + 10) = 0,\ a \neq 0$
c $a(x^2 + 4x + 29) = 0,\ a \neq 0$
d $a(x^2 - 2\sqrt{2}x + 3) = 0,\ a \neq 0$
e $a(x^2 - 4x + 1) = 0,\ a \neq 0$
f $a(3x^2 + 2x) = 0,\ a \neq 0$
g $a(x^2 + 2) = 0,\ a \neq 0$
h $a(x^2 + 12x + 37) = 0,\ a \neq 0$

2 a $a = -6,\ b = 10$ b $a = -2,\ b = -1$
c $a = -2,\ b = 8$ or $a = 0,\ b = 0$

EXERCISE 6B.4

2 z^* **4** a $\left(\frac{ac+bd}{c^2+d^2}\right) + \left(\frac{bc-ad}{c^2+d^2}\right)i$

7 **a** $z^2 = (a^2 - b^2) + (2ab)i$
c $z^3 = (a^3 - 3ab^2) + (3a^2b - b^3)i$

8 **a** $a = 0$ or $(b = 0,\ a \neq -1)$
b $a^2 - b^2 = 1$ and neither a nor b is 0, and $a \neq -1$

9 **c** $(z_1 z_2 z_3 z_4 \ldots . z_n)^* = {z_1}^* {z_2}^* {z_3}^* \ldots . z_n^*$ **d** $(z^n)^* = (z^*)^n$

EXERCISE 6C.1

1 **a** $3x^2 + 6x + 9$ **b** $5x^2 + 7x + 9$ **c** $-7x^2 - 8x - 9$
d $4x^4 + 13x^3 + 28x^2 + 27x + 18$

2 **a** $x^3 + x^2 - 4x + 7$ **b** $x^3 - x^2 - 2x + 3$
c $3x^3 + 2x^2 - 11x + 19$ **d** $2x^3 - x^2 - x + 5$
e $x^5 - x^4 - x^3 + 8x^2 - 11x + 10$
f $x^4 - 2x^3 + 5x^2 - 4x + 4$

3 **a** $2x^3 - 3x^2 + 4x + 3$ **b** $x^4 + x^3 - 7x^2 + 7x - 2$
c $x^3 + 6x^2 + 12x + 8$ **d** $4x^4 - 4x^3 + 13x^2 - 6x + 9$
e $16x^4 - 32x^3 + 24x^2 - 8x + 1$
f $18x^4 - 87x^3 + 56x^2 + 20x - 16$

4 **a** $6x^3 - 11x^2 + 18x - 5$ **b** $8x^3 + 18x^2 - x + 10$
c $-2x^3 + 7x^2 + 13x + 10$ **d** $2x^3 - 7x^2 + 4x + 4$
e $2x^4 - 2x^3 - 9x^2 + 11x - 2$
f $15x^4 + x^3 - x^2 + 7x - 6$ **g** $x^4 - 2x^3 + 7x^2 - 6x + 9$
h $4x^4 + 4x^3 - 15x^2 - 8x + 16$
i $8x^3 + 60x^2 + 150x + 125$
j $x^6 + 2x^5 + x^4 - 4x^3 - 4x^2 + 4$

EXERCISE 6C.2

1 **a** $Q(x) = x,\ R = -3,\ x^2 + 2x - 3 = x(x+2) - 3$
b $Q(x) = x - 4,\ R = -3,\ x^2 - 5x + 1 = (x-4)(x-1) - 3$
c $Q(x) = 2x^2 + 10x + 16,\ R = 35,$
$2x^3 + 6x^2 - 4x + 3 = (2x^2 + 10x + 16)(x - 2) + 35$

2 **a** $x^2 - 3x + 6 = (x+1)(x-4) + 10$
b $x^2 + 4x - 11 = (x+1)(x+3) - 14$
c $2x^2 - 7x + 2 = (2x-3)(x-2) - 4$
d $2x^3 + 3x^2 - 3x - 2 = (x^2 + x - 2)(2x + 1)$
e $3x^3 + 11x^2 + 8x + 7 = (x^2 + 4x + 4)(3x - 1) + 11$
f $2x^4 - x^3 - x^2 + 7x + 4 = (x^3 - 2x^2 + \frac{5}{2}x - \frac{1}{4})(2x+3) + \frac{19}{4}$

3 **a** $x + 2 + \dfrac{9}{x-2}$ **b** $2x + 1 - \dfrac{1}{x+1}$
c $3x - 4 + \dfrac{3}{x+2}$ **d** $x^2 + 3x - 2$
e $2x^2 - 8x + 31 - \dfrac{124}{x+4}$ **f** $x^2 + 3x + 6 + \dfrac{7}{x-2}$

EXERCISE 6C.3

1 **a** quotient is $x + 1$, remainder is $-x - 4$
b quotient is 3, remainder is $-x + 3$
c quotient is $3x$, remainder is $-2x - 1$
d quotient is 0, remainder is $x - 4$

2 **a** $1 - \dfrac{2x}{x^2 + x + 1}$, $x^2 - x + 1 = 1(x^2 + x + 1) - 2x$
b $x - \dfrac{2x}{x^2 + 2}$, $x^3 = x(x^2 + 2) - 2x$
c $x^2 + x + 3 + \dfrac{3x - 4}{x^2 - x + 1}$,
$x^4 + 3x^2 + x - 1 = (x^2 + x + 3)(x^2 - x + 1) + 3x - 4$
d $2x + 4 + \dfrac{5x + 2}{(x-1)^2}$, $2x^3 - x + 6 = (2x+4)(x-1)^2 + 5x + 2$
e $x^2 - 2x + 3 - \dfrac{4x + 3}{(x+1)^2}$, $x^4 = (x^2 - 2x + 3)(x+1)^2 - 4x - 3$
f $x^2 - 3x + 5 + \dfrac{15 - 10x}{(x-1)(x+2)}$,
$x^4 - 2x^3 + x + 5 = (x^2 - 3x + 5)(x-1)(x+2) + 15 - 10x$

3 quotient is $x^2 + 2x + 3$, remainder is 7

4 quotient is $x^2 - 3x + 5$, remainder is $15 - 10x$

EXERCISE 6D.1

1 **a** $4, -\frac{3}{2}$ **b** $-3 \pm i$ **c** $3 \pm \sqrt{3}$ **d** $0, \pm 2$
e $0, \pm i\sqrt{2}$ **f** $\pm 1, \pm i\sqrt{5}$

2 **a** $x = 1, -\frac{2}{5}$ **b** $x = -\frac{1}{2}, \pm i\sqrt{3}$ **c** $z = 0, 1 \pm i$
d $x = 0, \pm\sqrt{5}$ **e** $z = 0, \pm i\sqrt{5}$ **f** $z = \pm i\sqrt{2}, \pm\sqrt{5}$

3 **a** $(2x+3)(x-5)$ **b** $(z - 3 + i\sqrt{7})(z - 3 - i\sqrt{7})$
c $x(x + 1 + \sqrt{5})(x + 1 - \sqrt{5})$ **d** $z(3z - 2)(2z + 1)$
e $(z+1)(z-1)(z+\sqrt{5})(z-\sqrt{5})$
f $(z+i)(z-i)(z+\sqrt{2})(z-\sqrt{2})$

5 **a** $P(z) = a(z^2 - 4)(z - 3),\ a \neq 0$
b $P(z) = a(z+2)(z^2+1),\ a \neq 0$
c $P(z) = a(z-3)(z^2 + 2z + 2),\ a \neq 0$
d $P(z) = a(z+1)(z^2 + 4z + 2),\ a \neq 0$

6 **a** $P(z) = a(z^2 - 1)(z^2 - 2),\ a \neq 0$
b $P(z) = a(z-2)(z+1)(z^2+3),\ a \neq 0$
c $P(z) = a(z^2 - 3)(z^2 - 2z + 2),\ a \neq 0$
d $P(z) = a(z^2 - 4z - 1)(z^2 + 4z + 13),\ a \neq 0$

EXERCISE 6D.2

1 **a** $a = 2,\ b = 5,\ c = 5$ **b** $a = 3,\ b = 4,\ c = 3$

2 **a** $a = 2,\ b = -2$ or $a = -2,\ b = 2$ **b** $a = 3,\ b = -1$

4 $a = -2,\ b = 2,\ x = 1 \pm i$ or $-1 \pm \sqrt{3}$

5 **a** $a = -1$, zeros are $\frac{3}{2}, \dfrac{-1 \pm i\sqrt{3}}{2}$
b $a = 6$, zeros are $-\frac{2}{3}, \dfrac{1 \pm i\sqrt{11}}{2}$

6 **a** $a = -3,\ b = 6$ zeros are $-\frac{1}{2}, 2, \pm 2i$
b $a = 1,\ b = -15$ zeros are $-3, \frac{1}{2}, 1 \pm \sqrt{2}$

7 **a** $P(x) = (x+3)^2(x-3)$ or $P(x) = (x-1)^2(x+5)$
b If $m = -2$, zeros are -1 (repeated) and $\frac{2}{3}$.
If $m = \frac{14}{243}$, zeros are $\frac{1}{9}$ (repeated) and $-\frac{14}{9}$.

EXERCISE 6E.1

1 **a** $P(x) = Q(x)(x-2) + 7$, $P(x)$ divided by $x - 2$ leaves a remainder of 7.
b $P(-3) = -8$, $P(x)$ divided by $x + 3$ leaves a remainder of -8.
c $P(5) = 11$, $P(x) = Q(x)(x-5) + 11$

2 **a** 1 **b** 1 **3** **a** $a = 3$ **b** $a = 2$

4 $a = -5,\ b = 6$ **5** $a = -3,\ n = 4$

6 **a** -3 **b** 1 **7** $3z - 5$

EXERCISE 6E.2

1 **a** $k = -8,\ P(x) = (x+2)(2x+1)(x-2)$
b $k = 2,\ P(x) = x(x-3)(x+\sqrt{2})(x-\sqrt{2})$

2 $a = 7,\ b = -14$

3 **a** If $k = 1$, zeros are $3, -1 \pm i$.
If $k = -4$, zeros are $\pm 3, 1$.
b $m = -\frac{10}{7}$

4 **a** **i** $P(a) = 0$, $x-a$ is a factor **ii** $(x-a)(x^2+ax+a^2)$
b **i** $P(-a) = 0$, $x+a$ is a factor **ii** $(x+a)(x^2-ax+a^2)$

5 **b** $a = 2$

EXERCISE 6E.3

1 $P(x) = a(2x+1)(x^2-2x+10)$ $a \neq 0$

2 $p(x) = 4x^3 - 20x^2 + 36x - 20$

3 **a** $p = -3$, $q = 52$ other zeros are $2+3i$, -4

4 $a = -13$, $b = 34$ other zeros are $3-i$, $-2 \pm \sqrt{3}$

5 $a = 3$, $P(z) = (z+3)(z+i\sqrt{3})(z-i\sqrt{3})$

6 $k = 2$, $P(x) = (x+i\sqrt{5})(x-i\sqrt{5})(3x+2)$

EXERCISE 6E.4

1 **a** sum $= \frac{3}{2}$, product $= 2$ **b** sum $= \frac{4}{3}$, product $= \frac{5}{3}$
c sum $= 1$, product $= -4$ **d** sum $= \frac{3}{2}$, product $= 4$
e sum $= 0$, product $= 9$ **f** sum $= 0$, product $= -1$

2 **a** sum $= \frac{20}{3}$, product $= \frac{22}{3}$ **b** -40 **c** -44

3 **a** $k = \frac{1}{10}$ **b** $\frac{20+\sqrt{10}}{5}$ **4** $m = -2$, $n = 1$

5 $a = \pm 2$ **6** **b** $p = -1$, $q = -1$, $r = 1$

EXERCISE 6F.1

1 **a** cuts the x-axis at α **b** touches the x-axis at α
c cuts the x-axis at α with a change in shape

2 **a** $P(x) = 2(x+1)(x-2)(x-3)$
b $P(x) = -2(x+3)(2x+1)(2x-1)$
c $P(x) = \frac{1}{4}(x+4)^2(x-3)$
d $P(x) = \frac{1}{10}(x+5)(x+2)(x-5)$
e $P(x) = \frac{1}{4}(x+4)(x-3)^2$
f $P(x) = -2(x+3)(x+2)(2x+1)$

3 **a** $P(x) = (x-3)(x-1)(x+2)$
b $P(x) = x(x+2)(2x-1)$ **c** $P(x) = (x-1)^2(x+2)$
d $P(x) = (3x+2)^2(x-4)$

4 **a** **F** **b** **C** **c** **A** **d** **E** **e** **D** **f** **B**

5 **a** $P(x) = 5(2x-1)(x+3)(x-2)$
b $P(x) = -2(x+2)^2(x-1)$
c $P(x) = (x-2)(2x^2-3x+2)$

EXERCISE 6F.2

1 **a** $P(x) = 2(x+1)^2(x-1)^2$
b $P(x) = (x+3)(x+1)^2(3x-2)$
c $P(x) = -2(x+2)(x+1)(x-2)^2$
d $P(x) = -\frac{1}{3}(x+3)(x+1)(2x-3)(x-3)$
e $P(x) = \frac{1}{4}(x+1)(x-4)^3$ **f** $P(x) = x^2(x+2)(x-3)$

2 **a** **C** **b** **F** **c** **A** **d** **E** **e** **B** **f** **D**

3 **a** $P(x) = (x+4)(2x-1)(x-2)^2$
b $P(x) = \frac{1}{4}(3x-2)^2(x+3)^2$
c $P(x) = 2(x-2)(2x-1)(x+2)(2x+1)$
d $P(x) = (x-1)^2(\frac{8}{3}x^2 + \frac{8}{3}x - 1)$

EXERCISE 6F.3

1 **a** -1, $2 \pm \sqrt{3}$ **b** 1, $1 \pm i$ **c** $\frac{7}{2}$, $-1 \pm 2i$
d $\frac{1}{2}$, $\pm i\sqrt{10}$ **e** $\pm\frac{1}{2}$, 3, -2 **f** 2, $1 \pm 3i$

2 **a** $x = -2$, $\pm i\sqrt{3}$ **b** $x = -2$, $-\frac{1}{2}$, 1
c $x = 2$ (treble root) **d** $x = -2$, $\frac{3}{2}$, 3
e $x = -3$, 2, $1 \pm \sqrt{2}$ **f** $x = -\frac{1}{2}$, 3, $2 \pm i$

3 **a** $(x-1)(x-1+i)(x-1-i)$
b $(x+3)(x+2i)(x-2i)$
c $(2x-1)(x-2-\sqrt{3})(x-2+\sqrt{3})$
d $(x-2)(x-1+2i)(x-1-2i)$
e $(x-1)(2x-3)(2x+1)$
f $(x+2)(3x-2)(x-i\sqrt{3})(x+i\sqrt{3})$
g $(x+1)(2x-1)(x-1-i\sqrt{3})(x-1+i\sqrt{3})$
h $(2x+5)(x+2i)(x-2i)$

4 **a** -3.27, -0.860, 2.13 **b** -2.52, -1.18, 2.70

5 **a** $a = 700$, the time at which the barrier has returned to its original position
b $k = \frac{85}{36\,000\,000}$, $f(t) = \frac{85}{36\,000\,000}t(t-700)^2$

6 March

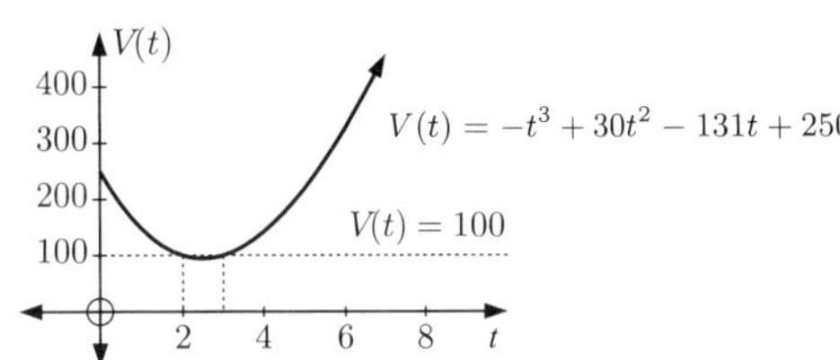

$V(t) = -t^3 + 30t^2 - 131t + 250$

7 9.938 m or 1.112 m

REVIEW SET 6A

1 **a** $a = 4$, $b = 0$ **b** $a = 3$, $b = -4$
c $a = 3$, $b = -7$ or $a = 14$, $b = -\frac{3}{2}$

2 **a** $12 + 5i$ **b** $-1 + i$ **c** $18 + 26i$

3 $\mathcal{R}e\,(z) = \frac{7\sqrt{3}}{4}$, $\mathcal{I}m\,(z) = -\frac{3}{4}$ **4** $z = \frac{3}{5} - \frac{1}{5}i$

6 $w = \dfrac{(a+1)^2 - b^2}{(a+1)^2 + b^2} + i\left(\dfrac{2(a+1)b}{(a+1)^2 + b^2}\right)$
w is purely imaginary if $b = \pm(a+1)$, $a \neq -1$, $b \neq 0$

7 **a** $12x^4 - 9x^3 + 8x^2 - 26x + 15$
b $4x^4 - 4x^3 + 13x^2 - 6x + 9$

8 **a** $x^2 - 2x + 4 - \dfrac{8}{x+2}$ **b** $x - 5 + \dfrac{19x + 30}{(x+2)(x+3)}$

9 **a** sum $= \frac{4}{3}$, product $= \frac{8}{3}$ **b** sum $= 0$, product $= -5$

10 "If a polynomial $P(x)$ is divided by $x - k$ until a constant remainder R is obtained then $R = P(k)$."

11 $a = 7$, $b = 0$ or $a = 4$, $b = \pm\sqrt{3}$ **12** $3x - 7$

13 $P(z) = z^4 - 6z^3 + 14z^2 - 10z - 7$

14 $k = 3$, $b = 27$, $x = 3, -3$; $k = -1$, $b = -5$, $x = -1, 5$

15 $P(x) = a(x^4 - 8x^3 + 22x^2 - 16x - 11)$, $a \neq 0$

16 $k = -2$, $n = 36$

17 **Another hint:** Show that $(\alpha\beta)^3 + (\alpha\beta)^2 - 1 = 0$

REVIEW SET 6B

1 $x = 2$, $y = -2$ **2** $z = 3 - 2i$ or $-3 + 2i$ **4** 7

5 $a = 0$, $b = -1$ **6** $a = 0$, $b = 0$ or $a = 3$, $b = 18$

7 2 **8** $(2z+1)(z+i\sqrt{5})(z-i\sqrt{5})$

9 $P(x) = (x+2)^2(x-1)(4x-3)$

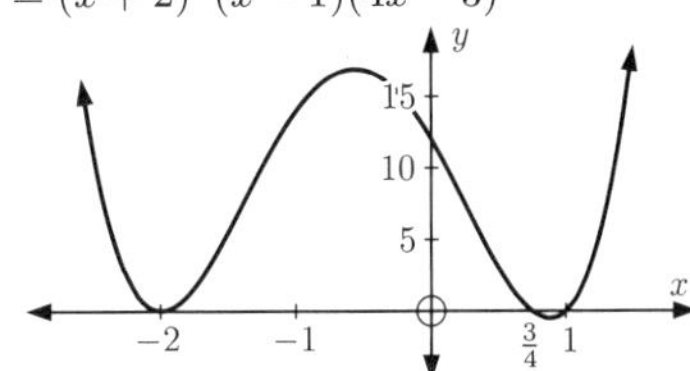

10 $1, -\frac{1}{2}, 1 \pm i\sqrt{5}$ **11** $(z+2)^2(z-1+i)(z-1-i)$

12 $P(z) = a(z^2-4z+5)(z^2+2z+10), \ a \neq 0$

13 $k=-4$, zeros are $3 \pm 2i, -1 \pm \sqrt{2}$ **14** $\pm 2i, -1 \pm i$

15 7 **16** $k \in]-\infty, 10-3\sqrt{10}[$ or $k \in]10+3\sqrt{10}, \infty[$

17 **a** $m=1, \ n=\pm 2$ **b**

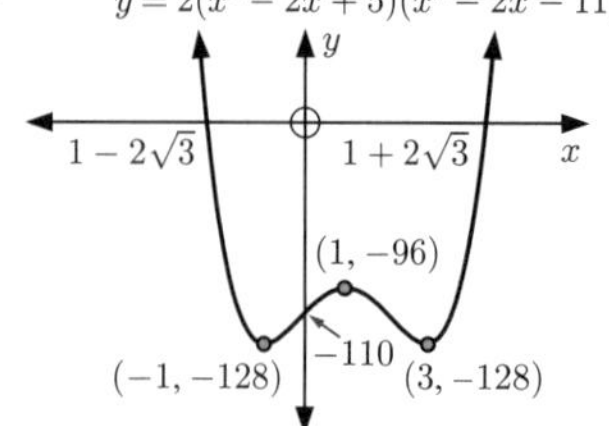

REVIEW SET 6C

1 $x=-1, \ y=2$ **2** $z=4-2i$ or $2+i$

4 **a** $x=0, \ y=0$ **b** $x=5, \ y=-7$
c $x=0, \ y=0,$ *or* $x=1, \ y=0,$
or $x=-\frac{1}{2}, \ y=\frac{\sqrt{3}}{2},$ *or* $x=-\frac{1}{2}, \ y=-\frac{\sqrt{3}}{2}$

6 **a** sum $= -\frac{3}{2}$, product $= -3$ **b** sum $= 0$, product $= \frac{3}{2}$

7 $z = -\frac{3737}{169} + \frac{4416}{169}i$ **8** 267 214

9 $a=-21$, other zeros are $5+i, \frac{1}{2}$

10 **a** $P(x) = a(2x-1)(x^2+2), \ a \neq 0$
b $P(x) = a(x^2-2x+2)(x^2+6x+10), \ a \neq 0$

11 **a** $k = 0, 4, -\frac{343}{8}$
b $P(x) = (x+2)^2(2x-1)$ when $k=4$

12 $z = -\frac{1}{2}, 2, \pm i\sqrt{2}$

13 **a** a is real, $k > 0$ **b** a is real, $k \leqslant 0$

14 $a=7, \ b=-20$ **15** $k = 5 \pm 2\sqrt{2}$

16 quotient is x^2+3x-9, remainder is $5x+17$, $a=4, \ b=-18$

17 **b** If $m=\sqrt{3}$, coefficient is $-4-2\sqrt{3}$
If $m=-\sqrt{3}$, coefficient is $-4+2\sqrt{3}$

EXERCISE 7A

1 **a** 4, 13, 22, 31 **b** 45, 39, 33, 27
c 2, 6, 18, 54 **d** 96, 48, 24, 12

2 **a** Starts at 8 and each term is 8 more than the previous term. Next two terms 40, 48.
b Starts at 2, each term is 3 more than the previous term; 14, 17.
c Starts at 36, each term is 5 less than the previous term; 16, 11.
d Starts at 96, each term is 7 less than the previous term; 68, 61.
e Starts at 1, each term is 4 times the previous term; 256, 1024.
f Starts at 2, each term is 3 times the previous term; 162, 486.
g Starts at 480, each term is half the previous term; 30, 15.
h Starts at 243, each term is $\frac{1}{3}$ of the previous term; 3, 1.
i Starts at 50 000, each term is $\frac{1}{5}$ of the previous term; 80, 16.

3 **a** Each term is the square of the term number; 25, 36, 49.
b Each term is the cube of the term number; 125, 216, 343.
c Each term is $n(n+1)$ where n is the term number; 30, 42, 56.

4 **a** 79, 75 **b** 1280, 5120 **c** 625, 1296
d 13, 17 **e** 16, 22 **f** 6, 12

EXERCISE 7B.1

1 **a** 1 **b** 13 **c** 79

2 **a** 7, 9, 11, 13 **b**

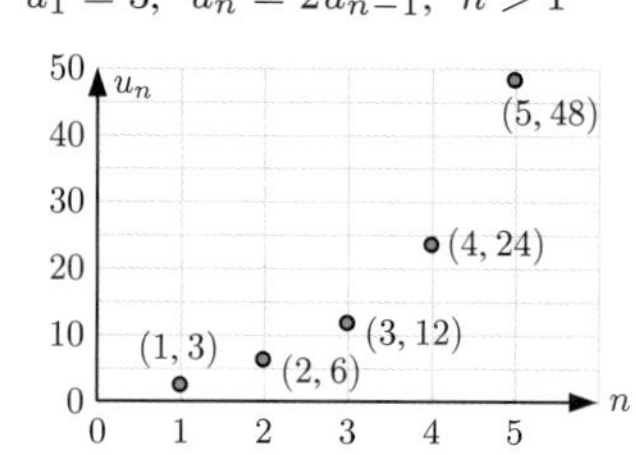

3 **a** 2, 4, 6, 8, 10 **b** 4, 6, 8, 10, 12
c 1, 3, 5, 7, 9 **d** −1, 1, 3, 5, 7
e 5, 7, 9, 11, 13 **f** 13, 15, 17, 19, 21
g 4, 7, 10, 13, 16 **h** 1, 5, 9, 13, 17

4 **a** 2, 4, 8, 16, 32 **b** 6, 12, 24, 48, 96
c $3, \frac{3}{2}, \frac{3}{4}, \frac{3}{8}, \frac{3}{16}$ **d** −2, 4, −8, 16, −32

5 17, 11, 23, −1, 47

EXERCISE 7B.2

1 **a** −1 **b** −5 **c** −9 **d** −13

2 **a** $u_1 = 3, \ u_n = 2u_{n-1}, \ n > 1$
b

3 **a** $\frac{1}{2}, \frac{2}{3}, \frac{3}{5}, \frac{5}{8}$
b

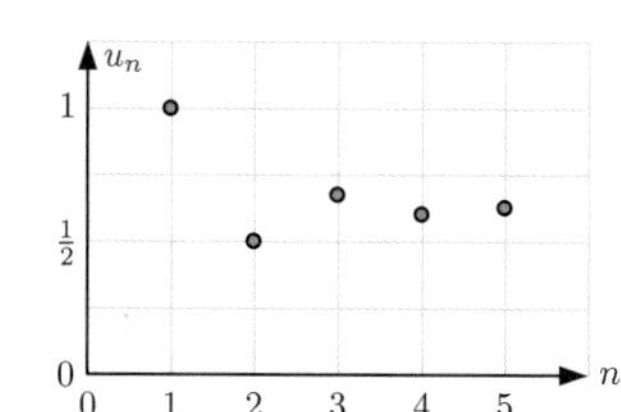

c **i** **Hint:** As $n \to \infty$, $u_{n+1} \to u$, and $u_n \to u$.

4 **a** 0.707 11, 0.765 37, 0.752 63, 0.755 36 **c** 0.754 88

5 $\approx 0.682\,33$

EXERCISE 7C

1 **a** 73 **b** 65 **c** $21\frac{1}{2}$

2 **a** 101 **b** −107 **c** $a+14d$

3 **a** $u_1=6, \ d=11$ **b** $u_n = 11n-5$ **c** 545
d yes, u_{30} **e** no

4 **a** $u_1 = 87, \ d=-4$ **b** $u_n = 91-4n$ **c** −69 **d** u_{97}

5 **b** $u_1 = 1, \ d=3$ **c** 169 **d** $u_{150} = 448$

6 **b** $u_1 = 32, \ d = -\frac{7}{2}$ **c** −227 **d** $n \geqslant 68$

7 **a** $k = 17\frac{1}{2}$ **b** $k=4$ **c** $k=4$ **d** $k=0$
e $k=-2$ or 3 **f** $k=-1$ or 3

8 **a** $u_n = 6n-1$ **b** $u_n = -\frac{3}{2}n + \frac{11}{2}$
c $u_n = -5n+36$ **d** $u_n = -\frac{3}{2}n + \frac{1}{2}$

9 **a** $6\frac{1}{4}, 7\frac{1}{2}, 8\frac{3}{4}$ **b** $3\frac{5}{7}, 8\frac{3}{7}, 13\frac{1}{7}, 17\frac{6}{7}, 22\frac{4}{7}, 27\frac{2}{7}$

10 a $u_1 = 36,\ d = -\frac{2}{3}$ b u_{100} **11** $u_{7692} = 100\,006$

12 a Month 1 = 5 cars, Month 2 = 18 cars, Month 3 = 31 cars, Month 4 = 44 cars, Month 5 = 57 cars, Month 6 = 70 cars
b The constant difference $d = 13$. c 148 cars
d 20 months

13 a $u_1 = 34,\ d = 7$ b 111 online friends c 18 weeks

14 a Day 1 = 97.3 tonnes, Day 2 = 94.6 tonnes, Day 3 = 91.9 tonnes
b $d = -2.7$, the cattle eat 2.7 tonnes of hay each day.
c $u_{25} = 32.5$. After 25 days (that is, July 25th) there will be 32.5 tonnes of hay left.
d 16.3 tonnes

EXERCISE 7D.1

1 a $b = 18,\ c = 54$ b $b = 2\frac{1}{2},\ c = 1\frac{1}{4}$
c $b = 3,\ c = -1\frac{1}{2}$

2 a 96 b 6250 c 16

3 a 6561 b $\frac{19\,683}{64}$ c 16 d ar^8

4 a $u_1 = 5,\ r = 2$ b $u_n = 5 \times 2^{n-1},\ u_{15} = 81\,920$

5 a $u_1 = 12,\ r = -\frac{1}{2}$
b $u_n = 12 \times (-\frac{1}{2})^{n-1},\ u_{13} = \frac{3}{1024}$

6 $u_1 = 8,\ r = -\frac{3}{4},\ u_{10} \approx -0.600\,677\,490\,2$

7 $u_1 = 8,\ r = \frac{1}{\sqrt{2}}$ **Hint:** $u_n = 2^3 \times (2^{-\frac{1}{2}})^{n-1}$

8 a $k = \pm 14$ b $k = 2$ c $k = -2$ or 4

9 a $u_n = 3 \times 2^{n-1}$ b $u_n = 32 \times (-\frac{1}{2})^{n-1}$
c $u_n = 3 \times (\pm\sqrt{2})^{n-1}$ d $u_n = 10 \times (\pm\sqrt{2})^{1-n}$

10 a $u_9 = 13\,122$ b $u_{14} = 2916\sqrt{3} \approx 5050.7$
c $u_{18} \approx 0.000\,091\,55$

EXERCISE 7D.2

1 a i ≈ 1550 ants ii ≈ 4820 ants b ≈ 12.2 weeks

2 a ≈ 278 b Year 2047

3 a i ≈ 73 ii ≈ 167 b ≈ 30.5 years

4 a i ≈ 2860 ii $\approx 184\,000$ b ≈ 14.5 years

EXERCISE 7D.3

1 a \$3993.00 b \$993.00 **2** €11 470.39

3 a ¥43 923 b ¥13 923 **4** \$23 602.32

5 ¥148 024.43 **6** £51 249.06 **7** \$14 976.01

8 £11 477.02 **9** €19 712.33 **10** ¥19 522.47

EXERCISE 7E

1 a i $S_n = \sum_{k=1}^{n} (8k - 5)$ ii 95

b i $S_n = \sum_{k=1}^{n} (47 - 5k)$ ii 160

c i $\sum_{k=1}^{n} 12\left(\frac{1}{2}\right)^{k-1}$ ii $23\frac{1}{4}$

d i $\sum_{k=1}^{n} 2\left(\frac{3}{2}\right)^{k-1}$ ii $26\frac{3}{8}$

e i $\sum_{k=1}^{n} \frac{1}{2^{k-1}}$ ii $1\frac{15}{16}$ f i $\sum_{k=1}^{n} k^3$ ii 225

2 a 24 b 27 c 10 d 25 e 168 f 310

3 $\sum_{n=1}^{20} (3n - 1) = 610$ **5** a 420 b ≈ 2232

6 a $1 + 2 + 3 + ... + (n-2) + (n-1) + n$

b
$$\begin{array}{rcccccccccc} & 1 & + & 2 & + & 3 & + ... + & (n-1) & + & n \\ + & n & + & (n-1) & + & (n-2) & + ... + & 2 & + & 1 \\ \hline & n+1 & + & n+1 & + & n+1 & + ... + & n+1 & + & n+1 \end{array}$$

c $S_n = \frac{n(n+1)}{2}$ d $a = 16,\ b = 3$

7 b $\sum_{k=1}^{n} (k+1)(k+2) = \frac{n(n^2 + 6n + 11)}{3}$,
$\sum_{k=1}^{10} (k+1)(k+2) = 570$

EXERCISE 7F

1 a 820 b $3087\frac{1}{2}$ c -1460 d -740

2 a 1749 b 2115 c $1410\frac{1}{2}$

3 a 160 b -630 c 135 **4** 203 **5** $-115\frac{1}{2}$

6 18 layers **7** a 65 b 1914 c 47 850

8 a 14 025 b 71 071 c 3367

9 a $u_n = 2n - 1$ c $S_1 = 1,\ S_2 = 4,\ S_3 = 9,\ S_4 = 16$

10 56, 49 **11** 10, 4, -2 or -2, 4, 10

12 2, 5, 8, 11, 14 or 14, 11, 8, 5, 2

13 34th week (total sold = 2057)

14 a $u_1 = 7,\ u_2 = 10$ b 64

EXERCISE 7G.1

1 a $23.9766 \approx 24.0$ b $\approx 189\,134$ c ≈ 4.000
d ≈ 0.5852

2 a $S_n = \frac{3 + \sqrt{3}}{2}\left((\sqrt{3})^n - 1\right)$ b $S_n = 24(1 - (\frac{1}{2})^n)$
c $S_n = 1 - (0.1)^n$ d $S_n = \frac{40}{3}(1 - (-\frac{1}{2})^n)$

3 a $u_1 = 3$ b $r = \frac{1}{3}$ c $u_5 = \frac{1}{27}$

4 a 3069 b $\frac{4095}{1024} \approx 4.00$ c $-134\,217\,732$

5 c \$26 361.59

6 a $\frac{1}{2}, \frac{3}{4}, \frac{7}{8}, \frac{15}{16}, \frac{31}{32}$ b $S_n = \frac{2^n - 1}{2^n}$
c $1 - (\frac{1}{2})^n = \frac{2^n - 1}{2^n}$ d as $n \to \infty,\ S_n \to 1$
e As $n \to \infty$, the sum of the fractions approaches the area of a 1×1 unit square.

7 54 or $\frac{2}{3}$

8 The 20th terms are: arithmetic 39, geometric 3^{19}
or arithmetic $7\frac{1}{3}$, geometric $(\frac{4}{3})^{19}$

9 a $n = 37$ b $n = 11$

11 a $A_3 = \$8000(1.03)^3 - (1.03)^2R - 1.03R - R$
b $A_8 = \$8000(1.03)^8 - (1.03)^7R - (1.03)^6R - (1.03)^5R - (1.03)^4R - (1.03)^3R - (1.03)^2R - (1.03)R - R = 0$
c $R = \$1139.65$

EXERCISE 7G.2

1 a i $u_1 = \frac{3}{10}$ ii $r = 0.1$ b $S = \frac{1}{3}$

2 a $\frac{4}{9}$ b $\frac{16}{99}$ c $\frac{104}{333}$ **4** a 54 b 14.175

5 **a** 1 **b** $4\frac{2}{7}$ **6** $u_1 = 9,\ r = \frac{2}{3}$

7 $u_1 = 8,\ r = \frac{1}{5}$ and $u_1 = 2,\ r = \frac{4}{5}$

8 **b** $S_n = 19 - 20 \times (0.9)^n$ **c** 19 seconds

9 **a** convergent, sum $= 12$ **b** not convergent, $n = 10$

10 70 cm **11** $x = \frac{1}{2}$

REVIEW SET 7A

1 **a** arithmetic, $d = -8$
b geometric, $r = 1$ *or* arithmetic, $d = 0$
c geometric, $r = -\frac{1}{2}$ **d** neither **e** arithmetic, $d = 4$

2 $k = -\frac{11}{2}$ **3** $u_n = 33 - 5n,\ S_n = \frac{n}{2}(61 - 5n)$

4 $k = \pm\frac{2\sqrt{3}}{3}$ **5** $u_n = \frac{1}{6} \times 2^{n-1}$ or $-\frac{1}{6} \times (-2)^{n-1}$

6 21, 19, 17, 15, 13, 11

7 **a** $u_n = 89 - 3n$ **b** $u_n = \dfrac{2n+1}{n+3}$
c $u_n = 100(0.9)^{n-1}$

8 **a** $1 + 4 + 9 + 16 + 25 + 36 + 49 = 140$
b $\frac{4}{3} + \frac{5}{4} + \frac{6}{5} + \frac{7}{6} = \frac{99}{20}$

9 **a** $10\frac{4}{5}$ **b** $16 + 8\sqrt{2}$ **10** 27 metres

11 **a** $u_n = 3n + 1$ **12** $a = b = c$

13 $x = 3,\ y = -1,\ z = \frac{1}{3}$ or $x = \frac{1}{3},\ y = -1,\ z = 3$

14 $x = \frac{3}{2}$ ($x = -\frac{6}{7}$ gives a divergent series)

REVIEW SET 7B

1 **a** $\frac{1}{3}$, 1, 3, 9 **b** $\frac{5}{4}, \frac{8}{5}, \frac{11}{6}, 2$ **c** 5, -5, 35, -65

2 **b** $u_1 = 6,\ r = \frac{1}{2}$ **c** $u_{16} \approx 0.000\,183$

3 **a** $n = 81$ **b** $-1\frac{1}{2}$ **c** 375

4 **a** 1587 **b** $47\frac{253}{256} \approx 48.0$ **5** $u_{12} = 10\,240$

6 **a** €8415.31 **b** €8488.67 **c** €8505.75

7 **a** 42 **b** $u_{n+1} - u_n = 5$ **c** $d = 5$ **d** 1672

8 $u_n = \left(\frac{3}{4}\right) 2^{n-1}$ **a** 49 152 **b** 24 575.25

9 $u_{11} = \frac{8}{19\,683} \approx 0.000\,406$ **10** **a** 17 **b** $255\frac{511}{512} \approx 256$

11 **a** $\frac{1331}{2100} \approx 0.634$ **b** $6\frac{8}{15}$ **12** $13 972.28

13 **a** ≈ 3470 **b** Year 2029 **14** $x > -\frac{1}{2}$

REVIEW SET 7C

1 **a** $d = -5$ **b** $u_1 = 63,\ d = -5$ **c** -117
d $u_{54} = -202$

2 **a** $u_1 = 3,\ r = 4$ **b** $u_n = 3 \times 4^{n-1},\ u_9 = 196\,608$

3 $u_n = 73 - 6n,\ u_{34} = -131$

4 **a** $\sum_{k=1}^{n} (7k - 3)$ **b** $\sum_{k=1}^{n} \left(\frac{1}{2}\right)^{k+1}$

5 **a** 70 **b** ≈ 241 **c** $\frac{64}{1875}$

6 12 **7** **a** £18 726.65 **b** £18 855.74

8 **a** $u_1 = 54,\ r = \frac{2}{3}$ and $u_1 = 150,\ r = -\frac{2}{5}$
b $|r| < 1$ in both cases, so the series will converge.
For $u_1 = 54,\ r = \frac{2}{3},\ S = 162$
For $u_1 = 150,\ r = -\frac{2}{5},\ S = 107\frac{1}{7}$

9 **a** 35.5 km **b** 1183 km **10** **a** $0 < x < 1$ **b** $35\frac{5}{7}$

12 $S = \dfrac{2 - 2^{\frac{1}{n+1}}}{2^{\frac{1}{n+1}} - 1}$

EXERCISE 8A

1 18 **2** **a** 4 **b** 8 **c** 24 **3** 6 **4** 42

5 1680 **6** **a** 125 **b** 60 **7** 17 576 000

8 **a** 4 **b** 9 **c** 81

EXERCISE 8B

1 **a** 13 **b** 20 **c** 19 **d** 32 **2** 13

EXERCISE 8C.1

1 1, 1, 2, 6, 24, 120, 720, 5040, 40 320, 362 880, 3 628 800

2 **a** 6 **b** 30 **c** $\frac{1}{7}$ **d** $\frac{1}{30}$ **e** 100 **f** 21

3 **a** $n,\ n \geqslant 1$ **b** $(n+2)(n+1),\ n \geqslant 0$
c $(n+1)n,\ n \geqslant 1$

4 **a** $\frac{7!}{4!}$ **b** $\frac{10!}{8!}$ **c** $\frac{11!}{6!}$ **d** $\frac{13!}{10!3!}$ **e** $\frac{3!}{6!}$ **f** $\frac{4!16!}{20!}$

5 **a** $6 \times 4!$ **b** $10 \times 10!$ **c** $57 \times 6!$ **d** $131 \times 10!$
e $81 \times 7!$ **f** $62 \times 6!$ **g** $10 \times 11!$ **h** $32 \times 8!$

6 **a** 11! **b** 9! **c** 8! **d** 9
e 34 **f** $n + 1$ **g** $(n-1)!$ **h** $(n+1)!$

EXERCISE 8C.2

1 **a** 3 **b** 6 **c** 35 **d** 210

2 **a** **i** 28 **ii** 28 **3** $k = 3$ or 6

EXERCISE 8D

1 **a** W, X, Y, Z
b WX, WY, WZ, XW, XY, XZ, YW, YX, YZ, ZW, ZX, ZY
c WXY, WXZ, WYX, WYZ, WZX, WZY, XWY, XWZ, XYW, XYZ, XZW, XZY, YWX, YWZ, YXW, YXZ, YZX, YZW, ZWX, ZWY, ZXW, ZXY, ZYW, ZYX

2 **a** AB, AC, AD, AE, BA, BC, BD, BE, CA, CB, CD, CE, DA, DB, DC, DE, EA, EB, EC, ED
b ABC, ABD, ABE, ACB, ACD, ACE, ADB, ADC, ADE, AEB, AEC, AED, BAC, BAD, BAE, BCA, BCD, BCE, BDA, BDC, BDE, BEA, BEC, BED, CAB, CAD, CAE, CBA, CBD, CBE, CDA, CDB, CDE, CEA, CEB, CED, DAB, DAC, DAE, DBA, DBC, DBE, DCA, DCB, DCE, DEA, DEB, DEC, EAB, EAC, EAD, EBA, EBC, EBD, ECA, ECB, ECD, EDA, EDB, EDC
2 at a time: 20 3 at a time: 60

3 **a** 120 **b** 336 **c** 5040 **4** **a** 12 **b** 24 **c** 36

5 720 **a** 24 **b** 24 **c** 48

6 **a** 343 **b** 210 **c** 120

7 **a** 648 **b** 64 **c** 72 **d** 136

8 **a** 120 **b** 48 **c** 72 **9** **a** 3 628 800 **b** 241 920

10 **a** 48 **b** 24 **c** 15 **11** **a** 360 **b** 336 **c** 288

12 **a** 15 120 **b** 720

13 **a** 3 628 800 **b** **i** 151 200 **ii** 33 600

EXERCISE 8E

1 ABCD, ABCE, ABCF, ABDE, ABDF, ABEF, ACDE, ACDF, ACEF, ADEF, BCDE, BCDF, BCEF, BDEF, CDEF, $\binom{6}{4} = 15$

2 $\binom{17}{11} = 12\,376$ **3** **a** $\binom{9}{5} = 126$ **b** $\binom{1}{1}\binom{8}{4} = 70$

4 **a** $\binom{13}{3} = 286$ **b** $\binom{1}{1}\binom{12}{2} = 66$

5 **a** $\binom{12}{5} = 792$
b **i** $\binom{2}{2}\binom{10}{3} = 120$ **ii** $\binom{2}{1}\binom{10}{4} = 420$

6 $\binom{3}{3}\binom{1}{0}\binom{11}{6} = 462$

7 a $\binom{1}{1}\binom{9}{3}=84$ b $\binom{2}{0}\binom{8}{4}=70$
c $\binom{2}{0}\binom{1}{1}\binom{7}{3}=35$

8 a $\binom{16}{5}=4368$ b $\binom{10}{3}\binom{6}{2}=1800$
c $\binom{10}{5}\binom{6}{0}=252$
d $\binom{10}{3}\binom{6}{2}+\binom{10}{4}\binom{6}{1}+\binom{10}{5}\binom{6}{0}=3312$
e $\binom{16}{5}-\binom{10}{5}\binom{6}{0}-\binom{10}{0}\binom{6}{5}=4110$

9 a $\binom{6}{2}\binom{3}{1}\binom{7}{2}=945$ b $\binom{6}{2}\binom{10}{3}=1800$
c $\binom{16}{5}-\binom{9}{0}\binom{7}{5}=4347$

10 $\binom{20}{2}-20=170$

11 a i $\binom{12}{2}=66$ ii $\binom{11}{1}=11$
b i $\binom{12}{3}=220$ ii $\binom{11}{2}=55$

12 $\binom{9}{4}=126$

13 a Selecting the different committees of 4 from 5 men and 6 women in all possible ways.
b $\binom{m+n}{r}$

14 a $\dfrac{\binom{12}{6}}{2}=462$ b $\dfrac{\binom{12}{4}\binom{8}{4}\binom{4}{4}}{3!}=5775$

15 $\binom{10}{2}\times\binom{7}{2}=945$

16 $\binom{10}{2}\binom{9}{2}+\binom{9}{2}\binom{8}{2}+\binom{10}{2}\binom{8}{2}+\binom{10}{2}\binom{9}{1}\binom{8}{1}$
$+\binom{10}{1}\binom{9}{2}\binom{8}{1}+\binom{10}{1}\binom{9}{1}\binom{8}{2}=12\,528$

17 a 45, Yes b 37 128 c 3 628 800

EXERCISE 8F

1 a $p^3+3p^2q+3pq^2+q^3$ b x^3+3x^2+3x+1
c $x^3-9x^2+27x-27$ d $8+12x+6x^2+x^3$
e $27x^3-27x^2+9x-1$ f $8x^3+60x^2+150x+125$
g $8a^3-12a^2b+6ab^2-b^3$ h $27x^3-9x^2+x-\frac{1}{27}$
i $8x^3+12x+\dfrac{6}{x}+\dfrac{1}{x^3}$

2 a $1+4x+6x^2+4x^3+x^4$
b $p^4-4p^3q+6p^2q^2-4pq^3+q^4$
c $x^4-8x^3+24x^2-32x+16$
d $81-108x+54x^2-12x^3+x^4$
e $1+8x+24x^2+32x^3+16x^4$
f $16x^4-96x^3+216x^2-216x+81$
g $16x^4+32x^3b+24x^2b^2+8xb^3+b^4$
h $x^4+4x^2+6+\dfrac{4}{x^2}+\dfrac{1}{x^4}$
i $16x^4-32x^2+24-\dfrac{8}{x^2}+\dfrac{1}{x^4}$

3 a $x^5+10x^4+40x^3+80x^2+80x+32$
b $x^5-10x^4y+40x^3y^2-80x^2y^3+80xy^4-32y^5$
c $1+10x+40x^2+80x^3+80x^4+32x^5$
d $x^5-5x^3+10x-\dfrac{10}{x}+\dfrac{5}{x^3}-\dfrac{1}{x^5}$

4 a 1 6 15 20 15 6 1
b i $x^6+12x^5+60x^4+160x^3+240x^2+192x+64$
ii $64x^6-192x^5+240x^4-160x^3+60x^2-12x+1$
iii $x^6+6x^4+15x^2+20+\dfrac{15}{x^2}+\dfrac{6}{x^4}+\dfrac{1}{x^6}$

5 a $7+5\sqrt{2}$ b $161+72\sqrt{5}$ c $232-164\sqrt{2}$

6 a $64+192x+240x^2+160x^3+60x^4+12x^5+x^6$
b 65.944 160 601 201

7 a $a=2$ and $b=e^x$ b $T_3=6e^{2x}$ and $T_4=e^{3x}$

8 $2x^5+11x^4+24x^3+26x^2+14x+3$ **9** a 270 b 4320

EXERCISE 8G

1 a $1^{11}+\binom{11}{1}(2x)^1+\binom{11}{2}(2x)^2+....+\binom{11}{10}(2x)^{10}+(2x)^{11}$
b $(3x)^{15}+\binom{15}{1}(3x)^{14}\left(\frac{2}{x}\right)^1+\binom{15}{2}(3x)^{13}\left(\frac{2}{x}\right)^2+....$
$....+\binom{15}{14}(3x)^1\left(\frac{2}{x}\right)^{14}+\left(\frac{2}{x}\right)^{15}$
c $(2x)^{20}+\binom{20}{1}(2x)^{19}\left(-\frac{3}{x}\right)^1+\binom{20}{2}(2x)^{18}\left(-\frac{3}{x}\right)^2+....$
$....+\binom{20}{19}(2x)^1\left(-\frac{3}{x}\right)^{19}+\left(-\frac{3}{x}\right)^{20}$

2 a $T_6=\binom{15}{5}(2x)^{10}5^5$ b $T_4=\binom{9}{3}(x^2)^6y^3$
c $T_{10}=\binom{17}{9}x^8\left(-\frac{2}{x}\right)^9$ d $T_9=\binom{21}{8}(2x^2)^{13}\left(-\frac{1}{x}\right)^8$

3 a $T_{r+1}=\binom{7}{r}x^{7-r}b^r$ b $b=-2$

4 a $\binom{15}{5}2^5$ b $\binom{9}{3}(-3)^3$

5 a
1 1
1 2 1
1 3 3 1
1 4 6 4 1
1 5 10 10 5 1

b i 2
ii 4
iii 8
iv 16
v 32

c The sum of the numbers in row n of Pascal's triangle is 2^n.
d Let $x=1$, in the expansion of $(1+x)^n$.

6 a $\binom{10}{5}3^52^5$ b $\binom{6}{3}2^3(-3)^3$ c $\binom{6}{3}2^3(-3)^3$
d $\binom{12}{4}2^8(-1)^4$

7 a $\binom{8}{6}=28$ b $2\binom{9}{3}3^6x^6-\binom{9}{4}3^5x^6=91\,854x^6$

8 $T_3=\binom{6}{2}(-2)^2x^8y^8$

9 a $84x^3$ b $n=6$ and $k=-2$ **10** $a=2$

11 $1+10x+35x^2+40x^3-30x^4$

14 $\displaystyle\sum_{r=0}^{n}2^r\binom{n}{r}=3^n$ **15** $(-1)^{100}=1$

17 a $(3+x)^n=3^n+\binom{n}{1}3^{n-1}x+\binom{n}{2}3^{n-2}x^2+$
$\binom{n}{3}3^{n-3}x^3+....+\binom{n}{n-1}3^1x^{n-1}+x^n$
b 4^n

REVIEW SET 8A

1 a $n(n-1)$, $n\geqslant 2$ b $n+2$ **2** 28

3 a 24 b 6 **4** a 900 b 180

5 a $a=e^x$ and $b=-e^{-x}$
b $(e^x-e^{-x})^4=e^{4x}-4e^{2x}+6-4e^{-2x}+e^{-4x}$

6 $362+209\sqrt{3}$ **7** It does not have one. **8** $c=3$

9 a $2^n+\binom{n}{1}2^{n-1}x^1+\binom{n}{2}2^{n-2}x^2+\binom{n}{3}2^{n-3}x^3+....$
$....+\binom{n}{n-1}2^1x^{n-1}+x^n$
b 3^n **Hint:** Let $x=1$ in a.

REVIEW SET 8B

1 a 45 b 120 **2** 64.964 808

3 $\binom{6}{2}\times 3^4\times(-2)^2=4860$

4 2500 **5** a 252 b 246

6 **a** $9 \times 9 \times 8 \times 7 = 4536$ numbers **b** 952 numbers

7 $(a+b)^6 = a^6 + 6a^5b + 15a^4b^2 + 20a^3b^3 + 15a^2b^4 + 6ab^5 + b^6$
a $x^6 - 18x^5 + 135x^4 - 540x^3 + 1215x^2 - 1458x + 729$
b $1 + \frac{6}{x} + \frac{15}{x^2} + \frac{20}{x^3} + \frac{15}{x^4} + \frac{6}{x^5} + \frac{1}{x^6}$

8 $\binom{12}{6} \times 2^6 \times (-3)^6$ **9** $8\binom{6}{2} - 6\binom{6}{1} = 84$ **10** $a = \pm 4$

REVIEW SET 8C

1 **a** $26^2 \times 10^4 = 6\,760\,000$ **b** $5 \times 26 \times 10^4 = 1\,300\,000$
c $26 \times 25 \times 10 \times 9 \times 8 \times 7 = 3\,276\,000$

2 **a** 3003 **b** 980 **c** 2982

3 **a** $x^3 - 6x^2y + 12xy^2 - 8y^3$
b $81x^4 + 216x^3 + 216x^2 + 96x + 16$

4 20 000 **5** 60 **6** $k = -\frac{1}{4}$, $n = 16$ **7** 4320

8 **a** 43 758 teams **b** 11 550 teams **c** 41 283 teams

9 $k = 180$ **10** $q = \pm\sqrt{\frac{3}{35}}$

EXERCISE 9A

1 $4n - 1$, $\mathbb{Z}^+$

2 **a** all $n \in \mathbb{Z}^+$, $n \geqslant 2$ **b** 10 for all $n \in \mathbb{Z}^+$
c 3 for all $n \in \mathbb{Z}^+$ **d** $\frac{1}{n+1}$ for all $n \in \mathbb{Z}^+$

3 **a** $\sum_{i=1}^{n} 2i = n(n+1)$ for all $n \in \mathbb{Z}^+$
b $\sum_{i=1}^{n} i \times i! = (n+1)! - 1$ for all $n \in \mathbb{Z}^+$
c $\sum_{i=1}^{n} \frac{i}{(i+1)!} = \frac{(n+1)! - 1}{(n+1)!}$ for all $n \in \mathbb{Z}^+$
d $\sum_{i=1}^{n} \frac{1}{(3i-1)(3i+2)} = \frac{n}{6n+4}$ for all $n \in \mathbb{Z}^+$

4 *Proposition*: The maximum number of triangles for n points within the original triangle is given by $T_n = 2n + 1$, $n \in \mathbb{Z}^+$.

5 **a** $n = 4$ $n = 5$

$C_4 = 8$ $C_5 = 16$

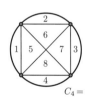

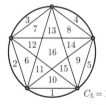

b *Conjecture*: The number of regions for n points placed around a circle is given by $C_n = 2^{n-1}$, $n \in \mathbb{Z}^+$.

c $n = 6$

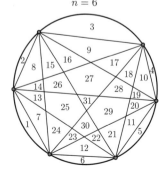

No. By the conjecture we expect $2^5 = 32$ regions, but there are only 31.

EXERCISE 9B.2

4 **b** $\frac{11}{210}$ **5** **c** $\frac{3\,628\,799}{3\,628\,800}$

8 **a** *Conjecture*: $u_n = n^2$ for $n \in \mathbb{Z}^+$

9 **a** *Conjecture*: $u_n = \frac{n}{2n+1}$ for $n \in \mathbb{Z}^+$

10 **a** $A_1 = 2$, $B_1 = 1$; $A_2 = 7$, $B_2 = 4$;
$A_3 = 26$, $B_3 = 15$; $A_4 = 97$, $B_4 = 56$
c For $n = 1, 2, 3, 4$, $(A_n)^2 - 3(B_n)^2 = 1$
Conjecture: $(A_n)^2 - 3(B_n)^2 = 1$ for $n \in \mathbb{Z}^+$

EXERCISE 10A

1 **a** $\frac{\pi}{2}^c$ **b** $\frac{\pi}{3}^c$ **c** $\frac{\pi}{6}^c$ **d** $\frac{\pi}{10}^c$ **e** $\frac{\pi}{20}^c$
f $\frac{3\pi}{4}^c$ **g** $\frac{5\pi}{4}^c$ **h** $\frac{3\pi}{2}^c$ **i** $2\pi^c$ **j** $4\pi^c$
k $\frac{7\pi}{4}^c$ **l** $3\pi^c$ **m** $\frac{\pi}{5}^c$ **n** $\frac{4\pi}{9}^c$ **o** $\frac{23\pi}{18}^c$

2 **a** 0.641^c **b** 2.39^c **c** 5.55^c **d** 3.83^c **e** 6.92^c

3 **a** $36°$ **b** $108°$ **c** $135°$ **d** $10°$ **e** $20°$
f $140°$ **g** $18°$ **h** $27°$ **i** $210°$ **j** $22.5°$

4 **a** $114.59°$ **b** $87.66°$ **c** $49.68°$ **d** $182.14°$
e $301.78°$

5 **a**

Degrees	0	45	90	135	180	225	270	315	360
Radians	0	$\frac{\pi}{4}$	$\frac{\pi}{2}$	$\frac{3\pi}{4}$	π	$\frac{5\pi}{4}$	$\frac{3\pi}{2}$	$\frac{7\pi}{4}$	2π

b

Deg.	0	30	60	90	120	150	180	210	240	270	300	330	360
Rad.	0	$\frac{\pi}{6}$	$\frac{\pi}{3}$	$\frac{\pi}{2}$	$\frac{2\pi}{3}$	$\frac{5\pi}{6}$	π	$\frac{7\pi}{6}$	$\frac{4\pi}{3}$	$\frac{3\pi}{2}$	$\frac{5\pi}{3}$	$\frac{11\pi}{6}$	2π

EXERCISE 10B

1 **a** 49.5 cm, 223 cm^2 **b** 23.0 cm, 56.8 cm^2

2 **a** 3.14 m **b** 9.30 m^2 **3** **a** 5.91 cm **b** 18.9 cm

4 **a** 0.686^c **b** 0.6^c

5 **a** 0.75^c, 24 cm^2 **b** 1.68^c, 21 cm^2 **c** 2.32^c, 126.8 cm^2

6 10 cm, 25 cm^2

8 **a** 11.7 cm **b** 11.7 **c** 37.7 cm **d** 3.23^c

9 **a** $\alpha \approx 18.43$ **b** $\theta \approx 143.1$ **c** 387 m^2

10 25.9 cm **11** **b** 2 h 24 min **12** 227 m^2

13 **a** $\alpha = 5.739$ **b** $\theta = 168.5$ **c** $\phi = 191.5$ **d** 71.62 cm

EXERCISE 10C

1 **a** **i** A(cos 26°, sin 26°), B(cos 146°, sin 146°), C(cos 199°, sin 199°)
ii A(0.899, 0.438), B(−0.829, 0.559), C(−0.946, −0.326)
b **i** A(cos 123°, sin 123°), B(cos 251°, sin 251°), C(cos(−35°), sin(−35°))
ii A(−0.545, 0.839), B(−0.326, −0.946), C(0.819, −0.574)

2

θ (degrees)	0°	90°	180°	270°	360°	450°
θ (radians)	0	$\frac{\pi}{2}$	π	$\frac{3\pi}{2}$	2π	$\frac{5\pi}{2}$
sine	0	1	0	−1	0	1
cosine	1	0	−1	0	1	0
tangent	0	undef	0	undef	0	undef

3 **a** **i** $\frac{1}{\sqrt{2}} \approx 0.707$ **ii** $\frac{\sqrt{3}}{2} \approx 0.866$

b

θ (degrees)	30°	45°	60°	135°	150°	240°	315°
θ (radians)	$\frac{\pi}{6}$	$\frac{\pi}{4}$	$\frac{\pi}{3}$	$\frac{3\pi}{4}$	$\frac{5\pi}{6}$	$\frac{4\pi}{3}$	$\frac{7\pi}{4}$
sine	$\frac{1}{2}$	$\frac{1}{\sqrt{2}}$	$\frac{\sqrt{3}}{2}$	$\frac{1}{\sqrt{2}}$	$\frac{1}{2}$	$-\frac{\sqrt{3}}{2}$	$-\frac{1}{\sqrt{2}}$
cosine	$\frac{\sqrt{3}}{2}$	$\frac{1}{\sqrt{2}}$	$\frac{1}{2}$	$-\frac{1}{\sqrt{2}}$	$-\frac{\sqrt{3}}{2}$	$-\frac{1}{2}$	$\frac{1}{\sqrt{2}}$
tangent	$\frac{1}{\sqrt{3}}$	1	$\sqrt{3}$	-1	$-\frac{1}{\sqrt{3}}$	$\sqrt{3}$	-1

4 **a**

Quadrant	*Degree measure*	*Radian measure*	$\cos\theta$	$\sin\theta$	$\tan\theta$
1	$0° < \theta < 90°$	$0 < \theta < \frac{\pi}{2}$	+ve	+ve	+ve
2	$90° < \theta < 180°$	$\frac{\pi}{2} < \theta < \pi$	−ve	+ve	−ve
3	$180° < \theta < 270°$	$\pi < \theta < \frac{3\pi}{2}$	−ve	−ve	+ve
4	$270° < \theta < 360°$	$\frac{3\pi}{2} < \theta < 2\pi$	+ve	−ve	−ve

b **i** 1 and 4 **ii** 2 and 3 **iii** 3 **iv** 2

5 **a** **i** 0.985 **ii** 0.985 **iii** 0.866 **iv** 0.866
v 0.5 **vi** 0.5 **vii** 0.707 **viii** 0.707
b $\sin(180° - \theta) = \sin\theta$ as the points have the same y-coordinate
d **i** 135° **ii** 129° **iii** $\frac{2\pi}{3}$ **iv** $\frac{5\pi}{6}$

6 **a** **i** 0.342 **ii** -0.342 **iii** 0.5 **iv** -0.5
v 0.906 **vi** -0.906 **vii** 0.174 **viii** -0.174
b $\cos(180° - \theta) = -\cos\theta$
d **i** 140° **ii** 161° **iii** $\frac{4\pi}{5}$ **iv** $\frac{3\pi}{5}$

8 **a** ≈ 0.6820 **b** ≈ 0.8572 **c** ≈ -0.7986
d ≈ 0.9135 **e** ≈ 0.9063 **f** ≈ -0.6691

9 **a** $\widehat{AOQ} = 180° - \theta$ or $\pi - \theta$ radians
b [OQ] is a reflection of [OP] in the y-axis and so Q has coordinates $(-\cos\theta, \sin\theta)$.
c $\cos(180° - \theta) = -\cos\theta$, $\sin(180° - \theta) = \sin\theta$

10 **a**

θ^c	$\sin\theta$	$\sin(-\theta)$	$\cos\theta$	$\cos(-\theta)$
0.75	0.682	-0.682	0.732	0.732
1.772	0.980	-0.980	-0.200	-0.200
3.414	-0.269	0.269	-0.963	-0.963
6.25	-0.0332	0.0332	0.999	0.999
-1.17	-0.921	0.921	0.390	0.390

b $\sin(-\theta) = -\sin\theta$, $\cos(-\theta) = \cos\theta$
c **i** Q has coordinates $(\cos(-\theta), \sin(-\theta))$ or $(\cos\theta, -\sin\theta)$ (since it is the reflection of P in the x-axis).
$\therefore$ $\cos(-\theta) = \cos\theta$ and $\sin(-\theta) = -\sin\theta$.
ii $\cos(2\pi - \theta) = \cos(-\theta) = \cos\theta$ {from **c i**}

EXERCISE 10D.1

1 **a** $\cos\theta = \pm\frac{\sqrt{3}}{2}$ **b** $\cos\theta = \pm\frac{2\sqrt{2}}{3}$ **c** $\cos\theta = \pm 1$
d $\cos\theta = 0$

2 **a** $\sin\theta = \pm\frac{3}{5}$ **b** $\sin\theta = \pm\frac{\sqrt{7}}{4}$ **c** $\sin\theta = 0$
d $\sin\theta = \pm 1$

3 **a** $\sin\theta = \frac{\sqrt{5}}{3}$ **b** $\cos\theta = -\frac{\sqrt{21}}{5}$ **c** $\cos\theta = \frac{4}{5}$
d $\sin\theta = -\frac{12}{13}$

4 **a** $-\frac{1}{2\sqrt{2}}$ **b** $-2\sqrt{6}$ **c** $\frac{1}{\sqrt{2}}$ **d** $-\frac{\sqrt{7}}{3}$

5 **a** $\sin x = \frac{2}{\sqrt{13}}$, $\cos x = \frac{3}{\sqrt{13}}$
b $\sin x = \frac{4}{5}$, $\cos x = -\frac{3}{5}$
c $\sin x = -\sqrt{\frac{5}{14}}$, $\cos x = -\frac{3}{\sqrt{14}}$
d $\sin x = -\frac{12}{13}$, $\cos x = \frac{5}{13}$

6 $\sin x = \dfrac{-k}{\sqrt{k^2+1}}$, $\cos x = \dfrac{-1}{\sqrt{k^2+1}}$

EXERCISE 10D.2

1 **a** $\theta \approx 1.33$ or 4.47 **b** $\theta \approx 0.592$ or 5.69
c $\theta \approx 0.644$ or 2.50 **d** $\theta = \frac{\pi}{2}$ or $\frac{3\pi}{2}$
e $\theta \approx 0.876$ or 4.02 **f** $\theta \approx 0.674$ or 5.61
g $\theta \approx 0.0910$ or 3.05 **h** $\theta \approx 1.52$ or 4.66
i $\theta \approx 1.35$ or 1.79

2 **a** $\theta \approx 1.82$ or 4.46 **b** $\theta = 0, \pi,$ or 2π
c $\theta \approx 1.88$ or 5.02 **d** $\theta \approx 3.58$ or 5.85
e $\theta \approx 1.72$ or 4.86 **f** $\theta \approx 1.69$ or 4.59
g $\theta \approx 1.99$ or 5.13 **h** $\theta \approx 2.19$ or 4.10
i $\theta \approx 3.83$ or 5.60

EXERCISE 10E

1 **a** 0 **b** $-2\tan\theta$ **c** $3\cos\theta$ **d** $4\sin\theta$
e $\cos^2\alpha$ **f** $\sin^2\alpha$ **g** 1

2 **a** $\sin\theta$ **b** $-2\sin\theta$ **c** 0 **d** $-\cos\theta$
e $4\cos\theta$ **f** $5\sin\theta$

3 **Hint:** $\theta - \phi = -(\phi - \theta)$

4 **a** $\tan\theta$ **b** $-\tan\theta$ **c** 1 **d** $\tan\theta$
e $\tan\theta$ **f** $\tan\theta$

EXERCISE 10F

1

	a	b	c	d	e
$\sin\theta$	$\frac{1}{\sqrt{2}}$	$\frac{1}{\sqrt{2}}$	$-\frac{1}{\sqrt{2}}$	0	$-\frac{1}{\sqrt{2}}$
$\cos\theta$	$\frac{1}{\sqrt{2}}$	$-\frac{1}{\sqrt{2}}$	$\frac{1}{\sqrt{2}}$	-1	$-\frac{1}{\sqrt{2}}$
$\tan\theta$	1	-1	-1	0	1

2

	a	b	c	d	e
$\sin\beta$	$\frac{1}{2}$	$\frac{\sqrt{3}}{2}$	$-\frac{1}{2}$	$-\frac{\sqrt{3}}{2}$	$-\frac{1}{2}$
$\cos\beta$	$\frac{\sqrt{3}}{2}$	$-\frac{1}{2}$	$-\frac{\sqrt{3}}{2}$	$\frac{1}{2}$	$\frac{\sqrt{3}}{2}$
$\tan\beta$	$\frac{1}{\sqrt{3}}$	$-\sqrt{3}$	$\frac{1}{\sqrt{3}}$	$-\sqrt{3}$	$-\frac{1}{\sqrt{3}}$

3 **a** $\cos 120° = -\frac{1}{2}$, $\sin 120° = \frac{\sqrt{3}}{2}$, $\tan 120° = -\sqrt{3}$
b $\cos(-45°) = \frac{1}{\sqrt{2}}$, $\sin(-45°) = -\frac{1}{\sqrt{2}}$, $\tan(-45°) = -1$

4 **a** $\cos 90° = 0$, $\sin 90° = 1$ **b** $\tan 90°$ is undefined

5 **a** $\frac{3}{4}$ **b** $\frac{1}{4}$ **c** 3 **d** $\frac{1}{4}$ **e** $-\frac{1}{4}$ **f** 1
g $\sqrt{2}$ **h** $\frac{1}{2}$ **i** $\frac{1}{2}$ **j** 2 **k** -1 **l** $-\sqrt{3}$

6 **a** 30°, 150° **b** 60°, 120° **c** 45°, 315°
d 120°, 240° **e** 135°, 225° **f** 240°, 300°

7 **a** $\frac{\pi}{4}, \frac{5\pi}{4}$ **b** $\frac{3\pi}{4}, \frac{7\pi}{4}$ **c** $\frac{\pi}{3}, \frac{4\pi}{3}$
d $0, \pi, 2\pi$ **e** $\frac{\pi}{6}, \frac{7\pi}{6}$ **f** $\frac{2\pi}{3}, \frac{5\pi}{3}$

8 **a** $\frac{\pi}{6}, \frac{11\pi}{6}, \frac{13\pi}{6}, \frac{23\pi}{6}$ **b** $\frac{7\pi}{6}, \frac{11\pi}{6}, \frac{19\pi}{6}, \frac{23\pi}{6}$ **c** $\frac{3\pi}{2}, \frac{7\pi}{2}$

9 **a** $\theta = \frac{\pi}{3}, \frac{5\pi}{3}$ **b** $\theta = \frac{\pi}{3}, \frac{2\pi}{3}$ **c** $\theta = \pi$
d $\theta = \frac{\pi}{2}$ **e** $\theta = \frac{3\pi}{4}, \frac{5\pi}{4}$ **f** $\theta = \frac{\pi}{2}, \frac{3\pi}{2}$
g $\theta = 0, \pi, 2\pi$ **h** $\theta = \frac{\pi}{4}, \frac{3\pi}{4}, \frac{5\pi}{4}, \frac{7\pi}{4}$
i $\theta = \frac{5\pi}{6}, \frac{11\pi}{6}$ **j** $\theta = \frac{\pi}{3}, \frac{2\pi}{3}, \frac{4\pi}{3}, \frac{5\pi}{3}$

10 **a** $\theta = k\pi, \ k \in \mathbb{Z}$ **b** $\theta = \frac{\pi}{2} + k\pi, \ k \in \mathbb{Z}$

REVIEW SET 10A

1 **a** $\frac{2\pi}{3}$ **b** $\frac{5\pi}{4}$ **c** $\frac{5\pi}{6}$ **d** 3π

2 **a** $\frac{\pi}{3}$ **b** $15°$ **c** $84°$

3 **a** 0.358 **b** -0.035 **c** 0.259 **d** -0.731

4 **a** $1, 0$ **b** $-1, 0$

6 **a** $\sin\left(\frac{2\pi}{3}\right) = \frac{\sqrt{3}}{2}, \ \cos\left(\frac{2\pi}{3}\right) = -\frac{1}{2}, \ \tan\left(\frac{2\pi}{3}\right) = -\sqrt{3}$
b $\sin\left(\frac{8\pi}{3}\right) = \frac{\sqrt{3}}{2}, \ \cos\left(\frac{8\pi}{3}\right) = -\frac{1}{2}, \ \tan\left(\frac{8\pi}{3}\right) = -\sqrt{3}$

7 $\frac{1}{\sqrt{15}}$ **8** $\pm\frac{\sqrt{7}}{4}$ **9** **a** $\frac{\sqrt{3}}{2}$ **b** 0 **c** $\frac{1}{2}$

10 **a** $-\frac{3}{\sqrt{13}}$ **b** $\frac{2}{\sqrt{13}}$

11 perimeter $= 12$ units, area $= 8$ units^2 **12** $\frac{\sqrt{6}}{\sqrt{11}}$

13 **a** 0 **b** $\sin\theta$ **c** $-\sin^2\alpha$

REVIEW SET 10B

1 **a** $(0.766, -0.643)$ **b** $(-0.956, 0.292)$

2 **a** 1.239^c **b** 2.175^c **c** -2.478^c

3 **a** $171.89°$ **b** $83.65°$ **c** $24.92°$ **d** $-302.01°$

4 111 cm^2

5 M$(\cos 73°, \sin 73°) \approx (0.292, 0.956)$
N$(\cos 190°, \sin 190°) \approx (-0.985, -0.174)$
P$(\cos 307°, \sin 307°) \approx (0.602, -0.799)$

6 $\approx 103°$

7 **a** $150°, 210°$ **b** $45°, 135°$ **c** $120°, 300°$

8 **a** $\theta = \pi$ **b** $\theta = \frac{\pi}{3}, \frac{2\pi}{3}, \frac{4\pi}{3}, \frac{5\pi}{3}$

9 **a** $133°$ **b** $\frac{14\pi}{15}$ **c** $174°$

10 perimeter ≈ 34.1 cm, area ≈ 66.5 cm^2

11 $r \approx 8.79$ cm, area ≈ 81.0 cm^2

12 **a** $\theta \approx 0.841$ or 5.44 **b** $\theta \approx 3.39$ or 6.03
c $\theta \approx 1.25$ or 4.39

REVIEW SET 10C

1 **a** $72°$ **b** $225°$ **c** $140°$ **d** $330°$

2

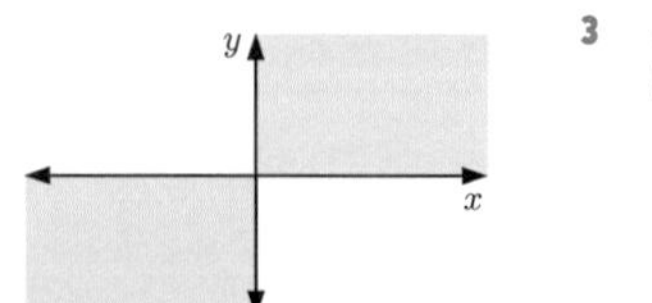

3 **a** $0, -1$
b $0, -1$

4 **a** $\sin(\pi - p) = m$ **b** $\sin(p + 2\pi) = m$
c $\cos p = \sqrt{1 - m^2}$ **d** $\tan p = \dfrac{m}{\sqrt{1 - m^2}}$

5 **a** **i** $60°$ **ii** $\frac{\pi}{3}$ **b** $\frac{\pi}{3}$ units **c** $\frac{\pi}{6}$ units^2 **6** 3

8 **a** $\frac{\sqrt{7}}{4}$ **b** $-\frac{\sqrt{7}}{3}$ **c** $-\frac{\sqrt{7}}{4}$ **9** **a** $2\frac{1}{2}$ **b** $1\frac{1}{2}$ **c** $-\frac{1}{2}$

10 **a** 0 **b** $-\cos\theta$

EXERCISE 11A

1 **a** 28.9 cm^2 **b** 384 km^2 **c** 28.3 cm^2 **2** $x \approx 19.0$

3 18.9 cm^2 **4** 137 cm^2 **5** 374 cm^2 **6** 7.49 cm

7 11.9 m **8** **a** $48.6°$ or $131.4°$ **b** $42.1°$ or $137.9°$

9 $\frac{1}{4}$ is not covered

10 **a** 36.2 cm^2 **b** 62.8 cm^2 **c** 40.4 mm^2 **11** 4.69 cm^2

EXERCISE 11B

1 **a** 28.8 cm **b** 3.38 km **c** 14.2 m

2 $\widehat{BAC} \approx 52.0°$, $\widehat{ABC} \approx 59.3°$, $\widehat{ACB} \approx 68.7°$

3 **a** $112°$ **b** 16.2 cm^2 **4** **a** $40.3°$ **b** $107°$

5 **a** $\cos\theta = 0.65$ **b** $x \approx 3.81$

6 **a** $x = 3 + \sqrt{22}$ **b** $x = \dfrac{-3 + \sqrt{73}}{2}$ **c** $x = \frac{5}{\sqrt{3}}$

7 **a** $\theta \approx 75.2$ **b** 6.30 m

8 **a** $x \approx 10.8$ **b** $x \approx 9.21$ **9** $x \approx 1.41$ or 7.78

10 **a** $x = 2$ **b** $4\sqrt{6}$ cm^2 **11** $63°, 117°, 36°, 144°$

EXERCISE 11C.1

1 **a** $x \approx 28.4$ **b** $x \approx 13.4$ **c** $x \approx 3.79$

2 **a** $a \approx 21.3$ cm **b** $b \approx 76.9$ cm **c** $c \approx 5.09$ cm

EXERCISE 11C.2

1 $C \approx 62.1°$ or $C \approx 117.9°$

2 **a** $\widehat{BAC} \approx 49.5°$ **b** $\widehat{ABC} \approx 72.0°$ or $108°$
c $\widehat{ACB} \approx 44.3°$

3 No, $\dfrac{\sin 85°}{11.4} \neq \dfrac{\sin 27°}{9.8}$ **4** $\widehat{ABC} = 66°$, BD ≈ 4.55 cm

5 $x \approx 17.7, \ y \approx 33.1$

6 **a** $\widehat{B} = 83°$ or $97°$ **b** $\widehat{B} = 83°$
c .... *cosine rule* as it avoids the *ambiguous case.*

7 Area ≈ 25.1 cm^2 **8** $\widehat{BAC} \approx 15.7°$ **9** $x = 8 + \frac{11}{2}\sqrt{2}$

EXERCISE 11D

1 17.7 m **2** 207 m **3** $23.9°$ **4** 77.5 m

5 **a** 5.63 km **b** $115°$
c **i** Esko **ii** 3.68 min **d** $295°$

6 $9.38°$ **7** 69.1 m **8** **a** 38.0 m **b** 94.0 m

9 $55.1°$ **10** AC ≈ 11.7 km, BC ≈ 8.49 km

11 **a** 74.9 km^2 **b** 7490 hectares **12** 9.12 km

13 85.0 mm **14** 10.1 km **15** 29.2 m **16** 37.6 km

REVIEW SET 11A

1 14 km^2

2 If the unknown is an angle, use the cosine rule to avoid an ambiguous case.

3 **a** $x = 3$ or 5 **b** Kady can draw 2 triangles, so she should ask for more information.

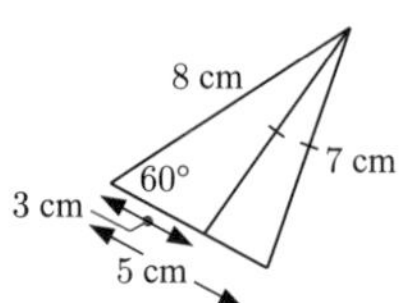

4 $\frac{12}{13}$ **6** 42 km

8 **a** $d^2 = x^2 - (10\cos 20°)x + 25$ **b** $x = 5\cos 20°$

9 **a** maximum value 16 when $x = 6$
b **i** $y = 12 - x$ **ii** $y^2 = x^2 - (16\cos\theta)x + 64$
d max. area $= 8\sqrt{5}$ units^2 when $x = y = 6$ {isosceles $\triangle$}

REVIEW SET 11B

1 a $x \approx 34.1$ b $x \approx 18.9$
2 $\text{AC} \approx 12.6$ cm, $A \approx 48.6°$, $C \approx 57.4°$ 3 113 cm^2
4 7.32 m 5 3.52 km 6 a 2:18 pm b $\approx 157°$
7 204 m 8 560 m, bearing 079.7°

REVIEW SET 11C

1 a $x \approx 41.5$ b $x \approx 15.4$
2 a $x \approx 47.5$ or 132.5 b $\text{AC} \approx 14.3$ cm or 28.1 cm
3 $\widehat{\text{EDG}} \approx 74.4°$ 4 a 10 600 m^2 b 1.06 ha
5 179 km, bearing 352°
6 a The information given could give two triangles:
b ≈ 2.23 m^3
7 b ii $b = 104$, $d = 76.2$ iii $a = 95.4$, $c = 84.6$
8 a $\text{QS} = \sqrt{45 - 36\cos\phi}$
b i $\widehat{\text{RSQ}} = 52.5°$ ii 24.8 units iii 23.2 units^2

EXERCISE 12A

1 a periodic b periodic c periodic d not periodic
e periodic f periodic g not periodic h not periodic
2 a

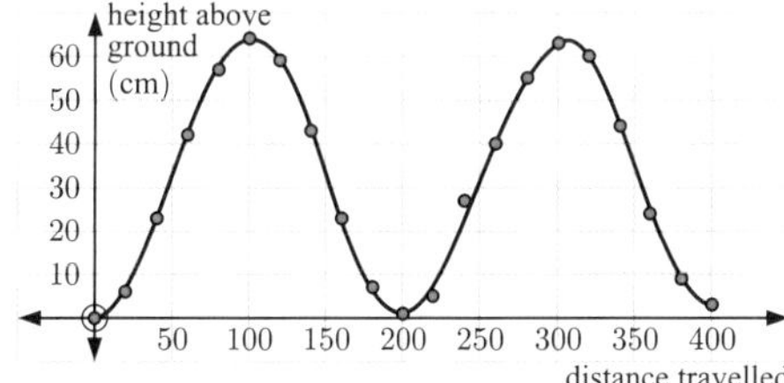

b A curve can be fitted to the data.
c The data is periodic.
i $y = 32$ (approx.) ii ≈ 64 cm
iii ≈ 200 cm iv ≈ 32 cm
3 a

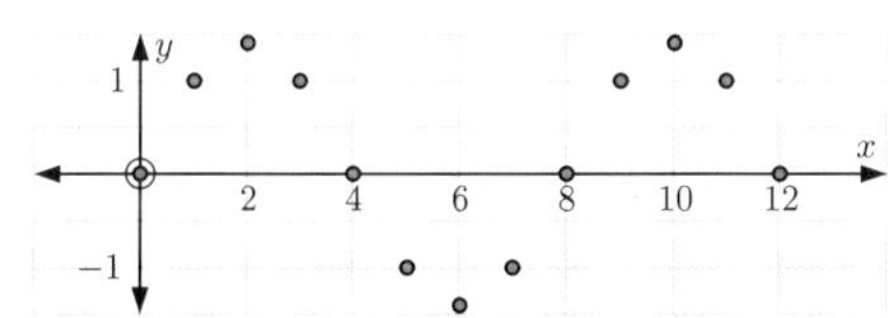

Data exhibits periodic behaviour.
b

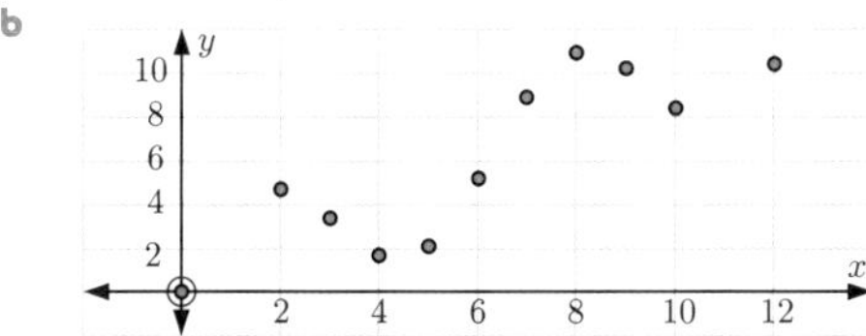

Not enough information to say data is periodic.

EXERCISE 12B.1

1 a

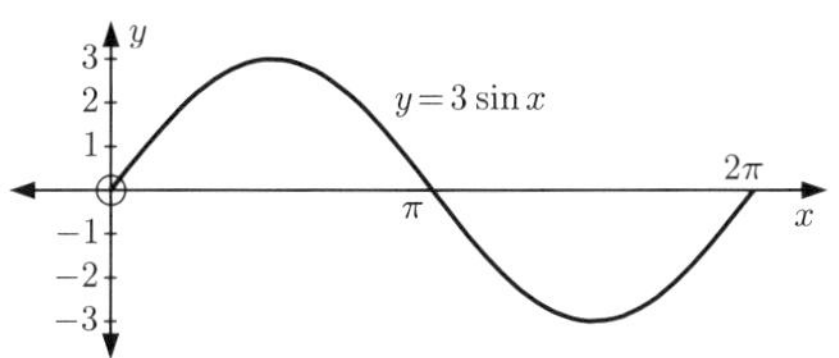

b

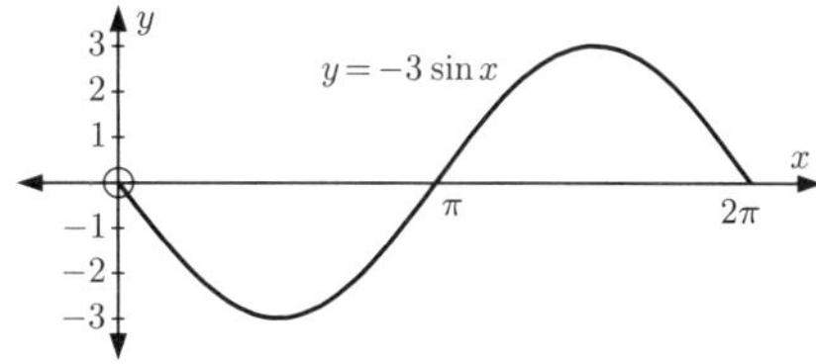

c

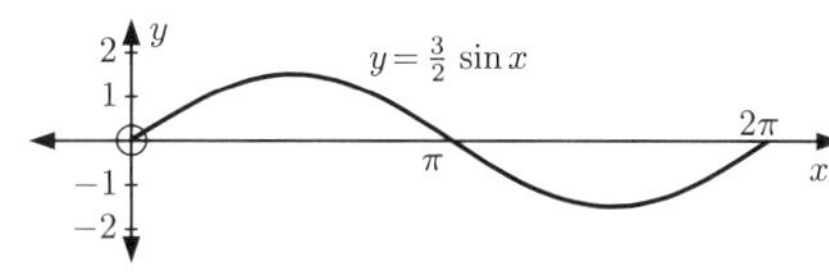

d

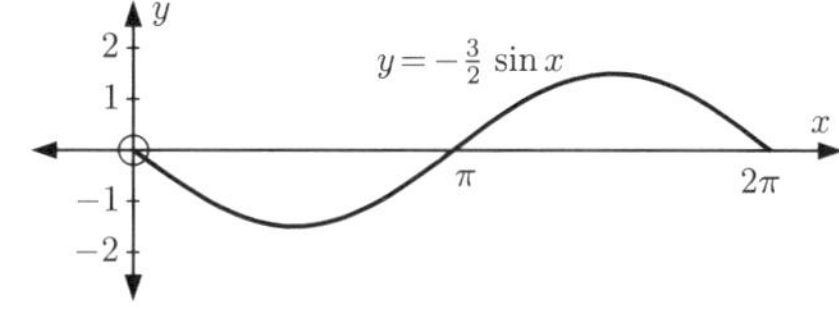

2 a

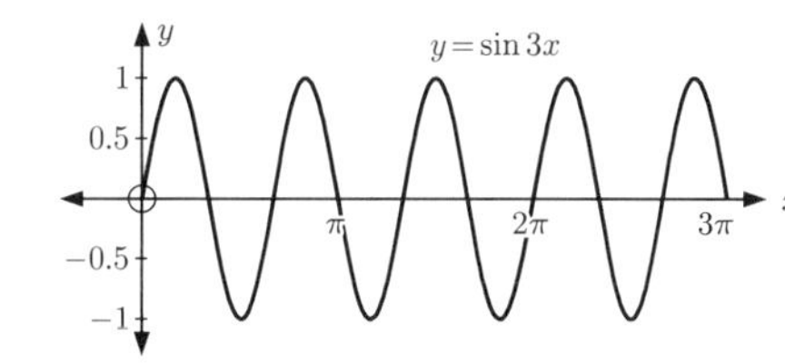

b

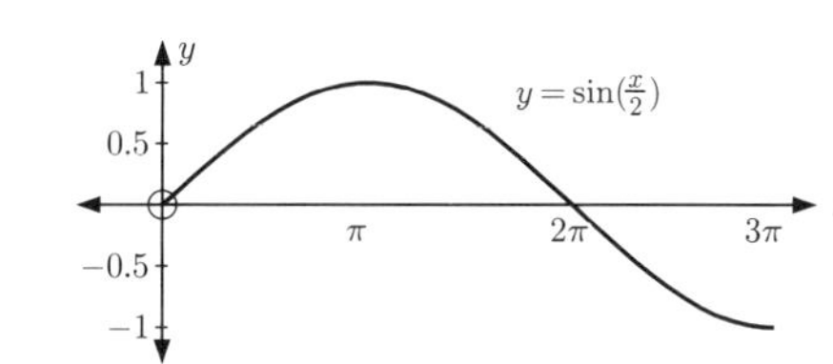

c

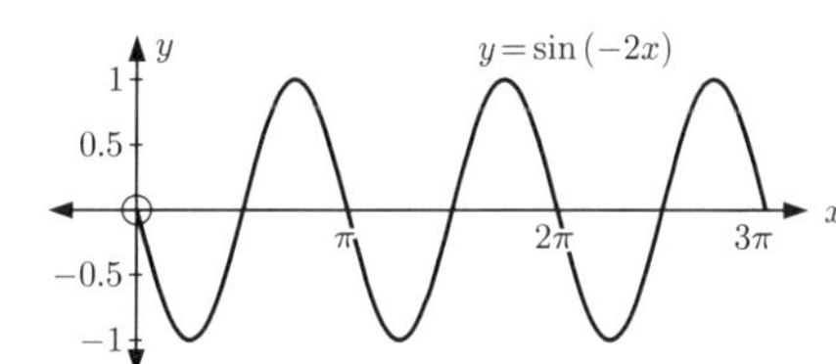

3 a $\frac{\pi}{2}$ b $\frac{\pi}{2}$ c 6π d $\frac{10\pi}{3}$
4 a $b = \frac{2}{5}$ b $b = 3$ c $b = \frac{1}{6}$ d $b = \frac{\pi}{2}$ e $b = \frac{\pi}{50}$
5 a

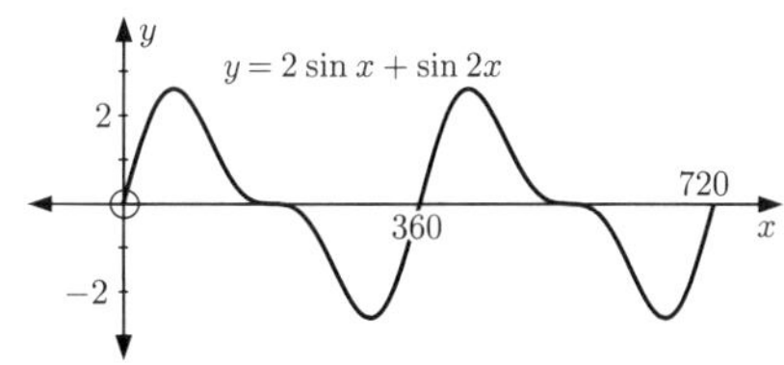

b

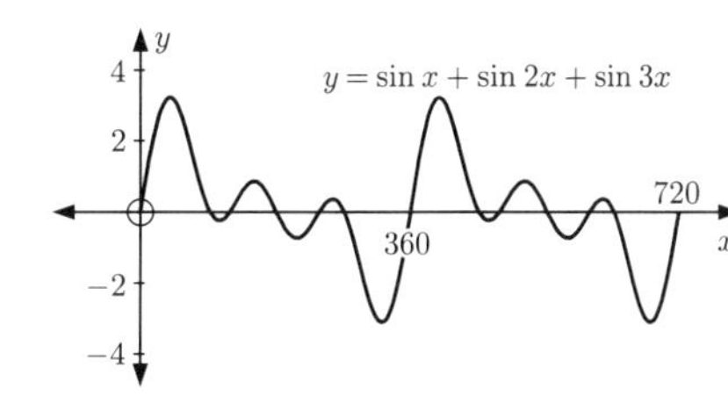

c

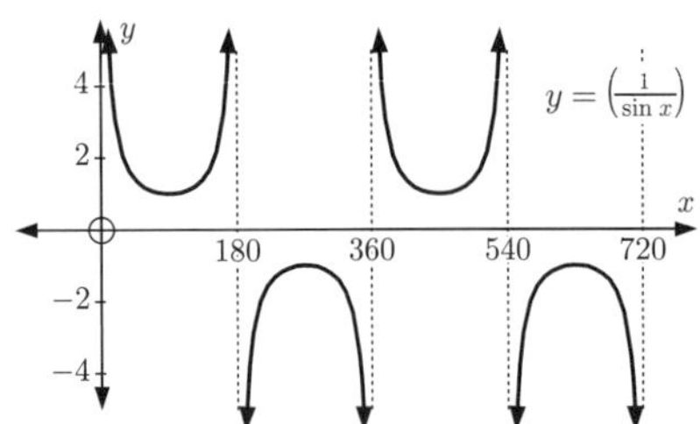

6 a i

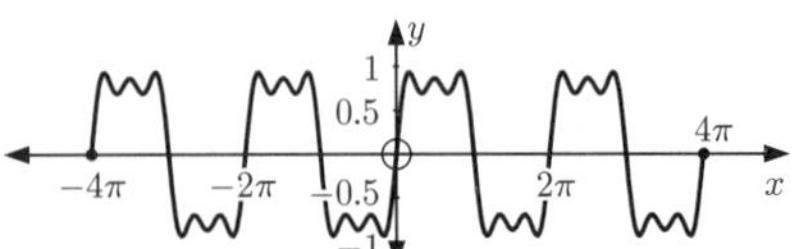

ii

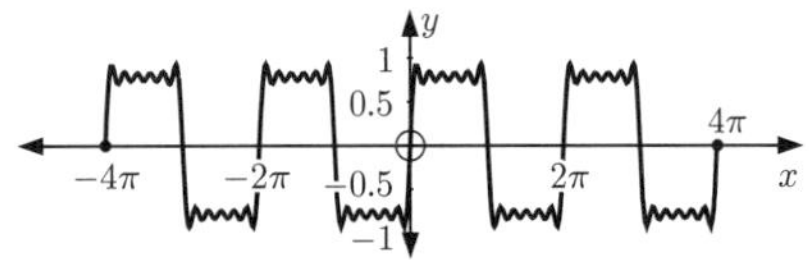

b

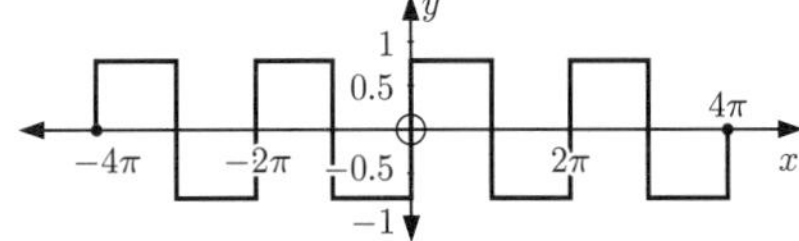

EXERCISE 12B.2

1 a

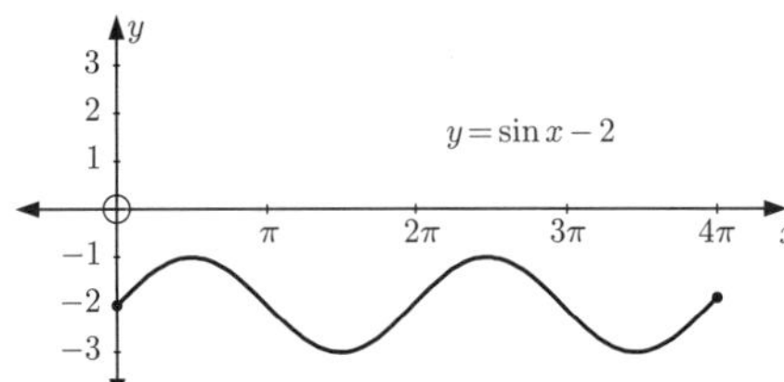

b

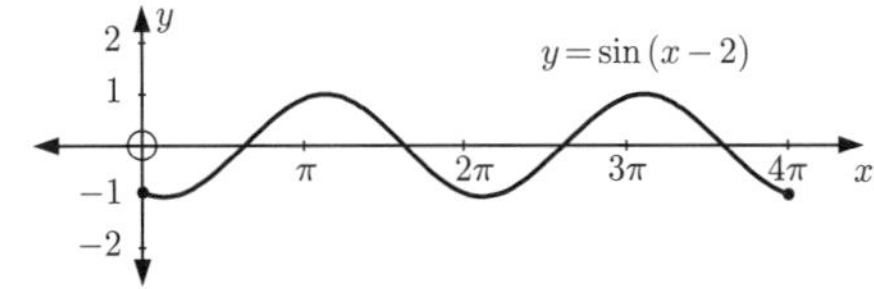

c

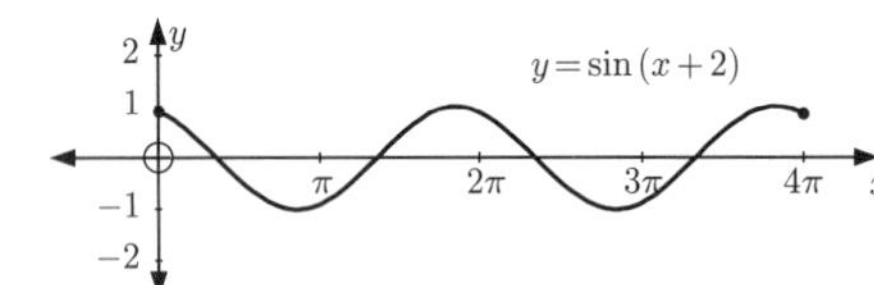

d

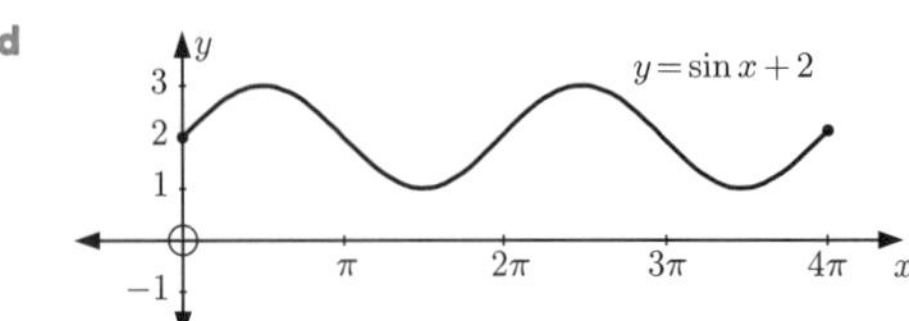

e

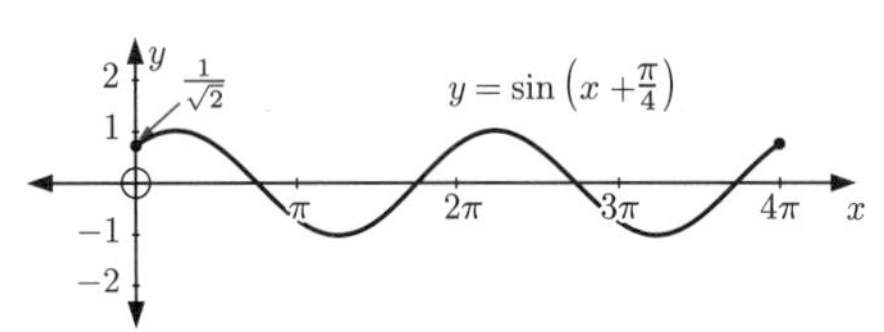

f

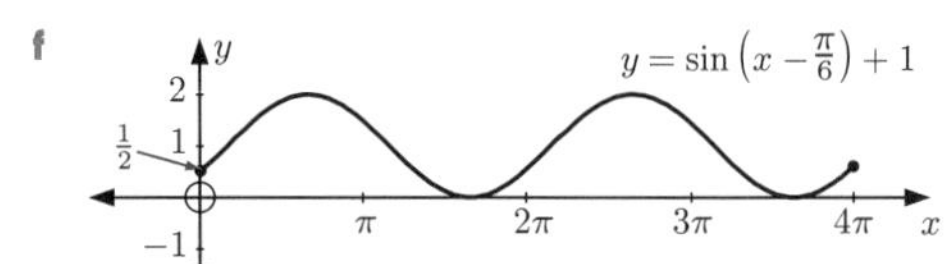

2 a $\frac{2\pi}{5}$ b 8π c π

3 a $\frac{2}{3}$ b 20 c $\frac{1}{50}$ d $\frac{\pi}{25}$

4 a vert. translation -1
b horiz. translation $\frac{\pi}{4}$ right
c vert. stretch, factor 2
d horiz. stretch, factor $\frac{1}{4}$
e vert. stretch, factor $\frac{1}{2}$
f horiz. stretch, factor 4
g reflection in the x-axis
h translation $\begin{pmatrix} -2 \\ -3 \end{pmatrix}$
i vert. stretch, factor 2, followed by a horiz. stretch, factor $\frac{1}{3}$
j translation $\begin{pmatrix} \frac{\pi}{3} \\ 2 \end{pmatrix}$

EXERCISE 12C

1 a $T \approx 6.5 \sin \frac{\pi}{6}(t - 4.5) + 20.5$

2 a $T \approx 4.5 \sin \frac{\pi}{6}(t - 10.5) + 11.5$

3 a $H \approx 7 \sin 0.507(t - 3.1)$
b

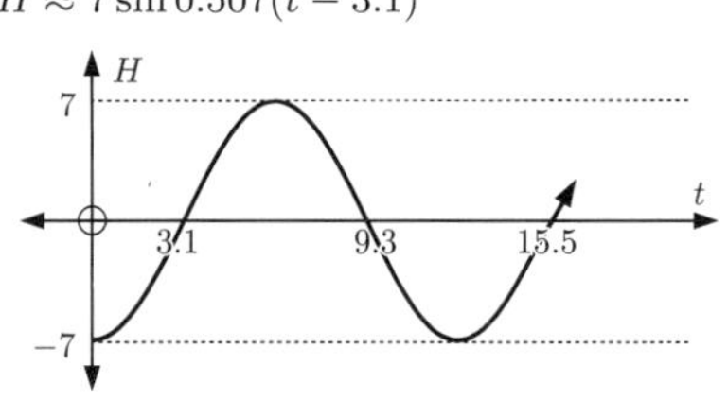

4 a $T \approx 9.5 \sin \frac{\pi}{6}(t - 10.5) - 9.5$
b A reasonable fit but not perfect.

5 $H = 10 \sin(\frac{\pi}{50}(t - 25)) + 12$

EXERCISE 12D

1 a $y = \cos x + 2$

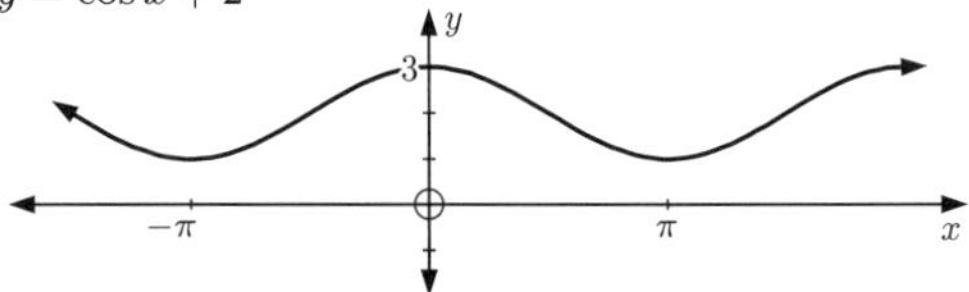

b $y = \cos x - 1$

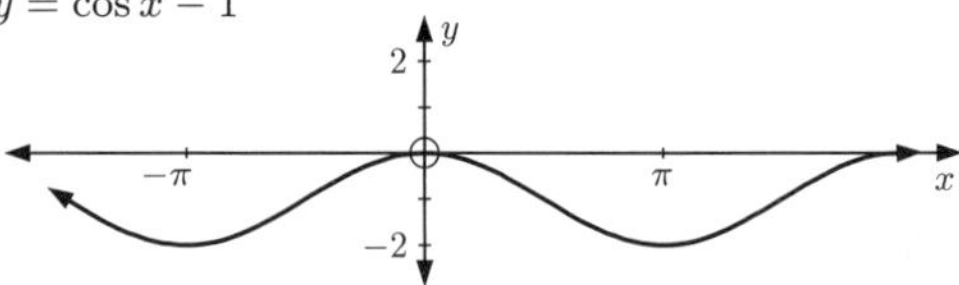

c $y = \cos\left(x - \frac{\pi}{4}\right)$

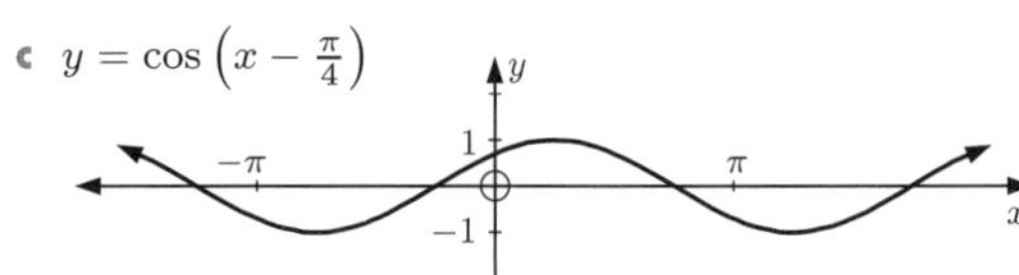

d $y = \cos\left(x + \frac{\pi}{6}\right)$

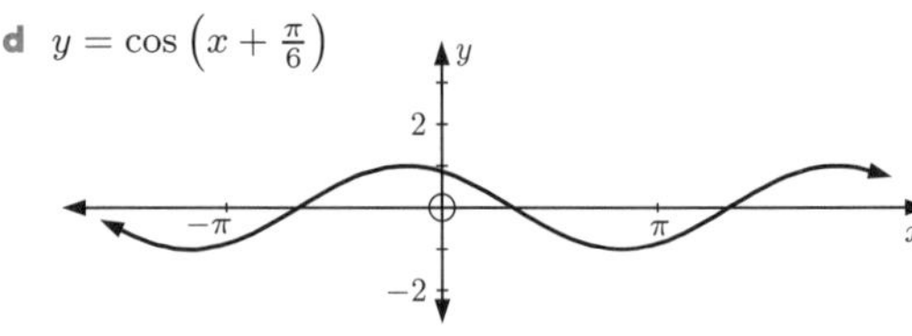

e $y = \frac{2}{3}\cos x$

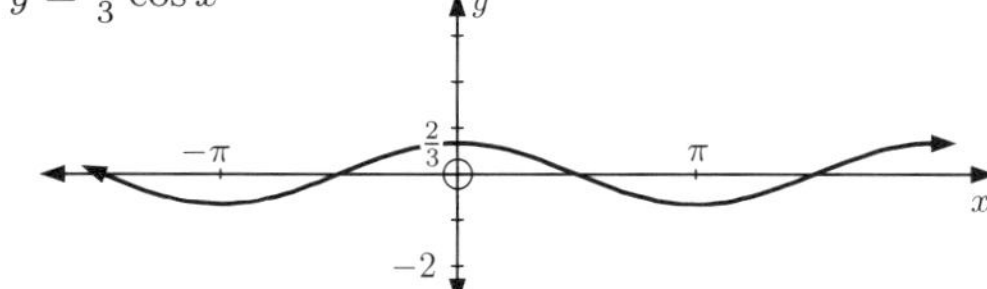

f $y = \frac{3}{2}\cos x$

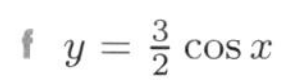

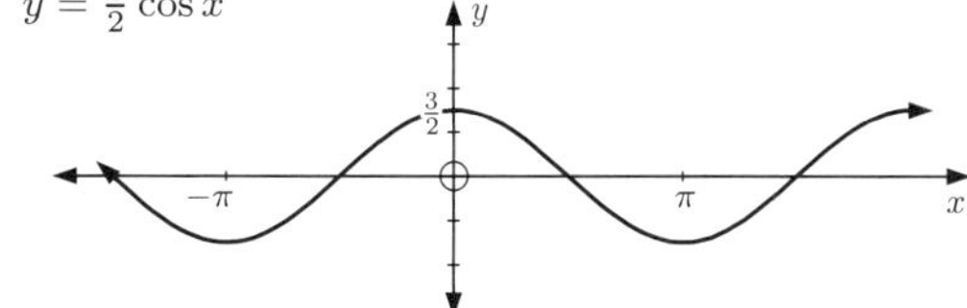

g $y = -\cos x$

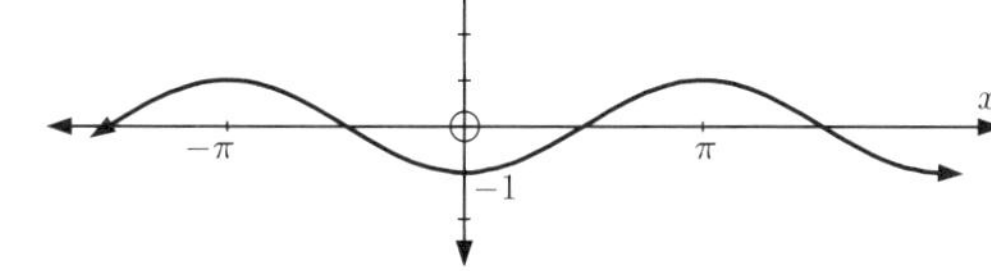

h $y = \cos\left(x - \frac{\pi}{6}\right) + 1$

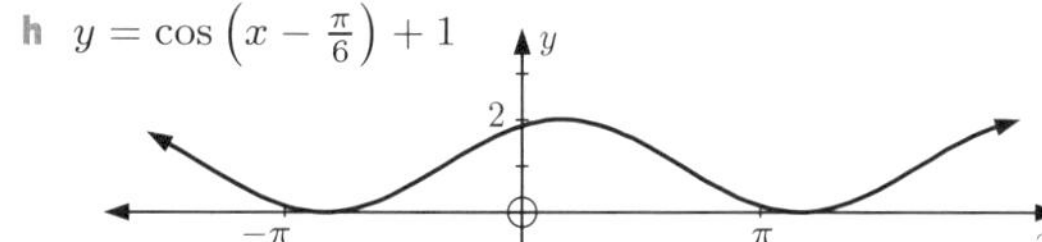

i $y = \cos\left(x + \frac{\pi}{4}\right) - 1$

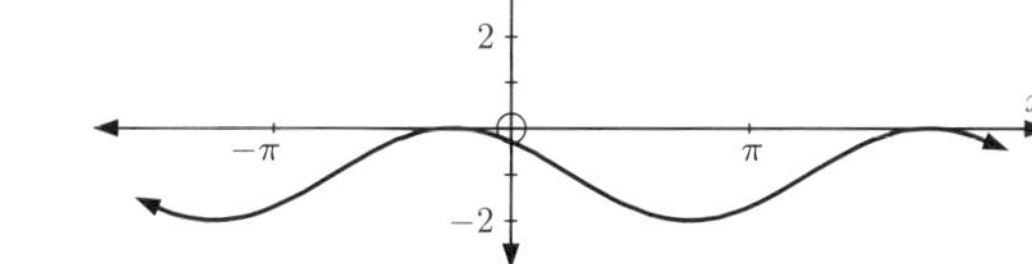

j $y = \cos 2x$

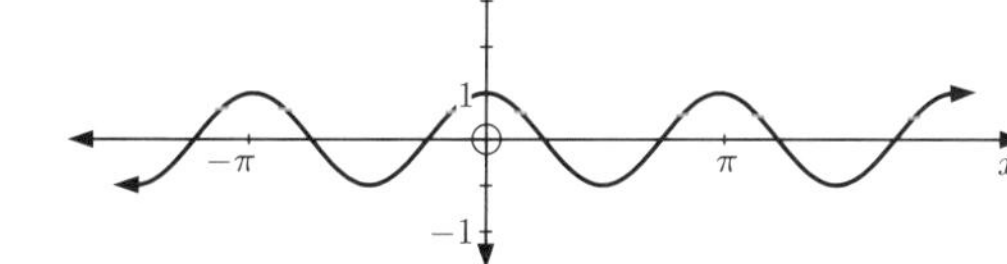

k $y = \cos\left(\frac{x}{2}\right)$

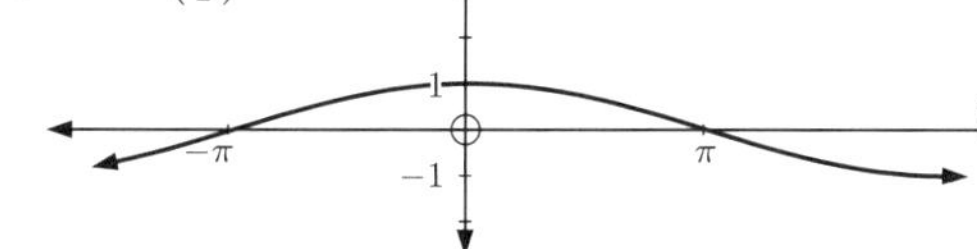

l $y = 3\cos 2x$

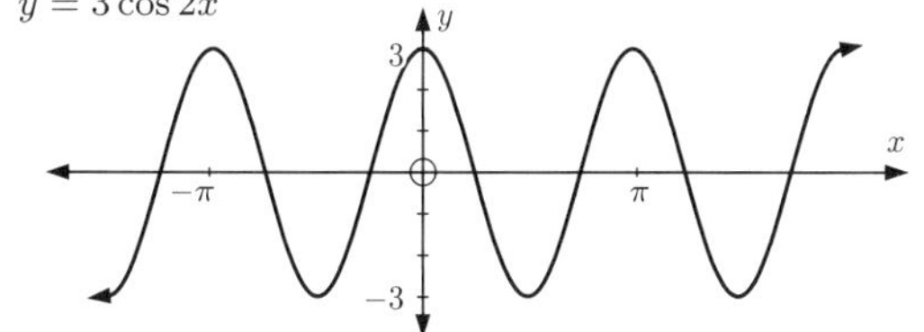

2 a $\frac{2\pi}{3}$ b 6π c 100

3 $|a| =$ amplitude, $b = \dfrac{2\pi}{\text{period}}$, $c =$ horizontal translation, $d =$ vertical translation

4 a $y = 2\cos 2x$ b $y = \cos\left(\frac{x}{2}\right) + 2$

c $y = -5\cos\left(\frac{\pi}{3}x\right)$

EXERCISE 12E

1 a i $y = \tan\left(x - \frac{\pi}{2}\right)$

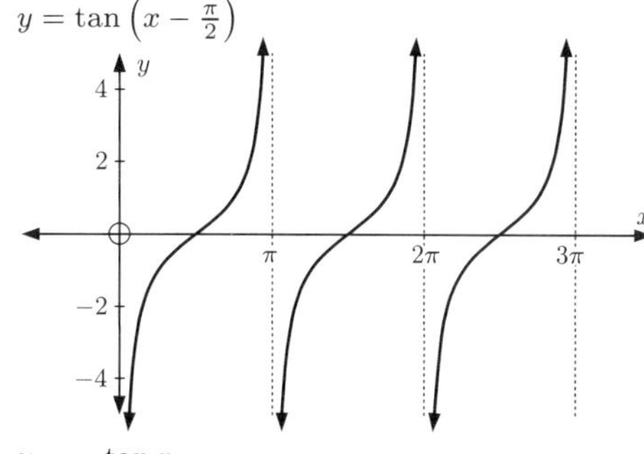

ii $y = -\tan x$

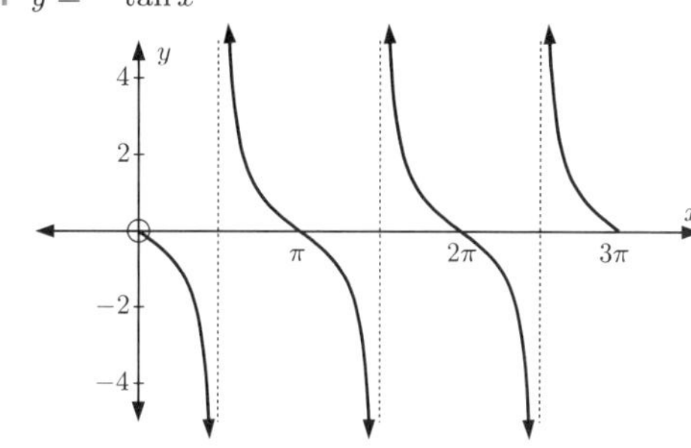

iii $y = \tan 3x$

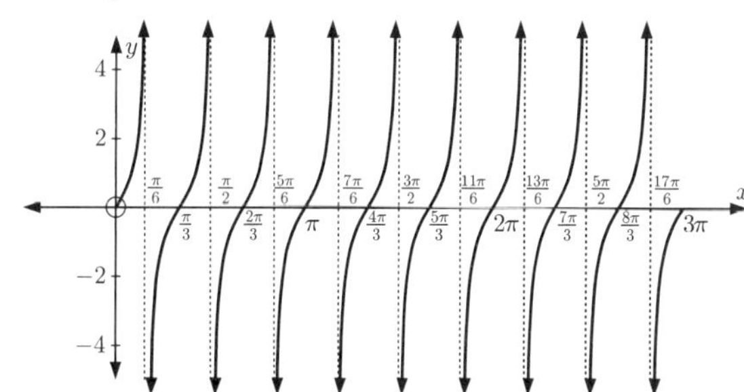

2 a translation through $\begin{pmatrix} \frac{1}{2} \\ 2 \end{pmatrix}$ b reflection in x-axis

c horizontal stretch, factor 2 and vertical stretch with factor 2

3 a π b $\frac{\pi}{3}$ c $\frac{\pi}{n}$

EXERCISE 12F

1 a 1 b undefined c 1

2 a π b 6π c π

3 a $b = 1$ b $b = 3$ c $b = 2$ d $b = \frac{\pi}{2}$

4 a

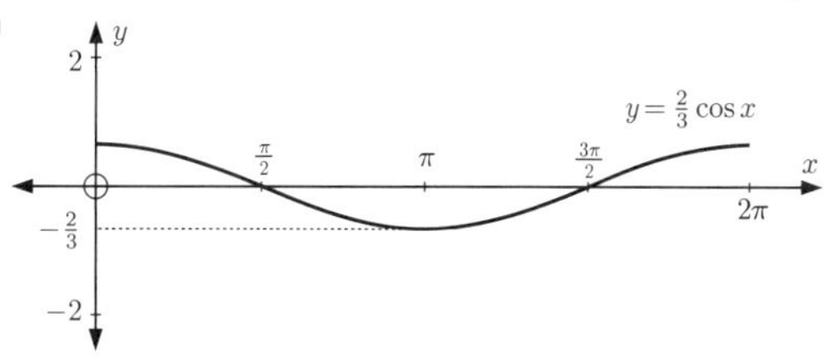

b

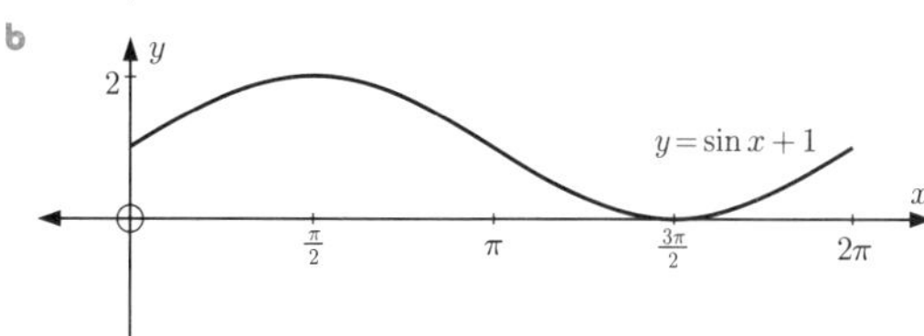

c 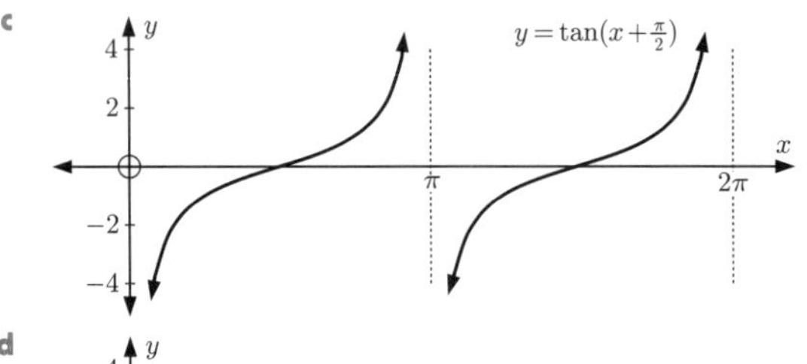

d 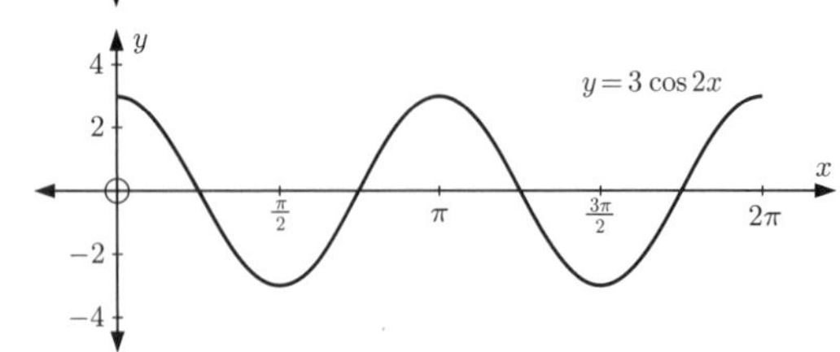

e 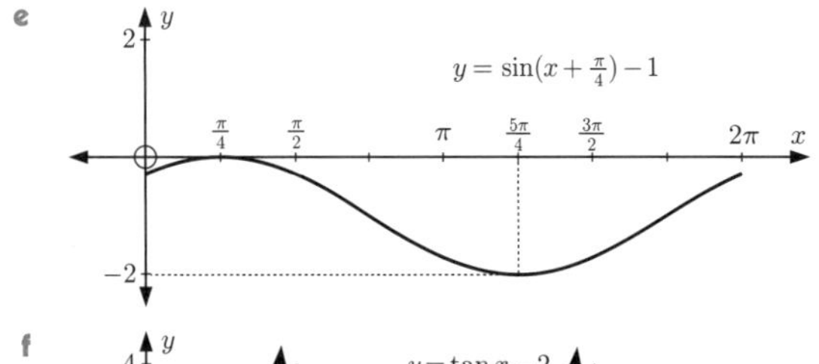

f 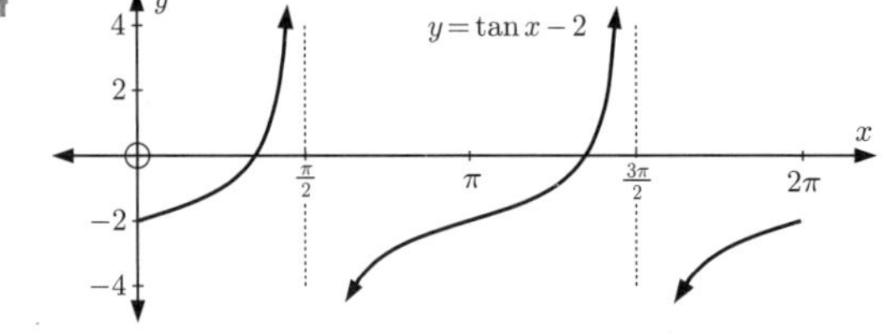

5

	a	b	c	d	e	f
maximum value	1	3	undef.	4	3	-2
minimum value	-1	-3	undef.	2	-1	-4

6 **a** vertical stretch, factor $\frac{1}{2}$ **b** horizontal stretch, factor 4
c reflection in the x-axis
d vertical translation down 2 units
e horizontally translate $\frac{\pi}{4}$ units to the left
f reflection in the y-axis

7 $m = 2,\ n = -3$ **8** $p = \frac{1}{2},\ q = 1$

EXERCISE 12G

1 **a** $\frac{2}{\sqrt{3}}$ **b** $-\frac{1}{\sqrt{3}}$ **c** $-\frac{2}{\sqrt{3}}$ **d** undefined

2 **a** $\csc x = \frac{5}{3},\ \sec x = \frac{5}{4},\ \cot x = \frac{4}{3}$
b $\csc x = -\frac{3}{\sqrt{5}},\ \sec x = \frac{3}{2},\ \cot x = -\frac{2}{\sqrt{5}}$

3 **a** $\sin x = -\frac{\sqrt{7}}{4},\ \tan x = -\frac{\sqrt{7}}{3},\ \csc x = -\frac{4}{\sqrt{7}},$
$\sec x = \frac{4}{3},\ \cot x = -\frac{3}{\sqrt{7}}$
b $\cos x = -\frac{\sqrt{5}}{3},\ \tan x = \frac{2}{\sqrt{5}},\ \csc x = -\frac{3}{2},$
$\sec x = -\frac{3}{\sqrt{5}},\ \cot x = \frac{\sqrt{5}}{2}$
c $\sin x = \frac{\sqrt{21}}{5},\ \cos x = \frac{2}{5},\ \tan x = \frac{\sqrt{21}}{2},$
$\csc x = \frac{5}{\sqrt{21}},\ \cot x = \frac{2}{\sqrt{21}}$
d $\sin x = \frac{1}{2},\ \cos x = -\frac{\sqrt{3}}{2},\ \tan x = -\frac{1}{\sqrt{3}},$
$\sec x = -\frac{2}{\sqrt{3}},\ \cot x = -\sqrt{3}$
e $\sin\beta = -\frac{1}{\sqrt{5}},\ \cos\beta = -\frac{2}{\sqrt{5}},\ \csc\beta = -\sqrt{5},$
$\sec\beta = -\frac{\sqrt{5}}{2},\ \cot\beta = 2$
f $\sin\theta = -\frac{3}{5},\ \cos\theta = -\frac{4}{5},\ \tan\theta = \frac{3}{4},$
$\csc\theta = -\frac{5}{3},\ \sec\theta = -\frac{5}{4}$

4 **a** 1 **b** 1 **c** $\frac{\cos x}{\sin^2 x}$ **d** $\cos x$ **e** $\cos x$ **f** $5\sin x$

5 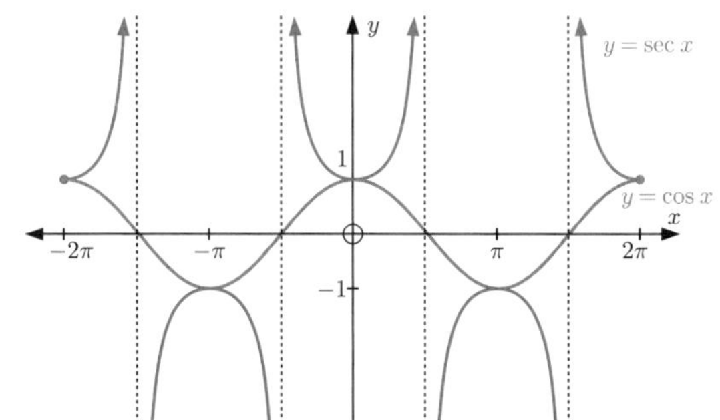

6 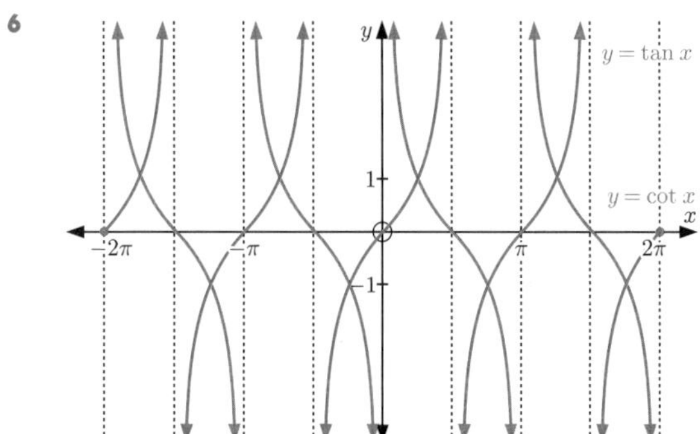

7 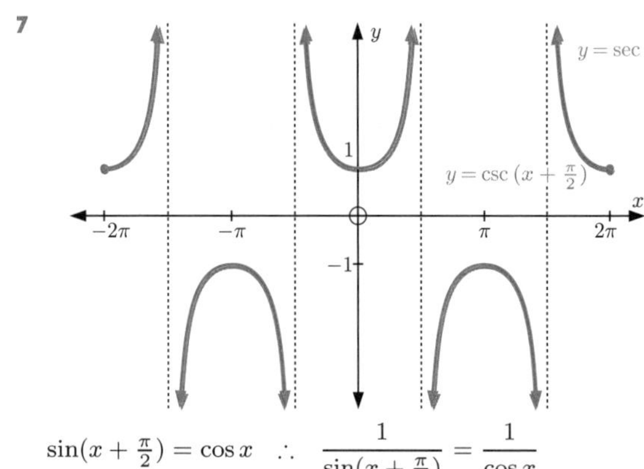

$\sin(x + \frac{\pi}{2}) = \cos x \quad \therefore \quad \frac{1}{\sin(x + \frac{\pi}{2})} = \frac{1}{\cos x}$

EXERCISE 12H

1

Function	*Restricted domain*	*Restricted range*
$y = \sin x$	$-\frac{\pi}{2} \leqslant x \leqslant \frac{\pi}{2}$	$-1 \leqslant y \leqslant 1$
$y = \cos x$	$0 \leqslant x \leqslant \pi$	$-1 \leqslant y \leqslant 1$
$y = \tan x$	$-\frac{\pi}{2} < x < \frac{\pi}{2}$	$y \in \mathbb{R}$

Inverse function	*Domain*	*Range*
$y = \arcsin x$	$-1 \leqslant x \leqslant 1$	$-\frac{\pi}{2} \leqslant y \leqslant \frac{\pi}{2}$
$y = \arccos x$	$-1 \leqslant x \leqslant 1$	$0 \leqslant y \leqslant \pi$
$y = \arctan x$	$x \in \mathbb{R}$	$-\frac{\pi}{2} < y < \frac{\pi}{2}$

2 **a** 0 **b** $-\frac{\pi}{2}$ **c** $\frac{\pi}{4}$ **d** $-\frac{\pi}{4}$ **e** $\frac{\pi}{6}$ **f** $\frac{5\pi}{6}$ **g** $\frac{\pi}{3}$
h $\frac{3\pi}{4}$ **i** $-\frac{\pi}{6}$ **j** ≈ -0.874 **k** ≈ 1.24 **l** ≈ -1.55

3 **a** $(0, 0)$ **b** $(0, 0)$ **c** $(0.739, 0.739)$

4 **a** horizontal asymptotes $y = -\frac{\pi}{2}$, $y = \frac{\pi}{2}$ **b** No

5 **a** $\arcsin(\sin\frac{\pi}{3}) = \frac{\pi}{3}$ **b** $\arccos(\cos(-\frac{\pi}{6})) = \frac{\pi}{6}$
c $\tan(\arctan(0.3)) = 0.3$ **d** $\cos(\arccos(-\frac{1}{2})) = -\frac{1}{2}$
e $\arctan(\tan\pi) = 0$ **f** $\arcsin(\sin\frac{4\pi}{3}) = -\frac{\pi}{3}$

REVIEW SET 12A

1 **a** no **b** yes

2

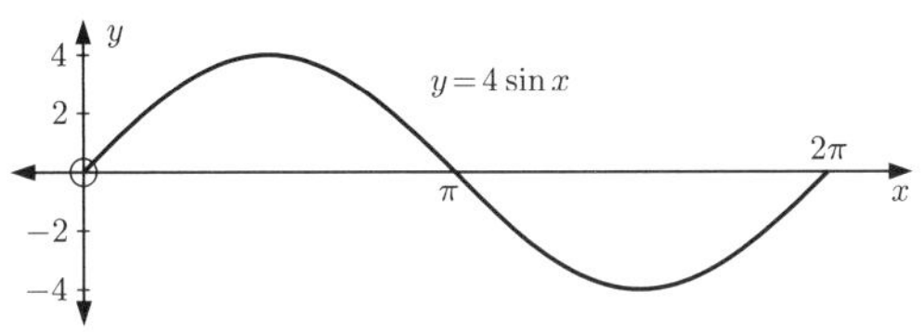

3 **a** minimum $= 0$, maximum $= 2$
b minimum $= -2$, maximum $= 2$

4 **a** 10π **b** $\frac{\pi}{2}$ **c** 4π **d** $\frac{\pi}{3}$

5

Function	*Period*	*Amplitude*
$y = -3\sin(\frac{x}{4}) + 1$	8π	3
$y = \tan 2x$	$\frac{\pi}{2}$	undefined
$y = 3\cos\pi x$	2	3

Function	*Domain*	*Range*
$y = -3\sin(\frac{x}{4}) + 1$	$x \in \mathbb{R}$	$-2 \leqslant y \leqslant 4$
$y = \tan 2x$	$x \neq \pm\frac{\pi}{4}, \pm\frac{3\pi}{4}, ...$	$y \in \mathbb{R}$
$y = 3\cos\pi x$	$x \in \mathbb{R}$	$-3 \leqslant y \leqslant 3$

6 **a** $y = -4\cos(2x)$ **b** $y = \cos\left(\frac{\pi}{4}x\right) + 2$

7 **a** vertical stretch with scale factor 3 and horizontal stretch with scale factor $\frac{1}{2}$
b translate $\begin{pmatrix} \frac{\pi}{3} \\ -1 \end{pmatrix}$

8 **a** $\sin x = \frac{2\sqrt{2}}{3}$, $\tan x = 2\sqrt{2}$, $\csc x = \frac{3}{2\sqrt{2}}$,
$\sec x = 3$, $\cot x = \frac{1}{2\sqrt{2}}$
b $\sin x = -\frac{4}{\sqrt{41}}$, $\cos x = -\frac{5}{\sqrt{41}}$, $\csc x = -\frac{\sqrt{41}}{4}$,
$\sec x = -\frac{\sqrt{41}}{5}$, $\cot x = \frac{5}{4}$

9 **a** $\arctan(\tan(-0.5)) = -0.5$
b $\arcsin(\sin(-\frac{\pi}{6})) = -\frac{\pi}{6}$ **c** $\arccos(\cos 2\pi) = 0$

REVIEW SET 12B

1 **a**

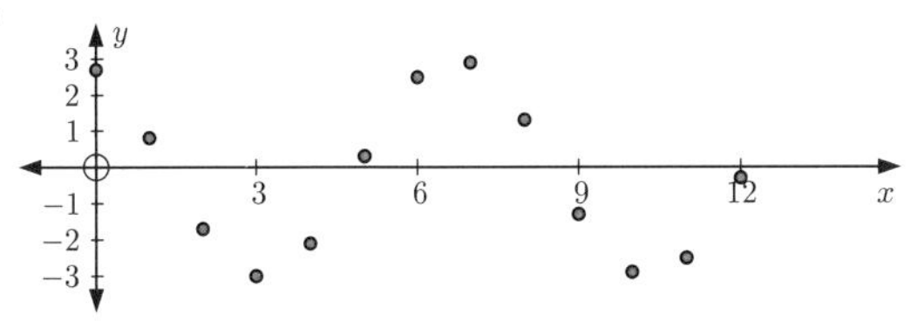

approximately periodic

b

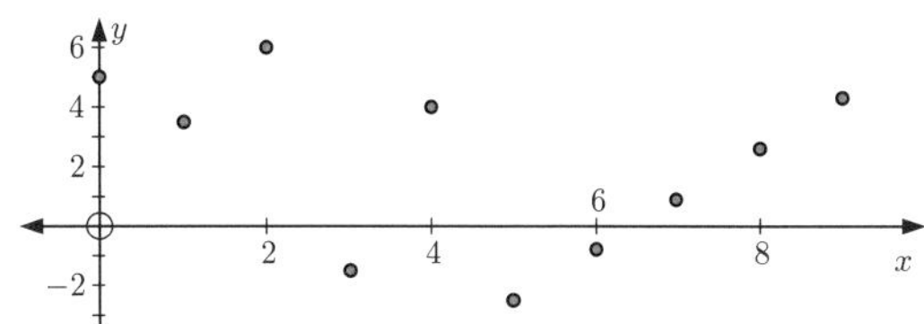

not periodic

2

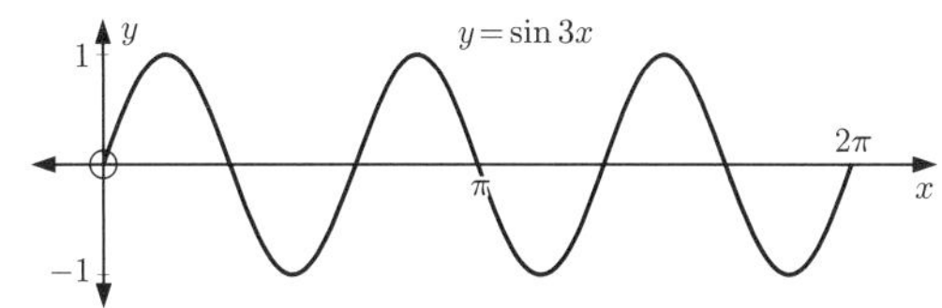

3 **a** 6π **b** $\frac{\pi}{4}$

4

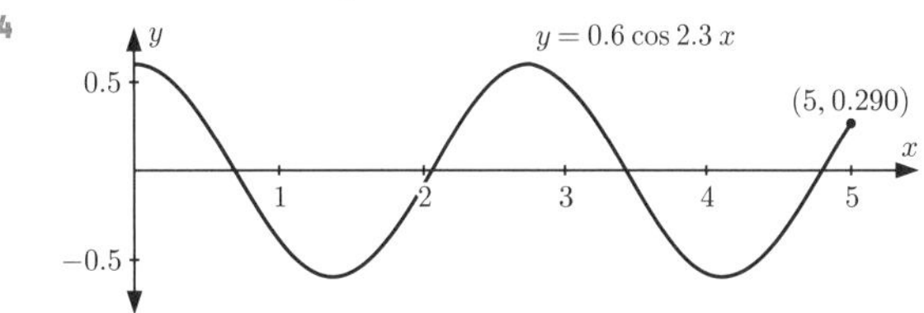

5 **a** maximum: $-5°C$, minimum: $-79°C$
b $T \approx 37\sin(0.008\,98n) - 42$ **c** ≈ 700 Mars days

6 **a** maximum $= 2$, minimum $= -8$
b maximum $= 1\frac{1}{3}$, minimum $= \frac{2}{3}$

7 **a** reflection in x-axis, horizontal stretch with scale factor $\frac{1}{2}$
b vertical stretch with scale factor 2, then translate $\begin{pmatrix} \frac{\pi}{4} \\ \frac{1}{2} \end{pmatrix}$, then horizontal stretch with scale factor 2

8 **a**

y = sec x
y = csc x

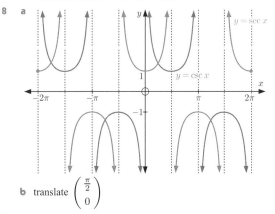

b translate $\begin{pmatrix} \frac{\pi}{2} \\ 0 \end{pmatrix}$

9 **a**

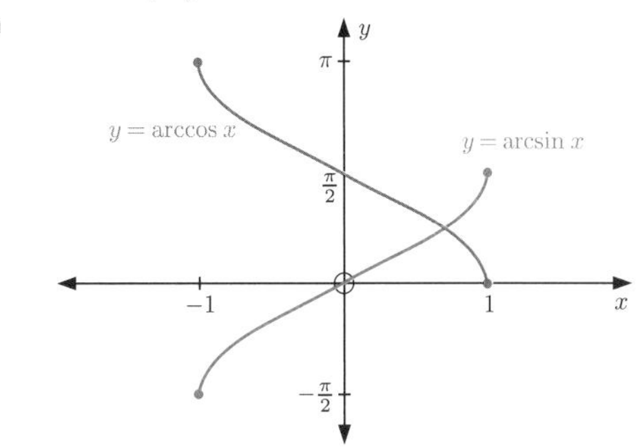

b $y = \arcsin x$: Domain $= \{x \mid -1 \leqslant x \leqslant 1\}$, Range $= \{y \mid -\frac{\pi}{2} \leqslant y \leqslant \frac{\pi}{2}\}$

$y = \arccos x$: Domain $= \{x \mid -1 \leqslant x \leqslant 1\}$, Range $= \{y \mid 0 \leqslant y \leqslant \pi\}$

c reflect in y-axis (*or* reflect in x-axis), translate $\begin{pmatrix} 0 \\ \frac{\pi}{2} \end{pmatrix}$

REVIEW SET 12C

1 **a** $b = \frac{1}{3}$ **b** $b = 24$ **c** $b = \frac{2\pi}{9}$

2 **a**

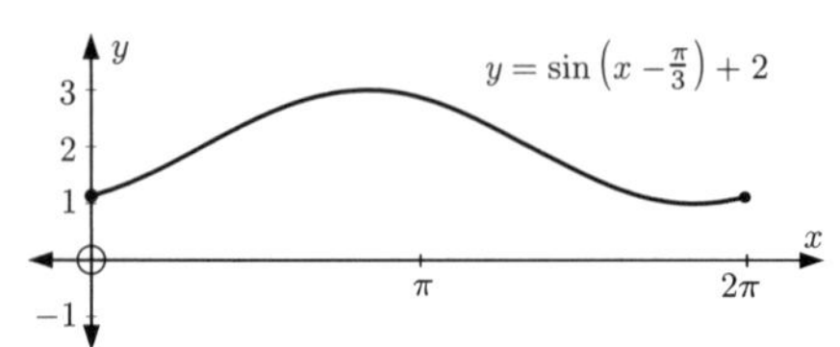

b $1 \leqslant k \leqslant 3$

3 **a** The function repeats itself over and over in a horizontal direction, in intervals of length 8 units.

b **i** 8 **ii** 5 **iii** -1

4 **a**

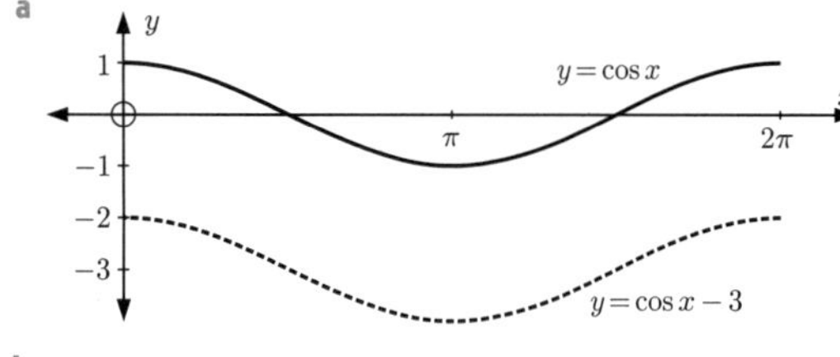

b

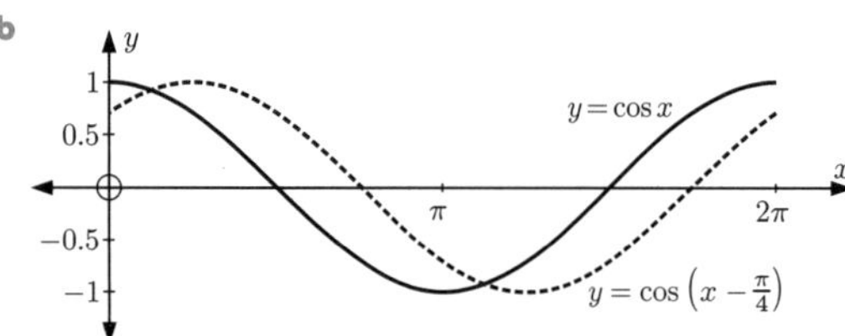

c

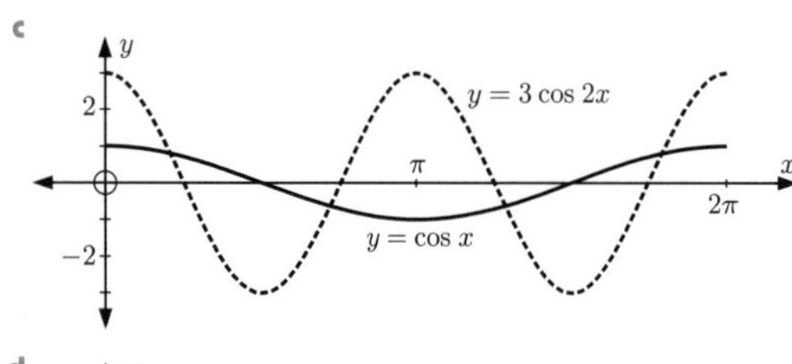

d

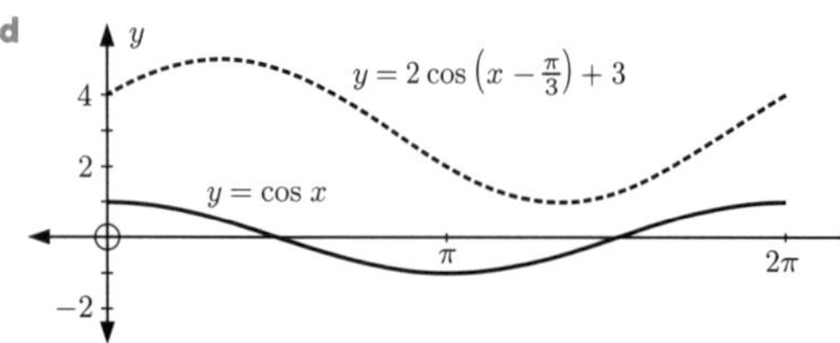

5 **a** $T \approx 7.05 \sin \frac{\pi}{6}(t - 10.5) + 24.75$

6 **a** translation through $\begin{pmatrix} \frac{\pi}{3} \\ 1 \end{pmatrix}$

b a vertical stretch with scale factor 2 followed by a reflection in the x-axis

c a horizontal stretch with scale factor $\frac{1}{3}$

7 **a** $y = \frac{3}{4}\sin(4x - 0.340) + \frac{1}{4}$ **b** $y = \tan(x - \frac{\pi}{2})$

8 **a** $\sec x$ **b** $\sin x$ **c** $\cos x$

9 **a** $-\frac{\pi}{2} < x < \frac{\pi}{2}$

b

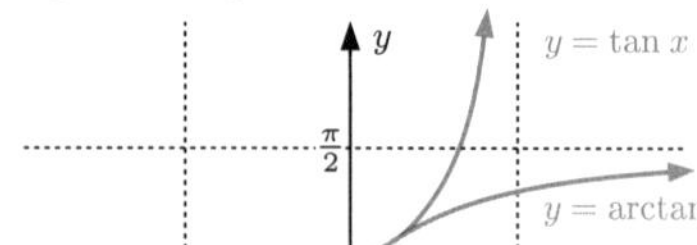
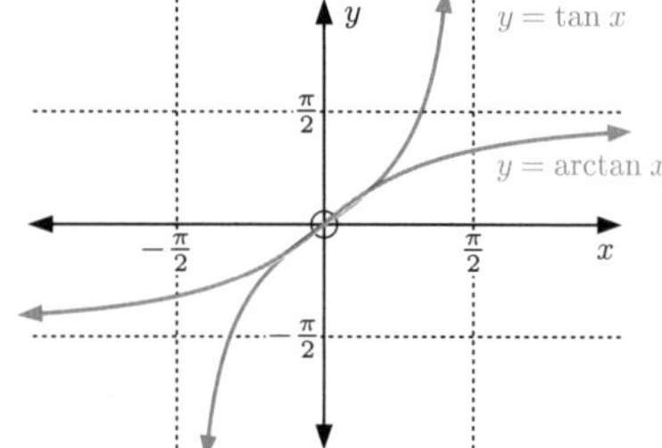

EXERCISE 13A.1

1 **a** $x \approx 0.3, 2.8, 6.6, 9.1, 12.9$ **b** $x \approx 5.9, 9.8, 12.2$

2 **a** $x \approx 1.2, 5.1, 7.4$ **b** $x \approx 4.4, 8.2, 10.7$

3 **a** $x \approx 0.4, 1.2, 3.5, 4.3, 6.7, 7.5, 9.8, 10.6, 13.0, 13.7$

b $x \approx 1.7, 3.0, 4.9, 6.1, 8.0, 9.3, 11.1, 12.4, 14.3, 15.6$

4 **a** **i** ≈ 1.6 **ii** ≈ -1.1

b **i** $x \approx 1.1, 4.2, 7.4$ **ii** $x \approx 2.2, 5.3$

EXERCISE 13A.2

1 **a** $x \approx 0.446, 2.70, 6.73, 8.98$

b $x \approx 2.52, 3.76, 8.80, 10.0$

c $x \approx 0.588, 3.73, 6.87, 10.0$

2 **a** $x \approx -0.644, 0.644$ **b** $x \approx -4.56, -1.42, 1.72, 4.87$

c $x \approx -2.76, -0.384, 3.53$

3 **a** $x \approx 1.08, 4.35$ **b** $x \approx 0.666, 2.48$

c $x \approx 0.171, 4.92$ **d** $x \approx 1.31, 2.03, 2.85$

4 $x \approx -0.951, 0.234, 5.98$

EXERCISE 13A.3

1 **a** $x = \frac{\pi}{3}, \frac{5\pi}{3}, \frac{7\pi}{3}, \frac{11\pi}{3}$ **b** $x = \frac{\pi}{4}, \frac{3\pi}{4}, \frac{9\pi}{4}, \frac{11\pi}{4}$

c $x = \frac{\pi}{4}, \frac{5\pi}{4}, \frac{9\pi}{4}, \frac{13\pi}{4}$

2 **a** $x = -\frac{5\pi}{3}, -\frac{4\pi}{3}, \frac{\pi}{3}, \frac{2\pi}{3}$ **b** $x = -\frac{5\pi}{4}, -\frac{3\pi}{4}, \frac{3\pi}{4}, \frac{5\pi}{4}$

c $x = -\frac{5\pi}{4}, -\frac{\pi}{4}, \frac{3\pi}{4}, \frac{7\pi}{4}$

3 **a** $0 \leqslant 2x \leqslant 4\pi$ **b** $0 \leqslant \dfrac{x}{3} \leqslant \frac{2\pi}{3}$

c $\frac{\pi}{2} \leqslant x + \frac{\pi}{2} \leqslant \frac{5\pi}{2}$ **d** $-\frac{\pi}{6} \leqslant x - \frac{\pi}{6} \leqslant \frac{11\pi}{6}$

e $-\frac{\pi}{2} \leqslant 2(x - \frac{\pi}{4}) \leqslant \frac{7\pi}{2}$ **f** $-2\pi \leqslant -x \leqslant 0$

4 **a** $-3\pi \leqslant 3x \leqslant 3\pi$ **b** $-\frac{\pi}{4} \leqslant \dfrac{x}{4} \leqslant \frac{\pi}{4}$

c $-\frac{3\pi}{2} \leqslant x - \frac{\pi}{2} \leqslant \frac{\pi}{2}$ **d** $-\frac{3\pi}{2} \leqslant 2x + \frac{\pi}{2} \leqslant \frac{5\pi}{2}$

e $-2\pi \leqslant -2x \leqslant 2\pi$ **f** $0 \leqslant \pi - x \leqslant 2\pi$

5 **a** $x = \frac{\pi}{3}, \frac{5\pi}{3}, \frac{7\pi}{3}$ **b** $x = \frac{\pi}{6}, \frac{5\pi}{6}, \frac{7\pi}{6}, \frac{11\pi}{6}, \frac{13\pi}{6}, \frac{17\pi}{6}$

c $x = 0, \frac{4\pi}{3}, 2\pi$

6 **a** $x = \frac{2\pi}{3}, \frac{4\pi}{3}, \frac{8\pi}{3}, \frac{10\pi}{3}, \frac{14\pi}{3}$

b $x = -330°, -210°, 30°, 150°$

c $x = \frac{5\pi}{6}, \frac{7\pi}{6}, \frac{17\pi}{6}$ **d** $x = -\frac{5\pi}{3}, -\pi, \frac{\pi}{3}, \pi$

e $x = -\frac{13\pi}{6}, -\frac{3\pi}{2}, -\frac{\pi}{6}, \frac{\pi}{2}, \frac{11\pi}{6}, \frac{5\pi}{2}$ **f** $x = 0, \frac{3\pi}{2}, 2\pi$

g $x = \frac{\pi}{2}, \frac{3\pi}{2}, \frac{5\pi}{2}$

h $x = -\frac{8\pi}{9}, -\frac{4\pi}{9}, -\frac{2\pi}{9}, \frac{2\pi}{9}, \frac{4\pi}{9}, \frac{8\pi}{9}$

i $x = 0, \frac{\pi}{4}, \frac{\pi}{2}, \frac{3\pi}{4}, \pi$ **j** $x = 0, \frac{\pi}{6}, \pi, \frac{7\pi}{6}, 2\pi$

7 **a** $x = \frac{\pi}{3}, \frac{5\pi}{3}$ **b** $x = \frac{5\pi}{4}, \frac{7\pi}{4}$
c $x = \frac{5\pi}{12}, \frac{7\pi}{12}, \frac{17\pi}{12}, \frac{19\pi}{12}$ **d** $x = \frac{13\pi}{12}, \frac{19\pi}{12}$

8 $\theta = \frac{\pi}{3}, \frac{4\pi}{3}$
a $x = \frac{\pi}{2}, \frac{3\pi}{2}$ **b** $x = \frac{\pi}{12}, \frac{\pi}{3}, \frac{7\pi}{12}, \frac{5\pi}{6}, \frac{13\pi}{12}, \frac{4\pi}{3}, \frac{19\pi}{12}, \frac{11\pi}{6}$
c $x = \frac{\pi}{3}, \frac{2\pi}{3}, \frac{4\pi}{3}, \frac{5\pi}{3}$

9 **a** $x = \frac{3\pi}{4}, \frac{7\pi}{4}$ **b** $x = \frac{5\pi}{24}, \frac{17\pi}{24}, \frac{29\pi}{24}, \frac{41\pi}{24}$

10 **a** $x = 0°, 90°, 180°$ **b** $x = \frac{\pi}{4}, \frac{5\pi}{4}, \frac{9\pi}{4}$

11 **a**

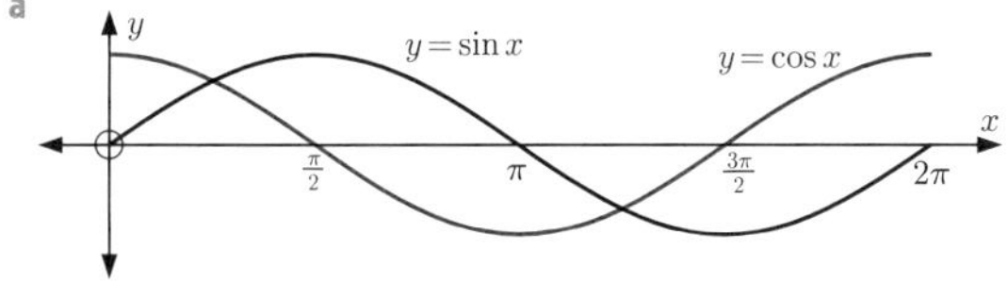

b $x = \frac{\pi}{4}$ or $\frac{5\pi}{4}$

12 **a** $x = \frac{3\pi}{4}$ or $\frac{7\pi}{4}$ **b** $x = \frac{\pi}{12}, \frac{5\pi}{12}, \frac{3\pi}{4}, \frac{13\pi}{12}, \frac{17\pi}{12}, \frac{7\pi}{4}$
c $x = \frac{\pi}{6}, \frac{2\pi}{3}, \frac{7\pi}{6}, \frac{5\pi}{3}$

13 **a** $x = 1$ **b** $x = -\frac{\sqrt{3}}{2}$ **c** $x = -\frac{1}{\sqrt{2}}$
d $x = -\frac{1}{2}$ **e** no solution exists **f** $x = 0$

EXERCISE 13B

1 **a** 22 m **b** 100 s **c** $t \approx 31.5$ s, 68.5 s, 132 s, 168 s

2 **a** **i** 7500 grasshoppers **ii** 10 300 grasshoppers
b 10 500 grasshoppers, when $t = 4$ weeks
c **i** at $t = 1\frac{1}{3}$ wks and $6\frac{2}{3}$ wks **ii** at $t = 9\frac{1}{3}$ wks
d $2.51 \leqslant t \leqslant 5.49$ weeks

3 **a** 20 m **b** at $t = \frac{3}{4}$ minute **c** 3 minutes
d

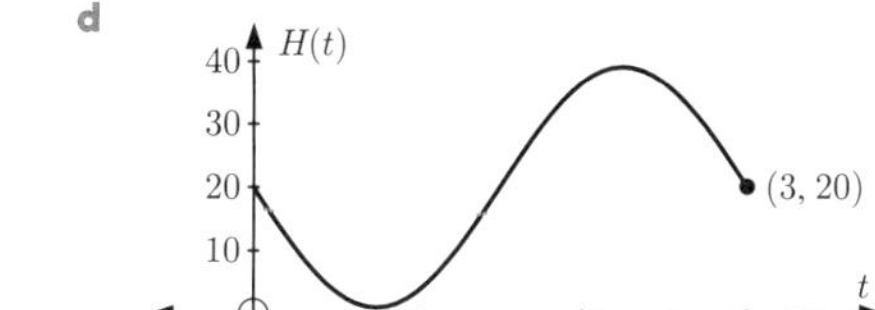

4 **a** 400 water buffalo
b **i** 577 water buffalo **ii** 400 water buffalo
c 650, which is the maximum population.
d 150, after 3 years **e** $t \approx 0.262$ years

5 **a** $H(t) = 3\cos(\frac{\pi t}{2}) + 4$ **b** $t \approx 1.46$ s

6 **a** **i** true **ii** true **b** 116.8 cents L^{-1}
c on the 5th, 11th, 19th, and 25th days
d 98.6 cents L^{-1} on the 1st and 15th days

EXERCISE 13C.1

1 **a** $2\sin\theta$ **b** $3\cos\theta$ **c** $2\sin\theta$ **d** $\sin\theta$
e $-2\tan\theta$ **f** $-3\cos^2\theta$

2 **a** 3 **b** -2 **c** -1 **d** $3\cos^2\theta$
e $4\sin^2\theta$ **f** $\cos\theta$ **g** $-\sin^2\theta$ **h** $-\cos^2\theta$
i $-2\sin^2\theta$ **j** 1 **k** $\sin\theta$ **l** $\sin\theta$

3 **a** $2\tan x$ **b** $\tan^2 x$ **c** $\sin x$ **d** $\cos x$
e $5\sin x$ **f** $\dfrac{2}{\cos x}$

4 **a** $1 + 2\sin\theta + \sin^2\theta$ **b** $\sin^2\alpha - 4\sin\alpha + 4$
c $\tan^2\alpha - 2\tan\alpha + 1$ **d** $1 + 2\sin\alpha\cos\alpha$
e $1 - 2\sin\beta\cos\beta$ **f** $-4 + 4\cos\alpha - \cos^2\alpha$

5 **a** $-\tan^2\beta$ **b** 1 **c** $\sin^2\alpha$
d $\sin^2 x - \tan^2 x$ **e** 13 **f** $\cos^2\theta$ **g** 0

EXERCISE 13C.2

1 **a** $(1 - \sin\theta)(1 + \sin\theta)$ **b** $(\sin\alpha + \cos\alpha)(\sin\alpha - \cos\alpha)$
c $(\tan\alpha + 1)(\tan\alpha - 1)$ **d** $\sin\beta(2\sin\beta - 1)$
e $\cos\phi(2 + 3\cos\phi)$ **f** $3\sin\theta(\sin\theta - 2)$
g $(\tan\theta + 3)(\tan\theta + 2)$ **h** $(2\cos\theta + 1)(\cos\theta + 3)$
i $(3\cos\alpha + 1)(2\cos\alpha - 1)$ **j** $\tan\alpha(3\tan\alpha - 2)$
k $(\sec\beta + \csc\beta)(\sec\beta - \csc\beta)$ **l** $(2\cot x - 1)(\cot x - 1)$
m $(2\sin x + \cos x)(\sin x + 3\cos x)$

2 **a** $1 + \sin\alpha$ **b** $\tan\beta - 1$ **c** $\cos\phi - \sin\phi$
d $\cos\phi + \sin\phi$ **e** $\dfrac{1}{\sin\alpha - \cos\alpha}$ **f** $\dfrac{\cos\theta}{2}$
g $\sin\theta$ **h** $\cos\theta$ **i** $\sec\theta + 1$

EXERCISE 13D

1 **a** $\frac{24}{25}$ **b** $-\frac{7}{25}$ **c** $-\frac{24}{7}$ **2** **a** $-\frac{7}{9}$ **b** $\frac{1}{9}$

3 **a** $\cos\alpha = \frac{-\sqrt{5}}{3}$ **b** $\sin 2\alpha = \frac{4\sqrt{5}}{9}$

4 **a** $\sin\beta = \frac{-\sqrt{21}}{5}$ **b** $\sin 2\beta = \frac{-4\sqrt{21}}{25}$

5 **a** $\frac{1}{3}$ **b** $\frac{2\sqrt{2}}{3}$ **6** **a** $\tan A = -\frac{7}{3}$ **b** $\tan A = \frac{3}{2}$

7 $\tan\left(\frac{\pi}{8}\right) = \sqrt{2} - 1$ **8** $\frac{3}{2}$

9 **a** $\sin 2\alpha$ **b** $2\sin 2\alpha$ **c** $\frac{1}{2}\sin 2\alpha$ **d** $\cos 2\beta$
e $-\cos 2\phi$ **f** $\cos 2N$ **g** $-\cos 2M$ **h** $\cos 2\alpha$
i $-\cos 2\alpha$ **j** $\sin 4A$ **k** $\sin 6\alpha$ **l** $\cos 8\theta$
m $-\cos 6\beta$ **n** $\cos 10\alpha$ **o** $-\cos 6D$ **p** $\cos 4A$
q $\cos\alpha$ **r** $-2\cos 6P$

11 **a** $x = 0, \frac{2\pi}{3}, \pi, \frac{4\pi}{3}, 2\pi$ **b** $x = \frac{\pi}{2}, \frac{3\pi}{2}$ **c** $x = 0, \pi, 2\pi$

13 **a** $\cos A = \frac{7}{10}$ **b** $\cos A = \frac{3}{4}$

EXERCISE 13E

1 **a** $\cos\theta$ **b** $-\sin\theta$ **c** $\sin\theta$ **d** $-\cos\alpha$ **e** $-\sin A$
f $-\sin\theta$ **g** $\dfrac{1 + \tan\theta}{1 - \tan\theta}$ **h** $\dfrac{1 + \tan\theta}{1 - \tan\theta}$ **i** $\tan\theta$

2 **a** $\frac{1}{2}\sin\theta + \frac{\sqrt{3}}{2}\cos\theta$ **b** $\frac{\sqrt{3}}{2}\sin\theta - \frac{1}{2}\cos\theta$
c $-\frac{1}{\sqrt{2}}\sin\theta + \frac{1}{\sqrt{2}}\cos\theta$ **d** $-\frac{\sqrt{3}}{2}\sin\theta + \frac{1}{2}\cos\theta$

3 **a** $\cos\theta$ **b** $\sin 3A$ **c** $\sin(B - A)$ **d** $\cos(\alpha - \beta)$
e $-\cos(\theta + \phi)$ **f** $2\sin(\alpha - \beta)$ **g** $\tan\theta$ **h** $\tan(3A)$

5 **a** $\cos 2\alpha$ **b** $-\sin 3\phi$ **c** $\cos\beta$

7 **a** $2 + \sqrt{3}$ **b** $-2 - \sqrt{3}$ **8** $\frac{7}{17}$ **9** 7

10 **a** -1 **b** $\tan(2A)$ **11** **a** $\frac{9+5\sqrt{2}}{2}$ **b** $\frac{4\sqrt{2}}{7}$

12 $\sqrt{3}$ **13** $\tan A = -\frac{1}{21}$ **14** $\tan A = \pm 1$

15 $\tan\alpha = \frac{25}{62}$ **16** $\tan\alpha = \frac{1}{8}$

17 $\tan(A + B + C)$
$$= \frac{\tan A + \tan B + \tan C - \tan A \tan B \tan C}{1 - \tan A \tan B - \tan A \tan C - \tan B \tan C}$$

20 $k = 2$, $b = \frac{\pi}{6}$ **21** **b** $\theta = -\frac{8\pi}{9}, -\frac{4\pi}{9}, -\frac{2\pi}{9}, \frac{2\pi}{9}, \frac{4\pi}{9}, \frac{8\pi}{9}$

22 **a** $\sin 3\theta = -4\sin^3\theta + 3\sin\theta$

b $\theta = 0, \frac{\pi}{4}, \frac{3\pi}{4}, \pi, \frac{5\pi}{4}, \frac{7\pi}{4}, 2\pi, \frac{9\pi}{4}, \frac{11\pi}{4}, 3\pi$

24 **a** $2\cos x - 5\sin x = \sqrt{29}\cos(x + 1.19)$ **b** $x \approx 0.761, \pi$
d $x \approx 0.761$ (the solution $x = \pi$ has been lost)

25 **Hint:** Let $\theta = \arctan(5)$ $\therefore$ $\tan\theta = 5$, etc. **27** $\frac{\pi}{4}$

28 **c** **i** $\frac{1}{2}\sin 4\theta + \frac{1}{2}\sin 2\theta$ **ii** $\frac{1}{2}\sin 7\alpha + \frac{1}{2}\sin 5\alpha$
iii $\sin 6\beta + \sin 4\beta$ **iv** $2\sin 5\theta + 2\sin 3\theta$
v $3\sin 7\alpha - 3\sin\alpha$ **vi** $\frac{1}{6}\sin 8A - \frac{1}{6}\sin 2A$

29 **c** **i** $\frac{1}{2}\cos 5\theta + \frac{1}{2}\cos 3\theta$ **ii** $\frac{1}{2}\cos 8\alpha + \frac{1}{2}\cos 6\alpha$
iii $\cos 4\beta + \cos 2\beta$ **iv** $3\cos 8x + 3\cos 6x$
v $\frac{3}{2}\cos 5P + \frac{3}{2}\cos 3P$ **vi** $\frac{1}{8}\cos 6x + \frac{1}{8}\cos 2x$

30 **c** **i** $\frac{1}{2}\cos 2\theta - \frac{1}{2}\cos 4\theta$ **ii** $\frac{1}{2}\cos 5\alpha - \frac{1}{2}\cos 7\alpha$
iii $\cos 4\beta - \cos 6\beta$ **iv** $2\cos 3\theta - 2\cos 5\theta$
v $5\cos 6A - 5\cos 10A$ **vi** $\frac{1}{10}\cos 4M - \frac{1}{10}\cos 10M$

31 **a** $\sin A\cos A = \frac{1}{2}\sin 2A$, $\cos^2 A = \frac{1}{2}\cos 2A + \frac{1}{2}$,
$\sin^2 A = \frac{1}{2} - \frac{1}{2}\cos 2A$
c $\cos\left(\frac{S+D}{2}\right)\cos\left(\frac{S-D}{2}\right) = \frac{1}{2}\cos S + \frac{1}{2}\cos D$
d $\sin\left(\frac{S+D}{2}\right)\sin\left(\frac{S-D}{2}\right) = \frac{1}{2}\cos D - \frac{1}{2}\cos S$

32 **a** $2\sin 3x\cos 2x$ **b** $2\cos 5A\cos 3A$ **c** $-2\sin 2\alpha\sin\alpha$
d $2\cos 4\theta\sin\theta$ **e** $-2\sin 4\alpha\sin 3\alpha$ **f** $2\sin 5\alpha\cos 2\alpha$
g $2\sin 3B\sin B$ **h** $2\cos\left(x + \frac{h}{2}\right)\sin\left(\frac{h}{2}\right)$
i $-2\sin\left(x + \frac{h}{2}\right)\sin\left(\frac{h}{2}\right)$

EXERCISE 13F

1 **a** $x = 0, \pi, \frac{7\pi}{6}, \frac{11\pi}{6}, 2\pi$ **b** $x = \frac{\pi}{3}, \frac{\pi}{2}, \frac{3\pi}{2}, \frac{5\pi}{3}$
c $x = \frac{\pi}{3}, \pi, \frac{5\pi}{3}$ **d** $x = \frac{7\pi}{6}, \frac{3\pi}{2}, \frac{11\pi}{6}$
e no solutions **f** $x = \frac{\pi}{6}, \frac{5\pi}{6}, \frac{7\pi}{6}, \frac{11\pi}{6}$

2 **a** $x = 0, \frac{2\pi}{3}, \frac{4\pi}{3}, 2\pi$ **b** $x = \frac{\pi}{3}, \frac{5\pi}{3}$ **c** $x = \frac{\pi}{2}, \frac{7\pi}{6}, \frac{11\pi}{6}$
d $x = 0, \frac{\pi}{6}, \frac{\pi}{2}, \frac{5\pi}{6}, \pi, \frac{7\pi}{6}, \frac{3\pi}{2}, \frac{11\pi}{6}, 2\pi$
e $x = \frac{\pi}{4}$ **f** $x = \frac{\pi}{6}, \frac{5\pi}{6}$

3 **a** $x = \frac{\pi}{6}, \frac{\pi}{2}, \frac{5\pi}{6}$ **b** $x = -\frac{5\pi}{6}, -\frac{\pi}{6}, 0$
c $x = -\frac{2\pi}{3}, -\frac{\pi}{3}, \frac{\pi}{3}, \frac{2\pi}{3}$

4 **a** $x \approx 0.896, 2.25$ **b** $x \approx 3.33, 6.10$
c $x \approx 0.730, 2.41, 3.87, 5.55$

EXERCISE 13G

1 **a** $\dfrac{1 - \sin^n x}{1 - \sin x}$ **b** $\dfrac{1}{1 - \sin x}$ if $-1 < \sin x < 1$,
c $\frac{7\pi}{6}$ or $\frac{11\pi}{6}$ not convergent for $\sin x = \pm 1$

2 **b** **i** $\sin 8x$ **ii** $\dfrac{\sin 20x}{2\sin x}$ **iii** $\dfrac{\sin 2nx}{2\sin x}$

3 **b** **i** $\dfrac{\sin(2^4 x)}{2^4}$ **ii** $\dfrac{\sin(2^6 x)}{2^6}$
c $\sin x\cos x\cos 2x \dots \cos 2^n x = \dfrac{\sin(2^{n+1}x)}{2^{n+1}}$

4 **b** 0

REVIEW SET 13A

1 **a** $x \approx 115°, 245°, 475°, 605°$ **b** $x \approx 25°, 335°, 385°$

2 **a** $x = \frac{7\pi}{6}, \frac{11\pi}{6}, \frac{19\pi}{6}, \frac{23\pi}{6}$ **b** $x = -\frac{7\pi}{4}, -\frac{5\pi}{4}, \frac{\pi}{4}, \frac{3\pi}{4}$

3 **a** $\frac{4\pi}{9}, \frac{5\pi}{9}, \frac{10\pi}{9}, \frac{11\pi}{9}, \frac{16\pi}{9}, \frac{17\pi}{9}$ **b** $\frac{3\pi}{4}, \frac{7\pi}{4}, \frac{11\pi}{4}$

4 **a** $x = \frac{\pi}{6}, \frac{7\pi}{6}$ **b** $x = 0, \frac{\pi}{4}, \pi, \frac{5\pi}{4}, 2\pi$

5 **a** $-\sin\theta$ **b** $\cos\theta$

6 **a** $1 - \cos\theta$ **b** $\dfrac{1}{\sin\alpha + \cos\alpha}$ **c** $\dfrac{-\cos\alpha}{2}$

7 **a** $-\frac{\sqrt{7}}{4}$ **b** $\frac{3\sqrt{7}}{8}$ **c** $-\frac{1}{8}$ **d** $-3\sqrt{7}$

9 **a** $\dfrac{-1 - \sqrt{3}}{2\sqrt{2}}$ **b** $2 - \sqrt{3}$

10 **a** $x = \frac{\pi}{3}, \frac{2\pi}{3}, \frac{4\pi}{3}$ or $\frac{5\pi}{3}$ **b** $x = \frac{\pi}{3}, \frac{5\pi}{6}, \frac{4\pi}{3}, \frac{11\pi}{6}$

12 **a** $x = 0, \frac{3\pi}{2}, 2\pi, \frac{7\pi}{2}, 4\pi$ **b** $x = \frac{\pi}{6}, \frac{2\pi}{3}, \frac{7\pi}{6}, \frac{5\pi}{3}$

13 **c** $x = \frac{16}{3}$ or 3

REVIEW SET 13B

1 **a** $x \approx 0.392, 2.75, 6.68$ **b** $x \approx 5.42$

2 **a** $x \approx 1.12, 5.17, 7.40$ **b** $x \approx 0.184, 4.62$

3 **a** $\frac{120}{169}$ **b** $\frac{119}{169}$ **c** $\frac{120}{119}$

4 **a** **i** $x \approx 1.33, 4.47, 7.61$ **ii** $x \approx 5.30$
iii $x \approx 2.83, 5.97, 9.11$
b **i** $x = -\frac{\pi}{2}, \frac{\pi}{2}$ **ii** $x = -\frac{2\pi}{3}, -\frac{\pi}{6}, \frac{\pi}{3}, \frac{5\pi}{6}$
iii $x = -\frac{2\pi}{3}, -\frac{\pi}{3}, \frac{\pi}{3}, \frac{2\pi}{3}$
c $x \approx 0.612, 3.75, 6.90$

6 $\sin\left(\frac{x}{2}\right) = \frac{\sqrt{7}}{2\sqrt{2}}$

7 **a** $x \approx 1.27, 5.02$ **b** $x = \frac{\pi}{12}, \frac{\pi}{4}, \frac{3\pi}{4}, \frac{11\pi}{12}, \frac{17\pi}{12}, \frac{19\pi}{12}$
c $x \approx 1.09, 2.05$

8 **a** 5000 **b** 3000, 7000
c $0.5 < t < 2.5$ and $6.5 < t \leqslant 8$

9 $x \approx 1.37, 5.44, 7.65$

10 $3\sin x + 4\cos x = 5\cos(x + 5.64)$ **11** 60 m

12 1.5 m **13** **a** 28 milligrams per m^3 **b** 8:00 am Monday

REVIEW SET 13C

1 **a** $x \approx -6.1, -3.4$ **b** $x \approx 0.8$

2 **a** $x = \frac{3\pi}{2}$ **b** $x = \frac{\pi}{6}, \frac{5\pi}{6}, \frac{7\pi}{6}, \frac{11\pi}{6}$

3 **a** $x = \frac{\pi}{2}, \frac{3\pi}{2}, \frac{5\pi}{2}, \frac{7\pi}{2}$ **b** $x = -\pi, -\frac{\pi}{3}, \pi, \frac{5\pi}{3}$

4 **a** $\cos\theta$ **b** $-\sin\theta$ **c** $5\cos^2\theta$ **d** $-\cos\theta$
e $\csc\theta$ **f** $\sin 2\theta$

5 $\sin\alpha = \frac{1}{\sqrt{5}}$

6 **a** $4\sin^2\alpha - 4\sin\alpha + 1$ **b** $1 - \sin 2\alpha$

8 **a** $x = \frac{\sqrt{3}}{2}$ **b** $x = 2 + \frac{1}{\sqrt{3}}$

EXERCISE 14A.1

1

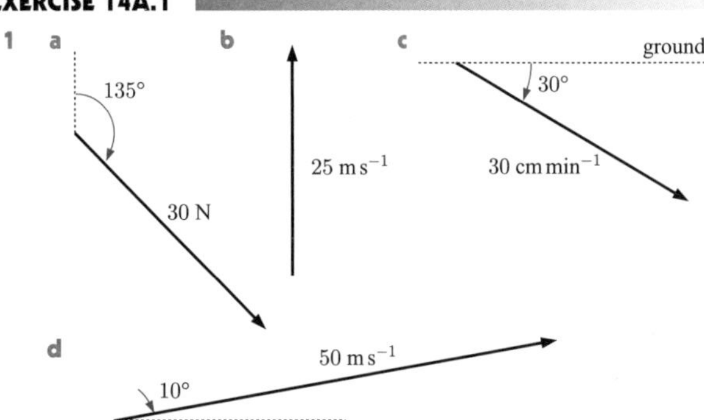

2 **a** 100 m s^{-1}

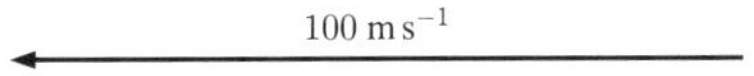

b 45°, 75 m s^{-1}

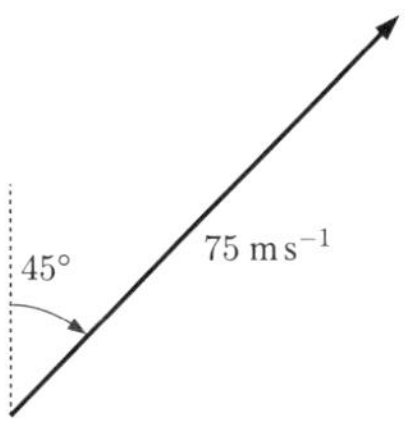

3 **a** 30 N, N, 45°, *Scale*: 1 cm ≡ 10 N

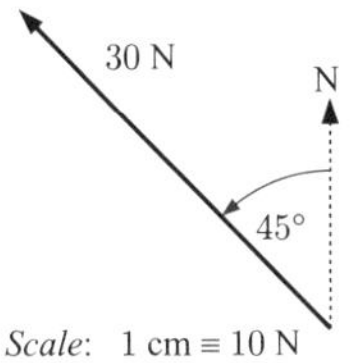

b 36 m s^{-1}, *Scale*: 1 cm ≡ 10 m s^{-1}

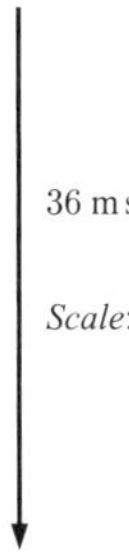

c y, 4 units, 15°, x, *Scale*: 1 cm ≡ 1 unit

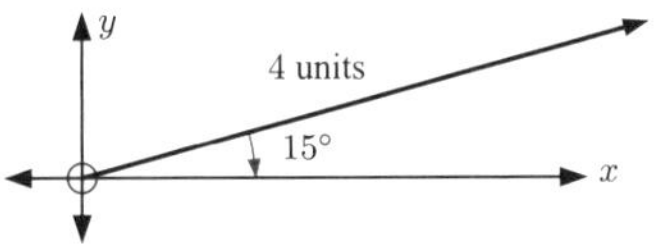

d *Scale*: 1 cm ≡ 30 km h^{-1}, 150 km h^{-1}, 8°

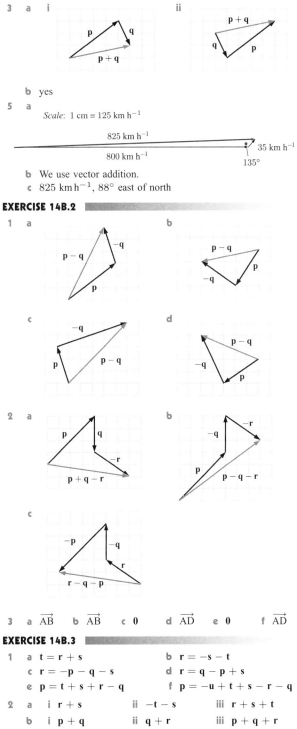

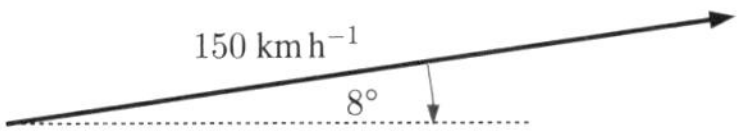

EXERCISE 14A.2

1 **a** **p**, **q**, **s**, **t** **b** **p**, **q**, **r**, **t** **c** **p** and **r**, **q** and **t**
d **q**, **t** **e** **p** and **q**, **p** and **t**

2 **a** true **b** true **c** false **d** false **e** true **f** false

3 **a** **i** $\overrightarrow{BC}$ **ii** $\overrightarrow{ED}$
b **i** $\overrightarrow{FE}$, $\overrightarrow{BC}$
ii $\overrightarrow{DE}$, $\overrightarrow{EF}$, $\overrightarrow{FE}$, $\overrightarrow{FA}$, $\overrightarrow{AF}$, $\overrightarrow{AB}$, $\overrightarrow{BA}$, $\overrightarrow{BC}$, $\overrightarrow{CB}$, $\overrightarrow{CD}$, $\overrightarrow{DC}$
c $\overrightarrow{FC}$ (or $\overrightarrow{CF}$)

EXERCISE 14B.1

1 **a** **q**, **p + q**, **p** **b** **p + q**, **p**, **q** **c** **p + q**, **q**, **p** **d** **p + q**, **q**, **p** **e** **q**, **p**, **p + q** **f** **p + q**, **q**, **p**

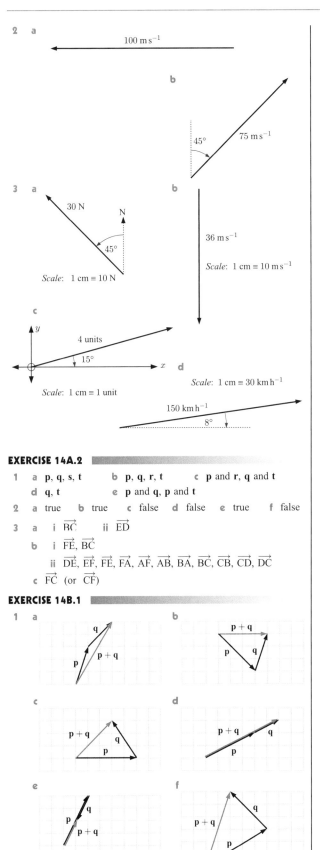

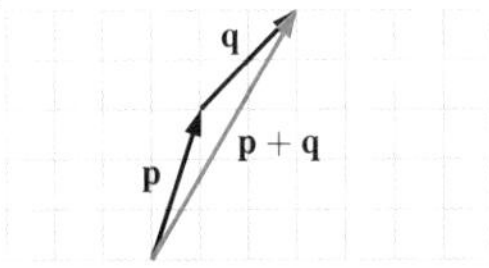

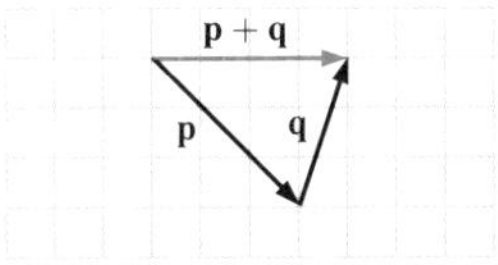

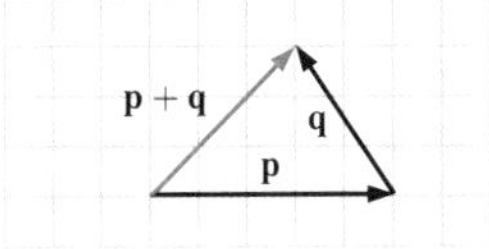

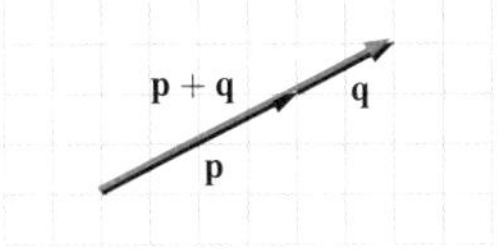

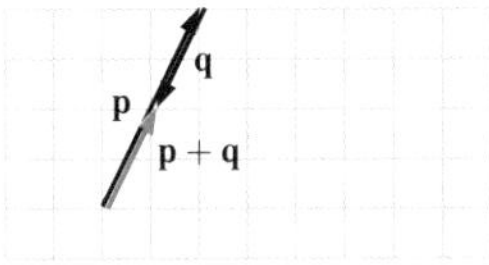

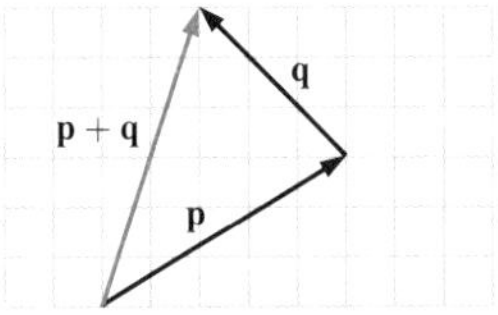

2 **a** $\overrightarrow{AC}$ **b** $\overrightarrow{BD}$ **c** **0** **d** $\overrightarrow{AD}$ **e** $\overrightarrow{AD}$ **f** **0**

3 **a** **i** **p**, **q**, **p + q** **ii** **p + q**, **q**, **p**
b yes

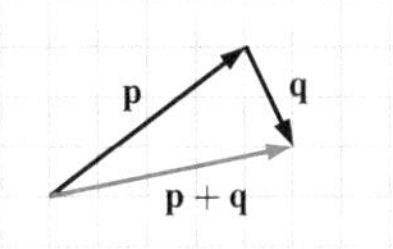

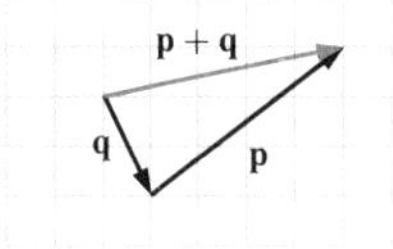

5 **a** *Scale*: 1 cm ≡ 125 km h^{-1} 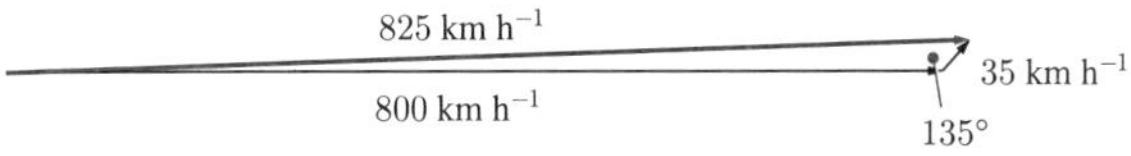
825 km h^{-1}, 35 km h^{-1}, 800 km h^{-1}, 135°
b We use vector addition.
c 825 km h^{-1}, 88° east of north

EXERCISE 14B.2

1 **a** **−q**, **p − q**, **p** **b** **p − q**, **p**, **−q** **c** **−q**, **p**, **p − q** **d** **p − q**, **−q**, **p**

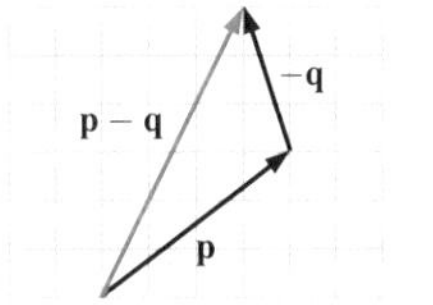

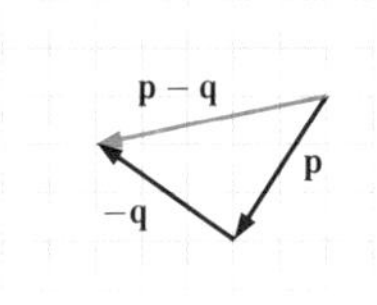

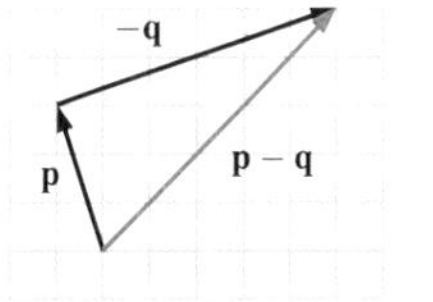

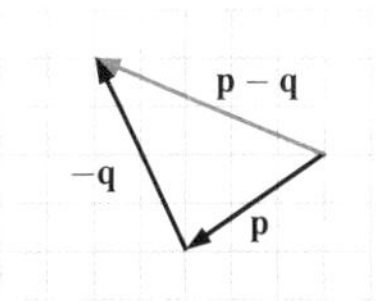

2 **a** **p**, **q**, **−r**, **p + q − r** **b** **−r**, **−q**, **p**, **p − q − r** **c** **−p**, **−q**, **r**, **r − q − p**

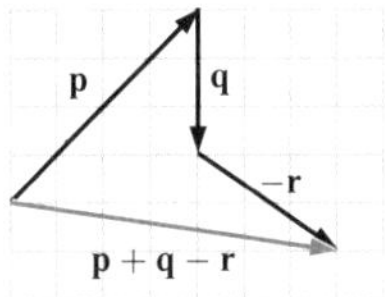

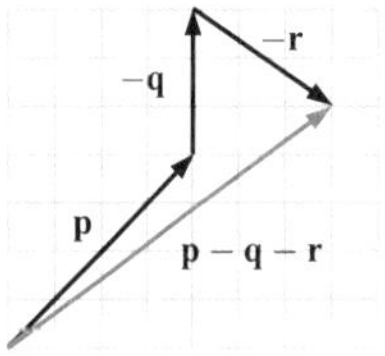

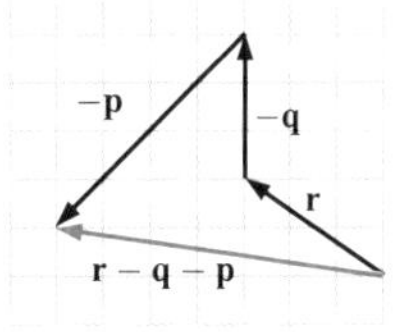

3 **a** $\overrightarrow{AB}$ **b** $\overrightarrow{AB}$ **c** **0** **d** $\overrightarrow{AD}$ **e** **0** **f** $\overrightarrow{AD}$

EXERCISE 14B.3

1 **a** $\mathbf{t}=\mathbf{r}+\mathbf{s}$ **b** $\mathbf{r}=-\mathbf{s}-\mathbf{t}$
c $\mathbf{r}=-\mathbf{p}-\mathbf{q}-\mathbf{s}$ **d** $\mathbf{r}=\mathbf{q}-\mathbf{p}+\mathbf{s}$
e $\mathbf{p}=\mathbf{t}+\mathbf{s}+\mathbf{r}-\mathbf{q}$ **f** $\mathbf{p}=-\mathbf{u}+\mathbf{t}+\mathbf{s}-\mathbf{r}-\mathbf{q}$

2 **a** **i** $\mathbf{r}+\mathbf{s}$ **ii** $-\mathbf{t}-\mathbf{s}$ **iii** $\mathbf{r}+\mathbf{s}+\mathbf{t}$
b **i** $\mathbf{p}+\mathbf{q}$ **ii** $\mathbf{q}+\mathbf{r}$ **iii** $\mathbf{p}+\mathbf{q}+\mathbf{r}$

EXERCISE 14B.4

1 **a** $-\mathbf{r}$ **b** $2\mathbf{s}$ **c** $\frac{1}{2}\mathbf{r}$ **d** $-\frac{3}{2}\mathbf{s}$

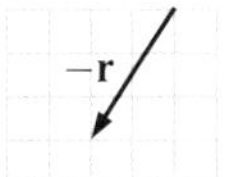

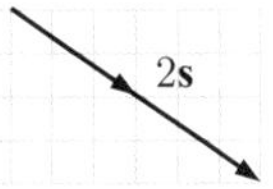

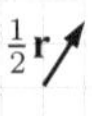

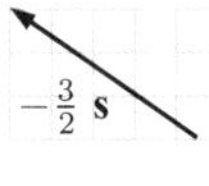

e

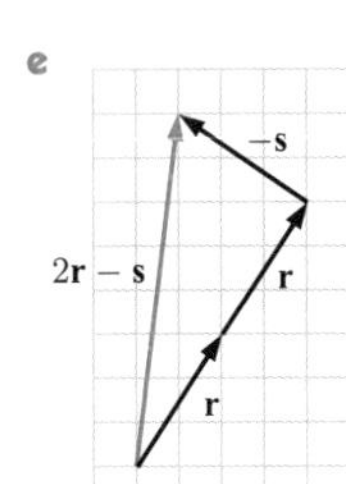

f

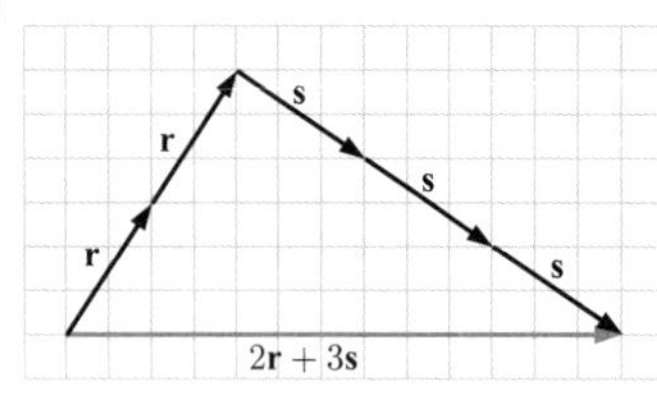

g

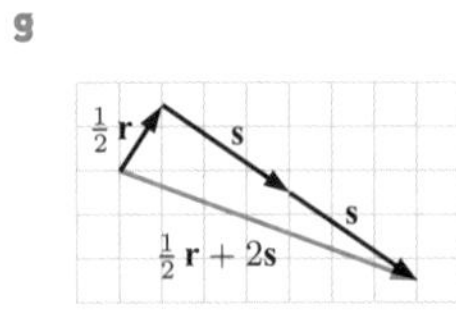

h

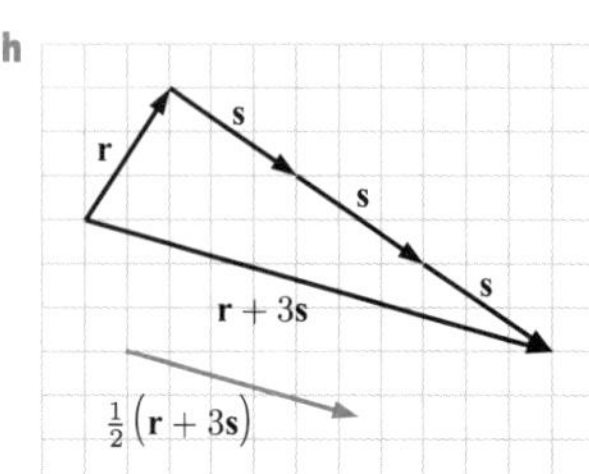

2 **a** p q **b** p q **c** p q $\mathbf{p} = 2\mathbf{q}$

d

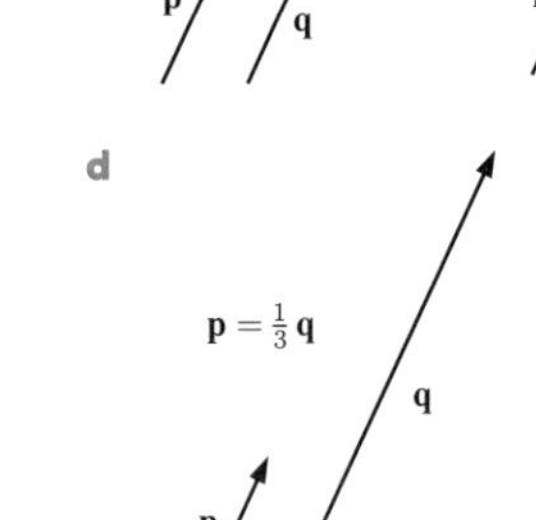

e

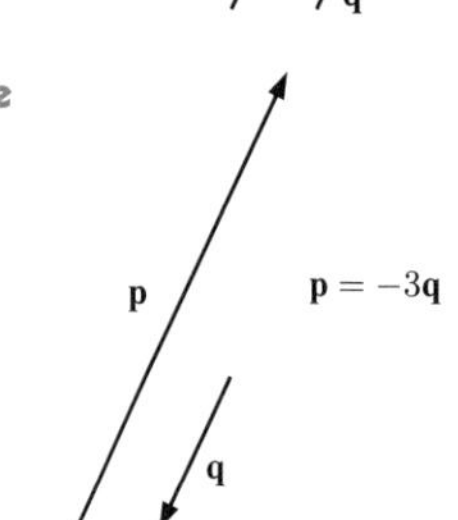

3 **a**

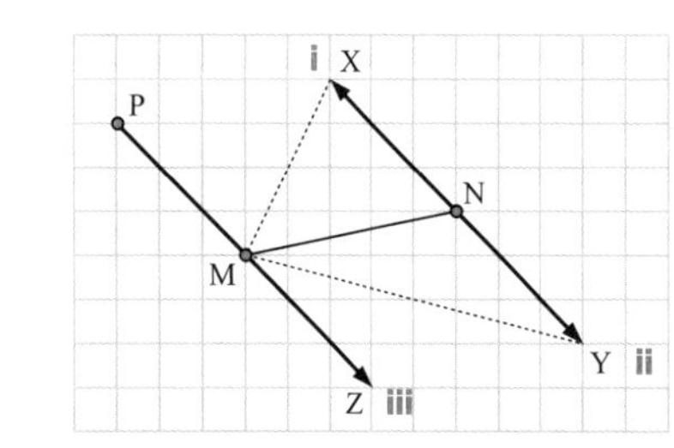

b a parallelogram

4 **a** $-\mathbf{p}$ **b** $\mathbf{p}+\mathbf{q}$ **c** $\frac{1}{2}(\mathbf{p}+\mathbf{q})$ **d** $\frac{1}{2}(\mathbf{q}-\mathbf{p})$

5 **a** $\mathbf{b}$ **b** $2\mathbf{b}$ **c** $\mathbf{b}-\mathbf{a}$ **d** $\mathbf{b}-\mathbf{a}$

EXERCISE 14C

1 **a** $\begin{pmatrix}7\\3\end{pmatrix}$, $7\mathbf{i}+3\mathbf{j}$ **b** $\begin{pmatrix}-6\\0\end{pmatrix}$, $-6\mathbf{i}$

c $\begin{pmatrix}2\\-5\end{pmatrix}$, $2\mathbf{i}-5\mathbf{j}$ **d** $\begin{pmatrix}0\\6\end{pmatrix}$, $6\mathbf{j}$

e $\begin{pmatrix}-6\\3\end{pmatrix}$, $-6\mathbf{i}+3\mathbf{j}$ **f** $\begin{pmatrix}-5\\-5\end{pmatrix}$, $-5\mathbf{i}-5\mathbf{j}$

2 **a** $3\mathbf{i}+4\mathbf{j}$ **b** $2\mathbf{i}$ **c** $2\mathbf{i}-5\mathbf{j}$ **d** $-\mathbf{i}-3\mathbf{j}$

4 3 2

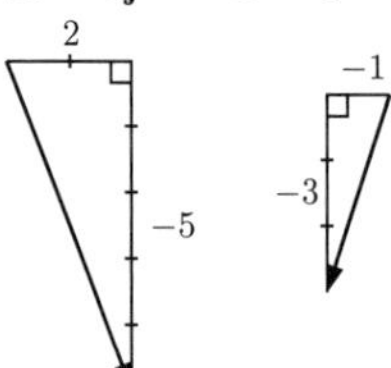

−1 −3

3 **a** $\begin{pmatrix}-4\\-1\end{pmatrix}$, $-4\mathbf{i}-\mathbf{j}$ **b** $\begin{pmatrix}-1\\-5\end{pmatrix}$, $-\mathbf{i}-5\mathbf{j}$

c $\begin{pmatrix}2\\0\end{pmatrix}$, $2\mathbf{i}$ **d** $\begin{pmatrix}3\\-4\end{pmatrix}$, $3\mathbf{i}-4\mathbf{j}$

e $\begin{pmatrix}-3\\4\end{pmatrix}$, $-3\mathbf{i}+4\mathbf{j}$ **f** $\begin{pmatrix}3\\5\end{pmatrix}$, $3\mathbf{i}+5\mathbf{j}$

4 **a** $\begin{pmatrix}1\\2\end{pmatrix}$ **b** $\begin{pmatrix}-1\\3\end{pmatrix}$ **c** $\begin{pmatrix}0\\-5\end{pmatrix}$ **d** $\begin{pmatrix}4\\-2\end{pmatrix}$

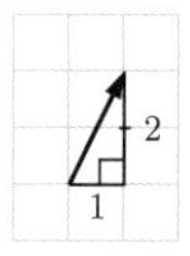

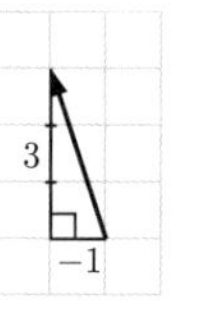

−5

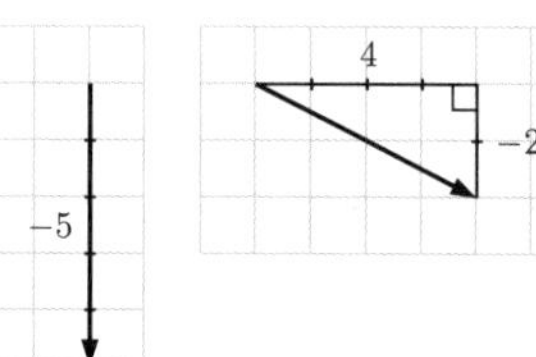

5 $\begin{pmatrix}0\\0\end{pmatrix}$

EXERCISE 14D

1 **a** 5 units **b** 5 units **c** 2 units **d** $\sqrt{8}$ units **e** 3 units

2 **a** $\sqrt{2}$ units **b** 13 units **c** $\sqrt{17}$ units **d** 3 units **e** $|k|$ units

3 **a** unit vector **b** unit vector **c** not a unit vector **d** unit vector **e** not a unit vector

4 **a** $k=\pm1$ **b** $k=\pm1$ **c** $k=0$ **d** $k=\pm\frac{1}{\sqrt{2}}$ **e** $k=\pm\frac{\sqrt{3}}{2}$

5 $p=\pm3$

EXERCISE 14E

1 **a** $\begin{pmatrix}-2\\6\end{pmatrix}$ **b** $\begin{pmatrix}-2\\6\end{pmatrix}$ **c** $\begin{pmatrix}-1\\-1\end{pmatrix}$ **d** $\begin{pmatrix}-1\\-1\end{pmatrix}$

e $\begin{pmatrix}-5\\-3\end{pmatrix}$ **f** $\begin{pmatrix}-5\\-3\end{pmatrix}$ **g** $\begin{pmatrix}-6\\4\end{pmatrix}$ **h** $\begin{pmatrix}-4\\1\end{pmatrix}$

2 **a** $\begin{pmatrix}-3\\7\end{pmatrix}$ **b** $\begin{pmatrix}-4\\-3\end{pmatrix}$ **c** $\begin{pmatrix}-8\\-1\end{pmatrix}$ **d** $\begin{pmatrix}-6\\9\end{pmatrix}$

e $\begin{pmatrix}0\\-5\end{pmatrix}$ **f** $\begin{pmatrix}6\\-9\end{pmatrix}$

3 **a** $\mathbf{a}+\mathbf{0}=\begin{pmatrix}a_1\\a_2\end{pmatrix}+\begin{pmatrix}0\\0\end{pmatrix}=\begin{pmatrix}a_1+0\\a_2+0\end{pmatrix}=\begin{pmatrix}a_1\\a_2\end{pmatrix}=\mathbf{a}$

b $\mathbf{a}-\mathbf{a}=\begin{pmatrix}a_1\\a_2\end{pmatrix}-\begin{pmatrix}a_1\\a_2\end{pmatrix}=\begin{pmatrix}a_1-a_1\\a_2-a_2\end{pmatrix}=\begin{pmatrix}0\\0\end{pmatrix}=\mathbf{0}$

4 **a** $\begin{pmatrix}-3\\-15\end{pmatrix}$ **b** $\begin{pmatrix}-1\\2\end{pmatrix}$ **c** $\begin{pmatrix}0\\14\end{pmatrix}$ **d** $\begin{pmatrix}5\\-3\end{pmatrix}$

e $\begin{pmatrix}\frac{5}{2}\\\frac{11}{2}\end{pmatrix}$ **f** $\begin{pmatrix}-7\\7\end{pmatrix}$ **g** $\begin{pmatrix}5\\11\end{pmatrix}$ **h** $\begin{pmatrix}3\\\frac{17}{3}\end{pmatrix}$

5 **a** $\begin{pmatrix}8\\-1\end{pmatrix}$ **b** $\begin{pmatrix}8\\-1\end{pmatrix}$ **c** $\begin{pmatrix}8\\-1\end{pmatrix}$

6 **a** $\sqrt{13}$ units **b** $\sqrt{17}$ units **c** $5\sqrt{2}$ units **d** $\sqrt{10}$ units **e** $\sqrt{29}$ units

7 **a** $\sqrt{10}$ units **b** $2\sqrt{10}$ units **c** $2\sqrt{10}$ units **d** $3\sqrt{10}$ units **e** $3\sqrt{10}$ units **f** $2\sqrt{5}$ units **g** $8\sqrt{5}$ units **h** $8\sqrt{5}$ units **i** $\sqrt{5}$ units **j** $\sqrt{5}$ units

EXERCISE 14F

1 a $\begin{pmatrix}2\\4\end{pmatrix}$ b $\begin{pmatrix}-2\\5\end{pmatrix}$ c $\begin{pmatrix}3\\-3\end{pmatrix}$ d $\begin{pmatrix}1\\-5\end{pmatrix}$
e $\begin{pmatrix}6\\-5\end{pmatrix}$ f $\begin{pmatrix}1\\3\end{pmatrix}$

2 a (4, 2) b (2, 2) 3 a $\begin{pmatrix}2\\1\end{pmatrix}$ b (3, 3)

4 a $\begin{pmatrix}5\\1\end{pmatrix}$ b $\begin{pmatrix}-5\\-1\end{pmatrix}$ c D(−1, −2)

5 a $\overrightarrow{AB} = \begin{pmatrix}4\\k-3\end{pmatrix}$, $|\overrightarrow{AB}| = \sqrt{16+(k-3)^2} = 5$ units
b $k = 0$ or 6

6 a $\overrightarrow{AB} = \begin{pmatrix}2\\3\end{pmatrix}$, $\overrightarrow{AC} = \begin{pmatrix}3\\-3\end{pmatrix}$
b $\overrightarrow{BC} = \overrightarrow{BA} + \overrightarrow{AC} = -\overrightarrow{AB} + \overrightarrow{AC}$ c $\overrightarrow{BC} = \begin{pmatrix}1\\-6\end{pmatrix}$

7 a $\begin{pmatrix}-5\\4\end{pmatrix}$ b $\begin{pmatrix}1\\2\end{pmatrix}$ c $\begin{pmatrix}6\\-5\end{pmatrix}$

8 a M(1, 4) b $\overrightarrow{CA} = \begin{pmatrix}7\\5\end{pmatrix}$, $\overrightarrow{CM} = \begin{pmatrix}5\\3\end{pmatrix}$, $\overrightarrow{CB} = \begin{pmatrix}3\\1\end{pmatrix}$

EXERCISE 14G

1 a
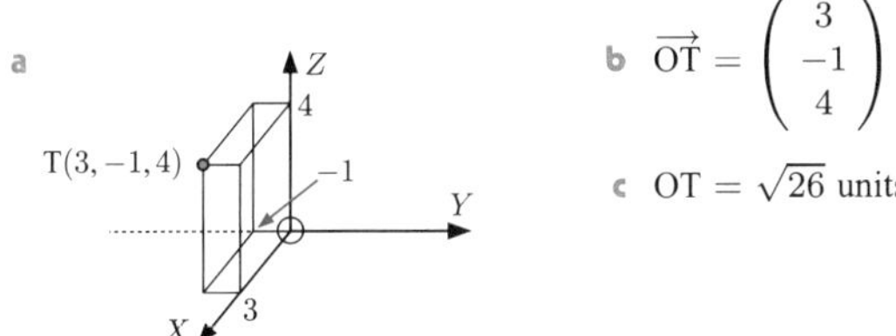

b $\overrightarrow{OT} = \begin{pmatrix}3\\-1\\4\end{pmatrix}$
c OT $= \sqrt{26}$ units

2 a
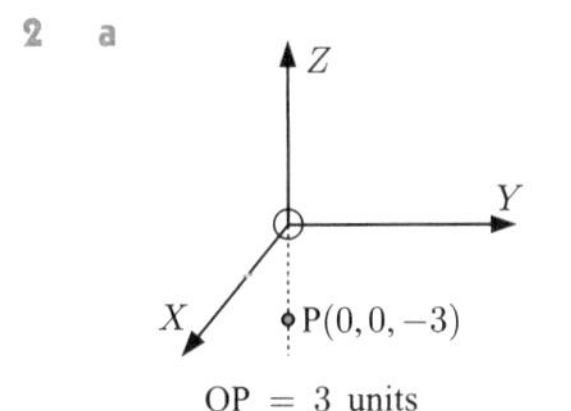

b
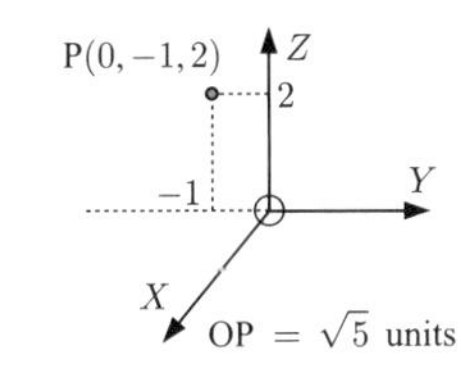

c
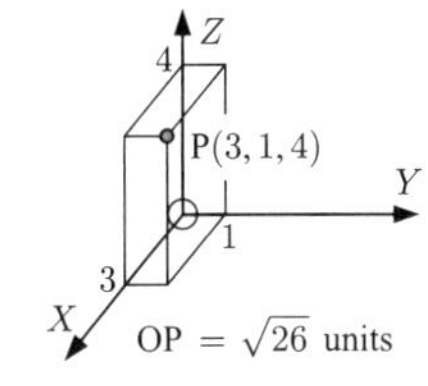

d
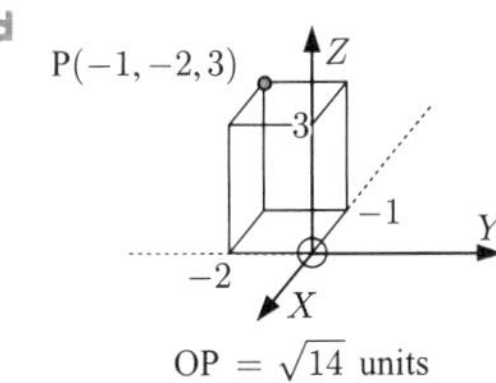

3 a $\overrightarrow{AB} = \begin{pmatrix}4\\-1\\-3\end{pmatrix}$, $\overrightarrow{BA} = \begin{pmatrix}-4\\1\\3\end{pmatrix}$
b AB $= \sqrt{26}$ units, BA $= \sqrt{26}$ units

4 $\overrightarrow{OA} = \begin{pmatrix}3\\1\\0\end{pmatrix}$, $\overrightarrow{OB} = \begin{pmatrix}-1\\1\\2\end{pmatrix}$, $\overrightarrow{AB} = \begin{pmatrix}-4\\0\\2\end{pmatrix}$

5 a $\overrightarrow{NM} = \begin{pmatrix}5\\-4\\-1\end{pmatrix}$ b $\overrightarrow{MN} = \begin{pmatrix}-5\\4\\1\end{pmatrix}$
c MN $= \sqrt{42}$ units

6 a $\overrightarrow{OA} = \begin{pmatrix}-1\\2\\5\end{pmatrix}$, OA $= \sqrt{30}$ units
b $\overrightarrow{AB} = \begin{pmatrix}3\\-2\\-2\end{pmatrix}$, AB $= \sqrt{17}$ units
c $\overrightarrow{AC} = \begin{pmatrix}-2\\-1\\-5\end{pmatrix}$, AC $= \sqrt{30}$ units
d $\overrightarrow{CB} = \begin{pmatrix}5\\-1\\3\end{pmatrix}$, CB $= \sqrt{35}$ units
e ABC is scalene, and not right angled.

7 a $\sqrt{13}$ units b $\sqrt{14}$ units c 3 units

9 a right angled b straight line (not a triangle)

10 a $|\overrightarrow{AB}| = \sqrt{158}$ units, $|\overrightarrow{BC}| = \sqrt{129}$ units,
$|\overrightarrow{AC}| = \sqrt{29}$ units, and $29 + 129 = 158$
b area ≈ 30.6 units^2

11 (0, 3, 5), $r = \sqrt{3}$ units

12 a $(0, y, 0)$ b (0, 2, 0) and (0, −4, 0)

13 a $a = 5$, $b = 6$, $c = -6$ b $a = 4$, $b = 2$, $c = 1$

14 a $k = \pm\frac{\sqrt{11}}{4}$ b $k = \pm\frac{2}{3}$

15 a $r = 2$, $s = 4$, $t = -7$ b $r = -4$, $s = 0$, $t = 3$

16 a $\overrightarrow{AB} = \begin{pmatrix}2\\-5\\-1\end{pmatrix}$, $\overrightarrow{DC} = \begin{pmatrix}2\\-5\\-1\end{pmatrix}$
b ABCD is a parallelogram.

17 a S(−2, 8, −3) b midpoints are at $(-\frac{1}{2}, 3, 1)$

EXERCISE 14H

1 a $\mathbf{x} = \frac{1}{2}\mathbf{q}$ b $\mathbf{x} = 2\mathbf{n}$ c $\mathbf{x} = -\frac{1}{3}\mathbf{p}$
d $\mathbf{x} = \frac{1}{2}(\mathbf{r} - \mathbf{q})$ e $\mathbf{x} = \frac{1}{5}(4\mathbf{s} - \mathbf{t})$ f $\mathbf{x} = 3(4\mathbf{m} - \mathbf{n})$

2 a $\mathbf{x} = \begin{pmatrix}4\\-6\\-5\end{pmatrix}$ b $\mathbf{x} = \begin{pmatrix}1\\-\frac{2}{3}\\\frac{5}{3}\end{pmatrix}$ c $\mathbf{x} = \begin{pmatrix}\frac{3}{2}\\-1\\\frac{5}{2}\end{pmatrix}$

3 $\overrightarrow{AB} = \begin{pmatrix}3\\4\\-2\end{pmatrix}$, AB $= \sqrt{29}$ units

4 a $\overrightarrow{AB} = 4\mathbf{i} - 5\mathbf{j} - \mathbf{k}$ b $\sqrt{42}$ units

5 a $\sqrt{10}$ b $\sqrt{6}$ c $2\sqrt{10}$ d $2\sqrt{10}$
e $-3\sqrt{6}$ f $3\sqrt{6}$ g $3\sqrt{2}$ h $\sqrt{14}$

6 $\overrightarrow{AC} = -\mathbf{i} - 2\mathbf{k}$

8 C(5, 1, −8), D(8, −1, −13), E(11, −3, −18)

9 a parallelogram b parallelogram c not parallelogram

10 a D(9, −1) b R(3, 1, 6) c X(2, −1, 0)

11 a $\overrightarrow{BD} = \frac{1}{2}\mathbf{a}$ b $\overrightarrow{AB} = \mathbf{b} - \mathbf{a}$ c $\overrightarrow{BA} = -\mathbf{b} + \mathbf{a}$
d $\overrightarrow{OD} = \mathbf{b} + \frac{1}{2}\mathbf{a}$ e $\overrightarrow{AD} = \mathbf{b} - \frac{1}{2}\mathbf{a}$ f $\overrightarrow{DA} = \frac{1}{2}\mathbf{a} - \mathbf{b}$

12 a $\begin{pmatrix}-1\\5\\-1\end{pmatrix}$ b $\begin{pmatrix}-3\\4\\-2\end{pmatrix}$ c $\begin{pmatrix}-3\\6\\-5\end{pmatrix}$

13 **a** $\begin{pmatrix}3\\1\\-2\end{pmatrix}$ **b** $\begin{pmatrix}1\\-3\\4\end{pmatrix}$ **c** $\begin{pmatrix}1\\4\\-9\end{pmatrix}$

d $\begin{pmatrix}-1\\\frac{3}{2}\\-\frac{7}{2}\end{pmatrix}$ **e** $\begin{pmatrix}1\\-4\\7\end{pmatrix}$ **f** $\begin{pmatrix}4\\2\\-2\end{pmatrix}$

14 **a** $\sqrt{11}$ units **b** $\sqrt{14}$ units **c** $\sqrt{38}$ units

d $\sqrt{3}$ units **e** $\begin{pmatrix}\sqrt{11}\\-3\sqrt{11}\\2\sqrt{11}\end{pmatrix}$ **f** $\begin{pmatrix}-\frac{1}{\sqrt{11}}\\\frac{1}{\sqrt{11}}\\\frac{3}{\sqrt{11}}\end{pmatrix}$

15 **a** $a=\frac{1}{3},\ b=2,\ c=1$ **b** $a=1,\ b=-1,\ c=2$

c $a=4,\ b=-1$

EXERCISE 14I

1 $r=3,\ s=-9$ **2** $a=-6,\ b=-4$

3 **a** $\overrightarrow{AB}\parallel\overrightarrow{CD}$, AB = 3CD

b $\overrightarrow{RS}\parallel\overrightarrow{KL}$, RS = $\frac{1}{2}$KL opposite direction

c A, B, and C are collinear and AB = 2BC

4 **a** $\overrightarrow{PR}=\begin{pmatrix}-1\\-3\\3\end{pmatrix}$, $\overrightarrow{QS}=\begin{pmatrix}-2\\-6\\6\end{pmatrix}$, $2\overrightarrow{PR}=\overrightarrow{QS}$

b PR = $\frac{1}{2}$QS

5 **a** $\begin{pmatrix}4\\8\end{pmatrix}$ **b** $\begin{pmatrix}-1\\-2\end{pmatrix}$

6 **a** $\frac{1}{\sqrt{5}}(\mathbf{i}+2\mathbf{j})$ **b** $\frac{1}{\sqrt{13}}(2\mathbf{i}-3\mathbf{k})$ **c** $\frac{1}{3}(2\mathbf{i}-2\mathbf{j}+\mathbf{k})$

7 **a** $\frac{3}{\sqrt{5}}\begin{pmatrix}2\\-1\end{pmatrix}$ **b** $\frac{2}{\sqrt{17}}\begin{pmatrix}1\\4\end{pmatrix}$

8 **a** $\overrightarrow{AB}=\begin{pmatrix}2\sqrt{2}\\-2\sqrt{2}\end{pmatrix}$ **b** $\overrightarrow{OB}=\begin{pmatrix}3+2\sqrt{2}\\2-2\sqrt{2}\end{pmatrix}$

c $B(3+2\sqrt{2},\ 2-2\sqrt{2})$

9 **a** $\pm\frac{1}{3}\begin{pmatrix}2\\-1\\-2\end{pmatrix}$ **b** $\pm\frac{2}{3}\begin{pmatrix}-2\\-1\\2\end{pmatrix}$

10 **a** $\sqrt{2}\begin{pmatrix}-1\\4\\1\end{pmatrix}$ **b** $\frac{5}{3}\begin{pmatrix}1\\2\\2\end{pmatrix}$

11 **c** $a=7,\ b=-1$ **d** $a=-\frac{7}{2},\ b=-\frac{21}{2}$

EXERCISE 14J

1 **a** 7 **b** 22 **c** 29 **d** 66 **e** 52 **f** 3 **g** 5 **h** 1

2 **a** 2 **b** 2 **c** 14 **d** 14 **e** 4 **f** 4

3 **a** -1 **b** $94.1°$ **4** **a** $\approx 140°$ **b** $\approx 114°$

5 **a** 1 **b** 1 **c** 0 **6** **a** 5 **b** -9

7 **a** **i** ± 12 **ii** 6

b **i** $\mathbf{a}\bullet\mathbf{b}$ is not 0 **ii** 12 units

c **i** $\mathbf{c}=\mathbf{d}$ **ii** $\mathbf{c}=-\mathbf{d}$

8 **a** $(\cos\theta,\ \sin\theta)$

b $\overrightarrow{BP}=\begin{pmatrix}\cos\theta+1\\\sin\theta\end{pmatrix}$, $\overrightarrow{AP}=\begin{pmatrix}\cos\theta-1\\\sin\theta\end{pmatrix}$

c $\overrightarrow{AP}\bullet\overrightarrow{BP}=\cos^2\theta+\sin^2\theta-1$

$\therefore\ \overrightarrow{AP}\bullet\overrightarrow{BP}=0$

d The angle in a semi-circle is a right angle.

10 **a** **i** $t=6$ **ii** $t=-\frac{3}{2}$

b **i** $t=-8$ **ii** $t=-\frac{6}{7}$

c **i** $t=0$ or 2 **ii** $t=\frac{-1\pm\sqrt{5}}{2}$

11 **b** **Hint:** Show $\mathbf{a}\bullet\mathbf{b}=\mathbf{b}\bullet\mathbf{c}=\mathbf{a}\bullet\mathbf{c}=0$

c **i** $t=-\frac{3}{2}$ **ii** $t=-\frac{5}{6}$

12 **a** $\widehat{BAC}$ is a right angle **b** not right angled

c $\widehat{BAC}$ is a right angle **d** $\widehat{ACB}$ is a right angle

13 $\overrightarrow{AB}\bullet\overrightarrow{AC}=0$, $\therefore$ $\widehat{BAC}$ is a right angle

14 **b** $|\overrightarrow{AB}|=\sqrt{14}$ units, $|\overrightarrow{BC}|=\sqrt{14}$ units, ABCD is a rhombus

c 0, the diagonals of a rhombus are perpendicular

15 **a** $k\begin{pmatrix}-2\\5\end{pmatrix},\ k\neq 0$ **b** $k\begin{pmatrix}2\\-1\end{pmatrix},\ k\neq 0$

c $k\begin{pmatrix}1\\3\end{pmatrix},\ k\neq 0$ **d** $k\begin{pmatrix}3\\4\end{pmatrix},\ k\neq 0$

e $k\begin{pmatrix}0\\1\end{pmatrix},\ k\neq 0$

16 **Hint:** Choose a vector $\begin{pmatrix}a\\b\\c\end{pmatrix}$, where a and b are integers.

Solve for c such that $\begin{pmatrix}a\\b\\c\end{pmatrix}\bullet\begin{pmatrix}1\\2\\-1\end{pmatrix}=0.$

17 $\widehat{ABC}\approx 62.5°$, the exterior angle $\approx 117.5°$

18 **a** $54.7°$ **b** $60°$ **c** $35.3°$

19 **a** $30.3°$ **b** $54.2°$ **20** **a** $M(\frac{3}{2},\frac{5}{2},\frac{3}{2})$ **b** $51.5°$

21 **a** $t=0$ or -3 **b** $r=-2,\ s=5,\ t=-4$

22 **a** $74.5°$ **b** $72.5°$

23 $\mathbf{a}=\begin{pmatrix}1\\0\\0\end{pmatrix}$, $\mathbf{b}=\begin{pmatrix}0\\1\\0\end{pmatrix}$, $\mathbf{c}=\begin{pmatrix}0\\0\\1\end{pmatrix}$ will do

$\mathbf{a}\bullet\mathbf{b}=\mathbf{a}\bullet\mathbf{c}$, but $\mathbf{b}\neq\mathbf{c}$

25 **a** **Hint:** Square both sides.

b Consider the parallelogram.

Find $\overrightarrow{AB}$ and $\overrightarrow{OC}$, etc.

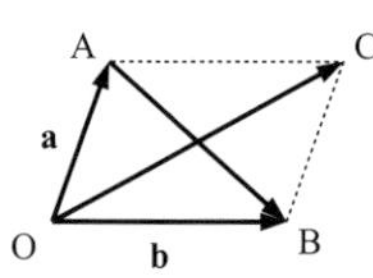

26 -7 **27** $\mathbf{a}\bullet\mathbf{b}$ is a scalar and so $\mathbf{a}\bullet\mathbf{b}\bullet\mathbf{c}$ is a scalar 'dotted' with a vector which is meaningless.

EXERCISE 14K.1

1 **a** $\begin{pmatrix}2\\5\\11\end{pmatrix}$ **b** $\begin{pmatrix}2\\4\\1\end{pmatrix}$ **c** $-\mathbf{i}-\mathbf{j}-\mathbf{k}$ **d** $\mathbf{i}-6\mathbf{j}+2\mathbf{k}$

2 **a** $\mathbf{a}\times\mathbf{b}=\begin{pmatrix}-11\\-2\\5\end{pmatrix}$ **b** $\mathbf{a}\bullet(\mathbf{a}\times\mathbf{b})=0=\mathbf{b}\bullet(\mathbf{a}\times\mathbf{b})$

c $\mathbf{a}\times\mathbf{b}$ is a vector perpendicular to both **a** and **b**

3 **a** $\mathbf{i}\times\mathbf{i}=\mathbf{0}$, $\mathbf{j}\times\mathbf{j}=\mathbf{0}$, $\mathbf{k}\times\mathbf{k}=\mathbf{0}$,

$\mathbf{a}\times\mathbf{a}=\mathbf{0}$ for all vectors **a**.

b **i** $\mathbf{i}\times\mathbf{j}=\mathbf{k}$, $\mathbf{j}\times\mathbf{i}=-\mathbf{k}$ **ii** $\mathbf{j}\times\mathbf{k}=\mathbf{i}$, $\mathbf{k}\times\mathbf{j}=-\mathbf{i}$

iii $\mathbf{i}\times\mathbf{k}=-\mathbf{j}$, $\mathbf{k}\times\mathbf{i}=\mathbf{j}$

$\mathbf{a}\times\mathbf{b}=-\mathbf{b}\times\mathbf{a}$ for all vectors **a** and **b**.

5 **a** $\begin{pmatrix}1\\4\\2\end{pmatrix}$ **b** 17

6 **a** $\begin{pmatrix}2\\-1\\-1\end{pmatrix}$ **b** $\begin{pmatrix}0\\5\\0\end{pmatrix}$ **c** $\begin{pmatrix}2\\4\\-1\end{pmatrix}$ **d** $\begin{pmatrix}2\\4\\-1\end{pmatrix}$

9 **a** $\mathbf{a}\times\mathbf{b}$ **b** $\mathbf{0}$ **c** $2(\mathbf{b}\times\mathbf{a})$ **d** $\mathbf{0}$

10 **a** $k\begin{pmatrix}-4\\1\\3\end{pmatrix}$ **b** $k\begin{pmatrix}6\\22\\-15\end{pmatrix}$ **c** $(-\mathbf{i}+\mathbf{j}-2\mathbf{k})n$

d $(5\mathbf{i}+\mathbf{j}+4\mathbf{k})n,\quad n, k\in\mathbb{R},\quad n, k\neq 0$

11 $k\begin{pmatrix}4\\-5\\-7\end{pmatrix},\ k\neq 0,\quad \frac{\sqrt{10}}{6}\begin{pmatrix}4\\-5\\-7\end{pmatrix}$ and $-\frac{\sqrt{10}}{6}\begin{pmatrix}4\\-5\\-7\end{pmatrix}$

12 **a** $\begin{pmatrix}2\\5\\-1\end{pmatrix}$ **b** $\begin{pmatrix}2\\0\\1\end{pmatrix}$

EXERCISE 14K.2

1 **a** $\mathbf{i}\times\mathbf{k}=-\mathbf{j},\quad \mathbf{k}\times\mathbf{i}=\mathbf{j}$

2 **a** $\mathbf{a}\bullet\mathbf{b}=-1,\quad \mathbf{a}\times\mathbf{b}=\begin{pmatrix}1\\5\\1\end{pmatrix}$

b $\cos\theta=-\frac{1}{\sqrt{28}}$ **c** $\sin\theta=\frac{\sqrt{27}}{\sqrt{28}}$ **d** $\sin\theta=\frac{\sqrt{27}}{\sqrt{28}}$

4 **a** **i** $\overrightarrow{OA}=\begin{pmatrix}2\\3\\-1\end{pmatrix},\quad \overrightarrow{OB}=\begin{pmatrix}-1\\1\\2\end{pmatrix}$

ii $\overrightarrow{OA}\times\overrightarrow{OB}=\begin{pmatrix}7\\-3\\5\end{pmatrix}$ **iii** $|\overrightarrow{OA}\times\overrightarrow{OB}|=\sqrt{83}$

b Area $\triangle OAB=\frac{1}{2}|\overrightarrow{OA}||\overrightarrow{OB}|\sin\theta$

$=\frac{1}{2}|\overrightarrow{OA}\times\overrightarrow{OB}|=\frac{\sqrt{83}}{2}$ units^2

5 **a** $\overrightarrow{OC}$ is parallel to $\overrightarrow{AB}$ **b** $\mathbf{a}\times\mathbf{b}=\mathbf{b}\times\mathbf{c}$

REVIEW SET 14A

1 **a**

b

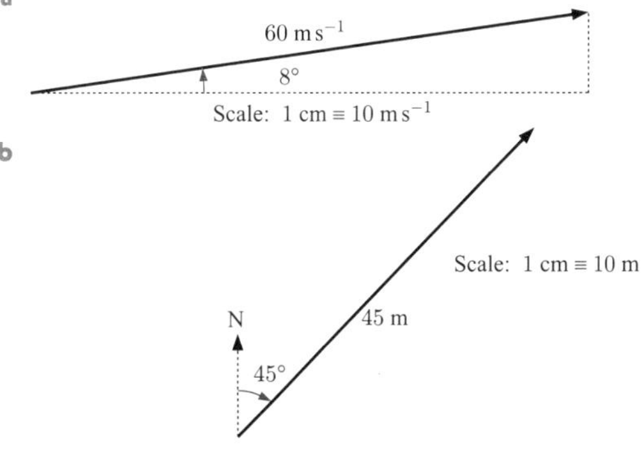

2 **a** $\overrightarrow{AC}$ **b** $\overrightarrow{AD}$

3 **a** $\mathbf{q}=\mathbf{p}+\mathbf{r}$ **b** $\mathbf{l}=\mathbf{k}-\mathbf{j}+\mathbf{n}-\mathbf{m}$ **4** $\begin{pmatrix}1\\4\end{pmatrix}$

5 **a** $\mathbf{p}+\mathbf{q}$ **b** $\frac{3}{2}\mathbf{p}+\frac{1}{2}\mathbf{q}$ **6** $m=5,\ n=-\frac{1}{2}$

7 $\begin{pmatrix}8\\-8\\7\end{pmatrix}$ **8** **a** -13 **b** -36 **10** $k=6$

11 **a** $\mathbf{a}\bullet\mathbf{b}$ is a scalar, so $\mathbf{a}\bullet\mathbf{b}\bullet\mathbf{c}$ is a scalar dotted with a vector, which is meaningless.

b $\mathbf{b}\times\mathbf{c}$ must be done first otherwise we have the cross product of a scalar with a vector, which is meaningless.

12 $t\begin{pmatrix}5\\4\end{pmatrix},\ t\neq 0$ **13** **a** **i** $\mathbf{p}+\mathbf{q}$ **ii** $\frac{1}{2}\mathbf{p}+\frac{1}{2}\mathbf{q}$

14 **a** $k=\pm\frac{7}{\sqrt{33}}$ **b** $k=\pm\frac{1}{\sqrt{2}}$

15 **a** $\mathbf{a}\bullet\mathbf{b}=-4$ **b** $\mathbf{b}\bullet\mathbf{c}=10$ **c** $\mathbf{a}\bullet\mathbf{c}=-10$

16 $a=-2,\ b=0$

17 If θ is acute, $\mathbf{u}\bullet\mathbf{v}=\sqrt{199}$; if θ is obtuse, $\mathbf{u}\bullet\mathbf{v}=-\sqrt{199}$.

18 **a** **i** $\mathbf{q}+\mathbf{r}$ **ii** $\mathbf{r}+\mathbf{q}$ **b** DB = AC, [DB] ∥ [AC]

19 **a** $t=-4$ **b** $\overrightarrow{LM}=\begin{pmatrix}5\\-3\\-4\end{pmatrix},\quad \overrightarrow{KM}=\begin{pmatrix}-2\\-2\\-1\end{pmatrix}$

$\therefore\ \widehat{M}=90^\circ$

REVIEW SET 14B

1 **a** **b**

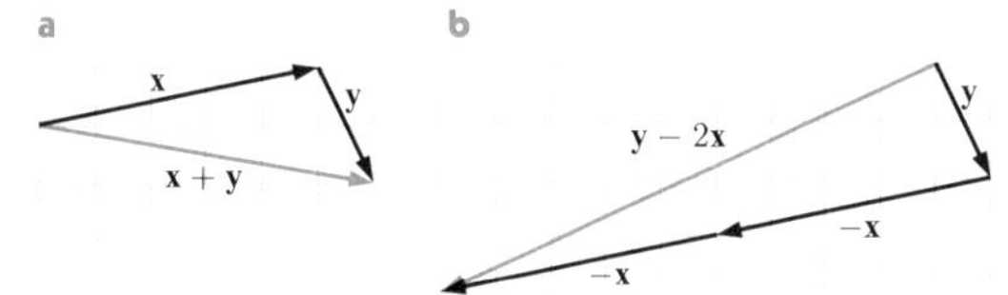

2 AB = AC = $\sqrt{53}$ units and BC = $\sqrt{46}$ units
$\therefore$ △ is isosceles

3 **a** $\sqrt{13}$ units **b** $\sqrt{10}$ units **c** $\sqrt{109}$ units

4 $r=4,\ s=7$

5 **a** $\begin{pmatrix}-6\\1\\3\end{pmatrix}$ **b** $\sqrt{46}$ units **c** $(-1,\ 3\frac{1}{2},\ \frac{1}{2})$ **6** $c=\frac{50}{3}$

7 $\begin{pmatrix}1\\-\frac{5}{3}\\-\frac{2}{3}\end{pmatrix}$ **8** 64.0° **9** (0, 0, 1) and (0, 0, 9)

10 $t=\frac{2}{3}$ or -3 **12** **a** 8 **b** $\approx 62.2^\circ$

13 **a** $\overrightarrow{AC}=-\mathbf{p}+\mathbf{r},\quad \overrightarrow{BC}=-\mathbf{q}+\mathbf{r}$

14 $\pm\frac{4}{\sqrt{14}}(3\mathbf{i}-2\mathbf{j}+\mathbf{k})$ **15** $\approx 16.1^\circ$ **16** $k\begin{pmatrix}0\\-2\\1\end{pmatrix}$

17 **a** $k=\pm\frac{1}{2}$ **b** $-\frac{5}{\sqrt{14}}\begin{pmatrix}3\\2\\-1\end{pmatrix}$

18 $\approx 80.3^\circ$ **19** $\approx 26.4^\circ$

REVIEW SET 14C

1 **a** $\overrightarrow{PQ}$ **b** $\overrightarrow{PR}$

2 **a** $\begin{pmatrix}3\\-3\\11\end{pmatrix}$ **b** $\begin{pmatrix}7\\-3\\-26\end{pmatrix}$ **c** $\sqrt{74}$ units

3 **a** AB = $\frac{1}{2}$CD, [AB] ∥ [CD] **b** C is the midpoint of [AB].

4 **a** $\overrightarrow{PQ}=\begin{pmatrix}-3\\12\\3\end{pmatrix}$ **b** $\sqrt{162}$ units **c** $\sqrt{61}$ units

5 **a** $\mathbf{r}+\mathbf{q}$ **b** $-\mathbf{p}+\mathbf{r}+\mathbf{q}$ **c** $\mathbf{r}+\frac{1}{2}\mathbf{q}$ **d** $-\frac{1}{2}\mathbf{p}+\frac{1}{2}\mathbf{r}$

6 **a** $\mathbf{x}=\begin{pmatrix}-1\\ \frac{1}{3}\\ \frac{2}{3}\end{pmatrix}$ **b** $\mathbf{x}=\begin{pmatrix}1\\ -10\\ 2\end{pmatrix}$ **7** $\mathbf{v}\bullet\mathbf{w}=\pm 6$

8 $\frac{1}{\sqrt{5}}\mathbf{i}+\frac{2}{\sqrt{5}}\mathbf{k}$ **9** $t=2\pm\sqrt{2}$

10 $\widehat{K}\approx 123.7°$, $\widehat{L}\approx 11.3°$, $\widehat{M}=45°$

11 **a** $k=\pm\frac{12}{13}$ **b** $k=\pm\frac{1}{\sqrt{3}}$ **12** $\approx 40.7°$

14 **a** $\overrightarrow{PQ}=\begin{pmatrix}5\\ -2\\ -4\end{pmatrix}$ **b** $\approx 41.81°$

15 $\overrightarrow{OT}=\begin{pmatrix}4\\ 8\end{pmatrix}$ or $\begin{pmatrix}2\\ -2\end{pmatrix}$ **16** $\approx 61.6°$

17 $\sin\theta=\frac{2}{\sqrt{5}}$ **18** $\pm\frac{3}{\sqrt{59}}\begin{pmatrix}3\\ 7\\ 1\end{pmatrix}$

EXERCISE 15A

1 **a** 6 m s^{-1}, 1 m s^{-1}, 7 m s^{-1} $\therefore 7\text{ m s}^{-1}$

b 6 m s^{-1}, 5 m s^{-1}, 1 m s^{-1} $\therefore 5\text{ m s}^{-1}$

2 **a** 1.34 m s^{-1} in the direction 26.6° to the left of her intended line.

b **i** 30° to the right of Q **ii** 1.04 m s^{-1}

3 **a** 24.6 km h^{-1} **b** $\approx 9.93°$ east of south

4 **a** 82.5 m **b** 23.3° to the left of straight across **c** 48.4 s

5 **a** The plane's speed in still air would be $\approx 437\text{ km h}^{-1}$. The wind slows the plane down to 400 km h^{-1}.

b 4.64° north of due east

EXERCISE 15B

1 **a** $\frac{\sqrt{101}}{2}$ units^2 **b** $\frac{\sqrt{133}}{2}$ units^2 **c** $\frac{\sqrt{69}}{2}$ units^2

2 $8\sqrt{2}$ units^2 **3** **a** D(−4, 1, 3) **b** $\sqrt{307}$ units^2

4 $k=2\pm 2\sqrt{33}$

5 $S=\frac{1}{2}\{|\mathbf{a}\times\mathbf{b}|+|\mathbf{a}\times\mathbf{c}|+|\mathbf{b}\times\mathbf{c}|+|(\mathbf{b}-\mathbf{a})\times(\mathbf{c}-\mathbf{a})|\}$

EXERCISE 15C

1 **a** **i** $\begin{pmatrix}x\\ y\end{pmatrix}=\begin{pmatrix}3\\ -4\end{pmatrix}+\lambda\begin{pmatrix}1\\ 4\end{pmatrix}$, $\lambda\in\mathbb{R}$

ii $x=3+\lambda$, $y=-4+4\lambda$, $\lambda\in\mathbb{R}$ **iii** $4x-y=16$

b **i** $\begin{pmatrix}x\\ y\end{pmatrix}=\begin{pmatrix}5\\ 2\end{pmatrix}+\lambda\begin{pmatrix}-2\\ 5\end{pmatrix}$, $\lambda\in\mathbb{R}$

ii $x=5-2\lambda$, $y=2+5\lambda$, $\lambda\in\mathbb{R}$ **iii** $5x+2y=29$

c **i** $\begin{pmatrix}x\\ y\end{pmatrix}=\begin{pmatrix}-6\\ 0\end{pmatrix}+\lambda\begin{pmatrix}3\\ 7\end{pmatrix}$, $\lambda\in\mathbb{R}$

ii $x=-6+3\lambda$, $y=7\lambda$, $\lambda\in\mathbb{R}$ **iii** $7x-3y=-42$

d **i** $\begin{pmatrix}x\\ y\end{pmatrix}=\begin{pmatrix}-1\\ 11\end{pmatrix}+\lambda\begin{pmatrix}-2\\ 1\end{pmatrix}$, $\lambda\in\mathbb{R}$

ii $x=-1-2\lambda$, $y=11+\lambda$, $\lambda\in\mathbb{R}$ **iii** $x+2y=21$

2 **a** $x=-1+2\lambda$, $y=4-\lambda$, $\lambda\in\mathbb{R}$

b Points are: (−1, 4), (1, 3), (5, 1), (−3, 5), (−9, 8)

3 **a** When $\lambda=1$, $x=3$, $y=-2$, $\therefore$ yes **b** $k=-5$

4 **a** (0, 8)

b It is parallel to $\begin{pmatrix}-1\\ 3\end{pmatrix}$ and in the opposite direction.

c $\begin{pmatrix}x\\ y\end{pmatrix}=\begin{pmatrix}0\\ 8\end{pmatrix}+\mu\begin{pmatrix}1\\ -3\end{pmatrix}$, $\mu\in\mathbb{R}$

5 **a** **i** $\begin{pmatrix}x\\ y\\ z\end{pmatrix}=\begin{pmatrix}1\\ 3\\ -7\end{pmatrix}+\lambda\begin{pmatrix}2\\ 1\\ 3\end{pmatrix}$, $\lambda\in\mathbb{R}$

ii $x=1+2\lambda$, $y=3+\lambda$, $z=-7+3\lambda$, $\lambda\in\mathbb{R}$

iii $\frac{x-1}{2}=y-3=\frac{z+7}{3}$

b **i** $\begin{pmatrix}x\\ y\\ z\end{pmatrix}=\begin{pmatrix}0\\ 1\\ 2\end{pmatrix}+\lambda\begin{pmatrix}1\\ 1\\ -2\end{pmatrix}$, $\lambda\in\mathbb{R}$

ii $x=\lambda$, $y=1+\lambda$, $z=2-2\lambda$, $\lambda\in\mathbb{R}$

iii $x=y-1=\frac{-z+2}{2}$

c **i** $\begin{pmatrix}x\\ y\\ z\end{pmatrix}=\begin{pmatrix}-2\\ 2\\ 1\end{pmatrix}+\lambda\begin{pmatrix}1\\ 0\\ 0\end{pmatrix}$, $\lambda\in\mathbb{R}$

ii $x=-2+\lambda$, $y=2$, $z=1$, $\lambda\in\mathbb{R}$

iii $y=2$, $z=1$

d **i** $\begin{pmatrix}x\\ y\\ z\end{pmatrix}=\begin{pmatrix}0\\ 2\\ -1\end{pmatrix}+\lambda\begin{pmatrix}2\\ -1\\ 3\end{pmatrix}$, $\lambda\in\mathbb{R}$

ii $x=2\lambda$, $y=2-\lambda$, $z=-1+3\lambda$, $\lambda\in\mathbb{R}$

iii $\frac{x}{2}=-y+2=\frac{z+1}{3}$

e **i** $\begin{pmatrix}x\\ y\\ z\end{pmatrix}=\begin{pmatrix}3\\ 2\\ -1\end{pmatrix}+\lambda\begin{pmatrix}0\\ 0\\ 1\end{pmatrix}$, $\lambda\in\mathbb{R}$

ii $x=3$, $y=2$, $z=-1+\lambda$, $\lambda\in\mathbb{R}$

iii $x=3$, $y=2$

6 **a** $\begin{pmatrix}x\\ y\\ z\end{pmatrix}=\begin{pmatrix}1\\ 2\\ 1\end{pmatrix}+\lambda\begin{pmatrix}-2\\ 1\\ 1\end{pmatrix}$, $\lambda\in\mathbb{R}$

b $\begin{pmatrix}x\\ y\\ z\end{pmatrix}=\begin{pmatrix}0\\ 1\\ 3\end{pmatrix}+\lambda\begin{pmatrix}3\\ 0\\ -4\end{pmatrix}$, $\lambda\in\mathbb{R}$

c $\begin{pmatrix}x\\ y\\ z\end{pmatrix}=\begin{pmatrix}1\\ 2\\ 5\end{pmatrix}+\lambda\begin{pmatrix}0\\ -3\\ 0\end{pmatrix}$, $\lambda\in\mathbb{R}$

d $\begin{pmatrix}x\\ y\\ z\end{pmatrix}=\begin{pmatrix}0\\ 1\\ -1\end{pmatrix}+\lambda\begin{pmatrix}5\\ -2\\ 4\end{pmatrix}$, $\lambda\in\mathbb{R}$

7 **a** $\begin{pmatrix}-2\\ 0\\ 3\end{pmatrix}$ **b** $\begin{pmatrix}-1\\ 1\\ -3\end{pmatrix}$ **c** $\begin{pmatrix}3\\ 2\\ 1\end{pmatrix}$ **d** $\begin{pmatrix}-2\\ 4\\ 3\end{pmatrix}$

8 **a** $(-\frac{1}{2}, \frac{9}{2}, 0)$ **b** (0, 4, 1) **c** (4, 0, 9)

9 **a** (x_0, y_0, z_0) **b** $\begin{pmatrix}l\\ m\\ n\end{pmatrix}$

c $\frac{x-x_0}{l}=\frac{y-y_0}{m}=\frac{z-z_0}{n}$, $l, m, n\neq 0$

10 (0, 7, 3) and $(\frac{20}{3}, -\frac{19}{3}, -\frac{11}{3})$

11 **a** (1, 2, 3) **b** $(\frac{7}{3}, \frac{2}{3}, \frac{8}{3})$

EXERCISE 15D

1 $75.7°$ 2 $\mathbf{b}_1 \bullet \mathbf{b}_2 = 0$ 3 $75.5°$

4 a $28.6°$ b $x = -\frac{48}{7}$

5 a $78.7°$ b $63.4°$ c $63.4°$ d $71.6°$

EXERCISE 15E

1 a $(1, 2)$ b

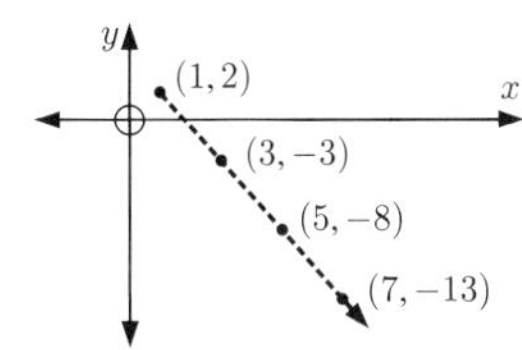

c $\begin{pmatrix} 2 \\ -5 \end{pmatrix}$

d $\sqrt{29}$ cm s^{-1}

2 a $\begin{pmatrix} x \\ y \end{pmatrix} = \begin{pmatrix} 2 \\ 3 \end{pmatrix} + t\begin{pmatrix} 4 \\ -5 \end{pmatrix}$, $t \geqslant 0$ b $(8, -4.5)$

c 45 minutes

3 a $\begin{pmatrix} -3+2t \\ -2+4t \end{pmatrix}$ d

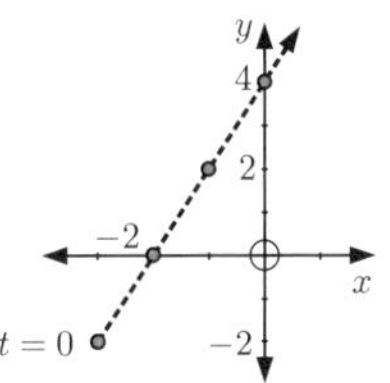

b $\begin{pmatrix} 2 \\ 8 \end{pmatrix}$

c i $t = 1.5$ s
ii $t = 0.5$ s

4 a i $(-4, 3)$ ii $\begin{pmatrix} 12 \\ 5 \end{pmatrix}$ iii 13 m s^{-1}

b i $(3, 0, 4)$ ii $\begin{pmatrix} 2 \\ -1 \\ -2 \end{pmatrix}$ iii 3 m s^{-1}

5 a $\begin{pmatrix} 120 \\ -90 \end{pmatrix}$ b $\begin{pmatrix} 20\sqrt{5} \\ 10\sqrt{5} \end{pmatrix}$ 6 $\begin{pmatrix} -12 \\ 30 \\ -84 \end{pmatrix}$

7 a A is at $(4, 5)$, B is at $(1, -8)$

b For A it is $\begin{pmatrix} 1 \\ -2 \end{pmatrix}$. For B it is $\begin{pmatrix} 2 \\ 1 \end{pmatrix}$.

c For A, speed is $\sqrt{5}$ km h^{-1}. For B, speed is $\sqrt{5}$ km h^{-1}.

d $\begin{pmatrix} 1 \\ -2 \end{pmatrix} \bullet \begin{pmatrix} 2 \\ 1 \end{pmatrix} = 0$

8 a $\begin{pmatrix} x_1 \\ y_1 \end{pmatrix} = \begin{pmatrix} -5 \\ 4 \end{pmatrix} + t\begin{pmatrix} 3 \\ -1 \end{pmatrix}$

$\therefore \ x_1(t) = -5 + 3t, \ \ y_1(t) = 4 - t$

b speed $= \sqrt{10}$ km min^{-1}

c a minutes later, $(t - a)$ min have elapsed.

$\therefore \ \begin{pmatrix} x_2 \\ y_2 \end{pmatrix} = \begin{pmatrix} 15 \\ 7 \end{pmatrix} + (t-a)\begin{pmatrix} -4 \\ -3 \end{pmatrix}$

$\therefore \ x_2(t) = 15 - 4(t-a), \ \ y_2(t) = 7 - 3(t-a)$

d Torpedo is fired at 1:35:28 pm and the explosion occurs at 1:37:42 pm.

9 a $\begin{pmatrix} -3 \\ 1 \\ -0.5 \end{pmatrix}$ b ≈ 19.2 km h^{-1}

c $\begin{pmatrix} x \\ y \\ z \end{pmatrix} = \begin{pmatrix} 6 \\ 9 \\ 3 \end{pmatrix} + t\begin{pmatrix} -3 \\ 1 \\ -0.5 \end{pmatrix}$, $t \in \mathbb{R}$ d 1 hour

EXERCISE 15F

1 a $\frac{3}{5}\sqrt{5}$ units b $\frac{1}{2}\sqrt{2}$ units c $\frac{9}{2}\sqrt{2}$ units d 0 units

2 a $6\mathbf{i} - 6\mathbf{j}$ b $\begin{pmatrix} 6-6t \\ -6+8t \end{pmatrix}$ c when $t = \frac{3}{4}$ hours

d $t = 0.84$ and position is $(0.96, 0.72)$

3 a $\begin{pmatrix} -120 \\ -40 \end{pmatrix}$ b $\begin{pmatrix} x \\ y \end{pmatrix} = \begin{pmatrix} 200 \\ 100 \end{pmatrix} + t\begin{pmatrix} -120 \\ -40 \end{pmatrix}$

c $(80, 60)$ d $|\overrightarrow{OP}| = \sqrt{80^2 + 60^2} = 100$ km

e at 1:45 pm and $d_{\min} \approx 31.6$ km f 2:30 pm

4 a A$(18, 0)$ and B$(0, 12)$ b R is at $\left(x, \dfrac{36-2x}{3}\right)$

c $\overrightarrow{PR} = \begin{pmatrix} x-4 \\ \frac{36-2x}{3} \end{pmatrix}$ and $\overrightarrow{AB} = \begin{pmatrix} -18 \\ 12 \end{pmatrix}$

d $\left(\frac{108}{13}, \frac{84}{13}\right)$ and distance ≈ 7.77 km

5 a A$(3, -4)$ and B$(4, 3)$ b for A $\begin{pmatrix} -1 \\ 2 \end{pmatrix}$, for B $\begin{pmatrix} -3 \\ -2 \end{pmatrix}$

c $82.9°$ d at $t = 1.5$ hours

6 a $(2, -1, 4)$ b $\sqrt{27}$ units

7 a $(2, \frac{1}{2}, \frac{5}{2})$ b $\sqrt{\frac{3}{2}}$ units

EXERCISE 15G

1 a

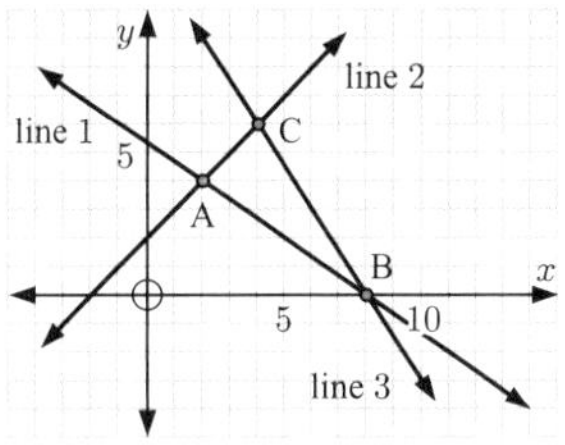

b A$(2, 4)$,
B$(8, 0)$,
C$(4, 6)$

c BC = BA
$= \sqrt{52}$ units
$\therefore$ isosceles $\triangle$

2 a

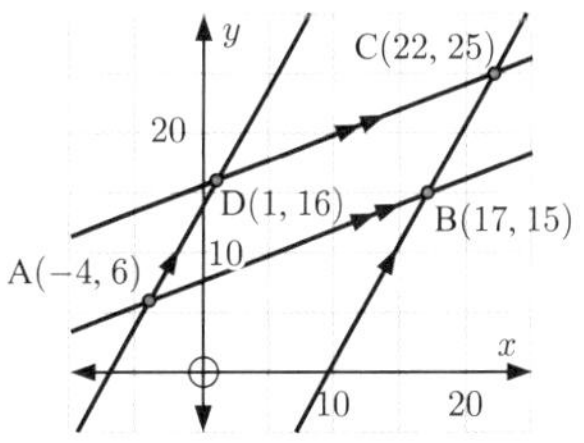

b A$(-4, 6)$,
B$(17, 15)$,
C$(22, 25)$,
D$(1, 16)$

3 a A$(2, 3)$, B$(8, 6)$, C$(5, 0)$

b AB = BC = $\sqrt{45}$ units, AC = $\sqrt{18}$ units

4 a P$(10, 4)$, Q$(3, -1)$, R$(20, -10)$

b $\overrightarrow{PQ} = \begin{pmatrix} -7 \\ -5 \end{pmatrix}$, $\overrightarrow{PR} = \begin{pmatrix} 10 \\ -14 \end{pmatrix}$, $\overrightarrow{PQ} \bullet \overrightarrow{PR} = 0$

c $\widehat{QPR} = 90°$ d 74 units^2

5 a A is at $(2, 5)$, B$(18, 9)$, C$(14, 25)$, D$(-2, 21)$

b $\overrightarrow{AC} = \begin{pmatrix} 12 \\ 20 \end{pmatrix}$ and $\overrightarrow{DB} = \begin{pmatrix} 20 \\ -12 \end{pmatrix}$

i $\sqrt{544}$ units ii $\sqrt{544}$ units iii 0

c Diagonals are perpendicular and equal in length, and as their midpoints are both $(8, 15)$, ABCD is a square.

EXERCISE 15H.1

1 a $x = 2,\ y = -3$ b $x = -1,\ y = 5$
c $x = -2,\ y = -4$

2 a intersecting b parallel c intersecting
d coincident e intersecting f parallel

3 a The second equation is the same as the first when divided throughout by 2. The lines are coincident.
b It gives no more information than the first. Gives the same solutions for x and y.
c i when $x = t,\ y = \dfrac{3-t}{2},\ t \in \mathbb{R}$
ii when $y = s,\ x = 3 - 2s,\ s \in \mathbb{R}$

4 b The system is inconsistent and so has no solutions.
c The lines are parallel.

5 b The lines are coincident. Infinitely many solutions exist of the form $x = t,\ y = \dfrac{5-2t}{3},\ t \in \mathbb{R}$.

6 b If $k \neq 4$, the system is inconsistent and so has no solutions. The lines are parallel.
If $k = 4$, the system has infinitely many solutions of the form $x = t,\ y = 3t - 2,\ t \in \mathbb{R}$. The lines are coincident.

7 a $\left[\begin{array}{cc|c} 3 & -1 & 8 \\ 0 & 0 & k-16 \end{array}\right]$
b i $k = 16$ ii $x = t,\ y = 3t - 8,\ t \in \mathbb{R}$
c i when $k \neq 16$ ii The lines are parallel.

8 a $\left[\begin{array}{cc|c} 4 & 8 & 1 \\ 0 & 2a+8 & -21 \end{array}\right]$ b $a \neq -4$
d When $a = -4$, last row is $0\ \ 0 \mid -21$. So, the system is inconsistent and $\therefore$ no solutions exist.

9 a A unique solution for $m \neq 2$ or -2.
b If $m \neq \pm 2$, $x = \dfrac{6}{m+2},\ y = \dfrac{6}{m+2}$;
if $m = 0$, unique solution is $x = 3,\ y = 3$.
c If $m = 2$, there are infinitely many solutions of the form $x = t,\ y = 3 - t$ (t is real). If $m = -2$, there are no solutions. The lines are parallel.

EXERCISE 15H.2

1 a They intersect at $(1, 2, 3)$, angle $\approx 10.9°$.
b Lines are skew, angle $\approx 62.7°$.
c They are parallel, $\therefore$ angle $= 0°$.
d They are skew, angle $\approx 11.4°$.
e They intersect at $(-4, 7, -7)$, angle $\approx 40.2°$.
f They are parallel, $\therefore$ angle $= 0°$.
g They are coincident, $\therefore$ angle $= 0°$

2 Line 1 and line 2 are parallel.
Line 1 and line 3 are skew with angle $\approx 48.2°$.
Line 2 and line 3 are skew with angle $\approx 48.2°$.

EXERCISE 15I

1 a $2x - y + 3z = 8$ b $3x + 4y + z = 19$
c $x + 3y + z = 10$

2 a $\begin{pmatrix} 2 \\ 3 \\ -1 \end{pmatrix}$ b $\begin{pmatrix} 3 \\ -1 \\ 0 \end{pmatrix}$ c $\begin{pmatrix} 0 \\ 0 \\ 1 \end{pmatrix}$ d $\begin{pmatrix} 1 \\ 0 \\ 0 \end{pmatrix}$

3 a $y = 0$ b $z = 4$

4 a ii $-2x + 6y + z = 18$ b ii $-5x + 3y + 12z = 12$
c ii $-y + z = 3$ {many vector forms exist}

5 a $x = 1 + \lambda,\ y = -2 - 3\lambda,\ z = 4\lambda,\ \lambda \in \mathbb{R}$
b $x = 3 + \lambda,\ y = 4 - \lambda,\ z = -1 - 2\lambda,\ \lambda \in \mathbb{R}$

6 $x = 2 - t,\ y = -1 + 3t,\ z = 3 - 3t,\ t \in \mathbb{R}$; $(1, 2, 0)$

7 a $x = 1 + t,\ y = -2 + 2t,\ z = 4 - 5t,\ t \in \mathbb{R}$
b i $(0, -4, 9)$ ii $(1, -2, 4)$ iii $(-5, -14, 34)$

8 a $(-1, -1, 4)$; 3 units b $(0, 1, -3)$; $2\sqrt{11}$ units
c $(-\frac{1}{7}, -\frac{26}{7}, -\frac{17}{7})$; $2\sqrt{\frac{3}{7}}$ units

9 $(1, -3, 0)$ 10 X axis at $(2, 0, 0)$

11 a $y - 3z = -7$ b $x - z = -2$ c $3x - y = 1$

12 $y - 2z = 8$

13 a $k = -\frac{3}{2}$ b B$(3, 6, -\frac{11}{2})$ or $(-1, -2, \frac{5}{2})$

14 a N$(3.4, 1.2, 1)$, $d = \frac{2}{\sqrt{5}}$ units
b N$(\frac{1}{6}, \frac{5}{6}, -\frac{1}{3})$, $d = \frac{5}{\sqrt{6}}$ units

16 d i $\frac{10}{\sqrt{6}}$ units ii $2\sqrt{3}$ units

17 a $\frac{19}{2\sqrt{6}}$ units b $\dfrac{|d_2 - d_1|}{\sqrt{a^2 + b^2 + c^2}}$ units 18 $\frac{26}{\sqrt{138}}$ units

19 $2x - y + 2z = -1$ and $2x - y + 2z = 11$

EXERCISE 15J

1 a $\approx 13.1°$ b $0°$ (the line and plane are parallel)
c $\approx 11.3°$ d $\approx 30.7°$

2 a $\approx 83.7°$ b $\approx 84.8°$ c $\approx 86.2°$ d $\approx 73.2°$
e $\approx 62.3°$

EXERCISE 15K

1 a $x = 1 + 2t,\ y = t,\ z = 0,\ t \in \mathbb{R}$
b $x = 4,\ y = -2,\ z = 1$ c $x = 4,\ y = -3,\ z = 2$
d no solution, system is inconsistent

2 a Either no solutions or an infinite number of solutions.
b i $a_1 = ka_2,\ b_1 = kb_2,\ c_1 = kc_2,\ d_1 \neq kd_2$ for some k
ii $a_1 = ka_2,\ b_1 = kb_2,\ c_1 = kc_2,\ d_1 = kd_2$ for some k
c i Planes meet in a line
$x = -2 + 3t,\ y = t,\ z = 5,\ t \in \mathbb{R}$
ii Planes meet in a line
$x = 2 - 2t,\ y = t,\ z = 1 + 3t,\ t \in \mathbb{R}$
iii Planes are coincident
$\therefore\ x = 6 - 2s + 3t,\ y = s,\ z = t,\ s, t \in \mathbb{R}$

3 a If $k = -2$, planes are coincident with infinitely many solutions.
If $k \neq -2$, planes meet in a line with infinitely many solutions.
b If $k = 16$, planes are coincident with infinitely many solutions.
If $k \neq 16$, planes are parallel with no solutions.

5 a Meet at a point $(1, -2, 4)$
b Meet in a line $x = \dfrac{9-t}{3},\ y = \dfrac{6+5t}{3},\ z = t,\ t \in \mathbb{R}$
c Meet in a line $x = 3t - 3,\ y = t,\ z = 5t - 11,\ t \in \mathbb{R}$
d No solutions as 2 planes are parallel and intersected by 3rd plane.
e Two planes are coincident and the other cuts obliquely at the line $x = \frac{5}{2} + \frac{1}{2}t,\ y = -\frac{3}{2} + \frac{3}{2}t,\ z = t,\ t \in \mathbb{R}$
f Meet at the point $(3, -2, 0)$

6 If $k = 5$ the planes meet in a line $x = -10t,\ y = -1 - 7t,\ z = t,\ t \in \mathbb{R}$. If $k \neq 5$, the line of intersection of any two planes is parallel to the third $\therefore$ no solutions.

7 **a** $\left[\begin{array}{ccc|c} 1 & 3 & 3 & a-1 \\ 0 & -7 & -5 & 9-2a \\ 0 & 0 & a+1 & a+1 \end{array}\right]$

b $x = \dfrac{19-6t}{7}$, $y = \dfrac{-5t-11}{7}$, $z = t$ (t is real)
Planes meet in a line.

c $x = \frac{1}{7}a + 2$, $y = \frac{2}{7}a - 2$, $z = 1$.
Planes meet at a point.

8 $\left[\begin{array}{ccc|c} 1 & 2 & m & -1 \\ 0 & -2(m+1) & 1-m^2 & 1+m \\ 0 & 0 & (m+1)(m+5) & -7(m+1) \end{array}\right]$

a If $m = -5$, no solution. The line of intersection of any two planes is parallel to the third.

b If $m = -1$, infinitely many solutions. Two planes are coincident and the third meets in a line.

c **i** If $m \neq -5$ or -1, there is a unique solution. The planes meet at a point.

9 They meet at the point $\left(\frac{94}{29}, \frac{-68}{29}, \frac{64}{29}\right)$.

REVIEW SET 15A

1 **a** $\begin{pmatrix} x \\ y \end{pmatrix} = \begin{pmatrix} -6 \\ 3 \end{pmatrix} + t\begin{pmatrix} 4 \\ -3 \end{pmatrix}$, $t \in \mathbb{R}$

b $x = -6 + 4t$, $y = 3 - 3t$, $t \in \mathbb{R}$ **c** $3x + 4y = -6$

2 $m = 10$

3 **a** $(5, 2)$ **b** $\begin{pmatrix} 4 \\ 10 \end{pmatrix}$ is a non-zero scalar multiple of $\begin{pmatrix} 2 \\ 5 \end{pmatrix}$

c $\begin{pmatrix} x \\ y \end{pmatrix} = \begin{pmatrix} 5 \\ 2 \end{pmatrix} + s\begin{pmatrix} 4 \\ 10 \end{pmatrix}$

4 **a** $x = 2 + t$, $y = 4t$, $z = 1 - 3t$, $t \in \mathbb{R}$

b Use $\cos\theta = \dfrac{|\overrightarrow{PQ} \bullet \overrightarrow{QR}|}{|\overrightarrow{PQ}||\overrightarrow{QR}|}$

5 **a** A(5, 2), B(6, 5), C(8, 3)

b $|\overrightarrow{AB}| = \sqrt{10}$ units, $|\overrightarrow{BC}| = \sqrt{8}$ units, $|\overrightarrow{AC}| = \sqrt{10}$ units

c isosceles

6 **a** OABC is a rhombus. So, its diagonals bisect its angles.

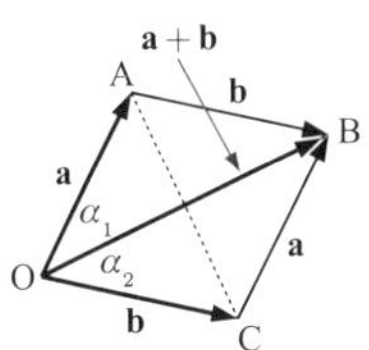

b $\begin{pmatrix} x \\ y \\ z \end{pmatrix} = \begin{pmatrix} 7 \\ 3 \\ -4 \end{pmatrix} + \lambda\begin{pmatrix} 0 \\ \frac{1}{3} \\ \frac{1}{3} \end{pmatrix}$, $\lambda \in \mathbb{R}$

c $(7, 3\frac{3}{4}, -3\frac{1}{4})$

7 **a** $\begin{pmatrix} x \\ y \\ z \end{pmatrix} = \begin{pmatrix} 3 \\ 2 \\ -1 \end{pmatrix} + \lambda\begin{pmatrix} -4 \\ 0 \\ 5 \end{pmatrix}$, $\lambda \in \mathbb{R}$

b $-4x + 5z = 24$ **c** $(-5, 2, 9)$ or $(11, 2, -11)$

8 $(6, -1, -10)$ **9** **a** ± 7 **b** $\frac{\sqrt{14}}{2}$ units^2

10 **a** $\frac{17}{3}$ units **b** $(\frac{8}{3}, \frac{7}{3}, \frac{4}{3})$

11 **a** They do not meet, the line is parallel to the plane.

b $\frac{16}{\sqrt{14}}$ units

12 **a** $(\frac{1}{5}, \frac{17}{5}, \frac{9}{5})$ **b** $(-1, 3, 1)$ **c** $6x - 8y - 5z = -35$

13 $\frac{31}{\sqrt{110}}$ units **14** $(4, 1, -3)$ and $(1, -5, 0)$

15 **a**

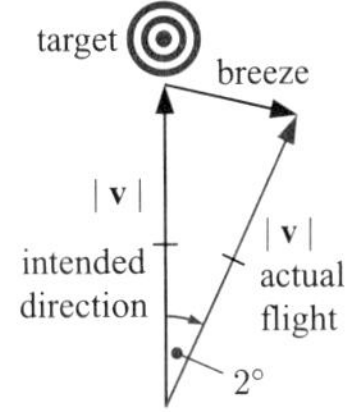

b **i** isosceles triangle $\therefore$ 2 remaining angles $= 89°$ each, breeze makes angle of $180 - 89 = 91°$ to intended direction of the arrow.

ii bisect angle $2°$ and use $\sin 1° = \dfrac{\frac{1}{2}\text{ speed}}{|\mathbf{v}|}$

$\therefore$ speed $= 2|\mathbf{v}|\sin 1°$

16 **a** X(7, 3, −1), D(7, 1, −2) **b** Y(5, 3, −2)

c $\overrightarrow{BD} = \begin{pmatrix} 3 \\ -3 \\ 0 \end{pmatrix}$ and $\overrightarrow{BY} = \begin{pmatrix} 1 \\ -1 \\ 0 \end{pmatrix}$ So, $\overrightarrow{BD} = 3\overrightarrow{BY}$, etc.

17 If $k = -2$, the planes meet in the line $x = \frac{4}{3}$, $y = -\frac{11}{3} + t$, $z = t$, $t \in \mathbb{R}$. If $k \neq -2$ the planes meet at the point $(\frac{4}{3}, -\frac{14}{3}, -1)$.

REVIEW SET 15B

1 $\begin{pmatrix} x \\ y \end{pmatrix} = \begin{pmatrix} 0 \\ 8 \end{pmatrix} + \lambda\begin{pmatrix} 5 \\ 4 \end{pmatrix}$, $\lambda \in \mathbb{R}$

2 **a** **i** $-6\mathbf{i} + 10\mathbf{j}$ **ii** $-5\mathbf{i} - 15\mathbf{j}$

iii $(-6 - 5t)\mathbf{i} + (10 - 15t)\mathbf{j}$, $t \geqslant 0$

b $t = 0.48$ h

c shortest distance ≈ 8.85 km, so will miss reef

3 **a** **i** $\begin{pmatrix} x \\ y \end{pmatrix} = \begin{pmatrix} 2 \\ -3 \end{pmatrix} + t\begin{pmatrix} 4 \\ -1 \end{pmatrix}$, $t \in \mathbb{R}$

ii $x = 2 + 4t$, $y = -3 - t$, $t \in \mathbb{R}$

b **i** $\begin{pmatrix} x \\ y \\ z \end{pmatrix} = \begin{pmatrix} -1 \\ 6 \\ 3 \end{pmatrix} + t\begin{pmatrix} 6 \\ -8 \\ -3 \end{pmatrix}$, $t \in \mathbb{R}$

ii $x = -1 + 6t$, $y = 6 - 8t$, $z = 3 - 3t$, $t \in \mathbb{R}$

4 **a** 11.5° east of due north **b** ≈ 343 km h^{-1} **5** 8.13°

6 **a** $x_1(t) = 2 + t$, $y_1(t) = 4 - 3t$, $t \geqslant 0$

b $x_2(t) = 13 - t$, $y_2(t) = [3 - 2a] + at$, $t \geqslant 2$

c interception occurred at 2:22:30 pm

d bearing $\approx 12.7°$ west of south, ≈ 4.54 units per minute

7 **a** $\approx 15.8°$ **b** $\approx 65.9°$

8 **b** $\approx 28.6°$ **c** $2x + 3y + 6z = 147$ **d** 14 units

9 **a** $4x + 2y + z = 3$ **b** $\approx 64.1°$

10 **a** $\begin{pmatrix} \sqrt{3} \\ -\sqrt{3} \\ \sqrt{3} \end{pmatrix}$ and $\begin{pmatrix} -\sqrt{3} \\ \sqrt{3} \\ -\sqrt{3} \end{pmatrix}$

b $\frac{1}{\sqrt{74}}\mathbf{i} + \frac{8}{\sqrt{74}}\mathbf{j} + \frac{3}{\sqrt{74}}\mathbf{k}$ or $-\frac{1}{\sqrt{74}}\mathbf{i} - \frac{8}{\sqrt{74}}\mathbf{j} - \frac{3}{\sqrt{74}}\mathbf{k}$

c $k = -7$ or 11

11 72.35° or 107.65°

12 **a** $\begin{pmatrix} x \\ y \\ z \end{pmatrix} = \begin{pmatrix} 2 \\ -1 \\ 3 \end{pmatrix} + \lambda\begin{pmatrix} -2 \\ 2 \\ -4 \end{pmatrix}$, $\lambda \in \mathbb{R}$

b $\left(2-\frac{2}{\sqrt{6}},\ -1+\frac{2}{\sqrt{6}},\ 3-\frac{4}{\sqrt{6}}\right)$ and $\left(2+\frac{2}{\sqrt{6}},\ -1-\frac{2}{\sqrt{6}},\ 3+\frac{4}{\sqrt{6}}\right)$

13 7.82° **14** $\frac{9\sqrt{2}}{2}$ units²

15 **a** intersecting at (4, 3, 1), angle ≈ 44.5°
b skew, angle ≈ 71.2°

16 **a** $5\begin{pmatrix}-1\\1\\1\end{pmatrix}$ **b** $m=1$ **c** $x-y-z=0$
d $t=2$ **e** $\frac{4}{\sqrt{114}}$

17 **b** $k=-1$
c $t(p+10)=q+2$ has infinitely many solutions for t when $p+10=0$ and $q+2=0$, $\therefore\ p=-10,\ q=-2$

18 **a** $\left[\begin{array}{ccc|c}1&-3&2&-5\\0&10&-4-k&25\\0&0&k+4&-5\end{array}\right]$

b No solutions if $k=-4$. Two planes are parallel and intersected by the third plane.
c **i** Unique solution when $k\neq -4$.
ii $x=1+\dfrac{10}{k+4},\ y=2,\ z=\dfrac{-5}{k+4}$ **iii** (3, 2, −1)
Planes meet at a point.

REVIEW SET 15C

1 $2\sqrt{10}(3\mathbf{i}-\mathbf{j})$

2 **a** (−4, 3) **b** (28, 27) **c** $\begin{pmatrix}8\\6\end{pmatrix}$ **d** 10 m s^{-1}

3 **a** (KL) is parallel to (MN) as $\begin{pmatrix}5\\-2\end{pmatrix}$ is parallel to $\begin{pmatrix}-5\\2\end{pmatrix}$
b (KL) is perpendicular to (NK) as $\begin{pmatrix}5\\-2\end{pmatrix}\bullet\begin{pmatrix}4\\10\end{pmatrix}=0$
and (NK) is perpendicular to (MN) as $\begin{pmatrix}4\\10\end{pmatrix}\bullet\begin{pmatrix}-5\\2\end{pmatrix}=0$
c K(7, 17), L(22, 11), M(33, −5), N(3, 7) **d** 261 units²

4 30.5°

5 **a** $|\overrightarrow{AB}|=\sqrt{27}$ units
b A lies on the line **r** where $\lambda=-3$ and B lies on **r** where $\lambda=0$ $\therefore$ the line between A and B is the same as line **r**, so it can be described by **r**.
c 70.5°

6 **a** Road A: $\begin{pmatrix}x\\y\end{pmatrix}=\begin{pmatrix}-9\\2\end{pmatrix}+\lambda\begin{pmatrix}4\\-3\end{pmatrix},\ \lambda\in\mathbb{R}$
Road B: $\begin{pmatrix}x\\y\end{pmatrix}=\begin{pmatrix}6\\-18\end{pmatrix}+\mu\begin{pmatrix}5\\12\end{pmatrix},\ \mu\in\mathbb{R}$
b Road B, 13 km

7 **a** $\overrightarrow{AB}\bullet\overrightarrow{AC}=\begin{pmatrix}-2\\-1\\6\end{pmatrix}\bullet\begin{pmatrix}5\\2\\2\end{pmatrix}=0$
b **i** $x=4-2t,\ y=2-t,\ z=-1+6t,\ t\in\mathbb{R}$
ii $x=4+5s,\ y=2+2s,\ z=-1+2s,\ s\in\mathbb{R}$

8 $a=4$ or $-\frac{36}{5}$

9 **a** $14x+29y-4z=32$ **b** ≈ 55.9° **c** $r=\frac{2\pm\sqrt{10}}{2}$

10 **a** $\mathbf{n}=\begin{pmatrix}5\\-1\\3\end{pmatrix}$ **b** D(−1, −1, 2) **c** ≈ 11.8 units²
d $\left(\frac{1}{6},\frac{5}{6},\frac{2}{3}\right)$

11 **a** $\overrightarrow{PQ}=\begin{pmatrix}1\\4\\-3\end{pmatrix}$, $|\overrightarrow{PQ}|=\sqrt{26}$ units, $\overrightarrow{QR}=\begin{pmatrix}-4\\-1\\4\end{pmatrix}$
b $x=2+\lambda,\ y=4\lambda,\ z=1-3\lambda,\ \lambda\in\mathbb{R}$
c $\begin{pmatrix}x\\y\\z\end{pmatrix}=\begin{pmatrix}2\\0\\1\end{pmatrix}+\lambda\begin{pmatrix}1\\4\\-3\end{pmatrix}+\mu\begin{pmatrix}-4\\-1\\4\end{pmatrix},\ \lambda,\mu\in\mathbb{R}$

12 **a** 3 units **b** (1, 2, 4) **c** $\sqrt{116}$ units

13 **a** $5x+y+4z=3$ **b** $x=5t,\ y=t,\ z=4t,\ t\in\mathbb{R}$
c $\left(\frac{5}{14},\frac{1}{14},\frac{2}{7}\right)$

14 **a** intersecting at (−1, 2, 3) **b** $\theta\approx 27.0°$

15 **b** $2x+3y+6z=147$ **c** 14 units
d (5, 7, 3) on line 1 and (9, 13, 15) on line 2

16 **a** A(2, −1, 0) **c** $\mathbf{r}=\begin{pmatrix}0\\-3\\2\end{pmatrix}+u\begin{pmatrix}3\\1\\-4\end{pmatrix}$
d $3x-y+2z=7$ **e** $\sqrt{14}$ units²
f normal is $\begin{pmatrix}x\\y\\z\end{pmatrix}=\begin{pmatrix}3\\-2\\-2\end{pmatrix}+\lambda\begin{pmatrix}3\\-1\\2\end{pmatrix}$

17 **e** The three planes have no common point of intersection. The line of intersection of any two planes is parallel to the third plane.

EXERCISE 16A

1

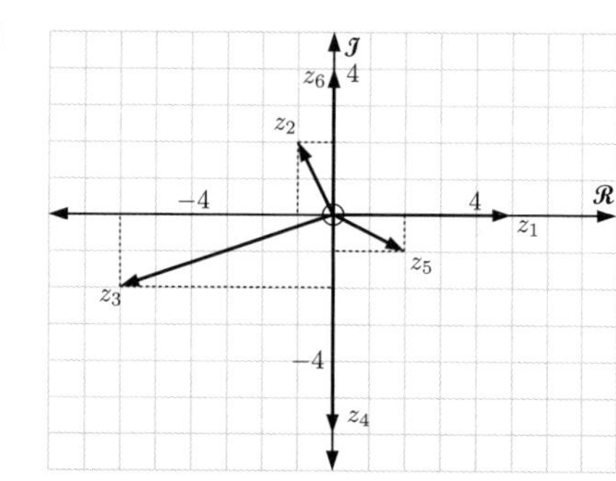

2 **a** $4+i$ **c** $-1+5i$ **d** $-7i$
b $-2+3i$

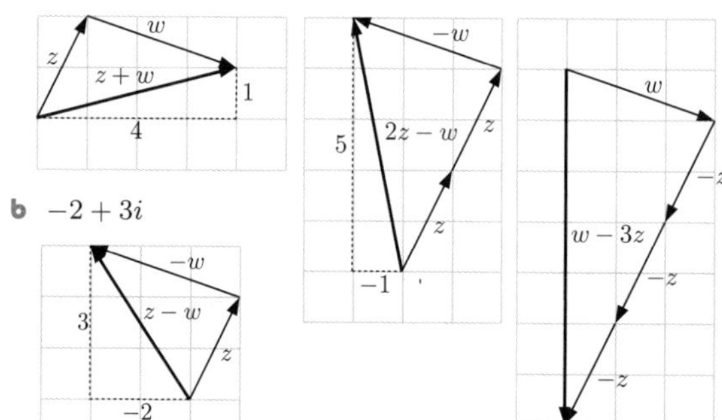

3 **a** $5-i$

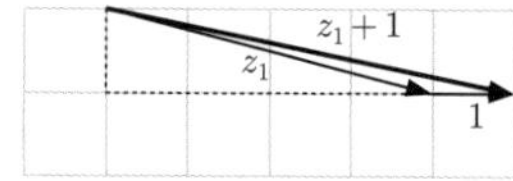

b $4+i$

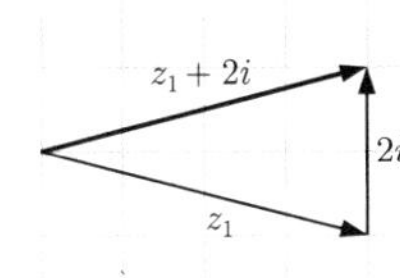

c $4+\frac{5}{2}i$

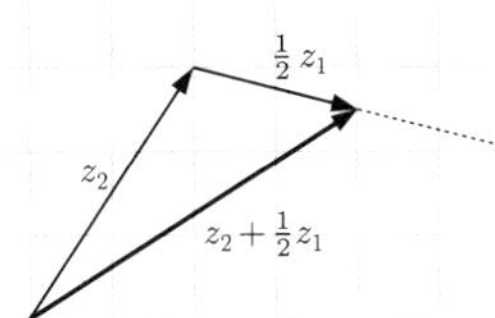

d $4-\frac{1}{2}i$

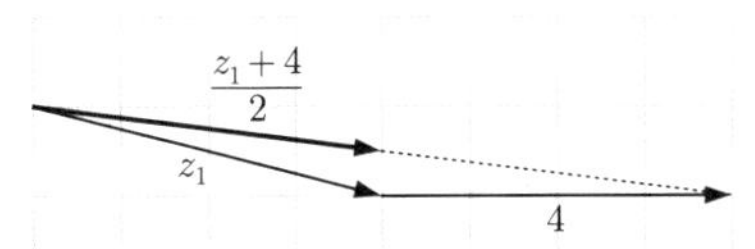

4 **a** **b** **c**

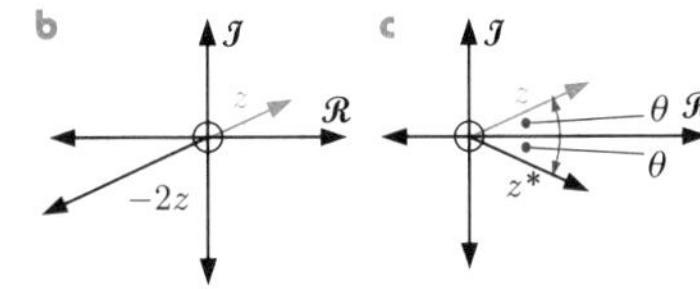

d **e** **f**

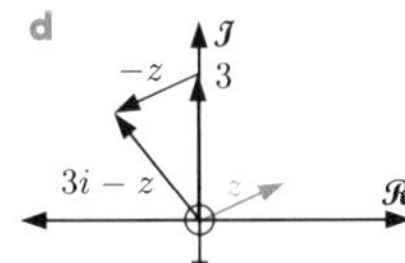

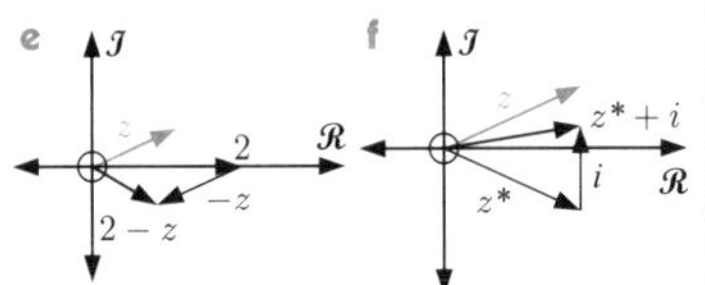

g **h**

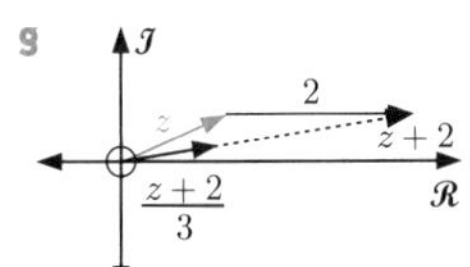

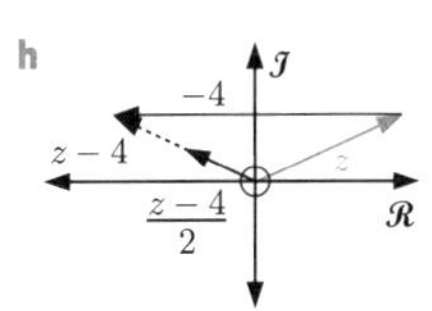

5 **a** **b**

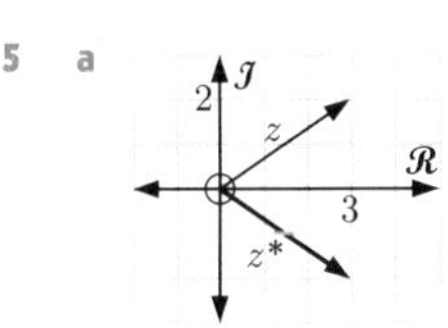

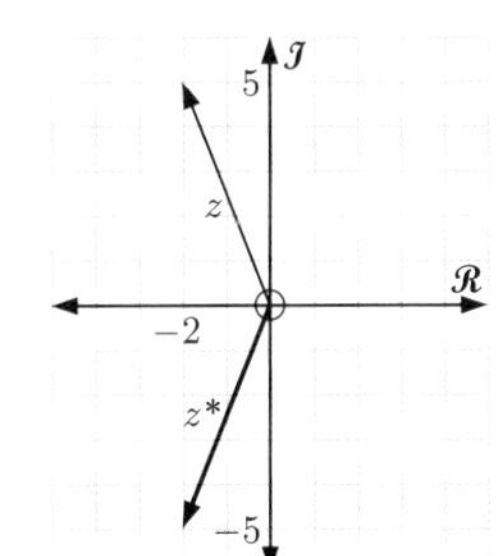

6

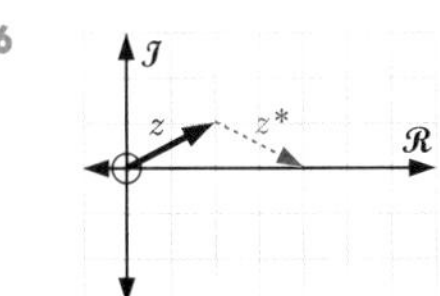

7

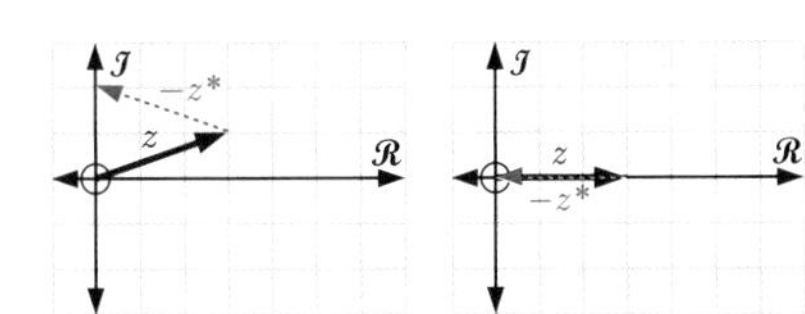

EXERCISE 16B.1

1 **a** 5 **b** 13 **c** $2\sqrt{17}$ **d** 3 **e** 4

2 **a** $\sqrt{5}$ **b** $\sqrt{5}$ **c** 5 **d** 5 **e** $5\sqrt{2}$ **f** $5\sqrt{2}$
g $\frac{1}{\sqrt{2}}$ **h** $\frac{1}{\sqrt{2}}$ **i** 5 **j** 5 **k** $5\sqrt{5}$ **l** $5\sqrt{5}$

4 **a** 1 **b** $|r|$

6 **b** $|z_1 z_2 z_3 .... z_n| = |z_1|\,|z_2| .... |z_n|$ **e** $2^{20} = 1\,048\,576$

7 **a** 6 **b** 9 **c** $3\sqrt{5}$ **d** 3 **e** $\frac{1}{3}$ **f** $\frac{2}{9}$

8 **a** $\left[\dfrac{a^2+b^2-1}{(a-1)^2+b^2}\right] + \left[\dfrac{-2b}{(a-1)^2+b^2}\right]i$ **b** 0

9 **a** 3 **b** 2

EXERCISE 16B.2

1 **a** **i** $4\sqrt{2}$ units **ii** $(1, 4)$
b **i** $5\sqrt{5}$ units **ii** $(-\frac{3}{2}, 2)$

2 **a** **i** $w+z$ **ii** $w-z$

4 **a** reflection in the $\mathcal{R}$-axis **b** rotation of π about O
c reflection in the $\mathcal{J}$-axis
d clockwise rotation of $\frac{\pi}{2}$ about O

5 $z = 2+6i$

EXERCISE 16C.1

1 **a** $4 \text{ cis } 0$ **b** $4 \text{ cis } \frac{\pi}{2}$ **c** $6 \text{ cis } \pi$
d $3 \text{ cis } \left(-\frac{\pi}{2}\right)$ **e** $\sqrt{2} \text{ cis } \frac{\pi}{4}$ **f** $2\sqrt{2} \text{ cis } \left(-\frac{\pi}{4}\right)$
g $2 \text{ cis } \left(\frac{5\pi}{6}\right)$ **h** $4 \text{ cis } \frac{\pi}{6}$

2 0

3 $k\sqrt{2} \text{ cis } \left(\frac{\pi}{4}\right)$ if $k > 0$, $-k\sqrt{2} \text{ cis } \left(-\frac{3\pi}{4}\right)$ if $k < 0$,
not possible if $k = 0$

4 **a** $2i$ **b** $4\sqrt{2}+4\sqrt{2}i$ **c** $2\sqrt{3}+2i$
d $1-i$ **e** $-\frac{\sqrt{3}}{2}+\frac{3}{2}i$ **f** -5

5 **a** 1 **b** 1

EXERCISE 16C.2

1 **a** $\text{cis } 3\theta$ **b** $\text{cis } 2\theta$ **c** $\text{cis } 3\theta$ **d** $\frac{\sqrt{3}}{2}+\frac{1}{2}i$
e $\sqrt{2}+i\sqrt{2}$ **f** 8 **g** $-2i$ **h** -4 **i** $4i$

2 **a** -1 **b** -1 **c** $\frac{1}{2}+\frac{\sqrt{3}}{2}i$

3 **a** $|z| = 2$, $\arg(z) = \theta$ **b** $2 \text{ cis } (-\theta)$
c $2 \text{ cis } (\theta+\pi)$ **d** $2 \text{ cis } (\pi-\theta)$

4 **a** $\text{cis } \left(\frac{\pi}{2}\right)$ **b** $r \text{ cis } (\theta+\frac{\pi}{2})$
c clockwise rotation of $\frac{\pi}{2}$ about O

5 **a** **i** $\text{cis } (-\theta)$ **ii** $\text{cis } (\theta-\frac{\pi}{2})$
b If $z = r \text{ cis } \theta$ then $z^* = r \text{ cis } (-\theta)$ in polar form.

EXERCISE 16C.3

1 **a** $z = 2\sqrt{2} \text{ cis } \left(-\frac{\pi}{4}\right)$ **b** $\frac{2\sqrt{2}}{3} \text{ cis } \left(\frac{\pi}{12}\right)$
c

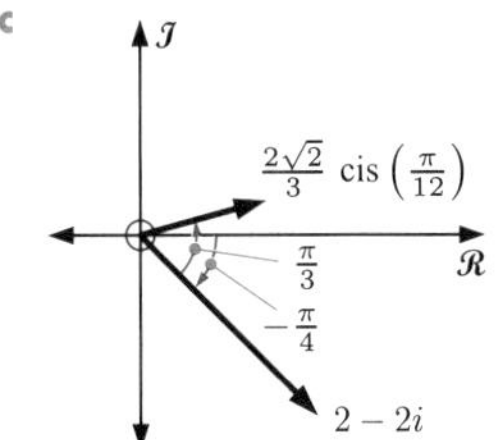

d The modulus of z has been multiplied by $\frac{1}{3}$, then rotated anti-clockwise through $\frac{\pi}{3}$ about the origin.

2 **a** $z = 2 \text{ cis } \left(\frac{5\pi}{6}\right)$ **b** $\text{cis } \left(\frac{\pi}{2}\right) = i$

c

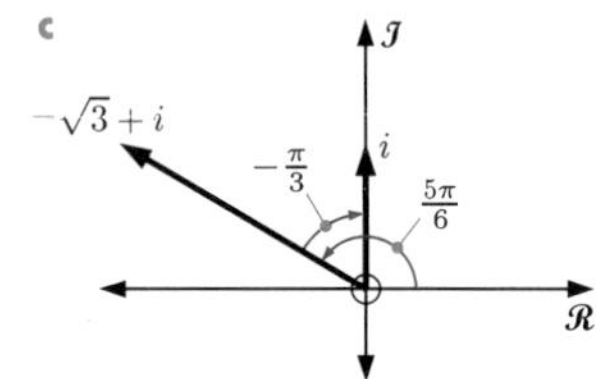

d The modulus of z has been halved, then rotated clockwise through $\frac{\pi}{3}$ about the origin.

3 **a** $\cos\left(\frac{\pi}{12}\right)=\frac{\sqrt{2}+\sqrt{6}}{4}$, $\sin\left(\frac{\pi}{12}\right)=\frac{\sqrt{6}-\sqrt{2}}{4}$

b $\cos\left(\frac{11\pi}{12}\right)=\frac{-\sqrt{2}-\sqrt{6}}{4}$, $\sin\left(\frac{11\pi}{12}\right)=\frac{\sqrt{6}-\sqrt{2}}{4}$

6 **a** $|-z|=3,\ \arg(-z)=\theta-\pi$

b $|z^*|=3,\ \arg(z^*)=-\theta$ **c** $|iz|=3,\ \arg(iz)=\theta+\frac{\pi}{2}$

d $|(1+i)z|=3\sqrt{2},\ \arg((1+i)z)=\theta+\frac{\pi}{4}$

e $\left|\frac{z}{i}\right|=3,\ \arg\left(\frac{z}{i}\right)=\theta-\frac{\pi}{2}$

f $\left|\frac{z}{1-i}\right|=\frac{3}{\sqrt{2}},\ \arg\left(\frac{z}{1-i}\right)=\theta+\frac{\pi}{4}$

7 **a** $|z-1|=2\sin\frac{\phi}{2},\ \arg(z-1)=\frac{\phi}{2}+\frac{\pi}{2}$

b $z-1=\left(2\sin(\frac{\phi}{2})\right)\operatorname{cis}\left(\frac{\phi}{2}+\frac{\pi}{2}\right)$

c $(z-1)^*=\left(2\sin(\frac{\phi}{2})\right)\operatorname{cis}\left(-\frac{\phi}{2}-\frac{\pi}{2}\right)$

8 **a** $z_2-z_1=\overrightarrow{AB},\quad z_3-z_2=\overrightarrow{BC}$ **b** $\left|\frac{z_2-z_1}{z_3-z_2}\right|=1$

c $\arg\left(\frac{z_2-z_1}{z_3-z_2}\right)=\frac{2\pi}{3}$ **d** 1

9 $a=-\sqrt{3}$

EXERCISE 16C.4

1 **a** $-1.41+1.01i$ **b** $1.27-3.06i$ **c** $-2.55-1.25i$

2 **a** $5\operatorname{cis}(-0.927)$ **b** $13\operatorname{cis}(-1.97)$ **c** $17.7\operatorname{cis}(2.29)$

3 **a** $2\operatorname{cis}\frac{\pi}{4}$ **b** $\sqrt{19}\operatorname{cis}(-2.50)$

4 **a** $a(x^2+2x+4)=0,\ a\neq 0$

b $a(x^2-2x+2)=0,\ a\neq 0$

EXERCISE 16D

1 **a** -1 **b** $\frac{1}{2}+\frac{\sqrt{3}}{2}i$ **c** $-i$

3 **a** $\frac{\theta}{2}$ **b** $\frac{\pi}{2}+\theta$ **c** $2\theta-\frac{\pi}{2}$ **d** $\frac{\pi}{2}-\theta$

4 **a** $\approx 0.540+0.841i$ **b** $\approx 0.455+0.891i$ **c** $e^{-\frac{\pi}{2}}$

d -1

EXERCISE 16E

2 **a** 32 **b** -1 **c** $-64i$

d $\sqrt{5}\operatorname{cis}\left(\frac{\pi}{14}\right)\approx(2.180+0.498i)$ **e** $\sqrt{3}+i$

f $16+16\sqrt{3}i$

3 **a** $128-128i$ **b** $1024+1024\sqrt{3}i$

c $\frac{1}{524\,288}\left(\frac{-1}{\sqrt{2}}+\frac{1}{\sqrt{2}}i\right)$ **d** $\frac{1}{64}(1-i)$

e $\sqrt{2}\cos\left(-\frac{\pi}{12}\right)+i\sqrt{2}\sin\left(-\frac{\pi}{12}\right)$ **f** $\frac{1}{64}(-\sqrt{3}-i)$

4 **a** $|z|^{\frac{1}{2}}\operatorname{cis}\left(\frac{\theta}{2}\right)$ **b** $-\frac{\pi}{2}<\phi\leqslant\frac{\pi}{2}$ **c** True **5** $\operatorname{cis}3\theta$

6 **a** $1+i=\sqrt{2}\operatorname{cis}\left(\frac{\pi}{4}\right),\ z^n=2^{\frac{n}{2}}\operatorname{cis}\left(\frac{n\pi}{4}\right)$

b **i** $n=4k$, k any integer **ii** $n=2+4k,\ k\in\mathbb{Z}$

7 **a** $\left|z^3\right|=8,\ \arg(z^3)=3\theta$

b $\left|iz^2\right|=4,\ \arg(iz^2)=\frac{\pi}{2}+2\theta$

c $\left|\frac{1}{z}\right|=\frac{1}{2},\ \arg\left(\frac{1}{z}\right)=-\theta$

d $\left|\frac{-i}{z^2}\right|=\frac{1}{4},\ \arg\left(\frac{-i}{z^2}\right)=-\frac{\pi}{2}-2\theta$

9 **b** $\tan 3\theta=\dfrac{3\tan\theta-\tan^3\theta}{1-3\tan^2\theta}$

c **i** $x=\frac{1}{\sqrt{2}}$, $\cos\left(\frac{5\pi}{12}\right)$, $\cos\left(\frac{11\pi}{12}\right)$

ii $x=\tan\left(\frac{\pi}{9}\right)$, $\tan\left(\frac{4\pi}{9}\right)$, $\tan\left(\frac{7\pi}{9}\right)$

10 **a** $\overrightarrow{AB}\equiv z_2-z_1$, $\overrightarrow{BC}\equiv z_3-z_2$

Hint: Notice that $\overrightarrow{BC}$ is a $90°$ rotation of $\overrightarrow{BA}$ about B.

b $\overrightarrow{OD}\equiv z_1+z_3-z_2$

11 **a** $\cos 4\theta=8\cos^4\theta-8\cos^2\theta+1$

b $\sin 4\theta=4\cos^3\theta\sin\theta-4\cos\theta\sin^3\theta$

12 **a** **iii** $\left(z+\frac{1}{z}\right)^3=z^3+3z+\frac{3}{z}+\frac{1}{z^3}$

b **Hint:** When $n=1$, $2i\sin\theta=z-\frac{1}{z}$. Now cube both sides.

EXERCISE 16F.1

1 $1,\ -\frac{1}{2}\pm i\frac{\sqrt{3}}{2}$

2 **a** $z=\sqrt{3}-i,\ 2i,\ -\sqrt{3}-i$

b $z=\frac{3\sqrt{3}}{2}-\frac{3}{2}i,\ 3i,\ -\frac{3\sqrt{3}}{2}-\frac{3}{2}i$

3 $-1,\ \frac{1}{2}\pm\frac{\sqrt{3}}{2}i$

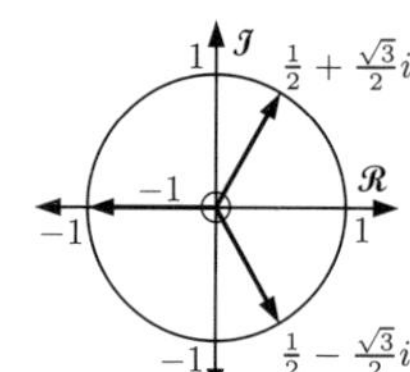

4 **a** $z=\pm 2,\ \pm 2i$

b $z=\sqrt{2}\pm i\sqrt{2},\ -\sqrt{2}\pm i\sqrt{2}$

5 $\operatorname{cis}\left(\frac{3\pi}{8}\right)$, $\operatorname{cis}\left(\frac{7\pi}{8}\right)$, $\operatorname{cis}\left(\frac{-\pi}{8}\right)$, $\operatorname{cis}\left(\frac{-5\pi}{8}\right)$

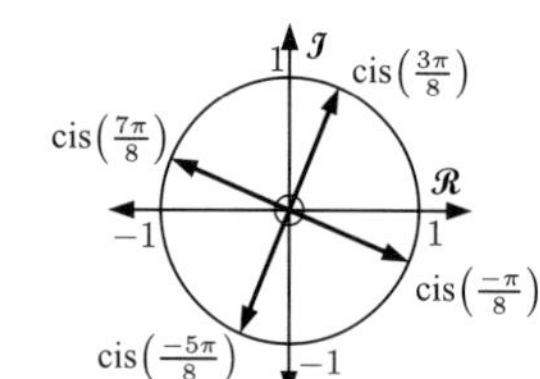

6 **a** $z=\sqrt{2}\operatorname{cis}\left(\frac{\pi}{12}\right)$, $\sqrt{2}\operatorname{cis}\left(\frac{3\pi}{4}\right)=-1+i$, $\sqrt{2}\operatorname{cis}\left(\frac{-7\pi}{12}\right)$

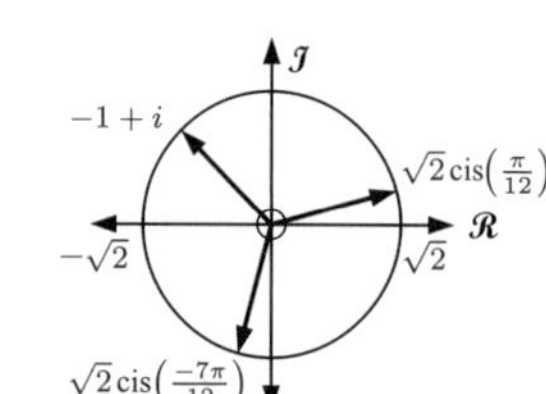

b $z=\sqrt{2}\operatorname{cis}\left(\frac{\pi}{4}\right)=1+i$, $\sqrt{2}\operatorname{cis}\left(\frac{11\pi}{12}\right)$, $\sqrt{2}\operatorname{cis}\left(\frac{-5\pi}{12}\right)$

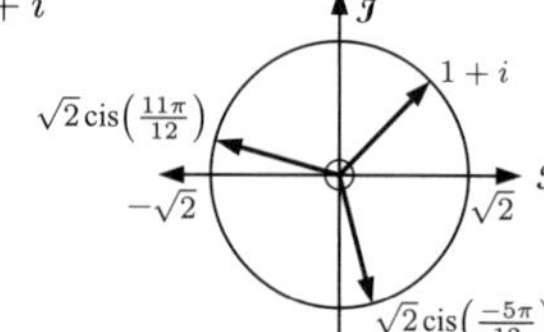

c $z = \frac{\sqrt{3}}{2} + \frac{1}{2}i$ or $-\frac{\sqrt{3}}{2} - \frac{1}{2}i$

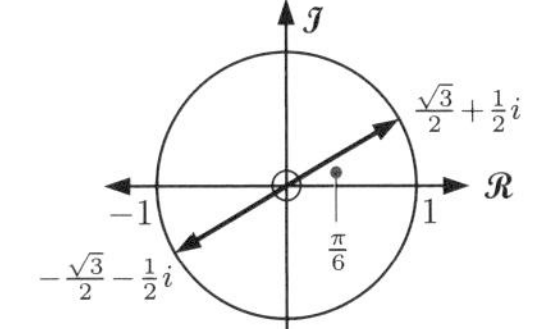

d $z = \sqrt[4]{2}\text{ cis}\left(\frac{\pi}{24}\right)$, $\sqrt[4]{2}\text{ cis}\left(\frac{13\pi}{24}\right)$, $\sqrt[4]{2}\text{ cis}\left(\frac{-11\pi}{24}\right)$, $\sqrt[4]{2}\text{ cis}\left(\frac{-23\pi}{24}\right)$

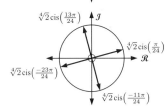

e $z = \sqrt{2}\text{ cis}\left(\frac{-3\pi}{20}\right)$, $\sqrt{2}\text{ cis}\left(\frac{\pi}{4}\right) = 1+i$, $\sqrt{2}\text{ cis}\left(\frac{13\pi}{20}\right)$, $\sqrt{2}\text{ cis}\left(\frac{-11\pi}{20}\right)$, $\sqrt{2}\text{ cis}\left(\frac{-19\pi}{20}\right)$

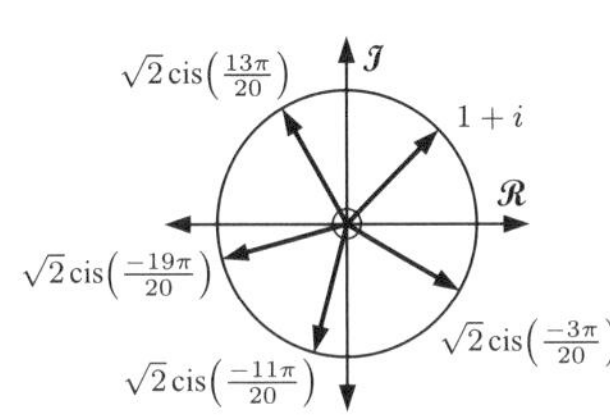

f $z = \sqrt[3]{4}\text{ cis}\left(\frac{-5\pi}{18}\right)$, $\sqrt[3]{4}\text{ cis}\left(\frac{7\pi}{18}\right)$, $\sqrt[3]{4}\text{ cis}\left(\frac{-17\pi}{18}\right)$

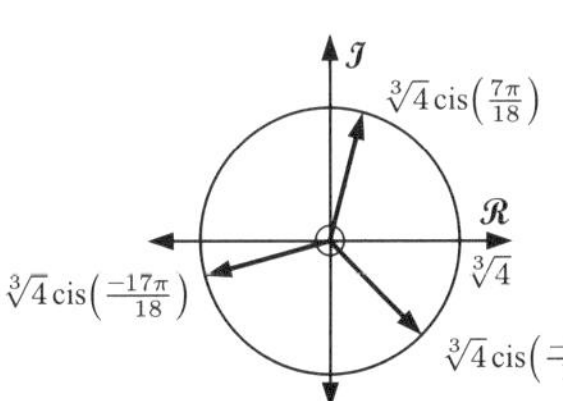

7 **a** $z = \frac{1}{\sqrt{2}} - \frac{1}{\sqrt{2}}i$, $\frac{1}{\sqrt{2}} + \frac{1}{\sqrt{2}}i$, $-\frac{1}{\sqrt{2}} + \frac{1}{\sqrt{2}}i$, or $-\frac{1}{\sqrt{2}} - \frac{1}{\sqrt{2}}i$

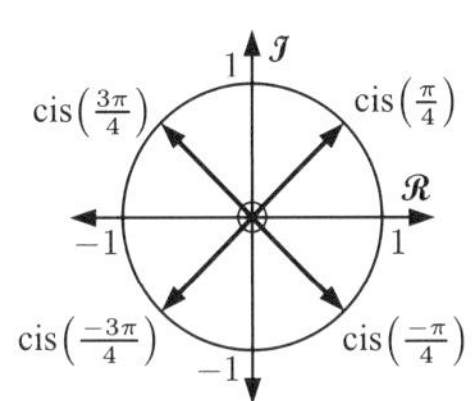

b $z^4 + 1 = (z^2 - \sqrt{2}z + 1)(z^2 + \sqrt{2}z + 1)$

8 **a** $|z| = 1$ and $\arg z = -\frac{2\pi}{3}$ **b** $z^3 = \text{cis}\,(-2\pi) = 1$

c Simplifies to $2(z + z^*) - 5$ where $z + z^*$ is always real.

9 **a** $16\text{ cis}\left(-\frac{\pi}{2}\right)$

b **i** $2\text{ cis}\left(\frac{7\pi}{8}\right)$ **ii** $2\cos\left(\frac{7\pi}{8}\right) + [2\sin\left(\frac{7\pi}{8}\right)]i$

EXERCISE 16F.2

1 **a** **i** $z = w^n - 3$ $(n = 0, 1, 2)$ and $w = \text{cis}\left(\frac{2\pi}{3}\right)$

ii $z = 2w^n + 1$ $(n = 0, 1, 2)$ and $w = \text{cis}\left(\frac{2\pi}{3}\right)$

iii $z = \dfrac{1 - w^n}{2}$ $(n = 0, 1, 2)$ and $w = \text{cis}\left(\frac{2\pi}{3}\right)$

3 **a** $z = \text{cis}\left(-\frac{4\pi}{5}\right)$, $\text{cis}\left(-\frac{2\pi}{5}\right)$, $\text{cis}\,0$, $\text{cis}\left(\frac{2\pi}{5}\right)$, $\text{cis}\left(\frac{4\pi}{5}\right)$

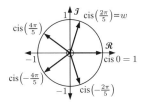

c $1 - w^5$

4 **b** **ii** **Hint:** The LHS is a geometric series.

EXERCISE 16G

1 $z = 1 + 2i$ or $1 - i$

2 **a** $y = x$ **b** $y = \frac{1}{\sqrt{3}}x + 1,\ x > 0$ **c** $7x^2 + 16y^2 = 112$

4 $\dfrac{\cos n\theta - \cos\theta - \cos[(n+1)\theta] + 1}{2 - 2\cos\theta}$

5 $2^n \cos^n\left(\frac{\theta}{2}\right)\cos\left(\frac{n\theta}{2}\right)$

REVIEW SET 16A

1 Real part is $16\sqrt{3}$. Imaginary part is 16.

2 **a** $2x + 4y = -1$ **b** $y = -x$ **3** $|z| = 4$

4 **a** reflection in $\mathcal{R}$-axis
b anti-clockwise rotation of π about O
c anti-clockwise rotation of $\frac{\pi}{2}$ about O

8 **b** $z = \dfrac{\alpha+2}{\alpha-1}, \dfrac{\alpha^2+2}{\alpha^2-1}, \dfrac{\alpha^3+2}{\alpha^3-1}, \dfrac{\alpha^4+2}{\alpha^4-1}$ where $\alpha = \text{cis}\left(\frac{2\pi}{5}\right)$

11 $2\sqrt{3} - 2i,\ -2\sqrt{3} - 2i,\ 4i$

12 **a** $\left|(2z)^{-1}\right| = \frac{1}{2}$, $\arg\left((2z)^{-1}\right) = -\theta$

b $|1 - z| = 2\sin\left(\frac{\theta}{2}\right)$, $\arg(1 - z) = \frac{\theta}{2} - \frac{\pi}{2}$

13 $-1 + i\sqrt{3} = 2\text{ cis}\left(\frac{2\pi}{3}\right)$, $m = \frac{3k}{2}$, k is an integer

REVIEW SET 16B

1 $\frac{1}{\sqrt{2}} - \frac{1}{\sqrt{2}}i$

2 **a** $5 + 2i$ **b** $2\sqrt{2}$ **c** 17^5 **d** ≈ -2.03

3 $a = 0,\ b = -1$

4 **a** $x = 0,\ y > 1$ **b** $3x^2 + 3y^2 - 20x + 12 = 0$

5 $4\text{ cis}\left(-\frac{\pi}{3}\right)$, $n = 3k$, k is an integer **6** $\frac{3}{2} \pm \frac{3\sqrt{3}}{2}i,\ -3$

7 **a** $\left|z^3\right| = 64$, $\arg\left(z^3\right) = 3\theta$ **b** $\left|\frac{1}{z}\right| = \frac{1}{4}$, $\arg\left(\frac{1}{z}\right) = -\theta$

c $|iz^*| = 4$, $\arg(iz^*) = \frac{\pi}{2} - \theta$

9 **a** $\frac{5\pi}{12}$ **b** $-\frac{11\pi}{12}$

10 $2 \leqslant |z| \leqslant 5$, $-\frac{\pi}{4} < \arg z \leqslant \frac{\pi}{2}$

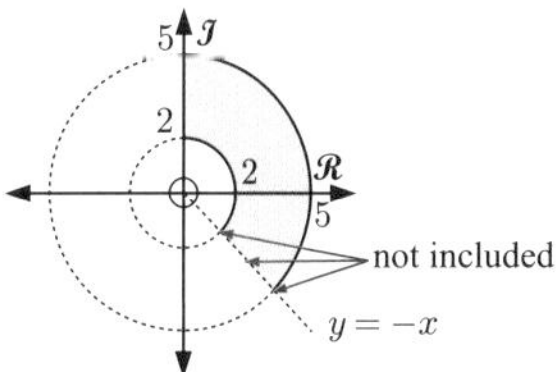

12 $1 + z = 2\cos\left(\frac{\alpha}{2}\right)\text{cis}\left(\frac{\alpha}{2}\right)$, $|1 + z| = 2\cos\left(\frac{\alpha}{2}\right)$, $\arg(1 + z) = \frac{\alpha}{2}$

13 **b** $\left|\dfrac{z_2 - z_1}{z_3 - z_2}\right| = 1$, $\arg\left(\dfrac{z_2 - z_1}{z_3 - z_2}\right) = \frac{2\pi}{3}$

14 $z = 1, w, w^2, w^3, w^4$ where $w = \text{cis}\left(\frac{2\pi}{5}\right)$

a $z = \frac{3}{2},\ w + \frac{1}{2},\ w^2 + \frac{1}{2},\ w^3 + \frac{1}{2},\ w^4 + \frac{1}{2}$, $w = \text{cis}\left(\frac{2\pi}{5}\right)$

b $z = 0,\ w - 1,\ w^2 - 1,\ w^3 - 1,\ w^4 - 1$, $w = \text{cis}\left(\frac{2\pi}{5}\right)$

c $z = \dfrac{w+1}{w-1}, \dfrac{w^2+1}{w^2-1}, \dfrac{w^3+1}{w^3-1}, \dfrac{w^4+1}{w^4-1}$, $w = \text{cis}\left(\frac{2\pi}{5}\right)$

REVIEW SET 16C

1 **a** $5\text{ cis}\left(-\frac{\pi}{2}\right)$ **b** $4\text{ cis}\left(-\frac{\pi}{3}\right)$ **c** $-k\sqrt{2}\text{ cis}\left(\frac{3\pi}{4}\right)$, $k < 0$

2 $b = \frac{1}{\sqrt{3}}$

3 **b** $(1 - i)z = 4\text{ cis}\left(\alpha - \frac{\pi}{4}\right)$, $\arg((1 - i)z) = \alpha - \frac{\pi}{4}$

4 **a** $\left|\dfrac{{z_1}^2}{{z_2}^2}\right| = 1, \quad \arg\left(\dfrac{{z_1}^2}{{z_2}^2}\right) = \pi$

5 $\mathcal{R}e\left(\left(\dfrac{z}{w}\right)^4\right) = \dfrac{a}{2b}, \quad \mathcal{I}m\left(\left(\dfrac{z}{w}\right)^4\right) = \dfrac{-a\sqrt{3}}{2b}$

6 **a** $z = 1, w, w^2, w^3$, and w^4 where $w = \text{cis}\left(\frac{2\pi}{5}\right)$

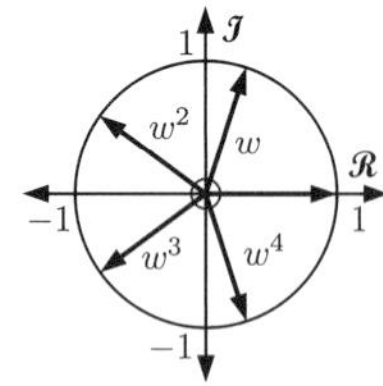

c 31

7 $\sqrt{3} - i, \ 2i, \ -\sqrt{3} - i$

8 **a** $(\text{cis}\,\theta)^3$ **b** $(\text{cis}\,\theta)^{-2}$ **c** $(\text{cis}\,\theta)^{-1}$

9 $z = 2^{0.3}\,\text{cis}\left(\frac{\pi}{20}\right), \ 2^{0.3}\,\text{cis}\left(\frac{9\pi}{20}\right), \ 2^{0.3}\,\text{cis}\left(\frac{17\pi}{20}\right),$
$2^{0.3}\,\text{cis}\left(\frac{-7\pi}{20}\right), \ 2^{0.3}\,\text{cis}\left(\frac{-3\pi}{4}\right)$

12 **a** $a(z^2 - 2\cos\left(\frac{2\pi}{5}\right)z + 1) = 0, \ a \neq 0$
b $a(z^2 + z - 1) = 0, \ a \neq 0$

14 **b** **i**

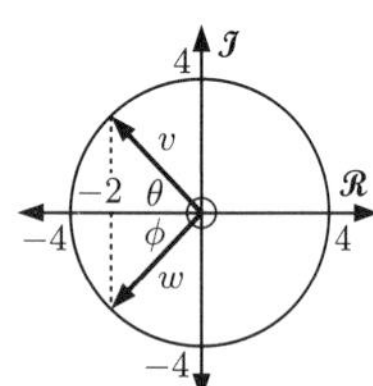

ii $\cos\theta = \frac{1}{2} \Rightarrow \theta = \frac{\pi}{3}$
$\therefore \ \arg v = \frac{2\pi}{3}$
iii $\arg w = -\frac{2\pi}{3}$
iv $\dfrac{\pi(4m - 7)}{6}$
v $m = \frac{7}{4}$

EXERCISE 17A

1 **a** 7 **b** 7 **c** 11 **d** 16 **e** 0 **f** 5
2 **a** 5 **b** 7 **c** c
3 **a** -2 **b** 7 **c** -1 **d** 1
4 **a** $x = 0$ **b** $x = 0$
5 **a** -3 **b** 5 **c** -1 **d** 6 **e** -4 **f** -8 **g** 1 **h** 2 **i** 5

EXERCISE 17B.1

1 $\lim\limits_{x\to\infty} \dfrac{1}{x^2} = 0$ **2** **a** 3 **b** $-\frac{2}{3}$ **c** -1 **d** 1 **e** 1

EXERCISE 17B.2

1 **a** **i** as $x \to 0^-$, $f(x) \to -\infty$
as $x \to 0^+$, $f(x) \to \infty$
as $x \to \infty$, $f(x) \to 0^+$
as $x \to -\infty$, $f(x) \to 0^-$
vertical asymptote $x = 0$, horizontal asymptote $y = 0$
ii $\lim\limits_{x\to-\infty} f(x) = 0, \quad \lim\limits_{x\to\infty} f(x) = 0$

b **i** as $x \to -3^-$, $f(x) \to \infty$
as $x \to -3^+$, $f(x) \to -\infty$
as $x \to \infty$, $f(x) \to 3^-$
as $x \to -\infty$, $f(x) \to 3^+$
vertical asymptote $x = -3$, horizontal asymptote $y = 3$
ii $\lim\limits_{x\to-\infty} f(x) = 3, \quad \lim\limits_{x\to\infty} f(x) = 3$

c **i** as $x \to -\frac{2}{3}^-$, $f(x) \to -\infty$
as $x \to -\frac{2}{3}^+$, $f(x) \to \infty$
as $x \to \infty$, $f(x) \to -\frac{2}{3}^+$
as $x \to -\infty$, $f(x) \to -\frac{2}{3}^-$
vertical asymptote $x = -\frac{2}{3}$,
horizontal asymptote $y = -\frac{2}{3}$
ii $\lim\limits_{x\to-\infty} f(x) = -\frac{2}{3}, \quad \lim\limits_{x\to\infty} f(x) = -\frac{2}{3}$

d **i** as $x \to 1^-$, $f(x) \to \infty$,
as $x \to 1^+$, $f(x) \to -\infty$,
as $x \to \infty$, $f(x) \to -1^-$
as $x \to -\infty$, $f(x) \to -1^+$
vertical asymptote $x = 1$,
horizontal asymptote $y = -1$
ii $\lim\limits_{x\to-\infty} f(x) = -1, \quad \lim\limits_{x\to\infty} f(x) = -1$

e **i** as $x \to \infty$, $f(x) \to 1^-$ horizontal asymptote $y = 1$
as $x \to -\infty$, $f(x) \to 1^-$ no vertical asymptote
ii $\lim\limits_{x\to-\infty} f(x) = 1, \quad \lim\limits_{x\to\infty} f(x) = 1$

f **i** as $x \to \infty$, $f(x) \to 0^+$
as $x \to -\infty$, $f(x) \to 0^-$
horizontal asymptote $y = 0$
no vertical asymptote
ii $\lim\limits_{x\to-\infty} f(x) = 0, \quad \lim\limits_{x\to\infty} f(x) = 0$

2 **a**

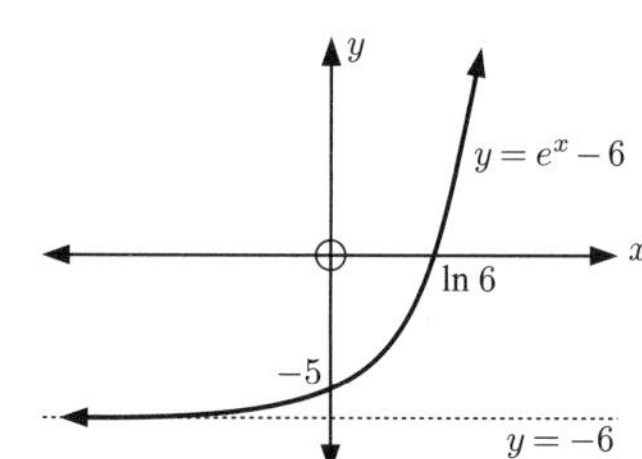

b **i** $\lim\limits_{x\to-\infty} (e^x - 6) = -6$
ii $\lim\limits_{x\to\infty} (e^x - 6)$ does not exist
$y = -6$ is a horizontal asymptote of $y = e^x - 6$.

3 $\lim\limits_{x\to-\infty} (2e^{-x} - 3)$ does not exist, $\lim\limits_{x\to\infty} (2e^{-x} - 3) = -3$

4 **a** vertical asymptote $x = 0$ as $x \to 0^+$, $y \to -\infty$
b vertical asymptote $x = 0$, horizontal asymptote $y = 0$
as $x \to 0^-$, $f(x) \to \infty$ as $x \to \infty$, $f(x) \to \infty$
as $x \to 0^+$, $f(x) \to 0^+$ as $x \to -\infty$, $f(x) \to 0^+$

5 **a** vertical asymptote $x = 0$ as $x \to 0^+$, $y \to -\infty$
b oblique asymptote $y = -x$ as $x \to \infty$, $y \to \infty$
as $x \to -\infty$, $y \to (-x)^+$
c oblique asymptote $y = x$ as $x \to \infty$, $y \to x^-$
as $x \to -\infty$, $y \to x^+$
d horizontal asymptote $y = 0$ as $x \to \infty$, $y \to 0^+$

EXERCISE 17C

1 **a** 2 **b** 1 **c** 1 **d** 4 **e** $\frac{1}{2}$ **f** 2π
2 **b** πr^2 **c** Area of circle $= \pi r^2$ **3** **c** $-\sin x$

EXERCISE 17D

1 **a**

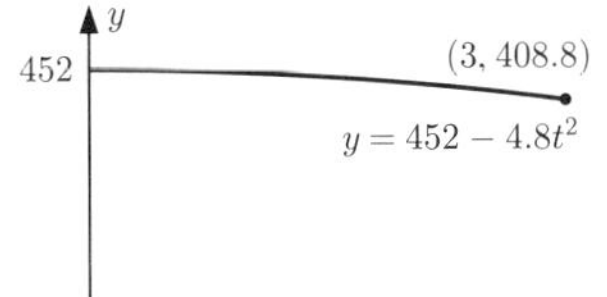

b No

c **i** 0 m s^{-1} **ii** 9.6 m s^{-1} **iii** 19.2 m s^{-1} **iv** 28.8 m s^{-1}

2 **a**

x	Point B	Gradient of [AB]
0	(0, 0)	2
1	(1, 1)	3
1.5	(1.5, 2.25)	3.5
1.9	(1.9, 3.61)	3.9
1.99	(1.99, 3.9601)	3.99
1.999	(1.999, 3.996 001)	3.999

x	Point B	Gradient of [AB]
5	(5, 25)	7
3	(3, 9)	5
2.5	(2.5, 6.25)	4.5
2.1	(2.1, 4.41)	4.1
2.01	(2.01, 4.0401)	4.01
2.001	(2.001, 4.004 001)	4.001

b $\lim\limits_{x \to 2} \dfrac{x^2 - 4}{x - 2} = 4$
The gradient of the tangent to $y = x^2$ at the point (2, 4) is 4.

EXERCISE 17E

1 **a** 3 **b** 0 **2** **a** 4 **b** -1 **3** $f(2) = 3$, $f'(2) = 1$

EXERCISE 17F

1 **a** **i** 1 **ii** 0 **iii** $3x^2$ **iv** $4x^3$ **b** $f'(x) = nx^{n-1}$

2 **a** 2 **b** $2x - 3$ **c** $-2x + 5$

3 **a** $\dfrac{dy}{dx} = -1$ **b** $\dfrac{dy}{dx} = 4x + 1$ **c** $\dfrac{dy}{dx} = 3x^2 - 4x$

4 **a** 12 **b** 108 **5** **a** 3 **b** -12 **c** 9 **d** 10

6 **a** $\dfrac{dy}{dx} = 3x^2 - 3$ **b** $(-1, 2)$ and $(1, -2)$

7 **a** $\dfrac{-4}{x^2}$ **b** $\dfrac{-9}{(x-2)^2}$

8 **a** $-\frac{2}{27}$ **b** $-\frac{45}{289}$ **c** $\frac{1}{4}$ **d** $-\frac{1}{2}$

REVIEW SET 17A

1 **a** -1 **b** -1 **c** 8

2 **a** $f'(x) = 2x + 2$ **b** $\dfrac{dy}{dx} = -6x$

3 **a** horizontal asymptote $y = -3$
as $x \to \infty$, $y \to \infty$ as $x \to -\infty$, $y \to -3^+$
b no asymptotes
c vertical asymptote $x = 0$ as $x \to 0^-$, $y \to -\infty$

4 **a** 4 **b** $\frac{2}{3}$ **c** π **5** 3

6 **a** $f'(t) = -9.6t \text{ m s}^{-1}$
b 19.2 m s^{-1} (the $-$ sign indicates travelling downwards)

REVIEW SET 17B

1 **a** as $x \to -\frac{2}{3}^-$, $y \to \infty$, as $x \to \infty$, $y \to \frac{1}{3}^-$
as $x \to -\frac{2}{3}^+$, $y \to -\infty$, as $x \to -\infty$, $y \to \frac{1}{3}^+$
vertical asymptote $x = -\frac{2}{3}$, horizontal asymptote $y = \frac{1}{3}$

b $\lim\limits_{x \to -\infty} \dfrac{x - 7}{3x + 2} = \frac{1}{3}$, $\lim\limits_{x \to \infty} \dfrac{x - 7}{3x + 2} = \frac{1}{3}$

2 **a**

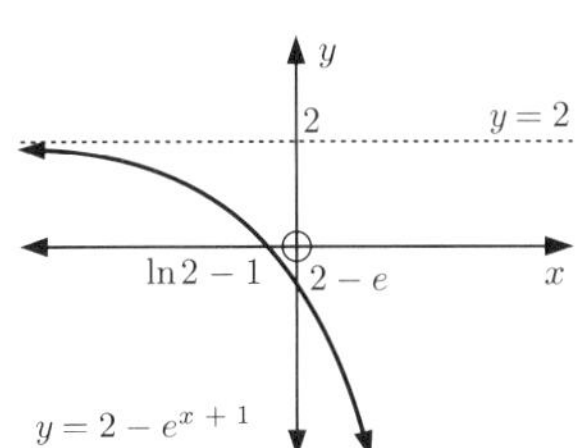

b $\lim\limits_{x \to -\infty} (2 - e^{x+1}) = 2$,
$\lim\limits_{x \to \infty} (2 - e^{x+1})$ does not exist

c $y = 2$

3 **a** $x = 0$ **b** $x = 1$ **4** $\cos x$, h and x are in radians

5 **b** **i** 12.2 **ii** 12.02 **iii** 12
c The gradient of the tangent to $y = 2x^2$ at $(3, 18)$ is 12.

6 $\lim\limits_{x \to \infty} \left(\dfrac{2x + 3}{4 - x}\right) = \lim\limits_{x \to -\infty} \left(\dfrac{2x + 3}{4 - x}\right) = -2$

REVIEW SET 17C

1 **a** -3 **b** 3 **c** -1

2 **a**

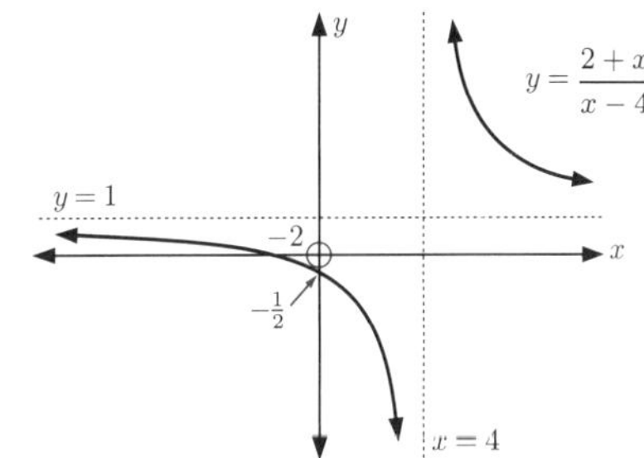

b as $x \to 4^-$, $y \to -\infty$, as $x \to \infty$, $y \to 1^+$,
as $x \to 4^+$, $y \to \infty$, as $x \to -\infty$, $y \to 1^-$
vertical asymptote $x = 4$, horizontal asymptote $y = 1$

c $\lim\limits_{x \to -\infty} \dfrac{2 + x}{x - 4} = 1$, $\lim\limits_{x \to \infty} \dfrac{2 + x}{x - 4} = 1$

3 $f'(1) = 2$ **4** **c** $\cos x$

5 **a** $\dfrac{dy}{dx} = 4x$ **b** gradient $= 16$ **c** when $x = -3$

EXERCISE 18A

1 **a** $3x^2$ **b** $6x^2$ **c** $14x$ **d** $\dfrac{3}{\sqrt{x}}$
e $\dfrac{1}{\sqrt[3]{x^2}}$ **f** $2x + 1$ **g** $-4x$ **h** $2x + 3$
i $2x^3 - 12x$ **j** $\dfrac{6}{x^2}$ **k** $-\dfrac{2}{x^2} + \dfrac{6}{x^3}$
l $2x - \dfrac{5}{x^2}$ **m** $2x + \dfrac{3}{x^2}$ **n** $-\dfrac{1}{2x\sqrt{x}}$
o $8x - 4$ **p** $3x^2 + 12x + 12$

2 **a** $7.5x^2 - 2.8x$ **b** $2\pi x$ **c** $-\dfrac{2}{5x^3}$
d 100 **e** 10 **f** $12\pi x^2$

3 **a** 6 **b** $\dfrac{3\sqrt{x}}{2}$ **c** $2x-10$ **d** $2-9x^2$ **e** $2x-1$

f $-\dfrac{2}{x^3}+\dfrac{3}{\sqrt{x}}$ **g** $4+\dfrac{1}{4x^2}$ **h** $6x^2-6x-5$

4 **a** 4 **b** $-\frac{16}{729}$ **c** -7 **d** $\frac{13}{4}$ **e** $\frac{1}{8}$ **f** -11

5 $b=3,\ c=-4$

6 **a** $\dfrac{2}{\sqrt{x}}+1$ **b** $\frac{1}{3}x^{-\frac{2}{3}}$ **c** $\dfrac{1}{x\sqrt{x}}$ **d** $2-\dfrac{1}{2\sqrt{x}}$

e $-\dfrac{2}{x\sqrt{x}}$ **f** $6x-\frac{3}{2}\sqrt{x}$ **g** $-\frac{25}{2}x^{-\frac{7}{2}}$ **h** $2+\frac{9}{2}x^{-\frac{5}{2}}$

7 **a** $\dfrac{dy}{dx}=4+\dfrac{3}{x^2}$, $\dfrac{dy}{dx}$ is the gradient function of $y=4x-\dfrac{3}{x}$ from which the gradient at any point can be found.

b $\dfrac{dS}{dt}=4t+4\ \text{m s}^{-1}$, $\dfrac{dS}{dt}$ is the instantaneous rate of change in position at the time t, or the velocity function.

c $\dfrac{dC}{dx}=3+0.004x$ \$ per toaster, $\dfrac{dC}{dx}$ is the instantaneous rate of change in cost as the number of toasters changes.

EXERCISE 18B.1

1 **a** $g(f(x))=(2x+7)^2$ **b** $g(f(x))=2x^2+7$

c $g(f(x))=\sqrt{3-4x}$ **d** $g(f(x))=3-4\sqrt{x}$

e $g(f(x))=\dfrac{2}{x^2+3}$ **f** $g(f(x))=\dfrac{4}{x^2}+3$

2 **a** $g(x)=x^3,\ \ f(x)=3x+10$

b $g(x)=\dfrac{1}{x},\ \ f(x)=2x+4$

c $g(x)=\sqrt{x},\ \ f(x)=x^2-3x$

d $g(x)=\dfrac{10}{x^3},\ \ f(x)=3x-x^2$

EXERCISE 18B.2

1 **a** $u^{-2},\ u=2x-1$ **b** $u^{\frac{1}{2}},\ u=x^2-3x$

c $2u^{-\frac{1}{2}},\ u=2-x^2$ **d** $u^{\frac{1}{3}},\ u=x^3-x^2$

e $4u^{-3},\ u=3-x$ **f** $10u^{-1},\ u=x^2-3$

2 **a** $8(4x-5)$ **b** $2(5-2x)^{-2}$ **c** $\frac{1}{2}(3x-x^2)^{-\frac{1}{2}}\times(3-2x)$

d $-12(1-3x)^3$ **e** $-18(5-x)^2$

f $\frac{1}{3}(2x^3-x^2)^{-\frac{2}{3}}\times(6x^2-2x)$ **g** $-60(5x-4)^{-3}$

h $-4(3x-x^2)^{-2}\times(3-2x)$ **i** $6(x^2-\dfrac{2}{x})^2\times(2x+\dfrac{2}{x^2})$

3 **a** $-\frac{1}{\sqrt{3}}$ **b** -18 **c** -8 **d** -4 **e** $-\frac{3}{32}$ **f** 0

4 $a=3,\ b=1$ **5** $a=2,\ b=1$

6 **a** $\dfrac{dy}{dx}=3x^2$, $\dfrac{dx}{dy}=\frac{1}{3}y^{-\frac{2}{3}}$ **Hint:** Substitute $y=x^3$

b $\dfrac{dy}{dx}\times\dfrac{dx}{dy}=\dfrac{dy}{dy}=1$

EXERCISE 18C

1 **a** $2x-1$ **b** $4x+2$ **c** $2x(x+1)^{\frac{1}{2}}+\frac{1}{2}x^2(x+1)^{-\frac{1}{2}}$

2 **a** $2x(2x-1)+2x^2$ **b** $4(2x+1)^3+24x(2x+1)^2$

c $2x(3-x)^{\frac{1}{2}}-\frac{1}{2}x^2(3-x)^{-\frac{1}{2}}$

d $\frac{1}{2}x^{-\frac{1}{2}}(x-3)^2+2\sqrt{x}(x-3)$

e $10x(3x^2-1)^2+60x^3(3x^2-1)$

f $\frac{1}{2}x^{-\frac{1}{2}}(x-x^2)^3+3\sqrt{x}(x-x^2)^2(1-2x)$

3 **a** -48 **b** $406\frac{1}{4}$ **c** $\frac{13}{3}$ **d** $\frac{11}{2}$

4 **b** $x=3$ or $\frac{3}{5}$ **c** $x\leqslant 0$ **d** $x=0$

e As we approach the point $x=0$ from the right, the curve has steeper and steeper gradient and approaches vertical.

5 $x=-1$ and $x=-\frac{5}{3}$

EXERCISE 18D

1 **a** $\dfrac{7}{(2-x)^2}$ **b** $\dfrac{2x(2x+1)-2x^2}{(2x+1)^2}$

c $\dfrac{(x^2-3)-2x^2}{(x^2-3)^2}$ **d** $\dfrac{\frac{1}{2}x^{-\frac{1}{2}}(1-2x)+2\sqrt{x}}{(1-2x)^2}$

e $\dfrac{2x(3x-x^2)-(x^2-3)(3-2x)}{(3x-x^2)^2}$

f $\dfrac{(1-3x)^{\frac{1}{2}}+\frac{3}{2}x(1-3x)^{-\frac{1}{2}}}{1-3x}$

2 **a** 1 **b** 1 **c** $-\frac{7}{324}$ **d** $-\frac{28}{27}$

3 **b** **i** never $\{\dfrac{dy}{dx}$ is undefined at $x=-1\}$

ii $x\leqslant 0$ and $x=1$

4 **b** **i** $x=-2\pm\sqrt{11}$ **ii** $x=-2$

c $\dfrac{dy}{dx}$ is zero when the tangent to the function is horizontal (gradient 0), at its turning points or points of horizontal inflection. $\dfrac{dy}{dx}$ is undefined at vertical asymptotes.

EXERCISE 18E

1 **a** $2\dfrac{dy}{dx}$ **b** $-3\dfrac{dy}{dx}$ **c** $3y^2\dfrac{dy}{dx}$ **d** $-y^{-2}\dfrac{dy}{dx}$

e $4y^3\dfrac{dy}{dx}$ **f** $\frac{1}{2}y^{-\frac{1}{2}}\dfrac{dy}{dx}$ **g** $-2y^{-3}\dfrac{dy}{dx}$

h $y+x\dfrac{dy}{dx}$ **i** $2xy+x^2\dfrac{dy}{dx}$ **j** $y^2+2xy\dfrac{dy}{dx}$

2 **a** $-\dfrac{x}{y}$ **b** $-\dfrac{x}{3y}$ **c** $\dfrac{x}{y}$ **d** $\dfrac{2x}{3y^2}$

e $\dfrac{-2x-y}{x}$ **f** $\dfrac{3x^2-2y}{2x}$

3 **a** 1 **b** $-\frac{1}{9}$

EXERCISE 18F

1 **a** $4e^{4x}$ **b** e^x **c** $-2e^{-2x}$ **d** $\frac{1}{2}e^{\frac{x}{2}}$

e $-e^{-\frac{x}{2}}$ **f** $2e^{-x}$ **g** $2e^{\frac{x}{2}}+3e^{-x}$

h $\dfrac{e^x-e^{-x}}{2}$ **i** $-2xe^{-x^2}$ **j** $\dfrac{-1}{x^2}e^{\frac{1}{x}}$

k $20e^{2x}$ **l** $40e^{-2x}$ **m** $2e^{2x+1}$

n $\frac{1}{4}e^{\frac{x}{4}}$ **o** $-4xe^{1-2x^2}$ **p** $-0.02e^{-0.02x}$

2 **a** e^x+xe^x **b** $3x^2e^{-x}-x^3e^{-x}$ **c** $\dfrac{xe^x-e^x}{x^2}$

d $\dfrac{1-x}{e^x}$ **e** $2xe^{3x}+3x^2e^{3x}$ **f** $\dfrac{xe^x-\frac{1}{2}e^x}{x\sqrt{x}}$

g $\frac{1}{2}x^{-\frac{1}{2}}e^{-x}-x^{\frac{1}{2}}e^{-x}$ **h** $\dfrac{e^x+2+2e^{-x}}{(e^{-x}+1)^2}$

3 **a** 108 **b** -1 **c** $\frac{9}{\sqrt{19}}$

4 a $\dfrac{6e^{3x}}{(1-e^{3x})^3}$ b $-\dfrac{e^{-x}}{2}(1-e^{-x})^{-\frac{3}{2}}$

c $\dfrac{1-2e^{-x}+xe^{-x}}{\sqrt{1-2e^{-x}}}$

5 $k=-9$ 6 a $\dfrac{dy}{dx}=2^x\ln 2$ 7 $(0, 0)$ or $(2, \dfrac{4}{e^2})$

8 a $2^x\ln 2$ b $5^x\ln 5$ c $2^x+x2^x\ln 2$

d $\dfrac{3x^2-x^3\ln 6}{6^x}$ e $\dfrac{x2^x\ln 2-2^x}{x^2}$ f $\dfrac{1-x\ln 3}{3^x}$

9 $\dfrac{dy}{dx}=\dfrac{-(54e^{-2x}+3x^2e^{3y}+8xy^3)}{3x^2(xe^{3y}+4y^2)}$

EXERCISE 18G

1 a $\dfrac{1}{x}$ b $\dfrac{2}{2x+1}$ c $\dfrac{1-2x}{x-x^2}$

d $-\dfrac{2}{x}$ e $2x\ln x+x$ f $\dfrac{1-\ln x}{2x^2}$

g $e^x\ln x+\dfrac{e^x}{x}$ h $\dfrac{2\ln x}{x}$ i $\dfrac{1}{2x\sqrt{\ln x}}$

j $\dfrac{e^{-x}}{x}-e^{-x}\ln x$ k $\dfrac{\ln(2x)}{2\sqrt{x}}+\dfrac{1}{\sqrt{x}}$ l $\dfrac{\ln x-2}{\sqrt{x}(\ln x)^2}$

m $\dfrac{4}{1-x}$ n $\ln(x^2+1)+\dfrac{2x^2}{x^2+1}$

2 a $\ln 5$ b $\dfrac{3}{x}$ c $\dfrac{4x^3+1}{x^4+x}$ d $\dfrac{1}{x-2}$

e $\dfrac{6}{2x+1}[\ln(2x+1)]^2$ f $\dfrac{1-\ln(4x)}{x^2}$ g $-\dfrac{1}{x}$

h $\dfrac{1}{x\ln x}$ i $\dfrac{-1}{x(\ln x)^2}$

3 a $\dfrac{-1}{1-2x}$ b $\dfrac{-2}{2x+3}$ c $1+\dfrac{1}{2x}$

d $\dfrac{1}{x}-\dfrac{1}{2(2-x)}$ e $\dfrac{1}{x+3}-\dfrac{1}{x-1}$ f $\dfrac{2}{x}+\dfrac{1}{3-x}$

g $\dfrac{9}{3x-4}$ h $\dfrac{1}{x}+\dfrac{2x}{x^2+1}$ i $\dfrac{2x+2}{x^2+2x}-\dfrac{1}{x-5}$

4 2 5 $a=3,\ b=-e$

6 a $\dfrac{1}{x\ln 2}$ b $\dfrac{1}{x\ln 10}$ c $\log_3 x+\dfrac{1}{\ln 3}$

7 $\dfrac{da}{db}=\dfrac{a^4b-2ae^{2a}-a}{4abe^{2a}\ln b-3a^3b^2+b}$

EXERCISE 18H

1 a $2\cos(2x)$ b $\cos x-\sin x$

c $-3\sin(3x)-\cos x$ d $\cos(x+1)$ e $2\sin(3-2x)$

f $\dfrac{5}{\cos^2(5x)}$ g $\frac{1}{2}\cos\left(\frac{x}{2}\right)+3\sin x$

h $\dfrac{3\pi}{\cos^2(\pi x)}$ i $4\cos x+2\sin(2x)$

2 a $2x-\sin x$ b $\dfrac{1}{\cos^2 x}-3\cos x$

c $e^x\cos x-e^x\sin x$ d $-e^{-x}\sin x+e^{-x}\cos x$

e $\dfrac{\cos x}{\sin x}$ f $2e^{2x}\tan x+\dfrac{e^{2x}}{\cos^2 x}$

g $3\cos(3x)$ h $-\frac{1}{2}\sin\left(\frac{x}{2}\right)$ i $\dfrac{6}{\cos^2(2x)}$

j $\cos x-x\sin x$ k $\dfrac{x\cos x-\sin x}{x^2}$ l $\tan x+\dfrac{x}{\cos^2 x}$

4 a $2x\cos(x^2)$ b $-\dfrac{1}{2\sqrt{x}}\sin(\sqrt{x})$ c $-\dfrac{\sin x}{2\sqrt{\cos x}}$

d $2\sin x\cos x$ e $-3\sin x\cos^2 x$

f $-\sin x\sin(2x)+2\cos x\cos(2x)$ g $\sin x\sin(\cos x)$

h $-12\sin(4x)\cos^2(4x)$ i $-4\csc(4x)\cot(4x)$

j $2\sec(2x)\tan(2x)$ k $-8\csc^2(2x)\cot(2x)$

l $-12\cot^2\left(\frac{x}{2}\right)\csc^2\left(\frac{x}{2}\right)$

5 a $-\frac{9}{8}$ b 0

6 a $\sec x+x\sec x\tan x$ b $e^x(\cot x-\csc^2 x)$

c $8\sec(2x)\tan(2x)$ d $-e^{-x}\left[\cot\left(\frac{x}{2}\right)+\frac{1}{2}\csc^2\left(\frac{x}{2}\right)\right]$

e $x\csc x\,[2-x\cot x]$ f $\sqrt{\csc x}\left[1-\frac{x}{2}\cot x\right]$

g $\tan x$ h $\csc(x^2)\left[1-2x^2\cot(x^2)\right]$

i $\dfrac{-\sqrt{x}\csc^2 x-\frac{1}{2}x^{-\frac{1}{2}}\cot x}{x}\equiv-\dfrac{\cos x\sin x+2x}{2x\sqrt{x}\sin^2 x}$

EXERCISE 18I

3 a $\dfrac{2}{1+4x^2}$ b $\dfrac{-3}{\sqrt{1-9x^2}}$ c $\dfrac{1}{\sqrt{16-x^2}}$

d $\dfrac{-1}{\sqrt{25-x^2}}$ e $\dfrac{2x}{1+x^4}$ f -1

4 a $\dfrac{dy}{dx}=\arcsin x+\dfrac{x}{\sqrt{1-x^2}}$

b $\dfrac{dy}{dx}=e^x\arccos x-\dfrac{e^x}{\sqrt{1-x^2}}$

c $\dfrac{dy}{dx}=-e^{-x}\arctan x+\dfrac{e^{-x}}{1+x^2}$

5 c $\dfrac{dy}{dx}=-\dfrac{1}{\sqrt{a^2-x^2}},\quad x\in\,]-a, a[$

EXERCISE 18J

1 a 6 b $\dfrac{3}{2x^{\frac{5}{2}}}$ c $12x-6$

d $\dfrac{12-6x}{x^4}$ e $24-48x$ f $\dfrac{20}{(2x-1)^3}$

2 a $-6x$ b $2-\dfrac{30}{x^4}$ c $-\frac{9}{4}x^{-\frac{5}{2}}$ d $\dfrac{8}{x^3}$

e $6(x^2-3x)(5x^2-15x+9)$ f $2+\dfrac{2}{(1-x)^3}$

3 a 9 b 10 c 12 d 6

5 a $x=1$
b $x=0, \pm\sqrt{6}$

6

x	-1	0	1
$f(x)$	$-$	0	$+$
$f'(x)$	$+$	$-$	$+$
$f''(x)$	$-$	0	$+$

8 b $f''(x)=3\sin x\cos 2x+6\cos x\sin 2x$

9 a $\dfrac{1}{x^2}$ b $\dfrac{1}{x}$ c $\dfrac{2}{x^2}(1-\ln x)$

10 a 0 b 3 c 0 d 6

11 **Hint:** Find $\dfrac{dy}{dx}$ and $\dfrac{d^2y}{dx^2}$ and substitute into the equation.

15 a $\dfrac{-y^2-x^2}{y^3}$ b $\dfrac{y^2-x^2}{y^3}$ c $\dfrac{2y}{x^2}$

16 a $\dfrac{dV}{dq}=\dfrac{V-1}{3V-q}$ b $\dfrac{d^2q}{dV^2}=\dfrac{2q-3V-3}{(1-V)^2}$

18 **a** **i** $-e^{-x}(x+1)$ **ii** $e^{-x}(x)$
iii $-e^{-x}(x-1)$ **iv** $e^{-x}(x-2)$
b $f^{(n)}(x) = (-1)^n e^{-x}(x-n+2)$

REVIEW SET 18A

1 **a** -17 **b** -17 **c** -6

2 **a** $6x - 4x^3$ **b** $1 + \frac{1}{x^2}$ **3** $(0, 0)$

4 **a** $3x^2 e^{x^3+2}$ **b** $\frac{1}{x+3} - \frac{2}{x}$, $x > -3$, $x \neq 0$
c $\frac{e^y(2y+1)}{2 - xe^y(2y+1)}$

6 **a** $5\cos(5x)\ln x + \frac{\sin(5x)}{x}$
b $\cos x \cos(2x) - 2\sin x \sin(2x)$
c $-2e^{-2x}\tan x + \frac{e^{-2x}}{\cos^2 x}$

7 $\frac{\sqrt{3}}{2}$ **8** $\frac{dy}{dx} = -1$, $\frac{d^2y}{dx^2} = 0$

9 **a** $\frac{dM}{dt} = 8t(t^2+3)^3$
b $\frac{dA}{dt} = \frac{\frac{1}{2}t(t+5)^{-\frac{1}{2}} - 2(t+5)^{\frac{1}{2}}}{t^3}$

10 **a** $-\frac{2}{x\sqrt{x}} - 3$ **b** $\frac{1}{2}(x^2-3x)^{-\frac{1}{2}}(2x-3)$

11 **a** $\frac{23}{4}$ **b** $-\frac{1}{8\sqrt{2}}$

REVIEW SET 18B

1 **a** $\frac{dy}{dx} = 5 + 3x^{-2}$ **b** $\frac{dy}{dx} = 4(3x^2+x)^3(6x+1)$
c $\frac{dy}{dx} = 2x(1-x^2)^3 - 6x(x^2+1)(1-x^2)^2$

2 $(-2, 19)$ and $(1, -2)$

3 **a** $-2(5-4x)^{-\frac{1}{2}}$ **b** $-4(5-4x)^{-\frac{3}{2}}$ **c** $-24(5-4x)^{-\frac{5}{2}}$

4 **a**

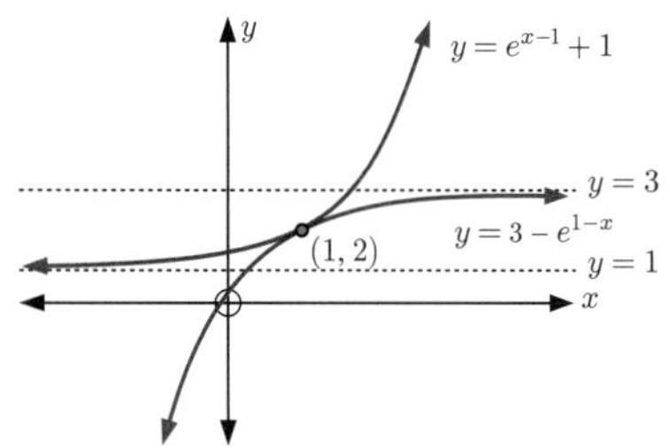

b $(1, 2)$ **c** gradient is $e^0 = 1$ for both
d The tangents to each of the curves at this point are the same line.

5 **a** $\frac{3x^2-3}{x^3-3x}$ **b** $\frac{e^x(x-2)}{x^3}$
c $\frac{dy}{dx} = \frac{e^{x+y}(y^2+1)}{2y - e^{x+y}(y^2+1)}$

6 $x = -\frac{1}{2}, \frac{3}{2}$ **7** **a** $\pi + 1$ **b** 2 **c** $-\frac{\sqrt{2}}{2}$

8 $\frac{dy}{dx} = 3b\cos(bx) + 2a\sin(2x)$, $a = 2$, $b = -1$ or 1

9 **a** $10 - 10\cos(10x)$ **b** $\tan x$
c $5\cos(5x)\ln(2x) + \frac{\sin(5x)}{x}$

10 **a** $f'(x) = \frac{3(x+3)^2\sqrt{x} - \frac{1}{2}x^{-\frac{1}{2}}(x+3)^3}{x}$
b $f'(x) = 4x^3\sqrt{x^2+3} + x^5(x^2+3)^{-\frac{1}{2}}$

11 **a** $\frac{dy}{dx} = \sqrt{\cos x} - \frac{x\sin x}{2\sqrt{\cos x}}$
b $\frac{dy}{dx} = e^x\left[\cot(2x) - 2\csc^2(2x)\right]$ **c** $\frac{dy}{dx} = \frac{-1}{\sqrt{4-x^2}}$

12 $A = -14$, $B = 21$ **13** $\frac{x}{3y}$

REVIEW SET 18C

1 **a** $\frac{dy}{dx} = 3x^2(1-x^2)^{\frac{1}{2}} - x^4(1-x^2)^{-\frac{1}{2}}$
b $\frac{dy}{dx} = \frac{(2x-3)(x+1)^{\frac{1}{2}} - \frac{1}{2}(x^2-3x)(x+1)^{-\frac{1}{2}}}{x+1}$

2 **a** $\frac{d^2y}{dx^2} = 36x^2 - \frac{4}{x^3}$ **b** $\frac{d^2y}{dx^2} = 6x + \frac{3}{4}x^{-\frac{5}{2}}$

3 $(1, e)$

4 **a** $f'(x) = \frac{e^x}{e^x+3}$ **b** $f'(x) = \frac{3}{x+2} - \frac{1}{x}$
c $x^{x^2+1}(2\ln x + 1)$

5 0

6 **a** $f'(x) = \frac{1}{2}x^{-\frac{1}{2}}\cos(4x) - 4x^{\frac{1}{2}}\sin(4x)$,
$f''(x) = -\frac{1}{4}x^{-\frac{3}{2}}\cos(4x) - 4x^{-\frac{1}{2}}\sin(4x)$
$- 16x^{\frac{1}{2}}\cos(4x)$
b $f'(\frac{\pi}{16}) = \frac{2-\pi}{\sqrt{2\pi}} \approx -0.455$, $f''(\frac{\pi}{8}) = \frac{-8\sqrt{2}}{\sqrt{\pi}} \approx -6.38$

8 **a** $x \approx -11.7$ or $x \approx -0.255$
b $x \approx -1.73$ or $x \approx 1.73$
c $x = -3$ or $x = 0$ or $x = 3$

9 **a** $f(x) = -5\sin(4x)$ **b** $x = \frac{\pi}{8}, \frac{3\pi}{8}, \frac{5\pi}{8}, \frac{7\pi}{8}$

10 **a** $\frac{y^2 - e^x y}{e^x - 2xy}$ **b** 0

13 **a** $e^{ax}(ax+1)$, $ae^{ax}(ax+2)$, $a^2e^{ax}(ax+3)$, $a^3e^{ax}(ax+4)$
b $f^{(n)}(x) = a^{n-1}e^{ax}(ax+n)$

EXERCISE 19A

1 **a** $y = -7x + 11$ **b** $4y = x + 8$ **c** $y = -2x - 2$
d $y = -2x + 6$ **e** $y = -5x - 9$ **f** $y = -5x - 1$

2 **a** $6y = -x + 57$ **b** $7y = -x + 26$ **c** $3y = x + 11$
d $6y = -x + 43$

3 **a** $y = 21$ and $y = -6$ **b** $(\frac{1}{2}, 2\sqrt{2})$
c $k = -5$ **d** $y = -3x + 1$

4 **a** $a = -4$, $b = 7$ **b** $a = 2$, $b = 4$

5 **a** $3y = x + 5$ **b** $9y = x + 4$
c $16y = x - 3$ **d** $y = -4$

6 **a** $y = 2x - \frac{7}{4}$ **b** $y = -27x - \frac{242}{3}$
c $57y = -4x + 1042$ **d** $2y = x + 1$

7 $a = 4$, $b = 3$

8 **a** Domain $\{x \mid x < 2\}$ **c** $8x + 3y = -19$

9 **a** $x + ey = 2$ **b** $x + 3y = 3\ln 3 - 1$
c $2x + e^2 y = \frac{2}{e^2} - e^2$

11 **a** $y = x$ **b** $y = x$ **c** $2x - y = \frac{\pi}{3} - \frac{\sqrt{3}}{2}$ **d** $x = \frac{\pi}{4}$

12 **a** $\sqrt{2}x - y = \sqrt{2}\left(\frac{\pi}{4} - 1\right)$ **b** $2x + y = \frac{2\pi}{3} + \sqrt{3}$

13 **a** $x - 2\sqrt{3}y = \frac{\pi}{6} - 4\sqrt{3}$ **b** $\sqrt{6}x + y = \pi\sqrt{6} + \sqrt{2}$

14 **a** $(-4, -64)$ **b** $(4, -31)$

15 **a** $f'(x) = 2x - \dfrac{8}{x^3}$ **b** $x = \pm\sqrt{2}$ **c** tangent is $y = 4$

16 A is $(\frac{2}{3}, 0)$, B is $(0, -2e)$

17 **a** $y = (2a-1)x - a^2 + 9$; $y = 5x$, contact at $(3, 15)$
$y = -7x$, contact at $(-3, 21)$

b $y = 0$, $y = 27x + 54$ **c** $y = -\sqrt{14}x + 4\sqrt{14}$

18 $y = e^a x + e^a(1-a)$ so $y = ex$ is the tangent to $y = e^x$ from the origin

19 **a**

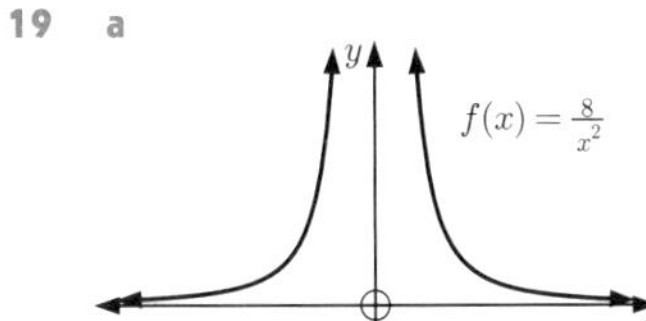

b $16x + a^3y = 24a$

c A is $(\frac{3}{2}a, 0)$, B is $(0, \dfrac{24}{a^2})$

d Area $= \dfrac{18}{a}$ units^2, area $\to 0$ as $a \to \infty$

20 $a = \frac{1}{4}$, point of intersection $(\frac{1}{2}, \frac{\sqrt{3}}{2})$ **21** $b = 3$

22 $\approx 63.43°$

23 **a** **Hint:** They must have the same y-coordinate at $x = b$ and the same gradient.

c $a = \frac{1}{2e}$ **d** $y = e^{-\frac{1}{2}}x - \frac{1}{2}$

24 **a** $x + \sqrt{3}y = 2$, $x - \sqrt{3}y = 2$ **b** $(2, 0)$

EXERCISE 19B

1 **a** i $x \geqslant 0$ ii never **b** i never ii $-2 < x \leqslant 3$
c i $x \leqslant 2$ ii $x \geqslant 2$ **d** i all real x ii never
e i $1 \leqslant x \leqslant 5$ ii $x \leqslant 1$, $x \geqslant 5$
f i $2 \leqslant x < 4$, $x > 4$ ii $x < 0$, $0 < x \leqslant 2$

2 **a** increasing for $x \geqslant 0$, decreasing for $x \leqslant 0$
b decreasing for all x
c increasing for $x \geqslant 0$, never decreasing
d increasing for $x \geqslant -\frac{3}{4}$, decreasing for $x \leqslant -\frac{3}{4}$
e decreasing for $x > 0$, never increasing
f incr. for $x \leqslant 0$ and $x \geqslant 4$, decr. for $0 \leqslant x \leqslant 4$
g increasing for all x
h increasing for $x > 0$, never decreasing
i increasing for $-\sqrt{\frac{2}{3}} \leqslant x \leqslant \sqrt{\frac{2}{3}}$,
decreasing for $x \leqslant -\sqrt{\frac{2}{3}}$, $x \geqslant \sqrt{\frac{2}{3}}$
j decreasing for all x
k decreasing for $x \leqslant -1$, increasing for $x \geqslant -1$
l increasing for $x \geqslant 1$, decreasing for $0 \leqslant x \leqslant 1$

3 **a** decr. for $x \leqslant -\frac{1}{2}$, $x \geqslant 3$, incr. for $-\frac{1}{2} \leqslant x \leqslant 3$
b increasing for $x \geqslant 0$, decreasing for $x \leqslant 0$
c increasing for $x \geqslant -\frac{3}{2} + \frac{\sqrt{5}}{2}$ and $x \leqslant -\frac{3}{2} - \frac{\sqrt{5}}{2}$
decreasing for $-\frac{3}{2} - \frac{\sqrt{5}}{2} \leqslant x \leqslant -\frac{3}{2} + \frac{\sqrt{5}}{2}$
d increasing for $x \leqslant 2 - \sqrt{3}$, $x \geqslant 2 + \sqrt{3}$
decreasing for $2 - \sqrt{3} \leqslant x \leqslant 2 + \sqrt{3}$

4 **a** sign diagram: $-$ | -1 | $+$ | 1 | $-$ (x)

b increasing for $-1 \leqslant x \leqslant 1$
decreasing for $x \leqslant -1$, $x \geqslant 1$

5 **a** sign diagram: $-$ | -1 | $+$ | 1 | $-$ (x)

b increasing for $-1 \leqslant x < 1$
decreasing for $x \leqslant -1$, $x > 1$

6 **a** sign diagram: $-$ | -1 | $+$ | 1 | $+$ | 3 | $-$ (x)

b increasing for $-1 \leqslant x < 1$, $1 < x \leqslant 3$
decreasing for $x \leqslant -1$, $x \geqslant 3$

7 **a** decreasing for $-1 < x < 1$, never increasing
b increasing for $-1 < x < 1$, never decreasing
c increasing for all x
d increasing for $(2k-1)\pi \leqslant x \leqslant 2k\pi$, $k \in \mathbb{Z}$,
decreasing for $2k\pi \leqslant x \leqslant (1 + 2k)\pi$, $k \in \mathbb{Z}$
e increasing for $(-\frac{1}{2} + 2k)\pi \leqslant x \leqslant (\frac{1}{2} + 2k)\pi$, $k \in \mathbb{Z}$,
decreasing for $(\frac{1}{2} + 2k)\pi \leqslant x \leqslant (\frac{3}{2} + 2k)\pi$, $k \in \mathbb{Z}$
f increasing for $(\frac{1}{2} + k)\pi < x < (\frac{3}{2} + k)\pi$, $k \in \mathbb{Z}$,
never decreasing

8 **a** increasing for $x \geqslant \sqrt{3}$ and $x \leqslant -\sqrt{3}$
decreasing for $-\sqrt{3} \leqslant x < -1$, $-1 < x < 1$, $1 < x \leqslant \sqrt{3}$
b increasing for $x \leqslant 0$, decreasing for $x \geqslant 0$
c increasing for $x \geqslant 2$, decreasing for $x < 1$, $1 < x \leqslant 2$
d increasing for $x \leqslant -1$,
decreasing for $-1 \leqslant x < 0$, $x > 0$
e increasing for all x

EXERCISE 19C

1 **a** A - local max, B - stationary inflection, C - local min.

b

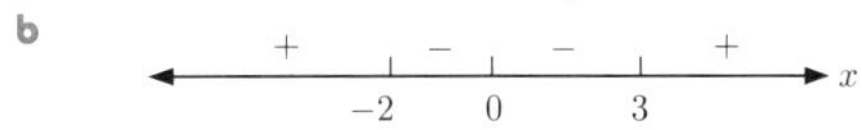

c i $x \leqslant -2$, $x \geqslant 3$ ii $-2 \leqslant x \leqslant 3$

d sign diagram: $-$ | -4 | $+$ | 0 | $-$ | 5 | $+$ (x)

2 **a** **b**

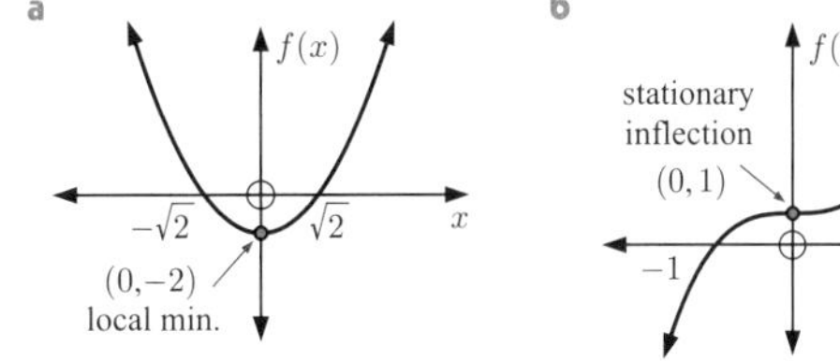

c

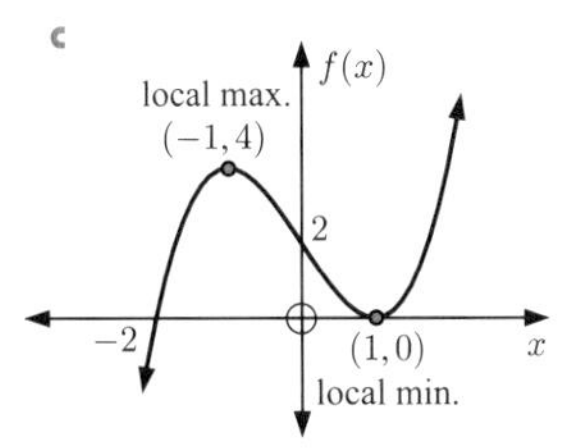

d

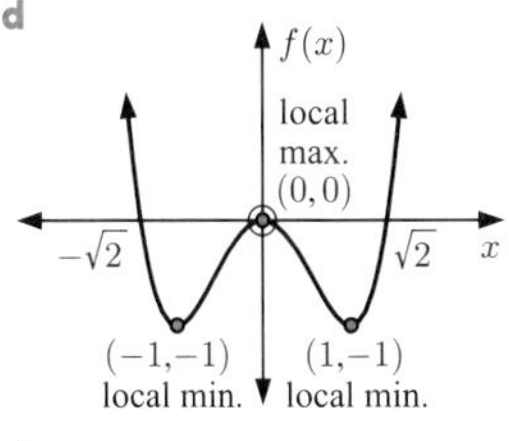

e

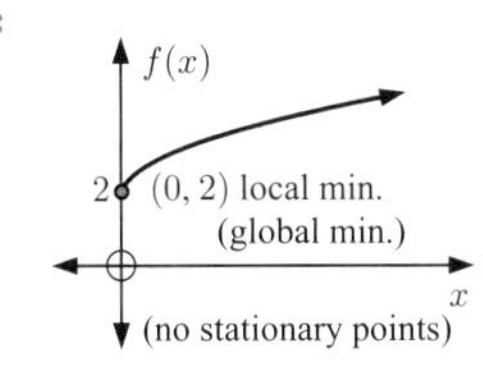

f

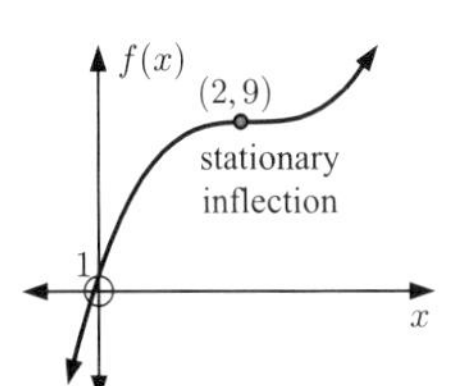

g

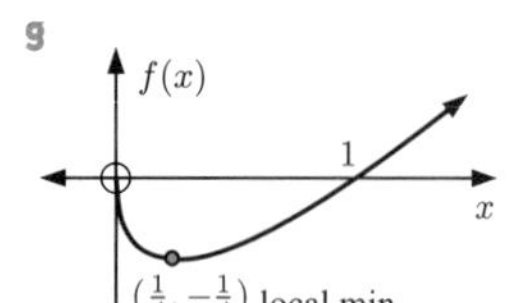

h

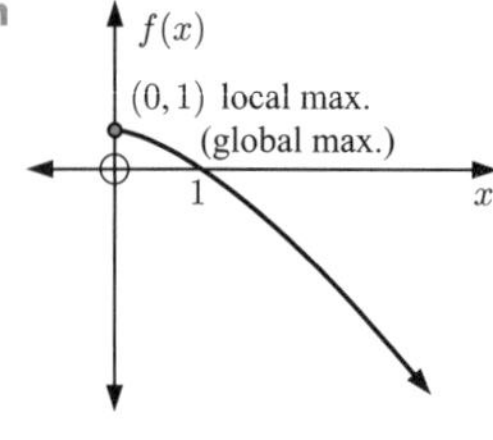

i

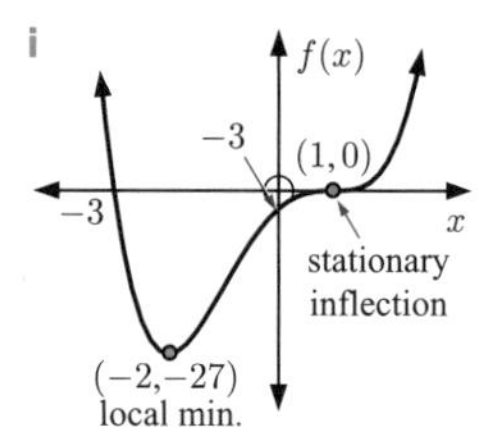

j

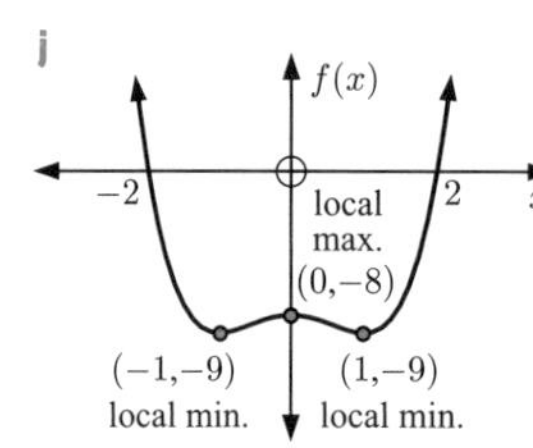

3 $x = -\dfrac{b}{2a}$, local min if $a > 0$, local max if $a < 0$

4 **a** local maximum at $(1, e^{-1})$
b local max. at $(-2, 4e^{-2})$, local min. at $(0, 0)$
c local minimum at $(1, e)$ **d** local maximum at $(-1, e)$

5 $a = 9$

6 **a** $a = -12$, $b = -13$
b local max at $(-2, 3)$, local min at $(2, -29)$

7 **a** $x > 0$

8 **a**

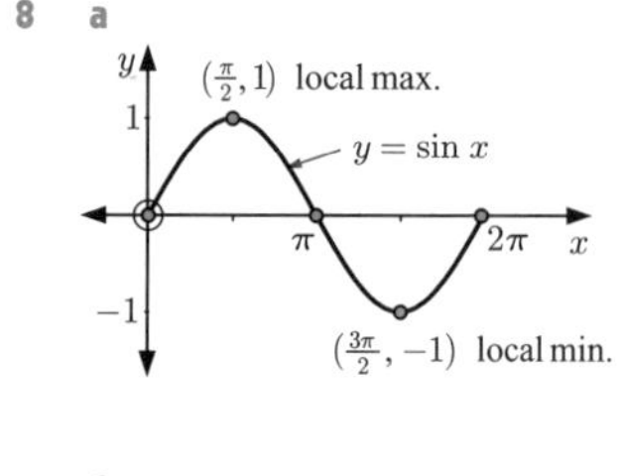

b

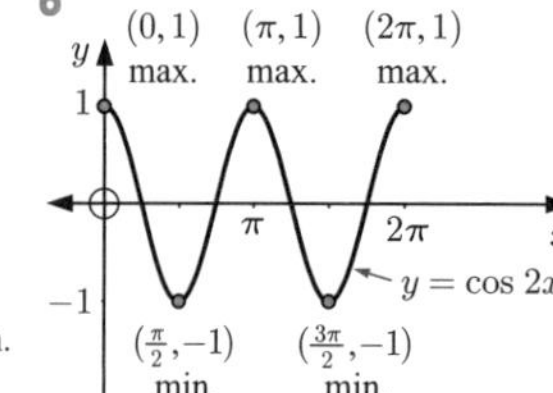

c **d**

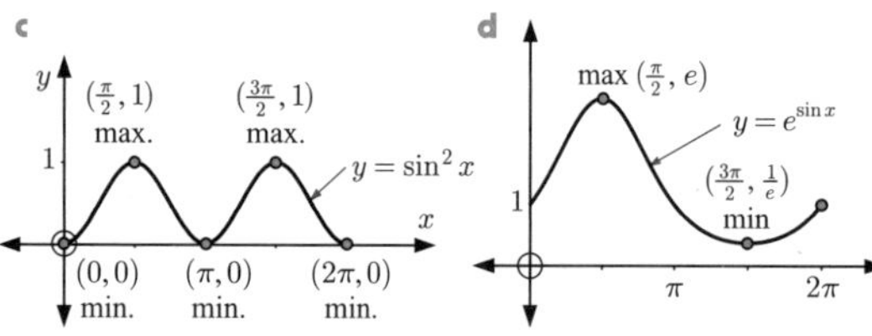

e

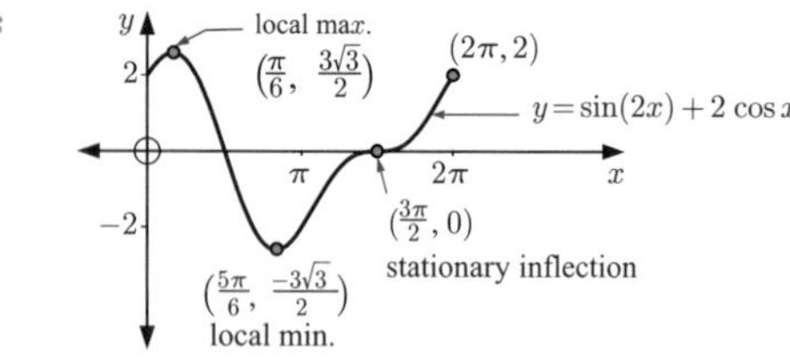

9 $P(x) = -9x^3 - 9x^2 + 9x + 2$

10 **a** greatest value is 63 when $x = 5$,
least value is -18 when $x = 2$
b greatest value is 4 when $x = 3$ and $x = 0$,
least value is -16 when $x = -2$.

12 $a = \frac{\sqrt{e}}{2}$, $b = -\frac{1}{8}$

13 **Hint:** Show that as $x \to 0$, $f(x) \to -\infty$,
and as $x \to \infty$, $f(x) \to 0$.

14 **Hint:** Show that $f(x) \geqslant 1$ for all $x > 0$.

15 **a** $x = \frac{\pi}{2}, \frac{3\pi}{2}$
b $(0, 1)$ is a local minimum, $(\pi, -1)$ is a local maximum
$(2\pi, 1)$ is a local minimum
c $f(x)$ has a period 2π
d

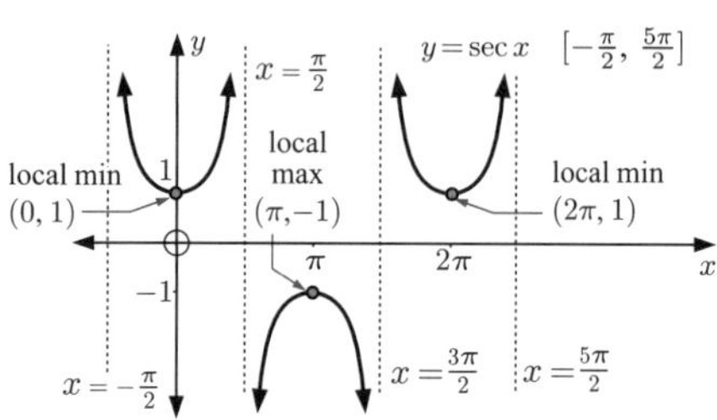

EXERCISE 19D.1

1 **a**

Point	$f(x)$	$f'(x)$	$f''(x)$
A	$+$	$-$	$+$
B	$-$	0	$+$
C	$+$	$+$	0
D	$+$	0	$-$
E	0	$-$	$-$

b B is a local minimum, D is a local maximum
c C is a non-stationary point of inflection

2 **a** no inflection **b** stationary inflection at $(0, 2)$
c non-stationary inflection at $(2, 3)$
d stationary inflection at $(0, 2)$
non-stationary inflection at $(-\frac{4}{3}, \frac{310}{27})$
e no inflection **f** stationary inflection at $(-2, -3)$

3 **a** **i** local minimum at $(0, 0)$
ii no points of inflection
iii decreasing for $x \leqslant 0$, increasing for $x \geqslant 0$
iv function is concave up for all x
v

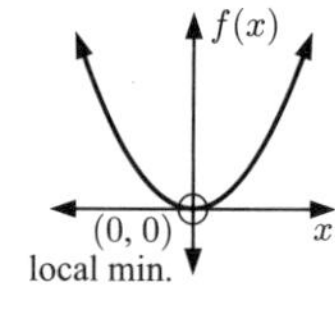

b **i** stationary inflection at $(0, 0)$
ii stationary inflection at $(0, 0)$
iii increasing for all real x
iv concave down for $x \leqslant 0$, concave up for $x \geqslant 0$
v

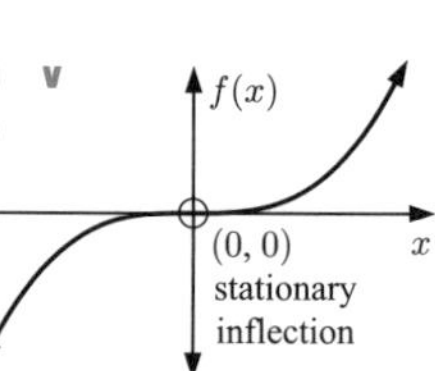

c **i** $f'(x) \neq 0$, no stationary points
ii no points of inflection
iii increasing for $x \geqslant 0$, never decreasing
iv concave down for $x \geqslant 0$, never concave up
v

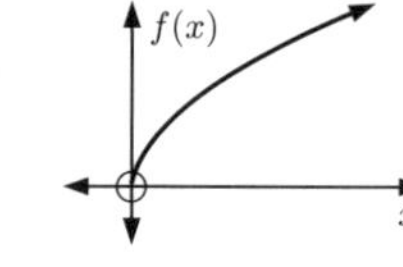

d **i** local max. at $(-2, 29)$
local min at $(4, -79)$
ii non-stationary inflection at $(1, -25)$
iii increasing for $x \leqslant -2$, $x \geqslant 4$
decreasing for $-2 \leqslant x \leqslant 4$

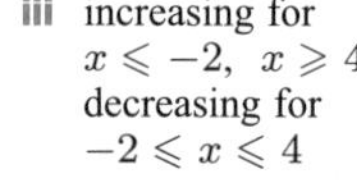

iv concave down for $x \leqslant 1$,
concave up for $x \geqslant 1$
v

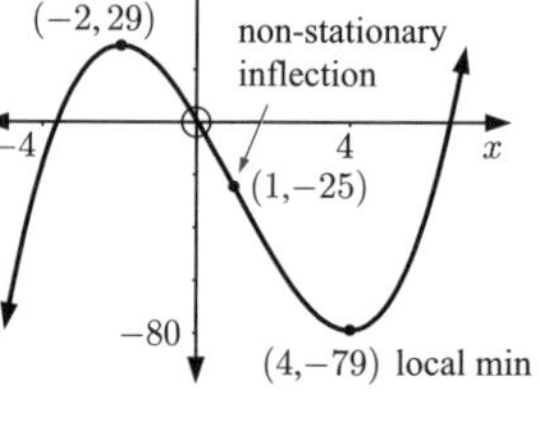

e **i** stat. inflect. at $(0, -2)$
local min. at $(-1, -3)$

ii stat. inflect. at $(0, -2)$
non-stationary inflection at $(-\frac{2}{3}, -\frac{70}{27})$

iii increasing for $x \geqslant -1$,
decreasing for $x \leqslant -1$

iv concave down for $-\frac{2}{3} \leqslant x \leqslant 0$
concave up for $x \leqslant -\frac{2}{3}$, $x \geqslant 0$

v

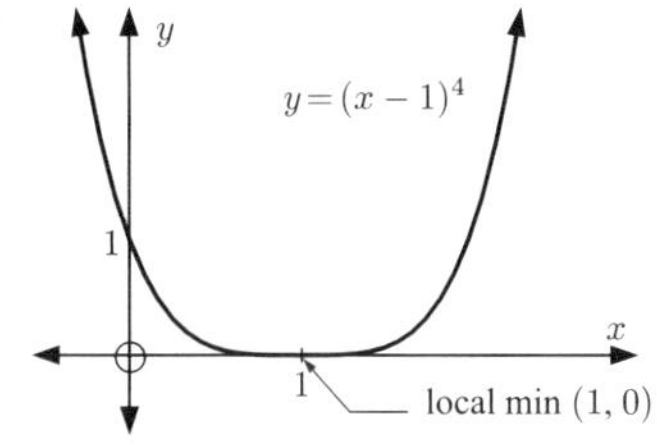

f **i** local min. at $(1, 0)$

ii no points of inflection

iii increasing for $x \geqslant 1$,
decreasing for $x \leqslant 1$

iv concave up for all x

v

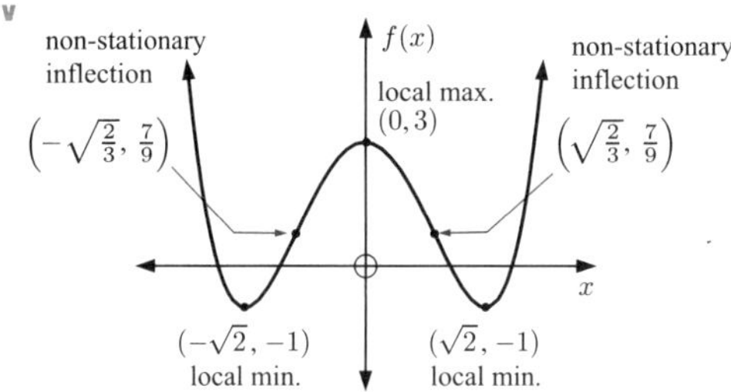

g **i** local minimum at $(-\sqrt{2}, -1)$ and $(\sqrt{2}, -1)$,
local maximum at $(0, 3)$

ii non-stationary inflection at $(\sqrt{\frac{2}{3}}, \frac{7}{9})$
non-stationary inflection at $(-\sqrt{\frac{2}{3}}, \frac{7}{9})$

iii increasing for $-\sqrt{2} \leqslant x \leqslant 0$, $x \geqslant \sqrt{2}$
decreasing for $x \leqslant -\sqrt{2}$, $0 \leqslant x \leqslant \sqrt{2}$

iv concave down for $-\sqrt{\frac{2}{3}} \leqslant x \leqslant \sqrt{\frac{2}{3}}$
concave up for $x \leqslant -\sqrt{\frac{2}{3}}$, $x \geqslant \sqrt{\frac{2}{3}}$

v

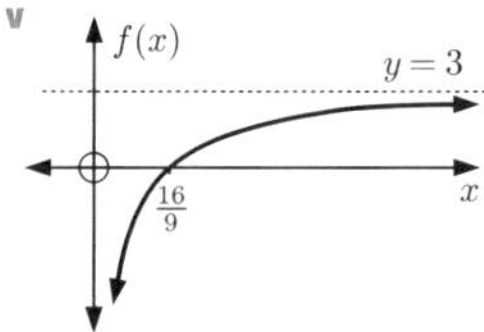

h **i** no stationary points

ii no inflections

iii increasing for $x > 0$,
never decreasing

iv concave down for $x > 0$,
never concave up

v

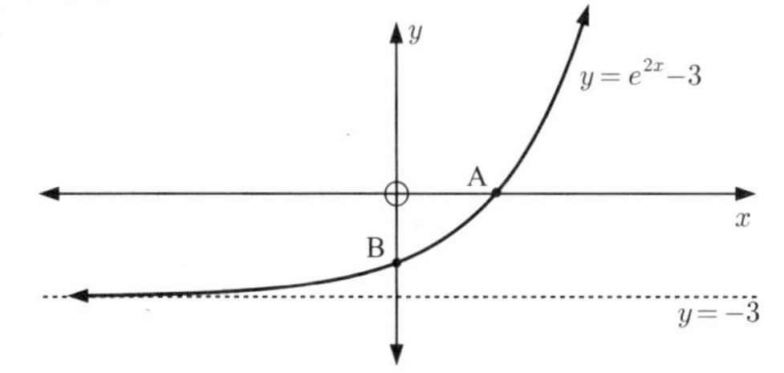

4 **a** $A(\frac{\ln 3}{2}, 0)$, $B(0, -2)$ **b** $f'(x) = 2e^{2x} > 0$ for all x

c $f''(x) = 4e^{2x} > 0$ for all x

d

e as $x \to -\infty$, $e^{2x} \to 0$ $\therefore$ $e^{2x} - 3 \to -3^+$

5 **a** $f(x)$: x-int. at $x = \ln 3$, y-int. at $y = -2$
$g(x)$: x-int. at $x = \ln(\frac{5}{3})$, y-int. at $y = -2$

b $f(x)$: as $x \to \infty$, $f(x) \to \infty$
as $x \to -\infty$, $f(x) \to -3^+$
$g(x)$: as $x \to \infty$, $g(x) \to 3^-$
as $x \to -\infty$, $g(x) \to -\infty$

c intersect at $(0, -2)$ and $(\ln 5, 2)$

d

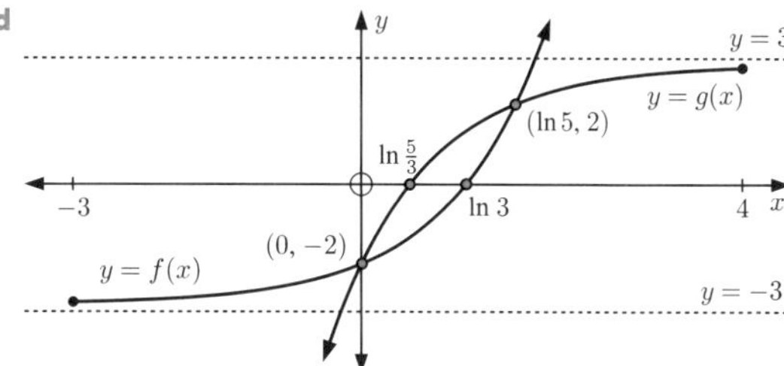

6 **a** $P(\frac{1}{2}\ln 3, 0)$,
$Q(0, -2)$

b $\dfrac{dy}{dx} = e^x + 3e^{-x}$
> 0 for all x

c y is concave down below the x-axis and concave up above the x-axis

d

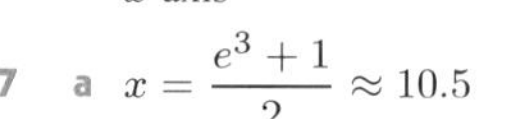

7 **a** $x = \dfrac{e^3 + 1}{2} \approx 10.5$ **b** no, $\therefore$ there is no y-int.

c gradient $= 2$ **d** Domain $= \{x \mid x > \frac{1}{2}\}$

e $f''(x) = \dfrac{-4}{(2x-1)^2} < 0$ for all $x > \frac{1}{2}$, so $f(x)$ is concave down

f

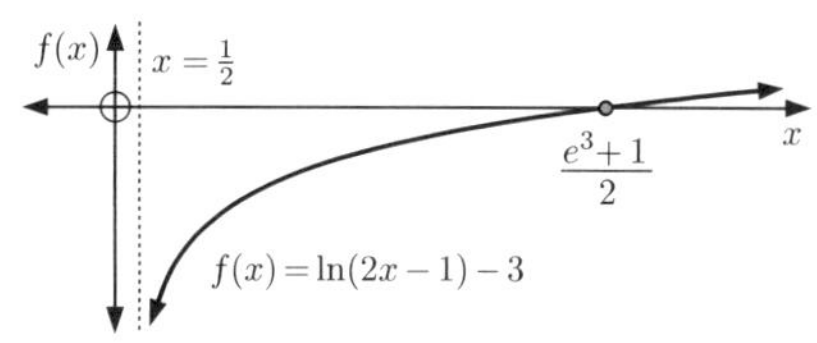

8 **a** $x > 0$

b $f'(x) > 0$ for all $x > 0$, so $f(x)$ is always increasing. Its gradient is always positive. $f''(x) < 0$ for all $x > 0$, so $f(x)$ is concave down for all $x > 0$.

c

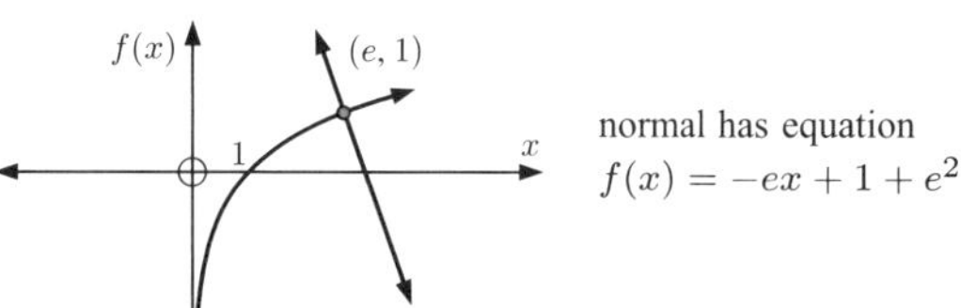

normal has equation
$f(x) = -ex + 1 + e^2$

9 **a** $f(x)$ does not have any x or y-intercepts

b as $x \to \infty$, $f(x) \to \infty$, as $x \to -\infty$, $f(x) \to 0^-$

c local minimum at $(1, e)$ **d** **i** $x > 0$ **ii** $x < 0$

e

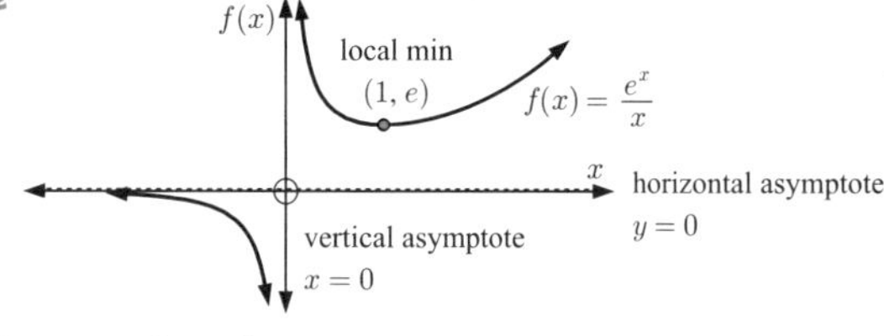

f $ey = -2x - 3$

10 **a** There is a local maximum at $\left(0, \frac{1}{\sqrt{2\pi}}\right)$.
$f(x)$ is incr. for all $x \leqslant 0$ and decr. for all $x \geqslant 0$.

b non-stationary inflections at $\left(-1, \frac{1}{\sqrt{2e\pi}}\right)$ and $\left(1, \frac{1}{\sqrt{2e\pi}}\right)$

c as $x \to \infty$, $f(x) \to 0^+$
as $x \to -\infty$, $f(x) \to 0^+$

d

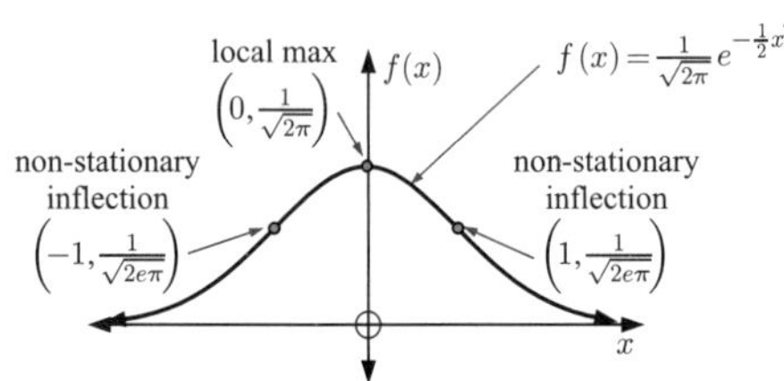

11 **a** x-intercept at $x = 0$, y-intercept at $y = 0$

b as $x \to \infty$, $y \to \infty$, as $x \to -\infty$, $y \to 0^-$

c local minimum at $(-1, -\frac{1}{4})$

d concave down for $x \leqslant -2$, concave up for $x \geqslant -2$

e

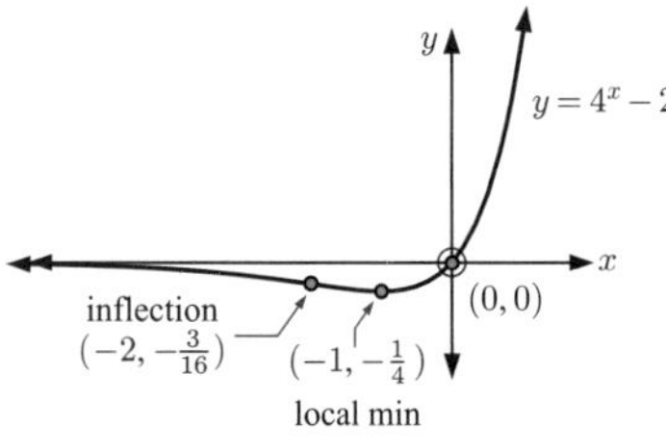

12 **a** **i** Show that $f'(t) = Ae^{-bt}(1 - bt)$

ii Show that $f''(t) = Abe^{-bt}(bt - 2)$

b

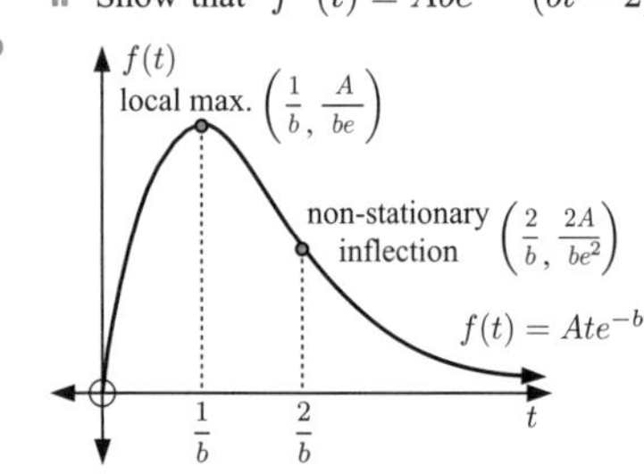

13 **a** y-intercept is $\dfrac{C}{1 + A}$

c

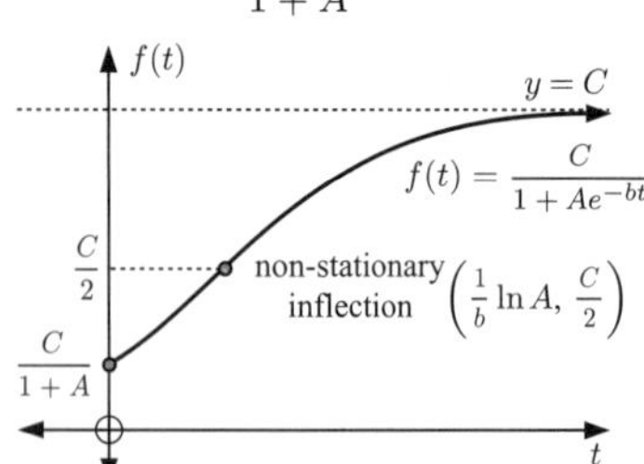

EXERCISE 19D.2

1 **a**

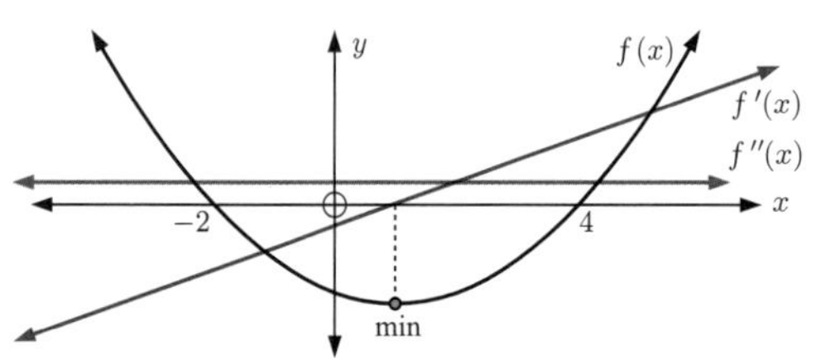

b

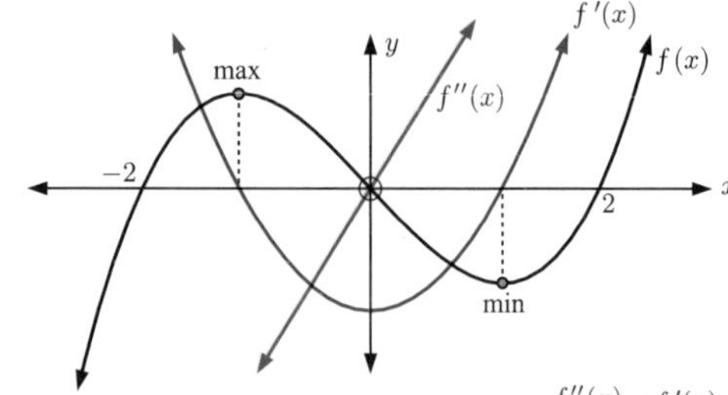

c

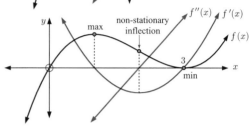

2 **a**

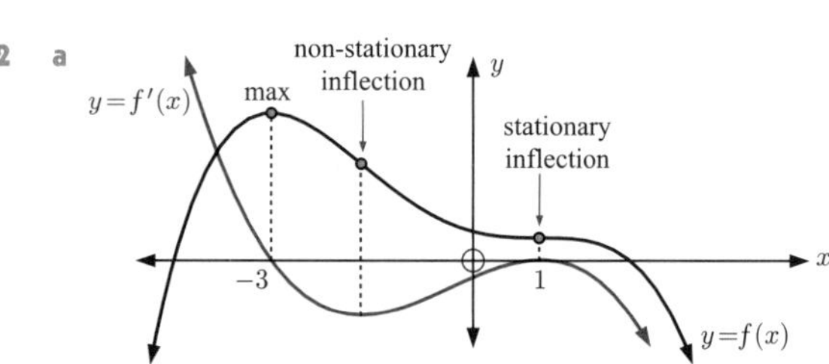

b

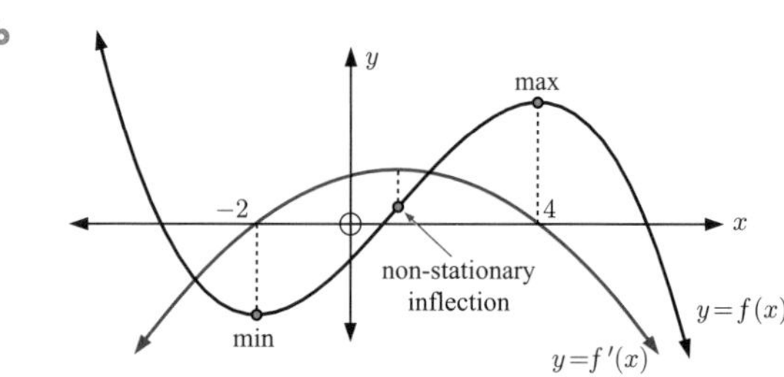

REVIEW SET 19A

1 $y = 4x + 2$ **2** $x = 1$

3 **a** $x = -3$ **b** x-int. $\frac{2}{3}$, y-int. $-\frac{2}{3}$

c $f'(x) = \dfrac{11}{(x+3)^2}$

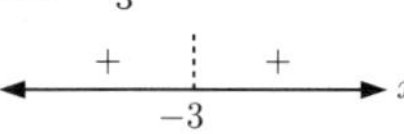

d no stationary points

4 $y = \dfrac{e}{2}x + \dfrac{1}{e} - \dfrac{e}{2}$ **6** $a = \frac{5}{2}$, $b = -\frac{3}{2}$

8 **a** y-intercept at $y = -1$, no x-intercept

b $f(x)$ is defined for all $x \neq 1$

c $f'(x) \leqslant 0$ for $x < 1$ and $1 < x \leqslant 2$
and $f'(x) \geqslant 0$ for $x \geqslant 2$
$f''(x) > 0$ for $x > 1$, $f''(x) < 0$ for $x < 1$
The function is decreasing for all defined values of $x \leqslant 2$, and increasing for all $x \geqslant 2$. The curve is concave down for $x < 1$ and concave up for $x > 1$.

d

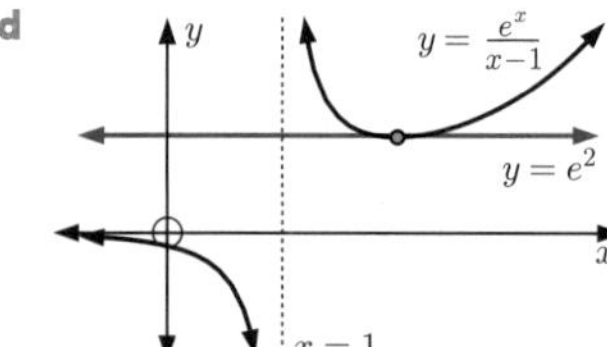

e tangent is $y = e^2$

9 $a = 64$ **10** $\text{P}(0, 7.5)$, $\text{Q}(3, 0)$

11

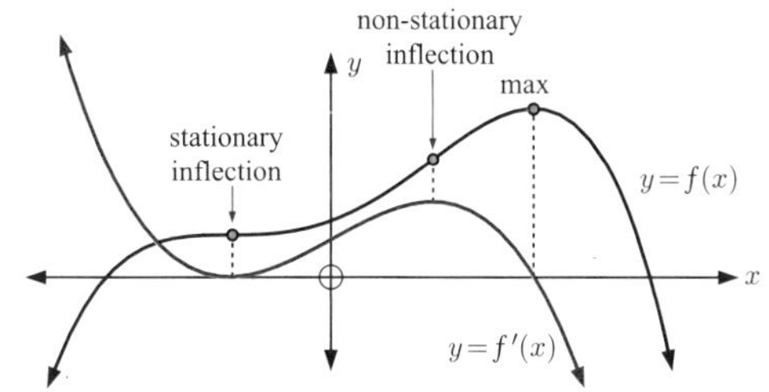

12 Tangent is $y = \ln 3$, so it never cuts the x-axis.

13 **a** $2\sqrt{3}x - y = \frac{2\sqrt{3}\pi}{3} - 2$ **b** $4x + y = 4\sqrt{3} + \frac{\pi}{3}$

REVIEW SET 19B

1 $y = 7$, $y = -25$ **2** $\frac{3267}{152}$ units^2

3 **a** $a = -6$

b local max. $(-\sqrt{2}, 4\sqrt{2})$, local min. $(\sqrt{2}, -4\sqrt{2})$

c

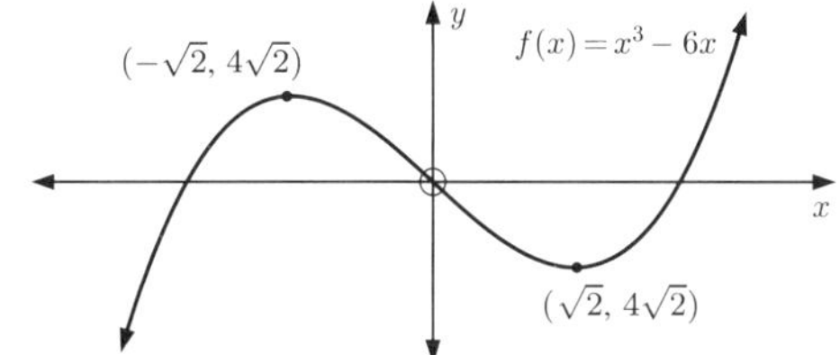

4 $p = 1$, $q = -8$ **5** $(-2, -25)$ **6** $a = \frac{1}{2}$

7 **a** local minimum at $(0, 1)$ **b** as $x \to \infty$, $f(x) \to \infty$

c $f''(x) = e^x$ [sign diagram: + , $f''(x)$]

thus $f(x)$ is concave up for all x.

d

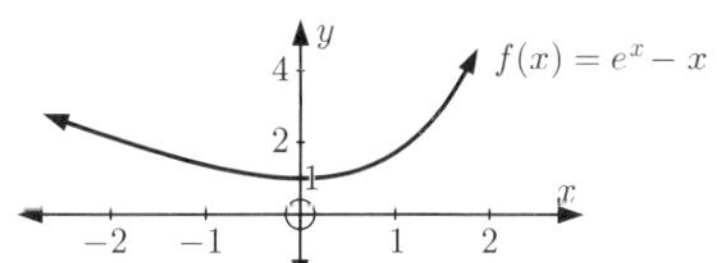

8 $y = \frac{1}{5}x - \frac{11}{5}$ (or $x - 5y = 11$)

10 $g(x) = -2x^2 + 6x + 3$

11 **a** for $0 \leqslant x \leqslant \frac{\pi}{2}$ and $\frac{3\pi}{2} \leqslant x \leqslant 2\pi$

b increasing for $\frac{3\pi}{2} \leqslant x \leqslant 2\pi$, decreasing for $0 \leqslant x \leqslant \frac{\pi}{2}$

c

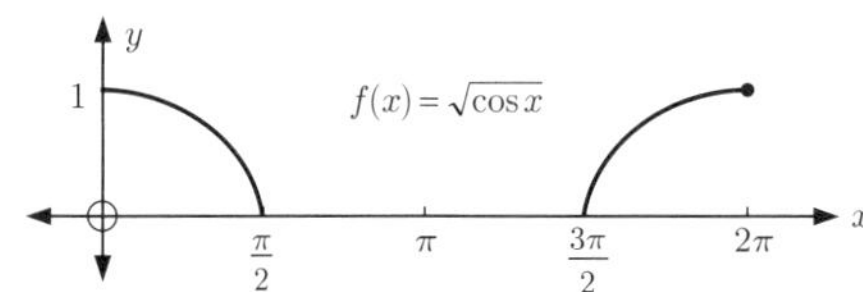

12

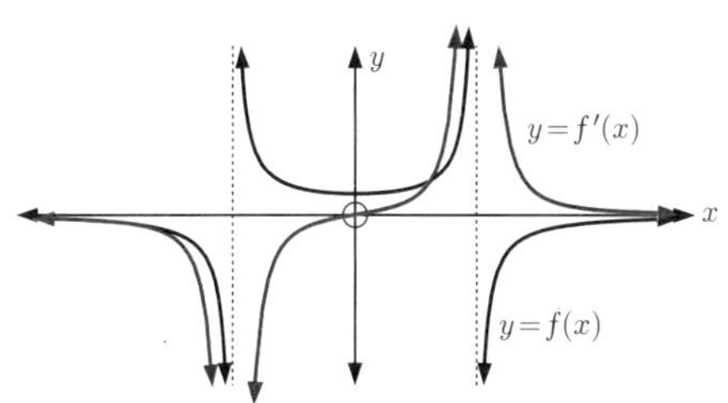

13 **a**

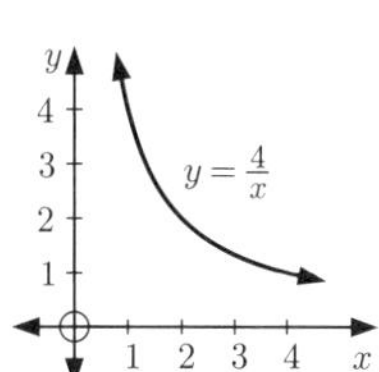

b $y = -\frac{4}{k^2}x + \frac{8}{k}$

c $\text{A}(2k, 0)$, $\text{B}\left(0, \frac{8}{k}\right)$

d Area $= 8$ units^2

e $k = 2$

14 $a = 9$, $b = -16$

REVIEW SET 19C

1 $y = 16x - \frac{127}{2}$

2 **a** $f(3) = 2$, $f'(3) = -1$ **b** $f(x) = x^2 - 7x + 14$

3 $a = -1$, $b = 2$

4 **a** $a = 2$ and the tangent is $y = 3x - 1$ **b** $(-4, -13)$

5 $(0, \ln 4 - 1)$

6 **a** local maximum at $(-2, 51)$, local minimum at $(3, -74)$ non-stationary inflection at $(\frac{1}{2}, -11\frac{1}{2})$

b increasing for $x \leqslant -2$, $x \geqslant 3$ decreasing for $-2 \leqslant x \leqslant 3$

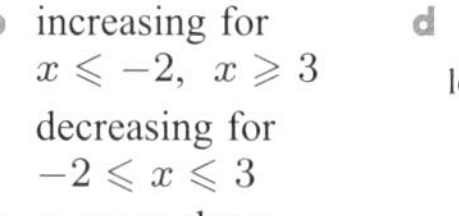

c concave down for $x \leqslant \frac{1}{2}$, concave up for $x \geqslant \frac{1}{2}$

d

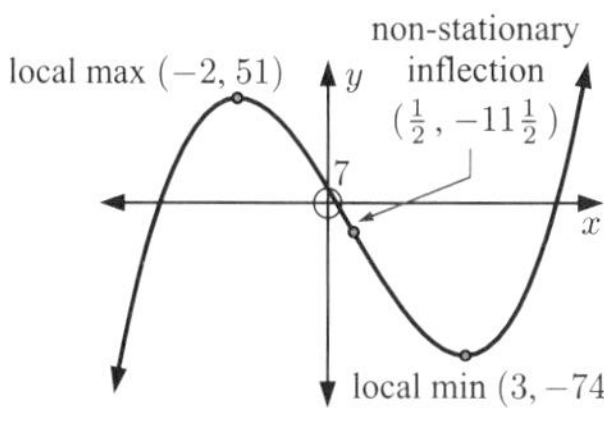

7 $\text{BC} = \frac{8\sqrt{10}}{3}$ **Hint:** The normal is $y = -3x + 8$.

8 **a** y-intercept at $y = 0$, x-intercept at $x = 0$ and $x = 2$

b local maximum at $\left(\frac{2}{3}, \frac{32}{27}\right)$, local minimum at $(2, 0)$, non-stationary inflection at $\left(\frac{4}{3}, \frac{16}{27}\right)$

c

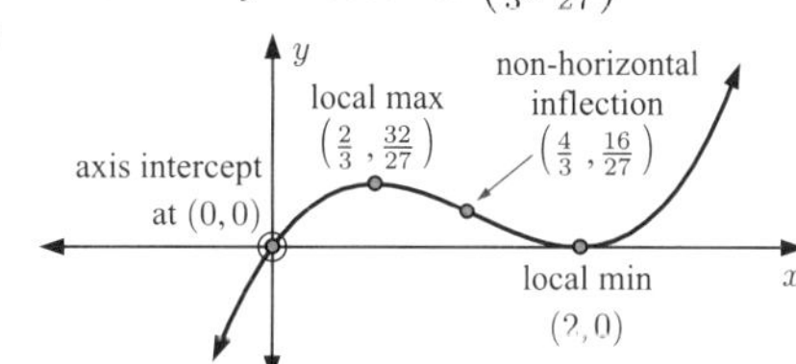

9 **a** $2x + 3y = \frac{2\pi}{3} + 2\sqrt{3}$ **b** $\sqrt{2}y - 4x = 1 - 2\pi$

10 $a = 9$, $b = 2$, $f''(-2) = -18$ **11** $4y = 3x + 5$

12 **a** $x > 0$ **b** Sign diag of $f'(x)$ Sign diag of $f''(x)$

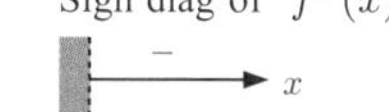

$f(x)$ is increasing for all $x > 0$ and is concave downwards for all $x > 0$.

c

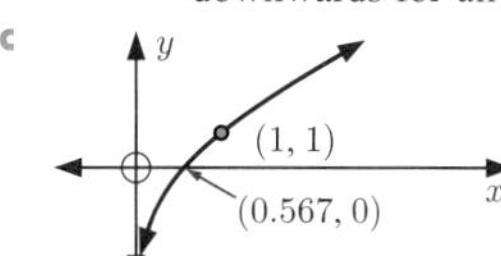

d normal is $x + 2y = 3$

13

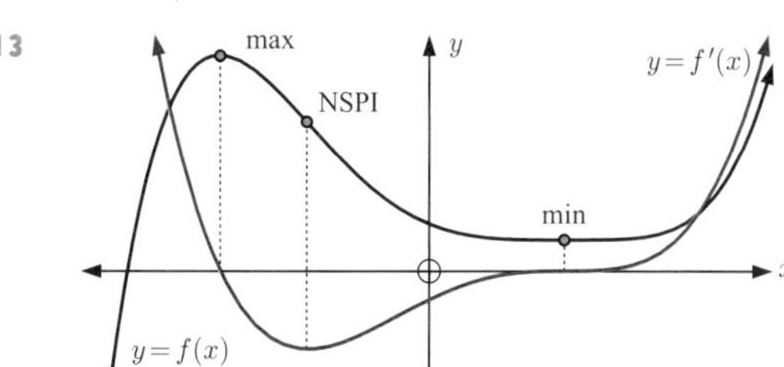

14 $f'(x) = 0$ for all x, so $f(x)$ is a constant $(\frac{\pi}{2})$.

15 **a** $\sqrt{3}x + y = 4$, $\sqrt{3}x - y = -4$ **b** $(0, 4)$

EXERCISE 20A.1

1 **a** 7 m s^{-1} **b** $(h+5) \text{ m s}^{-1}$
c $5 \text{ m s}^{-1} = s'(1)$ is the instantaneous velocity at $t = 1$ s
d average velocity $= (2t + h + 3) \text{ m s}^{-1}$,
$\lim_{h \to 0} (2t + h + 3) = 2t + 3 \text{ m s}^{-1}$ is the instantaneous velocity at time t seconds

2 **a** -14 cm s^{-1} **b** $(-8 - 2h) \text{ cm s}^{-1}$
c $-8 \text{ cm s}^{-1} = s'(2)$
$\therefore$ instantaneous velocity $= -8 \text{ cm s}^{-1}$ at $t = 2$
d $-4t = s'(t) = v(t)$ is the instantaneous velocity at time t seconds

3 **a** $\frac{2}{3} \text{ cm s}^{-2}$ **b** $\dfrac{2\sqrt{1+h} - 2}{h} \text{ cm s}^{-2}$
c $1 \text{ cm s}^{-2} = v'(1)$ is the instantaneous accn. at $t = 1$ s
d $\dfrac{1}{\sqrt{t}} \text{ cm s}^{-2} = v'(t)$, the instantaneous accn. at time t

4 **a** velocity at $t = 4$ **b** acceleration at $t = 4$

EXERCISE 20A.2

1 **a** $v(t) = 2t - 4 \text{ cm s}^{-1}$, $a(t) = 2 \text{ cm s}^{-2}$

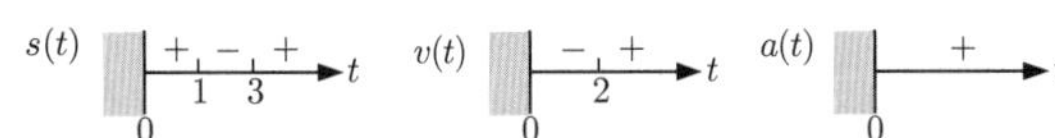

b $s(0) = 3$ cm, $v(0) = -4 \text{ cm s}^{-1}$, $a(0) = 2 \text{ cm s}^{-2}$
The object is initially 3 cm to the right of the origin and is moving to the left at 4 cm s^{-1}. It is accelerating at 2 cm s^{-2} to the right.
c $s(2) = -1$ cm, $v(2) = 0 \text{ cm s}^{-1}$, $a(2) = 2 \text{ cm s}^{-2}$
The object is instantaneously stationary, 1 cm to the left of the origin and is accelerating to the right at 2 cm s^{-2}.
d At $t = 2$, $s(2) = 1$ cm to the left of the origin.
e

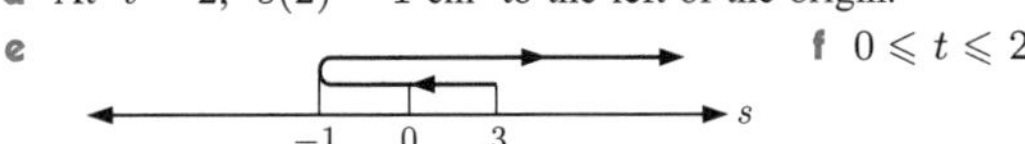

f $0 \leqslant t \leqslant 2$

2 **a** $v(t) = 98 - 9.8t \text{ m s}^{-1}$, $a(t) = -9.8 \text{ m s}^{-2}$

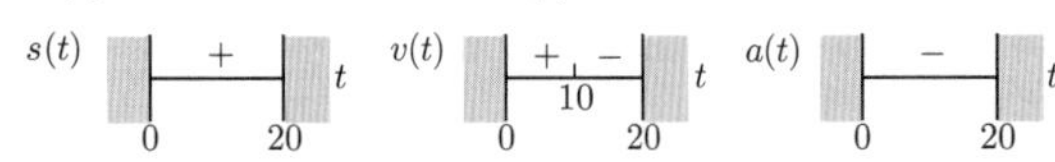

b $s(0) = 0$ m above the ground, $v(0) = 98 \text{ m s}^{-1}$ skyward
c $t = 5$ s Stone is 367.5 m above the ground and moving skyward at 49 m s^{-1}. Its speed is decreasing.
$t = 12$ s Stone is 470.4 m above the ground and moving groundward at 19.6 m s^{-1}. Its speed is increasing.
d 490 m **e** 20 seconds

3 **a** 1.2 m
b $s'(t) = 28.1 - 9.8t$ represents the instantaneous velocity of the ball.
c $t = 2.87$ s. The ball has reached its maximum height and is instantaneously at rest.
d 41.5 m
e **i** 28.1 m s^{-1} **ii** 8.5 m s^{-1} **iii** 20.9 m s^{-1}
$s'(t) \geqslant 0$ when the ball is travelling upwards.
$s'(t) \leqslant 0$ when the ball is travelling downwards.
f 5.78 s
g $s''(t)$ is the rate of change of $s'(t)$, or the instantaneous acceleration.

4 **b** 69.6 m s^{-1}

5 **a** $v(t) = 12 - 6t^2 \text{ cm s}^{-1}$, $a(t) = -12t \text{ cm s}^{-2}$
b $s(0) = -1$ cm, $v(0) = 12 \text{ cm s}^{-1}$, $a(0) = 0 \text{ cm s}^{-2}$
Particle started 1 cm to the left of the origin and was travelling to the right at a constant speed of 12 cm s^{-1}.
c $t = \sqrt{2}$ s, $s(\sqrt{2}) = 8\sqrt{2} - 1$ cm
d **i** $t \geqslant \sqrt{2}$ s **ii** never

6 **a** $v(t) = 3t^2 - 18t + 24 \text{ m s}^{-1}$ $a(t) = 6t - 18 \text{ m s}^{-2}$

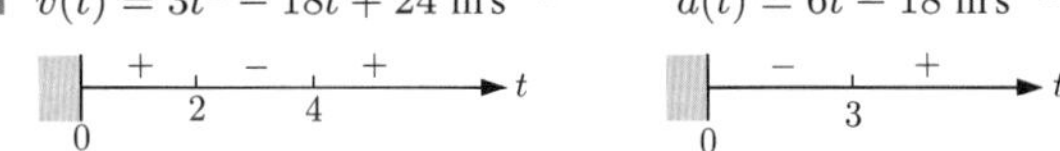

b $x(2) = 20$, $x(4) = 16$

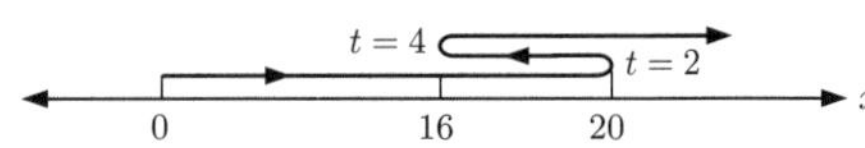

c **i** $0 \leqslant t \leqslant 2$ and $3 \leqslant t \leqslant 4$ **ii** $0 \leqslant t \leqslant 3$
d 28 m

7 **a** $v(t) = 100 - 40e^{-\frac{t}{5}} \text{ cm s}^{-1}$, $a(t) = 8e^{-\frac{t}{5}} \text{ cm s}^{-2}$
b $s(0) = 200$ cm on positive side of origin
$v(0) = 60 \text{ cm s}^{-1}$, $a(0) = 8 \text{ cm s}^{-2}$
c as $t \to \infty$, $v(t) \to 100 \text{ cm s}^{-1}$ (below)
d

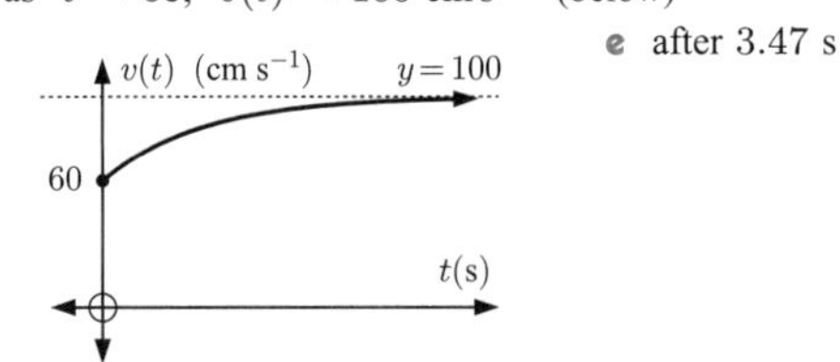

e after 3.47 s

8 **a** $x(0) = -1$ cm, $v(0) = 0 \text{ cm s}^{-1}$, $a(0) = 2 \text{ cm s}^{-2}$
b At $t = \frac{\pi}{4}$ seconds, the particle is $(\sqrt{2} - 1)$ cm left of the origin, moving right at $\sqrt{2} \text{ cm s}^{-1}$, with increasing speed.
c changes direction when $t = \pi$, $x(\pi) = 3$ cm
d increasing for $0 \leqslant t \leqslant \frac{\pi}{2}$ and $\pi \leqslant t \leqslant \frac{3\pi}{2}$

9 **Hint:** $s'(t) = v(t)$ and $s''(t) = a(t) = g$
Show that $a = \frac{1}{2}g$, $b = v(0)$, $c = 0$.

10 **a**

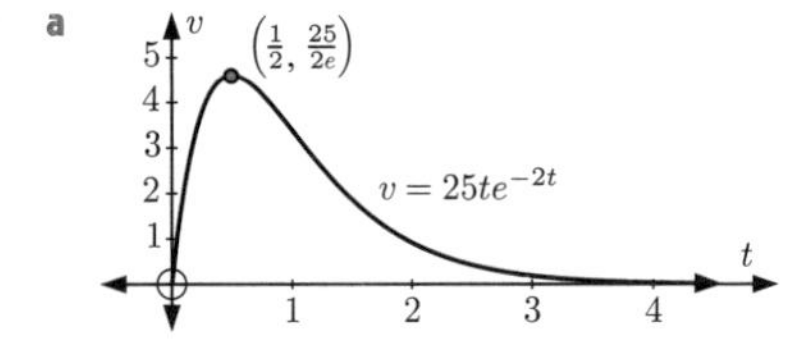

c $0 \leqslant t \leqslant \frac{1}{2}$

EXERCISE 20B

1 **a** \$118 000 **b** $\dfrac{dP}{dt} = 4t - 12$, \$1000s per year
c $\dfrac{dP}{dt}$ is the rate of change in profit with time
d **i** $0 \leqslant t \leqslant 3$ years **ii** $t > 3$ years
e minimum profit is \$100 000 when $t = 3$
f $\left.\dfrac{dP}{dt}\right|_{t=4} = 4$ Profit is increasing at \$4000 per year after 4 years.
$\left.\dfrac{dP}{dt}\right|_{t=10} = 28$ Profit is increasing at \$28 000 per year after 10 years.
$\left.\dfrac{dP}{dt}\right|_{t=25} = 88$ Profit is increasing at \$88 000 per year after 25 years.

2 **a** 19 000 m^3 per minute **b** 18 000 m^3 per minute

3 **a** **i** $Q(0) = 100$ **ii** $Q(25) = 50$ **iii** $Q(100) = 0$

b **i** decr. 1 unit per year **ii** decr. $\frac{1}{\sqrt{2}}$ units per year

c $Q'(t) = -\frac{5}{\sqrt{t}} < 0$

4 **a** 0.5 m

b $t = 4$: 9.17 m, $t = 8$: 12.5 m, $t = 12$: 14.3 m

c $t = 0$: 3.9 m year^{-1}, $t = 5$: 0.975 m year^{-1}, $t = 10$: 0.433 m year^{-1}

d as $\frac{dH}{dt} = \frac{97.5}{(t+5)^2} > 0$, for all $t \geqslant 0$, the tree is always growing, and $\frac{dH}{dt} \to 0$ as t increases

5 **a** $C'(x) = 0.0009x^2 + 0.04x + 4$ dollars per pair

b $C'(220) = \$56.36$ per pair. This estimates the additional cost of making one more pair of jeans if 220 pairs are currently being made.

c \$56.58 This is the actual increase in cost to make an extra pair of jeans (221 rather than 220).

d $C''(x) = 0.0018x + 0.04$.
$C''(x) = 0$ when $x = -22.2$. This is where the rate of change is a minimum, however it is out of the bounds of the model (you cannot make < 0 jeans!).

6 **a** **i** €4500 **ii** €4000

b **i** decrease of €210.22 per km h^{-1}

ii increase of €11.31 per km h^{-1}

c $\frac{dC}{dv} = 0$ at $v = \sqrt[3]{500\,000} \approx 79.4$ km h^{-1}

7 **a** $\frac{dV}{dt} = -1250\left(1 - \frac{t}{80}\right)$ L min^{-1}

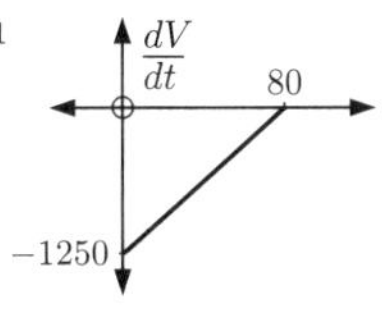

b at $t = 0$ when the tap was first opened

c $\frac{d^2V}{dt^2} = \frac{125}{8}$ L min^{-2}

This shows that the rate of change of V is constantly increasing, so the outflow is decreasing at a constant rate.

8 **a** The near part of the lake is 2 km from the sea, the furthest part is 3 km.

b $\frac{dy}{dx} = \frac{3}{10}x^2 - x + \frac{3}{5}$.

$\left.\frac{dy}{dx}\right|_{x=\frac{1}{2}} = 0.175$, height of hill is increasing as gradient is positive.

$\left.\frac{dy}{dx}\right|_{x=1\frac{1}{2}} = -0.225$, height of hill is decreasing as gradient is negative.

$\therefore$ top of the hill is between $x = \frac{1}{2}$ and $x = 1\frac{1}{2}$.

c 2.55 km from the sea, 63.1 m deep

9 **a** When $\frac{dP}{dt} = 0$, the population is not changing over time, so it is stable.

b 4000 fish **c** 8000 fish

10 **a** $k = \frac{1}{50}\ln 2 \approx 0.0139$

b **i** 20 grams **ii** 14.3 grams **iii** 1.95 grams

c 9 days and 6 minutes (216 hours)

d **i** -0.0693 g h^{-1} **ii** -2.64×10^{-7} g h^{-1}

e **Hint:** You should find $\frac{dW}{dt} = -\frac{1}{50}\ln 2 \times 20e^{-\frac{1}{50}\ln 2t}$

11 **a** $k = \frac{1}{15}\ln\left(\frac{19}{3}\right) \approx 0.123$ **b** 100°C

c $c = -k \approx -0.123$

d **i** decreasing at 11.7°C min^{-1}

ii decreasing at 3.42°C min^{-1}

iii decreasing at 0.998°C min^{-1}

12 **a** 43.9 cm **b** 10.4 years

c **i** growing at 5.45 cm per year

ii growing at 1.88 cm per year

13 **a** $A(0) = 0$

b **i** $k = \frac{\ln 2}{3}$ (≈ 0.231)

ii 0.728 litres of alcohol produced per hour

14 **a** $\frac{2C}{3}$ bees **b** 37.8% increase **c** Yes, C bees

d 3047 bees

e $B'(t) = \frac{0.865C}{e^{1.73t}(1 + 0.5e^{-1.73t})^2}$

and so $B'(t) > 0$ for all $t \geqslant 0$

$\therefore$ $B(t)$ is increasing over time.

f

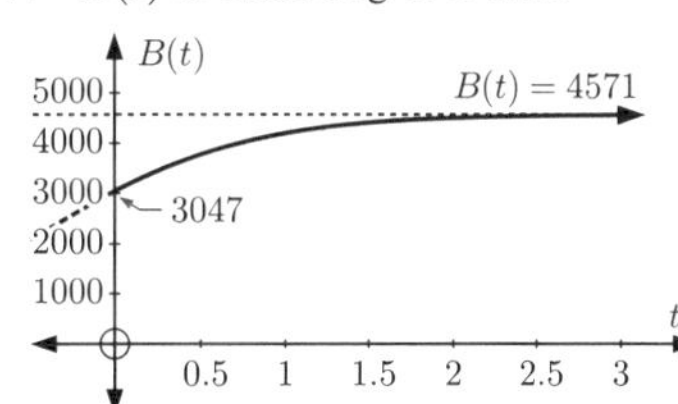

15 $\frac{21}{\sqrt{2}}$ cm^2 per radian

16 **a** rising at 2.73 m per hour **b** rising

17 **a** $-34\,000\pi$ units per second **b** $V'(t) = 0$

18 **b** **i** 0 **ii** 1 **iii** ≈ 1.11

EXERCISE 20C

1 250 items

2 **b**

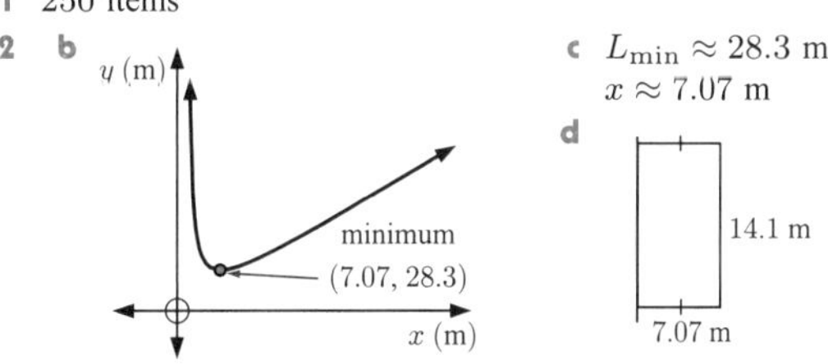

c $L_{min} \approx 28.3$ m, $x \approx 7.07$ m

d

3 50 fittings **4** 10 blankets **5** 14.8 km h^{-1}

6 **a** at 4.41 months old **b**

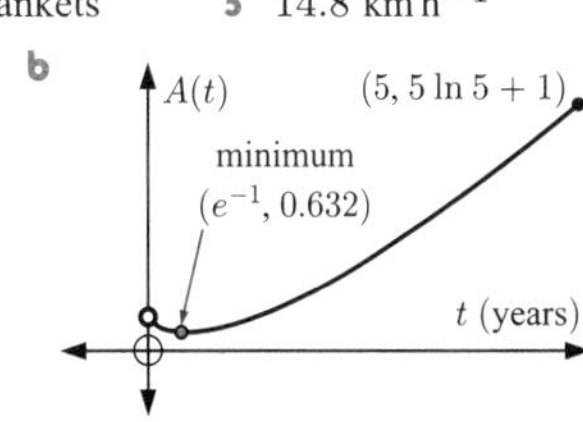

7 Maximum hourly cost is \$680.95 when 150 hinges are made per hour. Minimum hourly cost is \$529.80 when 104 hinges are made per hour.

8 20 kettles

9 **a** $2x$ cm **b** $V = 200 = 2x \times x \times h$

c **Hint**: Show $h = \frac{100}{x^2}$ and substitute into the surface area equation.

d

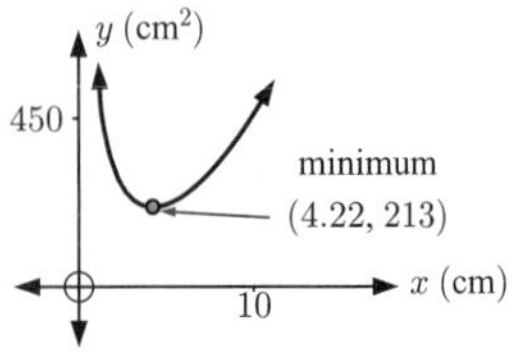

e $SA_{\min} \approx 213$ cm^2, $x \approx 4.22$ cm

f

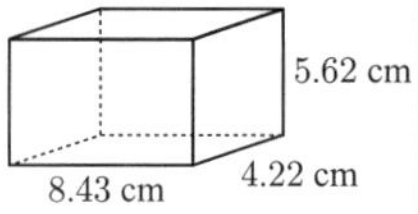

10 $C\left(\frac{1}{\sqrt{2}}, e^{(-\frac{1}{2})}\right)$

11 **a** recall that $V_{\text{cylinder}} = \pi r^2 h$ and that $1 \text{ L} = 1000 \text{ cm}^3$
b recall that $SA_{\text{cylinder}} = 2\pi r^2 + 2\pi r h$
c

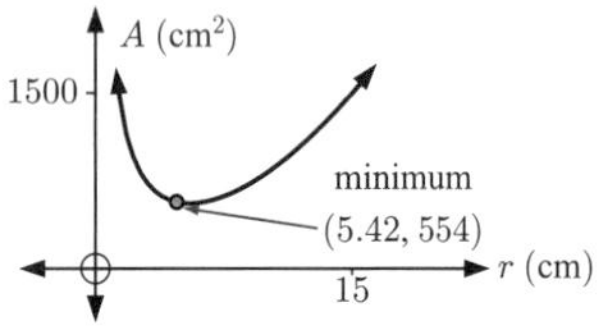

d $r \approx 5.42$ cm, $h \approx 10.84$ cm
e

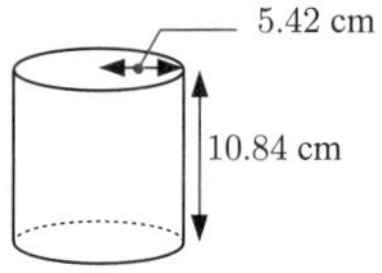

12 **b** $\theta \approx 1.91$, $A \approx 237$ cm^2 **13** **b** 6 cm × 6 cm

14 **a** $0 \leqslant x \leqslant 63.7$
b $l = 100$ m, $x = \frac{100}{\pi} \approx 31.83$ m, $A = \frac{20\,000}{\pi} \approx 6366$ m^2

15 after 13.8 weeks **16** after 40 minutes

17 **a** $D(x) = \sqrt{x^2 + (24 - x)^2}$
b $\dfrac{d[D(x)]^2}{dx} = 4x - 48$

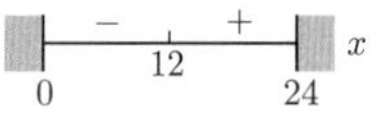

c Smallest $D(x) \approx 17.0$ m

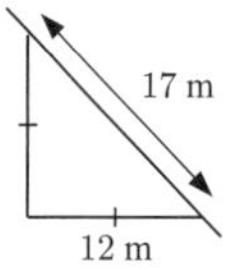

18 **a** $\text{QR} = \left(\dfrac{2 + x}{x}\right)$ m **c** **Hint:** All solutions < 0 can be discarded as $x \geqslant 0$.
d 416 cm

19 **c** $\theta = 30°$, $A \approx 130$ cm^2

20 **a** **Hint:** Show that $\text{AC} = \frac{\theta}{360} \times 2\pi \times 10$
b **Hint:** Show that $2\pi r = \text{AC}$
c **Hint:** Use the result from **b** and Pythagoras' theorem.
d $V = \frac{1}{3}\pi \left(\frac{\theta}{36}\right)^2 \sqrt{100 - \left(\frac{\theta}{36}\right)^2}$
e

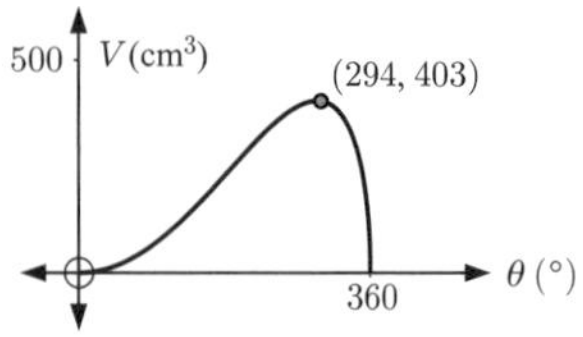

f $\theta \approx 294°$

21 **a** **Hint:** Use the cosine rule.
b 3553 km^2

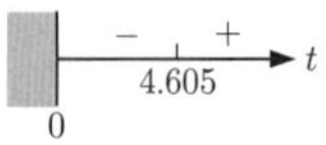

c 5:36 pm

22 **a** For $x < 0$ or $x > 6$, X is not on AC.
c $x = 2.67$ km This is the distance from A to X which minimises the time taken to get from B to C. (**Proof:** Use sign diagram or second derivative test. Be sure to check the end points.)

23 3.33 km **24** 4 m from the 40 cp lamp

25 9.87 m **26** **c** $4\sqrt{2}$ m **27** 1.34 m from A

28 1 hour 34 min 53 s when $\theta \approx 36.9°$

29 **a** $\tan\alpha = \dfrac{2}{x}$ and $\tan(\alpha + \theta) = \dfrac{3}{x}$
b $\theta = \arctan\left(\frac{3}{x}\right) - \arctan\left(\frac{2}{x}\right)$ **c** $x = \sqrt{6}$
d The maximum angle of view (θ) occurs when Sonia's eye is $\sqrt{6}$ m from the wall.

30 at grid reference (3.54, 8), ≈ 15.6 km

31 between A and N, 2.58 m from N

32 $\sqrt{\frac{3}{2}} : 1$ **33** **d** 63.7%

EXERCISE 20D

1 a is decreasing at 7.5 units per second

2 increasing at 1 cm per minute

3 **a** 4π m^2 per second **b** 8π m^2 per second

4 increasing at 6π m^2 per minute

5 decreasing at 0.16 m^3 per minute **6** $\frac{20}{3}$ cm per minute

7 $\frac{25\sqrt{3}}{6} \approx 7.22$ cm per minute

8 decreasing at $\frac{250}{13} \approx 19.2$ m s^{-1}

9 **a** 0.2 m s^{-1} **b** $\frac{8}{90}$ m s^{-1}

10 decreasing at $\frac{\sqrt{2}}{100}$ radians per second

11 **a** decr. at $\frac{3}{100}$ rad sec^{-1} **b** decr. at $\frac{1}{100}$ rad sec^{-1}

12 increasing at 0.128 radians per second

13 increasing at 0.12 radians per minute

14 increasing at 24.3 m s^{-1}

15 **a** $\frac{\sqrt{3}}{2}\pi$ cm s^{-1} **b** 0 cm s^{-1}

16 **a** $\frac{200}{\sqrt{13}}\pi$ rad sec^{-1} **b** 100π rad sec^{-1}

17 **b** $\frac{\sqrt{3}}{120}$ m min^{-1}

REVIEW SET 20A

1 **a** $x(0) = 3$ cm, $x'(0) = 2$ cm s^{-1}, $x''(0) = 0$ cm s^{-2}
b $t = \frac{\pi}{4}$ s and $\frac{3\pi}{4}$ s **c** 4 cm

2 **a** $v(t) = (6t^2 - 18t + 12)$ cm s^{-1}, $a(t) = (12t - 18)$ cm s^{-2}

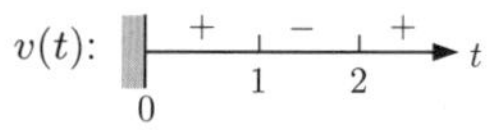

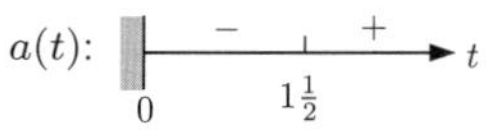

b $s(0) = 5$ cm to left of origin
$v(0) = 12$ cm s^{-1} towards origin
$a(0) = -18$ cm s^{-2} (reducing speed)
c At $t = 2$, particle is 1 cm to the left of the origin, is stationary and is accelerating towards the origin.
d $t = 1$, $s = 0$ and $t = 2$, $s = -1$
e

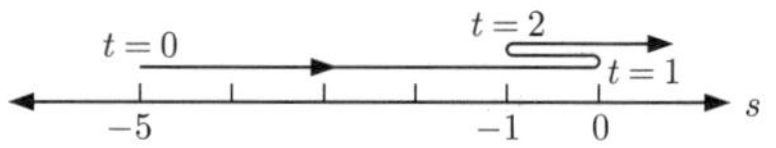

f Speed is increasing for $1 \leqslant t \leqslant 1\frac{1}{2}$ and $t \geqslant 2$.

3 **b** $k = 9$ **4** 6 cm from each end

5 **a** $v(t) = 2 + \dfrac{4}{(t+1)^2}$ cm s^{-1}

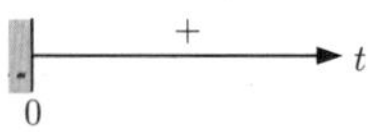

$a(t) = -\dfrac{8}{(t+1)^3}$ cm s^{-2}

b The particle is at O, moving right at 3 cm s^{-1}, and slowing down at 1 cm s^{-2}.

c The particle never changes direction.

d

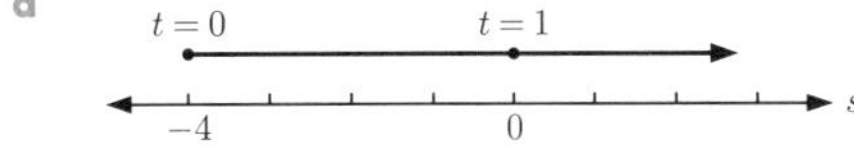

e **i** never **ii** never

6 $x = \frac{k}{2}\left(1 - \frac{1}{\sqrt{3}}\right)$

7 **a** $v(0) = 0\ \text{cm s}^{-1}$, $v(\frac{1}{2}) = -\pi\ \text{cm s}^{-1}$, $v(1) = 0\ \text{cm s}^{-1}$, $v(\frac{3}{2}) = \pi\ \text{cm s}^{-1}$, $v(2) = 0\ \text{cm s}^{-1}$

b $0 \leqslant t \leqslant 1,\ 2 \leqslant t \leqslant 3,\ 4 \leqslant t \leqslant 5,$ etc.
So, for $2n \leqslant t \leqslant 2n+1,\ n \in \{0, 1, 2, 3, ...\}$

8 **a** $a^2 + b^2 - 2ab\cos\theta = c^2 + d^2 - 2cd\cos\phi$

9 P is $(\ln a, 1)$ **10** **a** **i** 5 km **ii** $2\sqrt{10}$ km

REVIEW SET 20B

1 **a** 60 cm **b** **i** 4.24 years **ii** 201 years

c **i** 16 cm per year **ii** 1.95 cm per year

2 **a** $v(t) = -8e^{-\frac{t}{10}} - 40\ \text{m s}^{-1}$
$a(t) = \frac{4}{5}e^{-\frac{t}{10}}\ \text{m s}^{-2}\quad \{t \geqslant 0\}$

b $s(0) = 80$ m, $v(0) = -48\ \text{m s}^{-1}$, $a(0) = 0.8\ \text{m s}^{-2}$

c as $t \to \infty$, $v(t) \to -40\ \text{m s}^{-1}$ (below)

d $v(t)\ \text{m s}^{-1}$, t, -40, -48

e $t = 10\ln 2$ seconds

3 **a** **i** \$535 **ii** \$1385.79

b **i** $-\$0.267$ per km h^{-1} **ii** \$2.33 per km h^{-1}

c 51.3 km h^{-1}

4 **a** $v(t) = 3 - \dfrac{1}{2\sqrt{t+1}}$ $\quad a(t) = \dfrac{1}{4(t+1)^{\frac{3}{2}}}$

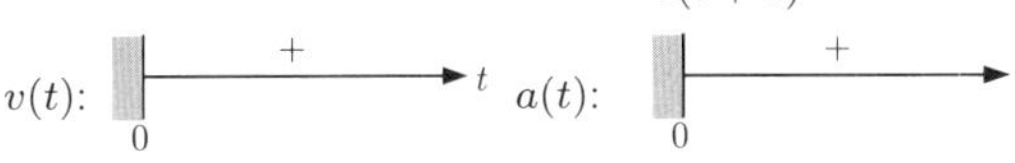

b $x(0) = -1,\ v(0) = 2.5,\ a(0) = 0.25$
Particle is 1 cm to the left of the origin, is travelling to the right at 2.5 cm s^{-1}, and accelerating at 0.25 cm s^{-2}.

c Particle is 21 cm to the right of the origin, is travelling to the right at 2.83 cm s^{-1}, and accelerating at 0.009 26 cm s^{-2}.

d never changes direction **e** never decreasing

5 **a** \$20 000 **b** \$146.53 per year

6 100 or 101 shirts, \$938.63 profit

7 **b** $A = 200x - 2x^2 - \frac{1}{2}\pi x^2$ **c**

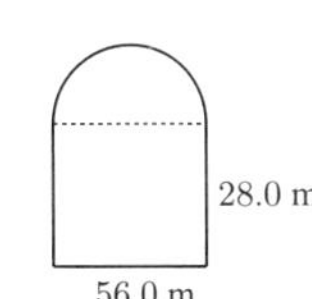

8 **a** $\text{LQ} = \dfrac{8}{x}$ km

b **Hint:** Show that $(\text{length of pipe})^2 = (\text{LQ}+1)^2 + (8+x)^2$ then simplify.

c $\dfrac{d\,[L(x)]^2}{dx} = 2\left(\dfrac{x+8}{x}\right)\left(\dfrac{x^3-8}{x^2}\right)$

d 11.2 km (when $x = 2$ km)

9 **a** $\dfrac{x^2}{9} + \dfrac{y^2}{4} = 1$ **b** $\dfrac{dy}{dx} = -\dfrac{2\cos\theta}{3\sin\theta}$

c 6 units^2, when $\theta = \frac{\pi}{4}, \frac{3\pi}{4}, \frac{5\pi}{4}, \frac{7\pi}{4}$

10 $\frac{20\sqrt{10}}{3} \approx 21.1$ m per minute

11 **a** $V(r) = \frac{8}{9}\pi r^3\ \text{m}^3$

b $\dfrac{dr}{dt} = -\dfrac{8}{375\pi} \approx -0.006\,79\ \text{m min}^{-1}$

REVIEW SET 20C

1 **a** $y = \dfrac{1}{x^2},\ x > 0$ **c** base is 1.26 m square, height 0.630 m

2 **a** $v(t) = 15 + \dfrac{120}{(t+1)^3}\ \text{cm s}^{-1}$, $a(t) = \dfrac{-360}{(t+1)^4}\ \text{cm s}^{-2}$

b At $t = 3$, particle is 41.25 cm to the right of the origin, moving to the right at 16.88 cm s^{-1} and decelerating at 1.41 cm s^{-2}.

c speed is never increasing

3 $\text{A}\left(\dfrac{1}{2}, \dfrac{1}{e}\right)$

4 **a** 2 m **b** $H(3) = 4$ m, $H(6) = 4\frac{2}{3}$ m, $H(9) = 5$ m

c $H'(0) = \frac{4}{3}\ \text{m year}^{-1}$, $H'(3) = \frac{1}{3}\ \text{m year}^{-1}$
$H'(6) = \frac{4}{27}\ \text{m year}^{-1}$, $H'(9) = \frac{1}{12}\ \text{m year}^{-1}$

d $H'(t) = \dfrac{12}{(t+3)^2} > 0$ for all $t \geqslant 0$

$\therefore$ the height of the tree is always increasing.

e

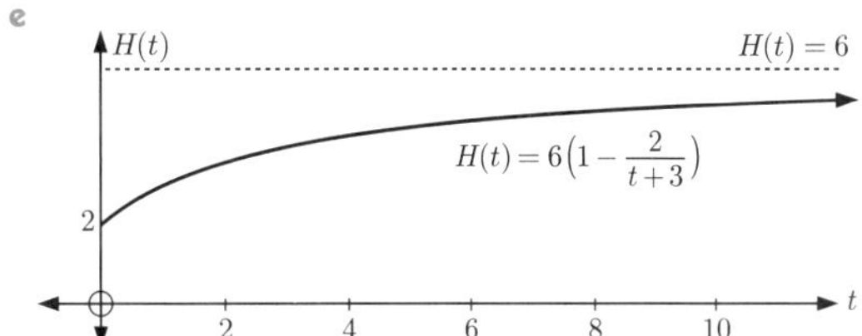

5 **a** $v(t) = 25 - \dfrac{10}{t}\ \text{cm min}^{-1}$, $a(t) = \dfrac{10}{t^2}\ \text{cm min}^{-2}$

b $s(e) = 25e - 10$ cm, $v(e) = 25 - \dfrac{10}{e}\ \text{cm min}^{-1}$,
$a(e) = \dfrac{10}{e^2}\ \text{cm min}^{-2}$

c As $t \to \infty$, $v(t) \to 25\ \text{cm min}^{-1}$ from below

d

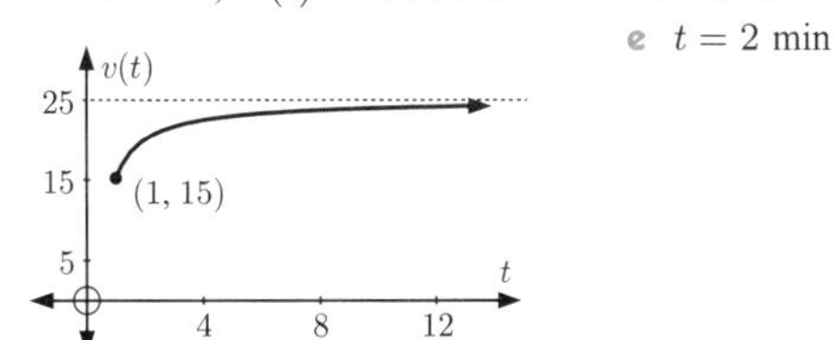

e $t = 2$ min

6 **b** $\dfrac{d[A(x)]^2}{dx} = 5000x - 4x^3$

c Area is a maximum when $x \approx 35.4$, $A = 1250\ \text{m}^2$.

7 **b** $\frac{1}{\sqrt{2}}$ m above the floor

8 **a** **Hint:** Use Pythagoras to find h as a function of x and then substitute into the equation for the volume of a cylinder.

b radius = 4.08 cm, height = 5.77 cm

9 3.60 m s^{-1}

EXERCISE 21A.1

1 a i 0.6 units^2 ii 0.4 units^2 b 0.5 units^2

2 a 0.737 units^2 b 0.653 units^2

3

n	A_L	A_U
10	2.1850	2.4850
25	2.2736	2.3936
50	2.3034	2.3634
100	2.3184	2.3484
500	2.3303	2.3363

converges to $\frac{7}{3}$

4 a i

n	A_L	A_U
5	0.160 00	0.360 00
10	0.202 50	0.302 50
50	0.240 10	0.260 10
100	0.245 03	0.255 03
500	0.249 00	0.251 00
1000	0.249 50	0.250 50
10 000	0.249 95	0.250 05

ii

n	A_L	A_U
5	0.400 00	0.600 00
10	0.450 00	0.550 00
50	0.490 00	0.510 00
100	0.495 00	0.505 00
500	0.499 00	0.501 00
1000	0.499 50	0.500 50
10 000	0.499 95	0.500 05

iii

n	A_L	A_U
5	0.549 74	0.749 74
10	0.610 51	0.710 51
50	0.656 10	0.676 10
100	0.661 46	0.671 46
500	0.665 65	0.667 65
1000	0.666 16	0.667 16
10 000	0.666 62	0.666 72

iv

n	A_L	A_U
5	0.618 67	0.818 67
10	0.687 40	0.787 40
50	0.738 51	0.758 51
100	0.744 41	0.754 41
500	0.748 93	0.750 93
1000	0.749 47	0.750 47
10 000	0.749 95	0.750 05

b i $\frac{1}{4}$ ii $\frac{1}{2}$ iii $\frac{2}{3}$ iv $\frac{3}{4}$ c area $= \dfrac{1}{a+1}$

5 a

n	Rational bounds for π
10	$2.9045 < \pi < 3.3045$
50	$3.0983 < \pi < 3.1783$
100	$3.1204 < \pi < 3.1604$
200	$3.1312 < \pi < 3.1512$
1000	$3.1396 < \pi < 3.1436$
10 000	$3.1414 < \pi < 3.1418$

b $n = 10\,000$

EXERCISE 21A.2

1 a

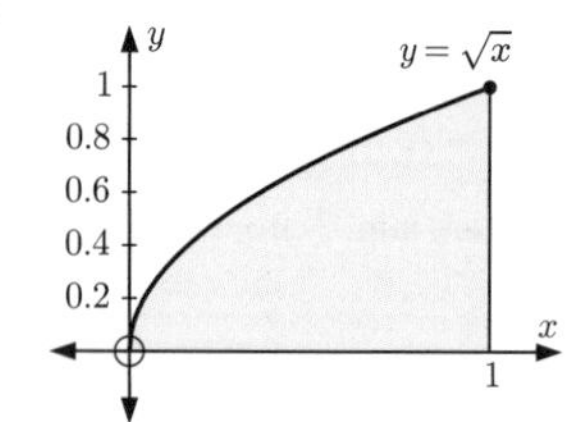

b

n	A_L	A_U
5	0.5497	0.7497
10	0.6105	0.7105
50	0.6561	0.6761
100	0.6615	0.6715
500	0.6656	0.6676

c $\int_0^1 \sqrt{x}\,dx \approx 0.67$

2 a

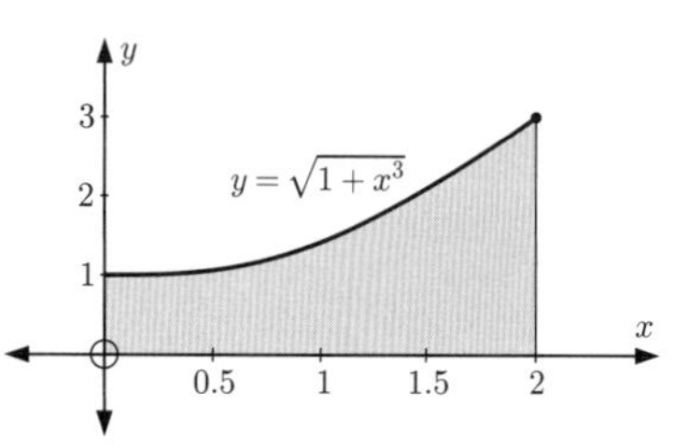

b $A_L = \dfrac{2}{n}\sum\limits_{i=0}^{n-1}\sqrt{1+x_i^3}$, $A_U = \dfrac{2}{n}\sum\limits_{i=1}^{n}\sqrt{1+x_i^3}$

c

n	A_L	A_U
50	3.2016	3.2816
100	3.2214	3.2614
500	3.2373	3.2453

d $\int_0^2 \sqrt{1+x^3}\,dx \approx 3.24$

3 a 18 b 4.5 c 2π

EXERCISE 21B

1 a i $\dfrac{x^2}{2}$ ii $\dfrac{x^3}{3}$ iii $\dfrac{x^6}{6}$ iv $-\dfrac{1}{x}$

v $-\dfrac{1}{3x^3}$ vi $\frac{3}{4}x^{\frac{4}{3}}$ vii $2\sqrt{x}$

b The antiderivative of x^n is $\dfrac{x^{n+1}}{n+1}$ $(n \neq -1)$.

2 a i $\frac{1}{2}e^{2x}$ ii $\frac{1}{5}e^{5x}$ iii $2e^{\frac{1}{2}x}$ iv $100e^{0.01x}$

v $\frac{1}{\pi}e^{\pi x}$ vi $3e^{\frac{x}{3}}$

b The antiderivative of e^{kx} is $\dfrac{1}{k}e^{kx}$.

3 a $\dfrac{d}{dx}(x^3+x^2) = 3x^2+2x$

$\therefore$ the antiderivative of $6x^2+4x = 2x^3+2x^2$

b $\dfrac{d}{dx}(e^{3x+1}) = 3e^{3x+1}$

$\therefore$ the antiderivative of $e^{3x+1} = \frac{1}{3}e^{3x+1}$

c $\dfrac{d}{dx}(x\sqrt{x}) = \frac{3}{2}\sqrt{x}$

$\therefore$ the antiderivative of $\sqrt{x} = \frac{2}{3}x\sqrt{x}$

d $\dfrac{d}{dx}(2x+1)^4 = 8(2x+1)^3$

$\therefore$ the antiderivative of $(2x+1)^3 = \frac{1}{8}(2x+1)^4$

EXERCISE 21C

1 a $\frac{1}{4}$ units^2 b $2\frac{1}{3}$ units^2 c $\frac{2}{3}$ units^2

3 a $3\frac{3}{4}$ units^2 b $24\frac{2}{3}$ units^2 c $\frac{-2+4\sqrt{2}}{3}$ units^2

d ≈ 3.48 units^2 e 2 units^2 f ≈ 3.96 units^2

4 $\frac{9\pi}{4}$ units^2

5 c i $\int_0^1(-x^2)dx = -\frac{1}{3}$, the area between $y=-x^2$ and the x-axis from $x=0$ to $x=1$ is $\frac{1}{3}$ units^2.

ii $\int_0^1(x^2-x)dx = -\frac{1}{6}$, the area between $y=x^2-x$ and the x-axis from $x=0$ to $x=1$ is $\frac{1}{6}$ units^2.

iii $\int_{-2}^0 3x\,dx = -6$, the area between $y=3x$ and the x-axis from $x=-2$ to $x=0$ is 6 units^2

d $-\pi$

EXERCISE 21D

1 $\dfrac{dy}{dx} = 7x^6$; $\int x^6\,dx = \frac{1}{7}x^7 + c$

2 $\dfrac{dy}{dx} = 3x^2+2x$; $\int(3x^2+2x)\,dx = x^3+x^2+c$

3 $\dfrac{dy}{dx} = 2e^{2x+1}$; $\int e^{2x+1}\,dx = \frac{1}{2}e^{2x+1}+c$

4 $\dfrac{dy}{dx} = 8(2x+1)^3$; $\int(2x+1)^3\,dx = \frac{1}{8}(2x+1)^4+c$

5 $\dfrac{dy}{dx} = \frac{3}{2}\sqrt{x}$; $\int\sqrt{x}\,dx = \frac{2}{3}x\sqrt{x}+c$

6 $\dfrac{dy}{dx} = -\dfrac{1}{2x\sqrt{x}}$; $\displaystyle\int \frac{1}{x\sqrt{x}}\,dx = -\frac{2}{\sqrt{x}} + c$

7 $\dfrac{dy}{dx} = -2\sin 2x$; $\int \sin 2x dx = -\frac{1}{2}\cos 2x + c$

8 $\dfrac{dy}{dx} = -5\cos(1-5x)$; $\int \cos(1-5x)dx = -\frac{1}{5}\sin(1-5x)+c$

9 $\int (2x-1)(x^2-x)^2\,dx = \frac{1}{3}(x^2-x)^3 + c$

11 $\dfrac{dy}{dx} = \dfrac{-2}{\sqrt{1-4x}}$; $\displaystyle\int \frac{1}{\sqrt{1-4x}}\,dx = -\frac{1}{2}\sqrt{1-4x} + c$

12 $2\ln(5-3x+x^2)+c$ {since $5-3x+x^2$ is >0}

13 $\dfrac{d}{dx}(\csc x) = -\dfrac{\cos x}{\sin^2 x}$; $\displaystyle\int \frac{\cos x}{\sin^2 x} = -\csc x + c$

14 $\displaystyle\int \frac{-3}{x^2+1}\,dx = -3\arctan x + c$ 15 $\dfrac{1}{\ln 2}2^x + c$

16 $x\ln x - x + c$

EXERCISE 21E.1

1 a $\dfrac{x^5}{5} - \dfrac{x^3}{3} - \dfrac{x^2}{2} + 2x + c$ b $\frac{2}{3}x^{\frac{3}{2}} + e^x + c$
c $3e^x - \ln|x| + c$ d $\frac{2}{5}x^{\frac{5}{2}} - 2\ln|x| + c$
e $-2x^{-\frac{1}{2}} + 4\ln|x| + c$ f $\frac{1}{8}x^4 - \frac{1}{5}x^5 + \frac{3}{4}x^{\frac{4}{3}} + c$
g $\frac{1}{3}x^3 + 3\ln|x| + c$ h $\frac{1}{2}\ln|x| + \frac{1}{3}x^3 - e^x + c$
i $5e^x + \frac{1}{12}x^4 - 4\ln|x| + c$

2 a $-3\cos x - 2x + c$ b $2x^2 - 2\sin x + c$
c $-\cos x - 2\sin x + e^x + c$ d $\frac{2}{7}x^3\sqrt{x} + 10\cos x + c$
e $\frac{1}{9}x^3 - \frac{1}{6}x^2 + \sin x + c$ f $\cos x + \frac{4}{3}x\sqrt{x} + c$

3 a $\frac{1}{3}x^3 + \frac{3}{2}x^2 - 2x + c$ b $\frac{2}{3}x^{\frac{3}{2}} - 2x^{\frac{1}{2}} + c$
c $2e^x + \dfrac{1}{x} + c$ d $-2x^{-\frac{1}{2}} - 8x^{\frac{1}{2}} + c$
e $\frac{4}{3}x^3 + 2x^2 + x + c$ f $\frac{1}{2}x^2 + x - 3\ln|x| + c$
g $\frac{4}{3}x^{\frac{3}{2}} - 2x^{\frac{1}{2}} + c$ h $2x^{\frac{1}{2}} + 8x^{\ \frac{1}{2}} - \frac{20}{3}x^{\ \frac{3}{2}} + c$
i $\frac{1}{4}x^4 + x^3 + \frac{3}{2}x^2 + x + c$

4 a $\frac{2}{3}x^{\frac{3}{2}} + \frac{1}{2}\sin x + c$ b $2e^t + 4\cos t + c$
c $3\sin t - \ln|t| + c$ d $\tan x - 2\cos x + c$
e $\frac{1}{2}\theta^2 + \cos\theta + c$ f $2\ln|\theta| - \tan\theta + c$

5 a $y = 6x + c$ b $y = \frac{4}{3}x^3 + c$
c $y = \frac{10}{3}x\sqrt{x} - \frac{1}{3}x^3 + c$ d $y = -\dfrac{1}{x} + c$
e $y = 2e^x - 5x + c$ f $y = x^4 + x^3 + c$

6 a $x - 2x^2 + \frac{4}{3}x^3 + c$ b $\frac{2}{3}x^{\frac{3}{2}} - 4\sqrt{x} + c$
c $x + 2\ln|x| + \dfrac{5}{x} + c$

7 a $f(x) = \frac{1}{4}x^4 - \frac{10}{3}x\sqrt{x} + 3x + c$
b $f(x) = \frac{4}{3}x^{\frac{3}{2}} - \frac{12}{5}x^{\frac{5}{2}} + c$ c $f(x) = 3e^x - 4\ln|x| + c$

8 $e^x\sin x + e^x\cos x$, $e^x\sin x + c$

9 $-e^{-x}\sin x + e^{-x}\cos x$, $e^{-x}\sin x + c$

10 $\cos x - x\sin x$, $\sin x - x\cos x + c$

11 $\tan x\sec x$, $\sec x + c$ 12 $1 - \dfrac{3}{1+x^2}$, $x - 3\arctan x + c$

13 a $\dfrac{d}{dx}(\arccos x) = \dfrac{-1}{\sqrt{1-x^2}}$, $\dfrac{d}{dx}(\arcsin x) = \dfrac{1}{\sqrt{1-x^2}}$
b $\arcsin x + c_1$ and $-\arccos x + c_2$

EXERCISE 21E.2

1 a $f(x) = x^2 - x + 3$ b $f(x) = x^3 + x^2 - 7$
c $f(x) = e^x + 2\sqrt{x} - 1 - e$ d $f(x) = \frac{1}{2}x^2 - 4\sqrt{x} + \frac{11}{2}$

2 a $f(x) = \dfrac{x^3}{3} - 4\sin x + 3$
b $f(x) = 2\sin x + 3\cos x - 2\sqrt{2}$
c $f(x) = \frac{2}{3}x^{\frac{3}{2}} - 2\tan x - \frac{2}{3}\pi^{\frac{3}{2}}$

3 a $f(x) = \frac{1}{3}x^3 + \frac{1}{2}x^2 + x + \frac{1}{3}$
b $f(x) = 4x^{\frac{5}{2}} + 4x^{\frac{3}{2}} - 4x + 5$
c $f(x) = -\cos x - x + 4$ d $f(x) = \frac{1}{3}x^3 - \frac{16}{3}x + 5$

EXERCISE 21F

1 a $\frac{1}{8}(2x+5)^4 + c$ b $\dfrac{1}{2(3-2x)} + c$ c $\dfrac{-2}{3(2x-1)^3} + c$
d $\frac{1}{32}(4x-3)^8 + c$ e $\frac{2}{9}(3x-4)^{\frac{3}{2}} + c$ f $-4\sqrt{1-5x} + c$
g $-\frac{3}{5}(1-x)^5 + c$ h $-2\sqrt{3-4x} + c$

2 a $-\frac{1}{3}\cos(3x) + c$ b $-\frac{1}{2}\sin(-4x) + x + c$
c $\frac{1}{2}\tan(2x) + c$ d $6\sin\left(\frac{x}{2}\right) + c$
e $-\frac{3}{2}\cos(2x) + e^{-x} + c$ f $\frac{1}{2}e^{2x} - 4\tan\left(\frac{x}{2}\right) + c$
g $-\cos\left(2x + \frac{\pi}{6}\right) + c$ h $3\sin\left(\frac{\pi}{4} - x\right) + c$
i $-2\tan\left(\frac{\pi}{3} - 2x\right) + c$ j $\frac{1}{2}\sin(2x) - \frac{1}{2}\cos(2x) + c$
k $-\frac{2}{3}\cos(3x) + \frac{5}{4}\sin(4x) + c$ l $\frac{1}{16}\sin(8x) + 3\cos x + c$

3 a $y = \frac{1}{3}(2x-7)^{\frac{3}{2}} + 2$ b $(-8, -19)$

4 a $\frac{1}{2}x + \frac{1}{4}\sin(2x) + c$ b $\frac{1}{2}x - \frac{1}{4}\sin(2x) + c$
c $\frac{3}{2}x + \frac{1}{8}\sin(4x) + c$ d $\frac{5}{2}x + \frac{1}{12}\sin(6x) + c$
e $\frac{1}{4}x + \frac{1}{32}\sin(8x) + c$ f $\frac{3}{2}x + 2\sin x + \frac{1}{4}\sin(2x) + c$

5 a $\frac{1}{2}(2x-1)^3 + c$ b $\frac{1}{5}x^5 - \frac{1}{2}x^4 + \frac{1}{3}x^3 + c$
c $-\frac{1}{12}(1-3x)^4 + c$ d $x - \frac{2}{3}x^3 + \frac{1}{5}x^5 + c$
e $-\frac{8}{3}(5-x)^{\frac{3}{2}} + c$ f $\frac{1}{7}x^7 + \frac{3}{5}x^5 + x^3 + x + c$

6 b $\frac{1}{32}\sin(4x) + \frac{1}{4}\sin(2x) + \frac{3}{8}x + c$

7 a $2e^x + \frac{5}{2}e^{2x} + c$ b $\frac{3}{5}e^{5x-2} + c$ c $-\frac{1}{3}e^{7-3x} + c$
d $\frac{1}{2}\ln|2x-1| + c$ e $-\frac{5}{3}\ln|1-3x| + c$
f $-e^{-x} - 2\ln|2x+1| + c$ g $\frac{1}{2}e^{2x} + 2x - \frac{1}{2}e^{-2x} + c$
h $-\frac{1}{2}e^{-2x} - 4e^{-x} + 4x + c$ i $\frac{1}{2}x^2 + 5\ln|1-x| + c$

8 a $y = x - 2e^x + \frac{1}{2}e^{2x} + c$
b $y = x - x^2 + 3\ln|x+2| + c$
c $y = -\frac{1}{2}e^{-2x} + 2\ln|2x-1| + c$

9 Both are correct. Recall that:
$$\frac{d}{dx}(\ln(Ax)) = \frac{d}{dx}(\ln A + \ln x) = \frac{1}{x}, \quad A, x > 0$$

10 $p = -\frac{1}{4}$, $f(x) = \frac{1}{2}\cos(\frac{1}{2}x) + \frac{1}{2}$

12 a $f(x) = -e^{-2x} + 4$
b $f(x) = x^2 + 2\ln|1-x| + 2 - 2\ln 2$
c $f(x) = \frac{2}{3}x^{\frac{3}{2}} - \frac{1}{8}e^{-4x} + \frac{1}{8}e^{-4} - \frac{2}{3}$

13 $x - \frac{1}{2}\cos(2x) + c$ **14** $\frac{1}{4}\sin 2x + 2\sin x + \frac{3}{2}x + c$

15 $\int \frac{2x-8}{x^2-4}\,dx = 3\ln|x+2| - \ln|x-2| + c$

16 $\int \frac{2}{4x^2-1}\,dx = \frac{1}{2}\ln|2x-1| - \frac{1}{2}\ln|2x+1| + c$

EXERCISE 21G.1

1 **a** $\frac{1}{5}(x^3+1)^5 + c$ **b** $\frac{1}{3}e^{x^3+1} + c$
c $\frac{1}{5}\sin^5 x + c$ **d** $e^{\frac{x-1}{x}} + c$

2 **a** $\frac{1}{4}(2+x^4)^4 + c$ **b** $2\sqrt{x^2+3} + c$
c $\frac{1}{8(1-x^2)^4} + c$ **d** $\frac{2}{3}(x^3+x)^{\frac{3}{2}} + c$
e $\frac{1}{5}(x^3+2x+1)^5 + c$ **f** $-\frac{1}{27(3x^3-1)^3} + c$
g $-\frac{1}{2(x^2+4x-3)} + c$ **h** $\frac{1}{5}(x^2+x)^5 + c$

3 **a** $e^{1-2x} + c$ **b** $e^{x^2} + c$ **c** $2e^{\sqrt{x}} + c$ **d** $-e^{x-x^2} + c$

4 **a** $\ln\left|x^2+1\right| + c$ **b** $-\frac{1}{2}\ln\left|2-x^2\right| + c$
c $\ln\left|x^2-3x\right| + c$ **d** $2\ln\left|x^3-x\right| + c$
e $-2\ln\left|5x-x^2\right| + c$ **f** $-\frac{1}{3}\ln\left|x^3-3x\right| + c$

5 **a** $f(x) = -\frac{1}{9}(3-x^3)^3 + c$ **b** $f(x) = 4\ln|\ln x| + c$
c $f(x) = -\frac{1}{3}(1-x^2)^{\frac{3}{2}} + c$ **d** $f(x) = -\frac{1}{2}e^{1-x^2} + c$
e $f(x) = -\ln\left|x^3-x\right| + c$ **f** $f(x) = \frac{1}{4}(\ln x)^4 + c$

6 **a** $\frac{1}{5}\sin^5 x + c$ **b** $-2(\cos x)^{\frac{1}{2}} + c$
c $-\ln|\cos x| + c$ **d** $\frac{2}{3}(\sin x)^{\frac{3}{2}} + c$
e $-(2+\sin x)^{-1} + c$ **f** $\frac{1}{2}(\cos x)^{-2} + c$
g $\ln|1-\cos x| + c$ **h** $\frac{1}{2}\ln|\sin(2x)-3| + c$
i $-\frac{1}{2}\cos(x^2) + c$ **j** $\frac{1}{4}\tan^4 x + c$
k $-\frac{1}{6}\csc^3(2x) + c$ **l** $-\frac{1}{3}\sin^3 x + \sin x + c$

7 **a** $-\cos x + \frac{2}{3}\cos^3 x - \frac{1}{5}\cos^5 x + c$
b $\frac{1}{5}\sin^5 x - \frac{1}{7}\sin^7 x + c$ **c** $\frac{1}{8}\sin^4(2x) + c$

8 **a** $-e^{\cos x} + c$ **b** $\ln|\sin x - \cos x| + c$ **c** $e^{\tan x} + c$

9 **a** $\ln|\sin x| + c$ **b** $\frac{1}{3}\ln|\sin(3x)| + c$ **c** $-\cot x + c$
d $\sec x + c$ **e** $-\csc x + c$ **f** $\frac{1}{3}\sec(3x) + c$
g $-2\csc\left(\frac{x}{2}\right) + c$ **h** $\frac{1}{2}\sec^2 x + c$ **i** $-2\sqrt{\cot x} + c$

10 **a** $\frac{1}{4}\left[\ln(x^2+7)\right]^2 + c$
b $\frac{2}{7}(x-16)^{\frac{7}{2}} + \frac{64}{5}(x-16)^{\frac{5}{2}} + \frac{512}{3}(x-16)^{\frac{3}{2}} + c$
c $\frac{1}{12}\arctan\left(\frac{x}{3}\right) + c$ **d** $2\sqrt{x-1} - 2\arctan\sqrt{x-1} + c$
e $\frac{1}{5}\sqrt{4x^2-1} - \frac{1}{5}\arccos\left(\frac{1}{2x}\right) + c$

11 **a** $\frac{d}{dx}(\arcsin x) = \frac{1}{\sqrt{1-x^2}}$,
$\int \frac{1}{\sqrt{1-x^2}}\,dx = \arcsin x + c$
b $\int \frac{1}{\sqrt{a^2-x^2}}\,dx = \arcsin\left(\frac{x}{a}\right) + c$
c Domain $= \{x \mid -a \leqslant x \leqslant a,\ \ a \neq 0\}$
d **i** $4\arcsin x + c$ **ii** $3\arcsin\left(\frac{x}{2}\right) + c$
iii $\frac{1}{2}\arcsin(2x) + c$ **iv** $\frac{2}{3}\arcsin\left(\frac{3x}{2}\right) + c$

12 **a** $\frac{d}{dx}(\arctan x) = \frac{1}{x^2+1}$, $\int \frac{1}{x^2+1}\,dx = \arctan x + c$
b $\int \frac{1}{x^2+a^2}\,dx = \frac{1}{a}\arctan\left(\frac{x}{a}\right) + c,\ \ a \neq 0$
c Domain $= \{x \mid x \in \mathbb{R}\}$
d **i** $\frac{1}{4}\arctan\left(\frac{x}{4}\right) + c$ **ii** $\frac{1}{2}\arctan(2x) + c$
iii $\frac{1}{2\sqrt{2}}\arctan\left(\frac{x}{\sqrt{2}}\right) + c$ **iv** $\frac{5}{6}\arctan\left(\frac{2x}{3}\right) + c$

13 **a** $\frac{1}{4\xi} - \frac{x}{(x+\xi)^2} + c$ **b** $\frac{1}{3\xi}\left(\frac{x-\xi}{x+\xi}\right)^{\frac{3}{2}} + c$

14 $2x^{\frac{1}{2}} - 3x^{\frac{1}{3}} + 6x^{\frac{1}{6}} - 6\ln(x^{\frac{1}{6}}+1) + c$

EXERCISE 21G.2

1 **a** $\frac{2}{5}(x-3)^{\frac{5}{2}} + 2(x-3)^{\frac{3}{2}} + c$
b $\frac{2}{7}(x+1)^{\frac{7}{2}} - \frac{4}{5}(x+1)^{\frac{5}{2}} + \frac{2}{3}(x+1)^{\frac{3}{2}} + c$
c $-(3-x^2)^{\frac{3}{2}} + \frac{1}{5}(3-x^2)^{\frac{5}{2}} + c$
d $\frac{1}{5}(t^2+2)^{\frac{5}{2}} - \frac{2}{3}(t^2+2)^{\frac{3}{2}} + c$

2 **a** $x - 3\arctan\left(\frac{x}{3}\right) + c$ **b** $\frac{1}{2}\arcsin x - \frac{1}{2}x\sqrt{1-x^2} + c$
c $\ln(x^2+9) + c$ **d** $2\ln(1+[\ln x]^2) + c$
e $\sqrt{x^2-4} - 2\arccos\left(\frac{2}{x}\right) + c$
f $\cos x - \frac{2}{3}\cos^3 x + c$ **g** $\frac{1}{2}\arcsin\left(\frac{2x}{3}\right) + c$
h $\frac{1}{2}x^2 - \frac{1}{2}\ln(1+x^2) + c$ **i** $\frac{1}{6}\arctan\left(\frac{2}{3}\ln x\right) + c$
j $\frac{1}{16}\ln\left(\frac{|x|}{\sqrt{x^2+16}}\right) + c$ **k** $\frac{-1}{16x}\sqrt{16-x^2} + c$
l $2\arcsin\left(\frac{x}{2}\right) - \frac{1}{4}x(2-x^2)\sqrt{4-x^2} + c$

EXERCISE 21H

1 **a** $xe^x - e^x + c$ **b** $-x\cos x + \sin x + c$
c $\frac{1}{3}x^3\ln x - \frac{1}{9}x^3 + c$ **d** $-\frac{1}{3}x\cos 3x + \frac{1}{9}\sin 3x + c$
e $\frac{1}{2}x\sin 2x + \frac{1}{4}\cos 2x + c$ **f** $x\tan x + \ln|\cos x| + c$

2 **a** $x\ln x - x + c$ **b** $x(\ln x)^2 - 2x\ln x + 2x + c$

3 $x\arctan x - \frac{1}{2}\ln(x^2+1) + c$

4 **a** $-x^2e^{-x} - 2xe^{-x} - 2e^{-x} + c$
b $\frac{1}{2}e^x(\sin x + \cos x) + c$ **c** $-\frac{1}{2}e^{-x}(\cos x + \sin x) + c$
d $-x^2\cos x + 2x\sin x + 2\cos x + c$

5 **a** $u^2e^u - 2ue^u + 2e^u + c$ **b** $x(\ln x)^2 - 2x\ln x + 2x + c$

6 **a** $-u\cos u + \sin u + c$ **b** $-\sqrt{2x}\cos\sqrt{2x} + \sin\sqrt{2x} + c$

7 $\frac{2}{3}\sqrt{3x}\sin\sqrt{3x} + \frac{2}{3}\cos\sqrt{3x} + c$

EXERCISE 21I

1 **a** $\ln\left|e^x - e^{-x}\right| + c$ **b** $\frac{7^x}{\ln 7} + c$ **c** $\frac{(3x+5)^6}{18} + c$
d $\ln(2-\cos x) + c$ **e** $x\tan x + \ln|\cos x| + c$
f $\frac{1}{2}\ln|\sin 2x| + c$ **g** $\frac{(x+3)^5}{5} - \frac{3}{4}(x+3)^4 + c$
h $\frac{x^3}{3} + \frac{3}{2}x^2 + 3x + \ln|x| + c$

2 **a** $-x^2e^{-x} - 2xe^{-x} - 2e^{-x} + c$
b $\frac{2}{5}(1-x)^{\frac{5}{2}} - \frac{2}{3}(1-x)^{\frac{3}{2}} + c$

c $\dfrac{x^3\sqrt{1-x^2}}{4} - \dfrac{x\sqrt{1-x^2}}{8} + \dfrac{\arcsin x}{8} + c$

d $\frac{3}{2}\arccos\left(\frac{2}{x}\right) + c$

e $\frac{2}{7}(x-3)^{\frac{7}{2}} + \frac{12}{5}(x-3)^{\frac{5}{2}} + 6(x-3)^{\frac{3}{2}} + c$

f $\ln|\cos x| + \dfrac{1}{2\cos^2 x} + c$ **g** $-\dfrac{\ln(x+2)+1}{x+2} + c$

h $\frac{1}{\sqrt{2}}\arctan\left(\frac{x+1}{\sqrt{2}}\right) + c$

3 **a** $\frac{1}{3}\arctan\left(\frac{x}{3}\right) + c$ **b** $8\arcsin\left(\sqrt{x}\right) + c$

c $x\ln(2x) - x + c$ **d** $\frac{1}{2}e^{-x}(\sin x - \cos x) + c$

e $\ln\left|\frac{x}{\sqrt{x^2+1}}\right| + c$ **f** $\dfrac{\arctan^2 x}{2} + c$

g $\frac{9}{2}\arcsin\left(\frac{x}{3}\right) + \frac{x\sqrt{9-x^2}}{2} + c$

h $-\left[\dfrac{(\ln x)^2 + 2\ln x + 2}{x}\right] + c$

i $\frac{2}{3}(x-3)^{\frac{3}{2}} + 6\sqrt{x-3} + c$

j $-\frac{1}{15}(\sin x \sin 4x + 4\cos x \cos 4x) + c$

k $\ln\left|x^2 - 2x + 5\right| + \frac{5}{2}\arctan\left(\frac{x-1}{2}\right) + c$

l $\sin x - \frac{1}{3}\sin^3 x + c$ **m** $\frac{1}{2}\ln\left|x^2+4\right| + 2\arctan\left(\frac{x}{2}\right) + c$

n $\arcsin\left(\frac{x}{2}\right) + 2\sqrt{4-x^2} + c$

o $2 - x - 6\ln|2-x| - \dfrac{12}{2-x} + \dfrac{4}{(2-x)^2} + c$

p $\dfrac{\sin^6 x}{6} - \dfrac{\sin^8 x}{4} + \dfrac{\sin^{10} x}{10} + c$

EXERCISE 21J.1

1 **a** $\frac{1}{4}$ **b** $\frac{2}{3}$ **c** $e - 1$ (≈ 1.72) **d** $1\frac{1}{2}$ **e** $6\frac{2}{3}$
f $\ln 3$ (≈ 1.10) **g** 1.52 **h** 2 **i** $e - 1$ (≈ 1.72)

2 **a** $\frac{1}{2}$ **b** $\frac{1}{2}$ **c** $\sqrt{3} - 1$ **d** $\frac{1}{3}$ **e** $\frac{\pi}{8} + \frac{1}{4}$ **f** $\frac{\pi}{4}$

4 **a** $m = -5$ **b** $m = -1$ or $\frac{4}{3}$ **5** ≈ 1.49

6 **a** $2e^{-1} - 1$ (≈ -0.264) **b** 1 **c** $3\ln 3 - 2$ (≈ 1.30)

EXERCISE 21J.2

1 **a** $\frac{1}{12}$ **b** 1.56 **c** $20\frac{1}{3}$ **d** 0.0337
e $\frac{1}{2}\ln\left(\frac{2}{7}\right)$ (≈ -0.626) **f** $\frac{1}{2}(\ln 2)^2$ (≈ 0.240)
g 0 **h** $2\ln 7$ (≈ 3.89)

2 **a** $2 - \sqrt{2}$ **b** $\frac{1}{24}$ **c** $\frac{1}{2}\ln 2$ **d** $\ln 2$ **e** $\ln 2$ **f** $\frac{1}{4}$

3 $\dfrac{3^{n+1}}{2n+2}$, $n \neq -1$, undefined for $n = -1$

4 **a** $\frac{28\sqrt{3}}{5} - \frac{44\sqrt{2}}{15}$ **b** $\frac{48\sqrt{6}-54}{5}$ **c** $\frac{1054\sqrt{3}}{35}$

EXERCISE 21J.3

1 **a** $\int_1^4 \sqrt{x}\,dx = \frac{14}{3}$, $\int_1^4 (-\sqrt{x})\,dx = -\frac{14}{3}$
b $\int_0^1 x^7\,dx = \frac{1}{8}$, $\int_0^1 (-x^7)\,dx = -\frac{1}{8}$

2 **a** $\frac{1}{3}$ **b** $\frac{7}{3}$ **c** $\frac{8}{3}$ **d** 1

3 **a** -4 **b** 6.25 **c** 2.25 **4** **a** $\frac{1}{3}$ **b** $\frac{2}{3}$ **c** 1

5 **a** 6.5 **b** -9 **c** 0 **d** -2.5

6 **a** 2π **b** -4 **c** $\frac{\pi}{2}$ **d** $\frac{5\pi}{2} - 4$

7 **a** $\int_2^7 f(x)\,dx$ **b** $\int_1^9 g(x)\,dx$ **8** **a** -5 **b** 4

9 **a** 4 **b** 0 **c** -8 **d** $k = -\frac{7}{4}$ **10** 0

REVIEW SET 21A

1 **a** 2π units^2 **b** 4 units^2

2 **a** $8\sqrt{x} + c$ **b** $-\frac{3}{2}\ln|1 - 2x| + c$
c $-\frac{1}{4}\cos(4x - 5) + c$ **d** $-\frac{1}{3}e^{4-3x} + c$

3 **a** $12\frac{4}{9}$ **b** $\sqrt{2}$ **c** $\frac{5}{54}$

4 $\dfrac{dy}{dx} = \dfrac{x}{\sqrt{x^2-4}}$; $\int$ is $\sqrt{x^2-4} + c$ **5** $b = \frac{\pi}{4}, \frac{3\pi}{4}$

6 **a** $\dfrac{9x}{2} - 4\sin x + \frac{1}{4}\sin(2x) + c$ **b** $\frac{1}{2}\sin(x^2) + c$

7 $\dfrac{d}{dx}(3x^2 + x)^3 = 3(3x^2 + x)^2(6x + 1)$
$\int (3x^2 + x)^2(6x + 1)\,dx = \frac{1}{3}(3x^2 + x)^3 + c$

8 **a** 6 **b** 3

9 **a** $\frac{1}{8}\sin^8 x + c$ **b** $-\frac{1}{2}\ln|\cos(2x)| + c$ **c** $e^{\sin x} + c$

10 $f\left(\frac{\pi}{2}\right) = 3 - \frac{\pi}{2}$ **11** **a** $\frac{\pi}{3} - 1$ **b** $\ln 2$ **c** $\frac{1}{2}\ln 3$

12 $-\frac{32}{3}(4-x)^{\frac{3}{2}} + \frac{16}{5}(4-x)^{\frac{5}{2}} - \frac{2}{7}(4-x)^{\frac{7}{2}} + c$

13 $x\arctan x - \frac{1}{2}\ln(x^2 + 1) + c$

14 **a** $\frac{1}{4}(x^2 + 1)^4 + c$ **b** **i** $\frac{15}{4}$ **ii** $-\frac{609}{8}$

15 **a** $A = 4$, $B = -2$, $C = -2$
b $4\ln|x| - 2\ln|x + 1| - 2\ln|x - 1| + c$ **c** $\ln\left(\frac{16}{25}\right)$

REVIEW SET 21B

1 **a**

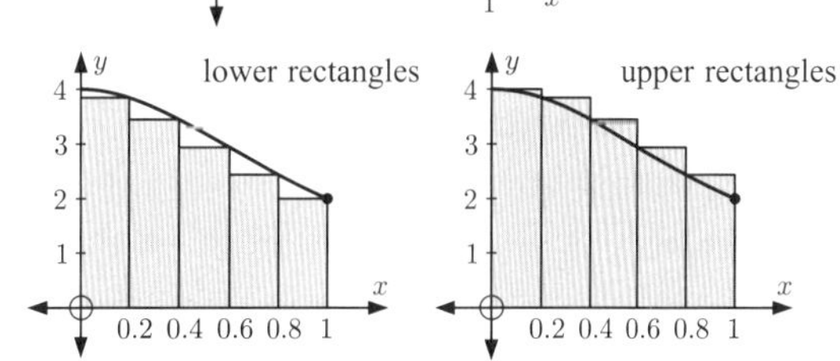

b

n	A_L	A_U
5	2.9349	3.3349
50	3.1215	3.1615
100	3.1316	3.1516
500	3.1396	3.1436

c $\displaystyle\int_0^1 \frac{4}{1+x^2}\,dx \approx 3.142$

2 **a** $y = \frac{1}{5}x^5 - \frac{2}{3}x^3 + x + c$ **b** $y = 400x + 40e^{-\frac{x}{2}} + c$
c $y = \frac{1}{6}x^6 - \frac{1}{2}x^4 + \frac{1}{2}x^2 + c$

3 **a** 3.528 **b** 2.963 **4** $\dfrac{2(\ln x)}{x}$, $\frac{1}{2}(\ln x)^2 + c$

5 $f(x) = 3x^3 + 5x^2 + 6x - 1$ **6** $a = \ln\sqrt{2}$

7 **a** $3 - \sqrt{7}$ **b** $e^2 - 2e^1$

8 **a** $f(x) = \frac{1}{4}x^4 + \frac{1}{3}x^3 - \frac{10}{3}x + 3$ **b** $3x + 26y = 84$

9 **a** $e^{3x} + 6e^{2x} + 12e^x + 8$ **b** $\frac{1}{3}e^3 + 3e^2 + 12e - 7\frac{1}{3}$

10 $\frac{19}{384}$ **11** $23\frac{1}{6}$ **12** $x\tan x + \ln|\cos x| + c$

13 **a** $2 + \dfrac{-5}{2x+1}$ **b** $\displaystyle\int_0^2 \frac{4x-3}{2x+1}\,dx = 4 - \tfrac{5}{2}\ln 5$ (≈ -0.0236)

14 **a** $\frac{1}{2}e^{-x}(\sin x - \cos x) + c$ **b** $e^x(x^2 - 2x + 2) + c$
c $\frac{1}{3}(9 - x^2)^{\frac{3}{2}} - 9\sqrt{9 - x^2} + c$

15 **a** $\ln\left(\frac{|x+2|}{(x-1)^2}\right) + c$ **b** $\ln\left(\frac{(x-1)^2}{|x+2|}\right) + c$

REVIEW SET 21C

1 **a** $-2e^{-x} - \ln|x| + 3x + c$ **b** $\frac{1}{2}x^2 - 2x + \ln|x| + c$
c $9x + 3e^{2x-1} + \frac{1}{4}e^{4x-2} + c$

2 $f(x) = \frac{1}{3}x^3 - \frac{3}{2}x^2 + 2x + 2\frac{1}{6}$

3 **a** $\frac{2}{3}(\sqrt{5} - \sqrt{2})$ **b** $\frac{\pi}{6} + \frac{\sqrt{3}}{4}$ **c** $\frac{1}{2}\ln 2$ **4** $e^{-\pi}$

5 if $n \neq -1$, $\dfrac{1}{2(n+1)}(2x+3)^{n+1} + c$
if $n = -1$, $\frac{1}{2}\ln|2x+3| + c$

6 $a = \frac{1}{3}$, $f'(x) = 2\sqrt{x} + \frac{1}{3\sqrt{x}}$ is never 0 as $\sqrt{x} \geqslant 0$ for all x
$\therefore$ $f'(x) > 0$ for all x

7 $a = 0$ or ± 3

8 **a** $A = \frac{17}{4}$, $B = \frac{25}{4}$ **b** $\int_0^2 (4 - x^2)\,dx \approx \frac{21}{4}$

9 **a** $2\sqrt{x^2 - 5} + c$ **b** $\dfrac{1}{3\cos^3 x} + c$ **c** $\frac{14\sqrt{7}}{3}$

10 $\tan x$, $\int \tan x\,dx = \ln(\sec x) + c$

11 **a** $5\arcsin\left(\frac{x}{3}\right) + c$ **b** $\frac{1}{6}\arctan\left(\frac{2x}{3}\right) + c$
c $\frac{80}{3}\sqrt{5} - \frac{124}{15}\sqrt{2}$

12 $A = 1$, $B = 2$, $C = 1$, $D = 4$, $\dfrac{(x+1)^3}{3} + 4\ln|x-2| + c$

13 **a** $\cos x + x\sin x + c$ **b** $\sqrt{x^2 - 4} + 2\arccos\left(\frac{2}{x}\right) + c$

14 $1 - \dfrac{1}{4e}$ **15** $\dfrac{\cos^{1-\frac{n}{2}} x}{\frac{n}{2} - 1} + c$, for $n \neq 2$,
$-\ln|\cos x| + c$, for $n = 2$

EXERCISE 22A

1 **a** 30 units^2 **b** $\frac{9}{2}$ units^2 **c** $\frac{27}{2}$ units^2 **d** 2 units^2

2 **a** $\frac{1}{3}$ units^2 **b** 2 units^2 **c** $63\frac{3}{4}$ units^2
d $(e - 1)$ units^2 **e** $20\frac{5}{6}$ units^2 **f** 18 units^2
g $\ln 4$ units^2 **h** $4\frac{1}{2}$ units^2 **i** $(2e - \dfrac{2}{e})$ units^2

3 **a** $3\frac{1}{3}$ units^2 **b** $12\frac{2}{3}$ units^2 **4** $\frac{2}{3}$ units^2

5 **b** $\frac{\pi}{2}$ units^2 **6** $(4\ln 4 - 3)$ units^2

7 **a** $-x\cos x + \sin x + c$ **b** $(1 + \cos 1 - \sin 1)$ units^2

EXERCISE 22B

1 **a** $4\frac{1}{2}$ units^2 **b** $(1 + e^{-2})$ units^2 **c** $\ln 3$ units^2
d $1\frac{5}{27}$ units^2 **e** 2 units^2 **f** $2\frac{1}{4}$ units^2 **g** $(\frac{\pi}{2} - 1)$ units^2

2 $10\frac{2}{3}$ units^2

3 **a, b**

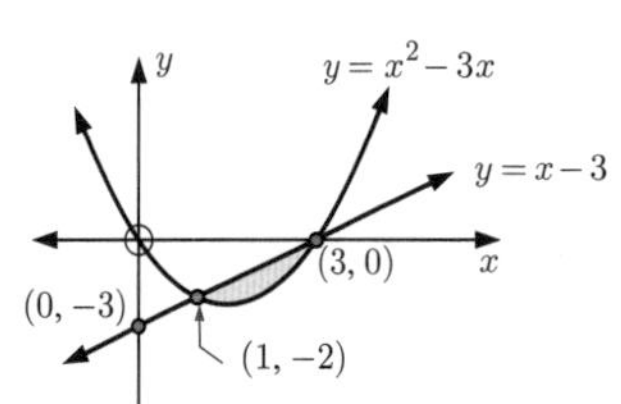

c $1\frac{1}{3}$ units^2

4 $\frac{1}{3}$ units^2

5 **a, b**

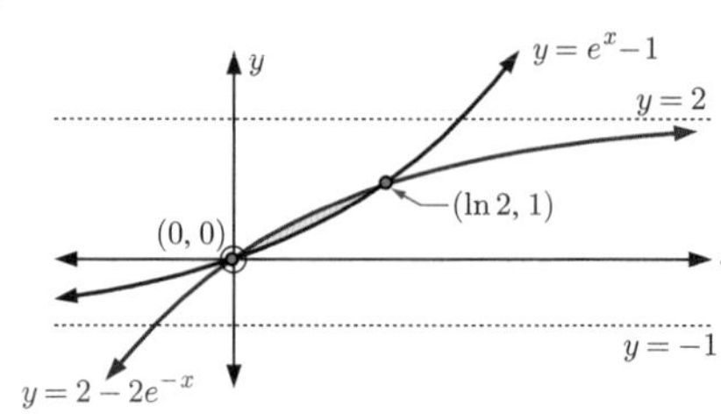

c enclosed area $= 3\ln 2 - 2$ (≈ 0.0794) units^2

6 $\frac{1}{2}$ units^2

7 **a**

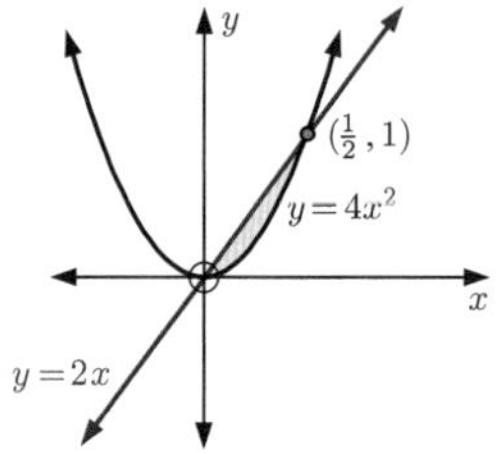

enclosed area $= \frac{1}{12}$ units^2

b

enclosed area $= \frac{1}{3}$ units^2

8 $\ln\left(\frac{27}{4}\right) - 1$ units^2 **9** $(\sqrt{2} - 1)$ units^2

10 **a** $(\frac{\pi}{4}, 1)$ **b** $\ln\sqrt{2}$ units^2

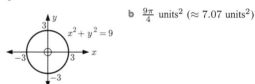

11 **a** **b** $\frac{9\pi}{4}$ units^2 $(\approx 7.07$ units$^2)$

12 **a** $40\frac{1}{2}$ units^2 **b** 8 units^2 **c** 8 units^2

13 **a** C_1 is $y = \sin(2x)$, C_2 is $y = \sin x$ **b** $A(\frac{\pi}{3}, \frac{\sqrt{3}}{2})$
c $2\frac{1}{2}$ units^2

14 **a** **i** $A = \int_{-2}^{-1} (x^3 - 7x - 6)\,dx + \int_{-1}^{3} (7x + 6 - x^3)\,dx$
ii $A = \int_{-2}^{3} \left|x^3 - 7x - 6\right|\,dx$
b Area $= 32\frac{3}{4}$ units^2

15 **a** $21\frac{1}{12}$ units^2 **b** 8 units^2 **c** $101\frac{3}{4}$ units^2

16 **a** $\int_3^5 f(x)\,dx = -$ (area between $x = 3$ and $x = 5$)
b $\int_1^3 f(x)\,dx - \int_3^5 f(x)\,dx + \int_5^7 f(x)\,dx$

17 **a** C_1 is $y = \cos^2 x$, C_2 is $y = \cos(2x)$
b $A(0, 1)$, $B(\frac{\pi}{4}, 0)$, $C(\frac{\pi}{2}, 0)$, $D(\frac{3\pi}{4}, 0)$, $E(\pi, 1)$
c Area $= \int_0^{\pi} (\cos^2 x - \cos(2x))\,dx$

18 **a** 2.88 units^2 **b** 4.97 units^2

19 **a** $k \approx 1.7377$ **b** $b \approx 1.3104$

20 **a** $k \approx 2.3489$ **b** $a = \sqrt{3}$

21 **a**

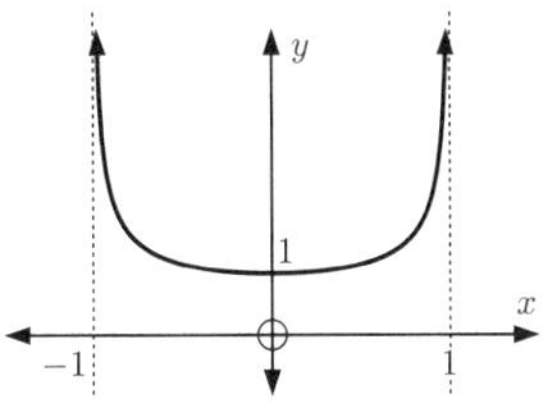

b **i** $f(-x) = \frac{1}{\sqrt{1-(-x)^2}}$
$= \frac{1}{\sqrt{1-x^2}}$
$= f(x)$ for all x

ii For y to have meaning, $1 - x^2 > 0$ which has solution $x \in]-1, 1[$.

c Area $= \frac{\pi}{6}$ units^2

EXERCISE 22C.1

1 110 m

2 **a** **i** travelling forwards
ii travelling backwards (opposite direction)
b 16 km **c** 8 km from starting point (on positive side)

3 **a**

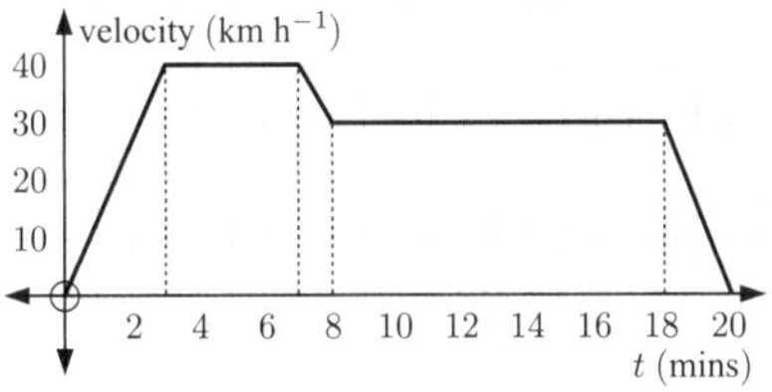

b 9.75 km

EXERCISE 22C.2

1 **a** $s(t) = t - t^2 + 2$ cm **b** $\frac{1}{2}$ cm **c** 0 cm

2 **a** $s(t) = \frac{1}{3}t^3 - \frac{1}{2}t^2 - 2t$ cm **b** $5\frac{1}{6}$ cm
c $1\frac{1}{2}$ cm left of its starting point

3 **a** $s(t) = 32t + 2t^2 + 16$ m
b no change of direction
so displacement $= s(t_1) - s(0) = \int_0^{t_1} (32 + 4t)\,dt$
c acceleration $= 4$ m s^{-2}

4 $\frac{\sqrt{3}+2}{4}$ m **5** **a** 41 units **b** 34 units **6** **b** 2 m

7 **a** 40 m s^{-1}
b 47.8 m s^{-1}
c 1.39 seconds
d as $t \to \infty$, $v(t) \to 50^-$
e $a(t) = 5e^{-0.5t}$ and as $e^x > 0$ for all x, $a(t) > 0$ for all t.
f

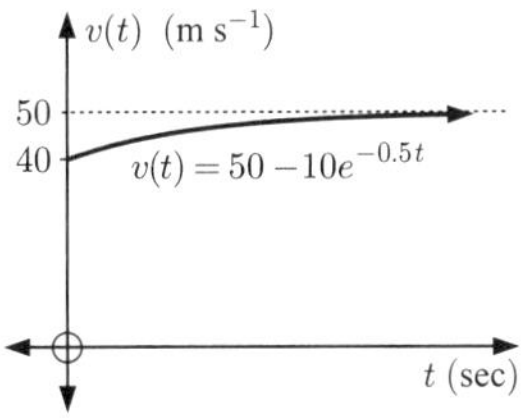

g 134.5 m

8 **a** $v(t) = \frac{1}{t+1} - 1$ m s^{-1} **b** $s(t) = \ln|t+1| - t$ metres
c $s(2) = \ln 3 - 2 \approx -0.901$ m, $v(2) = -\frac{2}{3}$ m s^{-1}, $a(2) = -\frac{1}{9}$ m s^{-2}
The object is approximately 0.901 m to the left of the origin, travelling left at $\frac{2}{3}$ m s^{-1}, with acceleration $-\frac{1}{9}$ m s^{-2}.

9 **a** $v(t) = \frac{t^2}{20} - 3t + 45$ m s^{-1}
b $\int_0^{60} v(t)\,dt = 900$. The train travels a total of 900 m in the first 60 seconds.

10 **a** Show that $v(t) = 100 - 80e^{-\frac{1}{20}t}$ m s^{-1} and as $t \to \infty$, $v(t) \to 100$ m s^{-1}. **b** 370.4 m

EXERCISE 22D

1 €4250

2 **a** $P(x) = 15x - 0.015x^2 - 650$ dollars
b maximum profit is $3100, when 500 plates are made
c $46 \leqslant x \leqslant 954$ plates (you cannot produce part of a plate)

3 $\approx 14\,400$ calories **4** 76.3°C

5 **a** When $x = 0$, $y = 0$ and $\frac{dy}{dx} = 0$
b $y = -\frac{1}{120}(1-x)^4 - \frac{x}{30} + \frac{1}{120}$ **c** 2.5 cm (at $x = 1$ m)

6 **a** $y = \left(\frac{0.01}{3}x^3 - \frac{0.005}{12}x^4 - \frac{0.08}{3}x\right)$ metres **b** 3.33 cm
c 2.375 cm **d** 1.05°

7 **Extra hint:** $\frac{dC}{dV} = \frac{1}{2}x^2 + 4$ and $\frac{dV}{dx} = \pi r^2$

8 3.820 20 units **9** **e** 0.974 km × 2.05 km

EXERCISE 22E.1

1 **a** 36π units^3 **b** 8π units^3 **c** $\frac{127\pi}{7}$ units^3
d $\frac{255\pi}{4}$ units^3 **e** $\frac{992\pi}{5}$ units^3 **f** $\frac{250\pi}{3}$ units^3
g $\frac{\pi}{2}$ units^3 **h** $\frac{40\pi}{3}$ units^3

2 **a** 18.6 units^3 **b** 30.2 units^3

3 **a** 186π units^3 **b** $\frac{146\pi}{5}$ units^3 **c** $\frac{\pi}{2}(e^8 - 1)$ units^3

4 **a** a circle **b** Volume of revolution $= \pi \int_a^b y^2\,dx$ $= \int_0^4 \pi(4\sqrt{x})^2\,dx$
c 128π units^3

5 **a** 63π units^3 **b** ≈ 198 cm^3

6 **a** a cone of base radius r and height h
b $y = -\left(\frac{r}{h}\right)x + r$ **c** $V = \frac{1}{3}\pi r^2 h$

7 **a** a sphere of radius r

8 **a** 8π units^3 **b** $\frac{1023\pi}{5}$ units^3 **c** $\frac{\pi}{2}(e^4 - 1)$ units^3
d $\frac{483\pi}{5}$ units^3 **e** $\frac{256\pi}{5}$ units^3

9 48π units^3

10 **a** $\frac{\pi^2}{4}$ units^3 **b** $\frac{\pi^2}{8}$ units^3 **c** 2π units^3
d $\pi\sqrt{3}$ units^3 **e** $\frac{\pi}{3}$ units^3 **f** $\pi\left(2 - \frac{\pi}{2}\right)$ units^3

11 **a**

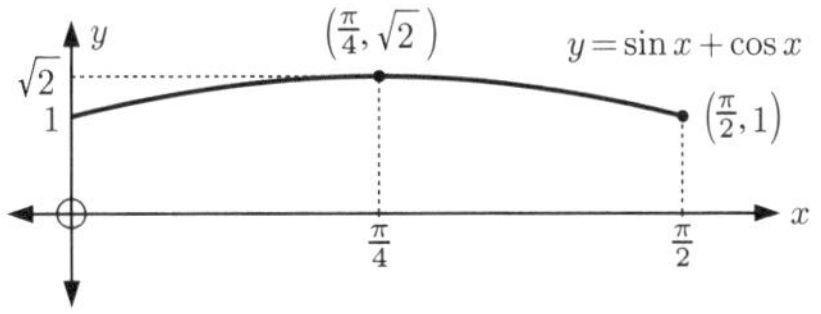

b $\pi\left(\frac{\pi}{4} + \frac{1}{2}\right)$ units^3

12 **a**

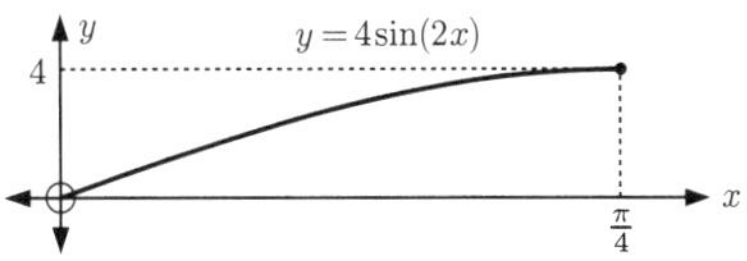

b $2\pi^2$ units^3

EXERCISE 22E.2

1 **a** A(−1, 3), B(1, 3) **b** $\frac{136\pi}{15}$ units^3

2 **a** A(2, e) **b** $\pi(e^2 + 1)$ units^3

3 **a** A(1, 1) **b** $\frac{11\pi}{6}$ units^3

4 **a**

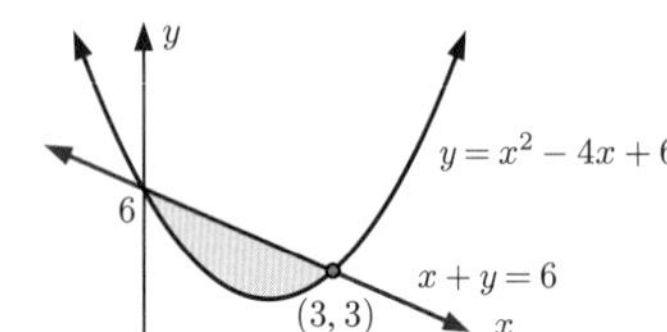

b $\frac{162\pi}{5}$ units^3

5 **a** A(5, 1) **b** $\frac{9\pi}{2}$ units^3

6 **b** see diagram alongside:

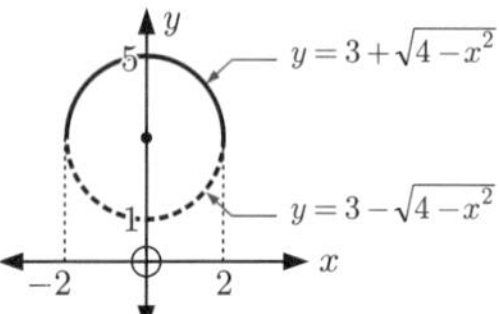

c $24\pi^2 \approx 237$ units^3

8 $V = 36\pi$ units^3, which is independent of r.

REVIEW SET 22A

1 $A = \int_a^b [f(x) - g(x)]\, dx + \int_b^c [g(x) - f(x)]\, dx$
$+ \int_c^d [f(x) - g(x)]\, dx$

2 **a** $2 + \pi$ units^2 **b** -2 units^2 **c** π units^2

3 no, total area shaded $= \int_{-1}^{1} f(x)\, dx - \int_1^3 f(x)\, dx$

4 $k = \sqrt[3]{16}$

5 **Hint:** Show that the areas represented by the integrals can be arranged to form a $1 \times e$ unit rectangle.

6 4.5 units^2 **7** $(3 - \ln 4)$ units^2

8 **a** $v(t)$: + (0 to 2), − (2 to 4), + (after 4); t (seconds)

b The particle moves in the positive direction initially, then at $t = 2$, $6\frac{2}{3}$ m from its starting point, it changes direction. It changes direction again at $t = 4$, $5\frac{1}{3}$ m from its starting point, and at $t = 5$, it is $6\frac{2}{3}$ m from its starting point again.

c $6\frac{2}{3}$ m **d** $9\frac{1}{3}$ m

9 **a** 312π units^3 **b** 402π units^3 **c** 2π units^3
d $\pi\left(\frac{3\pi - 8}{4}\right)$ units^3

10 $\frac{kL^4}{4}$ m **11** **a** $a = -3$ **b** A has x-coordinate $\sqrt[3]{4}$

12 **c** $\int_0^4 y_A\, dx = 2\pi$, $\int_4^6 y_B\, dx = -\frac{\pi}{2}$ **d** $\frac{3\pi}{2}$

13 2π units^3 **16** **a** $\frac{31\pi}{5}$ units^3 **b** $\frac{65\pi}{4}$ units^3

REVIEW SET 22B

1 **a** $a(t) = 2 - 6t$ m s^{-2} **b** $s(t) = t^2 - t^3 + c$ m
c -4 m (4 m to the left)

2 **a** local maximum at $(1, \frac{1}{2})$, local minimum at $(-1, -\frac{1}{2})$
b as $x \to \infty$, $f(x) \to 0^+$, as $x \to -\infty$, $f(x) \to 0^-$
c

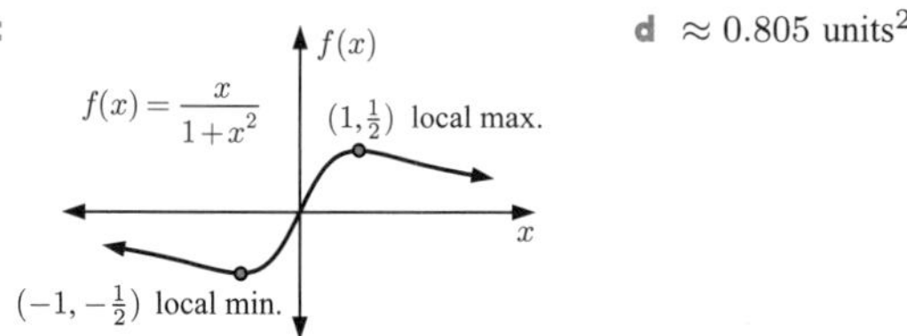

d ≈ 0.805 units^2

3 2.35 m

4 **a** $v(0) = 25$ m s^{-1}, $v(3) = 4$ m s^{-1}
b as $t \to \infty$, $v(t) \to 0^+$
c

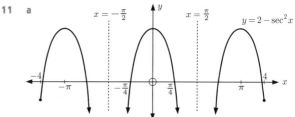

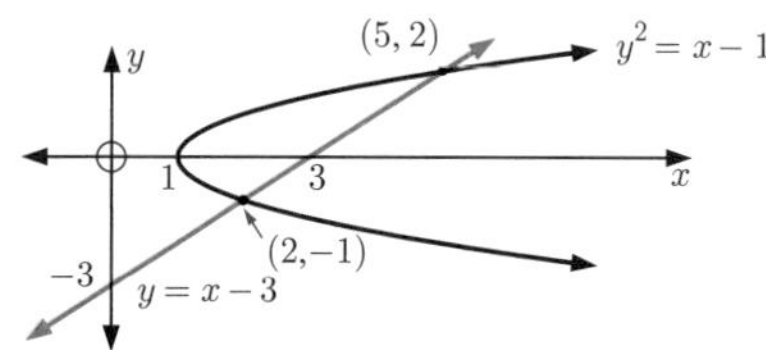

d 3 seconds **e** $a(t) = \frac{-200}{(t+2)^3}$ m s^{-2}, $t \geqslant 0$ **f** $k = \frac{1}{5}$

5 **a** 0 and -0.7292 **b** 0.2009 units^2

6 **a**

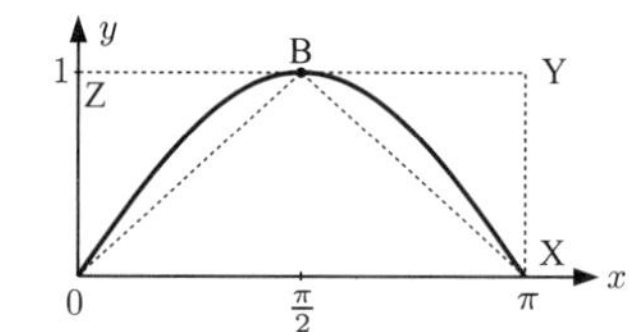

b $(2, -1)$ and $(5, 2)$ **c** 4.5 units^2

7 $m = \frac{\pi}{3}$ **8** **a** $a \approx 0.8767$ **b** ≈ 0.1357 units^2

9 **a** $\pi\left(\frac{3\pi}{32} - \frac{1}{8\sqrt{2}}\right)$ units^3 **b** ≈ 124 units^3

10 ≈ 31.2 units^2

11 **a**

b $x = -\frac{\pi}{2}$ and $x = \frac{\pi}{2}$
c x-intercepts are $x = -\frac{5\pi}{4}, -\frac{3\pi}{4}, -\frac{\pi}{4}, \frac{\pi}{4}, \frac{3\pi}{4}, \frac{5\pi}{4}$
y-intercept is $y = 1$
d $(\pi - 2)$ units^2

12 ≈ 2.59 units^2

13 **a**

From the graph,
area $\triangle$OBX < area under the curve < area OXYZ
$\therefore \frac{1}{2}\pi(1) < \int_0^\pi \sin x\, dx < \pi(1)$
$\therefore \frac{\pi}{2} < \int_0^\pi \sin x\, dx < \pi$

b partition as

c 2 units^2

14 **a** 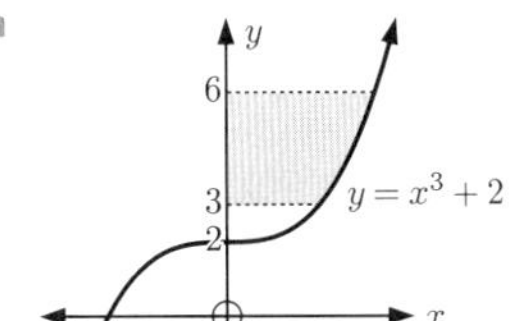

b $f(y)=\sqrt[3]{y-2}$
c $3\sqrt[3]{4}-\frac{3}{4}\approx 4.01$ units^2

15 $40\frac{1}{2}$ units^2

16 **Hint:** 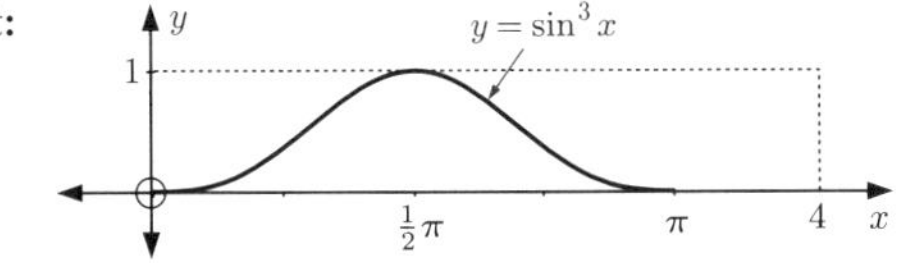

REVIEW SET 22C

1 **a** $v(t)=3t^2-30t+27$ cm s^{-1}
b $\int_0^6 (3t^2-30t+27)\,dt = s(6)-s(0) = -162$ cm (162 cm to the left of the origin)

2 **a** 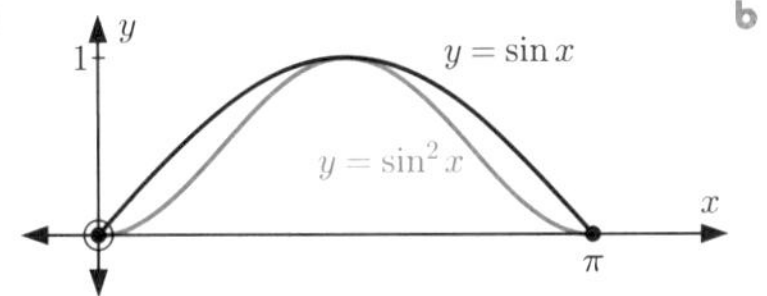

b $(1-\frac{\pi}{4})$ units^2

3 $a=\ln 3$, $b=\ln 5$ **4** $(\frac{\pi^2}{2}-2)$ units^2 **5** $(2-\frac{\pi}{2})$ units^2

6 $k=1\frac{1}{3}$ **7** π units^2 **8** **a** $\frac{\pi^2}{2}$ units^3 **b** 18π units^3

9 $21\frac{1}{12}$ units^2 **10** $\ln 2$ units^2

11 **a** $I(t)=\dfrac{100}{t}+100$ **b** **i** 105 milliamps **ii** as $t\to\infty$, $I\to 100$

12 **a** 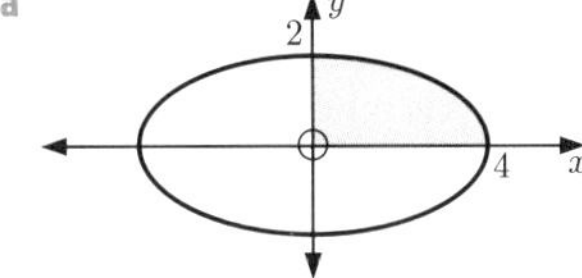

13 **a** $\frac{128\pi}{3}$ units^3

14 **a** $\sec x$, $\ln|\tan x+\sec x|+c$

b **i**

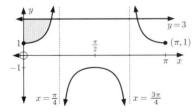

ii 0.965 units^2

15 $\frac{\pi}{2}$ units^3 **16** $\frac{96\pi}{5}$ units^3

EXERCISE 23A

1 **a** Heights can take any value from 170 cm to 205 cm, e.g., 181.37 cm.

b 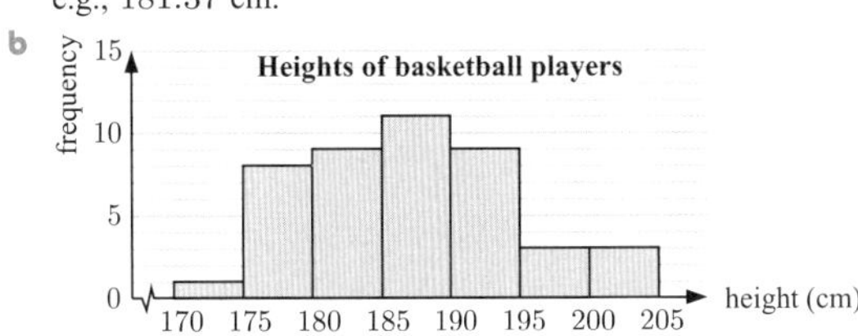

c The modal class is $185 \leqslant H < 190$ cm, as this occurred the most frequently.
d slightly positively skewed

2 **a** Continuous numerical, but has been rounded to become discrete numerical data.

b

Time (min)	*Tally*	*Frequency*
0 - 9	ⅠⅠⅠⅠ̸ Ⅰ	6
10 - 19	ⅠⅠⅠⅠ̸ ⅠⅠⅠⅠ̸ ⅠⅠⅠⅠ̸ ⅠⅠⅠⅠ̸ ⅠⅠⅠⅠ̸ Ⅰ	26
20 - 29	ⅠⅠⅠⅠ̸ ⅠⅠⅠⅠ̸ ⅠⅠⅠ	13
30 - 39	ⅠⅠⅠⅠ̸ ⅠⅠⅠⅠ	9
40 - 49	ⅠⅠⅠⅠ̸ Ⅰ	6

c positively skewed
d The modal travelling time class is 10 - 19 minutes.

3 **a** column graph **b** frequency histogram

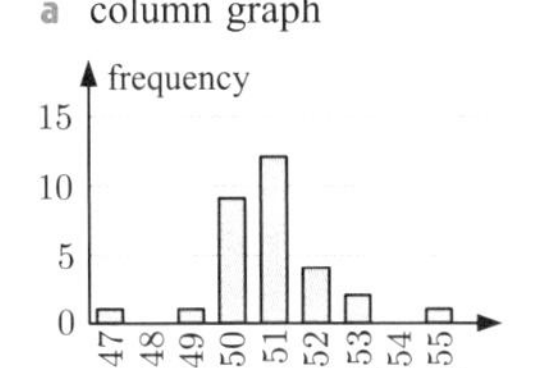

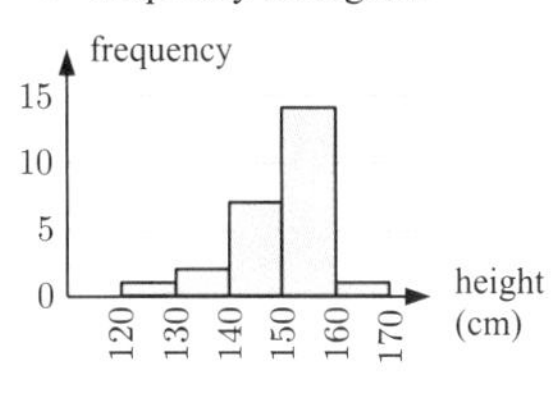

4 **a** 20 **b** 58.3% **c** **i** 1218 **ii** 512

EXERCISE 23B.1

1 **a** **i** 5.61 **ii** 6 **iii** 6
b **i** 16.3 **ii** 17 **iii** 18
c **i** 24.8 **ii** 24.9 **iii** 23.5

2 **a** A: 6.46, B: 6.85 **b** A: 7, B: 7
c The data sets are the same except for the last value, and the last value of A is less than the last value of B, so the mean of A is less than the mean of B.
d The middle value of the data sets is the same, so the median is the same.

3 **a** mean: \$29 300, median: \$23 500, mode: \$23 000
b The mode is the lowest value, so does not take the higher values into account.
c No, since the data is positively skewed, the median is not in the centre.

4 **a** mean: 3.19, median: 0, mode: 0
b The data is very positively skewed so the median is not in the centre.
c The mode is the lowest value so does not take the higher values into account.
d yes, 21 and 42 **e** no

5 **a** 44 **b** 44 **c** **i** 40.2 **ii** increase mean to 40.3

6 \$185 604 **7** 3144 km

8 **a** ≈ 70.9 g **b** ≈ 210 g **c** 139 g

9 116 **10** 17.25 goals per game

11 **a** mean = \$163 770, median = \$147 200 (differ by \$16 570)
b **i** mean selling price **ii** median selling price

12 $x=15$ **13** $a=5$ **14** 37

15 14.8 **16** 6 and 12 **17** 7 and 9

EXERCISE 23B.2

1 **a** 1 head **b** 1 head **c** 1.43 heads

2 **a** **i** 2.96 phone calls **ii** 2 phone calls **iii** 2 phone calls

c positively skewed
d The mean takes into account the larger numbers of phone calls.
e the mean

b 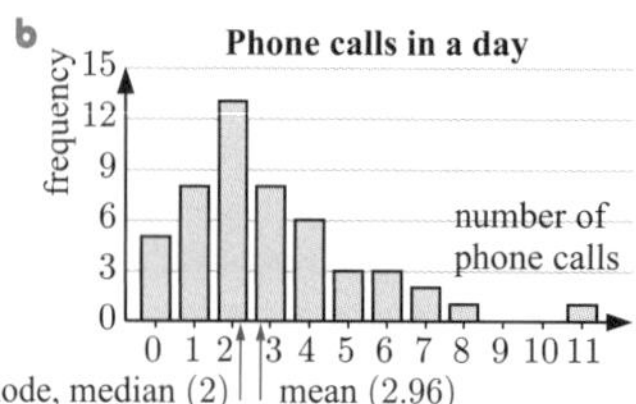

3 a i 49 matches ii 49 matches iii 49.0 matches
b no
c The sample of only 30 is not large enough. The company could have won its case by arguing that a larger sample would have found an average of 50 matches per box.

4 a i 2.61 children ii 2 children iii 2 children
b This school has more children per family than the average Australian family. c positive
d The mean is larger than the median and the mode.

5 a

Donation	*Frequency*
1	7
2	9
3	2
4	4
5	8

b 30
c i \$2.90 ii \$2 iii \$2
d the mode

6 a $x = 5$ b 75%
7 a i 5.63 ii 6 iii 6 b i 6.81 ii 7 iii 7
c the mean d yes
8 10.1 cm
9 a mean for A ≈ 50.7, mean for B ≈ 49.9
b No, as to the nearest match, A is 51 and B is 50.
10 a i €31 500 ii €28 000 iii €33 300 b The mode.

EXERCISE 23B.3

1 31.7
2 a 70 b $\approx 411\,000$ litres, ≈ 411 kL c ≈ 5870 L
3 a 125 people b ≈ 119 marks c $\frac{3}{25}$ d 28%

EXERCISE 23C.1

1 a Sample A

b, c

	A	B
μ	8	8
σ	2	1.06

Sample A has a larger standard deviation, which agrees with part **a**. That is, Sample A has a wider spread.

d The standard deviation is calculated using all data values, not just two.

2 a

	μ	σ
Andrew	25	4.97
Brad	30.5	12.6

b Andrew

3 a Rockets: range = 11, $\mu = 5.7$;
Bullets: range = 11, $\mu = 5.7$
b We suspect the Rockets, they have two zeros.
c Rockets: $\sigma = 3.9$ $\leftarrow$ greater variability
Bullets: $\sigma \approx 3.29$
d standard deviation

4 a We suspect variability in standard deviation since the factors may change every day.
b i mean ii standard deviation c less variability

5 a $\mu = 69$, $\sigma \approx 6.05$ b $\mu = 79$, $\sigma \approx 6.05$
c The distribution has simply shifted by 10 kg. The mean increases by 10 kg and the standard deviation remains the same.

6 a $\mu = 1.01$ kg; $\sigma = 0.17$ b $\mu = 2.02$ kg; $\sigma = 0.34$
c Doubling the values doubles the mean and the standard deviation.
7 a 0.809 b 0.150
c the extreme value greatly increases the standard deviation
8 $p = 6$, $q = 9$ 9 $a = 8$, $b = 6$ 10 b $\mu = 8.7$

EXERCISE 23C.2

1 $\mu \approx 1.72$ children, $\sigma \approx 1.67$ children
2 $\mu \approx 14.5$ years, $\sigma \approx 1.75$ years
3 $\mu \approx 37.3$ toothpicks, $\sigma \approx 1.45$ toothpicks
4 $\mu \approx 48.3$ cm, $\sigma \approx 2.66$ cm
5 $\mu \approx \$390.30$, $\sigma \approx \$15.87$

REVIEW SET 23A

1 a i 3.15 cm ii 4.5 cm
b 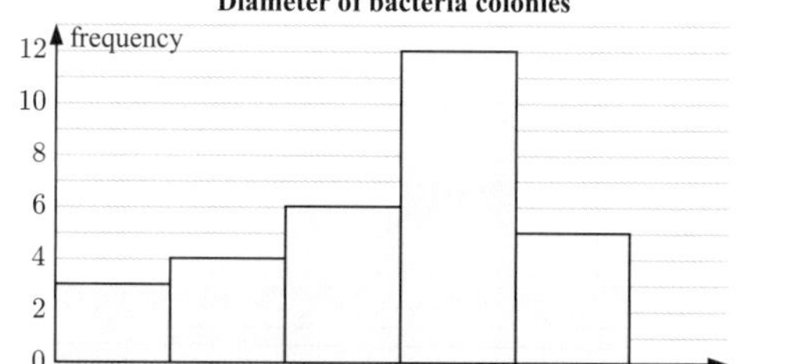

c The distribution is slightly negatively skewed.
2 $a = 8$, $b = 6$

3 a

Distribution	*Girls*	*Boys*
shape	pos. skewed	approx. symm.
median	36 s	34.5 s
mean	36 s	34.45 s
modal class	34.5 - 35.5 s	34.5 - 35.5 s

b The girls' distribution is positively skewed and the boys' distribution is approximately symmetrical. The median and mean swim times for boys are both about 1.5 seconds lower than for girls. Despite this, the distributions have the same modal class because of the skewness in the girls' distribution. The analysis supports the conjecture that boys generally swim faster than girls with less spread of times.
4 $k = 9$
5 a i 4 ii 5 iii ≈ 5.07 b positively skewed
6 a 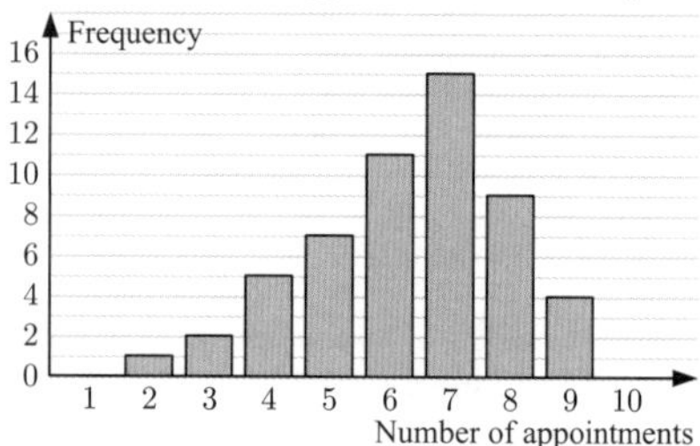

b negatively skewed c i 7 ii ≈ 6.33
7 $\mu = 14$ 8 $a = 5$, $b = 11$
9 a $\mu = 8 + \frac{1}{5}a$ b $a = 5$ or 15

REVIEW SET 23B

1 a highest = 97.5 m, lowest = 64.6 m
b use groups $60 \leqslant d < 65$, $65 \leqslant d < 70$, ...

c **Distances thrown by Thabiso**

Distance (m)	*Tally*	*Frequency*
$60 \leqslant d < 65$	|	1
$65 \leqslant d < 70$	|||	3
$70 \leqslant d < 75$	~~||||~~	5
$75 \leqslant d < 80$	||	2
$80 \leqslant d < 85$	~~||||~~ |||	8
$85 \leqslant d < 90$	~~||||~~ |	6
$90 \leqslant d < 95$	|||	3
$95 \leqslant d < 100$	||	2
	Total	30

d

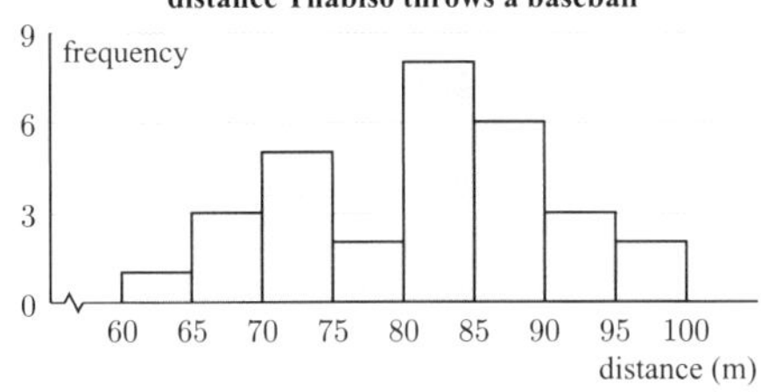

e i ≈ 81.1 m ii ≈ 83.1 m

2 b $\mu = k + 3$ 3 $\mu \approx 26.0$, $\sigma \approx 8.31$

4 $\mu \approx 33.6$ L, $\sigma \approx 7.63$ L

5 a $\mu \approx 49.6$, $\sigma \approx 1.60$
b Does not justify claim. Need a larger sample.

6 a

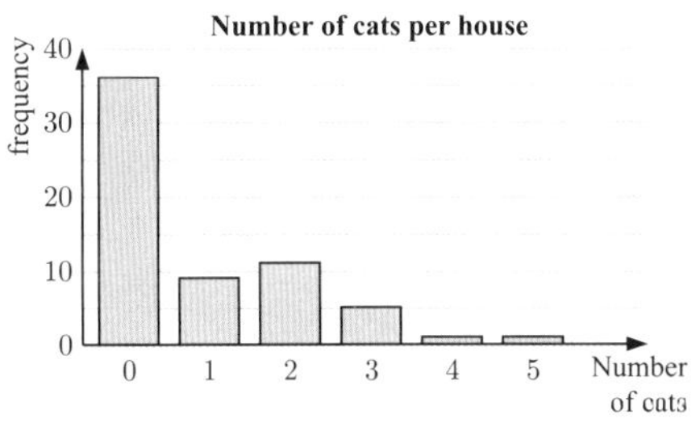

b positively skewed c i 0 ii 0.873 iii 0
d The mean, as it suggests that some people have cats. (The mode and median are both 0.)

7 a 426 cars b $\mu \approx 2.86$, $\sigma \approx 1.32$

8 a 60.9 b 60 c ≈ 12.8

9

		Kevin	Felicity
a	μ	41.2	39.5
b	σ	7.61	9.22

c Felicity d Kevin

10 22.6

REVIEW SET 23C

1 a discrete
b

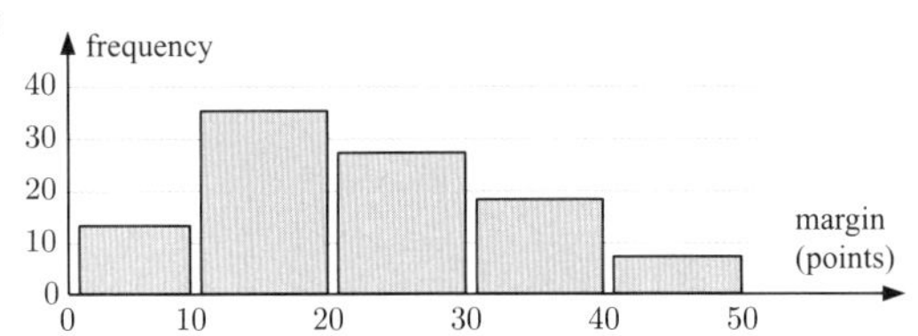

c No, as we do not know each individual data value, only the intervals they fall in.

2 a $x = 7$ b $\sigma^2 \approx 10.2$ 3 ≈ 414 customers

4 $m = 27$, $n = 36$

5 a median $= m + \frac{1}{2}$, mean $= m + 1$ b variance $= 10$

6 $\mu =$ €103.51, $\sigma \approx$ €19.40

7 a No, extreme values have less effect on the standard deviation of a larger population.
b i mean ii standard deviation
c A low standard deviation means that the weight of biscuits in each packet is, on average, close to 250 g.

8 $\mu \approx 8.56$, $\sigma \approx 0.970$

9 a

	Roger	Clinton
μ	84.1	76.8
σ	6.60	3.83

b Clinton c Roger

EXERCISE 24A

1 a ≈ 0.779 b ≈ 0.221

2 a ≈ 0.487 b ≈ 0.051 c ≈ 0.731

3 a 43 days b i ≈ 0.0465 ii ≈ 0.186 iii ≈ 0.465

4 a ≈ 0.0895 b ≈ 0.126

EXERCISE 24B

1 a {A, B, C, D} b {BB, BG, GB, GG}
c {ABCD, ABDC, ACBD, ACDB, ADBC, ADCB, BACD, BADC, BCAD, BCDA, BDAC, BDCA, CABD, CADB, CBAD, CBDA, CDAB, CDBA, DABC, DACB, DBAC, DBCA, DCAB, DCBA}
d {GGG, GGB, GBG, BGG, GBB, BGB, BBG, BBB}

2 a

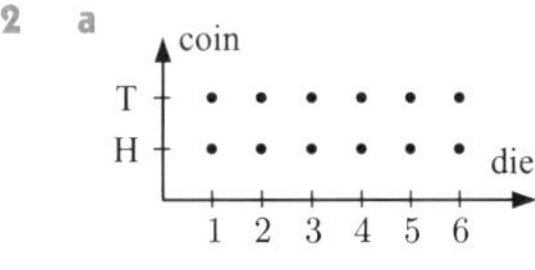

b

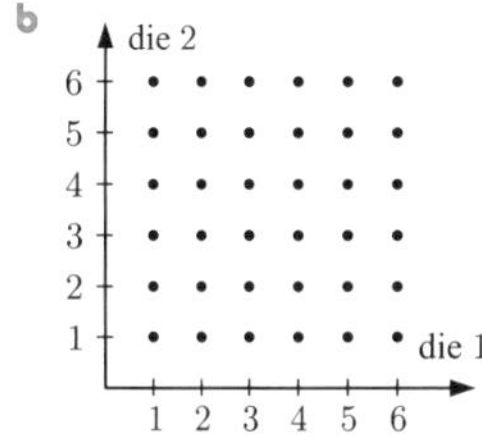

c

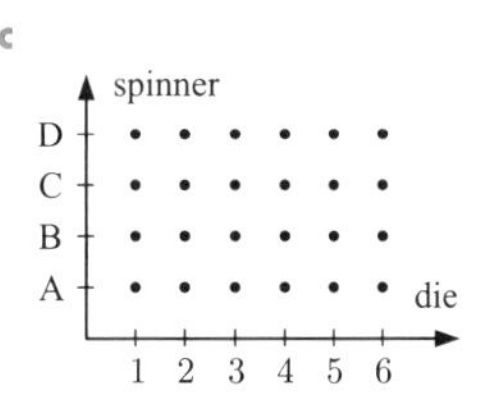

d

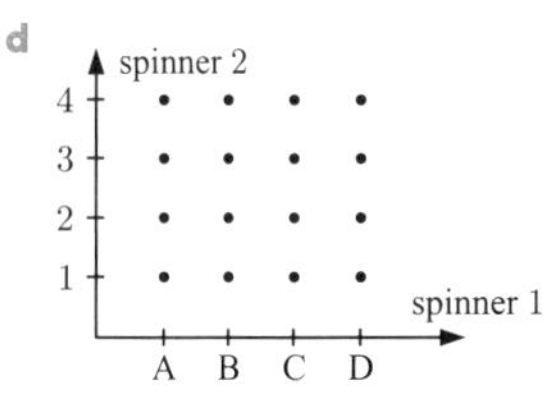

3 a

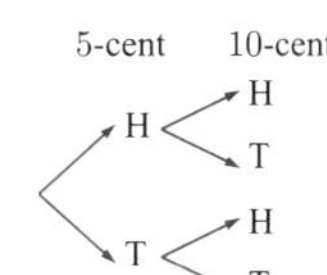

b

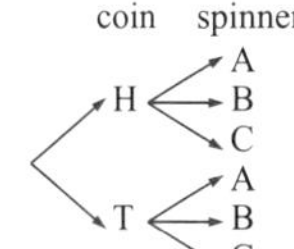

c

d

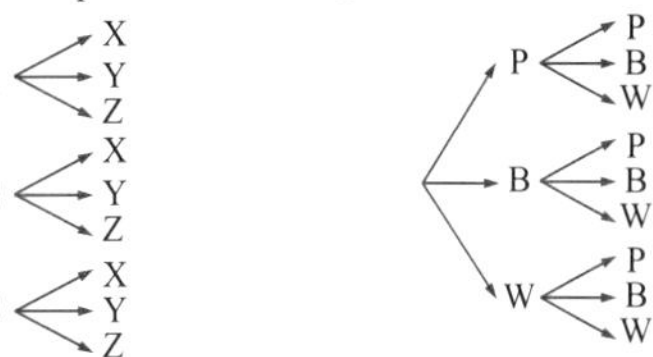

EXERCISE 24C.1

1 **a** $\frac{1}{5}$ **b** $\frac{1}{3}$ **c** $\frac{7}{15}$ **d** $\frac{4}{5}$ **e** $\frac{1}{5}$ **f** $\frac{8}{15}$

2 **a** 4 **b** **i** $\frac{2}{3}$ **ii** $\frac{1}{3}$

3 **a** $\frac{1}{4}$ **b** $\frac{1}{9}$ **c** $\frac{4}{9}$ **d** $\frac{1}{36}$ **e** $\frac{1}{18}$ **f** $\frac{1}{6}$ **g** $\frac{1}{12}$ **h** $\frac{1}{3}$

4 **a** $\frac{1}{7}$ **b** $\frac{2}{7}$ **c** $\frac{124}{1461}$ **d** $\frac{237}{1461}$ {remember leap years}

5 **a** {AKN, ANK, KAN, KNA, NAK, NKA}

b **i** $\frac{1}{3}$ **ii** $\frac{1}{3}$ **iii** $\frac{2}{3}$ **iv** $\frac{2}{3}$

6 **a** {GGG, GGB, GBG, BGG, GBB, BGB, BBG, BBB}

b **i** $\frac{1}{8}$ **ii** $\frac{1}{8}$ **iii** $\frac{1}{8}$ **iv** $\frac{3}{8}$ **v** $\frac{1}{2}$ **vi** $\frac{7}{8}$

7 **a** {ABCD, ABDC, ACBD, ACDB, ADBC, ADCB, BACD, BADC, BCAD, BCDA, BDAC, BDCA, CABD, CADB, CBAD, CBDA, CDAB, CDBA, DABC, DACB, DBAC, DBCA, DCAB, DCBA}

b **i** $\frac{1}{2}$ **ii** $\frac{1}{2}$ **iii** $\frac{1}{2}$ **iv** $\frac{1}{2}$

EXERCISE 24C.2

1 **2** **a**

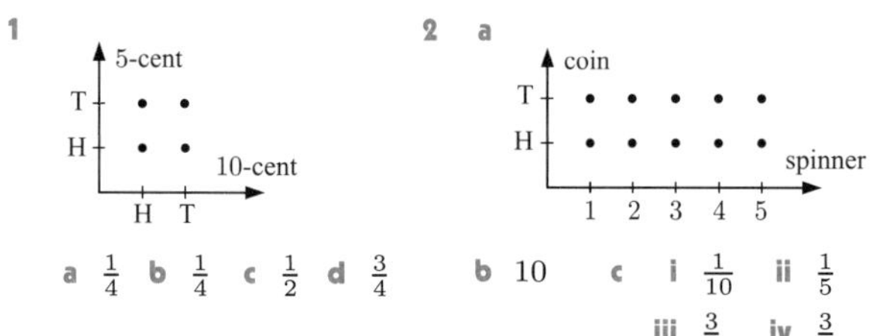

1 **a** $\frac{1}{4}$ **b** $\frac{1}{4}$ **c** $\frac{1}{2}$ **d** $\frac{3}{4}$

2 **b** 10 **c** **i** $\frac{1}{10}$ **ii** $\frac{1}{5}$ **iii** $\frac{3}{5}$ **iv** $\frac{3}{5}$

3 **a** $\frac{1}{36}$ **b** $\frac{1}{18}$ **c** $\frac{5}{9}$ **d** $\frac{11}{36}$ **e** $\frac{5}{18}$ **f** $\frac{25}{36}$ **g** $\frac{1}{6}$ **h** $\frac{5}{18}$ **i** $\frac{2}{9}$ **j** $\frac{13}{18}$

EXERCISE 24D

1 **a** $\frac{259}{537} \approx 0.482$ **b** $\frac{197}{537} \approx 0.367$ **c** $\frac{115}{537} \approx 0.214$ **d** $\frac{225}{259} \approx 0.869$ **e** $\frac{34}{115} \approx 0.296$

2 **a** 7510 **b** **i** ≈ 0.325 **ii** ≈ 0.653 **iii** ≈ 0.243

3 **a** ≈ 0.428 **b** ≈ 0.240 **c** ≈ 0.758 **d** ≈ 0.257 **e** ≈ 0.480

EXERCISE 24E.1

1 **a** $\frac{6}{7}$ **b** $\frac{36}{49}$ **c** $\frac{216}{343}$ **2** **a** $\frac{1}{8}$ **b** $\frac{1}{8}$

3 **a** 0.0096 **b** 0.8096 **4** **a** $\frac{1}{16}$ **b** $\frac{15}{16}$

5 **a** 0.56 **b** 0.06 **c** 0.14 **d** 0.24

6 **a** $\frac{8}{125}$ **b** $\frac{12}{125}$ **c** $\frac{27}{125}$

EXERCISE 24E.2

1 **a** $\frac{14}{55}$ **b** $\frac{1}{55}$ **2** **a** $\frac{7}{15}$ **b** $\frac{7}{30}$ **c** $\frac{7}{15}$

3 **a** $\frac{3}{100}$ **b** $\frac{3}{100} \times \frac{2}{99} \approx 0.000\,606$

c $\frac{3}{100} \times \frac{2}{99} \times \frac{1}{98} \approx 0.000\,006\,18$ **d** $\frac{97}{100} \times \frac{96}{99} \times \frac{95}{98} \approx 0.912$

4 **a** $\frac{4}{7}$ **b** $\frac{2}{7}$ **5** **a** $\frac{10}{21}$ **b** $\frac{1}{21}$

EXERCISE 24F

1 **a**

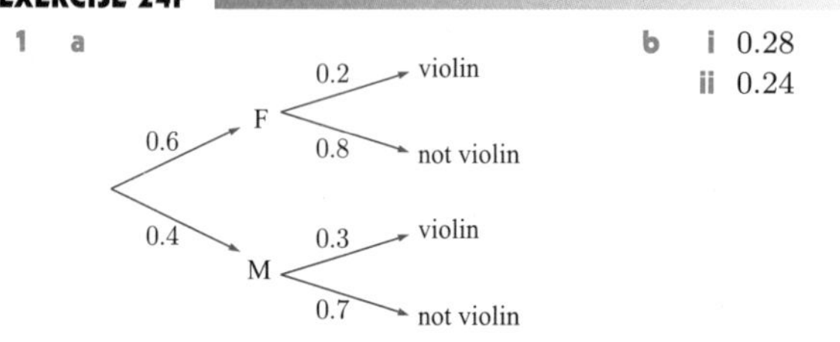

b **i** 0.28 **ii** 0.24

2 **a**

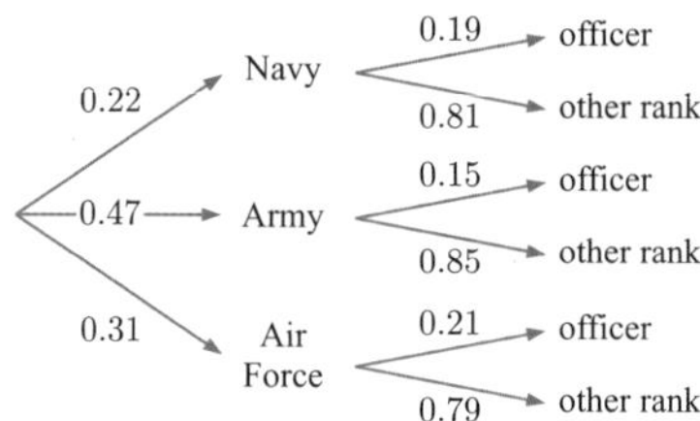

b **i** ≈ 0.177 **ii** ≈ 0.958 **iii** ≈ 0.864

3 **a**

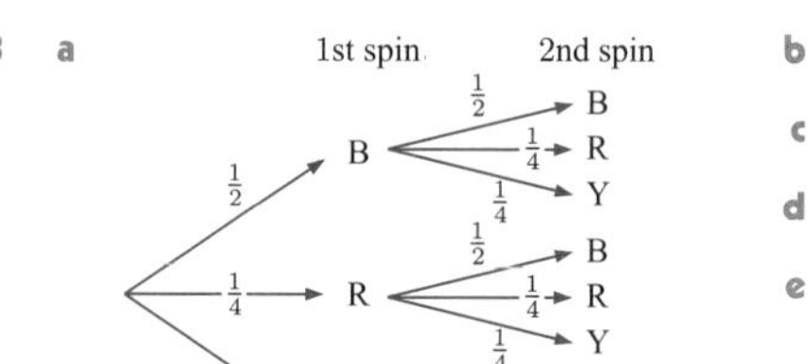

b $\frac{1}{4}$ **c** $\frac{1}{16}$ **d** $\frac{5}{8}$ **e** $\frac{3}{4}$

4

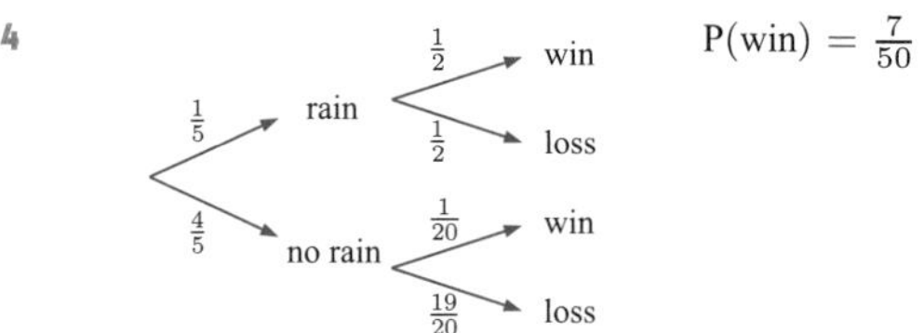

5 0.032 **6** $\frac{17}{40}$ **7** $\frac{9}{38}$ **8** **a** $\frac{11}{30}$ **b** $\frac{19}{30}$

EXERCISE 24G

1 **a** $\frac{20}{49}$ **b** $\frac{10}{21}$

2 **a**

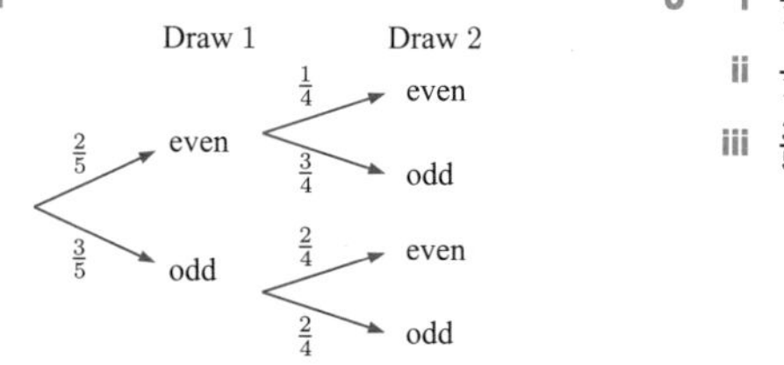

b **i** $\frac{3}{10}$ **ii** $\frac{1}{10}$ **iii** $\frac{3}{5}$

3 **a** $\frac{2}{9}$ **b** $\frac{5}{9}$

4 **a** **i** $\frac{1}{3}$ **ii** $\frac{2}{15}$ **iii** $\frac{4}{15}$ **iv** $\frac{4}{15}$

b These are all the possible outcomes, so their probabilities must sum to 1.

5 **a**

Spin 1 Spin 2

$\frac{1}{4}$ B

$\frac{3}{4}$ R: $\frac{1}{4}$ B, $\frac{3}{4}$ R

b $\frac{7}{16}$

6 **a** $\frac{1}{5}$ **b** $\frac{3}{5}$ **c** $\frac{4}{5}$ **7** $\frac{19}{45}$

8 **a** $\frac{2}{100} \times \frac{1}{99} \approx 0.000\,202$ **b** $\frac{98}{100} \times \frac{97}{99} \approx 0.960$

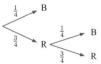

c $1 - \frac{98}{100} \times \frac{97}{99} \approx 0.0398$

9 $\frac{7}{33}$ **10** 7 to start with

EXERCISE 24H.1

1 **a** $A = \{1, 2, 3, 6\}$, $B = \{2, 4, 6, 8, 10\}$ **b**

c **i** $n(A) = 4$ **ii** $A \cup B = \{1, 2, 3, 4, 6, 8, 10\}$ **iii** $A \cap B = \{2, 6\}$

2 **a** **b**

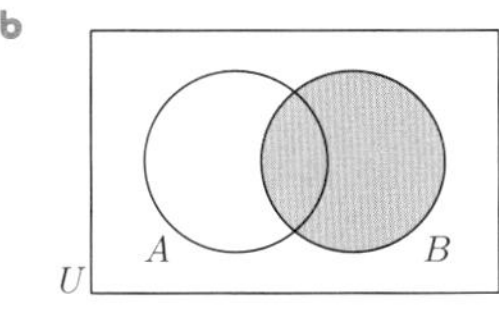

c

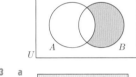

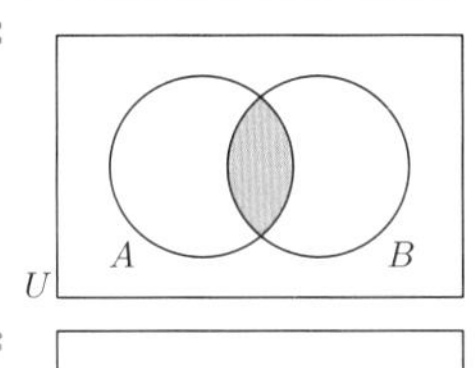

d

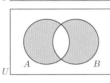

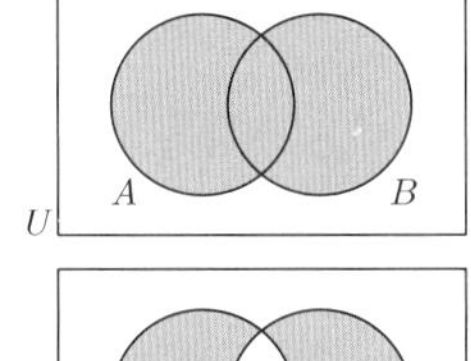

e **f**

3 **a**

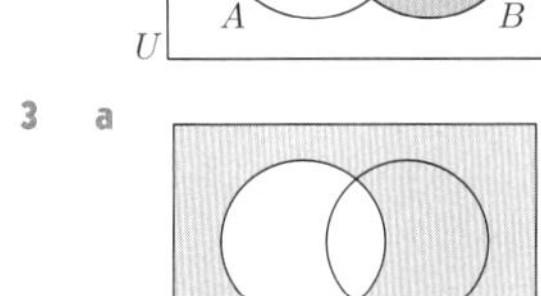

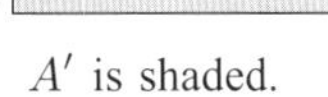

A' is shaded.

b

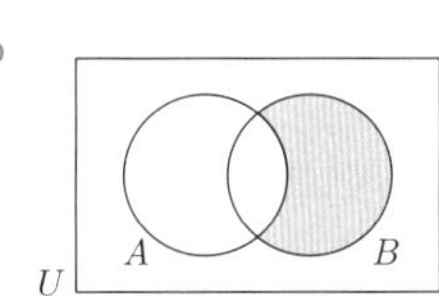

$A' \cap B$ is shaded.

c

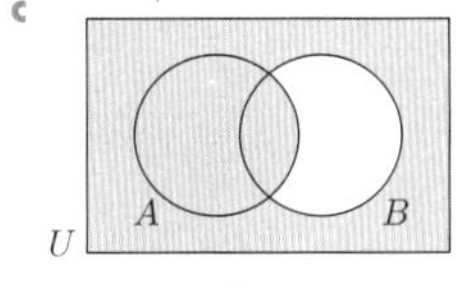

$A \cup B'$ is shaded.

d

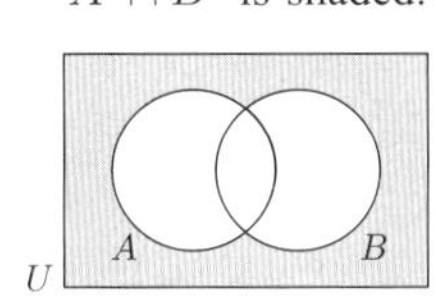

$A' \cap B'$ is shaded.

4 **a**

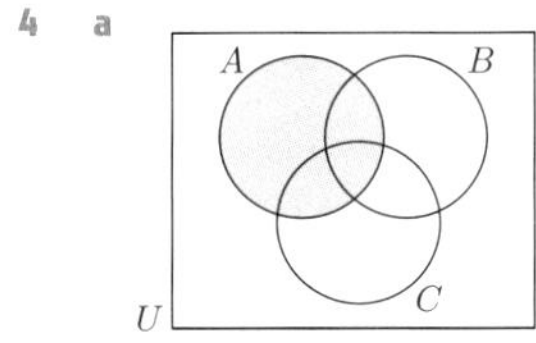

b

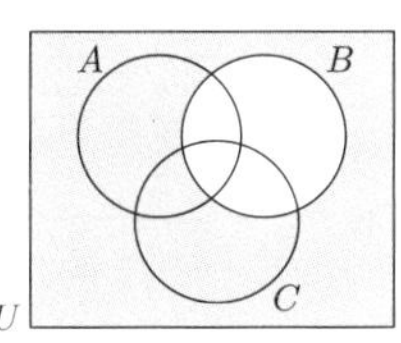

c

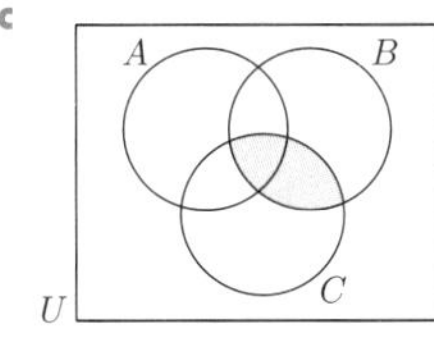

d

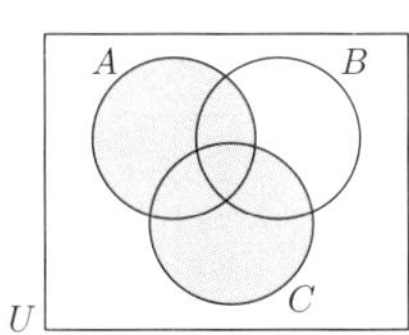

e

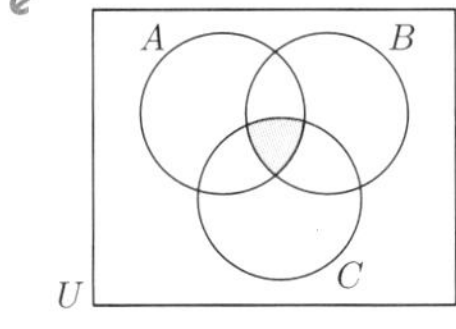

f

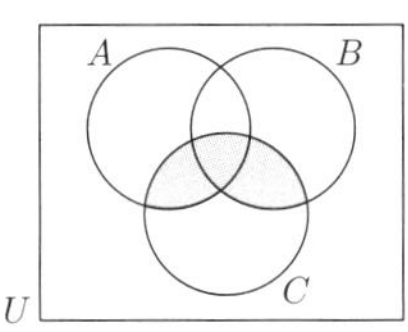

5 **a** 29 **b** 17 **c** 26 **d** 5

6 **a** 65 **b** 9 **c** 4 **d** 52

7 **a** **b** **i** $\frac{12}{14} = \frac{6}{7}$ **ii** $\frac{3}{14}$

8 **a** $\frac{19}{40}$ **b** $\frac{1}{2}$ **c** $\frac{4}{5}$ **d** $\frac{5}{8}$ **e** $\frac{13}{40}$

9 **a** $\frac{19}{25}$ **b** $\frac{13}{25}$ **c** $\frac{6}{25}$ **10** **a** $\frac{7}{15}$ **b** $\frac{1}{15}$ **c** $\frac{2}{15}$

11 **a** $a = 3$, $b = 3$

b **i** $\frac{12}{40} = \frac{3}{10}$ **ii** $\frac{4}{40} = \frac{1}{10}$ **iii** $\frac{7}{40}$ **iv** $\frac{15}{40} = \frac{3}{8}$ **v** $\frac{25}{40} = \frac{5}{8}$

EXERCISE 24H.2

1 For each of these, draw **two** diagrams. Shade the first with the LHS set and the second with the RHS set.

2 **a** **i** $n(A) = 14$ **ii** $n(B) = 19$ **iii** 2 **iv** 31

3 **a** **i** $\dfrac{b+c}{a+b+c+d}$ **ii** $\dfrac{b}{a+b+c+d}$ **iii** $\dfrac{a+b+c}{a+b+c+d}$ **iv** $\dfrac{a+b+c}{a+b+c+d}$

b $P(A \cup B) = P(A) + P(B) - P(A \cap B)$

EXERCISE 24I

1 0.6 **2** 0.2 **3** 0.35

4 **a** 22 study both **b** **i** $\frac{9}{25}$ **ii** $\frac{11}{20}$

5 **a** $\frac{3}{8}$ **b** $\frac{7}{20}$ **c** $\frac{1}{5}$ **d** $\frac{15}{23}$

6 **a** $\frac{14}{25}$ **b** $\frac{4}{5}$ **c** $\frac{1}{5}$ **d** $\frac{5}{23}$ **e** $\frac{9}{14}$ **7** $\frac{5}{6}$

8 **a** $\frac{13}{20}$ **b** $\frac{7}{20}$ **c** $\frac{11}{50}$ **d** $\frac{7}{25}$ **e** $\frac{4}{7}$ **f** $\frac{1}{4}$

9 **a** $\frac{3}{5}$ **b** $\frac{2}{3}$ **10** **a** 0.46 **b** $\frac{14}{23}$ **11** $\frac{70}{163}$

12 **a** 0.45 **b** 0.75 **c** 0.65

13 **a** 0.0484 **b** 0.393 **14** $\frac{2}{3}$

EXERCISE 24J

1 $P(R \cap S) = 0.4 + 0.5 - 0.7 = 0.2$ and $P(R) \times P(S) = 0.2$ $\therefore$ are independent events

2 **a** $\frac{7}{30}$ **b** $\frac{7}{12}$ **c** $\frac{7}{10}$ No, as $P(A \mid B) \neq P(A)$

3 **a** 0.35 **b** 0.85 **c** 0.15 **d** 0.15 **e** 0.5

4 $\frac{14}{15}$ **5** **a** $\frac{91}{216}$ **b** 26

6 **Hint:** Show $P(A' \cap B') = P(A')\,P(B')$ using a Venn diagram and $P(A \cap B)$

7 0.9

8 **a** **i** $\frac{13}{20}$ **ii** $\frac{7}{10}$ **b** No, as $P(C \mid D) \neq P(C)$

9 **a** **i** $\left(\frac{12}{13}\right)^4 \left(\frac{1}{13}\right)$ **iii** $\frac{13}{25}$ **b** ≈ 0.544 **10** $\frac{63}{125}$

EXERCISE 24K

1 $\frac{6}{55}$ **2** $\frac{1}{12}$ **3** ≈ 0.318 **4** ≈ 0.530

5 **a** $\frac{1}{10}$ **b** $\frac{2}{5}$

6 **a** ≈ 0.0288 **b** ≈ 0.635 **c** ≈ 0.966

7 $\frac{1}{5}$ **8** **a** ≈ 0.0962 **b** ≈ 0.0962

EXERCISE 24L

1 a 0.0435 b ≈ 0.598 **2** a ≈ 0.773 b ≈ 0.556

3 $\frac{10}{13}$ **4** ≈ 0.424 **5** 0.0137 **6** $\frac{15}{83}$ **7** $\frac{99}{148}$

8 a $\frac{9}{19}$ b $\frac{10}{19}$ **10** a 0.95 b ≈ 0.306 c 0.6

11 a 0.104 b ≈ 0.267 c ≈ 0.0168

REVIEW SET 24A

1 ABCD, ABDC, ACBD, ACDB, ADBC, ADCB, BACD, BADC, BCAD, BCDA, BDAC, BDCA, CABD, CADB, CBAD, CBDA, CDAB, CDBA, DABC, DACB, DBAC, DBCA, DCAB, DCBA

a $\frac{1}{2}$ b $\frac{1}{3}$

2 a $\text{P}(A') = 1 - m$ b $0 \leqslant m \leqslant 1$

c i $2m(1-m)$ ii $2m - m^2$

3 a $\frac{3}{8}$ b $\frac{1}{8}$ c $\frac{5}{8}$ **4** a $\frac{2}{5}$ b $\frac{13}{15}$ c $\frac{4}{15}$

5 a 0 b 0.45 c 0.8

6 a Two events are independent if the occurrence of either event does not affect the probability that the other occurs. For A and B independent, $\text{P}(A) \times \text{P}(B) = \text{P}(A \text{ and } B)$.

b Two events A and B are mutually exclusive if they have no common outcomes. $\text{P}(A \text{ or } B) = \text{P}(A) + \text{P}(B)$

7

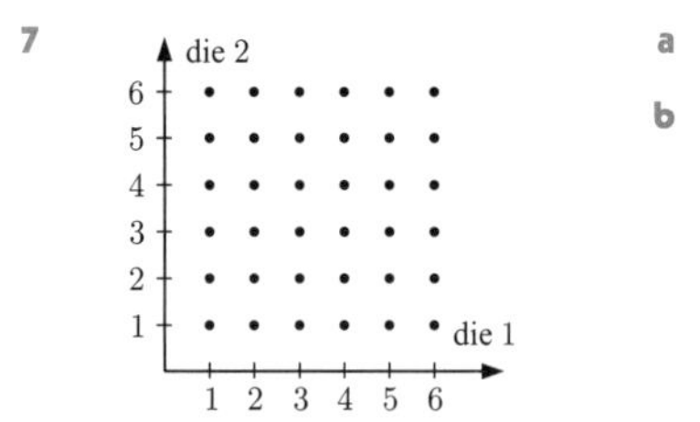

a $\frac{2}{9}$ b $\frac{5}{12}$

8 a $\frac{1}{4}$ b $\frac{37}{40}$ c $\frac{2}{5}$ **9** $\frac{5}{9}$

10 a $10(\frac{4}{5})^3(\frac{1}{5})^2 \approx 0.205$ b $5(\frac{4}{5})^4(\frac{1}{5}) + (\frac{4}{5})^5 \approx 0.737$

11 $\frac{4}{9}$ **12** $\frac{148}{243}$ **13** a $\frac{1}{50}$ b $\frac{13}{50}$

REVIEW SET 24B

1 P(N wins)

$= \frac{44}{125}$

$= 0.352$

2 a $\frac{4}{500} \times \frac{3}{499} \times \frac{2}{498} \approx 0.000\,000\,193$

b $1 - \frac{496}{500} \times \frac{495}{499} \times \frac{494}{498} \approx 0.0239$

3 a ≈ 0.259 b ≈ 0.703

4 a

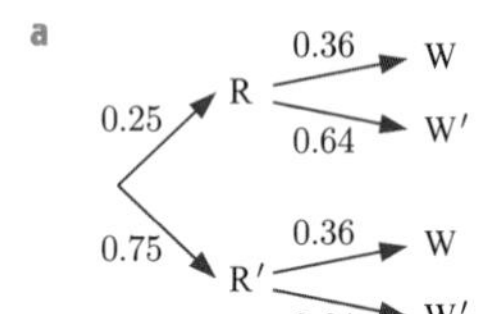

b i 0.09 ii 0.52

c That the two events (rain and wind) are independent.

5 $1 - 0.9 \times 0.8 \times 0.7 = 0.496$

6 a i 0.89 ii 20 b i 0.077 ii 0.11 c 0.81

7 a

b i $\frac{3}{28}$ ii $\frac{23}{35}$

8 a

	Female	Male	Total
Smoker	20	40	60
Non-smoker	70	70	140
Total	90	110	200

b i $\frac{7}{20}$ ii $\frac{1}{2}$ c i ≈ 0.121 ii ≈ 0.422

9 a $\text{P}(M \cap C) = 0.85$, $\text{P}(M)\,\text{P}(C) \approx 0.801$, so not independent b $\frac{3}{88}$

10 $\frac{3}{14}$ **11** ≈ 0.127 **12** a ≈ 0.0238 b ≈ 0.976

13 $\frac{1}{2}$ **14** a 0 b $\frac{3}{14}$ c $\frac{2}{15}$

REVIEW SET 24C

1 BBBB, BBBG, BBGB, BGBB, GBBB, BBGG, BGBG, BGGB, GGBB, GBBG, GBGB, BGGG, GBGG, GGBG, GGGB, GGGG.

P(2 children of each sex) $= \frac{3}{8}$

2 a $\frac{5}{33}$ b $\frac{19}{66}$ c $\frac{5}{11}$ d $\frac{16}{33}$

3 a i $\frac{3}{25}$ ii $\frac{24}{25}$ iii $\frac{11}{12}$ b i $\frac{11}{115}$ ii $\frac{104}{115}$

4 a $\frac{5}{8}$ b $\frac{1}{4}$ c $\frac{6}{13}$

5 a 0.93 b 0.8 c 0.2 d 0.65

6

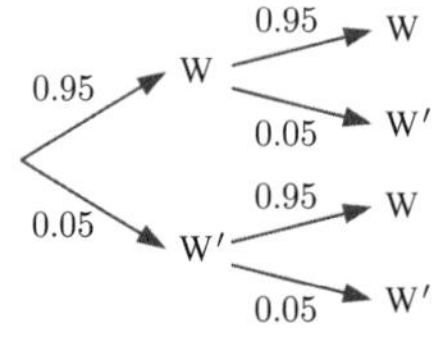

0.9975

7 a $\frac{31}{70}$ b $\frac{21}{31}$

8 a

	Men	Women	Total
Tea	15	24	39
Coffee	35	26	61
Total	50	50	100

b

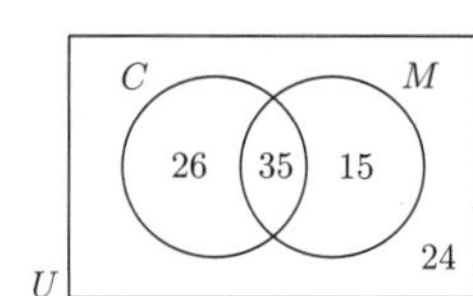

c i 0.39 ii $\frac{35}{61} \approx 0.574$

9 a

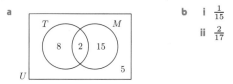

b i $\frac{1}{15}$ ii $\frac{2}{17}$

10 b ≈ 0.9876 c i ≈ 0.547 ii ≈ 0.266

d A 20 year old of either gender is expected to live for longer than 30 years, so it is unlikely the insurance company will have to pay out the policy.

11 a 0.078 31 b 0.076 63

12 **B** [probability $p(1-q)(1+q)$] is more likely than **A** [probability $p(1-q)(2-p)$]

13 $\frac{5}{8}$ **14** **a** 1.54×10^{-6} **b** 1.28×10^{-8}

EXERCISE 25A

1 **a** continuous **b** discrete **c** continuous **d** continuous **e** discrete **f** discrete **g** continuous **h** continuous

2 **a** **i** $X =$ the height of water in the rain gauge **ii** $0 \leqslant X \leqslant 400$ mm **iii** continuous

b **i** $X =$ stopping distance **ii** $0 \leqslant X \leqslant 50$ m **iii** continuous

c **i** number of switches until failure **ii** any integer $\geqslant 1$ **iii** discrete

3 **a** $X = 0, 1, 2, 3,$ or 4

b

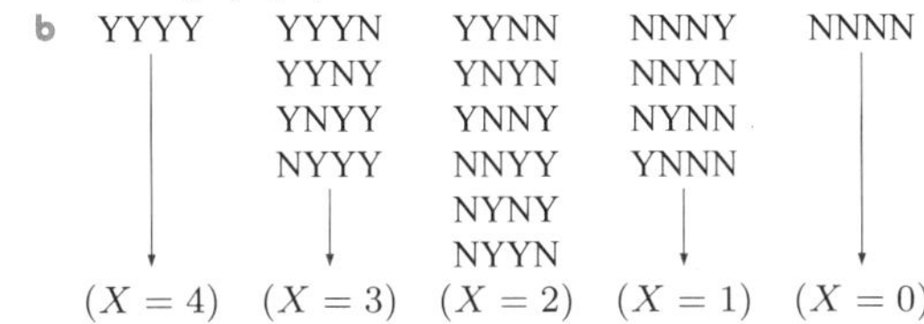

c **i** $X = 2$ **ii** $X = 2, 3,$ or 4

4 **a** $X = 0, 1, 2,$ or 3

b HHH → $(X=3)$; HHT, HTH, THH → $(X=2)$; TTH, THT, HTT → $(X=1)$; TTT → $(X=0)$

c $P(X=3) = \frac{1}{8}$, $P(X=2) = \frac{3}{8}$, $P(X=1) = \frac{3}{8}$, $P(X=0) = \frac{1}{8}$

d

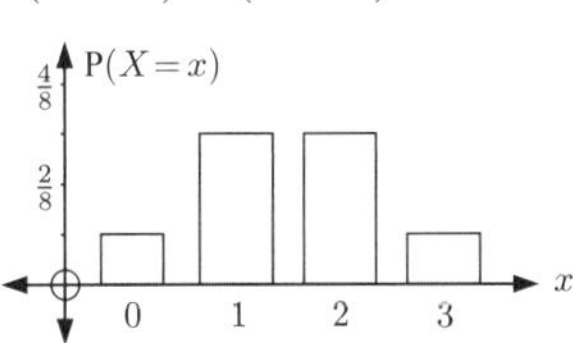

EXERCISE 25B

1 **a** $k = 0.2$ **b** $k = \frac{1}{7}$

2 **a** $P(2) = 0.1088$

b $a = 0.5488$ is the probability that Jason does not hit a home run in a game.

c $P(1) + P(2) + P(3) + P(4) + P(5) = 0.451\,2$ and is the probability that Jason will hit one or more home runs in a game.

d

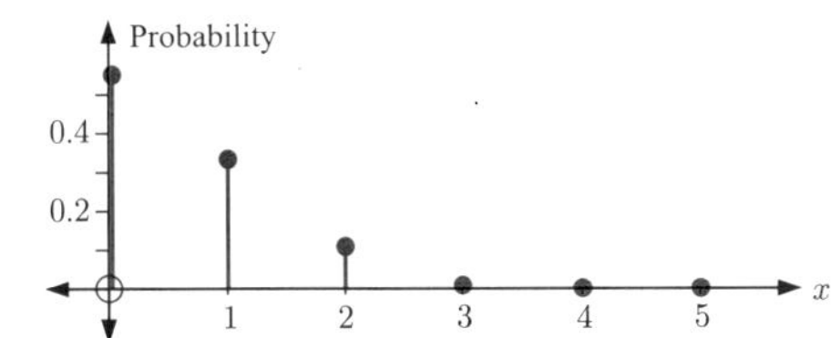

e mode $= 0$ home runs, median $= 0$ home runs

3 **a** $\sum P(x_i) > 1$ **b** $P(5) < 0$ which is not possible

4 **a** X is the number of hits that Sally has in each match. $X = 0, 1, 2, 3, 4,$ or 5

b **i** $k = 0.23$ **ii** $P(X \geqslant 2) = 0.79$ **iii** $P(1 \leqslant X \leqslant 3) = 0.83$

c mode $= 3$ hits, median $= 3$ hits

5 **a**

roll 1 \ roll 2	1	2	3	4	5	6
6	(6, 1)	(6, 2)	(6, 3)	(6, 4)	(6, 5)	(6, 6)
5	(5, 1)	(5, 2)	(5, 3)	(5, 4)	(5, 5)	(5, 6)
4	(4, 1)	(4, 2)	(4, 3)	(4, 4)	(4, 5)	(4, 6)
3	(3, 1)	(3, 2)	(3, 3)	(3, 4)	(3, 5)	(3, 6)
2	(2, 1)	(2, 2)	(2, 3)	(2, 4)	(2, 5)	(2, 6)
1	(1, 1)	(1, 2)	(1, 3)	(1, 4)	(1, 5)	(1, 6)

b **i** $P(0) = 0$, $P(1) = 0$, $P(2) = \frac{1}{36}$, $P(3) = \frac{2}{36}$, $P(4) = \frac{3}{36}$, $P(5) = \frac{4}{36}$, $P(6) = \frac{5}{36}$, $P(7) = \frac{6}{36}$, $P(8) = \frac{5}{36}$, $P(9) = \frac{4}{36}$, $P(10) = \frac{3}{36}$, $P(11) = \frac{2}{36}$, $P(12) = \frac{1}{36}$

ii

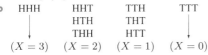

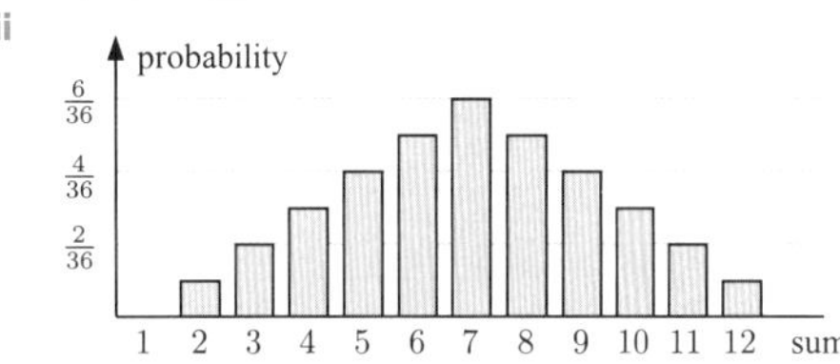

iii mode $= 7$, median $= 7$

6 **a** $k = \frac{1}{12}$ **b** $k = \frac{12}{25}$

7 **a** $P(0) = 0.1975k$; $P(1) = 0.0988k$; $P(2) = 0.0494k$; $P(3) = 0.0247k$; $P(4) = 0.0123k$

b $k = \frac{81}{31}$ (≈ 2.61), $P(X \geqslant 2) = \frac{7}{31} \approx 0.226$

8 **a** $P(0) = 0.665$ **b** $P(X \geqslant 1) = 0.335$

9 **a**

x	0	1	2
$P(X = x)$	$\frac{3}{28}$	$\frac{15}{28}$	$\frac{10}{28}$

b

x	0	1	2	3
$P(X = x)$	$\frac{1}{56}$	$\frac{15}{56}$	$\frac{30}{56}$	$\frac{10}{56}$

10 **a**

Die 1 \ Die 2	1	2	3	4	5	6
1	2	3	4	5	6	7
2	3	4	5	6	7	8
3	4	5	6	7	8	9
4	5	6	7	8	9	10
5	6	7	8	9	10	11
6	7	8	9	10	11	12

b $\frac{1}{6}$

c

d	2	3	4	5	6	7	8	9	10	11	12
$P(D = d)$	$\frac{1}{36}$	$\frac{2}{36}$	$\frac{3}{36}$	$\frac{4}{36}$	$\frac{5}{36}$	$\frac{6}{36}$	$\frac{5}{36}$	$\frac{4}{36}$	$\frac{3}{36}$	$\frac{2}{36}$	$\frac{1}{36}$

d $\frac{15}{26}$

11 **a**

Die 1 \ Die 2	1	2	3	4	5	6
1	0	1	2	3	4	5
2	1	0	1	2	3	4
3	2	1	0	1	2	3
4	3	2	1	0	1	2
5	4	3	2	1	0	1
6	5	4	3	2	1	0

b

N	0	1	2	3	4	5
$P(N = n)$	$\frac{6}{36}$	$\frac{10}{36}$	$\frac{8}{36}$	$\frac{6}{36}$	$\frac{4}{36}$	$\frac{2}{36}$

c **i** $\frac{1}{6}$ **ii** $\frac{2}{5}$

12 **a** $\displaystyle\sum_{n=0}^{\infty} \frac{(0.2)^n e^{-0.2}}{n!} = 1$

b **i** $\sum P(x_i) = 1$ **ii** $P(X = 0) = e^{-0.2} \approx 0.819$, $P(X = 1) = 0.2e^{-0.2} \approx 0.164$, $P(X = 2) = 0.02e^{-0.2} \approx 0.0164$
iii $P(X \geqslant 3) \approx 0.001\,15$

EXERCISE 25C

1 102 days **2** **a** $\frac{1}{8}$ **b** 25 **3** 30 times
4 \$1.50 **5** 15 days **6** 27
7 **a** **i** 0.55 **ii** 0.29 **iii** 0.16
b **i** 4125 **ii** 2175 **iii** 1200
8 **a** **i** $\frac{1}{6}$ **ii** $\frac{1}{3}$ **iii** $\frac{1}{2}$
b **i** \$1.33 **ii** \$0.50 **iii** \$3.50
c lose 50 cents **d** lose \$50
9 **a** €3.50 **b** −€0.50, no **c** **i** $k = 3.5$ **ii** $k > 3.5$
10 **a** £2.75 **b** £3.75
11 **a** $P(X \leqslant 3) = \frac{1}{12}$, $P(4 \leqslant X \leqslant 6) = \frac{1}{3}$, $P(7 \leqslant X \leqslant 9) = \frac{5}{12}$, $P(X \geqslant 10) = \frac{1}{6}$ **c** $a = 5$
d organisers would lose \$1.17 per game **e** \$2807

EXERCISE 25D

1 **a** $k = 0.03$ **b** $\mu = 0.74$ **c** $\sigma \approx 0.996$
2 **a** mode = 3 **b** median = 3 **c** $\mu = 2.5$ **d** $\sigma \approx 0.671$
3 **a**

x_i	0	1	2	3
$P(x_i)$	0.216	0.432	0.288	0.064

b $\mu = 1.2$, $\sigma \approx 0.849$

5 **a**

x_i	1	2	3	4	5
$P(x_i)$	0.1	0.2	0.4	0.2	0.1

b $\mu = 3.0$, $\sigma \approx 1.10$ **c** **i** $P(\mu - \sigma < x < \mu + \sigma) \approx 0.8$
ii $P(\mu - 2\sigma < x < \mu + 2\sigma) \approx 1$
6 \$390
7 **a**

M	1	2	3	4	5	6
$P(M_i)$	$\frac{1}{36}$	$\frac{3}{36}$	$\frac{5}{36}$	$\frac{7}{36}$	$\frac{9}{36}$	$\frac{11}{36}$

b **i** mode = 6 **ii** median = 5 **iii** $\mu \approx 4.47$
iv $\sigma \approx 1.40$
8 Tossing a coin P(head) = P(tail) = $\frac{1}{2}$ or rolling a die $P(1) = P(2) = P(3) = ... = P(6) = \frac{1}{6}$

EXERCISE 25E

1 **a** 3.4 **b** 1.64 **c** ≈ 1.28
2 **a** $k = 0.3$ **b** 6.4 **c** 0.84
3 **a** 2 **b** 5 **c** 1 **d** 1 **e** 3 **f** 1 **g** 9
4 **a** $a = 0.15$, $b = 0.35$ **5** **a** $a = -\frac{1}{84}$ **b** 4 **c** $\sqrt{3}$
6 **a**

x	0	1	2	3	4
p_x	$\frac{1}{16}$	$\frac{4}{16}$	$\frac{6}{16}$	$\frac{4}{16}$	$\frac{1}{16}$

b **i** 2 **ii** 1

7 **a**

x	0	1	2
p_x	$\frac{7}{15}$	$\frac{7}{15}$	$\frac{1}{15}$

b **i** 0.6 **ii** ≈ 0.611

8 **a** $a = 0.25$, $b = 0.35$ **b** 0.99
9 **a** $a = \frac{1}{4}$
b **i** $E(X) = 2\frac{11}{12}$, $Var(X) = \frac{275}{144} \approx 1.91$
ii $E(2X) = 5\frac{5}{6}$, $Var(2X) = 7\frac{23}{36} \approx 7.64$
iii $E(X - 1) = 1\frac{11}{12}$, $Var(X - 1) = \frac{275}{144} \approx 1.91$
10 17 and 4 respectively

11 **a** $E(aX + b) = E(aX) + E(b) = aE(X) + b$
b **i** 13 **ii** −5 **iii** $3\frac{1}{3}$
12 **a** $E(Y) = 13$, $Var(Y) = 16$ **b** $E(Y) = -7$, $Var(Y) = 16$
c $E(Y) = 0$, $Var(Y) = 1$
13 **a** $E(Y) = 2E(X) + 3$ **b** $E(Y^2) = 4E(X^2) + 12E(X) + 9$
c $Var(Y) = 4E(X^2) - 4\{E(X)\}^2$

EXERCISE 25F.1

1 **a** $(p+q)^4 = p^4 + 4p^3q + 6p^2q^2 + 4pq^3 + q^4$
b $4\left(\frac{1}{2}\right)^3\left(\frac{1}{2}\right) = \frac{1}{4}$
2 **a** $(p+q)^5 = p^5 + 5p^4q + 10p^3q^2 + 10p^2q^3 + 5pq^4 + q^5$
b **i** $5\left(\frac{1}{2}\right)^4\left(\frac{1}{2}\right) = \frac{5}{32}$ **ii** $10\left(\frac{1}{2}\right)^2\left(\frac{1}{2}\right)^3 = \frac{5}{16}$
iii $\left(\frac{1}{2}\right)^4\left(\frac{1}{2}\right) = \frac{1}{32}$
3 **a** $\left(\frac{2}{3} + \frac{1}{3}\right)^4 = \left(\frac{2}{3}\right)^4 + 4\left(\frac{2}{3}\right)^3\left(\frac{1}{3}\right) + 6\left(\frac{2}{3}\right)^2\left(\frac{1}{3}\right)^2 + 4\left(\frac{2}{3}\right)\left(\frac{1}{3}\right)^3 + \left(\frac{1}{3}\right)^4$
b **i** $\left(\frac{2}{3}\right)^4 = \frac{16}{81}$ **ii** $6\left(\frac{2}{3}\right)^2\left(\frac{1}{3}\right)^2 = \frac{8}{27}$ **iii** $\frac{8}{9}$
4 **a** $\left(\frac{3}{4} + \frac{1}{4}\right)^5 = \left(\frac{3}{4}\right)^5 + 5\left(\frac{3}{4}\right)^4\left(\frac{1}{4}\right)^1 + 10\left(\frac{3}{4}\right)^3\left(\frac{1}{4}\right)^2 + 10\left(\frac{3}{4}\right)^2\left(\frac{1}{4}\right)^3 + 5\left(\frac{3}{4}\right)\left(\frac{1}{4}\right)^4 + \left(\frac{1}{4}\right)^5$
b **i** $10\left(\frac{3}{4}\right)^3\left(\frac{1}{4}\right)^2 = \frac{135}{512}$ **ii** $\frac{53}{512}$ **iii** $\frac{47}{128}$
5 **a** ≈ 0.154 **b** ≈ 0.973

EXERCISE 25F.2

1 **a** The binomial distribution applies, as tossing a coin has two possible outcomes (H or T) and each toss is independent of every other toss.
b The binomial distribution applies, as this is equivalent to tossing one coin 100 times.
c The binomial distribution applies as we can draw out a red or a blue marble with the same chances each time.
d The binomial distribution does not apply as the result of each draw is dependent upon the results of previous draws.
e The binomial distribution does not apply, assuming that ten bolts are drawn without replacement. We do not have a repetition of independent trials.
2 **a** ≈ 0.0305 **b** ≈ 0.265 **3** $\approx 0.000\,864$
4 **a** ≈ 0.268 **b** ≈ 0.800 **c** ≈ 0.200
5 **a** ≈ 0.476 **b** ≈ 0.840 **c** ≈ 0.160 **d** ≈ 0.996
6 **a** ≈ 0.231 **b** ≈ 0.723 **c** 1.25 apples
7 **a** ≈ 0.0280 **b** $\approx 0.002\,46$ **c** ≈ 0.131 **d** ≈ 0.710
8 **a** **i** ≈ 0.998 **ii** ≈ 0.807 **b** 105 students
9 **a** **i** ≈ 0.290 **ii** ≈ 0.885 **b** 18.8
10 ≈ 0.0341 **11** at least 4 dice
12 **a** ≈ 0.863 **b** ≈ 0.475 **13** ≈ 0.837
14 **b** $P(0) = \frac{1}{32}$, $P(1) = \frac{5}{32}$, $P(2) = \frac{10}{32}$, $P(3) = \frac{10}{32}$, $P(4) = \frac{5}{32}$, $P(5) = \frac{1}{32}$

EXERCISE 25F.3

1 **a** **i** $\mu = 3$, $\sigma = 1.22$
ii

x_i	0	1	2	3
$P(x_i)$	0.0156	0.0938	0.2344	0.3125

x_i	4	5	6
$P(x_i)$	0.2344	0.0938	0.0156

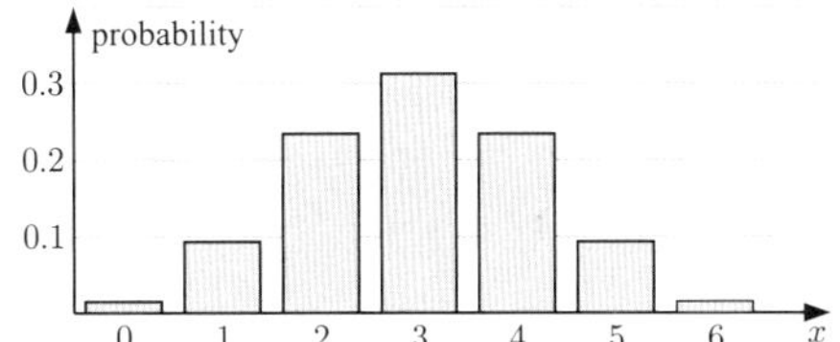

iii The distribution is bell-shaped.

b i $\mu = 1.2$, $\sigma = 0.980$

ii

x_i	0	1	2	3
$P(x_i)$	0.2621	0.3932	0.2458	0.0819

x_i	4	5	6
$P(x_i)$	0.0154	0.0015	0.0001

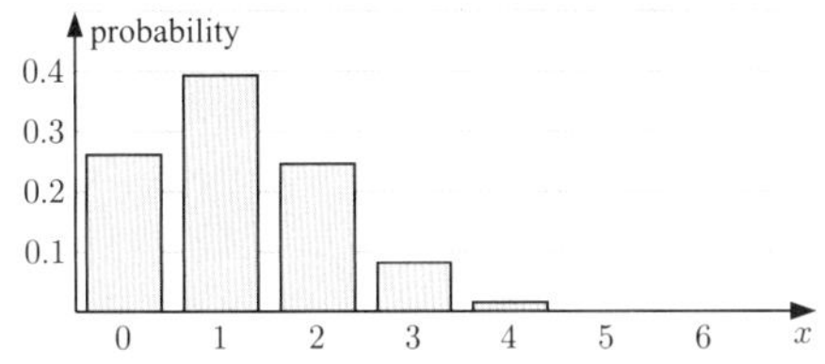

iii The distribution is positively skewed.

c i $\mu = 4.8$, $\sigma = 0.980$

ii

x_i	0	1	2	3
$P(x_i)$	0.0001	0.0015	0.0154	0.0819

x_i	4	5	6
$P(x_i)$	0.2458	0.3932	0.2621

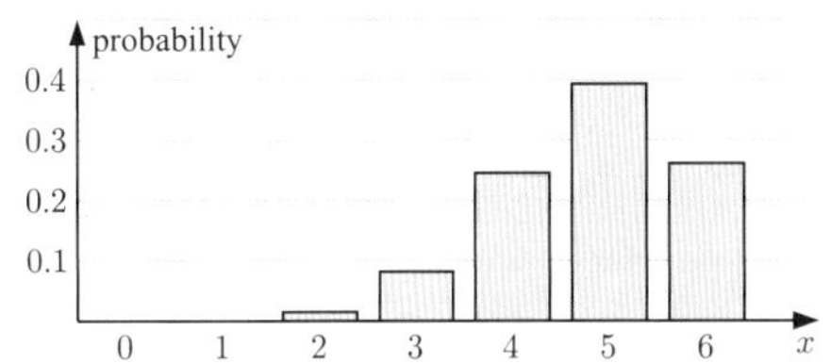

iii The distribution is negatively skewed and is the exact reflection of **b**.

2 $\mu = 5$, $\sigma^2 = 2.5$

3 **a**

x_i	0	1	2	3
p_i	$(1-p)^3$	$3p(1-p)^2$	$3p^2(1-p)$	p^3

4 **a** $\mu = 1.2$, $\sigma = 1.07$ **b** $\mu = 28.8$, $\sigma = 1.07$

5 $\mu = 3.9$, $\sigma \approx 1.84$

EXERCISE 25G

1 **a** 1.5 **b** $p_x = \dfrac{(1.5)^x e^{-1.5}}{x!}$ for $x = 0, 1, 2, 3, 4, 5, 6, \ldots$

x	0	1	2	3	4	5	6
f	12	18	12	6	3	0	1
$52p_x$	11.6	17.4	13.1	6.5	2.4	0.7	0.2

The fit is excellent.

2 **a** ≈ 7.13 **b** $p_x = \dfrac{(7.1289)^x e^{-7.1289}}{x!}$, $x = 0, 1, 2, 3, \ldots$

c i ≈ 0.0204 ii ≈ 0.0753 iii ≈ 0.838 iv ≈ 0.974

3 **a** ≈ 1.69, so $p_x = \dfrac{(1.694)^x e^{-1.694}}{x!}$, $x = 0, 1, 2, 3, \ldots$

b

x	0	1	2	3	4	5	6	7
f	91	156	132	75	33	9	3	1
$500p_x$	92	156	132	74	32	11	3	1

The fit is excellent.

c $\sigma \approx 1.29$ and $\sqrt{m} \approx 1.30$, so, σ is very close to $\sqrt{m}$ in value.

4 **a** ≈ 0.0498 **b** ≈ 0.577 **c** ≈ 0.185 **d** ≈ 0.440

5 **a** $m = \frac{3+\sqrt{33}}{2}$, $m > 0$ **b** i ≈ 0.506 ii ≈ 0.818

6 **a** ≈ 0.271 **b** ≈ 0.271 **c** ≈ 0.677

7 **a** 0.901 **b** 6 years

8 **a** ≈ 0.150 **b** ≈ 0.967 **c** mode = 1 flaw per metre

9 **a** $m \approx 6.8730$ **b** i ≈ 0.177 ii ≈ 0.417

10 **a** $y = e^{-x}\left(1 + x + \dfrac{x^2}{2}\right)$

b

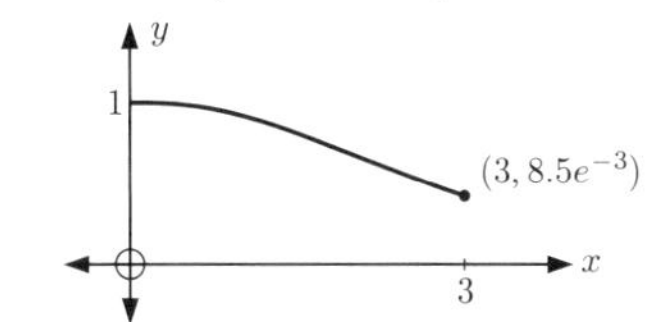

c $P'(x) = -\frac{1}{2}x^2 e^{-x}$ which is < 0 for all $x \in \mathbb{Z}^+$

$\therefore$ $P(x)$ decreases as x increases.

REVIEW SET 25A

1 **a** $a = \frac{5}{9}$ **b** $\frac{4}{9}$ **2** $\mu = 4.8$ defectives, $\sigma \approx 2.15$

3 **a** $k = 0.05$ **b** 0.15 **c** 1.7 **d** ≈ 0.954

4 **a** $\left(\frac{3}{5}+\frac{2}{5}\right)^4 = \left(\frac{3}{5}\right)^4 + 4\left(\frac{3}{5}\right)^3\left(\frac{2}{5}\right) + 6\left(\frac{3}{5}\right)^2\left(\frac{2}{5}\right)^2 + 4\left(\frac{3}{5}\right)\left(\frac{2}{5}\right)^3 + \left(\frac{2}{5}\right)^4$

b i $\frac{216}{625}$ ii $\frac{328}{625}$

5 **a** $\frac{1}{10}$ **b** $\frac{3}{5}$ **c** $\frac{3}{10}$ **d** $1\frac{1}{5}$

6 **a** £7 **b** No, she would lose £1 per game in the long run.

7 **a** $k = \binom{7}{x}$ **b** $\mu = \frac{7}{3} \approx 2.33$, $\sigma^2 = \frac{14}{9} \approx 1.56$

8 **a** $\left(\frac{4}{5}+\frac{1}{5}\right)^5 = \left(\frac{4}{5}\right)^5 + 5\left(\frac{4}{5}\right)^4\left(\frac{1}{5}\right) + 10\left(\frac{4}{5}\right)^3\left(\frac{1}{5}\right)^2 + 10\left(\frac{4}{5}\right)^2\left(\frac{1}{5}\right)^3 + 5\left(\frac{4}{5}\right)\left(\frac{1}{5}\right)^4 + \left(\frac{1}{5}\right)^5$

b i $\frac{64}{3125} \approx 0.0205$ ii $\frac{128}{625} \approx 0.205$

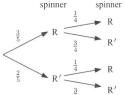

9 **a** \$4 **b** \$75

10 **a**

Pentagonal spinner — Square spinner

$\frac{3}{5}$ R: $\frac{1}{4}$ R, $\frac{3}{4}$ R′

$\frac{2}{5}$ R′: $\frac{1}{4}$ R, $\frac{3}{4}$ R′

b $\frac{11}{20}$

c i $P(X=1) = \binom{10}{1}\left(\frac{11}{20}\right)^1\left(\frac{9}{20}\right)^9$,

$P(X=9) = \binom{10}{9}\left(\frac{11}{20}\right)^9\left(\frac{9}{20}\right)^1$

ii It is more likely that exactly one red will occur 9 times.

11 **a** $y = \frac{1}{4}$ **b** $x = 16$ **c** median = 14, mode = 6

REVIEW SET 25B

1 a $k=\frac{8}{5}$ b 0.975 c 2.55 d ≈ 0.740

2 a 0.302 b 0.298 c 0.561 3 6.43 surgeries

4 a 42 donations b 0.334

5 a 0.849 b 2.56×10^{-6} c 0.991 d 0.000 246

6 a i 0.100 ii 0.114 b 3.41 games

7 a $n=120,\ \ p=\frac{1}{4}$

b i ≈ 0.0501 ii ≈ 0.878 iii ≈ 0.172

8 $a=\frac{1}{6}$

9 a $(0.6+0.4)^6$

$=(0.6)^6+6(0.6)^5(0.4)+15(0.6)^4(0.4)^2+20(0.6)^3(0.4)^3$

X wins 6 | X wins 5, Y wins 1 | X wins 4, Y wins 2 | X wins 3, Y wins 3

$+\,15(0.6)^2(0.4)^4+6(0.6)(0.4)^5+(0.4)^6$

X wins 2, Y wins 4 | X wins 1, Y wins 5 | Y wins 6

b i $20(0.6)^3(0.4)^3\approx 0.276$

ii $6(0.6)(0.4)^5+(0.4)^6\approx 0.0410$

10 0.156 11 a $p=0.3$ b 0.850

12 a $k=\frac{12}{145}$ b $\approx 2.81,\ \approx 1.19$

c median $=3$, mode $=4$

13 $m\approx 1.2,\ \approx 0.879$

REVIEW SET 25C

1 a $k=\frac{12}{11}$ b $k=\frac{1}{2}$

2 a

x	0	1	2	3	4
$\mathrm{P}(X=x)$	0.0625	0.25	0.375	0.25	0.0625

b $\mu=2$ c 1

3 $\mu=480,\ \sigma\approx 17.0$ 4 a 0.259 b 0.337 c 0.922

5 a i $\frac{2}{5}$ ii $\frac{1}{10}$ iii $\frac{1}{10}$ b \$2.70

6 a ≈ 0.0660 b ≈ 0.563 7 $n=11$

8 a ≈ 0.0388 b 25 of them

9 a $(\frac{1}{3}+\frac{2}{3})^5$ b ≈ 0.313

10 a i $\mu=1.28,\ \ \sigma=1.13$ b ≈ 0.366

ii $\mathrm{E}(Y)=1.14,\ \ \sigma_Y=0.566$

11 a $m=4,\ \ m>0$ b $\dfrac{13}{e^4}\approx 0.238$

12 ≈ 0.736 as $p\approx 0.1$ 13 a 0.0516 b No

EXERCISE 26A

1 a $a=-\frac{3}{32}$ b

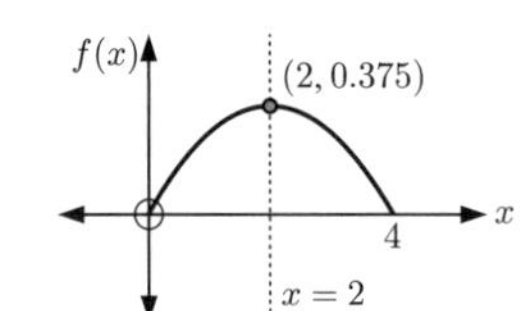

c i 2 ii 2 iii 2 iv 0.8

2 a $b=\sqrt[3]{30}$ b mean ≈ 1.55 c ≈ 0.483

3 a $k\approx 1.0524$ b median ≈ 0.645

4 a $k=-\frac{4}{375}$ b 4 c ≈ 3.46 d $3\frac{1}{3}$ e $1\frac{1}{9}$

5 a $k\leqslant\frac{5}{12}$ and $\int_0^k (5-12y)\,dy=1$

b $k=\frac{1}{3}$ $\{k=\frac{1}{2}$ does not satisfy $k\leqslant\frac{5}{12}\}$

c If $k=\frac{1}{2}$ the graph goes below the horizontal axis.

d $\mu=\frac{7}{54}$, median ≈ 0.116

6 a $k=\dfrac{1}{b-a}$

b $\mu=\dfrac{a+b}{2}$, median $=\dfrac{a+b}{2}$, mode is undefined

c $\mathrm{Var}(X)=\dfrac{(a-b)^2}{12}$, $\sigma=\dfrac{b-a}{\sqrt{12}}$

7 a median ≈ 0.347 b mode $=0$

8 a $a=\frac{\pi}{18}$ b $\mu\approx 0.0852$ c 0.0334 d 0.0501

9 $a=\frac{5}{32},\ \ k=2$

10 a

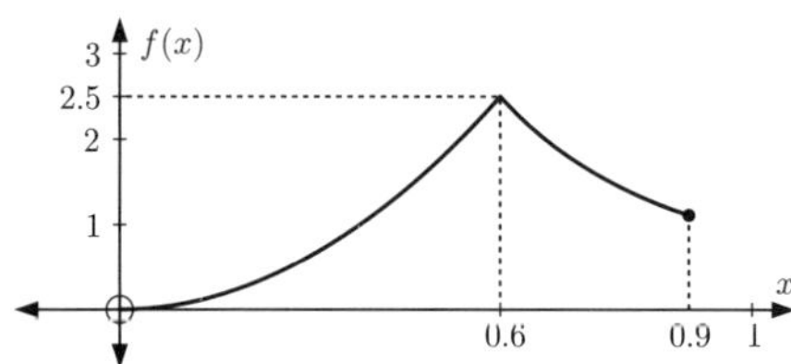

b $f(x)\geqslant 0$ for all $x\in[0,\,0.9]$ and area under curve $=1$

c $\mu\approx 0.590$, median $=0.6$, mode $=0.6$

d $\mathrm{Var}(X)\approx 0.0300$, $\sigma\approx 0.173$

e 0.652 The task can be performed between 18 minutes and 42 minutes 65.2% of the time.

EXERCISE 26B

1

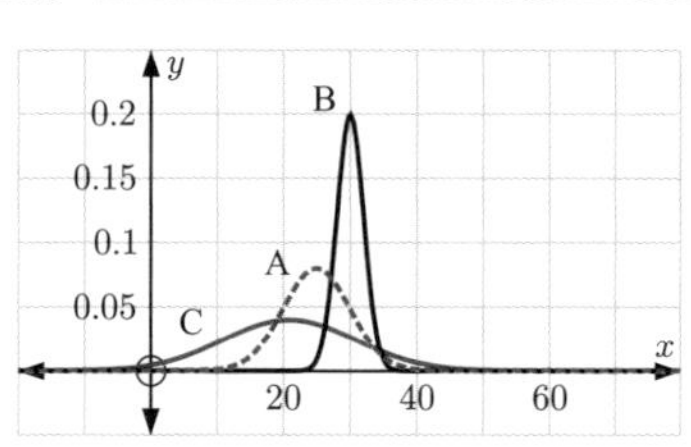

2 a, b The mean volume (or diameter) is likely to occur most often with variations around the mean occurring symmetrically as a result of random variations in the production process.

3 a 0.683 b 0.477

4 a 15.9% b 84.1% c 97.6% d 0.13%

5 3 times 6 a 5 b 32 c 137 competitors

7 a 459 babies b 446 babies 8 a 0.954 b 2.9

9 a $\mu=176$ g, $\sigma=24$ g b 81.9%

10 a i 34.1% ii 47.7%

b i 0.136 ii 0.159 iii 0.0228 iv 0.841

c $k=178$

11 a $\mu=155$ cm, $\sigma=12$ cm b 0.84

12 a 84.1% b 2.28% c i 2.15% ii 95.4%

d i 97.7% ii 2.28%

13 a ≈ 41 days b ≈ 254 days c ≈ 213 days

EXERCISE 26C

1 a 0.341 b 0.383 c 0.106

2 a 0.341 b 0.264 c 0.212 d 0.945

e 0.579 f 0.383

3 0.378 4 a 90.4% b 4.78%

5 a 0.003 33 b 61.5% c 23 eels

6 a 0.303 b 0.968 c 0.309

7 a 10.3% b 0.544 8 a 84.1% b 0.880

EXERCISE 26D

1 **a** z-scores:
English ≈ 1.82 Geography ≈ 1.61
Mandarin ≈ 2.33 Biology $= 0.9$
Maths ≈ 2.27

b Mandarin, Maths, English, Geography, Biology

2 **a** Physics -0.463, Chemistry 0.431, Maths 0.198, German 0.521, Biology -0.769

b German, Chemistry, Maths, Physics, Biology

3 **a**

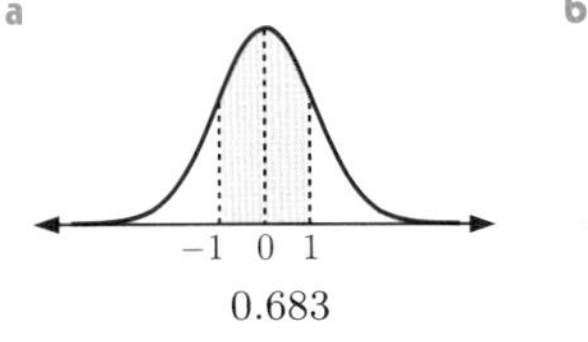

0.683

b

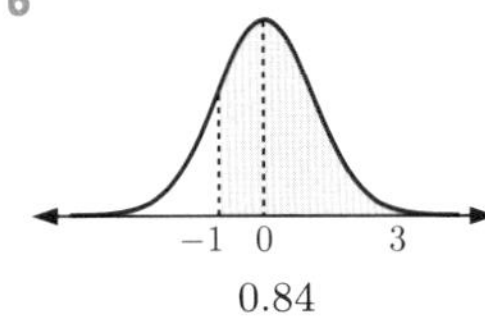

0.84

c

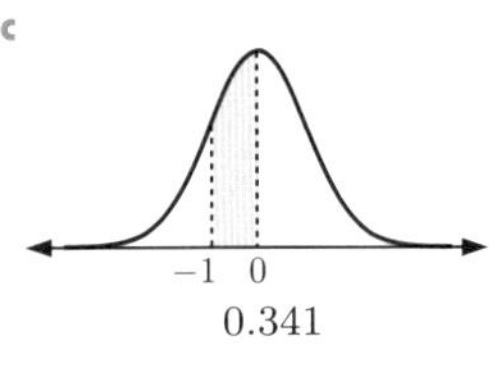

0.341

d

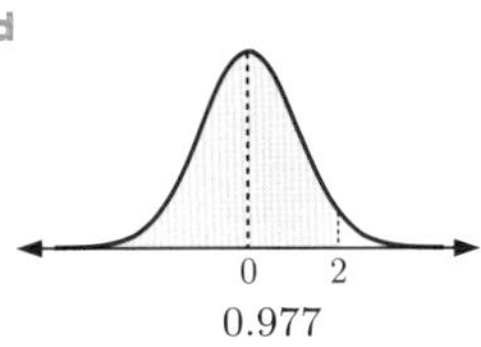

0.977

e

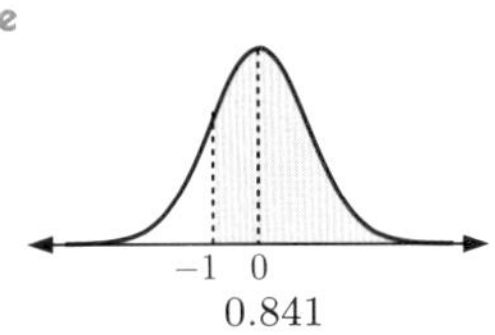

0.841

f

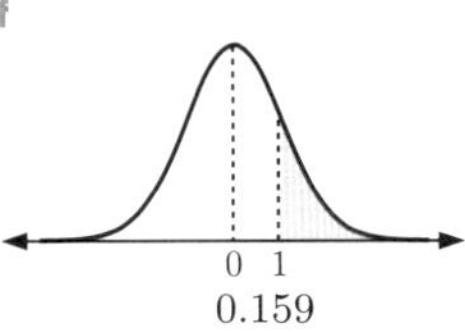

0.159

4 **a** $\text{E}\left(\dfrac{X-\mu}{\sigma}\right) = \text{E}\left(\frac{1}{\sigma}X - \frac{\mu}{\sigma}\right)$, etc.

b $\text{Var}\left(\dfrac{X-\mu}{\sigma}\right) = \text{Var}\left(\frac{1}{\sigma}X - \frac{\mu}{\sigma}\right)$, etc.

5 **a** $a = -1,\ b = 2$ **b** $a = -0.5,\ b = 0$ **c** $a = 0,\ b = 3$

6 **a**

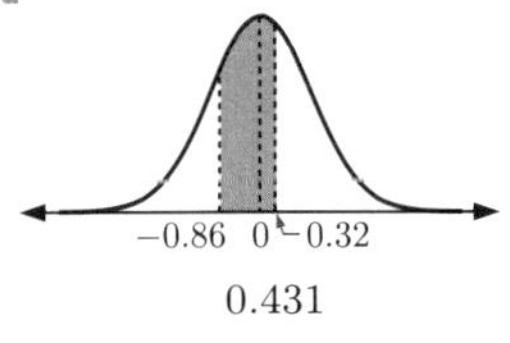

0.431

b

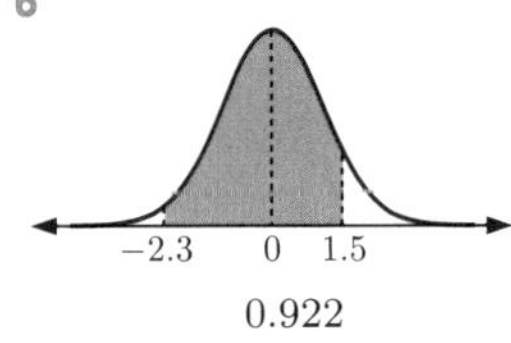

0.922

c

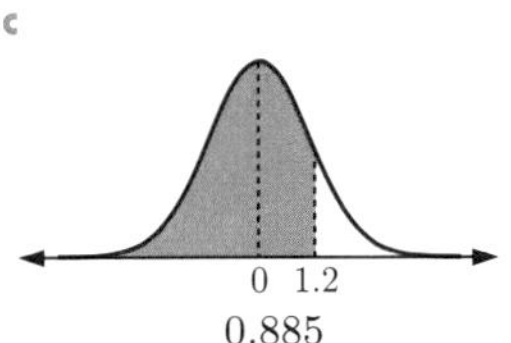

0.885

d

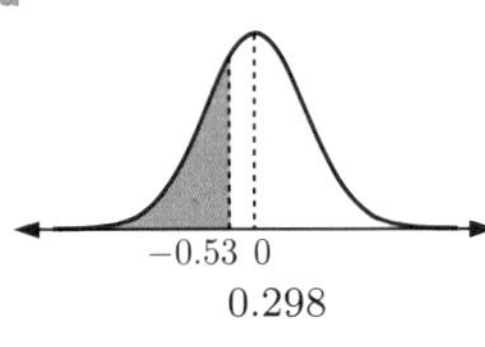

0.298

e

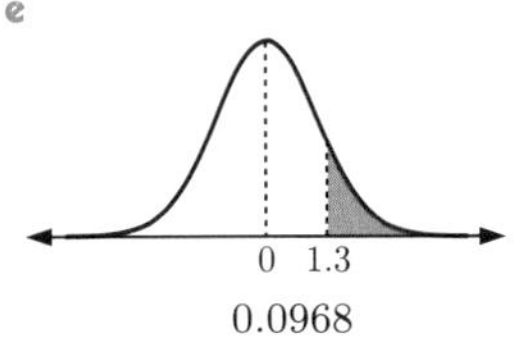

0.0968

f

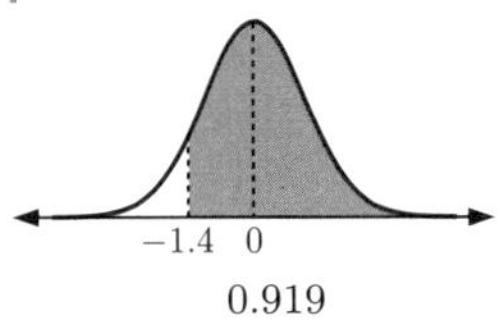

0.919

g

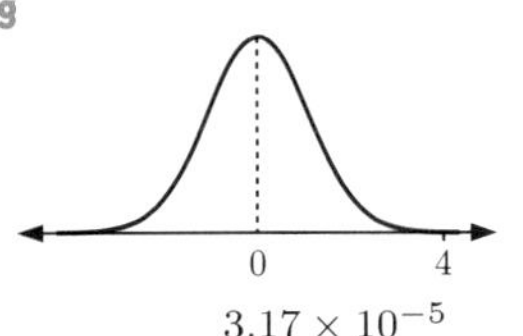

3.17×10^{-5}

h

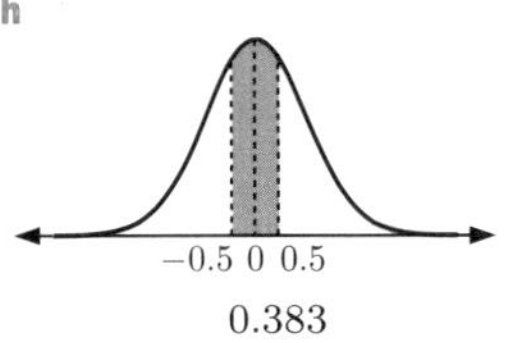

0.383

i

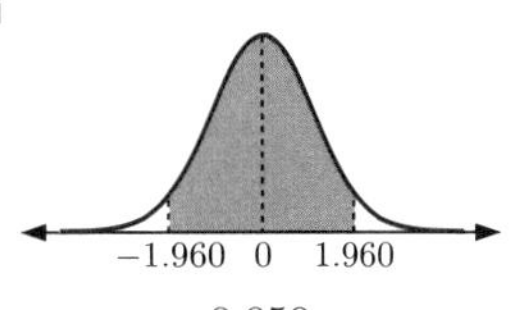

0.950

7 **a** **i** $z_1 \approx -0.859$, $z_2 \approx 1.18$ **ii** 0.687

EXERCISE 26E.1

1 **a** **b** **c** **d**

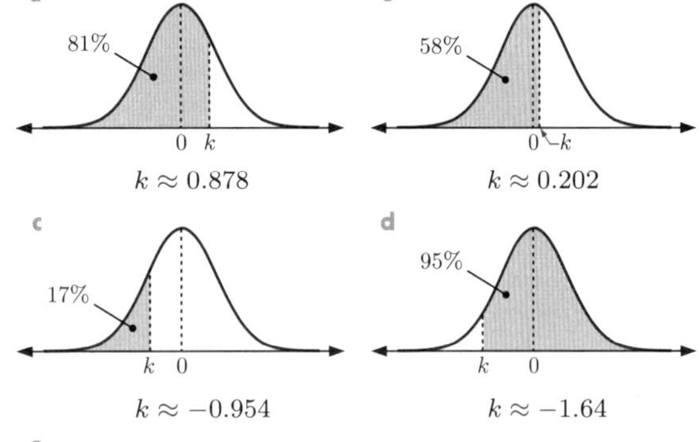

a $k \approx 0.878$ **b** $k \approx 0.202$
c $k \approx -0.954$ **d** $k \approx -1.64$

e

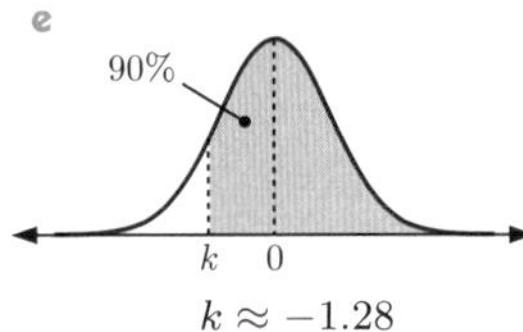

$k \approx -1.28$

2 **a**

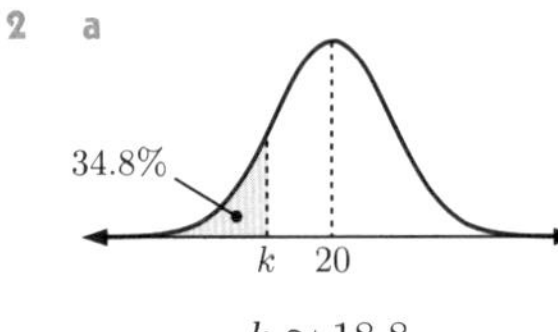

$k \approx 18.8$

b

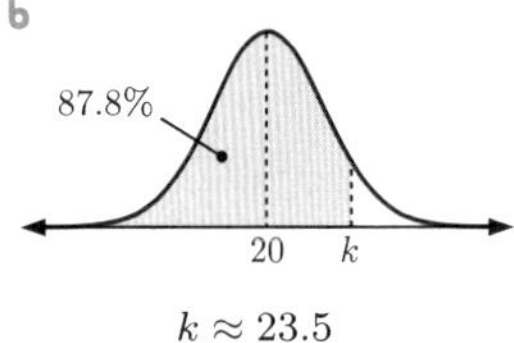

$k \approx 23.5$

c

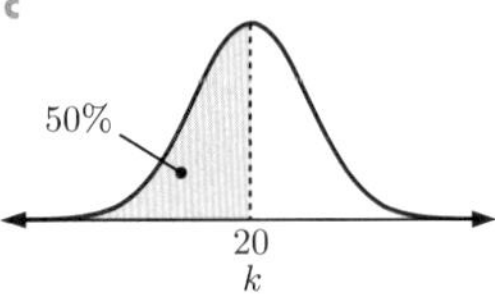

$k = 20$

3 **a**

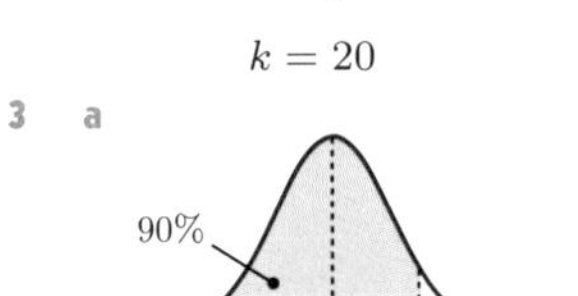

$k \approx 49.2$

b

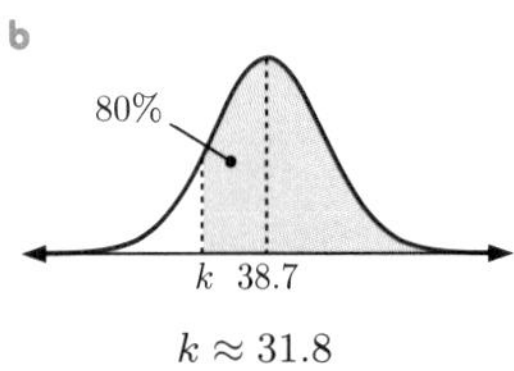

$k \approx 31.8$

4 **a** $a \approx 21.4$ **b** $a \approx 21.8$ **c** $a \approx 2.82$

5 82.9 **6** 24.7 cm **7** 75.2 mm **8** 65.6%

9 between 502 mL and 504 mL

EXERCISE 26E.2

1 112.4 **2** 0.193 m **3** $96.50 **4** 4:01:24 pm

5 $\mu \approx 23.6$, $\sigma \approx 24.3$

6 **a** $\mu \approx 52.4$, $\sigma \approx 21.6$ **b** $\mu \approx 52.4$, $\sigma \approx 21.6$, 54.4%

7 **a** $\mu \approx 4.00$ cm, $\sigma \approx 0.003\,53$ cm **b** 0.604

8 **a** $\mu = 2.00$ cm, $\sigma = 0.0305$ cm **b** 0.736

REVIEW SET 26A

1 a 2.28% b 95.4% c 68.3%

2 a i 2.28% ii 84% b 0.341

3 a $a = 6.3$ grams b $b \approx 32.3$ grams

4 a $\int_0^1 a(x+1)x(x-1)(x-2)\,dx = 1$ gives $a = \frac{30}{11}$

b mode is $\frac{1}{2}$ d median is $\frac{1}{2}$

5 a Harri's score is 2 standard deviations below the mean.

b 97.7% c 17

6 $k \approx 2$ 7 a $a = \frac{1}{4}$ b mode $= \frac{2}{\sqrt{3}}$ d $\mu = \frac{16}{15}$

8 29.5 m 9 a 0.136 b 0.341

10 a 2.28% b 84.1% c 81.9%

11 a $k = \ln 2$ b 0.3 c $\mu = 1 - \ln 2$

REVIEW SET 26B

1 a 68.3% b 95.4% c 81.9% d 13.6%

2 a i 81.9% ii 84.1% b 0.477 c $x \approx 61.9$

3 $\mu \approx 31.2$ mm 4 $k \approx 0.885$

5 a $a = -\frac{3}{10}$ b

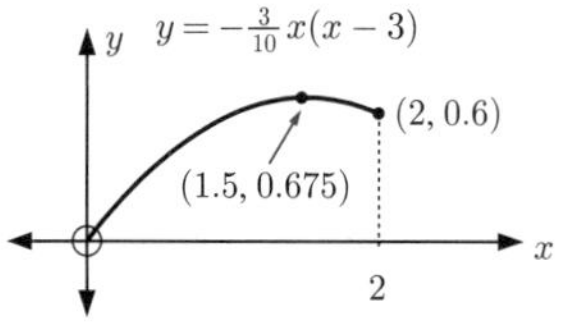

c i 1.2
ii 1.5
iii ≈ 1.24
iv 0.24

d $\frac{13}{20}$

6 a $\mu = 29.0$, $\sigma \approx 10.7$ b i 0.713 ii 0.250

7 a 6.68% b 0.854

8 a 1438 candidates b 71.1 marks c IQR ≈ 20.2 marks

9 a 0.260 b 29.3 weeks

10 a $\mu = 61.2$, $\sigma \approx 22.6$ b ≈ 0.244

11 a The relative difficulty of each test is not known.

b z-score for English $= 1$, z-score for Chemistry $= 1$

$\therefore$ Kerry's performance relative to the rest of the class is the same in both tests.

REVIEW SET 26C

1 a mean is 18.8, standard deviation is 2.6 b 13.6 to 24.0

2 a 0.364 b 0.356 c $k \approx 18.2$

3 0.207 4 $\mu \approx 80.0$ cm, median and mode are also 80.0 cm.

5 a ≈ 0.330 b ≈ 0.796 6 0.0708

7 a 68.3% b 0.0884

8 a $\int_0^1 \frac{4}{1+x^2}\,dx = \pi$ which is $\neq 1$

b $F(x) = \frac{1}{\pi} f(x)$, $\therefore$ $k = \frac{1}{\pi}$

c $\mu = \frac{2}{\pi}\ln 2$, $\text{Var}(X) = \frac{4}{\pi} - 1 - \left(\frac{2\ln 2}{\pi}\right)^2$

9 $\sigma \approx 0.501$ mL 10 0.403

11 a $k = 8$ b median $= 2\frac{2}{7}$ c $\mu \approx 2.75$, $\text{Var}(X) \approx 2.83$

EXERCISE 27A

1 a $(1-i)^2 = -2i$, $(1-i)^{4n} = (-4)^n$ b 256 c $1 \pm i$

2 $u_1 = 2$, $u_n = 3n^2 - 3n + 3$, $n > 1$ 3 $x > 3$

4 a $z = \frac{1}{2}\text{cis}\left(\frac{2\pi}{3}\right)$, $w = \frac{1}{2}\text{cis}\left(\frac{\pi}{4}\right)$

c $\cos\left(\frac{11\pi}{12}\right) = -\frac{\sqrt{2}+\sqrt{6}}{4}$, $\sin\left(\frac{11\pi}{12}\right) = \frac{\sqrt{6}-\sqrt{2}}{4}$

6 $x = -\frac{3\pi}{2}$ or $\frac{\pi}{2}$ 7 a $-e^2$ b $e^2 - 3$

8 a $\frac{12}{13}$ b $-\frac{5}{12}$ c $-\frac{120}{169}$ d $\frac{169}{119}$

9 $x = \frac{2\pi}{3}$ or $\frac{5\pi}{3}$ 10 $d = 2$ 11 $(0, -1, -1)$

12 a $y = 2x - 3$ b

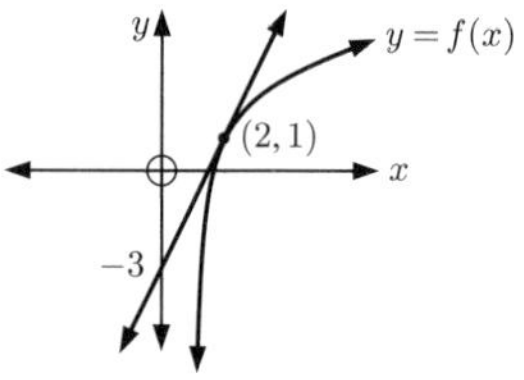

d $]\frac{3}{2}, 2[$

14 $\frac{1}{2}x^2 \arctan x - \frac{1}{2}x + \frac{1}{2}\arctan x + c$

15 a $x = 3$ b $x = \frac{\ln 2}{\ln 3}$ (or $\log_3 2$)

18 an arc of a circle, centre (0, 0), radius $\frac{\text{AB}}{2}$

19 a $A = 1$, $B = 0$, $C = -1$ b $\frac{1}{2}\ln 7 - \frac{3}{2}\ln 3$

20 c $z^4 = r^4 \text{cis}\,4\theta$, $\frac{1}{z} = \frac{1}{r}\text{cis}(-\theta)$, $iz^* = r\,\text{cis}\left(\frac{\pi}{2} - \theta\right)$

21 a i $A' \cap B$ ii B

22 a $-\sqrt{1-x^2} + c$ b $\arctan x + \frac{1}{2}\ln(1+x^2) + c$

c $\arcsin x + c$

24 a 9 b $\sum_{k=1}^{n} \frac{1}{\sqrt{k}+\sqrt{k+1}} = \sqrt{n+1} - 1$

26 $\frac{(m+2)(m+1)(n+2)(n+1)}{4}$ 27 $\frac{1}{2(n-1)}$

30 a $x = -\frac{10}{9}$ b $x = -1$ c $x = e^e$ d $x = \frac{1}{5}$

31 $\frac{1}{3}\tan^3 x + \tan x + c$

32 Either $A \cup B = U$, or, A and B are disjoint.

33 $-23 - 84\sqrt{2}i$ 34 $\theta = -\frac{11\pi}{12}, -\frac{7\pi}{12}, \frac{\pi}{12}, \frac{5\pi}{12}$

35 $a = -1$, $b = -7$

36 a $x < 2$ b $x < \frac{1}{2}$ c $x < 2$ 37 $\frac{3}{4}$

39 a $x \in\,]-\infty, 0[\, \cup\,]2, \infty[$ b $\frac{2x-2}{x(x-2)}$

c $4x - 3y = 12 - 3\ln 3$

40 b $\triangle$PQR is right angled at Q, and is isosceles as QR = QP.

43 $\frac{dy}{dx} = \frac{1 - y\cos(xy)}{x\cos(xy) + 2y}$ 45 $a = 2$, $n = 3$

46 a $\frac{24}{49}$ b $\frac{16}{25}$

47 a (0, 4). A translation of $\binom{2}{1}$.

b (0, 6). A translation of $\binom{2}{0}$ followed by a vertical stretch of factor 2.

c $(-2, -5)$

d $(\frac{1}{2}, 3)$. A horizontal stretch of factor $\frac{1}{2}$ followed by a translation of $\binom{1.5}{0}$.

e $(-2, \frac{1}{3})$ f $(3, -2)$. A reflection in $y = x$.

48 a $a = (r + \frac{1}{r})\cos\theta$, $b = (r - \frac{1}{r})\sin\theta$

b $r = 1$ or z is real and non-zero

51 a

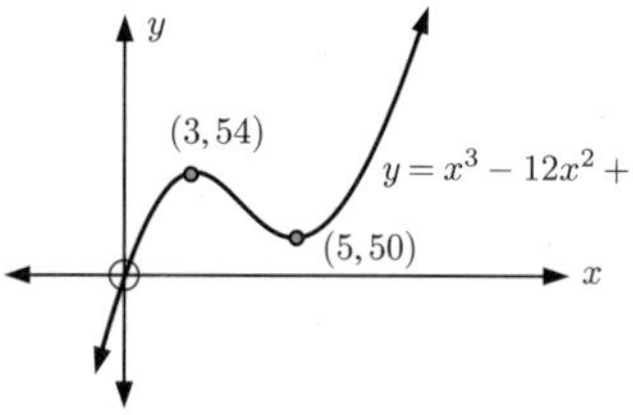

b $50 < k < 54$

$\therefore$ $k \in\,]50, 54[$

52 **a** $y^2 = x^2 + 64 - 16x\cos\theta, \ \cos\theta = \dfrac{3x-10}{2x}$
c $8\sqrt{5}$ cm^2 when $\triangle$ is isosceles

53 **a** $k = 2$ **b** $u_n = \dfrac{3n-8}{2}, \ n \in \mathbb{Z}^+$ **54** $\frac{5}{2}\sqrt{6}$ units^2

55 $\dfrac{8}{x}$ **56** $y = 4\sin(\frac{\pi}{2}x) - 1$ **57** $\frac{7}{10}$

58 $a = \frac{4}{3}, \ b = \frac{2}{3}$ or $a = -\frac{4}{3}, \ b = -\frac{2}{3}$ **59** $x = \frac{15}{7}, \ y = \frac{10}{7}$

60 $x \in [-1, 0] \cup [1, 2], \quad f'(x) = \dfrac{2x-1}{\sqrt{1-(1+x-x^2)^2}}$

61 $f(n) = 3n^2 - 3n + 4$ **62** **a** ≈ 0.34 **b** $\sigma \approx 5$

63 VAs $x = -\frac{\pi}{2}, \ x = -\frac{\pi}{4}, \ x = \frac{\pi}{2}, \ x = \frac{3\pi}{4}$ No HAs

64 **a** $5\mathbf{i} - 2\mathbf{j} + \mathbf{k}$ **b** $\frac{\sqrt{30}}{6}(5\mathbf{i} - 2\mathbf{j} + \mathbf{k})$

65 $2\sin\theta, \ \theta - \frac{\pi}{2}$ **66** $-x^2\cos x + 2x\sin x + 2\cos x + c$

67 $f'(x) = \dfrac{2x+29}{5}$ **68** $x = 4$ or 2 **70** **b** 1

71 **a** $\dfrac{dy}{dx} = \dfrac{3y-2x}{2y-3x}$ **b** $(\sqrt{7}, 0), \ (-\sqrt{7}, 0)$

72 **a** **i** $\frac{13}{21}$ **ii** $\frac{11}{21}$ **b** $\frac{2}{3}$

73 **a** $z = \frac{1}{\sqrt{2}}e^{i(-\frac{7\pi}{12})}$ **b** $n = 12$

74 **a** $t = \frac{2}{3}$ **b** $t = -\frac{1}{3}$

75 Translate $\begin{pmatrix} 1 \\ -10 \end{pmatrix}$ then reflect in x-axis.

76 $\frac{1}{2}$ unit^2 **77** $f(x) = \frac{1}{3}(x+3)(4x+1)(2x-3)^2$

78 **a** $x < \frac{1}{5}$ or $x > \frac{2}{7}$ **b** $-6 < x < 1$ or $x \geqslant 2$

79 $x = -\frac{11\pi}{12}, -\frac{3\pi}{4}, -\frac{7\pi}{12}, \frac{\pi}{12}, \frac{\pi}{4}, \frac{5\pi}{12}$

80 $\dfrac{dy}{dx} = \dfrac{-y^2 - ye^{xy}}{xe^{xy} + 2xy - \cos y}$ **81** $x = \sqrt{5} \quad (y = 0)$

82 $\frac{3}{4}$ unit^2 **83** $a = 5, \ b = 9$

84

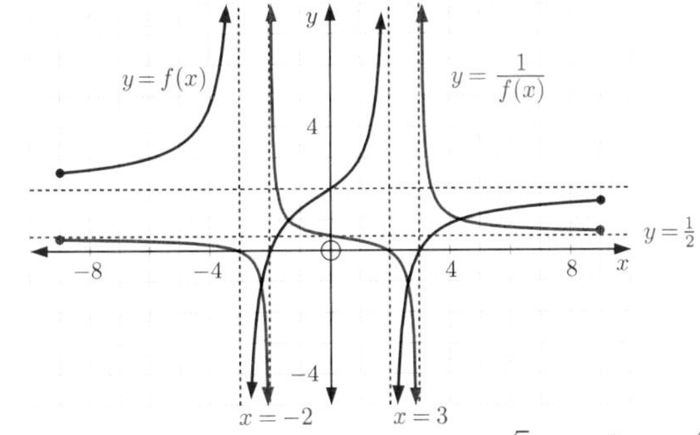

85 $\frac{1}{40\pi}$ cm s^{-1} **86** $a = 2$ **87** $(\frac{\pi}{4}, \frac{\sqrt{2}}{4}), \ (-\frac{3\pi}{4}, -\frac{\sqrt{2}}{4})$

88 $a = \frac{3}{5}$ **89** $x = \frac{\pi}{6}, \frac{5\pi}{6}, \frac{4\pi}{3}, \frac{5\pi}{3}$

90 **a** $\left(\begin{array}{ccc|c} 1 & -2 & 3 & 1 \\ 1 & p & 2 & 0 \\ -2 & p^2 & -4 & q \end{array}\right)$

c **i** $p \neq 0$, 3 planes meet at a point
ii $p = 0, \ q \neq 0$, 3 planes have no common point of intersection, planes 2 and 3 are parallel
iii $p = q = 0$, planes 2 and 3 are coincident
d $x = -2 - 4t, \quad y = t, \quad z = 1 + 2t, \quad t \in \mathbb{R}$

91 $P(z) = (z^2 + 2)(z^2 - 2z + 5)$ **92** $\frac{\sqrt{13}-2}{3}$

94 **a** $5 + i\sqrt{2}$
b $(\sqrt{3})^3 \operatorname{cis}(\arctan(\frac{\sqrt{2}}{5})); \quad a = \sqrt{3}, \quad \theta = \arctan(\frac{\sqrt{2}}{5})$
c $z = \sqrt{3}\operatorname{cis}(\frac{1}{3}\arctan(\frac{\sqrt{2}}{5}) + k\frac{2\pi}{3}), \quad k = 0, 1, 2$

95 **a** $P(x, \sqrt{a^2 - x^2})$
b **i** $\overrightarrow{AP} = \begin{pmatrix} x + a \\ \sqrt{a^2 - x^2} \end{pmatrix}, \quad \overrightarrow{AB} = \begin{pmatrix} 2a \\ 0 \end{pmatrix},$
$\overrightarrow{OP} = \begin{pmatrix} x \\ \sqrt{a^2 - x^2} \end{pmatrix}$
ii $\cos(\widehat{PAB}) = \sqrt{\dfrac{x+a}{2a}}, \quad \cos(\widehat{POB}) = \dfrac{x}{a}$
c angle at centre is twice angle on the circle subtended by the same arc.

97 **a** $|\mathbf{u} - \mathbf{v}| = \sqrt{2 - 2\cos\theta}, \quad |\mathbf{u} + \mathbf{v}| = \sqrt{2 + 2\cos\theta}$
b $\cos\theta = \frac{12}{13}$

98 $\cot(\frac{\theta}{2})$ **100** **b** πab

103 $x = 2, \ y = \frac{1}{8}$ or $x = 64, \ y = 4$

104 **a** $\sqrt{2}\operatorname{cis}\left(-\frac{\pi}{4}\right),$
$\sqrt{2}\operatorname{cis}\left(\frac{5\pi}{12}\right),$
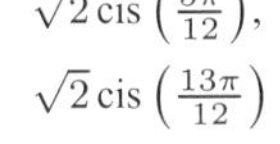
$\sqrt{2}\operatorname{cis}\left(\frac{13\pi}{12}\right)$
b

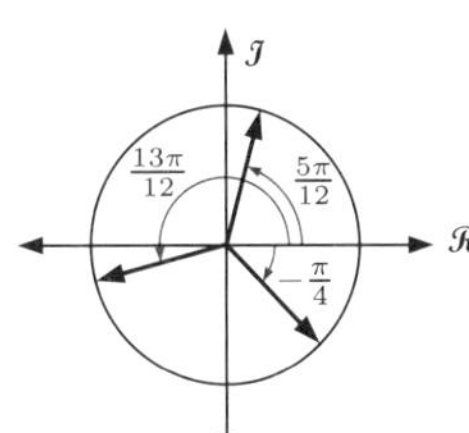

105 $m = -\frac{6}{19}$ or 6 **106** $\frac{\pi}{4}$ **110** $\sigma = \sqrt{1 + \sqrt{7}}$

111 **a** $A = \frac{1}{2}, \ B = -\frac{1}{2}$ **c** $\frac{3}{4}$

112 **a** $k = 2$ **b** $\frac{1}{2}\ln 3$ units^2

114 **a** $\dfrac{a^{n+1}(a\cos n\theta - \cos(n+1)\theta) - a\cos\theta + 1}{a^2 - 2a\cos\theta + 1}$
b $\dfrac{a^{n+1}(a\sin n\theta - \sin(n+1)\theta) + a\sin\theta}{a^2 - 2a\cos\theta + 1}$

115 $\frac{2}{3}(x+2)^{\frac{3}{2}} - x - 2 - 2\sqrt{x+2} + 2\ln(\sqrt{x+2} + 1) + c$

116 **a** $1 - (1-p)^n$ **b** $\dfrac{\binom{n}{k} p^k (1-p)^{n-k}}{\sum\limits_{r=k}^{n} \binom{n}{r} p^r (1-p)^{n-r}}$
c **i** $p = 1$ **ii** no real solutions

117 **b** $x = \tan\left(\frac{\pi}{16}\right), \ \tan\left(\frac{5\pi}{16}\right), \ \tan\left(\frac{9\pi}{16}\right), \ \tan\left(\frac{13\pi}{16}\right)$

118 **a** $\left(\begin{array}{ccc|c} 2 & -1 & 3 & 4 \\ 0 & 2 & a & 6 - a \\ 0 & 0 & a^2 - 4a & a^2 + 4a - 32 \end{array}\right)$
b **i** $a = 0$, 3 planes have no common point of intersection
ii $a = 4; \ x = \dfrac{5 - 5t}{2}, \ y = 1 - 2t, \ z = t, \ t \in \mathbb{R}$, 3 planes meet in a line
iii $a \neq 0$ or 4; $x = -\dfrac{a}{2} - \dfrac{12}{a}, \ y = -1 - a, \ z = \dfrac{a+8}{a}$, 3 planes meet at a point
When $a = 2$, the solution is $x = -7, \ y = -3, \ z = 5$.

119 **a** $f'(x) = e^{1-2x^2}(1 - 4x^2) \quad f''(x) = e^{1-2x^2}(16x^3 - 12x)$
b local min. at $(-\frac{1}{2}, -\frac{\sqrt{e}}{2})$ local max. at $(\frac{1}{2}, \frac{\sqrt{e}}{2})$
c $x = 0$ or $\pm\frac{\sqrt{3}}{2}$
d as $x \to \infty, \ f(x) \to 0^+$; as $x \to -\infty, \ f(x) \to 0^-$

e

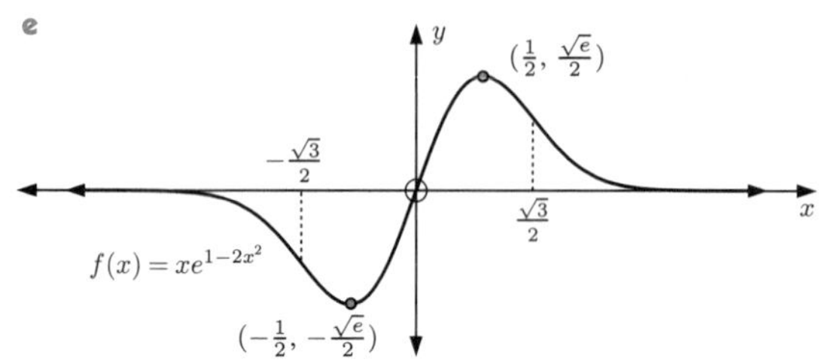

f $k = \frac{1}{\sqrt{2}}$

120 **b** $2\left|\cos(\frac{\theta-\phi}{2})\right|$, $\frac{\theta+\phi}{2}$ **121** $\frac{x}{2}(\sin(\ln x) - \cos(\ln x))$

122 **a** $x^2 - 2x + (1 + k^2)$ **b** $k = 0, \pm1, \pm2, \pm3$
c $pq = -1, -2, -5, -10$ **d** $1, -2, 1 \pm 2i$

123 **a**

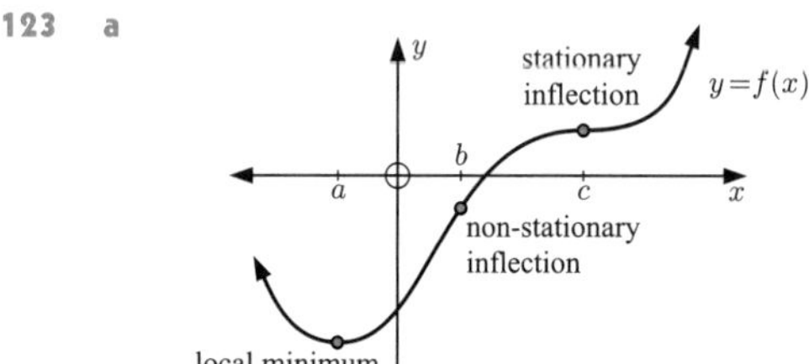

b $f(x) = f_1(x) + k, \ k \in \mathbb{R}$

124 **a** $\frac{25\pi}{2}$ units^2 **b** $25\pi\sin(\frac{\alpha}{2})$ units^2

125 **b** $y = e^{2x-1}$ **126** **b** $A + B + C = k\pi, \ k \in \mathbb{Z}$

127 **b** $2\sqrt[3]{3} - \sqrt{2}$ **130** **a** $P = Q = \frac{1}{2a}$

131 $x = \frac{5}{2}, \ y = -\frac{1}{2}$ **133** $\frac{19}{64}a = k$

EXERCISE 27B

1 $x = -3$ or ≈ -7.64

2 **a** ≈ 1574 trees **b** converges to 1000

3 $\frac{3-\sqrt{5}}{2} \approx 0.382$ **5** $-1 + 2i$ **7** $\alpha \approx 6.92°$

8 **b** $\sqrt{\frac{10}{11}} \approx 0.953$ units **9** $\frac{1}{2}\sqrt{63} \approx 3.97$ km

11 **a** $u_n = \cos\theta\tan^{n-2}\theta$
b $u_1 = 1$ and $u_{n+1} = u_n^2\cos\theta, \ n \in \mathbb{Z}^+$

12 $1 - (\frac{121}{128})^{10} \approx 0.430$ **13** **a** ≈ 1.48 units **b** ≈ 3.82 units

15 **a** $8\,\text{cis}\,(-\frac{\pi}{2})$ **b** $z = 2\,\text{cis}\,(-\frac{\pi}{6}), \ 2\,\text{cis}\,(\frac{\pi}{2}), \ 2\,\text{cis}\,(-\frac{5\pi}{6})$
c **e** $-8i$

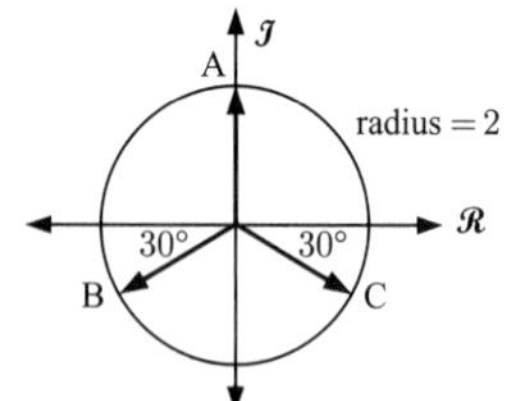

16

144°
25 cm
15 cm

17 **a** $\sigma \approx 3.599\,86$
b ≈ 0.781

18 **a** $\frac{3x}{x-2}$ **b** $\frac{2x+1}{x-1}$ **19** $\sqrt{3} \approx 1.73$ m

20 $x = \frac{2}{a^2 - 1}$ **22** **a** $2p^2 - p^4$ **b** $p \approx 0.541$

23 **b** $h(x) = (x - t)(x - 2t)(x - 3t)$
c $(-3, 27), \ (-2, 16), \ (-1, 5)$

24 **a** 66 **b** ≈ 264 **25** ≈ 6.40 cm

26 **a** $w = \frac{[a^2 - 1 - b^2] + i[2ab]}{(a+1)^2 + b^2}$
b purely imaginary if $a^2 - b^2 = 1$ and $ab \neq 0$

27 $f(x) = (x+1)^2(2x - 5), \ x = -1$ or $x \geqslant \frac{5}{2}$ **28** $\frac{17}{32}$

29 **b** 1, 1, 1 **c** $a_n^2 - 3b_n^2 = 1$ for all $n \in \mathbb{Z}^+$

30 **a** $H = A\sin\theta + B\cos\theta$

31 **a** -224 **b** 880 **c** -40 **32** ≈ 0.114

33 **a** $x^2 + 4x + 5$ **b** $a = -4$, real zero is 4 **34** $n = 3$ or 8

35 **a** $k = \frac{1}{3}$ **b** mean $= 800$, $\sigma = \frac{40\sqrt{3}}{3}$ **36** 3

37

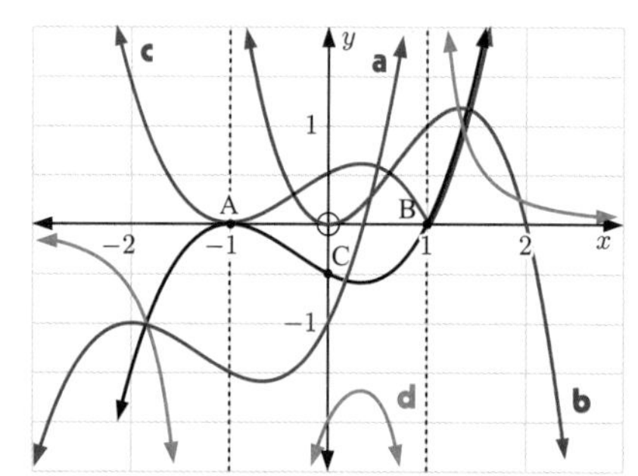

38 **a** ≈ 0.242 **b** ≈ 0.769

39 **a** D(7, 1, −2), X(7, 3, −1), Y(5, 3, −2)
b **Hint:** Show $\overrightarrow{BD} = k\overrightarrow{BY}$

40 $y = \frac{1}{2}x + \frac{1}{4}\sin 2x + 4$ **41** ≈ 0.842 **42** $a = \frac{1}{2}$

43 $1 + 2i, \ -1 \pm 3i$

44 **a** $n\binom{n}{2}$ committees **b** $\binom{n}{3}$ committees
d **i** $\frac{4}{11}$ **ii** $\frac{1}{8}$

45 **a** $a = 13, \ b = 12, \ c = \frac{\pi}{30}, \ d = 15$ **b** ≈ 24.9 m

46 $9b = 2a^2$ **47** $\frac{\pi}{6}(e^2 - 1)$ units^3

48 **a** $(\frac{1}{5}, \frac{17}{5}, \frac{9}{5})$ **c** $(-1, 3, 1)$ **d** $6x - 8y - 5z = -35$

49 **a** ≈ 0.549 **b** $\approx 0.001\,72$

50 **a** $x = 0$ **b** $x = 0.2$ or 0.3

51 **a** $f(x) = -x^2 + 6x - 13$ **b** $f(x) = -(x-3)^2 - 4$

52 **a** $\frac{\pi}{3}$ **b** $x = 1 - \frac{\pi}{2} \approx -0.571, \ t = 1 - \frac{\pi}{6} \approx 0.476$
c A horizontal stretch factor $\frac{1}{3}$, followed by a vertical stretch factor 2, followed by a translation $\binom{1}{4}$.
d $\{x \mid x \in [-1, 1], \ x \neq 1 - \frac{\pi}{2}, \ x \neq 1 - \frac{\pi}{6}\}$; range $y \in \mathbb{R}$

53 ≈ 20.6 cm **54** **a** 8008 **b** 5320 **c** 2211

55 **b** $k = -1$ **c** $p = -2, \ q = 6$ **56** $47\frac{6}{7}$ or $31\frac{1}{7}$

57 **a** ≈ 0.785 **b** ≈ 0.995

58 **a** when $x = \frac{\pi}{2}$ **b** $f''(x) = e^{\sin^2 x}(\sin^2 2x + 2\cos 2x)$
c (0.999, 2.03) and (2.14, 2.03)

59 **a** $P(x) = (ax + b)(2x^2 - 3x + 1)$
b $P(x) = (3x + 7)(2x^2 - 3x + 1)$

60 $a = \sqrt{e^6 - 1}$

61 $\sqrt{2}\,\text{cis}\left(-\frac{19\pi}{20}\right), \sqrt{2}\,\text{cis}\left(-\frac{11\pi}{20}\right), \sqrt{2}\,\text{cis}\left(-\frac{3\pi}{20}\right), \sqrt{2}\,\text{cis}\left(\frac{13\pi}{20}\right)$

62 $\approx 7.82°$ **63** **a** $2x\sqrt{1-x^2}$ **b** $-\frac{2}{3}(1-x^2)^{\frac{3}{2}}+c$ **c** $\frac{2}{3}$

64 **a** $a=\frac{3}{5}$, P(0, 3, −8) **b** $\theta \approx 18.8°$
c $5x-11y-7z=23$

65 **a** $t \approx 0.880$ and $t \approx 3.79$ **b** ≈ 24.5 m
c **i** $t \approx 2.33$
ii The speed is at a local maximum to the left, so the velocity is a minimum.

66 3.76 units^2 **67** **a** $a=8,\ b=25,\ c=26$ **b** $z \geqslant -2$

68 **a**

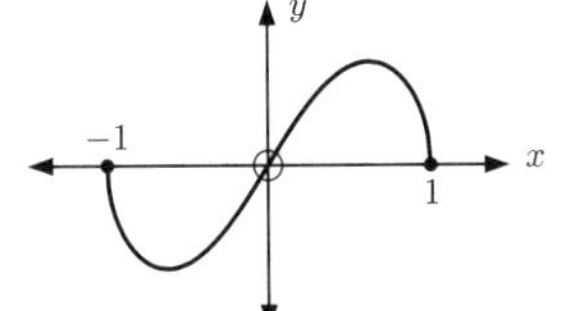

b ≈ 0.676 units^3

70 **b** $x=\sqrt[3]{4}+\sqrt[3]{2}$ **71** $f(x)=3e^{\frac{x}{3}}$

72 **a** \$865.25 **b** $R=\frac{r}{100m}P\left[\frac{(1+\frac{r}{100m})^{mn}}{(1+\frac{r}{100m})^{mn}-1}\right]$

73 **a** $11\binom{21}{10}=3\,879\,876$ choices

74 **a** $k=-\frac{1}{2e}$
b **i** $k=-\frac{1}{2e},\ k \geqslant 0$ **ii** $-\frac{1}{2e}<k<0$ **iii** $k<-\frac{1}{2e}$
c **i** (0, 0) **ii** $y=-x$, 45°

75 $n=3$ **76** ≈ 2.58 **77** $\frac{107}{576}$ **78** **a** $\frac{3}{5}$ **b** 2.2 cm

80 **a** 1 **b** $\frac{\pi}{2}-1$ **d** $(\frac{\pi}{2})^3-3\pi+6$

81 **b** $x=2\cos\left(\frac{2\pi}{9}\right),\ 2\cos\left(\frac{8\pi}{9}\right),\ 2\cos\left(\frac{14\pi}{9}\right)$

82 **b**

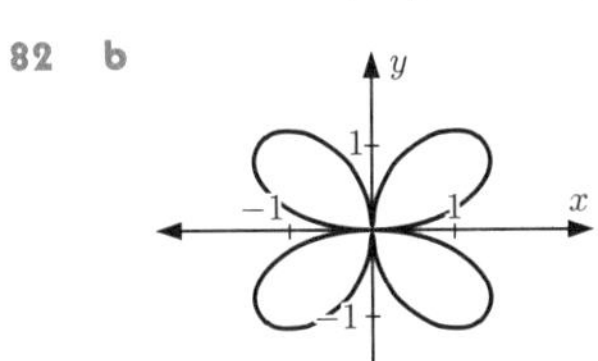

83 **a** $\frac{2n-m-p}{mp-n^2}$

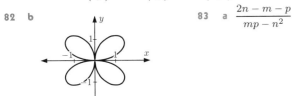

84 **a** A(2, −3, 1) **b** B(−1, 2, 3) **c** $p=0$ or $\frac{62}{371}$

85 **b** $e^x=\sum_{n=0}^{\infty}\frac{x^n}{n!}$ **86** 14.5

88 **a** (0, 0), (1, 0), (−1, 0)
b $\left(\frac{\sqrt{6}}{4},\frac{\sqrt{2}}{4}\right),\ \left(\frac{\sqrt{6}}{4},-\frac{\sqrt{2}}{4}\right),\ \left(-\frac{\sqrt{6}}{4},\frac{\sqrt{2}}{4}\right),$
$\left(-\frac{\sqrt{6}}{4},-\frac{\sqrt{2}}{4}\right)$

89 **a** $x\ln x-x+c$ **b** $k=e$ **c** ≈ 2.16

90 **b** r km **c** ≈ 1730 km h^{-1}

91 **b** $f(x)\approx\frac{1}{\sqrt{1-x^2}}+2$

92 **a** **i** Domain is $\{x \mid x \geqslant 0\}$ **ii** −8, at (8, −8)
iii

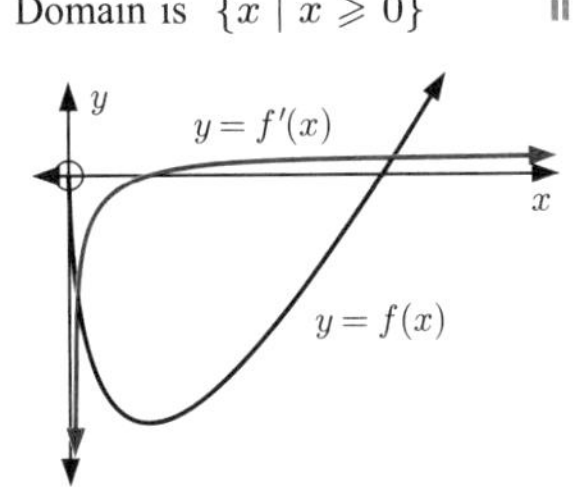

b **ii** **Hint:** Let $a=x,\ b=8$.

93 **a** $k=-2,\quad a=-1$
c $x=3+t,\quad y=-3-2t,\quad z=t,\quad t\in\mathbb{R}$
d (3, −3, 0) **e** $\approx 61.9°$

94 **a** $12!=479\,001\,600$
b **i** 43 545 600 **ii** 7 257 600 **iii** 159 667 200
iv 58 060 800
c **i** 5775 **ii** 1575

95 **b** $v=\frac{-2}{t+2}$ **c** 1.39 units

97 **a** dec at ≈ 322 cm^3/min **b** dec at ≈ 275 cm^3/min

98 **b** **i** $x=y^{\frac{7}{6}}$ **ii** $x=z^{\frac{14}{9}}$

99 **a** 2π
b

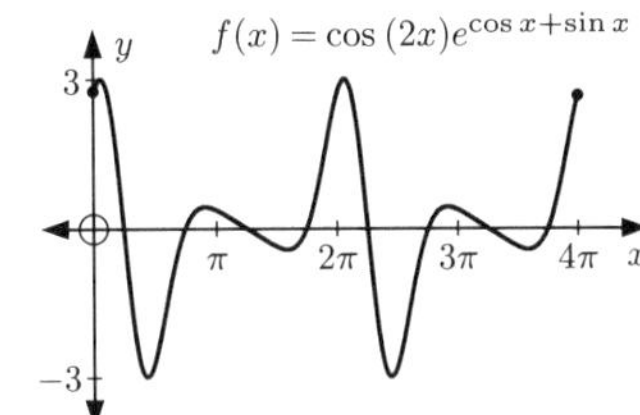

c **i** $e^{\cos x+\sin x}(\sin x+\cos x-1)+c$ **ii** $x=\frac{\pi}{4}$
d $e^{\sqrt{2}}(\sqrt{2}-1)\approx 1.70$ units^2

100 **a** **i**

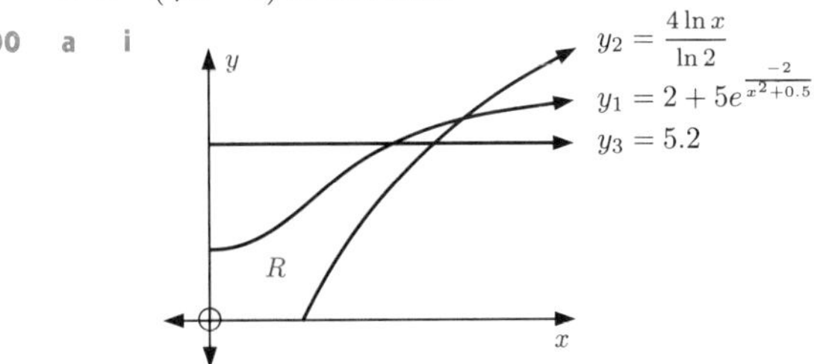

ii $2+5e^{-4}$
c **i**
$$V=\pi\left[\int_0^{5.2}2^{\frac{y}{2}}\,dy-\int_{2+\frac{5}{e^4}}^{5.2}\left(\frac{-2}{\ln\left(\frac{y-2}{5}\right)}-0.5\right)dy\right]\text{ cm}^3$$
ii ≈ 31.1 cm^3
d **i**

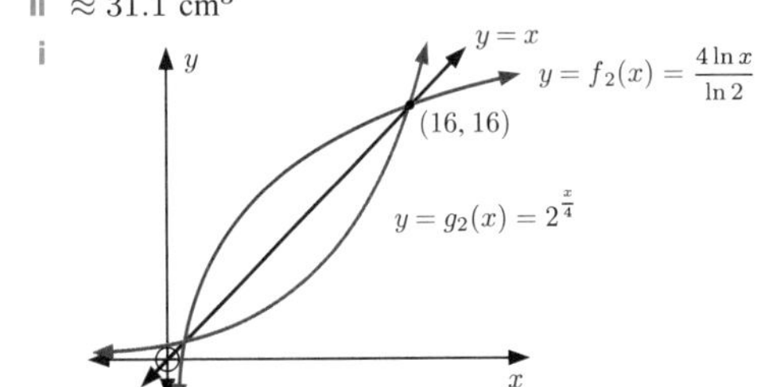

iii inverse functions
e **i** $g_1(x)=\sqrt{\frac{-2}{\ln\left(\frac{x-2}{5}\right)}-0.5}$
ii Domain is $\{x \mid 2+\frac{5}{e^4}\leqslant x<7\}$

INDEX